Values of fundamental constants

Quantity	Symbol	Value
Speed of light in vacuum	c_0	$3.00 \times 10^8 \,\mathrm{m/s}$
Gravitational constant	G	$6.6738 \times 10^{-11} \,\mathrm{N \cdot m^2/kg^2}$
Avogadro's number	N_A	$6.0221413 \times 10^{23} \,\mathrm{mol^{-1}}$
Boltzmann's constant	k_B	$1.380 \times 10^{-23} \,\mathrm{J/K}$
Charge on electron	e	$1.60 \times 10^{-19} \,\mathrm{C}$
Electric constant	ϵ_0	$8.85418782 \times 10^{-12} \,\mathrm{C^2/(N \cdot m^2)}$
Magnetic constant	μ_0	$4\pi \times 10^{-7} \,\mathrm{T \cdot m/A}$
Planck's constant	h	$6.626 \times 10^{-34} \,\mathrm{J \cdot s}$
Electron mass	m_e	$9.11 \times 10^{-31} \,\mathrm{kg}$
Proton mass	m_p	$1.6726 \times 10^{-27} \,\mathrm{kg}$
Neutron mass	m_n	$1.6749 \times 10^{-27} \,\mathrm{kg}$
Atomic mass unit	amu	$1.6605 \times 10^{-27} \,\mathrm{kg}$

Other useful numbers

Number or quantity	Value
π	3.1415927
e	2.7182818
1 radian	$57.2957795°$
Absolute zero ($T = 0$)	$-273.15\,°\mathrm{C}$
Average acceleration g due to gravity near Earth's surface	$9.8 \,\mathrm{m/s^2}$
Speed of sound in air at 20 °C	$343 \,\mathrm{m/s}$
Density of dry air at atmospheric pressure and 20 °C	$1.29 \,\mathrm{kg/m^3}$
Earth's mass	$5.97 \times 10^{24} \,\mathrm{kg}$
Earth's radius (mean)	$6.38 \times 10^6 \,\mathrm{m}$
Earth–Moon distance (mean)	$3.84 \times 10^8 \,\mathrm{m}$

PRINCIPLES & PRACTICE OF
PHYSICS

Eric Mazur
Harvard University

With contributions from

Catherine H. Crouch
Swarthmore College

Peter A. Dourmashkin
Massachusetts Institute of Technology

PEARSON

Boston   Columbus   Indianapolis   New York   San Francisco   Hoboken
Amsterdam   Cape Town   Dubai   London   Madrid   Milan   Munich   Paris   Montréal   Toronto
Delhi   Mexico City   São Paulo   Sydney   Hong Kong   Seoul   Singapore   Taipei   Tokyo

Executive Editor: Becky Ruden
Publisher: Jim Smith
Project Managers: Beth Collins and Martha Steele
Program Manager: Katie Conley
Managing Development Editor: Cathy Murphy
Senior Development Editor: Margot Otway
Development Editor: Irene Nunes
Editorial Assistant: Sarah Kaubisch
Text Permissions Project Manager: Liz Kincaid
Text Permissions Specialist: Paul Sarkis
Associate Content Producer: Megan Power
Production Management: Rose Kernan
Copyeditor: Carol Reitz
Compositor: Cenveo® Publisher Services
Design Manager: Mark Ong
Interior Designer: Hespenheide Design
Cover Designer: Tandem Creative, Inc.
Illustrators: Rolin Graphics
Photo Permissions Management: Maya Melenchuk
Photo Researcher: Eric Schrader
Manufacturing Buyer: Jeff Sargent
Vice-President of Marketing: Christy Lesko
Marketing Manager: Will Moore
Senior Marketing Development Manager: Michelle Cadden

Cover Photo Credit: Franklin Kappa

Credits and acknowledgments borrowed from other sources and reproduced, with permission, in this textbook appear on the appropriate page within the text or on p. C-1.

Library of Congress Cataloging-in-Publication Data on file.

ISBN 10: 0-321-94920-X; ISBN 13: 978-0-321-94920-2 (Student edition)
ISBN 10: 0-321-95836-5; ISBN 13: 978-0-321-95836-5 (Books a la carte)

4 16

Brief Contents

Volume 1 of *Principles of Physics* includes Chapters 1–21. Volume 2 of *Principles of Physics* includes Chapters 22–34.

About the Author

Eric Mazur is the Balkanski Professor of Physics and Applied Physics at Harvard University and Area Dean of Applied Physics. Dr. Mazur is a renowned scientist and researcher in optical physics and in education research, and a sought-after author and speaker.

Dr. Mazur joined the faculty at Harvard shortly after obtaining his Ph.D. at the University of Leiden in the Netherlands. In 2012 he was awarded an Honorary Doctorate from the École Polytechnique and the University of Montreal. He is a Member of the Royal Academy of Sciences of the Netherlands and holds honorary professorships at the Institute of Semiconductor Physics of the Chinese Academy of Sciences in Beijing, the Institute of Laser Engineering at the Beijing University of Technology, and the Beijing Normal University.

Dr. Mazur has held appointments as Visiting Professor or Distinguished Lecturer at Carnegie Mellon University, the Ohio State University, the Pennsylvania State University, Princeton University, Vanderbilt University, Hong Kong University, the University of Leuven in Belgium, and National Taiwan University in Taiwan, among others.

In addition to his work in optical physics, Dr. Mazur is interested in education, science policy, outreach, and the public perception of science. In 1990 he began developing peer instruction, a method for teaching large lecture classes interactively. This teaching method has developed a large following, both nationally and internationally, and has been adopted across many science disciplines.

Dr. Mazur is author or co-author of over 250 scientific publications and holds two dozen patents. He has also written on education and is the author of *Peer Instruction: A User's Manual* (Pearson, 1997), a book that explains how to teach large lecture classes interactively. In 2006 he helped produce the award-winning DVD *Interactive Teaching*. He is the co-founder of Learning Catalytics, a platform for promoting interactive problem solving in the classroom, which is available in MasteringPhysics®.

To the Student

Let me tell you a bit about myself.

I always knew exactly what I wanted to do. It just never worked out that way.

When I was seven years old, my grandfather gave me a book about astronomy. Growing up in the Netherlands I became fascinated by the structure of the solar system, the Milky Way, the universe. I remember struggling with the concept of infinite space and asking endless questions without getting satisfactory answers. I developed an early passion for space and space exploration. I knew I was going to be an astronomer. In high school I was good at physics, but when I entered university and had to choose a major, I chose astronomy.

It took only a few months for my romance with the heavens to unravel. Instead of teaching me about the mysteries and structure of the universe, astronomy had been reduced to a mind-numbing web of facts, from declinations and right ascensions to semi-major axes and eccentricities. Disillusioned about astronomy, I switched majors to physics. Physics initially turned out to be no better than astronomy, and I struggled to remain engaged. I managed to make it through my courses, often by rote memorization, but the beauty of science eluded me.

It wasn't until doing research in graduate school that I rediscovered the beauty of science. I knew one thing for sure, though: I was never going to be an academic. I was going to do something useful in my life. Just before obtaining my doctorate, I lined up my dream job working on the development of the compact disc, but I decided to spend one year doing postdoctoral research first.

It was a long year. After my postdoc, I accepted a junior faculty position and started teaching. That's when I discovered that the combination of doing research—uncovering the mysteries of the universe—and teaching—helping others to see the beauty of the universe—is a wonderful combination.

When I started teaching, I did what all teachers did at the time: lecture. It took almost a decade to discover that my award-winning lecturing did for my students exactly what the courses I took in college had done for me: It turned the subject that I was teaching into a collection of facts that my students memorized by rote. Instead of transmitting the beauty of my field, I was essentially regurgitating facts to my students.

When I discovered that my students were not mastering even the most basic principles, I decided to completely change my approach to teaching. Instead of lecturing, I asked students to read my lecture notes at home, and then, in class, I taught by questioning—by asking my students to reflect on concepts, discuss in pairs, and experience their own "aha!" moments.

Over the course of more than twenty years, the lecture notes have evolved into this book. Consider this book to be my best possible "lecturing" to you. But instead of listening to me without having the opportunity to reflect and think, this book will permit you to pause and think; to hopefully experience many "aha!" moments on your own.

I hope this book will help you develop the thinking skills that will make you successful in your career. And remember: your future may be—and likely will be—very different from what you imagine.

I welcome any feedback you have. Feel free to send me email or tweets.

I wrote this book for you.

Eric Mazur
@eric_mazur #MazurText
mazur@harvard.edu
Cambridge, MA

To the Instructor

They say that the person who teaches is the one who learns the most in the classroom. Indeed, teaching led me to many unexpected insights. So, also, with the writing of this book, which has been a formidably exciting intellectual journey.

Why write a new physics text?

In May 1993 I was driving to Troy, NY, to speak at a meeting held in honor of Robert Resnick's retirement. In the car with me was a dear friend and colleague, Albert Altman, professor at the University of Massachusetts, Lowell. He asked me if I was familiar with the approach to physics taken by Ernst Mach in his popular lectures. I wasn't. Mach treats conservation of momentum before discussing the laws of motion, and his formulation of mechanics had a profound influence on Einstein.

The idea of using conservation principles derived from experimental observations as the basis for a text—rather than Newton's laws and the concept of force—appealed to me immediately. After all, most physicists never use the concept of force because it relates only to mechanics. It has no role in quantum physics, for example. The conservation principles, however, hold throughout all of physics. In that sense they are much more fundamental than Newton's laws. Furthermore, conservation principles involve only algebra, whereas Newton's second law is a differential equation.

It occurred to me, however, that Mach's approach could be taken further. Wouldn't it be nice to start with conservation of both momentum *and* energy, and only *later* bring in the concept of force? After all, physics education research has shown that the concept of force is fraught with pitfalls. What's more, after tediously deriving many results using kinematics and dynamics, most physics textbooks show that you can derive the same results from conservation principles in just one or two lines. Why not do it the easy way first?

It took me many years to reorganize introductory physics around the conservation principles, but the resulting approach is one that is much more unified and modern—the conservation principles are the theme that runs throughout this entire book.

Additional motives for writing this text came from my own teaching. Most textbooks focus on the acquisition of information and on the development of procedural knowledge. This focus comes at the expense of conceptual understanding or the ability to transfer knowledge to a new context. As explained below, I have structured this text to redress that balance. I also have drawn deeply on the results of physics education research, including that of my own research group.

I have written this text to be accessible and easy for students to understand. My hope is that it can take on the burden of basic teaching, freeing class time for synthesis, discussion, and problem solving.

Setting a new standard

The tenacity of the standard approach in textbooks can be attributed to a combination of inertia and familiarity. Teaching large introductory courses is a major chore, and once a course is developed, changing it is not easy. Furthermore, the standard texts worked for *us*, so it's natural to feel that they should work for our students, too.

The fallacy in the latter line of reasoning is now well-known thanks to education research. Very few of our students are like us at all. Most take physics because they are required to do so; many will take no physics beyond the introductory course. Physics education research makes it clear that the standard approach fails these students.

Because of pressure on physics departments to deliver better education to non-majors, changes are occurring in the way physics is taught. These changes, in turn, create a need for a textbook that embodies a new educational philosophy in both format and presentation.

Organization of this book

As I considered the best way to convey the conceptual framework of mechanics, it became clear that the standard curriculum truly deserved to be rethought. For example, standard texts are forced to redefine certain concepts more than once—a strategy that we know befuddles students. (Examples are *work*, the standard definition of which is incompatible with the first law of thermodynamics, and *energy*, which is redefined when modern physics is discussed.)

Another point that has always bothered me is the arbitrary division between "modern" and "classical" physics. In most texts, the first thirty-odd chapters present physics essentially as it was known at the end of the 19th century; "modern physics" gets tacked on at the end. There's no need for this separation. Our goal should be to explain physics in the way that works best for students, using our full contemporary understanding. *All* physics is modern!

That is why my table of contents departs from the "standard organization" in the following specific ways.

Emphasis on conservation laws. As mentioned earlier, this book introduces the conservation laws early and treats them the way they should be: as the backbone of physics. The advantages of this shift are many. First, it avoids many of the standard pitfalls related to the concept of force, and it leads naturally to the two-body character of forces and the laws of motion. Second, the conservation laws enable students to solve a wide variety of problems without any calculus. Indeed, for complex systems, the conservation laws are often the natural (or only) way to solve problems. Third, the book deduces the conservation laws from observations, helping to make clear their connection with the world around us.

Table 1 Scheduling matrix

Topic	Chapters	Can be inserted after chapter...	Chapters that can be omitted without affecting continuity
Mechanics	1–14		6, 13–14
Waves	15–17	12	16–17
Fluids	18	9	
Thermal Physics	19–21	10	21
Electricity & Magnetism	22–30	12 (but 17 is needed for 29–30)	29–30
Circuits	31–32	26 (but 30 is needed for 32)	32
Optics	33–34	17	34

I and several other instructors have tested this approach extensively in our classes and found markedly improved performance on problems involving momentum and energy, with large gains on assessment instruments like the Force Concept Inventory.

Early emphasis on the concept of system. Fundamental to most physical models is the separation of a system from its environment. This separation is so basic that physicists tend to carry it out unconsciously, and traditional texts largely gloss over it. This text introduces the concept in the context of conservation principles and uses it consistently.

Postponement of vectors. Most introductory physics concerns phenomena that take place along one dimension. Problems that involve more than one dimension can be broken down into one-dimensional problems using vectorial notation. So a solid understanding of physics in one dimension is of fundamental importance. However, by introducing vectors in more than one dimension from the start, standard texts distract the student from the basic concepts of kinematics.

In this book, I develop the complete framework of mechanics for motions and interactions in one dimension. I introduce the second dimension when it is needed, starting with rotational motion. Hence, students are free to concentrate on the actual physics.

Just-in-time introduction of concepts. Wherever possible, I introduce concepts only when they are necessary. This approach allows students to put ideas into immediate practice, leading to better assimilation.

Integration of modern physics. A survey of syllabi shows that less than half the calculus-based courses in the United States cover modern physics. I have therefore integrated selected "modern" topics throughout the text. For example, special relativity is covered in Chapter 14, at the end of mechanics. Chapter 32, Electronics, includes sections on semiconductors and semiconductor devices. Chapter 34, Wave and Particle Optics, contains sections on quantization and photons.

Modularity. I have written the book in a modular fashion so it can accommodate a variety of curricula (See Table 1, "Scheduling matrix").

The book contains two major parts, Mechanics and Electricity and Magnetism, plus five shorter parts. The two major parts by themselves can support an in-depth two-semester or three-quarter course that presents a complete picture of physics embodying the fundamental ideas of modern physics. Additional parts can be added for a longer or faster-paced course. The five shorter parts are more or less self-contained, although they do build on previous material, so their placement is flexible. Within each part or chapter, more advanced or difficult material is placed at the end.

Pedagogy

This text draws on many models and techniques derived from my own teaching and from physics education research. The following are major themes that I have incorporated throughout.

Separation of conceptual and mathematical frameworks. Each chapter is divided into two parts: Concepts and Quantitative Tools. The first part, Concepts, develops the full conceptual framework of the topic and addresses many of the common questions students have. It concentrates on the underlying ideas and paints the big picture, whenever possible without equations. The second part of the chapter, Quantitative Tools, then develops the mathematical framework.

Deductive approach; focus on ideas before names and equations. To the extent possible, this text develops arguments deductively, starting from observations, rather than stating principles and then "deriving" them. This approach makes the material easier to assimilate for students. In the same vein, this text introduces and explains each idea before giving it a formal name or mathematical definition.

Stronger connection to experiment and experience. Physics stems from observations, and this text is structured so that it can do the same. As much as possible, I develop the material from experimental observations (and preferably those that students can make) rather than assertions. Most chapters use actual data in developing ideas, and new notions are always introduced by going from the specific to the general— whenever possible by interpreting everyday examples.

By contrast, standard texts often introduce laws in their most general form and then show that these laws are consistent with specific (and often highly idealized) cases. Consequently the world of physics and the "real" world remain two different things in the minds of students.

Addressing physical complications. I also strongly oppose presenting unnatural situations; real life complications must always be confronted head-on. For example, the use of unphysical words like *frictionless* or *massless* sends a message to the students that physics is unrealistic or, worse, that the world of physics and the *real* world are unrelated entities. This can easily be avoided by pointing out that friction or mass may be neglected under certain circumstances and pointing out *why* this may be done.

Engaging the student. Education is more than just transfer of information. Engaging the student's mind so the information can be assimilated is essential. To this end, the text is written as a dialog between author and reader (often invoking the reader—*you*—in examples) and is punctuated by Checkpoints—questions that require the reader to stop and think. The text following a Checkpoint often refers directly to its conclusions. Students will find complete solutions to all the Checkpoints at the back of the book; these solutions are written to emphasize physical reasoning and discovery.

Visualization. Visual representations are central to physics, so I developed each chapter by designing the figures before writing the text. Many figures use multiple representations to help students make connections (for example, a sketch may be combined with a graph and a bar diagram). Also, in accordance with research, the illustration style is spare and simple, putting the emphasis on the ideas and relationships rather than on irrelevant details. The figures do not use perspective unless it is needed, for instance.

Structure of this text

Division into *Principles* and *Practice* texts

I've divided this text into a *Principles* text, which teaches the physics, and a *Practice* text, which puts the physics into practice and develops problem-solving skills. This division helps address two separate intellectually demanding tasks: understanding the physics and learning to solve problems. When these two tasks are mixed together, as they are in standard texts, students are easily overwhelmed. Consequently many students focus disproportionately on worked examples and procedural knowledge, at the expense of the physics.

Structure of *Principles* chapters

As pointed out earlier, each *Principles* chapter is divided into two parts. The first part (Concepts) develops the conceptual framework in an accessible way, relying primarily on qualitative descriptions and illustrations. In addition to including Checkpoints, each Concepts section ends with a one-page Self-quiz consisting of qualitative questions.

The second part of each chapter (Quantitative Tools) formalizes the ideas developed in the first part in mathematical terms. While concise, it is relatively traditional in nature—teachers should be able to continue to use material developed for earlier courses. To avoid creating the impression that equations are more important than the concepts behind them, no equations are highlighted or boxed.

Both parts of the *Principles* chapters contain worked examples to help students develop problem-solving skills.

Structure of the *Practice* chapters

This text contains material to put into practice the concepts and principles developed in the corresponding chapters in the *Principles* text. Each chapter contains the following sections:

1. *Chapter Summary.* This section provides a brief tabular summary of the material presented in the corresponding *Principles* chapter.
2. *Review Questions.* The goal of this section is to allow students to quickly review the corresponding *Principles* chapter. The questions are straightforward one-liners starting with "what" and "how" (rather than "why" or "what if").
3. *Developing a Feel.* The goals of this section are to develop a quantitative feel for the quantities introduced in the chapter; to connect the subject of the chapter to the real world; to train students in making estimates and assumptions; to bolster students' confidence in dealing with unfamiliar material. It can be used for self-study or for a homework or recitation assignment. This section, which has no equivalent in existing books, combines a number of ideas (specifically, Fermi problems and tutoring in the style of the *Princeton Learning Guide*). The idea is to start with simple estimation problems and then build up to Fermi problems (in early chapters Fermi problems are hard to compose because few concepts have been introduced). Because students initially find these questions hard, the section provides many hints, which take the form of questions. A key then provides answers to these "hints."
4. *Worked and Guided Problems.* This section contains complex worked examples whose primary goal is to teach problem solving. The Worked Problems are fully solved; the Guided Problems have a list of questions and suggestions to help the student think about how to solve the problem. Typically, each Worked Problem is followed by a related Guided Problem.
5. *Questions and Problems.* This is the chapter's problem set. The problems 1) offer a range of levels, 2) include problems relating to client disciplines (life sciences, engineering, chemistry, astronomy, etc.), 3) use the second person as much as possible to draw in the student, and 4) do not spoon-feed the students with information and unnecessary diagrams. The problems are classified into three levels as follows: (•) application of single concept; numerical plug-and-chug; (••) nonobvious application of single concept or application of multiple concepts from current chapter; straightforward numerical or algebraic computation; (•••) application of multiple concepts, possibly spanning multiple chapters. Context-rich problems are designated CR.

As I was developing and class-testing this book, my students provided extensive feedback. I have endeavored to incorporate all of their feedback to make the book as useful as possible for future generations of students. In addition, the book was class-tested at a large number of institutions, and many of these institutions have reported significant increases in learning gains after switching to this manuscript. I am confident the book will help increase the learning gains in your class as well. It will help you, as the instructor, coach your students to be the best they can be.

Instructor supplements

The **Instructor Resource DVD** (ISBN 978-0-321-56175-6/0-321-56175-9) includes an Image Library, the Procedure and special topic boxes from *Principles*, and a library of presentation applets from **ActivPhysics**, PhET simulations, and PhET Clicker Questions. **Lecture Outlines** with embedded **Clicker Questions in PowerPoint®** are provided, as well as the *Instructor's Guide* and *Instructor's Solutions Manual*.

The *Instructor's Guide* (ISBN 978-0-321-94993-6/0-321-94993-5) provides chapter-by-chapter ideas for lesson planning using *Principles & Practice of Physics* in class, including strategies for addressing common student difficulties.

The *Instructor's Solutions Manual* (ISBN 978-0-321-95053-6/0-321-95053-4) is a comprehensive solutions manual containing complete answers and solutions to all Developing a Feel questions, Guided Problems, and Questions and Problems from the *Practice* text. The solutions to the Guided Problems use the book's four-step problem-solving strategy (Getting Started, Devise Plan, Execute Plan, Evaluate Result).

MasteringPhysics® is the leading online homework, tutorial, and assessment product designed to improve results by helping students quickly master concepts. Students benefit from self-paced tutorials that feature specific wrong-answer feedback, hints, and a wide variety of educationally effective content to keep them engaged and on track. Robust diagnostics and unrivalled gradebook reporting allow instructors to pinpoint the weaknesses and misconceptions of a student or class to provide timely intervention.

MasteringPhysics enables instructors to:

- Easily assign **tutorials** that provide individualized coaching.
- Mastering's hallmark **Hints** and **Feedback** offer scaffolded instruction similar to what students would experience in an office hour.

- **Hints** (declarative and Socratic) can provide problem-solving strategies or break the main problem into simpler exercises.
- **Feedback** lets the student know precisely what misconception or misunderstanding is evident from their answer and offers ideas to consider when attempting the problem again.

Learning Catalytics™ is a "bring your own device" student engagement, assessment, and classroom intelligence system available within MasteringPhysics. With Learning Catalytics you can:

- Assess students in real time, using open-ended tasks to probe student understanding.
- Understand immediately where students are and adjust your lecture accordingly.
- Improve your students' critical-thinking skills.
- Access rich analytics to understand student performance.
- Add your own questions to make Learning Catalytics fit your course exactly.
- Manage student interactions with intelligent grouping and timing.

The **Test Bank** (ISBN 978-0-130-64688-0/0-130-64688-1) contains more than 2000 high-quality problems, with a range of multiple-choice, true-false, short-answer, and conceptual questions correlated to *Principles & Practice of Physics* chapters. Test files are provided in both TestGen® and Microsoft® Word for Mac and PC.

Instructor supplements are available on the Instructor Resource DVD, the Instructor Resource Center at www. pearsonhighered.com/irc, and in the Instructor Resource area at www.masteringphysics.com.

Student supplements

MasteringPhysics (www.masteringphysics.com) is designed to provide students with customized coaching and individualized feedback to help improve problem-solving skills. Students complete homework efficiently and effectively with tutorials that provide targeted help.

Interactive eText allows you to highlight text, add your own study notes, and review your instructor's personalized notes, 24/7. The eText is available through MasteringPhysics, www.masteringphysics.com.

Eric Mazur
🐦 @eric_mazur #MazurText
mazur@harvard.edu
Cambridge, MA

Acknowledgments

This book would not exist without the contributions from many people. It was Tim Bozik, currently President, Higher Education at Pearson plc, who first approached me about writing a physics textbook. If it wasn't for his persuasion and his belief in me, I don't think I would have ever undertaken the writing of a textbook. Tim's suggestion to develop the art electronically also had a major impact on my approach to the development of the visual part of this book.

Albert Altman pointed out Ernst Mach's approach to developing mechanics starting with the law of conservation of momentum. Al encouraged me throughout the years as I struggled to reorganize the material around the conservation principles.

I am thankful to Irene Nunes, who served as Development Editor through several iterations of the manuscript. Irene forced me to continuously rethink what I had written and her insights in physics kept surprising me. Her incessant questioning taught me that one doesn't need to be a science major to obtain a deep understanding of how the world around us works and that it is possible to explain physics in a way that makes sense for non-physics majors.

Catherine Crouch helped write the final chapters of electricity and magnetism and the chapters on circuits and optics, permitting me to focus on the overall approach and the art program. Peter Dourmashkin helped me write the chapters on special relativity and thermodynamics. Without his help, I would not have been able to rethink how to introduce the ideas of modern physics in a consistent way.

Many people provided feedback during the development of the manuscript. I am particularly indebted to the late Ronald Newburgh and to Edward Ginsberg, who meticulously checked many of the chapters. I am also grateful to Edwin Taylor for his critical feedback on the special relativity chapter and to my colleague Gary Feldman for his suggestions for improving that chapter.

Lisa Morris provided material for many of the Self-quizzes and my graduate students James Carey, Mark Winkler, and Ben Franta helped with data analysis and the appendices. I would also like to thank my uncle, Erich Lessing, for letting me use some of his beautiful pictures as chapter openers.

Many people helped put together the *Practice* book. Without Daryl Pedigo's hard work authoring and editing content, as well as coordinating the contributions to that book, the manuscript would never have taken shape. Along with Daryl, the following people provided the material for the *Practice* book: Wayne Anderson, Bill Ashmanskas, Linda Barton, Ronald Bieniek, Michael Boss, Anthony Buffa, Catherine Crouch, Peter Dourmashkin, Paul Draper, Andrew Duffy, Edward Ginsberg, William Hogan, Gerd Kortemeyer, Rafael Lopez-Mobilia, Christopher Porter, David Rosengrant, Gay Stewart, Christopher Watts, Lawrence Weinstein, Fred Wietfeldt, and Michael Wofsey.

I would also like to thank the editorial and production staff at Pearson. Margot Otway helped realize my vision for the art program. Martha Steele and Beth Collins made sure the production stayed on track. In addition, I would like to thank Frank Chmely for his meticulous accuracy checking of the manuscript. I am indebted to Jim Smith and Becky Ruden for supporting me through the final stages of this process and to Carol Trueheart, Alison Reeves, and Christian Botting of Prentice Hall for keeping me on track during the early stages of the writing of this book. Finally, I am grateful to Will Moore for his enthusiasm in developing the marketing program for this book.

I am also grateful to the participants of the NSF Faculty Development Conference "Teaching Physics Conservation Laws First" held in Cambridge, MA, in 1997. This conference helped validate and cement the approach in this book.

In between these two stages, Carol Trueheart helped keep me motivated and facilitated the transition from Prentice Hall to Pearson.

Finally, I am indebted to the hundreds of students in Physics 1, Physics 11, and Applied Physics 50 who used early versions of this text in their course and provided the feedback that ended up turning my manuscript into a text that works not just for instructors but, more importantly, for students.

Reviewers of *Principles & Practice of Physics*

Over the years many people reviewed and class-tested the manuscript. The author and publisher are grateful for all of the feedback the reviewers provided, and we apologize if there are any names on this list that have been inadvertently omitted.

Edward Adelson, *Ohio State University*
Albert Altman, *University of Massachusetts, Lowell*
Susan Amador Kane, *Haverford College*
James Andrews, *Youngstown State University*
Arnold Arons, *University of Washington*
Robert Beichner, *North Carolina State University*
Bruce Birkett, *University of California, Berkeley*
David Branning, *Trinity College*
Bernard Chasan, *Boston University*
Stéphane Coutu, *Pennsylvania State University*
Corbin Covault, *Case Western Reserve University*
Catherine Crouch, *Swarthmore College*
Paul D'Alessandris, *Monroe Community College*
Paul Debevec, *University of Illinois at Urbana-Champaign*
N. John DiNardo, *Drexel University*
Margaret Dobrowolska-Furdyna, *Notre Dame University*
Paul Draper, *University of Texas, Arlington*
David Elmore, *Purdue University*
Robert Endorf, *University of Cincinnati*
Thomas Furtak, *Colorado School of Mines*
Ian Gatland, *Georgia Institute of Technology*
J. David Gavenda, *University of Texas, Austin*
Edward Ginsberg, *University of Massachusetts, Boston*
Gary Gladding, *University of Illinois*
Christopher Gould, *University of Southern California*
Victoria Greene, *Vanderbilt University*
Benjamin Grinstein, *University of California, San Diego*
Kenneth Hardy, *Florida International University*
Gregory Hassold, *Kettering University*
Peter Heller, *Brandeis University*
Laurent Hodges, *Iowa State University*
Mark Holtz, *Texas Tech University*
Zafar Ismail, *Daemen College*
Ramanathan Jambunathan, *University of Wisconsin Oshkosh*
Brad Johnson, *Western Washington University*
Dorina Kosztin, *University of Missouri Columbia*
Arthur Kovacs, *Rochester Institute of Technology* (deceased)
Dale Long, *Virginia Polytechnic Institute* (deceased)

John Lyon, *Dartmouth College*
Trecia Markes, *University of Nebraska, Kearney*
Peter Markowitz, *Florida International University*
Bruce Mason, *University of Oklahoma*
John McCullen, *University of Arizona*
James McGuire, *Tulane University*
Timothy McKay, *University of Michigan*
Carl Michal, *University of British Columbia*
Kimball Milton, *University of Oklahoma*
Charles Misner, *University of Maryland, College Park*
Sudipa Mitra-Kirtley, *Rose-Hulman Institute of Technology*
Delo Mook, *Dartmouth College*
Lisa Morris, *Washington State University*
Edmund Myers, *Florida State University*
Alan Nathan, *University of Illinois*
K.W. Nicholson, *Central Alabama Community College*
Fredrick Olness, *Southern Methodist University*
Dugan O'Neil, *Simon Fraser University*
Patrick Papin, *San Diego State University*
George Parker, *North Carolina State University*
Claude Penchina, *University of Massachusetts, Amherst*
William Pollard, *Valdosta State University*
Amy Pope, *Clemson University*
Joseph Priest, *Miami University* (deceased)
Joel Primack, *University of California, Santa Cruz*
Rex Ramsier, *University of Akron*
Steven Rauseo, *University of Pittsburgh*
Lawrence Rees, *Brigham Young University*
Carl Rotter, *West Virginia University*
Leonard Scarfone, *University of Vermont*
Michael Schatz, *Georgia Institute of Technology*
Cindy Schwarz, *Vassar College*
Hugh Scott, *Illinois Institute of Technology*
Janet Segar, *Creighton University*
Shahid Shaheen, *Florida State University*
David Sokoloff, *University of Oregon*
Gay Stewart, *University of Arkansas*
Roger Stockbauer, *Louisiana State University*
William Sturrus, *Youngstown State University*
Carl Tomizuka, *University of Illinois*
Mani Tripathi, *University of California–Davis*
Rebecca Trousil, *Skidmore College*
Christopher Watts, *Auburn University*
Robert Weidman, *Michigan Technological University*
Ranjith Wijesinghe, *Ball State University*
Augden Windelborn, *Northern Illinois University*

Detailed Contents

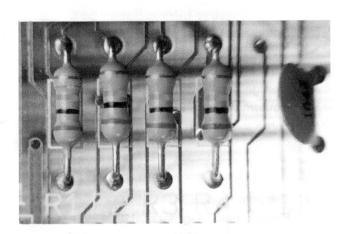

1 Foundations

CONCEPTS

QUANTITATIVE TOOLS

Chances are you are taking this course in physics because someone told you to take it, and it may not be clear to you *why* you should be taking it. One good reason for taking a physics course is that, first and foremost, physics provides a fundamental understanding of the world. Furthermore, whether you are majoring in psychology, engineering, biology, physics, or something else, this course offers you an opportunity to sharpen your reasoning skills. Knowing physics means becoming a better problem solver (and I mean *real* problems, not textbook problems that have already been solved), and becoming a better problem solver is empowering: It allows you to step into unknown territory with more confidence. Before we embark on this exciting journey, let's map out the territory we are going to explore so that you know where we are going.

1.1 The scientific method

Physics, from the Greek word for "nature," is commonly defined as the study of matter and motion. Physics is about discovering the wonderfully simple unifying patterns that underlie absolutely everything that happens around us, from the scale of subatomic particles, to the microscopic world of DNA molecules and cells, to the cosmic scale of stars, galaxies, and planets. Physics deals with atoms and molecules; gases, solids, and liquids; everyday objects, and black holes. Physics explores motion, light, and sound; the creation and annihilation of matter; evaporation and melting; electricity and magnetism. Physics is all around you: in the Sun that provides your daylight, in the structure of your bones, in your computer, in the motion of a ball you throw. In a sense, then, physics is the study of all there is in the universe. Indeed, biology, engineering, chemistry, astronomy, geology, and so many other disciplines you might name all use the principles of physics.

The many remarkable scientific accomplishments of ancient civilizations that survive to this day testify to the fact that curiosity about the world is part of human nature. Physics evolved from *natural philosophy*—a body of knowledge accumulated in ancient times in an attempt to explain the behavior of the universe through philosophical speculation—and became a distinct discipline during the scientific revolution that began in the 16th century. One of the main changes that occurred in that century was the development of the **scientific method,** an iterative process for going from observations to validated theories.

In its simplest form, the scientific method works as follows (**Figure 1.1**): A researcher makes a number of observations concerning either something happening in the natural world (a volcano erupting, for instance) or something happening during a laboratory experiment (a dropped brick and a dropped Styrofoam peanut travel to the floor at different speeds). These observations then lead the researcher to formulate a **hypothesis,** which is a tentative explanation of the observed phenomenon. The hypothesis is used to predict

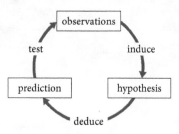

Figure 1.1 The scientific method is an iterative process in which a hypothesis, which is inferred from observations, is used to make a prediction, which is then tested by making new observations.

the outcome of some related natural occurrence (how a similarly shaped mountain near the erupting volcano will behave) or related laboratory experiment (what happens when a book and a sheet of paper are dropped at the same time). If the predictions prove inaccurate, the hypothesis must be modified. If the predictions prove accurate in test after test, the hypothesis is elevated to the status of either a **law** or a **theory.**

A law tells us *what* happens under certain circumstances. Laws are usually expressed in the form of relationships between observable quantities. A theory tells us *why* something happens and explains phenomena in terms of more basic processes and relationships. A scientific theory is not a mere conjecture or speculation. It is a thoroughly tested explanation of a natural phenomenon, one that is capable of making predictions that can be verified by experiment. The constant testing and retesting are what make the scientific method such a powerful tool for investigating the universe: The results obtained must be repeatable and verifiable by others.

Exercise 1.1 Hypothesis or not?

Which of the following statements are hypotheses? (*a*) Heavier objects fall to Earth faster than lighter ones. (*b*) The planet Mars is inhabited by invisible beings that are able to elude any type of observation. (*c*) Distant planets harbor forms of life. (*d*) Handling toads causes warts.

SOLUTION (*a*), (*c*), and (*d*). A hypothesis must be experimentally verifiable. (*a*) I can verify this statement by dropping a heavy object and a lighter one at the same instant and observing which one hits the ground first. (*b*) This statement asserts that the beings on Mars cannot be observed, which precludes any experimental verification and means this statement is not a valid hypothesis. (*c*) Although we humans currently have no means of exploring or closely observing distant planets, the statement is in principle testable. (*d*) Even though we know this statement is false, it *is* verifiable and therefore is a hypothesis.

Because of the constant reevaluation demanded by the scientific method, science is not a stale collection of facts but rather a living and changing body of knowledge. More important, any theory or law *always* remains tentative, and the testing never ends. In other words, it is not possible to prove any scientific theory or law to be absolutely true (or even absolutely false). Thus the material you will learn in this book does not represent some "ultimate truth"—it is true only to the extent that it has not been proved wrong.

A case in point is *classical mechanics,* a theory developed in the 17th century to describe the motion of everyday objects (and the subject of most of this book). Although this theory produces accurate results for most everyday phenomena, from balls thrown in the air to satellites orbiting Earth, observations made during the last hundred years have revealed that under certain circumstances, significant deviations from this theory occur. It is now clear that classical mechanics is applicable for only a limited (albeit important) range of phenomena, and new branches of physics— *quantum mechanics* and the theory of *special relativity* among them—are needed to describe the phenomena that fall outside the range of classical mechanics.

The formulation of a hypothesis almost always involves developing a **model,** which is a simplified conceptual representation of some phenomenon. You don't have to be trained as a scientist to develop models. Everyone develops mental models of how people behave, how events unfold, and how things work. Without such models, we would not be able to understand our experiences, decide what actions to take, or handle unexpected experiences. Examples of models we use in everyday life are that door handles and door hinges are on opposite sides of doors and that the + button on a TV remote increases the volume or the channel number. In everyday life, we base our models on whatever knowledge we have, real or imagined, complete or incomplete. In science we must build models based on careful observation and determine ways to fill in any missing information.

Let's look at the iterative process of developing models and hypotheses in physics, with an eye toward determining what skills are needed and what pitfalls are to be avoided (**Figure 1.2**). Developing a scientific hypothesis often begins

with recognizing patterns in a series of observations. Sometimes these observations are direct, but sometimes we must settle for indirect observations. (We cannot directly observe the nucleus of an atom, for instance, but a physicist can describe the structure of the nucleus and its behavior with great certainty and accuracy.) As Figure 1.2 indicates, the patterns that emerge from our observations must often be combined with simplifying assumptions to build a model. The combination of model and assumptions is what constitutes a hypothesis.

It may seem like a shaky proposition to build a hypothesis on assumptions that are accepted without proof, but making these assumptions—*consciously*—is a crucial step in making sense of the universe. All that is required is that, when formulating a hypothesis, we must be aware of these assumptions and be ready to revise or drop them if the predictions of our hypothesis are not validated. We should, in particular, watch out for what are called *hidden assumptions*— assumptions we make without being aware of them. As an example, try answering the following question. (Turn to the final section of the *Principles* volume, "Solutions to checkpoints," for the answer.)

1.1 I have two coins in my pocket, together worth 30 cents. If one of them is not a nickel, what coins are they?

Advertising agencies and magicians are masters at making us fall into the trap of hidden assumptions. Imagine a radio commercial for a new drug in which someone says, "Baroxan lowered my blood pressure tremendously." If you think that sounds good, you have made a number of assumptions without being aware of them—in other words, hidden assumptions. Who says, for instance, that lowering blood pressure "tremendously" is a good thing (dead people have tremendously low blood pressure) or that the speaker's blood pressure was too high to begin with?

Magic, too, involves hidden assumptions. The trick in some magic acts is to make you assume that something happens, often by planting a false assumption in your mind. A magician might ask, "How did I move the ball from here to there?" while in reality he is using two balls. I won't knowingly put false assumptions into *your* mind in this book, but on occasion you and I (or you and your instructor) may unknowingly make different assumptions during a given discussion, a situation that unavoidably leads to confusion and misunderstanding. Therefore it is important that we carefully analyze our thinking and watch for the assumptions that we build into our models.

If the prediction of a hypothesis fails to agree with observations made to test the hypothesis, there are several ways to address the discrepancy. One way is to rerun the test to see if it is reproducible. If the test keeps producing the same result, it becomes necessary to revise the hypothesis, rethink the assumptions that went into it, or reexamine the original observations that led to the hypothesis.

Figure 1.2 Iterative process for developing a scientific hypothesis.

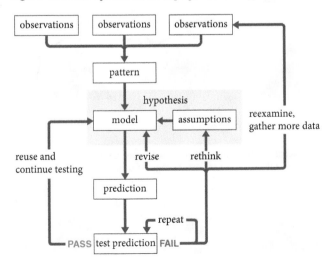

Exercise 1.2 Dead music player

A battery-operated portable music player fails to play when it is turned on. Develop a hypothesis explaining why it fails to play, and then make a prediction that permits you to test your hypothesis. Describe two possible outcomes of the test and what you conclude from the outcomes. (*Think* before you peek at the answer below.)

SOLUTION There are many reasons the player might not turn on. Here is one example. Hypothesis: The batteries are dead. Prediction: If I replace the batteries with new ones, the player should work. Possible outcomes: (1) The player works once the new batteries are installed, which means the hypothesis is supported; (2) the player doesn't work after the new batteries are installed, which means the hypothesis is not supported and must be either modified or discarded.

✋ **1.2** In Exercise 1.2, each of the conclusions drawn from the two possible outcomes contains a hidden assumption. What are the hidden assumptions?

The development of a scientific hypothesis is often more complicated than suggested by Figures 1.1 and 1.2. Hypotheses do not always start with observations; some are developed from incomplete information, vague ideas, assumptions, or even complete guesses. The refining process also has its limits. Each refinement adds complexity, and at some point the complexity outweighs the benefit of the increased accuracy. Because we like to think that the universe has an underlying simplicity, it might be better to scrap the hypothesis and start anew.

Figure 1.2 gives an idea of the skills that are useful in doing science: interpreting observations, recognizing patterns, making and recognizing assumptions, thinking logically, developing models, and using models to make predictions. It should not come as any surprise to you that many of these skills are useful in just about any context. Learning physics allows you to sharpen these skills in a very rigorous way. So, whether you become a financial analyst, a doctor, an engineer, or a research scientist (to name just a few possibilities), there is a good reason to take physics.

Figure 1.1 also shows that doing science—and physics in particular—involves two types of reasoning: *inductive*, which is arguing from the specific to the general, and *deductive*, arguing from the general to the specific. The most creative part of doing physics involves inductive reasoning, and this fact sheds light on how you might want to learn physics. One way, which is neither very useful nor very satisfying, is for me to simply tell you all the general principles physicists presently agree on and then for you to apply those principles in examples and exercises (**Figure 1.3a**). This approach involves deductive reasoning only and robs you of the opportunity to learn the skill that is the most likely to benefit your career: discovering underlying patterns. Another way is for me to present you with data and observations and make you part of the discovery and refinement of the physics principles (Figure 1.3b). This approach

Figure 1.3

(*a*) Learning science by applying established principles

| principles | — apply to → | examples, exercises |

(*b*) Learning science by discovering those principles for yourself before applying them

| observations, data | — discover → | principles | refine

is more time-consuming, and sometimes you may wonder why I'm not just *telling* you the final outcome. The reason is that discovery and refinement are at the heart of doing physics!

✋ **1.3** After reading this section, reflect on your goals for this course. Write down what you would like to accomplish and why you would like to accomplish this. Once you have done that, turn to the final section of the *Principles* volume, "Solutions to checkpoints," and compare what you have written with what I wrote.

1.2 Symmetry

One of the basic requirements of any law of the universe involves what physicists call *symmetry*, a concept often associated with order, beauty, and harmony. We can define **symmetry** as follows: An object exhibits symmetry when certain operations can be performed on it without changing its appearance. Consider the equilateral triangle in **Figure 1.4a**. If you close your eyes and someone rotates the triangle by 120° while you have your eyes closed, the triangle appears

Figure 1.4

(*a*) Rotational symmetry: Rotating an equilateral triangle by 120° doesn't change how it looks

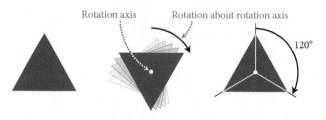

(*b*) Reflection symmetry: Across each reflection axis (labeled R), two sides of the triangle are mirror images of each other

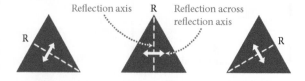

Figure 1.5 The symmetrical arrangement of atoms in a salt crystal gives these crystals their cubic shape.

(*a*) Micrograph of salt crystals

(*b*) Symmetrical arrangement of atoms in a salt crystal

(*c*) Da Vinci's Vitruvian Man shows the reflection symmetry of the human body

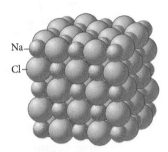

the same when you open your eyes, and you can't tell that it has been rotated. The triangle is said to have *rotational symmetry,* one of several types of geometrical symmetry.

Another common type of geometrical symmetry, *reflection symmetry,* occurs when one half of an object is the mirror image of the other half. The equilateral triangle in Figure 1.4 possesses reflection symmetry about the three axes shown in Figure 1.4*b*. If you imagine folding the triangle in half over each axis, you can see that the two halves are identical. Reflection symmetry occurs all around us: in the arrangement of atoms in crystals (**Figure 1.5*a*** and *b*) and in the anatomy of most life forms (Figure 1.5*c*), to name just two examples.

The ideas of symmetry—that something appears unchanged under certain operations—apply not only to the shape of objects but also to the more abstract realm of physics. If there are things we can do to an experiment that leave the result of the experiment unchanged, then the phenomenon tested by the experiment is said to possess certain symmetries. Suppose we build an apparatus, carry out a certain measurement in a certain location, then move the apparatus to another location, repeat the measurement, and get the same result in both locations.* By moving the apparatus to a new location (*translating* it) and obtaining the same result, we have shown that the observed phenomenon has *translational symmetry.* Any physical law that describes this phenomenon must therefore mathematically exhibit translational symmetry; that is, the mathematical expression of this law must be independent of the location.

Likewise, we expect any measurements we make with our apparatus to be the same at a later time as at an earlier time; that is, translation in time has no effect on the measurements. The laws describing the phenomenon we are

studying must therefore mathematically exhibit symmetry under translation in time; in other words, the mathematical expression of these laws must be independent of time.

Exercise 1.3 Change is no change

Figure 1.6 shows a snowflake. Does the snowflake have rotational symmetry? If yes, describe the ways in which the flake can be rotated without changing its appearance. Does it have reflection symmetry? If yes, describe the ways in which the flake can be split in two so that one half is the mirror image of the other.

Figure 1.6 Exercise 1.3.

SOLUTION I can rotate the snowflake by 60° or a multiple of 60° (120°, 180°, 240°, 300°, and 360°) in the plane of the photograph without changing its appearance (**Figure 1.7*a***). It therefore has rotational symmetry.

I can also fold the flake in half along any of the three blue axes and along any of the three red axes in Figure 1.7*b*. The flake therefore has reflection symmetry about all six of these axes.

Figure 1.7

(*a*) Rotational symmetry

60°

(*b*) Reflection symmetry

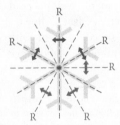

---

*In moving our apparatus, we must take care to move any relevant external influences along with it. For example, if Earth's gravity is of importance, then moving the apparatus to a location in space far from Earth does not yield the same result.

A number of such symmetries have been identified, and the basic laws that govern the inner workings of the physical world must reflect these symmetries. Some of these symmetries are familiar to you, such as translational symmetry in space or time. Others, like electrical charge or parity symmetry, are unfamiliar and surprising and go beyond the scope of this course. Whereas symmetry has always implicitly been applied to the description of the universe, it plays an increasingly important role in physics: In a sense the quest of physics in the 21st century is the search for (and test of) symmetries because these symmetries are the most fundamental principles that all physical laws must obey.

1.4 You always store your pencils in a cylindrical case. One day while traveling in the tropics, you discover that the cap, which you have placed back on the case day in, day out for years, doesn't fit over the case. What do you conclude?

1.3 Matter and the universe

The goal of physics is to describe all that happens in the universe. Simply put, the **universe** is the totality of matter and energy combined with the space and time in which all events happen—everything that is directly or indirectly observable. To describe the universe, we use *concepts,* which are ideas or general notions used to analyze natural phenomena.* To provide a quantitative description, these concepts must be expressed quantitatively, which requires defining a procedure for measuring them. Examples are the length or mass of an object, temperature, and time intervals. Such **physical quantities** are the cornerstones of physics. It is the accurate measurement of physical quantities that has led to the great discoveries of physics. Although many of the fundamental concepts we use in this book are familiar ones, quite a few are difficult to define in words, and we must often resort to defining these concepts in terms of the procedures used to measure them.

The fundamental physical quantity by which we map out the universe is **length**—a distance or an extent in space. The length of a straight or curved line is measured by comparing the length of the line with some standard length. In 1791, the French Academy of Sciences defined the standard unit for length, called the **meter** and abbreviated m, as one ten-millionth of the distance from the equator to the North Pole. For practical reasons, the standard was redefined in 1889 as the distance between two fine lines engraved on a bar of platinum-iridium alloy kept at the International Bureau of Weights and Measures near Paris. With the advent of lasers, however, it became possible to measure the speed of light with extraordinary accuracy, and so the meter was redefined in 1983 as the distance traveled by light in vacuum in a time interval of 1/299,792,458 of a second. This number is chosen so as to make the speed of light exactly 299,792,458 meters per second and yield a standard length for the meter that is very close to the length of the original platinum-iridium standard. This laser-based standard is final and will never need to be revised.

1.5 Based on the early definition of the meter, one ten-millionth of the distance from the equator to the North Pole, what is Earth's radius?

Now that we have defined a standard for length, let us use this standard to discuss the structure and size scales of the universe. Because of the extraordinary range of size scales in the universe, we shall round off any values to the nearest power of ten. Such a value is called an **order of magnitude.** For example, any number between 0.3 and 3 has an order of magnitude of 1 because it is within a factor of 3 of 1; any number greater than 3 and equal to or less than 30 has an order of magnitude of 10. You determine the order of magnitude of any quantity by writing it in scientific notation and rounding the coefficient in front of the power of ten to 1 if it is equal to or less than 3 or to 10 if it is greater than 3.[†] For example, 3 minutes is 180 s, which can be written as 1.8×10^2 s. The coefficient, 1.8, rounds to 1, and so the order of magnitude is 1×10^2 s $= 10^2$ s. The quantity 680, to take another example, can be written as 6.8×10^2; the coefficient 6.8 rounds to 10, and so the order of magnitude is $10 \times 10^2 = 10^3$. And Earth's circumference is 40,000,000 m, which can be written as 4×10^7 m; the order of magnitude of this number is 10^8 m. You may think that using order-of-magnitude approximations is not very scientific because of the lack of accuracy, but the ability to work effectively with orders of magnitude is a key skill not just in science but also in any other quantitative field of endeavor.

All ordinary matter in the universe is made up of basic building blocks called **atoms** (Figure 1.8). Nearly all the matter in an atom is contained in a dense central nucleus, which consists of **protons** and **neutrons,** two types of subatomic particles. A tenuous *cloud* of **electrons,** a third type of subatomic particle, surrounds this nucleus. Atoms are spherical and have a diameter of about 10^{-10} m. Atomic nuclei are also spherical, with a diameter of about 10^{-15} m, making atoms mostly empty space. Atoms attract one another when they are a small distance apart but resist being squeezed into one another. The arrangement of atoms in a material determines the properties of the material.

Figure 1.9 shows the relative size of some representative objects in the universe. The figure reveals a lot about the organization of matter in the universe and serves as a visual model of the structure of the universe. Roughly

---

*When an important concept is introduced in this book, the main word pertaining to the concept is printed in **boldface type.** All important concepts introduced in a chapter are listed at the end of the chapter, in the Chapter Glossary.

[†]The reason we use 3 in order-of-magnitude rounding, and not 5 as in ordinary rounding, is that orders of magnitude are logarithmic, and on this logarithmic scale log 3 = 0.48 lies nearly halfway between log 1 = 0 and log 10 = 1.

Figure 1.8 Scanning tunneling microscope image showing the individual atoms that make up a silicon surface. The size of each atom is about 1/50,000 the width of a human hair.

Figure 1.9 A survey of the size and structure of the universe.

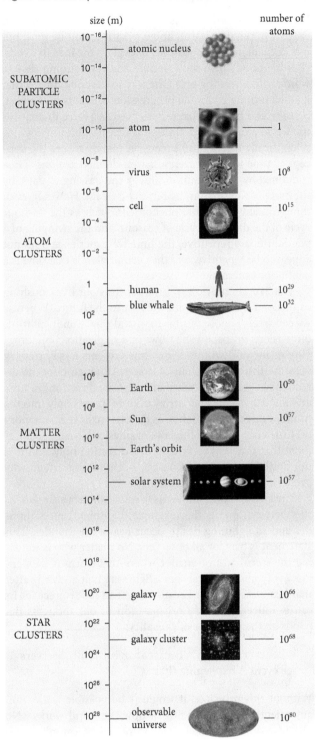

speaking, there is clustering of matter from smaller to larger at four length scales. At the subatomic scale, most of the matter in an atom is compressed into the tiny atomic nucleus, a cluster of subatomic particles. Atoms, in turn, cluster to form the objects and materials that surround us, from viruses to plants, animals, and other everyday objects. The next level is the clustering of matter in stars, some of which, such as the Sun, are surrounded by planets like Earth. Stars, in turn cluster to form galaxies. As we shall discuss in Chapter 7, this clustering of matter reveals a great deal about the way different objects interact with one another.

Exercise 1.4 Tiny universe

If all the matter in the observable universe were squeezed together as tightly as the matter in the nucleus of an atom, what order of magnitude would the diameter of the universe be?

SOLUTION From Figure 1.9 I see that there are about 10^{80} atoms in the universe. I can arrange these atoms in a cube that has 10^{27} atoms on one side because such a cube could accommodate $10^{27} \times 10^{27} \times 10^{27} = 10^{81}$ atoms. Given that the diameter of a nucleus is about 10^{-15} m, the length of a side of this cube would be

$$(10^{27} \text{ atoms})(10^{-15} \text{ m per atom}) = 10^{12} \text{ m,}$$

which is a bit larger than the diameter of Earth's orbit around the Sun.

An alternative method for obtaining the answer is to realize that the matter in a single nucleus occupies a cubic volume of about $(10^{-15} \text{ m})^3 = 10^{-45} \text{ m}^3$. If all the matter in the universe were squeezed together just as tightly, it would occupy a volume of about 10^{80} times the volume of an atomic nucleus, or $10^{80} \times 10^{-45} \text{ m}^3 = 10^{35} \text{ m}^3$. The side of a cube of this volume is equal to the cube root of 10^{35} m^3, or 4.6×10^{11} m, which is the same order of magnitude as my first answer.

1.6 Imagine magnifying each atom in an apple to the size of the apple. What would the diameter of the apple then be?

1.4 Time and change

Profound and mysterious, time is perhaps the greatest enigma in physics. We all know what is meant by *time,* but it is difficult, if not impossible, to explain the idea in words. (Put the book down for a minute and try defining *time* in words before reading on.) One way to describe time is that it is the infinite continual progression of events in the past, present, and future, often experienced as a force that moves the world along. This definition is neither illuminating nor scientifically meaningful because it merely relates the concept of time to other, even less well-defined notions. Time is defined by the rhythm of life, by the passing of days, by the cycle of the seasons, by birth and death. However, even though many individual phenomena, such as the 24-hour cycle of the days, the cycle of seasons, and the swinging of a pendulum, are repetitive, the time we experience does not appear to be repetitive, and the current view is that time is a continuous succession of events.

The irreversible flow of time controls our lives, pushing us inexorably forward from the past to the future. Whereas we can freely choose our location and direction in all three dimensions in space, time flows in a single direction, dragging us forward with it. Time thus presents less symmetry than the three dimensions of space: Although opposite directions in space are equivalent, opposite directions in time are not equivalent. The "arrow of time" points only into the future, a direction we define as the one we have no memory of. Curiously, most of the laws of physics have no requirement that time has to flow in one direction only, and it is not until Chapter 19 that we can begin to understand why events in time are irreversible.

The arrow of time allows us to establish a *causal relationship* between events. For example, lightning causes thunder and so lightning has to occur *before* the thunder. This statement is true for all observers: No matter who is watching the storm and no matter where that storm is happening, every observer first sees a lightning bolt and only after that hears the thunder because an effect never precedes its cause. Indeed, the very organization of our thoughts depends on the **principle of causality:**

> Whenever an event A causes an event B, all observers see event A happening first.

Without this principle, it wouldn't be possible to develop any scientific understanding of how the world works. (No physics course to take!) The principle of causality also makes it possible to state a definition: **Time** is a physical quantity that allows us to determine the sequence in which related events occur.

To apply the principle of causality and sort out causes and effects, it is necessary to develop devices—clocks—for keeping track of time. All clocks operate on the same principle: They repeatedly return to the same state. The rotation of Earth about its axis can serve as a clock if we note the instant the Sun reaches its highest position in the sky on successive days. A swinging pendulum, which repeatedly returns to the same vertical position, can also serve as a clock. The time interval between two events can be determined by counting the number of pendulum swings between the events. The accuracy of time measurements can be increased by using a clock that has a large number of repetitions in a given time interval.

1.7 (*a*) State a possible cause for the following events: (*i*) The light goes out in your room; (*ii*) you hear a loud, rumbling noise; (*iii*) a check you wrote at the bookstore bounces. (*b*) Could any of the causes you named have occurred after their associated event? (*c*) Describe how you feel when you experience an event but don't know what caused it—you hear a strange noise when camping, for instance, or an unexpected package is sitting on your doorstep.

The familiar standard unit for measuring time is the **second** (abbreviated s), originally defined as $1/86{,}400$ of a day but currently more accurately defined as the duration of 9,192,631,770 periods of certain radiation emitted by cesium atoms. **Figure 1.10** gives an idea of the vast range of time scales in the universe.

The English physicist Isaac Newton stated, "Absolute, true, and mathematical time, of itself and from its own nature, flows equably without relation to anything external." In other words, the notion of past, present, and future is universal—"now" for you, wherever you are at this instant, is also "now" everywhere else in the universe. Although this notion of the universality of time, which is given the name **absolute time,** is intuitive, experiments described in Chapter 14 have shown this notion to be false. Still, for many experiments and for the material we discuss in most of this book, the notion of absolute time remains an excellent approximation.

Now that we have introduced space and time, we can use these concepts to study events. Throughout this book, we focus on **change,** the transition from one state to another. Examples of change are the melting of an ice cube, motion (a change in location), the expansion of a piece of metal, the flow of a liquid. As you will see, one might well call physics the study of the changes that surround us and convey the passage of time. What is most remarkable about all this is that we shall discover that underneath all the changes we'll look at, certain properties remain *unchanged.* These properties give rise to what are called *conservation laws,* the most fundamental and universal laws of physics.

There is a profound aesthetic appeal in knowing that symmetry and conservation are the cornerstones of the laws that govern the universe. It is reassuring to know that an elegant simplicity underlies the structure of the universe and the relationship between space and time.

1.8 A single chemical reaction takes about 10^{-13} s. What order of magnitude is the number of sequential chemical reactions that could take place during a physics class?

Figure 1.10 A survey of the time scales on which events in the universe take place.

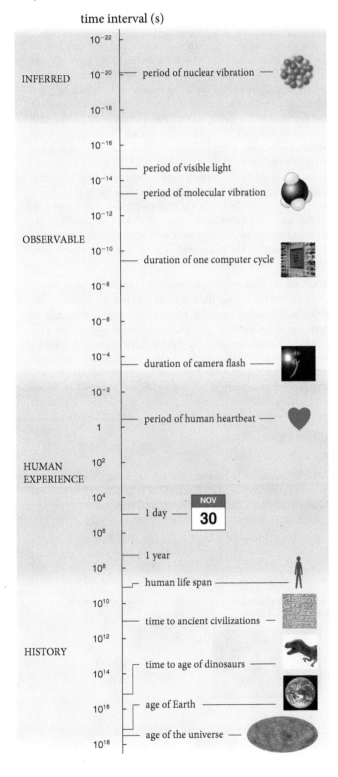

time interval (s)

1.5 Representations

Of all our senses, it is vision that most informs our mind. For this reason, expert problem solvers rarely start working on a problem without first making some sort of visual representation of the available information, and you should always do the same. Such a drawing helps you establish a clear mental image of the situation, interpret the problem in light of your own knowledge and experience, develop a qualitative understanding of the problem, and organize the information in a meaningful way. Without the drawing, you have to juggle all the information in your head. The drawing frees your brain to deal with the solution. As an example, consider the situation described in Checkpoint 1.9. First try solving the problem without a piece of paper and note the mental effort it takes; then make a sketch and work out the solution.

1.9 You and your spouse are working out a seating arrangement around a circular table for dinner with John and Mary Jones, Mike and Sylvia Masters, and Bob and Cyndi Ahlers. Mike is not fond of the Ahlers, and Sylvia asked that she not be seated next to John. You would like to alternate men and women and avoid seating spouses next to each other. Determine an arrangement that satisfies all the constraints.

Were you able to solve the problem without a sketch? Doing so would be difficult because there is more information than most people can comfortably keep in mind at once. When you represent the information visually, however, it is relatively easy to solve this problem (see the solution in the back of this book). The drawing breaks the problem down into small steps and helps you articulate in your mind what you are trying to accomplish.

Visual representations are not helpful only for making seating arrangements. They work just as well for solving physics problems, although it may not immediately be clear *how* to represent the available information. For this reason we shall develop a number of different ways—pictures, sketches, diagrams, graphs—and a number of different context-specific procedures to represent information visually in our study of physics. As you will see, these visual representations are an integral part of getting a grip on a problem and developing a model (Figure 1.11).

As we discussed in Section 1.1, descriptions of the physical world always begin with simplified representations. When you solve physics problems, elaborate drawings clutter your mind with irrelevant information and prevent you from getting a clear view of the important features. One of the most basic skills in physics, therefore, is to decide what to leave out of your drawings. If you leave out essential features, the representation is useless, but if you put in too many details, it becomes impossible to analyze the situation. Sometimes it is necessary to begin with an oversimplification in order to develop a feel for a given situation. Once this initial understanding has been gained, however,

CONCEPTS

Figure 1.11 Multiple representations help you solve problems. (*a*) Many problems start with a verbal representation. (*b*) Turning the words into a sketch helps you to grasp the problem. (*c*) The sketch can then give meaning to a mathematical representation of the problem.

(*a*)

Two collisions are carried out to crash-test a 1000-kg car: (*a*) While moving at 15 mph, the car strikes an identical car initially at rest. (*b*) While moving at 15 mph, the car strikes an identical car moving toward it and also traveling at 15 mph. For each collision, what is the amount of kinetic energy that can be converted to another form in the collision, and what fraction of the total initial (*a*) kinetic energy of the two-car system does this represent?

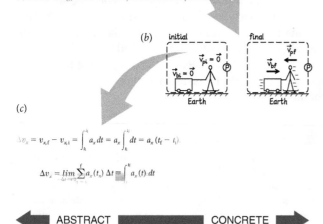

(*c*)

$$\Delta v_x = v_{x,f} - v_{x,i} = \int_{t_i}^{t_f} a_x\, dt = a_x \int_{t_i}^{t_f} dt = a_x\,(t_f - t_i)$$

$$\Delta v_x = \lim_{\Delta t \to 0} \sum_{i}^{f} a_x(t_n)\,\Delta t = \int_{t_i}^{t_f} a_x(t)\, dt$$

◀ ABSTRACT ▬▬▬▬▬ CONCRETE ▶

it becomes possible to construct less idealized models, with each successive model being a more realistic representation of the real-world situation.

In any drawings you make, therefore, you should treat everyday objects as simplifications that can be characterized by a minimum number of features or quantities. Some joke that physicists hold a grossly oversimplified view of the world, thinking in such terms as, for instance, "Consider a spherical cow…." The world around us, cows included, contains infinitely many details that may play a role in the grand scheme of things, but to get a grip on any problem it is important to begin by leaving out as much detail as possible. If the resulting model reproduces the main features of the real world, you know you have taken the essential

attributes into account. As an example, **Figure 1.12** shows a progression from a photograph of a cow to an abstract rendition of it. To study the pattern on the cow's hide, you need the photograph. If you are interested in only the position of a certain cow in a certain pasture, however, a simple dot suffices (Figure 1.12*d*). By reducing the cow to a dot, you have discarded any information about its size and shape, but as you will see as you continue with your study of physics, this information is often not relevant.

Exercise 1.5 Stretching a spring

For a physics laboratory assignment, one end of a spring is attached to a horizontal rod so that the spring hangs vertically, and a ruler is hung vertically alongside the spring. The stretching properties of the spring are to be measured by attaching eight identical beads to the spring's free end. With no beads attached, the free end of the spring is at a ruler reading of 23.4 mm. With one bead attached, the end of the spring drops to 25.2 mm. When the second, third, and fourth beads are attached one at a time, the end drops to ruler readings of 26.5 mm, 29.1 mm, and 30.8 mm, respectively. Adding the fifth and sixth beads together moves the spring end to 34.3 mm, and adding the last two beads moves the end to 38.2 mm. (*a*) Make a pictorial representation of this setup. (*b*) Tabulate the data. (*c*) Plot the data on a graph, showing the ruler readings on the vertical axis and the numbers of beads on the horizontal axis. (*d*) Describe what can be inferred from the data.

SOLUTION (*a*) The important items that should appear in my drawing are the spring, the rod from which it is suspended, the ruler, and at least one bead. My drawing should also indicate how the ruler readings are obtained. I can illustrate this procedure with just one bead, as **Figure 1.13a** shows. By drawing the spring first with no beads attached and then with one bead attached and showing the two ruler readings, I've represented the general procedure of adding beads one (or two) at a time and paying attention to how each addition changes the position of the spring end.

(*b*) See Figure 1.13*b*.

(*c*) See Figure 1.13*c*.

(*d*) The relationship between the ruler readings and the numbers of suspended beads is linear. That is to say, each additional bead stretches the spring by about the same amount.

Figure 1.12 Increasingly simplified and abstract representations of a cow.

(*a*) Photograph   (*b*) Simplified sketch   (*c*) Rectangle: represents cow's position and extent   (*d*) Dot: shows only cow's position

◀ CONCRETE ▬▬▬▬▬▬▬▬▬▬▬▬▬▬▬▬▬▬▬ ABSTRACT ▶

Figure 1.13

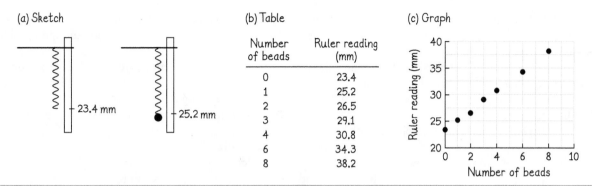

(a) Sketch

23.4 mm

25.2 mm

(b) Table

Number of beads	Ruler reading (mm)
0	23.4
1	25.2
2	26.5
3	29.1
4	30.8
6	34.3
8	38.2

(c) Graph

Exercise 1.5 demonstrates how various ways of representing information help you interpret data. The pictorial representation of Figure 1.13a helps you visualize how the measurements were taken. The table of Figure 1.13b organizes the data, and the graph of Figure 1.13c allows you to recognize the linear relationship between how far the spring stretches and the number of beads suspended from it—something that is not at all obvious from the text or the table.

Like the sketch, the graph is a simplified representation of the stretching of the spring. Each representation involves a loss of information and detail. The sketch is a simplified two-dimensional representation of a three-dimensional setup, and the graph shows only one piece of information for each measurement: the position of the bottom of the spring. All other information is left out in order to reveal one crucial point: How much the spring stretches is proportional to the number of beads suspended from it (something we look at in more detail in Chapter 9).

You may be beginning to wonder what the role of mathematics is in physics, given that we haven't used any thus far. One of the main roles of mathematics in physics is allowing us to express succinctly and unambiguously ideas that if expressed verbally would require a lot of words whose meaning may not be precise. Take this statement, for example:

The magnitude of the acceleration of an object is directly proportional to the magnitude of the vector sum of the forces exerted on the object and inversely proportional to the object's inertia. The direction of the acceleration is the same as the direction of the vector sum of the forces.

(Don't worry about the meaning of this statement for now; just note that it is quite a mouthful.) We can express this statement mathematically as

$$\vec{a} = \frac{\Sigma \vec{F}}{m}.$$

These two expressions—one verbal, the other mathematical—are identical. Neither one is more accurate than the other. Right now the symbols in the equation may make no more sense to you than hieroglyphs. Once you understand their meaning, however, the mathematical expression can be parsed much more quickly than the verbal expression can. In this respect, mathematics plays the same role as visual representations: It relieves the brain of having to keep track of many words. Another important benefit of mathematical representation is that we can use the techniques of mathematics to manipulate the symbols and obtain new insights.

Without an understanding of the meaning of the concepts in any expression (acceleration and force in our example here), however, verbal expressions and mathematical ones are both meaningless, and so it is important to focus first on the meanings of concepts. As you will notice, this book has been designed to develop concepts first and to emphasize the visual representation of these concepts before moving on to a mathematical treatment (see the box "Organization of this book").

1.10 Picture a long, straight corridor running east-west, with a water fountain located somewhere along it. Starting from the west end of the corridor, a woman walks a short distance east along the corridor and stops before reaching the water fountain. The distance from her to the fountain is twice the distance she has walked. She then continues walking east, passes the water fountain, and stops 60 m from her first stop. Now the distance from her to the fountain is twice the distance from her to the east end of the corridor. How long is the corridor?

CONCEPTS

Organization of this book

The material for this course is presented in two volumes. The first volume (the one you are reading now, *Principles*) is aimed at guiding you in developing a solid understanding of the principles of physics. The second volume, *Practice*, provides a broad variety of questions and problems that allow you to apply and sharpen your understanding of physics.

Each chapter in the *Principles* volume is divided into two parts: *Concepts* and *Quantitative Tools*. The *Concepts* part develops the conceptual framework for the subject of the chapter, concentrating on the underlying ideas and helping you develop a mental picture of the subject. The *Quantitative Tools* part develops the mathematical framework, building on the ideas developed in the *Concepts* part. Interspersed in the text of each chapter are several types of learning aids to guide you through the chapter:

1. **Checkpoints ( ).** These questions compel you to test yourself on how well you understand the material you just read. Do not skip them as you read the text. First of all, the answer to a checkpoint may be necessary to understand the text following that checkpoint. Second, and more important, I've put these checkpoints right in the text because working on them means learning the material. The answers to all the checkpoints are at the end of this volume.

2. **Exercises and examples.** The fully-worked-out exercises and examples help you develop and apply problem-solving strategies. It is generally a good idea to attempt to solve the problem by yourself before reading the solution. More information on general approaches to problem solving is given in Section 1.8.

3. **Procedures.** Approaches for analyzing specific situations are given in separate, highlighted boxes.

4. **Self-quiz.** The Self-quiz in each *Principles* chapter, which always comes at the end of the conceptual part of the chapter, allows you to assess your understanding of the concepts before you move on to the quantitative treatment. Complete each Self-quiz before working on the quantitative part of the chapter, even if you are already familiar with most of the material covered. Before tackling the quantitative material, be sure to resolve any difficulties you might have in answering a Self-quiz question, either by rereading the material in the conceptual part of the chapter or by consulting your instructor.

5. **Glossary.** At the end of each chapter is a list defining the important concepts in the chapter (which are the terms printed in bold).

CONCEPTS

Self-quiz

1. Two children in a playground swing on two swings of unequal length. The child on the shorter swing is considerably heavier than the child on the longer swing. You observe that the longer swing swings more slowly. Formulate a hypothesis that could explain your observation. How could you test your hypothesis?

2. What symmetries do you observe in the quilt patterns of **Figure 1.14**?

Figure 1.14

(a) 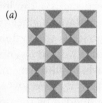   (b) 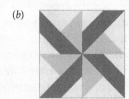   (c) 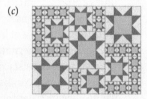   (d)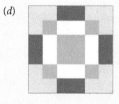

3. Give the order of magnitude of these quantities in meters or seconds: (*a*) length of a football field, (*b*) height of a mature tree, (*c*) one week, (*d*) one year.

4. Starting from the first floor, an elevator stops at floors 5, 2, 4, 3, 6, and 4 before returning to the first floor. (*a*) Represent this motion visually. (*b*) If the distance between the first and sixth floors is 15 m, what is the total distance traveled by the elevator?

Answers

1. One hypothesis is that longer swings swing more slowly than shorter swings. You can test this hypothesis by adjusting the length of either swing until the two lengths are the same and then asking the children to remount their respective swings and swing again. If the originally longer swing is still the slower one, your hypothesis is not correct. If the two swings now have the same speed, your hypothesis is correct. Another hypothesis is that heavier children swing faster than lighter ones. You can test this hypothesis by asking the children to trade places. If the longer swing now swings faster than the shorter swing, your hypothesis is correct. If the longer swing still swings more slowly, your hypothesis is incorrect.

2. See **Figure 1.15**. (*a*) Reflection symmetry about a horizontal line through the center. (*b*) Rotational symmetry by multiples of 90°. (*c*) Rotational symmetry by 180°. (*d*) Rotational symmetry by multiples of 90° and reflection symmetry about a horizontal, vertical, or diagonal line through the center.

Figure 1.15

(a) 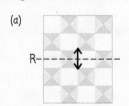   (b) 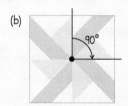   (c) 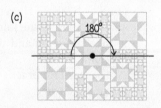   (d)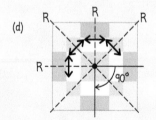

3. (*a*) 100 yards is about 100 m; the order of magnitude is thus 100 m $= 10^2$ m. (*b*) An average mature tree is between 5 and 20 m tall, for an average of 12 m $= 1.2 \times 10^1$ m. The coefficient 1.2 rounds to 1, and so the order of magnitude is 1×10^1 m $= 10$ m. (*c*) 1 week $=$ (1 week)(7 days/week)(24 h/day)(60 min/h)(60 s/min) $=$ 604,800 s $= 6 \times 10^5$ s; the coefficient 6 rounds to 10, and so the order of magnitude is 10×10^5 s $= 10^6$ s. (*d*) 1 year $=$ 52 weeks $=$ (52 weeks)(604,800 s/week) $=$ 31,449,600 s $= 3.1 \times 10^7$ s; the coefficient 3.1 rounds to 10, and so the order of magnitude is 10×10^7 s $= 10^8$ s.

4. (*a*) See **Figure 1.16** for one way to represent the motion. Note that the elevator itself is not represented because showing it would add nothing we need to the visual information. The only thing we are interested in is distances traveled, represented by the vertical lines. (*b*) If the distance between floors 1 and 6 is 15 m, one floor is (15 m)/5 $=$ 3.0 m. From the figure, I see that the numbers of floors traveled between successive stops are 4, 3, 2, 1, 3, 2, and 3, for a total of 18 floors, or 18(3.0 m) $=$ 54 m.

Figure 1.16   floor

Table 1.1 Physical quantities and their symbols

Physical quantity	Symbol
length	ℓ
time	t
mass	m
speed	v
volume	V
energy	E
temperature	T

1.6 Physical quantities and units

Because physics is a quantitative science, statements must be expressed in numbers, which requires either measuring or calculating numerical values for physical quantities. In this section we review some basic rules for dealing with physical quantities, which in this book are represented by italic symbols—typically letters from the Roman or Greek alphabet, such as t for time and σ for electrical conductivity. Table 1.1 gives the symbols for some of the physical quantities we use throughout the book.

Physical quantities are expressed as the product of a numerical value and a unit of measurement. For example, the length ℓ of an object that is 1.2 m long can be expressed as $\ell = 1.2$ m. The unit system used in science and engineering throughout the world and in everyday life in most countries is the Système International (International System), and the units are collectively called **SI units.** This system consists of seven base units (Table 1.2) from which all other units currently in use can be derived. For example, the physical quantity speed, which we discuss in Chapter 2, is defined as the distance traveled divided by the time interval over which the travel takes place. Thus the SI derived unit of speed is meters per second (m/s), the base unit of length divided by the base unit of time. A list of SI derived units and their relationship to the seven base units is given in Appendix C.

Be careful not to confuse abbreviations for units with symbols for physical quantities. Unit abbreviations are printed in roman (upright) type—m for meters, for instance—and symbols for physical quantities are printed in italic (slanted) type—t for time, say. Also, bear in mind that you can add and subtract quantities only if they have the same units; it is meaningless to add, say, 3 m to 4 kg.

To produce multiples of any SI unit and conveniently work with very large or very small numbers, we modify the unit name with prefixes representing integer powers of ten (Table 1.3). For example, a billionth of a second is denoted by 1 ns (pronounced "one nanosecond"):

$$1 \text{ ns} = 10^{-9} \text{ s}. \tag{1.1}$$

One thousand meters is denoted by 1 km, "one kilometer." Prefixes are never used without a unit and are never combined into compound prefixes. The unit *kilogram* contains a prefix (*kilo-*) because it is derived from the non-SI unit *gram* (1 kg = 1000 g). Therefore 10^{-6} kg never becomes 1 μkg. Instead, the names and multiples of the kilogram are constructed by adding the appropriate prefix to the word *gram* and the abbreviation g. For example, 10^{-6} kg becomes 1 mg, "one milligram."

The standard practice in engineering is to use only the powers of ten that are multiples of three.

Table 1.2 The seven SI base units

Name of unit	Abbreviation	Physical quantity
meter	m	length
kilogram	kg	mass
second	s	time
ampere	A	electric current
kelvin	K	thermodynamic temperature
mole	mol	amount of substance
candela	cd	luminous intensity

Table 1.3 SI prefixes

10^n	Prefix	Abbreviation	10^n	Prefix	Abbreviation
10^0	—	—			
10^3	kilo-	k	10^{-3}	milli-	m
10^6	mega-	M	10^{-6}	micro-	μ
10^9	giga-	G	10^{-9}	nano-	n
10^{12}	tera-	T	10^{-12}	pico-	p
10^{15}	peta-	P	10^{-15}	femto-	f
10^{18}	exa-	E	10^{-18}	atto-	a
10^{21}	zetta-	Z	10^{-21}	zepto-	z
10^{24}	yotta-	Y	10^{-24}	yocto-	y

1.11 Use prefixes from Table 1.3 to remove either all or almost all of the zeros in each expression. (*a*) ℓ = 150,000,000 m, (*b*) t = 0.000 000 000 012 s, (*c*) 1200 km/s, (*d*) 2300 kg.

Of the seven SI base units, we have already discussed two, the meter and the second. We discuss the base unit for mass, the kilogram, in Chapter 4, the base unit for electric current, the ampere, in Chapter 27, and the base unit for temperature, the kelvin, in Chapter 20.

The **mole** (abbreviated mol) is the SI base unit that measures the quantity of a given substance. A mole is currently defined as the number of atoms in 12×10^{-3} kg of carbon-12, the most common form of carbon. This number is called **Avogadro's number** N_A, after the 19th-century Italian physicist Amedeo Avogadro. The currently accepted experimental measurement of Avogadro's number is

$$N_A = 6.0221413 \times 10^{23}.$$

Note that the mole is simply a number: Just as *one dozen* means 12 of anything and *one gross* means 144 of anything, *one mole* means 6.022×10^{23} of anything. So 1 mol of helium atoms is 6.022×10^{23} helium atoms, and 1 mol of carbon dioxide molecules is 6.022×10^{23} carbon dioxide molecules.

The final SI base unit, the *candela*, measures luminous intensity. One candela (1 cd) is roughly the amount of light generated by a single candle; the light emitted by a 100-watt light bulb is about 120 cd. The definition of the candela takes into account how the human eye perceives the intensity of various colors and is therefore rather unwieldy. For this reason we do not use the candela in this book in the chapters dealing with light, concentrating instead on the amount of energy carried by light.

An important concept used throughout physics is **density,** the physical quantity that measures how much of some substance there is in a given volume. Depending on the quantity being measured, there are various types of density. For example, *number density* is the number of objects per unit volume. If there are N objects in a volume V, then the number density n of these objects is

$$n \equiv \frac{N}{V}. \tag{1.3}$$

(The symbol $\equiv$ means that the equality is either a definition or a convention.) If the objects in a given volume are packed together more tightly, the number density is higher (**Figure 1.17**). *Mass density* ρ (Greek rho) is the amount of mass m per unit volume:

$$\rho \equiv \frac{m}{V}. \tag{1.4}$$

Figure 1.17 Number density.

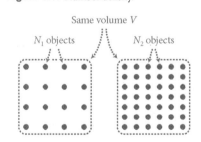

The greater the number N of objects in a given space V, the higher the number density $n = N/V$. In this case $N_2 > N_1$, so $n_2 > n_1$.

Exercise 1.6 Helium density

At room temperature and atmospheric pressure, 1 mol of helium gas has a volume of 24.5×10^{-3} m^3. The same amount of *liquid* helium has a volume of 32.0×10^{-6} m^3. What are the number and mass densities of (*a*) the gaseous helium and (*b*) the liquid helium? The mass of one helium atom is 6.647×10^{-27} kg.

SOLUTION (*a*) I know from Eq. 1.2 that 1 mol of helium contains 6.022×10^{23} atoms, and I can use this information in Eq. 1.3 to get the number density:

$$n = \frac{6.022 \times 10^{23} \text{ atoms}}{24.5 \times 10^{-3} \text{ m}^3} = 2.46 \times 10^{25} \text{ atoms/m}^3.$$

For the mass density, I must know the mass of 1 mol of helium atoms, and so I multiply the mass of one helium atom by the number of atoms in 1 mol:

$$m = (6.647 \times 10^{-27} \text{ kg/atom})(6.022 \times 10^{23} \text{ atoms/mol})$$

$$= 4.003 \times 10^{-3} \text{ kg/mol}.$$

Equation 1.4 then yields

$$\rho = \frac{4.003 \times 10^{-3} \text{ kg}}{24.5 \times 10^{-3} \text{ m}^3} = 0.163 \text{ kg/m}^3.$$

(*b*) For the liquid helium, the same reasoning gives me

$$n = \frac{6.022 \times 10^{23} \text{ atoms}}{32.0 \times 10^{-6} \text{ m}^3} = 1.88 \times 10^{28} \text{ atoms/m}^3$$

$$\rho = \frac{4.003 \times 10^{-3} \text{ kg}}{32.0 \times 10^{-6} \text{ m}^3} = 125 \text{ kg/m}^3.$$

Examples of non-SI units accepted for use along with SI units are the minute (1 min = 60 s), the hour (1 h = 3600 s), the liter (1 L = 10^{-3} m^3), and the metric ton (1 t = 10^3 kg).

A number of traditional, non-SI units are used in engineering; in various businesses, industries, sports, and trades; and in everyday life in the United States. Examples are inches, feet, yards, miles, acres, ounces, pints, gallons, and fluid ounces. These units are nondecimal, which makes it hard to interconvert them. When solving problems in this course, always begin by converting any quantities given in non-SI units to the SI equivalents. A conversion table is given in Appendix C.

The simplest way to convert from one unit to another is to write the conversion factor for the two units as a ratio. For, example, in Appendix C, we see that 1 in. = 25.4 mm. By bringing the 25.4 mm to the left side of the equals sign, we can write this as either

$$\frac{1 \text{ in.}}{25.4 \text{ mm}} = 1 \quad \text{or} \quad \frac{25.4 \text{ mm}}{1 \text{ in.}} = 1 \tag{1.5}$$

Note that you *must* write the units in these expressions because without them you obtain the incorrect expressions $\frac{1}{25.4} = 1$ and $\frac{25.4}{1} = 1$. Because multiplying any number by 1 doesn't change the number, you can use these ratios to convert units. For example, to express 4.5 in. in millimeters, you multiply by the ratio on the right in Eq. 1.5 and cancel out the inches:

$$4.5 \text{ in.} = (4.5 \text{ in.})\left(\frac{25.4 \text{ mm}}{1 \text{ in.}}\right) = 4.5 \times 25.4 \text{ mm} = 1.1 \times 10^2 \text{ mm.} \qquad (1.6)$$

1.12 Why is the ratio on the left in Eq. 1.5 not suitable for converting inches to millimeters?

Exercise 1.7 Unit conversions

Convert each quantity to a quantity expressed either in meters or in meters raised to some power: (*a*) 4.5 in., (*b*) 3.2 acres, (*c*) 32 mi, (*d*) 3.0 pints.

SOLUTION I obtain my conversions factors from Appendix C.

(*a*) $\qquad (4.5 \text{ in.})\left(\frac{2.54 \times 10^{-2} \text{ m}}{1 \text{ in.}}\right) = 1.1 \times 10^{-1} \text{ m.}$

(*b*) $\qquad (3.2 \text{ acres})\left(\frac{4.047 \times 10^3 \text{ m}^2}{1 \text{ acre}}\right) = 1.3 \times 10^4 \text{ m}^2.$

(*c*) $\qquad (32 \text{ mi})\left(\frac{1.609 \times 10^3 \text{ m}}{1 \text{ mi}}\right) = 5.1 \times 10^4 \text{ m.}$

(*d*) $\qquad (3.0 \text{ pints})\left(\frac{4.732 \times 10^{-4} \text{ m}^3}{1 \text{ pint}}\right) = 1.4 \times 10^{-3} \text{ m}^3.$

1.13 (*a*) Using what you know about the diameters of atoms from Section 1.3, estimate the length of one side of a cube made up of 1 mol of closely packed carbon atoms. (*b*) The mass density of graphite (a form of carbon) is $2.2 \times 10^3 \text{ kg/m}^3$. By how much does the length you calculated in part *a* change when you do your calculation with this mass density value? Remember that 1 mol is the number of atoms in $12 \times 10^{-3} \text{ kg}$ of carbon.

1.7 Significant digits

The numbers we deal with in physics fall into two categories: exact numbers that are known with complete certainty (integers, such as the 14 in "I have 14 books on my desk") and inexact numbers that result from measurements and are known to only within some finite precision. Consider, for example, using a ruler to measure the width of a piece of paper (Figure 1.18). The width falls between the ruler marks for 21 mm and 22 mm and is closer to 21 mm than 22 mm. We might guess that the width is about 21.3 mm, but we cannot be sure about the last digit without a better measurement method. By recording the width as 21 mm, we are indicating that the actual value lies between 20.5 mm and 21.5 mm. The value 21 mm is said to have two **significant digits**—digits that are known reliably.

By expressing a value with the proper number of significant digits, we can convey the precision to which that value is known. For numbers that don't contain any zeros, all digits shown are significant, which means that 21 has two significant digits, as just noted, and 21.3 has three significant digits (implying that the actual value lies between 21.25 and 21.35).

With numbers that contain zeros, the situation is more complicated. *Leading zeros,* which means any that come before the first nonzero digit, are never significant: 0.037 has two significant digits. Zeros that come between two nonzero digits,

Figure 1.18 If you measure the width of a sticky note with the ruler shown, you can reliably read off two digits.

as in 0.602, are always significant. *Trailing zeros* are those that come after the last nonzero digit in a number, as in 3.500 and 20. Trailing zeros to the right of the decimal point are always significant: 25.10 has four significant digits. However, trailing zeros in numbers that do not contain a decimal point are ambiguous. For example, in the number 7900, it is not clear whether the trailing zeros are significant or not. The number of significant digits is at least two but could be three or four. To accurately convey the precision for such numbers, we use scientific notation, which lets us place significant zeros to the right of the decimal point: 7.900×10^3 has four significant digits, 7.90×10^3 has three, and 7.9×10^3 has two.

Important note: To simplify the notation in this book, we consider all trailing zeros in numbers that do not contain a decimal point to be significant: four significant digits for 3400, for instance, and two significant digits for 30.

Do not confuse significant digits with the number of decimal places or the number of digits: 0.000584 has three significant digits, six decimal places, and seven digits; 58.4 has three significant digits, one decimal place, and three digits. With the exception of fundamental constants such as the speed of light, most values in this book are given to two or three significant digits.

When you use a calculator for the math involved in solving physics problems, the calculator display usually shows many more digits than the number of significant digits allowed by the problem data. In such cases, you need to round your answer to the correct number of significant digits. If the digit just to the right of the last significant digit you are allowed is less than 5, report your last significant digit as it appears on the calculator display. If the digit just to the right of the last significant digit is 5 or greater, increase your last significant digit by 1. For example, if you are allowed two significant digits, 1.356 rounds to 1.4, 2.5199 rounds to 2.5, and 7.95 rounds to 8.0.

Multiple roundings can yield an accumulation of errors, and therefore it is best to wait until you have obtained the final result in a multistep calculation before rounding. In intermediate results, therefore, retain a few more digits than what your input quantities have and then round off to the correct number of significant digits only in your final result.

Exercise 1.8 Significant digits

(*a*) How many significant digits are there in 403.54 kg, 3.010×10^{57} m, 2.43×10^{-3} s, 14.00 μm, 0.0140 s, 5300 kg? (*b*) Round 12,300 kg and 0.0125 s to two significant digits.

SOLUTION (*a*) 403.54 kg has five, 3.010×10^{57} m has four, 2.43×10^{-3} s has three, 14.00 μm has four, 0.0140 s has three, 5300 kg has four in this book (but is considered ambiguous in general).

(*b*) 1.2×10^4 kg (or 12 Mg); 0.013 s (or 1.3×10^{-2} or 13 ms).

Suppose you measure the mass and volume of some object to be $m = 1.2$ kg and $V = 0.123$ m^3, and you wish to use these values to compute the object's mass density. If you substitute the measured values into Eq. 1.4 and carry out the division on your calculator, you get

$$\rho \equiv \frac{m}{V} = \frac{1.2 \text{ kg}}{0.123 \text{ m}^3} = 9.756\,097\,56 \text{ kg/m}^3. \tag{1.7}$$

You measured the mass to two significant digits and the volume to three significant digits, but your calculator suggests you have determined the density to a precision of better than one part in a billion! The rigorous way to deal with uncertainties in measurements is to recalculate any computed value by using the high and low uncertainties for each value in the calculation, which is a time-consuming task. For this reason, we shall use two shortcuts. The first deals with multiplication and division:

> **When multiplying or dividing quantities, the number of significant digits in the result is the same as the number of significant digits in the input quantity that has the fewest significant digits.**

In your calculation of the mass density in Eq. 1.7, the input quantity that has the fewest significant digits is the mass (two significant digits), and so the answer should be rounded to two significant digits:

$$\rho \equiv \frac{m}{V} = \frac{1.2 \text{ kg}}{0.123 \text{ m}^3} = 9.8 \text{ kg/m}^3. \tag{1.8}$$

Now suppose you determine the mass of two parts of an object and obtain 105 kg for one part and 0.01 kg for the other (three and one significant digit, respectively). Adding the two measured values yields a value containing five significant digits:

$$105 \text{ kg} + 0.01 \text{ kg} = 105.01 \text{ kg}. \tag{1.9}$$

Given that the precision in the value 105 kg is ± 0.5 kg, it makes no sense to report the final result to five significant digits. Rounding so that the number of significant digits in the sum is the same as the number of significant digits in the input value that has the fewest significant digits would yield 100 kg, which deviates from the actual value (around 105 kg) by significantly more than the precision in the individual measurements (± 0.5 kg and ± 0.005 kg). To solve this problem, whenever we add or subtract numerical values, we focus on the number of *decimal places*:

> **When adding or subtracting quantities, the number of decimal places in the result is the same as the number of decimal places in the input quantity that has the fewest decimal places.**

For the addition in Eq. 1.9, the input quantity that has the fewest decimal places is 105 kg (zero decimal places), and so

$$105 \text{ kg} + 0.01 \text{ kg} = 105 \text{ kg}. \tag{1.10}$$

Bear in mind that *counting numbers* (that is, integers) are exact and so have an infinite number of significant digits. For example, the product of the width of two buttons each 15.7 mm wide is $2 \times 15.7 \text{ mm} = 31.4 \text{ mm}$ (not $3 \times 10^1 \text{ mm}$ because the 2 is a counting number).

Exercise 1.9 Significant digits and calculations

Calculate: (*a*) $f = a/(bc)$, where $a = 2.34 \text{ mm}^2$, $b = 54.26 \text{ m}$, and $c = 0.14 \, \mu\text{m}$; (*b*) $g = kt^3$, where $k = 1.208 \times 10^{-2} \text{ s}^{-3}$ and $t = 2.84$ s; (*c*) $f + g$; (*d*) the sum of $b = 54.26$ m and $c = 1.4$ mm; (*e*) $h = k(m - n)$, where $k = 1.252$, $m = 32.21$, and $n = 32.1$.

SOLUTION (*a*) I first need to convert the millimeters-squared in $a = 2.34 \text{ mm}^2$ to meters-squared and the micrometers in $c = 0.14 \, \mu\text{m}$ to meters:

$$2.34 \text{ mm}^2 \left(\frac{1 \text{ m}}{1 \times 10^3 \text{ mm}} \right)^2 = 2.34 \text{ mm}^2 \left(\frac{1 \text{ m}^2}{1 \times 10^6 \text{ mm}^2} \right) = 2.34 \times 10^{-6} \text{ m}^2$$

$$0.14 \, \mu\text{m} \left(\frac{1 \text{ m}}{1 \times 10^6 \, \mu\text{m}} \right) = 0.14 \times 10^{-6} \text{ m}.$$

(Continued)

Substituting the values given, I get

$$f = \frac{2.34 \times 10^{-6} \text{ m}^2}{(54.26 \text{ m})(0.14 \times 10^{-6} \text{ m})} = 0.30804,$$

which I must round to two significant digits, 0.31, because that is the number of significant digits in the value given for c.

(b) $g = (1.208 \times 10^{-2} \text{ s}^{-3})(2.84 \text{ s})^3 = 0.2767$, which I must round to three significant digits, 0.277, because of t.

(c) $f + g = 0.30804 + 0.2767 = 0.58474$, which I round to 0.58 because I must report only two decimal places, limited by $c = 0.14 \ \mu\text{m}$. Note that if I had added the rounded values for f and g, my result reported to two significant digits would not be 0.58: $f + g = 0.31 + 0.277 = 0.587$, which rounds to 0.59. Be careful—reporting the correct number of significant digits can be tricky.

(d) $b + c = 54.26 \text{ m} + 0.0014 \text{ m} = 54.26 \text{ m}$ (reported to two decimal places, limited by the 54.26 value).

(e) $h = 1.252(32.21 - 32.1) = 0.1$ (reported to one significant digit, limited by the 32.1 value).

The number zero requires special consideration. When some physical quantity is exactly zero, we denote this quantity by a zero *without units*. The speed of an object that is not moving is exactly zero and is therefore denoted by $v = 0$ (no units). However, if we measure a speed as zero to two significant digits, we write $v = 0.0 \text{ m/s}$ (note the units), implying that the actual value is zero to within 0.05 m/s.

1.14 (a) Express the circumference of a circle of radius $R = 27.3$ mm with the correct number of significant digits. (b) Let $a = 12.3$, $b = 3.241$, and $c = 55.74$. Compute $a + b + c$. (c) Let $m = 4.00$, $n = 3.00$, and $k = 7$ (exact). Compute $f = m^2/k$, $g = n^2/k$, and $f + g$.

1.8 Solving problems

You encounter problems every day. You are in a rush to get somewhere but can't find your car keys. You run out of flour while baking a birthday cake, and the supermarket is closed. Your flight is canceled on your way to a job interview. You want to buy those great new shoes, but there's no money in the bank. There is no formula for solving these or any other types of problems (if there were a formula, you wouldn't have a problem!). Because real-world problems defy one-shot solutions, we must break them down into smaller parts and solve them in steps. In this section we develop a four-step problem-solving strategy that will help you tackle a broad variety of problems (the procedure is summarized in the box "Solving problems").

Most physics problems are formulated in words. This is true not only of the problems in this book but also of questions you might have about the world around you. Standing on the rim of the Grand Canyon, you might wonder, If I drop a stone, how long does it take to hit the bottom?, or while watching the Olympics, you might wonder, Is there a physical limit to how high an athlete can jump?

❶ GETTING STARTED

Because it's not clear which road leads you most efficiently to the answer to a given problem, the first step of our problem-solving strategy, *getting started*, is the most difficult one. It is therefore useful to begin with something you *can* do: Organize the information given and be sure you are clear on exactly what is asked for in the problem. To clarify the goal of the problem in your mind, it is crucial to begin by visualizing the situation: Sketch, describe in your own words, list, and/or tabulate the main features of the problem to relieve yourself from having to hold all this information in memory. In making a sketch, be sure to follow the guidelines given in Section 1.5: Make your rendering as simple as possible and show all relevant numerical information in the sketch. Recasting your problem visually or verbally forces you to spell out what you want to accomplish and often automatically leads to ideas on how to solve the problem.

Once you have the information organized, ask yourself which concepts and principles apply. Throughout this book, we develop principle-specific procedures for solving problems. By determining which principles apply to a problem you are working on, you can identify which problem-solving procedures apply to your problem.

Finally, you must determine whether or not you have all the information necessary to solve the problem. Sometimes you may have to supplement the information provided with things you know about the situation at hand. Furthermore, because no problem can be formulated with absolute precision, you will often need to make a number of simplifying assumptions. This lack of precision is often a source of frustration for anyone beginning to solve physics problems. How do you know which assumptions are valid for a given problem? Do not let this question trouble you—as long as you are aware of the assumptions you make, you can always reexamine them once you have obtained an answer and then refine the assumptions and the solution.

❷ DEVISE PLAN

Next you must *devise a plan* for solving your problem, which means spelling out what you must do to solve the problem. Are there any physical relationships or equations that you can apply to determine the information you are trying to obtain? A good plan is to outline the steps you need to take to obtain a solution. For some (but certainly not all) problems, these steps are carried out mathematically.

❸ EXECUTE PLAN

In the third step, you *execute your plan* by following the steps you have outlined, substituting the information given and carrying out any mathematical operations necessary to isolate the quantity you wish to determine. In problems involving numerical values, you should solve for an algebraic answer first, waiting until the final step to substitute numerical values and obtain a numerical answer. The only exception to this algebraic-answer-first approach is that, in order to simplify your algebraic expressions, you should eliminate any quantities that are zero from them as soon as possible.

In the final step, substituting numerical values into your algebraic expression, do not forget to show the correct units on all numerical values and to carry these units through the calculation.

Once you have obtained an answer, you should check your work for these five important points (which you can remember via the acronym *VENUS*):

1. **Vectors/scalars used correctly?** As you will see in Chapter 2, the quantities we deal with in this book fall into two classes, vectors and scalars. Make sure you have expressed your answer in the appropriate form. (In this chapter all quantities are scalars.)

2. **Every question asked in problem statement answered?** Reread the problem statement to make sure you have completely answered every question.

3. **No unknown quantities in answers?** If your answer is an algebraic expression, be sure that every variable appearing in terms to the right of the equals sign has been given as a known quantity in the problem statement. Example: "A car moving at speed v . . . " means that v in this problem is a known quantity.

4. Units correct? Each numerical answer should be expressed as a number with the appropriate units for that quantity (Section 1.6). When your answer is an algebraic expression, be sure that the combined units of all terms to the right of the equals sign work out to be equivalent to the unit of the quantity to the left of the equals sign.

5. Significant digits justified? The number of significant digits reported in your answer should reflect the number of significant digits of the quantities given in the problem statement (Section 1.7).

As a reminder to yourself, put a checkmark beside each answer to indicate that you checked these five points.

❹ EVALUATE RESULT

You may think you are done, but there is one final—and very important—step left: *Evaluate your answer*. In this step, you should reflect on your work, examine the result, and determine whether your answer is reasonable (or, if you can't tell whether or not it is reasonable, at least make sure that it is not unreasonable). The first thing you might try is seeing whether your answer conforms to what you expect based on your situation sketch, diagram, or other information given. If any part of your answer is unexpected, go back and check your math and any assumptions you made.

If your result is an algebraic expression or if your numerical answer comes directly from an algebraic expression, check whether or not that expression behaves as you expect with changes in physical quantities. For example, say you obtain the expression $v = k/t$ for the speed of an object and you know from experience that for this object the speed is zero for large values of t. If you let t go to infinity, your expression should indeed give the right answer: $v = k/\infty = 0$. If it didn't, you must go back and check your math and any assumptions you made.

Some problems can be solved in more than one way. If you solve your problem in two ways and obtain the same answer, you can be pretty sure your answer is correct. If you don't get the same answer from the two ways, you'd better go back and check your work!

Procedure: Solving problems

Although there is no set approach when solving problems, it helps to break things down into several steps whenever you are working a physics problem. Throughout this book, we use the four-step procedure summarized here to solve problems. For a more detailed description of each step, see Section 1.8.

1. **Getting started.** Begin by carefully analyzing the information given and determining in your own words what question or task is being asked of you. Organize the information by making a sketch of the situation or putting data in tabular form. Determine which physics concepts apply, and note any assumptions you are making.
2. **Devise plan.** Decide what you must do to solve the problem. First determine which physical relationships or equations you need, and then determine the order in which you will use them. Make sure you have a sufficient number of equations to solve for all unknowns.
3. **Execute plan.** Execute your plan, and then check your work for the following five important points:

   Vectors/scalars used correctly?
   Every question asked in problem statement answered?
   No unknown quantities in answers?
   Units correct?
   Significant digits justified?

   As a reminder to yourself, put a checkmark beside each answer to indicate that you checked these five points.
4. **Evaluate result.** There are several ways to check whether an answer is reasonable. One way is to make sure your answer conforms to what you expect based on your sketch and the information given. If your answer is an algebraic expression, check to be sure the expression gives the correct trend or answer for special (limiting) cases for which you already know the answer. Sometimes there may be an alternative approach to solving the problem; if so, use it to see whether or not you get the same answer. If any of these tests yields an unexpected result, go back and check your math and any assumptions you made. If none of these checks can be applied to your problem, check the algebraic signs and order of magnitude.

If none of the evaluation methods can be applied to your answer, you should at least verify that there is nothing obviously wrong with it by checking that the algebraic signs you obtain make sense and that any numerical value has the correct order of magnitude. For example, if the algebraic sign of your solution tells you that an object moves to the left when you know it should move to the right, you know your answer can't be right. You can find some techniques for estimating orders of magnitude in the next section. Using these techniques, you should be able to obtain reasonable upper and lower bounds for your answer. Provided your answer does not lie outside this range, you know there is not something obviously wrong with it.

With the exception of one-step exercises that are applications of procedures, the examples in this book follow this four-step procedure.

Example 1.10 Length of string

A certain cylinder is 15 in. long and has a circumference of 4.0 in. When one end of a piece of string is attached to the left end of the cylinder and the string is then wound exactly five times around, the string just reaches to the cylinder's right end. What is the length of the string?

❶ GETTING STARTED I begin by making a sketch of the situation (**Figure 1.19a**). The shortest length is going to be obtained when the string is wound uniformly, with each of the five windings running one-fifth of the length of the cylinder. I need to find a way to determine the length of this spiraling string. This appears to be a geometry problem, but the surface of the cylinder is curved, not flat.

❷ DEVISE PLAN If I imagine drawing a straight line between the endpoints of the string, cutting the cylinder surface along that line, and then flattening the cylinder out, I obtain the situation shown in Figure 1.19b. Now the solution is clear: The total length of the string is the sum of the hypotenuses of five identical triangles.

❸ EXECUTE PLAN The width of each triangle is equal to one-fifth of the length of the cylinder: (15 in.)/5 = 3.0 in. The height of each triangle is equal to the circumference of the cylinder (4.0 in.). With these values, I can now use the Pythagorean theorem to determine the length ℓ of the hypotenuse of each triangle:

$$\ell = \sqrt{(3.0\ \text{in.})^2 + (4.0\ \text{in.})^2} = 5.0\ \text{in.}$$

The length of the string is equal to five times this value, or 25 in. ✔*

❹ EVALUATE RESULT The answer I obtain should be larger than the string length I would obtain if the string were wound around the cylinder in one place, which would give a string length of five times the circumference: 5 × 4.0 in. = 20 in. My answer should be smaller than what I would get if the string were wound five times around the cylinder at one end and then run along the cylinder length to the other end, which would give a string length of five times the circumference plus the length of the cylinder: (5 × 4.0 in.) + 15 in. = 35 in. The answer I obtained does indeed lie between these values, giving me confidence that it is correct.

Figure 1.19

(a)

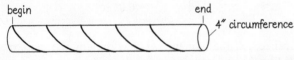

begin   end

4″ circumference

(b)

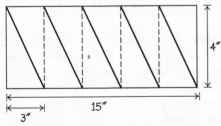

4″

15″

3″

*The checkmark shows that I checked my answer regarding the five important details listed on page 21: **Vector/scalar:** all quantities scalar; **Every question answered:** yes; **No unknown quantities in answers:** no; **Units correct:** yes, inches; **Significant digits:** everything given to two significant digits.

1.9 Developing a feel

Making order-of-magnitude estimates is possibly the most important skill to retain from a physics course because this is a skill you will use long after you leave your last classroom, no matter what you plan for a career. Economists, for instance, carry out order-of-magnitude estimates to predict market trends, and business owners might carry out such estimates to validate a business plan, to name just two possibilities.

Here in this physics course, to get back to the present, order-of-magnitude estimates allow you to develop a feel for a problem before you try to solve it. They also allow you to validate your answer: If your order-of-magnitude estimate and your answer disagree, you know to go back and reevaluate your work.

Order-of-magnitude estimates involve determining relationships between known and unknown quantities, making gross oversimplifications, combining assumptions and ideas, and even outright guessing. In general, there are multiple ways to obtain such estimates. The uncertainty in how to carry out an estimate is intimidating at first, and the lack of accuracy may appear unscientific to you. It isn't—estimation is an important aspect of science, and it is a great way to obtain physical insight and at the same time train your analytical and outside-the-box thinking skills.

For example, suppose we need to determine the number of cells in the human body. That we can't see individual cells tells us they must be very small, and therefore the number is probably very large. We could *guess* the number of cells, but the answer would probably be meaningless because we have no way of assessing its accuracy without looking it up somewhere. A billion? Ten billion? Hundred billion? What do you think?

To obtain an order-of-magnitude estimate, the first thing to do is to think of a strategy. For the number of cells in the human body, for example, if we knew the volume of a typical human body and the volume of a typical cell, we could divide the former by the latter to obtain our answer. Knowing neither volume, however, we need to estimate both.

There are several ways to estimate the volume of the human body. One way is to model the body as a cylinder (**Figure 1.20**). For the height of the cylinder, we take the height of a typical human, say $h_{human} \approx 1.7$ m. The symbol $\approx$ indicates that this value is an approximation. For the diameter, we take $d_{human} \approx 0.2$ m. That makes the diameter less than the left-to-right thickness of the body but greater than the front-to-back thickness, which should be in the right ballpark. This gives a volume of

$$V_{human} = \left(\frac{1}{4}\pi d_{human}^2\right)h_{human} \approx 0.05 \text{ m}^3.$$

Another way is to realize that the human body has roughly the same mass density as water—about 1 kg/L, which is

$$1 \text{ kg/L}\left(\frac{1 \text{ L}}{1 \times 10^{-3} \text{ m}^3}\right) = 1000 \text{ kg/m}^3.$$

The average human has a mass of about 70 kg, which corresponds to 70 L of water, and thus $V_{human} \approx (70 \text{ L})(1 \times 10^{-3} \text{ m}^3/1 \text{ L}) = 0.07 \text{ m}^3$, which is in the same ballpark as the estimate we got using the body-as-cylinder approach.

You may feel that there is too big a difference between 0.05 m³ and 0.07 m³. Indeed, if you were shopping for clothing, you'd want your clothing to fit your body better than that. However, here we are interested in only the order of magnitude. Either estimate is good enough for this purpose because for both the order of magnitude is the same: 0.05 m³ = 5×10^{-2} m³, which rounds to 10×10^{-2} m³, and so the order of magnitude is 10^{-1} m³; 0.07 m³ = 7×10^{-2} m³, which also rounds to 10×10^{-2} m³, giving the same order of magnitude: 10^{-1} m³.

Next we need to estimate the volume of a typical human cell. You may remember from biology class that the diameter of a typical cell is around $20 \,\mu\text{m} = 20 \times 10^{-6}$ m. Or, if you never learned cell sizes, you may recall that you can clearly see cells under a microscope and that a microscope can see objects as small as 10^{-6} m. To see the structure of a cell clearly under a microscope, the cell would need to be, say, at least 50 times the minimum size the microscope

Figure 1.20 To determine the volume of the human body, we approximate the body as a cylinder.

can handle, making our cell-diameter estimate 50×10^{-6} m. Although you may think there is a big difference between 20×10^{-6} m and 50×10^{-6} m, these two numbers are only a factor of 2.5 apart, which is much less than a factor of 10 ("one order of magnitude").

Cells viewed under a microscope tend to spread out on the glass and are thus less high than they are wide. Let's for simplicity assume they are disk-shaped and that their height is about one-quarter of their diameter. That means they have a diameter of 20×10^{-6} m and a height of 5×10^{-6} m, and so $V_{cell} = (\frac{1}{4}\pi d_{cell}^2)h_{cell} \approx 1.6 \times 10^{-15}$ m³. The number of cells is thus $V_{human}/V_{cell} \approx (0.05$ m³$)/(1.6 \times 10^{-15}$ m³$) \approx 3.1 \times 10^{13}$. Given the roughness of the estimates and assumptions, I give the final answer as an order of magnitude: 10^{14} (about 100 trillion cells).

I can be quite confident that the number of cells is somewhere between 10^{13} (10 trillion) and 10^{15} (1000 trillion), both substantially greater than the largest guess I offered earlier (100 billion $= 10^{11}$). The difference between that largest guess and the order-of-magnitude estimate is that I have no idea how much confidence to place in the guess, whereas I can feel reasonably certain about the result of my order-of-magnitude estimate.

You may think that this exercise is outrageous and that it is impossible to have faith in any order-of-magnitude value because it requires so many assumptions and estimates. However, the fact that we combine so many assumptions and estimates is precisely why the method works: We are unlikely to be either too high on every estimate or too low on every estimate. Instead, we unwittingly overestimate some values and underestimate others, and the over- and underestimates tend to cancel out. The result is that the final answer very likely is in the right ballpark. What is more important is that, in the process of obtaining an order-of-magnitude value, we develop insights and some feeling for the quantities involved.

Figure 1.21 shows in graphical form the process by which we obtained our estimate of the number of cells N_{cell} in the human body. We got to our answer through a combination of *devising a strategy* (using a ratio of volumes to determine the number of cells), *simplifying* (treating the human body as a cylinder and the cell as a disk), *estimating* (the diameter of the human body), and using snippets of *knowledge*. Figure 1.21 also shows why this approach can be intimidating at first: To determine a single target unknown, N_{cell}, we have to divide the problem into a larger number of questions, an approach that makes the number of unknowns increase initially! The point of this dividing, of course, is to arrive at values we know (or can guess).

The diagram in Figure 1.21a reduces the problem of obtaining N_{cell} to one of obtaining another value, d_{cell}. If you know the value of d_{cell} from your biology course, you are done. If you don't know the value, you can *translate* the problem into another one, as illustrated in Figure 1.21b. If you get stuck, guess upper and lower bounds for your unknowns, use the middle value to obtain an answer, and then reevaluate to see how your *guesstimate* affects the final answer.

Making order-of-magnitude estimates requires practice. As you go through the exercise more often, you'll gain confidence, and the process will come more naturally. An important point is not to worry too much about accuracy when working out orders of magnitude: A factor of ten doesn't matter. You'll be surprised how far you can get with this technique, some resourcefulness, and commonly available knowledge.

✋ **1.15** To work an order-of-magnitude problem, you need the value of the mass density of water, but you don't know what that value is. Design either a translation strategy (Figure 1.21b) or a division strategy (Figure 1.21a) for making an order-of-magnitude estimate of this mass density.

Figure 1.21 The process by which we estimated the number of cells in the human body.

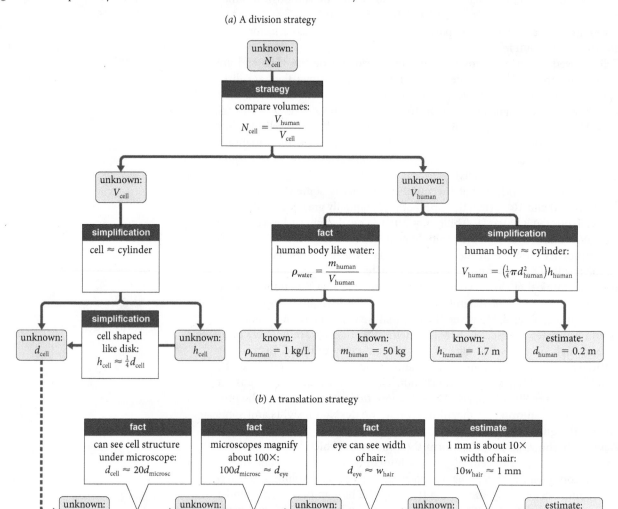

(a) A division strategy

(b) A translation strategy

Exercise 1.11 Selling books

Consider writing a book for high-school graduates in the United States who are planning to take a year off—sometimes called a gap year—before enrolling in college. There are currently no similar books. Estimate how many such books might be purchased annually.

SOLUTION I can estimate the number of books sold by estimating the percentage of college-bound high-school graduates who take a gap year. I know two things: (1) Not all graduates do this, and (2) there is a rising interest in doing it. I estimate that 5% of high-school graduates take a gap year before going to college, but I still have to estimate the number of college-bound high-school students in the United States. I know that approximately 50% of high-school graduates go on to college, but how many high-school graduates are there each year?

If I assume every person in the country graduates from high school when she or he is 18, I've reduced this part of the problem to determining the number of 18-year-olds in the country. I know that the total population is 300 million and that the average life span is about 75 years. If I assume that age is uniformly distributed, then 1/75 of the U.S. population should be 18 years old at any given instant (this is an oversimplification but is good enough for an order-of-magnitude estimate). So the number of high-school graduates in a given year is $(1/75)(3 \times 10^8) = 4 \times 10^6$, and 5% of this is 2×10^5. Thus the number of students taking a gap year each year is approximately 200,000, which strikes me as a reasonable number. If 5% of these students buy the proposed book, about 10,000 copies will be purchased each year.

1.16 Make an order-of-magnitude estimate of the mass of Earth in kilograms.

Chapter Glossary

SI units of physical quantities are given in parentheses.

Absolute time The notion that time is the same for all observers in the universe, regardless of their location or motion.

Atom A basic building block of matter.

Avogadro's number The number of atoms or molecules in 1 mol:

$$N_A = 6.022 \times 10^{23}. \qquad (1.2)$$

See also *mole*.

Change The transition from one state to another.

Density A measure of how much of some substance there is in a given volume. The *number density* (m^{-3}) is the number of objects per volume:

$$n \equiv \frac{N}{V}. \qquad (1.3)$$

The *mass density* (kg/m^3) is the amount of mass per volume:

$$\rho \equiv \frac{m}{V}. \qquad (1.4)$$

Electron A subatomic particle manifesting itself in a cloud around atomic nuclei.

Hypothesis A tentative explanation of observations to be used as a starting point for further investigation.

Law A description of a relationship between observable quantities that manifests itself in recurring patterns of events.

Length ℓ (m) A distance or extent in space.

Meter (m) The SI base unit of length, defined as the distance traveled by light in vacuum in $1/299{,}792{,}458$ of a second.

Model A simplified conceptual representation of a phenomenon.

Mole (mol) The SI base unit for measuring the quantity of a given substance, defined as the number of atoms in 12×10^{-3} kg of carbon-12, which is measured to be 6.022×10^{23}. See also *Avogadro's number*.

Order of magnitude A value rounded off to the nearest power of ten.

Physical quantity A physical property that can be measured and that is expressed as the product of a numerical value and a physical unit.

Principle of causality Whenever an event A causes an event B, all observers see event A happening first.

Proton, neutron Subatomic particles residing in atomic nuclei.

Scientific method An iterative process for going from observations to theories validated by experiments.

Second (s) The SI base unit of time, defined as the duration of $9{,}192{,}631{,}770$ periods of certain radiation emitted by cesium atoms.

SI units Units of measure in the International System, used in science and engineering.

Significant digits Digits in a numerical value that are reliably known.

Symmetry An object exhibits symmetry when certain operations can be performed on it without changing its appearance.

Theory A well-tested explanation of a natural phenomenon in terms of more basic processes and relationships.

Time t (s) A physical quantity that allows us to determine the sequence of related events.

Universe The totality of matter and energy combined with the space and time in which all events happen.

2 Motion in One Dimension

The branch of physics that deals with the quantitative representation of motion is called *kinematics.* Kinematics does not consider the causes and effects of motion; its goal is simply to provide a mathematical description of motion. In this chapter, we follow a kinematics approach; that is, we represent motion on a graph and quantify that motion without worrying about what causes it. In Chapter 8 we shall begin to look at the causes of motion, but for now all we care about is the motion while it is happening.

By describing the motion represented in a film clip, we begin to develop the most basic skill in physics: making simplified representations of real-world situations. You'll see how to use tables, graphs, and mathematical functions to represent quantitative data on motion. Finally, you'll learn the distinction between the terms *position* and *displacement* as well as the distinction between *speed* and *velocity*.

2.1 From reality to model

The opening picture of this chapter shows the *Shinkansen*—the Japanese bullet train—speeding over a bridge. Although the image conveys the idea of motion, it does not contain any quantitative representation of that motion. If I were to ask, "How fast is the train moving?" you wouldn't be able to answer if all you had to go on was this photograph. In order to answer my question, you need a more quantitative representation of the motion.

To analyze the motion of an object, we need to keep track of the object's position at different instants. If the position changes as time passes, the object is in motion. If the position is not changing, the object is said to be **at rest.** As a first step in measuring position, let's analyze the film clip reproduced in **Figure 2.1**, a record of me walking (and you thought physics was going to be dull!). As you can see from the clip, I first walked forward a certain distance, stood still for a short while, and then walked backward.

The imaginary straight line on the ground along which I took my steps is our *reference axis.* My position is determined by my location along this axis relative to some arbitrarily chosen reference point called the *origin.* Giving my position at different instants in time determines my motion in much the same way the film clip does. Choosing the left edge of the frame as the origin, we can say that in frame 1 the center of my body is 2.5 mm from that point; a few frames later it is 8.5 mm from the edge (frame 5); another few frames and it is 12.0 mm from the edge (frame 8). In other words, we can describe my motion by specifying my position for every frame, which is what **Table 2.1** does.

The same information is shown graphically in **Figure 2.2**, where my position relative to the left edge is plotted along the vertical axis and the horizontal axis represents the frame number. The formal way of describing this graph is to say that it shows my position (information on the vertical axis) *as a function of* the frame number

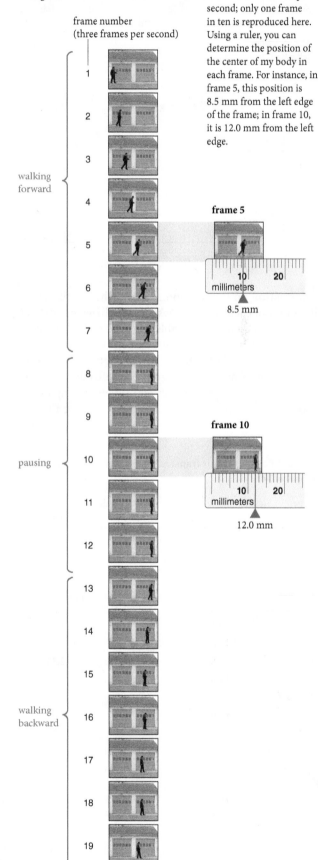

Figure 2.1 Film clip showing me walking to the right, pausing, then walking back to the left. The camera recorded this event at 30 frames per second; only one frame in ten is reproduced here. Using a ruler, you can determine the position of the center of my body in each frame. For instance, in frame 5, this position is 8.5 mm from the left edge of the frame; in frame 10, it is 12.0 mm from the left edge.

frame number
(three frames per second)

walking forward

frame 5

10   20
millimeters
8.5 mm

pausing

frame 10

10   20
millimeters
12.0 mm

walking backward

Figure 2.2 My distance from the left edge of each frame in Figure 2.1 versus frame number. We see that I start, in frame 1, 2.5 mm from the left edge of the frame, walk forward to 12.0 mm from the left edge at frame 8, stay there for five frames, then walk backward to 8.5 mm from the left edge.

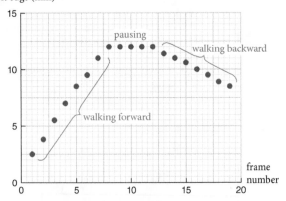

(information on the horizontal axis). The coordinates of the center of each circle correspond to the values listed in the first two columns in Table 2.1. The size of the circle is representative of the accuracy of the measurements. The upward trend of data points between frames 1 and 8 indicates that my distance from the left edge increases with increasing frame number and represents my motion walking from left to right. Then, representing the time

interval when I stood still, the data points stay flat from frame 8 to 12. Finally, from frame 12 onward, the data points show a downward trend representing my motion back to the left.

2.1 (*a*) Using a metric ruler, measure the position of the center of my body from the left edge in each frame of Figure 2.1. Compare your list of values with the positions given in Table 2.1. (*b*) Plot the positions you measured as a function of frame number—no need for graph paper, just draw your axes and plot your data by eye.

Taking information from the real world and creating a simplified abstract representation of it—making a model of a real-world happening—is one of the most important skills in physics. You'll be making such models hundreds of times throughout this course, some graphical like Figure 2.2, some in the form of tables as Table 2.1, some in the form of mathematical expressions.

2.2 (*a*) In Figure 2.2 the data points for frames 8–12 are all at position 12.0 mm. Explain what this alignment of data points means in terms of what is happening in the real world. (*b*) The data points for frames 7 and 14 are aligned horizontally. What does this alignment imply about the real-world motion? (*c*) What would it mean if two data points were aligned vertically? Could this happen?

2.2 Position and displacement

Figure 2.2 is of limited value as a model of a real-world happening. It tells us only that, in 19 frames of a certain video camera, my image moved from a position 2.5 mm from the left edge of the frame to a position 8.5 mm from the edge. It would be more useful to have this information in terms of real-world distances and time intervals. To know what distance I traveled and how long the trip took, I should have put markers on the ground at known intervals and a clock in the camera's field of view. Even though I did not do this, we can still determine position and time by *calibrating* the graph, as you can see in the next checkpoint.

2.3 (*a*) Take a reasonable guess at my height and calculate what real-world distance 1 mm in Figure 2.1 corresponds to. (*b*) Using this result, calculate the distance I walked between frames 1 and 10.

This checkpoint gives a procedure for calibrating the vertical axis in Figure 2.2 without taking any new pictures. Using 1.8 m as my height, I obtained the position values listed in the third column of Table 2.1. For motion in one dimension, position can be represented by the x coordinate along a reference axis (also called the x axis), and this coordinate is denoted by the symbol x. Because the x coordinate can be positive or negative, we always explicitly denote its

Table 2.1 Position versus frame number from the film sequence in Figure 2.1

Frame number	Distance from left edge (mm)	Position value x (m)	Time t (s)
1	2.5	+1.0	0
2	3.8	+1.5	0.33
3	5.5	+2.2	0.67
4	7.0	+2.8	1.00
5	8.5	+3.4	1.33
6	9.5	+3.8	1.67
7	11.0	+4.4	2.00
8	12.0	+4.8	2.33
9	12.0	+4.8	2.67
10	12.0	+4.8	3.00
11	12.0	+4.8	3.33
12	12.0	+4.8	3.67
13	11.4	+4.6	4.00
14	11.0	+4.4	4.33
15	10.6	+4.2	4.67
16	10.0	+4.0	5.00
17	9.5	+3.8	5.33
18	8.9	+3.6	5.67
19	8.5	+3.4	6.00

CONCEPTS

algebraic sign. In frame 4, for example, my position is given by $x = +2.8$ m. Note that by following this procedure, x values increase to the right.

We can do something similar for time. The camera takes 30 frames per second, and so the time interval between successive frames on the original film is $1/30$ s $= 0.033$ s. If we imagine having started a stopwatch at the first frame, that stopwatch would read 0.033 s at the second frame, 0.067 s at the third, 0.100 s at the fourth, and so on. Because only every tenth frame is shown in Figure 2.1, we must multiply these readings by 10. All these clock readings are listed in the fourth column in Table 2.1. Clock readings are represented by the symbol t; each value of t represents an instant in time. In everyday language *instant* can mean different things. For example, when you say, "I'll be back in an instant," the word *instant* refers to a short time interval. In physics, however, an instant has no duration at all; it represents a single value in time.

Figure 2.3 shows the resulting calibrated data in graphical form, with the position given by x on the vertical axis and time t on the horizontal axis. We refer to such a graph as a *position-versus-time graph*.

To obtain the data shown in Figures 2.2 and 2.3, we arbitrarily chose an origin. An origin is always needed in order to specify a position: The statement "the ball's position is $x = +3.2$ m" makes no sense unless we know from where to start measuring these 3.2 meters. The choice of origin is arbitrary. In analyzing Figure 2.1, for instance, I arbitrarily chose the left edge of the frame as the origin. To show that the choice of origin is arbitrary, suppose we replot our data, but now take my position in frame 1 to correspond to $x = 0$ and the clock reading at the first frame to be $t = 0.33$ s. Doing so produces the graph shown in **Figure 2.4**. Compare this graph with Figure 2.3 and you see that the change in origin and in the instant at which $t = 0$ changes the x and t values of each data point. However, as the next checkpoint shows, changing the origin and the

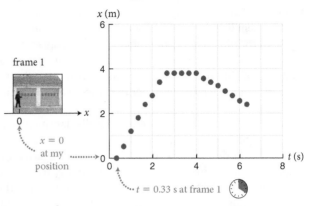

Figure 2.4 Position-versus-time graph for the data plotted in Figure 2.3, but with a different choice of origin.

instant at which $t = 0$ does not affect the relative positions of the data points.

2.4 (*a*) What are the final position values in Figures 2.3 and 2.4? (*b*) What values are obtained by subtracting the initial position value from the final position value in each figure?

We can represent the motion of an object that moves from one position to another by an arrow that points from the initial position to the final position (**Figure 2.5**). This arrow represents a physical quantity called the **displacement** of the object. The displacement does not depend in any way on the choice of reference axis or origin; you don't need any axis to draw the arrow in Figure 2.5. Numerically, we can represent the displacement by subtracting the initial x coordinate from the final x coordinate. The resulting value is called the *x component of the displacement*:

> The x component of an object's displacement is the change in its x coordinate. The value of the x component of the displacement is obtained by subtracting the initial x coordinate from the final x coordinate and is independent of the choice of origin.

Figure 2.3 Position-versus-time graph obtained by calibrating the data points in Figure 2.2. The vertical axis now represents my position, measured in meters, relative to a point on the left edge of the frame. The x and t values are listed also in Table 2.1.

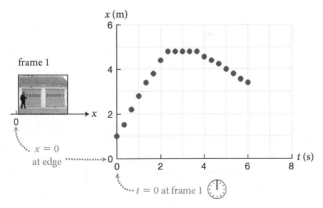

Figure 2.5 The arrow pointing from my initial position to my final position represents my displacement.

The term *x component* serves to remind us that this quantity is measured along some specific *x* axis. In the second part of this chapter you will see how to specify a displacement like the one in Figure 2.5 by choosing an *x* axis and determining the *x* component of the displacement. Later on, in Chapter 10, when we discuss motion that is not along a straight line, we shall need more than one axis and a correspondingly larger number of components to specify displacements.

Note that the *x* component of the displacement can be either positive or negative: It is positive for displacements in the direction of increasing *x* and negative for displacements in the direction of decreasing *x*. To remind ourselves of this fact, we shall always explicitly denote the algebraic sign of the *x* component of the displacement. The *x* component of the displacement between my initial and final positions in Figure 2.3, for example, is +2.4 m, indicating that I end up 2.4 m to the right of my starting position.

Returning to Checkpoint 2.4, we see that because every data point in Figure 2.4 is shifted down and to the right by exactly the same amount, the value obtained by subtracting the initial *x* coordinate from the final *x* coordinate is not affected by the change in origin. If it puzzles you that the *x* component of the displacement is independent of the choice of origin, consider the following: Suppose you tell a friend, "See you at the bakery in thirty minutes." Even if your watch says 3:30 at that instant and your friend's watch says 2:30, you will both arrive at the bakery at the same time: 30 minutes after you made the date. Your reference point is the clock reading 3:30 and your friend's reference point is the clock reading 2:30, but both of you add the same value—30 minutes—to your reference point. Just like the *x* component of the displacement (a *change in* the *x* coordinate), a time interval (a *change in* the clock reading) does not depend on the choice of reference point.

2.5 Suppose you walk in a straight line from a point P to a point Q, 2 m away from P, and then walk back along the same line to P. (*a*) What is the *x* component of your displacement for the round trip? (*b*) What distance did you travel during the round trip? (*c*) Is *distance traveled* the same thing as the *x component of the displacement*?

As this checkpoint shows, it is important to distinguish between the *x* component of the displacement and the *distance traveled*.

The distance traveled is the distance covered by a moving object along the path of its motion.

In contrast to the *x* component of the displacement, which can be either positive or negative, the distance traveled is always positive. Note also that, for an object that turns around, as in Checkpoint 2.5, the distance traveled is

greater than the distance between the endpoints. Only when the motion is always in the same direction is the distance traveled equal to the distance between the endpoints.

2.3 Representing motion

The graphs of Figures 2.3 and 2.4 specify position only at the specific instants represented by the frames in Figure 2.1. What about the positions at intermediate instants? To answer this question, we would need to look at all the frames taken between those shown in Figure 2.1. Doing so would give us ten times as many data points because the camera took ten times the number of frames shown in the figure. If we wanted information more detailed than that, we would need a camera that takes more than 30 frames per second.

As we repeat our measurements over smaller and smaller time intervals, the spacing between data points becomes increasingly smaller, eventually yielding what looks like a continuous curve instead of discrete points. If we assume the motion proceeds smoothly from frame to frame, we can obtain the same result by *interpolating* the data points in Figure 2.3—in other words, by drawing a smooth curve through the points, as in **Figure 2.6**.

2.6 (*a*) From Figure 2.6, how many seconds did it take me to go from *x* = +1.0 m to *x* = +4.0 m? (*b*) From *x* = +2.0 m to *x* = +3.0 m? (*c*) At what instant did I reach *x* = +2.5 m? (*d*) For how long was I at *x* = +2.5 m?

Figure 2.7 shows how the position-versus-time graph of Figure 2.6 relates to Figure 2.1. Figure 2.6 is a further simplification of our representation of the motion in the film clip. In reality, the motion may have been less smooth than the graph suggests. However, the film clip and Figure 2.6 both tell the same story: I moved forward about 3.8 m in 2.3 s, stood still for 1.3 s, and then moved 1.4 m backward in 2.3 s. Even though it gives less detail than the film clip, the simple curve in Figure 2.6 entirely describes the motion.

Figure 2.6 By interpolating the data points in Figure 2.3, we obtain a smooth curve representing my motion.

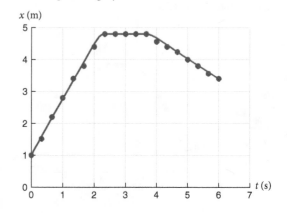

Figure 2.7 If you connect the images in sequential frames by a line, you obtain a curve representing the motion. Turn the page sideways and compare the curve through the images with the curve in Figure 2.6.

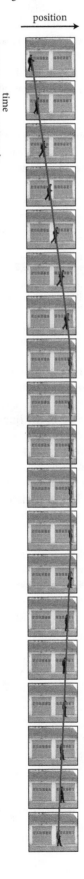

position

time

Example 2.1 Graphical representation

The curve in **Figure 2.8** is a graphical representation of the motion of a certain object. (*a*) What was the *x* coordinate of the object at $t = 0.50$ s? (*b*) At what instant(s) did the object reach $x = +0.80$ m? (*c*) What distance did the object travel between $t = 0.80$ s and $t = 1.2$ s? (*d*) How long did it take for the object to move from $x = +1.0$ m to $x = 0$?

Figure 2.8 A position-versus-time graph representing the motion of an object. The curve, which shows the position of the object at any instant between $t = 0$ and $t = 1.2$ s, can be represented by a mathematical function $x(t)$ (Examples 2.1 and 2.2).

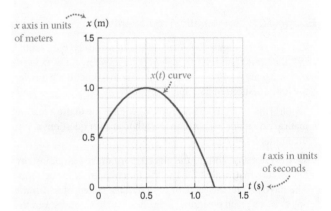

x axis in units of meters

x (m)

$x(t)$ curve

t axis in units of seconds

t (s)

❶ **GETTING STARTED** This problem requires me to use Figure 2.8 to answer questions about the motion of the object. The problem boils down to reading off values from the graph.

❷ **DEVISE PLAN** To find the object's position *x* at a given instant *t*, I draw (or imagine drawing) a vertical line at the value on the *t* axis corresponding to the given instant in time and read off the value of *x* at which that line intercepts the curve in the graph. Likewise, to find the instant *t* at which the object is at a given position *x*, I draw a horizontal line at the given value of *x* and read off the value of *t* at which that horizontal line intercepts the curve.

❸ **EXECUTE PLAN** (*a*) A vertical line drawn at $t = 0.50$ s in Figure 2.8 intercepts the curve at $x = +1.0$ m. ✔

(*b*) A horizontal line at $x = +0.80$ m intercepts the curve at $t = 0.20$ s and at $t = 0.80$ s. ✔

(*c*) At $t = 0.80$ s the object was at $x = +0.80$ m, and at $t = 1.2$ s it was at $x = 0$. The distance traveled was thus 0.80 m. ✔

(*d*) The object was at $x = +1.0$ m at $t = 0.50$ s and at $x = 0$ at $t = 1.2$ s. Thus it took the object 1.2 s − 0.50 s = 0.7 s to move between these points. ✔

✔ The checkmarks show that I carried out the following checks: **Vectors** and scalars used appropriately; **Everything** answered; **No** unknowns left; **Units** correct; **Significant** digits justified.

❹ **EVALUATE RESULT** That I obtain two instants in part *b* is OK because the object can return to a certain position. The answers to parts *c* and *d* (0.80 m and 0.7 s) are within the range of the curve (1.0 m and 1.2 s), so my answers are not unreasonable.

A curve in a position-versus-time graph like the one shown in Figures 2.6 and 2.8 is called an *x*(*t*) *curve* because it can be represented by a mathematical function *x*(*t*).

The curve and the function specify where the object is at any given instant. Note that $x(t)$ is a single symbol that means "x at instant t"; it is not the product of x and t and should not be confused with the label "x (m)" on the x axis, which means "x in units of meters."

To find the position of the object at a specific instant, we can either read off the position from the graph or substitute that instant into the function. For example, "$x(t = 0.2$ s)" gives the position of the object at the instant $t = 0.2$ s; this position can be obtained by substituting 0.2 s for t in the mathematical expression for $x(t)$.

Example 2.2 Mathematical representation

The curve in Figure 2.8 can be mathematically represented by the function $x(t) = a + bt + ct^2$, where $a = 0.50$ m, $b = +2.0$ m/s, and $c = -2.0$ m/s^2. Use this function to answer the four questions in Example 2.1.

❶ GETTING STARTED This problem requires me to use a mathematical expression rather than a graph to answer questions about the motion of an object.

❷ DEVISE PLAN To find the object's position at a given instant, I substitute that instant for t in the function given in the problem and compute the value of x. To find the instant at which the object is at a given position, I substitute that position for x in the function given in the problem and solve for t.

❸ EXECUTE PLAN (a) Substituting $t = 0.50$ s into the function, I find for the position at $t = 0.50$ s: $x(t = 0.50$ s) $= 0.50$ m $+ (+2.0$ m/s)(0.50 s) $+ (-2.0$ m/s^2)(0.50 s)$^2 = +1.0$ m. ✔

(b) To find the instants at which the object reached $x = +0.80$ m, I must set x in the mathematical function equal to $+0.80$ m: $+0.80$ m $= 0.50$ m $+ (+2.0$ m/s)$t + (-2.0$ m/s$^2)t^2$ or, simplifying the equation by omitting the units, $(2.0)t^2 - (2.0)t + 0.30 = 0$. This is a quadratic equation in t, which has two solutions: $t = [2.0 \pm \sqrt{4.0 - 4 \times 2.0 \times 0.30}]/(4.0)$. Working out this expression yields $t = 0.18$ s and $t = 0.82$ s. ✔

(c) At $t = 0.80$ s, the object's position is $x(t = 0.80$ s) $= 0.50$ m $+ (+2.0$ m/s)(0.80 s) $+ (-2.0$ m/s^2)(0.80 s)$^2 = +0.82$ m. At $t = 1.2$ s, the object is at $x = 0.50$ m $+ (+2.0$ m/s)(1.2 s) $+ (-2.0$ m/s^2)(1.2 s)$^2 = +0.02$ m. The distance it traveled is the difference between these two positions: $x(t = 0.80$ s) $- x(t = 1.2$ s) $= +0.82$ m $-(+0.02$ m) $= 0.80$ m. ✔

(d) I begin by calculating at what instant the object reached $x = +1.0$ m. Substituting this value into the mathematical expression yields 1.0 m $= 0.50$ m $+ (+2.0$ m/s)$t + (-2.0$ m/s$^2)t^2$ or, simplifying the equation by omitting the units, $-(2.0)t^2 + (2.0)t + 0.50 = 1.0$. Solving this quadratic equation yields $t = 0.50$ s. Next I must find the instant at which the object reached $x = 0$. Following the same procedure, I find two solutions: $t = -0.20$ s and $t = 1.2$ s. The first solution is before the starting point of the motion represented in Figure 2.8, and so I use the second solution to find that the object took 1.2 s $- 0.50$ s $= 0.7$ s to travel this distance. ✔

❹ EVALUATE RESULT To check my work, I can compare my answers with the ones I obtained in Example 2.1. Indeed, my answers are the same as (or very close to) the ones I obtained from reading the graph.

2.4 Average speed and average velocity

Now that you know how to represent motion in terms of position and time, let us use this information to calculate first the average speed and then the average velocity of the motion.

First, average speed. The more slowly I move, the larger the time interval I require to travel a given distance. Suppose, for instance, I had walked twice as slowly while the camera was rolling. I would have taken twice as much time to get to, say, the end of the forward part of my walk, which is the point $x = +4.8$ m in Figure 2.6. The distance traveled from my starting point would have been the same as before (4.8 m $- 1.0$ m $= 3.8$ m), but it would have taken 4.67 s to reach this point rather than the 2.33 s shown in Figure 2.6. This difference is shown in **Figure 2.9**.

The rate at which an $x(t)$ curve rises with increasing time is called the *slope* of the curve. Notice how the slope becomes less steep when the speed decreases.

2.7 For each speed in Figure 2.9, what are (a) the distance traveled between $t = 0$ and $t = 1.50$ s and (b) the ratio of the distance traveled to the corresponding time interval? For each speed, what are (c) the time interval required to move from $x = 2.00$ m to $x = 3.50$ m and (d) the ratio of the distance traveled to the corresponding time interval?

This checkpoint reveals how to extract speed from an $x(t)$ curve: Divide the distance traveled by the time interval required to travel this distance. The ratios of parts b and d correspond to the slopes of the lines in Figure 2.9 and bear out what I said before: The steeper the slope (the higher the number you got for your ratios), the higher the speed.

Suppose someone rode a bicycle 20 km in 2 h. Using the above procedure to calculate the cyclist's speed, we get $(20$ km)$/(2$ h) $= 10$ km/h. This doesn't necessarily mean the speed was constant—the person could have, for instance,

Figure 2.9 Forward motion at two speeds. Walking half as fast means taking twice as long to travel the same distance. To get to the arbitrary point $x = 3.5$ m, for example, takes about 3.0 s at half speed but only about 1.5 s at normal speed.

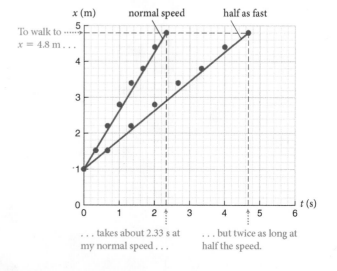

gone much faster than 10 km/h and taken a break somewhere in the middle of the trip. From the data provided, we can say only that the person's speed was 10 km/h *on average*. Hence:

An object's average speed is the distance traveled divided by the time interval required to travel that distance.

Note that this definition involves the *distance traveled* and not the *x component of the displacement*. To see why, let us again use the data from my walk. According to Table 2.1, the *x* component of my displacement was (final position)−(initial position) = (+3.4 m) − (+1.0 m) = +2.4 m over the 6.0-s duration of the trip. The distance I traveled, on the other hand, is the sum of my forward and backward distances: 4.8 m − 1.0 m = 3.8 m for the forward part and 4.8 − 3.4 = 1.4 m for the backward part. Thus the distance I traveled in 6 s was 5.2 m, so my average speed was (5.2 m)/(6.0 s) = 0.87 m/s, *not* (2.4 m)/(6.0 s) = 0.40 m/s.

If I had walked at a constant speed of 0.40 m/s for 6 s, my *x(t)* curve would be curve 1 in **Figure 2.10**. A constant speed of 0.40 m/s implies that I traveled 0.40 m every second. More generally, a constant speed during a given time interval means that when the time interval is divided into any number of equal-sized subintervals, the distance traveled is the same in all subintervals. A constant speed

Figure 2.10 Two of the infinite number of ways for getting from an initial position x_i to a final position x_f. Curve 1: walking in one direction only; curve 2: walking beyond the endpoint and then backtracking. Both ways result in a displacement of +2.4 m in 6 s.

curve 1:
Walk slowly forward from beginning to end.

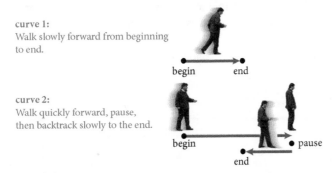

curve 2:
Walk quickly forward, pause, then backtrack slowly to the end.

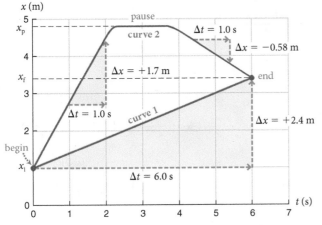

therefore shows up as a straight line in a position-versus-time graph.

Instead of walking at a constant speed from the initial position x_i directly to the final position x_f in Figure 2.10, however, I got to x_f by first walking beyond that point and then walking back, as illustrated by curve 2 in Figure 2.10.

2.8 (*a*) At what instant did I first pass the position $x_f = +3.4$ m (see curve 2 in Figure 2.10)? (*b*) What was my average speed up to that instant?

In Figure 2.10 let us use the symbol x_p to denote the point at which I paused in my walk. Walking forward, I traveled $x_p - x_i = 4.8$ m − 1.0 m = +3.8 m in 2.3 s, and so my average speed was (3.8 m)/(2.3 s) = 1.7 m/s. After pausing for 1.3 s, I traveled the distance $x_p - x_f =4.8$ m − 3.4 m = 1.4 m in 6.0 s − 3.6 s = 2.4 s. So the average speed at which I walked backward was (1.4 m)/(2.4 s) = 0.58 m/s. My higher forward speed is shown in Figure 2.10 by the steeper slope of curve 2 between x_i and x_p.

In a graph of position versus time, the steeper the slope of the curve, the higher the speed.

Another point to note about curve 2 in Figure 2.10 is that the curve slopes up from x_i to x_p and down from x_p to x_f. The reason is that I moved in the direction of increasing *x* from x_i to x_p and in the direction of decreasing *x* from x_p to x_f. Average speed, as we defined it, tells us only how fast something moves without indicating the direction in which the motion takes place. In contrast, the slope of an *x(t)* curve tells us both how fast and in what direction an object is moving. Motion in the positive *x* direction gives a rising *x(t)* curve and positive slope; motion in the negative *x* direction gives a falling *x(t)* curve and a negative slope.

The quantity that gives us both the speed and the direction of travel is **velocity**. Like displacement, velocity can be represented by an arrow (more on these arrows and their representation in the second part of this chapter). For now, we define the *x component of the average velocity* as follows:

The *x* component of an object's average velocity is the *x* component of its displacement divided by the length of the time interval taken to travel that displacement.

As before, the term *x component* serves to remind us that this quantity depends on the choice of axis and that it can be both positive and negative (in contrast to *average speed*, which is always positive). For motion in the positive *x* direction, the *x* component of the average velocity is positive; for motion in the negative *x* direction, it is negative. Just as we are doing for the *x* component of the displacement, we indicate the direction of motion along the *x* axis by using an algebraic sign. For example, in going from x_i to x_p along curve 2 in Figure 2.10, the *x* component of my average velocity is $+1.7$ m/s, meaning the motion is in the positive x direction; from x_p to x_f it is -0.58 m/s. (My *speed*, however, was 1.7 m/s between x_i and x_p and 0.58 m/s between x_p and x_f. Speed has no direction and therefore carries no algebraic sign.)

Example 2.3 Speed and velocity

With about 20 min to spare, you walk leisurely from your dorm to class, which is 1.0 km away. Halfway there, you realize you have forgotten your notebook and run back to your dorm. You walked 6.0 min before turning around, and you travel three times as far per unit of time running as you do when walking. What are (*a*) the x component of your average velocity and (*b*) your average speed over the entire trip?

❶ GETTING STARTED To begin this problem, it helps to make a sketch to visualize the situation (**Figure 2.11**).

Figure 2.11

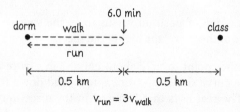

❷ DEVISE PLAN To find the x component of your average velocity, I need to divide the x component of your displacement by the time interval over which this displacement took place. To find your average speed, I must divide the distance traveled by the time interval.

❸ EXECUTE PLAN (*a*) When you are back at the dorm, the x component of your displacement is zero because your initial and final positions are the same. Therefore the x component of your average velocity, (x component of the displacement)/(time interval), is zero too. ✔

(*b*) Because you traveled halfway to class and back, and the distance between your dorm and the classroom is 1 km, the distance you traveled by the time you returned to your dorm is 1.0 km. You walked for 6.0 min before turning around. Because you run three times as fast as you walk, it takes you only 2.0 min to return to your dorm, and so the total round-trip time is 8.0 min, or 480 s. Your average speed therefore is (distance traveled)/(time interval) = $(1000 \text{ m})/(480 \text{ s}) = 2.1$ m/s. ✔

❹ EVALUATE RESULT If you start and end at the same place, your displacement is zero and so your average velocity is zero, regardless of how fast you move. The answer I obtained in part *b* means that you would cover about $(2.1 \text{ m/s})(3600 \text{ s}) = 7600 \text{ m} = 7.6$ km in one hour. This speed—$(7.6 \text{ km/h})/(1.6 \text{ km/mi}) = 4.7$ mi/h—is in the right ballpark for a person doing a combination of walking and running.

✋ **2.9** In Example 2.3, what was your average speed during the time interval that you walked (*a*) away from and (*b*) back toward your dorm? (*c*) Is the average of your answers to parts *a* and *b* the same as the average speed calculated in Example 2.3? Why or why not?

Let us now return to the motion illustrated in the film clip. In Figure 2.10 we determined the x component of the average velocity for my motion in each direction: $+1.7$ m/s walking forward, zero standing still, and -0.58 m/s walking backward. We can obtain more detailed information by calculating the x component of the average velocity from one frame to the next. We can do this by calculating the x components of the displacements between successive frames from the values of x in Table 2.1. The change in the value of x from one frame to the next gives the x component of the displacements between frames. If we then divide these displacement values by the time interval between frames (0.33 s), we get values for the x component of my average velocity that are averaged over much smaller time intervals than before. The result of this computation is shown in **Figure 2.12**.

The first thing that stands out in this graph is that the x component of the average velocity obtained from one frame to another varies quite a bit around the x component of the average velocity we calculated before. This variation is due to variations in my speed as I walked and to limits in the accuracy of determining the relatively small displacement values from one frame to the next.

Another striking feature in Figure 2.12 is that the x component of the average velocity is *negative* for $t \geq 4$ s. This is because the calculation for the x component of the displacement (final x coordinate minus initial x coordinate) automatically takes into account the direction in

Figure 2.12 Ratio of displacements to time intervals from frame to frame in Figure 2.1.

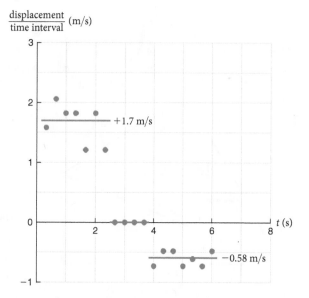

which the motion takes place. Between $t = 5.0$ s and 5.3 s, for instance, I moved from $x = +4.0$ m to $+3.8$ m (see Table 2.1), and so the x component of my displacement was $+3.8$ m $-$ $(+4.0$ m$) = -0.2$ m. Whenever x decreases, the x component of the displacement is *negative*. The time intervals, on the other hand, are always positive because time runs forward only. Therefore the ratio of the x component of the displacement to the time interval has the same algebraic sign as the x component of the displacement.

Figure 2.13 shows how the x component of the average velocity, denoted by v_x, changes as a function of time. The curve in this figure (which we'll refer to as a $v_x(t)$ *curve*) contains the same information as the $x(t)$ curve in Figure 2.6. Both are simplified representations of the motion shown in Figure 2.1.

2.10 (*a*) Suppose an object moves from an initial position $x_i = -1.2$ m to a final position $x_f = -2.3$ m without reversing its direction of travel. Is the x component of its average velocity positive or negative? (*b*) Repeat for an object moving from $x_i = +1.2$ m to $x_f = -1.2$ m. (*c*) Do your answers agree with the above statement about the algebraic sign of the x component of the average velocity being the same as the sign of the x component of the displacement?

Figure 2.13 Velocity-versus-time graph for the data shown in Figure 2.12.

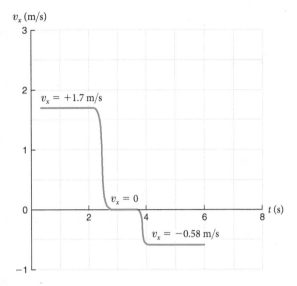

Self-quiz

1. Describe the motion represented by each graph in **Figure 2.14**.

Figure 2.14

(a)

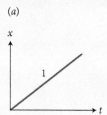

(b)

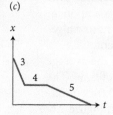

(c)

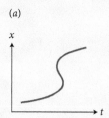

(d)

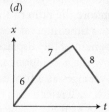

2. One or more of the graphs in **Figure 2.15** represent an impossible motion. Identify which ones and explain why the motion is not possible.

Figure 2.15

(a)

(b)

(c)

(d)

3. Two cars start from some point P, drive 10 km to a point Q, and then return to P. Car 1 completes the trip in 10 min; car 2 takes 11 min. For each car, what are (a) the x component of the average velocity and (b) the average speed?

4. Suppose you travel 10 km from P to Q at 10 km/h and another 10 km from Q to R at 20 km/h. What is your average speed during the entire trip from P to R?

5. (a) Which of the numbered lines 1–8 in the graphs in Figure 2.14 represents the greatest average speed? (b) For which of the numbered lines is the x component of the average velocity negative?

Answers

1. (a) An object moving away from the origin at constant speed; (b) an object at rest a certain distance from the origin; (c) an object moving toward the origin, then pausing halfway between its starting point and the origin, then continuing (more slowly) toward the origin; (d) an object moving away from the origin at constant speed for one-third of the time, then continuing in the same direction at a slower constant speed for another third of the time, and finally moving toward the origin at some intermediate constant speed.

2. (a) and (d). Put the left edge of a ruler along the x axis and slowly slide the ruler to the right. The point where this edge intersects the curve shows the position of the object at the instant indicated by the horizontal axis. For both (a) and (d) there are instants at which the ruler edge intersects the curve at more than one place for the same value of t. This tells you that these curves are meaningless because no object can be in two places at the same instant.

3. The x component of the displacement of both cars is zero because they end where they began. So the x component of their average velocities is zero. The distance they travel, however, is 20 km. So the average speed of car 1 is (20 km)/(10 min) = 2.0 km/min, and the average speed of car 2 is (20 km)/(11 min) = 1.8 km/min.

4. If you answered 15 km/h, think again—you cannot simply average the two speeds! To calculate the average speed, you need to know how long the entire trip from P to R took and the distance traveled. The 10 km from P to Q takes (10 km)/(10 km/h) = 1 h; the 10 km from Q to R takes (10 km)/(20 km/h) = 0.5 h, and so the total duration of the trip is 1.5 h. The average speed for the 20-km trip from P to R is thus (20 km)/(1.5 h) = 13 km/h.

5. (a) Line 3 because its slope is steepest. (b) Lines 3, 5, and 8 because the $x(t)$ curves slope downward with increasing time.

2.5 Scalars and vectors

The physical quantities we use in this book fall into two categories. The first type, called **scalars,** are completely specified by a number, which can be positive or negative, and a unit of measure. An example is temperature, a quantity that is specified completely by a number and a unit (as in 21 °C or 70 °F). The second type, called **vectors,** require, in order to be completely specified, both a magnitude (the "how much" and "of what," specified by a number and a unit) and a direction in space.

Exercise 2.4 Scalar or vector?

(*a*) Is the statement *Yesterday the car was parked 3 m from the telephone pole* sufficient to determine the position of the car? Is position a scalar or a vector? (*b*) Is the statement *New York is 200 miles from Boston* sufficient to determine the distance between the two cities? Is distance a vector or a scalar? (*c*) Is the statement *My friend walked 3 m due north from the telephone pole* sufficient to determine your friend's position at the end of her short walk? Is displacement a vector or a scalar?

SOLUTION (*a*) No. The car could have been anywhere on the perimeter of a circle of radius 3 m centered on the telephone pole. A direction is required in addition to a distance, and so position is a vector. ✔

(*b*) Yes. All that is required to specify distance is a number (200) and a unit of length (miles). Because distance does not depend on a direction in space, it is a scalar. ✔

(*c*) Yes. *3 m north* completely specifies your friend's displacement from the telephone pole because it gives both a distance (3 m) and a direction (north). Because it requires a direction, displacement is a vector. ✔

2.11 Are the following quantities vectors or scalars: (*i*) the price of a movie ticket, (*ii*) the average velocity of a ball launched vertically upward, (*iii*) the position of the corner of a rectangle, (*iv*) the length of a side of that rectangle?

Scalars follow the ordinary arithmetic laws of addition, subtraction, multiplication, and division. For example, 2 m + 3 m = 5 m and 2 × 5 s = 10 s. Vectors, on the other hand, obey special laws of addition, subtraction, and multiplication that we begin to discuss in the following sections and then learn about in more detail in Chapter 10.

The only type of motion we study through Chapter 9 is motion in one dimension, which means it occurs either forward or backward along a straight line. For now, therefore, the difference between scalars and vectors is not a major issue because in one dimension the direction of any vector can be specified by an algebraic sign. The difference between scalars and vectors becomes important when we discuss motion in more than one dimension in Chapter 10. It is therefore important that we be able to distinguish vectors from scalars, and so in this book we use arrows to denote vectors:*

$$\vec{a} \quad \text{vector}$$

$$b \quad \text{scalar.}$$

The arrow is a reminder that vectors have a direction in space and that they obey rules that are different from those for scalars. Throughout this book, we graphically represent vectors by thick arrows of appropriate length and direction and label these arrows with the corresponding symbol (see, for example, Figures 2.16 and 2.17).

QUANTITATIVE TOOLS

---

*Note that the direction of the arrow used in the vector notation (that is, the $\rightarrow$ in $\vec{a}$) does not represent the direction of the vector. In particular, it does not reverse direction when we multiply a vector by −1.

The **magnitude of a vector** tells us how much there is of that vector. In Exercise 2.4c, for instance, the magnitude of your friend's displacement is 3 m. When we draw one vector arrow longer than another in a diagram, we are saying that the vector represented by the longer arrow has a magnitude that is greater than the magnitude of the vector represented by the shorter arrow. A magnitude never has a sign associated with it.

There are two common ways of designating the magnitude of a vector symbolically: by drawing vertical lines on either side of its symbol or by removing the overarrow. Thus the magnitude of vector $\vec{b}$ is denoted by either $|\vec{b}|$ or b. In this book, we usually use b for simplicity.

To specify vectors mathematically, we introduce a **unit vector**—a vector whose sole purpose is to define a direction in space. Unit vectors have no units, and the number associated with them (their *magnitude*) is one. They are denoted by a caret or "hat" (^) to distinguish them from ordinary vectors, whose magnitude can have any value; for example, the unit vector pointing in the direction of increasing x along the x axis is denoted by $\hat{\imath}$ (pronounced *i-hat*). In **Figure 2.16a** this unit vector is represented by a black arrow that points in the direction of increasing x along the axis. Because unit vectors have a magnitude of 1 and no units, we have

$$|\hat{\imath}| \equiv 1. \tag{2.1}$$

Any vector along the x axis can be written in **unit vector notation** as the product of a scalar and this unit vector:

$$\vec{b} \equiv b_x \hat{\imath} \quad \text{(one dimension)} \tag{2.2}$$

where b_x is a scalar called the *x component of the vector $\vec{b}$*. The subscript x indicates that the quantity refers to the x axis and reminds you that the sign of b_x depends on our choice of axis direction. If the vector $\vec{b}$ points in the positive x direction (that is, in the direction of $\hat{\imath}$, the direction of increasing x), the scalar b_x is positive. If the vector $\vec{b}$ points in the negative x direction (that is, opposite the direction of $\hat{\imath}$, the direction of decreasing x), the scalar b_x is negative.

If we take the absolute value of Eq. 2.2 and substitute Eq. 2.1, we see that, for vectors in one dimension, the magnitude of the vector $\vec{b}$ is equal to the absolute value of b_x:

$$b \equiv |\vec{b}| = |b_x \hat{\imath}| = |b_x||\hat{\imath}| = |b_x| \quad \text{(one dimension).} \tag{2.3}$$

For practice, let us determine the x components of the vectors in Figure 2.16 and also write the vectors in unit vector notation. In Figure 2.16b, the vector $\vec{b}$ points in the positive x direction, which means that the scalar b_x (which is the x component of $\vec{b}$) is a positive number. The length of the arrow in Figure 2.16b tells us that the magnitude of $\vec{b}$ is 2.0 m (no sign; never a sign for a *magnitude*). From Eq. 2.3, therefore, the absolute value of b_x is 2.0 m. Consequently, the x component of $\vec{b}$ is $b_x = +2.0$ m. In unit vector notation, we write $(+2.0 \text{ m})\hat{\imath}$ for our vector $\vec{b}$.

In Figure 2.16c, the vector $\vec{c}$ points in the negative x direction, and so the scalar c_x (the x component of c) is a negative number. The magnitude of $\vec{c}$ along the x axis is 1.5 m, which means the absolute value of c_x is 1.5 m. The x component of $\vec{c}$ is thus $c_x = -1.5$ m. In unit vector notation, we write $(-1.5 \text{ m})\hat{\imath}$ for our vector $\vec{c}$.

Figure 2.17 summarizes how a vector $\vec{b}$ is represented graphically and mathematically. To remind ourselves that x components are signed numbers, we always show the algebraic sign of a component, even if it is positive (for instance, $b_x = +2.0$ m). Note also that the x component of a vector always has the same units as the quantity represented by that vector.

Figure 2.16 The product of a unit vector $\hat{\imath}$ and a scalar b_x yields a vector $\vec{b} = b_x \hat{\imath}$.

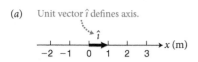

(a) Unit vector $\hat{\imath}$ defines axis.

(b) Multiplying $\hat{\imath}$ by $b_x = +2.0$ m gives $\vec{b}$.

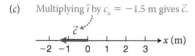

(c) Multiplying $\hat{\imath}$ by $c_x = -1.5$ m gives $\vec{c}$.

Figure 2.17 Graphical and mathematical representation of a vector.

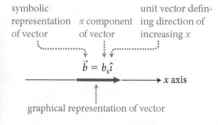

symbolic representation of vector x component of vector   unit vector defining direction of increasing x

$$\vec{b} = b_x \hat{\imath}$$

→ x axis

graphical representation of vector

- magnitude of vector: $b = |b_x|$
- If $b_x > 0$: $\vec{b}$ is in same direction as $\hat{\imath}$
- If $b_x < 0$: $\vec{b}$ is in opposite direction to $\hat{\imath}$

In Chapter 10 you'll learn how to mathematically describe vectors that do not lie along the x axis. For motion in one dimension, however, we can always align the axis along the line of motion, and so all vectors can be written in the form given by Eq. 2.2. Because x components are sufficient to completely specify vectors in one dimension, we shall for now work with mostly components rather than vectors.

✋🌀 **2.12** Consider the axis and unit vector shown in **Figure 2.18**. Let the vector $\vec{r}$ have its tail at the origin and its tip at my feet. (*a*) What is the x component of that vector? (*b*) Write the vector $\vec{r}$ in terms of its x component and the unit vector. (*c*) What does the vector $\vec{r}$ represent? (*d*) Does $r_x = +2.5$ m unambiguously determine a vector in Figure 2.18? If so, draw the vector, and state what it might represent.

2.6 Position and displacement vectors

The most fundamental vector is displacement, which, as noted in Section 2.2, can be represented graphically by an arrow that points from an object's initial position to its final position. **Figure 2.19**, for example, shows the displacement from my position in frame 3 in Figure 2.1 to my position in frame 10. This vector has a magnitude that is equal to the distance between my initial and final positions (2.6 m), and its direction is toward increasing values of x along the reference axis.

Mathematically we denote displacement by $\Delta\vec{r}$, where the Greek letter Δ (delta) means "the change in" and $\vec{r}$ is a vector specifying position (more on this vector in a bit). The change in a quantity is always taken to be the final value minus the initial value. If your height h increases from 1.75 m to 1.78 m, for instance, the change in your height is $\Delta h = 1.78$ m $- 1.75$ m $= +0.03$ m. If the number n of cookies in a jar is down to 10 from an initial 12, the change is $\Delta n = 10 - 12 = -2$ cookies.

As we saw in Section 2.2, the x component of the displacement is equal to the change in the x coordinate: $\Delta r_x \equiv \Delta x$. Because of this identity, which is specific to position and displacement, we henceforth denote the x component of the displacement by Δx rather than by Δr_x. So, if the x coordinate of an object changes from some initial value x_i at instant t_i to some final value x_f at some later instant t_f, the x component of the object's displacement is given by

$$\Delta x \equiv x_f - x_i. \tag{2.4}$$

As we saw in Section 2.2, the *x component of the displacement* and the *distance traveled* are not necessarily the same. In particular, if an object turns around during its motion, we need to add the distances between successive turning points (including the initial and final positions) in order to calculate the distance traveled. The **distance** d between two points is equal to the absolute value of the difference in the x coordinates of those two points:

$$d \equiv |x_1 - x_2| \quad \text{(one dimension)}. \tag{2.5}$$

Because distance is obtained by taking the absolute value of a quantity, it is always positive. Likewise, the distance traveled, which is the sum of the distances between successive points, is always positive.

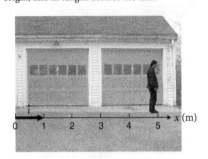

Figure 2.18 The unit vector $\hat{\imath}$ completely specifies a reference axis. The unit vector's direction determines the direction of increasing x along the axis. The tail of the vector locates the origin, and its length defines the unit.

Figure 2.19 Superposition of frames 3 and 10 of Figure 2.1. The arrow is my displacement.

Example 2.5 Displacement as a vector

On the reference axis shown in **Figure 2.20**, draw vectors representing my displacements between the following frames in Figure 2.1: (*a*) 4 and 8, (*b*) 11 and 19, (*c*) 6 and 17. Use the values listed in Table 2.1 to describe each vector in terms of its magnitude and direction.

Figure 2.20 Example 2.5.

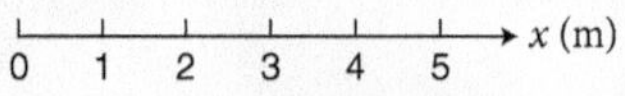

$$\xrightarrow{\quad|\quad|\quad|\quad|\quad|\quad|\quad} x\,(m)$$
$$0\quad1\quad2\quad3\quad4\quad5$$

❶ GETTING STARTED The pairs of frames given in each part of this example define the initial and final positions. For each pair, I need to determine the displacement vector.

❷ DEVISE PLAN I know that a displacement vector points from the initial position to the final position, and so I must draw the tail of the vector at the initial x coordinate and its head at the final x coordinate. Then I can use Eq. 2.3 to determine the magnitude of the displacement vector.

❸ EXECUTE PLAN (*a*) The displacement points from my initial position at $x_i = +2.8$ m to my final position at $x_f = +4.8$ m (see Table 2.1). The x component of the displacement is $\Delta x = x_f - x_i = +4.8\text{ m} - (+2.8\text{ m}) = +2.0$ m. The magnitude of this vector is $|\Delta \vec{r}| = |\Delta x| = 2.0$ m; its direction is to the right because the x component is positive (**Figure 2.21**). ✔

Figure 2.21

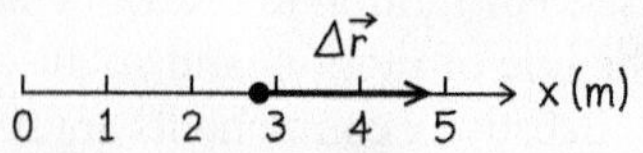

$$\xrightarrow{\quad|\quad|\quad|\quad|\quad|\quad|\quad} x\,(m)$$
$$0\quad1\quad2\quad3\quad4\quad5$$

(*b*) The displacement points from $x_i = +4.8$ m to $x_f = +3.4$ m. The x component of the displacement is $\Delta x = x_f - x_i = +3.4\text{ m} - (+4.8\text{ m}) = -1.4$ m. The magnitude of this vector is $|\Delta \vec{r}| = |\Delta x| = |-1.4\text{ m}| = 1.4$ m; its direction is to the left because the x component is negative (**Figure 2.22**). ✔

Figure 2.22

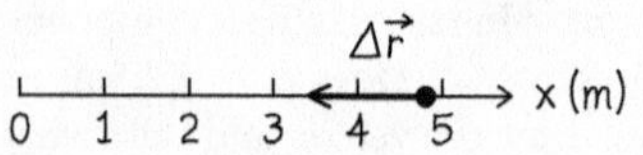

$$\xleftarrow{\quad|\quad|\quad|\quad|\quad|\quad|\quad} x\,(m)$$
$$0\quad1\quad2\quad3\quad4\quad5$$

(*c*) The displacement points from $x_i = +3.8$ m to $x_f = +3.8$ m, and thus it has zero magnitude (and no direction, of course; see **Figure 2.23**). ✔

Figure 2.23

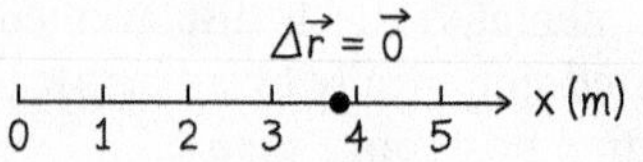

$$\Delta \vec{r} = \vec{0}$$
$$\xrightarrow{\quad|\quad|\quad|\quad|\quad|\quad|\quad} x\,(m)$$
$$0\quad1\quad2\quad3\quad4\quad5$$

❹ EVALUATE RESULT The arrow in part *a* points right, as I would expect because I moved forward (that is, to the right). In part *b* the arrow points left because I moved backward. Even though I moved between frames 6 and 17, I end at my initial position and so the displacement is zero.

When a vector has zero magnitude as in Example 2.5*c*, we say that it is equal to the *zero vector*, denoted by a zero with an arrow: $\Delta \vec{r} = \vec{0}$. In diagrams, we denote the zero vector by a dot (see Figure 2.23).

2.13 (*a*) What is the x component of the displacement shown in Figure 2.19? (*b*) Write the displacement in terms of its x component and the unit vector.

Rearranging Eq. 2.4, I get an expression for the x coordinate of the final position in terms of the x coordinate of the initial position and the x component of the displacement:

$$x_f = x_i + \Delta x. \tag{2.6}$$

This expression tells you that you can determine the x coordinate of an object's final position by adding the x coordinate of the object's initial position and the x component of its displacement.

Example 2.6 Working with displacements

(*a*) An object moves from an initial position at $x_i = +3.1$ m to a final position at $x_f = +1.4$ m. What is the x component of the object's displacement? (*b*) The x component of an object's displacement is $+2.3$ m. If the object's initial position is at $x_i = +1.6$ m, what is the x coordinate of its final position? (*c*) After undergoing a displacement of -1.3 m, an object is at $x_f = -0.4$ m. What is the x coordinate of the object's initial position?

❶ GETTING STARTED To visualize what is happening, I begin by making a sketch showing what is given for each situation and which quantities I need to determine (**Figure 2.24**).

❷ DEVISE PLAN The relationship among the initial x coordinate, the final x coordinate, and the x component of the

Figure 2.24

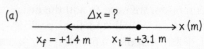

(*a*) $\Delta x = ?$
$$\xleftarrow{\hspace{3cm}\bullet\hspace{1cm}} x\,(m)$$
$x_f = +1.4$ m $\quad x_i = +3.1$ m

(*b*) $\Delta x = +2.3$ m
$$\bullet\xrightarrow{\hspace{3cm}} x\,(m)$$
$x_i = +1.6$ m $\quad x_f = ?$

(*c*) $\Delta x = -1.3$ m
$$\xleftarrow{\hspace{3cm}\bullet} x\,(m)$$
$x_f = -0.4$ m $\quad x_i = ?$

displacement is given by Eqs. 2.4 and 2.6. For each part of the problem, I know two of the three terms in each of these equations, and so I can use them to find the unknown quantity.

❸ EXECUTE PLAN (*a*) The *x* component of the displacement is given by Eq. 2.4:

$$\Delta x = x_f - x_i = +1.4 \text{ m} - (+3.1 \text{ m}) = -1.7 \text{ m}. ✔$$

(*b*) Using Eq. 2.6, I get

$$x_f = x_i + \Delta x = +1.6 \text{ m} + (+2.3 \text{ m}) = +3.9 \text{ m}. ✔$$

(*c*) Rearranging terms in Eq. 2.4, I have

$$x_i = x_f - \Delta x = -0.4 \text{ m} - (-1.3 \text{ m}) = +0.9 \text{ m}. ✔$$

❹ EVALUATE RESULT In the sketch I made, the displacement in part *a* points to the left, so the *x* component of the displacement should indeed be negative. In part *b* the displacement is to the right, so the final *x* coordinate should be larger than the initial *x* coordinate. For part *c* I expect the initial *x* coordinate to be positive because the 1.3-m displacement is larger than the distance from the final *x* coordinate to the origin (0.4 m).

Example 2.7 Displacement, distance, distance traveled

An object moves from point P at $x = +2.3$ m to point Q at $x = +4.1$ m and then to point R at $x = +1.5$ m. (*a*) What is the *x* component of the object's displacement after traveling from P to R? (*b*) What is the distance between the object's initial and final positions? (*c*) What is the distance traveled by the object?

❶ GETTING STARTED A simple sketch helps me visualize the motion and the relative locations of P, Q, and R (**Figure 2.25**).

Figure 2.25

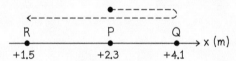

❷ DEVISE PLAN For part *a*, I can use Eq. 2.4 to calculate the *x* component of the object's displacement given x_i and x_f. In parts *b* and *c*, I need to calculate distances, which are given by Eq. 2.5.

❸ EXECUTE PLAN (*a*) The object's initial position is at $x_i = +2.3$ m and its final position is at $x_f = +1.5$ m, so the *x* component of its displacement is (see Eq. 2.4)

$$\Delta x = x_f - x_i = +1.5 \text{ m} - (+2.3 \text{ m}) = -0.8 \text{ m}. ✔$$

(*b*) The distance between the initial and final positions is $d = |x_P - x_R| = |+2.3 \text{ m} - (+1.5 \text{ m})| = 0.8$ m (see Eq. 2.5). ✔

(*c*) The distance between P and Q is $d = |x_P - x_Q| = |+2.3 \text{ m} - (+4.1 \text{ m})| = 1.8$ m; the distance between Q and R is $d = |x_Q - x_R| = |+4.1 \text{ m} - (+1.5 \text{ m})| = 2.6$ m. The distance traveled is thus 1.8 m + 2.6 m = 4.4 m. ✔

❹ EVALUATE RESULT The sketch shows that the displacement from P to R is in the negative *x* direction, and so my answer to part *a* should be negative. The distance between P and R is the absolute value of the displacement, and my answer to part *b* is indeed the absolute value of my answer to part *a*. In part *c*, the total distance traveled (4.4 m) is greater than the distance between P and R. The sketch I made readily confirms that this should be the case.

✋ **2.14** (*a*) Consider an *x* axis on which positive values of *x* are to the right of the origin. For an object initially located to the left of the origin, is the *x* component of a displacement to the right positive or negative? (*b*) Does the answer to part *a* depend on whether or not the object crosses the origin?

If displacement is a vector, then the position of an object can also be represented by a vector. To see this, multiply Eq. 2.4 through by $\hat{\imath}$:

$$\Delta x \,\hat{\imath} = (x_f - x_i)\hat{\imath} = x_f \,\hat{\imath} - x_i \,\hat{\imath}. \tag{2.7}$$

The left side represents the displacement

$$\Delta x \,\hat{\imath} = \Delta r_x \,\hat{\imath} = \Delta \vec{r}. \tag{2.8}$$

The vector $x_i \,\hat{\imath}$ in Eq. 2.7 is represented in **Figure 2.26** by an arrow running from the origin to my initial position. This arrow represents the *position vector* (or simply the **position**)—a vector that lets us determine the position of a point in space relative to some chosen origin. Mathematically, position is represented by the symbol $\vec{r}$. For one-dimensional motion we have

$$\vec{r} \equiv x \,\hat{\imath} \quad \text{(one dimension)}, \tag{2.9}$$

and so we see that the *x* component of position is equal to the *x* coordinate: $r_x \equiv x$.

Figure 2.26 Superposition of frames 3 and 10 of Figure 2.1. The arrows show my initial and final positions, $\vec{r}_i$ and $\vec{r}_f$, and the relationship of these vectors to the displacement $\Delta \vec{r}$.

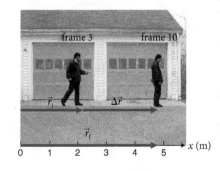

My initial position can be thus written as $x_i\,\hat{\imath} = \vec{r}_i$; my final position is $\vec{r}_f = x_f\,\hat{\imath}$, and so Eq. 2.7 becomes

$$\Delta\vec{r} = \vec{r}_f - \vec{r}_i. \tag{2.10}$$

This equation states that the displacement (the change in position) is equal to the final position minus the initial position. Figure 2.26 graphically shows the relationship between displacement $\Delta\vec{r}$ and the initial and final positions.

Exercise 2.8 Displacement and position vectors

(a) Write each displacement in Example 2.5 in terms of the unit vector $\hat{\imath}$ shown in Figure 2.18. (b) What is the magnitude of each displacement vector in part a? (c) Write my final position in Figure 2.26 in terms of the unit vector.

SOLUTION

(a) $\quad \Delta\vec{r} = (+2.0\text{ m})\hat{\imath}$

$\quad\quad \Delta\vec{r} = (-1.4\text{ m})\hat{\imath}$

$\quad\quad \Delta\vec{r} = (0)\hat{\imath} = \vec{0}.$ ✔

(b) $\quad |(+2.0\text{ m})| = 2.0\text{ m}$

$\quad\quad |(-1.4\text{ m})| = 1.4\text{ m}$

$\quad\quad |0| = 0.$ ✔

(c) $\quad \vec{r}_f = (+4.8\text{ m})\hat{\imath}.$ ✔

Rearranging Eq. 2.10, we get

$$\vec{r}_f = \vec{r}_i + \Delta\vec{r}, \tag{2.11}$$

which tells us that the final position is equal to the sum of the initial position and the displacement. Together with Figure 2.26, Eq. 2.11 gives a graphical prescription for adding two vectors:

> To add two vectors, place the tail of the second vector in the sum at the tip of the first vector; the vector representing the sum runs from the tail of the first vector to the tip of the second.

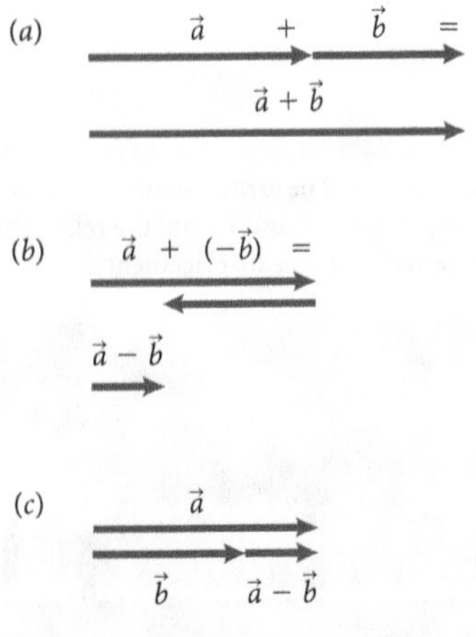

Figure 2.27*a* shows this procedure for two arbitrary vectors $\vec{a}$ and $\vec{b}$. Equation 2.10 also gives us a prescription for subtracting vectors:

$$\Delta\vec{r} = \vec{r}_f - \vec{r}_i = \vec{r}_f + (-1)\vec{r}_i. \tag{2.12}$$

This equation states that we can find the displacement $\Delta\vec{r}$ by subtracting the initial position $\vec{r}_i$ from the final position $\vec{r}_f$ or, alternatively, by adding the vector $(-1)\vec{r}_i$ to $\vec{r}_f$. The vector $(-1)\vec{r}_i$ has the same magnitude as $\vec{r}_i$ but points in the opposite direction. So the graphical prescription for subtracting two vectors is:

> To subtract a vector from another vector, reverse the direction of the vector being subtracted and add the reversed vector to the vector from which you are subtracting.

Figure 2.27*b* shows this procedure for two arbitrary vectors $\vec{a}$ and $\vec{b}$.

Figure 2.26 suggests another way of finding the difference between two vectors: The vectors $\vec{r}_i$ and $\vec{r}_f$ both have their tails at the origin, and the vector $\vec{r}_f - \vec{r}_i$ points from the tip of $\vec{r}_i$ to the tip of $\vec{r}_f$. So an alternative graphical prescription for subtracting two vectors is:

Figure 2.27 (a) Adding two vectors. (b, c) Two ways to subtract one vector from another.

To subtract a vector from another vector, place their tails at the same point and then draw a vector from the tip of the vector being subtracted to the tip of the vector from which you are subtracting.

or

The vector $\vec{a} - \vec{b}$ representing the difference between two vectors $\vec{a}$ and $\vec{b}$ points from the tip of $\vec{b}$ (the vector being subtracted) to the tip of $\vec{a}$.

This approach is illustrated in Figure 2.27c.

In later chapters we discuss rules for multiplying vectors. For now it suffices to know how to multiply a vector by a scalar. The effect of such a multiplication is the same as that of multiplying the unit vector by a scalar (see Figure 2.16): Multiplying a vector $\vec{a}$ by a scalar c yields a vector whose length equals the length of the original vector times the scalar:

$$c\,\vec{a} = c(a_x\hat{\imath}) = (ca_x)\hat{\imath}. \tag{2.13}$$

If the scalar c is positive, then the product vector points in the same direction as the original vector $\vec{a}$; if c is negative, the product vector points in the direction opposite that of $\vec{a}$.

Although our current treatment of vectors is limited to one dimension, we shall be able to use the same methods when dealing with vectors in more than one dimension in Chapter 10.

2.15 Consider the three vectors illustrated in **Figure 2.28**. (a) Make a sketch representing the vectors $\vec{a} + \vec{c}$, $\vec{a} - \vec{b}$, and $\vec{c} - \vec{b}$. (b) Are the vectors $\vec{a} + \vec{c}$ and $\vec{c} + \vec{a}$ identical? (c) Are the vectors $\vec{c} - \vec{b}$ and $\vec{b} - \vec{c}$ identical?

Figure 2.28 Checkpoint 2.15.

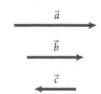

2.7 Velocity as a vector

Like displacement, average velocity is a vector. As we saw in Section 2.4, by dividing the x component of the displacement by the time interval required for that displacement, we obtain the x component of the average velocity during that time interval. Consider, for example, an object for which the x component of the displacement in the time interval between clock readings t_i and t_f is Δx. We can write the time interval as the change in the clock reading, $\Delta t = t_f - t_i$, and so the x component of the average velocity is

$$v_{x,\,av} \equiv \frac{\Delta x}{\Delta t} = \frac{x_f - x_i}{t_f - t_i}. \tag{2.14}$$

As before, the subscript x reminds us that this component is measured relative to our reference axis and that it can be either positive or negative, depending on the sign of Δx (Δt is always positive because $t_f > t_i$).

Equation 2.14 gives us the x component of the vector $\vec{v}_{av} = v_{x,\,av}\,\hat{\imath}$. Multiplying the first and middle terms of Eq. 2.14 by $\hat{\imath}$ and substituting $\Delta\vec{r} = \Delta x\,\hat{\imath}$ (Eq. 2.8), we see that average velocity can be written as

$$\vec{v}_{av} \equiv v_{x,\,av}\,\hat{\imath} = \frac{\Delta\vec{r}}{\Delta t}. \tag{2.15}$$

Because Δt is always positive ($t_f > t_i$), Eq. 2.15 shows that for motion in one dimension, the average velocity $\vec{v}_{av}$ always points in the same direction as the displacement.

Example 2.9 Average velocity and average speed

Consider my motion between frames 1 and 8 in Figure 2.1. Use the values listed in Table 2.1 to determine the answers to these questions: (a) What is my average speed over this time interval? (b) What is the x component of my average velocity over this time interval? (c) Write the average velocity in terms of the unit vector $\hat{\imath}$.

❶ GETTING STARTED Frames 1 through 8 correspond to my forward motion. To determine my average speed and velocity, I need to know both my displacement between these frames and the time interval between them.

❷ DEVISE PLAN I can get the average speed from its definition: the distance traveled divided by the time interval. Because I travel in only one direction between frames 1 and 8, the distance traveled is equal to the distance between my initial and final positions. For part b, I can use Eq. 2.14 to get the x component of the average velocity, and Eq. 2.15 shows me how to write the average velocity in unit vector form.

❸ EXECUTE PLAN (a) From Table 2.1, $|x_f - x_i| = |+4.8\text{ m} - (+1.0\text{ m})| = 3.8$ m. I traveled from x_i to x_f in time interval $\Delta t = 2.3$ s. My average speed was thus $(3.8\text{ m})/(2.3\text{ s}) = 1.7$ m/s. ✔

(b) The x component of my displacement is $\Delta x = x_f - x_i = +4.8\text{ m} - (+1.0\text{ m}) = +3.8$ m. Using Eq. 2.14, I find for the x component of average velocity $v_{x,av} = \Delta x/\Delta t = (+3.8\text{ m})/(2.3\text{ s}) = +1.7$ m/s. ✔

(c) $\vec{v}_{av} = \Delta\vec{r}/\Delta t = (+3.8\text{ m})\hat{\imath}/(2.3\text{ s}) = (+1.7\text{ m/s})\hat{\imath}$. ✔

❹ EVALUATE RESULT The answer to part a shows that I cover a distance roughly equal to my height each second, which is reasonable for a person walking. Because I walk in one direction only, the answer to part b has the same value as the answer to part a. The positive sign indicates that my motion is in the positive x direction, as I expect. My answer to part c expresses this result in unit vector form.

2.16 (a) Consider my motion between frames 6 and 17 in Figure 2.1. Use the values in Table 2.1 to determine the answers to these questions: What is my average speed over this time interval? What is the x component of my average velocity? What is the average velocity? (b) Repeat for the motion between frames 1 and 17.

2.8 Motion at constant velocity

If an object moves at constant velocity, we can drop the term *average* when referring to velocity. When we do so, the x component of the average velocity is simply v_x. According to Eq. 2.14, we thus have

$$\Delta x = v_x \Delta t \quad \text{(constant velocity).} \tag{2.16}$$

To give just one example, a car going at a constant velocity of $+10$ m/s in the x direction travels a distance of 10 m during each second of its motion. A plot of position versus time is therefore a straight line, as indicated in **Figure 2.29a**. Recall from Checkpoint 2.7 and the text immediately following it that the ratio $\Delta x/\Delta t$ is the slope of the $x(t)$ curve. Looking at Figure 2.29a, you can see that this ratio $\Delta x/\Delta t$ is the tangent of the angle that the $x(t)$ curve makes with any horizontal line drawn through the curve. So we see that for motion at constant velocity, the slope of the $x(t)$ curve is numerically equal to the x component of the velocity:

$$\frac{\Delta x}{\Delta t} = v_x \quad \text{(constant velocity).} \tag{2.17}$$

Another useful fact can be obtained by looking at the $v_x(t)$ curve for an object moving at constant velocity. Because the x component of the velocity always has the same value when the velocity is constant, this curve is a horizontal line, as shown in Figure 2.29b. Notice that the area $v_x \Delta t$ (height times width) of the shaded rectangle of width Δt under this line is, from Eq. 2.16, equal to the x component of the displacement, Δx, during the time interval.* This is true for any time interval Δt, and so:

Figure 2.29 Graphs of position and velocity as a function of time for an object moving at constant velocity.

(a) Position versus time: straight line with nonzero slope

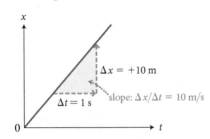

(b) Velocity versus time: horizontal line

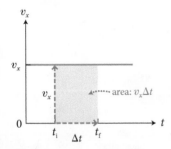

---

*The word *area* requires some caution here: The height of the rectangle represents the x component of a velocity (in meters per second), and the width represents a time interval (in seconds), and so the product—the "area"—has units of $(\text{m/s}) \times \text{s} = \text{m}$, not m^2 as would a surface area in the traditional sense. This is so because the area of the rectangle, $v_x \Delta t = \Delta x$, represents the x component of a displacement, which has units of meters.

The x component of the displacement during any time interval is given by the area under the $v_x(t)$ curve between the beginning and the end of the time interval.

Because the x component of the velocity can be positive or negative, the area can also be positive or negative, as it should given that it represents the x component of a displacement. As we shall see in Chapter 3, this correspondence between the x component of the displacement and the area under the $v_x(t)$ curve holds not just for motion at constant velocity but for any type of motion.

For an object moving at constant velocity, Eq. 2.16 gives us a way of determining the x coordinate of the position of the object at any time. Remembering that $\Delta x = x_f - x_i$, we can write Eq. 2.16 in the form

$$x_f - x_i = v_x \Delta t \quad \text{(constant velocity)},\qquad (2.18)$$

or, solving for x_f,

$$x_f = x_i + v_x \Delta t \quad \text{(constant velocity)}.\qquad (2.19)$$

This expression allows us to determine, at any time t_f, the x coordinate of the position of an object moving at constant velocity, provided we know the object's initial position coordinate x_i at time t_i.

Example 2.10 Motion at constant velocity

Figure 2.30 shows the position-versus-time graph for part of the motion of an object moving at constant velocity. (a) What is the x component of the object's velocity? (b) Write an expression for $x(t)$, the x coordinate of the position of the object at an arbitrary time t. (c) What is the x coordinate of the object's position at $t = 25$ s?

Figure 2.30 Example 2.10.

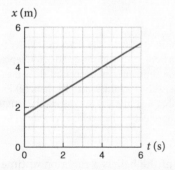

x (m)

❶ GETTING STARTED The $x(t)$ curve in Figure 2.30 is a straight line, as it should be for a motion at constant velocity. To answer the question, I have to read off information from the graph.

❷ DEVISE PLAN The x component of the object's velocity is given by Eq. 2.14. Once I have found this value, I can use Eq. 2.19 to answer part b. Although the graph doesn't show the position of the object at $t = 25$ s, I can use my solution to part b to find the position by substituting $t = 25$ s.

❸ EXECUTE PLAN (a) From the graph I see that the object's position changes from $x_i = +1.6$ m to $x_f = +5.2$ m in the time interval between $t_i = 0$ and $t_f = 6.0$ s. Substituting these values into Eq. 2.14, I find for the x component of the velocity:

$$v_x = \frac{x_f - x_i}{t_f - t_i} = \frac{+5.2\text{ m} - (+1.6\text{ m})}{6.0\text{ s} - 0} = \frac{+3.6\text{ m}}{6.0\text{ s}}$$
$$= +0.60 \text{ m/s.} ✔$$

(b) I take the object's position at $t = 0$ as the initial position. Substituting $t_i = 0$, $x_i = +1.6$ m, and $v_x = +0.60$ m/s from part a into Eq. 2.19, I get

$$x_f = +1.6\text{ m} + (+0.60\text{ m/s})(t_f - 0)$$
$$= +1.6\text{ m} + (+0.60\text{ m/s})t_f.$$

Because the choice of t_f is arbitrary, I may just as well write, for the x coordinate of the position at any instant of time t:

$$x(t) = +1.6\text{ m} + (+0.60\text{ m/s})t. ✔$$

(c) Substituting $t = 25$ s into the expression I obtained in part b yields

$$x(t = 25\text{ s}) = +1.6\text{ m} + (+0.60\text{ m/s})(25\text{ s})$$
$$= +1.6\text{ m} + 15\text{ m} = +17\text{ m.} ✔$$

❹ EVALUATE RESULT The x component of the object's velocity is positive, as I would expect from the positive slope of the $x(t)$ curve. I can double-check the value (0.60 m/s) by reading off the increase in x over a 1-s interval in the graph. The result is indeed about 0.6 m. I cannot read off the value for x for $t = 25$ s from the graph because the graph extends to only $t = 6$ s. However, from the graph I see that the object starts near $x = 2$ m and the x coordinate increases by 3 m every 5 s. In 25 s, the object displacement should therefore be 5(3 m) = 15 m, and so the object's final position should be near (2 m) + (15 m) = 17 m, which is what I got.

QUANTITATIVE TOOLS

2.17 In Example 2.10, suppose I had taken x_i to be the x coordinate of the object's position not at $t = 0$ but at $t = 2.0$ s. Show that the expression for the x coordinate of the position is the same as the one I obtained in Example 2.10b.

2.9 Instantaneous velocity

If you are driving and wonder How fast am I going *now*?, the answer to your question is provided by the speedometer. This gauge gives your *instantaneous speed*, which is generally not equal to your average speed during the entire trip. Likewise, if you drop a ball from a certain height, the ball speeds up as it falls, and so its velocity changes over time. In addition to being able to calculate the ball's average velocity as it falls over a certain distance, you might be interested in knowing its velocity at any given instant—its **instantaneous velocity.**

Figure 2.31 shows successive positions of a falling ball at 0.0300-s time intervals. Such a picture can be obtained using a strobe light to illuminate the ball at fixed intervals and thus record its position at fixed intervals. As the ball falls, the positions at which it is photographed get farther and farther apart. In other words, the magnitude of the ball's displacement per time interval (the magnitude of its average velocity, from Eq. 2.15) increases. Suppose we want to find the ball's velocity at position 2: We want to answer the question How fast is the ball moving at the instant it passes position 2? Using the ruler in the picture, we can measure how long the ball takes to fall a certain distance after position 2, but doing that gives the ball's *average* velocity over that distance. To get a value closer to the ball's instantaneous velocity at position 2, we have to make the time interval over which we measure the distance shorter and shorter.

From here on, the word *velocity* without any adjective is used to designate the instantaneous value of the velocity. When we are interested in the average value over a certain time interval, we shall always use the adjective *average*. In equations we drop the subscript av when we mean the instantaneous value—by convention, $\vec{v}_{av}$ = average velocity, $\vec{v}$ = instantaneous velocity. Likewise, *speed* always means *instantaneous speed*.

2.18 (*a*) Determine the x component of the ball's average velocity between positions 2 and 9 in Figure 2.31. (*b*) Repeat between positions 2 and 8. (*c*) Repeat for increasingly smaller time intervals. (*d*) As the time interval decreases, does the x component of the average velocity increase, decrease, or stay the same? Explain. (*e*) Plot the x components of the average velocity you obtained in parts *a–c* versus the length of the time interval. (*f*) Do you expect the x component of the average velocity to become zero as the time interval approaches zero?

Figure 2.32 shows the position of the ball as a function of time, as obtained from Figure 2.31. The x component of the average velocity measured over an interval Δt is given by Eq. 2.14, and in Checkpoint 2.18 you used that equation to evaluate the x component of the average velocity between positions 2 and 9:

$$v_{x,29} = \frac{x_9 - x_2}{t_9 - t_2}.$$ (2.20)

(I substituted 29 for the subscript av to indicate that we are averaging over the time interval between t_2 and t_9.) This calculation yields the slope of the hypotenuse of the right triangle between positions 2 and 9 in Figure 2.32a. As you reduced the size of Δt, you got closer and closer to the value of the velocity at position 2. As the checkpoint shows, the x component of the displacement gets smaller as Δt gets smaller, but the ratio of the two (the slope of the hypotenuse)

Figure 2.31 Successive positions of a falling ball recorded at 0.0300-s intervals.

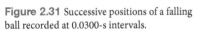
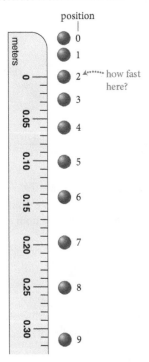

Figure 2.32 Procedure for finding the x component of the velocity of the ball in Figure 2.31 from its $x(t)$ curve. (a) As we reduce the time interval $\Delta t = t_f - t_2$ over which the x component of the velocity is averaged, Δx shrinks and the straight line between (x_2, t_2) and (x_f, t_f) approaches the tangent to the curve at (x_2, t_2). (b) The slope of the tangent at (x_2, t_2) gives the x component of the velocity at that point.

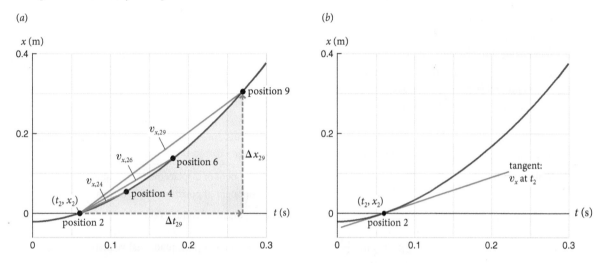

never becomes zero. This is not surprising because the x component of the ball's average velocity had some positive value while passing position 2. Thus that component should also be nonzero at position 2.

Letting the interval Δt approach zero, we obtain the x component of the velocity at instant t:

$$v_x \equiv \lim_{\Delta t \to 0} \frac{\Delta x}{\Delta t}. \tag{2.21}$$

The right side of this equation is the definition of the *derivative* of x with respect to time:

$$v_x \equiv \frac{dx}{dt}. \tag{2.22}$$

As you can see from Figure 2.32, the hypotenuse of the triangle spanned by the time interval Δt approaches the tangent to the $x(t)$ curve at time t_2, when Δt goes to zero. The x component of the velocity at instant t is thus equal to the slope of the tangent to the $x(t)$ curve at instant t.

The velocity (which is a vector) is then

$$\vec{v} = v_x \,\hat{\imath} = \left(\lim_{\Delta t \to 0} \frac{\Delta x}{\Delta t} \right) \hat{\imath}. \tag{2.23}$$

Because the unit vector is constant, we can pull it inside the parentheses and write

$$\left(\lim_{\Delta t \to 0} \frac{\Delta x}{\Delta t} \right) \hat{\imath} = \lim_{\Delta t \to 0} \frac{\Delta x \,\hat{\imath}}{\Delta t} = \lim_{\Delta t \to 0} \frac{\Delta \vec{r}}{\Delta t}, \tag{2.24}$$

where I have used the relationship $\Delta \vec{r} = \Delta x \,\hat{\imath}$ from Eq. 2.8. The term on the right in Eq. 2.24 is the derivative of the position vector, and so we can write

$$\vec{v} = \frac{d\vec{r}}{dt}. \tag{2.25}$$

The magnitude of this vector is the **speed** $v = |\vec{v}|$.

Table 2.2 Symbols and their meaning

Symbol	Meaning
t	clock reading = instant in time
$\Delta t \equiv t_f - t_i$	change in clock reading = interval of time
x	x coordinate
$d \equiv |x_1 - x_2|$	straight-line distance between two points (never has sign)
$\hat{\imath}$	unit vector defining direction of x axis (vector magnitude 1)
$\vec{b} = b_x \hat{\imath}$	vector and unit vector notation (see examples below)
b_x	x component of vector $\vec{b}$ (can be positive or negative)
$b \equiv |\vec{b}| = |b_x|$	magnitude of vector $\vec{b}$ (always positive)

Vector	Meaning	Component	Meaning
$\Delta \vec{r} = \Delta r_x \hat{\imath}$	displacement	$\Delta r_x \equiv \Delta x$	x component of displacement = change in x coordinate
$\vec{r} = r_x \hat{\imath}$	position	$r_x \equiv x$	x component of position = x coordinate
$\vec{v}_{av} = v_{x,av} \hat{\imath}$	average velocity	$v_{x,av} \equiv \dfrac{\Delta x}{\Delta t}$	x component of average velocity
$\vec{v} = v_x \hat{\imath}$	(instantaneous) velocity	$v_x \equiv \lim\limits_{\Delta t \to 0} \dfrac{\Delta x}{\Delta t} \equiv \dfrac{dx}{dt}$	x component of (instantaneous) velocity

Table 2.2 summarizes the various symbols we have introduced in this chapter and their meaning.

Example 2.11 Time intervals and derivatives

Suppose the x coordinate of the position of an object moving along the x axis varies in time according to the expression $x(t) = ct^3$, where c is a constant. Derive an expression for the x component of the velocity as a function of time.

❶ **GETTING STARTED** There are two ways to obtain the answer to this problem. One is to use differential calculus: The x component of the velocity is the derivative of the function $x(t)$ with respect to time. The other way is to evaluate the x component of the velocity over a finite time interval Δt and then let that interval go to zero. If I use both methods, I can easily verify my answer.

❷ **DEVISE PLAN** I'll use the (harder) second method first. To do so, I have to determine the change in the x coordinates between two instants t_i and $t_f = t_i + \Delta t$. The x component of the average velocity is then the ratio of this change in the x coordinate and Δt. Equation 2.21 then allows me to find the x component of the (instantaneous) velocity at any instant by taking the limit of this result as Δt goes to zero.

❸ **EXECUTE PLAN** The x coordinate of the object's initial position at time t_i is $x_i = ct_i^3$. At a later time $t_f = t_i + \Delta t$, the x coordinate of the position is

$$x_f = ct_f^3 = c(t_i + \Delta t)^3$$

$$= ct_i^3 + 3ct_i^2 \Delta t + 3ct_i(\Delta t)^2 + c(\Delta t)^3.$$

Therefore the x component of the displacement in the time interval Δt is

$$\Delta x = x_f - x_i$$

$$= \left[ct_i^3 + 3ct_i^2 \Delta t + 3ct_i(\Delta t)^2 + c(\Delta t)^3 \right] - ct_i^3$$

$$= 3ct_i^2 \Delta t + 3ct_i(\Delta t)^2 + c(\Delta t)^3.$$

The x component of the average velocity is thus

$$v_{x,av} = \frac{\Delta x}{\Delta t} = 3ct_i^2 + 3ct_i\Delta t + c(\Delta t)^2.$$

If I let Δt approach zero, the second and third terms on the right go to zero, so that I have

$$v_x = \lim_{\Delta t \to 0} \frac{\Delta x}{\Delta t} = 3ct^2, ✔$$

where I have dropped the subscript i because the result holds for any instant t. This equation gives the velocity at any time t. It tells me that v_x changes as the square of t.

❹ **EVALUATE RESULT** I can verify my result using differential calculus:

$$v_x(t) = \frac{d}{dt}x(t) = \frac{d}{dt}(ct^3) = 3ct^2.$$

Example 2.12 The limiting process

Consider again an object moving along the x axis according to the expression $x(t) = ct^3$, with $c = +0.12 \text{ m/s}^3$. (a) Determine the x component of the object's average velocity during the interval from $t_i = 1.0$ s to $t_f = 2.0$ s. (b) Repeat for the interval from $t_i = 1.0$ s to $t_f = 1.1$ s. (c) Show that your result approaches the x component of the velocity at $t = 1.0$ s if you continue to reduce the interval by factors of ten. Ignore the significant digits rules taught in Chapter 1. Instead, use all the digits your calculator shows in every step.

❶ GETTING STARTED I have to use a fixed initial instant ($t_i = 1.0$ s) and evaluate the change in the function $x(t)$ over increasingly smaller time intervals Δt, starting with $\Delta t = 1.0$ s.

❷ DEVISE PLAN Evaluating the change in $x(t)$ is just a matter of doing it, and so my plan is to roll up my sleeves and dig in!

❸ EXECUTE PLAN (a) At $t_i = 1.0$ s, the object is at $x_i = (+0.12 \text{ m/s}^3)(1.0 \text{ s})^3 = +0.12$ m. At $t_f = 2.0$ s, it is at $x_f = (+0.12 \text{ m/s}^3)(2.0 \text{ s})^3 = +0.96$ m, and so $\Delta x = +0.96$ m $-$ ($+0.12$ m) $= +0.84$ m. The x component of the average velocity therefore is $(\Delta x)/(\Delta t) = (+0.84 \text{ m})/(1.0 \text{ s}) = +0.84 \text{ m/s}$. ✔

(b) Between $t_i = 1.0$ s and $t_f = 1.1$ s, I have $\Delta x = (+0.12 \text{ m/s}^3)(1.1 \text{ s})^3 - (+0.12 \text{ m}) = +0.03972$ m, and $v_{x,\text{av}} = (+0.03972 \text{ m})/(0.10 \text{ s}) = +0.3972 \text{ m/s}$. ✔

Table 2.3 Example 2.12

Δt (s)	Δx (m)	$v_{x,\text{av}} = \Delta x / \Delta t$ (m/s)
1	+0.84	+0.84
0.1	+0.039 72	+0.3972
0.01	+0.003 636 12	+0.363 612
0.001	+0.000 360 360 12	+0.360 360 12
0.0001	+0.000 036 003 600 12	+0.360 036 0012
0.000 01	+0.000 003 600 036 00	+0.360 003 6000
0.000 001	+0.000 000 360 000 36	+0.360 000 3600
0.000 0001	+0.000 000 036 000 00	+0.360 000 0360

(c) Table 2.3 shows what happens as I continue to shorten the time interval by factors of ten. As Δt becomes smaller and smaller, $v_{x,\text{av}}$ approaches 0.36 m/s. ✔

❹ EVALUATE RESULT If I substitute $c = +0.12 \text{ m/s}^3$ and $t = 1.0$ s into the expression for the x component of the velocity obtained in Example 2.11, I should obtain the same answer as the limiting value for small intervals Δt. Indeed, I find

$$v_x(t) = 3ct^2 = 3(+0.12 \text{ m/s}^3)(1.0 \text{ s})^2 = +0.36 \text{ m/s}.$$

2.19 (a) Suppose the position for a certain object is constant. Determine the x component of its velocity by using Eq. 2.22 and by evaluating the limit in Eq. 2.21 of the ratio of the x component of the displacement over a finite time interval. (b) Repeat for the case where the x coordinate of the position changes with time according to the expression $x = ct$, where c is a constant. Explain your answer.

Chapter Glossary

SI units of physical quantities are given in parentheses.

At rest The condition of an object whose position is not changing.

Component of a vector The scalar by which the unit vector must be multiplied to generate a given vector along the axis defined by the unit vector. See also *unit vector notation*.

Displacement $\Delta \vec{r}$ (m) A vector pointing from an object's initial position to its final position:

$$\Delta \vec{r} = \vec{r}_f - \vec{r}_i. \tag{2.10}$$

Distance d (m) A distance d between two points is a scalar equal to the absolute value of the difference in the x coordinates of those two points:

$$d \equiv |x_1 - x_2| \quad \text{(one dimension)}. \tag{2.5}$$

Magnitude of a vector A positive scalar associated with a vector. The magnitude of a vector in one dimension is equal to the absolute value of the x component of the vector:

$$b \equiv |\vec{b}| = |b_x| \quad \text{(one dimension)}. \tag{2.3}$$

Position $\vec{r}$ (m) A vector that specifies the location of an object along a reference axis. The vector runs from the origin to the location of the object and can be written as the product of the x coordinate of the object's position and the unit vector $\hat{\imath}$:

$$\vec{r} \equiv x\,\hat{\imath} \quad \text{(one dimension)}. \tag{2.9}$$

Scalar A quantity that is entirely specified by a number and a unit of measure.

Speed v (m/s) The *average speed* v_{av} is a scalar equal to the distance traveled divided by the time interval it took to travel that distance. The *instantaneous speed* (or simply *speed*) is a scalar equal to the magnitude of the velocity: $v = |\vec{v}|$.

Unit vector A vector of magnitude one with no units used to specify a direction in space.

Unit vector notation A mathematical representation of a vector, the product of a scalar and a unit vector:

$$\vec{b} \equiv b_x\,\hat{\imath} \quad \text{(one dimension)}. \tag{2.2}$$

The scalar b_x is called the x component of vector $\vec{b}$. The subscript x reminds you that b_x is a signed number whose value depends on the choice of the unit vector.

Vector A quantity that requires a number, a unit, and a direction in space to be completely specified.

Velocity $\vec{v}$ (m/s) A vector that gives the time rate of change in the position of an object. The *average velocity* over any interval is obtained by dividing the displacement during that time interval by the duration of the interval:

$$\vec{v}_{av} \equiv \frac{\Delta \vec{r}}{\Delta t}. \tag{2.15}$$

The *instantaneous velocity* (or simply *velocity*) is the limit of the average velocity as the time interval goes to zero:

$$\vec{v} = \lim_{\Delta t \to 0} \frac{\Delta \vec{r}}{\Delta t} \equiv \frac{d\vec{r}}{dt}. \tag{2.25}$$

3 Acceleration

CONCEPTS

QUANTITATIVE TOOLS

In Section 2.8 we considered objects moving at constant velocity—unchanging speed and unchanging direction. Most motions, however, are not at constant velocity because things speed up and slow down and change direction all the time. Any change in velocity is caused by *acceleration*, the subject of this chapter.

3.1 Changes in velocity

Consider a car at rest at a traffic light. The light turns green, and the driver puts the car in motion. As long as the car's velocity is changing, the car is *accelerating*:

If an object's velocity is changing, the object is accelerating.

An object's velocity can change quickly or slowly: A sports car takes off much more quickly than a heavily loaded truck. Starting from rest, a sports car can reach a speed of 30 m/s in as little as 5 s; a truck might take minutes to reach the same speed. The difference between the two motions is the *rate* at which the speeds change. For example, if a sports car's speed increases from 0 to 10 m/s in 2.5 s, the rate at which the speed changes is $(10 \text{ m/s} - 0)/(2.5 \text{ s}) = 4.0 \text{ m/s}^2$. This rate, which is typically greater for a sports car than for a truck, is the magnitude of a quantity we call *acceleration*. Like velocity, acceleration is a vector, and so we need not only a magnitude but also a direction to specify it completely. We represent acceleration by the symbol $\vec{a}$.

For now, let's return to the example of the sports car and choose the x axis along its direction of travel. The rate at which the x component of the car's velocity changes is called the x component of its **average acceleration:**

The x component of the average acceleration of an object is the change in the x component of the velocity divided by the time interval during which this change took place.

I deliberately use the word *average* here because nothing prevents the x component of the car's velocity from increasing at different rates. For example, the x component of its velocity could increase from 0 to +5.0 m/s in 1.0 s, and then from +5.0 m/s to +10 m/s in 1.5 s.

The SI unit of acceleration, m/s² (meters per second squared), is a shorthand way of writing (m/s)/s (meters per second, per second). If something accelerates at +1 m/s², the x component of its velocity increases by +1 m/s each second. One of the units of time in the s² comes from the change in velocity; the other comes from the time interval during which this change occurs.

🖐 **3.1** The x component of a car's velocity increases from 0 to +5.0 m/s in 1.0 s, and then from +5.0 m/s to +10 m/s in the next 2.0 s. What is the x component of its average acceleration (*a*) during the first second, (*b*) during the last 2 seconds, and (*c*) during the entire 3.0-s interval?

Figure 3.1 The direction of the car's acceleration depends on the direction of the x axis and on whether the car is speeding up or slowing down.

(*a*) Car speeds up in positive x direction

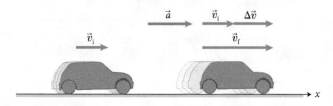

(*b*) Car slows down in positive x direction

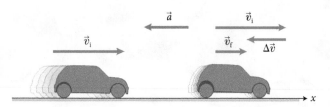

(*c*) Car slows down in negative x direction

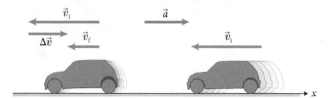

To see how the direction of acceleration relates to the motion, consider **Figure 3.1**. In Figure 3.1*a* a car traveling in the positive x direction (which means that its velocity points in the positive x direction) speeds up. The magnitude of the car's final velocity $\vec{v}_f$ is greater than the magnitude of the car's initial velocity $\vec{v}_i$, and so the change in the car's velocity $\Delta\vec{v} \equiv \vec{v}_f - \vec{v}_i$, which points from the tip of $\vec{v}_i$ to the tip of $\vec{v}_f$ (see Figure 2.27*a*), is in the positive x direction. By definition the acceleration $\vec{a}$ always points in the same direction as $\Delta\vec{v}$, and so here $\vec{a}$ points in the positive x direction. Consequently, the x component of the car's acceleration is positive: $a_x > 0$.

The car in Figure 3.1*b* also travels in the positive x direction, but this car slows down, and so $\Delta\vec{v}$ now points in the negative x direction—opposite the velocity vector. The acceleration vector, again pointing in the same direction as $\Delta\vec{v}$, also points in the negative x direction, giving a negative x component of the car's acceleration: $a_x < 0$.

Because its velocity is decreasing, you may be tempted to say that the car in Figure 3.1*b* is "decelerating"—an everyday expression used to indicate that something is slowing down. In physics, however, the word *deceleration* is not necessary and therefore best avoided; *acceleration* refers to *any* change in velocity, regardless of whether the object is speeding up or slowing down.

Figure 3.1*c* shows a surprising fact: A positive a_x does not have to mean speeding up. As you can see from the figure, the car slows down while traveling in the negative x direction. Because of the direction of travel, the car's velocity $\vec{v}$ now points in the negative x direction. The vector $\Delta\vec{v}$, however, points in the positive x direction (still from the tip of $\vec{v}_i$ to the tip of $\vec{v}_f$). The acceleration therefore also points in the positive x direction, and so the x component of the car's acceleration is *positive* even though the car slows down!

Whenever an object's velocity and acceleration vectors point in the same direction, the object speeds up. When they point in opposite directions, the object slows down.

✋ **3.2** The x component of the velocity of a car changes from -10 m/s to -2.0 m/s in 10 s. (*a*) Is the car traveling in the positive or negative x direction? (*b*) Does $\Delta\vec{v}$ point in the positive or negative x direction? (*c*) Is the x component of the car's acceleration positive or negative? (*d*) Is the car speeding up or slowing down?

When working with objects traveling in the negative x direction, don't confuse "positive x component of acceleration" with "speeding up." The x component of velocity in Figure 3.1*c*, for example, is increasing (it gets less negative), and so a_x is positive even though the car is slowing down. Something speeds up when the *magnitude* of its velocity (which, remember from Chapter 2, is called its *speed*) increases, regardless of which way the object is moving. Whether or not the x component of an object's acceleration is positive, however, depends on the direction in which the object is moving.

As mentioned in Section 2.8, the $x(t)$ curve for an object moving at constant velocity is a straight line (see Figure 2.29*a*). When the velocity is not constant, the $x(t)$ curve is not straight. In **Figure 3.2*a***, the x component of displacement Δx_2 during a certain time interval Δt beginning at t_2 is larger than the x component of displacement Δx_1 during the same time interval beginning at t_1. This means that the x component of the object's velocity is increasing. In Figure 3.2*b*, where the $x(t)$ curve bends downward, the x component of displacement Δx_2 beginning at t_2 is smaller than the x component of displacement Δx_1 beginning at t_1, telling us that the x component of the velocity is decreasing.

Figure 3.2 Position-versus-time graphs for objects accelerating along the positive x axis.

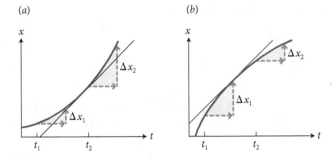

Figure 3.3 Position-versus-time graphs for objects accelerating along the positive x axis.

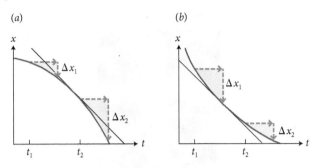

✋ **3.3** (*a*) In **Figure 3.3**, are the x components of the velocity represented by the $x(t)$ curves positive or negative? (*b*) Are the speeds increasing or decreasing? (*c*) Are the x components of $\Delta\vec{v}$ positive or negative? (*d*) Are the x components of the acceleration positive or negative?

If you answered Checkpoint 3.3 correctly, you are to be congratulated. Getting this right the first time around is not easy.

An interesting point to note is the following: In Figures 3.2*a* and 3.3*b*, the curves lie above the tangent to any point on them (they are said to *curve up* because they bend upward from the tangent), and the corresponding x components of acceleration are positive. The curves in Figures 3.2*b* and 3.3*a* lie below the tangent to any point; they are said to *curve down*, and the corresponding x components of acceleration are negative. So the sign of the x component of the acceleration is directly related to the curvature of an $x(t)$ curve:

The curvature of an $x(t)$ curve is a measure of the x component of acceleration. An upward curvature corresponds to a positive x component of acceleration. A downward curvature corresponds to a negative x component of acceleration.

✋ **3.4** (*a*) A car is speeding up in the negative x direction. In what direction do $\vec{a}$ and $\vec{v}$ point? (*b*) To which of the four graphs in Figures 3.2 and 3.3 does the situation represented in Figure 3.1*c* correspond?

3.2 Acceleration due to gravity

As an example of accelerated motion, think of an object falling in a straight line toward Earth's surface. We attribute the falling to *gravity* and say that there is a gravitational attraction between Earth and the object. Even though we don't know the ultimate cause of gravitational attraction, we can describe its effects.

Figure 3.4 Multiple-flash picture of a ball falling. The time interval between successive flashes is 0.05 s. Note how the distance traveled increases from one time interval to the next.

Figure 3.5 (*a*) Feathers and stones released from a certain height inside and outside an evacuated tube. (*b*) In air, the feather soon lags behind the stone; in vacuum, the two fall at the same speed.

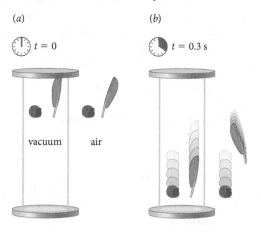

Figure 3.4 shows the positions of a falling ball recorded at equal time intervals of 0.05 s. Note how the spacing between successive positions increases as the ball drops. The picture tells us that the ball covers increasingly large distances in successive equal time intervals, and so we know that its speed increases as it falls. In other words, the ball accelerates. Two questions come to mind: *What is the magnitude of the acceleration as an object falls?* and *Does the acceleration depend on the physical characteristics of the object?*

Let us begin with the second question. If you drop a feather and a stone from the same height, there is no denying that the stone hits the ground first. This observation suggests that heavier objects fall faster than lighter ones—a common thought until the Italian mathematician, astronomer, and physicist Galileo Galilei (1564–1642) began to question its validity. Galileo observed that there is no difference between the amount of time taken by two stones to fall the same distance, even if the difference in weight is substantial and the stones are dropped from a considerable height. He argued that any observed difference in the amount of time it takes two objects to fall is a consequence of air resistance, which has a tendency to slow down lighter objects more than heavier ones.

3.5 Hold a book and a sheet of paper, cut to the same size as the book, side by side 1 m above the floor. Hold the paper parallel to the floor and the book with its covers parallel to the floor, and release them at the same instant. Which hits the floor first? Now put the paper on top of the book and repeat the experiment. What do you notice?

Figure 3.5 shows the effect of air resistance on a falling stone and a falling feather. One stone and one feather are dropped inside a glass tube from which the air has been pumped out. An identical stone and feather are released in air next to the tube. Figure 3.5*a* shows the situation at the

simultaneous release of both pairs; Figure 3.5*b* shows the situation 0.3 s after release. In air, the feather lags behind the stone. Inside the tube, however, both the stone and the feather fall the same distance. In vacuum, in other words, the feather "falls like a rock." Repeating this experiment with many different objects, we discover that, in the absence of air resistance, the acceleration due to gravity does not depend on the physical characteristics of the falling object.

The motion of objects moving under the influence of gravity only is called **free fall.** The feather inside the evacuated tube is in free fall, but the feather outside the tube is not because its motion is influenced not just by gravity but by air resistance, too.

Notice that the effect of air resistance on the stone outside the tube is hardly noticeable—it falls just as fast as the stone in vacuum. This is true for many objects as long as they are not too light and as long as they are not dropped from too great a height. So, unless you drop a light object from any height or a heavy one from the top of the Eiffel Tower, you can ignore air resistance and consider all dropped objects to be in free fall.

Let us now turn to the first question we posed earlier: What is the magnitude of the acceleration due to gravity? In **Figure 3.6a** I have plotted the positions of the ball in Figure 3.4; for each image of the ball, I measured the position from the center of the ball to the center of the topmost ball. Figure 3.6*b* shows the *x* component of the displacement between successive exposures; note how the magnitude of the displacement increases from one exposure to the next, telling us that the ball is accelerating.

3.6 What does Figure 3.6*b* tell you about the velocity of the ball from one exposure to the next?

Checkpoint 3.6 is tricky not because I wanted to mislead you but because it is important to see the difference

Figure 3.6 (*a*) Position-versus-time graph of the freely falling ball in Figure 3.4 for successive flashes $n = 1, 2, \ldots, 10$. (*b*) The change in the ball's x coordinate during the 10 exposures.

(*a*)

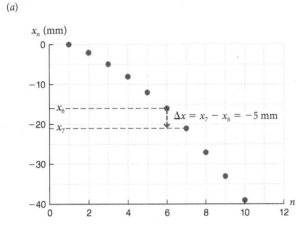

(*b*)

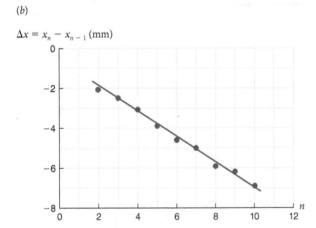

Figure 3.7 Gravity map of Earth's oceans compiled from Seasat satellite data. The coloration shows variations in the value of the acceleration due to gravity of a few tenths of a percent.

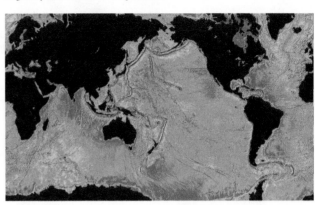

displacement increases at a constant rate. Because velocity is proportional to displacement, this graph tells us that the velocity of the ball increases at a constant rate; in other words, the ball's acceleration is constant. So, for an object in free fall, the acceleration remains *constant* throughout the object's fall toward Earth.

Experiments show that, near Earth's surface in the absence of air resistance, the magnitude of the acceleration of all objects due to gravity for all objects in free fall is 9.8 m/s^2.

In the absence of air resistance, the magnitude of the acceleration of all falling objects is 9.8 m/s^2, and consequently the amount of time it takes to fall from a certain height is the same for all falling objects.

So, each second an object falls, its speed increases by about 10 m/s. After 1 s, an object released from rest near Earth's surface has a speed of about 10 m/s. After 2 s, its speed is about 20 m/s, and so on. The value of the acceleration due to gravity varies a few tenths of a percent over Earth's surface (**Figure 3.7**).

✋ **3.7** (*a*) Does the speed of a falling object increase or decrease? (*b*) If the positive x axis points up, does v_x increase or decrease as the object falls? (*c*) Is the x component of the acceleration positive or negative?

between plotting position versus time and plotting displacement (*change* in position) versus time.*

The ball's position at a given instant is simply its distance from the starting position, which is what Figure 3.6*a* shows. What Figure 3.6*b* shows is the x component of the *displacement* $\Delta x = x_f - x_i$ from one exposure to the next. Because the motion is not at constant velocity, the displacement changes from one exposure to the next. I include this graph to show the following remarkable fact: The data points lie on a straight line, which means that the magnitude of the

---

*If you skipped the checkpoint, please do it now before I give you the answers. When you read a thriller, do you skip to the last page to find out how it ends? If you are puzzled by my questions, don't be discouraged and don't give up too soon. I wrote these questions precisely to arouse your curiosity. Discovering the answers to them is part of the learning process. Once you figure out the answers, you are unlikely to forget them. Taking the time to answer each checkpoint gives you something of lasting value. And if you like challenges, figuring out these checkpoints is without any doubt the easiest and most enjoyable way to learn physics. So take my advice to heart: Don't spoil the puzzle. You can do it—and you won't regret it!

3.3 Projectile motion

So far we have considered only the motion of freely falling objects released from rest. Let us now see what happens when we *launch* an object upward, so that it is already moving when we release it. An object that is launched but not self-propelled is called a *projectile,* and its motion is called **projectile motion.** The path the object follows is called its **trajectory.** The motion of an object in free fall is a special case of projectile motion.

Figure 3.8 A ball launched vertically upward travels to its highest position and then retraces its path; the vertical "ruler" in each frame allows us to determine the position of the ball as a function of time.

(a)

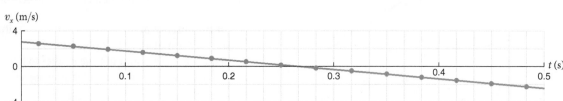

····Ball's downward path offset for clarity—ball actually falls straight down.

$\vec{v}$ x

(b) Film clip of ball's motion; frames are taken at 30 frames per second

1 2 3 4 5 6 7 8 9 10 11 12 13 14 15 16

(c) x(t) curve for ball

x (m)

0.7

0.5

0.3

0.1

0   0.1   0.2   0.3   0.4   0.5 t (s)

(d) $v_x(t)$ curve for ball

v_x (m/s)

4

0

−4

0.1   0.2   0.3   0.4   0.5 t (s)

Consider a ball launched straight up, as in **Figure 3.8a**. The ball first moves up and then reverses its direction of travel; its trajectory therefore is a straight line. Figure 3.8b shows a film clip of the motion of the ball; between frames 1 and 9 the ball moves vertically upward, and between frames 9 and 16 it moves vertically downward.

Figure 3.8c shows the ball's position for each frame in the film clip, and in Figure 3.8d I have plotted the ball's average velocity in each frame, obtained by dividing the displacement from one frame to the next by the 33-ms time interval between frames. As you can see, the velocity gets smaller and smaller as the ball nears the top of the trajectory. The ball then reverses direction, and so its velocity becomes negative. Beyond frame 9, the ball's speed—the magnitude of its velocity—increases again.

Because the ball slows down as it moves upward, its acceleration $\vec{a}$ and velocity $\vec{v}$ must point in opposite directions. Thus, because the velocity points upward, the acceleration points downward while the ball is moving upward. As the ball moves downward, it speeds up, and so now its acceleration and velocity both point downward. In other words, the ball's acceleration points downward both on the way up and on the way down.

Figure 3.8d shows that the ball's velocity changes *linearly*—it changes by equal amounts in equal time intervals—and so the acceleration is constant. The slope of the straight line in Figure 3.8d is about -10 m/s^2, which is the acceleration due to gravity. Therefore we see that, both on the way up and on the way down, the acceleration of a projectile is equal *both in magnitude and in direction* to that of a freely falling object.

If you throw a stone vertically upward and it has a velocity of $+20 \text{ m/s}$ right after leaving your hand, it slows down to $+10 \text{ m/s}$ after 1 s and reverses direction after 2 s. After 3 s, the stone is moving downward with a velocity of -10 m/s. Each second, gravity reduces the stone's velocity by another 10 m/s regardless of whether the stone is moving upward, moving downward, or instantaneously has zero velocity at the top. If you throw the stone harder, it leaves your hand with a higher speed, but once it leaves your hand, its motion is determined *solely* by gravity! In other words:

If an object is launched upward, the launch affects only the initial velocity. Once the object is released, the rest of its motion is determined by gravity alone (free fall).

🖐 **3.8** Imagine throwing a ball downward so that it has an initial speed of 10 m/s. What is its speed 1 s after you release it? 2 s after?

What happens at the very top of the trajectory of a ball launched upward? Because the x component of the ball's velocity goes from positive to negative at the top, there must be an instant at which the x component of the velocity is zero. The x component of the acceleration, however, is *not* zero at the top. To see why, consider the motion over a small time interval Δt during which the ball reaches its highest position. Just before the ball reaches the top, its velocity $\vec{v}$ points upward; just after the ball reaches the top, the velocity points downward. The vector $\Delta \vec{v}$ therefore points downward, and so does the acceleration $\vec{a}$. This holds true regardless of the size of the time interval: As we make Δt smaller, Δv_x becomes smaller, too, but the ratio $\Delta v_x / \Delta t$ remains equal to the slope of the $v_x(t)$ curve in Figure 3.8*d*. So the acceleration of the ball is the same *everywhere*: on the way up, on the way down, and at the top.

If it seems odd that the acceleration at the top is nonzero, remember that the velocity must change there from up to down, and any change in velocity—be it a change in magnitude or a change in direction—means a nonzero acceleration. That the x component of the velocity is zero doesn't matter: The x component of the velocity doesn't remain zero for any finite time interval; it is only *instantaneously* zero, and in order for it to change, there must be an acceleration. Note, also, that even though the x component of the velocity is instantaneously zero, it is not correct to say that the ball is "instantaneously at rest"; *at rest* means that the position of the ball is not changing (see Section 2.1), which is not the case at any position along the ball's trajectory, including the very top.

3.4 Motion diagrams

To help us visualize the motion of an object described in a problem, we make a **motion diagram** representing the position of the moving object at equally spaced time intervals. Motion diagrams look very similar to the multiple-flash picture shown in Figure 3.4 and provide a visual overview of all you know about the motion. The procedure for making such diagrams is explained in the box on page 60.

Suppose a bicycle rider going 8.0 m/s applies the brake and slows down to a stop over a distance of 20 m in 5.2 s. The motion diagram in **Figure 3.9** shows that same information in graphical form. Note how the diagram shows all

Figure 3.9 Motion diagram for bicycle rider slowing down to a stop.

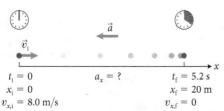

$t_i = 0$
$x_i = 0$
$v_{x,i} = 8.0 \text{ m/s}$

$a_x = ?$

$t_f = 5.2 \text{ s}$
$x_f = 20 \text{ m}$
$v_{x,f} = 0$

we know about the motion: the initial velocity, the distance it takes to come to a stop, and the time interval over which the stopping takes place. The spacing of the dots shows that the object of interest is slowing down. The question mark shows what information is missing. Making such a diagram helps you visualize the problem, organize the information given, and choose a convenient reference axis.

Nearly all problems in physics—and motion diagrams in particular—involve considering changes that occur over a certain time interval. Throughout this book we shall use clocks as in Figure 3.9 to mark the beginning and end of the time interval of interest. An important part of any physics problem is to record the parameters that characterize the situation under consideration at the beginning and end of that time interval (the *initial* and *final conditions*). When drawing motion diagrams, for example, we record an object's initial and final positions as well as its initial and final velocities. We begin applying this procedure in the next two examples.

Exercise 3.1 Accelerating car

Starting from rest, a car accelerates for 7.0 s at 3.0 m/s² and then travels at constant velocity for another 4.0 s. Make a motion diagram representing this motion.

SOLUTION The motion consists of two parts: constant acceleration during the first 7.0 s and constant velocity (zero acceleration) during the last 4.0 s. In my motion diagram (**Figure 3.10**) I show both parts of the motion: On the left the spacing between the dots increases to show the acceleration; on the right the spacing is constant to show motion at constant velocity.

Figure 3.10

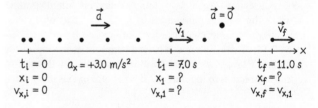

$t_i = 0$ $a_x = +3.0 \text{ m/s}^2$ $t_1 = 7.0 \text{ s}$ $t_f = 11.0 \text{ s}$
$x_i = 0$ $x_1 = ?$ $x_f = ?$
$v_{x,i} = 0$ $v_{x,1} = ?$ $v_{x,f} = v_{x,1}$

I chose a positive x axis pointing in the direction of travel and with the origin at the car's starting position. The initial conditions are $t_i = 0$, $x_i = 0$, and $v_{x,i} = 0$.

Before specifying the final conditions, I must consider the instant where the car stops accelerating. This instant, which I denote by t_1, occurs 7.0 s after the car begins to accelerate. Neither the car's position nor its velocity at that instant is known, so I denote these values by question marks in the diagram.

The car then moves at constant velocity for 4.0 s, so the final conditions are $t_f = 11.0 \text{ s}$, $x_f = ?$, and $v_{x,f} = v_{x,1}$. I complete the diagram by indicating that the x component of the car's acceleration is $+3.0 \text{ m/s}^2$ between t_i and t_1 and zero between t_1 and t_f. ✔

✔ The checkmark shows that I carried out the following checks: **Vectors and scalars used appropriately; Everything answered; No unknowns left; Units correct; Significant digits justified.**

Procedure: Analyzing motion using motion diagrams

When you solve motion problems, it is important to begin by making a diagram that summarizes what you know about the motion.

1. Use dots to represent the moving object at equally spaced time intervals. If the object moves at constant speed, the dots are evenly spaced; if the object speeds up, the spacing between the dots increases; if the object slows down, the spacing decreases.
2. Choose an x (position) axis that is convenient for the problem. Most often this is an axis that (a) has its origin at the initial or final position of the object and (b) is oriented in the direction of motion or acceleration.
3. Specify the position and velocity at all relevant instants. In particular, specify the *initial conditions*—position and velocity at the beginning of the time interval of interest—and the *final conditions*—position and velocity at the end of that time interval. Also specify all positions where the velocity reverses direction or the acceleration changes. Label any unknown parameters with a question mark.
4. Indicate the acceleration of the object between all the instants specified in step 3.
5. To consider the motion of more than one object, draw separate diagrams side by side, one for each object, using one common x axis.
6. If the object reverses direction, separate the motion diagram into two parts, one for each direction of travel.

Example 3.2 Can this be?

A newspaper article you read claims that by the time it reaches the ground, a stone dropped from the top of the Empire State Building (which has approximately 100 floors) "travels the length of a window faster than you can say *Watch out! A stone!*" Estimate whether this is true.

❶ GETTING STARTED At first this problem appears ill-defined. What am I supposed to calculate? Where to begin? If the answer to the first question is not immediately clear, the answer to the second is: with a motion diagram. I therefore begin by organizing the information following the procedure in the Procedure box "Analyzing motion using motion diagrams". I show the motion represented by a series of dots of increasing spacing to represent the acceleration due to gravity (**Figure 3.11**). I chose an x axis having its origin at the top of the building (the stone's initial position) and pointing downward (in the direction of the motion). This choice of axis means that downward corresponds to the *positive x* direction—x increases as the stone falls. The initial conditions are then

$$t_i = 0, \quad x_i = 0, \quad v_{x,i} = 0.$$

The clock reading t_i and position x_i are zero because of my arbitrary choice of reference point. I could have chosen any other values, but the choice I made simplifies things later on. That the x component of the initial velocity is zero is an assumption because the problem doesn't specify this detail. All I know is that the stone is dropped, and it seems reasonable to assume it is not thrown downward or upward, but simply released from rest.

Because I am interested in the stone's speed just as it reaches the ground, I choose that instant as the final instant. To estimate the distance between the initial and final positions, I assume that each floor is about 3 m high, making the building about 300 m high.* So the final conditions are

$$t_f = ?, \quad x_f = +300 \text{ m}, \quad v_{x,f} = ?$$

Between $x_i = 0$ and $x_f = +300$ m, the stone accelerates downward because of gravity. I make a simplifying assumption that air resistance doesn't affect the acceleration of the stone very much, so I write "$a_x = +9.8 \text{ m/s}^2$" between the initial and final positions.†

❷ DEVISE PLAN To complete the estimating task given in the problem statement, I need to compare the speed of the stone at the bottom of its fall to the speed it would have to have if it traveled the length of a window in the same time interval it takes to say "Watch out! A stone!" Because no values are given, I'll have to use estimates and then evaluate the speed using the relation among speed, distance traveled, and acceleration.

Figure 3.11

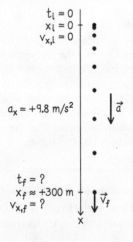

$t_i = 0$
$x_i = 0$
$v_{x,i} = 0$

$a_x = +9.8 \text{ m/s}^2$ $\vec{a}$

$t_f = ?$
$x_f \approx +300 \text{ m}$
$v_{x,f} = ?$ $\vec{v}_f$

x

---

* The Empire State Building is 381 m high, and so my estimate is off by about 30%, but you will soon see that this error is irrelevant.

† Note that a_x is positive because both the acceleration and the x axis point downward. This assumption is almost certainly an oversimplification because air resistance cannot be ignored for a fall from about 300 m. However, we have not yet developed the tools to include air resistance, and so I have no other choice.

③ EXECUTE PLAN I estimate that it takes roughly 2 s to say those words and a window is about 2 m tall. So, as long as the stone is traveling faster than $(2\text{ m})/(2\text{ s}) = 1\text{ m/s}$ just before it hits the ground, it passes by a window before the four words can be uttered. I also know that the acceleration in free fall is about 10 m/s^2, so the speed of the stone increases by 10 m/s each second. That means that the stone's speed is 10 m/s 1 s after it is released. I know from experience that it takes far longer than 1 s for an object to fall such a great distance, and so, even without further calculation, I see that the statement must be true. ✔

④ EVALUATE RESULT Even if I account for some slowing down due to air resistance, I know from experience that a stone dropped from even a much smaller height travels more than 1 m in 1 s just before it reaches the ground. (In the second part of this chapter, you will be able to use the motion diagram to calculate how fast the stone moves just before hitting the ground.)

3.9 Make a motion diagram for the following situation: A seaside cliff rises 30 m above the ocean surface, and a person standing at the edge of the cliff launches a rock vertically upward at a speed of 15 m/s. After reaching the top of its trajectory, the rock falls into the water.

Self-quiz

1. Two stones are released from rest at a certain height, one 1 s after the other. (*a*) Once the second stone is released, does the difference in their speeds increase, decrease, or stay the same? (*b*) Does their separation increase, decrease, or stay the same? (*c*) Is the time interval between the instants at which they hit the ground less than, equal to, or greater than 1 s? (Use $x(t)$ curves to help you visualize this problem.)

2. Which of the graphs in **Figure 3.12** depict(s) an object that starts from rest at the origin and then speeds up in the positive x direction?

Figure 3.12

(*a*) 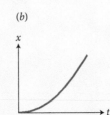 (*b*) 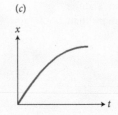 (*c*) 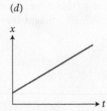 (*d*)

3. Which of the graphs in **Figure 3.13** depict(s) an object that starts from a positive position with a positive x component of velocity and accelerates in the negative x direction?

Figure 3.13

(*a*) 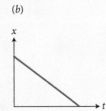 (*b*) 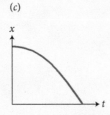 (*c*) 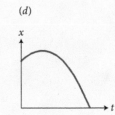 (*d*)

4. Draw a motion diagram for an object that initially has a negative x component of velocity but has a positive x component of acceleration.

Answers

1. (*a*) Both stones accelerate at about 10 m/s², and so the speed of each increases about 10 m/s each second. Because the speeds increase at the same rate, the difference in the speeds remains the same. (*b*) As indicated by the vertical lines between the $x(t)$ curves for both stones in **Figure 3.14**, the separation increases because the speed of the first stone is always greater. (*c*) Both stones execute the same motion—they start from rest and have the same acceleration and travel the same distance. So the second stone always remains 1 s behind (the $x(t)$ curves are separated horizontally by 1 s at all instants; see the horizontal lines separating the curves in Figure 3.14). Therefore the second stone hits the ground 1 s after the first one hits.

2. Choice *b* is the correct answer because its initial position is zero and the slope is initially zero but then increasing, indicating that the object speeds up. Choice *a* depicts a motion that does not begin at the origin and has the object speeding up in the negative x direction. Choice *c* depicts an object slowing down while moving in the positive x direction. Choice *d* depicts an object that is initially not at the origin and that moves at constant velocity in the positive x direction.

3. Choice *d*. Choice *a* does not have a positive initial position. Choice *b* represents zero acceleration. Choice *c* represents zero initial velocity.

4. An object that has a negative x component of velocity must slow down in order to have a positive x component of acceleration. **Figure 3.15** shows what a motion diagram for this situation looks like.

Figure 3.14

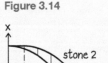

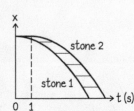

Figure 3.15

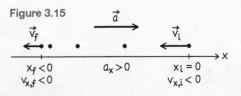

3.5 Motion with constant acceleration

As we saw in Section 3.1, the x component of the average acceleration is the change in the x component of the velocity divided by the time interval during which this change took place:

$$a_{x,\text{av}} \equiv \frac{\Delta v_x}{\Delta t} = \frac{v_{x,\text{f}} - v_{x,\text{i}}}{t_\text{f} - t_\text{i}}. \tag{3.1}$$

Notice the similarity between this definition and the definition of average velocity (Eq. 2.14):

$$v_{x,\text{av}} \equiv \frac{\Delta x}{\Delta t} = \frac{x_\text{f} - x_\text{i}}{t_\text{f} - t_\text{i}}. \tag{3.2}$$

Velocity is the rate of change in position, and acceleration is the rate of change in velocity.

Let us begin by analyzing motion with constant acceleration. The x component of the velocity of an object whose acceleration is constant increases by equal amounts Δv_x during equal time intervals Δt. For example, the x component of the velocity of an object that has a constant acceleration $a_x = +5\ \text{m/s}^2$ increases by $5\ \text{m/s}$ each second. If we plot the object's x component of velocity versus time, we therefore get a straight line with a nonzero slope, as in **Figure 3.16**.

To obtain an expression that allows us to determine the x component of the final velocity $v_{x,\text{f}}$ at some final instant t_f, we rewrite Eq. 3.1 in the form*

$$\Delta v_x \equiv v_{x,\text{f}} - v_{x,\text{i}} = a_x \Delta t \quad \text{(constant acceleration)} \tag{3.3}$$

or

$$v_{x,\text{f}} = v_{x,\text{i}} + a_x \Delta t \quad \text{(constant acceleration).} \tag{3.4}$$

So, if we know the x component of the initial velocity $v_{x,\text{i}}$ at some initial instant t_i for an object moving with constant acceleration, we add $a_x \Delta t$ to this x component of the velocity to obtain the x component of the final velocity $v_{x,\text{f}}$ at a later instant $t_\text{f} = t_\text{i} + \Delta t$.

How do we determine the displacement Δx for an accelerating object? In Section 2.8 we learned that the displacement of an object moving at constant velocity is equal to the area under the $v_x(t)$ curve. The motion we are considering now is not at constant velocity because acceleration is involved, but we can still apply what we know. We begin by approximating the $v_x(t)$ curve of Figure 3.16 by the stepped curve shown at the top of the shaded area in **Figure 3.17a**. The shaded area consists of many small rectangles of equal width δt adjusted in height to fit under the stepped $v_x(t)$ curve. (We use a lowercase Greek delta, δ, instead of an uppercase delta, Δ, to indicate that the time interval is small.) The area of each rectangle gives the displacement δx during an interval δt, and this area is the x component of the velocity at the beginning of the time interval multiplied by δt. The total displacement $\Delta x = x_\text{f} - x_\text{i}$ in the time interval from t_i to t_f is approximately equal to the sum of the areas of all the rectangles. The smaller the width of the rectangles, the more accurate the approximation.

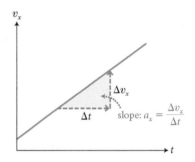

Figure 3.16 For an object moving with constant acceleration, the $v_x(t)$ curve is a straight line.

Figure 3.17 Determining the displacement for an object with constant acceleration.

(a)

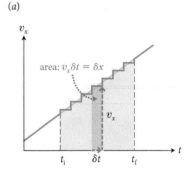

(b)

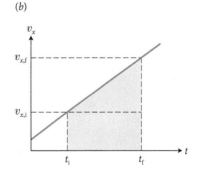

---

*Because the acceleration is constant, it always has the same (average) value, and so we can drop the subscript av.

In the limit that Δt approaches zero, the displacement $\Delta x = x_f - x_i$ is given exactly by the trapezoidal area under the $v_x(t)$ curve between t_i and t_f in Figure 3.17*b*. This area is equal to the sum of the areas of the shaded triangle and rectangle in Figure 3.17*b*. The area of the triangle is

$$\tfrac{1}{2}(v_{x,f} - v_{x,i})(t_f - t_i) = \tfrac{1}{2}\Delta v_x\,\Delta t = \tfrac{1}{2}a_x(\Delta t)^2, \tag{3.5}$$

where I have used Eq. 3.3 to substitute $a_x\Delta t$ for Δv_x. The area of the rectangle is

$$(v_{x,i} - 0)(t_f - t_i) = v_{x,i}\Delta t. \tag{3.6}$$

The displacement is thus the sum of Eqs. 3.5 and 3.6:

$$x_f - x_i = v_{x,i}\Delta t + \tfrac{1}{2}a_x(\Delta t)^2 \quad \text{(constant acceleration)}, \tag{3.7}$$

and the object's final position is

$$x_f = x_i + v_{x,i}\Delta t + \tfrac{1}{2}a_x(\Delta t)^2 \quad \text{(constant acceleration)}. \tag{3.8}$$

Equations 3.4 and 3.8 can be written without the delta notation when we reset our clock to zero at the beginning of a motion segment, so that $t_i = 0$ and $\Delta t = t_f - t_i = t_f$. Equations 3.8 and 3.4 then become, respectively,

$$x_f = x_i + v_{x,i}t_f + \tfrac{1}{2}a_x t_f^2 \quad \text{(constant acceleration)} \tag{3.9}$$

and
$$v_{x,f} = v_{x,i} + a_x t_f \quad \text{(constant acceleration)}, \tag{3.10}$$

giving us the position and the x component of the velocity of an object at any instant t_f.

Equations 3.9 and 3.10 are the basic equations for motion at constant acceleration when the initial conditions at $t_i = 0$ are known. Note that when $a_x = 0$, these equations become identical to the equations for motion at constant velocity: Eq. 3.10 yields an x component of the velocity v_x that is constant, and Eq. 3.9 becomes identical to Eq. 2.19.

Equation 3.9 gives the final position of an object at instant t_f. Because this instant is arbitrary, we can drop the subscript f to obtain an expression for the $x(t)$ curve of an object moving at constant acceleration:

$$x(t) = x_i + v_{x,i}t + \tfrac{1}{2}a_x t^2 \quad \text{(constant acceleration)}. \tag{3.11}$$

This expression shows that, for motion with constant acceleration, position depends on the square of t, and the $x(t)$ curve is thus a parabola. If a_x is positive, the parabola curves upward; if it is negative, the parabola curves downward (for an example, see Figure 3.8*c*). Likewise, from Eq. 3.10 we see that the $v_x(t)$ curve for motion at constant acceleration is a straight line:

$$v_x(t) = v_{x,i} + a_x t \quad \text{(constant acceleration)}. \tag{3.12}$$

For motion at constant velocity, we can set $a_x = 0$ in Eqs. 3.11 and 3.12, yielding straight lines for both $x(t)$ and $v_x(t)$. Table 3.1 summarizes the three basic types of motions we have studied so far: at rest, constant velocity, and constant acceleration.

Table 3.1 Kinematics graphs for three basic types of motion

	Motion diagram	Position versus time	Velocity versus time	Acceleration versus time
At rest		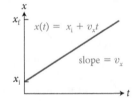	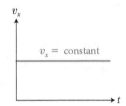	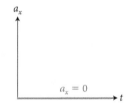
Constant velocity		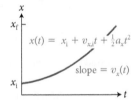	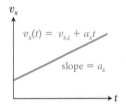	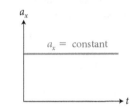
Constant acceleration	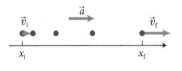			

Example 3.3 Accelerating car, again

Use Eqs. 3.11 and 3.12 to determine how far the car in Exercise 3.1 travels.

① GETTING STARTED To quantitatively analyze the situation in Exercise 3.1, I begin by separating the motion diagram in Figure 3.10 into two parts: one representing the acceleration of the car (**Figure 3.18a**) and one representing its motion at constant speed (Figure 3.18b). Note that the initial position and velocity in Figure 3.18b are not labeled with a question mark; even though they are not given, they are identical to the final position and velocity in Figure 3.18a, and so their values can be determined by solving the first part of the problem. The only unknown in the second part is the final position of the car.

② DEVISE PLAN Equation 3.11 allows me to determine the position of the accelerating car during the first part of the motion. To calculate the distance traveled during the second part of the motion, I need to know the velocity of the car. I can obtain this velocity from Eq. 3.12. From Eq. 2.18, I know that the distance traveled is then the product of this constant velocity and the duration of the time interval (4.0 s). Adding this distance to the position of the car at the end of the first part of the motion then gives me the total distance traveled by the car.

Figure 3.18

(a)

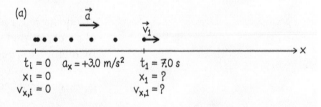

(b)

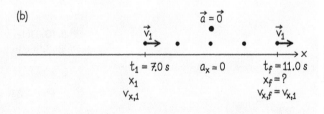

(Continued)

QUANTITATIVE TOOLS

❸ **EXECUTE PLAN** Given that $x_i = 0$ and $v_{x,i} = 0$, Eq. 3.11 reduces to $x_1 = x(t_1) = \frac{1}{2}a_x t_1^2$. Substituting $a_x = +3.0 \text{ m/s}^2$ and $t_1 = 7.0 \text{ s}$, I obtain for the position of the car at the end of the first part of the motion:

$$x_1 = \tfrac{1}{2}(+3.0 \text{ m/s}^2)(7.0 \text{ s})^2 = +74 \text{ m}.$$

From Eq. 3.12, I obtain the velocity at the beginning of the second interval:

$$v_1 = 0 + (+3.0 \text{ m/s}^2)(7.0 \text{ s}) = +21 \text{ m/s}.$$

At this constant velocity, the car's displacement during the 4.0 s of the second interval is

$$\Delta x = v_{x,1}\Delta t = (+21 \text{ m/s})(11.0 \text{ s} - 7.0 \text{ s}) = +84 \text{ m},$$

and so the x component of the car's total displacement is $(+74 \text{ m}) + (+84 \text{ m}) = +158 \text{ m}$. The car in Exercise 3.1 thus traveled a distance of 158 m. ✔

❹ **EVALUATE RESULT** With an acceleration of 3.0 m/s^2 (about one-third of the acceleration due to gravity), the car reaches a speed of

$$\frac{21 \text{ m}}{1 \text{ s}} \times \frac{1 \text{ mi}}{1609 \text{ m}} \times \frac{3600 \text{ s}}{1 \text{ h}} = 47 \text{ mi/h}$$

over the 7.0 s of the first interval and travels a distance of 74 m or about the length of a sports field. Both the car's final speed and the distance traveled are in line with what I would expect. During the second part of the motion, the car travels roughly the same distance in 4.0 s, which is also reasonable.

Example 3.4 Collision or not?

You are bicycling at a steady 6.0 m/s when someone suddenly walks into your path 2.5 m ahead. You immediately apply the brakes, which slow you down at 6.0 m/s^2. Do you stop in time to avoid a collision?

❶ **GETTING STARTED** In order to avoid a collision, you must come to a stop in less than 2.5 m. The problem therefore boils down to calculating the distance traveled under the given conditions and comparing this distance to 2.5 m.

❷ **DEVISE PLAN** Equation 3.7 gives me the displacement, but I don't know the time interval Δt. I can obtain this time interval from Eq. 3.4: $\Delta t = (v_{x,f} - v_{x,i})/a_x$, which contains no unknowns on the right side.

❸ **EXECUTE PLAN** Substituting the expression for the time interval from Eq. 3.4 into Eq. 3.7 gives the x component of the displacement necessary to stop:

$$\Delta x = v_{x,i}\frac{v_{x,f} - v_{x,i}}{a_x} + \tfrac{1}{2}a_x\left(\frac{v_{x,f} - v_{x,i}}{a_x}\right)^2 = \frac{v_{x,f}^2 - v_{x,i}^2}{2a_x}. \quad (1)$$

When I let the x axis point in the direction of the motion, the x component of the initial velocity is $v_{x,i} = +6.0 \text{ m/s}$, the x component of the final velocity is zero, $v_{x,f} = 0$, and the x component of the acceleration is $a_x = -6.0 \text{ m/s}^2$. Substituting these values into Eq. 1, I get

$$\Delta x = \frac{0 - (+6.0 \text{ m/s})^2}{2(-6.0 \text{ m/s}^2)} = +3.0 \text{ m},$$

which is more than the 2.5 m separating you from the person walking into your path. You will not be able to stop in time to avoid a collision. ✔

❹ **EVALUATE RESULT** Because $a_x = -6.0 \text{ m/s}^2$, your speed decreases by 6.0 m/s each 1 s, and so it takes 1 s to come to a stop from your initial speed of 6.0 m/s. Because your speed is decreasing, you travel less than 6.0 m, but without carrying out the calculation, I can't tell whether or not that distance is less than 2.5 m. All I can really say is that my answer should be less than 6.0 m, which it is.

Equation 3.4 allows us to determine the x component of the final velocity of an object undergoing constant acceleration after a certain time interval Δt. By multiplying both sides of Eq. 1 in Example 3.4 by $2a_x$ and rearranging terms, I get an expression for the x component of the final velocity of an object undergoing constant acceleration over a certain displacement Δx:

$$v_{x,f}^2 = v_{x,i}^2 + 2a_x\Delta x \quad \text{(constant acceleration).} \qquad (3.13)$$

3.10 Determine the velocity of the stone dropped from the top of the Empire State Building in Example 3.2 just before the stone hits the ground.

3.6 Free-fall equations

The magnitude of the **acceleration due to gravity** is designated by the letter g:

$$g \equiv |\vec{a}_{\text{free fall}}|. \qquad (3.14)$$

As noted in Section 3.2, near Earth's surface the magnitude of the acceleration due to gravity is $g = 9.8 \text{ m/s}^2$. The direction of this acceleration is always downward, and so if we choose a positive x axis that points upward, the x component of the acceleration due to gravity is given by $a_x = -g$, where the minus sign takes into account that the acceleration vector points downward.

If an object is dropped from a certain height with zero initial velocity along an upward-pointing x axis, then, according to Eqs. 3.9 and 3.10,

$$x_f = x_i - \tfrac{1}{2}gt_f^2 \qquad\qquad (3.15)$$

and

$$v_{x,f} = -gt_f. \qquad\qquad (3.16)$$

Example 3.5 Dropping the ball

Suppose a ball is dropped from height $h = 20$ m above the ground. How long does it take to hit the ground, and what is its velocity just before it hits?

❶ GETTING STARTED I begin by drawing a motion diagram, arbitrarily choosing an x axis that points upward and has its origin at the initial position of the ball (**Figure 3.19**). I make two simplifying assumptions: that the ball is released from rest ($v_{x,i} = 0$ at $t_i = 0$) and that I may ignore air resistance (so that the ball is in free fall).

Figure 3.19

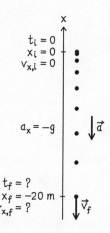

When the ball is released from rest at the origin, the initial conditions are

$$t_i = 0, \quad x_i = 0, \quad v_{x,i} = 0.$$

The final position x_f at instant t_f is a distance h below the initial position, and so, just before impact at instant t_f, the final conditions are

$$t_f = ?, \quad x_f = -h, \quad v_{x,f} = ?.$$

As the ball falls, its acceleration is negative (because $\vec{a}$ points toward increasingly negative values of x), and so $a_x = -g$.

❷ DEVISE PLAN Because the acceleration of the ball is constant, I can use Eqs. 3.9 and 3.10. Doing so gives me two equations to determine the two unknowns t_f and $v_{x,f}$.

❸ EXECUTE PLAN Substituting the initial and final conditions into Eq. 3.9 yields

$$-h = 0 + 0 - \tfrac{1}{2}gt_f^2 = -\tfrac{1}{2}gt_f^2,$$

and so

$$t_f = \sqrt{\frac{2h}{g}}.$$

Substituting $h = 20$ m and $g = 9.8$ m/s^2, I obtain for the amount of time the ball takes to fall 20 m

$$\Delta t = t_f - t_i = \sqrt{\frac{2h}{g}} - 0 = \sqrt{\frac{2(20\text{ m})}{9.8\text{ m/s}^2}} = \sqrt{4.0\text{ s}^2} = 2.0\text{ s.} ✔$$
$$(1)$$

Next I use Eq. 3.10. Because the ball starts from rest, I have

$$v_{x,f} = 0 - gt_f = -gt_f = -(9.8\text{ m/s}^2)(2.0\text{ s})$$

$$= -20\text{ m/s.} ✔$$

❹ EVALUATE RESULT The 2.0 s I obtained for the time interval it takes the ball to fall is not unreasonable: 20 m corresponds approximately to the height of a six-story building, and from experience I know that something dropped from such a height takes a few seconds to reach the ground. My answer for the final velocity $v_{x,f} = -20$ m/s also makes sense: The x component of the velocity is negative because it points in the negative x direction, and if the ball were moving always at a constant speed of 20 m/s, it would cover the 20-m distance in 1 s. Because it moves at that speed only at the end of the drop, it takes longer to fall, as Eq. 1 shows.

3.11 Repeat Example 3.5 for an x axis that points downward (leaving the origin at the ball's initial position).

Example 3.6 A stone launched up

A stone is launched straight up from ground level at a speed of 8.0 m/s. (a) How high does it rise? (b) How many seconds does it take for the stone to hit the ground?

① GETTING STARTED I am asked to consider two parts of the trajectory: part *a* from the launch position to the top of the trajectory and part *b* from the launch position to the top and back down to the launch position. I begin therefore by drawing two motion diagrams: one for the upward motion and the other for the downward motion. I (arbitrarily) choose an *x* axis pointing upward, put the origin at the launching position, and start the clock at $t_i = 0$ at the instant the stone is launched (Figure 3.20).

Figure 3.20

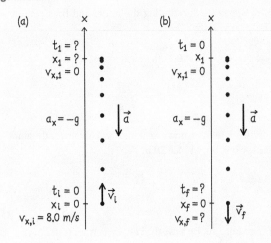

For the upward motion (Figure 3.20a), the initial conditions are $t_i = 0$, $x_i = 0$, and $v_{x,i} = +8.0$ m/s. The hard part of this problem is specifying the final conditions, and here is where the motion diagram helps. I know neither the instant t_1 at which the highest position is reached nor the location x_1 of that position. I do, however, know *something* about the stone at the highest position. (Put the book down and think for a second—you know it, too.)

The key is the word *highest*: The stone reverses direction at the highest position, and so its velocity must be zero there. So, for the upward part of the motion, the stone's final conditions are $t_f = t_1 = ?$, $x_f = x_1 = ?$ and $v_{x,f} = v_{x,1} = 0$.

For the downward motion, I reset the clock to zero at the top, as indicated by $t_1 = 0$ in Figure 3.20b. Doing this simplifies the calculation because it makes this part of the problem nearly identical to Example 3.5. I should not forget, however, that $\Delta t = t_f - t_i = t_f - 0 = t_f$ now represents the duration of the downward motion only, not the total duration for the upward and downward travel.

② DEVISE PLAN To determine the height reached by the stone, I need to obtain a value for x_1. The stone is undergoing constant acceleration, so its position and velocity are given by Eqs. 3.8 and 3.4. For the upward part of the motion I see from my

motion diagram that there are two unknowns (x_1 and t_1), and so the number of unknowns is equal to the number of equations. After I obtain x_1 and t_1, the downward part of the motion also contains two unknowns ($v_{x,f}$ and t_f), which I can calculate using the same two equations.

③ EXECUTE PLAN (a) For the upward portion of the motion I can use Eqs. 3.9 and 3.10 instead of Eqs. 3.8 and 3.4 because the clock reads zero at the beginning of the motion. With the *x* axis pointing upward, the acceleration of the stone is $a_x = -g$, and so these equations become

$$x_1 = 0 + v_{x,i}t_f - \tfrac{1}{2}gt_f^2 \tag{1}$$

and

$$0 = v_{x,i} - gt_f. \tag{2}$$

Solving Eq. 2 for t_f, I get

$$t_f = \frac{v_{x,i}}{g} = \frac{+8.0 \text{ m/s}}{9.8 \text{ m/s}^2} = 0.82 \text{ s}. \tag{3}$$

Substituting this result into Eq. 1, I determine the height to which the stone rises

$$x_1 = v_{x,i}\left(\frac{v_{x,i}}{g}\right) - \tfrac{1}{2}g\left(\frac{v_{x,i}}{g}\right)^2 = \frac{v_{x,i}^2}{g} - \frac{v_{x,i}^2}{2g} = \frac{v_{x,i}^2}{2g}, \tag{4}$$

and so

$$x_1 = \frac{(+8.0 \text{ m/s})^2}{2(9.8 \text{ m/s}^2)} = +3.3 \text{ m}. \checkmark$$

(b) For the downward part of the motion, Eq. 3.7 becomes

$$0 - x_1 = 0 - \tfrac{1}{2}g(t_f - t_1)^2.$$

Rearranging terms and substituting Eq. 4 for x_1, I have

$$(t_f - t_1)^2 = \frac{2x_1}{g} = \frac{2v_{x,i}^2}{2g^2} = \left(\frac{v_{x,i}}{g}\right)^2,$$

which shows that the amount of time it takes for the stone to fall back to the ground, $\Delta t = t_f - t_1 = v_{x,i}/g$, is identical to the amount of time it takes for the stone to reach its highest position (see Eq. 3).

The amount of time it takes for the round trip is

$$t_f = 2\left(\frac{v_{x,i}}{g}\right) = 2\left(\frac{8.0 \text{ m/s}}{9.8 \text{ m/s}^2}\right) = 1.6 \text{ s}. \checkmark$$

④ EVALUATE RESULT That it takes as much time to travel up as it takes to fall back down makes sense if you look at the position-versus-time graph for a projectile in Figure 3.8c. The parabola is symmetrical about the maximum, meaning that the acceleration is the same in both directions, and so both parts of the motion must take the same amount of time. The stone's initial velocity is 8.0 m/s, so at an acceleration of about 10 m/s², it should take 0.8 s to slow down to zero and reach its highest

position. The downward part of the motion is similar to the upward part, with the motion reversed, and so the total time should be twice as long. At a constant 8.0 m/s, the stone would travel (0.8 s)(8.0 m/s) = 6.4 m, but because it slows it travels only half that distance. My answers appear to make sense!

✋ **3.12** In Example 3.6, what is the x component of the velocity of the stone just before it hits the ground?

3.7 Inclined planes

Galileo concluded that the acceleration due to gravity is constant even though he was unable to study falling objects using high-speed video or multiple-flash photography. Instead, he used inclined planes at shallow angles of incline to reduce the acceleration. By rolling balls along such inclined planes, he came to the following remarkable conclusion:

> **When a ball rolls down an incline starting from rest, the ratio of the distance traveled to the square of the amount of time needed to travel that distance is constant.**

For instance, if a ball released from position $x_i = 0$ at $t_i = 0$ reaches position x_1 at instant t_1, x_2 at t_2, and x_3 at t_3 (**Figure 3.21**), Galileo found experimentally that

$$\frac{x_1}{t_1^2} = \frac{x_2}{t_2^2} = \frac{x_3}{t_3^2}. \tag{3.17}$$

Note that for an object moving with constant acceleration and starting from rest at $x_i = 0$ and $t_i = 0$, we have, from Eq. 3.9,

$$x_f = \tfrac{1}{2} a_x t_f^2, \tag{3.18}$$

so that

$$\frac{x_f}{t_f^2} = \tfrac{1}{2} a_x, \tag{3.19}$$

which tells us that Galileo's ratio (Eq. 3.17) is directly proportional to the x component of the acceleration. The fact that the ratio doesn't change as the ball rolls down the incline demonstrates that the acceleration is constant.

When Galileo increased the angle of the incline, the ratio x_f/t_f^2 increased, too, although for any fixed angle it remained constant regardless of the distance covered by the ball. Galileo was able to study only a limited range of angles, but he conjectured that for a given angle of incline the ratio x_f/t_f^2 would remain constant. For a vertical incline, he reasoned, the ball would be in free fall and still have a constant acceleration.

We saw in Section 3.2 that Galileo's conjecture was correct. **Figure 3.22a** shows a low-friction cart rolling down inclines of different angles. For each angle, the component of the velocity along the incline increases linearly with time, indicating that the component of the acceleration along the incline is constant for that angle. Moreover, the component of the acceleration along the incline (the slope of the lines in Figure 3.22b) becomes larger as the angle is increased. Figure 3.22c

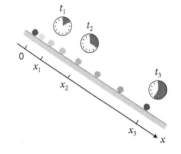

Figure 3.21 Galileo measured how long it takes a ball to roll down planes inclined at different angles and concluded that the acceleration due to gravity is constant.

Figure 3.22 The acceleration of a low-friction cart rolling down an incline increases as the angle of incline increases.

(a)

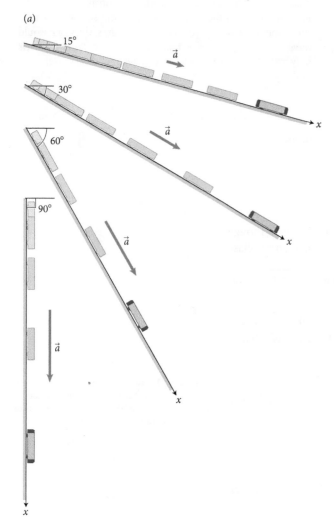

(b) Component of cart's velocity along incline as function of time

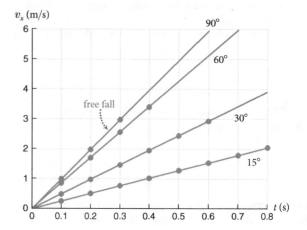

(c) Component of acceleration along incline as function of angle of incline

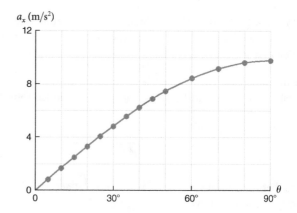

shows how the component of the acceleration along the incline varies with the angle of incline θ. Experimentally we determine that the component of the acceleration along the incline obeys the relationship

$$a_x = +g \sin \theta \tag{3.20}$$

when the x axis points downward along the incline. For a vertical incline, $\theta = 90°$ and Eq. 3.20 correctly yields the x component of the acceleration of a freely falling object.

3.13 As the angle θ of the incline used to collect the data of Figure 3.22 is increased beyond 90°, what happens to the acceleration? Does this result make sense (provided you always put the cart on the top side of the track)?

3.8 Instantaneous acceleration

So far, we have discussed only motion that has constant acceleration, but not all acceleration is constant. For example, when a car accelerates, the acceleration levels off as the car approaches top speed. This is illustrated in Figure 3.23. If the car's acceleration were constant, the $v_x(t)$ curve would be the dashed straight line in Figure 3.23a; instead, the x component of the velocity levels off at about $+50 \text{ m/s}$.

To determine the car's acceleration at a given instant, we follow the procedure of successive approximations we used in Section 2.9 to determine instantaneous velocity. Suppose we are interested in the x component of the car's acceleration 10 s after it begins accelerating. The x component of the *average* acceleration is obtained by dividing the change in the x component of the velocity during a certain time interval by the duration of that time interval (Eq. 3.1). For example, the x component of the car's average acceleration during the 25-s time interval following the instant we are interested in is given by the slope of the straight line segment PQ_1 in Figure 3.23a. To obtain the x component of the acceleration at $t = 10$ s, we make the time interval Δt over which we evaluate Δv_x successively shorter. As Δt approaches zero, the straight line segment PQ becomes tangent to the $v_x(t)$ curve at P, and the ratio $\Delta v_x / \Delta t$ approaches the x component of the **instantaneous acceleration** (or simply **acceleration**)* at point P:

$$a_x \equiv \lim_{\Delta t \to 0} \frac{\Delta v_x}{\Delta t}. \tag{3.21}$$

The right side of Eq. 3.21 is the definition of the derivative of v_x with respect to t, and so

$$a_x = \frac{dv_x}{dt}. \tag{3.22}$$

Note that the result is again a function of time. The acceleration-versus-time graph for the accelerating car of Figure 3.23 is shown in **Figure 3.24**. The slope of the $v_x(t)$ curve at a given point along the curve of Figure 3.23b gives the x component of the acceleration at that instant. For example, the value of the x component of the acceleration 10 s after the car begins accelerating is equal to the slope of the tangent to the $v_x(t)$ curve at point P (of Figure 3.23b).

In Section 2.9 we saw that the slope of the $x(t)$ curve for an object gives the x component of the object's velocity. Now we see that the slope of the $v_x(t)$ curve gives the x component of the object's acceleration:

$$\text{position } x \quad \xrightarrow{\begin{array}{c} \textit{slope of} \\ x(t) \textit{ curve} \end{array}} \quad \begin{array}{c} x \text{ component of} \\ \text{instantaneous} \\ \text{velocity } \vec{v} \end{array} \quad \xrightarrow{\begin{array}{c} \textit{slope of} \\ v_x(t) \textit{ curve} \end{array}} \quad \begin{array}{c} x \text{ component of} \\ \text{instantaneous} \\ \text{acceleration } \vec{a} \end{array}$$

Mathematically this can be expressed as

$$a_x = \frac{dv_x}{dt} = \frac{d}{dt}\left(\frac{dx}{dt}\right) \equiv \frac{d^2x}{dt^2}, \tag{3.23}$$

where the expression on the right side represents the *second derivative* (the "derivative of the derivative") of x with respect to t.

Equation 3.23 allows us to obtain the x component of the acceleration of an object from its position. Often we are interested in the inverse procedure, however: What is the displacement during a given time interval for, say, the varying

Figure 3.23 Velocity-versus-time graphs for an accelerating car. (*a*) As the time interval Δt over which the x component of the acceleration is averaged gets smaller, Δv_x gets smaller and the straight-line segment PQ approaches the tangent to the curve at P. (*b*) The slope of the tangent to the curve at P gives the x component of the instantaneous acceleration at that instant.

(*a*)

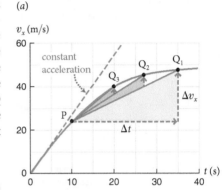

(*b*)

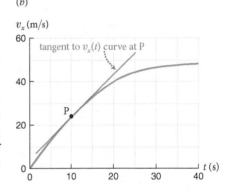

---

* As with velocity, the word *instantaneous* is implied whenever we say "acceleration." Only when we want to speak of *average* acceleration will we use the adjective.

Figure 3.24 Determining the change in the x component of the velocity for an accelerating object.

(a) To find change in velocity over time interval, we approximate acceleration curve by steps of constant acceleration.

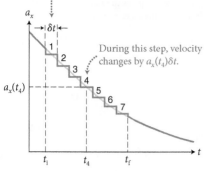

During this step, velocity changes by $a_x(t_4)\delta t$.

(b) Velocity change over entire interval is sum of velocity changes over steps.

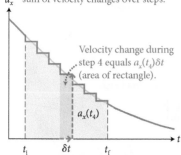

Velocity change during step 4 equals $a_x(t_4)\delta t$ (area of rectangle).

(c) As we make δt smaller, approximation improves and sum of rectangular areas approaches area under curve.

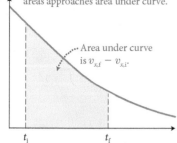

Area under curve is $v_{x,f} - v_{x,i}$.

acceleration shown in Figure 3.24? The trick is to apply what we know about constant-acceleration motion over time intervals small enough to make the acceleration approximately constant over that time interval.

Suppose we want to know the displacement during the time interval between t_i and t_f. We first obtain the change in velocity during that time interval and then calculate the displacement.

We begin by dividing up the time interval between t_i and t_f into a number of equal shorter intervals δt and approximating the acceleration by the stepped series of short segments of constant acceleration as shown in Figure 3.24a. Applying Eq. 3.3 to each shorter interval δt, we can write the x component of the change in velocity during each segment as

$$(\Delta v_x)_n = a_x(t_n)\delta t, \tag{3.24}$$

where $n = 1, 2, \ldots, 7$ and $a_x(t_n)$ is the x component of the acceleration at t_n (this instant is at the middle of each short interval, as shown in Figure 3.24a). The product $a_x(t_n)\delta t$ is the area of the dark shaded rectangle in Figure 3.24b. The x component of the velocity change during the entire interval from t_i to t_f is then approximately equal to the sum of the $(\Delta v_x)_n$ for all the short intervals δt between t_i and t_f:

$$\Delta v_x \approx \sum_n (\Delta v_x)_n = \sum_n a_x(t_n)\delta t, \tag{3.25}$$

where $\sum a_x(t_n)\delta t$ represents the sum of all the rectangular areas under the steps between t_i and t_f (the light shaded area in Figure 3.24b).

To get a better approximation, we can divide the time interval $t_f - t_i$ into a larger number of smaller intervals δt, so that the variation of the acceleration over each interval becomes even smaller and the red stepped curve is closer to the actual (light red) $a_x(t)$ curve.

An exact result is obtained if we let δt approach zero (and the number of small intervals approaches infinity), as in Figure 3.24c:

$$\Delta v_x = \lim_{\delta t \to 0} \sum_n a_x(t_n)\delta t, \tag{3.26}$$

which is precisely the definition of the *integral* of the x component of the acceleration with respect to t from t_i to t_f:

$$\Delta v_x = \int_{t_i}^{t_f} a_x(t)dt, \tag{3.27}$$

This integral represents the area under the $a_x(t)$ curve between t_i and t_f in Figure 3.24c. Integrals, like derivatives, can be evaluated mathematically, providing us with a convenient method for calculating the right side of Eq. 3.27 and allowing us to determine velocity from acceleration.

Once we know the velocity, we can use the same approach to obtain the displacement from the velocity. **Figure 3.25** graphically shows how to determine the x component of the displacement by evaluating the area under the $v_x(t)$ curve. In analogy to Eq. 3.27 we can thus write

$$\Delta x = \int_{t_i}^{t_f} v_x(t)dt. \tag{3.28}$$

This expression allows us to mathematically determine the x component of the displacement from a $v_x(t)$ curve. The x component of the velocity, in turn, can be obtained from the $a_x(t)$ curve (Eq. 3.27).

In Section 2.8, I showed that for motion at constant velocity the area under a $v_x(t)$ curve is equal to the x component of the object's displacement. Now we see that this holds true for accelerated motion as well. The geometrical relationship among acceleration, velocity change, and displacement is thus

$$\text{acceleration } a_x \quad \xrightarrow{\substack{\text{area under}\\ a_x(t) \text{ curve}}} \quad \substack{\text{velocity}\\ \text{change } \Delta v_x\\ \text{from } t_\text{i} \text{ to } t_\text{f}} \quad \xrightarrow{\substack{\text{area under}\\ v_x(t) \text{ curve}}} \quad \substack{\text{displacement } \Delta x\\ \text{from } t_\text{i} \text{ to } t_\text{f}}$$

Exercise 3.7 Using calculus to determine displacement

Suppose an object initially at x_i at $t_\text{i} = 0$ has a constant acceleration whose x component is a_x. Use calculus to show that the x component of the velocity and the x coordinate at some final instant t_f are given by Eqs. 3.10 and Eq. 3.9, respectively.

SOLUTION Because the acceleration is constant, I can pull a_x out of the integration in Eq. 3.27:

$$\Delta v_x = v_{x,\text{f}} - v_{x,\text{i}} = \int_{t_\text{i}}^{t_\text{f}} a_x dt = a_x \int_{t_\text{i}}^{t_\text{f}} dt = a_x(t_\text{f} - t_\text{i}).$$

Substituting $t_\text{i} = 0$ and rearranging terms, I obtain Eq. 3.10:

$$v_{x,\text{f}} = v_{x,\text{i}} + a_x t_\text{f}. \checkmark$$

For an arbitrary final instant t, I can drop the subscript f. Substituting this expression into Eq. 3.28, I get

$$\Delta x = \int_{t_\text{i}}^{t_\text{f}} (v_{x,\text{i}} + a_x t) dt$$

or, pulling constant terms out of the integration and then carrying out the integration,

$$\Delta x = x_\text{f} - x_\text{i} = v_{x,\text{i}} \int_{t_\text{i}}^{t_\text{f}} dt + a_x \int_{t_\text{i}}^{t_\text{f}} t \, dt = v_{x,\text{i}} t_\text{f} + a_x \left[\tfrac{1}{2} t^2 \right]_{t_\text{i}}^{t_\text{f}},$$

which yields Eq. 3.9:

$$x_\text{f} = x_\text{i} + v_{x,\text{i}} t_\text{f} + \tfrac{1}{2} a_x t_\text{f}^2. \checkmark$$

3.14 Take the first and second time derivatives of x_f in Eq. 3.9. What do you notice?

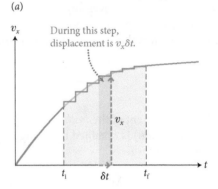
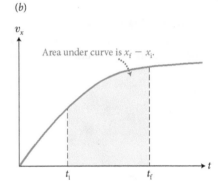

Figure 3.25 (*a*) To determine the x component of the displacement, we approximate the $v_x(t)$ curve by short segments of constant velocity. (*b*) When the segments are made infinitesimally small, the x component of the displacement becomes equal to the area under the $v_x(t)$ curve.

(*a*)

During this step, displacement is $v_x \delta t$.

(*b*)

Area under curve is $x_\text{f} - x_\text{i}$.

Chapter Glossary

SI units of physical quantities are given in parentheses.

Acceleration $\vec{a}$ (m/s^2) A vector that gives the time rate of change in the velocity of an object. The x component of the *average acceleration* over any interval is obtained by dividing the change in the x component of the velocity during that time interval by the duration of the interval:

$$a_x \equiv \frac{\Delta v_x}{\Delta t}. \tag{3.1}$$

The x component of the *instantaneous acceleration* (or simply *acceleration*) is obtained by taking the limit as the time interval goes to zero:

$$a_x = \lim_{\Delta t \to 0} \frac{\Delta v_x}{\Delta t} \equiv \frac{dv_x}{dt} = \frac{d}{dt}\left(\frac{dx}{dt}\right) = \frac{d^2x}{dt^2}. \tag{3.23}$$

Acceleration due to gravity (m/s^2) The acceleration of an object in free fall. Near Earth's surface and in the absence of air resistance, the magnitude of this acceleration is

$$g \equiv |\vec{a}_{\text{free fall}}| = 9.8 \text{ m/s}^2 \quad \text{(near Earth's surface)}.$$

Free fall The motion of an object subject to only the influence of gravity.

Motion diagram A diagram that shows the position of a moving object at equally spaced time intervals and summarizes what is known about the object's initial and final conditions (its position, velocity, and acceleration). See the procedure on page 60.

Projectile motion The motion of an object that is launched but not self-propelled. (Such an object is called a *projectile*.) The launch affects only the object's initial velocity. Once the object is launched, its motion is determined by gravity only, and hence the object is in free fall on the way up as well as on the way down.

Trajectory The path taken by a projectile. For motion at constant acceleration, the x coordinate and the x component of the velocity of an object are given by

$$x_f = x_i + v_{x,i}\Delta t + \tfrac{1}{2}a_x(\Delta t)^2 \text{ (constant acceleration)} \tag{3.8}$$

$$v_{x,f} = v_{x,i} + a_x\Delta t \qquad \text{(constant acceleration).} \tag{3.4}$$

4 Momentum

CONCEPTS

QUANTITATIVE TOOLS

In the preceding two chapters, we developed a mathematical framework for describing motion along a straight line. In this chapter, we continue our study of motion by investigating inertia, a property of objects that affects their motion. The experiments we carry out in studying inertia lead us to discover one of the most fundamental laws in physics—conservation of momentum.

4.1 Friction

Picture a block of wood sitting motionless on a smooth wooden surface. If you give the block a shove, it slides some distance but eventually comes to rest. Depending on the smoothness of the block and the smoothness of the wooden surface, this stopping may happen sooner or it may happen later. If the two surfaces in contact are very smooth and slippery, the block slides for a longer time interval than if the surfaces are rough or sticky. This you know from everyday experience: A hockey puck slides easily on ice but not on a rough road.

Figure 4.1 shows how the velocity of a wooden block decreases on three different surfaces. The slowing down is due to *friction*—the resistance to motion that one surface or object encounters when moving over another. Notice that, during the interval covered by the velocity-versus-time graph, the velocity decrease as the block slides over ice is hardly observable. The block slides easily over ice because there is very little friction between the two surfaces. The effect of friction is to bring two objects to rest with respect to each other—in this case the wooden block and the surface it is sliding on. The less friction there is, the longer it takes for the block to come to rest.

Figure 4.1 Velocity-versus-time graph for a wooden block sliding on three different surfaces. The rougher the surface, the more quickly the velocity decreases.

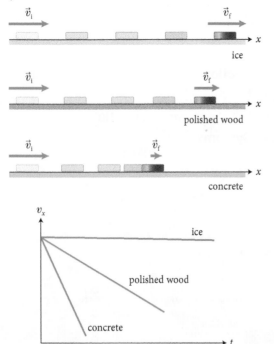

Figure 4.2 Low-friction track and carts used in the experiments described in this chapter.

You may wonder whether it is possible to make surfaces that have no friction at all, such that an object, once given a shove, continues to glide forever. There is no totally frictionless surface over which objects slide forever, but there are ways to minimize friction. You can, for instance, float an object on a cushion of air. This is most easily accomplished with a low-friction track—a track whose surface is dotted with little holes through which pressurized air blows. The air serves as a cushion on which a conveniently shaped object can float, with friction between the object and the track all but eliminated. Alternatively, one can use wheeled carts with low-friction bearings on an ordinary track. Figure 4.2 shows low-friction carts you may have encountered in your lab or class. Although there is still some friction both for low-friction tracks and for the track shown in Figure 4.2, this friction is so small that it can be ignored during an experiment. For example, if the track in Figure 4.2 is horizontal, carts move along its length without slowing down appreciably. In other words:

In the absence of friction, objects moving along a horizontal track keep moving without slowing down.

Another advantage of using such carts is that the track constrains the motion to being along a straight line. We can then use a high-speed camera to record the cart's position at various instants, and from that information determine its speed and acceleration.

4.1 (*a*) Are the accelerations of the motions shown in Figure 4.1 constant? (*b*) For which surface is the acceleration largest in magnitude?

4.2 Inertia

We can discover one of the most fundamental principles of physics by studying how the velocities of two low-friction carts change when the carts collide. Let's first see what happens with two identical carts. We call these *standard carts* because we'll use them as a standard against which to compare the motion of other carts. First we put one standard cart on the low-friction track and make sure it doesn't move. Next we place the second cart on the track some distance from the first one and give the second cart a shove toward the first. The two carts collide, and the collision alters the velocities of both.

Figure 4.3 shows a high-speed film sequence of such a collision. Cart 1 is initially at rest, and consequently its position doesn't change in the first seven frames. Cart 2 approaches cart 1 from the left and collides with it 60 ms after

Figure 4.3 (*a*) Two identical carts collide on a low-friction track. The length of track visible in each frame is 0.40 m. Cart 1 is initially at rest; cart 2 collides with it in the middle of the sequence. (*b*) Two curves, one marking the position of the rear of cart 1 and the other marking the position of the front of cart 2, superimposed on the film clip.

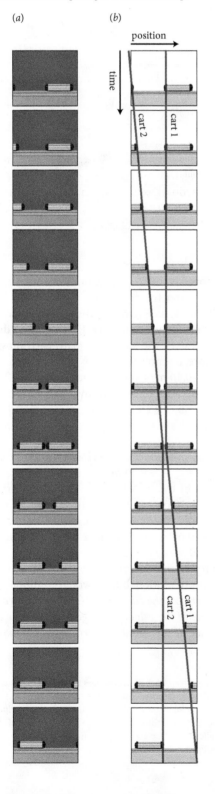

Figure 4.4 Position-versus-time graph for the two carts in Figure 4.3. The curves correspond to the curves of Figure 4.3*b*. All 120 original frames in the sequence were used to obtain this graph.

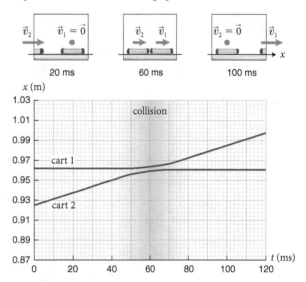

the beginning of the film sequence. The collision causes cart 1 to move to the right and cart 2 to stop dead in its tracks (or, rather, track).

Measuring the positions of the two carts at various instants gives us the $x(t)$ curves shown in **Figure 4.4**, where the shaded region shows the time interval during which the collision took place. Although the collision appears "instantaneous" to any observer, it takes about 10 ms for the motion of the carts to adjust.

Figure 4.5, which shows the velocities of the two carts, tells us that initially the velocities are constant: 0 for cart 1 and about $+0.58$ m/s (the plus sign indicating motion to the right) for cart 2. After the collision, the velocities are

Figure 4.5 Velocity-versus-time graph for the carts of Figure 4.3 before and after the collision. Before the collision, cart 1 is at rest; after the collision, cart 2 is at rest and cart 1 moves at the initial velocity of cart 2.

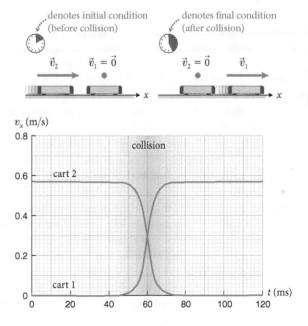

CONCEPTS

Figure 4.6 Velocity-versus-time graph for two identical carts before and after a collision on a low-friction track. Both carts are initially moving in the same direction.

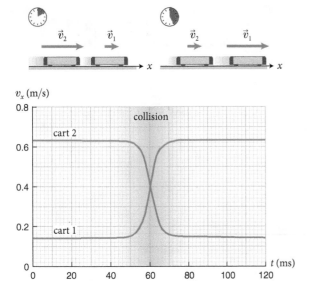

Figure 4.7 Velocity-versus-time graph for a standard cart and a double cart before and after the two collide on a low-friction track. The collision sets the double cart into motion and reverses the direction of travel of the standard cart.

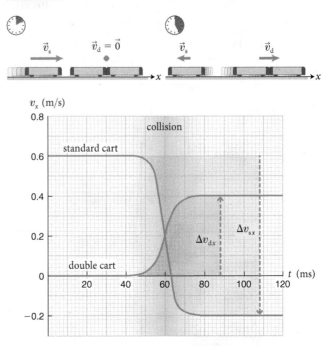

interchanged: Cart 1 now moves to the right at $+0.58$ m/s, and cart 2 has come to a complete stop. We can also see this interchanging of velocities in Figures 4.3 and 4.4, where the slopes of the two $x(t)$ curves are interchanged after the collision.

We can repeat the experiment, this time giving cart 2 a harder shove, and then repeat it again, this third time giving cart 2 just a gentle nudge. What we discover is that, *no matter what the initial velocity of cart 2 is,* the collision always interchanges the two velocities.

Further experiments show that it is not necessary that cart 1 be initially at rest. **Figure 4.6** shows what happens when both carts are moving in the same direction at the instant of collision: Once again the collision interchanges the two velocities.

4.2 What is the change in velocity of (*a*) cart 1 and (*b*) cart 2 in Figure 4.6? (*c*) What do you notice about your two answers?

You can repeat this experiment with many different initial velocities, with the carts moving in the same direction or in opposite directions, and you will always observe that the collision interchanges the velocities of the carts. Moreover, Checkpoint 4.2 shows that, if the velocity of one cart increases by a certain amount as a result of the collision, then the velocity of the other cart decreases by exactly the same amount.

To determine whether the amount of material that makes up each cart affects the motion, let's fasten two standard carts together so that the size of this unit is twice the size of the other cart we are going to use, which we continue to call the standard cart. With this double cart at rest on the track, we shove the standard cart toward it at a speed of 0.60 m/s. This time the moving (standard) cart does not come to a

complete stop; instead, the collision reverses its direction of travel, so that after the collision it moves to the left with a speed of 0.2 m/s, as **Figure 4.7** shows. After the collision, the initially stationary double cart, being more difficult to set in motion than the single standard cart, moves to the right, as before, but now its speed is lower than in the earlier experiments (0.40 m/s versus about 0.60 m/s). This result makes sense because we already know from experience that more massive objects are harder to set into motion than less massive ones—it's easier to throw a small stone than a large boulder.

The x component of the velocity of the double cart changed by $\Delta v_{dx} = +0.40$ m/s $- 0 = +0.40$ m/s, and the x component of the velocity of the standard cart changed by $\Delta v_{sx} = -0.20$ m/s $- (+0.60$ m/s$) = -0.80$ m/s.* The magnitude of the double cart's velocity change, $|\Delta v_{dx}| = 0.40$ m/s, is half that of the standard cart, $|\Delta v_{sx}| = 0.80$ m/s.

We can repeat this experiment and vary the initial speeds and directions of motion, but we would continue to observe that *no matter how the carts move (or do not move) initially, the magnitude of the velocity change of the double cart is always half that of the standard cart.* Also, the two velocity

---

*Because we must keep track of the changes in velocity of two colliding objects, we need an additional subscript. For example, $\Delta v_{dx} = v_{dx,f} - v_{dx,i}$ is the change in velocity for the double cart. The final subscript, separated by a comma, designates the instant: f for final and i for initial. The middle subscript indicates that we are dealing with the x component. The first subscript designates the object (d for double cart).

Figure 4.8 Velocity-versus-time graph for a standard cart and a half cart before and after the two collide on a low-friction track.

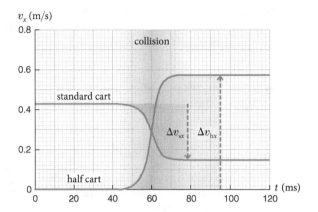

changes are always in opposite directions, which means the x component of the velocity of one cart increases and that of the other cart decreases.

Now let us repeat the collision once again, this time with the original cart 1 cut in half so that it is half the size of the other cart we'll be using. With the half cart at rest on the track, we push a standard cart toward it. As **Figure 4.8** shows, the magnitude of the velocity change of the half cart is twice the magnitude of the velocity change of the standard cart: $|\Delta v_{hx}| = 0.58 \text{ m/s}$ and $|\Delta v_{sx}| = 0.29 \text{ m/s}$.

A pattern is developing here: For a group of objects all made of the same material, the motion of larger objects is harder to change than the motion of smaller objects. The larger objects put up more resistance when we try to change their velocity. This tendency of an object to resist a change in its velocity is called **inertia.**

> **Inertia is a measure of an object's tendency to resist any change in its velocity.**

The results of our experiments are summarized in **Table 4.1**. As you can see, the *ratio of the inertias* of the two carts is equal to the inverse of the ratio of their velocity changes. In experiment 2, the double cart inertia is twice the standard cart inertia, and the ratio $|\Delta v_{dx}| : |\Delta v_{sx}|$ is 0.5. In experiment 3, the half cart inertia is half that of the standard cart, and the ratio $|\Delta v_{hx}| : |\Delta v_{sx}|$ is 2.

The data in Table 4.1 suggest that we could now take a new cart of unknown inertia, let a standard cart collide with it, and then *determine* from the changes in velocity the inertia of the new cart relative to that of the standard cart.

Table 4.1 Ratio of Velocity Changes in Collisions Between Two Carts

Experiment	Cart 1	Cart 2	$|\Delta v_{1x}| : |\Delta v_{2x}|$
1	standard	standard	1
2	double	standard	0.5
3	half	standard	2

An oil supertanker has an enormous inertia. Even at full throttle it can take as long as 10 min to bring a large ship to a halt, which explains why ships like the *Titanic* and the *Exxon Valdez* could not avoid a collision in spite of seeing the danger well ahead of time.

✋ **4.3** The x component of the final velocity of the standard cart in Figure 4.8 is positive. Can you make it negative by adjusting this cart's initial speed while still keeping the half cart initially at rest?

Figure 4.9 is a graph of the collision between some unknown cart and a standard cart—both solid and made of the same material. The magnitude of the unknown cart's change in velocity is one-third that of the standard cart (Checkpoint 4.4). We conclude from this result that the unknown cart's inertia is three times that of the standard cart. In other words, the amount of material in the unknown cart is three times that in the standard cart.

✋ **4.4** Verify that $|\Delta v_{ux}| / |\Delta v_{sx}| \approx 1/3$ for the two carts in Figure 4.9.

Figure 4.9 Velocity-versus-time graph for a standard cart and a cart of unknown inertia before and after the two collide on a low-friction track. The unknown inertia can be determined from the ratio of the carts' velocity changes.

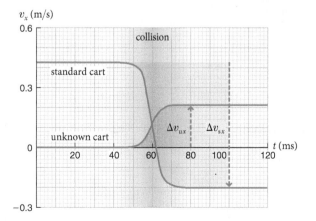

4.3 What determines inertia?

Having established that an object's inertia (its resistance to a change in velocity) is proportional to the amount of material in the object, we now come to the next important question: Is it only the amount of material that determines inertia? So far, all the carts we've been studying have been made of the same material. Would our results in experiment 1 be the same if we replaced one of the standard carts with a cart of the same volume of material but made of a different material—plastic instead of metal, say? As you probably know from experience, the answer is *no*: For example, a lead ball is more difficult to set in motion than a rubber ball of equal volume.

So let's make two carts of identical amounts of material, one of plastic and the other of metal, and see what happens when the two collide. The result, shown in **Figure 4.10**, is very different from the result for two identical carts (Figure 4.5).

4.5 What is the ratio of the x components of the change in velocity for the plastic and metal carts, $\Delta v_{px}/\Delta v_{mx}$, in Figure 4.10?

As we expect, the x component of the velocity of the plastic cart changes by a much larger amount than that of the metal cart (but it is still opposite in sign). In other words, it is easier to change the motion of the plastic cart than the motion of the metal cart. Given that inertia is a measure of an object's tendency to resist any change in its velocity, we conclude that an object's inertia is determined not just by the volume of material it contains but also by what that material is.

What about other physical properties—shape, say, or the smoothness of an object's surface? Experiments show that inertia does not depend on any of these other properties:

The inertia of an object is determined entirely by the type of material of which the object is made and by the amount of that material contained in the object.

Figure 4.10 Velocity-versus-time graph for two colliding carts that have the same volume but are made of different materials.

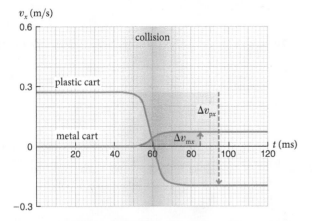

Once we have chosen one object as the standard for inertia, we can determine the inertia of any other object—whatever its size, shape, or composition—by letting it collide with our standard and measuring the ratio of the changes in the velocity of the two objects.

4.6 Is the inertia of the cart of unknown inertia in Figure 4.9 greater or less than that of the standard cart?

Example 4.1 Friction and collisions

Figure 4.11 shows the $v_x(t)$ curves for a collision between two identical carts moving not on a low-friction track but rather on a rough surface, so that friction affects their motion. Are the changes in the velocity of the carts caused by the collision still equal in magnitude?

Figure 4.11 Example 4.1.

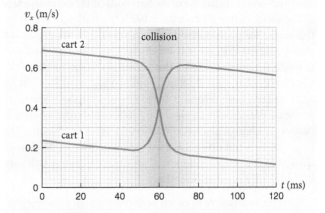

1 GETTING STARTED Looking at the graph, I'm tempted to say yes because if I tilt Figure 4.11 a bit, it looks very similar to Figure 4.6. Determining the change in velocity due to the collision, however, is not as straightforward because friction also affects the velocities. So, because the velocities of the carts before the collision decrease due to friction, it is not possible to read off a well-defined "initial" value of the velocity. The shape of the graph suggests that friction continues to decrease the velocities in the same way during and after the collision, so I have to determine a way to separate the change in velocity due to friction from the change due to the collision.

2 DEVISE PLAN Because the effect of friction is a steady decrease in velocity, I can extrapolate the $v_x(t)$ curve for cart 1 to calculate what its velocity would have been at $t = 80$ ms if the collision had not taken place. To determine the change in velocity due to the collision, I then read off the actual value of the velocity of cart 1 at $t = 80$ ms and subtract the two velocities. Then I repeat the procedure for cart 2 and compare the changes in velocity.

❸ **EXECUTE PLAN** By placing a ruler along the first part of the $v_x(t)$ curve for cart 1, I observe that the curve lies in line with the second part of the $v_x(t)$ curve for cart 2. Reading off the value of the curve at $t = 80$ ms gives a value of $+0.15$ m/s. Repeating this procedure for cart 2, I obtain a value of $+0.60$ m/s. Now I can determine the velocities the carts would have had at $t = 80$ ms if the collision had not taken place. As the graph shows, however, the carts do collide and their velocities at $t = 80$ ms are interchanged. The magnitudes of the changes in velocity caused by the collision are thus 0.45 m/s and 0.45 m/s. These velocity changes are equal in magnitude. ✔

❹ **EVALUATE RESULT** Extrapolation of the velocity curves in Figure 4.11 shows that the collision interchanges the velocities of the two carts, just as in a collision between identical carts in the absence of friction. The interchanging means that the changes in velocity must be of equal magnitude, as I found numerically.

As this example shows, the magnitude of the changes in velocity in a collision is the same even in the presence of friction. This means that even in the presence of friction, we can determine an object's inertia from the magnitude of the velocity changes in a collision.

4.4 Systems

In the preceding chapters we were always concerned with just a single object—a puck, a car, and so on. Starting with this chapter, however, most situations we encounter deal with any number of objects that interact with one another. To analyze such situations, we shall usually focus on one or more principal objects—a person climbing a mountain, two carts colliding, gas atoms in a container. The first step in any analysis therefore is to separate the object(s) of interest from the rest of the universe:

> **Any object or group of objects that we can separate, in our minds, from the surrounding environment is a system.**

For example, when considering a collision between two carts on a low-friction track, we might consider both carts together as the system. When someone throws a ball and we are interested in only the motion of the ball, a logical choice of system might be just the ball by itself. Once we have chosen the ball as our system, everything else—the thrower, the air around the ball, Earth—is outside the system and constitutes its **environment.** This imaginary separation of the universe into two parts—objects within the system and everything in the rest of the universe—makes it possible to develop simple but powerful accounting procedures.

When solving problems, you should imagine a boundary enclosing the objects of interest. What is inside this well-defined boundary constitutes your system and what is outside is the environment. The region of space enclosed in the system boundary may be quite large and contain within it a great many processes or activities, or it may be very small and contain an inert piece of material undergoing no change. The precise shape of the system boundary

Figure 4.12 Two choices of system for carts colliding on a track.

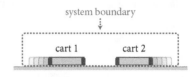

(*a*) Choice 1: system consists of both carts

(*b*) Choice 2: system consists of one cart

is unimportant; what counts is that it clearly separates the objects inside the system from those outside. To make this separation explicit, it will generally be helpful to make a pictorial representation of the objects within the system. In your sketch, you should draw a boundary around the objects that constitute your system.

There may be more than one way to separate the system and the environment in any given situation. For two carts colliding on a low-friction track, as illustrated in **Figure 4.12**, we can define our system as containing both carts (in which case the track is part of the environment) or just one cart (in which case the track and the other cart are part of the environment). Deciding what to include in the system will be dictated by the information you wish to learn. Often the choice is obvious; sometimes you will need to rely on experience to make this decision. More important, once you decide to include a certain object in the system, *it must stay that way throughout the analysis.* Failing to make a consistent choice of system is a frequent source of error.

Exercise 4.2 Choosing a system

Indicate at least two possible choices of system in each of the following two situations. For each choice, make a sketch showing the system boundary and state which objects are inside the system and which are outside. (*a*) After you throw a ball upward, it accelerates downward toward Earth. (*b*) A battery is connected to a light bulb that illuminates a room.

SOLUTION (*a*) The description of the situation mentions three objects: the ball, Earth, and you. One option is to include all three of them in the system (**Figure 4.13a**). As a second choice, I include you and the ball in the system (Figure 4.13*b*). ✔

Figure 4.13

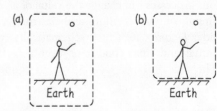

(b) Again I have three objects: the battery, the light bulb, and the room. I can choose just one of them—the battery—as my system (Figure 4.14a) or two of them—the battery and the light bulb (Figure 4.14b). ✔

Figure 4.14

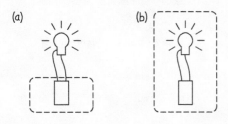

Note that my choices of system are arbitrary. Nothing in the problem prescribes the choice of system. If you tried this problem on your own before looking at my solution and you made different choices, then your answer is just as "correct" as mine!

Defining a system tells us nothing whatsoever about what is happening within it. It is simply a tool to help us set up an accounting scheme. Once we have chosen a system, we can study how certain quantities associated with the system change over time by determining the value of these quantities at the beginning and end of a time interval.

We shall in particular be interested in **extensive quantities**—that is, quantities whose value is proportional to the size or "extent" of the system. More specifically, if we divide the system into a number of pieces, then the sum of an extensive quantity for all the separate pieces is equal to the value of that quantity for the entire system. The number of trees in a park, for example, is extensive: If we divide the park into two parts and add the number of trees in each part, then we obtain the number of trees in the park. The price per gallon of gasoline is not an extensive quantity: If we divide a tankful of gas into two parts and add the price per gallon for the two parts, then we obtain twice the price per gallon for the entire tank. Quantities that do not depend on the extent of the system are **intensive quantities.**

✋ **4.7** Are the following quantities extensive or intensive: (a) inertia, (b) velocity, (c) the product of inertia and velocity?

Only four processes can change the value of an extensive quantity: input, output, creation, and destruction. To see this, consider Figure 4.15, which schematically represents a park's initial and final conditions over a certain time interval. The edge of the park is the system boundary, and

Figure 4.15 System diagram of a park. The number of trees in the park is an extensive quantity.

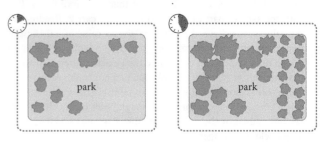

the park itself—trees and all—is the system. Diagrams that show a system's initial and final conditions are called **system diagrams;** we shall use such diagrams throughout this book. The change in the number of trees over the time interval is given by

change = final tree count − initial tree count.

The number of trees can change because new trees grow (creation) and old trees die or are taken down (destruction). Alternatively, new trees can be brought into the park (input) and some trees can be moved out of the park (output):

change = input − output + creation − destruction.

This equation is illustrated graphically in Figure 4.16a. If we understand how these four processes affect the trees, we can explain (or predict) any change in their number.

The accounting is often simplified by constraints put on the system. For example, if we build a fence around the park so that no trees can be transported into or out of the park, then the number of trees can change only because of creation and destruction. If there is no transfer of the extensive quantity under consideration across the boundary of the system, the change in that property can be due to only creation and destruction (Figure 4.16b):

change = creation − destruction.

Under certain circumstances we can also exclude creation and destruction. Suppose, for example, that we are counting the number of indestructible benches in the park rather than trees. The number of these benches changes only when benches are transported into or out of the park across the park boundary (unless there is a factory making indestructible benches in the park—an unlikely assumption). Any extensive quantity that cannot be created or destroyed is said to be **conserved.** The value of a

Figure 4.16 The effect of input, output, creation, and destruction on the extensive quantities of a system.

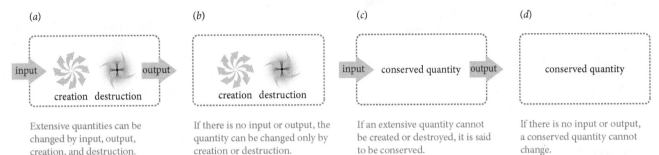

(a)

input · creation destruction · output

Extensive quantities can be changed by input, output, creation, and destruction.

(b)

creation destruction

If there is no input or output, the quantity can be changed only by creation or destruction.

(c)

input · conserved quantity · output

If an extensive quantity cannot be created or destroyed, it is said to be conserved.

(d)

conserved quantity

If there is no input or output, a conserved quantity cannot change.

conserved quantity can change only due to transfer of that quantity across the system boundary (Figure 4.16c):

$$\text{change} = \text{input} - \text{output}.$$

For a conserved quantity in a system for which there is no transfer, things are even more simple: Its value cannot change at all (Figure 4.16d). Conserved quantities play an important role in physics because their accounting is greatly simplified. In systems for which there is no transfer of a conserved quantity across the system boundary, this quantity does not change *regardless of the processes occurring inside or outside the system* (see the box "Conservation").

Exercise 4.3 Accounting principles

(*a*) Classify and give examples of the kinds of processes that can change (*i*) the number of loaves of bread in a bakery, (*ii*) the number of Lego pieces inside a house, and (*iii*) the number of coins in a safe that remains locked. (*b*) For each of these three cases, what is the system? Is the transfer of the quantity of interest across the system boundary possible? Is the quantity of interest conserved?

SOLUTION (*a*) (*i*) Input: a delivery of loaves of bread to the bakery; output: customers leaving the bakery with loaves they purchased; creation: the baking of a batch of loaves; destruction: the eating of loaves by patrons (inside the bakery; destruction outside the bakery falls under "output" because the loaves first leave the bakery). (*ii*) Input: Lego pieces are brought into the house;

Conservation

Imagine a sheet of paper of width *a* and length *b*. The circumference of this sheet is $2a + 2b$, and the surface area of one side is *ab*. Suppose we cut the sheet lengthwise into two equal parts. Are there any quantities that are unchanged by the cutting?

Circumference is not a conserved quantity: The sum of the circumferences of the two parts is not equal to the circumference of the single sheet. The circumference of each cut piece is $2(\frac{1}{2}a) + 2b$, and the combined circumference of the two pieces is $2(a + 2b)$, whereas the original circumference was $2(a + b)$.

Surface area is conserved: The surface area of each cut piece is $\frac{1}{2}ab$, and so the sum of the surface areas of the two pieces is *ab*, exactly as it was before we cut up the sheet.

Trivial? Well, suppose that instead of cutting the sheet in two, we cut it into 148 equal pieces. Can we say anything about the sum of the circumferences? Not without some brainwork! Conservation of surface area, on the other hand, allows us to state—with no calculations whatsoever—that the sum of the surface areas of all the pieces is still *ab*.

We have found a conserved quantity: surface area! No matter in how many pieces a sheet of paper is cut, the sum of the surface areas of the pieces is equal to the surface area of the original sheet.

If you are a skeptic (and in science you should always be skeptical), you could calculate the surface area of one of the 148 pieces and verify that the sum of the surface areas of all the pieces is indeed *ab*. To convince yourself that conservation of surface area is a general law, you might also want to repeat this procedure for a few more cases—say, 79 and 237 pieces.

Once you are convinced, however, it is easy to see the value of the conservation law. Suppose that instead of cutting the sheet neatly into equal pieces, someone tears it into ragged pieces, hands them to you, and asks, "What is the combined surface area of all of these pieces?" Even though an actual determination of the surface area would be difficult at best, you would answer, without hesitation, "*ab*"!

The idea that certain quantities are unaffected by events is appealing, not only because the notion of conservation intuitively makes sense (*some* things must remain unchanged) but also because it provides simple constraints on the behavior of a system.

Is surface area truly conserved? The answer is *no*: you can burn the sheet and make the entire surface area disappear. As we shall see, only a few quantities are truly conserved.

output: Lego pieces are taken out of the house; creation: assuming there is no manufacturing of Lego pieces in the house, there is no creation; destruction: for all practical purposes, I can consider the Lego pieces to be indestructible and so there is no destruction. (*iii*) Input: if the safe is locked it's impossible to add coins, so the input is zero; output: zero, too; creation: zero; destruction: zero. (Everything is zero, meaning the number of coins can't change inside a locked safe—which is precisely why people put valuables in safes!) ✔

(*b*) (*i*) The bakery and its contents make up the system. Loaves of bread can be taken into and out of the bakery, so transfer is possible. The number of loaves of bread is not conserved because loaves can be created and destroyed. (*ii*) The house and its contents are the system. Lego pieces can be taken into and out of it, so transfer is possible. Assuming that the Lego pieces cannot be fabricated or destroyed, I can consider their number to be conserved in this problem. (*iii*) The safe and its contents are the system. The safe remains locked so nothing can be put into it or taken out, and so transfer is not possible. Coins cannot be made or destroyed, so their number is conserved. ✔

Strictly speaking, indestructible benches in a park, Lego pieces, and coins are not truly conserved—all of them are manufactured somewhere (creation) and all of them can be destroyed (destruction). You may be wondering, then, whether *anything* is conserved. The answer is yes: Several quantities we shall encounter are conserved, giving rise to some of the most fundamental principles in physics.

4.8 Which of these quantities is extensive? (*a*) money, (*b*) temperature, (*c*) humidity, (*d*) volume.

Self-quiz

1. Two carts give the same velocity-versus-time graphs when they collide with the same standard cart. Can you conclude from this information that the two carts are identical?

2. The graphs in Figures 4.17a and 4.17b show the effects of carts A and B colliding (separately) with a standard cart S. List the three carts in order of increasing inertia.

Figure 4.17

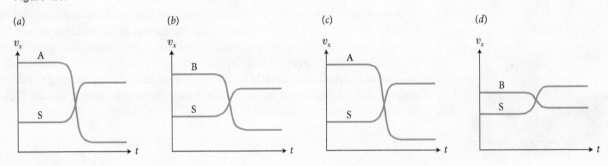

3. The graphs in Figures 4.17c and 4.17d show the effects of carts A and B colliding (separately) with a standard cart S. Which has greater inertia, A or B?

4. You do not know whether or not carts A and B of question 3 are made of the same material. Can you tell from the graphs which cart contains a larger quantity of material?

5. Imagine accounting for the number of cattle on a ranch. What do you choose as your system? What are the processes corresponding to input, output, creation, and destruction? Is transfer into and out of the system possible? Is the accountable quantity conserved?

Answers

1. No. Although the identical graphs tell you that the two carts have identical inertias, they need not be identical physically. A large plastic cart, for instance, can have the same inertia as a much smaller metal cart.

2. A and B (identical and less), S (greater). Because the magnitude of the velocity change for S is smaller than that for either A or B, cart S must have greater inertia. The *ratio* of the magnitudes of the velocity changes is about the same in both graphs, however, indicating that A and B have about the same inertias.

3. B. Don't be fooled by the bigger velocity change for S in part c; you must first consider each graph separately. Graph c tells you that A has less inertia than S because the magnitude of the velocity change for A is *larger* than that for S. Graph d tells you that B has greater inertia than S because the magnitude of the velocity change for B is *smaller* than that for S. Hence B must have greater inertia than A.

4. No. Although B has greater inertia, you cannot conclude anything about the quantity of material in it relative to A because you don't know from what materials the two carts are made.

5. The ranch is the system. Input: cattle transported into the ranch; output: cattle transported out of the ranch; creation: cattle born on the ranch; destruction: cattle expiring on the ranch. Transfer is possible because there is input and output, and the number of cattle is not conserved because cattle can be created and destroyed.

This replica of the standard kilogram is kept under two evacuated glass jars at the National Research Laboratory of Metrology in Japan. It is compared once every thirty years against the original international standard kept in France.

4.5 Inertial standard

The inertia of an object is represented by the symbol m (m is for *mass*, a concept that is related to inertia and that we shall examine later on). We shall use the term *inertia* rather than *mass* to underscore the physical meaning of this concept: An object's resistance to a change in its velocity. Inertia is a scalar, which means it does not depend on the orientation of any axis in space. By international agreement, the inertial standard is a platinum-iridium cylinder stored at the International Bureau of Weights and Measures in Sèvres, France. The inertia of this standard is defined as one **kilogram,** the basic SI unit of inertia ($m_s \equiv 1$ kg).*

Armed with a replica of this inertial standard and the knowledge gained in Sections 4.1 through 4.3, we can determine the inertia m of any other object. To do so, we make the object we are studying collide with the inertial standard and then measure the changes in velocity. In Section 4.2, we defined the ratio of the inertias of two colliding objects as the inverse of the ratio of the magnitude of their velocity changes, and so we can use this ratio for our inertial standard and our object of unknown inertia:

$$\frac{m_u}{m_s} \equiv -\frac{\Delta v_{sx}}{\Delta v_{ux}} \tag{4.1}$$

or

$$m_u \equiv -\frac{\Delta v_{sx}}{\Delta v_{ux}} m_s. \tag{4.2}$$

Why the minus sign? Well, suppose the object of unknown inertia is a replica of the standard. The left side of Eq. 4.1 then becomes $m_s/m_s = 1$. From the collision in Figure 4.5, we know, however, that the velocity of one of the carts increases (positive Δv_x) while that of the other decreases by the same amount (negative Δv_x). Without the minus sign, Eq. 4.1 would therefore become $1 = -1$. Notice that in all collisions we considered (Figures 4.5 through 4.9), the x components of the velocity change for the two colliding objects have opposite signs. Because this is true for *all* collisions, the minus sign in Eq. 4.1, the defining equation for inertia, means inertia is always a *positive* quantity.

In principle, this simple procedure makes it possible to determine the inertia of any object. Substituting the value $m_s \equiv 1$ kg in Eq. 4.2 allows us to determine the inertia of any other object in kilograms. In practice, it may not be a good idea to let objects collide in order to determine their inertia. (I don't recommend determining the inertia of a Ming vase in this way!) In Chapter 13 we shall discuss equivalent but more practical ways to determine the inertia of objects.

Example 4.4 Determining the inertia of a stone

A small stone is fastened to the top of a standard cart of inertia 1 kg to form a combination of unknown inertia m_u. A second standard cart is then launched with an initial velocity given by $v_{sx,i} = +0.46$ m/s toward the combination that is initially at rest. After the collision, the x component of the velocity of the cart with the stone is $v_{ux,f} = +0.38$ m/s and that of the standard cart is $v_{sx,f} = -0.08$ m/s. What is the inertia of the stone?

❶ GETTING STARTED This collision is much like the one depicted in Figure 4.7 with a standard cart moving toward a cart

of greater inertia that is initially at rest. The ratio of the carts' inertias follows from the ratio of the magnitude of their velocity changes.

❷ DEVISE PLAN To calculate the unknown inertia m_u, I can use Eq. 4.2. I know the x components of the two carts' initial and final velocities, so all the quantities on the right side of the equation are known.

---

*You may associate the unit kilogram with *weight*, a physical property that has to do with gravitation rather than inertia. More about the concept of weight in Chapter 13.

❸ EXECUTE PLAN I begin by determining the changes in the x components of the velocity for each cart. For the cart with the stone, $\Delta v_{\mathrm{u}x} = v_{\mathrm{u}x,\mathrm{f}} - v_{\mathrm{u}x,\mathrm{i}} = +0.38 \text{ m/s} - 0 = +0.38 \text{ m/s}$; for the standard cart $\Delta v_{\mathrm{s}x} = v_{\mathrm{s}x,\mathrm{f}} - v_{\mathrm{s}x,\mathrm{i}} = -0.08 \text{ m/s} - 0.46 \text{ m/s} = -0.54 \text{ m/s}$. Substituting these values into Eq. 4.2, I get

$$m_{\mathrm{u}} = -\frac{\Delta v_{\mathrm{s}x}}{\Delta v_{\mathrm{u}x}} m_{\mathrm{s}} = -\frac{-0.54 \text{ m/s}}{+0.38 \text{ m/s}} (1 \text{ kg}) = 1.4 \text{ kg}.$$

This represents the inertia of the stone and the cart as one unit. Given that the inertia of the standard cart is 1 kg, the inertia of the stone must be 0.4 kg. ✔

❹ EVALUATE RESULT The inertia of the cart-and-stone combination is greater than 1 kg, as I would expect. If the magnitude of the velocity change of this combination were half that of the standard cart ($\frac{1}{2}(0.54 \text{ m/s}) = 0.27 \text{ m/s}$), then the inertia of the cart-and-stone combination would be twice that of a standard cart, and so the inertia of the stone would be 1.0 kg. The magnitude of the velocity change is more than half, so I should expect the stone's inertia to be less than 1.0 kg, which agrees with the answer I found.

✋ⓒ **4.9** (*a*) Suppose that instead of a replica of the platinum-iridium cylinder kept in France, we had chosen another object as our inertial standard and defined the inertia of that object as being exactly equal to 1 kg. Would the inertia of our unknown object as measured against the new standard be different from that object's inertia measured against the French standard? (*b*) Would the outcome of a collision between two arbitrary objects be different?

4.6 Momentum

Our definition of inertia leads to the definition of another important physical quantity: *momentum.* Let's rewrite Eq. 4.2 in a slightly different form by multiplying both sides by $\Delta v_{\mathrm{u}x}$ and then bringing all terms to the left side:

$$m_{\mathrm{u}} \Delta v_{\mathrm{u}x} + m_{\mathrm{s}} \Delta v_{\mathrm{s}x} = 0. \tag{4.3}$$

Because the changes in velocity Δv_x can be written in the form $v_{x,\mathrm{f}} - v_{x,\mathrm{i}}$, we can rewrite this equation as

$$m_{\mathrm{u}}(v_{\mathrm{u}x,\mathrm{f}} - v_{\mathrm{u}x,\mathrm{i}}) + m_{\mathrm{s}}(v_{\mathrm{s}x,\mathrm{f}} - v_{\mathrm{s}x,\mathrm{i}}) = 0 \tag{4.4}$$

or

$$m_{\mathrm{u}}v_{\mathrm{u}x,\mathrm{f}} - m_{\mathrm{u}}v_{\mathrm{u}x,\mathrm{i}} + m_{\mathrm{s}}v_{\mathrm{s}x,\mathrm{f}} - m_{\mathrm{s}}v_{\mathrm{s}x,\mathrm{i}} = 0. \tag{4.5}$$

This form of Eq. 4.3, which contains nothing but products of inertias and velocities, is important because it suggests something new. The product of the inertia and the velocity of an object is called the **momentum** (plural: *momenta*) of that object. The conventional symbol for momentum is p, and so we can write

$$\vec{p} \equiv m\vec{v}. \tag{4.6}$$

As you learned in Section 2.6, the product of a vector and a scalar is a vector. Because inertia is a scalar and velocity is a vector, momentum is a vector. The direction of the momentum vector for any object is the same as the direction of the object's velocity vector. The x component of the momentum is the product of the inertia and the x component of the velocity:

$$p_x \equiv mv_x. \tag{4.7}$$

The SI units of momentum are the units of the product mv: $\text{kg} \cdot \text{m/s}$. For example, the magnitude of the momentum of an object of inertia 1 kg moving at a speed of 1 m/s is $1 \text{ kg} \cdot \text{m/s}$. Inertia is an intrinsic property of an object (you can't change it without changing the object), but the value of the momentum of that object, like its velocity, can change.

Example 4.5 Bullet and bowling ball

Compare the magnitude of the momenta of a 0.010-kg bullet fired from a rifle at 1300 m/s and a 6.5-kg bowling ball lumbering across the floor at 4.0 m/s.

❶ GETTING STARTED Momentum is the product of inertia and velocity. I have to calculate this quantity for both the bullet and the bowling ball and then compare the resulting values.

❷ DEVISE PLAN Equation 4.6 gives the momentum of an object. To determine the magnitude of the momentum of an object, I must take the product of the inertia m and the speed v: $p = mv$.

❸ EXECUTE PLAN Substituting the values given in the problem statement, I get

$$p_{bullet} = (0.010 \text{ kg})(1300 \text{ m/s}) = 13 \text{ kg} \cdot \text{m/s} ✔$$

$$p_{bowling} = (6.5 \text{ kg})(4.0 \text{ m/s}) = 26 \text{ kg} \cdot \text{m/s}. ✔$$

❹ EVALUATE RESULT Surprisingly, the magnitudes of the momenta are very close! I have no way of evaluating momenta because I don't have much experience yet with this quantity. However, the bullet has less inertia and a high speed and the bowling ball has greater inertia and a low speed, so it is not unreasonable that the product of these quantities is similar.

Momentum is a quantitative measure of "matter in motion" and depends on both the amount of matter in motion and how fast that matter is moving. Momentum is very different from inertia. A truck, for example, has greater inertia than a fly (it has a higher resistance to a change in its velocity), but if the truck is at rest and the fly is in motion, then the magnitude of the fly's momentum is larger than that of the truck, which is zero. In Example 4.5, the inertias of the bullet and the bowling ball are very different, yet their momenta are similar. Conceptually you can think of an object's momentum as its capacity to affect the motion of other objects in a collision.

With the definition of momentum, we can rewrite Eq. 4.5 in the form

$$p_{ux,f} - p_{ux,i} + p_{sx,f} - p_{sx,i} = 0. \tag{4.8}$$

If we write $\Delta p_{ux} \equiv p_{ux,f} - p_{ux,i}$ and $\Delta p_{sx} \equiv p_{sx,f} - p_{sx,i}$, Eq. 4.8 takes on the beautifully simple form

$$\Delta p_{ux} + \Delta p_{sx} = 0. \tag{4.9}$$

This equation means that, whenever an object of unknown inertia collides with the inertial standard, the changes in the x components of the momenta of the two objects add up to zero. In other words, the change in the x component of the momentum for one object is always the negative of the change for the other.

Example 4.6 Collisions and momentum changes

(a) A red cart with an initial speed of 0.35 m/s collides with a stationary standard cart ($m_s = 1.0$ kg). After the collision, the standard cart moves away at a speed of 0.38 m/s. What is the momentum change for each cart? (b) The experiment is repeated with a blue cart, and now the final speed of the standard cart is 0.31 m/s. What is the momentum change for each cart in this second collision? (c) If in the collisions $v_{rx,f} = +0.032$ m/s and $v_{bx,f} = -0.039$ m/s, what are the inertias of the red and the blue carts?

❶ GETTING STARTED I begin organizing the information given in the problem in a picture by showing the initial and final conditions for each of the two collisions (**Figure 4.18**).

Figure 4.18

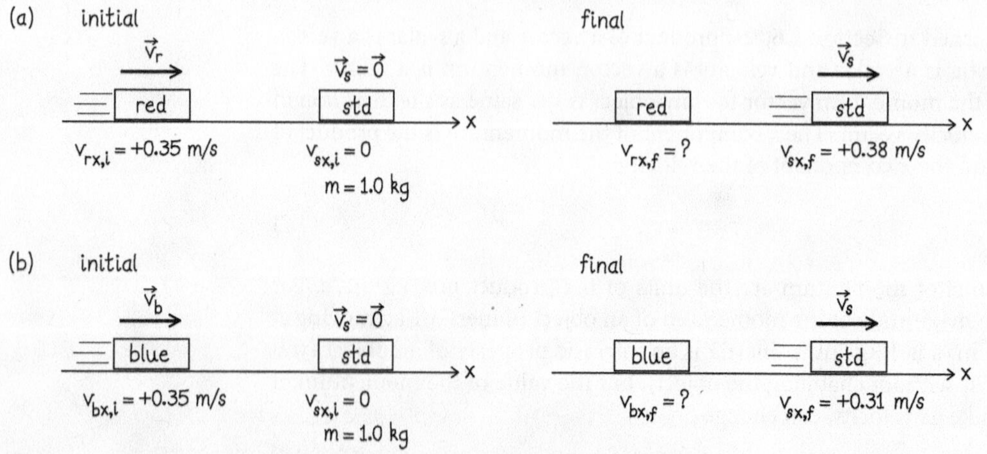

② DEVISE PLAN In each collision, I know the inertia and the x components of the initial and final velocities of the standard cart, so I can readily determine the change in the x component of its momentum. To determine the change in momentum for the other carts, I can then use Eq. 4.9, which tells me that for each collision the changes in the x components of the momenta of the two carts must add up to zero. For part c, I can use Eq. 4.7 together with the values for the red and blue carts' initial and final velocities and their changes in momenta calculated in parts a and b to determine the inertias.

③ EXECUTE PLAN (a) Because it initially is at rest, the standard cart's initial momentum is zero. The x component of its final momentum is $p_{sx,f} = (1.0 \text{ kg})(+0.38 \text{ m/s}) = +0.38 \text{ kg} \cdot \text{m/s}$, and so the x component of its change in momentum is

$$\Delta p_{sx} = +0.38 \text{ kg} \cdot \text{m/s} - 0 = +0.38 \text{ kg} \cdot \text{m/s}. ✔$$

Because the x components of the momentum changes of the two carts must add up to zero (Eq. 4.9), the corresponding x component of the red cart's momentum change is $\Delta p_{rx} = -0.38 \text{ kg} \cdot \text{m/s}. ✔$

(b) Applying the same procedure for this second collision, I get

$$\Delta p_{sx} = (1.0 \text{ kg})(+0.31 \text{ m/s}) - 0 = +0.31 \text{ kg} \cdot \text{m/s};$$

$$\Delta p_{bx} = -0.31 \text{ kg} \cdot \text{m/s}. ✔$$

(c) The change in momentum for the red cart is $\Delta p_{rx} = p_{rx,f} - p_{rx,i}$. Using Eq. 4.7, I can write this change in momentum as

$$\Delta p_{rx} = m_r v_{rx,f} - m_r v_{rx,i} = m_r(v_{rx,f} - v_{rx,i})$$

and so

$$m_r = \frac{\Delta p_{rx}}{v_{rx,f} - v_{rx,i}} = \frac{-0.38 \text{ kg} \cdot \text{m/s}}{+0.032 \text{ m/s} - 0.35 \text{ m/s}}$$

$$= 1.2 \text{ kg}. ✔$$

Likewise, for the blue cart:

$$m_b = \frac{\Delta p_{b,x}}{v_{bx,f} - v_{bx,i}} = \frac{-0.31 \text{ kg} \cdot \text{m/s}}{-0.039 \text{ m/s} - 0.35 \text{ m/s}}$$

$$= 0.80 \text{ kg}. ✔$$

④ EVALUATE RESULT If either the red or the blue cart had an inertia of 1.0 kg, then the cart would come to rest in the collision. My work shows that the inertia of the red cart is greater than that of a standard cart. Indeed, after the collision it still moves in the same direction it initially moved, as it should for a cart of greater inertia colliding with one of lesser inertia (see, for example, Figure 4.8). In contrast, the inertia of the blue cart is less than that of the standard cart. After the collision it bounces back and travels in the opposite direction, as I would expect (see Figure 4.7).

Equation 4.9 can be written in vectorial form:

$$\Delta \vec{p}_u + \Delta \vec{p}_s = \vec{0} \qquad (4.10)$$

and so, in a collision between an object of unknown inertia and the inertial standard, the changes in the momenta of the two objects add to zero.

4.10 In part a of Example 4.6, what are the directions of the changes in momentum, $\Delta \vec{p}_s$ and $\Delta \vec{p}_r$, of the two carts?

4.7 Isolated systems

As we saw in Checkpoint 4.7, momentum is an extensive property—you can add up the momenta of all the objects in a system to obtain the *momentum of the system*. In particular, the momentum of a system of two moving carts is the sum of the momenta of the two individual carts:

$$\vec{p} \equiv \vec{p}_1 + \vec{p}_2. \qquad (4.11)$$

With this definition we can begin to develop an accounting scheme for the momentum of a system. In particular, we shall be interested in determining what causes the momentum of a system to change. Let us begin by reexamining a number of situations we have encountered. In Figure 4.19a a puck slides to a stop on a wood floor. In Figure 4.19b a cart moves at constant velocity on a low-friction track. In Figure 4.19c a standard cart collides with another cart on a low-friction track, and in Figure 4.19d someone gives a cart moving at constant

QUANTITATIVE TOOLS

Figure 4.19 Velocity-versus-time and acceleration-versus-time graphs for four situations. The shading indicates nonzero acceleration.

(*a*) Puck slows to a halt

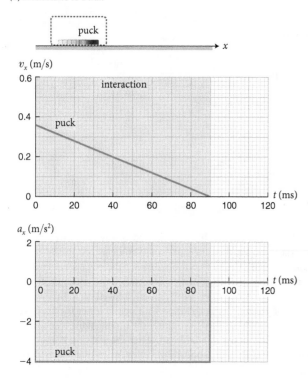

(*b*) Cart moves at constant velocity

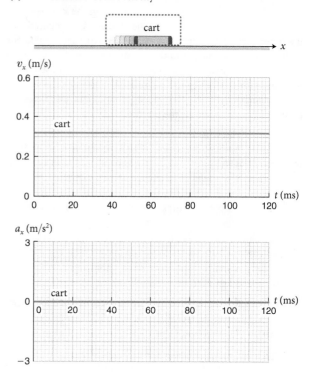

(*c*) Standard cart collides with cart of unknown inertia

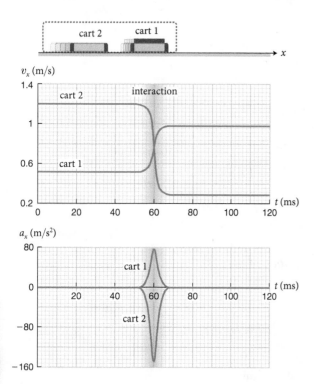

(*d*) Cart moving at constant velocity is given a shove

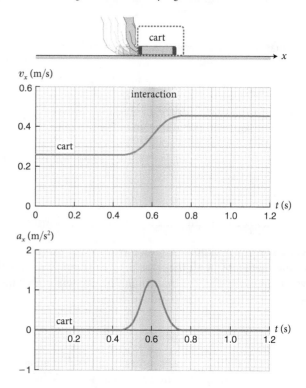

Table 4.2 Interactions and momentum changes in Checkpoint 4.11

Situation	Interacting objects	System	System interacting?	$\Delta\vec{p}$
a	floor ↔ puck	puck	yes	nonzero
b	none	cart	no	$\vec{0}$
c	cart 1 ↔ cart 2	cart 1	yes	nonzero
d	cart 1 ↔ cart 2	carts 1 & 2	no	$\vec{0}$
e	hand ↔ cart	cart	yes	nonzero

velocity on a low-friction track a shove. In some cases the velocity (and therefore the momentum) remains constant; in others it changes.

4.11 Is the change in the momentum $\Delta\vec{p}$ zero or nonzero for the following choices of system over the 120-ms time interval in Figure 4.19: (*a*) puck in Figure 4.19*a*, (*b*) cart in Figure 4.19*b*, (*c*) cart 1 in Figure 4.19*c*, (*d*) both carts in Figure 4.19*c*, (*e*) cart in Figure 4.19*d*?

In the graphs, I have highlighted the time intervals during which the acceleration of objects within the system is nonzero—that is, the time interval when the velocity and therefore the momentum of objects within the system change. Notice that this time interval always coincides with the time interval during which the object under consideration *interacts* with something else. In Figure 4.19*a* the puck slows down because the puck and the floor rub against (interact with) each other, in Figure 4.19*c* the two carts collide with (interact with) each other, and in Figure 4.19*d* the cart and the hand interact with each other. Conversely, in Figure 4.19*b* the interaction between the track and the cart has been virtually eliminated by reducing friction, and so the acceleration is zero.

In these examples, *interaction* refers to objects touching or rubbing against each other. Interactions are the subject in Chapter 7. For now, we'll use the word *interaction* to mean two objects acting on each other in such a way that one or both are accelerated.

Table 4.2 summarizes the observations from Checkpoint 4.11. Note that whenever the system interacts with something outside the system, the momentum of the system changes. Interactions across the boundary of a system are called *external interactions*. Interactions between two objects inside the system are called *internal interactions*. Looking at Figure 4.19 and Table 4.2, we see that external interactions transfer momentum into or out of the system.

A system for which there are no external interactions is said to be **isolated.** For such a system there is no input or output of momentum, and so the momentum of the system does not change: $\Delta\vec{p} = \vec{0}$. In Figure 4.19*b*, the cart's motion is not influenced by the track, and so the cart constitutes an isolated system. In Figure 4.19*c*, the two carts interact with each other but not with anything outside the system, and so the two carts together constitute a system that is isolated (but either cart by itself does not; see Table 4.3). In Figures 4.19*a* and 4.19*d*, where

Table 4.3 Two choices of system for carts colliding

	Choice 1	Choice 2
System:	carts 1 & 2	cart 1
Environment:	track	cart 2 & track
Interactions:	internal	external
System isolated?	yes	no
Momentum changing?	no	yes

Figure 4.20 The choice of system determines whether the interactions in question are internal or external and hence whether the system is isolated or not. We use the symbol $\textcircled{P}$ to denote an isolated system, indicating that no momentum crosses the system's boundary.

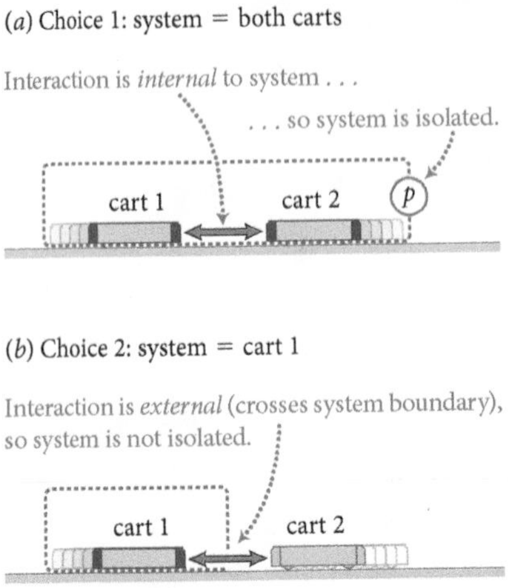

(*a*) Choice 1: system = both carts

Interaction is *internal* to system . . .

. . . so system is isolated.

cart 1   cart 2 $\textcircled{P}$

(*b*) Choice 2: system = cart 1

Interaction is *external* (crosses system boundary), so system is not isolated.

cart 1   cart 2

$\Delta\vec{p} \neq \vec{0}$, the system is not isolated. In Figure 4.19*a* the only object in the system (the puck) interacts with the floor, which is outside the system. In Figure 4.19*d*, the only object in the system (the cart) interacts with the hand, which is outside the system.

Figure 4.20 and Table 4.3 illustrate two system choices for studying a collision. The system containing both carts is isolated (the interaction between the two carts takes place inside the system; it is an internal interaction), and the momentum of the system is constant. The system containing just cart 1 is not isolated because it interacts with cart 2, which is now part of the environment, and the momentum of the system is not constant.

In isolated systems the analysis of momentum is simplified because we can concentrate on what happens inside the system. The procedure for choosing an isolated system is shown in the Procedure box on the next page and illustrated for a collision between two carts on a low-friction track in **Figure 4.21**. We begin by separating from one another all the objects that could play a role: the carts, the track, and the air. Next we identify all possible interactions: the collision between the two carts, the interaction between the air and the carts, and any possible interaction between the track and the carts. We know that each cart by itself rolls without slowing down on the track, so the interaction between the track and the carts and between the air and the carts does not have any significant effect on the carts' velocity. We therefore eliminate these and retain only the collision between the carts, which does change their velocities.*

Having eliminated all but the interaction between the carts, we can choose an isolated system by enclosing the two carts so that this interaction becomes internal. Finally, we complete the procedure by making a system diagram, showing the initial and final conditions of the isolated system (before and after the collision). In general, the interaction between two colliding objects affects their motion much more strongly than any other interaction, and so we can for all practical purposes always consider the system of two colliding objects to be isolated.

Figure 4.21 Applying the procedure for choosing an isolated system (see the Procedure box) to the collision between two carts on a low-friction track.

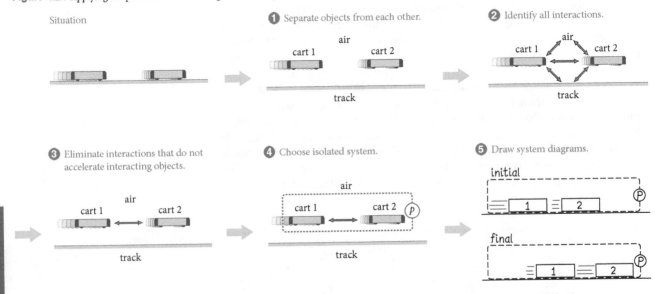

*Leaving out the air-cart and track-cart interactions is a simplification, because both *do* have a small effect on the carts' acceleration. However, you can ignore the effects of the these interactions because they are so much smaller than the effects of the collision. Never be afraid to make a simplifying assumption; you can always return to the beginning of a problem and refine your assumptions.

Procedure: Choosing an isolated system

When you analyze momentum changes in a problem, it is convenient to choose a system for which no momentum is transferred into or out of the system (an isolated system). To do so, follow these steps:

1. Separate all objects named in the problem from one another.
2. Identify all possible interactions among these objects and between these objects and their environment (the air, Earth, etc.).
3. Consider each interaction individually and determine whether it causes the interacting objects to accelerate. Eliminate any interaction that does not affect (or has only a negligible effect on) the objects' accelerations during the time interval of interest.
4. Choose a system that includes the object or objects that are the subject of the problem (for example, a cart whose momentum you are interested in) in such a way that none of the remaining interactions cross the system boundary. Draw a dashed line around the objects in your choice of system to represent the system boundary. None of the remaining interactions should cross this line.
5. Make a system diagram showing the initial and final states of the system and its environment.

Exercise 4.7 Who's pulling whom?

A person standing on a skateboard on horizontal ground pulls on a rope fastened to a cart. Both the person and the cart are initially at rest. Use the Procedure box to identify an isolated system and make a system diagram.

SOLUTION See **Figure 4.22**. I begin by separating the objects in the problem: the person, the cart, and Earth. (I could always go into more detail—include the air, the rope, and the skateboard—but it pays to keep things as simple as possible. For that reason, I consider the skateboard to be part of the person and the rope to be part of the cart.) The cart interacts with Earth and the person; the person interacts with the cart and Earth. Ignoring friction in the wheels of the cart, I know that the interaction between the cart and Earth has no effect on any motion, and so I can eliminate it from the analysis. The same holds for the interaction between the person (the skateboard) and Earth. I then draw a boundary around the person and the cart, making the interaction between the two internal. Because there are no external interactions, this system is isolated. Finally I draw a system diagram showing the initial and final conditions of the system with the person and cart initially at rest and then moving toward each other as the person pulls on the rope. ✔

Figure 4.22

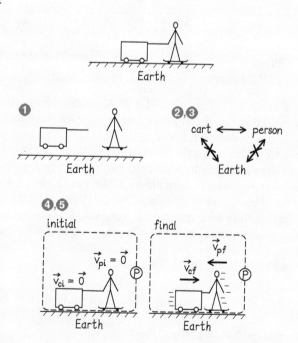

✋ **4.12** Imagine sitting on a sled on the slippery surface of a frozen lake. To reposition yourself closer to the back of the sled, you push with your legs against the front end. (*a*) Do you constitute an isolated system? (*b*) How about you and the sled? Ignore the interaction between sled and ice.

4.8 Conservation of momentum

In Section 4.4 we saw that the change in an extensive property can be due to transport into or out of the system and to creation or destruction. In Section 4.7 we saw that whenever we consider a system that is isolated, the momentum of the system is not changing:

$$\Delta \vec{p} = \vec{0} \quad \text{(isolated system).} \tag{4.12}$$

You may be tempted to conclude from this observation that momentum is conserved. All we can conclude, however, is that there is no creation or destruction of momentum *inside the systems we considered*—not that it is impossible to create or destroy momentum under any circumstances.

Because it is a direct consequence of the definition of momentum, Eq. 4.10 must—by definition—hold for *any* object that collides with the inertial standard. So if we take any two objects 1 and 2 and let each collide with the inertial standard on a level, low-friction track, we *always* obtain—no matter how we vary the initial velocities and no matter what the values of inertias m_1 and m_2 are—that

$$\Delta \vec{p}_1 + \Delta \vec{p}_s = \vec{0} \quad \text{(by definition)} \tag{4.13}$$

and

$$\Delta \vec{p}_2 + \Delta \vec{p}_s = \vec{0} \quad \text{(by definition).} \tag{4.14}$$

Suppose now that, instead of making objects 1 and 2 collide with the inertial standard, we make them collide with each other. Should we expect to obtain

$$\Delta \vec{p}_1 + \Delta \vec{p}_2 = \vec{0}? \tag{4.15}$$

You may be tempted to think that this equation follows from our earlier observations, but that is not the case. The value of $\Delta \vec{p}_s$ in Eqs. 4.13 and 4.14 need not be the same because Δv_{sx}, and therefore $\Delta \vec{p}_s$, vary from collision to collision (compare, for instance, Figures 4.8 and 4.9). So each of these equations holds for a separate collision and therefore we cannot combine them to draw any conclusions about the relationship between $\Delta \vec{p}_1$ and $\Delta \vec{p}_2$. The only way to verify whether Eq. 4.15 holds is to do the measurement.

Figure 4.23 shows the graphs for a collision between two arbitrary carts. To get the momentum values shown in Figure 4.23*b*, I multiplied the *x* component of the velocity of each cart by the cart's inertia determined by *independently* letting each cart collide with the inertial standard. The change in the *x* component of the momentum of cart 1 can be obtained by multiplying its inertia ($m_1 = 0.36$ kg) with the *x* components of the initial and final velocities read off from Figure 4.23*a* ($v_{1x,i} = 0$ and $v_{1x,f} = +0.17$ m/s):

$$\Delta p_{1x} = m_1(v_{1x,f} - v_{1x,i}) = (0.36 \text{ kg})(+0.17 \text{ m/s} - 0)$$

$$= 0.061 \text{ kg} \cdot \text{m/s.}$$

Figure 4.23 (a) Velocity-versus-time graph and (b) momentum-versus-time graph of two carts before and after the two collide on a low-friction track. The sum of the momenta is the same before and after the collision. The inertias are $m_1 = 0.36$ kg and $m_2 = 0.12$ kg.

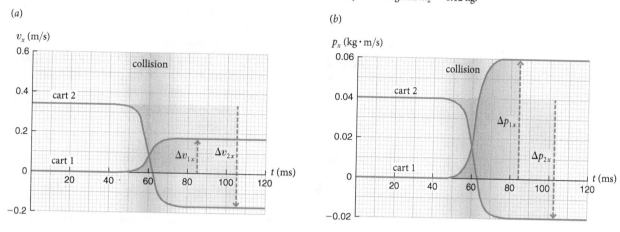

For cart 2 we have $v_{2x,i} = +0.34$ m/s, $v_{2x,f} = -0.17$ m/s, and $m_2 = 0.12$ kg, so

$$\Delta p_{2x} = m_2(v_{2x,f} - v_{2x,i}) = (0.12 \text{ kg})(-0.17 \text{ m/s} - 0.34 \text{ m/s})$$

$$= -0.061 \text{ kg} \cdot \text{m/s}.$$

Because the two changes in momentum are equal in magnitude and of opposite sign, they add up to zero: $\Delta p_{1x} + \Delta p_{2x} = 0$. Consequently the momentum of the system comprising the two carts does not change:

$$\Delta \vec{p} \equiv \Delta \vec{p}_1 + \Delta \vec{p}_2 = \vec{0}. \tag{4.16}$$

Repeating the experiment with another pair of carts or changing the initial velocities always yields the same result: The momentum of the system comprising any two colliding objects is not changed by the collision. The collision merely transfers some momentum from one cart to the other—one gains a certain amount of momentum, and the other loses the same amount.

✋ **4.13** (a) How much momentum is transferred in the collisions in Figure 4.23? (b) Is this momentum transferred from cart 1 to cart 2, or in the opposite direction?

Experiments show that *any* interaction between two objects—not just collisions—transfers momentum from one object to the other. The sum of the momenta of the interacting objects, however, never changes. No interactions or other phenomena have ever been observed where momentum is created out of nothing or destroyed. We must therefore conclude that momentum is conserved:

Momentum can be transferred from one object to another, but it cannot be created or destroyed.

This statement is one of the most fundamental principles of physics and is often referred to as the **conservation of momentum**.

For an isolated system, conservation of momentum means

$$\Delta \vec{p} = \vec{0} \quad \text{(isolated system)}. \tag{4.17}$$

QUANTITATIVE TOOLS

For a system that is not isolated, we have

$$\Delta \vec{p} = \vec{J}, \tag{4.18}$$

where $\vec{J}$ represents the transfer of momentum from the environment to the system. The quantity $\vec{J}$ is called the **impulse** delivered to the system. Like momentum, impulse is a vector and has SI units of kg·m/s. Depending on its direction relative to the momentum, the impulse can increase or decrease the magnitude of the momentum of the system. $\vec{J}$ therefore represents what we called "input" and "output" in Section 4.4.

✋ **4.14** (*a*) What is the magnitude of the impulse delivered to cart 1 in Figure 4.23? (*b*) Write the impulse delivered to cart 1 in vector form. (*c*) Does the fact that the change in the momentum of cart 1 is nonzero mean that momentum is not conserved?

Equation 4.18, which we shall use extensively throughout this book, is called the **momentum law.** This equation embodies conservation of momentum because it tells us that the momentum of a system can change only due to a transfer of momentum to the system ($\vec{J}$). If momentum were not conserved, then the right side of this equation would need to contain an additional term to account for changes in momentum due to the creation or destruction of momentum. For an isolated system the impulse is zero, $\vec{J} = 0$, and the momentum law takes on the form of Eq. 4.17.

Example 4.8 Bounce

A 0.20-kg rubber ball is dropped from a height of 1.0 m onto a hard floor and bounces straight up. Assuming the speed with which the ball rebounds from the floor is the same as the speed it has just before hitting the floor, determine the impulse delivered by the floor to the rubber ball.

❶ GETTING STARTED I define the ball to be my system in this problem. The impulse delivered to the ball is then given by the change in its momentum (Eq. 4.18). I need to develop a way to determine this change in momentum.

❷ DEVISE PLAN To solve this problem I need to first determine the velocity of the ball just before it hits the floor. I therefore break the problem into two parts: the downward fall of the ball and its collision with the floor. I can use Eq. 3.15 to determine the time interval it takes the ball to fall from its initial height (assuming the ball is initially at rest). As it falls, the ball undergoes constant acceleration, so I can use Eq. 3.10 to calculate its velocity just before it hits the floor and, because its speed is not changing as it rebounds, I also know its velocity after it bounces up. Knowing the velocities, I can calculate the ball's momentum before and after the bounce using Eq. 4.7. The change in momentum then gives the impulse according to Eq. 4.18.

❸ EXECUTE PLAN From Eq. 3.15 I see that it takes an object $t = \sqrt{2h/g} = \sqrt{2(1.0\ \text{m})/(9.8\ \text{m/s}^2)} = 0.45\ \text{s}$ to fall from a height of 1.0 m. Choosing the positive x axis pointing upward and substituting this result into Eq. 3.10, I obtain for the x component of the velocity of the ball just before it hits the floor:

$$v_{x,f} = 0 + (-9.8\ \text{m/s}^2)(0.45\ \text{s}) = -4.4\ \text{m/s}.$$

Now that I know the ball's velocity just before it hits the ground I can obtain the x component of the momentum of the ball just before it hits the ground by multiplying the ball's velocity by its inertia: $p_{x,i} = (0.20\ \text{kg})(-4.4\ \text{m/s}) = -0.88\ \text{kg·m/s}$. (I use the subscript i to indicate that this is the initial momentum of the ball before the collision with the floor.) If the ball rebounds with the same speed, then the x component of the momentum after the collision with the floor has the same magnitude but opposite sign: $p_{x,f} = +0.88\ \text{kg·m/s}$. The change in the ball's momentum is thus

$$\Delta p_x = p_{x,f} - p_{x,i} = +0.88\ \text{kg·m/s} - (-0.88\ \text{kg·m/s})$$

$$= +1.8\ \text{kg·m/s}.$$

The interaction with the ground changes the momentum of the ball, making it rebound. The ball does not constitute an isolated system, and the change in its momentum is due to an impulse delivered by Earth to the ball. To determine the impulse, I substitute the change in momentum of the ball into Eq. 4.18:

$$\vec{J} = \Delta \vec{p} = \Delta p_x \hat{\imath} = (+1.8\ \text{kg·m/s})\hat{\imath}. \checkmark$$

❹ EVALUATE RESULT The x component of the velocity of the ball just before it hits the floor is negative because the ball moves downward, in the negative x direction. After the collision it moves in the opposite direction, and consequently the x components of the changes in velocity and momentum are both positive, as I calculated. Because the x component of the change in the ball's momentum is positive, the impulse is directed upward (in the positive x direction). This makes sense because this impulse changes the direction of travel of the ball from downward to upward.

Because the momentum law takes on the simplest form for an isolated system, we shall usually try to choose an isolated system (unlike the above example). In such a case, we can write out the $\Delta\vec{p}$ term in Eq. 4.17:

$$\vec{p}_f = \vec{p}_i \quad \text{(isolated system)}. \tag{4.19}$$

In this form, the momentum law states that the initial and final values of the momentum of an isolated system are equal. For two colliding (or otherwise interacting) objects moving along the x axis, Eqs. 4.17 and 4.19 become

$$\Delta p_{1x} + \Delta p_{2x} = 0 \quad \text{(isolated system)} \tag{4.20}$$

and
$$p_{1x,f} + p_{2x,f} = p_{1x,i} + p_{2x,i} \quad \text{(isolated system)}. \tag{4.21}$$

Equations 4.20 and 4.21 are equivalent: Eq. 4.20 states that any change in momentum in one object is made up by a change of the same magnitude but in the opposite direction in the other object; Eq. 4.21 states that the sum of the momenta is the same before and after the collision. Which of these two equations you use in solving a problem is a matter of convenience.

4.15 (*a*) Are the changes in velocity in Figure 4.23*a* equal in magnitude? Why or why not? (*b*) Determine the velocity changes of the carts and verify that $m_1/m_2 = -\Delta v_{2x}/\Delta v_{1x}$ in Figure 4.23*a*. (*c*) Determine the initial and final momenta of the two carts. (*d*) What is the momentum of the system before the collision? (*e*) After the collision? (*f*) Are the momentum changes equal in magnitude and opposite in direction? Why or why not?

To convince you of the fundamental importance of the conservation of momentum, let me show you something else. Suppose we let carts 1 and 2 of Figure 4.23 collide with the same initial velocities as before, but this time we put something sticky between them, so that they remain locked together once they hit each other, as in **Figure 4.24** (next page). The velocities of the two carts *after* the collision must now be identical.

4.16 (*a*) Do the two carts in Figure 4.24 still constitute an isolated system like the carts in Figure 4.23? (*b*) What does your answer to part *a* imply about the momentum of the system comprising the two carts after the collision?

As you saw in the preceding checkpoint, the system of the two carts is isolated and so the momentum of the system should not change. **Figure 4.25a** on page 99 shows graphically that the final velocity of the combined carts is somewhere between the values of the two initial velocities. Before the collision, cart 1 is at rest and the x component of the velocity of cart 2 is $+0.34$ m/s; after the collision, both carts move at $v_{x,f} = +0.085$ m/s. The x component of the momentum of the system before the collision is thus $p_{1x,i} + p_{2x,i} = 0 + (0.12 \text{ kg})(0.34 \text{ m/s}) = 0.041 \text{ kg} \cdot \text{m/s}$. After the collision, the x component of the momentum of the system is $p_{1x,f} + p_{2x,f} = (0.36 \text{ kg})(0.085 \text{ m/s}) + (0.12 \text{ kg})(0.085 \text{ m/s}) = 0.031 \text{ kg} \cdot \text{m/s} + 0.010 \text{ kg} \cdot \text{m/s} = 0.041 \text{ kg} \cdot \text{m/s}$. The data indeed confirm that the momentum of the system does not change.

Note that the before-collision parts of Figures 4.23*a* and 4.25*a* are identical. The after-collision parts of these two figures are quite different from each other, however. Yet even with these two entirely different outcomes—the carts in Figure 4.23 flying off in opposite directions after colliding and those in Figure 4.25 sticking together—the momentum of the system does not change in both cases:

$$\vec{p}_i = \vec{p}_f \quad \text{(isolated system)}. \tag{4.22}$$

QUANTITATIVE TOOLS

Figure 4.24 (*a*) Two identical carts collide and stick together on a low-friction track. (*b*) High-speed film sequence of such a collision. The length of track visible in each frame is 0.40 m. (*c*) Curves indicating the positions of the rear of cart 1 and the front of cart 2 superimposed on the sequence.

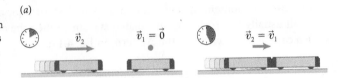

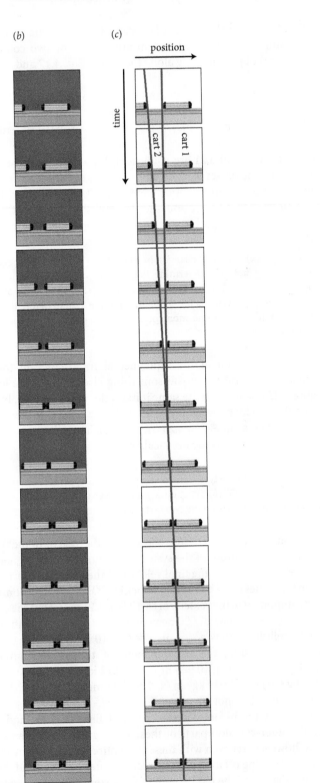

Figure 4.25 (*a*) Velocity-versus-time graph and (*b*) momentum-versus-time graph for the carts of Figure 4.24. Notice that the sum of the momenta is again the same before and after the collision. The inertia of cart 1 is 0.36 kg, and that of cart 2 is 0.12 kg.

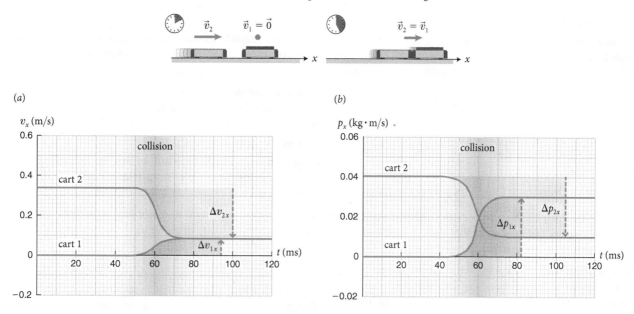

The momentum law is not confined to two objects. Suppose we put *three* objects in motion on a low-friction track such that they all end up colliding with one another. Let 1 first collide with 2 and then 2 with 3 (you can pick whatever sequence you like). In the first collision, $\vec{p}_1$ and $\vec{p}_2$ change but their sum remains unchanged. In addition, the sum of all three momenta remains unchanged because nothing has happened to $\vec{p}_3$ yet. When 2 and 3 collide, $\vec{p}_2$ and $\vec{p}_3$ change but the sum $\vec{p}_2 + \vec{p}_3$ does not change. So again, the momentum of the system comprising all three remains constant. In other words, throughout the two collisions, the momentum of the system comprising all three objects, $\vec{p}$, remains unchanged:

$$\vec{p}_i = \vec{p}_{1,i} + \vec{p}_{2,i} + \vec{p}_{3,i} = \vec{p}_{1,f} + \vec{p}_{2,f} + \vec{p}_{3,f} = \vec{p}_f \quad \text{(isolated system).} \quad (4.23)$$

This argument can be extended to any number of objects and any number of collisions or interactions.

That we can apply conservation of momentum to predict the final momentum of carts colliding on a low-friction track may seem like nothing special because you are indeed unlikely to encounter this situation outside the physics classroom. Conservation of momentum has much broader applications, however, because it governs *everything that happens in the universe.* Conservation of momentum applies to atoms and elementary particles at the subatomic scale, to stars and galaxies at the cosmic scale, and to everything in between. Next time you fly in an airplane consider this: you are moving forward because of conservation of momentum. The aircraft engines "throw air backward" so the airplane and the passengers move forward. Conservation of momentum is used to solve many science and engineering problems: from calculating the forces of impact during vehicle collisions, to the trajectories of satellites, to the load-carrying capabilities of artificial limbs.

4.17 A railroad car moves at velocity v_i on a track toward three other railroad cars at rest on the track. All the cars are identical, and each stationary car is some distance from the others. The moving car hits the first stationary car and locks onto it. Then the two hit the second stationary car, lock onto that one, and so on. When all four cars have locked onto one another, what is the velocity v_f of the four-car train?

Chapter Glossary

SI units of physical quantities are given in parentheses.

Conservation of momentum Momentum can be transferred from one object to another, but it cannot be created or destroyed. The momentum of an isolated system therefore cannot change:

$$\Delta \vec{p} = \vec{0} \quad \text{(isolated system).} \qquad (4.17)$$

Conserved quantity A quantity that cannot be created or destroyed.

Environment Everything that is not part of the system.

Extensive quantity If the value of a quantity for a system is equal to the sum of the values of that quantity for each part of the system, then the quantity is extensive.

Impulse $\vec{J}$ (kg·m/s) A vector defined as the transfer of momentum from the environment to the system due to an interaction between the two. For an isolated system, the impulse is zero: $\vec{J} = \vec{0}$.

Inertia m (kg) A scalar that is a measure of an object's tendency to resist any change in its velocity. The inertia m_u of any object can be found by letting that object collide with the inertial standard $m_s = 1$ kg along the x axis and taking the negative of the ratio of the changes in the x components of the velocities:

$$m_u \equiv -\frac{\Delta v_{sx}}{\Delta v_{ux}} m_s. \qquad (4.2)$$

Intensive quantity If the value of a quantity for a system is equal to the value of that quantity for any part of that system, then that quantity is intensive.

Isolated system A system for which there are no external interactions. For such a system there is no transfer of momentum into or out of the system. See the Procedure box on page 93.

Kilogram (kg) The SI base unit of inertia, defined as being equal to the inertia of a platinum-iridium cylinder stored in Sèvres, France.

Momentum $\vec{p}$ (kg·m/s) A vector that is the product of the inertia and the velocity of an object:

$$\vec{p} \equiv m\vec{v}. \qquad (4.6)$$

Momentum law The law that accounts for the change in the momentum of a system or object. Because momentum is conserved, the momentum of a system can change only due to a transfer of momentum between the environment and the system:

$$\Delta \vec{p} = \vec{J}. \qquad (4.18)$$

System Any object or group of objects that we can separate, in our minds, from the surrounding environment.

System diagram A schematic diagram that shows a system's initial and final conditions.

Gemasolar solar thermal plant, owned by Torresol Energy. ©SENER

5 Energy

CONCEPTS

QUANTITATIVE TOOLS

Now that we know about conservation of momentum, can we determine the final velocities of two colliding objects if the only things we know are the initial velocities and the fact that the momentum is conserved? The answer is *no* if our only tool is the momentum law. In the collisions represented in Figures 4.23 and 4.25, for instance, momentum is unchanged in both cases, and yet the two outcomes are definitely not the same. So knowing only that momentum remains constant isn't enough. We need additional information in order to predict future positions and velocities. In the process of looking for this additional information, we shall develop another conservation law—the law of conservation of *energy*.

5.1 Classification of collisions

If you look at the velocity-versus-time graphs in Chapter 4, you will notice that the velocity difference in the two carts, $\vec{v}_2 - \vec{v}_1$, in most cases has the same magnitude before and after the collision. Figure 5.1 shows two graphs of collisions we considered in Chapter 4 and highlights the velocity difference before and after the collision. This difference is the **relative velocity** of the carts: $\vec{v}_{12} \equiv \vec{v}_2 - \vec{v}_1$ is the velocity of cart 2 relative to cart 1 and $\vec{v}_{21} \equiv \vec{v}_1 - \vec{v}_2$ is the velocity of cart 1 relative to cart 2. The subscript in the relative velocity symbol is always shown with the object we are studying printed last: v_{12} is the velocity *of cart 2* and v_{21} is the velocity *of cart 1*. (We cannot use Δ here because we reserve that symbol to denote the change in a *single* quantity; now we are dealing with the difference between the same quantity for two different objects.)

The magnitude of the relative velocity is called the **relative speed.** Thus $v_{12} \equiv |\vec{v}_2 - \vec{v}_1|$ is the speed of cart 2 (last subscript in v_{12}) relative to cart 1. For motion along the x axis, the relative speed is the absolute value of the difference in the x components of the velocities: $v_{12} \equiv |v_{2x} - v_{1x}|$.

Note that the sequence of the subscripts for relative speeds does not matter: $v_{12} \equiv |v_{2x} - v_{1x}| = |v_{1x} - v_{2x}| = v_{21}$ (the speed of cart 2 relative to cart 1 is equal to the speed of cart 1 relative to cart 2).

A collision in which the relative speed before the collision is the same as the relative speed after the collision is called an **elastic collision.** Collisions between hard objects are generally elastic collisions. For instance, a superball bouncing off a hard floor bounces up with nearly the same speed with which it came down. Thus the relative speed of floor and ball does not change, and the collision is elastic.

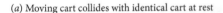 **5.1** Two cars are moving along a highway with neither one accelerating. Is their relative speed equal to the difference between their speeds? Why or why not?

Any collision for which the relative speed after the collision is lower than that before the collision is called an **inelastic collision.** If you drop a tennis ball to the floor, it rebounds with a lower speed than it had when it hit the floor. The relative speed is reduced, and the collision is inelastic.

A special type of inelastic collision is one in which the two objects move together after the collision so that their relative speed is reduced to zero. We call this special case a **totally inelastic collision.** Imagine dropping a ball of dough to the floor. Splat! The dough sticks to the floor, and the relative speed of dough and floor is reduced to zero. The collision in Figure 4.25 is another example of an inelastic collision.

As you might imagine, whether a collision is elastic, inelastic, or totally inelastic depends on the properties of the objects involved. However, if we know what happens to the relative speed in a collision, we can use that knowledge together with conservation of momentum to determine the final velocities.

Figure 5.1 Velocity-versus-time graphs of (*a*) two identical carts colliding on a low-friction track, with one of the carts initially at rest, and (*b*) a standard cart colliding with a cart of unknown inertia that is initially at rest. Notice that for both collisions the relative speeds of the two carts are the same before and after the collision.

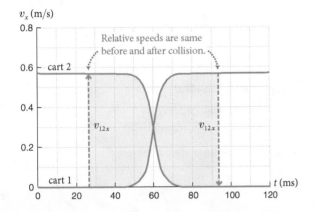

(*a*) Moving cart collides with identical cart at rest

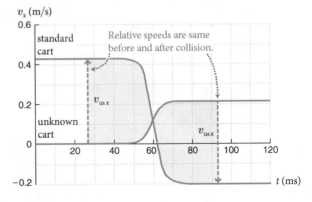

(*b*) Standard cart collides with cart of unknown inertia at rest

Exercise 5.1 Classifying collisions

Are the following collisions elastic, inelastic, or totally inelastic? (a) A red billiard ball moving at $v_{rx,i} = +2.2$ m/s hits a white billiard ball initially at rest. After the collision, the red ball is at rest and the white ball moves at $v_{wx,f} = +1.9$ m/s. (b) Cart 1 moving along a track at $v_{1x,i} = +1.2$ m/s hits cart 2 initially at rest. After the collision, the two carts move at $v_{1x,f} = +0.4$ m/s and $v_{2x,f} = +1.6$ m/s. (c) A piece of putty moving at $v_{px,i} = +22$ m/s hits a wooden block moving at $v_{bx,i} = +1.0$ m/s. After the collision, the two move at $v_{x,f} = +1.7$ m/s.

SOLUTION (a) The initial relative speed is $v_{wr,i} = |v_{rx,i} - v_{wx,i}| = |+2.2 \text{ m/s} - 0| = 2.2$ m/s; the final relative speed is $v_{wr,f} = |v_{rx,f} - v_{wx,f}| = |0 - 1.9 \text{ m/s}| = 1.9$ m/s, lower than the initial relative speed, which means the collision is inelastic. ✔

(b) $v_{12i} = |v_{2x,i} - v_{1x,i}| = |0 - (+1.2 \text{ m/s})| = 1.2$ m/s; $v_{12f} = |v_{2x,f} - v_{1x,f}| = |+1.6 \text{ m/s} - (+0.4 \text{ m/s})| = 1.2$ m/s. Because the relative speeds are the same, the collision is elastic. ✔

(c) After the collision, both the putty and the block travel at the same velocity, making their relative speed zero. The collision is totally inelastic. ✔

 5.2 (a) An outfielder catches a baseball. Is the collision between ball and glove elastic, inelastic, or totally inelastic? (b) When a moving steel ball 1 collides head-on with a steel ball 2 at rest, ball 1 comes to rest and ball 2 moves away at the initial speed of ball 1. Which type of collision is this? (c) Is the sum of the momenta of the two colliding objects constant in part a? In part b?

5.2 Kinetic energy

Relative speed is not an extensive quantity, and so we cannot develop an accounting scheme for it as we did for momentum in Chapter 4. The trick in studying elastic collisions therefore is to obtain a quantity that allows us to express the fact that the relative speed doesn't change in the form: (something of object 1) + (something of object 2) doesn't change. As I shall show you in Section 5.5, for an object of inertia m moving at

speed v this something is the quantity $K = \frac{1}{2}mv^2$, called the object's **kinetic energy** (literally "energy* associated with motion"). Unlike relative speed, which refers to two objects, kinetic energy is associated with a single object. Furthermore, kinetic energy is always positive and independent of the direction of motion (which means that it is a scalar). A ball moving at speed v to the left has exactly the same kinetic energy as the same ball moving at speed v to the right (or any other direction in space). Whenever the speed of an object changes, the kinetic energy changes.

 5.3 Is kinetic energy an extensive quantity?

To develop some feel for this new quantity and its relationship to relative speed, let us calculate the kinetic energies of the carts before and after the collisions shown in Figure 5.2. The two collisions have identical initial conditions, but one is elastic (Figure 5.2a) and the other is totally inelastic (Figure 5.2b).

Table 5.1 gives the initial and final kinetic energies of the carts, obtained by reading off the x components of the velocities from the figure. For the elastic collision, where the relative speed doesn't change, the sum of the two kinetic energies before the collision is equal to the sum after the collision: $(0.12 + 0) \text{ kg} \cdot \text{m}^2/\text{s}^2 = (0.086 + 0.029) \text{ kg} \cdot \text{m}^2/\text{s}^2$. For the totally inelastic collision, both the relative speed and the sum of the two kinetic energies change.

In general we observe:

In an elastic collision, the sum of the kinetic energies of the objects before the collision is the same as the sum of the kinetic energies after the collision.

---

*For now think of "energy" as the capacity to do things like accelerating and heating objects. We shall develop a more complete picture of what energy is in later sections and chapters.

Figure 5.2 Velocities for elastic and inelastic collisions between two carts. Cart 1 has inertia 0.12 kg, and cart 2 has inertia 0.36 kg.

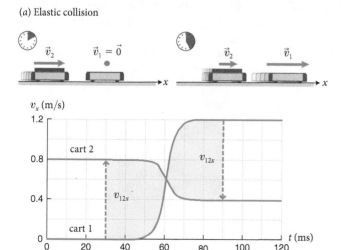

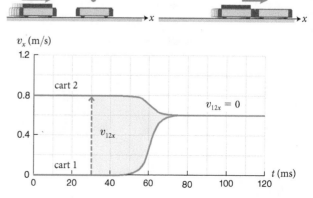

CONCEPTS

Table 5.1 Kinetic energy in elastic and totally inelastic collisions

	Inertia m (kg)	ELASTIC Velocity v_x (m/s)		Kinetic energy $\frac{1}{2}mv^2$ (kg · m²/s²)		TOTALLY INELASTIC Velocity v_x (m/s)		Kinetic energy $\frac{1}{2}mv^2$ (kg · m²/s²)	
		before	after	before	after	before	after	before	after
Cart 1	0.12	0	+1.2	0	0.086	0	+0.60	0	0.022
Cart 2	0.36	+0.80	+0.40	0.12	0.029	+0.80	+0.60	0.12	0.065
Relative speed		0.80	0.80			0.80	0		
Kinetic energy of system				0.12	0.12			0.12	0.087

In Section 5.5 we'll see that this statement is equivalent to the requirement that the relative speed remains the same.

For problems involving elastic collisions, we now have two tools: the momentum conservation law, which tells us that the momentum of an isolated system doesn't change, and the fact that the relative speed of the colliding objects doesn't change (or, alternatively, the fact that the kinetic energy of the system doesn't change). Let's now use both tools in an example.

Example 5.2 Carts colliding

(*a*) Is the collision in Figure 4.23 elastic, inelastic, or totally inelastic? How can you tell? (*b*) Verify your answer by comparing the initial kinetic energy of the two-cart system with the final kinetic energy.

❶ GETTING STARTED In Figure 4.23, it looks as if the initial and final relative speeds are the same, which makes the collision elastic. The problem asks me to confirm this fact by calculating the initial and final kinetic energies of the system.

❷ DEVISE PLAN To answer part *a*, I need to determine the initial and final relative speeds of the carts from the velocities, which I get from Figure 4.23a: $v_{1x,i} = 0$; $v_{2x,i} = +0.34$ m/s; $v_{1x,f} = +0.17$ m/s; $v_{2x,f} = -0.17$ m/s. To answer part *b*, I use $K = \frac{1}{2}mv^2$. The inertias of the carts are given in the figure caption: $m_1 = 0.36$ kg and $m_2 = 0.12$ kg.

❸ EXECUTE PLAN (*a*) $v_{12i} = |v_{2x,i} - v_{1x,i}| = |(+0.34 \text{ m/s}) - 0|$ $= 0.34$ m/s; $v_{12f} = |v_{2x,f} - v_{1x,f}| = |(-0.17 \text{ m/s}) - (+0.17 \text{ m/s})|$ $= 0.34$ m/s. The relative speed is unchanged, and thus the collision is elastic. ✔

(*b*) The initial values are

$$K_{1i} = \tfrac{1}{2}m_1 v_{1i}^2 = \tfrac{1}{2}(0.36 \text{ kg})(0)^2 = 0$$

$$K_{2i} = \tfrac{1}{2}m_2 v_{2i}^2 = \tfrac{1}{2}(0.12 \text{ kg})(0.34 \text{ m/s})^2 = 0.0069 \text{ kg · m}^2/\text{s}^2$$

so $K_i = K_{1i} + K_{2i} = 0.0069 \text{ kg · m}^2/\text{s}^2$.

The final values are

$$K_{1f} = \tfrac{1}{2}m_1 v_{1f}^2 = \tfrac{1}{2}(0.36 \text{ kg})(0.17 \text{ m/s})^2 = 0.0052 \text{ kg · m}^2/\text{s}^2$$

$$K_{2f} = \tfrac{1}{2}m_2 v_{2f}^2 = \tfrac{1}{2}(0.12 \text{ kg})(-0.17 \text{ m/s})^2 = 0.0017 \text{ kg · m}^2/\text{s}^2$$

so $K_f = (0.0052 \text{ kg · m}^2/\text{s}^2) + (0.0017 \text{ kg · m}^2/\text{s}^2)$

$= 0.0069 \text{ kg · m}^2/\text{s}^2,$

which is the same as before the collision, as it should be for an elastic collision. ✔

❹ EVALUATE RESULT Because I've reached the same conclusion—the collision is elastic—using two approaches, I can be pretty confident that my solution is correct.

Because kinetic energy is a scalar extensive quantity, bar diagrams are a good way to visually represent changes in this quantity. Figure 5.3, for example, shows the initial and final kinetic energies of the carts in the two collisions of Figure 5.2. Before the collision, only cart 2 has kinetic energy; after the collision, both carts have kinetic energy. For the elastic collision (Figure 5.3a), the sum of the kinetic

Figure 5.3 Bar diagrams representing the initial and final kinetic energies of the carts for the collisions shown in Figure 5.2. (*a*) The sum of the kinetic energies does not change for the elastic collision because the changes are equal and opposite. (*b*) The sum of the kinetic energies changes for the totally inelastic collision: The change in K_1 is smaller than the change in K_2.

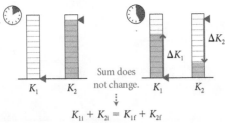

(*a*) Elastic collision

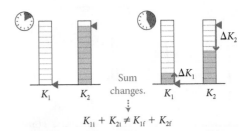

(*b*) Totally inelastic collision

CONCEPTS

energies after the collision is the same as the sum before the collision because the change in the kinetic energy of cart 1 is the negative of that of cart 2. For the totally inelastic collision (Figure 5.3*b*), the changes in the kinetic energies do not cancel, and so the system's kinetic energy after the collision is not equal to the system's kinetic energy before the collision.

✋ **5.4** A moving cart collides with an identical cart initially at rest on a low-friction track, and the two lock together. What fraction of the initial kinetic energy of the system remains in this totally inelastic collision?

5.3 Internal energy

In both inelastic and totally inelastic collisions, the relative speed changes and therefore the kinetic energy of the system changes. For example, the decrease in the kinetic energy of cart 2 in Figure 5.2*b*, represented by the downward pointing arrow labeled ΔK_2 in Figure 5.3*b*, is larger than the increase in the kinetic energy K_1 of cart 1, resulting in a decrease in the kinetic energy of the stuck-together carts. Where does this kinetic energy go? Does it simply vanish, or does it go elsewhere?

We can determine the answer to this question by looking more closely at inelastic collisions between objects and noticing that changes occur in the **state** of one or both objects. What I mean by *state* is the condition of the object as specified by some complete set of physical parameters: shape, temperature, whatever—*every possible physical variable that defines the object*. In inelastic collisions, objects deform (their shape changes) and heat up (their temperature changes): A ball of dough changes shape as it hits the ground, two cars deform as they collide, hands vigorously clapped together heat up and make sound.

The transformation of a system from an initial state to a final state is called a **process.** Processes cause change (see Section 1.4), and so in physics we aim to understand processes. Collisions that change either the motion or the state of objects are an example of a process. Other examples are the melting of an ice cube, the burning of fuel, the flow of a liquid, and an explosive separation. Initially we focus on collisions because they are easily visualized and we can study what happens in collisions using carts on tracks. As we gain experience, we shall replace collisions with other processes.

In inelastic collisions, the state of the objects after the collision is different from the state before the collision. If I were to make a movie of the ball of dough hitting the ground and then play the movie in reverse, you'd have no trouble telling that the movie was being played in the wrong direction. This is because this inelastic collision involves changes that cannot undo themselves—the dough is flat and at rest after the collision and round and moving before (and never the other way around). The same is true for a collision between two cars: The cars are damaged after the collision and not before, and it is not possible to repair

Figure 5.4 Before or after? (*a*) Inelastic collisions are an irreversible process. There is no question that the picture was taken *after* the accident because the state of the two cars has changed. (*b*) Elastic collisions are a reversible process. Can you tell whether this picture was taken before or after the ball collided with the racquet?

(*a*)

(*b*)

them simply by pulling them apart (**Figure 5.4a**). After the clapping, your hands are warmer than before, and you cannot cool your hands by "unclapping" them. All these inelastic collisions are **irreversible processes,** which means that the changes that occur in the state of the colliding objects cannot spontaneously undo themselves.

In contrast, viewing a movie of a superball bouncing off the floor or two carts colliding on a low-friction track (without sticking together), you'd be hard pressed to tell whether the movie is playing forward or in reverse. This is because elastic collisions are a **reversible process,** which means there are no permanent changes in the state of the colliding objects. The objects look the same before and after the collision (Figure 5.4*b*).

I have summarized these facts in **Table 5.2**. Notice how a change in the relative speed (and therefore a change in the sum of the kinetic energies) goes hand in hand with a change in the state: The sum of the kinetic energies of two colliding objects doesn't change unless their states change. Let us look at this connection with an eye to formulating a new conservation law.

Suppose we could associate some quantity having the same units as kinetic energy ($kg \cdot m^2/s^2$) with the state of an object—let's call this quantity the object's **internal energy** (denoted by E_{int}). Suppose further that we could arrange things in such a way that in an inelastic collision the increase in the sum of the internal energies of the colliding objects is equal to the decrease in their kinetic energies. This would mean that in an inelastic collision one form of energy is converted to another form (kinetic to internal) but the sum of the kinetic and internal energies—collectively called the **energy** of the system—doesn't change.

Table 5.2 Elastic and inelastic collisions

Collision type	Relative speed	State
elastic	unchanged	unchanged
inelastic	changed	changed
totally inelastic	changed (becomes zero)	changed

Figure 5.5 (*a*) Elastic collision. Because the relative speed of the carts does not change, the sum of the kinetic energies is also unchanged. The carts' states do not change, and so their internal energies are also unchanged. (*b*) Inelastic collision. Because the relative speed changes as a result of the collision, the kinetic energy changes. The cars' states also change, and so the internal energy changes too. For both collisions the sum of the kinetic and internal energies is constant.

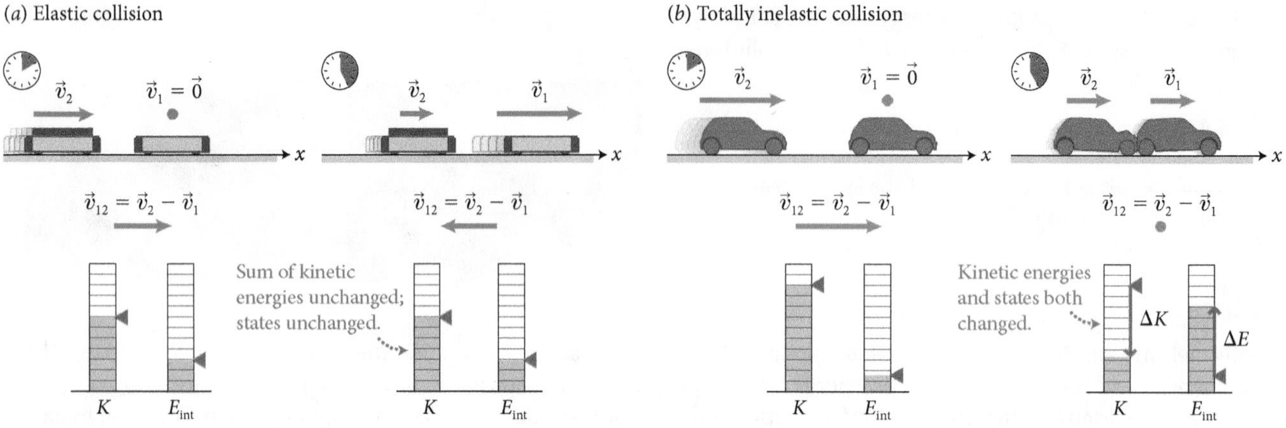

(*a*) Elastic collision

(*b*) Totally inelastic collision

In Chapter 7 we'll learn how to specify the state of an object and how to calculate the corresponding internal energy. For now, all we need to do is account for the missing kinetic energy in an inelastic collision by saying:

In any inelastic collision, the states of the colliding objects change, and the sum of their internal energies increases by an amount equal to the decrease in the sum of their kinetic energies.

Whenever some kinetic energy disappears in a collision, we can *always* determine changes in state to account for that loss. The big appeal of the relationship between state and internal energy is that we can now say:

The energy of a system of two colliding objects does not change during the collision.

This statement holds for all types of collisions: elastic, inelastic, and totally inelastic. In the elastic collision shown in **Figure 5.5a**, the collision alters the velocities and thus the kinetic energies of both carts. However, the sum of the kinetic energies before and after the collision remains unchanged, as shown by the kinetic energy bars. Because the collision is elastic, there are no changes in the states of the carts, which means their internal energies also remain unchanged, as shown by the internal energy bars. So the energy of the two-cart system does not change.

In the totally inelastic collision of Figure 5.5b, the sum of the kinetic energies decreases. Because changes occur in the states—the cars change shape—the internal energies of the carts change as well. The changes in motion and state are such that the energy of the system is the same before and after the collision. We account for the lost kinetic energy by saying that it has been converted to internal energy.

Note that in making the bar diagrams in Figure 5.5, I had to choose some initial value for the internal energies of the colliding objects. As we have no way (yet) of calculating internal energies, the values I chose are arbitrary. At present

we are interested only in *changes* in energy, however, and so the initial value is not important.

5.5 A piece of dough is thrown at a wall and sticks to it. Does the internal energy of the dough-wall system increase, decrease, or stay the same?

Example 5.3 Internal energy change

A 0.2-kg cart 1 initially at rest is struck by an identical cart 2 traveling at $v_{2x,i} = +0.5$ m/s along a low-friction track. After the collision, the velocity of cart 2 is reduced to $v_{2x,f} = +0.2$ m/s. (*a*) Is the collision elastic, inelastic, or totally inelastic? (*b*) By what amount does the internal energy of the two-cart system change? (*c*) Make a bar diagram showing the initial and final kinetic and internal energies of the two carts.

❶ GETTING STARTED I begin by organizing the information given in the problem in a sketch (**Figure 5.6**). To classify the collision, I need to determine the final relative speed, but the final velocity of cart 1 is not given.

Figure 5.6

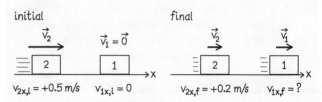

❷ DEVISE PLAN The two-cart system is isolated, and so the momentum of the system does not change, regardless of the type of collision. I can use this information to determine the final velocity of cart 1 and the final relative speed of the carts. By comparing the final and initial relative speeds, I can determine the type of collision. Once I know the initial and final velocities, I can calculate the kinetic energies using $K = \frac{1}{2}mv^2$ and determine what fraction of the initial kinetic energy has been converted to internal energy.

❸ EXECUTE PLAN (a) The initial relative speed is

$$|v_{2x,i} - v_{1x,i}| = |(+0.5 \text{ m/s}) - 0| = 0.5 \text{ m/s}.$$

To determine $v_{1x,f}$, I apply conservation of momentum to the system. The initial momentum of the system is

$$(0.2 \text{ kg})(+0.5 \text{ m/s}) + (0.2 \text{ kg})(0) = (0.2 \text{ kg})(+0.5 \text{ m/s})$$

and its final momentum is

$$(0.2 \text{ kg})(+0.2 \text{ m/s}) + (0.2 \text{ kg})(v_{1x,f}).$$

Conservation of momentum requires these two momenta to be equal:

$$(0.2 \text{ kg})(+0.5 \text{ m/s}) = (0.2 \text{ kg})(+0.2 \text{ m/s}) + (0.2 \text{ kg})(v_{1x,f})$$

$$(+0.5 \text{ m/s}) = (+0.2 \text{ m/s}) + v_{1x,f}$$

$$v_{1x,f} = +0.3 \text{ m/s}.$$

The final relative speed is thus

$$|v_{2x,f} - v_{1x,f}| = |(+0.2 \text{ m/s}) - (+0.3 \text{ m/s})| = 0.1 \text{ m/s},$$

which is different from the initial value. Thus the collision is inelastic. (I know that the collision is not *totally* inelastic because the relative speed has not been reduced to zero.) ✔

(b) The initial kinetic energies are

$$K_{1i} = 0$$

$$K_{2i} = \tfrac{1}{2}(0.2 \text{ kg})(0.5 \text{ m/s})^2 = 0.025 \text{ kg} \cdot \text{m}^2/\text{s}^2$$

so $\quad K_i = K_{1i} + K_{2i} = 0.025 \text{ kg} \cdot \text{m}^2/\text{s}^2.$

The final kinetic energies are

$$K_{1f} = \tfrac{1}{2}(0.2 \text{ kg})(0.3 \text{ m/s})^2 = 0.009 \text{ kg} \cdot \text{m}^2/\text{s}^2$$

$$K_{2f} = \tfrac{1}{2}(0.2 \text{ kg})(0.2 \text{ m/s})^2 = 0.004 \text{ kg} \cdot \text{m}^2/\text{s}^2$$

so $\quad K_f = K_{1f} + K_{2f} = 0.013 \text{ kg} \cdot \text{m}^2/\text{s}^2.$

The kinetic energy of the system has changed by an amount $(0.013 \text{ kg} \cdot \text{m}^2/\text{s}^2) - (0.025 \text{ kg} \cdot \text{m}^2/\text{s}^2) = -0.012 \text{ kg} \cdot \text{m}^2/\text{s}^2.$ To keep the energy of the system (the sum of its kinetic and internal energies) unchanged, the decrease in kinetic energy must be made up by an increase in internal energy. This tells me that the internal energy of the system increases by $0.012 \text{ kg} \cdot \text{m}^2/\text{s}^2.$ ✔

(c) My bar diagram is shown in **Figure 5.7**. The final kinetic energy bar is about half of the initial kinetic energy bar. Because I don't know the value of the initial internal energy, I set it to zero and make the final internal energy bar equal in height to the difference in the kinetic energy bars. ✔

Figure 5.7

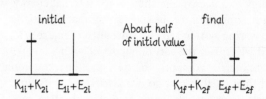

❹ EVALUATE RESULT If the collision were elastic, the velocities of the carts would be interchanged and cart 2 would come to a stop (see Figure 4.5). So the collision must be inelastic. Indeed, I found that both the relative speed and the sum of the kinetic energies change in the collision, as expected for an inelastic collision.

We are now in a position to extend the idea of internal energy to other interactions. Consider, for example, a cart initially at rest on a low-friction track set in motion by an expanding spring that is held fixed at one end, as shown in **Figure 5.8a**. As the cart is accelerated by the spring, the cart's kinetic energy increases but its state doesn't change, and so its energy increases (Figure 5.8b). Where did this energy come from? The spring sets the cart in motion, and so it makes sense to assume that the spring transfers energy to the cart. Indeed, the spring expands—its state changes—and so its internal energy changes. If we include the spring

Figure 5.8 Initial and final energies for two choices of system.

(a) Expanding spring accelerates cart from rest

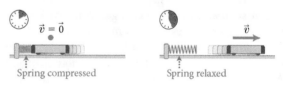

(b) Initial and final energies: system = cart only

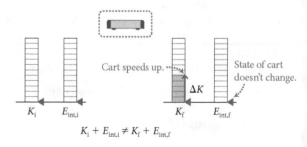

(c) Initial and final energies: system = cart + spring

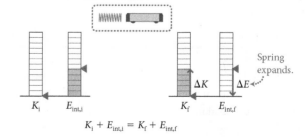

in the system and attribute the increase in the cart's kinetic energy to a decrease in the spring's internal energy, we can again arrange things in such a way that the energy of the cart-spring system remains unchanged (Figure 5.8c).

If we replace the cart in Figure 5.8 with a cart with a different inertia, compress the spring to the same initial state, and then let that cart go, we discover that its final kinetic energy is exactly the same as that of the first cart. So the spring always transfers the same amount of energy as it expands from a given compressed state to its relaxed state *regardless* of the object to which it transfers that energy.

5.6 Think of a few other ways to put the cart of Figure 5.8 in motion. In each case, can you account for the increase in the cart's kinetic energy by either a change in state or a change in motion of another object?

Checkpoint 5.6 suggests that a change in the cart's kinetic energy can always be attributed to either a change in state (and therefore a change in the internal energy) or a change in the motion of another object. Because the states of objects can change in many different ways, we associate different forms of internal energy with different kinds of state change. For example, the internal energy associated with a change in chemical state is called *chemical energy* and the internal energy associated with a change in the temperature of an object is called *thermal energy*. Table 5.3 lists additional forms of internal energy.

Absolutely everything happening around us entails changes in state and therefore changes in internal energy. Rivers flowing, air masses moving, machines lifting things, people walking, and atoms emitting light can all be expressed in terms of changes in state (and therefore changes in internal energy). More important, we discover that any change in state or motion is always accompanied by a compensating change in state or motion, and we can always attribute energy to these changes in such a way that the energy of the system remains unchanged. In other words, energy cannot be destroyed or created, and energy, like momentum, is a conserved quantity. Indeed, no observation has ever been found to violate the law of **conservation of energy:**

Energy can be transferred from one object to another or converted from one form to another, but it cannot be destroyed or created.

Table 5.3 Various state changes and their associated internal energy

State change	Internal energy
temperature change	thermal energy
chemical change	chemical energy
reversible change in shape	elastic energy
phase transformation	transformation energy

Kinetic energy and all forms of internal energy are thus different manifestations of the same conserved quantity: *energy.*

5.7 (a) Is the momentum of the cart-spring system in Figure 5.8 constant? (b) Is the system isolated?

5.4 Closed systems

As Checkpoint 5.7 shows, the system comprising the spring and cart in Figure 5.8 is not isolated. However, no energy is transferred to it, and therefore the energy of the system is constant.* Any system to or from which no energy is transferred is called a **closed system.** An important point to keep in mind is that a closed system need not be isolated (and likewise an isolated system is not necessarily closed). The procedure for choosing a closed system is described in the Procedure box. To see how this procedure works, let's look at the setup in **Figure 5.9**—some fuel (gasoline, say) being burned in a container open to the air.

I begin by making a sketch of the canister with the fuel before and after the combustion (Figure 5.9a). The combustion involves two changes in state: a change in the chemical state of the fuel and the air, and a change in the

Figure 5.9 (a) The burning of fuel involves a change in chemical composition and in temperature. (b) If we choose a closed system, the change in chemical energy must be compensated by a change in thermal energy.

(a) Sketch initial and final conditions, identify changes, choose system

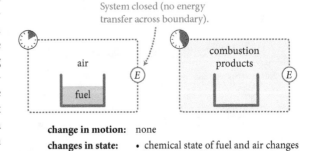

change in motion:	none
changes in state:	• chemical state of fuel and air changes
	• temperature rises

(b) Draw energy bar diagrams for initial and final conditions

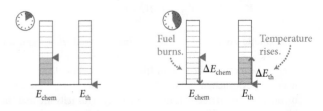

---

*How do I know that no energy is transferred to the cart-spring system? The expanding spring and the accelerating cart do not cause any changes in the state or motion of the environment (the track, Earth, and so on). Consequently the energy of the environment doesn't change, which means that no energy has been transferred from the environment to the system.

Procedure: Choosing a closed system

When we analyze energy changes, it is convenient to choose a system for which no energy is transferred to or from the system (a closed system). To do so, follow this procedure:

1. Make a sketch showing the initial and final conditions of the objects under consideration.
2. Identify all the changes in state or motion that occur during the time interval of interest.
3. Choose a system that includes all the objects undergoing these changes in state or motion. Draw a dashed line around the objects in your chosen system to represent the system boundary. Write "closed" near the system boundary to remind yourself that no energy is transferred to or from the system.
4. Verify that nothing in the surroundings of the system undergoes a change in motion or state that is related to what happens inside the system.

Once you have selected a closed system, you know that its energy remains constant.

temperature. The fuel and the air immediately surrounding it are what undergo these changes in state, and so I include them in my system. Because there are no other related changes in state, I know that my system is closed and its energy remains constant.

I can now make a bar diagram to represent the changes in energy that take place inside the system (Figure 5.9*b*). The change in the chemical state corresponds to a change in chemical energy, and the change in the temperature corresponds to a change in thermal energy. I therefore draw two bars. The combustion raises the temperature, and so the thermal energy increases. Because the system is closed and its energy remains constant, there must be a decrease in chemical energy that compensates for the increase in thermal energy.

Given that the energy of a closed system remains constant, we can focus on the energy conversions and transfers that happen inside the system. We speak of an *energy conversion* when energy is converted from one form to another. In the combustion illustrated in Figure 5.9, for example, an amount ΔE_{chem} of chemical energy is converted to an equal amount ΔE_{th} of thermal energy. When energy is transferred from one object to another, we speak of an *energy transfer*. An example is shown in Figure 5.3*a*, where energy is transferred from cart 2 to cart 1.

Because we are still lacking the quantitative tools, let me illustrate the transfer and conversion of energy qualitatively with a few examples. By converting some form of internal energy to kinetic energy, we can put objects in motion. For example, a car burns gasoline as it accelerates from rest. If we ignore the effects of air resistance, this situation represents a change in the chemical state of the gasoline and a change in the motion of the car. Chemical energy (a form of internal energy associated with the chemical state) stored in the gasoline is converted to kinetic energy of the car (Figure 5.10).

> ✋ **5.8** In describing what's going on in Figure 5.10, I ignored the change in temperature of the engine that accompanies the combustion of the fuel. Make a bar diagram that includes this change in temperature.

Figure 5.10 (*a*) An accelerating car constitutes a closed system because no changes occur in its environment. (*b*) The increase in the car's kinetic energy can be attributed to a decrease in chemical energy (due to the combustion of fuel).

(*a*) Initial and final conditions, changes in state and motion, system

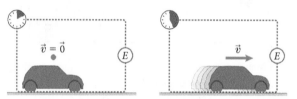

change in motion: car accelerates
change in state: chemical state of fuel changes

(*b*) Energy bar diagrams for initial and final conditions

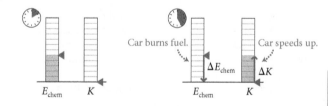

The generation and consumption of electrical power provide many other examples of energy conversions and transfers. In a coal-burning electrical power plant, coal is burned in a boiler to heat water and convert it to steam. The two complementary changes in state (a change in the chemical state of the coal and a change in the temperature and state of the water) correspond to a conversion of chemical energy to thermal energy. The steam drives a turbine, where the steam cools and a turbine blade spins, and so thermal energy is converted to kinetic energy. The moving turbine drives a generator, which converts the kinetic energy to electrical energy. Power lines transfer energy from the plant to our homes, where it is converted to many forms. Lamps convert electrical energy to light (a form of energy) and thermal energy (lamps get hot), a stereo converts electrical energy to sound energy, and on and on.

Exercise 5.4 State changes and internal energy

Choose an appropriate closed system and make a bar diagram representing the energy conversions and transfers that occur when (*a*) a pan of water is heated on a propane burner, (*b*) a cyclist accelerates from rest, and (*c*) a spring-loaded gun fires a ball of putty.

SOLUTION For each case, I apply the steps of the Procedure box on page 109. My sketches are shown in the figures.

(*a*) Changes in motion: none. Changes in state: the temperature of the water increases, and the chemical states of the propane and the air change. The bar diagram shows an increase in thermal energy and an equal decrease in chemical energy (**Figure 5.11**). Chemical energy is converted to thermal energy, and in the process energy is transferred from the propane to the water.

Figure 5.11

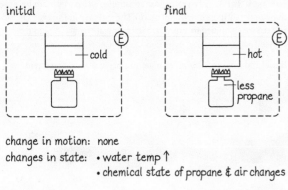

(*b*) Changes in motion: the bicycle and the cyclist accelerate. Changes in state: the chemical state of the cyclist changes because setting the bicycle in motion requires muscles to contract, a physiological process that involves a complex series of chemical reactions. The bar diagram shows an increase in kinetic energy and an equal decrease in chemical energy (**Figure 5.12**). Chemical energy is converted to kinetic energy of the bicycle and cyclist.

Figure 5.12

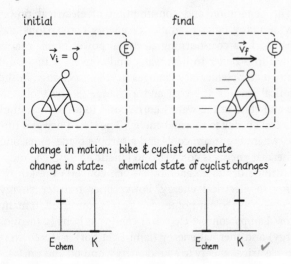

(*c*) Changes in motion: the putty is accelerated. Changes in state: the spring expands. The bar diagram shows a decrease in the elastic energy of the spring and an increase in the kinetic energy of the putty (**Figure 5.13**). As the spring expands, elastic energy is converted to kinetic energy of the putty.

Figure 5.13

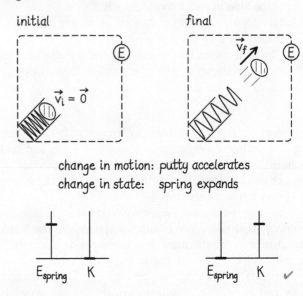

We started our study of physics by describing motion in terms of velocities, accelerations, and inertias. Now we have two new quantities, *momentum* and *energy*, and we have found two fundamental conservation laws that allow us to use simple accounting principles to describe changes in motion.

5.9 (*a*) Can the magnitude of the momentum of an object change without a change in the object's kinetic energy? (*b*) Without a change in the object's energy? (*c*) What are your answers to parts *a* and *b* for a system consisting of more than one object?

Self-quiz

1. Consider an isolated object at rest in space. The object contains internal energy in some form or another. Is it—in principle—possible to convert the internal energy to kinetic energy so that the object starts to move?

2. Imagine squeezing a piece of foam with your hands. Choose an appropriate closed system and make a bar diagram representing the energy conversions and transfers that occur during the squeezing.

3. When you heat a pot of water on a gas stove, the water temperature increases until the water begins to boil. This change in thermal state from cool water to hot water is due to chemical energy from the burning gas being converted to thermal energy of the water. Once boiling starts, the water temperature stays constant until all the water has turned to steam, even though the burning gas continues to transfer energy to the water. While the water is boiling off and becoming steam, what becomes of the energy released by the burning gas?

4. An electric fan turns electrical energy into wind energy (a form of kinetic energy because it involves moving air). Suppose a blowing fan is suddenly unplugged. Even though the fan no longer receives electrical energy, it continues to blow air while the blades slowly come to a stop. What type of energy is converted to wind energy after the fan is unplugged?

Answers

1. No. Getting the object to move would violate the law of conservation of momentum because the object would start with zero momentum ($\vec{p} = m\vec{v} = m\vec{0} = \vec{0}$) and end with nonzero momentum ($\vec{p} = m\vec{v}$).

2. Draw the foam before and after the squeezing. Two changes in state occur: The shape of the foam changes, and the chemical state of your muscles changes. These two changes in state correspond to changes in the foam's elastic energy and in chemical energy. In your closed system, include the foam and yourself (Figure 5.14).

Figure 5.14

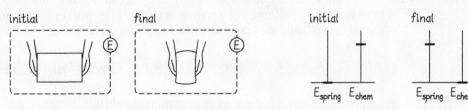

change in motion: none
changes in state: • shape of foam changes
                  • chemical state of muscles changes

3. It goes into changing the phase of the water from liquid to steam.

4. The kinetic energy of the fan blades is converted to wind energy.

5.5 Elastic collisions

As we saw in Section 5.1, we can classify collisions according to what happens to the relative velocity of the two colliding objects. The **relative velocity** of cart 2 relative to cart 1 is defined as

$$\vec{v}_{12} \equiv \vec{v}_2 - \vec{v}_1. \tag{5.1}$$

The relative velocity of cart 1 relative to cart 2 is the negative of this vector:

$$\vec{v}_{21} \equiv \vec{v}_1 - \vec{v}_2 = -\vec{v}_{12}. \tag{5.2}$$

If the magnitude of this relative velocity (the objects' **relative speed** $v_{12} \equiv |\vec{v}_2 - \vec{v}_1|$) is the same before and after the collision, the collision is elastic:

$$v_{12i} = v_{12f} \quad \text{(elastic collision).} \tag{5.3}$$

For two objects moving along the x axis, we can write this as

$$v_{2x,i} - v_{1x,i} = -(v_{2x,f} - v_{1x,f}) \quad \text{(elastic collision),} \tag{5.4}$$

where I have added a minus sign on the right because if object 2 initially moves faster than object 1, object 1 must move faster after the collision (**Figure 5.15**). In other words, in an elastic collision, if $v_{2x,i} > v_{1x,i}$, we must have $v_{2x,f} < v_{1x,f}$. The relative velocity in an elastic collision always changes sign after the collision: $v_{12x,i} = -v_{12x,f}$.

Because we can consider a system made up of two colliding objects to be isolated during the collision, conservation of momentum requires that the momentum of the colliding objects remains unchanged in the collision:

$$m_1 v_{1x,i} + m_2 v_{2x,i} = m_1 v_{1x,f} + m_2 v_{2x,f} \quad \text{(isolated system).} \tag{5.5}$$

Equations 5.4 and 5.5 allow us to determine the final velocities of the objects, given their initial velocities.

Figure 5.15 A collision between two objects moving in the same direction at different speeds.

If object 2 initially moves faster than object 1 . . .

. . . then after the collision, object 1 must move faster than object 2.

Example 5.5 Elastic collision

Two carts, one of inertia $m_1 = 0.25$ kg and the other of inertia $m_2 = 0.40$ kg, travel along a straight horizontal track with velocities $v_{1x,i} = +0.20$ m/s and $v_{2x,i} = -0.050$ m/s. What are the carts' velocities after they collide elastically?

❶ GETTING STARTED I begin by organizing the information in a sketch to help visualize the situation (**Figure 5.16**). I need to determine the velocities after the collision from the information given, knowing that the collision is elastic.

Figure 5.16

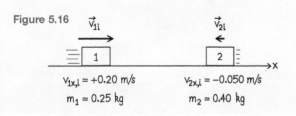

$v_{1x,i} = +0.20$ m/s $v_{2x,i} = -0.050$ m/s

$m_1 = 0.25$ kg $m_2 = 0.40$ kg

❷ DEVISE PLAN To determine the two unknowns $v_{1x,f}$ and $v_{2x,f}$, I need two equations. Because the collision is elastic, I can use Eqs. 5.4 and 5.5.

❸ EXECUTE PLAN I begin by solving Eq. 5.4 for $v_{2x,f}$:

$$v_{2x,f} = v_{1x,i} - v_{2x,i} + v_{1x,f}$$

so that I can eliminate $v_{2x,f}$ from Eq. 5.5:

$$m_1 v_{1x,i} + m_2 v_{2x,i} = m_1 v_{1x,f} + m_2 (v_{1x,i} - v_{2x,i} + v_{1x,f}).$$

Solving this expression for $v_{1x,f}$ yields

$$v_{1x,f} = \frac{m_1 - m_2}{m_1 + m_2} v_{1x,i} + \frac{2m_2}{m_1 + m_2} v_{2x,i}. \tag{1}$$

Repeating this procedure, I can eliminate $v_{1x,f}$ and solve for $v_{2x,f}$:

$$v_{2x,f} = \frac{2m_1}{m_1 + m_2} v_{1x,i} - \frac{m_1 - m_2}{m_1 + m_2} v_{2x,i}. \qquad (2)$$

Substituting the given values into Eqs. 1 and 2 yields $v_{1x,f} = -0.11$ m/s and $v_{2x,f} = +0.14$ m/s. ✔

❹ EVALUATE RESULT The final velocities yield the final relative speed $v_{12f} = |(+0.14 \text{ m/s}) - (-0.11 \text{ m/s})| = 0.25$ m/s. This value is the same as the initial relative speed $v_{12i} = |(-0.05 \text{ m/s}) - (+0.20 \text{ m/s})| = 0.25$ m/s, as required for an elastic collision.

Let me now show how Eqs. 5.4 and 5.5 lead to an expression that shows that the kinetic energy of two colliding objects doesn't change in an elastic collision. Rearranging the terms in Eq. 5.5, I get

$$m_1(v_{1x,i} - v_{1x,f}) = m_2(v_{2x,f} - v_{2x,i}), \qquad (5.6)$$

while Eq. 5.4 yields

$$v_{1x,i} + v_{1x,f} = v_{2x,i} + v_{2x,f}. \qquad (5.7)$$

Now I multiply the left side of Eq. 5.6 by the left side of Eq. 5.7 and likewise for the right sides to get

$$m_1(v_{1x,i} - v_{1x,f})(v_{1x,i} + v_{1x,f}) = m_2(v_{2x,f} - v_{2x,i})(v_{2x,i} + v_{2x,f}), \qquad (5.8)$$

which multiplies out to

$$m_1 v_{1x,i}^2 - m_1 v_{1x,f}^2 = m_2 v_{2x,f}^2 - m_2 v_{2x,i}^2. \qquad (5.9)$$

For motion along the x axis, the square of the x component of a velocity is the same as the square of the speed, $v_x^2 = v^2$, and so I can drop all the subscripts x. If I put all the v_i terms on the left and all the v_f terms on the right, I get

$$m_1 v_{1i}^2 + m_2 v_{2i}^2 = m_1 v_{1f}^2 + m_2 v_{2f}^2. \qquad (5.10)$$

I now divide both sides by 2 to get

$$\tfrac{1}{2} m_1 v_{1i}^2 + \tfrac{1}{2} m_2 v_{2i}^2 = \tfrac{1}{2} m_1 v_{1f}^2 + \tfrac{1}{2} m_2 v_{2f}^2. \qquad (5.11)$$

As you saw in Section 5.2, the quantity $\tfrac{1}{2} mv^2$ is the **kinetic energy**, which we denote by K:*

$$K \equiv \tfrac{1}{2} mv^2. \qquad (5.12)$$

With this definition, Eq. 5.11 simplifies to

$$K_{1i} + K_{2i} = K_{1f} + K_{2f} \quad \text{(elastic collision)}, \qquad (5.13)$$

where $K_{1i} = \tfrac{1}{2} m_1 v_{1i}^2$ is the initial kinetic energy of object 1, and so on. In words, in an elastic collision the sum of the initial kinetic energies is the same as the sum of the final kinetic energies.

---

*The factor $\tfrac{1}{2}$ in Eq. 5.12 is there for convenience. We could have defined kinetic energy as simply mv^2, but the definition given in Eq. 5.12 simplifies matters later on.

Note that there is no new physics in Eqs. 5.11 and 5.13—both express the same fact as Eq. 5.3. The great advantage of writing this information in the form of Eq. 5.13 is that this form expresses Eq. 5.3 in terms of an extensive quantity. Because kinetic energy is an extensive quantity, the kinetic energy of a system comprising two objects is the sum of the kinetic energies of the individual objects: $K = K_1 + K_2$. Equation 5.13 thus tells us that the kinetic energy K of a system of objects undergoing an elastic collision does not change:

$$K_i = K_f \quad \text{(elastic collision)}. \tag{5.14}$$

This result can also be written in the form

$$\Delta K = 0 \quad \text{(elastic collision)}. \tag{5.15}$$

As you know from studying Table 5.1, the SI unit for kinetic energy is $\text{kg} \cdot \text{m}^2/\text{s}^2$. Because energy is such an important quantity, we give this combination of units the name **joule** (rhymes with *pool*), after the British physicist James Prescott Joule (1818–1889), who studied energy:

$$1 \, \text{kg} \cdot \text{m}^2/\text{s}^2 \equiv 1 \, \text{J}.$$

Example 5.6 Collision and kinetic energy

A rubber ball of inertia $m_b = 0.050$ kg is fired along a track toward a stationary cart of inertia $m_c = 0.25$ kg. The kinetic energy of the system after the two collide elastically is 2.5 J. (*a*) What is the initial velocity of the ball? (*b*) What are the final velocities of the ball and the cart?

❶ **GETTING STARTED** I organize the problem graphically (**Figure 5.17**). I choose my x axis in the direction of the incoming rubber ball. Only one initial velocity is given. I need to determine the other initial velocity and both final velocities.

Figure 5.17

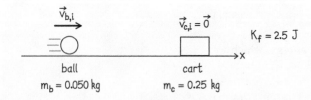

$K_f = 2.5$ J

ball
$m_b = 0.050$ kg

cart
$m_c = 0.25$ kg

❷ **DEVISE PLAN** Because the collision is elastic, I know that the kinetic energy of the system doesn't change, which means that the final value (2.5 J) is the same as the initial value. Because the cart is initially at rest, all of this kinetic energy belongs initially to the ball, and so I can use Eq. 5.12 to determine the initial velocity

of the ball. Once I have this information, I know the initial velocities of both colliding objects and I can use Eqs. 1 and 2 from Example 5.5 to calculate the final velocities.

❸ **EXECUTE PLAN** (*a*) From Eq. 5.12, I obtain the initial speed of the ball:

$$v_{b,i} = \sqrt{\frac{2K_{b,i}}{m_b}} = \sqrt{\frac{2(2.5 \, \text{J})}{0.05 \, \text{kg}}} = 10 \, \text{m/s}.$$

Because the ball is initially moving in the positive x direction, its initial velocity is given by $v_{bx,i} = +10$ m/s. ✔

(*b*) I can now substitute the two initial velocities and the inertias into Eqs. 1 and 2 of Example 5.5, which gives me $v_{bx,f} = -6.7$ m/s and $v_{cx,f} = +3.3$ m/s. ✔

❹ **EVALUATE RESULT** It makes sense that the velocity of the ball is reversed by the collision because the inertia of the cart is so much greater than that of the ball. Now that I know both the initial and final velocities, I can also check to make sure that the relative speed remains the same. Because the cart is initially at rest, the initial relative speed is 10 m/s; the final relative speed is $v_{bc,f} = |(+3.3 \, \text{m/s}) - (-6.7 \, \text{m/s})| = 10$ m/s, which is the same, as required for an elastic collision.

5.10 Conservation of momentum tells us that mv_x doesn't change for an isolated system. Now we see that in an elastic collision mv_x^2 also doesn't change. Does this mean that mv_x^3 remains unchanged, too?

5.6 Inelastic collisions

In a totally inelastic collision between two objects, the objects move together after the collision and so the two final velocities are identical. This means that the final relative speed is zero:

$$v_{12f} = 0 \quad \text{(totally inelastic collision).} \tag{5.16}$$

Therefore if you know the inertias m_1 and m_2 of the objects and the x components of their initial velocities $v_{1x,i}$ and $v_{2x,i}$, you can obtain the final velocities from Eq. 5.5 by setting

$$v_{1x,f} = v_{2x,f} \quad \text{(totally inelastic collision).} \tag{5.17}$$

The majority of collisions, however, fall somewhere between the two extremes of elastic and totally inelastic. For these collisions, the final relative speed is between zero and the value of the initial relative speed. For these cases, it is convenient to define the ratio of relative speeds as

$$e \equiv \frac{v_{12f}}{v_{12i}}. \tag{5.18}$$

The unitless quantity e is called the **coefficient of restitution** of the collision. It tells us how much of the initial relative speed is restituted (restored) after the collision. Because it is a ratio of speeds (which are always positive), e is always positive.

In general, it is more convenient to write Eq. 5.18 in terms of components:

$$e = -\frac{v_{2x,f} - v_{1x,f}}{v_{2x,i} - v_{1x,i}} = -\frac{v_{12x,f}}{v_{12x,i}}, \tag{5.19}$$

where the minus sign appears because the relative velocity changes sign after the collision (see Figure 5.15).

When $e = 1$, Eq. 5.19 is equal to Eq. 5.4 and the collision is elastic (all of the initial relative speed is restored). A value of $e = 0$ means that v_{12f} must be zero, and so the collision is totally inelastic. For values of e between 0 and 1, the collision is inelastic. This information is summarized in Table 5.4.

In an inelastic collision, therefore, we again have two equations—Eqs. 5.1 and 5.19—but we need to know the value of e to be able to calculate the final velocities. The details of how the states of the objects change in the collision (these state changes depend on the material and structural properties of the objects) are hidden in this coefficient e.

Table 5.4 Coefficient of restitution for various processes

Process	Relative speed	Coefficient of restitution
totally inelastic collision	$v_{12f} = 0$	$e = 0$
inelastic collision	$0 < v_{12f} < v_{12i}$	$0 < e < 1$
elastic collision	$v_{12f} = v_{12i}$	$e = 1$
explosive separation*	$v_{12f} > v_{12i}$	$e > 1$

*See Section 5.8.

Example 5.7 Restitution

A white car of inertia 1200 kg that is moving at a speed of 7.2 m/s rear-ends a blue car of inertia 1000 kg that is initially at rest. Immediately after the collision, the white car has a speed of 3.6 m/s. What is the coefficient of restitution for this collision?

❶ **GETTING STARTED** By putting all the information in graphical form (**Figure 5.18**), I see that I do not have a value for the final velocity of the blue car, which I need in order to calculate the coefficient of restitution.

Figure 5.18

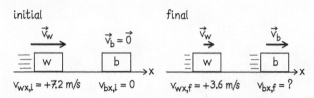

❷ **DEVISE PLAN** Because the system of the two colliding cars is isolated, I can use conservation of momentum (Eq. 5.5) to obtain the final velocity of the blue car, $v_{bx,f}$.

❸ **EXECUTE PLAN** I begin by writing Eq. 5.5 with subscripts appropriate to this particular collision:

$$m_w v_{wx,i} + m_b v_{bx,i} = m_w v_{wx,f} + m_b v_{bx,f}$$

or

$$m_w v_{wx,i} + 0 = m_w v_{wx,f} + m_b v_{bx,f}.$$

Solving for $v_{bx,f}$ and substituting the values given, I get

$$v_{bx,f} = \frac{m_w}{m_b}(v_{wx,i} - v_{wx,f})$$

$$= \frac{1200 \text{ kg}}{1000 \text{ kg}}(7.2 \text{ m/s} - 3.6 \text{ m/s}) = 4.3 \text{ m/s}.$$

Now I can substitute the initial and final velocities into Eq. 5.19 to obtain the coefficient of restitution:

$$e = -\frac{v_{wx,f} - v_{bx,f}}{v_{wx,i} - v_{bx,i}}$$

$$= -\frac{(+3.6 \text{ m/s}) - (+4.3 \text{ m/s})}{(7.2 \text{ m/s}) - 0} = 0.097. ✔$$

❹ **EVALUATE RESULT** The coefficient of restitution is rather small, but that is to be expected because cars don't bounce like superballs. Indeed, the blue car moves just a little bit faster than the white car after the collision (4.3 m/s versus 3.6 m/s), as I expect.

✋ **5.11** In a totally inelastic collision between two objects in an isolated system with one of the objects initially at rest, is it possible to lose *all* of the system's initial kinetic energy?

5.7 Conservation of energy

Because no energy is transferred to or from a closed system, conservation of energy requires that any change in the kinetic energy of a closed system be compensated by an equal change in the internal energy so that the sum of the kinetic and internal energies of the system does not change:

$$K_i + E_{int,i} = K_f + E_{int,f} \quad \text{(closed system).} \tag{5.20}$$

In contrast to Eq. 5.14, which holds for only an elastic collision, Eq. 5.20 holds for any closed system. If we write the sum of the kinetic and internal energies of an object or system as the **energy** of the object or system,

$$E \equiv K + E_{int}, \tag{5.21}$$

then we can rewrite Eq. 5.20 as

$$E_i = E_f \quad \text{(closed system)} \tag{5.22}$$

or

$$\Delta E = 0 \quad \text{(closed system).} \tag{5.23}$$

Note the parallel between Eq. 4.17, $\Delta \vec{p} = \vec{0}$, and Eq. 5.23. Equation 4.17 embodies conservation of momentum and states that the momentum of an isolated system cannot change. Equation 5.23 embodies conservation of energy: It states

Figure 5.19 Two examples of energy conservation in a closed system. (*a*) The ball's kinetic energy is converted to internal energy in the deformed mattress. (*b*) Chemical energy stored in the battery is converted to thermal energy (the battery gets hot). In both cases, as the bar graphs show, the system's energy remains the same.

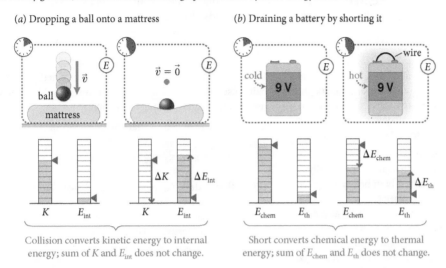

(*a*) Dropping a ball onto a mattress

Collision converts kinetic energy to internal energy; sum of K and E_{int} does not change.

(*b*) Draining a battery by shorting it

Short converts chemical energy to thermal energy; sum of E_{chem} and E_{th} does not change.

that the energy of a closed system cannot change; that is, energy cannot be created or destroyed.

Even though we cannot yet calculate the internal energy E_{int}, Eq. 5.23 allows us to draw a number of important conclusions. First, if the kinetic energy of a closed system changes, then the state of the system must change in such a way that its internal energy changes by an amount

$$\Delta E_{int} = -\Delta K \quad \text{(closed system).} \tag{5.24}$$

As an example, consider **Figure 5.19a**. A ball is dropped onto a mattress, where it comes to rest. During the time interval from the instant just before the ball hits the mattress to the instant at which the ball comes to rest, the motion of the ball changes and the shape of the mattress changes. The ball and mattress constitute a closed system, and so the decrease in kinetic energy must be accompanied by an increase in internal energy. Equation 5.24 requires the loss in kinetic energy to be equal to the gain in internal energy.

The second conclusion we can draw from Eq. 5.23 is that if the kinetic energy of a closed system does not change ($\Delta K = 0$), neither does the internal energy of the system ($\Delta E_{int} = 0$). Because E_{int} is the sum of the internal energies of all the parts that make up the system, however, $\Delta E_{int} = 0$ does *not* mean that no changes in state take place. For instance, E_{int} remains constant when internal energy is transferred from one part of a closed system to another:

$$\Delta E_{int,1} = -\Delta E_{int,2} \Rightarrow \Delta E_{int} = 0 \quad \text{(closed system, } \Delta K = 0) \tag{5.25}$$

or when one form of internal energy is converted to another:

$$\Delta E_{form\,1} = -\Delta E_{form\,2} \Rightarrow \Delta E_{int} = 0 \text{ (closed system, } \Delta K = 0). \tag{5.26}$$

As an example, consider Figure 5.19*b*. When the battery is drained rapidly, it becomes very hot. Because there is no motion before and after the draining, no change in kinetic energy occurs, but the chemical energy in the battery is converted to thermal energy. Equation 5.26 requires the loss in chemical energy to be equal to the gain in thermal energy:

$$\Delta E_{chem} + \Delta E_{th} = 0. \tag{5.27}$$

In practice there are often more than two simultaneous changes of state in a system (see Checkpoint 5.8, for instance), but regardless of how many energy conversions and transfers take place, the law of conservation of energy requires that the *amount of energy* in a closed system never changes.

Example 5.8 Making light

A 0.20-kg steel ball is dropped into a ball of dough, striking the dough at a speed of 2.3 m/s and coming to rest inside the dough. If it were possible to turn all of the energy converted in this totally inelastic collision into light, how long could you light a desk lamp? It takes 25 J to light a desk lamp for 1.0 s.

❶ GETTING STARTED I begin by applying the procedure for choosing a closed system. Although the problem doesn't specify it explicitly, I'm assuming the dough is at rest both before and after the steel ball is dropped in it; it could, for example, be at rest on a countertop. Only the steel ball has kinetic energy initially, and all of this energy is converted to internal energy as the ball comes to rest in the dough (**Figure 5.20**). So I have to calculate the initial kinetic energy of the ball and determine how long that amount of energy could light a lamp, given that 25 J lights a lamp for 1.0 s.

Figure 5.20

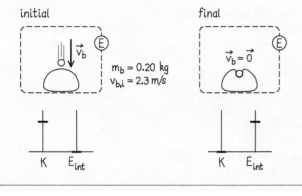

❷ DEVISE PLAN To determine the initial kinetic energy of the ball, I use Eq. 5.12. Then I divide this result by 25 J to determine how many seconds I can light a lamp.

❸ EXECUTE PLAN The initial kinetic energy of the ball is

$$K_{b,i} = \tfrac{1}{2} m_b v_{b,i}^2 = \tfrac{1}{2}(0.20 \text{ kg})(2.3 \text{ m/s})^2 = 0.53 \text{ J}.$$

Given that a desk lamp requires 25 J per second, this 0.53 J lights a lamp for

$$\frac{\text{energy available}}{\text{energy needed per second}} = \frac{0.53 \text{ J}}{25 \text{ J/s}} = 0.021 \text{ s}. ✔$$

❹ EVALUATE RESULT The length of time I obtained, two hundredths of a second, is not very much! However, a 0.20-kg steel ball moving at 2.3 m/s does not have much kinetic energy: I know from experience that a small steel ball's ability to induce state changes—to crumple or deform objects, for example—is very limited. So it makes sense that one can't light a desk lamp for very long.

✋ **5.12** A gallon of gasoline contains approximately 1.2×10^8 J of energy. If all of this energy were converted to kinetic energy in a 1200-kg car, how fast would the car go?

5.8 Explosive separations

Figure 5.21 When a cannon is fired, internal (chemical) energy is converted to kinetic energy.

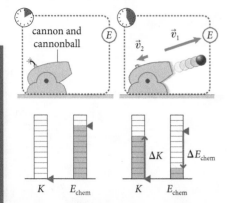

Is it possible to have a process in which kinetic energy is *gained* at the expense of internal energy? Yes, in an *explosion*, or any other type of **explosive separation,** where objects separate or break apart from each other, kinetic energy increases and internal energy decreases. The firing of a cannon is one example (**Figure 5.21**). Initially the cannon and cannonball are at rest. When the cannon is fired, the cannonball flies out of the barrel, and the cannon *recoils* in the opposite direction. (If it didn't recoil, the momentum of the system would not remain zero.) Both the cannon and cannonball therefore gain kinetic energy at the expense of chemical energy in the gun powder. This situation is the inverse of a totally inelastic collision: In this explosive separation, we start with the two objects together ($v_{12} = 0$) and end with them moving apart ($v_{12} > 0$).

Figure 5.22a shows an explosive separation that can be carried out on a low-friction track. Two carts, of inertias m and $3m$, are held against a compressed spring. When the carts are released, they move apart as the spring expands. As it expands, the spring's state changes and so does its internal energy; the decrease

Figure 5.22 Another example of an explosive separation.

(*a*) When carts are released, spring pushes them apart

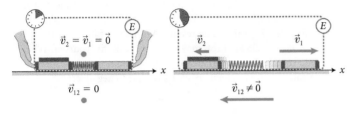

$\vec{v}_{12} = 0$

$\vec{v}_{12} \neq 0$

(*b*) Initial and final energies of system

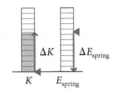

(*c*) Velocity-versus-time graph for the motion

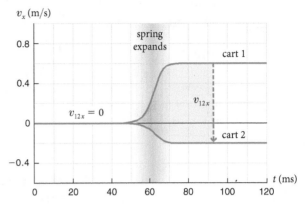

in internal energy of the spring causes an increase in the kinetic energy of the carts (Figure 5.22*b*). Notice how the velocity-versus-time graph for this explosive separation (Figure 5.22*c*) is the inverse of the one in Figure 5.2*b*.

To determine the final speeds in Figure 5.22, we need to know how much energy E_{int} the spring releases, a topic not covered until Chapter 9. Once we know ΔE_{int}, we have two equations that allow us to obtain the two final velocities, one a consequence of conservation of momentum:

$$0 = m_1 v_{1x,f} + m_2 v_{2x,f} \qquad (5.28)$$

(because $v_{1x,i} = v_{2x,i} = 0$) and the other a consequence of conservation of energy:

$$\Delta K + \Delta E_{int} = \tfrac{1}{2} m_1 v_{1f}^2 + \tfrac{1}{2} m_2 v_{2f}^2 + \Delta E_{int} = 0. \qquad (5.29)$$

Note that in this example the *initial* relative speed is zero, making e in Eq. 5.19 infinite. There is no restitution because there is no initial relative speed to be restituted. In the more general case where the two objects have a nonzero initial relative speed, $e > 1$.

Example 5.9 Spring energy

A 0.25-kg cart is held at rest against a compressed spring as in Figure 5.8*a* and then released. The cart's speed after it separates from the spring is 2.5 m/s. The spring is then compressed by the same amount between a 0.25-kg cart and a 0.50-kg cart, as shown in Figure 5.22*a*, and the carts are released from rest. What are the carts' speeds after separating from the spring?

❶ **GETTING STARTED** The key point in this problem is the identical compression of the spring in the two cases: The initial state of the spring is therefore the same before both releases. Because the spring ends in the same uncompressed state in both cases, the change in its internal energy must be the same in both cases. In the first case, all of this energy is transferred to the 0.25-kg cart. In the second case, the same amount of energy is distributed between the two carts.

❷ **DEVISE PLAN** To calculate the kinetic energy of the single cart in the first release, I use Eq. 5.12. This gives me the amount of energy stored in the compressed spring. The final velocities of the two carts in the second case are then given by Eqs. 5.28 and 5.29.

❸ **EXECUTE PLAN** From Eq. 5.12, I get

$$K = \tfrac{1}{2} m v^2 = \tfrac{1}{2}(0.25 \text{ kg})(2.5 \text{ m/s})^2 = 0.78 \text{ J}$$

and so the change in the spring's internal energy is $\Delta E_{int} = -0.78$ J. Next I rewrite Eq. 5.28 as $v_{1x,f} = -(m_2/m_1)v_{2x,f}$. Substituting this result in Eq. 5.29, I get

$$\tfrac{1}{2} m_1 \left(\frac{m_2}{m_1}\right)^2 v_{2x,f}^2 + \tfrac{1}{2} m_2 v_{2x,f}^2 = -\Delta E_{int}.$$

(Continued)

Solving for the final velocity of cart 2 gives

$$v_{2xf} = \sqrt{\frac{-2m_1 \Delta E_{int}}{m_2(m_1 + m_2)}}$$

$$v_{2xf} = \sqrt{\frac{-2(0.25\ \text{kg})(-0.78\ \text{J})}{(0.50\ \text{kg})(0.25\ \text{kg} + 0.50\ \text{kg})}} = 1.0\ \text{m/s.} ✔$$

Substituting this result into my rewritten Eq. 5.28, $v_{1x,f} = -(m_2/m_1)v_{2x,f}$, I get $v_{1x,f} = -2.0$ m/s. ✔

❹ **EVALUATE RESULT** The carts move in opposite directions, as expected. I also note that cart 1 moves at twice the speed of cart 2, as it should to keep the final momentum of the system zero. Finally, because my assignment of m_1 and m_2 is arbitrary, I verify that I get the same result when I substitute $m_1 = 0.50$ kg and $m_2 = 0.25$ kg. (You may want to check this yourself. When you reverse the inertias, why does the velocity of cart 1 reverse to positive and the velocity of cart 2 reverse to negative?)

 5.13 Does each cart in Example 5.9 get half of the spring's energy? Why or why not?

Chapter Glossary

SI units of physical quantities are given in parentheses.

Closed system A system to or from which no energy is transferred. See the Procedure box on page 109.

Coefficient of restitution e (unitless) A scalar equal to the ratio of relative speeds after and before a collision of two objects:

$$e \equiv \frac{v_{12f}}{v_{12i}} \tag{5.18}$$

Conservation of energy Energy can be transferred from one object to another or converted from one form to another, but it cannot be created or destroyed. The energy of a closed system cannot change:

$$\Delta E = 0 \quad \text{(closed system).} \tag{5.23}$$

Elastic, inelastic, totally inelastic collision Collisions between two objects are classified according to what happens to the relative speed $v_{12} = |\vec{v}_2 - \vec{v}_1|$ of the two objects, as summarized in Table 5.4.

Energy E (J) A scalar that provides a quantitative measure of the state or motion of an object or system. Energy appears in many different forms. The energy of an object or system always refers to the sum of all forms of energy in that object or system.

Explosive separation A process in which objects break apart from one another and the relative speed of the objects increases.

Internal energy E_{int} (J) Any energy not associated with the motion of an object or system. Internal energy is a quantitative measure of the state of the object or system.

Irreversible process A process involving changes that cannot undo themselves spontaneously.

Joule (J) The derived SI unit of energy, defined as $1\ \text{J} \equiv 1\ \text{kg} \cdot \text{m}^2/\text{s}^2$.

Kinetic energy K (J) Energy associated with the motion of an object. The kinetic energy K of an object of inertia m moving at speed v is

$$K \equiv \tfrac{1}{2}mv^2. \tag{5.12}$$

Process The transformation of a system from an initial state to a final state.

Relative velocity $\vec{v}_{12}$ (m/s) The velocity of one object relative to another:

$$\vec{v}_{12} \equiv \vec{v}_2 - \vec{v}_1. \tag{5.1}$$

The magnitude of this velocity is called the *relative speed* $v_{12} \equiv |\vec{v}_2 - \vec{v}_1|$.

Reversible process A process that can run backward so that the initial state is restored.

State The condition of an object (or a system) as specified by a complete set of variables.

6

Principle of Relativity

CONCEPTS

QUANTITATIVE TOOLS

Did you ever, sitting in a car at a red light and looking at the car next to you, slam on the brakes because you thought you were starting to roll but it turned out your car never moved? If you did, you experienced the relativity of motion, first described quantitatively by Galileo. You may have heard the term *relativity* in the context of Albert Einstein's famous theories of general relativity and special relativity. We'll touch upon one basic aspect of general relativity in Chapter 13 and discuss special relativity in Chapter 14. Underlying all of relativity is something we have so far ignored in our discussion of motion: the fact that the velocity measured for any object depends on the motion of the **observer** (the person doing the measuring). For example, to a person sitting in a moving train, a suitcase on the overhead rack is at rest, but to a person standing on a station platform watching the train speed by, the suitcase is not at rest. According to the person in the train, the suitcase has zero momentum ($m\vec{v}$) and zero kinetic energy ($\frac{1}{2}mv^2$). According to the person on the platform, the suitcase has nonzero momentum and nonzero kinetic energy. In this chapter we investigate whether or not the laws of conservation of momentum and conservation of energy depend on the velocity of the observer. In other words, if these laws are valid for one observer, are they also valid for an observer who is moving relative to the first observer?

6.1 Relativity of motion

Remember from Chapter 2 that whenever we talk about motion, we must specify a reference axis along which the motion occurs and an origin. Together, the axis and origin are called a **reference frame.** In our study of motion thus far, we have considered the motion of carts along a low-friction track, collecting our data using a ruler affixed to the track; that is to say, we collected our data while standing in a reference frame at rest relative to the surface of Earth (this is called an **Earth reference frame**).

What happens to the data if we are moving as we collect them—in other words, if we are in a reference frame that is moving relative to Earth? Imagine that we are moving alongside the track and want to describe a collision between two carts. Do the conservation laws hold in a reference

frame that is moving relative to Earth? Or, to put the question another way, does the behavior of the carts change if we are in motion as we study that behavior?

Consider, for example, the situation depicted in **Figure 6.1a**: two observers watching the motions of two cars. Car 1 is moving at constant velocity $\vec{v}_1$, and car 2 is at rest relative to Earth. Observer E is at rest relative to Earth, while observer M is inside car 1. The velocity $\vec{v}_M$ of observer M is the same as the velocity of car 1: $\vec{v}_M = \vec{v}_1$.

Relative to observer M, the position of car 1 is not changing; that is, car 1 is at rest relative to observer M. The distance to car 2, however, is steadily increasing, and so relative to observer M, car 2, observer E, and all the surroundings move to the left (Figure 6.1b). To distinguish quantities measured by different observers, we add a capital letter subscript to denote the observer. For instance, $\vec{v}_{M2}$ represents observer M's measurement of the velocity of car 2, also called the velocity of car 2 *relative to* observer M. When we are comparing quantities in various reference frames, quantities measured relative to Earth are given the subscript E. For example, relative to Earth the velocity of car 1 is represented by $\vec{v}_{E1}$ (When the Earth reference frame is the only reference frame, we drop the subscript E, recovering the notation we have used before: $\vec{v}_{E1} = \vec{v}_1$.)

To determine a relationship between the quantities measured by two observers moving relative to each other, let's start with a simple example. **Figure 6.2** shows five frames of a film of two carts on a low-friction track. In Figure 6.2a the positions of the two carts are measured with ruler A, which is affixed to the track (that is, in the Earth reference frame). Relative to ruler A, cart 1 is at rest and its position remains fixed at 22.5 mm. Cart 2 is moving to the right, and its position on ruler A changes from 0 to 14.4 mm.

In Figure 6.2b the positions of the two carts are measured with ruler B, which moves along with cart 2: The zero mark on ruler B remains aligned with the front of cart 2. The position of cart 2 therefore remains fixed at a ruler reading of 0, and so, relative to ruler B, cart 2 is at rest. On this ruler, the position of cart 1 does *not* remain fixed, however—the reading for this cart's position decreases from 22.5 mm to 8.1 mm over the five frames. That is to say, cart 1 moves 22.5 mm − 8.1 mm = 14.4 mm *to the left* along ruler B.

Figure 6.1 The motion you perceive depends on your reference frame.

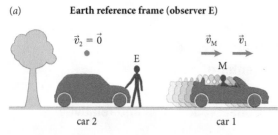

(*a*) **Earth reference frame (observer E)**

Relative to observer E, car 2 is at rest and car 1 moves to the right.

(*b*) **Reference frame of car 1 (observer M)**

But relative to observer M, car 1 is at rest while car 2, observer E, and Earth move to the left.

Figure 6.2 Two identical carts on a low-friction track. The positions of the carts are measured with (a) a ruler affixed to the track and (b) a ruler moving along with cart 2. In the Earth reference frame, cart 1 is at rest and cart 2 is moving to the right.

(a) Ruler is fixed to track (b) Ruler moves with cart 2

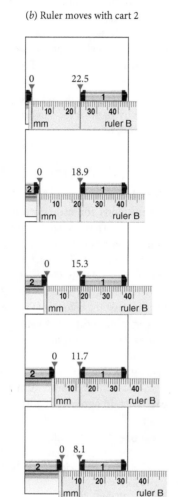

Figure 6.3a shows a position-versus-frame graph for the two carts drawn from the readings on ruler A in Figure 6.2a. What is shown in the graph corresponds to what an observer at rest alongside the track sees: cart 1 at rest and the x component of the velocity of cart 2 a constant $+3.6$ mm/frame. These are the observations taken in the Earth reference frame.

Figure 6.3b is the position-versus-frame graph for the two carts drawn from the readings on ruler B in Figure 6.2b. What is shown in the graph corresponds to what an observer M sitting on top of cart 2 sees: cart 2 at rest and the x component of the velocity of cart 1 a constant -3.6 mm/frame (the negative sign indicates that the cart moves to the left in the figure).

Example 6.1 Moving at my own speed

Suppose that in the situation shown in Figure 6.2, a third ruler is affixed to some device (not shown) that moves to the right along the track at a speed of 2.0 mm/frame. If in the Earth reference frame cart 2 again moves at $+3.6$ mm/frame and cart 1 is again at rest, what are the cart velocities according to an observer moving along with ruler C?

❶ **GETTING STARTED** If I arbitrarily let ruler C start in the position shown in frame 1 of Figure 6.2b—the zero mark on ruler C aligned with the front of cart 2—this ruler is shifted 2.0 mm to the right in the second frame and another 2.0 mm to the right in each successive frame. I have to use the readings from this ruler to determine the cart velocities.

❷ **DEVISE PLAN** I begin by making a table showing the position of each cart in each frame according to an observer moving along with ruler C. Because the ruler moves 2.0 mm to the right in each frame, I can obtain the readings on this ruler by subtracting 2.0 mm from each cart position in frame 2 of Figure 6.2a, subtracting 4.0 mm from each cart position in frame 3, subtracting 6.0 mm from each cart position in frame 4, and so on. I can then use these positions to make a position-versus-frame graph like the ones in Figure 6.3 and determine the velocities of the carts from the slopes of the curves.

Figure 6.3 Position-versus-frame graphs for the carts of Figure 6.2 as determined from readings on (a) ruler A and (b) ruler B.

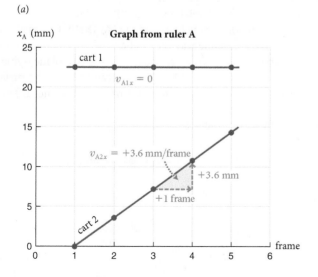

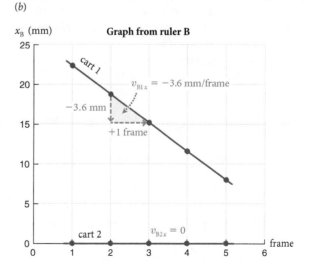

❸ **EXECUTE PLAN** My table of the cart positions measured with ruler C is shown in **Figure 6.4**. By plotting these data, I see that the x component of the velocity of cart 1 is -2.0 mm/frame and that of cart 2 is $+1.6$ mm/frame. ✔

Figure 6.4

x_{M2} (mm)	x_{M1} (mm)
0	22.5
1.6	20.5
3.2	18.5
4.8	16.5
6.4	14.5

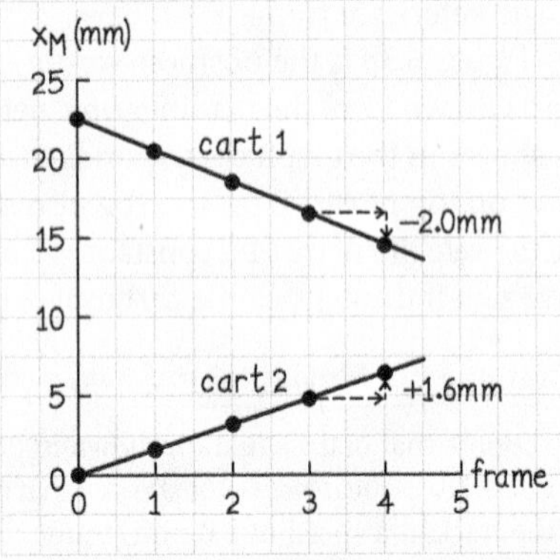

❹ **EVALUATE RESULT** The observer moving with ruler C sees cart 1 moving to the left, meaning that the x component of its velocity should be negative, as I found. This observer and cart 2 both move in the positive x direction in the Earth reference frame, but the speed of cart 2 is higher than that of ruler C, and so cart 2 moves to the right relative to the observer. Therefore the x component of its velocity should be positive, in agreement with what I found. Because the velocity of ruler C is different from the velocity of either cart, neither cart is at rest relative to ruler C.

This example and the text preceding it show that motion is a relative concept. In the Earth reference frame, cart 1 is at rest and cart 2 is moving. Relative to the moving ruler in Figure 6.2b, cart 1 is moving and cart 2 is at rest. And relative to the ruler of Example 6.1, both carts are moving. In all three reference frames, however, we discover that the carts move at constant velocity. We could just as well have used a ruler moving at $+1.0$ mm/frame, $+4.0$ mm/frame, -0.1 mm/frame, or any other constant velocity. In each case, we would reach the same conclusion: The carts move at constant velocity. So, if an object moves at constant velocity in the Earth reference frame, its motion observed from *any reference frame moving at constant velocity relative to Earth* is also at constant velocity.

Comparing Figures 6.3a and 6.3b, we see that the x component of each cart's velocity according to ruler A (Figure 6.3a) is 3.6 mm/frame higher than the velocity

according to ruler B (Figure 6.3b). So, an object's velocity $\vec{v}_{Ao}$ determined from the readings on ruler A is equal to the sum of the velocity $\vec{v}_{AB}$ of ruler B relative to ruler A and the velocity $\vec{v}_{Bo}$ of the object relative to ruler B: $\vec{v}_{Ao} = \vec{v}_{AB} + \vec{v}_{Bo}$. For cart 1, we have $0 = +3.6$ mm/frame $+ (-3.6$ mm/frame); for cart 2, $+3.6$ mm/frame $= +3.6$ mm/frame $+ 0$. You can readily verify that the same is true for the velocities determined with ruler C in Example 6.1.

✋ **6.1** What is the velocity of each cart in Figure 6.2 measured by an observer moving at -3.0 mm/frame in the Earth reference frame?

6.2 Inertial reference frames

How does the choice of reference frame affect our accounting methods for momentum and energy? To answer this question, let's consider the two carts of Figure 6.2 from the point of view of two observers. To an observer E, who is at rest relative to Earth, cart 1 is at rest and cart 2 is moving at constant velocity (**Figure 6.5a**). To an observer M, who is moving along with cart 2, cart 2 is at rest and cart 1 is moving at constant velocity (Figure 6.5b). Even though the observers obtain different values for the carts' velocities, these values are constant, and therefore both observers conclude that the momentum of each cart is constant. Both observers also agree that the two carts are isolated, which means that conservation of momentum yields $\Delta \vec{p} = \vec{0}$ (Eq. 4.17) for each cart, in agreement with their observations.

✋ **6.2** From the point of view of each observer in Figure 6.5, (*a*) is the energy of each cart constant? (*b*) Is the isolated system containing cart 1 closed? (*c*) Is the isolated system containing cart 2 closed?

Although we checked for only the simple case of carts moving at constant velocity on a low-friction track, Figure 6.5 and Checkpoint 6.2 suggest that the accounting procedures and principles we introduced in the preceding two chapters can be applied in any reference frame moving at constant velocity relative to Earth. Indeed, when we are moving at constant velocity relative to Earth, things around us behave the same way they behave when we are at rest on the ground. For example, imagine being in a plane flying at a constant speed of 260 m/s in a straight line. The surface of the coffee in the cup in front of you looks no different

Figure 6.5 The carts of Figure 6.2 seen by (*a*) an observer in the Earth reference frame and (*b*) an observer moving along with cart 2.

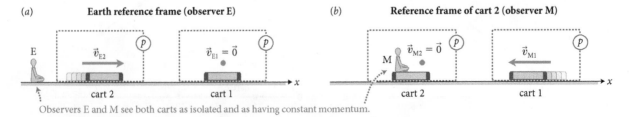

Observers E and M see both carts as isolated and as having constant momentum.

Figure 6.6 Motion of a marble in an accelerating car viewed from (*a*) the Earth reference frame and (*b*) a reference frame affixed to the car. $\vec{a}_{Ec}$ is the **E**arth observer's measurement of the **c**ar's acceleration, $\vec{a}_{ME}$ is the **m**oving observer's measurement of **E**arth's acceleration, and $\vec{a}_{Mc}$ is the **m**oving observer's measurement of the **c**ar's acceleration.

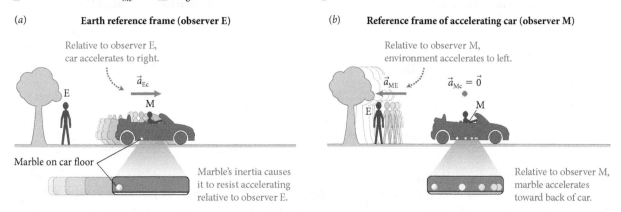

from the way it looks at home. If you drop your keys, they fall straight down just as they would if the plane were at rest on the ground. And when the flight attendant pours coffee into your cup while you all move at 260 m/s, you don't expect him to spill it on you any more than you do when a waiter pours your coffee in a restaurant.

In fact, if the engines are very quiet, the ride is smooth, and the window shades are down, it is impossible for you to determine whether or not the plane is moving. Here is the simplest possible experiment: Put a marble on the floor of the plane. Provided the floor is level relative to the ground, the marble remains at rest where you placed it. This is true whether the plane is at rest or cruising at a constant 260 m/s. When the plane is sitting at the gate, inertia keeps the marble at rest relative to the floor (which is at rest relative to the ground), and so $\vec{a} = \vec{0}$. In the air, inertia keeps the marble moving along with the plane at a constant velocity relative to the ground (or, put differently, inertia keeps the marble at rest relative to the floor), and so again $\vec{a} = \vec{0}$.

The Earth reference frame or any reference frame moving at constant velocity relative to Earth is called an **inertial reference frame.** We can tell whether or not a reference frame is inertial by testing whether or not the **law of inertia** holds:

> **In an inertial reference frame, any isolated object that is at rest remains at rest, and any isolated object in motion keeps moving at a constant velocity.**

The law of inertia does not hold in a reference frame that is accelerating relative to Earth. For example, imagine sitting in the passenger seat of a car moving at constant velocity with a marble on the floor near your feet. As long as the car is on a horizontal surface and keeps going straight at a constant speed, the marble remains at rest. When the driver suddenly accelerates forward, however, everything in the car lurches backward, with the marble probably ending up somewhere in the back of the car. According to an observer at rest in the Earth reference frame (**Figure 6.6a**), the marble resists being accelerated forward because of its inertia, and unless you have glued it to the floor or restrained it in some other way, it fails to keep up with the accelerating car. Seen

from inside the car, the isolated marble suddenly accelerates backward, and so the law of inertia does not hold in the reference frame of the accelerating car (Figure 6.6*b*). Reference frames in which the law of inertia does not hold are called *noninertial reference frames.*

We have singled out the Earth reference frame as the basic inertial reference frame. There is nothing fundamentally special about the Earth reference frame, however, other than that we perform most experiments in this frame. Strictly speaking, Earth is not an inertial reference frame because it revolves around a north-south axis and orbits the Sun in a nearly circular orbit. Because motion on a curved path means that the direction of the velocity is changing and so the velocity is not constant, Earth is accelerating and therefore is a noninertial reference frame. For most cases, the acceleration is too small to be noticeable, and so we may, for most practical purposes, consider the Earth reference frame to be inertial.

Exercise 6.2 Inertial or not?

Which of these reference frames are inertial: one affixed to (*a*) a merry-go-round, (*b*) the space shuttle orbiting Earth, (*c*) an airplane taking off, (*d*) a train moving at constant speed along a straight track?

SOLUTION (*b*) and (*d*). If I imagine placing a marble at rest in each reference frame and then observing the marble's motion, I can tell whether the law of inertia holds in that reference frame. From experience, I know that a marble remains at rest on the floor of a train moving at constant speed along a straight track, and so the train's reference frame is inertial. I also know that a marble moves away from where I place it on a merry-go-round and in an airplane taking off, so those reference frames are noninertial. Having never been on the space shuttle, I have no direct experience with such a reference frame. However, I have seen astronauts and objects floating in the space shuttle, and this motion suggests that objects stay at rest or keep moving as the law of inertia states, suggesting that the reference frame of the shuttle is inertial.* ✔

---

*Just like the Earth reference frame, the shuttle reference frame is only approximately inertial.

Figure 6.7 A stationary cart and an accelerating cart viewed from (a) the Earth reference frame and (b) a reference frame affixed to the accelerating cart.

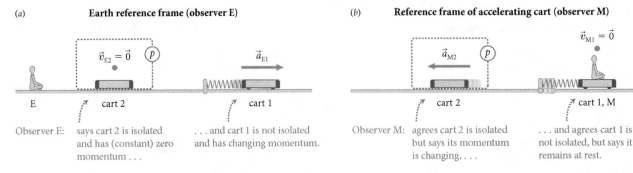

Our accounting procedures for momentum and energy cannot be used in noninertial reference frames. Consider, for example, the two carts shown in **Figure 6.7**. Cart 1 is being accelerated by a spring, and cart 2 is at rest in the Earth reference frame. Cart 2 constitutes an isolated system, but cart 1 is not isolated because it interacts with the spring. To observer E in the Earth reference frame (Figure 6.7a), the behavior of both carts is in agreement with the momentum law: The momentum of the nonisolated cart 1 changes, while the momentum of the isolated cart 2 is constant. For observer M, however, who is accelerating along with cart 1 (Figure 6.7b), things don't quite add up. From this observer's perspective, cart 1 remains at rest even though it interacts with the spring, and the momentum of cart 2 changes even though that cart is isolated. Equations 4.17 ($\Delta \vec{p} = \vec{0}$) and 5.23 ($\Delta E = 0$), which embody the laws of conservation of momentum and energy, do not hold in the noninertial reference frame of observer M in Figure 6.7b.

Are we going to run into problems because the laws of the universe are different in noninertial reference frames?

No, because nothing prescribes the reference frame; we get to choose it. So for now we just avoid using noninertial reference frames. For the accelerating car, for instance, we would choose not a reference frame affixed to the car but the Earth reference frame.

✋ **6.3** From the point of view of each observer in Figure 6.7, (a) is the energy of each cart constant? (b) Is the isolated system containing cart 1 closed? (c) Is the isolated system containing cart 2 closed? (d) Do the observations made by each observer agree with the conservation of energy law?

6.3 Principle of relativity

In the preceding section, we saw that the conservation laws apply for single objects in inertial reference frames. Let us now test the conservation laws for interacting objects.

Figure 6.8 shows velocity-versus-time graphs for two-cart collisions. The values in Figure 6.8a were measured by an observer in the Earth reference frame. As you saw in

Figure 6.8 Velocity-versus-time graphs for two carts colliding on a low-friction track as seen (a) from the Earth reference frame and (b) from a reference frame moving along the track at $v_{EMx} = -0.20$ m/s relative to Earth. The inertias are 0.36 kg for cart 1 and 0.12 kg for cart 2.

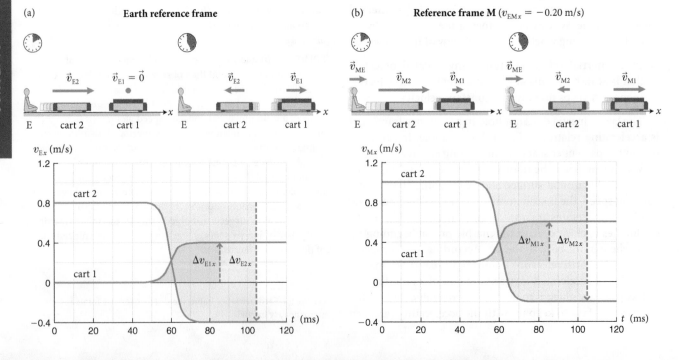

Checkpoint 6.1, we can obtain the velocities of the carts in reference frame M of Figure 6.8b by subtracting the velocities in reference frame M from the velocities in the Earth reference frame. Because reference frame M in Figure 6.8b moves at a constant −0.20 m/s relative to Earth, we must subtract −0.20 m/s (that is, add 0.20 m/s) from the x component of each velocity in the Earth reference frame to obtain the corresponding velocity in reference frame M.

The only difference between Figures 6.8a and 6.8b is that the curves in Figure 6.8b are shifted upward by 0.20 m/s, the relative speed of the two reference frames. However, the *changes* in the x components of the cart's velocities, Δv_{1x} and Δv_{2x}, are not affected by the velocity of reference frame M. Because the inertia of any object is not affected by the motion of an observer studying the object, we observe that the changes in the x component of the carts' momenta, $\Delta p_{1x} = m_1\Delta v_{1x}$ and $\Delta p_{2x} = m_2\Delta v_{2x}$, are the same in both reference frames. So, because conservation of momentum requires that $\Delta\vec{p} = 0$ in the Earth reference frame, we also obtain $\Delta\vec{p} = 0$ in reference frame M.

If reference frame M in Figure 6.8b had moved at any other constant velocity relative to Earth, the vertical shift of the curves would be different but the shape of the curves would be the same. Again we would obtain the same momentum changes for the two carts, and again we would conclude that the collision does not change the momentum of the system. So, if the change in the system momentum is zero in the Earth reference frame, it must be zero in all reference frames moving at a constant velocity relative to Earth. Experiments show that this statement can be extended to any inertial reference frame, not just the Earth reference frame.

Example 6.3 Momentum shove

A 0.12-kg cart moving on a straight, low-friction track gets a shove so that its speed changes. Its initial and final speeds measured in the Earth reference frame are 0.40 m/s and 0.80 m/s (Figure 6.9). Determine the change in the cart's momentum as seen from (a) the Earth reference frame and (b) a reference frame M moving in the same direction as the cart at a speed of 0.60 m/s relative to the track.

Figure 6.9 Example 6.3.

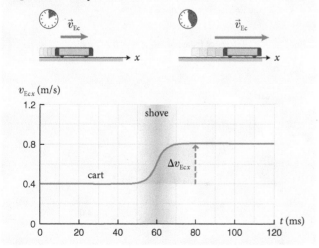

❶ **GETTING STARTED** To determine the momentum change in either case, I need to know the cart's initial and final velocities in the two reference frames. In the Earth reference frame, $v_{Ecx,i} = +0.40$ m/s and $v_{Ecx,f} = +0.80$ m/s. In reference frame M, which moves at $v_{MEx} = +0.60$ m/s relative to Earth, these values are 0.60 m/s lower: $v_{Mcx,i} = -0.20$ m/s and $v_{Mcx,f} = +0.20$ m/s. I then make a velocity-versus-time graph for the cart as viewed from reference frame M (**Figure 6.10**).

Figure 6.10

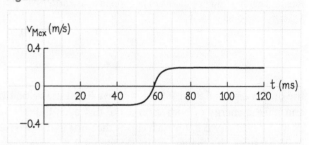

❷ **DEVISE PLAN** Now that I have the x components of the initial and final velocities in both reference frames, I can calculate the x components of the initial and final momenta and determine the change in momentum.

❸ **EXECUTE PLAN** (a) In the Earth reference frame, the x components of the initial and final momenta are $p_{Ecx,i} = mv_{Ecx,i} = (0.12$ kg$)(+0.40$ m/s$) = +0.048$ kg·m/s and $p_{Ecx,f} = mv_{Ecx,f} = (0.12$ kg$)(+0.80$ m/s$) = +0.096$ kg·m/s. So the x component of the change in momentum is $\Delta p_{Ecx} = (+0.096$ kg·m/s$) - (+0.048$ kg·m/s$) = +0.048$ kg·m/s. ✔

(b) In reference frame M, $p_{Mcx,i} = mv_{Mcx,i} = (0.12$ kg$) \times (-0.20$ m/s$) = -0.024$ kg·m/s and $p_{Mcx,f} = mv_{Mcx,f} = (0.12$ kg$)(+0.20$ m/s$) = +0.024$ kg·m/s. So $\Delta p_{Mcx} = (+0.024$ kg·m/s$) - (-0.024$ kg·m/s$) = +0.048$ kg·m/s. ✔

❹ **EVALUATE RESULT** The momentum change is the same in the two reference frames, as I expect. In reference frame M, the $v_x(t)$ curve shifts up or down, but its shape does not change, and therefore the values of $\Delta\vec{v}$ and $\Delta\vec{p}$ must also remain the same.

Example 6.3 reaffirms what we have already seen:

The change in a system's momentum is the same in any inertial reference frame.

✋ **6.4** In Example 6.3, what is the change in the cart's kinetic energy due to the shove (a) in the Earth reference frame, (b) in a reference frame moving in the same direction as the cart at 0.60 m/s relative to Earth, and (c) in a reference frame moving in the same direction as the cart at 0.80 m/s relative to Earth?

Unlike changes in momentum, changes in kinetic energy are not the same in two reference frames moving relative to each other. As Checkpoint 6.4 shows, the cart's kinetic energy increases, decreases, or stays the same depending on the reference frame.

The observation that changes in kinetic energy depend on the reference frame is disturbing. What does this result imply about conservation of energy? To begin addressing this question, let's look first at an elastic collision, such as

Table 6.1 An elastic collision seen from two reference frames

Cart	Inertia (kg)	v_x (m/s) before	v_x (m/s) after	Δv_x		Kinetic energy (10^{-3} J) before	Kinetic energy (10^{-3} J) after	ΔK
Earth reference frame								
1	0.36	0	+0.40	+0.40		0	29	+29
2	0.12	+0.80	−0.40	−1.2		38 +	9 +	−29 +
					K	38	38	0
Reference frame moving at −0.20 m/s relative to Earth								
1	0.36	+0.20	+0.60	+0.40		7	65	+58
2	0.12	+1.0	−0.20	−1.2		60 +	2 +	−58 +
					K	67	67	0

the one in Figure 6.8. In Table 6.1 I have listed the velocities and kinetic energy of both carts. Note that even though the change in velocity Δv_x for each cart is the same in the two reference frames, the change in kinetic energy ΔK for each cart is not. In a given reference frame, however, the absolute value of ΔK is the same for both carts, and so the change in the kinetic energy K of the two-cart system is zero in each frame.

A simple way of arguing that the kinetic energy of the two-cart system *has* to remain unchanged in elastic collisions in any inertial reference frame is to look at how we formulated kinetic energy in Chapter 5 in terms of the relative velocities of two objects in a collision. As we saw earlier, differences in velocity do not depend on the velocity of the reference frame (Figure 6.11). In particular, our experimental observation that the relative speed of two colliding objects is unchanged in an elastic collision holds when we measure that relative speed in a reference frame that is moving at constant velocity. More generally, we observe:

> **The kinetic energy of a system of two elastically colliding objects does not change in any inertial reference frame.**

Even though K does not change in elastic collisions, the fact that the individual ΔK values depend on the reference frame could be a big problem for conservation of energy

in inelastic collisions. This reference-frame dependence for the individual ΔK values could mean that ΔK for the system also depends on the reference frame. Recall from Section 5.3 that in inelastic collisions some kinetic energy is converted to internal energy, which is a measure of the state of the system. Because changes in kinetic energy are reference-frame-dependent, the amount that gets converted could be reference-frame-dependent as well and that would mean that the final state of the system would depend on the reference frame! To determine whether or not this is the case, we examine an inelastic collision in the next example.

Example 6.4 Energy conversion

Consider a collision between the two carts of Table 6.1, starting from the same initial velocities, but with $v_{E1x,f} = +0.30$ m/s. Make a table like Table 6.1 for this situation, and compare the amount of kinetic energy converted to internal energy in the Earth reference frame and in a reference frame M moving at −0.20 m/s relative to the track.

1 GETTING STARTED Because all the initial values are the same, I can copy all the "before" values from Table 6.1. I am given the final velocity of cart 1, and so I need to determine the final velocity of cart 2 in the Earth reference frame, then determine the final velocities in reference frame M, calculate the corresponding kinetic energies, and determine the change in kinetic energy in each reference frame.

Figure 6.11 The relative velocity of the two colliding carts of Figure 6.8 is the same in (a) the Earth reference frame and (b) a reference frame moving along the track at $v_{EMx} = -0.20$ m/s relative to Earth.

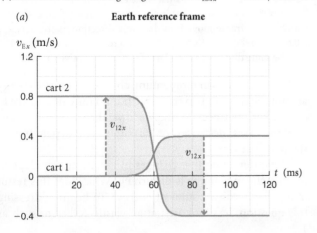

(a) **Earth reference frame**

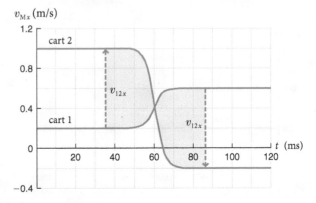

(b) **Reference frame M** ($v_{EMx} = -0.20$ m/s)

② **DEVISE PLAN** To determine $v_{E2x,f}$, I apply conservation of momentum to the isolated system of the two carts, which tells me that $\Delta \vec{p} = \vec{0}$. To calculate the two final velocities in reference frame M, I subtract the velocity of this reference frame from the carts' velocities in the Earth reference frame. Then I determine the kinetic energies from $K = \frac{1}{2}mv^2$ and complete the table by determining the changes in kinetic energy.

③ **EXECUTE PLAN** Because $\vec{p}_i = \vec{p}_f$, I have

$$m_1 v_{E1x,i} + m_2 v_{E2x,i} = m_1 v_{E1x,f} + m_2 v_{E2x,f}.$$

Substituting $v_{E1x,i} = 0$ and solving for $v_{E2x,f}$, I get

$$v_{E2x,f} = \frac{m_2 v_{E2x,i} - m_1 v_{E1x,f}}{m_2}.$$

Substituting the given values into this expression, I obtain $v_{E2x,f} = -0.10$ m/s.

Because $v_{EMx} = -0.20$ m/s, I must add 0.20 m/s to the x components of the carts' velocities in the Earth reference frame to determine the corresponding values in reference frame M:

$$v_{M1x,i} = 0 + (+0.20 \text{ m/s}) = +0.20 \text{ m/s}$$

$$v_{M2x,i} = (+0.80 \text{ m/s}) + (+0.20 \text{ m/s}) = +1.00 \text{ m/s}$$

$$v_{M1x,f} = (+0.30 \text{ m/s}) + (+0.20 \text{ m/s}) = +0.50 \text{ m/s}$$

$$v_{M2x,f} = (-0.10 \text{ m/s}) + (+0.20 \text{ m/s}) = +0.10 \text{ m/s}.$$

Now I can calculate all the kinetic energies and complete the table (**Table 6.2**). ✔

④ **EVALUATE RESULT** Each cart's kinetic energy change is different in the two reference frames because, as just noted in the text, changes in kinetic energy are reference-frame-dependent. However, the change in the *system's* kinetic energy is the same in the two reference frames: −21 mJ in both cases. This tells me that the amount of kinetic energy converted to internal energy—and therefore the final state of the two carts—does not depend on the reference frame, as I expect.

✋ **6.5** Repeat Example 6.4 but let the collision be totally inelastic.

The results of Example 6.4 and Checkpoint 6.5 are reassuring. Imagine that the converted kinetic energy becomes thermal energy, with the result that the temperature of each cart increases. You wouldn't expect these temperature changes to be different when you view the collision from reference frames that are moving at constant speed relative to each other. Likewise, in a collision between two cars, where the converted kinetic energy goes into crushing the fenders, you wouldn't expect the bending of the metal to depend on the reference frame. The amount of kinetic energy converted to internal energy during an inelastic collision does not depend on the velocity of the reference frame relative to the colliding objects.

These observations, together with the results we obtained in the examples in this section, tell us that the same laws and principles we discussed in Chapters 4 and 5 hold in any inertial reference frame. In particular, the momentum and energy laws can be used in any inertial reference frame. Earlier we saw that it is not possible to tell the difference between various inertial reference frames—that is, reference frames in which the law of inertia holds. It is an experimental fact that things behave the same way in all inertial reference frames. We should therefore expect the laws that describe that behavior to be the same too. This leads us to formulate the **principle of relativity:**

> **The laws of the universe are the same in all inertial reference frames moving at constant velocity relative to each other.**

The principle of relativity provides a criterion for judging theories: For a theory to be valid, it must prescribe the same behavior in all inertial reference frames. Although observers in different inertial reference frames record different values for the velocities of carts involved in a collision and therefore different momenta and kinetic energies, they agree that the momentum and energy of the two-cart system do not change.

It follows from the principle of relativity that no experiment carried out entirely in one inertial reference frame can tell what the motion of that reference frame is relative to another reference frame. Returning to the example of Section 6.2: Sitting in a very quiet airplane during a smooth, turbulence-free ride, it is impossible to tell if you are flying

CONCEPTS

Table 6.2 An inelastic collision seen from two reference frames

Cart	Inertia (kg)	v_x (m/s) before	after	Δv_x		Kinetic energy (10^{-3} J) before	after	ΔK
Earth reference frame								
1	0.36	0	+0.30	+0.30		0	16	+16
2	0.12	+0.80	−0.10	−0.90		$-\frac{38}{38}$ +	$-\frac{1}{17}$ +	$-\frac{37}{-21}$ +
					K	38	17	−21
Reference frame moving at −0.20 m/s relative to Earth								
1	0.36	+0.20	+0.50	+0.30		7	45	+38
2	0.12	+1.0	+0.10	−0.90		$-\frac{60}{67}$ +	$-\frac{1}{46}$ +	$-\frac{59}{-21}$ +
					K	67	46	−21

or if the airplane is sitting on the runway with its engines running. Of course, you can tell you are moving relative to Earth's surface by peeking out of the window and observing the landscape move by. By timing the motion of some reference points on Earth, you could even determine the velocity of the airplane relative to Earth. By peeking out of the window, however, your observations are no longer confined to a single reference frame. To determine your velocity relative to Earth's surface, you need to take measurements in two reference frames.

A consequence of the principle of relativity is that it is not possible to deduce from measurements taken entirely in one reference frame the motion of that reference frame relative to other reference frames.

This statement also means that no inertial reference frame is preferred over any other; it is not possible to determine the *absolute* velocity of any reference frame or object. The laws of the universe are the same in all inertial reference frames, and there is no reference frame that is "at rest" in some absolute sense.

When you are standing on Earth's surface, are you "at rest" or moving at 30 km/s (the speed of Earth in its orbit around the Sun), or is your speed higher still—say, the speed of the Sun in its galactic orbit? None of these statements is meaningful because no experiments can be done to determine your speed in empty space. You can only speak of your velocity or speed *relative to something else*.

When specifying the velocity of an object, we must therefore always state relative to what we measure that velocity. For simplicity, when the Earth reference frame is the only reference frame of interest, we will omit this relative statement and the subscript E. The phrase "his velocity was $\vec{v}$" will be understood to mean the velocity is measured relative to Earth.

🖐 **6.6** Is the coefficient of restitution *e* different in two inertial reference frames that are moving at constant velocity relative to each other? (See Eq. 5.18 if you have forgotten the definition of *e*.)

6.4 Zero-momentum reference frame

Let me end this first part of the chapter by addressing a practical question. If you can choose whichever inertial reference frame you wish, is any choice better than another? The reference frame you are in (usually one at rest relative to Earth) is a logical choice. By choosing some other reference frame, however, you could adjust the value of the momentum of a system up or down by some fixed amount without violating the conservation laws. You could, for instance, adjust the velocity of your reference frame in such a way that the momentum of the system you are observing becomes zero. Such a reference frame is called the system's **zero-momentum reference frame.**

To distinguish the zero-momentum reference frame from other reference frames, we'll refer to quantities measured in that reference frame with the capital letter Z. As we shall see, examining collisions from this reference frame often simplifies things.

The prescription for calculating the velocity of the zero-momentum reference frame is as follows:

The velocity $\vec{v}_{EZ}$ of a system's zero-momentum reference frame relative to Earth is equal to the momentum of the system in the Earth reference frame divided by the inertia of the system.

Example 6.5 Zeroing out the momentum

For the colliding carts in Figure 6.8, (*a*) determine the velocity of the zero-momentum reference frame relative to Earth and (*b*) show that the system momentum measured in the zero-momentum reference frame is zero both before and after the collision.

❶ **GETTING STARTED** This example requires me to apply the prescription for calculating the velocity of the zero-momentum reference frame and then verify that the system momentum is indeed zero in that reference frame.

❷ **DEVISE PLAN** To determine the velocity of the zero-momentum reference frame relative to Earth, I calculate the momentum of the two-cart system in the Earth reference frame and then divide this result by the sum of the inertias of the two carts.

Once I have determined the velocity of the zero-momentum reference frame, I subtract it from the carts' velocities in the Earth reference frame to obtain their velocities in the zero-momentum reference frame. I then use these velocities to calculate the carts' initial and final momenta.

❸ **EXECUTE PLAN** (*a*) In the Earth reference frame, cart 1 (inertia 0.36 kg) is initially at rest and cart 2 (inertia 0.12 kg) is moving at +0.80 m/s. The *x* component of the combined momentum of the two carts is therefore (0.12 kg)(+0.80 m/s) = +0.096 kg·m/s. The *x* component of the velocity of the zero-momentum reference frame is thus

$$\frac{0.096 \text{ kg}\cdot\text{m/s}}{0.36 \text{ kg} + 0.12 \text{ kg}} = +0.20 \text{ m/s.} ✔$$

(*b*) To obtain the *x* component of the velocities of the carts in the zero-momentum reference frame, I must subtract 0.20 m/s from their values in the Earth reference frame: $v_{Z1x,i} = -0.20$ m/s and $v_{Z2x,i} = +0.60$ m/s. Then $p_{Z1x,i} = -0.072$ kg·m/s and $p_{Z2x,i} = +0.072$ kg·m/s, making the initial momentum of the system measured from this reference frame zero. ✔

For the *x* component of the final velocities in the zero-momentum reference frame, I get $v_{Z1x,f} = +0.20$ m/s and $v_{Z2x,f} = -0.60$ m/s. So $p_{Z1x,f} = +0.072$ kg·m/s and $p_{Z2x,f} = -0.072$ kg·m/s, making the system's final momentum measured from this reference frame zero. ✔

❹ **EVALUATE RESULT** In the zero-momentum reference frame, the momentum of the two-cart system before the collision is zero, as it must be from the definition of the zero-momentum frame. I also observe that it is zero after the collision, as I expect given that the system is isolated.

Figure 6.12 Collision between two carts as seen (*a*) from the Earth reference frame and (*b*) from the zero-momentum reference frame. The inertias of the carts are $m_1 = 0.36$ kg and $m_2 = 0.12$ kg and the *x* axis points to the right in the diagrams.

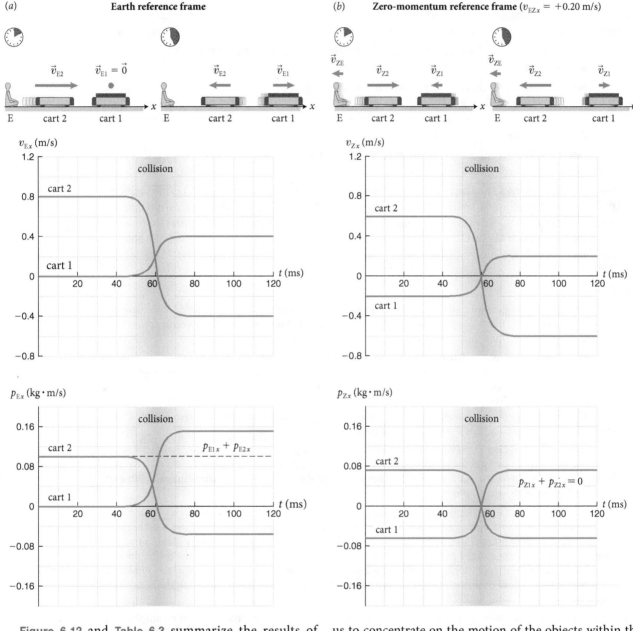

Figure 6.12 and Table 6.3 summarize the results of Example 6.5. Note how the shift from the Earth reference frame to the zero-momentum reference frame increases symmetry in the momentum-versus-time graph. In the zero-momentum reference frame, the velocity and momentum of each cart simply change sign in the collision. In this frame, the system as a whole is at rest, allowing us to concentrate on the motion of the objects within the system.

✋ **6.7** Is the kinetic energy of the two-cart system in Figure 6.12 in the zero-momentum reference frame less than, equal to, or greater than the system's kinetic energy in the Earth reference frame?

Table 6.3 Symmetry in a zero-momentum reference frame

| Cart | Velocity change (m/s) | | Momentum change (kg · m/s) | |
	Earth reference frame	Zero-momentum reference frame	Earth reference frame	Zero-momentum reference frame
1	$0 \rightarrow +0.40$	$-0.20 \rightarrow +0.20$	$0 \rightarrow +0.14$	$-0.072 \rightarrow +0.072$
2	$+0.80 \rightarrow -0.40$	$+0.60 \rightarrow -0.60$	$+0.096 \rightarrow -0.048$	$+0.072 \rightarrow -0.072$

Self-quiz

1. A space traveler discovers an object that accelerates in her reference frame. Which conclusion is correct? (*a*) Her reference frame is noninertial. (*b*) The object is not isolated. (*c*) You cannot tell.

2. A jogger starts from rest along a straight track. Consider the jogger-Earth system to be isolated. As the jogger's speed increases, does the speed of Earth change or remain constant?

3. A driver slams on the brakes of a car in which a marble is glued to the dashboard. To an observer standing by the side of the road, the marble slows down and has a nonzero acceleration. This happens because (*a*) the observer is in a noninertial reference frame, (*b*) the marble is not isolated, (*c*) some other reason.

4. Is it always possible to choose a zero-momentum reference frame for an isolated system that contains more than two objects?

5. On a low-friction track, two carts with unequal inertias move toward each other. After the collision, the two stick together and are at rest relative to the track. Before the collision, a zero-momentum reference frame would (*a*) move in the direction of the cart that has the greater inertia, (*b*) move in the direction of the faster cart, (*c*) not move relative to the track. (*d*) Can't tell—you need to know the carts' inertias and velocities.

Answers

1. (*c*). To claim that her reference frame is inertial, you need to know whether or not an *isolated* object remains at rest in that reference frame. From the information given, you don't know whether the object is accelerating because it is not isolated or because the reference frame is noninertial.

2. It changes. Strange as it sounds, Earth's velocity does change every time you start moving! Otherwise, conservation of momentum would not hold. Can you verify this statement experimentally? Not for the jogger-Earth case, because the ratio $\Delta v_{jogger}/\Delta v_{Earth}$ is equal to the inverse ratio of the inertias. Because Earth's inertia is enormous, its velocity change is infinitesimally small. This being so, how can you be sure you have the correct answer? Certainty comes from knowing that the law of conservation of momentum has been found to hold everywhere it has ever been tested—including on a cosmic scale.

3. (*b*). From the point of view of the observer, if the marble were not glued to the dashboard, the inertia of the marble would keep it moving at the velocity it had before the driver hit the brakes. The marble slows down only because the car's dashboard holds it in place. In other words, the dashboard and the marble interact with each other, and therefore the marble is not isolated.

4. Yes. The momentum of any system of objects is the sum of the individual momenta. To obtain the momentum of an object of inertia m_o in a reference frame moving at velocity $\vec{v}_{EM}$ relative to Earth, we must subtract $m_o\vec{v}_{EM}$ from the momentum of that object in the Earth reference frame. The system's momentum in that reference frame is then equal to the momentum of the system in the Earth reference frame minus the product $(m_1 + m_2 + m_3 + \cdots)\vec{v}_{EM}$. We can always choose a velocity $\vec{v}_{EM}$ such that the magnitude of this product is equal to the magnitude of the momentum of the system in the Earth reference frame and the algebraic signs are opposites.

5. (*c*). After the collision, both carts come to rest in the reference frame of the track, which means that the momentum of the two-cart system after the collision is zero. This means that the system's momentum before the collision was also zero. Therefore the reference frame of the track is a zero-momentum reference frame.

6.5 Galilean relativity

Consider two observers, A and B, moving at constant velocity relative to each other. Suppose they observe the same event and describe it relative to their respective reference frames and clocks (**Figure 6.13**). Let the origins of the two observers' reference frames coincide at $t = 0$ (Figure 6.13a). Observer A sees the event as happening at position $\vec{r}_{Ae}$ at clock reading t_{Ae} (Figure 6.13b).* Observer B sees the event at position $\vec{r}_{Be}$ at clock reading t_{Be}. What is the relationship between these clock readings and positions?

If, as we discussed in Chapter 1, we assume time is absolute—the same everywhere—and if the two observers have synchronized their (identical) clocks, they both observe the event at the same clock readings, which means

$$t_{Ae} = t_{Be}. \tag{6.1}$$

Because the clock readings of the two observers always agree, we can omit the subscripts referring to the reference frames:

$$t_A = t_B = t. \tag{6.2}$$

From Figure 6.13 we see that the position $\vec{r}_{AB}$ of observer B in reference frame A at instant t_e is equal to B's displacement over the time interval $\Delta t = t_e - 0 = t_e$, and so $\vec{r}_{AB} = \vec{v}_{AB} t_e$ because B moves at constant velocity $\vec{v}_{AB}$. Therefore

$$\vec{r}_{Ae} = \vec{r}_{AB} + \vec{r}_{Be} = \vec{v}_{AB} t_e + \vec{r}_{Be}. \tag{6.3}$$

Equations 6.2 and 6.3 allow us to relate event data collected in one reference frame to data on the same event e collected in a reference frame that moves at constant velocity relative to the first one (neither of these has to be at rest relative to Earth, but their origins must coincide at $t = 0$). To this end we rewrite these equations so that they give the values of time and position in reference frame B

Figure 6.13 Two observers moving relative to each other observe the same event. Observer B moves at constant velocity $\vec{v}_{AB}$ relative to observer A. (a) The origins O of the two reference frames overlap at instant $t = 0$. (b) At instant t_e, when the event occurs, the origin of observer B's reference frame has a displacement $\vec{v}_{AB} t_e$ relative to reference frame A.

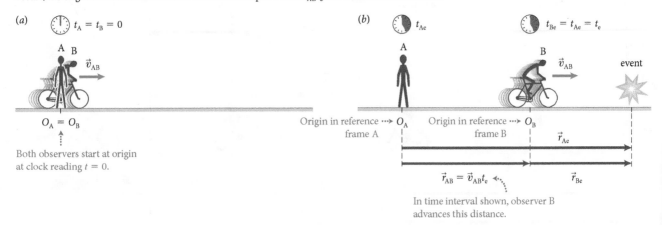

(a) $t_A = t_B = 0$

A B

$\vec{v}_{AB}$

$O_A = O_B$

Both observers start at origin at clock reading $t = 0$.

(b) t_{Ae}

$t_{Be} = t_{Ae} = t_e$

A

B

$\vec{v}_{AB}$

event

Origin in reference ⤳ O_A
frame A

Origin in reference ⤳ O_B
frame B

$\vec{r}_{Ae}$

$\vec{r}_{AB} = \vec{v}_{AB} t_e$

$\vec{r}_{Be}$

In time interval shown, observer B advances this distance.

*Remember our subscript form: The capital letter refers to the reference frame; the lowercase e is for "event." Thus the vector $\vec{r}_{Ae}$ represents observer **A**'s measurement of the position at which the **e**vent occurs.

in terms of quantities measured in reference frame A:

$$t_B = t_A = t \tag{6.4}$$

$$\vec{r}_{Be} = \vec{r}_{Ae} - \vec{v}_{AB} t_e. \tag{6.5}$$

These two equations, called the **Galilean transformation equations,** form the basis for *Galilean relativity*; they allow us to relate observations made in different reference frames. The Galilean transformation equations are predicated on the assumption that measurements of time intervals and lengths are not affected by motion. Even though this assumption sounds reasonable, it has been invalidated by experimental confirmation of the theory of special relativity (see Chapter 14). For most everyday phenomena, however, when speeds are much lower than the speed of light (about 3.0×10^8 m/s, see Section 1.3), Eqs. 6.4 and 6.5 remain excellent approximations.

6.8 A jogger runs in place on a treadmill whose belt moves at $v_{EBx} = +2.0$ m/s relative to Earth. Let the origins of the Earth reference frame and the reference frame B moving along with the top surface of the belt coincide at $t = 0$. (*a*) What is the jogger's position in the reference frame of the belt at $t = 10$ s? (*b*) Use Eq. 6.3 to show that an Earth observer's measurement of the jogger's position is $r_{Ejx} = 0$ at all instants.

Now that we have determined the relationship between positions in two reference frames, let us examine the relationship between the velocities of a cyclist relative to two observers, A and B, who are moving at constant velocity $\vec{v}_{AB}$ relative to each other (**Figure 6.14**). Each observer determines the velocity of the cyclist by measuring the cyclist's displacement during a certain time interval Δt.

Figure 6.14 The cyclist's displacement $\Delta \vec{r}_{Ac}$ measured by observer A is different from the displacement $\Delta \vec{r}_{Bc}$ measured by observer B, who is moving relative to observer A.

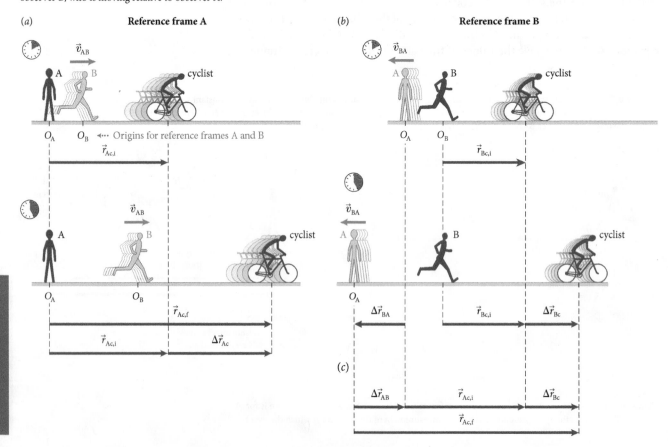

According to observer A, the bicycle undergoes a displacement $\Delta \vec{r}_{Ac}$ in the time interval Δt (Figure 6.14a). The bicycle's displacement $\Delta \vec{r}_{Bc}$ measured by observer B, however, is different because B undergoes a displacement $\Delta \vec{r}_{AB}$ relative to A (Figure 6.14b). Comparing the displacements, we obtain (Figure 6.14c)

$$\vec{r}_{Ac,f} - \vec{r}_{Ac,i} = \Delta \vec{r}_{Ac} = \Delta \vec{r}_{AB} + \Delta \vec{r}_{Bc}. \qquad (6.6)$$

If we divide the two sides of this equation by the time interval Δt, we obtain

$$\frac{\Delta \vec{r}_{Ac}}{\Delta t} = \frac{\Delta \vec{r}_{AB}}{\Delta t} + \frac{\Delta \vec{r}_{Bc}}{\Delta t}, \qquad (6.7)$$

and if we let Δt approach zero, we get

$$\vec{v}_{Ac} = \vec{v}_{AB} + \vec{v}_{Bc}. \qquad (6.8)$$

Rewriting this equation so that it gives the velocity of an object o in reference frame B in terms of quantities measured in reference frame A, we get

$$\vec{v}_{Bo} = \vec{v}_{Ao} - \vec{v}_{AB}. \qquad (6.9)$$

Equation 6.8 tells us that velocities obtained in different reference frames are additive. If an object moves at a velocity $\vec{v}_{Bo}$ in a reference frame that is moving at a velocity $\vec{v}_{AB}$ relative to some observer A, then the velocity $\vec{v}_{Ao}$ of that object in A's reference frame is the sum $\vec{v}_{AB} + \vec{v}_{Bo}$.

As an example, imagine walking in a train that moves along a straight track at 25 m/s relative to Earth. If you are walking forward at 1.0 m/s relative to the train, what is your velocity relative to Earth? In 1.0 s, the train moves 25 m relative to Earth, and in that same time interval you move that same 25 m plus an additional 1.0 m. Because you move (25 m) + (1.0 m) = 26 m each second, your velocity relative to Earth is 26 m/s.

To determine the relationship between an object's acceleration in the two reference frames, we begin by writing what a change in the object's velocity in reference frame B corresponds to in reference frame A. Using Eq. 6.8 and keeping in mind that the relative velocity $\vec{v}_{AB}$ of the two reference frames is constant, we get

$$\Delta \vec{v}_{Ao} = \vec{v}_{Ao,f} - \vec{v}_{Ao,i} = (\vec{v}_{AB} + \vec{v}_{Bo,f}) - (\vec{v}_{AB} + \vec{v}_{Bo,i})$$

$$= \vec{v}_{Bo,f} - \vec{v}_{Bo,i} = \Delta \vec{v}_{Bo}. \qquad (6.10)$$

In words, *changes in velocity are the same in any two reference frames moving at constant relative velocity*—a conclusion we reached in Section 6.3.

Dividing both sides of Eq. 6.10 by Δt and taking the limit as Δt approaches zero, we determine that the accelerations in the two reference frames are identical, too:

$$\vec{a}_{Ao} \equiv \lim_{\Delta t \to 0} \frac{\Delta \vec{v}_{Ao}}{\Delta t} = \lim_{\Delta t \to 0} \frac{\Delta \vec{v}_{Bo}}{\Delta t} \equiv \vec{a}_{Bo}. \qquad (6.11)$$

In particular, if an object moves at constant velocity in the Earth reference frame ($\vec{a}_{Eo} = \vec{0}$), then it moves at constant velocity ($\vec{a}_{Mo} = \vec{0}$) in any reference frame M that moves at constant velocity relative to Earth, confirming what we concluded in Section 6.1.

Two bullet trains passing each other. If each train has a speed of 50 m/s relative to the track, at what speed does the oncoming train approach the one from which the photo is taken?

Answer: 100 m/s.

To simplify working with quantities in various reference frames, note from Figure 6.13*b* that

$$\vec{r}_{Ae} = \vec{r}_{AB} + \vec{r}_{Be} \qquad (6.12)$$

The first and last subscripts on either side of this equation are the same, as indicated by the arrows. Moreover, if you imagine canceling the two identical subscripts B that follow each other on the right side and contracting the two vectors $\vec{r}$ into one, you get

$$\vec{r}_{A\cancel{B}} + \vec{r}_{\cancel{B}e} = \vec{r}_{Ae}, \qquad (6.13)$$

which is the same as Eq. 6.12. This "subscript cancellation" works with any relative quantity. For example, observer A's measurement of the velocity of an object is equal to observer A's measurement of the velocity of observer B plus observer B's measurement of the velocity of the object:

$$\vec{v}_{Ao} = \vec{v}_{A\cancel{B}} + \vec{v}_{\cancel{B}o}, \qquad (6.14)$$

which is the same as Eq. 6.8.

Figure 6.15 shows another useful subscript operation: Reversing the order of the subscripts reverses the direction of a relative vector $\vec{r}_{AB}$. The vector describing the position of observer B relative to A has the same magnitude as the vector $\vec{r}_{BA}$ describing the position of observer A relative to B, but it points in the opposite direction:

$$\vec{r}_{AB} = -\vec{r}_{BA}, \qquad (6.15)$$

and by taking the derivative with respect to time of this equation, we obtain a similar expression for the velocity:

$$\vec{v}_{AB} = -\vec{v}_{BA}. \qquad (6.16)$$

Figure 6.15 The position vectors of two observers relative to each other are of equal magnitude and point in opposite directions.

Position vectors are each other's opposites.

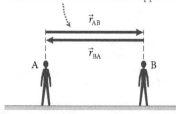

Procedure: Applying Galilean relativity

In problems dealing with more than one reference frame, you need to keep track not only of objects, but also of reference frames. For this reason, each quantity is labeled with two subscripts. The first subscript denotes the observer; the second denotes the object of interest. For example, if we have an observer on a train and also a car somewhere on the ground but in sight of the train, then $\vec{a}_{Tc}$ is the train observer's measurement of the acceleration of the car. Once you understand this notation and a few basic operations, working with relative quantities is easy and straightforward.

observer A's measurement of velocity of car:

$$\vec{v}_{Ac}$$

observer ⋯⋯⋮⋮⋯⋯ object of interest

1. Begin by listing the quantities given in the problem, using this double-subscript notation.
2. Write the quantities you need to determine in the same notation.

3. Use subscript cancellation (Eq. 6.13) to write an equation for each quantity you need to determine, keeping the first and the last subscripts on each side the same. For example, in a problem where you need to determine $\vec{v}_{Tc}$ involving a moving observer B, write

$$\vec{v}_{Tc} = \vec{v}_{TB} + \vec{v}_{Bc}.$$

4. If needed, use subscript reversal (Eq. 6.15) to eliminate any unknowns.
5. Use the kinematics relationships from Chapters 2 and 3 to solve for any remaining unknowns, making sure you stay in one reference frame.

You can use this procedure and the subscript operations for any of the three basic kinematic quantities (position, velocity, and acceleration).

Example 6.6 Who is moving?

You are driving at 25 m/s on a straight, horizontal road when a truck going 30 m/s in the same direction overtakes you. Let the positive x direction point in the direction of travel, and let the origins of the reference frames affixed to your car and the truck coincide at the instant the truck overtakes you. (*a*) What is your car's velocity as measured by someone in the truck? (*b*) What is the velocity of the truck relative to your car? (*c*) What is your car's position as measured by someone in the truck 60 s after overtaking you?

❶ GETTING STARTED This example involves two reference frames: the Earth reference frame (E) and the reference frame moving along with the truck (T). The velocities of the car and the truck are given in the Earth reference frame: $v_{\text{E}cx} = +25$ m/s and $v_{\text{E}Tx} = +30$ m/s. I need to determine the velocity of my car $\vec{v}_{\text{T}c}$ relative to the truck, the velocity of the truck $\vec{v}_{c\text{T}}$ relative to my car, and my car's position $\vec{r}_{\text{T}c}$ relative to the truck 60 s after the overtaking.

❷ DEVISE PLAN I begin by using subscript cancellation to write an equation that gives the velocity $\vec{v}_{\text{T}c}$ of my car relative to the truck in terms of velocities involving the Earth reference frame: $\vec{v}_{\text{T}c} = \vec{v}_{\text{TE}} + \vec{v}_{\text{E}c}$. The value of $\vec{v}_{\text{E}c}$ is given ($+25$ m/s), but I don't have a value for $\vec{v}_{\text{TE}}$. However, I can use subscript reversal to write $v_{\text{TE}x} = -v_{\text{E}Tx} = -30$ m/s. Once I have $\vec{v}_{\text{T}c}$, I can

use subscript reversal once more to determine $\vec{v}_{c\text{T}}$. The product of $\vec{v}_{\text{T}c}$ and the 60-s time interval gives me $\vec{r}_{\text{T}c}$ (from Eq. 2.16 $\Delta x = v_x \Delta t$).

❸ EXECUTE PLAN (*a*) Using subscript cancellation and reversal, I get $\vec{v}_{\text{T}c} = \vec{v}_{\text{TE}} + \vec{v}_{\text{E}c} = -\vec{v}_{\text{E}T} + \vec{v}_{\text{E}c}$ or, in terms of x components,

$$v_{\text{T}cx} = -v_{\text{E}Tx} + v_{\text{E}cx} = -(+30\text{ m/s}) + (+25\text{ m/s})$$

$$= -5.0\text{ m/s}. ✔$$

(*b*) $v_{c\text{T}x} = -v_{\text{T}cx} = +5.0$ m/s ✔

(*c*) My displacement in the truck's reference frame is $\Delta x_{\text{T}c} = (-5.0\text{ m/s})(60\text{ s}) = -300$ m. Thus 60 s after the truck passes you, a person in the truck measures your position as $r_{\text{T}cx} = 300$ m behind the truck. ✔

❹ EVALUATE RESULT The truck is moving faster than my car, and so relative to the truck my car moves in the negative x direction, confirming the minus sign on my answers to parts *a* and *c*. Relative to me, the truck moves in the positive x direction, in agreement with the plus sign of my answer to part *b*.

✋ **6.9** In a train moving due north at 3.1 m/s relative to Earth, a passenger carrying a suitcase walks forward down the aisle at 1.2 m/s relative to the train. A spider crawls along the bottom of the suitcase at 0.5 m/s southward relative to the suitcase. What is the velocity of the spider relative to Earth?

6.6 Center of mass

To relate momentum in two reference frames moving relative to each other, we must examine inertia. Equation 4.2 told us how to obtain the inertia m_0 of an object by letting the object collide with the standard inertia:

$$m_0 \equiv -\frac{\Delta v_{sx}}{\Delta v_{ox}}m_s. \tag{6.17}$$

Suppose this inertia m_0 is measured by two observers, A and B, who are moving at constant velocity $\vec{v}_{\text{AB}}$ relative to each other. Even though the velocity values measured by the two observers differ, Eq. 6.10 tells us that the moving observer obtains the same velocity changes Δv_{sx} and Δv_{ox} as those measured in the Earth reference frame. This means that we can omit the subscript referring to the reference frame and write

$$m_{\text{Ao}} = m_{\text{Bo}} = m_0. \tag{6.18}$$

Therefore the momenta of an object measured in two reference frames A and B are related by

$$\vec{p}_{\text{Ao}} \equiv m_0\vec{v}_{\text{Ao}} = m_0(\vec{v}_{\text{AB}} + \vec{v}_{\text{Bo}}) \equiv m_0\vec{v}_{\text{AB}} + \vec{p}_{\text{Bo}}, \tag{6.19}$$

where we have used Eq. 6.8 to relate the velocities in the two reference frames. Equation 6.19 shows that the momentum measured in reference frame B is different from the momentum measured in reference frame A. In particular, if the velocity of reference frame B relative to reference frame A is $\vec{v}_{AB} = \vec{p}_{Ao}/m_o$, the momentum measured in reference frame B is zero: $\vec{p}_{Bo} = \vec{0}$ (and so reference frame B is a zero-momentum reference frame for that object).

We can make the momentum of a system of objects zero in the same way. Suppose we have a system made up of objects that have inertias $m_1, m_2, \ldots$, and we let the momentum of the system in reference frame A be $\vec{p}_{A\,sys}$. Using Eq. 6.19, we can write for the momentum of the system

$$\vec{p}_{A\,sys} = \vec{p}_{A1} + \vec{p}_{A2} + \cdots$$

$$= (m_1\vec{v}_{AB} + \vec{p}_{B1}) + (m_2\vec{v}_{AB} + \vec{p}_{B2}) + \cdots$$

$$= (m_1 + m_2 + \cdots)\vec{v}_{AB} + (\vec{p}_{B1} + \vec{p}_{B2} + \cdots) = m\vec{v}_{AB} + \vec{p}_{B\,sys}, \quad (6.20)$$

where $m = m_1 + m_2 + \cdots$ is the inertia of the system. If we adjust the velocity $\vec{v}_{AB}$ of reference frame B relative to reference frame A such that

$$\vec{v}_{AB} = \frac{\vec{p}_{A\,sys}}{m}, \quad (6.21)$$

then, according to Eq. 6.20, reference frame B is a zero-momentum reference frame:

$$\vec{p}_{B\,sys} = \vec{p}_{A\,sys} - m\vec{v}_{AB} = \vec{p}_{A\,sys} - m\frac{\vec{p}_{A\,sys}}{m} = \vec{0}. \quad (6.22)$$

Equations 6.21 and 6.22 show that relative to Earth the velocity of the zero-momentum reference frame Z for a system of objects is equal to the system's momentum measured in the Earth reference frame divided by the inertia of the system (as we stated in Section 6.4):

$$\vec{v}_{EZ} = \frac{\vec{p}_{E\,sys}}{m} = \frac{m_1\vec{v}_{E1} + m_2\vec{v}_{E2} + \cdots}{m_1 + m_2 + \cdots}. \quad \begin{array}{l}\text{(zero-momentum} \\ \text{reference frame)}\end{array} \quad (6.23)$$

The velocity of the zero-momentum reference frame is related to the position of the **center of mass** of a system. This position is defined as

$$\vec{r}_{cm} \equiv \frac{m_1\vec{r}_1 + m_2\vec{r}_2 + \cdots}{m_1 + m_2 + \cdots}, \quad (6.24)$$

where $\vec{r}_1, \vec{r}_2, \ldots$ represent the positions of the objects of inertia $m_1, m_2, \ldots$ in the system.* Equation 6.24 applies to the values of the positions measured in *any* reference frame (and for this reason we can omit the reference-frame subscripts). To show the relationship between the velocity of the zero-momentum reference frame and the position of the center of mass, we take the derivative of the left and right sides of Eq. 6.24 with respect to time:

$$\frac{d\vec{r}_{cm}}{dt} = \frac{m_1(d\vec{r}_1/dt) + m_2(d\vec{r}_2/dt) + \cdots}{m_1 + m_2 + \cdots}. \quad (6.25)$$

---

*Even though I pointed out in Chapter 5 that *mass* and *inertia* are synonyms, we should—to be consistent with our use of the word *inertia* for the quantity m—call $\vec{r}_{cm}$ the position of the *center of inertia*. Historically, however, $\vec{r}_{cm}$ has always been called the position of the *center of mass*, a tradition we shall follow.

Because $d\vec{r}/dt = \vec{v}$ (Eq. 2.25), we can rewrite Eq. 6.25 in the form

$$\vec{v}_{cm} \equiv \frac{d\vec{r}_{cm}}{dt} = \frac{m_1\vec{v}_1 + m_2\vec{v}_2 + \cdots}{m_1 + m_2 + \cdots}, \qquad (6.26)$$

where $\vec{v}_{cm}$ is the system's **center-of-mass velocity**. Note that this velocity is precisely the velocity of the zero-momentum reference frame in Eq. 6.23. Thus we conclude that, when measured from a reference frame moving at the same velocity as the center of mass of a system, the system's momentum is automatically zero. (In other words, such a reference frame is a zero-momentum reference frame.) In figures we shall denote the center of mass by a circle with a cross in it: $\otimes$.

Example 6.7 Two-cart center of mass

The positions of two identical carts at rest on a low-friction track are measured relative to two axes oriented in the same direction along the track. On axis A, cart 1 is at x_{A1} and cart 2 is at x_{A2}. On axis B, cart 1 is at the origin and cart 2 is at x_{B2}. (a) Determine the position of the center of mass of the two-cart system relative to each axis. (b) If instead of identical carts you have $m_1 = 2m_2$, what is the position of the center of mass on each axis?

❶ GETTING STARTED I begin by making a sketch of the situation (**Figure 6.16**). Each axis corresponds to a reference frame that is at rest relative to Earth. Because the origins of the two axes don't coincide, the x components of the cart positions on axis A are different from those on axis B. I have to use these x components to determine the position of the system's center of mass on each axis for two cases: when the cart inertias are identical and when they are not.

Figure 6.16

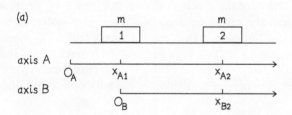

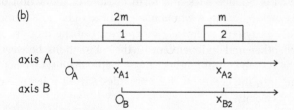

❷ DEVISE PLAN The position of the center of mass of a system of objects is given by Eq. 6.24. To use this equation in this problem, I must write it in terms of components. For axis A, I have

$$x_{A\,cm} = \frac{m_1 x_{A1} + m_2 x_{A2}}{m_1 + m_2}, \qquad (1)$$

and for axis B I replace the subscripts A with B. To solve this problem, I substitute the given values into this equation.

❸ EXECUTE PLAN (a) Substituting $m_1 = m_2 = m$ into Eq. 1, I get for axis A

$$x_{A\,cm} = \frac{m x_{A1} + m x_{A2}}{m + m} = \frac{m(x_{A1} + x_{A2})}{2m} = \frac{x_{A1} + x_{A2}}{2}. ✔$$

This position is halfway between the the two carts (see Figure 6.16). For axis B, I get

$$x_{B\,cm} = \frac{m x_{B1} + m x_{B2}}{m + m} = \frac{m(0 + x_{B2})}{2m} = \frac{x_{B2}}{2}, ✔$$

which is again the position midway between the two carts.

(b) Now I have $m_1 = 2m_2 = 2m$, and so for axis A,

$$x_{A\,cm} = \frac{2m x_{A1} + m x_{A2}}{2m + m} = \frac{2x_{A1} + x_{A2}}{3}. ✔$$

This position is one-third of the way from x_{A1} to x_{A2} on axis A because $(2x_{A1} + x_{A2})/3 = x_{A1} + (x_{A2} - x_{A1})/3$. For axis B,

$$x_{B\,cm} = \frac{2m \cdot (0) + m x_{B2}}{2m + m} = \frac{x_{B2}}{3}, ✔$$

which is one-third of the way between the origin of axis B and cart 2. If you go back to the carts in Figure 6.16b and mark the center-of-mass position on both axes, you'll see that the position $x_{B\,cm} = x_{B2}/3$ is aligned with the position $x_{A\,cm} = (2x_{A1} + x_{A2})/3$, as it must be. The center of mass of any given system can be at one position only, regardless of which reference frame is used to locate that position.

❹ EVALUATE RESULT When $m_1 = m_2$, the center of mass lies at the center of the system (midway between the carts) regardless of the choice of axis, which makes sense. When $m_1 = 2m_2$, the center of mass lies twice as close to cart 1 as to cart 2. This makes sense because I expect the center of mass to shift toward the object with the larger inertia. If, for example, the inertia of cart 1 were so large that the inertia of cart 2 was negligible, I could drop the second term in the numerator and denominator of Eq. 1, which would make the position of the center of mass be that of cart 1.

QUANTITATIVE TOOLS

✋ **6.10** In Example 6.7, let $m_1 = 3m_2$. (a) Where on axis A is the center of mass of the two-cart system? (b) Where on axis A would you need to place a third cart of inertia $m_3 = m_1$ so that the center of mass of the three-cart system is at the position of cart 2?

Example 6.7 demonstrates an important point: The position of the center of mass of a system is a property of the system that is independent of the choice of reference frame. Equation 6.24 suggests that the center of mass represents a sort of "average" position of a system, but the center of mass is more important than that. Back in Chapter 2 we began our study of motion by specifying the position of an object at each instant by the position vector $\vec{r}$. Real objects, however, are *extended* rather than *pointlike,* and so to say "the position" of a system or of a real object is not a precise statement unless we specify a fixed reference point in the system or on the object. For example, the statement "I parked my car 3 m from the fire hydrant" could mean that the front end of the car is 3 m away from the hydrant, or the midpoint of the car is 3 m from the hydrant, or any other point on the car is 3 m from the hydrant. The center of mass allows us to specify a fixed position in a system according to an exact prescription given by Eq. 6.24.

For extended objects, we can apply Eq. 6.24 by dividing the object into small pieces and substituting the position and inertia of each piece into the equation. This method is described in Appendix D. For symmetrical objects that have a uniformly distributed inertia, however, we don't need to carry out the calculation because the center of mass lies on an axis of symmetry and on any plane of symmetry. For example, the center of mass of a uniform rod lies at the center of the rod, midway between its ends, and the center of mass of a uniform sphere or cube lies at the geometric center of the sphere or cube.

Center of mass is also an important tool in our quest for simplification: It allows us to be precise even for systems where the motion is complex. To take a specific example, consider the motions shown in **Figure 6.17**. Because the two carts are identical, the center of mass of the two-cart system is always halfway between them. Notice that the center-of-mass motion is unaffected by the collision in Figure 6.17a. This is so because the momentum of the system doesn't change, and so the motion of the system must be constant. The center of mass of the system moves at velocity $\vec{v}_{cm}$ both before and after the collision.

The importance of the center of mass becomes even more evident in Figure 6.17b. Now the carts bounce back and forth because of the action of the spring while the two-cart system as a whole moves to the right. The motion of the center of mass of the system is entirely independent of the motion of the carts relative to each other and independent of the interactions between the carts. Even though the system continuously changes shape (the distance between the carts keeps changing), the center of mass provides a fixed reference point that moves at constant velocity.

✋ **6.11** (a) Determine the center-of-mass velocity of the two carts in Figure 6.8a before and after the collision, and verify that it is equal to the velocity of the carts at the point where the two $v_x(t)$ curves intersect. (b) Is the velocity of the carts at the point where the two $v_x(t)$ curves intersect always equal to the center-of-mass velocity? (c) Determine the velocities of the two carts in the zero-momentum reference frame, before and after the collision, and show that in this reference frame the momentum of the system is always zero.

Figure 6.17 Two identical carts moving on a low-friction track. (*a*) The carts collide elastically. (*b*) The carts are connected by a spring and bounce back and forth about the center of mass. In both cases, the center of mass moves at constant velocity, so that the $x(t)$ curve for the center of mass is always a straight line.

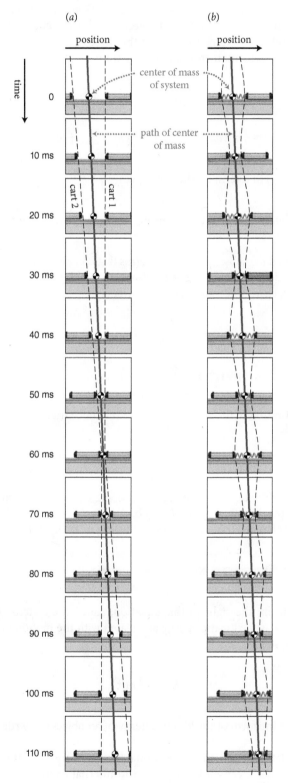

6.7 Convertible kinetic energy

We can derive some additional useful expressions involving the center-of-mass velocity. Keeping in mind that $\vec{v}_{Eo} = \vec{v}_{EZ} + \vec{v}_{Zo}$ (Eq. 6.8) and that for the zero-momentum reference frame $\vec{v}_{EZ} = \vec{v}_{cm}$ (Eqs. 6.23 and 6.26), we can derive an expression that gives, for a system of objects, the kinetic energy K_{Esys} measured in the Earth reference frame in terms of the corresponding kinetic energy K_{Zsys} measured in the zero-momentum reference frame:

$$K_{Esys} = \tfrac{1}{2}m_1 v_{E1x}^2 + \tfrac{1}{2}m_2 v_{E2x}^2 + \cdots$$

$$= \tfrac{1}{2}m_1(v_{cmx} + v_{Z1x})^2 + \tfrac{1}{2}m_2(v_{cmx} + v_{Z2x})^2 + \cdots. \quad (6.27)$$

Working out the first term on the right gives

$$\tfrac{1}{2}m_1(v_{cmx} + v_{Z1x})^2 = \tfrac{1}{2}m_1 v_{cm}^2 + m_1 v_{cmx} v_{Z1x} + \tfrac{1}{2}m_1 v_{Z1}^2, \quad (6.28)$$

where I dropped the subscripts x on the terms that contain squares of the x component of the velocity because $v_x^2 = v^2$. Working out the other terms in Eq. 6.27 and gathering terms that contain equal powers of v_{cmx}, we obtain

$$K_{Esys} = \tfrac{1}{2}(m_1 + m_2 + \cdots)v_{cm}^2 + (m_1 v_{Z1x} + m_2 v_{Z2x} + \cdots)v_{cmx}$$

$$+ (\tfrac{1}{2}m_1 v_{Z1}^2 + \tfrac{1}{2}m_2 v_{Z2}^2 + \cdots). \quad (6.29)$$

From the definition of the zero-momentum reference frame, we have

$$m_1 v_{Z1x} + m_2 v_{Z2x} + \cdots = p_{Z1x} + p_{Z2x} + \cdots = (p_{Zsys})_x = 0, \quad (6.30)$$

so that the middle term on the right in Eq. 6.29 disappears. The last term in Eq. 6.29 is the kinetic energy K_{Zsys} of the system measured in the zero-momentum reference frame, and so substituting $m = m_1 + m_2 + \cdots$, we have

$$K_{Esys} = \tfrac{1}{2}mv_{cm}^2 + K_{Zsys}. \quad (6.31)$$

The first term on the right in this equation, called the system's **translational kinetic energy,** is the kinetic energy associated with the motion of the center of mass of the system:

$$K_{cm} \equiv \tfrac{1}{2}mv_{cm}^2. \quad (6.32)$$

This term has the form of the kinetic energy of an object of inertia m moving at speed v_{cm}.

For an isolated system, the translational kinetic energy K_{cm} is *nonconvertible*, which means that it cannot be converted to internal energy. To see why this is true, note that this kinetic energy is a function of the system's center-of-mass velocity, which cannot change for an isolated system (see Section 6.8).

The remainder of the kinetic energy—the K_{Zsys} term in Eq. 6.31—is the system's **convertible kinetic energy,** the amount that can be converted to internal energy without changing the momentum of the system. It is equal to the system's kinetic energy minus the (nonconvertible) translational kinetic energy:

$$K_{conv} \equiv K_{Zsys} = K_{Esys} - \tfrac{1}{2}mv_{cm}^2$$

$$= (\tfrac{1}{2}m_1v_{E1}^2 + \tfrac{1}{2}m_2v_{E2}^2 + \cdots) - \tfrac{1}{2}(m_1 + m_2 + \cdots)v_{cm}^2. \qquad (6.33)$$

With this last equation, we can eliminate any reference to the zero-momentum reference frame in Eq. 6.31, leaving only quantities measured in the Earth reference frame:

$$K_{Esys} = K_{cm} + K_{conv}. \qquad (6.34)$$

If we had started not from an expression for the kinetic energy in the Earth reference frame (Eq. 6.27) but from the corresponding expression in any other inertial reference frame A, we would have obtained the same results, with A substituted for all the subscripts E. For this reason, we can omit the subscript E and rewrite Eqs. 6.31–6.34 in a form that holds in any inertial reference frame:

$$K = K_{cm} + K_{conv}, \qquad (6.35)$$

where

$$K_{conv} = (\tfrac{1}{2}m_1v_1^2 + \tfrac{1}{2}m_2v_2^2 + \cdots) - \tfrac{1}{2}mv_{cm}^2. \qquad (6.36)$$

The kinetic energy of a system can be split into a convertible part and a nonconvertible part. The nonconvertible part is the system's translational kinetic energy $K_{cm} = \tfrac{1}{2}mv_{cm}^2$. The remainder of the kinetic energy is convertible.

For a system of two colliding objects, the convertible kinetic energy is

$$K_{conv} = K - \tfrac{1}{2}mv_{cm}^2 = (\tfrac{1}{2}m_1v_1^2 + \tfrac{1}{2}m_2v_2^2) - \tfrac{1}{2}(m_1 + m_2)v_{cm}^2. \qquad (6.37)$$

With some algebra (see Checkpoint 6.12), substituting the Eq. 6.26 form of v_{cm} into this expression yields

$$K_{conv} = \tfrac{1}{2}\left(\frac{m_1m_2}{m_1 + m_2}\right)v_{12}^2 \quad \text{(two-object system).} \qquad (6.38)$$

If I now write

$$\mu \equiv \frac{m_1m_2}{m_1 + m_2}, \qquad (6.39)$$

then the convertible kinetic energy for two objects takes the simple form

$$K_{conv} = \tfrac{1}{2}\mu v_{12}^2 \quad \text{(two-object system),} \qquad (6.40)$$

where $\vec{v}_{12} = \vec{v}_2 - \vec{v}_1$ is the relative velocity of the two objects (see Eq. 5.1). The right side of this expression has the form of the kinetic energy of an object of inertia μ moving at a speed v_{12}. The quantity represented by the Greek letter μ (mu), which we introduced to simplify our notation, thus has the same units as inertia. It is called the **reduced inertia,** or *reduced mass*, because it is less than the inertia of either of the two colliding objects.

6.12 Verify that Eq. 6.38 is valid by substituting Eq. 6.26 for v_{cm} into Eq. 6.37 and working through the algebra.

Example 6.8 Crash test

Two collisions are carried out to crash-test a 1000-kg car. (*a*) While moving at 15 mi/h, the car strikes an identical car initially at rest. (*b*) While moving at 15 mi/h, the car strikes an identical car moving toward it and also traveling at 15 mi/h. For each collision, how much kinetic energy can be converted to internal energy, and what fraction of the initial kinetic energy of the two-car system does this represent?

1 GETTING STARTED I am given the inertias of two cars and their initial velocities in two collisions. I must calculate two values for each collision: the amount of kinetic energy converted during the collision and what fraction of the system's energy this converted amount is. I begin by making a sketch of the situation before each collision (**Figure 6.18**). I assume that the car that is initially at rest is free to roll (its parking brake is not on), and I ignore any friction so that I can consider the two-car system to be isolated. I also convert the speed from miles per hour to meters per second:

$$\frac{15 \text{ mile}}{1 \text{ hour}} \times \frac{1 \text{ hour}}{3600 \text{ s}} \times \frac{1609 \text{ m}}{1 \text{ mile}} = 6.7 \text{ m/s}.$$

Figure 6.18

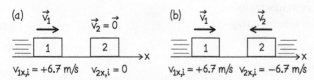

2 DEVISE PLAN The amount of energy that can be converted is given by Eq. 6.40, which requires me to calculate the reduced inertia of the two-car system (Eq. 6.39) and the relative speed of the two cars (Eq. 5.1). To determine the fraction of the initial kinetic energy that this converted amount of energy represents, I divide the converted amount by the sum of the initial kinetic energies $\frac{1}{2}mv^2$ of the two cars.

3 EXECUTE PLAN (*a*) Substituting $m_1 = m_2 = 1000$ kg into Eq. 6.39, I obtain the reduced inertia of the two-car system: $\mu = 500$ kg. The initial relative speed of the two cars is

$$v_{12} = |v_{2x} - v_{1x}| = |0 - (+6.7 \text{ m/s})| = 6.7 \text{ m/s}$$

and so

$$K_{conv} = \tfrac{1}{2}(500 \text{ kg})(6.7 \text{ m/s})^2 = 11 \text{ kJ}. \checkmark$$

The initial kinetic energy of the system is $K_i = K_{1,i} = \frac{1}{2}(1000 \text{ kg})(6.7 \text{ m/s})^2 = 22$ kJ, and so the fraction of energy converted is $K_{conv}/K_i = (11 \text{ kJ})/(22 \text{ kJ}) = 0.50$. ✔

(*b*) The initial relative speed of the two cars is

$$v_{12} = |v_{2x} - v_{1x}| = |6.7 \text{ m/s} - (-6.7 \text{ m/s})| = 13.4 \text{ m/s}$$

and so $\quad K_{conv} = \tfrac{1}{2}(500 \text{ kg})(13.4 \text{ m/s})^2 = 45 \text{ kJ}. \checkmark$

The initial kinetic energy of the system is $K_i = K_{1,i} + K_{2,i} = \frac{1}{2}(1000 \text{ kg})(6.7 \text{ m/s})^2 + \frac{1}{2}(1000 \text{ kg})(6.7 \text{ m/s})^2 = 45$ kJ, and so all of the initial kinetic energy is convertible: $K_{conv}/K_i = (45 \text{ kJ})/(45 \text{ kJ}) = 1.0$. ✔

4 EVALUATE RESULT When both cars move at 15 mi/h, the system has twice as much kinetic energy as when only one car moves at that speed, which makes sense. When only one car moves, however, the momentum of the system is nonzero, which means that not all the kinetic energy can be converted, as I found. In the second collision, the momentum of the system is zero, and so all of the kinetic energy can be converted, in agreement with my answer.

For an inelastic collision, v_{12} changes and so K changes, too. Using Eq. 6.35, we can write for the change in the kinetic energy of the system:

$$\Delta K = \Delta K_{cm} + \Delta K_{conv}. \tag{6.41}$$

The first term on the right is zero because the system is isolated, and so its translational kinetic energy cannot change. Substituting Eq. 6.40 into the last term, we have

$$\Delta K = \tfrac{1}{2}\mu v_{12f}^2 - \tfrac{1}{2}\mu v_{12i}^2 = \tfrac{1}{2}\mu(v_{12f}^2 - v_{12i}^2). \tag{6.42}$$

If we pull v_{12i}^2 out of the parentheses, we can write this expression in terms of the coefficient of restitution e (see Eq. 5.18):

$$\Delta K = \tfrac{1}{2}\mu v_{12i}^2 \left(\frac{v_{12f}^2}{v_{12i}^2} - 1\right) = \tfrac{1}{2}\mu v_{12i}^2(e^2 - 1). \tag{6.43}$$

This value represents the amount of kinetic energy converted to internal energy during the inelastic collision. The maximum change in the system's kinetic energy occurs in a totally inelastic collision ($e = 0$):

$$\Delta K = -\tfrac{1}{2}\mu v_{12i}^2 \quad \text{(totally inelastic collision).} \tag{6.44}$$

So, when two colliding objects stick together, their relative velocity becomes zero and all of the convertible kinetic energy is converted to internal energy (compare Eqs. 6.40 and 6.44). Even though the collision is totally inelastic, however, the two-object system still has some kinetic energy because the nonconvertible part of the system's kinetic energy ($\tfrac{1}{2}mv_{cm}^2$) is unaffected by the collision.

For an elastic collision, $e = 1$, and so Eq. 6.43 becomes $\Delta K = 0$: The kinetic energy is the same before and after the collision, as expected.

6.13 A moving object that has inertia m strikes a stationary object that has inertia $0.5m$. (*a*) What fraction of the initial kinetic energy of the system is convertible? (*b*) Why can't the rest be converted?

6.8 Conservation laws and relativity

In Section 6.3 we introduced the principle of relativity, which states that the laws of the universe are the same in any inertial reference frame. In this section we prove this statement for the momentum and energy laws, which embody the laws of conservation of momentum and energy.

Let us consider a system from two reference frames, A and B, that are moving at velocity $\vec{v}_{AB}$ relative to each other. In Section 6.6 we saw that the momentum of an object is different in two reference frames that are moving relative to each other (Eq. 6.19), but now we are interested in *changes* in momentum. For an object of inertia m_o, we can write

$$\Delta\vec{p}_{Ao} = m_o\Delta\vec{v}_{Ao}. \tag{6.45}$$

Because a change in velocity in reference frame A is the same as a change in reference frame B (Eq. 6.10), we have

$$\Delta\vec{p}_{Ao} = m_o\Delta\vec{v}_{Ao} = m_o\Delta\vec{v}_{Bo} = \Delta\vec{p}_{Bo}, \tag{6.46}$$

which tells us that the change in momentum of an object is the same in reference frames A and B. So, for the momentum of a system of objects, we have

$$\Delta\vec{p}_{A\,sys} = \Delta\vec{p}_{B\,sys}. \tag{6.47}$$

In words, *changes in the momentum of a system are the same in any two reference frames moving at constant velocity relative to each other.*

Equation 6.47 tells us that if the change in momentum of a system is zero in reference frame A, then it is also zero in reference frame B. So, if a system is isolated in reference frame A, it is also isolated in reference frame B, as required by the principle of relativity (**Figure 6.19**).

Next we examine changes in the energy of a system. Let us first consider a change in the system's internal energy. In Section 5.3 we introduced internal energy as a quantitative measure of a change in state. Given that the state of any object or system cannot depend on the motion of the observer, we must conclude that any change in internal energy must be independent of the reference frame:

$$\Delta E_{A\,int} = \Delta E_{B\,int}. \tag{6.48}$$

Figure 6.19 Whether or not a system is isolated does not depend on the motion of the observer, as long as the reference frames are inertial relative to each other.

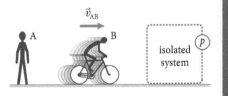

QUANTITATIVE TOOLS

Unfortunately, the situation is not so simple for a change in kinetic energy, as we saw in Section 6.3. If an object's velocity in reference frame A increases to $v_{Aox,i} + \Delta v_{Aox}$ from an initial value $v_{Aox,i}$, its kinetic energy increases by an amount

$$\Delta K_{Ao} = \tfrac{1}{2}m_o\big[(v_{Aox,i} + \Delta v_{Aox})^2 - (v_{Aox,i})^2\big]$$

$$= m_o v_{Aox,i}\Delta v_{Aox} + \tfrac{1}{2}m_o(\Delta v_{Aox})^2. \tag{6.49}$$

Substituting Eqs. 6.8 and 6.10 into the right side, we get

$$\Delta K_{Ao} = m_o(v_{ABx} + v_{Box,i})\Delta v_{Box} + \tfrac{1}{2}m_o(\Delta v_{Box})^2. \tag{6.50}$$

Applying Eq. 6.49 to reference frame B, we have

$$\Delta K_{Bo} = m_o v_{Box,i}\,\Delta v_{Box} + \tfrac{1}{2}m_o(\Delta v_{Box})^2. \tag{6.51}$$

When we expand Eq. 6.50, we get

$$\Delta K_{Ao} = m_o v_{ABx}\Delta v_{Box} + m_o v_{Box,i}\Delta v_{Box} + \tfrac{1}{2}m_o(\Delta v_{Box})^2. \tag{6.52}$$

Because the second and third terms on the right are exactly what we have on the right in Eq. 6.51, we see that

$$\Delta K_{Ao} = m_o v_{ABx}\Delta v_{Box} + \Delta K_{Bo} \neq \Delta K_{Bo} \quad \text{(single object),} \tag{6.53}$$

which means that the change in an object's kinetic energy depends on the reference frame in which that change is measured.

To investigate whether or not ΔK, the change in the kinetic energy of a system made up of two colliding objects, is independent of the reference frame in which the energy is measured, let us examine the kinetic energy converted to internal energy, given by Eq. 6.42. In reference frame A we have

$$\Delta K_A = \tfrac{1}{2}\mu(v_{A12,f}^2 - v_{A12,i}^2) \quad \text{(two-object system).} \tag{6.54}$$

Because the relative velocities are *differences* in velocities, Eq. 6.10 tells us that $v_{B12} = v_{A12}$, and so the change in the system's kinetic energy must also be the same in any two inertial reference frames:

$$\Delta K_B = \Delta K_A \tag{6.55}$$

(in spite of the fact that the changes in kinetic energy are different for the individual objects!). Combining Eqs. 6.55 and 6.48 and generalizing the result to a system of more than two objects, we obtain

$$\Delta K_B + \Delta E_{Bint} = \Delta K_A + \Delta E_{Aint} \tag{6.56}$$

or

$$\Delta E_{Asys} = \Delta E_{Bsys}. \tag{6.57}$$

In words, *changes in the energy of a system are the same in any two reference frames moving at constant velocity relative to each other.*

Equation 6.57 tells us that if the change in energy of a system is zero in reference frame A, then it is also zero in reference frame B. So, if a system is closed in reference frame A, it is also closed in reference frame B, as required by the principle of relativity.

✋ **6.14** Objects 1 ($m_1 = 1.0$ kg) and 2 ($m_2 = 3.0$ kg) collide inelastically. The velocities are $v_{1x,i} = +4.0$ m/s, $v_{2x,i} = 0$, $v_{1x,f} = -0.50$ m/s, and $v_{2x,f} = +1.5$ m/s. (*a*) What is the coefficient of restitution *e*? (*b*) Make a table like Table 6.1 showing the kinetic energy converted to internal energy both in the Earth reference frame and in a reference frame moving at $v_{EMx} = -1.0$ m/s relative to Earth.

Chapter Glossary

SI units of physical quantities are given in parentheses.

Center of mass A reference-frame-independent reference point on a system of objects, whose position is given by

$$\vec{r}_{cm} \equiv \frac{m_1\vec{r}_1 + m_2\vec{r}_2 + \cdots}{m_1 + m_2 + \cdots}. \tag{6.24}$$

Center-of-mass velocity $\vec{v}_{cm}$ (m/s) Velocity of the center of mass of a system of objects:

$$\vec{v}_{cm} \equiv \frac{m_1\vec{v}_1 + m_2\vec{v}_2 + \cdots}{m_1 + m_2 + \cdots}. \tag{6.26}$$

Convertible kinetic energy K_{conv} (J) The portion of the kinetic energy of a system that can be converted to internal energy:

$$K_{conv} \equiv K - \tfrac{1}{2}mv_{cm}^2. \tag{6.36}$$

The convertible kinetic energy measured for a system is the same in any inertial reference frame.

Earth reference frame A reference frame at rest relative to Earth.

Galilean transformation equations The relationships between the coordinates in time and space of an event in two reference frames A and B moving at constant velocity $\vec{v}_{AB}$ relative to each other:

$$t_B = t_A = t \tag{6.4}$$

$$\vec{r}_{Be} = \vec{r}_{Ae} - \vec{v}_{AB}t_e. \tag{6.5}$$

See also the Procedure box on page 136.

Inertial reference frame A reference frame in which the law of inertia holds.

Law of inertia In an inertial reference frame, any isolated object that is at rest remains at rest, and any isolated object that is in motion keeps moving at a constant velocity.

Observer A person (real or imaginary) carrying out measurements or observations in a reference frame.

Principle of relativity The laws of the universe are the same in any inertial reference frame.

Reduced inertia μ (kg) A scalar defined as

$$\mu \equiv \frac{m_1 m_2}{m_1 + m_2}. \tag{6.39}$$

Reference frame An axis and an origin that define a direction in space and a unit of length relative to which one can observe and measure motion.

Translational (nonconvertible) kinetic energy K_{cm} (J) The kinetic energy associated with the motion of the center of mass of a system:

$$K_{cm} \equiv \tfrac{1}{2}mv_{cm}^2. \tag{6.32}$$

The translational kinetic energy of an isolated system cannot be converted to internal energy because if it were, the system's momentum would not be constant.

Zero-momentum reference frame A reference frame in which the momentum of the system is zero. The velocity of this reference frame relative to another reference frame is equal to the system's center-of-mass velocity relative to that reference frame.

7 Interactions

In past chapters we saw how interactions can change the kinetic energy of objects as well as their internal energy. Every event that happens in this universe is the result of some interaction between objects. Interactions determine the structure of the universe, from the subatomic scale to the cosmological scale. In this chapter we study how interactions convert energy from one form to another. In the process we learn more about the concept of internal energy introduced in Chapter 5.

7.1 The effects of interactions

In the broadest sense, **interactions** are mutual influences between two objects that produce change (either physical change or a change in motion). As an example, consider a cart at rest on a horizontal track. The only way to get the cart to move is to make it *interact* with something else. We could, for instance, give it a shove or make it collide with another cart or fasten a magnet to it and push or pull it with another magnet.

Note that in each case we make the cart interact with something else. We can't make it change its state of motion without that something else: a hand, another cart, a magnet. If you look at Figure 4.19, you'll see that in each case where there is an interaction (*a*, *c*, and *d*), the interaction takes place between *two* objects: puck and floor, cart and cart, hand and cart.

An interaction that causes objects to accelerate can be either repulsive or attractive. A *repulsive interaction* is one in which the interacting bodies accelerate away from each other; an *attractive interaction* is one in which the interacting bodies accelerate toward each other. Some interactions are repulsive under some conditions and attractive under others, as illustrated in **Figure 7.1**.

7.1 (*a*) Imagine holding a ball a certain height above the ground. If you let the ball go, it accelerates downward. An interaction between the ball and what other object causes this acceleration? Is this interaction attractive or repulsive? (*b*) Once the ball hits the ground, its direction of travel reverses. Is this reversal the result of an attractive interaction or a repulsive one?

Figure 7.1 An interaction between two carts linked by a spring. When the spring is stretched, the interaction is attractive; when the spring is compressed, the interaction is repulsive.

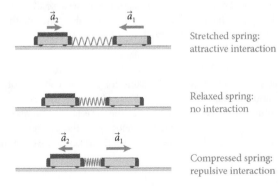

Stretched spring: attractive interaction

Relaxed spring: no interaction

Compressed spring: repulsive interaction

Figure 7.2 shows the x components of the velocities, momenta, and accelerations as well as the kinetic energies of two carts before and after an interaction. Cart 1, of inertia $m_1 = 0.12$ kg and initially at rest, is hit by cart 2, of inertia $m_2 = 0.24$ kg and initially moving at 0.55 m/s. The collision is elastic.

Figure 7.2a shows the x components of the velocities of the two carts. We've seen similar graphs before (Figure 4.8, for example). The x components of the final velocities are +0.73 m/s for cart 1 and +0.18 m/s for cart 2, and their relative speed remains 0.55 m/s, as it must for an elastic collision.

Figure 7.2b shows what happens to the x components of the momenta: One cart undergoes a momentum change in one direction, and the other cart undergoes an equal momentum change in the opposite direction. This is true at every instant: Even *during* the interaction, the momentum of the two-cart system is constant.

7.2 (*a*) Use Figure 7.2a to calculate the x components of the momenta of the two carts at $t = 30, 60,$ and 90 ms. (*b*) What is the x component of the momentum of the system at each of these three instants?

Notice in Figure 7.2 that the interaction that causes the changes in velocity and momentum is not instantaneous; instead, it takes place over a certain time interval (represented by the shaded area in the graphs). The x component of the average acceleration of each cart during the interaction is equal to the change in the x component of the velocity for that cart divided by the duration of the interaction. Because the change in speed for cart 1 is twice as big as that for cart 2 (Figure 7.2a) and because the time interval over which the interaction takes place is the same for both carts, the magnitude of the average acceleration of cart 1 is twice that of cart 2. This result is a direct consequence of conservation of momentum.

As we shall see in Section 7.7, this relationship between the average accelerations of colliding bodies is true for the instantaneous accelerations as well. So, whenever two objects interact, the ratio of the x components of their accelerations is equal to the negative inverse of the ratio of their inertias.

Example 7.1 Crash

A small car and a heavy truck moving at equal speeds in opposite directions collide head-on in a totally inelastic collision. Compare the magnitudes of (*a*) the changes in momentum and (*b*) the average accelerations of the car and the truck.

❶ **GETTING STARTED** I begin by making a sketch of the situation before and after the collision (**Figure 7.3**). Before the collision, the truck and car both move at the same speed. After the totally inelastic collision, the two move as one unit with zero relative velocity. Because the inertia m_t of the truck is greater than the inertia m_c of the car, the momentum of the system points in the same direction as the direction of travel of the truck. The combined wreck must therefore move in the same direction after the collision.

CONCEPTS

Figure 7.2 Conservation of momentum and kinetic energy in an elastic collision between two carts on a low-friction track. The inertia of cart 1 is 0.12 kg, and that of cart 2 is 0.24 kg.

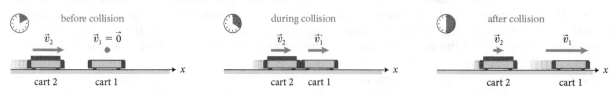

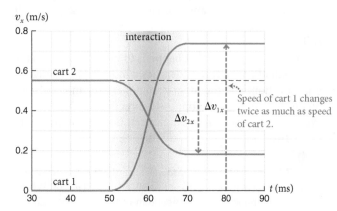

(a) Velocity

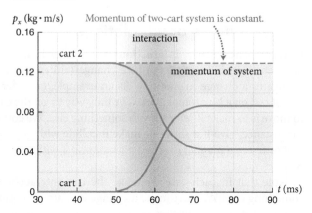

(b) Momentum

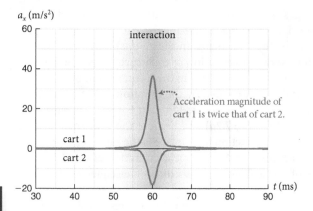

(c) Acceleration

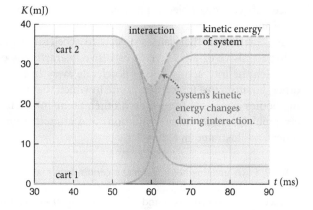

(d) Kinetic energy

---

❷ DEVISE PLAN To compare the changes in momentum, I can apply conservation of momentum to the isolated truck-car system. I can obtain the change in velocity by dividing the change in momentum by the inertia. Because the changes in velocity occur over the same time interval for both car and truck, and because the magnitude of the average acceleration is given by $\Delta v/\Delta t$, the ratio of the accelerations is the same as the ratio of the changes in velocity.

Figure 7.3 Example 7.1.

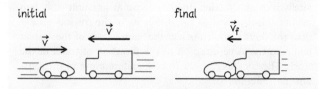

❸ EXECUTE PLAN (a) The momentum of the isolated truck-car system does not change in the collision, and so the magnitudes of the changes in momentum for the car and the truck are the same. ✔

(b) The change in the x component of the velocity of the truck is $\Delta p_{tx}/m_t$, and the change in the x component of the velocity of the car is $\Delta p_{cx}/m_c$. Because $m_t > m_c$ and because the magnitudes of the changes in momentum are equal, I conclude that the magnitude of the velocity change of the car is larger than that of the truck. ✔

❹ EVALUATE RESULT In any collision the magnitudes of the changes in momentum are the same for the two colliding objects, and so the answer to part *a* does not surprise me. That the magnitude of the velocity change for the car is larger also makes sense: As my sketch shows, the velocity of the car reverses, whereas the truck slows down somewhat but keeps traveling in the same direction.

Figure 7.2*d* shows the kinetic energy of the two carts. Just like the momentum, the kinetic energy of each cart changes, but because the collision is elastic, the kinetic energy of the system before the interaction is the same as the kinetic energy of the system afterward. Note, however, that the kinetic energy of the system does not remain constant *during* the collision! Unlike momentum, the kinetic energy of a system of colliding objects changes during the interaction—even when the interaction is an elastic collision.

To see why, remember from Section 5.5 that kinetic energy's being constant is mathematically equivalent to saying that the relative speed of two objects is the same before and after an interaction. Whenever two objects collide, however, their relative speed has to change during the interaction. If the objects were to maintain a constant relative speed during the interaction, changing their velocities would change the momentum of the system, and this can't happen if the system is isolated. Consequently the kinetic energy of any system of colliding objects has to change during the interaction, regardless of the type of collision—elastic, inelastic, or totally inelastic.

Figure 7.4 illustrates the instant of zero relative velocity for the elastic collision shown in Figure 7.2. At the instant at which the two $v(t)$ curves intersect ($t = 60$ ms), the velocities are equal, and hence the relative velocity is zero at that instant. If you cover up the rightmost diagram at the top of Figure 7.2 (representing the carts' motion after the collision), you have no way of distinguishing this elastic collision from a *totally inelastic* one. In other words, in an elastic collision and a totally inelastic collision that have the

same initial conditions, the same amount of kinetic energy is converted. The difference is that in the elastic collision any kinetic energy in the system before the collision that is converted to internal energy during the collision is all converted back to kinetic energy after the collision, whereas in the totally inelastic collision all of the initial kinetic energy converted to internal energy during the collision stays as internal energy and is not converted back to kinetic energy. The important point here is that, whenever two objects interact, their relative speed has to change, and therefore the kinetic energy of the system must change during the interaction.

✋ **7.3** (*a*) Use Figure 7.2*a* to calculate the kinetic energies of the two carts at $t = 30$, 60, and 90 ms. (*b*) What is the kinetic energy of the system at each instant?

This disappearance of kinetic energy doesn't jeopardize conservation of energy. The kinetic energy "missing" during the interaction has merely been temporarily converted to internal energy. Consider, for instance, a rubber ball thrown against a wall and bouncing off, as in **Figure 7.5**. At the instant of impact, the ball has zero velocity and has lost all the kinetic energy it had while moving toward the wall. Where has all this kinetic energy gone? It has gone into changing the shape of the ball. **Figure 7.6** shows a golf club striking a golf ball. The deformation of the ball at the instant of impact is clearly visible. After impact, both the rubber ball of Figure 7.5 and the golf ball of Figure 7.6 regain their original shape, and the kinetic energy converted to

Figure 7.4 Even in an elastic collision between two carts, there is an instant at which the carts have the same velocity, which means that at that instant their relative velocity is zero.

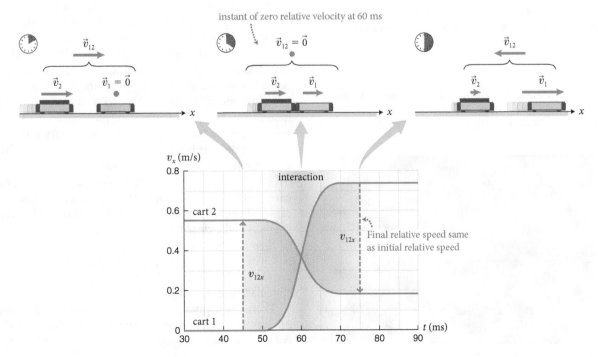

Figure 7.5 At the instant the ball in this elastic collision has zero velocity, all its kinetic energy has been used to deform the ball (and, to a lesser extent, the wall). Thus, this energy is stored temporarily as elastic potential energy.

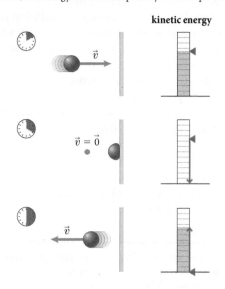

internal energy during the interaction reappears as kinetic energy. Because the system (ball and wall in Figure 7.5, ball and club in Figure 7.6) is closed, the energy of the system remains constant throughout the collision. In the elastic collision in Figure 7.4, similar changes in shape take place in the carts, but these changes occur too fast and are too small to be observed easily.

In inelastic collisions, the system's kinetic energy is not the same before and after the collision. As in an elastic collision, there is an instant at which the relative speed of the colliding objects is zero, but unlike in an elastic collision, not all of the kinetic energy converted to internal energy during the interaction reappears as kinetic energy after the collision. Some of the kinetic energy is converted irreversibly to internal energy (see Section 5.3).

Figure 7.6 At the instant of impact, a golf club and golf ball have zero relative speed, and much of the system's initial kinetic energy has gone into severe deformation of the ball (plus some deformation of the elastic club).

We can summarize the characteristics of an interaction:

1. *Two objects* are needed.
2. The *momentum* of a system of interacting objects is the same before, during, and after the interaction (provided the system is isolated).

Moreover, for interactions that affect the motion of objects:

1. The ratio of the *x* components of the accelerations of the interacting objects is the negative inverse ratio of their inertias. Because the velocities of both objects change in an interaction, the *individual* momenta and kinetic energies change.
2. The system's *kinetic energy changes during the interaction.* Part of it is converted to (or from) some internal energy. In an elastic collision, all of the converted energy reappears as kinetic energy *after* the collision. In an inelastic collision, some of the converted kinetic energy reappears as kinetic energy; in a totally inelastic collision, none of the converted kinetic energy reappears.

✋ **7.4** (*a*) In Figure 7.5, what is the momentum of the ball during the collision? (*b*) Is the momentum of the ball constant before, during, and after the collision? If so, why? If not, why not, and for what system is the momentum constant?

7.2 Potential energy

In any collision or interaction, the part of the converted kinetic energy that is temporarily stored in reversible changes in physical state and is then converted back to kinetic energy after the interaction is called **potential energy,** represented by *U*. Potential energy is a form of internal energy. The term *potential* refers to the fact that the energy has the potential to be converted back to kinetic energy—it is only temporarily stored. Just think of the rubber ball in Figure 7.5. At the instant it has zero velocity, all the energy that before impact was kinetic energy has gone into deforming the ball and is therefore now stored inside the ball. Because rubber is elastic, the stored energy has the potential to revert to kinetic energy. Similarly, if a cart on a track hits a spring mounted on a post (Figure 7.7), the cart compresses the spring, and all the energy that before impact was kinetic energy of the cart is temporarily converted to potential energy stored in the spring. As the spring returns to its original shape, this potential energy is converted back to kinetic energy of the cart. Throughout the interaction between the cart and the spring, the sum of the kinetic and potential energies of the closed cart-spring system is constant.

How and where is energy stored as potential energy? It is stored in reversible changes in the *configuration state* of the system, by which I mean the spatial arrangement of the system's interacting components.

There are many forms of potential energy, all related to the way interacting objects arrange themselves spatially. If you hoist a boulder up through a third-floor window, you

Figure 7.7 Energy conversion during the compression of a spring by a moving cart.

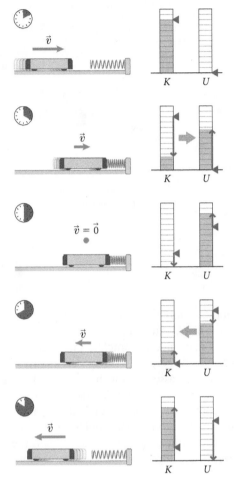

change the configuration state of the boulder-Earth system. As the boulder is hoisted upward, the form of potential energy called **gravitational potential energy** is stored in the Earth-boulder system. To see that the stored energy is potential energy, imagine cutting the rope that holds the boulder in place: The boulder accelerates as it tumbles back to Earth because of the gravitational interaction between it and Earth. As the boulder gains speed, its kinetic energy increases. Because energy is conserved and because the ball-Earth system is closed, the gain in kinetic energy has to be compensated by a loss of potential energy associated with the changed configuration state.

Moving to the atomic level, whenever you squeeze a ball or a spring, you change the configuration state of the atoms that make up the ball or spring. Reversible deformations correspond to changes in **elastic potential energy**. In Figure 7.7, for instance, the kinetic energy of the cart is temporarily converted to elastic potential energy in the spring.

> Potential energy is the form of internal energy associated with reversible changes in the configuration state of an object or system. Potential energy can be converted entirely to kinetic energy.

Exercise 7.2 Launch

A ball is pressed down on a spring and then released from rest. The spring launches the ball upward. Identify the energy conversions that occur between the instant the ball is released and the instant it reaches the highest point of its trajectory.

SOLUTION As the spring expands, elastic potential energy stored in the spring is converted to kinetic energy of the ball. As the ball travels upward, it slows down and the kinetic energy is converted to gravitational potential energy of the ball-Earth system. ✔

7.5 In Figure 7.7, the initial speed of the cart is v_i. Assuming no potential energy is initially stored in the spring, how much potential energy is stored in the spring at the instant depicted in the middle drawing and at the instant depicted in the bottom drawing? Give your answers in terms of m, v_i, and v.

7.3 Energy dissipation

The part of the converted kinetic energy that does not reappear after an inelastic collision is said to be *dissipated* (in other words, irreversibly converted). To understand what happens to energy that is dissipated, try the following simple experiment with a crisp sheet of paper lying flat on a horizontal surface. Without making any folds, bend one end of the sheet over the other, as illustrated in **Figure 7.8a**. If you remove your hand, the sheet straightens itself out. If you crumple up the sheet, however, as in Figure 7.8b, and then remove your hand, the crumpled sheet may expand a little, but it doesn't regain its original shape. There is an important difference between the two changes in shape. In Figure 7.8a the deformation takes place in a *coherent* manner, which means that at the atomic level, there is a pattern in the displacements of the atoms: They move orderly in rows, with each successive row experiencing a small displacement in the same direction as adjacent rows. As you bend the sheet in this fashion, you store potential energy in it. When you release the sheet, this potential energy appears as kinetic energy. In Figure 7.8b, the change in shape

Figure 7.8 Reversible and irreversible deformations of a sheet of paper. If the sheet is simply bent (a), it returns spontaneously to its original shape; if it is crumpled (b), it does not.

(*a*) Coherent deformation: reversible

(*b*) Incoherent deformation: irreversible

is *incoherent* because the atoms are displaced in random directions. The sheet can't straighten itself out because the randomly displaced atoms are in one another's way. Some of the smaller, less random deformations can restore themselves when you let go, but in general the energy used to do the crumpling has been irreversibly converted, producing a permanent deformation.

In addition to being dissipated to incoherent deformations, energy can be dissipated to incoherent motion, as shown in **Figure 7.9**. In Figure 7.9a a cart with a metal bar on top collides with a post that is anchored to the track. The collision is elastic, and so the cart's final kinetic energy is equal to the initial kinetic energy. In Figure 7.9b the metal bar has been replaced by a frame from which a large number of little balls are suspended. When this cart hits the post, the cart comes to a dead stop but the balls—initially at rest relative to the cart—chaotically jiggle back and forth in random directions. What has happened is that the initial kinetic energy of the cart has been converted to kinetic energy associated with the incoherent motion of the balls. We still have exactly the same amount of kinetic energy, but it has all become "internal." The initially coherent kinetic energy of the moving cart has been dissipated to the incoherent kinetic energy of the balls.

The experiment illustrated in Figure 7.9b is a large-scale model of what happens at the atomic scale in inelastic collisions. When a tennis ball bounces off a hard surface, part of the ball's kinetic energy is converted to incoherent energy of the atoms that make up the ball. (Imagine the atoms in the ball jiggling incoherently like the balls in Figure 7.9b.) Consequently the ball's kinetic energy after the bounce is

Figure 7.9 Reversible and irreversible collisions. In the totally inelastic collision in (b), the cart's coherent kinetic energy is converted completely to random energy of motion in the suspended balls.

(a) Cart rebounds elastically from barrier

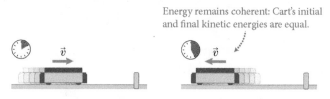

Energy remains coherent: Cart's initial and final kinetic energies are equal.

(b) Barrier stops cart, putting balls in random motion

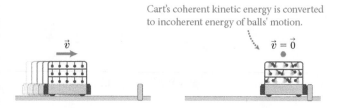

Cart's coherent kinetic energy is converted to incoherent energy of balls' motion.

less than that before the bounce, and so the rebound height is below the original height.

We can now give a complete classification of energy (**Figure 7.10**). All energy can be divided into two fundamental classes: energy associated with motion and energy associated with the configuration of interacting objects. Each class of energy comes in two forms: coherent and incoherent.

When all the atoms in an object move in a coherent fashion in the same direction—which means the entire object moves in that direction—the energy of motion is called

Figure 7.10 Classification of energy.

COHERENT (mechanical energy)	INCOHERENT (thermal energy, source energy)
kinetic energy $\vec{v}_{cm} \neq \vec{0}$	$\vec{v}_{cm} = \vec{0}$

ENERGY OF MOTION

Energy dissipation

Internal energy (all but kinetic energy)

ENERGY OF CONFIGURATION

potential energy

kinetic energy. Energy stored in coherent changes in configuration is called *potential energy*. The sum of a system's kinetic energy and potential energy is called the system's **mechanical energy** or **coherent energy.**

In addition to this coherent energy, a system can have **incoherent energy** associated with the incoherent motion and configuration of its parts. An important part of a system's incoherent energy is its **thermal energy.** (As we shall see in Checkpoint 7.7, not all incoherent energy is thermal energy.) The higher the thermal energy of an object, the higher its temperature—a concept we shall return to in Chapter 20. A rubber ball that has been squeezed repeatedly becomes noticeably hotter because the incoherent energy of the atoms that make up the ball has increased.

Here is a quick experiment you should try. Unfold a metal paper clip so that it takes on an S-shape. Holding the two ends, touch your lips with the central part of the clip—it should feel cold. Then vigorously bend the clip back and forth until it snaps in the middle and hold the two broken ends against your lips—the ends feel *hot*. The coherent energy you put into the back-and-forth bending has been converted to incoherent energy of the atoms in the metal (and some has gone into the breaking of the metal). When you hold the broken ends against your lips, you "feel" the atoms in the metal jiggling back and forth—they bounce against the atoms in your lips, transferring some of their energy to the atoms in your lips, thereby warming up your lips.

The sum of the system's incoherent energy and its potential energy is the system's *internal energy*.

Energy dissipation corresponds to the conversion of coherent energy into incoherent energy. This process is irreversible because incoherent energy cannot revert by itself to a coherent form. Consider the cart in Figure 7.9b, for example. If you shake it so that all the balls jiggle back and forth wildly, you are storing a certain amount of incoherent energy in the cart. If you then place the cart with the jiggling balls on the track against the post, this incoherent energy doesn't spontaneously turn into coherent kinetic energy of the cart as a whole—the cart does not suddenly start moving away from the post. What does happen, if you wait, is that the motion of the balls eventually stops. The kinetic energy of each ball gets dissipated, too—broken down into the incoherent jiggling of the individual atoms that make up each ball. (Because of the increased jiggling, the temperature of the balls increases but, as we shall see later, the temperature rise is imperceptibly small.)

This breaking down of the coherent energy of a large object into smaller units of incoherent thermal energy raises an interesting question: What happens to the energy once it has become thermal energy? Except for the motion of the Sun, the Moon, and other celestial objects, *all* motion we are familiar with comes to a stop unless something keeps it going. The stopping occurs because dissipation causes the coherent energy of the moving object to turn into incoherent energy. Does the energy of motion of the object's atoms dissipate, too? The answer to this question is *no*; at the

Figure 7.11 (*a*) Bouncing balls come to rest because their kinetic energy is converted to internal kinetic energy of the atoms (thermal energy). (*b*) Atoms have no internal structure to which kinetic energy can be dissipated, so they keep moving.

(*a*) Rubber balls bouncing in a box

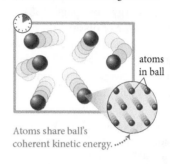
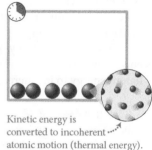

atoms in ball

Atoms share ball's coherent kinetic energy.

Kinetic energy is converted to incoherent atomic motion (thermal energy).

(*b*) Gas atoms moving in a container

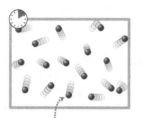
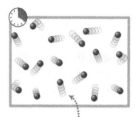
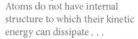

Atoms do not have internal structure to which their kinetic energy can dissipate . . .

. . . so they continue to move.

atomic level, there simply is no way to further break down and randomize the energy.

To see that this is so, suppose we throw a bunch of rubber balls into a box (**Figure 7.11a**). The balls initially bounce around, but if we look into the box a minute or so later, all of them are lying motionless on the bottom. All the energy is still there, but now it is inside the individual balls in the form of thermal energy. (The balls don't feel any warmer to you because the increase in thermal energy is so small as to be imperceptible.) Now suppose we observe a number of gas atoms randomly flying around in a container (Figure 7.11b). Provided the atoms do not interact with their environment, they keep flying around in the container *forever* because there are no smaller parts inside the atoms to which their kinetic energy can be dissipated.* And, because there is no dissipation at the atomic scale, nothing is required to keep atoms moving.

Friction, which occurs whenever two solid objects rub against each other or when an object moves through any fluid (air or water, say), is one of the causes of energy dissipation. Friction also causes wear and tear by removing microscopic pieces of material. This removal of material

---

*What about the particles that make up the atoms—electrons, protons, neutrons? As we shall see in Chapter 20, these parts can store energy only in certain minimum increments that are larger than the typical kinetic energy of an atom, and so the kinetic energy of an atom cannot be further dissipated inside the atom.

involves breaking the chemical bonds that hold atoms together. The incoherent configuration energy associated with the breaking of chemical bonds by friction or by crumpling, however, is many orders of magnitude smaller than the thermal energy generated in these processes (see the box "Coherent versus incoherent energy" on page 158). In general, we assume that the only type of dissipation caused by friction and crumpling is dissipation of mechanical energy into thermal energy. In Chapter 19 we shall examine energy dissipation and the accompanying conversion of coherent energy to incoherent energy in more detail.

7.6 Because of friction, a 0.10-kg hockey puck initially sliding over ice at 8.0 m/s slows down at a constant rate of 1.0 m/s² until it comes to a halt. (*a*) On separate graphs, sketch the puck's speed and its kinetic energy as functions of time. (*b*) To what form of energy is the kinetic energy of the puck converted?

7.4 Source energy

Because of the inexorable dissipation of mechanical energy, we need a source of energy—fuel, food, and so on—to either generate or maintain the mechanical energy of a system. For instance, the energy of gasoline is required to keep a car moving. Without this supply of energy, the car slows down as its kinetic energy dissipates because of friction. Likewise, to walk around the block, to send a message across the continent, to lift a basket, or to squeeze a spring, we need some source of energy.

Our current needs for mechanical energy are met using energy from a variety of sources—fossil and mineral fuels (oil, coal, natural gas), nuclear fuel, biomass fuel (food, wood, organic waste), water reservoirs, solar radiation, and wind. We refer to the energy obtained from these sources collectively as **source energy.** Broadly speaking, there are four kinds of source energy: *chemical energy* (energy associated with the configuration of atoms inside molecules) released in such chemical reactions as the burning of oil, coal, gas, and wood and the metabolizing of food; *nuclear energy* (energy

associated with the configuration of the nuclei of atoms) released in nuclear reactions; *solar energy* delivered by radiation from the Sun; and *stored solar energy* in the form of wind and hydroelectric energy.

Nearly all source energy can be traced back to energy delivered by the Sun. Nuclear reactions in the Sun create the solar radiation responsible for the energy content of biomass—the food we eat, the wood we burn. Fossil and mineral fuels were formed from the buried remains of plants and animals that lived and stored energy from the Sun millions of years ago. Wind energy results from the uneven heating of Earth's surface and atmosphere by solar radiation. Solar radiation evaporates the water that forms the clouds that deposit water in reservoirs behind dams.

7.7 How should chemical energy be classified in Figure 7.10?

To facilitate our accounting of energy in the rest of this book, we divide all energy into four categories: kinetic energy K, potential energy U, source energy E_s, and thermal energy E_{th} (**Figure 7.12a**). Interactions convert energy from one category to another, but conservation of energy requires that, for a closed system, the energy E of the system remains unchanged (Figure 7.12*b*).

Figure 7.13 illustrates the various types of energy conversions that can occur. Figure 7.13*a* shows the reversible conversion between the two forms of mechanical energy. Examples of this conversion are the (reversible) conversion between gravitational potential energy and kinetic energy for a ball thrown vertically upward and the (reversible) conversion between elastic potential energy and kinetic energy for a cart hitting a spring.

Figure 7.13*b* shows the (irreversible) dissipation of mechanical energy: Friction converts mechanical energy to thermal energy. Because of this dissipation, we need to convert source energy to maintain or generate mechanical energy (Figure 7.13*c*). For example, in a chemical reaction, the spatial arrangement of atoms inside molecules changes, and chemical energy is converted to kinetic energy of the

Figure 7.12 Energy accounting.

(*a*) The four categories of energy used for energy accounting

potential energy: coherent energy associated with configuration of interacting objects (gravitational and elastic potential energy go here)

source energy: incoherent energy used to produce other forms of energy (chemical, nuclear, solar, and stored solar energy go here)

kinetic energy: coherent energy associated with motion of objects

thermal energy: incoherent energy associated with chaotic motion of atoms making up objects

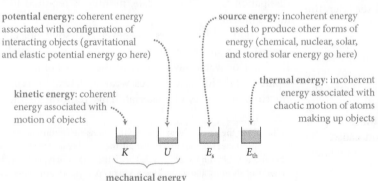

mechanical energy (all coherent energy)

(*b*) In any closed system, the sum of the four categories is constant

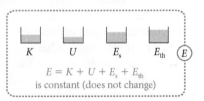

$$E = K + U + E_s + E_{th}$$
is constant (does not change)

Figure 7.13 Energy conversion processes.

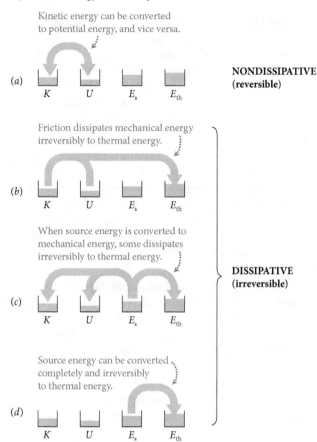

Kinetic energy can be converted to potential energy, and vice versa.

(a) K U E_s E_{th} NONDISSIPATIVE (**reversible**)

Friction dissipates mechanical energy irreversibly to thermal energy.

(b) K U E_s E_{th}

When source energy is converted to mechanical energy, some dissipates irreversibly to thermal energy.

(c) K U E_s E_{th} DISSIPATIVE (**irreversible**)

Source energy can be converted completely and irreversibly to thermal energy.

(d) K U E_s E_{th}

Figure 7.14 In the combustion of methane, each molecule of methane (CH_4) reacts with two oxygen molecules (O_2). The reaction rearranges the constituent atoms to form two water molecules (H_2O) and one molecule of carbon dioxide (CO_2). In the reaction, chemical energy originally in the reactant molecules is converted to kinetic energy of the products.

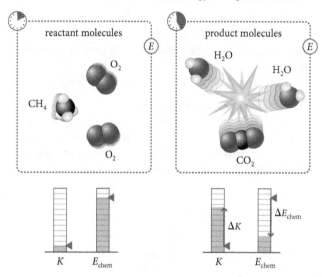

reactant molecules   product molecules

O_2 H_2O H_2O

CH_4 CO_2 O_2

K E_{chem} K E_{chem} ΔK ΔE_{chem}

reaction products (**Figure 7.14**). In the combustion of fuel in an engine, some of the kinetic energy of the combustion products can be converted to mechanical energy, but a large portion of it is converted to thermal energy (a car engine gets so hot that part of the mechanical energy generated by the engine must be used to turn a fan to *cool* the engine). Friction in the engine converts even more energy to thermal energy. Most of this thermal energy ends up either in the atmosphere or in the ground and cannot be easily recovered. We shall examine the generation of mechanical energy by *engines* in more detail in Chapter 21. For now, remember that the generation of mechanical energy from source energy is always accompanied by the generation of thermal energy, as illustrated in Figure 7.13*c*.

Figure 7.13*d* shows the conversion of source energy to thermal energy. An example of this process is the burning of gas or oil to heat our homes. The conversion of source energy to either mechanical or thermal energy is *irreversible*. The combustion of a methane-oxygen gas mixture, for example, causes permanent changes in the physical state of the gas mixture; it becomes a mixture of water and carbon dioxide. The reverse—obtaining methane from a mixture of water and carbon dioxide—is inconceivable even if energy is added to the system. Because of this irreversibility, the conversion of source energy to mechanical energy for transportation, heating, and so on—what we call *energy consumption*—decreases the available amount of source energy stored on Earth.

Because of dissipation, all forms of energy eventually are converted to thermal energy. As we shall see in Chapter 21, our ability to recover mechanical energy from thermal energy is very limited, however, making thermal energy not as useful as other forms of energy. Although some energy sources are renewable (solar, wind, water, biomass), they are not likely to meet our future demands for energy. Consequently the supply of source energy is limited, and a number of sources, such as oil and gas, are likely to run out during the 21st century.

7.8 Whenever you leave your room, you diligently turn off the lights to "conserve energy." Your friend tells you that energy is conserved regardless of whether or not your lights are off. Which of you is right?

As we have seen, interactions that convert source energy to either mechanical or thermal energy and interactions that convert mechanical energy to thermal energy are irreversible. Irreversible interactions change the thermal energy in a system and are called **dissipative interactions;** reversible interactions do not change the thermal energy in a system and are called **nondissipative interactions.**

To determine whether or not an interaction is dissipative, all you need to do is ask yourself whether the interaction is reversible:

Interactions that cause reversible changes are nondissipative; those that cause irreversible changes are dissipative.

Consider a puck sliding to a stop on a rough surface. If you were to make a movie of this action and then play the movie in reverse, you would immediately notice that something is not correct: A puck lying on a rough surface doesn't spontaneously start moving. So the interaction between the rough surface and the puck is irreversible and therefore dissipative.

CONCEPTS

Coherent versus incoherent energy

To get a feel for the different types of energy, let's look at a very ordinary object—a pencil. How many types of energy are (or can be) stored in a pencil?

Coherent (mechanical) energy

Kinetic energy: If the pencil is at rest on your desk, its kinetic energy is zero. If, however, you throw it across the room, you give it a kinetic energy of about **1 J.**

Potential energy: To store potential energy in the pencil, you must change its configuration by squeezing or bending it. The pencil, being stiff, does not bend or squeeze easily, and so, if you are lucky, you might be able to store **0.1 J** of elastic potential energy in it before it snaps.

Incoherent energy

Thermal energy: Thermal energy is the energy associated with the random jiggling of atoms. It is impossible to convert all of this energy to another form, and so instead let's imagine you take the pencil out of a drawer and carry it around in your pocket so that its temperature rises from room temperature to body temperature. This rise in temperature increases its thermal energy by **100 J.** If you could convert this 100 J of thermal energy to kinetic energy, your pencil would be moving at the speed of an airplane!

Chemical energy: Being made of wood, a pencil stores chemical energy, which can easily be released. If you burn the pencil, the wood turns to ashes, and the configuration energy stored in the chemical bonds is converted to thermal energy, which you feel as heat. The energy converted by burning the pencil is **100,000 J,** an amount equal to the kinetic energy of a medium-sized car moving at 35 mi/h.

Broken chemical bonds: If you bend the pencil enough, it breaks. The energy required to break the chemical bonds in the pencil is about **0.001 J.** This amount is only 1% of that 0.1 J of potential energy you added by bending the pencil to just before its breaking point; the remaining 99% ends up as thermal energy: When the pencil snaps, the stress created by the bending is relieved, and the snapping increases the jiggling of the atoms.

This comparison of energies makes two important points that are valid more generally. First, coherent forms of energy are insignificant compared with incoherent forms—most of the energy around us is in the form of incoherent energy. Second, when energy is dissipated, virtually all of it becomes thermal energy; the incoherent configuration energy associated with deformation, breaking, and abrasion is generally negligible compared to the energies required to cause these changes.

Exercise 7.3 Converting energy

For each of the following scenarios, choose a closed system, identify the energy conversions that take place, and classify each conversion according to the four processes shown in Figure 7.13. (*a*) A person lifts a suitcase. (*b*) A toy suspended from a spring bobs up and down. (*c*) A pan of water is brought to a boil on a propane burner. (*d*) A cyclist brakes and comes to a stop.

SOLUTION (*a*) Closed system: person, suitcase, and Earth. During the lifting, the potential energy of the Earth-suitcase system increases and the kinetic energy of the suitcase increases. The source energy is supplied by the person doing the lifting, who converts chemical (source) energy from food. In the process of converting this source energy, thermal energy is generated (the person gets hot). This process is represented in Figure 7.13*c*. ✔

(*b*) Closed system: toy, spring, and Earth. As the toy bobs, its height relative to the ground changes, its velocity changes, and the configuration of the spring changes. The bobbing thus involves conversions of gravitational and elastic potential energy and kinetic energy. This reversible process is represented in Figure 7.13*a*. ✔

(*c*) Closed system: pan of water and propane tank. As chemical (source) energy is released by burning the propane, the water is heated and its thermal energy increases. This process is represented in Figure 7.13*d*. ✔

(*d*) Closed system: Earth and cyclist. During the braking, the bicycle's kinetic energy is converted to thermal energy by friction (I'm ignoring the muscle source energy required to pull the brakes). This process is represented in Figure 7.13*b*. ✔

7.9 For each of the following processes, determine what energy conversion takes place and classify the interaction as dissipative or nondissipative. (Hint: Imagine what you would see if you played each situation in reverse.) (*a*) The launching of a ball by the expanding of a compressed spring, (*b*) the fall of a ball released a certain height above the ground, (*c*) the slowing down of a coasting bicycle, (*d*) the acceleration of a car.

7.5 Interaction range

The underlying mechanisms of interactions are still poorly understood, and as a result no one can answer the question, How do objects interact? From experience we know that not all objects interact with one another, but what determines when and why certain things do interact? It appears that each type of interaction occurs only between certain kinds of objects. So we assign to these objects certain *attributes* (magnetic strength, color, and so on) that are responsible for the interaction.

Note that it is the interaction that defines these attributes; we have no independent way of determining them. Consider a magnet, for instance. Is it a magnet because it behaves like one, or does it behave like a magnet because it *is* a magnet? In other words, do we say it has the attributes of a magnet because it interacts with other magnets? Or does it interact with other magnets because it has the attributes of a magnet?

Figure 7.15 Long-range interactions are perceptible over macroscopic distances; short-range ones, only over distances at the atomic scale or smaller. For instance, the contact interaction between two billiard balls is noticeable only when they are within nanometers of each other.

(*a*) Long-range interaction (magnetic)

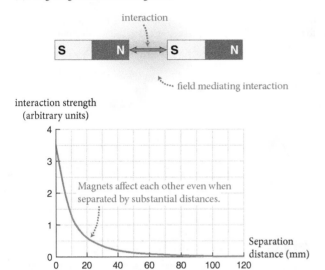

(*b*) Short-range interaction (contact)

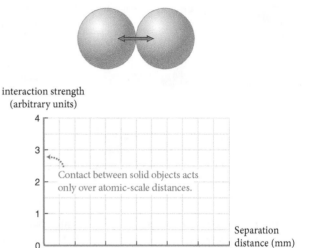

What all this boils down to is that we *define* the properties of matter (whether something is magnetic or nonmagnetic, heavy or light, red or green) according to the interactions the matter takes part in (it interacts with iron or doesn't, Earth pulls on it strongly or not, it interacts with green light or with red). Matter can therefore be classified according to its interactions, and the attributes we give to various types of matter are just a convenient way of indicating the types of interactions they take part in.

We are used to thinking of interactions as "things touching." You know, however, that a bar magnet can attract or repel another magnet located some distance away. The gravitational interaction works that way, too: Without any physical contact, Earth pulls downward on an object thrown up in the air. And so does electricity: On a day when the humidity is low, a plastic comb held near your head can attract your hair without physical contact.

To help us understand how things can interact without touching, we need to examine what we mean by *physical contact*. On the macroscopic level, *physical contact* means just that—two surfaces touching each other. On the atomic level, however, two atoms attract or repel each other even when they are separated by distances several times their size. Because all things are made up of atoms, there is no such thing as "touching" on the atomic level and therefore no such thing as physical contact. Your feet resting on the floor, for example, are supported by countless interactions between the atoms of your feet and those in the floor. So, on the atomic level, because atoms interact without physical contact, even feet "touching" floor is really an interaction without physical contact!

The strength of any interaction between two objects (in this context, think of *strength* in the sense of the magnitude of the acceleration caused by the interaction) is a function of the distance separating the objects. **Figure 7.15** shows how this strength varies with distance for magnets and for billiard

balls. Because the magnets "feel" each other over quite a large distance, their interaction is said to be a *long-range* interaction. The repulsion (or, depending on the orientation of the magnets, attraction) is greatest at small separation distances but continues to be felt even at relatively large distances. In contrast, there is no noticeable interaction between the billiard balls even when they are separated by as little as the width of a hair. The two balls interact only when they are brought sufficiently close for the individual atoms to "feel" one another (which we call "touching"). The physical contact between two objects is thus a *short-range* interaction.

As these examples illustrate, our notion of physical contact has to do with how rapidly the effects of an interaction change with distance. The difference between the magnets and billiard balls in Figure 7.15 is their **interaction range**—the distance over which the interaction is appreciable.

One model widely used to picture long-range interactions is the **field** (Figure 7.16*a*). In this model, every object is surrounded by a field, the properties of which are determined

Figure 7.16 (*a*) The field of a stationary spherical object occupies the space around the object. The shading indicates the strength of the field—a measure of how strongly the object interacts with other objects. The field shown here rapidly decreases in strength with distance. (*b*) An outward traveling ripple in the field created by shaking the object.

(*a*)   (*b*)

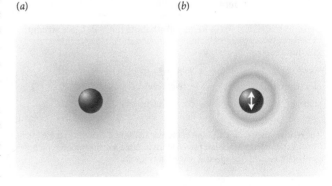

Figure 7.17 Protons and neutrons (the particles that make up the atomic nucleus) can interconvert by exchanging a particle called a pion. As a result of this pion exchange, an example of which is shown schematically below, protons and neutrons exchange places continually as though "pulled through each other." The effect is that of an attractive interaction between protons and neutrons.

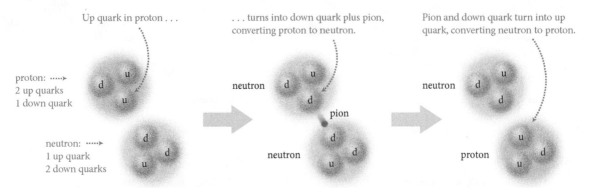

by the attributes of the object. An electrically charged object has an electric field, a magnetic object has a magnetic field, any object with what is called mass has a gravitational field, and so on. (We shall discuss mass in Chapter 13; for now just think of the mass of an object as a measure of how heavy that object is.) Interactions between objects are mediated by these fields.

Fields have an existence that is separate from the objects that create them. If I shake an object, the shaking creates a disturbance in the field that travels outward much like ripples on the surface of water (Figure 7.16*b*). The field of any object therefore contains information on the entire history of the object. As an example, think of watching the stars on a clear, moonless night. The light that reaches your eyes from some of the stars you are looking at was emitted as long as a million years ago. Some of the stars you see may not even exist anymore. Yet, via the ripples they once created in their fields—manifested as the light they once emitted—they are still interacting with you.

An alternative model explains interactions in terms of an exchange of fundamental particles called *gauge particles* (Figure 7.17 and Checkpoint 7.10). Physicists now believe that every interaction is caused by corresponding exchange of gauge particles. The existence of many of these particles has indeed been confirmed experimentally.

7.10 As an example of an interaction mediated by a "particle," imagine tossing a ball back and forth with a friend. You are both standing on an icy surface so slippery that friction is negligible. (*a*) Describe the effect the throwing and catching have on your momentum and your friend's momentum. (*b*) Is this an attractive or repulsive interaction? (*c*) Could you change the interaction into the other type?

7.6 Fundamental interactions

An interaction is **fundamental** if it cannot be explained in terms of other interactions. The interaction between, say, two colliding carts is not fundamental because it can be explained as being the result of the interactions between the atoms that make up the carts. Nor are the atomic interactions fundamental because they are the result of the interactions between the electrons and atomic nuclei that make up the atoms. And inside each atomic nucleus we observe yet another interaction responsible for holding the components of the nucleus together. This last interaction, the *strong interaction,* is today classified as fundamental because it cannot (yet) be explained in terms of other interactions.

All known interactions can be traced to just four fundamental interactions, two of them familiar from everyday life and two not (Table 7.1). The two familiar ones are the

Table 7.1 Fundamental Interactions

Type	Required attribute	Relative strength	Range	Gauge particle	Propagation speed
gravitational	mass	1	∞	graviton?	c?
weak	weak charge	10^{25}	10^{-18} m	vector bosons	varies
electromagnetic	electrical charge	10^{36}	∞	photon	c
strong	color charge	10^{38}	10^{-15} m	gluon	c

The relative strength is a measure of the magnitude of the effects of these interactions on two protons separated by about 10^{-15} m. The question marks in the last two columns indicate that the information provided has not yet been verified experimentally. The symbol c represents the speed at which lights travels.

gravitational and electromagnetic interactions. We know of the gravitational interaction as Earth pulling on objects. A nylon sock interacts with a silk blouse in the clothes dryer via the electric part of the electromagnetic interaction; a magnet interacts with a refrigerator door via the magnetic part. The two unfamiliar fundamental interactions are those that take place inside the nuclei of atoms: the weak interaction and the strong interaction. A brief overview of the four fundamental interactions follows.

1. Gravitational interaction. This long-range interaction manifests itself as an attraction between all objects that have *mass,* and it is mediated by a still undetected gauge particle called the *graviton.* It is the weakest fundamental interaction, as manifested by the observation that you can't feel it acting between your body and any nearby object even though both of you possess mass. Only for very massive objects does this interaction become appreciable: The gravitational interaction between Earth and your body, for instance, *is* noticeable.

Even though it is weak and decreases with increasing distance between the interacting objects, the gravitational interaction is what determines the structure of the universe on the cosmic scale. All other interactions cease to play a role on this scale. The existence of planetary systems and the shape of galaxies, such as the one shown in **Figure 7.18**, indicate the enormous range of the gravitational interaction.

2. Electromagnetic interaction. It is no exaggeration to say that this long-range interaction is responsible for most of what happens around us. It is responsible for the structure of atoms and molecules, for all chemical and biological processes, for the cohesion of matter into liquids and solids, for the repulsive interaction between objects such as a bat and ball, as well as for light and other electromagnetic radiation. The attribute of matter responsible for this interaction is called *electrical charge,* which comes in two varieties: *positive* and *negative,* each appearing in equal numbers in the universe. The gauge particle associated with the electromagnetic interaction is the *photon* (discussed in Chapter 34).

Figure 7.18 The matter in a galaxy is held together by the gravitational interaction—an indication of the tremendous range of this interaction.

Figure 7.19 Residual interaction. Even though a single magnet interacts strongly with a piece of iron, a pair of magnets joined together interacts only weakly with it.

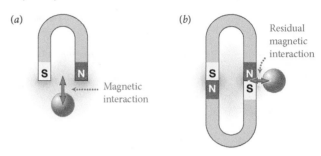

For moving charged particles, the electromagnetic interaction is complex (we'll treat it later in this book). For charged particles at rest, the interaction is either attractive or repulsive, depending on the combination of charged particles involved. Like the gravitational interaction, the strength of the electromagnetic interaction decreases rapidly with increasing distance between the interacting objects. However, the electromagnetic interaction is 36 orders of magnitude (that is, 10^{36} times) stronger than the gravitational interaction. To get a feel for just how large this difference in magnitude is, consider that the inertia of a grain of sand is "only" 33 orders of magnitude less than the inertia of Earth.

Because of the perfect balance in the amount of positive and negative charge in the universe and because the electromagnetic interaction is so strong, all matter tends to arrange itself in such a way that it is electrically neutral (which means it contains just as much positive as negative charge, and as a result most electromagnetic interactions cancel out). Consequently the electromagnetic interaction is most noticeable at the atomic scale. As noted in Section 7.5, what we call physical contact between objects is the result of the electromagnetic interactions between the individual charged parts of the atoms—interactions that average out at greater distances. The situation is much the same as when you let two horseshoe magnets attract each other until they stick together, as in **Figure 7.19**. The magnetic interaction between the pair and a piece of iron is considerably weaker than the interaction between that piece of iron and just one of the magnets. This is so because, in the pair, most of the effect of, say, the north pole of one magnet is canceled by the south pole of the other magnet. At short range, the canceling of the magnetic effect is not complete, and so there is some magnetic interaction—a *residual electromagnetic interaction*—left over, to interact weakly with the piece of iron.

3. Weak interaction. This repulsive interaction is responsible for some radioactive decay processes and for the conversion of hydrogen to helium in stars. The weak interaction acts inside the nucleus between subatomic particles that possess an attribute called *weak charge* and causes most subatomic particles to be unstable. The gauge particles associated with the weak interaction are called *vector bosons.*

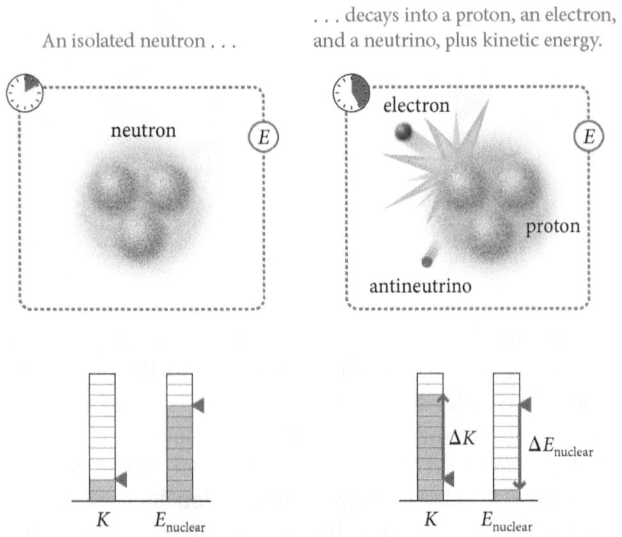

Figure 7.20 A neutron that is not part of an atomic nucleus disintegrates into a proton, an electron, and a subatomic particle called the antineutrino. This disintegration, called beta decay, is caused by the weak interaction.

An isolated neutron . . .

. . . decays into a proton, an electron, and a neutrino, plus kinetic energy.

neutron E

electron
proton
antineutrino E

K E_{nuclear}

ΔK $\Delta E_{\text{nuclear}}$

K E_{nuclear}

There is a strong parallel between the role of the electromagnetic interaction in chemical reactions and the role of the weak interaction in the decay of subatomic particles. Chemical reactions are caused by the rearrangement of electrons among atoms or groups of atoms (Figure 7.14). In the process, chemical energy is converted to (or from) kinetic energy of the reaction products. Likewise, when the weak interaction causes a subatomic particle to decay, internal energy of the original particle is converted to kinetic energy of the decay products (**Figure 7.20**).

There is a strong connection between the electromagnetic and weak interactions. Both can be considered manifestations of a single interaction called the *electroweak interaction*.

4. Strong interaction. The strong interaction acts between *quarks*, which are the building blocks of protons, neutrons, and other particles. This interaction, which can be either attractive or repulsive, is so strong that it completely overwhelms all other interactions among particles. The attribute required for this interaction is *color charge*, and the gauge particles are called *gluons*.

Quarks, like electrically charged particles and magnets, always arrange themselves in such a way that the effects of the strong interaction cancel. Only at very short range does some of the strong interaction remain. This *residual strong interaction* is responsible for holding the nucleus of an atom together.

In Table 7.1 note that the gravitational interaction—the fundamental interaction we are most directly aware of in our daily life—is orders of magnitude weaker than the other fundamental interactions. This is so because the stronger the interaction, the smaller the scale on which it affects the structure of the universe (**Figure 7.21**).

Figure 7.21 The fundamental interactions determine the structure of the universe at different scales. At the subatomic scale, the strong interaction is responsible for fusing quarks into the protons and neutrons that make up the atomic nucleus. The residual strong interaction and the weak interaction, along with electromagnetic repulsion between protons, are responsible for the formation and stability of atomic nuclei. Outside the atomic nucleus all effects of the strong interaction cancel, and weaker interactions become noticeable. Electromagnetic interactions between the positively charged atomic nucleus and negatively charged electrons are responsible for the structure of atoms. Interactions between atoms are due to residual electromagnetic interactions. At the human scale, the picture is dominated by the short-range residual electromagnetic interactions involved in contact between objects and by gravitational interactions. At the cosmic scale, the gravitational interaction plays the main role.

Size scale

10^{26} m

Gravitational interactions organize matter at planetary to cosmic scales.

Residual electromagnetic interactions assemble atoms into molecules and substances; these interactions are also responsible for contact interactions between objects.

electron cloud
nucleus

Electromagnetic interactions assemble electrons and nuclei into atoms.

Residual strong interactions assemble protons and neutrons into nuclei; **electromagnetic interactions** between protons limit nuclear size.

The **strong interaction** assembles quarks to form protons and neutrons.

10^{-15} m

CONCEPTS

Self-quiz

1. Two carts are about to collide head-on on a track. The inertia of cart 1 is greater than the inertia of cart 2, and the collision is elastic. The speed of cart 1 before the collision is higher than the speed of cart 2 before the collision. (*a*) Which cart experiences the greater acceleration during the collision? (*b*) Which cart has the greater change in momentum due to the collision? (*c*) Which cart has the greater change in kinetic energy during the collision?

2. Which of the following deformations are reversible and which are irreversible: (*a*) the deformation of a tennis ball against a racquet, (*b*) the deformation of a car fender during a traffic accident, (*c*) the deformation of a balloon as it is blown up, (*d*) the deformation of fresh snow as you walk through it?

3. Translate the kinetic energy graph in Figure 7.2 into three sets of energy bars: before the collision, during the collision, and after the collision. In each set, include a bar for K_1, a bar for K_2, and a bar for the internal energy of the system, and assume that the system is closed.

4. Describe a scenario to fit the energy bars shown in Figure 7.22. What happens during the interaction?

Figure 7.22

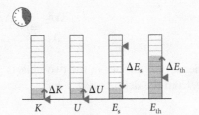

Figure 7.23

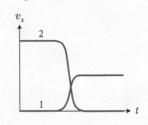

5. Describe a scenario to fit the velocity-versus-time curves for two colliding objects shown in Figure 7.23. What happens to the initial energy of the system of colliding objects during the interaction?

Answers

1. (*a*) The cart with the smaller inertia experiences the greater acceleration (see Figure 7.2). (*b*) The magnitude of $\Delta \vec{p}_1$ is the same as the magnitude of $\Delta \vec{p}_2$, but the changes are in opposite directions because the momentum of the system does not change during the collision. (*c*) $|\Delta K_1| = |\Delta K_2|$, but the changes are opposite in sign because the kinetic energy of the system before the elastic collision has to be the same as the kinetic energy of the system afterward.

2. (*a*) Reversible. The ball returns to its original shape. (*b*) Irreversible. The fender remains crumpled. (*c*) Irreversible. The balloon does not completely return to its original shape after deflation. (*d*) Irreversible. Your footprints remain.

3. See Figure 7.24. Before the collision $K_1 = 0$, K_2 is maximal, and $E_{int} = 0$; during the collision K_1, K_2, and E_{int} are all about one-third of the initial value of K_2; after the collision K_1 is about $7/8$ of the initial value of K_2, K_2 is about $1/8$ of the initial value of K_2, and $E_{int} = 0$. Because the system is closed, its energy is constant, which means the sum of the three bars is always the same.

Figure 7.24

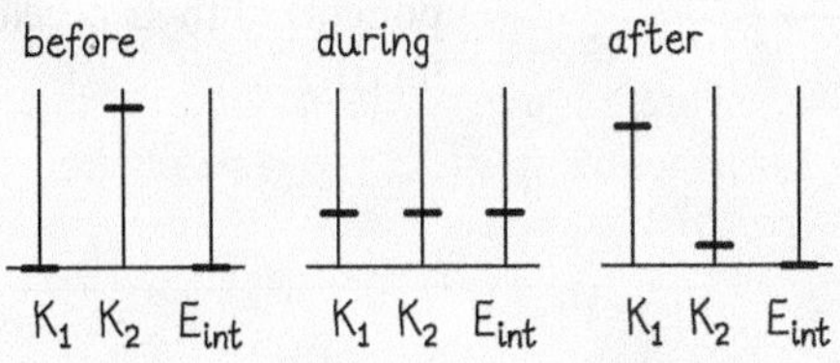

before   during   after

| | | | | | | | | | |

K_1 K_2 E_{int} K_1 K_2 E_{int} K_1 K_2 E_{int}

4. During the interaction, eight units of source energy is converted to two units of kinetic energy, two units of potential energy, and four units of thermal energy. One possible scenario is the vertical launching of a ball. Consider the system comprising you, the ball, and Earth from just before the ball is launched until after it has traveled some distance upward: The source energy goes down (you exert some effort), thermal energy goes up (in the process of exerting effort you heat up), kinetic energy goes up (the ball was at rest before the launch), and so does potential energy (the distance between the ground and the ball increases).

5. The graph represents an inelastic collision because the relative velocity of the two objects decreases to about half its initial value. In order for the momentum of the system to remain constant, the inertia of object 1 must be twice that of object 2. Possible scenario: Object 2, inertia m, collides inelastically with object 1, inertia $2m$. The collision brings object 2 to rest and sets object 1 in motion. The interaction converts the initial kinetic energy of object 2 to kinetic energy of cart 1 and to thermal energy and/or incoherent configuration energy of both carts.

7.7 Interactions and accelerations

Let us now prove a statement made in Section 7.1: For interactions that affect the motion of objects, the ratio of the x components of the accelerations of the interacting objects is equal to the negative inverse of the ratio of their inertias.

Whenever two objects interact, they exchange momentum and energy. If we consider the two-object system to be isolated during the interaction, conservation of momentum requires that, at any instant during the interaction, the momentum changes of the two objects be equal in magnitude and opposite in direction:

$$\Delta \vec{p}_1 = -\Delta \vec{p}_2. \tag{7.1}$$

Because the change in momentum for each object is caused by the interaction with the other object, the time interval over which the momenta change—the duration of the interaction—is the same for both objects. If we denote this time interval by Δt, we can write

$$\frac{\Delta \vec{p}_1}{\Delta t} = -\frac{\Delta \vec{p}_2}{\Delta t}. \tag{7.2}$$

If the inertias m_1 and m_2 of the two objects are not changed by the interaction, we have, from the definition $\vec{p} \equiv m\vec{v}$,

$$\frac{m_1 \Delta \vec{v}_1}{\Delta t} = -\frac{m_2 \Delta \vec{v}_2}{\Delta t}. \tag{7.3}$$

Using the definition of acceleration given as Eq. 3.21,

$$\lim_{\Delta t \to 0} \frac{\Delta v_x}{\Delta t} \equiv a_x, \tag{7.4}$$

we take the limit of both sides of Eq. 7.3 as Δt goes to zero and get for the x components of the accelerations:

$$m_1 a_{1x} = -m_2 a_{2x}. \tag{7.5}$$

Hence, both objects accelerate during the interaction, and the ratio of the x components of their accelerations is equal to the negative inverse of the ratio of their inertias:

$$\frac{a_{1x}}{a_{2x}} = -\frac{m_2}{m_1}. \tag{7.6}$$

This relationship between the accelerations of two interacting objects of constant inertia holds for all interactions in an isolated two-object system, regardless of what happens to the energy during the interaction.

7.11 A 1000-kg compact car and a 2000-kg van, each traveling at 25 m/s, collide head-on and remain locked together after the collision, which lasts 0.20 s. (*a*) According to Eq. 7.6, their accelerations during the collision are unequal. How can this be if both initially have the same speed and the time interval during which the collision takes place is the same amount of time for both? (*b*) Calculate the average acceleration in the direction of travel experienced by each vehicle during the collision.

7.8 Nondissipative interactions

Conservation of energy requires that the energy of any closed system remain unchanged. Using the four categories of energy introduced in Section 7.4, we can write this in the form

$$\Delta E = \Delta K + \Delta U + \Delta E_s + \Delta E_{th} = 0 \quad \text{(closed system)}. \qquad (7.7)$$

For nondissipative interactions, there are no changes in either the source energy or the thermal energy of the system: $\Delta E_s = 0$ and $\Delta E_{th} = 0$. Thus the only energy conversions allowed in nondissipative interactions are reversible transformations between kinetic energy K and potential energy U. For these types of interactions Eq. 7.7 simplifies to

$$\Delta E = \Delta K + \Delta U = 0 \quad \text{(closed system, nondissipative interaction)}. \qquad (7.8)$$

If we introduce the mechanical energy of a system

$$E_{mech} = K + U, \qquad (7.9)$$

we can write Eq. 7.8 as

$$\Delta E_{mech} = 0 \quad \text{(closed system, nondissipative interaction)}. \qquad (7.10)$$

In words, in a closed system in which only nondissipative interactions take place, the mechanical energy (the sum of the kinetic and potential energies) is constant.

We can deduce an important property of potential energy from the reversibility of nondissipative interactions. Consider the nondissipative interaction shown in Figure 7.25a. A cart, initially traveling to the right, bounces off a spring. The cart interacts with the spring, which is anchored to Earth. The system comprising the cart, spring, track, and Earth is closed. The interaction between the cart and Earth (which we'll consider to include the track and the spring) is nondissipative because it is reversible: Reversing the sequence yields the same result. From the motion of the cart, you cannot tell if time runs from t_1 to t_5 or the other way around (Figure 7.25b).

This reversibility means that the cart's kinetic energy at some given position x must be the same on the way in and on the way out. For example, K_{cart} at t_2—when the front end of the cart is at x_2—must be the same as K_{cart} at t_4—when the front end returns to x_2. Because K_{Earth} doesn't change (see Checkpoint 7.12), Eq. 7.8 yields

$$\Delta K_{cart} = -\Delta U_{spring}, \qquad (7.11)$$

where U_{spring} is the elastic potential energy associated with the shape of the spring. Consequently, the potential energy in the spring must also have the same value at any given position on the way in and on the way out. In other words, the potential energy stored in the spring must be a unique function of position: U_{spring} has a definite value at each position x of the end of the spring. More generally, the potential energy of any system (that is, the portion of the system's internal energy that is associated with reversible configuration changes) can be written in the form

$$U = U(x), \qquad (7.12)$$

where $U(x)$ is a unique function of a position variable x that quantifies the configuration of the system.

Figure 7.25 Reversible interaction between a cart and a spring anchored to a post.

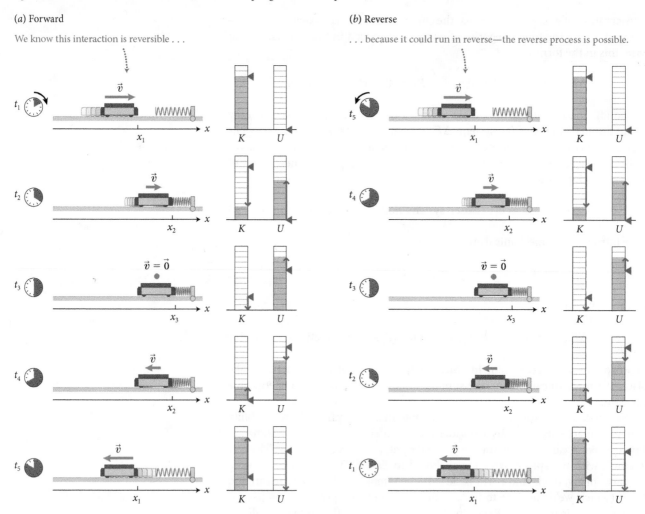

(*a*) Forward

We know this interaction is reversible . . .

(*b*) Reverse

. . . because it could run in reverse—the reverse process is possible.

A direct consequence of this dependence of potential energy on position is that the change in the kinetic energy of an object that moves from a position x_1 to a position x_2 during some nondissipative interaction depends *only* on those positions, not on the path taken by the object. Example 7.4 shows how this applies to the specific case of a cart bouncing off a spring.

7.12 Use the conservation laws to show that, when the spring in Figure 7.25 expands, the change in the kinetic energy of Earth is negligible and we are therefore justified in using Eq. 7.11.

Example 7.4 Path independence of change in potential energy

Figure 7.26 shows a cart striking a spring. In Figure 7.26*a*, consider the motion of the cart along the direct path from the initial position x_1, which is the position at which the cart makes initial contact with the free end of the spring, to the position x_2 (path A). In Figure 7.26*b*, consider the motion along the path from x_1 to the position of maximum compression x_3 and then back to x_2 (path B). Show that the change in the cart's kinetic energy is the same for both paths if the interaction caused by the spring is reversible.

❶ **GETTING STARTED** As the cart moves from x_1 to x_2 along path A, the spring is compressed and the cart's kinetic energy steadily decreases. Along path B, the spring is first compressed and the cart comes to a stop at x_3. The spring then expands, accelerating the cart back to x_2. To solve this problem, I'll consider the closed system made up of the cart, the spring, and Earth.

❷ **DEVISE PLAN** The change in the cart's kinetic energy is given by Eq. 7.11. Therefore if the change in the elastic potential energy

Figure 7.26 Example 7.4.

(*a*) Cart moves directly from x_1 to x_2

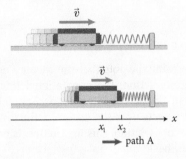

path A

(*b*) Cart moves from x_1 to x_2 via x_3

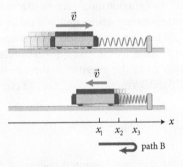

path B

of the spring is the same along both paths, then the change in the kinetic energy of the cart must also be the same along both paths. I'll first determine the change in potential energy between the initial and final positions of path A, then do the same between the initial and final positions of the two parts of path B.

❸ **EXECUTE PLAN** Because the potential energy associated with a reversible interaction is a function of x only (Eq. 7.12), the change in elastic potential energy along path A is

$$\Delta U_{\text{path A}} = U_{\text{f}} - U_{\text{i}} = U(x_2) - U(x_1).$$

Along path B I have two contributions to ΔU_{spring}: an increase in U from x_1 to x_3 and a decrease from x_3 to x_2:

$$\Delta U_{\text{path B}} = \Delta U_{13} + \Delta U_{32}$$

$$= \left[\, U(x_3) - U(x_1) \,\right] + \left[\, U(x_2) - U(x_3) \,\right]$$

$$= U(x_2) - U(x_1) = \Delta U_{\text{path A}}. ✔$$

So, the change in the cart's kinetic energy is the same for both paths.

❹ **EVALUATE RESULT** The change in the cart's kinetic energy is independent of the path joining x_1 and x_2 even though the cart ends up moving in opposite directions along the two paths. This is because the change in potential energy ΔU_{12} depends only on the coordinates of the initial and final positions x_1 and x_2.

Another way to see that this must be true is to consider only the motion from x_2 to x_3 and back. From x_2 to x_3 all the kinetic energy of the cart is converted to elastic potential energy of the spring. From x_3 to x_2 the elastic potential energy is converted back to kinetic energy, and so the motion from x_2 to x_3 and back to x_2 doesn't change the cart's kinetic energy. (Remember that kinetic energy is always positive and independent of the direction of travel.)

✋ **7.13** Show that, in Figure 7.26, the change in potential energy along a round trip from position x_1 to position x_2 and then back to x_1 is zero.

We can deduce a very important point from the fact that potential energy is a unique function of position. Consider, for example, a closed system with the potential energy function $U(x)$ shown in **Figure 7.27a**. If there is no dissipation, the sum of the kinetic and potential energies of this closed system is constant. Therefore, because the potential energy increases with increasing x, the kinetic energy must decrease with increasing x. So, if we consider two positions x_1 and $x_2 > x_1$, we see that the kinetic energy must be smaller at x_2 than at x_1. If the object whose kinetic energy is represented in the graph moves toward increasing x, as shown in Figure 7.27b, it slows down, and so its acceleration must point to the left. If, on the other hand, the object moves toward smaller x, as shown in Figure 7.27c, it speeds up, and so its acceleration again points to the left. This means that the object tends to be accelerated in the direction of lower potential energy, regardless of its original direction of motion. Quite generally we can state:

The parts of any closed system always tend to accelerate in the direction that lowers the system's potential energy.

In Figure 7.26, for example, the potential energy increases as the free end of the spring moves to the right. The free end therefore accelerates the cart to the left. This acceleration slows the cart down as it moves to the right and speeds it up as it moves to the left.

Figure 7.27 Acceleration in a closed system that has no energy dissipation.

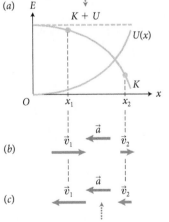

Acceleration always points toward positions of lower potential energy, regardless of direction of motion.

7.14 Consider a ball launched upward. Verify that its acceleration points in the direction that lowers the gravitational potential energy of the Earth-ball system.

7.9 Potential energy near Earth's surface

As a first example of how to determine the potential energy for a nondissipative interaction, consider a familiar interaction: the gravitational interaction near Earth's surface. As we discussed in Chapter 3, it is an experimental fact that, in the absence of air resistance, all objects near the ground fall with constant acceleration $g = 9.8 \text{ m/s}^2$. If we ignore friction, this interaction is nondissipative because free fall is reversible (see Checkpoint 7.9b).

If we release a ball from a certain height above the ground, the ball gains kinetic energy as it accelerates downward. As the ball falls, the distance between it and the ground decreases so that the (gravitational) configuration of the Earth-ball system and hence the potential energy associated with this configuration change. Because the ball and Earth constitute a closed system and because the gravitational interaction is nondissipative, we can use Eq. 7.8 to relate the change in the ball's kinetic energy to a change in the gravitational potential energy U^G (as we found in Checkpoint 7.12, Earth's kinetic energy doesn't change):*

$$\Delta U^G + \Delta K_b = 0, \tag{7.13}$$

where ΔU^G is the change in the gravitational potential energy of the Earth-ball system and ΔK_b is the change in the ball's kinetic energy.

Equation 7.13 allows us to obtain ΔU^G if we know ΔK_b. Using some of the kinematics from Chapter 3, we can obtain an expression that gives ΔU^G in terms of physical quantities other than the kinetic energy. Let's begin by looking at our ball as it falls from x_i to x_f in a time interval $\Delta t = t_f - t_i$. As illustrated in Figure 7.28, at t_i the ball is at a height x_i above the ground and the x component of its velocity is $v_{x,i}$; at t_f the ball is at x_f and the x component of its velocity is $v_{x,f}$. In that time interval, the ball accelerates with a constant acceleration given by Eqs. 3.1 and 3.14:

$$a_x = \frac{\Delta v_x}{\Delta t} = -g. \tag{7.14}$$

Next we turn to Eq. 3.7, which when we substitute $-g$ for a_x is

$$\Delta x = v_{x,i}\Delta t - \tfrac{1}{2}g(\Delta t)^2. \tag{7.15}$$

Rearranging Eq. 7.14 gives $\Delta t = -\Delta v_x/g$, and substituting this expression for Δt into Eq. 7.15 gives us

$$\Delta x = -v_{x,i}\frac{\Delta v_x}{g} - \tfrac{1}{2}g\left(\frac{\Delta v_x}{-g}\right)^2. \tag{7.16}$$

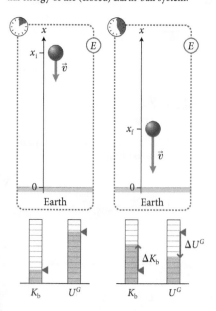

Figure 7.28 A freely falling ball gains kinetic energy at the expense of the gravitational potential energy of the (closed) Earth-ball system.

---

*We use superscripts to indicate the type of interaction associated with the potential energy: G for gravitational, E for electric, and so on.

Expanding the delta notations and working through the algebra, we get

$$x_f - x_i = -\frac{2v_{x,i}(v_{x,f} - v_{x,i}) + v_{x,f}^2 - 2v_{x,i}v_{x,f} + v_{x,i}^2}{2g} = -\frac{v_f^2 - v_i^2}{2g}, \quad (7.17)$$

where we have used the fact that $v_x^2 = v^2$. Multiplying through by g and by the inertia m_b of the ball, we get

$$m_b g(x_f - x_i) + \tfrac{1}{2}m_b(v_f^2 - v_i^2) = 0. \quad (7.18)$$

The second term on the left is precisely the change in the ball's kinetic energy, and so comparison with Eq. 7.13 tells us that the first term must be ΔU^G, the change in gravitational potential energy we are looking for. When the distance between the ball and the ground changes by an amount Δx, the gravitational potential energy changes by an amount $m_b g \Delta x$.

Because *all* objects falling near Earth's surface experience the same acceleration, this reasoning holds for any object. So, if the vertical coordinate of any object of inertia m changes by Δx, the change in the gravitational potential energy of the Earth-object system is

$$\Delta U^G = mg\Delta x \quad \text{(near Earth's surface).} \quad (7.19)$$

7.15 Suppose you raise this book (inertia $m = 3.4$ kg) from the floor to your desk, 1.0 m above the floor. (*a*) Does the gravitational potential energy of the Earth-book system increase or decrease? (*b*) By how much? (*c*) Conservation of energy requires that this change in potential energy be compensated for by a change in energy somewhere in the universe. Where?

Equation 7.19 gives the change in gravitational potential energy between two positions:

$$\Delta U^G = U_f^G - U_i^G = mg(x_f - x_i) = mgx_f - mgx_i, \quad (7.20)$$

from which we deduce that the gravitational potential energy of the Earth-object system is given by

$$U^G(x) = mgx \quad \text{(near Earth's surface),} \quad (7.21)$$

where x is the object's vertical coordinate with the x axis pointing up.

Implicit in this expression is that $U^G = 0$ at $x = 0$. Note, however, that if we write $U^G = mg(x + c)$, with c some arbitrary constant, Eq. 7.20 remains the same because the constant c, which appears in both terms, cancels out. This arbitrariness in the reference point for zero gravitational potential energy is no different from the arbitrariness in the reference point for measuring elevations or from the reference frame dependence of kinetic energy—physically, only *changes* in energy are relevant. Because of this arbitrariness, you can choose whichever reference point you like as the position where the gravitational potential energy is zero (for example, the ground at $x = 0$, as in Eq. 7.21).

Example 7.5 Path independence of gravitational potential energy

Ball A is released from rest at a height h above the ground. Ball B is launched upward from the same height at initial speed $v_{B,i}$. The two balls have the same inertia m. Consider this motion from the instant they are released to the instant they hit the ground. (a) Using kinematics, show that the change in kinetic energy is the same for both balls. (b) Show that this change in kinetic energy is equal to the negative of the change in the gravitational potential energy of the Earth-ball system between positions $x = +h$ and $x = 0$.

❶ GETTING STARTED I begin by representing these two motions graphically (**Figure 7.29**). Ball A drops straight down; ball B first rises a distance d above the launch position h and then falls from a height $h + d$.

Figure 7.29

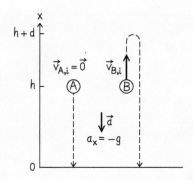

❷ DEVISE PLAN I can obtain the initial kinetic energies from the initial speeds, which are given. So all I need to do is calculate the final speeds using the kinematics equations for motion at constant acceleration from Chapter 3. From my sketch, I see that the x component of the acceleration of the balls is given by $a_x = -g$. For part b I can use Eq. 7.20 to calculate the change in gravitational potential energy.

❸ EXECUTE PLAN (a) The time interval ball A takes to fall to the ground can be obtained from Eq. 3.7:

$$\Delta x = x_f - x_i = 0 - h = v_{Ax,i}\Delta t - \tfrac{1}{2}g(\Delta t)^2$$

$$= -\tfrac{1}{2}g(\Delta t)^2$$

$$(\Delta t)^2 = \frac{2h}{g}. \tag{1}$$

The x component of the ball's final velocity is, from Eq. 7.14, $v_{Ax,f} = v_{Ax,i} - g\Delta t = -g\Delta t$, and so, using Eq. 1, I calculate that the kinetic energy of ball A just before it strikes the ground is

$$K_{A,f} = \tfrac{1}{2}mv_{A,f}^2 = \tfrac{1}{2}m[-g\Delta t]^2 = \tfrac{1}{2}mg^2\left(\frac{2h}{g}\right) = mgh. \tag{2}$$

Because the ball begins at rest, $K_{A,i} = 0$, and so the change in kinetic energy is $\Delta K_A = K_{A,f} - K_{A,i} = mgh$.

Equation 2 holds for any value of h. After reaching its highest position, ball B falls a distance $h + d$ to the ground, and so, substituting $h + d$ for h in Eq. 2, I obtain the final kinetic energy of ball B: $K_{B,f} = mg(h + d)$. To determine d, I examine ball B's motion from its initial position $x_i = h$ to the top of its path, where $x_f = h + d$, and again use Eq. 3.7:

$$\Delta x = (h + d) - h = d = v_{Bx,i}\Delta t_{top} - \tfrac{1}{2}g(\Delta t_{top})^2, \tag{3}$$

but the time interval Δt_{top} required to reach the top of the trajectory is not known. I do know, however, that the ball's velocity at the top is zero, and so I can say, again using Eq. 7.14, that $v_{Bx,top} = 0 = v_{Bx,i} - g\Delta t_{top}$, or $\Delta t_{top} = v_{Bx,i}/g$. Equation 3 then becomes

$$d = v_{Bx,i}\left(\frac{v_{Bx,i}}{g}\right) - \tfrac{1}{2}g\left(\frac{v_{Bx,i}}{g}\right)^2 = \frac{v_{B,i}^2}{g} - \tfrac{1}{2}\frac{v_{B,i}^2}{g} = \tfrac{1}{2}\frac{v_{B,i}^2}{g}.$$

The final kinetic energy of ball B is thus

$$K_{B,f} = mg(h + d) = mgh + mg\left(\tfrac{1}{2}\frac{v_{B,i}^2}{g}\right) = mgh + \tfrac{1}{2}mv_{B,i}^2.$$

Because $K_{B,i} = \tfrac{1}{2}mv_{B,i}^2$ is the initial kinetic energy of ball B, the gain in kinetic energy for ball B is $\Delta K_B = K_{B,f} - K_{B,i} = mgh$, which is indeed equal to that of ball A (Eq. 2). ✔

(b) According to Eq. 7.20, the change in gravitational potential energy is

$$\Delta U^G = mg(x_f - x_i) = mg(0 - h) = -mgh. ✔$$

This result is equal in magnitude to the gains in the balls' kinetic energies I found in part a.

❹ EVALUATE RESULT Because I obtained the same result in two ways, I can be quite confident of my answer. My result further confirms that the change in potential energy is independent of the path taken and depends on only the endpoints of the path.

7.16 Suppose that instead of choosing Earth and the ball as our system in the discussion leading up to Eq. 7.21, we had chosen to consider just the ball. Does it make sense to speak about the gravitational potential energy of the ball (the way we speak of its kinetic energy)?

7.10 Dissipative interactions

As noted in Section 7.3, a dissipative interaction is one in which there are changes in thermal energy. These interactions are *irreversible*. As an example of such an interaction consider Figure 7.30*a*, which shows a puck sliding along a wooden board and slowing to a stop. The friction between the board and the puck is dissipative: If we play the sequence in reverse, we obtain an impossible situation (Figure 7.30*b*). The friction acting on the puck in Figure 7.30*a* causes its kinetic energy to be converted to thermal energy—to be dissipated. The system comprising the puck, the board, and Earth (which holds the board in place) is closed, and so Eq. 7.7, $\Delta E = \Delta K + \Delta U + \Delta E_s + \Delta E_{th} = 0$, holds. We know that ΔU is zero because there is no change in the puck's height relative to Earth and there are no springs involved. The source energy has to be zero because no nuclear or chemical reactions are involved, and so Eq. 7.7 simplifies to

$$\Delta K + \Delta E_{th} = 0. \tag{7.22}$$

Because the only kinetic energy that changes is that of the puck (the changes in the kinetic energies of the board and Earth are negligible, see Checkpoint 7.12), we can say

$$\Delta K_{puck} = -\Delta E_{th}. \tag{7.23}$$

Figure 7.30 A puck that slides to a halt (*a*) dissipates energy through friction. Because this interaction is dissipative, the process is irreversible; the sequence shown in (*b*) cannot occur.

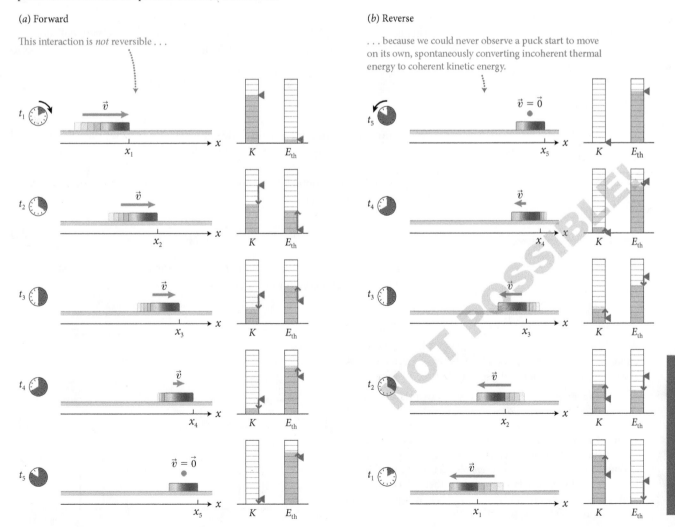

(*a*) Forward

This interaction is *not* reversible . . .

(*b*) Reverse

. . . because we could never observe a puck start to move on its own, spontaneously converting incoherent thermal energy to coherent kinetic energy.

Although this expression looks similar to Eq. 7.11, the right side is very different conceptually. In particular, thermal energy is not a function of position the way U is. The motion shown in Figure 7.30 can take place *anywhere* along the board (unlike the motion shown in Figure 7.25, which can take place only when the cart is at the spring).

Another way to see that the interaction between the puck and the board is dissipative is to look at what happens during a round trip. Imagine putting the puck down at some position x_A and then moving it back and forth between x_A and some other position x_B a dozen times. When the puck eventually returns to x_A, its state is changed: It is hotter because of the friction (and some wear has taken place). This means that energy has been dissipated irreversibly, and so $\Delta E_{th} \neq 0$ even though the puck is back at its starting position. Moving the cart of Figure 7.25 back and forth a dozen times, on the other hand, yields no change in the spring's potential energy: $\Delta U_{spring} = 0$ whenever the cart is back at its starting position.

Equation 7.7 also describes the changes in energy that take place during inelastic collisions. We know that ΔU is zero because any reversible configuration changes that occur during the first half of the collision are undone during the second half. Provided no source energy is converted during the collision, conservation of energy applied to the closed system of the colliding objects requires that the changes in kinetic energy and thermal energy add up to zero, as stated in Eq. 7.22, which we can restate in the form

$$\Delta K = -\Delta E_{th}. \tag{7.24}$$

This equation is applicable to elastic, inelastic, and totally inelastic collisions. In elastic collisions, no energy is dissipated. Thus $\Delta E_{th} = 0$ and so Eq. 7.24 reduces to

$$\Delta K = 0 \quad \text{(elastic collision)}, \tag{7.25}$$

which means that the kinetic energy of the system does not change, as we already know from Eq. 5.15. For inelastic collisions, we can compare Eq. 7.24 and Eq. 6.43, $\Delta K = \frac{1}{2}\mu v_{12i}^2(e^2-1)$, where the reduced inertia μ is $m_1 m_2/(m_1 + m_2)$ and the coefficient of restitution e (Eq. 5.18) is v_{12f}/v_{12i}. This comparison tells us that the amount of kinetic energy converted to thermal energy is determined by the coefficient of restitution:

$$\Delta E_{th} = -\Delta K = \tfrac{1}{2}\mu v_{12i}^2(1 - e^2). \tag{7.26}$$

The smaller the value of e, the larger the amount of energy dissipated from coherent energy to incoherent energy.

Example 7.6 Fender bender

A 1000-kg car traveling at 10 m/s collides with and attaches to an identical car that is initially at rest. (*a*) How much energy is dissipated in the collision? (*b*) Use Eq. 7.26 to verify your result.

❶ GETTING STARTED The collision is totally inelastic, and so the kinetic energy of the two-car system after the collision is less than it was before the collision. (It has to be nonzero, however, to maintain the momentum of the isolated two-car system.) Because the system can be considered closed during the collision and because no energy is stored reversibly ($\Delta U = 0$), the kinetic energy that disappears is converted to thermal energy (the crumpling heats the metal). I make the sketch in **Figure 7.31** to represent the energy changes in the system.

Figure 7.31

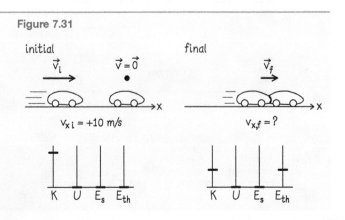

❷ DEVISE PLAN Equation 7.24 tells me that the increase in thermal energy in the system is equal to the decrease in kinetic energy. I can determine the initial kinetic energy from the data given. To obtain the final kinetic energy, I first use conservation of momentum to calculate the velocity of the two cars immediately after impact and then calculate the kinetic energy of the combined cars.

❸ EXECUTE PLAN (a) The initial kinetic energy is $\frac{1}{2}(1000 \text{ kg}) \times (10 \text{ m/s})^2 = 50$ kJ. After the collision, the cars, locked together and forming an object of inertia 2000 kg, must have the same combined momentum as the single car before the collision, and so

$$(2000 \text{ kg})v_{x,f} = (1000 \text{ kg})(+10 \text{ m/s})$$

$$v_{x,f} = +5.0 \text{ m/s}.$$

The final kinetic energy is thus

$$\frac{1}{2}(2000 \text{ kg})(5.0 \text{ m/s})^2 = 25 \text{ kJ}$$

so

$$\Delta K = 25 \text{ kJ} - 50 \text{ kJ} = -25 \text{ kJ}.$$

The amount of energy dissipated is 25 kJ. ✔

(b) For a totally inelastic collision, $e = 0$ and Eq. 7.26 becomes

$$\Delta E_{\text{th}} = \frac{1}{2}\mu v_{12i}^2 = \frac{1}{2}\left(\frac{m_1 m_2}{m_1 + m_2}\right)v_{12i}^2 = \frac{m^2}{4m}v_{12i}^2 = \frac{1}{4}m v_{12i}^2,$$

where I have used the fact that the two cars have the same inertia m. The initial relative speed v_{12i} is equal to the initial speed of car 1, and so $E_{\text{th}} = \frac{1}{4}(1000 \text{ kg})(10 \text{ m/s})^2 = 25$ kJ, as I found before. ✔

❹ EVALUATE RESULT The two methods yield the same result, which is reassuring. The energy dissipated is equal to the chemical energy stored in just one-fourth of a wooden pencil (see the box on page 158), and yet both cars are likely to be wrecked beyond repair. Although it is surprising that one could wreck two cars with the chemical energy stored in one pencil, the box on page 158 states that coherent forms of energy are insignificant compared with incoherent forms, leading me to believe that the answer I obtained is not unreasonable.

As we saw in Section 5.8, a system's kinetic energy increases in an explosive separation. Figure 7.32 shows two explosive separations. In Figure 7.32a the increase in kinetic energy is due to energy stored in a spring: One form of coherent energy (the elastic potential energy in the spring) is converted to another form of coherent energy (the kinetic energy of the moving carts). Because the compression of the spring is reversible, this interaction is nondissipative, and so this type of explosive separation is described by Eq. 7.8, $\Delta K + \Delta U = 0$.

The situation in Figure 7.32b is irreversible. Two carts are placed back to back against a firecracker, and when the firecracker is ignited, the explosive separation accelerates the carts away from each other. There is no change in potential energy because there is no change in the cart's height relative to Earth and there are no springs involved. There is, however, a change in source energy: During the explosive separation, chemical energy is converted partly to (coherent) kinetic energy of the carts and partly to (incoherent) thermal energy (things get hot during an explosion). For this dissipative interaction, we have, with $\Delta E_s = \Delta E_{\text{chem}}$,

$$\Delta K + \Delta E_{\text{chem}} + \Delta E_{\text{th}} = 0. \qquad (7.27)$$

Figure 7.32 Interactions that increase the kinetic energy of a system (a) nondissipatively and (b) dissipatively.

(a) Compressed spring pushes carts apart

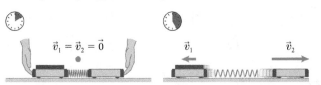

(b) Exploding firecracker pushes carts apart

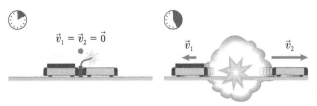

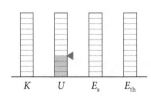

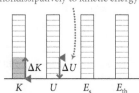

Elastic potential energy converted nondissipatively to kinetic energy

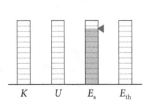

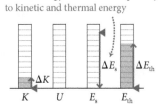

Source energy converted dissipatively to kinetic and thermal energy

More generally, for any interaction involving a rise in temperature, or a permanent deformation in one or more of the interacting objects, or the conversion of source energy—in short, for any dissipative interaction—we have

$$\Delta K + \Delta U + \Delta E_s + \Delta E_{th} = 0 \quad \text{(closed system, dissipative interaction).} \quad (7.28)$$

In general $\Delta E_s < 0$ because source energy is converted to another form of energy and $\Delta E_{th} > 0$ because dissipation increases the amount of thermal energy.

Exercise 7.7 Skateboard

Standing a few meters from a skateboard, a boarder begins running and then jumps onto the board. Make bar diagrams like the ones in Figure 7.32 to illustrate the energy conversions (*a*) while he accelerates and (*b*) while he jumps onto the board. Are the interactions dissipative or nondissipative?

SOLUTION (*a*) I choose a closed system consisting of the boarder, the skateboard, and Earth. The boarder's initial kinetic energy is zero. As he accelerates during his run, he gains kinetic energy while the kinetic energy of Earth and the board remain zero. As he accelerates, he converts source energy (chemical energy extracted from food) to kinetic energy (he speeds up) and thermal energy (he gets warmer). This conversion is irreversible and therefore dissipative. There are no reversible changes in potential energy, and therefore I draw the energy bars shown in **Figure 7.33**.

Figure 7.33

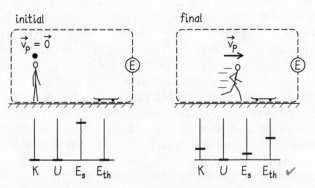

(*b*) The jump onto the skateboard is a totally inelastic collision. Initially the boarder moves while the skateboard is at rest; after he lands on the board, both move at the same velocity. During this "collision," there is no conversion of source energy and no change of potential energy. As in any other totally inelastic collision, however, part of the initial kinetic energy is dissipated to thermal energy between the boarder's feet and the board and between the wheels and the ground. My bar diagram is shown in **Figure 7.34**.

Figure 7.34

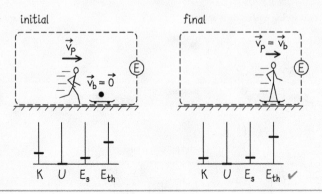

 7.17 How much energy is dissipated in the collision of Checkpoint 7.11?

Chapter Glossary

SI units of physical quantities are given in parentheses.

Coherent energy The sum of the coherent forms of energy (kinetic and potential energy), also called *mechanical energy*.

Dissipative interactions Interactions that involve changes in thermal energy. Such interactions are irreversible.

Elastic potential energy U(J) The potential energy associated with the reversible deformation of objects.

Field A way of describing the interaction of two objects that have a particular attribute, where each object produces in the space surrounding it a field that mediates the interaction.

Fundamental interaction An interaction that cannot (yet) be explained in terms of other interactions.

Gravitational potential energy U^G(J) The potential energy associated with the relative position of objects that are interacting gravitationally. Near Earth's surface, if the vertical coordinate of an object of inertia m is changed by an amount Δx, the gravitational potential energy changes by

$$\Delta U^G = mg\Delta x \quad \text{(near Earth's surface).} \quad (7.19)$$

Incoherent energy The internal energy associated with the incoherent motion and configuration of the parts of a system.

Interaction An event that produces physical change or a change in motion. An interaction that causes the interacting bodies to accelerate away from each other is *repulsive*. When the interacting bodies accelerate toward each other, the interaction is *attractive*. The ratio of the x components of the accelerations of two interacting objects is given by

$$\frac{a_{1x}}{a_{2x}} = -\frac{m_2}{m_1}. \quad (7.6)$$

Interaction range The distance over which the interaction between two objects is appreciable. The gravitational and electromagnetic interactions extend to infinite distances and are *long range*. The weak and strong nuclear interactions are *short range*.

Mechanical energy E_{mech}(J) The sum of the coherent forms of energy (kinetic and potential energy):

$$E_{\text{mech}} = K + U. \quad (7.9)$$

Nondissipative interactions Interactions that convert kinetic energy to potential energy, and vice versa. For these reversible interactions, the mechanical energy of a closed system does not change:

$$\Delta E_{\text{mech}} = \Delta K + \Delta U = 0 \quad \text{(nondissipative).} \quad (7.8)$$

Potential energy U(J) A coherent form of internal energy associated with reversible changes in the *configuration* of an object or system. Potential energy can be converted entirely to kinetic energy. The potential energy of a system of two interacting objects can always be written in the form

$$U = U(x), \quad (7.12)$$

where $U(x)$ is a unique function of a position variable x that quantifies the configuration of the system.

Source energy E_s(J) An incoherent form of internal energy associated with the random configuration of atoms. Source energy can be readily, but never entirely, converted to the two coherent forms of energy.

Thermal energy E_{th}(J) An incoherent form of internal energy associated with the random internal motion of the atoms that make up objects. Thermal energy cannot be entirely converted to coherent forms of energy.

8 Force

At this point in your study of physics, you can already solve a wide range of problems using just the conservation laws. Up to now we have applied these laws in only isolated or closed systems, however, and it is not always possible to identify such a system when working on problems. For this reason, you also need to know an approach to problem solving that is valid in systems that are interacting with their surroundings. If you look back at the material in earlier chapters, you will notice that many of the interactions we discussed involve pushing, pulling, or rubbing—all actions related to the everyday notion of *force*. So, in order to be able to deal with systems that are not isolated, you must know about the concept of force.

The term *force* is a familiar one—usually associated with the capacity to either move objects (for example, pushing a chair along the floor) or cause physical change (say, crushing an empty can). To which of the physical quantities I have defined so far does this notion of force best correspond, however, and how can we quantify it?

8.1 Momentum and force

To relate the intuitive concept of force to the quantities we have encountered so far, consider the following scenario. You are on a bicycle coming down a steep hill and approaching a sharp turn at 30 km/h. Suddenly you discover that your brakes are not working, leaving you no option but to slam into a concrete wall. The force of impact will be considerable, and if you were going faster—say, 40 km/h—it would be even greater. Conversely, if you were going at a mere 3 km/h, the impact might not be that bad. And it is not just your speed that matters. When you are carrying a backpack full of bricks and traveling at 30 km/h, the force of impact is greater than when you are traveling at the same speed but carrying nothing on your back.

Realizing that both your speed and your inertia govern the force of the impact you make with the wall, you may be tempted to believe that force is related solely to change in momentum, but things are not that simple. To see why, imagine that part of the concrete wall is padded with mattresses. Which part do you head for? You don't need any physics to realize that the mattresses offer you protection. When you hit a mattress, the force of impact is smaller than when you hit the concrete. In terms of momentum change and energy change, however, it doesn't make any difference which one you choose. In either case, your speed decreases by the same amount, 30 km/h, and so the decreases in your momentum and kinetic energy are the same no matter how you come to a stop.

The difference between hitting a mattress and hitting the concrete is that the mattress changes your momentum over a longer time interval. The momentum change therefore occurs at a slower rate. Evidently, the longer the interaction time interval, the smaller the force of impact. In other words, what determines the magnitude of the force of impact is not the absolute value of your momentum change but rather the rate at which this change happens. The faster your momentum changes, the larger the force of the impact on you.

This scenario illustrates two important points about forces. First, forces are manifestations of interactions—the force exerted by the mattress on you, for example, is one part of the interaction between you and the mattress (another part is the force exerted by you on the mattress). Second, for an object that is participating in one interaction only, we can quantitatively define **force:**

The force exerted on the object is the time rate of change in the object's momentum.

Because it is related to a change in momentum, force is a vector and so has direction as well as magnitude. The direction of any given force is the same as the direction of the momentum change the force causes. When you hit the wall in our bicycling example, for instance, the direction of the force exerted by the wall on you is opposite your direction of travel because the direction of your momentum change is opposite your direction of travel.

Even though the arguments leading up to the relationship between force and time rate of change in momentum all sound very plausible, I should warn you that the consequences of this relationship differ from your intuitive notions about forces in a number of ways.

✋ **8.1** Imagine pushing a crate in a straight line along a surface at a steady speed of 1 m/s. What is the time rate of change in the momentum of the crate?

The answer to this checkpoint should surprise you. After all, you *are* pushing the crate. So how can the time rate of change in the momentum of the crate be zero? The problem here is that the crate is interacting not just with you but also with the surface because there is friction between it and the surface. The directions and magnitudes of these two interactions (between the crate and your hands and between the crate and the surface) are such that their combined action causes no change in the momentum of the crate, and so it continues to move at constant speed in the direction of your push. In each interaction, there is a force exerted on the crate—exerted by you in one case and by the surface in the other case. The force exerted by your hands on the crate tends to move the crate forward, while the force exerted in the horizontal direction by the surface on the crate (the *force of friction* or *frictional force*) tends to resist the forward motion of the crate. If the two opposing forces exerted on the crate are equal in magnitude, their vector sum (sometimes called the *net force*) is zero. In other words, if the surface pushes on the crate just as hard as you push on it in the opposite direction, the acceleration effects of the

Figure 8.1 When you push a crate along a surface, two opposing forces are exerted on the crate: one by you (the push), the other by the surface (friction). If these forces are equal in magnitude, the crate moves at constant velocity.

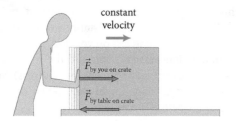

two opposing forces cancel and the crate moves at constant velocity (Figure 8.1). If the force exerted by the surface on the crate is smaller than the force you exert on it, the crate speeds up; if the force exerted by the surface is larger, the crate slows down.

For an object A interacting with more than one other object (such as a crate interacting both with you and with the surface), the resulting effect on A's momentum is due to the combined action of a number of forces, one for each interaction. Experimentally, we discover that forces obey the *principle of superposition*: The vector sum of the forces exerted on A is equal to the vector sum of the rate of change caused in A's momentum by the individual forces. Consequently:

> **The vector sum of all forces exerted on an object equals the time rate of change in the momentum of the object.**

For an object of fixed inertia, a change in momentum implies a change in velocity. So, when the vector sum of the forces exerted on an object is not zero, the object has a nonzero acceleration. The vector sum of the forces exerted on an object points in the same direction as the object's acceleration.

8.2 Imagine pushing on a crate initially at rest so that it begins to move along a floor. (*a*) While you are setting the crate in motion, increasing its speed in the desired direction of travel, what is the direction of the vector sum of the forces exerted on it? (*b*) Suppose you suddenly stop pushing and the crate slows to a stop. While the crate slows down, what is the direction of the vector sum of the forces exerted on it? (*c*) What is the direction of the vector sum of the forces once the crate comes to rest?

8.2 The reciprocity of forces

Another surprising aspect of the definition of force is that, because of the reciprocal nature of interactions, forces always come in pairs: When two objects interact, each exerts a force on the other. To see this, consider Figure 8.2, which illustrates two elastic collisions. In both collisions, cart 1 of inertia $m_1 = 0.12$ kg is initially at rest and cart 2 of inertia $m_2 = 0.24$ kg is initially moving at 0.60 m/s. For both

collisions, the final velocity of cart 1 is 0.80 m/s and that of cart 2 is 0.20 m/s. The only difference between the two collisions is that in Figure 8.2*a* the impact is softened by a spring, much like the cushioning of a fall by a mattress. As we did before, during the collision, we can ignore all interactions except that between the two carts.

8.3 (*a*) Verify that for both collisions in Figure 8.2 the momentum of the two-cart system remains constant. (*b*) Verify that both collisions are elastic.

As this checkpoint shows, in both collisions the momentum change of cart 1 is compensated for by an opposite change in the momentum of cart 2. Let's now relate these momentum changes to forces. For the soft collision, the changes in momentum take place during a time interval Δt_{soft} (about 10 ms). During this time interval, the *x* component of the momentum of cart 1 changes by an amount $\Delta p_{1x} = +0.096$ kg $\cdot$ m/s. Because force equals time rate of change in momentum, the *x* component of the average force exerted by cart 2 on cart 1 during the time interval Δt_{soft} is $(F_{by\ 2\ on\ 1})_x = \Delta p_{1x}/\Delta t_{soft} = (+0.096$ kg $\cdot$ m/s)/(0.010 s) = $+9.6$ kg $\cdot$ m/s^2.

Because the two-cart system is isolated, the change in the momentum of cart 2 must be equal in magnitude to that of cart 1 and in the opposite direction, and so $\Delta p_{2x} = -0.096$ kg $\cdot$ m/s. Because the time interval it takes for this change to occur is also Δt_{soft}, we calculate for the *x* component of the time rate of change in the momentum of cart 2, $(F_{by\ 1\ on\ 2})_x = \Delta p_{2x}/\Delta t_{soft} = -9.6$ kg $\cdot$ m/s^2. In words, the force exerted by cart 1 on cart 2 is equal in magnitude to the force exerted by cart 2 on cart 1 and in the opposite direction.

For the hard collision in Figure 8.2*b*, the momentum changes are the same as for the soft collision, but now the time interval during which these changes take place is much shorter, $\Delta t_{hard} \approx 1.0$ ms. As you might expect, the forces are much larger: For cart 1, we now have $(F_{by\ 2\ on\ 1})_x = \Delta p_{1x}/\Delta t_{hard} = (+0.096$ kg $\cdot$ m/s)/(0.0010 s) = $+96$ kg $\cdot$ m/s^2. The force exerted by cart 2 on cart 1 is again equal in magnitude to the force exerted by cart 1 on cart 2 and in the opposite direction because the two-cart system is isolated: $\Delta p_{1x} = -\Delta p_{2x}$. Now $(F_{by\ 1\ on\ 2})_x = \Delta p_{2x}/\Delta t_{hard} = -96$ kg $\cdot$ m/s^2. So we conclude:

> **Whenever two objects interact, they exert on each other forces that are equal in magnitude but opposite in direction.**

The pair of forces that two interacting objects exert on each other is called an **interaction pair.**

8.4 Does the conclusion just stated apply to inelastic collisions?

Figure 8.2 (a) Soft and (b) hard collisions between two carts on a low-friction track. The inertias are $m_1 = 0.12$ kg and $m_2 = 0.24$ kg.

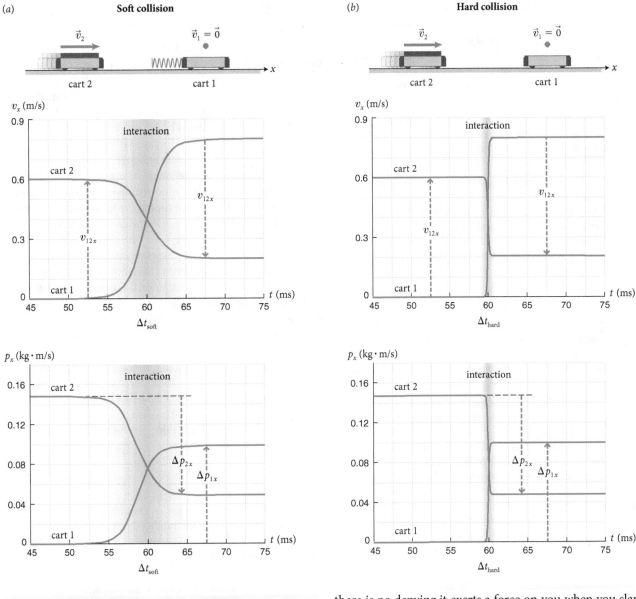

8.5 If you drop a book from a certain height, it falls (accelerating all the while) because of the gravitational force exerted by Earth on it. Because forces always come in interaction pairs, the book must exert a force on Earth. (a) How does the magnitude of $F_{\text{by book on Earth}}$ compare with the magnitude of $F_{\text{by Earth on book}}$? (b) Is Earth accelerating during the interaction? Why or why not?

The conclusion that interacting objects exert equal forces in opposite directions on each other is a direct result of the law of conservation of momentum and of our definition of force. This reciprocity of forces means, for instance, that if you push a crate along the floor, the crate pushes back on you with a force of equal magnitude. Can an inanimate object like a crate exert a force, and, if so, how do we know this force exists?

Let's settle the first question first, using the concrete wall of our earlier example. Even though the wall is inanimate,

there is no denying it exerts a force on you when you slam into it. I don't recommend trying it, but you can imagine feeling this force. So inanimate objects can exert forces.

Now for the second question: How do we know the crate exerts a force on you when you push it? To see that objects push back when you push on them, imagine leaning against a door when someone suddenly opens it away from you—you tumble into the door opening precisely because the force exerted by the door on you (a force that was keeping you upright) is suddenly removed. The force exerted by you on the door and the force exerted by the door on you form an *interaction pair*.

Here is another example: Imagine tipping over a chair by kicking it with your bare foot. The chair is tipped over by the force exerted by your foot on it. Your foot hurts because of the other half of the interaction pair—the force exerted by the chair on your foot.

Example 8.1 Splat!

A mosquito splatters onto the windshield of a moving bus and sticks to the glass. (*a*) During the collision, is the magnitude of the change in the mosquito's momentum smaller than, larger than, or equal to the magnitude of the change in the bus's momentum? (*b*) During the collision, is the magnitude of the average force exerted by the bus on the mosquito smaller than, larger than, or equal to the magnitude of the average force exerted by the mosquito on the bus?

❶ GETTING STARTED Because the mosquito sticks to the windshield, the bus and mosquito move together after impact and so this event is a totally inelastic collision. I therefore choose the mosquito plus the bus as my system.

❷ DEVISE PLAN During the collision, the system is isolated, and so its momentum doesn't change. I can use this fact to compare the magnitudes of the changes in momentum. The average forces exerted on the mosquito and the bus are each equal to the corresponding momentum change divided by the time interval over which that momentum change takes place. I also know that this time interval is the same for the mosquito and the bus.

❸ EXECUTE PLAN (*a*) Because the system's momentum is unchanged, I know that $\Delta \vec{p}_{mos} = -\Delta \vec{p}_{bus}$ and so $|\Delta \vec{p}_{mos}| = |\Delta \vec{p}_{bus}|$. ✔

(*b*) Because the changes in momentum are equal in magnitude and because the time interval over which the change takes place is the same for both objects, the magnitudes of the average forces must also be equal: $|\vec{F}_{by\ mos\ on\ bus,\ av}| = |\vec{F}_{by\ bus\ on\ mos,\ av}|$. ✔

❹ EVALUATE RESULT Although the answer to part *a* conforms to what I've seen before, the answer to part *b* is surprising. Intuitively I expect the magnitude of the average force exerted by the bus to be much larger than the magnitude of the average force exerted by the mosquito because the mosquito is crushed while nothing happens to the bus. However, I know that the force magnitude required to crush the mosquito is very small, and such a small force magnitude would have no effect on the bus. It therefore makes sense that the two forces are of equal magnitude.

✋ **8.6** In Example 8.1, the mosquito has an inertia of 0.1 g and is initially at rest, while the bus, with an inertia of 10,000 kg, has an initial speed of 25 m/s. The collision lasts 5 ms. (*a*) Calculate the final speeds of the mosquito and the bus. (*b*) What is the average acceleration of each during the collision? (*c*) By how much does the momentum of each change? (*d*) During the collision, how large is the average force exerted by the bus on the mosquito? (*e*) How large is the average force exerted by the mosquito on the bus?

8.3 Identifying forces

To identify the forces exerted on an object, it is convenient to distinguish between contact forces and field forces:

- **Contact forces** are forces that arise when objects physically touch each other. This category of forces includes forces due to pushing, pulling, and rubbing.
- **Field forces** are forces associated with what is called "action at a distance." In this case, the objects exerting forces

on each other do not need to be physically touching. For any object larger than atoms, gravitational and electromagnetic forces are the only field forces.

Identifying contact forces is easy: Whenever the surfaces of two objects touch, the objects exert contact forces on each other. Nearly all familiar forces fall into this category: the pull of a rope, the traction of a tire on a road, the force that keeps an airplane in the air, the push of a bumper, the hit of a hammer, the force that keeps you afloat when you are swimming, the drag caused by air resistance, and so on.

To determine which field forces to include in a given situation, gravitational and electromagnetic interactions must be considered. Gravitational interactions take place between any two objects. In our daily life, for instance, we each interact with Earth via the gravitational force, as does any other object—this book,* an apple, a house, and so on. The gravitational interaction we have with one another and with all other objects except Earth, however, is negligibly small. If you hold this book with your arms stretched out in front of you and let it go, it accelerates toward Earth, not toward you.

Because ordinary objects tend to be both electrically neutral and nonmagnetic, we usually do not need to consider field forces due to magnetic and electric interactions, unless a problem specifically mentions these types of interactions. We will treat electric and magnetic forces in detail in later chapters.

So, in most cases the only field force we need to consider is the gravitational force exerted by Earth. Only if a problem specifically mentions magnetic or electric interactions must we also take the field forces due to these interactions into account.

Exercise 8.2 What forces?

Identify all the forces exerted on the italicized object in each situation: (*a*) A *book* is lying on top of a magazine on a table. (*b*) A *ball* moves along a trajectory through the air. (*c*) A person is sitting on a *chair* on the floor of a room. (*d*) A *magnet* floats above another magnet that is lying on a table.

SOLUTION Because the interaction with the air is usually very small, I'm ignoring this interaction in this problem.

(*a*) Contact forces: The book is in contact with the magazine only, and so the magazine exerts a contact force on the book. Field forces: Earth exerts a gravitational force on the book. ✔

(*b*) Contact forces: The ball is not in contact with anything during its flight, and so there are no contact forces. Field forces: There is a gravitational interaction between Earth and the ball, so Earth exerts a gravitational force on the ball. ✔

(*c*) Contact forces: The chair is in contact with the floor and the person, and so both the floor and the person exert a force on the chair. Field forces: Earth exerts a gravitational force on the chair. ✔

---

*Hold a book with your arms stretched out in front of you and move it up and down. Do you feel the gravitational force exerted by Earth on the book? You can visualize this interaction as an invisible stretched spring reaching from the center of Earth to the book. Imagine stretching and relaxing this "gravitational spring" as you move the book up and down.

(*d*) Contact forces: The magnet floats, so it is not in contact with anything and thus there are no contact forces. Field forces: Both gravitational and magnetic forces occur in this problem: a gravitational force exerted by Earth on the magnet and a magnetic force exerted by the second magnet. ✔

8.7 A magnet lies on a table. You place a second magnet near the first one so that the two repel each other. Identify all the forces exerted on the first magnet.

8.4 Translational equilibrium

An object or system whose motion or state is not changing is said to be in **equilibrium.** For example, an object at rest or moving at constant velocity—in other words, its momentum is constant and its acceleration is zero—is said to be in **translational equilibrium.** According to the definition of force given in Section 8.1, constant momentum implies that the vector sum of the forces exerted on the object must be zero. Note that being in translational equilibrium doesn't necessarily mean that no forces are exerted on the object. It means only that all the force vectors add up to zero and therefore do not change the object's momentum.

As **Figure 8.3** illustrates, however, forces can also cause an object to rotate or to deform. Which combination of these effects occurs depends on where and how forces are exerted on the object. For now we consider only the acceleration caused by forces. Deformation is discussed in Chapters 7 and 8 and rotation in Chapter 11.

A book at rest on the floor is in translational equilibrium. So, too, is the crate pushed at constant velocity in Figure 8.1. In both of these cases, as with a coil spring squeezed between your hands, forces are exerted even though the object is in translational equilibrium.

> Whenever an object is at rest or moving at constant velocity, the vector sum of the forces exerted on the object is zero, and the object is said to be in translational equilibrium.

Figure 8.3 Forces can have various effects on an object.

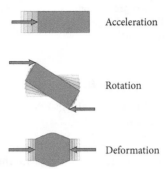

As an example of an object in translational equilibrium, consider a book lying on a table, as in **Figure 8.4a**. The book's momentum isn't changing, and therefore the vector sum of the forces exerted on it must be zero. What forces are exerted on the book? The only field force we need to consider is gravity. Even though the book is at rest, we know it is subject to gravity because if we suddenly remove the table from under it, the book accelerates downward (Figure 8.4b). Because the book is touching the table, we know there is a contact force exerted by the table on the book—the force that holds the book up and prevents it from falling through the table.

These are the only two forces exerted on the book. Because the vector sum of these forces must be zero, we conclude that the contact force exerted by the table on the book must be of the same magnitude as the gravitational force exerted by Earth on the book but in the opposite direction.

An object is also in translational equilibrium when it is moving at constant velocity. An example is the crate of Figure 8.1. Constant velocity means no change in momentum, and therefore the vector sum of the forces exerted on the crate must be zero. Considering only forces in the direction of motion, we determine two contact forces exerted on the crate: one by you pushing and one by the floor rubbing. In order for the vector sum of these forces to be zero, the frictional force exerted by the floor on the crate must be of the same magnitude as the force exerted by you on the crate but in the opposite direction.

Figure 8.4

(*a*) Book at rest on table (vector sum of forces exerted on book is zero)

(*b*) Book in free fall (vector sum of forces exerted on book is not zero)

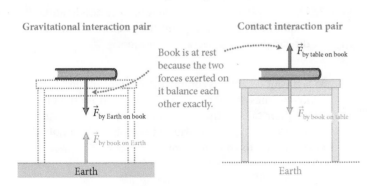

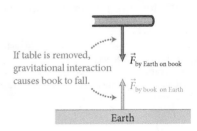

✋ **8.8** In Figure 8.4, are the contact force exerted by the table on the book and the gravitational force exerted by Earth on the book an interaction pair?

8.5 Free-body diagrams

Knowing about forces gives you a powerful tool for analyzing physical situations:

> If the vector sum of the forces exerted on an object is known, the time rate of change in the object's momentum is also known, and so the object's subsequent motion can be calculated.

Note the condition: To calculate the object's motion you need to know the vector sum of the forces exerted **on** the object only. The forces exerted **by** the object on its surroundings do not enter into the picture because they do not contribute to the change in the object's momentum. So, if you are considering an accelerating car and you need to calculate the time rate of change of its momentum, then you need to determine the vector sum of the forces exerted *on* the car (by the road, by the air, by gravity, and so on). The forces exerted *by* the car (on the road, on the air, . . .) determine what happens to those objects and not what happens to the car.

Whenever you are interested in one object in a collection of interacting objects, you must separate that one object and the forces exerted on it from the rest of the collection of objects and ignore the forces exerted by that object on anything else. To facilitate this separation, you should use what is called a **free-body diagram.** The procedure for drawing such diagrams is explained in "Drawing free-body diagrams." It is often also helpful to make a small sketch to identify the objects of interest. Be sure, however, to keep this "situation sketch" and your free-body diagram separate.

Until we begin to treat motion in more than one dimension in Chapter 10, we shall consider only the forces that are exerted in the direction of motion. For example, when

Procedure: Drawing free-body diagrams

1. Draw a center-of-mass symbol (a circle with a cross) to indicate the object you wish to consider*—this object is your system. Pretend the object is by itself in empty space (hence the name *free body*). If you need to consider more than one object in order to solve a problem, draw a separate free-body diagram for each.

2. List all the items in the object's environment that are in contact with the object. These are the items that exert *contact forces* on the object. *Do not add these items to your drawing*! If you do, you run the risk of confusing the forces exerted *on* the object with those exerted *by* the object.

3. Identify all the forces exerted *on* the object by objects in its environment. (For now, omit from consideration any force not exerted along the object's line of motion.) In general, you should consider (*a*) the *gravitational field force* exerted by Earth on the object and (*b*) the *contact force* exerted by each item listed in step 2.

4. Draw an arrow to represent each force identified in step 3. Point the arrow in the direction in which the force is exerted and place the tail at the center of mass. If possible, draw the lengths of the arrows so that they reflect the relative magnitudes of the forces. Finally, label each arrow in the form

$$\vec{F}^{\text{type}}_{\text{by on}},$$

where "type" is a single letter identifying the origin of the force (c for contact force, G for gravitational force), "by" is a single letter identifying the object exerting the force, and "on" is a single letter identifying the object subjected to that force (this object is the one represented by the center of mass you drew in step 1).

5. Verify that all forces you have drawn are exerted **on** and not **by** the object under consideration. Because the first letter of the subscript on $\vec{F}$ represents the object exerting the force and the second letter represents the object on which the force is exerted, every force label in your free-body diagram should have the same second letter in its subscript.

6. Draw a vector representing the object's acceleration *next to* the center of mass that represents the object. Check that the vector sum of your force vectors points in the direction of the acceleration. If you cannot make your forces add up to give you an acceleration in the correct direction, verify that you drew the correct forces in step 4. If the object is not accelerating (that is, if it is in translational equilibrium), write $\vec{a} = \vec{0}$ and make sure your force arrows add up to zero. If you do not know the direction of acceleration, choose a tentative direction for the acceleration.

7. Draw a reference axis. If the object is accelerating, it is often convenient to point the positive *x* axis in the direction of the object's acceleration.

When your diagram is complete it should contain only the center-of-mass symbol, the forces exerted *on* the object (with their tails at the center of mass), an axis, and an indication of the acceleration of the object. Do not add anything else to your diagram.

---

*Representing, say, a car by a single point might seem like an over-simplification, but because we are interested only in the motion of the car as a whole, any details (like its shape or orientation) do not matter. Any unnecessary details you add to your drawing only distract from the issue at hand.

pushing a cart on a horizontal track, you need only draw the forces along the horizontal direction. The forces along the vertical direction—an upward contact force exerted by the track on the cart and a downward gravitational force exerted by Earth on the cart—add up to zero and have no effect on the cart's motion.

The importance of free-body diagrams can hardly be exaggerated: Without a good free-body diagram, it is very difficult to analyze correctly any problem involving forces. In Section 4.4, I pointed out that separating a system from its environment is the most fundamental step in analyzing physical problems. Free-body diagrams help make this separation explicit. Let's therefore put the procedure into practice in a few examples.

Exercise 8.3 A book on the floor

Draw a free-body diagram for a book lying motionless on the floor.

SOLUTION I begin by drawing a small sketch of the situation. Next to it I draw a circle with a cross to represent the center of mass of the book—no line to represent the floor, just the center-of-mass symbol representing the book as if it were by itself in space (**Figure 8.5**). Of the objects in the book's environment, the only one I need to consider is the floor because that is the only thing in contact with the book.

Figure 8.5

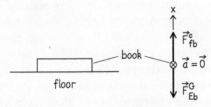

Next I name the forces exerted on the book. First there is a downward gravitational force exerted by Earth on the book. In addition, there is the one contact force exerted by the floor on the book. This force, which prevents the book from falling down through the floor, is directed upward.

Then I add arrows to the sketch to represent the forces I have identified. To represent the gravitational force, I draw an arrow pointing vertically downward with its tail at the center of mass. I label this force with a superscript G (for *gravity*) and the subscript Eb to indicate that it is a force exerted *by* Earth (E) *on* the book (b). For the contact force, I draw an upward-pointing arrow with its tail at the center of mass. I label this force with a superscript c (for *contact*) and the subscript fb to indicate that it is a force exerted *by* the floor (f) *on* the book (b). Finally, I verify that all forces in the diagram have subscripts ending in b because my diagram should contain only forces exerted *on* the book.

Because the book is at rest, its acceleration is zero, and therefore I write $\vec{a} = \vec{0}$ near the center of mass. From this information, I conclude that the vector sum of the forces exerted on the book (the sum of the two force vectors I have drawn) must be zero. I therefore adjust the lengths of the two arrows in the diagram so that they are of equal length, as shown in Figure 8.5. I complete the diagram by drawing an upward-pointing x axis. ✔

8.9 If Exercise 8.3 had asked about a book in free fall rather than one on the floor, what would the free-body diagram look like?

Exercise 8.4 Hanging around

Consider a person hanging motionless from a ring suspended from a cable, with the person's feet not touching the floor. Draw a free-body diagram for the ring.

SOLUTION I begin by drawing a sketch of the situation and then a center-of-mass symbol representing the ring (**Figure 8.6**). The things in the environment in contact with the ring are the cable and the person.

Figure 8.6

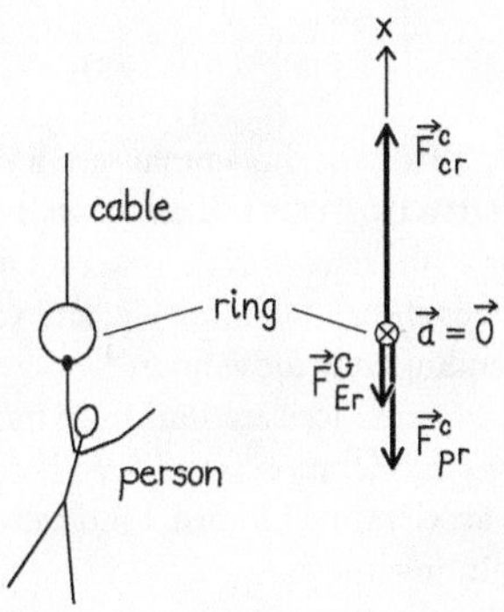

The next step is to identify all the forces exerted on the ring. As in the preceding example—and in *all* other examples concerning situations at or near Earth's surface—a vertically downward gravitational force is exerted by Earth (E) on the ring (r). I therefore draw an arrow from the center of mass down and label it $\vec{F}^G_{Er}$.

Next there are the contact forces: contact with the cable (c) and contact with the person (p). I represent the latter with a second arrow pointing downward. Because I don't know how this force compares in magnitude with the force of gravity, I choose its length arbitrarily. The other contact force is exerted by the cable on the ring. Because this force holds the ring up, I add an arrow pointing upward. Finally, I verify that the subscripts for each force end in r for ring because my diagram should contain only forces exerted *on* the ring.

Because the ring remains at rest, its acceleration is zero, and so I write $\vec{a} = \vec{0}$ near the center of mass. This tells me that the upward and downward forces exerted on the ring must add up to zero. I therefore adjust the length of the upward arrow so that it is equal to the sum of the lengths of the two downward arrows. I complete the diagram by adding an upward-pointing x axis. ✔

8.10 Draw a free-body diagram for the person in Exercise 8.4.

Exercise 8.5 Elevator woman

A woman stands in an elevator that is accelerating upward. Draw a free-body diagram for her.

SOLUTION After making a situation sketch, I draw a center-of-mass symbol to represent the woman (**Figure 8.7**). The only thing she is in contact with is the elevator floor—it is the floor that pushes her upward.

Figure 8.7

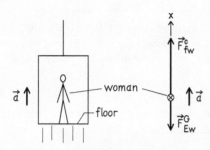

The forces exerted on the woman are a downward gravitational force exerted by Earth and an upward contact force exerted by the floor. I add arrows to the diagram to represent these forces, label them as shown in Figure 8.7, and verify that they all have subscripts ending in w for woman.*

What is the woman's acceleration? Even though she is at rest in the elevator, her acceleration is not zero because the elevator and she are both accelerating upward. I indicate this acceleration with an arrow pointing upward.

The upward acceleration means that the vector sum of the forces exerted on the woman is nonzero and directed upward. This can happen only if the upward force exerted by the elevator floor is larger than the gravitational force exerted by Earth. Therefore I adjust the lengths of the arrows as shown in Figure 8.7. Because the elevator has an upward acceleration, I add an upward-pointing x axis to my drawing. ✔

Warning: When drawing a free-body diagram, you may be tempted to add an arrow to show the vector sum of the forces (the "net force"), but this is not good practice. First, by drawing a vector showing the acceleration of the object, you have already indicated the direction of the vector sum of the forces. Second, this vector sum is the result of all the forces you have already drawn, not a separate force exerted by something on the object in your free-body diagram.

🖐 **8.11** You throw a ball straight up. Draw a free-body diagram for the ball (a) while it is still touching your hand and is accelerating upward, (b) at its highest point, and (c) on the way back down.

---

*If all the subscripts in a free-body diagram do not end in the same letter, do not merely change the subscripts! Repeat the identification of the forces and ask yourself if each force you drew points in the correct direction. For example, does the floor push up on the woman? Or did you draw the force that her feet exert *on the elevator floor*? Correct your diagram if necessary.

8.6 Springs and tension

To better understand the origin of contact forces like the upward push of a floor or the pull of a rope, let us examine the behavior of a spring. When a spring is neither stretched nor compressed, we say it is at its *relaxed length*. If you position a spring vertically and put a brick on top of it, as shown in **Figure 8.8a**, the brick *compresses* the spring and comes to rest in a position below the relaxed length of the spring. Figure 8.8 also shows the free-body diagram for the brick, which is subject to a downward force of gravity and an upward contact force—a *push*—exerted by the spring. Because the brick is at rest, the vector sum of the forces exerted on it must be zero. So the push by the spring and the gravitational force exerted by Earth are equal in magnitude and point in opposite directions. In general, whenever a spring holds an object in place, the spring must exert on the object an upward force, called a *support force*, that is equal in magnitude to the downward gravitational force exerted by Earth on the object.

Similarly, if we suspend the spring from a ceiling and fasten the brick at the bottom, the spring stretches, exerting on the brick an upward force—a *pull*—that counteracts the downward force of gravity (Figure 8.8b). The free-body diagram for this situation is identical to the one for the compressed spring; in both cases there is an upward force exerted by the spring on the brick. For the purpose of drawing a free-body diagram, it makes no difference at all whether it is a push or a pull that keeps the brick in place.

The force exerted by the spring always tends to return the spring to its relaxed length. When the spring is stretched, it exerts a pull (back toward the spring); when it is compressed, it exerts a push (away from the spring). Forces exerted by springs are therefore called *restoring forces*.

Now imagine adding additional identical bricks on top of the one shown in Figure 8.8a. Each additional brick increases the force exerted on the spring and further compresses it. If we repeat this experiment with the spring suspended from a ceiling, the spring stretches increasingly as the force exerted on the spring increases.

Over a certain range, called the *elastic range*, the compression or stretching of the spring is reversible; that is, when the force causing the compression or stretching is removed,

Figure 8.8 The free-body diagram for the brick is the same whether the spring (a) pushes or (b) pulls on it.

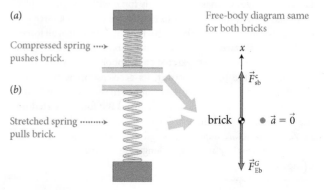

Figure 8.9 A soft spring exerts the same force on a given load as a stiff spring or a post or a rope, even though it deforms much more. In fact, the post and rope behave like very stiff springs, compressing and stretching very slightly in response to the load.

(*a*) Support pushes on brick

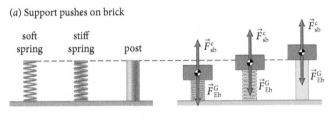

(*b*) Support pulls on brick

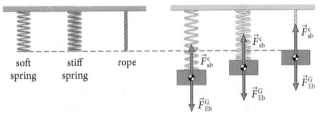

the spring returns to its original shape.* If stretched beyond a certain point, called the *elastic limit*, the spring no longer reverts to its original length when the load is removed. Additional load stretches the spring further until it eventually reaches the breaking point.

The amount of compression or stretching of a spring depends not only on the force exerted on the spring but also on the *stiffness* of the spring. A soft spring compresses or stretches more than a stiffer spring does under the same load (that is, when the same force is exerted on it). As **Figure 8.9** shows, however, the soft and stiff springs *exert exactly the same support force on the load.*

Now imagine putting a brick on a very stiff spring—one meant to support, say, a truck. You are unlikely to be able to see the spring compress (although you can measure the compression), but as with any other spring, compression is necessary in order to generate the upward force that holds the brick in place. With a very stiff spring, you need only a tiny compression to generate this support force. Even a solid object compresses a tiny bit when it supports another

---

*As you might have noticed, the word *elastic* is associated with reversible phenomena: Elastic collisions are reversible, elastic potential energy is associated with reversible deformations, and so on.

object, and as with a spring, the more load you pile on top of a solid object, the more it compresses in order to generate the force necessary to support its load.

Everything around you behaves this way: the chair you sit on, the bed you lie on, the floor you walk on, the road you drive on, and so on. Like a soft spring, an upholstered chair compresses noticeably when you sit on it. Sit on a table, though, and it "gives" a lot less. Even a concrete floor or a steel chair must compress ever so slightly to support you. The upholstered chair, the table, the concrete floor, and the steel chair all exert on you a support force that is equal in magnitude to the force of gravity exerted on you.

The same arguments can be applied to stretched objects (Figure 8.9*b*). In particular, the pull of a rope is like the pull of a stretched spring because the rope, like a very stiff spring, stretches ever so slightly while providing the support force that holds the object in place. As long as the deformation is reversible, the force exerted by a compressed or stretched material is called an **elastic force.**

Let us now turn our attention to the forces exerted *on* the object being compressed or stretched. **Figure 8.10a** shows a free-body diagram for a spring suspended from a ceiling and stretched by a brick. In addition to the two contact forces exerted by the ceiling and the brick on the spring, the spring is subject to a downward gravitational force. Because the spring is at rest, the upward force must be equal in magnitude to the sum of the downward forces exerted by gravity and the brick.

8.12 In Figure 8.10*a*, how does the magnitude of the downward force exerted *by the spring on the ceiling* compare with the magnitudes of the downward gravitational forces exerted by Earth on the spring and on the brick?

Be sure to check your answer to Checkpoint 8.12 before proceeding!

Figures 8.10*b* and 8.10*c* show free-body diagrams for a rope and a thread stretched by the same brick as in Figure 8.10*a*. The lower the inertia of the object being stretched, the smaller the force of gravity exerted on it. For objects of very low inertia, the magnitude of the gravitational force approaches zero, and the magnitudes of the two contact forces (exerted by the ceiling and the brick) become equal, as Figure 8.10*c* shows.

Because of the reciprocity of forces, the force exerted *by the thread on the ceiling* (not shown in the diagram) is equal

CONCEPTS

Figure 8.10 Free-body diagrams for a spring, rope, and thread that suspend a brick from the ceiling.

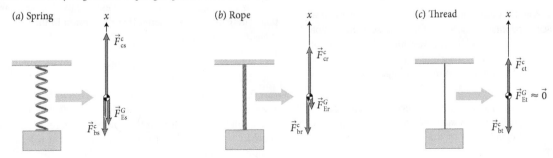

Figure 8.11 Transmission of a force by a rope to a box. Free-body diagrams for (*a*) the rope and (*b*) the box. (We assume that the force of gravity on the rope is much smaller than the forces exerted by the person and the box, so we can ignore it.)

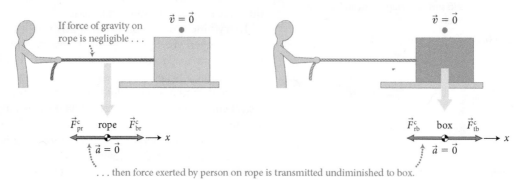

in magnitude to the force exerted by the brick on the thread. Because these two forces point in the same direction, the thread simply transmits the pulling force of the brick to the ceiling.

Of course, we don't usually use thread to pull on objects, but in situations in which a rope—or even a lightweight chain—is used to pull, we can generally ignore the gravitational force exerted on the rope or chain. The only requirement is that the pulling force be much greater than the gravitational force exerted on the rope or chain. So, if we ignore the gravitational force, the same transmission of forces described for our thread is seen when we use a rope. If you fasten a rope to a boat and pull on one end of the rope, the rope stretches a little and the opposite end of the rope transmits your force to the boat. In general, if the force of gravity exerted on any object being stretched—rope, spring, thread—is much smaller than the forces that cause the stretching, you can ignore the force of gravity. This is usually true when ropes, springs, or threads are used to pull or push on objects. Therefore:

> **The force exerted on one end of a rope, spring, or thread is transmitted undiminished to the other end, provided the force of gravity on the rope, spring, or thread is much smaller than the forces that cause the stretching.**

A rope being used to pull an object is subject to two equal outward forces, one on each end (**Figure 8.11**). The stress caused by this pair of forces is called the **tension** in the rope, and the forces causing the tension are called *tensile forces*. Tension is a scalar and is represented by $\mathcal{T}$. In an object subject to two equal tensile forces of magnitude F, one on each end, the tension is $\mathcal{T} = F$.

Example 8.6 Tug of war

If two people, A and B, pull on opposite ends of a rope that is at rest, each exerting a horizontal tensile force of magnitude F, the tension in the rope is $\mathcal{T} = F$. Suppose instead that one end of the rope is tied to a tree and A pulls on the other end by himself with the same force magnitude F. Is the tension in the rope larger than, equal to, or smaller than the tension when A and B pull on opposite ends?

❶ **GETTING STARTED** To determine the tension in each case, I need to know the magnitude of the tensile force exerted on each end of the rope. I begin by making a sketch of the two situations (**Figure 8.12**).

Figure 8.12

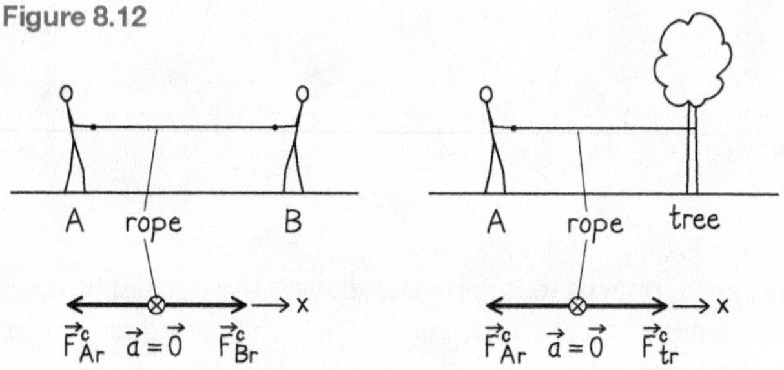

❷ **DEVISE PLAN** Because I must analyze various forces exerted on the rope, it is best to begin with a free-body diagram for the rope. In a tug of war only the horizontal forces matter, so I ignore the effect of the vertical gravitational force exerted by Earth and consider only the horizontal forces (Figure 8.12). Because the rope is always at rest, I know that the vector sum of the forces exerted on the rope must be zero. The magnitudes of the tensile forces exerted on either end are therefore equal to each other, and the tension in the rope is equal to the magnitude of the tensile force exerted on either end.

❸ **EXECUTE PLAN** The rope tied to the tree is subject to two contact forces, one exerted by A and the other exerted by the tree. Because the vector sum of the forces exerted on the rope must be zero, I make the two force arrows of equal length, as shown in my free-body diagram.

Because A pulls on the rope with a force of magnitude F, the tree must pull with a force of magnitude F in the opposite direction, and so the tension in the rope is $\mathcal{T} = F$. which is the same as when A and B pull on opposite ends. ✔

❹ **EVALUATE RESULT** The magnitude of the force exerted by A is F in both cases. Because the rope has zero acceleration in both cases, the magnitude of the force exerted by the tree must be the same as the magnitude of the force exerted by B when A and B pull. Therefore it makes sense that the tension in the rope is the same in the two cases.

✋ **8.13** In Example 8.6, suppose both people pull on the same end of the rope, each exerting a force F, while the other end is still tied to the tree. Is the tension in the rope larger than, equal to, or smaller than the tension when the two people pull on opposite ends?

Self-quiz

1. (*a*) Two forces, one twice the magnitude of the other, are exerted on identical objects during the same time interval. Which force causes the greater change in momentum? (*b*) Two identical forces are exerted on identical objects, one during a time interval twice as long as the other. Which force causes the greater change in momentum?

2. The free-body diagram of a box subjected to forces is shown in **Figure 8.13**. Because we are concerned about only the horizontal motion, the forces in the vertical direction have been omitted. (*a*) What is the direction of the change in the momentum of the box? (*b*) If the box is at rest before the forces shown in the diagram are exerted on it, what is the direction of its momentum once the forces are exerted?

Figure 8.13

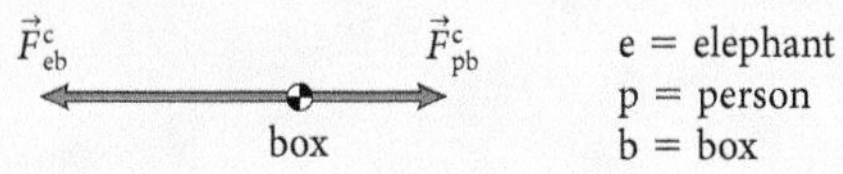

$\vec{F}_{eb}^c$ $\vec{F}_{pb}^c$

box

e = elephant
p = person
b = box

3. For the free-body diagram in question 2, list the objects in the environment of the box.

4. A block of wood rests on a shelf, and a second block of wood rests on top of the first. (*a*) List the objects in the environment of the top block. (*b*) List the objects in the environment of the bottom block. (*c*) Draw a free-body diagram for each block.

5. Complete the following sentences, applying the reciprocity of forces. (Example: If I push down on the seat of my chair, *the seat of my chair pushes up on me.*)

If I push to the right on the wall, . . .
If I pull to the right on a spring, . . .
If the floor pushes up on my feet, . . .
If Earth pulls down on me, . . .

Answers

1. (*a*) The larger force causes the greater change in momentum because force and change in momentum are proportional to each other over a given time interval. (*b*) The force exerted over the longer time interval causes the greater change in momentum because time interval and change in momentum are proportional to each other for a constant force.

2. (*a*) The vector sum of the forces exerted on the box points to the left because the force exerted on the box by the elephant is larger than the force exerted on the box by the person. The change in momentum is to the left, in the same direction as the vector sum of the forces. (*b*) The change in momentum is equal to the final momentum minus the initial momentum of zero. Therefore, the final momentum of the box must be to the left.

3. The objects in the environment of the box are an elephant and a person. If you also consider the vertical direction, you need to include the ground and Earth as being in the environment of the box.

4. (*a*) The objects in the environment of the top block are the bottom block and Earth. The shelf is not in contact with the top block and so is not part of the top block's environment. (*b*) The objects in the environment of the bottom block are the shelf, the top block, and Earth. (*c*) See **Figure 8.14**. Earth pulls downward on each block. The surface below each block pushes upward. If the bottom block pushes upward on the top block, the top block must push down on the bottom block.

Figure 8.14

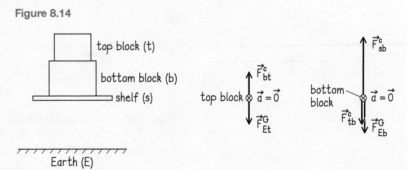

top block (t)
bottom block (b)
shelf (s)
Earth (E)

5. For each, swap the nouns and reverse the given direction.

If I push to the right on the wall, *the wall pushes to the left on me.*
If I pull to the right on a spring, *the spring pulls to the left on me.*
If the floor pushes up on my feet, *my feet push down on the floor.*
If Earth pulls down on me, *I pull up on Earth.**

---

*Many people find this last statement difficult to accept and are inclined to answer: If Earth pulls down on me, the ground pushes up on me. This statement, however, includes three objects (ground, Earth, me), not just the two required by the reciprocity of forces.

8.7 Equation of Motion

In the first part of this chapter, we determined that the vector sum of the forces exerted on an object is equal to the time rate of change in the object's momentum. If an object's momentum changes by an amount $\Delta\vec{p}_{object}$ in a time interval Δt, the *average* time rate of change in the object's momentum is

$$\frac{\Delta\vec{p}_{object}}{\Delta t}. \tag{8.1}$$

The *instantaneous* time rate of change in momentum is obtained by letting the time interval approach zero:

$$\lim_{\Delta t \to 0}\frac{\Delta\vec{p}_{object}}{\Delta t} = \frac{d\vec{p}_{object}}{dt} \equiv \Sigma\vec{F}_{object}, \tag{8.2}$$

where $\Sigma\vec{F}_{object}$ represents the vector sum of all the forces exerted on the object:*

$$\Sigma\vec{F}_{object} = \vec{F}_{1\,object} + \vec{F}_{2\,object} + \vec{F}_{3\,object} + \cdots, \tag{8.3}$$

where $\vec{F}_{1\,object}$ is the force exerted by object 1 on the object under consideration, and so on. Because we are concerned with only one object, we can omit the subscript on the left side of Eq. 8.3, and so Eq. 8.2 becomes

$$\Sigma\vec{F} \equiv \frac{d\vec{p}}{dt} \tag{8.4}$$

The vector sum of the forces points in the same direction as the change in momentum of the object.

Substituting $\vec{p} \equiv m\vec{v}$ in the term on the right in Eq. 8.4 gives

$$\frac{d\vec{p}}{dt} \equiv \frac{d(m\vec{v})}{dt} = m\frac{d\vec{v}}{dt}, \tag{8.5}$$

where I have used the fact that the inertia m of the object is constant. Because $d\vec{v}/dt$ is the acceleration $\vec{a}$ of the object, substituting Eq. 8.5 into Eq. 8.4 yields

$$\Sigma\vec{F} = m\vec{a} \tag{8.6}$$

or

$$\vec{a} = \frac{\Sigma\vec{F}}{m}. \tag{8.7}$$

In terms of components, Eqs. 8.6 and 8.7 become

$$\Sigma F_x = ma_x \quad \text{and} \quad a_x = \frac{\Sigma F_x}{m}. \tag{8.8}$$

Equation 8.7 is called the **equation of motion** for the object. For a given vector sum of forces, this equation allows us to obtain the acceleration and therefore to determine the motion of the object. Conversely, if we know the acceleration of an object, Eq. 8.6 allows us to determine the vector sum of the forces exerted on that object.

Because the SI unit of momentum is $kg \cdot m/s$ and the SI unit of time is seconds, the SI unit of force is $kg \cdot m/s^2$. This derived SI unit is given the special name **newton** (N) in honor of the English scientist Isaac Newton:

$$1\,N \equiv 1\,kg \cdot m/s^2. \tag{8.9}$$

---

*We drop the "by" subscript because the sum represents the forces exerted by *all* objects on the object under consideration.

Figure 8.15 Superposition of forces.

Effect of individual forces

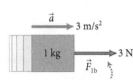

Exerted separately, two forces cause corresponding accelerations.

Effect of multiple forces

When exerted together, the two forces cause the same acceleration . . .

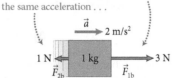

. . . as a single force equal to their vector sum.

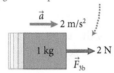

This definition means that a force of 1 N exerted on an object that has an inertia of 1 kg causes that object to accelerate at 1 m/s^2.

Note that while several forces can be exerted simultaneously on a single object, that object can have only one acceleration. For example, if a 1-kg object is simultaneously subject to two forces—say, a 3-N force directed to the right and a 1-N force directed to the left—the object accelerates at 2 m/s^2 to the right, just as if a single force of 2 N directed to the right had been exerted (see Figure 8.15). The reason forces can be added vectorially is that the effect of a force is a change in momentum, and momentum is a vector quantity. This property of adding forces vectorially is called the *superposition of forces*.

Example 8.7 Elevator stool

A person is sitting on a stool in an elevator. The forces exerted on the stool are a downward force of magnitude 60 N exerted by Earth, a downward force of magnitude 780 N exerted by the person, and an upward force of magnitude 850 N exerted by the elevator floor. If the inertia of the stool is 5.0 kg, what is the acceleration of the elevator?

❶ **GETTING STARTED** I know all the forces that are exerted on the stool, and I must determine its acceleration. I therefore begin by drawing a free-body diagram for the stool that includes the three forces mentioned in the problem statement (**Figure 8.16**). I arbitrarily choose an x axis pointing upward.

Figure 8.16

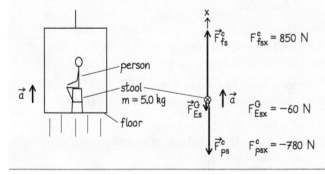

❷ **DEVISE PLAN** Equation 8.8 gives a relationship between the vector sum of the forces exerted on an object and the acceleration of the object. By substituting the forces given and the inertia of the stool, I can solve for a_x.

❸ **EXECUTE PLAN** From Eq. 8.8, I have

$$a_x = \frac{\sum F_x}{m} = \frac{1}{m}(F^c_{\text{fs}x} + F^c_{\text{ps}x} + F^G_{\text{Es}x})$$

$$= \frac{1}{5.0 \text{ kg}}[(+850 \text{ N}) + (-780 \text{ N}) + (-60 \text{ N})]$$

$$= +2.0 \text{ m/s}^2. ✔$$

❹ **EVALUATE RESULT** The acceleration I obtained is positive, which means that the elevator has an upward acceleration, which is what I expect because the force exerted by the floor on the stool is upward. The magnitude of the acceleration is about one-fifth the acceleration due to gravity, which strikes me as a reasonable value.

🖐 **8.14** (*a*) You exert a constant force of 200 N on a friend on roller skates. If she starts from rest, estimate how far she moves in 2.0 s. (*b*) When a person jumps off a wall, what is the magnitude of his acceleration? (*c*) Estimate the magnitude of the force exerted by Earth on the person during the jump in part *b*.

As we saw in Section 8.2, forces always come in interaction pairs. To see this quantitatively, consider an isolated system of two colliding carts. As we have seen before, the momentum of the two-cart system is constant, and so the changes in momentum caused by the collision must obey the relationship

$$\Delta \vec{p}_1 + \Delta \vec{p}_2 = \vec{0}, \tag{8.10}$$

where $\Delta \vec{p}_1$ is the change in momentum of cart 1 and $\Delta \vec{p}_2$ is that of cart 2. During an infinitesimal time interval dt, this relationship yields

$$d\vec{p}_1 + d\vec{p}_2 = \vec{0}. \tag{8.11}$$

Because dt is the same for both carts, the time rates of change in momentum must also add to zero:

$$\frac{d\vec{p}_1}{dt} + \frac{d\vec{p}_2}{dt} = \vec{0}. \tag{8.12}$$

According to Eq. 8.4, the time rate of change of momentum of an object is equal to the vector sum of the forces exerted on that object, and so

$$\Sigma \vec{F}_1 + \Sigma \vec{F}_2 = \vec{0}. \tag{8.13}$$

Because the carts are constrained to move in the horizontal direction, the forces in the vertical direction cancel, and so the only forces exerted on the carts we need to consider are those they exert on each other. The vector sum of the forces exerted on cart 1 is thus just the force exerted by cart 2 on cart 1, and the vector sum of the forces exerted on cart 2 is just the force exerted by cart 1 on cart 2:

$$\Sigma \vec{F}_1 = \vec{F}_{21} \quad \text{and} \quad \Sigma \vec{F}_2 = \vec{F}_{12}, \tag{8.14}$$

and thus $\vec{F}_{12} + \vec{F}_{21} = \vec{0}$, or

$$\vec{F}_{12} = -\vec{F}_{21}. \tag{8.15}$$

Newton's laws of motion

In the modern view, the conservation laws are the heart and soul of physics—they best represent the generally held belief that the universe has an underlying simplicity. In the 17th century, however, 200 years before the concept of energy and the conservation laws were formulated, the English scientist Isaac Newton published a work describing his ideas concerning the fundamental principles that govern forces and their effects on the motion of objects. Newton's three laws of motion, as they are known, have been the backbone of physics for more than two centuries.

We have already discussed each of these three laws—albeit under different, and in the current context more appropriate, names.

Newton's first law of motion, first formulated by Galileo Galilei, is what we called the *law of inertia* in Chapter 6:

In an inertial reference frame, any isolated object that is at rest remains at rest, and any isolated object that is in motion keeps moving at a constant velocity.

Newton's second law of motion corresponds to the *definition of force* given in Eq. 8.4:

The vector sum of the forces exerted on an object is equal to the time rate of change in the momentum of that object.

Equation 8.6, $\Sigma \vec{F} = m\vec{a}$, is frequently referred to as Newton's second law even though it holds only when m is constant.

Newton's third law of motion expresses the law of conservation of momentum in terms of forces (Eq. 8.15):

Whenever two objects interact, they exert on each other forces that are equal in magnitude but opposite in direction.

Newton is widely regarded as one of the greatest physicists of all time, not only for his formulation of the laws of motion but also for his law of gravity, for his development of the calculus, and for laying the foundation of modern optics.

QUANTITATIVE TOOLS

In words, whenever two objects interact, they exert forces on each other that are equal in magnitude and opposite in direction. The forces $\vec{F}_{12}$ and $\vec{F}_{21}$ form an **interaction pair.** Because interactions always involve two objects, every force that is exerted *on* an object has associated with it a force of equal magnitude in the opposite direction exerted *by* the object. Note, in Eq. 8.15, that the subscripts on the forces that form an interaction pair are interchanged. For that reason, both forces in an interaction pair can never be part of the same free-body diagram. Equation 8.15 is called **Newton's third law of motion** (see the box "Newton's laws of motion" on page 190); it is a direct consequence of the conservation of momentum expressed in Eq. 8.10.

8.15 If forces always come in interaction pairs and the forces in such a pair are equal in magnitude and opposite in direction (Eq. 8.15), how can the vector sum of the forces exerted on an object ever be nonzero?

8.8 Force of gravity

We have seen before that, when we ignore air resistance, all objects dropped from a certain height near Earth's surface have a downward acceleration given by $a_x = -g$ when we define up as being the positive x direction (see Chapter 3). For an object in free fall near Earth's surface, the only force exerted on the object is the gravitational force $\vec{F}_{Eo}^G$ exerted by Earth on the object. The left side of Eq. 8.8 is thus

$$\sum F_x = F_{Eox}^G. \tag{8.16}$$

The right side is $ma_x = -mg$, where m is the inertia of the object, and so the x component of the gravitational force exerted by Earth on the object is

$$F_{Eox}^G = -mg \text{ (near Earth's surface, } x \text{ axis vertically upward).} \tag{8.17}$$

The minus sign reflects the fact that the acceleration due to gravity (and therefore also the force of gravity) is directed downward and the positive x direction points upward. Because the acceleration due to gravity is the same for all objects, Eq. 8.17 holds for all objects in free fall near Earth's surface. Note that when additional forces are exerted on the object, the x component of the gravitational force exerted by Earth on the object is still given by Eq. 8.17, even though the x component of the object's acceleration is no longer equal to $-g$.

Example 8.8 Tennis ball launch

A tennis ball of inertia 0.20 kg is launched straight up in the air by hitting it with a racquet. If the magnitude of the acceleration of the ball while it is in contact with the racquet is $9g$, what are the magnitude and direction of the force exerted by the racquet on the ball?

❶ GETTING STARTED I am given the inertia and acceleration magnitude for a tennis ball hit by a racquet, and I must calculate the force—both magnitude and direction—responsible for this acceleration. Because this problem deals with forces exerted on the ball, I begin by drawing a situation sketch and a free-body diagram for the ball (Figure 8.17). While it is being hit, the ball is in contact with the racquet and so I include an upward contact force exerted by the racquet as well as a downward force of gravity exerted by Earth. Because the ball accelerates upward, I make the upward force vector longer than the downward force vector and point the positive x axis upward.

Figure 8.17

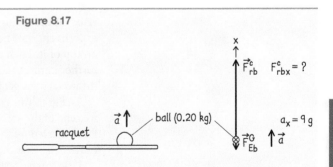

❷ DEVISE PLAN Equation 8.8 relates the ball's acceleration to the vector sum of the two forces exerted on it. I can use Eq. 8.17 to calculate one of the two forces and then solve for the other one in Eq. 8.8.

(Continued)

❸ **EXECUTE PLAN** Equation 8.8 yields

$$\Sigma F_x = F_{\mathrm{Eb}x}^{G} + F_{\mathrm{rb}x}^{c} = ma_x.$$

Because I've defined upward as the positive x direction, $a_x = +9g$. Substituting this value and Eq. 8.17 for the gravitational force exerted by Earth on the object, I get

$$(-mg) + F_{\mathrm{rb}x}^{c} = +9mg$$

or $\quad F_{\mathrm{rb}x}^{c} = +10mg = +10(0.20\text{ kg})(9.8\text{ m/s}^2)$

$$= +20\text{ N. } ✔$$

❹ **EVALUATE RESULT** The plus sign indicates that this force is directed upward, as expected. The magnitude of the force is ten times larger than the gravitational force exerted by Earth, which also makes sense if the ball is to be accelerated upward during the very short time interval that the racquet is in contact with the ball.

✋❂ **8.16** The magnitude of the gravitational force exerted by Earth on an object of inertia m_1 is m_1g. (*a*) What is the magnitude of the force exerted by the object on Earth (inertia m_E)? (*b*) What is the acceleration of Earth due to its gravitational interaction with the object?

Equation 8.17 shows that the magnitude of the force of gravity exerted on an object is proportional to the object's inertia, which is an intrinsic property of the object that has nothing to do with the presence of Earth or with gravity. This surprising conclusion is a direct consequence of the fact that, when air resistance is negligible, all objects experience the same acceleration due to gravity. It is more difficult to accelerate a large boulder than a feather, but the gravitational forces exerted by Earth on the boulder and on the feather are such that both have the same acceleration when they fall (ignoring air resistance). Equation 8.17 doesn't tell us much about the mechanism of gravity; it merely tells us what magnitude and direction the force of gravity must have to account properly for the free fall of objects near the surface of Earth. In Chapter 13, we'll return to the subject of gravity and its relationship to inertia. In the meantime, we can use Eq. 8.17 to calculate the magnitude of the force of gravity on any object near the surface of Earth.

✋❂ **8.17** Suppose you are in an elevator that is accelerating upward at 1 m/s^2. (*a*) Draw a free-body diagram for your body. (*b*) Determine the magnitude of the force exerted by the elevator floor on you.

8.9 Hooke's law

Now that we have established the relationship between the inertia of an object and the gravitational force exerted on it (Eq. 8.17), we can analyze the behavior of springs quantitatively. We begin by attaching bricks to the free end of a spring that is suspended from a ceiling. Each additional brick increases the force exerted on the spring and the amount by which the spring stretches. We repeat this experiment, but now we support the spring on a solid surface and stack bricks on top of it. Each additional brick increases the force exerted on the spring and further compresses it, up to the point where the spring "bottoms out"—which means its coils get in the way of further compression.

The graph in **Figure. 8.18** shows the x component of the displacement of the free end of the spring from its relaxed position at x_0 versus the x component of the force exerted *on* the spring *by* the bricks. The shaded region indicates the elastic range. The first thing to note is that, within the elastic range, the displacement and the force exerted by the load on the spring are linearly proportional to each other. The second thing to note is that the force exerted on the spring is always in the same direction as the displacement. We can express this quantitatively as

$$(F_{\text{by load on spring}})_x = k(x - x_0) \quad \text{(small displacement),} \tag{8.18}$$

where k is called the **spring constant**. The spring constant represents the inverse of the slope of the straight line through the data points in Figure 8.18; it tells us

Figure 8.18 The x component of the displacement of the free end of the spring from its relaxed position is proportional to the x component of the force causing the displacement.

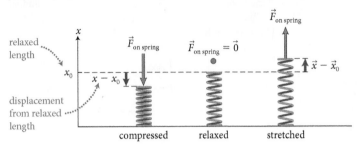

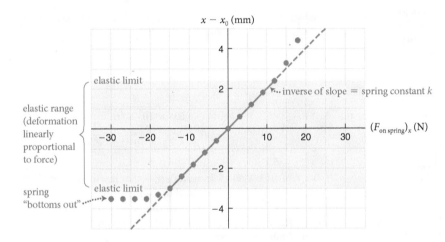

how large a force is required to give the free end of the spring a displacement $x - x_0$. The spring constant is therefore a quantitative measure of the stiffness of the spring discussed in Section 8.6. It is always positive, but different springs have different values of k, depending on how they respond to compression and stretching. The SI unit of the spring constant is N/m.

Often we are interested not in the force exerted *on* the spring, but in the force exerted *by* the spring. We know from Eq. 8.15 that

$$\vec{F}_{\text{by spring on load}} = -\vec{F}_{\text{by load on spring}} \tag{8.19}$$

and so the force exerted by the spring is given by

$$(F_{\text{by spring on load}})_x = -k(x - x_0) \quad \text{(small displacement).} \tag{8.20}$$

Note that the direction of the force of the spring on the load is opposite to the direction of the displacement. A stretched spring pulls back. A compressed spring pushes out. Equation 8.18 (or its equivalent, Eq. 8.20) is called **Hooke's law** after the British physicist Robert Hooke (1635–1703), who first discovered the linear relationship between a spring's stretching or compression and the force exerted on it.

Unlike the conservation principles, which are of universal validity, Eq. 8.20 is valid for only small displacement. As Figure 8.18 shows, Hooke's "law" fails if we compress or stretch the spring beyond certain limits. Still, Hooke's law is of great use because, for small deformations, many materials—from steel to foam—show a linear relationship between force and displacement when compressed or stretched. For this reason, the force that tends to return a stretched or compressed spring or material to its relaxed position is referred to as a *linear restoring force*.

Example 8.9 Spring compression

A book of inertia 1.2 kg is placed on top of the spring in Figure 8.18. What is the displacement of the top end of the spring from the relaxed position when the book is at rest on top of the spring?

❶ GETTING STARTED My task is to determine how much the top end of a spring is displaced when a 1.2-kg book is placed on the spring. Hooke's law tells me that this displacement is proportional to the force exerted by the spring on the book, so after making a sketch of the situation, I draw a free-body diagram for the book. The book is subject to a downward gravitational force exerted by Earth and an upward contact force exerted by the spring (**Figure 8.19**). Because the book is in translational equilibrium, the book's acceleration is zero and so the two forces are equal in magnitude. I arbitrarily point the x axis upward.

Figure 8.19

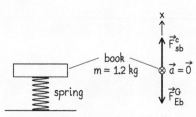

❷ DEVISE PLAN I can use Eq. 8.20 to determine the displacement, but in order to do so I need to know the spring constant k and the x component F^c_{sbx} of the force exerted by the spring on the book. The spring constant is given by the slope of the curve in Figure 8.18. From my free-body diagram I know that $\vec{F}^c_{sb}$ is equal to the negative of the force of gravity $\vec{F}^G_{Eb}$ exerted by Earth on the book.

❸ EXECUTE PLAN From the graph in Figure 8.18, I arbitrarily choose 25 N as my force value and see that it causes a displacement of 5.0 mm. The spring constant is thus, from Eq. 8.18:

$$k = \frac{(F_{\text{by load on spring}})_x}{x - x_0} = \frac{+25\ \text{N}}{+5.0 \times 10^{-3}\ \text{m}}$$

$$= 5.0 \times 10^3\ \text{N/m}.$$

Because the book is at rest, I know that $\vec{F}^c_{sb} + \vec{F}^G_{Eb} = \vec{0}$, and so $\vec{F}^G_{Eb} = -\vec{F}^c_{sb}$. The force of gravity is given by Eq. 8.17, and so I get for the force exerted by the spring on the book

$$F^c_{sbx} = -F^G_{Ebx}$$

$$= -(-mg)$$

$$= (1.2\ \text{kg})(9.8\ \text{m/s}^2)$$

$$= +12\ \text{N}.$$

Substituting this value and the spring constant into Eq. 8.20, I obtain for the displacement:

$$x - x_0 = \frac{-F^c_{sbx}}{k}$$

$$= \frac{-12\ \text{N}}{5.0 \times 10^3\ \text{N/m}}$$

$$= -2.4 \times 10^{-3}\ \text{m}. ✔$$

❹ EVALUATE RESULT It makes sense that the force exerted by the book on the spring is equal to the force exerted by Earth on the book ($\vec{F}^G_{Eb} = \vec{F}^c_{bs}$) because the spring has to keep the book from being pulled downward into the ground. I can thus also obtain my answer by reading off the displacement from Figure 8.18 for a force of magnitude 12 N (the magnitude of the force of gravity exerted on the book). This procedure yields a value that is very close to the result I obtained. The negative sign in my answer indicates that the top end of the spring is displaced downward, as I expect.

8.18 (a) Is a spring that has a large spring constant k stiffer or softer than a spring that has a small spring constant? (b) Which has a larger spring constant: steel or foam rubber?

8.10 Impulse

Back in Chapter 4 we defined the change in momentum of a system or object as the impulse delivered to it. In this section we determine a relationship between force and impulse, which permits us to determine the effect forces have on momentum.

We begin by considering a simple case: a single object subject to constant forces. To calculate what happens to the momentum of the object, we consider its motion during a time interval $\Delta t = t_f - t_i$. The constant forces exerted on the object give it a constant acceleration $\vec{a}$ given by Eq. 8.7. Because the acceleration is constant, we can write

$$\vec{a} = \frac{\Delta\vec{v}}{\Delta t}. \tag{8.21}$$

Multiplying both sides by the inertia m of the object, we get

$$m\vec{a} = m\frac{\Delta\vec{v}}{\Delta t} = \frac{\Delta\vec{p}}{\Delta t} \tag{8.22}$$

or

$$\Delta\vec{p} = m\vec{a}\,\Delta t. \tag{8.23}$$

From Eq. 8.6 we know that $\Sigma\vec{F} = m\vec{a}$, and so Eq. 8.23 becomes

$$\Delta\vec{p} = (\Sigma\vec{F})\,\Delta t \qquad \text{(constant force)}. \tag{8.24}$$

If we compare this to the momentum law ($\Delta\vec{p} = \vec{J}$, Eq. 4.18), we see that the impulse delivered by the forces to the object is

$$\vec{J} = (\Sigma\vec{F})\,\Delta t \qquad \text{(constant force)}. \tag{8.25}$$

This equation, called the **impulse equation,** states that the impulse delivered to an object—the change in an object's momentum—during a time interval Δt is equal to the product of the vector sum of the forces exerted on the object and the duration of the time interval. Equation 8.24 shows that the longer a force is exerted, the smaller it has to be to accomplish a given change in the momentum of the object. This is why, in the bicycling example discussed in Section 8.1, the force of impact is less when you hit the mattress instead of the concrete wall. Likewise, it hurts less to catch a ball (that is to say, bring the ball to rest) with a padded glove than without one. And, by holding out your hand in front of you and then drawing it back just as the ball strikes your glove, you can further increase Δt and thus reduce the force required to slow the ball to a stop.

Figure 8.20*a* shows a force-versus-time graph for an object subject to constant forces whose vector sum is equal to F_0. The impulse delivered during a time interval Δt is equal to the shaded rectangular area under the $F_x(t)$ curve—the width Δt times the height F_0.

So far in this discussion, the impulse-delivering forces we've looked at have been *constant* forces. However, the relationship between impulse and area under the $F_x(t)$ curve holds not only when $\Sigma\vec{F}$ is constant. Just as we found, in Section 3.8, that the area under the $a_x(t)$ curve of an object yields the change in velocity (Eq. 3.27, $\Delta v_x = \int_{t_i}^{t_f} a_x(t)dt$), we calculate for the impulse delivered by an arbitrary force $\Sigma\vec{F}$ that varies with time

$$\Delta\vec{p} = \vec{J} = \int_{t_i}^{t_f} \Sigma\vec{F}(t)\,dt \qquad \text{(time-varying force)}. \tag{8.26}$$

You can obtain this equation by integrating Eq. 8.4. You can also convince yourself of this result either by applying to force the procedure we applied to acceleration in Section 3.8 or by multiplying the left and right sides of Eq. 3.27 by the inertia m of the object.

Figure 8.20b shows the momentum of a cart getting a shove. The impulse delivered by the hand to the cart is equal to the change in momentum $\Delta\vec{p}$ shown in the figure. The force exerted by the hand on the cart is also shown in Figure 8.20b. This curve is obtained by applying what we know from Eq. 8.4 and taking the

Figure 8.20 The impulse delivered by a force is equal to the area under the $F_x(t)$ curve, as shown here for (a) a constant force and (b) a force that varies with time.

(a) Impulse delivered by constant force exerted on object

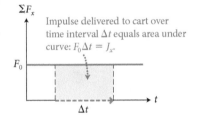

(b) Hand shoves cart, exerting force that varies with time

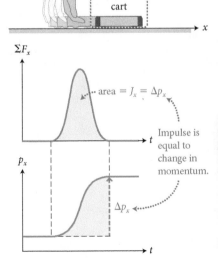

Figure 8.21 As a bat strikes a ball, the magnitude of the force exerted by the bat on the ball increases as the ball deforms, peaks at the instant the ball and bat have zero relative velocity, and finally decreases back to zero as the ball regains its shape and leaves the bat.

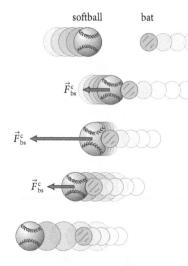

softball   bat

derivative of the $p_x(t)$ curve in Figure 8.20b. As the hand begins pushing, the force increases and peaks. Then, as the hand stops pushing, the force decreases to zero. The impact of a bat on a ball varies in a similar way with time, albeit much more rapidly (**Figure 8.21**). In general, the impulse delivered by a time-varying force during a given time interval is equal to the area under the $F_x(t)$ curve during that time interval (Eq. 8.26 and Figure 8.20b).

If the vector sum of the forces exerted on the object is zero, $\sum \vec{F} = \vec{0}$, the object is in translational equilibrium. Equation 8.26 then tells us that the momentum of the object does not change

$$\Delta \vec{p} = \vec{0} \qquad \text{(translational equilibrium).} \qquad (8.27)$$

In words, the momentum of an object in translational equilibrium does not change.

8.19 (a) A feather and a brick are falling freely in an evacuated tube. Is the magnitude of the gravitational force exerted by Earth on the feather larger than, smaller than, or equal to that exerted by Earth on the brick? (b) Suppose equal forces of 10 N are exerted on both objects for 2 s. Which object gains more momentum? (*Evacuated* means all the air has been sucked out of the tube, so that there is no air resistance to hinder the motion.)

8.11 Systems of two interacting objects

So far we have considered only the very simple case of a single object subject to a force. In this section we look at a system of two interacting objects. For such systems, it is useful to make a distinction between forces exerted from inside the system and forces exerted on the system from outside. The former are called *internal forces*, the latter *external forces*. In **Figure 8.22**, two carts equipped with repelling magnets move on a track while cart 1 gets a push. Because of the magnetic repulsion, the carts never touch each other; instead, cart 2 accelerates to the right when cart 1 approaches. If we choose the two carts as our system, the forces the carts exert on each other are internal, and the push exerted on cart 1 is external.

Figure 8.22 Two carts with repelling magnets moving under the influence of an external force exerted on cart 1.

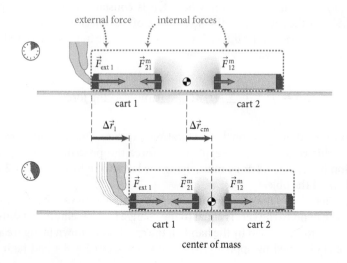

The momentum of the two-cart system is

$$\vec{p} = \vec{p}_1 + \vec{p}_2. \tag{8.28}$$

Differentiating this equation with respect to time and substituting the vector sum of the forces for each dp/dt term (Eq. 8.4) give

$$\frac{d\vec{p}}{dt} = \frac{d\vec{p}_1}{dt} + \frac{d\vec{p}_2}{dt} = \Sigma \vec{F}_1 + \Sigma \vec{F}_2. \tag{8.29}$$

The vector sum of the forces exerted on cart 1 consists of two parts: the external force exerted on cart 1 and an internal force exerted by cart 2 on cart 1 (due to the magnetic repulsion):

$$\Sigma \vec{F}_1 = \vec{F}_{\text{ext 1}} + \vec{F}_{21}^{\text{m}}. \tag{8.30}$$

Because no external force is exerted on cart 2, the vector sum of the forces exerted on it is

$$\Sigma \vec{F}_2 = \vec{F}_{12}^{\text{m}}. \tag{8.31}$$

Substituting Eqs. 8.30 and 8.31 into the right side of Eq. 8.29 gives

$$\frac{d\vec{p}}{dt} = \vec{F}_{\text{ext 1}} + \vec{F}_{21}^{\text{m}} + \vec{F}_{12}^{\text{m}}. \tag{8.32}$$

Because $\vec{F}_{12}^{\text{m}}$ and $\vec{F}_{21}^{\text{m}}$ form an interaction pair, we have $\vec{F}_{12}^{\text{m}} = -\vec{F}_{21}^{\text{m}}$, and so

$$\frac{d\vec{p}}{dt} = \vec{F}_{\text{ext 1}}. \tag{8.33}$$

Equation 8.33 shows that only the external force exerted on cart 1 changes the momentum of the system. The internal forces do not have any effect on the momentum of the system.

Using the definition of center-of-mass velocity from Section 6.6, we have $\vec{p} = m\vec{v}_{\text{cm}}$, with $m = m_1 + m_2$. Because the inertia m of the system remains constant in the interaction, we have

$$\frac{d\vec{p}}{dt} = \frac{d(m\vec{v}_{\text{cm}})}{dt} = m\frac{d\vec{v}_{\text{cm}}}{dt} \equiv m\vec{a}_{\text{cm}} \tag{8.34}$$

and hence we can rewrite Eq. 8.33 as simply

$$\vec{F}_{\text{ext 1}} = m\vec{a}_{\text{cm}} \tag{8.35}$$

or

$$\vec{a}_{\text{cm}} = \frac{\vec{F}_{\text{ext 1}}}{m}. \tag{8.36}$$

Equation 8.36 is the equation of motion for the center of mass of the system. In words:

> **The center of mass of a two-object system accelerates as though both objects were located at the center of mass and the external force were exerted at that point.**

You can now begin to see the power of the center-of-mass concept. Although we cannot determine the acceleration of the individual carts without knowing more about the magnetic interaction, Eq. 8.36 allows us to calculate the motion of their center of mass *regardless* of the details of the magnetic interaction. As you will see in the next section, this is true not just for a system of two carts subject to a single external force, but even for a system of many interacting objects subject to many external forces, regardless of the interactions within the system and regardless of where on the system the external forces are exerted.

8.12 Systems of many interacting objects

With the analysis of a system comprising two objects under our belt, we are ready to look at the more complicated case of a system in which there are more than two objects. As in the preceding section, we begin by differentiating the momentum of the system with respect to time:

$$\frac{d\vec{p}}{dt} = \frac{d\vec{p}_1}{dt} + \frac{d\vec{p}_2}{dt} + \cdots = \Sigma\vec{F}_1 + \Sigma\vec{F}_2 + \cdots, \tag{8.37}$$

where $\vec{p}_1, \vec{p}_2, \cdots$ are the momenta of the individual objects in the system and $\Sigma\vec{F}_1, \Sigma\vec{F}_2, \cdots$ are the vector sums of the forces exerted on each of them. Each of these sums has two parts: one due to internal interactions and one due to external interactions. For object 1, for example, we have a sum of internal forces and a sum of external forces exerted on it:

$$\Sigma\vec{F}_1 = \Sigma\vec{F}_{\text{ext 1}} + \Sigma\vec{F}_{\text{int 1}} \tag{8.38}$$

and we can write similar expressions for all the other objects. Therefore the right side of Eq. 8.37 consists of a sum of external and internal forces:

$$\frac{d\vec{p}}{dt} = \Sigma(\vec{F}_{\text{ext 1}} + \vec{F}_{\text{ext 2}} + \cdots) + \Sigma(\vec{F}_{\text{int 1}} + \vec{F}_{\text{int 2}} + \cdots). \tag{8.39}$$

All internal forces drop out of Eq. 8.39, however. To see why, consider the case in which there are no external forces. Then the right side of Eq. 8.39 reduces to the sum of all internal forces:

$$\frac{d\vec{p}}{dt} = \Sigma(\vec{F}_{\text{int 1}} + \vec{F}_{\text{int 2}} + \cdots). \tag{8.40}$$

In this case, however, the system is isolated (because no external forces are exerted on it), and so we know that its momentum does not change. Therefore $d\vec{p}/dt = \vec{0}$ and Eq. 8.40 becomes

$$\vec{0} = \Sigma(\vec{F}_{\text{int 1}} + \vec{F}_{\text{int 2}} + \cdots). \tag{8.41}$$

In words, all the internal forces add up to zero! Therefore all of them drop out of the right side of Eq. 8.39, and the time rate of change of the momentum of the system is the sum of the external forces:

$$\frac{d\vec{p}}{dt} = \Sigma(\vec{F}_{\text{ext 1}} + \vec{F}_{\text{ext 2}} + \cdots) \equiv \Sigma\vec{F}_{\text{ext}}, \tag{8.42}$$

where $\Sigma\vec{F}_{\text{ext}}$ is the vector sum of all of the external forces exerted on the system. (See also the box "Cancellation of internal forces" on page 199.)

Cancellation of internal forces

There is another way to see that internal forces always add to zero. We consider here a system consisting of three particles, but the proof can be extended to any number of particles.

Each particle is subject to forces from the other two particles and to an external force:

$$\Sigma \vec{F}_1 = \Sigma \vec{F}_{\text{ext}\,1} + \vec{F}_{21} + \vec{F}_{31}$$

$$\Sigma \vec{F}_2 = \Sigma \vec{F}_{\text{ext}\,2} + \vec{F}_{12} + \vec{F}_{32}$$

$$\Sigma \vec{F}_3 = \Sigma \vec{F}_{\text{ext}\,3} + \vec{F}_{13} + \vec{F}_{23}.$$

Summing up all the forces, we get

$$\Sigma \vec{F}_1 + \Sigma \vec{F}_2 + \Sigma \vec{F}_3$$
$$= \Sigma \vec{F}_{\text{ext}\,1} + \Sigma \vec{F}_{\text{ext}\,2} + \Sigma \vec{F}_{\text{ext}\,3} + (\vec{F}_{12} + \vec{F}_{21})$$
$$+ (\vec{F}_{13} + \vec{F}_{31}) + (\vec{F}_{32} + \vec{F}_{23}). \tag{1}$$

Remember, however, that two interacting particles exert on each other forces that are equal in magnitude but opposite in direction (Eq. 8.15). This means that the following equalities must hold:

$$\vec{F}_{12} = -\vec{F}_{21}$$

$$\vec{F}_{13} = -\vec{F}_{31}$$

$$\vec{F}_{32} = -\vec{F}_{23}.$$

So each term in parentheses in Eq. 1 vanishes, and only the external forces remain:

$$\Sigma \vec{F}_1 + \Sigma \vec{F}_2 + \Sigma \vec{F}_3$$
$$= \Sigma \vec{F}_{\text{ext}\,1} + \Sigma \vec{F}_{\text{ext}\,2} + \Sigma \vec{F}_{\text{ext}\,3}.$$

Using the definition of center-of-mass velocity again, we have $\vec{p} = m\vec{v}_{\text{cm}}$, where $m = m_1 + m_2 + \cdots$, so that

$$\frac{d\vec{p}}{dt} = \frac{d(m\vec{v}_{\text{cm}})}{dt} = m\frac{d\vec{v}_{\text{cm}}}{dt} \equiv m\vec{a}_{\text{cm}}. \tag{8.43}$$

After substituting Eq. 8.43 into Eq. 8.42, we obtain the same result as for the two-object nonisolated system:

$$\Sigma \vec{F}_{\text{ext}} = m\vec{a}_{\text{cm}}. \tag{8.44}$$

Now, however, this expression applies to many objects and many external forces.* What this result means is the following: If external forces are exerted on the objects in a many-object system, the system's center of mass moves as though all the objects in the system were concentrated at the center of mass and all external forces were exerted at that point.

The equation of motion for the center of mass of a nonisolated system is thus

$$\vec{a}_{\text{cm}} = \frac{\Sigma \vec{F}_{\text{ext}}}{m}. \tag{8.45}$$

Note the surprising similarities between the equations for a single object (Eqs. 8.4 and 8.6),

$$\frac{d\vec{p}}{dt} = \Sigma \vec{F} \quad \text{and} \quad \Sigma \vec{F} = m\vec{a}, \tag{8.46}$$

and those for the center of mass of a system of interacting objects (Eqs. 8.42 and 8.44),

$$\frac{d\vec{p}}{dt} = \Sigma \vec{F}_{\text{ext}} \quad \text{and} \quad \Sigma \vec{F}_{\text{ext}} = m\vec{a}_{\text{cm}}. \tag{8.47}$$

---

*Equation 8.44 is less general than Eq. 8.42. The latter is always true, but the former holds only if the inertia m of the system is not changing. Equation 8.44 cannot be used, for instance, to calculate the acceleration of a rocket where the burning and high-speed ejection of fuel result in an appreciable change in the rocket's inertia.

The most significant difference between Eqs. 8.6 and 8.44 is that instead of the acceleration $\vec{a}$ of a single object, Eq. 8.44 contains the center-of-mass acceleration $\vec{a}_{cm}$ of the system. This subtle but very important difference is what makes it possible to apply this analysis to everyday objects. Any object can be regarded as a system of interacting particles; after all, objects are made up of many interacting atoms. Thus all the relationships we just derived apply to macro-sized objects.

One interesting application is to macro-sized objects that deform easily. For example, imagine pushing a down pillow (**Figure 8.23**). When you begin pushing on one edge of the pillow, your hand deforms the side it touches, meaning that part of the pillow is accelerating. While your hand is accelerating that side of the pillow, the other side of the pillow doesn't move at all. In other words, different parts of the pillow have different instantaneous velocities and different instantaneous accelerations. It therefore makes no sense, while deformation is taking place, to speak of *the* acceleration $\vec{a}$ of the pillow, and Eqs. 8.46 cannot be applied. On the other hand, we can always use Eqs. 8.47 to describe the motion of the pillow's center of mass. In other words, the momentum $\vec{p}$ of the pillow, the acceleration $\vec{a}_{cm}$ of its center of mass, and the vector sum of the forces exerted on it are still well-defined quantities.

Perhaps the most remarkable aspect of this result is that we can always apply Eqs. 8.42 and 8.44 regardless of the internal interactions in a system. Even without any knowledge of these interactions, we can determine the motion of the center of mass of the pillow. What we give up by not considering the internal interactions is the ability to determine how the system deforms or what the motions of the different parts of the system are when the external force is exerted. In Figures 8.22 and 8.23, for example, we can determine only the displacement $\Delta \vec{r}_{cm}$ of the center of mass, not the displacement $\Delta \vec{r}_1$ of the point where the external force is exerted on the system, or the displacement of any other point.

The difference between Eqs. 8.46 and 8.47 vanishes when we consider a single object and ignore any deformation in it—that is to say, if we consider a **rigid object**. Although I argued in Section 8.6 that no object is truly rigid, most objects can be treated as being rigid for the purpose of describing their motion. For example, imagine pushing a stalled car toward the side of the road. For motion in one dimension, the front of the car moves the same way the back does—the minute compression of the car while your hands are pushing it is negligible. If you displace the back of the car by, say, 10 m, you can be pretty sure that the front of the car is also displaced by 10 m (give or take a fraction of the width of a hair). *All* parts of the car get displaced by 10 m in the same direction, and so the velocity and acceleration are the same for all parts of the car. For rigid objects we do not need to take into account any internal interactions or any difference in the displacement of various parts of the object.

Only for *deformable objects* (or for systems consisting of two or more rigid objects that can move relative to one another, like the two magnetically repelling carts in Figure 8.22) do you have to use the results derived in the last two sections.

8.20 (*a*) Show that Eqs. 8.47 reduce to Eqs. 8.46 when you consider a system of just one object. (*b*) Follow the procedure used to get from Eq. 8.21 to Eq. 8.24 for a system of many interacting objects.

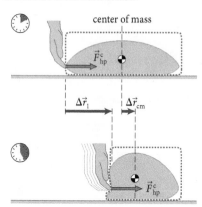

Figure 8.23 As the external force exerted by the hand displaces a down pillow, the pillow deforms, and so different parts of the pillow are displaced by different amounts. For objects that deform, it makes no sense to speak of "the displacement" or "the acceleration" of the object. However, it does makes sense to speak of the displacement or the acceleration of the center of mass.

center of mass

$\vec{F}_{hp}^{c}$

$\Delta \vec{r}_1$ $\Delta \vec{r}_{cm}$

$\vec{F}_{hp}^{c}$

Chapter Glossary

SI units of physical quantities are given in parentheses.

Contact force The force that an object exerts on a second object when the two are in physical contact with each other.

Elastic force The force caused by the reversible compression or stretching of an object.

Equation of motion The equation that relates the acceleration of an object to the vector sum of the forces exerted on it:

$$\vec{a} = \frac{\sum \vec{F}}{m}. \qquad (8.7)$$

For a system of more than one object, the equation of motion of the center of mass is

$$\vec{a}_{cm} = \frac{\sum \vec{F}_{ext}}{m}. \qquad (8.45)$$

Equilibrium Any object or system whose motion or state is not changing is in equilibrium.

Field force The force that an object exerts on a second object without the two being in physical contact with each other. The three types of field forces we consider are gravitational, electric, and magnetic forces.

Force $\vec{F}$ (N) When an object participates in an interaction, the force exerted on the object is given by the time rate of change caused in the object's momentum by that interaction. Forces obey the *principle of superposition*: The vector sum of the forces exerted on an object is equal to the vector sum of the rate of change caused in the object's momentum by the individual forces, and so

$$\sum \vec{F} \equiv \frac{d\vec{p}}{dt}. \qquad (8.4)$$

Free-body diagram A sketch showing a single object represented by its center of mass and all forces exerted *on* it. See the box on page 182.

Hooke's law The relationship between the force exerted on a spring (or other elastic object) and the displacement of the end of the spring from its relaxed value x_0 caused by that force:

$$(F_{\text{by load on spring}})_x = k(x - x_0) \quad \text{(small displacement)}, \qquad (8.18)$$

The constant k is the *spring constant*.

Impulse equation The equation that allows us to calculate the change in a system's momentum caused by forces exerted on the system during a time interval Δt. For a constant force,

$$\vec{J} = \left(\sum \vec{F} \right) \Delta t \qquad \text{(constant force)}. \qquad (8.25)$$

For a time-varying force,

$$\vec{J} = \int_{t_i}^{t_f} \sum \vec{F}(t) \, dt \qquad \text{(time-varying force)}. \qquad (8.26)$$

Interaction pair The forces that two interacting objects exert on each other. Conservation of momentum requires these two forces to be equal in magnitude and to point in opposite directions:

$$\vec{F}_{12} = -\vec{F}_{21}. \qquad (8.15)$$

Newton (N) A derived SI unit of force, defined as $1 \text{ N} \equiv 1 \text{ kg} \cdot \text{m/s}^2$.

Newton's laws of motion Three fundamental principles that govern forces and their effects on the motion of objects.

Rigid object An object whose deformation can be ignored when one or more forces are exerted on it. For such an object the distance between any two points on the object remains fixed.

Spring constant k (N/m) A scalar defined as the ratio of the magnitude of the force exerted on a spring (or other elastic object) and the displacement of the end of the spring from its relaxed value x_0 caused by that force. See also *Hooke's law*.

Tension $\mathcal{T}$ (N) A scalar representing the stress caused in an object subject to a pair of forces, called *tensile forces*, exerted so that they stretch the object. In an object subject to two equal tensile forces of magnitude F, one on each end of the object, the value of the tension is $\mathcal{T} = F$.

Translational equilibrium Any object whose momentum is not changing is in translational equilibrium.

9 Work

CONCEPTS

QUANTITATIVE TOOLS

As you learned in Chapter 8, a system subject to external forces—that is to say, forces exerted by objects outside the system on objects inside the system—is not isolated. The relationship between the momentum of a nonisolated system and the vector sum of the forces exerted on that system is simple: The vector sum of the forces is equal to the time rate of change in the system's momentum (Eq. 8.4). In addition to affecting a system's momentum, external forces can change the energy of the system. In the simplest case, a single force exerted on a system consisting of a single object causes that object to accelerate, and so the force changes the object's kinetic energy. Forces can also change the physical state of objects (think of compressing or stretching a spring or crumpling a piece of paper by exerting a force on it). Because the physical state of an object is related to the object's internal energy, we see that forces can change not only an object's kinetic energy but its internal energy as well.

To describe these changes in energy, physicists use the concept of **work:**

Work is the change in the energy of a system due to external forces.

When an external force $\vec{F}_{ext}$ causes a change in the energy of a system, physicists refer to this change either as "the *work done by* the force $\vec{F}_{ext}$ on the system" or as "the *work done by* the object or person exerting force $\vec{F}_{ext}$ on the system."

Because work is a change in energy, the unit for work is the same as that for energy—the joule.

Note that this definition of work is much more limited than the meaning of the word *work* in ordinary language, which refers to any physical or mental effort or activity. You may be "doing work" for your physics class while reading this book, but there is no work involved in the physics sense.

9.1 Force displacement

Work amounts to a mechanical transfer of energy either from a system to its environment or from the environment to the system. For instance, if you (part of the "environment") push a friend on a bicycle (system is rider plus bicycle) and your friend accelerates, the force exerted by you on your friend transfers energy from your body to hers,

increasing her kinetic energy. If you compress a spring, you transfer energy from your body to the spring, where it is stored in the form of elastic potential energy. In each case, the change in the system's energy is caused by an external force—a force exerted by you on the system.

If you wonder why a force has to be external to do work on a system—that is, to change the energy of the system—remember that forces represent interactions between objects. Any interaction between objects within the system rearranges the energy between these objects but doesn't change the overall amount of energy in the system. Only an interaction between an object inside the system and an object outside it—an external interaction—can transfer energy across the system boundary.

Do external forces *always* cause a change in the energy of a system? To answer this question, it is helpful to consider an example.

9.1 Imagine pushing against a brick wall as shown in Figure 9.1a. (a) Considering the wall as the system, is the force you exert on it internal or external? (b) Does this force accelerate the wall? Change its shape? Raise its temperature? (c) Does the energy of the wall change as a result of the force you exert on it? (d) Does the force you exert on the wall do work on the wall?

As this checkpoint illustrates, external forces exerted on a system do not always do work on the system. The other two examples in **Figure 9.1**, however, do involve changes in energy, and hence in these cases work is done by your external force on the system. If, as shown in Figure 9.1b, you push against something that has wheels and so can accelerate it, its kinetic energy changes. If you push against a deformable object, the object changes shape and so its potential energy changes (Figure 9.1c).

Why is the work zero in Figure 9.1a but nonzero in parts b and c? The key to answering this question lies in looking at the *point of application* of each force (the point at which the force is exerted on the system, as indicated by the black dots in Figure 9.1). When you push against the brick wall, it neither moves nor deforms, and so the point of application of the force does not move. When you push the cart, you must move along with the cart to accelerate it; in other words, the

Figure 9.1 An external force does work on a system only if it causes the point at which it is applied (represented by a black dot) to move.

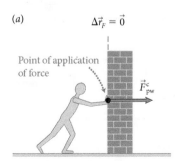

(a)

$\Delta \vec{r}_F = \vec{0}$

Point of application of force

$\vec{F}^c_{pw}$

Point of application does not move, so force does no work on wall.

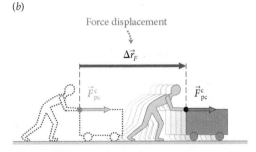

(b)

Force displacement

$\Delta \vec{r}_F$

$\vec{F}^c_{pc}$ $\vec{F}^c_{pc}$

Point of application moves, so force does work on cart.

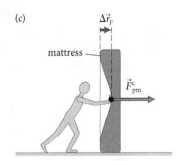

(c)

$\Delta \vec{r}_F$

mattress

$\vec{F}^c_{pm}$

Point of application moves, so force does work on mattress.

point of application of the force is displaced. Similarly, when your force changes the shape of the deformable mattress in Figure 9.1c, the point of application of the force is displaced.

In order for a force to do work, the point of application of the force must undergo a displacement.

Whenever an external force is exerted on a system and the displacement of the point of application of the force is zero, no work is done by the force on the system. Conversely, in one dimension, when the **force displacement** (shorthand for "the displacement of the point of application of the force") is nonzero, the external force does work on the system, and so the energy of the system changes. (In more than one dimension, the angle between the force and the force displacement also plays a role. We shall discuss this point in Chapter 10.)* To distinguish the force displacement from other displacements, we add a subscript F to the symbol for displacement: $\Delta \vec{r}_F$.

Exercise 9.1 Displaced forces

For which of the following forces is the force displacement nonzero: (a) the force exerted by a hand compressing a spring, (b) the force exerted by Earth on a ball thrown upward, (c) the force exerted by the ground on you at the instant you jump upward, (d) the force exerted by the floor of an elevator on you as the elevator moves downward at constant speed?

SOLUTION (a), (b), and (d).

(a) The point of application of the force is at the hand, which moves to compress the spring.

(b) The point of application of the force of gravity exerted by Earth on the ball is at the ball, which moves.

(c) The point of application is on the ground, which doesn't move.

(d) The point of application is on the floor of the elevator, which moves. ✔

9.2 You throw a ball straight up in the air. Which of the following forces do work on the ball while you throw it? Consider the interval from the instant the ball is at rest in your hand to the instant it leaves your hand at speed v. (a) The force of gravity exerted by Earth on the ball. (b) The contact force exerted by your hand on the ball.

9.2 Positive and negative work

When the energy of a system increases as a result of an external force exerted on the system, the change in energy is positive, and so the work done by that external force on

---

*Two additional external forces are exerted on the cart in Figure 9.1b that have nonzero force displacements: the force of gravity and the contact force exerted by the ground. However, these two forces are equal in magnitude and opposite in direction and so they "cancel out": Their net effect is the same as that of no force. Therefore no work is done by these forces on the cart.

the system is said to be positive. An external force can also decrease the energy of a system, however. For instance, imagine slowing down a friend who is coasting on a bicycle. In this case, the kinetic energy of the system decreases, the change in energy is negative, and so the work done by the external force (that is, by you on your friend) on the system is said to be negative.

9.3 A ball is thrown vertically upward. (a) As it moves upward, it slows down under the influence of gravity. Considering the changes in energy of the ball, is the work done by Earth on the ball positive or negative? (b) After reaching its highest position, the ball moves downward, gaining speed. Is the work done by the gravitational force exerted on the ball during this motion positive or negative?

Examples of positive and negative work are illustrated in **Figure 9.2**. In both cases the person exerts a force $\vec{F}^c_{pc}$ on the cart. What, then, is the difference between these two situations? The difference lies in the direction in which the force is applied. To speed the cart up, the person must push in the direction of travel of the cart; the force vector $\vec{F}^c_{pc}$ and the force displacement $\Delta \vec{r}_F$ point in the same direction. To slow the cart down, the force must be applied in the direction opposite the cart's direction of travel; the force $\vec{F}^c_{pc}$ and the force displacement $\Delta \vec{r}_F$ point in opposite directions.

The work done by a force on a system is positive when the force and the force displacement point in the same direction and negative when they point in opposite directions.

Figure 9.2 The work done on a system is positive if the system gains kinetic energy and negative if the system loses kinetic energy.

(a) Cart speeds up, so positive work is done on it

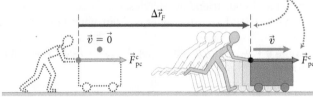

(b) Cart slows down, so negative work is done on it

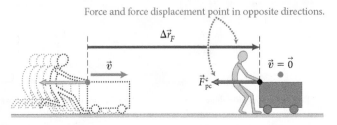

Figure 9.3 Only forces that are *external* to a system can do work on the system and change the system's energy.

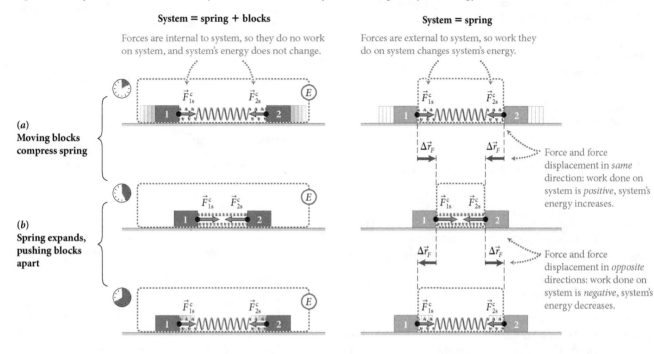

9.4 Go back to Checkpoint 9.3 and answer the same two questions in terms of the directions of the applied forces and the force displacements. Do your answers agree with those you gave in Checkpoint 9.3?

Let us next consider a situation involving changes in potential energy. **Figure 9.3a** shows a spring being compressed by two moving blocks. There are two ways to look at this problem. We can consider the spring and the blocks as a closed system whose energy must remain constant, or we can consider just the spring as a system on which the blocks do work.

For the closed blocks-spring system in Figure 9.3a, the potential energy of the spring increases during the compression by an amount equal to the decrease in the blocks' kinetic energy. Because no external forces are exerted on the system, no work is involved.

Considering just the spring as our system in Figure 9.3a, we note that the increase in the spring's potential energy results from the work done by the blocks on it. Because the spring's potential energy increases, the work done on the system must be positive. Indeed, the forces exerted by the blocks on the spring point in the same direction as the force displacements.

Figure 9.3b shows the reverse situation: Two blocks are held against a compressed spring and then released, so that they accelerate away from each other. For the closed blocks-spring system, the energy doesn't change, and we can account for the increase in kinetic energy of the blocks by a decrease in the potential energy of the expanding spring. If, however, we consider the spring by itself as our system, the decrease in energy of the system implies that the work done by the blocks on the spring is negative.

Let us see if we reach the same conclusion for the sign of the work in Figure 9.3b when we look at the directions of the forces and the force displacements. Because we are interested in the work done *on the spring*, we need to consider the forces exerted by the blocks on it. These forces still point toward the spring, as shown in Figure 9.3b.* The force displacements, however, now point in the opposite direction, away from the spring. Because each force and its force displacement point in opposite directions, the work done by the blocks on the spring is negative. Everything is consistent: Both energy considerations and the directions of the forces and force displacements tell us that the work done by the blocks on the spring is negative as the spring expands.

Exercise 9.2 Positive and negative work

Is the work done by the following forces considered in Exercise 9.1 positive, negative, or zero? In each case the system is the object on which the force is exerted. (a) The force exerted by a hand compressing a spring, (b) the force exerted by Earth on a ball thrown upward, (c) the force exerted by the ground on you at the instant you jump upward, (d) the force exerted by the floor of an elevator on you as the elevator moves downward at constant speed.

---

*Careful! You may think the spring exerts a force on the blocks and not the other way around. Don't forget, however, that the force exerted by the spring on each block is just one part of an interaction pair. If you are having trouble picturing the force exerted by the blocks on the spring, imagine the spring expanding without having to push against the blocks: The expansion would be easier because there would be no resistance from the blocks. So, in a sense, the force exerted by the blocks on the spring slows down the expansion of the spring.

SOLUTION (*a*) Positive. To compress a spring, I must move my hand in the same direction as I push. ✔

(*b*) Negative. The force exerted by Earth points downward; the point of application moves upward. ✔

(*c*) Zero, because the point of application is on the ground, which doesn't move. ✔

(*d*) Negative. The force exerted by the elevator floor points upward; the point of application moves downward. ✔

9.5 Suppose that instead of the two moving blocks in Figure 9.3*a*, just one block is used to compress the spring while the other end of the spring is held against a wall. (*a*) Is the system comprising the block and the spring closed? (*b*) When the system is defined as being only the spring, is the work done by the block on the spring positive, negative, or zero? How can you tell? (*c*) Is the work done by the wall on the spring positive, negative, or zero?

9.3 Energy diagrams

In analyzing situations involving work, we extend the bar diagrams introduced in Chapter 5 to include work done on the system. Consider, for example, the situation in **Figure 9.4*a*.** The energy of the system represented by this diagram increases from 10 J to 15 J. In order for this 5-J increase to occur, an amount of work equal to 5 J must be done by external forces on the system. The length of the vertical bar representing the final energy is equal to the sum of the lengths of the bars representing the initial energy and the work done by the external forces on the system.

Figure 9.4*b* represents a system whose energy decreases. In order for the energy to decrease, the work done by external forces on the system must be negative, and so the bar representing this work extends below the baseline.

Because any of the four kinds of energy discussed in Chapter 7 (kinetic energy, potential energy, source energy, and thermal energy) can change in situations involving work, our bar diagrams need to include more detail in order to show in what form the energy transferred to a system appears when some external force does work on it. The additional bars make these diagrams more cumbersome to draw, however, and therefore we introduce the following

Figure 9.4 Bar diagrams showing the work done on a system. Positive work adds to the system's energy; negative work subtracts from it.

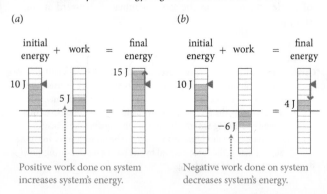

(*a*)

Positive work done on system increases system's energy.

(*b*)

Negative work done on system decreases system's energy.

Figure 9.5 (*a*) Changes in a system's energy can be shown (*b*) by initial and final bar diagrams or, more compactly, (*c*) by a single energy diagram, which shows the *changes* in energy in each category and a column showing work done on the system.

(*a*)

External work by person changes system's kinetic and potential energy.

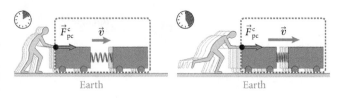

(*b*)

We can represent the changes in energy by initial and final bar diagrams . . .

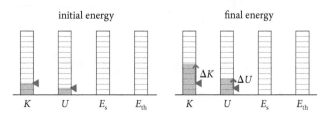

(*c*)   . . . or by a single **energy diagram.**

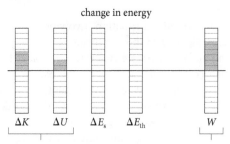

Changes in system's energy . . .   . . . equal work done on system.

simplification. Instead of drawing one set of bars for each of the initial and final individual energies, we draw just one set of bars representing the *change* in each category of energy, and we add a fifth bar, denoted by *W*, representing the work done by external forces on the system. We call these diagrams **energy diagrams.** Energy diagrams are a graphic representation of conservation of energy: They show that the energy of a system can change only due to the transfer of energy into or out of the system.

Just as with free-body diagrams, the first step in drawing an energy diagram is to define the system. In addition we should show any external force whose point of application undergoes a nonzero displacement; these are the forces that do work on the system. Consider, for example, the situation shown in **Figure 9.5*a*:** a person pushing a pair of carts connected by a spring. The pushing accelerates the carts and compresses the spring. Let us choose a system comprising the pair of carts and the spring; Figure 9.5*a* shows its initial and final states.

Procedure: Drawing energy diagrams

1. Specify the **system** under consideration by listing the components inside the system.
2. **Sketch** the system in its initial and final states. (The initial and final states may be defined for you by the problem, or you may have to choose the most helpful states to examine.) Include in your sketch any external forces exerted on the system that undergo a nonzero force displacement, and draw a dot at the point of application of each force.
3. Determine any nonzero **changes in energy** for each of the four categories of energy, taking into account the four basic energy-conversion processes illustrated in Figure 7.13:

   a. Did the speed of any components of the system change? If so, determine whether the system's kinetic energy increased or decreased and draw a bar representing ΔK for the system. For positive ΔK, the bar extends above the baseline; for negative ΔK, it extends below the baseline. (For some problems you may wish to draw separate ΔK bars for different objects in the system; if so, be sure to specify clearly the system component that corresponds to each bar and verify that the entire system is represented by the sum of the components.)

   b. Did the configuration of the system change in a reversible way? If so, draw a bar representing the change in potential energy ΔU for the system. If necessary, draw separate bars for the changes in different types of potential energy, such as changes in elastic and gravitational potential energy.

   c. Was any source energy consumed? If so, draw a bar showing ΔE_s. Source energy usually decreases, making ΔE_s negative, and so the bar extends below the baseline. Keep in mind that conversion of source energy is always accompanied by generation of thermal energy (Figure 7.13c and d).

   d. Does friction occur within the system, or is any source energy consumed? If so, draw a bar showing ΔE_{th}. In nearly all cases we consider, the amount of thermal energy increases, and so ΔE_{th} is positive.

4. Determine whether or not any **work** W is done by external forces on the system. Determine whether this work is negative or positive. Draw a bar representing this work, making the length of the bar equal to the sum of the lengths of the other bars in the diagram.

   If no work is done by external forces on the system, leave the bar for work blank, then go back and adjust the lengths of the other bars so that their sum is zero.

We also draw a vector representing the external force $\vec{F}^c_{pc}$ exerted by the person on the left cart and indicate the position of the point of application of that force with a black dot.

Because the carts speed up, the final kinetic energy of the system is greater than the initial kinetic energy and the change in kinetic energy $\Delta K = K_f - K_i$ is therefore positive. This change in kinetic energy corresponds to the upward-pointing arrow marked ΔK in Figure 9.5b, which is a set of the bar diagrams we used before now (no work values are shown). The compression of the spring causes the elastic potential energy stored in the spring to increase, and so the change in potential energy $\Delta U = U_f - U_i$ is also positive, as shown by the upward-pointing arrow in Figure 9.5b.

To represent this same information in an energy diagram (Figure 9.5c), we draw bars to represent the changes ΔK and ΔU. The heights of the bars, which are equal to the heights of the arrows in Figure 9.5b, now represent the *changes* in each energy category. The ΔE_s and ΔE_{th} bars are empty because there are no changes in these energies.

The change in the system's energy comes from the work done by external forces on the system. The length of the bar representing the work done on the system in Figure 9.5c must therefore equal the sum of the lengths of the ΔK and ΔU bars. Note that these energy arguments tell us that the work done by external forces on the system must be positive. We can verify this sign by checking the directions of the force exerted on the system and its force displacement. From Figure 9.5a, we see that both are to the right, and so the work done by external forces on the system is indeed positive.

The general procedure for making energy diagrams is summarized in the Procedure box "Drawing Energy Diagrams".

Exercise 9.3 Cart launch

A cart is at rest on a low-friction track. A person gives the cart a shove and, after moving down the track, the cart hits a spring, which slows down the cart as the spring compresses. Draw an energy diagram for the system that comprises the person and the cart over the time interval from the instant the cart is at rest until it has begun to slow down.

SOLUTION Following the steps in the Procedure box, I begin by listing the objects that make up the system: cart and person. Then I sketch the system in its initial and final states (**Figure 9.6**). The external forces exerted on the system are the force of gravity on person and cart, the contact force exerted by ground on person, the contact force exerted by track on cart, and the contact force exerted by spring on cart. The vertical forces add up to zero, and so the work done by them on the system is zero. The force exerted by the spring on the cart, however, undergoes a force displacement, so I include that force in my diagram and indicate its point of application, as shown in Figure 9.6.

Figure 9.6

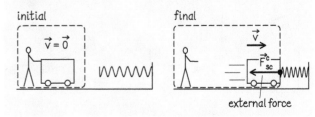

Next I determine whether there are any changes in energy. Kinetic energy: The cart begins at rest and ends with nonzero speed, which means a gain in kinetic energy (**Figure 9.7**). Potential energy: Neither the configuration of the cart nor that of the person changes in a reversible way, so the potential energy does not change. (The configuration of the spring does change in a reversible way, but the spring is not part of the system.) Source energy: The person has to exert effort to give the cart a shove, so source energy is consumed. The source energy of the system therefore decreases. Thermal energy: No friction is involved, but the conversion of source energy is always accompanied by the generation of thermal energy, so the thermal energy increases. (The person's muscles heat up.)

Figure 9.7

$$\Delta K \quad \Delta U \quad \Delta E_s \quad \Delta E_{th} \quad W$$

In Figure 9.6 the force exerted by the spring on the cart points to the left and the force displacement points to the right. This means that the work done by the spring on the cart is negative, and the W bar of my energy diagram must extend below the baseline. I adjust the lengths of the bars so that the length of the W bar is equal to the sum of the lengths of the other bars, yielding the energy diagram shown in Figure 9.7. ✔

 9.6 Draw an energy diagram for the cart in Figure 9.2b.

Exercise 9.4 Compressing a spring

A block initially at rest is released on an inclined surface. The block slides down, compressing a spring at the bottom of the incline; there is friction between the surface and the block. Consider the time interval from a little after the release, when the block is moving at some initial speed v, until it comes to rest against the spring. Draw an energy diagram for the system that comprises the block, surface, and Earth.

SOLUTION I begin by listing the objects that make up the system: block, surface, and Earth. Then I sketch the initial and final states of the system (**Figure 9.8**). The spring exerts external forces on the system. The bottom end of the spring exerts a

Figure 9.8

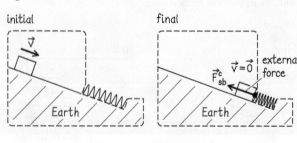

force on the surface edge, but this force has a force displacement of zero. The top end of the spring exerts a force $\vec{F}^c_{sb}$ on the block. Because this force undergoes a nonzero force displacement, I include it in my diagram and show a dot at its point of application.

Next I determine whether there are any energy changes. Kinetic energy: The block's kinetic energy goes to zero, and the kinetic energies of the surface and Earth do not change. Thus the kinetic energy of the system decreases, and ΔK is negative. Potential energy: As the block moves downward, the gravitational potential energy of the block-Earth system decreases, and so ΔU is negative. (Because the spring gets compressed, its elastic potential energy changes, but the spring is not part of the system.) Source energy: none (no fuel, food, or other source of energy is converted in this problem). Thermal energy: As the block slides, energy is dissipated by the friction between the surface and the block, so ΔE_{th} is positive.

To determine the work done on the system, I look at the external forces exerted on it. The point of application of the external force $\vec{F}^c_{sb}$ exerted by the spring on the block undergoes a force displacement opposite the direction of the force, so that force does negative work on the system. Thus the work done on the system by the external forces is negative, and the W bar extends below the baseline (**Figure 9.9**). I adjust the lengths of the bars so that the length of the W bar is equal to the sum of the lengths of the other three bars, yielding the energy diagram shown in Figure 9.9. ✔

Figure 9.9

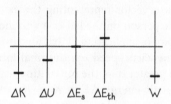

$$\Delta K \quad \Delta U \quad \Delta E_s \quad \Delta E_{th} \quad W$$

 9.7 Draw an energy diagram for the situation presented in Exercise 9.4, but choose the system that comprises block, spring, surface, and Earth.

9.4 Choice of system

For a given situation, different choices of system yield different energy diagrams, as you just saw if you did Checkpoint 9.7. To examine this point further, consider the situation shown in **Figure 9.10**. Using a long rope, a person lowers a basket from a balcony, smoothly putting the basket down on the ground. The basket initially moves downward at speed v_i, but friction between the person's hands and the rope slows the basket so that it ends up at rest on the ground. For simplicity we assume no source energy is consumed; the person just lets the rope slide, exerting no physical effort.

Figure 9.10 Different choices of system yield different energy diagrams.

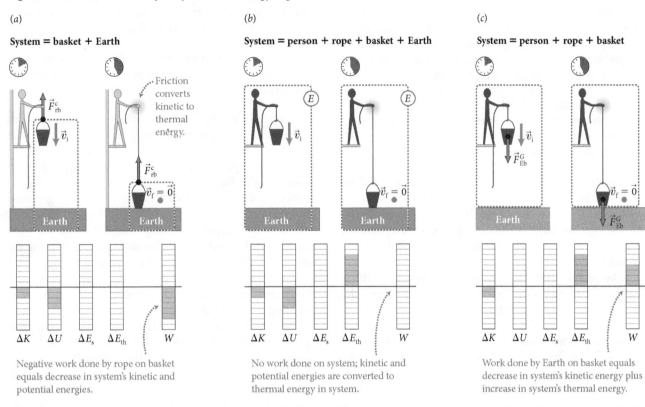

(a)

System = basket + Earth

Friction converts kinetic to thermal energy.

ΔK ΔU ΔE_s ΔE_{th} W

Negative work done by rope on basket equals decrease in system's kinetic and potential energies.

(b)

System = person + rope + basket + Earth

ΔK ΔU ΔE_s ΔE_{th} W

No work done on system; kinetic and potential energies are converted to thermal energy in system.

(c)

System = person + rope + basket

ΔK ΔU ΔE_s ΔE_{th} W

Work done by Earth on basket equals decrease in system's kinetic energy plus increase in system's thermal energy.

Let us begin by choosing the basket and Earth as our system, a choice that leads to the energy diagram shown in Figure 9.10a. Because the basket comes to rest, its kinetic energy decreases. As the distance between Earth and basket decreases, the potential energy of the Earth-basket system decreases. There is no conversion of source energy nor any dissipation of energy within the system, and so we leave the columns for ΔE_s and ΔE_{th} blank.

Next we examine the external forces exerted on the system. The person exerts forces both on Earth and, via the rope, on the basket. The latter involves a nonzero force displacement (the knot, which is the point of application for the force exerted by the rope on the basket, moves downward), so work is done on the system. This work is negative because the force exerted by the rope on the basket is upward and the force displacement is downward. In order to make the work equal to the change in the system's energy, the length of the bar representing work must be made equal to the sum of the lengths of the ΔK and ΔU bars.

What this energy diagram tells us is that the energy of the basket-Earth system decreases as energy is transferred out of the system. Because energy is transferred out of the system, the work done by the person on the system is negative.

What happens if we include the person and the rope in the system, as is Figure 9.10b? The changes in the kinetic and potential energies are the same as before, but now the person no longer exerts external forces and therefore cannot do any work on the system. In fact, because we have

included everything in the system, there are no external forces and so there is no work done by external forces on the system—the system is closed.

For this choice of system, we have to take into account the dissipation of energy because of friction between the person's hands and the rope, and so we have a positive ΔE_{th}. The resulting energy diagram tells us that the initial kinetic energy of the basket and the initial potential energy of the system are converted to thermal energy (the rope and the person's hands heat up). No work is done by external forces on the system because the system is closed. This conclusion is entirely consistent with the one we drew on the basis of Figure 9.10a, but in the process we've obtained more information on where the energy went.

Finally we consider the same situation for a system that comprises only basket, person, and rope, as in Figure 9.10c. The changes in the kinetic and thermal energies are as before. For this choice of system, however, there is no change in gravitational potential energy (remember that gravitational potential energy refers to the configuration of Earth and the basket *together*, and Earth is no longer part of the system). What was counted as gravitational potential energy in Figure 9.10a and b appears as work done by the gravitational force on the system here. Earth exerts a downward external force on the basket as the basket moves downward, so Earth does positive work on the basket. (Earth also exerts a downward force on the person, but this force involves no force displacement.)

In Section 9.5, I shall show that the positive work done by the gravitational force on the system in Figure 9.10c is exactly equal to the negative of the change in the gravitational potential energy in Figure 9.10a and b. As this example shows, in order to avoid double-counting, it is important always to remember the following point:

Gravitational potential energy always refers to the relative position of various parts within a system, never to the relative positions of one component of the system and its environment.

Whenever you are looking at the energy associated with the gravitational interaction between an object inside your system and an object outside the system, do not include that energy in the ΔU bar of your energy diagram. That energy must be accounted for in the W part of the diagram instead.

The energy diagram of Figure 9.10c shows how the initial kinetic energy of the basket and the positive work done by Earth on the system end up as thermal energy.

All three choices of system lead to a valid conclusion. By looking at a situation from various points of view, you gain a better understanding of what energy conversions and transfers take place. Also, by looking at a situation in various ways and verifying that your answers are consistent with one another, you can convince yourself that you have not overlooked something.

✋ **9.8** Draw an energy diagram for just the basket in Figure 9.10.

Although you are generally free to choose whichever system suits you best, there are certain choices of system that lead to complications. In particular:

When drawing an energy diagram, do not choose a system for which friction occurs at the boundary of the system.

The reason for this is simple: Wherever friction occurs, thermal energy is generated. The problem is that this thermal energy appears on both sides of the surface where friction occurs. If you pull on a rope to draw a large box toward yourself, as in **Figure 9.11**, friction between floor and box heats up both box and floor. The thermal energy generated ends up partly in the floor and partly in the box. How much ends up where depends, among other things, on the materials that rub against each other and is not something that can be determined easily. So, if you choose just the box as your system, an unknown amount of energy flows out of the system, and it is no longer possible to do the energy accounting for the system.

There is another reason friction at a system boundary leads to problems. Whenever friction occurs at a boundary, the frictional force is *external* to the system (it is exerted by the environment on the system) and hence has to be taken into account in the computation of the work done on the system. The force of friction, however, is not applied at a single, well-defined location. Instead, it is distributed over the surfaces that move relative to each other. Consequently it is not possible to determine the force displacement, which is the other piece of information necessary to determine the work done on the system.

Figure 9.11 If you need to do energy accounting, don't choose a system for which friction occurs at the system boundary.

(a)   **System = person + box**

Friction generates thermal energy in Earth and box . . . . . . but we don't know how much ends up in Earth and how much in box.

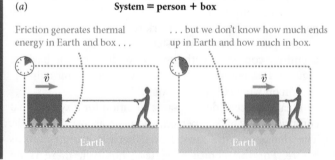

So, if system excludes Earth, we can't do energy accounting.

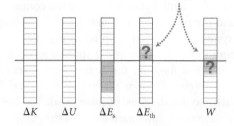

(b)   **System = person + box + Earth**

If system also includes Earth, all thermal energy remains in system . . .

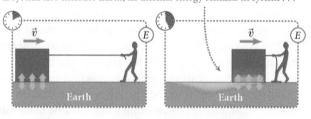

. . . so energy accounting is easy.

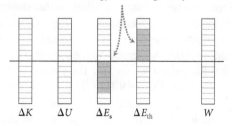

These problems with friction at a system boundary don't mean we can never treat any situations involving friction. Both our analyses of Figure 9.10 and Exercise 9.4 involve friction, but for all choices of system we considered, the two objects that rub against each other were kept together either inside or outside the system.

Example 9.5 Puck and spring

A puck sliding on a horizontal surface hits the free end of a spring that is held fixed at the other end. There is friction between puck and surface. Consider the interval from the instant before the puck comes in contact with the spring until the instant the puck has zero velocity and the spring reaches maximum compression. Draw an energy diagram for a system that contains the puck and for which the work done on the system is nonzero. (Your system may include objects in addition to the puck if you like.)

① GETTING STARTED I begin by making situation sketches at the beginning and end of the time interval of interest. The items in my sketches are the puck, the surface, and the spring (Figure 9.12).

Figure 9.12

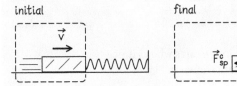

If I include all three items in my system, the system is closed, so the work done by external forces on the system is zero. Because there is friction between surface and puck, I must keep the two together. Given that the puck must be part of the system, I have no choice but to have my system be the puck plus the surface, so I draw my system boundary around these two objects. The only external force exerted on this system is the contact force exerted by the spring on the puck, so I add that force to my diagram and indicate its point of application.

② DEVISE PLAN All I have to do is follow the procedure for drawing an energy diagram for my system.

③ EXECUTE PLAN My energy diagram is shown in **Figure 9.13**. Because the puck slows to a stop, its kinetic energy decreases. Because there are no reversible changes in the system configuration and because no source energy is consumed, my bars for ΔU and ΔE_s must be zero. The friction causes dissipation of energy, so the ΔE_{th} bar shows an increase.

Figure 9.13

Because the external force exerted by the spring on the system points in the direction opposite the direction of the force displacement, the work done by the spring on the system is negative. I adjust the lengths of the bars so that the length of the W bar is equal to the sum of the lengths of the other bars. ✔

④ EVALUATE RESULT The puck slows down under the combined action of the frictional force and the force exerted by the spring. So part of the puck's initial kinetic energy is converted to thermal energy, and the rest goes into compressing the spring. Because the spring is not part of the system, that energy leaves the system as (negative) work.

✋ **9.9** (*a*) Draw an energy diagram for the situation shown in Figure 9.11 for the system that comprises Earth and the box. Assume the box keeps moving at constant velocity, and consider the rope to be part of the person. (*b*) Explain each bar in the energy diagram of Figure 9.11*b*.

Self-quiz

1. A stunt performer falls from the roof of a two-story building onto a mattress on the ground. The mattress compresses, bringing the performer to rest without his getting hurt. Is the work done by the mattress on the performer positive, negative, or zero? Is the work done by the performer on the mattress positive, negative, or zero?

2. Consider a weightlifter holding a barbell motionless above his head. (*a*) Is the sum of the forces exerted on the barbell zero? (*b*) Is the weightlifter exerting a force on the barbell? (*c*) If the weightlifter exerts a force, does this force do any work on the barbell? (*d*) Does the energy of the barbell change? (*e*) Are your answers to parts *c* and *d* consistent in light of the relationship between work and energy?

3. Do any of the systems in **Figure 9.14** undergo a change in potential energy? If yes, is the change positive or negative? Ignore any friction.

Figure 9.14

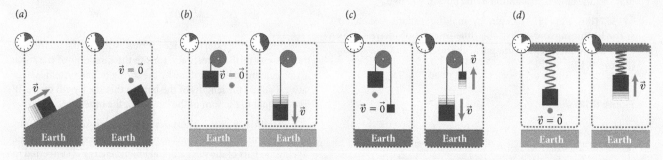

4. A person pushes a cart up a small hill, gaining speed along the way. Friction in the cart's wheels is negligible. Name the components of the system depicted in each energy diagram shown in **Figure 9.15**.

Figure 9.15

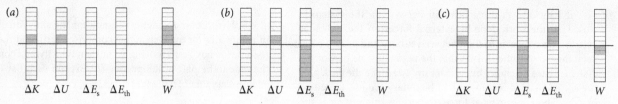

Answers

1. Because the mattress brings the performer to rest, $\vec{F}^c_{mp}$ must point upward. The point of application of this force moves downward as the mattress compresses, meaning the work done by the mattress on the performer is negative. The work done by the performer on the mattress is positive because $\vec{F}^c_{pm}$ points downward and its point of application moves downward.

2. (*a*) Yes, because the barbell remains motionless. (*b*) Yes. He exerts an upward force to counter the downward gravitational force. (*c*) No, because the point at which the lifter exerts a force on the barbell is not displaced. (*d*) No. (*e*) They are consistent. If no work is done on a system, the energy of the system does not change.

3. Yes. (*a*), (*c*), and (*d*). Because none of the situations involves friction or the conversion of source energy, the only things you need to consider are changes in the kinetic and potential energies and work done by external forces on the system. The system in part *a* is closed, and so the decrease in kinetic energy causes an increase in gravitational potential energy: $\Delta U > 0$. In part *b*, Earth is not included in the system, so there can be no potential energy component of the system's energy. In part *c*, the system is closed, so the increase in kinetic energy must be due to a decrease in gravitational potential energy: $\Delta U < 0$. In part *d*, Earth does negative work on the system, so the energy of the system must decrease. Because the kinetic energy increases, the elastic potential energy of the spring must decrease: $\Delta U < 0$.

4. If there is a change in potential energy, the cart and Earth must be in the system; if there is a change in source energy, the person must be in the system. (*a*) Cart and Earth. The person does positive work on the system. (*b*) Person, cart, and Earth. This system is closed; no work is done by external forces on it. Source energy is converted to increase the kinetic and potential energies (and some of it is dissipated). There is more kinetic and potential energy in this system than in part *a* because the person is included in the system. (*c*) Person and cart. Earth does negative work on the system.

9.5 Work done on a single particle

In the first part of this chapter we defined work as the change in the energy of a system due to external forces, and we represented work by the letter W. The SI unit of work is the same as that of energy, J. When work is done by external forces on a system, the energy of the system changes by an amount

$$\Delta E = W. \tag{9.1}$$

This equation is called the **energy law** and embodies conservation of energy: Because energy cannot be created or destroyed, the energy of a system can change only if energy is transferred into or out of the system.* For a system that is closed, $W = 0$ and so the energy law takes on the form we encountered in Chapter 5: $\Delta E = 0$.

To calculate the work done by an external force on a system, we start by considering a simple case: a single particle of inertia m subject to a constant force $\vec{F}$. The term **particle** refers to any object that has no internal structure and no extent in space. Because it lacks extent and internal structure, a particle cannot change shape and therefore any internal energy is fixed ($\Delta E_{int} = 0$). Only the kinetic energy of a particle can change, so

$$\Delta E = \Delta K \quad \text{(particle).} \tag{9.2}$$

To calculate what happens to the kinetic energy of a particle subject to a constant force $\vec{F}$, we consider the particle's motion over a time interval $\Delta t = t_f - t_i$ for the case where the particle's velocity at instant t_i is $\vec{v}_i$. The constant force gives the particle a constant acceleration given by the equation of motion (Eq. 8.8):

$$a_x = \frac{\sum F_x}{m} = \frac{F_x}{m}. \tag{9.3}$$

At the end of the time interval, the x component of the particle's velocity is given by Eq. 3.4,

$$v_{x,f} = v_{x,i} + a_x \Delta t, \tag{9.4}$$

and the x component of the particle's displacement is given by Eq. 3.7,

$$\Delta x = v_{x,i} \Delta t + \tfrac{1}{2} a_x (\Delta t)^2. \tag{9.5}$$

Because we are considering a particle, the displacement of the particle is also the force displacement.

Using these expressions, we can evaluate the change in the particle's kinetic energy:

$$\Delta K = K_f - K_i = \tfrac{1}{2} m (v_f^2 - v_i^2). \tag{9.6}$$

*For now, we consider only mechanical transfers of energy and ignore any other type of energy transfer. Later we'll also take into account transfer of energy by heating and radiation.

Substituting Eq. 9.4 into Eq. 9.6 and keeping in mind that $v^2 = v_x^2$ and $a^2 = a_x^2$, we get

$$\Delta K = \tfrac{1}{2}m\big[(v_{x,i} + a_x\Delta t)^2 - v_{x,i}^2\big]$$

$$= \tfrac{1}{2}m\big[2v_{x,i}a_x\Delta t + a_x^2(\Delta t)^2\big]$$

$$= ma_x\big[v_{x,i}\Delta t + \tfrac{1}{2}a_x(\Delta t)^2\big]$$

$$= ma_x\Delta x_F = F_x\Delta x_F, \qquad (9.7)$$

where the substitution $v_{x,i}\Delta t + \tfrac{1}{2}a_x(\Delta t)^2 = \Delta x_F$ comes from Eq. 9.5, the substitution $ma_x = F_x$ comes from Eq. 9.3, and Δx_F carries a subscript F to remind us that this is the **force displacement** (the displacement of the point of application of the force). Having found this result $\Delta K = F_x\Delta x_F$, we can use Eq. 9.2, $\Delta E = \Delta K$, to say that $\Delta E = F_x\Delta x_F$ and then Eq. 9.1, $\Delta E = W$, to finally say

$$W = F_x\Delta x_F \quad \text{(constant force exerted on particle, one dimension).} \qquad (9.8)$$

In words:

> For motion in one dimension, the work done by a constant force exerted on a particle equals the product of the x components of the force and the force displacement.

If the external force $\vec{F}$ and the force displacement $\Delta\vec{r}_F$ point in the same direction, F_x and Δx_F are either both positive or both negative, and so the product $F_x\Delta x_F$ is positive. Consequently, the kinetic energy increases. When the force and the particle's displacement point in opposite directions, F_x and Δx_F have opposite algebraic signs. The work done by the force on the particle is then negative, and so the kinetic energy of the particle decreases.

If more than one force is exerted on a particle, the x component of its acceleration is given by Eq. 8.8: $a_x = \sum F_x/m$. Substituting this expression for the x component of the acceleration into Eq. 9.7 gives $\Delta K = (\sum F_x)\Delta x_F$. Moving on to Eq. 9.8, we obtain the work done by these forces on the particle:

$$W = (\sum F_x)\Delta x_F \quad \text{(constant forces exerted on particle, one dimension).} \qquad (9.9)$$

Equation 9.9 is what we call the **work equation** for a particle. Together with the energy law (Eq. 9.1), this equation allows us to deal with systems that are not closed.

Notice the parallel between our treatments of momentum/impulse and energy/work (Figure 9.16). Conservation of momentum gives rise to the momentum law, Eq. 4.18, $\Delta\vec{p} = \vec{J}$, where $\vec{J}$ is the impulse delivered to the system. For an isolated system, the impulse is zero and so the momentum of the system does not change. If the system is not isolated, we can use the impulse equation, Eq. 8.25, $\vec{J} = (\sum\vec{F})\Delta t$, to calculate the change in the momentum of the system. Conservation of energy gives rise to the energy law, Eq. 9.1. The work done on a closed system is zero, and so the energy of the system does not change. If the system is not closed, we can use the work equation, Eq. 9.9, to calculate the change in the energy of the system. When we solve problems, it is possible to choose either a closed system or a system that is not closed. As the next example shows,

Figure 9.16 Energy and momentum changes for systems.

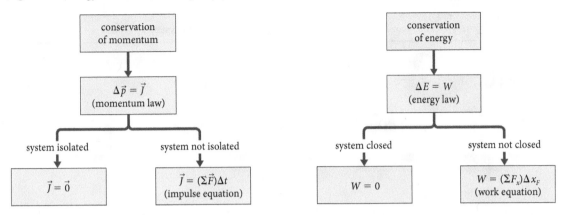

both choices lead to the same answer. If you are unable to obtain a solution with one approach, try the other.

Although we derived Eqs. 9.8 and 9.9 for a particle, they can also be used for any rigid object as long as only the kinetic energy of the object changes. In that case, we can use Eq. 9.2 to calculate the change in the energy of the object. Because all particles that make up the object undergo the same displacement $\Delta \vec{r}$, Eq. 9.8 gives the work done by a constant force exerted on the object.

Example 9.6 Work done by gravity

A ball of inertia m_b is released from rest and falls vertically. What is the ball's final kinetic energy after a displacement $\Delta x = x_f - x_i$?

❶ **GETTING STARTED** I begin by making a sketch of the initial and final conditions and drawing an energy diagram for the ball (**Figure 9.17**). I choose an x axis pointing upward. Because the ball's internal energy doesn't change as it falls (its shape and temperature do not change), I can treat the ball as a particle. Therefore only its kinetic energy changes. I can also assume air resistance is small enough to be ignored, so that the only

external force exerted on the ball is a constant gravitational force. This force has a nonzero force displacement and so does work on the ball. I therefore include this force in my diagram.

❷ **DEVISE PLAN** If I treat the ball as a particle, Eqs. 9.1 and 9.2 tell me that the change in the ball's kinetic energy is equal to the work done on it by the constant force of gravity, the x component of which is given by Eq. 8.17: $F^G_{Eb\,x} = -m_b g$. (The minus sign means that the force points in the negative x direction.) To calculate the work done by this force on the ball, I use Eq. 9.8.

❸ **EXECUTE PLAN** Substituting the x component of the gravitational force exerted on the ball and the force displacement $x_f - x_i$ into Eq. 9.8, I get

$$W = F^G_{Eb\,x}\Delta x_F = -m_b g(x_f - x_i).$$

Because the work is equal to the change in kinetic energy and the initial kinetic energy is zero, I have $W = \Delta K = K_f - 0 = K_f$, so

$$K_f = -m_b g(x_f - x_i). \; ✔$$

❹ **EVALUATE RESULT** Because the ball moves in the negative x direction, $\Delta x = x_f - x_i$ is negative and so the final kinetic energy is positive (as it should be).

An alternative approach is to consider the closed Earth-ball system. For that system, the sum of the gravitational potential energy and kinetic energy does not change, and so, from Eq. 7.13, $\Delta K + \Delta U^G = \frac{1}{2}m_b(v_f^2 - v_i^2) + m_b g(x_f - x_i) = 0$. Because the ball starts at rest, $v_i = 0$, and so I obtain the same result for the final kinetic energy: $\frac{1}{2}m_b v_f^2 = -m_b g(x_f - x_i)$.

Figure 9.17

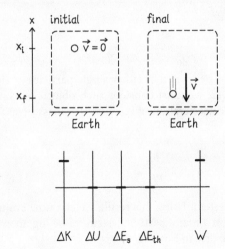

Figure 9.18 Two ways to determine the change in kinetic energy of a falling ball.

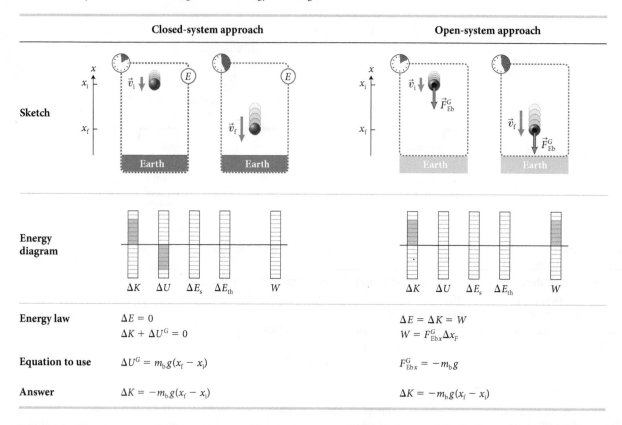

	Closed-system approach	Open-system approach

Energy law

Closed-system:
$$\Delta E = 0$$
$$\Delta K + \Delta U^G = 0$$

Open-system:
$$\Delta E = \Delta K = W$$
$$W = F^G_{Ebx}\,\Delta x_F$$

Equation to use

Closed-system:
$$\Delta U^G = m_b g (x_f - x_i)$$

Open-system:
$$F^G_{Ebx} = -m_b g$$

Answer

Closed-system:
$$\Delta K = -m_b g (x_f - x_i)$$

Open-system:
$$\Delta K = -m_b g (x_f - x_i)$$

The two approaches used in Example 9.6 are shown schematically in **Figure 9.18**. In the closed-system approach, the increase in kinetic energy is attributed to a decrease in the gravitational potential energy of the Earth-ball system. In the open-system approach, the increase in kinetic energy is attributed to the work done by the force of gravity on the ball. In this case it is meaningless to talk about gravitational potential energy because gravitational potential energy refers to the configuration of Earth and the ball *together*. Because Earth is not part of the system, it would not be clear whether to attribute this potential energy to Earth or to the ball.

9.10 Verify that in Example 9.6 the momentum transferred to the ball by the gravitational force is $\Delta p_x = m_b(v_{x,f} - v_{x,i})$.

9.6 Work done on a many-particle system

Everyday objects cannot always be treated as either single particles or rigid objects. As we saw in Section 8.12, we must consider such objects as systems of many interacting particles. The acceleration of the center of mass of such a system is given by Eq. 8.45:

$$\vec{a}_{cm} = \frac{\sum \vec{F}_{ext}}{m}, \tag{9.10}$$

where m is the inertia of the system. Using this result, we can now evaluate the work done by external forces on a many-particle system following an approach analogous to that of a single particle in Eqs. 9.4–9.8.

Consider the motion of a two-particle system over a time interval $\Delta t = t_f - t_i$. If the center of mass of the system moves at velocity $\vec{v}_{cm,i}$ at some initial instant t_i and has constant acceleration $\vec{a}_{cm}$, then from Eq. 9.4 the velocity of the center of mass at the end of the time interval is

$$\vec{v}_{cm,f} = \vec{v}_{cm,i} + \vec{a}_{cm}\Delta t \qquad (9.11)$$

and its displacement is (from Eq. 9.5)

$$\Delta \vec{x}_{cm} = \vec{v}_{cm,i}\Delta t + \tfrac{1}{2}\vec{a}_{cm}(\Delta t)^2. \qquad (9.12)$$

Using these expressions, we can evaluate the changes in the translational kinetic energy of the system from Eq. 6.32, $K_{cm} = \tfrac{1}{2}mv_{cm}^2$:

$$\Delta K_{cm} = K_{cm,f} - K_{cm,i} = \tfrac{1}{2}m(v_{cm,f}^2 - v_{cm,i}^2). \qquad (9.13)$$

Following the same derivation used to obtain Eq. 9.7, we calculate for the change in translational kinetic energy of the system

$$\Delta K_{cm} = ma_{cm\,x}\Delta x_{cm} = (\Sigma F_{ext\,x})\Delta x_{cm} \quad \text{(constant forces, one dimension).} \quad (9.14)$$

Even though this result looks very much like Eq. 9.7, ΔK_{cm} does not represent the work done by the external force on the system. The reasons are that the translational kinetic energy K_{cm} is only part of the kinetic energy of the system (which we know from Eq. 6.35, $K = K_{cm} + K_{conv}$) and for a system of many interacting particles changes can also occur in the other kinds of energy. Therefore $\Delta K_{cm} \neq \Delta E$. Knowing from Eq. 9.1 that $\Delta E = W$, we see that ΔK_{cm} from Eq. 9.14 cannot be equal to the work done by the external force on the many-particle system:

$$\Delta K_{cm} \neq W \quad \text{(many-particle system).} \qquad (9.15)$$

To see this more explicitly, consider the system consisting of two carts connected by a spring, illustrated in **Figure 9.19**. Cart 1 is subject to a single external force $\vec{F}_{ext\,1}$. If we ignore any changes in the carts' internal energies, we can consider the carts to be particles, which means that all of the system's internal energy is stored in the spring. The pushing on cart 1 both accelerates the system and compresses the spring, causing changes in the system's kinetic and internal energies. Unfortunately, we cannot calculate these changes because we don't know how much the spring compresses and we don't know anything about the instantaneous velocities of the carts. Because the work done by external forces on the system is equal to the change in the system's energy E—where E is the sum of all forms of energy—Eq. 9.14 does not provide an expression for the work done on the system.

So, how can we determine the work done by external forces on a many-particle system? Because work is energy transferred, we can use the fact that the work W_{env} done by a system on its environment has the same magnitude as the work W_{sys} done by the environment on the system but carries the opposite sign: $W_{env} = -W_{sys}$. If the system gains, say, 100 J, then conservation of energy dictates that something outside the system must lose 100 J. In **Figure 9.20a**, for example, the external force $\vec{F}_{h1}$ exerted by the hand on cart 1 gives the cart a displacement $\Delta \vec{r}_F$. According to Eq. 8.15, $\vec{F}_{12} = -\vec{F}_{21}$, cart 1 must exert a force $-\vec{F}_{h1}$ on the hand. The work done by cart 1 on the hand over the displacement $\Delta \vec{r}_F$ is, from Eq. 9.8, $-F_{h1\,x}\Delta x_F$ (Figure 9.20b). Because the work done by the two-cart system *on* the hand is $-F_{h1\,x}\Delta x_F$, the work done *by* the hand on the two-cart system must be $+F_{h1\,x}\Delta x_F$. In general, when an external

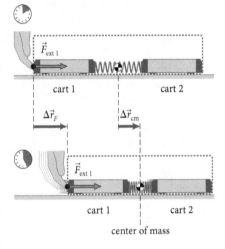

Figure 9.19 An external force does work on a system of two interacting carts.

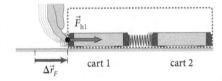

Figure 9.20 The work done by its environment on a system has the same magnitude as the work done by the system on its environment, but the two carry opposite signs: $W_{env} = -W_{sys}$.

(*a*) Hand does work on carts

(*b*) Carts do work on hand

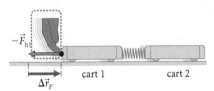

force is exerted on one of the particles in a two-particle system, the work equation is*

$$W = F_{\text{ext }1x}\Delta x_F \quad \text{(constant nondissipative force, one dimension).} \quad (9.16)$$

This work is equal to the product of the external force and Δx_F, the x component of the displacement *of the point of application of this force*, which is different from Δx_{cm}, the x component of the displacement *of the center of mass*, which appears in Eq. 9.14. Although we obtained this result for the specific case of two carts connected by a spring, the result does not depend on any of the specifics of the system and therefore is valid for any system of two interacting particles subject to a single, constant external force.

Example 9.7 Landing on his feet

A 60-kg person jumps off a chair and lands on the floor at a speed of 1.2 m/s. Once his feet touch the floor surface, he slows down with constant acceleration by bending his knees. During the slowing down, his center of mass travels 0.25 m. Determine the magnitude of the force exerted by the floor surface on the person and the work done by this force on him.

❶ **GETTING STARTED** I begin by making a sketch of the initial and final conditions, choosing my person as the system and assuming the motion to be entirely vertical (**Figure 9.21**). I point the x axis downward in the direction of motion, which means that the x components of both the displacement and the velocity of the center of mass are positive: $\Delta x_{\text{cm}} = +0.25$ m and $v_{\text{cm}\,x} = +1.2$ m/s. Two external forces are exerted on the person: a downward force of gravity $\vec{F}^G_{\text{Ep}}$ exerted by Earth and an upward contact force $\vec{F}^c_{\text{sp}}$ exerted by the floor surface. Only the point of application of the force of gravity undergoes a displacement, and so I need to include only that force in my diagram.

Figure 9.21

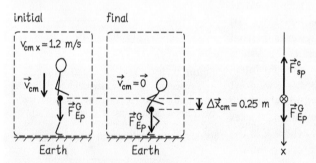

❷ **DEVISE PLAN** Knowing the initial center-of-mass velocity, I can use Eq. 9.13 to calculate the change in the person's translational kinetic energy ΔK_{cm}. To calculate the vector sum of the external forces exerted on the person, I can use the value I obtain for ΔK_{cm} and the displacement $\Delta x_{\text{cm}} = +0.25$ m in Eq. 9.14 to obtain the vector sum of the forces exerted on the person. If I then subtract the force of gravity, I obtain the force

exerted by the floor surface on the person. To determine the work done by this force on the person, I need to multiply it by the force displacement. Because the person slows down as he travels downward, his acceleration is upward and so the vector sum of the forces is upward too. To remind myself of this, I draw a free-body diagram in which the arrow for the upward contact force is longer than the arrow for the downward force of gravity.

❸ **EXECUTE PLAN** Because the person ends at rest, his final translational kinetic energy is zero, and Eq. 9.13 gives me

$$\Delta K_{\text{cm}} = 0 - \tfrac{1}{2}mv^2_{\text{cm,i}} = \tfrac{1}{2}(60 \text{ kg})(1.2 \text{ m/s})^2 = -43 \text{ J}.$$

Substituting this value and the displacement of the center of mass into Eq. 9.14 yields

$$\sum F_{\text{ext}\,x} = \frac{\Delta K_{\text{cm}}}{\Delta x_{\text{cm}}} = \frac{-43 \text{ J}}{0.25 \text{ m}} = -170 \text{ N}.$$

To obtain the force exerted by the floor from this vector sum, I add to the sketch a free-body diagram showing the two forces exerted on the person (Figure 9.21).

$$\sum F_{\text{ext}\,x} = F^G_{\text{Ep}\,x} + F^c_{\text{sp}\,x}$$

and so $F^c_{\text{sp}\,x} = \sum F_{\text{ext}\,x} - F^G_{\text{Ep}\,x}$. The x component of the force of gravity is $F^G_{\text{Ep}\,x} = mg = (60 \text{ kg})(9.8 \text{ m/s}^2) = +590$ N and so $F^c_{\text{sp}\,x} = -170$ N $- 590$ N $= -760$ N. ✔

To determine the work done by this force on the person, I must multiply the x component of the force by the force displacement. The point of application is at the floor, which doesn't move. This means that the force displacement is zero, and so the work done on the person is zero too: $W = 0$. ✔

❹ **EVALUATE RESULT** The contact force $F^c_{\text{sp}\,x}$ is negative because it is directed upward, as I expect. Its magnitude is larger than that of the force of gravity, as it should be in order to slow the person down.

---

*We can apply this line of reasoning only when the force $\vec{F}_{\text{ext }1}$ is nondissipative. As we saw in Section 9.4, if the force exerted at the boundary of a system is dissipative, we don't know how much of the dissipated energy ends up inside the system.

✋ **9.11** Compare the rightmost terms of Eqs. 9.14 and Eq. 9.16 to determine whether, in any two-particle system subject to a single, constant external force, ΔE is larger than, equal to, or smaller than ΔK_{cm}. Explain your answer.

We can easily generalize Eq. 9.16 to a many-particle system subject to several constant forces. When more than one external agent interacts with the system, we can follow the same argument to calculate the work done by each external agent on the system. If agent 1 exerts a force $\vec{F}_{ext\,1}$ on a system and the point of application of this force undergoes a displacement $\Delta\vec{r}_{F1}$, then the work done by that agent on the system is $W_1 = F_{ext\,1x}\Delta x_{F1}$.* Likewise, the work done by agent 2 is $W_2 = F_{ext\,2x}\Delta x_{F2}$. The work done by all the agents in the environment on the system is the sum of the work done by each on the system:

$$W = W_1 + W_2 + \cdots = F_{ext\,1x}\Delta x_{F1} + F_{ext\,2x}\Delta x_{F2} + \cdots \qquad (9.17)$$

or

$$W = \sum_n (F_{ext\,nx}\Delta x_{Fn}) \quad \text{(constant nondissipative forces, one dimension)}, \quad (9.18)$$

which is the work equation for a many-particle system subject to constant external forces.

Note that to calculate this work we must take the product of each external force and the force displacement *of that force*. If the system is an object that deforms, these displacement values are different in different parts of the system. For that reason both $F_{ext\,nx}$ and Δx_{Fn} in Eq. 9.18 carry a subscript n.

If the force is dissipative, we cannot use this expression because the dissipated energy is distributed on both sides of the system boundary. Dissipative forces are discussed in the next section. The results for a single particle and a system of particles are summarized in Table 9.1.

✋ **9.12** Show that for a one-particle system, Eqs. 9.14 and 9.18 both reduce to Eq. 9.9.

Table 9.1 Equation of motion and energy accounting for a particle and a system of particles

	Single particle or rigid object	System of particles or deformable object
Equation of motion	$\vec{a} = \dfrac{\sum\vec{F}}{m}$	$\vec{a}_{cm} = \dfrac{\sum\vec{F}_{ext}}{m}$
Energy law	$\Delta E = W$	$\Delta E = W$
Work equation	$W = (\sum F_x)\Delta x_F$	$W = \sum_n (F_{ext\,nx}\Delta x_{Fn})$
Kinetic energy change	$\Delta K = W$	$\Delta K_{cm} = (\sum F_{ext\,x})\Delta x_{cm}$

*When work is done by more than one force on an object or system, the work done by each force individually inherits the subscripts of the corresponding force.

QUANTITATIVE TOOLS

9.7 Variable and distributed forces

The results obtained in the preceding two sections apply only to forces that are constant and that are applied at well-defined locations in a system. If the forces are not constant or if we cannot determine the force displacement, then the results we obtained do not apply. Examples are the force exerted by a spring, which varies with the amount of compression or stretching, and the frictional force, which is distributed over a surface and for which it is not possible to determine the point of application. Let us examine these two cases in more detail.

What should one do if the force exerted on a system isn't constant? (We call such a force a *variable force*.) For this kind of force, the x component can be written as a function of x: $F_x = F_x(x)$. The trick is to apply the constant-force result over a force displacement small enough to make the force approximately constant over that displacement. Figure 9.22 shows how, for a force that varies with the position at which that force is applied, we can divide the force displacement into small segments.

Suppose we want to know the work done by the variable force on a system while the point of application of the force moves from x_i to x_f in Figure 9.22. We first divide the force displacement $x_f - x_i$ into N equal smaller force displacements δx_F such that the force is approximately constant across each δx_F (Figure 9.22a). Applying Eq. 9.8 to each smaller displacement δx_F, we can write the work done by the variable force on the system during each displacement as

$$W_n = F_x(x_n)\delta x_F, \tag{9.19}$$

where $F_x(x_n)$ is the x component of the force at x_n (this point x_n is located at the center of each interval, as shown in Figure 9.22a). This work is the area of the dark shaded rectangle in Figure 9.22a. The work done over the entire force displacement from x_i to x_f is then approximately equal to the sum of work W_n over all small force displacements δx_F between x_i and x_f:

$$W \approx \sum_n W_n = \sum_n F_x(x_n)\delta x_F, \tag{9.20}$$

where $\sum_n F_x(x_n)\delta x_F$ represents the sum of all the rectangular areas under the steps between x_i and x_f (the light shaded area in Figure 9.22a).

To get a better approximation, we can divide the displacement $x_f - x_i$ into a larger number of smaller force displacements δx_F, so that the variation of the external force over each displacement becomes even smaller and the red stepped curve is closer to the actual (light red) $F_x(x)$ curve.

An exact result is obtained if we let δx_F approach zero (and the number of small displacements approaches infinity), as in Figure 9.22b:

$$W = \lim_{\delta x_F \to 0} \sum_n F_x(x_n)\delta x_F, \tag{9.21}$$

which is precisely the definition of the integral of the x component of the external force with respect to x from x_i to x_f:

$$W = \int_{x_i}^{x_f} F_x(x)dx \quad \text{(nondissipative force, one dimension).} \tag{9.22}$$

 9.13 Show that Eq. 9.22 reduces to Eq. 9.8 when the force is constant.

An example of a force that varies with the position at which it is applied is the force exerted by a spring on a load: $\vec{F}_{sl} = -k(\vec{x} - \vec{x}_0)$ (Eq. 8.20).

Figure 9.22 Evaluating the work done by a variable force $\vec{F}(x)$ on a system.

(*a*) Work approximated by area under steps of constant force

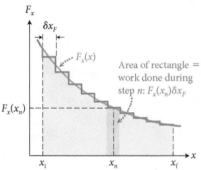

(*b*) As δx_F approaches zero, area under curve approaches work done

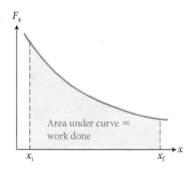

Example 9.8 Spring work

A brick of inertia m compresses a spring of spring constant k so that the free end of the spring is displaced from its relaxed position. What is the work done by the brick on the spring during the compression?

① GETTING STARTED I begin by making a sketch of the situation as the free end of the spring is compressed from its relaxed position x_0 to a position x (**Figure 9.23**). Because I need to calculate the work done by the brick on the spring, I choose the spring as my system. Three forces are exerted on the spring: contact forces exerted by the brick and by the floor, and the force of gravity. As usual when dealing with compressed springs, I ignore the force of gravity exerted on the spring (see Section 8.6). Only the force $\vec{F}^c_{bs}$ exerted by the brick on the spring undergoes a nonzero force displacement, so I need to show only that force in my diagram. Because the brick and spring do not exert any forces on each other when the spring is in the relaxed position, I do not draw this force in the initial state.

Figure 9.23

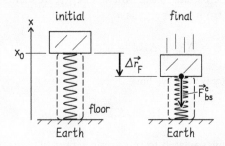

② DEVISE PLAN I need to calculate the work done by the brick on the spring. I could use Eq. 9.22 if I knew the force the brick exerts on the spring. That force is not given, but the force exerted

by the brick on the spring and the force exerted by the spring on the brick form an interaction pair: $\vec{F}^c_{bs} = -\vec{F}^c_{sb}$. I can use Eq. 8.20 to determine $\vec{F}^c_{sb}$ and then use it in Eq. 9.22.

③ EXECUTE PLAN Equation 8.20 tells me that the x component of the force exerted by the spring on the brick varies depending on how far the spring is compressed:

$$F_{sbx} = -k(x - x_0),$$

where x_0 is the coordinate of the relaxed position of the free end of the spring. The x component of the force exerted by the brick on the spring is thus

$$F_{bsx} = +k(x - x_0). \tag{1}$$

Because $x_0 > x$, F_{bsx} is negative, which means that $\vec{F}^c_{bs}$ points in the same direction as the force displacement. Thus the work done by the brick on the spring is positive. Now I substitute Eq. 1 into Eq. 9.22 and work out the integral to determine the work done by the brick on the spring:

$$W_{bs} = \int_{x_0}^{x} F_{bsx}(x)dx = \int_{x_0}^{x} k(x - x_0)dx$$

$$= \left[\tfrac{1}{2}kx^2 - kx_0x\right]_{x_0}^{x} = \tfrac{1}{2}k(x - x_0)^2. ✔ \tag{2}$$

④ EVALUATE RESULT Because the spring constant k is always positive (see Section 8.9 on Hooke's law), the work done by the brick on the spring is also positive. This is what I expect because the work done in compressing the spring is stored as potential energy in the spring.

9.14 In Example 9.8, consider the situation from the instant the brick is released to the instant at which it has zero velocity (when the spring compression is greatest). Draw an energy diagram for a system that comprises (a) the spring alone and (b) Earth, spring, and brick.

The energy law tells me that the change in the energy of the spring in Example 9.8 is equal to the work done by the brick on it: $\Delta E = W$. As you saw in part a of Checkpoint 9.14, only the potential energy of the spring (corresponding to the configuration of the spring) changes, and so $\Delta E = \Delta U_{spring} = W$. The change in the **potential energy of a spring** as its free end is displaced from its relaxed position x_0 to any position x is thus

$$\Delta U_{spring} = \tfrac{1}{2}k(x - x_0)^2. \tag{9.23}$$

This expression also describes the change in potential energy when the spring is stretched by some external force. To stretch the spring, the force exerted on it must point upward in Figure 9.23, which means that instead of $\Delta \vec{r}$ and $\vec{F}^c_{bs}$ both pointing downward as they do in compression, these two vectors both point upward. The expression for the work done by the stretching force on the spring (Eq. 2 in Example 9.8) remains the same because the displacement and the force exerted on the spring still both point in the same direction and so Eq. 9.23 also remains the same. Consequently the potential energy of a spring increases both when the spring is compressed and when it is stretched.

9.15 (Brain twister) As the brick in Example 9.8 moves downward, why is the magnitude of the force it exerts on the spring given by Eq. 1 in Example 9.8 and not by simply the magnitude *mg* of the gravitational force exerted by Earth on the brick? Hint: Draw a free-body diagram for the brick as it moves downward.

Next we turn to the force of friction. As we discussed in Section 9.4, friction poses a problem for two reasons. First, the energy dissipated by friction ends up being divided between two objects. Second, there is no single point of application for frictional forces—the force is distributed over two surfaces that move relative to each other.

Consider a block sliding across a rough surface. The block is initially moving at some speed v but eventually comes to rest because of friction between it and the surface. **Figure 9.24** shows an energy diagram for the closed block-surface system. Because there are no reversible changes in configuration and no source energy is used, we know that all of the block's kinetic energy is converted to thermal energy. So, if we know the block's initial kinetic energy, we can determine the thermal energy generated from the energy law:

$$\Delta E = \Delta K + \Delta E_{th} = 0. \qquad (9.24)$$

Because the final kinetic energy is zero, we have

$$\Delta E_{th} = -\Delta K = -(K_f - K_i) = K_i. \qquad (9.25)$$

How are these energy changes related to the displacement $\Delta \vec{r}_{cm}$ over which the block slides to a stop? Let the force of friction exerted by the surface on the block be $\vec{F}^f_{sb}$. According to Eq. 9.14, the change in the block's translational kinetic energy is equal to the product of the external force exerted on the block and the displacement of its center of mass:

$$\Delta K_{cm} = F^f_{sb x}\Delta x_{cm}. \qquad (9.26)$$

Because the frictional force and the block's displacement point in opposite directions, the right side of Eq. 9.26 is negative. Because the block is a rigid object, the translational kinetic energy is all of its kinetic energy, and so $\Delta K_{cm} = \Delta K$. Substituting Eq. 9.25, then yields an expression for the energy dissipated by friction:

$$\Delta E_{th} = -F^f_{sb x}\Delta x_{cm} \quad \text{(constant dissipative force, one dimension).} \qquad (9.27)$$

When you use this equation, keep in mind that the energy is dissipated on both sides of the surface where the friction occurs. So, even though the right sides of Eqs. 9.26 and 9.27 are the product of a constant force and a displacement, it would not be correct to refer to this product as "the work done by the friction force exerted on the block." Furthermore, Eq. 9.27 holds for travel in only one direction; if the block reverses the direction of travel, you have to apply Eq. 9.26 to each one-way segment separately. If the sum of the distances traveled over all the one-way segments is d_{path}, the change in thermal energy becomes

$$\Delta E_{th} = F^f_{sb} d_{path} \quad \text{(constant dissipative force, one dimension),} \qquad (9.28)$$

where F^f_{sb} is the magnitude of the force of friction and where we have dropped the minus sign of Eq. 9.27 because all quantities are positive.

Figure 9.24 Energy diagram for a system in which friction dissipates kinetic energy to thermal energy.

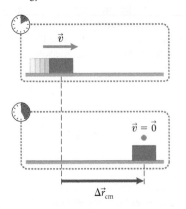

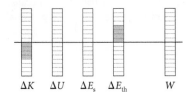

$$\Delta K \quad \Delta U \quad \Delta E_s \quad \Delta E_{th} \quad W$$

9.16 A 0.50-kg wood block slides 0.50 m on a horizontal floor before colliding elastically with a wall and reversing its direction of travel. If the block has an initial speed of 1.0 m/s and comes to rest exactly at its starting position, what is the magnitude of the frictional force between block and floor?

9.8 Power

Often we are interested not just in knowing how much energy is converted within a system or transferred to a system but also in knowing *how fast* the energy is converted or transferred. The rate at which energy is transferred or converted is called **power.** If the energy of a system changes by an amount ΔE over a time interval Δt, the average power is defined as

$$P_{av} = \frac{\Delta E}{\Delta t}. \tag{9.29}$$

The SI unit of power is the **watt** (W), which is equal to 1 J of energy per second: $1\ \mathrm{W} \equiv 1\ \mathrm{J/s}$.* A person in good physical condition can deliver about 75 W of average power over a prolonged time interval. This means a healthy person can do work—shovel snow, accelerate an object, and so on—at the rate of about 75 J/s. An athlete can deliver about 400 W over a sustained time interval and up to about 1000 W during short time intervals.

The instantaneous power is

$$P = \lim_{\Delta t \to 0} \frac{\Delta E}{\Delta t} = \frac{dE}{dt}. \tag{9.30}$$

Equation 9.30 gives the rate at which the energy of a system changes. The concept of power can be applied to the rate at which any type of energy changes. For example, the power at which thermal energy is generated is given by

$$P = \frac{dE_{th}}{dt}. \tag{9.31}$$

Example 9.9 Frictional power

A 0.50-kg wood block initially traveling at 1.0 m/s slides 0.50 m on a horizontal floor before coming to rest. What is the average rate at which thermal energy is generated?

1 GETTING STARTED I begin by making a sketch of the situation, listing all the given quantities and drawing an energy diagram for the closed block-floor system (**Figure 9.25**). I note that as the block slides to a stop, all of its kinetic energy is converted to thermal energy.

2 DEVISE PLAN To calculate the average power at which thermal energy is generated, I need to know the amount ΔE_{th} of thermal energy generated and the time interval Δt over which it is generated. The ratio of these two quantities gives the average power (Eq. 9.29). The amount of thermal energy is equal to the initial kinetic energy of the block. If I assume the block has constant acceleration as it comes to a stop, I can use equations from Chapter 3 to determine Δt from the block's initial speed and the distance over which it comes to a stop.

Figure 9.25

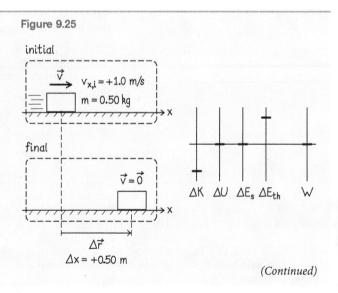

(Continued)

---

*Be careful about distinguishing between italic *W*, our symbol for work, and nonitalic W, the abbreviation for watt, the SI unit of power.

③ **EXECUTE PLAN** Because the block ends at rest, I know from Eq. 3.4, $v_{x,f} = v_{x,i} + a_x\Delta t$, that

$$a_x = -\frac{v_{x,i}}{\Delta t}.$$

Substituting this result into Eq. 3.8, $x_f = x_i + v_{x,i}\Delta t + \frac{1}{2}a_x(\Delta t)^2$, gives

$$\Delta x = v_{x,i}\Delta t - \frac{1}{2}\frac{v_{x,i}}{\Delta t}(\Delta t)^2 = \frac{1}{2}v_{x,i}\Delta t$$

or $\qquad \Delta t = \frac{2\Delta x}{v_{x,i}} = \frac{2(0.50\text{ m})}{1.0\text{ m/s}} = 1.0\text{ s}.$

The block's initial kinetic energy is $\frac{1}{2}mv_{x,i}^2 = \frac{1}{2}(0.50\text{ kg})(1.0\text{ m/s})^2 = 0.25$ J, so the thermal energy generated is $\Delta E_{th} = 0.25$ J. From Eq. 9.29, I then obtain for the average power at which this energy is generated: $P_{av} = (0.25\text{ J})/(1.0\text{ s}) = 0.25$ W. ✔

④ **EVALUATE RESULT** The value I obtain is rather small—as a gauge, consider that a 25-W light bulb consumes energy at 100 times this rate—but it's unlikely that the energy of a sliding block would be sufficient to generate any light.

Next consider a situation where a system's energy is changed by a constant external force $\vec{F}_{ext}$. One such instance might be when you push a cart. As long as the force you exert on the cart is constant, we can analyze the power being supplied by you to the cart in terms of the x components of the force $F_{ext\,x}$ and the force displacement Δx. We know from Eqs. 9.1 and 9.8 that

$$\Delta E = W = F_{ext\,x}\Delta x. \qquad (9.32)$$

Substituting this expression for ΔE into Eq. 9.29 yields

$$P_{av} = \frac{\Delta E}{\Delta t} = F_{ext\,x}\left(\frac{\Delta x}{\Delta t}\right) \qquad (9.33)$$

or, with the definition of average velocity,

$$P_{av} = F_{ext\,x}\,v_{x,av}, \qquad (9.34)$$

where $v_{x,av}$ is the x component of the average velocity of the point of application of the force (and because the cart is a rigid object it is also the x component of the average velocity of the cart). The power is positive when $\vec{F}_{ext}$ and $\vec{v}_{av}$ point in the same direction, and in this case the energy of the system increases. When $\vec{F}_{ext}$ and $\vec{v}_{av}$ point in opposite directions, the power is negative and the energy of the system decreases.

To obtain the instantaneous power, we take the limit as Δt in Eq. 9.33 goes to zero, yielding

$$P = F_{ext\,x}v_x \quad \text{(one dimension)}. \qquad (9.35)$$

Example 9.10 Car drag

A car requires 300 kJ of energy to overcome air resistance and maintain a constant speed of 20 m/s over a distance of 1.0 km. What is the force of air resistance exerted on the car?

❶ **GETTING STARTED** The air resistance causes dissipation of the car's kinetic energy. The problem statement tells me that 300 kJ is dissipated over a distance of 1.0 km when the car moves at 20 m/s.

❷ **DEVISE PLAN** From the speed and the distance traveled, I can calculate the time interval over which the 300 kJ is required.

I can then use Eq. 9.29 to calculate the rate at which the energy is converted. Knowing the velocity, I can then obtain the force of air resistance from Eq. 9.35.

❸ **EXECUTE PLAN** At 20 m/s it takes the car 50 s to cover 1.0 km. The rate at which energy is converted is thus $P = (300\text{ kJ})/(50\text{ s}) = 6.0$ kW. From Eq. 9.35, I obtain

$$F_x = \frac{P}{v_x} = \frac{6.0\text{ kW}}{20\text{ m/s}} = 300\text{ N}. ✔$$

❹ EVALUATE RESULT The answer I obtain—300 N—is the magnitude of the gravitational force exerted on a 30-kg (65-lb) object and therefore equal in magnitude to the force required to hold up such an object. I know, however, that if I stick my hand out a car window at highway speed, the force exerted by the air on my hand is not much larger than the force required to hold up a small object. The car, being much larger than my hand, intercepts a lot more air, but it is shaped more aerodynamically, and so it is reasonable that the force of air resistance is only 300 N.

✋ **9.17** (*a*) A gallon of gasoline contains about 1.4×10^8 J of chemical energy. A car consumes this amount of gasoline in approximately 30 min when cruising along a highway; a plane consumes the same amount in about 1 s when flying at cruising altitude. What is the average power of the energy release in each case? (*b*) A 10-kg load must be hoisted up the side of a building that is 50 m tall. How quickly can an athlete capable of delivering 500 W of power get the job done? (*c*) A 3-kW engine gets the job done six times faster than the athlete. How does the work done by the engine on the load compare with the work done by the athlete on the load?

Chapter Glossary

SI units of physical quantities are given in parentheses.

Energy diagram A diagram that accounts for changes in the energy of a system. When using such a diagram, avoid choosing a system in which friction occurs at the boundary and remember that potential energy involves an interacting pair of objects. See the Procedure box on page 207.

Energy law The law that accounts for the change in the energy of a system or object. Because energy is conserved, the energy of a system can change only due to a transfer of energy between the environment and the system:

$$\Delta E = W. \tag{9.1}$$

Force displacement $\Delta \vec{r}_F$ (m) The displacement of the point of application of a force.

Particle An object that does not have any internal structure or spatial extent. The internal energy of a particle is fixed: $\Delta E_{int} = 0$.

Potential energy of a spring U_{spring} (J) The change in the potential energy of a spring as its free end is displaced from its relaxed position x_0 to any position x is

$$\Delta U_{spring} = \tfrac{1}{2}k(x - x_0)^2. \tag{9.23}$$

Power P (W) A scalar equal to the rate at which energy is transferred or converted:

$$P = \frac{dE}{dt}. \tag{9.30}$$

In one dimension, the power delivered by a constant external force is

$$P = F_{ext\,x}v_x \quad \text{(one dimension)}, \tag{9.35}$$

where v_x is the velocity of the point of application of the force.

Watt (W) The derived SI unit of power, defined as $1\,\text{W} \equiv 1\,\text{J/s} = 1\,\text{kg}\cdot\text{m}^2/\text{s}^3$.

Work W (J) A scalar equal to the change in the energy of a system due to external forces. If the energy of a system increases, the work done by external forces on the system is positive; if the energy decreases, the work done by external forces on the system is negative.

Work equation The equation that gives the work done by external forces on a particle or system:

$$W = \left(\sum F_x \right)\Delta x_F \quad \begin{array}{l}\text{(constant forces exerted on}\\ \text{particle, one dimension)}\end{array} \tag{9.9}$$

and

$$W = \sum_n \left(F_{ext\,nx}\Delta x_{Fn} \right) \quad \begin{array}{l}\text{(constant nondissipative}\\ \text{forces, on many-particle}\\ \text{system, one dimension).}\end{array} \tag{9.18}$$

When the force and the force displacement point in the same direction, the work done by external forces on the system is positive; when they point in opposite directions, the work done by external forces on the system is negative. The work done by a variable nondissipative force exerted on a system is given by

$$W = \int_{x_i}^{x_f} F_x(x)\,dx \quad \begin{array}{l}\text{(nondissipative force,}\\ \text{one dimension).}\end{array} \tag{9.22}$$

10 Motion in a Plane

So far, we have restricted our study of physics to events that take place along a straight line—that is, in one dimension. In this chapter we develop the tools that allow us to deal with motion in a plane—in other words, motion that takes place in two dimensions. As you will see, any problem in two dimensions can be reduced to two one-dimensional problems (which you already know how to solve). After discussing the techniques that allow us to deal with vector quantities in two dimensions, we study the decomposition of two-dimensional force vectors. This decomposition allows us to examine more closely the contact forces that arise at the surface between two objects, which leads us to the subject of friction. Finally, we extend the concept of work so that it applies to situations where the force and force displacement vectors are not along the same line.

10.1 *Straight* is a relative term

To begin our discussion of motion in a plane, look at the film clip in **Figure 10.1**, which shows the motion of a ball dropped from the top of a pole attached to a cart that is moving at constant velocity along a horizontal track. You already know the resulting motion of the ball: It accelerates downward because of the force of gravity exerted on it (see Section 8.8). Indeed, the ball falls alongside the pole. As we discussed in Chapter 6, this is the way the motion must be because the laws of physics are the same in all inertial reference frames, regardless of whether the reference frames are at rest or moving at constant velocity with respect to the surface of Earth.

To extend this example to a more common situation, consider what happens when you drop an object inside the cabin of an airplane cruising at constant velocity. That object falls to the cabin floor in a straight line, exactly as it would if the airplane were at rest on the ground. Likewise, in the reference frame of the moving cart of Figure 10.1, the ball drops to the moving cart with a velocity that points vertically downward. In the cart's reference frame, the motion of the ball is free-fall motion, and a camera that moves along with the cart registers the motion shown in Figure 10.2*a*.

The picture is different, however, when observed from a stationary point located somewhere outside the cart—that is, from the Earth reference frame. Because the cart is moving to the right, the trajectory of the ball cannot be a straight vertical line in this reference frame. Figure 10.2*b* shows all the frames of the film clip of Figure 10.1 superimposed on one another. As you can see, in addition to its straight-downward motion relative to the cart, the ball is displaced horizontally from one frame to the next. Note that in spite of the rightward motion of the cart, the distance between ball and pole does not change. This constant distance is precisely what gives the illusion that the ball falls straight down! Your eyes tend to move along with the cart, and so you register only the vertical motion of the ball relative to the cart. The

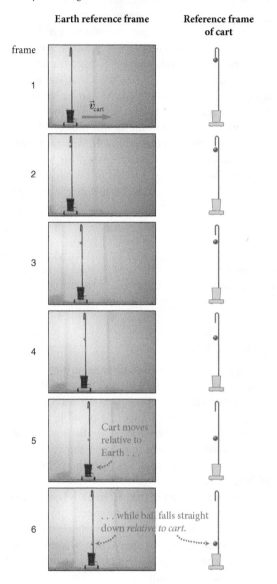

Figure 10.1 A ball dropped from a pole attached to a cart moving at constant velocity falls straight down *relative to the cart.*

Figure 10.2 Motion of the ball of Figure 10.1 as recorded by a camera that is (*a*) moving along with the cart and (*b*) stationary.

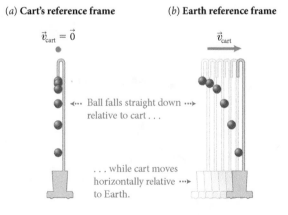

frames of the film clip, however, show that the horizontal displacement of the ball is identical to that of the cart: If the *x* component of the displacement of the cart is Δx from one frame to the next, then the *x* component of the displacement of the ball must also be Δx. So, we see that *straight* is a relative term, one that depends on the reference frame.

The analysis in Figure 10.2 tells us that the motion of the ball in the Earth reference frame can be broken down into two parts: free fall in the vertical direction (called the *vertical component* of the motion) and motion at constant velocity in the horizontal direction (the *horizontal component* of the motion). You already know how to treat each of these motions separately. All that is left to do, therefore, is to develop a way to describe both motions at the same time.

✋ **10.1** (*a*) In Figure 10.2, what is the ball's velocity the instant before it is released? (*b*) Is the ball's speed in the reference frame of the cart greater than, equal to, or smaller than its speed in the Earth reference frame?

10.2 Vectors in a plane

To describe the motion of the ball in Figure 10.1, we need to generalize our definition of vectors. First of all, we need not one but two reference axes—one for each component of the motion. **Figure 10.3** shows one possible choice for a set of perpendicular reference axes. To distinguish the two axes, we label the horizontal one *x* and the vertical one *y*. The origins of both axes coincide with the point at which the ball is released.

To specify the position of the ball in each frame, we need two coordinates: an *x coordinate* to specify position along the *x* axis and a *y coordinate* to specify position along the *y* axis. For example, in the coordinate system shown in Figure 10.3, the position of the ball in frame 6 of Figure 10.1 is specified by the coordinates $x = +0.22$ m, $y = -0.42$ m. These coordinates also determine the position vector of the ball (see Section 2.6), which points from the origin to the position of the ball, as shown in Figure 10.3.

Figure 10.3 Position vector for the ball in frame 6 of Figure 10.1, using reference axes in the Earth reference frame.

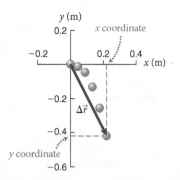

Figure 10.4 Displacement of the ball of Figure 10.1 in the reference frames of the cart and Earth.

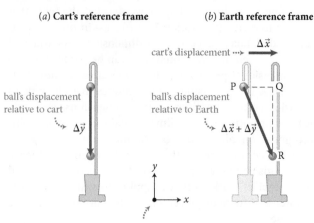

(*a*) **Cart's reference frame**   (*b*) **Earth reference frame**

"thumbnail axes" indicate coordinate system

Unlike vectors in one dimension (describing motion along a straight line), vectors in two dimensions (describing motion in a plane) do not necessarily point along the axes. Consequently, we need to reexamine the rules for vector addition and subtraction. We'll do this by looking at the vector that represents the ball's displacement between frames 2 and 6 in Figure 10.1. **Figure 10.4a** shows the ball's displacement in the reference frame of the moving cart. This vector is labeled $\Delta \vec{y}$ to reflect the fact that it is parallel to the *y* axis. Figure 10.4b shows the corresponding view in the Earth reference frame. The *cart's* displacement is labeled $\Delta \vec{x}$ because it is parallel to the *x* axis. The *ball's* displacement, however, now points diagonally from the ball's initial position P to its final position R. Note that the ball can also go from P to R by two successive displacements: first from P to Q and then from Q to R. Even if the ball follows this path, though, its displacement is still given by the vector pointing from P to R. In other words, the ball's displacement in the Earth reference frame is the *vector sum* of the horizontal displacement $\Delta \vec{x}$ and the vertical displacement $\Delta \vec{y}$.

As with vectors along a straight line (one dimension), the vector sum of two vectors in a plane (two dimensions) is obtained by placing the tail of the second vector at the head of the first vector. There is, however, one important difference between one-dimensional and two-dimensional vector addition (**Figure 10.5**). In one dimension, the sum of two

Figure 10.5 Vector addition in one and two dimensions.

(*a*) Adding vectors in one dimension

(*b*) Adding vectors in two dimensions

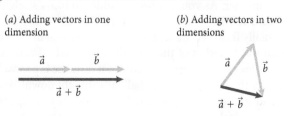

Figure 10.6 The order in which vectors are added does not affect the final answer. These two imaginary flights starting at LaGuardia airport consist of the same vectors in different orders; both end up at the same destination of John F. Kennedy airport.

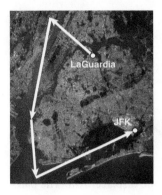
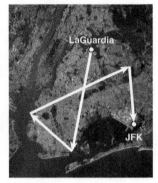

Figure 10.7 Vector addition and subtraction in two dimensions.

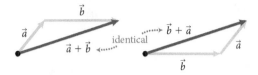

Vector addition: Vectors may be added in any order: $\vec{a} + \vec{b} = \vec{b} + \vec{a}$

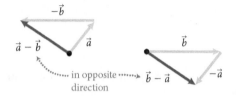

Vector subtraction: Order matters! $\vec{a} - \vec{b} \neq \vec{b} - \vec{a}$

vectors of equal length is either zero (when the vectors point in opposite directions) or a vector twice as long as each original vector (when they point in the same direction). In two dimensions, the length of the vector sum depends on the angle between the vectors being added. As Figure 10.5*b* shows, the length of the sum of two vectors can be smaller than the length of either original vector. When two or more vectors in a plane are added, the order in which the vectors are added is immaterial, as Figure 10.6 shows.

The procedure for subtracting vectors in a plane is identical to that for subtraction in one dimension (Figure 10.7):

> **To subtract a vector $\vec{b}$ from a vector $\vec{a}$, reverse the direction of $\vec{b}$ and then add the reversed $\vec{b}$ to $\vec{a}$.**

As in one dimension, the vector $\vec{a} - \vec{b}$ points from the tail of vector $\vec{a}$ to the tip of vector $-\vec{b}$. Note that vector addition is commutative but vector subtraction is not: $\vec{a} + \vec{b} = \vec{b} + \vec{a}$ but $\vec{a} - \vec{b} \neq \vec{b} - \vec{a}$.

We can use the principles of vector addition to develop an important procedure: the decomposition of a vector into components. Just as we can add any two vectors to form a new vector, we can decompose (break up) any vector into

two component vectors. For example, any vector $\vec{A}$ can be decomposed into **component vectors** $\vec{A}_x$ and $\vec{A}_y$ along the axes of some conveniently chosen set of mutually perpendicular axes, called a **rectangular coordinate system** (also called a *Cartesian coordinate system*).*

The procedure for decomposing a vector is shown in Figure 10.8. Note that if you add the resulting component vectors, you recover the original vector. When you decompose a vector, it is important to cross out the original vector so that you do not take both the original vector and the component vectors into account in any subsequent analysis.

---

*There is no formal requirement that the axes of the coordinate system be perpendicular to each other. Any set of two intersecting axes allows you to decompose and therefore uniquely specify all possible vectors in a plane. Calculations are considerably simpler, however, with axes that are perpendicular to each other.

Figure 10.8 Procedure for decomposing a vector using a set of mutually perpendicular axes.

To decompose a vector . . .

❶ Add axes with origin at vector tail.

❷ Drop lines from vector tip to axes; these determine lengths of component vectors.

❸ Replace original vector with component vectors along axes.

Original vector is sum of component vectors.

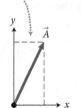
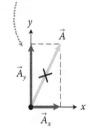

Figure 10.9 Decomposition of a vector $\vec{A}$ in three coordinate systems. In each coordinate system the component vectors $\vec{A}_x$ and $\vec{A}_y$ completely specify the position of point P relative to the origin. Which coordinate system is most convenient depends on the problem at hand.

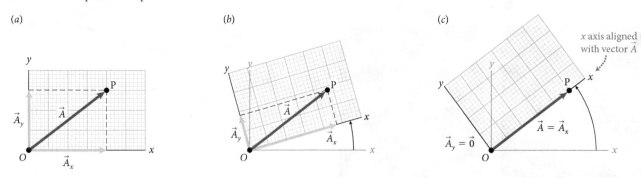

(a) *(b)* *(c)*

Figure 10.9 shows the decomposition of a vector $\vec{A}$ in three coordinate systems. In each coordinate system the component vectors $\vec{A}_x$ and $\vec{A}_y$ completely specify the position of point P relative to the origin O. In Figure 10.9c the coordinate system is such that the vector $\vec{A}$ lines up with the x axis. Now the component vector along the x axis is $\vec{A}$ and the component vector along the y axis is the zero vector. Which choice of coordinate system works best depends on the problem at hand (more on this later).

Exercise 10.1 Decompositions

If each major grid unit in Figure 10.9 corresponds to 1 m, specify the location of point P in terms of its x and y coordinates in each of the three coordinate systems.

SOLUTION (a) The vertical line dropped from P intercepts the x axis 4 grid units to the right of the y axis, so $x = +4.0$ m. The horizontal line dropped from P gives $y = +3.0$ m. The location of point P is thus given by $(x, y) = (+4.0$ m, $+3.0$ m$)$. ✔

(b) $(x, y) = (+4.7$ m, $+1.8$ m$)$. ✔

(c) $(x, y) = (+5.0$ m, $0)$. ✔

All the vector quantities we have encountered so far obey the rules for addition and subtraction outlined above, and all can be decomposed into component vectors. There is one significant difference, however, between the way vectors are treated in one-dimensional motion and the way they are treated in two-dimensional motion. As we saw in Section 2.8, the displacement and velocity of an object that moves along a straight line point in the same direction. Section 3.1 taught you that if the object speeds up, the acceleration also points in the same direction as the displacement, and if the object slows down, the acceleration points in the direction opposite that of the displacement and velocity. For motion in a plane, the relationship between the directions of these vectors is more complicated.

The displacement, instantaneous velocity, and acceleration of the ball in Figure 10.1 are shown in **Figure 10.10**. Because the ball's average velocity is its displacement divided by the time interval during which this displacement takes place, the average velocity (not shown in Figure 10.10)

points in the same direction as the displacement. The ball's instantaneous velocity is the average velocity evaluated over infinitesimally small time intervals, and so if the object does not move in a straight line, the instantaneous velocity does not necessarily point in the same direction as the displacement. In Figure 10.10, for instance, $\Delta \vec{r}$ points from the initial to the final position over a certain time interval, while $\vec{v}$ is tangent to the trajectory—that is, in the instantaneous direction of travel.

Let us next consider the acceleration. By definition, the ball's acceleration points in the direction of the change in velocity. Because the horizontal component of the ball's velocity is equal to the velocity of the cart and so is constant, there is no acceleration in the horizontal direction. Only the vertical component of the ball's velocity changes, and so the ball's acceleration points downward rather than in the direction of $\vec{v}$. If you recall the relationship between acceleration and force $(\Sigma \vec{F} = m\vec{a})$, this makes sense: The ball's acceleration is caused by the downward force of gravity exerted by Earth on the ball. Therefore both the force and the acceleration point straight down.

What does it mean when the instantaneous velocity and acceleration of an object point in different directions? To answer this question, I must decompose the acceleration into two components: one parallel to the instantaneous

Figure 10.10 Displacement, instantaneous velocity, and acceleration for the freely falling ball of Figure 10.1.

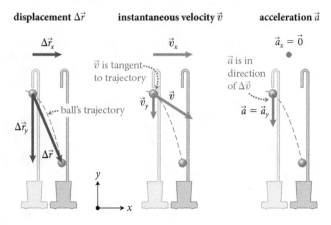

displacement $\Delta \vec{r}$ instantaneous velocity $\vec{v}$ acceleration $\vec{a}$

Figure 10.11 Velocity and acceleration vectors of the falling ball in frame 2 of Figure 10.1. Decomposition of $\vec{A}$ into component vectors parallel and perpendicular to $\vec{v}$.

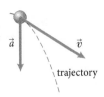

ball's acceleration and velocity

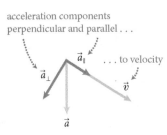

acceleration components perpendicular and parallel . . . . . . to velocity

Figure 10.13 Change in velocity caused by a perpendicular acceleration.

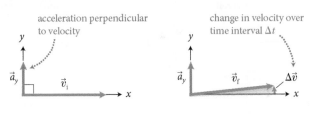

acceleration perpendicular to velocity

change in velocity over time interval Δt

velocity and one perpendicular to it (**Figure 10.11**). The parallel component causes the length of the instantaneous velocity—in other words, the ball's *speed*—to change. Just as for one-dimensional motion, the speed increases when the parallel component of the acceleration points in the same direction as the instantaneous velocity (as is the case with our ball-and-cart example) and decreases when they point in opposite directions. The perpendicular component of the acceleration causes the *direction* of the instantaneous velocity to change, which in our particular case of the ball and cart means that the ball does not move in a straight horizontal line the way the cart does. Instead, the ball's instantaneous velocity in any given frame of the film clip points more toward the ground than does its counterpart in the preceding frame. (In short, the ball curves downward.) In summary:

> **In two-dimensional motion, the component of the acceleration parallel to the instantaneous velocity changes the speed; the component of the acceleration perpendicular to the instantaneous velocity changes the direction of the instantaneous velocity but not its magnitude.**

These ideas are illustrated in **Figure 10.12**.

Figure 10.12 How the parallel and perpendicular components of acceleration affect velocity.

Acceleration parallel to velocity changes *speed:*

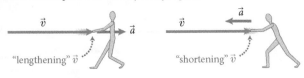

"lengthening" $\vec{v}$ "shortening" $\vec{v}$

Acceleration perpendicular to velocity changes *direction of motion:*

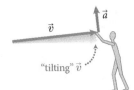

"tilting" $\vec{v}$

If you wonder how it can be that the perpendicular component changes *only* the direction of the velocity and does not affect its magnitude, examine **Figure 10.13**. An object initially moving in the positive x direction is given an acceleration in the positive y direction. Because of this acceleration, over a small but finite time interval Δt the object's initial velocity $\vec{v}_i$ gains a small component $\Delta \vec{v}$ in the y direction. The final velocity $\vec{v}_f = \vec{v}_i + \Delta \vec{v}$ therefore no longer points along the x axis—as we concluded before, $\vec{a}_y$, being perpendicular to $\vec{v}_i$, changes the direction of $\vec{v}_i$. Why then doesn't it also change the magnitude? The Pythagorean theorem tells us that the magnitude of $\vec{v}_f$ must be greater than that of $\vec{v}_i$, a result that appears to contradict what I just said about a perpendicular acceleration not affecting the magnitude of a horizontal velocity. The resolution of this contradiction lies in realizing that, as the direction of $\vec{v}$ changes, so does the angle between $\vec{v}$ and $\vec{a}_y$: The acceleration, because it is no longer exactly perpendicular to the velocity, now has a small component parallel to the velocity, and it is this component of the acceleration that causes the change in the magnitude of $\vec{v}$. We must therefore restrict our analysis to an infinitesimally small time interval. With this restriction, we can ignore the change in the magnitude of the velocity caused by $\Delta \vec{v}$: $v_i \approx v_f$.

✋ **10.2** In Figure 10.10, the ball's instantaneous velocity $\vec{v}$ does not point in the same direction as the displacement $\Delta \vec{r}$ (it points *above* the final position of the ball). Why?

10.3 Decomposition of forces

Now that you know how to decompose vectors into components, let us see how this technique works with forces. Consider, for example, putting a brick on a horizontal plank and then slowly tilting the plank, as illustrated in **Figure 10.14**. Initially, the brick remains at rest, but when the angle of incline θ exceeds some maximum value θ_{max}, the brick suddenly slides downward. The angle of incline at which an object on an inclined surface begins to slide depends on the materials out of which the surface and the object are made and on their smoothness. It is also strongly affected by the presence of dirt or lubricants.

Figure 10.14 The forces exerted on a brick as you raise one end of the plank on which it rests.

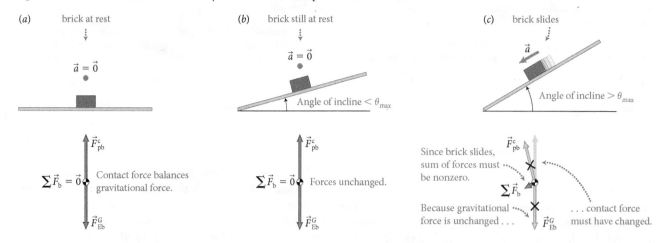

When the plank of Figure 10.14 is horizontal, the forces exerted on the brick are a downward gravitational force exerted by Earth and an upward contact force of equal magnitude exerted by the plank, as illustrated in the free-body diagram in Figure 10.14a. As long as the brick stays at rest, raising one end of the plank doesn't change the free-body diagram: The force of gravity on the brick does not depend on the inclination of the plank, and because the brick is still at rest we know that the force exerted by the plank must still be upward and equal in magnitude to the force of gravity, as shown in Figure 10.14b. When the angle of incline exceeds θ_{max} the situation changes, however. At that point, the contact force provided by the plank is no longer able to balance the downward force of gravity, and the brick accelerates down the incline. Although it is not immediately clear how to draw the free-body diagram, we know that the vector sum of the forces exerted on the brick

must point in the same direction as the brick's acceleration (Figure 10.14c).

Because the brick is constrained to move along the surface of the plank, its acceleration must always be parallel to the plank, as shown in Figure 10.14c. For this reason, it makes sense to analyze the forces in terms of components parallel and perpendicular to the plank, by pointing the x axis along the surface that constrains the motion, as shown in **Figure 10.15**. The force components perpendicular to the surface are called **normal components;** those parallel to the surface are called **tangential components.** Because the brick's acceleration must be parallel to the plank, the normal component vectors $\vec{F}^c_{pby}$ and $\vec{F}^G_{Eby}$ must always be equal in magnitude—whether the brick is at rest or sliding. (If these two forces were not of equal magnitude, the brick would accelerate in the direction perpendicular to the plank, and we know this doesn't happen.)

Figure 10.15 Free-body diagrams for a brick on a plank at different angles of inclination.

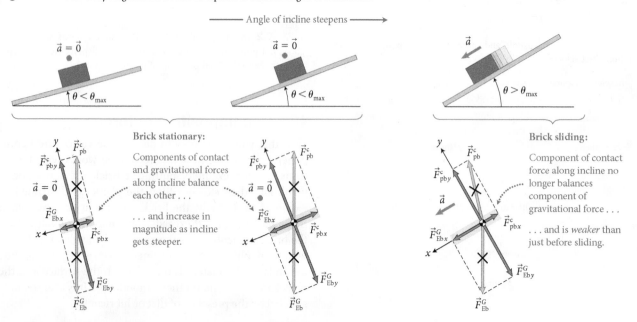

The tangential component vectors $\vec{F}_{\text{pb}x}^c$ and $\vec{F}_{\text{Eb}x}^G$, on the other hand, are equal in magnitude only when the brick is not accelerating down the incline. Note in Figure 10.15 that as the angle of incline is increased, the magnitude of the tangential component vector $\vec{F}_{\text{Eb}x}^G$ of the force of gravity increases. To prevent the brick from sliding down, the magnitude of the tangential component vector $\vec{F}_{\text{pb}x}^c$ of the contact force exerted by the plank on the brick must therefore also increase. When the angle of incline exceeds θ_{max}, the plank is no longer able to provide the sideways support necessary to hold the brick in place. The tangential component vectors of the forces no longer add up to zero, and the vector sum of the forces exerted on the brick is directed down the plank, as shown in Figure 10.14c.

✋ **10.3** A suitcase being loaded into an airplane moves at constant velocity on an inclined conveyor belt. Draw a free-body diagram for the suitcase as it moves up along with the belt. Show the normal and tangential components of the forces exerted on the suitcase.

Let us now return to a question raised in Section 10.2: What set of axes should you choose in a given problem? The example of the brick sliding down the inclined plank suggests an answer:

If possible, choose a coordinate system such that one of the axes lies along the direction of the acceleration of the object under consideration.

This choice of coordinate system allows you to break the problem neatly into two parts because only force components along the direction of acceleration can contribute to the motion.

When the acceleration of the object being considered is zero, choose your axes to coincide with the directions of as many forces as possible because no decomposition is necessary for any force that lies along an axis.

Example 10.2 Pulling a friend on a swing

Using a rope, you pull a friend sitting on a swing (**Figure 10.16**). (a) As you increase the angle θ, does the magnitude of the force $\vec{F}_{\text{rp}}^c$ required to hold your friend in place increase or decrease? (b) Is the magnitude of that force larger than, equal to, or smaller than the magnitude of the gravitational force $\vec{F}_{\text{Ep}}^G$ exerted by Earth on your friend? (Consider the situation for both small and large values of θ.) (c) Is the magnitude of the force $\vec{F}_{\text{sp}}^c$ exerted by the swing on your friend larger than, equal to, or smaller than F_{Ep}^G?

Figure 10.16 Example 10.2.

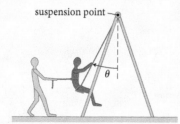

suspension point

θ

① **GETTING STARTED** I begin by drawing a free-body diagram of your friend (**Figure 10.17**). Three forces are exerted on him: $\vec{F}_{\text{Ep}}^G$, the force of gravity directed vertically downward, the horizontal force $\vec{F}_{\text{rp}}^c$ exerted by the rope, and a force $\vec{F}_{\text{sp}}^c$ exerted by the swing seat. This latter force $\vec{F}_{\text{sp}}^c$ is exerted by the suspension point via the chains of the swing and is thus directed along the chains. I therefore choose a horizontal x axis and a vertical y axis, so that two of the three forces lie along axes. Because your friend's acceleration is zero, the vector sum of the forces must be zero.

Figure 10.17

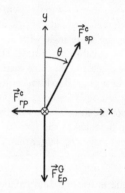

② **DEVISE PLAN** Because your friend is at rest, the vectors along the two axes must add up to zero. The best way to see how the magnitude of the force $\vec{F}_{\text{rp}}^c$ exerted by the rope must change as θ is increased is to draw free-body diagrams showing different values of θ. To answer parts b and c, I can compare the various forces in my free-body diagrams.

③ **EXECUTE PLAN** (a) I begin by decomposing $\vec{F}_{\text{sp}}^c$ into x and y components (**Figure 10.18a**). Because the forces must add to zero along both axes, I conclude from my diagram that $\vec{F}_{\text{sp}y}^c$ must be equal in magnitude to $\vec{F}_{\text{Ep}}^G$, the downward force of gravity. Likewise, $\vec{F}_{\text{sp}x}^c$ must be equal in magnitude to $\vec{F}_{\text{rp}}^c$, the horizontal force the rope exerts on your friend. Next, I draw a second free-body diagram for a larger angle θ (Figure 10.18b). As θ increases, $\vec{F}_{\text{sp}y}^c$ must remain equal in magnitude to $\vec{F}_{\text{Ep}}^G$ (otherwise your friend would accelerate vertically). As Figure 10.18b shows, increasing θ while keeping $\vec{F}_{\text{sp}y}^c$ constant requires the magnitude of $\vec{F}_{\text{sp}x}^c$ to increase. Because your friend is at rest, the forces in the horizontal direction must add up to zero and so $F_{\text{rp}}^c = |F_{\text{sp}x}^c|$. So if the magnitude of $\vec{F}_{\text{sp}x}^c$ increases, the magnitude of $\vec{F}_{\text{rp}}^c$ must increase, too. ✔

Figure 10.18

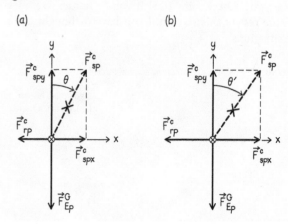

(a)　　　　　　　(b)

(b) From Figure 10.18 I see that $\tan\theta = |F^c_{spx}|/|F^c_{spy}|$. For $\theta < 45°$, $\tan\theta < 1$, and so $|F^c_{spx}| < |F^c_{spy}|$. Because $|F^c_{spy}| = F^G_{Ep}$ and $|F^c_{spx}| = F^c_{rp}$, I discover that for $\theta < 45°$, $F^c_{rp} < F^G_{Ep}$. When $\theta > 45°$, $\tan\theta > 1$, and so $|F^c_{spx}| > |F^c_{spy}|$ and $F^c_{rp} > F^G_{Ep}$. ✔

(c) $|\vec{F}^c_{spy}| = F^G_{Ep}$ and $F^c_{sp} = \sqrt{(F^c_{spx})^2 + (F^c_{spy})^2}$. Therefore, F^c_{sp} must always be larger than F^G_{Ep} when $\theta \neq 0$. ✔

❹ **EVALUATE RESULT** I know from experience that you have to pull harder to move a swing farther from its equilibrium position, and so my answer to part *a* makes sense. With regard to part *b*, when the swing is at rest at 45°, the forces $\vec{F}^c_{rp}$ and $\vec{F}^G_{Ep}$ on your friend make the same angle with the force $\vec{F}^c_{sp}$, and so $\vec{F}^c_{rp}$ and $\vec{F}^G_{Ep}$ should be equal in magnitude. The force of gravity is independent of the angle, but the force exerted by the rope increases with increasing angle, and so it makes sense that for angles larger than 45°, $\vec{F}^c_{rp}$ is larger than $\vec{F}^G_{Ep}$. In part *c*, because the vertical component of the force $\vec{F}^c_{sp}$ exerted by the seat on your friend always has to be equal to the force of gravity, adding a horizontal component makes $\vec{F}^c_{sp}$ larger than $\vec{F}^G_{Ep}$, as I found.

✋ **10.4** You decide to move a heavy file cabinet by sliding it across the floor. You push against the cabinet, but it doesn't budge. Draw a free-body diagram for it.

10.4 Friction

The force that opposes your push on the file cabinet in Checkpoint 10.4—the tangential component of the contact force exerted by the floor on the cabinet—has to do with friction. If the floor were very slick or if the cabinet had casters, there would be little friction and your push would easily move the cabinet. Instead, you have to lean against it with all your strength until, with a jerk, it suddenly begins to slide. Once you get the cabinet moving, you must keep pushing to keep it in motion. If you stop pushing, friction stops the motion.

✋ **10.5** *(a)* Suppose you push the file cabinet just enough to keep it moving at constant speed. Draw a free-body diagram for the cabinet while it slides at constant speed. *(b)* Suddenly you stop pushing. Draw a free-body diagram for the file cabinet at this instant.

Don't skip Checkpoint 10.5! It will be harder to understand the rest of this section if you haven't thought about these situations.

Even though the normal and tangential components of the contact force exerted by the floor on the cabinet belong to the same interaction, they behave differently and are usually treated as two separate forces: the normal component being called the **normal force** and the tangential component being called the **force of friction.**

To understand the difference between normal and frictional forces, consider a brick on a horizontal wooden plank supported at both ends (**Figure 10.19a**). Because the brick is at rest, the normal force $\vec{F}^c_{pb\perp}$ exerted by the plank on it is equal in magnitude to the gravitational force exerted on it. Now imagine using your hand to push down on the brick with a force $\vec{F}^c_{hb}$. Your downward push increases the total downward force exerted on the brick, and, like a spring under compression, the plank bends until the normal force it exerts on the brick balances the combined downward forces exerted by your hand and by Earth on the brick (Figure 10.19b). As you push down harder, the plank bends more, and the normal force continues to increase (Figure 10.19c) until you exceed the plank's capacity to provide support and it snaps, at which point the normal force suddenly disappears (Figure 10.19d). So, normal forces take on whatever value is required to prevent whatever is pushing down on a surface from moving through that surface—up to the breaking point of the supporting material.

Next imagine that instead of pushing down on the brick of Figure 10.19a, you gently push it to the right, as in **Figure 10.20**. As long as you don't push hard, the brick remains at rest. This tells you that the horizontal forces exerted on the brick add to zero, and so the plank must be exerting on the brick a horizontal frictional force that is equal in magnitude to your push but in the opposite direction. This horizontal force is caused by microscopic bonds between the surfaces in contact. Whenever two objects are placed in contact, such bonds form at the extremities of microscopic bumps on the surfaces of the objects. When you try to slide the surfaces past each other, these tiny bonds prevent sideways motion. As you push the brick to the right, the bumps resist bending and, like microscopic springs, each bump exerts a force to the left. The net effect of all these microscopic forces is to hold the brick in place. As you increase the force of your push, the bumps resist bending more and the tangential component of the contact force grows. This friction exerted by surfaces that are not moving relative to each other is called **static friction.**

Figure 10.19 A demonstration of the normal force.

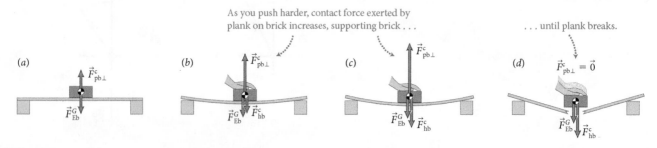

As you push harder, contact force exerted by plank on brick increases, supporting brick . . .

. . . until plank breaks.

Figure 10.20 The microscopic basis for forces of static and kinetic friction (b = brick, p = plank, h = hand).

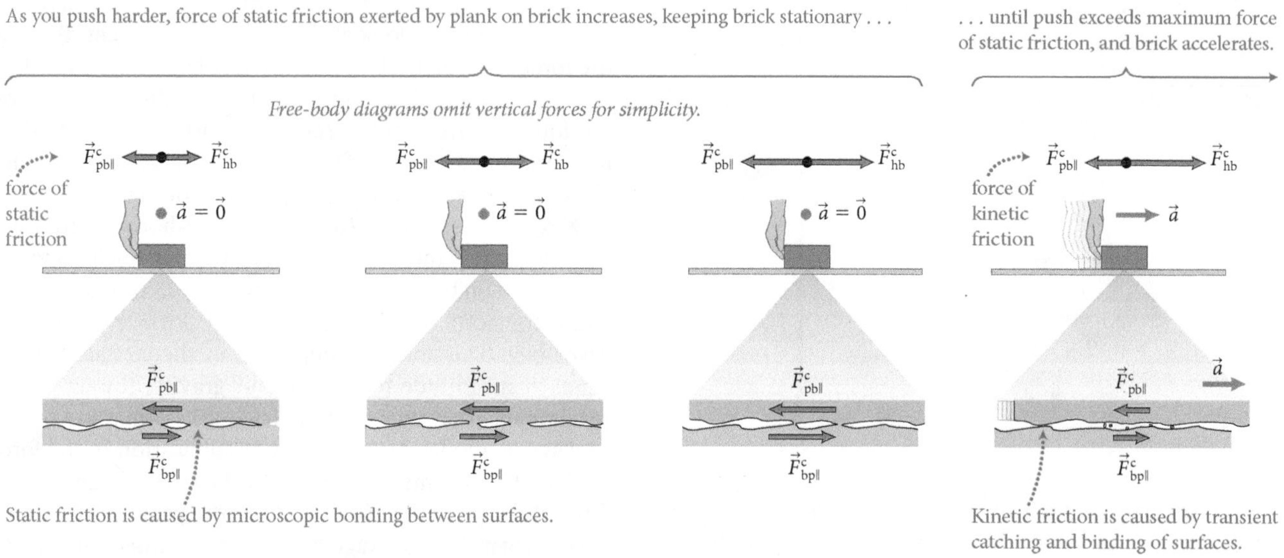

As you push harder, force of static friction exerted by plank on brick increases, keeping brick stationary . . .

. . . until push exceeds maximum force of static friction, and brick accelerates.

Free-body diagrams omit vertical forces for simplicity.

force of static friction

force of kinetic friction

Static friction is caused by microscopic bonding between surfaces.

Kinetic friction is caused by transient catching and binding of surfaces.

The tiny bonds cannot withstand large forces, however, and as you continue to increase your sideways push on the brick, the bonds get ripped apart, the tangential force they provide can no longer resist the push, and the brick starts to slide. The surface of the brick is no longer at rest relative to the surface on which the brick is resting—the two surfaces are said to *slip*.

Once the brick slips, bumps on the surfaces plow into each other, continually forming and breaking microscopic bonds, and microscopic pieces chip off from the surface. This constant bumping, making and breaking of bonds, and chipping cause what is called **kinetic friction.** Even though kinetic friction causes slipping surfaces to wear and to heat up, it is not always undesirable: Kinetic friction makes it possible to slow down the moving wheels of a car or bicycle, to use an eraser, and to warm up your hands by rubbing them, just to name a few examples.

So, just as the normal force takes on whatever value is necessary to prevent motion on a surface, the force of static friction takes on whatever value is necessary to prevent two surfaces from slipping past each other. The main differences between normal forces and forces of static friction are: (1) The maximum value of the force of static friction is generally much smaller than the maximum value of the normal force (it is relatively easy to push the brick across the plank but nearly impossible to push it through the plank), and (2) once the maximum value of the normal force is reached, the normal force disappears, but once the maximum value of the force of static friction is reached, there still is a smaller but nonzero force of kinetic friction.

To distinguish more easily among the various components of the contact force exerted by surface 1 on surface 2, we shall when necessary replace the superscript c by n for the normal force ($\vec{F}^c_{12\perp}$ becomes $\vec{F}^n_{12}$), by s for the force of static friction, and by k for the force of kinetic friction ($\vec{F}^c_{12\parallel}$ becomes $\vec{F}^s_{12}$ or $\vec{F}^k_{12}$).

10.6 Which type of friction—static or kinetic—plays a role in (a) holding a pencil, (b) chalk making marks on a chalkboard (c) skiing downhill, (d) polishing a metal surface, and (e) walking down an incline?

10.5 Work and friction

Like the normal force, the force of static friction is an elastic force: It causes no irreversible changes. The deformations that take place undo themselves when the force that caused them is removed. If you stop pushing on the file cabinet of Checkpoint 10.4 before it starts moving, for example, you will not scratch the floor. Kinetic friction, however, does cause damage, mainly because of the chipping mentioned earlier. Push the cabinet a short distance, no matter how carefully, and you will cause some damage, both to the bottom surface of the cabinet and to the floor surface. At best the damage is microscopic, but if you were to push the cabinet around the floor all day long, your landlady would probably not renew your lease. The heating and wear caused by kinetic friction turn coherent energy into incoherent energy. To sum up:

> **The force of kinetic friction is not an elastic force and so causes energy dissipation. The force of static friction is an elastic force and so causes no energy dissipation.**

In Section 9.4 I cautioned against choosing a system for which friction occurs at the boundary because you don't know how much of the dissipated energy ends up on each side of the boundary. Now you see that this caveat does not apply to static friction because static friction causes no energy dissipation.

This brings us to the question of whether or not the force of static friction can do work on a system. After all,

CONCEPTS

Figure 10.21 Two equivalent situations in which an object is accelerated by static friction.

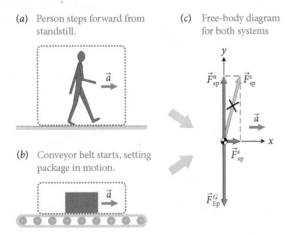

(a) Person steps forward from standstill.

(b) Conveyor belt starts, setting package in motion.

(c) Free-body diagram for both systems

static friction does not involve any motion of the two surfaces relative to each other, and work requires a nonzero force displacement. To answer this question, consider the two situations shown in **Figure 10.21**: A person starts to walk and a package is set in motion by a conveyor belt. In both cases the system (the person in Figure 10.21a and the package in Figure 10.21b) gains kinetic energy. Where does that energy come from, and what is the role of static friction in each case?

Let us first look at Figure 10.21a. To begin walking, the person leans slightly forward and raises one foot, with her other foot pressing on the floor downward and backward. The floor surface then supplies an upward contact force $\vec{F}^c_{sp}$ that is tilted forward, as shown in Figure 10.21c. The force $\vec{F}^c_{ps}$ of the person pressing down on the surface and the force $\vec{F}^c_{sp}$ of the surface pushing up form an interaction pair. Because she is not accelerating in the vertical direction, the forces in that direction must add to zero, and so the normal component of the contact force exerted on the person—the normal force $\vec{F}^n_{sp}$—is equal in magnitude to the force of gravity exerted on her. The tangential component $\vec{F}^s_{sp}$ is what propels her forward. As long as the foot doesn't slip, this tangential component is due to static friction; the forward acceleration is then equal to the force of static friction divided by the person's inertia.

To determine whether or not the force of static friction exerted by the floor surface on the person does work on the person, we need to look at the force displacement. Because the force of static friction is applied at the foot, which does not move in any direction while it is on the floor surface, the force displacement is zero. The surface therefore does no work on the person. That makes sense because *she* supplies the energy required to accelerate, not the surface.

Next, look at Figure 10.21b. The free-body diagram for this situation is the same as that for the person. Two forces are exerted on the package: a downward force of gravity $\vec{F}^G_{Ep}$ and a contact force $\vec{F}^c_{sp}$ exerted by the surface of the conveyor belt. The normal component of the contact force is equal in magnitude to the force of gravity; the tangential component (the force of static friction) is what causes the package to accelerate. The point of application of the force of static friction moves along with the package, and so the force of static friction does work on the package. This, too, makes sense: The package has no way of internally generating any kinetic energy; it is the conveyor that must do this. So, like any other elastic contact force, the force of static friction *can* do work on a system, and it is perfectly all right to choose a system in which *static* friction occurs at the boundary. It's only kinetic friction that causes problems because this type of friction dissipates energy, generating thermal energy that is distributed on both sides of the surface.

Figure 10.21 also dispels the often-heard myth that friction always *opposes* motion. In both cases shown in the figure, friction *causes* the motion. Without the forward pointing force of friction, neither the person nor the package would accelerate. Friction opposes the motion of two surfaces *relative to each other,* not motion per se. In Figure 10.21a, for instance, the person's rear foot would slip backward if there were no forward-pointing force of friction. The force of static friction prevents this slipping motion, and in doing so it pushes her forward.

10.7 Draw energy diagrams for the person and the package in Figure 10.21.

Self-quiz

1. In the diagrams in **Figure 10.22**, the velocity of an object is given along with the vector representing a force exerted on the object. For each case, determine whether the object's speed increases, decreases, or remains constant. Also determine whether the object's direction changes in the clockwise direction, changes in the counterclockwise direction, or doesn't change.

   Figure 10.22

2. Draw the sum $\vec{A} + \vec{B}$ and the difference $\vec{A} - \vec{B}$ of each vector pair in **Figure 10.23**.

   Figure 10.23

   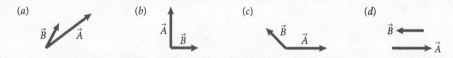

3. Each diagram in **Figure 10.24** indicates the momentum of an object before and after a force is exerted on it. For each case determine the direction of the force.

   Figure 10.24

4. While driving to school, you place your coffee cup on the dashboard. Which type of force of friction allows the cup to stay put on the dashboard when the car speeds up? Does this force do any work on the cup?

Answers

1. (a) $\vec{F}$ has component vectors parallel and perpendicular to $\vec{v}$. Because $\vec{F}_\parallel$ is in the same direction as $\vec{v}$, the speed increases. $\vec{F}_\perp$ tilts $\vec{v}$ clockwise from its current position. (b) Because the direction of $\vec{F}_\parallel$ is opposite the direction of $\vec{v}$, the speed decreases. $\vec{F}_\perp$ tilts $\vec{v}$ counterclockwise from its current position. (c) Because $\vec{F}_\parallel = 0$, the speed remains the same; $\vec{F}_\perp$ tilts $\vec{v}$ clockwise. (d) $\vec{F}_\parallel$ decreases the speed. $\vec{F}_\perp = 0$ and so the direction of $\vec{v}$ does not change.

2. The vector sum is equal to the diagonal of a parallelogram drawn using the two vectors placed tail to tail. The vector difference points from the tip of the second vector to the tip of the first vector (**Figure 10.25**).

   Figure 10.25

   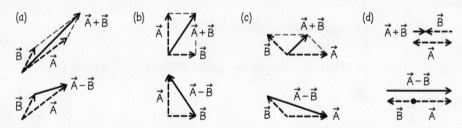

3. In each case the force is parallel to $\Delta\vec{p} = \vec{p}_f - \vec{p}_i$ (**Figure 10.26**).

   Figure 10.26

   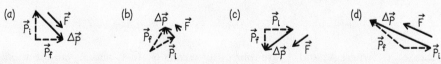

4. Because no sliding occurs, the force of friction must be static. Because the force displacement and the force point in the same direction when the car speeds up, this force does positive work on the cup.

Figure 10.27 Rectangular and polar coordinates.

(a) Rectangular coordinates

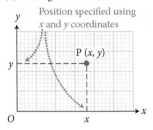

(b) Polar coordinates

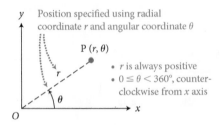

(c) Relationship between the two

10.6 Vector algebra

In a rectangular coordinate system, the position of a point P is specified by the rectangular coordinates x and y (**Figure 10.27a**). Alternatively, we can specify a point using **polar coordinates** (Figure 10.27b). The *radial* coordinate r gives the distance from the origin to the point and is always positive. The *angular* coordinate θ specifies the angle between r and the x axis and is measured counterclockwise from the positive x axis ($0 \leq \theta < 360°$). Any arbitrary point P is specified entirely by either (x, y) or (r, θ).

As can be seen from Figure 10.27c, the radial coordinate r is the hypotenuse of a right triangle with sides x and y, so

$$r = +\sqrt{x^2 + y^2}. \tag{10.1}$$

The plus sign in front of the square root indicates that r is never negative. Trigonometry tells us that $\tan \theta = y/x$, so

$$\theta = \tan^{-1}\left(\frac{y}{x}\right), \tag{10.2}$$

where the signs of x and y determine in which quadrant θ lies. Equations 10.1 and 10.2 allow us to determine polar coordinates from rectangular ones. Using trigonometry, we can also express rectangular coordinates in terms of polar coordinates:

$$x = r \cos \theta$$
$$y = r \sin \theta. \tag{10.3}$$

Exercise 10.3 Polar point

If each major grid unit in Figure 10.27 corresponds to 1 m, specify the location of point P in polar coordinates.

SOLUTION I begin by reading off the rectangular coordinates: $x = +3.6$ m, $y = +2.4$ m. Then I can use Eqs. 10.1 and 10.2 to convert to polar coordinates:

$$r = +\sqrt{x^2 + y^2} = \sqrt{(3.6 \text{ m})^2 + (2.4 \text{ m})^2} = 4.3 \text{ m}$$

and

$$\theta = \tan^{-1}\left(\frac{y}{x}\right) = \tan^{-1}\left(\frac{2.4 \text{ m}}{3.6 \text{ m}}\right) = 34°.$$

So, in polar coordinates point P is at $(r, \theta) = (4.3$ m, $34°)$. ✔

In rectangular coordinates, the position vector of a point P is

$$\vec{r} \equiv x\,\hat{\imath} + y\,\hat{\jmath}, \tag{10.4}$$

where $\hat{\imath}$ and $\hat{\jmath}$ represent unit vectors along the x and y axes, respectively. The component vectors of $\vec{r}$ are $x\,\hat{\imath}$ and $y\,\hat{\jmath}$. More generally, the decomposition of an arbitrary vector $\vec{A}$ can be written as follows (**Figure 10.28**):

$$\vec{A} = \vec{A}_x + \vec{A}_y = A_x\hat{\imath} + A_y\hat{\jmath}, \tag{10.5}$$

where $\vec{A}_x$ and $\vec{A}_y$ are *component vectors* of $\vec{A}$; A_x and A_y are the *x and y components* of $\vec{A}$. Keep in mind that the x and y components, A_x and A_y, are signed numbers—these symbols do *not* represent the magnitudes of the component

Figure 10.28 Decomposition of a general vector $\vec{A}$ into component vectors.

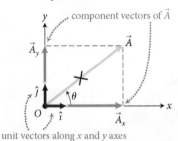

unit vectors along x and y axes

vectors $\vec{A}_x$ and $\vec{A}_y$. Whenever you see a subscript x or y (and later also z) on a symbol that does not have an arrow over it, the quantity represented by that symbol is a component of a vector and carries an algebraic sign.

Using the Pythagorean theorem, we can write the magnitude of $\vec{A}$ as

$$A \equiv |\vec{A}| = +\sqrt{A_x^2 + A_y^2}, \qquad (10.6)$$

where the plus sign in front of the square root indicates that only the positive square root applies because magnitude is always a positive quantity. The angle of $\vec{A}$ with respect to the x axis is specified by

$$\tan \theta = \frac{A_y}{A_x}, \qquad (10.7)$$

where the signs of A_x and A_y determine in which quadrant θ lies.

The decomposition of vectors is particularly useful when adding or subtracting them. As **Figure 10.29** shows, the x and y components of the vector sum $\vec{R}$ of two vectors $\vec{A}$ and $\vec{B}$ are equal to the sum of the corresponding components of $\vec{A}$ and $\vec{B}$:

$$\vec{R} = (A_x + B_x)\hat{\imath} + (A_y + B_y)\hat{\jmath} \qquad (10.8)$$

$$R_x = A_x + B_x$$
$$R_y = A_y + B_y. \qquad (10.9)$$

Figure 10.29 Adding vectors by components.

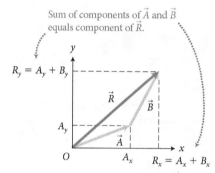

Sum of components of $\vec{A}$ and $\vec{B}$ equals component of $\vec{R}$.

Example 10.4 Speeding ball

A ball is thrown at an angle of 30° to the horizontal at a speed of 30 m/s. Write the ball's velocity in terms of rectangular unit vectors.

❶ **GETTING STARTED** I begin by making a sketch showing the velocity vector $\vec{v}$ and its decomposition in a rectangular coordinate system (**Figure 10.30**). I position the x axis along the horizontal in the direction of motion and the y axis along the vertical. I label the x and y components of the ball's velocity v_x and v_y, respectively.

❷ **DEVISE PLAN** Equation 10.5 tells me that the velocity vector can be written as $\vec{v} = v_x\hat{\imath} + v_y\hat{\jmath}$. To determine the x and y components, I apply trigonometry to the triangle made up by v, v_x, and v_y.

❸ **EXECUTE PLAN** From my sketch I see that $\cos \theta = v_x/v$ and $\sin \theta = v_y/v$, where $v = 30$ m/s and $\theta = 30°$. Therefore $v_x = v \cos \theta$ and $v_y = v \sin \theta$. Substituting the values given for v and θ, I calculate the rectangular components:

$$v_x = (30 \text{ m/s})(\cos 30°) = (30 \text{ m/s})(0.87) = +26 \text{ m/s}$$

$$v_y = (30 \text{ m/s})(\sin 30°) = (30 \text{ m/s})(0.50) = +15 \text{ m/s}.$$

The velocity in terms of unit vectors is thus

$$\vec{v} = v_x\hat{\imath} + v_y\hat{\jmath} = (+26 \text{ m/s})\hat{\imath} + (+15 \text{ m/s})\hat{\jmath}. ✔$$

❹ **EVALUATE RESULT** The x and y components are both positive, as I expect because I chose the direction of the axis in the direction that the ball is moving, and the magnitude of v_x is larger than that of v_y, as it should be for a launch angle that is smaller than 45°. I can quickly check my math by using Eq. 10.6 to calculate the magnitude of the velocity: $v = \sqrt{v_x^2 + v_y^2} = \sqrt{(26 \text{ m/s})^2 + (15 \text{ m/s})^2} = 30$ m/s, which is the speed at which the ball is launched.

Figure 10.30 Example 10.4.

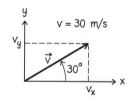

10.8 You navigate a ship from a pier to a buoy 1500 m northeast of the pier. There you sail to a point 300 m south and 700 m east of the buoy. What is your distance from the pier at that point?

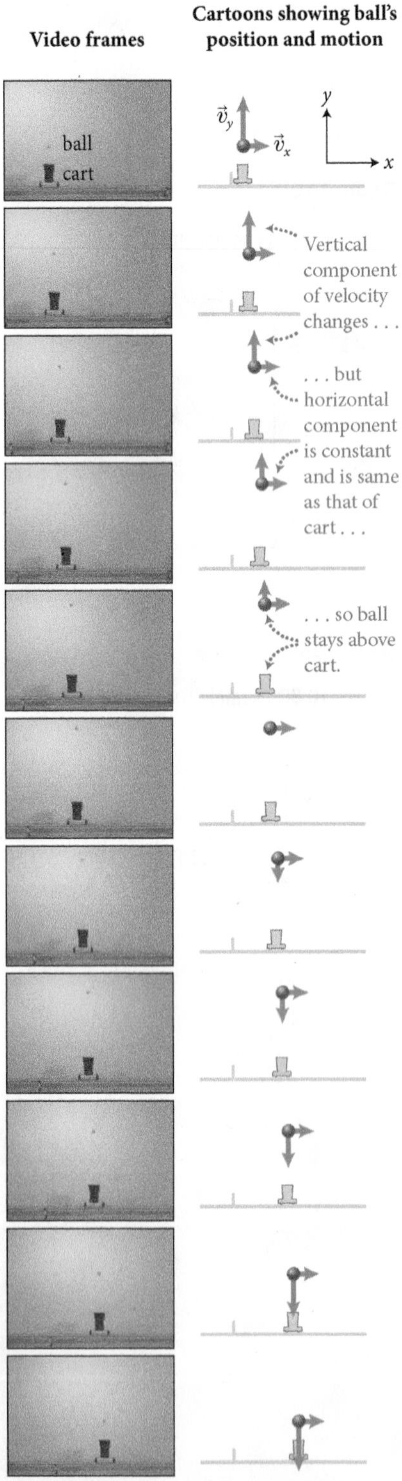

10.7 Projectile motion in two dimensions

We can now generalize the description of motion in a plane, starting with the position vector of an object:

$$\vec{r} = x\hat{\imath} + y\hat{\jmath}. \tag{10.10}$$

As in one dimension, the object's instantaneous velocity is

$$\vec{v} \equiv \frac{d\vec{r}}{dt} \equiv \lim_{\Delta t \to 0} \frac{\Delta \vec{r}}{\Delta t}. \tag{10.11}$$

Substituting Eq. 10.10 into Eq. 10.11, we obtain

$$\vec{v} = \lim_{\Delta t \to 0} \frac{\Delta x}{\Delta t}\hat{\imath} + \lim_{\Delta t \to 0} \frac{\Delta y}{\Delta t}\hat{\jmath} = v_x\hat{\imath} + v_y\hat{\jmath}, \tag{10.12}$$

so that

$$v_x = \frac{dx}{dt} \quad \text{and} \quad v_y = \frac{dy}{dt}. \tag{10.13}$$

Likewise, the components of the instantaneous acceleration are

$$a_x = \frac{dv_x}{dt} = \frac{d^2x}{dt^2} \quad \text{and} \quad a_y = \frac{dv_y}{dt} = \frac{d^2y}{dt^2}. \tag{10.14}$$

Knowing how to decompose vectors into components allows us to separate motion in a plane into two one-dimensional problems. Consider the motion of a ball launched straight up from a cart moving at constant velocity (**Figure 10.31**). Like the ball dropped from the moving cart in Figure 10.1, the launched ball remains vertically above the cart throughout its trajectory and so falls back into the cart. In the reference frame of the cart, the ball's motion is identical to the one-dimensional projectile motion we studied in Section 3.3. In the Earth reference frame, though, the ball's instantaneous velocity has a component in the horizontal direction in addition to the component in the vertical direction. The resulting two-dimensional projectile motion in the Earth reference frame is the curved trajectory shown in **Figure 10.32**.

How do we obtain the ball's position at some arbitrary instant $t_f = t_i + \Delta t$? Let us begin by writing the initial condition for the ball's motion in the Earth reference frame at instant t_i. Because the ball's acceleration is always vertically downward, we choose a coordinate system that has a horizontal x axis and a vertical y axis. In the Earth reference frame, the ball's initial velocity is thus

$$\vec{v}_i = v_{x,i}\hat{\imath} + v_{y,i}\hat{\jmath}. \tag{10.15}$$

As viewed from the Earth reference frame, the ball's initial horizontal velocity is equal to the cart's constant velocity. The presence of the horizontal velocity in Eq. 10.15 tells us that, as viewed from the Earth reference frame,

Figure 10.31 Ball launched vertically up from a cart moving at constant velocity in the horizontal direction.

the ball is launched not straight up but at an angle θ relative to the x axis given by

$$\tan \theta = \frac{v_{y,i}}{v_{x,i}}. \qquad (10.16)$$

The ball's velocity in the horizontal direction is constant; in the vertical direction, the ball has a constant downward acceleration of magnitude $a_y = -g$. The x component of the ball's instantaneous velocity is not changing. We can obtain the y component at instant t_f from Eq. 3.4:

$$v_{x,f} = v_{x,i} \quad \text{(constant velocity)} \qquad (10.17)$$

$$v_{y,f} = v_{y,i} - g\Delta t \quad \text{(constant acceleration)}. \qquad (10.18)$$

We can now use the results we derived for one-dimensional projectile motion to determine the position of the ball at any instant. The ball's horizontal position at instant t_f is given by Eq. 2.19, $x_f = x_i + v_{x,i}\Delta t$. For the vertical position we use Eq. 3.8, $x_f = x_i + v_{x,i}\Delta t + \frac{1}{2}a_x(\Delta t)^2$, replacing x by y and then setting $a_y = -g$:

$$x_f = x_i + v_{x,i}\Delta t \quad \text{(constant velocity)} \qquad (10.19)$$

$$y_f = y_i + v_{y,i}\Delta t - \frac{1}{2}g(\Delta t)^2 \quad \text{(constant acceleration)}. \qquad (10.20)$$

These two equations completely determine the trajectory of a projectile launched with initial velocity $\vec{v}_i$ from point (x_i, y_i).

Equations 10.19 and 10.20 give us x_f and y_f as functions of Δt because $x \propto \Delta t$ and $y \propto (\Delta t)^2$. By eliminating Δt from these equations, we obtain a function $y(x)$ that depends quadratically on x and describes a parabola. Hence the trajectory of a projectile whose initial velocity is not vertical is a parabola.

Figure 10.32 Trajectory of the ball in Figure 10.31 as seen from the Earth reference frame.

Earth reference frame

Example 10.5 Position of highest point

The ball of Figure 10.32 is launched from the origin of an xy coordinate system. Write expressions giving, at the top of its trajectory, the ball's rectangular coordinates in terms of its initial speed v_i and the acceleration due to gravity g.

❶ **GETTING STARTED** Because the ball is launched from the origin, $x_i = 0$ and $y_i = 0$. As the ball moves upward, the vertical component of its velocity, v_y, is positive. After crossing its highest position, the ball moves downward, and so now v_y is negative. As the ball passes through its highest position, therefore, v_y reverses sign, so at that position $v_y = 0$.

❷ **DEVISE PLAN** Taking the highest position as my final position and then substituting $v_{y,f} = 0$ into Eq. 10.18, I can determine the time interval Δt_{top} it takes the ball to travel to this position. Once I know Δt_{top}, I can use Eqs. 10.19 and 10.20 to obtain the ball's coordinates at the top.

❸ **EXECUTE PLAN** Substituting $v_{y,f} = 0$ into Eq. 10.18, I get

$$0 = v_{y,i} - g\Delta t_{top}.$$

Solving for Δt_{top} then yields

$$\Delta t_{top} = \frac{v_{y,i}}{g}. \qquad (1)$$

Using $x_i = 0$, $y_i = 0$ in Eqs. 10.19 and 10.20, I then calculate the location of the highest position:

$$x_{top} = 0 + v_{x,i}\left(\frac{v_{y,i}}{g}\right) = \frac{v_{x,i}v_{y,i}}{g}$$

and

$$y_{top} = 0 + v_{y,i}\left(\frac{v_{y,i}}{g}\right) - \frac{1}{2}g\left(\frac{v_{y,i}}{g}\right)^2$$

$$= \frac{v_{y,i}^2}{g} - \frac{1}{2}\frac{v_{y,i}^2}{g} = \frac{1}{2}\frac{v_{y,i}^2}{g}. \checkmark$$

❹ **EVALUATE RESULT** Because the ball moves at constant velocity in the horizontal direction, the x coordinate of the highest position is the horizontal velocity component $v_{x,i}$ multiplied by the time interval it takes to reach the top (given by Eq. 1). The y coordinate of the highest position does not depend on $v_{x,i}$ as I expect because the vertical and horizontal motions are independent of each other.

Example 10.6 Range of a projectile

How far from the launch position is the position at which the ball of Figure 10.32 is once again back in the cart? (This distance is called the *horizontal range* of the projectile.)

❶ **GETTING STARTED** As we saw in Example 3.6, the time interval taken by a projectile to return to its launch position from the top of the trajectory is equal to the time interval Δt_{top} it takes to travel from the launch position to the top. I can therefore use the Δt_{top} expression I found in Example 10.5 to solve this problem.

❷ **DEVISE PLAN** If the time interval it takes the ball to travel from its launch position to the top is Δt_{top}, then the time interval the ball is in the air is $2\Delta t_{top}$. Because the ball travels at constant velocity in the horizontal direction, I can use Eq. 10.19 to obtain the ball's range.

❸ **EXECUTE PLAN** From Example 10.5, I have $\Delta t_{top} = v_{y,i}/g$, and so the time interval spent in the air is $\Delta t_{flight} = 2v_{y,i}/g$. Substituting this value into Eq. 10.19 yields

$$x_f = 0 + v_{x,i}\left(\frac{2v_{y,i}}{g}\right) = \frac{2v_{x,i}v_{y,i}}{g}. ✔$$

❹ **EVALUATE RESULT** Because the trajectory is an inverted parabola, the top of the parabola lies midway between the two locations where the parabola intercepts the horizontal axis. So the location at which the parabola returns to the horizontal axis lies a horizontal distance twice as far from the origin as the horizontal distance at the top. In Example 10.5 I found that $x_{top} = v_{x,i}v_{y,i}/g$. The answer I get for the horizontal range is indeed twice this value.

✋ **10.9** Suppose a projectile's initial velocity is specified by the initial speed v_i and launch angle θ instead of by its rectangular components as in Eq. 10.15. (*a*) Using Eqs. 10.3, write expressions for the projectile's maximum height and horizontal range in terms of v_i and θ. (*b*) What angle θ gives the greatest value of y_{max}? (*c*) Of x_{max}?

10.8 Collisions and momentum in two dimensions

As we saw in Chapter 5, collisions in one dimension are entirely described by two equations: Eq. 4.17, $\Delta\vec{p} = \vec{0}$, which states that the momentum of the isolated system of colliding objects does not change, and Eq. 5.18, $e \equiv v_{12f}/v_{12i}$, which defines the coefficient of restitution. The same two equations also apply to collisions in two dimensions, but because momentum is a vector, Eq. 4.17 now yields two separate equations: one for the momentum in the x direction and the other for the momentum in the y direction. Consider, for example, a collision in an isolated system of two objects of inertia m_1 and m_2. For this system, Eq. 4.17 yields

$$\Delta p_x = \Delta p_{1x} + \Delta p_{2x} = m_1(v_{1x,f} - v_{1x,i}) + m_2(v_{2x,f} - v_{2x,i}) = 0 \quad (10.21)$$

$$\Delta p_y = \Delta p_{1y} + \Delta p_{2y} = m_1(v_{1y,f} - v_{1y,i}) + m_2(v_{2y,f} - v_{2y,i}) = 0. \quad (10.22)$$

If we know the initial velocities of the colliding objects, we have values for $v_{1x,i}, v_{1y,i}, v_{2x,i}$, and $v_{2y,i}$, but these two equations still contain four unknown quantities: $v_{1x,f}, v_{1y,f}, v_{2x,f}$, and $v_{2y,f}$. To obtain values for these four unknown quantities, we need four equations. Even if we know the coefficient of restitution of the colliding objects, we have only three equations (Eqs. 10.21, 10.22, and 5.18) and so, unlike collisions in one dimension, we cannot determine the outcome of the collision without some additional information about the final velocities.

Example 10.7 Pucks colliding

Pucks 1 and 2 slide on ice and collide. The inertia of puck 2 is twice that of puck 1. Puck 1 initially moves at 1.8 m/s; puck 2 initially moves at 0.20 m/s in a direction that makes an angle of 45° with the direction of puck 1. After the collision, puck 1 moves at 0.80 m/s in a direction that makes an angle of 60° with its original direction. What are the speed and direction of puck 2 after the collision?

❶ **GETTING STARTED** Because it takes place on a surface, the collision is two-dimensional. I begin by organizing the given information in a sketch (**Figure 10.33**). I let puck 1 move along the x axis before the collision. Then I add puck 2, which initially moves at an angle of 45° to the x axis. I draw puck 1 after the collision moving upward and to the right along a line that makes an angle of 60° with the x axis. I don't know anything about the motion of puck 2 after the collision, and so I arbitrarily draw it moving to the right and upward. I label the pucks and indicate the speeds I know. I need to determine the final speed v_{2f} of puck 2 and the angle θ between its direction of motion after the collision and the x axis.

Figure 10.33

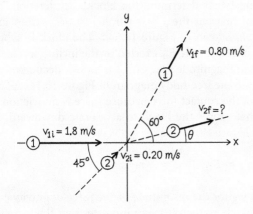

② DEVISE PLAN Equations 10.21 and 10.22 give the relationship between the initial and final velocities of the two pucks. I know the x and y components of both initial velocities and, after decomposing $\vec{v}_{1f}$, the x and y components of the final velocity of puck 1. Thus I can use Eqs. 10.21 and 10.22 to obtain values for the two components of $\vec{v}_{2f}$. Using these component values, I can calculate v_{2f} and θ.

③ EXECUTE PLAN The two unknown components are $v_{2x,f}$ and $v_{2y,f}$. Solving Eq. 10.21, for $v_{2x,f}$, I get

$$v_{2x,f} = -\frac{m_1}{m_2}(v_{1x,f} - v_{1x,i}) + v_{2x,i}$$

$$= -\tfrac{1}{2}[(0.80 \text{ m/s}) \cos 60° - (1.8 \text{ m/s})]$$

$$+ (0.20 \text{ m/s}) \cos 45°$$

$$= 0.84 \text{ m/s}.$$

Solving Eq. 10.22 for $v_{2y,f}$, I get

$$v_{2y,f} = -\frac{m_1}{m_2}(v_{1y,f} - v_{1y,i}) + v_{2y,i} = -\frac{m_1}{m_2}v_{1y,f} + v_{2y,i}$$

$$= -\tfrac{1}{2}(0.80 \text{ m/s}) \sin 60° + (0.20 \text{ m/s}) \sin 45°$$

$$= -0.21 \text{ m/s}.$$

Using these values, I obtain the final speed of puck 2 and its direction of motion:

$$v_{2f} = \sqrt{v_{2x,f}^2 + v_{2y,f}^2} = \sqrt{(0.84 \text{ m/s})^2 + (-0.21 \text{ m/s})^2}$$

$$= 0.87 \text{ m/s} \; ✔$$

$$\tan \theta = \frac{v_{2y,f}}{v_{2x,f}} = \frac{-0.21 \text{ m/s}}{0.84 \text{ m/s}} = -0.24$$

or $\quad \theta = -14°. \; ✔$

④ EVALUATE RESULT My positive result for $v_{2x,f}$ and negative result for $v_{2y,f}$ tell me that after the collision, puck 2 moves in the positive x direction and the negative y direction. Both of these directions make sense: I can see from my sketch that the x component of the velocity of puck 1 decreases in the collision. The corresponding decrease in the x component of its momentum must be made up by an increase in the x component of the momentum of puck 2. Given that puck 2 initially moves in the positive x direction, it must continue to do so after the collision. In the y direction, I note that because $v_{1y,i}$ is zero, the y component of the momentum of puck 1 increases in the positive y direction by an amount $m_1 v_{1y,f} = m_1[(0.80 \text{ m/s}) \sin 60°] = m_1(0.69 \text{ m/s})$. Therefore the y component of the momentum of puck 2 must undergo a change of equal amount in the negative y direction. Because the initial y component of the momentum of puck 2 is smaller than this value, $m_2 v_{2y,i} = 2m_1 v_{2y,i} = 2m_1[(0.20 \text{ m/s}) \sin 45°] = m_1(0.28 \text{ m/s})$, puck 2 must reverse its direction of motion in the y direction, yielding a negative value for $v_{2y,f}$, in agreement with what I found.

 10.10 Is the collision in Example 10.7 elastic?

10.9 Work as the product of two vectors

In Section 9.5, I showed that the work done by a constant force on a rigid object is equal to the product of the force and the force displacement:

$$W = F_x \Delta x_F \quad \text{(constant nondissipative force,}$$
$$\text{one dimension).} \qquad\qquad (10.23)$$

Work is a scalar, but force and displacement are vectors. What form does the right side of Eq. 10.23 take if the displacement and the force do not point along the same line? Consider the block sliding down an incline shown in

Figure 10.34 Forces exerted on a block sliding down an incline with negligible friction.

(a) Block slides with negligible friction.

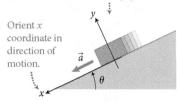

Orient x coordinate in direction of motion.

(b)

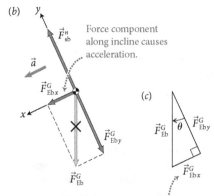

Force component along incline causes acceleration.

(c)

Use trigonometry to find magnitude of force component along incline:

$$F_{Ebx}^G = F_{Eb}^G \sin \theta = +mg \sin \theta$$

Figure 10.34a. The friction between the block and the surface of the incline is negligible. The (nondissipative) force of gravity exerted on the block does not line up with the block's displacement along the incline. In order to calculate the work done on the block, we evaluate the block's change in kinetic energy ΔK while it slides down the incline. Because there are no changes in the internal energy of the block, the work done on it is equal to ΔK.

To calculate ΔK, we must first determine the block's acceleration. Let us choose a coordinate system that has the x axis parallel to the incline and the y axis perpendicular to it, as shown in Figure 10.34a. The block is subject to two forces: a normal force of magnitude F_{sb}^n exerted by the incline surface and a downward force of gravity of magnitude F_{Eb}^G, which is shown decomposed into components F_{Ebx}^G and F_{Eby}^G in the free-body diagram in Figure 10.34b. (There is no tangential component of the contact force because there is no friction.) The x component of $\vec{F}_{Eb}^G$ is what causes the block to accelerate downward. Using trigonometry, we see from Figure 10.34c that

$$F_{Ebx}^G = F_{Eb}^G \sin \theta = +mg \sin \theta, \tag{10.24}$$

where we have substituted mg for the magnitude of the force of gravity exerted on the block (Eq. 8.17) and where the plus sign indicates that this component points in the positive x direction. This equation tells us that the magnitude of the downward acceleration of the block is

$$a_x = \frac{\sum F_x}{m} = \frac{F_{Ebx}^G}{m} = \frac{+mg \sin \theta}{m} = +g \sin \theta. \tag{10.25}$$

This is exactly what we found to be the acceleration of a cart rolling down an incline in Section 3.7 (Eq. 3.20). (We now see that the factor $\sin \theta$ in Eq. 3.20 comes from the decomposition of the force of gravity into components parallel and perpendicular to the incline.)

Suppose the block begins at rest on the incline and drops a vertical distance h, as illustrated in **Figure 10.35a.** If we let the origin coincide with the block's initial position, the block's initial condition is $x_i = 0$, $v_{x,i} = 0$, and its acceleration is given by Eq. 10.25. In order for the block to drop a vertical distance h, its displacement along the incline must be

$$\Delta x = +\frac{h}{\sin \theta}. \tag{10.26}$$

The time the block takes to cover this distance is obtained from Eq. 3.7:

$$\Delta x = v_{x,i}\Delta t + \tfrac{1}{2}a_x(\Delta t)^2 = 0 + \tfrac{1}{2}g \sin \theta (\Delta t)^2, \tag{10.27}$$

Figure 10.35 These blocks gain the same amount of kinetic energy because they descend (without friction) through the same vertical distance h.

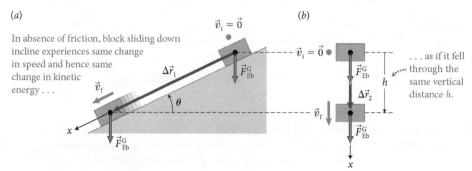

(a)

In absence of friction, block sliding down incline experiences same change in speed and hence same change in kinetic energy . . .

(b)

. . . as if it fell through the same vertical distance h.

and so, from Eqs. 10.26 and 10.27, we obtain

$$(\Delta t)^2 = \frac{2h}{g\sin^2\theta}. \tag{10.28}$$

The block's final kinetic energy is then

$$\tfrac{1}{2}mv_{x,f}^2 = \tfrac{1}{2}m(v_{x,i} + a_x\Delta t)^2 = \tfrac{1}{2}m(0 + g\sin\theta\,\Delta t)^2$$

$$= \tfrac{1}{2}mg^2\sin^2\theta(\Delta t)^2, \tag{10.29}$$

where the substitution for $v_{x,f}$ comes from Eq. 3.4 and that for a_x from Eq. 10.25. Substituting Eq. 10.28 into Eq. 10.29, we obtain

$$\tfrac{1}{2}mv_{x,f}^2 = \tfrac{1}{2}mg^2\sin^2\theta\left(\frac{2h}{g\sin^2\theta}\right) = mgh. \tag{10.30}$$

So, because the block's initial kinetic energy is zero, the change in kinetic energy, which is equal to the work done on the block, is

$$W = \Delta K = K_f - K_i = \tfrac{1}{2}mv_{x,f}^2 - 0 = mgh. \tag{10.31}$$

This is the same result we found in Section 7.9 for the change in kinetic energy of a freely falling object! So the work done on the freely falling block in Figure 10.35*b* when it drops a vertical distance *h* is the same as the work done on the sliding block in Figure 10.35*a* when it drops over the same vertical distance.

Note that we could have obtained the same result for the work done on the block along the incline by calculating the work done on it by *the component of the force of gravity along the incline*:

$$W = F^G_{Ebx}\Delta x = (+mg\sin\theta)\left(+\frac{h}{\sin\theta}\right) = mgh. \tag{10.32}$$

This means that neither the normal force nor the *y* component of the force of gravity—both of which are perpendicular to the displacement of the block—does any work on the block. This makes sense because it is the *x* component of the force of gravity that causes the displacement of the block and that does all the work on it. The forces in the *y* direction cause no displacement and thus do no work on it.

These results can be stated more compactly by defining a **scalar product** of two vectors. If the angle between two vectors $\vec{A}$ and $\vec{B}$ is ϕ, the scalar product is

$$\vec{A}\cdot\vec{B} \equiv AB\cos\phi. \tag{10.33}$$

The right side of Eq. 10.33 shows that the scalar product of vectors $\vec{A}$ and $\vec{B}$ is a scalar and is equal to the product of the magnitude of vector $\vec{A}$ ($|\vec{A}| \equiv A$) and the component of $\vec{B}$ along $\vec{A}$ ($B\cos\phi$). Note that the scalar product is commutative, so that the result is independent of the order of multiplication:

$$\vec{A}\cdot\vec{B} = \vec{B}\cdot\vec{A} = AB\cos\phi = BA\cos\phi, \tag{10.34}$$

which is shown graphically in **Figure 10.36**. Also, as the name suggests, the scalar product of two vectors is a number (a scalar), not a vector.

Figure 10.36 The scalar product of two vectors.

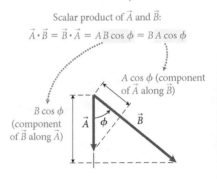

Scalar product of $\vec{A}$ and $\vec{B}$:
$\vec{A}\cdot\vec{B} = \vec{B}\cdot\vec{A} = AB\cos\phi = BA\cos\phi$

$A\cos\phi$ (component of $\vec{A}$ along $\vec{B}$)

$B\cos\phi$ (component of $\vec{B}$ along $\vec{A}$)

Using the definition of scalar product, we can write **work as a scalar product:**

$$W = \vec{F} \cdot \Delta \vec{r}_F \quad \text{(constant nondissipative force).} \quad (10.35)$$

In one dimension, when the force and the force displacement both point along the same line, Eq. 10.35 reduces to our old expression for work, $W = F_x \Delta x_F$ (Eq. 9.8). Note that if the force and the displacement are in opposite directions, $\phi = 180°$, and so $\cos \phi = -1$ and the work is negative.

Because $\cos 90° = 0$, the work done by a force perpendicular to $\Delta \vec{r}_F$ is zero. So for a block sliding down an incline, the normal force does no work and the work done by the force of gravity on the block is

$$W = \vec{F}_{Eb}^G \cdot \Delta \vec{r}_F = (mg)\left(\frac{h}{\sin \theta}\right) \cos \phi, \quad (10.36)$$

where θ is the angle of the incline and ϕ the angle between $\vec{F}_{Eb}^G$ and $\Delta \vec{r}_F$. Because $\phi = \frac{\pi}{2} - \theta$ and $\cos(\frac{\pi}{2} - \theta) = \sin \theta$, we see as before, that the work done by the force of gravity on the block is $W = mgh$.

Figure 10.37 shows how we can generalize this result to determine the work done by the gravitational force on an object moving across a surface of any shape. If we approximate the surface by, say, four inclined surfaces, we can apply Eq. 10.36 to each one. The total work done on a block sliding on the surface in Figure 10.37 as it descends from the top to the bottom is

$$W = mgh_1 + mgh_2 + mgh_3 + mgh_4 = mgh. \quad (10.37)$$

To obtain a better approximation we can divide the surface into more, smaller inclined surfaces, but each time we see that the total work done on the block is equal to mgh.

This result also makes sense from an energy perspective. Consider, for example a ball sliding down the surface of Figure 10.37, and let's assume that there is no friction between the ball and the surface. **Figure 10.38** shows two energy diagrams for this situation. For the closed Earth-ball system (Figure 10.38a)

Figure 10.37 An arbitrary incline can be broken into small straight line segments.

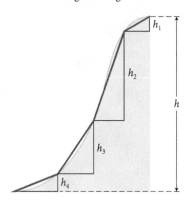

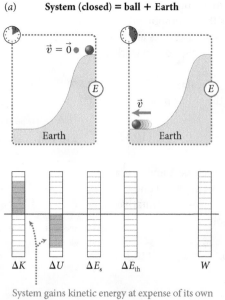

Figure 10.38 Two views of the energy changes that occur when a ball slides down an arbitrary incline with negligible friction.

(*a*) **System (closed) = ball + Earth**

System gains kinetic energy at expense of its own potential energy; no net change in system's energy.

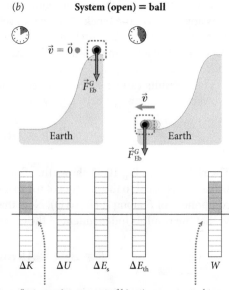

(*b*) **System (open) = ball**

System gains amount of kinetic energy equal to work done on it by gravitational force.

the change in the ball's kinetic energy ΔK_b and the change in the potential energy of the system add up to zero:

$$\Delta K_b + \Delta U^G = 0. \tag{10.38}$$

From Eq. 7.19, we know that the change in the gravitational potential energy is given by $\Delta U^G = -mgh$, and so

$$\Delta K_b = mgh. \tag{10.39}$$

Next let's consider just the ball by itself as our system (Figure 10.38b). There is no change in the ball's internal energy, so the change in the ball's kinetic energy is due to work done on it by the force of gravity: $\Delta K_b = W$. The change in the ball's kinetic energy does not depend on the system chosen, so we see from Eq. 10.39 that the work done on the ball is

$$W = mgh, \tag{10.40}$$

in agreement with Eqs. 10.36 and 10.37. In the process of deriving Eqs. 10.36 and 10.37, however, we derived an expression for the work done by a force that does not point in the same direction as the force displacement (Eq. 10.35).

Let us end by generalizing the result in Eq. 10.35 to a nondissipative force $\vec{F}(\vec{r})$ that is not constant. For such a force we follow the same procedure as in Section 9.7: we divide the force displacement $\Delta\vec{r}_F$ into many small force displacements $\delta\vec{r}_F$, such that the force is approximately constant over each small displacement, and then apply Eq. 10.35 to each small displacement (Figure 10.39):

$$W_n \approx \vec{F}(\vec{r}_n) \cdot \delta\vec{r}_{Fn}, \tag{10.41}$$

where $\vec{F}(\vec{r}_n)$ is the force at position $\vec{r}_n$ (at the beginning of each interval, as shown in Figure 10.39). The work done on the system over the entire force displacement $\Delta\vec{r}_F$ is then equal to the sum of the work done over all of the small displacements:

$$W \approx \sum_n W_n = \sum_n \vec{F}(\vec{r}_n) \cdot \delta\vec{r}_{Fn}. \tag{10.42}$$

To get a better approximation, we can divide the force displacement $\Delta\vec{r}_F$ into a larger number of smaller displacements $\delta\vec{r}_{Fn}$, so that the variation of the force over each displacement becomes even smaller. An exact result is obtained if we let $\delta\vec{r}_{Fn}$ approach zero (and the number of small displacements approaches infinity):

$$W = \lim_{\delta r_{Fn} \to 0} \sum_n \vec{F}(\vec{r}_n) \cdot \delta\vec{r}_{Fn}, \tag{10.43}$$

which is the definition of the line integral (Appendix B) of the force:

$$W = \int_{\vec{r}_i}^{\vec{r}_f} \vec{F}(\vec{r}) \cdot d\vec{r} \quad \text{(variable nondissipative force)}, \tag{10.44}$$

where the line integral should be evaluated over the path traced out by the point of application of the force.

For dissipative forces, we can follow the same procedure of dividing up the displacement to our approach of Section 9.7. Equation 9.27 then replaces Eq. 10.35, yielding for the change in thermal energy caused by the dissipative force:

$$\Delta E_{th} = -\int_{\vec{r}_i}^{\vec{r}_f} \vec{F}(\vec{r}_{cm}) \cdot d\vec{r}_{cm} \quad \text{(variable dissipative force)}. \tag{10.45}$$

Note that, in contrast to Eq. 10.44, this expression contains the position of the center of mass of the system rather than the point of application of the force.

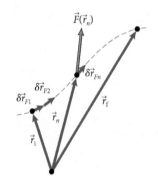

Figure 10.39 To evaluate the work done on a system by a variable force, we divide the force displacement into small pieces.

✋ **10.11** When is W in Eq. 10.40 positive, and when is it negative? Is your answer consistent with the definition of positive and negative work given in Section 9.2?

10.10 Coefficients of friction

To describe the motion of objects along surfaces, we need a quantitative expression for the force of friction. Like the normal force, the force of static friction can take on a range of values, depending on the situation.

Suppose you are still struggling with the file cabinet of Checkpoint 10.4. As you increase the strength of your push on the cabinet, the force of static friction opposing your push grows. If you change the direction of your push, the force of friction changes direction as well, always opposing your force. The force of static friction has a maximum value, however, evidenced by the fact that if you push hard enough, the cabinet starts sliding. So the question is: What is the *maximum* force of static friction in a given situation?

The answer to this question depends on many variables, but we can establish a number of basic facts. For example—this you know from experience—the maximum force of static friction depends on the force with which the surfaces press against each other. An empty file cabinet slides more easily than a full one, for instance, and the double brick in **Figure 10.40** requires twice as large a force to begin sliding as the single brick does. The reason for this dependence on load is shown in Figure 10.40. If you put a brick down on a large flat plate, the *contact area* between the brick and the plate is given by the surface area of the face of the brick that is touching the plate. At the microscopic level, however, surfaces are never completely flat, so the surfaces of the brick and the plate touch only at the tops of protrusions over a combined *effective contact area* that is much smaller than the contact area (typically only about $1/10,000$ of it). When the load increases, protrusions on the surfaces flatten, and the effective contact area increases, strengthening the bond between the surfaces.

Measurements show that the angle θ_{max} at which objects begin to slide down an incline (in other words, the angle at which the force of static friction reaches its maximum value) does not depend on the contact area. As illustrated in **Figure 10.41**, two identical bricks begin sliding at the same incline angle regardless of the contact area between the brick and the incline surface. I know it is counterintuitive that the contact area does not determine when the bricks begin to slide, but to understand what is going on, let's go to the microscopic level again. When the brick is oriented so that its largest face is in contact with the inclined surface, the force exerted by the brick on the inclined surface is spread out over a large number of contact points, with the effective contact area having a certain value. When the brick is on its side, there are fewer contact points, but each point now carries a greater share of the load and so flattens out more. This flattening increases the contact area available at each point, with the result that the effective contact area has the same value as when the brick was in the other orientation. Thus the effective contact area doesn't depend on the orientation of the brick—it depends only on the force with which the brick presses against the surface. These observations establish the following important point:

> **The maximum force of static friction exerted by a surface on an object is proportional to the force with which the object presses on the surface and does not depend on the contact area.**

To obtain an expression for the force with which an object presses on an inclined surface, consider **Figure 10.42**. The normal force $\vec{F}_{sb}^{n}$ exerted by the surface

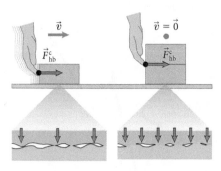

Figure 10.40 On a rough surface, a single brick slides more easily than a double brick. The harder two surfaces press against each other, the larger the effective contact area, and consequently the stronger the bond between the surfaces.

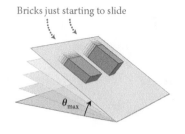

Figure 10.41 Regardless of its orientation, a brick starts sliding at the same angle of incline.

Bricks just starting to slide

θ_{max}

on the block and the normal force $\vec{F}_{bs}^n$ with which the surface of the block presses against the inclined surface (not shown in Figure 10.42) are an interaction pair and therefore equal in magnitude. As we have just determined, the maximum magnitude of the force of static friction is proportional to F_{bs}^c and therefore must also be proportional to F_{sb}^n, $(F_{sb}^s)_{max} = \mu_s F_{sb}^n$. For any two surfaces 1 and 2, we can therefore write

$$(F_{12}^s)_{max} = \mu_s F_{12}^n. \qquad (10.46)$$

The maximum magnitude of the force of static friction between two surfaces is proportional to the magnitude of the normal force exerted by the surfaces on each other. The unitless proportionality constant μ_s is called the **coefficient of static friction.** Its value depends on the two materials in contact and the condition of their surfaces. Values for a few representative materials are given in Table 10.1.

10.12 (*a*) A brick slides more easily on ice than on wood. For which combination (brick on ice or brick on wood) is μ_s larger? (*b*) For rubber on dry asphalt, μ_s is about 1. Ignoring any limits due to the runner's skill and condition, what maximum acceleration without slipping can a runner achieve using rubber soles on a dry, horizontal asphalt surface? [Hint: See Figure 10.21.]

The coefficient of static friction expressed in terms of the angle of incline can readily be determined experimentally by raising the incline of Figure 10.42 to the maximum angle at which the block remains at rest. At that angle θ_{max}, the force of static friction $\vec{F}_{sb}^s$ reaches the maximum value given by Eq. 10.46, and because the block remains at rest, the forces along the x axis must add to zero:

$$\sum F_x = F_{Ebx}^G + (F_{sb}^s)_{max} = 0 \qquad (10.47)$$

and so

$$(F_{sb}^s)_{max} = -F_{Ebx}^G. \qquad (10.48)$$

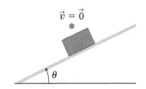

Figure 10.42 Free-body diagram for a block at rest on an inclined surface.

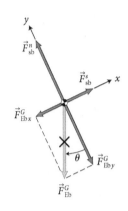

Table 10.1 Coefficients of friction

Material 1	Material 2	μ_s	μ_k
aluminum	aluminum	1.1–1.4	1.4
aluminum	steel	0.6	0.5
glass	glass	0.9–1.0	0.4
glass	nickel	0.8	0.6
ice	ice	0.1	0.03
rubber	aluminum	0.5	—
rubber	concrete	1.0–4.0	0.8
rubber	rubber	0.84	—
steel	steel	0.8	0.4
steel	brass	0.5	0.4
steel	copper	0.5	0.4
steel	lead	0.95	0.95
wood	wood	0.25–0.5	0.2

The values given are for clean, dry, smooth surfaces.

Because the block has no acceleration in the y direction, we also have

$$\sum F_y = F_{sb}^n + F_{Eby}^G = 0, \qquad (10.49)$$

and so

$$F_{sb}^n = -F_{Eby}^G. \qquad (10.50)$$

Substituting Eqs. 10.48 and 10.50 into 10.46, we get

$$-F_{Ebx}^G = \mu_s(-F_{Eby}^G) \qquad (10.51)$$

$$\mu_s = \frac{F_{Ebx}^G}{F_{Eby}^G}. \qquad (10.52)$$

Looking at Figure 10.42, we see that F_{Ebx}^G/F_{Eby}^G is the tangent of θ and so

$$\mu_s = \tan \theta_{max}. \qquad (10.53)$$

10.13 (*a*) To what angle θ_{max} does a coefficient of static friction of 1 correspond? (*b*) Put, in turn, a coin, a paper clip, a CD case, and a comb on the cover of this book and tilt the book to obtain the angle at which each object begins to slide. In each case, how close do you get to the angle calculated in part *a*? What do you conclude from these observations about the coefficients of static friction for these combinations of surfaces?

Equation 10.46 gives only the *maximum* value of the force of static friction. All we know in any situation in which this force is present but has not reached its maximum value is that the magnitude of the force of static friction exerted by surface 1 on surface 2 must obey the condition

$$F_{12}^s \leq \mu_s F_{12}^n, \qquad (10.54)$$

where F_{12}^n is the magnitude of the normal force exerted by surface 1 on surface 2. How, then, can you determine the magnitude of the frictional force exerted on an object in a given situation? Inequality 10.54 is not of much help, especially because the normal force can also take on almost any value! The key is to use what you know about the object's motion: If you know its acceleration and inertia, you can use the equation of motion, $\vec{a} = \sum\vec{F}/m$, to determine the vector sum of the forces exerted on the object. If you know what other forces are exerted on it, you can extract the values of the normal and frictional forces from this sum. The general procedure for dealing with situations involving frictional forces at surfaces is explained in the Procedure box "Working with frictional forces."

Procedure: Working with frictional forces

1. Draw a free-body diagram for the object of interest. Choose your x axis parallel to the surface and the y axis perpendicular to it, then decompose your forces along these axes. Indicate the acceleration of the object.
2. The equation of motion in the y direction allows you to determine the sum of the y components of the forces in that direction:

$$\sum F_y = ma_y.$$

Unless the object is accelerating in the normal direction, $a_y = 0$. Substitute the y components of the forces from your free-body diagram. The resulting equation allows you to determine the normal force.

3. The equation of motion in the x direction is

$$\sum F_x = ma_x.$$

If the object is not accelerating along the surface, $a_x = 0$. Substitute the x components of the forces from your free-body diagram. The resulting equation allows you to determine the frictional force.

4. If the object is not slipping, the normal force and the force of static friction should obey Inequality 10.54.

Example 10.8 Pulling a friend up a hill

A hiker is helping a friend up a hill that makes an angle of 30° with level ground. The hiker, who is farther up the hill, is pulling on a cable attached to his friend. The cable is parallel to the hill so that it also makes an angle of 30° with the horizontal. If the coefficient of static friction between the soles of the hiker's boots and the surface of the hill is 0.80 and his inertia is 65 kg, what is the maximum magnitude of the force he can exert on the cable without slipping?

❶ GETTING STARTED I begin by making a sketch and drawing the free-body diagram for the hiker (**Figure 10.43**). I chose an x axis that points up the hill and the y axis perpendicular to it. The forces exerted on the hiker are a force $\vec{F}_{ch}^c$ exerted by the cable and directed down the hill (this force forms an interaction pair with the force the hiker exerts on the cable), the force of gravity $\vec{F}_{Eh}^G$, and the contact force exerted by the surface of the hill on the hiker. This last force has two components: the normal force $\vec{F}_{sh}^n$ and the force of static friction $\vec{F}_{sh}^s$. The force of static friction is directed up the hill. If the hiker is not to slip, his acceleration must be zero, and so the forces in the normal (y) and tangential (x) directions must add up to zero.

Figure 10.43

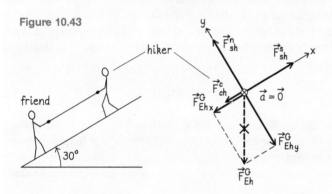

❷ DEVISE PLAN As suggested in the Procedure box on page 250, I'll write the equation of motion along the two axes, setting the acceleration in each direction to zero. Then I use Inequality 10.54 to determine the maximum force that the hiker can exert without slipping.

❸ EXECUTE PLAN Because the magnitude of the force of gravity exerted on the hiker is mg, I have in the y direction

$$\sum F_y = F_{Ehy}^G + F_{sh}^n = -mg\cos\theta + F_{sh}^n = ma_y = 0$$

$$F_{sh}^n = mg\cos\theta, \tag{1}$$

where the minus sign in $-mg\cos\theta$ indicates that F_{Ehy}^G is in the negative y direction. Likewise, I have in the x direction

$$\sum F_x = F_{Ehx}^G + F_{sh}^s - F_{ch}^c$$

$$= -mg\sin\theta + F_{sh}^s - F_{ch}^c = ma_x = 0$$

$$F_{sh}^s = mg\sin\theta + F_{ch}^c. \tag{2}$$

If the hiker is not to slip, the force of static friction must not exceed its maximum value. Substituting Eqs. 1 and 2 into Inequality 10.54, I get

$$mg\sin\theta + F_{ch}^c \leq \mu_s(mg\cos\theta)$$

$$F_{ch}^c \leq mg(\mu_s\cos\theta - \sin\theta).$$

The problem asks me about the magnitude F_{hc}^c of the force exerted *by* the hiker *on* the cable, which is the same as the magnitude F_{ch}^c of the force exerted *by* the cable *on* the hiker, so

$$F_{hc}^c \leq mg(\mu_s\cos\theta - \sin\theta).$$

Substituting the values given, I obtain

$$F_{hc}^c \leq (65\text{ kg})(9.8\text{ m/s}^2)(0.80\cos 30° - \sin 30°)$$

$$= 1.2 \times 10^2\text{ N.} ✔$$

❹ EVALUATE RESULT A force of 1.2×10^2 N corresponds to the gravitational force exerted by Earth on an object that has an inertia of only 12 kg, and so the hiker cannot pull very hard before slipping. I know from experience, however, that unless I can lock a foot behind some solid object, it is impossible to pull hard on an inclined surface, and thus the answer I obtained makes sense.

Example 10.9 Object accelerating on a conveyor belt

While designing a conveyor belt system for a new airport, you determine that, on an incline of 20°, the magnitude of the maximum acceleration a rubber belt can give a typical suitcase before the suitcase begins slipping is 4.0 m/s². What is the coefficient of static friction for a typical suitcase on rubber?

❶ GETTING STARTED: I begin by making a sketch and drawing a free-body diagram for the suitcase, choosing the x axis along the conveyor belt in the direction of acceleration and the y axis upward and perpendicular to it (**Figure 10.44**). I also draw the upward acceleration of the suitcase along the incline.

❷ DEVISE PLAN As in Example 10.8, I write out the equation of motion along both axes. The x component of the acceleration is $a_x = +a$, where the magnitude of the acceleration is

Figure 10.44

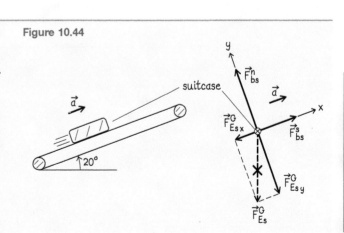

$a = 4.0 \text{ m/s}^2$. I can then use Inequality 10.54 to determine the coefficient of static friction.

❸ EXECUTE PLAN The equations of motion in the x and y directions are

$$\Sigma F_x = F_{Esx}^G + F_{bs}^s = -mg \sin \theta + F_{bs}^s = ma_x \qquad (1)$$

$$\Sigma F_y = F_{Esy}^G + F_{bs}^n = -mg \cos \theta + F_{bs}^n = ma_y = 0, \qquad (2)$$

with m the inertia of the suitcase. If a_x represents the maximum acceleration at which the suitcase does not slip, the force of static friction must be maximum. Substituting Eq. 10.46 for $(F_{bs}^s)_{max}$ into Eq. 1, I get

$$-mg \sin \theta + \mu_s F_{bs}^n = m(+a).$$

Solving this equation for μ_s and substituting $F_{bs}^n = mg \cos \theta$ from Eq. 2, I obtain

$$\mu_s = \frac{ma + mg \sin \theta}{mg \cos \theta} = \frac{a + g \sin \theta}{g \cos \theta}$$

$$= \frac{(4.0 \text{ m/s}^2) + (9.8 \text{ m/s}^2) \sin 20°}{(9.8 \text{ m/s}^2)(\cos 20°)} = 0.80. \checkmark$$

❹ EVALUATE RESULT From Table 10.1, I see that the value I obtained for the coefficient for static friction is close to the coefficient for rubber against rubber and therefore reasonable. I also note that the inertia m of the suitcase, which is not given, drops out of the final result.

Once surfaces slip relative to each other, the force of kinetic friction, like its static counterpart, is proportional to the normal force: It takes a greater force to keep a heavily loaded file cabinet moving than it does to keep an empty one moving. Experiments also show that objects sliding to a stop on a horizontal surface under the influence of kinetic friction undergo a constant acceleration; in other words, the force of kinetic friction between two slipping surfaces is constant and independent of the relative velocity of the two surfaces. In analogy to Eq. 10.46, we can say

$$F_{12}^k = \mu_k F_{12}^n, \qquad (10.55)$$

where μ_k is called the **coefficient of kinetic friction.** The kinetic coefficient μ_k is always smaller than μ_s (or else an object would never slip because it would "stop slipping the instant it starts slipping"). For most surface combinations, the value of μ_k is somewhere between 25% and 100% of the corresponding value of μ_s (Table 10.1).

Note that, unlike the force of static friction, which can take on any value below the maximum limit dictated by μ_s, the force of kinetic friction is constant and therefore always equal to the value given in Eq. 10.55.

10.14 Imagine putting a single and a double brick on a flat board, as in Figure 10.40, and then slowly raising one end of the board. (*a*) Does the single brick begin to slide before, after, or at the same time as the double brick? (*b*) Once the bricks start sliding, is the acceleration of the double brick larger than, equal to, or smaller than that of the single brick?

Chapter Glossary

SI units of physical quantities are given in parentheses.

Coefficient of friction μ_s, μ_k (unitless): A scalar equal to the ratio of the force of friction to the normal force. The maximum force of static friction exerted by surface 1 on surface 2 is given by

$$(F_{12}^s)_{\max} = \mu_s F_{12}^n, \qquad (10.46)$$

where μ_s is the *coefficient of static friction* and F_{12}^n is the normal force exerted by surface 1 on surface 2. The force of kinetic friction is given by

$$F_{12}^k = \mu_k F_{12}^n, \qquad (10.55)$$

where μ_k is the *coefficient of kinetic friction.*

Component vector Any vector $\vec{A}$ can be decomposed into component vectors $\vec{A}_x$ and $\vec{A}_y$, along some conveniently chosen axes, such that $\vec{A} = \vec{A}_x + \vec{A}_y$.

Frictional force The tangential component of the contact force exerted by a surface on an object. When the two surfaces in contact are not moving relative to each other, the tangential component is called the *force of static friction*; when they are moving, it is called the *force of kinetic friction.*

Normal component The component of a vector perpendicular to a surface.

Normal force The normal component of the contact force exerted by a surface on an object.

Polar coordinate system A coordinate system in which the two coordinates that locate a point are its distance from the origin and the angle between its position vector and the x axis.

Rectangular coordinate system A coordinate system in which two coordinates that locate a point are its perpendicular distances from each of two perpendicular axes. Convenient choices of coordinate systems: For an object that is accelerating, choose one of the axes along the direction of the acceleration; for an object in translational equilibrium, choose your axes such that the largest possible number of forces lie along one of the axes.

Scalar product The product of two vectors yielding a scalar quantity. For two vectors $\vec{A}$ and $\vec{B}$ that make an angle ϕ, the scalar product is defined as

$$\vec{A} \cdot \vec{B} \equiv AB \cos \phi. \qquad (10.33)$$

Tangential component The component of a vector parallel to a surface.

Work as scalar product W (J) The work done on an object by a constant nondissipative force $\vec{F}$ exerted on that object during a displacement $\Delta \vec{r}_F$ of the point of application of $\vec{F}$ is equal to the scalar product of $\vec{F}$ and $\Delta \vec{r}_F$:

$$W = \vec{F} \cdot \Delta \vec{r}_F \quad \text{(constant nondissipative force).} \qquad (10.35)$$

For a force that is not constant, the work done on a system is

$$W = \int_{\vec{r}_i}^{\vec{r}_f} \vec{F}(\vec{r}) \cdot d\vec{r} \quad \text{(variable nondissipative force).} \qquad (10.44)$$

11 Motion in a Circle

The motion we have been dealing with so far in this text is called **translational motion** (Figure 11.1*a*). This type of motion involves no change in an object's orientation; in other words, all the particles in the object move along identical parallel trajectories. During **rotational motion,** which we begin to study in this chapter, the orientation of the object changes, and the particles in an object follow different circular paths centered on a straight line called the *axis of rotation* (Figure 11.1*b*). Generally, the motion of rigid objects is a combination of these two types of motion (Figure 11.1*c*), but as we shall see in Chapter 12 this combined motion can be broken down into translational and rotational parts that can be analyzed separately. Because we already know how to describe translational motion, knowing how to describe rotational motion will complete our description of the motion of rigid objects.

As Figure 11.1*b* shows, each particle in a rotating object traces out a circular path, moving in what we call *circular*

motion. We therefore begin our analysis of rotational motion by describing circular motion. Circular motion occurs all around us. A speck of dust stuck to a spinning CD, a stone being whirled around on a string, a person on a Ferris wheel—all travel along the perimeter of a circle, repeating their motion over and over. Circular motion takes place in a plane, and so in principle we have already developed all the tools required to describe it. To describe circular and rotational motion we shall follow an approach that is analogous to the one we followed for the description of translational motion. Exploiting this analogy, we can then use the same results and insights gained in earlier chapters to introduce a third conservation law.

11.1 Circular motion at constant speed

Figure 11.2 shows two examples of circular motion: a block dragged along a circle by a rotating turntable and a puck constrained by a string to move in a circle. The block and puck are said to *revolve* around the vertical axis through the center of each circular path. Note that the axis about which they revolve is external to the block and puck and perpendicular to the plane of rotation. This is the definition of *revolve*—to move in circular motion around an *external* center. Objects that turn about an *internal* axis, such as the turntable in Figure 11.2*a*, are said to *rotate*. These two types of motion are closely related because a rotating object can be considered as a system of an enormous number of particles, each revolving around the axis of rotation.

Figure 11.1 Translational and rotational motion of a rigid object.

(*a*) Translational motion

All points on object follow identical trajectories.

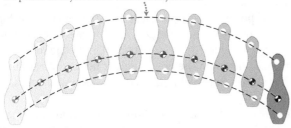

(*b*) Rotational motion

All points on object trace circles centered on axis of rotation.

axis of rotation

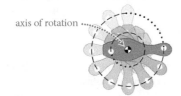

(*c*) Combined translation and rotation

Different points on object follow different trajectories.

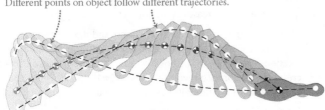

Figure 11.2 Examples of circular motion.

(*a*) Block revolves on rotating turntable

Block *revolves* because axis is external to it.

axis of rotation

Turntable *rotates* because axis is internal to it.

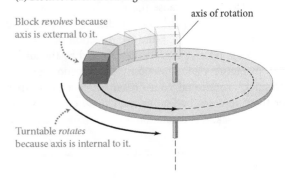

(*b*) Tethered puck revolves on air table

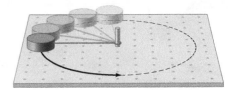

Figure 11.3 Average and instantaneous velocity for an object in circular motion.

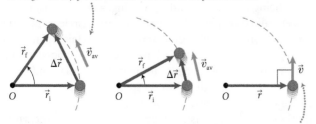

Average velocity points in same direction as displacement $\Delta\vec{r}$.

As time interval approaches zero, average velocity approaches instantaneous velocity, which is tangent to circular trajectory.

Figure 11.3 shows an overhead view of a puck moving counterclockwise along the arc of a circle. The puck's average velocity points in the same direction as the puck's displacement $\Delta\vec{r}$, which points from the puck's initial position to its final position. As we consider the motion for increasingly smaller time intervals, the angle between the position vectors $\vec{r}_i$ and $\vec{r}_f$ measured from the center of the circular trajectory approaches zero and $\Delta\vec{r}$ becomes tangent to the circular trajectory. This is true all along the trajectory, and so we can say:

> The instantaneous velocity $\vec{v}$ of an object in circular motion is always perpendicular to the object's position $\vec{r}$ measured from the center of the circular trajectory.

In the first part of this chapter, we study only objects in **circular motion at constant speed;** that is to say, the magnitude of their instantaneous velocity doesn't change.

Because the motion lies entirely in a plane, we can use Cartesian coordinates to describe the puck's position and velocity at any instant. In **Figure 11.4a**, for example, the puck's position is specified by (x_1, y_1) at instant t_1 and by (x_2, y_2) at instant t_2. However, because the magnitude of the position vector $\vec{r}$ is always equal to the radius of the circle, we can also give the magnitude r and the direction of this vector to specify the puck's position, as indicated by the angles θ_1 and θ_2 in Figure 11.4b. To specify the direction, we define the puck's **rotational coordinate,** which is a unitless number that increases by 2π for each circle completed by the revolving puck.* The rotational coordinate is denoted

Figure 11.4 The position of an object in circular motion can be given in Cartesian or polar coordinates.

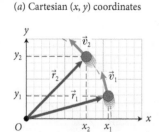

(a) Cartesian (x, y) coordinates   (b) Polar (r, θ) coordinates

*The adjective *rotational* is used for both rotation and revolution.

Figure 11.5 The rotational coordinate specifies the location of an object on a circle.

(a) Relationship between rotational coordinate and polar angle

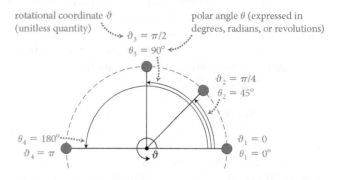

rotational coordinate ϑ (unitless quantity)

polar angle θ (expressed in degrees, radians, or revolutions)

$\vartheta_3 = \pi/2$
$\theta_3 = 90°$

$\vartheta_2 = \pi/4$
$\theta_2 = 45°$

$\theta_4 = 180°$
$\vartheta_4 = \pi$

$\vartheta_1 = 0$
$\theta_1 = 0°$

(b) Symbol used in this text to specify rotational coordinate system

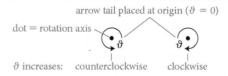

arrow tail placed at origin ($\vartheta = 0$)

dot = rotation axis

ϑ increases: counterclockwise   clockwise

by the Greek letter ϑ (script theta); for example, $\vartheta = \pi$ refers to the position of the puck half a revolution after starting from the origin. The rotational coordinate is related to *polar angle* θ as shown in **Figure 11.5a**. Unlike the rotational coordinate, however, the polar angle is not a unitless number but rather expressed in units of revolutions, degrees, or radians.

To measure ϑ we need to choose a direction of increasing ϑ and a zero, just as we need to specify a direction of increasing x and an origin to measure position along an axis. We shall indicate our choice of rotational coordinate system, which will generally be dictated by the motion under consideration, by drawing a curved arrow around the axis of rotation, as shown in Figure 11.5b.

Exercise 11.1 Rotational coordinates

Determine the rotational coordinate of the following objects for the indicated choices of rotational coordinate system. (a) The tip of the hour hand of a clock at 4:30 p.m., when zero is at noon on the same day and ϑ increases in the clockwise direction. (b) A faucet after being turned 3/4 of a turn in a clockwise direction, starting from zero and with ϑ increasing in the counterclockwise direction. (c) The object in Figure 11.4 if $\theta = 78.3°$, when zero is at $\theta = 0°$ and ϑ increases in the counterclockwise direction.

SOLUTION (a) The angle between noon and 4:30 p.m. is 135°, which is $(135°)/(360°) = 0.375$ of one revolution. Because the rotation takes place in the positive direction and because ϑ increases by 2π for each revolution, $\vartheta = (2\pi)(+0.375) = +2.36$. Note that there are no units because ϑ is a unitless quantity. ✔

(b) $\vartheta = (2\pi)(-3/4) = -\frac{3}{2}\pi.$ ✔

(c) $\vartheta = (2\pi)(+78.3°)/(360°) = +1.37.$ ✔

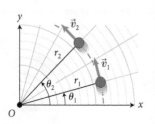

Figure 11.6 Two pucks in circular motion at constant speed can have different speeds even though their rotational velocity (the rate at which ϑ changes) is the same.

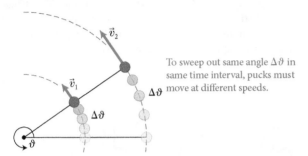

Figure 11.7 Instantaneous and rotational velocities for points on a rotating disk.

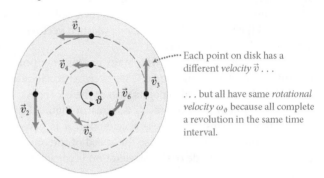

Each point on disk has a different *velocity* $\vec{v}$ . . .

. . . but all have same *rotational velocity* ω_ϑ because all complete a revolution in the same time interval.

We shall represent the rate at which an object's rotational coordinate changes by ω_ϑ (Greek omega). The subscript ϑ indicates that the sign of this quantity depends on our choice of direction for increasing ϑ, just as the sign of v_x (the rate at which an object's x coordinate changes) depends on our choice of x axis. Indeed, just as v_x is the component of a velocity vector $\vec{v}$, ω_ϑ is a component of a *rotational velocity vector* $\vec{\omega}$, which we shall discuss in Chapter 12. For convenience we shall refer to ω_ϑ as the **rotational velocity,** although strictly speaking it is the ϑ component of $\vec{\omega}$.

The magnitude of the rotational velocity is the **rotational speed,** which we denote by ω just as v denotes speed: $\omega = |\vec{\omega}| = |\omega_\vartheta|$. The time interval it takes an object in circular motion at constant speed to complete one revolution is called the **period** and is denoted by the letter T.

For example, if the puck in Figure 11.4 completes three revolutions in the negative ϑ direction in 2.0 s, its average rotational velocity for those three revolutions ($\Delta\vartheta = -3(2\pi) = -6\pi$) is $\omega_{\vartheta,\text{av}} = \Delta\vartheta / \Delta t = (-6\pi)/(2.0 \text{ s}) = -9.4 \text{ s}^{-1}$; its average rotational speed is 9.4 s^{-1}; and its period is $T = (2.0 \text{ s})/3 = 0.67$ s. Note that the units of rotational velocity and rotational speed are s^{-1} (*inverse second*).

The rotational coordinate of an object in circular motion at constant speed increases linearly with time. If it takes a puck in circular motion at constant speed 1.0 s to go from $\vartheta = 0$ to $\vartheta = +0.80$, then it also takes the puck 1.0 s to go from $\vartheta = +0.80$ to $\vartheta = +1.6$, and the object's rotational velocity is constant at $\omega_\vartheta = (+1.6 - 0.80)/(1.0 \text{ s}) = +0.80 \text{ s}^{-1}$.

What is the relationship between an object's speed and its rotational velocity? **Figure 11.6** shows two pucks in circular motion. The rotational coordinate of both pucks changes by the same amount $\Delta\vartheta$ during a time interval Δt, and so both have the same rotational velocity. Their speeds, however, are not the same because the puck that moves along the larger circle travels a greater distance along its trajectory than does the puck that moves along the smaller circle.

The advantage of using rotational velocity to describe rotational motion over using the ordinary velocity $\vec{v}$ becomes clear when we look at the rotating disk in **Figure 11.7**. As the velocity vectors $\vec{v}$ show, each point on the disk has a different velocity $\vec{v}$, with points farther from the axis of revolution having a greater speed than points closer to the axis. Yet all points on the disk execute the same number of revolutions per unit time, and so they all have the same rotational velocity ω_ϑ.

The velocity of an object in circular motion constantly changes direction. As **Figure 11.8a** shows, even though the initial and final *speeds* of the object are the same (indicated by the equal magnitudes of the velocity vectors), the vectors $\vec{v}_i$ and $\vec{v}_f$ point in different directions. The object therefore undergoes a change in velocity $\Delta\vec{v}$, which, by our definition of acceleration, means that the object accelerates.

Objects in circular motion have a nonzero acceleration even if they are moving at constant speed.

Because $\vec{r}$ and $\vec{v}$ are always perpendicular, when $\vec{r}$ rotates through an angle θ, $\vec{v}$ rotates through the same angle,

Figure 11.8 Average and instantaneous acceleration of an object in circular motion.

(a)

Puck accelerates because its velocity changes direction.

(b)

To find puck's average acceleration:

Shaded triangles are similar.

$\vec{a}_{\text{av}}$ has same direction as change in velocity $\Delta\vec{v} = \vec{v}_f - \vec{v}_i$.

$\vec{a}_{\text{av}} = \dfrac{\Delta\vec{v}}{\Delta t}$

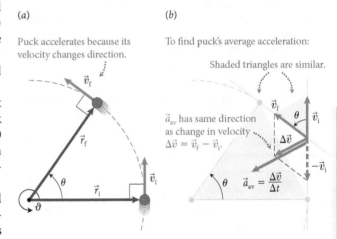

(c)

As time interval approaches zero, average acceleration approaches instantaneous acceleration.

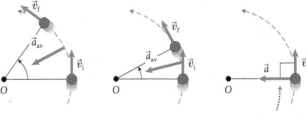

Instantaneous acceleration is perpendicular to $\vec{v}$.

which means the shaded triangles in Figure 11.8*b* are similar. To determine the change in velocity $\Delta\vec{v} = \vec{v}_f - \vec{v}_i$, we slide $\vec{v}_i$ and $\vec{v}_f$ along their tangents until their tails meet, as shown in Figure 11.8*b*, then reverse $\vec{v}_i$ (dashed vector) and add the reversed vector to $\vec{v}_f$.

Note that the vector $\Delta\vec{v}$ we obtain in this way points toward the center of the circle. The average acceleration $\vec{a}_{av} \equiv \Delta\vec{v}/\Delta t$ therefore also points toward the center. As we let the time interval Δt approach zero, as in Figure 11.8*c*, we see that the (instantaneous) acceleration $\vec{a}$ is perpendicular to the velocity $\vec{v}$. Because the speed of the object is constant, the magnitude of the acceleration is constant too.

> **An object executing circular motion at constant speed has an acceleration of constant magnitude that is directed toward the center of its circular path.**

This acceleration is called the **centripetal acceleration** (*centripetal* is Latin for "center-seeking").

In Chapter 10 we saw that the component of the acceleration parallel to an object's instantaneous velocity causes the speed of the object to change, and the perpendicular component causes the direction of the instantaneous velocity to change. For circular motion at constant speed, there is only a perpendicular component of acceleration (the *centripetal acceleration* we just described), and so only the direction of the velocity changes. Nothing pushes or pulls the object in the direction of motion, and so its speed doesn't change.

Exercise 11.2 Accelerations

Determine the direction of the average acceleration in each of the following situations: (*a*) A car goes over the top of a hill at constant speed. (*b*) A runner slows down after crossing a finish line on level ground. (*c*) A cyclist makes a left turn while coasting at constant speed on a horizontal road. (*d*) A roller-coaster car is pulled up a straight incline at constant speed.

SOLUTION For each situation I make a before-and-after sketch showing the initial and final velocities (**Figure 11.9**). The acceleration is nonzero if the direction of the velocity or the magnitude of the velocity changes.

Figure 11.9

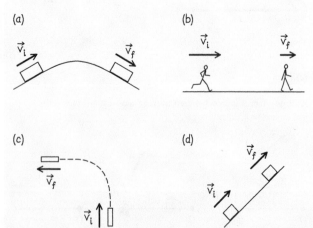

The average acceleration points in the same direction as the change in velocity $\Delta\vec{v}$. To determine the direction of this vector, I draw the vectors $\vec{v}_i$ and $\vec{v}_f$ with their tails together. The change in velocity then points from the tip of $\vec{v}_i$ to the tip of $\vec{v}_f$, as shown in **Figure 11.10**. For each situation, the direction of the average acceleration is given by the direction of $\Delta\vec{v}$. ✔

Figure 11.10

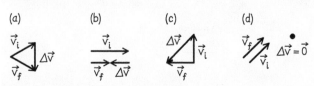

To describe circular motion it is convenient to choose a coordinate system that rotates around with the object under consideration, as illustrated in **Figure 11.11**. The *radial axis,* denoted by the letter *r*, points in the direction of the radius of the circular trajectory, away from the axis of rotation. The *tangential axis,* denoted by the letter *t*, points in the direction of increasing ϑ, tangent to the trajectory. Because the tangent to a circle is perpendicular to the radius drawn to the point of contact of the tangent, the *r* and *t* axes are always perpendicular to each other. Often we'll need a third axis, denoted by the letter *z*, to describe the direction of forces exerted on the object. This axis is chosen perpendicular to the plane of motion (and therefore perpendicular to both the *r* and *t* axes).

With this choice of coordinate system, the velocity of an object executing circular motion at constant speed is always directed along the *t* axis and the acceleration of the object is always directed inward along the *r* axis.

🖐 ⟳ **11.1** Suppose the object in Figure 11.8 is in *accelerated* circular motion, so that $|\vec{v}_f| > |\vec{v}_i|$. In which direction does the object's average acceleration point?

Figure 11.11 Rotating coordinate system used to describe circular motion.

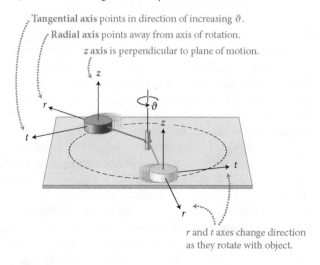

Tangential axis points in direction of increasing ϑ.

Radial axis points away from axis of rotation.

z axis is perpendicular to plane of motion.

r and *t* axes change direction as they rotate with object.

11.2 Forces and circular motion

The centripetal acceleration of an object in circular motion at constant speed tells us that the vector sum of the forces exerted on the object must be directed toward the center of the circle, continuously adjusting the object's direction. Without this inward-pointing vector sum of forces, the object would move in a straight line.

> **An object that executes circular motion at constant speed is subject to a force (or vector sum of forces) of constant magnitude directed toward the center of the circular trajectory.**

But wait! Aren't you pushed *outward* when a car makes a sharp turn? Say you are in a car that takes a right turn as illustrated in **Figure 11.12**. Even though the center of the arc describing the car's trajectory is on the right, there is no denying that your left shoulder is pressed against the left door! Don't forget, however, that because the direction of the car's velocity is changing, it has a nonzero acceleration, and so the reference frame of the car is noninertial.

In Chapter 6 we discussed how things are different in noninertial reference frames. Riding in a car that is traveling on a straight road when it makes an emergency stop, you feel as if you are being pushed forward. There is no force pushing you forward, however. Instead, the "push" you think you feel is nothing more than your body obeying the law of inertia and continuing at its original velocity. As your body tries to continue, it collides with the seat belt because the belt, being part of the car, begins to slow down. You exert a forward force on the belt, but it exerts a rearward force on you. Likewise, when a car initially traveling straight ahead makes a sharp right turn, your body has a tendency to continue in a straight line. As the car begins to curve rightward, the force exerted by the left door on you forces your body into the turn. So even though you feel you are being pushed outward, an inward contact force $\vec{F}_{cp}^{\,c}$ is exerted by the car on you, as illustrated in Figure 11.12. The feeling of being pushed outward

(leftward) arises only from the noninertial nature of the car's reference frame; in reality, your body simply tends to continue moving in a straight line. The moral is:

> **Avoid analyzing forces from a rotating frame of reference because such a frame is accelerating and therefore noninertial.**

The force required to make an object go around a curve can be a push, a pull, or a combination of the two. For example, the puck in Figure 11.2b is pulled inward by the string; in Figure 11.12, the person is pushed inward by the car.

The magnitude of the force required to make an object move in circular motion at constant speed depends on the speed of the object and the radius of the trajectory. To see how this force depends on speed, refer to Figure 11.8b. By mentally increasing the magnitude of the vectors $\vec{v}_i$ and $\vec{v}_f$ (which is equivalent to increasing the speed of the revolving object), you see how the vectors $\Delta\vec{v}$ and $\vec{a}$ also get longer. Because $\sum\vec{F} = m\vec{a}$, an object traveling at higher speed requires a larger inward force.

To see how the inward force depends on radius, look at Figure 11.13, which shows two pucks moving at the same speed along circles of different radii. During a given time interval, the rotational coordinate ϑ of the puck moving along the larger circle changes less than that of the puck moving along the smaller circle. Consequently the puck on the larger circle has a smaller change in velocity and a smaller centripetal acceleration. Because force is directly proportional to acceleration, we can say that the larger the radius, the smaller the force required to make the object go

Figure 11.13 Change in velocity for pucks moving along circles of different radius.

(a)

When two pucks move at same speed on circles of different radius . . .

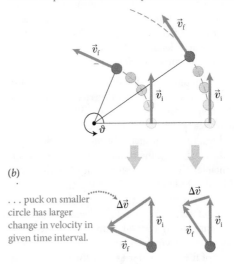

(b)

. . . puck on smaller circle has larger change in velocity in given time interval.

Figure 11.12 As you round a curve in a car, are you pushed inward or outward?

path of person in absence of forces

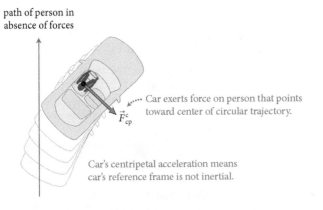

Car exerts force on person that points toward center of circular trajectory.

$\vec{F}_{cp}^{\,c}$

Car's centripetal acceleration means car's reference frame is not inertial.

Figure 11.14 The smaller the radius, the sharper the curve and the larger the force required to make the object follow the curve.

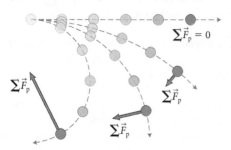

in a circle (**Figure 11.14**). As the radius approaches infinity, the object goes in a straight line and no force is required to change the direction of its velocity.

> **The inward force required to make an object move in circular motion increases with increasing speed and decreases with increasing radius.**

Example 11.3 Cube on turntable

A cube lies on a turntable initially rotating at constant speed. The rotational speed of the turntable is slowly increased, and at some instant the cube slides off the turntable. Explain why this happens.

❶ **GETTING STARTED** I'm not given much information, so I begin by making a sketch of the situation (**Figure 11.15a**). As the turntable rotates, the cube executes a circular motion. My task is to explain why the cube does not remain on the turntable as the turntable rotates faster.

Figure 11.15

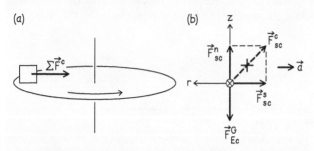

❷ **DEVISE PLAN** Because the cube executes circular motion, it has a centripetal acceleration, and so the vector sum of the forces exerted on it must point toward the center of the turntable. I therefore need to make a free-body diagram that reflects this combination of forces and determine how the forces change as the rotational speed of the turntable increases.

❸ **EXECUTE PLAN** To draw my free-body diagram, I must answer the question What are the forces exerted on the cube? First there is $\vec{F}^G_{Ec}$, the gravitational force exerted by Earth. This force is directed vertically downward and so cannot contribute to a force directed toward the turntable's center. The only other force exerted on the cube is $\vec{F}^c_{sc}$, the contact force exerted by the surface of the turntable. The normal component $\vec{F}^n_{sc}$ of this force must be equal in magnitude to $\vec{F}^G_{Ec}$ because the cube does not accelerate in the vertical direction. The horizontal component, which is the force of static friction $\vec{F}^s_{sc}$, is what forces the cube toward the center of its circular path. I therefore draw the free-body diagram shown in Figure 11.15b. Because the vertical component of the acceleration is zero, the forces in that

direction add to zero. Thus the vector sum of the forces exerted on the cube equals the force of static friction.

As the rotational speed increases, the magnitude of the centripetal acceleration of the cube also increases. This means that the magnitude of the force of static friction must get larger. At some instant, this force reaches its maximum value and so can no longer increase even though the rotational speed continues to increase. Consequently, the vector sum of the forces exerted on the cube is no longer large enough to give it the centripetal acceleration required for its circular trajectory. When this happens, the distance between the cube and the axis of rotation increases until the cube slides off the edge. ✔

❹ **EVALUATE RESULT** What makes the cube slide off the turntable is its tendency to continue in a straight line (that is, on a trajectory tangent to its circular trajectory). Up to a certain speed the force of static friction is large enough to overcome this tendency and keep the cube moving in a circle. Once the force of static friction reaches its maximum value, the cube begins to slide.

It is all too easy to think that there must be an *outward* force pushing the cube off the turntable as the rotational velocity increases in Example 11.3. As in a car rounding a corner, this thought arises because, without realizing it, you tend to view the situation from the rotating frame of the turntable. Because this is an easy mistake to make in situations involving rotation, you must always make sure you choose an inertial frame of reference.

If you can't imagine the cube moving off the turntable without being *pushed* off, try the following: Put some object—an eraser or your keys will do—on top and near the edge of a book as in **Figure 11.16a** and then gently,

Figure 11.16 The eraser in (*b*) falls off the book because of inertia, not because of a force "pushing" it off the book.

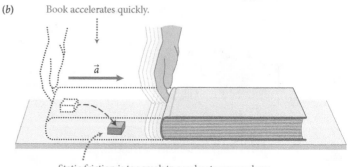

without touching the object, push the book along the surface of a table. The object moves along because it is being pushed forward by the force of static friction exerted by the book on it. If you give the book a hard shove, however, the object lags behind and falls off the rear edge of the book. Did it get pushed off? Someone moving along with the book might claim it did—just as someone in an accelerating car feels "pushed backward"—but such a person is not in an inertial reference frame. From the point of view of someone in the Earth reference frame, the book got yanked from under the object. The force of static friction exerted by the book on the object is not large enough to give the object the same acceleration as the book, and so the object falls off the rear edge without anything physically pushing it off. If you mark its starting position, you will notice that the object always ends up in front of its starting position, as shown in Figure 11.16*b*. This observation tells you that, even though it falls off the rear edge of the book, the object must have been pushed in the *forward* direction. It falls off the book not because it has been pushed off, but because the forward push exerted by the book on the object is not

large enough for the object to keep up with the accelerating book.

The same is true for the cube on the turntable: It falls off the edge not because it is pushed off but because the force of static friction between turntable and cube is not large enough to force the cube to go around in a circle.

11.2 Suppose I have two cubes on a turntable at equal distances from the axis of rotation. The inertia of cube 1 is twice that of cube 2. Do both cubes begin sliding at the same instant if I slowly increase the rotational velocity?

Figure 11.17 shows free-body diagrams for three objects moving in circular motion at constant speed. In each case the vector sum of the forces exerted on the object points toward the center of the circular trajectory.

11.3 (*a*) Does a bicycle always have to lean into a curve as illustrated in Figure 11.17*a*? (*b*) The rope holding the bucket in Figure 11.17*b* makes a small angle with the horizontal. Is it possible to swing the bucket around so that the rope is exactly horizontal?

Figure 11.17 Free-body diagrams for three objects in circular motion at constant speed.

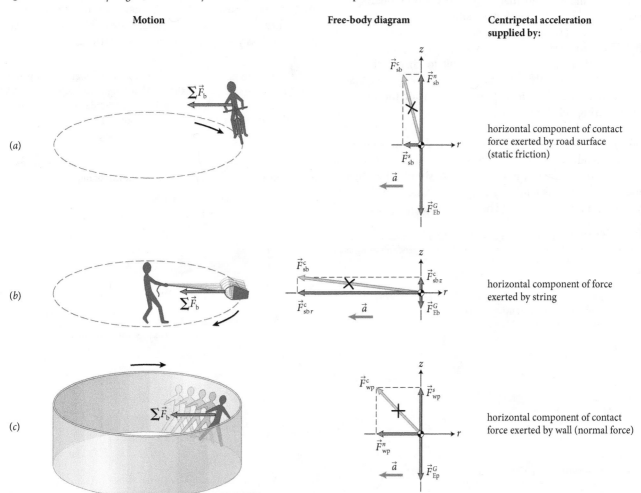

11.3 Rotational inertia

Having established the principles of circular motion at constant speed, let us now examine *changes* in rotational velocity. Imagine two identical pucks A and B on a horizontal air table (**Figure 11.18**). Puck B is initially at rest and fastened to a vertical post in the middle of the table by a short string that constrains the puck's motion to a circle of radius r_B. Suppose we give puck A a shove so that it strikes stationary puck B at speed v, as illustrated in Figure 11.18a. The two pucks collide elastically, and so immediately after the collision puck A is at rest and puck B has speed v. Puck B then begins to move around the post in circular motion at constant speed v. If the change in its rotational coordinate during a time interval Δt is $\Delta\vartheta_B$, then its rotational velocity after the collision is $\omega_{B\vartheta} = \Delta\vartheta_B/\Delta t$.

Now suppose we shove puck A at the same initial speed v toward a third identical puck C initially at rest and attached to a string of length $r_C > r_B$, as in Figure 11.18b. As before, puck A comes to rest after the collision and puck C moves away with speed v. Because both pucks B and C have the same speed v after the collision, they travel the same arc length in the same time interval Δt. However, their *rotational* speeds are different because the radii of their trajectories are different. The magnitude of the change $\Delta\vartheta_C$ in the rotational coordinate of C during the time interval Δt is smaller than that of B, and C's rotational speed, $\omega_C = |\Delta\vartheta_C/\Delta t|$, is therefore smaller than that of B.

If puck C is to attain the same final rotational velocity as B, it needs to be struck by a considerably faster moving puck A. Put differently, it is more difficult to change the rotational velocity of C than it is to change that of B. Note that C and B have the same inertia. If it weren't for the presence of the strings of different lengths, we would not be able to distinguish them: Once hit by A, both C and B would move off in a straight line with the same speed v. Apparently an object's tendency to resist a change in rotational velocity, its **rotational inertia,** is not given by simply the object's inertia—the object's tendency to resist a change in velocity as we defined it in Chapter 4. (The inertia surely plays a role: If I

increase the inertia of puck B, its rotational speed after the collision is lower and so is its change in rotational velocity.) That the longer string on C causes C's rotational velocity to change less than B's suggests that an object's rotational inertia depends on its location relative to the axis of rotation. The greater the distance between the bulk of an object's material and the axis of rotation, the larger its rotational inertia. You can easily verify this using a hammer (**Figure 11.19**). If you hold the hammer near its head, it is easily rotated. If you hold the hammer at the handle, as you would normally, it is more difficult to rotate.

There are many examples of the effect of rotational inertia. Skaters reduce their rotational inertia, and therefore spin more easily, by holding their arms close to their bodies; you run faster with your legs bent than with them extended. Conversely, when trying to keep your balance on a beam, you spread out your arms. This action moves more of your inertia away from the rotation axis and so increases your rotational inertia, which in turn increases your resistance to rotating off the beam. Tightrope walkers use a long pole for the same reason: The pole puts more inertia farther from the rotation axis and so increases the walkers' rotational inertia and gives them more time to adjust their balance.

One final important point: Unlike inertia, which has a single value, rotational inertia depends on the location of the rotation axis. An object can have a small rotational inertia about one axis and a large rotational inertia about another axis—just think of the hammer in Figure 11.19. Therefore the value of rotational inertia cannot be determined unless you specify the axis of rotation.

✋ **11.4** About which axis is the rotational inertia of a pencil (*a*) largest and (*b*) smallest: (1) a lengthwise axis through the core of the pencil; (2) an axis perpendicular to the pencil's length and passing through its midpoint; (3) an axis perpendicular to the pencil's length and passing through its tip?

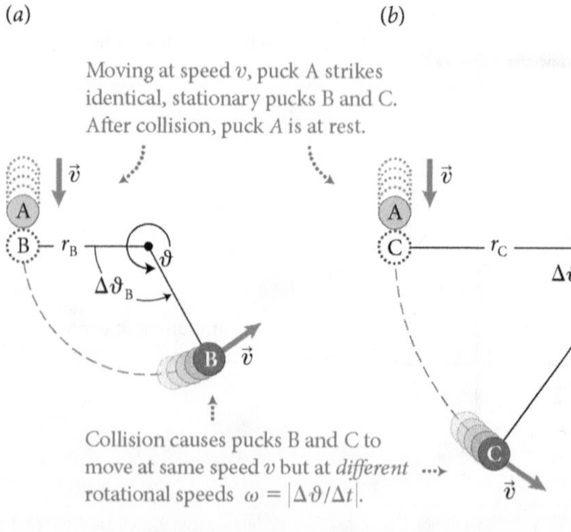

Figure 11.19 The rotational inertia of a hammer—its resistance to rotational motion—depends on the location of the axis of rotation.

Most of hammer's mass is in head.

center of mass

Axis of rotation far from center of mass: Hammer is hard to rotate (has high rotational inertia).

Axis of rotation at center of mass: Hammer is easy to rotate (has low rotational inertia).

Figure 11.18 When a moving puck strikes a stationary puck attached to a string, the rotational velocity of the puck on the string depends on string length.

(*a*)　　　　　　　　　(*b*)

Moving at speed v, puck A strikes identical, stationary pucks B and C. After collision, puck A is at rest.

$\vec{v}$

A

B — r_B — ϑ

$\Delta\vartheta_B$

B $\vec{v}$

$\vec{v}$

A

C — r_C — ϑ

$\Delta\vartheta_C$

Collision causes pucks B and C to move at same speed v but at *different* rotational speeds $\omega = |\Delta\vartheta/\Delta t|$.

C

$\vec{v}$

Self-quiz

1. Which of the following is in translational equilibrium? (*a*) An object whose center of mass is undergoing circular motion at constant speed. (*b*) A wheel spinning about an axis through its center of mass.

2. **Figure 11.20** shows the velocities $\vec{v}_i$ and $\vec{v}_f$ of a point on the rim of an object at two instants t_i and t_f for four different motions. For each case, determine the change in velocity $\Delta\vec{v}$ and the direction of the average acceleration during the time interval between t_i and t_f.

Figure 11.20

  (a)

  (b)

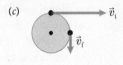 (c)

  (d)

3. Why doesn't the acceleration in question 2 point toward the center of the circle in all cases?

4. Describe the interaction responsible for providing a centripetal acceleration for (*a*) a car rounding a level curve, (*b*) a car rounding a banked curve, (*c*) a coin rotating along with a turntable, (*d*) a ball swung through a horizontal circle by a string that sweeps out a cone, (*e*) the Moon orbiting Earth, (*f*) clothes spinning in a dryer, (*g*) a marble rolling along the inside of a horizontal hoop, and (*h*) a ball on a string rolling in a horizontal circle.

5. A ball attached to a string (the far end of which is fixed) rolls in a horizontal circle. Under which conditions is the string more likely to break: (*a*) when the speed of the ball is increased for a given radius or (*b*) when the length of the string is increased for a given speed?

Answers

1. (*a*) An object undergoing circular motion at constant speed is accelerated because the direction of its center-of-mass velocity changes continuously. If it is accelerated, the vector sum of the forces exerted on the object is not zero, and the object cannot be in translational equilibrium. (*b*) If the center of mass of a spinning wheel is fixed, $a_{cm} = 0$ and so the object is in translational equilibrium.

2. For each case, $\Delta\vec{v} = \vec{v}_f - \vec{v}_i$ and the average acceleration is parallel to $\Delta\vec{v}$ (**Figure 11.21**).

Figure 11.21

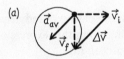 (a)

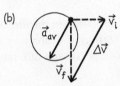 (b)

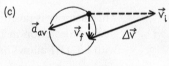 (c)

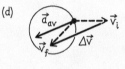 (d)

3. Notice that as the object travels around the circle in parts *b* and *c*, its speed changes. To change the speed, there must be a tangential component to the acceleration as well as a centripetal component. The tangential component requires that the acceleration not point directly toward the center.

4. (*a*) Force of static friction exerted by road on car, (*b*) horizontal component of contact force exerted by road on car, (*c*) force of static friction exerted by turntable on coin, (*d*) horizontal component of tensile force exerted by string on ball, (*e*) gravitational force exerted by Earth on the Moon, (*f*) centripetal component of contact force exerted by drum of dryer on clothes, (*g*) contact force exerted by hoop on marble, and (*h*) tensile force exerted by string on ball.

5. In case *a*, for a given radius, a greater speed means the ball travels through a larger angle during a specific time interval. If the angle is larger, the magnitude of $\Delta\vec{v}$ is larger. A larger magnitude of $\Delta\vec{v}$ requires that the acceleration and force also be larger. The force providing the acceleration is due to the tension in the string. Therefore, the greater speed requires more tension and the string is more likely to break. In case *b*, for a given speed, a larger radius means that the ball travels through a smaller angle in a specific time interval. If the angle is smaller, the magnitude of $\Delta\vec{v}$ is smaller. A smaller $\Delta\vec{v}$ requires a smaller acceleration and a smaller force. Therefore, a large radius requires a smaller tension in the string and the string is less likely to break.

Figure 11.22 Relationships between rotational coordinate ϑ, radius r, and arc length s.

(a)

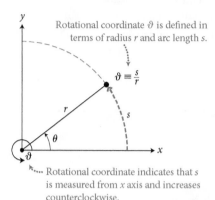

Rotational coordinate ϑ is defined in terms of radius r and arc length s.

$\vartheta \equiv \dfrac{s}{r}$

Rotational coordinate indicates that s is measured from x axis and increases counterclockwise.

(b)

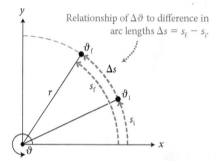

Relationship of $\Delta\vartheta$ to difference in arc lengths $\Delta s = s_f - s_i$.

11.4 Rotational kinematics

The **rotational coordinate** of an object moving along a circle of radius r is defined as the length of the arc s over which the object has moved divided by the radius:

$$\vartheta \equiv \frac{s}{r}. \tag{11.1}$$

The sign and magnitude of s (and therefore the sign and magnitude of ϑ) depend on our choice of rotational coordinate system. With the choice of rotational coordinate system shown in **Figure 11.22a**, s is measured along the perimeter of the circle, beginning at the positive x axis and ending at the location of the object, and the counterclockwise direction is taken to be positive. Because the rotational coordinate ϑ is a ratio of two lengths, it is a unitless quantity. In contrast, the polar angle θ swept out by the revolving object can be expressed in *radians* (rad), *degrees* (°), or *revolutions* (rev). The relationship between these units is

$$2\pi \text{ rad} = 360° = 1 \text{ rev.} \tag{11.2}$$

Note that, unlike the units for length, time, and inertia, no physical standard is required to specify the units for polar angle.

Rotational coordinate and polar angle are directly related to each other. After one revolution, the arc length s traveled by the object is equal to the circumference of the circle, and so $s = 2\pi r$. Substituting this value for s in Eq. 11.1 yields $\vartheta = 2\pi$ after one revolution. Knowing from Eq. 11.2 that 2π rad = 1 rev, we see that the rotational coordinate corresponds to the *number of radians in the polar angle*: $\vartheta = \theta/(1 \text{ rad})$.* In other words, given the polar angle θ of the object's position, we can obtain its rotational coordinate by expressing the polar angle in radians and dividing by 1 rad.

The change in the rotational coordinate of an object, the object's final rotational coordinate minus its initial rotational coordinate, is

$$\Delta\vartheta \equiv \vartheta_f - \vartheta_i. \tag{11.3}$$

This quantity is positive for a rotation in the direction of increasing ϑ and negative for a rotation in the direction of decreasing ϑ. Combining this expression with Eq. 11.1 allows us to express $\Delta\vartheta$ as a difference in arc lengths:

$$\Delta\vartheta = \vartheta_f - \vartheta_i = \frac{s_f}{r} - \frac{s_i}{r} = \frac{\Delta s}{r}, \tag{11.4}$$

where Δs is defined as the length of the arc between the final and initial locations of the object, as shown in Figure 11.22b.

11.5 Starting from a position with rotational coordinate zero, an object moves in the positive ϑ direction at a constant speed of 3.0 m/s along the perimeter of a circle of radius 2.0 m. (*a*) What is the object's rotational coordinate after 1.5 s? (*b*) How long does it take the object to complete one revolution? (*c*) What is its rotational coordinate 1.0 s after passing a point on the circle whose polar angle is 90°?

---

*The reason for this division by 1 rad is to cancel out the units of radian. The rotational coordinate ϑ is unitless, but the polar angle θ has units of radians. So for an object that has rotated to an angle of 6.1 rad, the rotational coordinate is (6.1 rad)/(1 rad) = 6.1; for an object that has rotated to an angle of 265°, the rotational coordinate is (265°)(2π rad/360°)/(1 rad) = 4.63.

In analogy to the definition of average velocity, we can define *average rotational velocity* $\omega_{\vartheta,\text{av}}$ as being equal to the change in the rotational coordinate divided by the time interval during which this change takes place:

$$\omega_{\vartheta,\text{av}} \equiv \frac{\Delta\vartheta}{\Delta t}. \tag{11.5}$$

The SI unit of average rotational velocity is s^{-1} (inverse second). The subscript ϑ is there to remind us that the average rotational velocity is a signed quantity that can be positive or negative, depending on the choice of positive direction of rotation for ϑ.

The **rotational velocity** ω_ϑ is obtained by letting the time interval during which the change in the rotational coordinate is measured approach zero:

$$\omega_\vartheta \equiv \lim_{\Delta t \to 0} \frac{\Delta\vartheta}{\Delta t} = \frac{d\vartheta}{dt}. \tag{11.6}$$

The instantaneous rotational velocity is equal to the time derivative of the rotational coordinate just as the instantaneous velocity is equal to the time derivative of position. The **rotational speed** is the absolute value of the rotational velocity:

$$\omega = |\omega_\vartheta|. \tag{11.7}$$

The relationship between instantaneous rotational velocity and instantaneous velocity can be obtained by substituting Eq. 11.4 into Eq. 11.5:

$$\omega_\vartheta = \frac{d\vartheta}{dt} = \frac{1}{r}\frac{ds}{dt}. \tag{11.8}$$

Because the rate at which the arc length s is swept out is the tangential component of the velocity of the object, $v_t = ds/dt$, we have

$$\omega_\vartheta = \frac{v_t}{r} \tag{11.9}$$

or

$$v_t = r\omega_\vartheta. \tag{11.10}$$

Both the tangential component of the velocity v_t and the rotational velocity ω_ϑ are signed quantities that are positive for motion in the direction of increasing ϑ and negative in the direction of decreasing ϑ. We can also express Eq. 11.10 in terms of speeds:

$$v = r\omega. \tag{11.11}$$

In words, the speed of an object in circular motion equals its rotational speed multiplied by the radius of the circular path. As we concluded when we analyzed Figure 11.6, the speed increases with increasing radius.

We can also define a **rotational acceleration** α_ϑ (Greek letter alpha):

$$\alpha_\vartheta \equiv \lim_{\Delta t \to 0} \frac{\Delta\omega_\vartheta}{\Delta t} = \frac{d\omega_\vartheta}{dt} = \frac{d^2\vartheta}{dt^2}. \tag{11.12}$$

The rotational acceleration has SI units s^{-2} (inverse seconds squared). It is a measure of the rate at which the rotational velocity increases.

For *circular motion at constant speed*, the rotational speed $\omega = v/r$ is constant, and so the rotational acceleration is zero. How can rotational acceleration be zero if, as we saw in Section 11.1, circular motion at constant speed requires a nonzero centripetal acceleration? The answer is that rotational acceleration measures the change in the magnitude of the velocity $\vec{v}$ with which an object travels around a circle, while centripetal acceleration measures the change in the *direction* of $\vec{v}$. Even if the object's speed v is constant, meaning its rotational acceleration is zero, the changing direction of $\vec{v}$ means there is a nonzero centripetal acceleration.

To obtain the magnitude of the centripetal acceleration, remember that if the position vector of a revolving object sweeps out an angle θ, the velocity vector $\vec{v}$ sweeps out the same angle, as illustrated in Figure 11.23. This means that the two shaded triangles in Figure 11.23 are similar and so

$$\frac{|\Delta \vec{r}|}{r} = \frac{|\Delta \vec{v}|}{v} \tag{11.13}$$

or, rearranging terms and dividing left and right by the time interval Δt during which the change in velocity direction took place,

$$\frac{|\Delta \vec{v}|}{\Delta t} = \frac{v}{r} \frac{|\Delta \vec{r}|}{\Delta t}. \tag{11.14}$$

In the limit for $\Delta t \to 0$, the term $|\Delta \vec{v}|/\Delta t$ approaches the magnitude of the (centripetal) acceleration, which we shall denote by a_c, and the term $|\Delta \vec{r}|/\Delta t$ approaches the speed v, and so the magnitude of the **centripetal acceleration** is

$$a_c = \frac{v^2}{r} \quad \text{(circular motion).} \tag{11.15}$$

This is the acceleration required to keep an object moving at speed v along the perimeter of a circle of radius r. As we saw in Section 11.1, the centripetal acceleration of an object executing a circular motion at constant speed is directed radially inward.

Although we derived Eq. 11.15 for circular motion at constant speed, *any* object constrained to move along a circle or part of a circle of radius r is subject to a radially inward acceleration of the magnitude given by Eq. 11.15. Because we defined the radial axis to be directed outward from the rotation axis (Figure 11.11), we can write

$$a_r = -\frac{v^2}{r} \quad \text{(any motion along arc of radius } r\text{).} \tag{11.16}$$

When the object's speed is not constant, we have, in addition to this radial component of the acceleration, a component parallel to the velocity (see Checkpoint 11.1). This component is called the **tangential acceleration** because it is always tangent to the trajectory:

$$a_t = \frac{dv}{dt}. \tag{11.17}$$

The tangential component is positive when the speed increases and negative when it decreases.

Figure 11.24 summarizes a number of important facts about circular motion. Figure 11.24a illustrates circular motion at constant speed and Figure 11.24b accelerated circular motion. Note that the velocity is always tangential to the

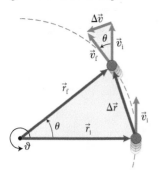

Figure 11.23 Change in the direction of the position $\vec{r}$ and the velocity $\vec{v}$ for an object moving in a circle at constant speed.

trajectory: that is, the radial component of the velocity is always zero. This is true for any type of circular motion, be it at constant speed or accelerated:

$$v_r = 0 \quad \text{(any type of circular motion).} \tag{11.18}$$

When the circular motion is at constant speed, the radial component of the acceleration is nonzero (Eq. 11.16) and the tangential acceleration is zero:

$$a_t = 0 \quad \text{(circular motion at constant speed).} \tag{11.19}$$

In that case, the **period** T (the time interval it takes the object to complete one revolution) is

$$T = \frac{2\pi r}{v} = \frac{2\pi}{\omega} \quad \text{(circular motion at constant speed),} \tag{11.20}$$

where we have used Eq. 11.11 relating the speed v to the rotational speed ω.

If the object speeds up or slows down, both components of the acceleration are nonzero. The acceleration no longer points to the center of the circle, as shown in Figure 11.24c, and the magnitude of the acceleration is

$$a = \sqrt{a_r^2 + a_t^2}. \tag{11.21}$$

When the tangential acceleration is nonzero, the rotational acceleration is nonzero too. To relate these two quantities, we substitute Eq. 11.10 for v_t:

$$a_t = \frac{dv_t}{dt} = r\frac{d\omega_\vartheta}{dt}. \tag{11.22}$$

Noting that $d\omega_\vartheta/dt$ is the rotational acceleration α_ϑ (Eq. 11.12), we can write Eq. 11.22 in the form

$$a_t = r\alpha_\vartheta. \tag{11.23}$$

Thus tangential acceleration is related to rotational acceleration in the same way that velocity and rotational velocity are related (Eq. 11.10) and, similarly, arc length and rotational coordinate (Eq. 11.1). For all three we can write (Table 11.1):

$$\text{translational motion quantity} = r \times \text{rotational motion quantity.}$$

If a_t is constant (and so the rotational acceleration is constant too), then the arc length traveled and the tangential velocity are given by the kinematics equations for motion at constant acceleration (Eqs. 3.8 and 3.4):

$$s_f = s_i + v_{t,i}\Delta t + \tfrac{1}{2}a_t(\Delta t)^2 \tag{11.24}$$

$$v_{t,f} = v_{t,i} + a_t\Delta t. \tag{11.25}$$

Dividing both sides of each of these equations by r and substituting the expressions from Table 11.1, we obtain the equivalent kinematics equations for rotational motion:

$$\vartheta_f = \vartheta_i + \omega_{\vartheta,i}\Delta t + \tfrac{1}{2}\alpha_\vartheta(\Delta t)^2 \quad \text{(constant rotational acceleration)} \tag{11.26}$$

$$\omega_{\vartheta,f} = \omega_{\vartheta,i} + \alpha_\vartheta\Delta t \quad \text{(constant rotational acceleration).} \tag{11.27}$$

Table 11.2 gives an overview of the correspondence between translational and rotational kinematic quantities and equations.

Figure 11.24 Circular motion and acceleration.

(*a*) Circular motion at constant speed

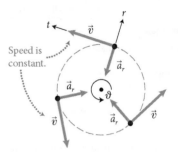

(*b*) Circular motion at increasing speed

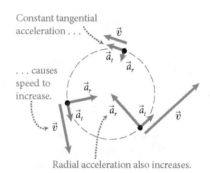

(*c*) Acceleration during circular motion

Table 11.1

$s = r\vartheta$	(11.1)
$v_t = r\omega_\vartheta$	(11.10)
$a_t = r\alpha_\vartheta$	(11.23)

QUANTITATIVE TOOLS

Table 11.2 Translational and rotational kinematics

Translational motion (constant acceleration)		Rotational motion (constant rotational acceleration)	
coordinate	x	rotational coordinate	ϑ
x component of displacement	$\Delta x = x_f - x_i$	change in rotational coordinate	$\Delta\vartheta = \vartheta_f - \vartheta_i$
x component of velocity	$v_x = \dfrac{dx}{dt}$	rotational velocity	$\omega_\vartheta = \dfrac{d\vartheta}{dt}$
x component of acceleration	$a_x = \dfrac{dv_x}{dt} = \dfrac{d^2x}{dt^2}$	rotational acceleration	$\alpha_\vartheta = \dfrac{d\omega_\vartheta}{dt} = \dfrac{d^2\vartheta}{dt^2}$
kinematics relationships (constant a_x):		rotational kinematics relationships (constant α_ϑ):	
	$v_{x,f} = v_{x,i} + a_x\Delta t$		$\omega_{\vartheta,f} = \omega_{\vartheta,i} + \alpha_\vartheta\Delta t$
	$x_f = x_i + v_{x,i}\Delta t + \frac{1}{2}a_x(\Delta t)^2$		$\vartheta_f = \vartheta_i + \omega_{\vartheta,i}\Delta t + \frac{1}{2}\alpha_\vartheta(\Delta t)^2$
		radial acceleration	$a_r = -\dfrac{v^2}{r} = -r\omega^2$
		tangential acceleration	$a_t = 0$

Example 11.4 Leaning into a curve

A woman is rollerblading to work and, running late, rounds a corner at full speed, sharply leaning into the curve (**Figure 11.25**). If, during the turn, she goes along the arc of a circle of radius 4.5 m at a constant speed of 5.0 m/s, what angle θ must her body make with the vertical in order to round the curve without falling?

Figure 11.25 Example 11.4.

Figure 11.26

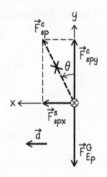

❶ **GETTING STARTED** As she rounds the circular arc at constant speed, the woman executes circular motion at constant speed. She must therefore undergo a centripetal acceleration as a result of the forces exerted on her. To help me determine the forces exerted on her, I draw a free-body diagram (**Figure 11.26**). The forces exerted on the rollerblader are the gravitational force $\vec{F}_{Ep}^{G}$ and a contact force $\vec{F}_{sp}^{c}$ exerted by the surface of the road. Now I see why she must lean into the turn: When she stands straight, the contact force is directed straight up, but as she leans, this force develops a component that pushes her toward the center of the circular arc and provides the necessary centripetal acceleration. I indicate the direction of this centripetal acceleration in my drawing and choose a set of axes—the x axis in the direction of the centripetal acceleration and the y axis upward. (Note that the resulting free-body diagram is similar to that for the bicyclist in Figure 11.17a.) I must determine the angle θ that $\vec{F}_{sp}^{c}$ makes with the vertical.

❷ **DEVISE PLAN** From my free-body diagram, I can draw two conclusions. First, the forces in the y direction must add to zero: $\sum F_y = 0$. Second, the x component of the contact force provides the centripetal acceleration given by Eq. 11.15. This gives me two equations from which I should be able to determine θ.

❸ **EXECUTE PLAN** Substituting the centripetal acceleration from Eq. 11.15 into the equation of motion in the x direction, I get

$$\sum F_x = ma_x = m(+a_c) = +m\frac{v^2}{r}, \qquad (1)$$

where m is the inertia of the rollerblader. (The rollerblader's inertia is not given, but I hope it will drop out and I won't need it.) From my diagram, I see that

$$\sum F_x = F_{spx}^{c} = F_{sp}^{c}\sin\theta. \qquad (2)$$

In the y direction I have $F^c_{\text{sp}y} = F^c_{\text{sp}} \cos\theta$ and $F^G_{\text{Ep}y} = -mg$. The equation of motion in the y direction then yields

$$\sum F_y = F^c_{\text{sp}y} + F^G_{\text{Ep}y} = F^c_{\text{sp}} \cos\theta - mg = 0.$$

Solving this equation for F^c_{sr} and substituting the result into Eq. 2, I get

$$\sum F_x = F^c_{\text{sp}} \sin\theta = \left(\frac{mg}{\cos\theta}\right)\sin\theta = mg\tan\theta.$$

Substituting this result into Eq. 1 then yields

$$mg\tan\theta = m\frac{v^2}{r}$$

$$\tan\theta = \frac{v^2}{gr} = \frac{(5.0 \text{ m/s})^2}{(9.8 \text{ m/s}^2)(4.5 \text{ m})} = 0.57.$$

This gives an angle $\theta = \tan^{-1}(0.57) = 0.52$ rad, or about 30°. ✔

❹ EVALUATE RESULT The angle of the skater in Figure 11.25 is about 30°, and so my answer appears to be reasonable.

🖐 **11.6** Estimate the maximum speed with which you could take the curve in Example 11.4. [Hint: What is the maximum coefficient of static friction?]

11.5 Angular momentum

Let us now return to the experiment discussed in Section 11.3: A stationary puck C fastened to a string of length r is struck by an identical puck A initially moving at speed v. For simplicity we'll ignore the pucks' extent and treat them as particles. After the elastic collision, the puck C moves in circular motion at constant speed v. Calling this puck's inertia m, we can say that its kinetic energy is

$$K = \tfrac{1}{2}mv^2. \tag{11.28}$$

Substituting the expression given by Eq. 11.10 for v, we can rewrite this kinetic energy in the form

$$K = \tfrac{1}{2}mv^2 = \tfrac{1}{2}m(r\omega)^2 = \tfrac{1}{2}(mr^2)\omega^2 \tag{11.29}$$

or, defining the term in parentheses as the **rotational inertia** I of puck C about the axis of rotation

$$I \equiv mr^2 \quad \text{(particle)}, \tag{11.30}$$

we get

$$K_{\text{rot}} = \tfrac{1}{2}I\omega^2. \tag{11.31}$$

The rotational inertia of an object about an axis opposes a change in rotational motion about that axis just as inertia opposes a change in translational motion. The SI unit of rotational inertia is $\text{kg}\cdot\text{m}^2$. Because the SI unit of ω is s^{-1}, we get, from Eq. 11.31 SI units of $\text{kg}\cdot\text{m}^2/\text{s}^2$ for the kinetic energy, which is the joule.

Note the remarkable similarity between Eqs. 11.31 and 11.28—inertia m and speed v of Eq. 11.28 are replaced by their rotational equivalents, rotational inertia I and rotational speed ω. The main advantage of writing the kinetic energy in the form of Eq. 11.31 will become clear when we deal with rotating extended objects in Section 11.6: The rotational velocity is the same in all parts of a rotating object, whereas the speed depends on the distance to the axis of rotation (recall Figure 11.7). Because the kinetic energy in Eq. 11.31 is due to rotational motion, it is called the **rotational kinetic energy** of the object.

🖐 **11.7** Suppose that in Figure 11.18 $r_C = 2r_B$. (a) What is the ratio of the rotational velocity of B to that of C? (b) What is the ratio of the rotational inertia of B to that of C?

The answers to this checkpoint might surprise you. That the rotational inertia of B is only one-fourth that of C means that B has only one-*fourth* the tendency to resist a change in its rotational motion. So why, when the same force is exerted, is B's rotational velocity only *twice* as large as C's rather than four times as much?

The reason the rotational velocity of B is not as great as you might expect is that the rotational velocity an object acquires depends on how far from the rotation axis it is struck. Increasing the string's length in Figure 11.18 not only increases the puck's rotational inertia but also increases the effectiveness of the blow from the incoming puck. Doubling r for puck C quadruples C's rotational inertia because of the factor r^2 in Eq. 11.30 but also increases puck A's ability to set C in rotational motion. As a result, B's rotational velocity is only twice that of C rather than four times as much.

Figure 11.27 shows two collisions between a bar that is free to rotate about its right end and a puck. In Figure 11.27a the puck strikes the end of the rod; in Figure 11.27b the rod is struck closer to the axis. Although the rotational inertia of the rod is fixed and although the puck has the same initial momentum, the rod rotates more slowly when struck closer to the rotation axis. (In the extreme case that it is hit right on the axis, the rod doesn't move at all.) So the incoming puck's ability to set other objects in rotational motion depends not only on its momentum but also on how far from the axis it strikes the other object.

Returning to the collision of Section 11.3, we have seen that substituting ω for v and I for m in the expression $\frac{1}{2}mv^2$ yields the final kinetic energy of the struck puck (C in our case), which is equal to the initial kinetic energy of the incoming puck (A):

$$\tfrac{1}{2}mv^2 \rightarrow \tfrac{1}{2}I\omega^2 = \tfrac{1}{2}(mr^2)\left(\frac{v}{r}\right)^2 = \tfrac{1}{2}mv^2. \tag{11.32}$$

This relationship tells us that the elastic collision causes all the (translational) kinetic energy initially in puck A to be converted to the rotational kinetic energy of puck C. The sum of the kinetic energy of puck A and the rotational kinetic energy of puck C (that is to say, the kinetic energy of the two-puck system) remains constant, as it should.

What happens if we make the same $m \rightarrow I$, $v \rightarrow \omega$ substitution in mv, the magnitude of the momentum of puck A? Let's try it:

$$mv \rightarrow I\omega = (mr^2)\left(\frac{v}{r}\right) = rmv. \tag{11.33}$$

The result is different this time: The quantity $I\omega$ is not equal to the magnitude mv of the momentum of the incoming puck A. Instead $I\omega$ is equal to the product of mv and the distance r from A to the axis of rotation at the instant of impact. This "rotational momentum" $I\omega$ thus represents precisely what we called the "ability to set in rotational motion" a moment ago: The greater the distance r

Figure 11.27 A rod free to rotate about one end is struck by an incoming puck at different distances from its axis of rotation.

(*a*) Puck strikes rod far from rotation axis

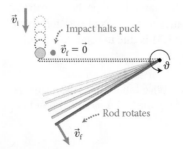

(*b*) Puck strikes rod closer to rotation axis

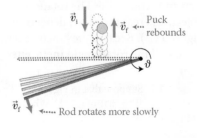

between the location of impact and the axis of rotation, the larger rmv and the easier it is for puck A to make puck C rotate.

The quantity $I\omega$ plays a fundamental role in physics. The larger the value of $I\omega$ for a moving object, the more easily the object can set another object in rotational motion. Because of the analogy to momentum, which measures an object's capacity to set other objects in motion along a straight line, this quantity is called the **angular momentum** and is denoted by the letter L:

$$L_\vartheta \equiv I\omega_\vartheta. \tag{11.34}$$

The SI units of angular momentum are $kg \cdot m^2/s$. As before, the subscript ϑ means that the angular momentum, like the rotational velocity, is a signed quantity that can be positive or negative, depending on the choice of the positive direction of rotation for ϑ. Substituting Eqs. 11.10 and 11.30 into Eq. 11.34, we get

$$L_\vartheta \equiv I\omega_\vartheta = (mr^2)\left(\frac{v_t}{r}\right) = rmv_t. \tag{11.35}$$

An object does not need to rotate or revolve in order to possess a nonzero angular momentum. For example, even though it moves in a straight line, the incoming puck A in Figure 11.18 has a nonzero angular momentum—it has the ability to set another puck in rotational motion. You can use Eqs. 11.34 and 11.35 to calculate the angular momentum of objects in circular motion, but what is the angular momentum of an object that moves along a straight line? The first thing to note is that it doesn't make sense to talk about the angular momentum of an object unless we specify an axis of rotation. Figure 11.28 shows how to determine the angular momentum for a particle moving along a straight line. For a given axis of rotation (real or imagined), the value of r in Eq. 11.35 is given by the perpendicular distance $r_\perp$ between the axis of rotation and the straight line defined by the momentum of the object (the radius of the circle for which this line, called the *line of action* of the momentum, is a tangent; see Figure 11.28), while the value of v_t is the speed of the object. The distance $r_\perp$ is called the **lever arm distance,** or simply **lever arm,** of the momentum relative to the axis of rotation. The magnitude of the angular momentum of a particle that moves along a straight line is thus

$$L = r_\perp mv \quad \text{(particle).} \tag{11.36}$$

The sign of the angular momentum is determined by the choice of the positive direction of rotation for ϑ. For example, the rotational coordinate of the particle in Figure 11.28 is increasing as it moves along its straight trajectory, and so the angular momentum of the particle is positive.

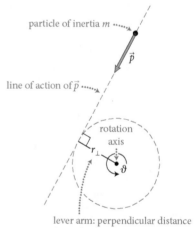

Figure 11.28 Relative to an axis of rotation, a particle of inertia m moving along a straight line has an angular momentum of magnitude $r_\perp mv$, where $r_\perp$ is the lever arm distance and v is the speed of the object.

particle of inertia m
line of action of $\vec{p}$
rotation axis
$r_\perp$

lever arm: perpendicular distance from rotation axis to line of action

Exercise 11.5 Angular momenta

Determine the angular momenta of objects A and B in **Figure 11.29** relative to (a) axis 1 and (b) axis 2. The inertia of the objects are $m_A = 2.0$ kg and $m_B = 0.50$ kg, and their speeds are $v_A = 0.60$ m/s and $v_B = 1.0$ m/s. The tall side of the rectangle is 2.0 m long and object B travels along the diagonal, which makes an angle of 30° with the long side of the rectangle.

SOLUTION (a) Object A: The lever arm distance relative to axis 1 is the long side of the rectangle: $r_\perp = 2.0$ m. The angular momentum is therefore

$$L_{A\vartheta} = +(2.0 \text{ m})(2.0 \text{ kg})(0.60 \text{ m/s}) = +2.4 \text{ kg} \cdot m^2/s.$$

The plus sign indicates that the rotational coordinate of A relative to axis 1 increases. ✔

Object B: B's line of motion is the diagonal of the rectangle and so goes through axis 1. Therefore $r_\perp = 0$ and thus $L_{B\vartheta} = 0$. ✔

Figure 11.29 Exercise 11.5.

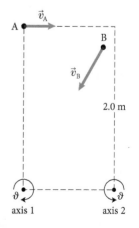

Figure 11.30

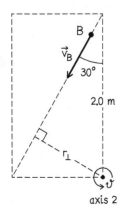

(b) Object A: The lever arm distance relative to axis 2 is the long side of the rectangle: $r_\perp = 2.0$ m. The angular momentum is therefore $L_{A\vartheta} = -(2.0\ \text{m})(2.0\ \text{kg})(0.60\ \text{m/s}) = -2.4\ \text{kg} \cdot \text{m}^2/\text{s}$. The minus sign indicates that the rotational coordinate of A relative to axis 2 decreases. ✔

Object B: To determine $r_\perp$ relative to axis 2, I draw a straight line perpendicular to B's line of motion (**Figure 11.30**). Using trigonometry, I see that $\sin 30° = r_\perp/(2.0\ \text{m})$, or $r_\perp = (0.5)(2.0\ \text{m}) = 1.0$ m, and so $L_{B\vartheta} = +(1.0\ \text{m})(0.50\ \text{kg})(1.0\ \text{m/s}) = +0.50\ \text{kg} \cdot \text{m}^2/\text{s}$. The plus sign indicates that the rotational coordinate of B relative to axis 2 increases. ✔

Going back to the experiment we discussed in Section 11.3, if we substitute the values of m, $r_\perp$, and v for puck A in Figure 11.18 just before impact, we obtain a value for the angular momentum of puck A just before the impact that is the same as that of puck B after the impact:

$$L_{A\vartheta,i} = L_{B\vartheta,f} = +r_B mv. \tag{11.37}$$

Because puck A comes to rest after the impact, the change in its angular momentum is $\Delta L_{A\vartheta} = L_{A\vartheta,f} - L_{A\vartheta,i} = 0 - r_B mv = -r_B mv$. The change in the angular momentum of puck B is $\Delta L_{B\vartheta} = L_{B\vartheta,f} - L_{B\vartheta,i} = +r_B mv - 0 = +r_B mv$. As you can readily verify, angular momentum is an extensive quantity (see Section 4.4), and so we can define the angular momentum of the system comprising both pucks as $L_\vartheta = L_{A\vartheta} + L_{B\vartheta}$. Because the change in the angular momentum of one puck is compensated for by the change in the angular momentum of the other puck, the angular momentum of the system does not change in the impact: $\Delta L_\vartheta = 0$.

This observation suggests that angular momentum, like momentum, is a conserved quantity. Indeed, if it weren't for friction, objects in circular motion would tend to continue their circular motion, just as objects in translational motion tend to continue to move along a straight line (see Section 4.1). The more we reduce friction, the longer a spinning wheel keeps turning. In the absence of friction, it would keep turning forever and its angular momentum would be constant. Consider, for example, the particle in circular motion shown in Figure 11.31. The particle is subject to a force $\vec{F}$ that has both a radial and a tangential component. The radial component keeps the particle moving in circular motion. The tangential component, which is tangent to the circular trajectory and points in the direction opposite the direction of motion, causes the particle to slow down. In the absence of any tangential forces, however, the angular momentum remains constant.

Strictly speaking, the fact that the angular momentum of the colliding pucks in Figure 11.18 remains constant is a consequence of conservation of momentum—the change in the momentum of B is compensated for by a change in the momentum of A. It turns out, however, that conservation of angular momentum is just as fundamental as conservation of momentum and conservation of energy. We can generalize this new conservation law to situations where it cannot be deduced from other conservation principles, which tells us that **conservation of angular momentum** is a fundamental law of physics:

> **Angular momentum can be transferred from one object to another, but it cannot be created or destroyed.**

This means that whenever the angular momentum of an object changes, there must be a compensating change of the opposite sign in the environment.

The angular momentum of an object or system on which no tangential forces are exerted is therefore constant:

$$\Delta L_\vartheta = 0 \quad \text{(no tangential forces).} \tag{11.38}$$

Figure 11.31 Tangential forces change the angular momentum.

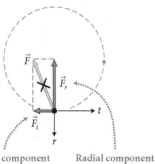

Tangential component changes speed and angular momentum.

Radial component causes centripetal acceleration.

Exercise 11.6 Spinning faster

Divers increase their spin by tucking in their arms and legs (**Figure 11.32**). Suppose the outstretched body of a diver rotates at 1.2 revolutions per second before he pulls his arms and knees

Figure 11.32 Exercise 11.6.

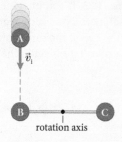

into his chest, reducing his rotational inertia from $9.4 \ \text{kg} \cdot \text{m}^2$ to $3.1 \ \text{kg} \cdot \text{m}^2$. What is his rotational velocity after he tucks in his arms and legs?

SOLUTION Once the diver is off the board, the only force exerted on his is the gravitational force exerted by Earth. This force does not affect the angular momentum of the diver (dropped objects do not spontaneously start to rotate), and so his angular momentum must remain constant. If his angular momentum before he tucks is $L_{\vartheta,i}$ and that after is $L_{\vartheta,f}$, then

$$L_{\vartheta,f} = L_{\vartheta,i}$$

and so, from Eq. 11.34,

$$I_f \omega_{\vartheta,f} = I_i \omega_{\vartheta,i}$$

$$\omega_{\vartheta,f} = \frac{I_i}{I_f} \omega_{\vartheta,i}$$

$$= \frac{9.4 \ \text{kg} \cdot \text{m}^2}{3.1 \ \text{kg} \cdot \text{m}^2} (1.2 \ \text{s}^{-1}) = 3.6 \ \text{s}^{-1}. ✔$$

11.8 Does the rotational kinetic energy of the diver in Exercise 11.6 change as he pulls her arms in? Explain.

Example 11.7 Dumbbell collision

In **Figure 11.33**, two identical pucks B and C, each of inertia m, are connected by a rod of negligible inertia and length ℓ that is free to rotate about a fixed axis through its center. A third identical puck A, initially moving at speed v_i, strikes the combination as shown. After the elastic collision, what are the rotational velocity of the dumbbell and the velocity of puck A?

Figure 11.33 Example 11.7

rotation axis

1 GETTING STARTED I begin with a two-part sketch (**Figure 11.34**), choosing an x axis in the direction of A's initial motion and choosing counterclockwise as the positive direction of rotation (this is the direction in which I expect the dumbbell to rotate after the collision. Because A hits B head-on and because the inertia of the dumbbell is twice that of A, I expect A

to bounce back and move in the negative x direction after the collision, as my after-collision sketch shows.

Figure 11.34

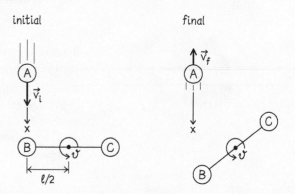

initial   final

2 DEVISE PLAN In elastic collisions the kinetic energy of the system remains constant (see Section 5.5). In this collision I need to consider the kinetic energy of puck A and the rotational kinetic energy of the dumbbell. Because there are two unknowns—A's final velocity and the dumbbell's final rotational velocity—I need an additional law to determine both. To this end I apply conservation of angular momentum (Eq. 11.38) to the system comprising puck A and the dumbbell.

(Continued)

③ EXECUTE PLAN The initial kinetic energy of the system is that of puck A, $\frac{1}{2}mv_i^2$. The final kinetic energy is the sum of the (translational) final kinetic energy of A and the rotational kinetic energy of the dumbbell, $K_f = \frac{1}{2}mv_f^2 + \frac{1}{2}I\omega_f^2$, where I is the rotational inertia of the dumbbell. Ignoring the negligible inertia of the rod, I can say that each puck in the dumbbell contributes a rotational inertia $m(\ell/2)^2$ given by Eq. 11.30, so that the rotational inertia of the dumbbell is

$$I = 2m(\ell/2)^2 = \tfrac{1}{2}m\ell^2. \tag{1}$$

Because the collision is elastic, the final kinetic energy must equal the initial kinetic energy, and so

$$\tfrac{1}{2}mv_i^2 = \tfrac{1}{2}mv_f^2 + \tfrac{1}{2}I\omega_f^2 = \tfrac{1}{2}mv_f^2 + \tfrac{1}{2}(\tfrac{1}{2}m\ell^2)\omega_f^2,$$

where I have substituted for I the expression I obtained in Eq. 1. Dividing both sides by $\frac{1}{2}m$ gives $v_i^2 = v_f^2 + \frac{1}{2}\ell^2\omega_f^2$. Because puck A moves along the x axis, $v_i^2 = v_{x,i}^2$ and $v_f^2 = v_{x,f}^2$. Because $\omega_f^2 = \omega_{\vartheta,f}^2$ I get

$$v_{x,i}^2 = v_{x,f}^2 + \tfrac{1}{2}\ell^2\omega_{\vartheta,f}^2. \tag{2}$$

Next I turn to conservation of angular momentum. The change in A's angular momentum is, from Eq. 11.36,

$$\Delta L_{A\vartheta} = L_{A\vartheta,f} - L_{A\vartheta,i} = (\ell/2)mv_{x,f} - (\ell/2)mv_{x,i}$$

$$= (\ell/2)m(v_{x,f} - v_{x,i}).$$

The initial angular momentum of the dumbbell $L_{d\vartheta,i}$ about the rotation axis is zero; its final angular momentum about this axis is, from Eq. 11.34, $L_{d\vartheta,f} = I\omega_{\vartheta,f}$. The change in the dumbbell's angular momentum is thus

$$\Delta L_{d\vartheta} = I\omega_{\vartheta,f} - 0 = \tfrac{1}{2}m\ell^2\omega_{\vartheta,f}.$$

Because the system is isolated, its angular momentum doesn't change (Eq. 11.38), and so

$$\Delta L_\vartheta = \Delta L_{A\vartheta} + \Delta L_{d\vartheta}$$

$$= (\ell/2)m(v_{x,f} - v_{x,i}) + \tfrac{1}{2}m\ell^2\omega_{\vartheta,f} = 0$$

or

$$v_{x,i} = v_{x,f} + \ell\omega_{\vartheta,f}. \tag{3}$$

Solving Eq. 3 for $v_{x,f}$, substituting the result into Eq. 2, and using the quadratic equation to solve for the rotational velocity of the dumbbell, I get

$$\omega_{\vartheta,f} = \frac{4v_{x,i}}{3\ell} = +\frac{4v_i}{3\ell}. \; ✔$$

Substituting this result back into Eq. 3, I obtain for the final velocity of puck A $v_{x,f} = -\frac{1}{3}v_{x,i} = -\frac{1}{3}v_i$. ✔

④ EVALUATE RESULT The final rotational velocity is positive, indicating that the dumbbell in Figure 11.34 rotates counterclockwise, in agreement with my drawing. The x component of the final velocity of puck A is negative, indicating that it bounces back, as I expected.

Figure 11.35 Procedure for determining the rotational inertia of an extended object.

(a)   To determine rotational inertia of an extended object . . .

(b)   . . . divide object into small segments of inertia δm and add up their rotational inertias.

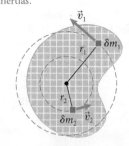

✋ 🖐 **11.9** (a) Follow the line of reasoning used in working from Eq. 11.28 to Eq. 11.31 to verify that the rotational inertia of the dumbbell in Example 11.7 is indeed $\frac{1}{2}m\ell^2$. (b) Does the momentum of the system remain constant in the collision of Figure 11.33?

11.6 Rotational inertia of extended objects

We can apply the concept of rotational inertia to extended rigid objects. Consider, for example, the rotating object in **Figure 11.35**. If you imagine breaking down the object into many small segments of equal inertia δm, each segment has a different velocity $\vec{v}$, but all move in circles about the axis at the same rotational speed ω.

The rotational kinetic energy of the rotating object is the sum of the kinetic energies of the segments:

$$K_{rot} = \tfrac{1}{2}\delta m_1 v_1^2 + \tfrac{1}{2}\delta m_2 v_2^2 + \cdots = \sum_n (\tfrac{1}{2}\delta m_n v_n^2). \tag{11.39}$$

Because each segment has a different velocity, it is difficult to evaluate this expression. Substituting Eq. 11.11, however, we obtain

$$K_{rot} = \sum_n \left[\tfrac{1}{2}\delta m_n (\omega r_n)^2\right] = \tfrac{1}{2}\left[\sum_n \delta m_n r_n^2\right]\omega^2, \tag{11.40}$$

where r_n measures the distance from segment n to the rotation axis. Because

$\delta m_n r_n^2$ is the rotational inertia I_n of segment n (Eq. 11.30), we can write this expression in the form

$$K_{\text{rot}} = \tfrac{1}{2}\left[\sum_n I_n\right]\omega^2 = \tfrac{1}{2}I\omega^2, \tag{11.41}$$

where I is the rotational inertia of the entire object. So, once we know the object's rotational inertia, it is straightforward to calculate its rotational kinetic energy.

Equations 11.40 and 11.41 show that the rotational inertia of an extended object can be obtained by dividing the object into small segments and computing the sum of the contributions of the small segment:

$$I = \sum_n \delta m_n r_n^2. \tag{11.42}$$

To evaluate this sum for the extended object, we take the limit of this expression as $\delta m_n \to 0$. In this limit, the sum becomes an integral:

$$I = \lim_{\delta m_n \to 0} \sum_n \delta m_n r_n^2 \equiv \int r^2 dm \quad \text{(extended object).} \tag{11.43}$$

This integral is difficult to evaluate for an arbitrarily shaped object. As the next few examples will show, however, the integral is relatively easy to calculate for objects that exhibit some symmetry and are *uniform* (that is to say, the inertia is uniformly distributed over the volume of the object). Table 11.3 lists the results of additional calculations.

Table 11.3 Rotational inertia of uniform objects of inertia M about axes through their center of mass

Rotation axes oriented so that object could roll on surface: For these axes, rotational inertia has the form cMR^2, where $c = I/MR^2$ is called the *shape factor*. The farther the object's material from the rotation axis, the larger the shape factor and hence the rotational inertia.

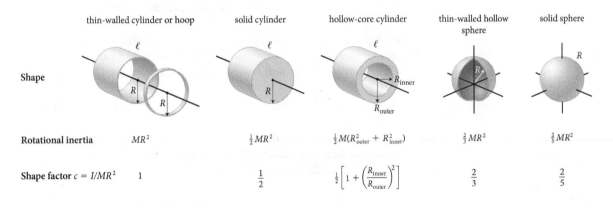

	thin-walled cylinder or hoop	solid cylinder	hollow-core cylinder	thin-walled hollow sphere	solid sphere
Shape					
Rotational inertia	MR^2	$\tfrac{1}{2}MR^2$	$\tfrac{1}{2}M(R_{\text{outer}}^2 + R_{\text{inner}}^2)$	$\tfrac{2}{3}MR^2$	$\tfrac{2}{5}MR^2$
Shape factor $c = I/MR^2$	1	$\dfrac{1}{2}$	$\dfrac{1}{2}\left[1 + \left(\dfrac{R_{\text{inner}}}{R_{\text{outer}}}\right)^2\right]$	$\dfrac{2}{3}$	$\dfrac{2}{5}$

Other axis orientations

	thin-walled hoop	solid cylinder	thin rod	rectangular plate
Shape				
Rotational inertia	$\tfrac{1}{2}MR^2$	$\tfrac{1}{4}MR^2 + \tfrac{1}{12}M\ell^2$	$\tfrac{1}{12}M\ell^2$	$\tfrac{1}{12}M(a^2 + b^2)$

Example 11.8 Rotational inertia of a hoop about an axis through its center

Calculate the rotational inertia of a hoop of inertia m and radius R about an axis perpendicular to the plane of the hoop and passing through its center.

❶ GETTING STARTED I begin by drawing the hoop and a coordinate system (**Figure 11.36**). Because the axis goes through the center of the hoop, I let the origin be at that location. The axis of rotation is perpendicular to the plane of the drawing and passes through the origin.

Figure 11.36

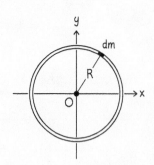

❷ DEVISE PLAN Equation 11.43 gives the rotational inertia of an object as the sum of the contributions from many small segments. If I divide the hoop into infinitesimally small segments each of inertia dm, I see that each segment lies the same distance $r = R$ from the rotation axis (one such segment is shown in Figure 11.36). This means I can pull the constant $r^2 = R^2$ out of the integral in Eq. 11.43, making it easy to calculate.

❸ EXECUTE PLAN Substituting $r = R$ in Eq. 11.43, I obtain

$$I = \int r^2 \, dm = R^2 \int dm = mR^2. \checkmark$$

❹ EVALUATE RESULT This result makes sense because all the material contained in the hoop lies at the same distance R from the rotation axis. Therefore the rotational inertia of the hoop is the same as that of a particle of inertia m located a distance R from the rotation axis, which I know from Eq. 11.30: $I = mR^2$.

For objects in which not all the material is the same distance from the rotation axis, we need to express dm in terms of either Cartesian or polar coordinates in order to evaluate the integral in Eq. 11.43. For uniform one-dimensional objects, we place the object along a positive x axis and introduce a new quantity: *inertia per unit length* $\lambda \equiv dm/dx$. Each small segment dm of the object can then be written as the product of λ and an infinitesimal length dx: $dm = \lambda \, dx$. Equation 11.43 thus becomes, after substituting x for r,

$$I = \lambda \int x^2 dx \quad \text{(uniform one-dimensional object)}, \qquad (11.44)$$

where the integration is carried out over the length of the object.

Likewise, for a uniform two-dimensional object, we convert dm to a surface area by defining an *inertia per unit area* $\sigma \equiv dm/dA$, so that Eq. 11.43 becomes

$$I = \sigma \int r^2 dA \quad \text{(uniform two-dimensional object)}, \qquad (11.45)$$

where the integration is carried out over the surface area of the object.

For a uniform three-dimensional object, we introduce an *inertia per unit volume* $\rho \equiv dm/dV$, so that Eq. 11.43 becomes

$$I = \rho \int r^2 dV \quad \text{(uniform three-dimensional object)}, \qquad (11.46)$$

where the integration is carried out over the volume of the object.

Example 11.9 Rotational inertia of a rod about an axis through its center

Calculate the rotational inertia of a uniform solid rod of inertia m and length ℓ about an axis perpendicular to the long axis of the rod and passing through its center.

❶ GETTING STARTED I begin with a sketch of the rod. For this one-dimensional object, I choose an x axis that lies along the rod's long axis, and because the rotation being analyzed is about a rotation axis located through the rod's center, I choose this point for the origin of my x axis (**Figure 11.37**).

Figure 11.37

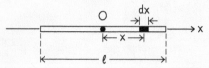

❷ DEVISE PLAN Because the rod is a uniform one-dimensional object, I can use Eq. 11.44 to calculate its rotational inertia. First

I determine the inertia per unit length λ. Then I carry out the integration from one end of the rod ($x = -\ell/2$) to the other ($x = +\ell/2$).

❸ **EXECUTE PLAN** The inertia per unit length is $\lambda = m/\ell$. Substituting this expression and the integration boundaries into Eq. 11.44, I get

$$I = \lambda \int x^2 dx = \frac{m}{\ell} \int_{-\ell/2}^{+\ell/2} x^2 dx = \frac{m}{\ell}\left[\frac{x^3}{3}\right]_{-\ell/2}^{+\ell/2} = \frac{1}{12}m\ell^2. \checkmark$$

❹ **EVALUATE RESULT** If I approximate each half of the rod as a particle located a distance $\ell/4$ from the origin I chose in Figure 11.37, the rotational inertia of the rod would be, from Eq. 11.30, $2(\frac{1}{2}m)(\frac{1}{4}\ell)^2 = \frac{1}{16}m\ell^2$. This is not too far from the value I obtained, so my answer appears to be reasonable.

✋ **11.10** Calculate the rotational inertia of the rod in Example 11.9 about an axis perpendicular to the long axis of the rod and passing through one end.

Example 11.10 Rotational inertia of hollow-core cylinder

Calculate the rotational inertia of a uniform hollow-core cylinder of inner radius R_{inner}, outer radius R_{outer}, length ℓ, and inertia m about an axis parallel to the cylinder's length and passing through its center, as in **Figure 11.38**.

Figure 11.38 Example 11.10.

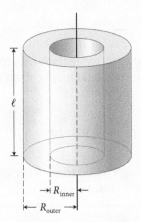

❶ **GETTING STARTED** As in Example 11.9, I will divide this cylinder into segments and integrate the contributions of all the segments over the volume of the cylinder. There are many ways to divide the cylinder into small segments dm, but I can simplify the integration by exploiting the cylindrical symmetry. Starting at the inner face of the wall, at R_{inner}, and moving toward the outer face, at R_{outer}, I divide the wall into a series of many thin-walled cylindrical shells, each of thickness dr and length ℓ and all concentric with the original cylinder, as shown in my top-down sketch of the cylinder (**Figure 11.39**). Because each shell is infinitely thin, all the material in each shell is the same distance r from the axis of rotation, which simplifies the calculation.

❷ **DEVISE PLAN** The hollow-core cylinder is a uniform, three-dimensional object, so I should use Eq. 11.46. First, I must determine the inertia per unit volume, m/V, for it. Next, I must express the infinitesimal volume dV of each shell in terms of r and dr. Finally, I should carry out the integration for values of r from R_{inner} to R_{outer}.

❸ **EXECUTE PLAN** Each thin-walled shell has an inertia $dm = \rho\, dV$, where ρ is inertia per unit volume and dV is the

Figure 11.39

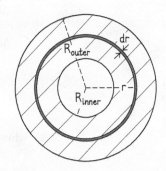

volume of the shell. To determine $\rho = m/V$ for the cylinder, I divide its inertia m by its volume V. The volume of the solid part of the cylinder plus the empty space that forms its core is $\pi R_{outer}^2 \ell$ and that of the core is $\pi R_{inner}^2 \ell$. So the volume of the cylinder is $V = \pi(R_{outer}^2 - R_{inner}^2)\ell$, and $\rho = m/\pi(R_{outer}^2 - R_{inner}^2)\ell$.

Each shell has an outer surface area of $2\pi r\ell$ and thickness dr, and so its volume is $dV = 2\pi r\ell\, dr$. The rotational inertia of the entire cylinder is thus

$$I = \rho \int r^2 dV = \rho \int_{R_{inner}}^{R_{outer}} r^2(2\pi r\ell)dr = 2\pi\rho\ell \int_{R_{inner}}^{R_{outer}} r^3 dr.$$

Working out the integral, I get

$$I = 2\pi\rho\ell\left[\frac{r^4}{4}\right]_{R_{inner}}^{R_{outer}} = \frac{\pi\rho\ell(R_{outer}^4 - R_{inner}^4)}{2}$$

$$= \frac{\pi\ell(R_{outer}^4 - R_{inner}^4)}{2}\frac{m}{\pi\ell(R_{outer}^2 - R_{inner}^2)}.$$

Factoring $R_{outer}^4 - R_{inner}^4 = (R_{outer}^2 + R_{inner}^2)(R_{outer}^2 - R_{inner}^2)$, I get for the rotational inertia of the hollow-core cylinder

$$I_{\text{hollow-core cylinder}} = \frac{1}{2}m(R_{outer}^2 + R_{inner}^2). \checkmark$$

❹ **EVALUATE RESULT** In the limit $R_{inner} = R_{outer}$, the cylinder becomes a thin-walled cylindrical shell of radius R_{outer}, and my result becomes $I = mR_{outer}^2$, as I expect for an object that has all its material a distance R from the axis of rotation.

The result from Example 11.10 allows us to determine the rotational inertia of a solid cylinder by setting R_{inner} equal to zero and R_{outer} equal to the radius R of the cylinder. About the long axis of the cylinder, we have

$$I_{\text{solid cylinder}} = \tfrac{1}{2}mR^2. \tag{11.47}$$

👆 **11.11** Note in Example 11.10 that the rotational inertia of a thick cylindrical shell is $\tfrac{1}{2}m(R_{outer}^2 + R_{inner}^2)$ while that of a solid cylinder is $\tfrac{1}{2}mR_{outer}^2$. Does the fact that the factor $R_{outer}^2 + R_{inner}^2$ is larger than the factor R_{outer}^2 mean that you can increase the rotational inertia of a solid cylinder by drilling a hole through it?

Figure 11.40 The parallel-axis theorem relates the rotational inertia about an axis through the center of mass to the rotational inertia about other parallel axes.

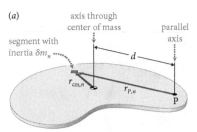

(a)

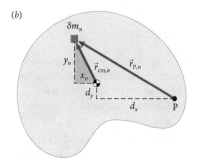

(b)

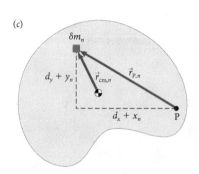

(c)

Because rotational inertia depends on the location of the rotation axis, tables such as Table 11.3 may seem to be of limited utility. However, there is a simple relationship between the rotational inertia of an object about an axis through its center of mass and its rotational inertia about any axis parallel to the center-of-mass axis. Consider the object shown in **Figure 11.40**. Let us derive an expression relating the rotational inertia of the object about an axis through its center of mass to the rotational inertia about a parallel axis through an arbitrary point P.

According to Eq. 11.42, the rotational inertia of the object about an axis through its center of mass is

$$I_{cm} = \sum_n \delta m_n r_{cm,n}^2 = \sum_n \delta m_n (x_n^2 + y_n^2), \tag{11.48}$$

where I have used the Pythagorean theorem for the shaded right triangle in Figure 11.40b to substitute for $r_{cm,n}^2$. The rotational inertia about a parallel axis through P a distance d away from the center of mass is

$$I = \sum_n \delta m_n r_{P,n}^2 = \sum_n \delta m_n[(d_x + x_n)^2 + (d_y + y_n)^2], \tag{11.49}$$

where now the Pythagorean substituting comes from the right triangle in Figure 11.40c. Working out the right side, I obtain

$$I = \sum_n \delta m_n(d_x^2 + 2d_x x_n + x_n^2 + d_y^2 + 2d_y y_n + y_n^2). \tag{11.50}$$

From the definition of the center of mass (Eq. 6.24), I know that if the center of mass is at the origin, then

$$\sum_n \delta m_n x_n = \sum_n \delta m_n y_n = 0, \tag{11.51}$$

and because d_x and d_y are constant, the terms that contain $2d_x x_n$ and $2d_y y_n$ drop out of Eq. 11.50, leaving me with

$$I = \sum_n \delta m_n[(d_x^2 + d_y^2) + (x_n^2 + y_n^2)] = \sum_n \delta m_n d^2 + I_{cm}. \tag{11.52}$$

The factor d^2 can be pulled out of the summation, and so we see that the rotational inertia about the parallel axis is

$$I = I_{cm} + md^2. \tag{11.53}$$

This relationship is called the **parallel-axis theorem.**

Example 11.11 Rotational inertia of a rod about an axis through one end

Use the parallel-axis theorem to calculate the rotational inertia of a uniform solid rod of inertia m and length ℓ about an axis perpendicular to the length of the rod and passing through one end.

❶ GETTING STARTED I first make a sketch of the rod, showing its center of mass and the location of the rotational axis (**Figure 11.41**). Because I am told to use the parallel-axis theorem, I know I have to work with the rod's center of mass. I know that for a uniform rod, the center of mass coincides with the geometric center, and so I mark that location in my sketch.

Figure 11.41 A solid rod of length ℓ rotating about an axis at one end.

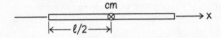

❷ DEVISE PLAN In Example 11.9, I determined that the rotational inertia about an axis through the rod's center is $I = \frac{1}{12}m\ell^2$. For a uniform rod, the center of mass coincides with the geometric center, so I can use Eq. 11.53 to determine the rotational inertia about a parallel axis through one end.

❸ EXECUTE PLAN The distance between the rotation axis and the center of mass is $d = \frac{1}{2}\ell$, and so, with $I_{cm} = \frac{1}{12}m\ell^2$ from Example 11.9, Eq. 11.53 yields

$$I = I_{cm} + md^2 = \tfrac{1}{12}m\ell^2 + m\left(\frac{\ell}{2}\right)^2 = \tfrac{1}{3}m\ell^2. \checkmark$$

❹ EVALUATE RESULT I obtained the same answer in Checkpoint 11.10 by directly working out the integral.

 11.12 About what axis is the rotational inertia of an extended object smallest?

Chapter Glossary

SI units of physical quantities are given in parentheses.

Angular momentum L_ϑ $(\text{kg} \cdot \text{m}^2/\text{s})$ The capacity of an object to make other objects rotate:

$$L_\vartheta \equiv I\omega_\vartheta. \qquad (11.34)$$

Centripetal acceleration a_c (m/s^2) The acceleration required to make an object go around a circular trajectory. An object going at speed v through an arc of radius r has a radially inward acceleration of magnitude

$$a_c = \frac{v^2}{r} \quad \text{(circular motion).} \qquad (11.15)$$

Circular motion at constant speed Motion at constant speed along a circular path. Any object in circular motion at constant speed is subject to a force (or vector sum of forces) directed toward the center of the circular path.

Conservation of angular momentum Angular momentum can be transferred from one object to another, but it cannot be created or destroyed.

Lever arm distance $r_\perp$ (m) The perpendicular distance between an axis of rotation and the line of action of a vector.

Parallel-axis theorem The relationship between the rotational inertia of an object about an axis through its center of mass and its rotational inertia about any parallel axis a distance d away:

$$I = I_{cm} + md^2. \qquad (11.53)$$

Period T (s) The time interval it takes an object to complete one revolution:

$$T = \frac{2\pi r}{v} = \frac{2\pi}{\omega} \quad \text{(circular motion at constant speed).} \qquad (11.20)$$

Rotational acceleration α_ϑ (s^{-2}) The rate at which an object's rotational velocity changes:

$$\alpha_\vartheta = \frac{d\omega_\vartheta}{dt} = \frac{d^2\vartheta}{dt^2}. \qquad (11.12)$$

The rotational acceleration changes the speed of an object moving in a circle of radius r. It is related to the tangential acceleration by

$$a_t = r\alpha_\vartheta. \qquad (11.23)$$

Rotational coordinate ϑ (unitless) A scalar that specifies the position of an object on a circle. It is defined as the length of arc s measured from the positive x axis divided by the radius r of the circle:

$$\vartheta \equiv \frac{s}{r}. \qquad (11.1)$$

Rotational inertia I $(\text{kg} \cdot \text{m}^2)$ A scalar that is a measure of an object's tendency to resist changes in rotational velocity. For a particle of inertia m, the rotational inertia about an axis a distance r from the particle is

$$I \equiv mr^2 \quad \text{(particle).} \qquad (11.30)$$

For an extended object,

$$I \equiv \int r^2 dm \quad \text{(object).} \qquad (11.43)$$

Rotational kinetic energy K_{rot} (J) The kinetic energy associated with rotational motion. For an object of rotational inertia I with a rotational speed ω:

$$K_{rot} = \tfrac{1}{2}I\omega^2. \qquad (11.31)$$

Rotational motion The type of motion for which the particles in an object follow different circular paths centered on a straight line called the *axis of rotation*.

Rotational speed ω (s^{-1}) The magnitude of the rotational velocity, $\omega = |\omega_\vartheta|$.

Rotational velocity ω_ϑ (s^{-1}) The rate at which an object's rotational coordinate changes:

$$\omega_\vartheta = \frac{d\vartheta}{dt}. \qquad (11.6)$$

For an object moving along a circle of radius r, the rotational velocity ω_ϑ and the tangential velocity v_t are related by

$$v_t = r\omega_\vartheta. \qquad (11.10)$$

Tangential acceleration The component of the acceleration parallel to a circular trajectory; for circular motion at constant speed, $a_t = 0$.

Translational motion The type of motion that involves no change in an object's orientation. For such motion, all the particles in the object move along identical parallel trajectories.

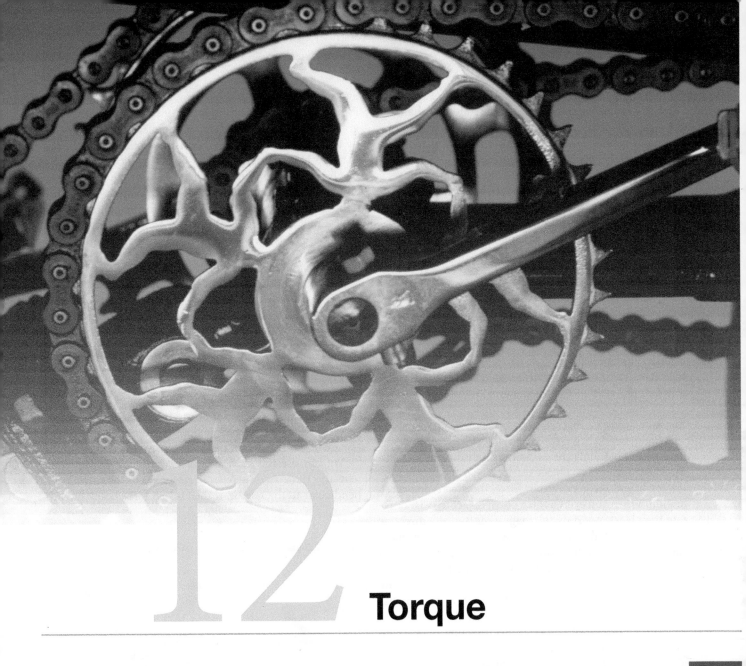

12 Torque

CONCEPTS

QUANTITATIVE TOOLS

Having described the basic principles of rotational motion, we now turn our attention to the cause of changes in this motion. Continuing to exploit the analogy between translational and rotational motion, we introduce a rotational analog of force called *torque*. A torque causes an object's angular momentum to change, in the same way that a force causes an object's momentum to change. We also study how to describe rotations that take place about an axis that is not fixed. To complete our description of rotations, we examine rotations in more than one dimension.

12.1 Torque and angular momentum

In the preceding chapter we studied how objects can be made to rotate. As I mentioned in Section 11.5, tangential components of forces cause the angular momentum of an object to change (Figure 11.31). If you exert a force on the edge of a stationary wheel in a direction tangential to the rim, as illustrated in **Figure 12.1a**, the wheel starts to rotate. To twist the cap off a jar, you exert a tangential force at the edge of the cap, as in Figure 12.1b.

When you exert a force on an object to set it in translational motion, only the magnitude and direction of that force play a role: The object accelerates in the direction of the force, and the greater your force, the greater the acceleration. When you exert a force on an object to twist or rotate the object, not only the magnitude and the direction of the force but also the point at which you exert it determine how effective that force is in rotating the object. For example, if you push on a seesaw to lift a child seated on the opposite end, as illustrated in **Figure 12.2**, it is best to push as far as possible from the pivot of the seesaw and to push in a direction perpendicular to the seesaw. If you push very close to the pivot or if you push in a direction not perpendicular to the seesaw, it is more difficult to rotate the seesaw about its pivot. For a force of a given magnitude, the ability to rotate the seesaw is greatest when the force is exerted as indicated in Figure 12.2a.

The reason that the ability of a force to rotate objects depends on the point of application is most easily understood in terms of energy and work. Suppose you raise the child of Figure 12.2 by exerting a force $\vec{F}$ straight down

Figure 12.1 To (a) make a wheel rotate and (b) twist a cap off a jar, you must exert a force tangential to their rims.

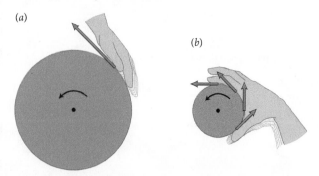

on the opposite end of the seesaw, as in **Figure 12.3a**. Raising the child a vertical distance Δy means increasing the gravitational potential energy of the Earth-child system by an amount $mg\,\Delta y$, where m is the child's inertia. This increase in potential energy is due to work done by you on the system. As you raise the child a distance Δy, your end of the seesaw goes down an equal amount. The work done by you on the system is equal to the magnitude of the force times the force displacement, or $F\,\Delta y_F$. If the child begins and ends at rest, so that there is no change in kinetic energy, the work done by you on the system must equal the change in potential energy of the Earth-child system: $mg\,\Delta y = F\,\Delta y_F$. So we see that to lift the child by pushing on the opposite end of the seesaw, the magnitude of the force $\vec{F}$ must be equal to the magnitude mg of the force of gravity on the child.

Suppose, however, you had pushed not on the far end of the seesaw but rather halfway between the pivot and the end. As Figure 12.3b shows, that point of the seesaw moves down only half the original force displacement, and so to do the same amount of work, the force you must exert is twice as great as before. The closer to the pivot you push, the greater the force required to raise the child.

We can use the same line of reasoning to understand why the force is most effective when it is oriented perpendicular to the seesaw. If the exerted force points in any

Figure 12.2 Easier and harder ways to rotate a seesaw.

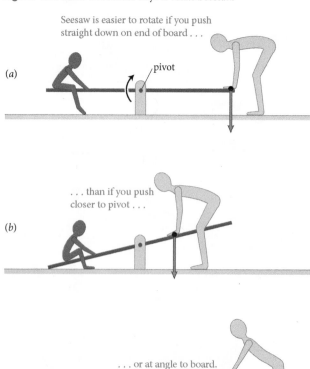

Seesaw is easier to rotate if you push straight down on end of board . . .

(a)

pivot

. . . than if you push closer to pivot . . .

(b)

. . . or at angle to board.

(c)

other direction, we can decompose it into components parallel and perpendicular to the seesaw (Figure 12.3c). The component parallel to the seesaw merely pushes the seesaw against the pivot and doesn't contribute to any rotational displacement of the seesaw: In the extreme case that the force is directed entirely along the seesaw, nothing happens. So it is only the component perpendicular to the seesaw, $\vec{F}_\perp$, that causes the seesaw to rotate.

To find out more accurately what determines the turning ability of a force, consider the rod in **Figure 12.4**. The rod is suspended by a pivot attached at a point that is off-center, and the rod is free to rotate about that point. Two objects of unequal inertias m_1 and m_2 are suspended on either end of the rod, keeping the rod in balance so that it is at rest. For simplicity we assume that the inertia of the rod is negligible compared to m_1 and m_2. Before reading on, complete the following checkpoint.

12.1 (*a*) Draw a free-body diagram for the rod of Figure 12.4. Let the inertia of the rod be negligible compared to m_1 and m_2. (*b*) Would the free-body diagram change if you slide object 2 to the left? (*c*) Experiments show that when $m_1 = 2m_2$ the rod is balanced for $r_2 = 2r_1$. How is the ratio r_1/r_2 related to the ratio m_1/m_2? (*d*) What would happen if you remove object 1 from the rod? (*e*) What would happen if you double the inertia of object 1? (*f*) What would happen if you carefully place a penny on top of object 1? Let the inertia of the two objects be significantly larger than that of the penny. (*g*) Is there a difference between what would happen in parts *e* and *f*?

The forces exerted by the two objects in Figure 12.4 each have the ability to rotate the rod—object 1 tends to rotate the rod in a counterclockwise direction, object 2 in a clockwise direction. This ability of a force to rotate an object about an axis is called the **torque** about that axis. Torque is the rotational analog of force: Forces cause (translational) acceleration; torques cause rotational acceleration about an axis. The rod in Figure 12.4 is balanced because the torques caused by the two forces—their ability to rotate the rod—are equal in magnitude, but the rotations they tend to cause are in opposite directions, and so they sum to zero.

Note that the magnitudes of the two forces that cause these equal torques in opposite directions are not themselves equal: $F_1 = m_1g$ and $F_2 = m_2g$, with $m_2 \neq m_1$. From experiment we know, however, that the rod is balanced when $r_1F_1 = r_2F_2$. This suggests that the torque caused by each force is $r_\perp F$, the product of the magnitude of that force and the lever arm distance of the force relative to the axis of rotation (the perpendicular distance between the line of action of the force and the pivot; see Section 11.5). The line of action of the upward force exerted by the pivot on the rod (see Checkpoint 12.1) goes through the axis of rotation. Consequently, the corresponding lever arm distance is zero, and so this force does not cause any torque about the pivot.

In Figure 12.4, the forces exerted on the rod are perpendicular to it. **Figure 12.5** shows two equivalent ways to determine the torque caused by a force that is not perpendicular to the radial line from the pivot to the point of application of the force. One way is to decompose the force into components parallel and perpendicular to this axis (Figure 12.5a). Only the perpendicular component causes the lever to rotate; the parallel component simply pulls on the pivot. The torque caused by $\vec{F}$ is thus $rF_\perp = r(F \sin \theta)$, where r is the length of the vector pointing from the axis of rotation to the point of application of the force. Note that we can also write this torque as $(r \sin \theta)F$, which is precisely $r_\perp F$, as we surmised. Figure 12.5b shows how to determine $r_\perp$: It is the perpendicular distance between the line of

Figure 12.3 The reason you can rotate the seesaw most easily by pushing downward on the end of the board.

(*a*) Pushing downward on board's end minimizes force you must exert on child + seesaw system because . . .

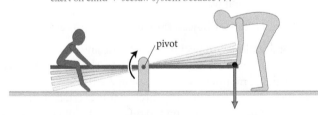

(*b*) . . . work done on system is proportional to force displacement Δy_F, which increases with distance from pivot.

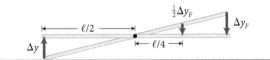

(*c*) Only the component of $\vec{F}$ perpendicular to board does work on seesaw.

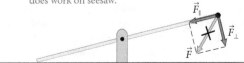

Figure 12.4 Unequal weights balance a rod that is suspended off-center.

(*a*)

(*b*)

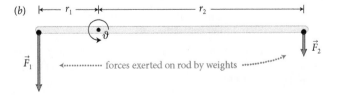

forces exerted on rod by weights

Figure 12.5 Two ways to determine the torque caused by a force exerted on a lever at an angle θ.

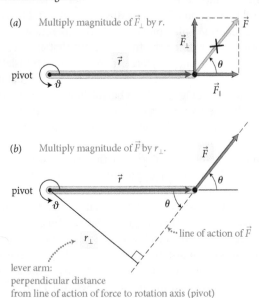

(a) Multiply magnitude of $\vec{F}_\perp$ by r.

pivot

(b) Multiply magnitude of $\vec{F}$ by $r_\perp$.

pivot

line of action of $\vec{F}$

lever arm:
perpendicular distance
from line of action of force to rotation axis (pivot)

action of the force and the axis of rotation. So, the torque caused by a force exerted on an object is the product of the magnitude of the force and its lever arm distance. It can be written equivalently as $rF_\perp$ and as $r_\perp F$.

Like other rotational quantities, torque carries a sign that depends on the choice of direction for increasing ϑ. In Figure 12.4, for example, the torque caused by $\vec{F}_1$ about the pivot tends to rotate the rod in the direction of increasing ϑ and so is positive; the torque caused by $\vec{F}_2$ is negative. The sum of the two torques about the pivot is then $r_1F_1 + (-r_2F_2)$. As we've seen, the two torques are equal in magnitude when the rod is balanced, and so the sum of the torques is zero. When the sum of the torques is not zero, the rod's rotational acceleration is nonzero, and so its rotational velocity and angular momentum change.

In the situations depicted in Figures 12.4 and 12.5 we used the pivot to calculate the lever arm distances. This is a natural choice because that is the point about which the object under consideration is free to rotate. However, torques also play a role for stationary objects that are suspended or supported at several different points and that are not free to rotate—for example, a plank or bridge supported at either end. To determine what reference point to use in such cases, complete the following exercise.

Exercise 12.1 Reference point

Consider again the rod in Figure 12.4. Calculate the sum of the torques about the left end of the rod.

SOLUTION I begin by making a sketch of the rod and the three forces exerted on it, showing their points of application on the rod (**Figure 12.6**).

Figure 12.6

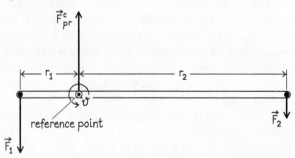

reference point

The lever arm distances must now be determined relative to the left end of the rod. The lever arm distance of force $\vec{F}_1$ to this point is zero, and so the torque caused by that force about the left end of the rod is zero. If I choose counterclockwise as the positive direction of rotation, $\vec{F}_2$ causes a negative torque about the left end of the rod; the force $\vec{F}_{pr}^c$ exerted by the pivot causes a positive torque about the left end of the rod. The lever arm distance of $\vec{F}_2$ about the left end of the rod is $r_1 + r_2$; that of $\vec{F}_{pr}^c$ is r_1. Because the rod is at rest, the magnitude of the force exerted by the pivot is equal to the sum of the forces $\vec{F}_1$ and $\vec{F}_2$. Taking into account the signs of the torques, we find that the sum of the torques about the left end of the rod is $r_1(F_1 + F_2) - (r_1 + r_2)F_2 = r_1F_1 - r_2F_2$. This is the same result we obtained for the torques about the pivot, and so the sum of the torques about the left end is zero. ✔

Exercise 12.1 shows that the sum of the torques about the left end of the rod is zero, just like the sum of the torques about the pivot. You can repeat the calculation for the torques about the right end of the rod or any other point, and each time you will find that the sum of the torques is zero. The reason is that the rod is not rotating about any point, and so the sum of the torques must be zero about any point. In general we can say:

For a stationary object, the sum of the torques is zero.

For a stationary object we can choose any reference point we like to calculate torques. It pays to choose a reference point that simplifies the calculation. As you have seen, we do not need to consider any force that is exerted at the reference point. So, by putting the reference point at the point of application of a force, we can eliminate that force from the calculation.

✋ **12.2** In the situation depicted in Figure 12.2a, you must continue to exert a force on the seesaw to keep the child off the ground. The force you exert causes a torque on the seesaw, and yet the seesaw's rotational acceleration is zero. How can this be if torques cause objects to accelerate rotationally?

Example 12.2 Torques on lever

Three forces are exerted on the lever of **Figure 12.7**. Forces $\vec{F}_1$ and $\vec{F}_3$ are equal in magnitude, and the magnitude of $\vec{F}_2$ is half as great. Force $\vec{F}_1$ is horizontal, $\vec{F}_2$ and $\vec{F}_3$ are vertical, and the lever makes an angle of 45° with the horizontal. Do these forces cause the lever to rotate about the pivot? If so, in which direction?

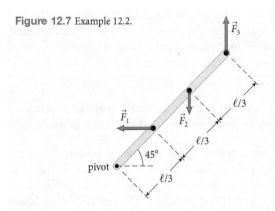

Figure 12.7 Example 12.2.

① **GETTING STARTED** I begin by arbitrarily choosing counterclockwise as the direction of increasing ϑ. With that choice of ϑ, $\vec{F}_1$ and $\vec{F}_3$ cause positive torques about the pivot, while $\vec{F}_2$ causes a negative torque. To answer the question, I need to determine the magnitude and sign of the sum of these three torques about the pivot.

② **DEVISE PLAN** The forces are not perpendicular to the long axis of the lever, and so I need to follow one of the two procedures shown in Figure 12.5 to determine the torques about the pivot. I arbitrarily choose to determine the lever arm distances. To determine these distances relative to the pivot, I make a sketch showing the forces and the perpendicular distance from the pivot to the line of action of each force (**Figure 12.8**). I can then get the magnitude of each torque by multiplying each force magnitude by the corresponding lever arm distance. Knowing the sign and magnitude of each torque, I can determine the combined effect of the three torques.

Figure 12.8

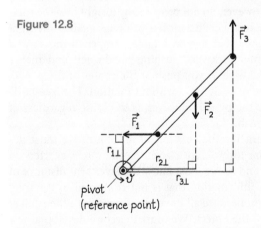

③ **EXECUTE PLAN** I know that $F_1 = 2F_2$. My sketch tells me that $r_{1\perp} = (\ell/3)\sin 45°$ and $r_{2\perp} = (2\ell/3)\cos 45°$, and so $r_{1\perp} = \frac{1}{2}r_{2\perp}$ because $\sin 45° = \cos 45°$. Therefore the torques caused by these two forces about the pivot are equal in magnitude: $r_{1\perp}F_1 = (\frac{1}{2}r_{2\perp})(2F_2) = r_{2\perp}F_2$. Because the two torques carry opposite signs, their sum is thus zero and their effects cancel. This means that the torque caused by $\vec{F}_3$ determines whether or not the lever rotates and, if so, in which direction. Because this torque is nonzero and counterclockwise, the lever rotates in a counterclockwise direction. ✔

④ **EVALUATE RESULT** Looking at Figure 12.7, I see that the two larger forces ($\vec{F}_1$ and $\vec{F}_3$) cause counterclockwise torques about the pivot, and only the smaller force $\vec{F}_2$ causes a clockwise torque. Thus it makes sense that the lever rotates in the counterclockwise direction.

✋ **12.3** (*a*) Without changing the magnitude of any of the forces in Example 12.2, how must you adjust the direction of $\vec{F}_3$ to prevent the lever from rotating? (*b*) If, instead of adjusting the direction of $\vec{F}_3$, you adjust the magnitude of $\vec{F}_2$, by what factor must you change it?

12.2 Free rotation

So far we have considered only rotations about a *fixed axis*, one that constrains the object to rotate about that axis and prevents it from moving away from the axis. Examples of such physical axes are axles, pivots, fulcrums, joints, and hinges. Objects don't need a physical axis to rotate, however. Put a pencil on your desk and flick one end with your finger. The pencil will move away from your hand while executing a rotating motion. A Frisbee flies through the air spinning about its center—without a physical axis through it. The rotation of objects that are not constrained by a physical axis or by other external constraints is called **free rotation.** About what point or axis do freely rotating objects like the pencil and the Frisbee rotate? Can an object be made to rotate about any point if nothing constrains it to rotate about some particular point?

Figure 12.9 shows the motion of a wrench after being launched as shown at the bottom of the figure. Hold the wrench by the end of the handle and sharply flip your wrist clockwise while launching the wrench upward. As the wrench moves upward, it executes a free rotation. The strobe photograph in Figure 12.9 allows us to analyze this motion. As you can see, the trajectory of the end of the handle is neither linear nor circular. Note, however, that the motion of the white dot on the wrench is remarkably simple: The dot moves straight up. What point does this white dot correspond to?

In Section 8.12, I showed that when external forces are exerted on the particles in a many-particle system, the system's center of mass moves as though all the particles in the system were concentrated at the center of mass and all external forces were exerted at that point. This statement is a direct result of conservation of momentum and of the definition of center of mass. The wrench, which consists of many atoms, is a many-particle system, and as it moves upward, Earth exerts a downward gravitational force on each atom in the wrench. The result we obtained in Section 8.12 thus states that the center of mass of the wrench should move upward as if it were a particle moving under the influence of gravity. Indeed, if you closely examine the spacing of the dots in Figure 12.9, you can see that the spacing decreases. This tells you that the dot slows down

Figure 12.9 Motion of a wrench that is launched upward so that it rotates.

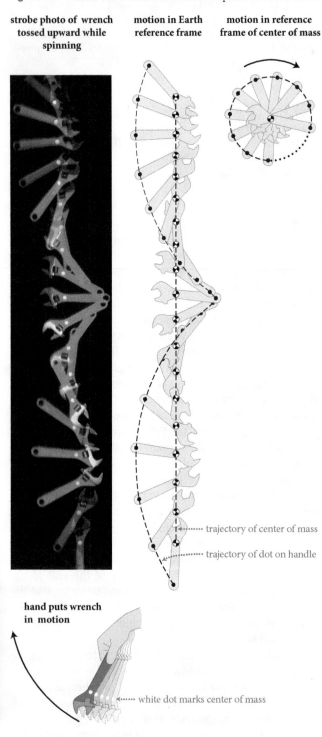

strobe photo of wrench tossed upward while spinning

motion in Earth reference frame

motion in reference frame of center of mass

·········· trajectory of center of mass

·········· trajectory of dot on handle

hand puts wrench in motion

·········· white dot marks center of mass

So, the white dot on the wrench is the center of mass of the wrench, and other points on the wrench rotate around it. The center of mass is in translational motion at constant acceleration and thus cannot be rotating. This is true in general:

> **Objects that are made to rotate without external constraints always rotate about the center of mass.**

This statement shows how remarkably powerful the concept of center of mass is. It allows us to break down the complex motion of the freely rotating wrench into two separate parts: the translational motion of the center of mass and a rotational motion of the wrench about the center of mass. This separation greatly simplifies the analysis of the motion because we can treat each part separately with the tools we have already developed.

12.4 As the wrench in Figure 12.9 moves upward, the upward translational motion of its center of mass slows down. Does the rotation about the center of mass also slow down? Which way does the wrench rotate when it falls back down after reaching its highest position?

12.3 Extended free-body diagrams

In Chapter 9 we developed a procedure for making free-body diagrams. These diagrams help us determine the *vector sum of the forces exerted on* an object, which, in turn, determines the motion of the object's center of mass. Free-body diagrams, however, do not provide any insight into the rotation of an object. For instance, if you push against the right end of a pencil that lies on a table, the pencil moves and rotates counterclockwise; pushing on the left end makes it rotate clockwise. If you push at the center of the pencil, you can even make it move without rotation. The free-body diagram for the pencil, however, is the same regardless of where you push.

To account for the rotation of an object, we must determine what torques are caused by the forces exerted on that object, and so we must know the lever arm distance of each force—information that is not contained in the free-body diagram because all forces are drawn with their tail at the center of the object. We must therefore develop a new kind of diagram that shows not only the forces exerted on an object but also the location of the point of application of each force relative to some chosen reference point. Such a diagram is called an **extended free-body diagram.** The procedure for making these diagrams is described in the Procedure box on page 7.

Drawing an extended free-body diagram and choosing a reference point help you determine the lever arm distances to that point so you can calculate the magnitude and direction of the torques about that point.

as it moves upward. Quantitative analysis of the motion of the dot shows that the magnitude of the dot's downward acceleration is exactly *g*—that is, equal to that of an object launched without rotation.

Procedure: Extended free-body diagrams

1. Begin by making a standard free-body diagram for the object of interest (the *system*) to determine what forces are exerted on it. Determine the direction of the acceleration of the center of mass of the object, and draw an arrow to represent this acceleration.
2. Draw a cross section of the object in the plane of rotation (that is, a plane perpendicular to the rotation axis) or, if the object is stationary, in the plane in which the forces of interest lie.
3. Choose a reference point. If the object is rotating about a hinge, pivot, or axle, choose that point. If the object is rotating freely, choose the center of mass. If the object is stationary, you can choose any reference point. Because forces exerted at the reference point cause no torque, it is most convenient to choose the point where the largest number of forces is exerted or where an unknown force is exerted. Mark the location of your reference point and choose a positive direction

of rotation. Indicate the reference point in your diagram by the symbol ⊙.
4. Draw vectors to represent the forces that are exerted on the object and that lie in the plane of the drawing. Place the tail of each force vector at the point where the force is exerted on the object. Place the tail of the gravitational force exerted by Earth on the object at the object's center of mass.* Label each force.
5. Indicate the object's rotational acceleration in the diagram (for example, if the object accelerates in the positive ϑ direction, write $\alpha_\vartheta > 0$ near the rotation axis). If the rotational acceleration is zero, write $\alpha_\vartheta = 0$.

---

*Is the gravitational force really exerted at the center of mass? Suppose the gravitational force exerted by Earth were exerted at some other point. The force would then cause a permanent torque about the center of mass, and any object dropped from rest would begin spinning, which is not true.

Exercise 12.3 Holding a ball

You hold a ball in the palm of your hand, as shown in **Figure 12.10.** The bones in your forearm act like a horizontal lever pivoted at the elbow. The bones are held up by the biceps muscle, which makes an angle of about 15° with the vertical. Draw an extended free-body diagram for your forearm.

Figure 12.10 Exercise 12.3.

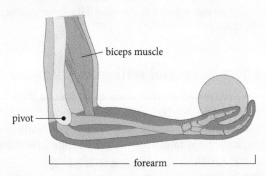

SOLUTION I begin by making a standard free-body diagram for the forearm, which is in contact with the upper-arm bone (the *humerus*), the biceps muscle, and the ball, all of which exert contact forces on it. In addition, there is the gravitational force $\vec{F}_{Ef}^G$ to show in my diagram. I know the direction of $\vec{F}_{bf}^c$ (down), $\vec{F}_{mf}^c$ (up and to the left), and $\vec{F}_{Ef}^G$ (down) but not the direction of $\vec{F}_{hf}^c$.

How do I come up with a reasonable direction for $\vec{F}_{hf}^c$ in my diagram? I know that the forearm is at rest, which means its linear acceleration is zero, and so the vector sum of the forces exerted on the forearm must be zero. This means that the component of $\vec{F}_{hf}^c$ parallel to the forearm must be equal in magnitude to the parallel component of $\vec{F}_{mf}^c$ but pointing in the opposite direction. Because I cannot determine off-hand what the perpendicular component of $\vec{F}_{hf}^c$ is, I draw a force vector pointing down at some arbitrary angle and adjust its length so that the two parallel components are equal in length. Finally I adjust all the perpendicular components so that their

sum is zero. This gives me the free-body diagram shown in **Figure 12.11a.** Because the arm is at rest, I write $\vec{a}_{cm} = \vec{0}$ next to the diagram. The zero acceleration tells me that the vector sum of the forces exerted on the arm is zero.

Figure 12.11

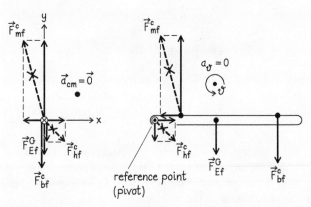

(a) Free-body diagram   (b) Extended free-body diagram

Next I turn to the extended free-body diagram. I represent the forearm as a horizontal rod, with the pivot on the left, as shown in Figure 12.11b. Because the forearm is at rest, I can choose any point as the reference point. Because I know the least about $\vec{F}_{hf}^c$, however, it makes sense to choose my reference point at the pivot where that force is exerted. I choose the positive direction of rotation to be counterclockwise and write $\alpha_\vartheta = 0$ next to it because the forearm is not rotating.

Finally I draw the same forces as in the free-body diagram, placing their tails at their points of application. The result is shown in Figure 12.11b. The perpendicular component of $\vec{F}_{mf}^c$ causes a positive torque, $\vec{F}_{Ef}^G$ and $\vec{F}_{bf}^c$ cause negative torques, and $\vec{F}_{hf}^c$ causes no torque because its lever arm distance is zero. ✔

You will work out a quantitative solution to this problem in Checkpoint 12.8.

12.5 (*a*) If the biceps muscle in Figure 12.10 were attached farther out toward the wrist, would the torque generated by the muscle about the pivot get greater, get smaller, or stay the same? (*b*) As the hand is raised above the level of the elbow, so that the forearm makes an angle of 15° with the horizontal, does the arm's capacity to lift objects increase, decrease, or stay the same?

As you may have noticed, the information in the extended free-body diagram overlaps with that in the free-body diagram. Is it then necessary to draw both? I strongly recommend doing so until you have gained some practice. The standard free-body diagram helps you find the vector sum of the forces exerted on the object, which determines the object's translational motion. In the extended free-body diagram, it is more cumbersome to decompose vectors and add components, and so it is easier to make errors. The extended free-body diagram, however, is necessary to determine the torques caused by the forces that are exerted on the object. The next exercise helps clarify the purpose of each diagram.

Exercise 12.4 Pushing a large crate

You are moving a large crate mounted on swivel wheels, exerting an off-center force $\vec{F}^c_{pc}$ as shown in **Figure 12.12**. Draw an extended free-body diagram for the crate. You can ignore any friction in the wheels.

Figure 12.12 Exercise 12.4.

$\vec{F}^c_{pc}$

SOLUTION The forces exerted on the crate are a downward force of gravity, an upward contact force exerted by the floor, and your horizontal push. I therefore draw the free-body diagram shown in **Figure 12.13a**.

Figure 12.13

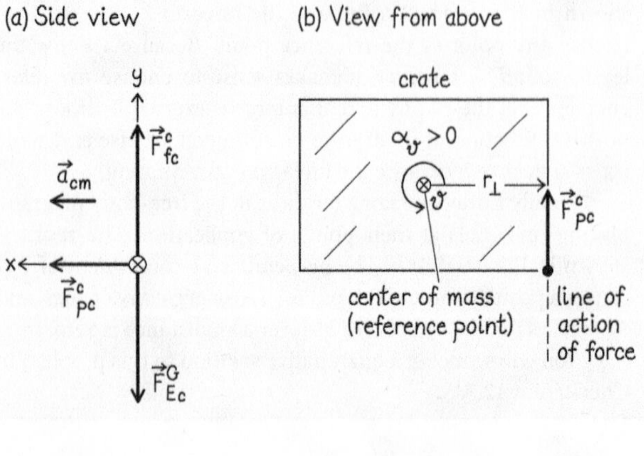

(a) Side view

y
$\vec{F}^c_{fc}$
$\vec{a}_{cm}$
x
$\vec{F}^c_{pc}$
$\vec{F}^G_{Ec}$

(b) View from above

crate
$\alpha_\vartheta > 0$
$r_\perp$
ϑ
$\vec{F}^c_{pc}$
center of mass (reference point)
line of action of force

The free-body diagram shows that the vector sum of the forces exerted on the crate is nonzero. From this information I conclude that the push causes the center of mass of the crate to accelerate in the direction of the push. I therefore draw an arrow next to the free-body diagram to indicate the direction of this center-of-mass acceleration.

The crate is not constrained to rotate about a fixed axis. The floor and the wheels it is resting on, however, do constrain it to rotate in the horizontal plane. In addition to accelerating the center of mass of the crate, the off-center push therefore causes the crate to rotate about a vertical axis through the center of mass. Because the rotation is in the horizontal plane, I draw an outline of the crate as seen from above for the extended free-body diagram.

The crate is not constrained to rotate about any fixed axis in the plane I have drawn, so the center of mass is the reference point I must use. I assume the crate to be uniform, so its center of mass is at the center. I mark the reference point in my diagram and choose counterclockwise as the direction of positive rotation.

The only force that lies in this plane is $\vec{F}^c_{pc}$, so the extended free-body diagram contains just one force (Figure 12.11*b*). The lever arm distance of $\vec{F}^c_{pc}$ is the distance between its line of action and the center of mass of the crate. From the diagram, I thus conclude that $\vec{F}^c_{pc}$ causes a counterclockwise torque. This torque causes the crate to rotate counterclockwise, and I write $a_\vartheta > 0$ next to the diagram to indicate that the rotational acceleration of the crate is positive. ✔

12.6 Suppose the force $\vec{F}^c_{pc}$ in Exercise 12.4 gives the center of mass of the crate an acceleration $\vec{a}_{cm}$. Would this acceleration be greater, smaller, or the same if the force were exerted exactly at the center of the crate?

12.4 The vectorial nature of rotation

So far we've treated all rotational quantities as if they were scalars. There are two reasons a scalar treatment of rotational quantities cannot be complete, however: Rotations have direction, and the rotation axis has a definite orientation in space. For rotations that lie in a single plane, an algebraic sign is sufficient to indicate the direction of rotation in the same way that an algebraic sign indicates direction in one-dimensional translational motion, as shown in **Figure 12.14**.

For rotations in three dimensions, the algebraic sign alone is not sufficient to determine the direction of rotation. Consider, for example, the two spinning disks in **Figure 12.15**. If you view them from the angle shown in Figure 12.15*a*, you could claim that the disk A spins counterclockwise and disk B clockwise. Seen from below, as in Figure 12.15*b*, however, disk A spins clockwise and the rotation of disk B, now seen sideways, is neither clockwise nor counterclockwise. So, in three dimensions the terms *clockwise* and *counterclockwise* are insufficient to specify the direction of a rotation, just as *positive* and *negative* are not enough to specify a displacement vector in more than one dimension.

Can a rotation be characterized by a vector? The vectors $\vec{r}$, $\vec{v}$, and $\vec{a}$ of an object in circular motion at constant

Figure 12.14 For both translation in one dimension and rotation in a plane, the two possible directions of motion can be distinguished by their algebraic sign.

(*a*) Translational motion in one dimension

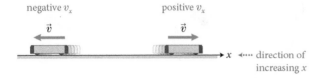

negative v_x positive v_x

$\vec{v}$ $\vec{v}$

$\longrightarrow x$ $\longleftarrow$···· direction of increasing x

(*b*) Rotational motion in a plane

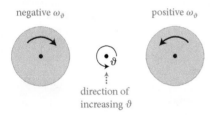

negative ω_ϑ positive ω_ϑ

ϑ

direction of increasing ϑ

speed all lie in the plane of the trajectory, as shown in **Figure 12.16a**. Their directions change every instant, and because all directions in the plane are equivalent, it makes no sense to associate a vector in this plane with the rotation. How would we agree on which direction to pick? The direction of the rotation axis, a direction in which the vectors $\vec{r}$, $\vec{v}$, and $\vec{a}$ never point, is the only direction associated with the rotation that remains fixed. We can specify the orientation of the rotation axis by a vector. Given that vector, you know the orientation of the plane in which the rotation takes place.

If we use a vector to specify the orientation of the rotation axis in space, we can use the two possible directions along this axis to represent the two possible directions of rotation about the axis. The convention for associating a vector along the rotation axis with the direction of rotation is illustrated in Figure 12.16*b* and is called the **right-hand rule:**

> **When you curl the fingers of your right hand along the direction of rotation, your thumb points in the direction of the vector that specifies the direction of rotation.**

Figure 12.15 For rotations in three dimensions, it is no longer possible to specify the direction of rotation unambiguously by algebraic sign or by the terms *clockwise* and *counterclockwise*.

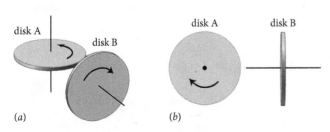

Whether a disk rotates clockwise or counterclockwise depends on your angle of view.

disk A   disk B   disk A   disk B

(*a*)   (*b*)

Figure 12.16 Right-hand rule and directions associated with rotations.

(*a*) Vectors $\vec{r}$, $\vec{v}$, and $\vec{a}$ cannot be associated with a direction of rotation

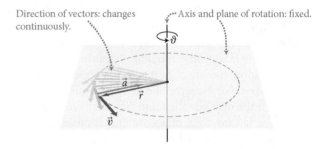

Direction of vectors: changes continuously.   ···· Axis and plane of rotation: fixed.

ϑ

$\vec{a}$

$\vec{r}$

$\vec{v}$

(*b*) The right-hand rule connects direction of rotation about an axis with a direction along the rotation axis

Right-hand rule: Curl fingers of right hand in direction of rotation; thumb will point in direction of vector representing rotation.

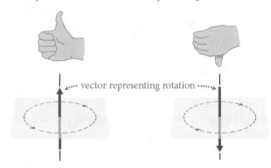

···· vector representing rotation ····

Armed with this rule, you can now specify the direction of any rotation and the orientation of the axis about which that rotation takes place with a vector and avoid the ambiguity of terms like *clockwise* and *counterclockwise*. For example, for the two disks in Figure 12.15*a* the right-hand rule yields the directions given by the two blue vectors in **Figure 12.17**. The advantage of the vectors is that we

Figure 12.17 The right-hand rule applied to the spinning disks of Figure 12.15.

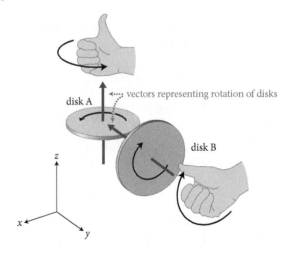

disk A   ···· vectors representing rotation of disks

disk B

z

x y

can specify them by giving their x, y, and z components. The vector representing the rotation of disk A points in the positive z direction, and so it has the form $(0, 0, +A)$. The vector representing the rotation of disk B has the form $(0, -B, 0)$, where A and B are positive numbers.

Note that the vector associated with a rotation always lies along the axis of that rotation. If you invert the direction of rotation, the vector points in the opposite direction along the axis. For example, if disk A in Figure 12.17 were rotating in the opposite direction from the one illustrated, you'd have to curl your fingers in the opposite direction, and so your thumb would point in the negative z direction, opposite the direction shown.

Regardless of your observation angle, the right-hand rule always yields the same direction for a given rotation. Suppose, for example, that you are tightening bolts into a plate as in **Figure 12.18**. When you tighten a bolt, it is natural to consider the bolt's rotation as being clockwise because you usually look at the head of the bolt when tightening it. Now imagine tightening a bolt on the backside of the plate by reaching around it. From where you stand, you must turn in a *counterclockwise* direction to tighten that bolt! The statement "to tighten a bolt, you must rotate it in a clockwise direction" holds only when you are facing the head of the bolt. As Figure 12.18 shows, however, when you align your right-hand thumb in the direction you want the bolt to move, your fingers curl in the correct direction, regardless of where you stand relative to the plate.

The right-hand rule can be used to determine not only the direction of a vector representing a rotation but also the other way around. Given a vector, we can use the right-hand rule to determine the corresponding rotation (**Figure 12.19**). Either one uniquely determines the other.

What kind of a vector is the vector we have just introduced? All the vector quantities we have encountered up to now are derived from the displacement vector—an arrow that points from an object's initial position to its final position. As we've seen, displacement vectors obey a commutation principle: The sum of several displacements is

Figure 12.18 The right-hand rule tells you how to tighten a bolt regardless of your position relative to the bolt.

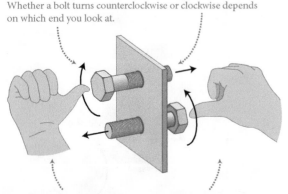

Whether a bolt turns counterclockwise or clockwise depends on which end you look at.

But if you point thumb of right hand in direction bolt should move, curled fingers *always* show which way to turn it.

Figure 12.19 The right-hand rule used two ways.

Given rotating object . . .   . . . curl right-hand fingers in direction of rotation . . .   . . . thumb specifies unique direction in space.

Given direction in space . . .   . . . point right-hand thumb in that direction . . .   . . . curled fingers specify unique rotation.

independent of the sequence in which the displacements are added. Displacement vectors commute in two and in three dimensions. Do rotational displacements commute? Rotations in a plane do, as you can easily verify by adding two angles. The result is independent of the order in which the two angles are added: $15° + 23° = 23° + 15° = 38°$. As **Figure 12.20** illustrates, however, rotations in three dimensions do not commute: If you give a beach ball two successive rotations over 90° angles about perpendicular axes, the final orientation of the ball depends on the sequence of the rotations. So, rotational displacements cannot be vectors because in three dimensions $\vartheta_1 + \vartheta_2 \neq \vartheta_2 + \vartheta_1$.

Figure 12.20 Rotational displacements are not commutative: The order in which you perform them matters.

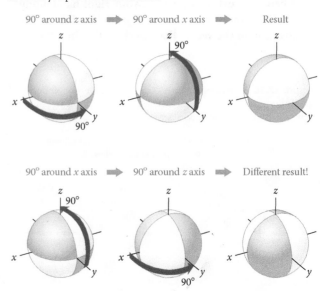

Figure 12.21 Successive rotations over very small angles are commutative. The dashed lines show the initial positions of the ball's seams.

For successive rotations over very small angles, order doesn't matter.

If, instead of rotating the ball twice over 90° angles, you rotate it over two small angles θ_1 and θ_2, the difference in the ball's final orientation is much smaller. As **Figure 12.21** shows, for two successive rotations over 6° angles, the difference is no longer noticeable. Why small angles commute can be understood by looking at **Figure 12.22**, which shows a close-up of the displacement of the seams on the ball in Figure 12.21. Each rotational displacement $\Delta\vartheta$ corresponds to a displacement over an arc of length $\Delta s = r\,\Delta\vartheta$, where r is the radius of the ball. For very small rotational displacements $d\vartheta_1$ and $d\vartheta_2$, the corresponding arcs ds_1 and ds_2 are approximately straight lines on a nearly flat surface.

Figure 12.22 Detail of the balls of Figure 12.21 in their final orientation. Because the displacements along the arcs corresponding to each rotation are nearly straight lines, the rotations commute.

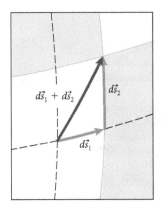
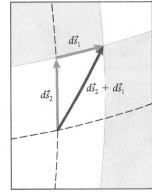

The smaller the angles θ_1 and θ_2, the more the arcs behave like vectors in two dimensions and so commute.

So, infinitesimally small rotational displacements do commute: $d\vartheta_1 + d\vartheta_2 = d\vartheta_2 + d\vartheta_1$. This means that rotational velocity—the rate at which the rotational displacement changes, $\omega = d\vartheta/dt$—obeys the commutation law: $\omega_1 + \omega_2 = \omega_2 + \omega_1$. We can therefore introduce a *rotational velocity vector* $\vec{\omega}$ whose direction is given by the right-hand rule and whose magnitude is ω. The rotational velocity vector is the basic vector describing rotational motion, much as displacement is the basic vector for translational motion. **Figure 12.23** shows two spinning tops with their rotational velocity vectors. Each vector determines the orientation of the top's axis of rotation, the direction in which the top rotates around this axis, and the rotational speed. The arrow representing the rotational velocity of top A is longer than the arrow of top B, indicating that A spins faster than B. We shall examine the vector nature of other rotational quantities in Section 12.8.

Vectors associated with a direction of rotation are called **axial vectors;** vectors derived from the displacement vector are called **polar vectors.** Axial vectors play an important role in physics, not only in rotation but also in electricity and magnetism.

✋ **12.7** Suppose the rotation of top A in Figure 12.23 slows down without a change in the direction of its axis of rotation. (*a*) In which direction does the vector $\Delta\vec{\omega}$ point? (*b*) Can the top's rotational acceleration be represented by a vector? If so, in which direction does this vector point?

Figure 12.23 Representing rotational velocity vectors.

Rotational velocity vector:

- *Direction* indicates rotation direction (via right-hand rule).
- *Length* indicates speed of rotation.

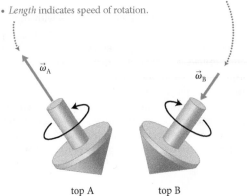

top A   top B

Self-quiz

1. A rope supports one end of a beam as shown in **Figure 12.24**. Draw the lever arm distance for the torque caused by the rope about the pivot.

Figure 12.24

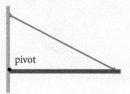

2. Draw a free-body diagram and an extended free-body diagram for (a) a door hanging on two hinges and (b) a bridge supported from each end, with a car positioned at one-quarter of the bridge's length from one support.

3. Which diagram in **Figure 12.25**—1, 2, or 3—shows the alarm clock on the left after it has been rotated in the directions indicated by (a) 90° about the x axis and then 90° about the y axis and (b) 90° about the y axis and then 90° about the x axis? Does the order of the rotation change your answer?

Figure 12.25

4. Give the direction of the rotational velocity vector associated with each spinning object shown in **Figure 12.26**.

Figure 12.26

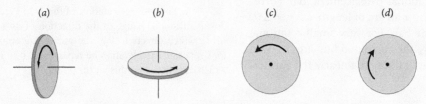

Answers

1. The lever arm distance $r_\perp$ is the perpendicular distance between the pivot and the line of action of the force exerted by the rope on the beam, as shown in **Figure 12.27**.

Figure 12.27

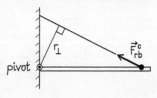

2. See **Figure 12.28**. (a) The door interacts with three objects: Earth, the top hinge, and the bottom hinge. Without the top hinge, the force of gravity would tend to rotate the door about an axis perpendicular to the door through the bottom hinge. The force exerted by the top hinge must balance the clockwise torque caused by the force of gravity about the axis through the bottom hinge. The horizontal components of the forces exerted by the hinges must cancel each other. (b) The bridge interacts with four objects: Earth, the right support, the left support, and the car. The upward forces from the supports must balance the downward gravitational forces of the car and the bridge. Because these forces must also counteract the counterclockwise torque caused by the car, the force exerted by the support closer to the car must be greater than the force exerted by the other support.

Figure 12.28

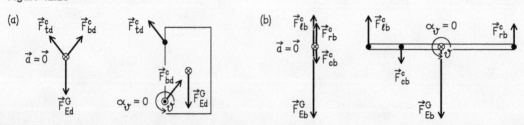

3. (a) 3; (b) 2. The order of rotation does make a difference.

4. Wrapping the fingers of your right hand in the direction of spin gives rotational velocity vectors that point (a) to the right, (b) up, (c) out of the page, and (d) into the page.

12.5 Conservation of angular momentum

In Section 11.5 we saw that angular momentum is a conserved quantity and that the angular momentum of a system rotating in a plane on which no tangential forces are exerted remains constant. Let us now examine how tangential forces are related to the concept of torque introduced in the first part of this chapter and derive mathematical expressions that relate the various rotational quantities and torque.

Consider the situation shown in **Figure 12.29**. A force $\vec{F}$ is exerted on a particle of inertia m that is constrained to move along a circular trajectory of radius r about a pivot. Because $\vec{F}$ has a component that is tangential to the circular trajectory, the particle speeds up. In the language we developed in the first part of this chapter, the particle's rotational acceleration is nonzero because $\vec{F}$ causes a torque on the particle about the pivot. The magnitude of this torque, which we denote by the Greek letter τ (*tau*), is given by

$$\tau \equiv rF \sin \theta = r_\perp F = rF_\perp, \tag{12.1}$$

where θ is the angle between the radius vector $\vec{r}$ and $\vec{F}$ and where $F_\perp$ is the magnitude of the tangential component of the force: $F_\perp = |F_t|$. Because torque is the product of a force and a distance, it has SI units of newton-meters (N · m). Work, being the product of displacement and force, has the same units. Torque and work represent two very different and unrelated quantities, however. Work corresponds to the transfer of energy by a force; torque is the ability of a force to change the rotational motion of an object. To avoid confusion, we shall always use joules for work and newton-meters for torque.

Let us now relate this torque to the particle's tangential acceleration a_t (which is perpendicular to the radius vector $\vec{r}$). This acceleration is given by

$$F_t = ma_t. \tag{12.2}$$

Combining Eqs. 12.1 and 12.2, we obtain

$$\tau_\vartheta = r(ma_t), \tag{12.3}$$

where the subscript ϑ means that τ_ϑ is a signed quantity. Although strictly speaking τ_ϑ is the component of a torque *vector*, we shall refer to it simply as torque. The algebraic sign of the torque is determined by the sign of the rotational acceleration it causes. The torque caused by the force $\vec{F}$ in Figure 12.29, for example, is positive because it causes the particle to accelerate in the positive ϑ direction ($a_t > 0$).

As we saw in Chapter 11, the tangential acceleration is related to the rotational acceleration by $a_t = r\alpha_\vartheta$, so we can write

$$\tau_\vartheta = rm(r\alpha_\vartheta) = mr^2 \alpha_\vartheta. \tag{12.4}$$

The quantity mr^2 is the rotational inertia of the particle, so Eq. 12.4 becomes

$$\tau_\vartheta = I\alpha_\vartheta. \tag{12.5}$$

Equation 12.5 shows how the torque caused by a force exerted on a particle relates to the particle's rotational acceleration. Notice the remarkable parallel between Eqs. 12.2 and 12.5: Instead of "force equals inertia times acceleration," we have "torque equals rotational inertia times rotational acceleration." Torque is to rotation what force is to translation, and therefore Eq. 12.5 is sometimes called the *rotational equation of motion*.

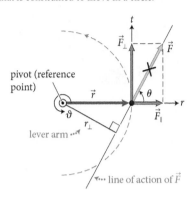

Figure 12.29 A force $\vec{F}$ is exerted on a particle that is constrained to move in a circle.

Figure 12.30 A rigid object constrained to rotate about an axis.

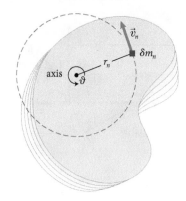

To find a similar expression for extended objects, consider the rotating extended object illustrated in **Figure 12.30**. Imagine breaking down the object into small particles of inertia δm_n, with each particle located a distance r_n from the axis and having velocity $\vec{v}_n$. Let each particle be subject to a torque $\tau_{n\vartheta}$ about the axis; this torque is caused by forces exerted by surrounding particles and by external forces. For each particle we can write an expression analogous to Eq. 12.4:

$$\tau_{n\vartheta} = \delta m_n \, r_n^2 \, \alpha_{n\vartheta}. \tag{12.6}$$

The sum of the torques on all particles is then

$$\sum_n \tau_{n\vartheta} = \sum_n (\delta m_n r_n^2 \alpha_{n\vartheta}) = \left(\sum_n \delta m_n r_n^2\right)\alpha_\vartheta, \tag{12.7}$$

where we have used the fact that all particles in a rigid rotating object must have the same rotational acceleration α_ϑ. The sum on the right side is the rotational inertia I of the extended object (Eq. 11.42), and so Eq. 12.7 becomes

$$\sum_n \tau_{n\vartheta} = I\alpha_\vartheta. \tag{12.8}$$

Figure 12.31 The internal torques caused by two interacting particles within a rotating rigid object have the same lever arm distance, so they cancel.

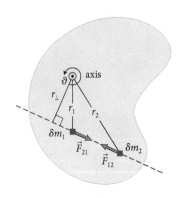

The sum on the left side contains torques caused by external forces (*external torques* for short) and internal torques. To find the contribution from internal torques, consider two interacting particles within the extended object, as illustrated in **Figure 12.31**. Because particles 1 and 2 interact with each other, they exert forces on each other that are equal in magnitude: $F_{12} = F_{21}$. As Figure 12.31 shows, the lever arm distance for the two forces is the same ($r_\perp$), and so the torque caused by $\vec{F}_{12}$ is equal in magnitude to the torque caused by $\vec{F}_{21}$ but opposite in direction. Their sum is therefore zero:

$$\tau_{21\vartheta} + \tau_{12\vartheta} = r_\perp F_{21} - r_\perp F_{12} = r_\perp(F_{21} - F_{12}) = 0. \tag{12.9}$$

Because this cancellation holds for all pairs of interacting particles within the object, all internal torques drop out of the sum on the left side of Eq. 12.8. Only torques caused by external forces remain. The expression we obtain for an extended object is therefore identical to that for a particle:

$$\sum \tau_{\text{ext}\,\vartheta} = I\alpha_\vartheta. \tag{12.10}$$

Another useful expression can be obtained by recognizing that the particle's rotational velocity and rotational acceleration are related by

$$\alpha_\vartheta = \frac{d\omega_\vartheta}{dt}. \tag{12.11}$$

Substituting this expression into Eq. 12.10 and recalling that angular momentum $L_\vartheta = I\omega_\vartheta$ (Eq. 11.34), we get

$$\sum \tau_{\text{ext}\,\vartheta} = I\frac{d\omega_\vartheta}{dt} = \frac{d}{dt}(I\omega_\vartheta) = \frac{dL_\vartheta}{dt}, \tag{12.12}$$

where we have used the fact that the rotational inertia I of a rigid object is constant. So torque is the time rate of change of angular momentum, much as force is the time rate of change of momentum. Equation 12.12 shows that if the sum of the torques caused by external forces exerted on an object is zero, the object's angular momentum does not change:

$$\sum \tau_{\text{ext}\,\vartheta} = \frac{dL_\vartheta}{dt} = 0 \quad \Rightarrow \quad \Delta L_\vartheta = 0. \tag{12.13}$$

Whenever the sum of the torques caused by the forces exerted on an object is zero—in other words, whenever its angular momentum is constant—the object is said to be in **rotational equilibrium.** In Chapter 8 we learned that an object is in *translational equilibrium* when its momentum is constant and that this condition is met when the vector sum of the external forces exerted on the object is zero. Now we can expand that notion to include angular momentum, which remains constant when the sum of the torques caused by external forces exerted on it is zero. An object that is in both translational and rotational equilibrium is said to be in **mechanical equilibrium:**

$$\sum \tau_{\text{ext}\,\vartheta} = 0 \quad \text{and} \quad \sum \vec{F}_{\text{ext}} = \vec{0} \quad \Leftrightarrow \quad \text{mechanical equilibrium.} \tag{12.14}$$

For a system that is not in rotational equilibrium, we have

$$\Delta L_\vartheta = J_\vartheta, \tag{12.15}$$

where J_ϑ represents the transfer of angular momentum from the environment to the system. The quantity J_ϑ is called the **rotational impulse** delivered to the system. Like angular momentum it has SI units of $\text{kg} \cdot \text{m}^2/\text{s}$. In analogy to the momentum law (Eq. 4.18) and the energy law (Eq. 9.1), which embody conservation of momentum and energy, respectively, Eq. 12.15 embodies conservation of angular momentum and is called the **angular momentum law.**

To calculate the rotational impulse, we can use Eq. 12.12. For a system that is subject to constant torques, the time rate of change of the angular momentum is constant, so over a time interval Δt,

$$\sum \tau_{\text{ext}\,\vartheta} = \frac{\Delta L_\vartheta}{\Delta t}$$

$$\Delta L_\vartheta = \left(\sum \tau_{\text{ext}\,\vartheta}\right)\Delta t. \tag{12.16}$$

Comparing Eqs. 12.15 and 12.16, we see that the rotational impulse is

$$J_\vartheta = \left(\sum \tau_{\text{ext}\,\vartheta}\right)\Delta t \quad \text{(constant torques)}. \tag{12.17}$$

Equation 12.17, called the **rotational impulse equation,** allows us to calculate the change in angular momentum for a system subject to constant external torques. **Figure 12.32** illustrates how conservation of angular momentum gives rise to the angular momentum law and how to treat isolated and nonisolated systems. Notice that the structure of this figure is identical to that of Figure 9.16, which illustrates the same principles for momentum and energy.

✋ **12.8** Use the conditions of mechanical equilibrium to express $\vec{F}^c_{\text{hfy}}$ in Figure 12.11b in terms of $\vec{F}^G_{\text{Ef}}$ and $\vec{F}^c_{\text{bf}}$. Let the distance between the ball and the pivot be ℓ and the distance from the point where the biceps muscle is attached to the pivot be $\ell/5$.

Figure 12.32 Angular momentum change for isolated and nonisolated systems.

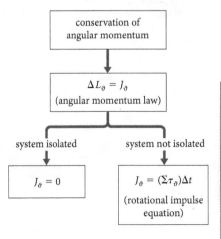

Example 12.5 Flywheel

A motor exerts a constant force of 120 N tangential to the rim of a 20-kg cylindrical flywheel of radius 0.50 m. The flywheel is free to rotate about an axis through its center and runs perpendicular to its face. If the flywheel is initially at rest and the motor is turned on for 2.0 s, how much work does the motor do on the flywheel?

❶ **GETTING STARTED** I begin by making a sketch of the situation to organize the information (**Figure 12.33**). The force exerted by the motor causes the flywheel to start spinning, which means the wheel's rotational kinetic energy changes. This is the only energy change in the system, and I know from the energy law (Eq. 9.1) that $\Delta E = W$. Therefore $\Delta K_{rot} = W$, and so to calculate the work done by the motor, I need to determine this change in rotational kinetic energy. Because the flywheel is at rest initially, I know that $\Delta K_{rot} = K_{rot,f}$.

Figure 12.33

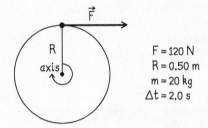

$F = 120$ N
$R = 0.50$ m
$m = 20$ kg
$\Delta t = 2.0$ s

❷ **DEVISE PLAN** To obtain $K_{rot,f}$, I can use Eq. 11.31, $K_{rot} = \frac{1}{2}I\omega^2$. The rotational inertia I of the flywheel (which is a solid cylinder) is $\frac{1}{2}mR^2$ (see Table 11.3). Because I'm interested in $K_{rot,f}$, I need the final value for ω, the wheel's rotational speed. How can I connect ω_f to anything I know in this problem? The relationship between ω and angular momentum is $L_\vartheta \equiv I\omega_\vartheta$ (Eq. 11.34), and I know from Eq. 12.16 that $\Delta L_\vartheta = \tau_\vartheta \Delta t$. I know Δt, but do I know anything about τ in terms of the information given—a force, an inertia, and a wheel radius? Yes, Eq. 12.1:

$\tau = rF_\perp$. Thus my plan is to express ω_f in terms of R and F and then use that expression for ω_f in $K_{rot} = \frac{1}{2}I\omega^2$ to calculate $K_{rot,f} = W$.

❸ **EXECUTE PLAN** The magnitude of the torque caused by the motor is $\tau = RF$, where R is the radius of the wheel and F is the magnitude of the force exerted by the motor. Equation 12.16 then gives

$$\Delta L_\vartheta = (\Sigma\tau_{ext\,\vartheta})\Delta t = +RF\Delta t.$$

Because the initial angular momentum is zero, I know that $\Delta L_\vartheta = I\omega_{\vartheta,f} = +I\omega_f$, and so I have

$$\Delta L_\vartheta = RF\Delta t = I\omega_f$$

$$\omega_f = \frac{RF\Delta t}{I}.$$

The final rotational kinetic energy is thus

$$K_{rot,f} = \frac{1}{2}I\omega_f^2 = \frac{1}{2}I\left(\frac{RF\Delta t}{I}\right)^2$$

$$= \frac{(RF\Delta t)^2}{2I} = \frac{(RF\Delta t)^2}{mR^2} = \frac{(F\Delta t)^2}{m},$$

and the work done on the flywheel is

$$W = K_{rot,f} = \frac{[(120\text{ N})(2.0\text{ s})]^2}{20\text{ kg}} = 2880\text{ J} = 2.9\text{ kJ.} ✔$$

❹ **EVALUATE RESULT** Delivering 2.9 kJ in 2.0 s corresponds to a power of $(2.9\text{ kJ})/(2.0\text{ s}) = 1.4$ kW, which is not an unreasonable amount for a large motor.

Example 12.6 Spinning up a compact disc

When you load a compact disc into a drive, a spinning conical shaft rises up into the opening in the center of the disc, and the disc begins to spin (**Figure 12.34**). Suppose the disc's rotational inertia is I_d, that of the shaft is I_s, and the shaft's initial rotational speed is ω_i. Does the rotational kinetic energy of the disc-shaft

Figure 12.34 Example 12.6.

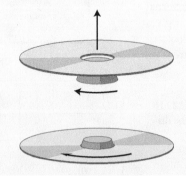

system remain constant in this process? Assume for simplicity that no external forces cause torques on the shaft.*

❶ **GETTING STARTED** I am given rotational inertias for a disc and a shaft, plus the initial rotational speed of the shaft, and my task is to determine whether or not the system's rotational kinetic energy changes when these two units interact: $\Delta K_{rot} \stackrel{?}{=} 0$. The disc is initially at rest, but as the shaft comes in contact with it, the two exert on each other forces that cause torques. The shaft speeds up the rotation of the disc, and the disc slows down the rotation of the shaft. The disc has no rotational velocity before the shaft touches it, and so initially all the system's rotational kinetic energy is in the shaft. After they reach a common rotational speed ω_f, both have rotational kinetic energy. I need to calculate the initial and final rotational kinetic energies of the shaft-disc system to answer the question.

---

*In reality, the shaft is connected to a motor that keeps it rotating at the correct rotational speed, and thus there is a force that causes a torque on the shaft.

② DEVISE PLAN The rotational kinetic energy of a rotating object is given by Eq. 11.31, $K = \frac{1}{2}I\omega^2$. I know the initial rotational speed of the shaft and its rotational inertia, so I can use this equation to calculate the initial rotational kinetic energy of the shaft-disc system. I do not know the final rotational speed of the system, but I do know that because there are no external torques on the system, Eq. 12.13 tells me that the angular momentum must remain constant: $\Delta L_\vartheta = 0$. Expressing ΔL_ϑ as the difference between the final and initial values gives me an expression containing ω_f and ω_i, which means I can probably get an expression for ω_f/ω_i that I can then use to compare the ratio $K_{rot,f}/K_{rot,i}$ and thereby determine whether or not $\Delta K_{rot} = 0$. Because the problem is stated in symbols rather than numerical values, my comparison will be between two algebraic expressions.

③ EXECUTE PLAN Because the torques that the disc and shaft cause on each other are internal and because there are no external torques, I have for the system's angular momentum

$$\Delta L_\vartheta = (I_s + I_d)\omega_{\vartheta,f} - I_s\,\omega_{\vartheta,i} = 0. \qquad (1)$$

If I let the initial direction of rotation of the shaft be positive, $\omega_{\vartheta,i} = +\omega_i$ and so $\omega_{\vartheta,f}$ is also positive. Rearranging terms

in Eq. 1, I find that the ratio of the final and initial rotational speeds is

$$\frac{\omega_f}{\omega_i} = \frac{I_s}{I_s + I_d}.$$

The system's initial rotational kinetic energy is $K_i = \frac{1}{2}I_s\,\omega_i^2$, its final rotational kinetic energy is $K_f = \frac{1}{2}(I_s + I_d)\omega_f^2$, and the ratio of the two is

$$\frac{K_f}{K_i} = \frac{\frac{1}{2}(I_s + I_d)}{\frac{1}{2}I_s}\frac{\omega_f^2}{\omega_i^2} = \frac{I_s}{I_s + I_d} < 1,$$

so $K_f < K_i$, or $\Delta K \neq 0$. The rotational kinetic energy of the system is not constant. ✔

④ EVALUATE RESULT The spinning up of the disc is like an inelastic "rotational collision": The disc initially at rest comes in contact with the spinning shaft, and the two reach a common rotational speed. While the disc is spinning up, some of the system's initial rotational kinetic energy is converted to thermal energy because of friction between disc and shaft, and so it makes sense that the system's rotational kinetic energy decreases.

12.9 Consider the situation in Example 12.6. (*a*) Is the vector sum of the forces exerted by the shaft on the compact disc nonzero while the disc is spinning up? (*b*) Is the disc isolated? ·

12.6 Rolling motion

The rotations we have considered so far fall into two classes: rotations about a fixed axis and free rotations. Forces exerted on objects that are constrained to rotate about a fixed axis cause rotational motion only because the axis prevents any translational motion. Forces exerted on freely rotating objects cause both translational motion of the center of mass and rotation about that point. These two motions are *independent* of each other. When you throw a Frisbee, for example, its translational speed v_{cm} is independent of its rotational speed ω (**Figure 12.35**). The Frisbee can have any combination of spin and translational speed: You can throw it so that it spins fast and moves away from you slowly, spins slowly and moves away from you fast, and so on.

The rolling motion of a wheel or ball represents an intermediate case between fixed and free rotations because rolling puts a constraint on the relationship between the translational and rotational motions. Furthermore, a rolling wheel or ball always rotates about its geometric center. We shall restrict our discussion to symmetric objects for which the center of mass is at the geometric center.

Figure 12.35 The translational and rotational motions of a thrown Frisbee are independent of each other.

Motion of flying, spinning disk . . .

. . . combines translation of center of mass with rotation around center of mass. In a free rotation, these two motions are *independent* of each other.

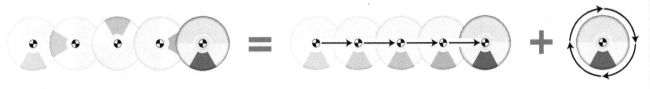

Earth reference frame

disk reference frame

Figure 12.36 When an object rolls without slipping, its translational and rotational motions are coupled.

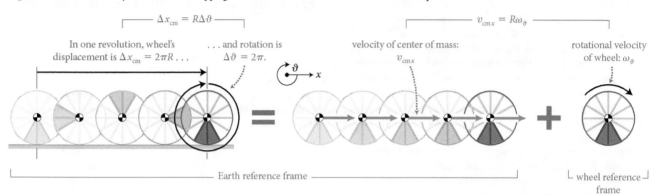

Consider the wheel of radius R rolling along a straight line on a flat surface in Figure 12.36. To describe this rolling motion we need to choose both a direction of rotation and a direction of the x axis. Usually it is advisable to choose these directions such that a positive change in ϑ corresponds to a positive change in x, as shown in Figure 12.36. If the wheel does not slip, its center advances a distance equal to its circumference, $2\pi R$, during each full revolution. Because the wheel's rotational displacement in one full revolution is 2π, the center-of-mass displacement Δx_{cm} caused by the rolling is related to the rotational displacement $\Delta\vartheta$ by

$$\Delta x_{cm} = R\Delta\vartheta. \tag{12.18}$$

Note that this expression is valid for any displacement; for example, for a quarter revolution $\Delta\vartheta = 2\pi/4 = \pi/2$ and $\Delta x_{cm} = 2\pi R/4 = (\pi/2)R = R\Delta\vartheta$. Dividing the left and right sides by the time interval Δt it takes to advance this distance and taking the limit of both sides for $\Delta t \to 0$ yield

$$v_{cmx} = R\omega_\vartheta \quad \text{(rolling motion without slipping).} \tag{12.19}$$

This equation expresses the **condition for rolling motion without slipping.** By coupling the translational and rotational speeds, this condition puts a constraint on the motion of any object that rolls without slipping. Unlike a Frisbee, whose (translational) speed is independent of its rotational speed, the speed of a moving car that is not skidding is proportional to the rotational speed of its wheels.

Equation 12.19 resembles Eq. 11.10 relating the translational and rotational speeds of an object in circular motion:

$$v_t = r\omega_\vartheta. \tag{12.20}$$

These two equations mean very different things, however. Equation 12.19 describes the translational motion of the wheel in the Earth reference frame, whereas Eq. 12.20 gives the tangential speed of any point on the rotating wheel in a reference frame moving along with the center of mass of the wheel. The velocity of a point on the wheel in the Earth reference frame is obtained by adding these two velocities, as shown in Figure 12.37. A point on the rim moves with a continuously

Figure 12.37 The trajectory of a point on the rim of a rolling wheel. Such a trajectory is called a *cycloid*.

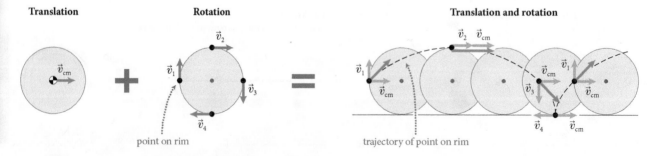

changing velocity along a trajectory resembling that of the handle of the wrench in Figure 12.9. Note in particular that whenever a point on the rim is in contact with the surface over which the wheel is rolling, that point has zero instantaneous velocity (on the rim $r = R$ and $v_t = r\omega_\vartheta$ points in the direction opposite the direction of $v_{cm} = R\omega$). This zero instantaneous velocity of the rim point in contact with the surface is a direct consequence of the requirement that the object rolls without slipping: The relative velocity of the two surfaces in contact must be zero for there to be no slipping. Figure 12.38 shows that points on the rim move in a direction perpendicular to the surface just before and just after reaching the bottom.

Let us next turn to the dynamics of rolling motion. Suppose a round object of inertia m and radius R is released from rest on a ramp as shown in Figure 12.39a. Why does the object roll down instead of sliding down as would a block of wood? The reason is that the force of static friction, which forces the instantaneous point of contact to be motionless, causes a torque about the center of the object. To understand the motion of the object, we must find the vector sum of the forces exerted on it and the sum of the torques caused by these forces. Figure 12.39b shows a free-body diagram for the object, which is subject to two forces: the force of gravity and a contact force exerted by the ramp, which has a normal component $\vec{F}_{ro}^n$ in the y direction and a tangential component $\vec{F}_{ro}^s$ in the x direction due to static friction. As we have done with other problems involving ramps, we chose the x axis to be along the ramp. For the object to accelerate down the ramp, the vector sum of the forces must point in the positive x direction, so $\sum F_y = F_{Eoy}^G + F_{ro}^n = 0$ and $\sum F_x = F_{Eox}^G - F_{ro}^s > 0$. The vector sum of the forces and the center-of-mass acceleration are related by

$$\sum F_x = F_{Eox}^G - F_{ro}^s = mg \sin\theta - F_{ro}^s = ma_{cm\,x}. \tag{12.21}$$

This equation contains two unknowns: the magnitude of the force of static friction F_{ro}^s and $a_{cm\,x}$. All we know about F_{ro}^s is that it must be less than or equal to its maximum value $(F_{ro}^s)_{max}$, but this fact doesn't help us solve Eq. 12.21. Let us therefore turn to the extended free-body diagram shown in Figure 12.39c, which shows the object with the forces exerted on it at their points of application. Because the object rotates about its center of mass, we determine the sum of the torques about the center of mass. Neither $\vec{F}_{Eo}^G$ nor $\vec{F}_{ro}^n$ causes a torque because their lines of action go through the center of mass, and so the lever arm distances for these forces are zero. Only $\vec{F}_{ro}^s$ causes a torque; its lever arm distance is the radius R of the object. Equation 12.10 then yields

$$\sum \tau_{ext\,\vartheta} = +F_{ro}^s R = I\alpha_\vartheta, \tag{12.22}$$

where I is the object's rotational inertia and α_ϑ is its rotational acceleration. Although this yields another equation, we have now added a third unknown: α_ϑ.

Figure 12.38 The point of contact of a wheel rolling without slipping has zero instantaneous velocity.

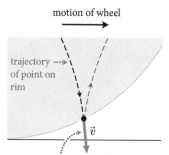

motion of wheel

trajectory ···▸ of point on rim

$\vec{v}$

Point on rim moves straight down just before touching road . . .

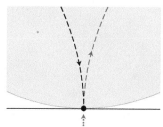

. . . has zero velocity the instant it touches road . . .

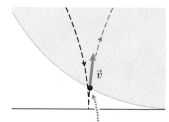

$\vec{v}$

. . . and moves straight up just after touching road.

Figure 12.39 A round object rolls down a ramp.

(a) Object rolls down ramp

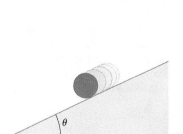

θ

(b) Free-body diagram

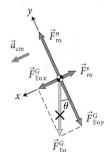

y

$\vec{F}_{ro}^n$

$\vec{a}_{cm}$

$\vec{F}_{Eox}^G$

$\vec{F}_{ro}^s$

x

θ

$\vec{F}_{Eoy}^G$

$\vec{F}_{Eo}^G$

(c) Extended free-body diagram

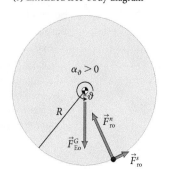

$\alpha_\vartheta > 0$

ϑ

R

$\vec{F}_{ro}^n$

$\vec{F}_{Eo}^G$

$\vec{F}_{ro}^s$

Because the object is rolling without slipping, its rotational and translational motions are coupled, and so its rotational and center-of-mass accelerations must be related. Differentiating both sides of Eq. 12.19, $v_{cmx} = R\omega_\vartheta$, with respect to time yields

$$a_{cmx} = R\alpha_\vartheta. \tag{12.23}$$

Substituting α_ϑ from Eq. 12.23 into Eq. 12.22 and solving for F_{ro}^s, we find

$$F_{ro}^s = \frac{I}{R^2}a_{cmx}. \tag{12.24}$$

Substituting this result for F_{ro}^s into Eq. 12.21 and solving for a_{cmx}, we obtain

$$a_{cmx} = +\frac{g\sin\theta}{1 + \dfrac{I}{mR^2}} = +\frac{g\sin\theta}{1 + c}, \tag{12.25}$$

where $c \equiv I/mR^2$ is the *shape factor* of the object ($0 < c \leq 1$; see Table 11.3). The numerator in Eq. 12.25 is the acceleration of an object sliding down a ramp with negligible friction (see Eq. 3.20). The denominator shows by what factor this acceleration is reduced for rolling motion. Because c is independent of m and R, Eq. 12.25 shows that the acceleration of the object does not depend on the values of m and R. Only the shape of the object matters because it affects the shape factor c. For all objects of a given shape, the time interval to roll down a ramp is thus the same. A large cylinder takes the same time interval as a small one, but a thin cylindrical shell takes longer than a solid cylinder because the shape factor of the shell is greater.

We can now also find an expression for the magnitude of the force of static friction $\vec{F}_{ro}^s$. Substituting a_{cmx} from Eq. 12.25 into Eq. 12.24 and eliminating the rotational inertia using $c \equiv I/mR^2$, which gives $I = cmR^2$, we obtain

$$F_{ro}^s = \frac{I}{R^2}a_{cmx} = \frac{cmR^2}{R^2}\frac{g\sin\theta}{1+c} = \frac{mg\sin\theta}{c^{-1}+1}. \tag{12.26}$$

Note that the force of static friction plays a dual role. It decreases the center-of-mass speed and acceleration of the rolling object by reducing the magnitude of the vector sum of the forces exerted on it: $\sum F_x = F_{Eox}^G - F_{ro}^s$. It also causes the torque that gives the object a rotational acceleration. In the absence of static friction, there would be no torque and objects would never roll—they would only slide.

Wheels certainly make it easier to push objects, but as Eq. 12.26 shows, they do not eliminate friction. In fact, the force of static friction to which wheels are subject slows them down, and as we have seen, the force of static friction can be *greater* than the force of kinetic friction. The next examples help clarify this paradox.

12.10 (*a*) A cylindrical shell and a solid cylinder, made of the same material and having the same inertia m and radius R, roll down a ramp. Is the force of static friction exerted on the shell greater than, smaller than, or equal to that on the solid cylinder? Explain. (*b*) For $\mu_s = 1$, what is the maximum angle of incline θ on which these objects can roll without slipping?

Example 12.7 Bicycle friction

As you accelerate from rest on a bicycle, how does the magnitude of the force of friction exerted by the road surface on the rear wheel compare with the magnitude of the force of friction exerted by the road surface on the front wheel? Ignore air resistance, assume both wheels have the same inertia m_w and the same radius R, and let m_{comb} be the combined inertia of you and your bicycle (including both wheels).

❶ GETTING STARTED I begin by making a sketch of my bicycle (Figure 12.40a). As I push on the pedals, the chain causes a torque on the rear wheel that makes the wheel rotate so that it exerts a rearward force on the road. Consequently the road surface exerts on the rear wheel a contact force in the opposite direction, and the tangential component of this force—the force of static friction—is what makes the bicycle go forward. This force, which I label $\vec{F}_{sr}^s$, is one of the forces I need for my comparison. (Note the subscripts here: r stands for "rear," not "road"; this is the frictional force exerted by the (road) surface on the rear wheel.)

The front wheel, because it is not connected to the chain, is merely pushed along with the rest of the bicycle. To make this wheel rotate in the right direction, the road surface must exert on it a rearward-pointing force of static friction $\vec{F}_{sf}^s$ (subscript f for "front" here). The torque caused by this frictional force is what makes the front wheel rotate, and this is the second force in my comparison.

Armed with this information, I make a free-body diagram for the bicycle (Figure 12.40b), showing the two frictional forces, the upward component of the contact force $\vec{F}_{sb}^n$ exerted by the road surface, and the downward force of gravity $\vec{F}_{Eb}^G$. Because the bicycle accelerates forward, I draw the vector arrow for $\vec{F}_{sr}^s$ longer than the vector arrow for $\vec{F}_{sf}^s$. Finally, I show the direction in which the bicycle's center of mass accelerates and choose a set of axes, letting the x axis point in the direction of motion. I also choose clockwise as the direction of increasing ϑ because that is the direction in which the wheels rotate.

Figure 12.40

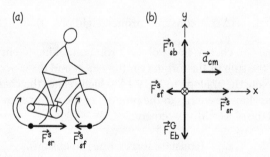

(a) (b)

❷ DEVISE PLAN The equation of motion in the form given by Eq. 8.45, $\vec{a}_{cm} = \Sigma\vec{F}_{ext}/m$, relates the vector sum of the forces shown in my free-body diagram to the acceleration $\vec{a}_{cm}$ of the bicycle's center of mass. Thus this equation gives me an expression that contains the two forces I'm after and seems a good place to begin. This expression contains the translational

acceleration, however, whereas the variables in this problem are rotational, so I will probably need Eq. 12.23 relating α_ϑ and a_{cmx}. Focusing on the front wheel first, I can use the rotational equation of motion, Eq. 12.5, $\tau_\vartheta = I\alpha_\vartheta$, to relate the torque caused by $\vec{F}_{sf}^s$ to α_ϑ for this wheel and then use Table 11.3 to express everything in terms of the given variables m_w and R. Once I have my expression for the magnitude F_{sf}^s, I can work with Eq. 8.45 to obtain an expression for F_{sr}^s in terms of F_{sf}^s and make my comparison.

❸ EXECUTE PLAN The vector sum of the forces in the x direction is, from Eq. 8.45

$$\Sigma F_x = F_{sr}^s - F_{sf}^s = m_{comb}\, a_{cmx}, \qquad (1)$$

so I have an expression that contains both of the forces I'm after.

According to Eq. 12.23, the rotational acceleration of the rear wheel is $\alpha_\vartheta = a_{cmx}/R$, and the rotational acceleration of the front wheel must be the same as that of the rear wheel. To give the front wheel rotational acceleration α_ϑ, the force $\vec{F}_{sf}^s$ must cause a torque $\tau_\vartheta = I\alpha_\vartheta$ on the wheel, where I is the wheel's rotational inertia. If I treat the wheel as a thin hoop of inertia m_w, I have, from Table 11.3, $I = m_w R^2$, which means the torque on the front wheel is $\tau_\vartheta = m_w R^2\alpha_\vartheta$. The lever arm distance for this torque is equal to the radius R of the wheel, and so for this wheel $\tau_\vartheta = F_{sf}^s R = m_w R^2\alpha_\vartheta$. I know that for both wheels $\alpha_\vartheta = a_{cmx}/R$, however, so I can write

$$F_{sf}^s R = (m_w R^2)\left(\frac{a_{cmx}}{R}\right).$$

Dividing both sides by R yields one of the two forces I must compare:

$$F_{sf}^s = m_w a_{cmx}. \qquad (2)$$

This is the magnitude of the force of static friction required to make the front wheel roll when the bicycle has acceleration a_{cmx}. The smaller the inertia of the wheel, the smaller this force magnitude.

Turning to the other force I need for my comparison—the force exerted on the rear wheel—I begin by eliminating a_{cmx} from Eq. 1 using Eq. 2 to get

$$F_{sr}^s - F_{sf}^s = \frac{m_{comb}}{m_w}F_{sf}^s$$

or

$$F_{sr}^s = F_{sf}^s\left(1 + \frac{m_{comb}}{m_w}\right) = F_{sf}^s\left(\frac{m_{comb} + m_w}{m_w}\right).$$

The magnitude of the force of static friction exerted on the rear wheel is greater than that exerted on the front wheel by the factor $(m_{comb} + m_w)/m_w$. ✔

❹ EVALUATE RESULT Because $m_{comb} \gg m_w$, the magnitude of the force of static friction exerted on the front wheel is much smaller than that on the rear wheel, as I expect.

Figure 12.41 A rigid object subject to a constant torque caused by a force $\vec{F}$ exerted on it undergoes a rotational displacement $\Delta\vartheta$.

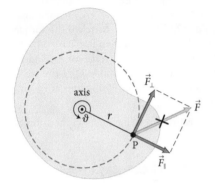

🖐 **12.11** (*a*) In Example 12.7, does the force of static friction exerted on the rear wheel cause a torque on the wheel? If so, in which direction? If not, why not? (*b*) In what direction is the sum of the torques on the rear wheel? What does this tell you about the relative magnitudes of the individual torques on the wheel?

12.7 Torque and energy

Torques cause objects to accelerate rotationally and thus cause a change in their rotational kinetic energy. To calculate this energy change, consider the object shown in Figure 12.41. A force $\vec{F}$ is exerted at point P on the object, a distance r from the axis of rotation. Let this force be such that the magnitude of its component perpendicular to r remains constant as the object rotates. This component thus causes a constant torque $\tau_\vartheta = +rF_\perp$ about the axis. Equation 12.10 relates the sum of the torques to the rotational acceleration. By applying the chain rule (Appendix B), we can write this equation in the form

$$\sum \tau_{\text{ext}\,\vartheta} = I\alpha_\vartheta = I\frac{d\omega_\vartheta}{dt} = I\frac{d\omega_\vartheta}{d\vartheta}\frac{d\vartheta}{dt} = I\frac{d\omega_\vartheta}{d\vartheta}\omega_\vartheta, \qquad (12.27)$$

so

$$\left(\sum \tau_{\text{ext}\,\vartheta}\right)d\vartheta = I\omega_\vartheta\,d\omega_\vartheta. \qquad (12.28)$$

Integrating the left side of this equation yields

$$\int_{\vartheta_i}^{\vartheta_f}\left(\sum\tau_{\text{ext}\,\vartheta}\right)d\vartheta = \left(\sum\tau_{\text{ext}\,\vartheta}\right)\int_{\vartheta_i}^{\vartheta_f}d\vartheta = \left(\sum\tau_{\text{ext}\,\vartheta}\right)\Delta\vartheta. \qquad (12.29)$$

Integrating the right side after pulling the (constant) rotational inertia I out of the integral gives

$$\int_{\omega_{\vartheta,i}}^{\omega_{\vartheta,f}} I\omega_\vartheta\,d\omega_\vartheta = I\int_{\omega_{\vartheta,i}}^{\omega_{\vartheta,f}}\omega_\vartheta\,d\omega_\vartheta = I\left[\tfrac{1}{2}\omega_\vartheta^2\right]_{\omega_{\vartheta,i}}^{\omega_{\vartheta,f}}$$

$$= \tfrac{1}{2}I\omega_{\vartheta,f}^2 - \tfrac{1}{2}I\omega_{\vartheta,i}^2 = K_{\text{rot,f}} - K_{\text{rot,i}}. \qquad (12.30)$$

Equations 12.28–12.30 thus show that the change in the rotational kinetic energy, $\Delta K_{\text{rot}} \equiv K_{\text{rot,f}} - K_{\text{rot,i}}$, is

$$\Delta K_{\text{rot}} = \left(\sum\tau_{\text{ext}\,\vartheta}\right)\Delta\vartheta \quad \text{(constant torques, rigid object).} \qquad (12.31)$$

This equation states that the change in an object's rotational kinetic energy is equal to the product of the sum of the torques on the object and the object's rotational displacement. Note the analogy to Eq. 9.14, which relates the change in an object's center-of-mass kinetic energy to the product of the vector sum of the forces exerted on it and the object's displacement:

$$\Delta K_{\text{cm}} = \left(\sum F_{\text{ext}}\right)\Delta x_{\text{cm}} \quad \text{(constant forces, one dimension).} \qquad (12.32)$$

The kinetic energy of an object or system that is in both translational and rotational motion is equal to the sum of its center-of-mass and rotational kinetic energies:

$$K = K_{\text{cm}} + K_{\text{rot}} = \tfrac{1}{2}mv_{\text{cm}}^2 + \tfrac{1}{2}I\omega^2, \qquad (12.33)$$

and the change in this kinetic energy is given by the sum of the changes in Eqs. 12.31 and 12.32:

$$\Delta K = \Delta K_{\text{cm}} + \Delta K_{\text{rot}}. \qquad (12.34)$$

Example 12.8 Rolling down a ramp

A solid cylindrical object of inertia m, rotational inertia I, and radius R rolls down a ramp that makes an angle θ with the horizontal. By how much does the cylinder's energy increase if it is released from rest and its center of mass drops a vertical distance h?

❶ **GETTING STARTED** I am given information about an object in the shape of a solid cylinder—inertia, rotational inertia, radius, and initial speed—and my task is to find out how much the object's energy has increased once it has rolled down a ramp such that its center of mass has traveled a vertical distance h. The object accelerates down the incline under the influence of the force of gravity. I therefore begin by making a sketch of the situation and drawing both free-body and extended free-body diagrams (Figure 12.42). The object is subject to a gravitational force exerted by Earth and a contact force exerted by the ramp. If I choose my axes as shown in my sketch, the contact force exerted by the ramp has a normal component $\vec{F}^n_{ro}$ in the y direction and a tangential component $\vec{F}^s_{ro}$ in the negative x direction due to static friction.

❷ **DEVISE PLAN** As the object rolls, both its translational and rotational kinetic energies increase. Because the shape of the object does not change and because static friction is nondissipative, the object's internal energy does not change. I can use Eq. 12.32 for the change in translational kinetic energy and Eq. 12.31 for the change in rotational kinetic energy. To express the two factors on the right in Eq. 12.32 in terms of my given variables, I use the geometry of the situation to express both factors in terms of $\sin \theta$.

❸ **EXECUTE PLAN** The change in translational kinetic energy is given by Eq. 12.32, $\Delta K_{cm} = (\sum F_{ext\,x})\Delta x_{cm}$, and the vector sum of the forces exerted on the object in the x direction is

$$\sum F_{ext\,x} = mg \sin \theta - F^s_{ro}$$

or, using Eq. 12.26,

$$\sum F_{ext\,x} = +mg \sin \theta \left(1 - \frac{1}{1 + c^{-1}}\right),$$

where c is the object's shape factor. Because the displacement of the object's center of mass along the plane is $\Delta x_{cm} = h/\sin \theta$, the change in its translational kinetic energy is

$$\Delta K_{cm} = (\sum F_{ext\,x})\Delta x_{cm} = mgh \left(1 - \frac{1}{1 + c^{-1}}\right).$$

Next, I use Eq. 12.31 to calculate the change in the object's rotational kinetic energy. From my extended free-body diagram I see that only the force of static friction causes a (positive) torque, so $\sum \tau_{ext\,\vartheta} = +F^s_{ro}R$. I can find the object's rotational displacement $\Delta \vartheta$ from Eq. 12.18:

$$\Delta \vartheta = \frac{\Delta x_{cm}}{R} = \left(\frac{h}{\sin \theta}\right)\left(\frac{1}{R}\right) = +\frac{h}{R \sin \theta},$$

so, from Eq. 12.31,

$$\Delta K_{rot} = (F^s_{ro}R)\left(\frac{h}{R \sin \theta}\right) = \left(\frac{mg \sin \theta}{1 + c^{-1}}\right)\left(\frac{h}{\sin \theta}\right) = \frac{mgh}{1 + c^{-1}},$$

where I have again used Eq. 12.26 to substitute for F^s_{ro}. Adding the two changes in kinetic energy, I obtain

$$\Delta E = \Delta K_{cm} + \Delta K_{rot} = mgh \left(1 - \frac{1}{1 + c^{-1}}\right)$$
$$+ mgh \left(\frac{1}{1 + c^{-1}}\right) = mgh. ✔$$

❹ **EVALUATE RESULT** My result indicates that the object's energy changes by the same amount it would change if it were simply in free fall! In other words, the only work done on the object is the work done by the gravitational force: $\vec{F}^G_{Eo} \cdot \Delta \vec{r} = mgh$ (see Section 10.9). This implies that the work done by all other forces on the object is zero. The normal force does no work on the object because it is perpendicular to the displacement of the object, but what is the work done by the force of static friction on the object? The object's displacement, $h/\sin \theta$, lies along the line of action of the force of static friction, and so it is tempting to write $-F^s_{ro}(h/\sin \theta)$ for the work done by the force of static friction on the object. However, the point of application for $\vec{F}^s_{ro}$ has zero velocity. At each instant, a different point on the object's surface touches the ramp, but the instantaneous velocity of that point is zero. The force displacement for $\vec{F}^s_{ro}$ is thus zero, so the work done by this force on the object is zero as well.

Figure 12.42

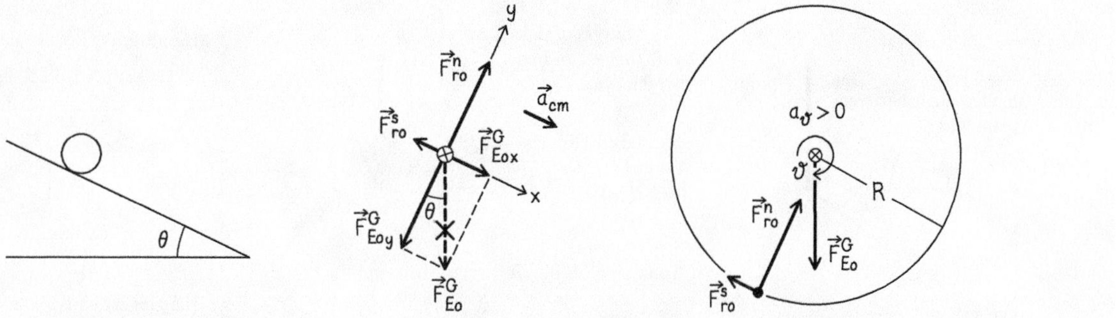

Figure 12.43 For a rotation about a fixed axis, the rotational velocity, rotational acceleration, and torque vectors all point along the rotation axis.

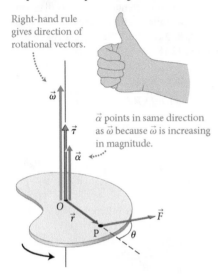

Right-hand rule gives direction of rotational vectors.

$\vec{\alpha}$ points in same direction as $\vec{\omega}$ because $\vec{\omega}$ is increasing in magnitude.

Example 12.8 shows that *the force of static friction does no work on a rolling object!* The only thing it does is take away some of the translational kinetic energy and generate an equal amount of rotational kinetic energy. It takes away translational kinetic energy by reducing the magnitude of the vector sum of the forces and converts it to rotational kinetic energy by causing a torque.

12.12 In Figure 12.41, does the force exerted on the rigid object do work on the object?

12.8 The vector product

To complete our treatment of rotation, we must examine the vectorial nature of the rotational components. As we saw in the first part of this chapter, an object's rotational velocity can be described by a vector $\vec{\omega}$. Angular momentum, which is the product of a scalar and $\vec{\omega}$, $\vec{L} = I\vec{\omega}$, is therefore also a vector and points in the same direction as $\vec{\omega}$. Because the rotational acceleration $\vec{\alpha}$ is the time derivative of $\vec{\omega}$, it must be a vector, too. Torque, which is the product of a scalar and $\vec{\alpha}$, $\vec{\tau} = I\vec{\alpha}$, therefore points in the same direction as $\vec{\alpha}$.

Consider the object in **Figure 12.43**. A force $\vec{F}$ exerted at point P increases the object's rotational velocity about the rotation axis at O. As the rotational velocity increases, the length of the vector representing $\vec{\omega}$ increases. Because the rotation takes place in a plane, the direction of $\vec{\omega}$ remains along the axis of rotation. The direction of $\vec{\alpha}$ is related to the direction of the vector representing the change in rotational velocity $\Delta\vec{\omega}$. Because the direction of $\vec{\omega}$ remains along the axis of rotation, $\Delta\vec{\omega}$, $\vec{\alpha}$, and $\vec{\tau}$ also lie along the rotation axis.

The magnitude of the torque is $rF \sin \theta$, where θ is the angle between $\vec{r}$ and $\vec{F}$. In Section 10.9, I showed that work on a system can be written as a scalar product of two vectors. To account for the vectorial nature of torque, we must introduce a new kind of product of two vectors: one that generates a new vector. Specifically, this product of the vectors $\vec{r}$ and $\vec{F}$ must generate a vector whose magnitude is $rF \sin \theta$ and whose direction is as indicated in **Figure 12.44**. Such a product is called the **vector product.** The vector product of two vectors $\vec{A}$ and $\vec{B}$, written $\vec{A} \times \vec{B}$, is a vector whose magnitude is

$$|\vec{A} \times \vec{B}| \equiv AB \sin \theta, \tag{12.35}$$

where θ is the angle between $\vec{A}$ and $\vec{B}$ when they are placed tail to tail (and $\theta \le 180°$). The direction of the vector $\vec{A} \times \vec{B}$ is determined as illustrated in Figure 12.44a: Place vectors $\vec{A}$ and $\vec{B}$ with their tails together and apply the right-hand rule to the direction of rotation obtained by rotating $\vec{A}$ into $\vec{B}$ through the

Figure 12.44 The vector product of two vectors.

(a) Vector product $\vec{A} \times \vec{B}$

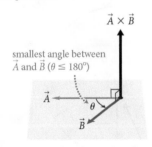

smallest angle between $\vec{A}$ and $\vec{B}$ ($\theta \le 180°$)

(b) Finding the direction of a vector product

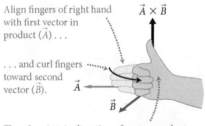

Align fingers of right hand with first vector in product ($\vec{A}$) . . .

. . . and curl fingers toward second vector ($\vec{B}$).

Thumb points in direction of vector product.

(c) Magnitude of a vector product

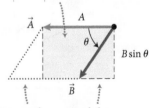

Magnitude of $\vec{A} \times \vec{B}$ equals area of rectangle: $|\vec{A} \times \vec{B}| = AB \sin \theta$.

$B \sin \theta$

Triangles are equal, so rectangle has same area as dotted parallelogram.

smallest angle. When you point the fingers of your right hand along the direction of the first vector in the product and curl them toward the direction of the second vector, your thumb points in the direction of the vector product (Figure 12.44b).

Figure 12.44c shows a geometric interpretation of the magnitude of the vector product. The area of the shaded rectangle is equal to $AB \sin \theta$, and because the two triangles are identical, the area of the parallelogram defined by the vectors $\vec{A}$ and $\vec{B}$ is equal to the area of the shaded rectangle. Therefore:

The magnitude of the vector product of two vectors is equal to the area of the parallelogram defined by them.

The magnitude of the vector product is greatest when the two vectors are perpendicular to each other and $\sin \theta = \sin 90° = 1$. When the two vectors are parallel, the parallelogram is reduced to a straight line and the vector product is zero. In particular, the vector product of a vector with itself is zero:

$$\vec{A} \times \vec{A} = \vec{0}. \tag{12.36}$$

Another interesting property of the vector product is that the multiplication sequence is important. The vector product $\vec{B} \times \vec{A}$ is equal in magnitude to $\vec{A} \times \vec{B}$ but points in the opposite direction, as shown in Figure 12.45:

$$\vec{B} \times \vec{A} = -(\vec{A} \times \vec{B}). \tag{12.37}$$

Using the definition of vector product, we can write the torque about the origin caused by a force $\vec{F}$ as

$$\vec{\tau} = \vec{r} \times \vec{F}. \tag{12.38}$$

You can easily verify that this product has the correct magnitude and direction.

Because torque is the time derivative of angular momentum (Eq. 12.12) and force is the time derivative of momentum (Eq. 8.4), we can write Eq. 12.38 as

$$\frac{d\vec{L}}{dt} = \vec{r} \times \frac{d\vec{p}}{dt}. \tag{12.39}$$

To see what this expression means, we use the product rule (Appendix B) to find the time derivative of the vector product $\vec{r} \times \vec{p}$:

$$\frac{d}{dt}(\vec{r} \times \vec{p}) = \frac{d\vec{r}}{dt} \times \vec{p} + \vec{r} \times \frac{d\vec{p}}{dt}. \tag{12.40}$$

The first term on the right side vanishes because the vector product of a vector and itself is zero:

$$\frac{d\vec{r}}{dt} \times \vec{p} = \vec{v} \times (m\vec{v}) = m(\vec{v} \times \vec{v}) = 0. \tag{12.41}$$

Therefore the right side of Eq. 12.39 is equal to the time derivative of the vector product $\vec{r} \times \vec{p}$:

$$\frac{d\vec{L}}{dt} = \frac{d}{dt}(\vec{r} \times \vec{p}). \tag{12.42}$$

The angular momentum vector can therefore be written as

$$\vec{L} = \vec{r} \times \vec{p}. \tag{12.43}$$

Figure 12.45

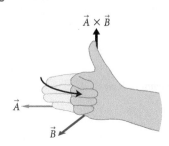

$\vec{A} \times \vec{B}$

$\vec{A}$

$\vec{B}$

Vector products $\vec{A} \times \vec{B}$ and $\vec{B} \times \vec{A}$ point in opposite directions.

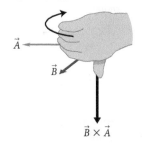

$\vec{A}$

$\vec{B}$

$\vec{B} \times \vec{A}$

Figure 12.46 The angular momentum of an isolated particle about an arbitrary point (called origin O here) is constant because $r \sin \theta = r_\perp$ is the same regardless of the position of the particle.

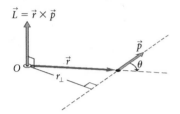

Like the rotational velocity vector, $\vec{L}$ is perpendicular to the plane defined by $\vec{r}$ and $\vec{p}$, a direction in which nothing moves—there is certainly no momentum in this direction.

Because it applies to any type of motion and not just rotational motion, the expression $\vec{L} = \vec{r} \times \vec{p}$ is much more general than $\vec{L} = I\vec{\omega}$. Consider, for example, the motion of the isolated particle in **Figure 12.46**. Because the particle is isolated, its trajectory is a straight line and its momentum is constant. From Eq. 12.35 we see that the magnitude of the particle's angular momentum with respect to the origin O is

$$L = r\,mv\,\sin\theta = r_\perp mv, \tag{12.44}$$

where θ is the angle between $\vec{r}$ and $\vec{p}$ and $r_\perp$ is the lever arm distance of $\vec{p}$ relative to O. This is the same result we obtained in Chapter 11 (Eq. 11.36). The right side of Eq. 12.44 is indeed constant. Even though $\vec{r}$ changes in direction and magnitude, the lever arm distance $r_\perp$ remains constant. Consequently the product $\vec{r} \times \vec{p}$, perpendicular to the plane defined by O and the particle's trajectory, is unchanged in direction and magnitude.

Table 12.1 Translational and rotational dynamics

	Translation	Rotation	Relationship
Momentum	$\vec{p} \equiv m\vec{v}$	$\vec{L} \equiv I\vec{\omega}$	$\vec{L} = \vec{r} \times \vec{p}$
Momentum change	$\Delta\vec{p} = \Sigma\vec{F}_{\text{ext}}\Delta t$	$\Delta\vec{L} = \Sigma\vec{\tau}_{\text{ext}}\Delta t$	
Force	$\Sigma\vec{F}_{\text{ext}} \equiv \dfrac{d\vec{p}}{dt} = m\vec{a}_{\text{cm}}$	$\Sigma\vec{\tau}_{\text{ext}} = \dfrac{d\vec{L}}{dt} = I\vec{\alpha}$	$\vec{\tau} = \vec{r} \times \vec{F}$
Kinetic energy	$K_{\text{cm}} = \frac{1}{2}mv^2$	$K_{\text{rot}} = \frac{1}{2}I\omega^2$	
Kinetic energy change	$\Delta K_{\text{cm}} = (\Sigma\vec{F}_{\text{ext}}) \cdot \Delta\vec{r}_{\text{cm}}$	$\Delta K_{\text{rot}} = (\Sigma\tau_{\text{ext}\,\vartheta})\Delta\vartheta$	

Table 12.2 Corresponding translational and rotational quantities

position	$\vec{r}$	$\leftrightarrow$	ϑ
velocity	$\vec{v}$	$\leftrightarrow$	$\vec{\omega}$
acceleration	$\vec{a}$	$\leftrightarrow$	$\vec{\alpha}$
inertia	m	$\leftrightarrow$	I

Table 12.1 summarizes the results obtained in this chapter and the remarkable parallels between the equations that describe the dynamics of translational and rotational motion. The two sets of equations are mathematically identical given the correspondences in **Table 12.2**. The only new mathematical tool we have introduced in this chapter is the vector product. Apart from this product, which is required to account for the vectorial nature of rotation, the mathematical framework for the two types of motion is identical. Finally, note that—unlike their translational counterparts—rotational inertia, angular momentum, and torque depend on the choice of origin and axis. You should therefore always specify your choice of axis when dealing with these quantities and—most important—consistently use the same axis for the different quantities throughout your calculations.

12.13 Consider a particle moving at constant speed along a circular trajectory centered on the origin. Write the relationship of the particle's velocity $\vec{v}$, rotational velocity $\vec{\omega}$, and position vector $\vec{r}$ in the form of a vector product.

Chapter Glossary

SI units of physical quantities are given in parentheses.

Angular momentum law The law that accounts for the change in the angular momentum of a system or object. Because angular momentum is conserved, the angular momentum of a system can change only if there is a transfer of angular momentum between the environment and the system:

$$\Delta L_\vartheta = J_\vartheta. \tag{12.15}$$

Axial vector A vector associated with a direction of rotation. The vector points along the rotation axis, with its direction given by the right-hand rule.

Condition for rolling motion without slipping The translational and rotational motions of a round object of radius R that rolls along a surface without slipping are coupled. The center-of-mass velocity and the rotational velocity of a rolling object are related by

$$v_{cm\,x} = R\omega_\vartheta. \tag{12.19}$$

Rolling motion occurs because of the force of static friction between the rolling object and the surface on which it moves. This force of static friction does no work on the object.

Extended free-body diagram A sketch showing a single object with all the forces exerted on it at their points of application. See the Procedure box on page 7.

Free rotation A rotation that is not constrained by a physical axis or by other external constraints. Freely rotating objects rotate about their center of mass. Their translational and rotational motions are independent of each other—the first determined by the vector sum of the forces exerted on the object and the second by the sum of the torques caused by the forces exerted on it.

Mechanical equilibrium Any object that is simultaneously in translational equilibrium and in rotational equilibrium is in mechanical equilibrium.

Polar vector A vector associated with a displacement or derived from it.

Right-hand rule The rule that relates the direction of a vector to a direction of rotation: When you curl the fingers of your right hand along the direction of rotation, your outstretched thumb points in the direction of the vector that specifies the direction of rotation.

Rotational equilibrium Any object or system whose angular momentum is not changing is in rotational equilibrium. This state requires the sum of the torques on the object to be zero:

$$\sum \vec{\tau}_{ext} = \vec{0} \implies \Delta \vec{L} = \vec{0} \quad \text{(rotational equilibrium).} \tag{12.13}$$

Rotational impulse J_ϑ (kg·m²/s) The amount of angular momentum transferred from the environment to a system.

Rotational impulse equation The equation that allows us to calculate the change in a system's angular momentum caused by torques on the system during a time interval Δt. For constant torques;

$$J_\vartheta = \left(\sum \tau_{ext\,\vartheta}\right)\Delta t \quad \text{(constant torques).} \tag{12.17}$$

Torque $\vec{\tau}$ (N·m) An axial vector that describes the tendency of a force to give an object a rotational acceleration. The torque caused by a force $\vec{F}$ exerted on an object about a point O is

$$\vec{\tau} = \vec{r} \times \vec{F}, \tag{12.38}$$

where $\vec{r}$ is the vector pointing from O to the point of application of $\vec{F}$. The direction of the torque is given by the right-hand rule; the torque magnitude is

$$\tau \equiv rF\sin\theta = r_\perp F = rF_\perp. \tag{12.1}$$

Vector product The product of two vectors that forms a new vector. The vector product of vectors $\vec{A}$ and $\vec{B}$ that make an angle θ is written as $\vec{A} \times \vec{B}$. This product is another vector whose magnitude is equal to the area $AB\sin\theta$ of the parallelogram defined by $\vec{A}$ and $\vec{B}$ and whose direction can be obtained by applying the right-hand rule to the direction of rotation obtained by rotating $\vec{A}$ into $\vec{B}$ through the smallest angle.

13

Gravity

The motion of celestial bodies has fascinated human-kind since antiquity. What keeps Earth's Moon, the planets, and other heavenly objects moving along their orbits? Motivated largely by aesthetics, Greek philosophers proposed that celestial bodies move in circles because they believed that circular motion was the most "natural" and the most perfect form of motion. As we have seen, however, objects travel in a straight line when they are not subject to any external forces. In the early 17th century, the German astronomer Johannes Kepler suggested that the Sun was somehow driving the planets around their orbits.

If the laws of physics are truly universal, then phenomena on a cosmic scale should follow the same laws that govern objects around us. In this chapter we see that we can indeed apply our notions of mechanics to describe the motion of celestial bodies.

13.1 Universal gravity

Careful analysis of the motion of the Moon and the planets shows that the Moon revolves around Earth and that Earth and the other planets revolve around the Sun at roughly constant speed in nearly circular orbits.* In other words, to a good approximation they move in circular motion at constant speed. These bodies generally rotate about an axis through their centers in addition to revolving around the Sun, but in this chapter we are concerned with their *orbital* motion—that is, rotational quantities associated with the revolving motion of objects.

As we saw in Chapter 11, any object that is in circular motion at constant speed has a centripetal acceleration. As illustrated in **Figure 13.1**, in the reference frame of the Sun,

Earth has an acceleration that points toward the Sun; in the reference frame of Earth, the Moon has an acceleration that points toward Earth.

To get some feel for the quantities involved in celestial mechanics, do the following checkpoint before proceeding.

13.1 The orbital period of the Moon around Earth is 27.32 days; that of Earth around the Sun is 365.26 days. (*a*) Which has the greater rotational speed: Earth or the Moon? Are the rotational speeds of Earth and the Moon around the Sun great or small relative to the rotational speeds you encounter in daily life? (*b*) The radius of the Moon's orbit around Earth is $R_M = 3.84 \times 10^8$ m; that of Earth's orbit around the Sun is $R_E = 1.50 \times 10^{11}$ m. Which has the greater speed? (*c*) Calculate the centripetal acceleration of Earth in the reference frame of the Sun and the centripetal acceleration of the Moon in the reference frame of Earth. (*d*) How do these accelerations compare with the acceleration due to gravity near Earth's surface?

As we saw in Chapter 11, some force is required to supply a centripetal acceleration. All examples we have considered so far involve contact forces. Neither the Moon nor Earth is in contact with anything, however, and so the force that holds them in orbit must be a field force. In the late 17th century, Isaac Newton made the bold proposal that the force that holds celestial bodies in orbit is a gravitational force—the same force that causes objects near Earth's surface to fall (**Figure 13.2**). Every piece of matter in the universe—feather or planet—attracts every other piece of matter. In other words, a gravitational attraction between Earth and the Moon is responsible for the centripetal acceleration of the Moon, just as a gravitational attraction between a terrestrial object and Earth is responsible for the free fall of that object. The reason the Moon does not fall to Earth like a freely falling ball is that the Moon has just the right speed to keep moving in a circle around Earth.

Figure 13.1 Both Earth and the Moon have nearly circular orbits and thus have a centripetal acceleration.

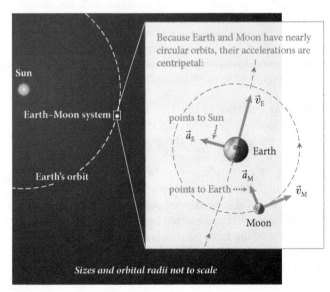

Because Earth and Moon have nearly circular orbits, their accelerations are centripetal:

$\vec{v}_E$
points to Sun
$\vec{a}_E$
Earth
$\vec{a}_M$
points to Earth
$\vec{v}_M$
Moon

Sizes and orbital radii not to scale

Sun
Earth–Moon system
Earth's orbit

Figure 13.2 The invisible tether that holds the Moon in orbit is, in reality, the same kind of gravitational interaction that causes a ball to fall near Earth's surface. The Moon's orbit is not to scale.

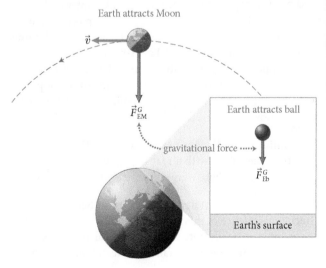

Earth attracts Moon
$\vec{v}$
$\vec{F}^G_{EM}$
gravitational force
Earth attracts ball
$\vec{F}^G_{Eb}$
Earth's surface

*Earth, for example, moves along an orbit that, to the eye, is indistinguishable from a circle. The speed of Earth along its orbit is constant to within 3%.

Figure 13.3 Light that strikes a surface area a^2 located a distance r from the source strikes a surface of area $2a \times 2a = 4a^2$ at a distance $2r$ and a surface of area $3a \times 3a = 9a^2$ at a distance $3r$. The intensity—the amount of light per unit area—decreases with distance r as $1/r^2$.

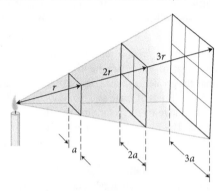

Figure 13.4 (a) A solid sphere exerts a gravitational force as if all of the sphere's matter were concentrated at the center of the sphere. (b) Because the magnitude of the gravitational force between two objects depends on the distance between them, the gravitational forces exerted by the various "chunks" of Earth's interior represented by δm have different magnitudes.

(a) Gravitational interaction between Earth and ball near Earth's surface

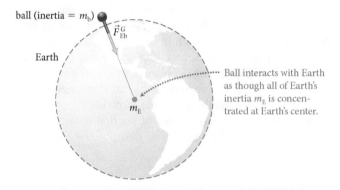

(b) Reason gravitation acts as though from Earth's center

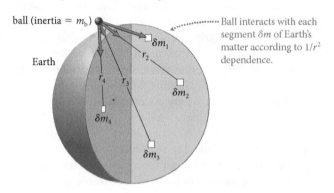

Newton furthermore postulated that the effect of gravity weakens with distance, much as the effect of a magnet weakens with distance. If gravity spreads out uniformly from an object, as the rays from a light source do, then the effect of gravity should fall off as $1/r^2$—the inverse of the square of the distance (**Figure 13.3**). The radius of the Moon's orbit is 60 times the radius of Earth, so the gravitational force exerted by Earth on the Moon should be smaller by a factor of $(\frac{1}{60})^2$ than the gravitational force exerted by Earth on an object near the ground. An object at the same distance from Earth as the Moon should therefore experience an acceleration due to gravity of magnitude $\frac{1}{3600}g = (9.8 \text{ m/s}^2)/3600 = 0.00272 \text{ m/s}^2$. This value is in excellent agreement with the Moon's centripetal acceleration (see Checkpoint 13.1), confirming the $1/r^2$ dependence of the gravitational force.

There is one catch in this reasoning: The dilution of the gravitational force by a factor of $1/r^2$ assumes that the object near Earth's surface is one Earth radius away from the point of attraction. In other words, a ball falling near the surface of Earth interacts with Earth as if all the matter were concentrated at its center—6378 km below the surface—as illustrated in **Figure 13.4a**. Yet if all matter interacts gravitationally with a $1/r^2$ dependence, then a piece of Earth-matter near the surface of Earth attracts the ball much more strongly than does a piece of Earth-matter deep inside or on the opposite side of Earth (Figure 13.4b). It is certainly not obvious that the forces exerted by all the different pieces of Earth add up to give the same result as in Figure 13.4a. Newton, however, proved mathematically that:

A uniform solid sphere exerts a gravitational force outside the sphere with a $1/r^2$ dependence as if all the matter in the sphere were concentrated at its center.

The proof of this statement is given in Section 13.8. A direct consequence of this statement is that the distance between gravitationally interacting spherical objects should always be taken to be the distance between their centers.

13.2 If the force of gravity decreases with the inverse square of the distance, why were we allowed, in all our earlier work on the gravitational force, to say that an object sitting on the ground, an object sitting in a tree 10 m above the ground, and an object flying at an altitude of 10 km all experience the same 9.8-m/s² acceleration due to gravity?

Further evidence of the $1/r^2$ dependence of the gravitational force is provided by the observed relationship between the radii R and the periods T of the planetary orbits. As it orbits the Sun, each planet has centripetal acceleration $a_c = v^2/R$, where $v = 2\pi R/T$ is the planet's linear speed and R is the radius of its orbit. So $a_c = 4\pi^2 R/T^2$ or $a_c \propto R/T^2$.* If this centripetal acceleration is provided by a gravitational attraction exerted by the Sun on the planet, the magnitude of the force (and hence the centripetal acceleration) should fall off as $1/R^2$. So, the $1/r^2$ dependence of the gravitational force requires that $a_c \propto 1/R^2$. Combining this

---

*The symbol $\propto$ means "is proportional to." That is, $a_c \propto R/T^2$ means that there is some number k such that $a_c = kR/T^2$.

Figure 13.5 The planetary period T versus the radius R of the orbit. The data are plotted on a logarithmic grid to accommodate the large range of values. To convince yourself of the $T^2 \propto R^3$ dependence, compare the values of T at $R = 1$ Earth radius and $R = 100$ Earth radii.

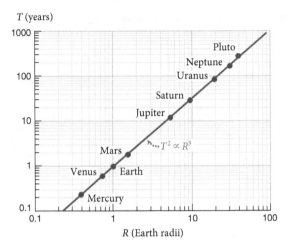

result with $a_c \propto R/T^2$, we get $1/R^2 \propto R/T^2$, or $T^2 \propto R^3$. In words:

The square of the period of a planetary orbit is proportional to the cube of the orbit's radius.

Kepler had established this fact from astronomical data in 1619, providing support for the $1/r^2$ dependence of the gravitational force proposed later by Newton. The excellent agreement with astronomical observations is shown in **Figure 13.5**.

✋ **13.3** Suppose the universe were two dimensional rather than three dimensional. (*a*) Following the line of reasoning illustrated in Figure 13.3, describe how the strength of the gravitational force would depend on distance in this flat universe. (*b*) How would the periods of the planetary orbits be related to their radii in this universe?

What determines the strength of the gravitational attraction? To answer this question, consider two identical lumps of clay, each of inertia m. If we let the lumps fall freely, each is subject to a single downward force of gravity and undergoes an acceleration g. To give each lump this acceleration, Earth must exert a force mg on it. Now imagine combining the two lumps. The double lump has twice the inertia, which means it is twice as hard to accelerate as either single lump. Yet, like any other object near Earth's surface, the double lump undergoes an acceleration g. From this observation, we conclude that Earth pulls twice as hard on the double lump as it does on either single lump. Apparently the gravitational pull on an object is proportional to the quantity of material in it, a quantity called the **mass** of the object. The double lump has twice the mass of either single lump.

The gravitational pull exerted by Earth on an object is proportional to the object's mass.

The acceleration of a freely falling object near Earth's surface does not depend on the type of material: In the absence of air resistance, all objects, regardless of their composition, configuration, or any other property of the object, have the same acceleration in free fall. This observation suggests that the inertia of an object and the force of gravity exerted on it are proportional to each other. Given that the force of gravity is proportional to the object's mass, we can consider mass and inertia to be equal and let the SI unit of mass be the kilogram (kg). As we shall see in the next chapter, this equality between mass and inertia breaks down for motion at very high speeds. In everyday situations, however, we can say:

The mass of an object is equal to the object's inertia.

For the description of everyday situations, therefore, we can denote both by the same symbol m and express each in units of kilograms. Unless we need to refer explicitly to an object's resistance to a change in velocity or account for effects that occur at very high speeds, we henceforth shall call both simply the *mass* of the object.

The equality of inertia and mass may seem obvious, but consider the two experiments shown in **Figure 13.6**. The collision experiment in Figure 13.6*a* allows us to determine the ratio of the objects' inertias. Now suppose we suspend the same two objects, from a beam that is supported at its center of mass. We arrange the objects' suspension points so that the beam is balanced as shown in Figure 13.6*b*. Experimentally we discover that for *any* two objects, the ratio m_1/m_2 determined in the first experiment is exactly equal to the ratio r_2/r_1 determined in the second experiment. As we saw in Chapter 12, when the beam is balanced, the torques on the beam add up to zero, so $r_1 F_{E1}^G = r_2 F_{E2}^G$, which means that

$$\frac{F_{E1}^G}{F_{E2}^G} = \frac{r_2}{r_1}.$$

Figure 13.6 Establishing the equivalence between inertia and mass.

(*a*) Collision between objects gives ratio of inertias

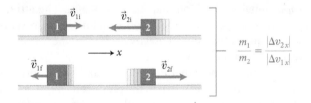

(*b*) Beam in equilibrium gives ratio of gravitational forces

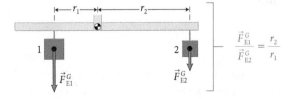

Figure 13.7 Tracing an ellipse. The greater the separation of the foci, the more eccentric (elongated) the ellipse ($0 < e < 1$). In the special case where the two foci coincide ($e = 0$), the ellipse becomes a circle of radius a.

(*a*) Drawing an ellipse with two pins and a string

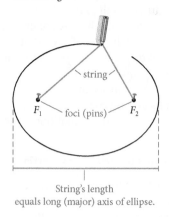

String's length equals long (major) axis of ellipse.

(*b*) Eccentricity of ellipse

Eccentricity *e* is ratio of distance between foci to long (major) axis of ellipse.

When foci coincide, $e = 0$ and ellipse is a circle.

(*c*) Axes of ellipse

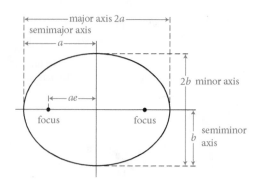

Combining this with the experimental observation that

$$\frac{m_1}{m_2} = \frac{r_2}{r_1},$$

we see that

$$\frac{F^G_{E1}}{F^G_{E2}} = \frac{m_1}{m_2}.$$

In words, for any two objects, the ratio of the gravitational forces they experience is equal to the ratio of their inertias. If you think about it, there is no reason to expect this result. The two experiments are completely different: The collision measures the objects' tendencies to resist a change in velocity—their *inertias*—and involves no gravity; the balancing experiment measures the gravitational pull on the objects—their *mass* — and involves no motion. This equivalence is therefore truly remarkable.

Because the gravitational force exerted by Earth on an object is proportional to that object's mass, we can expect the gravitational force exerted by Earth on the Moon to be proportional to the Moon's mass: $F^G_{EM} \propto m_M$. The attraction between Earth and the Moon is mutual, and so symmetry requires the force of gravity exerted by the Moon on Earth to be proportional to the mass of Earth: $F^G_{ME} \propto m_E$. In addition, we know that $F^G_{EM} = F^G_{ME}$, and so the force of gravity must be proportional to the product $m_M m_E$.

Combining this conclusion with the $1/r^2$ dependence, we can now write for the gravitational attraction between any two material objects of masses m_1 and m_2 separated by a distance r:

$$F^G_{12} \propto \frac{m_1 m_2}{r^2}.$$

This is Newton's **law of gravity.** It applies to all the mass in the universe: Every piece of matter attracts every other piece of matter in the universe according to this law.

The $1/r^2$ dependence also explains the shape of the planetary orbits. Using conservation of energy and angular momentum, we can show that the orbit of a body moving under the influence of gravity must be an ellipse (**Figure 13.7**), a circle, a parabola, or a hyperbola. A comet, for example, moves in an elliptical orbit with the Sun at one focus of the ellipse. The planets also move in elliptical orbits with the Sun at one focus. As you can see from the data in **Table 13.1**, however, the planetary orbits are close to circular, even for dwarf planet Pluto, whose orbit has the greatest eccentricity of all. Even though the effect of the eccentricity on the shape of the orbit is minimal, its effect on the speed is considerable. The Sun is appreciably off-center for Pluto's orbit, and so the distance to the Sun and the speed vary as the planet moves between the closest point to the Sun, the *perihelion*, and the farthest point, the *aphelion* (**Figure 13.8**).

Figure 13.8 In spite of having the greatest eccentricity in Table 13.1, Pluto's orbit (dashed line) is remarkably close to a circle (solid line). Because the Sun is off-center, Pluto's speed varies appreciably along the orbit.

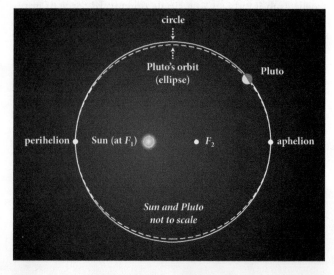

Table 13.1 Solar system data (in SI units and relative to Earth)

| | Mass | | Equatorial radius | | Orbit[†] | | | | |
	(kg)	(m_E)	(m)	(R_E)	semimajor axis (m)	(a_E)	eccentricity	period (s)	(years)
Sun	2.0×10^{30}	3.3×10^5	7×10^8	110	–	–	–	–	–
Mercury	3.30×10^{23}	0.06	2.440×10^6	0.38	5.79×10^{10}	0.39	0.206	7.60×10^6	0.24
Venus	4.87×10^{24}	0.81	6.052×10^6	0.95	1.082×10^{11}	0.72	0.007	1.94×10^7	0.62
Earth	5.97×10^{24}	1	6.378×10^6	1	1.496×10^{11}	1	0.017	3.16×10^7	1
Mars	6.42×10^{23}	0.11	3.396×10^6	0.53	2.279×10^{11}	1.52	0.09	5.94×10^7	1.88
Jupiter	1.90×10^{27}	318	7.149×10^7	11.2	7.783×10^{11}	5.20	0.05	3.74×10^8	11.86
Saturn	5.68×10^{26}	95.2	6.027×10^7	9.45	1.427×10^{12}	9.54	0.05	9.29×10^8	29.45
Uranus	8.68×10^{25}	14.5	2.556×10^7	4.01	2.871×10^{12}	19.2	0.05	2.65×10^9	84.02
Neptune	1.02×10^{26}	17.1	2.476×10^7	3.88	4.498×10^{12}	30.1	0.01	5.20×10^9	164.8
Pluto	1.31×10^{22}	0.002	1.151×10^6	0.18	5.906×10^{12}	39.5	0.25	7.82×10^9	247.9
Moon	7.3×10^{22}	0.012	1.737×10^6	0.27	3.84×10^8	0.0026	0.055	2.36×10^6	0.075

[†]The elliptical orbits of the planets and the Moon are specified by their *semimajor axis a* (half the major axis) and eccentricity *e*; see Figure 13.7. With the exception of Mercury and Pluto, the eccentricity is small and so the orbits are close to being circular.

13.4 Suppose the force of gravity between two objects 1 and 2 of masses m_1 and m_2 were proportional to the sum $m_1 + m_2$ rather than to the product $m_1 m_2$. (a) Would this dependence be consistent with the following two requirements: (i) The force exerted by 1 on 2 is equal in magnitude to the force exerted by 2 on 1. (ii) The force of gravity exerted by Earth on two identical objects is equal to the sum of the forces exerted on the individual objects. (b) Can you think of any other combination of m_1 and m_2 that satisfies requirements (i) and (ii)? (c) Suppose you are given a lump of matter of mass M. How should you divide this lump into two parts to maximize the force of gravity between the two parts?

Example 13.1 Comparing gravitational pulls

Compare the gravitational force exerted by Earth on you with (a) that exerted by a person standing 1 m away from you and (b) that exerted by Earth on Pluto.

❶ **GETTING STARTED** Although I don't know yet how to calculate a numerical value for the gravitational force, I know that the gravitational force between two objects with masses m_1 and m_2 separated by a distance r is proportional to the factor $m_1 m_2/r^2$. To compare gravitational forces, I should therefore compare these factors for each given situation.

❷ **DEVISE PLAN** My mass is about 70 kg, and I can get the masses of Earth and Pluto from Table 13.1. When I stand on the surface of Earth, the distance between me and the center of the planet is its radius, which is given in Table 13.1. To do part b, I need to know the distance between Pluto and Earth, which is not given in the table. Planetary orbits are very nearly circular, however, so I can consider the semimajor axis a to be the radius of each (nearly) circular orbit. Table 13.1 shows that the "radius" of Pluto's orbit is about 40 times greater than that of Earth's orbit, and so I make only a small error by taking the Sun-Pluto distance as a measure of the (average) Earth-Pluto distance. Armed with this information, I'll calculate the factor

$m_1 m_2/r^2$ first for Earth and me, then for the two situations described in the two parts of this problem.

❸ **EXECUTE PLAN** When I stand on the surface of Earth, $r = R_E$, and so the factor $m_1 m_2/r^2$ is

$$\frac{m_1 m_E}{R_E^2} = \frac{(70 \text{ kg})(5.97 \times 10^{24} \text{ kg})}{(6.38 \times 10^6 \text{ m})^2} = 1.0 \times 10^{13} \text{ kg}^2/\text{m}^2.$$

(a) For two 70-kg people separated by 1 m, I get

$$\frac{m_1 m_2}{r_{12}^2} = \frac{(70 \text{ kg})^2}{(1 \text{ m})^2} = 4.9 \times 10^3 \text{ kg}^2/\text{m}^2.$$

Thus the gravitational force exerted by a person standing 1 m from me is $(4.9 \times 10^3 \text{ kg}^2/\text{m}^2)/(1.0 \times 10^{13} \text{ kg}^2/\text{m}^2) = 4.9 \times 10^{-10}$ times the gravitational force exerted by Earth on me. ✔

(b) For Earth and Pluto, I have

$$\frac{m_E m_P}{r_{EP}^2} = \frac{(5.97 \times 10^{24} \text{ kg})(1.36 \times 10^{22} \text{ kg})}{(5.9 \times 10^{12} \text{ m})^2}$$

$$= 2.3 \times 10^{21} \text{ kg}^2/\text{m}^2.$$

This is $(2.3 \times 10^{21} \text{ kg}^2/\text{m}^2)/(1.0 \times 10^{13} \text{ kg}^2/\text{m}^2) = 2.3 \times 10^8$ times greater than the attraction between Earth and me. ✔

❹ **EVALUATE RESULT** That the gravitational force exerted by another person on me is more than a billion times smaller than that exerted by Earth on me makes sense: Only the gravitational attraction between Earth and objects on Earth is noticeable. That the gravitational force exerted by Earth on Pluto is 200,000,000 times greater than that exerted by Earth on me, even though Pluto is 1,000,000 times farther from Earth's center than I am, is amazing. Of course, Pluto's mass is about 10^{20} times greater than mine, and that factor more than makes up for the large difference between the Earth-me distance and the Earth-Pluto distance.

CONCEPTS

13.5 In part *a* of Example 13.1, what would the distance between the other person and you have to be for the gravitational force between the two of you to match the gravitational force exerted by Earth on you? Is it possible to verify your prediction?

13.2 Gravity and angular momentum

The force of gravity is a **central force**—a force whose line of action always lies along the straight line that connects the two interacting objects. A planet orbiting the Sun, for example, is subject to a force that is always directed toward the Sun. The fact that the force is central has an important consequence for angular momentum.

For example, consider an isolated system of two gravitationally interacting stars of masses m_1 and m_2 (**Figure 13.9**).

Figure 13.9 Gravitation is a central force, as illustrated by two stars orbiting their common center of mass.

(a) Two stars follow elliptical orbits around common center of mass

star 1   center of mass   star 2

(b) Gravitational forces stars exert on each other point along straight line connecting star centers. (Stars are shown at two instants in orbit.)

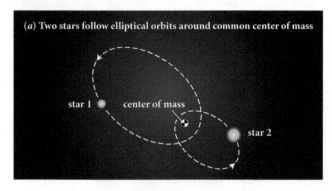

(c) Ratio r_1/r_2 of distances between each star and center of mass remains constant.

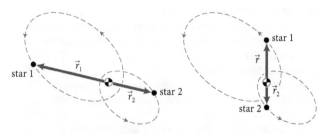

We choose a coordinate system such that the origin is at the center of mass of the system. Because the system is isolated, the center of mass remains fixed at the origin. This means that the ratio r_1/r_2 of the objects' distances from the center of mass must remain the same regardless of the objects' motion (when $\vec{r}_{cm} = 0$, $m_1\vec{r}_1 + m_2\vec{r}_2 = 0$ and so $r_1/r_2 = m_2/m_1$; see Section 6.6). Note how the ratio r_1/r_2 in Figure 13.9 remains constant and how the orbits of the two stars have the same shape (scaled by the factor r_1/r_2.)

Because the center of mass always lies on the line of action of the gravitational forces, the torques caused by each of these forces about the center of mass are zero and therefore the angular momentum of each star about the center of mass doesn't change. So, not only is the angular momentum of the two-star system constant, but the angular momentum of each star about the center of mass is constant, too.

> **In an isolated system of two objects interacting through a central force, each object has a constant angular momentum about the center of mass.**

Often we consider situations in which one of the two interacting objects is much more massive than the other, as shown in **Figure 13.10**. In that case, the center of mass lies nearly at the center of the more massive object. This permits us to view the situation from the position of the more massive object, which we can consider fixed at the origin. In Figure 13.10*a*, for example, the Earth moves in a nearly circular orbit about the Sun. Because the Sun is so much more massive than Earth, the center of the Sun is very close to the center of mass of the Sun-Earth system. We therefore can ignore the small difference and consider the two centers to coincide. If we consider the orbit to be circular, the center of mass also coincides with the center of the orbit. Figure 13.10*b* shows a comet moving along an elliptical orbit under the influence of a gravitational force exerted by the Sun. The center of mass again lies at the center of the Sun, which is at one of the foci of the ellipse (not at the center of the ellipse).

13.6 (*a*) If we ignore the influence of other celestial objects, the Earth-Sun system is isolated, and so it must rotate about the system's center of mass. Using the data in Table 13.1, determine how far from the center of the Sun the center of mass of the system is and what fraction of the Sun's radius this distance is. (*b*) As it passes the point shown in Figure 13.10*b*, does the comet speed up, slow down, or travel at constant speed? If the trajectory shown is part of an ellipse, where is the comet's velocity greatest? Where is it smallest?

Because the force is central and because the center of mass is at the Sun in Figure 13.10, the angular momentum

Figure 13.10 Whether an orbit is circular or elliptical, the gravitational force on the orbiting body points toward the Sun. If the orbit is elliptical, the Sun is at one focus, not at the center of the ellipse.

(*a*) Orbits of Earth and hypothetical comet (not to scale)

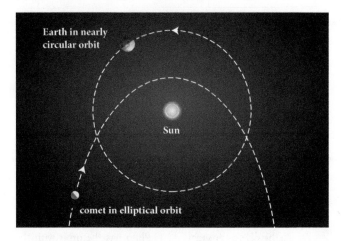

Earth in nearly
circular orbit

Sun

comet in elliptical orbit

(*b*) Force exerted by Sun on Earth and comet points toward Sun

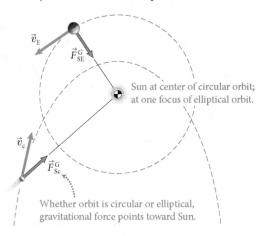

$\vec{v}_E$

$\vec{F}^G_{SE}$

$\vec{v}_c$

$\vec{F}^G_{Sc}$

Sun at center of circular orbit; at one focus of elliptical orbit.

Whether orbit is circular or elliptical, gravitational force points toward Sun.

Figure 13.11 The position vectors of an object moving (*a*) with constant momentum along a straight line and (*b*) with constant angular momentum along an arbitrary path sweep out equal areas in equal time intervals.

(*a*) Object moving with constant momentum along straight line

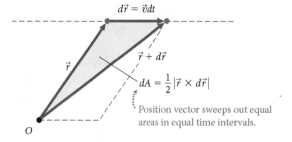

$d\vec{r} = \vec{v}dt$

$\vec{r}$

$\vec{r} + d\vec{r}$

$dA = \frac{1}{2}|\vec{r} \times d\vec{r}|$

Position vector sweeps out equal areas in equal time intervals.

O

(*b*) Object moving with constant angular momentum along curve

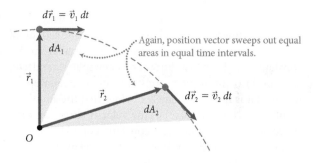

$d\vec{r}_1 = \vec{v}_1 \, dt$

dA_1

Again, position vector sweeps out equal areas in equal time intervals.

$\vec{r}_1$

$\vec{r}_2$

$d\vec{r}_2 = \vec{v}_2 \, dt$

dA_2

O

area of the dashed parallelogram in Figure 13.11*a*—twice the area dA of the shaded triangle. Because the particle's displacement can be written as $d\vec{r} = \vec{v}\,dt$, we have $dA = \frac{1}{2}|\vec{r} \times \vec{v}|dt = \frac{1}{2}L/m\,dt$. The rate at which the area A is swept out by the position vector is thus $dA/dt = \frac{1}{2}L/m$. In words:

> **The angular momentum of a particle about an origin is proportional to the rate at which area is swept out by the particle's position vector.**

So, when the angular momentum of a particle is constant, the rate dA/dt is also constant. Figure 13.11*b* shows the motion of an object along a curved trajectory. In two equal time intervals dt, the position vector sweeps out areas dA_1 and dA_2. If the angular momentum of the object about O is constant, then $dA_1 = dA_2$. The motion caused by a central force directed to the origin is such that the radius vector from the origin sweeps out equal areas in equal time intervals *regardless of the shape of the trajectory*. In the early 17th century, before Newton formulated the law of gravity and long before the concept of angular momentum was introduced, Johannes Kepler had already deduced from astronomical observations that the straight line from the Sun to each planet sweeps out equal areas in equal time intervals. (See the box "Kepler's laws of planetary motion.")

of each of the orbiting objects about the Sun is constant. The meaning of this statement for a circular orbit is simple: If the angular momentum is constant, then the object's speed must be constant (that is, we are dealing with circular motion at constant speed).

To understand the meaning for any other type of orbit, we must first introduce a geometric interpretation of angular momentum. Consider first the particle moving along a straight line illustrated in **Figure 13.11*a***. As we saw in Section 12.8 the angular momentum of any moving particle about the origin can be written as $\vec{L} = \vec{r} \times \vec{p}$, where $\vec{r}$ is the position of the particle (Eq. 12.43). In a time interval dt, the position vector $\vec{r}$ sweeps out the area of the shaded triangle in Figure 13.11*a*. As we saw in Section 12.8, the magnitude of the vector product of two vectors is equal to the area of the parallelogram defined by them. So $|\vec{r} \times d\vec{r}|$ is the

Figure 13.12 Three forms of motion with constant angular momentum.

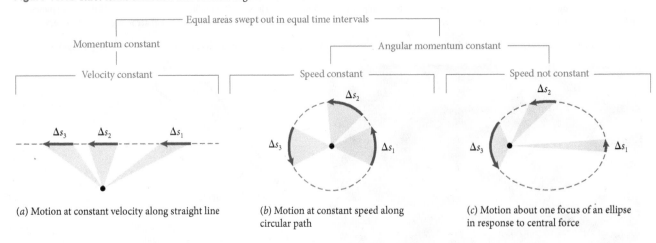

(a) Motion at constant velocity along straight line

(b) Motion at constant speed along circular path

(c) Motion about one focus of an ellipse in response to central force

Figure 13.12 shows three types of motion with constant angular momentum: motion at constant speed along a straight line (no force; constant velocity), circular motion at constant speed (central force; constant speed), and elliptical motion (central force; constant rate of area swept). For the third it is not immediately obvious that the angular momentum is constant: Neither the velocity direction nor the speed of the moving object is constant. For all three types of motion, however, the position vector sweeps out the same area in equal time intervals. To do so an object in elliptical motion (Figure 13.12c) must move fastest when it is closest to the object it is orbiting and slowest when it is the farthest away, in agreement with the answer to Checkpoint 13.6, part b.

Example 13.2 Pluto

Of all the objects in Table 13.1, Pluto has the orbit with the greatest eccentricity. (a) What is the ratio of the orbit's semiminor axis b to its semimajor axis a? (b) What is the ratio of Pluto's speed at perihelion to its speed at aphelion?

❶ **GETTING STARTED** I begin by making a sketch of Pluto's orbit, using the Sun as one focus (**Figure 13.13**). From Figure 13.8 I know that the orbit is very nearly circular but the Sun is off-center, making the orbit slightly elliptical. I mark the perihelion (shortest Sun-orbit distance) and aphelion (longest Sun-orbit distance) and show velocity vectors at those locations because I am asked about speeds there. In the ellipse as I have oriented it, with the line defining the foci horizontal, the semiminor axis b extends from the center of the major axis to point P at the top of the ellipse.

Kepler's laws of planetary motion

The German mathematician and astrologer Johannes Kepler devoted much of his life to analyzing celestial motion data collected by Tycho Brahe. Kepler discovered three laws of planetary motion.

Kepler's first law describes the shape of the planetary orbits:

> **All planets move in elliptical orbits with the Sun at one focus.**

Although the deviation from circular orbits is small, this statement was a radical departure from the accepted wisdom, dating back to Plato, that the planets, being heavenly bodies, were perfect and therefore could move in only perfect circles or combinations of circles.

Kepler's second law reveals that, even if planets are not in circular motion at constant speed, their motions obey the following requirement:

> **The straight line from any planet to the Sun sweeps out equal areas in equal time intervals.**

Kepler's third law relates the planetary orbits to one another:

> **The squares of the periods of the planets are proportional to the cubes of the semimajor axes of their orbits around the Sun.**

Kepler discovered this third law by painstakingly examining, over a period of many years, countless combinations of planetary data.

In keeping with Aristotelian notions, Kepler believed that a force was necessary to drive the planets along their orbits, not to keep them in orbit. Consequently, Kepler was unable to provide a correct explanation for these three laws. It was not until Newton that the single unifying reason for these laws was established. As we've seen, Kepler's three laws follow directly from the law of gravity: The first and third laws are a consequence of the $1/r^2$ dependence, and the second law reflects the central nature of the gravitational force.

Figure 13.13

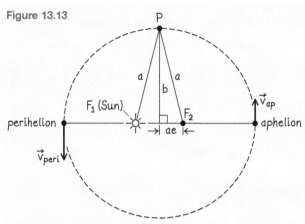

The ratio of the semimajor and semiminor axes can be obtained from either right triangle in Figure 13.13. Because the sum of the distances from any point on an ellipse to the two foci is $2a$ (see Figure 13.7), I know that $\overline{F_1P} = \overline{F_2P} = a$.

❷ **DEVISE PLAN** For part a, I can use the Pythagorean theorem and the values in Table 13.1 to calculate the ratio b/a. For part b I know that the planet's angular momentum about the Sun must be constant because the force exerted by the Sun on Pluto is a central force. I can therefore calculate the angular momentum at perihelion and at aphelion and set the two equal to obtain the ratio $v_{\text{peri}}/v_{\text{ap}}$.

❸ **EXECUTE PLAN** (a) Applying the Pythagorean theorem to the right triangle containing F_2 gives $b = \sqrt{a^2 - (ae)^2} = a\sqrt{1 - e^2}$. Using the value of e from Table 13.1, I get $b/a = \sqrt{1 - (0.25)^2} = 0.97$. ✔

(b) At perihelion, the Sun-Pluto distance is $(a - ae)$, and the angular momentum is $L_{\text{peri}} = m(a - ae)v_{\text{peri}} = ma(1 - e)v_{\text{peri}}$ (Eq. 11.36); at aphelion, $L_{\text{ap}} = ma(1 + e)v_{\text{ap}}$. Because $L_{\text{peri}} = L_{\text{ap}}$, I get

$$ma(1 - e)v_{\text{peri}} = ma(1 + e)v_{\text{ap}},$$

so

$$\frac{v_{\text{peri}}}{v_{\text{ap}}} = \frac{1 + e}{1 - e} = \frac{1 + 0.25}{1 - 0.25} = 1.7. ✔$$

❹ **EVALUATE RESULT** The semiminor and semimajor axes are equal to within 3%, which agrees with what is shown in Figure 13.8. That v_{peri} is greater than v_{ap} agrees with my answer to Checkpoint 13.6, part b.

Conservation of angular momentum is also responsible for the flat, disk shape of most galaxies (like the one seen from the side in the opening picture of this chapter) and of our solar system. Figure 13.14 illustrates the gravitational contraction of a slowly rotating cloud of interstellar gas and dust. Because the cloud's rotational inertia decreases as it contracts, its rate of rotation must increase, just as a skater spins faster with her arms pulled in. In the plane perpendicular to the rotation axis, part of the gravitational force directed toward the center of the cloud goes into providing the centripetal acceleration. Regions near this plane therefore undergo less contraction than do regions along the axis. This differential contraction results in the characteristic disk shape of many galaxies.

Figure 13.14 (a) Gravitational contraction of a slowly rotating cloud of interstellar gas and dust. (b) Conservation of angular momentum requires the rotational velocity to increase as the cloud contracts and its rotational inertia decreases. (c) Contraction is less efficient in the plane perpendicular to the rotation axis because part of the gravitational interaction goes into supplying the centripetal acceleration.

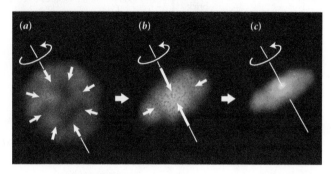

13.3 Weight

The law of gravity states that the force of gravity depends on location. Figure 13.15 shows astronauts orbiting Earth at an altitude of about 400 km. Are they floating "weightlessly" inside their spacecraft because the force of gravity at their location has been reduced to a negligible value? The calculation is easy and worth doing, as the next checkpoint shows.

13.7 The space shuttle typically orbits Earth at an altitude of about 300 km. (a) By what factor is the shuttle's distance to the center of Earth increased over that of an object on the ground? (b) The gravitational force exerted by Earth on an object in the orbiting shuttle is how much smaller than the gravitational force exerted by Earth on the same object when it is sitting on the ground? (c) What is the acceleration due to gravity at an altitude of about 300 km? (d) While in orbit, the shuttle's engines are off. Why doesn't the shuttle fall to Earth?

After answering this checkpoint, you may be wondering in what sense objects inside the space shuttle are "weightless." To resolve this issue, we must define what we mean by *weight*.

Figure 13.15 Astronauts Shane Kimbrough and Sandra Magnus float "weightlessly" in Space Shuttle Endeavour during mission STS-126.

CONCEPTS

Figure 13.16 Two methods for weighing a brick.

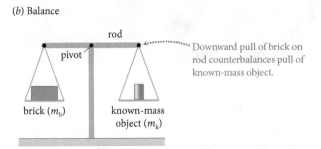

(a) Spring scale

Compression measures force exerted by brick on spring.

(b) Balance

rod
pivot
Downward pull of brick on rod counterbalances pull of known-mass object.

brick (m_b)

known-mass object (m_k)

Figure 13.17 Using a balance at Earth's surface and on the Moon. If the rod is balanced on Earth, the same objects balance the rod on the Moon.

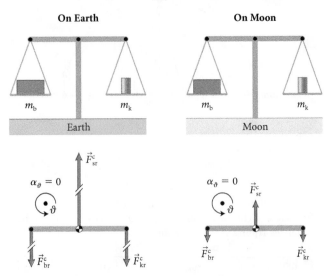

On Earth

m_b m_k

Earth

On Moon

m_b m_k

Moon

Figure 13.16 illustrates two types of devices used for weighing objects. When a load is put on a **spring scale,** the load either compresses or stretches a spring inside the scale (Figure 13.16a). The change in the spring's length is proportional to the force exerted on it by the load: $F^c_{1s\,x} = k(x - x_0)$ (Eq. 8.18). The deformation of the spring therefore provides a measure of the downward force exerted on it by the load. When you stand on a bathroom scale, for example, the downward force exerted by your body on the scale compresses a spring inside it, which, in turn, rotates a dial, and you read your weight off the dial. The rotation of the dial is proportional to the force exerted by your body on the scale. (The scale was calibrated at the factory by putting objects of known mass on it.)

✋ **13.8** A 1.0-kg brick is placed on a spring scale. (a) What is the magnitude of the force exerted by Earth on the brick? (b) What is the magnitude of the force exerted by the brick on the scale? (c) How many steps are required to answer part b? (d) Does the spring scale measure the pull of gravity on the brick?

Another type of weighing device, the **balance,** illustrated in Figure 13.16b, works as follows. A brick whose mass m_b is to be determined is suspended from one end of a rod pivoted about its center, and objects of known masses are suspended from the other end until the rod is in mechanical equilibrium. In mechanical equilibrium, the downward forces exerted by the loads on the two ends of the rod are equal, and so m_b must be equal to the sum m_k of the known masses.

Don't both devices measure the same thing? You may be surprised to learn that the answer to this question is *no*. To see why, suppose we repeat the experiments illustrated in Figure 13.16 on the surface of the Moon. The gravitational force exerted by the Moon on an object sitting on the lunar surface is about six times smaller than the gravitational force exerted by Earth on the same object sitting on Earth's

surface. If you were carrying a heavy load on the Moon, you would discover that the force needed to support it is less than on Earth; on the Moon the load feels a lot lighter. As the extended free-body diagrams in **Figure 13.17** show, however, the smaller gravitational pull of the Moon does not affect the mechanical equilibrium of the balance: The downward pull diminishes on *both* sides, and so the rod remains in mechanical equilibrium. With the balance, we still obtain $m_b = m_k$, just as we do on Earth.

The reading of the spring scale on the Moon, however, is not the same as it is on Earth. Because of the weaker pull of gravity on the Moon, the deformation of the spring is smaller, as shown in **Figure 13.18**. Spring scale readings on the Moon are about six times smaller than on Earth. So the spring scale tells us that the load weighs less on the Moon than on Earth, while the balance tells us that it weighs the same. This difference arises because:

A spring scale measures the downward force exerted on it by its load, but a balance compares gravitational forces and measures mass.

The result obtained with a spring scale depends on the strength of the gravitational pull on the load; the result obtained with a balance does not depend on the strength of the gravitational pull.*

✋ **13.9** A 1.0-kg brick is placed on a spring scale inside the space shuttle orbiting Earth at 300 km altitude. (a) What is the magnitude of the force exerted by Earth on the brick? (b) What is the magnitude of the force exerted by the brick on the spring scale? (Hint: See the astronauts in Figure 13.15.) (c) Does the spring scale measure the pull of gravity on the brick? (d) Why is your answer different from your answer to Checkpoint 13.8, part d?

---

*Many modern weighing devices are called "balances" even though in reality they operate like spring scales. Typically, these devices employ a pivoted rod, but instead of comparing gravitational forces, one end of the rod is connected to a spring or other mechanism for generating a calibrated force.

Figure 13.18 Free-body diagrams for a brick on a spring scale on the surface of Earth and the Moon. Because of the smaller gravitational attraction by the Moon, the spring's compression on the Moon is less than it is on Earth.

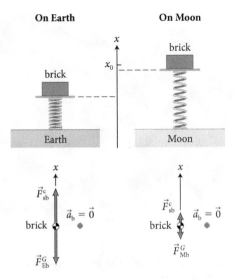

The answers to this checkpoint show that the reading of a spring scale depends not only on the gravitational pull on its load but also on the acceleration of the scale. To see this point more clearly, consider the situation shown in **Figure 13.19a**. Two bricks of identical mass m are placed on two identical spring scales, one on solid ground and the other in an elevator that is accelerating downward. Figure 13.19b shows the free-body diagrams for both bricks.

Figure 13.19 A spring scale holding a brick gives a different reading when it is stationary and when it is in an elevator accelerating downward. Because of the acceleration of the elevator, $\Delta x_{\text{elev}} < \Delta x_{\text{stat}}$.

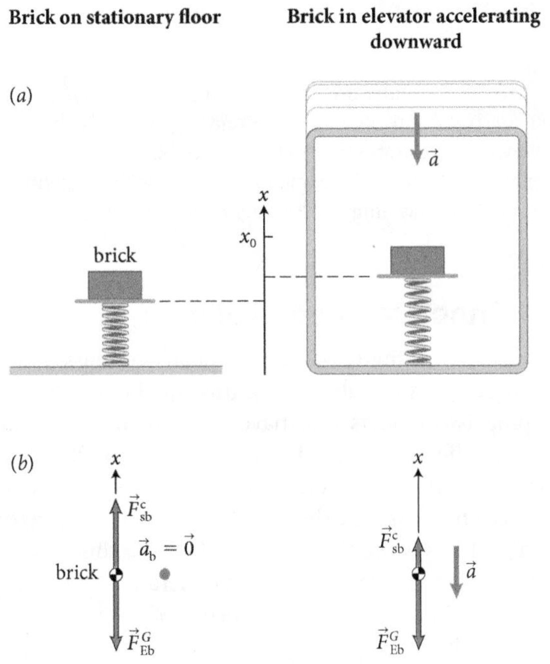

Because the brick in the elevator has a downward acceleration, we know that the vector sum of the forces exerted on it has to be nonzero and directed downward. The downward force of gravity has not changed, however. Consequently, the upward force exerted by the scale on the brick must be *smaller* than the magnitude of the downward force of gravity acting on it. Because the force exerted by the scale on the brick and the force exerted by the brick on the scale form an interaction pair, we conclude that the force exerted by the brick on the scale in the elevator is smaller than that exerted by the identical brick on the stationary spring scale. The spring scale in the elevator therefore gives a lower reading than the stationary scale.

If you are surprised that the result of weighing an object with a spring scale depends on acceleration, try the following experiment. Put an object in the palm of your hand with your arm stretched out in front of you, then rapidly move your hand downward. While you accelerate your hand downward, you can feel that the object presses less on your hand than it does when you hold your hand still. In other words, while your hand accelerates downward, the object in it feels lighter—it feels as though it "weighs" less.*

13.10 (a) An object is placed on a spring scale in an elevator moving upward at a constant speed. Is the reading of the scale greater than, smaller than, or the same as the reading obtained in a stationary elevator? (b) How does the reading of a scale in an elevator that has upward acceleration $a = 0.5g$ compare with the reading obtained in a stationary elevator? (c) What if the elevator has downward acceleration $a = g$?

As this checkpoint shows, objects in an elevator that is in free fall produce a zero spring scale reading. No force is necessary to support them: They, too, fall freely along with the elevator. Inside a freely falling elevator, a spring scale under your feet—falling away underneath you, together with the elevator and everything inside it—reads zero. If you were to hold a briefcase in the freely falling elevator, you would discover that the briefcase does not need to be supported by your hand because the force of gravity gives it the same downward acceleration g as you. The briefcase feels weightless—you could even let go and the briefcase would continue to hover next to you.

Any object in free fall—that is, any object subject to only a force of gravity—experiences weightlessness.

---

*You can also try this by accelerating your hand upward, in which case the object feels heavier. The effect, however, is less pronounced because it is more difficult to accelerate your arm upward. In the downward direction, gravity and your muscles work together to provide the acceleration; in the upward direction, your muscles have to not only accelerate your arm but also counteract the effect of gravity.

Figure 13.20 As a spacecraft orbits Earth a distance h above the ground, it falls a distance Δh. Because the craft moves so fast in a direction tangential to the (curved) ground, the ground "falls away" at the same rate, and the distance between the craft and the ground remains constant.

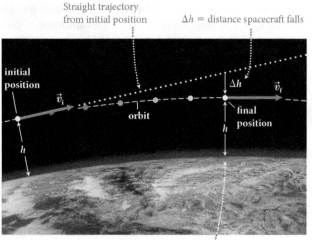

Height h of orbit remains constant because of distance Δh spacecraft falls.

Figure 13.21 If an airplane briefly follows a parabolic path such that it has a downward acceleration g, the airplane and the people inside it are in free fall. In this picture, astronaut candidates aboard a KC-135 Stratotanker briefly experience weightlessness as they and the plane are in a short free fall.

If this appears odd, consider once again the astronauts in orbit in Figure 13.15. The only force exerted on them is the force of gravity, and so they *are* in free fall! This free fall is precisely what makes them weightless. Because of their large speed and the curvature of Earth, however, they do not come closer to the ground. As they fall, the ground below them recedes at the same speed as they fall (Figure 13.20).

13.11 (*a*) How far does the space shuttle in Figure 13.20, 300 km above Earth, fall in 1.0 s? (*b*) If the radius of Earth is 6400 km, what is the shuttle's speed?

You don't need to be an astronaut to experience weightlessness. When you jump off a chair, you are momentarily weightless; during your short free fall, an object in the palm of your hand hovers weightlessly above your hand. A plane flying across a large parabolic arc may be in a controlled free fall for short time intervals. As shown in Figure 13.21, people inside such an airplane float weightlessly, just as astronauts in orbit do.

13.12 Planes can be put (safely) into free fall for periods up to 40 s. (*a*) If the plane shown in Figure 13.21 was flying horizontally before being put into free fall, how much altitude does it lose in those 40 s? (*b*) What is the plane's final vertical speed of descent? (*c*) At the instant shown, is the vertical component of the plane's velocity directed upward, directed downward, or zero?

The discussion in this section shows that our notion of *weight* corresponds most closely to the reading of a spring scale: The perceived "heaviness" of an object depends, as does the reading on the scale, not only on the pull of gravity on the object but also on its acceleration. The "weightless" astronauts in the space shuttle would indeed register a zero

scale reading, even though the pull of gravity on them is almost identical to the pull at Earth's surface.

Because spring scale readings depend on acceleration, however, weight is often defined as the force of gravity exerted *on* an object. Spring scale readings are then called *apparent* or *effective weight*. With this definition, the astronauts in Figure 13.15 and the people in Figure 13.21 are not weightless because the force of gravity on them (their "weight") is the same as it is on Earth's surface, and so they must be called *apparently weightless*. To avoid all these difficulties and the conflict between our intuitive notion of weight (spring scale reading) and the common definition of weight (force of gravity), we therefore avoid using the word *weight* in this book. Instead we shall use the more precise "force of gravity" to refer to the downward pull of gravity on an object and "scale reading" to refer to the force exerted by an object on a spring scale.

13.13 Consider an airplane flying at a constant speed of 900 km/h, 10 km above the ground. (*a*) What is the plane's downward acceleration? (*b*) Draw a free-body diagram for the airplane. (*c*) Why don't people in the airplane feel weightless as astronauts do traveling 290 km higher up?

13.4 Principle of equivalence

Let us return to the remarkable equality of mass and inertia. The simplest possible test of this equality is obtained by dropping two objects that have different inertias and are made of different materials and measuring their accelerations. As Galileo discovered, *all* objects fall with the same acceleration g. It is true that Earth pulls twice as hard on an object that has twice the mass, but because this object also has twice the inertia, it doesn't accelerate any faster. The equality of mass and inertia has been tested experimentally to one part in 10^{12}.

The equality of mass and inertia may seem obvious because we are so accustomed to the fact that heavier objects not only take more effort to support against gravity but also take more effort to be set in motion. Intuitively, both properties simply go by the amount of matter in the object. Yet this equality suggests a much deeper connection between gravity and acceleration.

In Section 13.3 we encountered another manifestation of this equivalence between gravity and inertia: The reading on a spring scale depends not only on the gravitational pull on the load but also on the acceleration of the scale. We can change the reading of the scale—our notion of weight—simply by accelerating the scale. This raises an interesting question: Can we design an experiment that distinguishes between gravity and an accelerated reference frame?

Consider a brick placed on a spring scale inside a container at rest on the surface of Earth, as illustrated in **Figure 13.22a**. Because of the gravitational attraction exerted by Earth on the brick, the brick compresses the spring to a scale reading x_E. To an observer in the Earth

Figure 13.22 If you observe a brick compressing a spring scale in an enclosed container, you cannot tell whether the compression is due to a gravitational interaction or to acceleration of the container.

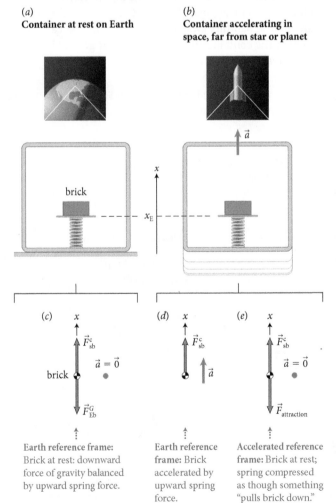

reference frame the brick is at rest, and so the upward force exerted by the scale is equal in magnitude to the downward force exerted by Earth (see Figure 13.22c). Now picture the same arrangement in space, far away from any planet or star, where the influence of gravity is negligible and so we can say that $g = 0$. Suppose the container is towed by a spaceship; if the spaceship fires its rockets, the container accelerates, as illustrated in Figure 13.22b. Objects inside the container no longer float weightlessly, but being dragged along by the container, they rest against the bottom of the container. If the brick is placed on the scale, the scale registers a reading proportional to the force exerted by the spring on the brick. This force is required to give the brick the same acceleration $\vec{a}$ as the container. From the point of view of an observer in the Earth reference frame, the brick has an upward acceleration $\vec{a}$ (see Figure 13.22d).

If the rockets are fired such that $a = g$, could you, from inside the container, tell the difference between being accelerated in space and being on Earth? To you, the scale and the scale reading appear just as they do on Earth. An object released from your hand falls with an acceleration g, just as it does on Earth. From inside the container, with both of your feet solidly on the floor, everything looks and feels just as it does on Earth—all objects appear to be attracted to the container floor.

Having observed all this, you draw the free-body diagram shown in Figure 13.22e: From your point of view, the brick is at rest, and so the vector sum of the forces exerted on it must be zero. So, in addition to an upward force exerted by the compressed spring on the brick, you draw a downward force, equal in magnitude to that of the upward force. To you, it appears as if this force is due to a downward attraction experienced by all objects inside the container. Without peeking outside, you cannot determine whether this attraction is present because underneath the container there is a planet or because the container is accelerating upward in space.

Notice that the fictitious attractive force experienced by objects inside the container is no less mysterious than the force of gravity. Both forces have the same effect: Objects are pulled toward the bottom of the container according to their mass or inertia (the distinction has now disappeared). Although the gravitational force is more familiar, and although Newton's law of gravity describes its behavior, we really have no idea how gravity "works." The conclusion is thus:

> One cannot distinguish locally between the effects of a constant gravitational acceleration of magnitude g and the effects of an acceleration of the reference frame of magnitude $a = g$.

This statement, called the **principle of equivalence,** is the starting point for the theory of general relativity developed by Albert Einstein in the early 1900s. The word *locally* is necessary because on a large scale, the $1/r^2$ dependence of the gravitational force becomes apparent: For a very tall

CONCEPTS

A Boeing 747-8 flight simulator.

container in Figure 13.22, the gravitational acceleration at the top of the container would be smaller than at the bottom; for a very wide container, it would become noticeable that objects are falling not parallel to one another but toward a single point—the attracting object's center. In the theory of general relativity, the effects of gravity are explained in an entirely geometric way. Gravity and acceleration are one.

On a practical note, our inability to distinguish between gravity and acceleration is exploited in aircraft flight simulators and motion simulators in the amusement industry. These devices are programmed to tilt in concert with a film simulating a ride in a vehicle, airplane, or spaceship. The effect is a startlingly realistic sensation of motion. The principle of operation is shown in **Figure 13.23**. Suppose you are in a closed container, with a ball suspended from the ceiling.

If the container rolls at constant speed along a track, as in Figure 13.23a, you cannot *feel* the motion and the string hangs vertically, although sounds from the wheels and little bumps in the track might give away that the container is in motion. The law of inertia (see Section 6.2) tells us, however, that you cannot determine the difference between moving at constant speed and sitting at rest. So, with the container on a stationary base, as in Figure 13.23b, if the container shakes a bit to simulate bumps and you hear the same sounds and see a movie of a landscape that flows by at constant speed, the illusion of moving at constant speed is complete.

Now suppose the container speeds up, as in Figure 13.23c. The chair presses against your back to accelerate you along with the container, and as a result you feel pushed backward against the chair. As Figure 13.23d shows, the same effect can be obtained by tilting a stationary container backward (together with appropriate sounds and visual clues). In both cases, also, the ball suspended from the ceiling tilts toward you, and objects don't fall straight to the floor. From inside the container, you cannot tell whether these things happen because the container accelerates or because gravity is in a different direction.

To simulate slowing down (see Figure 13.23e), the container needs to be tilted forward (see Figure 13.23f). The result is that you feel pushed forward, as in a sharply braking car, and the ball tilts away from you. Again you cannot tell whether the container is slowing down or gravity is pulling in a different direction.

13.14 How should the simulator in Figure 13.23 be tilted to simulate a right turn?

Figure 13.23 The principle of equivalence states that an observer inside a closed container cannot distinguish between motion at constant speed and rest, between speeding up and tilting backward, or between slowing down and tilting forward.

Vehicle: Effects caused by acceleration

(a) Constant velocity
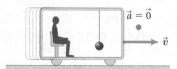

(c) Forward acceleration (speeding up)

(e) Rearward acceleration (slowing down)

Simulator: Identical effects caused by tilting

(b) No tilt

(d) Backward tilt

(f) Forward tilt

Figure 13.24 A light pulse traverses an accelerating container.

(a)

Earth reference frame

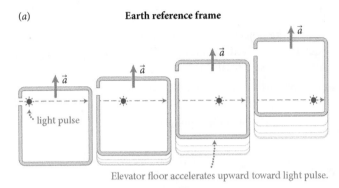

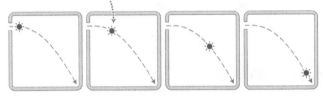

Elevator floor accelerates upward toward light pulse.

(b)

Reference frame of elevator

Pulse behaves as though falling in response to gravity.

The simulator example may give you the impression that the principle of equivalence has more to do with our senses than with the properties of the universe. This is not so, however. Because anything that moves along a straight line in an inertial reference frame must move along a curved path in an accelerated reference frame, the principle of equivalence tells us that *anything* that moves near an object that has mass must move along a curved path. Figure 13.24 shows that a light pulse that travels along a straight line in the Earth reference frame but then enters an accelerating elevator must move along a curved path as

viewed from inside the elevator. This effect is purely kinematic, and so given the equivalence between gravity and acceleration, the bending should also occur if the elevator, instead of accelerating upward, were placed on top of an object with a large mass.

13.15 Light travels at approximately 3.0×10^8 m/s. (a) How long does it take for a light pulse to cross an elevator 2.0 m wide? (b) How great an acceleration is necessary to make the pulse deviate from a straight-line path by 1.0 mm? Is it likely that this effect can be observed? (c) If light is bent by the gravitational pull of an object, light should "fall" when traveling parallel to the surface of Earth. How far does a beam of light travel in 0.0010 s, and how much does it fall over that distance? Is it likely that this effect can be observed?

Although the calculations in this checkpoint show that detection of the gravitational bending of light is difficult, the effect was observed within a few years of its prediction in 1915. Because of the large mass of the Sun, a ray that passes close to the Sun should undergo a deflection. Although it is normally impossible for us earthlings to see any stars close to the Sun, the light from the Sun is blocked by the Moon during a solar eclipse (Figure 13.25). By comparing a picture of stars near the Sun taken during an eclipse with a picture of the same stars taken when the Sun is far away from them (in other words, at night), it is possible to observe the bending of light caused by the Sun's gravitational attraction. The measured values of such deflections are in excellent agreement with the theoretical values.

13.16 (a) What would the scale reading be if the container in Figure 13.22b were traveling at constant speed instead of accelerating? (b) Describe the path of the light pulse in the reference frame of the elevator in Figure 13.24b if the elevator were traveling at constant speed.

Figure 13.25 When the Sun passes close to our line of sight to a star, the Sun's gravity bends the star's light, slightly shifting the star's apparent location in the sky. This effect is visible only during a solar eclipse, when the Moon blocks the Sun's glare. Diagram not to scale.

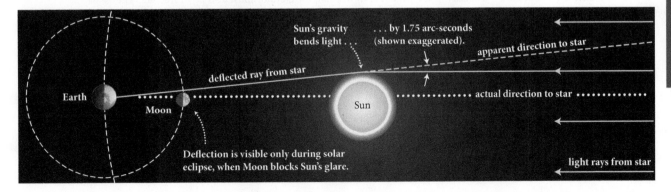

Self-quiz

1. Determine the ratio of Earth's gravitational force exerted on a 1.0-kg object at the following positions to that at sea level: (*i*) the top of Mt. Everest, altitude 8848 m, (*ii*) a shuttle orbiting at 200 km; (*iii*) the orbit of the Moon. Is the 1.0-kg object weightless in any of these circumstances?

2. The orbital period of Jupiter around the Sun is 11.86 times that of Earth. What is the ratio of their distances to the Sun?

3. If the speed of a planet in its orbit around the Sun changes, how can the planet's angular momentum be constant?

4. A brick of unknown mass is placed on a spring scale. When in equilibrium, the scale reads 13.2 N. Which of the following statements is/are true? (*i*) Earth always exerts a gravitational force of 13.2 N on the brick. (*ii*) The normal force the scale exerts on the brick is 13.2 N. (*iii*) The brick has an inertia of 1.32 kg.

5. Draw a free-body diagram of an individual experiencing weightlessness in an elevator. What must the elevator's acceleration be for this to occur?

6. If you were to travel in a vertical circle, at what point in the circle would you be most likely to experience weightlessness?

Answers

1. The gravitational force is proportional to the inverse square of the distance between Earth and the object. For each case, we have $F_{pos}/F_{sea\ level} = r_{sea\ level}^2/r_{pos}^2$, where $r_{sea\ level} = 6.378 \times 10^6$ m (Earth's radius) and r_{pos} is the distance between the position in question and Earth's center. (*i*) The distance to Earth's center is the sum of Earth's radius and the altitude of Mt. Everest: $r_{pos} = (6.378 \times 10^6$ m$) + (8848$ m$) = 6.387 \times 10^6$ m; $F_{pos}/F_{sea\ level} = 0.997$. (*ii*) $r_{pos} = (6.378 \times 10^6$ m$) + (200,000$ m$) = 6.578 \times 10^6$ m; $F_{pos}/F_{sea\ level} = 0.940$. (*iii*) The distance to Earth's center is the radius of the Moon's orbit. Given that the Moon's orbit is nearly circular, I use the value of the semimajor axis in Table 13.1: $r_{pos} = 3.84 \times 10^8$ m; $F_{pos}/F_{sea\ level} = 0.000276$. The object is never weightless because, in each case, it experiences a gravitational force and, in all but the last case, the magnitude of this force is very close to the magnitude at sea level on Earth.

2. The period squared is proportional to the cube of the orbital radius. Therefore, the square of the ratio of the periods ($11.86^2 = 141$) is equal to the cube of the ratio of the orbital radii. The ratio of the radii is then $141^{1/3} = 5.2$. Check this value against the data in Table 13.1.

3. If the angular momentum is constant but the speed changes, the rotational inertia must change. When a planet's speed increases, either its inertia must decrease or it must move nearer the object it orbits to keep the angular momentum constant. Conversely, when a planet's speed decreases, either its inertia must increase or it must move away from the object it orbits.

4. Only statement (*ii*) is true. The scale reading may change from location to location, and the inertia of the brick is 1.32 kg only when the gravitational acceleration at the specific location is 10 m/s².

5. If a person experiences weightlessness, the only force acting on her or him is the gravitational force. Contact with the elevator floor is not necessary. If the person is in free fall, the elevator must also be in free fall. The free-body diagram consists of just one force vector, $\vec{F}_{Ep}^G$ downward.

6. To experience weightlessness, the only force acting on you is the gravitational force exerted by Earth, which points downward. This downward force provides the centripetal acceleration needed for traveling in a circle. The only position from which the center of the circle is downward is the top of the circle.

13.5 Gravitational constant

As we saw in Section 13.1, Newton's **law of gravity** states that the gravitational force exerted by a object of mass m_1 on a object of mass m_2 is an attractive, central force that is proportional to the product of the masses and inversely proportional to the square of the distance between them. If the positions of the two objects are given by the vectors $\vec{r}_1$ and $\vec{r}_2$, then the distance between them is $r_{12} = |\vec{r}_2 - \vec{r}_1|$, as shown in Figure 13.26. We can thus write the magnitude of the gravitational force as

$$F_{12}^G = G\frac{m_1 m_2}{r_{12}^2}, \qquad (13.1)$$

where the proportionality constant G is called the **gravitational constant.** The force is attractive and directed along the straight line joining the objects.

What is the relationship between Eq. 13.1 and the magnitude mg of the gravitational force near Earth's surface? To answer this question, let object 1 in Eq. 13.1 be Earth with mass m_E and let object 2 be an arbitrary object of mass m_o near Earth's surface:

$$F_{Eo}^G = G\frac{m_E m_o}{r_{Eo}^2}. \qquad (13.2)$$

For an object a distance h above the ground, the distance between the object and Earth's center is $r_{Eo} = R_E + h$, where R_E is the radius of Earth. Near the ground, $h \ll R_E$, $r_{Eo} \approx R_E$, and Eq. 13.2 becomes

$$F_{Eo}^G = \frac{G m_E m_o}{r_{Eo}^2} \approx G\frac{m_E m_o}{R_E^2} = m_o\left(\frac{G m_E}{R_E^2}\right) \quad \text{(near Earth's surface).} \quad (13.3)$$

The term in parentheses represents the acceleration of the object due to the force F_{Eo}^G. Knowing from our study of the gravitational force that $F_{Eo}^G = m_o g$, we see that

$$g = \frac{G m_E}{R_E^2} \quad \text{(near Earth's surface).} \qquad (13.4)$$

Because G, m_E, and R_E are all constant, the law of gravity does indeed give a constant acceleration near the surface of Earth.

✋ **13.17** For an object released from a height $h \approx R_E$ above the ground, does the acceleration due to gravity decrease, increase, or stay the same as the object falls to Earth?

Working in about 1677, Newton knew the radius of Earth but not its mass, and so he could get only a rough value of G by estimating m_E (as you did in Checkpoint 1.16). The magnitude of G was determined experimentally by Henry Cavendish in 1798. The basic principle of Cavendish's measurement is shown in Figure 13.27. Two small lead spheres connected by a light rod are suspended from a long thin fiber. When two large lead spheres are placed near the small spheres, the gravitational attraction between the large and small spheres causes the rod to rotate, thus twisting the fiber. Equilibrium is reached when the torque caused by the gravitational forces is equal in magnitude to the opposing torque caused by the twisted fiber. By calibrating the degree to which known forces twist the fiber, Cavendish was able to measure the gravitational force between the lead spheres. Knowing this force, he then found the value of G by measuring the separation and inertias of the spheres. The value determined with modern instruments is

$$G = 6.6738 \times 10^{-11}\ \text{N} \cdot \text{m}^2/\text{kg}^2. \qquad (13.5)$$

Figure 13.26 The distance between two objects whose positions are given by the vectors $\vec{r}_1$ and $\vec{r}_2$ is equal to the magnitude of the vector $\vec{r}_2 - \vec{r}_1$.

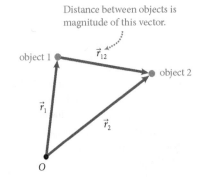

Distance between objects is magnitude of this vector.

Figure 13.27 Schematic representation of Cavendish's experimental setup to determine the gravitational constant. The entire setup is enclosed in a box to shield it from air currents.

Magnitude of gravitational forces between spheres can be deduced from torsion in fiber.

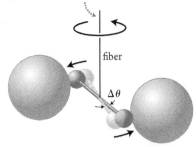

Because G is exceedingly small, the force exerted by two 1-kg objects that are 1 m apart is a mere 6.7×10^{-11} N. For comparison, the force of gravity exerted by Earth on a hair is around 10^{-3} N—15 million times stronger! This comparison shows how exceedingly small the gravitational force is. Only when at least one of the two interacting objects has a very large mass (10^{11} kg at the minimum) does the gravitational force become sizeable.

Example 13.3 Weighing Earth

Cavendish is said to have "weighed Earth" because his determination of G provided the first value for the planet's mass m_E. Given that the radius of Earth is about 6400 km and given the value of G in Eq. 13.5, determine m_E.

❶ GETTING STARTED I am given Earth's radius and G and must use these values to determine Earth's mass. An expression containing G and the acceleration due to gravity g seems a good place to begin.

❷ DEVISE PLAN Equation 13.4 gives the acceleration due to gravity near Earth's surface, g, in terms of G, R_E, and m_E. Knowing the values of g, G, and R_E, I can determine m_E.

❸ EXECUTE PLAN From Eq. 13.4, I obtain $m_E = gR_E^2/G$,

so $m_E = \dfrac{(9.8 \text{ m/s}^2)(6.4 \times 10^6 \text{ m})^2}{(6.6738 \times 10^{-11} \text{ N} \cdot \text{m}^2/\text{kg}^2)} = 6.0 \times 10^{24}$ kg. ✔

❹ EVALUATE RESULT Table 13.1 gives 5.97×10^{24} kg, which agrees with my answer to within less than 1%. Instead of looking up the value, I can also estimate Earth's mass to check my answer. Remember that we carried out that estimate in Checkpoint 1.16 and obtained 5×10^{24} kg, which is within 20% of my answer.

13.6 Gravitational potential energy

In Section 7.9 we determined the potential energy associated with the force of gravity near Earth's surface. Let us now obtain general expressions for the gravitational potential energy from the law of gravity. To do so, we consider two objects 1 and 2 of masses m_1 and m_2. Suppose the mass of object 1 is much greater than the mass of object 2, $m_1 \gg m_2$, so that if we place object 1 at the origin, we can consider it being fixed there. Now let object 2 move along the x axis from x_i to x_f under the influence of the gravitational attraction between the two (**Figure 13.28**). What is the work done by the gravitational force exerted by object 1 on object 2?

If we choose the x axis pointing to the right as indicated in Figure 13.28, the x component of the force exerted by object 1 on object 2 when object 2 is at some arbitrary position x is

$$F_{12x}^G = -G\frac{m_1 m_2}{x^2}. \tag{13.6}$$

Figure 13.28 An object of mass m_2 moves toward an object of mass $m_1 \gg m_2$ held fixed at the origin.

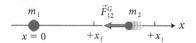

The x component of the force is negative because the force points in the negative x direction. The gravitational force varies with distance, and so to evaluate the work done by object 1 on object 2, we must evaluate the integral (see Eq. 9.22):

$$W = \int_{x_i}^{x_f} F_x(x)\,dx. \tag{13.7}$$

Substituting Eq. 13.6 into Eq. 13.7, we obtain

$$W = -Gm_1m_2 \int_{x_i}^{x_f} \left(\frac{1}{x^2}\right) dx = Gm_1m_2 \left[\frac{1}{x}\right]_{x_i}^{x_f} = Gm_1m_2 \left(\frac{1}{x_f} - \frac{1}{x_i}\right). \tag{13.8}$$

This is the work done by the gravitational force exerted by object 1 on the system consisting of object 2 only (**Figure 13.29a**). Because the force and the displacement point in the same direction, we know that the work must be positive. Indeed, the right side of Eq. 13.8 is positive because $x_f < x_i$.

As object 2 moves from x_i to x_f, it accelerates, and so its kinetic energy increases. Because no other energy associated with it changes, the increase in the kinetic energy must be equal to the work done on it by object 1, $\Delta K = W$, as illustrated in the energy diagram in Figure 13.29a.

If, instead of object 2 by itself, we now consider the (closed) system of the two interacting objects as a whole, then the increase in the kinetic energy of object 2 is due not to the work done by the external gravitational force exerted by object 1 on it but to a decrease in the gravitational potential energy of the system (Figure 13.29b and compare with Figure 9.18). This gravitational potential energy is a measure of the configuration of the system. The change in the gravitational potential energy of the system is thus $\Delta U^G = -\Delta K$. Because the change in kinetic energy does not depend on the choice of system, we see that the change in potential energy of the two-object system is the negative of W in Eq. 13.8:

$$\Delta U^G = Gm_1m_2\left(\frac{1}{x_i} - \frac{1}{x_f}\right). \qquad (13.9)$$

It is customary to choose the gravitational potential energy of a system to be zero at infinite separation ($x = \infty$), which is the separation distance at which the gravitational force is zero: $U^G(\infty) = 0$. So, if we let object 2 move in from $x = \infty$ to an arbitrary position x under the influence of the gravitational acceleration due to object 1, Eq. 13.9 yields

$$\Delta U^G = U^G(x) - U^G(\infty) = U^G(x) - 0 = Gm_1m_2\left(0 - \frac{1}{x}\right), \quad (13.10)$$

so the **gravitational potential energy** is

$$U^G(x) = -G\frac{m_1m_2}{x}, \qquad (13.11)$$

which is zero at infinite x. As object 2 moves in from infinitely far away, its kinetic energy increases, and so the potential energy of the two-object system decreases. Therefore setting $U^G(\infty) = 0$ at $x = \infty$ causes the potential energy to be negative for all values of $x < \infty$.

Figure 13.29 Energy diagrams for the situation shown in Figure 13.28.

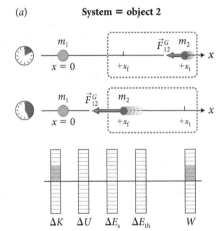

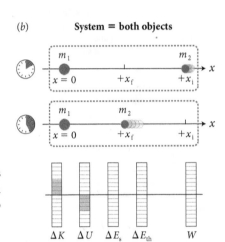

Example 13.4 Gravitational potential energy near Earth's surface

The distance between Earth's surface and an object of mass m is changed by an amount Δx. Show that when $x \approx R_E$ and $\Delta x \ll R_E$, where R_E is the radius of Earth, the general expression for gravitational potential energy, Eq. 13.9, reduces to the expression $\Delta U^G = mg\Delta x$ (Eq. 7.19).

❶ **GETTING STARTED** I begin by choosing an x axis that is perpendicular to Earth's surface and has the origin $x = 0$ at Earth's center. I then mark off a distance $x = R_E$ on this axis (**Figure 13.30a**). Because I'm told that $x \approx R_E$ and $\Delta x \ll R_E$, I let the initial position x_i be right at Earth's surface, $x = R_E$, and the final position x_f be just to the right of x_i.

To get some feel for the situation, I also graph the gravitational potential energy of Earth from Eq. 13.11, marking the object's initial and final positions (Figure 13.30b). Basically, the problem requires me to show that when x_i and x_f lie sufficiently close to each other, ΔU^G is linearly proportional to Δx. Graphically this is already clear from my graph: If I blow up the region around $x_i = R_E$, the $U^G(x)$ curve is linear.

❷ **DEVISE PLAN** If I substitute $x_i = R_E$ and $x_f = R_E + \Delta x$ into Eq. 13.9, I can obtain an expression showing how ΔU^G depends on Δx. I can then approximate the resulting expression for $\Delta x \ll R_E$.

Figure 13.30

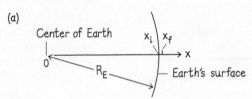

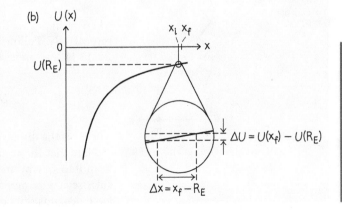

❸ **EXECUTE PLAN** After I substitute the masses of Earth and the object, $x_i = R_E$, and $x_f = R_E + \Delta x$, Eq. 13.9 becomes

$$\Delta U^G = Gm_E m \left(\frac{1}{R_E} - \frac{1}{R_E + \Delta x} \right). \qquad (1)$$

I know that $\Delta x \ll R_E$ in this problem, so I can write the second term in the parentheses as

$$\frac{1}{R_E + \Delta x} = \frac{1}{R_E} \left(\frac{1}{1 + \Delta x/R_E} \right) \approx \frac{1}{R_E} \left(1 - \frac{\Delta x}{R_E} \right) \qquad (2)$$

because, for any small number $\epsilon \ll 1$, $1/(1 + \epsilon) \approx 1 - \epsilon$ (see Appendix B). Substituting Eq. 2 into Eq. 1, I obtain

$$\Delta U^G = Gm_E m \left[\frac{1}{R_E} - \frac{1}{R_E} \left(1 - \frac{\Delta x}{R_E} \right) \right]$$

$$= Gm_E m \frac{\Delta x}{R_E^2} = m \frac{Gm_E}{R_E^2} \Delta x.$$

When I substitute, from Eq. 13.4, $g = Gm_E/R_E^2$, this expression becomes $\Delta U^G = mg\Delta x$. ✔

❹ **EVALUATE RESULT** This result confirms the conclusion I drew from my sketch: $U^G(x)$ is curved, but for sufficiently small displacements Δx, the curve can be approximated by a straight line. Near Earth's surface the slope of this line is mg.

Figure 13.31 To calculate the work done by object 1 on object 2 while object 2 moves along the dashed path from P to Q, we approximate the path by (*a*) one radial and one circular displacement or (*b*) a number of successive such displacements.

(*a*) Trajectory approximated by one radial segment PR and one circular segment RQ

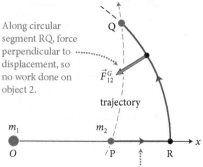

Along circular segment RQ, force perpendicular to displacement, so no work done on object 2.

Along radial segment PR, force not perpendicular to displacement, so work done on object 2.

(*b*) Trajectory approximated by multiple radial and circular segments

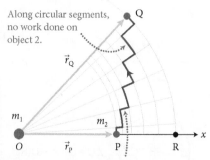

Along circular segments, no work done on object 2.

Radial segments add up to length of PR.

Because the force of gravity is a central force, we can easily generalize Eq. 13.11 to more than one dimension. Suppose object 2 moves from P to Q in **Figure 13.31a** along part of an elliptical orbit. Let us first approximate the displacement of object 2 by two successive displacements: one along the radial direction from P to R and one along a circular arc from R to Q. The work done by the force of gravity on object 2 over the radial displacement is given by Eq. 13.8. Along the circular arc RQ, the magnitude of the force of gravity is constant but always oriented perpendicular to the displacement. Because the work in more than one dimension is given by the scalar product of the force and the force displacement (Eq. 10.35), the work done by the force of gravity on object 2 along the circular arc is zero:

$$W_{R \to Q} = \int_R^Q \vec{F}_{12}^G \cdot d\vec{r} = 0. \qquad (13.12)$$

So the work done by object 1 on object 2 as it moves from P to Q along the path PRQ is equal to the work done along the radial displacement only.

A better approximation of the actual path of object 2 can be obtained by breaking down the motion into small segments and approximating each segment by the sum of radial and circular displacements, as shown in Figure 13.31b. Again, the work done on object 2 along each circular displacement is zero. Furthermore, all the radial displacements simply add up to the same radial displacement as in Figure 13.31a. So only the change in radial distance matters and the gravitational work done by object 1 on object 2 while object 2 moves from P to Q is

$$W_{P \to Q} = Gm_1 m_2 \left(\frac{1}{r_Q} - \frac{1}{r_P} \right), \qquad (13.13)$$

where r_P and r_Q are the magnitudes of the position vectors of object 2 at P and Q, respectively. Following the same line of reasoning that led from Eq. 13.8 to Eq. 13.11, we can write for the gravitational potential energy of the system

$$U^G(\vec{r}) = -G \frac{m_1 m_2}{r} \quad \text{(zero at infinite } r), \qquad (13.14)$$

where r is the distance between the two objects.

Note that the result we obtained in Eq. 13.13 holds for *any* path that leads from P to Q. Any arbitrary path can be broken down into small radial and circular segments, and as Figure 13.31b shows, the work done by the gravitational force depends only on the position of the endpoints relative to m_1, not on the path taken. Consider for example paths 1 and 2 from P to Q in Figure 13.32a. The work done is the same along both paths:

$$W_{P \to Q, \text{path 1}} = W_{P \to Q, \text{path 2}}. \qquad (13.15)$$

This path independence is characteristic of nondissipative interactions (see Section 7.8).

We also know that the work done along a path moving in one direction is the negative of the work done along that path moving in the opposite direction. This can readily be seen in Figure 13.31b: Reversing the direction of travel inverts the sign of the displacement (and therefore the sign of the work) along all radial components. So if we let object 2 move from Q to P along path 2 in Figure 13.32a, the work done on it is

$$W_{Q \to P, \text{path 2}} = -W_{P \to Q, \text{path 2}}. \qquad (13.16)$$

A consequence of Eq. 13.15 and 13.16 is that if an object moves from P to Q along path 1 and then back to P along path 2, as shown in Figure 13.32b, then the work done by the gravitational force on the object is zero:

$$W = W_{P \to Q, \text{path 1}} + W_{Q \to P, \text{path 2}} = W_{P \to Q, \text{path 2}} + (-W_{P \to Q, \text{path 2}}) = 0. \quad (13.17)$$

A path like the one shown in Figure 13.32b, in which any object following the path returns to its original position, is called a **closed path.** Because the gravitational potential energy depends only on the position of the endpoints and the endpoints are the same, the change in the potential energy of the two-object system is zero. Consequently the work done by the gravitational force exerted by one object on the other as it moves along a closed path is zero, as Eq. 13.17 shows.

✋ **13.18** Near Earth's surface, when $\Delta U^G = mg\Delta x$, it is customary to let the gravitational potential energy be zero at Earth's surface, instead of at infinite separation as we did in Eq. 13.10. Does this choice of zero yield greater or smaller values for the gravitational potential energy? By what amount?

13.7 Celestial mechanics

Now that we have derived an expression for the gravitational potential energy, we can write an expression for the energy of a system of two celestial objects. Consider a system consisting of two objects of which one has a much greater mass than the other ($M \gg m$), so that we can consider the center of mass of the two-object system to be fixed at the center of the larger object (see Section 13.2). Examples of such a system are a planet orbiting a star, a moon orbiting a planet, and a satellite orbiting Earth (Figure 13.33). For convenience we'll refer to the object with the greater mass M as the *star* and the object with the smaller mass m as the *satellite,* but the results apply to any pair of celestial objects with $M \gg m$. We shall consider the system to be closed and isolated so that neither its energy nor its angular momentum can change:

$$\Delta E = 0 \quad \text{and} \quad \Delta \vec{L} = \vec{0}. \qquad (13.18)$$

The energy of the system consists of the gravitational potential energy and the kinetic energy. Because only the satellite is in motion, we have

$$E = K + U^G = \tfrac{1}{2}mv^2 - \frac{GMm}{r}, \qquad (13.19)$$

where v is the speed of the satellite and r is the distance between the satellite and the star.

Figure 13.32 The gravitational work done on object 2 depends only on the endpoints.

(a)

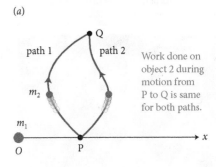

Work done on object 2 during motion from P to Q is same for both paths.

(b)

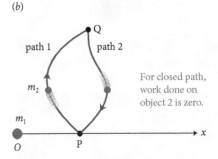

For closed path, work done on object 2 is zero.

Figure 13.33 When one object orbits another that has much greater mass, we can consider the center of mass of the system to be fixed at the center of the larger object.

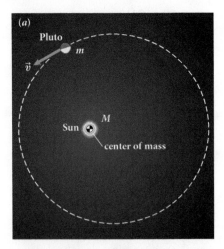

(a)

Pluto

$\vec{v}$

m

M

Sun

center of mass

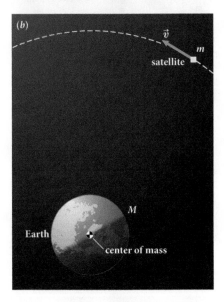

(b)

$\vec{v}$

m

satellite

M

Earth

center of mass

Figure 13.34 When the mechanical energy E_{mech} of a system is negative, the separation distance r between two objects in the system is limited to the region where $U^G < E_{mech}$.

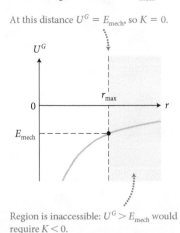

At this distance $U^G = E_{mech}$, so $K = 0$.

U^G

0

r_{max}

r

E_{mech}

Region is inaccessible: $U^G > E_{mech}$ would require $K < 0$.

Because the force of gravity is central, the angular momentum of each object is constant. For the satellite we have

$$L = r_{\perp} mv. \tag{13.20}$$

Equations 13.18–13.20 allow us to determine the shape of the satellite's orbit given the values of $\vec{r}$ and $\vec{v}$ at any given instant.

13.19 (a) If an object of mass m is released from rest a distance r from a star of mass $M \gg m$ and radius R_s, is the mechanical energy of the star-object system E_{mech} positive, negative, or zero? (b) Describe the motion of the object, and determine the maximum and minimum values of its kinetic energy during its motion. (c) Does the motion you described in part b satisfy conservation of angular momentum about the center of the star? (d) Suppose the object is now launched with nonzero velocity in a direction perpendicular to the straight line joining it and the center of the star. Is E_{mech} positive, negative, or zero?

As this checkpoint shows, the mechanical energy E_{mech} of a star-satellite system can be negative because U^G is always negative when we let the gravitational potential energy be zero at infinite separation. To understand the consequences of the energy being negative, look at **Figure 13.34**, which shows the gravitational potential energy U^G as a function of the distance r between satellite and star. As we saw in Section 13.6, U^G has a $1/r$ dependence and approaches zero as r goes to infinity. If the energy E of the system is negative, then there is a value $r = r_{max}$ for which $U^G = E_{mech}$—all of the mechanical energy is in the form of potential energy, and thus the satellite's kinetic energy is zero. Any separation distance r greater than r_{max} would yield a negative kinetic energy because $U^G > E_{mech}$ for $r > r_{max}$. Because the kinetic energy cannot be negative, the satellite's motion for negative values of E_{mech} is restricted to values of $r < r_{max}$ for which $U^G < E_{mech}$. The maximum separation distance r_{max} occurs when $U^G = E_{mech}$:

$$-\frac{GMm}{r_{max}} = E_{mech} \tag{13.21}$$

$$r_{max} = \frac{GMm}{-E_{mech}}. \tag{13.22}$$

In words, we say that for negative E_{mech}, the satellite is *bound* to the star: It cannot escape to infinity because it does not have enough kinetic energy. When the mechanical energy E_{mech} is positive, the gravitational potential energy is *always* smaller than E_{mech}, and so even at infinite separation, where the gravitational potential energy is maximum, there still is a positive amount of kinetic energy. Therefore there is no limit on the separation distance r for $E_{mech} > 0$; in this case, we say the satellite is *unbound* because it has enough energy to "escape" to infinity.

13.20 (a) If our satellite of mass m were to reach the position r_{max} given by Eq. 13.22, what would its angular momentum be? (b) What would its trajectory have to be to satisfy conservation of angular momentum? (c) If its angular momentum is nonzero, can the satellite ever reach r_{max}?

To determine the shape of the satellite's orbit, we must solve Eqs. 13.18–13.20. The solution of these simultaneous equations yields orbits that correspond to the *conic sections* illustrated in **Figure 13.35**, with the star at the focus of the section. The sections shown in the figure fall into two categories: *ellipses* (including the

Conic sections

Second-degree equations in two variables of the form

$$Ax^2 + Bxy + Cy^2 + Dx + Ey + F = 0$$

are called **conic sections** because the curves they represent can be obtained by intersecting a plane and a circular cone (Figure 13.35). These curves can also be written in the form

$$\frac{x^2}{a^2} + \frac{y^2}{a^2(1 - e^2)} = 1, \tag{1}$$

where a is the semimajor axis (see Figure 13.7) of the section and e is the eccentricity.

Ellipse: For $0 \leq e < 1$, we obtain an ellipse with a semimajor axis a and a semiminor axis b with

$b^2 = a^2(1 - e^2)$, where e is the eccentricity of the ellipse. For each point on the ellipse, the sum of the distances to the two foci at $(\pm ae, 0)$ is equal to $2a$. A special type of ellipse is obtained for $e = 0$: The two foci coincide, and the ellipse becomes a **circle** of radius a.

Hyperbola: When $e > 1$, the term that contains y^2 becomes negative, and Eq. 1 yields the two branches of a hyperbola with foci at $(\pm ae, 0)$ and oblique asymptotes $y = \pm(b/a)x$, with $b^2 = a^2(e^2 - 1)$. For each point on the hyperbola, the difference of the distances to the foci is equal to $2a$.

The limiting case between an ellipse and a hyperbola, which occurs when $e = 1$, yields a **parabola** of the form $y^2 = 4px$. The parabola can be thought of as an elongated ellipse with one focus at $(p, 0)$ and the other at $(\infty, 0)$.

Figure 13.35 Conic sections.

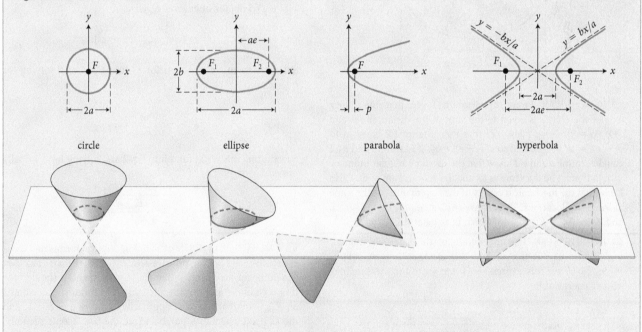

circle   ellipse   parabola   hyperbola

circle, which is a special type of ellipse) and *hyperbolas*. The limiting section between these two categories is the *parabola*. Ellipses are bound and therefore correspond to negative mechanical energy; hyperbolas are unbound and thus correspond to positive mechanical energy. The limiting case between bound and unbound motion occurs for $E_{mech} = 0$, when the satellite has just enough kinetic energy to approach infinity (with zero kinetic energy). The corresponding orbit is the parabola. Summarizing, we have

$$E_{mech} < 0: \quad \text{elliptical orbit}$$

$$E_{mech} = 0: \quad \text{parabolic orbit}$$

$$E_{mech} > 0: \quad \text{hyperbolic orbit}$$

Example 13.5 Rocket science

A satellite of mass m_{sat} is in an elliptical orbit around a star of mass $m_{star} \gg m_{sat}$. If the mechanical energy of the star-satellite system is E_{mech} and the magnitude of the angular momentum of the satellite about the star is L, what are the semimajor axis a and the eccentricity e of the satellite's orbit?

❶ **GETTING STARTED** I begin by making a drawing of the system (Figure 13.36), placing the star at one focus (F) of the ellipse. I also mark the positions of smallest (P) and and greatest (Q) separation between star and satellite, because at those locations the position vector $\vec{r}$ from star to satellite is perpendicular to the satellite's velocity $\vec{v}$, and therefore it is relatively easy to calculate the angular momentum.

Figure 13.36

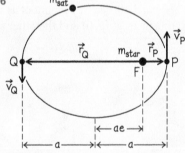

❷ **DEVISE PLAN** From my diagram I see that the major axis of the ellipse is equal to the sum of the distances FP and FQ: $2a = r_P + r_Q$. I also see that the distance FP is equal to $a - ae = a(1 - e)$, and so $a(1 - e) = r_P$. If I know r_P and r_Q, I can determine a and e. I know that the satellite's angular momentum at P must be the same as its angular momentum at Q. I also know that the mechanical energy of the system must be the same at these two locations. Thus I have two equations, $L_P = L_Q = L$ and $E_P = E_Q = E_{mech}$, from which I can determine r_P and r_Q.

❸ **EXECUTE PLAN** The magnitude of the satellite's angular momentum at P is the same as that at Q, so $L = m_{sat}r_P v_P = m_{sat}r_Q v_Q$. Using this expression, I can write for the satellite's kinetic energy at P:

$$K_P = \tfrac{1}{2}m_{sat}v_P^2 = \frac{m_{sat}^2 v_P^2 r_P^2}{2m_{sat}r_P^2} = \frac{L^2}{2m_{sat}r_P^2}.$$

The energy of the system at P is then

$$E_{mech} = K_P + U_P^G = \frac{L^2}{2m_{sat}r_P^2} - \frac{Gm_{sat}m_{star}}{r_P}. \tag{1}$$

After writing corresponding expressions for position Q and setting the system energy at P equal to that at Q, I get

$$\frac{L^2}{2m_{sat}r_P^2} - \frac{Gm_{sat}m_{star}}{r_P} = \frac{L^2}{2m_{sat}r_Q^2} - \frac{Gm_{sat}m_{star}}{r_Q}.$$

I now put both terms containing L on the left and combine them over their common denominator, put both terms containing G on the right and combine them over their common denominator, and work through the algebra to get

$$\frac{L^2}{2m_{sat}}(r_P + r_Q) = Gm_{sat}m_{star}r_P r_Q.$$

Substituting $r_P + r_Q = 2a$, $r_P = a(1 - e)$, and $r_Q = 2a - r_P = a(1 + e)$, I obtain, after some algebra,

$$L^2 = Gm_{sat}^2 m_{star}a(1 - e^2). \tag{2}$$

Substituting this expression for L^2 into Eq. 1 and solving for a give

$$a = -\frac{Gm_{sat}m_{star}}{2E_{mech}}. \checkmark$$

Substituting this result for a into Eq. 2 and solving for e, I finally obtain

$$e = \sqrt{1 + \frac{2E_{mech}L^2}{G^2 m_{sat}^3 m_{star}^2}}. \checkmark$$

❹ **EVALUATE RESULT** The orbit being elliptical means that E_{mech} is negative, and so a is positive and e is smaller than 1, as they should be for an ellipse. When $E_{mech} = 0$, the orbit should turn into a parabola and the satellite can go to infinity. Indeed, as I let E_{mech} approach zero, a approaches infinity and $e = 1$, and the ellipse turns into a parabola (see the box "Conic sections" on page 331).

13.21 Consider a planet of mass m moving at constant speed v in a circular orbit of radius R under the influence of the gravitational attraction of a star of mass M. (*a*) What is the planet's kinetic energy, in terms of m, M, G, and R? (*b*) What is the energy of the star-planet system? (*c*) Using the expression for the eccentricity obtained in Example 13.5, obtain an expression for E_{mech} for the circular orbit. (*d*) Compare your answers to parts *b* and *c*.

The results obtained in Example 13.5 show that the length $2a$ of the major axis of the ellipse depends on only the mechanical energy E_{mech} of the system. Because E_{mech} is negative, the greater E_{mech}, the closer it is to zero and the greater a.

Figure 13.37 shows five orbits with the same fixed energy E_{mech} but different values of L. The energy $E_{mech} = U^G + K$ is fixed because the satellite is launched

Figure 13.37 Orbits of a satellite of mass m launched in different directions with fixed speed and a fixed distance from an object of mass M. Because the mechanical energy E_{mech} of the system is always the same, the length $2a$ of the orbit's major axis (dashed line) is constant. As the angle between $\vec{v}_i$ and $\vec{r}_i$ is decreased, the angular momentum decreases and the eccentricity of the orbit increases.

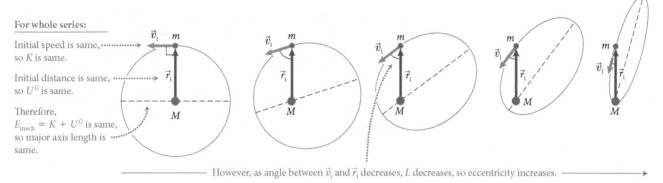

For whole series:

Initial speed is same, so K is same.

Initial distance is same, so U^G is same.

Therefore, $E_{\text{mech}} = K + U^G$ is same, so major axis length is same.

However, as angle between $\vec{v}_i$ and $\vec{r}_i$ decreases, L decreases, so eccentricity increases.

at a fixed initial speed (so K is fixed) from a point P that is a fixed distance r_i from the center of the larger object of mass M (so U^G is fixed). Because E_{mech} is fixed, the length $2a$ of the major axis is the same for all orbits. As the direction of $\vec{v}_i$ is changed (without changing the speed v_i and therefore E_{mech} and $2a$), the magnitude of the angular momentum changes, affecting the eccentricity of the orbit, which depends on both L and E_{mech}. For a fixed mechanical energy E_{mech} (that is to say, for a fixed value of the major axis), e increases as L decreases. So, as the angle between $\vec{v}_i$ and $\vec{r}_i$ gets smaller, the ellipse becomes more elongated. In all cases, the object of mass M is at one of the foci of the ellipse.

Figure 13.38 shows the orbits of an object launched multiple times from a fixed location that is a distance r_i from Earth's center. All launches are in the same direction, perpendicular to the straight line connecting the object to Earth's center, but the launch speed is increased for each successive launch. Because the initial gravitational potential energy U^G is always the same, the value of the mechanical energy E_{mech} is determined by the initial kinetic energy: $E_{\text{mech}} = U^G + K$.

If the object is released from rest, $K = 0$ and so $E_{\text{mech}} = U^G$. As you saw in Checkpoint 13.19, the object falls in a straight line toward the surface of Earth. When the object is given a small initial speed, however, the orbit is an ellipse

Figure 13.38 Orbits of an object launched at different speeds from a fixed location that is a fixed distance r_i from Earth's center.

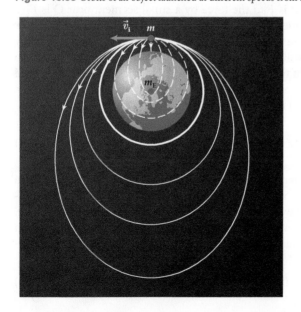

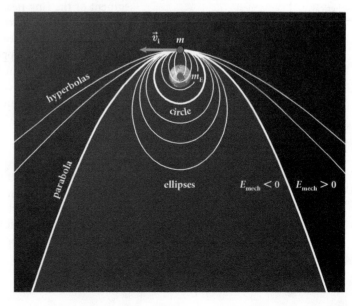

with Earth's center at the focus farther from the launch point. For small v_i, the orbit intersects Earth and the object crashes. As v_i is increased, the ellipse becomes rounder and longer, and for sufficiently great v_i, the object makes it around the planet.

As we continue to increase v_i, we reach a point when v_i has just the right magnitude to make v_i^2/r_i equal to the centripetal acceleration caused by Earth's gravitational force, and the object goes in a circular orbit. For greater v_i, the orbit is again elliptical, but Earth's center now coincides with the nearer focus of the ellipse.

As we continue to increase v_i, we reach a point where the mechanical energy E_{mech} of the Earth-object system is no longer negative and the orbit is unbound. This happens when

$$E_{mech} = \tfrac{1}{2}mv_{esc}^2 - \frac{Gmm_E}{r_i} = 0, \tag{13.23}$$

where v_{esc} is the object's *escape speed*, m is the mass of the launched object, m_E is Earth's mass, and r_i is the distance from the launch point to Earth's center. An object launched at the escape speed and moving without any further propulsion travels out to infinity along a parabolic path. At launch speeds greater than v_{esc} the orbit corresponds to one branch of a hyperbola.

13.22 (*a*) Determine an expression for the escape speed at Earth's surface. (*b*) What is the value of this escape speed? (*c*) Does it matter in which direction an object is fired at the escape speed?

13.8 Gravitational force exerted by a sphere

Central to Newton's law of gravity is the statement that a solid sphere exerts a gravitational force on an object outside the sphere as if all its mass were concentrated at the center of the sphere. To prove this statement, we'll calculate the force exerted on a particle of mass m by a thin spherical shell of mass M and radius R; the distance between the particle and the center of the shell is r. Once we have a result for the shell, we can obtain the result for a solid sphere by treating the solid sphere as a large number of concentric spherical shells, like the layers of an onion.

Consider first a vertical ring-shaped piece of the shell—the portion contained between the angles θ and $\theta + d\theta$ in **Figure 13.39a**. The force exerted by a small segment of the ring near P is $\vec{F}_{Pp}^G$; the force by a small segment near Q is $\vec{F}_{Qp}^G$. Only the components of these forces along the x axis contribute to the vector sum of the forces exerted on the particle. The same cancellation occurs for every pair of diametrically opposite small segments that make up the entire ring. To calculate the vector sum of the forces exerted on the particle, we therefore need to consider only the x components. The x component of the force exerted by a small segment of the ring of mass dm near P on the particle is

$$dF_{Ppx}^G = -G\frac{m\,dm}{s^2}\cos\alpha, \tag{13.24}$$

where α is the angle between the force and the straight line from the particle to the center of the shell and s is the distance from the particle to P. The x components of the forces exerted by all segments of the ring have the same magnitude. The vector sum of the forces exerted by the whole ring is thus equal to the sum of the contributions from all segments of the ring:

$$dF_{rpx}^G = -G\frac{m\,dM}{s^2}\cos\alpha, \tag{13.25}$$

Figure 13.39 To calculate the gravitational force exerted by a spherical shell of mass M on an object of mass m, the shell is divided into vertical rings.

(*a*)

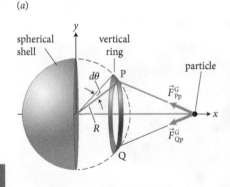

(*b*) Geometry of red triangle in part *a*.

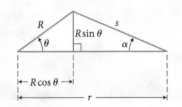

where dM is the mass of the entire ring (the sum of the masses of all segments). The width of the ring being equal to the arc length $R\,d\theta$, the area of the ring is (length) $\times$ (width) $= (2\pi R \sin \theta)(R\,d\theta) = 2\pi R^2 \sin \theta\,d\theta$. If the mass M of the entire shell is uniformly distributed over the shell, then the mass per unit area is $M/(4\pi R^2)$, so

$$dM = 2\pi R^2 \sin \theta\,d\theta\,\frac{M}{4\pi R^2} = \frac{M}{2}\sin \theta\,d\theta \tag{13.26}$$

or, combining Eqs. 13.25 and 13.26,

$$dF_{\mathrm{rp}x}^{G} = -\frac{GMm}{2}\frac{\cos \alpha \sin \theta\,d\theta}{s^2}. \tag{13.27}$$

Equation 13.27 gives the force exerted by one vertical ring of our spherical shell on the particle. Now we must integrate this result over the entire shell. Before we can carry out this integration, we must reduce the variables α, s, and θ, which are all related, to one variable. From Figure 13.39b we see that

$$\cos \alpha = \frac{r - R \cos \theta}{s}, \tag{13.28}$$

and from the law of cosines (see Appendix B), we have

$$s^2 = R^2 + r^2 - 2rR \cos \theta. \tag{13.29}$$

Differentiating the left side of this equation with respect to s yields $2s$. For the right side we get

$$\frac{d}{ds}(R^2 + r^2 - 2rR \cos \theta) = \frac{d}{d\theta}(R^2 + r^2 - 2rR \cos \theta)\frac{d\theta}{ds}$$

$$= 2rR \sin \theta\,\frac{d\theta}{ds}. \tag{13.30}$$

Equating the derivatives on the two sides, we get

$$2s = 2rR \sin \theta\,\frac{d\theta}{ds} \tag{13.31}$$

$$\sin \theta\,d\theta = \frac{s\,ds}{rR}. \tag{13.32}$$

Substituting Eq. 13.32 and $R \cos \theta$ from Eq. 13.29 into Eq. 13.28 and then into Eq. 13.27 yields

$$dF_{\mathrm{rp}x}^{G} = -\frac{GMm}{4r^2 R}\frac{(r^2 + s^2 - R^2)ds}{s^2} = -\frac{GMm}{4r^2 R}\left(\frac{r^2 - R^2}{s^2} + 1\right)ds. \tag{13.33}$$

To obtain the force exerted by the entire shell, we must integrate this expression from $s = r - R$ (at $\theta = 0$) to $s = r + R$ (at $\theta = \pi$):

$$F_{\mathrm{sp}x}^{G} = \int_{r-R}^{r+R} dF_{\mathrm{rp}x}^{G} = -\frac{GMm}{4r^2 R}\int_{r-R}^{r+R}\left(\frac{r^2 - R^2}{s^2} + 1\right)ds. \tag{13.34}$$

Working out the integral (see Checkpoint 13.23), we get

$$\int_{r-R}^{r+R}\left(\frac{r^2 - R^2}{s^2} + 1\right)ds = 4R \qquad (13.35)$$

so

$$F_{\text{sp}\,x}^G = -G\frac{Mm}{r^2} \quad \text{(shell)}, \qquad (13.36)$$

where the minus sign indicates that the force points toward the center of the shell. So we see that the shell exerts a force as if its entire mass M were concentrated at the center.

We can now easily extend this result to a solid sphere by considering many concentric shells. For each shell, we obtain an expression of the form given by Eq. 13.36; adding all terms (one for each shell) then yields

$$F_{\text{sp}}^G = G\frac{mM_{\text{sphere}}}{r^2} \quad \text{(sphere)} \qquad (13.37)$$

for the magnitude of the gravitational force exerted by a solid sphere of mass M_{sphere} on a particle of mass m placed a distance r from the sphere's center. Equation 13.37 shows that the solid sphere exerts a force as if its entire mass M were concentrated at the center.

13.23 (*a*) Work out the integral in Eq. 13.35. (*b*) Show that Eq. 13.33 still holds if the particle of mass m is inside the shell. (*c*) What are integration limits with this particle inside the shell? What is the value of the integral in Eq. 13.35 for these limits?

Chapter Glossary

SI units of physical quantities are given in parentheses.

Balance A device that uses a pivoted rod to compare the gravitational force exerted on an object of unknown mass with that exerted on objects of known mass.

Central force A force whose line of action always lies along the straight line connecting the two interacting objects.

Closed path A path in which any object that follows the path returns to its original position.

Conic sections Mathematical curves that are obtained by intersecting a plane and a circular cone. The shape of the orbit of two gravitationally interacting objects is a conic section.

Gravitational constant A fundamental constant that determines the strength of the gravitational interaction:

$$G = 6.6738 \times 10^{-11}\,\text{N} \cdot \text{m}^2/\text{kg}^2. \qquad (13.5)$$

See also *law of gravity*.

Gravitational potential energy U^G (J) The potential energy associated with a system of interacting objects. If the gravitational potential energy of a system of two objects with masses m_1 and m_2 is chosen to be zero at infinite separation, it is

$$U^G(\vec{r}) = -G\frac{m_1 m_2}{r} \qquad (13.14)$$

when the separation of the objects is r.

Kepler's laws Laws that describe celestial orbits deduced from measurements of the motions of the planets. Kepler's laws are a direct consequence of the law of gravity.

Law of gravity The law that describes the gravitational attraction between all objects that have mass. An object of mass m_1 exerts on an object of mass m_2 a distance r away an attractive central force of magnitude

$$F_{12}^G = G\frac{m_1 m_2}{r_{12}^2}, \qquad (13.1)$$

where G is the gravitational constant.

Mass m (kg) A scalar property of every piece of matter that determines the strength of its gravitational interaction with other pieces of matter. Except at very high speeds, the mass of an object is equal to the object's inertia.

Principle of equivalence A basic principle of the theory of general relativity that states that one cannot distinguish locally between the effects of a constant gravitational acceleration of magnitude g and an acceleration of the reference frame of magnitude $a = g$.

Spring scale A device that measures the downward force exerted on it by a load, by determining the deformation of a calibrated spring.

14

Special Relativity

CONCEPTS

QUANTITATIVE TOOLS

Time is the basic fabric of life because living means *changing* and change—a concept central to all we have discussed up to now—cannot exist without time. Time is so basic that we take for granted that we all agree on what it is. If, for instance, you and I arrange to meet here tomorrow at three o'clock and we make sure our watches agree now, we expect to see each other here tomorrow when our watches read three o'clock.

As we see in this chapter, however, our commonsense experience of time fails to explain certain experimental observations. In 1905 Albert Einstein realized these shortcomings and revised the basic notions of space and time with his *theory of special relativity*. The basic principles of this theory are the principle of relativity, which we discussed in Chapter 6, and the experimental fact that the measured value of the speed of light is independent of the motion of the source relative to the observer making the measurement. The revised notions of space and time require us to adjust our understanding of inertia, momentum, and energy. However, the conservation principles—the unifying principles of this text and the most fundamental laws of physics—continue to hold, as required by the principle of relativity.

14.1 Time measurements

Since the advent of clocks, the measurement of time has been so straightforward that it hardly requires much thought. Just as spatial coordinates determine *where* something happens, time tells us *when* it happens. Let's define an **event** as something that happens at a specific location at a specific instant. Take, for example, your act of reading this sentence. If you and I agree on a coordinate grid (for example, longitude, latitude, and altitude) and a time zone, you can specify the location and time of this event in such a way that it is uniquely and precisely defined. You are reading this sentence, say, in Helena, Montana (46.595°N, 112.030°W, alt. 1138 m), at 9:33 p.m. today.

Let's concentrate for now on the time part of describing events. By referring to the watch on your wrist, you can measure the instant at which any event near you occurs. There are two reasons you can count on this procedure to work. First, it is possible to construct a device—a clock—that measures equal, calibrated time intervals (say, seconds). Clocks serve as a "ruler" for time, just as an ordinary ruler helps us measure space. Second, we can *synchronize* clocks; that is, we can make sure two clocks run at the same rate and their readings agree.

Keeping time helps us sort out the sequence of events and establish a possible causal relationship between events (see Section 1.4). If you are at the location of two successive events, it's straightforward to determine which event happened first and what the time interval between them is. How, though, can you determine the sequence for two events that occur at different locations?

We need to be able to detect some sort of signal (such as sound or light) to observe an event, and all such signals

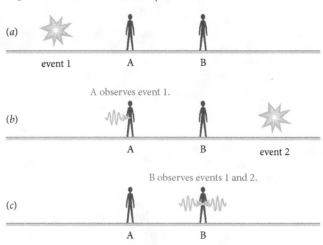

Figure 14.1 Two observers with synchronized clocks observe two events.

(a) event 1   A   B

A observes event 1.

(b) A   B   event 2

B observes events 1 and 2.

(c) A   B

travel at some finite speed. Of all signals known, light travels the fastest and therefore is the most convenient for timing events.

Figure 14.1 shows two events—the emission of light signals at two locations—and their observation by two observers using synchronized clocks. The light signals are of short duration, like a flash, so that we can consider each emission to take place at a definite instant. The signals travel at the speed of light (3.00×10^8 m/s in vacuum or air) and carry the message that the event has occurred. Observer A is closer to event 1 than to event 2; observer B is closer to event 2 than to event 1. In Figure 14.1a, event 1 occurs but the signal has not yet reached either observer. In Figure 14.1b, the signal from event 1 reaches observer A and event 2 occurs. Note that the signal from event 1 has not yet reached observer B and that the signal from event 2 has reached neither observer. A little later, the signals from both events reach observer B (Figure 14.1c) but the signal from event 2 has not yet reached observer A.

✋ **14.1** In Figure 14.1, based on the time of arrival of the light signals, which event is observed first according to (a) observer A and (b) observer B?

As this checkpoint shows, if you go by the instants at which the signals reach the two observers, the observers do not agree on the sequence of the two events. Observer A sees the events as occurring at two different instants, but observer B sees them as occurring at the same instant. The instant of *observation*, however, is not the instant at which the light signals *are emitted*. For events that occur around you in daily life, the difference between the instant each event happens and the instant you observe it is negligible because light travels so fast. Light emitted by the Sun, however, takes about 8 minutes to reach us. Thus, if a solar flare occurs at the surface of the Sun, that event takes place 8 minutes before we observe it. Light emitted by the Andromeda Galaxy takes 2 million years to reach Earth. If you look at the Andromeda Galaxy on a dark, moonless

Figure 14.2 One way to measure the time interval between two events is to travel from one event to the other, using a single clock to record the instant at which each occurs.

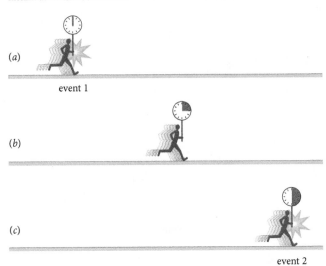

(a)

event 1

(b)

(c)

event 2

night, you see the light it emitted 2 million years ago. A lot could have happened to that galaxy in those 2 million years—for all you know, the entire galaxy might not be there anymore!

Even though we do not deal with these astronomical distances here in our discussion of time measurements, we still need to take into account the *signal delay* between event and observation. This signal delay is the time interval it takes for the signal to travel from the location of the event to the location of the observer. For nearby events, the signal delay is negligible. For events separated by an appreciable distance, we can avoid signal-delay complications in several ways. One straightforward, if not always practical, way is shown in **Figure 14.2**: An observer travels from the location of event 1 to the location of event 2 and records the two events on the same clock. In the observer's reference frame, the two events take place *at the same location* (the origin if you imagine axes attached to the observer) but at different instants. To avoid the complications of a noninertial reference frame (see Chapter 6), the observer should travel at constant velocity. The time interval obtained this way is called the **proper time interval:***

> The time interval between any two events that occur at the same position in a particular reference frame is called the proper time interval between the two events.

Not all pairs of events are separated by a proper time interval: If an observer does not have enough time to travel from the first event to the second event to observe both events at his location, there is no proper time interval for those events.

---

* From the French *propre* meaning *own*, as in one's "own time."

Example 14.1 Proper time interval

You are flying in a plane, and a friend is on the ground. Which of you measures the proper time interval between (a) the beginning and end of a brief in-flight video program and (b) two successive signals from a warning light on a tall building along your flight path?

1 GETTING STARTED The problem involves two observers (my friend and I) and two pairs of related events. My friend is in the Earth reference frame (E); I am in the reference frame of the plane (P). To simplify the discussion, I denote the beginning and end of the movie as events 1 and 2, respectively, and the two successive flashes of the warning light as events 3 and 4.

2 DEVISE PLAN To see who measures the proper time interval, I should make a sketch for each pair of events, showing the positions of the E and P reference frames. If either observer sees either pair of events as occurring at the same location, that observer measures the proper time interval.

3 EXECUTE PLAN (a) **Figure 14.3** shows that the beginning (event 1) and end (event 2) of the video program take place at the same location in reference frame P. Because that is my reference frame, I measure the proper time interval between these events. Figure 14.3 also shows that to my friend, standing at one location in the Earth reference frame, the two events happen at different locations (event 1 high and to his left, event 2 high and to his right). Thus he cannot measure the proper time interval. ✔

Figure 14.3

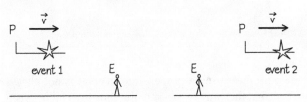

(b) **Figure 14.4** shows that the two flashes take place at the same place in reference frame E. Because this is my friend's reference frame, he measures the proper time interval between the two flashes. As for me, I see the first flash as coming from right below the plane but the second flash as coming from behind the plane. Two different locations from my point of view mean that I cannot measure the proper time interval between the flashes. ✔

Figure 14.4

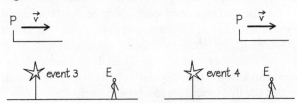

4 EVALUATE RESULT The video "happens" in the plane, so the plane is the reference frame in which to measure the proper time interval—the screen on which the video is projected is at rest in the plane. Likewise, the warning light is at rest in the Earth reference frame. My answers make sense.

CONCEPTS

Figure 14.5 Synchronized clocks can be used to measure the time interval between events that occur in different locations. (*a*) Each observer records the instant at which a nearby event occurs. (*b*) To synchronize their clocks, the two observers record the instant at which a single event occurs.

(*a*)

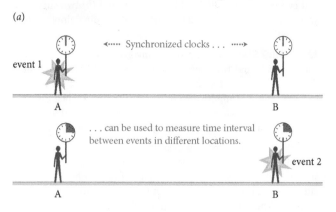

(*b*)

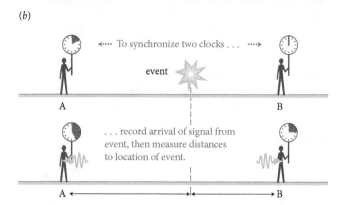

Figure 14.6 The scaffold of time: a reference frame made up of a grid of synchronized clocks. An event can occur anywhere in the grid, and the instant at which the event occurs is the instant shown on the clock nearest the event.

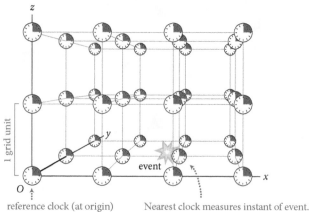

Because measuring a proper time interval is not always practical, another way to measure the time interval between events that occur in two locations is to place observers with identical clocks near each event, so that each observer can record the nearby event according to his clock (**Figure 14.5a**). Once the events are over, the two observers can communicate their clock readings and determine which event occurred first.

To synchronize the clocks, the observers can use the procedure illustrated in Figure 14.5*b*. An event sends out light signals to both observers, and each observer records the instant at which the signal reaches his clock. They then measure the distance from each clock to the location of the event and correct for the signal delay. If the clocks are synchronized, the observers obtain the same instant for the event after they subtract the signal delay from their recorded clock reading. If they don't obtain the same instant, they can use this procedure to adjust their clocks.

Because we don't always know where events are going to occur, it makes sense to construct a reference frame that has identical clocks arranged in a grid at regular fixed spacings, as shown in **Figure 14.6**. We can synchronize the clocks by sending a signal from the clock at the origin (the *reference clock*) to all the other clocks and correcting each

clock's reading for the signal delay. We can then define the instant in that reference frame at which any event occurs as the reading on the clock whose position coincides with that event. From now on, when we say "measuring the instant at which an event occurs in a given reference frame," we shall understand this to mean that we are using this procedure and that *an observer* in this reference frame is the observer stationed at the clock nearest the event.

Now that we have established a procedure for determining the instant at which an event occurs in a certain reference frame, we can determine the order of events in that reference frame. Two events that occur at different positions in the reference frame are **simultaneous** when the readings of the two clocks at the two locations of the events are identical. When the events are not simultaneous, the clock readings at the locations of the events allow us to determine the sequence of the events.

Exercise 14.2 Synchronizing clocks

Suppose you want to synchronize the network of clocks in Figure 14.6 by emitting a single light pulse at the origin. This signal triggers each clock to start running from some initial setting. If it takes the signal 1 s to travel along one grid unit and if the reference clock is set initially to 12 o'clock, what should the initial settings be for each clock (*a*) along the *x* axis and (*b*) along the diagonal in the *y* = 0 plane?

SOLUTION (*a*) Along the *x* axis, the signal arrives at the first clock to the right of the reference clock 1 s after the emission, when the reference clock reads 12:00:01, and so I set the first clock to 12:00:01. The next clock along the *x* axis starts running when the signal triggers it at 12:00:02, and so I set each clock on the *x* axis 1 s ahead of the clock immediately to its left. ✔

(*b*) The light must travel $\sqrt{2}$ grid units to reach the first clock along the diagonal. We must therefore set that clock ahead of the reference clock by $\sqrt{2}$ s = 1.414 s. The next clock along the diagonal is set ahead by $2\sqrt{2}$ s, and so on. ✔

Figure 14.7 A detector located midway between two light sources can detect whether signals are emitted (*a*) simultaneously or (*b*) not simultaneously.

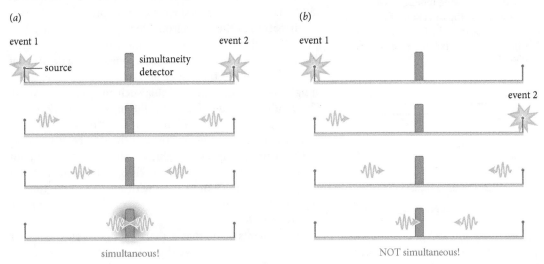

(a)

(b)

simultaneous!

NOT simultaneous!

✋ **14.2** Suppose an observer is standing at the origin in Figure 14.6 and all the clocks are synchronized. (*a*) Which of the following is true? The observer sees that (*i*) all the clocks read the same time, (*ii*) nearby clocks display later times than more distant clocks, (*iii*) nearby clocks display earlier times than more distant clocks. (*b*) Does the observer see all the clocks running at the same rate?

14.2 *Simultaneous* is a relative term

Suppose we determine that two events—for example, the emission of light signals by two sources—are simultaneous in a certain reference frame. Are these events also simultaneous in another reference frame that is in motion relative to the first? To help answer this question, we use a *simultaneity detector*: a device that emits a blue light when it receives two signals at the same instant.

When we position the detector midway between the positions where the two events take place, as illustrated in **Figure 14.7**, the time interval it takes to travel to the detector is the same for both signals. This means that, if the two signals reach the detector at the same instant and the blue light comes on, the events were simultaneous in the reference frame of the detector (Figure 14.7*a*). If the events occur at different instants, the two signals do not arrive at the detector at the same instant and the blue light does not come on (Figure 14.7*b*). The blue light thus tells us whether or not the two emission events were simultaneous *in the reference frame of the detector*. One advantage of this device is that we do not need to be in the reference frame of the detector to determine whether the two events are simultaneous in its reference frame. As long as the simultaneity detector is placed midway between the two events, the blue light tells

us whether the events are simultaneous in the detector's reference frame.

Up to now we have discussed only measurements made by an observer (or detector) at rest with respect to the sources of the signals. What happens to the transmission of signals when the source and observer are moving relative to each other? To find out, let's start by reviewing the relativity of speeds for everyday objects that move much more slowly than light.

Example 14.3 Fast ball

Consider a spring-loaded device that ejects a ball at some unknown everyday speed v_b. To measure this speed, you measure the time interval it takes the ball to pass through a stationary tube of known length ℓ (**Figure 14.8**). Now suppose the tube is moving at speed v_t toward the ejection device. What speed do you obtain by dividing the length ℓ of the tube by the time interval Δt it takes the ball to pass through the tube? Express your answer in terms of v_b and v_t.

Figure 14.8 Example 14.3.

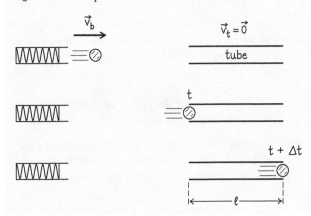

CONCEPTS

❶ GETTING STARTED I begin by drawing the ball at the instant it enters and the instant it exits the moving tube (**Figure 14.9**). I choose my positive x axis to be in the direction of the ball's motion and use d to denote the distance the tube moves while the ball passes through. The distance the ball travels as it passes through the tube is thus $\ell - d$.

Figure 14.9

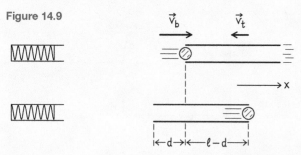

❷ DEVISE PLAN From my sketch I see that during the time interval Δt, two things happen: The tube advances a distance d, and the ball travels a distance $\ell - d$. Because speed equals distance traveled divided by the time interval needed to travel this distance, I can relate these two distances to the speeds v_b and v_t. This gives me two equations, from which I can solve for the two unknowns d and Δt. I also note that dividing ℓ by Δt gives me the speed v_{Mb}, which is the speed of the ball measured by an observer M moving along with the tube.

❸ EXECUTE PLAN Because the tube advances a distance d in the time interval Δt, I have $v_t = d/\Delta t$. Likewise, for the ball: $v_b = (\ell - d)/\Delta t = \ell/\Delta t - d/\Delta t$. The first term equals the speed v_{Mb} I am asked to determine. The second term is the speed of the tube v_t, and so $v_b = v_{Mb} - v_t$. Rearranging terms, I get $v_{Mb} = v_b + v_t$. The speed of the tube is added to that of the ball. ✔

❹ EVALUATE RESULT Using the subscript cancellation rule (Eq. 6.14), I can write $\vec{v}_{Mb} = \vec{v}_{ME} + \vec{v}_{Eb}$. With my choice of positive x axis to the right in Figure 14.9, I get $v_{Mbx} = v_{MEx} + v_{Ebx}$. Because $\vec{v}_{ME}$ is the velocity of Earth relative to an observer moving with the tube, I have $v_{MEx} = v_t$. Likewise $\vec{v}_{Eb}$ is the velocity of the ball relative to Earth, and so $v_{Ebx} = v_b$. Therefore, $v_{Mbx} = v_{MEx} + v_{Ebx} = v_t + v_b$, which confirms the result I obtained.

🖑 **14.3** In Example 14.3 what is the time interval Δt it takes the ball to pass through the moving tube if $v_b = v_t = v$? Express your answer in terms of ℓ and v.

Example 14.3 reaffirms what we learned in Chapter 6: The speed of an everyday object relative to an observer depends on the motion of both the object and the observer. Relative to an observer moving along with the tube in Figure 14.9, the ball moves not at speed v_b but at speed $v_b + v_t$. Measurements of the **speed of light in vacuum,** denoted by c_0, however, show something very different:*

The measured value of the speed of light propagating in vacuum is c_0, independent of the motion of the source relative to the observer making the measurement.

---

*The subscript zero on c_0 is there to remind us that this is the value obtained in vacuum. The zero indicates the absence of any medium through which the light propagates.

This counterintuitive result was first suspected after an experiment done by A.A. Michelson and E.W. Morley in 1887. Many experiments since then have provided extensive confirmation of the Michelson-Morley results. To help you grasp the full meaning of this remarkable result, imagine that we shine a beam of light from a source into a device that is at rest relative to the source and that measures the speed of light in vacuum (**Figure 14.10a**; such an apparatus exists). The result is a value of 299,792,458 m/s (see Section 1.3), which we shall round off to 3.00×10^8 m/s. If we now move the measuring device toward the source at speed v_d and measure the speed of light, we obtain not $c_0 + v_d$ but again c_0 (Figure 14.10b)! This baffling result is like discovering that the speed of the ball with respect to the tube in Figure 14.8 is the same regardless of the motion of the tube relative to Earth.

Experiments show that it makes no difference whether we move the source, move the measuring device, or move both. In all cases, the device measures the speed of light in vacuum to be c_0 (Figure 14.10c and d). This result holds even when the source or detector moves at a substantial fraction of c_0: If a source moves toward you at a speed of $0.2c_0$, you still measure the light traveling toward you at c_0, not $c_0 + 0.2c_0 = 1.2c_0$. After more than 100 years of repeated experiments, we can most certainly rule out an error in measurement.

Any quantity whose value does not depend on the choice of reference frame is called an **invariant.** Because the speed of light is the same in all inertial reference frames, it is an invariant. Invariants play an important role in physics because we want to develop theories and laws that are the same in different reference frames. To construct these theories and laws, we first need to establish which physical quantities are invariants. The invariance of these quantities tells us something fundamental about the underlying laws of the universe.

Measuring the speed of light involves nothing more complicated than measuring the distance traveled by light

Figure 14.10 The measured value of the speed of light in vacuum does not depend on the speeds of the source and the measuring device.

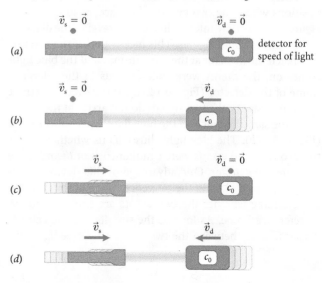

in a certain time interval and dividing that distance by the time interval. The unsettling conclusion from Figure 14.10 is that measured values of distances and time intervals must somehow depend on the motions of the source and the reference frame. This conclusion forces us to reexamine what we mean by *distances* and *time intervals,* two of the most basic quantities we introduced back in Chapter 2.

Let us start with the experimental fact that the speed of light in vacuum is always exactly c_0 and examine the consequences this result has on the experiment of Figure 14.7a. Suppose the entire setup—the two light sources and the simultaneity detector midway between them—now moves at a very high speed v past an observer at rest on the ground, as in **Figure 14.11**. (By *very high* I mean a speed large enough for the detector to move an appreciable distance in the time interval it takes the light signals to reach the detector.) Suppose further that the observer sees the two emission events occur at the same instant; that is, he is midway between the two sources at the instant they emit a light signal, and so the two signals reach him at the same instant (middle panel in Figure 14.11a). In other words, *the events are simultaneous in the observer's reference frame.*

As you can see from Figure 14.11a, however, the simultaneity detector moves to the right as the signals travel toward

it, and so it receives signal 2 before it receives signal 1. Therefore the blue light does not go on. In the reference frame of the detector, the two signals were *not* emitted simultaneously! Meanwhile, the observer is confused: In his reference frame, the two signals *were* emitted simultaneously, and so why didn't the blue detector light come on?

✋ **14.4** Suppose the observer in Figure 14.11a, at rest relative to the ground and standing midway between the two sources when they emit signals, is joined by another observer who moves along with the simultaneity detector. (a) Based on what the detector tells them, do the observers agree that the light signals do not reach the detector at the same instant? (b) How does each observer explain why the signals don't reach the detector at the same instant?

Suppose the sources in Figure 14.11 emit the signals in such a way that the fast-moving detector receives them at the same instant and the blue light turns on. To the observer at rest on the ground, source 1 must have emitted its light signal before source 2 emitted its signal because signal 1 has to catch up with the detector moving away from it (Figure 14.11b). In his reference frame, the two events are *not* simultaneous, and so he is once again confused by what

Figure 14.11 If the sources and the simultaneity detector move together at high speed relative to an observer, then (a) events that are simultaneous to the observer will not be so to the detector. (b) To make the signals strike the simultaneity detector at the same instant, the source behind the detector must emit its signal before the source in front of the detector.

(a) Events are simultaneous to observer

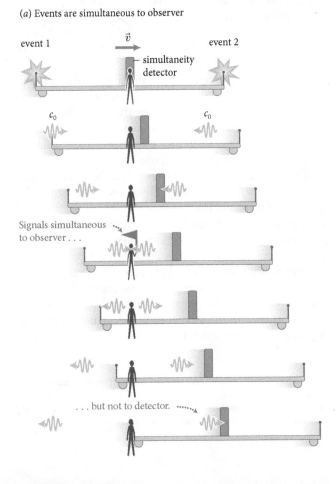

(b) Events are simultaneous to simultaneity detector

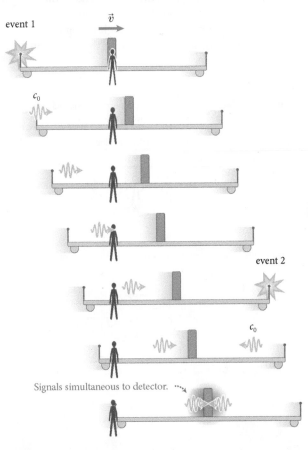

the detector is "saying": Blue light on means the events were simultaneous. The blue light tells us that the detector sees exactly the situation depicted in Figure 14.7a, and so in the reference frame of the detector the emission events are simultaneous. From these experiments we infer that:

Events that are simultaneous in one reference frame need not be simultaneous in another reference frame that is moving relative to the first.

What does this relativity of simultaneity imply for the grid of synchronized clocks of Figure 14.6? To answer this question, consider a pair of synchronized clocks at rest relative to observer A midway between the two clocks in **Figure 14.12**. The clocks are mounted on opposite ends of a rod that is one grid unit long. Because the clocks are synchronized, they strike 3 o'clock at the same instant according to observer A. Now consider the situation from the point of view of observer B, who is moving to the left at constant speed relative to A.

From B's perspective, the two-clock unit moves to the right, just like the simultaneity detector in Figure 14.11. Suppose events 1 and 2 in Figure 14.11 are our two clocks striking 3 o'clock. As Figure 14.11b shows, these two events occur at the same instant in the reference frame of the detector. Relative to observer B, however, the left clock strikes 3 o'clock (event 1) before the right clock does (event 2). So, in Figure 14.12b, too, the left clock strikes 3 o'clock *before* the right clock in the reference frame of observer B. In other words, the clock on the trailing end of the grid unit runs ahead of the clock on the leading end. This statement must hold true for any pair of neighboring clocks along the line of motion, which means that the clocks on a grid

Figure 14.12 Two synchronized clocks are unsynchronized in the reference frame of an observer moving relative to the clocks.

(a) Observer at rest relative to clocks sees clocks as synchronized.

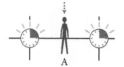

A

(b) However, to observer in motion relative to frame of clocks . . .

$\vec{v}_{BA}$

B

. . . clocks are *not* synchronized; trailing clock shows later time.

$\vec{v}_{BA}$

B

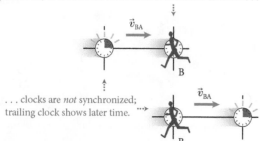

Figure 14.13 Desynchronization of time in a moving reference frame.

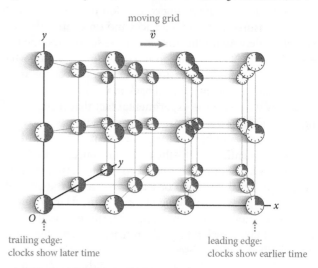

moving grid

$\vec{v}$

y

y

O

x

trailing edge: clocks show later time

leading edge: clocks show earlier time

moving relative to an observer desynchronize as shown in **Figure 14.13**. If you were to move along with the grid, all clocks would strike, say, 3 o'clock at the same instant. If the grid moves relative to you, however, the clocks at the trailing end of the grid strike 3 o'clock before those at the leading end.

What is one instant in one reference frame is a multitude of different instants in a reference frame that is moving relative to the first reference frame. As we saw in analyzing Figures 14.11 and 14.12, events that are simultaneous in one reference frame are not simultaneous in a reference frame that is moving relative to the first one.* Simultaneity is a relative concept, and time intervals are therefore not invariants.

The reason this notion of the relative nature of simultaneity defies common sense is that these effects are observable only at speeds that are much greater than everyday speeds. In Figure 14.11a, for example, in the time interval it takes the light to travel half the distance between the two sources, the simultaneity detector travels one-fourth of that distance. In other words, the detector moves at half the speed of light, or 1.5×10^8 m/s, close to a million times faster than the speed of an airplane! Had it instead moved at the speed of an airplane, the displacement between the top and bottom pictures in Figure 14.11 would have been negligible, and so, for all practical purposes, the signals would have arrived simultaneously at the detector both according to an observer on the ground and according to one on the moving assembly. It is only when observers (or detectors) move at speeds close to c_0 relative to each other that deviations from our everyday experience cannot be ignored.

---

* The only events we consider here are ones that occur at different positions along the line of motion of two reference frames in motion relative to each other. Events occurring along an axis perpendicular to the line of motion can be simultaneous in both reference frames.

14.5 Suppose the setup in Figure 14.11 is 10 m long. (*a*) How long does it take a light signal to travel from source to detector? (*b*) If the setup moves at 20 m/s, how far does it travel in the time interval you calculated in part *a*? (*c*) By how much does the displacement of the detector shorten (or lengthen) the travel time of the light signals? (*d*) Do you expect the time interval you calculated in part *c* to be measurable? (*e*) Repeat parts *b–d* for a detector moving at 1.0×10^8 m/s.

14.3 Space-time

The fact that simultaneity is relative rather than absolute makes us rethink the whole concept of time. If simultaneity depends on the motion of observers relative to each other, shouldn't the rate at which a clock is observed to tick depend on the motion of the observer relative to the clock? The answer is yes, and here is the easiest way to see that. Consider the device shown in **Figure 14.14**, called a *light clock*. At the bottom is a source that emits a very short light signal; at the top is a mirror that reflects the signal back down. At the instant the reflected signal reaches the bottom, the source is triggered to emit a new light signal. The up-and-down motion of the signals determines the period of this clock. If the mirror is a distance h away from the source, it takes each light signal a time interval h/c_0 to travel from the source up to the mirror and the same time interval to travel back down to the source. The period of this clock is thus $2h/c_0$. Just as the ticking of a mechanical clock defines the unit of time for the clock, this period defines the unit of time for the light clock.

Consider this clock from the perspective of two observers: one at rest relative to the clock (observer A, Figure 14.14*a*) and one moving to the left at very high speed v relative to the clock (observer B, Figure 14.14*b*). To observer A the clock signals move straight up and down. To observer B, however, the signals move along a longer diagonal path. Because the signals travel at exactly c_0 according to both observers, observer B sees the signals take longer to travel from source to mirror and from mirror to source.

14.6 (*a*) In Figure 14.14, is the period of the clock according to observer B greater than, equal to, or less than the period according to A? (*b*) Does the period of the clock according to B increase, decrease, or stay the same if we increase v?

Because the distance the signal travels from source to mirror and back to source is greater according to B than according to A, it takes longer for the clock to complete a cycle according to observer B. Therefore, the clock period is longer for B than it is for A. Observer B sees the clock as running more slowly than observer A. The greater the speed at which the clock moves relative to B, the more slowly she observes it ticking off its units of time. This

Figure 14.14 Because the speed of light in vacuum is always c_0, the light signal in a light clock takes longer to reach the mirror according to an observer moving relative to the clock than to an observer at rest relative to the clock.

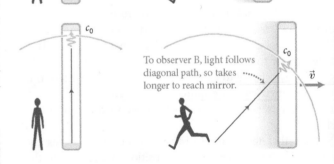

(*a*) Light clock as seen by observer A at rest relative to clock

(*b*) Light clock as seen by observer B moving left relative to clock

effect is called **time dilation** because the units of time appear stretched out (dilated):

When a clock is in motion relative to an observer, that observer sees the time units of the clock as being longer than the time units of an identical clock at rest relative to her. Consequently, she measures the clock in motion as running slow relative to the clock at rest.

Figure 14.15 Each observer sees the other observer's clock run more slowly.

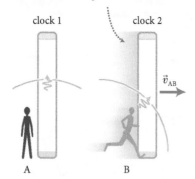

(a) **Reference frame of observer A**

Relative to A, clock 2 moving and therefore running slow

clock 1   clock 2

$\vec{v}_{AB}$

A   B

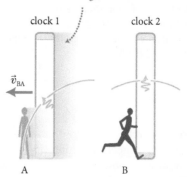

(b) **Reference frame of observer B**

Relative to B, clock 1 moving and therefore running slow

clock 1   clock 2

$\vec{v}_{BA}$

A   B

14.7 On your desk in front of you are two watches, one mechanical and one electronic. For each minute on the electronic watch, the mechanical watch advances by only 59 s. (a) Is the mechanical watch fast or slow relative to the electronic one? (b) Using the mechanical watch as the standard, is the electronic one fast or slow?

Ordinarily, if your watch runs slow relative to some clock, that clock runs fast relative to your watch, as you just saw in Checkpoint 14.7. Things are different when dealing with time dilation. Consider, for example, two identical clocks: clock 1, which is at rest relative to observer A, and clock 2, which is traveling along with observer B. To observer A, clock 2 is in motion and therefore running slow compared to clock 1 (**Figure 14.15a**). To observer B, however, clock 2 is at rest and clock 1 is moving (Figure 14.15b). From B's perspective, it is now the signal in clock 1 that takes longer to reach its mirror. So to B, clock 1 appears to run slowly relative to clock 2. The effect of time dilation is thus symmetrical, as required by the principle of relativity: Each observer sees the other observer's clock run more slowly.

Note that in Figure 14.14b the light signal leaves the source when the clock is at one position and returns to the source when the clock is at a different position. This means that the time interval between these events measured in B's reference frame is not the proper time interval. Observer A measures the proper time interval because the emission and return take place at the same position relative to A. This means that the proper time interval between the two events is the shortest, and we can say:

> **An observer who measures the proper time interval between two events (that is, an observer for whom the events occur at the same location) obtains the shortest time interval between those two events.**

Example 14.4 My nap was shorter

You are flying in a plane, and a friend is on the ground. Who measures a shorter duration of (a) a nap you take and (b) a nap your friend takes?

❶ **GETTING STARTED** Like Example 14.1, this problem involves two observers (my friend and I) and two pairs of events (the beginning and end of each person's nap). My friend is in the Earth reference frame (E); I am in the reference frame of the plane (P). To simplify the discussion, I denote the beginning and end of my nap as events 1 and 2, respectively, and the beginning and end of my friend's nap as events 3 and 4, respectively.

❷ **DEVISE PLAN** The proper time interval between two events is the shortest time interval between them. Therefore this problem boils down to determining who measures the proper time interval for each nap. I should therefore make a sketch for each pair of events.

❸ **EXECUTE PLAN** (a) **Figure 14.16** shows the beginning (event 1) and end (event 2) of my nap on the plane. To me, the two events take place at the same location, which means that the time interval I measure is the proper time interval. To my friend, the two events take place at different locations (I fall asleep far to his left and wake up directly overhead), so the interval he measures is not the proper one. Thus I measure the shorter duration for my nap. ✔

Figure 14.16

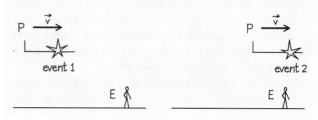

(b) **Figure 14.17** shows the beginning (event 3) and end (event 4) of my friend's nap on the ground. To my friend, the two events take place at the same location, which means that the time interval he measures is the proper time interval. To me, the two events take place at different locations (my friend falls asleep in front and below the plane and he wakes up directly underneath), so the interval I measure is not the proper one. Thus my friend measures the shorter duration for his nap. ✔

Figure 14.17

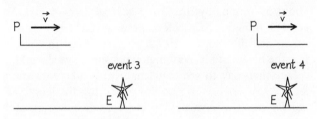

④ EVALUATE RESULT Each nap happens in the reference frame of the napper, and so it makes sense that each person measures the proper time interval of her or his nap.

You may ask yourself if this slowing down of moving clocks is just a peculiarity of the type of clock we have constructed. Who measures time with light clocks anyway? Suppose you have two identical mechanical clocks adjusted so that their periods are the same as those of a pair of identical light clocks. Now suppose you pair each mechanical clock with one of the light clocks and set one pair in motion. The stationary pair remains synchronized. Likewise, the moving pair must remain synchronized for this reason: If the motion of the clocks caused them to become unsynchronized, an observer moving along with them would note the difference in ticking rates and from this difference be able to determine his state of motion, which would violate the principle of relativity (see Section 6.3). Indeed, no one has ever observed a pair of initially synchronized clocks becoming unsynchronized because the pair's state of motion changes. So, we know that (1) moving light clocks run more slowly and (2) mechanical clocks behave just as light clocks do. We therefore conclude that moving mechanical clocks—or any other clock, for that matter, including your biological clock—must exhibit the same time-dilation effects as a light clock. It is time itself that dilates!

Even though time dilation is extremely small at everyday speeds, it has been measured by comparing two atomic clocks (which are not light clocks), one at rest on the surface of Earth and the other aboard an aircraft (see the box "Desynchronized twins").

Direct evidence that the time interval between two events depends on the motion of the observer is provided

Desynchronized twins

Suppose one member of a set of twins embarks on a space trip, flying away from Earth at a speed of $0.98c_0$. Ten years after her departure, she turns around and returns to Earth, again traveling at speed $0.98c_0$. To observers on Earth monitoring the trip, clocks on the space ship run five times more slowly than clocks on Earth. Likewise, the twin's heart—a clock—beats five times more slowly. Consequently, by the time she returns to Earth, she has aged not 20 years but 4 years (20 years/5)! When she meets her twin brother again, who has remained Earthbound this whole time interval, he's 16 years older than she is. According to him, she's been gone 20 years, but to her, the trip lasted only 4 years.

This scenario is disturbing because it defies our commonsense notion of time. The experiment has been carried out with a pair of identical ("twin") atomic clocks, one flown Eastward twice around the world while the other sat at the airport. Upon its return, the flown clock was behind the airport clock, and the time difference matched what theory predicts.

This desynchronization of twins (or twin clocks) is sometimes called the *twin paradox* because viewing the situation from the perspective of the traveling twin appears to yield a contradiction: To her it is her brother on Earth who is moving at $0.98c_0$, and so *he* should be the one aging more slowly! The paradox is that, as explained earlier, she does not see him as being 16 years younger—both agree that she is the younger one once the trip is over. In other words, even though she sees Earth/brother moving at $0.98c_0$ and so should see Earth clocks running *slow*, these Earth clocks measure a *longer* duration for her trip than her own clocks. How can this be?

The key to resolving this paradox is to remember that the brother remains in an inertial reference frame during the entire trip, but the sister doesn't. The spaceship has to turn around to return to Earth, and this change in direction means acceleration. During this acceleration, a marble initially at rest on the floor of the spaceship does not remain at rest, and thus there is a time interval during the trip when she is in a noninertial reference frame. In contrast, a marble initially at rest on Earth's surface is not affected by the spaceship's acceleration: in other words, the Earth remains an inertial reference frame all the while the traveling twin is away. Thus the physical situation is different in the two reference frames: the Earth reference frame always inertial, the reference frame of the spaceship not always inertial. The theory of special relativity shows that during the time interval in which she is accelerating, the sister would see Earth clocks speed *up*, so that when she gets back to Earth, she is not at all surprised to see that her brother has aged much more than she has.

by cosmic-ray muons, fast-moving elementary particles created when highly energetic protons from deep space collide with atoms some 10 km up in Earth's atmosphere. Laboratory experiments show that muons at rest have a *half-life* of 1.5×10^{-6} s. That is, if you have 100 muons at some instant $t = 0$, half of them will decay into some other type of particle during one half-life. Thus you will have only 50 muons left at $t = 1.5$ μs, 25 left at $t = 3.0$ μs, and so on.

Traveling at a speed nearly equal to c_0, a muon travels 450 m in 1.5×10^{-6} s, and so the number of muons should decrease by a factor of 2 each 450 m they travel through the atmosphere. Experiments done at the top of Mt. Washington at an altitude of 1900 m showed that 568 muons strike a certain detector per hour. At sea level the muons have traveled an additional 1900 m, or about four times 450 m, and so the count should be diminished by a factor of $(\frac{1}{2})^4$, yielding approximately 35 muons per hour. When the same detector is placed at sea level, however, not 35 but 412 muons per hour are detected—more than ten times as many as expected. How can this be?

Because of the time dilation that occurs at the speed they travel, the muons' half-life of 1.5×10^{-6} s is stretched to 1.4×10^{-5} s in our reference frame at rest on the ground. So, in the Earth reference frame the muons exist long enough for a substantial fraction of them to reach the ground before disintegrating.

✋ **14.8** Could you lengthen your life by traveling at high speed?

Motion is a relationship between time and distance. Thus, now that we understand how relative motion changes our ideas about time measurements, a logical next question to investigate might be whether motion changes the measurement of *distance*. In other words, is there a length change analogous to time dilation? An intriguing thought

experiment shows us that the answer is yes. Picture yourself traveling along with some muons as they move from the summit of Mt. Washington to sea level. They are at rest relative to you, and so there is no time dilation in your reference frame: According to your clock, the muons have a (proper) half-life of 1.5×10^{-6} s. If you are traveling with the muons (at rest in their reference frame), you see the ground rushing toward you at a speed of nearly c_0. Thus it should still take 4 half-lives for the surface of the sea to travel the 1900 m toward you, and as a result you should again see about 35 muons by the time sea level reaches you. You see 412, though, just as before. Because you "know" that the muon half-life is 1.5×10^{-6} s, you are forced to conclude that the height of Mt. Washington has somehow shrunk!

Another way to see that length measurements are affected by motion is to imagine measuring the length of a *moving* car. One way would be to record the positions of the front and back of the car on the pavement and then measure the distance between those marks. You need to make sure that you record the two positions simultaneously, however, because the car is moving relative to you. If you record the back position at one instant and the front position some time later, you won't get a correct result. Remember, however, that simultaneity is a relative concept, and therefore we can expect length measurements also to depend on the motion of an observer relative to the object whose length the observer is measuring.

To see how length measurements are affected by motion, consider again a grid unit with a pair of clocks. The unit is at rest and the clocks are synchronized in the reference frame of observer A (**Figure 14.18a**). Suppose a second observer B speeds up leftward when the clocks read 3 o'clock in A's reference frame. Because the clocks are synchronized, both of them read 3 o'clock in A's reference frame when observer B begins speeding up. From observer B's perspective, observer A and the clocks speed up to the right (Figure 14.18b). Because observer B is moving relative to the grid unit, the clocks are desynchronized, with the clock on the trailing

Figure 14.18 Length contraction of an accelerating grid.

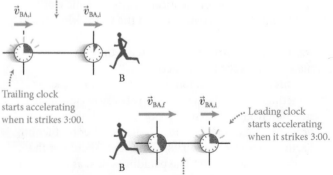

(*a*) **Reference frame of observer A**

Observer A sees clocks synchronized.

At 3:00, A sees B begin accelerating.

(*b*) **Reference frame of observer B**

Grid moves relative to B, so clocks not synchronized. Trailing clock shows later time.

Trailing clock starts accelerating when it strikes 3:00.

Leading clock starts accelerating when it strikes 3:00.

Because trailing clock accelerates first, B sees grid unit shortened.

end being ahead of the clock on the leading end (see Figure 14.13). In observer B's reference frame, therefore, the clock on the trailing end strikes 3 o'clock and begins speeding up *before* the clock on the leading end does. Consequently B sees the distance between the clocks (and the length of the rod between them) decrease. Whenever the unit speeds up in B's reference frame, B sees the length of the rod decrease. The faster the unit moves, the shorter the rod appears to B.

When the unit slows down relative to observer B, the opposite happens: The clock on the trailing end begins slowing down *before* the clock on the leading end begins slowing down, and so B sees the rod become longer again. When the unit is at rest relative to B, she measures the rod to have its **proper length:**

> **The length of an object measured by an observer at rest relative to the object is called the object's proper length.**

An observer measuring the length of an object moving relative to the observer always measures a value that is shorter than the object's proper length.

14.9 Suppose observer B in Figure 14.18 carries a unit of two synchronized clocks that is identical to the unit that is at rest relative to A. Observer B begins speeding up when the clocks in her reference frame read 3 o'clock. What happens to the length of B's unit according to observer A?

As this checkpoint shows, the shortening of the rod's length, like time dilation, is symmetrical, in agreement with the principle of relativity. Each observer sees the other observer's grid move at the same relative speed and therefore shortened by the same amount.

Note that the shortening occurs only in the direction of motion. To see why, consider our clock grid unit aligned as in

Figure 14.19. With the unit at rest, pegs are placed above and below it so that it fits snugly between them (Figure 14.19a). If the unit were to get shorter when it is in motion, it would fit more easily between the pegs (Figure 14.19b). Now imagine observing the same situation while traveling along with the unit. The unit is now at rest from your point of view and so does not shorten. The pegs, however, are in motion relative to you, and so the distance between them becomes smaller and the clocks hit the pegs. However, whether or not the clocks hit the pegs cannot depend on the reference frame of the observer, and so there cannot be any shortening in a direction perpendicular to the direction of motion.

When we discussed time dilation and considered the moving light clock in Figure 14.14b, we tacitly assumed its height to be fixed. Now we see that this assumption was justified.

The shortening of objects in the direction of motion is called **length contraction:**

> **When an object moves relative to an observer, the observer measures the length of the object in the direction of its motion to be shorter than the proper length of the object.**

Length contraction is a direct consequence of the relativity of simultaneity. Returning to the cosmic-ray muons, we can now see why they reach Earth's surface. To them, Earth's atmosphere flies by at nearly the speed of light. Length contraction reduces the thickness of the atmosphere so much that the atmosphere passes by the muons before they decay. To us on the ground, time dilation lengthens the half-life of the muons, and so they can cross the proper length of the atmosphere. To them, length contraction decreases the distance to the ground, and so they reach the ground during their proper lifetime.

Figure 14.19 Length contraction cannot occur in a direction perpendicular to the direction of motion.

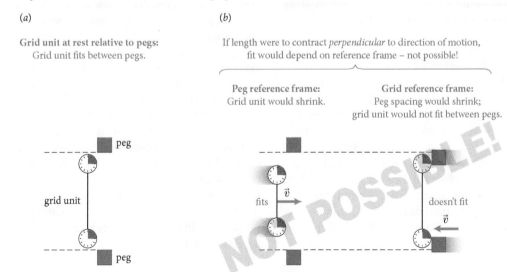

(a)
Grid unit at rest relative to pegs:
Grid unit fits between pegs.

(b)
If length were to contract *perpendicular* to direction of motion, fit would depend on reference frame – not possible!

Peg reference frame:
Grid unit would shrink.

Grid reference frame:
Peg spacing would shrink;
grid unit would not fit between pegs.

peg

grid unit

peg

fits $\vec{v}$

doesn't fit

$\vec{v}$

NOT POSSIBLE!

Common sense and special relativity

That observers who are in motion relative to each other assign different values to the length of an object and to the time interval between two events clashes with common sense. Even though experiments confirm the predictions of special relativity, it is hard to believe that, say, the length of a rod and the time interval between your lunch and dinner do not have fixed, absolute values. This difficulty arises from the fact that in everyday circumstances we can get away with treating lengths and time intervals as invariants.

How "real" then is length contraction or the slowing down of clocks? It is just as real as the pitch of an approaching car horn sounding higher than that of a stationary or receding car. (This phenomenon is called the *Doppler effect*; see Section 17.7.) Riding along in a car with a stuck car horn, you measure the same pitch a roadside observer would measure when the car is at

rest. Yet there is no denying that the pitch is perceived to change when car and observer are in motion relative to each other. These observations don't mean that the horn changes when it moves relative to an observer. What changes is the pitch *perceived by the observer*.

The same statements can be made about a moving rod. Any observer measuring the rod's length when the rod is at rest relative to him obtains the same value as you would (the proper length). Just as the pitch of a moving car horn is perceived to be changing, so is the measured length of a moving rod. The rod itself doesn't change—the *measurement of its length* yields a different value. In that sense relativistic effects are every bit as real as the changing pitch of a horn. For example, few muons would make it across what we measure to be 10 km of atmosphere if that length did not appear to be shortened to less than 1 km to them.

Time and space are connected. If we consider time and space to constitute a four-dimensional *space-time* entity, we see that, depending on the reference frame, the time dimension stretches (dilates) and one of the three spatial dimensions contracts. As we shall see in Section 14.5, a quantity called the *space-time interval* remains unaffected by this distortion of space-time geometry. At everyday speeds, these distortions are so small that they are not perceived. For motions at very high speeds, such as that of cosmic-ray muons, however, the consequences of these effects can be significant (see the box "Common sense and special relativity").

✋ **14.10** Consider a long pole moving through a tunnel, with the direction of motion being along the long axis of the pole. An observer on the ground determines that the moving pole has the same length as the tunnel. Does an observer moving along with the pole agree that the pole fits exactly in the tunnel?

14.4 Matter and energy

We developed the entire framework of mechanics from our everyday notions of space and time. In Chapters 2 and 3, we introduced the definitions of displacement, velocity, and acceleration. Then we introduced inertia, which depends on a ratio of changes in velocity (see Section 4.5). Many other quantities, such as momentum, energy, and force, in turn depend on inertia. So all the physics we have developed so far hinges on our concepts of space and time.

As we just saw in the preceding sections, however, everyday notions of space and time fail at very high speed. Additionally, there is plenty of experimental evidence that

our notion of energy fails at very high speeds. For example, experiments show that quadrupling the kinetic energy of an electron initially moving at $0.8c_0$ by doing work on it increases its speed by just 20% rather than doubling it as we would expect from the expression $K = \frac{1}{2}mv^2$.

Figure 14.20*a* shows the energy diagram for accelerating an electron from rest to $0.8c_0$. (The diagram is simplified by lumping all internal energy into one bar.) Because electrons have no known internal structure, $\Delta E_{\text{int}} = 0$, and so only their kinetic energy can change by doing work on them. The change in kinetic energy must thus be equal to the work done on the electron.

Now suppose we do additional work on the electron after it has attained a speed of $0.8c_0$. If the additional work is three times the amount of work done in accelerating the electron from rest to $0.8c_0$, our expression for kinetic energy, $K = \frac{1}{2}mv^2$, predicts a new speed that is twice the original speed: $v_f = 1.6c_0$. Experiments show, however, that the speed increases by just 20% and the value we obtain for ΔK using the expression $K = \frac{1}{2}mv^2$ is very much smaller than expected (Figure 14.20*b*). We must therefore conclude that our expression for kinetic energy is not valid at high speeds.

This experiment shows that it is more difficult to accelerate an electron when it is moving at $0.8c_0$ than when it is at rest. Now remember that an object's inertia is a measure of its resistance to acceleration (Section 4.2). We have so far assumed that an object's inertia is constant—an intrinsic property that is independent of the object's physical state or motion relative to an observer. Experimental data show that this assumption does not hold as an object's speed approaches c_0. In particular, as its kinetic energy increases, the object's inertia increases.

Figure 14.20 Simplified energy diagrams for accelerating an electron.

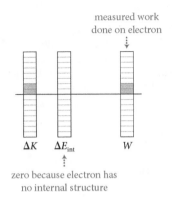

(a) Electron accelerated from 0 to $0.8c_0$

measured work
done on electron

ΔK ΔE_{int} W

zero because electron has
no internal structure

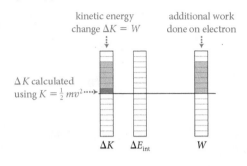

(b) Doing three times as much work increases electron's
speed by just 20%

kinetic energy
change $\Delta K = W$

additional work
done on electron

ΔK calculated
using $K = \frac{1}{2}mv^2$

ΔK ΔE_{int} W

14.11 Do two observers, one moving at very high speed relative to the other, agree on the inertia of an object? Why or why not?

As Checkpoint 14.11 shows, inertia depends on the reference frame and therefore is not an invariant. In Chapter 13 we introduced mass—the property that determines the gravitational interaction of objects. The way in which objects interact gravitationally cannot depend on the motion of an observer relative to the interacting objects, which means that, in contrast to inertia, mass must be an invariant. That is to say, all observers should agree on the value of the mass of an object, regardless of their motion relative to the object. That mass is an invariant and inertia is not implies that there *is* a difference between mass and inertia, contrary to what we established in Chapter 13. As we shall see in Section 14.7, the difference between the two becomes apparent only at very high speeds.

Let us now turn to the nonkinetic part of energy: internal energy. Remember that the internal energy of an object or system is a quantitative measure of the state of that object or system. Because the state of an object cannot depend on the motion of the observer relative to the object, internal energy also must be an invariant.

So, we see that kinetic energy and inertia are not invariants, but internal energy and mass are. We also found that inertia depends on kinetic energy. Experiments show a similar connection between mass and internal energy. The primary example is provided by nuclear reactions, such as those taking place in the Sun. One of these reactions is the fusion of a proton and a deuteron D (the nucleus of a hydrogen isotope consisting of one neutron and one proton) to form a helium-3 nucleus (two protons and one neutron), while 8.80×10^{-13} J is released as radiation energy:

$$p + D \rightarrow {}^3He + radiation.$$

Figure 14.21 shows a simplified energy diagram for this reaction. Given that this is a nuclear reaction, nuclear energy (a form of internal energy) must be converted to radiation energy. Because the radiation energy leaves the system, the internal energy decreases.

Measurements show that the sum of the mass of the proton and the mass of the deuteron ($m_p + m_D = 5.01621 \times 10^{-27}$ kg) is greater than the mass of the helium-3 nucleus (5.00641×10^{-27} kg). In other words, the mass of the system decreases as its internal energy decreases.

Figure 14.21 Simplified energy diagram for the fusion of a proton and a deuteron. During the reaction, nuclear energy is emitted in the form of radiation. Experiments show that the amount of energy released is 8.80×10^{-13} J.

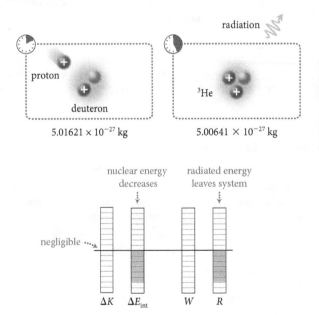

radiation

proton

deuteron

5.01621×10^{-27} kg

${}^3He$

5.00641×10^{-27} kg

nuclear energy
decreases

radiated energy
leaves system

negligible

ΔK ΔE_{int} W R

CONCEPTS

Figure 14.22 An ionized deuterium molecule and a helium ion contain the same elementary particles, yet have different masses. (Nuclei not to scale.)

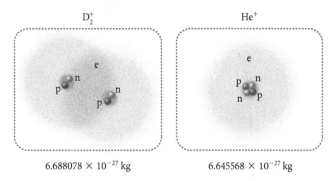

D_2^+ He^+

6.688078×10^{-27} kg 6.645568×10^{-27} kg

Another example of the connection between mass and internal energy is provided by comparing the masses of an ionized deuterium molecule and an ionized helium atom (Figure 14.22). Both consist of two protons, two neutrons, and an electron, but the two particles have different configurations. Accurate measurements show that this different in configuration—which means a difference in internal energy—results in a small mass difference.

These observations indicate a relationship between mass and internal energy that is parallel to the one between inertia and kinetic energy:

Increasing the internal energy of a system increases the mass of the system.

Mass is a property we associate with particles or objects that have physical substance—made of what we call "matter." Internal energy is a quantitative measure of the object's state. Now we see that the two are proportional to each other and that whenever a system's state changes, both its internal energy and its mass change.

✋ **14.12** In each of the following situations, does the mass of the system increase, decrease, or stay the same? (*a*) A spring is compressed (system: spring). (*b*) A cup of coffee cools down (system: coffee).

Exercise 14.5 The mass of energy

(*a*) In the nuclear reaction illustrated in Figure 14.21, what amount of mass is "lost"? (*b*) What is the ratio of energy decrease to mass loss as a result of the reaction? (*c*) Show that the units of your answer to part *b* are equivalent to m^2/s^2.

SOLUTION (*a*) The mass decreases by $(5.00641 \times 10^{-27}$ kg$) - (5.01621 \times 10^{-27}$ kg$) = -9.80 \times 10^{-30}$ kg. ✔

(*b*) The text states that the energy released in the reaction is 8.80×10^{-13} J. Therefore the ratio of energy decrease to mass loss is $(8.80 \times 10^{-13}$ J$)/(9.80 \times 10^{-30}$ kg$) = 8.98 \times 10^{16}$ J/kg. ✔

(*c*) J/kg = (kg · m²/s²)/kg = m²/s². ✔

Example 14.6 Spring mass

A spring of spring constant $k = 0.16$ N/m is compressed to half its 50-mm relaxed length. Using the result of Exercise 14.5, determine the change in the spring's mass.

❶ **GETTING STARTED** I begin by drawing a simplified energy diagram for the compressed spring (Figure 14.23). The work done on the spring during the compression increases the spring's potential energy and therefore the spring's internal energy. I need to determine the increase in mass corresponding to this energy increase.

Figure 14.23

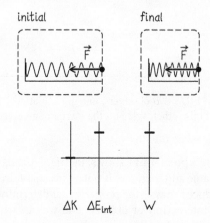

❷ **DEVISE PLAN** To calculate the increase in the spring's potential energy, I can use Eq. 9.23, $\Delta U_{spring} = \frac{1}{2}k(x - x_0)^2$. In part *b* of Exercise 14.5, I determined the ratio of energy change to mass change. I can use this ratio to convert the spring's energy change to a mass change.

❸ **EXECUTE PLAN** The spring has been compressed by an amount $x - x_0 = 50$ mm $- 25$ mm $= 2.5 \times 10^{-2}$ m. From Eq. 9.23, I get $\Delta U_{spring} = \frac{1}{2}(0.16$ N/m$)(2.5 \times 10^{-2}$ m$)^2 = 5.0 \times 10^{-5}$ J. In Exercise 14.5*b*, I found that the ratio of energy change to mass change is 8.98×10^{16} J/kg, so the change in the spring's mass is $(5.0 \times 10^{-5}$ J$)/(8.98 \times 10^{16}$ J/kg$) = 5.6 \times 10^{-22}$ kg. ✔

❹ **EVALUATE RESULT** The change in the spring's mass is much too small to be measured. That is reassuring because in my experience I've never been able to observe any change in mass when compressing springs.

Exercise 14.5 and Example 14.6 show that the ratio of mass change to energy change is very small; that is, a tiny amount of mass corresponds to an enormous amount of energy. Table 14.1 lists the fractional changes in mass for a number of changes in state (corresponding to changes in energy). As a rule of thumb, nuclear reactions release an amount of energy corresponding to about 10^{-3} of the mass; chemical reactions release an amount of energy corresponding to about 10^{-9} of the mass. As you can see from the table, other processes cause even smaller fractional changes in mass. For example, no one has been able to

Table 14.1 Changes in energy and fractional mass changes due to changes in state

System before	System after	Energy change (J)	Fractional change in system's mass
1-kg book and Earth	1-kg book and Earth separated by an additional 1 m	10	2×10^{-41}
cup of tea	cup of tea heated by 10 °C	4000	5×10^{-13}
1 mole of H_2 molecules	2 moles of H atoms	5×10^5	3×10^{-9}
p + D	^{3}He	9×10^{-13}	2×10^{-3}

observe the change in mass that occurs when we heat a cup of tea because this change is on the order of 10^{-13} kg, much too small to measure.

In Chapter 13 we established that mass and inertia are equivalent at everyday speeds. Consequently increasing the internal energy of an object at rest must increase not only the object's mass but also its inertia. So, whereas mass depends on internal energy only, inertia depends on both internal energy and kinetic energy:

Increasing the energy $E = K + E_{int}$ of a system increases its inertia. The increase in inertia is the same whether the added energy goes into kinetic energy K, internal energy E_{int}, or a combination of the two.

Table 14.2 summarizes the correspondence between matter and energy. The mass and internal energy of a system are invariants—that is, independent of the motion of observer and system relative to each other. Inertia and energy are not invariants. In addition, energy is a conserved quantity. In the second part of this chapter, we'll use the conservation principles to derive quantitative expressions for the relationships between mass and internal energy and between inertia and energy.

Table 14.2 Matter and energy

Quantity	Invariant?	Conserved?	Related to
E_{int}	yes	no	mass
E	no	yes	inertia

14.13 Do the inertia and mass of the system increase, decrease, or remain the same when (*a*) an elementary particle is accelerated (system: particle) and (*b*) an elementary particle slams into a metal target (system: particle and target)?

CONCEPTS

Self-quiz

1. Imagine you are at the origin of Figure 14.6, where, as in Exercise 14.2, a light pulse takes 1 s to travel one grid unit. When your clock reads 12:00:05, what do you observe as the reading on the clocks at positions (3, 0) and (0, 2)?

2. Which of the following statements are true when we compare a moving light clock and one at rest in the Earth reference frame? (*i*) The signal in the moving clock travels a greater distance during one cycle. (*ii*) The signal in the moving clock takes more time to complete one cycle. (*iii*) The cycle on the moving clock is longer. (*iv*) Time on the moving clock runs more slowly. (*v*) The moving clock completes an hour first.

3. Standing near the finish line at a race track, you measure the time interval between two cars as they cross the finish line. Is the time interval between these events as measured by the driver in the second car the same as, greater than, or less than the time interval you measure?

4. A particle of mass m is moving at speed v toward an identical particle that is at rest in the Earth reference frame. Is the inertia of the particles in their zero-momentum reference frame greater than, equal to, or less than m?

5. An observer on the ground observes a car moving past a gate and notes that the car is longer than the gate. For an observer in the car, which of the following are possibilities? (*i*) The gate is longer than the car. (*ii*) The gate is the same length as the car. (*iii*) The gate is shorter than the car.

Answers

1. Because it takes 1 s for light to travel one grid unit, it takes 3 s for the clock at (3, 0) to be seen by you. Its time is thus 3 s earlier than that of the origin clock, or 12:00:02. The light from the clock at (0, 2) takes 2 s to reach you, and so that clock reads 12:00:03.

2. All but (*v*) are true. Because the signal in the moving clock travels an equal vertical distance but a greater horizontal distance, the overall distance is greater. At constant speed, a greater distance requires a longer time interval. This longer time interval makes the duration of the clock's cycle longer, and the clock runs slow. If the clock runs slow, the hour passes less quickly.

3. Because you measure both events at the same position, you obtain the proper time interval between them. The driver in the second car moves relative to you, and so, according to the argument used in the solution to Example 14.4, measures a longer time interval.

4. In the zero-momentum reference frame both particles are moving, and therefore each particle has an inertia that is greater than the mass m.

5. Because moving objects contract in the direction of motion, the gate gets shorter to the observer in the car. Because the gate is shorter than the car in the Earth reference frame, it is also shorter in the car's reference frame. Thus only (*iii*) is possible.

14.5 Time dilation

Let us determine how the period of a light clock depends on the motion of the clock relative to the observer. The time interval between the instant a light signal is emitted at the bottom of a light clock and the instant the signal returns to the bottom is the period of that clock. In the reference frame of observer A in Figure 14.24a, the clock of height h is at rest, and so the period measured by observer A is $T_A = 2h/c_0$. In the reference frame of observer B, the clock is in motion and the signal follows the inverted-V path shown in Figure 14.24b. Because this path is longer than $2h$, the period T_B measured by observer B is longer than T_A.

During one round trip of its light signal, the clock moves a distance vT_B to the right in B's reference frame. From the right triangle in Figure 14.24b, we see that the distance d traveled by the signal during that time interval must satisfy

$$\left(\frac{d}{2}\right)^2 = h^2 + \left(v\frac{T_B}{2}\right)^2. \tag{14.1}$$

Because the signal travels at speed c_0, we have $d = c_0 T_B$. Substituting this relationship into Eq. 14.1 and multiplying the result by 4 to remove any fractions, we get

$$(c_0 T_B)^2 = 4h^2 + (vT_B)^2 \tag{14.2}$$

or

$$(c_0^2 - v^2)T_B^2 = 4h^2. \tag{14.3}$$

Dividing both sides of this equation by c_0^2 gives

$$\frac{c_0^2 - v^2}{c_0^2}T_B^2 = \frac{4h^2}{c_0^2} = T_A^2, \tag{14.4}$$

where we have used $T_A = 2h/c_0$. Rearranging terms, we get

$$T_B = \sqrt{\frac{c_0^2}{c_0^2 - v^2}}\, T_A = \frac{1}{\sqrt{1 - v^2/c_0^2}}\, T_A. \tag{14.5}$$

Because the factor in front of T_A is greater than 1 for $0 < v < c_0$, we see that $T_B > T_A$, as expected.

Figure 14.24 The period of a light clock moving relative to an observer is determined by the time interval it takes the light signal to travel the two diagonal distances $d/2$.

(*a*) **Reference frame of observer A:** clock at rest relative to observer

(*b*) **Reference frame of observer B:** clock moving relative to observer

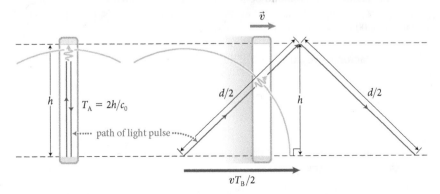

Figure 14.25 As the speed v at which any object travels approaches c_0, the Lorentz factor γ approaches infinity. Up to a speed of about $0.1c_0$, γ is very close to 1.

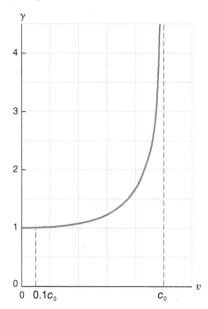

Table 14.3 Gamma factor

v/c_0	γ
0	1
0.1	1.005
0.5	1.15
0.9	2.3
0.98	5.0
0.99	7.1
0.999	22.4
0.9999	70.71
0.99999	223.61
0.999999	707.107

The factor on the right side of Eq. 14.5 is often represented by the Greek letter gamma, γ, and called the **Lorentz factor:**

$$\gamma \equiv \frac{1}{\sqrt{1 - v^2/c_0^2}}. \tag{14.6}$$

This factor is a measure of how pronounced relativistic deviations are from our everyday notions of time and space. At ordinary speeds, $v \ll c_0$, and so $\gamma \approx 1$; at high speeds, γ approaches infinity, as shown in **Figure 14.25**.

Table 14.3 lists values of γ for some values of v/c_0. Note that γ remains very close to 1 even up to very high speeds. For example, an object traveling at $0.1c_0$ could circle Earth in about 1 s, and yet the deviation of γ from 1 is less than 1%, meaning that the effects of time dilation are very small. As a rule of thumb, we can say that for speeds less than $0.1c_0$ the deviation of γ from 1 is negligible. We'll refer to speeds less than $0.1c_0$ as *nonrelativistic speeds* and speeds greater than $0.1c_0$ as *relativistic speeds*.

Conversely, we can obtain the ratio v^2/c_0^2 from γ by first squaring Eq. 14.6:

$$\gamma^2 = \frac{1}{1 - v^2/c_0^2} \tag{14.7}$$

and then inverting this equation:

$$\frac{1}{\gamma^2} = 1 - \frac{v^2}{c_0^2}, \tag{14.8}$$

so

$$\frac{v^2}{c_0^2} = 1 - \frac{1}{\gamma^2}. \tag{14.9}$$

For nonrelativistic speeds, we can approximate γ as*

$$\gamma = \frac{1}{\sqrt{1 - v^2/c_0^2}} \approx 1 + \tfrac{1}{2}\frac{v^2}{c_0^2} \quad (v/c_0 \ll 1), \tag{14.10}$$

so

$$\gamma - 1 \approx \tfrac{1}{2}\frac{v^2}{c_0^2} \quad \text{(within two significant digits when } v < 0.1c_0.) \tag{14.11}$$

As you will see shortly, this equation is useful for calculating γ when $v < 0.1c_0$.

14.14 Taking each motion to be relative to the Earth reference frame, calculate γ for (*a*) a car moving at 30 m/s, (*b*) an airplane moving at 250 m/s, (*c*) a spacecraft moving at 10,000 m/s, and (*d*) an electron moving at $0.60c_0$.

Using the factor γ, we can simplify Eq. 14.5 to

$$T_B = \gamma T_A. \tag{14.12}$$

---

*From the binomial expansion $\left[\sqrt{1 + \epsilon}\,\right]^{-1} = 1 - \tfrac{1}{2}\epsilon + \tfrac{3}{8}\epsilon^2 + \cdots$ (see Appendix B). For small ϵ, we can ignore terms of order ϵ^2 and higher.

This equation provides an exact quantitative expression for time dilation. In words, the period T_B measured by an observer in motion relative to the clock is lengthened by a factor γ relative to the period T_A measured by an observer at rest relative to the clock. Note that T_B is not a proper time interval between the emission of successive light signals because these events occur at different positions in the reference frame of the observer. Only an observer at rest relative to the clock can measure the proper time interval between the emission of successive light signals. To this observer, the clock is at rest, and so the light signal travels a distance $2h$ between successive emission events. The proper time interval between the emission of successive signals in the clock is thus $2h/c_0$.

Time dilation does not occur in light clocks only. As we pointed out in Section 14.3, the principle of relativity tells us that any clock should exhibit the same effects. Time dilation affects the measurement of the time interval between any two events and therefore time itself. Generalizing Eq. 14.12, we can say that the time intervals between two events as measured by two observers in motion relative to each other are related by

$$\Delta t_v = \gamma \Delta t_{\text{proper}}, \tag{14.13}$$

where Δt_{proper} is the time interval measured by an observer for whom both events occur at the same position, Δt_v is the time interval between the same two events measured by a second observer moving at speed v relative to the first observer, and γ is the factor given by Eq. 14.6.

Note that Eq. 14.13 involves two observers measuring the time interval Δt between two events. When applying this equation, you must therefore always begin by identifying the two events and the two observers. Next you should determine which observer measures the proper time interval and what speed v to use in Eq. 14.13. As an example, consider Figure 14.26. In Figure 14.26a two events occur at the same position relative to observer A, and so A measures the proper time interval Δt_{proper}. Observer B moves at speed v_{AB} relative to A, and so she measures a time interval $\Delta t_v = \gamma \Delta t_{\text{proper}}$, where the v in γ is equal to v_{AB}. In Figure 14.26b two events occur at the same position relative to B, and so B measures the proper time interval Δt_{proper}. Observer A moves at speed $v_{BA} = v_{AB}$ relative to B and so he measures a time interval $\Delta t_v = \gamma \Delta t_{\text{proper}}$, where the v in γ is equal to v_{AB}.

Figure 14.26 Proper time measurements.

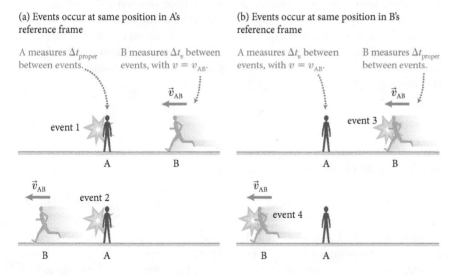

(a) Events occur at same position in A's reference frame

A measures Δt_{proper} between events.   B measures Δt_v between events, with $v = v_{AB}$.

$\vec{v}_{AB}$

event 1

A   B

$\vec{v}_{AB}$ event 2

B   A

(b) Events occur at same position in B's reference frame

A measures Δt_v between events, with $v = v_{AB}$.   B measures Δt_{proper} between events.

$\vec{v}_{AB}$

event 3

A   B

$\vec{v}_{AB}$ event 4

B   A

Example 14.7 Flight delay

You are on an airplane flying at a steady 250 m/s at cruising altitude. At noon the pilot announces that you are passing over the center of London and exactly 6.0 h later, according to your watch, comes on the address system to say that you are passing over Abu-Dhabi. (*a*) Do observers on the ground measure a longer, equal, or shorter time interval between these two events? (*b*) Is the difference between the time interval measured on the ground and that measured by your watch detectable?

❶ **GETTING STARTED** This problem involves two reference frames (the Earth reference frame and the reference frame of the plane) and two events (plane passing over London and plane passing over Abu-Dhabi). I denote the plane passing over the city as event 1 and the plane passing over Abu-Dhabi as event 2. The time interval between these events is 6.0 h in the reference frame of the plane.

❷ **DEVISE PLAN** The proper time interval is the shortest time interval between the events, and so this problem boils down to determining who measures the proper time interval. Once I have determined that, I can use Eq. 14.13 to determine the time interval measured by the ground observers and determine the difference between the two intervals.

❸ **EXECUTE PLAN** (*a*) Relative to the plane, events 1 and 2 occur at the same position. Therefore the 6.0-h time interval I measure is the proper time interval in the plane reference frame. The ground observers move at 250 m/s relative to my reference frame, and so, according to Eq. 14.13, they measure a longer time interval. ✔

(*b*) To determine the difference, I subtract the proper time interval from both sides of Eq. 14.13:

$$\Delta t_v - \Delta t_{\text{proper}} = \gamma \Delta t_{\text{proper}} - \Delta t_{\text{proper}} = (\gamma - 1)\Delta t_{\text{proper}}.$$

From Eq. 14.11 I have

$$\gamma - 1 \approx \tfrac{1}{2}\frac{v^2}{c_0^2} = \frac{(250 \text{ m/s})^2}{2(3.00 \times 10^8 \text{ m/s})^2} = 3.47 \times 10^{-13}.$$

The difference in the time intervals is thus $(3.47 \times 10^{-13})(6.0 \text{ h})$ $(3600 \text{ s/1 h}) = 7.5 \times 10^{-9}$ s, which is orders of magnitude smaller than can be detected with ordinary clocks. ✔

❹ **EVALUATE RESULT** That the difference in time intervals is too small to be easily measured is reassuring in two ways. First, from experience I know that my watch doesn't stop telling the correct time while flying (other than time zone adjustments). Second, the speed of the airplane is just a small fraction of the speed of light $(250 \text{ m/s})/(3.00 \times 10^8 \text{ m/s}) \approx 10^{-6}$, and so, for all practical purposes, $\gamma = 1$.

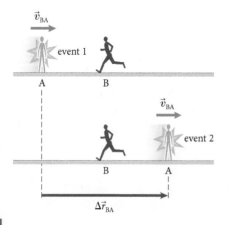

Figure 14.27 Observer B can determine the proper time interval between any pair of events by measuring the time interval between them and the distance separating them.

Reference frame of observer B

$\vec{v}_{\text{BA}}$

event 1

A   B

$\vec{v}_{\text{BA}}$

event 2

B   A

$\Delta \vec{r}_{\text{BA}}$

Let's examine Figure 14.26*a* from observer B's point of view and see if there is any way that B can determine the proper time interval between the two events without the need to refer to observer A's reference frame. In B's reference frame, the events occur at two locations (**Figure 14.27**). Let's choose the *x* axis such that we can denote A's displacement in B's reference frame by Δx. Relative to B, A moves at velocity $\vec{v}_{\text{BA}}$. If B measures a time interval Δt between the two events, then the *x* component of A's displacement is

$$\Delta x = v_{\text{BA}x}\Delta t = v_{\text{BA}x}\gamma \Delta t_{\text{proper}}, \quad (14.14)$$

where we have used Eq. 14.13 to convert Δt to the proper time interval between the events. Equation 14.14 relates the proper time interval to measurements taken entirely in B's reference frame (Δx and Δt), but the equation also contains the velocity of A's reference frame relative to B. To obtain an expression that does not contain this velocity requires a bit of algebraic manipulation on the expression $(c_0 \Delta t_{\text{proper}})^2$. We begin by multiplying and dividing this quantity by γ^2, substituting Eq. 14.8 for $1/\gamma^2$:

$$(c_0 \Delta t_{\text{proper}})^2 = \gamma^2 (c_0 \Delta t_{\text{proper}})^2 \left(1 - \frac{v^2}{c_0^2}\right). \quad (14.15)$$

Multiplying out the term in parentheses, we get

$$(c_0 \Delta t_{\text{proper}})^2 = c_0^2 \gamma^2 (\Delta t_{\text{proper}})^2 - v^2 \gamma^2 (\Delta t_{\text{proper}})^2. \quad (14.16)$$

Because $\Delta t = \gamma \Delta t_{\text{proper}}$ (Eq. 14.13), the first term on the right is equal to $(c_0 \Delta t)^2$. The magnitude of velocity $\vec{v}_{\text{BA}}$ is equal to the speed v that appears in the Lorentz factor, and so we can write $v_{\text{BA}x}^2 = v^2$. Equation 14.14 then tells us that the second term on the right in Eq. 14.16 is equal to $(\Delta x)^2$, and so we can rewrite this equation as

$$(c_0 \Delta t_{\text{proper}})^2 = (c_0 \Delta t)^2 - (\Delta x)^2. \tag{14.17}$$

Note that the right side of this expression contains only measurements made in B's reference frame: the time interval Δt and the spatial separation Δx of the events. The quantity $c_0 \Delta t$ represents the distance traveled by a light signal in the time interval between the events. The quantity

$$s^2 \equiv (c_0 \Delta t)^2 - (\Delta x)^2 \tag{14.18}$$

is called the **space-time interval** between two events. Because both terms on the right in Eq. 14.18 have units m^2, we see that s^2 represents a distance squared.

Depending on how $c_0 \Delta t$ compares to Δx, we can distinguish between different types of event pairs. When the distance $c_0 \Delta t$ traveled by a light signal in the time interval between the events is greater than the distance between the events, $(c_0 \Delta t)^2 > (\Delta x)^2$, and Eq. 14.18 tells us that $s^2 > 0$. For such a pair of events, there is an inertial reference frame for which the two events occur at the same position ($\Delta x = 0$) and the events are separated in time only. For this reason the space-time interval of a pair of events for which $s^2 > 0$ is said to be *timelike*. A clock in the reference frame in which the two events occur at the same position measures the proper time interval Δt_{proper} between the events and uniquely determines their sequence in time. Because the speed of light c_0 and the proper time interval Δt_{proper} between two events have definite values, any observer who measures the distance and time interval between the two events must obtain the same space-time interval $s^2 = (c_0 \Delta t_{\text{proper}})^2$ between them. In other words, the space-time interval is an invariant, even though neither Δx nor Δt is an invariant.

In a reference frame where the two events do not occur at the same position, $\Delta x > 0$. When $s^2 > 0$, Eq. 14.17 tells us that, for any reference frame in which $\Delta x > 0$, the time interval Δt between the events is greater than Δt_{proper}, as we concluded earlier. When Δt_{proper} is positive, Δt must be positive in all reference frames, and so a sequence of events that has a timelike space-time interval is the same in all inertial reference frames.

When the distance $c_0 \Delta t$ traveled by a light signal in the time interval between the events is less than the distance Δx between them, Eq. 14.18 tells us that $s^2 < 0$. For such a pair of events, there is no inertial reference frame for which the two events occur at the same position and therefore the events have no proper time interval. There is, however, a reference frame for which the events are simultaneous ($\Delta t = 0$) and separated in space only: $(\Delta x)^2 = -s^2$. The space-time interval for a pair of events for which $s^2 < 0$ is therefore said to be *spacelike*. In any other reference frame, the events are not simultaneous, and we cannot uniquely determine their sequence. Depending on the reference frame, Δt can be positive, zero, or negative, and so either event may precede the other.

When two events are exactly separated by the distance that light travels in the time interval between them, $s^2 = 0$ and the space-time interval is said to be *lightlike*. When you look at a source of light (your desk lamp or a star, say), the emission of the light and your observation of that light are separated by a lightlike space-time interval.

Because the space-time interval s^2 is an invariant, it is timelike, spacelike, or lightlike regardless of the reference frame. How does the nature of the space-time interval affect the principle of causality we discussed in Chapter 1? To distinguish cause and effect, it is necessary to determine the sequence of events. For a pair of events that have a timelike space-time interval ($s^2 > 0$), there is a reference frame in which we can study both events at the same position, and so we can determine their sequence with certainty and establish their causal relationship. For a pair of events that has a spacelike space-time interval ($s^2 < 0$), we cannot ascertain the time order of events and therefore cannot determine any causal relationship. For two events to be causally related, they must be able to interact physically. When $s^2 < 0$, the events occur in such rapid succession that a light signal cannot travel from one event to the other in the time interval between them. Because no information can travel faster than the speed of light, such events cannot interact and so cannot be causally related. Only events that have a timelike space-time interval can be causally related, and for such events the time sequence can be uniquely determined.

14.15 An elementary particle is observed to disintegrate after traveling 120 m in 5.0×10^{-7} s from the location where it was created. How long does the particle exist before disintegrating according to an observer at rest relative to the particle?

14.6 Length contraction

In Section 14.3 we saw that the length of an object measured by an observer moving relative to the object is shortened. Remember also that time dilation and length contraction are two manifestations of the same effect. To work out the amount of shortening, we now turn the clocks from Figure 14.24 on their side as in **Figure 14.28**, which shows the clock from the perspective of an observer A at rest relative to the clock and from the perspective of an observer B in motion relative to the clock.

We denote the proper length of the clocks by ℓ_{proper}. Our goal is to see what shortened length ℓ_v observer B obtains. According to A, the period of the clock is

$$T_A = \frac{2\ell_{\text{proper}}}{c_0}. \tag{14.19}$$

The period of the clock measured by B is given by Eq. 14.12: $T_B = \gamma T_A$. This period is the sum of two time intervals: the time interval Δt_{right} during which the light signal travels to the right from source to mirror and the time interval

Figure 14.28 By calculating the distance traveled by the light pulse in the reference frame of the observer moving relative to the clock, we can work out by how much the length of the clock is shortened.

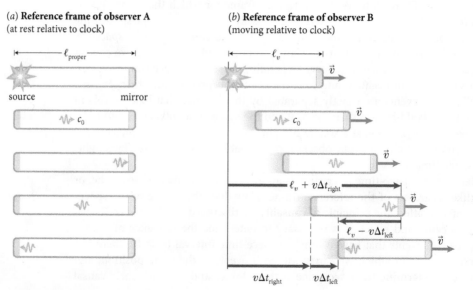

(a) **Reference frame of observer A**
(at rest relative to clock)

(b) **Reference frame of observer B**
(moving relative to clock)

Δt_{left} during which the signal travels from mirror to source. During Δt_{right}, the clock moves a distance $v\Delta t_{\text{right}}$ as measured by B. The distance traveled by the signal during that time interval is thus $d_{\text{right}} = \ell_v + v\Delta t_{\text{right}}$. Because the signal travels at speed c_0 independent of the motion of the source or the motion of the observer, the distance traveled by the signal in the time interval Δt_{right} must be $d_{\text{right}} = c_0\Delta t_{\text{right}}$. Equating the two expressions for d_{right}, we have

$$c_0\Delta t_{\text{right}} = \ell_v + v\Delta t_{\text{right}}. \tag{14.20}$$

Solving for Δt_{right} yields

$$\Delta t_{\text{right}} = \frac{\ell_v}{c_0 - v}. \tag{14.21}$$

While the signal in the clock returns from the mirror to the source, the clock moves an additional distance $v\Delta t_{\text{left}}$.* The signal therefore travels a distance $d_{\text{left}} = \ell_v - v\Delta t_{\text{left}}$ again as measured by observer B. Because the signal travels at speed c_0 independent of the motion of the source or observer, the distance traveled by the signal in the time interval Δt_{left} must be $d_{\text{left}} = c_0\Delta t_{\text{left}}$. Equating the two expressions for d_{left}, we have

$$c_0\Delta t_{\text{left}} = \ell_v - v\Delta t_{\text{left}}. \tag{14.22}$$

Solving for Δt_{left} yields

$$\Delta t_{\text{left}} = \frac{\ell_v}{c_0 + v}. \tag{14.23}$$

Combining this result with Eq. 14.21, we get for the sum of the two time intervals

$$T_{\text{B}} = \Delta t_{\text{right}} + \Delta t_{\text{left}}$$

$$= \ell_v\left(\frac{1}{c_0 - v} + \frac{1}{c_0 + v}\right). \tag{14.24}$$

Combining terms, we obtain

$$T_{\text{B}} = \frac{2\ell_v c_0}{c_0^2 - v^2}. \tag{14.25}$$

Dividing the numerator and denominator on the right by c_0^2, we get

$$T_{\text{B}} = \frac{2\ell_v}{c_0}\left(\frac{1}{1 - v^2/c_0^2}\right) = \frac{2\ell_v}{c_0}\gamma^2, \tag{14.26}$$

where we have substituted γ^2 from Eq. 14.7. Substituting this expression for T_{B} and Eq. 14.19 for T_{A} into Eq. 14.12, we get

$$\frac{2\ell_v}{c_0}\gamma^2 = \gamma\frac{2\ell_{\text{proper}}}{c_0} \tag{14.27}$$

or

$$\ell_v = \frac{\ell_{\text{proper}}}{\gamma}. \tag{14.28}$$

---

*The subscript *left* refers to the motion of the light, not the motion of the clock.

Figure 14.29 Length measurements.

(a) Object is at rest in A's reference frame

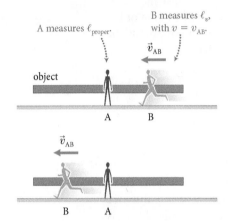

(b) Object moves with observer B

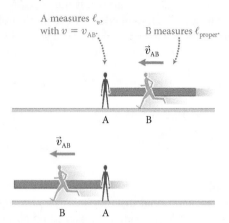

Because $\gamma > 1$, the measured length ℓ_v of the clock moving relative to us (or any other object moving relative to us) is shorter than its proper length ℓ_{proper}.

Note that Eq. 14.28 involves two observers measuring the length of an object: observer A at rest relative to the object and observer B in motion relative to it. When applying this equation, you must therefore always begin by identifying (1) the observers and (2) the observer's reference frame, if any, in which the object is at rest. The observer in this reference frame measures the object's proper length. As an example, consider the two observers in Figure 14.29. In Figure 14.29*a* the object is at rest in A's reference frame, and so A measures the proper length ℓ_{proper}. Observer B moves at speed v_{AB} relative to A, and so he measures a length $\ell_v = \ell_{\text{proper}}/\gamma$, where the v in γ is equal to v_{AB}. In Figure 14.29*b* the object is at rest relative to B, and so B measures the proper length ℓ_{proper}. Observer A moves at speed $v_{BA} = v_{AB}$ relative to B, and so she measures a length $\ell_v = \ell_{\text{proper}}/\gamma$, where the v in γ is equal to v_{AB}.

The Procedure box "Time and length measurements at very high speeds" summarizes how to treat time intervals and distances when dealing with motion at very high speeds.

Procedure: Time and length measurements at very high speeds

At very high speeds, measurements of lengths and time intervals depend on the reference frame. Problems involving objects or observers moving at very high speeds therefore require extra care when we consider lengths and time intervals.

1. Identify each observer or reference frame with a letter and determine in which reference frame each numerical quantity given in the problem statement is measured.
2. If the problem involves any time intervals, identify the events that define that time interval. Determine in which observer's reference frame, if any, the two events occur at the same position. This observer measures the proper time interval between the events (Figure 14.26).
3. Identify any distances or lengths mentioned in the problem. Determine for which observer, if any, the object defining that distance or length remains at rest. This observer measures the proper length (Figure 14.29).
4. Use Eqs. 14.13 and 14.28 to determine the time intervals and lengths in reference frames other than the ones in which the intervals and lengths are given.

Example 14.8 Electron dash

An electron moving at $0.80c_0$ relative to the Earth reference frame travels the 100-m length of a building. What is the length of the building according to an observer moving along with the electron?

❶ GETTING STARTED This problem involves two reference frames: the reference frame of the building (which is the Earth reference frame) and the reference frame of the electron. Although the problem doesn't mention this explicitly, the 100-m length of the building is likely to be the length measured in Earth's reference frame, and so it is the building's proper length. I summarize this information in a table:

	Electron reference frame	Earth reference frame
$v_{electron}$	0	$0.8c_0$
$\ell_{building}$	?	100 m

❷ DEVISE PLAN Equation 14.28 relates the building length measured by the observer in the electron's reference frame to the proper length. I know the building's proper length (100 m) and the speed of the observer relative to the reference frame in which the proper length is measured ($0.80c_0$). I can calculate γ from Eq. 14.6 and then use Eq. 14.28 to obtain the length of the building in the electron's reference frame.

❸ EXECUTE PLAN $\gamma = 1/\sqrt{1 - v^2/c_0^2} = 1/\sqrt{1 - (0.80)^2} = 1.67$, and so $\ell_v = \ell_{building} = (100 \text{ m})/1.67 = 60 \text{ m}.$ ✔

❹ EVALUATE RESULT From our length contraction discussion in Section 14.3, I know that an observer measuring the length of an object that is moving relative to her measures a length shorter than the proper length. In this example, the observer is in the electron's reference frame, and the building being measured is moving relative to that reference frame. Therefore I expect the measured length to be shorter than the proper length.

Example 14.9 Cosmic ray longevity

A proton strikes an air molecule in Earth's atmosphere, creating a muon 6.0 km above the ground. The muon travels toward Earth at nearly c_0 and decays just before reaching the ground. An observer traveling along with the muon determines that it lives for 2.2×10^{-6} s. According to this observer, what is the distance between the ground and the muon at the instant the muon is created?

❶ GETTING STARTED The problem mentions two reference frames: the Earth reference frame and the reference frame of the muon. The altitude is measured in the Earth reference frame, whereas the muon's lifetime is measured in the reference frame of the muon. The muon's lifetime is the time interval between its creation at an altitude of 6.0 km (event 1) and its decay at the ground (event 2). The speed of the muon in the Earth reference frame is (nearly) c_0. I summarize this information in tabular form:

	Earth reference frame	Muon reference frame
Δt		2.2×10^{-6} s (proper time interval)
h_{atm}	6.0 km (proper length)	?
v_{muon}	c_0	

❷ DEVISE PLAN The 6.0-km distance traveled by the muon in the Earth reference frame is the proper length of the atmosphere. To calculate the (contracted) distance traveled by the muon in the muon reference frame, I can use Eq. 14.28, but I first need to determine γ for the motion of the muon. I can determine the time interval between events 1 and 2 in the Earth reference frame by dividing the distance traveled in the Earth reference frame (6.0 km) by the muon speed, which I take to be c_0. This time interval is the lifetime of the muon in the Earth reference frame. Because I know the muon lifetime in the reference frame of the muon (2.2×10^{-6} s), which is the proper time interval, I can use Eq. 14.13 to determine γ.

❸ EXECUTE PLAN The distance traveled in the Earth reference frame is 6.0×10^3 m, and so in this reference frame the time interval between events 1 and 2 is $\Delta t_v = (6.0 \times 10^3 \text{ m})/(3.00 \times 10^8 \text{ m/s}) = 2.0 \times 10^{-5}$ s. In the muon reference frame, the muon is at rest and exists for the proper time interval $\Delta t_{proper} = 2.2 \times 10^{-6}$ s. I can obtain γ from Eq. 14.13:

$$\gamma = \frac{\Delta t_v}{\Delta t_{proper}} = \frac{2.0 \times 10^{-5} \text{ s}}{2.2 \times 10^{-6} \text{ s}} = 9.1.$$

Using this value for γ, I determine the distance ℓ_v traveled by the muon in its reference frame from Eq. 14.28:

$$\ell_v = \frac{\ell_{proper}}{\gamma} = \frac{6.0 \times 10^3 \text{ m}}{9.1} = 6.6 \times 10^2 \text{ m}.$$ ✔

❹ EVALUATE RESULT The muon lifetime in the Earth reference frame is longer than the muon's proper lifetime, as it should be. Because the muon lives longer in the Earth reference frame, it can travel a greater distance (6.0 km instead of 660 m). I can also check my result by considering the situation from the muon's reference frame. In that reference frame, the atmosphere flies by at nearly the speed of light for a time interval corresponding to the muon's proper lifetime, and so $\ell_v = c_0 \Delta t_{proper} = (3.00 \times 10^8 \text{ m/s})(2.2 \times 10^{-6} \text{ s}) = 6.6 \times 10^2 \text{ m}$, which is the result I got.

QUANTITATIVE TOOLS

Figure 14.30 A muon created at the top of the atmosphere is able to reach the ground before decaying—but observers in the Earth reference frame and the muon reference frame would explain this result differently.

(*a*) **Earth reference frame**

Muon reaches ground because its clock runs slow.

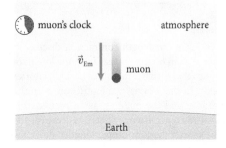

(*b*) **Muon reference frame**

Muon reaches ground because atmosphere moves relative to it and hence is contracted.

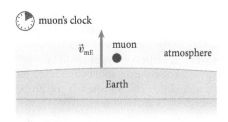

Table 14.4 Muon math

Reference frame	ℓ (km)	Δt (s)
Earth	6.0	2.0×10^{-5}
muon	0.66	2.2×10^{-6}

Figure 14.30 is a graphical representation of Example 14.9. To the observer in the Earth reference frame, the length of the muon's path is equal to the thickness (proper length) of the atmosphere, but the particle's lifetime is lengthened by time dilation (Figure 14.30*a*). Figure 14.30*b* shows the same decay viewed by the observer attached to the muon. To this observer, the muon is at rest and decays in the proper time interval, but the moving atmosphere is contracted. What appears as time dilation of the muon's lifetime to one observer appears as length contraction of the atmosphere to the other (Table 14.4). Time dilation and length contraction are thus two manifestations of the same effect.

In Chapter 6 we derived the Galilean transformation equations (Eqs. 6.4 and 6.5) to relate the coordinates of an event e in any two inertial reference frames A and B moving at constant velocity relative to each other. At very high speeds, these equations have to be replaced with equations that take into account the effects of time dilation and length contraction. For two reference frames A and B that have origins that coincide at $t_A = t_B = 0$ and that are moving at constant velocity in the x direction relative to each other, the new equations, called the **Lorentz transformation equations,** are (see Appendix E):

$$t_{Be} = \gamma\left(t_{Ae} - \frac{v_{ABx}}{c_0^2}x_{Ae}\right) \tag{14.29}$$

$$x_{Be} = \gamma(x_{Ae} - v_{ABx}t_{Ae}) \tag{14.30}$$

$$y_{Be} = y_{Ae} \tag{14.31}$$

$$z_{Be} = z_{Ae}. \tag{14.32}$$

At nonrelativistic speeds, these equations reduce to the Galilean transformation equations.

From the Lorentz transformation equations, we obtain that the x component of the velocity of an object o in reference frame A, v_{Aox}, is related to that in reference frame B, v_{Box}, by the expression (see Appendix E)

$$v_{Box} = \frac{v_{Aox} - v_{ABx}}{1 - \frac{v_{ABx}}{c_0^2}v_{Aox}}. \tag{14.33}$$

For nonrelativistic speeds, the denominator is approximately equal to 1 and we obtain the usual addition of velocities: $v_{Box} = v_{Aox} - v_{ABx}$ (see Eq. 6.9).

14.16 A car travels 1.0 km down a straight road at 100 km/h. How much shorter is the road in the car's reference frame?

14.7 Conservation of momentum

In Chapter 4 we introduced inertia to account for velocity changes in collisions and took the momentum of a particle to be the product of its inertia and its velocity. This approach led us to conservation of momentum, one of the fundamental principles that we have used throughout this text. In Section 14.4 we found that inertia depends on energy and therefore on speed. Because conservation principles are fundamental and because they embody an appealing simplicity, we would like to preserve them in light of the modifications required by special relativity. To this end, we begin again by taking the momentum of a particle as the product of its inertia and its velocity but this time allowing the inertia to depend on velocity:

$$\vec{p} \equiv m_v\vec{v}. \tag{14.34}$$

The subscript v on the inertia symbol m_v is there to remind us that inertia is no longer an invariant. We now examine what modifications are needed to our concept of inertia to preserve conservation of momentum.

To obtain an expression for m_v, let us look at an elastic collision between two identical particles and require the system to be isolated so that the momentum of the two particles remains constant. The particles are projected at each other at speed u (Figure 14.31). Their velocities are equal in magnitude but opposite in direction, both before and after the collision. The position of the center of mass of this system is at the location where the particles collide and it remains at rest. In other words, Figure 14.31 represents the collision viewed from the zero-momentum reference frame. We begin our analysis of this collision by choosing our axes as shown in Figure 14.32a.

In the zero-momentum reference frame, the speeds of the particles are the same before and after the collision, and so we gain no information on how inertia depends on speed. Let us therefore consider the same collision from the point of view of an observer A moving parallel to the x axis of Figure 14.32 with a velocity equal to the component of the velocity of particle 1 along the x axis. To this observer, particle 1 moves up and down parallel to the y axis (Figure 14.32b). For convenience, let's denote the event when particle 1 leaves observer A as the *throw* and its return to observer A as the *catch*.

To an observer B moving along the x axis with a velocity equal to the component of the velocity of particle 2 along the x axis, the collision looks exactly the same except that particles 1 and 2 are interchanged (Figure 14.32c).

14.17 In Figure 14.32, is the time interval between the throw and the catch of particle 1 in reference frame A greater than, equal to, or less than the time interval between the throw and the catch of particle 2 in reference frame B?

Note that in all three reference frames only the y components of $\vec{v}_1$ and $\vec{v}_2$ change. Because the observers' motion is always in the x direction, there is no length contraction in the y direction, and so to both observers the particles travel the same distance d in the y direction between throw and catch.

In reference frame A, the y component of the velocities of the two particles have the same magnitudes, but the x components are different. Therefore particles 1 and 2 move at different speeds v_1 and v_2, respectively, and so even though we began by declaring the two particles to be identical, the inertia m_{v1} of particle 1

Figure 14.31 Collision between two identical particles as seen in the zero-momentum reference frame.

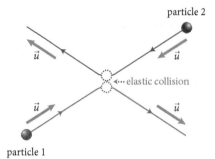

Figure 14.32 Analysis of the trajectories in Figure 14.31 from three reference frames, showing our choice of axes.

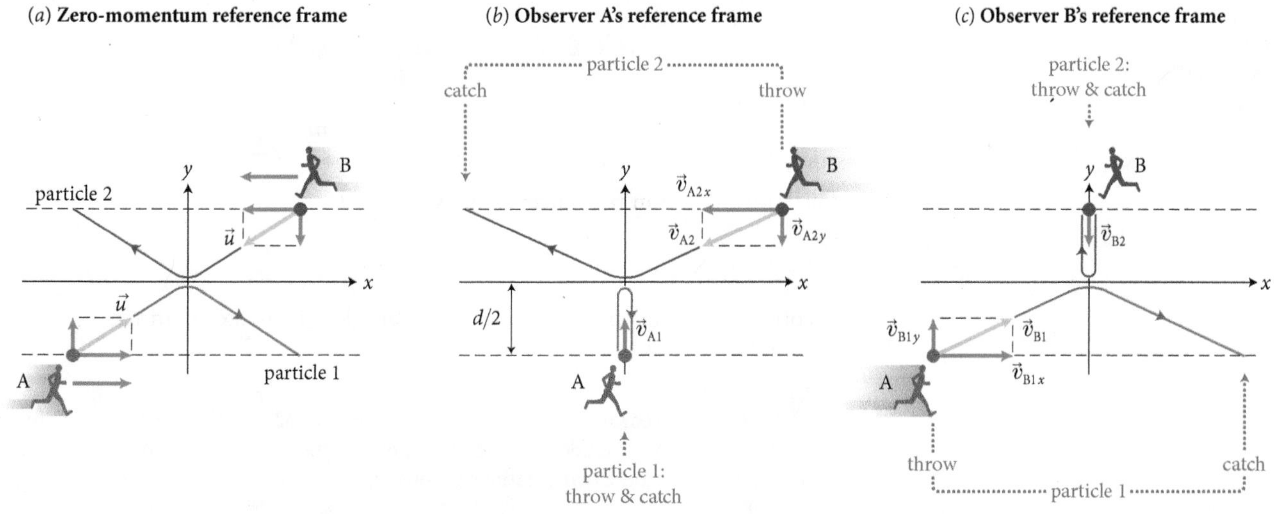

(a) Zero-momentum reference frame

(b) Observer A's reference frame

(c) Observer B's reference frame

measured in reference frame A must be different from the inertia m_{v2} of particle 2 measured in this frame. We can obtain the ratio of these inertias by requiring the momentum of the two-particle system to remain constant. To this end, let us evaluate the velocity changes caused by the collision in reference frame A. Let the time intervals between the throw and catch of particles 1 and 2 in reference frame A be Δt_{A1} and Δt_{A2}, respectively. The changes in the y component of the velocities in reference frame A are thus

$$\Delta v_{A1y} = -\frac{d}{\Delta t_{A1}} \quad \text{and} \quad \Delta v_{A2y} = \frac{d}{\Delta t_{A2}}. \tag{14.35}$$

The time interval Δt_{A1} measured by observer A between the throw and catch of particle 1 in reference frame A is the proper time interval between these two events because they occur at the same position according to observer A: $\Delta t_{A1} = \Delta t_{\text{proper}}$. The time interval Δt_{A2} observer A measures between the throw and catch of particle 2, however, is not a proper time interval because for observer A these two events occur at different positions (Figure 14.32b).

Relative to observer B, the throw and catch of particle 2 occur at the same position (Figure 14.32c), and so the time interval measured by B is the proper time interval. Because of symmetry, the collision seen by observer A is identical to that seen by observer B, and so the proper time interval measured by A between the throw and catch of particle 1 is equal to the proper time interval measured by B between the throw and catch of particle 2, as you saw in Checkpoint 14.17:

$$\Delta t_{A1} = \Delta t_{B2} = \Delta t_{\text{proper}}. \tag{14.36}$$

Equation 14.13, $\Delta t_v = \gamma \Delta t_{\text{proper}}$, tells us that if observer B measures the proper time interval between the throw and catch of particle 2, observer A moving relative to B obtains that the time interval between those events is $\gamma \Delta t_{\text{proper}}$, where γ depends on the relative velocity of A and B. Therefore

$$\Delta t_{A2} = \gamma \Delta t_{\text{proper}}, \tag{14.37}$$

where v in $\gamma = 1/\sqrt{1 - v^2/c^2}$ is the speed of B relative to A: $v = |\vec{v}_{A2x}|$. Substituting Eq. 14.36 into Eq. 14.37 yields $\Delta t_{A2} = \gamma \Delta t_{A1}$, and so Eq. 14.35 becomes

$$\Delta v_{A1y} = -\frac{d}{\Delta t_{A1}} \quad \text{and} \quad \Delta v_{A2y} = \frac{d}{\gamma \Delta t_{A1}}. \tag{14.38}$$

Because the two-particle system is isolated, the sum of the particles' momentum changes in reference frame A must add to zero:

$$\Delta p_{A1y} + \Delta p_{A2y} = m_{v1} \Delta v_{A1y} + m_{v2} \Delta v_{A2y}$$

$$= -m_{v1} \frac{d}{\Delta t_{A1}} + m_{v2} \frac{d}{\gamma \Delta t_{A1}} = 0, \tag{14.39}$$

which means that in reference frame A,

$$m_{v2} = \gamma m_{v1}. \tag{14.40}$$

According to observer A, the inertia of particle 2 is greater than the inertia of particle 1.

14.18 (a) According to observer A in Figure 14.32, which particle is moving faster? (b) Does A conclude that the inertia of a particle increases or decreases with speed? (c) Answer these two questions for observer B.

Equation 14.40 gives us the relationship between the inertias of two particles moving relative to each other and relative to us. To determine the relationship between the inertia of a particle moving relative to us and that of an identical particle at rest relative to us, consider the limit in which the y components of the velocities are very small. In that case, particle 1 in Figure 14.32b is virtually at rest in reference frame A, $v_{A1} \approx 0$, and particle 2 moves horizontally along with observer B at speed $v_{A2} \approx |\vec{v}_{A2x}| = v$. When the particle is at rest relative to us, its inertia and mass are equivalent and so $m_v \approx m$, where m is the mass of the particle. If we denote the inertia of the moving particle by m_v, Eq. 14.40 shows us how a particle's **inertia** m_v, which depends on speed, is related to its mass m, which is an invariant:

$$m_v = \gamma m. \qquad (14.41)$$

The first thing to note is that for speeds up to about $0.1c_0$, $\gamma \approx 1$, and so mass and inertia are virtually indistinguishable, in agreement with what we established in Chapter 13. As the speed approaches c_0, however, m_v increases to infinity. Because inertia is a measure of an object's resistance to being accelerated, m_v approaching infinity means it becomes more and more difficult to speed up the particle as its speed approaches c_0. We therefore must conclude that no object or particle that has mass can be accelerated to speed c_0.

Exercise 14.10 Sluggish electron

By what factor does the inertia of an electron increase (a) as the electron is accelerated to $\frac{1}{3}c_0$ and (b) as its speed is increased by another 20%? (c) Does the mass of the electron change as it is accelerated?

SOLUTION From Eq. 14.41, I know that the factor by which the inertia increases is $m_v/m = \gamma$. (a) At $\frac{1}{3}c_0$, $\gamma = 1/\sqrt{1 - (\frac{1}{3})^2} = 1.06$. The inertia increases by 6%. ✔

(b) Increasing the speed by another 20% means adding

$$\tfrac{1}{5}\left(\tfrac{1}{3}c_0\right) = \tfrac{1}{15}c_0$$

to the speed $\frac{1}{3}c_0$, giving

$$v = \tfrac{1}{3}c_0 + \tfrac{1}{15}c_0 = \tfrac{2}{5}c_0$$

$$\gamma = \frac{1}{\sqrt{1 - (\frac{2}{5})^2}} = 1.09.$$

The inertia of the electron increases by another 3%. ✔

(c) No. Mass is an invariant. Thus the mass of an object is independent of the motion of either the observer or the object. ✔

🖐 **14.19** To what fraction of c_0 must an electron at rest in the Earth reference frame be accelerated in order for its inertia to increase by a factor of 1000?

Now that we have an expression for the inertia of a moving particle (Eq. 14.41), we can obtain an expression for the particle's momentum in terms of its mass and velocity. Substituting Eq. 14.41 into Eq. 14.34 gives

$$\vec{p} = \gamma m\vec{v}. \qquad (14.42)$$

Unlike the expression for momentum you learned in Chapter 4, $\vec{p} = m\vec{v}$, Eq. 14.42 does not show the momentum increasing linearly with increasing velocity. At ordinary speeds, Eq. 14.42 reduces to the familiar form $\vec{p} = m\vec{v}$. As the speed approaches c_0, the momentum increases rapidly to infinity (Figure 14.33). Experiments such as the one described in the next example have conclusively shown that Eq. 14.42 correctly accounts for the momentum of a particle at all speeds up to c_0.

Figure 14.33 As the speed of a particle approaches c_0, the magnitude of its momentum approaches infinity. Up to a speed of about $0.1c_0$, the magnitude of the momentum is very close to mv.

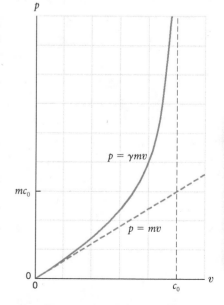

Example 14.11 Fast collision

A fast-moving electron collides head-on with a proton moving at $0.60c_0$ as measured in the Earth reference frame. What is the electron's initial speed if the two particles combine to form a system that after the collision is at rest in the Earth reference frame? The mass of the proton is 1836 times that of the electron.

❶ GETTING STARTED I begin by making a sketch showing the electron-proton system before and after the collision (**Figure 14.34**). I choose the positive x axis to be in the initial direction of motion of the electron. I can consider the electron-proton system to be isolated, and so the initial momentum of the system must be equal to the final momentum (which is zero).

Figure 14.34

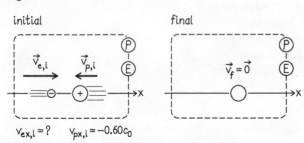

❷ DEVISE PLAN I can use Eq. 14.42 to calculate the initial momentum of each particle and then set the sum of these momenta equal to zero. The resulting expression contains four quantities: the unknown initial velocity component $v_{ex,i}$ of the electron, the initial velocity component $v_{px,i}$ of the proton $(-0.60c_0)$, the mass m_p of the proton, and the mass m_e of the electron. Because I am given that $m_p/m_e = 1836$, I can solve for the unknown initial velocity of the electron.

❸ EXECUTE PLAN The initial momentum of the system must be zero, so

$$p_{px,i} + p_{ex,i} = \gamma_{p,i}\, m_p v_{px,i} + \gamma_{e,i}\, m_e v_{ex,i} = 0, \quad (1)$$

where I have added subscripts to the Lorentz factors to keep the factor for the proton separate from the factor for the electron. Dividing the two terms containing γ in Eq. 1 by m_e, and using Eq. 14.6 to write out the Lorentz factors, I can rewrite Eq. 1 as

$$\frac{m_p}{m_e} \frac{v_{px,i}}{\sqrt{1 - v_{px,i}^2/c_0^2}} = -\frac{v_{ex,i}}{\sqrt{1 - v_{ex,i}^2/c_0^2}}.$$

When I square both sides and then substitute the values for the mass ratio and the speed of the proton, I get

$$(1836)^2 \frac{(-0.60c_0)^2}{1 - (-0.60)^2} = \frac{v_{ex,i}^2}{1 - (v_{ex,i}/c_0)^2},$$

which I can solve for $v_{ex,i}$:

$$v_{ex,i} = (1 + 5.3 \times 10^{-7})^{-1/2} c_0 = 0.99999974\, c_0. ✔$$

I show eight significant digits because the 1 in the above expression is exact (infinite number of significant digits) and so the sum $1 + 5.3 \times 10^{-7}$ is known to the eighth decimal place (1.00000053).

❹ EVALUATE RESULT I note that this speed is close to c_0 (but not greater than or equal to it). The fact that the electron needs to travel at such a high speed is not surprising because the proton's mass is nearly 2000 times that of the electron.

14.20 In Example 14.11, (*a*) is the collision elastic, inelastic, or totally inelastic (see Section 5.1)? (*b*) How fast would the electron have to move if the expression for momentum were $\vec{p} = m\vec{v}$ instead of Eq. 14.42?

14.8 Conservation of energy

Now we turn our attention to energy. Suppose we accelerate a particle from rest to a speed approaching c_0.* To determine the increase in its kinetic energy, let us calculate the work done on the particle. From Chapter 9, we know that the work done on the particle is given by Eq. 9.22:

$$W = \int_{x_i}^{x_f} F(x)\,dx = \int_{x_i}^{x_f} \frac{dp}{dt}\,dx, \quad (14.43)$$

where we have substituted the definition of force (Eq. 8.4). Substituting $p = \gamma m v$ and noting that the mass m is constant, we can write the integrand in Eq. 14.43 as

$$\frac{dp}{dt}dx = \frac{d}{dt}(\gamma m v)dx = m\frac{d(\gamma v)}{dt}dx. \quad (14.44)$$

QUANTITATIVE TOOLS

*From here on, speeds are assumed to be relative to the Earth reference frame unless stated otherwise.

Note that because γ is a function of v, it varies with time and cannot be pulled out of the differentiation. Using $dx = v\,dt$ and canceling the dt factors, we get

$$\frac{dp}{dt}dx = m\frac{d(\gamma v)}{dt}dx = m\frac{d(\gamma v)}{dt}v\,dt = mv\,d(\gamma v), \qquad (14.45)$$

and so the work done on the particle is

$$W = m\int_{i}^{f} v\,d(\gamma v) = m\int_{i}^{f}(v^2\,d\gamma + v\gamma\,dv), \qquad (14.46)$$

where we have used the chain rule $d(\gamma v) = \gamma\,dv + v\,d\gamma$ (Appendix B).

The relationship between v^2 and γ is given by Eq. 14.9: $v^2/c_0^2 = 1 - 1/\gamma^2$. Differentiating both sides of Eq. 14.9 with respect to time yields

$$\frac{2v}{c_0^2}\frac{dv}{dt} = 0 - \frac{-2}{\gamma^3}\frac{d\gamma}{dt}, \qquad (14.47)$$

and then multiplying both sides by $\frac{1}{2}c_0^2\gamma\,dt$ yields

$$\gamma v\,dv = \frac{c_0^2}{\gamma^2}d\gamma. \qquad (14.48)$$

From Eq. 14.8, $1/\gamma^2 = 1 - v^2/c_0^2$, we can say

$$\frac{1}{\gamma^2} = \frac{c_0^2 - v^2}{c_0^2} \quad \text{and} \quad \frac{c_0^2}{\gamma^2} = c_0^2 - v^2, \qquad (14.49)$$

and so c_0^2/γ^2 in Eq. 14.48 can be replaced by $c_0^2 - v^2$, yielding $\gamma v\,dv = (c_0^2 - v^2)d\gamma$. Substituting this result into Eq. 14.46 and noting that $\gamma = 1$ when $v = 0$ at the lower integration boundary, we get

$$W = m\int_{1}^{\gamma} [v^2\,d\gamma + (c_0^2 - v^2)d\gamma] = mc_0^2\int_{1}^{\gamma}d\gamma = (\gamma - 1)mc_0^2. \quad (14.50)$$

All of this work done on the particle goes toward increasing its kinetic energy, and so the particle's final **kinetic energy** is

$$K = (\gamma - 1)mc_0^2 = \gamma mc_0^2 - mc_0^2 = m_v c_0^2 - mc_0^2, \qquad (14.51)$$

where we have used the expression for the inertia, $m_v = \gamma m$, from Eq. 14.41.

This expression for the kinetic energy of a particle looks completely different from the expression we have used so far, $K = \frac{1}{2}mv^2$ (see Section 5.2). In particular, as the speed approaches c_0, the particle's kinetic energy becomes infinitely great (Figure 14.35). Therefore it would take an infinite amount of energy to accelerate a particle to speed c_0, so c_0 is the limiting speed at which any object can travel relative to another other object.

✋ **14.21** Use Eq. 14.11, $\gamma - 1 \approx 0.5(v^2/c_0^2)$, to show that Eq. 14.51 reduces to the familiar $K = \frac{1}{2}mv^2$ at ordinary speeds ($v \ll c_0$).

Each term on the right of Eq. 14.51 represents a quantity of energy, so Eq. 14.51 gives us the kinetic energy of a particle as the difference between two amounts of energy. In Chapter 5 we defined the energy of a system as the sum of that

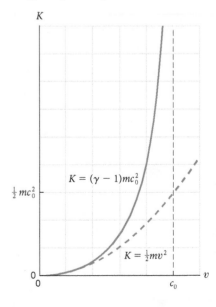

Figure 14.35 As the speed of a particle approaches c_0, its kinetic energy approaches infinity. Up to speeds of about $0.25c_0$, the kinetic energy is very close to $\frac{1}{2}mv^2$.

$K = (\gamma - 1)mc_0^2$

$K = \frac{1}{2}mv^2$

system's kinetic and internal energies, $E \equiv K + E_{int}$ (Eq. 5.21), and so we see that the kinetic energy of a system is obtained by subtracting the internal energy from the system's energy:

$$K = E - E_{int}. \tag{14.52}$$

In the right side of Eq. 14.51, note that the first term in the difference is proportional to inertia m_v and the second is proportional to mass m. In Section 14.4 we established that there is a relationship between inertia and energy and a relationship between mass and internal energy. Comparing Eqs. 14.51 and 14.52 thus suggests that the energy is given by

$$E = m_v c_0^2 \tag{14.53}$$

and the **internal energy** is given by

$$E_{int} = mc_0^2. \tag{14.54}$$

The assignment made in Eqs. 14.53 and 14.54 may look like a leap of faith, but in Section 14.4 we found that in a nuclear reaction between a proton and a deuteron, both the internal energy and the mass of the system decrease. Equation 14.54 tells us that the changes in internal energy and mass should be related as

$$\Delta E_{int} = E_{int,f} - E_{int,i}$$
$$= m_f c_0^2 - m_i c_0^2 = \Delta m c_0^2, \tag{14.55}$$

or $\Delta E_{int}/\Delta m = c_0^2$. The experimental data show that the ratio $\Delta E_{int}/\Delta m = c_0^2$ is 8.98×10^{16} m^2/s^2 (see Example 14.5), which is precisely c_0^2, confirming Eq. 14.54. The term E_{int} in Eq. 14.54 thus represents the energy that corresponds to *all* of the mass of the particle.

All forms of internal energy we have discussed so far require an arbitrary choice of reference point (see, for example, Eq. 7.21), and so we usually considered only differences in internal energy. Equation 14.54, however, provides us with a measure of internal energy that does not depend on any choice of reference point. What is more, we can calculate E_{int} from the mass m, which is a quantity we can determine experimentally.

For example, a proton, as mentioned in Chapter 7, is made up of quarks held together by the strong nuclear interaction. If we add up the values of the masses of these quarks (which are determined from a combination of theory and experiment and known only approximately), we discover that their sum is just 1% of the mass of the proton. The remaining 99% of the mass of the proton must therefore come from the nuclear potential energy that binds the quarks together. Because protons (and neutrons, which also consist of three quarks) are the basic building blocks of all matter, it follows that most of what we consider matter really *is* energy! Even though we have no mathematical expression for nuclear potential energy, we can determine its value from the masses of the proton and the quarks, and the relationship $E_{int} = mc_0^2$.

14.22 (*a*) Draw a simplified energy diagram for the totally inelastic collision between two identical objects moving at speed v toward each other. After the collision, the two objects form a compound object that is at rest. (*b*) Which of the following quantities are constant in the collision: kinetic energy, mass, energy?

Example 14.12 Alpha emission

A uranium nucleus ^{238}U at rest decays to a thorium nucleus ^{234}Th by emitting an alpha particle (which is a helium nucleus ^{4}He):

$$^{238}\text{U} \rightarrow {}^{234}\text{Th} + {}^{4}\text{He}.$$

The mass of the uranium atom is $m_U = 395.292599 \times 10^{-27}$ kg, the mass of the thorium atom is $m_{Th} = 388.638509 \times 10^{-27}$ kg, and the mass of the helium atom is $m_{He} = 6.646478 \times 10^{-27}$ kg. What is the combined kinetic energy of the fragments?

❶ GETTING STARTED I begin by drawing a simplified energy diagram for the alpha emission (**Figure 14.36**), choosing the

Figure 14.36

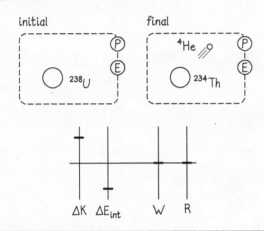

uranium nucleus and the fragments as my (closed) system. Because the system is closed, the change in kinetic energy is the negative of the change in internal energy: $\Delta K = -\Delta E_{int}$.

❷ DEVISE PLAN Because the initial kinetic energy is zero, the final kinetic energy of the fragments is equal to $-\Delta E_{int}$. To determine this change in internal energy, I can calculate the change in the mass of the system and then use Eq. 14.55.

❸ EXECUTE PLAN

$$\Delta m = (m_{Th} + m_{He}) - m_U = -7.61 \times 10^{-30} \text{ kg}.$$

Substituting this value into Eq. 14.55 yields

$$\Delta E_{int} = \Delta m c_0^2 = (-7.61 \times 10^{-30} \text{ kg})(3.00 \times 10^8 \text{ m/s})^2$$

$$= -6.85 \times 10^{-13} \text{ J},$$

so the final kinetic energy of the fragments is 6.85×10^{-13} J. ✔

❹ EVALUATE RESULT The answer I obtain is a tiny amount of energy, contrary to my expectations for nuclear energy! However, this is the energy liberated by the nucleus of a single atom. One atom is about 10^{-10} m in diameter, which means a cube of matter 1 mm on a side contains about $[(10^{-3} \text{ m})/(10^{-10} \text{ m})]^3 = 10^{21}$ atoms. If each atom liberates an amount of energy equal to the answer I got, my cube yields $(10^{21})(6.85 \times 10^{-13} \text{ J}) = 7 \times 10^8$ J of energy, which is a very large amount and more in line with what I'd expect.

14.23 An atom in the filament of a light bulb emits energy in the form of light. (*a*) Draw a simplified energy diagram representing this emission. (*b*) Does the mass of the atom increase, decrease, or stay the same? (*c*) Does the inertia of the atom increase, decrease, or stay the same? Assume the atom remains at rest in the Earth reference frame.

We can obtain a useful relationship from our expressions for momentum, $\vec{p} = \gamma m \vec{v}$ (Eq. 14.42), and energy, $E = m_v c_0^2$ (Eq. 14.53), by evaluating the quantity $E^2 - (c_0 p)^2$. Substituting γm for m_v in Eq. 14.53, we have

$$E^2 - (c_0 p)^2 = \gamma^2 m^2 c_0^4 - c_0^2 \gamma^2 m^2 v^2 = \gamma^2 m^2 c_0^2 (c_0^2 - v^2). \quad (14.56)$$

From Eq. 14.49, we see that $c_0^2 - v^2 = c_0^2/\gamma^2$, and so

$$E^2 - (c_0 p)^2 = (m c_0^2)^2. \quad (14.57)$$

The right side of this expression is an invariant, and so its value does not depend on the reference frame. The left side, however, contains the energy E and the momentum magnitude p of the system under consideration, both of which depend on the reference frame! Again, we have constructed an invariant from two quantities that are not invariants themselves. Equation 14.57 tells us that if we measure the energy and the magnitude of the momentum of a system in a certain reference frame and then compute the quantity $E^2 - (c_0 p)^2$, we obtain the same value as the corresponding combination in any other reference frame. In each case, the result is equal to $(m c_0^2)^2$, and so the quantity $E^2 - (c_0 p)^2$ is an invariant.

Energy and momentum are conserved quantities when evaluated with respect to each reference frame, and so for a system that is both isolated and closed, the energy and momentum laws tell us that $E_i = E_f$ and $\vec{p}_i = \vec{p}_f$. We have used these equations extensively to determine E_f and p_f given E_i and p_i. Equation 14.57 relates values of E and p to values of m. Because values of m are known from separate measurements, Eq. 14.57 allows us to eliminate momentum from the problem. The next example and checkpoint illustrate the usefulness of Eq. 14.57.

Example 14.13 Proton collision

Two protons, each of mass $m_p = 1.67 \times 10^{-27}$ kg, collide to produce a particle that has a mass 300 times the mass of each proton. What is the minimum kinetic energy that one of the protons must have in order to create this particle if the other proton is initially at rest?

❶ GETTING STARTED I begin by organizing the information in a sketch (**Figure 14.37**). I denote the incoming proton by 1 and the proton initially at rest by 2. The two protons form an isolated, closed system, and so the energy law requires $E_{1i} + E_{2i} = E_f$ and the momentum law requires $\vec{p}_{1i} = \vec{p}_f$ (because $\vec{p}_{2i} = 0$).

Figure 14.37

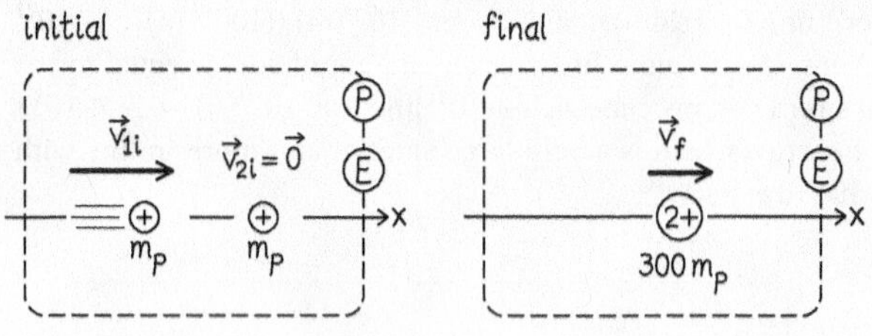

❷ DEVISE PLAN The initial kinetic energy K_{1i} of proton 1 is given by Eq. 14.52, with the internal energy E_{1int} given by Eq. 14.54. I must therefore determine the initial energy E_{1i} of proton 1. I can use Eq. 14.57 to relate the energy E of each particle to its momentum p and then combine this with the energy and momentum law to solve for the initial energy of proton 1.

❸ EXECUTE PLAN Applying Eq. 14.57 to proton 1, proton 2, and the final particle, I get

$$E_{1i}^2 - (c_0 p_{1i})^2 = (m_p c_0^2)^2 \tag{1}$$

$$E_{2i}^2 = (m_p c_0^2)^2 \tag{2}$$

$$E_f^2 - (c_0 p_f)^2 = (300 m_p c_0^2)^2. \tag{3}$$

Substituting $E_{1i} + E_{2i} = E_f$ and $p_{1i} = p_f$ into Eq. 3 yields

$$(E_{1i} + E_{2i})^2 - (c_0 p_{1i})^2 = (300 m_p c_0^2)^2.$$

Working out the left side, I get

$$E_{1i}^2 + 2E_{1i}E_{2i} + E_{2i}^2 - (c_0 p_{1i})^2 = E_{1i}^2 - (c_0 p_{1i})^2 + E_{2i}^2 + 2E_{1i}E_{2i}$$
$$= (300 m_p c_0^2)^2,$$

so, after substituting from Eqs. 1 and 2, I have

$$(m_p c_0^2)^2 + (m_p c_0^2)^2 + 2E_{1i}(m_p c_0^2) = (300 m_p c_0^2)^2.$$

Solving for E_{1i} gives

$$E_{1i} = \frac{(300 m_p c_0^2)^2 - 2(m_p c_0^2)^2}{2(m_p c_0^2)} = \tfrac{1}{2}(300^2 - 2)m_p c_0^2. \tag{4}$$

From Eqs. 14.52 and 14.54, I have $K_{1i} = E_{1i} - m_p c_0^2$, so

$$K_{1i} = \left[\tfrac{1}{2}(300^2 - 2)m_p c_0^2\right] - m_p c_0^2 = \tfrac{1}{2}(300^2 - 4)m_p c_0^2$$
$$= \tfrac{1}{2}(300^2 - 4)(1.67 \times 10^{-27}\text{ kg})(3.00 \times 10^8\text{ m/s})^2$$
$$= 6.76 \times 10^{-6}\text{ J. } ✔$$

❹ EVALUATE RESULT My answer is large for a particle that has a mass on the order of 10^{-27} kg. Comparing my result in Eq. 4 with Eq. 14.51, however, I see that $\gamma - 1 = (300^2 - 4)/2$, or $\gamma = 4.5 \times 10^4$, and so the proton has to be accelerated to very close to the speed of light. It is therefore not unreasonable that the energy of the proton is on the order of microjoules.

✋ **14.24** In Example 14.13, what is the minimum kinetic energy of each proton if they have the same initial speed and collide head-on?

Chapter Glossary

SI units of physical quantities are given in parentheses.

Event A physical occurrence that takes place at a specific location at a specific instant.

Inertia m_v (kg) A scalar that is a quantitative measure of an object's tendency to resist any change in its velocity. Inertia is related to the object's mass m:

$$m_v = \gamma m. \tag{14.41}$$

At nonrelativistic speeds, the inertia of an object is virtually indistinguishable from the object's mass: $m_v \approx m$.

Internal energy E_{int} (J) Any energy not associated with the motion of an object or system. Internal energy is a quantitative measure of the state of the object or system. The internal energy of an object is an invariant and is related to the mass m of the object by

$$E_{int} = mc_0^2. \tag{14.54}$$

Invariant A quantity whose value does not depend on the choice of reference frame.

Kinetic energy K (J) The energy associated with the motion of an object. The kinetic energy of a particle moving at speed v is

$$K = (\gamma - 1)mc_0^2. \tag{14.51}$$

At nonrelativistic speeds, this expression simplifies to $K = \frac{1}{2}mv^2$.

Length contraction When an object moves at speed v relative to an observer, the observer measures the length ℓ_v of the object in the direction of its motion to be shorter than the proper length ℓ_{proper} of the object:

$$\ell_v = \frac{\ell_{proper}}{\gamma}. \tag{14.28}$$

Lorentz factor γ (unitless) A scalar factor that appears frequently in special relativity:

$$\gamma \equiv \frac{1}{\sqrt{1 - v^2/c_0^2}}. \tag{14.6}$$

This factor is a measure of how pronounced relativistic deviations are from our everyday notions of time and space. At nonrelativistic speeds, $\gamma \approx 1$; at relativistic speeds, $\gamma > 1$ and γ approaches infinity as v approaches c_0.

Lorentz transformation equations The relationships between the coordinates of an event e in two reference frames A and B that are coinciding at $t_A = t_B = 0$ and that are moving at constant velocity $\vec{v}_{AB}$ in the x direction relative to each other:

$$t_{Be} = \gamma\left(t_{Ae} - \frac{v_{ABx}}{c_0^2}x_{Ae}\right) \tag{14.29}$$

$$x_{Be} = \gamma(x_{Ae} - v_{ABx}t_{Ae}) \tag{14.30}$$

$$y_{Be} = y_{Ae} \tag{14.31}$$

$$z_{Be} = z_{Ae}. \tag{14.32}$$

Proper length ℓ_{proper} (m) The length of an object measured by an observer at rest relative to the object.

Proper time interval Δt_{proper} (s) The time interval between two events that occur at the same position (and can therefore be measured by a single clock at rest). An observer measuring the proper time interval between two events records the shortest time interval between those events.

Simultaneity Events are *simultaneous* when they are observed to occur at the same instant in a reference frame made up of a grid of synchronized clocks. Events that are simultaneous in one reference frame need not be simultaneous in a different reference frame.

Space-time interval s^2 (m^2) An invariant given by

$$s^2 \equiv (c_0\Delta t)^2 - (\Delta x)^2. \tag{14.18}$$

For events that have a proper time interval, the space-time interval is equal to the square of the distance traveled by light in the proper time interval.

Speed of light The speed at which light travels. The value of this speed is independent of the motion of the source or observer and is the limiting speed at which any object can travel relative to any other object. In vacuum the speed of light is $c_0 = 299{,}792{,}458$ m/s.

Time dilation The effect that causes a clock moving relative to an observer to be measured as running more slowly than an identical clock at rest with respect to that observer. If Δt_{proper} is the proper time interval between two events measured by an observer for whom both events occur at the same position, then the time interval between those events measured by an observer moving at speed v relative to the first observer is

$$\Delta t_v = \gamma \Delta t_{proper}. \tag{14.13}$$

15 Periodic Motion

ny motion that repeats itself at regular time intervals is called **periodic motion.** The motion of, say, the Moon revolving around Earth or Earth revolving around the Sun is one type of familiar periodic motion, as is the rotation of the hands of a clock. In this chapter, we are interested in a particular type of periodic motion—namely, back-and-forth periodic motion. Such motion, called either **vibration** or **oscillation** (the two words mean the same thing), is common at all scales in the universe. A rocking chair, a swing, the pendulum of a grandfather clock, the strings on a guitar, the wings of a mosquito all oscillate. At the atomic level, atoms oscillate inside solids. At the cosmic level, the entire universe may oscillate in an ever-repeating cycle of expansion and contraction. The physical natures of these periodic motions are very different from one another, and yet they are all closely related and their mathematical descriptions are always of the same form, differing only in the quantities involved.

In this chapter, we first develop a model for the simplest type of oscillation, called *simple harmonic motion,* and relate it to a periodic motion we are already familiar with: rotational motion. Then we show that any oscillation can be described in terms of simple harmonic motions. After presenting a number of examples, we discuss the effects of energy dissipation during oscillations.

15.1 Periodic motion and energy

Figure 15.1 shows the periodic motion of a spring-cart system. The system, initially at rest, is compressed and then released from rest. After being released, the cart oscillates back and forth along the low-friction track, alternately compressing and stretching the spring.

✋ **15.1** (*a*) List the forces exerted on the spring-cart system of Figure 15.1 right after it is released, and draw a free-body diagram for each object in the system. (*b*) Which of these forces do work on the system as it oscillates? (*c*) As the hand pushes on the cart and compresses the spring, is the work done by the cart on the spring positive, negative, or zero? (*d*) How does the work done by the cart on the spring compare with the work done by the spring on the cart?

To understand why the motion of the cart in Figure 15.1 is periodic and why the cart doesn't simply return to its initial position and remain there, let us look at the mechanical energy in the cart-spring system. As the spring is compressed by the hand, elastic potential energy is stored in the system. Just before the cart is released, the system is at rest, so the mechanical energy of the system consists entirely of potential energy (**Figure 15.2a**). After the cart is released, the spring-cart system is closed—none of the forces exerted on it does any work on it—and so the energy of the system must remain constant.

In order to analyze the system's energy changes in more detail, we choose an *x* axis that points in the initial direction of motion of the cart and that has its origin at the

Figure 15.1 Compressing and releasing the spring cause the cart to oscillate back and forth on the track.

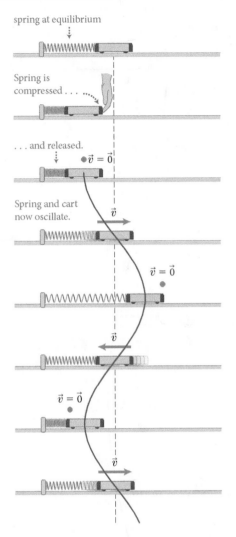

center of the cart's equilibrium position. As the expanding spring speeds up the cart in the positive *x* direction, the kinetic energy increases and the potential energy decreases (Figure 15.2*b*). When the cart reaches the equilibrium position at *x* = 0, the potential energy is zero (Figure 15.2*c*). The cart, however, does not stop at this position because its kinetic energy is not zero. The cart therefore overshoots the equilibrium position and begins stretching the spring. As the spring stretches, it exerts on the cart a restoring force that slows the cart down and converts kinetic energy back to elastic potential energy (Figure 15.2*d*). This continues until the kinetic energy falls to zero, at which point the cart reverses its direction of travel (Figure 15.2*e*). The potential energy is converted back to kinetic energy (Figure 15.2*f*), and the cart again returns to the equilibrium position, this time traveling leftward (Figure 15.2*g*). The potential energy is zero again, but with the cart's kinetic energy being nonzero, it overshoots the equilibrium position once again. Provided no energy is dissipated, the system returns to its initial state (compare Figure 15.2*a* and 15.2*i*). A new cycle begins, and the cart repeats its back-and-forth motion.

Figure 15.2 Periodic motion is the result of a continuous exchange between kinetic and potential energy.

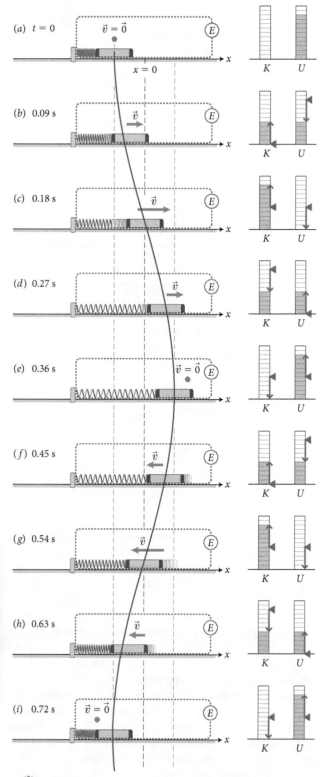

The time interval it takes to complete a full cycle of the motion is the *period* T (see Section 11.1). For the motion shown in Figure 15.2 the period is equal to the time elapsed between (*a*) and (*i*): $T = 0.72$ s. The inverse of the period, called the **frequency** of the motion, $f = 1/T$, gives the number of cycles completed per second. In Figure 15.2, $f = 1/(0.72 \text{ s}) = 1.4 \text{ s}^{-1}$ and so the cart completes 1.4 cycles each second.

The magnitude of the cart's maximum displacement from the equilibrium position is called the **amplitude** A of the periodic motion. The amplitude is related to the mechanical energy of the system. In Figure 15.2, for example, the greater the initial compression of the spring, the greater the amplitude and the greater the initial potential energy.

In practice, periodic motion in mechanical systems does not continue indefinitely because some energy is always dissipated (friction in the cart's bearings, air resistance, heating of the spring). This dying out of periodic motion due to dissipation of energy is called *damping*. Initially, we shall ignore damping and, with this simplification, we find that:

Periodic motion is characterized by a continuous conversion between potential and kinetic energy in a closed system.

All systems that exhibit this behavior have one common feature: a restoring interaction that tends to return the system to equilibrium. Some examples are illustrated in **Figure 15.3**. Each of these systems executes periodic motion when disturbed from its equilibrium position, and in each case the amplitude of the motion is a measure of the mechanical energy of the system.

✋ **15.3** For each system in Figure 15.3, identify (*a*) the restoring force and (*b*) the type of potential energy associated with the motion.

Figure 15.3 Examples of oscillating systems.

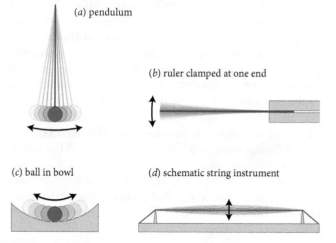

(*a*) pendulum

(*b*) ruler clamped at one end

(*c*) ball in bowl

(*d*) schematic string instrument

✋ **15.2** (*a*) In Figure 15.2*e*, the cart's displacement from the equilibrium position is maximum. Is the *x* component of the cart's acceleration at that instant positive, negative, or zero? (*b*) At which instant(s) in Figure 15.2 is the magnitude of the cart's acceleration greatest? At which instant(s) is it smallest? (Use the blue position curve to answer these questions.)

Figure 15.4 The oscillation of an object suspended from a spring is isochronous: The period T does not depend on the amplitude of the oscillation.

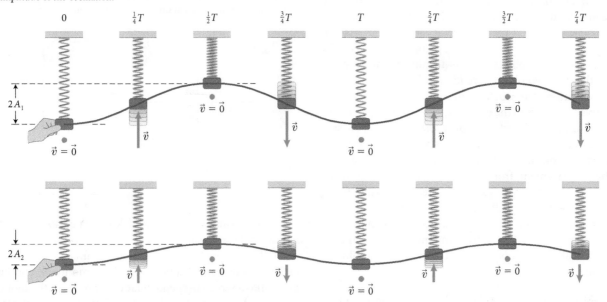

15.2 Simple harmonic motion

Measuring the period of any oscillating system reveals a surprising fact: When the amplitude of the motion is not too great, the period is independent of the amplitude. For example, an object suspended from a spring has a fixed period of oscillation regardless of its amplitude (**Figure 15.4**). The period of a pendulum, too, is independent of the amplitude of its swing, which is why pendulums are used to regulate the movement of mechanical clocks. Likewise, when you pluck a guitar string, the pitch of the sound, which is determined by the string's period of oscillation, is the same regardless of how far back you pull the string.

A system that exhibits equal periods for all amplitudes is called *isochronous*. Provided their amplitudes are not too great, all the systems shown in Figure 15.3 are isochronous. Individual systems have different periods that are determined by the properties of the system, but any given system has a period that is essentially independent of amplitude.

The graph of displacement versus time for an isochronous oscillating system is shown in **Figure 15.5**. The $x(t)$ curves for an isochronous oscillation are all sine functions. For this reason they are referred to as *sinusoidal curves*. Any periodic motion that yields a sinusoidal $x(t)$ curve is called **simple harmonic motion,** and a system that executes such motion is called a *simple harmonic oscillator*. For small amplitudes, all the systems shown in Figure 15.3 yield this type of $x(t)$ curve, and so all are examples of simple harmonic motion. The motion is called harmonic because in musical instruments harmonic motion is responsible for the sound emitted by the instruments.

The fact that simple harmonic motion is isochronous allows us to draw some important conclusions. Consider, for example, the two isochronous $x(t)$ curves in Figure 15.5, one with twice the amplitude of the other. These could be the curves of two identical pendulums or any other of the systems shown in Figure 15.3, with one oscillating object moving twice as far out as the other. In one period, object 1 always covers twice the distance that object 2 covers. Because the periods are the same, the velocity of object 1 must always be twice as great as that of object 2 at the same instant in the periodic motion. Neither motion has constant velocity or constant acceleration—the velocities and accelerations change constantly. However, in order for the velocity of object 1 always to be twice that of object 2, the acceleration of object 1 must, at each instant, be twice as great as the acceleration of object 2.

Figure 15.5 Positions-versus-time graphs for two isochronous objects.

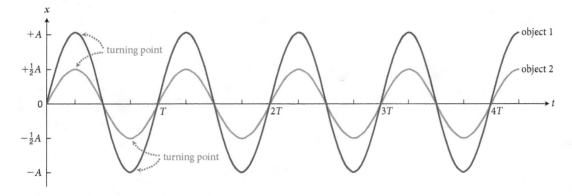

As you saw in Checkpoint 15.2, the acceleration is greatest when the displacement from the equilibrium position is greatest (that is, at the *turning points,* which are the positions where the object turns around). This tells us that the acceleration of object 1 at its turning point $x_1 = A$ is twice as great as that of object 2 at its turning point $x_2 = \frac{1}{2}A$. Given that the two objects are identical, we conclude that the acceleration of an isochronous object is proportional to the displacement from the equilibrium position. Because the acceleration of an object is proportional to the vector sum of the forces exerted on it, this tells us that the restoring force exerted on the oscillating object is also proportional to the displacement from the equilibrium position. We already know that springs exert linear restoring forces (Hooke's law, Section 8.9). The observation that all systems in Figure 15.3 are isochronous tells us, then, that the restoring forces in all these systems must also be linearly proportional to the displacement from the equilibrium position.

A object executing simple harmonic motion is subject to a linear restoring force that tends to return the object to its equilibrium position and is linearly proportional to the object's displacement from its equilibrium position.

15.4 Suppose the spring in Figure 15.1 is compressed twice as much. (*a*) By how much does the mechanical energy of the spring-cart system increase? (*b*) What is the relationship between the amplitude of the oscillation and the mechanical energy in the oscillating system?

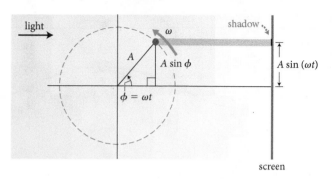

Figure 15.6 If we project the shadow of a ball moving in a circle at constant speed onto a screen perpendicular to the plane of the motion, the shadow moves up and down in simple harmonic motion.

Simple harmonic motion is closely related to a periodic motion we have already studied: circular motion at constant speed. To see how the two are related, imagine projecting the shadow of a small sphere in circular motion at constant speed onto a screen perpendicular to the plane of the motion (**Figures 15.6** and **15.7**). While the sphere sweeps out a circular trajectory, its shadow oscillates up and down on the screen. The position of the shadow at any instant is given by the vertical component of the sphere's position vector at that instant. If the sphere begins on the horizontal axis and moves with a constant rotational speed ω, the angle ϕ between the sphere's position vector and the horizontal axis grows steadily with time: $\phi = \omega t$. The position of the sphere's shadow on the screen is then described by $A \sin(\omega t)$, where A is the radius of the sphere's trajectory. Because it is described by a sinusoidal function, the projection of circular motion at constant speed is simple harmonic motion.

Figure 15.7 Correspondence between circular motion at constant speed and simple harmonic motion for (*a*) an object suspended from a spring and (*b*) a pendulum.

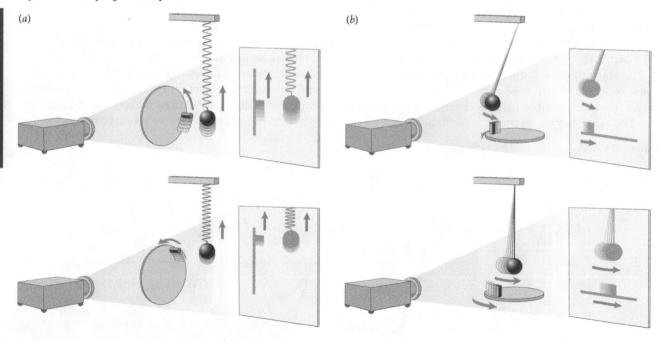

Figure 15.8 Fourier analysis of a periodic function into sinusoidal functions. The period of each function is indicated by the shading. (*a*) Original periodic function. (*b*) Set of harmonics used to describe the function in (*a*). (*c*) Successive sum of harmonics showing an increasingly improving match to the original periodic function.

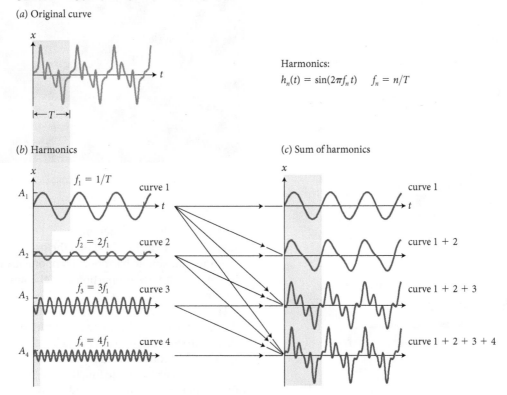

(*a*) Original curve

Harmonics:
$$h_n(t) = \sin(2\pi f_n t) \qquad f_n = n/T$$

(*b*) Harmonics

(*c*) Sum of harmonics

The correspondence between circular motion at constant speed and simple harmonic motion can be demonstrated experimentally by projecting the shadows of an object in simple harmonic motion and an object in circular motion at constant speed on a screen, as in Figure 15.7. If the rotational speed is adjusted so that its period equals that of the oscillation, the shadows of the two objects move synchronously.

15.5 What are (*a*) the direction of the velocity of the shadow on the screen in Figure 15.6 and (*b*) the direction of the shadow's acceleration?

15.3 Fourier's theorem

Simple harmonic motion is important because, as Figure 15.3 suggests, it is very common. Another reason it is important is that, according to a rule known as **Fourier's theorem,**[*] any periodic function can be written as a sum of sinusoidal functions of frequency $f_n = n/T$, where $n \geq 1$ is an integer and T is the period of the function. This means that *any* periodic motion can be treated as a superposition of simple

harmonic motions. So, if we understand simple harmonic motion, we can deal with any other type of periodic motion by breaking it down into its simple harmonic components. Because any motion that is nonrepetitive can be thought of as periodic motion that has an infinitely long period, Fourier's theorem has broad implications: The time-dependent motion of *any* system can be expressed in terms of sinusoidal functions.

Consider, for example, the periodic motion represented by the nonsinusoidal curve in **Figure 15.8a**. Although very different from simple harmonic motion, this motion can be expressed as the sum of a few simple harmonic components, shown in the sinusoidal curves of Figure 15.8*b*. The first component has the same frequency $f_1 = 1/T$ as the original periodic function; this frequency, the lowest in the sum, is called the **fundamental frequency** or **first harmonic.** All other components, called **higher harmonics,** are at integer multiples of the fundamental frequency: $f_n = nf_1 = n/T$.[†] When these higher harmonics are added to the first harmonic, the shape of the resulting periodic function is not sinusoidal, but the period remains the same (Figure 15.8*c*).

---

[*]After the French mathematician Jean-Baptiste Joseph Fourier (1768–1830).

[†]In general, an infinite sum of sines and cosines is required to break down arbitrary periodic functions. Only periodic functions for which $f(t) = -f(-t)$ (that is, they are odd in t, such as the one shown in Figure 15.8) can be analyzed with only sines (which are also odd in t).

Figure 15.9 (*a*) Periodic function and (*b*) amplitudes of the harmonic functions in the Fourier series of the function shown in part *a*. (*c*) Spectrum of the periodic function, obtained by plotting the squares of the amplitudes.

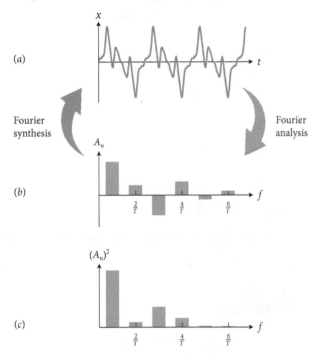

Figure 15.10 Spectrum analyzer display on a stereo component.

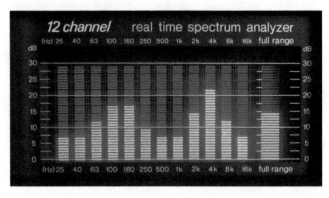

By adjusting the amplitude of each harmonic, we can make the sum of harmonics fit the original periodic function.

The breaking down of a function into harmonic components is called *Fourier analysis,* and the resulting sum of sinusoidal functions is called a *Fourier series.* **Figure 15.9** shows one way of graphically representing the Fourier series: The amplitudes A_n of all the harmonics are plotted against the frequencies of the harmonics. As you saw in Checkpoint 15.4, the energy of a simple harmonic oscillator is proportional to the square of the amplitude. For this reason, it is customary to plot A_n^2 as a function of frequency, as shown in Figure 15.9*c*; such a plot is called the **spectrum** of the periodic function.

The spectrum shows the contribution of each harmonic to the original periodic function. A similar type of visual display of frequency content is found on some stereophonic components. Displays such as the one shown in **Figure 15.10** provide a visual representation of the frequency content of the sound being played.

Figure 15.11 shows some important relationships between a periodic function and its spectrum. The period T of the function determines the frequency of the first harmonic. In other words, the duration of the slowest component of the function determines the lowest frequency in the spectrum. Conversely, the duration T_{min} of the fastest component in the function determines the highest frequency in the spectrum.

Fourier analysis has many applications, ranging from electronic signal processing to chemical analysis and voice

recognition. The ear functions as a Fourier analyzer because the inner ear breaks down the oscillations produced by sound into simple harmonic components, each producing a nerve impulse proportional to the amplitude of that component.

The inverse of Fourier analysis—the creation of periodic functions by adding sinusoidal functions together—is called *Fourier synthesis.* A musician's electronic synthesizer, for example, produces a sum of harmonics that add up to produce different periodic functions. When the amplitudes of the different harmonics are adjusted so that they match the frequency content of real instruments, the synthesizer simulates the sounds of many musical instruments.

15.6 (*a*) What does the spectrum of a single sinusoidal function of period T look like? (*b*) As T is increased, what change occurs in the spectrum?

Figure 15.11 Relationship between a periodic function and its spectrum.

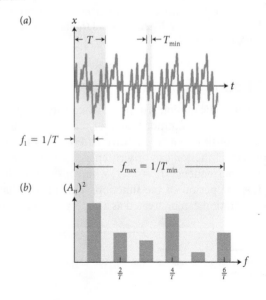

15.4 Restoring forces in simple harmonic motion

Periodic motion takes place about a position of translational or rotational equilibrium (see Section 12.5) and requires a restoring force that tends to return the object to the equilibrium position. Consider, for example, an object moving along an x axis and subject to the vector sum of forces whose x component is shown graphically in **Figure 15.12**. The object is in translational equilibrium at each position where the vector sum is zero: at x_1, x_2, and all values of $x > x_3$. Can the object oscillate about any of these positions? From the graph we see that $\sum F_x$ is positive to the left of x_1, and so the vector sum $\sum \vec{F}$ points toward x_1. To the right of x_1 $\sum F_x$ is negative, and so the vector sum of forces again points toward x_1. The forces exerted on the object tend to return the object to x_1 whenever it moves away from x_1. Around x_1 the vector sum of forces "restores" the translational equilibrium so that the object can oscillate around x_1.

Because the vector sum of the forces always pushes the object back to x_1, this equilibrium position is said to be *stable*. You can visualize this position by imagining a ball at the bottom of a bowl: When the ball is displaced in any direction, a restoring force accelerates it back toward the bottom.

On either side of equilibrium position x_2, the direction of the vector sum of the forces is such that the object accelerates away from the equilibrium position. Such an equilibrium position is called *unstable* because the forces tend to amplify any disturbance from equilibrium. Imagine balancing a ball on top of a hill: The slightest displacement of the ball in any direction is enough to make it accelerate away from the equilibrium position.

Figure 15.12 Whenever the vector sum of the forces exerted on an object is zero, the object is in translational equilibrium. This graph shows stable, unstable, and neutral equilibria.

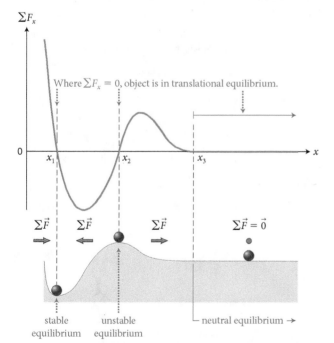

Positions for which $x > x_3$ represent a *neutral* equilibrium: Moving the ball in any direction has no effect on its subsequent acceleration.

What all this tells us is that objects tend to oscillate about *stable* equilibrium positions because the forces about such positions tend to restore the equilibrium. A consequence of the restoring forces about stable equilibrium positions is:

In the absence of friction, a small displacement of a system from a position of stable equilibrium causes the system to oscillate.

✋ **15.7** (*a*) Are there any equilibrium positions in the force-versus-distance graph in **Figure 15.13**? If so, is the equilibrium stable, unstable, or neutral? (*b*) Compare the magnitude of the restoring force for equal displacements on either side of x_0 in Figure 15.13.

Figure 15.13 shows why so many systems move in simple harmonic motion. As we saw in Section 15.2, simple harmonic motion requires *linear* restoring forces—that is, forces that are linearly proportional to the displacement from the equilibrium position. In Checkpoint 15.7*b* you saw that the restoring force about the stable equilibrium position x_0 in Figure 15.13 is not linear. However, as illustrated by the successive enlargements in Figure 15.13, the curve, which gives the relationship between force and position, becomes indistinguishable from a straight line for small displacements from the equilibrium position x_0 (say, within the shaded region). In other words:

For sufficiently small displacements away from the equilibrium position, restoring forces are always linearly proportional to the displacement. Consequently, for small displacements, objects execute simple harmonic motion about any stable equilibrium position.

Figure 15.13 For sufficiently small displacements near a stable equilibrium position, the restoring force is linear in the displacement.

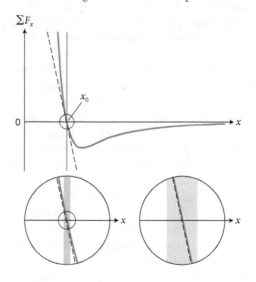

CONCEPTS

Figure 15.14 Cause of the restoring force for a taut string displaced from equilibrium.

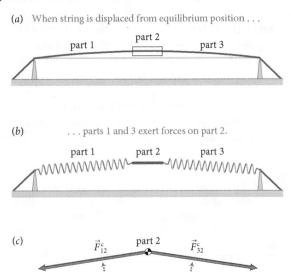

(*a*)   When string is displaced from equilibrium position . . .

part 2
part 1   part 3

(*b*)   . . . parts 1 and 3 exert forces on part 2.

part 1   part 2   part 3

(*c*)   part 2

$\vec{F}^c_{12}$ $\vec{F}^c_{32}$

Forces add up to nonzero restoring force.

Figure 15.15 Cause of the restoring torque on a beam bent away from its equilibrium shape.

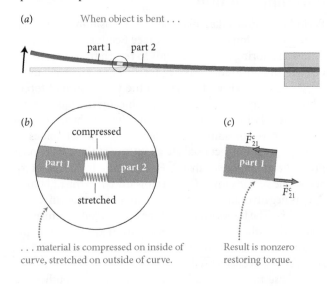

(*a*)   When object is bent . . .

part 1   part 2

(*b*)

compressed

part 1   part 2

stretched

. . . material is compressed on inside of curve, stretched on outside of curve.

(*c*)

$\vec{F}^c_{21}$

part 1

$\vec{F}^c_{21}$

Result is nonzero restoring torque.

The restoring forces in the examples in Figure 15.3 fall into two categories: elastic (*b*, *d*) and gravitational (*a*, *c*). We saw in Chapter 8 that, for small displacements from the equilibrium position, the elastic restoring force of a spring is linearly proportional to the displacement. The same is true for other elastic restoring forces. **Figure 15.14**, for example, shows a string displaced from its equilibrium position. Each part of the stretched string is subject to elastic (contact) forces exerted by the parts immediately adjacent to it. Part 2 in the middle of the string in Figure 15.14*b*, for instance, is subject to a force exerted by part 1 and a force exerted by part 3. In the equilibrium position, when the string is straight, the elastic forces exerted by parts 1 and 3 on part 2 add to zero. As the string is displaced, however, the elastic forces exerted by the two neighboring parts no longer line up (Figure 15.14*c*) and their vector sum provides a restoring force that tends to return the string to its equilibrium position. For small displacements, as we just saw, this restoring force is linearly proportional to the displacement from the equilibrium position.

The same line of reasoning applies to the restoring forces that arise when materials are twisted or bent. **Figure 15.15** shows that bending an object compresses the material that lies on the inside of the curve and stretches the material that lies on the outside of the curve. The resulting internal elastic forces cause torques on each part of the object, torques that tend to return the object to its equilibrium shape. For small displacements, the internal elastic forces are linearly proportional to the displacement, and consequently the restoring torque, too, is linearly proportional to the (rotational) displacement.

Let us next examine a pendulum, where the restoring force is due to gravity. **Figure 15.16***a* shows a pendulum bob of mass *m* suspended from a string of length ℓ. If the bob is much more massive and much smaller than the string that supports the bob, we can ignore the mass of the string and treat the bob as a particle. Such a pendulum is called

a *simple pendulum*. The bob is pulled back over an arc of length *s*. As the free-body diagram in Figure 15.16*b* shows, the bob is subject to two forces: a contact force exerted by the string and a gravitational force. The restoring force is provided by the component of the gravitational force perpendicular to the string, $\vec{F}^G_{Eb\perp}$. From Figure 15.16*b*, we see that the magnitude of $\vec{F}^G_{Eb\perp}$ is equal to $mg \sin\vartheta$ and from Figure 15.16*a* that $\sin\vartheta = x/\ell$, and so the magnitude of the restoring force exerted on the bob is $mg \sin\vartheta = (mg/\ell)x$. Provided the rotational displacement is small, the bob's displacement from its equilibrium position along the circular arc *s* is approximately equal to the bob's horizontal displacement *x*, $s \approx x$. For small angles, therefore, the restoring force is linearly proportional to the bob's displacement .

Figure 15.16 For small angles the horizontal displacement *x* of a pendulum is almost the same as the length of the arc *s* along which it is displaced.

(*a*) Pendulum displaced from equilibrium

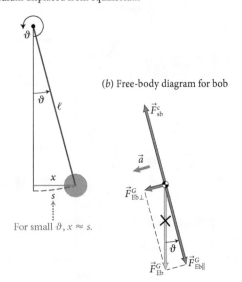

ϑ

ϑ ℓ

x

s

For small ϑ, $x \approx s$.

(*b*) Free-body diagram for bob

$\vec{F}^c_{sb}$

$\vec{a}$

$\vec{F}^G_{Eb\perp}$

ϑ

$\vec{F}^G_{Eb}$ $\vec{F}^G_{Eb\|}$

Example 15.1 Displaced string

Show that for small displacements the restoring force exerted on part 2 of the displaced string in Figure 15.14 is linearly proportional to the displacement of that part from its equilibrium position.

❶ GETTING STARTED Figure 15.14c shows the forces exerted by parts 1 and 3 on part 2 when the string is displaced from its equilibrium position. I'll assume that these forces are much greater than the force of gravity exerted on part 2 so that I can ignore gravity in this problem. The force that pulls the string away from the equilibrium position is not shown, which means the string has been released after being pulled away from the equilibrium position.

I begin by making a free-body diagram and choosing a set of axes (**Figure 15.17**). The x components of the forces $\vec{F}_{12}^c$ and $\vec{F}_{32}^c$ cancel; the sum of the y components is the restoring force. The magnitude of the y components is determined by the angle θ between the x axis and either $\vec{F}_{12}^c$ or $\vec{F}_{32}^c$.

Figure 15.17

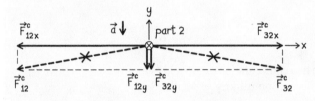

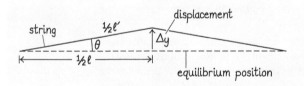

I also make a sketch of the displaced string, showing the displacement Δy of part 2 from its equilibrium position. I denote the length of the string in its equilibrium position by ℓ and the length of the displaced string by ℓ'.

❷ DEVISE PLAN The forces $\vec{F}_{12}^c$ and $\vec{F}_{32}^c$ are equal in magnitude and their y components are determined by $\sin\theta$, which is equal to $\Delta y/(\frac{1}{2}\ell')$. If the displacement is small, I can assume that the length of the string doesn't change much from its equilibrium value, so that $\ell \approx \ell'$. Because the forces $\vec{F}_{12}^c$ and $\vec{F}_{32}^c$ are proportional to the tension in the string, I can also consider these forces to be constant. Using this information, I can express the restoring force in terms of the displacement Δy.

❸ EXECUTE PLAN From my sketch I see that the restoring force is $F_{12y}^c + F_{32y}^c$. Because $\vec{F}_{12}^c$ and $\vec{F}_{32}^c$ are equal in magnitude, I can write the sum of the y components as $2F_{12y}^c = 2F_{12}^c \sin\theta$. I also know that $\sin\theta = \Delta y/(\frac{1}{2}\ell') \approx \Delta y/(\frac{1}{2}\ell)$. Combining these two relationships, I obtain for the magnitude of the restoring force: $F_{restoring} = 2F_{12y}^c \approx (4F_{12}^c/\ell)\Delta y$. For small displacements, the term in parentheses is constant and so the restoring force is, indeed, proportional to the displacement Δy. ✔

❹ EVALUATE RESULT I made two assumptions to derive my answer. The first is that gravity can be ignored. Indeed, taut strings tend to be straight, indicating that gravity (which would make the strings sag) doesn't play an appreciable role. The other

assumption I made was that the length of the string doesn't change much when it is displaced from its equilibrium position. This assumption is also justified because the displacement of a string tends to be several orders of magnitude smaller than the string length.

✋ **15.8** Using a calculator, determine the percent error in the approximation $\sin\theta \approx \theta$ for polar angles of 1°, 5°, 10°, and 20°.

In Section 15.1, I showed how oscillations arise from a continuous conversion between kinetic energy and potential energy. Another way to look at oscillations is to say:

Oscillations arise from an interplay between inertia and a restoring force.

When an object that is subject to a restoring force is displaced from its equilibrium position, the restoring force accelerates it back toward the equilibrium position. Once it gets there, however, it has a nonzero velocity, and its inertia causes it to overshoot the equilibrium position.

This picture of an interplay between inertia and restoring force allows us to make some qualitative predictions about the period of an oscillator. The greater the mass of the oscillating object, the harder it is to accelerate and so the longer the period.* Increasing the restoring force has the opposite effect: The greater the restoring force, the greater the object's acceleration and the shorter the period—a stiff spring causes a much more rapid oscillation than a soft, sloppy one.

The period of an oscillating object increases when its mass is increased and decreases when the magnitude of the restoring force is increased.

This relationship does not hold for pendulums, however, because for a pendulum the restoring force and mass are proportional to each other. Increasing the mass of a pendulum bob makes the pendulum harder to accelerate but increases the magnitude of the gravitational force exerted on the pendulum bob—the restoring force—in proportion (see Figure 15.16b). The two effects cancel, and so:

The period of a pendulum is independent of the mass of the pendulum.

✋ **15.9** (a) What effect, if any, does increasing the length ℓ of a simple pendulum have on its period? (b) A pendulum and an object suspended from a spring are taken to the Moon, where the acceleration due to gravity is smaller than that on Earth. How does the period of the pendulum on the Moon compare with its period on Earth? (c) How does the period of the suspended object on the Moon compare with its period on Earth?

---

*You can demonstrate this as follows: Position a plastic or thin wooden ruler on a table so that about half its length overhangs the edge of the table. While pressing the ruler down on the table, pull the free end down and release, noting the rapid oscillations. Next attach an eraser or other small object on top of the overhanging part of the ruler and repeat the experiment. Because of the increased mass, the period is now much longer.

Self-quiz

1. Is an oscillating object in translational equilibrium?

2. A pendulum bob swings through a circular arc defined by positions A and D in Figure 15.18. (*a*) At A, B, and C, what are the possible directions of the velocity of the bob and of the restoring force exerted on it? (*b*) What is the direction of the bob's acceleration at A and C?

3. (*a*) In the cart-spring system shown in Figure 15.19, is the cart's speed greater at A or at B? (*b*) At which of these two positions is the restoring force acting on the cart greater? (*c*) At which of these two positions is the cart's acceleration greater?

Figure 15.18

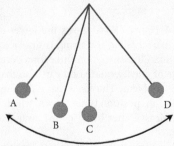

Figure 15.19

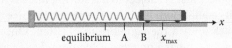

Figure 15.20

4. A T-shaped wooden structure is balanced on a pivot in the two configurations shown in Figure 15.20. Which configuration is in stable equilibrium? Which configuration is likely to oscillate?

Answers

1. No. When an object is in translational equilibrium, the sum of the forces exerted on it is zero. An oscillating object is subject to a variable force that accelerates the object back and forth.

2. (*a*) The velocity is always tangent to the bob's trajectory; the restoring force always points along the arc toward the equilibrium position C. At A, the bob reaches its maximum displacement, so its velocity must be zero; the restoring force points along the arc toward C. At B, the velocity can be in either direction along the arc as the bob swings back and forth, but the restoring force always points along the arc towards C. At C, the velocity can be to the right or left, but the restoring force is zero because this is the equilibrium position. (*b*) At A, the acceleration is tangent along the arc toward C. At C, the bob does have a centripetal radial acceleration directed vertically upward, but no tangential acceleration.

3. (*a*) Once released from the stretched position shown, the cart speeds up as it moves left toward the equilibrium position, so its speed is greater at A than at B. (*b*) The restoring force is proportional to the displacement from the equilibrium position, so it is greater at B than at A. (*c*) The acceleration is proportional to the restoring force, so it, too, is greater at B.

4. If the structure in configuration A is tipped slightly to the left as shown in Figure 15.21, the gravitational force exerted on the part to the left of the vertical line through the pivot is greater than that exerted on the part to the right of the pivot, and so the torques caused by these forces make it tip farther in the same direction, away from the equilibrium position. Conversely, if it is tipped to the right, it tips farther to the right. This configuration is unstable, and so the structure does not oscillate. If the structure in configuration B is tipped slightly either way, the torques caused by gravitational forces exerted on the parts on either side of the pivot cause it to move in the opposite direction, back toward the equilibrium position. Configuration B is stable, and so the structure oscillates.

Figure 15.21

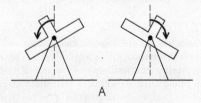

15.5 Energy of a simple harmonic oscillator

In Section 15.2, you saw that there is a correspondence between simple harmonic motion and circular motion at constant speed. We begin our quantitative description of simple harmonic motion by exploiting this correspondence.

Figure 15.22 shows a sinusoidally varying, time-dependent function describing some simple harmonic motion represented by a rotating arrow of length A whose tip traces out a circle. The circle is called a **reference circle,** and the arrow is called a **phasor.** The phasor rotates in a counterclockwise direction with a constant rotational speed ω. The vertical component of the phasor varies sinusoidally with time (see Figures 15.6 and 15.7).

Let us first establish a relationship between the rotational speed of the phasor and the frequency f of the corresponding simple harmonic motion. If the phasor completes one revolution in a period T, its rotational speed is

$$\omega \equiv \frac{\Delta \vartheta}{\Delta t} = \frac{2\pi}{T}, \tag{15.1}$$

and the frequency of the corresponding simple harmonic motion is, as noted in Section 15.1,

$$f \equiv \frac{1}{T}. \tag{15.2}$$

The derived SI unit of frequency is the **hertz** (Hz) after the 19th-century German physicist Heinrich Hertz, who produced the first radio waves:

$$1\ \text{Hz} \equiv 1\ \text{s}^{-1}. \tag{15.3}$$

Substituting $1/T$ from Eq. 15.2 into Eq. 15.1, we find

$$\omega = 2\pi f. \tag{15.4}$$

So the rotational speed ω of the phasor is greater than the frequency f of the simple harmonic motion by a factor of 2π. The reason for this factor is that ω measures the change in rotational position per unit time, and in one revolution the rotational position ϑ changes by 2π. Frequency, on the other hand, measures the number of cycles per unit time, and one revolution on the reference circle corresponds to one oscillation cycle (Figure 15.22).

In the context of oscillations, ω is often called the **angular frequency** because it has the same unit (s^{-1} or inverse second) as the frequency f. Keep in mind, however, that the frequency f and the angular frequency ω are not equal (see Eq. 15.4). To facilitate remembering the distinction between the two, we use the SI unit hertz for f and the SI unit s^{-1} for ω.

Figure 15.22 Reference circle and $x(t)$ curve for a simple harmonic oscillator.

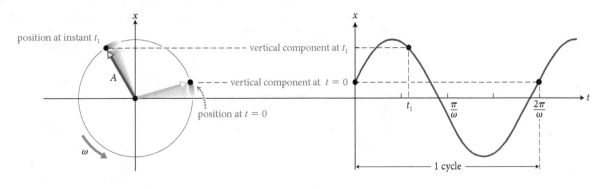

Figure 15.23 The phase of a simple harmonic oscillator at any instant is given by the angle between the phasor and the positive horizontal axis measured in the counterclockwise direction.

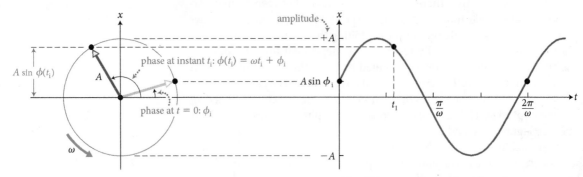

The rotational position of the tip of the phasor relative to the positive horizontal axis is called the **phase** of the motion (**Figure 15.23**). The phase is represented by the symbol $\phi(t)$. For constant angular frequency, the phase increases linearly with time:

$$\phi(t) = \omega t + \phi_i \quad \text{(constant angular frequency),} \qquad (15.5)$$

where the constant ϕ_i is the *initial phase* at $t = 0$. The phase $\phi(t)$ tells us at what point in the cycle the motion is; the initial phase ϕ_i is the phase at $t = 0$. Because rotational position is a unitless number, the phase is unitless too. You determine the phase by expressing the angle between the phasor and the positive horizontal axis in radians and dividing by 1 rad.

The vertical component of A in Figure 15.23 can now be written in the form

$$x(t) = A \sin \phi(t) = A \sin(\omega t + \phi_i) \quad \text{(simple harmonic motion),} \quad (15.6)$$

where the **amplitude** A of the oscillation is equal to the length of the phasor. Equation 15.6 gives a general expression for the position of a simple harmonic oscillator as a function of time when the equilibrium position is at $x = 0$—a pendulum bob, an object suspended from a spring, a taut string. Because the motion is periodic, $x(t)$ has the same value each time the phase increases by 2π. For a crest (where $x(t) = A$ and the sine function in Eq. 15.6 has the value 1), the phase could be $\frac{\pi}{2}$, or $\frac{\pi}{2}$ plus an integer multiple of 2π; for a trough (where $x(t) = -A$ and the sine function in Eq. 15.6 has the value -1), the phase could be $\frac{3\pi}{2}$, or $\frac{3\pi}{2}$ plus an integer multiple of 2π.

Exercise 15.2 Oscillation phases

Consider the spring-cart system of Figure 15.1 again. The cart is pulled away from its equilibrium position in the positive x direction and then released at $t = 0$. (*a*) What is the initial phase of the cart's oscillation? What is the phase of the cart's oscillation (*b*) half a period later and (*c*) one period later? What are the cart's initial phases in the following two situations? (*d*) The cart is at its equilibrium position when, at $t = 0$, it is given a sharp blow in the negative x direction so that it starts oscillating. (*e*) The cart is first pulled back a distance d in the positive x direction, and then at $t = 0$ it is given a sharp blow, also in the positive x direction, so that it carries out an oscillation of amplitude $A = 2d$.

SOLUTION (*a*) The cart starts at maximum amplitude in the positive x direction, so the phasor is pointing straight up and the initial phase of the cart's oscillation is $\phi_i = (90°)(2\pi \text{ rad}/360°)/(1 \text{ rad}) = \pi/2.$ ✔

(*b*) In one period, the phasor completes a full cycle, so half a period later the phasor has traversed half a cycle (180°). The angle between the phasor and the horizontal axis is thus $90° + 180° = 270°$, and thus $\phi(t) = (270°)(2\pi \text{ rad}/360°)/(1 \text{ rad}) = 3\pi/2.$ ✔

(*c*) In one period the phasor has completed a full cycle, so the phase is equal to the initial phase $\phi_i + 2\pi = \frac{\pi}{2} + 2\pi = \frac{5\pi}{2}$. ✔

(*d*) If the cart is in the equilibrium position, the phasor must be parallel to the horizontal axis and so the phasor is at 0° or 180°. When the phasor is at 0°, the cart moves in the positive *x* direction; when it is at 180°, the cart moves in the negative *x* direction, so the initial phase is $\phi_i = (180°)(2\pi \text{ rad}/360°)/(1 \text{ rad}) = \pi$. ✔

(*e*) The displacement of the cart is given by Eq. 15.6. At $t = 0$: $x(0) = A \sin \phi_i$. I am given that $x(0) = +d$ and $A = 2d$, so $+d = 2d \sin \phi_i$, or $\sin \phi_i = \frac{1}{2}$, and so $\phi_i = \pi/6$ or $5\pi/6$. Because the cart moves in the positive *x* direction at $t = 0$, the initial phase must be $\phi_i = \pi/6$. ✔

By differentiating Eq. 15.6 with respect to time, we can obtain general expressions for the *x* components of the velocity and acceleration of a harmonic oscillator:

$$v_x \equiv \frac{dx}{dt} = \omega A \cos(\omega t + \phi_i) \qquad \text{(simple harmonic motion)} \qquad (15.7)$$

$$a_x \equiv \frac{d^2x}{dt^2} = -\omega^2 A \sin(\omega t + \phi_i) \quad \text{(simple harmonic motion)}. \qquad (15.8)$$

Equations 15.6–15.8 are the basic kinematic equations for simple harmonic motion. We derived them without considering any particular system—everything followed from the fact that simple harmonic motion corresponds to the projection onto a screen of some object moving in circular motion at constant speed.

15.10 (*a*) What are the algebraic signs of *x*, v_x, and a_x when the phase $\phi(t) = \omega t + \phi_i$ is between 0 and $\frac{\pi}{2}$? (*b*) Repeat for $\pi < \phi(t) < \frac{3\pi}{2}$.

Substituting Eq. 15.6 into Eq. 15.8, we find that the *x* component of the acceleration of an object in simple harmonic motion is proportional to the displacement and points in the direction opposite the displacement (that is, toward the equilibrium position):

$$a_x = -\omega^2 x \quad \text{(simple harmonic motion)}. \qquad (15.9)$$

Substituting the definition of the acceleration into Eq. 15.9, we obtain a second-order differential equation in the variable *x*:

$$\frac{d^2x}{dt^2} = -\omega^2 x \quad \text{(simple harmonic motion)}. \qquad (15.10)$$

This equation is called the **simple harmonic oscillator equation** because any system that satisfies this equation undergoes simple harmonic motion. As we shall see in the next two sections, the symbols may take on different meanings for different types of simple harmonic motion, but the mathematical form is always that given in Eq. 15.10.

Multiplying both sides of Eq. 15.9 by the mass of the oscillating object, we have

$$ma_x = -m\omega^2 x \quad \text{(simple harmonic motion)}. \qquad (15.11)$$

The left side of Eq. 15.11 is equal to the *x* component of the vector sum of the forces acting on the oscillating object, $ma_x = \sum F_x$, and so

$$\sum F_x = -m\omega^2 x \quad \text{(simple harmonic motion)}. \qquad (15.12)$$

Because m and ω are positive constants, Eq. 15.12 tells us that an object in simple harmonic motion is subject to a force that always points in a direction opposite the object's displacement x from the equilibrium position and is linearly proportional to x—in other words, a *linear restoring force* as we already concluded in Section 15.4.

Let us next examine the mechanical energy of a simple harmonic oscillator, using kinematic Equations 15.6–15.8. The work done by the forces exerted on the oscillating object when it is moving from the equilibrium position to position x is found by integrating the x component of the position-dependent force exerted on the oscillator (Eq. 9.22):

$$W = \int_{x_0}^{x} \Sigma F_x(x)dx = -\int_{x_0}^{x} m\omega^2 x \, dx, \tag{15.13}$$

where we have substituted the restoring force given in Eq. 15.12. This work causes the kinetic energy of the oscillating object to change, and so Eq. 15.13 gives us the change in kinetic energy ΔK. Working out the integration thus yields

$$\Delta K = -m\omega^2 \int_{x_0}^{x} x dx = -m\omega^2 \left[\tfrac{1}{2}x^2 \right]_{x_0}^{x} = \tfrac{1}{2}m\omega^2 x_0^2 - \tfrac{1}{2}m\omega^2 x^2. \tag{15.14}$$

Because the interaction that causes the oscillation is reversible, this change in kinetic energy must be provided by some form of potential energy. (In the examples we discussed, the energy was provided by gravitational potential energy or the elastic potential energy of a spring). If we include the object that causes the oscillation in our system, the system must be closed, so $\Delta E = \Delta K + \Delta U = 0$. Therefore the change in potential energy of the (closed) system is $\Delta U = -\Delta K$, or, substituting Eq. 15.14,

$$\Delta U = U(x) - U(x_0) = \tfrac{1}{2}m\omega^2 x^2 - \tfrac{1}{2}m\omega^2 x_0^2. \tag{15.15}$$

If we let the potential energy at the equilibrium position be zero, $U(x_0) = 0$, then we see that the potential energy of the oscillating system is $U(x) = \tfrac{1}{2}m\omega^2 x^2$, and so the mechanical energy is

$$E = K + U = \tfrac{1}{2}mv^2 + \tfrac{1}{2}m\omega^2 x^2 \tag{15.16}$$

or, substituting for x from Eq. 15.6 and for v_x from Eq. 15.7,

$$E = \tfrac{1}{2}m\omega^2 A^2 \cos^2(\omega t + \phi_i) + \tfrac{1}{2}m\omega^2 A^2 \sin^2(\omega t + \phi_i)$$

$$= \tfrac{1}{2}m\omega^2 A^2 \quad \text{(simple harmonic motion),} \tag{15.17}$$

where we have used the fact that $\cos^2 \phi + \sin^2 \phi = 1$ for all ϕ. Equation 15.17 shows that the mechanical energy of a simple harmonic oscillator is constant and proportional to the square of the amplitude, as we concluded earlier.

15.11 (*a*) Use Eq. 15.15 to determine the maximum potential energy of a simple harmonic oscillator. (*b*) Does your answer to part *a* agree with Eq. 15.17?

15.6 Simple harmonic motion and springs

Whenever an object undergoes reversible deformation as a result of stretching, compressing, bending, or twisting, it exerts an elastic restoring force. Consider, for example, the spring-cart system in **Figure 15.24**. The spring exerts on the oscillating cart a force $F_{scx}^c = -k(x - x_0)$ (Eq. 8.20), where k is the spring constant and $x - x_0$ is the displacement of the cart from its equilibrium position at x_0, where the spring is relaxed.

We can use this information about the restoring force to obtain an equation for the cart's motion in terms of the amplitude, angular frequency, and phase of the motion. If we let the origin be at the equilibrium position, $x_0 = 0$ and so

$$F_{scx}^c = -kx. \tag{15.18}$$

The equation of motion for the cart in the horizontal direction is then

$$\sum F_{cx} = ma_x = -kx, \tag{15.19}$$

where m is the cart's mass, or

$$ma_x = m\frac{d^2x}{dt^2} = -kx \tag{15.20}$$

or

$$\frac{d^2x}{dt^2} = -\frac{k}{m}x. \tag{15.21}$$

The solution to this simple harmonic oscillator equation is a sinusoidal function $x(t)$, and comparison with Eq. 15.10 tells us that $k/m = \omega^2$. Therefore the angular frequency of the oscillation is

$$\omega = +\sqrt{\frac{k}{m}}. \tag{15.22}$$

The plus sign in front of the square root means that we must take the positive square root because ω is always positive.

The motion of the cart is given by Eq. 15.6:

$$x(t) = A\sin\left(\sqrt{\frac{k}{m}}\,t + \phi_i\right). \tag{15.23}$$

Figure 15.25 shows a number of different sinusoidal functions of the form given by Eq. 15.23 that satisfy Eq. 15.21. All of the functions $x(t)$ shown in Figure 15.25 complete one cycle in a period T, and so they all have the same angular frequency ω, whose value is fixed by properties of the system—the spring constant k and the cart's mass m (Eq. 15.22). However, the peak heights tell you that the curves do not all have the same amplitude A, and the fact that each curve reaches its maximum at a different instant tells you that they have different initial phases ϕ_i. Without information in addition to the values of k and m, we cannot determine the amplitude A and initial phase ϕ_i.

In order to determine A and ϕ_i, we have to know the position and x component of the velocity at some instant. The next examples illustrate how to determine these quantities for some specific situations.

Figure 15.24 Restoring force exerted by spring on cart when cart is displaced from its equilibrium position x_0.

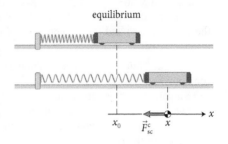

Figure 15.25 Four of the infinite number of solutions that satisfy Eq. 15.21.

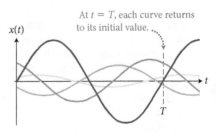

Example 15.3 Cart pulled away and released from rest

A cart of mass $m = 0.50$ kg fastened to a spring of spring constant $k = 14$ N/m is pulled 30 mm away from its equilibrium position and then released with zero initial velocity. What are the cart's position and the x component of velocity 2.0 s after being released?

❶ **GETTING STARTED** I begin by making a sketch of the initial condition and representing the oscillation by a reference circle and a graph of position versus time (Figure 15.26). I choose the positive x axis to be in the same direction as the cart's initial displacement, with the origin of the axis at the equilibrium position. Because the cart is released with zero initial velocity, its initial kinetic energy at $x = +30$ mm is $K = 0$, and so each time it returns to that position it again has $K = 0$. This tells me that the cart's initial displacement is also its maximum displacement, and so $A = 30$ mm. Because the cart begins at its maximum positive displacement, I know that the oscillation begins with the phasor straight up, and so the initial phase is $\phi_i = +\pi/2$.

Figure 15.26

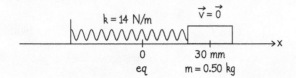

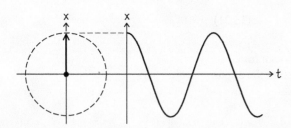

❷ **DEVISE PLAN** To determine the position of the cart at $t = 2.0$ s, I can use Eq. 15.6. This equation contains the angular frequency ω, which I can determine from Eq. 15.22, and the amplitude A and initial phase ϕ_i, which I already determined. To calculate the x component of the cart's velocity, I can use Eq. 15.7.

❸ **EXECUTE PLAN** From Eq. 15.22, I get for the angular frequency of the oscillation:

$$\omega = \sqrt{\frac{k}{m}} = \sqrt{\frac{14 \text{ N/m}}{0.50 \text{ kg}}} = \sqrt{28 \text{ s}^{-2}} = 5.3 \text{ s}^{-1}.$$

Substituting this angular frequency, $t = 2.0$ s, $A = 30$ mm, and $\phi_i = +\pi/2$ into Eq. 15.6, I obtain for the cart's position at $t = 2.0$ s:

$$x(2.0 \text{ s}) = (30 \text{ mm}) \sin\left[(5.3 \text{ s}^{-1})(2.0 \text{ s}) + \frac{\pi}{2}\right] = -12 \text{ mm.} ✔$$

Substituting the same values into Eq. 15.7, I obtain for the x component of the cart's velocity:

$$v_x(2.0 \text{ s}) = (5.3 \text{ s}^{-1})(30 \text{ mm}) \cos\left[(5.3 \text{ s}^{-1})(2.0 \text{ s}) + \frac{\pi}{2}\right]$$

$$= +1.5 \times 10^2 \text{ mm/s.} ✔$$

❹ **EVALUATE RESULT** The magnitude of the value I obtained for x is smaller than the amplitude, as it should be. The phase at $t = 2.0$ s is $\phi(t) = (5.3 \text{ s}^{-1})(2.0 \text{ s}) + \pi/2 = 12$, which corresponds to $12/(2\pi) = 1.9$ cycles, so the phasor is in the fourth quadrant, which confirms that x should be negative. The cart moves a distance of $4A = 0.12$ m in one cycle, which takes $2\pi/\omega = 1.2$ s, and so the cart's average speed is $(0.12 \text{ m})/(1.2 \text{ s}) = 0.10$ m/s, indicating that the magnitude of the velocity component I obtain is reasonable. Because the cart is moving in the positive x direction when the phasor is in the fourth quadrant, the x component of the velocity should be positive, as I found.

✋ **15.12** (*a*) In Example 15.3 is the initial condition that $v_x = 0$ satisfied by Eq. 15.7? (*b*) Is it correct to say that the x component of the velocity given by Eq. 15.7 must be positive whenever x is negative so as to make sure that the cart moves back to its equilibrium position?

Example 15.4 Cart struck with spring already compressed

Cart 1 of mass $m = 0.50$ kg fastened to a spring of spring constant $k = 14$ N/m is pushed 15 mm in from its equilibrium position and held in place by a ratchet (Figure 15.27). An identical cart 2 is launched at a speed of 0.10 m/s toward cart 1. The carts collide elastically, releasing the ratchet and setting cart 1 in motion. After the collision, cart 2 is immediately removed from the track. (*a*) What is the maximum compression of the spring? (*b*) How many seconds elapse between the instant the carts collide and the instant the spring reaches maximum compression?

Figure 15.27 Example 15.4

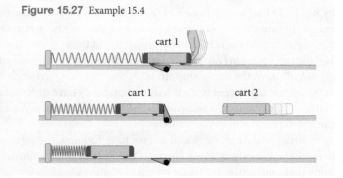

❶ GETTING STARTED If I ignore the effect of the spring during the collision, I can say that the elastic collision interchanges the velocities of the two carts. I make a sketch of the initial condition of cart 1 just before the collision, choosing the positive x axis to the right, the equilibrium position at $x = 0$, and the initial displacement of the cart 15 mm to the left of $x = 0$. I also draw a reference circle and sketch the oscillation resulting from the collision (Figure 15.28). With this choice of axis, the x component of the initial velocity of cart 1 is $v_{x,i} = -0.10$ m/s. Because m and k are the same as in Example 15.3, the angular frequency is again $\omega = 5.3$ s^{-1}.

Figure 15.28

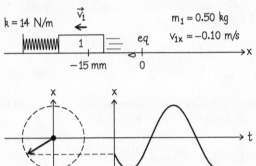

In contrast to the situation in Example 15.3, the initial displacement of cart 1 is not equal to the amplitude of the oscillation because the collision increases the cart's displacement from the equilibrium position. In other words, cart 1 moves leftward immediately after the collision. It continues moving in this direction until the elastic restoring force building up in the compressing spring causes the cart to stop. This maximum-compression position gives the amplitude of the oscillation.

❷ DEVISE PLAN I can determine the value of the amplitude from the mechanical energy of the cart-spring system, which is the sum of the initial potential energy in the spring and the initial kinetic energy of the cart. The potential energy in the spring is given by Eq. 9.23; because the equilibrium position x_0 is at the origin, this equation reduces to $U_{spring} = \frac{1}{2}kx^2$. The kinetic energy is given by $K = \frac{1}{2}mv^2$. At the position of maximum compression, all of the mechanical energy is stored in the spring, $x = -A$, and so $E_{mech} = U_{spring} = \frac{1}{2}kA^2$.

Once I know A, I can determine the initial phase from Eq. 15.6. I can then use that same equation to solve for t at the position of maximum compression, where $x = -A$.

❸ EXECUTE PLAN (a) The initial kinetic and potential energies of the cart-spring system are

$K = \frac{1}{2}(0.50 \text{ kg})(-0.10 \text{ m/s})^2 = 0.0025$ J

$U = \frac{1}{2}kx^2 = \frac{1}{2}(14 \text{ N/m})(-15 \text{ mm})^2 \left(\dfrac{1.0 \text{ m}}{1000 \text{ mm}}\right)^2 = 0.0016$ J

so

$E_{mech} = K + U = (0.0025 \text{ J}) + (0.0016 \text{ J}) = 0.0041$ J.

At the position of maximum compression, all of this energy is stored in the spring, and so $\frac{1}{2}kA^2 = 0.0041$ J, or with the k value given,

$$A = \sqrt{\dfrac{2(0.0041 \text{ J})}{14 \text{ N/m}}} = 0.024 \text{ m} = 24 \text{ mm}. \checkmark$$

(b) Substituting the value for A determined in part a and the initial condition $x_i = -15$ mm at $t = 0$ into Eq. 15.6, I obtain

$$x(0) = A \sin(0 + \phi_i) = (24 \text{ mm}) \sin \phi_i = -15 \text{ mm}$$

$$\sin \phi_i = \dfrac{-15 \text{ mm}}{24 \text{ mm}} = -0.63 \quad \text{or} \quad \phi_i = \sin^{-1}(-0.63).$$

Two initial phases satisfy this relationship, $\phi_i = -0.68$ and $\phi_i = -\pi + 0.68 = -2.5$, but only the latter gives a negative x component of the velocity (see Eq. 15.7), as required by the initial condition.

At the first instant of maximum compression, $\sin(\omega t + \phi_i) = -1$, which means $\omega t + \phi_i = -\frac{1}{2}\pi$. Solving for t yields $t = (-\frac{1}{2}\pi - \phi_i)/\omega = \left[-\frac{1}{2}\pi - (-2.5)\right]/(5.3 \text{ s}^{-1}) = 0.17$ s. $\checkmark$

❹ EVALUATE RESULT At 24 mm, the amplitude is greater than the 15-mm initial displacement from the equilibrium position, as I would expect. From the reference circle part of my sketch I see that it takes about one-eighth of a cycle to go from the position of impact to the position of maximum compression. From Eq. 15.1 I see that the number of seconds needed to complete one cycle is $T = 2\pi/\omega = 2\pi/(5.3 \text{ s}^{-1}) = 1.2$ s, and so the 0.17-s value I obtained for seconds elapsed between collision and maximum compression is indeed close to one-eighth of a cycle. The assumption that the influence of the spring can be ignored during the collision is justified because the force exerted by the spring is small relative to the force of the impact: At a compression of 15 mm, the magnitude of the force exerted by the spring is $(14 \text{ N/m})(0.015 \text{ m}) = 0.21$ N. The force of impact is given by the time rate of change in the cart's momentum, $\Delta \vec{p}/\Delta t$. The magnitude of the momentum change is $\Delta p = m\Delta v = (0.50 \text{ kg})(0.10 \text{ m/s}) = 0.050$ kg·m/s. If the collision takes place in, say, 20 ms, the magnitude of the force of impact is $(0.050 \text{ kg·m/s})/(0.020 \text{ s}) = 2.5$ N, which is more than 10 times greater than the magnitude of the force exerted by the spring.

✋ **15.13** Suppose cart 2 were not removed from the track of Figure 15.27 immediately after the collision but instead were left stationary at the collision point. (a) At what instant do the two carts collide again? (b) Describe the motion of the two carts after this second collision.

Example 15.5 Vertical oscillations

A block of mass $m = 0.50$ kg is suspended from a spring of spring constant $k = 100$ N/m. (a) How far below the end of the relaxed spring at x_0 is the equilibrium position x_{eq} of the suspended block (Figure 15.29a)? (b) Is the frequency f with which the block oscillates about this equilibrium position x_{eq} greater than, smaller than, or equal to that of an identical system that oscillates horizontally about x_0 on a surface for which friction can be ignored (Figure 15.29b)?

Figure 15.29 Example 15.5.

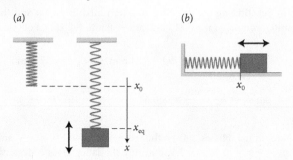

(a)   (b)

x_0

x_{eq}

x

① GETTING STARTED I begin by making a free-body diagram for the suspended block, choosing the positive x axis pointing downward. Two forces are exerted on the block: an upward force $\vec{F}^c_{sb}$ exerted by the spring and a downward gravitational force $\vec{F}^G_{Eb}$ exerted by Earth (**Figure 15.30a**). When the suspended block is in translational equilibrium at x_{eq} (which lies below x_0), the vector sum of these forces must be zero. With the block at x_{eq}, the spring is stretched such that the end attached to the block is also at x_{eq}. When the block is below x_{eq}, the spring is stretched farther, causing the magnitude of $\vec{F}^c_{sb}$ to increase, and so now the vector sum of the forces exerted on the block points upward. When the block is above x_{eq} (but below the position x_0 of the end of the spring when it is relaxed), the spring is stretched less than when the block is at x_{eq}, causing the magnitude of $\vec{F}^c_{sb}$ to decrease, and so the vector sum of the forces exerted on the block points downward. The vector sum of $\vec{F}^c_{sb}$ and $\vec{F}^G_{Eb}$ thus serves as a restoring force.

Figure 15.30

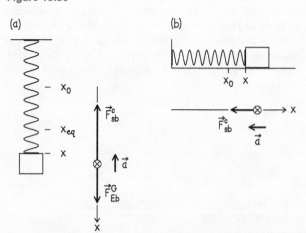

(a)   (b)

x_0

x_{eq}

x

$\vec{F}^c_{sb}$

$\vec{a}$

$\vec{F}^G_{Eb}$

x

x_0 x

$\vec{F}^c_{sb}$

$\vec{a}$

x

I also make a free-body diagram for the horizontal arrangement (Figure 15.30b), showing only the horizontal forces (the force of gravity and the normal force exerted by the surface cancel out). I let the positive x axis point to the right. In this case, the restoring force is $\vec{F}^c_{sb}$ only.

② DEVISE PLAN In translational equilibrium, the vector sum of the forces exerted on the suspended block is zero, and so I can determine the magnitude of the force exerted by the spring. I can then use Hooke's law (Eq. 8.20) to determine the distance between the equilibrium position and x_0. To compare the oscillation frequencies of the two systems, I should write the simple harmonic oscillator equation for each system in the form given by Eq. 15.12.

③ EXECUTE PLAN (a) For translational equilibrium, I have

$$\Sigma F_x = F^c_{sbx} + F^G_{Ebx} = -k(x_{eq} - x_0) + mg = 0,$$

where $x_{eq} - x_0$ is the displacement of the spring's end from its relaxed position. Solving for $x_{eq} - x_0$, I obtain

$$x_{eq} - x_0 = \frac{mg}{k} = \frac{(0.50 \text{ kg})(9.8 \text{ m/s}^2)}{100 \text{ N/m}} = 0.049 \text{ m.} ✔$$

(b) For the horizontal arrangement, I have in the situation depicted in my sketch (Figure 15.30b)

$$\Sigma F_x = F^c_{sbx} = -k(x - x_0), \tag{1}$$

so if I let the origin of my x axis be at the position of the end of the relaxed spring, $x_0 = 0$, the rightmost factor in Eq. 1 reduces to $-kx$ and thus $\Sigma F_x = -kx$. Next I turn to the vertical arrangement. In the position illustrated in Figure 15.30a, the x component of the upward force exerted by the spring is

$$F^c_{sbx} = -k(x - x_0) = -k(x - x_{eq}) - k(x_{eq} - x_0). \tag{2}$$

From part a I know that $k(x_{eq} - x_0)$ is equal to mg. The x component of the vector sum of the forces exerted on the block at position x is thus

$$\Sigma F_x = F^G_{Ebx} + F^c_{sbx} = mg - k(x - x_{eq}) - mg$$

$$= -k(x - x_{eq}).$$

If, as usual, I let the origin be at the equilibrium position, $x_{eq} = 0$, then this result for ΣF_x is identical to Eq. 1. Comparing these results to Eq. 15.12, I see that in both cases $k = m\omega^2$ and so the oscillation frequencies $f = \omega/2\pi$ of the two systems are the same: $f_{vert} = f_{hor}$. ✔

④ EVALUATE RESULT The two oscillations take place about different equilibrium positions (x_0 for the horizontal case, x_{eq} for the vertical case), but the effect of the combined gravitational and elastic forces in the vertical arrangement is the same as that of just the elastic force in the horizontal arrangement because the force exerted by the spring is linear in the displacement.

15.14 Convince yourself that the argument presented in the solution to Example 15.5 is also valid for displacements of the block above x_{eq}.

15.7 Restoring torques

Some simple harmonic oscillators involve rotational rather than translational displacements. As an example, consider the disk suspended from a thin fiber shown in Figure 15.31. Such a setup is called a *torsional oscillator*. If the disk is rotated in the horizontal plane over an angle ϑ_{max}, the fiber is twisted, and elastic potential energy is stored in the fiber. When the disk is released, it oscillates as energy is converted back and forth between the elastic potential energy in the fiber and the rotational kinetic energy of the disk. Translational oscillations, as in our simple pendulum and block-spring setup, are due to the interplay between a restoring force and inertia; rotational oscillations are due to an interplay between a restoring torque and rotational inertia.

For a quantitative description of the disk's oscillation, we use the rotational form of the equation of motion for the disk, which relates the sum of the torques caused by the forces acting on the disk to its rotational acceleration α_ϑ (Eq. 12.10):

$$\sum \tau_\vartheta = I\alpha_\vartheta, \tag{15.24}$$

where I is the rotational inertia of the disk.

To complete the description, we must relate the torque to the rotational displacement—just as we related force to translational displacement for translational oscillations. In the case of the rotating disk, the torque is caused by the twisted fiber: A clockwise rotation of the fiber causes a counterclockwise torque, and vice versa; greater rotational displacements of the disk cause greater torques. For small rotational displacements, we might expect that the relationship between rotational displacement and torque is linear:

$$\tau_\vartheta = -\kappa(\vartheta - \vartheta_0). \tag{15.25}$$

The minus sign indicates that the direction of the torque is opposite the direction of the rotational displacement: The torque tends to rotate the disk back toward its equilibrium position. This rotational Form of Hooke's law is found to hold for a wide variety of materials. The constant κ (the Greek letter kappa), called the *torsional constant*, depends on the properties of the material being twisted. Just as stiff springs have large spring constants, stiff materials have large torsional constants.

If we let the rotational position at equilibrium be zero, $\vartheta_0 = 0$, we can combine Eqs. 15.24 and 15.25, yielding

$$I\alpha_\vartheta = I\frac{d^2\vartheta}{dt^2} = -\kappa\vartheta \tag{15.26}$$

or

$$\frac{d^2\vartheta}{dt^2} = -\frac{\kappa}{I}\vartheta. \tag{15.27}$$

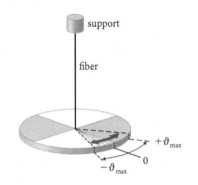

Figure 15.31 Torsional oscillator.

support

fiber

$+\vartheta_{max}$

0

$-\vartheta_{max}$

Notice the similarity between this equation and two earlier ones: Eq. 15.10 for a generic harmonic oscillator and Eq. 15.21 for a mass oscillating on a spring. The meaning of the symbols is different, but the mathematical form is identical. The disk therefore executes simple harmonic motion, and the rotational position $\vartheta(t)$ is a sinusoidal function of time.

To find the position as a function of time, we exploit the correspondence between Eqs. 15.21 and 15.27, substituting rotational displacement ϑ for translational displacement x in Eq. 15.23 and rotational inertia I for mass m and the torsional constant κ for the spring constant k in Eq. 15.22:

$$\vartheta = \vartheta_{max} \sin(\omega t + \phi_i) \qquad (15.28)$$

$$\omega = \sqrt{\frac{\kappa}{I}} \quad \text{(torsional oscillator)}, \qquad (15.29)$$

where ϑ_{max} in Eq. 15.28 is the maximum rotational displacement (the amplitude of the rotational oscillation). Note that the ω in Eq. 15.29 represents the angular frequency of the torsional oscillator, not its rotational speed.

15.15 For the torsional oscillator shown in Figure 15.31, what effect, if any, does a decrease in the radius of the disk have on the oscillation frequency f? Assume the disk's mass is kept the same.

The pendulum is another example of a rotational oscillator. For a pendulum, however, the restoring torque is not caused by an elastic force, and so we cannot use Eq. 15.25. Instead, we must write an expression for the restoring torque due to gravity and relate the magnitude of this torque to the rotational displacement of the pendulum. Consider an object suspended about a rotation axis located a distance ℓ_{cm} from the object's center of mass, as illustrated in **Figure 15.32a**. The restoring torque is provided by the component of the force of gravity perpendicular to ℓ_{cm}: $F^G_{Eo\perp} = -mg \sin \vartheta$ (Figure 15.32b). The lever arm of this force relative to the rotation axis is ℓ_{cm}, and so for a rotational displacement ϑ, the torque caused by the force of gravity about the axis is

$$\tau_\vartheta = -\ell_{cm}(mg \sin \vartheta), \qquad (15.30)$$

where m is the object's mass. For small rotational displacements, $\sin \vartheta \approx \vartheta$, and so once again we obtain a restoring torque that is linearly proportional to the rotational displacement:

$$\tau_\vartheta = -(m\ell_{cm} g)\vartheta. \qquad (15.31)$$

After $\tau_\vartheta = I\alpha_\vartheta = I\, d^2\vartheta/dt^2$ is substituted for the left side of Eq. 15.31, we get

$$\frac{d^2\vartheta}{dt^2} = -\frac{m\ell_{cm} g}{I}\vartheta, \qquad (15.32)$$

and so the angular frequency of oscillation is

$$\omega = \sqrt{\frac{m\ell_{cm} g}{I}} \quad \text{(pendulum)}. \qquad (15.33)$$

Figure 15.32 (a) Oscillating pendulum. (b) Extended free-body diagram for the pendulum.

(a)

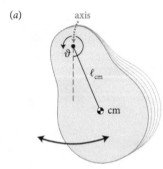

(b)

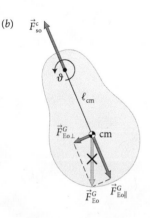

Example 15.6 The simple pendulum

Suppose a simple pendulum consisting of a bob of mass m suspended from a string of length ℓ is pulled back and released. What is the period of oscillation of the bob?

❶ GETTING STARTED I begin by making a sketch of the simple pendulum (**Figure 15.33**), indicating the equilibrium position by a vertical dashed line.

Figure 15.33

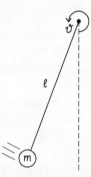

❷ DEVISE PLAN The period of the pendulum is related to the angular frequency (Eq. 15.1). To calculate the angular frequency, I can use Eq. 15.33 with $\ell_{cm} = \ell$. If I treat the bob as a particle, I can use Eq. 11.30, $I = mr^2$, with $r = \ell$ to calculate the bob's rotational inertia about the point of suspension.

❸ EXECUTE PLAN Substituting the bob's rotational inertia into Eq. 15.33, I get

$$\omega = \sqrt{\frac{m\ell g}{m\ell^2}} = \sqrt{\frac{g}{\ell}},$$

so, from Eq. 15.1, $\omega = 2\pi/T$, I obtain

$$T = \frac{2\pi}{\omega} = 2\pi\sqrt{\frac{\ell}{g}}. \checkmark$$

❹ EVALUATE RESULT My result says that T is independent of the mass m of the bob, in agreement with what is stated in Section 15.4. Increasing g decreases the period as it should: A greater g means a greater restoring force, and so the bob is pulled back to the equilibrium position faster. It also makes sense that increasing ℓ increases the period: As my sketch shows, for greater ℓ the bob has to move a greater distance to return to the equilibrium position.

Example 15.7 Measuring g

The oscillations of a thin rod can be used to determine the value of the acceleration due to gravity. A rod that is 0.800 m long and suspended from one end is observed to complete 100 oscillations in 147 s. What is the value of g at the location of this experiment?

❶ GETTING STARTED I begin by making a sketch of the situation. To simplify the problem, I assume that the rod pivots about the fixed end (**Figure 15.34**).

Figure 15.34

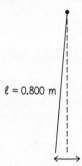

$\ell = 0.800$ m

❷ DEVISE PLAN I calculated the rotational inertia of a thin rod about one end in Example 11.11: $I = \frac{1}{3}m\ell^2$, where m is the rod's mass and ℓ its length. I can use this equation and Eq. 15.33 to solve for g, noting that ℓ_{cm} in Eq. 15.33 is half the length of the rod ($\ell_{cm} = \frac{1}{2}\ell$).

❸ EXECUTE PLAN Substituting the rod's rotational inertia and the center-of-mass distance $\ell_{cm} = \frac{1}{2}\ell$ into Eq. 15.33, I get

$$\omega = \sqrt{\frac{m\frac{1}{2}\ell g}{\frac{1}{3}m\ell^2}} = \sqrt{\frac{3g}{2\ell}}.$$

Solving for g yields $g = \frac{2}{3}\omega^2\ell$. The period of the rod is $(147\text{ s})/(100) = 1.47$ s, and the angular frequency is $\omega = 2\pi f = 2\pi(1/T) = 2\pi/(1.47\text{ s}) = 4.27\text{ s}^{-1}$. Substituting this information into the expression for g yields $g = \frac{2}{3}(4.27\text{ s}^{-1})^2(0.800\text{ m}) = 9.74\text{ m/s}^2$. $\checkmark$

❹ EVALUATE RESULT I note that the rod's mass does not appear in my expression for ω. Because ω is proportional to $1/T$, this again confirms what Section 15.4 says about the period of a pendulum being independent of the pendulum's mass because the restoring force is proportional to the pendulum's mass. The value I obtained, 9.74 m/s², is reasonably close to the acceleration due to gravity near Earth's surface, giving me confidence in my solution.

15.16 Can you use the experiment described in Example 15.7 to measure the gravitational constant G?

Example 15.8 The simple pendulum as a one-dimensional oscillator

The angular frequency ω of a simple pendulum can be calculated by treating the pendulum as a one-dimensional oscillator. In Section 15.4, we used this approach to analyze the restoring force exerted on a pendulum, considering the effect of the force of gravity on the horizontal displacement of the bob and ignoring the slight difference between the horizontal displacement x and the arc length s along which the bob swings. Show that this treatment yields for ω the same expression I obtained in Example 15.6, $\omega = \sqrt{g/\ell}$.

❶ GETTING STARTED I make a sketch of the pendulum, showing the bob displaced a horizontal distance x from its equilibrium position, the distance I must concentrate on in this problem (**Figure 15.35***a*). I also draw a free-body diagram for the

Figure 15.35

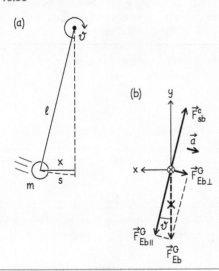

bob (Figure 15.35*b*), showing $\vec{F}^G_{\text{Eb}\parallel}$ and $\vec{F}^G_{\text{Eb}\perp}$, with the component vectors of $\vec{F}^G_{\text{Eb}}$ parallel to and perpendicular to the string in its displaced orientation.

❷ DEVISE PLAN The component of the gravitational force perpendicular to the string in its displaced orientation, $\vec{F}^G_{\text{Eb}\perp}$, provides the restoring force. For small displacements, this component is nearly parallel to the x axis, and so I can take $F^G_{\text{Eb}\perp}$ to be the restoring force in the x direction. This expression will permit me to use the simple harmonic oscillator equation (Eq. 15.10) to obtain an express for ω.

❸ EXECUTE PLAN From my free-body diagram, I see that

$$F^G_{\text{Eb}\perp} = -mg \sin \theta = -mg\frac{x}{\ell},$$

The x component of the force exerted on the bob is equal to $F^G_{\text{Eb}\perp} \cos \theta$. In the small-angle approximation, $F^G_{\text{Eb}\perp} \cos \theta \approx F^G_{\text{Eb}\perp}$. Thus I can write Eq. 15.11 as

$$ma_x \approx -\frac{mg}{\ell}x,$$

or

$$a_x \equiv \frac{d^2x}{dt^2} \approx -\frac{g}{\ell}x.$$

Comparing this result with Eq. 15.10, $d^2x/dt^2 = -\omega^2 x$, tells me that $-\omega^2 x = -(g/\ell)x$, an equality that yields the same angular frequency of oscillation as in Example 15.6: $\omega = \sqrt{g/\ell}$. ✔

❹ EVALUATE RESULT Given that my result is the same as that obtained using a different method, I feel confident that it is correct.

Figure 15.36 Damped oscillating systems.

(a) Oscillating block is slowed by friction

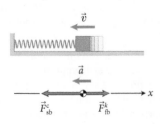

(b) Oscillating cart is slowed by air drag on vane

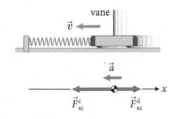

🌀 15.17 Imagine placing a pendulum in an elevator. While the elevator accelerates upward, is the frequency f of the pendulum greater than, smaller than, or equal to the frequency when the elevator is at rest?

15.8 Damped oscillations

Up to this point, we have ignored any dissipation of energy in simple harmonic motion. Mechanical oscillations always involve some friction, however, and so the energy of the oscillation is slowly converted to thermal energy. As this conversion takes place, the oscillating object slows down. The general term for this slowing down because of energy dissipation is *damping*, and the system is said to execute a **damped oscillation.**

Consider, for example, the two systems in **Figure 15.36**. Figure 15.36*a* shows a block oscillating under the action of a spring. In Figure 15.36*b* the block is replaced by a cart with a vane on top. As the block oscillates, it rubs against the floor, and its motion is slowed down by friction with the floor. The motion of the cart is hindered by air drag caused by the vane.

The figure also shows the free-body diagrams for the block and the cart. Each object is subject to two forces in the horizontal direction: a restoring force exerted by the spring and a *damping force*—a force due to friction or drag that dissipates the energy of the oscillation. Damping forces can take many forms. For the oscillating block rubbing on the floor, the magnitude of the force of kinetic friction is $F^k_{\text{fb}} = \mu_k(mg)$, with μ_k the coefficient of kinetic friction. The direction of this force is always opposite the direction of motion (that is, in the direction opposite the direction of the velocity vector).

Drag forces exerted by air or liquids at low speeds tend to be proportional to the velocity of the object: The faster you ride your bicycle, for instance, the greater the drag force exerted on you by the air. So

$$\vec{F}^{d}_{ac} = -b\vec{v} \quad \text{(low speed)}, \tag{15.34}$$

where the coefficient b, called the *damping coefficient*, depends on the shape of the object on which the drag force acts and on the properties of the gas or liquid exerting the force. The SI units of the damping coefficient are kg/s. The minus sign tells you that the direction of the drag force is opposite the direction of the velocity.

For most oscillators, velocity-dependent drag forces are the main cause of energy dissipation. In the presence of drag, the equation of motion for the cart in Figure 15.36 becomes

$$\Sigma F_x = F^{c}_{scx} + F^{d}_{acx} = -kx - bv_x = ma_x, \tag{15.35}$$

where k is the spring constant and m the mass of the cart. When we write the x components of the velocity and the acceleration as time derivatives of x, Eq. 15.35 can be written in the form

$$m\frac{d^2x}{dt^2} + b\frac{dx}{dt} + kx = 0. \tag{15.36}$$

As long as the damping is not too great, the solution to this differential equation is a sinusoidally varying function whose amplitude decreases with time. The solution, given here without proof, is

$$x(t) = Ae^{-bt/2m}\sin(\omega_d t + \phi_i), \tag{15.37}$$

where the angular frequency ω_d of the damped oscillator is given by

$$\omega_d = \sqrt{\frac{k}{m} - \frac{b^2}{4m^2}} = \sqrt{\omega^2 - \left(\frac{b}{2m}\right)^2}. \tag{15.38}$$

Figure 15.37 shows oscillations for various values of the damping coefficient b. For $b = 0$, there is no damping, which means $\omega_d = \omega$ and $e^{-bt/2m} = 1$. In this case,

Figure 15.37 Damped simple harmonic motion for various values of the damping coefficient b not so great as to prevent oscillations.

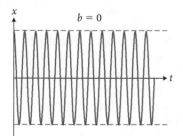

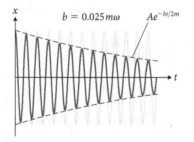

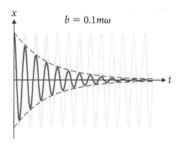

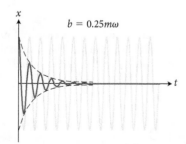

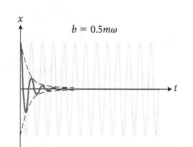

Eq. 15.37 is identical to Eq. 15.6, the equation for the position of an undamped simple harmonic oscillator. For light damping (small b), the second term in the square root in Eq. 15.38 is negligible, and the damping does not affect the angular frequency of the oscillation: $\omega_d \approx \omega_0$. As Figure 15.37 shows, however, damping decreases the amplitude of the oscillation with each swing. Mathematically, this is expressed by the exponential (see Appendix B) factor $e^{-bt/2m}$ in Eq. 15.37, which makes the amplitude at any given instant equal to $x_{max} = Ae^{-bt/2m}$, where x_{max} is the time-dependent amplitude of the damped oscillator and A is the (constant) amplitude of the undamped system. In the figure, this decreasing amplitude x_{max} is indicated by the dashed exponential curves, called the *envelope* of the oscillation. The stronger the damping, the faster the amplitude decreases. For very large damping, oscillations cease to occur and the system just returns to the equilibrium position. Such a system is said to be *overdamped*.

The ratio m/b has units of time, and it is customary to call this ratio the **time constant** $\tau \equiv m/b$. (Even though the symbol for the time constant is the same as that for torque, there is no relationship between the two.) With this simplification, the decreasing amplitude can be written as

$$x_{max}(t) = Ae^{-t/2\tau}. \tag{15.39}$$

The mechanical energy of the oscillator, which is proportional to the square of the amplitude (Eq. 15.17), thus decreases as

$$E(t) = \tfrac{1}{2}m\omega^2 x_{max}^2 = (\tfrac{1}{2}m\omega^2 A^2)e^{-t/\tau} = E_0\,e^{-t/\tau}, \tag{15.40}$$

where $E_0 = \tfrac{1}{2}m\omega^2 A_0^2$ is the initial mechanical energy. Equation 15.40 shows that after a time interval equal to the time constant, so that $t = \tau$, the energy has decreased to $e^{-\tau/\tau} = e^{-1} \approx 0.37$ of its initial value; only about one-third of the initial energy remains. During each time interval of duration τ, the energy again decreases by the factor e^{-1}. After two time constants, for instance, only about 10% of the initial energy remains.

Because oscillation periods vary widely, it is often convenient to express the time constant τ in terms of the period T so that we know after how many *cycles* (rather than seconds) the energy has decreased by a certain factor. In terms of cycles, the decay of the oscillation can be characterized by the *quality factor*, defined as

$$Q \equiv \omega\tau = 2\pi\frac{\tau}{T}. \tag{15.41}$$

The quality factor is a unitless ratio similar to the rotational coordinate ϑ—if $Q = 2\pi$, then $\tau = T$, and so the energy decays by a factor of e^{-1} in one cycle. When $Q = 1000(2\pi)$, it takes 1000 cycles for the energy to decay by a factor of e^{-1}. A high Q implies that $\omega_d \approx \omega_0$. For mechanical oscillators with low friction (piano strings, tuning forks), Q can be as large as a few thousand, meaning the oscillator executes thousands of oscillations before its energy decreases by a factor of e^{-1}. A bell with a large Q rings for a longer time interval than one with a small Q.

15.18 A tuning fork that sounds the tone musicians call middle C oscillates at frequency $f = 262$ Hz. If the amplitude of the fork's oscillation decreases by a factor of 3 in 4.0 s, what are (*a*) the time constant of the oscillation and (*b*) the quality factor?

Chapter Glossary

SI units of physical quantities are given in parentheses.

Angular frequency $\omega\,(\text{s}^{-1})$ A positive scalar equal to the rotational speed of a rotating object whose projection produces an oscillation of frequency f, so that

$$\omega = 2\pi f. \tag{15.4}$$

Amplitude $A\,(\text{m})$ A positive scalar equal to the magnitude of the maximum displacement in a periodic motion.

Damped oscillation An oscillation whose amplitude decreases over time due to dissipation of energy.

Fourier's theorem Any periodic function with period T can be written as a sum of sinusoidal functions of frequency $f_n = n/T$, where n is a positive integer.

Frequency $f\,(\text{Hz})$ A positive scalar equal to the number of cycles per second of a periodic motion. The frequency is the inverse of the period:

$$f \equiv 1/T. \tag{15.2}$$

Fundamental frequency $f_1\,(\text{Hz})$ The frequency of the first term ($n = 1$) in the sum of sinusoidal functions of frequency $f_n = n/T$ that add up to describe a periodic function.

Harmonic A term in a sum of sinusoidal functions used to describe periodic functions. The lowest frequency term is called the *first harmonic*. The other terms are called *higher harmonics*.

Hertz (Hz) The derived SI unit of frequency: $1\text{ Hz} \equiv 1\text{ s}^{-1}$.

Oscillation Periodic back-and-forth motion. Also called *vibration*.

Periodic motion Any motion that repeats itself at regular time intervals, called the *period T* of the motion.

Phase ϕ (unitless) A scalar that specifies the time-dependent argument of the sine function describing simple harmonic motion:

$$\phi(t) = \omega t + \phi_i. \tag{15.5}$$

The constant ϕ_i, the *initial phase*, specifies the phase at $t = 0$.

Phasor A rotating arrow whose component on a vertical axis traces out a simple harmonic motion.

Reference circle A circle traced out by the tip of a phasor.

Simple harmonic motion Motion for which the displacement is a sinusoidally varying function of time given by

$$x(t) = A\sin(\omega t + \phi_i), \tag{15.6}$$

where A is the *amplitude,* ω the *angular frequency,* and ϕ_i the *initial phase* of the motion. The period T of a simple harmonic oscillator is independent of the amplitude A.

Simple harmonic oscillator equation An equation satisfied by any system that undergoes simple harmonic motion:

$$\frac{d^2x}{dt^2} = -\omega^2 x, \tag{15.10}$$

where x represents some sort of displacement and ω is the angular frequency of the oscillation. Any such system is called a *simple harmonic oscillator.*

Spectrum A graph that shows the squares of the amplitude A of the simple harmonic components of a periodic function versus frequency f. The spectrum provides a visual display of the frequency content of the function.

Time constant $\tau\,(\text{s})$ A scalar that gives the interval over which the energy of a damped simple harmonic oscillator is reduced by a factor of $1/e$.

Vibration See *oscillation.*

16 Waves in One Dimension

Many phenomena around us involve waves. Throw a stone into a pond and ripples move outward along the surface of the water. Clap your hands and a sound wave travels outward. A large earthquake is detected all over the world because the earthquake causes a wave to propagate through Earth's interior. In general, a **wave** is a *disturbance* propagating (moving) either through a material or through empty space.

The most important feature of waves is that they do not carry matter from one place to another. After a wave has passed through a medium, the medium looks just as it did before it was disturbed. You prove this every time you and a stadium full of other fans work together to create a wave (**Figure 16.1**). Once you've done your small part—stand up, sit down—you are sitting in the same seat; in other words, you haven't moved from one location to another in the stadium. Or throw a rock into a still pond and watch the motion of a floating leaf: The leaf rises and falls as the wave created by the rock passes by but never moves away from its original position. Once the pond is still again, the leaf is just where it was before. A little thought will tell you that the same is true for every molecule of water in the pond. As we shall see, momentum and energy—not matter—are what move along with any wave.

Waves are collective phenomena: A stadium full of fans can create a wave that travels around the stadium; a single person, on the other hand, cannot create such a wave. The person can move up and down or run from one place to another, but neither motion is a wave. A wave can thus never exist at a single position in space: A single water molecule does not make a wave—a water wave requires an extended volume of water.

The examples with which I opened this chapter—water waves, sound waves, seismic waves—are all *mechanical waves,* which are waves that involve the (temporary) displacement of particles. Such waves require a *medium* to propagate along

or through. The propagation is caused by the interaction of the particles in the medium. The medium for water ripples is the water; for the sound of clapping hands, the medium is the air; for seismic waves, it is the material that makes up Earth's interior. Some waves, such as radio waves and light, do not require any medium; they can propagate through empty space as well as through a medium. Even though such *nonmechanical waves* are very different in origin and nature from the mechanical waves we discuss in this chapter and the next, they share many of the properties of mechanical waves. For this reason, when we deal with nonmechanical waves in later chapters, we shall carry over many results derived for mechanical waves.

16.1 Representing waves graphically

Suppose one end of a long, taut string is rapidly displaced up and down once, as shown in **Figure 16.2**. As the end of the string is raised, it pulls up the neighboring part of the string. As that neighboring part begins moving up,

Figure 16.2 Generation of a wave pulse that propagates along a string.

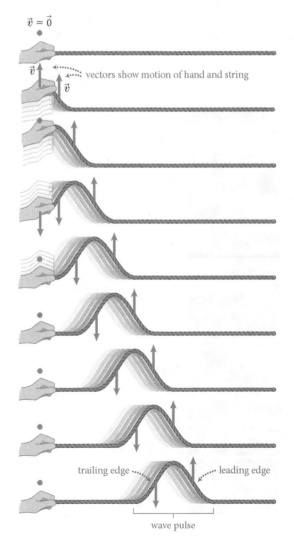

Figure 16.1 Fans at a stadium creating a wave. People get up and raise their arms as soon as the person next to them does so. Because of the small delay between one person and the next, not everyone gets up at the same instant, and the wave travels around the stadium at some finite speed.

it pulls up the next part of the string, which, in turn, pulls up the next part, and so forth. The upward displacement of the end of the string is thus transmitted along the string because neighboring pieces of the string pull on each other. As the end of the string is moved back down, a similar chain of events occurs in the reverse direction. This pulling along of neighboring pieces of string, first up and then down, produces a **wave pulse,** which travels along the string at constant speed without changing shape. It is important that you do not confuse the horizontal motion of the pulse with the vertical-only motion of the string. The medium through which the pulse passes does not move in the horizontal direction. Each particle of the string moves only once up and once down as the pulse passes through it. **Figure 16.3** shows this vertical-only motion for a string of beads.

16.1 (*a*) Using the choice of axes shown in Figure 16.3, draw a position-versus-time graph showing how the *x* and *y* components of the position of the large bead change with time. (*b*) Are the *x* and *y* components of the velocity of the large bead positive, zero, or negative while the bead is on the leading edge of the pulse? Repeat for the trailing edge.

Figure 16.3 shows that the wave motion is different from the motion of the particles of the medium through

Figure 16.3 Video frames sequence of a wave pulse propagating along a string of beads.

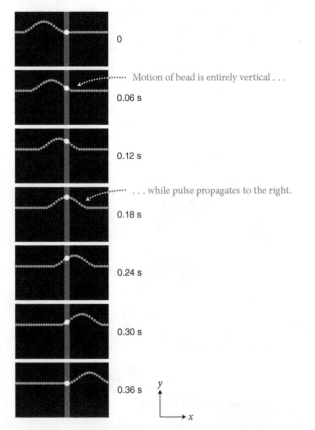

0

Motion of bead is entirely vertical . . .

0.06 s

0.12 s

. . . while pulse propagates to the right.

0.18 s

0.24 s

0.30 s

0.36 s

which the pulse travels. The beads move in the vertical direction only, with each bead executing the same motion—first up and then down. Each bead moves at a different instant, however: first bead 1, then bead 2, then bead 3, and so on. At the leading edge of the pulse, each bead's motion is upward; at the trailing edge, it is downward. After the pulse has passed, all the beads are back in their initial positions.

An important point to keep in mind therefore is:

The motion of a wave (or of a single wave pulse) is distinct from the motion of the particles in the medium that transmits the wave (or pulse).

To emphasize this point, we shall use the letter *c* to denote the **wave speed** of a wave through a medium (along a string in this case) and *v* to denote the speed of the particles in the medium (the vertical motion of the beads in this case).*

16.2 (*a*) Is the wave speed *c* of the pulse in Figure 16.3 constant? (*b*) Determine that speed if the distance between adjacent beads is 5.0 mm.

As this checkpoint shows:

The wave speed *c* of a wave pulse along a string is constant.

The motion of the various parts of the string, however, is not so simple. As the pulse propagates down the string, the displacements of various particles depend on time *and* on position along the string. At any given instant, different particles on the string have different displacements. Figures 16.2 and 16.3 show "snapshots" of a wave pulse along a string. Each snapshot shows the displacements of the particles on the string from their equilibrium positions as a function of position at one instant. For example, **Figure 16.4a** shows a snapshot of a triangular wave pulse traveling along a string. Such a triangular wave pulse is created by moving the end of the string first upward and then more rapidly downward. The snapshot is superimposed on a calibrated grid so that we can read off the displacements of various particles of the string from their equilibrium positions. The vector $\vec{D}$ in Figure 16.4a represents the **displacement** of the particle at *x* = 1.0 m at instant *t* = 0.94 s. In the context of waves, displacement is denoted by the symbol $\vec{D}$, and the initial position is always taken to be the equilibrium position of the particle.

The graphic representation of all the particle displacements at this instant is shown in Figure 16.4b, which gives the *y* components of the displacements of the particles

---

*We use *c* for the speed of *any* wave, not just light (which is also a wave). The important point is never to confuse the speed of a wave with the speed of the particles being disturbed by the wave.

Figure 16.4 Distinction between the wave function and displacement curves for a triangular wave pulse propagating along a string parallel to the x axis.

(a) Snapshot of wave at $t = 0.94$ s

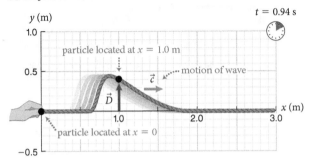

(b) Wave function at $t = 0.94$ s

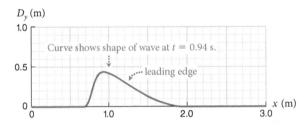

(c) Displacement curve for particle located at $x = 0$

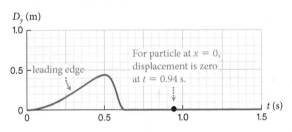

(d) Displacement curve for particle located at $x = 1.0$ m

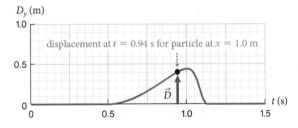

of the string as a function of position x along the string (the left end of the string being at $x = 0$). The function shown in Figure 16.4b is called the **wave function.** The wave function represents the shape of the wave at a given instant and changes with time as the wave travels along the string.

16.3 From Figure 16.4a, determine the displacement of a point (a) 1.0 m, (b) 1.5 m, and (c) 2.0 m from the end of the string. (d) On a snapshot taken a short time interval later, is the displacement at $x = 1.5$ m greater than, equal to, or smaller than that shown in Figure 16.4a?

A complementary way of representing the wave pulse of Figure 16.4a is to plot the displacement of one particle on the string as a function of time. Figure 16.4c, for example, shows the y component of the displacement of the end of the string at $x = 0$ as a function of time. The function shown in Figure 16.4c is called the **displacement curve.** The graph shows that the end of the string started moving up at $t = 0$, reached a maximum displacement of about $+0.50$ m at about $t = 0.50$ s, and then quickly moved back down to its initial position, which it reached at about 0.63 s. This is the motion that generated the triangular wave pulse propagating along the string.

The displacement curve for any string through which a wave pulse moves is different for different particles on the string, as you can see by comparing Figure 16.4c and d. Figure 16.4d shows the displacement curve for the particle at $x = 1.0$ m. The displacement remains zero until the leading edge of the pulse has traveled from $x = 0$ to $x = 1.0$ m. This happens at $t = 0.50$ s, and at that instant the leading edge of the pulse begins raising the $x = 1.0$ m particle until it reaches its maximum displacement at $t = 1.0$ s. The much steeper trailing end of the pulse shown in Figure 16.4a then quickly returns the particle to its equilibrium position between $t = 1.0$ s and $t = 1.1$ s.

Why is the displacement curve the mirror image of the wave function? Because the wave pulse travels to the right, the leading edge of the pulse is on the right on the wave function graph. The displacement curve, on the other hand, has small values of t (earlier instants) on the left and large values of t (later instants) on the right. So in Figure 16.4c the leading edge of the pulse is on the *left*. (To convince yourself, look at the particle at $x = 1.0$ m in Figure 16.4a and visualize the motion of that particle as the pulse travels past it. Sketch the displacement of the particle as a function of time and compare your result with Figure 16.4d).

Note that the displacement curves in Figure 16.4c and d have the same form: The one for $x = 1.0$ m is simply shifted to the right. This tells us that both particles execute the same motion at different instants.

16.4 (a) From Figure 16.4c and d, determine how long it takes the pulse to travel from $x = 0$ to $x = 1.0$ m. (b) Using your answer to part a, determine the wave speed.

Mechanical waves are divided into two categories, depending on how the medium moves relative to the wave motion. For a wave propagating along a string, the medium movement is perpendicular to the pulse movement. Such waves are called **transverse waves.** In the other type of

Figure 16.5 Longitudinal wave pulse propagating along a spring.

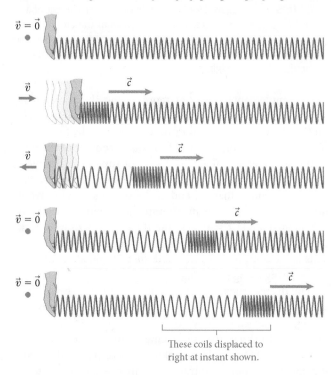

These coils displaced to right at instant shown.

Figure 16.6 Wave function and displacement curve for a longitudinal wave pulse propagating along a spring.

(*a*) Snapshot of wave pulse at instant t_1

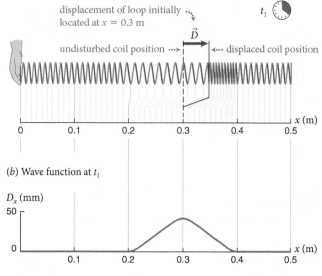

displacement of loop initially located at $x = 0.3$ m

$\vec{D}$

undisturbed coil position ···→ ←··· displaced coil position

(*b*) Wave function at t_1

(*c*) Displacement curve for coil located at $x = 0$

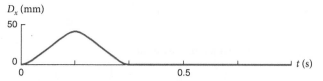

mechanical wave, the medium movement is parallel to the pulse movement, and waves that behave this way are called **longitudinal waves.** Let's now look at this second type of wave in more detail.

Consider a long coil spring whose end is rapidly displaced back and forth as in Figure 16.5. As the left end of the spring is displaced to the right, each coil pushes on the next, displacing the coils one after the other to the right. When the left end is pulled back to its original position, each coil pulls the next one back. The result is again a wave pulse propagating to the right, with each coil temporarily being displaced in the direction of propagation of the wave pulse.

Even though the direction of motion of the individual coils coincides with the direction of propagation of the wave pulse, the motion of the coils and that of the pulse are still distinct. The individual coils move back and forth, but the wave pulse travels at constant speed along the spring.

Longitudinal waves can also be represented using wave functions and displacement curves, but now the wave functions no longer look like a snapshot of the propagating wave. Figure 16.6*a* shows a snapshot of a pulse propagating along a spring at some instant t_1, overlaid on a calibrated grid. The corresponding wave function is shown in Figure 16.6*b*—the wave function gives the *x* component of the displacement of each coil as a function of the coil's position. The coils between $x = 0$ and $x = +0.20$ m are undisturbed, and hence the displacement $\vec{D}$ is zero between $x = 0$ and $x = +0.20$ m. The coils between $x = +0.20$ m and $x = +0.40$ m are all displaced rightward. The *x* component of the displacement is positive because the

displacement is in the positive *x* direction. To the right of $x = +0.40$ m, the displacement is again zero because the pulse has not yet reached that part of the spring.

Figure 16.6*c* shows the displacement curve for the end of the spring at $x = 0$.

16.5 (*a*) The pulse in Figure 16.6*b* is symmetrical even though Figure 16.6*a* shows that the length of the stretched portion of the spring is 0.15 m and the length of the compressed portion of the spring is only 0.05 m. Explain how the symmetrical curve of Figure 16.6*b* is a correct representation of the asymmetrical situation shown in Figure 16.6*a*. (*b*) Suppose that instead of the end of the spring moving to the right, as illustrated in Figure 16.6, the end is moved to the left. Would this disturbance still cause a wave pulse to travel to the right? If yes, sketch the wave function for this wave pulse at some instant $t > 0$. If no, what happens instead?

16.2 Wave propagation

A wave pulse is not an object—it has no mass—and so the description of wave motion is very different from the description of the motion of objects. To study the propagation of a wave pulse along a string, let us consider a collection of beads connected by short strings. The mass of the strings is negligible relative to that of the beads. Figure 16.7*a* shows

Figure 16.7 Forces and accelerations that cause a wave pulse to propagate along a string of beads.

(a) Wave pulse propagates along string of beads

(b) Free-body diagrams for bead 3

(c) Velocity and acceleration vectors for bead 3

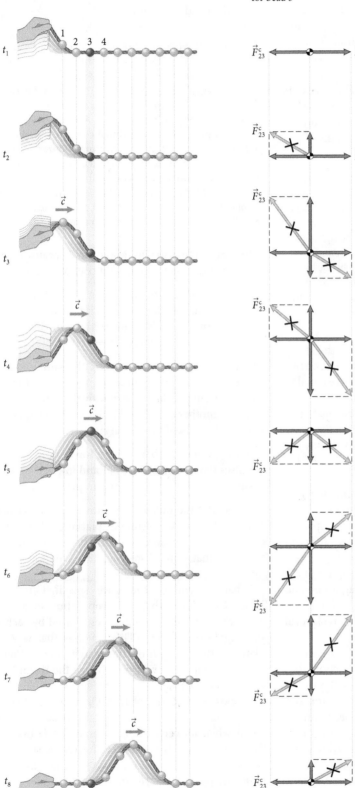

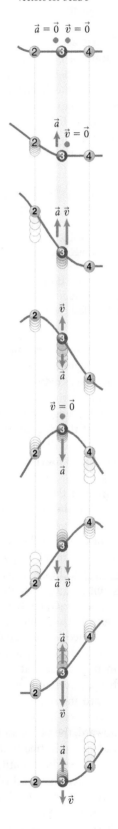

the first ten of these beads as a pulse travels along the string. The vertical gray lines show that the motion of the beads is entirely vertical.

Let us focus on the third bead along the string. At t_1 the left end of the string has just begun moving up, and only bead 1 is displaced from its initial position. The top free-body diagram in Figure 16.7b shows the two contact forces exerted on bead 3. If the string is held taut, $\vec{F}_{23}^c$ and $\vec{F}_{43}^c$ point in opposite horizontal directions. (This amounts to saying that the force of gravity on the beads is negligible relative to the contact forces exerted by the strings, and because the bead does not accelerate, the vector sum of the forces exerted on it must be zero.)

✋ **16.6** Answer these five questions about the situation at t_1 in Figure 16.7, assuming the right end of the string of beads is attached to a wall that is not shown. (a) If $F_{43}^c = 5$ N, what is the magnitude of the force exerted by the wall on the rightmost bead? (b) What is the magnitude of the horizontal component of $\vec{F}_{12}^c$? (c) What is the direction of the force exerted by the hand on the string? (d) How does the magnitude of this force compare with the magnitude F_{43}^c? (e) What are the effects of the horizontal and vertical components of this force?

This checkpoint makes an important point: If the string is held taut, the magnitudes of the horizontal components of all the contact forces between adjacent beads are equal to the tension $\mathcal{T}$ in the string.

At t_2 the leading edge of the pulse has propagated to bead 3. Because of the displacement of bead 2, $\vec{F}_{23}^c$ is no longer horizontal. The displacement of 2 also causes the string between 2 and 3 to stretch, and so the magnitude F_{23}^c increases.

The magnitude of the horizontal component of $\vec{F}_{23}^c$ is still equal to $\mathcal{T}$, but the vector sum of the forces exerted on bead 3 is now nonzero and points vertically upward. Consequently, bead 3 now has a nonzero upward acceleration and so begins moving upward (Figure 16.7c). As it moves upward, it stretches and changes the direction of the string connecting it to bead 4, and so the magnitude and direction of $\vec{F}_{43}^c$ change. These changes in $\vec{F}_{43}^c$ have two effects: $\vec{F}_{43}^c$ begins to slow down bead 3, and the reciprocal force $\vec{F}_{34}^c$ begins to accelerate bead 4 upward.

At some instant between t_3 and t_4, $\vec{F}_{23}^c$ and $\vec{F}_{43}^c$ again become equal in magnitude and opposite in direction (but this time with $\vec{F}_{23}^c$ directed upward and to the left and $\vec{F}_{43}^c$ directed downward and to the right), and so the vector sum of the forces exerted on bead 3 is zero. As the bead continues to move upward, the vector sum of the forces begins to oppose the bead's motion (free-body diagram at t_4). Consequently the bead slows down until it comes to a stop at t_5. At that instant, the bead has reached its maximum displacement, and the vector sum of the forces exerted on it points downward. As the bead moves back down, the vector sum of the forces exerted on it decreases. At t_7 the vector sum points upward again, slowing the downward motion of the

bead until it reaches its initial position again at t_8. Once the pulse has moved beyond bead 3 and the string around it is straight again, the vector sum of $\vec{F}_{23}^c$ and $\vec{F}_{43}^c$ is again zero, and so the bead remains at rest.

✋ **16.7** (a) Note that the displacement of bead 4 at t_4 in Figure 16.7a is the same as that of bead 3 at t_3. How do the velocity and acceleration of bead 4 at t_4 compare with those of bead 3 at t_3? (b) You move one end of two different strings, A and B, up and down in the same way. Suppose the resulting pulse travels twice as fast on string A as on string B. How does the velocity of a particle of string A compare with the velocity of a particle of string B that has the same nonzero displacement? (c) Sketch the pulses as they propagate along the two strings and point out any differences between the two. (d) For each string, consider a particle located 0.20 m from the left end. Sketch a displacement-versus-time graph for each particle, and point out any differences between your two graphs.

Be sure not to skip Checkpoint 16.7 because it makes two very important—and somewhat counter-intuitive—points:

> When a particle of the string is displaced from its equilibrium position, its velocity $\vec{v}$ and acceleration $\vec{a}$ are determined only by the initial disturbance and are independent of the wave speed c.

All particles of the string execute the same motion as that caused by the initial disturbance, *regardless of wave speed c* (which determines only how quickly they execute this motion after one another). As Checkpoint 16.7 shows, however, the *shape* of the pulse depends on c:

> For a given disturbance, high wave speeds yield wave pulses that are stretched out and low wave speeds result in pulses that are more compressed.

Regardless of the pulse shape, however, any particle of the string that reaches a certain displacement has the same velocity and acceleration as any other particle at the instant it reaches that same displacement in the same part of the pulse.

What determines the wave speed? Our analysis of Figure 16.7 shows that the propagation of a wave pulse along a string is due to the forces exerted by each piece on the neighboring pieces. This suggests that wave speed is related to the strength of these forces. Let us therefore investigate what happens if we change the magnitude of the tension in the string.

Suppose that at instant t_3 in Figure 16.7, when bead 3 is on the leading edge of the pulse, this bead has an acceleration $\vec{a}$ when its vertical displacement is $\vec{D}$ (**Figure 16.8a**). This acceleration is caused by the vector sum $\sum \vec{F}$ of the forces exerted on the bead. If we now increase the tension in the string (Figure 16.8b), the magnitudes of the horizontal components of the contact forces increase by the same amount. Because the contact forces are now greater, the vector sum of the forces exerted on bead 3 reaches the same value as in Figure 16.8a with a smaller displacement

Figure 16.8 Effects of tension and mass on wave speed. (*a*) The free-body diagrams show the angle between $\vec{F}_{23}^c$ and $\vec{F}_{43}^c$ that is required to cause an acceleration $\vec{a}$ at a displacement $\vec{D}$. (*b*) When the tension is greater, a shallower angle is required to produce the same condition so the wave propagates faster. (*c*) When the mass is greater, a sharper angle is required to produce the same condition so the wave propagates more slowly.

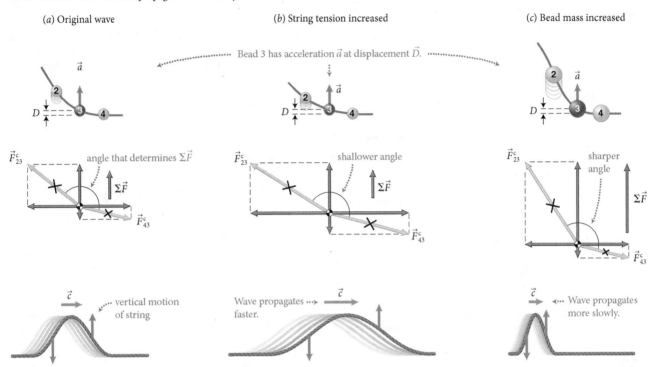

(*a*) Original wave

(*b*) String tension increased

(*c*) Bead mass increased

of bead 2. In other words, it takes a shorter time interval for the pulse to propagate from one bead to the next. A greater tension in the string thus causes a higher wave speed.

Increasing the mass of the beads accomplishes the opposite effect: For a given tension, increasing the mass of each bead decreases the beads' accelerations. To give this larger bead 3 the same acceleration $\vec{a}$ that the smaller bead 3 in Figure 16.8*a* has, bead 2 must now be displaced farther. Thus now it takes a longer time interval for the pulse to propagate from one bead to the next (Figure 16.4*c*). Increasing the mass of the beads corresponds to increasing the mass per unit length of a continuous string.

The speed c of a wave propagating along a string increases with increasing tension in the string and decreases with increasing mass per unit length along the string.

This result makes sense: A greater tension means a greater restoring force and therefore a greater tendency to pass the displacement along, resulting in a higher wave speed. If the mass is greater, there is a greater resistance to passing the displacement along, and so the wave speed is lower.

16.8 Two strings, A and B, are identical except that the tension in A is greater than that in B. Suppose you move the left end of each string rapidly up and down once. For each string, sketch (*a*) the wave functions at the instant the pulse on A has traveled halfway down the length of the string and (*b*) the displacement curve for a particle midway down the string.

If the end of a string is made to execute a periodic motion, the resulting wave is called a **periodic wave**. **Figure 16.9** shows a special type of periodic wave, called a **harmonic wave**, obtained by moving the end of the string so that it oscillates harmonically. The result is a sinusoidally shaped wave traveling along the string.

16.9 Plot the displacement of the left end of the string in Figure 16.9 as a function of time. What similarities and differences exist between your graph and the shape of the wave in Figure 16.9?

Note how the shape of the string in Figure 16.9 resembles the curves representing simple harmonic oscillations from Chapter 15. The shape in Figure 16.9, however, represents the displacement of the medium as a function of *position*, whereas the simple harmonic oscillator curves in Chapter 15 represented displacement as a function of *time*. If you focus on one particle of the string as the wave moves through it, however, you would see that its up-and-down motion is a delayed version of the hand's periodic motion. So a number of quantities we have used to describe periodic motion are also useful to describe waves (*amplitude*: maximum displacement of any given particle of the medium from its equilibrium position; *frequency*: number of cycles per second executed by each particle of the medium; *period*: time interval taken to complete one cycle). A periodic wave repeats itself over a distance called the **wavelength,** denoted by the symbol λ (Figure 16.9). Each time the left end of the

CONCEPTS

Figure 16.9 If the end of a string is made to execute a simple harmonic motion, the resulting traveling wave is sinusoidal in shape. During one period of the motion, the wave advances by one wavelength.

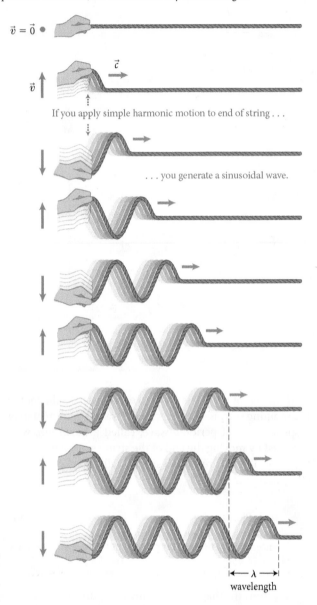

If you apply simple harmonic motion to end of string . . .

. . . you generate a sinusoidal wave.

λ
wavelength

Figure 16.10 The effect on a wave pulse of the speed with which the hand moves up and down. In each case the leading edge of the pulse propagates the same distance during the same time interval, showing that the wave speed c is independent of the velocities $\vec{v}$ of the particles in the string.

(*a*) String moved slowly   (*b*) String moved quickly

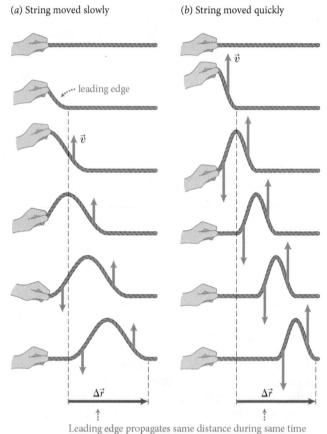

leading edge

$\Delta\vec{r}$ $\Delta\vec{r}$

Leading edge propagates same distance during same time interval, so wave speed c is same.

string executes a complete oscillation, the wave advances by one wavelength. If the speed at which the wave propagates is c, then the wave advances a distance cT during the period T of the sinusoidal motion. Therefore:

The wavelength of a periodic wave is equal to the product of the wave speed and the period of the wave motion.

Contrary to what you may expect, moving your hand up and down more quickly does not generate a faster-traveling pulse. Figure 16.10 shows two pulses generated on identical strings held under identical tensions. The end of the string in Figure 16.10*b* is displaced up and down more quickly than that in Figure 16.10*a*. Both pulses have the same

displacement $\Delta\vec{r}$ during a time interval Δt, so the wave speed c is the same for both pulses. To a very good approximation, we determine experimentally that:

The speed c of a wave propagating along a string is independent of the velocities $\vec{v}$ of the individual pieces of string. The value of c is determined entirely by the properties of the medium.

If wave speed depended on the velocity of the particles, the shape of a wave pulse would change as it propagates because some parts of it would propagate more quickly than others. As long as the displacement caused by the wave is not too large, however, the shape of the waves remains unchanged (see Figure 16.3).

16.10 Consider a harmonic wave traveling along a string. Are the following quantities determined by the source of the wave, by the properties of the string, or by both: (*a*) the period of oscillation of a particle of the string; (*b*) the speed c at which the wave travels along the string; (*c*) the wavelength; (*d*) the maximum speed v of a particle of the string?

Figure 16.11 A wave pulse carries kinetic and potential energy. If there is no energy dissipation, the amount of energy in (*b*) is the same as that in (*a*).

(*a*)

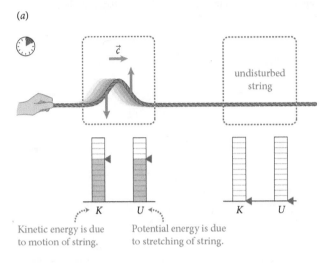

Kinetic energy is due to motion of string.

Potential energy is due to stretching of string.

(*b*)

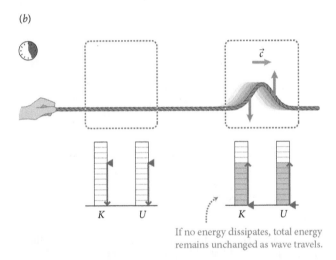

If no energy dissipates, total energy remains unchanged as wave travels.

Figure 16.11 illustrates how a propagating wave pulse carries two forms of energy along with it: kinetic energy associated with the motion of individual particles and elastic potential energy associated with the stretching of the string as the pulse is passing through. In Section 16.3 we shall see that a mechanical wave always carries equal amounts of these two forms of energy.

✋ **16.11** Does the wave pulse in Figure 16.11 also carry along momentum? Justify your answer.

16.3 Superposition of waves

Waves have a remarkable property: Two waves can pass straight through each other without changing each other's shape. **Figure 16.12** shows a series of snapshots of two pulses propagating toward each other along a string. When the two pulses overlap, the displacements caused by the two waves

Figure 16.12 Constructive interference. Two pulses propagating in opposite directions interfere constructively if they displace the string particles in the same direction (here, upward). When the pulses cross, their displacements add algebraically, giving rise to a displacement greater than that caused by either pulse individually. The pulses do not interact with each other: After they separate, their shapes are unaltered.

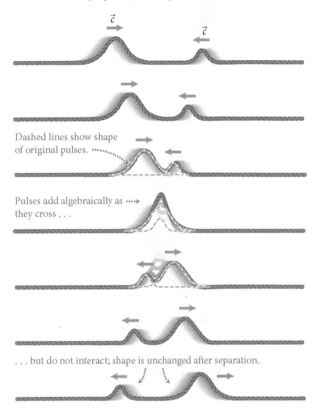

Dashed lines show shape of original pulses.

Pulses add algebraically as they cross . . .

. . . but do not interact; shape is unchanged after separation.

add algebraically, provided the medium obeys Hooke's law (see Section 8.9) for the combined displacement. This phenomenon is called the **superposition of waves:**

> If two or more waves overlap in a medium that obeys Hooke's law, then the resulting wave function at any instant is the algebraic sum of the individual wave functions.

The resulting wave may have a very complicated shape when the waves overlap, but when they separate again, each wave still has its original shape.

The superposition of waves further distinguishes waves from objects: Two objects cannot occupy the same location in space at the same instant, but two waves *can* occupy the same location at the same instant.

Because the wave functions of two overlapping waves add algebraically, the resultant displacement is greater than that of either wave when the individual displacements have the same sign (Figure 16.12). When they have opposite signs (**Figure 16.13**), the wave with the smaller displacement decreases the displacement caused by the other wave.

When two waves overlap, we say that the waves *interfere* with each other, and the phenomenon is called **interference.** The adding of waves of the same sign, as in Figure 16.12,

Figure 16.13 Destructive interference. Two pulses propagating in opposite directions interfere destructively if they displace the string particles in opposite directions. When the pulses cross, their displacements add algebraically, giving rise to a decreased displacement. After the pulses separate, their shapes are unaltered.

Figure 16.14 Complete destructive interference of two pulses of identical shape propagating in opposite directions while displacing string particles in opposite directions. At the instant the pulses overlap, their displacements add to zero.

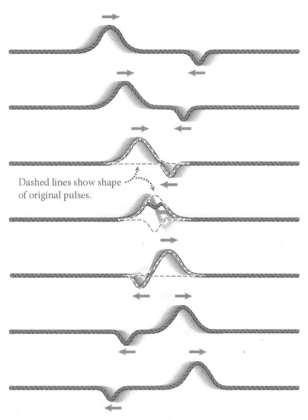

Dashed lines show shape of original pulses.

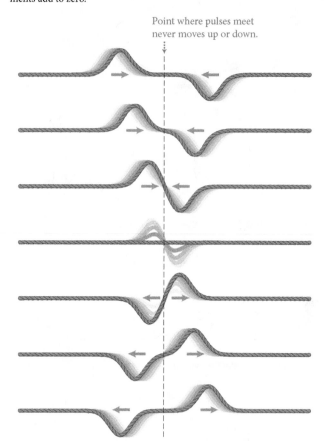

Point where pulses meet never moves up or down.

is called *constructive interference,* and the adding of waves of opposite signs, as in Figure 16.13, is called *destructive interference.* This terminology is somewhat misleading because the waves do not *interact* in any way. As you can see from Figures 16.12 and 16.13, when the two waves separate, their shapes are unchanged. Although there is no destruction or construction, interference of waves does affect the displacement that occurs at specific locations, as the next checkpoint shows.

✋ **16.12** (*a*) Is the maximum displacement the same for all particles of the string in Figure 16.12? For all particles of the string in Figure 16.13? (*b*) For each string, sketch a displacement curve for a point near the left end of the string. (*c*) Repeat part *b* for the point of each string at which the pulses meet. (*d*) Reconcile your answer to part *a* with your sketches for parts *b* and *c*.

An extreme form of destructive interference occurs when the two pulses crossing each other are the same size and shape but of opposite algebraic signs, as in **Figure 16.14**. At the instant the two pulses overlap, the displacements cancel and the string is flat. Even though it appears as though both pulses have disappeared, they continue unchanged after the superposition.

✋ **16.13** (*a*) Sketch a displacement curve for the point on the string in Figure 16.14 at which the two pulses meet. (*b*) When two wave pulses overlap, how is the velocity of a point in the overlap region related to the velocities of the corresponding points of the individual pulses?

As Checkpoint 16.13 shows, the point at which the two pulses in Figure 16.14 meet never moves. When the two pulses overlap, they cause equal displacements in opposite directions, and so the vector sum of the displacements is zero. A point that remains stationary in a medium through which waves move is called a **node.**

We can deduce a useful fact by examining what happens to the energy in the pulses when they overlap exactly. Because the two pulses are identical, each carries the same amount of energy $E_1 = K_1 + U_1$, and the total energy in the two pulses is thus $2K_1 + 2U_1$ (**Figure 16.15a**). When the string displacements caused by the two pulses cancel each other exactly (the pulses interfere destructively), the string is straight, and so the elastic potential energy is zero. The kinetic energy, however, is not zero. The upward-moving leading edge of pulse 1 adds to the upward-moving trailing edge of pulse 2, and likewise for the downward-moving edges. The result is that the points within the overlapping

Figure 16.15 Interference of two pulses traveling in opposite directions. At the instant the pulses overlap, the string displacement is zero everywhere, and so the potential energy is zero; all the energy is kinetic.

(*a*) Two pulses of identical shape traveling in opposite directions and having displacements in opposite directions

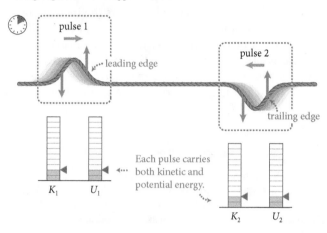

(*b*) At the instant the pulses overlap, the string displacement is zero everywhere, and so the potential energy is zero; all the energy is kinetic

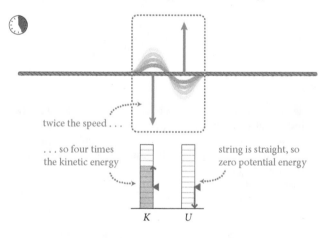

pulses move twice as fast as those within the individual pulses before and after the overlap. Because kinetic energy depends on the square of the velocities, the kinetic energy of points in the overlap region is four times as great as that of an individual pulse. The energy during the instant shown in Figure 16.15*b* is thus $4K_1$, which must be equal to the initial energy $2K_1 + 2U_1$ of the two separate pulses. So we determine that $2K_1 + 2U_1 = 4K_1$, which can be true only if $K_1 = U_1$—in other words, if each pulse has an equal amount of kinetic and potential energy. Because these arguments do not depend on the shape of the pulse, it follows that:

> **A wave contains equal amounts of kinetic and potential energy.**

16.4 Boundary effects

When a wave pulse reaches a boundary where the transmitting medium ends, the pulse is *reflected,* which means that its direction of propagation is reversed. After reaching

Figure 16.16 Reflection of a pulse at a boundary where the string end is fixed and so cannot move.

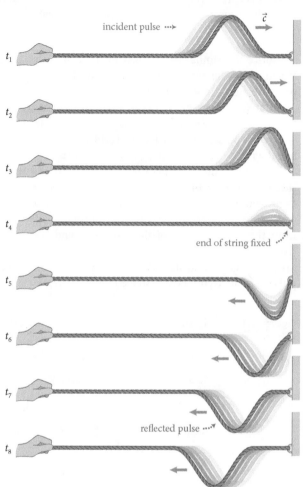

the boundary, a pulse originally traveling to the right starts moving to the left through the medium, and a pulse originally traveling to the left starts moving to the right. Before the pulse reaches the boundary, we refer to it as the *incident pulse;* after reflection, it is the *reflected pulse.* Whether or not a reflected pulse has the same orientation as the incident pulse depends on the properties of the boundary. Consider first a pulse propagating along a string toward the end of the string that has been anchored so that it cannot move (**Figure 16.16**). As the leading edge of the pulse reaches the fixed end, the pulse pulls the end upward. The effect of this upward pull is that the fixed end exerts a downward force on the string.* This downward force exerted on the string produces an *inverted* reflected pulse. If no energy is lost in the reflection, the reflected pulse is identical in shape to the incident pulse.

---

*If you think it is hard to imagine that the fixed end pulls the string downward, imagine pulling yourself up from a chinning bar. When you do so, you exert a downward force on the bar, and the bar exerts an upward force on you. Because the bar is fixed, you end up moving up even though you pull down on the bar.

✋ **16.14** The string in Figure 16.16 is perfectly straight at t_4. What has happened to the energy in the incident pulse at that instant?

Notice the similarity between Figures 16.16 and 16.14. If you cover the right half of Figure 16.14 with a piece of paper so that the left edge of the paper goes through the point at which the pulses meet, the two figures are identical. Indeed, as we saw in Checkpoint 16.13, the point at which the pulses meet in Figure 16.14 is a node, precisely like the fixed end of the string in Figure 16.16.

The correspondence between Figure 16.16 and the left side of Figure 16.14 suggests a procedure for determining the shape of a reflected wave pulse: On the incident pulse superpose an identical, but inverted, pulse that approaches from the opposite side of the fixed end and reaches the fixed end at the same instant as the incident pulse. This procedure is best illustrated using an example involving a pulse with an asymmetrical shape (**Figure 16.17**).

Figure 16.17a shows a triangular wave pulse approaching the fixed end of a string. To construct the reflected pulse we first draw an inverted triangular pulse on the opposite side of the fixed point and moving in the opposite direction. This inverted "reflected pulse" reaches the fixed point at the same instant as the incident pulse, and the two interfere (Figure 16.17b and c). While the two pulses overlap in space, the resulting shape of the pulse is obtained by adding the incident and reflected waves. Once the reflected pulse separates from the fixed end, its shape is the inverse of the incident wave pulse (Figure 16.17d).

Example 16.1 Reflection from a fixed end

Consider a triangular wave pulse approaching the fixed end of a string (**Figure 16.18**). Sketch the shape of the string (a) when a point halfway up the leading edge of the pulse has reached the fixed end and (b) when the peak of the pulse has reached the fixed end.

Figure 16.17 Reflection of a triangular pulse at a boundary where the string end cannot move.

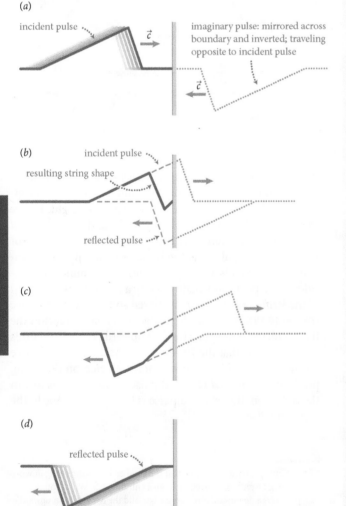

(a)

incident pulse

$\vec{c}$

imaginary pulse: mirrored across boundary and inverted; traveling opposite to incident pulse

$\vec{c}$

(b)

incident pulse

resulting string shape

reflected pulse

(c)

(d)

reflected pulse

Figure 16.18 Examples 16.1 and 16.2.

$\vec{c}$

❶ **GETTING STARTED** I begin by sketching the pulse at the two instants given in the problem, pretending that the fixed end is not there and no reflection has occurred (**Figure 16.19**). I use a dashed line for the pulse to indicate that this is not the actual shape of the string at these instants (because reflection *has* occurred), and I draw a vertical line to indicate the (horizontal) position of the fixed end.

Figure 16.19

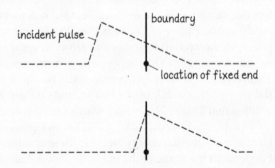

incident pulse

boundary

location of fixed end

❷ **DEVISE PLAN** Once any part of the pulse reaches the fixed end, reflection begins. To determine the pulse shape during reflection, I must draw a reflected pulse that is both inverted and reversed relative to the incident pulse and then add that reflected pulse to my incident pulses in Figure 16.19.

❸ **EXECUTE PLAN** (*a*) At the fixed end, the sum of the displacement of the incident pulse and the displacement of the reflected pulse is zero. In the region where the two pulses do not overlap, the sum of the displacements of the two pulses is equal to the displacement of the trailing edge of the incident pulse. Between this no-overlap region and the fixed end, I add the dashed and dotted lines algebraically to determine the shape of the string. The resulting shape is shown in **Figure 16.20a**. ✔

(*b*) Following the same procedure yields Figure 16.20*b*. ✔

Figure 16.20

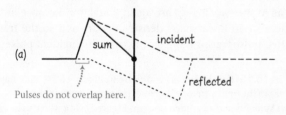

(*a*)   sum   incident

Pulses do not overlap here.   reflected

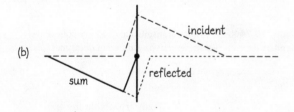

(*b*)   incident

sum   reflected

❹ **EVALUATE RESULT** The pulses I drew terminate at the node at the fixed end, as expected. In Figure 16.20*b* the string shape resembles the reflected pulse more than the incident pulse because most of the incident pulse has traveled past the fixed end.

Next consider reflection from a free end. To maintain the tension in the string at the free end, we connect the end to a light ring that can slide freely along a vertical rod (**Figure 16.21**). When it reaches a height equal to the maximum height of the pulse, however, the free end keeps moving upward, as you can see in Figure 16.21. The reason for this overshooting is that there are no string particles to the right of the free end exerting a downward force. As it overshoots, the free end exerts an upward force on the string to its left, pulling the string above the maximum height of the pulse. After reaching its maximum height, the ring moves back down to the equilibrium level. This up and down motion of the ring creates a left-moving pulse in the string, just as the up and down motion of the hand at the left created the original pulse. This leftward-moving pulse is the reflection of the original pulse. Although the reflected pulse is left–right reversed, it is not inverted.

To construct the reflected pulse, we follow the same procedure we used to construct a pulse reflected from a fixed end, but we do not invert the reflected pulse.

Figure 16.21 Reflection of a wave pulse at a boundary where the string end is free to move vertically.

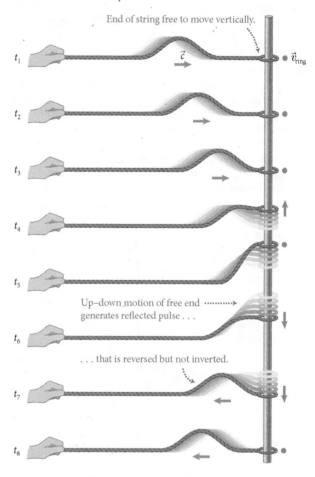

End of string free to move vertically.

t_1 $\vec{c}$ $\vec{v}_{ring}$

t_2

t_3

t_4

t_5

Up–down motion of free end generates reflected pulse . . .

t_6

. . . that is reversed but not inverted.

t_7

t_8

Example 16.2 Reflection from a free end

Let the triangular wave pulse of Figure 16.18 approach the free end of a string. Sketch the shape of the string (*a*) when a point halfway up the leading edge of the incident pulse has reached the fixed end and (*b*) when the peak of the pulse has reached the fixed end.

❶ **GETTING STARTED** I begin by sketching the pulse at the two instants given in the problem, ignoring the free end (**Figure 16.22**). I use a dashed line for the pulse to indicate that this is not the actual shape of the string at these instants, and I draw a vertical line to indicate the (horizontal) position of the free end of the string.

Figure 16.22

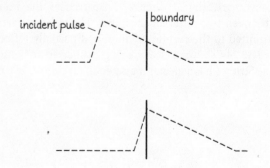

incident pulse   boundary

❷ **DEVISE PLAN** Once any part of the pulse reaches the free end, reflection begins. To determine the pulse shape during reflection, I must draw a reflected pulse that is reversed but not inverted and then add that reflected pulse to my incident pulses in Figure 16.22.

❸ **EXECUTE PLAN** (*a*) At the free end, the displacement of the reflected pulse is the same as that of the incident pulse, and so the string displacement is double what it was before the incident pulse arrived. In the region where the two pulses do not overlap, the sum of the displacements of the two pulses is equal to the displacement of the trailing edge of the incident pulse. Between this no-overlap region and the free end, I add the dashed and dotted lines algebraically to determine the shape of the string shown in **Figure 16.23a**. ✔

(*b*) Following the same procedure yields Figure 16.23*b*. ✔

Figure 16.23

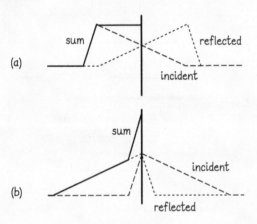

❹ **EVALUATE RESULT** Once the incident pulse arrives at the free end, the displacement of the free end at any given instant is twice that of the incident pulse, and therefore the free end swings far up in Figure 16.23*b*, in agreement with what I expect from Figure 16.21.

✋ **16.15** (*a*) When the free end of the string in Figure 16.21 reaches its maximum displacement at t_5, what is the kinetic energy in the pulse? (*b*) What is the potential energy in the pulse? (*c*) Is the energy in the pulse at this instant the same as the energy at t_1?

A case intermediate between a boundary where the end of the string is fixed and a boundary where it is free to move vertically is when a pulse reaches the boundary between two different mediums. The pulse is partially transmitted to the second medium and partially reflected. **Figure 16.24a** shows a heavy string connected to a lighter string. String 1 has a greater mass per unit length—a quantity

called the **linear mass density** and represented by the symbol μ—than string 2. The wave speed along string 1 is c_1, and that along string 2 is c_2. The tensions in the two strings are equal, but because of the difference in mass density, $c_2 > c_1$. Because string 2 is much less massive than string 1, the end of string 1 behaves somewhat like a free end, and so the reflected pulse is not inverted.

Figure 16.24*b* illustrates a pulse propagating along the same string 1 toward a string 2 that has a greater linear mass density. String 2 restricts the movement of the end of string 1, so now the boundary behaves like a fixed end, reflecting an inverted pulse back along string 1. The tensions in the two strings are again equal, but because of the difference in linear mass density, $c_2 < c_1$ and so the transmitted pulse travels more slowly than the incident pulse.

✋ **16.16** For the strings and pulse in Figure 16.24, what happens in the limit where (*a*) $\mu_2 \to 0$, (*b*) $\mu_2 \to \infty$, and (*c*) $\mu_2 = \mu_1$? (*d*) Why is the transmitted pulse in Figure 16.24*a* wider than the incident pulse?

Figure 16.24 Reflection and transmission of a wave pulse at the boundary between two strings that have different linear mass densities.

(*a*) Pulse propagates into string of smaller mass density: $\mu_1 > \mu_2$, so $c_2 > c_1$

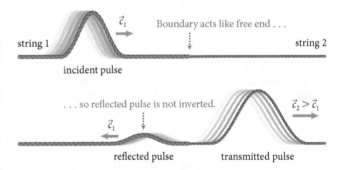

(*b*) Pulse propagates into string of greater mass density: $\mu_1 < \mu_2$, so $c_2 < c_1$

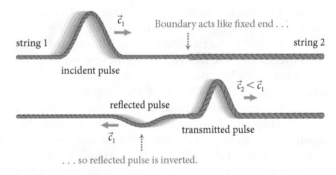

Self-quiz

1. Given the wave function for $t = 0.50$ s shown in **Figure 16.25**, draw the displacement curves of particles at $x = 0$ and $x = 1.0$ m as the wave passes them. If the wave pulse was generated at $x = 0$, what is the speed of the wave?

Figure 16.25

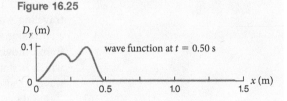

2. Given the displacement curve shown in **Figure 16.26** for a bead at $x = 0$ on a string as a wave with a speed of 2.0 m/s passes, draw the wave function at $t = 2.0$ s.

Figure 16.26

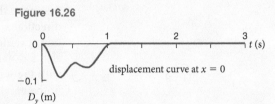

3. The two pulses shown in **Figure 16.27** are traveling in opposite directions with a speed of 1.0 m/s. Draw the sum of the two pulses at $t = 1.0$ s, $t = 2.0$ s, and $t = 3.0$ s.

Figure 16.27

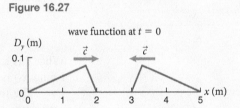

Answers

1. See **Figure 16.28**. The leading edge of the wave travels 0.50 m in 0.50 s, giving a wave speed of 1.0 m/s. In the displacement for the particle at $x = 0$, the leading edge passes $x = 0$ at $t = 0$ and the trailing edge passes $x = 0$ at 0.50 s. In the displacement curve for the particle at $x = 1.0$ m, the leading edge passes $x = 1.0$ m at $t = 1.0$ s and the trailing edge passes $x = 1.0$ m at 1.5 s.

Figure 16.28

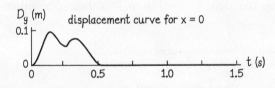

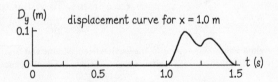

2. See **Figure 16.29**. With a wave speed of 2.0 m/s and 1.0 s needed for the whole wave to pass $x = 0$, the length of the wave must be 2.0 m. At $t = 1.0$ s, the leading edge of the wave passes $x = 2.0$ m. At $t = 2.0$ s, the leading edge passes $x = 4.0$ m and the trailing edge passes 2.0 m.

Figure 16.29

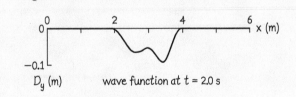

3. See **Figure 16.30**. The two pulses start out 1.0 m apart, and each has a width of 2.0 m. With a relative speed of 2.0 m/s, they begin to interfere at $t = 0.5$ s. At $t = 1.0$ s, the pulses overlap by half their width, with the trailing edges sticking out. At $t = 2.0$ s, the pulses also overlap by half their width, but now the leading edges are sticking out. At $t = 3.0$ s, the waves are separated again.

Figure 16.30

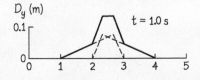

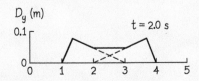

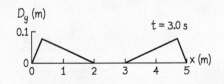

16.5 Wave functions

The displacement of the medium caused by a traveling wave is a function of both space and time, and so the mathematical description of a one-dimensional traveling wave requires a function in two variables. Let us begin by examining a traveling wave of constant shape, looking at the pulses first from a reference frame moving along with the wave and then from the Earth reference frame, which is stationary relative to the medium. In the moving reference frame, the wave is stationary and so its description is simple.

Figure 16.31*a* shows a transverse wave on a string at instant $t = 0$ when the origins of the two reference frames coincide. The wave travels in the positive x direction at speed c without changing its shape. Figure 16.31*b* shows the situation at some instant t, when the wave and the moving reference frame have traveled a distance ct to the right. As seen from the reference frame moving along with the wave, the wave remains stationary (Figure 16.31*c* and *d*). If the curve in the moving reference frame is described by some function $f(x_M)$, then the y component of the displacement $\vec{D}_M$ of any particle of the string as seen from the moving reference frame is given by

$$D_{My} = f(x_M). \tag{16.1}$$

What this tells us is that when we view the situation from the moving reference frame, x_M is the only variable we need to specify to determine the displacement D_{My}. The function $f(x_M)$, which specifies the (constant) shape of the wave as seen from the moving reference frame, is called the *time-independent wave function*.

When we view from the Earth reference frame, the wave has the same shape as in the moving reference frame but it is moving in the positive direction along the x axis. We must therefore specify both x and t in order to determine the displacement $\vec{D}$, and so we write $D_y = f(x, t)$. The function $f(x, t)$, called the *time-dependent wave function*, completely specifies the changing shape of the wave disturbance in the medium as seen from the Earth reference frame. To visualize this function, we need to make a three-dimensional

Figure 16.31 Traveling wave pulses seen from the Earth reference frame and from a reference frame moving along with the pulse at instant $t = 0$ (when the origins of the two reference frames overlap) and at some instant t later.

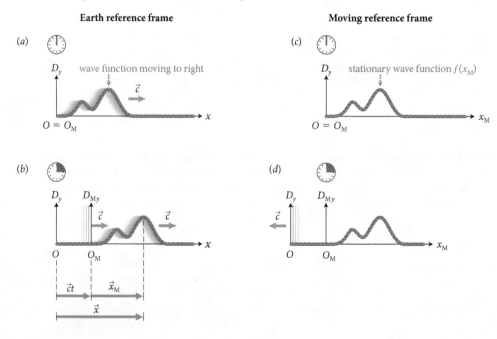

Figure 16.32 Time-dependent wave function.

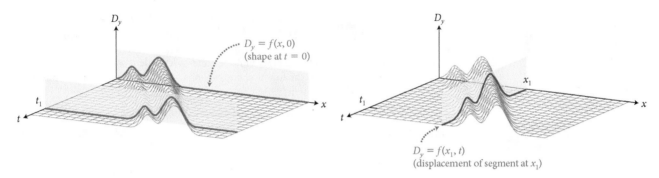

(a) Shape of the medium at instants $t = 0$ and $t = t_1$

(b) Displacement of the medium at a fixed position as a function of time

$D_y = f(x, 0)$
(shape at $t = 0$)

$D_y = f(x_1, t)$
(displacement of segment at x_1)

graph of $f(x, t)$ versus the variables x and t (**Figure 16.32a**). The surface shown in the graph gives the value of $D_y = f(x, t)$ for each pair of values of x and t. A vertical cut through the surface along the x axis yields the curve labeled $D_y = f(x, 0)$. This curve gives the shape of the wave disturbance at $t = 0$. A vertical cut through the surface parallel to the x axis intercepting the t axis at t_1 gives $D_y = f(x, t_1)$, the shape of the wave disturbance at t_1. As you can see, the wave has advanced in the positive x direction. Vertical cuts parallel to the t axis give the displacement at a fixed position in the medium as a function of time (Figure 16.32b). The curve $f(x_1, t)$, for example, shows how the displacement at x_1 changes with time.

The surface in Figure 16.32 is completely determined by the constant shape shown in Figure 16.31c and the wave speed c. To see how this is true, imagine sliding the curve in Figure 16.31c diagonally across the xt plane in Figure 16.32a. Because three-dimensional graphs are difficult to draw and to read, we shall use two-dimensional graphs to represent traveling waves that are constant in shape.

Let us now relate the displacement viewed from a moving reference frame to the displacement viewed from the Earth reference frame. At instant t, the y component of the displacement of any point P on the three-dimensional surface of an $f(x, t)$ graph is given by $D_y = f(x, t)$ in the Earth reference frame and by $D_{My} = f(x_M)$ in the moving reference frame (Figure 16.31b). Because the displacement of the string particles does not depend on which reference frame we use, we must have $D_y = D_{My}$. From Figure 16.31b we see that $x_M = x - ct$, so

$$f(x_M) = f(x - ct). \tag{16.2}$$

Substituting these relationships into Eq. 16.1, we discover that the y component of the displacement viewed from the Earth reference frame is given by

$$D_y = f(x - ct). \tag{16.3}$$

This equation describes a wave of constant form $f(x)$ traveling in the positive x direction at speed c. For a wave traveling in the negative x direction at speed c, we have

$$D_y = f(x + ct). \tag{16.4}$$

Any wave traveling along the x axis has either the form $f(x - ct)$ or the form $f(x + ct)$. Provided the quantities x, c, and t appear in the combination $(x - ct)$ or $(x + ct)$ in a function, that function can represent a traveling wave. Thus, the function $\sin[k(x - ct)]$, where k is a constant, represents a harmonic wave that travels in the positive direction along the x axis. The function $\sin[k^2(x^2 - c^2t^2)]$, however, does not represent a traveling harmonic wave.

QUANTITATIVE TOOLS

Exercise 16.3 Traveling waves

(a) Consider the time-dependent wave function

$$D_y = f(x, t) = \begin{cases} b(x - ct) & \text{for } 0 < x - ct \leq 1.0 \text{ m} \\ 0 \text{ for} & x - ct \leq 0 \quad \text{or} \quad x - ct > 1.0 \text{ m}, \end{cases}$$

where $b = 0.80$ and $c = 2.0$ m/s. Plot the time-independent wave function for a few values of t to verify that the function corresponds to a wave traveling in the positive x direction at a speed of 2.0 m/s.

(b) Let the shape of a wave at $t = 0$ be described by the function

$$f(x, 0) = \frac{a}{x^2 + b}.$$

If the wave travels in the negative x direction at a speed c, what is the mathematical form of the time-dependent wave function $f(x, t)$?

SOLUTION (a) Because I need to plot the function versus x, I first add ct to each term in the inequality $0 < x - ct \leq 1.0$ m and get $ct < x \leq (1.0 \text{ m}) + ct$. For $t = 0$, the function is nonzero when $0 < x \leq 1.0$ m; for $t = 1.0$ s when $2.0 \text{ m} < x \leq 3.0$ m; for $t = 2.0$ s when $4.0 \text{ m} < x \leq 5.0$ m; and so on. So, for $t = 0$ I plot the function $(0.80)x$ between $x = 0$ and $x = 1.0$ m; for $t = 1.0$ s I plot the function $0.80(x - 2.0 \text{ m})$ between $x = 2.0$ m and $x = 3.0$ m; and so on (**Figure 16.33**). My graphs show a triangular wave that is constant in shape and displaced in the positive x direction by $+2.0$ m each second. ✔

Figure 16.33

Region satisfying inequality $0 < x - (2.0 \text{ m/s})t \leq 1.0 \text{ m}$

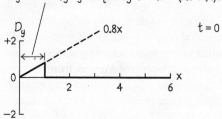

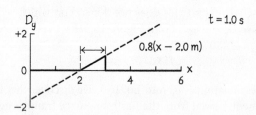

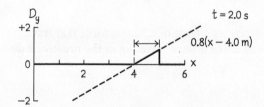

(b) When viewed from a reference frame traveling along with the wave, the wave function does not depend on time and has the form

$$f(x_M) = \frac{a}{x_M^2 + b}.$$

Equation 16.2 gives me the relationship between the displacement viewed from the Earth reference frame and one moving along with the pulse in the positive x direction. Because the pulse in part b of this exercise moves in the negative direction, I have to invert the sign in front of the speed c, so $f(x + ct) = f(x_M)$ and $x_M = x + ct$. Substituting these relationships into $f(x_M)$, I obtain the time-dependent wave function

$$f(x, t) = \frac{a}{(x + ct)^2 + b}. \checkmark$$

Let us consider a transverse harmonic wave traveling along a string aligned with an x axis. Such a wave can be generated by making the end of the string execute a simple harmonic motion. **Figure 16.34a** shows the shape of the string at an instant at which the left end, at $x = 0$, is in the equilibrium position. For convenience, we choose this instant to be $t = 0$. The time-independent wave function is

$$f(x) = A \sin(kx), \tag{16.5}$$

where A is the amplitude of the wave and k is a constant. The wave repeats itself over a distance λ (the wavelength), so the displacement $f(x)$ is the same at both ends of a wavelength—that is, at x and $x + \lambda$:

$$A \sin(kx) = A \sin(kx + k\lambda), \tag{16.6}$$

Because a sine function repeats itself when its argument is increased by 2π, it follows from Eq. 16.6 that $k\lambda = 2\pi$, so

$$k = \frac{2\pi}{\lambda}. \tag{16.7}$$

The constant k, called the **wave number** of the motion, thus gives the number of wavelengths in a length of 2π m. Because wavelength has units of meters, the wave number k has SI units of m^{-1}.

If the wave travels at speed c in the positive x direction, the time-dependent wave function is obtained by replacing x in Eq. 16.5 by $(x - ct)$:

$$D_y = f(x, t) = A \sin[k(x - ct)]. \tag{16.8}$$

Figure 16.34 Wave function of a harmonic wave (a) at $t = 0$ and (b) after one period. (c) The corresponding displacement curve for $x = 0$.

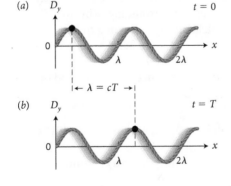

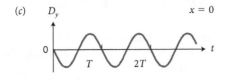

Figure 16.34*b* shows the wave function after the string end at $x = 0$ has executed one complete cycle (that is, at $t = T$, where T is the period of the motion). During this cycle, the wave has advanced by a distance corresponding to one wavelength. To see what this means, compare the positions of the black dots in Figure 16.34*a* and *b*. Think of these two dots as representing the same point on the traveling wave. In the time interval from $t = 0$ to $t = T$, this point of the wave has moved a distance λ. (Do *not* think of the dots as representing the same particle *of the string* because, remember, the medium does not move horizontally in this wave motion; only the momentum and energy associated with the wave move horizontally.) Because the wave moves at wave speed c, this distance corresponds to

$$\lambda = cT \tag{16.9}$$

or, substituting the frequency $f \equiv 1/T$,

$$\lambda f = c. \tag{16.10}$$

Because the period of the motion is T, we have for $x = 0$ $\sin(kct) = \sin[kc(t + T)]$. This can be true only if $kcT = 2\pi$, or

$$kc = \frac{2\pi}{T} \equiv \omega, \tag{16.11}$$

where ω is the angular frequency as defined in Section 15.5: $\omega = 2\pi f = 2\pi/T$. Using the angular frequency, we can rewrite the time-dependent wave function in Eq. 16.8 in the form:

$$D_y = f(x, t) = A \sin(kx - \omega t). \tag{16.12}$$

Let us now analyze the vertical displacement of the string as the wave passes through. We can use Eq. 16.12 to examine the vertical displacement of any particle of the string as a function of time. For instance, substituting $x = 0$ into Eq. 16.12 and using the trigonometric identity $\sin(-\alpha) = -\sin \alpha$, we obtain

$$D_y = f(0, t) = A \sin(-\omega t) = -A \sin(\omega t). \tag{16.13}$$

The displacement curve for this function $D_y(0, t)$ is shown in Figure 16.34*c*. (If you are having trouble understanding why the displacement curve dips downward initially, think about what is going to happen in Figure 16.34*a* to the string particle located at $x = 0$ just after $t = 0$: It is going to be pulled down below the *x* axis.)

Next we consider a point a distance *d* from the end of the string. Substituting $x = d$ into Eq. 16.12 yields

$$D_y = f(d, t) = A \sin(kd - \omega t) = A \sin(-\omega t + kd) \tag{16.14}$$

or, using the trigonometric identity $\sin(-\alpha) = \sin(\alpha - \pi)$,

$$D_y = f(d, t) = A \sin(\omega t - kd - \pi) = A \sin(\omega t + \text{constant}). \tag{16.15}$$

This expression, which gives the *y* component of the displacement of the string at any arbitrary position $x = d$, shows that all points along the string execute a simple harmonic motion with a period $2\pi/\omega = T$.

Figure 16.35 Any wave can be expressed in terms of sinusoidally varying (harmonic) waves.

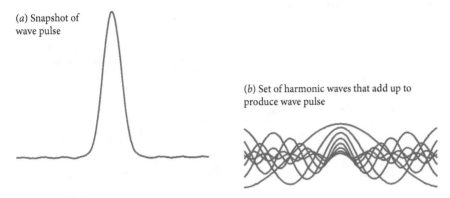

(*a*) Snapshot of wave pulse

(*b*) Set of harmonic waves that add up to produce wave pulse

In deriving Eq. 16.8, we assumed that $f = 0$ at $x = 0$ and $t = 0$. To obtain a more general expression that allows for a nonzero initial displacement, we add to the argument of the sine in Eq. 16.12 (the *phase* of the wave, see Section 15.5) an initial phase ϕ_i:

$$D_y = f(x, t) = A \sin(kx - \omega t + \phi_i). \qquad (16.16)$$

The initial phase specifies the initial condition: At $x = 0$ and $t = 0$, Eq. 16.16 becomes $D_y = f(0, 0) = A \sin \phi_i$, and so, for a given amplitude, ϕ_i determines the initial value of the vertical displacement at $x = 0$.

In Section 15.3 we discussed Fourier's theorem, which states that the time dependence of any function can be expressed in terms of sinusoidally varying functions of time. Because of the superposition principle, the same theorem can be applied to waves: Any wave can be expressed in terms of sinusoidally varying waves (harmonic waves). **Figure 16.35a**, for example, shows a wave pulse obtained by adding together the harmonic waves shown in Figure 16.35*b*. At the center, all of the harmonic waves add up to form the pulse; everywhere else, they add up to nearly zero (surprising, but true!). So, if we understand harmonic waves, we can deal with more complicated waves by determining the harmonic waves that add up to give these waves.

 16.17 (*a*) Which of the following functions could represent a traveling wave?

    (*i*) $A \cos(kx + \omega t)$ (*ii*) $e^{-k|x - ct|^2}$

    (*iii*) $b(x - ct)^2 e^{-x}$ (*iv*) $-(b^2 t - x)^2$

(*b*) Which of the following functions can be made into a traveling wave?

    (*i*) $x/(1 + bx^2)$ (*ii*) xe^{-kx} (*iii*) x^2

16.6 Standing waves

When a harmonic wave travels along a string that has a fixed end, the reflected wave interferes with the incident wave. **Figure 16.36** shows the striking pattern that results from this interference. Initially the wave travels to the right, but as the reflected wave begins traveling back at t_3, the two waves traveling in opposite directions create a pulsating *stationary* pattern. Some particles of the string—the nodes (see Section 16.3)—do not move at all, while the sinusoidal loops between the nodes move up and down rather than along the string. The motion has its greatest amplitude at points halfway between the nodes; these points are called **antinodes.** The pulsating stationary pattern caused by harmonic waves of the same amplitude traveling in opposite directions is called a **standing wave.**

Figure 16.36 When a harmonic wave is reflected from the fixed end of a string, the reflected wave interferes with the incident wave, forming a standing wave.

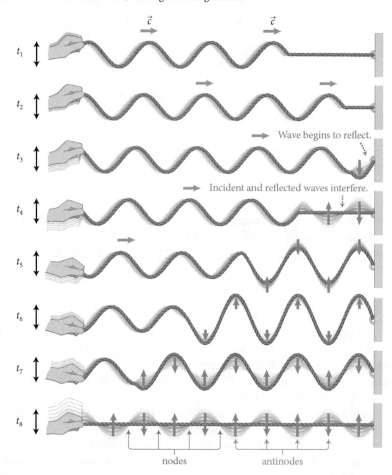

Figure 16.37 illustrates how standing waves come about. Each diagram shows a blue wave propagating to the right and a red wave propagating to the left. The two waves have the same amplitude and wavelength. The speed at which they propagate is determined by the medium and is therefore the same for both waves. The bottom part of each diagram shows the sum of the two waves, obtained by adding, for each value of x, the vertical displacements of the two waves at that value of x. In the diagram labeled $t = 0$, the crests of the red wave have the same x coordinate as the crests of the blue wave, and the same is true for the troughs: Each trough in the red wave aligns with a trough in the blue wave. Waves like this, where comparable points align, are said to be *in phase* (their phases are the same). The two waves reinforce each other—they interfere constructively—and adding them yields a wave that has twice the amplitude. A quarter period later, at $t = \frac{1}{4}T$, one has a crest when the other has a trough and the waves are 180° *out of phase*. The two waves now interfere destructively, and the string is flat at this instant.

16.18 (*a*) In the standing wave pattern of Figure 16.37, how is the energy distributed between kinetic and potential at $t = 0$, $t = \frac{1}{8}T$, and $t = \frac{1}{4}T$? (*b*) Is the energy in a length of the string corresponding to one wavelength constant? (*c*) Does the standing wave transport energy? If so, in which direction? If not, why not?

To see why certain points are nodes, let us denote the y components of the displacements of the red and blue waves by $D_{1y} = f_1(x, t)$ and $D_{2y} = f_2(x, t)$, respectively, and look at what happens at the points $x = \lambda/2$ and $x = \lambda$. At

Figure 16.37 Standing wave created by two counterpropagating waves f_1 and f_2.

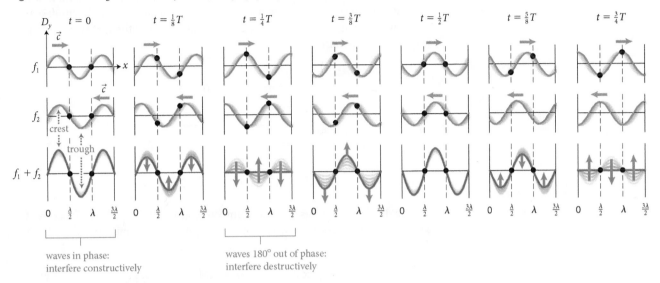

waves in phase:
interfere constructively

waves 180° out of phase:
interfere destructively

$t = 0$, we have $D_{1y} = D_{2y} = 0$, and so $D_{1y} + D_{2y} = 0$. Now consider the situation at $t = \frac{1}{8}T$. Wave f_1 has advanced to the right, while wave f_2 has moved an equal distance to the left. At $\lambda/2$, the vertical displacement of the string caused by wave f_1 is now positive, but that caused by wave f_2 is negative and equal in magnitude, so the two add to zero and the string does not move from its equilibrium position. At $x = \lambda$ in the $t = \frac{1}{8}T$ diagram, we see the same thing but with the displacement directions reversed: f_1 displaces the string in the negative y direction, and f_2 displaces it by the same amount in the positive y direction. Because both waves always travel equal distances in opposite directions, the displacements caused by them at the points $x = \lambda/2$ and $x = \lambda$ are *always* equal in magnitude and opposite in sign. Consequently $D_{1y} + D_{2y}$ remains zero at these points for all t.

We can analyze the situation quantitatively by writing the wave traveling to the right as

$$D_{1y} = f_1(x, t) = A \sin(kx - \omega t) \qquad (16.17)$$

and the wave traveling to the left as

$$D_{2y} = f_2(x, t) = A \sin(kx + \omega t). \qquad (16.18)$$

The combined wave then is

$$D_y = f_1(x, t) + f_2(x, t) = A[\sin(kx - \omega t) + \sin(kx + \omega t)]. \qquad (16.19)$$

Using the trigonometric identities $\sin \alpha + \sin \beta = 2 \sin\frac{1}{2}(\alpha + \beta) \cos\frac{1}{2}(\alpha - \beta)$ and $\cos \alpha = \cos(-\alpha)$ (see Appendix B), we can write Eq. 16.19 as

$$D_y = f_1(x, t) + f_2(x, t) = 2A \sin kx \cos \omega t = [2A \sin kx]\cos \omega t. \qquad (16.20)$$

This is *not* a traveling wave because it is not a function of $x - ct$ or of $x + ct$ (see the text immediately following Eq. 16.4). Instead this function describes a standing wave on a string, one in which the string displacement oscillates harmonically as $\cos \omega t$ with an amplitude that varies along x as $2A \sin kx$ (Figure 16.38).

Figure 16.38 A standing wave is a stationary harmonic wave with an amplitude that varies like a sine. The spacing between adjacent nodes is equal to $\lambda/2$.

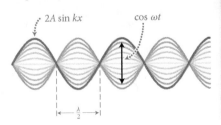

2A sin kx cos ωt

For a standing wave of the form given in Eq. 16.20 the displacement is *always* zero at the locations where $\sin kx$ is zero. These points are the nodes; they occur when kx is a whole-number multiple of π. Using the fact that $k = 2\pi/\lambda$ (Eq. 16.7), we obtain

$$\frac{2\pi}{\lambda}x = n\pi \qquad n = 0, \pm 1, \pm 2, \ldots, \tag{16.21}$$

and so nodes occur at

$$x = 0, \pm\frac{\lambda}{2}, \pm\lambda, \pm\frac{3\lambda}{2}, \ldots. \tag{16.22}$$

The displacement has its maximum value at the points where $\sin kx = \pm 1$. These antinodes occur when kx is an odd whole-number multiple of $\pi/2$:

$$\frac{2\pi}{\lambda}x = \frac{n\pi}{2} \qquad n = \pm 1, \pm 3, \pm 5, \ldots, \tag{16.23}$$

and so antinodes occur at

$$x = \pm\frac{\lambda}{4}, \pm\frac{3\lambda}{4}, \pm\frac{5\lambda}{4}, \ldots. \tag{16.24}$$

Note that, as shown in Figure 16.38, adjacent nodes are separated by a distance $\lambda/2$. Adjacent antinodes are also spaced by a distance $\lambda/2$ and lie midway between the nodes.

16.19 (*a*) Do two counterpropagating waves that have the same wavelength but different amplitudes cause standing waves? (*b*) Do two counterpropagating waves that have the same amplitude but different wavelengths cause standing waves?

16.7 Wave speed

To determine a quantitative expression for the wave speed c, consider the situation shown in **Figure 16.39**. The end of a taut horizontal string is moved vertically upward at a constant velocity $\vec{v}$. The displacement of the end causes a triangular wave pulse to propagate along the string. The bend on the leading edge travels in the horizontal direction with wave speed c. Just as with the string of beads in Section 16.2, all segments of the string execute the same motion as the end of the string: Each segment begins at rest and then, as the pulse reaches it, moves vertically upward at speed v.

Let us consider the motion of the string during a small time interval Δt beginning at some instant t. At the beginning of the interval, the end of the string has been raised a distance vt and the bend has traveled a distance ct (Figure 16.39b). The slanted segment of the string moves vertically up at constant speed v; at the instant represented in Figure 16.39b, segment A moves up while the rest of the string remains horizontal. A short time interval Δt later, segment A has moved farther up and is still moving at v, so the momentum of segment A has not changed.

Let us choose the x axis in the direction of the wave and the y axis in the upward direction. During the time interval Δt, the bend has advanced by a distance $\Delta x = c\Delta t$, and so an additional length $c\Delta t$ of string (segment B) now also moves vertically upward with speed v. The mass of this segment is $\mu(c\Delta t)$,

Figure 16.39 Procedure for obtaining a quantitative expression for the wave speed *c*.

(*a*) Triangular wave pulse generated by lifting end of string at constant velocity

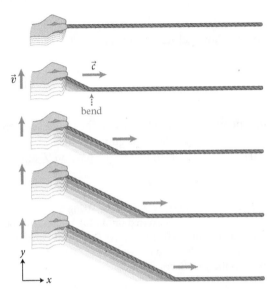

(*b*) In time interval Δ*t*, bend advances distance *c*Δ*t*

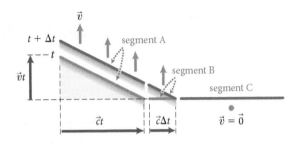

(*c*) Free-body diagram for segment B

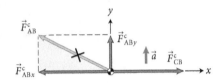

where μ is the mass per unit length, or **linear mass density,** of the string. For a uniform string of length ℓ and mass m, the linear mass density is

$$\mu \equiv \frac{m}{\ell} \quad \text{(uniform linear object).} \qquad (16.25)$$

The SI units of linear mass density are kg/m. The change in the momentum of segment B is equal to the mass of B times its change in velocity $\Delta\vec{v}_B = \vec{v} - 0 = \vec{v}$. The mass of B is equal to the linear mass density of the string times the equilibrium length of B, $m_B = \mu(c\Delta t)$, and so $\Delta\vec{p}_B = m_B\Delta\vec{v}_B = \mu(c\Delta t)\vec{v}$. This change in momentum occurs during the time interval Δt, and so the rate of change is

$$\frac{\Delta\vec{p}_B}{\Delta t} = \frac{\mu(c\Delta t)\vec{v}}{\Delta t} = \mu c\vec{v}. \qquad (16.26)$$

The rate of change in momentum of segment B of the string is by definition equal to the vector sum of the forces exerted on it:

$$\sum\vec{F}_B = \frac{\Delta\vec{p}_B}{\Delta t} = \mu c\vec{v}. \qquad (6.27)$$

Figure 16.39*c* shows the free-body diagram for segment B. Segment A exerts on B a force slanting upward along the direction of segment A; segment C exerts on B a force in the horizontal direction. The *x* components of these two forces must balance because segment B is not accelerating in the horizontal direction. We also know that the *x* components of these two forces are equal in magnitude to the tension $\mathcal{T}$ in the string. The vector sum of the forces exerted on B is thus equal to the vertical component of the force exerted by A on B. Finally, we note that the shaded triangles in Figure 16.39*b* and *c* are similar, so

$$\frac{F^c_{ABy}}{F^c_{ABx}} = \frac{F^c_{ABy}}{\mathcal{T}} = \frac{vt}{ct} = \frac{v}{c}, \qquad (16.28)$$

and thus the vector sum of the forces acting on segment B of the string is

$$\Sigma F_y = F^c_{ABy} = \frac{Tv}{c}.$$ (16.29)

Combining this result with Eq. 16.27, we get $\mu cv = Tv/c$, or

$$c = \sqrt{\frac{T}{\mu}}.$$ (16.30)

Note that the wave speed is independent of the velocity v of the string segments. The result in Eq. 16.30 agrees with the qualitative arguments in Section 16.2: The speed of a pulse along a string increases with increasing tension T and with decreasing linear mass density.

16.20 (a) Draw a free-body diagram for segment A of the string at instant t in Figure 16.39. What is the vector sum of the forces exerted on A? (b) What does your answer to part a tell you about the change in momentum of A? Does your conclusion about A's momentum make sense? (c) Does the hand do work on the string? (d) What happens to the energy of the string as the end moves upward?

Example 16.4 Measuring mass

You use a hammer to give a sharp horizontal blow to a 10-kg lead brick suspended from the ceiling by a wire that is 5.0 m long. It takes 70 ms for the pulse generated by the sudden displacement of the brick to reach the ceiling. What is the mass of the wire?

❶ **GETTING STARTED** I begin by making a sketch of the situation (**Figure 16.40**a). Because I know the pulse travels 5.0 m in 70 ms, I can determine the wave speed along the wire: $c = (5.0\text{ m})/(0.070\text{ s}) = 71\text{ m/s}$. This wave speed depends on

Figure 16.40

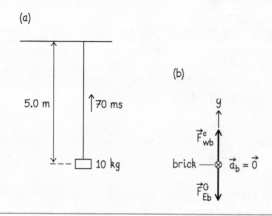

(a)

5.0 m   ↑70 ms

□ 10 kg

(b)

y

$\vec{F}^c_{wb}$ ↑

brick — ⊗ $\vec{a}_b = \vec{0}$

$\vec{F}^G_{Eb}$ ↓

the linear mass density of the wire and the tension in the wire. The latter is determined by the force exerted by the brick on the wire. Because the force exerted by the brick on the wire and the force exerted by the wire on the brick are equal in magnitude, I draw a free-body diagram for the brick (Figure 16.40b) to help me determine that magnitude.

❷ **DEVISE PLAN** To determine the mass of the wire, I can use Eq. 16.25. To use this equation, I must know the length of the wire, which is given, and the linear mass density μ, which I can calculate from Eq. 16.30. To obtain μ from Eq. 16.30, I need to know c (which I already calculated) and the tension T in the wire. This tension is equal to F^c_{bw}, which in turn is equal to F^c_{wb}. From my free-body diagram, I see that $F^c_{wb} = F^G_{Eb}$, and so I know F^c_{bw} and therefore T.

❸ **EXECUTE PLAN** The tension is equal to the downward force of gravity exerted on the brick: $T = F^G_{Eb} = (10\text{ kg})(9.8\text{ m/s}^2) = 98\text{ N}$. Substituting the values for T and c into Eq. 16.30, I calculate the linear mass density of the wire $\mu = T/c^2 = (98\text{ N})/(71\text{ m/s})^2 = 0.019\text{ kg/m}$. The mass of the wire is thus $m = (0.019\text{ kg/m})(5.0\text{ m}) = 9.5 \times 10^{-2}\text{ kg}.$ ✔

❹ **EVALUATE RESULT** The value I obtain—about 0.1 kg—is not unreasonable for a 5.0-m-long wire that can support a lead brick.

16.8 Energy transport in waves

Imagine moving the end of a very long, taut string up and down so that a harmonic wave begins traveling along the string at the wave speed c given in Eq. 16.30. Energy moves along with the wave, and so, as you shake the string end, you must supply energy to the string. At what rate must you supply this energy? To answer this question, let us first consider the energy involved in generating a simple triangular wave as shown in **Figure 16.41**.

As it raises the end of the string, the hand does work on the string. The work done by the hand on the string during a time interval Δt is equal to the scalar product of the force $\vec{F}_{hA}^c$ exerted by the hand on segment A of the string and the force displacement. From Figure 16.41a and b, we see that

$$W = F_{hAy}^c(v\Delta t), \tag{16.31}$$

where F_{hAy}^c is the y component of $\vec{F}_{hA}^c$ and $v\Delta t$ is the y component of the force displacement. If we ignore the force of gravity on segment A of the string, the force exerted by the hand on A is transmitted undiminished to B, and so $\vec{F}_{hA}^c$ is equal in magnitude to the force $\vec{F}_{AB}^c$ exerted by A on B. Using Eq. 16.29, we then have

$$F_{hAy}^c = \mathcal{T}\frac{v}{c}. \tag{16.32}$$

Substituting this result and $\mathcal{T} = \mu c^2$ from Eq. 16.30 into Eq. 16.31, we discover that the work done by the hand on the string during the time interval Δt is

$$W = \left(\mathcal{T}\frac{v}{c}\right)v\Delta t = \mu c\, v^2\Delta t. \tag{16.33}$$

The energy law, $\Delta E = W$ (Eq. 9.1), tells us that the work done by the hand on the string changes the energy of the string. The average rate at which the hand supplies energy to the string (the *average power*) is equal to the change in the energy of the string divided by the time interval over which the energy changes (Eq. 9.29), and so $P_{av} \equiv \Delta E/\Delta t = \mu c v^2$.

As the hand does work on the string, the kinetic energy and the potential energy of the string change: The kinetic energy changes because more of the string is set in motion, and the potential energy changes because the raising stretches the string, storing elastic potential energy. Let us first calculate the change in kinetic energy. After a time interval Δt, segment A has a velocity v, and so its kinetic energy is $\frac{1}{2}m_A v^2$, where m_A is the mass of A. The original length of segment A is $c\Delta t$ (Figure 16.41a), and so $m_A = \mu c\Delta t$. The change in kinetic energy is thus

$$\Delta K = \tfrac{1}{2}(\mu c\Delta t)v^2 - 0 = \tfrac{1}{2}\mu c v^2\Delta t, \tag{16.34}$$

which, Eq. 16.33 tells us, is exactly half the energy supplied to the string!

The other half must be potential energy associated with the stretching of segment A (the rest of the string is undisturbed). Let the length of A at the instant shown in Figure 16.41a be ℓ_A. The change in the length of the string is then $\Delta \ell = \ell_A - c\Delta t$. If $\Delta \ell$ is small, we can treat the string as a spring of spring constant k_s, and so Eq. 9.23 tells us the change in potential energy stored in the stretched string is

$$\Delta U = \tfrac{1}{2}k_s\Delta\ell^2 = \tfrac{1}{2}(k_s\Delta\ell)\Delta\ell = \tfrac{1}{2}F_{hA}^c\Delta\ell, \tag{16.35}$$

where we have used Hooke's law, Eq. 8.18, to substitute $F_{hA}^c = k_s\Delta\ell$. Because segment A is not accelerating, we see from the free-body diagram for segment A (Figure 16.41b) that $\vec{F}_{hA}^c = -\vec{F}_{BA}^c$. Because the force exerted by segment B on segment A and the force exerted by segment A on segment B form an interaction pair, we have $\vec{F}_{hA}^c = \vec{F}_{AB}^c$. Comparing the similar shaded triangles in Figure 16.41a and the free-body diagram for segment B (Figure 16.41c), we see that

$$\frac{F_{AB}^c}{F_{ABx}^c} = \frac{F_{hA}^c}{\mathcal{T}} = \frac{\ell_A}{c\Delta t} \tag{16.36}$$

Figure 16.41 Procedure for determining the energy involved in generating a simple triangular wave.

(*a*) Triangular wave pulse generated by lifting end of string at constant velocity

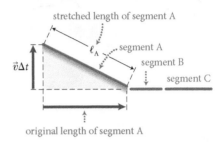

(*b*) Free-body diagram for segment A

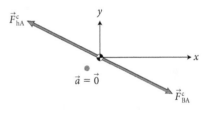

(*c*) Free-body diagram for segment B

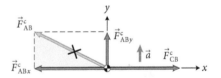

or $F_{hA}^c = T\ell_A/(c\Delta t)$, and so Eq. 16.35 becomes

$$\Delta U_A = \tfrac{1}{2}\frac{T\ell_A}{c\Delta t}\Delta\ell = \tfrac{1}{2}(\mu c^2)\frac{\ell_A}{c\Delta t}\Delta\ell = \tfrac{1}{2}\frac{\mu c}{\Delta t}\ell_A\Delta\ell, \qquad (16.37)$$

where we have used Eq. 16.30 to eliminate T. As long as the vertical displacement of the end of the string is small, $\ell_A \approx c\Delta t$, and so

$$\ell_A\Delta\ell = \ell_A(\ell_A - c\Delta t) = \ell_A^2 - \ell_A(c\Delta t) \approx \ell_A^2 - (c\Delta t)^2 \qquad (16.38)$$

or, when we apply the Pythagorean theorem to the shaded triangle in Figure 16.41a, $\ell_A\Delta\ell = (v\Delta t)^2$, and therefore

$$\Delta U = \tfrac{1}{2}\frac{\mu c}{\Delta t}(v\Delta t)^2 = \tfrac{1}{2}\mu c v^2\Delta t. \qquad (16.39)$$

So, in the limit of small displacement, the change in elastic potential energy is indeed equal to the change in kinetic energy, and the triangular pulse contains equal amounts of kinetic and potential energy.

Let us next consider the energy carried in a harmonic wave described by $y(x, t) = A\sin(kx - \omega t)$. We begin by calculating the amount of energy E_λ stored in a length of string corresponding to one wavelength of the harmonic wave. Because during each period T the wave advances by one wavelength, E_λ is the amount of energy that must be supplied in one period. The average power is then

$$P_{av} \equiv \frac{\Delta E}{\Delta t} = \frac{E_\lambda}{T}. \qquad (16.40)$$

To calculate E_λ, we first divide the string into very small segments, each of length dx. If the linear mass density of the string is μ, the mass of each segment is $dm = \mu\, dx$. Each segment carries out a simple harmonic motion, and the energy associated with this simple harmonic motion is $\tfrac{1}{2}(dm)\omega^2 A^2$ (Eq. 15.17). At any given instant, part of this energy is in the form of kinetic energy associated with the motion of the segment and part is in the form of potential energy associated with its deformation. The sum of the two, however, always equals $\tfrac{1}{2}(dm)\omega^2 A^2$. All the segments in a length of string one wavelength long add up to a mass of $\sum dm = m$, and from the definition of linear mass density, we have $m = \mu\lambda$. In a length of string one wavelength long, therefore, the combined energy contributed by all the segments of length dx is

$$E_\lambda = \tfrac{1}{2}(\mu\lambda)\omega^2 A^2. \qquad (16.41)$$

The average power that must be supplied is thus

$$P_{av} = \tfrac{1}{2}\mu\lambda A^2\omega^2/T = \tfrac{1}{2}\mu A^2\omega^2 c, \qquad (16.42)$$

where we have used the fact that $\lambda/T = c$ (Eq. 16.9). This result makes sense: It takes more effort to generate a wave of greater (angular) frequency or amplitude (in each case you must shake the end with a greater velocity). Furthermore, as we would expect, a string of greater linear mass density requires more energy. Finally, if the wave speed increases, the energy is carried away more quickly, and so the rate at which it needs to be supplied increases.

Example 16.5 Delivering energy

A wire with linear mass density $\mu = 0.0500$ kg/m is held taut with a tension of 100 N. At what rate must energy be supplied to the wire to generate a traveling harmonic wave that has a frequency of 500 Hz and an amplitude of 5.00 mm?

①　GETTING STARTED The rate at which energy must be supplied to the wire is the power. Because none of the quantities given varies with time, the (instantaneous) power is equal to the average power, which is given by Eq. 16.42.

②　DEVISE PLAN To calculate the average power, I need to know μ and A, both of which are given, as well as ω and c. The angular frequency ω is related to the frequency f by Eq. 15.4, $\omega = 2\pi f$, and I can obtain the wave speed c from Eq. 16.30.

③　EXECUTE PLAN From Eq. 16.30 I obtain $c = \sqrt{T/\mu} = \sqrt{(100\ \text{N})/(0.0500\ \text{kg/m})} = 44.7$ m/s. Equation 15.4 yields $\omega = 2\pi f = 2\pi(500\ \text{Hz}) = 3.14 \times 10^3\ \text{s}^{-1}$. Substituting these values into Eq. 16.42 gives

$$P_{\text{av}} = \tfrac{1}{2}(0.0500\ \text{kg/m})(0.00500\ \text{m})^2(3.14 \times 10^3\ \text{s}^{-1})^2(44.7\ \text{m/s})$$

$$= 275\ \text{W}.\ ✔$$

④　EVALUATE RESULT The answer I obtain is a fairly large power—comparable to the power delivered by a 250-W light bulb or by a person exercising. The wave travels very fast, though, and the frequency is high, so the answer is not unreasonable.

16.21 Suppose that instead of shaking the end of a very long, taut string, you shake a point in the center of the string, keeping the amplitude and the frequency the same. Is the rate at which you must supply energy smaller than, equal to, or greater than if you shake the end?

16.9 The wave equation

In Chapter 15 I argued that simple harmonic oscillations are common because linear restoring forces are common. Linear restoring forces give rise to the equation of motion for a simple harmonic oscillator, which in turn gives rise to sinusoidally oscillating solutions. Waves are equally common because whenever a medium is disturbed it tends to oscillate sinusoidally and the oscillation couples to neighboring regions of the medium. As we shall show in this section, systems of coupled simple harmonic oscillators give rise to an equation called the wave equation, whose solutions are traveling waves. Here we derive this equation for the special case of a transverse wave traveling along a string; we shall encounter the same equation again when discussing electromagnetic waves.

Figure 16.42a shows a piece of a string that has been displaced from equilibrium by a passing wave. Let us focus on segment B between the positions x_i and $x_f = x_i + \Delta x$ and write the equation of motion for that segment. The segment is subject to two forces, one from each of the two adjacent pieces of string (we again ignore the force of gravity because it is much smaller than the tension). Figure 16.42b shows the free-body diagram for segment B. As we saw in Section 16.2, the x components of the forces are equal in magnitude to the tension T in the string and add to zero. The vector sum of the forces exerted on B is thus equal to the sum of the y components:

$$\sum F_{By} = F^c_{ABy} + F^c_{CBy} = T \tan \theta_f - T \tan \theta_i, \tag{16.43}$$

where θ_i and θ_f are the directions of the tangents to the string at x_i and x_f. For small displacements, these angles are small, and so we can use the small-angle approximation $\tan \theta \approx \theta$ and rewrite Eq. 16.43 as

$$\sum F_{By} \approx T(\theta_f - \theta_i) = T\Delta\theta. \tag{16.44}$$

The equation of motion for segment B is

$$\sum F_{By} = m_B a_{By} = (\mu \Delta x)a_{By}, \tag{16.45}$$

where m_B is the mass of segment B, a_{By} is the y component of its acceleration,

Figure 16.42 Segment of a string transmitting a wave pulse.

(a) String transmitting wave pulse

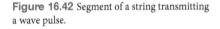
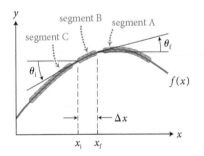

(b) Free-body diagram for segment B

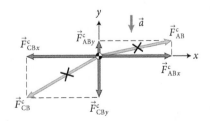

and μ is its linear mass density. Substituting $T\Delta\theta$ from Eq. 16.44 into the left side of Eq. 16.45 and rearranging terms yields

$$\frac{\Delta\theta}{\Delta x} = \frac{\mu}{T}a_{By}. \tag{16.46}$$

In the limit $\Delta x \to 0$, the left side of this expression yields the derivative of the angle θ with respect to x. The angle θ is determined by the function f that describes the shape of the string in Figure 16.42a—the derivative of f with respect to x is equal to the slope, $\tan\theta$. A complication is that the shape of the string changes with time and so is a function of both x and t. We therefore have to use the *partial derivative* of $f(x, t)$ with respect to x, which means that we hold t constant while taking the derivative with respect to x (see Appendix B). The partial derivative of $f(x, t)$ with respect to x is written $\partial f/\partial x$, so we have

$$\tan\theta = \frac{\partial f}{\partial x}. \tag{16.47}$$

For small θ we can write $\tan\theta \approx \theta$, and so, in the limit $\Delta x \to 0$, the left side of Eq. 16.46 becomes the second partial derivative of $f(x, t)$ with respect to x:

$$\lim_{\Delta x \to 0} \frac{\Delta\theta}{\Delta x} \equiv \frac{\partial\theta}{\partial x} \approx \frac{\partial}{\partial x}\left(\frac{\partial f}{\partial x}\right) \equiv \frac{\partial^2 f}{\partial x^2}. \tag{16.48}$$

Next we consider the right side of Eq. 16.46. The acceleration a_{By} is given by the second derivative with respect to time of the vertical position $y = f(x, t)$ of segment B. So the first partial derivative with respect to time $\partial f/\partial t$ gives the y component of the velocity of segment B, and the second partial derivative gives the y component of the acceleration:

$$a_{By} = \frac{\partial^2 f}{\partial t^2}. \tag{16.49}$$

Substituting Eqs. 16.48 and 16.49 into Eq. 16.46, we get

$$\frac{\partial^2 f}{\partial x^2} = \frac{\mu}{T}\frac{\partial^2 f}{\partial t^2}, \tag{16.50}$$

and substituting Eq. 16.30, $c = \sqrt{T/\mu}$, yields

$$\frac{\partial^2 f}{\partial x^2} = \frac{1}{c^2}\frac{\partial^2 f}{\partial t^2}. \tag{16.51}$$

This is the one-dimensional **wave equation.** Any function of the form $f(x - ct)$ or $f(x + ct)$ is a solution to this equation (the proof of this is left as a homework exercise). All waves, mechanical and nonmechanical, result from equations of this form (but only for a taut string do we get Eq. 16.50 with $c = \sqrt{T/\mu}$).

It is helpful to keep in mind what the terms in the wave equation tell us about the wave. The second partial derivative of f with respect to x on the left is related to the curvature of the string carrying the wave—the greater $\partial^2 f/\partial x^2$, the sharper the curvature of the string. The right side is the acceleration of particles of the string divided by the square of the wave speed. Thus the wave equation tells us that the curvature of the wave increases as the medium acceleration increases and decreases as the wave speed increases.

For a fixed wave speed, the greater the acceleration of the particles in the medium, the greater the curvature. A downward curvature (concave down or $\partial^2 f/\partial x^2 < 0$) corresponds to a downward acceleration ($\partial^2 f/\partial t^2 < 0$). An upward curvature (concave up or $\partial^2 f/\partial x^2 > 0$) corresponds to an upward acceleration

$(\partial^2 f/\partial t^2 > 0)$. This is shown graphically in **Figure 16.43**. In Figure 16.43a, the curvature of the string is greatest (and downward) at the top. As we have seen, the acceleration is also greatest (and downward) at that point (compare with Figure 16.7c). At point Q, the curvature is upward; indeed, as the pulse travels to the right, point Q has to accelerate in the upward direction. At point R, the curvature and acceleration are both zero. Another illustration of the acceleration-curvature relationship at fixed wave speed is seen by comparing Figure 16.43a and b: If we move the end of the string up and down more quickly, the acceleration at point P increases and the pulse becomes narrower, giving a greater curvature at P.

The inverse relationship between curvature and wave speed is seen by comparing Figure 16.43a and c. With P (as well as all other points) having the same acceleration in the two drawings, the greater wave speed in part c stretches the string horizontally, decreasing the curvature.

Exercise 16.6 Sinusoidal solution to the wave equation

Show that a sinusoidal traveling wave of the form $f(x, t) = A \sin(kx - \omega t)$ satisfies the wave equation for any value of k and ω.

SOLUTION The first partial derivative of $f(x, t)$ with respect to x is

$$\frac{\partial f}{\partial x} = kA \cos(kx - \omega t),$$

and so the second partial derivative with respect to x is

$$\frac{\partial^2 f}{\partial x^2} = -k^2 A \sin(kx - \omega t). \tag{1}$$

The first partial derivative of $f(x, t)$ with respect to t is

$$\frac{\partial f}{\partial t} = -\omega A \cos(kx - \omega t).$$

Differentiating again gives

$$\frac{\partial^2 f}{\partial t^2} = -\omega^2 A \sin(kx - \omega t).$$

Multiplying both sides of this equation by $1/c^2$ and using $k = \omega/c$ (Eq. 16.11), I obtain

$$\frac{1}{c^2}\frac{\partial^2 f}{\partial t^2} = -\frac{\omega^2}{c^2} A \sin(kx - \omega t) = -k^2 A \sin(kx - \omega t). \tag{2}$$

Because the right sides of Eqs. 1 and 2 are equal, the left sides must also be equal, so

$$\frac{\partial^2 f}{\partial x^2} = \frac{1}{c^2}\frac{\partial^2 f}{\partial t^2},$$

which is Eq. 16.51, and so the wave equation is satisfied. ✔

16.22 If any time-dependent sinusoidal harmonic function, $f(x, t) = A \sin(kx - \omega t)$, satisfies the wave equation, what determines the values of the wave number k and the angular frequency ω for, say, a wave on a string?

Figure 16.43 Relationships among curvature, acceleration, and wave speed.

One-dimensional wave equation

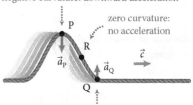

$$\underbrace{\frac{\partial^2 f}{\partial x^2}}_{\text{curvature}} = \frac{1}{c^2} \underbrace{\frac{\partial^2 f}{\partial t^2}}_{\text{acceleration}}$$

(a)

negative curvature: downward acceleration

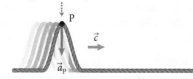

zero curvature: no acceleration

positive curvature: upward acceleration

(b)

strong curvature: great acceleration

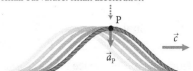

(c)

small curvature: small acceleration

Chapter Glossary

SI units of physical quantities are given in parentheses.

Antinode A point in space crossed by two or more waves simultaneously, where the amplitude of the displacement is greatest.

Displacement (of particle in wave) $\vec{D}$ (m) A vector that points from the equilibrium position to the actual position of a particle in a medium disturbed by a wave.

Displacement curve A curve that shows how the displacement of a specific particle in a medium disturbed by a wave varies as a function of time.

Harmonic wave A periodic wave whose displacement can be represented by a sinusoidally varying function of space and time:

$$f(x, t) = A \sin[k(x - ct) + \phi_i], \qquad (16.16)$$

where k is the wave number, c the wave speed, and ϕ_i the initial phase.

Interference When two or more waves are simultaneously present at a certain point, they interfere with each other. If the displacements are in the same direction, the interference is said to be *constructive;* if they are in opposite directions, the interference is said to be *destructive.*

Linear mass density μ (kg/m) A scalar that represents the amount of mass per unit length of uniform strings and other uniform one-dimensional objects:

$$\mu \equiv \frac{m}{\ell}. \qquad (16.25)$$

Longitudinal waves Waves for which the displacement is parallel to the direction of propagation.

Node A point in space that is crossed by two or more waves simultaneously, where the displacement remains zero.

Periodic wave A wave for which the displacement of particles in the medium at any fixed position is a periodic function of time.

Standing wave A pulsating stationary pattern caused by two counterpropagating waves of identical amplitude A and wavelength λ. The pattern is characterized by a series of *nodes* separated by $\lambda/2$.

Superposition of waves In a medium that obeys Hooke's law, the resultant displacement of two or more overlapping waves is the algebraic sum of the displacements of the individual waves. Consequently two waves can pass through each other without changing shape.

Transverse waves Waves for which the displacement is perpendicular to the direction of propagation.

Wave A disturbance that propagates through a material or through space. Waves involve transfer of energy without transfer of matter. *Mechanical waves* require a medium in order to be transmitted and are due to the coupling of the particles that make up the medium.

Wave equation The equation that is satisfied by any function that represents a wave traveling at wave speed c:

$$\frac{\partial^2 f}{\partial x^2} = \frac{1}{c^2} \frac{\partial^2 f}{\partial t^2}. \qquad (16.51)$$

All such functions are of the form $f(x - ct)$ or $f(x + ct)$.

Wave function A function that describes the displacement caused by a wave. The *time-dependent wave function* is a function in two variables (x and t) that gives the displacement as a function of space and time. The *time-independent wave function* is a function in one variable (x) that gives the displacement as a function of space at one specific instant.

Wave number k (m^{-1}) A scalar that represents the number of wavelengths that fit in a length of 2π m:

$$k = \frac{2\pi}{\lambda} = \frac{\omega}{c}. \qquad (16.7, 16.11)$$

Wave pulse A single isolated propagating disturbance.

Wave speed c (m/s) The speed at which a wave propagates. For mechanical waves, this speed, denoted by the letter c, is determined by the properties of the medium. The wave speed is distinct from the velocity of the particles in the medium.

Wavelength λ (m) A scalar that gives the minimum distance over which a periodic wave repeats itself.

$$\lambda = cT. \qquad (16.9)$$

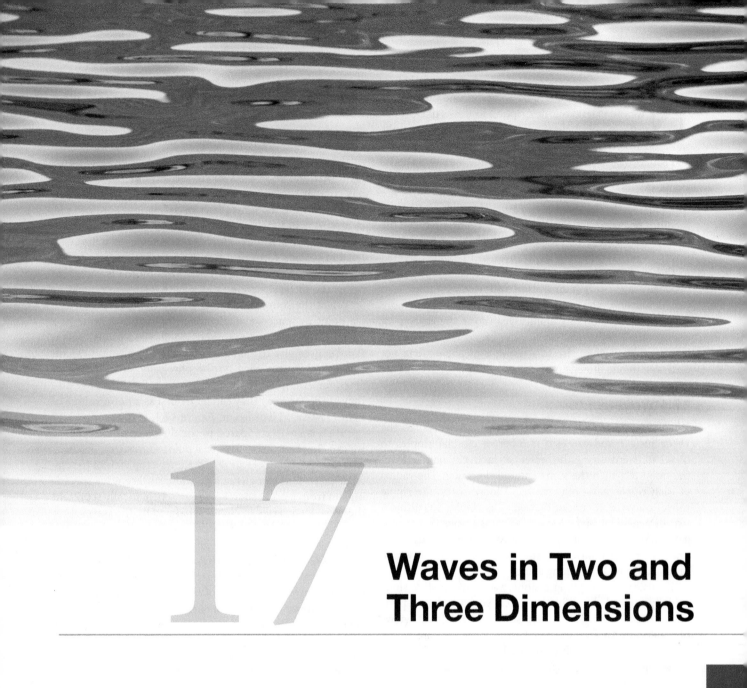

17

Waves in Two and Three Dimensions

CONCEPTS

QUANTITATIVE TOOLS

ow that we have treated waves in one dimension, we can extend our description to waves that propagate in more than one dimension—for example, waves on a water surface, which propagate in two dimensions, and sound waves, which propagate in three dimensions. The basic mechanism for these waves is the same as that for one-dimensional waves: When particles in a medium are coupled together, a disturbance caused at one point in a medium propagates to the surrounding points. Because the basic mechanism is the same, most of what we know about waves in one dimension carries over to the waves treated in this chapter. However, there are important differences between waves in one dimension and waves in more than one dimension. For example, waves propagating in more than one dimension lose amplitude even in the absence of energy dissipation, and wave interference in more than one dimension gives rise to a number of phenomena not observed in one dimension.

17.1 Wavefronts

When you drop a stone in water, the stone disturbs the water surface and causes a number of circular ripples to travel outward along the water surface, emanating from the point at which the stone entered the water. The radius of each ripple starts out small and increases as the ripple moves outward at constant speed. The ripples constitute a two-dimensional **surface wave,** which is a wave that propagates in two dimensions. Surface waves are analogous to the one-dimensional waves we studied in Chapter 16, where propagation is always along a line.

If, instead of dropping a stone, you move a stick sinusoidally up and down in the water, you generate a periodic surface wave. **Figure 17.1** shows two cutaway views of a periodic surface wave at two instants that are half a period apart. The graph below each view shows the vertical component of the displacement of the water surface as

a function of the radial distance r to the source—in other words, the *wave function* of the wave. Because the wave is circular, the wave function is the same in all radial directions. Note the similarity between the shape of the curves in these graphs and the shape of the wave in Figure 16.9 for a periodic wave on a string. Because the surface wave has circular symmetry, all points that are equidistant from the source have the same phase of oscillation. For example, at instant t_i all points that are a distance λ from the source have reached their maximum displacement above the equilibrium surface, forming a circular crest. Half a period later, the same points have reached their maximum displacement below the surface, forming a circular trough.

When the source of the wave can be localized to a single point in space—a small stone dropped in a pond, the stick in Figure 17.1—the source is said to be a **point source.** Point sources are treated as if they have no physical extent. The actual source need not be physically small, however. Stars, for example, are much larger than Earth, yet because they are so far away they can be treated as a point source of light (a wave). Likewise, in Figure 17.1, if the diameter of the stick is small compared to the diameter of the ripples, we can consider the stick to be a point source.

The blue region of **Figure 17.2** shows a top view of the circular crests and troughs of a periodic surface wave spreading out from a point source. We can schematically represent the wave by a series of circles separated by the wavelength λ (white region). Each circle corresponds to points on the surface that have the same phase of oscillation. Curves or surfaces on which all points have the same phase constitute a **wavefront.** As the wave travels away from the source at the wave speed c, the radius of each wavefront increases.

This expansion of circular surface waves causes the wave energy to be spread out over a larger and larger region.

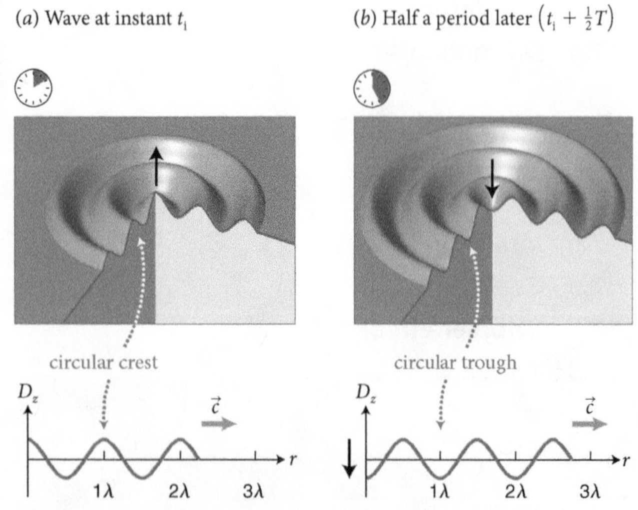

Figure 17.2 Top view of the circular crests and troughs formed in a periodic surface wave. We represent the wave schematically as a series of wavefronts.

Figure 17.1 Cutaway views of a wave on the surface of a liquid and the corresponding vertical displacement of the surface as a function of distance r from the oscillating wave source, which is located at $r = 0$.

(*a*) Wave at instant t_i

(*b*) Half a period later $\left(t_i + \frac{1}{2}T\right)$

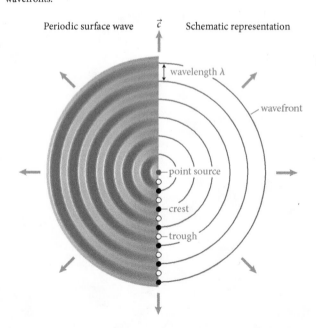

Figure 17.3 A surface wave moving at speed c. Because surface waves expand in two dimensions, the wave energy is distributed over a larger and larger circumference.

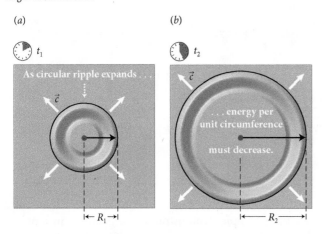

(a)

t_1

As circular ripple expands . . .

$\vec{c}$

$\vdash R_1 \dashv$

(b)

t_2

$\vec{c}$

. . . energy per unit circumference must decrease.

$\vdash\!\!\!-\!\!\!- R_2 \!\!-\!\!\!-\!\dashv$

Figure 17.5 The wavefronts from a point source emitting waves in three dimensions are uniformly expanding, concentric spheres.

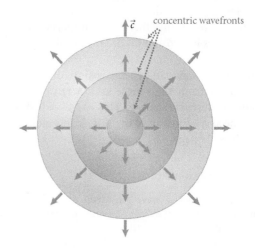

$\vec{c}$ concentric wavefronts

Consider, for example, the spreading ripples in **Figure 17.3**. At instant t_1 the ripples cross a circle of radius R_1 centered on the wave source; at instant t_2 they are about to cross a circle of radius R_2. If there is no loss of wave energy as the wave moves from R_1 to R_2—that is, if there is no dissipation of energy—the amount of energy that crosses the circle of radius R_2 must be the same as the amount that crosses the circle of radius R_1. Because the circumference increases, there is less energy per unit length along the circumference at t_2 than there is at t_1.

🖐 **17.1** Let $t_2 = 2t_1$ in Figure 17.3. (a) How does R_1 compare with R_2? (b) If the energy in the wave is E and there is no dissipation of energy, what is the energy per unit length along the circumference at R_1? At R_2? (c) How does the energy per unit length along a wavefront vary with radial distance r?

As Checkpoint 17.1 shows, the expansion of circular wavefronts causes the energy per unit length along a wavefront to decrease as $1/r$. Because the energy in the wave is proportional to the square of the wave amplitude (Eq. 16.41, $E_\lambda = \frac{1}{2}\mu\lambda\omega^2 A^2$), it follows that the amplitude is proportional to $1/\sqrt{r}$. (You can see this amplitude decrease in the cutaway view in **Figure 17.4**.) Keep in mind that

this amplitude decrease has nothing to do with energy dissipation: It is due entirely to the spreading out of the wave, which causes the energy to be distributed over a larger circumference.

Some waves—sound waves and light waves are two examples—spread out in three dimensions. If the source is a point source and the spreading is uniform in all three dimensions, the wavefronts are spherical (**Figure 17.5**). The resulting waves are called **spherical waves.** As a three-dimensional wave propagates, each wavefront expands outward at the wave speed c, and the energy carried by the wavefront is spread out over a larger and larger spherical area. When a wavefront reaches a distance r from the source, the energy in the wavefront is spread out over a surface area of $A = 4\pi r^2$, and so the energy per unit area carried by the wavefront is proportional to $1/r^2$ (**Figure 17.6**). The amplitude of a spherical wave is therefore proportional to $1/r$ (Eq. 16.41 again).

Figure 17.4 Cutaway view of a periodic surface wave. Because the energy in a given wavefront is spread out over an increasingly large circumference as the wavefront moves away from the source, the amplitude decreases as the wavefront moves away from the source.

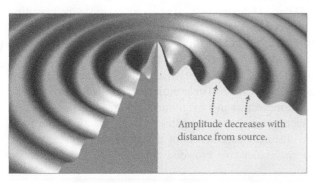

Amplitude decreases with distance from source.

Figure 17.6 The inverse-square relationship between energy and radius for a wave expanding uniformly in three dimensions without dissipation. All the energy that crosses area A_1 also crosses A_2 and A_3. Because the area of a sphere is proportional to $r^2 (A_{sphere} = 4\pi r^2)$, the energy per unit of surface area decreases as $1/r^2$.

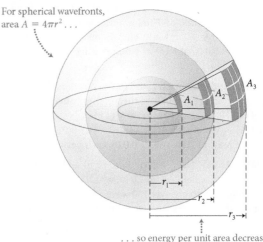

For spherical wavefronts, area $A = 4\pi r^2$ . . .

A_3

A_1

A_2

r_1

r_2

r_3

. . . so energy per unit area decreases as $1/r^2$ (inverse-square relationship).

Because it is difficult to represent wavefronts in three dimensions, we usually represent spherical wavefronts by a series of circles, just as we do for two-dimensional waves. Keep in mind, however, that there is an important difference between waves in two and in three dimensions:

> The amplitude of waves in two dimensions decreases with distance r from the source as $1/\sqrt{r}$; in three dimensions it decreases as $1/r$. This decrease is due purely to the spreading out of the wavefronts and involves no loss of energy.

Example 17.1 Ripple amplitude

The amplitude of a surface wave for which $\lambda = 0.050$ m is 5.0 mm at a distance of 1.0 m from a point source. What is the amplitude of the wave (a) 10 m from the source and (b) 100 m from the source? During a time interval equal to 100 periods, by what percentage does the amplitude of a wavefront decrease (c) after the wavefront passes a position 1.0 m from the source and (d) after it passes a position 100 m from the source?

❶ GETTING STARTED I am given that the amplitude $A = 5.0$ mm at $r = 1.0$ m. As the wave spreads out, its amplitude diminishes, and I need to calculate the amplitude at $r = 10$ m and $r = 100$ m. In addition I need to determine by how much the wave attenuates as it propagates over a 100-period time interval past these two positions.

❷ DEVISE PLAN Because the wave is a two-dimensional surface wave, the amplitude is proportional to $1/\sqrt{r}$. I know the amplitude $A_{1.0\,m}$ at $r = 1.0$ m, so I can use this dependence to determine the amplitude at other distances from the source. For parts a and b, I need to determine $A_{10\,m}$ and $A_{100\,m}$ at $r = 10$ m and $r = 100$ m. In 100 periods, the wavefront travels a distance equal to 100 wavelengths, which is $100 \times (0.050$ m$) = 5.0$ m, so for parts c and d, I also need to determine the amplitudes 5 m beyond those positions.

❸ EXECUTE PLAN (a) The ratio of the amplitudes at 1.0 m and 10 m is $\sqrt{(1.0\,\text{m})}/\sqrt{(10\,\text{m})} = \sqrt{(1.0/10)} = 0.32$, and so the amplitude at 10 m is $0.32 \times (5.0$ mm$) = 1.6$ mm. ✔

(b) At 100 m, $\sqrt{(1.0\,\text{m})}/\sqrt{(100\,\text{m})} = 0.10$, and so the amplitude is $0.10 \times (5.0$ mm$) = 0.50$ mm. ✔

(c) 100 periods after passing the 1.0-m position, the wavefront is 6.0 m from the source and the amplitude decreases by a factor of $\sqrt{(1.0\,\text{m})}/\sqrt{(6.0\,\text{m})} = 0.41$. This means $A_{6.0\,m}$ is 41% of $A_{1.0\,m}$: $A_{6.0\,m} = 0.41(5.0$ mm$) = 2.0$ mm, a decrease of about 60%. ✔

(d) Then 100 periods after passing the 100-m position, the wavefront is 105 m from the source and the amplitude decreases by a factor of $\sqrt{(100\,\text{m})}/\sqrt{(105\,\text{m})} = \sqrt{(1/1.05)} = 0.98$. This means that $A_{105\,m}$ is 98% of $A_{100\,m}$: $A_{105\,m} = 0.98(0.50$ mm$) = 0.49$ mm, a decrease of 2%—a mere 0.02% each period. ✔

❹ EVALUATE RESULT The amplitudes at 10 m and 100 m are both smaller than the amplitude at 1.0 m, which is what I expect. That the amplitude decrease over 100 wavelengths is less farther away from the source agrees with what I expect from the $1/\sqrt{r}$ dependence.

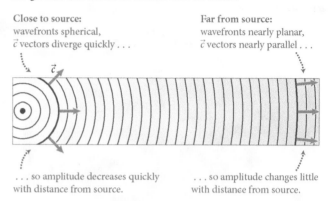

Figure 17.7 Far from the source, wavefronts are nearly flat, so amplitude changes little with additional distance from the source.

Close to source: wavefronts spherical, $\vec{c}$ vectors diverge quickly . . .

Far from source: wavefronts nearly planar, $\vec{c}$ vectors nearly parallel . . .

. . . so amplitude decreases quickly with distance from source.

. . . so amplitude changes little with distance from source.

As this example demonstrates, the decrease in amplitude is most pronounced close to the source; far from the source, the decrease is hardly noticeable. Figure 17.7 illustrates the same point in a different way: Far from the source, the wavefronts are nearly flat rather than curved, and as a wavefront straightens, a fixed length of it diverges less. Far from a point source, a spherical wavefront essentially becomes a two-dimensional flat wavefront called a **planar wavefront**. Likewise, far from a point source, a circular surface wavefront essentially becomes a one-dimensional straight wavefront.

✋ **17.2** Notice that in the views of the surface wave in Figure 17.1 the amplitude does not decrease with increasing radial distance r. How could such waves be generated?

17.2 Sound

The human ear can detect longitudinal waves at frequencies in the range from about 20 Hz to 20 kHz. Such waves, which propagate through any kind of material—solid, liquid, or gas—constitute what we call **sound.** Sound propagates through any medium that can exert elastic restoring forces. If you find it hard to think of a gas as being able to exert an elastic force, squeeze an empty plastic bottle first with a tightly sealed cap and then without it. Without the cap, squeezing the bottle simply pushes some air out of the bottle. With the cap, the bottle exerts an elastic force because of the air in it. It is this elastic force that is responsible for transmitting sound waves.

Figure 17.8 shows a simple wave pulse propagating through a long, air-filled tube. One end of the tube is fitted with a piston. Starting at t_2, the piston is rapidly moved back and forth, first pushing the molecules in the layer of air adjoining the piston closer together and then letting them expand again. These molecules, in turn, push and pull on the molecules in a neighboring layer, and so on, creating a disturbance that propagates through the air in the tube. The disturbance consists of a *compression* (region where the molecules are crowded together) and a *rarefaction*

Figure 17.8 A piston generates a longitudinal sound pulse that propagates along an air-filled tube. The disturbance consists of a compression and a rarefaction. The wave function is similar to the one for a longitudinal wave in a spring (see Figure 16.6).

(a) Longitudinal wave pulse created by in-and-out movement of piston

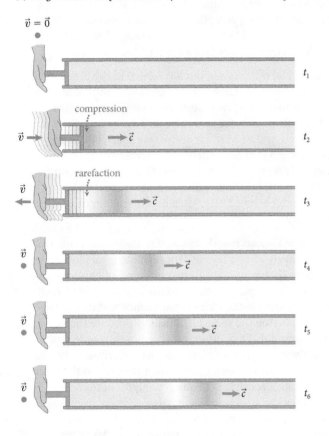

(b) Wave function of pulse at instant t_6

(region where they are spaced far apart). The resulting wave is called a *compressional wave*. The wave speed c at which the disturbance travels through the tube depends on the density and the elastic properties of the medium. For dry air at a temperature of 20 °C, the speed of sound waves is 343 m/s.

Exercise 17.2 Wavelength of audible sound

Given that the speed of sound waves in dry air is 343 m/s, determine the wavelengths at the lower and upper ends of the audible frequency range (20 Hz–20 kHz).

SOLUTION The wavelength is equal to the distance traveled in one period. At 20 Hz, the period is $1/(20 \text{ Hz}) = 1/(20 \text{ s}^{-1}) = 0.050$ s, so the wavelength is $(343 \text{ m/s})(0.050 \text{ s}) = 17$ m. ✔
The period of a wave of 20 kHz is $1/(20{,}000 \text{ Hz}) = 5 \times 10^{-5}$ s, so the wavelength is $(343 \text{ m/s})(5.0 \times 10^{-5} \text{ s}) = 17$ mm. ✔

Figure 17.9 A simple mechanical model for longitudinal waves.

(a) Identical beads coupled by springs

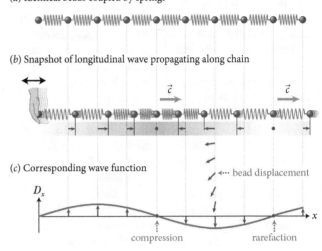

(b) Snapshot of longitudinal wave propagating along chain

(c) Corresponding wave function

The wave pulse in Figure 17.8 is longitudinal—the air molecules are displaced in the same direction as that in which the wave propagates. In bulk gases and liquids that have no internal friction, there can be no transverse waves because such gases and liquids do not exert elastic forces in the transverse direction. Notice the similarity between the wave pulse shown in Figure 17.8 and the longitudinal wave pulse propagating through a coil spring shown in Figure 16.5.

Figure 17.9a shows a chain of identical beads coupled together by springs, which can serve as a simple mechanical model for studying longitudinal waves. A snapshot of a harmonic longitudinal wave propagating through the chain is shown in Figure 17.9b. The leftmost bead is made to oscillate back and forth harmonically. Because each bead executes, in turn, the same longitudinal oscillation, a compressional wave propagates along the chain. Although the beads move in the same direction as the wave, there is no permanent displacement of beads in that direction. What are transported along the chain, however, are energy (as kinetic energy of the beads and elastic potential energy of the springs) and momentum.

✋ **17.3** Does the wave speed along the chain shown in Figure 17.9 increase or decrease when (a) the spring constant of the springs is increased and (b) the mass of the beads is increased?

From Section 16.1, you know that a snapshot of a transverse wave gives us the wave function directly. Things are not so simple with longitudinal waves, however. Because with these waves the displacement of the medium is along the axis of wave propagation, it is cumbersome to interpret any wave snapshots. The wave function shown in Figure 17.9c, however, provides an alternative representation of the wave and reveals its harmonic nature, which is not immediately obvious in Figure 17.9b. This graph is obtained by plotting the displacement of each bead from

The perception of sound

The human ear is capable of detecting longitudinal waves transmitted through a gas, liquid, or solid medium in what is called the *audible frequency range*. This range, which varies with age and also from person to person, extends from about 20 Hz to 20 kHz. The sensitivity of the ear varies greatly over this range, however, peaking at about 3 kHz. Waves outside this range cause no auditory nerve signals in humans, although large-amplitude waves of very low frequency, called *subsonic waves* (such as earthquakes), can be felt. Some animals, such as bats, can detect longitudinal waves of frequencies up to 120 kHz, called *ultrasonic waves* because they are above the audible frequency range for humans.

The perception of sound is a complicated and subjective matter. The mechanical part of this process, which is determined by the physiology of the ear, is well understood. A sound wave that reaches the ear propagates through a complex system of solid and liquid parts of the outer and middle ear to the inner ear, where the wave causes *hair cells* to oscillate. These oscillations cause the auditory nerve to send signals to the brain.

The neurological processing of these signals by the brain is not yet fully understood. The properties of sound are typically described using terms such as *loudness, pitch,* and *timbre* or *tonal quality*. These terms are subjective, however, and so there is no simple correlation between them and the parameters used to describe waves in physics (amplitude, frequency, harmonic content).

Loudness: The perception of loudness is connected to sound wave energy and therefore amplitude. Because the ear is capable of detecting waves over a large range of wave energies—from the very faint sound of a pin drop to the loud roar of an airliner taking off—the response is highly nonlinear (discussed in Section 17.5). The correlation between loudness and energy is complicated by the fact that the perception of loudness also depends on frequency: A 1-kHz wave sounds louder than a 200-Hz wave of equal energy (see Figure 17.26).

Pitch: The sensation of pitch is related to the physical property of frequency. The correlation is nonlinear; if the frequencies of two sound waves differ by a factor of 2, their pitches are said to be an *octave* apart. Sounds at 256 Hz and 512 Hz are one octave apart, but so are sounds at 3 kHz and 6 kHz. The perceived pitch also depends on the amplitude of the sound wave. Above 3 kHz, pitch increases with increasing sound wave amplitude; below 2 kHz, pitch decreases with increasing amplitude.

Timbre: We can readily distinguish two musical instruments even when they emit tones of identical pitch because they produce different mixtures of harmonics. The pitch is typically related to the lowest, or fundamental, frequency in the series of harmonics, and two instruments playing the same pitch usually produce the same fundamental. They differ, however, in the content of the higher harmonics, which is what determines the timbre of the instrument. Even though the ear and the brain constitute a remarkable Fourier analyzer in the sense that they can distinguish sounds of different harmonic contents, we cannot characterize timbre with a precise statement about measurable parameters. (An alternative term you might run across for timbre is *tonal quality*.)

---

its equilibrium position versus the bead's equilibrium position. To create Figure 17.9c, the displacement vector of each bead in Figure 17.9b is rotated 90° counterclockwise and its tail placed on the x axis at the *equilibrium* position of the bead given in Figure 17.9a.

✋ **17.4** (*a*) Plot the velocity of the beads along the chain in Figure 17.9b as a function of their equilibrium position x. (*b*) Plot the linear density (number of beads per unit length) as a function of x.

As Checkpoint 17.4 suggests, we can also represent the wave in Figure 17.9 by plotting the linear density versus position. Comparing the linear density curve of Figure S17.4b with the $D_x(x)$ curve of Figure S17.4a gives us some important information:

The compressions and rarefactions in longitudinal waves occur at the locations where the medium displacement is zero.

(You can get this same information by comparing Figure 17.9b and c.)

The propagation of sound through a medium is very similar to that of a compressional wave through the chain in Figure 17.9. Disturbed molecules in a solid, liquid, or gas are subject to intermolecular forces that tend to return the molecules to their equilibrium positions. For small displacements, the restoring force is proportional to the displacement. It is this restoring force that is responsible for the propagation of the disturbance caused by the piston in Figure 17.8.

When not confined to a tube, a compressional wave traveling through air spreads out in three dimensions because, unlike beads along a string, air molecules interact with the surrounding molecules in all three directions. Consequently, sound waves tend to form spherical wavefronts. **Figure 17.10** shows a sound wave generated by an oscillating tuning fork. A series of spherical shells of compression and rarefaction travel outward from the fork. Each molecule in the air executes a sinusoidal back-and-forth movement

Figure 17.10 The periodic wave emitted by a tuning fork causes a series of outward-traveling spherical wavefronts. The dot represents an air molecule that oscillates about its equilibrium position as the wavefronts pass by.

Figure 17.11 Image of the spherical compressional wavefronts produced by sound waves emanating from a telephone. This image was obtained using laser illumination. Reprinted with the permission of Alcatel-Lucent USA, Inc.

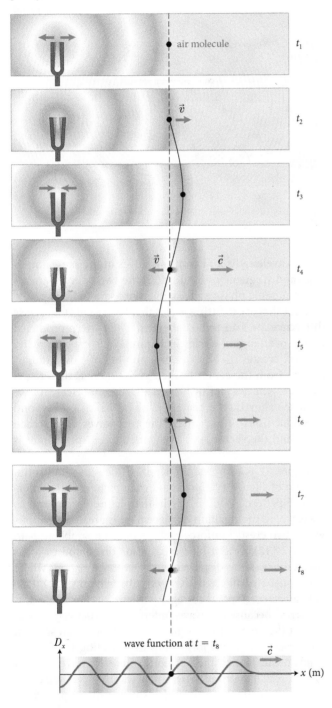

as successive shells travels by; the sine curve in the figure shows the motion of one molecule.

The displacements of air molecules disturbed by a sound wave are very small. At a distance of a few hundred meters, for example, the deafening roar of an airliner taking off causes displacements of only 10^{-6} m—about one-hundredth of the width of a hair. These small displacements coupled with the high speed at which sound waves travel make it impossible to directly see the compressional wavefronts that accompany sound. Only under special laser illumination is it possible to observe these wavefronts (**Figure 17.11**).

Notice how the wavefronts shown in Figures 17.10 and 17.11 resemble the circular wavefronts of a surface wave, such as the one represented in Figure 17.2. The compressions correspond to crests, and the rarefactions correspond to troughs. When we draw circles to represent the spherical wavefronts of sound waves, which are longitudinal waves, the circles represent not crests and troughs but rather compressions and rarefactions.

✋ **17.5** Even though the sound of a loudspeaker carries in all directions, the loudness in front of the speaker is considerably greater than the loudness behind it. What is the shape of the wavefronts produced by a loudspeaker?

17.3 Interference

The superposition of overlapping waves in two and three dimensions leads to a number of interesting phenomena. Consider first the situation illustrated in **Figure 17.12** on the next page, where two identical circular wave pulses traveling on a liquid surface cross as they expand. At t_2 the two crests meet at a point midway between the two sources, giving rise to constructive interference. At this point the displacement of the liquid surface is twice as great as that due to the individual pulses. As the pulses continue to expand, this point of overlap splits into two points, one moving upward and the other moving downward. At t_3 the crest of each pulse meets the trough of the other pulse, leading to destructive interference. Because the height of

Figure 17.12 Interference of two identical sets of circular wave pulses expanding on a liquid surface.

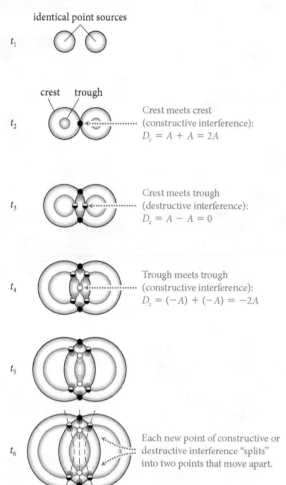

Figure 17.13 Moiré pattern formed by the overlapping wavefronts of two closely spaced, coherent wave sources (located at the centers of the two concentric sets of circles). The center of each band in the pattern corresponds to a nodal line.

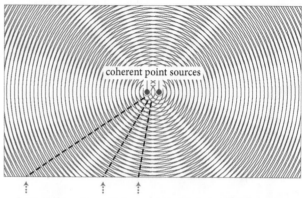

nodal lines (lines along which waves interfere destructively)

the crest is equal to the depth of the trough, the vector sum of the medium displacement is zero.* Each of these points splits into two points that move vertically away from each other. At t_4 the two troughs meet and reinforce each other, giving rise to constructive interference again. At this point the depression of the medium is doubled. Note that, as shown at t_6, the points of constructive interference move along a straight line perpendicular to the line that connects the two sources, while the points of destructive interference move along curved paths.

Let us now apply these ideas to the overlapping of periodic surface waves. When periodic circular waves from two nearby sources cross each other, the crests and troughs overlap in many places, giving rise to a complex interference pattern. **Figure 17.13** shows the circular wavefronts for two sources separated by a distance equal to

*The waves do not cancel exactly close to the sources because the amplitudes of the two waves differ due to the $1/\sqrt{r}$ factor in the wave amplitude. Far from the sources, however, this difference is negligible.

three wavelengths and emitting waves of the same amplitude and frequency. In addition, the two sources always have the same phase and are therefore said to be *in phase*. Sources that are emitting waves that have a constant phase difference are said to be **coherent.** The pattern produced by the overlapping circles in Figure 17.13, called a *moiré pattern*, creates the illusion of a number of bands radiating out from the center of the figure (*moiré* is the French word for the wavy finish seen in some fabrics). As we shall see shortly, the centers of these bands correspond to **nodal lines**—lines where the two waves cancel each other and the vector sum of the medium displacement is always zero.

Figure 17.14a shows a magnified view of the central portion of Figure 17.13 overlaid on a picture of the interference pattern. White areas in the picture correspond to areas where the medium displacement is greatest—regions where two crests reinforce each other (comparable to t_2 in Figure 17.12). Bright blue areas also correspond to areas where the medium displacement is greatest—regions where two troughs reinforce each other (t_4 in Figure 17.12). Pale blue areas correspond to regions where the surface is undisturbed because the waves interfere destructively. If you look at the picture from a glancing angle, you can see that these pale blue areas form outward radiating bands. The bands coincide with the positions where the wavefronts of one source fall halfway between the wavefronts of the other (which gives the illusion of bands in the moiré pattern in Figure 17.13).

Figure 17.14b shows a detail of the interference pattern. In addition to the solid dark blue arcs that indicate wavefronts marking the crests of the waves generated by each source, bright blue arcs indicate the locations of the troughs halfway between each pair of neighboring crests. Figure 17.14c shows the same region half a period later. All wavefronts have advanced by half a wavelength, and the nodes (•) and

Figure 17.14 Details of the interference pattern in the central region of Figure 17.13. White and blue regions represent, respectively, areas of maximum positive and maximum negative medium displacement. The intermediate regions of zero displacement line up to form the gray nodal bands seen in Figure 17.13.

(a)

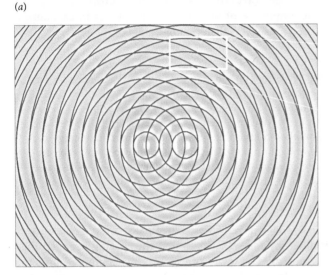

(b)

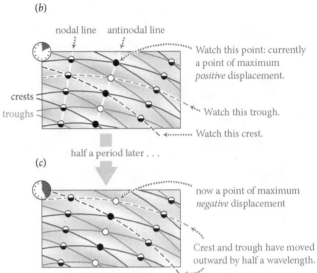

nodal line   antinodal line

Watch this point: currently a point of maximum *positive* displacement.

crests
troughs

Watch this trough.

Watch this crest.

half a period later . . .

(c)

now a point of maximum *negative* displacement

Crest and trough have moved outward by half a wavelength.

antinodes (● and ○) have moved along with the wavefronts. See, for example, the node at the intersection of the dashed dark blue crest from the left source and the dashed bright blue trough from the right source—in Figure 17.14*b* it is the second node from the top; in Figure 17.14*c* it is the uppermost. In addition, all white and bright blue areas are interchanged, as are all filled and unfilled circles (see, for example, the highlighted filled and open circles in Figure 17.14*b* and *c*). Where two crests initially reinforced each other, two troughs now meet. (Of course, the nodal lines stay in place, by definition.)

17.6 (*a*) What is the medium displacement at a point halfway between a filled circle and a neighboring open circle in Figure 17.14*b*? (*b*) Sketch the variation of this displacement as a function of time. (*c*) What is the medium displacement at all points between two neighboring half-filled circles? (*d*) Sketch the variation of this displacement as a function of time at the midpoint between two neighboring half-filled circles.

The half-filled circles line up to form a path, and as Checkpoint 17.6 demonstrates,* *all* points along this path remain undisturbed and are therefore nodes. The line that connects the half-filled circles is thus a nodal line, and the pale blue bands that appear in Figure 17.13 are thus centered on nodal lines. The filled and open circles line up to form what are called **antinodal lines.** Points on these antinodal lines have the greatest amplitude of oscillation; in other words, these points are antinodes.

17.7 Is a point halfway between a filled and a neighboring open circle in Figure 17.14*b* a node?

The interference pattern formed by overlapping waves in two dimensions can readily be observed for waves generated on a water surface (**Figure 17.15**). The nodal lines are clearly visible in the picture. All waves in more than one dimension—sound waves, light waves, x-ray waves, radio waves—set up interference patterns that contain nodal and antinodal lines. For example, two loudspeakers that are some distance apart and that emit sound waves of the same frequency set up an interference pattern that is not very different from the ones for circular water waves. By adjusting

Figure 17.15 Photograph of the interference pattern caused by overlapping circular waves on a water surface.

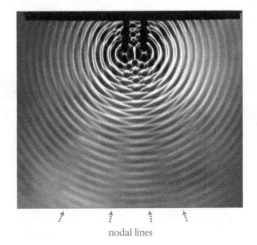

nodal lines

---

*I cannot overstress the importance of doing Checkpoint 17.6. By completing this checkpoint, you will be teaching yourself the key feature of interference of waves in more than one dimension.

Figure 17.16 Less can be more. (*a*) When both sources generate waves, point Q lies on a nodal line and thus never experiences displacement of the medium (because the waves from the two sources interfere destructively there). If these were sound waves, there would be silence at Q. (*b*) However, when only one source generates waves, Q experiences their full effect.

(*a*) Both sources generate waves

(*b*) Only S_2 generates waves

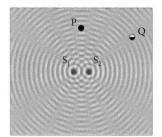
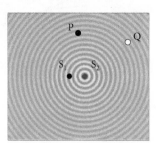

your position in front of the speakers, you can find regions where the destructive interference occurs, causing the loudness of the sound to be low.

The existence of nodal regions has a number of consequences. Consider, for example, the interference pattern shown in **Figure 17.16*a***. Two coherent sources S_1 and S_2 emit circular waves. The waves cause the medium displacement at point P to vary, but at point Q, located on a nodal line, the medium displacement *remains zero*. If you were standing at that location, you could not tell whether the sources were emitting waves or not! Each source separately, however, does emit waves in the direction of Q. Figure 17.16*b* shows that, with S_1 is turned off, the wavefronts from S_2 that reach Q are not interfered with. In other words, if you were standing at Q, you would hear sound when S_1 is turned off but not when both sources are on. So, *removing* a source *increases* the amplitude of the wave at Q! Conversely, adding a second source (in other words, returning to the situation in Figure 17.16*a*)

decreases (or even eliminates) the wave amplitude in certain regions. So:

> **When the waves from two coherent sources interfere, the amplitude of the sum of these waves in certain directions is less than that of a single source.**

17.8 (*a*) If the amplitude of the wave generated by each source in Figure 17.16 is A at points P and Q, what is the overall amplitude at P in Figure 17.16*a*? In Figure 17.16*b*? (*b*) Given that the wave energy that passes Q in Figure 17.16*b* becomes zero when S_1 is turned on, do you expect the energy that passes P to increase or decrease? (*c*) Use your answer to part *a* to determine by what factor the amount of energy that passes P changes, and reconcile your answer with your answer to part *b*.

Just what a given interference pattern looks like is determined by the wavelength and the spacing between the two sources. The greater the source separation relative to the wavelength, the greater the number of nodal lines. We determine the number of nodal lines in a pattern by imagining a straight line running through the centers of the two sources and then counting the number of nodal lines on either side of this line:

> **If two coherent sources located a distance d apart emit identical waves of wavelength λ, then the number of nodal lines on either side of a straight line running through the centers of the sources is the greatest integer smaller than or equal to $2(d/\lambda)$.**

You can study the appearance of nodal lines as the separation between the sources is varied in **Figure 17.17**.

17.9 By how many wavelengths are the sources in Figure 17.16*a* separated?

Figure 17.17 The effect of changing the separation between two point sources.

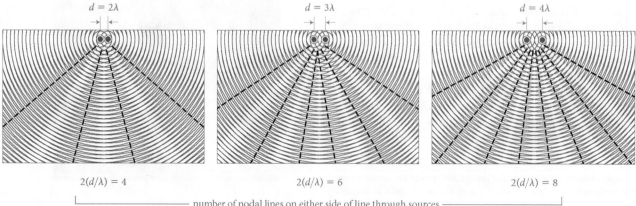

$d = 2\lambda$ $d = 3\lambda$ $d = 4\lambda$

$2(d/\lambda) = 4$ $2(d/\lambda) = 6$ $2(d/\lambda) = 8$

number of nodal lines on either side of line through sources

Example 17.3 Nodes

For the situation shown in Figure 17.16a, how many nodes are there along the line segment S_1S_2 that connects the centers of the two sources?

1 GETTING STARTED The sources are coherent and so send identical waves in opposite directions along S_1S_2. This means there is a standing wave (see Section 16.6) along that segment. From Figure 17.16b I can tell that exactly three wavelengths fit between S_1 and S_2.

2 DEVISE PLAN The two sources are in phase, so I can determine the number of nodes by plotting the amplitudes of the waves due to each source along S_1S_2 at two instants and then adding those amplitudes. The nodes occur wherever the amplitude of the combined wave remains zero.

3 EXECUTE PLAN Figure 17.18a shows, for the instant at which the medium displacement at both sources is maximum, the medium displacement caused by S_1 and S_2 and the sum of the two. Figure 17.18b shows the displacements one-quarter cycle later. Comparing the two graphs, I see that six positions along line segment S_1S_2 (denoted by the black dots in the bottom graphs) remain stationary, indicating six nodes. ✔

Figure 17.18

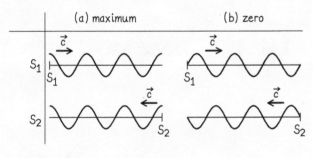

displacement of medium at sources is:

(a) maximum   (b) zero

resultant standing wave:

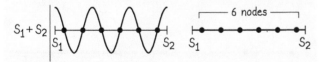

4 EVALUATE RESULT Because the sources are separated by a distance equal to three wavelengths and because there is one node for each half-wavelength (see Section 16.6), my answer makes sense.

Example 17.3 shows us why there are twice as many nodes and therefore twice as many nodal lines as there are wavelengths between the sources. Each node corresponds to a nodal line, and so you now see why there is a connection between the number of nodal lines and the source spacing.

17.10 Along the leftward and rightward extensions of the straight line joining S_1 and S_2 in Figure 17.16, why are there no nodes to the left of S_1 and to the right of S_2?

Figure 17.19 on the next page shows what happens when additional coherent sources are added between the two sources of Figure 17.14. Let us focus our attention on the two points P and Q. When just two sources are present (Figure 17.19a), P is on a nodal line and Q is on an antinodal line. The path length diagram in Figure 17.19a shows the waves arriving at P and Q. The distance from source 1 to P is 7λ; the distance from source 2 to P is $9\frac{1}{2}\lambda$. Consequently the two waves *always* arrive at P out of phase—they cause medium displacements in opposite directions and so always cancel. The distance from source 1 to Q is 8λ; the distance from source 2 is 9λ. So the two waves arrive in phase at Q and always reinforce each other.

Figure 17.19b shows how the addition of a single source halfway between the two original sources redistributes the flow of wave energy and concentrates the waves in fewer triangular areas. The phase of the wave arriving from the additional source is approximately halfway between that from sources 1 and 2. In other words, source 3 is about $8\frac{1}{4}\lambda$ from P and $8\frac{1}{2}\lambda$ from Q. Because the waves from 1 and 2 cancel each other at P, the wave from 3 arrives undisturbed at P, so P is no longer on a nodal line. At Q, the additional wave from 3 arrives out of phase with those from 1 and 2, so the additional source diminishes the wave amplitude at Q.

If two additional sources are added, almost all the wave energy becomes concentrated in a narrow cone perpendicular to the straight line that connects the sources (Figure 17.19c). The phases of the waves are such that, even though there are five sources instead of two, the amplitude of the waves arriving at Q is *smaller* than before and that of the waves arriving at P is very small.

With ten coherent sources, all the waves are concentrated in a cone along the direction perpendicular to the straight line that joins the sources (Figure 17.19d). Outside of this cone, the waves—all arriving at any given point, including P and Q, with a different phase—cancel each other. **Figure 17.20** on page 445 shows what happens when 100 coherent sources are placed close together, now forming a row 10λ wide: A *beam* of nearly straight wavefronts emanates in a direction perpendicular to the row of sources. There are hardly any waves outside the beam because the waves from all the different sources tend to cancel each other. Summarizing:

When many coherent point sources are placed close together along a straight line, the waves nearly cancel out in all directions except the direction perpendicular to the axis of the sources.

17.11 How does the wave amplitude along the beam of wavefronts in Figure 17.20 change with distance from the row of sources?

Figure 17.19 When additional coherent sources are added between the two sources spaced by three wavelengths, the interference pattern changes: The greater the number of sources, the more the waves are concentrated along the axis perpendicular to the sources. The displacement curves show the displacements at P and Q for each source and the sum of these displacements (bold curve).

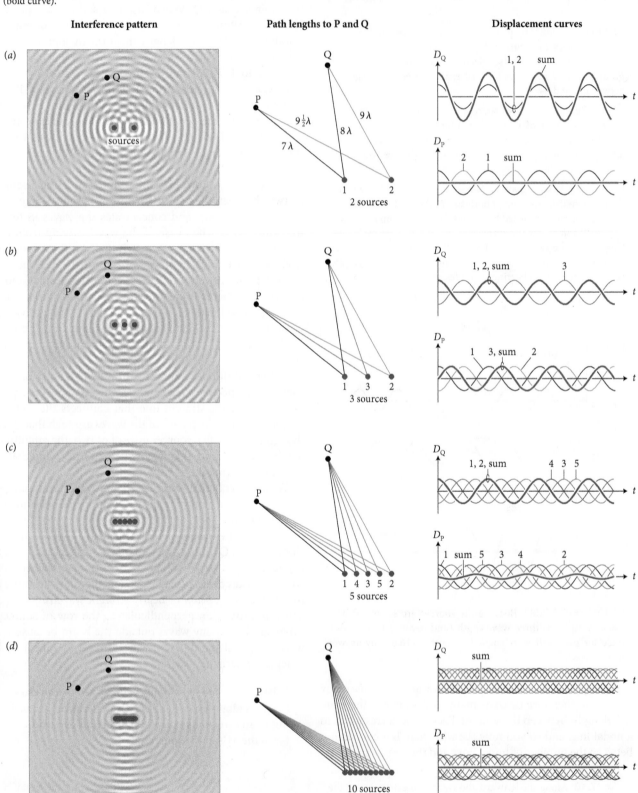

Figure 17.20 A 10λ-wide row of 100 coherent sources produces nearly planar wavefronts. The flow of energy is entirely along a beam perpendicular to the row of sources.

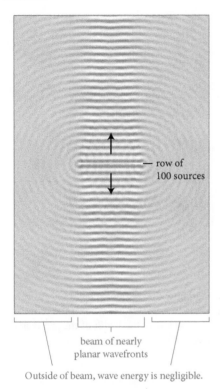

— row of
100 sources

beam of nearly
planar wavefronts

Outside of beam, wave energy is negligible.

Figure 17.21 According to Huygens' principle, any wavefront can be regarded as a collection of closely spaced, coherent point sources. Advancing wavefronts can then be constructed by drawing circular wavelets centered at points along an earlier wavefront.

old wavefront

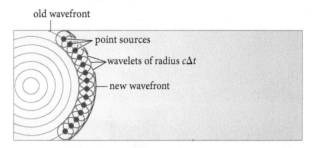

point sources
wavelets of radius $c\Delta t$
new wavefront

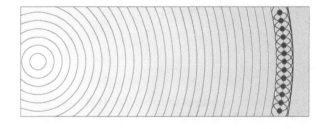

17.4 Diffraction

The nearly straight wavefronts that emanate from continuous rows of coherent sources suggest that *any* wavefront may be regarded as having originated from a collection of many closely spaced, coherent point sources emitting waves in all directions. The waves cancel in all directions except the direction perpendicular to the wavefront. This picture of wavefronts consisting of closely spaced, coherent point sources is called **Huygens' principle** after the 17th-century Dutch physicist Christiaan Huygens.

Figure 17.21 illustrates how Huygens' principle can be used to construct a future wavefront given the wavefront at a certain instant. We begin by dividing the original wavefront into a series of closely spaced point sources. All those point sources emit in phase because they are all on the same wavefront. After a short time interval Δt, each source has emitted a small circular wavefront, called a *wavelet*, of radius $c\Delta t$. Wavelets of neighboring sources overlap in such a way that they interfere destructively with each other except in the forward direction. All these forward-moving wavelets combine to form the wavefront we aim to construct, labeled "new wavefront" in Figure 17.21.*

---

*For a *single* row of coherent point sources, wavefronts emanate in *two* directions along a line perpendicular to the straight line connecting the sources (Figure 17.20). When additional rows of coherent point sources are added in front of or behind this row and these sources are given the proper phases, propagation in one of the two directions is suppressed by interference.

Huygens' principle also helps explain how waves spread around corners. If you stand near an open window, you hear sounds from the street even if you are not standing directly in front of the window. Likewise, if you stand behind a tree, you can hear sounds coming from in front of the tree because the sound waves spread around the tree.

Figure 17.22 on the next page shows nearly planar wavefronts incident on gaps of various size. If the gap is much larger than the wavelength, the gap lets through a *beam* of wavefronts with very little spreading. As the gap gets narrower, the waves beyond the gap spread out more and more, so that what gets through the gap no longer constitutes a beam. When the gap width is smaller than the wavelength, the wavefronts beyond the gap spread out in all directions—in essence, the gap now acts like a single point source. This spreading, called **diffraction,** is another characteristic feature that distinguishes waves from objects: Waves spread out when going through a narrow opening, but objects do not. If you throw a handful of buckshot through an opening, for instance, the "beam" of buckshot doesn't spread out after passing the opening.

The reason waves diffract is directly related to the observation that many coherent point sources in a row give rise to waves in just one direction because the waves emitted by each source cancel in all but the forward direction (Figure 17.20). Here we see the opposite occurring: As the number of point sources is reduced (the gap is narrowed in our example), the cancellation of waves along directions away from the straight line perpendicular to the line of sources becomes less complete and spreading occurs. As the width of the gap becomes comparable to the wavelength, the gap acts like a point source, leading to spreading of the wave in all directions.

CONCEPTS

Figure 17.22 When a planar wavefront passes through a gap, some spreading, called diffraction, occurs. The smaller the gap, the greater the amount of spreading. When the width d of the gap is smaller than the wavelength λ, the gap acts like a point source.

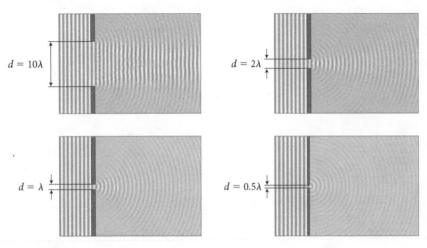

Obstacles or apertures whose width is smaller than the wavelength of an incident wave give rise to considerable spreading of that wave.

The amount of spreading depends strongly on the wavelength. For a given aperture, low-frequency sounds diffract more strongly than higher-frequency ones. If a band marches by below your window and you are not standing directly in the window opening, you hear the low-frequency drone of drums and tubas more than the high-frequency sounds of cymbals and piccolos. Move near the window and the sound of the band turns from dull to sharp as the high-frequency sounds emitted by it now also reach your ears.

A direct consequence of wave diffraction is that it is impossible to create a beam of waves that is narrower than the wavelength of the waves. This means that diffraction limits the ability of waves to locate objects. For example, it is impossible for a person to locate small objects by detecting reflected sound waves because, with their relatively long wavelengths (0.02 m to 20 m), these waves diffract around most objects. We therefore locate objects using visible light, whose wavelength is in the range of 4×10^{-7} m to 7×10^{-7} m (less than one-hundredth the width of a human hair). Only objects and apertures that are smaller than the wavelength of visible light diffract light.

Bats can detect ultrasonic waves of frequencies up to 120 kHz. The wavelength of these waves is about 3 mm. By emitting such waves and detecting their reflections, bats can locate small insects. Ultrasonic waves are also used for diagnostic purposes in medicine. Because of their short wavelength, it is possible to emit a narrow beam of ultrasonic waves. The beam is rapidly scanned across a sample to create an image from the reflections of the beam that occur inside the sample (**Figure 17.23**).

17.12 Suppose the barriers in Figure 17.22 were held at an angle to the incident wavefronts. Sketch the transmitted wavefronts for the case where the width of the gap is much smaller than the wavelength of the incident waves.

Figure 17.23 Ultrasound image of a three-month-old fetus in utero. The 2-MHz waves (1 megahertz = 1 MHz = 10^6 Hz) used to obtain this picture have a wavelength of 0.7 mm.

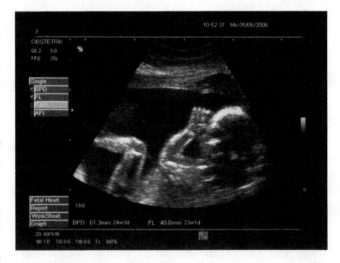

Self-quiz

1. A circular wave travels such that a crest occurs at radius r_1 at instant t_1. The same crest has a radius $r_2 = 2r_1$ at $t_2 = 2t_1$. (*a*) What is the wave speed? (*b*) Where is the crest at $t_3 = 3t_1$? (*c*) What is the ratio (energy per unit length at t_3)/(energy per unit length at t_1)?

2. A spherical wave travels such that a crest occurs at radius r_1 at instant t_1. The same crest has a radius $r_2 = 2r_1$ at $t_2 = 2t_1$. (*a*) What is the wave speed? (*b*) Where is the crest at $t_3 = 3t_1$? (*c*) What is the ratio (energy per unit area at t_3)/(energy per unit area at t_1)?

3. Two circular wavefronts interfere as shown in **Figure 17.24**. (*a*) Draw the antinodal lines. (*b*) Is the medium displacement along these lines constant?

Figure 17.24

  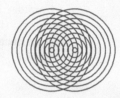

4. Two coherent sources are placed two wavelengths apart. Point P lies on a nodal line in the interference pattern of the two sources, and point Q lies on an antinodal line. (*a*) If one source is turned off, does the energy arriving at P increase, decrease, or remain the same? (*b*) If one source is turned off, does the energy arriving at Q increase, decrease, or remain the same? (*c*) Do your answers depend on which source is turned off?

5. Because sound waves diffract around an open doorway, you can hear sounds coming from outside the doorway. You cannot, however, see objects outside the doorway unless you are directly in line with them. What does this observation imply about the wavelength of light?

Answers

1. (*a*) The speed is the distance traveled divided by the time interval over which the motion occurs: r_1/t_1. (*b*) At $3t_1$ the crest is at $3r_1$. (*c*) The energy per unit length at r_1 is proportional to $1/r_1$. That at r_3 is proportional to $1/r_3 = 1/3r_1$. The ratio of the energies per unit length at the two instants is thus $1/3$.

2. (*a*) The speed is r_1/t_1. (*b*) The crest is at $3r_1$. The answers to parts *a* and *b* to this and the preceding question have to do with only the speed of the waves, not their dimensionality. (*c*) The energy per unit area at r_1 is proportional to $1/r_1^2$. That at r_3 is proportional to $1/r_3^2 = 1/(3r_1)^2$. The ratio of the energies per unit area at the two instants is $1/9$.

3. (*a*) See **Figure 17.25**. (*b*) No. The medium displacement at any point on an antinodal line varies as a function of time.

Figure 17.25

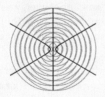 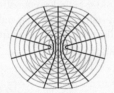 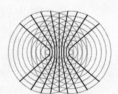

4. (*a*) Being on a nodal line, P initially receives no energy. After one source is turned off, it receives energy from the other source, and so the energy P receives increases. (*b*) Being on an antinodal line, Q receives four times the energy of a single source because the amplitude at Q is twice that of a single source. When one source is turned off, Q receives only the energy from a single source, and so the energy received at Q decreases. (*c*) It does not matter which source is turned off. The interference pattern exists only when both sources are on, and each point receives the energy from only a single source when the second source is turned off.

5. Because light does not diffract as it travels through the doorway, this observation implies that the wavelength of the light must be smaller than the width of the doorway. Given that visible light has wavelengths between 4×10^{-7} m and 7×10^{-7} m and most doorways are about 1 m wide and 2 m tall, this is indeed the case.

17.5 Intensity

As we saw in Section 17.1, the propagation of waves in more than one dimension causes the wave energy to be diluted over a larger and larger region. For waves in three dimensions, we define the **intensity** I as the energy delivered by the wave per unit time per unit area normal to the direction of propagation. Because the energy delivered per unit time is the definition of power, we can say

$$I \equiv \frac{P}{A}, \tag{17.1}$$

where P is the power delivered by the wave over an area A. For three-dimensional waves, the SI unit of intensity is thus W/m^2.

Consider a point source sending out waves uniformly in all directions. If there is no dissipation of energy, all the energy generated by the source must cross an imaginary sphere centered on the source. If the power delivered by the source is P_s, the intensity is

$$I = \frac{P_s}{A_{sphere}} = \frac{P_s}{4\pi r^2} \quad \text{(uniformly radiating point source)}, \tag{17.2}$$

and so we see that the intensity of a wave in three dimensions is proportional to $1/r^2$, as we concluded in Section 17.1.

For a two-dimensional surface wave, the intensity is defined as the energy incident per unit time per unit *length* normal to the direction of propagation. For a surface wave that delivers a power P over a length L,

$$I_{surf} \equiv \frac{P}{L}. \tag{17.3}$$

Consequently the SI unit of the intensity of a surface wave is W/m because we must replace the area of a sphere in Eq. 17.2 by the circumference of a circle: $I = P_s/(2\pi r)$. The intensity of two-dimensional waves is thus proportional to $1/r$.

Example 17.4 Sound wave energy

The minimum intensity audible to the human ear is called the *threshold of hearing*. For a 1.0-kHz sound wave, this threshold is approximately $10^{-12}\ W/m^2$. The maximum tolerable intensity, called the *threshold of pain,* is about $1.0\ W/m^2$ for a 1.0-kHz sound wave. (*a*) For a 1.0-kHz sound wave, estimate the amount of energy delivered to the ear in 1.0 s at the threshold of hearing and at the threshold of pain. (*b*) If the sound is produced by a loudspeaker 1.0 m away from the ear, what is, in each case, the power emitted by the loudspeaker?

❶ GETTING STARTED I am given two intensities, which give the power per unit area, and asked in part *a* to calculate the energy arriving at the ear in 1.0 s for each intensity. Because energy per unit time is power, part *a* reduces to identifying the area over which the energy is delivered. For part *b*, I am given a loudspeaker-to-ear distance, and I need to calculate the power for each loudspeaker at that distance.

❷ DEVISE PLAN For part *a*, to estimate the energy delivered to the ear in 1.0 s, I have to estimate the surface area of the ear. I can then use Eq. 17.1 to calculate the power delivered to the ear. To obtain the energy delivered, I multiply the power by the time interval. For part *b*, I assume that the loudspeaker is a point source that radiates sound uniformly in all directions. With this assumption, I can use the given intensities and source-to-ear distance in Eq. 17.2 to calculate the power emitted by the speaker.

❸ EXECUTE PLAN (*a*) The surface area of the outer ear is approximately $(0.030 \text{ m})^2 \approx 9.0 \times 10^{-4} \text{ m}^2$, so the power delivered to the ear at the threshold of hearing is about

$$P = IA = (10^{-12} \text{ W/m}^2)(9.0 \times 10^{-4} \text{ m}^2) = 9.0 \times 10^{-16} \text{ W}$$

$$= 9.0 \times 10^{-16} \text{ J/s}.$$

In 1.0 s, a mere 9.0×10^{-16} J—about a millionth of a billionth of a joule—gets delivered to the ear when a sound source emits at an intensity equal to the threshold of hearing. ✔

At the threshold of pain, the power delivered to the ear is

$$(1.0 \text{ W/m}^2)(9.0 \times 10^{-4} \text{ m}^2) = 9.0 \times 10^{-4} \text{ W},$$

and so 0.90 mJ is delivered to the ear in 1.0 s. ✔

(*b*) For the power emitted by the loudspeaker at the two thresholds, Eq. 17.2 gives

threshold of hearing:

$$P_s = I(4\pi r^2) = (10^{-12} \text{ W/m}^2)[4\pi(1.0 \text{ m})^2]$$

$$= 1.3 \times 10^{-11} \text{ W} ✔$$

threshold of pain:

$$P_s = (1.0 \text{ W/m}^2)[4\pi(1.0 \text{ m})^2] = 13 \text{ W}. ✔$$

❹ EVALUATE RESULT The energies delivered at the threshold of hearing are very small, and I have no good way of evaluating those. I do know that the ear can detect sounds made by the tiniest of movements—the rustle of a leaf, for example—and these movements cannot transmit much energy, so at least I know that the values I obtained are not unreasonably large. I know that headphones and loudspeakers can easily emit power levels at the threshold of pain. The fact that headphones can be powered with tiny batteries tells me that the 0.9-mW answer I obtained in part *a* is not unreasonable. Most stereo systems are rated at tens to hundreds of watts, so the 13 W in part *b* is also in the right range.

As Example 17.4 illustrates, the human ear is capable of handling sound waves over a tremendous range of intensities—intensities that differ by as much as a factor of 10^{12}. To deal conveniently with such an enormous intensity range, it is common to use a logarithmic scale for intensity. For example, if the intensity *I* increases by a factor of 10^{12}, as it does when going from the threshold of hearing to the threshold of pain, the common logarithm (see Appendix B), log *I*, increases by only 12. The intensity of a sound wave can then be expressed as part of a ratio that also includes the intensity I_{th} at the threshold of hearing:

$$\log\left(\frac{I}{I_{th}}\right). \tag{17.4}$$

Because the reference level ($I_{th} = 10^{-12} \text{ W/m}^2$) is chosen arbitrarily, the quantity $\log(I/I_{th})$, even though it is unitless, is given an artificial unit. This unit is the *bel* (B), in honor of the American scientist Alexander Bell (1847–1922). It is customary to express intensity levels of sound in **decibels** (dB), one-tenth of a bel: 1 dB = 0.1 B. The **intensity level** β of a sound, measured in decibels, is thus defined as

$$\beta \equiv (10 \text{ dB}) \log\left(\frac{I}{I_{th}}\right). \tag{17.5}$$

Figure 17.26 Average auditory response of the human ear. The ear is sensitive to longitudinal waves in air ranging in frequency from 20 Hz to 20 kHz. The sensitivity is greatest at 3 kHz, where the threshold of hearing is lowest. The curves give the intensity levels required to produce harmonic tones of the same perceived loudness.

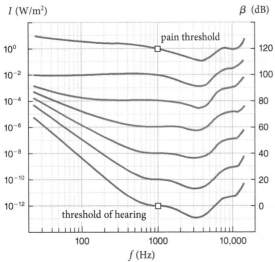

Table 17.1 Approximate intensity levels

Source	distance (m)	β (dB)	Description
Jet engine	50	140	pain
Pneumatic hammer	10	110	
Shout	1.5	100	very loud
Car horn	10	90	
Hair dryer	0.2	80	loud
Automobile interior		70	
Conversation	1	60	moderate
Office background		50	
Library background		40	
Suburban bedroom		30	quiet
Whisper	1	20	
Normal breathing	5	10	barely audible

At the threshold of hearing, where the intensity is $I = 10^{-12}$ W/m², we have $\log(I/I_{th}) = \log(1) = 0$, and so the intensity level $\beta = (10\ \text{dB})(0) = 0$ dB. At the threshold of pain, where the intensity is $I = 1$ W/m², we have $\log[(1\ \text{W/m}^2)/(10^{-12}\ \text{W/m}^2)] = 12$, and so the intensity level is $\beta = (10\ \text{dB})(12) = 120$ dB.

As **Figure 17.26** illustrates, the thresholds for hearing and pain vary with frequency. The five curves between the two threshold curves give the intensity required to produce harmonic sounds that a listener perceives as all being equally loud. For example, a sinusoidal 200-Hz tone of intensity 10^{-8} W/m² and a sinusoidal 1-kHz tone of intensity 10^{-10} W/m² are perceived to have the same loudness by the average listener. The right vertical scale gives the corresponding intensity level in decibels. **Table 17.1** gives the intensity levels in decibels of a number of common sound sources.

Exercise 17.5 Doubling the intensity

A clarinet can produce about 70 dB of sound. By how much does the intensity level increase if a second clarinet is played at the same time?

SOLUTION If the intensity of the sound produced by one clarinet is I_c, the intensity level of one clarinet is

$$\beta_1 = (10\ \text{dB}) \log\left(\frac{I_c}{I_{th}}\right) = 70\ \text{dB}.$$

The second clarinet doubles the intensity, so the intensity level becomes

$$\beta_2 = (10\ \text{dB}) \log\left(\frac{2I_c}{I_{th}}\right) = (10\ \text{dB})\left[\log 2 + \log\left(\frac{I_c}{I_{th}}\right)\right]$$

$$= (10\ \text{dB})\log 2 + \beta_1,$$

where I have used the logarithmic relationship $\log AB = \log A + \log B$. Because $\log 2 \approx 0.3$, the intensity level increases to $\beta_2 \approx (10\ \text{dB})(0.3) + 70\ \text{dB} = 73$ dB. So, even though the intensity doubles, the intensity *level* increases by only 3 dB. ✔

17.13 In Exercise 17.5, how many clarinets must play at the same time in order to increase the intensity level from 70 dB to 80 dB?

QUANTITATIVE TOOLS

Figure 17.27 When two sources emit waves of equal amplitude but slightly different frequencies, the waves superpose to produce a wave that has a pulsating amplitude (dashed line). We call this phenomenon *beating*.

(*a*) Displacement curves for two waves of equal amplitude but slightly different frequencies

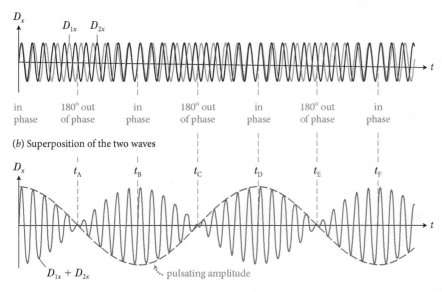

(*b*) Superposition of the two waves

17.6 Beats

In the first part of this chapter, the only type of interference we considered was interference between sources that emit waves of the same frequency. Let us now consider two sources that emit waves of equal amplitude but slightly different frequencies f_1 and f_2. Let the medium displacements caused by the individual waves at some fixed location be given by

$$D_{1x} = A \sin(2\pi f_1 t) \qquad (17.6)$$

$$D_{2x} = A \sin(2\pi f_2 t). \qquad (17.7)$$

These displacements are plotted in **Figure 17.27a**. Note that because the frequencies are different, the phase relationship between the two sources is not constant. At $t = 0$, the displacements are in phase, but over time they get out of phase, with the phase of one wave getting ahead of the other. At some instants the two waves are in phase $(0, t_B, t_D, \dots)$, and at these instants their crests coincide, giving rise to oscillations of double amplitude. At other instants, however, the waves are 180° out of phase $(t_A, t_C, \dots)$, and at these instants the crests and troughs of the two waves cancel each other.

The medium displacement that results from the superposition of the two waves is shown in Figure 17.27b. The resultant wave is an oscillation that changes amplitude with time. At instants $0, t_B, t_D,$ and t_F the amplitude is maximal; at instants $t_A, t_C,$ and t_E the amplitude is zero. This phenomenon, called **beating,** can be heard clearly with two tuning forks that oscillate at slightly different frequencies. Each maximum in the amplitude (at $0, t_B, t_D, \dots$) corresponds to what sounds like a *beat*. When both tuning forks simultaneously produce sound, you hear a tone whose intensity periodically fades in and out.

This phenomenon is similar to the moiré pattern in **Figure 17.28**, produced by two striped bands that have slightly different line spacings. At some places the lines of the two bands overlap, producing bright regions (labeled P, Q, R, S), while at others the lines fall precisely in between one another, producing dark regions.

Figure 17.28 Moiré pattern showing beating between two patterns with slightly different line spacings.

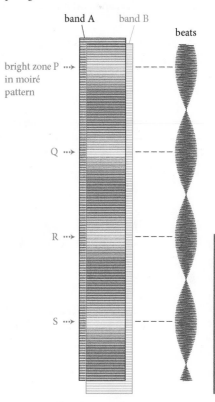

✋ **17.14** The lines in band A in Figure 17.28 are more closely spaced than those in band B. (*a*) How many more lines does band A have between points P and S? (*b*) Describe what happens to the pattern if band B is displaced downward by one-half the spacing between its lines.

Example 17.6 Lapping rounds

You and a friend are running laps around a race track. You begin together, and both of you run at constant speed, but you run faster than she does. Because the time interval T_y you take to complete one lap is smaller than the time interval T_f your friend takes to complete one lap, you finish the first lap ahead of her. Each lap you get farther ahead, until eventually you overtake your friend after she has run n laps (**Figure 17.29**). What is the frequency $f_{overtake}$ at which you overtake her? Express your answer in two ways: first in terms of T_y and T_f and then in terms of the frequencies f_y and f_f at which each of you completes one lap.

❶ **GETTING STARTED** Regardless of how many laps my friend has run, when I overtake her I have completed exactly one lap more than she has. So, if she has completed n laps, I have completed $(n + 1)$ laps. At that instant, the process starts over, and I overtake her again when I have completed another $(n + 1)$ laps. Note that n need not be a whole number of laps; for example, if I first overtake my friend after 8.2 laps (she having completed 7.2 laps) and we both keep running at constant speed, the next overtaking occurs when I have completed 16.4 laps.

❷ **DEVISE PLAN** The frequency at which I overtake my friend is the reciprocal of the time interval it takes me to overtake her. I know that she completes n laps, each of duration T_f, at the same instant I complete $(n + 1)$ laps, each lap of duration T_y. From this information I can determine n and the instant at which I first overtake her. The frequency at which I overtake her is the reciprocal of the period between that instant and the beginning of the race.

❸ **EXECUTE PLAN** At the instant I first overtake my friend, I have run for a time interval $(n + 1)T_y$ and she has run for a time interval nT_f. Because we started at the same instant, I have $nT_f = (n + 1)T_y$, giving me

$$n = \frac{T_y}{T_f - T_y}.$$

The instant at which the first overtake occurs is

$$t_{overtake} = nT_f = \frac{T_y T_f}{T_f - T_y},$$

and so the overtaking frequency is

$$f_{overtake} = \frac{1}{t_{overtake}} = \frac{T_f - T_y}{T_y T_f}. ✔$$

To express $f_{overtake}$ in terms of f_y and f_f, I use $f_y = 1/T_y$ and $f_f = 1/T_f$ and write

$$f_{overtake} = \frac{T_f - T_y}{T_y T_f} = \frac{1}{T_y} - \frac{1}{T_f} = f_y - f_f. ✔$$

❹ **EVALUATE RESULT** Because $f_y > f_f$, the overtaking frequency is positive, as I expect. My result shows that if both of us run laps at the same frequency, $f_y = f_f$, the overtaking frequency is zero. That makes sense because, running at the same rate, we never overtake each other. Another limiting case is when my friend remains at rest. In that case f_f is zero and so the overtaking frequency is equal to the frequency at which I complete laps. My answer makes sense in the limiting cases, which gives me confidence it is correct.

Figure 17.29 Example 17.6.

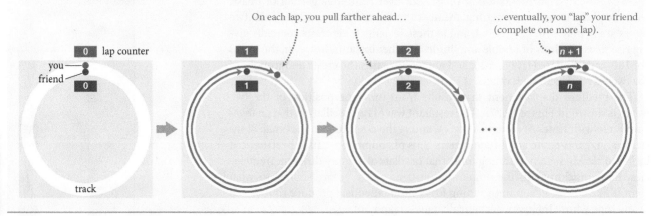

On each lap, you pull farther ahead... ...eventually, you "lap" your friend (complete one more lap).

you — friend —

lap counter
track

Suppose that in Example 17.6 your friend had run faster than you. In that case we could apply the same reasoning, but the indices y and f would now have been switched and the end result would have been $f_{overtake} = f_f - f_y$. (Because $f_y < f_f$, the overtaking frequency is again positive.) The general result, in case we do not know which frequency is greater, is thus $f_{overtake} = |f_f - f_y|$.

We can apply the reasoning I used in solving this example to the beating of two waves: The phases of the two waves catch up each time the wave of higher frequency has completed one more full oscillation than the other wave. If the wave frequencies are f_1 and f_2, then the frequency at which the beats occur, called the **beat frequency**, must, in analogy to the result from Example 17.6, be

$$f_{beat} \equiv |f_1 - f_2|. \tag{17.8}$$

We can reach the same conclusion by adding the expressions for the displacements of the two waves given by Eqs. 17.6 and 17.7:

$$D_x = D_{1x} + D_{2x} = A(\sin 2\pi f_1 t + \sin 2\pi f_2 t). \tag{17.9}$$

The sum of two sine functions can be written using the identity

$$\sin \alpha + \sin \beta = 2\cos\tfrac{1}{2}(\alpha - \beta)\sin\tfrac{1}{2}(\alpha + \beta), \tag{17.10}$$

so

$$D_x = 2A\cos\tfrac{1}{2}[2\pi(f_1 - f_2)t]\sin\tfrac{1}{2}[2\pi(f_1 + f_2)t]. \tag{17.11}$$

Because $\cos\theta = \cos(-\theta)$, we can replace the term $(f_1 - f_2)$ in Eq. 17.11 by $(f_2 - f_1)$ when $f_1 < f_2$. If we write $\Delta f = |f_1 - f_2|$ for the absolute value of the difference in frequency and $f_{av} = \tfrac{1}{2}(f_1 + f_2)$ for the average frequency, we can rewrite Eq. 17.11 in the form

$$D_x = 2A\cos\left[2\pi\left(\tfrac{1}{2}\Delta f\right)t\right]\sin(2\pi f_{av}t). \tag{17.12}$$

This expression tells us that the resulting wave has a frequency that is the average of the two original frequencies and an amplitude that varies with time. The factor $2A\cos\left[2\pi(\tfrac{1}{2}\Delta f)t\right]$ corresponds to the dashed curve in Figure 17.27b. It represents the slow variation in the amplitude of the combined wave.

The frequency of the amplitude variation is $\tfrac{1}{2}\Delta f = \tfrac{1}{2}|f_1 - f_2|$. However, *two* beats occur in each cycle of this amplitude variation. In Figure 17.27, one cycle in the amplitude variation lasts from $t = 0$ to $t = t_D$ (look at the dashed curve in Figure 17.27b and notice how it returns to its starting value at t_D). During this period *two* beats occur, one at t_A and the other at t_C. The beat frequency is therefore *twice* the amplitude variation frequency $\tfrac{1}{2}|f_1 - f_2|$, and so we see that the beat frequency is indeed equal to the absolute value of the difference in the frequencies of the two waves: $f_{beat} = |f_1 - f_2|$.

Exercise 17.7 Tuning a piano

Your middle-C tuning fork oscillates at 261.6 Hz. When you play the middle-C key on your piano together with the tuning fork, you hear 15 beats in 10 s. What are the possible frequencies emitted by this key?

SOLUTION The beat frequency—the number of beats per second—is equal to the difference between the two frequencies (Eq. 17.8). I am given the frequency of the tuning fork, $f_t = 261.6$ Hz, and the beat frequency, $f_B = (15 \text{ beats})/(10 \text{ s}) = 1.5$ Hz. I do not know, however, whether the frequency f_p of the struck middle-C piano key is higher or lower than that of the tuning fork.

If it is higher, I have $f_B = f_p - f_t$.

If it is lower, then $f_B = f_t - f_p$.

So

$$f_p = f_t \pm f_B = 261.6 \text{ Hz} \pm 1.5 \text{ Hz}$$

and the possible frequencies emitted by the out-of-tune middle-C key are 260.1 Hz and 263.1 Hz. ✔

17.15 Do two sound waves of slightly different frequencies and *different* amplitudes cause beats?

17.7 Doppler effect

When an observer and a source of waves move relative to each other, the observed wave frequency is not the same as the frequency emitted by the source. When observer and source move toward each other, the observed frequency is higher than the emitted frequency; when they move away from each other, the observed frequency is lower than the emitted frequency. This phenomenon is called the **Doppler effect.**

Let us determine how the observed frequency deviates from the actual frequency of the source, depending on the speeds of source and observer relative to the medium through which the waves propagate. In **Figure 17.30a,** a loudspeaker that is stationary relative to a medium emits a sound wave that is detected by a microphone that is also stationary relative to the medium. The period of the wave is T, and so one wavefront of a given phase (a crest, for example) leaves the loudspeaker each period T. Once the first wavefront reaches the microphone, it thereafter receives one wavefront each period T, and so the frequency received

Figure 17.30 The Doppler effect demonstrated by a speaker and a microphone. Notice that motion of the speaker toward the microphone and motion of the microphone toward the speaker both cause the frequency received by the microphone to increase—but not by the same amount. The two situations are not symmetrical because motion of the speaker shortens the wavelength of the propagating sound, whereas motion of the microphone does not.

(*a*) Speaker and microphone are stationary relative to medium

(*b*) Speaker moves toward microphone

(*c*) Microphone moves toward speaker

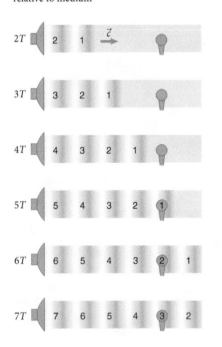

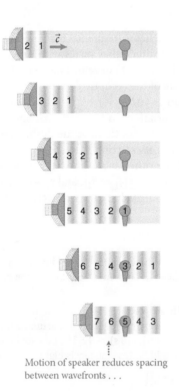

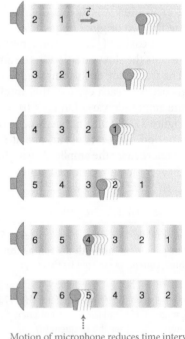

Motion of speaker reduces spacing between wavefronts . . .

Motion of microphone reduces time interval between arrival of successive wavefronts . . .

$D_x(t)$ graphs for waves received by microphone

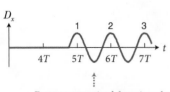

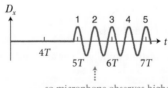

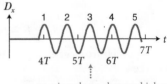

Frequency received by microphone is same as frequency emitted by speaker.

. . . so microphone observes higher frequency.

. . . so microphone observes higher frequency (but not as high as in part *b*).

by the microphone is identical to the frequency emitted by the loudspeaker. The graph at the bottom of Figure 17.30*a* shows the longitudinal medium displacement caused by the sound wave at the location of the microphone as a function of time.

Figure 17.30*b* shows what happens when the loudspeaker emits the same sound wave while moving at constant velocity toward the microphone. Because of the motion of the loudspeaker, each wavefront is emitted closer to the previous wavefront than if the loudspeaker were stationary. Because the speed of the wavefronts, which is determined by the properties of the medium (air in this case), does not depend on the motion of the loudspeaker, the more closely spaced wavefronts generated by the moving loudspeaker reach the microphone at shorter time intervals than those of the stationary loudspeaker. For example, between $t = 5T$ and $t = 7T$ only two cycles between wavefronts 1 and 3 pass the microphone when the source is stationary (Figure 17.30*a*), but four wave cycles between wavefronts 1 and 5 pass the microphone when the source is moving (Figure 17.30*b*). So the frequency detected by the microphone for the moving source is $(4 \text{ cycles})/(7T - 5T) = (4 \text{ cycles})/(2T)$, or two cycles per period instead of one cycle per period; in other words, the frequency heard by the microphone is twice the frequency emitted by the moving loudspeaker.

To obtain a general expression for the frequency received from a source S moving at speed v_s, let us evaluate the spacing between wavefronts. **Figure 17.31** shows the motion of the source during one period. (For now we consider the case when the source moves more slowly than the wave, $v_\mathrm{s} < c$; in the next section we consider what happens when $v_\mathrm{s} > c$.) The labeled wavefront was emitted when the source was at $x = 0$. During the time interval T, the source has moved from $x = 0$ to $x_1 = +v_\mathrm{s}T$, at which point it emits the next wavefront. The first wavefront meanwhile has moved from $x = 0$ out to a radius $x_2 = cT$. To an observer somewhere to the right of Figure 17.31, the distance between the two wavefronts is decreased: $\Delta x = x_2 - x_1 = cT - v_\mathrm{s}T = (c - v_\mathrm{s})T$. The period T_0 of the wave that reaches this observer is equal to the time interval between the

Figure 17.31 Doppler effect for a source moving toward an observer. The observer is off to the right and not shown.

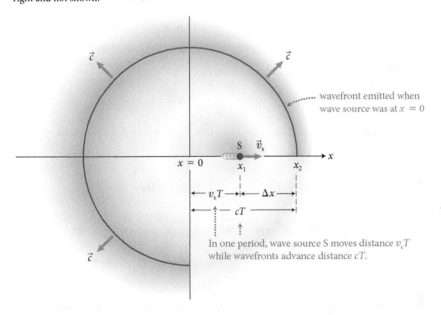

wavefront emitted when wave source was at $x = 0$

In one period, wave source S moves distance $v_\mathrm{s}T$ while wavefronts advance distance cT.

arrival of two successive wavefronts and is thus equal to the distance Δx divided by the speed of the waves:

$$T_0 = \frac{\Delta x}{c} = \frac{c - v_s}{c}T \quad \text{(source approaching observer, } v_s < c\text{).} \quad (17.13)$$

The frequency of the approaching source heard by the observer is thus

$$f_0 \equiv \frac{1}{T_0} = \frac{c}{c - v_s}\frac{1}{T}$$

$$= \frac{c}{c - v_s}f_s \quad \text{(source approaching observer, } v_s < c\text{),} \quad (17.14)$$

where f_s is the frequency emitted by the source. Note that because v_s is positive, $[c/(c - v_s)] > 1$, so the observed frequency is higher than the emitted frequency, as expected.

Now let us consider an observer on the opposite side of Figure 17.31. To this observer, the source is receding and the distance between wavefronts is increased: $\Delta x = (c + v_s)T$. Consequently the minus sign in front of v_s becomes a plus in Eqs. 17.13 and 17.14:

$$f_0 \equiv \frac{1}{T_0} = \frac{c}{c + v_s}\frac{1}{T} = \frac{c}{c + v_s}f_s \quad \text{(source receding from observer),} \quad (17.15)$$

and the observed frequency is now lower than the emitted frequency, as we expect. Contrary to the situation described by Eq. 17.14, there is no limit on the speed of the source in this case.

Example 17.8 Moving train

Standing alongside a straight section of railroad track as a train passes, you record the sound of the train's horn. By analyzing the recording, you determine that the frequency of the sound was 483 Hz as the train approached and 405 Hz after it passed. At what speed was the train moving? Use 343 m/s for the speed of sound in air.

❶ GETTING STARTED I am given two frequencies for one sound source, so I know this problem is about a Doppler shift. I assume the train moved at constant velocity and denote its unknown speed by v. Because the horn—that is, the sound source—was on the train, v is also the speed at which the horn moved, and I know I can use some Doppler equation to calculate this speed. The wave speed is given as $c = 343$ m/s.

❷ DEVISE PLAN As the train approaches, the frequency I hear is given by Eq. 17.14. After it passes, the frequency is given by Eq. 17.15. This gives me two equations with two unknowns, f_s (the frequency of the source) and v, so I can solve for v.

❸ EXECUTE PLAN As the train approaches, the observed frequency is given by Eq. 17.14:

$$f_1 = \frac{c}{c - v_s}f_s = \frac{c}{c - v}f_s. \quad (1)$$

After the train passes, the observed frequency is given by Eq. 17.15:

$$f_2 = \frac{c}{c + v_s}f_s = \frac{c}{c + v}f_s. \quad (2)$$

Solving Eq. 2 for f_s and substituting the result into Eq. 1, I obtain

$$f_1 = \frac{c}{c - v}\left(\frac{c + v}{c}f_2\right) = \frac{c + v}{c - v}f_2.$$

Multiplying both sides by $(c - v)$ gives

$$f_1(c - v) = f_2(c + v)$$

$$v(f_1 + f_2) = c(f_1 - f_2).$$

Therefore

$$v = \frac{f_1 - f_2}{f_1 + f_2}c = \frac{483 \text{ Hz} - 405 \text{ Hz}}{483 \text{ Hz} + 405 \text{ Hz}}(343 \text{ m/s}) = 30 \text{ m/s.} ✔$$

❹ EVALUATE RESULT This is about 110 km/h or 70 mi/h, not unreasonable for a train.

17.16 In Example 17.8, what is the frequency of the sound the horn makes?

Figure 17.32 Doppler effect for an observer moving toward the source.

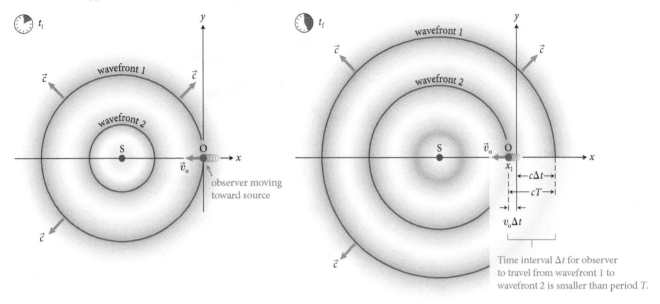

Time interval Δt for observer to travel from wavefront 1 to wavefront 2 is smaller than period T.

Applications of the Doppler effect range from measuring blood flow by reflecting a beam of ultrasound waves from red blood cells to measuring the speed of a storm using radar. The latter uses electromagnetic waves (see Chapter 30), which, like sound waves, exhibit a Doppler effect.

Any movement by an observer also affects the frequency of sound waves. When an observer is moving toward a stationary source, as in Figure 17.30c, the observer travels through the oncoming wavefronts and consequently observes more of them per unit time than if she or he were at rest. Between $t = 4T$ and $t = 6T$ in Figure 17.30c, for example, three wave cycles pass the microphone, and so the observed frequency is $(3 \text{ cycles})/(6T - 4T) = (1.5 \text{ cycle})/T = 1.5f_s$. Because the microphone moves toward the loudspeaker, each subsequent wavefront has to travel a shorter distance to reach the microphone, and so the observed frequency is higher than the emitted frequency.

Figure 17.32 allows us to relate the increase in frequency to the speed v_o of a observer O who is moving toward the source S. At t_i the observer is at $x = 0$ just as wavefront 1 passes her. At t_f the observer meets wavefront 2 at position x_1, and so she detects one complete cycle (that is, two wavefronts) in $\Delta t = t_f - t_i$. During that time interval, wavefronts 1 and 2 have moved a distance $c\Delta t$. From the figure we see that the sum of the distances traveled by the wavefront and the moving observer is equal to the distance between wavefronts, so $c\Delta t + v_o\Delta t = cT$. The time interval at which successive wavefronts reach the observer is thus

$$\Delta t = \frac{cT}{c + v_o}. \tag{17.16}$$

The frequency she observes is

$$f_o \equiv \frac{1}{\Delta t} = \frac{c + v_o}{c}\frac{1}{T} = \frac{c + v_o}{c}f_s \quad \text{(observer approaching source).} \tag{17.17}$$

Because $(c + v_o) > c$, the observed frequency is higher, as expected.

Let us now consider an observer who moves away from the source at a speed that is smaller than the speed of sound, $v_o < c$. (If the observer were to move

Figure 17.33 Doppler effect for an observer moving away from the source at a speed $v_o < c$.

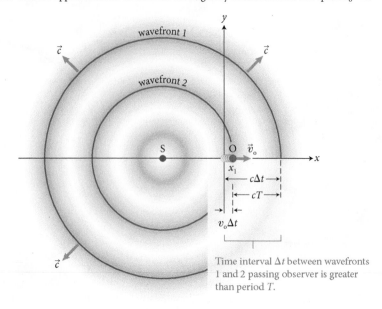

Time interval Δt between wavefronts 1 and 2 passing observer is greater than period T.

away at a greater speed, the sound would never catch up with the observer.) Again, we take the observer to be at $x = 0$ just as wavefront 1 passes him at t_i. Figure 17.33 shows the situation at t_f when wavefront 2 catches up with the observer at position x_1. In the time interval $\Delta t = t_f - t_i$ the observer hears a complete cycle. During that time interval, wavefronts 1 and 2 have moved a distance $c\Delta t$. From the figure we see that the distance between wavefronts is equal to the difference between the distances traveled by the wavefront and the moving observer, so $c\Delta t - v_o\Delta t = cT$. The time interval at which successive wavefronts reach the observer is thus

$$\Delta t = \frac{cT}{c - v_o}. \tag{17.18}$$

Consequently the plus sign in front of v_o in Eq. 17.17 becomes a minus, and the observed frequency is thus

$$f_o \equiv \frac{1}{\Delta t} = \frac{c - v_o}{c}\frac{1}{T}$$

$$= \frac{c - v_o}{c}f_s \quad \text{(observer receding from source, } v_o < c). \tag{17.19}$$

Because $(c - v_o) < c$, the observed frequency is lower, as expected.

If both source and observer are moving, the two effects multiply—the moving source causes wavefronts to be spaced differently, giving rise to the observed frequency given in Eqs. 17.14 and 17.15, and the moving observer detects the wavefronts at a different rate, as indicated by Eqs. 17.17 and 17.19. The two effects can be combined in one equation:

$$f_o = \left(\frac{c \pm v_o}{c}\right)\left(\frac{c}{c \pm v_s}\right)f_s = \frac{c \pm v_o}{c \pm v_s}f_s, \tag{17.20}$$

which can be written in the more symmetrical form

$$\frac{f_o}{f_s} = \frac{c \pm v_o}{c \pm v_s}. \tag{17.21}$$

Figure 17.34 To use the general Doppler equation, the key thing you need to remember is that the observed frequency increases when source and observer move toward each other and decreases when they move apart.

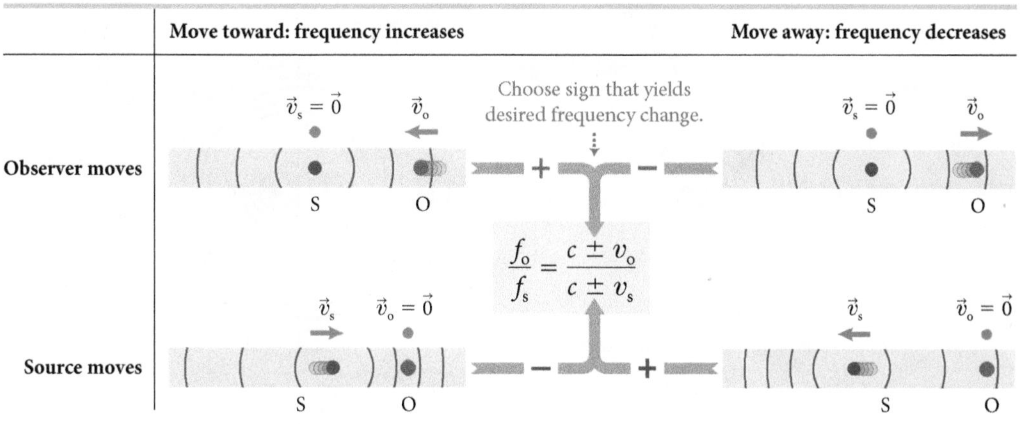

Keep in mind that in Eqs. 17.20 and 17.21 the speeds of the observer and source are measured relative to the medium. Also, these expressions are subject to the same conditions as in Eqs. 17.14 and 17.19: when the source is approaching the observer $v_s < c$, and when the observer is receding from the source $v_o < c$.

Rather than trying to remember the correct choices for the signs in front of v_o and v_s in this expression, it is easiest to remember that the observed frequency increases when source and observer move toward each other and decreases when they move apart. For example, the top left of Figure 17.34 shows an observer approaching a stationary source ($v_s = 0$). The observed frequency is higher and therefore you select the plus sign in front of v_o in Eq. 17.21. When the observer moves away from the stationary source (top right of Figure 17.34), the observed frequency is lower and so you select the minus sign in front of v_o in Eq. 17.21. The bottom left of Figure 17.34 shows a source moving toward an observer who is at rest relative to the medium ($v_o = 0$). The observed frequency is higher, and therefore you select the minus sign in front of v_s in Eq. 17.21. In the bottom right of Figure 17.34 the source moves away from the stationary observer; the observed frequency is now lower and you select the plus sign in front of v_s in Eq. 17.21.

17.17 (*a*) Suppose an observer moves at half the speed of sound toward a stationary source of sound waves. By what factor does the observed frequency differ from the emitted frequency? (*b*) By what factor does the observed frequency differ if, instead of the observer moving, the source moves toward the stationary observer at half the speed of sound? (*c*) Why is there a difference?

17.8 Shock waves

The observed frequency f_o given by Eq. 17.14 becomes infinite when the speed of the source approaches the speed of sound. As can be seen in Figure 17.35a, the wavefronts in front of a source moving at a speed $v_s < c$ crowd together. When a source moves at the speed of sound, *all* wavefronts in the forward direction travel with the source, and so the spacing between them is reduced to zero. A source traveling *faster* than the the speed of sound overtakes the wavefronts it has generated. The wavefronts then pile up to form a wavefront that is wedge-shaped in two dimensions and cone-shaped in three dimensions.

This wedge- or cone-shaped wavefront, called a **shock wave,** occurs whenever the speed of any object moving through a medium is higher than the the speed of sound in the medium, *even if the object emits no sound.* A bullet moving faster

Figure 17.35 The formation of a shock wave as the speed of a source reaches and then exceeds the speed of sound in the medium.

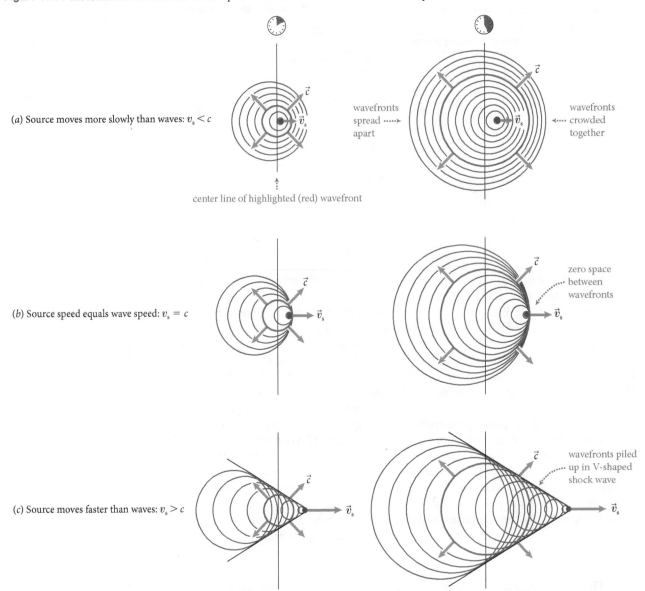

(a) Source moves more slowly than waves: $v_s < c$

wavefronts spread ······▸ apart

wavefronts ◂····· crowded together

center line of highlighted (red) wavefront

(b) Source speed equals wave speed: $v_s = c$

zero space ···· between wavefronts

(c) Source moves faster than waves: $v_s > c$

wavefronts piled up in V-shaped shock wave

than the speed of sound in air, for example, creates a shock wave (**Figure 17.36**) even though the bullet emits no sound. Air piles up in front of the speeding bullet like snow in front of a snowplow, and this piling up of the air creates a compression in front of the bullet. This compression then moves outward at the speed of sound, just as the compression in Figure 17.8 does, and forms the wedge- or cone-shaped shock wave, which trails the bullet like a water wake trailing behind a ship.

When a shock wave sweeps over your ears, the piled-up wavefronts sound like an explosion, just as a small boat gets jolted when struck by the wake of a passing speedboat. For a small object, such as a bullet, the shock wave produces a sharp, cracking sound. The crack of a long whip is a shock wave produced when the tip of the whip is made to move faster than the speed of sound. For large objects, the shock wave sounds like a thunderous boom called a *sonic boom*. The sonic boom of a supersonic airplane reaches the pain threshold of the human ear even at a distance of 20 km. For this reason, supersonic transport is not feasible over land. The cone-shaped shock wave dragging behind the plane would break windows on the ground even at its maximum flying altitude of about 20 km.

QUANTITATIVE TOOLS

Figure 17.36 Shock wave produced by a bullet traveling through air.

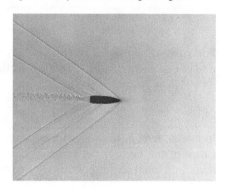

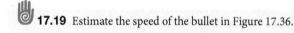 **17.18** Does an observer moving at twice the speed of sound toward a stationary sound source detect a sonic boom?

As can be seen in **Figure 17.37**, the higher the source speed, the narrower the wedge or cone formed by the shock wave. From the right-angle triangle PQR, we see that the angle the shock wave makes with the direction of motion is given by

$$\sin \theta = \frac{c}{v_s} \quad (v_s > c). \tag{17.22}$$

For motion through air, the ratio v_s/c is called the *Mach number*. A plane flying at Mach 1.8, for example, has a speed of 1.8 times the speed of sound.

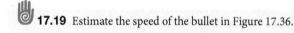 **17.19** Estimate the speed of the bullet in Figure 17.36.

Figure 17.37 The shape of a shock wave depends on the speed of the source. The higher the source speed, the smaller the angle θ and so the sharper the cone or wedge.

(*a*) Lower source speed v_s: wider angle θ

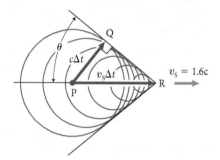

(*b*) Higher source speed v_s: narrower angle θ

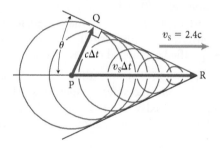

Chapter Glossary

SI units of physical quantities are given in parentheses.

Antinodal line A line in the interference pattern that is formed by the overlapping of two or more wave sources where the displacement of the medium has the maximum amplitude.

Beat frequency f_{beat} (Hz) The frequency at which beats occur between two waves of slightly different frequencies:

$$f_{beat} = |f_1 - f_2|. \qquad (17.8)$$

Beating Oscillating amplitude variation due to the superposition of two waves of equal amplitude but slightly different frequencies.

Coherent sources Two or more sources that are emitting waves that have a constant phase difference.

Decibel (dB) The logarithmic unit of measurement (not part of the SI system) that expresses the magnitude of a physical quantity (usually power or intensity) relative to a specified or implied reference level. See also *intensity level*.

Diffraction The spreading out of waves around obstacles or beyond the sides of apertures. The spreading is more pronounced when the size of the obstacle or aperture is about equal to or smaller than the wavelength of the wave.

Doppler effect A change in the observed wave frequency due to the relative motion of a wave source and an observer. The observed frequency f_o is given by

$$\frac{f_o}{f_s} = \frac{c \pm v_o}{c \pm v_s}, \qquad (17.21)$$

where v_s and v_o are the speeds relative to the medium of the source and the observer, respectively. The signs are to be chosen such that the frequency $f_o > f_s$ when source and observer approach each other and $f_o < f_s$ when they move apart.

Huygens' principle Any wavefront may be regarded as a collection of many closely spaced, coherent point sources.

Intensity I (W/m^2) For a three-dimensional wave: the energy incident per second per unit area normal to the direction of propagation:

$$I \equiv \frac{P}{A}. \qquad (17.1)$$

For a two-dimensional wave: the energy incident per second per unit *length* normal to the direction of propagation:

$$I_{surf} \equiv \frac{P}{L}. \qquad (17.3)$$

Intensity level β (dB) A quantity proportional to the logarithm of the ratio of the intensity of a sound wave to the threshold intensity for hearing ($I_{th} = 10^{-12}$ W/m^2):

$$\beta \equiv (10 \text{ dB}) \log\left(\frac{I}{I_{th}}\right). \qquad (17.5)$$

Nodal line A line in the interference pattern where the displacement of the medium is always zero.

Planar wavefront A two-dimensional, flat wavefront. All wavefronts are approximately planar when the distance to the source is large relative to the wavelength. The amplitude of a planar wavefront does not decrease as the wavefront propagates.

Point source A single identifiable source of a wave or other disturbance that can be treated as if it has no physical extent and is located at a single point in space.

Shock wave A conical or wedge-shaped disturbance caused by the piling up of wavefronts from a source moving at a speed greater than or equal to the wave speed. The higher the source speed, the narrower the angle of the shock wave:

$$\sin \theta = \frac{c}{v_s} \quad (v_s > c). \qquad (17.22)$$

Sound Longitudinal waves that propagate through a solid, liquid, or gas in the frequency range 20 Hz–20 kHz (called the *audible* frequency range).

Spherical wave A three-dimensional wave whose wavefronts are spheres. In the absence of energy dissipation, the amplitude of a spherical wave decreases with distance r to the source as $1/r$.

Surface wave A wave that spreads out in two dimensions. In the absence of energy dissipation, the amplitude of a surface wave decreases with distance r to the source as $1/\sqrt{r}$.

Wavefront A curve or surface in a medium on which all points of a propagating wave have the same phase of oscillation.

18

Fluids

CONCEPTS

QUANTITATIVE TOOLS

In earlier chapters we dealt with the motion and behavior of solid objects only. In the next few chapters we extend our description to include two other states of matter: liquids and gases (**Figure 18.1**). The interior forces exerted on the atoms in a solid keep each atom at a particular location in the solid. Because the atoms are not free to move relative to one another, a piece of solid material has a definite size and shape. In a liquid, the forces exerted on the particles* are not strong enough to keep them from moving relative to one another. Consequently a liquid sample has a definite volume but adjusts to the shape of its container. In a gas, the particles are far apart. They move about randomly and freely, and so a sample of gas has neither a definite volume nor a definite shape.

Liquids and gases are collectively called **fluids** because they flow.

Because fluids flow and change shape, we cannot describe their behavior in the same way we described the behavior of solids. When you look at the flow of water in a river, for example, it makes no sense to speak of "the velocity of the water" because the water may move fast in one spot and more slowly in another, or it may be flowing in one direction at one location and in another direction a bit farther downstream. Unlike a solid object, in other words, water in a river cannot be characterized by a simple set of kinematic quantities—the motion of its center of mass and rotation about that center. However, you can specify its velocity at a specific location.

In general, therefore, describing the motion of fluids requires an approach in which we associate quantities such as velocity not with a body of fluid as a whole but with a position within the body of fluid.

18.1 Forces in a fluid

The result of exerting forces on fluids is very different from the result when forces are exerted on solids. To better understand this difference, let us begin by examining how a solid object at rest responds to a set of forces whose vector sum is zero. Because the vector sum is zero, the object remains at rest, but because no solid material is perfectly rigid, the object deforms when opposing forces are exerted on it. Such opposing forces are said to cause a *stress*. A solid object that returns to its original shape when the stress is removed is said to be *elastic* (see Section 8.6). Beyond the elastic limit, the object deforms permanently, and still greater forces may cause the object to fracture. Within the elastic range, the amount of deformation caused by a given stress depends on the properties of the material and typically obeys Hooke's law (see Section 8.9) for small deformations; that is, the deformation is linearly proportional to the magnitude of the applied forces.

Because the atoms in a solid are at fixed positions, a solid object can withstand forces applied to its surface in a number of ways. A set of opposing forces along the same line of action (**Figure 18.2a**) causes either *tensile stress*, which stretches the object, or *compressive stress*, which compresses it. Under such stress, the object's dimensions perpendicular to the line of action of the forces also change to compensate for the stretching or compressing along the line of action.

Figure 18.1 States of matter.

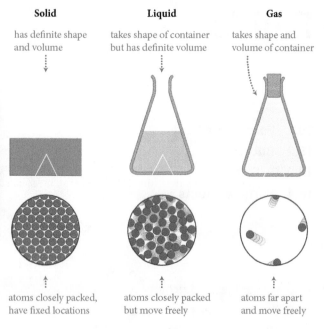

Solid	Liquid	Gas
has definite shape and volume	takes shape of container but has definite volume	takes shape and volume of container

atoms closely packed, have fixed locations

atoms closely packed but move freely

atoms far apart and move freely

Figure 18.2 Different types of stresses caused by equal forces applied in opposite directions to solid objects.

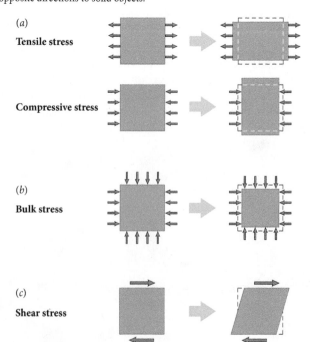

(a)
Tensile stress

Compressive stress

(b)
Bulk stress

(c)
Shear stress

*Both liquids and gases can be made up of either atoms or molecules. From now on, we'll use the generic term *particle* to refer to either atoms or molecules in a liquid or gas.

In spite of this compensating change, however, precise measurements show that the object's volume increases by a small amount when the object is stretched and decreases by a small amount when the object is compressed.

When opposing forces are exerted perpendicular to all surfaces of a solid object (Figure 18.2b), the atoms are pushed closer together in all directions, resulting in what is called *bulk stress*. Such a stress causes the volume of the object to decrease.

A third type of stress, *shear stress*, occurs when opposing forces are exerted tangent to the surfaces of a solid object (Figure 18.2c). For small forces, the object's shape changes—turning a rectangular face into a parallelogram, say—but the volume of the object does not change.

Let us now examine how a fluid responds to these same three types of stresses: tensile/compressive, bulk, and shear. If you put your hands under water parallel to each other and some distance apart and then move them toward each other, you feel resistance. Your moving hands exert a force on the water and, because forces always come in interaction pairs, the water, in turn, exerts a force on your hands. Once your hands stop moving, however, this force exerted by the water on them disappears. In addition, when you remove your hands, the water molecules don't flow back to where they were before you moved your hands. In other words, water is not elastic the way some solid materials are. If you compress an elastic solid—a foam rubber ball, say—the way you "compress" the water by bringing your hands together, you must keep exerting the force if you want to keep the ball compressed. Once you remove your hands, the ball returns to its original shape. This is not true for most liquids (water, rubbing alcohol, and gasoline, for instance) or for all gases under ordinary circumstances; these fluids are not elastic. You can easily verify that they also cannot support static tensile or shear stresses. Such fluids are said to be *nonviscous*. Liquids such as jelly or tar, which do not behave the way water does in response to a shear stress, can support a shear stress and are said to be *viscous*.

Let us next look at how fluids behave when subjected to bulk stress. If you exert forces on a fluid so as to cause a bulk stress, as in Figure 18.2b, the fluid resists being compressed. Imagine putting water in a cylinder fitted with a piston and pushing down on the piston (**Figure 18.3a**). Because the water has nowhere to go and because the water molecules are already as close together as they can get, a very large force is required to change the volume of the water, and so you feel a rapidly growing resistive force exerted by the water on the piston. If, instead of water, you put a gas in the cylinder, the gas compresses easily as you push on the piston because gas particles are far apart and can easily be pushed closer together. As the gas is compressed, however, it exerts on the piston a force of increasing magnitude. Once that force is equal in magnitude to the force you exert on the piston, the system reaches mechanical equilibrium (Figure 18.3b). The greater the force you exert, the more the gas is compressed.

Figure 18.3 All fluids resist bulk stresses, but liquids don't compress much, whereas gases are easily compressed.

(*a*) Liquids oppose compression

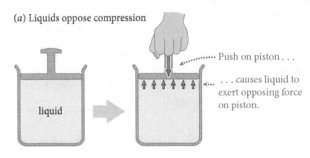

(*b*) Gases compress more easily

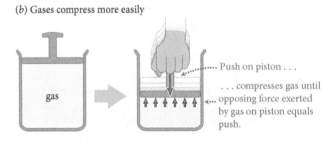

(*c*) Fluids exert forces on all surfaces they contact

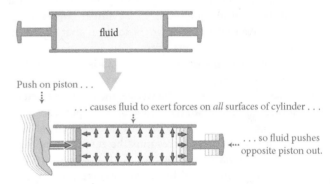

So, a fluid can sustain forces that are perpendicular to its surfaces, provided the forces are exerted simultaneously on *all* its surfaces. If you pushed the left piston in Figure 18.3c inward without also pushing the right piston inward, for example, you would simply push the water and the right piston out of the tube. Likewise, if you replaced a section of either the top or bottom wall of the cylinder in Figure 18.3c with a piston and again pushed on the left piston, you would also need to push both the right piston and the newly added piston inward in order to keep the water contained. From these observations we conclude that, in mechanical equilibrium, a fluid in a container exerts a (normal) force that is distributed along *all* surfaces of the container that are in contact with the fluid, and those surfaces exert on the fluid a force that is of equal magnitude but in the opposite direction.

The main difference between gases and liquids is that gases tend to compress easily, often greatly changing the volume and therefore the mass density (mass per unit volume, see Section 1.6), whereas the volume and mass density of a liquid do not change easily. For simplicity, from now on we shall assume that the liquids we deal with are *incompressible*; that is, their mass density and volume are fixed.

> **When a gas is subjected to bulk stress, its mass density and volume change easily. When a liquid is subjected to bulk stress, its mass density and volume can be considered unchanged.**

18.1 In Figure 18.3c, suppose you push the left piston rightward with a force of magnitude 10 N. (*a*) To keep the water in the cylinder at rest, what force must you exert on the (identical) right piston? (*b*) How does the force $\vec{F}_{wr}^c$ exerted by the water on the right piston compare to the force $\vec{F}_{pl}^c$ exerted by you on the left piston?

The two pistons in Checkpoint 18.1 are identical. Suppose we divide the right piston in Figure 18.3c into two pistons, each having half the surface area of the original piston. As you might expect, to keep the water in the cylinder at rest while pushing the left piston rightward with a force of magnitude 10 N, you would need to exert on each half-piston a force of magnitude 5 N, which suggests a linear proportionality between surface area and force.

To examine this relationship in more detail, consider a fluid at mechanical equilibrium in a cylinder (**Figure 18.4**). Forces are exerted on the two pistons so that the fluid is subject to a compressive stress. Now imagine a wedge-shaped volume of the fluid. Because liquids cannot sustain any shear stresses (resulting from forces tangent to the wedge surfaces) exerted on the wedge, the forces exerted by the surrounding fluid on the wedge must be perpendicular to the three wedge surfaces. Because the wedge is at rest, the vector sum of these forces must be zero. Let the three wedge surfaces have areas A_1, A_2, and A_3. Each of these areas is proportional to the length of the corresponding side of the dashed triangle in Figure 18.4a, and each is subject to a normal force exerted by the surrounding fluid. Let us denote these forces by $\vec{F}_1$, $\vec{F}_2$, and $\vec{F}_3$. Because these vectors add up to zero, we can arrange them in a triangle as shown in Figure 18.4b. Because the vectors are perpendicular

to the sides of the dashed triangle in Figure 18.4a, the triangle in Figure 18.4b is similar to the dashed triangle, and so the ratio of the force magnitudes $F_1:F_2:F_3$ is the same as the ratio of the lengths of the sides of the dashed triangle. Because the dimension of the wedge in the direction perpendicular to the plane of the drawing is fixed, the area A of each side of the wedge is proportional to its length. Therefore $F_1:F_2:F_3 = A_1:A_2:A_3$, or $F_1/A_1 = F_2/A_2 = F_3/A_3$. In other words, the ratio of the magnitude of the force exerted by the fluid on a surface and the area of that surface is constant. This ratio is called **pressure,** denoted by the symbol P. Pressure is a scalar and is taken to be positive when a fluid is under compression.

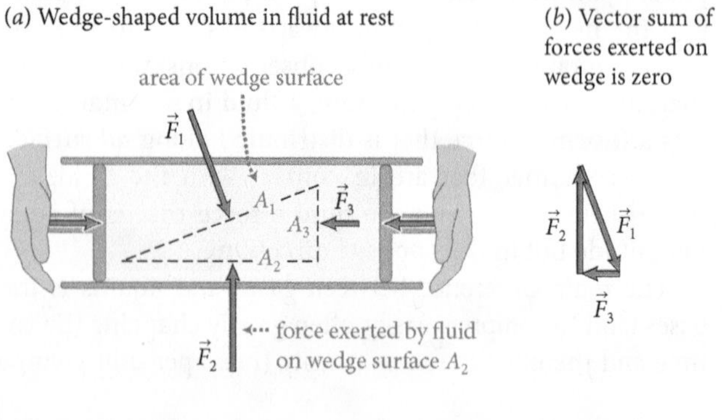

It is important to distinguish pressure from force. The two are related because forces exerted on the fluid cause the pressure in a fluid. Force, however, is a vector, while pressure is a scalar. In addition, forces of the same magnitude can cause very different pressures depending on the surface area over which these forces are exerted. Imagine, for example, exerting a force of 10 N on your skin first with the tip of a pin and then with the cover of a book. Even though the force is the same in both cases, the effect is very different. At the tip of the pin, the pressure caused by the 10-N force is very great because the surface area of the pin is small. The pressure caused by the pin is likely to cause pain, and the pin might penetrate into your skin. The book, however, spreads that same 10-N force out over a much greater surface area, and so the resulting pressure is much smaller.

Another important point to keep in mind is that because forces always come in interaction pairs, the force exerted by a container wall on a fluid (which causes the pressure in the fluid) and the force exerted by the fluid on the container wall form an interaction pair. The pressure in a fluid therefore causes the fluid to exert forces on anything that is in contact with it. Because of the mobility of the particles in a fluid, pressure is transmitted in all directions to all parts of the fluid. However, because fluids cannot sustain shear stresses, the forces exerted by the fluid on any surfaces that contain the fluid are always exerted perpendicular to the surface.

In gases, pressure can be only positive. Liquids, however, can sustain negative pressures. For example, if you hold the container of Figure 18.3a upside down as shown in **Figure 18.5a** and pull down on the piston, you feel a large resistive force, which tells you that the liquid pulls the piston inward. The water molecules exert an attractive force on one another and on the surfaces of the container, pulling them inward as you pull the piston down. So, just as a liquid under compression causes an outward push that we associate with a positive pressure, we find that a "stretched" liquid causes an inward pull that we associate with a negative pressure.

(You can easily try this yourself. Fill a glass to the top with water, place an index card on top of the glass, making sure no air is left between the water surface and the card, and then turn the glass upside down while holding the card in place with the palm of your hand. If you slowly remove your hand from the card, the water remains in the glass.)

Figure 18.4 Forces exerted on the sides of a wedge-shaped volume of fluid.

(*a*) Wedge-shaped volume in fluid at rest

(*b*) Vector sum of forces exerted on wedge is zero

Figure 18.5 Liquids can sustain negative pressures, whereas pressures in a gas can be positive only.

(*a*) Because liquid particles exert attractive forces on each other and on the container walls, liquid can hold piston up against force of gravity

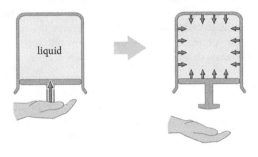

(*b*) Gases expand to fill any volume and therefore cannot sustain negative pressures

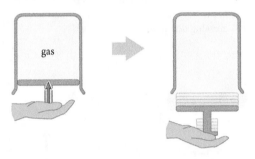

Gas particles are so far apart that they interact only when they collide, and so a gas always expands to fill whatever container it is placed in. A gas can therefore not sustain negative pressure, and a piston just falls out of a gas container that is held upside down (Figure 18.5*b*).

18.2 Consider a plastic cup full of water. (*a*) Is the water exerting a force on the bottom of the cup? (*b*) On the sides of the cup? (*c*) Is the water pressure negative, zero, or positive?

The answers to Checkpoint 18.2 may have surprised you. What causes the pressure in the water if the top surface of the water is open to the environment and no piston is pushing down on that top surface? The cause is gravity. Imagine the column of water in the glass as being a stack of thin "layers" of water. Gravity exerts a force on each layer, which means that each layer pushes down on the layer underneath it, just as a book presses down on the surface of a table because of gravity. At the top of the water column, the atmosphere presses down on the surface, our imaginary top layer.

To examine the effect of gravity on the pressure in a fluid, consider again a water-filled tube fitted with identical pistons on either end, but now we orient the tube vertically (Figure 18.6*a*). Even with no force exerted on the top piston, you have to exert an upward force on the bottom piston to prevent the water from falling out of the tube. As the free-body diagram in Figure 18.6*b* shows, when you are not pressing down on the top piston, the water is subject to two forces: a downward force of gravity and an upward

Figure 18.6 Gravity exerts a downward force on a volume of fluid.

(*a*) You must exert upward force on bottom piston to keep water from falling out

(*b*) Free-body diagram for water

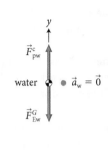

force of equal magnitude exerted by the bottom piston.* So the pressure at the top of the water is zero (no force exerted on the surface), whereas the pressure at the bottom of the tube is nonzero.

Now imagine dividing a liquid in a container into two equal parts, as shown in **Figure 18.7a**. Ignoring any force exerted at the top surface (we'll return to this issue in a bit), we see from the free-body diagrams in Figure 18.7*b* that the top half must be supported against the pull of gravity by the bottom half. The bottom of the container, in turn, must support both halves. Consequently the pressure at the bottom of the container is greater than the pressure at the center.

Figure 18.7 Pressure increases with distance below the surface because liquid layers at the top press down on lower layers of liquid.

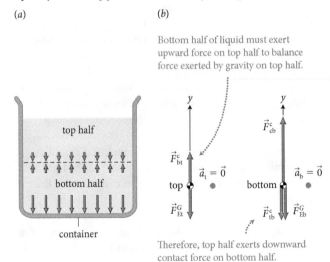

Bottom half of liquid must exert upward force on top half to balance force exerted by gravity on top half.

Therefore, top half exerts downward contact force on bottom half.

---

*As we shall see later, the atmosphere exerts a force on each piston, but the combined effect of these two forces is negligible. For simplicity, we also ignore the mass of the pistons, which means there is no gravitational force exerted on them and so the top piston exerts no force on the water.

18.3 How does the pressure in the liquid vary with height above the bottom of the container in Figure 18.7?

The conclusion you reached in Checkpoint 18.3 holds for any liquid that is at rest in a container of any shape:

The pressure in a liquid at rest in a container decreases linearly with height, regardless of the shape of the container

In a gas, the pressure also decreases with height, but the effect is much less pronounced because gases are much less dense than liquids. Also, because gases are compressible, gas density varies with height and so the decrease in pressure with height is not linear. Air, for example, is densest at sea level, where it is compressed by all the air above. At the cruising altitude of most airplanes (10 km), the mass density of the air is one-third of its value at sea level, and the atmospheric pressure is just one-quarter of its value at sea level.

At sea level, you are at the bottom of an "ocean" of air. The mass of a column of air that has a cross section the size of a fingernail is approximately 1 kg. In other words, at sea level the atmosphere exerts on each fingernail-sized area of your body a force that is equivalent to the force of gravity on a 1-kg object. If we were to add up the magnitudes of the forces exerted on the entire surface area of your body, you would obtain a force equal to the force of gravity exerted on 2×10^4 kg! The force with which the atmosphere pushes down on the surface of your orange juice at breakfast is more than four times the force with which this book presses down on your lap.

Given that the average atmospheric pressure at sea level is $1 \times 10^5 \text{ N/m}^2$, it is clear that we cannot ignore the pressure at the top of the liquid in Figure 18.7. As you saw in Checkpoint 18.1, any force exerted at one surface of a liquid is transmitted to any other part of the liquid. This means that if we change the pressure in a container of liquid at any location, the change in pressure is the same throughout the liquid. This is called **Pascal's principle:**

A pressure change applied to an enclosed liquid is transmitted undiminished to every part of the liquid and to the walls of the container in contact with the liquid.

Consequently, if atmospheric pressure at the top of a column of water in a container open to the atmosphere is $1 \times 10^5 \text{ N/m}^2$, the pressure in the water near that surface must also be $1 \times 10^5 \text{ N/m}^2$. Below the surface, the pressure in the water increases with greater depth as more water pushes down due to gravity. The next example gives you some feel for pressure variation with height in both water and air.

Example 18.1 Water and air pressure

Consider a cylindrical container that has a radius of 30.0 mm and is filled with water to a height of 0.150 m. If the container is at sea level, what is the pressure (*a*) at the water surface and (*b*) at the bottom of the container? (*c*) How far above the water surface is the decrease in atmospheric pressure the same as the pressure decrease between the bottom and top of the water? The mass densities of air and water are 1.20 kg/m³ and 1.00×10^3 kg/m³, respectively.

❶ **GETTING STARTED** I begin by making a sketch of the situation (Figure 18.8a). The problem deals with the pressure at three locations: at the bottom of the container (point A), at the water surface (B), and at some location above the surface (C). Because 0.150 m is a negligible distance in the context of this problem, the pressure at B is essentially equal to the atmospheric pressure at sea level. At A the pressure should be greater than at B because of the force of gravity exerted on the water. At C the pressure should be smaller than at B because the volume of air above C pressing down on C is smaller than the volume of air above B pressing down on B.

Figure 18.8

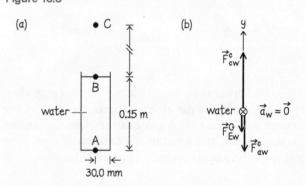

❷ **DEVISE PLAN** The pressure at A is equal to the magnitude of the force $\vec{F}^c_{cw}$ exerted by the container bottom on the water divided by the area of the container bottom. To determine the magnitude of this force, I make a free-body diagram for the water (Figure 18.8b). The forces exerted on the water are the force of gravity $\vec{F}^G_{Ew}$, the downward force $\vec{F}^c_{aw}$ exerted by the atmosphere, and the upward force exerted by the container bottom, which I need to determine. Because the water is at rest, the vector sum of these forces must be zero, and so the magnitude of the force exerted by the container bottom on the water is $F^c_{cw} = F^G_{Ew} + F^c_{aw}$. The force $\vec{F}^c_{aw}$ is determined by atmospheric pressure. To calculate F^G_{Ew} I need to know the mass of the water, which I can obtain by multiplying the volume of the water by the mass density of water.

Because the difference in pressure between A and B is the same as that between B and C, I know that the mass of the air column between B and C must be equal to the mass of the water column between A and B. The ratio of the distances between these locations must therefore be equal to the ratio of the volumes, which is equal to the inverse of the ratio of the mass densities (assuming that the density of air is approximately constant over that column of air).

③ EXECUTE PLAN (a) The pressure at B is essentially equal to atmospheric pressure at sea level, which I can look up: $P_B = 1.01325 \times 10^5 \, \text{N/m}^2$. ✔

(b) Dividing both sides of $F^c_{cw} = F^G_{Ew} + F^c_{aw}$ by the bottom surface area A, I get $P_A = F^G_{Ew}/A + P_{atm}$ because $P_A = F^c_{cw}/A$ and $P_{atm} = P_B = F^c_{aw}/A$. The volume of the water is $V = Ah$, and so the mass of the water is this volume times the mass density of water: $m_{water} = Ah\rho$. The magnitude of the force of gravity exerted on the water is $F^G_{Ew} = m_{water}g = Ah\rho g$, and this force is exerted over the bottom surface area A of the container. The pressure at point A is thus

$$P_A = \frac{F^G_{Ew}}{A} + P_{atm} = \frac{Ah\rho g}{A} + P_{atm} = h\rho g + P_{atm}$$

$$= (0.150 \, \text{m})(1.0 \times 10^3 \, \text{kg/m}^3)(9.8 \, \text{m/s}^2)$$

$$+ \, 1.01325 \times 10^5 \, \text{N/m}^2 = 1.03 \times 10^5 \, \text{N/m}^2. \checkmark$$

(c) The ratio of the distances is $\overline{AB}/\overline{BC} = \rho_{air}/\rho_{water} = (1.20 \, \text{kg/m}^3)/(1.00 \times 10^3 \, \text{kg/m}^3) = 1.20 \times 10^{-3}$. I am given that $\overline{AB} = 0.150 \, \text{m}$, so $\overline{BC} = (0.150 \, \text{m})/(1.20 \times 10^{-3}) = 1.25 \times 10^2 \, \text{m}$. ✔

④ EVALUATE RESULT The answer I obtained in part b shows that the pressure difference between the top and bottom of the container is very small. That makes sense—I know from experience that my ears, which are very sensitive to pressure, don't experience any pressure difference when I swim a mere 0.15 m under the surface of water.

The answer to part c tells me that the distance above sea level I would have to climb in order to obtain the same pressure difference is about three orders of magnitude greater than the distance from B to A (0.15 m versus 120 m). Again, I know from experience that I don't experience any detectable pressure difference when going up 120 m in air, so my answer is not unreasonable.

18.4 How far below the surface of a deep body of water is the pressure twice what it is at the surface?

18.2 Buoyancy

The decrease in pressure with height is responsible for the **buoyant force,** an upward force experienced by objects when they are either fully or partially submerged in a fluid. The tendency of certain objects to float in air or water, called *buoyancy,* is due to this force.

If you hold a brick under water with its top surface parallel to the surface of the water, as illustrated in **Figure 18.9a,** the pressure in the water at the bottom of the brick is greater than the pressure at the top, so the upward force exerted by the water on the bottom surface of the brick is greater than the downward force the water exerts on the top surface of the brick. The free-body diagram for the brick in Figure 18.9b shows the forces exerted on the brick: a downward gravitational force, an upward force exerted by your hand, an upward force exerted by the water on the bottom surface, and a downward force exerted by the water on the top surface. Because the pressure is greater at the bottom of the brick than at the top, the two forces exerted by the water on the brick add up to an upward buoyant force (Figure 18.9c). Buoyant forces are denoted by a superscript b. (We can ignore the sideways forces exerted by the water on the brick because they cancel out.)

18.5 (a) What happens to the pressure difference between the top and bottom of the brick in Figure 18.9a when the brick is held deeper in the water? (b) What happens to the pressure difference when the pressure at the surface of the water is increased? (c) What is the effect of the pressure in the water on the vertical sides of the brick? (d) Consider the same brick held in the air. Does the air exert a buoyant force on the brick? If so, in which direction is this force?

Figure 18.9 Cause of the buoyant force experienced by objects that are partly or fully submerged in a fluid.

(a) Brick held under water

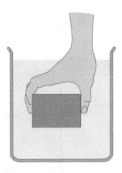

(b) Free-body diagram that separates force exerted by water on brick's top and bottom surfaces

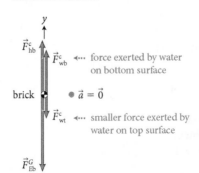

(c) Free-body diagram showing buoyant force exerted by water on brick

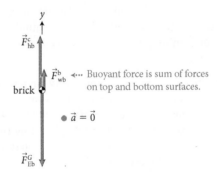

Figure 18.10 The force exerted by gravity on the volume of liquid *displaced* by a submerged object is equal in magnitude to the buoyant force exerted by the liquid on the object.

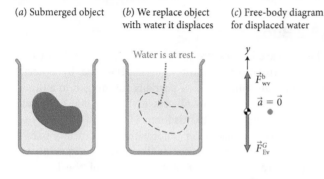

(*a*) Submerged object   (*b*) We replace object with water it displaces   (*c*) Free-body diagram for displaced water

How large is the buoyant force that is exerted on an object submerged either fully or partially in a fluid? For an object that has a rectangular cross section, such as the brick in Figure 18.9*a*, we can determine the pressures at the top and bottom of the object, multiply the pressures by the surface area to get the two force magnitudes, and then take the difference of these magnitudes to obtain the buoyant force. However, many objects do not have a rectangular cross section. Consider, for example, the fully submerged object in Figure 18.10*a*. Even if we know the pressure at every location on the surface of the object, calculating the buoyant force exerted by the water on the object is a complicated matter. We know, however, that if we remove the object from the water, water fills the space previously occupied by the object (Figure 18.10*b*). This volume of water within the boundary defined by the object is said to be *displaced* by the object when the object is fully submerged. Figure 18.10*c* shows the free-body diagram for this displaced volume of water. Because all the water is at rest, we know that the vector sum of the forces exerted on the volume of water within the boundary must be zero. Two forces are exerted on this volume of water: a downward force of gravity exerted by Earth and a buoyant force exerted by the surrounding water. Given that the vector sum of these two forces is zero, we know that the buoyant force exerted on the water within the boundary must be upward and equal in magnitude to the magnitude of the force of gravity exerted by Earth on the water within the boundary.

Because the buoyant force exerted by the surrounding water on the boundary in Figure 18.10*a* cannot depend on the nature of the material inside the boundary, we have a way of determining the buoyant force exerted on an object of arbitrary shape:

An object submerged either fully or partially in a fluid experiences an upward buoyant force that is equal in magnitude to the magnitude of the force of gravity exerted on the fluid displaced by the object. The volume of the displaced fluid is equal to the volume of the submerged portion of the object.

This statement is called **Archimedes' principle.** Note that there is no reference to the shape or composition of the submerged object: The only factor that comes into play is the volume of the displaced fluid.

18.6 If the buoyant force exerted on an object is always equal in magnitude to the force of gravity exerted on the fluid displaced by the object, why does a brick placed in a barrel of water sink?

Now that we know how to determine the buoyant force exerted on an object, we can understand why some objects float and others sink. As Checkpoint 18.6 shows, a brick sinks because the force of gravity exerted by Earth on it is greater than the force of gravity exerted by Earth on an equal volume of water (equal in magnitude to the buoyant force). Because the vector sum of these two forces is directed downward, the brick sinks unless it is held up.

When the buoyant force is greater in magnitude than the force of gravity exerted on an object fully submerged in water, the object accelerates toward the surface of the water when you release it. A block of wood released under water, for example, moves up to the surface, and so the force of gravity exerted by Earth on the block is smaller than the force of gravity exerted on an equal volume of water (Figure 18.11*a*). Once it floats at the surface, only part of the block is submerged, and so now the volume of displaced water is smaller than the volume of the block (Figure 18.11*b*). Because the block is at rest at the surface, the force of gravity exerted by Earth on it and the buoyant force exerted by the water must be equal in magnitude, as shown in Figure 18.11*c*.

Objects made out of materials that normally sink can be made to float by designing the objects in such a way that the volume of water they displace is great enough to offset the force of gravity exerted on them. For example, a steel ship floats even though a block of steel sinks. The reason

Figure 18.11 The concept of displaced volume can be used to determine the buoyant force on floating as well as submerged objects.

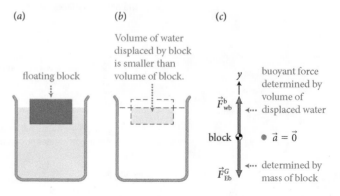

Figure 18.12 Steel ships can float even though steel sinks. The ship's hollow interior means that the volume of water displaced by the ship is larger than the volume of steel. Thus, the upward buoyant force exerted on the ship is greater than the downward gravitational force exerted on the steel.

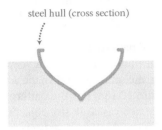

steel hull (cross section)

Volume of water displaced by hull exceeds volume of steel.

for this is illustrated in **Figure 18.12**: As long as the force of gravity exerted on the ship is smaller than the force of gravity exerted on the volume of water displaced by the ship, the buoyant force is great enough to keep the ship floating. Because the hull of a ship is hollow, the volume of displaced water is much greater than the volume of steel.

Example 18.2 Mass densities

When a piece of cork floats in water, three-quarters of its volume is above the water surface. How does the mass density of cork compare with the mass density of water?

❶ **GETTING STARTED** I begin by making a sketch of the situation. Because one-quarter of the cork's volume is under water, I know that the cork displaces a volume of water equal to one-quarter of the volume of the cork. This also tells me that the force of gravity exerted on the whole cork $m_{cork}g$ is equal to the force of gravity exerted on the displaced water $m_{disp}g$ (**Figure 18.13**), which means that the mass of the displaced water is equal to the mass of the cork.

Figure 18.13

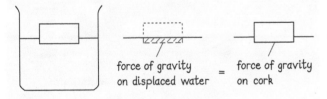

force of gravity on displaced water = force of gravity on cork

❷ **DEVISE PLAN** The mass density of an object is equal to the amount of mass per unit volume. Because I know how the volumes of equal masses of cork and water compare (Figure 18.13), I can compare the mass densities of cork and water.

❸ **EXECUTE PLAN** The volume of a piece of cork of a certain mass is four times greater than the volume of an equal mass of water, and so the mass density of cork is one-quarter that of water. ✔

❹ **EVALUATE RESULT** A "light" material such as cork has less mass per unit volume and therefore a smaller mass density than water, as I expect.

Figure 18.14 Checkpoint 18.7.

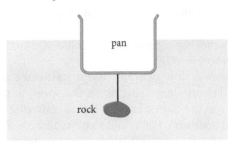

pan

rock

✋ **18.7** A pan with a rock in it is floating in water. Suppose you remove the rock and use a very lightweight string to suspend it from the bottom of the pan (**Figure 18.14**). If the combination still floats, is the volume of water displaced when the rock is suspended greater than, equal to, or smaller than the volume of water displaced when the rock is in the pan?

18.3 Fluid flow

We have so far restricted our study to fluids at rest. Now we turn our attention to fluids in motion; such motion is usually referred to as *fluid flow*. When the flow is such that the velocity of the fluid at any given location is constant, such as when water moves through a pipe, the flow is said to be **laminar.** Not all fluid flow is laminar, however. For example, the motion of rising smoke or the motion of water in the wake of a large ship is characterized by chaotic changes. Such flow is said to be **turbulent.**

To better understand fluid flow, we introduce the concept of **streamlines,** which are lines drawn to represent the paths taken by the particles in a fluid in motion. Streamlines can be made visible by injecting smoke into a flow of gas (**Figure 18.15**) or ink into a flow of liquid (**Figure 18.16**).

Figure 18.15 Streamline flow over a car.

Figure 18.16 Streamline flow around a cylinder.

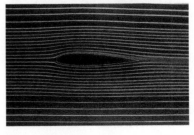

When flow is laminar, as in Figures 18.15 and 18.16, streamlines maintain their shape and position. In turbulent flow, streamlines are erratic and often curl into whirling, circularly moving regions of fluid called *vortices* (singular *vortex*, Figure 18.17).

Whether a flow of fluid past a stationary object is laminar or turbulent depends on the speed of the flow, the shape of the object, and the fluid's resistance to shear stress (its *viscosity*). For a fluid that has zero viscosity and that has no friction when flowing past surfaces, streamlines just bend around a blunt object (**Figure 18.18a**). Except where the fluid has to move out of the object's

Figure 18.17 Turbulent flow around a cylinder.

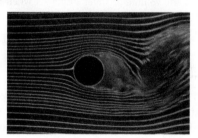

way, the flow speed is the same everywhere. In such situations, the streamline pattern in regions downstream from the object is the same as the streamline pattern upstream of the object. (Although such "totally nonviscous" fluids don't exist, examining these models provides useful insights.)

Figure 18.18b shows the flow pattern for a fluid that has nonzero viscosity when that fluid is moving at a very low speed. The streamlines have the same general shape as for the nonviscous fluid of Figure 18.18a, but the pattern of speeds is very different in the two cases. At the object's surface, the viscous fluid in Figure 18.18b is stationary because of friction between the fluid and the surface. Far above and below the object, the flow is unaffected by the presence of the object. Between those two extremes, the flow speed increases from layer to layer. Because the viscosity is nonzero, adjacent layers of fluid exert shear forces on each other, with faster-moving layers trying to speed up slower-moving layers and slower-moving layers slowing down faster-moving ones. The resulting "rubbing" of layers of fluid causes energy dissipation.

As flow speed is increased, the torques caused by the forces that adjacent layers exert on each other become so

Figure 18.18 Whether the flow of fluid past a stationary object is laminar or turbulent depends on the fluid's speed and viscosity and on the object's shape.

(a) Low flow speed, zero viscosity

(b) Low flow speed, nonzero viscosity

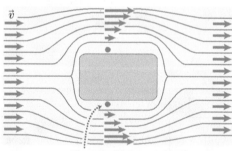

Flow speed is zero at object's surface.

(c) High flow speed, turbulence

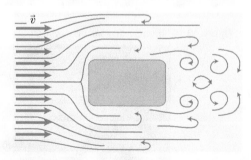

(d) High flow speed, streamlined object

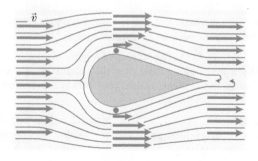

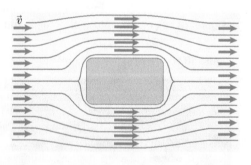

great that the flow develops vortices and becomes turbulent (Figure 18.18c). The turbulent wake alongside and downstream from the object is erratic and causes significant energy dissipation.

✋ **18.8** Instead of laminar fluid flow past a stationary object as in Figure 18.18a, consider the motion of an object moving at constant velocity through a stationary fluid. Do you expect the flow pattern surrounding the object to be the same as in Figure 18.18a or different?

To prevent turbulence and minimize energy dissipation at high flow speeds, it helps to *streamline* objects, which means to adjust their shape so as to maintain streamline flow around them. For example, when an object has a rounded upstream end and a long, tapered downstream tail, as in Figure 18.18d, the turbulent wake is very narrow and the flow pattern is similar to the low-speed flows illustrated in Figure 18.18a and b. At the surface of a streamlined object such as the one in Figure 18.18d, the fluid is still stationary, but the flow speed increases rapidly over a thin boundary layer. Outside this boundary layer, the flow resembles that of a nonviscous fluid.

Now that we have some general feeling for fluid flow, let us look at flow through a pipe. In a pipe of fixed diameter, the streamlines are all parallel to the walls of the pipe. For an (ideal) nonviscous fluid flowing through such a pipe, the flow speed is the same everywhere and the pipe offers no resistance to the flow (Figure 18.19a). For any viscous fluid at low flow speed, the flow speed is zero at the pipe walls and highest at the center (Figure 18.19b). At high flow speed, the speed is more constant across the diameter of the pipe, dropping off only in a boundary layer (Figure 18.19c). The higher the speed, the thinner the boundary layer.

Let us next consider flow through a constriction in a pipe (Figure 18.20a). To simplify matters, we ignore any turbulence or boundary-layer effects. The first thing to note is that the streamlines must crowd together in the narrow part of the pipe. (A streamline cannot have an end because that would mean a fluid particle moving along that streamline would come to a stop.) Therefore the density of streamlines is greater in the narrow part than it is in the wide part.

Figure 18.20 What happens when a fluid flows from a wider region of a pipe into a narrower region.

(a) When a pipe carrying a fluid narrows, the streamlines get closer together

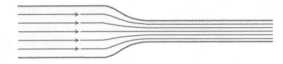

(b) Where the pipe is narrower, the fluid must flow faster

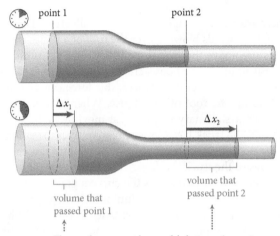

These volumes must be equal (what goes in must come out), so fluid must move faster where pipe is narrow.

If the flow is laminar, the mass of the fluid passing point 1 in a given time interval in the wide part must be equal to the mass of the fluid passing point 2 in the narrow part in that time interval (Figure 18.20b). (If that were not the case, mass would accumulate between those two points and the flow would not be laminar.) Consequently, in a given time interval, the displacement Δx_2 of a fluid particle moving through the narrow part is greater than the displacement Δx_1 of a fluid particle moving through the wide part. This means that the overall fluid flows faster in the narrow part of the pipe where the density of streamlines is greater than in the wide part.

A flowing fluid speeds up when the region through which it flows narrows and slows down when the region through which it flows widens. The density of streamlines reflects the changes in flow speed: When streamlines in a laminar flow get closer together, the flow speed increases; when they get farther apart, the flow speed decreases.

This principle applies to every particle of a moving fluid. Thus each particle must accelerate as it moves along a streamline from the wide part of the pipe to the narrow part. This means that the surrounding fluid must exert a force to the right on the particle while it is in the tapered

Figure 18.19 Flow speed patterns through a pipe of fixed diameter.

(a) Nonviscous fluid   (b) Viscous fluid at low speed   (c) Viscous fluid at high speed

$\vec{v}$ $\vec{v}$ $\vec{v}$

boundary layer

region of the tube in Figure 18.20. Because the force exerted on fluid particles is a measure of the pressure in the fluid, the pressure is greater in the wide part of the pipe than in the narrow part.

When the flow speed in a laminar flow increases, the pressure in the fluid decreases.

This effect is sometimes called the *Bernoulli effect* after the Swiss physicist who first described the pressure difference quantitatively. The pressure difference occurs along the streamlines. Because there is no flow in the direction perpendicular to a streamline, the pressure must be the same across neighboring streamlines.

There are many examples of the Bernoulli effect. During a storm, high winds blowing over the roof of a house cause the pressure above the roof to be lower than the pressure in the still air inside the house, creating an upward force on the roof. In extreme circumstances, this force can be great enough to lift the roof off the house. When a large truck speeds past a stationary car, the fast-moving air between the car and the truck has a lower pressure than the still air on the other side of the car, and consequently the car is pulled toward the truck.

Another familiar example is the curved, gravity-defying motion of a rapidly moving, spinning ball. **Figure 18.21a** shows a top view of a curveball—a baseball thrown with a counterclockwise spin. Because of friction between the surface of the ball and the air, air is dragged around with the spinning ball as the ball moves through the air. Figure 18.21*b* shows the air movement viewed from a reference frame moving along with the ball. The motion of the air caused by the spinning of the ball adds to the flow of air past the ball, causing the air behind the ball to be deflected to the right and the air on the left of the ball to move faster than the air on the right. The deflection of the air means that the direction of the momentum of the air shifts rightward, and

that change must be accompanied by an equal-magnitude leftward momentum change for the ball. Alternatively, according to the Bernoulli effect, the pressure in the air to the left of the ball is lower than the pressure to the right, causing a leftward force on the ball. Both of these effects make the ball veer to the left. Likewise a tennis ball that has "top spin" (meaning the top of the ball spins in the direction of the ball's translational motion) curves sharply downward.

The dimples on the surface of golf balls increase the friction between the air and the ball surface to enhance the effect of spin. A golf ball struck with a club is given a rapid backspin because of the slant of the club's face. The backspin forces the air behind the ball downward, causing the ball to experience a lift that carries it farther than it would travel without spin. The lift gained by the backspin more than compensates for the loss of range resulting from air drag.

Example 18.3 Magic pull

When air is blown downward into the narrow opening of a funnel (Figure 18.22), a lightweight ball initially positioned below the wide opening is pulled up into the funnel and held in place. Explain how the ball can be pulled upward even though the airflow and gravity are both directed downward.

Figure 18.22 Example 18.3.

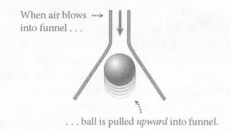

When air blows ⋯▸
into funnel . . .

. . . ball is pulled *upward* into funnel.

❶ **GETTING STARTED** I know that if I were to blow against the ball, it would move away from me. So I reason that to pull the ball upward, there must be exerted on it an upward force that is greater than the combined effect of the downward gravitational and airflow forces. The ball is subject to an upward buoyant force, but I know this force is smaller than that of gravity alone because a ball—even a lightweight one—does not float in air. So there must be an additional force pulling the ball upward. Because this force is absent when no air flows past the ball, that flow must be causing the upward force.

❷ **DEVISE PLAN** To understand the effects of the flowing air, I should sketch the streamlines (I assume laminar flow). The density of the streamlines tells me how the speed of the flow varies, and from this information I can determine how the air pressure varies around the ball, which in turn tells me in which direction the air exerts a force on the ball.

❸ **EXECUTE PLAN Figure 18.23** shows how the streamlines must bend around the ball. In order to pass around the ball, they must be squeezed together between the ball and the sides of the funnel. At the bottom of the funnel, the diameter of the funnel is very wide and so the streamlines are spaced far apart.

Figure 18.21 Motion of a spinning ball as seen (*a*) from the Earth reference frame and (*b*) from a reference frame moving along with the ball. Because the air is dragged along with the ball surface, the air moves faster to the left of the ball and more slowly to the right of the ball.

(*a*)　　　　　　　　(*b*)

Earth reference　　Reference frame moving with ball
frame

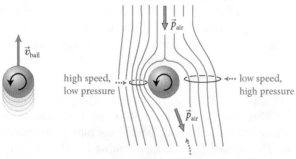

high speed,　　　　　　　⋯◂ low speed,
low pressure　　　　　　　high pressure

Air behind ball is deflected by ball's spin.

Figure 18.23

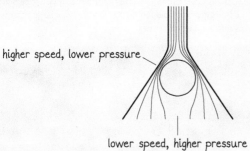

This tells me that the flow speed is smaller below the ball than above it and consequently that the pressure is greater below the ball than above it. This pressure difference causes an upward force on the ball. If the pressure difference is great enough, this force can exceed the combined effect of the force of gravity and the downward force of the airflow on the ball. ✔

❹ **EVALUATE RESULT** My answer is consistent with the observation given in the problem statement, even though the ball's being pulled *upward* by a downward flow of air seems nothing short of magic! I assumed the flow to be laminar, which may not be entirely correct for an object like a ball (which is not streamlined). However, because air is a gas and because gases have very low viscosity, the assumption that the flow remains laminar is not unreasonable.

✋ **18.9** The cloth roof of a convertible car often bulges at high speed, even when the top fits tightly and no wind is getting caught under it. Explain what causes the bulging.

18.4 Surface effects

A distinguishing feature between liquids and gases is that liquids have a well-defined surface but gases do not. If you put a gas in an open container, the gas escapes. A liquid, however, remains in the container it is placed in, and small quantities of liquid placed on a solid surface form drops. In each case, the liquid has a well-defined surface that separates it from the surrounding air. On a small scale—that is, for small quantities of liquid and for small objects—this surface behaves like an elastic skin, pulling small volumes of liquid into drops and supporting small objects such as dust particles or water spiders.

We can understand the behavior of liquid surfaces by considering what happens at the molecular scale. When the particles in a fluid are far apart, they exert no appreciable force on one another (**Figure 18.24**). As they get closer together, however, the interaction between particles has both an attractive and a repulsive component. At a distance of a few particle diameters, particles exert attractive forces, called *cohesive forces*, on one another. When the separation distance becomes less than a particle diameter, however, the particles strongly resist being pushed closer together, resulting in very strong *repulsive forces* between particles. In a gas, the particles move at such great speed and their

Figure 18.24 Interaction between fluid particles in a gas when the particles are far apart and in a liquid when they are closer together.

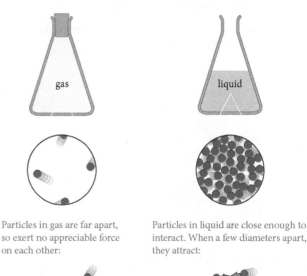

Particles in gas are far apart, so exert no appreciable force on each other:

Particles in liquid are close enough to interact. When a few diameters apart, they attract:

When closer than a diameter apart, they repel strongly:

separation distances are so great that the cohesive forces are negligible. In other words, the gas particles essentially don't interact except when they collide and exert repulsive forces on one another.

Cohesive forces play an important role in liquids, however. The piston in Figure 18.5a stays in position because of the cohesive forces in the liquid, for instance. Likewise, if you press pure water with no air dissolved in it between two smooth plates, an enormously large force is required to separate the plates.

The cohesive forces between particles cause a liquid surface to act like a stretched elastic sheet that tends to minimize the surface area. To see why this happens, compare the forces exerted on a particle in the interior of a liquid with those exerted on a particle at the surface. The interior particle experiences attractive forces from neighbors in all directions, and the vector sum of these forces is zero. At the surface, however, there are no neighbors above the particle, and so the vector sum of the cohesive forces points into the liquid perpendicular to the surface. The surface particles cannot move into the liquid, however, because as they are pulled inward, they experience a strong repulsive outward force from neighbors.

Because of the inward pull on all surface particles, a small volume of liquid tends to arrange itself in the shape of a sphere (**Figure 18.25**). In a gas there is no attraction between neighboring particles, and therefore a collection of gas particles tends to fly apart, expanding into all available surrounding space. To keep gas particles together, you can

Figure 18.25 The cohesive forces between molecules in a liquid cause a liquid surface to act like a stretched elastic sheet. Consequently, small volumes of liquid tend to adopt a spherical shape.

(*a*) Difference between vector sum of cohesive force on molecules at the surface and in the interior of a liquid

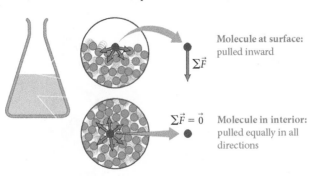

Molecule at surface: pulled inward

$\Sigma\vec{F}$

$\Sigma\vec{F} = \vec{0}$ Molecule in interior: pulled equally in all directions

(*b*) Balloon as analogy for surface forces on liquid droplet

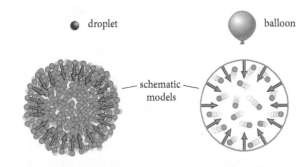

droplet

balloon

schematic models

Inward pull on surface molecules tends to make droplet spherical.

Stretched balloon skin exerts analogous inward force on gas.

put them inside a balloon. The stretched balloon pushes inward on the gas particles, exerting a force on them and keeping them together. The effect of the inward pull on the surface particles in a liquid is much like the force exerted by a balloon on the particles in a gas. The liquid surface thus acts like a membrane that is under tension and that pulls the liquid together in a spherical shape.

Just as an inflated balloon is said to be under tension, so do we say that a liquid surface has *surface tension*. In the absence of gravity, surface tension causes even large volumes of liquid to take on a spherical shape (**Figure 18.26**). In the presence of gravity, however, drops of liquid are not perfectly spherical, with larger drops flattening out more than small ones. The reason for the difference in how much large and small drops lose their spherical shape is that only particles on the drop surface are pulled into the spherical shape. Gravity, in contrast, pulls on all particles in a volume of liquid, not just those in the surface layer. To state it another way, surface tension is a surface effect but gravity is a volume effect. As the next exercise shows, the relative importance of these two effects changes with drop size.

Figure 18.26 Astronaut Kevin Ford watches a water bubble float freely between him and the camera aboard the International Space Station. When no large gravitational force is involved, even a large volume of liquid coalesces into a spherical shape.

Exercise 18.4 Scaling a drop

Suppose surface tension increases in direct proportion to surface area (an oversimplification!). How does the ratio of surface tension to the force of gravity exerted on a drop change as the drop diameter increases tenfold?

SOLUTION The force of gravity is proportional to the mass of each drop, which is proportional to each drop's volume. The ratio of surface tension to the force of gravity on each drop is therefore proportional to the surface-area-to-volume ratio. For simplicity, I approximate the smaller drop as a sphere with diameter d. This means that I should compare the surface-area-to-volume ratio of this sphere with the surface-area-to-volume ratio of a sphere of diameter $10d$.

The surface area of a sphere of radius $R = \frac{1}{2}d$ is $4\pi R^2 = \pi d^2$, and its volume is $\frac{4}{3}\pi R^3 = \frac{4}{3}\pi(d/2)^3 = \frac{1}{6}\pi d^3$, and so the surface-area-to-volume ratio of a sphere is $(\pi d^2)/(\frac{1}{6}\pi d^3) = 6/d$. The surface-area-to-volume ratio thus decreases by a factor of 10 as the diameter of the drop increases by a factor of 10. ✔

The analogy between surface tension and the tension in the rubber of a balloon allows us to draw some important conclusions. Consider, for example, the air-filled balloon of **Figure 18.27**. The rubber in the uninflated end of the balloon is less stretched than the rubber in the inflated part,

Figure 18.27 A partially inflated balloon. The tension is greatest where the rubber is stretched the most.

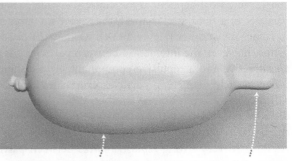

more stretch, so more tension

little stretch, so little tension

and so there is less tension in the uninflated rubber than in the inflated rubber. (You may have noticed that balloons are softest and most easily squeezed where they are least inflated.) To understand how the tension in a balloon relates to its shape and to the pressure in the air inside the balloon, answer the following checkpoint first.

18.10 How does the air pressure in the uninflated part of the balloon in Figure 18.27 compare with the air pressure in the inflated part?

Did the answer to Checkpoint 18.10 surprise you? How can one part of the balloon be soft and another part hard if the air pressure is the same everywhere inside the balloon? The reason has to do with *radius of curvature,* which is the radius of the circle that approximates a curve at the point under consideration. As **Figure 18.28** shows, the radius of curvature is different in the two parts. The air pressure must be the same in the two parts, as you learned in Checkpoint 18.10, and the inward force exerted by the rubber surface of the balloon on the air inside is caused by the tension in the rubber. As you can see by comparing the free-body

diagrams for a small surface segment in Figure 18.28a and b, the smaller the radius of curvature, the smaller the tensile force needed to provide a given inward force. In the part of the balloon where the radius of curvature is smaller, the angle of the tensile forces is large and so smaller tensile forces add up to the same vector sum. Indeed when the radius is smaller, the rubber is stretched less and so the tensile force is smaller.

In an elastic membrane enclosing a gas, tension in the membrane increases with increasing radius of curvature, but the pressure is the same everywhere inside the volume enclosed by the membrane, as required by Pascal's principle.

The situation is different with liquids because a liquid surface is not exactly like a stretched rubber skin. In particular, surface tension in a liquid is determined entirely by the cohesive forces between the liquid particles and is constant regardless of the shape of the liquid surface. So the tensile forces exerted on a given segment of the infinitesimally thin surface layer separating liquid and gas are the same no matter what the radius of curvature of the surface is. As shown in **Figure 18.29**, these constant tensile forces exerted on the surface segment of a liquid droplet cause a greater inward force when the radius of curvature is small. Because the segment is at rest, this inward force on the segment must be balanced by the forces exerted by the liquid and air on

Figure 18.28 The reason why the balloon in Figure 18.27 has an uninflated end. Because the air pressure must be the same everywhere inside a balloon, areas with a small radius of curvature must have smaller-magnitude tensile forces than areas with a larger radius of curvature.

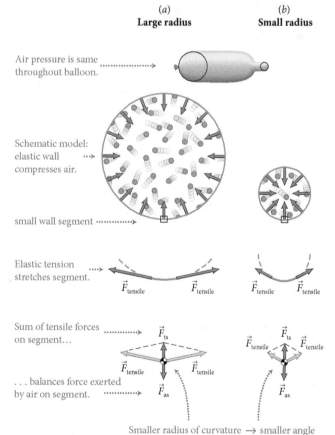

Figure 18.29 Because surface tension in a liquid is independent of the shape of the liquid surface, a small liquid sphere has a greater internal pressure than a larger liquid sphere.

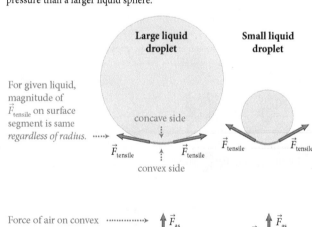

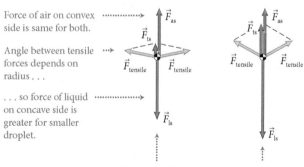

either side of the surface layer. The outward force exerted by the liquid on the segment is therefore greater than the inward force exerted by the air on the segment. Because these forces result from the pressures in the fluids on either side of the surface, we can conclude that the pressure in the liquid on the concave side of the surface must be greater than the pressure in the air outside the droplet. We will refer to this difference as the *pressure difference across the surface:*

> **As the radius of curvature of a liquid surface decreases, the pressure difference across the surface increases, with the pressure being greatest on the concave side of the surface.**

This statement and the earlier boldfaced statement describing pressure in a membrane-enclosed gas are consequences of **Laplace's law,** which relates surface curvature to surface tension and to the pressure in a liquid.

Surface tension makes it possible for small insects to walk on a water surface and for small objects to float on a liquid surface. For example, if you carefully place a steel paperclip on a water surface, it stays there even though a thicker steel cylinder would sink (**Figure 18.30a**). There are two ways to explain this. The first is to invoke Laplace's law: The water surface beneath the paperclip is concave up and so the pressure above the surface is greater than the pressure just below it. The pressure just below the surface where the paperclip rests is atmospheric pressure, while the pressure above the surface where the paperclip rests is atmospheric pressure plus the pressure caused by the force exerted by the paperclip on the surface.

The second way to explain why the paperclip floats is illustrated in Figure 18.30b, which shows free-body diagrams for the segment of the surface just below the paperclip and for the paperclip. The tensile forces due to surface tension balance the downward force $\vec{F}_{ns}$ exerted by the paperclip on the surface segment. This downward force $\vec{F}_{ns}$ forms an interaction pair with the upward force $\vec{F}_{sn}$ exerted by the surface segment on the paperclip, which balances the downward force of gravity on the paperclip and makes the paperclip float.

✋ **18.11** (*a*) What is the difference between the pressure just below the water surface in a pool and the pressure in the air just above it? (*b*) Is the pressure inside a raindrop greater than, equal to, or smaller than the pressure in the surrounding air? (*c*) How does the pressure inside a small raindrop compare with the pressure inside a larger raindrop?

A liquid surface that comes in contact with a solid surface forms an angle with the solid surface (**Figure 18.31**). This angle, called the *contact angle,* is the same for a given liquid-solid combination (Figure 18.31*a*).* The contact angle is defined as the angle between the solid surface and the tangent to the liquid surface at the point where liquid and solid meet; the contact angle is always measured in the liquid. If the contact angle is smaller than 90°, the liquid is

Figure 18.31 Small drops are nearly spherical; larger ones flatten out. For a given combination of liquid and solid surface, the contact angle is fixed and does not depend on the drop size or shape.

(*a*)

Contact angle θ_c is fixed for given combination of liquid and solid.

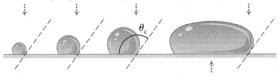

Larger droplets flatten out.

(*b*)

When contact angle < 90°, we say that liquid wets surface.

Figure 18.30 Surface tension enables paperclips to float on water.

(*a*)

(*b*)

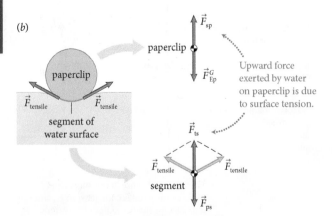

---

*The gas surrounding the liquid and solid also affects the contact angle. Because this effect is generally very small, we shall ignore it.

Figure 18.32 Free-body diagrams for a particle at the gas-liquid-solid junction for a liquid that (*a*) wets and (*b*) does not wet a horizontal surface.

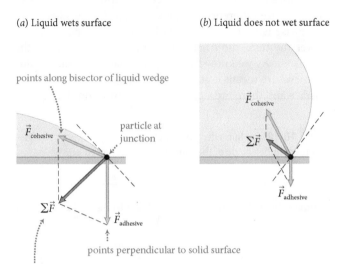

(*a*) Liquid wets surface

(*b*) Liquid does not wet surface

points along bisector of liquid wedge

particle at junction

$\vec{F}_{\text{cohesive}}$

$\vec{F}_{\text{cohesive}}$

$\Sigma\vec{F}$

$\Sigma\vec{F}$

$\vec{F}_{\text{adhesive}}$

$\vec{F}_{\text{adhesive}}$

points perpendicular to solid surface

Vector sum must point perpendicular to liquid surface (or particle would move along surface).

Figure 18.33 The concave or convex shape of the surface of a liquid in a capillary tube is determined by the contact angle.

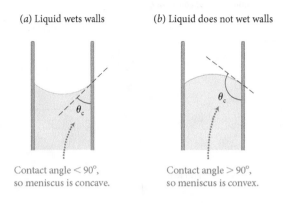

(*a*) Liquid wets walls

(*b*) Liquid does not wet walls

θ_{c}

θ_{c}

Contact angle < 90°, so meniscus is concave.

Contact angle > 90°, so meniscus is convex.

said to *wet* the solid surface (Figure 18.31*b*). If the contact angle is greater than 90°, as in Figure 18.31*a*, the liquid does not wet the surface; it forms drops instead of spreading out and wetting the surface.

The contact angle is strongly influenced by how smooth and clean the solid surface is and by the presence of impurities in the liquid. For example, water beads up into drops on a paraffin wax surface, but adding a bit of soap to the water makes the drop spread out and wet the paraffin surface.

The contact angle for a liquid on a solid surface is due to the interplay between the cohesive forces between the particles in the liquid and the *adhesive forces** between the liquid particles and the atoms in the solid. Consider, for example, a particle at the point where the surface meets the solid (**Figure 18.32**). This particle is subject to a cohesive force exerted by a wedge of liquid and an adhesive force exerted by the solid surface. Symmetry requires the cohesive force to be directed along the bisector of the wedge. Likewise, the adhesive force must be perpendicular to the solid surface. (If the adhesive force were not perpendicular to the solid surface, there would be a flow tangential to the surface in response to any tangential component of that force.) Ignoring the effects of gravity (which are very small on a single particle), we can say that the vector sum of the forces exerted on the particle is the sum of the cohesive force and the adhesive force. Because the liquid cannot sustain any shear force, the vector sum of the forces on any particle at

its surface must always be perpendicular to the surface (see Section 18.1). Given the directions of these two forces and their vector sum, their relative magnitudes are fixed. For example, for a liquid that wets a solid with a contact angle of 60° or smaller, the adhesive force is always greater than the cohesive force (Figure 18.32*a*), while for a nonwetting combination, the cohesive force is always greater than the adhesive force (Figure 18.32*b*).

The contact angle also determines the shape of the edges of a liquid surface at any container walls (**Figure 18.33**). In a very narrow tube, called a *capillary* tube (after the Latin word for *hair*), the surface of a liquid is curved. This curved surface, called a **meniscus,** is concave when the liquid wets the walls of the capillary tube (Figure 18.33*a*) and convex when the liquid is nonwetting (Figure 18.33*b*). By looking closely at the edge of the surface of a liquid in a glass, you can determine whether or not the liquid wets the glass.

The free-body diagrams for a particle at the point where the liquid surface meets the glass are shown in **Figure 18.34** for a concave meniscus and a convex meniscus. Notice the similarity with the free-body diagrams in Figure 18.32.

Figure 18.34 Free-body diagrams for a particle at the gas-liquid-solid junction for a liquid that (*a*) wets and (*b*) does not wet a vertical surface.

(*a*) Liquid wets wall

(*b*) Liquid does not wet wall

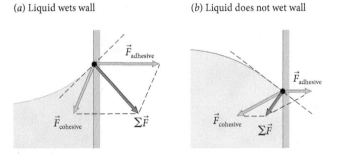

$\vec{F}_{\text{adhesive}}$

$\vec{F}_{\text{adhesive}}$

$\vec{F}_{\text{cohesive}}$

$\vec{F}_{\text{cohesive}}$

$\Sigma\vec{F}$

$\Sigma\vec{F}$

* Cohesive forces are exerted between like particles. Adhesive forces are exerted between unlike particles.

CONCEPTS

Figure 18.35 The pressure difference across the concave surface of a liquid that wets the walls of a capillary tube causes the liquid to rise in the tube, a phenomenon called *capillary rise*.

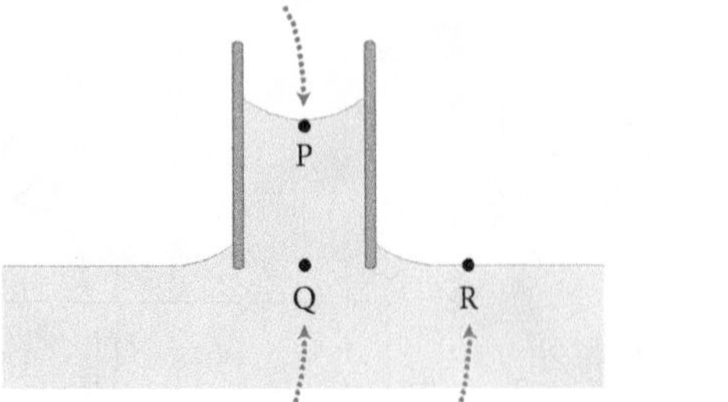

Curvature means pressure drops across surface, so $P_P < P_{atm}$.

P

Q   R

Points Q and R are at same height, so $P_Q = P_R = P_{atm}$.

The formation of a meniscus is responsible for the rise of liquid in a capillary tube, as **Figure 18.35** shows. Because the meniscus is concave, the pressure at point P in the liquid, just below the meniscus, must be smaller than the atmospheric pressure just above the meniscus. The pressure at Q, a point directly below P and at the same height as the surface of the liquid surrounding the capillary tube, must be the same as the pressure at R (otherwise the liquid would flow so as to eliminate any horizontal pressure difference). Because the surface at R is flat, there is no pressure difference across the surface, and so we know that the pressure at R (and therefore the pressure at Q) is equal to atmospheric pressure. The pressure difference between points P and Q causes an upward buoyant force to be exerted by the liquid

in the tube between P and Q, causing a *capillary rise* until the downward force of gravity exerted on the liquid that has been raised into the tube balances the upward force due to the pressure difference. As illustrated in **Figure 18.36**, capillary rise is more pronounced in a narrow tube. The narrower the tube, the smaller the radius of curvature of the meniscus. As we know from Laplace's law, a smaller radius of curvature means a greater pressure difference across the surface and therefore a greater upward force on the liquid.

18.12 Explain why surface effects cannot support a large solid cylinder made of the same material as a needle.

Figure 18.36 The narrower the tube, the greater the capillary rise.

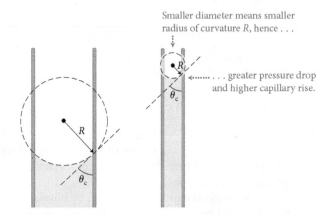

Smaller diameter means smaller radius of curvature R, hence . . .

. . . greater pressure drop and higher capillary rise.

R

θ_c

R

θ_c

Self-quiz

1. Masons use a hose open at both ends and filled with water to ensure that the blocks they lay remain level (Figure 18.37). How does such a hose help them do this?

Figure 18.37

2. (*a*) Imagine a rock and a balloon of the same volume completely submerged in water. How do the buoyant forces exerted on them compare? (*b*) How do the buoyant forces compare when the balloon volume is half the rock volume? (*c*) Which object floats and which sinks in part *a* and in part *b*?

3. For a thrown ball to rise, what type of spin should the thrower place on it?

4. (*a*) Draw a free-body diagram analogous to Figure 18.32 for a gas-liquid-solid boundary at which the contact angle is 90°. Make your solid surface horizontal. (*b*) How do the magnitudes and directions of the adhesive and cohesive forces exerted on your particle compare? (*c*) How does the magnitude of the cohesive force between two liquid particles compare with the magnitude of the adhesive force between a liquid particle and an atom in the solid?

5. (*a*) Imagine a capillary tube placed in a nonwetting liquid (as opposed to the wetting-liquid situation of Figure 18.35). How does the pressure at a point P just below the now convex surface compare with atmospheric pressure? (*b*) How does the level in the tube compare with the level of the liquid outside the tube?

Answers:

1. The air pressure at both water surfaces is atmospheric pressure. Therefore the height of the water surface at the left end of the hose must be the same as the height of the water surface at the right end, in order to balance the pressure throughout the water. If the masons align the two ends of the hose with the two ends of a row of blocks and no water bubbles out of either end of the hose, the two ends of the wall are at the same height, meaning the row is level.

2. (*a*) Because the rock and balloon occupy the same volume, they displace the same volume of water, and so the surrounding water exerts equal buoyant forces on them. (*b*) The smaller volume of the balloon displaces a smaller volume of water, and so the buoyant force exerted on the balloon is half the buoyant force exerted on the rock. (*c*) In either case, the balloon rises and the rock sinks. The buoyant force on a rock is always smaller than the force of gravity exerted on it. Likewise the buoyant force on a balloon is always greater than the force of gravity exerted on it, regardless of the volume of the balloon.

3. The thrower needs to place a "backspin" on the ball, which is a spin in which the top of the ball moves toward the thrower. This direction of spin causes the speed of the air relative to the surface of the ball to be greater at the top of the ball than at the bottom. Consequently the air pressure is smaller at the top of the ball than at the bottom. As a result, the ball is subject to an upward force due to this pressure difference.

4. (*a*) See Figure 18.38. A contact angle of 90° means that the tangent line to the drop at the contact point is vertical. (*b*) The cohesive force makes an angle of 45° with the horizontal, and the adhesive force is directed vertically downward. The vector sum of the forces must be horizontal, perpendicular to the tangent to the drop at the contact point. The cohesive force is greater than the adhesive force because the vertical component of the cohesive force must cancel the vertically downward adhesive force. (*c*) They are equal. For a particle on the liquid-solid boundary, there are solid atoms below the boundary, liquid particles above the boundary, and liquid particles to the left and the right. The vector sum of the forces exerted on the liquid particle at the boundary is zero and so the magnitude of the cohesive forces between two liquid particles must be the same as the magnitude of the adhesive force between a liquid particle and an atom in the solid.

Figure 18.38

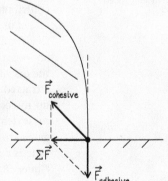

5. (*a*) Because the surface is convex, the pressure at P is greater than atmospheric pressure. (*b*) Because the pressure at R is still atmospheric pressure, the difference in pressure causes a downward force on the liquid in the tube and a resulting *capillary fall*.

18.5 Pressure and gravity

In the first part of this chapter we defined the ratio of the magnitude of the force $\vec{F}_{fs}^c$ exerted by a fluid on a surface and the area A of that surface as the **pressure** in a fluid:

$$P \equiv \frac{F_{fs}^c}{A}. \tag{18.1}$$

Pressure is a scalar, and the SI unit of pressure is newtons per meter-squared, N/m^2. This derived unit is called the **pascal** (Pa), in honor of the French scientist Blaise Pascal:

$$1 \text{ Pa} \equiv 1 \text{ N}/m^2. \tag{18.2}$$

This definition means that a fluid in which the pressure is 1 Pa exerts a force of magnitude 1 N on a 1-m^2 area of the container that holds the fluid. Because a pressure of 1 Pa is very small, pressures are often given in kilopascals: 1 kPa = 1000 Pa. For example, the average pressure in the atmosphere at sea level is

$$P_{atm} = 101.3 \text{ kPa}. \tag{18.3}$$

This pressure is sometimes referred to as 1 atmosphere (1 atm). *Atmosphere* is not an SI unit, however, and so any pressure given in atmospheres must be converted to an SI unit (pascals or kilopascals) before you can do most calculations.

Inside a box that contains absolutely no atoms or molecules, the pressure is zero because nothing exerts a force on the walls of the box. The box is said to contain a *perfect vacuum*. In practice, a perfect vacuum does not exist because it is impossible to remove every particle from a volume of space. Even in deep outer space, the pressure, though very small, is nonzero. In practice, a *vacuum* is any space in which the pressure is less than about 1 Pa. In such a vacuum, $P = 0$ is a very good approximation.

There are several contributions to pressure in a fluid. One contribution is from the gravitational force exerted on the fluid. The pressure in the atmosphere at sea level, for example, is due to the gravitational force exerted by Earth on all the air surrounding the planet. Other contributions are from the motion of the fluid and, when the fluid is a liquid, from the curvature of the liquid surface. A fourth contribution is from the collisions that take place between the fluid particles, and between the fluid particles and the container walls—the pressure in an inflated balloon is an example of this fourth contribution. We'll discuss this contribution to the pressure in a gas in more detail in Chapter 19. In the remainder of this chapter, we concentrate on the gravitational, motion, and surface contributions to pressure.

Let us begin by looking at a familiar type of pressure, the *atmospheric pressure* that results from the air surrounding Earth. The atmosphere can be modeled as an "ocean" of air extending from the ground up. Because gases are compressible, the distribution of gas molecules in the air is not uniform. All the molecules are attracted to the ground by Earth's gravitational pull, which means that higher-up molecules press down on lower ones, compressing the air and making the mass density high near sea level and lower at altitudes above sea level. Because gases are compressible, the relationship between atmospheric pressure and altitude is complicated by the fact that the mass density of air varies with altitude. Figure 18.39 shows how the average atmospheric pressure varies with altitude. If the mass density of air were constant, the curve would be a straight line with a negative slope. If the extremes were as in Figure 18.39—100 kPa at sea level and essentially zero at 30 km—the atmospheric pressure at the midpoint between

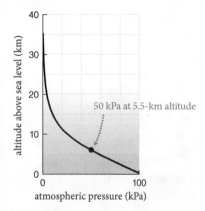

Figure 18.39 Variation in atmospheric pressure with altitude above sea level.

50 kPa at 5.5-km altitude

altitude above sea level (km)

atmospheric pressure (kPa)

these two altitudes, 15 km, would be half of its sea-level value. Instead we see a nonlinear relationship in Figure 18.39, with the pressure at 50 kPa, half the sea-level value, not at an altitude of 15 km but at a mere 5.5 km. This means that half of the atmosphere is below 5.5 km. At the cruising altitude of a commercial airplane (10 km), the atmospheric pressure is about one-quarter of what it is at sea level, meaning that 75% of the atmosphere is below the airplane. At an altitude of 100 km, the pressure in the atmosphere drops to about 1 Pa, close to vacuum.

Even though the changes in atmospheric pressure are considerable over large distances, they are usually negligible over small vertical distances because, even at its densest, at sea level, the mass density of air is so small. Near sea level, for example, the atmospheric pressure drops 12 Pa for every meter of altitude. The pressure difference between the air near your feet and the air near your head is therefore about 20 Pa, which is negligibly small relative to atmospheric pressure, 1×10^5 Pa.

Example 18.5 Altimeter

Airplane cabins are pressurized to prevent passengers from developing altitude sickness during flight. Your altimeter watch, which uses pressure to determine altitude, indicates a cabin altitude of 2500 m even though the captain has just announced that the airplane is flying at an altitude of 10,000 m. Armed with this information, and ignoring the motion of the airplane, you calculate the integral of the magnitude of the force exerted by the air on the fuselage of the airplane, which you estimate to be a cylinder of length 60 m and diameter 6.0 m.

❶ GETTING STARTED Pressure decreases with altitude as shown in Figure 18.39. Because the cabin is pressurized, the pressure inside the cabin is greater than it would be if it were allowed to equalize with the exterior atmospheric pressure at 10,000 m. That the altimeter indicates an altitude of 2500 m tells me that the cabin pressure is equal not to atmospheric pressure at sea level (about 100 kPa) but to a value somewhat lower than 100 kPa. (Because I can feel my ears "pop" as the airplane takes off and lands, I know that the cabin pressure must indeed change somewhat.)

❷ DEVISE PLAN The force exerted by the air on the fuselage is the vector sum of two forces: an outward force due to the pressure P_{in} inside the cabin and an inward force due to the pressure P_{out} outside the cabin. According to Eq. 18.1, the magnitude of each of these forces is PdA, where dA is the surface area of a segment of the fuselage. Because $P_{in} > P_{out}$, the air exerts on each

fuselage segment an outward force of magnitude $(P_{in} - P_{out})dA$. Figure 18.39 gives me the values for P_{out} at 10,000 m and for what the cabin pressure P_{in} must be given that it corresponds to atmospheric pressure at 2500 m. Ignoring the tiny variation in air pressure over the height of the airplane, the values of P_{in} and P_{out} are identical over the entire fuselage, and so I just need to multiply the pressure difference by the surface area $A = \int dA$ of the entire fuselage to compute the integral of the magnitude of the forces exerted on the fuselage.

❸ EXECUTE PLAN From Figure 18.39, I see that $P_{out} = P_{10\,km} = 24$ kPa and $P_{in} = P_{2.5\,km} = 80$ kPa, so $P_{in} - P_{out} = 80$ kPa $-$ 24 kPa $= 56$ kPa. The surface area of the (assumed) cylindrical fuselage is equal to circumference times length plus the surface area of the two circular ends: $A = 2\pi R\ell + 2\pi R^2 = 2\pi(3.0\text{ m})(60\text{ m}) + 2\pi(3.0\text{ m})^2 = 1.2 \times 10^3 \text{ m}^2$. The sum of the magnitudes of the forces exerted by the air on the fuselage is thus $(56 \text{ kPa})(1000 \text{ Pa}/1 \text{ kPa})(1.2 \times 10^3 \text{ m}^2) = 6.7 \times 10^7$ N. ✔

❹ EVALUATE RESULT This force magnitude is very large, corresponding to the force of gravity exerted on 6.8×10^6 kg, or 6800 metric tons! The fuselage does not rupture because the force is distributed over a very large area. For comparison, if the pressure inside the cabin were maintained at a pressure equal to the atmospheric pressure at sea level, the pressure difference across the fuselage would be $P_{in} - P_{out} = 101$ kPa $-$ 24 kPa $= 77$ kPa and the force would be some 40% greater.

✋ **18.13** Consider a book lying on a table at sea level with the front cover facing up. The book is 0.28 m tall, 0.22 m wide, and 50 mm thick; its mass is 3.0 kg. How does the force exerted by the atmosphere on the front cover compare with the force of gravity exerted on the book?

The answer to Checkpoint 18.13 is astonishing. With the book under so much pressure, how can anyone lift it? The reason you can is that the air is pushing on the book from all directions. Not only does it push down on the front cover, it also pushes up on the bottom cover. Because the surface areas of the top and bottom covers are the same and because the air pressure below the book is virtually the same as that above the book, the magnitudes of the upward and downward forces exerted by the atmosphere on the book are the same, and so their vector sum is zero.

You may wonder how the atmosphere can exert an upward force on the bottom cover if the book is lying flat on a table. The reason is that there is always a tiny bit of air between the cover and the table surface, and the air pressure in this layer has to be the same as the air pressure everywhere else around the book (otherwise air would rush into or out of the space between table and cover to eliminate the pressure difference). If you were to remove the air from underneath the book—say, by putting the book on top of the opening of a chamber from which the air has been evacuated—it would be nearly impossible to lift the book off the table.

18.14 A dart that has a suction cup at one end sticks to a ceiling. Describe what force holds the dart against the ceiling.

Figure 18.40 Model for describing how the pressure in a liquid varies with height.

(a) Imaginary cylindrical volume of liquid at rest inside larger volume of same liquid

(b) Free-body diagram for cylinder

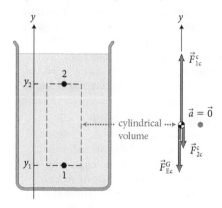

In contrast to gases, liquids are nearly incompressible, which means that we can take the mass density in a liquid to be uniform. Therefore we can derive a simple expression to describe how the pressure in a liquid varies with height. Consider a cylindrical volume of liquid inside a large stationary body of identical liquid (**Figure 18.40a**). The cylindrical volume has volume V and horizontal cross-sectional area A, and Figure 18.40b shows that three forces are exerted on it in the vertical direction: an upward force exerted by the column of liquid below the bottom of the cylindrical volume (point 1), a downward force exerted by the column of liquid above the top of the volume (point 2), and a downward force of gravity. Because our volume is at rest, the vector sum of the three forces must be zero, and so, with the positive y axis pointing upward, we can say for all the forces exerted on the cylindrical volume in the vertical direction,

$$\sum F_{cy} = F^c_{1cy} + F^c_{2cy} + F^G_{Ecy} = P_1 A - P_2 A - mg = 0, \qquad (18.4)$$

where m is the mass of the liquid in the cylindrical volume. Because the mass density ρ of the liquid is the same everywhere, we can write for the mass of the cylindrical volume $m = \rho V = \rho A (y_2 - y_1)$, and so

$$P_1 A - P_2 A - \rho A (y_2 - y_1) g = 0 \qquad (18.5)$$

or, dividing all terms by A and bringing terms with the same subscripts to the same side,

$$P_1 + \rho g y_1 = P_2 + \rho g y_2 \quad \text{(stationary liquid)}. \qquad (18.6)$$

Equation 18.6 tells us that the quantity $P + \rho g y$ is the same at the top and bottom of the cylindrical volume. Because the top and bottom are at arbitrary positions, the fact that $(P + \rho g y)_{\text{top}} = (P + \rho g y)_{\text{bottom}}$ means that the quantity $P + \rho g y$ is the same *everywhere* in the *entire* liquid. In other words, the pressure in a stationary liquid is the same at all points in any given horizontal plane through the liquid (that is, at some fixed value of y).

Rearranging Eq. 18.6 to bring all the terms containing y to one side ($P_1 = P_2 + \rho g y_2 - \rho g y_1$), we see that the pressure at all points in a horizontal plane 1 located a vertical distance $d = y_2 - y_1$ below another horizontal plane 2 is given by

$$P_1 = P_2 + \rho g d \quad \text{(stationary liquid, 1 below 2)}. \qquad (18.7)$$

If we let point 2 be at the surface of the liquid (Figure 18.41), P_2 is the pressure at the surface, and Eq. 18.7 tells us that the pressure at a depth $d = y_2 - y_1$ below the surface is given by

$$P = P_{surface} + \rho g d \quad \text{(stationary liquid).} \qquad (18.8)$$

This expression shows that the pressure P in a stationary liquid increases with depth or, put differently, P decreases linearly with the height in the liquid, just as we concluded in Section 18.2.

The term that contains g is the gravitational contribution to the pressure and is sometimes referred to as *hydrostatic pressure*.

Now consider two liquid-filled tubes, one wide and one narrow and both open at one end, connected as shown in Figure 18.42. In each tube, the pressure at the surface of the liquid is equal to atmospheric pressure. This means that the downward force $P_{atm}A_{wide}$ exerted by the atmosphere on the surface of the liquid in the wide tube is significantly greater than the downward force $P_{atm}A_{narrow}$ exerted by the atmosphere on the surface of the liquid in the narrow tube. How does this difference in force affect the levels of the water in each tube?

To answer this question, let's consider the pressures at points 1 and 2 in Figure 18.42. Because these points are in the same horizontal plane, the pressures must be equal, $P_1 = P_2$. Equation 18.8 tells us that these pressures are related to the (atmospheric) pressure at the surface by

$$P_1 = P_{atm} + \rho g d_1 \quad \text{and} \quad P_2 = P_{atm} + \rho g d_2. \qquad (18.9)$$

Because $P_1 = P_2$, we must have $d_1 = d_2$: The liquid rises to the same height in both tubes, even though the forces exerted by the atmosphere on the surfaces of the liquid are different. The reason is that, despite the difference in these downward forces, the *pressure* at the bottom of each tube must be the same ($P_1 = P_2$). If the pressure were not the same, the pressure difference would cause the liquid to flow until the pressure difference is evened out, which occurs only when $d_1 = d_2$. The equal height of the liquid surfaces in the two tubes means that these surfaces are in the same horizontal plane. So the pressure at two points in a horizontal plane is the same as long as the two points are connected by an uninterrupted path through the liquid.

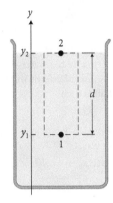

Figure 18.41 If the cylindrical volume is defined such that its top is at the liquid surface, the pressure at the top is equal to atmospheric pressure.

Figure 18.42 Liquid rises to the same level in connected tubes, regardless of their width or shape.

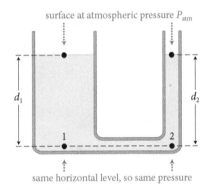

surface at atmospheric pressure P_{atm}

same horizontal level, so same pressure

Example 18.6 Dam

A dam of horizontal length ℓ holds water of mass density ρ to a height h. What is the magnitude of the force exerted by the water on the dam?

❶ GETTING STARTED I begin with a sketch (Figure 18.43a). The pressure in the water behind the dam is equal to atmospheric pressure at the water surface and increases linearly with depth as given by Eq. 18.8. If I choose my y axis pointing upward, with the origin at the bottom of the dam, the pressure varies with y as shown in Figure 18.43b. For convenience, I call the water side of the dam face A and the opposite side face B.

❷ DEVISE PLAN Because the pressure varies with y, the magnitude of the force $\vec{F}_{wd}$ also varies with y. I therefore divide the area of face A into thin horizontal strips each of width dy (Figure 18.43a). Because I know the pressure at a given value of y, I can calculate the force dF_{wd} exerted on the strip. To calculate the force exerted by the water on the whole dam, I integrate my expression over y from $y = 0$ at the bottom of the dam to $y = h$ at the water surface.

Figure 18.43

(a)

(b)

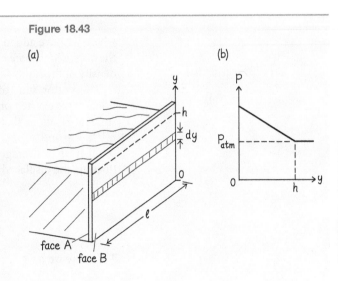

❸ EXECUTE PLAN Because a point in the water located at some arbitrary value of y is at a depth $d = h - y$ below the water surface and because $P_{\text{surface}} = P_{\text{atm}}$, I write Eq. 18.8 in the form

$$P(y) = P_{\text{atm}} + \rho g(h - y).$$

The area of each strip is $dA = \ell\, dy$, and so the magnitude of the force exerted by the water on each strip is $dF = P(y)dA = P(y)\ell\, dy$. The magnitude of the force exerted by the water on the entire dam is then

$$F_{\text{wd}} = \int dF_{\text{ws}} = \ell \int_0^h P(y)dy = \ell\left[P_{\text{atm}}y + \rho ghy - \tfrac{1}{2}\rho gy^2 \right]_0^h$$

$$= \ell h\left(P_{\text{atm}} + \tfrac{1}{2}\rho gh\right).\ ✔$$

❹ EVALUATE RESULT The product ℓh is the surface area of the dam, and so the term $\ell h P_{\text{atm}}$ is the magnitude of the force exerted by the atmosphere on the dam. Because the atmosphere exerts a force of the same magnitude on face B in the opposite direction, the first term in my answer drops out of the vector sum of the forces exerted by the water on face A of the dam and by the atmosphere on face B of the dam. The remaining term, $\ell h(\tfrac{1}{2}\rho gh)$, is the product of the area of the face of the dam and $\tfrac{1}{2}\rho gh$, which is the pressure increase halfway down in the water. Because the pressure increases linearly with depth, this term represents the average value of the pressure increase between surface and bottom, and so my answer makes sense.

Let us now turn to buoyancy. The volume of liquid in Figure 18.40 is held in place by the buoyant force $\vec{F}_{lv}^{b}$ exerted by the surrounding liquid. The vector sum of the forces exerted on the volume of liquid is thus

$$\sum \vec{F}_v = \vec{F}_{lv}^{b} + \vec{F}_{Ev}^{G} = \vec{0}. \tag{18.10}$$

Because the magnitude of the force of gravity on the volume of liquid is ρVg, we obtain for the magnitude of the **buoyant force** exerted by the surrounding liquid on the volume of liquid

$$F_{lv}^{b} = \rho Vg. \tag{18.11}$$

The direction of the buoyant force is always opposite the direction of the gravitational acceleration. Because this expression holds for any object in any fluid, we can write for the buoyant force on any object o in a fluid f:

$$F_{fo}^{b} = \rho_f V_{\text{disp}}g, \tag{18.12}$$

where we have added a subscript to V to remind us that this volume represents the *displaced volume* of fluid and a subscript on ρ to remind us that it is the mass density of the fluid.

The vector sum of the upward buoyant force and the downward gravitational force exerted on an object in a given fluid determines whether the object sinks or floats in that fluid. When they are equal in magnitude, $F_{fo}^{b} = F_{Eo}^{G}$ or $\rho_f V_{\text{disp}}g = mg = \rho_{o,\text{av}} V_o g$, where we have written the mass of the object as the product of its average mass density $\rho_{o,\text{av}}$ and volume V_o. For an object that is floating, the displaced volume is less than the volume of that object:

$$V_{\text{disp}} = \frac{\rho_{o,\text{av}}}{\rho_f} V_o \quad \text{or} \quad \frac{\rho_{o,\text{av}}}{\rho_f} < 1. \tag{18.13}$$

Therefore we have

$$\rho_{o,\text{av}} < \rho_f \quad \text{(object floats).} \tag{18.14}$$

An object placed in a fluid floats whenever the average mass density of the object is smaller than the mass density of the fluid.

Conversely, an object sinks when the magnitude of the buoyant force is smaller than that of the gravitational force. Working through the same thought sequence that got us from forces to Eq. 18.14, we see that

$$\rho_{o,av} > \rho_f \quad \text{(object sinks).} \tag{18.15}$$

An object placed in a fluid sinks whenever the average mass density of the object exceeds the mass density of the fluid.

When the object's average mass density is equal to the mass density of the fluid, $\rho_{o,av} = \rho_f$, the object is said to have *neutral buoyancy*—it can "hang" motionless anywhere in the fluid.

✋ **18.15** An object floats with 80% of its volume submerged in water. How does the average mass density of the object compare with that of water?

18.6 Working with pressure

The results in Section 18.5 allow us to understand a number of applications of pressure. Let's first consider measurements of pressure. A *pressure gauge* is used to measure pressures that are near or greater than atmospheric pressure; a *vacuum gauge* is used to measure pressures that are smaller than atmospheric pressure.

The most common pressure gauges have a flexible, sealed chamber that changes shape when the pressure inside or outside the chamber changes. The deformation of the chamber is measured electrically, optically, or mechanically. For example, in a simple mechanical pressure gauge (**Figure 18.44**), the chamber deformation is coupled via a gear to an indicating needle on a calibrated dial. The reading on a pressure gauge is the pressure in excess of atmospheric pressure. Thus when you use a pressure gauge to check the pressure in, say, a bicycle tire, the reading on the gauge, called *gauge pressure* P_{gauge}, is related to the actual pressure P in the tire by

$$P = P_{gauge} + P_{atm}. \tag{18.16}$$

For example, when a tire pressure gauge reads 2.4 atm, the pressure inside the tire is 3.4 atm. If the pressure inside a container is smaller than atmospheric pressure, the pressure gauge reading is negative.

The simplest pressure gauge is the open-tube *manometer* (**Figure 18.45**). A manometer is essentially a liquid-filled, U-shaped tube, one leg of which is connected to the container where the pressure P is to be measured and the other leg of which is open to the atmosphere. The pressure in the liquid at the bottom of the U is P_{bottom}, while that at the surface of the liquid in the right leg is atmospheric pressure. Applying Eq. 18.7 to these two points, we get $P_{bottom} = P_{atm} + \rho g y_{right}$. Applying the same procedure to the left leg, we have $P_{bottom} = P + \rho g y_{left}$, and so $P = P_{atm} + \rho g(y_{right} - y_{left})$ or, if we denote the difference in the heights of the liquid in the two legs by h, the pressure measured by the manometer is

$$P = P_{atm} + \rho g h. \tag{18.17}$$

Equation 18.17 shows that using a manometer to measure the pressure in some fluid-filled container requires knowing atmospheric pressure, which means we cannot use a manometer to measure atmospheric pressure. We can

Figure 18.44 Simple mechanical pressure gauge.

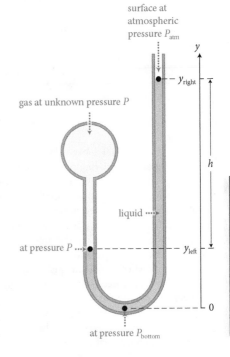

Figure 18.45 An open-tube manometer. The difference in the levels in the two legs of the manometer gives the pressure P inside the bulb.

QUANTITATIVE TOOLS

Figure 18.46 A barometer. The height of the column of liquid in the tube gives the atmospheric pressure.

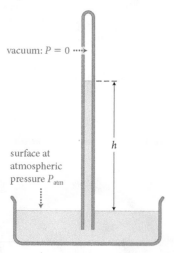

vacuum: $P = 0$

h

surface at atmospheric pressure P_{atm}

use another device, though, a *barometer*, illustrated in **Figure 18.46**. This device consists of a shallow reservoir; a long, thin tube closed at one end and open at the other; and a dense liquid, usually mercury. Some of the liquid sits in the reservoir. The tube is filled completely with the liquid, and its open end is covered temporarily. The tube is then inverted and its open end plunged into the liquid in the reservoir. As the temporary cover is removed, some of the liquid drains from the tube into the reservoir, leaving a vacuum above the liquid in the closed end of the tube. The level of the liquid in the tube remains above the level of the liquid in the reservoir so as to satisfy Eq. 18.7 when we define the liquid surface in the reservoir as point 1 and the liquid surface in the tube as point 2. When we let the reservoir liquid surface be at $y_1 = 0$, the left side of Eq. 18.6 becomes $P_{atm} + 0$. Because there is a vacuum above the tube liquid surface, the right side of Eq. 18.6 is $0 + \rho gh$, and so we have

$$P_{atm} = \rho gh. \tag{18.18}$$

If we know the mass density ρ of the liquid, we can determine atmospheric pressure by measuring the height h of the liquid in the tube.

Example 18.7 Liquid support

In **Figure 18.47**, the top tank, which is open to the atmosphere, contains water ($\rho_{water} = 1.0 \times 10^3 \text{ kg/m}^3$) and the bottom tank contains oil ($\rho_{oil} = 0.92 \times 10^3 \text{ kg/m}^3$) covered by a piston. The tank on the right has a freely movable partition that keeps the oil and water separate. The partition is a vertical distance 0.10 m below the open surface of the water. If the piston in the bottom tank is 0.50 m below the open surface of the water and has a surface area of $8.3 \times 10^{-3} \text{ m}^2$, what must the mass of the piston be to keep the system in mechanical equilibrium? For simplicity, ignore the mass of the partition.

Figure 18.47 Example 18.7.

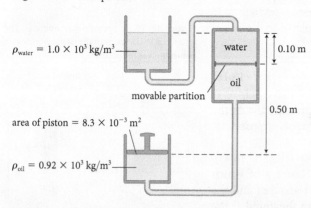

$\rho_{water} = 1.0 \times 10^3 \text{ kg/m}^3$

water

oil

0.10 m

movable partition

0.50 m

area of piston = $8.3 \times 10^{-3} \text{ m}^2$

$\rho_{oil} = 0.92 \times 10^3 \text{ kg/m}^3$

❶ **GETTING STARTED** I begin by making a sketch of the arrangement and identifying points of interest (**Figure 18.48a**). Point 1 is at the water surface in the top tank, and the pressure here is atmospheric pressure. Point 2 is on the partition in the right tank separating the water from the oil. In mechanical equilibrium, the partition is not accelerating, and so, because I ignore the partition mass, the pressure must be the same on the two sides of the partition. Point 3 is at the position in the bottom tank where the oil and piston meet.

I also draw a free-body diagram for the piston, which also has zero acceleration in mechanical equilibrium (**Figure 18.48b**). The piston is subject to an upward force $\vec{F}^c_{op}$ exerted by the oil, a downward force $\vec{F}^c_{ap}$ exerted by the air above the piston, and a downward gravitational force $\vec{F}^G_{Ep}$.

Figure 18.48

(a)

(b)

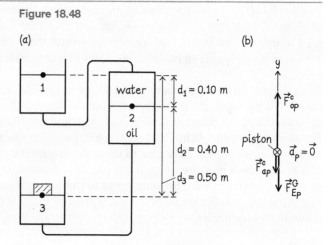

1

water

2

oil

$d_1 = 0.10$ m

$d_2 = 0.40$ m

$d_3 = 0.50$ m

3

y

$\vec{F}^c_{op}$

piston

$\vec{a}_p = 0$

$\vec{F}^c_{ap}$

$\vec{F}^G_{Ep}$

❷ **DEVISE PLAN** When the system is in mechanical equilibrium, the piston is stationary. Therefore the vector sum of the forces exerted on it must be zero. From my free-body diagram, I see that the magnitude of the upward force exerted by the oil on the piston is equal in magnitude to the sum of the downward gravitational force exerted on the piston and the downward force exerted by the air on the piston: $F^c_{op} = F^G_{Ep} + F^c_{ap} = m_p g + F^c_{ap}$.

The magnitudes F^c_{ap} and F^c_{op} are given by a rearranged form of Eq. 18.1: $F^c_{ap} = P_{air} A_{piston}$, and $F^c_{op} = P_3 A_{piston}$. The problem therefore reduces to determining the pressure P_3 at point 3, where the oil and piston meet. That point is 0.40 m below the partition in the tank on the right, where the pressure is P_2. Because points 2 and 3 are in a connected liquid (the oil), I can use Eq. 18.7 to relate P_2 and P_3. Points 1 and 2 are also in a connected liquid (the water), so I can use Eq. 18.7 to relate P_2 and $P_1 = P_{atm}$.

❸ **EXECUTE PLAN** Because the vector sum of the forces exerted on the piston is zero, I have $F^c_{op} - F^c_{ap} - F^G_{Ep} = P_3 A_{piston} - P_{air} A_{piston} - m_{piston} g = 0$, or

$$m_{piston} = \frac{(P_3 - P_{air}) A_{piston}}{g}. \tag{1}$$

To obtain P_3, I first apply Eq. 18.7 to points 1 and 2, remembering that the lower point always goes on the left side of the equation: $P_2 = P_1 + \rho_{water}gd_1 = P_{air} + \rho_{water}gd_1$. Then I apply Eq. 18.7 to points 2 and 3: $P_3 = P_2 + \rho_{oil}gd_2 = P_{air} + \rho_{water}gd_1 + \rho_{oil}gd_2$. Substituting this expression into Eq. 1, I get

$$m_{piston} = \frac{(P_{air} + \rho_{water}gd_1 + \rho_{oil}gd_2 - P_{air})A_{piston}}{g}$$

$$= (\rho_{water}d_1 + \rho_{oil}d_2)A_{piston}$$

$$= [(1.0 \times 10^3 \text{ kg/m}^3)(0.10 \text{ m}) + (0.92 \times 10^3 \text{ kg/m}^3)$$

$$\times (0.40 \text{ m})](8.3 \times 10^{-3} \text{ m}^2) = 3.9 \text{ kg.} ✔$$

④ **EVALUATE RESULT** The atmospheric pressure term drops out because it affects both the water surface and the piston. Points 1 and 3 are separated by 0.5 m. Because I know that the pressure difference due to 10 m of water is equal to atmospheric pressure at sea level, or about 100 kPa (see Checkpoint 18.4; these are handy numbers to remember), I know that 0.5 m of water should correspond to a pressure difference of about 5 kPa. The mass density of oil is a bit smaller than that of water, but I can ignore the difference. The surface area of the piston is about 10^{-2} m^2, and so the force giving rise to the 5-kPa pressure difference is about 50 N, which is equal to the gravitational force exerted on a mass of 5 kg, which is close to the answer I obtained.

✋ **18.16** Oil and water don't mix, and the mass density of oil is smaller than that of water. Suppose water is poured into a U-shaped tube that is open at both ends until the water surface is halfway up each leg of the tube, and then some oil is poured on top of the water in the right leg. Once the system comes to equilibrium, are the top of the oil column in the right leg and the top of the water column in the left leg at the same height? If not, which is higher?

The principles we have discussed in this section allow us to understand the working of *hydraulic systems*, which are used for lifting heavy loads or exerting large forces. The basic idea behind a hydraulic system is simple: A liquid, usually oil, is used to transmit a force applied at one point to another point. **Figure 18.49a** shows a schematic hydraulic system: two pistons fitted into two connected, oil-filled cylinders. Because the oil is incompressible, a force exerted on one piston increases the pressure everywhere in the oil, and so the force is transmitted to the other piston.

The tube that connects the two cylinders can be any length or shape, which means it can snake through tight spaces and can also fork, allowing an applied force to be distributed. In the brake system of a car, for example, the force your foot exerts on the brake pedal is transmitted to the piston in an oil-filled master cylinder. The pressure increase in the oil is transmitted through steel tubes to the four brakes in the wheels, where the oil exerts a force on the brake pads, pushing them against a disk attached to the wheel, which slows down the car (Figure 18.49b).

Another great advantage of hydraulic systems is that they permit us to "multiply" a force by varying the size of the pistons. In **Figure 18.50**, a force of magnitude F_1 is exerted on a small piston of area A_1, causing a gauge pressure

Figure 18.49 Hydraulic systems make it possible to lift heavy loads or exert large forces.

(a) Basic hydraulic system

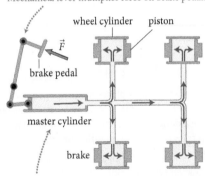

(b) Schematic model of hydraulic brake system

Mechanical lever multiplies force on brake pedal.

Hydraulic system multiplies and distributes force.

Figure 18.50 Hydraulic system used to "multiply" a force.

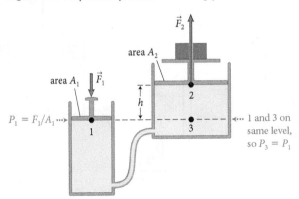

Procedure: Working with pressure in liquids at rest

The branch of physics that deals with pressure in a liquid at rest is called *hydrostatics*. The pressure in a liquid at rest is determined by gravity and by what happens at the boundary of the liquid. To determine the pressure in such liquids:

1. Begin by making a sketch showing all the **boundaries** and identifying all the factors that affect pressure: pistons, gases at surfaces open to the atmosphere, and so on. Note the known vertical heights of liquid surfaces, the areas of these surfaces, the surface areas of pistons, and the liquid mass densities.
2. Determine the **pressure at each surface.** The pressure at a liquid surface open to the air is equal to atmospheric pressure P_{atm}. The pressure at a liquid surface

bordering a vacuum is zero: $P = 0$. The pressure at a liquid surface open to a gas other than the atmosphere is equal to the pressure in the gas: $P = P_{gas}$. The pressure at a liquid surface that is in contact with a solid, such as a piston, is $P = F^c_{sl}/A$, where F^c_{sl} is the magnitude of the force exerted by the solid on the liquid and A is the area over which that force is exerted.
3. Use **horizontal planes.** The pressure is the same at all points on a horizontal plane in a connected liquid. The pressure difference between two horizontal planes 1 and 2 is given by $P_1 = P_2 + \rho g d$ (Eq. 18.7), where d is the vertical distance between the horizontal planes and 1 is below 2.

$P_1 = F_1/A_1$ in the liquid at the small piston. Because $P_3 = P_2 + \rho g h$, the gauge pressure P_2 at the large piston is then

$$P_2 = P_3 - \rho g h = P_1 - \rho g h, \tag{18.19}$$

where h is the vertical distance between the pistons. When the liquid in a hydraulic system is oil, we can ignore the term $\rho g h$ because the mass density of oil is not very great and h is generally small, so that $P_1 \approx P_2$. The gauge pressure P_2 at point 2 causes the liquid to exert a force of magnitude F_2 on the piston. If the area of that piston is A_2, we have $P_2 = F_2/A_2$, so that

$$\frac{F_1}{A_1} = \frac{F_2}{A_2} \tag{18.20}$$

or

$$F_2 = \frac{A_2}{A_1} F_1. \tag{18.21}$$

Because $A_2 > A_1$, we have $F_2 > F_1$. Therefore a small force applied to the small piston causes the liquid to exert a large force on the large piston (even though the *pressure* is the same at both points). The factor A_2/A_1 is what permits you to exert large forces on the brakes of a car even though the force your foot exerts on the brake pedal is relatively small.

Example 18.8 Liquid work

Suppose a load is placed on the large piston in Figure 18.50 and we wish to raise that load by a certain distance. How does the work done on the small piston compare with the work required to raise the load? Ignore the masses of the pistons.

❶ GETTING STARTED I begin by making a sketch (**Figure 18.51**). To keep my subscripts simple, I call the small piston 1 and the large piston 2. I denote the displacements of the two pistons by $\Delta \vec{x}_1$ and $\Delta \vec{x}_2$, and the forces exerted on them by $\vec{F}_1$ and $\vec{F}_2$.

❷ DEVISE PLAN The magnitudes F_1 and F_2 are related by Eq. 18.21, $F_2 = (A_2/A_1)F_1$. Each force does an amount of work given by the scalar product of the force and the force displacement, which is equal to the displacement of the piston on which the force is exerted. Both forces do positive work because the

Figure 18.51

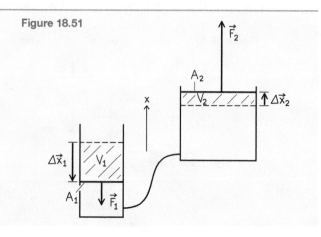

direction of each force is the same as the direction of the force displacement. Because the liquid in the cylinders is incompressible, I know that the volume V_1 of liquid pushed out of the small cylinder must be equal to the volume V_2 of liquid pushed into the large cylinder: $V_1 = V_2$. This expression will help me relate the piston displacements to the piston surface areas.

❸ **EXECUTE PLAN** Raising the load a distance $d = \Delta x_2$ requires work $W_2 = \vec{F}_2 \cdot \Delta \vec{x}_2 = F_2 d$. Using Eq. 18.21, I can write for the work done on the small piston

$$W_1 = \vec{F}_1 \cdot \Delta \vec{x}_1 = F_1 \Delta x_1 = F_2 \frac{A_1}{A_2} \Delta x_1. \quad (1)$$

The volume of liquid pushed out of the small cylinder is $V_1 = A_1 \Delta x_1$, and that pushed into the large cylinder is $V_2 = A_2 \Delta x_2 = A_2 d$. Because these two volumes are equal, I have $A_1 \Delta x_1 = A_2 d$, or $\Delta x_1 = (A_2/A_1)d$. Substituting this expression into Eq. 1, I get $W_1 = F_2 d$, telling me that the same work is done on both pistons. ✔

❹ **EVALUATE RESULT** My answer tells me that no energy is gained by using the hydraulic lift, which makes sense. The force exerted on the small piston is increased by the factor A_2/A_1 when it is transmitted to the large piston, but the displacement decreases by the inverse factor A_1/A_2, and so the work done is the same.

✋ **18.17** In the derivation of Eq. 18.21, we ignored the effect of atmospheric pressure on the two pistons. How is the result we obtained affected by atmospheric pressure?

18.7 Bernoulli's equation

Next we turn our attention to fluids in motion. Consider the laminar flow of a nonviscous fluid through the tapered tube in **Figure 18.52**. In a time interval Δt, a thin "slice" of fluid in the wide part of the tube undergoes a displacement Δx_1. The volume of fluid that enters the tube at point 1 is thus $A_1 |\Delta x_1|$, where A_1 is the cross-sectional area of the wide part of the tube. If the mass density of the fluid at point 1 is ρ_1, the amount of mass entering the tube at point 1 is $m_1 = \rho_1 A_1 |\Delta x_1|$. Likewise, if a thin slice of fluid in the narrow part of the tube has a displacement Δx_2, and the cross-sectional area of the narrow part is A_2, then the mass of the fluid emerging at point 2 is $\rho_2 A_2 |\Delta x_2|$. If the flow is laminar, the flow speed at any location doesn't change with time. This means that no mass can accumulate in the tube, and so the mass of the fluid emerging from the narrow part at point 2 during a time interval Δt must be the same as the mass entering at point 1:

$$\rho_1 A_1 |\Delta x_1| = \rho_2 A_2 |\Delta x_2| \quad \text{(laminar flow of nonviscous fluid).} \quad (18.22)$$

If the flow speed of the fluid is v_1 at point 1 and v_2 at point 2, we have $|\Delta x_1| = v_1 \Delta t$ and $|\Delta x_2| = v_2 \Delta t$. Substituting these expressions into Eq. 18.22 and dividing both sides by Δt, we obtain

$$\rho_1 A_1 v_1 = \rho_2 A_2 v_2 \quad \text{(laminar flow of nonviscous fluid).} \quad (18.23)$$

Equation 18.23 is called the **continuity equation.** This equation quantitatively expresses the idea that the mass of the fluid emerging from the narrow part at point 2 during a time interval Δt is the same as the mass entering at point 1.

For an incompressible fluid, the mass density is the same in all parts of the tube and so

$$A_1 v_1 = A_2 v_2 \quad \text{(laminar flow of nonviscous, incompressible fluid).} \quad (18.24)$$

This equation expresses quantitatively what we concluded in Section 18.3: A fluid flows faster in a narrow part of a tube than in a wide part.

Figure 18.52 A fluid flows through a tapered tube.

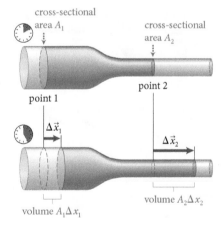

We can write Eq. 18.24 in another form by introducing a quantity Q, called the **volume flow rate,** the rate at which a volume of fluid crosses a section of a tube. If a volume V crosses a section in a time interval Δt, the volume flow rate is

$$Q \equiv \frac{V}{\Delta t}. \tag{18.25}$$

The SI unit of volume flow rate is meters cubed per second (m^3/s). For a nonviscous fluid flowing at speed v through a pipe of cross-sectional area A, we have $V = A|\Delta x| = Av\Delta t$, and so the volume flow rate is

$$Q = Av. \tag{18.26}$$

Equation 18.24 thus tells us that the volume flow rate is the same at all locations in an incompressible fluid: $Q_1 = Q_2$.

We also concluded in Section 18.3 that in order for a fluid to accelerate in the region where a tube narrows, the pressure must be greater in the wide part. Let us therefore now relate pressure to flow speed. Consider the tapered tube in **Figure 18.53**. At point 1 the tube's cross-sectional area is A_1, the fluid's speed is v_1, the pressure is P_1, and the vertical height above some arbitrarily chosen origin is y_1. The corresponding quantities at point 2 are A_2, v_2, P_2, and y_2. Let us consider the system made up of Earth and the fluid initially between points 1 and 2. How does the energy of this system change?

To answer this question, let us apply the energy law (Eq. 9.1, $\Delta E = W$) to this system over a time interval Δt. As Figure 18.53 shows, during the time interval Δt the system has shifted to the right. The difference between the initial and final states of the system is a "missing" volume of fluid on the left and an "added" volume of fluid on the right. The continuity equation requires the masses of these two volumes to be the same: $m_1 = m_2 = m$. Because the mass of the fluid between points 1 and 2 doesn't change, the only forms of energy we need to consider are the kinetic and potential energies associated with m_1 and m_2. The change in kinetic energy is

$$\Delta K = \tfrac{1}{2}m_2 v_2^2 - \tfrac{1}{2}m_1 v_1^2 = \tfrac{1}{2}m(v_2^2 - v_1^2), \tag{18.27}$$

and the change in gravitational potential energy is

$$\Delta U^G = m_2 g y_2 - m_1 g y_1 = mg(y_2 - y_1). \tag{18.28}$$

Substituting Eqs. 18.27 and 18.28 into the left side of the energy law yields

$$\Delta E = \Delta K + \Delta U^G = \tfrac{1}{2}m(v_2^2 - v_1^2) + mg(y_2 - y_1). \tag{18.29}$$

Our system is not closed because the fluid that is in contact with the left and right termini of the system exerts forces $\vec{F}_1$ and $\vec{F}_2$ on the system and the points of application of these forces undergo nonzero force displacements $\Delta \vec{x}_{F1}$ and $\Delta \vec{x}_{F2}$,* respectively. So let us now evaluate the work W done on the system. We take the time interval Δt to be small enough so that we can consider the forces to be constant. Because $\vec{F}_1$ and the force displacement of $\vec{F}_1$ are in the same direction, the work done on the system at point 1 is positive:

$$W_1 = F_{1x}\Delta x_{F1} = +P_1 A_1 v_1 \Delta t. \tag{18.30}$$

Figure 18.53 A volume of fluid flows through a tapered tube that changes altitude.

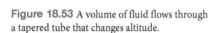

Between these points, mass of fluid does not change.

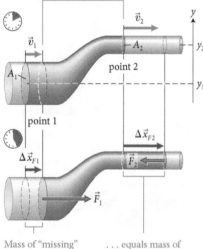

Mass of "missing" fluid here . . .

. . . equals mass of "added" fluid here.

---

*The subscript F on the displacement means we are considering the displacement of the point of application of the force (see Section 9.1).

At point 2, negative work is done on the system because the force and force displacement point in opposite directions:

$$W_2 = F_{2x}\Delta x_{F2} = -P_2 A_2 v_2 \Delta t, \tag{18.31}$$

and so the work done on the system is

$$W = P_1 A_1 v_1 \Delta t - P_2 A_2 v_2 \Delta t. \tag{18.32}$$

Using Eq. 18.24, we can write $A_1 v_1 \Delta t = A_2 v_2 \Delta t$. The quantity $A_1 v_1 \Delta t$ is the volume of the mass of fluid m that flows into (or out of) the section of tube we are considering. From the definition of mass density, we know that $V = m/\rho$, which means that

$$\frac{m}{\rho} = A_1 v_1 \Delta t. \tag{18.33}$$

Substituting this expression into Eq. 18.32, we get

$$W = \frac{m}{\rho}(P_1 - P_2). \tag{18.34}$$

Because the work done on the system must be equal to the change in the system's energy, we have, from Eqs. 18.29 and 18.34:

$$\tfrac{1}{2}m(v_2^2 - v_1^2) + mg(y_2 - y_1) = \frac{m}{\rho}(P_1 - P_2) \tag{18.35}$$

or, dividing both sides by m, rearranging terms, and multiplying through by ρ,

$$P_1 + \rho g y_1 + \tfrac{1}{2}\rho v_1^2 = P_2 + \rho g y_2 + \tfrac{1}{2}\rho v_2^2 \quad \begin{array}{l}\text{(laminar flow} \\ \text{of incompressible,} \\ \text{nonviscous fluid).}\end{array} \tag{18.36}$$

Together with the continuity equation (Eq. 18.23), Eq. 18.36, known as **Bernoulli's equation,** forms the foundation of the hydrodynamics of incompressible fluids. Strictly speaking, Bernoulli's equation applies only to liquids because we used Eq. 18.24, which is valid only for incompressible fluids. However, for the flow of a gas at modest velocities (and for height differences $y_2 - y_1$ that are not too great), the pressure variations have little effect on the mass density and Bernoulli's equation is still a reasonable approximation.

Because the points 1 and 2 are arbitrary, Eq. 18.36 tells us that the quantity $P + \rho g y + \tfrac{1}{2}\rho v^2$ is the same at different locations in a flow stream. If the fluid is stationary, $v = 0$ everywhere, and Eq. 18.36 takes on the same form as Eq. 18.6. If points 1 and 2 are at the same height, then $y_1 = y_2$, and Eq. 18.36 becomes

$$P_1 + \tfrac{1}{2}\rho v_1^2 = P_2 + \tfrac{1}{2}\rho v_2^2 \quad \begin{array}{l}\text{(horizontal laminar flow} \\ \text{of incompressible,} \\ \text{nonviscous fluid).}\end{array} \tag{18.37}$$

This equation confirms what we concluded in Section 18.3: The pressure P is greater where the flow speed v is smaller, and vice versa.

Example 18.9 Leaky bucket

Water leaks out of a small hole in the side of a bucket. The hole is a distance d below the surface of the water, and the cross section of the hole is much smaller than the diameter of the bucket. At what speed does the water emerge from the hole?

❶ GETTING STARTED I begin by making a sketch of the bucket (Figure 18.54), defining my point 1 at the water surface and my point 2 at the hole. The bucket can be regarded as a tube open at both ends, with the top opening, cross-sectional area A_1, as one end and the hole, cross-sectional area A_2, as the other end.

Figure 18.54

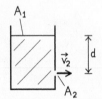

❷ DEVISE PLAN Because $A_1 \gg A_2$, the continuity equation (Eq. 18.23, $\rho_1 A_1 v_1 = \rho_2 A_2 v_2$) tells me that the speed v_1 at which the surface of the water moves downward is much smaller than the speed v_2 at which the water emerges from the hole, $v_1 \ll v_2$. For all practical purposes, therefore, I can assume $v_1 \approx 0$. I also know that the pressure both at the water surface and at the hole is equal to atmospheric pressure. I can then use Bernoulli's equation (Eq. 18.36) to determine v_2.

❸ EXECUTE PLAN Because $P_1 = P_2 = P_{atm}$, Bernoulli's equation reduces to $\rho g y_1 + \frac{1}{2}\rho v_1^2 = \rho g y_2 + \frac{1}{2}\rho v_2^2$. Dividing through by $\frac{1}{2}\rho$, setting $v_1 = 0$, and bringing the terms containing y to the left side, I get $v_2^2 = 2g(y_1 - y_2) = 2gd$, and so $v_2 = \sqrt{2gd}$. ✔

❹ EVALUATE RESULT My answer is equal to the speed acquired over a vertical distance d by a freely falling object starting from rest. Because the Earth-object system is closed, $\Delta K + \Delta U^G = \frac{1}{2}mv^2 - mgd = 0$, and so $v^2 = 2gd$. My result makes sense: For each water drop that emerges from the hole, the water at the surface is reduced by an equal amount and the kinetic energy of the emerging drop must be equal to the decrease in potential energy.

18.18 Suppose that the diameter of the hole in Example 18.9 is 10.0 mm, the bucket diameter is 0.50 m, and the distance from the hole to the water surface is 0.30 m. At what speed does the water emerge from the hole?

18.8 Viscosity and surface tension

So far in our discussion of fluid flow we have ignored energy dissipation. As pointed out in Section 18.3, viscosity is a measure of a fluid's resistance to shear stress. Dissipation of energy occurs in both liquids and gases, and so both have a nonzero viscosity. Consider, for example, two flat plates separated by a fluid (**Figure 18.55**). The lower plate is held fixed while the upper plate is translated in the plane of the plate. Because of friction between adjacent fluid layers, a force $\vec{F}_{p,ext}$ is required to move the upper plate at constant speed v. Experiments show that the magnitude of this force is proportional to the surface area A of the upper plate and to the speed v, and inversely proportional to the distance d between the plates. Because the force is directed in the direction opposite that of the velocity of the upper plate, we can write for a plate moving in the positive x direction:

$$F_{fpx} \equiv -\eta A \frac{v_x}{d}. \tag{18.38}$$

The proportionality constant η (Greek lowercase eta) is the **viscosity** of the fluid. The SI unit for viscosity is the pascal-second, with $1 \text{ Pa} \cdot \text{s} = 1 \text{ kg}/(\text{m} \cdot \text{s})$. The greater the viscosity of a fluid, the greater the force needed to move the plate at constant speed. Because the work done by the plate on the fluid ends up as thermal energy in the fluid, a greater viscosity means more energy dissipation in the fluid as fluid layers rub past one another.

The ratio v_x/d in Eq. 18.38 can also be written as dv_x/dy because both v_x/d and dv_x/dy express the change in the x component of the velocity divided by the distance over which the change occurs. The derivative dv_x/dy, called the *velocity gradient,* is a measure of the rate at which one fluid layer moves relative to an adjacent layer. The advantage of expressing Eq. 18.38 in terms of the velocity gradient is that the resulting equation can be applied more generally:

$$F_{fpx} = -\eta A \frac{dv_x}{dy}. \tag{18.39}$$

Figure 18.55 To measure viscosity, a fluid is placed between two parallel plates. The bottom plate is held fixed while the top plate is translated. The force required to keep the top plate moving at constant speed gives the viscosity of the fluid.

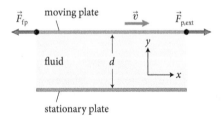

Table 18.1 Viscosity of fluids

Fluid	Viscosity (Pa · s)	Temperature (°C)
Air	1.71×10^{-5}	0
	1.81×10^{-5}	20
	2.18×10^{-5}	100
Blood	3.0×10^{-3}	20
	2.0×10^{-3}	37
Castor oil	5.3	0
	0.986	20
	0.017	100
Water	1.79×10^{-3}	0
	1.00×10^{-3}	20
	0.282×10^{-3}	100

Because viscous forces are generally small, fluids are often used as lubricants to reduce friction between solids. Table 18.1 lists some typical viscosities. The viscosity of gases is typically several orders of magnitude smaller than that of liquids. The viscosity of liquids typically decreases as temperature increases. Honey, for example, becomes more runny as it is heated. In contrast, gases typically get more viscous at higher temperatures.

Many instances of fluid flow, such as the velocity distribution of water through pipes or the flow of blood through arteries, involve laminar flow through cylindrical tubes. Let us therefore consider the flow of a fluid through a horizontal cylindrical pipe of length ℓ and radius R. Because the boundary layer of the fluid sticks to the interior wall of the pipe, this layer is at rest. The layer of fluid adjacent to the boundary layer moves slowly in the direction of the flow, and adjacent layers move at increasing speeds, with the fluid at the center of the pipe having the maximum speed (see Figure 18.19b).

Consider a cylinder of fluid of radius r flowing in the positive x direction through a tube of radius $R > r$. The dashed outline in Figure 18.56 shows such a cylinder of fluid. Because of friction between the fluid in the cylinder and the more slowly moving layer of fluid adjacent to the cylinder boundary, a viscous force $\vec{F}_{\text{fc}}^{\text{viscous}}$ is exerted on the cylinder in the negative x direction. To keep the cylinder moving at constant velocity through the tube, a force of equal magnitude must be exerted on it in the positive x direction. This force requires a pressure difference ΔP between the ends of the cylinder, and so we see that for a fluid with nonzero viscosity, pressure must decrease as the fluid moves along the tube. For motion at constant speed, we have

$$\sum \vec{F}_{\text{c}} = \vec{F}_{\text{fc}}^{\text{pressure}} + \vec{F}_{\text{fc}}^{\text{viscous}} = 0. \qquad (18.40)$$

If the pressure at the left end of the cylinder is P_1, the fluid to the left of this end exerts a force of magnitude $F_1^{\text{pressure}} = P_1(\pi r^2)$ in the positive x direction, where πr^2 is the cross-sectional area of the cylinder. If the pressure at the right end of the cylinder is P_2, the fluid to the right of this end exerts a force of magnitude $F_2^{\text{pressure}} = P_2(\pi r^2)$ in the negative x direction. The vector sum of these two forces is $(P_1 - P_2)(\pi r^2)$. The viscous force F^{viscous} on

Figure 18.56 In mechanical equilibrium, the force due to the pressure on the two ends of the fluid in the cylinder defined by the dashed lines must balance the viscous force opposing the motion of the cylinder.

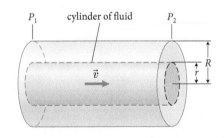

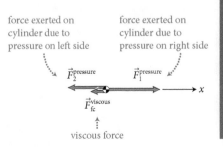

QUANTITATIVE TOOLS

the cylinder is given by Eq. 18.39, with A equal to the surface area $2\pi r\ell$ of the cylinder of fluid and $dv_x/dy = -dv_x/dr$.* Substituting this information into Eq. 18.40, we get

$$(P_1 - P_2)(\pi r^2) + \eta(2\pi r\ell)\frac{dv_x}{dr} = 0 \tag{18.41}$$

or, rearranging terms,

$$-dv_x = \frac{(P_1 - P_2)(\pi r^2)}{\eta(2\pi r\ell)}dr = \frac{(P_1 - P_2)}{2\eta\ell}rdr. \tag{18.42}$$

If we integrate the right side from r to R, the x component of velocity goes from v_x (the value at r) to 0 (the value at R, where the fluid is at rest because its boundary layer sticks to the tube wall):

$$-\int_{v_x}^{0} dv_x = \frac{P_1 - P_2}{2\eta\ell}\int_{r}^{R} rdr = \frac{P_1 - P_2}{2\eta\ell}\left[\tfrac{1}{2}r^2\right]_{r}^{R} \tag{18.43}$$

or $\qquad v_x = \dfrac{P_1 - P_2}{4\eta\ell}(R^2 - r^2)$ (laminar flow in cylindrical tube). (18.44)

This equation shows that the velocity varies with the square of r, and so its curve has a parabolic shape (see Figure 18.19b) with a maximum at the center of the tube, where $r = 0$.

Because the fluid velocity varies with r, the volume flow rate is no longer simply Av. To calculate the volume flow rate, we consider a thin cylindrical shell of thickness dr (Figure 18.57). The shell has a cross-sectional area $dA = 2\pi rdr$ and contributes to the overall volume flow rate an amount

$$dQ = v_x dA = v_x(2\pi rdr) = \frac{P_1 - P_2}{4\eta\ell}(R^2 - r^2)(2\pi rdr)$$

$$= \frac{\pi(P_1 - P_2)}{2\eta\ell}(R^2r - r^3)dr, \tag{18.45}$$

where we have substituted Eq. 18.44 for v_x. To obtain the volume flow rate Q through the tube, we must integrate this expression from $r = 0$ to $r = R$. Integrating the factor $(R^2r - r^3)$ in Eq. 18.45 yields

$$\int_{0}^{R}(R^2r - r^3)dr = \left[\tfrac{1}{2}R^2r^2 - \tfrac{1}{4}r^4\right]_{0}^{R} = \tfrac{1}{2}R^4 - \tfrac{1}{4}R^4 = \tfrac{1}{4}R^4, \tag{18.46}$$

and so the volume flow rate through a cylindrical tube is

$$Q = \frac{\pi R^4}{8\eta\ell}(P_1 - P_2) \quad \text{(laminar flow in cylindrical tube).} \tag{18.47}$$

Equation 18.47, called **Poiseuille's law,** shows that a high viscosity leads to a low-volume flow rate, which is reasonable. The strong dependence on the tube radius R means that a small adjustment in radius can lead to a large increase in volume flow rate. For example, increasing R by just 20% increases the volume flow rate by a factor of $(1.2)^4 = 2.1$.

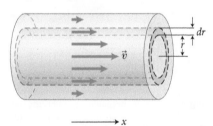

Figure 18.57 Each cylindrical shell (dashed lines) contributes an amount dQ to the volume flow rate through the tube.

---

*In Figure 18.19b v_x decreases with increasing r, and so dv_x/dr is negative. In Figure 18.55 v_x increases with increasing y, and so dv_x/dy is positive.

Example 18.10 Blood flow

Blood flows through an artery that has a radius of 3.0 mm. What is the maximum flow speed if the volume flow rate through the artery is $2.0 \times 10^{-6} \text{ m}^3/\text{s}$?

❶ **GETTING STARTED** I assume that the artery is cylindrical, which means I can use the results from this section.

❷ **DEVISE PLAN** I can obtain the maximum flow speed by setting $r = 0$ in Eq. 18.44. This equation contains the unknown factor $(P_1 - P_2)/(4\eta\ell)$. The expression for the volume flow rate, Eq. 18.47, however, contains the same factor, which gives me two equations with two unknowns—the maximum flow speed v and the factor $(P_1 - P_2)/(4\eta\ell)$—so I can solve for both unknowns.

❸ **EXECUTE PLAN** Setting $r = 0$ in Eq. 18.44 gives $v_{\max} = (P_1 - P_2)R^2/(4\eta\ell)$. Next, I rewrite Eq. 18.47 in the form

$(P_1 - P_2)/(4\eta\ell) = 2Q/(\pi R^4)$, which I substitute into my expression for $v_{\max}$:

$$v_{\max} = \frac{(P_1 - P_2)R^2}{4\eta\ell} = \frac{2Q}{\pi R^4}R^2 = \frac{2Q}{\pi R^2}$$

$$= \frac{2(2.0 \times 10^{-6} \text{ m}^3/\text{s})}{\pi(3.0 \times 10^{-3} \text{ m})^2} = 0.14 \text{ m/s.} ✔$$

❹ **EVALUATE RESULT** I can rewrite the equality $v_{\max} = (2Q)/(\pi R^2)$ from my final equation in the form $Q = \frac{1}{2}\pi R^2 v_{\max} = \frac{1}{2}Av_{\max}$, where A is the cross-sectional area of the artery. This equation is very similar to Eq. 18.26, the definition of volume flow rate. Indeed, $\frac{1}{2}v_{\max}$ is the average flow speed. The value I obtain also appears to be reasonable—a given volume of blood travels about 1 m in 7 s.

Having dealt with forces in the bulk of a fluid, let us now turn to forces at liquid surfaces. As we saw in Section 18.4, these forces cause surface tension at the surface and are responsible for the shape of drops of liquid and capillary rise.

Figure 18.58 shows a device commonly used to measure surface tension, a length of metal bent into a U shape and attached to a loosely fitted wire to form a rectangular loop. The wire fits so loosely that it slides off when the loop is held with the U inverted. When the loop is dipped into a soap solution, however, the soap film that forms inside the loop holds the wire stationary, as Figure 18.58 shows. An upward force must be exerted on the wire to hold it in place by balancing the downward force of gravity. The only likely candidate for this upward force is the one exerted by the film, so that $F_{\text{fw}} = F^G_{\text{Ew}}$. Experiments show that F_{fw} remains constant as we move the wire up or down. Because the thickness of the film changes as the wire is moved, the observation that the force magnitude F_{fw} does not change means that the film thickness is unimportant. Therefore we must attribute the upward force exerted on the wire to the front and back surfaces of the film. Experimentally we also find that the magnitude of this force is proportional to the length ℓ of the wire. The constant of proportionality

$$\gamma \equiv \frac{F_{\text{fw}}}{2\ell} \tag{18.48}$$

is called the **surface tension** of the liquid that makes up the film. The factor 2 in the denominator accounts for the two surfaces of the film.

The SI unit for surface tension is newtons per meter. Table 18.2 gives the surface tension for a number of liquids in contact with air.

We can also understand surface tension in terms of the energy associated with the surface of a liquid. Moving the wire in Figure 18.58 causes a force displacement Δx_F, and so work $W = F_{\text{fw}x}\Delta x_F$ is done by the film on the wire. Because $\vec{F}_{\text{fw}}$ points in the positive x direction, we have $F_{\text{fw}x} = +2\ell\gamma$, and so $W = 2\ell\gamma\Delta x_F = -2\gamma\Delta A$, where ΔA is the change in the surface area A of the film. (When the wire moves, Δx_F and ΔA have opposite algebraic signs, so $\ell\Delta x_F = -\Delta A$.) As we move the wire down, we increase the surface area A of the film, and so $\Delta A > 0$ and the work done by the film on the wire is negative. Consequently the work done by the wire on the film is positive (the energy of the film increases). Where does the energy associated with this work go? As we

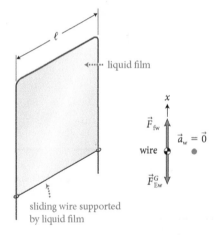

Figure 18.58 The force exerted by a liquid film on a wire is due to surface tension in the film.

Table 18.2 Surface tensions of liquids in contact with air.

Liquid	Surface tension (N/m)	Temperature (°C)
alcohol	2.23×10^{-2}	20
copper	1.100	1130
glycerin	6.31×10^{-2}	20
gold	1.000	1070
mercury	0.465	20
neon*	5.15×10^{-3}	−247
olive oil	3.20×10^{-2}	20
oxygen*	1.57×10^{-2}	−193
silver	0.800	970
water	7.57×10^{-2}	0
	7.28×10^{-2}	20
	5.89×10^{-2}	100

*Values given for liquid in contact with vapor.

increase the surface area, more liquid particles must be brought to the surface. For each particle brought to the surface, we reduce the number of bonds between particles (see Figure 18.25), and so the energy associated with the work we do in moving the wire downward goes into breaking bonds between neighboring particles in order to bring them to the surface. So, another way of understanding surface tension is to see it as the energy required to increase the surface area of a liquid. The greater the surface tension of a liquid, the more energy required to increase the surface area of a sample of the liquid by a certain amount.

As we saw in Section 18.4, surface tension is responsible for the pressure difference across a curved liquid surface. To derive the relationship between surface tension and pressure difference, consider the spherical body of liquid in **Figure 18.59a**. The radius of the sphere is R, and the liquid has a surface tension γ. The pressure inside the liquid is P_{in}, and the pressure outside is P_{out}. Let's consider the forces exerted on one hemisphere, as illustrated in Figure 18.59b. Because the hemisphere remains at rest, the vector sum of the

Figure 18.59 Model for deriving the relationship between surface tension and pressure difference across a curved liquid surface. The surface tension forces are all directed perpendicular to the circular cross section, and the pressure-difference forces are all directed outward.

(a) Sphere of liquid

(b) Forces exerted on half of sphere due to surface tension and to pressure difference across surface

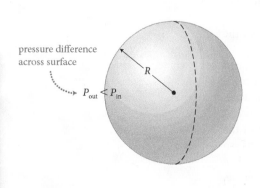

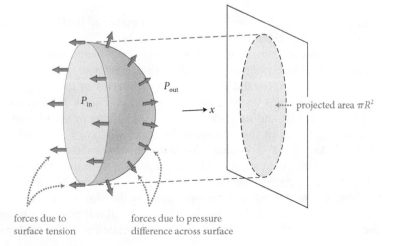

forces due to surface tension

forces due to pressure difference across surface

outward force due to the pressure difference across the surface and the inward force due the surface tension must be zero:

$$\sum F_{\text{hemi } x} = F_x^{\text{pressure}} + F_x^{\text{surf tension}} = 0. \qquad (18.49)$$

The force due to the pressure difference is perpendicular to the hemisphere surface. The components of the forces that are exerted on each segment of the surface but are not in the x direction cancel, and the x components of all these forces sum to the pressure difference $P_{\text{in}} - P_{\text{out}}$ times πR^2, which is the area of the circle projected on the screen in Figure 18.59b (see Guided Problem 18.4 in the *Practice* volume). The force due to the surface tension is exerted along the perimeter of the hemisphere, which has length $\ell = 2\pi R$. Therefore the magnitude of the force due to surface tension is $F_x^{\text{surf tension}} = 2\pi R\gamma$ (from Eq. 18.48, $\gamma = F/2\ell$; using this equation would give us $F = 2\ell\gamma = 2(2\pi R)\gamma = 4\pi R\gamma$, but we have to drop the factor 2 from Eq. 18.48 because here we are working with only one surface). Substituting these forces into Eq. 18.49 gives

$$(P_{\text{in}} - P_{\text{out}})\pi R^2 - 2\pi R\gamma = 0 \qquad (18.50)$$

or, rearranging terms,

$$P_{\text{in}} - P_{\text{out}} = \frac{2\gamma}{R} \quad \text{(spherical surface).} \qquad (18.51)$$

Equation 18.51 is a quantitative form of **Laplace's law** for a spherical surface.
For a cylindrical surface, it can be shown that Eq. 18.51 takes the form

$$P_{\text{in}} - P_{\text{out}} = \frac{\gamma}{R} \quad \text{(cylindrical surface).} \qquad (18.52)$$

Example 18.11 Transmural pressure

The pressure difference across the wall of a blood vessel is called *transmural pressure*. Suppose an aorta of radius 11 mm has a transmural pressure of 12 kPa and a capillary of radius 4.0 μm has a transmural pressure of 5.0 kPa. Which blood vessel requires a greater tensile stress (tensile force per unit length) in its wall?

❶ **GETTING STARTED** I consider the aorta and the capillary to be elastic cylindrical tubes. Because the wall of a blood vessel keeps the blood confined within the tube, the tension in the wall plays the role of the surface tension γ of an unconfined liquid.

❷ **DEVISE PLAN** To determine γ for each vessel, I can use Eq. 18.52, which relates the pressure difference across the wall to the radius of the cylinder.

❸ **EXECUTE PLAN** I first rewrite Eq. 18.52 in the form $\gamma = (P_{\text{in}} - P_{\text{out}})R$. For the aorta,

$$\gamma = (12 \text{ kPa})\left(\frac{1000 \text{ Pa}}{1 \text{ kPa}}\right)(11 \text{ mm})\left(\frac{1 \text{ m}}{1000 \text{ mm}}\right)$$

$$= 1.3 \times 10^2 \text{ N/m.}$$

(You should be able to confirm that the units in this expression are indeed newtons per meter.) For the capillary,

$$\gamma = (5.0 \text{ kPa})\left(\frac{1000 \text{ Pa}}{1 \text{ kPa}}\right)(4.0 \text{ } \mu\text{m})\left(\frac{1 \text{ m}}{10^6 \text{ } \mu\text{m}}\right)$$

$$= 0.020 \text{ N/m.}$$

The tensile force per unit length in the aorta is much greater than that in the capillary. ✔

❹ **EVALUATE RESULT** The pressure differences across the aorta and capillary are of the same order of magnitude, and so less tension is required when the radius is small. This is the same situation as in Figure 18.27: The pressure difference is the same everywhere on the balloon surface, but the tension is less where the radius of curvature is smaller.

✋ **18.19** What is Laplace's law, Eq. 18.51, for a soap bubble?

Chapter Glossary

SI units of physical quantities are given in parentheses.

Archimedes' principle An object submerged either fully or partially in a fluid experiences an upward buoyant force that is equal in magnitude to the force of gravity exerted on the fluid displaced by the object. The volume of the displaced fluid is equal to the volume of the submerged portion of the object.

Bernoulli's equation The equation that relates pressure, laminar flow speed, and elevation in an incompressible fluid:

$$P_1 + \rho g y_1 + \tfrac{1}{2}\rho v_1^2 = P_2 + \rho g y_2 + \tfrac{1}{2}\rho v_2^2 \quad (18.36)$$

(laminar flow of incompressible, nonviscous fluid).

Buoyant force The upward force exerted by fluids on objects that are fully or partially submerged in a fluid. For an object that displaces a volume V_{disp} of a fluid of uniform mass density ρ_f, the magnitude of the buoyant force is

$$F_{\text{fo}}^b = \rho_f V_{\text{disp}} g. \quad (18.12)$$

Continuity equation The equation that relates the laminar flow speed of a fluid to the cross-sectional area of the tube through which it flows:

$$\rho_1 A_1 v_1 = \rho_2 A_2 v_2 \quad (18.23)$$

(laminar flow of nonviscous fluid).

Fluid Any substance that has the ability to flow; any liquid or gas.

Laminar flow The flow of a fluid for which the velocity of the fluid at any location is constant.

Laplace's law The relationship among surface curvature, surface tension, and the pressure difference across a surface. For a spherical surface,

$$P_{\text{in}} - P_{\text{out}} = \frac{2\gamma}{R} \quad \text{(spherical surface)}, \quad (18.51)$$

and for a cylindrical surface,

$$P_{\text{in}} - P_{\text{out}} = \frac{\gamma}{R} \quad \text{(cylindrical surface)}. \quad (18.52)$$

Meniscus The curved surface at the top of a liquid in a narrow tube.

Pascal (Pa) The derived SI unit of pressure, defined as $1\,\text{Pa} = 1\,\text{N/m}^2 = 1\,\text{kg}/(\text{m}\cdot\text{s}^2)$.

Pascal's principle A pressure change applied to an enclosed liquid is transmitted undiminished to every part of the liquid and to the walls of the container in contact with the liquid.

Poiseuille's law The relationship between the pressure difference across a cylindrical tube and the volume flow rate Q through the tube:

$$Q = \frac{\pi R^4}{8\eta \ell}(P_1 - P_2) \quad (18.47)$$

(laminar flow in cylindrical tube).

Pressure P (Pa) A scalar that gives the ratio of the magnitude of the force exerted by a fluid perpendicular to an area and the surface area:

$$P \equiv \frac{F_{\text{fs}}^c}{A}. \quad (18.1)$$

Streamline A path taken by particles in a fluid in motion.

Surface tension γ (N/m) A scalar that gives the magnitude of the force per unit length exerted by a liquid surface.

Turbulent flow Chaotically changing flow characterized by whirling, circularly moving regions of fluid.

Viscosity η (Pa·s) A scalar measure of the resistance of a fluid to shear deformation. For a layer of fluid between two flat plates, the viscosity is defined by the equation that relates the force required to keep the top plate moving at constant speed and the surface area A of the plate, the difference in the x components of the velocities between the two plates, $v_{\text{top},x} - v_{\text{bottom},x}$, and the distance $y_{\text{top}} - y_{\text{bottom}}$ between the plates:

$$F_{\text{fp}\,x} \equiv -\eta A \frac{v_{\text{top},x} - v_{\text{bottom},x}}{y_{\text{top}} - y_{\text{bottom}}}. \quad (18.38)$$

Volume flow rate Q (m³/s) A scalar that gives the rate at which a volume of fluid crosses a section of a tube:

$$Q \equiv \frac{V}{\Delta t}. \quad (18.25)$$

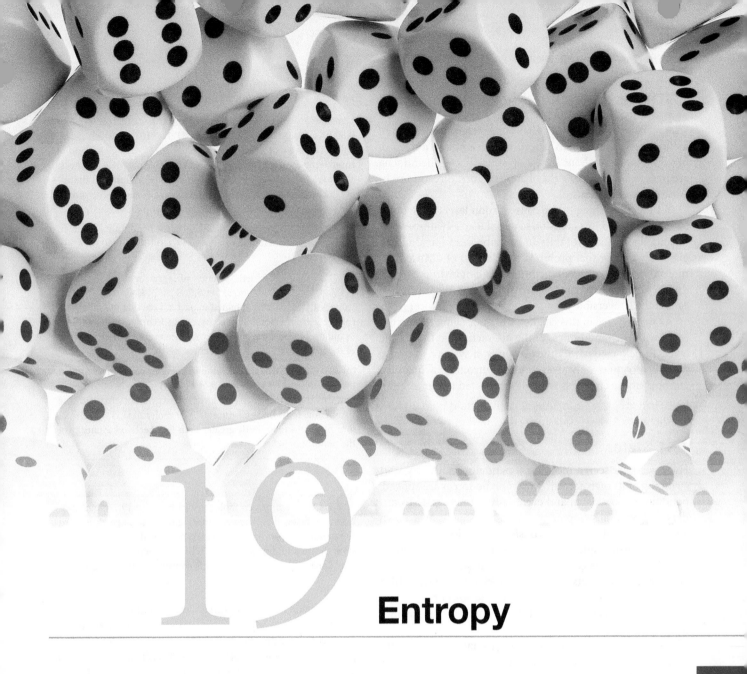

19 Entropy

CONCEPTS

QUANTITATIVE TOOLS

The conservation laws allow us to analyze a broad range of phenomena, yet they cannot explain the *irreversibility* of time. Cars are damaged in a collision, a cube of ice melts when placed in a drink, and a drop of ink diffuses throughout a glass of water. These phenomena are *irreversible,* meaning they happen in one direction only. Always, no exceptions.

Yet there's nothing in the conservation laws that prohibits cars damaged in a collision to spontaneously move apart "undamaged," that prohibits an ice cube from forming in a glass of water, or that prohibits a collection of diffused ink molecules from coming together into a single drop. So why do phenomena such as these never happen? What is it that causes the irreversible flow of time toward the future?

As we shall see in this chapter, the underlying reason has to do with the randomness of the interactions at the atomic level. To help us quantify the tendency of systems to evolve in one direction only over time, I introduce a quantity, *entropy,* whose change is strongly connected to our sense of irreversibility.

19.1 States

Have you ever seen a motionless pendulum suddenly start swinging of its own accord? No; when no forces are applied, pendulums slow down and eventually come to rest. To understand why a swinging pendulum always slows down and never speeds up, let us consider a pendulum swinging back and forth inside a container. As the pendulum swings, kinetic energy gets converted to gravitational potential energy, which then gets converted to kinetic energy, which then . . . . If no energy is dissipated, the sum of these two energies remains constant and the pendulum keeps swinging forever.

Over time, however, the swings get smaller because with each swing a small part of the coherent mechanical energy of the pendulum is converted to incoherent thermal energy (see Section 7.3). This conversion is due to friction in the pendulum suspension and to collisions with the air molecules surrounding the pendulum. When the pendulum has come to rest, all of its mechanical energy has been dissipated. This energy is not gone—it still exists but now in the form of incoherent thermal energy associated with the motion of the molecules in the air surrounding the pendulum. So the universal tendency for the mechanical energy to dissipate appears to be related to the incoherent nature of the motion of atoms and molecules at the atomic scale.

Not one of the conservation laws we have studied forbids the conversion of that thermal energy back to mechanical energy. The pendulum could start swinging again if the air molecules would supply the kinetic energy. To understand why this transfer of energy from air molecules to pendulum is never observed, we need to look at molecular motion on the atomic scale. Let's begin by getting some feel for how much energy the pendulum and the air molecules have.

Exercise 19.1 Pendulum energy

Consider a 0.10-kg pendulum swinging at a maximum speed of 0.80 m/s inside a box that contains 1.0×10^{23} nitrogen molecules. The mass of a nitrogen molecule is 4.7×10^{-26} kg, and at room temperature a typical nitrogen molecule moves at 500 m/s. What are (a) the mechanical energy of the pendulum, (b) the average kinetic energy of one nitrogen molecule, and (c) the sum of the average kinetic energies of all the nitrogen molecules?

SOLUTION (a) The mechanical energy of the pendulum is equal to its maximum kinetic energy: $\frac{1}{2}mv^2 = \frac{1}{2}(0.10 \text{ kg})(0.80 \text{ m/s})^2 = 3.2 \times 10^{-2}$ J. ✔

(b) The average kinetic energy of a nitrogen molecule is $\frac{1}{2}(4.7 \times 10^{-26} \text{ kg})(500 \text{ m/s})^2 = 5.9 \times 10^{-21}$ J. Comparing this value with the result I obtained in part *a*, I see that the mechanical energy of the pendulum is more than 18 orders of magnitude (a billion billion times) greater than the average kinetic energy of the individual nitrogen molecules. ✔

(c) The sum of the average kinetic energies of all the nitrogen molecules is $(1.0 \times 10^{23})(5.9 \times 10^{-21} \text{ J}) = 5.9 \times 10^2$ J. Note that this thermal energy (the kinetic energy associated with the incoherent motion of all the nitrogen molecules) is more than four orders of magnitude greater than the mechanical energy of the pendulum. ✔

19.1 The nitrogen molecules in Exercise 19.1 move considerably faster than the pendulum and continuously bombard the pendulum from all sides. (*a*) Why does this bombardment slow the pendulum down? (*b*) When the pendulum is at rest, why don't we see the effects of this continuous bombardment?

The inertia of a pendulum at rest is so much greater than that of the surrounding air molecules that the effect of collisions between air molecules and the pendulum is hardly noticeable. To see the effect of such collisions, we would have to greatly reduce the inertia of the pendulum. In 1827 the Scottish botanist Robert Brown was the first to observe the effect of this random molecular bombardment. Using a microscope to observe grains of pollen suspended in water, Brown noticed that the grains bounced around, following a zigzag path like the one shown in Figure 19.1.

This random motion, called *Brownian motion,* is due to the random bombardment of the grains by the surrounding

Figure 19.1 A grain of pollen suspended in water undergoes a random zigzag motion due to collisions with surrounding water molecules.

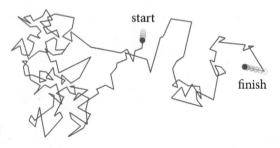

water molecules. Because the grains are small, the bombardment is not always equal on all sides, causing the grains to zigzag. The small random movements are what makes up Brownian motion.

19.2 How would Brownian motion change if the mass of the grain in Figure 19.1 were increased?

An important conclusion we can draw from the observation of Brownian motion in pollen grains suspended in water is that the bombardment is incoherent. As a result of this lack of coherence, the displacement of each grain over a long time interval is nearly zero and the grain never gains or loses any kinetic energy. If the grain were constantly kicked in the same direction, it would keep accelerating in that direction and gain kinetic energy. The same is true for a pendulum suspended in the air and hanging at rest: If it were constantly hit from one direction, it *would* start swinging.

Because it is impossible to treat the details of the incoherent atomic motion quantitatively, we need to use a very different approach to connect the atomic world of atoms and molecules to the *macroscopic* world we have described so far (*macroscopic* means "large relative to the size of a molecule"). In particular, we need to use probability theory to determine the likelihood of certain events taking place. As we shall see, the odds strongly favor slowing down a pendulum over speeding it up. To begin, let's review the essentials of probability theory using the throwing of two dice as an example.

Exercise 19.2 Throwing dice

When you throw two dice, one red and one blue, many times, what is the fraction of throws for which the dots on the two dice add to 4?

SOLUTION There are 36 combinations for the pair of dice (**Figure 19.2**), three of which add to 4. On every throw, each of the 36 combinations is equally likely to occur. If you throw the dice many times, the fraction of throws that add to 4 approaches the ratio $3/36 = 1/12$. ✔

19.3 What sum of dots are you most likely to throw with two dice?

The **probability** that a certain event will occur is the fraction of times that event occurs in a large number of repetitions. Exercise 19.2 tells you that the probability of throwing a 4 is $1/12$. If you throw two dice 12 times, you won't necessarily throw a 4 exactly once. The more times you throw the dice, however, the closer the fraction of times you throw a 4 will be to $1/12$.

If we consider the two dice as a system, there are several ways we can describe the state of this system. We can characterize it by the sum of the dots shown on the two

Figure 19.2 The 36 possible combinations for two dice.

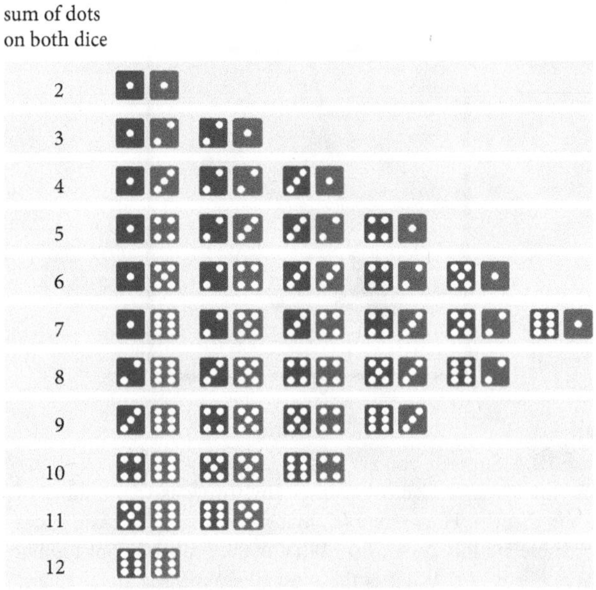

sum of dots
on both dice

dice after any given throw, or we can give a more detailed description by specifying the number of dots showing on each die. For systems with very large numbers of particles, we often care only about large-scale properties, not the detailed specifications of each particle. When we describe a system in terms of its large-scale properties, we are describing the system's **macrostate.** When we give a detailed specification of all the particles that make up the system, we are describing the system's **basic state.*** For a volume of gas, for example, the macrostate is described by large-scale properties that can be readily measured, such as temperature, pressure, and volume. The basic state is described by specifying the position and velocity of each individual gas particle.

Before describing a gas, let us return to the simpler system of two dice. The macrostate of the system after any throw is given by the sum thrown; the basic state is given by the number of dots shown on each die. As you can see in Figure 19.2, there are 36 basic states for the system and 11 macrostates. Because any basic state is equally likely to occur, the probability of any one basic state occurring is $1/36$. To calculate the probability of, say, the macrostate 9, we divide the number of basic states associated with this macrostate by the number of basic states associated with all the macrostates. Four out of the 36 basic states correspond to the macrostate 9. Therefore the probability of throwing a 9 is $(1/36)(4) = 4/36$. Because there are more basic states for the macrostate 7 than there are for the macrostate 9, you are more likely to throw, on average, a 7 than a 9.

---

*Often the term *microstate* is used instead of *basic state*. Because *microstate* and *macrostate* sound alike and are often hard to tell apart in spoken language, we'll use the term *basic state*.

CONCEPTS

Figure 19.3 As a pendulum in a box swings, it collides with three particles. The pendulum slows down as its energy is redistributed into the particles.

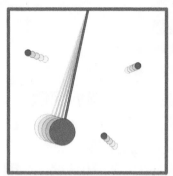

19.4 Suppose you are tossing four coins and decide to define the system's macrostate after any throw as the number of heads thrown. How many (*a*) macrostates and (*b*) basic states are there for this system? (*c*) What is the probability of tossing three heads? (*d*) What is the most likely macrostate after any toss?

Let us now see how all this applies to a swinging pendulum. Consider one swinging inside a box containing three gas particles (**Figure 19.3**). The particles are *distinguishable*—that is, we can tell them apart—and they collide randomly, but elastically, with one another, with the walls of the box, and with the pendulum. According to *quantum mechanics*, the energy of confined systems can only change by very small, indivisible, discrete units called *quanta* (singular: *quantum*), and many experiments confirm this discrete nature of energy units. The collisions in the box thus transfer energy units between the colliding objects.

To get a feel for how energy units are redistributed as the particles collide with the pendulum, we ignore any dissipation of energy in the pendulum's suspension and we assume that there are six identical (and therefore *indistinguishable*) energy units to be distributed among the pendulum and the three particles. (We'll drop the restriction that the energy units are indistinguishable later.) We further assume that the collisions distribute the energy units randomly throughout the system: that is, the redistribution jumbles up both the speeds of the particles and the directions in which they move. This assumption is justified not only by the observation of Brownian motion but also by the fact that the assumption leads to a behavior that accurately describes our observations.

Let us investigate how the six energy units can be distributed among the pendulum and the three gas particles. If we describe the system's macrostate by the number of energy units in the pendulum, there are seven macrostates. **Figure 19.4** shows how, for each of these seven macrostates, the energy units can be distributed over the pendulum and the three particles. Note that as the amount of energy in the particles increases, the number of basic states associated with that macrostate increases.

19.5 What are the six ways in which you can distribute three energy units over three particles such that one of the particles has two energy units?

Because collisions completely randomize the energy distribution, each basic state is equally likely to occur, and we can use statistics to determine the probability of any given macrostate. The probability of any one basic state is 1/84 (because there are 84 equally likely basic states). The probability of any macrostate is the number of basic states

Figure 19.4 The ways in which six energy units can be distributed among the pendulum and three particles of Figure 19.3. Distribution of six energy units among pendulum and three particles; each solid-colored box corresponds to one of the six units of energy. The number of basic states associated with all macrostates is 84. As the number of energy units in the pendulum decreases (left), the fraction of basic states associated with that macrostate increases (right).

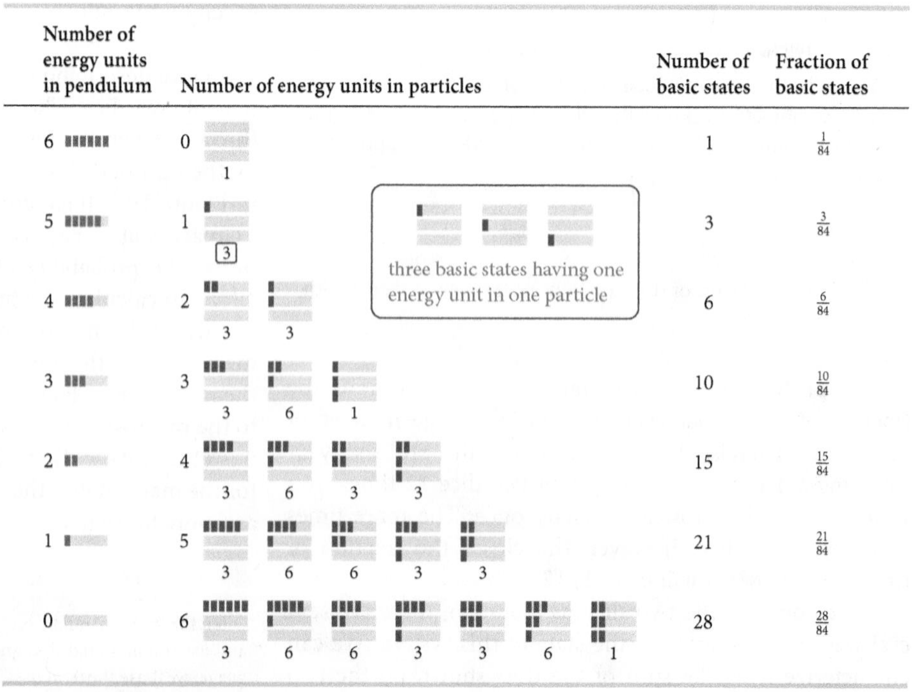

Number of energy units in pendulum	Number of energy units in particles	Number of basic states	Fraction of basic states
6	0	1	$\frac{1}{84}$
5	1	3	$\frac{3}{84}$
4	2	6	$\frac{6}{84}$
3	3	10	$\frac{10}{84}$
2	4	15	$\frac{15}{84}$
1	5	21	$\frac{21}{84}$
0	6	28	$\frac{28}{84}$

three basic states having one energy unit in one particle

Figure 19.5 As the fraction of energy in the pendulum increases, the probability of that corresponding macrostate decreases.

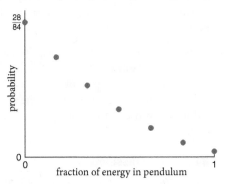

associated with that macrostate divided by the number of basic states associated with all macrostates, which is the number given in the last column of Figure 19.4. Figure 19.5 shows how this probability depends on the fraction of energy in the pendulum.

19.6 What is the probability of finding (*a*) all of the energy in the pendulum and (*b*) either zero or one unit of energy in the pendulum?

We "digitized" the energy by arbitrarily dividing it into six units. However, collisions could redistribute smaller units of energy, and so the mechanical energy of the pendulum could change by smaller amounts. Figure 19.6 shows how the probability of finding a certain fraction of energy in the pendulum relative to that of finding zero energy in it changes as we divide the energy into smaller units. Note by comparing Figures 19.5 and 19.6 that the division into finer units has little effect on the relative probabilities. In other words, the picture we obtain by dividing the energy into very coarse energy units is a remarkably good approximation of the much more realistic fine division predicted by quantum mechanics.

What happens if we increase the number of particles? Figure 19.7 shows that as the number of particles increases, the spike in the relative probability near the origin sharpens

Figure 19.6 As the energy is divided into more units, the probability of finding a certain fraction of the system's energy in the pendulum relative to that of finding zero energy in it approaches the solid red line.

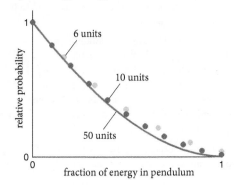

Figure 19.7 As the number of particles increases from 3 to 100, the probability of finding a certain fraction of the system's 50 energy units in the pendulum relative to that of finding zero energy in it decreases significantly.

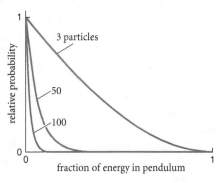

and the relative probability of all 50 energy units in the pendulum drops precipitously: With 3 particles, it is on the order of 10^{-4}; with 50 particles, it is on the order of 10^{-29}; with 100 particles, it drops to 10^{-40}.

Even with a very large number of particles, however, the relative probability of the pendulum containing all the energy is nonzero. So, if we release a pendulum in a box that contains many particles and the mechanical energy of the pendulum is gradually transferred to the particles, there is a very small but nonzero probability of the pendulum gaining all its energy back again.

We can get a feeling for how long it would take for this to happen by multiplying the average time interval between collisions by the number of basic states for the system. This gives us the length of time it would take on average to cycle randomly through all the basic states. This time interval is called the *recurrence time* of the system.

In a box of volume 1 m³ containing 3 air molecules at room temperature, the average time interval between collisions is approximately 10^{15} s. So the recurrence time is approximately $(84)(10^{15}\text{ s}) = 10^{17}$ s. If there are 10^{20} particles, the average time interval between collisions is approximately 10^{-19} s. For a large number of particles, the number of basic states is equal to the number of particles raised to the number of energy units: $(10^{20})^6 = 10^{120}$ (Section 19.6). Thus the recurrence time is approximately $(10^{120})(10^{-19}\text{ s}) = 10^{101}$ s. The age of the universe is only about 10^{18} s, so don't hold your breath waiting for the pendulum to recover all its energy!

19.7 (*a*) Averaged over a long time interval, for what fraction of the time interval does the pendulum in Figure 19.3 have more than 50% of the energy of the system? (*b*) Does this fraction increase, decrease, or stay the same as we increase the number of particles?

19.2 Equipartition of energy

In our description of a pendulum swinging among gas particles, we saw that as the number of particles increases, the probability of nonzero energy in the pendulum goes to zero (Figure 19.7). Suppose now that we have a gas of four particles instead of a pendulum and three particles. The bars for the energy units in the pendulum in Figure 19.4 then represent the energy of the fourth particle, and we see that the probability is highest for that particle to have little or no energy. We can use the same argument for each of the four particles: Each particle tends to have the smallest amount of energy possible. Because the thermal energy E_{th} associated with the incoherent motion of the particles is fixed, the system tends to distribute this energy equally over all the particles, with each particle having, on average, an amount of energy E_{th}/N, where N is the number of particles. If at any instant one of the particles has a greater fraction of the thermal energy E_{th}, this energy is rapidly redistributed over the rest of the system.

Each particle is like the pendulum. Therefore if we exchange the role of the pendulum with the role of one of the particles in Figure 19.4, we obtain the macrostates and basic states shown in **Figure 19.8**. Again, the odds are highest for that particle to have little or no energy, and as we saw in Figure 19.7, these odds become even more pronounced for large numbers of particles. Each particle tends to minimize its energy.

The situation is the same for the pendulum and the three particles. On average, each part of this system has an equal share $E/4$ of the system's energy E. If we replace either the pendulum or one of the particles with an oscillating spring, with a heavier particle, or with some other object, we can go through the same arguments and in each case we see that:

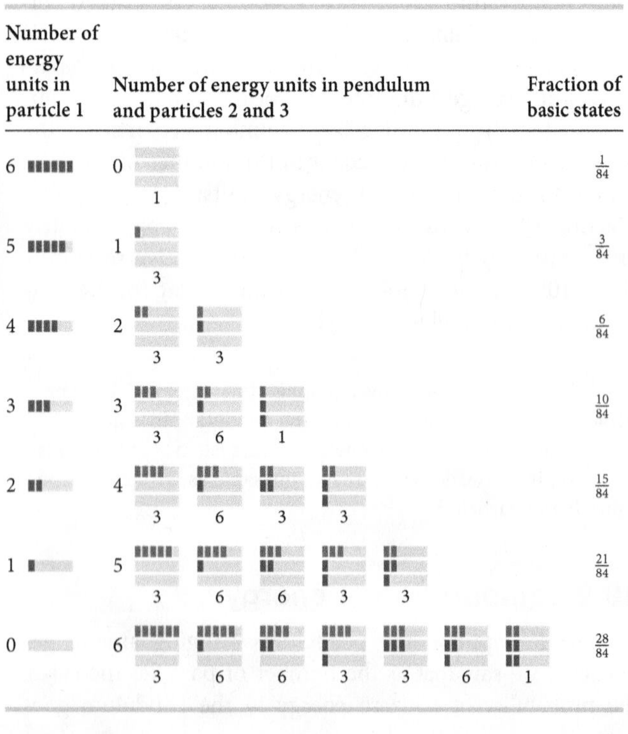

As long as the interactions between different parts of a system randomize the distribution of energy, each part of the system tends to have an equal share of the system's energy.

This equal distribution of energy among the parts of a system is called **equipartition of energy.** In Section 20.3 we shall define more precisely what "part of the system" means. For now, consider each indivisible element of the system—a particle, a pendulum, an oscillating spring, whatever—as a "container" for energy.

Example 19.3 Pendulum dissipation

Consider again the 0.10-kg pendulum of Exercise 19.1. The pendulum starts out swinging at a maximum speed of 0.80 m/s and stops swinging in 5.0 min. It is in a box that contains 1.0×10^{23} nitrogen molecules, each having a mass of 4.652×10^{-26} kg and moving at 500 m/s. The pendulum collides, on average, 1.0×10^{22} times per second with the nitrogen molecules. (a) In one collision what amount of mechanical energy is transferred on average to a nitrogen molecule, and what is the ratio of this transferred energy to the average kinetic energy of the nitrogen molecule? (b) By what fraction has the average kinetic energy of a nitrogen molecule increased when the pendulum stops swinging? (c) If the final mechanical energy of the pendulum is equal to the average kinetic energy of a nitrogen molecule, what is the pendulum's average final speed?

❶ GETTING STARTED The collisions gradually transfer all the pendulum's initial kinetic energy to the nitrogen molecules. For simplicity, I assume that each collision transfers the same amount of energy.

❷ DEVISE PLAN The number of collisions that occur during the 5.0 min it takes the pendulum to stop swinging is $(3.0 \times 10^2 \text{ s})(1.0 \times 10^{22} \text{ collisions/s}) = 3.0 \times 10^{24}$ collisions. Dividing the initial kinetic energy of the pendulum $(3.2 \times 10^{-2}$ J; Exercise 19.1a) by this number gives me the average energy transferred in each collision. I can then compare this energy with the average kinetic energy of a nitrogen molecule calculated in Exercise 19.1b $(5.9 \times 10^{-21}$ J).

When the pendulum stops, all its mechanical energy has been transferred to the molecules. In Exercise 19.1, I determined the initial mechanical energy of the pendulum $(3.2 \times 10^{-2}$ J) and the sum of the average kinetic energies of all the nitrogen molecules $(5.9 \times 10^2$ J). The ratio of these quantities gives me the fractional increase in the kinetic energy of each molecule.

The final average kinetic energy of the pendulum is half of its final mechanical energy: $\frac{1}{2} m_{pend}(v_{pend,f}^2)_{av} = \frac{1}{2} E_{pend,f}$. To calculate the pendulum's average final speed, I set $E_{pend,f}$ equal to the final average kinetic energy of a nitrogen molecule from Exercise 19.1b.

❸ EXECUTE PLAN (a) The average amount of mechanical energy transferred from pendulum to molecule in each collision is $(3.2 \times 10^{-2}$ J)/$(3.0 \times 10^{24}$ collisions) $= 1.1 \times 10^{-26}$ J/collision. From Exercise 19.1b, the average kinetic energy of each molecule is 5.9×10^{-21} J. So the ratio of the energy transferred per collision to the average molecular kinetic energy is $(1.1 \times 10^{-26}$ J)/$(5.9 \times 10^{-21}$ J) $= 1.9 \times 10^{-6}$. ✔

(b) The fractional increase in the average kinetic energy of each nitrogen molecule is the initial kinetic energy of the

Figure 19.8 If we interchange the pendulum with one of the particles in Figure 19.4, we obtain an identical configuration of macrostates and basic states.

Number of energy units in particle 1	Number of energy units in pendulum and particles 2 and 3	Fraction of basic states
6 ▮▮▮▮▮▮	0 / 1	$\frac{1}{84}$
5 ▮▮▮▮▮	1 / 3	$\frac{3}{84}$
4 ▮▮▮▮	2 / 3 3	$\frac{6}{84}$
3 ▮▮▮	3 / 3 6 1	$\frac{10}{84}$
2 ▮▮	4 / 3 6 3 3	$\frac{15}{84}$
1 ▮	5 / 3 6 6 3 3	$\frac{21}{84}$
0	6 / 3 6 6 3 3 6 1	$\frac{28}{84}$

pendulum (which all ends up in the molecules) divided by the combined kinetic energy of the molecules from Exercise 19.1: $(3.2 \times 10^{-2} \text{ J})/(5.9 \times 10^{2} \text{ J}) = 5.4 \times 10^{-5}$. ✔

(c) Because the final mechanical energy of the pendulum is equal to the average kinetic energy of a nitrogen molecule, I have $\frac{1}{2} m_{\text{pend}}(v^2_{\text{pend,f}})_{\text{av}} = \frac{1}{2}(5.9 \times 10^{-21} \text{ J})$, or $(0.10 \text{ kg})(v^2_{\text{pend,f}})_{\text{av}} = 5.9 \times 10^{21}$ J. Therefore the final average speed of the pendulum is $(v_{\text{pend,f}})_{\text{av}} = [(5.9 \times 10^{-21} \text{ J})/(0.10 \text{ kg})]^{1/2} = 2.4 \times 10^{-10}$ m/s. ✔

❹ **EVALUATE RESULT** Part *a* tells me that each collision transfers only a very small fraction of the pendulum's energy. This makes sense because it takes many collisions for the pendulum to come to rest. The small value I obtained in part *b* for the fractional increase in the molecules' energy makes sense because I know from experience that the air temperature (a quantity related to the kinetic energy of the molecules in the air) in the region around a swinging pendulum does not increase appreciably as the pendulum moves. Finally, my answer to part *c* shows that, for all practical purposes, the final average speed of the pendulum indeed is zero, as I had assumed in part *a*.

As Example 19.3 illustrates, the pendulum begins with much more mechanical energy than each individual gas molecule. Each collision between the pendulum and a nitrogen molecule can increase or decrease the energy of the pendulum. On average, however, these energy exchanges gradually decrease the energy of the pendulum until all nitrogen molecules and the pendulum have roughly the same energy. The pendulum's excess energy has been dissipated over all the molecules, raising their average energy by a minuscule amount.

19.8 Suppose a container holds a gas made up of two kinds of particles 1 and 2, with $m_2 = 2m_1$. Assume the particles interact via elastic collisions. (*a*) How do the average kinetic energies of the two kinds of particles compare? (*b*) How do their average speeds compare?

19.3 Equipartition of space

Collisions not only randomize the energy distribution in a gas, they also randomize spatial distributions. Consider, for example, a container divided into two compartments by a removable partition (**Figure 19.9**). In the left compartment is a gas, and the right compartment is empty. If we remove the partition, the gas diffuses to fill the entire volume. Likewise, if we put one type of gas on one side and another type of gas on the other side and then remove the partition, the two gases diffuse and mix. After some length of time has passed, both types of particles are evenly distributed throughout the container.

Both the diffusion of a single gas and the mixing of two different gases are irreversible processes. In Figure 19.9, for example, once the particles have expanded into the right side of the container, they do not spontaneously contract again into the left side.

Figure 19.9 (*a*) A container has gas in one half and vacuum in the other, divided by a partition. (*b*) When the partition is removed, the gas expands irreversibly into the vacuum. After the expansion is complete, there is an equal likelihood of finding a gas particle anywhere in the container.

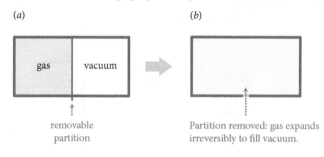

(*a*)   (*b*)

removable partition

Partition removed: gas expands irreversibly to fill vacuum.

Like dissipation of energy, this irreversible diffusion is due to the incoherent motion of particles at the atomic level. Collisions between the particles jumble up both the speeds of the particles and the directions in which they move. Consequently the likelihood at any instant of finding a particle at any given position in the container is the same as at any other position or instant—there is no "preferred" position for the particles.

19.9 Suppose there are initially two indistinguishable particles in the left compartment of the container in Figure 19.9a. After the partition is removed, what is the probability at any given instant of finding one particle in each half of the container?

Checkpoint 19.9 suggests that we can use the statistical approach of Section 19.2 to determine the most likely locations of particles in a container. Consider, for example, a box that contains six indistinguishable particles. In order to count basic states, we need to "digitize" space just as we digitized energy in Section 19.2. So let's divide the container into four equal compartments as shown in **Figure 19.10** (there are no real partitions).

The basic state of this system is determined by specifying how many particles are in each of the four compartments. There are many ways to characterize the system. Let's arbitrarily define the macrostate of the system to be the number

Figure 19.10   an instant when all particles are in top left compartment

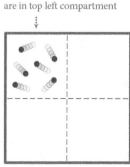

Figure 19.11

(a)

(b)

Figure 19.12 Distribution of six particles in four equal compartments of a box.

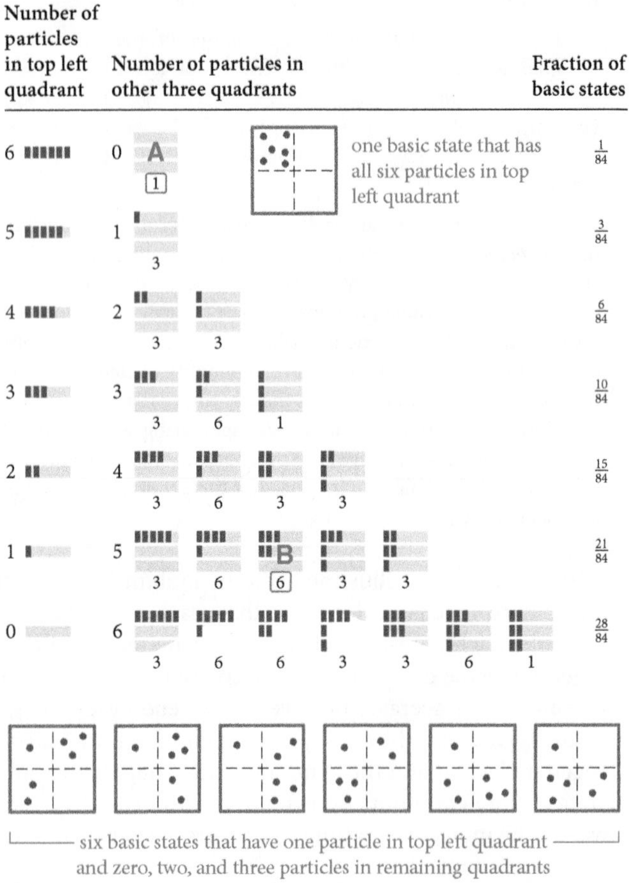

Number of particles in top left quadrant	Number of particles in other three quadrants		Fraction of basic states
6 ▮▮▮▮▮▮	0 A ①	one basic state that has all six particles in top left quadrant	$\frac{1}{84}$
5 ▮▮▮▮▮	1 ▮ 3		$\frac{3}{84}$
4 ▮▮▮▮	2 ▮▮ ▮ 3 3		$\frac{6}{84}$
3 ▮▮▮	3 ▮▮▮ ▮▮ ▮ 3 6 1		$\frac{10}{84}$
2 ▮▮	4 ▮▮▮▮ ▮▮▮ ▮▮ ▮▮ 3 6 3 3		$\frac{15}{84}$
1 ▮	5 ▮▮▮▮▮ ▮▮▮▮ ▮▮▮ B ▮▮▮ ▮▮ 3 6 ⑥ 3 3		$\frac{21}{84}$
0	6 ▮▮▮▮▮▮ ▮▮▮▮▮ ▮▮▮▮ ▮▮▮▮ ▮▮▮ ▮▮▮ ▮▮ 3 6 6 3 3 6 1		$\frac{28}{84}$

—— six basic states that have one particle in top left quadrant —— and zero, two, and three particles in remaining quadrants

of particles in the upper left compartment. (An alternative macrostate would be the number of particles in the lower right compartment, say, or the number in the lower half.) In Figure 19.10, all six particles are in the upper left compartment. The basic state is thus specified by the set of numbers (6, 0, 0, 0), representing the number of particles in each compartment, and the macrostate by the number 6 (for the number of particles in the upper left compartment). To determine the probability of having this distribution rather than any other, we need to determine the number of basic states and then the probability of each macrostate.

Exercise 19.4 Spatial macrostates and basic states

In Figure 19.10, how many basic states correspond to the macrostate in which (*a*) all six particles are in the top left compartment and (*b*) five particles are in the upper left compartment?

SOLUTION (*a*) One basic state because there is only one way to put all six particles in the upper left compartment (Figure 19.11*a*). ✔

(*b*) When five particles are in the upper left compartment, the last particle can be placed in any of the three other compartments. These three basic states are shown in Figure 19.11*b*. ✔

As you may have noticed, the problem of determining the distribution of six particles in four equal compartments of a container is identical to the energy distribution problem of Section 19.2, where we distributed six indistinguishable units of energy among four distinguishable "energy containers." Here we are distributing six indistinguishable "units of matter" among four distinguishable "spatial containers." The correspondence between the two problems is summarized in Table 19.1.

Using the mapping from Table 19.1, we can adapt Figure 19.4 to our current discussion, as shown in Figure 19.12. The bar in the leftmost column, which represented the energy in the pendulum in Figure 19.4, now represents the number of particles in the top left compartment of the box. All the bars arranged in groups of three represent the number of particles in each of the other three compartments. The distribution labeled A thus represents the only basic state that corresponds to having all six particles in the top left compartment: The other three compartments are empty. The distribution labeled B represents six of the 21 basic states associated with the macrostate in which one particle is in the top left compartment. These basic states have three

Table 19.1 Two problems, one idea

	Section 19.2	Section 19.3
phenomenon	distribution of energy	distribution of particles
indistinguishable items to be distributed	units of energy	particles
over distinguishable	pendulum and particles	compartments of container
macrostate defined by	energy units in pendulum	particles in top left compartment
basic state defined by	energy units in pendulum and each particle	particles in each compartment

particles in any one of the other three compartments, two particles in another, and zero particles in the last remaining compartment.

19.10 (*a*) What is the probability of finding all six particles in one compartment? (*b*) Of finding five particles in one compartment?

Example 19.5 Particles in box

Use the data in Figure 19.12 to determine the average number of particles in the top left compartment of the four-compartment box shown there.

❶ GETTING STARTED The rightmost column in Figure 19.12 gives the fraction of basic states associated with each of the seven macrostates. This fraction represents the probability of finding the indicated number of particles in the top left compartment. For example, the probability of finding three particles there is $\frac{10}{84}$. In other words, if I observe the system for a long time interval, I'll see three particles in the top left compartment $\frac{10}{84}$th of the time.

❷ DEVISE PLAN To calculate the average number of particles in the top left compartment, I should take the sum over all macrostates of the product of the number of particles in that compartment (leftmost column in Figure 19.12) and the probability of finding that macrostate (rightmost column).

❸ EXECUTE PLAN Summing the products of the values in the leftmost and rightmost columns gives $6(\frac{1}{84}) + 5(\frac{3}{84}) + 4(\frac{6}{84}) + 3(\frac{10}{84}) + 2(\frac{15}{84}) + 1(\frac{21}{84}) + 0(\frac{28}{84}) = 1.5$. So, averaged over time, there are 1.5 particles in the top left compartment. ✔

❹ EVALUATE RESULT My answer makes sense because there are six particles in the box to be divided over four compartments: $6/4 = 1.5$.

Example 19.5 shows that even though the most probable macrostate has no particles in the upper left compartment, the average number of particles in that compartment is nonzero. Let us generalize our result by dividing the box into a large number M of equal compartments and increasing the number of particles N. On average each compartment tends to have its "share" N/M of the particles, yielding a uniform distribution of all the particles over space.

This tendency to distribute particles uniformly over space is called **equipartition of space** in analogy to the concept of equipartition of energy we encountered in Section 19.2. The diffusion of ink in water or smoke in air and the uniform distribution of gas particles in a container are all manifestations of this equipartitioning of space.

19.11 Suppose the partition in Figure 19.9 is designed to slide freely to the left and right. It is positioned in the center so that the container is divided into two equal compartments and clamped in place. Then $2N$ particles of a gas are added to the left compartment and N particles are added to the right compartment. When the clamp on the partition is removed, does the partition move to the right, move to the left, or stay where it is?

19.4 Evolution toward the most probable macrostate

In Sections 19.2 and 19.3, we assumed that the distribution of energy and the spatial distribution of particles are entirely random, like the throw of dice. However, there is an important difference between the throwing of dice and, say, the slowing down of a pendulum.

Each time you throw two dice, the combination you get is completely independent of the combinations you threw before. After throwing a sum of 7, for example, you can throw 12, 2, another 7, or any other sum. The dissipation of the energy in a swinging pendulum, on the other hand, takes many collisions. Because a single collision redistributes only a very small portion of the system's energy, the way the energy is distributed after any given collision is only slightly different from the way it was distributed just before the collision. In other words, the energy distribution in each post-collision macrostate is not independent of the energy distribution in the pre-collision macrostate.

19.12 In the system shown in Figure 19.4, suppose each collision transfers exactly one unit of energy from one colliding object to the other. (*a*) If the system is in the macrostate that has six energy units in the pendulum, which basic state(s) can the system be in after one collision? (*b*) If the system is in the macrostate that has five energy units in the pendulum, which *macrostates* can the system be in after one collision? (*c*) How many basic states do the macrostates in part *b* correspond to?

Checkpoint 19.12 shows that only neighboring macrostates are accessible to the system after each collision. So, if we start with all the energy in the pendulum, the system can only gradually evolve to the most probable macrostate (the one with all the energy in the gas particles). It takes a minimum of six collisions to reach the most probable macrostate. The number is likely to be much greater because, as you saw in doing the checkpoint, collisions can take place between two particles, leaving the energy in the pendulum unchanged, or a collision may *add* energy to the pendulum.

19.13 Suppose the system in Figure 19.4 is in the macrostate in which two energy units are in the pendulum and one collision takes place. (*a*) What are the accessible macrostates? (*b*) What is the probability for each of these macrostates?

Note that the probability of reducing the energy in the pendulum is always higher than that of increasing the energy, and so the system gradually evolves to macrostates of increasing probability. Once the system is in the most probable macrostate, collisions continue to shuffle energy units around, but the system remains near this most probable macrostate. The system is therefore at equilibrium, and the most probable macrostate is the **equilibrium state.** If the pendulum and particles do temporarily occupy a macrostate that is less probable than the equilibrium state, then any subsequent collisions tend to move the system back toward equilibrium.

CONCEPTS

Figure 19.13 Box containing 20 particles (14 on the left, 6 on the right) and 10 units of energy. The particles in the two compartments can exchange energy with one another through collisions with the partition.

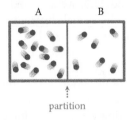

partition

As we have seen, the probability of all the energy moving from the particles to the pendulum is vanishingly small, even for a gas of just 100 particles. This one-way flow of energy from pendulum to gas particles is the essence of *irreversibility* (see Section 5.3). Irreversibility is strongly connected to our sense of time. The direction of time is clear because everything around us is evolving irreversibly toward equilibrium. Now you can see that the underlying reason for irreversibility is statistics: Any system that is in a low-probability macrostate tends to evolve toward the macrostate that has the maximum number of basic states. Once the system has reached that equilibrium state, its probability of spontaneously returning to a macrostate that has a small number of basic states is vanishingly small.

To get a better sense of this evolution toward equilibrium, let's examine the particles in the closed box of **Figure 19.13**. The box contains 20 particles and is divided into compartments A and B. The partition prevents the particles from traveling from one compartment to the other but still allows particles from the two compartments to exchange energy with one another via collisions with the partition.

The energy of the system is constant—no energy can leak out because the system is closed. Suppose A contains 14 particles, B contains six particles, and the box contains ten units of energy. **Table 19.2** shows how these ten units can be distributed among the 20 particles. Let's define the macrostate by the energy units in compartment A. The number of basic states associated with each compartment is represented by the symbol Ω (the Greek capital omega) plus a subscript identifying the compartment. For example, there is only one basic state for compartment A ($\Omega_A = 1$) for macrostate $E_A = 0$. With nine energy units in A ($E_A = 9$) and the other one in B, there are 497,420 basic states for A ($\Omega_A = 497,420$) and six basic states for B ($\Omega_B = 6$).

✋ **19.14** How many ways are there to distribute two indistinguishable energy units among six distinguishable particles?

Note in Table 19.2 that as the number of energy units in each compartment of the box increases, the number of basic states in that compartment increases rapidly. Because the number of energy units in the box is constant, the number of basic states in one compartment decreases as the number of basic states in the other compartment increases. What then is the most probable macrostate for the box as a whole?

To answer this question, consider the macrostate $E_A = 1$ in Table 19.2. There are 14 ways to distribute this 1 energy unit in A (it can go into any one of the 14 particles of A), so there are 14 basic states associated with compartment A for $E_A = 1$. One energy unit in A leaves nine energy units to be distributed in B, and as you can see from the table, there are 2002 ways to distribute these nine units over the six particles in B. For each of the 14 basic states associated with compartment A, there are 2002 basic states associated with compartment B. The number of basic states associated with the macrostate $E_A = 1$ is thus $14 \times 2002 = 28,028$. More generally, for a given macrostate, the number Ω of

Table 19.2 Number of basic states Ω associated with macrostates having 10 energy units distributed over 20 particles

E_A	E_B	Ω_A	Ω_B	$\Omega = \Omega_A \Omega_B$	$\ln \Omega$
Energy units in A	Energy units in B	Corresponding number of basic states for A	Corresponding number of basic states for B	Corresponding number of basic states for macrostate	
0	10	1	3003	3.00×10^3	8.01
1	9	14	2002	2.80×10^4	10.2
2	8	105	1287	1.35×10^5	11.8
3	7	560	792	4.44×10^5	13.0
4	6	2380	462	1.10×10^6	13.9
5	5	8568	252	2.16×10^6	14.6
6	4	27,132	126	3.42×10^6	15.0
7	3	77,520	56	4.34×10^6	15.3
8	2	203,490	21	4.27×10^6	15.3
9	1	497,420	6	2.98×10^6	14.9
10	0	1,144,066	1	1.14×10^6	14.0
Total		1,961,256	8008	$\Omega_{\text{tot}} = 2.00 \times 10^7$	

Figure 19.14 Probability of finding a given fraction of the system's energy in compartment A of the box in Figure 19.13. As the number of energy units increases from 10 to 1000, the probability distribution becomes narrower but remains centered about the mean energy.

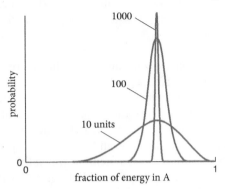

basic states available to the system is obtained by multiplying Ω_A by Ω_B: $\Omega = \Omega_A\Omega_B$.

The probability of each macrostate is obtained by dividing Ω, the number of basic states associated with that macrostate, by Ω_{tot}, the number of basic states associated with all macrostates (2.00×10^7; see Table 19.2). The table shows you that this probability is greatest for the macrostate $E_A = 7$, as you would expect. Given that there are 14 particles in A and six in B, on average each particle has half an energy unit, and so the $E_A = 7$ macrostate corresponds to an equipartitioning of the energy. The curve labeled 10 units in **Figure 19.14** shows this probability as a function of the fraction of energy contained in A.

Example 19.6 Probability of macrostates

In Figure 19.13, after a very large number of particle-partition collisions have occurred, what is the probability of finding the system in (a) the macrostate $E_A = 1$ and (b) the macrostate $E_A = 7$?

❶ **GETTING STARTED** Because all basic states are equally likely, the probability of finding the system in macrostate E_A is equal to the fraction Ω/Ω_{tot}, where Ω is the number of basic states of the system associated with the macrostate E_A and Ω_{tot} is the number of basic states associated with all macrostates (2.00×10^7; Table 19.2).

❷ **DEVISE PLAN** To calculate the probability of a given macrostate E_A, I divide the value of Ω for that macrostate given in Table 19.2 by $\Omega_{tot} = 2.00 \times 10^7$.

❸ **EXECUTE PLAN** (a) For $E_A = 1$, Table 19.2 tells me that $\Omega = 2.80 \times 10^4$. The probability of macrostate $E_A = 1$ is thus $(2.80 \times 10^4)/(2.00 \times 10^7) = 1.40 \times 10^{-3}$. ✔

(b) For the macrostate $E_A = 7$, $\Omega = 4.34 \times 10^6$. So the probability of this macrostate occurring is $(4.34 \times 10^6)/(2.00 \times 10^7) = 2.17 \times 10^{-1}$. ✔

❹ **EVALUATE RESULT** My result shows that the macrostate $E_A = 7$ is more than 150 times more probable than the macrostate $E_A = 1$. This makes sense because, as we saw earlier, the macrostate $E_A = 7$ is the equilibrium state for which there is an equipartition of energy.

If we increase the number of energy units in the box of Figure 19.13 to 100 or 1000, the number of basic states grows exponentially, and if we plot the probability of each macrostate as a function of the fraction of energy in A, we obtain the two curves labeled 100 and 1000 in Figure 19.14. Just as we saw in Figure 19.7, the most probable macrostate doesn't change, but the probability peaks much more narrowly around this state. In other words, the most probable macrostate—the equilibrium state—is now even more likely than any other macrostate.

Note that the number of basic states is very large, even with just ten energy units and 20 particles. In a box of volume 1 m^3 containing air at atmospheric pressure and room temperature, there are on the order of 10^{25} particles and 10^{20} energy units per particle, and so the number of basic states becomes unimaginably large—on the order of ten raised to the power 10^{21}! Because the number of basic states is so large, it is more convenient to work with the natural logarithm of that number. As you can see from the rightmost column in Table 19.2, the natural logarithm of the number of basic states is indeed much more manageable.

Figure 19.15 shows how the natural logarithms of Ω_A, Ω_B, and Ω vary with the number of energy units in compartment A in Figure 19.13. As you can see, the natural logarithm of the number of basic states changes much less rapidly than the number of basic states. Note that as E_A increases, the number of basic states Ω_A increases. As E_A increases, however, E_B decreases and so Ω_B decreases. The number of basic states Ω is maximum when $E_A = 7$ and $E_B = 3$, representing an equipartition of energy. The most probable macrostate (equilibrium) is achieved when there is equipartition of energy.

✋ **19.15** What is the average energy per particle in compartments A and B in Figure 19.13 (a) when there is one energy unit in A and (b) when the system is at equilibrium?

As you can see from Table 19.2, with $E_A = 1$ the number of basic states for the system (2.80×10^4) is more than 100 times smaller than it is at equilibrium ($E_A = 7$, $\Omega = 4.34 \times 10^6$). Collisions between the particles and the partition redistribute

Figure 19.15 Natural logarithm of the number of basic states for compartment A, for compartment B, and for the two compartments in Figure 19.13 combined. The number of basic states is maximal when the energy is equipartitioned (seven energy units in A).

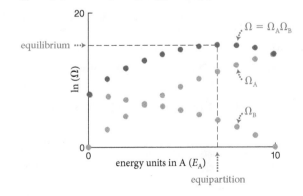

Figure 19.16 If the system evolves toward equilibrium from an initial macrostate with 1 energy unit in A, the number of basic states increases in A and decreases in B. The decrease for B is smaller than the increase for A and so the number of basic states still increases.

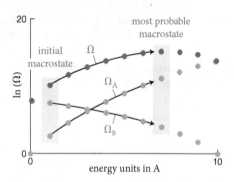

the energy among the particles, and over time the system evolves toward the most probable macrostate—the one for which each particle has its "fair share" of the system's energy ($10/20 = 0.5$ unit), just as you found in Checkpoint 19.15b. As the system evolves toward equilibrium from the $E_A = 1$ macrostate (Figure 19.16), the number of basic states in B *decreases;* that is, B moves toward a macrostate of *lower* probability. However, the rate at which Ω_A increases is greater than the rate at which Ω_B decreases, and so Ω still increases.

Example 19.7 Equilibrium

Suppose the system depicted in Figure 19.13 evolves toward equilibrium from the macrostate $E_A = 4$, $E_B = 6$. Determine the changes in (a) the number of energy units and (b) the number of basic states for compartment A, for compartment B, and for the system.

❶ **GETTING STARTED** From Table 19.2 I see that the equilibrium macrostate (the one that has the greatest number of basic states associated with it) has $E_A = 7$ and $E_B = 3$. This is the macrostate toward which the system evolves.

❷ **DEVISE PLAN** The change in the number of energy units in A is $\Delta E_A = E_{A,f} - E_{A,i}$. The change in the number of energy units in B is $\Delta E_B = E_{B,f} - E_{B,i}$. The change for this system is the sum of these two values. To determine the changes in the number of basic states, I obtain the number of basic states for the initial and final macrostates from Table 19.2 and compute the differences.

❸ **EXECUTE PLAN** (a)

$$\Delta E_A = E_{A,f} - E_{A,i} = 7 - 4 = +3;$$

$$\Delta E_B = E_{B,f} - E_{B,i} = 3 - 6 = -3;$$

$$\Delta E = \Delta E_A + \Delta E_B = 3 - 3 = 0. ✔$$

(b) $\Delta \Omega_A = \Omega_{A,f} - \Omega_{A,i} = 77{,}520 - 2380 = 75{,}140$

$\Delta \Omega_B = \Omega_{B,f} - \Omega_{B,i} = 56 - 462 = -406$

$\Delta \Omega = \Omega_f - \Omega_i = (\Omega_{A,f})(\Omega_{B,f}) - (\Omega_{A,i})(\Omega_{B,i})$

$\quad = (4.34 \times 10^6) - (1.10 \times 10^6) = 3.24 \times 10^6. ✔$

❹ **EVALUATE RESULT** My answer to part a shows that the change in the number of energy units in the system is zero, as I expect because the system is closed. My result in b shows that even though

Ω_B goes down, Ω_A increases significantly more, and so there is a gain in the number of basic states, as should happen when a system evolves toward equilibrium. Because Ω is the product of Ω_A and Ω_B, $\Delta\Omega$ is not simply the sum of $\Delta\Omega_A$ and $\Delta\Omega_B$. My result for $\Delta\Omega$, however, shows an increase in the number of basic states, as I expect.

The evolution toward equilibrium in Figure 19.13 (or in any other system) is a gradual process that seeks out the most probable macrostates through a very large number of random collisions. Once the system has reached equilibrium, the probability of returning to the macrostate $E_A = 1$ energy unit is very small—even for a system of just 20 particles. For a more realistic system, in which the number of particles is much greater, the odds favor equipartitioning so strongly that the probability of observing a system away from equilibrium is virtually zero unless the system is disturbed by outside influences.

The irreversible evolution of a system toward equilibrium can be quantified by the number of basic states Ω:

A closed system always evolves so as to maximize the number of basic states Ω. When this number has reached a maximum, the system is at equilibrium.

This irreversible tendency of closed systems to maximize Ω is called the **second law of thermodynamics.***

Often we cannot directly determine the number of basic states Ω for a system. The second law of thermodynamics, however, allows us to reach conclusions about Ω in a closed system without having to consider any details at the atomic level. Consider, for example, a closed system consisting of a drop of ink placed in a beaker of water. We know from experience that the ink diffuses until it is uniformly distributed in the water (equipartition of space). In Section 19.3 we saw that the spreading out of particles increases the number of basic states—Ω for six particles in a container is much greater at equilibrium than when all the particles are clustered in a corner. For the ink, too, Ω increases as the ink spreads out, and so we can use our intuitive sense of the direction of time to reach conclusions about the number of basic states: If a process in a closed system runs only one way (the process is irreversible), the number of basic states of that system must be increasing. Conversely, the number of basic states of a closed system undergoing a reversible process remains constant.

✋ **19.16** For which of these systems does Ω increase over time: (a) a thermos bottle containing warm water with an ice cube in it, (b) a cup of tea that is at the same temperature as the air in an insulated room, (c) a well-shaken bottle containing oil and water, (d) a seed germinating in soil inside a closed container, (e) the universe.

---

*Thermodynamics is the branch of physics that deals with thermal energy and engines, subjects we discuss in the next two chapters. The second law of thermodynamics was first formulated (in a different form) in the context of this branch of physics. The formulation given here represents the modern and more general view of this law.

Self-quiz

1. Figure 19.4 shows that there are 15 basic states for four energy units distributed over three particles. List all 15.

2. Suppose a container holds three types of gas. The masses of the gas particles are m, $4m$, and $9m$. How do the average speeds of the particles compare?

3. In Figure 19.12, what is the probability of finding three particles in any compartment of the box?

4. The two partitions in the container shown in **Figure 19.17** can slide freely. Where will they be positioned when the space is equipartitioned?

Figure 19.17

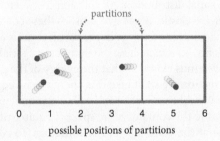

partitions

possible positions of partitions

5. A box with a fixed partition through which energy can be exchanged contains 50 particles, 40 on the left and 10 on the right. The system contains 250 energy units. As the system approaches equilibrium, how many energy units end up on the left? On the right?

6. If a closed system tends to change in a specific direction, such as ice melting in an insulated glass of water, what can you say about the number of basic states for this system?

Answers

1. 400, 040, 004, 310, 301, 031, 130, 013, 103, 220, 202, 022, 112, 121, 211.

2. According to the concept of equipartition of energy, the energy needs to be shared equally among all the gas particles, $\frac{1}{2}mv_1^2 = \frac{1}{2}(4m)v_2^2 = \frac{1}{2}(9m)v_3^2$, or $v_1^2 = 4v_2^2 = 9v_3^2$. Thus $v_1 = 2v_2 = 3v_3$.

3. Figure 19.12, shows that there are ten basic states with three particles in the top left compartment, regardless of the number of particles in the other compartments. When there are two particles in the top left compartment, there are six basic states with three particles in any other compartment. With one particle in the top left compartment, there are $6 + 3 = 9$ basic states with three particles in any other compartment. With no particles in the top left compartment, there are $6 + 3 = 9$ basic states with three particles in any other compartment. Therefore there are $10 + 6 + 9 + 9 = 34$ basic states with three particles in any compartment. The probability of finding three particles in any compartment is thus $34/84$.

4. The partition shown at position 2 moves to position 4, and the partition shown at 4 moves to 5, so that there is 1 particle per unit length of the box.

5. Because at equilibrium the energy must be equipartitioned, the 250 units are evenly distributed across the 50 particles; with five units/per particle. The 40 particles on the left have 200 units of energy, and the 10 particles on the right have 50 units.

6. According to the second law of thermodynamics, the system evolves to the macrostate that has the maximum number of basic states. Because the process is irreversible, the final macrostate must have a greater number of basic states than the initial macrostate.

Figure 19.18 Each of the N particles inside the container can be placed in any one of the M equal-sized compartments.

N particles in volume V

volume V divided into M compartments . . .

. . . each of which has volume δV

19.5 Dependence of entropy on volume

In Sections 19.2 and 19.3, we saw that the equipartition of energy and space describes the most probable macrostate of any system—that is, the state that has the greatest number of basic states. The number of *atomic parameters,* such as the position and velocity of each particle, is intractably large for any real system, and so it is generally impossible to specify the basic state of the system. The macrostate, on the other hand, is readily specified by a few measurable *macroscopic parameters,* such as volume, pressure, or temperature.

To make the connection between macroscopic and atomic parameters, we consider a gas that consists of a very large number N of particles in a container of volume V. The particles occupy a negligible fraction of the volume V and interact with one another only during collisions that randomize both the energy distribution and the spatial distribution. All collisions between the particles and the walls of the container are elastic. A gas that satisfies these conditions is called an **ideal gas.** At high temperature and low pressure, an ideal gas is a good model for a real gas.

Let us now calculate how the number of spatial basic states depends on V and N. The system is closed so that the energy of the gas remains constant. Throughout this section we shall take this energy to be equipartitioned over the gas particles. At equilibrium each particle has an equal probability of occupying any location inside the container; the space is equipartitioned and each particle has its fair share of the volume V of the container. To count basic states, we divide the container into equal-sized compartments as we did in Section 19.3. If there are M equal-sized compartments and just one particle, that particle can be placed in any one of the M compartments. Thus there are M basic states for that one particle (**Figure 19.18**). A second particle can also be placed in any of the M compartments, and so, for two particles, there are M^2 basic states. Continuing this process for N particles, we see that Ω, the number of basic states, is*

$$\Omega = M^N \quad \text{(closed system).} \tag{19.1}$$

Exercise 19.8 Number of basic states

Suppose ten distinguishable particles are equipartitioned in a container that is divided into 100 equal-sized compartments. (*a*) How many basic states are associated with this system? (*b*) What is the natural logarithm of this number of basic states?

SOLUTION (*a*) The number of basic states is $\Omega = M^N = 100^{10} = (10^2)^{10} = 10^{20}$, an unimaginably large number. ✔

(*b*) The natural logarithm of the number of basic states is $\ln \Omega = \ln(10^{20}) = 46$, a much more manageable number. ✔

19.17 If you decrease the size of the compartments in Exercise 19.8 by a factor of 10, what is the change (*a*) in the number of basic states and (*b*) in the natural logarithm of that number?

Exercise 19.8 and Checkpoint 19.17 illustrate that the natural logarithm of the number of basic states is much more manageable and less sensitive to a change in compartment size than the number itself. The natural logarithm of the number of basic states from Eq. 19.1 is

$$\ln \Omega = \ln(M^N) = N \ln M \quad \text{(closed system at equilibrium),} \tag{19.2}$$

---

*We assume here that the particles are distinguishable from one another. In reality, the particles of an ideal gas are indistinguishable. In the case of indistinguishable particles and when $M \gg N \gg 1$, the number of basic states is found to be $M^N/N!$ instead of the result shown in Eq. 19.1. For simplicity, we shall limit the treatment in this section to systems of distinguishable particles.

where we have used the identity $\ln(a^b) = b \ln a$. If the volume of the equal-sized compartments is δV, the number of compartments is $M = V/\delta V$, and so Eq. 19.2 becomes

$$\ln \Omega = N \ln \left(\frac{V}{\delta V} \right)$$

$$= N \ln V - N \ln \delta V \quad \text{(closed system at equilibrium),} \qquad (19.3)$$

where we have used the identify $\ln(a/b) = \ln a - \ln b$.

Because the quantity $\ln \Omega$—the natural logarithm of the number of basic states—is so important, we define a quantity called **entropy:**[*]

$$S \equiv \ln \Omega. \qquad (19.4)$$

Entropy is a unitless quantity that is a measure of the number of basic states in a system. The greater the number of basic states, the greater the entropy. Entropy allows us to quantify the spreading out of energy or particles over space.

As we saw in Section 19.4, a closed system always evolves so as to maximize the number of basic states. As the number of basic states increases, so does the entropy: $S_f > S_i$. Therefore,

$$\Delta S > 0 \quad \text{(closed system evolving toward equilibrium).} \qquad (19.5)$$

At equilibrium, the number of basic states reaches its maximum value Ω_{max} and the entropy of the system no longer changes:

$$\Delta S = 0 \quad \text{(closed system at equilibrium).} \qquad (19.6)$$

Equations 19.5 and 19.6 constitute the **entropy law,** a mathematical expression of the second law of thermodynamics. The entropy law implies that the entropy of a closed system never decreases (it can only increase or remain constant).

It is important to note that Eqs. 19.5 and 19.6 apply to only closed systems. For a system that is not closed, the entropy can increase, remain the same, or decrease. Back in Figure 19.16, for example, the particles in B transfer some of their energy to the particles in A, so B does not constitute a closed system. As we saw, Ω_B decreases and so the entropy of B decreases:

For systems that are not closed, the entropy can increase, decrease, or stay the same.

For a system of N gas particles equipartitioned over a volume V, we can obtain an expression for how entropy depends on volume by substituting Eq. 19.3 into Eq. 19.4:

$$S = N \ln V - N \ln \delta V \quad \text{(closed system at equilibrium).} \qquad (19.7)$$

This expression describes a gas that is equipartitioned over a volume V—that is, at equilibrium. The value for S given by this expression thus represents the *maximum* value for the entropy of a gas of N particles in a volume V. If the gas is not at equilibrium, its entropy is smaller than that given by Eq. 19.7.

---

[*]Historically, entropy has been defined as $S \equiv k_B \ln \Omega$, where k_B is a constant that makes the entropy a much smaller number and gives it units of J/K (see Eq. 19.39). In this book, however, we shall use the definition given in Eq. 19.3. This definition is called a *statistical definition* of entropy.

The first term on the right in Eq. 19.7 shows that for a fixed number of particles N, the entropy (that is, the natural logarithm of the number of basic states) increases as we increase the volume V of the container. The second term on the right depends on the size of the compartments: The smaller the compartments, the smaller this term and therefore the greater the entropy. That the entropy depends on the volume δV of the compartments may seem disturbing—after all, the partitioning is completely arbitrary. In general, however, $\delta V \ll V$, and so the second term on the right in Eq. 19.7 is much smaller than the first one and can be ignored.

In general, we shall be concerned only with *changes* in entropy, in which case the second term in Eq. 19.7 drops out entirely. Consider, for example, the expansion of a gas made up of a fixed number of particles from an initial volume V_i to a final volume $V_f > V_i$. According to Eq. 19.7, the change in entropy is

$$\Delta S = S_f - S_i = (N \ln V_f - N \ln \delta V) - (N \ln V_i - N \ln \delta V)$$

$$= N \ln V_f - N \ln V_i = N \ln\left(\frac{V_f}{V_i}\right) \quad \text{(closed system at equilibrium).} \quad (19.8)$$

Because the second term on the right in Eq. 19.7 cancels out, the entropy change does not depend on the size of the compartments. Note that because $V_f > V_i$ the change in entropy is positive.

19.18 (*a*) What does a positive entropy change in a closed system imply about the change in the number of basic states? (*b*) How does Eq. 19.8 show that a gas expands if given more room? (*c*) How does it show that the gas does not contract spontaneously into a subvolume?

Let us next examine how entropy changes when a system of two interacting gases evolves to equilibrium. Consider, for example, two gases inside a container with a partition separating the gases into two compartments, one of volume V_A containing N_A particles and the other of volume V_B containing N_B particles (**Figure 19.19a**).* If we let the partition move freely, the system evolves to the macrostate that has the maximum number of basic states. If the partition moves to the right, V_A increases, and so the number of basic states for gas A increases. As V_A increases, V_B decreases, and so the number of basic states for gas B decreases. Where does the partition come to rest? In other words, what value of V_A maximizes the number of basic states for the system as a whole?

In Section 19.4, we determined that the number of basic states for a system comprising two gases is the product of the number of basic states for the two separate gases:

$$\Omega = \Omega_A \Omega_B, \quad (19.9)$$

Using the definition of entropy (Eq. 19.4), we obtain

$$S = \ln(\Omega_A \Omega_B) = \ln \Omega_A + \ln \Omega_B = S_A + S_B, \quad (19.10)$$

where we have used the fact that $\ln(ab) = \ln a + \ln b$. Note that, unlike Ω, the natural logarithm of Ω and thus entropy are extensive properties:

The entropy of a combined system is the sum of the entropies of the individual systems.

Figure 19.19 (*a*) Two gases of unequal densities are placed on either side of a partition. (*b*) When the partition is free to move, the two gases evolve to the macrostate corresponding to the maximum number of basic states.

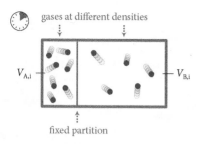

gases at different densities

$V_{A,i}$ — $V_{B,i}$

fixed partition

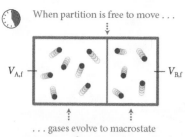

When partition is free to move . . .

$V_{A,f}$ — $V_{B,f}$

. . . gases evolve to macrostate corresponding to maximum number of basic states.

*We assume energy is equipartitioned over the system made up of both compartments and focus on only volume changes here. In the next section we shall examine the equipartitioning of energy.

Example 19.9 Partitioned gas

A partition divides a container into two equal compartments, A and B. Initially compartment A contains a gas of distinguishable particles and compartment B is empty. (*a*) What is the change in the entropy of the gas if the partition is removed and the gas is allowed to expand into the volume V of the container? (*b*) If the partition is reinserted after the gas has reached equilibrium in the volume V, what is the change in the entropy of the gas due to the reinsertion?

❶ GETTING STARTED When the partition is removed, the gas expands to fill the container, so the final volume occupied by the gas is twice the initial volume: $V_f = 2V_i$. Reinserting the partition divides the container into two equal compartments A and B again. The only difference is that now B is not empty; it contains half the particles, and A contains the other half.

❷ DEVISE PLAN To obtain the change in entropy due to the removal of the partition, I substitute $V_f = 2V_i$ into Eq. 19.8. For part *b*, the entropy change due to the reinsertion is equal to the difference between the entropy of the gas after the partition is reinserted and its entropy before the partition is reinserted. To determine the entropy before the partition is reinserted, I use Eq. 19.7. Equation 19.10 tells me that after reinsertion the entropy of the gas is equal to the sum of the entropies in the two compartments. Because each compartment has half the volume of the container and contains half the gas particles, the entropies in the two compartments are equal to each other. To determine these entropies, I use Eq. 19.7 again.

❸ EXECUTE PLAN (*a*) Equation 19.8 gives
$$\Delta S = N \ln(2V_i/V_i) = N \ln 2. ✔$$

(*b*) From Eq. 19.7, I obtain for the entropy before reinsertion of the partition: $S_i = N \ln V - N \ln \delta V$. Combining Eqs. 19.10 and 19.7, I obtain for the final entropy,

$$S_f = S_A + S_B = 2\left[\frac{N}{2}\ln\left(\frac{V/2}{\delta V}\right)\right] = N \ln \frac{V}{2} - N \ln \delta V$$
$$= N \ln V - N \ln 2 - N \ln \delta V.$$

The change in entropy is thus $\Delta S = S_f - S_i = -N \ln 2. ✔$

❹ EVALUATE RESULT My answer to part *a* shows that the removal of the partition and subsequent expansion of the gas result in a positive change in entropy. This is what I expect given that the system evolves toward equilibrium. My answer to part *b* is the negative of the answer to part *a*, which tells me that the change in entropy due to the combined removal and reinsertion of the partition is zero: $\Delta S = N \ln 2 - N \ln 2 = 0$. That is surprising because the gas has expanded into a larger volume. However, the particles in this problem are distinguishable, and in reality gas particles are indistinguishable. For distinguishable particles, the reinsertion reduces the number of basic states (a particle in compartment B can no longer be in compartment A, and vice versa). If the particles are indistinguishable, reinserting the partition causes no entropy change.

Assuming that the gases on each side of the partition in Figure 19.19 remain at equilibrium, we have from Eq. 19.7

$$S_A = N_A \ln V_A - N_A \ln \delta V \qquad (19.11)$$

$$S_B = N_B \ln V_B - N_B \ln \delta V = N_B \ln(V - V_A) - N_B \ln \delta V, \qquad (19.12)$$

where $V = V_A + V_B$ is the combined (constant) volume of the system. Increasing V_A thus increases S_A but decreases S_B. Figure 19.20 shows how the entropies of the two compartments and the entropy of the combined system depend on V_A. When the number of basic states, and thus the entropy in Eq. 19.10, are maximum, Eq. 19.6 applies, so the partition comes to rest and the combined system is at equilibrium. We therefore would like to determine for what value of V_A the system's entropy S is maximized.

At the maximum value of S, the slope of the S curve in Figure 19.20 is zero. This slope is given by the derivative of S with respect to V_A, and so at equilibrium,

$$\frac{dS}{dV_A} = \frac{dS_A}{dV_A} + \frac{dS_B}{dV_A} = 0 \quad \text{(equilibrium)} \qquad (19.13)$$

or
$$\frac{dS_A}{dV_A} = -\frac{dS_B}{dV_A} \quad \text{(equilibrium)}. \qquad (19.14)$$

The quantities dS_A/dV_A and dS_B/dV_A represent the slopes of the S_A and S_B curves in Figure 19.20. Equation 19.14 thus tells us that these two curves must have opposite slopes at the equilibrium value of V_A.

Figure 19.20 As the partition in Figure 19.19 moves to the right and V_A increases, S_A increases and S_B decreases until the entropy of the system is maximum. At this value of V_A, the partition comes to rest, and the system is at equilibrium.

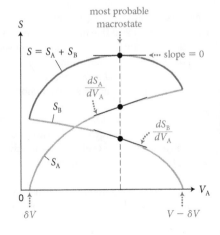

QUANTITATIVE TOOLS

We can now show that the condition expressed in Eq. 19.14 leads to equipartitioning of the space by calculating the derivatives. Because V is constant,

$$\frac{dV_B}{dV_A} = \frac{d}{dV_A}(V - V_A) = -1, \qquad (19.15)$$

we have $dV_A = -dV_B$ and so can rewrite Eq. 19.14 in the form

$$\frac{dS_A}{dV_A} = -\frac{dS_B}{(-dV_B)} = \frac{dS_B}{dV_B} \quad \text{(equilibrium)}. \qquad (19.16)$$

Equation 19.16 tells us that at equilibrium the quantity dS/dV evaluated on either side of the partition is the same.

Taking the derivative of Eq. 19.7 with respect to V, we get

$$\frac{dS}{dV} = \frac{d}{dV}(N \ln V - N \ln \delta V) = \frac{d}{dV}(N \ln V) \qquad (19.17)$$

because $N \ln \delta V$ does not depend on V (N is fixed and δV is a constant). Recalling that

$$\frac{d}{dx} \ln x = \frac{1}{x}, \qquad (19.18)$$

we get

$$\frac{dS}{dV} = \frac{N}{V}. \qquad (19.19)$$

The quantity N/V represents the number of particles per unit volume and is the number density of the gas (Eq. 1.3). Substituting Eq. 19.19 into Eq. 19.16, we obtain

$$\frac{N_A}{V_A} = \frac{N_B}{V_B} \quad \text{(equilibrium)}, \qquad (19.20)$$

which is exactly the requirement for the equipartition of space: At equilibrium the number densities are the same on both sides of the partition, and so each particle has the same amount of space available to it.

Example 19.10 Sliding partition

The box in Figure 19.19 contains seven gas particles in compartment A and five in compartment B, and the partition separating the compartments is free to move. Let the volume of the box be V and the initial ratio of the compartment volumes be $V_{B,i}/V_{A,i} = 3$. What is the change in the entropy of this gas as it evolves from this macrostate to equilibrium?

❶ GETTING STARTED Once the equilibrium state is reached, the number densities are equal on the two sides of the partition and the change in the entropy of the gas—which is my system—is the sum of the entropy changes in the two compartments.

❷ DEVISE PLAN Equation 19.20 gives me the volume ratio at equilibrium, which I call the final volume ratio. Knowing the initial and final ratios, I can use Eq. 19.8 to determine the entropy change in each compartment and then use Eq. 19.10 to determine the entropy change of the system.

❸ EXECUTE PLAN From the given initial volume ratio, I know that $V_{B,i} = 3V_{A,i}$. The volume of the box is the sum of the volumes of the compartments: $V = V_{A,i} + V_{B,i} = V_{A,i} + 3V_{A,i} = 4V_{A,i}$, and so $V_{A,i} = V/4$ and $V_{B,i} = 3V/4$. From Eq. 19.20 the ratio of the final volumes is equal to the ratio of the numbers of particles. I know that the number of particles in each compartment cannot change, so I can say that $V_{B,f}/V_{A,f} = N_B/N_A = 5/7$, which means $V_{B,f} = 5V_{A,f}/7$.

After the partition moves, the sum of the volumes is still V, and so I know $V = V_{A,f} + V_{B,f} = V_{A,f} + 5V_{A,f}/7 = 12V_{A,f}/7$.

Thus $V_{A,f} = 7V/12$ and $V_{B,f} = 5V/12$. Substituting these volumes into Eq. 19.8 gives

$$\Delta S_A = N_A \ln\frac{V_{A,f}}{V_{A,i}} = 7 \ln\frac{7V/12}{V/4} = 7 \ln\tfrac{7}{3} = 5.93$$

$$\Delta S_B = N_B \ln\frac{V_{B,f}}{V_{B,i}} = 5 \ln\frac{5V/12}{3V/4} = 5 \ln\tfrac{5}{9} = -2.94.$$

Because the entropy of the system is the sum of the entropies of the particles in the two compartments (Eq. 19.10), I have

$$\Delta S = \Delta S_A + \Delta S_B = 5.93 - 2.94 = 2.99. ✔$$

❹ **EVALUATE RESULT** The volume of compartment A increases and the volume of compartment B decreases, so I expect the entropy of the particles in A to increase and the entropy of the particles in B to decrease, as my results show. Because the magnitude of the change in entropy is greater for A than for B, the entropy change of the system is positive, as it should be for a closed system moving toward equilibrium.

✋ **19.19** Explain why the changes in entropy in Example 19.10 do not depend on the volume V of the box.

19.6 Dependence of entropy on energy

Let us now turn our attention to how entropy depends on the energy of a system. For simplicity we restrict ourselves to a *monatomic* ideal gas (that is, a gas made up of atoms for which we can ignore any internal structure), so that none of the atoms' kinetic energy can be converted to internal energy. The incoherent kinetic energy of all the atoms makes up the thermal energy of the gas. If the center of mass of the gas is at rest, the energy of the system is entirely in the form of thermal energy.

Let us now calculate how the number of basic states depends on the thermal energy E_{th} and the number of atoms N of a closed system. Throughout this section we shall take the atoms to be distributed uniformly over the (constant) volume V of the system. At equilibrium each atom has its fair share of the thermal energy E_{th} of the system. A single atom moving at speed v contributes an amount of energy $\frac{1}{2}mv^2$ to E_{th}. Because the atom is free to move in any direction, many velocities correspond to a given speed v (**Figure 19.21**) and thus to a given share of the thermal energy E_{th}. A basic state is specified by the components v_x, v_y, and v_z of all N atoms. To count the basic states, we can use the procedure from Section 19.5: We divide the available "velocity space" in Figure 19.21 into small compartments and count the number of ways we can distribute N atoms among these compartments.

What is the available velocity space? To determine this velocity space, we introduce the **root-mean-square speed,**[*] or *rms speed* for short:

$$v_{rms} \equiv \sqrt{(v^2)_{av}}, \tag{19.21}$$

where $(v^2)_{av}$ is the average of the squares of the speeds of all the atoms in a volume of gas. Applying the Pythagorean theorem to the two shaded triangles in Figure 19.21, we have

$$v^2 = v_x^2 + v_y^2 + v_z^2. \tag{19.22}$$

Figure 19.21 Velocity space determined by the components of velocity. All velocity vectors ending on the surface of the sphere correspond to the same speed and therefore the same kinetic energy.

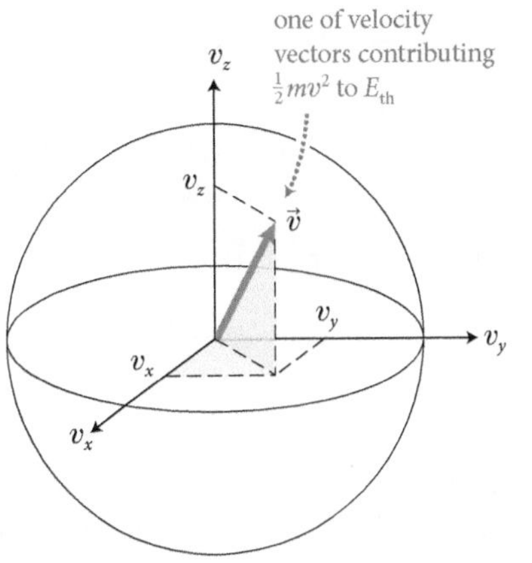

---

[*]This value is obtained by taking the square *root* of the *mean* (average) of the *squares* of the speeds.

If we average over all atoms, this expression becomes

$$v_{rms}^2 = (v^2)_{av} = (v_x^2)_{av} + (v_y^2)_{av} + (v_z^2)_{av}. \qquad (19.23)$$

Given that the gas is the same in all directions, there is no difference between the x, y, and z directions, and so

$$(v_x^2)_{av} = (v_y^2)_{av} = (v_z^2)_{av} \qquad (19.24)$$

or

$$v_{rms}^2 = 3(v_x^2)_{av}. \qquad (19.25)$$

19.20 Five atoms moving along the x axis have x components of velocities $+2.1$ m/s, -3.2 m/s, $+4.5$ m/s, -0.3 m/s, and -3.1 m/s. Determine (a) the x component of the average velocity, (b) the average speed, and (c) the rms speed.

To return to determining the volume of the velocity space, let us now assume that the components of the velocities of the atoms are limited to a range of a few times the rms speed. That is, let us assume that

$$-av_{rms} \leq v_x \leq av_{rms}, \qquad (19.26)$$

Figure 19.22 Calculating the volume $\mathcal{V}$ of the velocity space.

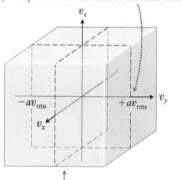

y components of velocities lie between $\pm av_{rms}$

Cube defines "velocity space."

and likewise for the y and z components, and where a is a number between, say, 1 and 10. This may seem like a totally arbitrary restriction, but you will see in a bit that the end result is not affected by this restriction. With this simplifying assumption the velocities of all the atoms in our system lie in a cube having sides of length $2av_{rms}$ (**Figure 19.22**). The "volume" $\mathcal{V}$ of the velocity space of the atoms in a gas-filled container is then $\mathcal{V} = (2av_{rms})^3$.

We can relate this volume to the thermal energy of the gas because we know that the average kinetic energy K_{av} of one atom is equal to the thermal energy of the gas divided by the number N of gas atoms in the container:

$$K_{av} \equiv \tfrac{1}{2}m(v^2)_{av} = \tfrac{1}{2}mv_{rms}^2 = \frac{E_{th}}{N} \quad \text{(monatomic ideal gas)}, \qquad (19.27)$$

where we have used the definition of rms speed (Eq. 19.21). Now we can rewrite the rms speed as

$$v_{rms} = \left(\frac{2E_{th}}{mN}\right)^{1/2}, \qquad (19.28)$$

and so we see that the volume $\mathcal{V}$ of the velocity space is proportional to $(E_{th})^{3/2}$:

$$\mathcal{V} = (2av_{rms})^3 = \left[2a\left(\frac{2E_{th}}{mN}\right)^{1/2}\right]^3 = \left(\frac{8a^2}{mN}\right)^{3/2}(E_{th})^{3/2} = b(E_{th})^{3/2}, \qquad (19.29)$$

where $b = (8a^2/mN)^{3/2}$ is a constant.

19.21 Work out the algebra between the third and fourth terms in Eq. 19.29.

In analogy to what we did in Section 19.5, we now divide this volume $\mathcal{V}$ of the velocity space into smaller velocity-space compartments, each of volume $\delta\mathcal{V}$. The number of velocity-space compartments is

$$M_v = \frac{\mathcal{V}}{\delta\mathcal{V}} = \frac{b}{\delta\mathcal{V}}(E_{th})^{3/2}, \qquad (19.30)$$

and the number of ways we can distribute N distinguishable atoms in these M_v compartments is (see Eq. 19.1):

$$\Omega_v = (M_v)^N = \left[\frac{b}{\delta\mathcal{V}} (E_{th})^{3/2} \right]^N. \qquad (19.31)$$

For a monatomic ideal gas in which the energy is equipartitioned, the entropy is maximum and given by

$$S \equiv \ln \Omega_v = \tfrac{3}{2} N \ln E_{th} - N \ln \frac{\delta\mathcal{V}}{b} \quad \text{(monatomic ideal gas at equilibrium).} \qquad (19.32)$$

The first term to the right of the equals sign shows that for a fixed number of atoms N, the entropy increases as we increase the thermal energy E_{th} of the gas. The factors $\delta\mathcal{V}$ and b in Eq. 19.32 tell us that the entropy depends on the size of the velocity-space compartments in which the atoms are distributed and on the restriction we imposed in Eq. 19.26. However, the term that contains these factors drops out when we consider changes in entropy:

$$\Delta S = S_f - S_i = \tfrac{3}{2} N \ln E_{th,f} - \tfrac{3}{2} N \ln E_{th,i} = \tfrac{3}{2} N \ln \left(\frac{E_{th,f}}{E_{th,i}} \right). \qquad (19.33)$$

Now we can examine how entropy changes when a system of two interacting gases initially having different amounts of thermal energy evolves toward a system-wide equilibrium state. Consider a closed system of two gases that can exchange energy with each other, as illustrated in Figure 19.23. The volume of each compartment and the number of atoms in each are fixed. In Section 19.4 we saw that such a system tends to evolve toward an equipartitioning of the system's thermal energy. If thermal energy is exchanged across the partition, one gas gains thermal energy and the other one loses thermal energy. Because the entropy of the gas depends on its thermal energy, the exchange of thermal energy causes the entropy to increase on one side of the partition and decrease on the other side. Figure 19.24 shows how the entropies of the two gases and the entropy of the combined system depend on the thermal energy $E_{th,A}$ of gas A. When the number of basic states, and thus the entropy, are maximum, Eq. 19.6 applies and the exchange of thermal energy stops and the system is said to be in **thermal equilibrium.** We therefore would like to determine for what value of $E_{th,A}$ the system's entropy $S = S_A + S_B$ is maximized.

At the maximum entropy level, the slope of curve S in Figure 19.24 is zero, and so

$$\frac{dS}{dE_{th,A}} = \frac{dS_A}{dE_{th,A}} + \frac{dS_B}{dE_{th,A}} = 0 \quad \text{(closed system at thermal equilibrium)} \qquad (19.34)$$

or

$$\frac{dS_A}{dE_{th,A}} = -\frac{dS_B}{dE_{th,A}} \quad \text{(closed system at thermal equilibrium).} \qquad (19.35)$$

The quantities $dS_A/dE_{th,A}$ and $dS_B/dE_{th,A}$ represent the slopes of curves S_A and S_B in Figure 19.24. Eq. 19.35 thus tells us that these two curves must have opposite slopes at thermal equilibrium. Because E_{th} is constant,

$$\frac{dE_{th,B}}{dE_{th,A}} = \frac{d}{dE_{th,A}} [E_{th} - E_{th,A}] = -1, \qquad (19.36)$$

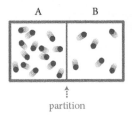

Figure 19.23 Container holding two gases separated by a partition. The gases can exchange energy with each other through collisions with the partition.

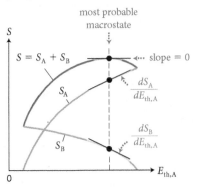

Figure 19.24 How the entropies of the gases in Figure 19.23 depend on the thermal energy $E_{th,A}$ of gas A.

we have $dE_{th,A} = -dE_{th,B}$ and so Eq. 19.35 becomes

$$\frac{dS_A}{dE_{th,A}} = -\frac{dS_B}{d(-E_{th,B})} = \frac{dS_B}{dE_{th,B}} \quad \text{(closed system at thermal equilibrium).} \quad (19.37)$$

As you have probably noticed, there is a strong parallel between the quantity dS/dE_{th} in the above derivation and the quantity dS/dV we discussed in the preceding section (compare, also, Figures 19.20 and 19.24). In Eq. 19.19 we saw that the quantity dS/dV represents the number density of the gas and that at equilibrium the number density tends to be the same in all compartments of the system (Eq. 19.20).

To discover what physical concept dS/dE_{th} corresponds to, answer the following checkpoint.

19.22 (*a*) Using Figure 19.24, compare the kinetic energy of an atom in gas A with that of an atom in gas B in the thermal equilibrium state (that is, on the dashed line) and in a macrostate to the left of the dashed line. (*b*) In each of these two macrostates, which gas is "hotter"? (*c*) For the same two macrostates, compare the quantities $dS_A/dE_{th,A}$ and $dS_B/dE_{th,B}$.

As Checkpoint 19.22 suggests, the quantity dS/dE_{th} is smaller for a hot gas than for a cooler gas. We therefore define this quantity to be inversely proportional to a quantity called the **absolute temperature** T:

$$\frac{1}{k_B T} \equiv \frac{dS}{dE_{th}}, \quad (19.38)$$

The unit for absolute temperature, the **kelvin** (K), is one of the SI base units. In Section 20.2 we'll see how this unit is defined. The proportionality constant k_B, called the **Boltzmann constant,** is

$$k_B = 1.381 \times 10^{-23} \, \text{J/K}. \quad (19.39)$$

The constant k_B is chosen in such a way as to reconcile the units for absolute temperature with common temperature measurements, which we shall discuss in the next chapter. For now, consider T to be a quantitative measure of the temperature obtained by adding 273.15 to the reading on a thermometer calibrated in degrees Celsius.

With this definition of absolute temperature, we see that thermal equilibrium (Eq. 19.37) implies that the absolute temperatures of the two gases are the same:

$$T_A = T_B \quad \text{(thermal equilibrium).} \quad (19.40)$$

We can obtain another consequence of the thermal equilibrium condition by taking the derivative of Eq. 19.32 with respect to E_{th}:

$$\frac{dS}{dE_{th}} = \frac{d}{dE_{th}}\left[\tfrac{3}{2}N \ln E_{th} - N \ln \frac{\delta \mathcal{V}}{b}\right] = \frac{d}{dE_{th}}\left[\tfrac{3}{2}N \ln E_{th}\right] = \tfrac{3}{2}\frac{N}{E_{th}}, \quad (19.41)$$

where the derivative of the term $N \ln(\delta \mathcal{V}/b)$ is zero because N, $\delta \mathcal{V}$, and b are constant. Substituting Eq. 19.41 into Eq. 19.37 and inverting the result yield

$$\frac{E_{th,A}}{N_A} = \frac{E_{th,B}}{N_B} \quad \text{(thermal equilibrium),} \quad (19.42)$$

which is exactly the requirement for equipartition of energy: At thermal equilibrium the amount of thermal energy per atom is the same on either side of the partition.

19.23 Suppose 14 atoms are in compartment A and six atoms are in compartment B, as shown in Figure 19.23. When thermal equilibrium energy is reached, what percentage of the thermal energy do you expect to find in each compartment?

19.7 Properties of a monatomic ideal gas

Let us next connect some of the atomic parameters that describe an ideal gas to its macroscopic parameters. Consider a monatomic ideal gas that consists of a very large number N of indistinguishable atoms of mass m inside a closed rigid container of volume V. The collisions between the atoms and between the atoms and the walls of the container are all elastic. **Figure 19.25** shows one atom-wall collision. Because the collision is elastic, the speed v of the atom does not change; the collision simply reverses the direction of the velocity component v_x perpendicular to the wall, leaving the velocity component v_y parallel to the wall unchanged.

What is true for the atom in Figure 19.25 is true for all the other gas atoms. Therefore the forces exerted by the wall and by the atoms are perpendicular to the wall. The collisions between the wall and the gas atoms are the origin of the pressure in the gas.

19.24 Let the system comprising the atoms and wall in Figure 19.25 be isolated. (*a*) What is the direction of the atom's change in momentum? (*b*) What is the direction of the change in momentum of the wall? (*c*) How are these momentum changes related to the forces that wall and atom exert on each other?

As we saw in Section 18.5, the pressure P in a gas is equal to the magnitude of the force $\vec{F}^c_{gA}$ exerted by the gas on a surface per unit area. Because the force exerted by the gas on the surface and the force exerted by the surface on the gas form a interaction pair, we have $F^c_{gA} = F^c_{Ag}$. Over a section of wall of area A, Eq. 18.1 then yields

$$P \equiv \frac{F^c_{gA}}{A} = \frac{F^c_{Ag}}{A}. \tag{19.43}$$

The force exerted by the wall on the gas is equal to the rate of change in the momentum of the gas (see Section 8.7), and our task is to evaluate the magnitude of this rate of change. To simplify matters, we initially assume that the x component of the velocities of all the atoms in the gas have the same magnitude $|v_x|$. (We shall see shortly that our result doesn't depend on this oversimplification.) Let's begin by evaluating the change in momentum of a single gas atom when it collides with the wall. Each collision like the one in Figure 19.25 changes the sign of the x component of the atom's velocity, and so the magnitude of the atom's change in momentum is $2m|v_x|$.

To calculate the rate at which the magnitude of the momentum of the gas changes, we need to determine how many collisions occur within a finite time interval Δt. In order for an atom, for which the x component of its velocity is v_x, to collide with the wall in a time interval Δt, the atom has to be within a distance $|v_x|\Delta t$ of the wall. So let us consider all the atoms in a cylinder of cross-sectional area A and length $|v_x|\Delta t$ (**Figure 19.26**). At equilibrium, the atoms are uniformly distributed over the volume of the container, and so the number of atoms per unit volume is N/V. The volume of the cylinder is

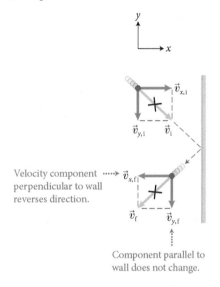

Figure 19.25 An atom collides elastically with a wall, reversing the perpendicular component of the incoming velocity. Its speed remains unchanged.

Velocity component perpendicular to wall reverses direction.

Component parallel to wall does not change.

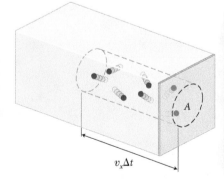

Figure 19.26 On average, one half of the gas atoms contained in the cylinder strike the wall during a time interval Δt.

$A|v_x|\Delta t$, and so the number of atoms in the cylinder is $N_{cyl} = (N/V)A|v_x|\Delta t$. At any given instant, there should be as many atoms moving in the $+x$ direction as there are atoms moving in the $-x$ direction, and so half of the atoms in the cylinder in Figure 19.26 collide with the wall in the time interval Δt. The magnitude of the change in momentum for all gas atoms is thus equal to the number of collisions $N_{cyl}/2$ times the magnitude of the change in momentum $2m|v_x|$ for a single collision:

$$|\Delta p| = \tfrac{1}{2}\frac{N}{V}(A|v_x|\Delta t)(2m|v_x|) = \frac{NAmv_x^2}{V}\Delta t. \tag{19.44}$$

To calculate the magnitude of the rate of change of the momentum of the gas, we divide the left and right sides of the equation by Δt:

$$\frac{|\Delta p|}{\Delta t} = \frac{NAmv_x^2}{V}. \tag{19.45}$$

The quantity on the left, the magnitude of the rate of change of the momentum of the gas, is equal to the magnitude of the force exerted by the wall on the gas in the cylinder in Figure 19.26, and so, substituting Eq. 19.45 into Eq. 19.43, we obtain for the pressure in the gas

$$P = \frac{|\Delta p|/\Delta t}{A} = \frac{Nmv_x^2}{V}. \tag{19.46}$$

In deriving Eq. 19.46 we used the unlikely assumption that the x components of the velocities of all atoms have the same magnitude $|v_x|$. When we use this assumption, each atom contributes a force $(Amv_x^2)/V$, and the force in Eq. 19.45 is the sum of N such forces. If the atoms move at different speeds, we should replace v_x^2 in Eq. 19.46 by the average value $(v_x^2)_{av}$. (We can do this because N times the average value of a quantity is equal to the sum of N actual values.)

When we use the relationship between $(v_x^2)_{av}$ and v_{rms}^2 expressed in Eq. 19.25, $v_{rms}^2 = 3(v_x^2)_{av}$, Eq. 19.46 becomes

$$P = \frac{Nm\tfrac{1}{3}v_{rms}^2}{V} = \tfrac{2}{3}\frac{N(\tfrac{1}{2}mv_{rms}^2)}{V} \tag{19.47}$$

or, substituting $N(\tfrac{1}{2}mv_{rms}^2) = E_{th}$ (see Eq. 19.27),

$$P = \tfrac{2}{3}\frac{E_{th}}{V} \quad \text{(monatomic ideal gas).} \tag{19.48}$$

Equation. 19.47 shows that the pressure in a monatomic ideal gas is determined by two atomic parameters: the number density N/V and the average kinetic energy $\tfrac{1}{2}mv_{rms}^2$ of an atom.

We can obtain an expression for the thermal energy of a monatomic ideal gas by substituting Eq. 19.41, $dS/dE_{th} = \tfrac{3}{2}N/E_{th}$, into Eq. 19.38, $1/k_BT = dS/dE_{th}$. Doing this we obtain

$$\frac{1}{k_BT} = \tfrac{3}{2}\frac{N}{E_{th}} \tag{19.49}$$

or $\qquad\qquad E_{th} = \tfrac{3}{2}Nk_BT \quad \text{(monatomic ideal gas).} \tag{19.50}$

Substituting Eq. 19.50 into Eq. 19.48, we obtain

$$P = \frac{N}{V}k_BT \quad \text{(ideal gas system at equilibrium)}. \quad (19.51)$$

This relationship among pressure, volume, temperature, and number of atoms is called the **ideal gas law.** We obtained this result among macroscopic parameters (pressure, volume, temperature) by considering what happens at the atomic scale for a monatomic ideal gas. Experiments show that, at low pressure and high temperature, the ideal gas law describes the behavior of *any* gas at equilibrium, not just monatomic gases.

Because the thermal energy of a gas is the sum of the average kinetic energies of its atoms, $E_{th} = NK_{av}$ (Eq. 19.27), we can use the Eq. 19.50 expression for E_{th} and write

$$K_{av} = \tfrac{1}{2}mv_{rms}^2 = \tfrac{3}{2}k_BT. \quad (19.52)$$

Equation 19.52 allows us to obtain an expression for the rms speed of an atom in a monatomic ideal gas:

$$v_{rms} = \sqrt{\frac{3k_BT}{m}}. \quad (19.53)$$

The rms speed of a monatomic ideal gas atom is inversely proportional to the square root of the mass of the atom and proportional to the square root of the temperature of the gas. If the temperature of a gas goes up, the rms speed of the atoms goes up, too (Figure 19.27).

19.25 Use the ideal gas law to express the equilibrium condition in Eq. 19.20 in terms of pressure.

Figure 19.27 When the temperature of a monatomic ideal gas in a fixed volume is increased, the rms speed of the gas atoms increases.

T_1 $\qquad$ $T_2 > T_1$

Example 19.11 Helium speed

A sample of helium gas containing 6.02×10^{23} atoms at atmospheric pressure $(1.01 \times 10^5$ Pa$)$ occupies a volume of 22.4×10^{-3} m³. The mass of a helium atom is 6.646×10^{-27} kg. What is the rms speed of the helium atoms?

❶ **GETTING STARTED** I can treat the helium gas as a monatomic ideal gas. The rms speed of the atoms is determined by the temperature of the sample, which is not given.

❷ **DEVISE PLAN** Equation 19.53 gives the rms speed of atoms in a monatomic ideal gas. I am given the mass m of one helium atom, but not the temperature T. To determine T I can use the ideal gas law (Eq. 19.51).

❸ **EXECUTE PLAN** From Eq. 19.51, I obtain $T = PV/(Nk_B)$. Substituting this expression for the temperature into Eq. 19.53 yields

$$v_{rms} = \sqrt{\frac{3k_B PV/(Nk_B)}{m}} = \sqrt{\frac{3PV}{mN}} = 1.30 \times 10^3 \text{ m/s.} ✔$$

❹ **EVALUATE RESULT** Those helium atoms sure are zipping around fast! At 1.30×10^3 m/s, however, they are moving about ten times more slowly than the escape speed from Earth $(1.12 \times 10^4$ m/s; see Checkpoint 13.22), and so my result is at least not unreasonable. I also note that using my calculated rms speed in Eq. 19.47 yields a value of 1.01×10^5 Pa for P, giving me confidence in my answer.

19.26 A certain number of helium atoms are placed in one container, and an equal number of argon atoms are placed in a separate identical container. The pressure is the same in both containers. Helium atoms have a smaller mass than argon atoms. For which gas, if either, (a) is the thermal energy smaller, (b) is the average kinetic energy of an atom smaller, (c) is the rms speed of an atom smaller?

19.8 Entropy of a monatomic ideal gas

Now that we have derived an expression that gives the thermal energy of a gas in terms of its temperature (Eq. 19.50), we can express the entropy in terms of temperature. Substituting Eq. 19.50 into Eq. 19.32 gives

$$S = \tfrac{3}{2}N \ln\left(\tfrac{3}{2}Nk_{\mathrm{B}}T\right) - N \ln\frac{\delta\mathcal{V}}{b}$$

$$= \tfrac{3}{2}N \ln T + \tfrac{3}{2}N \ln\left(\tfrac{3}{2}Nk_{\mathrm{B}}\right) - N \ln\frac{\delta\mathcal{V}}{b} \tag{19.54}$$

or, taking all constant terms together,

$$S = \tfrac{3}{2}N \ln T + \text{constant (monatomic ideal gas in thermal equilibrium).}$$

$$\tag{19.55}$$

This expression describes a gas for which the energy is equipartitioned over the N atoms—that is, in thermal equilibrium. As before, the constant drops out when we consider changes in entropy (Eq. 19.8). For example, if the temperature of the gas is raised from T_i to T_f while the volume is kept constant, the change in entropy is

$$\Delta S = \tfrac{3}{2}N \ln\left(\frac{T_f}{T_i}\right) \quad \text{(constant volume).} \tag{19.56}$$

Example 19.12 Entropy change for a cooling gas

A sample of a monatomic ideal gas consisting of 1.00×10^{23} atoms in a volume of 4.00×10^{-3} m^3 is initially at a temperature of 290 K. The gas is cooled until it reaches a pressure of 5.00×10^4 N/m^2. (a) What is the change in entropy for the gas and for each atom? (b) What is the ratio of the final rms speed of each atom to its initial rms speed? (c) How does the change in the entropy of the gas depend on this ratio $v_{\mathrm{rms,f}}/v_{\mathrm{rms,i}}$?

❶ GETTING STARTED Both rms speed and entropy depend on temperature. As the gas cools, its temperature decreases, and so the rms speed of the atoms and the entropy of the gas both change.

❷ DEVISE PLAN I can use the ideal gas law, Eq. 19.51, to determine the final temperature of the gas. Once I know the final temperature, I can calculate the entropy change from Eq. 19.56 and the ratio of final to initial rms speeds from Eq. 19.53. To determine how the change in entropy depends on the ratio of rms speeds, I combine Eqs. 19.53 and Eq. 19.56.

❸ EXECUTE PLAN (a) From Eq. 19.51, the final temperature of the gas is

$$T_f = \frac{P_f V}{Nk_{\mathrm{B}}} = \frac{(5.00 \times 10^4 \ \text{N/m}^2)(4.00 \times 10^{-3} \ \text{m}^3)}{(1.00 \times 10^{23})(1.38 \times 10^{-23} \ \text{J/K})} = 145 \ \text{K.}$$

Equation 19.56 then yields for the change in the entropy of the gas:

$$\Delta S = S_f - S_i = \tfrac{3}{2}N \ln\left(\frac{T_f}{T_i}\right) = \tfrac{3}{2}(1.00 \times 10^{23}) \ln\left(\frac{145 \ \text{K}}{290 \ \text{K}}\right)$$

$$= -1.04 \times 10^{23}. \ ✔$$

Because there are 1.00×10^{23} atoms, the change in entropy per atom is -1.04. ✔

(b) From Eq. 19.53 I see that the ratio of final to initial rms speeds is $v_{\mathrm{rms,f}}/v_{\mathrm{rms,i}} = (T_f/T_i)^{1/2} = [(145 \ \text{K})/(290 \ \text{K})]^{1/2} = 0.707.$ ✔

(c) Because $a \ln N = \ln N^a$, I can write Eq. 19.56 as $\Delta S = N \ln(T_f/T_i)^{3/2}$. From part b I know that $v_{\mathrm{rms,f}}/v_{\mathrm{rms,i}} = (T_f/T_i)^{1/2}$, so I can write $T_f/T_i = v_{\mathrm{rms,f}}^2/v_{\mathrm{rms,i}}^2$. Combining these two expressions, I see that the relationship between the entropy change of the gas and the rms speed ratio of the atoms is $\Delta S = N \ln(v_{\mathrm{rms,f}}/v_{\mathrm{rms,i}})^3.$ ✔

❹ EVALUATE RESULT As the gas cools, the rms speed of the atoms decreases, as my answer to part b shows. As the rms speed decreases, the number of basic states possible decreases, and so the entropy should decrease, in agreement with my answer to part a. Because the system is not closed, there is no reason its entropy cannot decrease, as I have found here.

Equation 19.50, $E_{\mathrm{th}} = \tfrac{3}{2}Nk_{\mathrm{B}}T$, tells me that the term on the right in Eq. 19.52 is equal to E_{th}/N, and from this it follows that E_{th} is proportional to v_{rms}^2. Equation 19.32 shows that the entropy is proportional to $\tfrac{3}{2}\ln E_{\mathrm{th}} = \ln E_{\mathrm{th}}^{3/2}$ and so to $\ln(v_{\mathrm{rms}}^2)^{3/2} = \ln v_{\mathrm{rms}}^3$, in agreement with my result in part c.

19.27 (a) Does the negative change in entropy in Example 19.12 violate the Eq. 19.5 part of the entropy law? (b) What is the change in the thermal energy of the gas in Example 19.12?

We can combine Eq. 19.55 with Eq. 19.7 to obtain an expression that tells us how entropy depends on both volume and temperature. To do so, remember that if we consider both the basic states associated with the spatial distribution of the atoms in a gas and those associated with the velocity distribution, the number of basic states associated with both types of distributions is the product of the number of each type of basic state (Eq. 19.9):

$$\Omega = \Omega_{space}\,\Omega_{vel}. \tag{19.57}$$

Taking the natural logarithm of this equation, we have, from $S \equiv \ln \Omega$,

$$S = S_{space} + S_{vel}. \tag{19.58}$$

To obtain an expression for the entropy of a gas that provides the dependence on both the volume and the temperature of the gas, we must therefore substitute Eqs. 19.7 and 19.55 into Eq. 19.58:

$$S = N \ln V + \tfrac{3}{2} N \ln T + constant \quad (equilibrium), \tag{19.59}$$

where the constant term $-N \ln \delta V$ in Eqs. 19.7 has been combined with constant terms in Eq. 19.55 to give a new constant term. Equation 19.59 can also be written as

$$S = N \ln(T^{3/2}V) + constant \quad (equilibrium). \tag{19.60}$$

Like Eqs. 19.7 and 19.55, Eq. 19.60 describes a system at equilibrium: If different compartments of the system are not at equilibrium with each other, the entropy of the system is smaller than that given by Eq. 19.60.

Equation 19.60 is in agreement with what we saw in the conceptual part of this chapter: As we increase either the temperature or the volume, the number of basic states—and therefore the entropy—increases. If the system changes from an equilibrium state with temperature T_i and volume V_i to a new equilibrium state with temperature T_f and volume V_f, the change in entropy is

$$\Delta S = \tfrac{3}{2} N \ln\!\left(\frac{T_f}{T_i}\right) + N \ln\!\left(\frac{V_f}{V_i}\right). \tag{19.61}$$

Most important, the change in entropy is independent of *how* the system is changed: ΔS depends on only the initial and final equilibrium states of the system. Note that for a system that is not closed, there are no constraints on ΔS: It can be positive, negative, or zero.

19.28 Does the entropy of a gas whose temperature is halved and volume doubled increase, decrease, or stay the same?

Chapter Glossary

SI units of physical quantities are given in parentheses.

Absolute temperature T (K) A quantity related to the rate of change of entropy with respect to thermal energy:

$$\frac{1}{k_B T} \equiv \frac{dS}{dE_{th}}, \qquad (19.38)$$

where k_B is the Boltzmann constant.

Basic state The state of a system that is described using a complete specification of all the constituent particles.

Boltzmann constant k_B (J/K) A constant of proportionality that reconciles the units for absolute temperature with common temperature measurements:

$$k_B = 1.380 \times 10^{-23} \text{ J/K}. \qquad (19.39)$$

Entropy S (unitless) A quantity equal to the natural logarithm of the number of basic states Ω for a system:

$$S \equiv \ln \Omega. \qquad (19.4)$$

For a system of N particles contained in a volume V at a temperature T at equilibrium, the entropy is given by

$$S = N \ln(T^{3/2}V) + \text{constant} \quad \text{(equilibrium). (19.60)}$$

Entropy law (also called the *second law of thermodynamics*) A closed system always evolves so as to maximize the number of basic states. As the number of basic states increases, so does the entropy:

$$\Delta S > 0 \quad \text{(closed system evolving toward equilibrium).} \qquad (19.5)$$

At equilibrium, the number of basic states reaches its maximum value Ω_{max} and the entropy of the system no longer changes:

$$\Delta S = 0 \quad \text{(closed system at equilibrium).} \qquad (19.6)$$

Equilibrium state The macrostate of a system that has the greatest number of basic states and therefore is the most probable macrostate.

Equipartition of energy The equal distribution of energy among all constituent parts of a system.

Equipartition of space The equal distribution of particles over all available volume.

Ideal gas A gas that consists of a very large number of particles that move incoherently, occupy a negligible fraction of the volume occupied by the gas, and interact with one another and the walls of the container only via elastic collisions.

Ideal gas law A relationship among the pressure, volume, temperature, and number of atoms in an ideal gas system at equilibrium:

$$P = \frac{N}{V} k_B T \quad \text{(ideal gas system at equilibrium).} \quad (19.51)$$

Kelvin (K) The SI base unit of absolute temperature.

Macrostate The state of a system that is described using only quantities that can be determined by measurements made on scales much larger than molecular sizes.

Probability The fraction of times that an event occurs in a large number of repetitions.

Root-mean-square speed v_{rms} (m/s) The square root of the average of the squares of the speeds of all the atoms in a volume of gas. For an ideal gas made up of atoms of mass m at temperature T, the rms speed is

$$v_{rms} = \sqrt{\frac{3k_B T}{m}}. \qquad (19.53)$$

Second law of thermodynamics See *Entropy law*.

Thermal equilibrium The condition in which all parts of a system are at the same temperature. In this condition thermal energy is equipartitioned over the system and the entropy of the system is at its maximum.

20

Energy Transferred Thermally

CONCEPTS

QUANTITATIVE TOOLS

We use terms like *hot* and *cold, heating* and *temperature* based on our everyday interpretations of these words. We know from experience that to raise the temperature of water to the boiling point for a cup of tea, we need to heat the water, which requires energy. The branch of physics that deals with these concepts, called *thermodynamics* (from the Greek *therme,* meaning "heat," and *dunamis,* meaning "power"), was developed in the early 19th century to describe the working of steam engines. The concepts of entropy and absolute temperature introduced in Chapter 19 were developed as a consequence of that work. In this chapter we connect the statistical ideas from Chapter 19 to the macroscopic concepts of temperature, pressure, and volume.

20.1 Thermal interactions

The concept of temperature is familiar: It's a measure of "hotness." According to what we saw in Chapter 19, if two objects that are not at the same temperature are placed in contact with each other, energy is transferred from the object at the higher temperature to the object at the lower temperature until the temperatures of the two are equal. We attribute this transfer of energy to a **thermal interaction** between the two objects at different temperatures. As we saw in Chapter 19, this thermal interaction is the overall effect of countless interactions at the atomic scale.

Let us begin by considering the familiar situation shown in **Figure 20.1**: heating a pot of water on a stove. As we heat the water, energy flows from the flame to the pot and then from the pot to the water. The flame, which is hotter (that is, at a higher temperature) than the pot, heats the pot, and the pot, which is hotter than the water, heats the water. If we consider the water as our system and make an energy

diagram for that system, we obtain the result shown in **Figure 20.2**. As the water gets hotter, its thermal energy increases. The burning flame provides that energy, but it is not part of the system, and so there is no change in our system's source energy. The system's kinetic energy and potential energy do not change either, and so the system's energy increases because of the increase in thermal energy.

How do we account for this increase in the energy of the system? Although there are external forces exerted on the water—the force of gravity and a normal force exerted by the pot on the water—neither force causes any displacement, and so does no work on the system.

Consequently the energy diagram of Figure 20.2 is unbalanced: The energy changes are not equal to the work done on the system. The reason for this unbalance is that this energy diagram does not account for the energy transferred into the system by the heating. To make the diagram complete, we need to add a bar that represents the energy transferred into the system by the thermal interactions among the flame, pot, and water. The amount of energy transferred by thermal interactions ("thermally") is denoted by the letter Q (**Figure 20.3**). The law of conservation of energy requires that the change in the energy of the system be equal to the energy transferred into it, so the height of the bar representing Q in Figure 20.3 has to be equal to the height of the bar representing the increase in thermal energy.

20.1 (*a*) What are the SI units of Q? (*b*) For the process depicted in Figure 20.2*a*, make an energy diagram for each of these systems: (*i*) water, pot, and flame; (*ii*) pot and flame; (*iii*) pot.

Figure 20.1 When a gas burner heats a pot of water, energy flows from higher temperature to lower temperature.

(*a*) Flame heats pot of water

(*b*) Flow of energy from higher to lower temperature

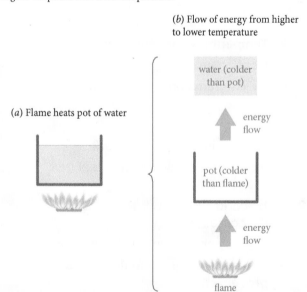

Figure 20.2 When a pot of water is heated, the energy diagram for a system comprising only the water is unbalanced.

(*a*) Initial and final states

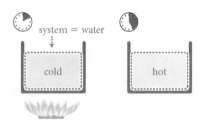

system = water

cold   hot

(*b*) Energy diagram for system

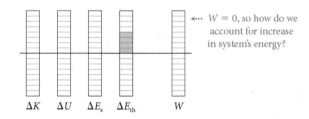

ΔK ΔU ΔE_s ΔE_{th} W

$W = 0$, so how do we account for increase in system's energy?

Figure 20.3 To complete the energy diagram of Figure 20.2, we must add a bar to account for the energy transferred thermally into the system (represented by the symbol Q).

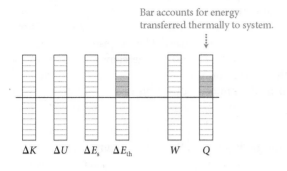

Bar accounts for energy transferred thermally to system.

ΔK ΔU ΔE_s ΔE_{th} W Q

The quantity Q is frequently called *heat*, a term we shall avoid because it is easy to confuse the verb *to heat* with the noun and to think of "heat" as a property of an object instead of a quantity of energy transferred across a boundary. When it is necessary to refer to Q by name, we shall use the term **energy transferred thermally,** reserving the verb *to heat* to refer to the process of increasing the thermal energy of a system. (For example, *I am heating a pot of water* means that I am increasing its thermal energy.)

Now consider the situation in **Figure 20.4a**, where two gases are separated by a fixed partition that is not thermally insulating. Initially gas A in the left compartment is at a lower temperature than gas B in the right compartment. Because the partition is not thermally insulating, the gases interact thermally, and energy is transferred thermally from the hotter gas to the colder one until their temperatures are the same. The energy transferred thermally is represented by the arrow labeled Q. The energy diagram for gas A in Figure 20.4b consequently shows an increase in thermal energy and the energy transferred thermally, Q, is positive.

 20.2 Make an energy diagram for gas B in Figure 20.4.

As Checkpoint 20.2 shows, the sign of Q depends on the direction of the energy transfer. Just as for work, if the transfer is into the system (that is, the energy of the system increases), Q is positive (Figure 20.4b); if the transfer is out of the system, Q is negative (Checkpoint 20.2).

Let us examine in more detail the difference between transferring energy by doing work on a system (that is, through mechanical interactions) and by transferring energy thermally to it. In **Figure 20.5a** a spring-loaded piston compresses an ideal gas into a smaller volume, doing work on the gas. We shall ignore any friction between the piston and the container walls. The ideal gas is *thermally insulated;* that is, it is not interacting thermally with its surroundings. A process that does not involve any thermal transfer of energy is called an **adiabatic process** ($Q = 0$).

Because the process depicted in Figure 20.5a is adiabatic, the amount of energy transferred to the gas is equal to the work done on the gas by the piston and is determined by the force $\vec{F}^c_{pg}$ exerted by the piston on the gas and the force displacement $\Delta \vec{x}_F$.

How is the energy of the gas raised? There are no changes in the kinetic energy of the center of mass of the

Figure 20.4 When gases of different temperatures are placed in thermal contact, energy is transferred thermally from the hotter to the cooler gas until the temperatures of the two gases are the same.

(a) Gases at different temperatures are placed in thermal contact

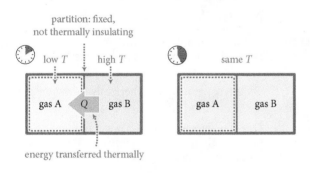

partition: fixed, not thermally insulating

low T high T same T

gas A Q gas B   gas A   gas B

energy transferred thermally

(b) Energy diagram for gas A

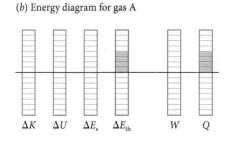

ΔK ΔU ΔE_s ΔE_{th} W Q

Figure 20.5 A piston does work on a thermally insulated ideal gas, increasing the gas's temperature.

(a) Spring-loaded piston compresses thermally insulated ideal gas

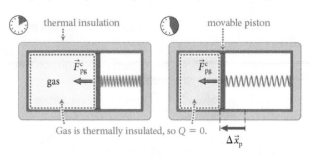

thermal insulation   movable piston

$\vec{F}^c_{pg}$ $\vec{F}^c_{pg}$

gas

Gas is thermally insulated, so $Q = 0$. $\Delta \vec{x}_p$

(b) Energy diagram for gas

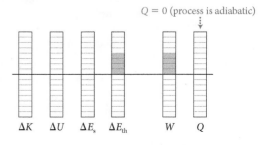

$Q = 0$ (process is adiabatic)

ΔK ΔU ΔE_s ΔE_{th} W Q

Figure 20.6 A stationary piston does not change the speed of a gas particle that collides with it, but a moving piston does.

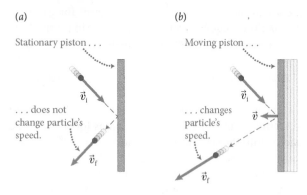

(a)

Stationary piston . . .

$\vec{v}_i$

. . . does not change particle's speed.

$\vec{v}_f$

(b)

Moving piston . . .

$\vec{v}_i$

. . . changes particle's speed.

$\vec{v}$

$\vec{v}_f$

gas because the gas as a whole remains at rest. There are also no changes in the potential energy of the gas because there is no potential energy associated with the interactions between particles in an ideal gas. Nor are there any changes in the source energy. Therefore the work W done on the gas must raise the thermal energy of the gas (Figure 20.5b). We can understand how this happens by considering the (elastic) collisions of the gas particles with the piston. If a particle collides with the stationary piston (**Figure 20.6a**), the particle's speed is unaltered by the collision. When the piston is moving toward the incoming particle, however, the particle gains speed because of the collision and therefore its kinetic energy increases, just as a tennis ball bounces

faster off your racquet when you swing it toward the ball. So, as the piston moves in, the particles colliding with it speed up (Figure 20.6b), increasing their incoherent kinetic energy and thus the thermal energy of the gas.

Note that in Figure 20.5a the volume decreases. During this decrease, the force exerted by the piston on the gas and the force displacement are in the same direction. Consequently the work done on the gas is positive. Had the piston moved in the opposite direction (expanding the volume), the work done on the gas would have been negative.

> When the volume of a gas decreases, the work done on the gas is positive. When the volume increases, the work done on the gas is negative. When there is no change in volume, the work done on the gas is zero.

Summarizing the information from Figures 20.4 and 20.5, we see that there are two ways to transfer energy to a system: mechanically (by doing work on the system) and thermally (see the box "Transferring energy to a system"). Both types of transfer can occur simultaneously. In that case the change in the energy of the system ΔE is equal to the sum of the work done on the system, W, and the energy transferred thermally to it, Q.

Let us now turn our attention to the transformation of a system from one macrostate to another. Such a transformation is a called a **process,** and if the process is carried out in such a way that the system remains very close to equilibrium at all instants during the process, the process is said to be **quasistatic.**

Transferring energy to a system

	DOING WORK	TRANSFERRING ENERGY THERMALLY
Technical term:	Mechanical interaction between system and its environment	Thermal interaction between system and its environment
Mediated by:	Macroscopic pushes and pulls	Atomic collisions
Requires:	A nonzero external force $\vec{F}_{ext}$ exerted on the system and a nonzero force displacement $\Delta \vec{x}_F$	A nonzero temperature difference ΔT between the system and its environment.
Amount of energy transferred *into* system:	W ("work done on system")	Q ("energy transferred thermally to system")
Sign:	$W > 0$ when external force and force displacement are in the same direction (energy of system increases in absence of other transfers)	$Q > 0$ when environment is at a higher temperature than system (energy of system increases in absence of other transfers)
	$W < 0$ when external force and force displacement are in opposite directions (energy of system decreases in absence of other transfers)	$Q < 0$ when environment is at a lower temperature than system (energy of system decreases in absence of other transfers)
Equilibrium:	The vector sum of the external forces is zero, and the force displacement of each external force is zero (mechanical equilibrium)	System is at same temperature as its environment (thermal equilibrium)

Figure 20.7 Quasistatic versus non-quasistatic expansion of a cylinder containing a gas.

(*a*) Quasistatic expansion: piston rises gradually

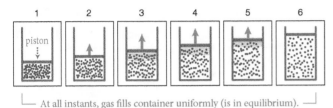

└── At all instants, gas fills container uniformly (is in equilibrium). ──┘

(*b*) Non-quasistatic expansion: piston rises suddenly

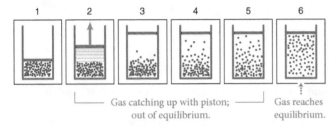

└── Gas catching up with piston; ──┘   Gas reaches
        out of equilibrium.   equilibrium.

Figure 20.7a shows the expansion of a volume of gas in a cylinder. To make the expansion quasistatic, the piston is moved slowly so that the gas always uniformly fills all the available space, and each particle has its equipartition share of the thermal energy. In each frame shown in the drawing, the gas is in a different state, but each state is an equilibrium state. In a quasistatic process, therefore, the pressure may change during the process, but at any given instant the pressure has a uniform, well-defined value at all locations in the volume of gas. The same is true of the temperature: The final temperature may differ from the initial temperature, but at any given instant the temperature has a uniform, well-defined value at all locations in the volume of gas. There are no locations where the pressure is greater and no spots that are hotter than others. In Figure 20.7a, the entropy changes as the gas expands (because the entropy of an ideal gas changes with volume and temperature, Eq. 19.60), but at each instant it has the maximum value for that equilibrium state.

Suppose, however, that the piston moves at a rapid speed, as illustrated in Figure 20.7b. The motion of the piston is so fast that the gas can't keep up with it. In frame 2, for example, the gas still mostly occupies the original volume. Consequently, the pressure at the bottom of the cylinder is still nearly the same as it was in frame 1, while it is nearly zero at the top, where we still have a vacuum. Although the piston reaches its final position in frame 3, it takes the gas another three frames before it is evenly distributed and in equilibrium. In frames 2–5 of Figure 20.7b, the gas is not in equilibrium, and so the entropy is not maximum. Because

particles in a gas typically move at very great speeds, they redistribute themselves rapidly. Thus, in practice, even if a piston moves rapidly as in Figure 20.5, the resulting process can still be quasistatic.

✋ **20.3** Suppose you were to play the two film clips shown in Figure 20.7 backward. Would the resulting processes be possible?

Let us now examine how the entropy changes when energy is transferred to a system in a quasistatic process. First consider the process shown in Figure 20.4, where energy is transferred thermally between two gases. The energy transferred thermally to gas A is positive ($Q > 0$). The temperature of gas A increases. Therefore by Eq. 19.56 the entropy of the gas increases. Conversely, Q is negative for gas B, the temperature of B decreases, and so its entropy decreases. Consequently, the sign of the entropy change of a system is the same as the sign of Q for that system:

Transferring energy thermally to a system increases its entropy. Transferring energy thermally from a system decreases its entropy.

Next, let us determine what happens to the entropy of a system when work is done on it in the absence of any thermal interaction ($Q = 0$). Consider, for example, the quasistatic compression of the thermally insulated gas in Figure 20.5 again. As the piston does work on the gas, the thermal energy of the gas goes up, but at the same time the volume goes down. In Chapter 19 we saw that an increase in the thermal energy of a system raises its entropy. A decrease in the system volume, however, lowers the entropy, so we can't say offhand what happens to the entropy of the gas as it is compressed.

Let us therefore consider the closed system consisting of the contents of the box in Figure 20.5a. In contrast to the equilibration of temperatures in Figure 20.4, which is an irreversible process, the compression of the gas and the expansion of the spring are reversible. Because the process is quasistatic, the system remains in equilibrium at all instants during the compression, and therefore the entropy law tells us that the entropy of the (closed) system doesn't change: $\Delta S = 0$ (Eq. 19.6).

Because we are interested in the process (which takes place in the gas), we divide the system into two parts: the gas and the spring-loaded piston ("piston" for short). Because entropy is an extensive quantity, we can write the change in the system's entropy as $\Delta S = \Delta S_{gas} + \Delta S_{piston} = 0$.

In the spring-loaded piston, potential energy is converted to kinetic energy, both of which are coherent forms of energy (see Figure 7.10). Regardless of the speed of the piston and the configuration of the spring, there is only one way to distribute the energy and atoms such that all the energy is coherent. At any instant, therefore, the number of basic states of the spring-loaded piston is 1, and so the entropy of the spring-loaded piston does not change: $\Delta S_{piston} = 0$.

If $\Delta S_{\text{piston}} = 0$ and $\Delta S = 0$, then the entropy of the gas cannot change either: $\Delta S_{\text{gas}} = 0$. So we can conclude that in a quasistatic process during which we do work on a gas without thermally transferring any energy to it ($Q = 0$), the entropy of the gas does not change. Generalizing this idea to any type of system undergoing a quasistatic adiabatic process, we can state:

The entropy of a system undergoing a quasistatic adiabatic process does not change.

In a quasistatic adiabatic process, the temperature and volume changes are coordinated in such a way as to keep the entropy constant. This conclusion holds for only *quasistatic* adiabatic processes. For instance, in Example 19.9, a gas adiabatically expands into a larger volume, increasing the entropy of the gas. The expansion is adiabatic, but not quasistatic. Therefore the entropy of a system undergoing a non-quasistatic adiabatic process can change.

20.4 Which of the following systems undergo a quasistatic process? (*a*) A gas-containing balloon is popped inside a vacuum container. System: container, gas, and balloon. (*b*) A cup of coffee cools down. System: coffee. (*c*) A sample of compressed gas in a cylinder fitted with a heavy piston is allowed to expand, pushing the piston outward. System: gas. (*d*) An ice cube melts in a glass of hot water. System: water and ice cube.

20.2 Temperature measurement

In Chapter 19 we defined the absolute temperature as a rather abstract concept related to the rate at which entropy changes with changes in thermal energy. Entropy is a measure of the number of available basic states in a system, and thermal energy is the energy associated with the incoherent internal motion of the system. Neither of these quantities can be measured directly. We must therefore figure out some convenient way to measure the temperature of an object.

Most of the methods that have been developed to measure temperature rely on measuring some physical quantity that varies with temperature. For example, in a glass thermometer, illustrated in Figure 20.8, the thermal expansion of a liquid provides a means of measuring temperature. A glass tube is connected to a bulb filled with glycerine, mercury or some other liquid whose mass density decreases with rising temperature. Because the mass of the liquid is constant, its volume expands and therefore the temperature can be determined by measuring the height of the liquid in the glass tube. To calibrate this type of thermometer, it is placed in melting ice and allowed to come to thermal equilibrium. Once equilibrium is reached, a mark is made on the glass tube at the height of the liquid in the tube. The thermometer is then placed in boiling water and allowed to come to thermal equilibrium. The tube is marked at the new height of the liquid in the tube. The two marks on the tube are assigned some standard values representing the freezing

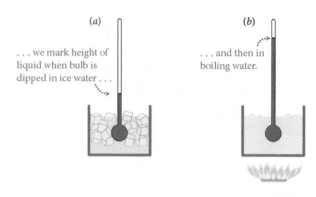

Figure 20.8 Calibrating a gas thermometer by immersing its bulb in ice water and then in boiling water.

To calibrate thermometer . . .

(*a*)

. . . we mark height of liquid when bulb is dipped in ice water . . .

(*b*)

. . . and then in boiling water.

and boiling points of water—historically 0 °C and 100 °C, respectively.

Now, if the expansion of the liquid is proportional to temperature, the temperature of any substance can be measured by immersing the thermometer in the substance and noting how far up the glass tube the enclosed liquid rises. If the thermometer is immersed in a glass of milk and the liquid in the tube rises, say, 27/100 of the distance between the 0 °C mark and the 100 °C mark, the temperature of the milk is 27 °C. Of course, in commercial thermometers there is no need to measure heights in fractions because the glass tube is marked all along its length and the temperature is read directly.

Is the expansion of a liquid proportional to the absolute temperature we defined in Chapter 19? No—as a matter of fact, the expansion of one liquid isn't even proportional to that of another liquid. For this reason, glass thermometers are accurate over only a relatively narrow range of temperatures.

How, then can we measure temperature accurately? In Section 19.7 we derived the ideal gas law, $P = Nk_{\text{B}}T/V$, which relates the absolute temperature to two macroscopic quantities that are easily measured: pressure and volume. This law is a direct consequence of our definition of absolute temperature and is valid for an ideal gas, that is, it is valid as long as the pressure, volume, and absolute temperature are such that we can ignore the internal structure of the gas particles and the interactions between particles (except for collisions between them). No gas is ideal, but many gases can approximate ideal behavior. For a dilute gas of helium atoms, for instance, the conditions for ideal behavior are satisfied down to temperatures that are quite low.

Figure 20.9 illustrates the basic principle of an **ideal gas thermometer**, which measures the pressure in a gas sample kept enclosed in a fixed volume. A bulb filled with a gas at low pressure is connected to a thin, flexible U-shaped tube that contains mercury. The end of the tube opposite

Figure 20.9 An ideal gas thermometer.

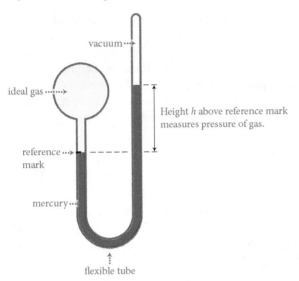

we raise the right arm of the flexible U-tube until the mercury level in the left arm is back at the reference mark (Figure 20.10b).

Provided the mercury level in the left arm is always kept at the reference mark, the volume V of the gas is fixed and the height h of the mercury in the right arm measures the pressure in the gas at a constant volume. Under these conditions, the height h is a direct measure of the pressure and therefore the absolute temperature T. The ideal gas law tells us that at $T = 0$ the pressure is zero, which means there is a vacuum above both arms of the U-shaped tube. So we know that at $T = 0$ the height h should be zero.

To calibrate the thermometer and define the unit of absolute temperature, we need to find some reproducible thermal phenomenon and assign it a fixed value. The current standard of absolute temperature is based on the **triple point of water**—the unique conditions under which liquid water, solid ice, and water vapor (a gas) coexist in thermal equilibrium. To relate absolute temperature to historical definitions of temperature, the absolute temperature at the triple point of water is set at 273.16 K:*

$$T_{tp} \equiv 273.16 \text{ K.}$$

the gas-filled bulb is capped and evacuated. The difference in height between the levels of mercury in the two arms of the tube determines the pressure in the gas in the bulb (see Section 18.6). For a fixed value of V, the ideal gas law, $P = Nk_BT/V$, tells us that the pressure P is proportional to the absolute temperature T.

Suppose now that the absolute temperature of the gas in the bulb is raised. The pressure in the gas increases, pushing the mercury in the left arm down below the reference mark and increasing the volume V of the gas, as illustrated in **Figure 20.10a**. To restore the gas to its original volume,

Once the height of the mercury column in an ideal gas thermometer has been measured at the triple point of water, we can draw a straight line between the origin at $T = 0$, where we know h should be zero, and the triple point. We can then determine any other absolute temperature simply by reading off the absolute temperature corresponding to the measured height (**Figure 20.11**). The important point is that the curve is linear: As long as the gas behaves like an ideal gas and is kept at constant volume, the pressure (and therefore the height of the mercury column) is directly proportional to the absolute temperature.

Figure 20.10 Using an ideal gas thermometer to measure temperature.

(a) (b)

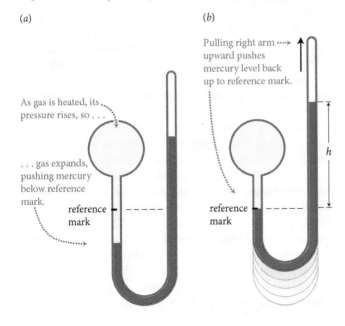

Figure 20.11 The height-versus-temperature graph is linear for an ideal gas thermometer. The thermometer is calibrated at the triple point of water. Any temperature can then be read from the graph.

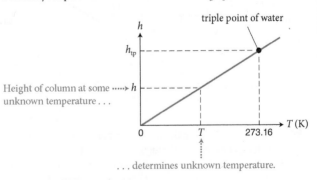

---

*The triple point of water occurs at a temperature of 0.01 °C (roughly 32 °F) and a pressure of 610 Pa (about one-hundredth of atmospheric pressure).

Temperature measurement scales

Methods for measuring and defining temperature were developed long before the development of the gas thermometer and the concept of absolute temperature. Starting in the early 18th century, several scales were defined for measuring temperature, two of which are still in use. The most widely used scale is the *Celsius scale*. Temperatures on this scale are measured in units that are called *degrees Celsius* (°C) and that have the same magnitude as the kelvin (a temperature change of 1 °C is the same as a temperature change of 1 K). The zero of the Celsius scale is the freezing point of water, and the boiling point of water (at a pressure of 101.325 kPa) is set at 100 °C.

We can convert between the Celsius and Kelvin scales as follows. If T_C represents the temperature expressed in degrees Celsius and T the absolute temperature expressed in kelvins, we have

$$T_C = \frac{1\,°C}{1\,K}T - 273.15\,°C. \tag{1}$$

The older *Fahrenheit scale* is used in the United States. Temperatures on this scale are measured in units called *degrees Fahrenheit* (°F). On this temperature scale, the freezing point of water is 32 °F and the boiling point is 212 °F, placing the boiling and freezing points of water exactly 180 °F apart. If T_F is the temperature expressed in degrees Fahrenheit, the relationship between the Celsius and the Fahrenheit scales is

$$T_C = \frac{5\,°C}{9\,°F}(T_F - 32\,°F). \tag{2}$$

Notice that no degree symbol, °, is used with kelvins: 13.0 °C, 55.4 °F, but 286 K. Readings in degrees Celsius or Fahrenheit can be negative, but readings in kelvins are always positive.

Example 20.1 Measuring absolute temperature

When the bulb of an ideal gas thermometer is immersed in water at the triple point, the height of the mercury column above the reference mark is 760 mm. The same thermometer is then immersed in a sample of unknown temperature, and the height of the mercury column moves to 1167 mm above the reference mark. What is the absolute temperature of the sample?

❶ GETTING STARTED The height of the mercury column is linearly proportional to the absolute temperature. I therefore begin by making a graph like Figure 20.11, summarizing the information given in the problem (**Figure 20.12**). I denote the unknown absolute temperature by T and the height of the mercury at this unknown temperature by $h = 1167$ mm.

Figure 20.12

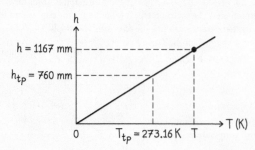

❷ DEVISE PLAN Because the line giving height as a function of absolute temperature goes through the origin, its slope is equal to the ratio h_{tp}/T_{tp}. From my graph I see that the ratio h/T must have the same value: $h/T = h_{tp}/T_{tp}$. From this expression I can calculate the unknown temperature.

❸ EXECUTE PLAN Solving for T, I get $T = T_{tp}h/h_{tp} = (273.16\ \text{K})(1167\ \text{mm})/(760\ \text{mm}) = 419\ \text{K}.$ ✔

❹ EVALUATE RESULT The absolute temperature I obtain is about 50% higher than T_{tp}, which is what I would expect, given that the mercury height is about 50% higher than that at the triple point.

In principle, the operation of the ideal gas thermometer as described above is independent of the gas used—we can use *any* gas we want. In practice, deviations from the ideal gas law complicate matters a bit, and the science and technology of gas thermometers are more complex than presented here. Nonetheless, gas thermometers serve as the standard reference for all other types of thermometers. Commercial thermometers typically give the temperature on a different scale (see the box "Temperature measurement scales" and Table 20.1). In thermodynamics, and in physics in general, it makes the most sense to use absolute temperatures. For simplicity we shall henceforth refer to the absolute temperature T expressed in kelvins simply as *temperature*.

In drawing Figure 20.11 we assumed that we can extrapolate the linear relationship between pressure and temperature from the ideal gas law down to zero. In practice, it is impossible for any gas to reach this temperature because as the temperature decreases, the gas condenses into a liquid. Of all gases, helium becomes liquid at the lowest temperature (4.22 K, which corresponds to −269 °C or −452 °F), and so its useful range extends below that of any other gas. Zero temperature ($T = 0$) is called **absolute zero.** Any system at absolute zero has zero entropy (that is, it has only one available basic state), and its thermal energy is a minimum. Absolute zero therefore is the limit of low temperature.

Table 20.1 Comparison of temperature scales (defining points in bold)

Temperature	Kelvin (K)	Celsius (°C)	Fahrenheit (°F)
Absolute zero	0	−273.15	−459.67
Water freezes (at a pressure of 101.325 kPa)	273.15	**0**	**32**
Triple point of water	**273.16**	0.01	32
Room temperature	293	20	68
Average human body temperature	310	37.0	98.6
Water boils (at a pressure of 101.325 kPa)	373.12	**100**	**212**
Titanium melts	1941	1668	3034
Surface of the Sun	5778	5505	9941

Exercise 20.2 Temperature conversions

Derive an expression for converting a temperature (a) from the Celsius scale to the Fahrenheit scale, (b) from the Kelvin scale to the Fahrenheit scale, and (c) from the Fahrenheit scale to the Kelvin scale.

SOLUTION (a) To obtain an expression for the temperature T_F measured in degrees Fahrenheit in terms of the temperature T_C measured in degrees Celsius, I use Eq. 2 from the "Temperature measurement scales" box, solving this expression for T_F:

$$T_F = \frac{9\,°F}{5\,°C}\,T_C + 32\,°F.\ ✔$$

(b) Substituting the expression for converting degrees Celsius to kelvins yields

$$T_F = \frac{9\,°F}{5\,°C}\left(\frac{1\,°C}{1\,K}\,T - 273.15\,°C\right) + 32\,°F.\ ✔$$

(c) I can use the expression for T_F I obtained in part b. Solving for T yields:

$$T = \left[\frac{1\,K}{1\,°C}\frac{5\,°C}{9\,°F}(T_F - 32\,°F)\right] + 273.15\,K.\ ✔$$

🖐 **20.5** (a) During a certain day the outdoor temperature ranges between 60 °F and 80 °F. What is the corresponding range in degrees Celsius and in kelvins? (b) The filament temperature of an incandescent light bulb is about 2800 K. What is the corresponding temperature in degrees Celsius and in degrees Fahrenheit? (Assume only two significant digits in 2800 K.)

20.3 Heat capacity

The amount of energy it takes to raise the temperature of an object depends on the amount and kind of material in the object. Indeed, we know from experience that it takes longer to heat a large pot full of water than to heat the same pot containing only a cup of water. The larger quantity of water requires more energy, and therefore more gas needs to be burned.

We can characterize how much energy it takes to raise the temperature of an object by a scalar quantity called the object's **heat capacity.** The heat capacity of an object is equal to the amount of energy transferred thermally to the object divided by the resulting change in temperature.* The SI unit of heat capacity is J/K.

🖐 **20.6** Which is greater in each pair: (a) the thermal energy of a pool of water or that of a glass of water at the same temperature, (b) the heat capacity of a pool of water or that of a glass of water at the same temperature, (c) the thermal energy of a cold glass of water or that of an equal amount of warm water?

Because heat capacity is an extensive quantity (it depends on the amount of material in an object), it makes sense to introduce a related intensive quantity so that we can compare the thermal properties of different materials. Traditionally, heat capacity is turned into an intensive quantity by dividing it by the mass of the object. The resulting quantity, called **specific heat capacity** and denoted by the lowercase letter c, is a measure of how much energy is required to raise the temperature of 1 kg of a certain material by 1 K. The adjective *specific* means "per unit mass." The SI unit of specific heat capacity is J/K · kg.

Table 20.2 gives the specific heat capacities of a variety of materials. As you can see, different materials have very different specific heat capacities. For example, it takes about ten times less energy to raise the temperature of a 1-kg block of copper by 1 K (385 J) than it takes to raise the temperature of an equal mass of water by 1 K (4181 J).

Why are these values so different from one another, and what determines the specific heat capacity of a material? Remember from Chapter 19 that when an object is in thermal equilibrium, its thermal energy is equally distributed among all the parts of the object that can store energy. Clearly, different materials have different abilities to store thermal energy. To understand these differences, let us first examine the heat capacity of an ideal gas.

*Strictly speaking this quantity is the average heat capacity, but because the heat capacity does not depend much on temperature, we can omit the word "average".

Table 20.2 Specific heat capacities

Material	c (J/K $\cdot$ kg)
alcohol	2436
aluminum	897
brass	385
carbon graphite	685
copper	385
glass	837
gold	130
helium	3116
ice (268 K)	2090
iron	449
lead	130
marble	860
mercury	140
nickel	443
platinum	133
silver	235
steam (373 K)	1556
titanium	527
water	4181
wood	1700
zinc	388

All values are reported at a temperature of 298 K (unless otherwise noted) and atmospheric pressure and with the gases held at constant volume.

Table 20.3 Heat capacities per particle in units of k_B

Class	Material	C_V/k_B
Monatomic gases	helium	1.50
	argon	1.50
	krypton	1.50
	xenon	1.51
	radon	1.51
Polyatomic gases	ammonia	3.37
	hydrogen	2.47
	nitrogen	2.50
Liquids	mercury	3.38
	water	8.97
Solids	aluminum	2.91
	copper	2.94
	gold	3.08
	iron	3.02
	silver	3.05
	titanium	3.04
	zinc	3.05

All values are reported at a temperature of 298 K and atmospheric pressure and with gases held at constant volume (as indicated by the subscript V on C_V).

In Chapter 19 we found that the thermal energy of a monatomic ideal gas is given by $E_{th} = \frac{3}{2} N k_B T$ (Eq. 19.50). Each of the N ideal gas atoms thus has an average amount of thermal energy equal to $\frac{3}{2} k_B T$, where $k_B = 1.381 \times 10^{-23}$ J/K is the Boltzmann constant. We arrived at this result by considering the kinetic energy associated with the motion of an ideal gas atom along the x, y, and z directions (see Section 19.6). The motion along each axis is called a **degree of freedom.** Because there are three independent axes for translation, there are three "translational degrees of freedom." In thermal equilibrium, the equipartition of energy provides each translational degree of freedom of each atom with an equal share of the thermal energy—the **equipartition energy share** $\frac{1}{2} k_B T$—for a total of $\frac{3}{2} k_B T$ per atom.

If we raise the temperature of an ideal gas by 1 K, the thermal energy of the gas goes up by $\frac{3}{2} N k_B (1 \text{ K})$, and that of each atom goes up by $\frac{3}{2} k_B (1 \text{ K})$. If this energy is transferred thermally to the gas, we expect the heat capacity per atom to be equal to $\frac{3}{2} k_B$.

Table 20.3 lists the heat capacities per particle, denoted by the uppercase letter C, in units of k_B for a number of substances. As you can see, it is indeed $\frac{3}{2}$ for monatomic gases, such as helium and argon, but deviates from this value for polyatomic gases. For hydrogen, for example, the value is closer to $\frac{5}{2}$ than to $\frac{3}{2}$, suggesting that a hydrogen molecule has two additional degrees of freedom that can store energy.

The reason for this discrepancy is that we treated each ideal gas particle as an atom that has only kinetic energy associated with translational motion along three independent axes.

A dumbbell-shaped hydrogen molecule (**Figure 20.13a**), however, also has energy associated with rotational motion. The molecule can rotate about three independent axes (Figure 20.13b). Rotation about the z axis does not store any appreciable energy, because the rotational inertia about that axis is negligibly small.* So each hydrogen molecule has two additional "rotational degrees of freedom" for a total of five (three translational and two rotational). If each degree of freedom stores its equipartition share of $\frac{1}{2} k_B T$, we should obtain an energy per molecule of $\frac{5}{2} k_B T$. Consequently we expect a heat capacity per particle of $\frac{5}{2} k_B$, in good agreement with the experimental value of $2.47 k_B$.

20.7 Consider the four-atom ammonia molecule illustrated in **Figure 20.14**. (a) Which of the three rotational degrees of freedom can store thermal energy? (b) Based on your answer to part a, what is the thermal energy per molecule? (c) What is the heat capacity per molecule? (d) Does your prediction in part c agree with the experimental value listed in Table 20.3?

---

*Although the atoms have a diameter of about 10^{-10} m and are separated by about the same distance, nearly all the mass is concentrated in the nucleus, which has a smaller diameter of about 10^{-15} m. Therefore the "bar" connecting the two atoms at the end of the H_2 dumbbell is about 10^5 times greater than the diameter of the atoms, and so the rotational inertia about the axis running through the length of the bar of the dumbbell is essentially zero (see Checkpoint 11.4).

Figure 20.13 A diatomic molecule has energy associated with rotation about two axes, as well as energy associated with translation along three axes.

(*a*) Diatomic molecule

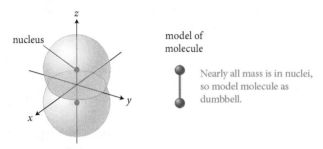

(*b*) Rotational degrees of fredom for diatomic molecule

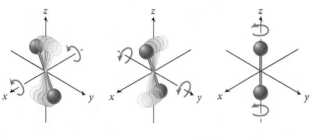

Rotation about *x* and *y* axes stores rotational energy.

Rotation about *z* axis stores negligible energy (because rotational inertia is negligible).

Figure 20.15 shows how, for a gas of diatomic hydrogen molecules, the heat capacity per particle in units of k_B varies with temperature. Notice how the heat capacity per particle in units of k_B is close to 2.5 (the value listed in Table 20.3) only around room temperature. As the temperature is lowered to about 50 K, the heat capacity per particle in units of k_B decreases to 1.5 (or $\frac{3}{2}$). Why does hydrogen gas behave like a monatomic ideal gas at low temperatures? To ask the same question another way, why doesn't the rotational-degrees-of-freedom argument presented above for hydrogen hold at low temperatures? The experimental data suggest that, at low temperatures, the rotational motions of the molecules do not contribute to their thermal energy.

Figure 20.14 Ammonia molecule.

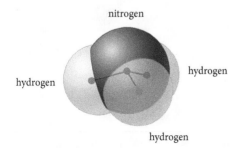

Figure 20.15 Temperature dependence of the heat capacity per particle *C* in units of k_B for hydrogen gas molecules.

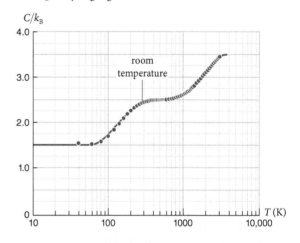

The reason for this behavior is that, according to quantum mechanics, vibrational and rotational energy comes in small fixed amounts, called quanta. Below about 100 K, the equipartition share $\frac{1}{2}k_B T$ is smaller than the quantum of energy needed to make a molecule rotate about one of the possible rotation axes. Therefore when one molecule collides with another, the amount of energy transferred during the collision is not great enough to set either molecule rotating (**Figure 20.16***a*). Above 100 K, the hydrogen molecules have sufficient kinetic energy to set each other spinning (Figure 20.16*b*). At room temperature (about 298 K), the value $\frac{1}{2}k_B T$ is equal to several rotational quanta, and so the equipartition of energy includes the two independent rotations of the hydrogen molecules, as described in the text preceding Checkpoint 20.7.

Figure 20.16 Collisions between hydrogen molecules. (*a*) At low temperatures (*T* < 100 K), the molecules have insufficient kinetic energy to cause any changes in rotational motion. (*b*) At higher temperatures, molecules can be set into rotational motion.

(*a*)

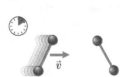 

If kinetic energy < quantum of rotational energy . . .

. . . collision cannot change rotational kinetic energy.

(*b*)

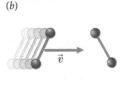 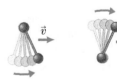

If kinetic energy > quantum of rotational energy . . .

. . . collision can change rotational kinetic energy.

Figure 20.17 At temperatures above 1000 K, diatomic molecules begin to vibrate along the axis connecting the two atoms.

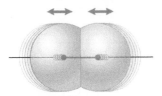

Figure 20.15 shows that at temperatures above 1000 K, the heat capacity per particle increases further. At such temperatures, the equipartition share becomes comparable to the quantum associated with another type of repetitive motion: vibration of the two hydrogen atoms along the axis of the molecule (**Figure 20.17**). Although there is only one axis for this type of motion in hydrogen, this vibrational motion contributes two degrees of freedom because it involves both potential and kinetic energy, each of which can store an amount of energy $\frac{1}{2}k_BT$. So the number of degrees of freedom for hydrogen is seven (three translational, two rotational, and two vibrational). If each degree of freedom contributes $\frac{1}{2}k_BT$, we expect a thermal energy per molecule of $\frac{7}{2}k_BT$. Indeed, experiments show that the heat capacity per hydrogen molecule approaches $\frac{7}{2}k_B$ around 3000 K. To summarize:

Only degrees of freedom for which $\frac{1}{2}k_BT$ is greater than the quantum of energy associated with that degree of freedom contribute to the heat capacity of a gas. In thermal equilibrium, each contributing degree of freedom of a particle in an ideal gas stores $\frac{1}{2}k_BT$ of thermal energy.

So, for a gas in which each particle has d contributing degrees of freedom, we expect the heat capacity per particle to be equal to $\frac{d}{2}k_B$.

Example 20.3 Energy quanta for nitrogen

At room temperature, do collisions between nitrogen molecules have enough energy to set a nitrogen molecule rotating? Enough to set it vibrating? The quantum of energy associated with rotation is $E_{rot} = 4.0 \times 10^{-23}$ J and the quantum of energy associated with vibration is $E_{vib} = 4.7 \times 10^{-20}$ J.

❶ GETTING STARTED At a given absolute temperature T each molecule stores $\frac{1}{2}k_BT$ of thermal energy per contributing degree of freedom. If the quantum of energy associated with a certain degree of freedom is greater than the energy stored per degree of freedom, then that degree of freedom cannot store any energy.

❷ DEVISE PLAN To solve this problem, I need to compare the equipartition energy share at room temperature with the energy quanta associated with rotation and vibration. I can obtain the equipartition energy share at room temperature by substituting $T = 300$ K (room temperature) into $\frac{1}{2}k_BT$.

❸ EXECUTE PLAN At 300 K the equipartition energy share is $\frac{1}{2}k_BT = \frac{1}{2}(1.38 \times 10^{-23}$ J/K$)(300$ K$) = 2.07 \times 10^{-21}$ J. This value is greater than E_{rot} but smaller than E_{vib}, which means that

at room temperature the nitrogen molecules rotate but do not vibrate. ✔

❹ EVALUATE RESULT Table 20.3 confirms my conclusion: At room temperature the heat capacity per particle of nitrogen molecules is $2.50 \approx \frac{5}{2}$, indicating that rotations contribute to the storing of thermal energy but vibrations do not.

We can apply the same treatment to the vibrating atoms in a solid crystal. Consider, for example, the simple cubic arrangements of atoms shown in **Figure 20.18**. Each atom is "tied" in place by elastic bonds with neighboring atoms, so there are no translational or rotational degrees of freedom. However, each atom can vibrate along the three independent axes of the crystal. As we have seen, each vibration contributes two degrees of freedom, and so each atom should have six degrees of freedom, each storing $\frac{1}{2}k_BT$ of thermal energy. We therefore expect a heat capacity per atom of $3k_B$, which is in good agreement with the values for a number of simple crystalline solids (see Table 20.3).

Figure 20.19 shows the temperature dependence of the heat capacity per atom for a number of solids. Again we see that quantum effects affect the heat capacity. Below certain temperatures, the heat capacity of each solid decreases to zero because $\frac{1}{2}k_BT$ is smaller than the quantum of energy associated with the vibrations.

The interplay between the various degrees of freedom and the number of basic states available for the equipartition of energy explains the heat capacities for different materials over a broad range of temperatures. For a more accurate and complete description, however, we need *quantum mechanics*—a branch of physics that accounts for the discrete nature of energy and is beyond the scope of this text.

In all of this discussion we have considered materials that do not undergo any phase transition from one state of matter to another. During a phase transition, adding energy thermally to a system does not change the temperature, but instead goes to changing the state of matter. (See the box "Phase transitions.")

Figure 20.18 Each atom in a crystal can vibrate along three independent axes.

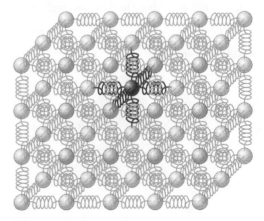

Figure 20.19 Temperature dependence of the heat capacity per atom for a number of solids.

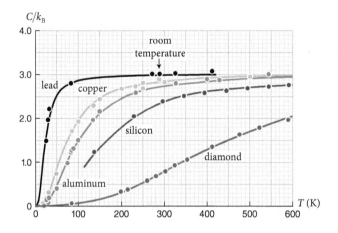

🖐 **20.8** Two identical volumes of gases—one monatomic and the other diatomic—are at the same pressure at room temperature. The temperature of each is raised by the same amount by thermally transferring energy to each of them. Is the energy transferred to the monatomic gas smaller than, equal to, or greater than the energy transferred to the diatomic gas?

20.4 *PV* diagrams and processes

We have seen that a system can exchange energy with its surroundings in two ways: by work done on it and by a thermal transfer of energy. The exchange of energy changes the state of the system. As we saw in Section 20.1, the system

might get hotter, or expand, or the pressure might increase. In this section we examine the interplay between energy exchanges and changes in state for an ideal gas.

The equilibrium state of a gas can be completely specified by the four macroscopic measurable quantities we discussed in Chapter 19: the pressure P of the gas, its volume V, its temperature T, and the number of molecules N in that volume. Generally we shall consider samples that consist of a fixed number N of particles, so we can represent the equilibrium state of a gas by the triplet of values (P, V, T).

As we saw in Chapter 19, for an ideal gas the four macroscopic variables are related to one another by Eq. 19.51: $P = (N/V)k_BT$.* This means that if the number of particles in a gas is fixed, we need only two variables to completely specify the equilibrium state of the gas. If we choose these two variables to be P and V, then the equilibrium state of a fixed amount of ideal gas can be represented by a point on a *PV diagram* such as the ones shown in **Figure 20.20** on the next page.

We can read the corresponding pair of values (P, V) from the axes and determine the temperature from the ideal gas law (assuming we know N). Once we know the temperature, we can calculate the thermal energy and the entropy of the gas (see Eqs. 19.50 and 19.59 for monatomic ideal gases).

> **Any point on a *PV* diagram represents an equilibrium state in which the gas has a well-defined temperature, thermal energy, and entropy.**

---

*For a nonideal gas the four macroscopic variables are also related to one another, but the relationship is more complicated than that expressed by the ideal gas law.

Phase transitions

Most substances occur in several *states of matter* (also called *phases*): the solid state, the liquid state, and the gaseous state. H_2O, for example, exists in the solid state as ice, in the liquid state as water, and in the gaseous state as water vapor. The transition from one state of matter to another is called a *phase transition*. During a phase transition any energy transferred thermally does not change the temperature, but instead goes toward changing the state. At a given pressure, therefore, phase transitions occur at a fixed temperature and are accompanied by a change in volume.

For example, at normal atmospheric pressure ice changes to water at a temperature of 0 °C. To melt 1 kg of ice requires a thermal transfer of energy of 3.34×10^5 J. This energy is required to break the bonds that hold the water molecules together in ice.

Because the energy transferred thermally to a substance undergoing a phase transition is an extensive quantity, we introduce an intensive quantity by dividing the energy transferred thermally by the mass of the substance undergoing the transition. The resulting quantity, $L \equiv Q/m$, is called the *specific transformation energy*

(traditionally often called *latent heat,* although the term *heat,* which is an energy and not a specific energy, is a misnomer). Because there are two phase transitions between the three states of matter, each substance has two specific transformation energies: one associated with the solid-liquid transition (*specific transformation energy for melting L_m*) and one with the liquid-gas transition (*specific transformation energy for vaporization L_v*).

Phase transitions are reversible. Energy must be added to a substance to make it undergo a transition from a lower-temperature phase to a higher-temperature phase. In the other direction, energy must be removed from the system. For example, to change 1 kg of water at 100 °C to water vapor at 100 °C at normal atmospheric pressure requires thermally transferring 2.256×10^6 J to the water. When that water vapor condenses back to water at 100 °C and normal atmospheric pressure, the same 2.256×10^6 J must be removed from the vapor.

Table 20.4 on the next page lists the specific transformation energies and associated phase transition temperatures for the melting and vaporization of a number of substances at normal atmospheric pressure.

Figure 20.20 Isotherms and isentropes in a *PV* diagram for an ideal gas. Points 1 and 2 represent two equilibrium states, both at *T* = 300 K. In general, on a *PV* diagram, the isentrope passing through a given point is steeper than the isotherm passing through that point.

(*a*) *PV* diagram showing isotherms (curves of constant temperature *T*) for an ideal gas

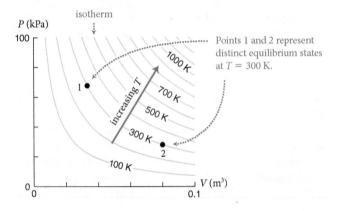

(*b*) **I**sentropes (curves of constant entropy *S*) superimposed on isotherms

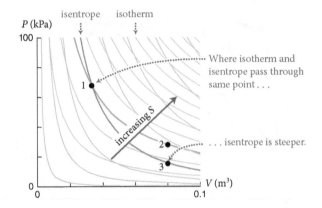

The ideal gas law implies that all equilibrium states (*P*, *V*) at a fixed temperature lie on a curve such that the product *PV* is constant or, to express this information another way, *P* ∝ 1/*V*. These curves are called **isotherms**. Figure 20.20*a* shows a number of isotherms, each corresponding to a fixed temperature and therefore a fixed thermal energy of the gas. Isotherms that are higher on the *PV* diagram represent higher temperatures and greater thermal energy.

20.9 Consider the equilibrium states represented by points 1 and 2 in Figure 20.20*a*. Compare these quantities for an ideal gas in these two states: pressure *P*, volume *V*, temperature *T*, entropy *S*, and thermal energy E_{th}.

Another important family of curves associated with the *PV* diagram for an ideal gas is shown in Figure 20.20*b*. Each

Table 20.4 Phase transition temperatures and specific transformation energies for melting (m) and vaporization (v)

Substance	T_m (K)	L_m (10^3 J/kg)	T_v (K)	L_v (10^3 J/kg)
Helium	–	–	4.230	20.8
Hydrogen	13.84	58.6	20.37	449
Nitrogen	63.18	25.5	77.36	199
Oxygen	54.36	13.8	90.19	213
Ammonia	195	331	239.8	1369
Carbon tetrachloride	250.3	16	349.8	194
Ethanol	159	100.7	351	837
Water	273.15	334	373.12	2256
Mercury	234	11.8	630	272
Aluminum	933.6	395	2740	10,500
Lead	600.5	24.5	2023	871
Sulfur	392	38.1	717.75	326
Gold	1336.15	64.5	2933	1578
Copper	1356	134	2840	5069

curve in this family connects all points that correspond to equilibrium states that have the same entropy. Such curves are called **isentropes.**

An isentrope passing through a given point on a *PV* diagram is steeper than the isotherm passing through that point. To see why this is so, consider the isentrope and isotherm passing through point 1 in Figure 20.20a. As we saw in Checkpoint 20.9, the entropy of state 2 on the isotherm is greater than the entropy of state 1.* Equation 19.60, $S = N\ln(T^{3/2}V) + \text{constant}$, tells us that in order to return the entropy to its original value, the temperature must decrease as we go at constant volume from state 2 to state 3, which lies on the isentrope. So the isentrope passing through state 1 has to be steeper than the isotherm (Figure 20.20b).

Isentropes that are higher on the *PV* diagram represent greater entropies. You can see this by going from state 3 to state 2 in Figure 20.20b. The volume is the same for both states, but the fact that you move from one isotherm to a higher isotherm means that the temperature and thus the entropy (see Eq. 19.60) increase.

Let us now turn our attention to the transformation of a system from one equilibrium state to another. If the process is quasistatic, it is a succession of near-equilibrium states. Provided these states are very close to equilibrium, we can represent them by points on a *PV* diagram, and the succession of points then constitutes a path on the *PV* diagram, such as the one shown in **Figure 20.21**.

> *During a quasistatic process, the system remains near equilibrium at all instants. Such a process can be represented by a continuous path on a PV diagram.*

Processes that are not quasistatic cannot be represented on a *PV* diagram. For such processes we can only show the points that correspond to the initial and final equilibrium states.

There are infinitely many ways to change the state of a gas. **Figure 20.22** shows two quasistatic processes that transform a gas from the same initial state i to the same final state f. Because the two processes have the same initial and final equilibrium states and because each state has a well-defined thermal energy and entropy, the change in these two quantities is independent of the process. For an ideal gas whose center of mass is at rest, there are no changes in the kinetic, potential, or source energy. Thus the change in the energy of the gas is equal to the change in the thermal energy, and we can state:

> *When an ideal gas is transformed from one equilibrium state to another, the changes in the thermal energy and entropy of the gas are independent of the process.*

Note that this statement holds even for non-quasistatic, irreversible processes between equilibrium states. The changes in thermal energy and entropy depend on only

Figure 20.21 A path on a *PV* diagram represents a succession of equilibrium states that takes the system from some initial state to some final state. Such a transformation is called a quasistatic process.

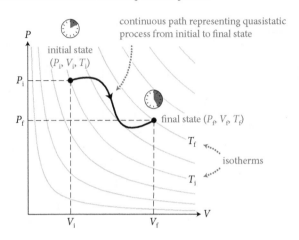

the initial and final states—what happens in between does not matter. If the change in thermal energy is independent of the process, then the energy transferred to the system ($W + Q$) must also be independent of the process. Are W and Q each independent of the process, too?

As we shall see in Section 20.5, the magnitude of the work done on a gas during any quasistatic process carried out on the gas is given by the area under the path that represents the process in a *PV* diagram. **Figure 20.23** on the next page shows a number of quasistatic processes, with the area under the path representing each process shaded; the larger the shaded area, the greater the magnitude of the work done on the gas. If the path is in the direction of increasing volume (Figure 20.23a and b), the work done on the gas is negative. If the path is in the direction of decreasing volume (Figure 20.23c), the work done on the gas is positive. For a vertical path (Figure 20.23d), the work is zero.

Comparing the sizes of the hatched areas in parts a and b of Figure 20.23 tells us that the amount of work done on the gas is different in the two graphs. These graphs represent processes A and B of Figure 20.22, telling us that the

Figure 20.22 Two quasistatic processes that transform a gas from an initial state i to a final state f.

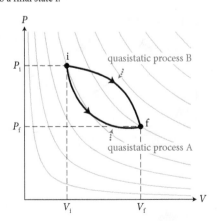

---

*For simplicity, we shall omit the word *equilibrium* when referring to states presented by specific points on *PV* diagrams. For example, "state 1" stands for "the *equilibrium* state represented by point 1."

Figure 20.23 The work done on a gas during a quasistatic process is equal to the area under the path on the *PV* diagram representing that process.

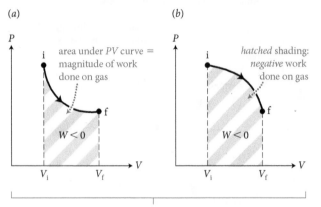

(a)

area under *PV* curve = magnitude of work done on gas

$W < 0$

$V_f > V_i$: Gas expands, so work done on gas is negative.

(b)

hatched shading: *negative* work done on gas

$W < 0$

(c)

solid shading: *positive* work done on gas

$W > 0$

$V_f < V_i$: Gas compressed, so work done on gas is positive.

(d)

$V_i = V_f$

$V_f = V_i$: Volume does not change, so no work done on gas.

magnitude of the work needed to carry out process B is greater than the work needed to carry out process A. Unlike changes in thermal energy that occur during a process, the amount of work done does depend on the path taken.

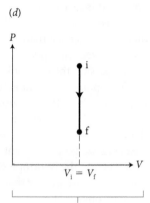 **20.10** (*a*) For which of the two quasistatic processes shown in Figure 20.22, if either, is the change in entropy greater? (*b*) Is the energy transferred thermally to the gas the same along both paths?

As Checkpoint 20.10 makes clear, the quantity of energy *Q* transferred thermally to the system, like the work *W* done on it, depends on the path taken.

Example 20.4 Different process, same change?

In quasistatic process A of **Figure 20.24**, an ideal gas is brought from an initial to a final state while its temperature is held fixed. In quasistatic process B, an identical gas is brought in two steps from the same initial state as in process A to the same final state.

Figure 20.24 States i and f are on the same isotherm.

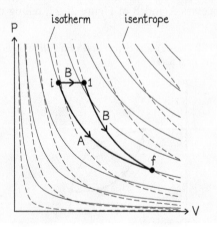

In the first step, the gas expands at constant pressure from the initial state to an intermediate state 1, and in the second step, the gas is brought adiabatically to the final state. Is the quantity of energy transferred thermally to the gas during process B greater than, equal to, or smaller than during process A?

❶ **GETTING STARTED** During both processes, the gas expands, and so the work done on the gas is negative in both cases. The quantity of energy transferred thermally depends on both the work done on the gas and the change in the thermal energy of the gas.

❷ **DEVISE PLAN** The initial and final states lie on the same isotherm, which means the temperature of the gas is the same in the two states. Therefore the thermal energy of the gas is the same in the two states, and so the change in thermal energy is zero for both processes. This means that the quantity of energy transferred thermally to the gas is the negative of the work done on the gas.

❸ **EXECUTE PLAN** The area under the path for process B in Figure 20.24 is larger than the area under the path for process A, and so the magnitude of the work done on the gas is greater for process B. This means that the quantity of energy transferred thermally must be greater during process B. ✔

❹ **EVALUATE RESULT** I am told that the second step in process B is adiabatic, which means that $Q = 0$. The problem therefore boils down to comparing the energy transferred thermally during process A with the energy transferred thermally during the first step of process B. During process A, the change in the thermal energy is zero, and so all the energy transferred thermally to the gas can be used by the gas to do work on its environment. That state 1 is on a higher isotherm than state i tells me that the temperature of the gas increases during the first step of process B. Therefore part of the energy transferred thermally is required to raise the thermal energy of the gas. It therefore makes sense that *Q* is greater for process B.

Figure 20.25 Four types of constrained quasistatic processes.

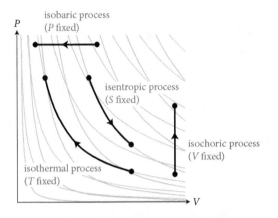

Table 20.5 Constrained quasistatic processes

Process	Constraint	Consequence
isochoric	V fixed	$W = 0$
isentropic (adiabatic)	S fixed	$Q = 0.$
isobaric	P fixed	
isothermal	T fixed	$\Delta E_{th} = 0$

The quasistatic processes in Example 20.4 are examples of *constrained processes*: Their path on a *PV* diagram is determined by a constraint placed on the system.

Figure 20.25 shows four important constrained quasistatic processes in ideal gases that we shall study in the second part of this chapter. When the pressure in the gas is held fixed during a quasistatic process, the path in the *PV* diagram is a horizontal line at the fixed pressure *P*; such a process is called isobaric (Greek for "same weight"). When the volume is held fixed, the path is a vertical line, and the process is called isochoric ("same volume"). Because the volume doesn't change, no work is done on (or by) the gas in an isochoric process (Figure 20.23*d*).

When the temperature is held fixed, the path lies along an isotherm, and the process is called isothermal ("same temperature"). As we saw in Chapter 19, the thermal energy of an ideal gas is proportional to the temperature of the gas,

and so in an isothermal process, the thermal energy does not change.

When the entropy is constant during a process, the path lies along an isentrope and the process is called isentropic ("same entropy"). As we saw in Section 20.1, the entropy of a system remains constant during quasistatic adiabatic processes, and therefore these processes are isentropic. Paths along isentropes thus represent processes for which no energy is transferred thermally to the system ($Q = 0$). Isentropes are sometimes also called *adiabats*, and isentropic processes are also called *adiabatic processes*.*

Table 20.5 summarizes these constrained quasistatic processes and the consequences of the constraints imposed.

20.11 For each process in Figure 20.25, determine whether the following quantities are positive, negative, or zero: W, Q, ΔE_{th}, ΔS.

---

*Note, however, that an adiabatic process need not be isentropic. Only a *quasistatic* adiabatic process is isentropic.

Self-quiz

1. Figure 20.26 shows the energy changes for a certain system over a short time interval. Describe what is happening to the system. From the diagram, what can you conclude about the temperature of the environment at the beginning of the time interval?

Figure 20.26

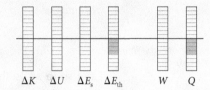

ΔK ΔU ΔE_s ΔE_{th} W Q

2. On a really hot day, which temperature scale (°F or °C) registers the greater number? On a really cold day, which temperature scale registers the smaller number? At what temperature do the two scales register the same number?

3. Which material in Table 20.2 has the greatest specific heat capacity? The smallest? Imagine that objects made of these materials are placed on your hand. The objects have the same mass and are initially at the same temperature. Each object equilibrates thermally with your hand, which is at a higher temperature. Describe any differences in the equilibration process.

4. What is the approximate thermal energy of a molecule that has three degrees of translational freedom, two degrees of rotational freedom, and two degrees of vibrational freedom? The molecule is in a gas at temperature T.

5. Which types of processes are represented in Figure 20.24? For each process, determine whether the following quantities are positive, negative, or zero: ΔT, ΔV, ΔP.

Answers

1. The thermal energy of the system is decreasing ($\Delta E_{th} < 0$) because energy is leaving the system ($Q < 0$) and entering the environment. That energy leaves the system means that initially the temperature of the environment must be lower than the temperature of the system.

2. The Fahrenheit scale registers the greater number on a really hot day. If the Fahrenheit temperature is 110 °F, the Celsius temperature is 43.3 °C. The Celsius scale registers the smaller number on a really cold day. If the Fahrenheit temperature is −20 °F, the Celsius temperature is −29 °C. The scales read the same temperature at −40 °C = −40 °F.

3. Water is the highest, gold and lead the lowest. Because water has such a high specific heat capacity, a large amount of energy must leave your hand and enter the water as your hand and the water equilibrate, much more than the gold or lead would. This, in part, is why it is so hard to keep warm if you are wet and why gold jewelry so quickly comes to the same temperature as your skin.

4. With seven degrees of freedom and an equipartition energy share of $\frac{1}{2}k_BT$, the molecule has roughly $\frac{7}{2}k_BT$ of thermal energy.

5. Process A in Figure 20.24 is isothermal because it follows an isotherm. $\Delta T = 0$ because the process is isothermal, $\Delta V > 0$ because $V_f > V_i$, and $\Delta P < 0$ because $P_f < P_i$. The first step of process B is isobaric because the pressure is constant. $\Delta T > 0$ because T_i is on a lower isotherm than T_1, $\Delta V > 0$ because $V_1 > V_i$, and $\Delta P = 0$ because $P_1 = P_i$. The second step of process B is isentropic because it follows an isentrope. $\Delta T < 0$ because T_f is on a lower isotherm than T_1, $\Delta V > 0$ because $V_f > V_1$, and $\Delta P < 0$ because $P_f < P_1$.

20.5 Change in energy and work

If we take into account both the transfer of mechanical energy (work) W and the energy transferred thermally Q, the energy law (Eq. 9.1) becomes

$$\Delta E = W + Q. \tag{20.1}$$

Remember that this law is a direct consequence of conservation of energy: The only way the energy of a system can change is by a transfer of some form of energy to or from the surroundings.

Let's first consider only situations where the system is an ideal gas and where the translational kinetic energy, potential energy, and source energy of the gas remain constant ($\Delta K = \Delta U = \Delta E_s = 0$). The energy law then reduces to

$$\Delta E_{th} = W + Q \quad \text{(ideal gas).} \tag{20.2}$$

Equation 19.50 tells us that the thermal energy of a monatomic ideal gas consisting of N atoms is

$$E_{th} = \tfrac{3}{2} N k_B T \quad \text{(monatomic ideal gas).} \tag{20.3}$$

As we saw in Section 20.3, at thermal equilibrium each degree of freedom of an ideal gas particle contributes $\tfrac{1}{2} k_B T$ to the particle's thermal energy. For a monatomic ideal gas, each atom has three degrees of freedom, yielding the factor 3 in Eq. 20.3. Generalizing this result for an ideal gas with d contributing degrees of freedom, we have

$$E_{th} = \frac{d}{2} N k_B T \quad \text{(ideal gas).} \tag{20.4}$$

20.12 What is the thermal energy associated with a gas sample that contains N hydrogen molecules at room temperature ($T = 300$ K)?

When the temperature of an ideal gas sample that contains a fixed number N of particles changes, Eq. 20.4 tells us that the thermal energy of the sample changes by an amount

$$\Delta E_{th} = \frac{d}{2} N k_B \Delta T. \tag{20.5}$$

As we noted in Section 20.4, this change in thermal energy depends on only the initial and final equilibrium states and is independent of the path taken from initial state to final state.

To calculate the work done on a gas, consider the situation shown in Figure 20.27: A piston with surface area A quasistatically compresses a gas in a cylinder. The x component of the force exerted by the piston on the gas is F^c_{pgx}. The work done by this external force on the gas is (Eq. 9.22)

$$W = \int_{x_i}^{x_f} F^c_{pgx} \, dx.$$

The magnitude of the force exerted by the piston on the gas is related to the gas pressure P by Eq. 19.43:

$$P = \frac{F^c_{pg}}{A} \tag{20.7}$$

Figure 20.27 A piston that has a surface area A exerts a force on a gas, compressing it.

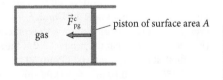

Figure 20.28 Checkpoint 20.13.

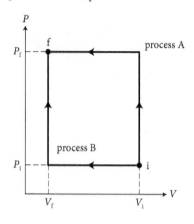

or $F_{pg}^c = PA$. If we take the positive x axis to the right as in Figure 20.27, we have $F_{pgx}^c = -F_{pg}^c = -PA$. When the piston moves a distance dx, the volume of the gas changes by an amount $dV = A\,dx$, and so the work done on the gas is

$$W = -\int_{x_i}^{x_f} PA\,dx = -\int_{V_i}^{V_f} P\,dV. \tag{20.8}$$

The integral to the right of the minus sign in this expression represents the area under the path taken in the PV diagram, as we discussed in Section 20.4 (see Figure 20.23). If the process is in the direction of increasing volume (expansion), the integral in Eq. 20.8 is positive, making the work done on the gas negative (Figure 20.23a and b). If the process is in the direction of decreasing volume (compression), the work done on the gas is positive (Figure 20.23c).

When the pressure is constant (that is, during a quasistatic isobaric process), we can pull the pressure out of the integral, and the work done on the gas is

$$W = -P\int_{V_i}^{V_f} dV = -P(V_f - V_i) = -P\Delta V \quad \text{(quasistatic isobaric process).} \tag{20.9}$$

20.13 Which quasistatic process from the initial state i to the final state f in Figure 20.28 requires more work done on the gas?

From Checkpoint 20.13 we see that the work done on a gas depends on the manner in which the gas is transformed from its initial state to its final state.

Example 20.5 Same change, different process

In Figure 20.28, an ideal gas is brought from an initial state i to a final state f by two processes. The initial volume and pressure are $0.50\ \text{m}^3$ and 100 kPa, and the final values for these variables are $0.10\ \text{m}^3$ and 500 kPa. How much energy is transferred thermally to the gas during (a) process A and (b) process B?

❶ GETTING STARTED Both processes are represented by a continuous path on a PV diagram and so are quasistatic. From the graph and the data given, I know that $V_i = 0.50\ \text{m}^3$, $V_f = 0.10\ \text{m}^3$, $P_i = 100$ kPa, and $P_f = 500$ kPa.

❷ DEVISE PLAN To calculate the quantity of energy transferred thermally, I can use the energy law as given in Eq. 20.2, $\Delta E_{th} = W + Q$, which gives me $Q = \Delta E_{th} - W$. From Checkpoint 20.13, I know that work is done on the gas only during the constant-pressure legs. To calculate W during these legs, I can use Eq. 20.9. The change in the thermal energy ΔE_{th} is given by Eq. 20.5, but I do not know d, N, or ΔT. The ideal gas law, however, relates pressure and volume to temperature: $PV = Nk_BT$ (Eq. 19.51). I note that the product PV is the same for the initial and final states: $P_iV_i = (100\ \text{kPa})(0.50\ \text{m}^3) = 50\ \text{kPa} \cdot \text{m}^3$ and $P_fV_f = (500\ \text{kPa})(0.10\ \text{m}^3) = 50\ \text{kPa} \cdot \text{m}^3$. This tells me that the

initial and final states lie on an isotherm, so $\Delta T = 0$. Therefore Eq. 20.5 becomes $\Delta E_{th} = 0$, so Eq. 20.2 yields $Q = -W$.

❸ EXECUTE PLAN The change in volume for both processes is $\Delta V = V_f - V_i = (0.10\ \text{m}^3) - (0.50\ \text{m}^3) = -0.40\ \text{m}^3$.

(a) From Eq. 20.9, the work done on the gas for process A is $W = -P\Delta V = -(500\ \text{kPa})(-0.40\ \text{m}^3) = 2.0 \times 10^2$ kJ. Therefore $Q_A = -2.0 \times 10^2$ kJ. ✔

(b) $W = -P\Delta V = -(100\ \text{kPa})(-0.40\ \text{m}^3) = 40$ kJ, which leads to $Q_B = -40$ kJ. ✔

❹ EVALUATE RESULT The work done on the gas during process A is five times greater than the work done during process B. This is not surprising because in A the compression from V_i to V_f is done at a pressure that is five times greater than the pressure at which the compression is done in B. Because the thermal energy of the gas does not change, any energy acquired in the form of work done on the gas has to leave as energy transferred thermally out of the gas. Both Q_A and Q_B should therefore be negative, and Q_A should be greater than Q_B, which is what I found.

As we have seen, the change in thermal energy for an ideal gas undergoing some quasistatic process is independent of the process and is given by Eq. 20.5. Checkpoint 20.13 and Example 20.5, however, show that the work done on the gas and the energy transferred thermally depend on the process.

✋ **20.14** Consider the three processes shown in **Figure 20.29**. Process B is isothermal. (*a*) For which of the three processes is the energy transferred thermally greatest? (*b*) For which of the three processes is the energy transferred thermally smallest?

Figure 20.29 Checkpoint 20.14.

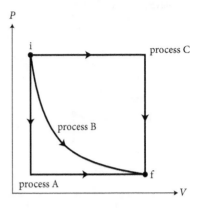

20.6 Isochoric and isentropic ideal gas processes

Let us now take a deeper look at what the energy law tells us about the transfers of mechanical and thermal energy that occur during constrained quasistatic processes. We begin by examining isochoric (constant volume) and isentropic (constant entropy) quasistatic processes in this section and then isobaric (constant pressure) and isothermal (constant temperature) quasistatic processes in the next section.

An example of an isochoric quasistatic process is shown in Figure 20.30. An ideal gas sample consisting of N particles enclosed in a fixed volume is heated. The energy diagram for the volume of ideal gas during this process is shown in Figure 20.30*b*. Because the gas is heated, its temperature and therefore its thermal energy increase. Because the volume is fixed, no work is done on the gas. The amount Q of energy transferred thermally to the gas raises the thermal energy of the gas.

Figure 20.30*c* shows the corresponding PV diagram. The ideal gas law, $P = Nk_BT/V$ (Eq. 19.51), tells us that if the volume is kept fixed, the pressure in the gas increases as the temperature increases. Because the volume is fixed and the pressure increases, the path that represents the heating process is straight up. The arrow representing Q has solid shading because energy is transferred to the gas ($Q > 0$).

Because $V_i = V_f$, the integral in Eq. 20.8 is zero, and so the work done on the gas is zero:

$$W = 0 \quad \text{(isochoric process).} \tag{20.10}$$

The energy law for an ideal gas, $\Delta E_{th} = W + Q$, then reduces to

$$\Delta E_{th} = Q \quad \text{(isochoric process).} \tag{20.11}$$

Substituting the expression given for ΔE_{th} in Eq. 20.5, we see that

$$Q = \frac{d}{2}Nk_B\Delta T \quad \text{(isochoric process).} \tag{20.12}$$

In Section 20.3, we introduced the quantity **heat capacity per particle**. Mathematically this quantity is defined as the energy transferred thermally divided by the number of particles and by the temperature change:

$$C_V \equiv \frac{Q}{N\Delta T} \quad \text{(constant volume),} \tag{20.13}$$

where here we add a subscript V to remind ourselves that this result is obtained at constant volume. Substituting Eq. 20.12 into this definition for C_V yields

$$C_V = \frac{d}{2}k_B \quad \text{(constant volume).} \tag{20.14}$$

Note that this result predicts that the heat capacity per particle is determined by the number of degrees of freedom, in agreement with the experimental results we discussed in Section 20.3.

Figure 20.30 An example of an isochoric quasistatic process.

(*a*) Fixed volume of gas is heated, raising T and P

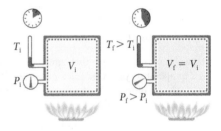

(*b*) Energy diagram for gas

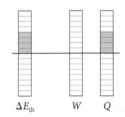

(*c*) PV diagram showing the isochoric process

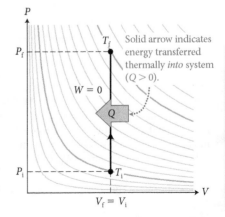

QUANTITATIVE TOOLS

Figure 20.31 Two ideal gas processes. Process A is isochoric; process B is not.

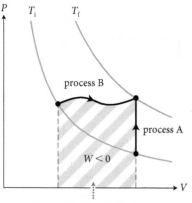

Hatched shading: Work done on gas in process B is *negative*.

Equation 20.14 allows us to rewrite Eq. 20.5 in the form

$$\Delta E_{th} = N C_V \Delta T \quad \text{(any ideal gas process).} \qquad (20.15)$$

Even though we used Eq. 20.5 to arrive at Eq. 20.15 for ΔE_{th} under isochoric conditions, Eq. 20.15 holds for all ideal gas processes, not just isochoric ones. To see why, consider two samples of N ideal gas particles undergoing the two processes in **Figure 20.31**. Both samples are initially at temperature T_i and end at temperature T_f. Process A is isochoric; process B is some arbitrary process. Because the temperature change is the same for the two processes, Eq. 20.5 tells us the thermal energy change must also be the same. This means we can use Eq. 20.15 to calculate ΔE_{th} for any process, given the change in temperature ΔT in the process and knowing C_V for the gas. In contrast to Eq. 20.5, which expresses the thermal-energy change in terms of the number of degrees of freedom of each ideal gas particle, Eq. 20.15 gives the thermal-energy change in terms of macroscopically measured quantities.

 20.15 Is Eq. 20.15 valid for an isothermal process ($\Delta T = 0$)?

Example 20.6 Energy transferred thermally

A sample of hydrogen gas that contains 1.00 mol of H_2 molecules at a temperature of 300 K and a pressure of 60.0 kPa undergoes an isochoric process. The final pressure is 80.0 kPa. How much energy is transferred thermally to the gas?

❶ GETTING STARTED I begin by making a PV diagram of the process, treating the hydrogen as an ideal gas (**Figure 20.32**). Because the process is isochoric, I represent it by a vertical line. Because $P_f > P_i$, I indicate that the path representing the process points upward.

Figure 20.32

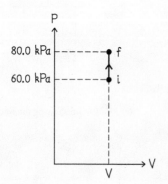

❷ DEVISE PLAN The quantity of energy transferred thermally during an isochoric process is given by Eq. 20.12. To use this equation, I need to know N, d, and ΔT. I know from Avogadro's number that my 1.00-mol sample contains $N = 6.02 \times 10^{23}$ molecules, and I can obtain d from Eq. 20.14 if I use Table 20.3 to get the value of C_V/k_B for H_2 molecules. For ΔT, I am given the initial temperature but not the final temperature. I know the initial and final pressures, however, and because the

volume is fixed in this isochoric process, I can use the ideal gas law to determine the final temperature and so have a value for ΔT.

❸ EXECUTE PLAN From Table 20.3 I see that, for H_2 molecules, $C_V/k_B = 2.47$ at room temperature (about 300 K). From Figure 20.15 I see that C_V/k_B stays at that value through 700 K, and so, from Eq. 20.14, I know that $d/2 = C_V/k_B = 2.47$.

Next I rewrite the ideal gas law (Eq. 19.51) in the form $V = N k_B T/P$. Because V, N, and k_B are constant, I know that T/P is constant, and so $T_i/P_i = T_f/P_f$. Solving for T_f, I get

$$T_f = \frac{T_i P_f}{P_i} = \frac{(300\ \text{K})(80.0\ \text{kPa})}{60.0\ \text{kPa}} = 400\ \text{K},$$

so $\Delta T = 400\ \text{K} - 300\ \text{K} = 100\ \text{K}$. Substituting my values for N, d, and ΔT in Eq. 20.12 gives me

$$Q = \frac{d}{2} N k_B \Delta T = 2.47(6.02 \times 10^{23})(1.38 \times 10^{-23}\ \text{J/K})(100\ \text{K})$$

$$= 2.05 \times 10^3\ \text{J.} \checkmark$$

❹ EVALUATE RESULT Because the temperature of the gas increases, its thermal energy must increase. The process is isochoric, so no work is done on the gas. Therefore the value of the quantity of energy transferred thermally must be positive, as my result is.

The quantity of energy transferred is significant—approximately equal to the kinetic energy of a 1-kg object moving at about 65 m/s—but I know that thermal energies tend to be large relative to kinetic energies. (See, for example, the box "Coherent versus incoherent energy" in Chapter 7.)

Next we consider a quasistatic isentropic process, a process for which the entropy remains unchanged:

$$\Delta S = 0 \quad \text{(isentropic process).} \qquad (20.16)$$

As we saw in Section 20.1, no energy is transferred thermally in an isentropic quasistatic process:

$$Q = 0 \quad \text{(isentropic quasistatic process).} \qquad (20.17)$$

An example of an isentropic quasistatic process is shown in **Figure 20.33a**. An ideal gas sample consisting of N particles is compressed very slowly by grains of sand dropped on the piston to ensure that the process is quasistatic. The gas is thermally insulated so that $Q = 0$. The increasing force exerted on the piston increases the pressure in the gas and decreases its volume. The corresponding energy diagram for the volume of ideal gas is shown in Figure 20.33b. The work done by the piston on the gas is positive, increasing the thermal energy of the gas. Because $Q = 0$, Eq. 20.2 becomes

$$\Delta E_{th} = W \quad \text{(isentropic quasistatic process).} \qquad (20.18)$$

The increase in thermal energy means that the temperature of the gas goes up even though the gas is not being heated. During an isentropic compression, work done on the gas raises the temperature of the gas. You may have noticed this effect while using a bicycle pump: If you pump vigorously, the air in the pump—and therefore the pump and the tire tube—becomes noticeably hotter.

Figure 20.33c shows the PV diagram for the isentropic process. The shaded area under the path is the work done by the piston on the gas. As the solid shading indicates, this work is positive ($W > 0$). Because the entropy of the gas doesn't change when $Q = 0$, the path in Figure 20.33c lies along an isentrope, and because the gas is compressed, the final volume is smaller than the initial volume. As you can see from the PV diagram, the path crosses the isotherms and ends on an isotherm at a higher temperature, indicating that the process raises the temperature of the gas, as we saw before.

If we substitute Eq. 20.15 into Eq. 20.18, we obtain an expression that relates the work done on the gas W to the temperature change:

$$\Delta E_{th} = W = NC_V \Delta T \quad \text{(isentropic process).} \qquad (20.19)$$

✋ **20.16** Is the product PV in the final state in Figure 20.33c greater than, equal to, or smaller than the product PV in the initial state?

Figure 20.33 An example of an isentropic quasistatic process.

(a) Thermally insulated volume of ideal gas is compressed slowly

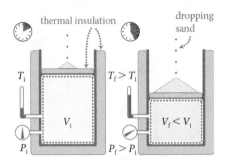

(b) Energy diagram for gas

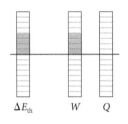

(c) PV diagram showing the isentropic process

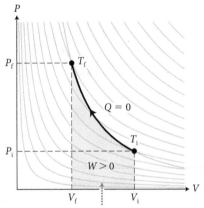

Solid shading: work done on gas is *positive*.

20.7 Isobaric and isothermal ideal gas processes

Next we turn to an isobaric (constant pressure) quasistatic process. An example is shown in **Figure 20.34a** on the next page: An ideal gas sample consisting of N particles is heated while it is being held at constant pressure by a piston that exerts a constant force on the gas. If the piston were held in place, the heating would raise the temperature of the gas and, according to the ideal gas law, the pressure would go up. Because the piston is free to move, however, the pressure stays constant and the gas expands instead. Therefore the gas does work on its environment. The energy diagram for the volume of the ideal gas during this process is shown in Figure 20.34b. The heating transfers energy to the gas ($Q > 0$), part of which goes into raising the temperature (and therefore the thermal energy

Figure 20.34 An example of an isobaric quasistatic process.

(*a*) Ideal gas heated at constant pressure

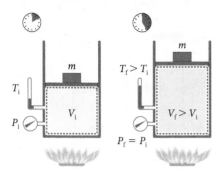

(*b*) Energy diagram for gas

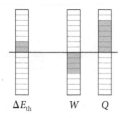

ΔE_{th} W Q

(*c*) *PV* diagram showing the isobaric process

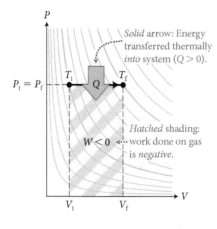

Solid arrow: Energy transferred thermally *into* system ($Q > 0$).

Hatched shading: work done on gas is *negative*.

of the gas: $\Delta E_{\text{th}} > 0$), and part of which is used by the gas to do work on its environment ($W < 0$).

Figure 20.34*c* shows the *PV* diagram for the process. Because the pressure is constant, the path lies on a horizontal line. Because the gas expands, the direction of the path is to the right, toward increasing volume. The shaded area under the path is the negative of the work done on the gas ($W < 0$). (Remember our convention: A hatched area signifies that the work done *on the gas* is negative.) We draw an arrow representing Q with solid shading to represent the fact that we transfer energy thermally to the gas ($Q > 0$).

Because this process is at constant pressure, we can define heat capacity per particle at constant pressure in analogy to Eq. 20.13:

$$C_P \equiv \frac{Q}{N\Delta T} \quad \text{(constant pressure)}, \tag{20.20}$$

and so in an isobaric process the energy transferred thermally is

$$Q = NC_P\Delta T \quad \text{(isobaric process)}. \tag{20.21}$$

From the ideal gas law (Eq. 19.51), $PV = Nk_{\text{B}}T$, we have for the initial and final states $PV_{\text{i}} = Nk_{\text{B}}T_{\text{i}}$ and $PV_{\text{f}} = Nk_{\text{B}}T_{\text{f}}$. The change in volume is therefore $\Delta V = V_{\text{f}} - V_{\text{i}} = Nk_{\text{B}}\Delta T/P$, which means the work done on the gas is, from Eq. 20.9,

$$W = -P\Delta V = -Nk_{\text{B}}\Delta T \quad \text{(isobaric process)}. \tag{20.22}$$

Now we have all three terms in the energy law for an ideal gas, $\Delta E_{\text{th}} = W + Q$. Equation 20.15 gives the change in the thermal energy ΔE_{th}, Eq. 20.22 gives the work done on the gas W, and Eq. 20.21 gives the energy Q transferred thermally into the gas. We can use this information to obtain an expression for C_P for ideal gases. Substituting Eqs. 20.15, 20.21, and 20.22 into the energy law for an ideal gas, we obtain

$$NC_V\Delta T = -Nk_{\text{B}}\Delta T + NC_P\Delta T \quad \text{(ideal gas)}. \tag{20.23}$$

Dividing by $N\Delta T$, we see that for an ideal gas, the heat capacities per particle are related by the expression

$$C_P = C_V + k_{\text{B}} \quad \text{(ideal gas)} \tag{20.24}$$

or, substituting from Eq. 20.14 for C_V,

$$C_P = \frac{d}{2}k_{\text{B}} + k_{\text{B}} = \left(\frac{d}{2} + 1\right)k_{\text{B}} \quad \text{(ideal gas)}. \tag{20.25}$$

Using Eqs. 20.14 and 20.25, we can now write the **heat capacity ratio** as

$$\gamma \equiv \frac{C_P}{C_V} = \frac{\left(\dfrac{d}{2} + 1\right)k_{\text{B}}}{\dfrac{d}{2}k_{\text{B}}} = 1 + \frac{2}{d} \quad \text{(ideal gas)}. \tag{20.26}$$

For a monatomic ideal gas, for which $d = 3$, the heat capacity ratio is $\gamma = \frac{5}{3}$.

Example 20.7 Heat capacity ratio

Experimentally one obtains $C_P = 4.844 \times 10^{-23}$ J/K for nitrogen gas at room temperature. (a) Calculate the heat capacity ratio γ for a gas of nitrogen molecules, N_2, at room temperature. (b) Which degrees of freedom of the N_2 molecules contribute to the heat capacity?

❶ GETTING STARTED I am given the room-temperature C_P value for nitrogen gas and asked to calculate γ and then to determine which of the degrees of freedom of the N_2 molecules contribute to the heat capacity. I assume the gas behaves ideally so that I can apply the results of this section.

❷ DEVISE PLAN The heat capacity ratio γ is given by Eq. 20.26. To calculate this ratio I need to know d, the number of degrees of freedom of the N_2 molecules that contribute to storing energy at room temperature. Equation 20.25 relates d to C_P, which is given. Once I determine d using this relationship, I can calculate γ and establish which degrees of freedom contribute to the heat capacity.

❸ EXECUTE PLAN (a) From Eq. 20.25, I get $\frac{d}{2} + 1 = C_P/k_B$, or $d = 2(C_P/k_B - 1)$. Substituting the value given for C_P, I get

$$d = 2\left(\frac{4.844 \times 10^{-23} \text{ J/K}}{1.381 \times 10^{-23} \text{ J/K}} - 1\right) = 5.015.$$

Substituting this value for d into Eq. 20.26, I get $\gamma = 1 + \frac{2}{5.015} = 1.399.$ ✔

(b) Because the nitrogen molecule is diatomic, it has three translational, two rotational, and two vibrational degrees of freedom. I know from Section 20.3 that vibrational motion does not begin to contribute to the heat capacity until the temperature of a gas is about 1000 K. Because my sample is at room temperature, and given that my result for d is very close to 5, I conclude that the translational and rotational degrees of freedom are the ones that contribute to the heat capacity ratio. ✔

❹ EVALUATE RESULT I can check the reasonableness of my calculated value $\gamma = 1.399$ by using it in Eq. 20.26 to determine C_V for the nitrogen molecule and then comparing what I get with the value obtained using data from Table 20.3. From Eq. 20.26: $C_V = C_P/\gamma = (4.844 \times 10^{-23} \text{ J/K})/1.40 = 3.46 \times 10^{-23}$ J/K; from Table 20.3: $C_V = 2.50k_B = 2.50(1.381 \times 10^{-23} \text{ J/K}) = 3.45 \times 10^{-23}$ J/K. That these two calculations result in about the same value for C_V gives me confidence in the value I obtained for γ.

The nitrogen molecule, N_2, is similar to the hydrogen molecule, H_2. I therefore expect the contribution of the various degrees of freedom to the heat capacity of the two molecules to be similar. I know from Section 20.3 that it is translational and rotational motions that contribute in H_2, which agrees with the motion types I chose in part b.

✋ 20.17 Suppose the isobaric process in Figure 20.34c begins and ends on the same isotherms as the isochoric process in Figure 20.30c. Which process requires a greater amount Q of energy transferred thermally?

An example of an isothermal (constant temperature) quasistatic process is shown in Figure 20.35a on the next page. Sand is dropped onto a piston, which then moves downward, compressing an ideal gas sample. The gas sample consists of N particles and is in thermal contact with a large volume of water. To keep the gas at constant temperature, the water volume must be large enough that adding (or removing) energy thermally to it does not noticeably change the water temperature. Such a body of water is called a *thermal reservoir*. The thermal reservoir is not part of the system (the gas by itself). The purpose of the thermal reservoir is to keep the temperature of our system constant by thermally exchanging energy with it.

The work done by the piston on the gas is converted to thermal energy. For the process to be isothermal, the work has to be done sufficiently slowly for the gas and the thermal reservoir to remain in thermal equilibrium. A quasistatic isothermal process is thus a slow interplay between work done on the gas and thermal transfer of energy out of the gas.

The temperature of the gas remains constant, and hence the thermal energy does not change:

$$\Delta E_{th} = 0 \quad \text{(isothermal process)}. \qquad (20.27)$$

The energy diagram for the volume of ideal gas during this isothermal process is shown in Figure 20.35b. The piston is doing work on the gas ($W > 0$), and

Figure 20.35 An example of an isothermal quasistatic process.

(a) Ideal gas compressed slowly at constant temperature

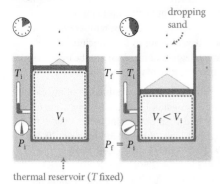

thermal reservoir (*T* fixed)

(b) Energy diagram for gas

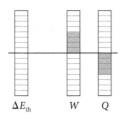

ΔE_{th} W Q

(c) PV diagram showing the isothermal process

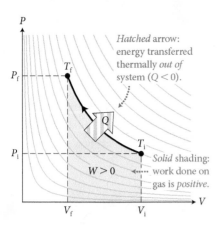

energy is transferred thermally from the gas ($Q < 0$). Because the temperature is constant, there is no change in the thermal energy. Hence the energy law for an ideal gas is $0 = W + Q$, which means

$$Q = -W \quad \text{(isothermal process).} \quad (20.28)$$

The work done on the gas can be found by substituting $P = Nk_BT/V$ from the ideal gas law into Eq. 20.8:

$$W = -\int_{V_i}^{V_f} \frac{Nk_BT}{V}dV = -Nk_BT\int_{V_i}^{V_f}\frac{dV}{V} \quad \text{(isothermal process),} \quad (20.29)$$

where we have pulled the (constant) temperature T out of the integral. Solving the integral, we have

$$W = -Nk_BT(\ln V_f - \ln V_i) = -Nk_BT\ln\left(\frac{V_f}{V_i}\right) \quad \text{(isothermal process).} \quad (20.30)$$

The energy transferred thermally to the gas is then, from Eq. 20.28,

$$Q = -W = Nk_BT\ln\left(\frac{V_f}{V_i}\right) \quad \text{(isothermal process).} \quad (20.31)$$

Figure 20.35*c* shows the *PV* diagram for the isothermal process. The path lies along an isotherm. The shading for the arrow representing Q is hatched because energy is transferred thermally out of the gas. Note that $V_f < V_i$, and so $Q < 0$, as we concluded earlier (see Checkpoint 20.11).

Table 20.6 summarizes the expressions we have obtained for the various terms in the energy law for an ideal gas for each of the four constrained ideal gas processes.

Table 20.6 Energy law for an ideal gas and constrained ideal gas processes

Process	Constraint	W	Q	ΔE_{th}	Energy law for ideal gas
isochoric	*V* fixed	0	$NC_V\Delta T$		$\Delta E_{th} = Q$
isentropic	*S* fixed		0	$NC_V\Delta T$	$W = \Delta E_{th}$
isobaric	*P* fixed	$-Nk_B\Delta T$	$NC_P\Delta T$	$NC_V\Delta T$	$\Delta E_{th} = W + Q$
isothermal	*T* fixed	$-Nk_BT\ln\left(\frac{V_f}{V_i}\right)$		0	$Q = -W$

Example 20.8 Isothermal compression

An ideal gas sample is compressed quasistatically at constant temperature. The initial pressure and volume are $P_i = 1.01 \times 10^5$ Pa and $V_i = 3.00 \times 10^{-2}$ m^3, and the final volume is $V_f = 2.00 \times 10^{-2}$ m^3. How much work is done on the gas during the compression?

❶ **GETTING STARTED** I begin by making a *PV* diagram of the process (**Figure 20.36**). The process is an isothermal compression, so it is represented by a path going up along an isotherm toward the left.

Figure 20.36

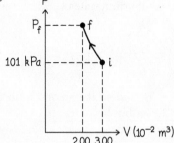

② DEVISE PLAN The work done on an ideal gas in an isothermal process is given by Eq. 20.30. I know the initial and final volumes, but not the factor Nk_BT. I can determine this factor from the ideal gas law, $PV = Nk_BT$; it does not matter whether I use initial values or final values in my calculation because the temperature is the same in both cases. Because I do not know the final pressure, however, I should use initial values.

③ EXECUTE PLAN From the ideal gas law, I have $Nk_BT = P_iV_i$. From Eq. 20.30, the work done on the gas is

$$W = -Nk_BT \ln\left(\frac{V_f}{V_i}\right) = -P_iV_i \ln\left(\frac{V_f}{V_i}\right).$$

Substituting values for the variables, I get

$$W = -(1.01 \times 10^5 \, \text{Pa})(3.00 \times 10^{-2} \, \text{m}^3) \ln\left(\frac{2.00 \times 10^{-2} \, \text{m}^3}{3.00 \times 10^{-2} \, \text{m}^3}\right)$$

$$= 1.23 \times 10^3 \, \text{J}. \; ✔$$

④ EVALUATE RESULT My answer is positive, as I expect for the work done on a gas in compressing it. To see whether the magnitude of my answer is in the right ballpark, I assume that the volume change is achieved by moving a piston of surface area $1 \times 10^{-2} \, \text{m}^2$ over a distance of 1 m. At a pressure of 1.01×10^5 Pa, the gas exerts on the piston a force of magnitude $(1 \times 10^{-2} \, \text{m}^2)(1.01 \times 10^5 \, \text{Pa}) \approx 10^3$ N. The piston exerts a force of equal magnitude on the gas. If this force were constant, the work done on the gas during a force displacement of 1 m would be $W = (10^3 \, \text{N})(1 \, \text{m}) = 10^3$ J, which is of the same order of magnitude as the answer I obtained. (This estimate doesn't take into account the increase in pressure and so is a bit lower than the actual value.)

✋ 20.18 Consider an ideal gas undergoing the two processes depicted in **Figure 20.37** between the same initial and final states that lie on an isotherm. In process A the gas is taken via an isochoric process to an intermediate state 1 and then via an isobaric process to the final state. In process B the gas undergoes an isentropic transformation to an intermediate state 2 and then an isobaric transformation to the final state. Which process has the greater value for (a) the work done on the gas and (b) the magnitude of the energy transferred thermally to the gas?

20.8 Entropy change in ideal gas processes

In this section we examine the entropy changes that occur for the four constrained quasistatic processes discussed in the preceding two sections. Recall from Chapter 19 that the entropy of a monatomic ideal gas is determined completely by the volume and the temperature of the gas (Eq. 19.59):

$$S = N \ln V + \tfrac{3}{2} N \ln T + \text{constant} \quad \text{(monatomic ideal gas).} \quad (20.32)$$

We obtained this expression for a monatomic ideal gas where each atom has three degrees of freedom (Section 20.3). The factor 3 in the $\frac{3}{2}$ in this expression comes from the three dimensions of the velocity space among which we equipartitioned the energy (Eq. 19.29). In the more general case of an ideal gas with d degrees of freedom, we replace the 3 by d:

$$S = N \ln V + \frac{d}{2} N \ln T + \text{constant} \quad \text{(ideal gas).} \quad (20.33)$$

Figure 20.37 Checkpoint 20.18.

(a) Process A: isochoric step, then isobaric step

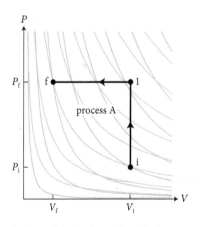

(b) Process B: isentropic step, then isobaric step

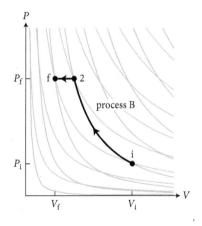

During a process in which the gas undergoes a transformation from an initial equilibrium state (V_i, T_i) to a final equilibrium state (V_f, T_f), the entropy changes by an amount (see Eqs. 19.8 and 19.56)

$$\Delta S = N \ln\left(\frac{V_f}{V_i}\right) + \frac{d}{2} N \ln\left(\frac{T_f}{T_i}\right) \quad \text{(ideal gas).} \tag{20.34}$$

For an isothermal process, $T_i = T_f$, and so the second term on the right is zero because $\ln(1) = 0$. Equation 20.34 thus becomes

$$\Delta S = N \ln\left(\frac{V_f}{V_i}\right) \quad \text{(isothermal process),} \tag{20.35}$$

a result we already obtained in Eq. 19.8. Comparing this result with Eq. 20.31, we see that we can express the entropy change for an isothermal process as

$$\Delta S = \frac{Q}{k_B T} \quad \text{(isothermal process).} \tag{20.36}$$

This expression allows us to calculate the change in entropy directly once we know the amount of energy transferred thermally, allowing us to bypass Eq. 20.34. Equation 20.36 is applicable only when the temperature is constant, however. For any other process, Eq. 20.34 provides the simplest approach for calculating the entropy change.

For an isochoric process, $V_f = V_i$, and so Eq. 20.34 reduces to

$$\Delta S = \frac{d}{2} N \ln\left(\frac{T_f}{T_i}\right) \quad \text{(isochoric process)} \tag{20.37}$$

or, substituting from Eq. 20.14,

$$\Delta S = \frac{N C_V}{k_B} \ln\left(\frac{T_f}{T_i}\right) \quad \text{(isochoric process).} \tag{20.38}$$

✋ **20.19** Can you use Eq. 20.36 to calculate the change in entropy for an isochoric process?

In an isobaric process, both V and T change, but the two are related by the ideal gas law, $PV = N k_B T$. Because P is constant in any isobaric process, we have $V_i = (N k_B/P) T_i$ and $V_f = (N k_B/P) T_f$. Therefore

$$\frac{V_f}{V_i} = \frac{(N k_B/P) T_f}{(N k_B/P) T_i} = \frac{T_f}{T_i} \quad \text{(isobaric process).} \tag{20.39}$$

With this result, Eq. 20.34 becomes

$$\Delta S = N \ln\left(\frac{T_f}{T_i}\right) + \frac{d}{2} N \ln\left(\frac{T_f}{T_i}\right)$$

$$= N\left(1 + \frac{d}{2}\right) \ln\left(\frac{T_f}{T_i}\right) \quad \text{(isobaric process).} \tag{20.40}$$

Substituting from Eq. 20.25, we can rewrite this expression in the form

$$\Delta S = \frac{N C_P}{k_B} \ln\left(\frac{T_f}{T_i}\right) \quad \text{(isobaric process).} \tag{20.41}$$

Note the similarity between the expression for the entropy change in an isochoric process (Eq. 20.38) and the expression for the entropy change in an isobaric process (Eq. 20.41). The only difference is the type of heat capacity per particle: For the isochoric process we use C_V; for the isobaric process we use C_P.

As an example, consider the isobaric expansion shown in Figure 20.34. The final temperature is higher than the initial temperature, and so, because $\ln(T_f/T_i) > 0$, Eq. 20.41 tells us that the entropy increases. Indeed, from the PV diagram in Figure 20.34c we see that the final state must lie on a higher isentrope than the initial state, indicating that the entropy of the final state is greater than the entropy of the initial state.

In an isentropic process, the entropy doesn't change, and so

$$\Delta S = 0 \quad \text{(isentropic process)}. \tag{20.42}$$

This result implies that during an isentropic process the changes in volume and temperature are such that the two terms in Eq. 20.34 cancel each other. In other words, the volume and temperature changes are correlated in such a way as to keep the entropy of the system constant.

✋ **20.20** Suppose two different combinations of constrained quasistatic processes take two identical ideal gases from the same initial equilibrium state to the same final equilibrium state. Is the change in entropy the same or different in the two cases?

Table 20.7 summarizes the entropy changes for the four constrained ideal gas processes. As you saw in Checkpoint 20.20, the value of the entropy change is independent of the process and depends on only the temperatures and volumes of the initial and final equilibrium states. For *any* process, we can use Eq. 20.34 to calculate the entropy change. The expressions summarized in Table 20.7, however, provide a quick way to obtain the entropy changes for the four basic constrained ideal gas processes. Keep in mind that in these processes the system is not closed, and so the entropy change can be positive, negative, or zero.

Equation 20.33 can be rewritten in a form that allows us to derive useful relationships among the pressure, volume and temperature of any two states, (P_i, V_i, T_i) and (P_f, V_f, T_f), that lie on an isentrope. We begin by combining the first and second terms on the right in Eq. 20.33, which gives us

$$S = N \ln(T^{d/2}V) + \text{constant}. \tag{20.43}$$

Because the entropy is constant, the quantity $T^{d/2}V$ must be constant as well for a sample that consists of a fixed number N of particles: $T_i^{d/2}V_i = T_f^{d/2}V_f$. Raising

Table 20.7 Entropy changes in ideal gas processes

Process	ΔS	Equation
isochoric	$\dfrac{NC_V}{k_B} \ln\left(\dfrac{T_f}{T_i}\right)$	20.38
isothermal	$N \ln\left(\dfrac{V_f}{V_i}\right) = \dfrac{Q}{k_B T}$	20.35, 20.36
isobaric	$\dfrac{NC_P}{k_B} \ln\left(\dfrac{T_f}{T_i}\right)$	20.41
isentropic	0	20.42

the constant quantity $T^{d/2}V$ to the power $\frac{2}{d}$, and substituting $\frac{2}{d} = \gamma - 1$ from Eq. 20.26, we see that $(T^{d/2}V)^{2/d} = TV^{\gamma-1}$, or

$$T_i V_i^{\gamma-1} = T_f V_f^{\gamma-1} \quad \text{(isentropic process).} \tag{20.44}$$

This expression provides the relationship between volume and temperature along an isentrope. Using the ideal gas law, we can obtain expressions relating other quantities along an isentrope. Multiplying the ideal gas law, $PV = Nk_B T$, by $V^{\gamma-1}$, we obtain:

$$PV^\gamma = Nk_B TV^{\gamma-1}. \tag{20.45}$$

From Eq. 20.44 we see that the right side of Eq. 20.45 is constant for an isentropic process, and thus the quantity PV^γ must also be constant:

$$P_i V_i^\gamma = P_f V_f^\gamma \quad \text{(isentropic process).} \tag{20.46}$$

From the ideal gas law, we know that V is proportional to T/P, which means that $PV^\gamma \propto P(T/P)^\gamma = P^{1-\gamma}T^\gamma$ is constant for an isentropic process. Raising this quantity to the power $1/\gamma$, we get $(P^{1-\gamma}T^\gamma)^{1/\gamma} = P^{(1/\gamma)-1}T$, and so in an isentropic process,

$$P_i^{(1/\gamma)-1}T_i = P_f^{(1/\gamma)-1}T_f \quad \text{(isentropic process).} \tag{20.47}$$

Equations 20.44, 20.46 and 20.47 relate the pressure, volume, and temperature for the initial and final equilibrium states of an ideal gas sample that consists of N particles undergoing an isentropic process.

Example 20.9 Entropy changes

Figure 20.38 shows two quasistatic processes that take a sample of an ideal gas containing N particles from an initial state (P_i, V_i) to a final state (P_f, V_f). Process A is a one-step isentropic compression. Process B has two steps: an isochoric pressure increase to an intermediate state 1 followed by an isobaric decrease in volume. Express the changes in entropy for process A and process B in terms of the given quantities.

Figure 20.38

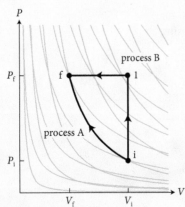

❶ **GETTING STARTED** The fact that the initial and final states are on different isotherms (red curves) in Figure 20.38 tells me

that the two are at different temperatures, which I call T_i and T_f. These temperatures are related to the initial and final volume and pressure by the ideal gas law. For the intermediate state, I can write $(P_1, V_1) = (P_f, V_i)$. I'll denote the temperature of this state by T_1.

❷ **DEVISE PLAN** Table 20.7 gives expressions for the entropy changes in the four constrained processes. I can eliminate the temperature from these expressions by using the ideal gas law, $PV = Nk_B T$, and Eq. 20.39.

❸ **EXECUTE PLAN** Process A is isentropic, which means that its entropy doesn't change: $\Delta S_A = 0$. ✔

I first obtain the entropy change for the isochoric step of process B from Eq. 20.38:

$$\Delta S_{B,\text{isochoric}} = \frac{NC_V}{k_B}\ln\!\left(\frac{T_1}{T_i}\right).$$

From the ideal gas law, I get for the initial and intermediate states, respectively:

$$P_i V_i = Nk_B T_i \tag{1}$$

$$P_f V_i = Nk_B T_1. \tag{2}$$

Dividing Eq. 2 by Eq. 1 gives me $T_1/T_i = P_f/P_i$, and so for the first step of process B I have

$$\Delta S_{B,isochoric} = \frac{NC_V}{k_B}\ln\left(\frac{P_f}{P_i}\right). \qquad (3)$$

The entropy change for the isobaric step of process B is given by Eq. 20.41:

$$\Delta S_{B,isobaric} = \frac{NC_P}{k_B}\ln\left(\frac{T_f}{T_1}\right).$$

Equation 20.39 yields $T_f/T_1 = V_f/V_i$, so I can write the entropy change as

$$\Delta S_{B,isobaric} = \frac{NC_P}{k_B}\ln\left(\frac{V_f}{V_i}\right). \qquad (4)$$

The entropy change for process B is the sum of the entropy changes for the two steps, so, adding Eqs. 3 and 4, I get

$$\Delta S_B = \frac{NC_V}{k_B}\ln\left(\frac{P_f}{P_i}\right) + \frac{NC_P}{k_B}\ln\left(\frac{V_f}{V_i}\right). \checkmark$$

④ **EVALUATE RESULT** The two processes connect the same initial and final states, so the entropy change should be the same for the two processes. Therefore the expression I obtained for process B should be zero: $\Delta S_B = 0$. Because the process that takes the gas from the initial to the final state is isentropic, Eq. 20.46 tells me that $P_f/P_i = (V_i/V_f)^\gamma$. Summing my expressions for $\Delta S_{B,isochoric}$ and $\Delta S_{B,isobaric}$ and making this substitution for P_f/P_i yield

$$\Delta S_B = \frac{NC_V}{k_B}\ln\left(\frac{V_i}{V_f}\right)^\gamma + \frac{NC_P}{k_B}\ln\left(\frac{V_f}{V_i}\right)$$

$$= \frac{NC_V}{k_B}\gamma\ln\left(\frac{V_i}{V_f}\right) + \frac{NC_P}{k_B}\ln\left(\frac{V_f}{V_i}\right).$$

Because $\gamma \equiv C_P/C_V$ (Eq. 20.26), I obtain, as expected

$$\Delta S_B = \frac{NC_P}{k_B}\ln\left(\frac{V_i}{V_f}\right) + \frac{NC_P}{k_B}\ln\left(\frac{V_f}{V_i}\right)$$

$$= \frac{NC_P}{k_B}\ln\left(\frac{V_i}{V_f}\right) - \frac{NC_P}{k_B}\ln\left(\frac{V_i}{V_f}\right) = 0.$$

✋ **20.21** For a quasistatic ideal gas process that ends on a higher isentrope, are W, Q, ΔE_{th}, and ΔS positive, negative, or zero when the gas is (a) expanded and (b) compressed?

20.9 Entropy change in nonideal gas processes

Let us now briefly consider what happens to the energy and entropy when the temperature of a liquid or gas changes. When we deal with a solid or liquid of given mass and constant volume, it is more convenient to work with specific heat capacity (see Section 20.3 and Table 20.2) rather than heat capacity per particle. The specific heat capacity is obtained by dividing the energy transferred thermally by the mass of the object or system and by the temperature change:*

$$c_V \equiv \frac{Q}{m\Delta T}. \qquad (20.48)$$

And so for a given change in temperature, the energy transferred thermally is

$$Q = mc_V\Delta T. \qquad (20.49)$$

Because the volume is constant, no work is done and the energy law reduces to $\Delta E_{th} = Q$, so the system's energy change is

$$\Delta E_{th} = mc_V\Delta T. \qquad (20.50)$$

---

*We are assuming here that the specific heat capacity does not depend on temperature. As Figure 20.15 shows, however, this is only approximately true over limited temperature ranges.

To calculate the corresponding entropy change, we rewrite the definition of absolute temperature (Eq. 19.38), which gives the relationship between changes in thermal energy and entropy, as follows:

$$dS = \frac{dE_{th}}{k_B T}. \tag{20.51}$$

For an infinitesimally small change in temperature, Eq. 20.50 becomes $dE_{th} = mc_V dT$, and so Eq. 20.51 becomes

$$dS = \frac{mc_V}{k_B} \frac{dT}{T}. \tag{20.52}$$

To determine the entropy change for a finite temperature change, we integrate this expression for a temperature change $\Delta T = T_f - T_i$:

$$\Delta S = \frac{mc_V}{k_B} \int_{T_i}^{T_f} \frac{dT}{T} = \frac{mc_V}{k_B} \ln\left(\frac{T_f}{T_i}\right). \tag{20.53}$$

We can also use Eq. 20.51 to obtain a more general expression for the entropy change in an isothermal process. Let the energy thermally transferred be Q. At constant temperature the entropy change is then

$$\Delta S = \frac{Q}{k_B T} \quad \text{(isothermal process)}. \tag{20.54}$$

Note that this expression is identical to the one we obtained earlier for an ideal gas (Eq. 20.36).

Let us next determine what happens to the energy and entropy of a system during a phase transition. In Section 20.3 we defined the **specific transformation energy** as the energy transferred thermally to a substance undergoing a phase transition divided by the mass of the substance:

$$L \equiv \frac{Q}{m} \quad \text{(phase transition with } Q > 0\text{)}. \tag{20.55}$$

Because the quantity L is always positive and Q can be negative or positive, Eq. 20.55 holds only for melting and vaporization. For solidification and condensation, Q is negative and so we have

$$L \equiv \frac{-Q}{m} \quad \text{(phase transition with } Q < 0\text{)}. \tag{20.56}$$

Because the temperature is constant, we can use Eq. 20.54 to calculate the corresponding entropy change

$$\Delta S = \frac{Q}{k_B T} = \frac{mL}{k_B T} \quad \text{(melting or vaporization)}. \tag{20.57}$$

For solidification and condensation the entropy change is of the opposite sign:

$$\Delta S = -\frac{mL}{k_B T} \quad \text{(solidification or condensation)}. \tag{20.58}$$

20.22 An amount of water with a mass of 0.010 kg at a temperature of 20 °C is placed in a freezer and turns into an ice cube at 0.0 °C. What is the corresponding entropy change? Ignore any volume changes.

QUANTITATIVE TOOLS

Chapter Glossary

SI units of physical quantities are given in parentheses.

Absolute zero The temperature at which the thermal energy of a system is minimum. A system at absolute zero has only a single microstate and zero entropy.

Adiabatic process A process that does not involve any thermal transfer of energy ($Q = 0$).

Degree of freedom Any independent way in which a particle can store energy.

Energy transferred thermally Q (J) A scalar equal to the change in the energy of a system due to thermal interactions. If the energy of the system increases, the energy transferred thermally is positive; if the energy decreases, the energy transferred thermally is negative.

Equipartition energy share In thermal equilibrium, each degree of freedom of a system at temperature T stores an amount of energy equal to $\frac{1}{2}k_B T$ per particle.

Heat capacity (J/K) A scalar defined as the amount of energy transferred thermally to an object divided by the resulting change in temperature.

Heat capacity per particle C (J/K) A scalar equal to the heat capacity of an object divided by the number of particles making up that object:

$$C \equiv \frac{Q}{N\Delta T}. \qquad (20.13, 20.20)$$

A subscript V or P is added when the heating takes place at constant volume or at constant pressure.

Heat capacity ratio γ (unitless) A scalar equal to the ratio of the heat capacity per particle at constant pressure to the heat capacity per particle at constant volume. For an ideal gas:

$$\gamma \equiv \frac{C_P}{C_V} = \frac{\left(\frac{d}{2} + 1\right)k_B}{\frac{d}{2}k_B} = 1 + \frac{2}{d} \quad \text{(ideal gas). (20.26)}$$

Ideal gas thermometer A device for measuring temperature using an ideal gas.

Isentrope A curve on a PV diagram that represents states having the same entropy. These curves are sometimes called *adiabats*.

Isentropic process A process for which the entropy of the system is constant. Quasistatic adiabatic processes are isentropic. An isentropic process is an adiabatic process.

Isobaric process A process for which the pressure in the system is held constant.

Isochoric process A process for which the volume of the system is held constant.

Isotherm A curve on a PV diagram that represents states having the same temperature.

Isothermal process A process for which the temperature of the system is held constant.

Process The transformation of a system from one macrostate to another.

Quasistatic process A process for which the system always remains very close to equilibrium as the transformation takes place. A path on a PV diagram always represents a quasistatic process.

Specific heat capacity c (J/K·kg) A scalar equal to the heat capacity of an object divided by the mass of the object.

Specific transformation energy L (J/kg) The energy transferred thermally to a substance undergoing a phase transition divided by the mass of the substance:

$$L \equiv \frac{Q}{m} \quad \text{(phase transition with } Q > 0\text{).} \qquad (20.55)$$

$$L \equiv \frac{-Q}{m} \quad \text{(phase transition with } Q < 0\text{).} \qquad (20.56)$$

Thermal interaction An interaction between two objects at different temperatures that results in a thermal transfer of energy from the object at the higher temperature to the object at the lower temperature.

Thermal reservoir A body of material large enough so that its temperature does not change in a thermal interaction with the system under consideration.

Triple point of water The temperature and pressure at which the solid, liquid, and gas phases of water coexist in thermal equilibrium. By international convention, the temperature of the triple point of water is 273.16 K.

21 Degradation of Energy

Some textbooks define energy as the *capacity to do work*. In the two preceding chapters, however, we saw that energy can do more than that. It can also be used to heat objects, and in this chapter we ask (and answer) two important questions about energy: To what extent can we use mechanical energy for heating and thermal energy for doing work? And, more specifically, to what extent can we convert energy from forms provided by nature to forms useful for technology and society?

The oceans of the world, for example, contain a very large amount of thermal energy. If we could convert just a small fraction of that energy to useful forms, we could satisfy the world's energy needs for thousands of years. Why can't a ship take in some of the thermal energy in the ocean and use it for propulsion? Such a scheme does not violate the energy law: Thermal energy would simply be converted to an equal amount of mechanical energy. As we see in this chapter, however, the entropy law imposes constraints on conversions of energy.

21.1 Converting energy

In Chapter 5 we saw that energy is a quantitative measure of the state of an object and that the law of conservation of energy implies that any change of state must be accompanied by a compensating change of state (or by a set of compensating changes of state). Whenever a pair of complementary changes of different types take place, energy is converted from one form to another. For example, burning propane gas involves two changes of state: a change in the molecular chemical bonds of the reagents (which we associate with chemical energy) and a change in the temperature of the gases (which we associate with thermal energy). The combustion is irreversible: It would be pretty startling if the hot gases would suddenly "unburn," cooling down and reconstituting the original propane and oxygen. Because the process is irreversible, we can conclude that the entropy of the gas increases in the process of burning (see Section 19.4).

A ball accelerating in free fall is another example of a complementary pair of changes of state: a change in the state of motion of the ball (which we associate with kinetic energy) and a change in the configuration of the Earth-ball system (which we associate with gravitational potential energy). This process is reversible: A movie clip of the falling ball played backward shows an upward moving ball slowing down as if it had been thrown upward (which *can* happen). We can thus conclude that the entropy of the Earth-ball system does not change.

21.1 For each system, identify the complementary changes of state and associated energies, and determine whether or not the entropy of the system increases. (*a*) A puck slides to a stop on a surface. System: puck and surface. (*b*) A compressed spring is released and accelerates a cart on a low-friction track. System: spring and cart. (*c*) A projectile is fired from a cannon. System: projectile, cannon, and explosive chemicals. (*d*) A hot object and a cold object are placed in contact and reach thermal equilibrium. System: both objects.

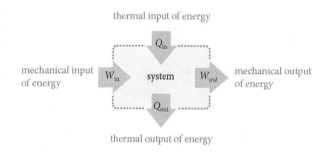

Figure 21.1 Energy input-output diagram.

To analyze the energy conversions that occur in a process, such as the ones in Checkpoint 21.1, we must choose a system and separately analyze all the energy transfers into and out of the system. As we saw in Chapter 20, energy can be transferred in two ways: mechanically and thermally. **Figure 21.1** shows for each of these two types of transfers an arrow representing the transfer of energy *into* the system (called *energy input*) and an arrow representing the transfer of energy *out of* the system (called *energy output*). Diagrams that show the energy inputs and energy outputs of a system are called **energy input-output diagrams.**

Energy inputs and energy outputs are always taken to be positive. The work done on a system is thus the difference between the **mechanical input of energy** and the **mechanical output of energy**: $W = W_{in} - W_{out}$. Likewise, the energy transferred thermally into a system is the difference between the **thermal input of energy** and the **thermal output of energy**: $Q = Q_{in} - Q_{out}$.

Figure 21.1 can represent the functioning of any device or industrial process. **Figure 21.2** shows the energy input-output diagrams for two devices: a car engine and a refrigerator. In a car, the combustion of gasoline provides a thermal input of energy to the engine (the system). The engine has two energy outputs: a mechanical output of energy to the wheels of the car and a thermal output of energy to the environment (Figure 21.2*a*). In a refrigerator there are two energy inputs (Figure 21.2*b*): a thermal input of energy from the refrigerator contents (which are part of the environment) and a mechanical input of energy from the compressor motor to the rest of the refrigerator (the system). The refrigerator also has a thermal output of energy to the environment. (You may have noticed that the back of a refrigerator gets hot.)

Each energy input or energy output in Figures 21.1 and Figure 21.2 is accompanied by changes in entropy in the environment and in the system. To gain some practice determining energy and entropy changes, let us analyze a few specific situations.

Exercise 21.1 Converting mechanical energy to thermal energy

Consider the device illustrated in **Figure 21.3**. A string is attached to a paddle wheel that is immersed in a viscous liquid in a container. As the string is pulled, the paddle wheel rotates at

CONCEPTS

Figure 21.2 Examples of energy input-output diagrams.

(a) Energy input-output diagram for car engine

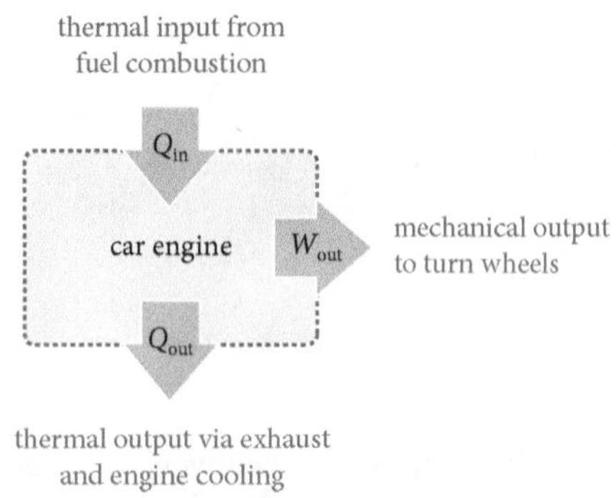

thermal input from
fuel combustion

Q_{in}

car engine W_{out} mechanical output
to turn wheels

Q_{out}

thermal output via exhaust
and engine cooling

(b) Energy input-output diagram for refrigerator

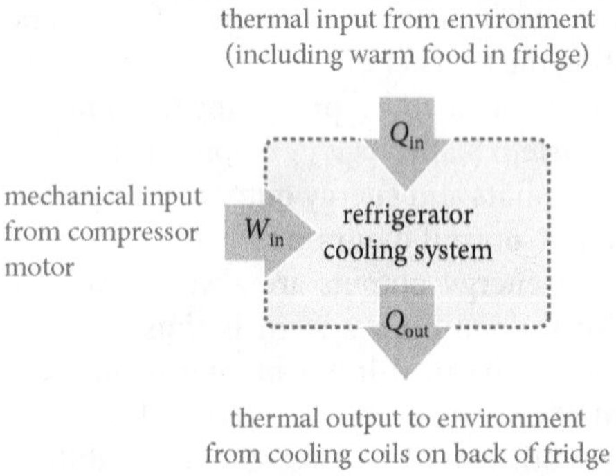

thermal input from environment
(including warm food in fridge)

Q_{in}

mechanical input
from compressor W_{in} refrigerator
motor   cooling system

Q_{out}

thermal output to environment
from cooling coils on back of fridge

constant speed, increasing the temperature of the liquid from T_i to T_f. (a) Draw an energy input-output diagram for the system comprising the paddle wheel and the liquid. Do the energy and entropy of the system comprising the paddle wheel and the liquid increase, decrease, or stay the same? (b) Assume the paddle wheel stops turning at the instant shown in Figure 21.3b, and we now

Figure 21.3

Hand does mechanical work on system by pulling string.

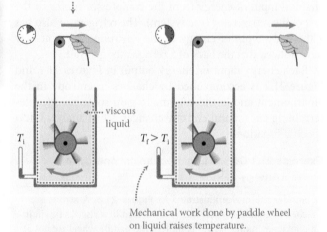

T_i

$T_f > T_i$

viscous
liquid

Mechanical work done by paddle wheel
on liquid raises temperature.

Figure 21.4

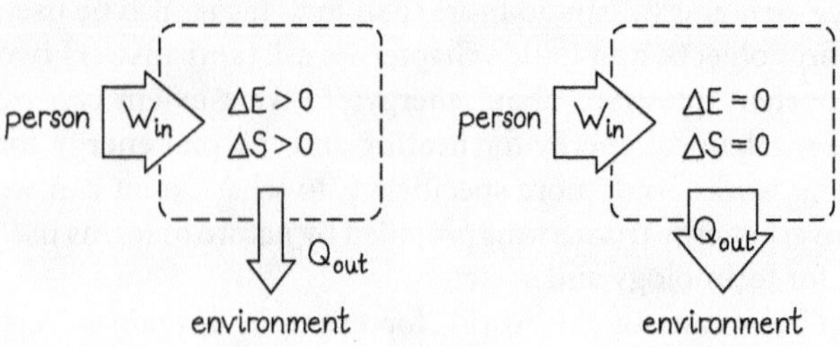

(a)

person W_{in} $\Delta E > 0$
                    $\Delta S > 0$

Q_{out}

environment

(b)

person W_{in} $\Delta E = 0$
                    $\Delta S = 0$

Q_{out}

environment

wait until the liquid cools back down to its initial temperature T_i. Draw an energy input-output diagram for the system, but this time over the time interval from the instant the wheel begins turning to the instant the liquid temperature is again T_i. Do the energy and entropy of the system increase, decrease, or stay the same over this new time interval?

SOLUTION (a) The work done on the system by pulling the string corresponds to an energy input. Because the system is not thermally insulated, energy is transferred thermally to the environment as the temperature of the system increases. My energy input-diagram therefore has one input and one output (**Figure 21.4a**). Because the temperature of the system increases, I know that the energy of the system increases and so the energy input must be greater than the energy output. The increase in thermal energy increases the entropy of the system. ✔

(b) If the liquid cools back to its initial temperature, the initial and final states of the system are the same, so the energy change and the entropy change are both zero. The input of energy is the same as in part a, but because the energy of the system is not changing, the thermal output of energy must now be equal in magnitude to the mechanical input (Figure 21.4b). ✔

Note that the energy law imposes no restrictions on energy conversions: The energy law permits any form of energy to be converted entirely to any other form of energy. The device in Exercise 21.1, for example, converts mechanical energy entirely to thermal energy. Let us now look at the conversion of thermal energy to mechanical energy.

✋ **21.2** (a) For the situation illustrated in **Figure 21.5**, draw an energy input-output diagram for the gas. (b) Do the energy and entropy of the gas change? Ignore any friction in the piston and any energy transferred thermally out of the gas.

As you saw in Checkpoint 21.2, energy transferred thermally to the gas does work on the piston, and so the device converts thermal energy to mechanical energy. Because the temperature of the gas goes up, however, the thermal energy of the gas increases. Thus not all thermal energy is converted to mechanical energy in this device.

To prevent the thermal energy of the gas from increasing, we could immerse the setup in a thermal reservoir held at some fixed temperature T_R, as illustrated in **Figure 21.6**. If we then slowly remove grains of sand lying on top of the piston, the upward force exerted by the gas on the piston is

Figure 21.5 When a volume of gas is heated at constant pressure in a cylinder fitted with a movable piston, both its temperature and its volume increase.

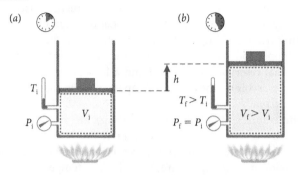

a tiny bit greater than the downward force exerted by the remaining sand on the piston, and so the piston moves up. As the gas expands, it does work on the piston. Because the temperature of the gas remains constant, its thermal energy doesn't change, and so the energy required to raise the piston must be transferred thermally from the reservoir. Here is a situation where thermal energy is converted entirely to mechanical energy.

As a device, the setup of Figure 21.6 is not very practical because to maintain a constant gas temperature, the expansion would have to proceed *very* slowly. Moreover, in order for the gas to continue to do work on the piston, its volume would have to keep expanding.

When designing practical devices for converting energy, we usually want them to be able to operate indefinitely. If the device were to keep expanding in volume as in Figure 21.6 or keep getting hotter or building up pressure over time, it could not be operated for very long. For this reason we want the state of the device that converts energy to remain the same or to return to the same state over some repeating time interval. Such a device is called a **steady device.** When a steady device evolves from an initial state through a series of processes back to its initial state, the series of processes is called a *cycle*. During the operation of the device, this cycle is repeated continuously. For simplicity, we shall assume that all steady devices involve only quasistatic processes and that no energy is dissipated due to friction.

Figure 21.6 A gas held at constant temperature in a cylinder fitted with a movable piston expands slowly as the load on the piston is diminished.

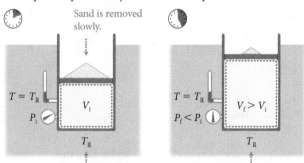

thermal reservoir at constant temperature T_R

Because the initial and final states of a steady device are the same over one cycle, there is no change in the energy of the system: $\Delta E = 0$. Over one cycle, therefore, the energy input must be equal to the energy output: $W_{in} + Q_{in} = W_{out} + Q_{out}$.

To see what constraints are imposed by the entropy law (Eq. 19.5; $\Delta S > 0$ for a closed system moving toward equilibrium) on the energy inputs and outputs of a steady device, we examine the entropy changes associated with these energy inputs and outputs. The device is not a closed system and so we cannot apply the entropy law to it. The device and its environment constitute a closed system, however, so we can write $\Delta S = \Delta S_{dev} + \Delta S_{env} \geq 0$ (Eqs. 19.5, 19.6). Because the initial and final states of the cycle are the same, we know that there cannot be any change in the entropy of the device: $\Delta S_{dev} = 0$ for one cycle. Therefore, we need to consider only entropy changes in the environment when we deal with steady devices and make sure these changes satisfy $\Delta S_{env} \geq 0$.

As we saw in Section 20.1, no entropy change is associated with a quasistatic adiabatic process. Because $Q = 0$ for mechanical inputs and outputs of energy and because we assume that the processes in a steady device are quasistatic, we have $\Delta S_{env} = 0$ for W_{in} and W_{out}. Therefore the only changes in the entropy of the environment we need to consider are those caused by thermal inputs and outputs of energy:

Steady devices operate under two constraints: (1) The energy input must be equal to the energy output. (2) The device's thermal inputs and outputs must increase the entropy of the environment (or leave it unaltered).

In Chapter 20, we saw that the entropy change associated with energy transferred thermally has the same sign as Q (Eq. 20.54). Therefore, a thermal output of energy increases the entropy of the environment. Conversely, a thermal input of energy decreases the entropy of the environment. **Figure 21.7** summarizes these entropy changes due to the various kinds of energy inputs and energy outputs for a steady device. We use a circular system boundary to indicate that the system is a steady device. The diagram represents the energy inputs and outputs during one complete cycle.

Figure 21.7 Energy and entropy changes associated with energy inputs and energy outputs for a steady device.

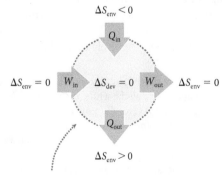

Circular system boundary indicates *steady device* ($\Delta E = 0$, $\Delta S_{dev} = 0$).

Figure 21.8 Energy input-output diagrams for two types of steady device.

(*a*) Device that converts mechanical energy to thermal energy

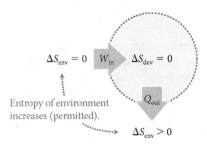

$\Delta S_{\text{env}} = 0$ W_{in} $\Delta S_{\text{dev}} = 0$

Q_{out}

Entropy of environment
increases (permitted).

$\Delta S_{\text{env}} > 0$

(*b*) Attempt to convert thermal energy entirely to mechanical energy

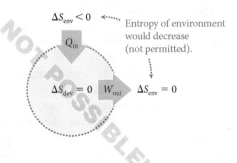

$\Delta S_{\text{env}} < 0$ Entropy of environment
would decrease
(not permitted).

Q_{in}

$\Delta S_{\text{dev}} = 0$ W_{out} $\Delta S_{\text{env}} = 0$

NOT POSSIBLE!

We are now in a position to analyze steady devices that convert energy. First consider a steady device that converts mechanical energy to thermal energy. Such a device has a mechanical input of energy and a thermal output of energy (**Figure 21.8a**). The mechanical input of energy involves no change in the entropy of the environment; the thermal output of energy causes an increase in the entropy of the environment. The entropy of the environment therefore increases, which is permitted by the entropy law. This means that a steady device can convert mechanical energy entirely to thermal energy.

Next consider a steady device that converts thermal energy to mechanical energy (Figure 21.8*b*). The thermal input of energy causes a decrease in the entropy of the environment; the mechanical output of energy involves no change in the entropy of the environment. The entropy of the environment therefore decreases, which is not permitted by the entropy law for a steady device. Therefore no steady device can convert thermal energy entirely to mechanical energy.

✋ **21.3** Suppose we slowly push the piston back to its original position in Figure 21.6, all the while maintaining a constant temperature in the gas. (*a*) Over the time interval that includes the expansion and the compression of the gas, has any energy been converted? (*b*) Does the entropy of the gas increase, stay the same, or decrease over that time interval? (*c*) If we keep repeating the expansion and compression, is this device a steady device?

21.2 Quality of energy

In this section we develop a graphical approach to keep track of entropy changes in the environment caused by steady devices that convert energy. Let us begin by considering the very simple situation illustrated in **Figure 21.9a**: A copper rod is in thermal contact with a high-temperature thermal reservoir on one end and a low-temperature thermal reservoir on the other end. Apart from its ends, the rod is thermally isolated from its environment. Because the temperatures of the thermal reservoirs are fixed, the rod remains in a steady state: One end remains at the temperature T_{H} of the high-temperature reservoir, and the other end remains at the temperature T_{L} of the low-temperature reservoir. (Remember that thermal reservoirs are large enough so that any energy transferred thermally to them does not change their temperature; see Section 20.7.) As a result of the temperature difference, energy is transferred thermally through the rod from the high-temperature reservoir to the low-temperature reservoir. Although the rod hardly qualifies as a device, its state doesn't change and so it satisfies the requirements that the entropy and energy of a steady device do not change over time. (The cycle for this device can be taken to be any arbitrary time interval.)

Figure 21.9*b* shows the energy input-output diagrams for the rod and for the two thermal reservoirs. The change in the energy of the high-temperature reservoir is equal to the thermal input of energy of the rod, $\Delta E_{\text{H}} = -Q_{\text{in}}$, while the change in the energy of the low-temperature reservoir is equal to the thermal output of energy of the rod, $\Delta E_{\text{L}} = Q_{\text{out}}$. Because the state of the rod is not changing, we know that $\Delta E_{\text{rod}} = Q_{\text{in}} - Q_{\text{out}} = 0$, so $Q_{\text{in}} = Q_{\text{out}}$ and therefore $\Delta E_{\text{H}} = -\Delta E_{\text{L}}$. The amount of energy transferred out of the high-temperature reservoir is equal to the amount of energy transferred into the low-temperature reservoir.

Figure 21.9 (*a*) A copper rod in thermal contact with a high-temperature reservoir and a low-temperature reservoir. (*b*) Energy input-output diagrams for the rod and the reservoirs.

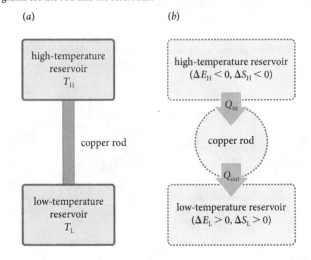

(*a*)

high-temperature
reservoir
T_{H}

copper rod

low-temperature
reservoir
T_{L}

(*b*)

high-temperature reservoir
($\Delta E_{\text{H}} < 0, \Delta S_{\text{H}} < 0$)

Q_{in}

copper rod

Q_{out}

low-temperature reservoir
($\Delta E_{\text{L}} > 0, \Delta S_{\text{L}} > 0$)

Figure 21.10 Entropy gradient. The shaded bar represents the rod's environment, which consists of the high-temperature reservoir (H) and the low-temperature reservoir (L).

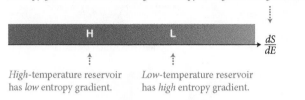

Entropy gradient measures change in entropy for a given change in energy.

$\dfrac{dS}{dE}$

High-temperature reservoir has *low* entropy gradient.　*Low*-temperature reservoir has *high* entropy gradient.

The thermal transfer of energy through the rod causes entropy changes in the two reservoirs (both part of the rod's environment) even though their temperatures remain unchanged. Because energy is transferred out of the high-temperature reservoir, its energy decreases, and so its entropy decreases too: $\Delta S_H < 0$. The increase of energy in the low-temperature reservoir causes its entropy to increase: $\Delta S_L > 0$. Because the transfer of energy is irreversible, we know that the entropy of the environment must increase, so that $\Delta S_H + \Delta S_L > 0$.

The reason entropy increases can be understood by looking at the way we defined temperature in Chapter 19: $dS/dE \equiv 1/(k_B T)$ (Eq. 19.38).* The quantity dS/dE, called the **entropy gradient**, is a measure of the change in entropy caused by a given change in energy. Equation 19.38 tells us that the magnitude of the change in entropy caused by transferring a given amount of energy to or from a reservoir is inversely proportional to the temperature of the reservoir. The entropy change ΔS caused by a given change in energy ΔE is therefore greater in magnitude for the low-temperature reservoir than for the high-temperature reservoir. Because ΔS_L is positive and ΔS_H is negative, it follows that $\Delta S_H + \Delta S_L > 0$.

✋ 21.4 What are the signs of ΔE_R, ΔS_R, and $\Delta S_R/\Delta E_R$ for a thermal transfer of energy (*a*) to and (*b*) from a thermal reservoir?

You can view the entropy gradient as the "entropy cost of an energy change." The greater the entropy gradient (or the lower the temperature), the greater the change in entropy caused by a given energy change. As you saw in Checkpoint 21.4, the entropy gradient is always positive, regardless of the direction of the energy transfer, because ΔS and ΔE always have the same algebraic sign.

We can arrange the rod's environment by its entropy gradient. Because the high-temperature reservoir has the smaller entropy gradient, we place it to the left of the low-temperature reservoir on the entropy-gradient scale in **Figure 21.10**. If we represent the thermal transfer of energy through the rod in **Figure 21.11a**, by a curved arrow, we obtain the diagram shown in Figure 21.11*b*. The tail of the

---

*As we did in Chapter 20, we assume that the kinetic, potential, and source energies of the reservoirs remain constant, so that $\Delta E = \Delta E_{th}$.

Figure 21.11 (*a*) We rearrange the device of Figure 21.9 to separate the system (the steady device) from its environment (the thermal reservoirs). (*b*) Using this geometry, we construct an entropy diagram showing the energy conversions and transfers in the steady device. The thermal input at low dS/dE causes entropy change in the environment of smaller magnitude than thermal output at high dS/dE, so the entropy of the environment increases: $\Delta S_H + \Delta S_L > 0$.

(*a*) We separate the device (the system) from its environment

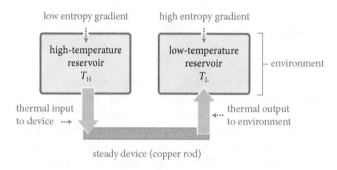

(*b*) Entropy diagram

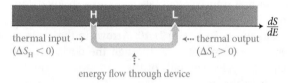

arrow represents the thermal input of energy from the high-temperature reservoir, and the head represents the thermal output of energy to the low-temperature reservoir.

Note that we can associate the sign of the entropy changes in the reservoirs with the direction of the arrow. The tail of the arrow (the energy input) points down and $\Delta S_H < 0$; the head (the energy output) points up and $\Delta S_L > 0$. The diagram in Figure 21.11*b* therefore allows us to see that the transfer of energy through the rod increases the entropy of the environment: The rod takes a certain amount of energy out of the environment at low entropy cost (small dS/dE) and returns the same amount of energy to the environment at a greater entropy cost (large dS/dE). The magnitude of the entropy change caused by the output is therefore greater than the magnitude of the entropy change caused by the input. Because the input causes a decrease in entropy and the output causes an increase in entropy, the combined effect is an increase in the entropy of the environment.

Now suppose the copper rod somehow transfers energy from the low-temperature reservoir to the high-temperature reservoir, as illustrated in **Figure 21.12**. The curved arrow representing the energy transfer now points from the low-temperature reservoir to the high-temperature reservoir and, as you can readily verify, the signs of the entropy changes are the opposite of the signs in Figure 21.11. The rod takes a certain amount of energy out of the environment at large dS/dE and returns the same amount of energy to the

Figure 21.12 Entropy diagram for a thermal transfer of energy from a low-temperature reservoir (L) to a high-temperature reservoir (H), a process forbidden by the entropy law.

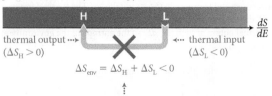

Input at high dS/dE would cause greater entropy change than output at low dS/dE, so entropy of environment would *decrease* — not possible!

environment at a smaller dS/dE. The entropy change of the environment in this process is thus negative, which is not allowed by the entropy law. Indeed, no one has ever observed energy being transferred thermally from a cold object to a hot object with no other changes in the system or the environment (that is, the cold object spontaneously getting colder and the hot object getting hotter).

The diagrams in Figures 21.11 and 21.12 are called **entropy diagrams.** An entropy diagram shows the energy conversions and transfers that take place in a steady device with the environment arranged according to an entropy gradient, as shown in Figure 21.10. Arrows in the diagram represent the energy transfers through the device during one complete cycle. The tails of the arrows represent energy inputs to the device, and the heads represent energy outputs. Thermal inputs of energy (tails) cause the entropy of the environment to decrease, while thermal outputs of energy (arrowheads) cause the entropy of the environment to increase.

Entropy diagrams permit us to keep track of entropy changes that result from the energy conversions by a steady device. In particular:

A transfer of energy that points to the right in an entropy diagram causes an increase in the entropy of the environment; a transfer that points to the left causes a decrease in the entropy of the environment.

As we shall see in Section 21.5, entropy diagrams can be used to derive quantitative information about entropy changes in the environment.

Let's now return to the example of a paddle wheel in a viscous liquid. This time, however, we immerse the container of viscous liquid in a large thermal reservoir that keeps the viscous liquid at a constant temperature T_R (Figure 21.13). The moving parts and the viscous liquid now constitute a steady device. A person does work on the device by pulling on the string and then the device transfers energy to the reservoir. The device thus converts mechanical energy to thermal energy.

To draw an entropy diagram for this energy conversion process, we must first determine where to put mechanical energy on the entropy-gradient scale. Remember that doing work—in other words, transferring energy mechanically—does not cause any change in entropy. Therefore $\Delta S = 0$

Figure 21.13 Paddle-wheel device from Exercise 21.1 is held at a constant temperature T_R by a large thermal reservoir of water.

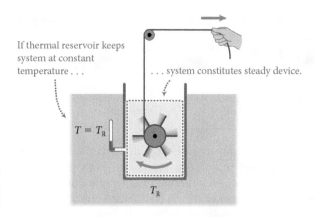

and so $\Delta S/\Delta E = 0$. The entropy gradient associated with mechanical energy is thus zero. Mechanical energy transfers therefore always appear on the far left at the origin of the dS/dE scale (Figure 21.14).

The only entropy change in Figure 21.14 is thus caused by the thermal output of energy. This entropy change is positive (arrow points up), and so the entropy of the environment increases.

The inverse of the process in Figure 21.14, illustrated in Figure 21.15—conversion of thermal energy entirely to mechanical energy by a steady device—cannot occur. In this process, the device would transfer energy thermally from the environment and mechanically to the environment. The curved arrow now points to the left, causing a decrease in the entropy of the environment, which is impossible. This process would mean that the paddle wheel in Figure 21.13 could rewind the string and pull the hand to the left by transferring energy thermally from the reservoir. The entropy law prevents this from happening.

Summarizing, we see that converting energy with a low entropy gradient to energy with a high entropy gradient with a steady device (arrow points right in entropy diagram) increases the entropy of the environment and is permitted by the entropy law. Conversely, converting energy with a high entropy gradient to energy with a low entropy gradient with a steady device (arrow points left in entropy diagram)

Figure 21.14 Entropy diagram for the process depicted in Figure 21.13.

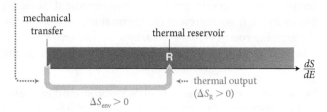

Figure 21.15 Entropy diagram for the conversion of thermal energy entirely to mechanical energy by a steady device.

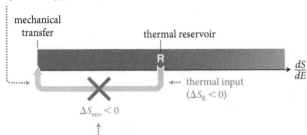

proposed mechanical output of energy ($\Delta S = 0$)

Entropy of environment would decrease: not possible.

decreases the entropy of the environment and is not permitted by the entropy law.

As energy shifts toward the right in an entropy diagram, the energy becomes less useful for, say, heating things or powering machines. You can do less with a tub of lukewarm water than with a bucket of boiling hot water that contains the same amount of thermal energy as the tub. We can therefore associate the entropy gradient with the **quality of the energy:** The smaller dS/dE, the lower the entropy cost and the higher the quality of energy (**Figure 21.16a**). The tendency of the entropy of the environment to increase and energy to shift from higher quality to lower quality (that is, to the right in the entropy diagram) corresponds to a *degradation* of energy quality (Figure 21.16b). In the device of Figure 21.13, for instance, high-quality mechanical energy

Figure 21.16 Energy with a large entropy gradient is less useful and therefore of lower quality than energy with a small entropy gradient.

Relationship of entropy gradient to quality of energy

(a)

Process that degrades energy quality

(b)
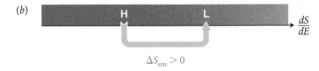

Process that upgrades energy quality (not possible on its own)

(c)
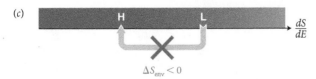

(which allows us to do useful work, such as lifting or accelerating things) is irreversibly turned into less useful, lower-quality thermal energy.

A conversion of lower-quality energy to higher-quality energy (an *upgrade* in energy quality) corresponds to a left-pointing curved arrow in entropy diagrams and produces a decrease in the entropy of the environment (Figure 21.16c). Such upgrades cannot occur in steady devices that have a single energy input and a single energy output. For example, it's impossible to convert low-quality thermal energy entirely to high-quality mechanical energy. This is why ships can't be powered by converting thermal energy from the ocean entirely to mechanical energy.

✋ **21.5** For a steady device that has a single input of energy and a single output of energy, does the quality of the energy increase, decrease, or stay the same when the entropy change in the environment is (*a*) positive and (*b*) zero?

Example 21.2 Energy and entropy changes in a thermal reservoir

Suppose the person pulling the string in Figure 21.13 does work W on the device immersed in a thermal reservoir R at temperature T_R. (*a*) Do the energy and entropy of the reservoir increase, decrease, or stay the same? (*b*) If the same experiment is repeated with the reservoir at a lower temperature, are the changes in the energy and entropy of the reservoir greater than, the same as, or smaller than in part *a*?

❶ **GETTING STARTED** Because the device is immersed in a thermal reservoir, the state of the device doesn't change. The device is therefore a steady device, and its energy and entropy don't change. Because the reservoir temperature is lower in part *b*, the states of the device and reservoir are different from those in part *a*. In particular, a lower reservoir temperature means that the reservoir's thermal energy is smaller.

❷ **DEVISE PLAN** I can draw an energy input-output diagram for the device and determine its energy inputs and outputs. From the diagram, I can deduce whether there is any thermal transfer of energy to (or from) the reservoir. To evaluate any entropy changes in the reservoir, I draw an entropy diagram for each part of the problem.

❸ **EXECUTE PLAN** (*a*) I know that the person does work on the device and so $W_{in} = W > 0$. Because the energy of the device doesn't change, the energy law tells me that $\Delta E = W + Q = 0$, so $Q = -W < 0$, which means that energy must be transferred thermally out of the device: $Q_{out} = -Q = W$. My energy input-output diagram is shown in **Figure 21.17a**.

The energy transferred thermally out of the device goes into the reservoir, which tells me that the energy of the reservoir increases by an amount W. ✔

To see what happens to the reservoir entropy, I need an entropy diagram. Mechanical energy initially in the device is converted to thermal energy. I represent this conversion by drawing a right-pointing arrow in my entropy diagram, running from the origin ($dS/dE = 0$ for mechanical energy) to the R representing the reservoir (Figure 21.17b). That the arrow points to the right tells me that the entropy of the reservoir increases. ✔

CONCEPTS

Figure 21.17

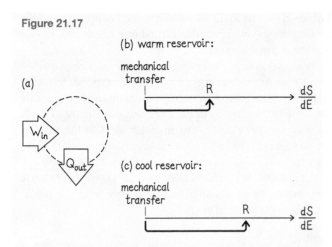

Figure 21.18 Entropy diagram for a process that upgrades and degrades energy.

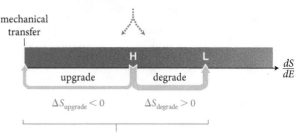

(b) Even though the temperature of the reservoir is lowered, the amount of work done on the device by the person is still W, so the temperature change does not affect my energy input-output diagram. This tells me that the energy of the reservoir increases by the same amount W as in part a. ✔

Because the reservoir temperature is lower than in part a, however, on my entropy diagram for this situation (Figure 21.17c), I show R farther to the right than in Figure 21.17b. The right-pointing arrow representing the conversion is longer, meaning a greater dS/dE value at this temperature. A greater entropy gradient means a higher entropy cost of energy change, and so the entropy change associated with the energy change W must be greater than the entropy change of part a. ✔

❹ **EVALUATE RESULT** I can also evaluate the entropy changes using Eq. 20.54, $\Delta S_R = Q_R/k_B T_R$. The quantity of energy transferred thermally to the reservoir is the same for both parts of the problem: $Q_R = W$. Because the reservoir temperature T_R is lower in part b, the entropy change must be greater here than in part a.

✋ **21.6** In an entropy diagram, which of these statements are true? (a) The quality of the energy is proportional to the entropy gradient. (b) The quality of the energy is inversely proportional to the entropy gradient. (c) High-temperature energy is of high quality. (d) High-temperature energy is of low quality. (e) Low-temperature energy has a small entropy gradient. (f) Low-temperature energy has a large entropy gradient.

21.3 Heat engines and heat pumps

As we just saw, it is not possible for a steady device that has a single energy input and a single energy output to upgrade energy quality. The only way for a steady device to upgrade energy quality is to also degrade energy quality in such a way that the change in entropy of the environment that results from both processes is either positive or zero: $\Delta S_{env} = \Delta S_{upgrade} + \Delta S_{degrade} \geq 0$.

The process illustrated in **Figure 21.18** is one way to combine two compensating processes. A steady device takes in energy from a high-temperature reservoir, converts part of this energy to mechanical energy, and thermally transfers the rest to a low-temperature reservoir. This process is allowed as long as the increase in entropy associated with the degradation of the energy (the thermal output) equals or exceeds the decrease in entropy associated with the upgrade (the mechanical output). A steady device of this type is called a **heat engine**.

One familiar heat engine that carries out the process illustrated in Figure 21.18 is the automobile engine we discussed briefly in Section 21.1. As shown in **Figure 21.19**, the thermal input of energy is provided by the combustion of gasoline at high temperature (about 2700 K). Roughly two-thirds of this input energy is discarded to the environment by the car's exhaust and cooling system, both of which are roughly at 400 K. The remaining one-third of the input energy is turned into mechanical output of energy (part of which is lost due to friction in the moving parts of the car). As you can see by the relative sizes of the Q_{out} and W_{out} arrows in Figure 21.19, in a typical car engine more energy is degraded than upgraded in order to satisfy the entropy law and keep the engine's final state the same as its initial state.

Because we arrange energy inputs and outputs on the horizontal scale of an entropy diagram according to their entropy gradient, the length of an arrow in the diagram is

Figure 21.19 Energy input-output diagram for an automobile engine, which is one type of heat engine.

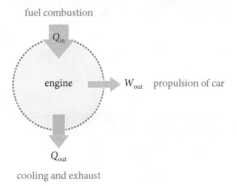

Figure 21.20 Obtaining entropy changes from entropy diagrams.

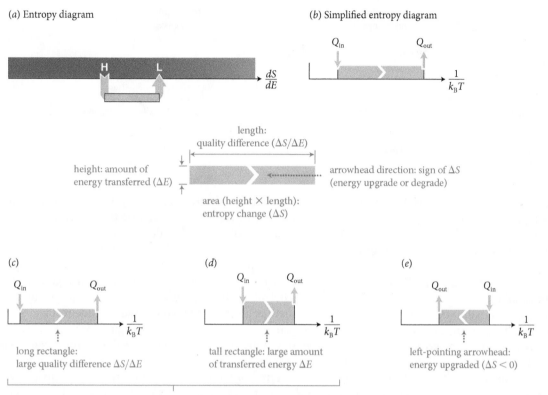

(*a*) Entropy diagram

(*b*) Simplified entropy diagram

length:
quality difference ($\Delta S/\Delta E$)

height: amount of
energy transferred (ΔE)

arrowhead direction: sign of ΔS
(energy upgrade or degrade)

area (height × length):
entropy change (ΔS)

(*c*)

long rectangle:
large quality difference $\Delta S/\Delta E$

(*d*)

tall rectangle: large amount
of transferred energy ΔE

(*e*)

left-pointing arrowhead:
energy upgraded ($\Delta S < 0$)

same area: same entropy change ΔS

a measure of the quality difference $\Delta S/\Delta E$ between the input and output. If we take the width of each arrow to represent the amount of energy ΔE transferred, then the area of the horizontal part of the arrow (the product of the arrow's width and length) represents the magnitude of the change in entropy (**Figure 21.20a**).

Because it is difficult to draw curved arrows and determine their area, we simplify our procedure for drawing entropy diagrams, replacing the arrows with rectangles (Figure 21.20b). The length of the rectangle corresponds to the quality difference between the energy output and input, and the height of the rectangle corresponds to the amount of energy transferred. The area of the rectangle thus represents the entropy change in the environment caused by the energy conversion or transfer. The direction of the energy transfer is indicated by an arrow inside the rectangle. In addition we mark the energy inputs and outputs by vertical arrows: down for input and up for output. Finally, because the entropy gradient dS/dE is equal to $1/k_BT$ (Eq. 19.38), we label the axis with $1/k_BT$.

There are two ways to increase the magnitude of the entropy change. The greater the quality difference between the input and output, the greater the entropy change. For example, in Figure 21.20c the energy input is of higher quality than in Figure 21.20b. Consequently, the entropy change produced by the conversion of this higher-quality energy is greater. Shifting the input of the arrow to the left (higher quality) increases its length and therefore its area

(which represents the entropy change). As you can see from Eq. 20.54, $\Delta S = Q/k_BT$, the entropy change is also proportional to the quantity of energy Q being transferred. So, a second way to induce a greater entropy change between two fixed-temperature reservoirs is to increase the amount of energy being transferred (represented by the greater height of the rectangle in Figure 21.20d). Finally, if the direction of an energy transfer is reversed, the sign of the entropy change is reversed but its magnitude remains the same (compare Figure 21.20e with 21.20b). **Figure 21.21** shows the entropy diagram for a process that both upgrades and degrades energy, with rectangles used to represent the transfers.

Because thermal inputs and outputs of energy occur at different temperatures, they are separated on the dS/dE scale. Mechanical inputs and outputs of energy, however, are always located at $dS/dE = 0$, and so they overlap in an entropy

Figure 21.21 Entropy diagram for a process that upgrades and degrades energy (compare with Figure 21.18).

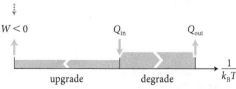

mechanical *output* of energy from system (*upward* arrow):
$W < 0$ because W = work done *on* system

Figure 21.22 For a steady device, the entropy change in the environment must be greater than or equal to zero.

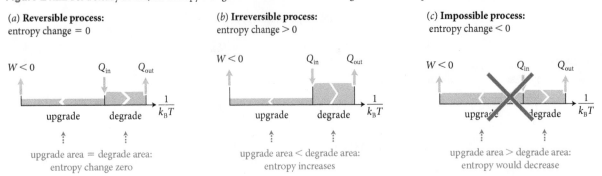

(a) **Reversible process:** entropy change = 0

upgrade area = degrade area: entropy change zero

(b) **Irreversible process:** entropy change > 0

upgrade area < degrade area: entropy increases

(c) **Impossible process:** entropy change < 0

upgrade area > degrade area: entropy would decrease

diagram. The end of an arrow (or rectangle) at $dS/dE = 0$ therefore does not represent W_{in} or W_{out} but rather the difference between these two, $W = W_{in} - W_{out}$, which is the work done on the system. When the arrow points up, energy is transferred mechanically from the system to the environment and hence $W < 0$ (Figure 21.21). When the arrow points down, work is done on the system, indicating a mechanical transfer of energy into the system, $W > 0$.

For a steady device, the entropy law requires the entropy change in the environment to be greater than or equal to zero, and so the area of the degrade rectangle must be larger than or equal to the area of the upgrade rectangle. **Figure 21.22a** shows a process for which the energy upgrade and energy degradation cause equal-magnitude changes in entropy. For such a process, the entropy change is zero, so the process is reversible and allowed. Figure 21.22b shows a process for which the magnitude of the entropy change caused by the energy upgrade is smaller than the magnitude of the entropy change caused by the energy degradation. For such a process, the entropy change is positive, so the process is irreversible and allowed. Figure 21.22c shows a process for which the magnitude of the entropy change caused by the energy upgrade is greater than the magnitude of the entropy change caused by the energy degradation. For such a process, the entropy change is negative, so the process is forbidden by the entropy law. The procedure for drawing entropy diagrams is shown in the Procedure box on page 573.

21.7 Draw an entropy diagram for the refrigerator in Figure 21.2b.

The entropy diagram for an irreversible process such as the one in Figure 21.22b can be treated as the entropy diagram for an irreversible degradation of energy (that is, hot to cold) superimposed on the entropy diagram for a reversible process (**Figure 21.23**). The upgrade rectangles in Figure 21.23a and b are the same, but in Figure 21.23b the height of the degrade rectangle has been adjusted so that the areas of the upgrade and degrade rectangles are the same. The entropy change in Figure 21.23b is therefore zero, and the process represented by this entropy diagram is reversible. Figure 21.23c shows the remainder of the energy transferred to the low-temperature reservoir. This process increases the entropy of the environment and is irreversible. The mechanical energy generated in Figure 21.23b is retrievable. Because the process is reversible, the mechanical energy can be converted back to thermal energy. The energy in Figure 21.23c, on the other hand, has been degraded and is irretrievable. It cannot be transferred back to the high-temperature reservoir without some other associated energy degradation. For now, the only processes we consider are reversible ones for which the entropy change is zero.

21.8 Why must the degrade rectangle in Figure 21.22a have a greater height than the upgrade rectangle?

Example 21.3 Hot or cold?

A reversible heat engine converts energy taken from a thermal reservoir at temperature T_{in} by doing work on the environment and thermally transferring energy to a thermal reservoir at temperature $T_{out} < T_{in}$. How do the thermal input and output of energy change if the temperature of the output reservoir is lowered without changing the amount of work done by the engine on the environment?

Figure 21.23 Entropy diagram for an irreversible process.

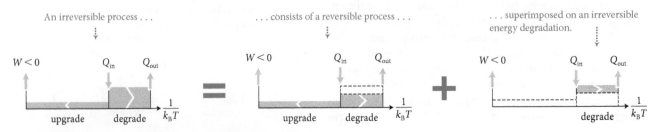

An irreversible process . . .

. . . consists of a reversible process . . .

. . . superimposed on an irreversible energy degradation.

Procedure: Drawing entropy diagrams for a steady device

1. Draw an entropy gradient axis pointing right. Add a short vertical axis to indicate the origin of the axis. Label the axis $1/k_BT$ to remind yourself of the relationship between entropy gradient and temperature.
2. Mark the positions along the axis of the thermal inputs and outputs of energy of the device. Transfers at higher temperature are on the left; transfers at lower temperature are on the right. Use downward-pointing arrows for inputs, use upward-pointing arrows for outputs, and label these Q_{in} and Q_{out}.
3. Mechanical transfers of energy are placed at the origin of the axis. Use a downward-pointing arrow when the work done on the system is positive and an upward-

pointing arrow when the work done on the system is negative. Label this arrow by W and indicate whether $W > 0$ or $W < 0$.

4. Draw rectangles to represent each energy transfer. Make the height of each rectangle proportional to the amount of energy transferred (but see point 5 below). Use an arrowhead to indicate the direction of each transfer.
5. For the operation of the device to be permitted by the entropy law, the combined area of the rectangles in which the arrows point right must be larger than the combined area of the rectangles in which the arrows point left.

① **GETTING STARTED** I begin by drawing an entropy diagram for the heat engine (**Figure 21.24a**). Because the engine is reversible, the area of the upgrade rectangle must equal the area of the degrade rectangle. The height of the upgrade rectangle is equal to the magnitude of W and is fixed. The height of the degrade rectangle is equal to Q_{out}.

Figure 21.24 Example 21.3.

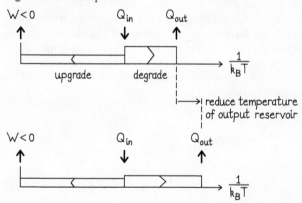

② **DEVISE PLAN** To determine how the thermal input and output of energy change as the temperature of the output reservoir is lowered, I need to draw another entropy diagram (Figure 21.24b), keeping the upgrade rectangle the same as in Figure 21.24a and shifting the right side of the degrade rectangle to the right to represent the lower temperature of the output reservoir. When I make this change, I must remember that the upgrade and degrade rectangles must still have the same area, and so now my degrade rectangle is not as tall as in Figure 21.24a. Also, I know that the sum of the outputs must equal the sum of the inputs, and so $W_{out} + Q_{out} = Q_{in}$.

③ **EXECUTE PLAN** In any entropy diagram, the height of the degrade rectangle represents Q_{out}. Because the height of the degrade rectangle is less in Figure 21.24b than in Figure 21.24a, I conclude that Q_{out} decreases when the temperature of the output reservoir is lowered. Because $W_{out} + Q_{out} = Q_{in}$ and W_{out} is not changed, Q_{in} must also decrease. ✔

④ **EVALUATE RESULT** My result means that if the temperature difference between the reservoirs increases, I need less energy

Q_{in} to do the same amount of work on the environment. The reason is that I can compensate for the energy upgrade from thermal to mechanical with less energy: At low temperature a given amount of energy causes a greater entropy change and so my answer makes sense.

✋ **21.9** Suppose you keep W and T_{out} in the reversible process of Figure 21.24a fixed but increase T_{in}. Do the magnitudes of these quantities increase, decrease, or stay the same: (a) ΔS associated with the upgrade, (b) ΔS associated with the degradation, (c) Q_{in}?

Example 21.3 and Checkpoint 21.9 show that it is beneficial to take energy in at the highest possible temperature and discard energy at the lowest possible temperature. Doing so minimizes the thermal input of energy for a fixed amount of work done on the environment.

We can define the **efficiency** of a heat engine as follows:

The efficiency of a heat engine is the ratio of the work done by the engine on the environment to the thermal input of energy into the engine. For a heat engine operating between two fixed-temperature reservoirs, the efficiency is greatest when the thermal input of energy takes place at the highest possible temperature and the thermal output of energy takes place at the lowest possible temperature.

We see now that no heat engine—even a reversible one—can have an efficiency of 1.00 because some fraction of the thermal input of energy has to be degraded to make up for the entropy decrease that occurs in the process of upgrading the energy. The imperfect efficiency of heat engines has nothing to do with friction in the moving parts or other irreversible phenomena; rather, it is a direct consequence of the entropy law.

✋ **21.10** When the temperature of the output reservoir in Example 21.3 is lowered, does the efficiency increase, remain the same, or decrease?

Figure 21.25 Energy conversions in a heat pump.

(a) Entropy diagrams for energy conversions in heat pump

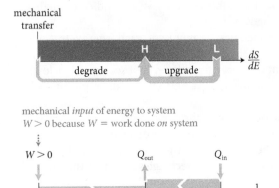

Figure 21.26 A heat pump can serve as an air-conditioner in summer by removing energy from inside the house and as a heater in winter by transferring energy thermally from outside to inside the house.

(a) Summer   (b) Winter

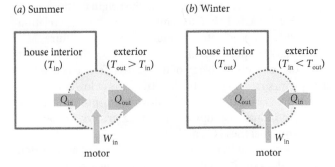

on the heat pump. The **coefficient of performance of cooling** of a heat pump is defined as the ratio of the thermal input of energy (the energy transferred thermally from the space that must be cooled) to the work done on the heat pump.

Example 21.4 Cooling performance

For a set temperature T_{in} of a refrigerator interior, does the coefficient of performance of cooling increase, decrease, or stay the same as the temperature T_{out} of the room that contains the refrigerator increases? Assume the refrigerator is a reversible heat pump.

❶ GETTING STARTED I begin by drawing an entropy diagram for the refrigerator (**Figure 21.27a**). Because the refrigerator is reversible, the areas of the upgrade and degrade rectangles are equal. The coefficient of performance of cooling is the ratio of the thermal input of energy Q_{in} (the height of the upgrade rectangle) to the work W done on the refrigerator (the height of the degrade rectangle).

Figure 21.27

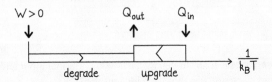

(b) Energy input-output diagram for refrigerator (a type of heat pump)

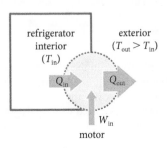

As you saw in Checkpoint 21.7, there are more ways to cause a degradation of energy that compensates for an upgrade than the one shown in Figure 21.21. For example, you can swap the roles of the upgrade and degradation shown in Figure 21.21. The resulting entropy diagram, shown in **Figure 21.25a**, represents the operation of a **heat pump.** In a heat pump, mechanical energy is degraded in order to thermally transfer energy from a lower-temperature region of the environment to a higher-temperature region of the environment—a direction opposite the natural direction in which energy is transferred thermally. Note that work is done on the device, so $W > 0$.

The refrigerator is one familiar example of a heat pump (Figure 21.25b). The interior of the refrigerator constitutes the reservoir at T_{in}; the air surrounding the refrigerator is the reservoir at T_{out} (coils at the back or bottom of the refrigerator transfer the energy Q_{out} to the air).

Heat pumps are also used to heat and cool houses. In the summer, a heat pump operates as an air-conditioner and cools a house by moving energy from inside the house to outside (from cold to hot; **Figure 21.26a**). In the winter, the same heat pump heats by moving energy from outside the house to inside (again from cold to hot, Figure 21.26b).

In designing a heat pump for cooling, we want to maximize the amount of energy transferred thermally from the low-temperature reservoir for a given amount of work done

❷ DEVISE PLAN To determine how the coefficient of performance of cooling changes as T_{out} increases, I must draw another entropy diagram, shifting the thermal output (the point where the two rectangles meet) to the left (Figure 21.27b). I must still have the area of the upgrade rectangle equal the area of the degrade rectangle, so in this diagram, I increase the height of the degrade rectangle and decrease that of the upgrade rectangle.

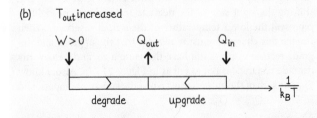

❸ EXECUTE PLAN The height of the upgrade rectangle corresponds to Q_{in}, and the height of the degrade rectangle corresponds to W. Shifting Q_{out} leftward to represent the increased T_{out} made the height of the upgrade rectangle (Q_{in}) smaller and the height of the degrade rectangle (W) greater. The ratio Q_{in}/W therefore decreases, which means that the coefficient of performance of cooling decreases. ✔

❹ EVALUATE RESULT The fact that the coefficient of performance of cooling decreases means that more energy is required to cool the refrigerator interior when the air outside the refrigerator is at a higher temperature, which makes sense intuitively. The reason is that when the refrigerator discards energy to the environment at a higher temperature, a given amount of energy causes a smaller entropy increase than when the energy is discarded at a lower temperature. This degradation of energy therefore compensates a smaller upgrade, reducing the coefficient of performance of cooling.

In designing a heat pump for heating, we want to maximize the thermal output of energy to the high-temperature reservoir for a given amount of work done on the heat pump. The **coefficient of performance of heating** of a heat pump is defined as the ratio of the thermal output of energy (the energy delivered thermally to the space that must be heated) to the work done on the heat pump.

Ordinary heating transfers energy from a high-temperature reservoir (say, burning fuel) to a low-temperature reservoir (a room to be heated, say), and the thermal input of energy Q_{in} is equal to the thermal output of energy Q_{out} (Figure 21.11). Equivalently, in Figure 21.14, $W = W_{in} > 0$ is converted to an identical amount of energy Q_{out}. In a heat pump, however, $Q_{out} > Q_{in}$, as Figure 21.26 indicates, because $Q_{out} = Q_{in} + W$. Note that while heat engines have an efficiency less than 1, the coefficient of performance of heating always exceeds 1. A typical heat pump has a coefficient of performance of heating of 3 to 4.

✋ **21.11** Estimate these quantities for the heat pump in Figure 21.26: (*a*) coefficient of performance of cooling and (*b*) coefficient of performance of heating.

21.4 Thermodynamic cycles

Heat engines and heat pumps are made up of three important components: a collection of moving parts, thermal reservoirs, and a *working substance* that undergoes a series of thermodynamic processes. For example, in the (heat) engine of an automobile, a mixture of air and gasoline is the working substance that produces the motion of the pistons that propels the automobile. Thermal energy generated in the combustion is transferred thermally to the environment through the engine's exhaust, and so the environment serves as a thermal reservoir. In a power plant, the working substance is water, which drives the blades of steam turbines. The steam boiler and the outside air serve as thermal reservoirs that transfer energy thermally to and from the water.

In this section we look at the thermodynamic processes that the working substance undergoes during the operation of a heat engine or heat pump. As we have seen, the engine or pump is a steady device: It must periodically return to a certain initial state. The working substance must therefore undergo a *cyclic process*—a sequence of steps in which the working substance begins and ends in the same thermal equilibrium state. Under those circumstances, the energy and the entropy of the working substance return to their initial values in one cycle.

Figure 21.28a illustrates a cyclic process that involves an expansion from state 1 to state 2 and then a compression back to state 1 via another path. We saw in Chapter 20 that during an expansion the work done on the system is

Figure 21.28 The work done on a system during a thermodynamic cyclic process is equal in magnitude to the area enclosed by the pathways making up the cycle.

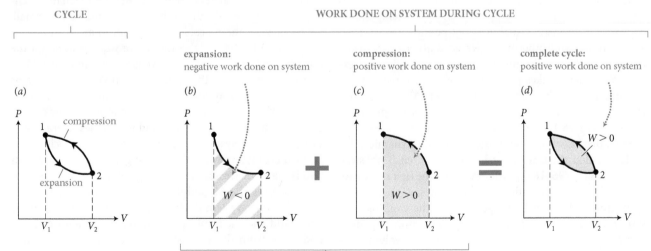

Magnitude of work done on system is smaller during expansion than compression because expansion takes place at lower pressure.

negative and during a compression the work done on the system is positive. If the expansion takes place at a lower pressure than the compression (as it does in Figure 21.28a), the work done on the system during the expansion is smaller in magnitude than the work done during the compression. Therefore the work done on the system during a complete cycle from 1 to 2 and back to 1 is positive.

As we saw in Chapter 20, the magnitude of the negative work done on the system during the expansion $1 \rightarrow 2$ is equal to the hatched area in Figure 21.28b. The magnitude of the positive work done on the system during the compression $2 \rightarrow 1$ is equal to the shaded area in Figure 21.28c. Adding these two areas and taking their signs into account, we see that the work done on the system during a complete cycle is equal to the shaded area enclosed by the path that represents the cycle, as illustrated in 21.28d.

✋ **21.12** If you begin at state 1 again but this time go through the cycle in Figure 21.28 clockwise rather than counterclockwise, what are the magnitude and sign of (a) the work done on the system and (b) the work done on the environment?

Figure 21.28 and Checkpoint 21.12 demonstrate an important point:

The magnitude of the work done on a system during a cycle is equal to the area enclosed by the path that represents the cycle in a PV diagram. The work done on the system is negative for a clockwise cycle and positive for a counterclockwise cycle.

Because heat engines do positive work on the environment, their cycles are clockwise. Conversely, the environment does positive work on heat pumps, and so their cycles are counterclockwise.

What thermodynamic processes occur in a heat engine? To answer this question, let us consider first the thermal input of energy from the high-temperature reservoir. This process takes place at a constant temperature T_{in} and so is isothermal. The thermal output of energy at temperature $T_{out} < T_{in}$ is also isothermal. It is not possible to construct a cycle from two isotherms, so we need other processes to connect the two isotherms. Because no other thermal exchange of energy takes place, these other processes we are looking for must be isentropic. **Figure 21.29** shows a thermodynamic cycle that consists of two isothermal processes ($1 \rightarrow 2$ and $3 \rightarrow 4$) and two isentropic processes ($2 \rightarrow 3$ and $4 \rightarrow 1$). Because of its importance in thermodynamics, this cycle is called the **Carnot cycle** after the French physicist Sadi Carnot (1796–1832), who produced the first successful description of the operation of heat engines.

Beginning in state 1 in Figure 21.29, the working substance expands isothermally at a temperature T_{in} until it reaches some state 2 (at which point its volume has increased). The working substance then undergoes an isentropic expansion during which its temperature drops from T_{in} to T_{out}. From this state 3, the working substance is

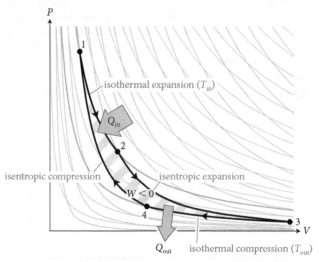

Figure 21.29 *PV* diagram for the thermodynamic process known as the Carnot cycle.

compressed isothermally to state 4, from where it undergoes an isentropic compression back to state 1, completing the cycle.

Energy is transferred thermally into and out of the system only during the two isothermal processes, but all four processes involve work. The work done on the system during the cycle is equal to the area enclosed by the cycle and has been shaded in Figure 21.29. Because W is negative, the shading is hatched. (Remember that is what we want: The heat engine must do positive work *on its environment*, and so the work done *on the system* is negative.)

The Carnot cycle has a number of deficiencies that makes it impractical for a heat engine. First of all, as you can see from the small hatched area in Figure 21.29, the amount of work done on the environment in each cycle is small. More important, the presence of the isotherms means that the cycle cannot be repeated quickly. As we saw in Section 20.7, isothermal processes are very slow, and so an engine that operates on the Carnot cycle would put out mechanical energy at a very low rate; that is, its power would be very small.

To alleviate these problems, we want to avoid isotherms and "open up" the cycle so that more work is done on the environment. If we replace the two isotherms in Figure 21.29 by isobars, we obtain the cycle shown in **Figure 21.30a**. This cycle, called the **Brayton cycle**, does substantially more work on the environment than the Carnot cycle and can be carried out fairly quickly. Gas turbines and jet engines operate on a cycle that is closely approximated by the Brayton cycle.

In the Brayton cycle, the working substance undergoes two isobaric processes ($1 \rightarrow 2$ and $3 \rightarrow 4$) and two isentropic processes ($2 \rightarrow 3$ and $4 \rightarrow 1$). Energy thermally enters and leaves the system during the isobaric processes. If we begin and end the isentropic compression in Figures 21.29 and 21.30 from the same states 4 and 1, the hatched area enclosed by the Brayton cycle is much larger than the area enclosed by the Carnot cycle. The larger hatched area means that a steady device that operates on the Brayton cycle has

Figure 21.30 The thermodynamic process known as the Brayton cycle.

(a) *PV* diagram for Brayton cycle

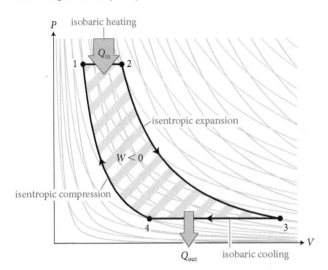

(b) Flow chart for gas turbine operating on Brayton cycle

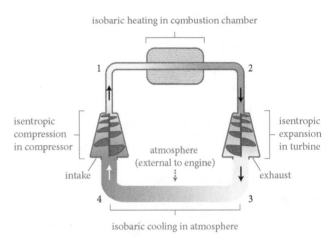

a much greater mechanical output of energy than a steady device that operates on the Carnot cycle. Also, the isobaric and isentropic processes can be carried out rapidly, which means the power output in a Brayton cycle can be high.

A gas turbine engine that operates on the Brayton cycle has three main components: a gas compressor, a combustion chamber, and an expansion turbine. In practice, such an engine is an open system: Fresh air flows in at one end, and hot gases are expelled at the other end. For the purpose of thermodynamic analysis, however, we can pretend that the exhaust gases are reused in the intake.

Figure 21.30*b* shows a flow chart for a gas turbine engine. Gas—typically ambient air—is drawn into the compressor, which isentropically compresses the air to high pressure $4 \rightarrow 1$. The compressed air then flows into the combustion chamber, where it is heated to temperatures as high as 2000 K at constant pressure ($1 \rightarrow 2$). The heated, pressurized air then expands isentropically ($2 \rightarrow 3$), and the hot gases are expelled at high speed from the rear of the engine, driving the engine forward. The exhaust gases then cool

back down to the intake temperature—another constant-pressure process ($3 \rightarrow 4$)—and the cycle is complete.

To power the compressor, a turbine is placed in the expansion chamber. The expanding gases drive this turbine, which in turn feeds some energy from the expanding gases to the compressor. In a jet engine, for example, the turbine and the compressor are placed on a single drive shaft, so the turbine directly feeds the compressor (**Figure 21.31**).

Although the Brayton cycle delivers more energy and power than the Carnot cycle, it is not as efficient, which means that the work done on the environment divided by the thermal input of energy is smaller. The reason for this diminished efficiency is illustrated in **Figure 21.32a**. In the Brayton cycle, the thermal input and output of energy do not take place at fixed temperatures as in the Carnot cycle. The Brayton thermal input of energy starts at the temperature T_1 reached during compression and then moves up to

Figure 21.31 Jet engine.

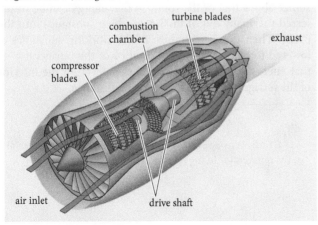

Figure 21.32 Entropy diagrams for Brayton and Carnot cycles.

(a) Brayton cycle

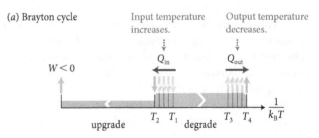

(b) Carnot cycle

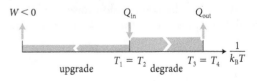

Figure 21.33 A Brayton cycle run in reverse is the cycle for a heat pump.

(a) *PV* diagram for Brayton cycle run in reverse

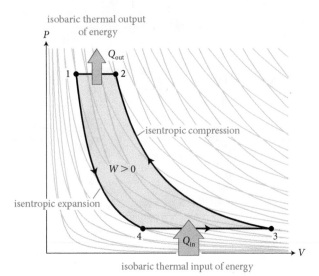

(b) Flow chart for refrigerator operating on reverse Brayton cycle

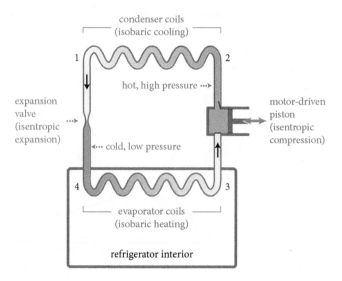

a higher temperature T_2 during combustion. The thermal output of energy starts at temperature T_3 (the temperature at the end of the isentropic expansion) and then slides down to the lower temperature T_4 (the temperature at the beginning of the isentropic compression).

As we saw in Section 21.3, the efficiency of a heat engine is highest when the thermal input of energy takes place at the highest possible temperature and the output at the lowest possible temperature. So the part of the cycle where thermal input of energy occurs at T_2 and thermal output of energy occurs at T_4 has the greatest efficiency. All other parts of the cycle have lower efficiency, which contributes to an overall efficiency that is lower than the efficiency we would see if *all* energy exchange had taken place at only the highest and lowest temperatures, T_2 and T_4, respectively. By contrast, thermal input of energy at one fixed high temperature and output at one fixed low temperature is exactly what happens in a Carnot cycle (Figure 21.32b).

By running the Brayton cycle in reverse (**Figure 21.33a**), we obtain a heat pump. The cycle used in refrigerators, air-conditioners, and home heat pumps is a bit more complex because the working substance in these devices changes from liquid to vapor and back. However, if we assume the working substance is a gas, the basic thermodynamic processes are the isentropic ($3 \rightarrow 2$ and $1 \rightarrow 4$) and isobaric ($2 \rightarrow 1$ and $4 \rightarrow 3$) ones illustrated in Figure 21.33a.

Figure 21.33b shows a flow chart of the reverse Brayton cycle that takes place in a refrigerator. The working substance is first isentropically compressed by a motor-driven piston ($3 \rightarrow 2$). The compression raises both the temperature and the pressure in the working substance, which is

then cooled down at constant pressure in a long coiled tube called the *condenser* ($2 \rightarrow 1$). This tube is exposed to the ambient temperature. After passing through the condenser, the high-pressure and now somewhat cooler working substance passes through an expansion valve ($1 \rightarrow 4$) and expands isentropically. In the process of expanding, the working substance cools significantly. It then passes through a second set of coils called the *evaporator* ($4 \rightarrow 3$). Because the temperature of the space around the evaporator is higher than the temperature of the working substance in the evaporator, energy is thermally transferred from the surrounding space to the evaporator. As Figure 21.33b shows, this space surrounding the evaporator is the interior of the refrigerator, where food is stored and thus kept cold. After passing through the evaporator, the working substance is returned to the compressor, where the cycle repeats.

Note that a heat pump transfers energy from the interior of the refrigerator at low temperature to the room at a higher temperature. To accomplish this, it needs mechanical energy to first raise the temperature of the working substance above that of the room so that energy can be transferred thermally out of the hot working substance into the room. Then the temperature of the working substance is lowered to below that of the interior of the refrigerator so that energy can be transferred thermally from the interior of the refrigerator into the working substance.

21.13 Why doesn't the transfer of energy from the interior of the refrigerator to the room violate the entropy law?

Self-quiz

1. Describe the energy transfers in Figure 21.34. What is implied about the energy of the system?

Figure 21.34

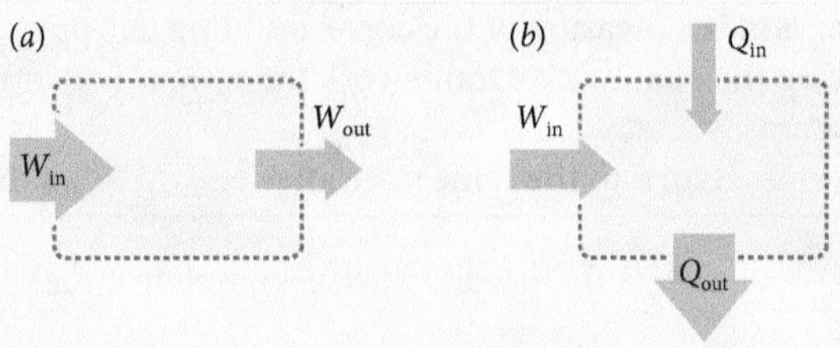

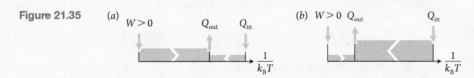

2. Describe the energy conversions in Figure 21.35. Which are possible?

Figure 21.35

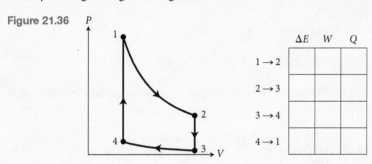

3. Write the ratios for the efficiency of a heat engine, the coefficients of performance of cooling for a refrigerator, and the coefficients of performance of heating a room using a heat pump in terms of W, Q_{in}, and Q_{out}. Make a generalization about the quantity on the top of each ratio and the quantity on the bottom of each ratio.

4. Figure 21.36 shows an ideal gas cycle consisting of two isotherms and two isochores. Complete the table by writing the algebraic signs of each of the terms ΔE, W, and Q. If a term is zero, write zero.

Figure 21.36

	ΔE	W	Q
$1 \rightarrow 2$			
$2 \rightarrow 3$			
$3 \rightarrow 4$			
$4 \rightarrow 1$			

Answers

1. (*a*) More work is done on the system than the system does on the environment, so the energy of the system increases. (*b*) More energy is transferred thermally out of the system than into it, and work is done on the system. Depending on how the work done on the system compares to the difference between the thermal transfers of energy, the energy of the system can increase, decrease, or stay the same.

2. In both cases mechanical energy is degraded into thermal energy of a high-temperature reservoir while thermal energy of a low-temperature reservoir is upgraded to thermal energy of a high-temperature reservoir. (*a*) Because the area of the degrade rectangle is larger than the area of the upgrade rectangle, entropy is increasing, so this process is possible. (*b*) Because the area of the upgrade rectangle is larger than the area of the degrade rectangle, entropy is decreasing, so this process is not possible.

3. Efficiency $= -W/Q_{in}$, the coefficient of performance of cooling $= Q_{in}/W$, and the coefficient of performance of heating $= Q_{out}/W$. The quantities in the numerator are the desired results: work done by the engine on the environment, energy transferred thermally out of the contents of the refrigerator, and energy transferred thermally into the room to be heated. The quantities in the denominator are the energies required to obtain the desired result: the thermal transfer of energy required to run the engine and the work required to run the refrigerator or heat pump.

4. ΔE: $1 \rightarrow 2 = 0$ (isotherm); $2 \rightarrow 3 = -$(temperature decreases); $3 \rightarrow 4 = 0$ (isotherm); $4 \rightarrow 1 = +$(temperature increases). W: $1 \rightarrow 2 = -$(expansion); $2 \rightarrow 3 = 0$ (isochoric process); $3 \rightarrow 4 = +$(compression); $4 \rightarrow 1 = 0$ (isochoric process). Q (sign follows from signs of ΔE and W): $1 \rightarrow 2 = +$; $2 \rightarrow 3 = -$; $3 \rightarrow 4 = -$; $4 \rightarrow 1 = +$.

21.5 Entropy constraints on energy transfers

In the first part of this chapter, we used entropy diagrams to describe the energy conversions that occur in heat engines and pumps. We stated, without proof, that the area of the rectangle for any energy conversion or transfer represented in such a diagram is a measure of the corresponding entropy change. In this section we develop the quantitative framework for computing entropy changes from entropy diagrams.

Steady devices return to the same state after one cycle, so that

$$\Delta E = 0 \quad \text{(steady device, complete cycle).} \tag{21.1}$$

The energy law (Eq. 20.1) then tells us that for a steady device $Q + W = 0$. Because $Q = Q_{in} - Q_{out}$, we see that over one cycle the work done on a steady device must be equal to the difference in the thermal output and input of energy:

$$W = Q_{out} - Q_{in} \quad \text{(steady device, complete cycle).} \tag{21.2}$$

Because the initial and final states are the same, the entropy of a steady device doesn't change either:

$$\Delta S_{dev} = 0 \quad \text{(steady device, complete cycle).} \tag{21.3}$$

From Chapter 19 we know that only processes in which the entropy either increases or remains unchanged can occur:

$$\Delta S_{dev} + \Delta S_{env} \geq 0. \tag{21.4}$$

Equation 21.3 thus means that we must have

$$\Delta S_{env} \geq 0 \quad \text{(steady device, complete cycle).} \tag{21.5}$$

As we saw in the first part of this chapter, the entropy constraint in Equation 21.5 limits the type of energy transfers that can occur in a steady device.

✋ **21.14** What is the change in the energy of the environment in a device for which Eqs. 21.1–21.5 hold?

We begin by considering a steady device that transfers energy thermally from a reservoir at temperature T_{in} to a reservoir at a lower temperature $T_{out} < T_{in}$ (Figure 21.37a).

In Chapter 20 we saw that the entropy change associated with energy transferred thermally at a fixed temperature is given by (Eq. 20.54)

$$\Delta S = \frac{Q}{k_B T} \quad \text{(isothermal process).} \tag{21.6}$$

The entropy diagram in Figure 21.37a shows two thermal transfers of energy—a transfer out of the environment at the high-temperature reservoir and a transfer into the environment at the low-temperature reservoir. Let us first apply Eq. 21.6 to the high-temperature reservoir. Because energy is transferred thermally out of that reservoir, $Q_H = -Q_{in} < 0$. Equation 21.6 thus yields a negative entropy change of the high-temperature reservoir, which is at temperature T_{in}:

$$\Delta S_H = -\frac{Q_{in}}{k_B T_{in}}. \tag{21.7}$$

Figure 21.37 Entropy diagram for thermal transfer of energy (a) from a high-temperature reservoir to a low-temperature reservoir and (b) from a low-temperature reservoir to a high-temperature reservoir.

(a) Transfer to lower-temperature reservoir

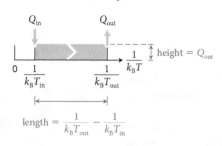

(b) Transfer to higher-temperature reservoir

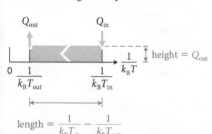

Likewise, for the low-temperature reservoir, which is at temperature T_{out}, $Q_L = Q_{out} > 0$, so the entropy change is positive:

$$\Delta S_L = \frac{Q_{out}}{k_B T_{out}}. \tag{21.8}$$

Adding the entropy changes in the environment, we get for the environment entropy change during the complete cycle represented in the entropy diagram

$$\Delta S_{env} = \Delta S_H + \Delta S_L = -\frac{Q_{in}}{k_B T_{in}} + \frac{Q_{out}}{k_B T_{out}} \quad \text{(steady state).} \tag{21.9}$$

Because $\Delta E = 0$ (Eq. 21.1) and because no work is involved, the energy law tells us that $\Delta E = Q_{in} - Q_{out} = 0$, so $Q_{in} = Q_{out}$. Substituting this result in Eq. 21.9, we obtain for the entropy change in the environment

$$\Delta S_{env} = -\frac{Q_{out}}{k_B T_{in}} + \frac{Q_{out}}{k_B T_{out}} = Q_{out}\left(\frac{1}{k_B T_{out}} - \frac{1}{k_B T_{in}}\right) \quad \text{(steady state).} \tag{21.10}$$

The right side of this equation gives exactly what we have been calling the area of the rectangle representing the energy transfer in the entropy diagram. As Figure 21.37a shows, the height of this rectangle is Q_{out} (a positive quantity):

$$\text{height of rectangle} = Q_{out}. \tag{21.11}$$

If we use the quantity $1/k_B T$ to quantify the horizontal scale in the diagram, we see that the left edge of the rectangle is at $1/k_B T_{in}$ and the right edge is at $1/k_B T_{out}$. The length of the rectangle is thus

$$\text{length of rectangle} = \frac{1}{k_B T_{out}} - \frac{1}{k_B T_{in}}. \tag{21.12}$$

Because $T_{out} < T_{in}$, this quantity is positive. Note that the length is exactly equal to the term in parentheses in Eq. 21.10, so the product of the height and the length—the area—of the rectangle indeed represents the entropy change of the environment ΔS_{env} due to the transfer of energy from the high-temperature reservoir to the low-temperature reservoir.

Let us next consider the reverse process: a thermal transfer of energy from a low-temperature reservoir to a high-temperature reservoir. As we saw in Section 21.2, such an upgrade of energy is not possible. The entropy diagram for this impossible process is shown in Figure 21.37b. Note that the rectangle has the same length and height as the one in Figure 21.37a. It points to the left (from low temperature to high temperature), indicating that the entropy change is negative. Therefore we take the negative of the area of the rectangle to compute the entropy change in the environment:

$$\Delta S_{env} = -Q_{out}\left(\frac{1}{k_B T_{in}} - \frac{1}{k_B T_{out}}\right) \quad \text{(steady state).} \tag{21.13}$$

Because the entropy change in the environment must be positive (Eq. 21.5), we see that this type of transfer cannot occur by itself.

The procedure for computing the entropy change from an entropy diagram is presented in the Procedure box on page 582.

Procedure: Computing entropy changes from entropy diagrams

To determine the entropy change in the environment from an entropy diagram, compute the area of the rectangles that represent the energy transfers as follows:

1. Determine the length of the horizontal part of each rectangle by subtracting the position of the left end of the rectangle on the $1/k_B T$ scale from the position of the right end on the scale. The position of mechanical energy is always at the zero point on this scale. Verify that the result is a positive quantity. (If it isn't, you made a mistake somewhere.)

2. Express the height of each rectangle in terms of Q_{in}, Q_{out}, and W. Remember that the height must be a positive quantity; Q_{in} and Q_{out} are always positive, but W can be positive or negative. Avoid using the input or output where two rectangles meet; if necessary, you can

use Eq. 21.2, $W = Q_{out} - Q_{in}$, to express the height in terms of the quantities that are given in the problem.

3. Determine the sign of the entropy change. If the rectangle points left (energy upgrade), the entropy change is negative. If the rectangle points right (energy degradation), the entropy change is positive.

4. The entropy change for each rectangle is given by the area of the rectangle (the length multiplied by the height computed in steps 1 and 2) preceded by the algebraic sign determined in step 3:

$$\Delta S = (\text{sign}) \times (\text{length}) \times (\text{width}).$$

5. The entropy change for the process is the sum of the individual entropy changes you have calculated.

Example 21.5 Thermal transfer of energy

A steel bar transfers energy from a thermal reservoir at 750 K to one at 520 K. If the bar transfers 1.5×10^6 J of energy during a certain time interval over which the temperature distribution of the bar does not change, what is the entropy change of the environment?

❶ GETTING STARTED Because the temperature distribution does not change, the state of the bar remains constant and I can consider the bar to be a steady device. I begin by drawing an entropy diagram (**Figure 21.38**), showing a right-pointing arrow in the rectangle because the direction of the energy transfer is from the higher-temperature reservoir to the lower-temperature one. I am given $T_{in} = 750$ K, $T_{out} = 520$ K, and the amount of energy transferred $Q_{in} = Q_{out} = 1.5 \times 10^6$ J.

Figure 21.38

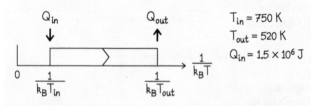

$T_{in} = 750$ K
$T_{out} = 520$ K
$Q_{in} = 1.5 \times 10^6$ J

❷ DEVISE PLAN To determine the entropy change of the environment, I follow the procedure shown above that tells me how to compute the entropy change from my entropy diagram.

❸ EXECUTE PLAN Because the rectangle arrow points to the right, I know that the entropy of the environment increases and thus the entropy change is positive. The height of the rectangle is Q_{in}. The length is $[1/k_B T_{out} - 1/k_B T_{in}]$. The entropy change is thus

$$\Delta S_{env} = +Q_{in}\left[\frac{1}{k_B T_{out}} - \frac{1}{k_B T_{in}}\right] = \frac{+Q_{in}}{k_B}\left[\frac{1}{T_{out}} - \frac{1}{T_{in}}\right].$$

Substituting the values given, I obtain

$$\Delta S_{env} = \frac{1.5 \times 10^6 \text{ J}}{1.38 \times 10^{-23} \text{ J/K}}\left[\frac{1}{520 \text{ K}} - \frac{1}{750 \text{ K}}\right]$$

$$= 6.4 \times 10^{25}. ✔$$

❹ EVALUATE RESULT Energy is degraded (hot to cold), so I expect a positive entropy change. The number I obtained for the magnitude of the entropy change is very large, but I know from Chapter 19 that entropy values are very large, and so my answer is not unreasonable.

✋ **21.15** Suppose the bar in Example 21.5 is disconnected from the high-temperature reservoir and allowed to thermally equilibrate with the low-temperature reservoir. (*a*) Are the changes in the energy and entropy of the bar during this process positive, negative, or zero? (*b*) Is there any change in the entropy of the environment? (*c*) Does the sum $S_{bar} + S_{env}$ change?

Next we consider a device that converts mechanical energy (work) to thermal energy. In Section 21.1 we discussed a paddle wheel immersed in water. Now we consider a somewhat different situation: a motor doing work on an electric generator that delivers electrical energy to an electric blanket, which converts the electrical energy to thermal energy (**Figure 21.39a**). We define the generator, blanket, and wires that connect the two as our system. The motor (which is part of the environment) does work on the system, and energy is transferred thermally by the blanket (part of the system) to the environment. We assume that the temperatures of the blanket and the environment remain constant. This assumption ensures that the initial and final states of the system are the same.

Figure 21.39b shows the entropy diagram for this process. Because the transfer is to the right, the entropy change is positive. The height of the rectangle is Q_{out}. (Because $W = Q_{out}$ we could also write the height of the rectangle as W.) The rectangle stretches from the origin (0 on the $1/k_B T$ scale) to $1/k_B T_{out}$, so its length is $1/k_B T_{out}$. The entropy change is thus

$$\Delta S_{env} = +Q_{out}\frac{1}{k_B T_{out}} = \frac{Q_{out}}{k_B T_{out}} \quad \text{(steady state).} \quad (21.14)$$

This is indeed the result we would expect: No entropy change is associated with the work done on the system, and the entropy change caused by the thermal transfer of energy to the environment is $Q_{out}/k_B T_{out}$, which is exactly the result we obtain in Eq. 21.14.

21.16 (*a*) Draw an entropy diagram for a steady device that converts thermal energy from a thermal reservoir to mechanical energy. (*b*) From the diagram, compute the entropy change that occurs during the process and show that such a device is impossible.

21.6 Heat engine performance

Consider a heat engine that takes in energy thermally from a high-temperature reservoir, converts part of this energy to mechanical energy by doing work on the environment ($W < 0$), and discards the remainder to a low-temperature reservoir. As we saw in Section 21.3, the entropy diagram has two rectangles: one for the energy upgrade and the other for the degradation (**Figure 21.40**). To determine the entropy change in the environment over one complete cycle, we add the entropy changes of the two rectangles:

$$\Delta S_{env} = \Delta S_{upgrade} + \Delta S_{degrade} \quad \text{(complete cycle).} \quad (21.15)$$

The entropy change for the degradation is given by Eq. 21.10, so we need only calculate the change for the upgrade. The height of the upgrade rectangle is $-W$. Because $W = Q_{out} - Q_{in}$ (Eq. 21.2), the height of the rectangle can also be written as $Q_{in} - Q_{out}$. The rectangle stretches from the origin (0 on the $1/k_B T$ scale) to $1/k_B T_{in}$, so its length is $1/k_B T_{in}$. Because the transfer is to the left, the entropy change is negative. The entropy change associated with the upgrade is thus

$$\Delta S_{upgrade} = -(Q_{in} - Q_{out})\left[\frac{1}{k_B T_{in}}\right]. \quad (21.16)$$

Figure 21.40 Entropy diagram for a heat engine that upgrades a portion of the thermal energy to mechanical energy and degrades the rest.

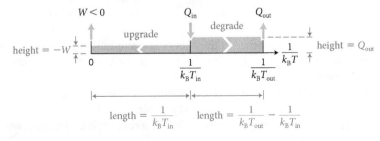

Figure 21.39 A setup that converts mechanical energy (from the motor) to thermal energy.

(*a*)

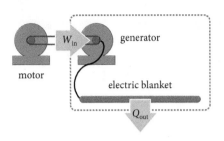

(*b*)

Substituting Eqs. 21.16 and 21.10 into Eq. 21.15 gives

$$\Delta S_{env} = -(Q_{in} - Q_{out})\left[\frac{1}{k_B T_{in}}\right] + Q_{out}\left[\frac{1}{k_B T_{out}} - \frac{1}{k_B T_{in}}\right], \quad (21.17)$$

or
$$\Delta S_{env} = -\frac{Q_{in}}{k_B T_{in}} + \frac{Q_{out}}{k_B T_{out}} \quad \text{(complete cycle)}, \quad (21.18)$$

which is exactly Eq. 21.9, the expression we derived for the entropy change for a simple device that thermally transfers energy from a high-temperature reservoir to a low-temperature reservoir. The reason we obtain the same expression for the more complicated case of a heat engine is that no entropy change is associated with mechanical transfers of energy. Only the thermal transfers of energy to and from the thermal reservoirs contribute to the entropy change.

For a reversible heat engine, the entropy change in the environment must be equal to zero over one cycle (everything is back to what it was before), $\Delta S_{env} = 0$, so

$$-\frac{Q_{in}}{k_B T_{in}} + \frac{Q_{out}}{k_B T_{out}} = 0 \quad \text{(reversible heat engine)}. \quad (21.19)$$

Rearranging this equation, we get

$$\frac{Q_{out}}{Q_{in}} = \frac{T_{out}}{T_{in}} \quad \text{(reversible heat engine)}. \quad (21.20)$$

For a reversible heat engine that transfers energy from a thermal reservoir at T_{in}, the amount of energy discarded is proportional to T_{out}. The amount of energy discarded is independent of the type of working substance used and depends on only the temperatures of the two reservoirs. If we lower T_{out}, we decrease the amount of discarded energy, which means a greater portion of the input energy can be converted to mechanical energy. So we can maximize the work done on the environment by minimizing the temperature of the low-temperature reservoir.

21.17 Is it possible to reduce the amount of thermally discarded energy to zero by bringing T_{out} down to absolute zero?

Figure 21.41 shows the energy input-output diagram for a heat engine. In Section 21.3 we defined the efficiency of such an engine as the ratio of the work done on the environment $(-W)$ to the thermal input of energy (Q_{in}). Using the symbol η (lowercase Greek eta) for efficiency, we can write

$$\eta \equiv \frac{-W}{Q_{in}}. \quad (21.21)$$

Substituting W from Eq. 21.2 yields

$$\eta = \frac{Q_{in} - Q_{out}}{Q_{in}} = 1 - \frac{Q_{out}}{Q_{in}}. \quad (21.22)$$

Equation 21.22 is a consequence of the energy law and holds for both reversible and irreversible heat engines that involve transfers of energy to and from two thermal reservoirs.

Figure 21.23 shows that for a given amount of work $-W$ done by a heat engine, Q_{in} is smallest when the engine is reversible. Therefore the efficiency of a heat engine is maximum when it is reversible. For a reversible heat engine, we can

Figure 21.41 Energy input-output diagram for a heat engine.

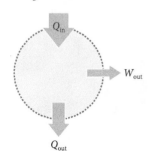

substitute Eq. 21.20 into Eq. Eq. 21.22 to obtain an expression for the maximum efficiency:

$$\eta_{max} = 1 - \frac{T_{out}}{T_{in}} = \frac{T_{in} - T_{out}}{T_{in}} \quad \text{(reversible heat engine).} \quad (21.23)$$

In practice, the efficiency of a heat engine is always smaller than this maximum efficiency because it is impossible to completely eliminate irreversible processes. The important point to remember, however, is that even for a reversible heat engine, the efficiency is less than 1. The maximum efficiency is obtained with a minimum transfer of thermal energy to the low-temperature reservoir and has *nothing* to do with the details of the workings of the heat engine. *Only* the temperatures of the two reservoirs have any effect on the maximum efficiency: It is greatest when the ratio T_{out}/T_{in} is smallest.

Example 21.6 Environmental impact of power plant

A power plant burns fossil fuel to produce steam at 650 K. The pressurized steam drives a steam turbine, where the steam condenses to water and is discarded at 310 K. What is the minimum quantity of energy discarded thermally by a 1.0-GW power plant to the environment during one day?

❶ GETTING STARTED Because I am interested in the minimum quantity of thermal energy discarded, I can treat the turbine as a reversible heat engine. Energy is thermally transferred to the turbine from the steam, which serves as the high-temperature reservoir at $T_{in} = 650$ K. Energy is thermally discarded along with the condensed steam at $T_{out} = 310$ K.

❷ DEVISE PLAN To calculate the minimum quantity of energy discarded to the environment, $Q_{out,min}$, I can use Eq. 21.2, which tells me that for a given value of W, Q_{out} is minimum when Q_{in} is minimum. No value is given for Q_{in}, but Eq. 21.21 tells me that Q_{in} is minimum when η has its maximum value, which I can calculate using Eq. 21.23. In one day, a 1.0-GW power plant delivers an amount of energy equal to $(24 \text{ h})(3600 \text{ s}/1 \text{ h})$ $(1.0 \times 10^9 \text{ J/s}) = 8.6 \times 10^{13}$ J, and so $W = -8.6 \times 10^{13}$ J.

❸ EXECUTE PLAN The maximum efficiency is (Eq. 21.23)

$$\eta_{max} = \frac{650 \text{ K} - 310 \text{ K}}{650 \text{ K}} = 0.523,$$

and so the minimum quantity of energy transferred thermally to the turbine is

$$Q_{in,min} = \frac{-W}{\eta_{max}} = \frac{8.6 \times 10^{13} \text{ J}}{0.523} = 1.6 \times 10^{14} \text{ J}.$$

From Eq. 21.2 I thus calculate the minimum quantity of energy discarded thermally:

$$Q_{out,min} = Q_{in,min} + W = (1.6 \times 10^{14} \text{ J}) - (8.6 \times 10^{13} \text{ J})$$
$$= 7.8 \times 10^{13} \text{ J.} \checkmark$$

❹ EVALUATE RESULT The quantity of thermal energy leaving the turbine is smaller than the quantity of thermal energy entering it (as it should be) and approximately equal to the quantity of energy delivered by the plant. This makes sense because the maximum efficiency is close to $\frac{1}{2}$. I assumed the turbine to be reversible, which is unlikely. However, if the turbine is not reversible, more energy is dissipated, and so my result for the minimum amount of energy discarded still holds.

21.18 A hurricane thermally takes in energy from ocean water at 26 °C, discards some of it to the upper atmosphere at −52 °C, and converts the remainder to mechanical energy in the form of wind. (*a*) What is the maximum possible efficiency for the thermal-to-mechanical conversion? (*b*) If the temperature of the water increases by 1 °C, by what amount does the maximum efficiency increase?

If we run a heat engine in reverse, we obtain a heat pump (**Figure 21.42**). When the heat pump is used as a heater, we define the **coefficient of performance of heating** as

$$\text{COP}_{heating} \equiv \frac{Q_{out}}{W}. \quad (21.24)$$

Figure 21.42 Energy input-output diagram for a heat pump.

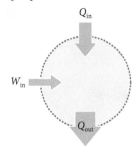

Because $W = W_{in} > 0$ and $W = Q_{out} - Q_{in}$, Q_{out} is always greater than or equal to W, and so $COP_{heating}$ for a heat pump is always greater than or equal to 1.

Substituting Eq. 21.2 into Eq. 21.24 gives

$$COP_{heating} = \frac{Q_{out}}{Q_{out} - Q_{in}} = \frac{1}{1 - Q_{in}/Q_{out}}. \quad (21.25)$$

For a reversible heat pump that transfers energy from a lower-temperature thermal reservoir at T_{in} to a higher-temperature reservoir at T_{out}, we can substitute Eq. 21.20 into Eq. 21.25 to obtain the maximum coefficient of performance of heating:

$$COP_{heating,max} = \frac{1}{1 - T_{in}/T_{out}} = \frac{T_{out}}{T_{out} - T_{in}} \quad \text{(reversible heat pump).} \quad (21.26)$$

This expression tells us that the maximum coefficient of performance of heating is greatest when the ratio T_{in}/T_{out} is close to 1.

21.19 What is the coefficient of performance of heating for the electric blanket in Figure 21.39?

When a heat pump is used to cool (as a refrigerator or air conditioner), we define the **coefficient of performance of cooling** as

$$COP_{cooling} \equiv \frac{Q_{in}}{W}. \quad (21.27)$$

Substituting Q_{in} from Eq. 21.2 gives

$$COP_{cooling} = \frac{Q_{out} - W}{W} = \frac{Q_{out}}{W} - 1 = COP_{heating} - 1. \quad (21.28)$$

For a reversible heat pump, we can substitute Eq. 21.26 into Eq. 21.28 to obtain an expression for the maximum coefficient of performance of cooling:

$$COP_{cooling,max} = \frac{T_{out}}{T_{out} - T_{in}} - 1$$

$$= \frac{T_{in}}{T_{out} - T_{in}} \quad \text{(reversible heat pump).} \quad (21.29)$$

For a refrigerator, the coefficient of performance of cooling is usually greater than 1 because the temperature of the interior T_{in} is around 255 K, and the temperature difference $T_{out} - T_{in}$ between the outside air and the interior is around 40 K.

Example 21.7 Cost of cooling off

The thermal insulation in a refrigerator is never perfect. Typically energy is transferred thermally from the room to the refrigerator interior at a rate of 60 W with the refrigerator door closed. Suppose the interior needs to be held at 4.0 °C and the temperature of the room is 20 °C. At what minimum rate must mechanical energy be delivered to the refrigerator to maintain the interior at 4.0 °C?

❶ **GETTING STARTED** Because I am interested in the minimum rate at which mechanical energy must be delivered, I can treat the refrigerator as a reversible heat pump. The room is the high-temperature reservoir, so $T_{out} = 20\ °C = 293\ K$. The refrigerator interior is the low-temperature reservoir, so $T_{in} = 4.0\ °C = 277\ K$. To maintain the temperature of the interior, the energy Q_{in} that leaks through the thermal insulation and moves from the room into the refrigerator must be transferred out. I am given that the rate at which this energy is transferred thermally into the refrigerator is $dQ_{in}/dt = 60$ W.

❷ DEVISE PLAN To calculate the rate at which mechanical energy must be delivered to balance a given quantity Q_{in} moving from the room to the refrigerator interior, I should first calculate how much work needs to be done to transfer that quantity of energy out of the refrigerator. To calculate the work required to transfer this quantity of energy Q_{in} I can use Eq. 21.27: $W = Q_{in}/COP_{cooling}$. This expression tells me that the amount of work needed is minimum when $COP_{cooling}$ has its maximum value, which I can determine from Eq. 21.29. To obtain the rate at which mechanical energy must be delivered, I take the time derivative of W.

❸ EXECUTE PLAN The maximum coefficient of performance of cooling is (Eq. 21.29)

$$COP_{cooling,max} = \frac{277\ K}{293\ K - 277\ K} = 17,$$

so the minimum work required is $W_{min} = Q_{in}/COP_{cooling,max}$. Taking the derivative of this expression, I obtain

$$\left(\frac{dW}{dt}\right)_{min} = \frac{dQ_{in}/dt}{COP_{cooling,max}} = \frac{60\ W}{17} = 3.5\ W. ✔$$

❹ EVALUATE RESULT My calculated value for the rate at which mechanical energy must be delivered to cancel the thermal energy that leaks from the room to the refrigerator interior is $1/17$ the rate at which that leaking occurs. This makes sense because the coefficient of performance of cooling is 17. This rate is small for what I expect for typical electrical appliances (several hundred watts), but I assumed the refrigerator was reversible. In reality, it is not reversible, and additional energy is dissipated, which means my calculated value is too low. Also, refrigerator doors are opened many times a day, thus increasing the rate of energy transferred thermally to the interior.

✋ **21.20** (*a*) If the heat pump from the refrigerator in Example 21.7 is used during the winter to heat a room to 20 °C when it's 4 °C outside, what is $COP_{heating}$? (*b*) At what rate must work be done on the heat pump if you want to heat the room at a rate of 500 W?

21.7 Carnot cycle

In Section 21.4, we identified the four processes that make up the Carnot cycle: two isothermal processes and two isentropic processes (**Figure 21.43**). Part of the energy taken in from a high-temperature reservoir is upgraded to mechanical energy, while the rest is degraded and discarded into a low-temperature reservoir (Figure 21.32*b*). The thermal transfer of energy occurs only during the isothermal processes at the two temperatures T_{in} and T_{out}.

As we study the Carnot cycle further, let us use an ideal gas as the working substance so that we can use our results for the entropy and thermal transfer of energy for each process from Tables 20.6 and 20.7, which are gathered together in **Table 21.1**.

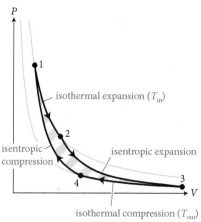

Figure 21.43 *PV* diagram for a Carnot cycle.

Table 21.1 Entropy and energy changes for the working substance in a Carnot cycle (see Tables 20.6 and 20.7)

Process	Definition	ΔS_{dev}	ΔE	W	Q
$1 \rightarrow 2$ (*isothermal*)	$\Delta T = 0$	$N \ln\left(\dfrac{V_2}{V_1}\right)$	0	$-Nk_B T_{in} \ln\left(\dfrac{V_2}{V_1}\right)$	$Nk_B T_{in} \ln\left(\dfrac{V_2}{V_1}\right)$
$2 \rightarrow 3$ (*isentropic*)	$Q = 0$	0	$NC_V(T_{out} - T_{in})$	$NC_V(T_{out} - T_{in})$	0
$3 \rightarrow 4$ (*isothermal*)	$\Delta T = 0$	$N \ln\left(\dfrac{V_4}{V_3}\right)$	0	$-Nk_B T_{out} \ln\left(\dfrac{V_4}{V_3}\right)$	$Nk_B T_{out} \ln\left(\dfrac{V_4}{V_3}\right)$
$4 \rightarrow 1$ (*isentropic*)	$Q = 0$	0	$NC_V(T_{in} - T_{out})$	$NC_V(T_{in} - T_{out})$	0
cycle	reversible	0	0	$-Nk_B(T_{in} - T_{out}) \ln\left(\dfrac{V_2}{V_1}\right)$	$-W$

To calculate the efficiency of a heat engine operating on this cycle from Eq. 21.21, we need to determine W and Q_{in}. We can obtain the work done on the working substance (our system) during the cycle from Table 21.1:

$$W = -Nk_B T_{in} \ln\left(\frac{V_2}{V_1}\right) - Nk_B T_{out} \ln\left(\frac{V_4}{V_3}\right) \quad \text{(Carnot cycle).} \qquad (21.30)$$

We do not need to write the two isentropic work terms because they cancel out.

For a complete cycle the entropy change must be zero, and so summing the entropy changes in Table 21.1, we get

$$N \ln\left(\frac{V_2}{V_1}\right) + N \ln\left(\frac{V_4}{V_3}\right) = 0 \qquad (21.31)$$

or

$$\ln\left(\frac{V_2}{V_1}\right) = -\ln\left(\frac{V_4}{V_3}\right), \qquad (21.32)$$

which means that

$$\frac{V_2}{V_1} = \frac{V_3}{V_4} \quad \text{(Carnot cycle).} \qquad (21.33)$$

Substituting Eq. 21.32 into Eq. 21.30, we obtain

$$W = -Nk_B(T_{in} - T_{out}) \ln\left(\frac{V_2}{V_1}\right) \quad \text{(Carnot cycle).} \qquad (21.34)$$

Because the volume ratio V_2/V_1 is greater than 1, $\ln(V_2/V_1)$ is positive. Because $T_{in} - T_{out}$ is also positive, the work done on the system is negative, which is consistent with our observation in Section 21.4: The work done on the system for a clockwise cycle is negative. Hence for a clockwise Carnot cycle, the work done on the environment is positive.

The high-temperature reservoir transfers energy thermally to the working substance during the first isothermal process (the one at T_{in}). The quantity Q_{in} is thus equal to the energy transferred thermally during the first isothermal process (see Table 21.1):

$$Q_{in} = Q_{1\to 2} = Nk_B T_{in} \ln\left(\frac{V_2}{V_1}\right). \qquad (21.35)$$

Remember that not all of this energy can be converted to mechanical energy because some of it must be transferred to the low-temperature reservoir during the second isothermal process to satisfy the entropy law (Figure 21.32b).

Substituting Eqs. 21.34 and 21.35 into Eq. 21.21 thus yields

$$\eta = \frac{-W}{Q_{in}} = \frac{Nk_B(T_{in} - T_{out}) \ln\left(\frac{V_2}{V_1}\right)}{Nk_B T_{in} \ln\left(\frac{V_2}{V_1}\right)} = \frac{T_{in} - T_{out}}{T_{in}} \quad \text{(Carnot cycle),} \quad (21.36)$$

which is exactly the maximum efficiency we obtained in Eq. 21.23 for any heat engine, regardless of the working substance.

Example 21.8 Carnot engine

In a reversible heat engine, 6.02×10^{23} molecules of nitrogen gas are used as the working substance. The gas undergoes a Carnot cycle as illustrated in Figure 21.43. The volume of the gas is 6.00×10^{-3} m³ in state 1 and 14.0×10^{-3} m³ in state 2. The low-temperature reservoir is at 320 K. How much energy is transferred thermally to the low-temperature reservoir during one cycle?

❶ **GETTING STARTED** The temperature of the low-temperature reservoir is $T_{out} = 320$ K. I am also given that $V_1 = 6.00 \times 10^{-3}$ m³ and $V_2 = 14.0 \times 10^{-3}$ m³. I need to determine Q_{out}, the amount of energy transferred thermally to the low-temperature reservoir in each cycle.

❷ **DEVISE PLAN** From Eq. 21.2 I know that $Q_{out} = W + Q_{in}$. Therefore this problem boils down to determining W and Q_{in}, which are given by Eq. 21.34 and 21.35, respectively. These equations both contain T_{in}, which is not given, but the terms containing T_{in} cancel and so T_{in} drops out.

❸ **EXECUTE PLAN** Substituting Eqs. 21.34 and 21.35 into Eq. 21.2 after solving the latter for Q_{out}, I get

$$Q_{out} = W + Q_{in} = -Nk_B(T_{in} - T_{out})\ln\left(\frac{V_2}{V_1}\right)$$

$$+ Nk_B T_{in}\ln\left(\frac{V_2}{V_1}\right) = Nk_B T_{out}\ln\left(\frac{V_2}{V_1}\right). \quad (1)$$

Substituting the values given yields

$$Q_{out} = (6.02 \times 10^{23})(1.38 \times 10^{-23} \text{ J/K})(320 \text{ K})$$

$$\ln\left(\frac{14.0 \times 10^{-3} \text{ m}^3}{6.00 \times 10^{-3} \text{ m}^3}\right)$$

$$= 2.25 \times 10^3 \text{ J. } ✔$$

❹ **EVALUATE RESULT** The quantity I obtain for Q_{out} is positive, as it should be. I also note that the expression I obtain for Q_{out} in Eq. 1 can be written as $-Nk_B T_{out}\ln(V_4/V_3)$ because $\ln(V_2/V_1) = -\ln(V_4/V_3)$ (Eq. 21.32). Therefore $Q_{out} = -Q_{3\to4}$ (Table 21.1). This makes sense because energy is transferred thermally to the low-temperature reservoir only during the second isothermal process ($3 \to 4$). The minus sign in $-Q_{3\to4}$ accounts for the fact that $Q < 0$ (energy transferred out of the system) and, by definition, Q_{out} is always positive.

21.21 What is the change in entropy during each of the four legs of the cycle in Example 21.8?

21.8 Brayton cycle

As we saw in Section 21.4, the Brayton cycle (Figure 21.44) produces more mechanical energy than the Carnot cycle. More importantly, because the Brayton cycle does not involve any isothermal processes, the rate at which work can be done is greater (yielding more power).

To determine the efficiency of the Brayton cycle, we begin by collecting the appropriate values for the entropy and energy changes from Tables 20.6 and 20.7. The result is shown in Table 21.2 on the next page.

Figure 21.44 *PV* diagram for a Brayton cycle.

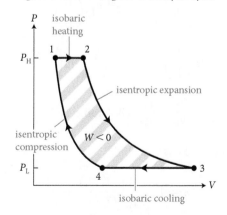

Table 21.2 Entropy and energy for the working substance in a Brayton cycle

Process	Definition	ΔS_{dev}	ΔE	W	Q
$1 \rightarrow 2$ (isobaric)	$\Delta P = 0$	$\dfrac{NC_P}{k_B}\ln\!\left(\dfrac{T_2}{T_1}\right)$	$NC_V(T_2 - T_1)$	$-Nk_B(T_2 - T_1)$	$NC_P(T_2 - T_1)$
$2 \rightarrow 3$ (isentropic)	$Q = 0$	0	$NC_V(T_3 - T_2)$	$NC_V(T_3 - T_2)$	0
$3 \rightarrow 4$ (isobaric)	$\Delta P = 0$	$\dfrac{NC_P}{k_B}\ln\!\left(\dfrac{T_4}{T_3}\right)$	$NC_V(T_4 - T_3)$	$-Nk_B(T_4 - T_3)$	$NC_P(T_4 - T_3)$
$4 \rightarrow 1$ (isentropic)	$Q = 0$	0	$NC_V(T_1 - T_4)$	$NC_V(T_1 - T_4)$	0
cycle	reversible	0	0	$-Q$	$NC_P(T_2 - T_1 + T_4 - T_3)$

Because the energy change over a cycle is zero, we have $W = -Q$. To obtain the work done during a cycle, therefore, we can sum the two terms in the rightmost column of Table 21.2 (which is simpler than taking the sum of the four W terms):

$$W = -Q = -NC_P(T_2 - T_1 + T_4 - T_3) \quad \text{(Brayton cycle).} \quad (21.37)$$

To calculate the efficiency (Eq. 21.21; $\eta = -W/Q_{\text{in}}$), we must obtain expressions for $-W$ and Q_{in}, both of which must be positive. Because $T_2 > T_1$, Q is positive during the first isobaric process ($1 \rightarrow 2$) at the high pressure P_H. Likewise, because $T_4 < T_3$, Q is negative during the second isobaric process ($3 \rightarrow 4$) at the low pressure P_L. The energy transferred thermally from the environment to the working substance is thus equal to the energy transferred thermally during the isobaric process $1 \rightarrow 2$:

$$Q_{\text{in}} = NC_P(T_2 - T_1). \quad (21.38)$$

Substituting Eqs. 21.37 and 21.38 into Eq. 21.21 and dividing out the factor NC_P, we obtain

$$\eta \equiv \frac{-W}{Q_{\text{in}}} = \frac{T_2 - T_1 + T_4 - T_3}{T_2 - T_1} = 1 - \frac{T_3 - T_4}{T_2 - T_1}. \quad (21.39)$$

We can simplify this expression for the efficiency of the Brayton cycle by using our knowledge about entropy. The entropy change for the cycle is zero (Table 21.2). This ΔS_{dev} for the cycle is the sum of the two nonzero ΔS_{dev} terms in Table 21.2, so setting the sum of these two terms equal to zero gives

$$\frac{NC_P}{k_B}\ln\!\left(\frac{T_2}{T_1}\right) + \frac{NC_P}{k_B}\ln\!\left(\frac{T_4}{T_3}\right) = 0 \quad (21.40)$$

$$\ln\!\left(\frac{T_2}{T_1}\right) = -\ln\!\left(\frac{T_4}{T_3}\right) = \ln\!\left(\frac{T_3}{T_4}\right), \quad (21.41)$$

which means that $\qquad \dfrac{T_2}{T_1} = \dfrac{T_3}{T_4} \quad$ (Brayton cycle). $\qquad (21.42)$

We now rewrite Eq. 21.39 as

$$\eta = 1 - \frac{T_4}{T_1}\left[\frac{(T_3/T_4) - 1}{(T_2/T_1) - 1}\right]. \quad (21.43)$$

By Eq. 21.42, the term inside the square brackets is equal to 1, so the efficiency is

$$\eta = 1 - \frac{T_4}{T_1}. \quad \text{(Brayton cycle)} \qquad (21.44)$$

The temperature T_4 is the lowest temperature in the cycle. The temperature T_1 is the temperature of the working substance at the completion of the isentropic compression, when the working substance exits the compressor (shown in Figure 21.30b).

During the isentropic compression, we have, from Eq. 20.47,

$$(P_H)^{(1/\gamma)-1}T_1 = (P_L)^{(1/\gamma)-1}T_4, \qquad (21.45)$$

where γ is the heat capacity ratio C_P/C_V given by Eq. 20.26. The temperature ratio in Eq. 21.44 then becomes

$$\frac{T_4}{T_1} = \left(\frac{P_H}{P_L}\right)^{(1/\gamma)-1}. \qquad (21.46)$$

The efficiency in Eq. 21.44 can thus be expressed in terms of the pressure increase accomplished by the compressor in the process $4 \to 1$:

$$\eta = 1 - \left(\frac{P_H}{P_L}\right)^{(1/\gamma)-1}. \quad \text{(Brayton cycle)} \qquad (21.47)$$

The ratio P_H/P_L is called the *pressure ratio*. For aircraft engines, this ratio typically is between 4 and 30. The higher the pressure ratio of an engine running on the Brayton cycle, the more efficient the engine.

Example 21.9 Performance and pressure ratio

Consider a jet engine operating on the Brayton cycle. The working substance is air, which consists primarily of diatomic molecules. The air is drawn in at atmospheric pressure, 1.01×10^5 Pa, at 288 K. In the combustion chamber, the temperature of the air is raised to 1520 K, and the engine has a pressure ratio of 30. What are the temperatures at the end of the two isentropic processes?

❶ GETTING STARTED Air enters the engine at point 4 in the cycle in Figure 21.44. Therefore I know that $T_4 = 288$ K. The air is compressed between state 4 and state 1 and the combustion takes place between state 1 and state 2, so I also know that $T_2 = 1520$ K. The isentropic processes end at states 1 and 3, so I need to determine T_1 and T_3.

❷ DEVISE PLAN To determine T_1 I can use Eq. 21.46, which relates T_1 to T_4 and the pressure ratio, both of which are given. This equation also contains the heat capacity ratio γ, which I can determine from Eq. 20.26, $\gamma = C_P/C_V = 1 + 2/d$, because I know that $d = 5$ for diatomic molecules. Once I have T_1, I can use Eq. 21.42 to determine T_3.

❸ EXECUTE PLAN From Eq. 20.26, I get

$$\gamma = 1 + \frac{2}{5} = \frac{7}{5},$$

so the exponent in Eq. 21.46 is $\frac{1}{\gamma} - 1 = \frac{5}{7} - 1 = -\frac{2}{7}$. I can then solve Eq. 21.46 for T_1:

$$T_1 = \frac{T_4}{(P_H/P_L)^{(1/\gamma)-1}} = \frac{288 \text{ K}}{30^{-2/7}} = 760 \text{ K}. ✔ \qquad (1)$$

Now that I have T_1 and T_2, I can get T_3 from Eq. 21.42:

$$T_3 = \frac{T_2 T_4}{T_1} = \frac{(1520 \text{ K})(288 \text{ K})}{760 \text{ K}} = 575 \text{ K}. ✔$$

❹ EVALUATE RESULT I know from Section 20.4 that on a given PV diagram, isentropes are steeper than isotherms, so Figure 21.44 tells me that T_1 and T_3 should lie between $T_4 = 288$ K and $T_2 = 1520$ K. The values I obtain indeed satisfy this condition.

21.22 (a) What is the efficiency of the jet engine in Example 21.9? (b) If the jet engine were to operate on a Carnot cycle between the same highest and lowest temperatures, what would its efficiency be?

Chapter Glossary

SI units of physical quantities are given in parentheses.

Brayton cycle A reversible thermodynamic cycle that consists of two isobaric processes and two isentropic processes.

Carnot cycle A reversible thermodynamic cycle that consists of two isothermal processes and two isentropic processes. This cycle operates at the maximum possible efficiency between two thermal reservoirs at fixed temperatures.

Coefficient of performance of cooling COP (unitless) The ratio of the thermal input of energy to the work done on a heat pump:

$$\text{COP}_{\text{cooling}} \equiv \frac{Q_{\text{in}}}{W}. \qquad (21.27)$$

Coefficient of performance of heating COP (unitless) The ratio of the thermal output of energy to the work done on a heat pump:

$$\text{COP}_{\text{heating}} \equiv \frac{Q_{\text{out}}}{W}. \qquad (21.24)$$

Efficiency η (unitless) The ratio of the work done by a heat engine on the environment to the thermal input of energy:

$$\eta \equiv \frac{-W}{Q_{\text{in}}}. \qquad (21.21)$$

Energy input-output diagram A diagram that shows the energy inputs and energy outputs of a system or device. Energy inputs and energy outputs can be due to mechanical interactions or thermal interactions.

Entropy diagram A diagram that shows the changes in entropy in the environment of a steady device due to the energy conversions and transfers caused by the device. See the Procedure box on page. 573.

Entropy gradient dS/dE (1/J) A scalar that represents the rate at which the entropy of a system changes with a change in thermal energy of the system. The entropy gradient is inversely proportional to the absolute temperature (Eq. 19.38).

Heat engine A steady device that does work on the environment by taking in energy thermally at a high temperature and putting out energy thermally at a lower temperature.

Heat pump A steady device that uses mechanical energy to take in energy thermally at a low temperature and put out energy thermally at a higher temperature.

Mechanical input of energy W_{in} (J) A positive scalar equal to the amount of energy transferred into a system due to a mechanical interaction.

Mechanical output of energy W_{out} (J) A positive scalar equal to the amount of energy transferred out of a system due to a mechanical interaction.

Quality of energy A measure of the ratio of energy change to entropy change. The higher the quality of the energy transferred, the smaller the associated entropy change. Mechanical energy represents the highest-quality energy.

Steady device A device that can be operated indefinitely because its state does not change more and more over time. The changes in energy and entropy of such a device must be zero either continuously or over some repeating time interval.

Thermal input of energy Q_{in} (J) A positive scalar equal to the amount of energy transferred into a system due to a thermal interaction.

Thermal output of energy Q_{out} (J) A positive scalar equal to the amount of energy transferred out of a system due to a thermal interaction.

22

Electric Interactions

CONCEPTS

QUANTITATIVE TOOLS

*E*lectricity is a familiar term—outlets, batteries, light bulbs, computers all involve electricity. It is no understatement to say that modern life depends on electricity, but what exactly *is* electricity? We all know what electricity does, but it's not that easy to explain what electricity *is*.

Electricity manifests itself in many ways: from the sparks that fly when you scuff your feet across a carpet on a dry winter day to the electricity we use in our homes to the transmission of radio and television programs. Even the attraction between magnets has to do with electricity. In this chapter, we begin our treatment of electricity with a discussion of static electricity.

22.1 Static electricity

When you tear off some plastic wrap from its roll, the wrap is attracted to anything that gets close: your hand, the countertop, a dish. This interaction between the plastic wrap and other objects doesn't have to involve any physical contact. For example, you can feel the presence of a piece of freshly torn-off plastic wrap with your cheek or the back of your hand even when your face or hand is held some distance away from the piece. You may have experienced many similar interactions: Styrofoam peanuts are attracted to your arms when you unpack a box full of them (Figure 22.1). Running a comb through your hair on a dry day causes the comb to attract your hair. After rubbing a balloon against a woolen sweater, you can hold the balloon close to a wall and *see* the attraction as the balloon moves toward the wall. In all these instances, the mass of the objects is too small for the interactions to be gravitational. What, then, is this interaction?

You may never have thought of these interactions as being particularly strong, but consider this: If you rub a comb through your hair and then pass the comb over some small bits of paper, the bits of paper jump up to your comb and stick to it. In other words, the bits of paper accelerate upward, which means the force exerted by your comb on them must be *greater* than the gravitational force exerted on them by Earth!

Now try this: Quickly pull a 20-cm strip of transparent tape* out of a dispenser and suspend it from the edge of a

Figure 22.1 Styrofoam peanuts cling to the cat's fur because of static electricity.

table (just be sure the table is not metal). Notice how the tape is attracted to anything brought nearby. It might even take some practice to prevent the tape from curling up and sticking to the underside of the table or to your hand. Bring a few objects near the suspended tape and notice the attractive interaction between them.† Go ahead—experiment!

22.1 Suspend a freshly pulled piece of transparent tape from the edge of your desk. (*a*) What happens when you hold a battery near the tape? Does it matter whether you point the + side or the − side of the battery toward the tape? Does a spent battery yield a different result? Does a wooden object yield a different result? (*b*) What happens when you hold a strip of freshly pulled tape near the power cord of a lamp? Does it make any difference if the lamp is on or off?

All these interactions involving static electricity are examples of **electric interactions.** The experiment you just did tells you there is no obvious connection between electric interactions and the electricity we think of as "flowing" in electric circuits and batteries. In Chapter 31 we shall see, however, that the two are connected.

Objects that participate in electric interactions exert an **electric force** on each other. The electric force is a field force (see Section 8.3): Objects exerting electric forces on each other need not be physically touching. As you may have noticed from the interaction between the strips of tape and various nearby objects, the magnitude of the electric force depends on distance: It decreases as you increase the separation.

22.2 Suspend a freshly pulled strip of transparent tape from the edge of your desk. (*a*) Pull a second strip of tape out of the dispenser and hold it near the first strip. What do you notice? (*b*) Does it matter which sides of the strips you orient toward each other?

As Checkpoint 22.2 makes clear, not all electric interactions are attractive. Even if you increase the mass of the strips by suspending paper clips from them, the repulsion between the strips is great enough to keep the paper clips apart (Figure 22.2). Now place your hand between two repelling strips and notice how both strips fly toward your hand! Then run each strip of tape several times between your fingers and notice how the electric interaction diminishes or even disappears.

22.3 Suspend two freshly pulled 20-cm strips of transparent tape from the edge of your desk. Cut two 20-cm strips of paper, making each strip the same width as the tape, and investigate the interactions between the paper strips and the tape by bringing them near each other. Which of the following combinations display an electric interaction: paper-paper, tape-paper, tape-tape?

---

*For best results, use the type called "magic" tape.
†If you find something that *repels* the tape, wipe the entire surface of the object with your hand and see if it still repels—it shouldn't. Mystified? Hang on! We'll soon be able to resolve your questions.

Figure 22.2 Strips of tape just pulled out of a dispenser repel each other. The repulsive force is great enough to keep the strips apart even when they are weighted down by paper clips.

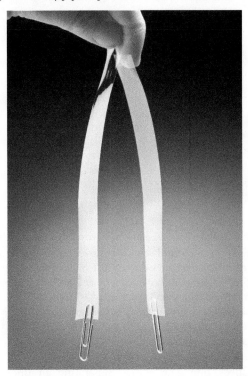

22.2 Electrical charge

As we saw in the preceding section, electric interactions are sometimes attractive and sometimes repulsive. In addition, the experiment you performed in Checkpoint 22.3 demonstrates that paper strips, which do not interact electrically with each other, do interact electrically with transparent tape. What causes these interactions? To answer this question, we need to carry out a systematic sequence of experiments.

Figure 22.3 illustrates a simple procedure for reproducibly creating strips of tape that interact electrically. A suspended strip created according to this procedure interacts in the following ways: It repels another strip created in the same manner, and it attracts any other object that does not itself interact electrically with other objects (**Figure 22.4**).

Figure 22.4

Tape strips prepared according to Figure 22.3 repel each other . . .

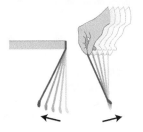

. . . but are attracted to your hand.

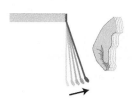

Let us call the attribute responsible for the electric interaction **electrical charge,** or simply **charge.** Saying that something carries an electrical charge is just another way of saying that that object interacts electrically with other objects that carry electrical charge. Freshly pulled strips of tape carry electrical charge, and two such strips interact because each possesses an electrical charge, just as your body and Earth interact because each possesses mass. The general term for any microscopic object that carries an electrical charge, such as an electron or ion, is **charge carrier.**

It is not immediately clear what attributes to assign to objects that do not interact electrically with each other but do interact with a charged tape strip—a strip of paper, your hand, an eraser, you name it. All we know for now is that the interaction between these objects and a charged tape is attractive rather than repulsive.

The electric charge on an object is not a permanent property; if you let a charged strip of tape hang for a while, it loses its ability to interact electrically. In other words, the strip is no longer charged—it is *discharged.* Depending on the humidity of the air, the discharging can take minutes or hours, but you can speed up the discharging by rubbing your fingers a few times over the entire length of a suspended charged strip of tape.* (The rubbing allows the charge to "leak away" from the tape by distributing itself over your body.)

---

*If rubbing your fingers along the tape doesn't do the job, try licking them before rubbing them over the tape.

Figure 22.3 Procedure for making strips of transparent tape that interact electrically. The purpose of the foundation strip is simply to provide a standard surface.

❶ On flat surface, stick down tape strip as foundation; flatten with thumb.

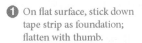

foundation strip

❷ Fold end of second strip to make handle; smooth onto foundation strip.

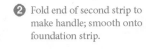
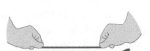

❸ Pull second strip off in one quick motion.

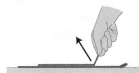

❹ Holding both ends of strip to prevent curling, hang strip on table edge.

Figure 22.5 Procedure for making strips of transparent tape that carry opposite charges.

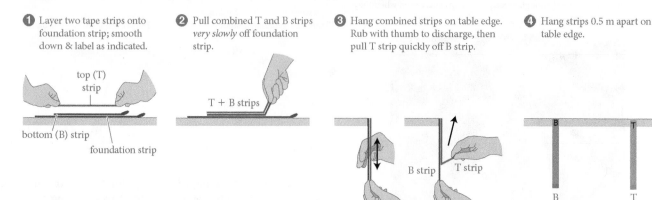

❶ Layer two tape strips onto foundation strip; smooth down & label as indicated.

top (T) strip

bottom (B) strip

foundation strip

❷ Pull combined T and B strips *very slowly* off foundation strip.

T + B strips

❸ Hang combined strips on table edge. Rub with thumb to discharge, then pull T strip quickly off B strip.

B strip   T strip

❹ Hang strips 0.5 m apart on table edge.

B   T

B   T

22.4 (*a*) Prepare a charged strip of transparent tape as described in Figure 22.3 and then suspend the strip from the edge of your desk. Verify that the tape interacts as you would expect with your hand, with a strip of paper, and with another charged strip of tape. (*b*) Rub your fingers along the hanging strip to remove all the charge from it, and then verify that it no longer interacts with your hand. If it does interact, rub again until it no longer interacts. (*c*) Predict and then verify experimentally how the uncharged suspended strip interacts with a strip of paper and with a charged strip of tape.

To restore the charge on a discharged strip, stick the strip on top of the foundation strip from which you pulled it off (step 1 in Figure 22.3), smooth it out, and then quickly pull it off again. You can recharge a strip quite a few times before it loses its adhesive properties. Once the tape does lose its adhesiveness, however, recharging it becomes impossible. It is generally a good idea to rub your finger over the foundation strip before you reuse it to make sure that it, too, is uncharged.

22.5 Recharge the discharged strip from Checkpoint 22.4 and verify that it interacts as before with your hand, with a strip of paper, and with another charged strip of tape.

A discharged tape strip interacts in the same way as objects that carry no charge. Such objects are said to be electrically **neutral.** They do not interact electrically with other neutral objects, but they do interact electrically with charged objects. We shall examine this surprising fact in more detail in Section 22.4.

Where does the electrical charge on a charged tape strip come from? Is charge *created* when two strips are pulled apart as in Figure 22.3? This is something we can check by sticking two strips of tape together, rubbing with our fingers to remove all charge from the combination, and then quickly separating the two strips (**Figure 22.5**).

22.6 Follow the procedure illustrated in Figure 22.5 to separate a pair of charged strips. (*a*) How does strip B interact with a neutral object? How does strip T interact with a neutral object? (*b*) Create a third charged strip and see how it interacts with strip B and with strip T. (*c*) Is strip T charged? (*d*) Is strip B charged? (*e*) Check what happens to the interactions with B and T strips when you discharge a B or a T strip by rubbing your fingers along its length.

As Checkpoint 22.6 shows, separating an uncharged pair of strips produces two charged ones, but the behavior of strip B is different from that of the other strips we have encountered so far!

22.7 Make two charged pairs of strips (B and T) following the procedure illustrated in Figure 22.5. Investigate the interaction of B with T, T with T, and B with B.

The interactions between the B and T strips are illustrated in **Figure 22.6**: Strips of the same type repel each other, while strips of different types attract each other. This series of experiments leads us to conclude that there are two types of charge on the tapes, one type on B strips and another type on T strips. Strips that carry the same type of charge, called *like charges,* exert repulsive forces on each other; strips that carry different types of charge, called *opposite charges,* exert attractive forces on each other.

Having determined that two types of electrical charge exist, a logical next question is: Are there even more types?

22.8 (*a*) Prepare one charged strip of tape according to Figure 22.3 and hang it from the edge of your desk. Hang a narrow strip of paper from the desk edge also, about 0.5 m away from the tape strip. Pass a *plastic* comb six times quickly through your hair and then show that the comb is charged. Be sure to use a plastic comb; combs made from other materials do not acquire a charge when passed through hair. The cheapest type of comb usually works best. (*b*) Make a pair of oppositely charged B and T strips (Figure 22.5) and investigate how they interact with a charged comb. (*c*) Does your comb behave like a B strip, a T strip, or neither?

CONCEPTS

Figure 22.6 Interactions of B and T charged strips.

Strips of same type repel each other.

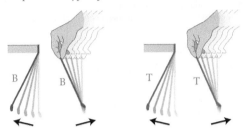

Strips of different types attract each other.

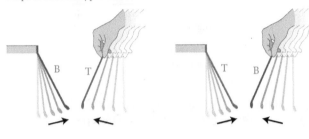

Experiments show that *any* charged object—obtained by rubbing objects together or otherwise—always attracts either a B strip or a T strip and repels the other. No one has ever found a charged object that repels or attracts *both* types of strips. In other words:

> There are two and only two types of charge. Objects that carry like charges repel each other; objects that carry opposite charges attract each other.

The two types of charge never appear independently of each other: Whenever two neutral objects are either rubbed together and then separated or, if an adhesive surface is involved, stuck together and then separated and one of them acquires a charge of one type, the other object always acquires a charge of the other type. The generation of opposite charges is obvious when you separate a neutral pair of tape strips. When you pass a comb through your hair, the comb acquires a charge of one type and your hair acquires a charge of the other type. On a dry day, you may have noticed that some hair strands stand up away from your head. Each charged strand is being repelled by the other charged strands, and so they are all getting as far away from one another as possible.

It can be shown that when two tape strips are separated, the forces exerted by the B strip and the T strip on a third charged strip are equal in magnitude, although one is attractive and the other repulsive. Furthermore, when the B and T strips are recombined, the combination is neutral again. These observations suggest that after you rub and then separate a pair of objects, the objects carry equal amounts of opposite charge. Combining these equal amounts of opposite charge produces zero charge. These observations indicate that all neutral matter contains equal amounts of

positive and negative charge. The two types of charge are called **positive** and **negative charges.** The definition of negative charge is as follows:*

> Negative charge is the type of charge acquired by a plastic comb that has been passed through hair a few times.

✋ **22.9** Does the B strip you created in Checkpoint 22.8 carry a positive charge or a negative charge?

When two neutral objects touch, some charge can be transferred from one object to the other, with the result that one object ends up with a surplus of one type of charge and the other object ends up with an equal surplus of the other type of charge. For example, when a neutral piece of styrofoam is rubbed with a neutral piece of plastic wrap, the styrofoam acquires a positive charge (meaning it contains more positive than negative charge) and the plastic wrap acquires a negative charge (it has a surplus of negative charge). Without further information, however, we cannot tell whether positive charge has been transferred from the wrap to the styrofoam, or negative charge has been transferred from the styrofoam to the plastic wrap, or a combination of these two. (See **Figure 22.7** on the next page.) Summarizing:

> All neutral matter contains equal amounts of positive and negative charge; charged objects contain unequal amounts of positive and negative charge.

In illustrations, surplus charge is represented by plus or minus signs. Keep in mind, however, that these signs never represent the only type of charge in an object. The plus signs on the positively charged styrofoam in Figure 22.7, for example, mean only that the styrofoam contains more positive than negative charge, either because some of its negative charge has been removed or because some positive charge has been added. In addition to the 12 positive charge carriers shown in Figure 22.7, the styrofoam contains millions and millions of positive charge carriers paired with millions and millions of negative charge carriers. A drawing such as Figure 22.7, shows only *unpaired* charge carriers (usually referred to as *surplus charge*).

As our observations in Figure 22.6 show, oppositely charged B and T strips attract each other. The interaction between positive and negative charge tends to bring positive and negative charge carriers as close together as possible. Because combining equal amounts of positive and negative charge results in zero charge, we can say that charge carriers always tend to arrange themselves in such a way as to produce uncharged objects—indeed, all matter around us tends to be neutral.

---

*Historically, negative charge was (arbitrarily) defined by Benjamin Franklin (1706–1790) as the charge acquired by a rubber rod rubbed with cat fur. Because plastic combs and hair are more easily accessible than rubber rods and cat fur, the definition of negative charge given here is more convenient.

Figure 22.7 Rubbing neutral styrofoam with neutral plastic wrap leaves the two objects with equal charges of opposite types.

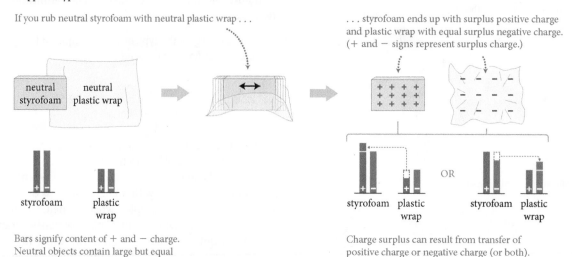

If you rub neutral styrofoam with neutral plastic wrap . . .

. . . styrofoam ends up with surplus positive charge and plastic wrap with equal surplus negative charge. (+ and − signs represent surplus charge.)

neutral styrofoam   neutral plastic wrap

OR

styrofoam   plastic wrap   styrofoam   plastic wrap   styrofoam   plastic wrap

Bars signify content of + and − charge. Neutral objects contain large but equal amounts of + and − charge.

Charge surplus can result from transfer of positive charge or negative charge (or both).

22.10 Imagine having a collection of charged marbles that retain their charge even when they touch other objects. Red marbles are positively charged, and blue marbles are negatively charged. (a) What happens if you place a bunch of red marbles close together on a flat horizontal surface? (b) What happens if you do the same with a bunch of blue marbles? (c) What happens if you do the same with an equal mixture of red and blue marbles? (d) What happens in part c if you have a few more red marbles than blue ones? (e) As a whole, is the collection of marbles in part d positively charged, negatively charged, or neither?

22.3 Mobility of charge carriers

To gain a better understanding of electrical charge, many additional experiments are required, most of which require items not easily found at home. A rubber rod rubbed with a piece of cat fur acquires—by Benjamin Franklin's original definition—a negative charge (and the fur acquires a positive charge). A glass rod rubbed with silk acquires a positive charge (and the silk a negative charge). Other materials also acquire a charge upon contact or rubbing, but these two combinations of rubber/fur and glass/silk provide the most convenient means of generating relatively large amounts of charge.

Interesting things happen when a charged rubber rod is brought into contact with an uncharged pith ball.* As the rod is brought near the ball, the ball moves toward the rod because of the attraction between the charged rod and the neutral ball (Figure 22.8a). As the ball touches the rod, however (Figure 22.8b), the crackling sounds of tiny sparks may be heard. The ball suddenly jumps away from the rod (Figure 22.8c), indicating that the interaction between rod and ball has become repulsive. This repulsive interaction indicates that the ball has acquired the same type of charge

*Pith is the soft, lightweight, spongelike material that makes up the interior of the stems of flowering plants.

as the rod (negative). In other words, some of the surplus negative charge on the rod has been transferred to the ball.

Charge can be transferred from one object to another by bringing the two into contact.

We can use this phenomenon to investigate the electrical behavior of different kinds of materials. For example, if we transfer some charge to one end of an uncharged rubber rod and then extend the charged end toward an uncharged pith ball, the two interact electrically, as shown in Figure 22.9a. If we hold the *uncharged* end near the pith ball, as in Figure 22.9b, however, no interaction occurs. This tells us that the charge does not flow from one end of the rubber rod to the other; instead, it remains near the spot where it has been deposited. Any material in which charge doesn't flow (or moves only with great difficulty) is called an **electrical insulator.**

Electrical insulators are materials through which charge carriers cannot flow easily. Any charge transferred to an insulator remains near the spot at which it was deposited.

Figure 22.8 A charged rubber rod can transfer charge to a neutral pith ball.

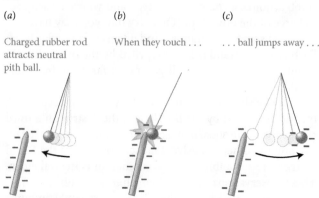

(a)   (b)   (c)

Charged rubber rod attracts neutral pith ball.

When they touch . . .

. . . ball jumps away . . .

. . . which tells us that rod & ball have same type of charge.

Figure 22.9 A rubber rod is an example of an electrical insulator.

(*a*)   (*b*)

Charged end of rubber
rod attracts pith ball . . .   . . . uncharged end doesn't.

So, rubber is electrical insulator:
Charge does not flow through it.

Glass, rubber, wood, and plastic are examples of electrical insulators. Air, particularly dry air, is also an insulator, although the presence of large amounts of charge can cause charge carriers to "jump" from one object to another, causing sparks.

Exercise 22.1 Electric forces

(*a*) Draw a free-body diagram for the pith ball in Figure 22.9*a*. (*b*) Two identical neutral pith balls A and B are suspended side by side from two vertical strings. After some charge is transferred from a charged rod to A, A and B interact. (B remains neutral because the two balls never come into contact with each other.) Sketch the orientation of A and B after the charge has been transferred to A. (*c*) Draw a free-body diagram for each ball.

SOLUTION (*a*) The ball is subject to three forces: a gravitational force, a contact force exerted by the string, and an attractive electric force exerted by the charged particles in the rod. This last force is directed horizontally toward the rod. Because the ball is at rest, I know that the vector sum of these three forces is zero. So the horizontal component of the force exerted by the string on the ball must be equal in magnitude to the electric force exerted by the rod on the ball, and the vertical component of the force exerted by the string on the ball must be equal in magnitude to the gravitational force exerted by Earth on the ball (**Figure 22.10***a*).

Figure 22.10

(a)   (b)

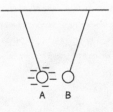

(c)

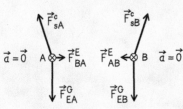

Figure 22.11 A metal rod is an electrical conductor.

(*a*)   (*b*)

If we transfer charge to
one end of metal rod . . .

. . . both ends of charged rod attract pith
balls equally . . .

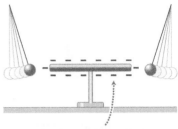

. . . meaning that metal is conductor
(charge can spread over it).

(*b*) As we saw in Section 22.2, a neutral object interacts electrically with a charged object. The electric force exerted by A on B and that exerted by B on A form an interaction pair and so their magnitudes are equal. Because the masses of the pith balls are the same, each is pulled in by the same distance. Thus my sketch is as shown in Figure 22.10*b*. (*c*) See Figure 22.10*c*.

In **Figure 22.11**, a charged rod is brought into contact with an uncharged metal rod supported on an electrically insulating stand. Once the charged rod has touched the metal rod, *all* points on the surface of the metal rod interact electrically with other objects, indicating that the charge spreads out over the metal rod. The tendency of charge to spread out over metal objects can be demonstrated with an *electroscope* (**Figure 22.12***a*). Two strips made of metal foil are suspended from a small metal rod in an electrically insulating enclosure; the rod is connected to a metal ball on top of the enclosure. When the metal ball is charged by an exterior source, the strips move away from each other. The explanation for this movement is that the added charge quickly moves from the metal ball through the metal rod and onto the two metal strips. Once the strips carry the same type of charge, they repel each other (Figure 22.12*b*).

Figure 22.12 An electroscope depends on electrical conduction.

(*a*)   (*b*)

Charge from rod spreads over conducting
elements, causing leaves to spring apart.

Leaves remain apart
after rod is removed.

metal stem

hinged
metal leaves

CONCEPTS

Figure 22.13 A conducting wire distributes charge between two conducting spheres.

(*a*)

Use charged rod . . .

(*b*)

. . . to charge one metal sphere.

(*c*)

Connect spheres with wire.

(*d*)

Charge distributes equally.

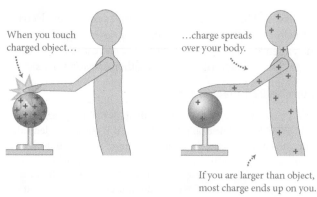

Another demonstration of the free motion of charges through metals is shown in **Figure 22.13**: When a long wire is used to connect a charged metal sphere to an uncharged metal sphere, charged particles flow from the charged sphere to the uncharged one. Because wires are made of metal, this experiment shows that, in contrast to what happens with electrical insulators, charge moves easily through a metal and across a metal-to-metal contact. Materials through which charge carriers can flow are called electrical **conductors,** and the flow of charge through conductors is called **conduction.**

> **Electrical conductors are materials through which charge carriers can flow easily. Any charge transferred to a conductor spreads out over the conductor and over any other conductors in contact with it.**

Metals are the only solid materials that are conductors at room temperature. (As noted earlier, glass, plastic, and most other solids are electrical insulators.) Although charge does not flow easily through pure water, minute amounts of impurities turn water into a fairly good conductor. Because most water contains some impurities, water is therefore usually considered a conductor.

Except for the outer layer of soil, Earth is also a good conductor. Consequently, when a charged, conducting object is connected to Earth by a wire, a process called **grounding,** charge carriers can flow between Earth ("ground") and the object. Because Earth is so large, it can supply or absorb a nearly unlimited number of charge carriers. In the absence of other nearby electrical influences, the grounded object is left with no surplus of either type of charge.

Because of its high water content, the human body is a conductor. Consequently, any time you touch a charged object, as in **Figure 22.14**, some of the charge moves into you—you act like a grounding agent just the way Earth does. As long as you keep touching the object, charge flows into your body, reducing the charge on the object (Figure 22.14). If the charge on the object is large, the charge that accumulates on your hair makes your hair stand up and separate as far as possible, like the leaves of an electroscope (**Figure 22.15**).

🖐 **22.11** (*a*) Why is it impossible to charge a metal rod held in your hand by rubbing the rod with other materials? (*b*) Why can you charge a rubber rod even when you hold it in your hand?

Figure 22.14 Because of its water content, the human body is a conductor.

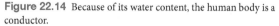

When you touch charged object...

...charge spreads over your body.

If you are larger than object, most charge ends up on you.

Is charge some sort of fluid that flows from one object to another, or is it composed of small particles that can be peeled off or stuck onto objects? To answer this question, we must look at the atomic structure of matter. All matter consists of atoms (see Section 1.3), the structure of which is schematically illustrated in **Figure 22.16**. Nearly all the atom's mass is concentrated in the extremely small nucleus at the center. The nucleus is composed of protons and neutrons. The region surrounding the nucleus, representing most of the atom's volume, is a cloud of electrons.

Figure 22.15 Charge spreads over the human body, so a large charge will cause your hairs to repel one another and stand on end.

Figure 22.16 Structure of the atom (not to scale).

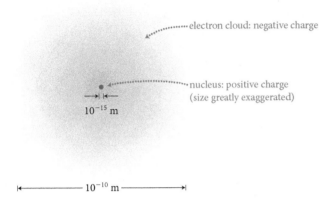

electron cloud: negative charge

nucleus: positive charge
(size greatly exaggerated)

10^{-15} m

10^{-10} m

Experiments show that electrons have a negative electrical charge—they are repelled by a charged comb* and by other electrons. Protons carry a positive charge, and neutrons carry no charge. The protons and neutrons in the nucleus are held together tightly by the strong interaction (see Chapter 7), which is great enough to overcome the electrical repulsion between the positively charged protons. The electrical attraction between the positively charged nucleus and the negatively charged electrons is responsible for keeping the electrons bound to the nucleus. The electron cloud does not collapse on the nucleus because of additional constraints imposed on the electrons by the laws of quantum mechanics.

Charge is an inherent property of the electron, which means it is impossible to remove the charge from an electron—there is no such thing as a discharged electron. Experiments show that

All electrical charge comes in whole-number multiples of the electrical charge on the electron.

For this reason, the magnitude of the charge on the electron, designated by the letter e, is called the **elementary charge.**

Every atom contains equal numbers of electrons and protons. Because atoms are neutral, the fact that they contain equal numbers of electrons and protons tells us that the magnitude of the positive charge on the proton is also e. The charge on an electron is $-e$, and that on a proton $+e$. As with electrons, the charge cannot be removed from a proton.

Given that macroscopic objects contain an immense number of atoms and that each atom can contain dozens of electrons and protons, we see that ordinary objects contain an immense number of positively charged protons, exactly balanced by an equal number of negatively charged electrons. A surplus of just a minute fraction of these numbers

is sufficient to give rise to a noticeable macroscopic charge. For example, when you pull apart two strips of transparent tape, the separation causes a surplus of less than one in a trillion (10^{12}) electrons. (Because there are about 10^{22} electrons in the strip, that fraction represents some 10^{10}, or ten billion electrons.)

When two atoms are brought close together, they may form a chemical bond by transferring one or more electrons from one atom to the other. Once such an electron transfer takes place, both atoms contain unequal numbers of electrons and protons and are now called **ions** instead of atoms. One of the two ions has gained one or more electrons, meaning it contains more electrons than protons and therefore carries a negative charge. The other ion, the one that lost electrons, contains more protons than electrons and so carries a positive charge.

Ions in solids are always immobile, but ions in liquids can move freely. For instance, in table salt, a compound made of pairs of sodium (Na^+) and chloride (Cl^-) ions, the charged ions hardly move at all, meaning that solid table salt is an electrical insulator. Dissolve table salt in water, however, and the solution contains large quantities of positively charged sodium ions and negatively charged chloride ions. Because these ions can move freely, the solution is an electrical conductor.

Some solids are made not of paired ions the way sodium chloride is but rather of individual atoms. In atomic solids that are electrical insulators, the electrons in the atoms are unable to move because each electron is bound to a specific atom. Diamond (made of the element carbon) and glass are two familiar examples. Metals are also atomic solids rather than ionic solids, but in metals, each atom gives up one or more electrons to a shared "gas" of electrons that spreads throughout the volume of the metal. The metal as a whole is still neutral: The negative charge of the electron gas is exactly balanced by the positive charge of the ions. The electrons in the gas are called *free electrons* because they can move freely inside the metal; these electrons are responsible for the easy flow of charge through a metal.

Nearly all electrical phenomena are due to the transfer of electrons—and therefore charge—from one atom to another. For example, when the sticky side of one strip of transparent tape is applied on top of the nonsticky side of a second strip, atoms in the adhesive from the top strip form chemical bonds with atoms in the nonsticky surface of the bottom strip by transferring electrons, as shown in **Figure 22.17** on the next page. These bonds are responsible for the adhesion of one strip to the other. When the strips are pulled apart quickly, the bonds are broken, but not all electrons manage to get back to the top strip. The bottom strip thus ends up with a surplus of electrons, making it negatively charged, and the top one with a deficit of electrons, making it positively charged.*

*Recall from our discussion of positive and negative charge carriers in Section 22.2 that a plastic comb carries a negative charge.

*Depending on the type of adhesive and the material of the backing, the transfer of electrons can also be in the other direction.

Figure 22.17 How strips of tape can acquire opposite charges when pulled apart.

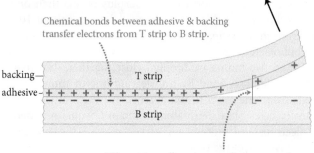

Chemical bonds between adhesive & backing transfer electrons from T strip to B strip.

backing—
adhesive—

T strip

B strip

When strips pull apart, some electrons remain on B strip, leaving surplus positive charge on T strip.

Figure 22.18 Because charge is conserved, the charge of a closed system does not change even when particles are created or destroyed.

(*a*) Electron-positron annihilation

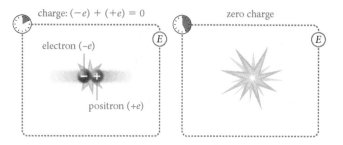

charge: $(-e) + (+e) = 0$

zero charge

electron $(-e)$

positron $(+e)$

(*b*) Decay of a free neutron

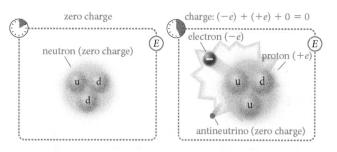

zero charge

charge: $(-e) + (+e) + 0 = 0$

neutron (zero charge)

electron $(-e)$

proton $(+e)$

antineutrino (zero charge)

As we saw in Section 10.4, friction between two surfaces also involves the breaking of chemical bonds. As with the separation of the tape strips, this bond breaking sometimes leaves surplus charge on the surfaces. When you touch a piece of plastic food wrap to a piece of styrofoam, for instance, chemical bonds form between atoms on the two surfaces. In these bonds, electrons from the styrofoam move to the wrap. If you then rub the surfaces against each other, these bonds are broken, and some of the electrons originally on the styrofoam stay on the wrap. If the breaking of the chemical bonds occurs slowly, the electrons migrate back and no surplus charge builds up. For that reason it is necessary to rub vigorously or to separate strips of tape quickly. The key point is:

Any two dissimilar materials become charged when brought into contact with each other. When they are separated rapidly, small amounts of opposite charge may be left behind on each material.

Because charging by breaking of chemical bonds is due to a transfer of charge, we now see that for every surplus of negative charge that appears in one place, an equal surplus of positive charge appears somewhere else. After the two strips in Figure 22.17 are separated, the sum of the positive charge on the T strip and the negative charge on the B strip is still zero. No creation or destruction of charge is involved, suggesting that electrical charge—like momentum, energy, and angular momentum—is a conserved quantity. The principle of **conservation of charge** states:

Electrical charge can be created or destroyed only in identical positive-negative pairs such that the charge of a closed system always remains constant.

No process has ever been found to violate this principle. Even when charged subatomic particles, such as electrons and protons, are created or destroyed—a process that can be observed in high-energy particle accelerators—charge is conserved. For example, when an electron (charge $-e$) collides with a subatomic particle called the *positron* (charge $+e$), both particles are destroyed, leaving nothing but a

flash of highly energetic radiation (**Figure 22.18a**). The charge of the electron-positron system before the collision is $(-e) + (+e) = 0$, and it is still zero after the collision. Likewise, in a process called beta decay, when a free neutron (carrying zero charge and made up of two down quarks and one up quark) decays into a proton (charge $+e$, one down and two up quarks), an electron (charge $-e$), and a neutral subatomic particle called the antineutrino (zero charge), the charge of the system comprising the neutron and the particles into which it decays remains zero (Figure 22.18*b* and Section 7.6).

22.12 When two objects made of the same material are rubbed together, friction occurs but neither material acquires surplus charge. Why?

22.4 Charge polarization

Let us now reexamine the interaction between a charged object and a neutral one. **Figure 22.19** shows the interaction between a charged rubber rod and an uncharged electroscope. With the rod far away (Figure 22.19*a*), the leaves of the electroscope hang straight down. When the rod is brought near the ball of the electroscope (Figure 22.19*b*), the leaves separate even without any contact between the rod and the electroscope. As the distance between the rod and the electroscope is increased again, the leaves drop down, showing that no charge has been transferred from the rod to the electroscope.

Figure 22.19 In (*a*) and (*b*), a charged rod induces polarization in an electroscope. (*c*) A schematic atom-level view.

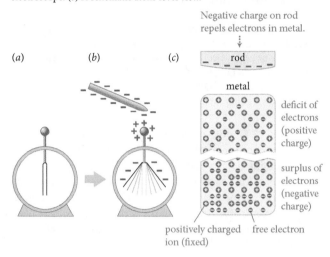

Why do the leaves separate even though the electroscope remains neutral? They separate because the negative charge on the rod repels the free electrons in the metallic parts of the electroscope: The free electrons are pushed as far away as possible from the rod (Figure 22.19*c*) and pile up in the leaves. This redistribution of charge is nearly instantaneous. The top of the electroscope thus ends up with a deficit of electrons—a positive charge—and the leaves end up with a surplus of electrons—a negative charge. The negative charge on the leaves is responsible for the repulsion between them. When the rod is removed, the electrons, being repelled by one another and attracted to the positive charge on the electroscope ball, immediately flow back to their normal positions, evening out the distribution of positive and negative charge.

✋ **22.13** (*a*) In Figure 22.19*b*, is the electroscope as a whole positively charged, negatively charged, or neutral? (*b*) How does the magnitude of the positive charge on the electroscope ball compare with the magnitude of the negative charge on the leaves? (*c*) Is the force exerted by the rod on the electroscope ball attractive or repulsive? Is the force exerted by the rod on the leaves attractive or repulsive? (*d*) How do you expect the magnitude of the force the rod exerts on the ball to compare with the magnitude of the force the rod exerts on the leaves?

Any separation of charge carriers in an object is called **charge polarization,** or simply **polarization,** and an object in which charge polarization occurs is said to be *polarized.* The electroscope of Figure 22.19*b*, for instance, is polarized by the nearby charged rod. In any object in which charge is polarized, there are two charged *poles*, one positive and the other negative. In the electroscope of Figure 22.19*b*, the positive pole is at the ball and the negative pole is in the foil strips.

In metals, the polarization induced by the presence of a nearby charged object is very great because the free electrons

in the metal move easily in response to the presence of the charged object. Even in electrical insulators, however, where there are no free electrons moving about, a nearby charged object induces some polarization. The basic reason for the polarization of insulators is illustrated in **Figure 22.20**: In the presence of an external charge, the center of the electron cloud and the nucleus of an atom shift away from each other, causing the atom to become polarized. So, when a negatively charged comb is brought near a small piece of paper, each atom in the paper becomes polarized—the electron clouds are pushed away from the comb, and the nuclei are pulled toward the comb. If we consider the paper as consisting of two overlapping parts that have the same shape but carry opposite charges, the positively charged part is pulled a bit toward the comb and the negatively charged part is pushed away, as shown in **Figure 22.21a** on the next page. This leaves the central part of the paper neutral but creates a sliver of surplus positive charge on the side facing the comb and an equal amount of surplus negative charge on the opposite side, and so the paper is polarized.

✋ **22.14** In an atom, what limits the separation between the electron cloud and the nucleus in the presence of an external charge? Why, for example, isn't the electron cloud in Figure 22.20*b* pulled all the way to the location of the external positive charge?

The polarization of atoms is responsible for the attraction between charged and neutral objects. In Figure 22.21, for example, the positively charged side of the paper is closer to the comb than the negatively charged side. Because the electric force decreases with increasing distance, the magnitude of the attractive force exerted on the positive side is greater than the magnitude of the repulsive force exerted on the negatively charged side (Figure 22.21*b*). Consequently, the vector sum of the electric forces exerted by the comb on the neutral piece of paper points toward the comb and the paper is pulled toward the comb.

Figure 22.20 Polarization of a neutral atom.

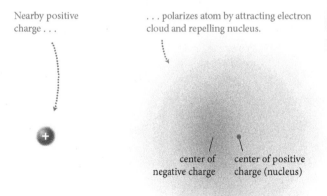

Figure 22.21 Polarization of a neutral insulator (bits of paper) by a charged comb. In (*a*), a single bit of paper is modeled as two offset sheets with opposite charges.

Charged comb picks up neutral paper

(*a*) Schematic model of interaction between comb and paper

(*b*) Free-body diagram for paper

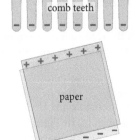

comb teeth

paper

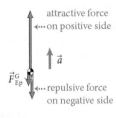

attractive force
on positive side

$\vec{a}$

$\vec{F}_{Ep}^{G}$

repulsive force
on negative side

22.15 (*a*) When a positively charged object is brought near a neutral piece of paper, is the vector sum of the forces exerted by the charged object on the paper attractive or repulsive? (*b*) Describe what would happen when a negatively charged comb is brought near an electroscope if protons, not electrons, were mobile in metals. (*c*) Can you deduce from the experiment illustrated in Figure 22.19 which charge carriers—electrons or protons—are mobile in metals?

Figure 22.22 illustrates how you can exploit polarization to charge neutral conducting objects. A negatively charged rod brought near a metal sphere induces polarization in the sphere. To get as far away as possible from the negative charge on the rod, the electrons in the metal of the sphere move to the right surface. Thus the surface of the sphere nearer the rod becomes the positive pole, and the surface farther from the rod becomes the negative pole. When you touch the metal, the electrons can move into your body, thereby getting even farther away from the negative charge on the rod. In essence, you have become the negative pole, while the sphere is the positive pole. If you remove your hand from the sphere, the sphere is left with a deficit of electrons (they stayed inside you!) and so carries a positive charge—it has become charged without ever touching a charged object. (Likewise, you now have a surplus of negative charge and carry a negative charge, which is spread so thin that it is hardly noticeable.) This process is called **charging by induction.**

Figure 22.22 Polarization can be exploited to charge neutral conducting objects.

❶ Charged rod induces polarization in metal sphere.

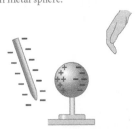

❷ When you touch sphere, negative charge gets farther from rod by spreading onto you.

❸ When you let go, you retain surplus of one type of charge and sphere retains surplus of opposite type of charge.

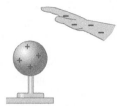

CONCEPTS

Self-quiz

1. You can use a positively charged object to charge a neutral object (*i*) by conduction or (*ii*) by induction. For each process, which type of charge (positive, negative, or neither) does the neutral object acquire?

2. Because we observe two types of electric interactions, attractive and repulsive, we postulate two types of charge. Do you think there are also two types of mass? Why do you think this? Do you think there are two types of magnetic pole? Why do you think this?

3. Is the statement *A plastic comb that has been passed through the hair a few times carries a negative charge* a physical law or a definition? What are some of the differences between a law and a definition?

4. A balloon rubbed in your hair or on your clothes sticks to a wall. If you place the rubbed side of the balloon against the wall, it sticks to the wall immediately. Try this, however: After rubbing the balloon in your hair, place the side of the balloon you did not rub against the wall and notice how the balloon turns until the rubbed side is touching the wall (*a*) Draw a free-body diagram for the balloon sticking to the wall. (*b*) Given that the balloon rotates so that the rubbed area is against the wall, do you think the balloon is an electrical insulator or conductor? (*c*) Was charge created in either the balloon or the wall in order for the sticking to occur? (*d*) Is any charge transferred from the balloon to the wall? Why or why not?

5. Air can act as both an insulator and a conductor. Consider reaching for a metal doorknob after scuffing your feet over a carpet. As your hand approaches the knob, a spark jumps between your hand and the knob. Explain how air acts as both insulator and conductor in this situation.

Answers

1. (*i*) Positive. When the two objects are brought into contact with each other (a necessary condition for conduction), surplus positive charge moves from the charged object to the neutral one. Thus the neutral object acquires the same type of charge as the charged object. (*ii*) Negative. Because the objects don't touch during charging by induction, charge carriers of the same type as the charged object escape from the neutral object during grounding.

2. Mass is the quantity responsible for gravitational interactions. Because all gravitational interactions are attractive, we can assume there is only one type of mass. Magnetic poles are responsible for magnetic interactions, which can be attractive or repulsive. Therefore there must be two types of magnetic pole (called north and south).

3. Definition. A law arises from observable phenomena and is found to be true in all cases that have been tested or observed (see Section 1.1). A definition cannot be tested. There is no test that allows us to observe that the charge on a comb passed through hair is negative. All we can do is show that the charged comb behaves in a fashion similar to other objects whose charge we call negative.

4. (*a*) See Figure 22.23. (*b*) Insulator. Because the only portion of the balloon that is attracted to the wall is the rubbed area, we know that the charge created by the rubbing does not spread out over the balloon surface. (*c*) No. Electrical charge can never be created. The surplus charge on the balloon was transferred from your hair or clothing. (*d*) No. If change were transferred, the balloon and the wall would repel each other.

Figure 22.23

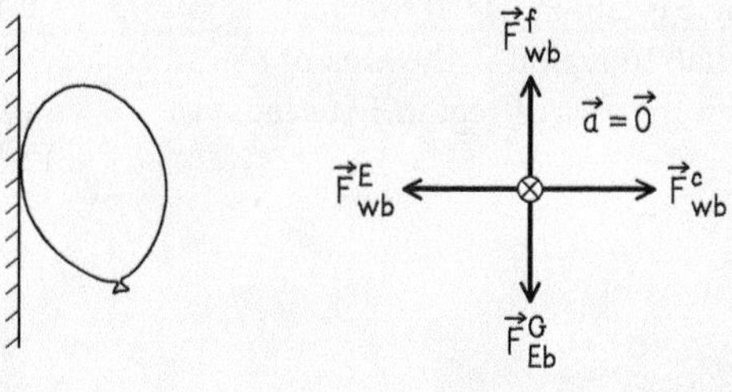

5. Scuffing your feet on the carpet transfers charge from the carpet to you. Before you get near the knob, the air insulates your charged body from the knob. As you move nearer and nearer to the knob, the magnitude of the electric force between the charge carriers in your hand and those in the knob increases until the forces are so great that the air molecules are ionized, thereby producing a conducting pathway between your hand and the knob. Now the ionized air acts as a conductor for the jumping charge.

22.5 Coulomb's law

Quantitative experiments with electrical charge are difficult to carry out because objects lose their charge and because charge carriers on objects tend to rearrange themselves in the presence of other charged objects. In the 18th century, however, the English clergyman and scientist Joseph Priestley carried out a remarkable experiment: He charged a hollow sphere and showed that no electric force was exerted on a small piece of charged cork placed inside the sphere. Remembering that Newton had proven that no gravitational force exists inside a hollow sphere (see the box "Zero gravitational force inside a spherical shell" and **Figure 22.24**, below) because the gravitational force decreases with the square of distance, Priestley proposed that the electric force, too, decreases as $1/r^2$.

In 1785, Charles Coulomb, a French physicist, provided direct evidence for an inverse-square law by measuring how the electric force between two charged spheres changes as the distance between the spheres changes. The basic apparatus for Coulomb's experiment is shown in **Figure 22.25**. A small dumbbell is suspended from a long fiber. When spheres A and B are charged, the electric

Zero gravitational force inside a spherical shell

A consequence of the inverse-square dependence on distance of the gravitational force is that a uniform spherical shell exerts *no force at all* on a mass placed anywhere inside it. This result is very important in electrostatics because it provides a strong test of the law describing the electric force between charged objects.

Consider the uniform spherical shell in Figure 22.24a. A particle of mass m is placed off-center inside the shell. To determine the force exerted by the shell on the particle, consider first the force exerted by region 1, defined as a very small region of the shell surface. Let the mass of this region be m_1 and its distance from the particle be d_1, which means the magnitude of the gravitational force exerted by this region on the particle is Gm_1m/d_1^2. Extending the cone defined by the particle and region 1 to the opposite side of the shell gives us a small region 2, of mass m_2 and at distance d_2 from the particle. The magnitude of the gravitational force exerted by this region on the particle is Gm_2m/d_2^2. If the particle is closer to region 1 than to region 2, the area of region 2 must be greater than the area of region 1 (because of

how we defined region 2 as being an extension of the cone formed by region 1 and the particle), which means region 2 contains more mass: $m_2 > m_1$.

Because the mass of the shell is distributed uniformly over the shell, the mass of each of our two regions is proportional to its area: $m_1/m_2 = a_1/a_2$. Because they are marked out by similar cones, the areas of the two regions are proportional to the squares of the distances to the particle: $a_1/a_2 = d_1^2/d_2^2$, which means that

$$\frac{m_1}{m_2} = \frac{d_1^2}{d_2^2}.$$

Rearranging terms, we get $m_1/d_1^2 = m_2/d_2^2$ *regardless* of the position of the particle, and so we see that the forces cancel: The two regions exert forces of equal magnitude in opposite directions on the particle (Figure 22.24b). We can now apply the same arguments to other pairs of small regions on either side of the particle, each yielding equal and opposite forces on the particle (Figure 22.24c). The vector sum of all the forces exerted on the particle by all the small regions making up the spherical shell is thus zero.

Figure 22.24

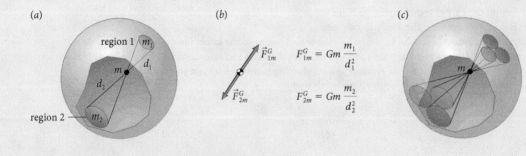

(a)

region 1 m_1

m d_1

d_2

region 2 — m_2

(b)

$\vec{F}_{1m}^G$

$\vec{F}_{2m}^G$

$F_{1m}^G = Gm\dfrac{m_1}{d_1^2}$

$F_{2m}^G = Gm\dfrac{m_2}{d_2^2}$

(c)

m

force between them twists the fiber. The amount of twist is a measure of the magnitude of the force between the two spheres (see also Section 15.7). A similar arrangement was used a few years later by Cavendish to study gravitational interactions (see Section 13.5).

Coulomb also devised a method for systematically varying the "quantity of charge" q on a metal sphere. He found that when a charged metal sphere is brought into contact with an identical uncharged metal sphere, the final charge is the same on each sphere—both exert a force of equal magnitude on a third charged object. In other words, each sphere gets half the original charge. By sharing charge among several identical metal spheres, Coulomb could produce spheres whose charge was one-half, one-quarter, one-eighth, and so on of the original charge (**Figure 22.26**).

By thus varying the charges on spheres A and B of his apparatus, Coulomb found that the **electric force** is proportional to the charge on each sphere. We can summarize these findings in one equation, called **Coulomb's law,** which gives the magnitude of the electric force exerted by two charged particles separated by a distance r_{12} and carrying charges q_1 and q_2:

$$F_{12}^E = k \frac{|q_1|\,|q_2|}{r_{12}^2}. \tag{22.1}$$

As we shall see in Chapter 27, the interaction between charged particles becomes more complicated when the particles are not at rest. For this reason the force in Coulomb's law is sometimes called the *electrostatic* force and the branch of physics that deals with stationary distributions of charge is called *electrostatics*.

If the positions of the two charged particles are given by the vectors $\vec{r}_1$ and $\vec{r}_2$, respectively, then the distance between them is $r_{12} = |\vec{r}_2 - \vec{r}_1|$. The value of the constant of proportionality k depends on the units used for charge, force, and length. The absolute-value signs around the charges in Eq. 22.1 are necessary because q_1 and q_2 can be negative but the *magnitude* F_{12}^E of the electric force must always be positive.

Coulomb's law bears a striking resemblance to Newton's law of gravity (Eq. 13.1):

$$F_{12}^G = G \frac{m_1 m_2}{r_{12}^2}. \tag{22.2}$$

Why these two laws have the same mathematical form remains a mystery. The main differences between the two are that mass is always positive but electrical charge can be positive or negative, which means that the gravitational force is always attractive but the electric force can be attractive or repulsive.

The derived SI unit of charge, called the **coulomb** (C), is defined as the quantity of electrical charge transported in 1 s by a current of 1 ampere, a quantity and unit we shall define in Chapter 27. One coulomb is equal to the magnitude of the charge on about 6.24×10^{18} electrons. Conversely, the magnitude of the charge of the electron is

$$e = 1/6.24 \times 10^{18}\,\text{C} = 1.60 \times 10^{-19}\,\text{C}. \tag{22.3}$$

The charge on any object comes in only whole-number multiples of this elementary charge:

$$q = ne, \quad n = 0,\ \pm 1,\ \pm 2,\ \pm 3, \dots . \tag{22.4}$$

This means that an object can have charge $q = 0$, $q = +7e$, $q = -4e$, and so forth, but not, for instance, $+1.2e$. Because the elementary charge is very small,

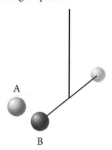

Figure 22.25 Schematic diagram of Coulomb's apparatus for measuring the electric force between two charged spheres.

A

B

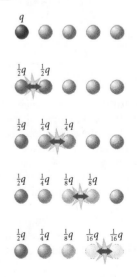

Figure 22.26 By successively allowing a charged sphere to touch an initially uncharged neighbor, we can distribute an amount of charge in ever-lessening amounts over a number of spheres.

the fact that charge exists only as whole-number multiples of the elementary charge isn't noticeable under ordinary circumstances. For example, running a comb through your hair easily gives the comb a surplus of about 10^{12} electrons, and the quantity of electrons flowing through a 100-W light bulb each second is about 10^{19}. These are such large numbers that the fact that charge comes in only whole-number multiples of the elementary charge normally remains unnoticed.

Using the coulomb as the unit of charge, we can determine the value of k in Eq. 22.1 experimentally by measuring the force between two known charged particles separated by a known distance:

$$k = 9.0 \times 10^9 \text{ N} \cdot \text{m}^2/\text{C}^2. \tag{22.5}$$

The value of this constant shows how large a unit the coulomb is: Two particles, each carrying a charge of 1 C, separated by 1 m exert on each other a force of 9 billion newtons—equal to the gravitational force exerted by Earth on several dozen loaded supertankers! It is very difficult to build up a charge of this magnitude on all but very large objects because things get ripped apart by the enormous forces. The largest accumulations of charge we know of occur in the atmosphere: Large clouds that accumulate a charge of about 50 C discharge through the air to Earth, causing lightning.

Example 22.2 Gravity versus electricity

Compare the magnitudes of the gravitational and electric forces exerted by the nucleus of a hydrogen atom—a single proton ($m_p = 1.7 \times 10^{-27}$ kg)—on an electron ($m_e = 9.1 \times 10^{-31}$ kg) when the two are 0.50×10^{-10} m apart.

❶ GETTING STARTED For simplicity, I assume I can treat the proton and electron as particles. I also assume they are at rest so that I can use the principles of electrostatics.

❷ DEVISE PLAN I can use Eq. 22.2 to calculate the magnitude of the gravitational force and Eq. 22.1 to calculate the magnitude of the electric force.

❸ EXECUTE PLAN

$$F_{pe}^G = G \frac{m_p m_e}{r_{pe}^2}$$

$$= (6.7 \times 10^{-11} \text{ N} \cdot \text{m}^2/\text{kg}^2) \frac{(9.1 \times 10^{-31} \text{ kg})(1.7 \times 10^{-27} \text{ kg})}{(0.50 \times 10^{-10} \text{ m})^2}$$

$$= 4.1 \times 10^{-47} \text{ N}$$

and

$$F_{pe}^E = k \frac{|q_p| \, |q_e|}{r_{pe}^2}$$

$$= (9.0 \times 10^9 \text{ N} \cdot \text{m}^2/\text{C}^2) \frac{(1.6 \times 10^{-19} \text{ C})(1.6 \times 10^{-19} \text{ C})}{(0.50 \times 10^{-10} \text{ m})^2}$$

$$= 9.2 \times 10^{-8} \text{ N}.$$

The electric force exerted by the proton on the electron is $(9.2 \times 10^{-8} \text{ N})/(4.1 \times 10^{-47} \text{ N}) \approx 10^{39}$ times greater than the gravitational force exerted by the proton on the electron. ✔

❹ EVALUATE RESULT The difference in magnitudes is in agreement with the information given in Table 7.1.

Example 22.3 Comb electricity

(a) A 0.020-kg plastic comb acquires a charge of about -1.0×10^{-8} C when passed through your hair. What is the magnitude of the electric force between two such combs held 1.0 m apart after being passed through your hair? (b) If two identical 0.020-kg combs carry one surplus electron for every 10^{11} electrons in the combs, what is the magnitude of the electric force between these combs held 1.0 m apart?

❶ GETTING STARTED Both parts of the problem require me to calculate the magnitude of the electric force between the combs. If I treat the combs as particles, I can use Eq. 22.1 to calculate this force.

❷ DEVISE PLAN To calculate the magnitude of the electric force between two charged objects, I need to know the charge on each object and their separation distance. I know these data for part a: $q_1 = q_2 = -1.0 \times 10^{-8}$ C and $r_{12} = 1.0$ m, where the subscripts 1 and 2 denote the two combs. For part b I am given only the separation distance, and so I need to determine the charge on each comb.

I am given the fraction of electrons added, and I know the charge on one electron. So, to determine the charge on each comb, I need to determine how many electrons each comb contains. The number of electrons in each comb is equal to the number of protons in the comb: $N_e = N_p$. I am given the mass of the comb and I know that the mass is determined by the protons and neutrons in all the atoms making up the comb (the electrons contribute very little). Given that the protons and neutrons have almost identical mass ($m_p = m_n = 1.7 \times 10^{-27}$ kg), I can determine the number N of protons and neutrons by dividing the mass of the comb by m_p. Given that most atoms contain roughly equal numbers of protons and neutrons, I can say that the number of protons is $N_p \approx N/2$.

❸ **EXECUTE PLAN** (a) Substituting the values given into Eq. 22.1, I get

$$F_{12}^E = k\frac{|q_1|\,|q_2|}{r_{12}^2}$$

$$= (9.0 \times 10^9 \, \text{N} \cdot \text{m}^2/\, \text{C}^2)\frac{(1.0 \times 10^{-8} \, \text{C})(1.0 \times 10^{-8} \, \text{C})}{(1.0 \, \text{m})^2}$$

$$= 9.0 \times 10^{-7} \, \text{N}. ✔$$

(b) The number of protons plus neutrons in the comb is

$$N = \frac{0.020 \, \text{kg}}{1.7 \times 10^{-27} \, \text{kg}} = 1.2 \times 10^{25},$$

and so $N_p \approx N/2 = 6 \times 10^{24}$. The number of electrons is equal to the number of protons, and so there are 6×10^{24} electrons in

each comb to begin with. Adding one surplus electron for every 10^{11} electrons means adding $(6 \times 10^{24})/(1 \times 10^{11}) = 6 \times 10^{13}$ electrons to each comb; these electrons carry a combined charge of $(6 \times 10^{13})(-1.6 \times 10^{-19} \, \text{C}) = -9.6 \times 10^{-6} \, \text{C}$. The magnitude of the repulsive electric force between the combs is then

$$F_{12}^E = k\frac{|q_1|\,|q_2|}{r_{12}^2}$$

$$= (9.0 \times 10^9 \, \text{N} \cdot \text{m}^2/\text{C}^2)\frac{(9.6 \times 10^{-6} \, \text{C})(9.6 \times 10^{-6} \, \text{C})}{(1.0 \, \text{m})^2}$$

$$\approx 1 \, \text{N}. ✔$$

❹ **EVALUATE RESULT** My answer to part a is a force too small to be felt, which is what I expect based on experience (two combs passed through hair don't exert an appreciable force on each other). In contrast, my answer to part b is phenomenally large for an electric force. The magnitude of the initial acceleration acquired by the combs would be $a_1 = F_{21}^E/m_1 = (1 \, \text{N})/(0.020 \, \text{kg}) = 50 \, \text{m/s}^2$, or about five times the acceleration due to gravity! Even though the fraction of electrons removed—one in 100 billion—is very small, the factor k in Eq. 22.1 is so great that the resulting force is also great. Indeed, I learned in Table 7.1 that the electromagnetic interaction is 36 orders of magnitude stronger than the gravitational interaction, so my answer is not unreasonable.

✋ **22.16** Two identical conducting spheres, one carrying charge $+q$ and the other carrying charge $+3q$, are initially held a distance d apart. The spheres are allowed to touch briefly and then returned to separation distance d. Is the magnitude of the force they exert on each other after the touching greater than, smaller than, or the same as the magnitude of the force they exerted on each other before the touching?

Like the gravitational force, the electric force is *central*; that is, its line of action is along the line connecting the two interacting charged particles. Consider, for example, the two particles carrying charges q_1 and q_2 shown in Figure 22.27a. The vector $\vec{r}_{12} \equiv \vec{r}_2 - \vec{r}_1$ gives the position of particle 2 relative to particle 1; this vector points from particle 1 to particle 2. We can define a unit vector pointing in this direction by dividing the vector $\vec{r}_{12}$ by its magnitude:

$$\hat{r}_{12} \equiv \frac{\vec{r}_2 - \vec{r}_1}{r_{12}}. \tag{22.6}$$

Depending on the algebraic sign of the charges, the electric force can be attractive or repulsive. For like charges ($q_1q_2 > 0$), the force is repulsive. In this case, the force $\vec{F}_{12}^E$ exerted by particle 1 on particle 2 points in the same direction as the unit vector $\hat{r}_{12}$ (Figure 22.27b). For opposite charges ($q_1q_2 < 0$), the force is attractive, and so $\vec{F}_{12}^E$ points in the direction opposite the direction of $\hat{r}_{12}$ (Figure 22.27c). In either case, $\vec{F}_{12}^E$ can be written in the form

$$\vec{F}_{12}^E = k\frac{q_1q_2}{r_{12}^2}\hat{r}_{12}. \tag{22.7}$$

Because $r_{12} = r_{21}$ and because $\hat{r}_{21}$ points in the direction opposite the direction of $\hat{r}_{12}$, the force $\vec{F}_{21}^E$ exerted *by* particle 2 *on* particle 1, which points in the direction

Figure 22.27 (a) Position vectors for two charged particles. (b) Repulsive forces exerted on each other by two particles carrying like charges. (c) Attractive forces exerted on each other by two particles carrying opposite charges.

(a)
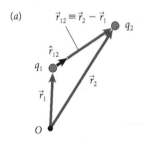

(b)
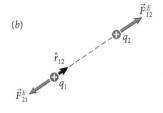

(c)

Figure 22.28 The reason Coulomb's law does not apply in a strict sense to macroscopic charged objects. The law is approximately correct if the objects are far apart relative to their radii.

(*a*) Charged spheres separated by a distance large compared to the sphere radii

Distance between centers of charge distributions same as distance between sphere centers

(*b*) Charged spheres separated by a distance small compared to the sphere radii

Charge repulsion causes distance between centers of charge distributions to differ from distance between sphere centers.

opposite the direction of $\vec{F}^E_{12}$, is obtained by simply switching the indices 1 and 2 in Eq. 22.7:

$$\vec{F}^E_{21} = k\frac{q_2 q_1}{r^2_{21}} \hat{r}_{21} = k\frac{q_1 q_2}{r^2_{12}}(-\hat{r}_{12}) = -\vec{F}^E_{12}, \qquad (22.8)$$

as we would expect for an interaction pair (see Eq. 8.15). Equation 22.7 is the vectorial form of Eq. 22.1.

✋ **22.17** Using your knowledge about work and potential energy, determine whether the potential energy of a closed system of two charged particles carrying like charge increases, decreases, or stays the same when the distance between the two is increased. Repeat for two particles carrying opposite charge.

Before going on, I should mention a limitation to Coulomb's law. Strictly speaking, it is applicable only to charged particles. This is so because the distance r_{12} is well defined only when the size of the charged objects is negligibly small compared with their separation distance. When the charged objects are not particles, the distance r_{12} is not equal to the center-to-center distance. You can see why with the help of Figure 22.28. In **Figure 22.28a**, the charge is distributed uniformly over the surface of each of two widely separated metal spheres. The way in which a collection of charge carriers is spread out over a macroscopic object is called a **charge distribution.** Because the charge distributions over the metal spheres in Figure 22.28*a* are uniform, the center of each charge distribution coincides with the center of the sphere, and so r_{12} is well defined. When we bring the spheres close together, as in Figure 22.28*b*, the like charge carriers repel one another and move to the far side of each sphere. Now the centers of the charge distributions no longer coincide with the spheres' centers, so r_{12} (the center-to-center distance of the two charge distributions) is not simply the distance separating the centers of the two conductors.

✋ **22.18** (*a*) Is the magnitude of the electric force between the two conducting spheres in Figure 22.28*b* greater or smaller than that obtained from Coulomb's law, which assumes the charge is concentrated at the center of each sphere? (*b*) Is the answer to part *a* the same if the charge on one of the conductors is negative instead of positive?

22.6 Forces exerted by distributions of charge carriers

Coulomb's law deals only with *pairs* of charged objects. To calculate the force exerted by an assembly of objects carrying charges q_2, q_3, q_4, ... on an object 1 carrying a charge q_1, we take the vector sum of all the forces exerted on object 1 by each of the other charged objects independently:

$$\sum \vec{F}^E_1 = \vec{F}^E_{21} + \vec{F}^E_{31} + \vec{F}^E_{41} + \cdots, \qquad (22.9)$$

where each term is given by Coulomb's law:

$$\sum \vec{F}^E_1 = k\frac{q_2 q_1}{r^2_{21}} \hat{r}_{21} + k\frac{q_3 q_1}{r^2_{31}} \hat{r}_{31} + k\frac{q_4 q_1}{r^2_{41}} \hat{r}_{41} + \cdots. \qquad (22.10)$$

In other words, we calculate the force exerted by object 2 on object 1, then calculate the force exerted by object 3 on object 1, and so forth, and then add the forces. This means that if we know the details of some distribution of charged objects, we can calculate the force exerted by this distribution of charged objects on a single charged particle. For distributions that contain large numbers of charged objects, the summation can be accomplished via an integration. We will limit our discussion here to simple cases involving only a few charged objects.

✋ **22.19** Figure 22.29 shows how a charged particle 1 interacts with two other charged particles 2 and 3. Determine the direction of the vector sum of the electric forces exerted on particle 2.

The basic limitation of Coulomb's law continues to apply when we are analyzing collections of charged objects: Eq. 22.10 is valid only for charged *particles,* not for charged extended bodies. Suppose, for example, that we replace each charged particle in Figure 22.29 by a conducting sphere carrying the same charge as each particle. Let us first consider just the interaction between spheres 1 and 2. When these oppositely charged spheres are placed near each other, the charge carriers on the two spheres rearrange themselves to be as close as possible to each other (**Figure 22.30a**). A similar type of rearrangement takes place when just spheres 1 and 3 are placed near each other (Figure 22.30b). When all three spheres are placed near one another, the positive charge on sphere 3 pushes the positive charge on sphere 2 up and pulls the negative charge on sphere 1 down (Figure 22.30c). Consequently, the forces that the spheres exert on one another are not the same as the forces exerted by the individual pairs (compare the forces in Figures 22.29 and 22.30).

Figure 22.29 Forces exerted by two charged particles 2 and 3 on charged particle 1.

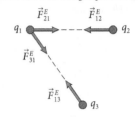

Figure 22.30

(*a*) Sphere 1 interacts with just sphere 2   (*b*) Sphere 1 interacts with just sphere 3   (*c*) Sphere 1 interacts with both spheres

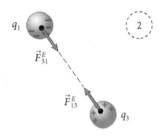

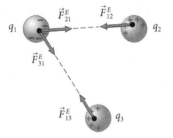

Example 22.4 Electric tug of war

You are given three charged particles. Particles 1 and 2 carry charge $+q$ and particle 3 carries charge $-4q$. (*a*) Determine the relative values of the separation distances r_{12} and r_{13} when the three particles are arranged along a straight line in such a way that the vector sum of the forces exerted on particle 1 is zero. (*b*) With the particles arranged this way, are the vector sums of the forces exerted on particles 2 and 3 also zero?

❶ **GETTING STARTED** Particle 2 exerts a repulsive force on particle 1; particle 3 exerts an attractive force on particle 1. In order for the two forces exerted on particle 1 to cancel, particles 2 and 3 must be on the same side of particle 1. Because the magnitude of the charge on particle 2 is smaller than that of the charge on particle 3, the distance r_{12} must be shorter than the distance r_{13}.

❷ **DEVISE PLAN** I choose my x axis vertically up along the line defined by the particles (**Figure 22.31a**). To determine the relative values of the separation distances r_{12} and r_{13} when the vector sum of the forces exerted on particle 1 is zero, I draw a free-body diagram for particle 1 (Figure 22.31b). The magnitude of each force exerted on this particle is given by Eq. 22.1, and so by

setting the sum of the x components of the forces equal to zero, I have an expression containing r_{12} and r_{13} and I can manipulate the expression to get the relative values of r_{12} and r_{13}.

Figure 22.31

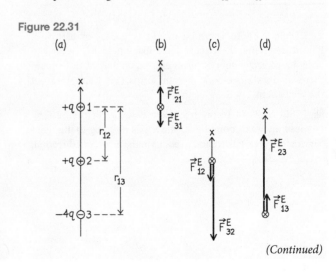

(Continued)

❸ EXECUTE PLAN (a) The x components of the two forces in Figure 22.31b must add to zero, so

$$\Sigma F_{1x} = F^E_{21x} + F^E_{31x}$$

$$= +k\frac{|q_1|\,|q_2|}{r^2_{12}} - k\frac{|q_1|\,|q_3|}{r^2_{13}} = k\frac{qq}{r^2_{12}} - k\frac{(q)(4q)}{r^2_{13}}$$

$$= +k\frac{q^2}{r^2_{12}} - k\frac{4q^2}{r^2_{13}} = 0.$$

I therefore must have $\dfrac{q^2}{r^2_{12}} = \dfrac{4q^2}{r^2_{13}}$,

or $r^2_{13} = 4r^2_{12}$, and so $r_{13} = 2r_{12}$. ✔

(b) The forces exerted by particles 1 and 3 on particle 2 both point in the negative x direction (Figure 22.31c) and so cannot sum to zero. The forces exerted by particles 1 and 2 on particle 3 both point in the positive x direction (Figure 22.31d) and so cannot sum to zero. ✔

❹ EVALUATE RESULT My answer for part a makes sense because the force exerted by each particle is inversely proportional to the square of the separation distance, and this force varies directly with the quantity of charge. Because the charge on particle 3 is four times greater than that on particle 2, the square of the separation distance r_{13} must be four times r_{12}, and so $r_{13} = 2r_{12}$, as I found.

Example 22.5 Electrostatic equilibrium

Consider four charged particles placed at the corners of a square whose sides have length d. Particles 1, 2, and 4 carry identical positive charges. In order for the vector sum of the forces exerted on particle 1 to be zero, what charge must be given to particle 3, which is in the corner diametrically opposite particle 1?

❶ GETTING STARTED I begin by making a sketch of the situation (**Figure 22.32a**). Because particles 1, 2, and 4 all carry identical positive charge, I write $q_1 = q_2 = q_4 = +q$. The separation between neighboring particles on the square is d. The separation between particles 1 and 3 is $\sqrt{2}d$.

Figure 22.32

(a)

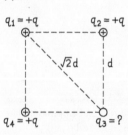

(b)

❷ DEVISE PLAN To determine the charge needed on particle 3, I must determine the electric force magnitude F^E_{31} needed to yield a zero vector sum of forces exerted on particle 1. I should therefore draw a free-body diagram for particle 1 and work out the vector sum.

❸ EXECUTE PLAN In my free-body diagram (Figure 22.32b), I choose my y axis pointing up and x axis pointing to the right. The force $\vec{F}^E_{21}$ is repulsive and so points in the negative x direction;

the force $\vec{F}^E_{41}$ is repulsive and points in the positive y direction. To make the vector sum of the forces exerted on 1 zero, $\vec{F}^E_{31}$ must be such that its components cancel $\vec{F}^E_{21}$ and $\vec{F}^E_{41}$. Along the x axis, I therefore have

$$\Sigma F_{1x} = F^E_{21x} + F^E_{31x} + F^E_{41x} = -F^E_{21} + F^E_{31}\cos 45° + 0 = 0,$$

so $F^E_{21} = F^E_{31}\cos 45°$. Substituting the Coulomb's law expressions for these two force magnitudes, I have

$$.k\frac{|q_2|\,|q_1|}{d^2} = k\frac{|q_3|\,|q_1|}{(d\sqrt{2})^2}\cos 45°$$

$$\frac{q^2}{d^2} = \frac{|q_3|q}{2\sqrt{2}d^2}$$

$$|q_3| = 2\sqrt{2}\,q.$$

From Figure 22.32b, I also know that $\vec{F}^E_{31}$ must be an attractive force, so

$$q_3 = -2\sqrt{2}q. ✔$$

❹ EVALUATE RESULT My result indicates that the charge on particle 3 is greater than q. This makes sense because the attractive force exerted by this particle must balance the forces exerted by particles 2 and 4, which are both closer to particle 1. To obtain my answer I solved only for the x component of $\vec{F}^E_{31}$ and did not consider the y component. However, my free-body diagram shows that the magnitudes of the components along the y axis are the same as those along the x axis, so analyzing the y components would have given me the same result.

Example 22.6 Electric trajectory

Consider the arrangement of charged particles shown in **Figure 22.33**. The charge magnitudes are the same in all three cases, but q_1 and q_2 are positive and q_3 is negative. Sketch the trajectory of particle 1 if it is released while particles 2 and 3 are held fixed. Ignore any gravitational force exerted by Earth on the particle.

Figure 22.33 Example 22.6

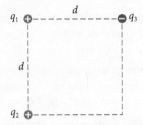

① GETTING STARTED I begin by drawing a free-body diagram for particle 1, choosing the x axis to the right and the y axis up (**Figure 22.34a**). Because the charges on the two particles have the same magnitude and because the separation distances are the same, the magnitudes F_{21}^E and F_{31}^E are the same.

Figure 22.34

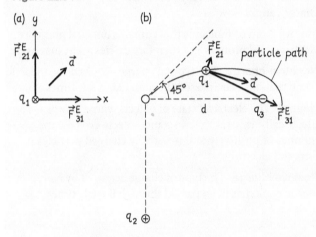

② DEVISE PLAN The direction in which particle 1 accelerates is determined by the vector sum of the forces exerted on it, and I can determine this direction from my free-body diagram. As the particle moves along its trajectory, however, the forces exerted on it change in both direction and magnitude, and so I must consider these changes in formulating my answer.

③ EXECUTE PLAN Because the magnitudes F_{21}^E and F_{31}^E are the same, the vector sum of the two forces exerted on particle 1 bisects the angle between the two forces, and so the initial acceleration of 1 (which points in the same direction as the vector sum of forces) points up and to the right at an angle of 45°, as illustrated in Figure 22.34a. As particle 1 moves in the direction indicated in Figure 22.34a, $\vec{F}_{21}^E$ and $\vec{F}_{31}^E$ change in both direction and magnitude. The magnitude of $\vec{F}_{21}^E$ decreases because the distance between 1 and 2 increases, and the magnitude of $\vec{F}_{31}^E$ increases because the distance between 1 and 3 decreases. The direction of the acceleration of particle 1 is the same as the direction of the vector sum of these two forces. The resultant motion is qualitatively illustrated in Figure 22.34b. ✔

④ EVALUATE RESULT My sketch makes sense: Particle 1 first moves up and to the right because it is repelled by particle 2 and attracted by particle 3. As it moves away from 2 and approaches 3, the effect of the attraction increases and so the trajectory curves and heads toward 3.

✋ **22.20** Seven small metal spheres are arranged in a hexagonal pattern as illustrated in **Figure 22.35**. Spheres 1 and 7 carry equal amounts of positive charge; the other spheres are uncharged. (*a*) To give sphere 7 an acceleration $\vec{a}$ that points to the right, what (single) other sphere must be charged? There may be more than one possibility. (*b*) What are the sign and magnitude of that charge?

Figure 22.35 Checkpoint 22.20.

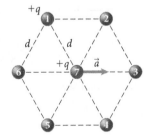

Chapter Glossary

SI units of physical quantities are given in parentheses.

Charge (electrical) q (C) A scalar that represents the attribute responsible for electromagnetic interactions, including electric interactions. There are two types of charge: *positive* ($q > 0$) and *negative* ($q < 0$). Two objects that carry the same type of charge exert repulsive forces on each other; objects that carry different types of charge exert attractive forces on each other.

Charge carrier Any microscopic object that carries an electrical charge.

Charge distribution The way in which a collection of charge carriers is distributed in space.

Charge polarization A spatial separation of the positive and negative charge carriers in an object. The polarization of neutral objects induced by the presence of external charged objects is responsible for the electric interaction between charged and neutral objects.

Charging by induction A method of charging a neutral object using a charged object, with no physical contact between them.

Conduction The flow of charge carriers through a material.

Conductor (electrical) Any material or object through which charge carriers can flow easily.

Conservation of charge The principle that the charge of a closed system cannot change. Thus charge can be transferred from one object to another and can be created or destroyed only in identical positive-negative pairs.

Coulomb (C) The derived SI unit of charge equal to the magnitude of the charge on about 6.24×10^{18} electrons. (The coulomb is defined as the quantity of electrical charge transported in 1 s by a current of 1 ampere, a unit we shall define in Chapter 27).

Coulomb's law The force law that gives the direction and magnitude of the electric force between two particles at rest carrying charges q_1 and q_2 separated by a distance r_{12}:

$$\vec{F}_{12}^{E} = k \frac{q_1 q_2}{r_{12}^2} \hat{r}_{12}. \tag{22.7}$$

The constant k has the value $k = 9.0 \times 10^9 \, \text{N} \cdot \text{m}^2/\text{C}^2$.

Electric force $\vec{F}^E$ (N) The force that charge carriers (and macroscopic objects that carry a surplus electrical charge) exert on each other. The magnitude and direction of this force are given by Coulomb's law.

Electric interaction A long-range interaction between charged particles or objects that carry a surplus electrical charge and that are at rest relative to the observer.

Elementary charge The smallest observed quantity of charge, corresponding to the magnitude of the charge of the electron: $e = 1.60 \times 10^{-19}$ C. See also *Coulomb*.

Grounding The process of electrically connecting an object to Earth ("ground"). Grounding permits the exchange of charge carriers with Earth, a huge reservoir of charge carriers. A charged, conducting object that is grounded will retain no surplus of either type of charge, assuming no other nearby electrical influences.

Insulator (electrical) Any material or object through which charge cannot flow easily.

Ion An atom or molecule that contains unequal numbers of electrons and protons and therefore carries a surplus charge.

Negative charge The type of charge acquired by a plastic comb that has been passed through hair a few times.

Neutral The electrical state of objects whose charge is zero. Electrically neutral macroscopic objects contain the same number of positively and negatively charged particles (protons and electrons).

Positive charge The type of charge acquired by hair after a plastic comb has been passed through it a few times.

The Electric Field

CONCEPTS

QUANTITATIVE TOOLS

In this chapter we revisit an issue we discussed briefly in Chapter 7: the long-range nature of electric and gravitational interactions. How does one charged object "reach out" and affect another charged object? What are the invisible "springs" that pull us—and everything around us—toward Earth's surface? You can describe these long-range interactions by saying that every charged object and every object that has mass has a "sphere of influence" surrounding it. The modern word for this sphere of influence is *field*. Fields are not imaginary. That sensation you felt when, while reading the beginning of Chapter 22, you held a piece of plastic food wrap near your face was the sensation of a field created by the charged particles in the wrap. The closer the wrap is to your skin, the stronger the sensation.

The concept of field is important for two reasons. First, it is impossible to describe the interaction between moving charged particles without it. Second, as you will soon see, it is often easier to deal with fields than with distributions of charge because frequently more is known about fields than about the way charge is distributed.

23.1 The field model

Newton's law of gravity and Coulomb's law describing the electric force between charged particles successfully account for the magnitudes of the gravitational and electric forces between stationary objects. However, they do not address the fundamental puzzle of how objects separated in space can interact without any mediator of the interaction (such an interaction is called *action at a distance*). Worse, they share a fundamental flaw: Both imply that the action of one object on another is instantaneous everywhere throughout space. Consider, for example, the two metal rods in Figure 23.1. Even if both are electrically neutral, their electrons interact. Suppose you quickly drive the electrons in rod A down to the bottom, as in Figure 23.1*b*. According to Coulomb's law, doing this will instantly change the force exerted by the electrons in A on those in B, regardless of how far apart the rods are. This means that it would be possible—in principle—to be standing at one position in space and instantly detect a change that occurs at some far distant position.

The idea that an object can directly and instantly influence another object regardless of their separation was troubling in the 19th century but became untenable in the early 20th century when it was demonstrated experimentally that the interaction between charged objects is not instantaneous. The principle illustrated in Figure 23.1, for example, is what makes possible the transmission of radio signals from a transmitting antenna (rod A) to a receiving antenna (rod B). We know from many experiments that such transmission is not instantaneous. As just one example, a radio signal takes about 0.1 s to travel from Earth's surface to an orbiting communications satellite. In other words, the picture conveyed by Newton's law of gravity and Coulomb's law—that one object directly and instantly affects other objects regardless of the distance between them—cannot be correct.

Instead we must adopt another model of long-range interactions, a model in which interactions take place through the intermediary of an **interaction field** (or simply a **field**). In the field model, an interacting object fills the space around itself with a field. When an object A is placed in the field of an object B, A can "feel" the presence of B's field. Instead of the two objects interacting directly as illustrated in Figure 23.2*a*, it is the field created by each object that acts on the other object (Figure 23.2*b*). The stronger the field, the greater the magnitude of the force resulting from the interaction.

In Figure 23.1, the electrons in rod B feel the field set up by the electrons in rod A. When the electrons in A accelerate, their motion causes a disturbance in A's field, and this disturbance propagates outward through space like the ripples on the surface of a pond. Only when these ripples in the field reach rod B can the motion of A's electrons be detected by those in rod B. In Chapter 30 we shall study the propagation of disturbances in fields due to accelerating charge carriers. For now, we shall concentrate on fields created by stationary objects.

The field model applies equally well to gravitational and electric interactions, with each interaction having its own type of field. The space around any object that has mass is filled with a *gravitational field,* and the space around any electrically charged object is filled with an *electric field*. Gravitational fields exert forces on objects that have mass, and electric fields exert forces on objects that either carry a charge or can be polarized. Let's begin by developing the concept of gravitational field.

Before we attempt to obtain a physical quantity we can use to describe any gravitational field, we should note a number of things. First, for any object A located in a gravitational field created by an object S (S is called the *source* of the field), the magnitude of the field felt by A depends only on the properties of S and on the position of A relative to S; the field magnitude does not depend in any way on the properties of A. Second, the field of an object is always there, even when the object is not interacting with anything else. A field therefore must be represented by a set of numerical

Figure 23.1 Newton's and Coulomb's laws imply that forces are exerted instantaneously across a distance—but experiments show that they do not.

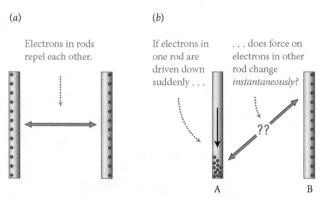

(a)

Electrons in rods repel each other.

(b)

If electrons in one rod are driven down suddenly . . .

. . . does force on electrons in other rod change *instantaneously?*

??

A   B

CONCEPTS

Figure 23.2 The field model for interaction at a distance.

(*a*) Model of direct interaction at a distance

(*b*) Field model of interaction at a distance

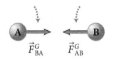

We model A and B as exerting forces directly on each other.

$\vec{F}^G_{BA}$ $\vec{F}^G_{AB}$

field of A

force due to field of A at location of B

$\vec{F}^G_{AB}$

AND

force due to field of B at location of A

$\vec{F}^G_{BA}$

field of B

Fields of A and B shown separately for clarity; both are present at same time.

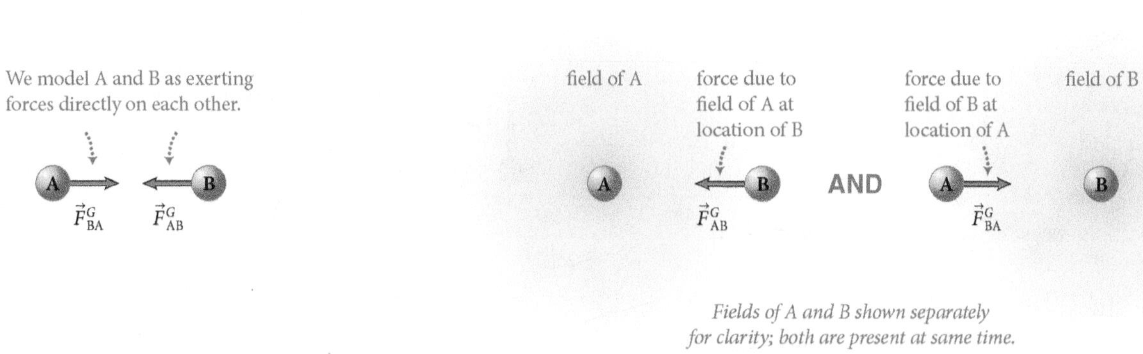

values that cover all the space outside the field source, with every point of this space having a different numerical value. One field representation you are already familiar with is that of a "temperature field," where the temperature across the surface of a region has a specific value at each location (**Figure 23.3**). Third, for stationary objects the field model must give the same forces as Newton's law of gravity and Coulomb's law. In particular, the field model must still yield forces between the two objects that are equal in magnitude and opposite in direction. It is not immediately obvious that the field model preserves this symmetry in the forces because a field is not something shared by two interacting objects—each object has its own field.

For example, in the gravitational interaction between a ball and Earth, the gravitational force exerted by Earth on the ball is due to Earth's gravitational field and the gravitational force exerted by the ball on Earth is due to the ball's field. These two gravitational fields are very different from each other: Earth's field pulls strongly on, say, a paper clip, while the effect of the ball's field on that paper clip is unmeasurably small. As you will see shortly, however, the symmetry of the interaction is preserved despite the asymmetry in the fields (see Checkpoint 23.3).

What physical quantity can we use to describe the gravitational field of an object? How about the gravitational force exerted by the object? Let's examine this possibility using Earth as our object. As we saw in Section 8.8, the magnitude of the gravitational force exerted by Earth on an object of mass m near Earth's surface is $F^G_{Eo} = mg$. This force is not a good quantity for describing Earth's field because the force depends not only on the source—Earth (which determines g)—but also on the mass m of the object placed in the field. As illustrated in **Figure 23.4**, two objects that have different masses m_1 and m_2 but are placed at the same height above Earth's surface are subject to different gravitational forces m_1g and m_2g. The quantity $g = F^G_E/m$, however—the gravitational force per unit of mass—is the same for any object.* This quantity is determined solely by the properties of Earth and is independent of those of any object that experiences a gravitational force exerted by Earth.

23.1 Two objects 1 and 2, of mass m_1 and m_2, are released from rest far from Earth, at a location where the magnitude of the acceleration due to gravity is much less than $g = 9.8 \text{ m/s}^2$. (*a*) What is the ratio F^G_{E1}/F^G_{E2}? (*b*) For these two objects, is the magnitude of the gravitational force exerted by Earth per unit mass independent of the properties of the objects?

Figure 23.3 The temperature across a region is specified by a set of values, with a specific temperature value for every position in that region. Such a set of values is called a *field*.

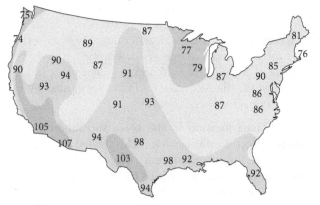

Figure 23.4 Comparison between gravitational force and gravitational acceleration on objects of different mass at the same distance from Earth.

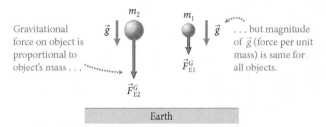

Gravitational force on object is proportional to object's mass . . .

m_2 $\vec{g}$

$\vec{F}^G_{E2}$

m_1 $\vec{g}$ . . . but magnitude of $\vec{g}$ (force per unit mass) is same for all objects.

$\vec{F}^G_{E1}$

Earth

---

*This is so because mass and inertia are equivalent (see Section 13.1), and so the force per unit of mass is equal to force divided by inertia, which is acceleration.

Figure 23.5 Vector field diagram for the gravitational field in a region near Earth.

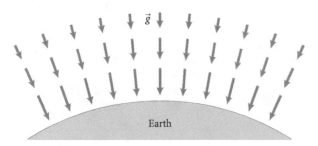

We can use the gravitational force per unit mass exerted by an object as a measure of the magnitude of the object's gravitational field. For example, near Earth's surface, because $g = 9.8 \text{ m/s}^2$, the magnitude of Earth's gravitational field is 9.8 N/kg; near the surface of the Moon, where $g_{\text{moon}} = 1.6 \text{ m/s}^2$, the magnitude of the Moon's gravitational field is 1.6 N/kg. (Remember that 1 N/kg = $1 \text{ (kg} \cdot \text{m/s}^2)/\text{kg} = 1 \text{ m/s}^2$.)

At any given location in the space surrounding a source object S, the magnitude of the gravitational field created by S is the magnitude of the gravitational force exerted on an object B placed at that location divided by the mass of B.

Unlike the temperature field in Figure 23.3, which is a *scalar field,* the gravitational field is a *vector field:* At every position, it has both a magnitude and a direction. **Figure 23.5,** for example, shows a **vector field diagram** representing the gravitational field near Earth. You can determine the magnitude and direction of this field in the space surrounding Earth using a **test particle** (an idealized particle whose mass is small enough that its presence does not perturb the object whose gravitational field we are measuring). Measure, at each location, the gravitational force exerted by Earth on the test particle, and then divide that force by the mass of the test particle to obtain the direction and the magnitude of the gravitational field at that location. As you can see from Figure 23.5, Earth's gravitational field, which can be represented at each position by a vector $\vec{g}$, always points toward the center of Earth, and its magnitude decreases with increasing distance away from Earth. Near Earth's surface, the magnitude of these vectors is $g = 9.8 \text{ N/kg}$.

✋ **23.2** A communications satellite orbits $1.4 \times 10^7 \text{ m}$ from Earth's center, at a location where the magnitude of Earth's gravitational field is 2.0 N/kg. (*a*) If the mass of the satellite is $m_s = 2000 \text{ kg}$, what is the magnitude of $\vec{F}^G_{Es}$? (*b*) If you place a 0.20-kg ball at the satellite's location, what is the magnitude of $\vec{F}^G_{Eb}$?

Checkpoint 23.2 illustrates that if you know the gravitational field at a certain position, you can easily calculate the gravitational force exerted by the source of that field on any object at that position by taking the product of the magnitude of the gravitational field at the location of the object and the mass of the object.

Figure 23.6 Electric force exerted on two objects of different inertia m and charge q by the electric fields created by two identical charged particles.

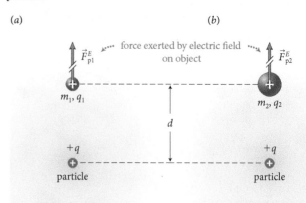

✋ **23.3** (*a*) Is the magnitude of the gravitational force exerted by Earth on a ball greater than, equal to, or smaller than the magnitude of the gravitational force exerted by the ball on Earth? (*b*) Is the magnitude of Earth's gravitational field at the position of the ball greater than, equal to, or smaller than the magnitude of the gravitational field of the ball at a distance equal to Earth's radius? (*c*) Explain how the answers to parts *a* and *b* can both be correct.

23.2 Electric field diagrams

Let us now apply the same ideas to electric interactions. **Figure 23.6a** shows object 1 of mass m_1 and charge q_1 a distance d from a particle that carries a charge $+q$. What is the electric field $\vec{E}$ created by the particle at the position of object 1? Before answering this question, answer the next checkpoint, which concerns the interactions of the particle with object 1 and with an object 2 of mass m_2 and charge q_2 (Figure 23.6b; $m_2 \neq m_1$ and $q_2 \neq q_1$).

✋ **23.4** (*a*) Are the electric forces $\vec{F}^E_{p1}$ and $\vec{F}^E_{p2}$ in Figure 23.6 equal? (*b*) What does the quantity $\vec{F}^E_{pi}/m_i$ represent? (*c*) Is this quantity the same for objects 1 and 2? If not, what quantity is the same for both of these objects?

As Checkpoint 23.4 makes clear, the quantity $\vec{F}^E_{pi}/q_i$—the electric force per unit charge—is determined entirely by the source of the electric field and is independent of the object on which the field exerts a force. So, in analogy to the gravitational field, we can say:

At any given location in the space surrounding a source object S, the electric field created by S is the electric force exerted on a charged test particle placed at that location divided by the charge of the test particle: $\vec{E}_S \equiv \vec{F}^E_{St}/q_t$.

Like gravitational fields, electric fields are vector fields. There is one difference between the two types of fields, however. Electric interactions can be either repulsive or attractive, and so the direction of the electric force—and hence the direction of the electric field—depends on the sign of the charge. Our rule is that:

The direction of the electric field at a given location is the same as the direction of the electric force exerted on a positively charged object at that location.

23.5 (*a*) If the particle in Figure 23.6 carries a negative charge $q < 0$ and q_1 and q_2 are positive, what are the directions of $\vec{F}^E_{p1}$ and $\vec{F}^E_{p2}$? (*b*) Does the electric field created by the particle point toward or away from the particle? (*c*) If q and q_2 are negative, what are the direction of $\vec{F}^E_{p2}$ and the direction of the electric field created by the particle at the location of object 2? (*d*) If q is positive, does the electric field created by the particle point toward or away from the particle? (*e*) How does the magnitude of the electric field created by a particle that carries a charge $+q$ ($q > 0$) compare with the magnitude of the electric field created by a particle that carries a charge $-q$ of identical magnitude at a distance d from each particle?

Figure 23.7 shows the vector field diagrams for the electric fields of particles that carry positive and negative charges. Because it is impossible to draw electric field vectors at *all* locations, the diagrams show vectors at only

Figure 23.7 Vector field diagrams for positively and negatively charged particles. The lengths of the vectors show that the electric field magnitude decreases with increasing distance from the source.

(*a*) Electric field of positively charged particle

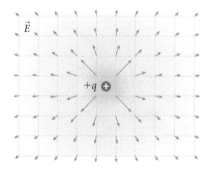

Electric field is directed away from positive source . . .

(*b*) Electric field of negatively charged particle

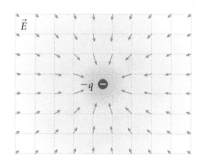

. . . and toward negative source.

Figure 23.8 Electric field pattern created by a small charged object in a solution that contains plastic fibers. The fibers align with the direction of the electric field created by the charged object.

certain positions; from these representative vectors you can get an idea of how the electric field looks as a whole. In addition, the drawing is limited to two dimensions, but you should visualize the electric field as spreading out in all three dimensions.

Electric fields can be made visible by putting charged objects in a (nonconducting) liquid that contains small uncharged plastic fibers or grass seed. Each fiber aligns itself in the direction of the electric field at the fiber's location (**Figure 23.8**).

23.6 If you know the electric field $\vec{E}$ at some location, how can you determine the magnitude and direction of the electric force exerted by that field on an object carrying a charge q and placed at that location?

23.3 Superposition of electric fields

The concept of electric field becomes especially useful when we consider the combined electric field that results from more than one charged object. Suppose we are interested in the electric field created by two particles that carry charges of equal magnitude but opposite sign. To determine the electric field created by the particles at a point P, we place a test particle* carrying a positive charge q_t at P and measure the vector sum of the forces exerted on it by the two charged source particles (**Figure 23.9** on the next page). The electric field at P is then equal to this vector sum divided by q_t.

Figure 23.9 illustrates the **superposition of electric fields:**

The combined electric field created by a collection of charged objects is equal to the vector sum of the electric fields created by the individual objects.

*When measuring electric fields, we assume that the charge q_t on the test particle is so small that the particle does not perturb the particles or objects that generate the electric field we are measuring.

Figure 23.9 The electric field due to multiple charged objects (here, a pair of charged particles) is the vector sum of the fields created by the individual objects.

(a)

To find electric field at P . . .

(b)

. . . we start with forces exerted by fields on test particle placed at P.

(c)

Just as vector sum of forces $\Sigma\vec{F}_t$ on test particle is sum of individual forces . . .

(d)

. . . so electric field $\vec{E}$ at P is sum of fields due to individual particles. It points in same direction as $\Sigma\vec{F}_t$.

The superposition principle holds regardless of the number of sources. Because of the vectorial nature of the electric interaction, electric forces add vectorially (Eq. 22.9). Consequently the electric field at P is equal to $(\vec{F}_{1t}^E + \vec{F}_{2t}^E)/q_t = \vec{F}_{1t}^E/q_t + \vec{F}_{2t}^E/q_t$, which is the vector sum of the electric fields created by the two sources individually.

The only caveat is the one I pointed out in Figure 22.30. When we deal with conductors, the distribution of charge on the individual conductors in isolation might be different from what it is when the conductors are placed close together.

Exercise 23.1 Electric field of two positively charged particles

Consider two identical particles 1 and 2 carrying charges $q_1 = q_2 > 0$ (Figure 23.10). What is the direction of the combined electric field at points P_1 through P_4?

Figure 23.10 Exercise 23.1.

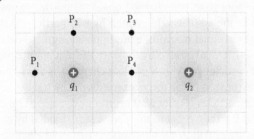

SOLUTION I place a positively charged test particle at each location and determine the vector sum of the (repulsive) forces exerted by 1 and 2 on each test particle. Because $\vec{E} = \Sigma\vec{F}/q_{\text{test}}$, the direction of $\vec{E}$ is the same as the direction of $\Sigma\vec{F}$ (Figure 23.11).

Figure 23.11

(a) Electric forces on test particles

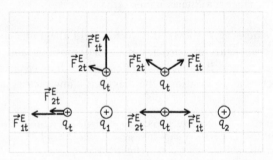

(b) Vector sum of forces on each test particle

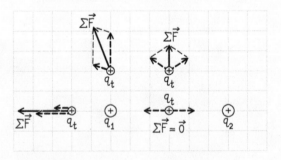

(c) Electric field at each tested point

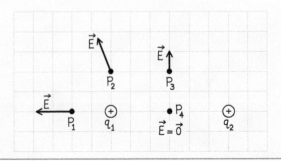

Figure 23.12 Vector field diagrams showing the superposition of the electric fields of the two charged particles of Figure 23.10.

(a) Electric field of particle 1

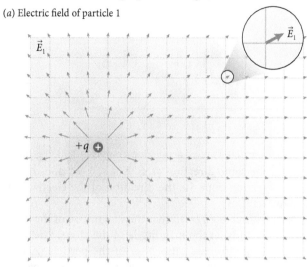

(b) Electric field of particle 2

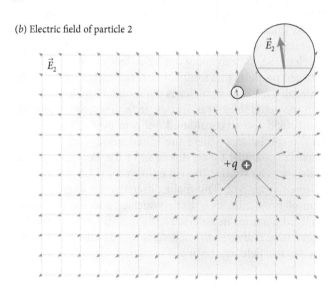

(c) Electric field of both particles

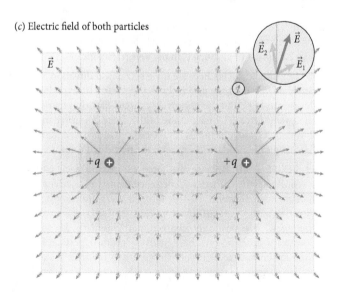

Figure 23.13 Pattern created by two identical charged particles in a liquid containing plastic fibers. Compare with the vector field diagram in Figure 23.12c.

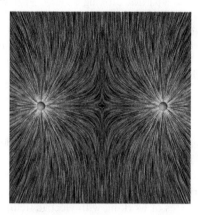

23.7 (a) If the charge on particle 2 in Exercise 23.1 is doubled so that $q_2 = 2_{q1}$, what happens to the direction of the electric field at points P_1 through P_4? (b) If the charge on particle 2 is negative so that $q_2 = -q_1$, what is the direction of the electric field at points P_1 through P_4?

Figure 23.11 provides a limited view of the electric field created by the two particles. A more complete view is given in **Figure 23.12**. This diagram is obtained by vectorially adding, for each grid point, the electric field vectors for the individual particles. Note how the pattern of vectors resembles the pattern created by two identically charged particles in a solution of plastic fibers (**Figure 23.13**).

Using the superposition principle, we can determine the electric field produced by any system of charged particles. **Figure 23.14**, for example, shows a vector diagram for the electric field generated by three charged particles. Because every charged object is made up of charged particles—electrons and protons—we can determine the electric field of any object at any position in space. For a real object, the calculation might be very tedious or even intractable because of the large number of charged particles, but the basic principle is as given above.

Figure 23.14 Vector field diagram of the electric field created by three charged objects.

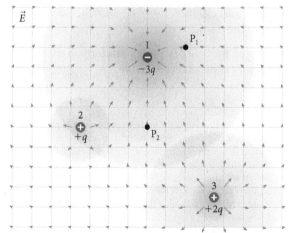

23.8 (*a*) In Figure 23.14, what is the direction of the force $\sum \vec{F}^E$ exerted on a particle carrying a charge $+q$ and placed at P_1? (*b*) How does $\sum \vec{F}^E$ change when the particle at P_1 carries a charge $+3q$? (*c*) Do the magnitude and direction of $\sum \vec{F}^E$ on a charged particle at P_1 change if the $+2q$ charge on object 3 is halved? (*d*) What is the direction of $\sum \vec{F}^E$ exerted on a particle carrying a charge $-q$ and placed at point P_2?

23.4 Electric fields and forces

Before developing additional techniques for determining the electric fields created by systems of charged particles, let us consider this question: What are the forces exerted by an electric field on charged or polarized objects? One of the advantages of working with electric fields is that, for any system of charged particles, once we know the electric field the system creates at some point P in space, we can determine the force exerted by the system on any other charged particle placed at P without worrying about any of the individual source particles in the system anymore.* In Chapter 22 we used the action-at-a-distance model to discuss the forces exerted by charged objects either on other charged objects or on polarized objects. Now we can use the field model to do the same thing. For stationary charged particles, both methods must yield the same result.

When we study the forces exerted by electric fields, it is useful to distinguish between uniform and nonuniform. In a *uniform electric field*, the direction and magnitude of the electric field are the same everywhere. No electric field is ever uniform throughout all space, but as we shall see in Section 23.7 it is possible to create regions of space where the electric field is uniform. In a *nonuniform electric field*, the direction and magnitude of the electric field vary from position to position. (All the electric fields we have considered so far are nonuniform.)

Let us first consider what happens to a charged particle placed in a uniform electric field. Because the electric field is defined as the electric force per unit of charge, the force $\vec{F}_p^E$ exerted by an electric field $\vec{E}$ on a particle carrying a charge q is $\vec{F}_p^E = q\vec{E}$.† Because $\vec{E}$ is the same everywhere, the force $\vec{F}_p^E$ exerted on the particle is constant and so it undergoes a constant acceleration $\vec{a} = \vec{F}_p^E/m = q\vec{E}/m = (q/m)\vec{E}$, where m is the particle's mass.

A charged particle placed in a uniform electric field undergoes constant acceleration.

---

*This procedure is valid only if the presence of the charged particle at P does not alter the way charge is distributed over the system. We shall refer to the particle or system of particles that creates the electric field at P as being "fixed."

†When dealing with forces exerted by fields, we drop the subscript representing the object that exerts the force (the "by" subscript) because the field is due to *all* objects other than the object on which the force is exerted. The superscript E reminds us that we are dealing with the force exerted by an electric field.

Figure 23.15 Forces exerted by a uniform electric field on a positively and a negatively charged particle.

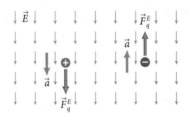

If the particle carries a positive charge, $q > 0$, $\vec{F}_p^E$ and $\vec{a}$ point in the same direction as $\vec{E}$. If $q < 0$, $\vec{F}_p^E$ and $\vec{a}$ point in the direction opposite the direction of the electric field (Figure 23.15).

Note from $\vec{a} = (q/m)\vec{E}$ that the magnitude of the acceleration depends on the magnitude of the electric field and on the charge-to-mass ratio q/m of the particle. A large charge q causes a greater force to be exerted on the particle and therefore a greater acceleration; a larger mass m means the particle has greater inertia and therefore the acceleration is smaller.

Because we have already studied motion with constant acceleration, we can apply our knowledge to the motion of charged particles in a uniform electric field. In general, the trajectory of these particles is parabolic, like the trajectory of a projectile fired near Earth's surface, where the gravitational field can be considered uniform over a limited area. In the special case where the initial velocity of a charged particle is parallel to the direction of the electric field, the trajectory is a straight line, like the vertical fall of an object released from rest. The main difference between the motion of projectiles near Earth's surface and the motion of charged particles in an electric field is that Earth's gravitational field is always directed vertically downward, whereas the electric field can be in any direction.

Example 23.2 Charged particle trajectories

Four charged particles are fired with a horizontal initial velocity $\vec{v}$ into a uniform electric field that is directed vertically downward. The effect of gravity is negligible. The particles have the following charges and masses: particle 1 $(+q, m)$; 2 $(+q, 2m)$; 3 $(+2q, 2m)$; 4 $(-q, m)$. Sketch the four trajectories.

❶ GETTING STARTED Because the electric field direction is vertically down, the three positively charged particles experience a downward force and the negatively charged particle experiences an upward force. The magnitude of this force doesn't change as the particles move through the electric field because both the field magnitude and the charges on the particles are constant.

❷ DEVISE PLAN Because the force exerted on each particle is constant, the particles experience constant accelerations. Because the direction of the force is perpendicular to the direction of the particles' initial motion, they all have a parabolic trajectory. The positively charged particles have a constant downward acceleration; the negatively charged particle has a constant upward acceleration. Because $\vec{a} = (q/m)\vec{E}$, the acceleration magnitude is greatest when q is large and/or m is small.

③ **EXECUTE PLAN** I draw trajectories that curve down for 1, 2, and 3 and up for 4 (**Figure 23.16**). The magnitude of the electric force exerted on particle 2 is the same as that exerted on particle 1, but the acceleration of particle 2 is smaller because this particle has the greater mass. I indicate this difference in acceleration by making trajectory 1 more curved than trajectory 2. The magnitude of the electric force exerted on particle 3 is twice as great as that exerted on particle 1, but 3's mass is also twice as great, and so the two particles have the same charge-to-mass ratio and therefore the same acceleration and trajectory. The magnitude of the electric force exerted on particle 4 is the same as that exerted on particle 1 but points in the opposite direction, and so trajectories 1 and 4 are identical in shape but curve in opposite directions. ✔

Figure 23.16

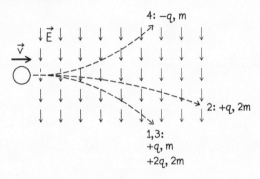

④ **EVALUATE RESULT** My sketch indicates that particles with increasingly positive charge-to-mass ratios curve increasingly downward. Conversely, particles with increasingly negative charge-to-mass ratios curve increasingly upward. This is what I expect because a particle's deflection is a function of both its charge, which determines the magnitude of the force exerted by the electric field on it (greater charge, greater deflection), and its mass, which relates the particle's acceleration to the force exerted on it (greater mass, smaller deflection).

✋ⓒ **23.9** A water droplet carrying a positive charge is released from rest in a uniform horizontal electric field near Earth's surface. The horizontal electric force is comparable in magnitude to the gravitational force exerted by Earth. Describe the droplet's trajectory.

In a nonuniform electric field, the force exerted on a charged particle varies from one position to another, so we cannot easily specify the particle's trajectory without knowing more about the electric field. As in a uniform electric field, however:

A positively charged particle placed in a nonuniform electric field has an acceleration in the same direction as the electric field; a negatively charged particle placed in such a field has an acceleration in the opposite direction.

Figure 23.17 Extended free-body diagram for a permanent dipole placed in a uniform electric field.

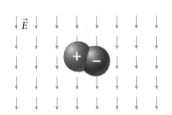
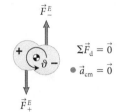

In Chapter 22 we found that charged objects can polarize electrically neutral objects by separating the centers of positive and negative charge in the latter. The resulting configuration of charge—equal amounts of positive and negative charge separated by a small distance—is called an **electric dipole** or simply **dipole.** Many molecules, such as water molecules, are *permanent dipoles;* that is to say, the centers of positive and negative charge are kept separated by some internal mechanism. **Figure 23.17** illustrates the forces exerted on a permanent electric dipole in a uniform electric field. Because the electric field is uniform and the magnitude of the charge on the positive end of the dipole is equal to the magnitude of the charge on the negative end, the forces exerted on the two ends are equal in magnitude but opposite in direction, making their vector sum zero. However, the forces exerted on the two ends cause a torque (see Chapter 12).

✋ⓒ **23.10** (a) What effect does the torque caused by the electric field have on the electric dipole in Figure 23.17? (b) Is the torque the same for every orientation of the molecule?

The orientation of an electric dipole can be characterized by a vector, the **dipole moment,** that, by definition, points from the center of negative charge to the center of positive charge, as shown in **Figure 23.18** on the next page. As Checkpoint 23.10 illustrates, the electric forces create a torque on the dipole that tends to align the dipole moment with the electric field.

In a nonuniform electric field, the situation is more complicated because the two ends of the dipole are now subject to forces that have different magnitudes as well as different directions. Consider, for example, the nonuniform electric field in Figure 23.18a, which is due to a positively charged particle to the left side of the figure. The magnitude of $\vec{F}_{-}^{E}$ is greater than the magnitude of $\vec{F}_{+}^{E}$ because the negative end of the dipole is closer to the positively charged particle. Thus the vector sum of the forces exerted on the

Figure 23.18 Extended free-body diagrams for permanent dipoles in nonuniform electric fields. The electric field shown is due to a positively charged particle to the left of the figure.

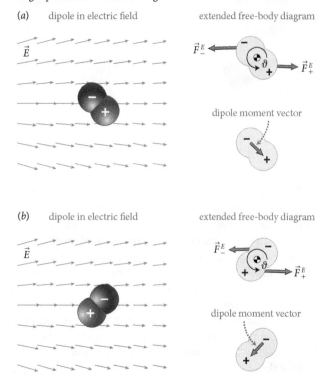

(*a*)   dipole in electric field   extended free-body diagram

dipole moment vector

(*b*)   dipole in electric field   extended free-body diagram

dipole moment vector

two ends is nonzero, and so the dipole experiences an acceleration whose magnitude and direction depend on its orientation with respect to the electric field. In addition, the forces create a torque about the dipole's center of mass. As in a uniform electric field:

> A permanent electric dipole placed in an electric field is subject to a torque that tends to align the dipole moment with the direction of the electric field. If the field is uniform, the dipole has zero acceleration; if the electric field is nonuniform, the dipole has a nonzero acceleration.

23.11 (*a*) Draw a free-body diagram for the dipole in Figure 23.18*a* and determine the direction of the dipole's center-of-mass acceleration. (*b*) Draw a free-body diagram for the dipole in Figure 23.18*b* and qualitatively describe the dipole's motion.

Self-quiz

1. Suppose someone discovers that blue and yellow objects attract each other, that two blue objects repel each other, and that two yellow objects repel each other. The strength of this "chromatic interaction" is found to depend on color depth: The deeper the color, the greater the magnitude of the interaction. How would you define the magnitude and direction of the "chromatic field" of an object?

2. (a) Does an electrically neutral particle that has mass interact with an electric field? (b) Does a charged particle interact with a gravitational field?

3. The two particles in **Figure 23.19** have the same mass, carry charges of the same magnitude ($q_1 = -q_2 > 0$), and are equidistant from point P. (a) What is the electric field direction at P? (b) At P, what is the direction of the gravitational field due to the two particles? Ignore Earth's gravitational field.

4. Can electric and gravitational fields exist in the same place at the same time?

5. What are the directions of the acceleration of each particle in **Figure 23.20**? Describe the resulting motions.

Figure 23.19

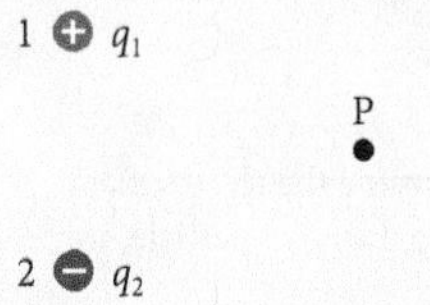

Figure 23.20

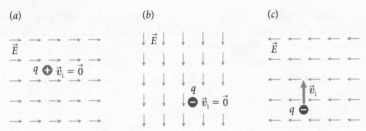

Answers

1. The gravitational field is defined as the gravitational force per unit of mass, with the field direction the same as the direction of the force. The electric field is defined as the electric force per unit of charge, with the field direction parallel to that of the force exerted on a positively charged particle. Therefore, the chromatic field can be defined as the chromatic force per unit of color, with the field direction parallel to that of the force exerted on a particle carrying some chosen color.

2. (a) No. Uncharged particles don't interact with electric fields. (Remember that a particle has no extent and therefore cannot be polarized.) (b) Yes, because any particle or object, charged or uncharged, interacts with a gravitational field.

3. (a) The electric field of particle 1 points away from the particle, which means that at P it points to the right and down. The electric field of particle 2 points toward the particle, meaning to the left and down at P. The vector sum of the electric fields at P therefore points straight down (**Figure 23.21a**). (b) The gravitational fields of the two particles point toward them from P, and so their vector sum points to the left (Figure 23.21b).

Figure 23.21

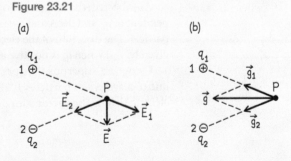

4. Yes. Consider a charged object near Earth's surface. This object is surrounded by an electric field, but it is also surrounded by Earth's (and to a lesser extent its own) gravitational field. These electric and gravitational fields exist in the same place at the same time.

5. (a) Recall from the discussion following Checkpoint 23.4 that the direction of the electric field at a given location is the same as the direction of the electric force exerted on a positively charged particle at that location. In Figure 23.20a, therefore, the positively charged particle experiences a force directed to the right. Because its initial velocity is zero, the particle moves in a straight line in the direction of the electric field. (b) The negatively charged particle in Figure 23.20b experiences an acceleration up the page, opposite the direction of the electric field. This particle moves in a straight line up the page. (c) The negatively charged particle in Figure 23.20c experiences an acceleration to the right, in the direction opposite the direction of the electric field. Because its initial velocity is perpendicular to the direction of the electric field, the particle travels in a parabolic trajectory up the page and curving to the right.

23.5 Electric field of a charged particle

In Section 23.2 we defined the **electric field** at a certain point P in space as the electric force experienced at P by a test particle carrying a charge q_t divided by the charge of the test particle:

$$\vec{E} \equiv \frac{\vec{F}_t^E}{q_t}. \tag{23.1}$$

The SI unit of electric field is the newton per coulomb (N/C).

Equation 23.1 requires no knowledge of the charge distribution that causes the electric field: It gives a prescription for determining the electric field at a given position in space. We can use Coulomb's law, however, to derive an expression for the electric field created at some point P due to a source particle carrying a charge q_s at position $\vec{r}_s$ (**Figure 23.22**). If we place a test particle carrying a charge q_t at P, Coulomb's law (Eq. 22.7) tells us that the force exerted on the test particle is

$$\vec{F}_{st}^E = k\frac{q_s q_t}{r_{st}^2}\hat{r}_{st}, \tag{23.2}$$

where $k = 9.0 \times 10^9 \text{ N} \cdot \text{m}^2/\text{C}^2$ is the proportionality constant that appears in Coulomb's law (Eq. 22.5), r_{st} is the distance between the two particles, and $\hat{r}_{st}$ is a unit vector pointing from the source particle to the test particle. If we divide the electric force exerted by the source particle on the test particle by the charge q_t on the test particle, we obtain an expression for the electric field created by the source particle at P:

$$\vec{E}_s = \frac{\vec{F}_{st}^E}{q_t} = k\frac{q_s}{r_{st}^2}\hat{r}_{st}. \tag{23.3}$$

Because the test particle has nothing to do with this electric field, we can omit any reference to it by writing $\vec{r}_{st} = \vec{r}_{sP}$ and referring only to the position of point P:

$$\vec{E}_s(P) = k\frac{q_s}{r_{sP}^2}\hat{r}_{sP}. \tag{23.4}$$

This expression represents the electric field at P due to a source particle carrying a charge q_s at position $\vec{r}_s$.

As expected, the magnitude of the electric field at P is proportional to q_s, is independent of q_t, and decreases as the inverse square of the distance r_{sP} from the source particle. The direction of the electric field is outward (that is to say, in the direction given by $\hat{r}_{sP}$) when q_s is positive and inward (antiparallel to $\hat{r}_{sP}$) when q_s is negative.

Using the superposition principle, we now can determine the electric field due to a system of particles 1, 2, . . . carrying charges q_1, q_2, ... The combined electric field is the vector sum of the individual electric fields:

$$\vec{E} = \vec{E}_1 + \vec{E}_2 + \cdots = \sum k\frac{q_i\hat{r}_{iP}}{r_{iP}^2}. \tag{23.5}$$

Once the electric field at a certain position is known, the force exerted on any particle carrying charge q placed at that position can be found from

$$\vec{F}_P^E = q\vec{E}. \tag{23.6}$$

(Remember that we omit the "by" subscript on the force when the force is exerted by a field.) If q is positive, the force exerted on the particle is in the same direction as the electric field; if q is negative, the force exerted on the particle is in the direction opposite the direction of the electric field.

Figure 23.22 To determine the electric field at P generated by a charged source particle, we place a test particle at P.

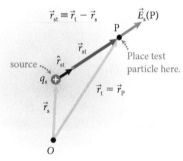

QUANTITATIVE TOOLS

Example 23.3 Electric field due to two charged particles

A point P is located at $x_P = 2.0$ m, $y_P = 3.0$ m. What are the magnitude and direction of the electric field at P due to a particle 1 carrying charge $q_1 = +10\ \mu C$ and located at $x_1 = 1.0$ m, $y_1 = 0$ and a particle 2 carrying charge $q_2 = +20\ \mu C$ and located at $x_2 = -1.0$ m, $y_2 = 0$?

❶ **GETTING STARTED** I begin by making a sketch of the situation (Figure 23.23). Each particle carries a positive charge, so the electric field due to each particle points away from the particle.

Figure 23.23

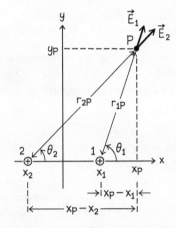

❷ **DEVISE PLAN** To determine the electric field $\vec{E}_P$ at P, I must take the vector sum of $\vec{E}_1$ and $\vec{E}_2$ at P. I can use Eq. 23.4 to calculate the magnitudes E_1 and E_2. To obtain the vector sum of the two fields, I add their x and y components.

❸ **EXECUTE PLAN** The distances from the particles to P are $r_{1P} = \sqrt{(x_P - x_1)^2 + y_P^2} = \sqrt{10\ \text{m}^2} = 3.2$ m and $r_{2P} = \sqrt{(x_P - x_2)^2 + y_P^2} = \sqrt{18\ \text{m}^2} = 4.2$ m. The magnitudes of the electric fields created by the particles at P are thus

$$E_1 = k\frac{|q_1|}{r_{1P}^2} = (9.0 \times 10^9\ \text{N} \cdot \text{m}^2/\text{C}^2)\frac{(1.0 \times 10^{-5}\ \text{C})}{10\ \text{m}^2}$$
$$= 0.90 \times 10^4\ \text{N/C}$$

$$E_2 = k\frac{|q_2|}{r_{2P}^2} = (9.0 \times 10^9\ \text{N} \cdot \text{m}^2/\text{C}^2)\frac{(2.0 \times 10^{-5}\ \text{C})}{18\ \text{m}^2}$$
$$= 1.0 \times 10^4\ \text{N/C}.$$

To calculate $\vec{E}_P$, I take the vector sum of $\vec{E}_1$ and $\vec{E}_2$ at P. In component form, I have

$$E_{Px} = E_{1x} + E_{2x} = E_1 \cos \theta_1 + E_2 \cos \theta_2$$
$$= E_1\frac{(x_P - x_1)}{r_{1P}} + E_2\frac{(x_P - x_2)}{r_{2P}}$$

$$E_{Py} = E_{1y} + E_{2y} = E_1 \sin \theta_1 + E_2 \sin \theta_2$$
$$= E_1\frac{y_P}{r_{1P}} + E_2\frac{y_P}{r_{2P}}.$$

Substituting the values given, I have

$$E_{Px} = (0.9 \times 10^4\ \text{N/C})\frac{1.0\ \text{m}}{3.2\ \text{m}} + (1.0 \times 10^4\ \text{N/C})\frac{3.0\ \text{m}}{4.2\ \text{m}}$$
$$= +1.0 \times 10^4\ \text{N/C}$$

$$E_{Py} = (0.9 \times 10^4\ \text{N/C})\frac{3.0\ \text{m}}{3.2\ \text{m}} + (1.0 \times 10^4\ \text{N/C})\frac{3.0\ \text{m}}{4.2\ \text{m}}$$
$$= +1.6 \times 10^4\ \text{N/C}.$$

Finally, I write this in vector form as

$$\vec{E}_P = (+1.0 \times 10^4\ \text{N/C})\hat{\imath} + (+1.6 \times 10^4\ \text{N/C})\hat{\jmath}.\ ✔$$

❹ **EVALUATE RESULT** Both E_{Px} and E_{Py} are positive, as I expect based on my sketch. The magnitudes of $\vec{E}_1$ and $\vec{E}_2$ are comparable, which is what I would expect: Particle 2 carries twice the charge of particle 1, but the square of its distance to P is greater by a factor of 1.8.

✋ **23.12** What is the magnitude of the electric force exerted by the electric field on an electron placed at point P in Figure 23.23? What is the initial acceleration of the electron if it is released from rest from that point? [$e = 1.6 \times 10^{-19}$ C; $m_e = 9.1 \times 10^{-31}$ kg]

23.6 Dipole field

Next we examine the electric field due to a permanent electric dipole. **Figure 23.24** on the next page shows a dipole that consists of a particle carrying a charge $+q_p$ at $x = 0$, $y = +\frac{1}{2}d$, and another particle carrying a charge $-q_p$ at $x = 0$, $y = -\frac{1}{2}d$, where d is the distance between the two particles. The charge q_p of the positively charged pole is called the *dipole charge*, and the distance d is called the *dipole separation*. Each particle creates an electric field at all positions in space, so the two fields overlap everywhere. We can determine the combined electric field at any position by adding the two fields vectorially. Let us do this for two general locations: anywhere along the x axis and anywhere along the y axis.

Figure 23.24 Calculating the electric field due to a dipole.

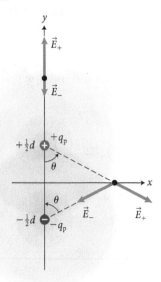

Along the x axis, which bisects the dipole, the magnitudes of the electric fields due to the two ends of the dipole are equal:

$$E_+ = E_- = k\frac{q_p}{x^2 + (d/2)^2}. \qquad (23.7)$$

The x components of these two electric fields point in opposite directions and so add to zero. The magnitude of the combined electric field is thus equal to the sum of the y components:

$$E_y = E_{+y} + E_{-y} = -(E_+ + E_-)\cos\theta$$

$$= -\left(2k\frac{q_p}{x^2 + (d/2)^2}\right)\left(\frac{d/2}{[x^2 + (d/2)^2]^{1/2}}\right) = -k\frac{q_p d}{[x^2 + (d/2)^2]^{3/2}}. \qquad (23.8)$$

The product $q_p d$ is a measure of the strength of the dipole and is the magnitude of the dipole moment introduced in Section 23.4. To specify both the strength and the orientation of the dipole we can write this quantity as a vector, called the **dipole moment:**

$$\vec{p} \equiv q_p \vec{r}_p, \qquad (23.9)$$

where $\vec{r}_p \equiv \vec{r}_{-+} = \vec{r}_+ - \vec{r}_-$ is the position of the positively charged particle relative to the negatively charged particle (and so $d = |\vec{r}_p|$). Because q_p is always taken to be positive, the dipole moment $\vec{p}$ points in the same direction as $\vec{r}_p$: along the axis of the dipole (the line that passes through the center of each particle), in the direction from the negative to the positive pole (**Figure 23.25**). Large permanent dipole moments can be caused either by a large dipole separation d or by a large dipole charge q_p. Conceptually you can think of the magnitude of the dipole moment as a measure of how strongly the dipole wants to align itself in the direction of an electric field. The SI unit of dipole moment is the C · m.

For distances far from the dipole ($x \gg d/2$), we may ignore $d/2$, and so $[x^2 + (d/2)^2]^{3/2} \to x^3$. Equation 23.8 thus becomes

$$E_y \approx -k\frac{p}{|x^3|} \quad \text{(far from dipole along the positive } x \text{ axis).} \qquad (23.10)$$

Figure 23.25 The dipole moment $\vec{p}$ points along the axis of the dipole from the negative to the positive pole.

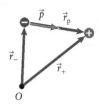

The right side of this equation is negative for both positive and negative x, and so anywhere along the x axis the dipole's electric field $\vec{E}$ points in the negative y direction, opposite the direction of the dipole moment. Equation 23.10 also shows that the magnitude of the electric field is inversely proportional to x^3, in contrast to the electric field of a charged particle, which is inversely proportional to x^2 (Eq. 23.4). The reason the electric field of a dipole approaches zero faster as x increases is that the angle between $\vec{E}_+$ and $\vec{E}_-$ in Figure 23.24 approaches 180° as x increases, and so the electric fields of the two poles tend to cancel each other more and more.

Along the y axis, the electric field created by either end of the dipole is directed along the y axis. Thus to determine the y component of the electric field of the dipole at any position along the y axis, we must add the y components of the fields from each particle. For $y > +d/2$:

$$E_y = E_{+y} + E_{-y} = k\frac{q_p}{[y - (d/2)]^2} - k\frac{q_p}{[y + (d/2)]^2}. \qquad (23.11)$$

After some algebra, this can be rewritten in the form

$$E_y = k\frac{q_p}{y^2}\left[\left(1 - \frac{d}{2y}\right)^{-2} - \left(1 + \frac{d}{2y}\right)^{-2}\right] \quad (y > +d/2). \qquad (23.12)$$

For distances far from the dipole, $y \gg +d/2$, so we can use the binomial series expansion, which states that for $x \ll 1$, $(1 + x)^n \approx 1 + nx$ (see Appendix B). Applying this expansion to the two terms inside the square brackets in Eq. 23.12, we get

$$E_y \approx k\frac{q_p}{y^2}\left[\left(1 + 2\frac{d}{2y}\right) - \left(1 - 2\frac{d}{2y}\right)\right]$$

$$= k\frac{q_p}{y^2}\left[\frac{2d}{y}\right] = 2k\frac{q_p d}{y^3} = 2k\frac{p}{y^3} \quad (y \gg d/2). \quad (23.13)$$

The right side of this equation has the same algebraic sign as y, so the dipole's electric field $\vec{E}$ points in the positive y direction. (Carrying out the same calculation for $y < -d/2$, you can show that the electric field still points in the positive y direction. In between the two charged particles, the electric field points in the negative y direction.) The magnitude of the electric field is inversely proportional to y^3—just as along the x axis, the electric fields of each of the two poles tend to cancel each other more and more as the distance from the dipole increases. One can show that the electric field of the dipole depends on $1/r^3$ for all positions far from the dipole (where r is the distance between the point under consideration and the center of the dipole). The reason is that the electric fields of the positive and negative ends of the dipole partially cancel each other, and this cancellation becomes more complete far from the dipole: The farther you are from the dipole, the smaller the separation between the charged particles appears to be.

23.13 The magnitude of the electric field created by dipole A at a certain point P is E_A. If the dipole is replaced with another dipole B that has its dipole moment oriented in the same direction, the magnitude of the electric field at point P is found to be greater: $E_B > E_A$. Which dipole has the greater dipole moment? For which of these two dipoles is the dipole charge q_p greater?

23.7 Electric fields of continuous charge distributions

So far we have dealt with only charged particles because Coulomb's law applies only to charged particles. However, most charged objects of interest—from charged combs to electrical components—are not particles. Instead, they are extended bodies. Although every macroscopic object consists of very large numbers of charged particles—protons and electrons—it is not practical to calculate the individual field of each of these particles and then add them vectorially. Instead, we shall treat any macroscopic charged object as having a continuous charge distribution and calculate the electric field created by the object by dividing the charge distribution on the object into infinitesimally small segments that may be considered charged source particles carrying a charge dq_s. For the charged macroscopic object shown in **Figure 23.26**, for example, we can use Coulomb's law to obtain the infinitesimal portion of the electric field at point P contributed by a segment:

$$d\vec{E}_s(P) = k\frac{dq_s}{r_{sP}^2}\hat{r}_{sP}. \quad (23.14)$$

Using the principle of superposition, we can then sum the contributions of all the segments that make up the object. Because the segments are infinitesimally small, this sum corresponds to an integral:

$$\vec{E} = \int d\vec{E}_s = k\int\frac{dq_s}{r_{sP}^2}\hat{r}_{sP}. \quad (23.15)$$

Figure 23.26 To calculate the electric field created at P by a continuous charge distribution, we divide the distribution into infinitesimally small segments that can be treated as charged source particles carrying charge dq_s.

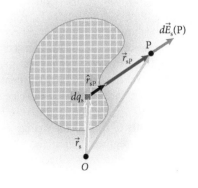

In order to evaluate this integral, we must express dq_s, $1/r_{sP}^2$, and $\hat{r}_{sP}$ in terms of the same coordinate(s). To do so, it is necessary to express the charge on the object in terms of a **charge density**—the amount of charge per unit of length, per unit of surface area, or per unit of volume. For a one-dimensional object, such as a thin charged wire of length ℓ carrying a charge q uniformly distributed along the wire, the *linear charge density*—the amount of charge per unit of length (in coulombs per meter)—is given by

$$\lambda \equiv \frac{q}{\ell} \quad \text{(uniform charge distribution).} \tag{23.16}$$

For uniformly charged two-dimensional objects, we use the *surface charge density*—the amount of charge per unit of area (in coulombs per square meter). For example, the surface charge density of a flat plate of area A carrying a uniformly distributed charge q is

$$\sigma \equiv \frac{q}{A} \quad \text{(uniform charge distribution).} \tag{23.17}$$

For a uniformly charged three-dimensional object, we use the *volume charge density*:

$$\rho \equiv \frac{q}{V} \quad \text{(uniform charge distribution),} \tag{23.18}$$

which gives the amount of charge per cubic meter.

The procedure on this page provides some helpful steps for carrying out the integral in Eq. 23.15, and the next four examples show how to put the procedure into practice.

Procedure: Calculating the electric field of continuous charge distributions by integration

To calculate the electric field of a continuous charge distribution, you need to evaluate the integral in Eq. 23.15. The following steps will help you evaluate the integral.

1. Begin by making a sketch of the charge distribution. Mentally divide the distribution into small segments. Indicate one such segment that carries a charge dq_s in your drawing.

2. Choose a coordinate system that allows you to express the position of the segment in terms of a minimum number of coordinates $(x, y, z, r, \text{ or } \theta)$. These coordinates are the integration variables. For example, use a radial coordinate system for a charge distribution with radial symmetry. Unless the problem specifies otherwise, let the origin be at the center of the object.

3. Draw a vector showing the electric field caused by the segment at the point of interest. Examine how the components of this vector change as you vary the position of the segment along the charge distribution. Some components may cancel, which greatly simplifies the

calculation. If you can determine the direction of the resulting electric field, you may need to calculate only one component. Otherwise express $\hat{r}_{sP}$ in terms of your integration variable(s) and evaluate the integrals for each component of the field separately.

4. Determine whether the charge distribution is one-dimensional (a straight or curved wire), two-dimensional (a flat or curved surface), or three-dimensional (any bulk object). Express dq_s in terms of the corresponding charge density of the object and the integration variable(s).

5. Express the factor $1/r_{sP}^2$, where r_{sP} is the distance between dq_s and the point of interest, in terms of the integration variable(s).

At this point you can substitute your expressions for dq_s and $1/r_{sP}^2$ into Eq. 23.15 and carry out the integral (or component integrals), using what you determined about the direction of the electric field (or substituting your expression for $\hat{r}_{sP}$).

Example 23.4 Electric field created by a uniformly charged thin rod

A thin rod of length ℓ carries a uniformly distributed charge q. What is the electric field at a point P along a line that is perpendicular to the long axis of the rod and passes through the rod's midpoint?

❶ GETTING STARTED I begin by making a sketch of the situation. After drawing a set of axes, I place the rod along the y axis, with the origin at the rod center and point P on the positive x axis (Figure 23.27).

Figure 23.27

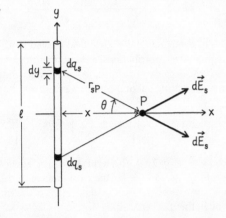

❷ DEVISE PLAN The word *thin* implies that I can treat the rod as a one-dimensional object. Because the rod is uniformly charged, I can thus use Eq. 23.16 to determine the linear charge density along the rod. To determine the electric field at P, I divide the rod lengthwise into a large number of infinitesimally small segments, each of length dy. Each segment contributes to the electric field at P an amount given by Eq. 23.14. For each segment above the x axis there is a corresponding segment below the axis at the same distance from P. The y components of the electric fields $d\vec{E}_s$ due to these two segments add up to zero, so I need to calculate only the x component dE_{sx}. To get the electric field created by the entire rod, I use Eq. 23.15 to integrate my result over the length of the rod.

❸ EXECUTE PLAN The charge dq_s on each segment dy is $dq_s = \lambda\, dy = (q/\ell)\, dy$. The x component of the electric field created by each segment at P is thus

$$dE_{sx} = k\frac{dq_s}{r_{sP}^2}\cos\theta = k\frac{q}{\ell r_{sP}^2}\cos\theta\, dy, \qquad (1)$$

where θ is the angle between the x axis and the line that connects the segment dy with P. Both θ and r_{sP} depend on the position y of the segment, so I must choose one integration variable and express the others in terms of that variable. I choose θ as the integration variable, which means I must express the factor dy/r_{sP}^2 in Eq. 1 in terms of θ. Using trigonometry, I have

$$\cos\theta = \frac{x}{r_{sP}} \qquad (2)$$

and

$$\tan\theta = \frac{y}{x}, \qquad (3)$$

where x is the x coordinate of point P. Differentiating Eq. 3 yields

$$dy = x\, d(\tan\theta) = \frac{x}{\cos^2\theta}\, d\theta. \qquad (4)$$

Next I divide Eq. 4 by r_{sP}^2 to obtain the factor dy/r_{sP}^2 I need. I use r_{sP}^2 on the left, but on the right I use Eq. 2 to write r_{sP}^2 in the form $x^2/\cos^2\theta$, yielding

$$\frac{dy}{r_{sP}^2} = \left(\frac{\cos^2\theta}{x^2}\right)\left(\frac{x}{\cos^2\theta}\, d\theta\right) = \frac{1}{x}\, d\theta.$$

Substituting this result into Eq. 1 and integrating over the entire rod yield

$$E_x = k\frac{q}{\ell}\int_{-\theta_{max}}^{+\theta_{max}}\frac{\cos\theta}{x}\, d\theta = \frac{kq}{\ell x}\int_{-\theta_{max}}^{+\theta_{max}}\cos\theta\, d\theta$$

$$= \frac{kq}{\ell x}\sin\theta\,\Big|_{-\theta_{max}}^{+\theta_{max}} = \frac{2kq}{\ell x}\sin\theta_{max},$$

where θ_{max}, the maximum value of θ, is the angle between the x axis and the line that connects the top end of the rod with P. Substituting $\sin\theta_{max} = y/r_{sP} = \frac{1}{2}\ell/\sqrt{(\ell/2)^2 + x^2}$ finally yields

$$E_x = \frac{kq}{x\sqrt{\ell^2/4 + x^2}}; E_y = 0; E_z = 0. ✔$$

❹ EVALUATE RESULT Very far from the rod along the positive x axis, $x \ll \ell$, so the rod looks like a particle. In this case, I can ignore the ℓ^2 term in the denominator and my result becomes identical to Eq. 23.4, the equation for a particle that carries a charge q ($E = kq/r^2$, or using the symbols from this problem, $E_x = kq/x^2$).

When P is very close to the rod, $x \ll \ell$, I can ignore the x^2 term in the denominator and write

$$E_x = \frac{2k(q/\ell)}{x} = \frac{2k\lambda}{x}. \qquad (5)$$

In this $x \ll \ell$ case, the distance from P to either end of the rod is much greater than the distance from P to the closest point on the rod (which is the rod midpoint, located at the origin), and thus the rod essentially looks "infinitely long" to an observer at P. Indeed, Eq. 5 shows that my result no longer depends on ℓ.

I also note that the rod's electric field is now inversely proportional to x rather than x^2. I saw in Chapter 17 that the amplitudes of waves that spread out in three dimensions are inversely proportional to x^2, whereas the amplitudes of waves that spread out in two dimensions are inversely proportional to x. My result therefore make sense because the electric field that emanates from a charged particle "spreads out" in three dimensions, but the field that emanates from an infinitely long charged rod spreads out in just two dimensions.

Example 23.5 Electric field created by a uniformly charged thin ring

A thin ring of radius R carries a uniformly distributed charge q. What is the electric field at point P along an axis that is perpendicular to the plane of the ring and passes through its center?

❶ **GETTING STARTED** I begin by making a sketch of the situation. I let the ring be in the xy plane, with the origin at the center of the ring, and I place P on the positive z axis (**Figure 23.28**).

Figure 23.28

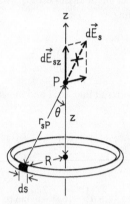

❷ **DEVISE PLAN** Because the ring is thin, I can use Eq. 23.16 to determine its linear charge density. To determine the electric field at P, I divide the ring into a large number of infinitesimally small segments, each of arc length ds. Each segment contributes to the electric field an amount given by Eq. 23.14. Because all segments are at the same distance from P, all contribute an electric field $d\vec{E}_s$ of the same magnitude. As shown in my sketch, the contribution $d\vec{E}_s$ makes an angle θ with the z axis, and so each segment ds produces a component parallel to the z axis and a component perpendicular to it. For each pair of segments on opposite sides of the ring, the components of $d\vec{E}_s$ perpendicular to the z axis add up to zero, and so I am concerned only with the z components.

To get the electric field created by the ring, I can use Eq. 23.15 to integrate my $d\vec{E}_s$ result over the circumference of the ring.

❸ **EXECUTE PLAN** Each segment carries a charge $dq_s = \lambda\, ds$, where $\lambda = q/2\pi R$ is the linear charge density along the ring. The magnitude of each segment's contribution to the electric field is, from Eq. 23.14,

$$dE_s = k\frac{dq_s}{r_{sP}^2} = k\frac{\lambda\, ds}{r_{sP}^2} = k\frac{\lambda\, ds}{z^2 + R^2}.$$

For the z component of $d\vec{E}_s$, I see from Figure 23.28 that the angle between the vectors $d\vec{E}_{sz}$ and $d\vec{E}_s$ is also θ, and so $\cos\theta = dE_{sz}/dE_s$. Then, combining this expression with the relationship

$$\cos\theta = \frac{z}{r_{sP}} = \frac{z}{\sqrt{z^2 + R^2}},$$

I have

$$dE_{sz} = \cos\theta\, dE_s = \left[\frac{z}{\sqrt{z^2 + R^2}}\right]\left[k\frac{\lambda\, ds}{z^2 + R^2}\right] = k\frac{z\lambda}{[z^2 + R^2]^{3/2}}\, ds.$$

To determine the electric field created by the ring, I must integrate the contributions around the ring, from $s = 0$ to $s = 2\pi R$. Because k, z, R, and λ are all independent of s, I can move everything out of the integral except ds:

$$E_z = \int dE_{sz} = k\frac{z\lambda}{[z^2 + R^2]^{3/2}}\int_0^{2\pi R} ds = k\frac{z\lambda(2\pi R)}{[z^2 + R^2]^{3/2}}.$$

Because $\lambda = q/2\pi R$, the term $\lambda(2\pi R)$ is equal to the charge q on the ring, so I get for the z component of the electric field along the axis perpendicular to the plane of the ring and passing through the ring center

$$E_x = 0;\ E_y = 0;\ E_z = k\frac{qz}{[z^2 + R^2]^{3/2}}.\ ✔$$

❹ **EVALUATE RESULT** Very far from the ring my expression for E_z should become the same as that for a charged particle. Indeed, when $z \gg R$, I can ignore the R^2 term in my result, and so

$$E_z \approx k\frac{qz}{[z^2]^{3/2}} = k\frac{qz}{z^3} = k\frac{q}{z^2},$$

as I expect.

At the center of the ring, $z = 0$ and so my expression yields $E_z = 0$. This result is reasonable because at the center of the ring the electric forces exerted by segments on opposite sides of the ring on a charged test particle add to zero. When the vector sum of these forces is zero, the electric field must be zero also.

Example 23.6 Electric field created by a uniformly charged disk

A thin disk of radius R carries a uniformly distributed charge. The surface charge density on the disk is σ. What is the electric field at a point P along the perpendicular axis through the disk center?

❶ **GETTING STARTED** I begin with a sketch, placing the disk in the xy plane, with the disk center at the origin. I let point P lie on the positive z axis (**Figure 23.29**).

❷ **DEVISE PLAN** Because of the circular symmetry of the disk, I divide it into a large number of ring-shaped segments, each of radius r and width dr. The charge on each ring is the product of the ring surface area (circumference times width) and the surface charge density: $dq_s = (2\pi r)dr\,\sigma$. The contribution of each

Figure 23.29

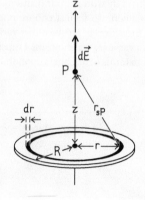

ring to the electric field at P is given by the expression for E_z I obtained for a uniformly charged thin ring in Example 23.5 with $2\pi r \sigma\, dr$ substituted for q and r substituted for R. So all I need to do is integrate over the entire disk.

3 EXECUTE PLAN Substituting $q = 2\pi r \sigma\, dr$ and $R = r$ into the Example 23.5 expression for E_z and integrating the result over the disk from $r = 0$ to $r = R$, I have

$$E_z = \int dE_z = k \int_0^R \frac{2\pi r \sigma z}{(z^2 + r^2)^{3/2}} dr.$$

Because σ and z are independent of r, I can move them out of the integral:

$$E_x = 0;\ E_y = 0;\ E_z = k\pi\sigma z \int_0^R \frac{2r\, dr}{(z^2 + r^2)^{3/2}}$$

$$= k\pi\sigma z \int_0^R \frac{d(r^2)}{(z^2 + r^2)^{3/2}} = k\pi\sigma z \frac{-2}{(z^2 + r^2)^{1/2}}\bigg|_0^R$$

$$= 2k\pi\sigma z \left[\frac{1}{(z^2)^{1/2}} - \frac{1}{(z^2 + R^2)^{1/2}} \right]. \checkmark \quad (1)$$

4 EVALUATE RESULT Let me evaluate my result for $z \gg R$, where, to an observer at P, the disk looks like a particle. For positive z, I can write the E_z in Eq. 1 as

$$E_z = 2k\pi\sigma \left[1 - \frac{z}{(z^2 + R^2)^{1/2}} \right]. \quad (2)$$

From the binomial series expansion (see Appendix B), I get in the case that $z \gg R$,

$$\frac{z}{(z^2 + R^2)^{1/2}} = \left(1 + \frac{R^2}{z^2} \right)^{-1/2} \approx 1 - \tfrac{1}{2}\frac{R^2}{z^2}.$$

Substituting this result into Eq. 2 and writing $q = \sigma(\pi R^2)$ for the charge on the disk, I get

$$E_z = 2k\pi\sigma \left(\tfrac{1}{2}\frac{R^2}{z^2} \right) = k\frac{\sigma\pi R^2}{z^2} = k\frac{q}{z^2},$$

which is the result for a charged particle, as I expect.

I can also evaluate my result for $z \approx 0$, where, to an observer at that location, the disk looks like it has an infinite radius (that is to say, it looks like an infinite flat sheet). In that case, the second term inside the brackets in Eq. 2 vanishes and $E_z = 2k\pi\sigma$. This tells me that the electric field of an infinite flat charged sheet is independent of z and constant throughout space, as I have sketched in **Figure 23.30**. (In other words, it is uniform.) While this lack of dependence on z is somewhat counterintuitive, it agrees with what I concluded earlier: The electric field that emanates from a charged particle "spreads out" in three dimensions and its amplitude is inversely proportional to the square of the distance from the particle, whereas the electric field that emanates from an infinitely long charged rod spreads out in just two dimensions and its amplitude is inversely proportional to the distance from the rod. As my sketch shows, the electric field that emanates from an infinite plane can't spread out at all (if the plane is truly infinite), and therefore its amplitude is independent of distance.

Figure 23.30

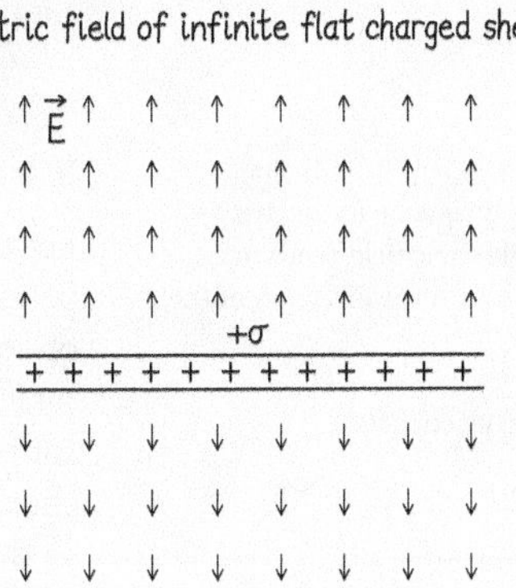

electric field of infinite flat charged sheet

23.14 (a) Describe the electric field between two infinitely large parallel charged sheets if the charge density of one sheet is $+\sigma$ and that of the other is $-\sigma$. (b) Describe the electric field outside the sheets.

Example 23.7 Electric field created by a uniformly charged sphere

A solid sphere of radius R carries a fixed, uniformly distributed charge q. Exploiting the analogy between Newton's law of gravity and Coulomb's law, use the result obtained in Section 13.8 to obtain an expression for the magnitude of the electric field created by the sphere at a point P outside the sphere.

1 GETTING STARTED To determine the electric field magnitude at point P, I need to determine the magnitude of the electric force exerted by the sphere on a test particle carrying a charge q_t at P and then divide that force magnitude by q_t.

2 DEVISE PLAN I can follow the same procedure as in Section 13.8 to calculate the gravitational force of a spherical object: I first divide the sphere into a series of thin concentric shells that resemble the layers in an onion and then divide each shell into a series of vertical rings (**Figure 23.31**). I then calculate the contribution of each ring to the electric field at P and integrate first over each shell and then over the sphere. The expression I get for F^E_{sphere} must be of the same form as that for the gravitational sphere, F^G_{sphere} (Eq. 13.37), because the gravitational force and the electric force are both inversely proportional to the square of the distance between the interacting particles. So all I need to do is replace G in Eq. 13.37 by k, M_{sphere} by q, and m by q_t to obtain the magnitude of the electric force exerted by the sphere on the test particle. To obtain an expression for the electric field, I then divide the result by q_t.

(Continued)

Figure 23.31

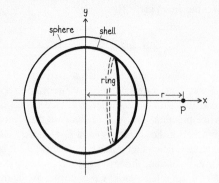

③ EXECUTE PLAN In analogy to Eq. 13.37, I write

$$F_{sp}^G = G \frac{m M_{sphere}}{r^2} \rightarrow F_{st}^E = k \frac{q_t q}{r^2},$$

where r is the distance from the center of the sphere to P. To obtain the magnitude of the electric field, I divide the electric force by q_t:

$$E_{sphere} = k \frac{q}{r^2}. \checkmark$$

④ EVALUATE RESULT Comparing my result with Eq. 23.3 (in scalar form), $E_s = k q_s / r_{st}^2$, shows that outside a uniformly charged solid sphere the magnitude of the electric field is the same as that surrounding a particle carrying the same charge and located at the center of the sphere. The result is independent of the radius R of the sphere and similar to the result obtained in Section 13.8: A solid sphere exerts a gravitational force as if the entire mass of the sphere were concentrated at the center. It makes sense that I obtain a similar result for the charged sphere because the gravitational force and the electric force are both proportional to $1/r^2$.

23.15 How does the electric field inside a uniformly charged sphere vary with distance from the sphere center? [Hint: What is the electric field inside a hollow uniformly charged sphere?]

23.8 Dipoles in electric fields

Let us end this chapter by considering the forces exerted by electric fields on dipoles. **Figure 23.32a** shows a dipole consisting of two particles that carry charges of equal magnitude but opposite sign connected by a rod of length d; the dipole makes an angle θ with a uniform electric field $\vec{E}$ created by some unseen distant source. As we saw in Section 23.4, the forces exerted by the electric field on the charged ends of the dipole are equal in magnitude but opposite in direction, and so the vector sum of the forces exerted on the dipole is zero. Consequently the acceleration of the center of mass of the dipole is zero. Because the forces are exerted on opposite ends of the dipole, however, they create torques that cause the dipole to rotate counterclockwise about its center of mass. Figure 23.32b shows that the force exerted on the positive end causes a counterclockwise torque of magnitude

$$\tau_+ = r_\perp F_+^E = (\tfrac{1}{2} d \sin \theta)(q_p E), \quad (23.19)$$

where $r_\perp$ is the lever arm of the force. The force exerted on the negative end causes an identical torque because the lever arm and the magnitude of the force are the same. The electric field thus causes a torque on the dipole equal to

$$\Sigma \tau_\vartheta = 2 (\tfrac{1}{2} d \sin \theta)(q_p E) = (q_p d) E \sin \theta \equiv p E \sin \theta. \quad (23.20)$$

This can be written in vectorial form as

$$\Sigma \vec{\tau} = \vec{p} \times \vec{E}, \quad (23.21)$$

where $\vec{p}$ is the dipole moment, which by definition points from the negative end of the dipole to the positive end and whose magnitude is given by Eq. 23.9. According to the right-hand rule (see Section 12.4), the vector product $\vec{p} \times \vec{E}$ in Eq. 23.21 gives a torque that points out of the plane of the drawing in

Figure 23.32 The torque on an electric dipole caused by an electric field tends to align the dipole moment $\vec{p}$ with the direction of the electric field.

(a) Electric dipole in electric field

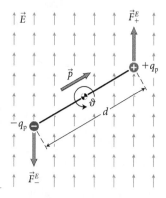

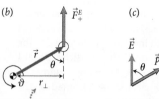

(b) Lever arm $\vec{r}_\perp$ depends on angle θ of dipole with respect to electric field.

(c) Torque on dipole is vector product $\Sigma \vec{\tau} = \vec{p} \times \vec{E}$.

Figure 23.33 A dipole interacts with a charged particle.

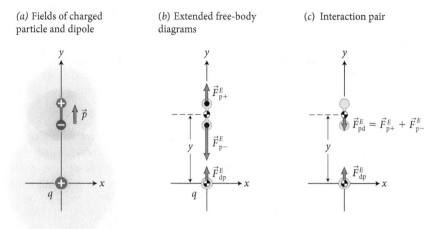

(a) Fields of charged particle and dipole

(b) Extended free-body diagrams

(c) Interaction pair

Figure 23.32*c*. As we saw in Section 12.8, such a torque indeed causes a counterclockwise rotation. The torque on the dipole is maximum when the dipole moment is perpendicular to the electric field and zero when it is parallel or antiparallel to the electric field.

 23.16 Is Eq. 23.20 valid if the center of mass is not in the middle of the dipole?

As we saw in Section 23.4, the vector sum of the forces exerted on dipoles in nonuniform electric fields is not zero. Consider, for example, the situation illustrated in **Figure 23.33***a*. A dipole with its dipole moment $\vec{p}$ aligned along the y axis is placed in the nonuniform electric field generated by a particle carrying a charge q and located at the origin. Because the distance between the negative end of the dipole and the particle is smaller than the distance between the positive end and the particle, the magnitude of the attractive force $\vec{F}_{p-}^{E}$ on the negative end is greater than the repulsive force $\vec{F}_{p+}^{E}$ on the positive end. Consequently the vector sum of the forces exerted by the nonuniform field on the dipole is nonzero, and the dipole is attracted to the particle. How does this attraction vary with the position y of the dipole?

To answer this question, we can write an expression for the vector sum of the forces $\sum \vec{F}_{d}^{E}$ exerted on the two ends of the dipole and examine how this sum varies with y. Alternatively, we can calculate the force $\vec{F}_{dp}^{E}$ exerted by the dipole on the particle, using our results from Section 23.6. This force and the vector sum of the forces exerted by the particle on the dipole form an interaction pair, and so their magnitudes are the same. Equation 23.13 tells us that, along the dipole axis, the magnitude of the electric field created by the dipole is $2k(p/y^3)$, and so the magnitude of the force exerted by the dipole on the particle is

$$F_{dp}^{E} = qE_{d} = 2k\frac{pq}{y^3}. \tag{23.22}$$

The magnitude of the force exerted by the particle on the dipole, being equal in magnitude to F_{dp}^{E}, is thus

$$F_{pd}^{E} = F_{p-}^{E} - F_{p+}^{E} = 2k\frac{pq}{y^3}. \tag{23.23}$$

Like the electric field of a dipole, the forces between a charged object carrying charge q and a dipole is inversely proportional to the cube of the distance between them.

✋ **23.17** How does doubling each of the following quantities affect the force between a dipole and a particle placed near the dipole and carrying charge q? (a) the charge q, (b) the dipole separation d of the dipole, (c) the dipole charge q_p, (d) the distance between the dipole and the charged particle

As we saw in Chapter 22, electrically neutral objects interact with a charged object because they become polarized in the presence of the charged object. Consider an isolated neutral atom. The centers of the atom's positive and negative charge distributions coincide, and therefore the atom's dipole moment is zero: $d = 0$ and so $\vec{p} = \vec{0}$ (**Figure 23.34a**). The presence of an external electric field—that is, an electric field created by some other charged object—causes a separation between the positive and negative charge centers and so induces a dipole moment (Figure 23.34b). To understand the interaction between charged objects and neutral ones, we must therefore study the interaction between a charged particle and what is called an **induced dipole**. The first question to ask is: How does the magnitude of the induced dipole moment depend on the presence of a charged particle?

When a neutral atom is placed in an electric field $\vec{E}$, it is found that, as long as the electric forces exerted by that field on the charged particles in the atom are not too large, the induced dipole separation d_{ind} in the atom obeys Hooke's law. In other words, the induced dipole separation is proportional to the magnitude of the applied electric force, $F_d^E = cd_{ind}$, with c being the "spring constant" of the atom. We can rewrite this as $d_{ind} = (1/c)F_d^E$, and because d_{ind} is proportional to the magnitude of the induced dipole moment p_{ind} and the magnitude of the force exerted on the dipole is proportional to the magnitude E of the electric field at the position of the dipole, the **induced dipole moment** is proportional to the field at the position of the dipole:

$$\vec{p}_{ind} = \alpha\vec{E} \quad (\vec{E} \text{ not too large}), \tag{23.24}$$

where α, the **polarizability** of the atom, is a constant that expresses how easily the charge distributions in the atom are displaced from each other. The SI unit of polarizability is $C^2 \cdot m/N$.

Figure 23.34 A charged particle induces a dipole in an electrically neutral atom.

(a) Neutral atom

(b) Charged particle induces dipole in atom

(c) Charged particle and dipole interact

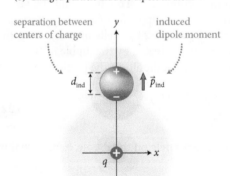

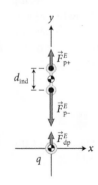

23.18 Given that the induced dipole moment $\vec{p}_{ind}$ points from the negative to the positive end of an induced dipole and the electric field $\vec{E}$ displaces the positive charge center in the direction of the electric field and the negative charge center in the opposite direction, do you expect the polarizability α to be positive or negative?

The electric field of a charged particle is given by Eq. 23.4, so the magnitude of the induced dipole moment is proportional to the inverse square of the distance between the particle and the dipole:

$$p_{ind} = \alpha E = \alpha k \frac{q}{y^2}. \tag{23.25}$$

In contrast, the dipole moment of a permanent dipole is constant.

We can now substitute the induced-dipole result of Eq. 23.25 into Eq. 23.23 to determine the force exerted by a charged particle on an induced dipole:

$$F_{pd}^E = 2k\frac{p_{ind}q}{y^3} = \alpha \frac{2k^2q^2}{y^5}. \tag{23.26}$$

This result shows that the interaction between a charged particle and a polarized object depends much more strongly on the distance between them $(1/y^5)$ than does the interaction between two charged objects $(1/y^2)$. You may have noticed this in Chapter 22 when comparing the attraction between two charged strips of tape with the attraction between a charged strip and a neutral object.* As the neutral object approaches the charged strip, the force varies so fast with distance that it is often difficult to prevent the tape from sticking to the neutral object.

23.19 (a) How does doubling the charge q_A carried by an object A affect the force exerted by A on another charged particle? (b) How does doubling q_A affect the force exerted by A on an induced dipole? (c) Explain why your answers to parts a and b are the same or different. (d) Can the force exerted by a charged particle cause a torque on an induced dipole?

---

*Try it! Pull two strips of transparent tape out of a dispenser, suspend one from the edge of a table and then move the other slowly toward it. Notice how the interaction between the strips varies relatively smoothly as a function of separation. Next, move your hand slowly toward the suspended strip and note how the force increases rapidly.

Chapter Glossary

SI units of physical quantities are given in parentheses.

Charge density, linear λ (C/m), surface σ (C/m^2), or volume ρ (C/m^3): A scalar that is a measure of the amount of charge per unit of length, area, or volume on a one-, two-, or three-dimensional object, respectively.

Dipole (electric) A neutral charge configuration in which the center of positive charge is separated from the center of negative charge by a small distance. Dipoles can be *permanent*, or they can be *induced* by an external electric field.

Dipole moment (electric) $\vec{p}$ (C·m) A vector defined as the product of the *dipole charge* q_p (the positive charge of the dipole) and the vector $\vec{r}_p$ that points from the center of negative charge to the center of positive charge:

$$\vec{p} \equiv q_p \vec{r}_p. \tag{23.9}$$

Electric field $\vec{E}$ (N/C) A vector equal to the electric force exerted on a charged test particle divided by the charge on the test particle:

$$\vec{E} \equiv \frac{\vec{F}_t^E}{q_t}. \tag{23.1}$$

Induced dipole A separation of the positive and negative charge centers in an electrically neutral object caused by an external electric field.

Induced dipole moment $\vec{p}_{ind}$ (C·m) A dipole moment induced by an external electric field in an electrically neutral object. For small electric fields, the induced dipole moment in an atom is proportional to the applied electric field:

$$\vec{p}_{ind} = \alpha \vec{E}, \tag{23.24}$$

where α is the *polarizability* of the atom.

Interaction field or **field** A physical quantity surrounding objects that mediates an interaction. Objects that have mass are surrounded by a *gravitational field*; those that carry an electrical charge are surrounded by an *electric field*. Both are *vector fields* specified by a direction and a magnitude at each position in space.

Polarizability α (C^2·m/N) A scalar measure of the amount of charge separation that occurs in an atom or molecule in the presence of an externally applied electric field.

Superposition of electric fields The electric field of a collection of charged particles is equal to the vector sum of the electric fields created by the individual charged particles:

$$\vec{E} = \vec{E}_1 + \vec{E}_2 + \cdots. \tag{23.5}$$

Test particle An idealized particle whose physical properties (mass or charge) are so small that the particle does not perturb the particles or objects generating the field we are measuring.

Vector field diagram A diagram that represents a vector field, obtained by plotting field vectors at a series of locations.

24 Gauss's Law

CONCEPTS

QUANTITATIVE TOOLS

In principle, Coulomb's law allows us to calculate the electric field produced by any discrete or continuous distribution of charged objects. In practice, however, the calculation is often so complicated that the sums or integrals that arise might require numerical evaluation on a computer. For this reason, it pays to search for additional methods to determine the electric field produced by a charge distribution. In this chapter we develop a relationship between an electric field and its source, known as *Gauss's law*, that can be used to determine the electric fields due to charge distributions that exhibit certain simple symmetries. These symmetries appear in many common applications, which makes Gauss's law an important tool in calculating electric fields. As we shall see in Chapter 30, Gauss's law is one of the fundamental equations of *electromagnetism*—the theory that describes electromagnetic interactions and electromagnetic waves.

24.1 Electric field lines

In Chapter 23 we used vector field diagrams to visualize electric fields. Another way to visualize electric fields, which will help us reach some new insights, is to draw **electric field lines.** These lines are drawn so that at any location the electric field $\vec{E}$ is tangent to them. Because the electric field is a vector, we assign to field lines a direction that corresponds with the direction of the electric field.

To draw an electric field line, imagine placing a test particle carrying a positive charge q_t somewhere near a charge distribution. Then move the test particle a small distance in the direction of the electric force exerted on it. (Remember from Chapter 23 that the electric field points in the same direction as the electric force exerted on a positively charged test particle.) Repeat the procedure to trace out a line (**Figure 24.1**). We label the field lines with the symbol E to remind us that they represent an electric field.

✋ **24.1** Draw several field lines representing the electric field of an isolated positively charged particle. Repeat for a negatively charged particle.

As Checkpoint 24.1 illustrates, the field line diagrams for a positive and for a negative isolated charged particle are similar, even though the electric fields point in opposite directions. They point radially outward from a positively charged particle and point radially inward toward a negatively charged particle. This direction means that electric field lines always start from a positively charged object and always end on a negatively charged object, never the other way around.

Because an electric field is present everywhere around a charged object, a field line passes through every location in space. In practice, we draw only a finite number of field lines to represent the entire field. **Figure 24.2**, for example, shows the pattern of field lines created by a pair of oppositely charged particles (that is, a dipole). Sixteen field lines emanate from the positively charged particle on the

Figure 24.1 Using a positively charged test particle to trace out an electric field line.

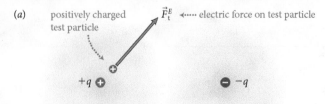

(a)

positively charged test particle $\vec{F}_t^E$ ⬅····· electric force on test particle

$+q$ ⊕   ⊖ $-q$

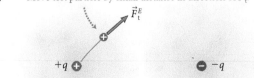

(b)   Move test particle by small distance in direction of $\vec{F}_t^E$

$\vec{F}_t^E$

$+q$ ⊕   ⊖ $-q$

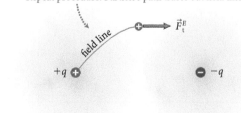

(c)   Repeat procedure. Particle's path traces out field line.

field line   ⊕ $\vec{F}_t^E$

$+q$ ⊕   ⊖ $-q$

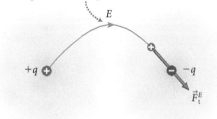

(d)   At each point, field line points in direction of electric force.

E

$+q$ ⊕   ⊖ $-q$

$\vec{F}_t^E$

left, and 16 field lines terminate on the negatively charged particle on the right. Notice the correspondence between this pattern and the corresponding vector field diagram in **Figure 24.3a** and the pattern created by the fibers in Figure 24.3b.

The number of field lines that emanate from a positively charged object is arbitrary; we could have chosen some number other than 16 for Figure 24.2. However, in a given field line diagram, the number of field lines is always proportional to the magnitude of the charge carried by the

Figure 24.2 Electric field line diagram for an electric dipole.

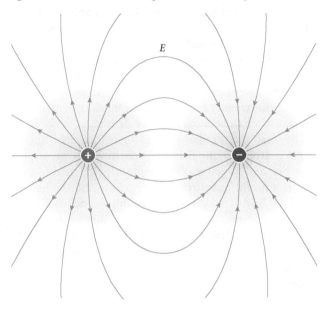

object. If, for example, 16 field lines emanate from an object that carries a charge $+q$, then 32 lines emanate from an object that carries a charge $+2q$ and eight lines terminate on an object that carries a charge $-q/2$.

> **The number of field lines that emanate from a positively charged object or terminate on a negatively charged object is proportional to the charge carried by the object.**

Exercise 24.1 Field lines of infinite charged plate

Draw a field line diagram for an infinite plate that carries a uniform positive charge distribution.

SOLUTION I know from Chapter 23 that the electric field produced by a charged plate of infinite area is always perpendicular

Figure 24.3 Two representations of the electric field of an electric dipole.

(*a*) Vector field diagram of an electric dipole

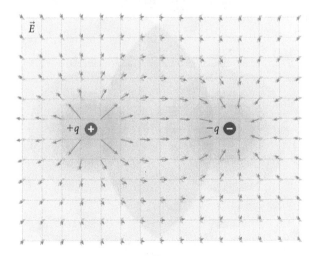

to the plate (see Figure 23.30). Thus, the field lines must be straight lines perpendicular to the plate. Because the plate is positively charged, I draw the field lines perpendicular to and away from the plate on either side (**Figure 24.4**). ✔

Figure 24.4

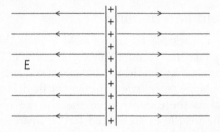

Like vector field diagrams, field line diagrams provide an incomplete view of the electric field and are awkward to draw for all but the simplest charge distributions. Both types of diagram are limited by the two-dimensional nature of illustrations. In particular, you should keep in mind that field lines emanate in three dimensions (**Figure 24.5**), not just in the plane of the drawing.

Figure 24.5 Although we generally use two-dimensional representations of field line diagrams, field lines emanate in three dimensions.

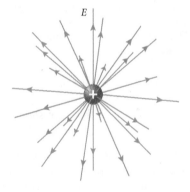

(*b*) Pattern created by electric dipole in a suspension of plastic fibers that align with the electric field

CONCEPTS

24.2 (*a*) Is it possible for two electric field lines to cross? (Hint: What is the direction of the electric field at the point of intersection?) (*b*) Can two electric field lines touch?

24.2 Field line density

The most remarkable feature of field lines is this: Even though we take into account only the *direction* of the electric field when drawing field lines, they also contain information about the *magnitude* of the electric field. Figure 24.5 shows that as the distance from the charged object increases, the field lines are spaced farther apart from one another. To see whether there is a quantitative correspondence between field line spacing and electric field magnitude, complete the next checkpoint.

24.3 Imagine a hollow sphere enclosing the charged object in Figure 24.5, centered on the object. (*a*) Given that 26 field lines emanate from the charged object, how many field lines cross the surface of the hollow sphere? (*b*) If the radius of the hollow sphere is *R*, what is the number of field line crossings per unit surface area? (*c*) Now consider a second sphere with radius 2*R*, also centered on the charged object. How many field lines cross the surface of this second sphere? (*d*) How does the number of field line crossings per unit area on the second sphere compare with that on the first sphere? (*e*) How does the electric field at a location on the second sphere compare with the field at a location on the first sphere?

From Checkpoint 24.3, we see that the electric field and the number of field line crossings per unit area both decrease as $1/r^2$. To express this correspondence quantitatively, we define a new quantity, the **field line density:**

> The field line density at a given position is the number of field lines per unit area that cross a surface perpendicular to the field lines at that position.

Figure 24.6 illustrates why the surface through which the field lines pass must be perpendicular to the field

lines. The field represented by the field lines in the figure is uniform—its magnitude and direction are the same everywhere. As you can see in the figure, the number of field lines that cross the surface depends on the orientation of the surface. The number of field lines that cross the surface is maximum when the surface is perpendicular to the field lines and decreases for any other orientation. We shall see later in this chapter how to account for the orientation of a surface when calculating field line density.

Because the number of field lines in a field line diagram is arbitrary, the field line density is also an arbitrary number, and so you may be wondering why field line density is a useful quantity. As you will see shortly, however, the field line density allows us to draw conclusions about electric field magnitudes. The only condition we make is that, in a given field line diagram, the number of field lines emanating from or terminating on charged objects is proportional to the magnitude of the charge carried by these objects.

24.4 (*a*) In Figure 24.6, for what orientation is the number of field lines that cross the surface a minimum? (*b*) How many field lines cross a plane surface of area 0.5 m² placed perpendicular to the field lines in Figure 24.6? (*c*) Using your answer to part *b*, what is the number of field line crossings *per unit area* through the 0.5-m² surface? (*d*) How does this compare to the number of field line crossings per unit area for the 1-m² surface in Figure 24.6*a*?

For the spherical surfaces of Checkpoint 24.3, the field lines are all perpendicular to the surface because the field lines are radial. The number of field line crossings you calculated per unit area *is* the field line density. These results lead us to conclude:

> At every position in a field line diagram, the magnitude of the electric field is proportional to the field line density at that position.

The box "Properties of electric field lines" on page 643 summarizes the properties of electric field lines.

Figure 24.6 The number of field lines that cross a given surface depends on the orientation of the surface relative to the field lines.

(*a*)

Plane perpendicular to field lines intersects maximum number of field lines.

(*b*)

Same plane at any other orientation intersects fewer field lines.

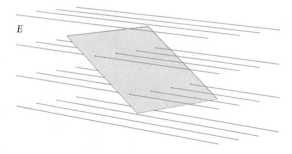

Properties of electric field lines

When working with electric field lines, keep the following points in mind:

1. Field lines emanate from positively charged objects and terminate on negatively charged objects.
2. At every position, the direction of the electric field is given by the direction of the tangent to the electric field line through that position.
3. Field lines never intersect or touch.
4. The number of field lines emanating from or terminating on a charged object is proportional to the magnitude of the charge on the object.
5. At every position, the magnitude of the electric field is proportional to the field line density.

Exercise 24.2 Field strength from field lines

Consider the field line diagram shown in **Figure 24.7**. (*a*) What are the signs of the charges on the two small spherical objects? (*b*) What are the relative magnitudes of these charges? (*c*) What is the ratio of the magnitudes of the electric fields at points P and R? (*d*) Is the electric field zero anywhere in the region shown?

Figure 24.7 Exercise 24.2.

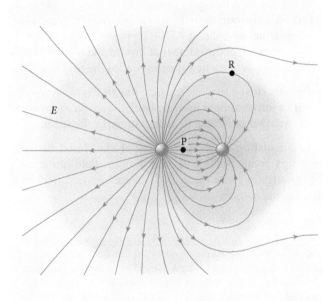

SOLUTION (*a*) Because field lines leave the left object and terminate on the right one, the left object carries a positive charge and the right object carries a negative charge. ✔

(*b*) From Figure 24.7 I see that about twice as many field lines leave the left object as end on the right one. Thus, the charge on the left object is about twice that on the right object.* ✔

(*c*) The magnitude of the electric field at each position is proportional to the field line density at that position. The field line density is equal to the number of field lines per unit length,

which is proportional to the inverse of the distance between adjacent field lines. Measuring with a ruler, I see that the distances between adjacent field lines at points P and R are $d_P \approx 1$ mm and $d_R \approx 6$ mm, so

$$\frac{E_P}{E_R} = \frac{d_R}{d_P} \approx \frac{6 \text{ mm}}{1 \text{ mm}} = 6. \checkmark$$

(*d*) The absence of field lines on the right suggests that the electric field is small (or even zero). Indeed, if a test particle carrying a positive charge is placed in that region, it is subject to a repulsive force exerted by the positively charged object on the left and an attractive force exerted by the negatively charged object on the right. If the test particle is to the right of the particles and $\sqrt{2}$ as far from the positively charged particle as it is from the negatively charged particle, the vector sum of the two forces is zero and so the electric field at that position is zero. ✔

Note that in Exercise 24.2, half of the field lines leave the area of interest. These field lines either eventually terminate on a negatively charged object (not shown) or continue out to "infinity."

✋ **24.5** Imagine moving the hollow sphere of radius R of Checkpoint 24.3*a* sideways so that the charged object is no longer at the center of the sphere (but still within it). (*a*) How does the number of field line crossings through the surface of the sphere change as it is moved? (*b*) How does the average number of field line crossings per unit surface area of the sphere change? (*c*) Does the electric field at a fixed position on the surface of the sphere change or remain the same as the sphere is moved? (*d*) Are your answers to parts *b* and *c* in contradiction, given the relationship between the electric field magnitude and field line crossings per unit area?

24.3 Closed surfaces

Checkpoint 24.5 leads us to another result that will be important in deriving Gauss's law: Whenever a charged particle is placed inside a hollow spherical surface, the number of field lines that pierce the surface is the same *regardless of where inside the surface the particle is placed.* This is true simply because so long as the charged particle is inside the surface, all the field lines emanating from the particle must go through the spherical surface. In fact, we don't even need to use a spherical surface—a

---

*If you answered that the magnitude of the charge on the left is four times that on the right because in three dimensions there must be four times as many field lines radiating outward from the left object, don't worry—we've hit on one of the shortcomings of field line representations. In general we shall go by the number of dimensions represented in the drawing (in this case, two).

Figure 24.8 Any surface that encloses a positively charged particle is pierced by all the field lines that emanate from that particle, regardless of the shape of the surface and the position of the particle within the surface.

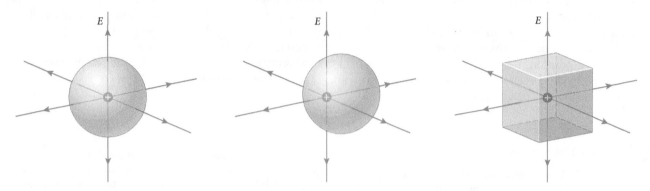

cube-shaped surface or any other surface enclosing the charged particle will do (**Figure 24.8**). In each case, the number of field lines that pierce the surface is equal to the number of field lines that emanate from or terminate on the charged particle enclosed by the surface.

✋ **24.6** Suppose eight field lines emanate from an object carrying a charge $+q$. How many field lines pierce the surface of a hollow sphere if the sphere contains (*a*) a single object carrying a charge $+2q$ and (*b*) two separate objects, each carrying a charge $+q$? (*c*) If the sphere is pierced by 20 field lines, what can you deduce about the combined charge on objects inside the sphere?

A surface that completely encloses a volume is called a **closed surface**. Checkpoint 24.6 suggests that a direct relationship exists between the number of field lines that cross a closed surface and the **enclosed charge**—the sum of all charge enclosed by that surface. However, what happens if a field line reenters the closed surface, as illustrated in **Figure 24.9a**? Field line 4 now crosses the closed surface *three* times, so the number of field line crossings is not six but eight. If you look closely at the figure, however, you will

discover that not all the crossings are the same. For seven of the crossings the field line goes outward (from the inside of the closed surface to the outside), while for the eighth crossing the field line goes inward. If we assign a value of $+1$ to each outward crossing and a value of -1 to each inward crossing, we obtain $(+7) + (-1) = 6$.

To keep track of the number of inward and outward field crossings, we define a new quantity called the *field line flux*:

> For any closed surface, the *field line flux* is the number of outward field lines crossing the surface minus the number of inward field lines crossing the surface.

In calculating the field line flux for any closed surface, we assign a value of $+1$ to each outward field line crossing the surface and a value of -1 to each inward field line crossing the surface.

✋ **24.7** (*a*) If more than one field line reenters the donut in Figure 24.9a, what happens to the field line flux? (*b*) Are there any closed surfaces enclosing a charged particle through which the field line flux is different from that through a simple sphere around that particle?

Figure 24.9 The number of field lines exiting a closed surface minus the number entering it is always equal to the number of field lines generated inside the surface.

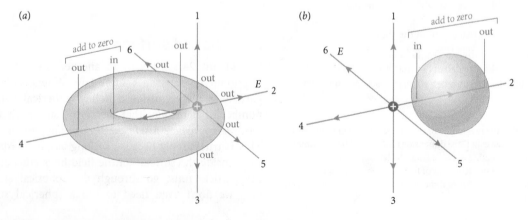

The field line flux through any closed surface is always equal to the number of field lines that originate from within that surface minus the number of field lines that terminate on charged objects within that surface. So far, we have drawn an arbitrary number of field lines, but we can make our statement more precise:

The *field line flux* through a closed surface is equal to the charge enclosed by the surface multiplied by the number of field lines per unit charge.

What about charged objects outside the closed surface? To see what effect such charged objects have, complete the following checkpoint.

24.8 (*a*) What is the field line flux through the closed spherical surface in Figure 24.9*b* due to a charged particle outside the sphere? (*b*) Does your answer to part *a* change if we move the particle around (but keep it outside the volume enclosed by the surface)?

Checkpoint 24.8 demonstrates a very important point:

The field line flux through a closed surface due to charged objects outside the volume enclosed by that surface is always zero.

This means that if we know the field line flux through a closed surface, then we can determine the charge enclosed by that surface, regardless of the distribution of charge outside the surface. This statement is a form of Gauss's law, which we shall describe mathematically in Section 24.7.

Exercise 24.3 Flux of an electric dipole

Consider the three-dimensional dipole field line diagram shown in Figure 24.10. Six field lines emanate from the positively charged end, and six terminate on the negatively charged end. (*a*) What is the field line flux through the surface of the cube that encloses the positively charged end shown in the figure? (*b*) What is the field line flux through the surface of a similar cube that encloses the negatively charged end?

Figure 24.10 Example 24.3.

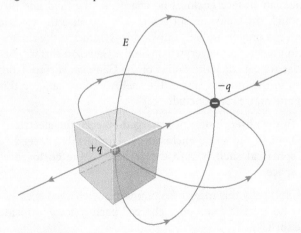

SOLUTION (*a*) Six field lines emanate from the positively charged particle. Each line crosses the surface of the cube in the outward direction and thus contributes a value of +1 to the field line flux. The field line flux is +6. ✔

(*b*) Six field lines terminate on the negatively charged particle, so there are again six field line crossings. However, these field lines are directed inward and so the field line flux is −6. ✔

24.9 What is the field line flux through the surface of a rectangular box that encloses *both* ends of an electric dipole?

The relationship between the field line flux through a closed surface and the enclosed charge is important because it can help us determine one from a knowledge of the other. For example, in the next section we shall use this relationship to derive two important theorems about isolated conducting objects.

24.10 Consider the two-dimensional field line diagram in Figure 24.11, part of which is hidden from view. (*a*) If the object in the top left carries a charge of +1 C, what is the charge enclosed in the region that is hidden? (*b*) What is the field line flux through a surface that encloses the entire area represented by the diagram?

Figure 24.11 Checkpoint 24.10: What is the charge inside the dashed region?

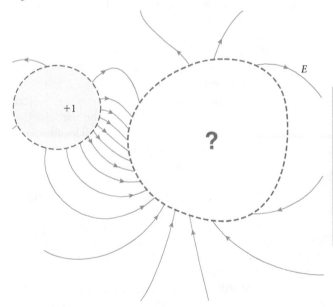

24.4 Symmetry and Gaussian surfaces

The relationship between field line flux and enclosed charge allows us to reach several important conclusions about charged objects and their electric fields without having to do

Figure 24.12 Using spherical Gaussian surfaces to examine the electric fields of a charged particle and a uniformly charged spherical shell. The electric fields, Gaussian surfaces, and charged shell are spherical and are shown here in cross section.

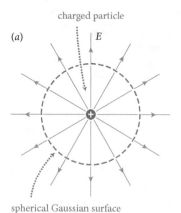

charged particle

(a) E

spherical Gaussian surface

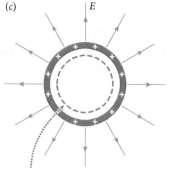

hollow shell with same charge as particle

(b) E

Gaussian surface outside shell: electric field is same as for particle.

(c) E

Gaussian surface inside shell: encloses no charge, so electric field must be zero.

any calculations. To apply this relationship in a given situation, we first need to select a closed surface. This surface need not correspond to a real object—any surface, real or imagined, will do. We'll refer to these closed surfaces as **Gaussian surfaces.** The choice of surface is dictated by the symmetry of the situation at hand. As a rule of thumb, we choose a surface such that the electric field is the same (and possibly zero) everywhere along as many regions of the surface as possible, because such a choice makes it easy to determine the field line flux through the surface.

Consider, for example, the charged particle shown in **Figure 24.12a.** As we have seen, the field lines for the particle radiate outward from it. (The figure shows only a two-dimensional cross section of the three-dimensional situation.) The field is symmetrical in all three dimensions— it has the same magnitude at the same distance from the center in any direction. Therefore, if we draw a spherical Gaussian surface that is concentric with the particle, the magnitude of the electric field is the same at all locations on the sphere. In other words, the field line density is the same all over the surface of the sphere. As we have seen in the preceding section, the field line flux through the Gaussian surface is proportional to the charge enclosed by the sphere.

Now suppose we replace the charged particle by a spherical shell that carries the same charge as the particle and still fits within our Gaussian surface (Figure 24.12b). If the charge is uniformly distributed over the shell, then the electric field should still be the same in all directions. Like the charged particle, the charged shell has *spherical symmetry* (see the box "Symmetry and Gauss's law" on page 647): Reorienting the spherical shell by rotating it over an angle about any axis does not change the charge configuration and so should not change the electric field at a given location. This means that the field lines should again be straight lines radiating uniformly outward. Also, the field line flux

through the Gaussian surface should still be the same because the surface encloses the same amount of charge. The only way that the field line fluxes through the Gaussian surfaces in Figures 24.12a and b can be uniform *and* equal in magnitude is if the electric fields are the same at every position on the spherical Gaussian surface. Because this argument holds for a spherical Gaussian surface of any radius, we can conclude:

> **The electric field outside a uniformly charged spherical shell is the same as the electric field due to a particle that carries an equal charge located at the center of the shell.**

This means that a uniformly charged shell exerts a electric force on a charged particle outside the shell as if all the shell's charge were concentrated at the center of the shell. Because a sphere may be viewed as a collection of shells, we can extend this statement to uniformly charged spheres.

Let us now turn our attention to the electric field in the space enclosed by the shell. We draw a spherical Gaussian surface that fits within the shell (Figure 24.12c). This Gaussian surface encloses no charge, so the field line flux through the Gaussian surface is zero. Because the electric field can only be radially outward by symmetry, the electric field must be zero everywhere on the Gaussian surface. Because we can vary the radius of the Gaussian surface from zero to the inner radius of the shell without changing this argument, we can conclude:

> **In the absence of other charged objects, the electric field in the space enclosed by a uniformly charged spherical shell is zero everywhere in the enclosed space.**

Physically this means that a uniformly charged shell exerts no electric force on a charged particle located inside the shell.

Symmetry and Gauss's law

The symmetry of an object is determined by its *symmetry operations*—manipulations that leave its appearance unchanged (see Section 1.2). A sphere, for example, looks the same if we reorient it by rotating it about any axis (**Figure 24.13a**). This type of symmetry is called **spherical symmetry.** An infinitely long, cylindrical rod does not look any different if we rotate it, reverse it, or translate it about its long axis (Figure 24.13b). The rod is said to have **cylindrical symmetry.** An infinite flat sheet has **planar symmetry:** It remains unchanged if it is rotated about an axis perpendicular to the sheet or translated along either of the two axes perpendicular to this axis (Figure 24.13c).

Many other types of symmetry may occur, but these three types play an important role in electrostatics. For charge configurations that exhibit any of these three symmetries, we can calculate the electric field due to the charge distribution directly using Gauss's law.

Because objects are never infinite, they cannot exhibit true cylindrical or planar symmetry. However, for a long straight wire or a large flat sheet we can often obtain good results by assuming they have cylindrical or planar symmetry. When we work problems, the words *long* and *large* imply that you may assume the object has infinite dimensions compared to other length scales of interest.

Figure 24.13 Three symmetries important for applications of Gauss's law.

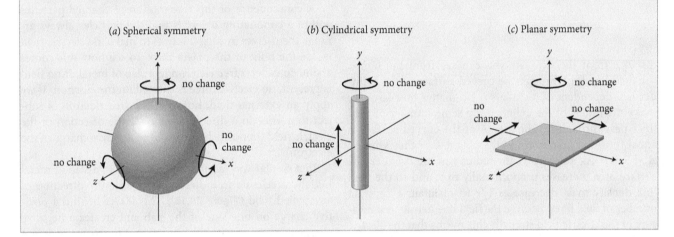

(a) Spherical symmetry (b) Cylindrical symmetry (c) Planar symmetry

24.11 There are two reasons the field line flux through a closed surface may be zero: because the field is zero everywhere or because the outward flux is balanced by an equal inward flux. Why can't the latter situation be true for the Gaussian surface in Figure 24.12c?

Particles, shells, and spheres are the only objects that exhibit spherical symmetry. **Figure 24.14** illustrates a different type of symmetry: the *cylindrical symmetry* of an infinitely long, uniformly charged straight wire.* Because of this symmetry, rotating the wire about its axis or moving it along the axis should not have any effect on the electric field at any position in space. For this to be the case, the field lines must be arranged radially along planes that are perpendicular to the wire (Figure 24.14). We can take advantage of this symmetry by drawing a Gaussian surface in the shape of a cylinder that is concentric with the wire, as shown in Figure 24.14.

Figure 24.14 The electric field of an infinite uniformly charged wire exhibits cylindrical symmetry. We can examine this field by surrounding the wire with a concentric cylindrical Gaussian surface.

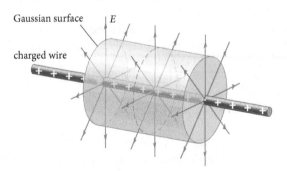

24.12 Consider a point on the curved part of the Gaussian surface in Figure 24.14. Does the magnitude of the electric field at that point increase, decrease, or stay the same if you (a) change the location of the point on the curved surface or (b) increase the radius of the Gaussian surface? (c) What is the field line flux through the left and right surfaces of the Gaussian surface?

_____
*For the wire to exhibit cylindrical symmetry, it has to be infinitely long. If the wire has finite length, you can tell when it is moved along its axis.

We can use the cylindrical Gaussian surface to determine how the electric field due to the charged wire decreases with

Figure 24.15 The electric field of a uniformly charged sheet exhibits planar symmetry. We examine its field by drawing a cylindrical Gaussian surface that straddles the sheet.

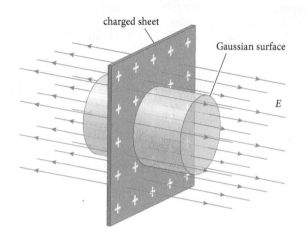

charged sheet

Gaussian surface

E

distance from the wire. As you can see from the figure, the field line flux through the curved surface of the cylinder is independent of its radius r. No matter how large we make the radius of the cylinder, the same number of field lines pass through it. The area A of the curved surface is equal to the perimeter times the height h of the cylinder: $A = 2\pi rh$. As we increase the radius r of the cylinder, the surface area increases proportionally to r, and so the field line density must decrease as $1/r$ to maintain a constant number of field lines. Because the field line density is a measure of the electric field strength, this means that the electric field due to the wire decreases as $1/r$, as we established in Example 23.4. A quantitative expression for the electric field due to a charged rod is given in Section 24.8.

Figure 24.15 shows a situation with a different symmetry: a charged sheet. If the sheet is very large and the charge is uniformly distributed along it, then the electric field lines must be perpendicular to the sheet and also uniformly distributed along it. The one-dimensional symmetry exhibited by the electric field due to the charged sheet is an example of *planar symmetry*. To take advantage of this symmetry, we draw a cylindrical Gaussian surface that straddles the sheet, as illustrated in Figure 24.15.

24.13 Consider a point on the right surface of the Gaussian surface in Figure 24.15. Does the magnitude of the electric field at that point increase, decrease, or stay the same if you (*a*) change the location of the point on the right surface, or (*b*) increase the height h of the Gaussian surface? (*c*) Is the field line flux through the right surface of the Gaussian surface positive, negative, or zero? (*d*) How does the field line flux through the right surface compare to that through the left surface? (*e*) Is the field line flux through the curved surface of the Gaussian surface positive, negative, or zero?

Because the area of the right surface and the field line flux through that surface don't change as we change the

height of the cylinder, we conclude that the electric field line density doesn't change with distance to the plane. Hence the magnitude of the electric field due to the charged sheet is the same everywhere (see also Example 23.6).

24.5 Charged conducting objects

Let us now apply the relationship between field line flux and enclosed charge to charged conducting objects. As we saw in Chapter 22, conducting materials permit the free flow of charge carriers within the bulk of the material. Conducting objects typically contain many charge carriers that are free to move, such as electrons (in a metal) or ions (in a liquid conductor). The material as a whole can still be electrically neutral; a neutral piece of metal, for example, contains as many positively charged protons as negatively charged electrons.

A consequence of this free motion of charged particles within a conducting object is that the particles always arrange themselves in such a way as to make the electric field inside the bulk of the object zero. To see how this comes about, consider a free electron in a slab of metal. If no field is present, no electric force is exerted on the electron. If we apply an external field, however, the free electron is subject to a force in a direction opposite the direction of the electric field (opposite because of the negative charge of the electron).

In a similar way, all the free electrons in a slab of metal initially accelerate in a direction opposite the direction of an applied field (**Figure 24.16**). This leaves behind a positive charge on one side of the slab and creates a negative charge on the opposite side. Because of this rearrangement of charge, an induced electric field builds up in a direction opposite the direction of the external field. As a result, the electric field inside the slab, which is the sum of the external electric field and the induced electric field, decreases. As this field decreases, so does the force exerted on the free electrons in the slab. When enough charge carriers have accumulated on each side of the slab to make the electric field inside the slab zero, the electric force exerted on the free electrons in the metal becomes zero and the material reaches **electrostatic equilibrium**—the condition in which the distribution of charge in a system does not change. The time interval it takes for a metal to reach electrostatic equilibrium is very short (about 10^{-16} s), so the rearrangement of charge carriers is virtually instantaneous. The important point to remember is:

> **The electric field inside a conducting object that is in electrostatic equilibrium is zero.**

Keep in mind that this statement holds *only* in electrostatic equilibrium. When charge carriers are made to flow through a conducting object—as in any electric or electronic device, like your stereo or a refrigerator—the electric field is *not* zero inside the object!

Suppose now we add charge to a conducting object. It makes sense to assume that the charged particles will

Figure 24.16 Why the electric field inside the bulk of a conducting object is zero when the object is in electrostatic equilibrium.

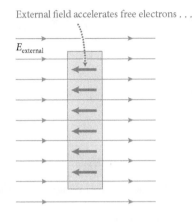

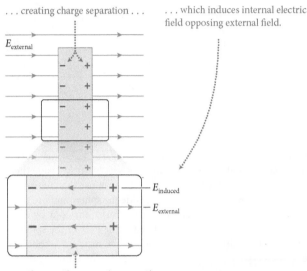

(*a*) No electric field

neutral metal

(*b*) Electric field just switched on

External field accelerates free electrons . . .

$E_{external}$

(*c*) Electrostatic equilibrium established

. . . creating charge separation . . .

. . . which induces internal electric field opposing external field.

$E_{external}$

$E_{induced}$

$E_{external}$

Once field within metal is zero, $\vec{E}_{inside} = \vec{E}_{external} + \vec{E}_{induced} = \vec{0}$, equilibrium is reached.

arrange themselves over the object in such a way as to spread out as far as possible from one another, given that particles carrying like charges repel. We can use a Gaussian surface to obtain a better understanding of where the charged particles go.

✋ **24.14** Consider a spherical Gaussian surface inside a positively charged conducting object that has reached electrostatic equilibrium. (*a*) Is the field line flux through the Gaussian surface positive, negative, or zero? (*b*) What can you conclude from your answer to part *a* about the charge enclosed by the Gaussian surface?

We can extend the result of Checkpoint 24.14 to conducting objects of any shape. Consider, for example, the irregularly shaped, charged conducting object shown in Figure 24.17. Draw a Gaussian surface of the same shape as the object, just below its surface. Given that the field is zero

Figure 24.17 Because the electric field inside a conducting object in electrostatic equilibrium is zero, we conclude that there cannot be any surplus charge inside the object.

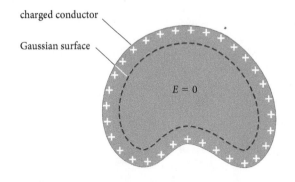

charged conductor

Gaussian surface

$E = 0$

everywhere inside the conducting object, the field line flux through the Gaussian surface is zero and the charge enclosed by the Gaussian surface is also zero. Because we can choose the Gaussian surface arbitrarily close to the surface of the object, we conclude:

> **Any surplus charge placed on an isolated conducting object arranges itself at the surface of the object. No surplus charge remains in the body of the conducting object once it has reached electrostatic equilibrium.**

✋ **24.15** Suppose the charged conducting object in Figure 24.17 contains an empty cavity. Does any surplus charge reside on the inner surface of the cavity?

Example 24.4 Charged particle in a cavity

An electrically neutral, conducting sphere contains an irregularly shaped cavity. Inside the cavity is a particle carrying a positive charge $+q$. What are the sign and magnitude of the charge on the sphere's outer surface?

❶ **GETTING STARTED** I am told that a sphere made of material that is an electrical conductor has a cavity in its interior and that a particle in the cavity carries charge $+q$. My task is to determine the sign and magnitude of any charge residing on the sphere's outer surface. I begin by sketching a vertical cross section through the sphere showing the cavity and the charged particle inside it (**Figure 24.18a** on the next page). The problem states that the sphere is electrically neutral, but the question posed implies that some charge resides on its outer surface. Because $\vec{E} = \vec{0}$ inside the conducting material, I know that an equal quantity of the opposite charge must accumulate somewhere else on the sphere.

Figure 24.18

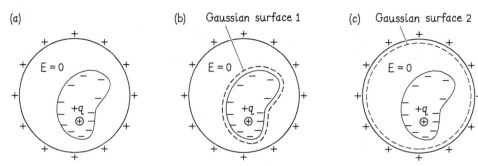

(a)

$E = 0$

$+q$

(b) Gaussian surface 1

$E = 0$

$+q$

(c) Gaussian surface 2

$E = 0$

$+q$

② DEVISE PLAN The sphere is conducting, so I know that once electrostatic equilibrium is reached, the electric field inside the bulk of the sphere must be zero: $\vec{E} = \vec{0}$. I can use this information to draw any Gaussian surface inside the sphere and use the following reasoning to determine the charge enclosed by my Gaussian surface: Because $\vec{E} = \vec{0}$ inside the bulk of the sphere and because I draw my Gaussian surface inside the sphere, $\vec{E} = \vec{0}$ everywhere on the Gaussian surface. Therefore the field line flux through the Gaussian surface is zero, which means the charge enclosed by this surface must be zero. Because I can draw my Gaussian surface anywhere inside the bulk of the conductor, I can use this information to determine the distribution of charge on the sphere.

③ EXECUTE PLAN I begin by drawing a Gaussian surface 1 enclosing the cavity (Figure 24.18*b*). Because the field line flux through this surface is zero, the charge enclosed by the surface must be zero. There is charge $+q$ inside the surface (in the charged particle), however, and so in order for the charge enclosed by the surface to be zero, a quantity of charge $-q$ must have migrated from someplace in the region surrounding the cavity and accumulated on the inner cavity surface.

Because the sphere is electrically neutral, the charge $-q$ that migrated to the inner cavity surface must leave a charge $+q$ behind somewhere else on the sphere. If I now draw Gaussian surface 2 just inside the sphere's outer surface (Figure 24.18*c*), I see that, because the field line flux through this Gaussian surface has to be zero, the charge enclosed by this surface is also zero. The positive charge $+q$ that results from the migration of charge $-q$ to the cavity surface must therefore reside outside Gaussian surface 2. Because I can draw surface 2 arbitrarily close to the sphere's outer surface, I conclude that the sphere's outer surface carries a charge $+q$. ✔

④ EVALUATE RESULT The negative charge that migrates from the region outside the cavity to the cavity surface arranges itself in such a way as to cancel the electric field that the charged particle creates in the region outside the cavity. Therefore all the field lines that start on the charged particle must end on the negative charge at the cavity surface. In order for all the field lines to end here, the quantity of negative charge on the cavity surface must be equal to the quantity of charge on the particle. The sphere is electrically neutral, and my choice for where I draw Gaussian surface 2 requires that all the positive charge resulting from the migration of negative charge to the cavity surface must accumulate outside Gaussian surface 2, which means right at the sphere's outer surface. Thus my answer makes sense.

That the electric field inside any conductor is zero in electrostatic equilibrium allows us to draw one additional important conclusion. Because the electric fields must be zero everywhere, including at the surface of a conducting object, there cannot be any component of the electric field parallel to the surface of the object, and therefore we can conclude:

> **In electrostatic equilibrium, the electric field at the surface of a conducting object is perpendicular to that surface.**

If there were a component of the electric field parallel to the surface, that component would cause any free charge carrier to move along the surface, which means the conductor is not in electrostatic equilibrium.

24.16 In Example 24.4, is the electric field inside the cavity zero?

Self-quiz

(For this self-quiz assume all situations are two-dimensional.)

1. In **Figure 24.19**, which of the two charged spheres carries a charge of greater magnitude?

2. Consider Gaussian surfaces 1–3 in Figure 24.19. Determine the field line flux through each surface.

3. In Figure 24.19, is the field line density greater at point A or point B? At which of these locations is the magnitude of the electric field greater? Is the field line density at point C zero or nonzero?

4. The electric field lines in **Figure 24.20** tell you there must be one or more charged particles inside the Gaussian surface defined by the dashed line. Could the electric field shown be due to a single particle inside the Gaussian surface? What must the signs and relative magnitudes of the charged particle(s) be in order to create the electric field lines shown?

5. **Figure 24.21** shows a small ball that carries a charge of $+q$ inside a conducting metal shell that carries a charge of $+2q$. (*a*) What are the sign and magnitude of the charge on the inner surface of the shell? (*b*) What are the sign and magnitude of the charge on the outer surface of the shell?

Figure 24.19

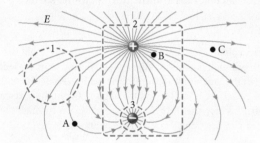

Figure 24.20

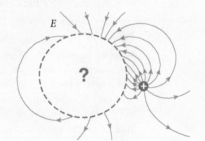

Figure 24.21

shell in cross section
(charge = $+2q$)

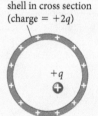

Answers

1. The number of field lines is proportional to the charge on the object. Because more lines emanate from the charged object on the top, that object must carry a greater charge.

2. For surface 1, all lines that enter the surface also exit the surface, so the field line flux is zero. For surface 2, 25 lines exit the surface and 6 lines enter the surface. The field line flux through surface 2 is thus $25 - 6 = 19$. Fifteen field lines cross surface 3, with all lines entering from the outside. The field line flux for surface 3 is -15.

3. Point B has the greater field line density because the lines are closer together at B than they are at A. The magnitude of the electric field is greater at point B because electric field strength is proportional to the field line density. Even though C is not on a field line, the field line density, which is represented by the spacing of the field lines *around* point C, is nonzero.

4. Because electric field lines converge on one point near the top of the area enclosed by the surface and diverge from a point near the bottom, there must be objects that carry both negative and positive charges inside the surface. Because more field lines enter the surface than exit the surface, the negatively charged object(s) must carry a charge of greater magnitude than the positively charged object(s).

5. (*a*) Because the electric field in the conducting shell is zero, the field line flux through a Gaussian surface drawn inside the material of the conducting shell must be zero. According to Gauss's law, the charge enclosed in the surface must also be zero. The charge on the inner surface of the shell must therefore be $-q$, which added to $+q$ gives zero. (*b*) For a neutral shell, a charge of $-q$ on the inside surface of the shell would leave a surplus of $+q$ on the outside surface. The surplus charge of $+2q$ that was placed on the shell also resides on the outside surface. The charge residing on the outside surface of the shell is thus $+3q$.

24.6 Electric flux

We introduced two important concepts in the first part of this chapter: the field line density, which is proportional to the strength of the electric field, and the field line flux, which represents the number of field lines going outward through a closed surface minus the number of field lines going inward. In this section we'll turn these concepts into quantities we can calculate.

Consider, for example, a trapezoidal box in a uniform electric field $\vec{E}$ (Figure 24.22a). The field line flux through the closed surface of the trapezoidal box is zero: Twenty field lines go into the back surface and 20 come out through the front surface. Another way of putting this is to say that the field line flux into the back surface is equal in magnitude to the field line flux out of the front surface. Instead of using field lines, however, whose number is chosen arbitrarily, we'll work with a quantity called the **electric flux,** represented by the symbol Φ_E (Φ is the Greek capital phi). The magnitude of the electric flux through a surface with area A in a uniform electric field of magnitude E is defined as

$$\Phi_E \equiv EA \cos \theta \quad \text{(uniform electric field),} \qquad (24.1)$$

where θ is the angle between the electric field and the normal to the surface.

✋ **24.17** (*a*) Consider the front surface of the trapezoidal box in Figure 24.22a, detached from the rest of the trapezoidal box. Does the field line flux through that surface increase, decrease, or stay the same if any of the following quantities is increased: (*i*) the area of the front surface, (*ii*) the magnitude of the electric field (keeping the area constant), (*iii*) the slope of the front surface (that is to say, the angle between the surface and the direction of the electric field is increased)? (*b*) Does the *electric flux* through the front surface increase, decrease, or stay the same if any of these quantities is changed?

As Checkpoint 24.17 shows, electric flux, as we've defined it in Eq. 24.1, behaves like the field line flux. To make the correspondence more precise, we define an *area vector* $\vec{A}$ for a flat surface area as a vector whose magnitude is equal to the surface area A and whose direction is normal to the plane of the area. On closed surfaces we choose $\vec{A}$ to point outward (we'll deal with open surfaces later). Figure 24.22b shows the area vectors associated with the front and back surfaces of the (closed) trapezoidal box. With this definition, Eq. 24.1 can be written as a scalar product (Eq. 10.33):

$$\Phi_E \equiv EA \cos \theta = \vec{E} \cdot \vec{A} \quad \text{(uniform electric field),} \qquad (24.2)$$

where θ is the angle between $\vec{E}$ and $\vec{A}$. Electric flux is a scalar, and the SI unit of electric flux is $N \cdot m^2 / C$.

✋ **24.18** Let the area of the back surface of the trapezoidal box in Figure 24.22 be 1.0 m^2, the magnitude of the electric field be 1.0 N/C, and $\theta = 30°$ for the front surface. (*a*) What are the magnitudes of the area vectors for the front and back surfaces of the trapezoidal box? (*b*) What are the electric fluxes through the front and back surfaces?

The above definition of electric flux applies only for uniform electric fields and flat surfaces. Let us therefore consider the more general case of an irregular surface in a nonuniform field (Figure 24.23). To calculate the electric flux through that surface, we divide the entire surface into small segments of surface area δA_i, with each segment being small enough that we can consider it to be essentially flat, and we can define an area vector $\delta \vec{A}_i$ whose magnitude is equal to the surface area δA_i of the segment and whose direction is normal to the segment. This allows us to apply Eq. 24.2 to each individual segment. The electric

Figure 24.22 Determining the electric flux through a flat surface.

(*a*) Trapezoidal box in uniform electric field

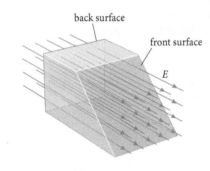

back surface

front surface

E

(*b*) Vector areas of front and back surfaces

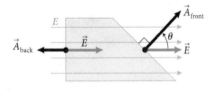

Figure 24.23 To obtain the electric flux through an irregularly shaped, nonplanar surface and/or for a nonuniform electric field, we divide the surface into small segments. For very small segments, each segment is essentially flat and the field through each segment is essentially uniform. The flux through the entire surface is then given by the sum of all of the contributions through each segment.

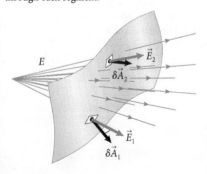

flux through a single segment is then $\Phi_{Ei} = \vec{E}_i \cdot \delta\vec{A}_i$, where $\vec{E}_i$ is the electric field vector at the location of the segment. To calculate the electric flux through the entire surface we must sum the electric flux through all the surface segments:

$$\Phi_E = \sum \vec{E}_i \cdot \delta\vec{A}_i. \qquad (24.3)$$

If we let the area of each segment approach zero, then the number of segments approaches infinity and the sum is replaced by an integral:

$$\Phi_E = \lim_{\delta A_i \to 0} \sum \vec{E}_i \cdot \delta\vec{A}_i = \int \vec{E} \cdot d\vec{A}. \qquad (24.4)$$

The integral in Eq. 24.4 is called a *surface integral* (see Appendix B); $d\vec{A}$ is the area vector of an infinitesimally small surface segment. If the surface is closed, this surface integral is written as

$$\Phi_E = \oint \vec{E} \cdot d\vec{A}, \qquad (24.5)$$

where the circle through the integral sign indicates that the integration is to be taken over the entire closed surface and $d\vec{A}$ is chosen to point outward. Because evaluating a surface integral is mathematically more complicated than single-variable integration, it is important to exploit any symmetry to simplify the calculation.

Example 24.5 Cylindrical Gaussian surface in a uniform electric field

Consider a cylindrical Gaussian surface of radius r and length ℓ in a uniform electric field $\vec{E}$, with the length axis of the cylinder parallel to the electric field (**Figure 24.24**). What is the electric flux Φ_E through this Gaussian surface?

Figure 24.24 Example 24.5.

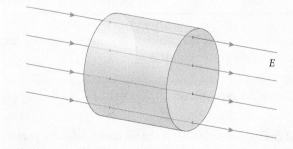

Figure 24.25

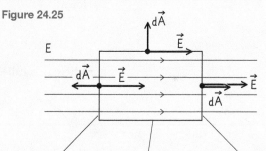

back flat surface   curved surface   front flat surface

❶ **GETTING STARTED** From Figure 24.24 I see that a cylindrical Gaussian surface consists of three regions: front and back flat surfaces and a curved surface joining them. The electric field is perpendicular to the front and back flat surfaces and parallel to the curved surface.

❷ **DEVISE PLAN** The electric flux is given by Eq. 24.5, so I can calculate the electric flux through the Gaussian surface by applying Eq. 24.5 to each of the three regions and then summing the three contributions to the electric flux. In order to evaluate the scalar product $\vec{E} \cdot d\vec{A}$ for each region, I sketch a side view of the Gaussian surface showing the vectors $\vec{E}$ and $d\vec{A}$ for each of the three regions (**Figure 24.25**).

❸ **EXECUTE PLAN** Applying Eq. 24.5 to the three regions, I can write the flux through the Gaussian surface as the sum of three surface integrals: one over the back surface, one over the curved surface, and one over the front surface:

$$\Phi_E = \oint \vec{E} \cdot d\vec{A} = \underbrace{\int \vec{E} \cdot d\vec{A}}_{\substack{\text{back flat} \\ \text{surface}}} + \underbrace{\int \vec{E} \cdot d\vec{A}}_{\substack{\text{curved} \\ \text{surface}}} + \underbrace{\int \vec{E} \cdot d\vec{A}}_{\substack{\text{front flat} \\ \text{surface}}}. \qquad (1)$$

From my sketch, I see that on the back flat surface the angle between $\vec{E}$ and $d\vec{A}$ is 180°, so $\vec{E} \cdot d\vec{A} = E(\cos 180°) \, dA = -E \, dA$. Because the magnitude of the electric field is the same everywhere, I can pull E out of the integral:

$$\underbrace{\int \vec{E} \cdot d\vec{A}}_{\substack{\text{back flat} \\ \text{surface}}} = \int (-E) \, dA = -E \int dA = -E(\pi r^2),$$

where πr^2 is the area of the back flat surface.

(Continued)

The integral over the curved region of the Gaussian surface yields a value of zero because the angle between between $\vec{E}$ and $d\vec{A}$ is 90° everywhere on the curved region, so $\vec{E} \cdot d\vec{A} = E(\cos 90°)\, dA = 0$.

Finally, for the front flat surface I have $\vec{E} \cdot d\vec{A} = E(\cos 0°)\, dA = E\, dA$, so

$$\int_{\substack{\text{front flat} \\ \text{surface}}} \vec{E} \cdot d\vec{A} = \int E\, dA = E \int dA = E(\pi r^2).$$

Adding up the three terms in Eq. 1 yields

$$\Phi_E = -E(\pi r^2) + 0 + E(\pi r^2) = 0. \checkmark$$

❹ EVALUATE RESULT Because there is no charge enclosed by the Gaussian surface, I know that the field line flux through the surface must be zero, a fact I confirm by looking at Figure 24.24: The four field lines shown contribute a flux of −4 on the back flat surface, +4 on the front flat surface, and 0 along the curved surface, for a total of −4 + 4 + 0 = 0. It therefore makes sense that the electric flux through this Gaussian surface also is zero.

24.19 Consider a spherical Gaussian surface of radius r with a particle that carries a charge $+q$ at its center. (a) What is the magnitude of the electric field due to the particle at the Gaussian surface? (b) What is the electric flux through the sphere due to the charged particle? (c) Combining your answers to parts a and b, what is the relationship between the electric flux through the sphere and the enclosed charge q_{enc}? (d) Would this relationship change if you doubled the radius r of the sphere?

24.7 Deriving Gauss's Law

Checkpoint 24.19 shows that the electric flux through a spherical Gaussian surface is equal to the charge q enclosed by the sphere times $4\pi k$, where $k = 9.0 \times 10^9\ \text{N} \cdot \text{m}^2/\text{C}^2$ is the proportionality constant that appears in Coulomb's law (see Eqs. 22.1 and 22.5). This relationship is usually written in the form

$$\Phi_E = 4\pi k q = \frac{q}{\epsilon_0}, \tag{24.6}$$

where ϵ_0 is called the **electric constant** (sometimes called *permittivity constant*):

$$\epsilon_0 \equiv \frac{1}{4\pi k} = 8.85418782 \times 10^{-12}\ \text{C}^2/(\text{N} \cdot \text{m}^2). \tag{24.7}$$

Equation 24.6 is a special case of **Gauss's law,** which states that the electric flux through the closed surface of an arbitrary volume is

$$\Phi_E = \oint \vec{E} \cdot d\vec{A} = \frac{q_{enc}}{\epsilon_0}, \tag{24.8}$$

where q_{enc} is the **enclosed charge**—the sum of all charge on an object or portion of an object enclosed by the closed surface. The formal proof of Gauss's law is an extension of the calculation you performed in Checkpoint 24.19 and is shown in the box "Electric flux though an arbitrary closed surface".

Gauss's law is a direct consequence of Coulomb's law with its $1/r^2$ dependence and the superposition of electric fields. In that respect, it contains nothing new. However, as the next section shows, Gauss's law greatly simplifies the calculation of electric fields due to charge distributions that exhibit one of the three symmetries we discussed in Section 24.4.

24.20 Suppose Coulomb's law showed a $1/r^{2.00001}$ dependence instead of a $1/r^2$ dependence. (a) Calculate the electric flux through a spherical Gaussian surface of radius R centered on a particle carrying a charge $+q$. (b) Substitute your result in Eq. 24.8. What do you notice?

Electric flux through an arbitrary closed surface

Consider a particle that carries a positive charge $+q$, surrounded by the irregularly shaped closed surface shown in Figure 24.26*a*.

1. To determine the electric flux through the irregular surface, we divide the volume enclosed by the surface into small square wedges that taper to a point at the charged particle, one of which is shown. We calculate the electric flux through the surface segment dA cut out by each wedge and then sum the contributions from all the wedges.

2. To determine the electric flux through dA in Figure 24.26*a*, we draw two spherical Gaussian surfaces around q: one with a radius r_1 equal to the distance between dA and q, the other with an arbitrary radius r_2. Our wedge from step 1 now defines two other small surface segments dA_1 and dA_2 on these two spheres.

3. If the segment dA is made very small, then the field lines in the wedge are nearly parallel to one another. Therefore, according to what we found in Section 24.6, the electric flux through dA is equal to that through dA_1 (Figure 24.26*b*). In addition, as you showed in Checkpoint 24.19, this flux is also equal to that through surface segment dA_2. In fact, the electric flux is the same through *any* surface that cuts through the wedge. Put differently, any surface that cuts through the wedge intercepts the same number of field lines.

4. We can repeat this procedure for each wedge. Each time, we see that the electric flux through a segment dA on the irregular surface is equal to that through a corresponding segment dA_2. As we add the contributions of all the wedges, we conclude that the electric flux through the closed irregular surface is equal to that through a sphere of arbitrary radius r centered on q.

5. If there is more than one charged particle inside the irregularly shaped surface, we can use the above arguments for each particle individually and then use the superposition of electric fields (Section 23.3):

$$\vec{E} = \sum \vec{E}_i,$$

where $\vec{E}_i$ is the electric field due to particle i alone. The electric flux due to all charged particles is then the sum of the electric fluxes due to the individual electric fields:

$$\Phi_E = \sum \Phi_{Ei} = \sum \frac{q_i}{\epsilon_0} = \frac{q_{enc}}{\epsilon_0}.$$

✋ **24.21** Suppose q is *outside* the irregularly shaped surface in Figure 24.26. Show that the electric flux due to q through the closed surface is zero. (Hint: Draw a small wedge from q through the surface and determine the electric flux through the two intersections between the wedge and the surface.)

Figure 24.26 Formal proof of Gauss's law.

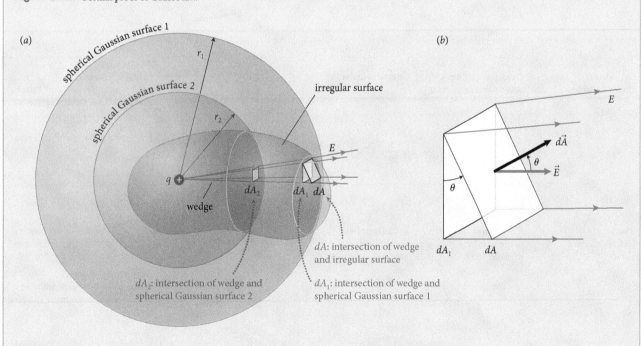

(a) spherical Gaussian surface 1
spherical Gaussian surface 2
r_1
r_2
irregular surface
E
q
dA_2 dA_1 dA
wedge
dA: intersection of wedge and irregular surface
dA_2: intersection of wedge and spherical Gaussian surface 2
dA_1: intersection of wedge and spherical Gaussian surface 1

(b)
E
$d\vec{A}$
θ
$\vec{E}$
θ
dA_1 dA

24.8 Applying Gauss's Law

Gauss's law relates the electric flux through a closed surface to the charge enclosed by it. In Section 24.5 we encountered one important application of Gauss's law: If we can choose a Gaussian surface such that the electric field is zero everywhere on that surface (inside a conducting material, for example), then we know that the charge enclosed by that surface must be zero. In this section we show how Gauss's law can be used to avoid having to carry out any integrations to calculate the electric field. In principle one can calculate the electric flux through any surface, but the calculation is not trivial in general. For the surfaces and charged objects shown in Section 24.4 and summarized in Figure 24.27, however, the electric flux is easy to calculate because the field lines are either parallel to the surface (in which case the electric flux is zero) or perpendicular to it and the magnitude of the electric field is constant (in which case the electric flux is simply the product of the magnitude of the electric field E and the surface area A).

To see the benefit of Gauss's law, consider a charged spherical shell of radius R that carries a uniformly distributed positive charge q. The electric field due to this charged shell can be calculated using the procedure outlined in Section 23.7: Divide the shell into infinitesimally small segments that carry a charge dq, apply Coulomb's law to each small segment, and integrate over the entire shell (Eq. 23.15):

$$\vec{E}_{sP} = k \int \frac{dq_s}{r_{sP}^2} \hat{r}_{sP}. \tag{24.9}$$

Figure 24.27 Applying Gauss's law to determine the electric fields of symmetrical charge distributions.

Symmetry of charge distribution	Electric field geometry	Gaussian surface		To find electric flux
Spherical (charged sphere)	$\vec{E}$ radiates uniformly outward in three dimensions.	Concentric sphere	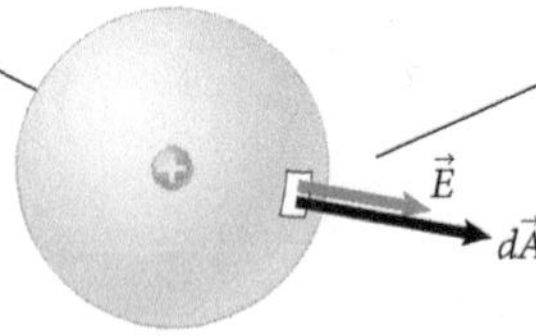	At all points, $\vec{E}$ is perpendicular to surface and has same magnitude.
Cylindrical (infinite charged rod)	$\vec{E}$ radiates uniformly outward perpendicular to axis.	Coaxial cylinder	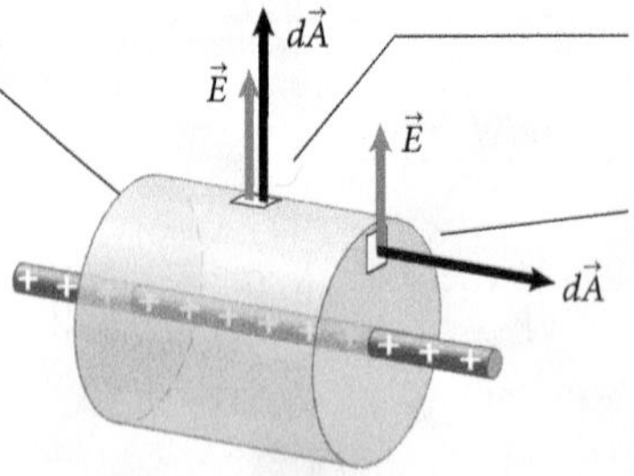	*Cylindrical surface:* At all points, $\vec{E}$ is perpendicular to surface and has same magnitude. *End faces:* $\vec{E}$ is parallel to face, so flux is zero.
Planar (infinite charged sheet)	$\vec{E}$ is uniform and perpendicular to plane.	Cylinder or box perpendicular to plane	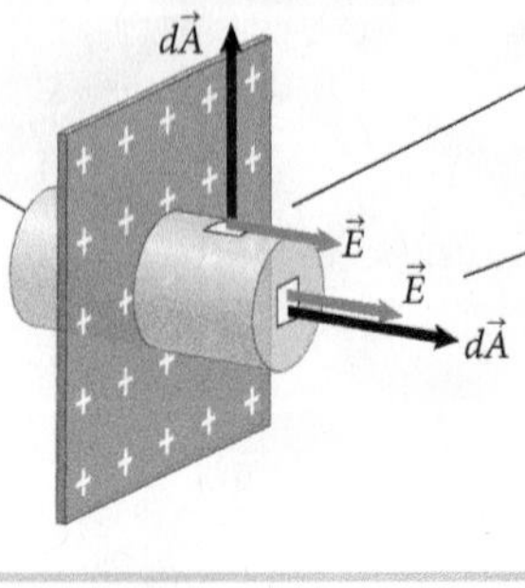	*Surface perpendicular to plane:* $\vec{E}$ is parallel to face, so flux is zero. *Faces parallel to plane:* At all points, $\vec{E}$ is perpendicular to surface and has same magnitude.

As you may imagine, this so-called *direct integration* is no simple matter (see Section 13.8 for a similar integral), even though the result, which we derived qualitatively in Section 24.4, is surprisingly simple. Taking advantage of the symmetry of the problem, however, we can use Gauss's law to calculate the answer in just two steps.

We begin by drawing a concentric spherical Gaussian surface of radius $r > R$ around the shell (**Figure 24.28a**). According to Gauss's law, the flux through the Gaussian surface is equal to the enclosed charge divided by ϵ_0:

$$\Phi_E = \frac{q_{enc}}{\epsilon_0} = \frac{q}{\epsilon_0}. \tag{24.10}$$

In addition, we know that because of the spherical symmetry the electric field has the same magnitude E at each position on the Gaussian surface and the field is perpendicular to the surface. Because the electric field is perpendicular, we have $\vec{E} \cdot d\vec{A} = E\, dA$, and because E has the same value everywhere on the Gaussian surface, we can pull the electric field out of the integral in Eq. 24.5:

$$\Phi_E = \oint \vec{E} \cdot d\vec{A} = \oint E\, dA = E \oint dA = EA, \tag{24.11}$$

where A is the area of the spherical Gaussian surface:

$$A = 4\pi r^2. \tag{24.12}$$

Substituting Eq. 24.12 into Eq. 24.11, we obtain

$$\Phi_E = 4\pi r^2 E. \tag{24.13}$$

Combining Eqs. 24.10 and 24.13, we obtain

$$4\pi r^2 E = \frac{q}{\epsilon_0} \tag{24.14}$$

or

$$E = \frac{1}{4\pi\epsilon_0}\frac{q}{r^2} = k\frac{q}{r^2}. \tag{24.15}$$

This is exactly the magnitude of the electric field due to a particle carrying a charge q located at the center of the shell, as we concluded in Example 23.7 for a solid sphere.

We can also use Gauss's law to determine the electric field inside the shell by drawing a concentric spherical Gaussian surface with radius $r < R$ (Figure 24.28b). For this surface the enclosed charge is zero, $q_{enc} = 0$, so the right side of Eq. 24.14 becomes zero. Consequently, the electric field inside the uniformly charged spherical shell must be zero.

Note that our calculation did not involve working out any integrals, even though Eq. 24.5 does contain an integral—the symmetry of the problem allows us to bypass the integration. The procedure box on page 658 shows how to calculate the electric field using Gauss's law for charge distributions that exhibit one of the symmetries listed in Figure 24.27. In the next three exercises we apply this procedure to calculate the electric field of a number of different charge distributions.

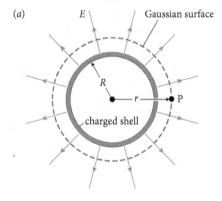

Figure 24.28 Applying Gauss's law to a charged spherical shell.

(a)

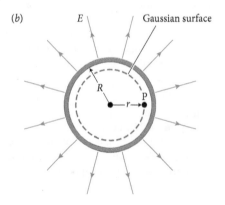

(b)

Procedure: Calculating the electric field using Gauss's Law

Gauss's law allows you to calculate the electric field for charge distributions that exhibit spherical, cylindrical, or planar symmetry without having to carry out any integrations.

1. Identify the symmetry of the charge distribution. This symmetry determines the general pattern of the electric field and the type of Gaussian surface you should use (see Figure 24.27).
2. Sketch the charge distribution and the electric field by drawing a number of field lines, remembering that the field lines start on positively charged objects and end on negatively charged ones. A two-dimensional drawing should suffice.
3. Draw a Gaussian surface such that the electric field is either parallel or perpendicular (and constant) to each face of the surface. If the charge distribution divides space into distinct regions, draw a Gaussian surface in each region where you wish to calculate the electric field.
4. For each Gaussian surface determine the charge q_{enc} enclosed by the surface.
5. For each Gaussian surface calculate the electric flux Φ_E through the surface. Express the electric flux in terms of the unknown electric field E.
6. Use Gauss's law (Eq. 24.8) to relate q_{enc} and Φ_E and solve for E.

You can use the same general approach to determine the charge carried by a charge distribution given the electric field of a charge distribution exhibiting one of the three symmetries in Figure 24.27. Follow the same procedure, but in steps 4–6, express q_{enc} in terms of the unknown charge q and solve for q.

Exercise 24.6 Electric field inside uniformly charged sphere

Consider a charged sphere of radius R carrying a positive charge q that is uniformly distributed over the volume of the sphere. What is the magnitude of the electric field a radial distance $r < R$ from the center of the sphere?

SOLUTION The sphere has spherical symmetry, so I know that the field must point radially outward in all directions. I therefore draw a concentric spherical Gaussian surface with a radius $r < R$ (**Figure 24.29**). Because the sphere carries a uniformly distributed charge q, the amount of charge enclosed by the Gaussian surface is determined by the ratio of the volumes of the Gaussian surface and the charged sphere:

$$q_{enc} = \frac{\frac{4}{3}\pi r^3}{\frac{4}{3}\pi R^3}q = \frac{r^3}{R^3}q. \tag{1}$$

The electric flux is given by the product of the magnitude of the electric field $E(r)$ and the surface area A of the Gaussian surface (Eq. 24.13):

$$\Phi_E = 4\pi r^2 E. \tag{2}$$

Substituting Eqs. 1 and 2 into Eq. 24.8, I obtain

$$4\pi r^2 E = \frac{r^3}{R^3}\frac{q}{\epsilon_0}$$

or

$$E = \frac{1}{4\pi\epsilon_0}\frac{q}{R^3}r = k\frac{q}{R^3}r, \checkmark$$

the same result I obtained in Checkpoint 23.15.

Figure 24.29

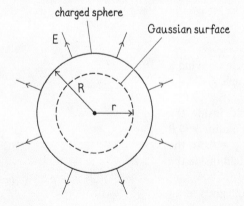

charged sphere

Gaussian surface

E

R

r

24.22 What is the electric field outside a solid sphere carrying a charge $+q$ uniformly distributed throughout its volume?

Exercise 24.7 Electric field of an infinitely long charged thin rod

What is the electric field magnitude a radial distance r from the central length axis of an infinitely long thin rod carrying a positive charge per unit length λ?

SOLUTION An infinitely long rod has cylindrical symmetry. I assume the rod's diameter is vanishingly small. From the symmetry, I know that the electric field points radially outward (see Section 24.4). I therefore make a sketch showing the rod and a few representative field lines. I then draw a cylindrical Gaussian surface of radius r and height h around the rod (**Figure 24.30**). The cylinder encloses a length h of the rod, so the enclosed charge is

$$q_{enc} = \lambda h. \qquad (1)$$

Figure 24.30

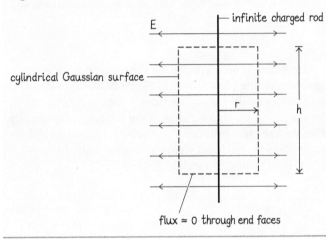

cylindrical Gaussian surface

infinite charged rod

flux = 0 through end faces

The electric flux through the top and bottom faces of the Gaussian surface is zero because the electric field is parallel to those faces. I also know that symmetry requires the electric field to have the same magnitude E at each position on the cylindrical part of the surface. I can therefore pull the electric field out of the integral. The electric flux through that part of the surface is

$$\Phi_E = \int \vec{E} \cdot d\vec{A} = \int E\, dA = E \int dA. \qquad (2)$$

cyl surface

The area of the cylindrical surface is equal to the circumference of the cylinder, $2\pi r$, times its height h, so Eq. 2 becomes

$$\Phi_E = E(2\pi r h). \qquad (3)$$

Substituting Eqs. 1 and 3 into Eq. 24.8, I obtain

$$E(2\pi r h) = \frac{\lambda h}{\epsilon_0}$$

or

$$E = \frac{\lambda}{2\pi \epsilon_0 r} = \frac{2k\lambda}{r}. \checkmark$$

This is the same result I obtained by direct integration in Example 23.4 for a finite charged rod in the limit that I am close to the rod (which makes the rod appear infinitely long). The direct integration, however, took almost two pages of work!

24.23 The direct integration procedure also yields an expression for a rod of *finite* length (see Example 23.4). Can you use Gauss's law to derive this expression as well? Why or why not?

Exercise 24.8 Electric field of an infinite charged sheet

What is the electric field a distance d from a thin, infinite nonconducting sheet with a uniform positive surface charge density σ?

SOLUTION An infinite sheet has planar symmetry. From the symmetry I know that the electric field points away from the sheet and that the magnitude of the electric field is the same everywhere (see Section 24.4). I make a sketch of the sheet and the electric field and then draw a Gaussian surface in the form of a cylinder that straddles the sheet (**Figure 24.31**). If the cross section of the cylinder has area A, then the cylinder encloses a piece of the sheet of area A and the enclosed charge is

$$q_{enc} = \sigma A. \qquad (1)$$

Figure 24.31

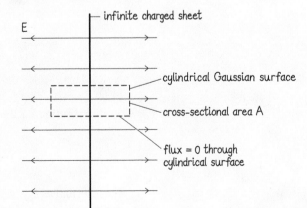

infinite charged sheet

cylindrical Gaussian surface

cross-sectional area A

flux = 0 through cylindrical surface

(Continued)

The electric flux through the cylindrical part of the Gaussian surface is zero because the field lines are parallel to that surface. The field is perpendicular to the two ends and points outward, however, so the electric flux through those ends is the product of the area of each end, A, and the magnitude of the electric field:

$$\Phi_E = 2EA. \qquad (2)$$

Substituting Eqs. 1 and 2 into Eq. 24.8, I get

$$2EA = \frac{\sigma A}{\epsilon_0}$$

or

$$E = \frac{\sigma}{2\epsilon_0}. \checkmark$$

Because $1/(2\epsilon_0) = 2\pi k$, I can also write this as $E = 2\pi k\sigma$, which is the same result I obtained by direct integration in Example 23.6 for a uniformly charged disk in the limit that I am very close to the disk (which makes the disk appear like an infinite sheet).

The situation is a little different for a *conducting* plate. Consider, for example, the infinite charged conducting plate shown in **Figure 24.32a**. As we saw in Section 24.5, any charge resides on the outside surfaces of the conducting object, so we have to consider the charge on *both* surfaces. If the surface charge density on the plate is σ and we use the same cylindrical Gaussian surface as we did for the nonconducting sheet, then the enclosed charge is not σA but $2\sigma A$ (σA for each of the two surfaces). Then Gauss's law yields

$$\Phi_E = \frac{q_{enc}}{\epsilon_0} = \frac{2\sigma A}{\epsilon_0}. \qquad (24.16)$$

Figure 24.32 Applying Gauss's law to an infinite charged conducting plate.

(a)

(b)

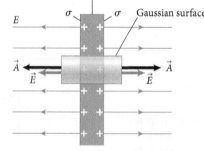 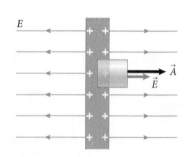

Substituting Eq. 24.16 into Eq. 2 of Exercise 24.8 yields

$$E = \frac{\sigma}{\epsilon_0} \quad \text{(infinite conducting plate)}. \qquad (24.17)$$

Alternatively, you can choose a cylindrical Gaussian surface that has one end buried in the plate (Figure 24.32b). In that case only one of the two surfaces is enclosed and so the enclosed charge is σA. However, now the electric flux through the left end is zero because $E = 0$ inside the bulk of the plate, so the electric flux through the Gaussian surface is not $2EA$ but EA. Substituting these values into Gauss's law yields again Eq. 24.17.

24.24 (*a*) A very large metal plate of surface area A carries a positive charge q. What is the surface charge density of the plate? What is the magnitude of the field created by the plate? (*b*) A very large, thin nonconducting sheet of surface area A carries a fixed, uniformly distributed positive charge q. What is the surface charge density of the sheet? What is the magnitude of the field created by the sheet?

Chapter Glossary

SI units of physical quantities are given in parentheses.

Closed surface Any surface that completely encloses a volume.

Cylindrical symmetry A configuration that remains unchanged if rotated or translated about one axis exhibits cylindrical symmetry.

Electric constant ϵ_0 ($C^2/(N \cdot m^2)$) A constant that relates the electric flux to the enclosed charge in Gauss's law:

$$\epsilon_0 \equiv \frac{1}{4\pi k} = 8.85418782 \times 10^{-12} \, C^2/(N \cdot m^2). \quad (24.7)$$

Electric field lines A representation of electric fields using lines of which the tangent to the line at every position gives the direction of the electric field at that position.

Electric flux Φ_E ($N \cdot m^2/C$) A scalar that provides a quantitative measure of the number of electric field lines that pass through an area. The electric flux through a surface is given by the surface integral

$$\Phi_E \equiv \int \vec{E} \cdot d\vec{A}. \quad (24.4)$$

Electrostatic equilibrium The condition in which the distribution of charge in a system does not change.

Enclosed charge The sum of all the charge within a given closed surface.

Field line density The number of field lines per unit area that cross a surface perpendicular to the field lines at that position. The field line density at a given position in a field line diagram is proportional to the magnitude of the electric field.

Field line flux The number of outward field line crossings through a closed surface minus the number of inward field line crossings. The field line flux through a closed surface is equal to the charge enclosed by the surface multiplied by the number of field lines per unit charge.

Gauss's law The relationship between the electric flux through a closed surface and the charge enclosed by that surface:

$$\Phi_E = \oint \vec{E} \cdot d\vec{A} = \frac{q_{enc}}{\epsilon_0}. \quad (24.8)$$

Gaussian surface Any closed surface used to apply Gauss's law.

Planar symmetry A configuration that remains unchanged if rotated about one axis or translated about any axis perpendicular to the axis of rotation exhibits planar symmetry.

Spherical symmetry A configuration that remains unchanged when rotated about any axis through its center exhibits spherical symmetry.

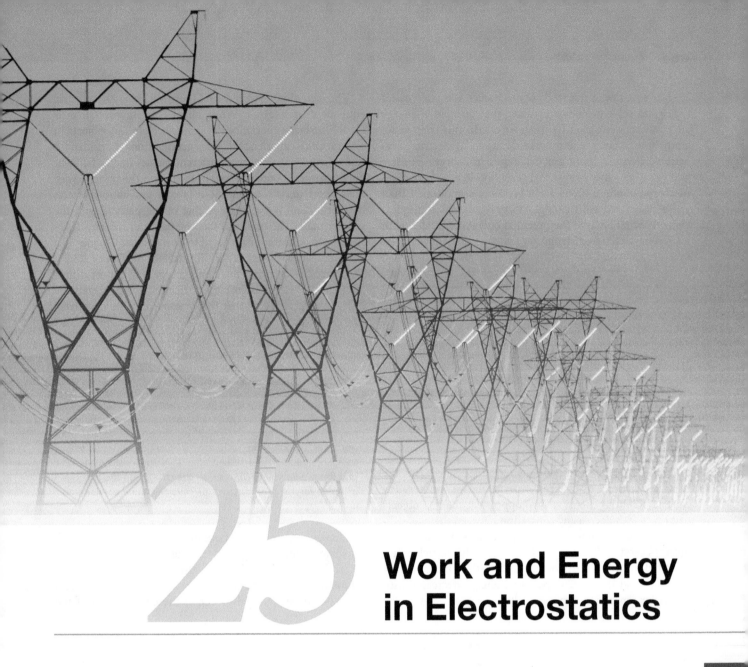

Work and Energy in Electrostatics

CONCEPTS

QUANTITATIVE TOOLS

As we saw in our study of mechanics, it is often easier to solve problems by using the concepts of energy and work than by using forces. In this chapter we study how to apply energy considerations to electric interactions. Because there are two types of charge—positive and negative—the energy changes associated with changes in charge configurations are a bit more complicated than those associated with changes in gravitational configurations. We first analyze the potential energy associated with a stationary charge distribution and then introduce a new quantity, *potential difference,* that is related to potential energy and that plays an important role in electronics because, unlike potential energy, it can be measured directly.

25.1 Electric potential energy

Figure 25.1 shows the energy changes that occur in closed systems of two charged objects. In Figure 25.1a, a positively charged particle is released from rest in the constant electric field of a large stationary object carrying a negative charge. The attractive electric interaction between the particle and the object accelerates the particle toward the object, and so the kinetic energy of the system increases. This increase in kinetic energy must be due to a decrease in **electric potential energy,** the potential energy associated with the relative positions of charged objects. As we can see, the electric potential energy of two oppositely charged objects decreases with decreasing separation between the two.

Figure 25.1b shows what happens in a system of two objects carrying like charges. In this case the particle is accelerated away from the object, and so the kinetic energy increases and the electric potential energy decreases with *increasing* separation between the two.

The situation depicted in Figure 25.1a is the electric equivalent of free fall. While all objects in free fall near Earth's surface experience the same acceleration, objects in electric fields experience *different* accelerations. Consider, for example, two particles with different masses in free fall near Earth's surface (**Figure 25.2a**). Even though the particle with the greater mass is subject to a greater gravitational force, its acceleration is the same as that of the particle with the smaller mass. The reason is that the particle's inertia (its resistance to acceleration) is equal to its mass (see Chapter 13). Now consider the situation illustrated in Figure 25.2b. Two particles carrying the same charge $+q$, but of different mass, are released near the surface of a large negatively charged object. The electric forces exerted by the negatively charged object on the two particles are equal in magnitude, but because the masses of the two particles are different, the particles' accelerations are different as well.

✋ **25.1** Suppose both particles in Figure 25.2b are released from rest. Let $m_2 > m_1$ and consider only electric interactions. (a) How do their kinetic energies compare after they have both undergone the same displacement? (b) How do their momenta compare? (c) How do their kinetic energies and momenta compare at some fixed instant after they have been released? (d) How would you need to adjust the charges on the particles in order for the two particles to have the same acceleration upon release?

Changes in electric potential energy can also be associated with changes in the orientation of charged objects. Consider, for example, the situation illustrated in **Figure 25.3**. An electric dipole is held near the surface of a large, positively charged object. If the electric field of the

Figure 25.1 Energy diagrams for closed systems in which a positively charged particle is released from rest near a large stationary object that carries (a) a negative or (b) positive charge.

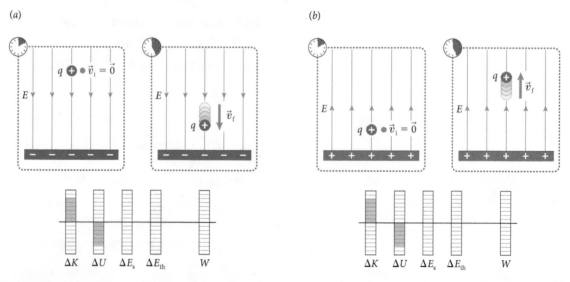

(a)

(b)

ΔK ΔU ΔE_s ΔE_{th} W

ΔK ΔU ΔE_s ΔE_{th} W

Figure 25.2 The free motion of charged particles in an electric field is different from the free fall of objects in a gravitational field.

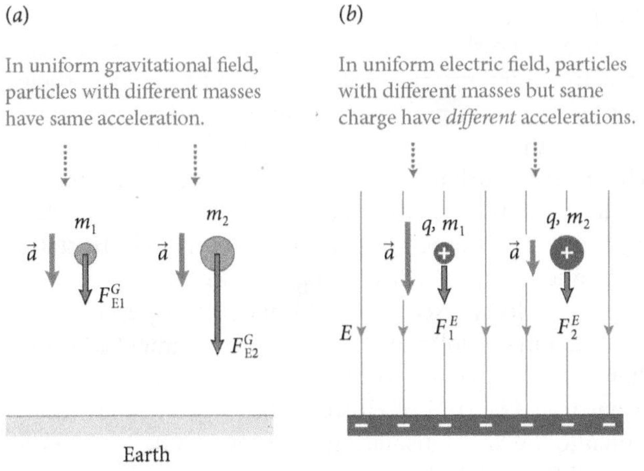

(*a*)

In uniform gravitational field, particles with different masses have same acceleration.

(*b*)

In uniform electric field, particles with different masses but same charge have *different* accelerations.

Figure 25.4 Energy diagram for a positively charged particle in the uniform electric field of a stationary, negatively charged object, that is not part of the system.

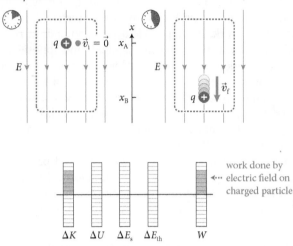

large charged object is uniform, then the dipole begins rotating as shown (see Sections 23.4 and 23.8). As it begins rotating, the dipole gains rotational kinetic energy. This means that the electric potential energy of the system must be changing as the orientation of the dipole changes. This change occurs because the positive side of the dipole gets farther away from the positively charged object, while the negative side gets closer to it; as we've seen, both of these motions correspond to a decrease of electric potential energy.

✋ **25.2** As the dipole in Figure 25.3 continues to rotate, it reaches the point where its axis is aligned with the electric field of the large object. (*a*) What happens to the electric potential energy as the dipole moves beyond that point? (*b*) Describe the motion of the dipole beyond that point. (*c*) How would the motion of the dipole change if it were released with a different orientation from the one shown in Figure 25.3?

Figure 25.3 Energy diagram for a system in which a dipole is released from rest near a positively charged stationary object.

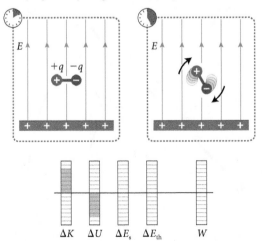

25.2 Electrostatic work

In general, we shall be considering the motion of a charged object through the constant electric field created by other stationary charged objects (such a field is sometimes called an *electrostatic field*). Therefore, our system—the charged object whose motion we are considering—is not closed. The energy of the system is not constant, and we must take into account the work done by the electric field on the system. For example, considering just the particle in Figure 25.1*a* as our system, we obtain the energy diagram shown in **Figure 25.4**. The particle can have only kinetic energy, so its increase in kinetic energy is now due to work done by the electric field on it.

✋ **25.3** (*a*) Suppose the particle in Figure 25.4 moves along the *x* axis from point A at $x = x_A$ to point B at $x = x_B$. How much work is done by the electric field $\vec{E}$ on it? (*b*) Suppose now that an external agent moves the particle back from B to A, starting and ending at rest as shown in **Figure 25.5**. How much work does the electric field do on the particle as it is moved? (*c*) How much work does the agent do on the particle while it is moved? (*d*) What is the combined work done by the agent and by the electric field on the particle as it is moved? (*e*) Draw an energy diagram to illustrate the energy changes of the particle as the agent moves it from B to A.

Figure 25.5 Checkpoint 25.3.

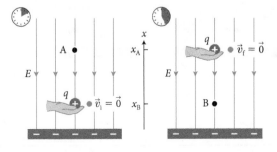

Checkpoint 25.3 illustrates the importance of distinguishing between the work done by the electric field on a charged particle and the work done by the agent moving it. We shall refer to the work done by an electrostatic field as **electrostatic work.** If the particle begins and ends at rest, the electrostatic work is equal in magnitude and opposite in sign to the work done by the agent doing the moving. Then the total work done on the particle—the sum of the electrostatic and mechanical work—is zero. Indeed, the kinetic energy of the particle does not change, and because the particle possesses no other form of energy, its energy remains unchanged: $\Delta E = W = 0$.

Note that the electric force between charged particles, just like the gravitational force, is a central force: Its line of action always lies along the line connecting the two interacting particles (see Section 13.2). A direct consequence of this fact is:

> **The electrostatic work done on a charged particle as it moves from one point to another is independent of the path taken by the particle and depends on only the positions of the endpoints of the path.**

The proof of this statement parallels the one for gravitational forces in Section 13.6. Imagine, for example, lifting a particle from A to B along curved path 2 in **Figure 25.6a** instead of along straight path 1. As shown in Figure 25.6b, the path can be approximated by small straight horizontal and vertical segments. Along the horizontal segments, the electric force is perpendicular to the force displacement, so the electrostatic work on the particle along these segments is zero. Along the vertical segments, the electrostatic work on the particle is nonzero, but note that each vertical segment corresponds to an equivalent vertical segment on path 1. Thus, the displacements along all the vertical segments of path 2 add up to precisely the displacement along

path 1. In other words, the electrostatic work on the particle along path 2 (or any other path from A to B) is equal to that along path 1.

Figure 25.6c shows how this argument can be generalized to a nonuniform electric field. Imagine a particle being moved from point A to point C along the gray trajectory in the electric field caused by an object carrying a charge at the origin. We can approximate the trajectory by a succession of small circular arcs centered about the origin and small straight radial segments. The electrostatic work done on the particle along the circular arcs is zero because the force exerted on the particle is perpendicular to the force displacement. The radial segments, on the other hand, contribute to the electrostatic work done on the particle. The sum of all the radial segments, however, is equal to the radial displacement from A to B. Because no electrostatic work is done on the particle along the circular path from B to C, we thus see that the electrostatic work done on the particle along the gray trajectory from A to C is equal to the electrostatic work done along the path from A to B. The electrostatic work done on the particle along *any* path from A to C is thus the same as the electrostatic work done from A to B.

✋ **25.4** Suppose the electrostatic work done on a charged particle as it moves along the gray path from A to C in Figure 25.6c is W. What is the electrostatic work done on the particle (a) along the path from C to B to A and (b) along the closed path from A to C to B and back to A?

Checkpoint 25.4 shows that the electrostatic work done on a charged particle that moves around a closed path—*any* closed path—in an electrostatic field is zero. We obtained a similar result for the work done by the gravitational force (Eq. 13.17). The physical reason for this result is that the

Figure 25.6 The electrostatic work done on a charged particle as the particle moves from point A to point B is independent of the path taken; it depends only on the positions of the endpoints of the path.

(a) Two paths by which particle can move from A to B

(b) Path 2 approximated by vertical & horizontal segments

(c) Same argument applied to nonuniform electric field

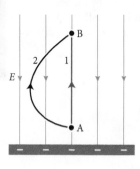

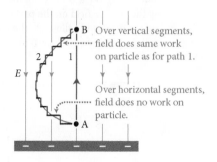

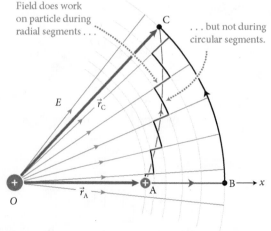

electric force between two charged particles, like the gravitational force between two objects that have mass, is nondissipative: When charged particles are moved around in electrostatic fields, no energy is irreversibly converted to other forms of energy. This is so because the electric interaction is nondissipative.

25.5 (*a*) If the electrostatic work done on a charged particle as it is moves around a closed path is zero, does this also mean that the electrostatic work done on the particle is zero as it moves along a piece of the closed path? (*b*) In Checkpoint 25.3a we found that the electrostatic work done on the particle is the product of the *x* components of the electric force and the force displacement. Is the same true for the electrostatic work done on a particle that is moved along path AB in Figure 25.6c? (*c*) What happens to this electrostatic work if (*i*) the charge on the particle and (*ii*) the mass of the particle is doubled?

The third part of Checkpoint 25.5 demonstrates a very important point: The electrostatic work done on a charged particle is proportional to the charge carried by that particle. This means that once we have calculated the electrostatic work done on a particle carrying a charge *q* along some path in an electrostatic field, we don't need to carry out the whole calculation again if we are interested in the electrostatic work done on another particle carrying a charge 2*q*. We know that the electrostatic work done on the second particle is twice the electrostatic work done on the first. Thus, if we know the electrostatic work done on a particle carrying a unit positive charge along some path, then we know the electrostatic work done on a particle carrying *any* charge between the same two points! We therefore introduce a new quantity, called **electrostatic potential difference** (or simply **potential difference**), defined as:

> **The potential difference between point A and point B in an electrostatic field is equal to the negative of the electrostatic work per unit charge done on a charged particle as it moves from A to B.**

The potential difference is a scalar, and because the electrostatic work done on a charged particle as it moves from one position to another can be positive or negative, the potential difference can also be positive or negative; the potential difference between any two points B and A is the negative of that between points A and B. It is important to keep in mind that potential difference *is not a form of energy*—it is electrostatic work done per unit charge and therefore has SI units of J/C.

You may be wondering why the potential difference is defined in terms of the *negative* of the electrostatic work done on a particle and not in terms of the energy required to move the particle, which has the opposite sign. The reason is that the energy required to move the particle depends on the change in the particle's kinetic energy. If the particle starts at rest and ends at a nonzero speed, the particle's kinetic energy increases and so the energy required is greater than when it starts and ends at rest. The electrostatic work

done on a particle, on the other hand, is independent of any change in the particle's kinetic energy.

25.6 (*a*) Is the potential difference along any path from A to C in Figure 25.6c positive, negative, or zero? (*b*) Along any path from C to B? (*c*) Along the straight path from B to A? (*d*) In Figure 25.4, is the potential difference between the particle's initial and final positions positive, negative, or zero? (*e*) Express this potential difference in terms of the change in the particle's kinetic energy ΔK and its charge *q*.

In principle, only the potential *difference* between the endpoints of a path is meaningful. We can, however, assign a value to the **potential** at each of these endpoints by choosing a reference point. Specifically, if there is a positive potential difference between points A and B, then A is at a lower potential than B; by assigning a value to the potential at one of the two points, the value of the potential at the other point is fixed. Potential and potential difference, which are immensely useful in solving problems in electrostatics, are discussed in more detail in Section 25.5.

25.3 Equipotentials

As we have seen, the electrostatic work done on a charged particle along the horizontal segments in Figure 25.6b and the circular arcs in Figure 25.6c is zero. Consequently, the potential difference between any two points on such an arc or horizontal segment is zero. In other words, the potential has the same value at all points along these arcs or segments. Such paths are said to be **equipotential lines:**

> **An equipotential line is a line along which the value of the electrostatic potential does not change. The electrostatic work done on a charged particle as it moves along an equipotential line is zero.**

The equipotential lines in Figure 25.6 are much like contour lines on a topographical map. For example, the contour lines in **Figure 25.7** on the next page connect points of equal elevation. If you follow a contour line, your elevation remains the same, and so the gravitational potential energy associated with the separation between Earth and your body remains constant. Consequently, no gravitational work is done by Earth on you while you follow contour lines—these lines represent "gravitational equipotential lines."

Returning to the electrostatic case, note that Figure 25.6 is a two-dimensional representation of a three-dimensional situation. Thus, the equipotential line segments shown are really parts of **equipotential surfaces.** In Figure 25.6a, for example, the electrostatic work done on a charged particle is zero along *any* displacement parallel to the surface of the negatively charged object, into or out of the plane of the drawing. Consequently, as shown in **Figure 25.8** on the next page, any surface parallel to the surface of the charged sheet causing the electric field is an equipotential surface. Often we'll use the term *equipotential* to denote an equipotential line or surface.

Figure 25.7 Contour lines on a map are analogous to equipotential lines. If you hike along a contour line, you neither gain nor lose gravitational potential energy.

(a)

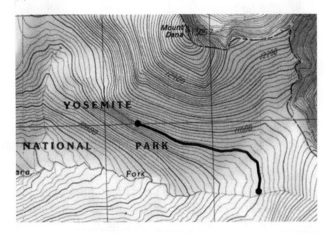

(b)

Just as with vector fields and field line diagrams, it is impossible to draw equipotential surfaces at all locations. In general, equipotentials are drawn with some fixed potential difference between them, just as contour lines represent a fixed difference in altitude. Keep in mind, however, that at any point between the equipotential surfaces shown in a figure we can, in principle, draw another equipotential surface.

✋ **25.7** Consider a single charged particle. Are there any equipotential lines or equipotential surfaces surrounding this particle?

The equipotential surfaces in Figure 25.8 and those in Checkpoint 25.7 are perpendicular to the electric field lines. This is true for *any* stationary charge distribution:

The equipotential surfaces of a stationary charge distribution are everywhere perpendicular to the corresponding electric field lines.

The proof of this statement is straightforward: If the electric field line were not perpendicular to the equipotential surface, then the electric field would have a nonzero component along the surface. This means there would be a nonzero component of electric force along the surface. By definition, however, the electrostatic work done on a charged particle is zero along an equipotential surface, and so there cannot be such a component.

Figure 25.9 shows a two-dimensional view of the equipotential surfaces of a more complicated stationary charge distribution. Note how, at every point in the diagram, the equipotentials are, indeed, perpendicular to the field lines.

Recall from Section 24.5 that in electrostatic equilibrium, the electric field inside the bulk of a conducting object is zero, regardless of the shape of the object or any charge carried by it. This means that no electrostatic work is done on a charged particle inside a charged or uncharged conducting object. Thus, the entire volume of the conducting object

Figure 25.9 Field lines and equipotentials for three stationary charged particles.

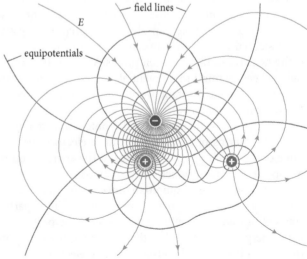

Figure 25.8 Equipotential surfaces in a uniform electric field in (a) two dimensions; and (b) three dimensions.

(a)   (b)

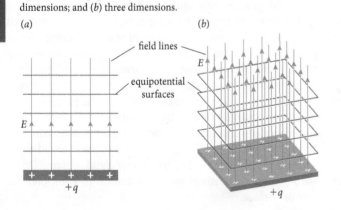

is an *equipotential volume.* In electrostatic equilibrium, all points within a conducting object are at the same electrostatic potential.

Example 25.1 Potential differences

Two metallic spheres A and B are placed on nonconducting stands. Sphere A carries a positive charge, and sphere B is electrically neutral. The two spheres are connected to each other via a wire, and the charge carriers reach a new electrostatic equilibrium. (*a*) Is the electric potential energy of the charge configuration after the spheres are connected greater than, smaller than, or equal to that of the original configuration? (*b*) Before the spheres are connected, is the potential difference between A and B positive, negative, or zero? Is it positive, negative, or zero after the spheres are connected?

❶ GETTING STARTED I need to evaluate the electric potential energy—the energy associated with the configuration of charge carriers—and the potential difference—the negative of the electrostatic work per unit charge done on a charge carrier. I begin with two sketches: one showing the charge distribution before the spheres are connected, and one showing the distribution after they are connected (**Figure 25.10a** and *b*). Once the spheres are connected, they form one conducting object, and the positive charge initially on A spreads out over both spheres.

Figure 25.10.

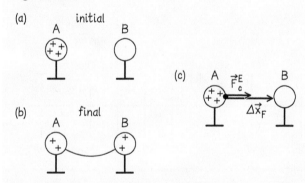

❷ DEVISE PLAN To evaluate the potential difference between A and B, I need to determine the negative of the electrostatic work done on a charged particle as it is moved from A to B and then divide the result by the charge on the particle.

❸ EXECUTE PLAN (*a*) After the two spheres are connected, the charge carriers spread out over both spheres, so they are farther apart than they were before the connection was made. Because I know that energy is required to push positively

charged particles together, I conclude that the electric potential energy associated with the charge configuration in Figure 25.10*b* is smaller than that associated with the configuration in Figure 25.10*a.* ✔

(*b*) When a positive charge carrier is moved from A to B, the electrostatic work done on the carrier is *positive* because the electric force exerted on the carrier (directed away from sphere A) and the force displacement (from A to B) point in the same direction (Figure 25.10*c*). Consequently, the potential difference between A and B must be negative. ✔

Once the spheres are connected, they form one conducting object. Because the electric field is zero inside the entire object, no energy is required to move charge carriers around inside A and B, or from A to B, or vice versa. Thus, the potential difference between A and B after they are connected is zero. ✔

❹ EVALUATE RESULT I know that a closed system always tends to arrange itself so as to lower the system's potential energy (see Section 7.8). It therefore makes sense that the electric potential energy of the system is smaller after the spheres are connected. That the potential difference between A and B is negative before they are connected is a direct consequence of the definition of potential difference. After A and B are connected, they form one large equipotential volume and so it makes sense that the potential difference is zero.

Example 25.1 allows us to make another very useful observation:

> **An electrostatic field is directed from points of higher potential to points of lower potential.**

Because the electric field gives the direction of the force exerted on a positively charged particle, a direct consequence of this fact is:

> **In an electrostatic field, positively charged particles tend to move toward regions of lower potential, whereas negatively charged particles tend to move toward regions of higher potential.**

25.8 When you hold a positively charged rod above a metallic sphere without touching it, a surplus of negative charge carriers accumulates at the top of the sphere, leaving a surplus of positive charge carriers at the bottom. Is the potential difference between the top and the bottom of the sphere positive, negative, or zero?

Self-quiz

1. Consider the situation illustrated in **Figure 25.11**. A positively charged particle is lifted against the uniform electric field of a negatively charged plate. Ignoring any gravitational interactions, draw energy diagrams for the following choices of systems: (*a*) particle and plate, $v_f = 0$; (*b*) particle only, $v_f = 0$; (*c*) particle, plate, and person lifting, $v_f = 0$; (*d*) particle and plate, $v_f \neq 0$.

2. A positively charged particle is moved from point A to point B in the electric field of the large, stationary, positively charged object in **Figure 25.12**. (*a*) Is the electrostatic work done on the particle positive, negative or zero? (*b*) How is the electrostatic work done on the particle along the straight path from A to B different from the electrostatic work done on the particle along a path from A to B via C?

3. **Figure 25.13** shows both the electric field lines and the equipotentials associated with the given charge distribution. (*a*) Is the potential at point A higher than, lower than, or the same as the potential at point B? (*b*) Is the potential at point C higher than, lower than, or the same as the potential at point B? (*c*) Is the potential at point C higher than, lower than, or the same as the potential at point A?

Figure 25.11

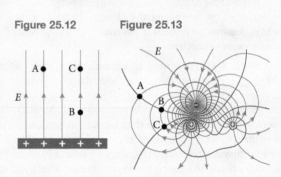

Figure 25.12

Figure 25.13

Figure 25.14

(a) ΔK ΔU ΔE_s ΔE_{th} W

(b) ΔK ΔU ΔE_s ΔE_{th} W

(c) ΔK ΔU ΔE_s ΔE_{th} W

Answers

1. Only the person involves the conversion of source energy, so $\Delta E_s = 0$ in cases *a*, *b*, and *d*. When $v_f = 0$, $\Delta K = 0$; in *d*, $v_f \neq 0$ and so $\Delta K > 0$. For case *a*, the electric potential energy of the system increases as a result of positive work done by the agent (hand) on the particle, see **Figure 25.14a**. For case *b*, the (kinetic) energy of the particle alone does not change because the positive work done by the agent on the particle and the negative electrostatic work done on it are equal in magnitude, so all bars are zero. For case *c*, a decrease in the source energy (provided by the agent) is responsible for the increase in electric potential energy of the system (Figure 25.14*b*). For case *d*, the positive work done by the agent increases both the kinetic energy of the particle and the electric potential energy of the system (Figure 25.14*c*).

2. (*a*) The electrostatic work done on the particle while it moves from point A to point B is negative because the angle between the force exerted on the particle and the force displacement is between 90° and 180° (see Eqs. 10.35 and 10.33). (*b*) The electrostatic work done along the two paths is the same because the electrostatic work done on a particle between two points is independent of the path taken.

3. (*a*) The same, because these points lie along an equipotential surface. (*b*) Higher. The potential increases in a direction opposite to the direction of the electric field, so point C is at a higher potential than point B. (*c*) Higher. Points A and B are at the same potential, and point C is at a higher potential than point B.

25.4 Calculating work and energy in electrostatics

To quantify the electrostatic work done on a charged particle, consider charged particles 1 and 2 in **Figure 25.15**. Particle 2 is moved from point A to point B through the nonuniform electric field of particle 1, which is held stationary. The electrostatic work done by particle 1 on particle 2 as it is moved along the solid path from A to B is given by (Eq. 10.44)

$$W_{12}(A \rightarrow B) = \int_A^B \vec{F}_{12}^E \cdot d\vec{\ell}, \qquad (25.1)$$

where $\vec{F}_{12}^E$ is the electric force exerted by particle 1 on particle 2 and $d\vec{\ell}$ is an infinitesimal segment of the path. This line integral (see Appendix B) is generally not easy to calculate because the magnitude of the electric force and the angle between the force and the path segment $d\vec{\ell}$ vary along the path. However, as we saw in Section 25.2, the electrostatic work done on particle 2 along the solid path from A to B in Figure 25.15 is the same as that done along the dashed path from A to C to B. The circular path from A to C is along an equipotential (see Checkpoint 25.7), so the electrostatic work done on the particle along that path is zero. The electric force is given by Coulomb's law (Eq. 22.7). If we take particle 1 to be at the origin, we can write $r_{12} = r$ for the distance between the two particles and $\hat{r}_{12} = \hat{r}$ for the unit vector pointing from particle 1 to particle 2. Along the radial path from C to B, an infinitesimal segment of the path can be written as $d\vec{\ell} = dr\,\hat{r}$, and so with $k = 1/(4\pi\epsilon_0)$ (Eq. 24.7), the integrand in Eq. 25.1 becomes

$$\vec{F}_{12}^E \cdot d\vec{\ell} = \frac{1}{4\pi\epsilon_0} \frac{q_1 q_2}{r^2} \hat{r} \cdot (dr\,\hat{r}) = \frac{1}{4\pi\epsilon_0} \frac{q_1 q_2}{r^2} dr. \qquad (25.2)$$

The line integral in Eq. 25.1 thus becomes

$$W_{12}(A \rightarrow B) = W_{12}(C \rightarrow B) = \frac{q_1 q_2}{4\pi\epsilon_0} \int_{r_C}^{r_B} \frac{1}{r^2} dr$$

$$= -\frac{q_1 q_2}{4\pi\epsilon_0} \left[\frac{1}{r}\right]_{r_C}^{r_B} = \frac{q_1 q_2}{4\pi\epsilon_0} \left[\frac{1}{r_C} - \frac{1}{r_B}\right]. \qquad (25.3)$$

The distance from particle 1 to point A is the same as the distance from particle 1 to point C, so we have $r_A = r_C$ and

$$W_{12}(A \rightarrow B) = \frac{q_1 q_2}{4\pi\epsilon_0} \left[\frac{1}{r_A} - \frac{1}{r_B}\right]. \qquad (25.4)$$

Generalizing this result to arbitrary initial and final points, we get

$$W_{12} = \frac{q_1 q_2}{4\pi\epsilon_0} \left[\frac{1}{r_{12,i}} - \frac{1}{r_{12,f}}\right], \qquad (25.5)$$

where $r_{12,i}$ and $r_{12,f}$ are the initial and final values of the distance separating particles 1 and 2. Note that this expression is independent of the path taken: The electrostatic work done on particle 2 depends on only the distance between the two particles at the endpoints. Equation 25.5 also does not require particle 1 to be at the origin.

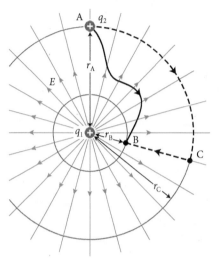

Figure 25.15 The electrostatic work done by particle 1 on particle 2 as the latter is moved from A to B is the same for the meandering solid path and for the dashed path ACB.

✋ **25.9** (a) Using Eq. 25.5, determine whether the electrostatic work done on particle 2 along path CB in Figure 25.15 is positive, negative, or zero. (b) Does moving the particle carrying charge q_2 along path CB involve positive, negative, or zero mechanical work done on the particle? Assume the particle begins and ends at rest; verify the consistency of your answer with part a. (c) By how much does the electric potential energy of the two charged particles change as particle 2 is moved from C to B? Is this change positive, negative, or zero?

As Checkpoint 25.9 demonstrates, the change in electric potential energy of the system that comprises the two particles in Figure 25.15 is the negative of the electrostatic work done by particle 1 on the system that comprises particle 2 only. (This relationship between electrostatic work and change in electric potential energy can also be seen in Figures 25.1 and Figure 25.4.) Taking the negative of Eq. 25.5, we thus obtain

$$\Delta U^E = \frac{q_1 q_2}{4\pi\epsilon_0}\left[\frac{1}{r_{12,\text{f}}} - \frac{1}{r_{12,\text{i}}}\right]. \tag{25.6}$$

Equation 25.6 gives only the *change* in electric potential energy, not *the* potential energy for a given configuration of charge. To obtain such an expression, we must first choose a zero of potential energy. It is customary to choose this zero to correspond to the configuration for which the force between the interacting particles is zero—that is to say, for infinite separation: $U^E = 0$ when $r_{12} = \infty$. Substituting this choice of reference point for the initial point in Eq. 25.6, we obtain

$$\Delta U^E \equiv U_{\text{f}}^E - U_{\text{i}}^E = U_{\text{f}}^E - 0 = \frac{q_1 q_2}{4\pi\epsilon_0}\left[\frac{1}{r_{12,\text{f}}} - 0\right]. \tag{25.7}$$

Thus, the **electric potential energy** for two particles carrying charges q_1 and q_2 and separated by distance r_{12} is

$$U^E = \frac{q_1 q_2}{4\pi\epsilon_0}\frac{1}{r_{12}} \quad (U^E \text{ zero at infinite separation}). \tag{25.8}$$

As this expression shows, the electric potential energy associated with two positively charged particles is positive: If they are brought together starting from infinite separation, the electric potential energy of the two particles increases. This is as it should be, because the two particles repel each other and so energy must be added to the system to bring it together ($W > 0$). The same idea applies to two negatively charged particles because the product $q_1 q_2$ is still positive. For a system of oppositely charged particles, on the other hand, $q_1 q_2 < 0$ and so $U^E < 0$. Indeed, the interaction is attractive and the potential energy decreases as the two are brought together.

✋ **25.10** Suppose we keep particle 2 stationary and move particle 1 in so that their final separation is the same as that in Figure 25.15. Is the electrostatic work done on particle 1 as it is moved in also given by the right-hand side of Eq. 25.4?

Example 25.2 Putting two charged particles together

Two small pith balls, initially separated by a large distance, are each given a positive charge of 5.0 nC. By how much does the electric potential energy of the two-ball system change if the balls are brought together to a separation distance of 2.0 mm?

❶ **GETTING STARTED** I am asked to calculate a change in the electric potential energy of a system. I take *large* to mean a separation distance great enough that the balls don't interact. So the initial state is one in which the two pith balls have infinite separation. In the final state they are 2.0 mm apart.

② **DEVISE PLAN** To calculate the change in electric potential energy I can use Eq. 25.6.

③ **EXECUTE PLAN** Substituting $q_1 = q_2 = 5.0 \times 10^{-9}$ C, $r_{12,i} \approx \infty$, $r_{12,f} = 2.0 \times 10^{-3}$ m, and $k = 1/(4\pi\epsilon_0) = 9.0 \times 10^9$ N·m²/C² into Eq. 25.6, I get

$$\Delta U^E = (9.0 \times 10^9 \text{ N·m}^2/\text{C}^2)(5.0 \times 10^{-9} \text{ C})^2 \left[\frac{1}{2.0 \times 10^{-3} \text{ m}} - 0 \right]$$

$$= 1.1 \times 10^{-4} \text{ J.} \checkmark$$

④ **EVALUATE RESULT** My answer is positive, which makes sense because the balls repel each other and so work must be done on the system as they are brought together. This work increases the potential energy of the system. The magnitude of the potential energy change is small, but so is the magnitude of the force between the balls: Substituting q_1, q_2, and $r_{12,f}$ into Coulomb's law (Eq. 22.1), I obtain a force magnitude of 0.056 N. This is the maximum force between the two balls, so it makes sense that the energy associated with this interaction is small.

We can readily generalize the expressions for electrostatic work and electric potential energy for situations involving more than two charged particles. To determine the electric potential energy of a system of three charged particles in a certain configuration, for example, we calculate the electrostatic work done while assembling the system in its final configuration, starting from a situation where all three particles are far apart. Placing the first particle, carrying charge q_1, in its final position involves no electrostatic work because the other two particles are far away and therefore not interacting with particle 1. Next we bring in particle 2, carrying charge q_2, as illustrated in **Figure 25.16a**. The electrostatic work done while moving particle 2 can be found by substituting infinity for $r_{12,i}$ and r_{12} for $r_{12,f}$ in Eq. 25.5:

$$W_{12} = -\frac{q_1 q_2}{4\pi\epsilon_0} \frac{1}{r_{12}}. \tag{25.9}$$

Finally we bring in particle 3, carrying charge q_3, as shown in Figure 25.16b. The electrostatic work done while moving particle 3 is given by Eq. 25.1:

$$W_3 = \int_i^f (\Sigma \vec{F}_3^E) \cdot d\vec{\ell}, \tag{25.10}$$

where $\Sigma \vec{F}_3^E$ is the vector sum of the forces exerted on particle 3. Particle 3 is subject to two forces, one exerted by particle 1 and one by particle 2, so

$$\int_i^f (\Sigma \vec{F}_3^E) \cdot d\vec{\ell} = \int_i^f (\vec{F}_{13}^E + \vec{F}_{23}^E) \cdot d\vec{\ell}$$

$$= \int_i^f \vec{F}_{13}^E \cdot d\vec{\ell} + \int_i^f \vec{F}_{23}^E \cdot d\vec{\ell} = W_{13} + W_{23}. \tag{25.11}$$

In other words, the electrostatic work done as 3 is moved to its final position is the sum of the electrostatic work done when only 1 is present plus that done when only 2 is present. Now we can apply Eq. 25.5 to each term in Eq. 25.11:

$$W_{13} + W_{23} = -\frac{q_1 q_3}{4\pi\epsilon_0} \frac{1}{r_{13}} - \frac{q_2 q_3}{4\pi\epsilon_0} \frac{1}{r_{23}}. \tag{25.12}$$

The total electrostatic work done on the three-particle system while assembling the charge configuration is the sum of Eqs. 25.9 and Eq. 25.12:

$$W = -\frac{q_1 q_2}{4\pi\epsilon_0} \frac{1}{r_{12}} - \frac{q_1 q_3}{4\pi\epsilon_0} \frac{1}{r_{13}} - \frac{q_2 q_3}{4\pi\epsilon_0} \frac{1}{r_{23}}. \tag{25.13}$$

Figure 25.16 To obtain the electrostatic potential energy of a system of three charged particles, we assemble the system one particle at a time.

(*a*) We bring in second charged particle

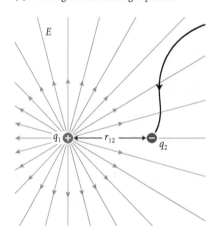

(*b*) We bring in third charged particle

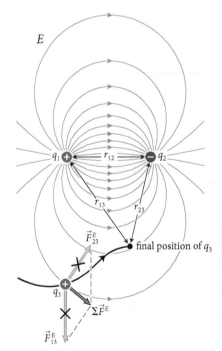

QUANTITATIVE TOOLS

Alternatively, with our choice of zero at infinity, the electric potential energy of the system is

$$U^E = \frac{q_1 q_2}{4\pi\epsilon_0}\frac{1}{r_{12}} + \frac{q_1 q_3}{4\pi\epsilon_0}\frac{1}{r_{13}} + \frac{q_2 q_3}{4\pi\epsilon_0}\frac{1}{r_{23}}$$

$$(U^E \text{ zero at infinite separation}). \quad (25.14)$$

In words, to determine the electric potential energy of a system of charged particles, we need to add the electric potential energies for each pair of particles.

25.11 Suppose the charged particles in Figure 25.16 are assembled in a different order—say, 3 first, then 1, and finally 2. Do you obtain the same result as in Eq. 25.13 and Eq. 25.14?

25.5 Potential difference

The negative of the electrostatic work per unit charge done on a particle that carries a positive charge q from one point to another is defined as the **potential difference** between those points:

$$V_{AB} \equiv V_B - V_A \equiv \frac{-W_q(A \to B)}{q}. \quad (25.15)$$

Potential difference is a scalar, and the SI units of potential difference are joules per coulomb (J/C). In honor of Alessandro Volta (1745–1827), who developed the first battery, this derived unit is given the name **volt** (V):

$$1\ V \equiv 1\ J/C. \quad (25.16)$$

For example, if the electric field in **Figure 25.17** does -12 J of electrostatic work on a particle carrying charge $q_2 = +2.0$ C as it is moved from point A to point B (in other words, it requires $+12$ J of work by an external agent without the particle gaining any kinetic energy), then the potential difference V_{AB} between A and B is $-(-12\ J)/(2.0\ C) = +6.0$ V. That is, the potential at B is 6.0 V higher than that at A.

Once we know the potential difference V_{AB} between A and B, we can obtain the electrostatic work done on *any* object carrying a charge q as it is moved along any path from A to B:

$$W_q(A \to B) = -q V_{AB}. \quad (25.17)$$

Keep in mind that the subscripts AB mean "from A to B." Because B is the final position, we write $V_{AB} \equiv V_B - V_A$. So when we refer to the "potential difference between A and B," we always mean the potential at B minus the potential at A.

Potential is important in practical applications because the potential difference between two points can be measured readily with a device called a *voltmeter* (we'll encounter these devices when we discuss electrical circuits in Chapters 31 and 32.) In the next chapter we shall discuss the operation of another familiar device, the *battery*, which allows one to maintain a constant potential difference between two points. A 9-V battery, for example, maintains a $+9$-V-potential difference between its negative and positive terminals. (The positive terminal is at the higher potential, and thus the potential difference is *positive* when going from $-$ to $+$.) For example, when a particle carrying charge $+1$ C is moved

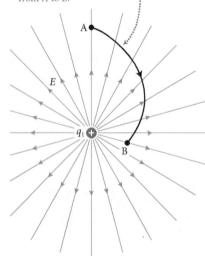

Figure 25.17 Once we know the electrostatic work done by a particle carrying a charge q_1 on a particle carrying a charge q_2 as the latter is moved from A to B in Figure 25.15, we can determine the potential difference between A and B in the electric field of particle 1.

Potential difference between A and B, $V_{AB} = V_B - V_A$, is negative of electrostatic work per unit charge done along a path from A to B.

from the negative terminal of a 9-V battery to the positive terminal, the particle undergoes a potential difference $V_{-+} = V_+ - V_- = +9$ V. Equation 25.17 tells us that the electrostatic work done on the particle is equal to

$$W_q(- \rightarrow +) = -q\, V_{-+} = -(+1\text{ C})(+9\text{ V}) = -9\text{ J}. \qquad (25.18)$$

That this quantity is negative indicates that the agent moving the particle must do a positive amount of work on the particle. Likewise, when a particle carrying a charge of -2 C is moved from the $-$ terminal to the $+$ terminal of a 9-V battery, the electrostatic work done on the particle is $-(-2\text{ C})(+9\text{ V}) = +18$ J.

For simplicity, we shall denote the magnitude of the potential difference between the terminals of a battery by V_{batt}. Therefore

$$V_{\text{batt}} = V_+ - V_-. \qquad (25.19)$$

Just as with potential energy, only potential *differences* are physically relevant. If we choose a reference point, however, we can determine the value for the potential at any other point. In the preceding section we chose infinity as the reference point for the electric potential energy for charged particles because they do not interact at infinite distance $U^E(\infty) = 0$. The same choice can be made for the potential, but when we deal with electrical circuits it is customary to assign zero potential to Earth (ground) because Earth is a good and very large conducting object through which the motion of charge carriers requires negligible energy.

Exercise 25.3 Potential and potential difference

The negative terminal of a 9-V battery is connected to ground via a wire. (a) What is the potential of the negative terminal? (b) What is the potential of the positive terminal? (c) What is the potential of the negative terminal if the positive terminal is connected to ground?

SOLUTION (a) I know from Section 25.3 that any conducting objects that are in electrical contact with each other form an equipotential. Once they are in contact, therefore, the ground (which is conducting), the wire, and the negative terminal are all at the same potential. If the potential of the ground is zero (an arbitrary but customary choice), then the potential of the negative terminal is also zero. ✔

(b) The potential difference between the negative and positive terminals is $+9$ V, meaning that the potential of the positive terminal is 9 V higher than that of the negative terminal: $V_{\text{batt}} = V_+ - V_- = +9$ V. If V_- is zero, then the potential of the positive terminal is $V_+ = +9$ V $+ V_- = (+9\text{ V}) + (0\text{ V}) = +9$ V. ✔

(c) With the positive terminal connected to ground, that terminal's potential becomes zero. Because the battery maintains a potential difference of $+9$ V between the negative and positive terminals, I now have $V_{\text{batt}} = V_+ - V_- = 0 - V_- = +9$ V and so the negative terminal is at a potential of -9 V. ✔

To obtain an explicit expression for the potential difference between two points A and B in the electric field of particle 1 carrying charge q_1, we start with the expression for the electrostatic work done by particle 1 on a particle 2 carrying charge q_2 as it is moved from A to B (Eq. 25.4). All we need to do is add a minus sign and divide by q_2:

$$V_{AB} \equiv \frac{-W_{12}(A \rightarrow B)}{q_2} = \frac{q_1}{4\pi\epsilon_0}\left[\frac{1}{r_B} - \frac{1}{r_A}\right]. \qquad (25.20)$$

Figure 25.18 Equipotentials, field lines, and graph of potential for a charged particle.

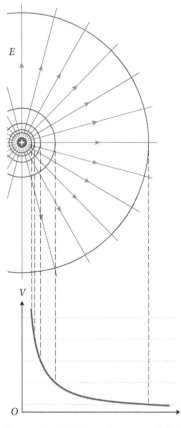

For charged particles, the choice of ground as a zero for the potential is not very meaningful. However, if we set the zero for potential at infinity and let r_A be at infinity, we obtain for the potential at a distance $r = r_B$ from a single charged particle

$$V(r) = \frac{1}{4\pi\epsilon_0}\frac{q_1}{r} \quad \text{(potential zero at infinity).} \quad (25.21)$$

Equation 25.21 confirms that the potential is constant on any spherical surface centered on the charged particle (that is, any surface for which r is constant), as we concluded in Checkpoint 25.7. **Figure 25.18** shows the potential for a charged particle in graphical form. The bottom of the figure shows the $1/r$ dependence of the potential; the top shows a two-dimensional view of the equipotentials around the charged particle. The $1/r$ dependence of the potential is reflected by the increasing spacing of the equipotentials (with a fixed potential difference between them) as r increases.

✋ **25.12** (a) Using Eq. 25.20, determine whether the potential difference between A and B in Figure 25.17 is positive, negative, or zero. (b) From the directions of the electric force and the force displacement, is the electrostatic work done on a positively charged particle as it is moved along a straight path from A to B positive, negative, or zero? Verify that your answer is consistent with the answer in part a.

Example 25.4 Atomic potential difference

A (simplistic) model of the hydrogen atom treats the electron as a particle carrying a charge $-e$ orbiting a proton (a particle carrying a charge $+e$) in a circle of radius $r_H = 0.53 \times 10^{-10}$ m. (a) How much energy is required to completely separate the electron from the proton? For simplicity, ignore the electron's kinetic energy. (b) Across what potential difference does the electron travel as it is separated from the proton?

❶ **GETTING STARTED** To completely separate the electron from the proton, I must increase the distance between them from r_H to infinity (at which point their interaction is reduced to zero). By increasing the separation, I increase the potential energy of the electron-proton system. The energy I must add to the system in order to separate the particles is equal to the increase in potential energy of the system.

❷ **DEVISE PLAN** I can calculate the potential energy increase from Eq. 25.6, setting $r_i = r_H$ and $r_f = \infty$. I can obtain the potential difference between the initial and final positions of the electron by taking the negative of the electrostatic work done by the electric field of the proton on the electron during the separation and dividing this value by the charge on the electron.

❸ **EXECUTE PLAN** (a) The change in the electric potential energy of the two-particle system is

$$\Delta U^E = \frac{q_1 q_2}{4\pi\epsilon_0}\left[\frac{1}{r_f} - \frac{1}{r_i}\right] = \frac{-e^2}{4\pi\epsilon_0}\left[0 - \frac{1}{r_H}\right]$$

$$= (9.0 \times 10^9 \text{ N}\cdot\text{m}^2/\text{C}^2)\frac{(1.6 \times 10^{-19}\text{C})^2}{0.53 \times 10^{-10}\text{ m}} = 4.3 \times 10^{-18} \text{ J}.$$

Because there are no changes in any other forms of energy in the system, the energy required to separate the electron and the proton in a hydrogen atom is 4.3×10^{-18} J. ✔

(b) The answer I obtained in part a is the mechanical work an external agent must do on the electron to separate it from the proton. I know that the energy of the electron does not change because the electron gains no kinetic energy. Considering just the electron as my system, I thus have $\Delta E = 0$ and so the work done on the electron is zero. This work has two parts: mechanical work done by the external agent and electrostatic work done by the electric field of the proton. Because the sum of these two parts is zero, I know that the electrostatic work done by the electric field of the proton on the electron must be the negative of the mechanical work done by the external agent, so $W_{pe}(r_H \to \infty) = -4.3 \times 10^{-18}$ J. The potential difference is thus (Eq. 25.15)

$$V_{H\infty} = \frac{-W_{pe}(r_H \to \infty)}{q} = \frac{-(-4.3 \times 10^{-18}\text{ J})}{-1.6 \times 10^{-19}\text{ C}} = -27 \text{ V}. ✔$$

❹ **EVALUATE RESULT** The positive sign on ΔU^E in part a means that the potential energy of the proton-electron system increases when the two are moved apart. Therefore mechanical work must be done to pull them apart, as I expect because the electron and proton attract each other. The value I obtain is extremely small, but I know that 1 m³ of matter contains about 10^{29} atoms (see Exercise 1.6), so the electric potential energy in a cubic meter of matter is on the order of $(10^{28})(4.3 \times 10^{-18} \text{ J}) \approx 10^{11}$ J. In the box "Coherent versus incoherent energy" on page 158 I learned that the amount of chemical energy in a pencil is on the order of 10^5 J. Given that chemical energy is derived from electric potential energy stored in chemical bonds and that the

volume of a pencil is about 10^{-5} m^3, my answer for part a is not unreasonable.

For part b, I could have used Eq. 25.20 directly. Substituting the values given into that equation, I obtain the same answer, which gives me confidence in the answer I obtained.

To obtain a more general result for the potential difference between one point and another in an arbitrary electric field $\vec{E}$, consider the situation illustrated in Figure 25.19. A particle carrying a charge q is moved from point A to point B in an electric field due to some charge distribution (not visible in the illustration). The electrostatic work done on the particle is

$$W_q(\text{A} \rightarrow \text{B}) = \int_\text{A}^\text{B} \vec{F}_q^E \cdot d\vec{\ell}. \qquad (25.22)$$

The vector sum of the forces exerted on the particle is equal to the product of the electric field and the charge q (Eq. 23.6):

$$\vec{F}_q^E = q\vec{E}, \qquad (25.23)$$

so

$$W_q(\text{A} \rightarrow \text{B}) = q\int_\text{A}^\text{B} \vec{E} \cdot d\vec{\ell}. \qquad (25.24)$$

The potential difference between point A and point B is therefore

$$V_{\text{AB}} \equiv \frac{-W_q(\text{A} \rightarrow \text{B})}{q} = -\int_\text{A}^\text{B} \vec{E} \cdot d\vec{\ell}. \qquad (25.25)$$

In evaluating this line integral, we must keep in mind that the integral does not depend on the path taken but only on the endpoints. It therefore pays to choose a path that facilitates evaluating the integral. The Procedure box below and the next example will help you gain practice calculating potential differences between two points.

Figure 25.19 To determine the potential difference between two points in an electric field, we must evaluate the electrostatic work done on a charged particle as the particle is moved along a path between those points.

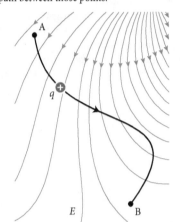

Procedure: Calculating the potential difference between two points in an electric field

The potential difference between two points in an electric field is given by Eq. 25.25. The following steps will help you evaluate the integral:

1. Begin by making a sketch of the electric field, indicating the points corresponding to the two points between which you wish to determine the potential difference.
2. To facilitate evaluating the scalar product $\vec{E} \cdot d\vec{\ell}$, choose a path between the two points so that $\vec{E}$ is either parallel or perpendicular to the path. If necessary, break the path into segments. If $\vec{E}$ has a constant value along the path (or a segment of the path), you can pull it out of the integral; the remaining integral is then equal to the length of the corresponding path (or the segment of the path).

3. Remember that to determine V_{AB} ("the potential difference between points A and B"), your path begins at A and ends at B. The vector $d\vec{\ell}$ therefore is tangent to the path, in the direction that leads from A to B (see also Appendix B).

At this point you can substitute the expression for the electric field and carry out the integral. Once you are done, you may want to verify the algebraic sign of the result you obtained: *negative* when a positively charged particle moves along the path in the direction of the electric field, and *positive* when it moves in the opposite direction.

Example 25.5 Electrostatic potential in a uniform field

Consider a uniform electric field of magnitude E between two parallel charged plates separated by a distance d. (a) What is the potential difference between the positive plate and the negative plate? (b) What is the value of the potential at a point P that lies between the plates and is a distance $a < d$ from the positive plate, if the potential of the negative plate is zero?

❶ **GETTING STARTED** I begin by making a sketch of the two parallel plates and the electric field (**Figure 25.20**). For part a I must calculate the potential difference between the positive and negative plates, so I choose a path that begins at the positive plate and ends at the negative plate. For part b I am asked to determine, relative to the potential at the negative plate, the potential at a point P located between the plates a distance a from the positive plate, so I choose a path that runs from P to the negative plate. I indicate the endpoints for both paths in my drawing and choose my x axis to be parallel with the electric field with the positive plate at $x = 0$.

Figure 25.20

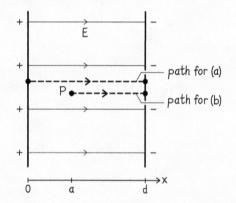

❷ **DEVISE PLAN** To obtain the potential difference between the endpoints of each path, I apply Eq. 25.25. With my choice of x axis the $d\vec{\ell}$ factor in Eq. 25.25 becomes $d\vec{\ell} = dx\,\hat{\imath}$. Because $\vec{E} = E\,\hat{\imath}$ I thus have $\vec{E} \cdot d\vec{\ell} = E\,dx$, and because the field is uniform, E is

constant. For part b, once I have the potential difference between the endpoints of the path from P to the negative plate, I can determine the potential at P because I am told that the potential at the negative plate is zero.

❸ **EXECUTE PLAN** (a) Applying Eq. 25.25 to the path from $x_i = 0$ (positive plate) to $x_f = d$ (negative plate), I get

$$V_{0d} = -\int_0^d E\,dx = -E\int_0^d dx = -Ed, \checkmark \qquad (1)$$

where I have pulled E out of the integral because it is constant.

(b) If I start the line integral in Eq. 1 at point P, I obtain

$$V(d) - V(a) = -E\int_a^d dx = -E(d - a)$$

or, because $V(d) = 0$,

$$V(a) = E(d - a). \checkmark$$

❹ **EVALUATE RESULT** The negative sign in Eq. 1 means that the potential at the end of the path (that is, at the negative plate) is *lower* than the potential at the beginning (the positive plate). This negative potential difference is in agreement with the sign of the electrostatic work done on a positively charged particle: Positive electrostatic work is done on the particle as it is moved from the positive plate to the negative plate, and so, according to Eq. 25.15, the potential difference should indeed be negative.

My result for part b indicates that the potential is positive at $x = a$ ($a < d$), decreases linearly with the distance a to the positive plate, and goes to zero at $a = d$. That makes sense, because I know that the potential of the positive plate must be higher than that of the negative plate and if $a = d$, I get $V(d) = 0$, as expected.

With an appropriate choice of reference point for the potential, Eq. 25.25 allows us to assign values to the potential at every point surrounding a charge distribution. This "potential field" is related to the electric field: Each can be determined from the other. A drawing that shows a set of equipotentials for a charge distribution is equivalent to a drawing that shows a set of field lines for that charge distribution. In Section 25.7 we shall show how the electric field can be derived from the potential field. The potential field, however, has advantages over the electric field. First, it is a scalar field whereas the electric field is a vector field, and so calculations involving the potential are generally simpler. Second, while no devices exist to measure electric field strength directly, we can measure the potential difference between two points with a voltmeter.

✋ **25.13** Verify that Eq. 25.25 is consistent with Eq. 25.20 by substituting the expression for the electric field of a charged particle.

By following the same procedure as in Checkpoint 25.13, we can now obtain the potential for a group of charged particles. Recall that the electric field due to a group of charged particles is equal to the sum of the electric fields of the individual charged particles (Eq. 23.5),

$$\vec{E} = \sum_n \vec{E}_n.$$ (25.26)

This gives us

$$V_{AB} = -\int_A^B \vec{E} \cdot d\vec{\ell} = -\int_A^B \left(\sum_n \vec{E}_n \right) \cdot d\vec{\ell}.$$ (25.27)

Because the integral of a sum is equal to a sum of integrals, we have

$$-\int_A^B \left(\sum_n \vec{E}_n \right) \cdot d\vec{\ell} = -\sum_n \int_A^B \vec{E}_n \cdot d\vec{\ell}.$$ (25.28)

The line integral after the summation sign is the negative of the potential difference due to the field $\vec{E}_n$, so we see that the total potential difference is the sum of the potential differences caused by the individual particles:

$$V_{AB} = \sum_n V_{AB,n},$$ (25.29)

where $V_{AB,n}$ is the potential difference caused by particle n. Substituting the potential of a single particle, Eq. 25.21, and again letting the potential at infinity be zero, we get

$$V_P = \frac{1}{4\pi\epsilon_0} \sum_n \frac{q_n}{r_{nP}} \quad \text{(potential zero at infinity)},$$ (25.30)

where q_n is the charge carried by particle n and r_{nP} is the distance of particle n from the point P at which we are evaluating the potential (**Figure 25.21**).

The line integral on the right in Eq. 25.25 has an important significance. As we argued in Section 25.2, the electrostatic work done on a charged particle moving around a closed path (as in **Figure 25.22**) is zero. Therefore, Eq. 25.24 must yield zero for a closed path:

$$W_q(\text{closed path}) = q \oint \vec{E} \cdot d\vec{\ell} = 0,$$ (25.31)

where the circle through the integral sign indicates that the integration is to be taken around a closed path. Because Eq. 25.31 holds for any value of q, we have

$$\oint \vec{E} \cdot d\vec{\ell} = 0 \quad \text{(electrostatic field).}$$ (25.32)

In other words, for any electrostatic field, the line integral of the electric field around a closed path is zero. This is equivalent to saying that you cannot extract energy from an electrostatic field by moving a charged particle around a closed path. In terms of potential, this means that if we start at some point P on the closed path and the potential has a value V_P at that point, then the potential can take on other values as we go around the closed path, but as we return to P, the value of the potential must once again be V_P. As we shall see later in Chapter 30, it *is* possible to get energy out of electric fields, however. Here we have shown only that it is *not* possible to extract energy by moving a charged particle around a closed path in *electrostatic* fields—that is, those due to stationary charge distributions.

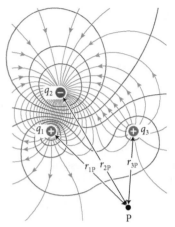

Figure 25.21 The potential at point P due to a group of charged particles is the algebraic sum of the potentials at P due to the individual particles.

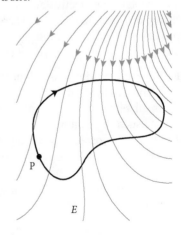

Figure 25.22 The electrostatic work done on a charged particle as the particle is moved around a closed path starting and ending at some point P is zero.

25.14 Describe how the potential varies as you go around the closed path in Figure 25.22 in the direction shown, starting from a potential V_P at P.

25.6 Electrostatic potentials of continuous charge distributions

For extended objects with continuous charge distributions, we cannot use Eq. 25.30 directly to calculate the potential. Instead we must divide the object into infinitesimally small segments, each carrying charge dq_s (which we can treat as a charged particle), and then integrate over the entire object.

Consider, for example the object shown in Figure 25.23. Let the zero of potential again be at infinity. Treating each segment as a charged particle, we calculate its contribution to the potential at P (Eq. 25.21):

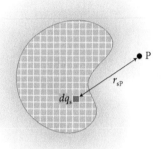

Figure 25.23 The potential due to an extended object is the algebraic sum of the potentials of all the infinitesimally small segments that make up the object.

$$dV_s = \frac{1}{4\pi\epsilon_0}\frac{dq_s}{r_{sP}}, \tag{25.33}$$

where r_{sP} is the distance between P and dq_s. The potential due to the entire object is then given by the sum over all the segments that make up the object. For infinitesimally small segments, this yields the integral

$$V_P = \int dV_s = \frac{1}{4\pi\epsilon_0}\int\frac{dq_s}{r_{sP}}, \quad \text{(potential zero at infinity)}, \tag{25.34}$$

where the integral is taken over the entire object. Note the parallel between this integral and the integral in Eq. 23.15 for calculating the electric field of a continuous charge distribution. Indeed, the procedure for calculating the potential of a continuous charge distribution is very similar to that for calculating the electric field of a continuous charge distribution (see Section 23.7). However, the potential in Eq. 25.34 is much easier to evaluate because it is a scalar and does not involve any unit vectors. The Procedure box below provides some helpful steps in evaluating Eq. 25.34, and the next two examples show how to put the procedure into practice.

PROCEDURE: Calculating the electrostatic potentials of continuous charge distributions

To calculate the electrostatic potential of a continuous charge distribution (relative to zero potential at infinity), you need to evaluate the integral in Eq. 25.34. The following steps will help you work out the integral:

1. Begin by making a sketch of the charge distribution. Mentally divide the distribution into infinitesimally small segments, each carrying a charge dq_s. Indicate one such segment in your drawing.

2. Choose a coordinate system that allows you to express the position of dq_s in the charge distribution in terms of a minimum number of coordinates (x, y, z, r, or θ). These coordinates are the integration variables. For example, use a radial coordinate system for a charge

distribution with radial symmetry. Never place the representative segment dq_s at the origin.

3. Indicate the point at which you wish to determine the potential. Express the factor $1/r_{sP}$, where r_{sP} is the distance between dq_s and the point of interest, in terms of the integration variable(s).

4. Determine whether the charge distribution is one dimensional (a straight or curved wire), two dimensional (a flat or curved surface), or three dimensional (any bulk object). Express dq_s in terms of the corresponding charge density of the object and the integration variable(s).

At this point you can substitute your expressions for dq_s and $1/r_{sP}$ into Eq. 25.34 and work out the integral.

Example 25.6 Electrostatic potential of a uniformly charged thin rod

A thin rod of length ℓ carries a uniformly distributed charge q. What is the potential V_P at point P a distance d from the rod along a line that runs perpendicular to the long axis of the rod and passes through one end of the rod?

❶ **GETTING STARTED** I begin by making a sketch. I let the y axis be along the rod, with the origin at the bottom of the rod. Point P then lies on the x axis (**Figure 25.24**). What I must calculate is the electrostatic potential V_P at P.

Figure 25.24

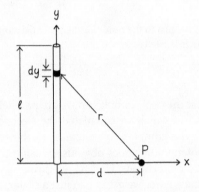

❷ **DEVISE PLAN** Because the rod is thin and uniformly charged, I can treat it as a one-dimensional object that has a linear charge density $\lambda = q/\ell$. To determine the electrostatic potential at point P, I divide the rod lengthwise into a large number of infinitesimally small segments, each of length dy. Each segment contributes to the potential at P an amount given by Eq. 25.33. To calculate the potential at P due to the entire rod, I can then use Eq. 25.34 to integrate my result over the entire length of the rod.

❸ **EXECUTE PLAN** Each length segment dy carries a charge $dq_s = \lambda dy = (q/\ell)dy$. To calculate the potential at P due to the

whole rod, I substitute the distance between each segment dy and P, $r_{sP} = \sqrt{y^2 + d^2}$, and my expression for dq_s into Eq. 25.34 and integrate from the bottom of the rod at $y = 0$ to the top at $y = \ell$:

$$V_P = \frac{1}{4\pi\epsilon_0} \frac{q}{\ell} \int_0^\ell \frac{dy}{\sqrt{y^2 + d^2}}.$$

Looking up the solution of the integral, I obtain

$$V_P = \frac{1}{4\pi\epsilon_0} \frac{q}{\ell} \left[\ln(y + \sqrt{y^2 + d^2}) \right]_0^\ell$$

$$= \frac{1}{4\pi\epsilon_0} \frac{q}{\ell} \left[\ln(\ell + \sqrt{\ell^2 + d^2}) - \ln d \right]$$

$$= \frac{1}{4\pi\epsilon_0} \frac{q}{\ell} \ln\left(\frac{\ell + \sqrt{\ell^2 + d^2}}{d} \right). ✔$$

❹ **EVALUATE RESULT:** If I let ℓ go to zero, my answer should become the same as that for a particle (Eq. 25.21). When $\ell \ll d$, I can ignore the term ℓ^2 in my answer and so the argument of the logarithm becomes $(\ell + d)/d = 1 + \ell/d$. Because $\ln(1 + \epsilon) \approx \epsilon$ when $\epsilon \ll 1$ (see Appendix B), my answer becomes

$$V_P \approx \frac{1}{4\pi\epsilon_0} \frac{q}{\ell} \ln\left(\frac{\ell + d}{d} \right) \approx \frac{1}{4\pi\epsilon_0} \frac{q}{\ell} \frac{\ell}{d} = \frac{1}{4\pi\epsilon_0} \frac{q}{d},$$

which is indeed equal to the electrostatic potential at a distance $r = d$ from a particle.

Example 25.7 Electrostatic potential of a uniformly charged disk

A thin disk of radius R carries a uniformly distributed charge. The surface charge density on the disk is σ. What is the electrostatic potential due to the disk at point P that lies a distance z from the plane of the disk along an axis that runs through the disk center and is perpendicular to the plane of the disk?

❶ **GETTING STARTED** I begin by making a sketch of the disk (**Figure 25.25**). I let the disk be in the xy plane, with the origin at the center of the disk and point P on the z axis.

Figure 25.25

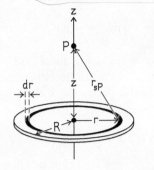

❷ **DEVISE PLAN** Because of the circular symmetry of the disk, I divide it into a large number of thin circular ring segments, each of radius r and thickness dr. All parts of a given ring are the same distance r_{sP} from point P, so each part makes the same contribution to the potential at P. I can therefore calculate the contribution of an entire ring segment to the potential at P using Eq. 25.33, substituting the charge dq_s on the segment and the distance from a point on the segment to P. The charge dq_s on the ring segment is given by the product of its area (circumference times thickness) and the surface charge density: $dq_s = (2\pi r) dr\, \sigma$. To calculate the potential at point P due to the entire disk, I can use Eq. 25.34 to integrate my result over the radius of the disk, using r as my integration variable.

❸ **EXECUTE PLAN:** The distance from a point on any ring segment to P is $r_{sP} = \sqrt{z^2 + r^2}$, and so each segment's contribution to the potential is given by

$$dV_P = \frac{1}{4\pi\epsilon_0} \frac{(2\pi r)\sigma\, dr}{\sqrt{z^2 + r^2}}.$$

(Continued)

To calculate the potential at P due to the whole disk, I integrate this expression from $r = 0$ to $r = R$:

$$V_P = \int dV_P = \frac{1}{4\pi\epsilon_0} \int_0^R \frac{2\pi r\sigma}{\sqrt{z^2 + r^2}}\, dr.$$

Looking up the solution of the integral, I get

$$V_P = \frac{2\pi\sigma}{4\pi\epsilon_0} \int_0^R \frac{r\, dr}{\sqrt{z^2 + r^2}} = \frac{\sigma}{2\epsilon_0}(\sqrt{z^2 + R^2} - |z|). ✔$$

❹ EVALUATE RESULT When z is very large relative to R, the disk should resemble a particle and my result should reduce to that

for a particle (Eq. 25.21). For large $z > 0$ I can use the binomial expansion (see Appendix B) to write the factor in parentheses as

$$\sqrt{z^2(1 + R^2/z^2)} - z \approx z\left(1 + \tfrac{1}{2}\frac{R^2}{z^2}\right) - z = \tfrac{1}{2}\frac{R^2}{z}.$$

Because $\sigma\pi R^2$ is equal to the charge q on the disk, my expression for V becomes

$$V_P \approx \frac{\sigma}{2\epsilon_0}\left(\tfrac{1}{2}\frac{R^2}{z}\right) = \frac{\sigma\pi R^2}{4\pi\epsilon_0 z} = \frac{1}{4\pi\epsilon_0}\frac{q}{z},$$

which is equal to the result for the potential along a z axis due to a particle located at the origin.

You may have noticed that the two examples above are parallel to corresponding examples in Section 23.7 in which we calculated the electric fields due to a thin charged rod or disk. If you compare the calculations, however, the advantage of working with the potential becomes obvious: Because the potential is a scalar you don't need to take vector sums. Haven't we thrown away some information, though? After all, the answer we get is also just a scalar, not a vector like the electric field. Figure 25.9 gives us some idea of the answer to this question: Because field lines and equipotentials are always perpendicular to each other, you can draw equipotentials if you know the field line pattern or, conversely, draw field lines if you know the equipotentials. It turns out that even though the potential is a scalar and the electric field is a vector, it is possible to determine one from the other.

✋ **25.15** Verify that the potentials obtained in Examples 25.6 and 25.7 have the correct sign.

25.7 Obtaining the electric field from the potential

For the potential to be a useful quantity, we must be able to determine the electric field (and therefore the forces exerted by this field) from the potential. For example, let the equipotentials in **Figure 25.26** represent the potential of some charge distribution. How can we use the known potential of the charge distribution to determine the value of the electric field at any point P?

We know that the electric field is perpendicular to the equipotentials, and so the electric field at P must be along the direction indicated in the figure. To determine the magnitude of the electric field, imagine moving a particle carrying a charge q over an infinitesimally small displacement $d\vec{s}$ along some arbitrary axis s. Let the particle be displaced from P, where the potential is V, to a point P′ where the potential is $V + dV$. According to Eq. 25.17, the electrostatic work done on the particle is

$$W_q(P \to P') = -qV_{PP'} = -q(V_{P'} - V_P) = -q\, dV \qquad (25.35)$$

because the potential difference between P and P′ is $(V + dV) - V = dV$. On the other hand, we also know that the electrostatic work done on the particle is equal to the scalar product of the electric force exerted on the particle and the force displacement $d\vec{r}_F = d\vec{s}$:

$$W_q(P \to P') = \vec{F}_q^E \cdot d\vec{s} = (q\vec{E}) \cdot d\vec{s}$$
$$= q(\vec{E} \cdot d\vec{s}) = qE\cos\theta\, ds, \qquad (25.36)$$

Figure 25.26 To determine the component of electric field along an axis, we calculate the electrostatic work done on a charged particle as the particle is moved over a short segment along that axis.

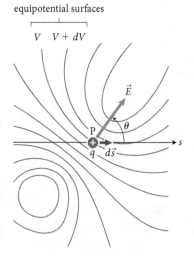

equipotential surfaces

where we have assumed that the force displacement is small enough that $\vec{E}$ can be considered constant between P and P'. Note that θ is the angle between $\vec{E}$ and the s axis, so $E \cos \theta$ is the component of the electric field along the s axis. We can write $E \cos \theta = E_s$, and equating the two expressions for the electrostatic work done on the particle, Eqs. 25.35 and 25. 36, we get

$$-q \, dV = qE_s \, ds \qquad (25.37)$$

or

$$E_s = -\frac{dV}{ds}. \qquad (25.38)$$

In other words, the component of the electric field along the s axis is given by the negative of the derivative of the potential with respect to s. The faster V varies (the more closely spaced the equipotentials), the greater the magnitude of the electric field.

Equation 25.38 gives only the component of the electric field along the (arbitrary) axis s. To determine all the components of the electric field, we must repeat this procedure for each of the three Cartesian coordinates:

$$E_x = -\frac{\partial V}{\partial x}, \quad E_y = -\frac{\partial V}{\partial y}, \quad E_z = -\frac{\partial V}{\partial z}. \qquad (25.39)$$

Note the partial derivatives; these are necessary because the function $V(x, y, z)$ depends on all three Cartesian coordinates. When you take the partial derivative with respect to one coordinate, just remember to keep the other coordinates constant. (For example, if $V = x^2 y$, then $\partial V/\partial x = 2xy$, $\partial V/\partial y = x^2$, and $\partial V/\partial z = 0$.)

Once the components of the electric field are determined, we can write the electric field in vectorial form:

$$\vec{E} = -\frac{\partial V}{\partial x}\hat{\imath} - \frac{\partial V}{\partial y}\hat{\jmath} - \frac{\partial V}{\partial z}\hat{k}. \qquad (25.40)$$

Equation 25.40 then tells us how to obtain the electric field from the potential.

25.16 Apply Eq. 25.40 to the potential you obtained for the uniform field between two parallel charged plates in Example 25.5, and verify that you get the correct expression for the electric field between the plates.

Example 25.8 The electrostatic potential and electric field due to a dipole

A permanent dipole consists of a particle carrying a charge $+q_p$ at $x = 0, y = +\frac{1}{2}d$ and another particle carrying a charge $-q_p$ at $x = 0, y = -\frac{1}{2}d$. Use the electrostatic potential at a point P on the axis of the dipole to determine the electric field at that point.

❶ GETTING STARTED I begin by making a sketch of the dipole (Figure 25.27). I place the dipole along my y axis, with the midpoint of the dipole length at the origin, so that the particles are at the coordinates given in the problem. I choose a point P on the positive y axis, and I let the y coordinate of P be y. Because the two charged particles lie on the y axis, the electric field they create at P must be directed along the y axis.

Figure 25.27

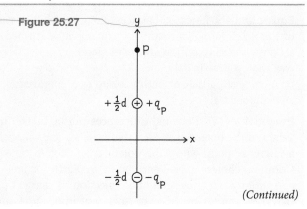

(Continued)

QUANTITATIVE TOOLS

❷ DEVISE PLAN The potential at P is the sum of the potentials due to the individual charged particles. To calculate these potentials, I can use Eq. 25.30. Once I have calculated the potential at P, I can use Eq. 25.40 to determine the electric field.

❸ EXECUTE PLAN The potential at P is

$$V_P = \frac{1}{4\pi\epsilon_0} \sum \frac{q_n}{r_{nP}} = \frac{1}{4\pi\epsilon_0}\left[\frac{+q_P}{y - \frac{1}{2}d} + \frac{-q_P}{y + \frac{1}{2}d}\right]. \quad (1)$$

Now I use this result to determine the electric field. Because I am working with the y axis, I work with the y component

of Eq. 25.40:

$$E_y = -\frac{\partial V}{\partial y} = \frac{1}{4\pi\epsilon_0}\left[\frac{q_P}{(y - \frac{1}{2}d)^2} - \frac{q_P}{(y + \frac{1}{2}d)^2}\right]. ✔$$

Because the electric field at P is directed along the y axis, the other components of the electric field are zero: $E_x = E_z = 0.$ ✔

❹ EVALUATE RESULT: This is the same result as in Eq. 23.11. (Remember $k = 1/(4\pi\epsilon_0)$.)

In comparing the derivation in Example 25.8 with the derivation of the electric field in Section 23.6, note that the calculation of the electric field via the potential does not involve any vector addition. The scalar nature of the potential therefore greatly simplifies calculations.

✋ 25.17 Calculate the electric field at any point on the axis of a thin charged disk from the potential we obtained in Example 25.7. Compare your answer to the result we obtained by direct integration in Section 23.7.

Chapter Glossary

SI units of physical quantities are given in parentheses.

Electric potential energy U^E (J) The form of potential energy associated with the configuration of stationary objects that carry electrical charge. When the reference point for the electric potential energy is set at infinity, the potential energy for two particles carrying charges q_1 and q_2 and separated by a distance r_{12} is

$$U^E = \frac{q_1 q_2}{4\pi\epsilon_0}\frac{1}{r_{12}} \quad (U^E \text{ zero at infinite separation}). \quad (25.8)$$

Electrostatic work W_q (J) Work done by an electrostatic field on a charged particle or object moving through that field. The electrostatic work depends on only the endpoints of the path. For a particle of charge q that is moved from point A to point B in an electric field, the electrostatic work is

$$W_q(A \rightarrow B) = q\int_A^B \vec{E} \cdot d\vec{\ell}. \quad (25.24)$$

Equipotentials Lines or surfaces along which the value of the potential is constant. The equipotential surfaces of a charge distribution are always perpendicular to the corresponding electric field lines. The electrostatic work done on a charged particle or object is zero as it is moved along an equipotential.

Potential V_P (V) Potential differences can be turned into values of the potential at every point in space by choosing a reference point where the potential is taken to be zero. Common choices of reference point are Earth (or *ground*) and infinity. The potential of a collection of charged particles (measured with respect to zero at infinity) at some

point P can be found by taking the algebraic sum of the potentials due to the individual particles at P:

$$V_P = \frac{1}{4\pi\epsilon_0}\sum \frac{q_n}{r_{nP}} \quad (\text{potential zero at infinity}), \quad (25.30)$$

where q_n is the charge carried by particle n and r_{nP} is the distance from P to that particle. For continuous charge distributions, the sum can be replaced by an integral:

$$V_P = \frac{1}{4\pi\epsilon_0}\int \frac{dq_s}{r_{sP}} \quad (\text{potential zero at infinity}). \quad (25.34)$$

The electric field can be obtained from the potential by taking the partial derivatives:

$$\vec{E} = -\frac{\partial V}{\partial x}\hat{\imath} - \frac{\partial V}{\partial y}\hat{\jmath} - \frac{\partial V}{\partial z}\hat{k}. \quad (25.40)$$

Potential difference V_{AB} (V) The potential difference between points A and B is equal to the negative of the electrostatic work per unit charge done on a charged particle as it is moved along a path from A to B:

$$V_{AB} \equiv \frac{-W_q(A \rightarrow B)}{q} = -\int_A^B \vec{E} \cdot d\vec{\ell}. \quad (25.25)$$

For electrostatic fields, the potential difference around a closed path is zero:

$$\oint \vec{E} \cdot d\vec{\ell} = 0 \quad (\text{electrostatic field}). \quad (25.32)$$

Volt (V) The derived SI unit of potential defined as $1\,\text{V} \equiv 1\,\text{J/C}.$

26

Charge Separation and Storage

CONCEPTS

QUANTITATIVE TOOLS

This chapter deals with generating and storing electric potential energy. To produce charged objects, positive and negative charge carriers must be pulled apart and then kept separate. Work is required to pull apart charge carriers, just as work is required to stretch a spring. In each case, this work results in energy storage in the system. We now look at what kind of changes in energy are involved in the separation of positive and negative charge carriers and how charge carriers that have been separated can be stored in simple arrangements of conductors.

26.1 Charge separation

Whenever objects are "charged" (by separating strips of Scotch tape, rubbing objects against each other, using batteries, etc.), the basic phenomenon is the same: Some process (pulling, rubbing, chemical reactions) separates positive and negative charge carriers from one another. As a concrete example, consider a rubber rod and a piece of fur. If you rub the two together and then separate them, they become oppositely charged because the rod pulls electrons away from the fur: The rod ends up with a surplus of electrons and the fur with a deficit. Provided none of the electrons on the rod leak away (to the air, your hand, etc.), the magnitude of the negative charge on the rod is equal to that of the positive charge on the fur.

What is the change in energy associated with this charge separation? Consider the rubber-fur system in its initial and final states (**Figure 26.1a**). To separate the positive and negative charge carriers, they must be pulled apart against an attractive electric force, just as the ends of a stretched spring are pulled apart against an elastic force. Because work must be done on the rod-fur system, the electric potential energy of the system is greater in the final state. This energy is supplied by you while you rub the two objects together and then increase their separation. Not all of the energy you put into the system goes into electric potential energy; the friction involved in the rubbing not only produces charge separation but also heats up the rod and the fur, so part of the work you do on the system increases the thermal energy. An energy diagram for the rod-fur system is shown in Figure 26.1b.

🖐 **26.1** Suppose you repeat the charging (starting again with uncharged rod and fur), but this time you rub longer and twice as much charge accumulates at each point on the two objects. How do the following quantities compare to what they were after the first charging: (*i*) the direction and magnitude of the electric field at point P in Figure 26.1a; (*ii*) the potential difference between two fixed points on the rod and the fur; and (*iii*) the electric potential energy in the rod-fur system?

Checkpoint 26.1 highlights the essence of this chapter. Be sure not to confuse *potential difference* and *electric potential energy*:

- The system's electric potential energy depends on the configuration of the positive and negative charge carriers in the system.

Figure 26.1 When we charge a rubber rod and a piece of fur by rubbing them together, we do work to separate charge and hence increase the electric potential energy of the system comprising the rod and fur.

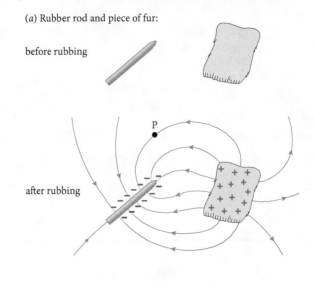

(*a*) Rubber rod and piece of fur:

before rubbing

P

after rubbing

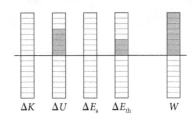

(*b*) Energy diagram for system of rod and fur

$\Delta K \quad \Delta U \quad \Delta E_s \quad \Delta E_{th} \quad W$

- The potential difference between points on the rod and the fur is a measure of the electrostatic work done on a particle carrying a unit of charge (not part of the system) while moving between those points.

In the next section we examine the proportionality between potential difference and charge separation in detail. In this section we concentrate on the relationship between charge separation and electric potential energy.

The crucial point to take away from Checkpoint 26.1 is:

Positive work must be done on a system to cause a charge separation of the positive and negative charge carriers in the system. This work increases the system's electric potential energy.

🖐 **26.2** If you double the separation between the charged rod and fur in Figure 26.1a, does the electric potential energy of the rod-fur system increase, decrease, or stay the same?

The amount of stored electric potential energy depends on the amount of charge that is separated and the distance that separates the charge carriers. More charge or a greater separation means more electric potential energy is stored. These arguments apply to *all* devices that separate charge, such as Van de Graaff generators (see the box below) and batteries (see Section 26.4). Once electric potential energy has been generated by separating charge carriers, this energy can be used for other processes, such as lighting a lamp, operating a radio, and so on.

Every **charge-separating device** (or **charging device**) has some mechanism that moves charge carriers *against* an electric field—a process that requires work to be done on the system of charge carriers. For the rod-fur system, this work is mechanical and is supplied by the person doing the rubbing and separating the objects. In a battery, chemical reactions drive charge carriers through a region where the electric field opposes their motion.

26.3 If you include the person doing the rubbing in the system considered in Figure 26.1, what is the resulting energy diagram?

Where is the electric potential energy of the rod-fur system stored? As you may recall from Section 7.2, potential energy is stored in reversible changes in the configuration of interacting components of a system. Electric potential energy, therefore, is associated with the configuration of the charge carriers in a system. A look at the electric field pattern suggests an alternative view, however. **Figure 26.2** shows how the electric field line pattern changes as the distance between the charged rod and the fur increases. Note how more of the space around the system becomes filled with field lines (indicating that the magnitude of the electric field there increases), while the density of the field lines between

Figure 26.2 Change in the electric field pattern as the distance between the rod and fur increases.

As rod and fur get farther apart:
Magnitude of electric field *between* them decreases . . .

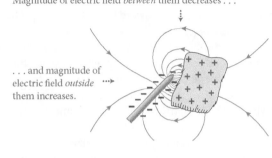

. . . and magnitude of electric field *outside* them increases.

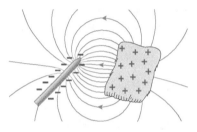

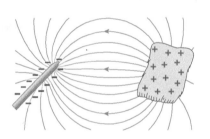

Van de Graaff generator

Figure 26.3 shows a schematic diagram of a Van de Graaff generator—a mechanical device invented in the 1930s by Robert J. Van de Graaff to separate large amounts of electrical charge. The basic principle is extremely simple: A nonconducting belt delivers charge carriers to a hollow conducting dome that rests on a nonconducting support. Machines of this type are used to generate the very large potential differences required in particle accelerators and for the generation of x rays.

Operation of the generator involves three important steps. The first step is a transfer of charge carriers to the belt at A. This transfer can be done by literally "spraying" charged particles onto the belt or simply by rubbing the rubber belt against some appropriate material.

The second step transports the charge carriers to the dome. This step is possible because the belt is nonconducting, so the charge carriers are not mobile—they are stuck to the belt, which is driven by a motor around a pulley inside the dome. The motor must do work on the charge carriers to move them against the electric field of the dome. (In the example shown in the diagram, positive charge carriers at B must be transported upward against the downward electric field of the positively charged dome.)

Figure 26.3 Schematic diagram of a Van de Graaff generator.

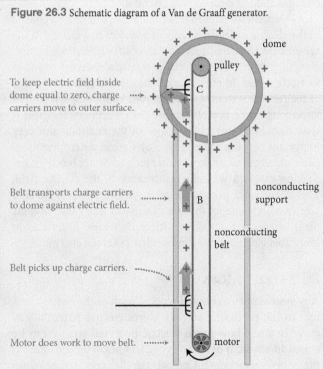

To keep electric field inside dome equal to zero, charge carriers move to outer surface.

Belt transports charge carriers to dome against electric field.

Belt picks up charge carriers.

Motor does work to move belt.

The third step transfers the charge carriers from the belt onto the dome. As we saw in Section 24.5, the electric field inside a hollow conductor is always zero and any charge inside a conductor moves toward the outer surface. Therefore, once the charge carriers are inside the dome, they tend to move to the outer surface of the dome. For this purpose, a comb of conducting needles is placed close to the belt at C. If the charge carriers on the belt are electrons, the electrons hop onto the comb and move via the connecting wire toward the outside of the metal dome, causing the dome to acquire a negative charge. Alternatively, the charge carriers on the belt can be positively ionized air molecules, in which case electrons in the comb are attracted toward the ions. These electrons then jump from the comb onto the belt, neutralizing the ions on the belt while leaving a positive charge behind on the outside of the dome.

Construction of a huge double Van de Graaff generator for the MIT Physics Department in South Dartmouth, Massachusetts, in 1933. These generators, currently at the Boston Museum of Science, generated opposite charge and were able to produce potential differences of 10,000,000 V between the two 4.5-m domes.

the rod and the fur decreases. Thus, as the distance between the rod and the fur changes, the electric potential energy changes and the electric field changes.

As we shall see later in this chapter, we can relate a change in the electric potential energy of a system to a change in the system's electric field (integrated over all of space), which suggests that the electric potential energy of a system is stored in its electric field. In other words, the electric potential energy of the rod-fur system is spread throughout the space around it. As long as the two objects are held stationary, the field is stationary, so the exact "location" of the energy is not very important because there is no way we can determine it experimentally. If we shake the charged rod or fur, however, the shaking causes a wavelike disturbance in the electric field. This disturbance propagates away from the rod or fur and carries with it energy that we can detect. For now we don't need to concern ourselves with such waves. It suffices to know that electric fields store electric potential energy.

26.2 Capacitors

Any system of two charged objects, such as the rod-fur system in the preceding section, stores electric potential energy. To study how much electric potential energy can be stored in a system of two objects, let's begin by considering the simple arrangement of two parallel conducting plates

shown in Figure 26.4a. A system for storing electric potential energy that consists of two conductors is called a **capacitor**; the arrangement in Figure 26.4a is called a *parallel-plate capacitor*.

Figure 26.4b illustrates a simple method for charging such a capacitor. Each plate is connected by a wire to a terminal of a battery, which maintains a fixed potential difference between its terminals. Figure 26.5 shows what happens when the connection is made between the battery and the capacitor. If the capacitor plates are far enough away from

Figure 26.4 A parallel-plate capacitor.

(*a*) Not charged

(*b*) Charged

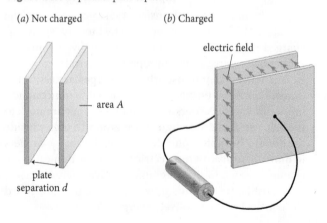

electric field

area *A*

plate separation *d*

Figure 26.5 Charging a capacitor.

(*a*) Capacitor not connected to battery

Zero potential difference
between uncharged plates.

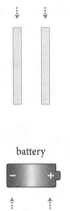

battery

Battery maintains potential
difference between terminals.

(*b*) Capacitor being charged

Electrons flow along wires in
direction of higher potential.

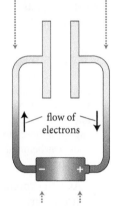

flow of
electrons

Chemical reactions in battery supply
charge to terminals, keeping potential
difference fixed.

(*c*) Capacitor fully charged

Potential of each plate now identical
to that of corresponding battery terminal.

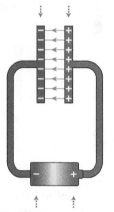

Potential difference between terminals
stays the same.

the battery, the potential difference between the plates initially is zero (Figure 26.5*a*). Immediately after the wires are connected, there is a potential difference between the ends of each wire. This difference in potential causes electrons (which are mobile in metal) in the wires to flow as indicated by the arrows in Figure 26.5*b*. A positive charge builds up on the plate connected to the positive terminal, and a negative charge of equal magnitude builds up on the other plate. As electrons leave one plate and accumulate on the other, the potential difference between the plates changes. This process continues until the potential is the same at both ends of each wire—that is, when the potential difference between the plates is equal to that between the terminals of the battery (Figure 26.5*c*). Because there is no longer any potential difference from one end of the wire to the other, the flow of electrons stops and the capacitor is said to be *fully charged*. In the process of achieving this state, the battery has done work on the electrons; this work has now become electric potential energy stored in the capacitor.

The time interval it takes to fully charge a capacitor depends on the properties of the capacitor, the battery, and the way the capacitor is connected to the battery. Typically, only a fraction of a second is needed for charging, although the time interval it takes to charge very large capacitors can be minutes (more on this in Chapter 32).

26.4 (*a*) Suppose that we disconnect the wires from the plates after the capacitor is charged as shown in Figure 26.5*c*. How does the potential difference between the plates after the wires are disconnected compare to that just before they are disconnected? (*b*) If we replace the battery in Figure 26.5 by a battery that maintains a greater potential difference between its terminals, is the magnitude of the charge on the plates greater than, smaller than, or the same as when the first battery is connected?

When a capacitor is not connected to anything, as in Checkpoint 26.4*a*, it is said to be *isolated*. For an isolated capacitor, the *quantity of charge on each plate* is fixed because the charge carriers have nowhere to go. In contrast, for a capacitor that is connected to a battery, the *potential difference across the capacitor* is fixed—the charge carriers on the plates always adjust themselves in such a way as to ensure that the potential difference across the capacitor is equal to that across the battery.

Figure 26.6 shows the electric fields of two isolated charged parallel-plate capacitors. The field is nearly uniform in the region between the plates, but it is nonuniform at the edges. When the spacing between the plates is small compared to the

Figure 26.6 Effect of plate separation in relation to plate area on the field of a parallel-plate capacitor.

Plate separation small
compared to plate area:

As plate separation becomes greater
compared to plate area . . .

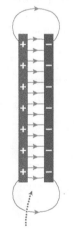

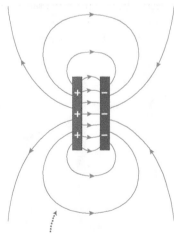

Electric field nearly uniform,
localized mainly between plates.

. . . more electric field "escapes"
from between plates.

Figure 26.7 Doubling the charge on a parallel-plate capacitor.

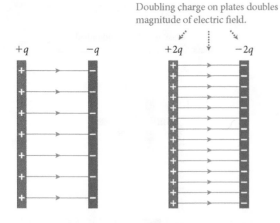

Doubling charge on plates doubles magnitude of electric field.

area of the plates, the effect of the nonuniform field is negligible: The electric field is confined almost entirely to the region between the plates, and for all practical purposes we can consider this field to be uniform. When discussing parallel-plate capacitors, we shall ignore the nonuniform fields at the edges and assume the electric field is entirely uniform between the plates. This simplification is justified by the geometry of most capacitors used in electronic applications.

Let us now examine the relationship between the magnitude of the charge on the plates of a parallel-plate capacitor and the potential difference between them. **Figure 26.7** shows an isolated parallel-plate capacitor carrying a positive charge $+q$ on one plate and a negative charge $-q$ on the other. If we double the magnitude of the charge on each plate, then the electric field between the plates doubles, too (see Checkpoint 23.14). Consequently, the electric force exerted on a charged particle between the plates doubles, so the electrostatic work done in moving a charged particle from one plate to the other also doubles. According to Eq. 25.15, the potential difference between the plates doubles as well. In other words, the potential difference between the plates is proportional to the magnitude of the charge on the plates.

What happens if we increase the plate separation of an isolated parallel-plate capacitor, as illustrated in **Figure 26.8**? The electric field remains the same because it is determined by the

Figure 26.8 Doubling the plate separation of a parallel-plate capacitor.

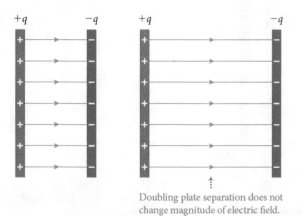

Doubling plate separation does not change magnitude of electric field.

surface charge density on the plates, which doesn't change (see Checkpoint 23.14). Because the distance between the plates increases, however, the electrostatic work done in moving a charged particle from one plate to the other increases—more work is required to move the particle over a greater distance—so the potential difference between the plates increases too, confirming the result we obtained in Example 25.5.

26.5 Suppose the two capacitors in Figure 26.8 are each connected to a 9-V battery. (*a*) Which of the two capacitors stores the greater amount of charge? (*b*) If, instead of the separation increasing, the area of the plates of the capacitor is halved and then the capacitor is connected to a 9-V battery, does the capacitor store more charge, less charge, or the same amount of charge as before the area of the plates was halved?

As Checkpoint 26.5 illustrates, the geometry of the capacitor determines its capacity to store charge. In general:

For a given potential difference between the plates of a parallel-plate capacitor, the amount of charge stored on its plates increases with increasing plate area and decreases with increasing plate separation.

Does this mean that we can increase the amount of charge stored on a parallel-plate capacitor indefinitely simply by making the plate separation infinitesimally small? The answer is *no*, because if the plate spacing is decreased while the potential difference between the capacitor plates is fixed, the charge on each plate increases and thus the magnitude of the electric field in the capacitor increases. When the electric field is about 3×10^6 V/m, the air molecules between the plates become *ionized* and the air becomes conducting, allowing a direct transfer of charge carriers between the plates. Once such a so-called **electrical breakdown** occurs, the capacitor loses all its stored energy in the form of a spark.

The opening page of this chapter shows an electrical breakdown of air between the charged dome of a very large Van de Graaff generator and a nearby metal object. The breakdown limits the maximum potential difference across a capacitor and thus the maximum amount of charge that can be stored on it. The electric field at which electrical breakdown occurs is called the *breakdown threshold*.

The breakdown threshold can be raised by inserting a nonconducting material between the capacitor plates. As we shall see in the next section, such a nonconducting material also greatly increases the amount of charge that can be stored by a capacitor. To understand why this is so, we begin by considering a simpler situation: the insertion of a conductor between the plates of a parallel-plate capacitor.

Figure 26.9a shows an isolated charged capacitor. Suppose we now insert a conducting slab between the plates of this capacitor (Figure 26.9b). As we saw in Section 24.5, the charge carriers in the conductor rearrange themselves in such a fashion as to eliminate the field inside the bulk of the conductor.

Figure 26.9 Inserting a conducting slab between the plates of a parallel-plate capacitor.

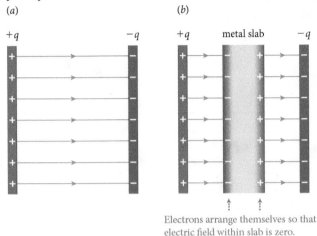

Electrons arrange themselves so that electric field within slab is zero.

🖐 **26.6** Suppose the capacitor in Figure 26.9a is charged and then disconnected from the battery. (a) As the conducting slab is inserted in the capacitor, as in Figure 26.9b, does the amount of charge on the capacitor plates increase, decrease, or stay the same? (b) How much charge accumulates on each side of the slab, once it is inserted? (c) What is the potential difference across the metal slab? (d) As the slab is inserted, does the magnitude of the potential difference between the capacitor plates increase, decrease, or stay the same?

As Checkpoint 26.6 demonstrates, the effect of the slab is to make the electric field in part of the space between the capacitor plates zero, thus reducing the magnitude of the potential difference between the plates for a given amount of charge on them. As the next example shows, the converse of this fact is that, for a given potential difference between its plates, the capacitor can store more energy with the slab inserted than it can without the slab. In other words, the slab increases the capacitor's capacity to store charge.

Example 26.1 Metal-slab capacitor

Suppose the capacitor in Figure 26.9 has a plate separation distance d and the plates carry charges $+q$ and $-q$ when the capacitor is connected to a battery that maintains a potential difference V_{batt} between its terminals. If a metal slab of thickness $d/2$ is inserted midway between the plates while the battery remains connected, what happens to (a) the magnitude of the electric field between the plates and (b) the quantity of charge on the plates?

❶ **GETTING STARTED** I am given the plate separation distance and plate charges for a capacitor connected to a battery, and I must determine how the electric field magnitude between the plates and the quantity of charge on each plate change when a metal slab is inserted. Because the capacitor remains connected to the battery, the potential difference across the capacitor is fixed. Because the slab is conducting, the electric field inside the slab is always zero.

❷ **DEVISE PLAN** The electric field magnitude determines the electrostatic work, which in turn determines the potential difference between the plates (which I know). To determine how the electric field magnitude changes when the metal slab is inserted, therefore, I must first determine how the potential between the plates changes when the slab is inserted. Once I know the electric field, I can determine how the slab affects the charge on the plates because I know that the magnitude of the electric field between the plates is proportional to the surface charge density σ (see Checkpoint 23.14, $E = 4k\pi\sigma$).

❸ **EXECUTE PLAN** To compare the potentials before and after the slab is inserted, I must plot V as a function of position between the plates (**Figure 26.10**). I choose my x axis to be parallel to the electric field, with the positive plate at $x = 0$ and the negative plate at $x = d$ as the zero of potential. In the absence of the metal slab, the field is uniform between the plates, and so the potential decreases linearly from $x = 0$ to $x = d$ (Figure 26.10a). Because the electric field inside the slab is zero, the potential does not vary across the slab (it is an equipotential; see Section 25.3). Because the battery keeps the potential at each plate constant, the potential-versus-distance curve must take on the zigzag form shown in Figure 26.10b.

Figure 26.10

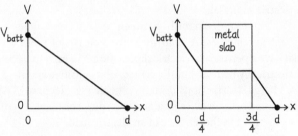

(a) Potential without slab   (b) Potential with metal slab

(a) Insertion of the metal slab affects the electric field between the plates because the slab is an electrical conductor and so the electric field inside it must always be zero. However, because V_{batt} is constant, I know that the electrostatic work, $W_q = F^E d = qV_{batt}$, done on a particle carrying a charge q to move the particle from one plate to the other must be the same whether or not the slab is in place there. Because no electrostatic work is done to move the particle through the slab where the electric field is zero, the field outside the slab must be greater to make up for the smaller distance over which the particle is moved. More precisely, because the distance over which the electric field is nonzero is reduced to $d/2$, the magnitude of the field must be twice what it was before the slab was inserted. ✔

(b) If the field doubles, then the charge per unit area must also double. Given that the area of the plates does not change, this means that the charge on the plates must double. ✔

❹ **EVALUATE RESULT** Inserting the metal slab with the battery connected is equivalent to halving the separation distance between the plates while keeping the potential difference constant. I know from Checkpoint 26.5 that, for a constant potential difference, the quantity of charge stored on a capacitor increases with decreasing plate separation, as I found.

✋ **26.7** (*a*) Does the position of the slab in Figure 26.9 affect the potential difference across the capacitor? Consider, in particular, the case in which the slab is moved all the way to one side and makes electrical contact with one of the plates. (*b*) Sketch the potential $V(x)$ as a function of x, with the slab off-center.

26.3 Dielectrics

As we just saw, decreasing the space inside an isolated charged capacitor where the electric field is nonzero increases its capacity to store electrical charge for a given potential difference across its plates. With a conducting slab inserted, however, the gap between either plate and the slab face nearest it is smaller than the plate-to-plate gap before the slab was inserted. Because a decreased gap with the potential difference held constant means E increases, we still have the problem of electrical breakdown.

Suppose, however, that we insert a nonconducting material—a **dielectric**—between the plates of a capacitor. As we discussed in Section 22.4, the electric field between the plates of the capacitor polarizes the dielectric. What effect does this polarization have? To answer this question we must first look in more detail at what happens in a polarized dielectric material.

We should distinguish between two general types of dielectric materials. A *polar* dielectric consists of molecules that have a permanent electric dipole moment; each molecule is electrically neutral, but the centers of its positive and negative charge distributions do not coincide (see Section 23.4). The atoms or molecules in a *nonpolar* dielectric have no dipole moment in the absence of an electric field.

Figure 26.11a shows the polarization of a nonpolar dielectric. In the presence of an electric field, the electrons in a nonpolar dielectric are displaced in the direction opposite to $\vec{E}$, inducing a dipole moment on each molecule.

✋ **26.8** Why are the electrons displaced in a direction *opposite* the electric field?

The polarization of a polar dielectric is shown in Figure 26.11*b*. In the absence of an electric field, the individual molecules' dipole moments are randomly aligned, so the material as a whole is not polarized. In the presence of an electric field, however, the molecular dipoles are subject to a torque (see Section 23.8) that tends to align the dipoles with the electric field, giving rise to a macroscopic polarization. In general, the polarization of polar dielectrics is much greater than that of nonpolar ones because the permanent dipole moments of the molecules in a polar dielectric are much greater than the induced dipole moments in a nonpolar dielectric.

Figure 26.12 illustrates the effect of the uniform polarization of the atoms or molecules in a polar or nonpolar dielectric. The charge enclosed by any volume that lies

Figure 26.11 Polarization of nonpolar and polar molecules in an electric field.

(a) Polarization of nonpolar molecules

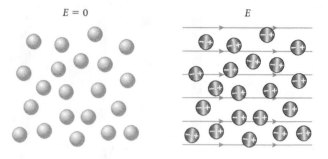

(b) Polarization of polar molecules

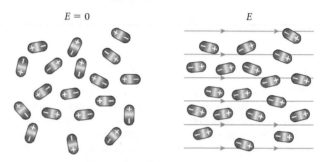

entirely inside the polarized dielectric is zero: The positive and negative charge carriers may not coincide exactly, but on average they occur in equal numbers. However, this is not true for a small volume at the surface of the dielectric. In the volume on the right in Figure 26.12, for example, a surplus of positive charge appears at the surface of the material. Thus, a polarized dielectric has a very thin sliver of surplus positive charge on one side and a sliver containing an equal amount of surplus negative charge on the other side (see also Figure 22.21). The surface charge on either side of the polarized dielectric is said to be **bound** because the charge carriers that cause it are not free to roam around in the material. In contrast, the charge on the capacitor plates is **free**

Figure 26.12 The reason a polarized dielectric exhibits a macroscopic polarization.

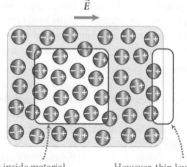

For volume inside material, enclosed charge sums to zero.

However, thin layer at surface has surplus of positive charge.

Figure 26.13 The polarization induced on a dielectric in a parallel-plate capacitor is equivalent to two thin sheets carrying opposite charge.

(*a*) Dielectric sandwiched between capacitor plates

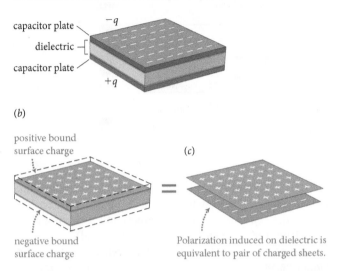

(*b*)

positive bound
surface charge

(*c*)

negative bound
surface charge

Polarization induced on dielectric is
equivalent to pair of charged sheets.

Figure 26.14 The presence of a polarized dielectric reduces the strength of the electric field between the plates of a capacitor.

(*a*)   (*b*)

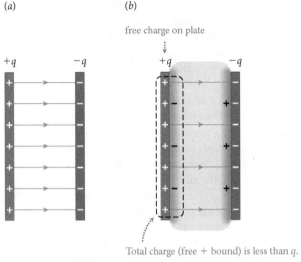

free charge on plate

Total charge (free + bound) is less than *q*.

because the charge carriers that cause it can move around freely. From a macroscopic point of view, a uniformly polarized dielectric differs from an unpolarized dielectric only by the presence of this bound surface charge.

What is the effect of the bound surface charge on the electric field inside the capacitor? Consider a dielectric slab inside an isolated charged capacitor (**Figure 26.13a**). The dielectric is polarized by the electric field of the capacitor; that is, a positive bound surface charge appears on the surface near the negatively charged capacitor plate, and a negative bound surface charge appears on the other side (Figure 26.13*b*). Imagine now that we could "freeze in" the polarization and consider just the slab by itself. Except for two sheets of charge at the top and bottom, the bulk of the dielectric is neutral. Thus, for all practical purposes, the polarized dielectric is equivalent to two very thin sheets carrying opposite charge (Figure 26.13*c*).

26.9 (*a*) In which direction does the electric field due to the bound surface charge point at a location above the top surface in Figure 26.13*c*? (*b*) In which direction does it point at a location between the top and bottom surfaces?

We can now obtain the electric field of the capacitor with the dielectric by superposition: It is equal to the electric field of the capacitor without the dielectric, plus the electric field of the polarized dielectric by itself. As you found in Checkpoint 26.9, the direction of the electric field due to the polarized dielectric is opposite that of the capacitor, so the presence of the dielectric decreases the electric field strength in the capacitor. Alternatively, we can say that each of the bound surface charges compensates for part of the free charge on the adjoining capacitor plate, so, in effect, the total charge (free and bound) on each side of the capacitor

is reduced (**Figure 26.14**). This reduction in charge, in turn, gives rise to a smaller electric field inside the capacitor. For some materials, the field inside can be reduced by a factor of several thousand.

26.10 (*a*) If the magnitude of the bound surface charge on the dielectric slab in Figure 26.14*b* were equal to the magnitude of the free charge on the capacitor plates, what would be the electric field inside the capacitor? (*b*) Could the magnitude of the bound surface charge ever be *greater* than the magnitude of the free charge on the plates?

Figure 26.15 shows what happens when a dielectric-filled capacitor is connected to a battery. The battery keeps the potential difference between the capacitor plates the same regardless of the presence of the dielectric. Because

Figure 26.15 The presence of a polarized dielectric increases the charge on the plates of a capacitor connected to a battery.

Battery keeps electric field between plates same in both cases:

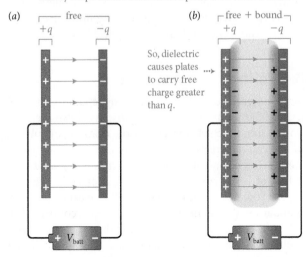

(*a*) —— free ——
$+q$ $-q$

(*b*) ┌ free + bound ┐
$+q$ $-q$

So, dielectric causes plates to carry free charge greater than *q*.

V_{batt} V_{batt}

the electric field in the capacitor is equal to the potential difference divided by the distance between the plates (see Example 25.5), it follows that, as long as the capacitor is connected to the battery, the electric field must be the same regardless of the dielectric. The electric field can be the same only if the distribution of charge causing the electric field is the same. In other words, regardless of the presence of the dielectric, we must have the same amount of total charge (free and bound) on each side of the capacitor. Let the magnitude of the free charge on the capacitor plates without the dielectric be q. As shown in Figure 26.15b, the polarization of the dielectric causes a negative bound surface charge next to the positive capacitor plate; the sum of the free charge on the conductor and the adjoining bound surface charge must still be equal to $+q$. Similarly, the sum of the negative free charge on the opposite plate and the positive bound surface charge on the adjoining dielectric still must be $-q$. Consequently, the magnitude of the free charge on each plate by itself can be much greater than without the dielectric. This extra charge is supplied by the battery.

26.11 Given that the electric field is the same in both capacitors in Figure 26.15, which stores the greater amount of electric potential energy?

If the answer to Checkpoint 26.11 surprises you—after all, the electric fields are the same in the two capacitors—remember that the amount of *charge separation* is not the same. The charge on the capacitor plates polarizes the dielectric, and this polarization is the result of charge separation in the molecules of the dielectric. Thus, instead of empty space without charge separation between the capacitor plates, we now have (in addition to a greater charge on the plates) a lot of additional charge separation on the microscopic scale. Most of the energy stored in the capacitor is not due to the separation of charge on the plates, but to the separation of charge in the dielectric between the plates. The pulling apart of the positive and negative charge distributions in the dielectric increases the electric potential energy stored in the dielectric, much like stretching a spring by pulling its ends apart stores elastic potential energy. This tells us that an electric field of a given magnitude in a dielectric stores more energy than an equal field in vacuum.

26.4 Voltaic cells and batteries

Electric potential energy is generated by separating charged particles. Earlier in this chapter we discussed two means of accomplishing such charge separation: charging by rubbing and the Van de Graaff generator. Another common way to generate electric potential energy is by means of a *voltaic cell*, the first of which was constructed by Alessandro Volta in around 1800. Assemblies of voltaic cells are called *batteries*. A standard 9-V alkaline battery, for example, consists of six 1.5-V cells connected together. While there are many types of voltaic cells and batteries, all have a common operating

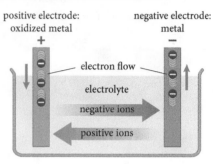

Figure 26.16 General operating principle of a voltaic cell. Electrons flow when the cell is connected to an electronic device.

principle: Chemical reactions turn chemical energy into electric potential energy by accumulating electrons on one side of the cell (the negative terminal) and removing electrons from the other side (the positive terminal).

The general principle of a voltaic cell is illustrated in **Figure 26.16**. Two conducting terminals, or *electrodes*, are submerged in an *electrolyte*—a solvent that contains mobile ions. One electrode is usually made from an oxidized metal; the oxidized metal reacts by accepting positive ions from the electrolyte and electrons from the electrode. The other electrode is generally metallic; it oxidizes by taking in negative ions and giving up electrons. Because of these reactions, a surplus of electrons builds up on the metallic terminal and a deficit of electrons builds up on the oxidized-metal terminal, causing a potential difference between the two. The reactions stop when the potential difference between the electrodes reaches a certain value called the *cell potential difference*. This value is determined by the type of chemicals in the cell and is typically on the order of a few volts. As we saw in Section 25.5, this potential difference can be used to do electrostatic work on charge carriers when a battery is connected to some device—such as a light bulb, a motor, or a capacitor.

As long as the cell is not connected to anything and its chemicals do not deteriorate, the cell remains in the same state indefinitely. When the cell is connected to a capacitor or to some other device, however, the surplus of electrons is removed from the negative electrode and electrons are supplied to the positive electrode. Then the chemical reactions resume in order to maintain the cell potential difference between the terminals. As the reactions proceed, the electrolyte becomes more dilute and the compositions of the electrodes change. The cell is exhausted when all the ions in the electrolyte have been depleted. For some types of cells, the chemical reactions can be reversed by supplying a potential difference to the terminals of the cell; electric potential energy is then converted back to chemical energy. Such cells are used in rechargeable batteries, which can be reused repeatedly to store and recover electric potential energy.

As we noted in Section 26.1, any charge-separating device involves the motion of charge carriers against the direction of an electric field. This is where work is done on the charge

Figure 26.17 Schematic diagram of a lead-acid cell and of the reactions taking place at the positive and negative electrodes.

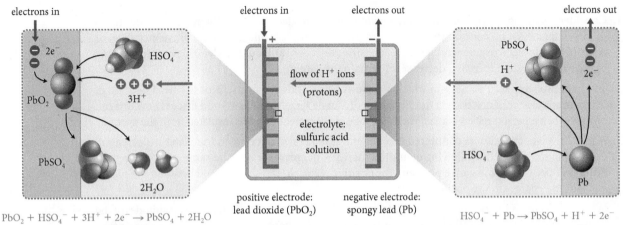

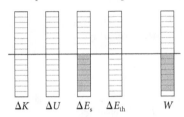

$$PbO_2 + HSO_4^- + 3H^+ + 2e^- \rightarrow PbSO_4 + 2H_2O$$

Reaction pulls electrons from positive terminal:

$\Delta K \quad \Delta U \quad \Delta E_s \quad \Delta E_{th} \quad W$

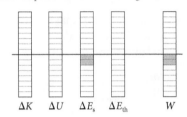

$$HSO_4^- + Pb \rightarrow PbSO_4 + H^+ + 2e^-$$

Reaction pushes electrons onto negative terminal:

$\Delta K \quad \Delta U \quad \Delta E_s \quad \Delta E_{th} \quad W$

carriers and where some form of energy is converted to electric potential energy. Inside a voltaic cell, electrons must be pulled away from the positively charged terminal and deposited onto the negatively charged terminal. This process occurs at the surface of the electrodes where the chemical reactions take place. The chemical reactions move charge carriers against strong opposing electric forces. The chemical energy released in the reactions provides the energy necessary to move the charge carriers against the electric field. The work done per unit charge is called the **emf** (pronounced e-m-f)* of the device:

The emf of a charge-separating device is the work per unit charge done by nonelectrostatic interactions in separating positive and negative charge carriers inside the device.

All charge-separating devices—batteries, voltaic cells, generators, solar cells—have some *nonelectrostatic* means to separate charge carriers and thereby create a potential difference across the terminals of the device.

26.12 As electrons leave one terminal and are added to the other, ions in the electrolyte must flow in the direction indicated in Figure 26.16 to maintain an even distribution of charge. What must be the direction of the electric field in the bulk of the electrolyte to cause this flow?

Figure 26.17 illustrates the operation of a lead-acid cell used in automobile batteries. A 12-V automobile battery consists of six such cells, each producing a potential difference of 2.1 V. The negative electrode of a lead-acid cell is composed of spongy lead (Pb) packed on a metal grid; the positive electrode contains lead dioxide (PbO_2) packed on a metal grid. The electrodes are immersed in sulfuric acid and chemical reactions convert the lead, the lead dioxide, and the sulfuric acid into lead sulfate ($PbSO_4$) and water. For every molecule of lead sulfate that is produced in these reactions, one electron is removed from the positive terminal and one is added to the negative terminal. The left and right sides of Figure 26.17 show energy diagrams for the species undergoing chemical reactions at each of the electrodes. For each reaction, the chemical energy of the species involved in the reaction decreases; this energy is used to do work on the electrons in the electrodes.

26.13 Given that the cell does *positive* work on the electrons, why is it that the work in both energy diagrams in Figure 26.17 is *negative*?

---

*emf stands for *electromotive force*, a misnomer because this quantity bears no relation to the concept of force. For this reason we shall always refer to this quantity by its abbreviation, rather than its original meaning.

Self-quiz

1. Consider again Figure 26.2 and imagine moving one more electron from the fur to the rod. (*a*) Is the work that must be done on the rod-fur system to accomplish this transfer positive, negative, or zero? (*b*) Is the electrostatic work positive, negative, or zero? (*c*) Does the electric potential energy of the rod-fur system increase, decrease, or remain the same?

2. You have probably seen pictures in which a person's hair stands out from his or her head because of "electrostatic charge." Look back at the discussion of Van de Graaff generators and discuss how this can happen when a person makes contact with the globe of the generator but is insulated from the ground.

3. A parallel-plate capacitor is connected to a battery. If the distance between the plates is decreased, do the magnitudes of the following quantities increase, decrease, or stay the same: (*i*) the potential difference between the negative plate and the positive plate, (*ii*) the electric field between the plates, and (*iii*) the charge on the plates?

4. When a dielectric is inserted between the plates of an isolated charged capacitor, do the magnitudes of the following quantities increase, decrease, or stay the same: (*i*) the charge on the plates, (*ii*) the electric field between the plates, and (*iii*) the potential difference between the negative plate and the positive plate?

5. Draw an energy diagram for the process of charging a capacitor with a dielectric as shown in Figure 26.15*b* for the following systems: (*a*) battery, capacitor, and dielectric; (*b*) dielectric only; (*c*) battery and capacitor. Ignore any dissipation of energy.

ANSWERS:

1. (*a*) To displace the electron toward the rod, you must apply a force directed toward the rod. Because the force and force displacement are in the same direction, you (an external agent) must do positive work. (*b*) The electric force exerted on the electron is directed toward the fur, opposite the direction of the force displacement, so the electrostatic work is negative. (*c*) The electric potential energy of the system increases because separating charge carriers increases a system's electric potential energy.

2. If a person is in contact with the globe of the generator but insulated from the ground, then the person acts as an extension of the globe. Electrical charge spreads out over the surface of the person, including the surface of each hair as well. Because each hair has a surplus of the same type of charge, the hairs repel each other and stand out, getting as far away from each other as possible.

3. (*i*) Stays the same. The battery keeps the potential difference across the capacitor constant. (*ii*) To keep a constant potential difference when the distance between the plates decreases, the magnitude of the electric field between the plates must increase because $Ed = V_{batt}$ (see Example 25.5). (*iii*) For the magnitude of the electric field to increase, the charge on the plates must increase.

4. (*i*) Because the capacitor is isolated, the charge on the plates must remain the same—there is no path for the charge to travel elsewhere. (*ii*) When the dielectric is inserted, the electric field due to the bound surface charge is in the opposite direction of the electric field due to the free charge on the plates and decreases the magnitude of the electric field between the plates. (*iii*) Because the magnitude of the electric field decreases and the separation between the plates is constant, the magnitude of the potential difference between the negative plate and the positive plate must also decrease.

5. See **Figure 26.18**. (*a*) During charging, a decrease in source energy (from the battery) increases the electric potential energy (more charge separation in the dielectric and on the capacitor plates). (*b*) The electric potential energy of the dielectric increases due to work done on it by the battery and the capacitor. The electric potential energy stored in the dielectric is smaller than that stored in part *a* because some electric potential energy is stored on the capacitor plates. (*c*) The decrease in source energy is the same as in part *a*. The electric potential energy stored on the capacitor is smaller than in part *a* because most of the converted source energy ends up in the dielectric, which is not part of the system considered. This energy leaves the system as negative work.

Figure 26.18

(a)

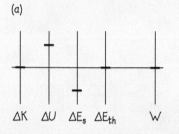

ΔK ΔU ΔE_s ΔE_{th} W

(b)

ΔK ΔU ΔE_s ΔE_{th} W

(c)

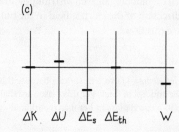

ΔK ΔU ΔE_s ΔE_{th} W

26.5 Capacitance

Figure 26.19 shows three capacitors, each one consisting of a pair of conducting objects carrying opposite charges of magnitude q. For each arrangement, the potential difference between the objects is proportional to q; that is, doubling q doubles the potential difference across the capacitor. The ratio of the magnitude of the charge on one of the objects to the magnitude of the potential difference across them is defined as the **capacitance** of the arrangement:

$$C \equiv \frac{q}{V_{\text{cap}}}. \qquad (26.1)$$

In Eq. 26.1, q represents *the magnitude of the charge on each conducting object* and V_{cap} is *the magnitude of the potential difference between the conducting objects.* Because both these quantities are positive, C is always positive.

The value of C depends on the size, shape, and separation of the conductors. In Figure 26.19, for example, the values of V would typically be different for the three capacitors, even though q is the same for each. Below we'll examine how to determine C for a given set of conductors.

✋ **26.14** Two capacitors, A and B, are each connected to a 9-V battery. If $C_A > C_B$, which capacitor stores the greater amount of charge?

The answer to Checkpoint 26.14 suggests a simple interpretation of C. As its name suggests, C represents the capacitor's *capacity to store charge*: The greater C, the greater the amount of charge stored for a given value of V_{cap}.

As you can see from Eq. 26.1, capacitance has SI units of coulomb per volt. This derived unit is given the name **farad** (F), in honor of the English physicist Michael Faraday:

$$1\,\text{F} \equiv 1\,\text{C/V}.$$

As you will see in Checkpoint 26.15, a capacitance of 1 F is enormous. The capacitance of capacitors commonly found in electronic devices is expressed in microfarads ($1\,\mu\text{F} = 1 \times 10^{-6}\,\text{F}$) and picofarads ($1\,\text{pF} = 1 \times 10^{-12}\,\text{F}$).

Figure 26.19 suggests a simple procedure for determining the capacitance of a given set of conductors: Determine the potential difference V_{cap} between the two conductors when they carry some given charge q, and use Eq. 26.1 to calculate C. Note that because conductors are equipotentials, V_{cap} represents the potential difference between *any* two points on the conductors measured along *any* path. The Procedure box below gives one procedure for determining the capacitance of a given set of conductors. In the next examples we apply this procedure to some simple configurations of conductors.

Figure 26.19 The electric fields and potential differences of three different capacitors.

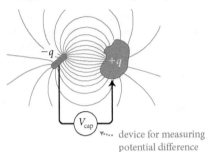

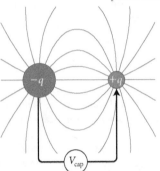

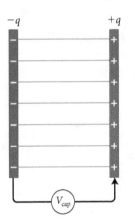

Procedure: Calculating the capacitance of a pair of conductors

To calculate the capacitance of a pair of conductors:

1. Let the conductors carry opposite charges of magnitude q.
2. Use Gauss's law, Coulomb's law, or direct integration to determine the electric field along a path leading from the negatively charged conductor to the positively charged conductor.

3. Calculate the electrostatic work W done on a test particle carrying a charge q_t along this path (Eq. 25.24) and determine the potential difference across the capacitor from Eq. 25.15:

$$V_{\text{cap}} = -W_{q_t}(- \rightarrow +)/q_t.$$

4. Use Eq. 26.1, $C \equiv q/V_{\text{cap}}$, to determine C.

Example 26.2 Parallel-plate capacitor

What is the capacitance of a parallel-plate capacitor that has a plate area A and a plate separation distance d?

❶ GETTING STARTED I begin by making a sketch of the capacitor, showing the electric field between the plates (**Figure 26.20a**). The problem doesn't specify the plate shape, so I simply show the capacitor from the side, representing each plate by a horizontal line. If I assume that the separation distance d is small, then the electric field is uniform and confined between the plates.

Figure 26.20

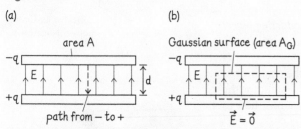

(a)

area A

$-q$

E

d

$+q$

path from $-$ to $+$

(b)

Gaussian surface (area A_G)

$-q$

E

$+q$

$\vec{E} = \vec{0}$

❷ DEVISE PLAN I can use the steps of the Procedure box on page 697 to determine the capacitance. The first step is to determine the electric field between the capacitor plates when they carry opposite charges of magnitude q. The second step is to obtain the electrostatic work done on a test particle moved from one plate to the other; I can use Eq. 25.24 to obtain this work. Because the field is uniform, it is most convenient to choose a path along a field line for the path over which the electrostatic work is done. As specified in the Procedure box, the path runs from the negatively charged plate to the positively charged plate. Once I know the electrostatic work, I know the potential difference across the capacitor and so can calculate the capacitance.

❸ EXECUTE PLAN Because of the planar symmetry, I can use Gauss's law to determine the electric field. I choose a cylindrical Gaussian surface straddling the surface of the positively charged plate. The cylinder height is less than d, and the area of the end surfaces is A_G (**Figure 26.20b**). The electric flux through this Gaussian surface is zero everywhere except through the top surface, where $\Phi = EA_G$. (The bottom surface is inside the conducting metal plate, where the electric field is zero.)

To apply Gauss's law, I also need to know the charge enclosed by the Gaussian surface. The positive plate carries a charge $+q$ distributed over a surface of area A, so the surface charge density is $\sigma = +q/A$ and the charge enclosed by the Gaussian surface is $q_{enc} = \sigma A_G = (q/A)A_G$. Applying Gauss's law, $\Phi = q_{enc}/\epsilon_0$ (Eq. 24.8), I get

$$EA_G = \frac{q}{\epsilon_0 A} A_G \quad \text{or} \quad E = \frac{q}{\epsilon_0 A} = \frac{\sigma}{\epsilon_0} \quad (1)$$

in agreement with Eq. 24.17.

Now that I know E, I can calculate the electrostatic work required to move a test particle carrying a charge $+q_t$ from the negatively charged to the positively charged plate. The electric force exerted on the test particle is upward in Figure 26.20, and the force displacement is downward because the particle moves from negative plate to positive plate. Because these two vectors point in opposite directions, the electrostatic work done on the test particle is negative, $W_{q_t} = -q_t Ed$, and the potential difference across the capacitor is, from Eq. 25.15,

$$V_{cap} \equiv \frac{-W_{q_t}(- \to +)}{q_t} = \frac{q_t Ed}{q_t} = Ed$$

or, substituting E from Eq. 1,

$$V_{cap} = \frac{qd}{\epsilon_0 A}.$$

Note that the potential difference is proportional to the magnitude of the charge on each plate, q. Using the definition of capacitance, I obtain

$$C \equiv \frac{q}{V_{cap}} = \frac{q}{qd/(\epsilon_0 A)} = \frac{\epsilon_0 A}{d}. \checkmark$$

❹ EVALUATE RESULT My result agrees with the conclusions we drew in Section 26.2: The capacitance (or quantity of charge stored for a given potential difference) increases with increasing plate area A and decreasing plate separation distance d. Also, I note that the electric field—and therefore the capacitance—do not depend on the plate: Circular or square plates give the same result.

✋ 26.15 The plate spacing in a typical parallel-plate capacitor is about 50 μm. (a) What is the plate area in a 1-μF capacitor? (b) Given that the electric field at which electrical breakdown occurs in air is about 3×10^6 V/m, what is the maximum charge that this capacitor can hold? (c) How many electrons does this charge correspond to? (The electron's charge is $e = 1.6 \times 10^{-19}$ C.)

Figure 26.21

(a) One way to design a compact capacitor with a large surface area

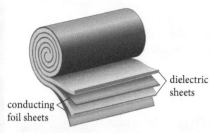

dielectric sheets

conducting foil sheets

(b) Some capacitors used in electronic circuits

As Checkpoint 26.15 shows, even modest capacitances require very large plate areas. Various techniques are used to keep the overall size of capacitors small, one of which involves rolling up two thin conducting sheets that are separated by thin sheets of a dielectric material (**Figure 26.21a**). Figure 26.21b shows a number of different capacitors used in electronic circuits.

Example 26.3 Coaxial cylindrical capacitor

Figure 26.22 shows a *coaxial capacitor* consisting of two concentric metal cylinders 1 and 2, of radii R_1 and $R_2 > R_1$, and both of length $\ell \gg R_2$. Both cylinders are made of metal. What is the capacitance of this arrangement?

Figure 26.22 Example 26.3.

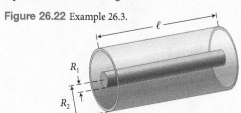

❶ **GETTING STARTED** To determine the capacitance, I must let the two cylinders carry opposite charges of magnitude q, which I assume to be uniformly distributed over each cylinder. If I let cylinder 1 carry a charge $+q$ and cylinder 2 carry a charge $-q$, the electric field points radially outward from cylinder 1 to cylinder 2 (**Figure 26.23**). Because the cylinders are very long relative to their separation distance $R_2 - R_1$, I assume that the electric field is confined to the volume between the cylinders.

Figure 26.23

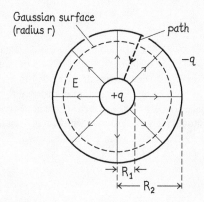

❷ **DEVISE PLAN** Again I refer to the Procedure box on page 697 to calculate the capacitance. For the path over which electrostatic work is done, I choose a straight path that goes radially from cylinder 2 to cylinder 1.

❸ **EXECUTE PLAN** Because of the cylindrical symmetry, I choose a cylindrical Gaussian surface (Figure 26.23). The length of the Gaussian surface is ℓ_G, and its radius is r ($R_2 > r > R_1$). The electric flux Φ through the curved portion of the Gaussian surface is equal to the product of the electric field strength E_r at a distance r from the common axis of cylinders 1 and 2 and the surface area of the Gaussian surface $A_G = (2\pi r)\ell_G$. Therefore $\Phi = 2\pi r \ell_G E_r$. Because the linear charge density on cylinder 1 is $+q/\ell$, the quantity of charge enclosed by the Gaussian surface is given by the product of the linear charge density and the length

of the Gaussian surface: $q_{enc} = +(q/\ell)\ell_G$. Applying Gauss's law, I get

$$2\pi r \ell_G E_r = \frac{q\ell_G}{\epsilon_0 \ell},$$

or $E_r = q/(2\pi\epsilon_0 \ell r)$, in agreement with the result we obtained in Exercise 24.7, $E = 2k\lambda/r$, because $k = 1/(4\pi\epsilon_0)$ and $q/\ell = \lambda$.

Now that I know E_r, I can calculate the electrostatic work required to move a test particle carrying a charge q_t from cylinder 2 to cylinder 1. Integrating the electric force exerted on the test particle over the force displacement from cylinder 2 (negatively charged) to cylinder 1 (positively charged), I get

$$W_{q_t} = \int_{R_2}^{R_1} \frac{qq_t}{2\pi\epsilon_0 \ell r}\, dr.$$

Working out the integral, I obtain

$$W_{q_t} = \frac{qq_t}{2\pi\epsilon_0 \ell}\left[\ln r\right]_{R_2}^{R_1} = \frac{qq_t}{2\pi\epsilon_0 \ell}\ln\!\left(\frac{R_1}{R_2}\right).$$

The potential difference between the negative cylinder 2 and the positive cylinder 1 is thus

$$V_{cap} \equiv \frac{-W_{q_t}}{q_t} = -\frac{q}{2\pi\epsilon_0 \ell}\ln\!\left(\frac{R_1}{R_2}\right) = \frac{q}{2\pi\epsilon_0 \ell}\ln\!\left(\frac{R_2}{R_1}\right).$$

Because $R_2 > R_1$, the logarithm is positive and therefore V_{cap} is positive, as it should be because I am bringing a quantity q_t of positive charge from a location of low potential on negatively charged cylinder 2 to a location of high potential on positively charged cylinder 1.

According to Eq. 26.1, the capacitance of the coaxial capacitor is thus

$$C \equiv \frac{q}{V_{cap}} = \frac{2\pi\epsilon_0 \ell}{\ln(R_2/R_1)}. \checkmark$$

❹ **EVALUATE RESULT** My result shows that the capacitance is proportional to ℓ, which makes sense: The longer the coaxial cylinders, the greater the quantity of charge that can be stored on them. Decreasing R_1 or increasing R_2 is equivalent to increasing the plate separation distance d in a parallel-plate capacitor, which decreases the capacitance. Indeed, my result shows a decreasing capacitance for decreasing R_1 or increasing R_2. (The dependence on R_1 and R_2 is a bit more complicated than the dependence on d in a parallel-plate capacitor because the electric field in the coaxial capacitor is nonuniform and because changing the radii of the cylinders affects their surface areas.)

🖐 **26.16** Coaxial cables used for cable television typically have a central metallic core of 0.20-mm radius, surrounded by a cylindrical metallic sheath of 2.0-mm radius. The two are separated by a plastic spacer. If the effect of the spacer can be ignored (that is, assuming the two conductors are separated by air), what is the capacitance of a 100-m-long cable?

Example 26.4 Spherical capacitor

What is the capacitance of a spherical capacitor consisting of two concentric conducting spherical shells of radii R_1 and $R_2 > R_1$?

❶ GETTING STARTED If I let the inner sphere carry a positive charge $+q$ and the outer one a negative charge $-q$, my sketch of the capacitor looks identical to the sketch of the coaxial capacitor of Example 26.3 (Figure 26.24). The calculation, however, will not be the same because now the electric field has spherical, not cylindrical, symmetry.

Figure 26.24

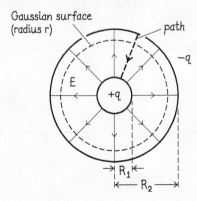

Gaussian surface (radius r)

path

$-q$

E

$+q$

R_1

R_2

❷ DEVISE PLAN For the path over which electrostatic work is done, I choose a straight path that goes radially from the outer sphere to the inner sphere. The outer sphere does not contribute to the electric field between the spheres because the field inside a hollow conductor is always zero (see Section 24.5). The electric field created by the inner sphere is the same as that created by a charged particle (Eq. 24.15, $E = kq/r^2$), so I can use this expression to follow steps 3 and 4 in the Procedure box on page 697.

❸ EXECUTE PLAN The electrostatic work done in moving a test particle carrying a charge q_t from the outer sphere to the inner sphere is

$$W_{q_t}(- \rightarrow +) = \int_{R_2}^{R_1} \frac{qq_t}{4\pi\epsilon_0 r^2} \, dr.$$

Working out the integral, I get

$$W_{q_t}(- \rightarrow +) = -\frac{qq_t}{4\pi\epsilon_0}\left[\frac{1}{r}\right]_{R_2}^{R_1} = -\frac{qq_t}{4\pi\epsilon_0}\left[\frac{1}{R_1} - \frac{1}{R_2}\right].$$

The potential difference between the outer and inner spheres is thus

$$V_{cap} \equiv \frac{-W_{q_t}(- \rightarrow +)}{q_t} = \frac{q}{4\pi\epsilon_0}\left[\frac{1}{R_1} - \frac{1}{R_2}\right],$$

so the capacitance is

$$C \equiv \frac{q}{V_{cap}} = 4\pi\epsilon_0\left[\frac{1}{R_1} - \frac{1}{R_2}\right]^{-1} = 4\pi\epsilon_0\left[\frac{R_2 - R_1}{R_1 R_2}\right]^{-1}$$

$$= 4\pi\epsilon_0\left[\frac{R_1 R_2}{R_2 - R_1}\right]. ✔$$

❹ EVALUATE RESULT I expect the capacitance to go up as the separation distance $R_2 - R_1$ between the spheres decreases, and this is just what my result shows. If I increase the spheres' radii while keeping their separation distance $R_2 - R_1$ fixed, the surface area $A = 4\pi R^2$ of each sphere increases and the capacitance should increase, in agreement with my result.

Figure 26.25 To determine the electric potential energy stored in a capacitor, we calculate the energy required to transfer charge from one conductor to the other.

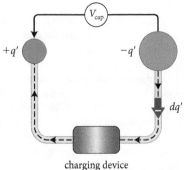

V_{cap}

$+q'$

$-q'$

dq'

charging device

✋ **26.17** (a) To calculate the "capacitance" of an isolated sphere, evaluate the result we obtained in Example 26.4 in the limit that R_2 goes to infinity. (b) What is the capacitance of the spherical metal dome of a Van de Graaff generator like the one shown in the chapter-opening photo, which has a radius of about 2.5 m. (c) Given that air breaks down in an electric field with a magnitude of about 3.0×10^6 V/m, what is the maximum amount of charge that can be stored on the dome before the air breaks down?

26.6 Electric field energy and emf

How much electric potential energy is stored in a charged capacitor? To answer this question, consider a simple capacitor consisting of two conducting objects. In order to charge the capacitor, some charge-separating device must transfer charge from one conductor to the other (Figure 26.25). During this transfer, the charge-separating device does work on the capacitor and this energy ends up as electric potential energy "stored in the capacitor."* One complication in the calculation of the work done by the charge-separating device is that as the magnitude of the charge on each conductor increases, the potential difference increases too, so the work required to transfer a unit of charge increases.

---

*We are assuming that there is no dissipation of energy, so that all the work done on the system ends up as electric potential energy. In practice this is a reasonable assumption.

Let us therefore break down the transfer of charge from one conductor to the other into small increments of charge dq', so that the potential difference is essentially constant during the transfer of a single increment. Consider some instant during the charging when the magnitude of the charge on each conductor is q'. The potential difference between the negative and positive conductors is then given by Eq. 26.1: $V_{cap} = q'/C$. As an additional increment of charge dq' is moved from the negative to the positive conductor, the electrostatic work done on it is $dW = -dq'V_{cap}$ (Eq. 25.17). Because the charge-separating device must do work on charge carriers against the electric force, the work done by the charge-separating device on the charge carriers is the negative of the electrostatic work, so the change in electric potential energy of the capacitor during the transfer is

$$dU^E = -dW = V_{cap}\, dq' = \frac{q'}{C}\, dq'. \tag{26.2}$$

When the magnitude of the charge on each conductor has increased from zero to its final value q, the electric potential energy stored in the capacitor is

$$U^E = \int dU^E = \int_0^q \frac{q'}{C}\, dq' = \frac{1}{C}\int_0^q q'\, dq' = \tfrac{1}{2}\frac{q^2}{C}. \tag{26.3}$$

Often it is more convenient to express the electric potential energy not in terms of the magnitude of the charge q on the capacitor, but in terms of the potential difference across it:

$$U^E = \tfrac{1}{2}\frac{q^2}{C} = \tfrac{1}{2}\,CV_{cap}^2 = \tfrac{1}{2}\,qV_{cap}. \tag{26.4}$$

Note that Eqs. 26.3 and 26.4 hold for any type of capacitor, regardless of the configuration of the conductors. All that enters into these expressions besides the charge or the potential difference is the capacitance, which depends on the size, shape, and the separation of the conductors.

26.18 A 1.0-μF parallel-plate capacitor with a plate spacing of 50 μm is charged up to the breakdown threshold. (*a*) If the electric field in the air between the capacitor plates is 3.0×10^6 V/m, how much energy is stored in the capacitor? Express your answer in joules. (*b*) How high must you raise this book ($m \approx 2$ kg) to increase the gravitational potential energy of the Earth-book system by the same amount?

As we discussed in Section 26.1, we can imagine electric potential energy to be stored either in the configuration of charge in the capacitor or in the electric field. We can use Eq. 26.4 and our knowledge about the electric field in a capacitor to relate electric potential energy to the electric field. From Example 25.5 we know that the magnitude of the potential difference between the plates of a parallel-plate capacitor is given by Ed, so, using the expression for C in Example 26.2 and Eq. 26.4, we can write for the electric potential energy stored in a parallel-plate capacitor

$$U^E = \tfrac{1}{2}\,CV_{cap}^2 = \tfrac{1}{2}\left(\frac{\epsilon_0 A}{d}\right)(Ed)^2 = \tfrac{1}{2}\,\epsilon_0 E^2(Ad). \tag{26.5}$$

The term in parentheses on the right side, Ad, is equal to the volume of the space between the capacitor plates—that is, the region to which the electric field is confined. Therefore, the energy per unit volume stored in the electric field—the **energy density** of the electric field—is

$$u_E \equiv \frac{U^E}{\text{volume}} = \tfrac{1}{2}\,\epsilon_0 E^2. \tag{26.6}$$

Although we derived this expression for the special case of a parallel-plate capacitor, it holds true for any electric field in vacuum. Any given region of space where a uniform electric field is present can be viewed as containing an amount of electric potential energy equal to $\tfrac{1}{2}\,\epsilon_0$ times the square of the magnitude of the electric field in that region times the volume. If the electric field is nonuniform, we must subdivide the volume of interest into small enough segments that E can be considered uniform within each segment, then apply Eq. 26.6 to each segment and take the sum of all the contributions. (This corresponds to integrating the energy density over the volume of the region that contains the electric field.)

26.19 A parallel-plate capacitor has plates of area A separated by a distance d. The magnitude of the charge on each plate is q. (*a*) Determine the magnitude of the force exerted by the positively charged plate on the negatively charged one. (*b*) Suppose you increase the separation between the plates by an amount Δx. How much work do you need to do on the capacitor to achieve this increase? (*c*) What is the change in the electric potential energy of the capacitor? (*d*) Moving the plate adds additional space with electric field between the plates. Show that the energy stored in the electric field in this additional space is equal to the work done on the capacitor.

The energy stored in a capacitor is supplied to it by the charging device—such as a generator, a battery, or a solar cell. Inside this device, nonelectrostatic interactions cause a separation of charge by doing work on charged particles. The work per unit charge done by the nonelectrostatic interactions on the charge carriers inside the device is called the emf and is denoted by $\mathcal{E}$:

$$\mathcal{E} \equiv \frac{W_{\text{nonelectrostatic}}}{q}. \tag{26.7}$$

The SI unit of emf is the same as that of potential: the volt. The rating of a battery—1.5 V or 9 V—gives its emf.*

If no energy is dissipated inside the charging device, *all* of the energy can be transferred to charge carriers outside the device. This transfer takes place through electric interactions. In Figure 26.25, for example, electric forces remove electrons from one object and push them onto the other, charging the capacitor. In the absence of any energy dissipation, the nonelectrostatic work done on charge carriers inside the device is equal to the electrostatic work done on charge carriers outside it. Because the electrostatic work per unit charge is the potential difference between the negative and positive terminals of the charging device, we have, for an ideal charging device,

$$V_{\text{device}} = \mathcal{E} \quad \text{(ideal device)}. \tag{26.8}$$

In practice, some energy is always dissipated inside the device, so not all of the nonelectrostatic work done on charge carriers inside the device can be turned into electrostatic work. Consequently, for most devices, $V_{\text{device}} < \mathcal{E}$.

---

*The term *voltage* is sometimes used to refer to a potential difference or to an emf (such as the rating of a battery). Potential difference, however, is related to *electrostatic* work done on charge carriers, whereas emf deals with *nonelectrostatic* work done on them. Thus, electrostatic work brings opposite charge carriers together, while nonelectrostatic work causes charge separation. To maintain this important distinction, we shall avoid the term *voltage*.

Example 26.5 Van de Graaff energy

The radius of the dome on the Van de Graaff generator shown on the opening page of this chapter is about 2.5 m, and air breaks down when the field magnitude is about 3.0×10^6 V/m. How much electric potential energy is stored in the electric field surrounding the dome just before the air there breaks down?

❶ GETTING STARTED I am given the radius of a Van de Graaff dome and asked to calculate how much potential energy is in the electric field surrounding the dome just before the field causes the air to break down. If I approximate the dome as a uniformly charged spherical shell, then the electric field surrounding it is the same as that surrounding a particle carrying the same charge [Eq. 24.15, $E = q/(4\pi\epsilon_0 r^2)$]. The magnitude of the field around the dome is greatest at the dome surface, so I take this E_{surf} value as the maximum value just before the air breaks down.

❷ DEVISE PLAN Equation 26.6 gives me the energy density of the electric field around the dome. I can substitute the Eq. 24.15 expression for the electric field of a sphere carrying a charge q into Eq. 26.6 to obtain an expression for the energy density of the electric field at an arbitrary distance r from the dome center. Because the electric field has the same magnitude at any location a distance r from the dome center, I can divide the space outside the dome into a series of thin spherical shells, each of thickness dr and all concentric with the dome (**Figure 26.26**), and then integrate over all shells from $r = R$ to $r = \infty$ to obtain the energy stored in the electric field surrounding the dome in terms of the charge q. I can then use Eq. 24.15 to eliminate q from my result and express the energy stored in terms of E_{surf}, which is given.

Figure 26.26

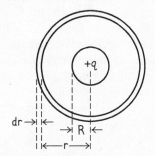

❸ EXECUTE PLAN Substituting Eq. 24.15 into Eq. 26.6, I get

$$u_E = \tfrac{1}{2}\epsilon_0 E^2 = \tfrac{1}{2}\epsilon_0\left(\frac{q}{4\pi\epsilon_0 r^2}\right)^2.$$

The volume of a thin spherical shell of radius r and thickness dr centered on the dome is equal to the surface area of a sphere of radius r times the shell thickness: $(4\pi r^2)dr$. The energy in that volume is thus $dU^E = u_E(4\pi r^2)dr$, and so the electric potential energy in the space around the dome is

$$U^E = \int_R^\infty dU^E = \int_R^\infty u_E(4\pi r^2)\,dr = \tfrac{1}{2}\epsilon_0\int_R^\infty\left(\frac{q}{4\pi\epsilon_0 r^2}\right)^2(4\pi r^2)dr$$

$$= \frac{q^2}{8\pi\epsilon_0}\int_R^\infty\frac{1}{r^2}\,dr = \frac{q^2}{8\pi\epsilon_0}\left[\frac{-1}{r}\right]_R^\infty = \frac{q^2}{8\pi\epsilon_0 R}. \quad (1)$$

Now I have an expression for U^E, the quantity I must determine, but I have no value for q. Given that the electric field at the dome surface is given by Eq. 24.15, $E_{surf} = q/(4\pi\epsilon_0 R^2)$, I can rearrange this expression to $q = E_{surf}(4\pi\epsilon_0 R^2)$ and rewrite Eq. 1 as $U^E = 2\pi\epsilon_0 E_{surf}^2 R^3$. Because E_{surf} is the maximum electric field magnitude around the dome, I know that this magnitude must be the breakdown value for air. Substituting the values given, I get

$$U^E = 2\pi(8.85 \times 10^{-12}\,\mathrm{C}^2/(\mathrm{N\cdot m}^2)(3.0 \times 10^6\,\mathrm{V/m})^2(2.5\,\mathrm{m})^3$$

$$= 7.8\,\mathrm{kJ}. \checkmark$$

❹ EVALUATE RESULT As a check on my work, I can calculate the electric potential energy of the charge stored on the dome using Eq. 26.4. In Checkpoint 26.17, you found that the capacitance of an isolated sphere is $C_{sphere} = 4\pi\epsilon_0 R$ and that the potential of a sphere is related to the electric field at its surface by $V_{cap} = ER$. Therefore $U^E = \tfrac{1}{2}CV_{cap}^2 = 2\pi\epsilon_0 R^3 E_{surf}^2$, which is the same result I obtained.

26.20 The flash unit on a typical camera uses a 100-μF capacitor to store electric potential energy. The capacitor is charged to a potential of 300 V. When the flash is fired, the energy in the capacitor is released to a bulb in a burst of about 1.0-ms duration. (a) How much energy is stored in the fully charged capacitor before it is fired? (b) What is the average power of the flash firing?

26.7 Dielectric constant

As we saw in Section 26.3, the capacitance of a capacitor can be increased by inserting a dielectric between the two conductors. For example, inserting a slab of mica between the plates of an isolated charged capacitor (**Figure 26.27**) decreases the potential difference across the capacitor by a factor of 5. This tells us that the mica reduces the electric field inside the isolated capacitor by a factor of 5. By definition, the magnitude of the potential difference V_0 across the isolated

Figure 26.27 The potential difference across an isolated parallel-plate capacitor is greater (a) without a dielectric between the plates than it is (b) with a dielectric.

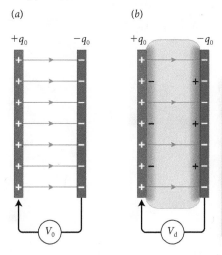

capacitor without a dielectric divided by the magnitude of the potential difference V_d with the dielectric is called the **dielectric constant** κ:

$$\kappa \equiv \frac{V_0}{V_d}. \tag{26.9}$$

Given that the magnitude q_0 of the charge on each plate of the isolated capacitor is not affected by the dielectric, we see from Eqs. 26.9 and 26.1 that

$$\kappa \equiv \frac{V_0}{V_d} = \frac{V_0/q_0}{V_d/q_0} = \frac{1/C_0}{1/C_d} = \frac{C_d}{C_0}, \tag{26.10}$$

where C_d is the capacitance of the capacitor with the dielectric and C_0 that without a dielectric. Therefore, the capacitance changes by the factor κ when a dielectric is inserted:

$$C_d = \kappa C_0. \tag{26.11}$$

The dielectric constant is always greater than 1 ($\kappa > 1$) because the presence of a dielectric decreases the electric field inside the capacitor. The greater the polarization of the dielectric material, the more reduced the electric field inside the dielectric and the greater the dielectric constant κ. Table 26.1 gives the dielectric constants for several commonly used dielectric materials. The dielectric constant for vacuum—that is, no material between the plates—is unity by definition. Because air is very dilute, the dielectric constant of air is nearly unity as well. If the dielectric is composed of polar molecules that can align themselves (such as the water molecules in liquid water), then the overall polarization is much greater than in nonpolar dielectrics, so the dielectric constant is large. For some polar materials, the dielectric constant can be in the thousands.

The electric field $\vec{E}$ inside the dielectric is the superposition of the electric field due to the free charge on the plates, $\vec{E}_{\text{free}}$, and the electric field due to the bound surface charge on the dielectric, $\vec{E}_{\text{bound}}$: $\vec{E} = \vec{E}_{\text{free}} + \vec{E}_{\text{bound}}$ (Figure 26.28a). We designate the magnitude of the free charge on the capacitor plates by q_{free} and the magnitude of the bound charge on the surfaces of the dielectric by q_{bound} (Figure 26.28b and c). With this notation, both q_{free} and q_{bound} are always positive. Using the expression for the electric field of a sheet of charge, we can thus write for the magnitude of the electric field inside the capacitor in the absence of a dielectric

$$E_{\text{free}} = \frac{\sigma_{\text{free}}}{\epsilon_0} = \frac{q_{\text{free}}}{\epsilon_0 A}, \tag{26.12}$$

Figure 26.28 (a) The electric field inside a dielectric-filled capacitor is the vector sum of the electric field due to the charged plates and that due to the polarized dielectric. (b) and (c) Bound and free charge on a vacuum-filled and a dielectric-filled isolated parallel-plate capacitor.

(a)

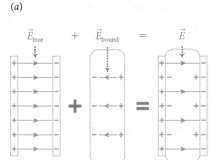

(b) vacuum-filled   (c) dielectric-filled

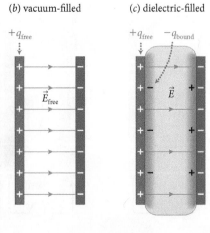

Table 26.1 Dielectric properties

Material	Dielectric constant κ	Breakdown threshold $E_{\max}$ (V/m)
Air (1 atm)	1.00059	3.0×10^6
Paper	1.5–3	4.0×10^7
Mylar (polyester)	3.3	4.3×10^8
Quartz	4.3	8×10^6
Mica	5	2×10^8
Oil	2.2–2.7	
Porcelain	6–8	
Water (distilled, 20 °C)	80.2	$6.5–7 \times 10^7$
Titania ceramic	126	8×10^6
Strontium titanate	322	
Barium titanate	1200	8×10^7

where $\sigma_{\text{free}} = q_{\text{free}}/A$ is the magnitude of the free surface charge density and A is the area of either capacitor plate. Likewise, the magnitude of the electric field due to the bound surface charge is

$$E_{\text{bound}} = \frac{\sigma_{\text{bound}}}{\epsilon_0} = \frac{q_{\text{bound}}}{\epsilon_0 A}, \tag{26.13}$$

where σ_{bound} is the magnitude of the bound surface charge density. Because $\vec{E}_{\text{free}}$ and $\vec{E}_{\text{bound}}$ point in opposite directions, the magnitude of the electric field $\vec{E}$ inside the dielectric is then

$$E = E_{\text{free}} - E_{\text{bound}} = \frac{\sigma_{\text{free}} - \sigma_{\text{bound}}}{\epsilon_0} = \frac{q_{\text{free}} - q_{\text{bound}}}{\epsilon_0 A}. \tag{26.14}$$

Let us determine the magnitude of the bound charge q_{bound}. If the plate separation is d, we can write Eq. 26.10 in the form

$$\kappa \equiv \frac{V_0}{V_{\text{d}}} = \frac{E_{\text{free}}\, d}{E\, d} = \frac{E_{\text{free}}}{E}, \tag{26.15}$$

where E_{free} is the magnitude of the electric field due to the free charge only, and E is the magnitude of the electric field inside the dielectric. In other words, the dielectric reduces the electric field by the factor κ: $E = E_{\text{free}}/\kappa$. Substituting Eqs. 26.12 and 26.14 into this expression, we get

$$\frac{q_{\text{free}}}{\kappa \epsilon_0 A} = \frac{q_{\text{free}} - q_{\text{bound}}}{\epsilon_0 A} \tag{26.16}$$

or

$$\frac{q_{\text{free}}}{\kappa} = q_{\text{free}} - q_{\text{bound}}. \tag{26.17}$$

Solving this expression for q_{bound}:

$$q_{\text{bound}} = \frac{\kappa - 1}{\kappa}\, q_{\text{free}}. \tag{26.18}$$

Because κ is always greater than 1, we see that the magnitude of the bound surface charge is always smaller than the magnitude of the free charge that causes it.

Next we consider the situation of a capacitor connected to a battery (Figure 26.29). In this situation, the potential difference across the capacitor is constant, but the charge on the plates changes when the dielectric is inserted. As we have seen in Section 26.3, the electric field inside the dielectric must be the same as before the dielectric was inserted. In other words, the sum of the free and bound charges must still be equal to q_0; that is, $q_0 = q_{\text{free}} - q_{\text{bound}}$ (Figure 26.29). Because the definition of capacitance involves only the charge on the capacitor plates, we can write

$$C_{\text{d}} \equiv \frac{q_{\text{free}}}{V_{\text{d}}} \tag{26.19}$$

and likewise

$$C_0 \equiv \frac{q_0}{V_0}. \tag{26.20}$$

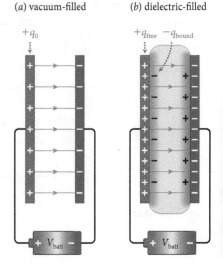

Figure 26.29 Bound and free charge on a vacuum-filled and a dielectric-filled parallel-plate capacitor connected to a battery.

(*a*) vacuum-filled   (*b*) dielectric-filled

Note the difference between q_{free} and q_0. Even though both represent free charge, they are not equal because when the dielectric is inserted, the battery increases the charge on the plate in order to maintain a constant potential difference across the capacitor, so $q_{\text{free}} > q_0$. Because $C_d = \kappa C_0$ (Eq. 26.11), we see from Eqs. 26.19 and 26.20 that the dielectric increases the charge on the capacitor plates by the factor κ:

$$q_{\text{free}} = \kappa q_0. \tag{26.21}$$

26.21 (a) In Figure 26.29, what is the magnitude of q_{bound}? Express your answer in terms of q_0 and the properties of the dielectric. (b) What is the bound surface charge density on the dielectric? Express your answer in terms of the electric field E.

Example 26.6 Capacitor with dielectric

A parallel-plate capacitor consists of two conducting plates with a surface area of 1.0 m^2 and a plate separation distance of 50 μm. (a) Determine the capacitance and the energy stored in the capacitor when it is charged by connecting it to a 9.0-V battery. (b) With the capacitor fully charged and disconnected from the battery, a 50-μm-thick sheet of Mylar is inserted between the plates. Determine the potential difference across the capacitor and the energy stored in it. (c) If the Mylar-filled capacitor is connected to the battery, how much work does the battery do to fully charge the capacitor?

❶ **GETTING STARTED** I am given information about a capacitor and a battery used to charge it. From this information, I must determine the capacitance and the energy stored in the capacitor with and without a sheet of Mylar between the plates connected to the battery, and determine what happens to the potential with and without a sheet of Mylar between the plates and with and without the battery connected to the capacitor. When connected to the capacitor, the battery maintains a constant potential across the capacitor. When the battery is not connected to the capacitor, the charge on the plates remains constant.

❷ **DEVISE PLAN** To calculate the energy stored in the capacitor I can use Eq. 26.4; I calculated the capacitance of a parallel-plate capacitor in Example 26.2. When the dielectric is added, the potential and the capacitance are given by Eqs. 26.9 and 26.11. From Table 26.1, I see that the dielectric constant of Mylar is $\kappa = 3.3$.

❸ **EXECUTE PLAN** (a) Using the result of Example 26.2, I get

$$C = \frac{[8.85 \times 10^{-12}\,C^2/(N \cdot m^2)](1.0\,m^2)}{50 \times 10^{-6}\,m}$$

$$= 0.18 \times 10^{-6}\frac{C^2}{N \cdot m} = 0.18\,\mu F, ✔$$

and Eq. 26.4 gives

$$U^E = \tfrac{1}{2}C(V_0)^2 = \tfrac{1}{2}(0.18\,\mu F)(9.0\,V)^2 = 7.2\,\mu J. ✔$$

(b) Because of the bound surface charge on the dielectric, the electric field between the capacitor plates decreases, and so the potential difference across the capacitor decreases, too. From the definition of the dielectric constant (Eq. 26.9), I have

$$V_d = \frac{V_0}{\kappa} = \frac{9.0\,V}{3.3} = 2.7\,V, ✔$$

where I obtained my value for κ from Table 26.1. To calculate the energy in the presence of the dielectric, I must first obtain an expression for the capacitance of the dielectric-filled capacitor. Substituting the expression for the capacitance of a parallel-plate capacitor (see Example 26.2) into Eq. 26.11 yields

$$C_d = \kappa C_0 = \frac{\kappa \epsilon_0 A}{d}$$

$$= \frac{(3.3)[8.85 \times 10^{-12}\,C^2/(N \cdot m^2)](1.0\,m^2)}{50 \times 10^{-6}\,m}$$

$$= 0.58\,\mu F.$$

The stored energy is thus

$$U^E = \tfrac{1}{2}CV_d^2 = \tfrac{1}{2}(0.58\,\mu F)(2.9\,V)^2 = 2.2\,\mu J. ✔$$

(c) The energy stored in the fully charged dielectric-filled capacitor is

$$U^E = \tfrac{1}{2}CV_{\text{batt}}^2 = \tfrac{1}{2}(0.58\,\mu F)(9.0\,V)^2 = 24\,\mu J.$$

From part b I know that before it was connected to the battery, the capacitor stored 2.2 μJ, and so the work done by the battery in charging the capacitor must be 24 μJ $-$ 2.2 μJ $=$ 22 μJ. ✔

❹ **EVALUATE RESULT** My answers to parts a and b show that the amount of energy stored decreases when the dielectric is inserted. That makes sense because, as the dielectric is brought near the plates, the charged plates induce a polarization on the dielectric and consequently it is pulled into the space between the plates (**Figure 26.30**). Therefore, the capacitor does *positive* work on whoever is holding the dielectric, and the energy in the capacitor decreases as the dielectric enters the space between the plates. This work is equal to the difference in energy between parts *a* and *b*: $W = 7.2\ \mu J - 2.2\ \mu J = 5.0\ \mu J$. My answer to part *c* is about three times greater than the value I calculated for U^E in part *a*, which is what I expect given that the capacitance is increased by the factor $\kappa = 3.3$ once the dielectric is inserted.

Figure 26.30

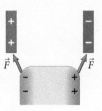

26.22 Verify that in the solution to part *a* of Example 26.6, (*a*) the ratio of units $C^2/(N \cdot m)$ is equivalent to the unit F and (*b*) the product of units $F \cdot V^2$ is equivalent to the unit J.

26.8 Gauss's law in dielectrics

Can we apply Gauss's law to calculate the electric fields inside dielectric materials? The answer is *yes,* because Gauss's law is a fundamental law that follows directly from the $1/r^2$ dependence of Coulomb's law. Thus, the presence of a dielectric cannot affect its validity.

Consider the situation illustrated in **Figure 26.31**. To determine the magnitude of the electric field E inside the dielectric, we consider the cylindrical Gaussian surface with cross-sectional area A shown in the figure. The electric flux is zero except through the right flat surface of the cylinder, so

$$\oint \vec{E} \cdot d\vec{A} = EA. \tag{26.22}$$

The charge enclosed by the Gaussian surface is not just the enclosed charge on the plate—we must also take into account the enclosed bound charge on the dielectric. The enclosed charge is thus $q_{enc} = q_{free,\ enc} - q_{bound,\ enc}$, and Gauss's law then gives

$$\oint \vec{E} \cdot d\vec{A} = EA = \frac{q_{free,\ enc} - q_{bound,\ enc}}{\epsilon_0}. \tag{26.23}$$

In this form, Gauss's law is not very useful, because in order to extract E from Eq. 26.23, we need to know the magnitude of the bound surface charge. Generally, we don't know the contribution from the bound charge in a given situation.

Substituting the relationship between the free and bound charges (Eq. 26.18), however, we can rewrite Eq. 26.23 in the form

$$\oint \vec{E} \cdot d\vec{A} = \frac{q_{free,\ enc}}{\epsilon_0 \kappa}. \tag{26.24}$$

This result—Gauss's law in dielectrics—is remarkable. The left side contains the electric flux of the electric field *inside the dielectric*. We can obtain this field, however, just by accounting for the enclosed *free* charge (and we already know how to deal with that charge). This relationship is valid because the effect of the bound charge is completely accounted for by the dielectric constant in the denominator. As Eq. 26.17 shows, dividing q_{free} by κ gives the difference of the free and bound charges.

Figure 26.31 A cylindrical Gaussian surface used to calculate the electric field inside a dielectric-filled parallel-plate capacitor.

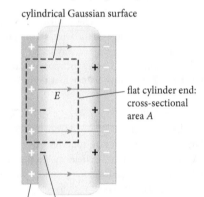

Because the dielectric constant affects the value of the electric field, **Gauss's law in matter** is usually written in the form

$$\oint \kappa \vec{E} \cdot d\vec{A} = \frac{q_{\text{free, enc}}}{\epsilon_0}. \tag{26.25}$$

This form of Gauss's law is very general: Even though we derived it for the special case of a parallel-plate capacitor, it holds in any situation, even one without a dielectric. In the absence of matter (that is, in vacuum), $\kappa = 1$, and because there is no bound charge we have $q_{\text{free, enc}} = q_{\text{enc}}$. Then Eq. 26.25 becomes identical to the familiar form of Gauss's law (Eq. 24.8).

Example 26.7 Electric field surrounding a charged insulated wire

A thin, long, straight wire is surrounded by plastic insulation of radius R and dielectric constant κ (**Figure 26.32**). The wire carries a uniform distribution of charge with a positive linear charge density λ. If the wire has a diameter d, what is the potential difference between the outer surface of the wire and the outer surface of the insulation?

Figure 26.32 Example 26.7.

plastic insulation

wire

R

1 GETTING STARTED: The insulation reduces the electric field created by the charge in the wire, so the potential difference between the wire surface and any location a distance R from the wire (in other words, any location on the outer surface of the insulation) is smaller than when there is no insulation around the wire.

2 DEVISE PLAN: The potential difference between two locations A and B can be obtained from Eq. 25.25, $V_{AB} = -\int_A^B \vec{E} \cdot d\vec{\ell}$, but using this expression requires me to know the electric field. To calculate the electric field inside the insulation, I can apply Eq. 26.25 to a cylindrical Gaussian surface that has radius r and length L and is concentric with the wire, as shown in **Figure 26.33**.

Figure 26.33

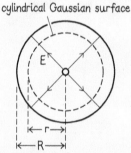

cylindrical Gaussian surface

E

r

R

3 EXECUTE PLAN: Because of the cylindrical symmetry of the wire, the electric field has the same magnitude E everywhere on the curved region of the Gaussian surface. The electric flux through that region of the Gaussian surface is equal to the product of the electric field at a distance r from the wire and the surface area: $\Phi = EA = E(2\pi rL)$ (Eq. 3 in Exercise 24.7). The free charge enclosed by the cylinder is λL, and with these substitutions Eq. 26.25 becomes

$$\kappa E(2\pi rL) = \frac{\lambda L}{\epsilon_0}$$

and

$$E = \frac{\lambda}{2\pi\kappa\epsilon_0 r}. \tag{1}$$

Substituting this expression for E into Eq. 25.25, I obtain for the potential difference between the outer surface of the wire and the outer surface of the insulation

$$V_{dR} = -\int_{d/2}^{R} \vec{E} \cdot d\vec{r} = -\frac{\lambda}{2\pi\kappa\epsilon_0} \int_{d/2}^{R} \frac{1}{r} dr$$

$$= -\frac{\lambda}{2\pi\kappa\epsilon_0} \ln \frac{2R}{d}. ✓$$

4 EVALUATE RESULT: Because $\ln(2R/d)$ is positive, the potential difference is negative, as it should be because I am moving away from the positively charged wire. As an additional check, if I set $\kappa = 1$ in Eq. 1, my result for the electric field becomes identical to the result I obtained for a thin wire without insulation in Exercise 24.7.

26.23 Show that, if you account for the free and bound charges, Gauss's law in vacuum (Eq. 24.8) yields the same result for the electric field outside the insulation as Gauss's law in matter (Eq. 26.25) does.

Chapter Glossary

SI units of physical quantities are given in parentheses.

Bound charge A surplus of charge in polarized matter due to charge carriers that are bound to atoms and cannot move freely within the bulk of the material.

Capacitance C (F) The ratio of the magnitude of the charge q on a pair of oppositely charged conductors and the magnitude V_{cap} of the potential difference between them:

$$C \equiv \frac{q}{V_{cap}}. \tag{26.1}$$

The capacitance is a measure of a capacitor's capacity to store charge (or, equivalently, electric potential energy).

Capacitor A pair of conducting objects separated by a nonconducting material or vacuum. Any such pair of objects stores electric potential energy when charge has been transferred from one object to the other.

Charge-separating device A device that transfers charge from one object to another. To achieve this charge transfer, the device must move charge carriers against an electric field, requiring the device to do work on the charge carriers. This work can be supplied from a variety of sources, such as mechanical or chemical energy. Examples of charge-separating devices are voltaic cells, batteries, and Van de Graaff generators.

Dielectric A nonconducting material inserted between the plates of a capacitor. Often used more broadly to describe any nonconducting material. *Polar* dielectrics are made up of molecules that have a nonzero dipole moment, whereas *nonpolar* dielectrics consist of nonpolar molecules.

Dielectric constant κ (unitless) The factor by which the potential across an isolated capacitor is reduced by the insertion of a dielectric:

$$\kappa \equiv \frac{V_0}{V_d}. \tag{26.9}$$

Electrical breakdown When a dielectric material is subject to a very large electric field, the molecules in the material may ionize, temporarily turning the dielectric into a conductor. The electric field magnitude at which breakdown occurs is called the *breakdown threshold*.

Emf $\mathscr{E}$ (V) The emf of a charge-separating device is the work per unit charge done by nonelectrostatic interactions in separating positive and negative charge carriers inside the device:

$$\mathscr{E} \equiv \frac{W_{nonelectrostatic}}{q}. \tag{26.7}$$

Energy density of the electric field u_E (J/m^3) The energy per unit volume contained in an electric field. In vacuum:

$$u_E = \tfrac{1}{2}\epsilon_0 E^2. \tag{26.6}$$

Farad (F) The derived SI unit of capacitance:

$$1\,\mathrm{F} \equiv 1\,\mathrm{C/V}.$$

Free charge A surplus of charge due to charge carriers that can move freely within the bulk of a material.

Gauss's law in matter For electric fields inside matter, Gauss's law can be written in the form

$$\oint \kappa \vec{E} \cdot d\vec{A} = \frac{q_{free,\,enc}}{\epsilon_0}. \tag{26.25}$$

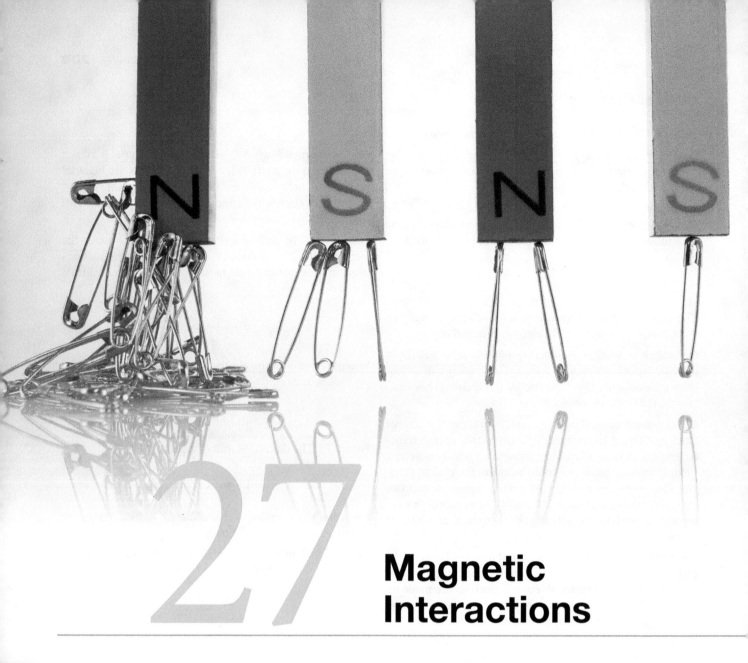

27 Magnetic Interactions

The word *magnetism* is derived from Magnesia, a province where the ancient Greeks mined *magnetite,* a mineral that attracts iron. The interactions between a magnet and a paper clip or a refrigerator door are familiar ones. These interactions may seem to have nothing to do with the subject of earlier chapters, but as we shall see, electricity and magnetism are closely related phenomena. They are two manifestations of one interaction, called the *electromagnetic interaction.* The discovery of the connection between electricity and magnetism in the 19th century opened the door to important technological breakthroughs, such as electric motors and generators, the transmission of radio signals, and the electronics and telecommunications industries.

In this chapter, we discuss interactions between magnets and introduce the concept of a magnetic field. We also begin to explore the connection between electricity and magnetism.

27.1 Magnetism

One simple definition of **magnet** is any object that attracts pieces of iron, such as iron filings or paper clips. Magnets come in many shapes and sizes, as **Figure 27.1** shows.

If you examine the surface of a magnet interacting with an object that contains iron, such as a paper clip, you will discover that certain parts of the magnets, called *magnetic poles,* interact more strongly with the paper clip than do other parts of the magnet. Disk-shaped magnets (Figure 27.1*b*) usually have poles on the two faces; paper clips stick to the flat faces but not to the curved surface. Bar magnets (Figure 27.1*c*) have poles at the ends, and most of the length of the bar doesn't interact very strongly with a paper clip. A horseshoe magnet (Figure 27.1*d*) is simply a bent bar magnet with poles at the ends of the bent bar.

Figure 27.1 Magnets come in many shapes and sizes.

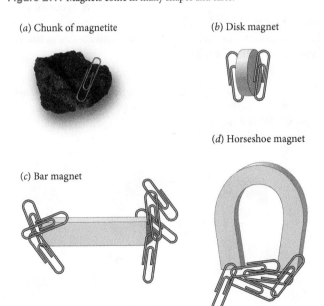

(*a*) Chunk of magnetite

(*b*) Disk magnet

(*d*) Horseshoe magnet

(*c*) Bar magnet

Figure 27.2 The needle of a compass is a small bar magnet that is free to rotate. When the compass is held horizontally, the needle points in the direction of Earth's North Pole. This end of the needle is defined as its north pole N.

Depending on which of their poles you hold near each another, two magnets can attract or repel without touching. This tells us that these interactions are long-range and that there are two types of magnetic poles. The interaction between magnets is one example of what are called **magnetic interactions.**

When a bar magnet is suspended and free to rotate, it aligns itself so that its poles lie on a line roughly north-south along Earth's surface. This north-south alignment of a freely rotating magnet is the basic operational principle of a compass needle, which is simply a freely rotating magnetic needle (**Figure 27.2**). This alignment provides a means of distinguishing between the two types of poles:

> **The pole of a freely suspended bar magnet that settles toward north is defined as being the north pole of the magnet (denoted N); the opposite pole is defined as being the south pole of the magnet (denoted S).**

To understand what causes this alignment, we must first examine how one magnetic pole interacts with another. If you hold the poles of two magnets near each other, you discover that the magnetic interaction between the two types of poles follows a rule very similar to that between the two types of charge (**Figure 27.3** on the next page):

> **Like magnetic poles repel each other; opposite magnetic poles attract each other.**

✋ **27.1** Because we cannot see any obvious difference between the ends of a bar magnet, could it be that *like* poles attract each other and *unlike* poles repel each other?

How do the poles of a magnet interact with objects that are not magnets? For most materials that are not magnets, the answer to this question is: not at all. Try picking up a penny, a wooden tooth pick, a piece of aluminum, or a piece

Figure 27.3 Interactions between magnetic poles.

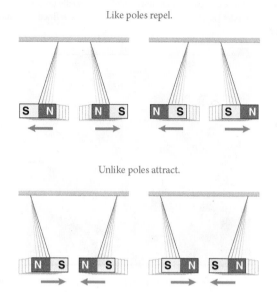

Like poles repel.

Unlike poles attract.

of plastic with a magnet. It won't work. However, both the north pole and the south pole of a magnet attract an iron paper clip (**Figure 27.4**).

Materials attracted by both types of magnetic poles include iron, nickel, cobalt, and certain alloys, such as steel. These materials are called *magnetic materials*. Normally, two objects made from these materials do not interact with each other. For example, one iron paper clip does not attract another one. However, just as a charged object can induce charge polarization on a neutral object, the presence of a magnet induces a **magnetic polarization** in a paper clip or any other object made from a magnetic material (**Figure 27.5**). The poles of the magnet then interact with the induced poles of the magnetized paper clip.

27.2 (*a*) Is the interaction between a charged object and an electrically neutral object always attractive? Why or why not? (*b*) In Figure 27.4, which type of magnetic pole is induced at the top of each paper clip?

Figure 27.4 Both the north and south poles of a magnet attract an unmagnetized iron object.

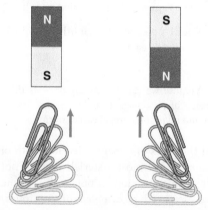

Figure 27.5 Comparing induced electric and magnetic polarization. Magnetic polarization can be induced only in an object made from magnetic material.

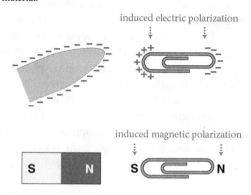

induced electric polarization

induced magnetic polarization

Unlike induced electric polarization, which vanishes as soon as the charged object is removed, some magnetic materials retain their induced magnetic polarization. For example, sewing needles left unmoved for long periods become magnetized by Earth's magnetic field. If you stroke a paper clip several times in the same direction with a magnet, the paper clip remains magnetically polarized even after you remove the magnet. In fact, you may even be able to pick up one paper clip with another (**Figure 27.6**). The clip can be *demagnetized* again—its magnetic polarization undone—by heating it, dropping it, or stroking it with a magnet in random directions.

In addition to the retention of magnetic polarization, there is another fundamental difference between magnetic and electric interactions: It is not possible to isolate one pole of a magnet the way we can separate a positively charged particle and a negatively charged particle from each other. For example, if we cut a bar magnet in two, we see that each of the resulting pieces has two opposite poles (**Figure 27.7**). The cutting has created an additional *pair* of opposite poles. Remarkably, if we carefully place the two pieces together again, the two new poles seem to vanish. A piece of iron held close to the cut is either not at all or only weakly attracted, as if the two newly formed opposite poles at the cut have "neutralized" each other.

Figure 27.6 If you stroke a paper clip several times in the same direction with a magnet, the clip retains some magnetic polarization.

Paper clip can be magnetized by stroking on magnet several times in one direction.

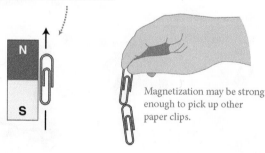

Magnetization may be strong enough to pick up other paper clips.

Figure 27.7 When a magnet is cut in two, each piece retains both an N and S pole.

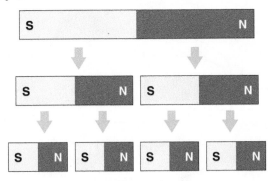

Each of the two halves can again be cut in half, but regardless of how small the pieces into which we divide the bar, each piece always has two poles. As we continue to divide the bar into smaller and smaller pieces, we come to a point where each part is a single atom. Experiments show that the iron, nickel, or cobalt atoms that make up magnets still behave like magnets. Even elementary particles behave like magnets—we shall refer to these as *elementary magnets.* Thus, a magnet consists of a large number of elementary magnets whose alignment is such that their combined effect is reinforced at the poles (**Figure 27.8a**). Because we cannot cut the elementary magnets, a pair of new poles appears every time we cut through the magnet (Figure 27.8b).

In spite of extensive searches, isolated magnetic poles, called *magnetic monopoles* (say, a north pole in the absence of a south pole), have never been found. Magnetism is therefore not due to magnetic monopoles, the way electricity is due to electrical charge. Although the pole of a magnet is not a monopole, we will occasionally rely on the picture of a bar magnet as a pair of opposite monopoles separated by a small distance to get an intuitive feel for magnetism. We call such an arrangement a **magnetic dipole.**

In an object made of magnetic material, the elementary magnets are randomly oriented relative to one another (**Figure 27.9a**). Some push or pull in one direction, while others push or pull in other directions. As a result, their effects cancel and the object as a whole does not act like a magnet. When a magnet is brought nearby, however, the elementary magnets in the object align themselves with

Figure 27.8 The concept of elementary magnets explains why cutting a magnet in two reveals new N and S poles.

In magnetized material, elementary magnets (◑) are aligned . . .

. . . so splitting magnet exposes two new poles.

Figure 27.9 (a) Unmagnetized and (b) magnetized pieces of magnetic material.

(a) Unmagnetized material: atoms oriented randomly

(b) Magnetized material: atoms aligned

the poles of the magnet, and the object becomes magnetized (Figure 27.9b). With some magnetic materials, the alignment is (partially) maintained even after the nearby magnet is removed, leaving the object magnetized.

You can use this model to visualize demagnetization. If a magnet is heated or handled roughly, the elementary magnets are jarred around and lose their alignment.

✋ **27.3** (a) Draw the elementary magnets inside a bar magnet and a horseshoe magnet, using the half-filled-circle format shown in Figures 27.8 and 27.9. (b) How many poles does the magnetized ring in **Figure 27.10** have? (c) If someone gave you such a ring, how could you verify that it is indeed magnetized as illustrated?

Figure 27.10 A magnetized ring (Checkpoint 27.3).

27.2 Magnetic fields

The long-range nature of magnetic interactions suggests that we can introduce the concept of a **magnetic field,** denoted by $\vec{B}$.* In analogy to the field model for electric interactions, a magnet is surrounded by a magnetic field. This magnetic field exerts a force on the poles of another magnet. Because there is no such thing as a "magnetic charge," it is not possible to map out a magnetic field using a "test magnetic

---

*The unintuitive symbol $\vec{B}$ was introduced early on in the description of magnetism. Whenever you see a *B* in an illustration or force superscript, remember that it stands for a magnetic field or force. The force exerted by a magnet on a paper clip, for example, is denoted by $\vec{F}^{B}_{mc}$.

CONCEPTS

Figure 27.11 Effect of a bar magnet on a compass needle.

(*a*) Forces exerted by magnet on poles of compass needle cause a torque.

(*b*) North pole of compass needle points toward south pole of magnet

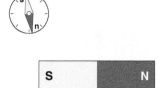

charge." We can, however, use a compass to map out the direction of the magnetic field.

27.4 (*a*) Which end of a compass needle is a north pole: the end that points toward Earth's North Pole or the other end? (*b*) If you place a compass near the north pole of a magnet, what happens to the compass needle? (*c*) Is Earth's geographic North Pole a magnetic north pole?

When a compass is placed near a bar magnet as illustrated in **Figure 27.11a**, the south pole of the bar magnet exerts an attractive force on the north pole of the compass needle and a repulsive force on the needle's south pole. These two forces cause a torque that tends to align the needle with its north pole pointing toward the bar magnet's south pole.

27.5 (*a*) What is the effect of the north pole of the bar magnet on the compass needle in Figure 27.11a? (*b*) What is the combined effect of the bar magnet's north and south poles on the needle? (*c*) A compass placed in the position shown in Figure 27.11b aligns itself in the direction indicated. What is the direction of the vector sum of the forces exerted by the bar magnet's north and south poles on the north pole of the needle? On the south pole of the needle?

We can now use the alignment of a compass needle to map out **magnetic field lines** by placing a compass somewhere near a bar magnet, moving the compass a small distance in the direction of the needle, and repeating this procedure to trace out a line, as illustrated in **Figure 27.12**. At

Figure 27.13 Magnetic field line pattern surrounding a bar magnet.

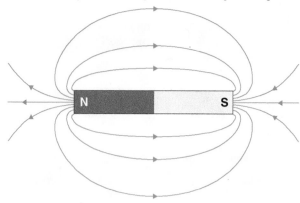

any location, the direction of the magnetic field $\vec{B}$ is tangent to the magnetic field line passing through the location. By convention:

Near a magnet, magnetic field lines point away from north poles and toward south poles.

A more complete diagram of the magnetic field pattern of a bar magnet is shown in **Figure 27.13**. Notice the similarity with an electric dipole field (Figure 24.2).

The field line patterns around a magnet can be made visible by sprinkling iron filings on a piece of paper placed over the magnetic. The little pieces of iron become magnetized and act like compass needles that align themselves along field lines (**Figure 27.14**).

The similarity between electric and magnetic field patterns suggests that we may be able to carry over many of the concepts we developed for electric fields into our study of magnetism. In particular, we can associate a magnetic field magnitude with the density of magnetic field lines:

At every location in a magnetic field line diagram, the magnitude of the magnetic field is proportional to the field line density at that location.

In the field pattern shown in **Figure 27.15a**, for example, the magnitude of the magnetic field is greatest near the bottom and smallest near the top. Occasionally we shall work with magnetic fields that are perpendicular to the plane of the drawing, and so we need to be able to represent them. Figure 27.15b and c show conventions for representing field lines perpendicular to the page.

In electrostatics, the concept of field line flux led to Gauss's law, and so we may ask ourselves if there is an analogous law

Figure 27.12 A magnetic field line can be traced by moving a compass small distances in the direction in which its needle points.

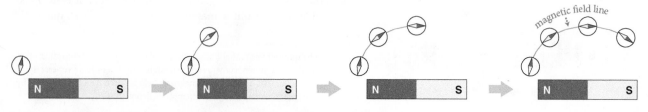

Figure 27.14 The magnetic field line pattern surrounding a bar magnet made visible by sprinkled iron filings around the magnet.

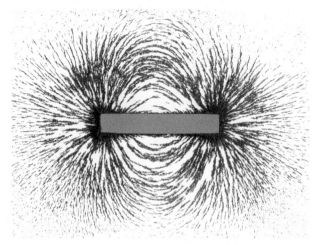

for magnetism. The answer is *yes,* but the result is somewhat different from that in electrostatics. To see why, complete this checkpoint.

✋ **27.6** (*a*) Consider a single elementary magnet inside a closed surface. Given that elementary magnets are particles without spatial extent, is the magnetic field line flux through the closed surface positive, negative, or zero? (*b*) Does adding a second elementary magnet inside the closed surface change your answer? (*c*) Consider a bar magnet inside a closed surface. Is the magnetic field line flux through the closed surface positive, negative, or zero? (*d*) Does your answer change if the closed surface cuts through the magnet?

The fundamental point of Checkpoint 27.6 is that each field line leaving a closed surface that encloses an elementary magnet reenters the surface somewhere else, so the magnetic field line flux due to an elementary magnet inside a closed surface is zero. Generalizing this statement to a collection of many elementary magnets, we say:

Figure 27.15 Conventions for representing a magnetic field.

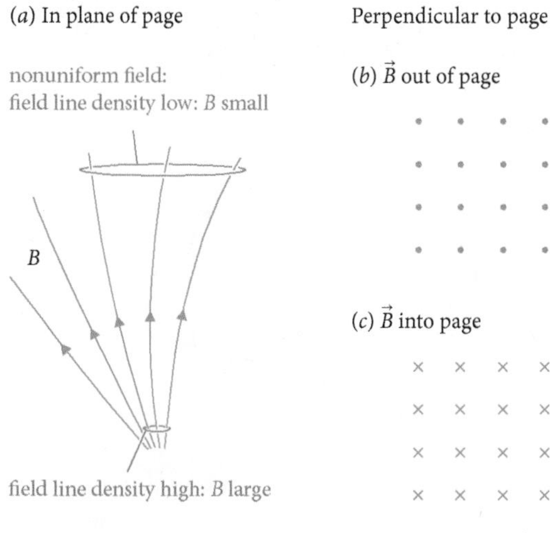

(*a*) In plane of page   Perpendicular to page

nonuniform field:
field line density low: *B* small

(*b*) $\vec{B}$ out of page

. . . . .

. . . . .

. . . . .

. . . . .

B

(*c*) $\vec{B}$ into page

× × × ×

× × × ×

× × × ×

field line density high: *B* large   × × × ×

The magnetic field line flux through a closed surface is always zero.

In electrostatics, electric field lines originate or terminate on electrical charge. Because of the absence of magnetic monopoles, however, we have no "magnetic charge" on which magnetic field lines could originate or terminate. Magnetic field lines must therefore always form loops that close on themselves. The statement that the magnetic field line flux through a closed surface is always zero is therefore a direct consequence of the absence of magnetic monopoles.

✋ **27.7** What is the direction of the magnetic field lines *inside* the bar magnet of Figure 27.13?

27.3 Charge flow and magnetism

The first indication of a connection between electricity and magnetism came in 1820, when the Danish physicist Hans Christian Ørsted discovered that a flow of charge carriers deflects the needle of a compass. This effect is illustrated in **Figure 27.16**, which shows battery terminals connected to a conducting rod.

The top of the rod in Figure 27.16 is connected to the negative battery terminal, and the bottom is connected to the positive terminal. As we saw in Chapter 26, this means that the bottom of the rod is at a higher potential than the top and an upward-pointing electric field arises inside the rod.*

The upward electric field in the rod exerts a downward electric force on the negatively charged electrons in the rod, causing a flow of charge carriers, called **current,** through the rod. The battery maintains a constant potential difference across the rod by adding electrons at the top of the rod and removing them at the bottom. Consequently, the battery maintains a constant current through the rod.

---

*In Section 24.5 we concluded that the electric field inside a conductor in electrostatic equilibrium is zero. The rod in Figure 27.16, however, is not in electrostatic equilibrium because the battery maintains a potential difference across it, causing free charge carriers in the rod to move.

Figure 27.16 A flow of charge carriers through a conducting rod causes a circular alignment of compass needles.

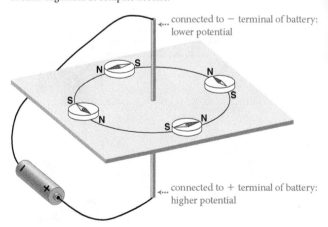

connected to − terminal of battery: lower potential

connected to + terminal of battery: higher potential

27.8 (*a*) Is the rod in Figure 27.16 electrically charged while connected to the battery? (*b*) Is there an electric field due to the rod at the positions of the compasses in Figure 27.16?

When the electrons flow through the rod in Figure 27.16, the compass needles align themselves in a circular pattern around the rod. This observation leads to a conclusion:

A flow of charged particles causes a magnetic field.

If the direction of the flow is reversed by reversing the battery, the compass needles turn around until they point in the opposite direction. If the flow of charge carriers is stopped, the needles point toward Earth's geographic North Pole, as usual.

27.9 Sketch the magnetic field line pattern in the horizontal plane around the rod in Figure 27.16.

It is important to note that a flow of positive charge carriers in one direction is equivalent to a flow of negative charge carriers in the opposite direction. In **Figure 27.17**, for example, the charge on the right increases and the charge on the left decreases regardless of which flow occurs. Experiments also show that the two types of flow are equivalent in terms of the magnetic field:

A flow of positive charge in one direction produces the same magnetic field as an equal flow of negative charge in the opposite direction.

Therefore, regardless of the actual movement of charge carriers through a current-carrying wire, we shall always denote the direction of decreasing potential (that is, the direction in which positive charge carriers would flow) with an arrow labeled with the symbol for current (*I*) next to the wire, as shown in **Figure 27.18**. We'll call this direction the *direction of current.*

Figure 27.17 A flow of positive charge carriers in one direction is equivalent to a flow of negative charge carriers in the other direction.

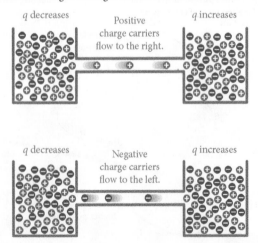

Figure 27.18 By definition, current has the direction in which positive charge carriers would flow, even if it is actually carried by negative charge carriers moving in the opposite direction.

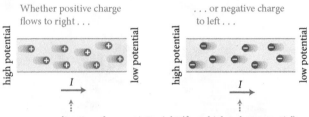

... we say *direction of current* is to right (from high to low potential).

We can now connect the direction of a magnetic field to that of current. **Figure 27.19** shows a single circular magnetic field line for the current-carrying rod from Figure 27.16. As seen from the top, the magnetic field curls in a counterclockwise direction around the rod. The downward flow of negative electrons through the rod corresponds to an upward current, and so the direction of the magnetic field is connected to that of the current by the *right-hand current rule*:

If you point the thumb of your right hand in the direction of current, your fingers curl in the direction of the magnetic field produced by that current.

27.10 Can you replace the current-carrying rod of Figure 27.16 by a magnet and get the same magnetic field?

The relationship between the magnetic field produced by a magnet and that produced by a current-carrying wire is not immediately obvious. We shall study this relationship in more detail in the next chapter. Before doing so, however, we still need to answer the question: How does a bar magnet interact with a straight current-carrying wire? The answer, which follows from the experiment illustrated in **Figure 27.20**, describes an interaction that is very different

Figure 27.19 The right-hand current rule relates the direction of a current to the direction of the resulting magnetic field.

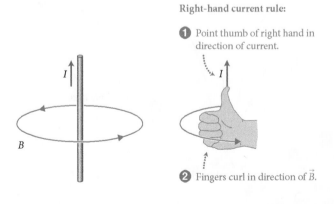

Right-hand current rule:

❶ Point thumb of right hand in direction of current.

❷ Fingers curl in direction of $\vec{B}$.

Figure 27.20 The magnetic force exerted by a bar magnet on a current-carrying wire depends on the magnet's orientation.

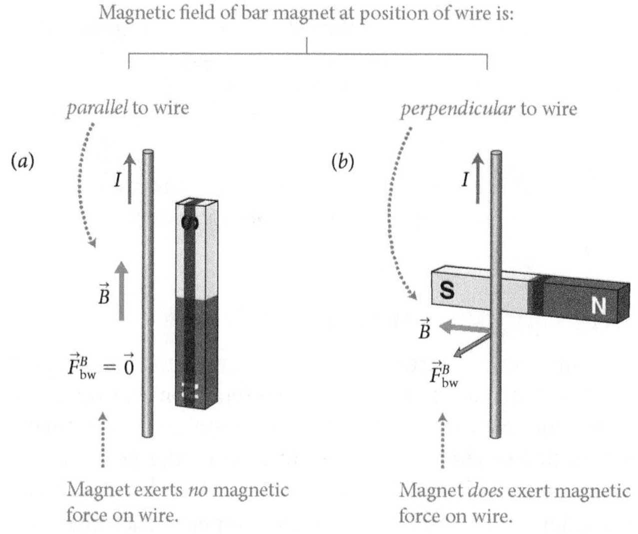

Magnetic field of bar magnet at position of wire is:

parallel to wire

(a)

$\vec{F}^B_{bw} = \vec{0}$

Magnet exerts *no* magnetic force on wire.

perpendicular to wire

(b)

$\vec{B}$

$\vec{F}^B_{bw}$

Magnet *does* exert magnetic force on wire.

Figure 27.22 When a magnet exerts a force on a current-carrying wire, the right-hand force rule relates the direction of the force to that of the current and the magnet's magnetic field.

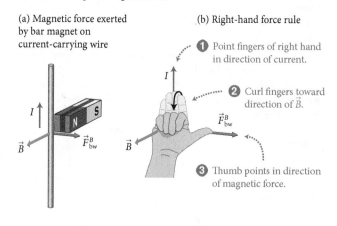

(a) Magnetic force exerted by bar magnet on current-carrying wire

(b) Right-hand force rule

❶ Point fingers of right hand in direction of current.

❷ Curl fingers toward direction of $\vec{B}$.

❸ Thumb points in direction of magnetic force.

from all the interactions we have encountered so far. When a bar magnet is held parallel to a current-carrying wire, the magnetic field of the magnet at the position of the wire is parallel to the wire (Figure 27.20a). In this position, the magnet exerts *no force* on the wire! When we rotate the magnet 90°, so that its long axis is perpendicular to the long axis of the wire, the magnet exerts a repulsive force on the wire (Figure 27.20b). The most striking aspect of this interaction is that the direction of the magnetic force is *perpendicular* to the magnetic field.

27.11 Use Figure 27.20 to determine the direction of the magnetic force exerted by the magnet on the wire when the magnet is in the orientations shown (a) in **Figure 27.21a** and (b) in Figure 27.21b. (Hint: First determine the direction of the magnetic field due to the magnet at the wire; then use the observations from Figure 27.20 to determine the direction of the magnetic force.)

The surprising *sideways* force exerted by the bar magnet on the wire in Figure 27.21b (see Checkpoint 27.11) is a result of the fact that the magnetic force exerted by a bar magnet on a current-carrying wire is always at right angles to both the magnetic field and the direction of current. This

observation suggests that these three directions can be connected by a right-hand rule. **Figure 27.22** shows that if you orient the fingers of your right hand in the direction of current in such a way that you can curl them so that the fingertips point in the direction of the magnetic field, then your right thumb points in the direction of the magnetic force exerted on the wire. Take a minute to verify the direction of the force in Figure 27.20b using this procedure, commonly called the *right-hand force rule:**

> The direction of the magnetic force exerted by a magnetic field on a current-carrying wire is given by the direction of the right-hand thumb when the fingers of that hand are placed along the direction of current in such a way that they can be curled toward the magnetic field.

Table 27.1 summarizes the two right-hand rules we have encountered so far.

Table 27.1 Right-hand rules in magnetism

Right-hand rule	thumb points along	fingers curl
current rule	current	along B-field
force rule	magnetic force	from current to B-field

Figure 27.21 Checkpoint 27.11.

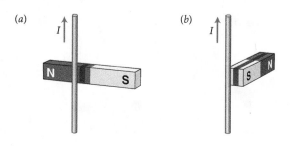

(a)

(b)

*There is nothing magical about right hands. You could use your left hand and change the rule to read, "The direction . . . given by the direction of the left-hand thumb when the fingers of that hand are placed along the direction of the magnetic field in such a way that they can be curled toward the direction of current." The same applies to determining the direction of a magnetic field surrounding a current-carrying rod. In Figure 27.19, for instance, orienting your left hand so that the thumb points in the direction *opposite* the direction of current automatically curls the fingers in the direction of the magnetic field. A consistent convention is what is important.

Example 27.1 Current-carrying rods

Two parallel rods carry currents in opposite directions. Determine the direction of the magnetic force exerted by each rod on the other rod.

❶ **GETTING STARTED** I begin by making a sketch of the rods, labeling them 1 and 2 and showing the currents in opposite directions (**Figure 27.23**). The current through rod 2 creates a magnetic field at the location of rod 1. This magnetic field exerts a force $\vec{F}_{21}^{B}$ on rod 1. Likewise, the magnetic field created at rod 2 by the current through rod 1 exerts a force $\vec{F}_{12}^{B}$ on rod 2.

Figure 27.23

(a)

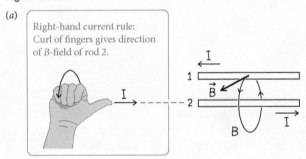

(b)

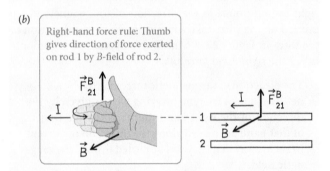

❷ **DEVISE PLAN** To determine the direction of each force, I need to apply the right-hand force rule. To apply this rule, I need to know the direction of the magnetic field created by each rod at the location of the other. I can determine these directions using the right-hand current rule.

❸ **EXECUTE PLAN** I begin with the right-hand current rule to determine the direction of the magnetic field due to the current through rod 2 at the location of rod 1. When I align my right thumb with the direction of current through rod 2, my fingers curl out of the page (Figure 27.23a), telling me that this is the direction of the magnetic field due to rod 2. Next I apply the right-hand force rule to determine the direction of the magnetic force $\vec{F}_{21}^{B}$ exerted by rod 2 on rod 1. I align the fingers of my right hand with the current through rod 1 in such a way that they can curl toward $\vec{B}$ (Figure 27.23b). My right thumb now points up in the plane of the page, telling me that this is the direction of $\vec{F}_{21}^{B}$. ✔

To determine the direction of the magnetic force $\vec{F}_{12}^{B}$ exerted by rod 1 on rod 2, I again begin with the right-hand current rule, which tells me that, at the location of rod 2, the magnetic field due to rod 1 is directed out of the page. Applying the force rule, I place my fingers along the direction of I_2 and see that curling them in the direction of $\vec{B}$ makes my thumb point downward in the plane of the page, so this is the direction of $\vec{F}_{12}^{B}$. ✔

The force $\vec{F}_{21}^{B}$ points upward, the force $\vec{F}_{12}^{B}$ points downward, and therefore the rods repel each other.

❹ **EVALUATE RESULT** It makes sense that the two rods exert forces on each other that point in opposite directions because the forces $\vec{F}_{12}^{B}$ and $\vec{F}_{21}^{B}$ form an interaction pair.

✋ **27.12** Determine the directions of the forces exerted by two parallel rods with currents in the same direction.

27.4 Magnetism and relativity

The magnetic interaction between two current-carrying rods is baffling. How can any interaction between the rods depend on the motion of the charge carriers in them? **Figure 27.24a** shows a schematic view of the positive and negative charge carriers that make up two parallel metal rods. The electrons in the two rods repel one another, but this repulsion is perfectly balanced by the attraction between the electrons in one rod and the positively charged ions in the other rod. Once the electrons are set in motion, however, this balance appears to be disturbed, and, depending on the signs of the currents, the rods either attract (Figure 27.24b) or repel each other.

To analyze this interaction, consider the simpler situation of two pairs of electrons. Relative to Earth, pair A is at rest and pair B is moving at constant velocity $\vec{v}$ (**Figure 27.25a**). To observer E, who is also at rest relative to Earth, the interaction between the two electrons of pair A is completely described by Coulomb's law: As we saw in Chapter 22, the

Figure 27.24 Schematic view of the interaction between two current-carrying rods.

(a) Metal rods don't interact when they carry no current

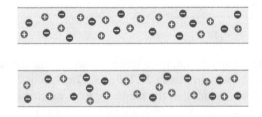

(b) When carrying a current, they do interact

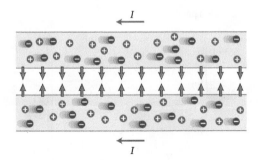

Figure 27.25 A relativistic view of the interaction between two charged particles.

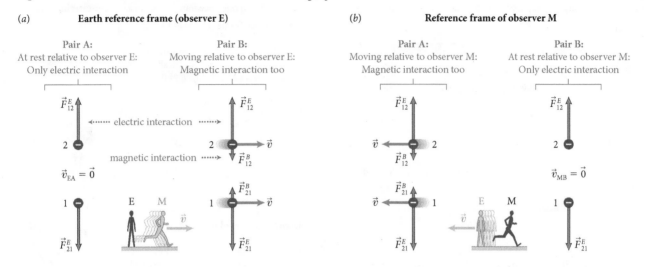

(a)   **Earth reference frame (observer E)**

Pair A:
At rest relative to observer E:
Only electric interaction

Pair B:
Moving relative to observer E:
Magnetic interaction too

(b)   **Reference frame of observer M**

Pair A:
Moving relative to observer M:
Magnetic interaction too

Pair B:
At rest relative to observer M:
Only electric interaction

electrons repel each other due to an electric interaction. According to observer E, the interaction between the electrons of pair B is *different* from that between the pair A electrons. In addition to the Coulomb repulsion, observer E perceives a magnetic interaction between the electrons of pair B because the moving electrons constitute a current.

The conclusions drawn by an observer M moving along with pair B are startlingly different (Figure 27.25*b*). To this observer, pair B is at rest and pair A is moving to the left at constant velocity. Consequently, to M the interaction between the electrons of pair B is purely electric, while that between the electrons of pair A must have a magnetic component as well. This odd conclusion is a direct consequence of the fact that magnetism is motion-dependent and therefore, like velocity, must be relative:

> **The observed interaction between charge carriers depends on their motion relative to the observer: The interaction can be purely electric, purely magnetic, or a combination of the two.**

At first sight, this statement appears to be a violation of the principle of relativity. If reference frame M moves at constant velocity relative to reference frame E, then both observers should agree on their observations (see Section 6.3). The resolution of this problem lies in the fact that the magnetic interaction between two moving charge carriers is a direct consequence of special relativity: *Magnetism is a relativistic effect.*

As explained in Chapter 14, relativistic effects tend to be so extraordinarily small that they cannot be observed under most ordinary conditions. So why is it that we can feel this relativistic effect every time we stick a note to a refrigerator with a magnet? To see why, let us return to the two current-carrying rods in Figure 27.24. Two 1-m rods, each 0.5 mm in diameter, contain an immense number of free electrons—more than 10^{22} of them. To appreciate how enormous this number is, imagine that these electrons

were not accompanied by an equal number of positively charged ions. The force with which two such collections of unbalanced electrons separated by a distance of 5 mm would repel each other is a phenomenal 10^{19} N—enough to lift up thousands of average-sized mountains. The fact that two electrically neutral metal rods do not exert any force on each other shows how incredibly accurate is the balance between positive and negative charge carriers in matter. In contrast, if we run large currents through the rods—by applying and maintaining a potential difference of, say, several volts across the length of each rod—then the magnetic force exerted by each rod on the other is a mere 10^{-2} N. This small but measurable magnetic force is 10^{21} times smaller than the electric repulsion between the electrons in the rods! Thus, if magnetism is indeed a relativistic effect (see below and Section 27.8 for details), we can now understand why we can so readily observe it. First, the electric force is incredibly large, and so even a small relativistic correction becomes measurable. Second, because electrical charge is so well balanced in all matter, electric forces are balanced, leaving only the small "magnetic" correction. In a sense, therefore, magnetism provides the most direct observation of a relativistic effect!

Let us now try to understand qualitatively how special relativity requires a magnetic interaction between charge carriers in motion. The treatment that follows relies on very little knowledge of special relativity—only the concept of length contraction is needed (see Section 14.6): When an object moves relative to an observer, the observer measures the length of the object in the direction of its motion to be shorter than the proper length of the object.

A direct result of length contraction is that the charge density of an object depends on its motion relative to the observer. Consider, for example, a rod carrying a positive charge q. At rest in the Earth reference frame, the rod has length ℓ_{proper} and its "proper" charge density is $\lambda_{proper} = q/\ell_{proper}$. If the rod is set in motion along its

Figure 27.26 Because a rod appears shorter when it is moving relative to the observer, its charge density appears larger than when the rod is at rest.

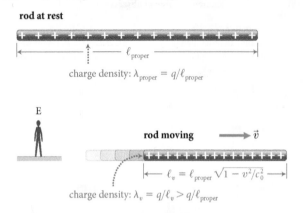

rod at rest

ℓ_{proper}

charge density: $\lambda_{\text{proper}} = q/\ell_{\text{proper}}$

E

rod moving ⟶ $\vec{v}$

$\ell_v = \ell_{\text{proper}} \sqrt{1 - v^2/c_0^2}$

charge density: $\lambda_v = q/\ell_v > q/\ell_{\text{proper}}$

lengthwise axis, its length measured by an observer at rest is smaller than ℓ_{proper} (**Figure 27.26**). Because the charge on the rod is still q, the observer at rest sees a greater charge density on the moving rod.*

27.13 Consider the two identical rods in Figure 27.26, one moving and the other at rest relative to observer E, at rest in the Earth reference frame. Suppose a second observer M moves along with the moving rod. (*a*) Which rod has the greater charge density according to observer M? (*b*) Suppose the charge on the rod moving in the Earth reference frame is adjusted so that its charge *density* as seen by observer E is the same as the charge density of the rod at rest in the Earth reference frame. Is the charge density on each rod as seen by observer M greater than, equal to, or smaller than $\lambda_{\text{proper}} = q/\ell_{\text{proper}}$?

Understanding the changes in charge density that occur when charge carriers move relative to the observer is the key point for the remainder of this section. Thus if you did not complete the preceding checkpoint, please go back and do it now before reading on.

Let us now apply these ideas to the interaction between a current-carrying wire and a charged particle. **Figure 27.27a** schematically shows a current-carrying wire consisting of fixed positively charged ions (red) and negatively charged electrons (blue) traveling to the right at velocity $\vec{v}$. (For clarity, the ions and electrons are drawn side by side rather than intermingled as they are in a real conductor.)

To an observer E at rest relative to the ions, the wire does not carry a surplus charge, and so any length of wire contains the same number of ions and electrons. That is, to E the linear charge density of the ions, λ_{proper}, is equal in magnitude to that of the electrons, $-\lambda_{\text{proper}}$. Consequently, there is no electric field outside the wire, and the wire cannot exert an electric force on a charged particle placed near the wire.

Suppose, however, that a positively charged particle moves alongside the wire—for simplicity, we let it move at the same velocity $\vec{v}$ relative to the wire as the electrons in the wire. First we consider the particle and the wire from

Figure 27.27 Observers E (at rest in the Earth reference frame) and M (moving in the Earth reference frame) observe a current-carrying wire.

(*a*) **Earth reference frame (observer E)**

To observer E, ions and electrons in wire have same charge density, so wire is electrically neutral and has no electric field: $\vec{E} = \vec{0}$.

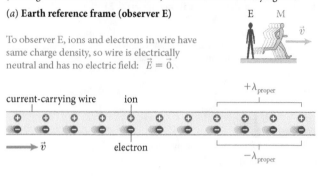

current-carrying wire   ion

$+\lambda_{\text{proper}}$

$\vec{v}$ electron

$-\lambda_{\text{proper}}$

(*b*) **Reference frame of observer M**

To observer M, charge density of ions in wire is greater than λ_{proper} and that of electrons is smaller than λ_{proper}, so wire is positively charged, and $\vec{E} \neq \vec{0}$.

$\vec{v}$ ⟵

$+\lambda > +\lambda_{\text{proper}}$

$|\lambda| < |\lambda_{\text{proper}}|$

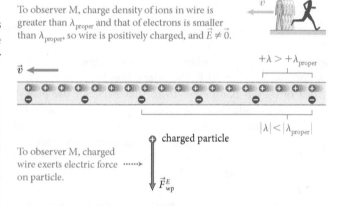

To observer M, charged wire exerts electric force on particle.

charged particle

$\vec{F}_{\text{wp}}^E$

(*c*) **Earth reference frame**

To observer E, wire cannot exert electric force on particle because it is electrically neutral.

$+\lambda_{\text{proper}}$

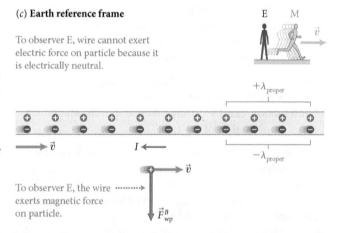

$\vec{v}$ I ⟵

$-\lambda_{\text{proper}}$

$\vec{v}$

To observer E, the wire exerts magnetic force on particle.

$\vec{F}_{\text{wp}}^B$

the point of view of an observer M moving along with the particle. Because the particle is at rest relative to M, it experiences no magnetic force according to this observer. From M's point of view, the electrons in the wire are at rest and the ions move at velocity $\vec{v}$ to the left (Figure 27.27*b*). In addition, as you saw in Checkpoint 27.13, the charge densities

---

*It is not immediately obvious that the charge of a system is not affected by the motion of the charge carriers. Experiments, however, show that charge is, indeed, an invariant (that is, its value does not depend on the choice of reference frame).

seen by M are different from those seen by E. To M, the ions are closer together because they are moving, which means their charge density is greater than λ_{proper}. The electrons, on the other hand, are at rest relative to M and so farther apart. According to observer M the magnitude of their linear charge density therefore must be smaller than that observed by E relative to whom they are moving.

The different charge densities mean that although the wire appears electrically neutral to E, it cannot also appear neutral to M. According to M, the magnitude of the electron density is smaller than λ_{proper}, and the magnitude of the ion density is greater than λ_{proper}. Thus, the wire appears positively charged to observer M. Observer M therefore sees a downward electric field due to the wire at the location of the charged particle, with the wire and the particle repelling each other.

In the reference frame of the fixed ions, a very different picture emerges. Observer E, seeing no electric field (Figure 27.27a), cannot attribute the repulsion between the wire and the particle to an electric interaction. Instead, E attributes this repulsion to a magnetic interaction between two currents, one in the wire and the other caused by the moving charged particle (Figure 27.27c). As we shall see in

Section 27.8, this magnetic interaction can be completely accounted for by the Coulomb interaction and special relativity, and in principle any other magnetic interaction can also be explained this way. However, because transforming back and forth from one moving reference frame to another is cumbersome, it is easier to develop a separate treatment for magnetism that does not require reference-frame transformations and that ignores any relativistic effects. It is important to keep in mind, however, that magnetic and electric interactions are two aspects of one *electromagnetic* interaction, with magnetism being a relativistic correction to the electric interaction.

27.14 What is the direction of current through the rod in (a) Figure 27.27b and (b) Figure 27.27c? (c) Is the direction of the force in Figure 27.27c in agreement with what we learned about the interaction of two parallel current-carrying wires in Section 27.3? (d) If the particle moving alongside the wire in Figure 27.27 carries a negative charge, is the force exerted by the wire on the particle attractive or repulsive? (e) Is this direction consistent with the direction of the forces exerted on each other by two parallel current-carrying wires?

CONCEPTS

Self-quiz

1. A compass sits on a table with its needle pointing to Earth's North Pole. A bar magnet with its long axis oriented along an east-west line is brought toward the compass from the right. If the needle turns clockwise, which pole of the bar magnet is nearer the compass?

2. Draw the magnetic field lines associated with the magnets in Figure 27.1*b–d* (both outside and inside the magnets). Assume that the pole on the left of each magnet is the north pole.

3. For each situation shown in Figure 27.28, apply the appropriate right-hand rule to determine the direction at position P of the magnetic field generated by the current-carrying wire.

Figure 27.28

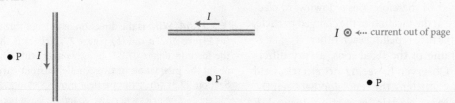

4. For each situation in Figures 27.20*a*, 27.20*b*, and 27.21*a*, reverse the polarity of the magnet and then apply the appropriate right-hand rule to determine the direction of the force exerted by the bar magnet on the current-carrying wire.

Answers

1. South pole. The clockwise rotation means the needle tip (which is a north pole by definition) moves toward the bar magnet. Because opposite poles attract, the bar magnet's south pole must be the closer one. (Remember that Earth's geographic North Pole is a magnetic south pole and so attracts the north pole of any compass needle.)

2. See Figure 27.29.

Figure 27.29

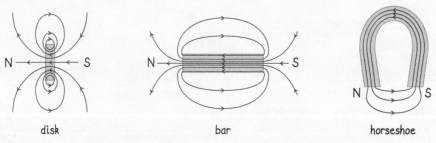

disk   bar   horseshoe

3. See Figure 27.30. (See Checkpoint 27.7 if you do not understand this answer.)

Figure 27.30

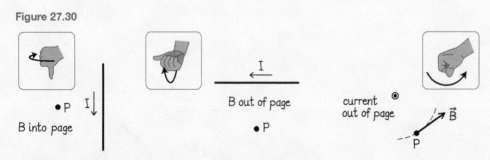

4. Reversing the magnet polarity reverses the direction of the force exerted on the wire, unless the force is zero, in which case it remains zero.

27.5 Current and magnetism

As we saw in the first part of this chapter, currents exert magnetic forces on one another. That is, a current creates a magnetic field, and magnetic fields exert forces on currents. Let us therefore begin this part of the chapter by introducing a quantitative definition of current. As mentioned in Section 27.3, **current** I is the rate at which charged particles cross a section of a conductor in a given direction. For a constant current, we have

$$I \equiv \frac{q}{\Delta t} \quad \text{(constant current),} \tag{27.1}$$

where q is the quantity of charge passing a given position in a time interval Δt. If the current is not constant, we evaluate the flow of charged particles over infinitesimally small time intervals, yielding

$$I \equiv \frac{dq}{dt}, \tag{27.2}$$

where dq is the infinitesimal quantity of charge crossing a given section of a conductor in an infinitesimally small time interval dt. Note that I can be positive or negative; by definition it is positive in the direction from high to low potential because that is the direction in which positive charge carriers flow (which we defined earlier as the direction of current).

The SI unit of current is the **ampere** (A). This base unit is defined to be the current through two parallel straight thin wires of infinite length separated by 1 m in vacuum when the wires exert a force of 2×10^{-7} N per meter of length on each other. As we saw in Chapter 22, the coulomb is derived from the ampere: 1 C corresponds to the quantity of charge transported by a current of 1 A through a chosen section in a time interval of 1 s, or

$$1\,\text{C} \equiv 1\,\text{A} \cdot \text{s}. \tag{27.3}$$

Even though a charge of 1 C is extremely large by ordinary standards, currents of several amperes are quite common. Simple devices, which we shall discuss in Chapter 31, allow us to measure the current through a conductor quite readily. In the remainder of this chapter, we shall discuss constant currents—that is, currents whose magnitude does not change with time.

Experiments show that when a straight wire carrying a current I is placed in a uniform external magnetic field,* the magnetic force $\vec{F}_w^B$ exerted by the magnetic field on the wire is proportional to the length ℓ of wire in the magnetic field and to the magnitude of the current I. The force also depends on the angle θ between the direction of the current through the wire and the magnetic field (**Figure 27.31**). When the wire is parallel to the magnetic field ($\theta = 0$), the magnetic force exerted on the wire is zero. When the wire and the magnetic field are perpendicular to each other ($\theta = 90°$), the force is maximum:

$$F_{w,\,max}^B = |I|\ell B \quad \begin{array}{l}\text{(straight wire perpendicular} \\ \text{to uniform magnetic field).}\end{array} \tag{27.4}$$

Because I is a signed quantity, we need to put absolute-value symbols around it to indicate its magnitude.

Equation 27.4 defines the magnitude B of the magnetic field. If we measure the magnetic force exerted by a magnetic field on a wire of known length carrying a known current, we can determine B from Eq. 27.4:

$$B \equiv \frac{F_{w,\,max}^B}{|I|\ell} \quad \begin{array}{l}\text{(straight wire perpendicular} \\ \text{to uniform magnetic field).}\end{array} \tag{27.5}$$

*This magnetic field does not include the magnetic field of the current-carrying wire.

Figure 27.31 A current-carrying wire in an external magnetic field (that is, a magnetic field created by an object other than the wire).

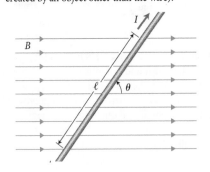

Table 27.2 Magnetic fields

Object	B (T)
Earth's surface	5×10^{-5}
small bar magnet	0.01
neodymium magnet	0.2
laboratory magnet	10
neutron star surface	10^8

It follows from this expression that the magnetic field has SI units of $N/(A \cdot m)$, a derived unit called the **tesla**:

$$1\,T \equiv 1\,N/(A \cdot m) = 1\,kg/(s^2 \cdot A). \qquad (27.6)$$

A magnetic field of 1 T is relatively large. Earth's magnetic field at Earth's surface varies between 3×10^{-5} T and 6×10^{-5} T. **Table 27.2** provides some examples of the magnitudes of various magnetic fields.

At intermediate angles θ, the magnitude of the magnetic force exerted on the wire is proportional to $\sin \theta$, and so for an arbitrary angle θ between 0 and 180°, the magnitude of the magnetic force exerted on the wire is

$$F_w^B = |I|\ell B \sin \theta \quad (0 < \theta < 180°). \qquad (27.7)$$

As we saw in Figure 27.22, the direction of the magnetic force exerted by a magnet on a current-carrying wire is always perpendicular to both the direction of the current I and the magnetic field $\vec{B}$, and is given by the right-hand force rule. In Figure 27.31, for example, the force is directed into the plane of the page. (If you line up the fingers of your right hand with the direction of current and then curl them, over the smallest angle, toward the direction of the magnetic field, your thumb points into the page.) If we define a vector $I\vec{\ell}$, whose magnitude is given by the product of the magnitude of the current and the length ℓ of the wire, and whose direction is given by the direction of current through the wire, we can write the **magnetic force** in Eq 27.7 as the vector product of two vectors (see Section 12.8):

$$\vec{F}_w^B = I\vec{\ell} \times \vec{B} \quad \text{(straight wire in uniform magnetic field).} \qquad (27.8)$$

This vector product represents a force that has the magnitude given in Eq. 27.7. The direction of $\vec{F}_w^B$ is always perpendicular to both $I\vec{\ell}$ (the direction of current) and $\vec{B}$ and is obtained by curling the fingers of the right hand from the direction of current to $\vec{B}$ (**Figure 27.32**).

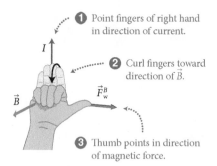

Figure 27.32 The right-hand rule for the direction of the vector product $I\vec{\ell} \times \vec{B}$.

1. Point fingers of right hand in direction of current.
2. Curl fingers toward direction of $\vec{B}$.
3. Thumb points in direction of magnetic force.

Example 27.2 Magnetic field meter

A metal bar 0.20 m long is suspended from two springs, each with spring constant $k = 0.10$ N/m, and the bar is in an external magnetic field directed perpendicular to the bar length (**Figure 27.33**). With a current of 0.45 A in the bar, the bar rises a distance $d = 1.5$ mm. (a) In what direction is the current? (b) What is the magnitude B of the external magnetic field?

Figure 27.33 Example 27.2.

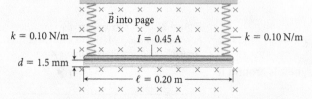

$k = 0.10$ N/m $\qquad$ $\vec{B}$ into page $\qquad$ $I = 0.45$ A $\qquad$ $k = 0.10$ N/m

$d = 1.5$ mm $\qquad$ $\ell = 0.20$ m

❶ **GETTING STARTED** I begin by making a free-body diagram for the bar with and without the current through the bar. Without the current, the bar is subject to two upward forces exerted by the springs and a downward force of gravity (**Figure 27.34a**). With the current turned on, an upward magnetic force lifts the bar, and so, according to Hooke's law (see Section 8.9), the springs exert a smaller force on the bar (**Figure 27.34b**). The amount by which

the bar rises determines by how much the force exerted by the springs is reduced, which in turn gives me the magnitude of the upward magnetic force exerted on the bar.

Figure 27.34

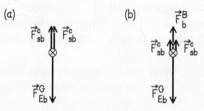

(a) $\qquad$ (b)

$\vec{F}_{sb}^c \qquad \vec{F}_{sb}^c \qquad \qquad \vec{F}_b^B$

$\qquad\qquad\qquad \vec{F}_{sb}^c \qquad \vec{F}_{sb}^c$

$\vec{F}_{Eb}^G \qquad\qquad\qquad \vec{F}_{Eb}^G$

❷ **DEVISE PLAN** I know the directions of the external magnetic field (into the page) and of the force (upward) exerted by that field, so I can determine the direction of the current from the right-hand force rule. Because the bar is straight and perpendicular to the external magnetic field, I can use Eq. 27.5 to relate the magnitude of the magnetic field to the magnitude of the magnetic force. I know both the current and the length of the bar but not the magnitude of the force exerted on the bar. I know, however, that this force magnitude is equal to the change in the magnitude of the force exerted by the springs, which I can calculate using Hooke's law (Eq. 8.20).

❸ EXECUTE PLAN (*a*) The external magnetic field is into the plane of the page (Figure 27.33) and the magnetic force is directed upward. I therefore lay my right hand on top of Figure 27.33 with my thumb pointing up to represent the upward force. I can lay my hand on the page palm up or palm down. In the palm-up position, curling my fingers draws the tips away from the direction of $\vec{B}$. In the palm-down position, curling them draws the tips toward the direction of $\vec{B}$. Therefore that's the position I want, telling me the current is to the right, in the direction my fingers point before I curl them. ✔

(*b*) From Eq. 8.20, I know that raising the bar a distance *d* reduces the spring force by an amount *kd*, and so the magnitude of the magnetic force is $F_b^B = 2kd$. Substituting this result into Eq. 27.5 gives

$$B = \frac{2kd}{|I|\ell} = \frac{2(0.10 \text{ N/m})(1.5 \times 10^{-3} \text{ m})}{(0.45 \text{ A})(0.20 \text{ m})} = 3.3 \times 10^{-3} \text{ T.} ✔$$

❹ EVALUATE RESULT: The magnitude of the magnetic field I obtain is about 100 times greater than Earth's magnetic field and of the same order of magnitude as that of a bar magnet and therefore not unreasonable.

27.15 Suppose the charge carriers flowing through the horizontal bar in Example 27.2 are negatively charged. In which direction must they flow so that the magnetic force exerted on the wire is still directed upward?

27.6 Magnetic flux

From the field line picture for electrostatics, we obtained Gauss's law, which states that the electric flux Φ_E through any closed surface is proportional to the enclosed charge. Gauss's law allows us to determine the electric field of certain symmetrical charge distributions with great ease.

Because magnetic fields can also be described by field lines, we can follow a similar treatment for magnetism. Consider a surface of area A in a uniform magnetic field B (**Figure 27.35**). If a line normal to the surface makes an angle θ with the field, we can define a **magnetic flux** in analogy to the electric flux defined in Eq. 24.2:

$$\Phi_B \equiv BA \cos \theta = \vec{B} \cdot \vec{A} \quad \text{(uniform magnetic field)}, \quad (27.9)$$

where $\vec{A}$ is an area vector (see Section 24.6) whose magnitude is equal to the area A of the surface and whose direction is normal to the surface. In the case of a closed surface, the direction of $\vec{A}$ is always chosen to be outward. Magnetic flux is a scalar, and as you can see from Eq. 27.9, magnetic flux has SI units of $\text{T} \cdot \text{m}^2$. This derived unit is given the name **weber** (Wb), in honor of the German physicist Wilhelm Weber:

$$1 \text{ Wb} \equiv 1 \text{ T} \cdot \text{m}^2 = 1 \text{ m}^2 \cdot \text{kg}/(\text{s}^2 \cdot \text{A}).$$

If the field is nonuniform or the surface is not flat, we follow the same procedure as for the electric flux and divide the surface into small surface elements, apply Eq. 27.9 to each surface element, and then sum the magnetic flux through all the elements. In the limit that the area of each element approaches zero, the sum is replaced by a surface integral (Eq. 24.4):

$$\Phi_B \equiv \int \vec{B} \cdot d\vec{A}, \quad (27.10)$$

where $d\vec{A}$ is the area vector of an infinitesimally small segment of the surface. The meaning of magnetic flux Φ_B is similar to that of electric flux. It is a quantitative measure of the number of magnetic field lines crossing the surface specified in the integration.

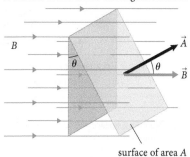

Figure 27.35 The magnetic flux through a surface of area A is given by the scalar product of the area vector $\vec{A}$ and the magnetic field $\vec{B}$.

As we saw in Section 27.2, magnetic field lines always form loops. Thus, the magnetic flux through any closed surface must always be zero because every field line that exits the surface must enter it somewhere else:

$$\Phi_B = \oint \vec{B} \cdot d\vec{A} = 0. \tag{27.11}$$

The zero reflects the fact that magnetic field lines always form loops or, equivalently, that there is no magnetic equivalent of an isolated charged particle. Equation 27.11 is called **Gauss's law for magnetism.** However, because of the zero result and because magnetic fields generally do not exhibit the same type of symmetry as electric fields, this expression does not allow us to determine magnetic fields in the same way that Gauss's law allows us to determine electric fields. As we shall see in Chapter 29, however, the magnetic flux is an important quantity when we consider changing magnetic fields.

Example 27.3 Magnetic flux through a loop

A square loop 0.20 m on each side is placed in a uniform magnetic field of magnitude 0.50 T. The plane of the loop makes a 30° angle with the magnetic field. What is the magnetic flux through the loop?

❶ GETTING STARTED I begin by drawing a side view to visualize the situation (**Figure 27.36**). I have to calculate the magnetic flux through the flat surface defined by the loop.

Figure 27.36

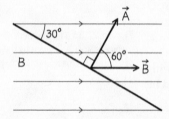

❷ DEVISE PLAN Because the magnetic field is uniform, I can use Eq. 27.9 to calculate the magnetic flux through the loop.

❸ EXECUTE PLAN The area of the loop is $A = (0.20 \text{ m})^2 = 0.040 \text{ m}^2$. Because the loop makes a 30° angle with the magnetic field, the angle between the magnetic field and a normal to the plane of the loop is 60°. Therefore

$$\Phi_B = AB \cos\theta = (0.040 \text{ m}^2)(0.50 \text{ T}) \cos(60°)$$

$$= 1.0 \times 10^{-2} \text{ Wb.} ✔$$

❹ EVALUATE RESULT I arbitrarily chose the area vector $\vec{A}$ to point upward and to the right. Had I chosen to point it downward and to the left, the angle between $\vec{A}$ and $\vec{B}$ would have been 120° and, given that $\cos 120° = -\frac{1}{2}$, I would have obtained $\Phi_B = -0.010$ Wb. The flat surface defined by the loop doesn't constitute a closed surface, however, and so there is no unique direction for $\vec{A}$. The sign of the magnetic flux is therefore not determined by the information given in the question. I have no means of evaluating the magnitude of the flux, so I carefully check my calculations one more time.

Figure 27.37 Any surface bounded by loop L yields the same magnetic flux.

(a)

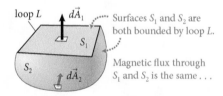

Surfaces S_1 and S_2 are both bounded by loop L.

Magnetic flux through S_1 and S_2 is the same . . .

(b)

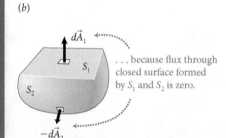

. . . because flux through closed surface formed by S_1 and S_2 is zero.

Because we can draw an infinite number of surfaces that are bounded by a loop, the expression "flux through the loop" in Example 27.3 appears open to interpretation. **Figure 27.37a**, for example, shows two surfaces S_1 and S_2 bounded by the same loop L. Which surface should we use when computing the magnetic flux? The answer is: *any* surface bounded by L yields the same magnetic flux. To see why, let us examine the magnetic fluxes through S_1 and S_2. In both cases we use a surface normal that points up through loop L.

The magnetic flux through S_2 is

$$\Phi_B = \int_{S_2} \vec{B} \cdot d\vec{A}_2. \tag{27.12}$$

To see how this flux relates to that through S_1, consider the magnetic flux through the closed surface made up of S_1 and S_2 (**Figure 27.37b**). Because this surface is closed, we know the magnetic flux through it is zero (Eq. 27.11):

$$\oint_{S_1+S_2} \vec{B} \cdot d\vec{A} = \int_{S_1} \vec{B} \cdot d\vec{A} + \int_{S_2} \vec{B} \cdot d\vec{A} = 0, \tag{27.13}$$

where the area vector $d\vec{A}$ points *outward* from the closed surface. In terms of the surface normals $d\vec{A}_1$ and $d\vec{A}_2$ shown in Figure 27.37a, Eq. 27.13 becomes

$$\oint_{S_1+S_2} \vec{B} \cdot d\vec{A} = \int_{S_1} \vec{B} \cdot d\vec{A}_1 + \int_{S_2} \vec{B} \cdot (-d\vec{A}_2)$$

$$= \int_{S_1} \vec{B} \cdot d\vec{A}_1 - \int_{S_2} \vec{B} \cdot d\vec{A}_2. \qquad (27.14)$$

The term $\int_{S_1} \vec{B} \cdot d\vec{A}_1$ represents the magnetic flux through the flat surface S_1; the term $\int_{S_2} \vec{B} \cdot d\vec{A}_2$ represents the magnetic flux through the curved surface S_2. Because the magnetic flux through the closed surface is zero (Eq. 27.11), the right side of Eq. 27.14 must also be zero, and so

$$\int_{S_1} \vec{B} \cdot d\vec{A}_1 = \int_{S_2} \vec{B} \cdot d\vec{A}_2. \qquad (27.15)$$

✋ **27.16** A cube 1.0 m on each side is placed in a 1.0-T magnetic field with the field perpendicular to one surface of the cube. What are (*a*) the magnetic flux through the side through which the field enters the cube and (*b*) the magnetic flux through the entire surface of the cube?

27.7 Moving particles in electric and magnetic fields

The magnetic force exerted by a magnetic field on a straight current-carrying wire is really the sum of the magnetic forces exerted on many individual charge carriers moving through the wire. By examining how a current I through a conductor is related to the properties of the charge carriers causing that current, we can therefore deduce the magnetic force acting on a single charge carrier.

Consider a constant current caused by a flow of charge carriers, each carrying a quantity of charge q through a wire of cross-sectional area A (Figure 27.38). If the charge carriers flow at speed v, then in a time interval Δt each advances a distance $\ell = v\Delta t$. Thus all the charge carriers in the shaded volume $V = A\ell$ pass through a cross section of the wire in a time interval Δt. If the wire contains n charge carriers per unit volume, the shaded volume contains nV charge carriers and the charge flowing through the cross section is $Q = nVq = nA\ell q = nA(v\Delta t)q$. The current through the wire is thus

$$I \equiv \frac{Q}{\Delta t} = \frac{nA(v\Delta t)q}{\Delta t} = nAqv. \qquad (27.16)$$

Substituting this result into Eq. 27.7, we can write for the magnitude of the magnetic force exerted on the current-carrying wire

$$F_{\text{w}}^B = |nA\ell qv|B \sin\theta = nA\ell|q|vB \sin\theta, \qquad (27.17)$$

where θ is the angle between the velocity of the charge carriers and the magnetic field. Because n is the number of charge carriers per unit volume and $A\ell$ is the volume of a length ℓ of the wire, the quantity $N = nA\ell$ represents the number of charge carriers in a length ℓ of the wire. The magnetic force exerted on that length of the wire can thus be written $F_{\text{w}}^B = N|q|vB \sin\theta$. Because there are N charge carriers in the length ℓ, the magnitude of the **magnetic force** exerted on a single particle carrying a charge q moving at velocity $\vec{v}$ is

$$F_{\text{p}}^B = |q|vB \sin\theta \qquad (27.18)$$

or, in vector form,

$$\vec{F}_{\text{p}}^B = q\vec{v} \times \vec{B}. \qquad (27.19)$$

Figure 27.38 If the charge carriers in a straight current-carrying wire move at speed v, they advance a distance $\ell = v\Delta t$ in a time interval Δt. All the charge carriers in the shaded volume pass through the cross-sectional area A in that time interval.

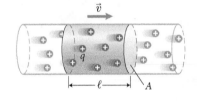

QUANTITATIVE TOOLS

Note that the vector $q\vec{v}$ always points in the direction of current—that is, from high potential to low potential—regardless of the sign of q. As we saw in Section 25.3, a negatively charged particle moves from low to high potential, so $\vec{v}$ points opposite the direction of current, but because $q < 0$, the vector $q\vec{v}$ points in the direction of current.

This is the fundamental expression for the magnetic force acting on a moving charge carrier. Even though we derived Eq. 27.19 for a positively charged particle, it holds for any kind of charge. Because of its velocity dependence, this force is very different from any of the other forces we have encountered so far. In particular, if the charge carrier is at rest *relative to the reference frame in which the magnetic field is measured*, the magnetic force vanishes. (In contrast, the electric force exerted on a charged particle is independent of the motion of the charged particle relative to the reference frame in which the electric field is measured.)

In the presence of both electric and magnetic fields, the **electromagnetic force** exerted on each charge carrier is

$$\vec{F}_p^{EB} = q\vec{E} + q\vec{v} \times \vec{B} = q(\vec{E} + \vec{v} \times \vec{B}). \qquad (27.20)$$

Figure 27.39 A charged particle moving in a uniform magnetic field travels (*a*) in a straight line when its velocity is parallel to the field and (*b*) in a circle when the two are perpendicular.

(*a*) Particle's velocity is parallel to field

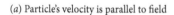
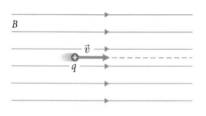

(*b*) Particle's velocity is perpendicular to field

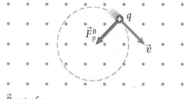

$\vec{B}$ out of page

Let us now examine what kind of trajectory a charged particle follows when it travels through a region of uniform magnetic and electric fields. We begin by examining the two special cases of a positive charge carrier traveling into a region of uniform magnetic field. When the carrier travels parallel to the direction of $\vec{B}$, as in **Figure 27.39a**, the magnetic force is zero because the vector product in Eq. 27.19 is zero when $\vec{v}$ is parallel to $\vec{B}$. (Alternatively, you can visualize the moving charge carrier as a current. As we have seen in Section 27.3, the magnetic force is zero when the current is parallel to the magnetic field.)

When the charge carrier moves perpendicular to a uniform magnetic field, as in Figure 27.39b, the magnetic force is nonzero. Because the magnetic force is always perpendicular to both $\vec{B}$ and $\vec{v}$, it lies in the plane of the drawing and is perpendicular to $\vec{v}$. Because the force acting on the change carrier always remains perpendicular to the direction of motion, the speed of the charge carrier does not change and we have the condition for circular motion at constant speed. The magnetic force, always directed toward the center of the circular path in which the charge carrier moves, provides the centripetal acceleration.

The equation of motion for the charge carrier is

$$\sum \vec{F} = m\vec{a}. \qquad (27.21)$$

If we ignore the force of gravity exerted on the charge carrier, the only force exerted on it is the magnetic force. Because $\vec{B}$ and $\vec{v}$ are perpendicular, the magnitude of this force is $F_p^B = |q|vB$ (Eq. 27.18). The magnitude of the charge carrier's centripetal acceleration is given by Eq. 11.15, $a_c = v^2/R$, where R is the radius of its circular trajectory. Therefore

$$|q|vB = \frac{mv^2}{R}. \qquad (27.22)$$

Solving for R, we obtain for the radius of the trajectory

$$R = \frac{mv}{|q|B}. \qquad (27.23)$$

Because the ratio $m/|q|$ is fixed for a given charge carrier, we see that the radius of its trajectory depends on the charge carrier's speed. In a given magnetic field, fast charge carriers move in larger circles than slow ones of the same type. Interestingly, the time interval needed for one full revolution is independent of the carrier's speed. This time interval is equal to the circumference of the trajectory divided by the carrier's speed:

$$T = \frac{2\pi R}{v} = \frac{2\pi}{v} \frac{mv}{|q|B} = \frac{2\pi m}{|q|B}. \qquad (27.24)$$

The corresponding angular frequency is

$$\omega = 2\pi f = \frac{2\pi}{T} = \frac{|q|B}{m}. \qquad (27.25)$$

This angular frequency is sometimes called the *cyclotron frequency* after a type of particle accelerator, called a *cyclotron*, in which particles are accelerated between successive semicircular trajectories.

27.17 A proton and an electron travel through a region of uniform magnetic field B. If their speeds are the same, what is the ratio R_p/R_e of the radii of their circular paths through the field?

Example 27.4 Mass spectrometer

Figure 27.40 shows a schematic of a device, called a *mass spectrometer*, for determining the mass of ions or other charged particles. The ions that enter the mass spectrometer are first accelerated by an electric field and then deflected by a magnetic field. The mass of the ions is obtained from the position at which they hit a detector after being deflected by the magnetic field.

Figure 27.40 Example 27.4.

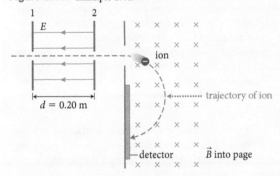

d = 0.20 m

trajectory of ion

detector $\vec{B}$ into page

In a certain mass spectrometer, the electric field is caused by a potential difference of 10 kV across a distance $d = 0.20$ m between plates 1 and 2, and the magnitude of the magnetic field is $B = 0.20$ T. An oxygen ion with a charge q of $-2e$ ($1e = 1.6022 \times 10^{-19}$ C) and a mass m of 16 atomic mass units (1 atomic mass unit $= 1$ u $= 1.6605 \times 10^{-27}$ kg) enters the electric field with negligible initial velocity. At what distance from the point of entry into the magnetic field does the ion hit the detector?

❶ GETTING STARTED From the schematic I see that the ion's motion consists of two parts: a linear motion at constant acceleration in the uniform electric field, followed by circular motion at constant speed in the uniform magnetic field. Because the drawing shows the ion entering the magnetic field perpendicular to the plane of the detector, the ion traces out a half circle before hitting the detector. The distance we must determine is therefore equal to the diameter of the circular trajectory of the ion.

❷ DEVISE PLAN Equation 27.23 gives the radius of the circular trajectory. I am given B, m, and q, but I don't know the speed v at which the ion enters the magnetic field. This speed is equal to the final speed the ion acquires after accelerating in the electric field. I know from Eq. 25.17 how much electrostatic work is done on the ion as it traverses the potential difference between plates 1 and 2 that set up the electric field. According to Eq. 25.17, I have $W_{ion}(1 \to 2) = -qV_{12}$. This work increases the kinetic energy of the ion. Because the initial kinetic energy is zero, I have $W_{ion}(1 \to 2) = \Delta K = K_f$. I know q, V_{12}, and m, so I can now calculate v from $K_f = \frac{1}{2}mv^2$ and then use that value in Eq. 27.23 to obtain the answer to this question.

❸ EXECUTE PLAN The electrostatic work done on the ion is

$$W_{ion}(1 \to 2) = -qV_{12} = 2eV_{12}.$$

From the expression for kinetic energy, I obtain for the ion's speed as it enters the magnetic field

$$v = \sqrt{\frac{K_f}{\frac{1}{2}m}} = \sqrt{\frac{4eV_{12}}{m}}.$$

Substituting this speed and the magnitude of the charge q on the oxygen ion into Eq 27.23, I obtain for the diameter of the ion's circular trajectory

$$2R = \frac{2mv}{|q|B} = \frac{2m}{2eB}\sqrt{\frac{4eV_{12}}{m}} = \frac{2}{B}\sqrt{\frac{mV_{12}}{e}}$$

$$= \frac{2}{0.20 \text{ T}} \sqrt{\frac{16(1.66 \times 10^{-27} \text{ kg})(1.0 \times 10^4 \text{ V})}{1.60 \times 10^{-19} \text{ C}}}$$

$$= 0.41 \text{ m}. ✔$$

❹ EVALUATE RESULT To be measurable in a laboratory, the distance from the point of entry at which the ions hit the detector must be neither too small nor too great. Given these constraints, the distance I obtained appears reasonable.

Figure 27.41 Charged particles whose speed satisfies Eq. 27.26 move in a straight line through magnetic and electric fields that are oriented perpendicular to each other.

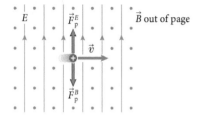

Let us next consider the trajectory of a charged particle in combined uniform electric and magnetic fields. If the charged particle has a velocity component perpendicular to the magnetic field, it is subject to a magnetic force in addition to an electric force. With an appropriate choice of fields, it is possible for the two forces to add up to zero, leaving the charged particle undisturbed. This arrangement is illustrated in **Figure 27.41**: A positive charged particle moves at right angles to both $\vec{E}$ and $\vec{B}$, which are perpendicular to each other. If the two forces are of equal magnitude, we have $F_p^B = F_p^E$, and so from Eqs. 27.18 and 23.6 $|q|vB = |q|E$ or

$$v = \frac{E}{B} \quad \text{(electric and magnetic force cancel).} \quad (27.26)$$

Thus, if we adjust the magnitudes of the electric and magnetic fields so that one charged particle goes through undeflected, then any other charged particle moving at the same velocity also passes through the fields undeflected, *regardless of its mass or charge*. Consequently, the setup shown in Figure 27.41 serves as a *velocity selector*: If charge carriers traveling at different speeds are injected from the left into a region where an electric field and a magnetic field are perpendicular to each other, only those whose speed satisfies Eq. 27.26 make it through without being deviated. Another important application of the cancellation of electric and magnetic forces is discussed in the box "The Hall effect" on page 731.

27.18 In Figure 27.41, do $\vec{F}_p^B$ and $\vec{F}_p^E$ still cancel if the charged particle (*a*) carries a negative charge, (*b*) travels in the opposite direction, (*c*) travels at a slight angle to the two fields?

Example 27.5 The mass of the electron

Figure 27.42 shows schematically part of the apparatus used in 1897 by J. J. Thomson to determine the charge-to-mass ratio of the electron. A beam of electrons, all moving at the same speed v, enters a region of electric and/or magnetic fields. When an electric field of magnitude 1.0 kV/m and a magnetic field of magnitude 1.2×10^{-4} T are turned on, the electrons go through the device undeflected. When the magnetic field is turned off, the electrons are deflected by 3.2 mm in the negative y direction after traveling the length $\ell = 0.050$ m of the apparatus. Given that the charge of the electron is $-e = -1.60 \times 10^{-19}$ C, what is the mass of each electron?

Figure 27.42 Example 27.5.

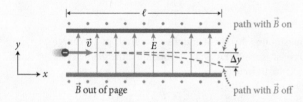

❶ **GETTING STARTED** When the magnetic field is turned on, the electrons travel in a straight line, which means that the forces exerted by the electric and magnetic fields on the electron cancel and the electrons travel at constant speed in the $+x$ direction. When the magnetic field is turned off, the electrons undergo a constant acceleration in the $-y$ direction due to the electric field while continuing to travel at constant speed in the $+x$ direction. I know from the expression $\vec{a} = (q/m)\vec{E}$ we derived in Section 23.4 that the acceleration of the electrons in the electric field depends on their charge-to-mass ratio. If I can determine this ratio from the electrons' trajectory, I can obtain their mass.

❷ **DEVISE PLAN** Because I know the magnitudes of the electric and magnetic fields that yield a straight trajectory, I can obtain the speed v of the electrons as they enter the apparatus from Eq. 27.26. Because the electrons move in the $+x$ direction, I know that the x component of their velocity is given by $v_x = +v$ (regardless of whether the magnetic field is on or off). When the magnetic field is off, their acceleration in the y direction is given by $a_y = F_y^E/m_e = -eE/m_e$. I can solve this expression for m_e, but I don't know a_y. I do know, however, that I can use kinematics to determine a_y.

❸ **EXECUTE PLAN** Because the electron's initial velocity in the y direction is zero, the amount of deflection is given by $\Delta y = \frac{1}{2}a_y\Delta t^2$, where Δt is the time interval during which the electrons travel in the electric field. This time interval is equal to $\Delta t = \ell/v_x = \ell/v$, so that

$$a_y = \frac{2\Delta y}{(\Delta t)^2} = \frac{2\Delta y}{(\ell/v)^2} = \frac{2\Delta y\, v^2}{\ell^2} = \frac{2\Delta y(E/B)^2}{\ell^2}, \quad (1)$$

where I have substituted Eq. 27.26 for the speed v of the electrons. Substituting $a_y = -eE/m_e$ into Eq. 1 and solving for m_e, I get

$$m_e = \frac{-eE}{a_y} = \frac{-e\ell^2 B^2}{2\Delta y E}$$

$$= \frac{(-1.60 \times 10^{-19}\,\text{C})(0.050\,\text{m})^2(1.2 \times 10^{-4}\,\text{T})^2}{2(-3.2 \times 10^{-3}\,\text{m})(1.0 \times 10^3\,\text{V/m})}$$

$$= 9.0 \times 10^{-31}\,\text{kg.} ✔$$

❹ **EVALUATE RESULT** The value I obtain is close to the published value of the electron mass ($m_e = 9.10938291 \times 10^{-31}$ kg), giving me confidence in my calculation.

The Hall effect

The canceling of electric and magnetic forces exerted on a charge carrier makes it possible to determine whether the mobile charge carriers in a conductor are positively or negatively charged. This determination is done using a phenomenon called the *Hall effect*.

Consider the rectangular conducting strip carrying an upward current illustrated in **Figure 27.43** (that is, the bottom of the strip is at a higher potential than the top). The strip is placed in a magnetic field directed into the page. If the current is caused by positive charge carriers (Figure 27.43*a*), the carriers move upward and so, according to the right-hand rule, the magnetic force acting on them is to the left. This force deflects the carriers toward the left side of the strip, where they pile up. This accumulation of positive charge carriers on the left causes an electric field to the right across the strip that exerts on the carriers an electric force that pulls them to the right. As more carriers accumulate on the left,

the magnitude of this electric force increases until it becomes equal in magnitude to the magnetic force. Once the two forces are equal in magnitude, the carriers are no longer deflected and so no more carriers accumulate on the left. Any subsequent charge carriers travel in a straight line.

The accumulation of charge carriers can be determined by measuring the potential difference between the left (L) and right (R) sides of the strip. If, as illustrated in Figure 27.43*b*, positive carriers accumulate on the left, this potential difference is positive:

$$V_{RL} = V_L - V_R > 0.$$

As illustrated in Figure 27.43*c*, however, if the mobile charge carriers causing the current are negatively charged, V_{RL} is *negative*. This is so because an upward current means that negative charge carriers move downward and, as you can verify using the right-hand rule, the magnetic force exerted on these carriers is also to the left. Thus, regardless of the sign of their charge, the mobile charge carriers pile up on the left. If they are positively charged, V_{RL} is positive; if they are negatively charged, V_{RL} is negative. Experiments on strips of common metals always yield a negative V_{RL}, showing that the mobile charge carriers in these metals are negatively charged, as we stated earlier (see Section 22.3 and Checkpoint 22.15).

27.19 (*a*) Express the magnitude of the electric field inside the strip in Figure 27.43 in terms of the width *w* of the strip and the potential difference V_{RL}. (*b*) Given the magnitude *B* of the magnetic field, what is the speed at which the charge carriers travel? (*c*) Show how this information, together with Eq. 27.16, can be used to determine the number density *n* of the charge carriers.

Figure 27.43

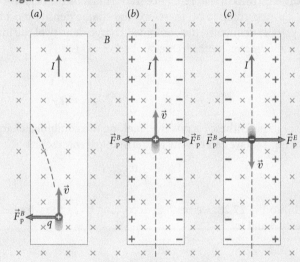

27.8 Magnetism and electricity unified

In this section we quantitatively examine the situation discussed in the latter part of Section 27.4—the interaction between a current-carrying wire and a positively charged particle moving parallel to the wire. For simplicity, we again let the charged particle move at the same speed *v* as the electrons in the wire.* In the Earth reference frame in Figure 27.27 the wire is electrically neutral: The linear charge density of the fixed positively charged ions $\lambda_{proper} > 0$ is equal in magnitude to that of the electrons, $-\lambda_{proper}$. The interaction between the wire and the charged particle is thus purely magnetic.

As we saw in Section 27.4, the wire appears to be positively charged to observer M, who moves along with the electrons. According to M, the average distance between the positively charged ions has decreased by the factor

$$\gamma = \frac{1}{\sqrt{1 - v^2/c_0^2}} \qquad (27.27)$$

---

*In the more general case where the charged particle does not move at the same speed as the electrons in the wire, the algebra becomes more complicated but the conclusions remain the same.

Figure 27.44 A positively charged particle moving parallel to a current-carrying wire, as seen by two observers in motion relative to each other.

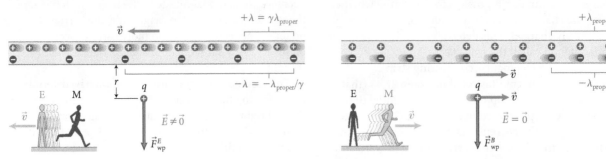

(a) **Reference frame of electrons in wire (observer M)**

(b) **Earth reference frame (observer E)**

due to length contraction (Eqs. 14.28 and 14.6; $\gamma > 1$). Thus the charge density of the ions is increased and has a magnitude $\lambda_{\text{M ions}} = \gamma\lambda_{\text{proper}}$ (**Figure 27.44a**). The electrons, on the other hand, are at rest relative to M and moving relative to E. Therefore the average distance between the electrons according to E must be decreased by the factor γ in Eq. 27.27 relative to the distance observed by M. Because to observer E in the Earth reference frame the wire is electrically neutral, the charge density of the electrons in that reference frame must be the negative of that of the ions: $-\lambda_{\text{proper}}$. According to M, therefore, the charge density must be $\lambda_{\text{M electrons}} = -\lambda_{\text{proper}}/\gamma$. Consequently, the combined charge density of the ions and the electrons in the moving reference frame of observer M is

$$\lambda_{\text{M}} = \lambda_{\text{M ions}} + \lambda_{\text{M electrons}} = \gamma\lambda_{\text{proper}} - \frac{\lambda_{\text{proper}}}{\gamma} = \lambda_{\text{proper}}\left(\gamma - \frac{1}{\gamma}\right). \quad (27.28)$$

Because $\gamma > 1$, the term in parentheses is positive, making the combined charge density positive. Using Eq. 14.9 we can rewrite the term in parentheses as

$$\gamma - \frac{1}{\gamma} = \gamma\left(1 - \frac{1}{\gamma^2}\right) = \gamma\frac{v^2}{c_0^2}. \quad (27.29)$$

Substituting Eq. 27.29 into Eq. 27.28, we see that according to M the wire has a (nonzero) charge density equal to

$$\lambda_{\text{M}} = \lambda_{\text{proper}}\gamma\frac{v^2}{c_0^2}. \quad (27.30)$$

Using the expression $E = 2k\lambda/r$ for the magnitude of the electric field of a charged wire (see Exercise 24.7), we obtain for the magnitude of the electric field at a distance r from the wire according to M

$$E_{\text{M}} = \frac{2k\lambda_{\text{M}}}{r} = \frac{2k\lambda_{\text{proper}}\gamma v^2}{rc_0^2}. \quad (27.31)$$

(Because the distance to the wire is perpendicular to the direction of motion, both observers measure the same distance r and so we need no subscript on r.)

According to observer M, the electric force exerted by the wire on the charged particle therefore has a magnitude

$$F_{\text{Mwp}}^E = |q|E_{\text{M}} = |q|\frac{2k\lambda_{\text{proper}}\gamma v^2}{rc_0^2} \quad (27.32)$$

and points perpendicular to and away from the wire.

Now we return to the Earth reference frame, as illustrated in Figure 27.44b. In this reference frame, the combined charge density of the ions and electrons is

$$\lambda_{\text{ions}} + \lambda_{\text{electrons}} = \lambda_{\text{proper}} - \lambda_{\text{proper}} = 0, \quad (27.33)$$

and so the electric field outside the wire is zero. Consequently, the force between the wire and the charged particle cannot be due to an electric interaction. We know experimentally, however, that according to observer E the current-carrying wire exerts a *magnetic* force on the particle. The direction of this force is also perpendicular to and away from the wire. Is this magnetic force the same as the electric force seen by M?

To answer this question we first need to determine the relationship between the forces measured by two observers in motion relative to each other. To determine this relationship, we shall use the definition of force given in Eq. 8.4, $\sum \vec{F} \equiv d\vec{p}/dt$, where $\sum \vec{F} = \vec{F}_{wp}$. The vector $d\vec{p}$ points in the same direction as the force—perpendicular to the relative velocity of the two frames—and so must be the same for both observers: $d\vec{p} = d\vec{p}_M$. Because of time dilation, however, we have from Eq. 14.13, $dt = \gamma dt_M$, where dt is the infinitesimal time interval measured by observer E and dt_M is the corresponding time interval measured by observer M (because M moves along with the particle, this is a proper time interval). The magnitude of the force exerted by the wire on the particle measured by observer E is thus

$$F_{wp} = \frac{dp}{dt} = \frac{dp_M}{\gamma dt_M} = \frac{F_{Mwp}}{\gamma}, \tag{27.34}$$

where F_{Mwp} is the magnitude of the force exerted by the wire on the particle measured by observer M. To observer M this force is electric in nature and its magnitude is given by Eq. 27.32. Substituting Eq. 27.32 on the right in Eq. 27.34 thus yields

$$F_{wp} = |q| \frac{2k\lambda_{proper}v^2}{rc_0^2}. \tag{27.35}$$

The quantity $\lambda_{proper}v$ is just the current I because if the electrons advance by a distance d in a time interval Δt, a quantity of charge $|q| = \lambda_{proper}d$ flows through the wire in that time interval. So the rate at which the charge carriers flow is

$$I \equiv \left| \frac{q}{\Delta t} \right| = \lambda_{proper} \frac{d}{\Delta t} = \lambda_{proper}v. \tag{27.36}$$

This result means that we can write Eq. 27.35 as

$$F_{wp} = |q|v\frac{2kI}{rc_0^2}. \tag{27.37}$$

Equation 27.37 gives the magnitude of the force exerted by the current-carrying wire on the particle measured by observer E. The magnitude depends on the current I through the wire, on the speed v and charge q of the particle, and on the distance r between the wire and the particle. Observer E interprets this force as a magnetic force due to the magnetic field caused by the current through the wire. From Eq. 27.18 we see that the magnitude of the magnetic force exerted by the wire on the moving particle is $F_{wp}^B = |q|vB$ (because $\vec{v}$ and $\vec{B}$ are at right angles, $\sin\theta = 1$). Substituting this expression into Eq. 27.37 and dividing both sides by qv, we see that the magnitude of the magnetic field according to observer E is

$$B = \frac{2kI}{rc_0^2} = \frac{2k}{c_0^2}\frac{I}{r}. \tag{27.38}$$

In words, the magnitude B of the magnetic field is proportional to the current I and inversely proportional to the distance r to the wire, with the proportionality constant being $2k/c_0^2$. As we shall see in the next chapter, experiments confirm this dependence. Here we see that Coulomb's law, together with special relativity, *requires* a magnetic field of this form. In other words, electricity and magnetism are two aspects of the same interaction, not two different interactions.

27.20 Explain why the $1/r$ dependence expressed in Eq. 27.38 is consistent with the symmetry of the wire causing the magnetic field.

Chapter Glossary

SI units of physical quantities are given in parentheses.

Ampere (A) The SI base unit of current, defined as the constant current through two straight parallel thin wires of infinite length placed 1 meter apart that produces a force between the wires of magnitude 2×10^{-7} N for each meter length of the wires. The coulomb and the ampere are related by

$$1 \text{ C} \equiv 1 \text{ A} \cdot \text{s}. \tag{27.3}$$

Current I (A) A scalar that gives the rate at which charge carriers cross a section of a conductor in a given direction:

$$I \equiv \frac{dq}{dt}. \tag{27.2}$$

The *direction of current* through a conductor is the direction in which the potential decreases. In this direction I is positive.

Electromagnetic force $\vec{F}^{EB}$ (N) The force exerted on a moving charged particle in the presence of both an electric field and a magnetic field:

$$\vec{F}_{\text{p}}^{EB} = q(\vec{E} + \vec{v} \times \vec{B}), \tag{27.20}$$

Gauss's law for magnetism The magnetic flux through a closed surface is always zero:

$$\Phi_B = \oint \vec{B} \cdot d\vec{A} = 0. \tag{27.11}$$

Magnet Any object that attracts pieces of iron, such as iron filings or paper clips.

Magnetic dipole An object with a pair of opposite magnetic poles separated by a small distance.

Magnetic field $\vec{B}$ (T) A vector that provides a measure of the magnetic interaction of objects. The magnitude of a uniform magnetic field can be determined by measuring the force exerted by that magnetic field on a straight wire of length ℓ carrying a current I:

$$B \equiv \frac{F_{\text{w, max}}^{B}}{|I|\ell} \quad \begin{array}{l} \text{(straight wire perpendicular} \\ \text{to uniform magnetic field).} \end{array} \tag{27.5}$$

The direction of the magnetic field at a certain location is given by the direction in which a compass needle points at that location.

Magnetic field line A representation of magnetic fields using lines of which the tangent at every position gives the direction of the magnetic field at that position. Near a magnet, magnetic field lines point away from north poles and toward south poles.

Magnetic flux Φ_B (Wb) A scalar that provides a quantitative measure of the number of magnetic field lines passing through an area. The magnetic flux through a surface is given by the surface integral

$$\Phi_B \equiv \int \vec{B} \cdot d\vec{A}. \tag{27.10}$$

Magnetic force $\vec{F}^B$ (N) The force exerted by magnets, current-carrying wires, and moving charged particles on each other. The magnetic force exerted on a straight wire of length ℓ carrying a current I and placed in a uniform magnetic field B is

$$\vec{F}_{\text{w}}^{B} = I\vec{\ell} \times \vec{B}, \tag{27.8}$$

where $I\vec{\ell}$ is a vector whose magnitude is given by the product of the magnitude of the current and the length ℓ of the wire, and whose direction is given by the direction of current through the wire. The magnetic force exerted on a particle carrying a charge q and moving at a velocity $\vec{v}$ through a magnetic field $\vec{B}$ is

$$\vec{F}_{\text{p}}^{B} = q\vec{v} \times \vec{B}. \tag{27.19}$$

Magnetic interaction The long-range interaction between magnets and/or current-carrying wires that are at rest relative to the observer.

Magnetic polarization The magnetic state induced in a piece of magnetic material because of the presence of a magnet.

Tesla (T) The SI derived unit of magnetic field:

$$1 \text{ T} \equiv 1 \text{ N}/(\text{A} \cdot \text{m}) = 1 \text{ kg}/(\text{s}^2 \cdot \text{A}). \tag{27.6}$$

Weber (Wb) The SI derived unit of magnetic flux:

$$1 \text{ Wb} \equiv 1 \text{ T} \cdot \text{m}^2 = 1 \text{ m}^2 \cdot \text{kg}/(\text{s}^2 \cdot \text{A}).$$

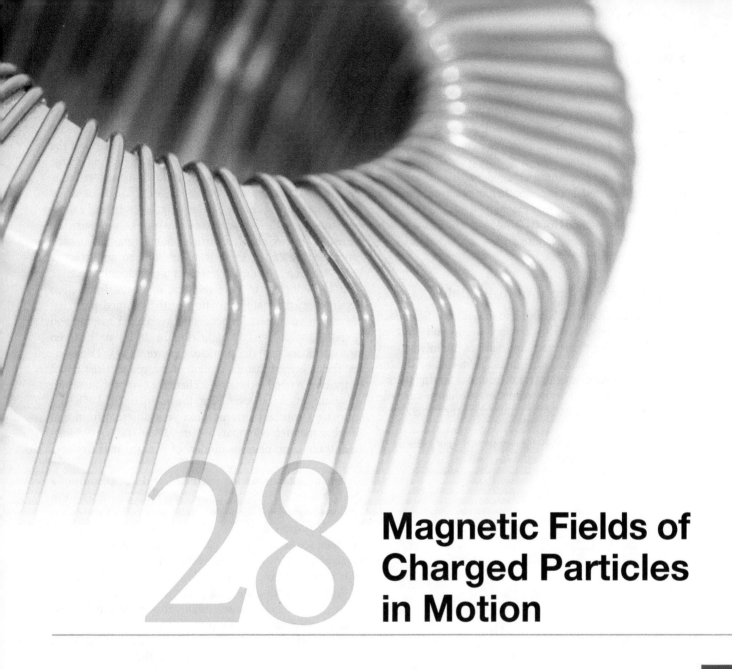

28 Magnetic Fields of Charged Particles in Motion

CONCEPTS

QUANTITATIVE TOOLS

*I*n this chapter we investigate further the relationship between the motion of charged particles and the occurrence of magnetic fields. As we shall see, *all* magnetism is due to charged particles in motion—whether moving along a straight line or spinning about an axis. It takes a moving or spinning charged particle to create a magnetic field, and it takes another moving or spinning charged particle to "feel" that magnetic field. We shall also discuss various methods for creating magnetic fields, which have wide-ranging applications in electromechanical machines and instruments.

28.1 Source of the magnetic field

As we saw in Chapter 27, magnetic interactions take place between magnets, current-carrying wires, and moving charged particles. **Figure 28.1** summarizes the interactions we have encountered so far. Figures 28.1*a–c* show the interactions between magnets and current-carrying wires. The sideways interaction between a magnet and a current-carrying wire (Figure 28.1*b*) is unlike any other interaction

Figure 28.1 Summary of magnetic interactions. Notice that stationary charged particles do not engage in magnetic interactions.

(*a*)

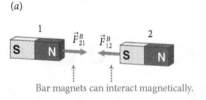

Bar magnets can interact magnetically.

(*b*)   (*c*)

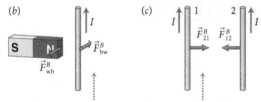

Current-carrying wires can interact magnetically.

(*d*)   (*e*)

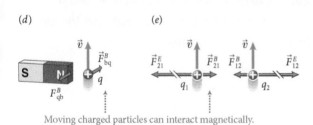

Moving charged particles can interact magnetically.

(*f*)   (*g*)

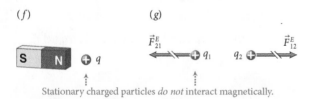

Stationary charged particles *do not* interact magnetically.

we have encountered. The forces between the wire and the magnet are not central; they do not point directly from one object to the other. As we saw in Section 27.7, the magnetic force exerted on a current-carrying wire is the sum of the magnetic forces exerted on many individual moving charge carriers. Similarly the magnetic field due to a current-carrying wire is the sum of the magnetic fields of many individual moving charge carriers. Figure 28.1*d* and *e* illustrate the magnetic interactions of moving charged particles. Note that for two charged particles moving parallel to each other (Figure 28.1*e*), there is, in addition to an attractive magnetic force, a (much greater) repulsive electric force.

It is important to note that the magnetic interaction depends on the state of motion of the charged particles. No magnetic interaction occurs between a bar magnet and a stationary charged particle (Figure 28.1*f*) or between two stationary charged particles (Figure 28.1*g*). These observations suggest that the motion of charged particles might be the origin of *all* magnetism. There are two problems with this assumption, however. First, the magnetic field of a wire carrying a constant current looks very different from that of a bar magnet. (Compare Figures 27.13 and 27.19.) Second, there is no obvious motion of charged particles in a piece of magnetic material.

Figure 28.2*a* shows the magnetic field line pattern of a straight wire carrying a constant current. The lines form circles centered on the wire, circles that reflect the cylindrical symmetry of the wire (the symmetry of an infinite cylinder, see Section 24.4). The horizontal distance between adjacent circles is smaller near the wire, where the magnitude of the field is greater.

A single moving charged particle does not have cylindrical symmetry because, unlike for an infinitely long straight wire, moving the particle up or down along its line of motion changes the physical situation. Because of the *circular* symmetry of the situation, the field still forms circles around the line of motion, but the magnitude of the field at a fixed distance from the particle's line of motion

Figure 28.2 Comparing the magnetic fields of a current-carrying wire and a moving charged particle.

(*a*) Magnetic field of a wire carrying a constant current

(*b*) Magnetic field of a moving charged particle

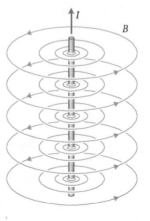

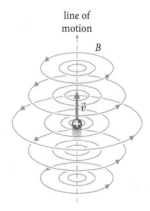

decreases if one moves away from the particle (Figure 28.2b). It is not at all obvious how the field pattern shown in Figure 28.2b could give rise to the magnetic field of a bar magnet; there are certainly no poles in the magnetic field of the moving charged particle.

✋📵 **28.1** Make a sketch showing the directions of the magnetic forces exerted on each other by (a) an electron moving in the same direction as the current through a wire, (b) a moving charged particle and a stationary charged particle, and (c) two current-carrying wires at right angle to each other as illustrated in **Figure 28.3**. (Hint: Determine the forces exerted at points P_1 through P_5).

Figure 28.3 Checkpoint 28.1c.

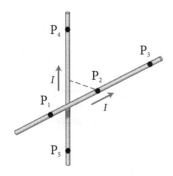

28.2 Current loops and spin magnetism

The circular pattern of magnetic field lines around a wire carrying a constant current suggests a method for generating a strong magnetic field: If a wire carrying a constant current is bent into a loop as shown in **Figure 28.4**, all the magnetic field lines inside the loop point in the same direction, reinforcing one another.*

Figure 28.4 The magnetic field of a wire loop carrying a constant current. The magnetic fields from all parts of the loop reinforce one another in the center of the loop.

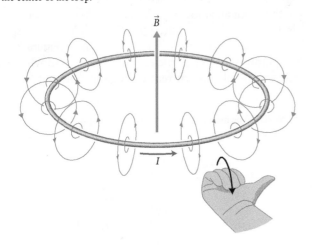

---

*For now, we'll ignore how to make charge carriers flow through such a loop. In Section 28.6 we'll discuss physical arrangements that accomplish the situation illustrated in Figure 28.4.

What does the magnetic field of such a current-carrying loop, called a **current loop** for short, look like? To answer this question, we treat the current loop as a collection of small segments of a current-carrying wire and determine the direction of the magnetic field at various points around the loop. As we did in Chapter 27, we shall assume all currents to be constant in the remainder of this chapter.

We begin by considering the magnetic field due to a small segment of the current loop at a point on the central axis that passes perpendicularly through the face of the loop (point A in **Figure 28.5** on the next page). Segment 1 carries a current that points into the page. The magnetic field lines of this segment are concentric circles centered on the segment. Using the right-hand current rule, we see that the magnetic field curls clockwise in the plane of the drawing, and so the magnetic field due to segment 1 at A points up and to the right. Figure 28.5b shows a magnetic field line through A due to segment 2. Because the current through segment 2 points out of the page, this field line curls counterclockwise, and so the magnetic field due to segment 2 at A points up and to the left.

Figure 28.5c shows the contributions from segments 1 and 2 together; because their horizontal components cancel, the vector sum of $\vec{B}_1$ and $\vec{B}_2$ points vertically up. The same arguments can be applied to any other pair of segments lying on opposite sides of the current loop. Therefore the magnetic field due to the entire current loop points vertically up at A.

Now consider point C at the center of the current loop. As illustrated in Figure 28.5d, the magnetic fields due to segments 1 and 2 point straight up there, too. The same is true for all other segments, and so the field at point C also points straight up. We can repeat the procedure for point D below the current loop, and as shown in Figure 28.5e, the magnetic field there also points up.

To determine the direction of the magnetic field at a point outside the current loop, consider point G in Figure 28.5f. Using the right-hand current rule, you can verify that the field due to segment 1 points vertically down and the field due to segment 2 points vertically up. Because G is closer to 1 than it is to 2, the magnetic field due to 1 is stronger, and so the sum of the two fields points down.

✋📵 **28.2** What is the direction of the magnetic field at a point vertically (a) above and (b) below segment 1 in Figure 28.5?

The complete magnetic field pattern of the current loop, obtained by determining the magnetic fields for many more points, is shown in **Figure 28.6a** on the next page. Close to the wire, the field lines are circular, but as you move farther away from the wire, the circles get squashed inside the loop and stretched outside due to the contributions of other parts of the loop to the magnetic field.

Figure 28.5 Mapping the magnetic field of a current loop. The magnetic field contributions from (*a*) segment 1 and (*b*) segment 2 at A (*c*) add up to a vertical field. Magnetic fields at (*d*) point C at the center of the ring, (*e*) point D below the ring, and (*f*) point G to the right of the ring. Note that in all cases the magnetic field of each segment is perpendicular to the line connecting that segment to the point at which we are determining the field.

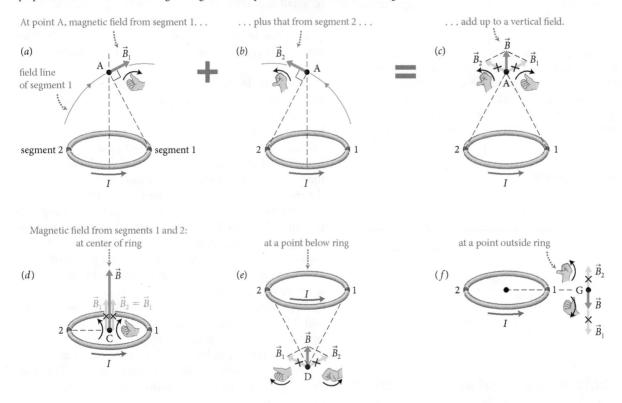

As you may have noticed, the field line pattern of a current loop resembles that of a bar magnet (Figure 28.6*b*). Indeed, if you shrink the size of both the current loop and the bar magnet, their magnetic field patterns become identical

Figure 28.6 The magnetic field of a current loop (*a*) resembles that of a dipole (*b, c*).

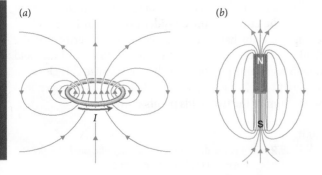

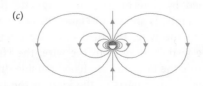

(Figure 28.6*c*). The magnetic field pattern in Figure 28.6*c* is that of an infinitesimally small magnetic dipole.

Given that a current loop produces a magnetic field similar to that of a bar magnet, is the magnetic field of a bar magnet then perhaps due to tiny current loops inside the magnet? More precisely, are elementary magnets (see Section 27.1) simply tiny current loops?

The connection between current loops and elementary magnets becomes clearer once we realize that a current loop is nothing but an amount of charge that revolves around an axis. Consider, for example, a positively charged ring spinning around a vertical axis through its center (**Figure 28.7**). The spinning charged particles cause a current moving in a circle exactly like the current through the circular loop in

Figure 28.7 The magnetic field of a charged spinning ring is identical to that of a current loop.

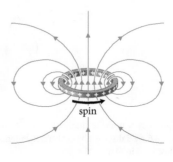

Figure 28.5. If we let the radius of the ring approach zero, the ring becomes a spinning charged particle, and its magnetic field pattern approaches that of the magnetic dipole illustrated in Figure 28.6c. This surprising result tells us:

> **A spinning charged particle has a magnetic field identical to that of an infinitesimally small magnetic dipole.**

Experiments show that most elementary particles, such as electrons and protons, possess an intrinsic angular momentum—as if they permanently spin around—and such spinning motion indeed would produce a magnetic field of the form shown in Figure 28.6c. Because of the combined intrinsic angular momentum of these elementary particles inside atoms, certain atoms have a magnetic field, causing them to be the elementary magnets we discussed earlier. The reason we cannot separate north and south magnetic poles is therefore a direct consequence of the fact that the magnetic field of a particle with intrinsic angular momentum is that of an infinitesimally small magnetic dipole.

✋ 28.3 Suppose a negatively charged ring is placed directly above the positively charged ring in Figure 28.7. If both rings spin in the same direction, is the magnetic interaction between them attractive or repulsive?

28.3 Magnetic dipole moment and torque

To specify the orientation of a magnetic dipole we introduce the **magnetic dipole moment.** This vector, represented by the Greek letter μ (mu), is defined to point, like a compass needle, along the direction of the magnetic field through the center of the dipole (Figure 28.8). For a bar magnet the magnetic dipole moment points from the south pole to the north pole. To determine the direction of $\vec{\mu}$ for a current loop, you can use a right-hand rule: When you curl the

Figure 28.8 The magnetic dipole moment points in the direction of the magnetic field through the center of the dipole.

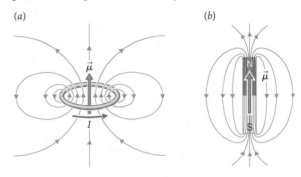

fingers of your right hand along the direction of the current, your thumb points in the direction of $\vec{\mu}$.

We now have three right-hand rules in magnetism, illustrated in **Figure 28.9** and summarized in **Table 28.1** together with the right-hand vector product rule. What helps distinguish these right-hand rules from one another is looking at which quantity curls: For a current-carrying wire, it is the magnetic field that curls, whereas for a current loop, it is the current that curls. The only additional thing you need to remember is that in the case of the magnetic force exerted on a current-carrying wire, the fingers are associated with the curl from the direction of the current to the direction of the magnetic field. Note, also, that the order of application

Table 28.1 Right-hand rules

Right-hand rule	thumb points along	fingers curl
vector product	$\vec{C} = \vec{A} \times \vec{B}$	from $\vec{A}$ to $\vec{B}$
current rule	current	along B-field
force rule	magnetic force	from current to B-field
dipole rule	$\vec{\mu}$ (parallel to $\vec{B}$)	along current loop

Figure 28.9 Right-hand rules in magnetism.

(a) Right-hand current rule

① Point right thumb in direction of current. ⋯⋯▶

② Fingers curl in direction of $\vec{B}$.

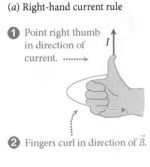

(b) Right-hand force rule

① Point fingers of right hand in direction of current.

③ Thumb points in direction of magnetic force.

② Curl fingers toward direction of $\vec{B}$.

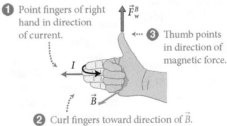

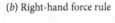

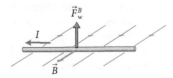

(c) Right-hand dipole rule

① Point right thumb in direction of magnetic dipole moment.

② Fingers curl in direction of current.

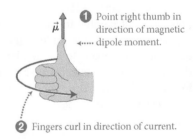

Figure 28.10 Magnetic forces exerted on a square current loop that is oriented so that its magnetic dipole moment is perpendicular to an external magnetic field.

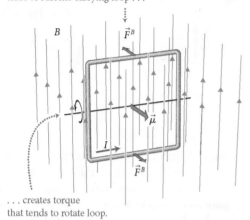

Force exerted by magnetic field on top and bottom sides of current-carrying loop . . .

. . . creates torque that tends to rotate loop.

Figure 28.11 Magnetic forces exerted on a square current loop that is oriented so that its magnetic dipole moment is parallel to an external magnetic field.

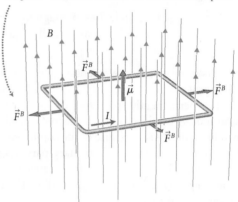

Magnetic field exerts force on all four sides of loop . . .

. . . but forces all lie in plane of loop, so cause no torque.

of each rule in Figure 28.9 can be reversed to solve for a different variable. If you know the direction in which the magnetic field lines curl, for example, you can use the right-hand current rule to determine the current direction.

28.4 Does the direction of the electric field along the axis inside an electric dipole coincide with the direction of the electric dipole moment?

What happens when a current loop is placed in a magnetic field? To find out, consider a square current loop of wire placed in a uniform magnetic field with its magnetic dipole moment perpendicular to the magnetic field (**Figure 28.10**). The current loop experiences magnetic forces on the top and bottom sides but not on the vertical sides because they are parallel to the direction of the magnetic field (see Section 27.3). Using the right-hand force rule of Figure 28.9b, we see that the magnetic forces exerted on the top and bottom sides cause a torque that tends to rotate the loop as indicated in Figure 28.10.

When the current loop is oriented with its magnetic dipole moment parallel to the magnetic field, all four sides experience a magnetic force (**Figure 28.11**). However, because all four forces lie in the plane of the loop, none of them causes any torque. Because the magnitudes of the four forces are the same, their vector sum is zero and the loop is not accelerated sideways.

28.5 As the current loop in Figure 28.10 rotates over the first 90°, do the magnitudes of the (a) magnetic force exerted on the horizontal sides and (b) the torque caused by these forces increase, decrease, or stay the same? (c) As the loop rotates, do the two vertical sides experience any force, and, if so, do these forces cause any torque? (d) What happens to the torque as the loop rotates beyond 90°?

Summarizing the results of Checkpoint 28.5:

A current loop placed in a magnetic field tends to rotate such that the magnetic dipole moment of the loop becomes aligned with the magnetic field.

This alignment is completely analogous to the alignment of the electric dipole moment in the direction of an external electric field, which we studied in Section 23.8 (see Figure 23.32).

28.6 Suppose the square current loop in Figure 28.10 is replaced by a circular loop with a diameter equal to the width of the square loop and with the same current. Does the circular loop experience a torque? If not, why not? If so, how does this torque compare with that on the square loop?

Example 28.1 Current loop torque

When placed between the poles of a horseshoe magnet as shown in Figure 28.12, does a rectangular current loop experience a torque? If so, in which direction does the loop rotate?

Figure 28.12 Example 28.1

❶ GETTING STARTED I know that a current loop placed in a magnetic field tends to rotate such that the magnetic dipole moment of the loop becomes aligned with the magnetic field. If the loop tends to rotate, it experiences a torque.

❷ DEVISE PLAN The simplest way to answer this question is to look at the directions of the magnetic field $\vec{B}$ and the magnetic dipole moment $\vec{\mu}$. By definition, the magnetic field between the poles of the magnet points from the north pole to the south pole. To determine the direction of $\vec{\mu}$, I can use the right-hand dipole rule.

③ EXECUTE PLAN I begin by sketching the loop and indicating the direction of the magnetic field (Figure 28.13*a*). To determine the direction of $\vec{\mu}$, I curl the fingers of my right hand along the direction of the current through the loop. My thumb shows that $\vec{\mu}$ points straight up. To align $\vec{\mu}$ with $\vec{B}$, therefore, the loop rotates in the direction shown by the curved arrow in Figure 28.13*a*. The loop must experience a torque in order for this rotation to occur. ✔

Figure 28.13

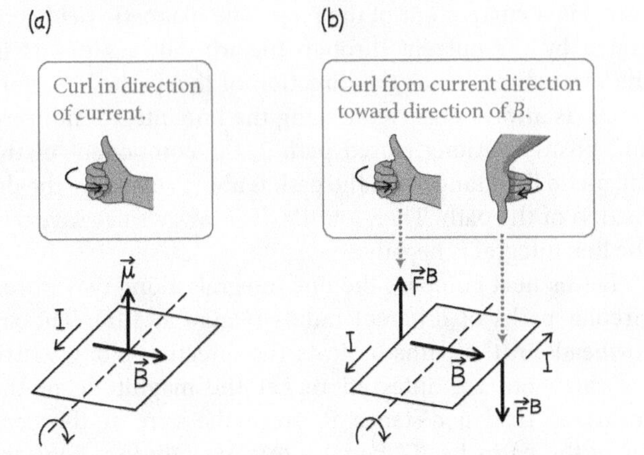

(a)

Curl in direction of current.

$\vec{\mu}$

I

$\vec{B}$

(b)

Curl from current direction toward direction of B.

$\vec{F}^B$

I

$\vec{B}$

I

$\vec{F}^B$

④ EVALUATE RESULT I can verify my answer by determining the force exerted by the magnetic field on each side of the loop and seeing if these forces cause a torque (Figure 28.13*b*). The front and rear sides experience no force because the current through them is either parallel or antiparallel to $\vec{B}$. To determine the direction of the force exerted on the left side of the loop, I point my right-hand fingers along the direction of the current through that side and curl them toward the direction of the magnetic field. My upward-pointing thumb indicates that the force exerted on the left side is upward. Applying the right-hand force rule to the right side of the loop tells me that the magnetic force exerted on that side is directed downward. The magnetic forces exerted on the left and right sides thus cause a torque that makes the loop rotate in the same direction I determined earlier.

The alignment of the magnetic dipole moment of current loops in magnetic fields is responsible for the operation of any device involving an electric motor (see the box below titled Electric motors).

✋ **28.7** Describe the motion of the current loop in Figure 28.12 if the magnitude of the magnetic field between the poles of the magnet is greater on the left than it is on the right.

Electric motors

The torque caused by the forces exerted on a current loop in a magnetic field is the basic operating mechanism of an electric motor. A problem with the arrangement shown in Figure 28.12, however, is that once the magnetic dipole moment is aligned with the magnetic field, the torque disappears. Also if the current loop overshoots this equilibrium position, the torque reverses direction (see Checkpoint 28.5*d*). The most common way to overcome this problem is with a *commutator*—an arrangement of two curved plates that reverses the current through the current loop each half turn. The result is that the current loop keeps rotating.

The basic operation is illustrated in Figure 28.14. Each half of the commutator is connected to one terminal of a battery and is in contact with one end of the current loop. In the position illustrated in Figure 28.14*a*, the black end is in contact with the negative commutator (that is, the half of the commutator connected to the negative battery

terminal). Consequently, the current direction is counterclockwise as seen from above through the current loop, and the magnetic dipole moment $\vec{\mu}$ points up and to the left. The vertical magnetic field therefore causes a torque that turns the loop clockwise to align $\vec{\mu}$ with $\vec{B}$.

Once the loop has rotated to the position shown in Figure 28.14*b*, $\vec{\mu}$ is aligned with $\vec{B}$ but the contact between the ends of the current loop and the commutator is broken, which means there is no current in the loop. The loop overshoots this equilibrium position, but as soon as it does so, the current direction reverses because now the black end of the loop is in contact with the positive commutator (Figure 28.14*c*). The current direction is now clockwise as seen from above, and so $\vec{\mu}$ is reversed. Consequently, the current loop continues to rotate clockwise, as illustrated in Figure 28.14*c–e*. After half a revolution, we reach the initial situation again and the sequence repeats.

Figure 28.14 Operating principle of an electric motor. At instant (b), the current direction through the current loop is reversed.

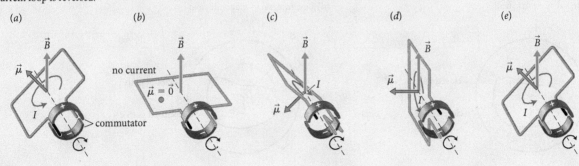

(a)   (b)   (c)   (d)   (e)

28.4 Ampèrian paths

In electrostatics, Gauss's law provides a powerful tool for determining the electric field of a charge distribution: The electric flux through a closed surface is determined by the amount of charge enclosed by that surface (see Section 24.3). The basic reason for Gauss's law is illustrated in **Figure 28.15a**: Electric field lines originate or terminate on charged particles, and the number of "field line piercings" through a closed surface is proportional to the amount of charge enclosed by that surface. As you saw in Checkpoint 27.6, however, Gauss's law for magnetism is not as helpful for determining magnetic fields. The reason is illustrated in Figure 28.15b: Magnetic field lines form loops, so if they exit a closed surface, they must reenter it at some other point. Consequently, the magnetic flux through a closed surface is always zero, regardless of whether or not the surface encloses any magnets or current-carrying wires. In mathematical language, the surface integral of the magnetic field over a closed surface is always zero.

Let us next consider line integrals of electric and magnetic fields. As we saw in Section 25.5, the line integral of the electrostatic field around a closed path is always zero. Consider, for example, the line integral of the electrostatic field around closed path 1 in Figure 28.15c. Because the electrostatic field generated by the charged particle at the center of path 1 is always perpendicular to the path, $\vec{E}$ is perpendicular to $d\vec{\ell}$. Therefore $\vec{E} \cdot d\vec{\ell} = 0$ and so the line integral is always zero. Any other path, such as closed path 2

in Figure 28.15c, can be broken down into small radial and circular segments. As we saw in Section 25.5, going once around a closed path, all the nonzero contributions along the radial segments add up to zero. Physically this means that as you move a charged particle around a closed path through an electrostatic field, the work done by the electrostatic field on the particle is zero.

Consider now, however, the line integral around closed path 1 in Figure 28.15d, which is concentric with a wire carrying a current out of the page. The magnetic field generated by the current through the wire always points in the same direction as the direction of this path. Therefore $\vec{B} \cdot d\vec{\ell}$ is always positive, making the line integral nonzero and positive. Along closed path 2, the component of the magnetic field tangent to the path is always opposite the direction of the path. Therefore $\vec{B} \cdot d\vec{\ell}$ is always negative and the line integral is negative.

Let us next compare the line integrals along two closed circular paths of different radius (**Figure 28.16a**). The arrowheads in the paths indicate the direction along which we carry out the integration. Let the magnitude of the magnetic field a distance R_1 from the wire at the center of the paths be B_1. Because this magnitude is constant along the entire circular path, the line integral along this path is the product of the field magnitude and the length of the path: $B_1(2\pi R_1) = 2\pi B_1 R_1$. Along closed path 2 we obtain $2\pi B_2 R_2$, where B_2 is the field magnitude a distance R_2 from the wire. As you may suspect from the cylindrical symmetry of the wire, the magnitude of the magnetic field

Figure 28.15 Surface and line integrals of electric and magnetic fields.

(*a*) Surface integral of electric field (Gauss's law)

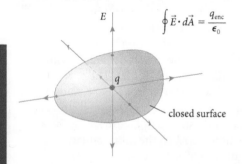

$$\oint \vec{E} \cdot d\vec{A} = \frac{q_{enc}}{\epsilon_0}$$

(*b*) Surface integral of magnetic field (Gauss's law for magnetism)

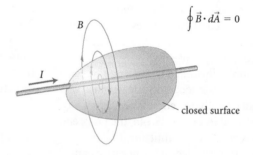

$$\oint \vec{B} \cdot d\vec{A} = 0$$

(*c*) Line integral of electric field

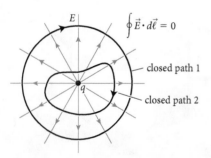

$$\oint \vec{E} \cdot d\vec{\ell} = 0$$

(*d*) Line integral of magnetic field

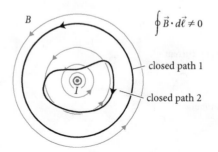

$$\oint \vec{B} \cdot d\vec{\ell} \neq 0$$

Figure 28.16 (*a*) Two closed circular paths concentric with a wire that carries a current directed out of the page. (*b*) A noncircular path encircling the current-carrying wire. The two arcs each represent one-eighth of a circle.

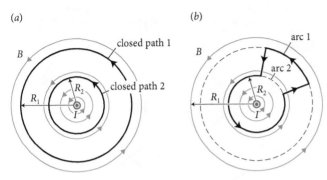

Figure 28.17 (*a*) A noncircular closed path encircling a current-carrying wire. (*b*) We can approximate the path by using small arcs and radial segments.

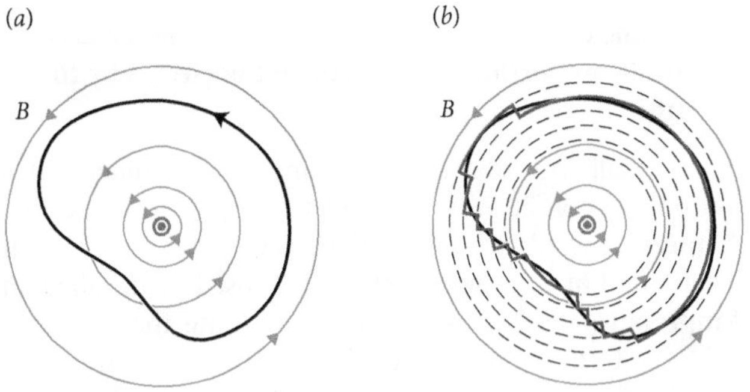

decreases as $1/r$ with distance r from the wire (we'll confirm this dependence in Example 28.6). This means that as we go from R_2 to R_1, the field decreases by a factor R_2/R_1. In other words, $B_1 = B_2(R_2/R_1)$ or $B_1R_1 = B_2R_2$. Thus we see that the line integrals along the two paths are equal: $2\pi B_1R_1 = 2\pi B_2R_2$. The same argument can be applied to any other closed circular path centered on the wire, from which we conclude that the line integral of the magnetic field of a straight current-carrying wire over any circular path centered on the wire has the same value.

28.8 If the magnitude of the current I through a wire is increased, do you expect the line integral of the magnetic field around a closed path around the wire to increase, decrease, or stay the same?

Now consider the noncircular path illustrated in Figure 28.16*b*. Most of the path lies along a circle of radius R_2, with the exception of one-eighth of a revolution, which is along two radial segments and an arc of radius R_1. Because the radial segments are perpendicular to the magnetic field, they do not contribute to the line integral. How do the line integrals along arcs 1 and 2 compare? As we just saw, the line integrals along the two closed circular paths in Figure 28.16*a* are equal, so the line integrals along one-eighth of each closed path in Figure 28.16*a* must also be equal. The same must be true in Figure 28.16*b*, which means the line integrals along arcs 1 and 2 are identical, and so the line integral along the noncircular path in Figure 28.16*b* is equal to that along any circular path centered on the wire.

We can make the deviations from a circular path progressively more complicated, but as illustrated in Figure 28.17, any path can always be broken down into small segments that are either radial or circular and concentric with the wire. The radial segments never contribute to the line integral because the magnetic field is always perpendicular to them, while the circular segments always add up to a single complete revolution. So, in conclusion:

The value of the line integral of the magnetic field along a closed path encircling a current-carrying wire is independent of the shape of the path.

28.9 What happens to the value of the line integral along the closed path in Figure 28.17*a* when (*a*) the direction of the current through the wire is reversed; (*b*) a second wire carrying an identical current is added parallel to and to the right of the first one (but still inside the path); and (*c*) the current through the second wire is reversed?

Next let's examine the line integral along a closed path near a current-carrying wire lying outside the path. One such path is shown in Figure 28.18 as two arcs joined by two radial segments. As before, the radial segments do not contribute to the line integral. The magnitudes of the line integrals along arcs 1 and 2 are equal, but because the direction of arc 2 is opposite the direction of the magnetic field, the line integral along that arc is negative. Consequently, the line integral along the entire path adds up to zero. We can again extend this statement to a path of a different form, but as long as the path does not encircle the current-carrying wire, the line integral is zero:

The line integral of the magnetic field along a closed path that does not encircle any current-carrying wire is zero.

Figure 28.18 A noncircular closed path not encircling a current-carrying wire.

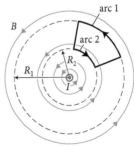

Putting the above results together, we see that the line integral of the magnetic field along a closed path tells us something about the amount of current encircled by the path:

The line integral of the magnetic field along a closed path is proportional to the current encircled by the path.

We shall put this statement in a more quantitative form in Section 28.5. As we shall see there, this law, called **Ampère's law,** plays a role analogous to Gauss's law: Given the amount of current encircled by a closed path, called an **Ampèrian path,** we can readily determine the magnetic field due to the current, provided the current distribution exhibits certain simple symmetries. Because the line integral along a closed path depends on the direction of integration, we must always choose a direction along the path when specifying an Ampèrian path. Exercise 28.2 illustrates the importance of the direction of the Ampèrian path.

28.10 Suppose the path in Figure 28.17 were tilted instead of being in a plane perpendicular to the current-carrying wire. Would this tilt change the value of the line integral of the magnetic field around the path?

Exercise 28.2 Crossed wires

Consider the Ampèrian path going through the collection of current-carrying wires in **Figure 28.19.** If the magnitude of the current is the same in all the wires, is the line integral of the magnetic field along the Ampèrian path positive, negative, or zero?

Figure 28.19 Exercise 28.2

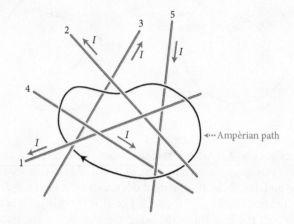

SOLUTION For each wire, I must determine whether or not the path encircles the wire and, if it does, whether the component of that wire's magnetic field tangent to the path points in the same direction as the path. I see that wires 1 and 3 go through the path but the other three wires lie either on top of the path or beneath it.

The direction of current through wire 1 is forward out of the plane of the page, so the magnetic field lines around this wire curl counterclockwise—opposite the direction of the Ampèrian path—giving a negative contribution to the line integral. The direction of current through wire 3 is down into the plane of the page, so it yields a positive contribution to the line integral. Because the two currents are equal in magnitude, the contributions to the line integral add up to zero. ✔

28.11 How do the following changes affect the answer to Exercise 28.2: (*a*) reversing the current through wire 1, (*b*) reversing the current through wire 2, (*c*) reversing the direction of the Ampèrian path?

Self-quiz

1. Determine the direction of the magnetic force exerted at the center of the wire or on the particles in **Figure 28.20**.

Figure 28.20

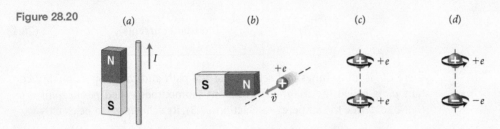

2. Determine the direction of the magnetic field at P due to (a) the current loop in **Figure 28.21a** and (b) segments A and C of the current loop in Figure 28.21b.

Figure 28.21

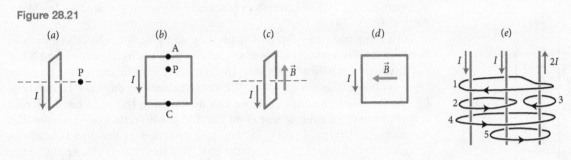

3. Determine in which direction the current loop rotates (a) in Figure 28.21c and (b) in Figure 28.21d.

4. (a) Determine the currents encircled by the five Ampèrian paths in Figure 28.21e. (b) Rank the paths according to the magnitudes of the line integral of the magnetic field along each path, greatest first.

Answers:

1. (a) No magnetic force is exerted by the magnet on the wire because the magnetic field at the location of the wire and the current are antiparallel. (b) The magnetic force acting on the particle is upward. To see this, consider the moving positively charged particle to be current in the direction of the velocity of the particle and then use the right-hand force rule, which makes your thumb point upward. (c) The magnetic dipole moments of the two spinning particles both point up and so the particles attract each other, just like two bar magnets oriented the way the spinning particles are (**Figure 28.22a**). (d) The magnetic dipole moment of the negative particle points down, that of the positive particle points up, and so the two particles repel each other. The comparable bar magnet orientation is shown in Figure 28.22b.

Figure 28.22

(a)

N
S	$\uparrow \vec{\mu}$

N
S	$\uparrow \vec{\mu}$

(b)

N
S	$\uparrow \vec{\mu}$

S
N	$\downarrow \vec{\mu}$

2. (a) The right-hand dipole rule tells you that the magnetic dipole of the loop and the magnetic field produced by the loop point to the right at P. (b) Segments A and C both contribute a magnetic field that points out of the page at P, so the magnetic field due to both segments also points out of the page at P.

3. (a) $\vec{\mu}$ for the current loop points to the right, so the current loop rotates counterclockwise about an axis perpendicular to the page and through the center of the loop. (b) $\vec{\mu}$ for the current loop points out of the page, so the loop rotates about an axis aligned with the vertical sides of the loop. The right side of the loop moves up out of the page, and the left side moves down into the page.

4. (a) Path 1 encircles two currents I in the same direction as the magnetic field: $+2I$. Path 2 encircles the same two currents in the opposite direction: $-2I$. Path 3 encircles $2I$ in the direction opposite the magnetic field direction: $-2I$. Path 4 encircles all three currents, which add up to zero. Path 5 encircles I in the direction opposite the magnetic field direction and $2I$ in the same direction as the magnetic field: $+I$. (b) Each line integral is proportional to the current encircled, making the ranking $1 = 2 = 3 > 5 > 4$.

28.5 Ampère's law

In the preceding section, we saw that the line integral of the magnetic field around a closed path, called an Ampèrian path, is proportional to the current encircled by the path, I_{enc}. This can be expressed mathematically as

$$\oint \vec{B} \cdot d\vec{\ell} = \mu_0 I_{enc} \quad \text{(constant currents),} \tag{28.1}$$

where $d\vec{\ell}$ is an infinitesimal segment of the path and the proportionality constant μ_0 is called the **magnetic constant** (sometimes called *permeability constant*). To define the ampere (see Section 27.5), its value is set to be exactly

$$\mu_0 = 4\pi \times 10^{-7} \text{ T} \cdot \text{m/A.}$$

Figure 28.23 A closed path encircling two of three straight current-carrying wires.

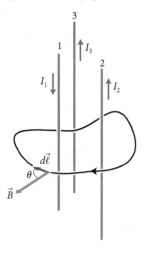

Equation 28.1 is called **Ampère's law**, after the French physicist André-Marie Ampère (1775–1836).

Let us illustrate how Eq. 28.1 is used by applying it to the closed path in **Figure 28.23**. From the figure we see that the path encircles wires 1 and 2 but not 3, which lies behind the path. To calculate the right side of Eq. 28.1, we must assign an algebraic sign to the contributions of currents I_1 and I_2 to I_{enc}. As we saw in Section 28.4, we can do so using the right-hand current rule. Putting the thumb of our right hand in the direction of I_1, we see that the magnetic field of I_1 curls in the same direction as the direction of the integration path (the Ampèrian path) indicated in the diagram. Applying the same rule to I_2 tells us that the magnetic field of this current curls in the opposite direction. Thus I_1 yields a positive contribution and I_2 yields a negative contribution to the line integral. The right side of Eq. 28.1 thus becomes $\mu_0(I_1 - I_2)$.

Procedure: Calculating the magnetic field using Ampère's law

For magnetic fields with straight or circular field lines, Ampère's law allows you to calculate the magnitude of the magnetic field without having to carry out any integrations.

1. Sketch the current distribution and the magnetic field by drawing one or more field lines using the right-hand current rule. A two-dimensional drawing should suffice.
2. If the field lines form circles, the Ampèrian path should be a circle. If the field lines are straight, the path should be rectangular.
3. Position the Ampèrian path in your drawing such that the magnetic field is either perpendicular or tangent to the path and constant in magnitude. Choose the direction of the Ampèrian path so that, where it runs parallel to the magnetic field lines, it points in the same direction as the field. If the current distribution divides space into distinct regions, draw an Ampèrian path in each region where you wish to calculate the magnetic field.

4. Use the right-hand current rule to determine the direction of the magnetic field of each current encircled by the path. If this magnetic field and the Ampèrian path have the same direction, the contribution of the current to I_{enc} is positive. If they have opposite directions, the contribution is negative.
5. For each Ampèrian path, calculate the line integral of the magnetic field along the path. Express your result in terms of the unknown magnitude of the magnetic field B along the Ampèrian path.
6. Use Ampère's law (Eq. 28.1) to relate I_{enc} and the line integral of the magnetic field and solve for B. (If your calculation yields a negative value for B, then the magnetic field points in the direction opposite the direction you assumed in step 1.)

You can use the same general approach to determine the current given the magnetic field of a current distribution. Follow the same procedure, but in steps 4–6, express I_{enc} in terms of the unknown current I and solve for I.

To calculate the line integral on the left side of Eq. 28.1, we first divide the closed path into infinitesimally small segments $d\vec{\ell}$, one of which is shown in Figure 28.23. The segments are directed tangentially along the path in the direction of integration. For each segment $d\vec{\ell}$, we take the scalar product of $d\vec{\ell}$ with the magnetic field $\vec{B}$ at the location of that segment, $\vec{B} \cdot d\vec{\ell}$, and then we add up the scalar products for all segments of the closed path. In the limit that the segment lengths approach zero, this summation becomes the line integral on the left in Eq. 28.1. In the situation illustrated in Figure 28.23, we cannot easily carry out this integration because of the irregular shape of the path. Just like Gauss's law in electrostatics, however, Ampère's law allows us to easily determine the magnetic field for highly symmetrical current configurations. The general procedure is outlined in the Procedure box on the previous page. The next two examples illustrate how this procedure can be applied to simplify calculating the line integral in Ampère's law.

Example 28.3 Magnetic field generated by a long straight current-carrying wire

A long straight wire carries a current of magnitude I, and this current creates a magnetic field $\vec{B}$. Derive an expression for the magnitude of the magnetic field a radial distance r from the wire.

❶ GETTING STARTED I begin by making a sketch of the wire, arbitrarily orienting it vertically (**Figure 28.24**). I know that the magnetic field is circular, so I draw one circular field line, centering it on the wire and giving it a radius r. Using the right-hand current rule, I determine the direction in which the magnetic field points along the circle and indicate that with an arrowhead in my drawing.

Figure 28.24

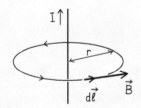

❷ DEVISE PLAN To begin my derivation I can use Ampère's law (Eq. 28.1). If I let the field line in my sketch be the Ampèrian path, the magnetic field is constant in magnitude and tangential all along the circular path, simplifying the integral on the left in Eq. 28.1. I let the direction of the Ampèrian path be the same as that of the magnetic field. That is to say, the infinitesimal path segment $d\vec{\ell}$ points in the same direction as $\vec{B}$ all along the Ampèrian path.

❸ EXECUTE PLAN With my choice of Ampèrian path, $d\vec{\ell}$ and $\vec{B}$ are always pointing in the same direction, and so $\vec{B} \cdot d\vec{\ell} = B \, dl$.

Because the magnitude B is the same all around the path, I can write for the left side of Eq. 28.1

$$\oint \vec{B} \cdot d\vec{\ell} = \oint B \, d\ell = B \oint d\ell.$$

The line integral on the right here is the sum of the lengths of all the segments $d\vec{\ell}$ around the Ampèrian path; that is to say, it is equal to the circumference of the circle. Therefore I have

$$\oint \vec{B} \cdot d\vec{\ell} = B \oint d\ell = B(2\pi r). \tag{1}$$

Because the direction in which the Ampèrian path encircles the current I is the same as the direction of the magnetic field generated by the current, the right side of Eq. 28.1 yields

$$\mu_0 I_{enc} = +\mu_0 I. \tag{2}$$

Substituting the right side of Eq. 2 and Eq. 1 into Eq. 28.1, I get $B(2\pi r) = \mu_0 I$ or

$$B = \frac{\mu_0 I}{2\pi r}. \checkmark$$

❹ EVALUATE RESULT My result shows that the magnitude of the magnetic field is proportional to I, as I expect (doubling the current should double the magnetic field), and inversely proportional to the radial distance r from the wire. I know from Chapter 24 that the electric field is also inversely proportional to r in cases that exhibit cylindrical symmetry, another indication that my result here makes sense.

✋ **28.12** Suppose the wire in Example 28.3 has a radius R and the current is uniformly distributed throughout the volume of the wire. Follow the procedure of Example 28.3 to calculate the magnitude of the magnetic field inside ($r < R$) and outside ($r > R$) the wire.

Example 28.4 Magnetic field generated by a large current-carrying sheet

A large flat metal sheet carries a current. The magnitude of the current per unit of sheet width is K. What is the magnitude of the magnetic field a distance d above the sheet?

❶ GETTING STARTED I begin by drawing the sheet and indicating a point P a distance d above it where I am to determine the magnetic field magnitude (**Figure 28.25a**). I draw a lengthwise arrow to show the current through the sheet.

Figure 28.25

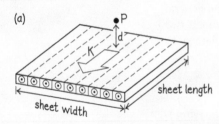
(a)

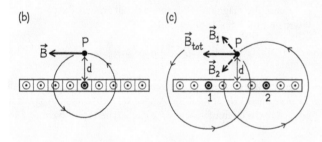
(b)   (c)

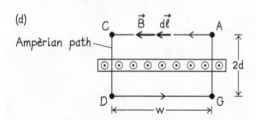
(d)

Ampèrian path

❷ DEVISE PLAN I'll solve this problem using Ampère's law (Eq. 28.1). I first need to determine the direction of the magnetic field on either side of the sheet. To this end, I divide the sheet into thin parallel strips, as indicated by the dashed lines in Figure 28.25a. I can then treat each strip as a current-carrying wire.

Figure 28.25b is a cross-sectional view of the sheet after I have divided it into strips. The perspective here is looking into the sheet in the direction opposite the direction of the current. Using the right-hand current rule, I determine that the strip right underneath point P contributes a magnetic field at P that points parallel to the sheet and to the left. Next I look at the contributions from the two strips labeled 1 and 2 in Figure 28.25c, equidistant on either side of P. The two large circles show the magnetic field lines from these two strips that go through P. Because the magnitudes of the contributions from the two strips are equal, the vector sum of their contributions to the magnetic field also points parallel to the sheet and to the left. The same argument can be applied to any other pair of strips that are equidistant on either side of P. Thus, at P the magnetic field due to the entire sheet must be parallel to the sheet and to the left. Similar reasoning shows that the magnetic field below the sheet is also parallel to the sheet and points to the right.

To exploit what I know about the magnetic field, I choose the rectangular path ACDG in Figure 28.25d as my Ampèrian path, making the direction of the path the same as that of the magnetic field. The width of this path is w, and its height is $2d$.

❸ EXECUTE PLAN I can write the line integral around the Ampèrian path as the sum of four line integrals, one over each side of the path:

$$\oint \vec{B} \cdot d\vec{\ell} = \int_A^C \vec{B} \cdot d\vec{\ell} + \int_C^D \vec{B} \cdot d\vec{\ell} + \int_D^G \vec{B} \cdot d\vec{\ell} + \int_G^A \vec{B} \cdot d\vec{\ell}.$$

Along sides CD and GA the magnetic field is perpendicular to the Ampèrian path and so $\vec{B} \cdot d\vec{\ell}$ is zero. Along each of the two horizontal sides, $d\vec{\ell}$ and $\vec{B}$ point in the same direction, and so $\vec{B} \cdot d\vec{\ell} = B\, d\ell$.

Symmetry requires that the magnitude of the magnetic field a distance d below the sheet be the same as the magnitude a distance d above it (because flipping the sheet upside down does not alter the physical situation). Therefore I can take the magnitude B out of the integral:

$$\oint \vec{B} \cdot d\vec{\ell} = \int_A^C B\, d\ell + \int_D^G B\, d\ell = B \int_A^C d\ell + B \int_D^G d\ell.$$

These two line integrals $\int d\ell$ yield the length of the two horizontal sides, which is the Ampèrian path width w, and so the left side of Eq. 28.1 becomes

$$\oint \vec{B} \cdot d\vec{\ell} = B(w + w) = 2Bw. \tag{1}$$

To get an expression for the right side of Eq. 28.1, I must first determine the amount of current encircled by the Ampèrian path. Because the magnitude of the current per unit of width through the sheet is K, the magnitude of the current through the Ampèrian path of width w is Kw. The right side of Eq. 28.1 thus becomes

$$\mu_0 I_{enc} = \mu_0 Kw. \tag{2}$$

Substituting the right sides of Eqs. 1 and 2 into Ampère's law, I get $2Bw = \mu_0 Kw$, or

$$B = \tfrac{1}{2}\mu_0 K. \checkmark$$

❹ EVALUATE RESULT Because the sheet has planar symmetry (see Section 24.4), I expect that, by analogy with the electric field around a flat charged sheet, the magnetic field on either side of my sheet here is uniform. That is, the field magnitude does not depend on the distance from the sheet, and this is just what my result shows. It also makes sense that my result shows that the magnitude of the magnetic field is proportional to the current per unit width K through the sheet.

Figure 28.26 Checkpoint 28.13.

(a)

✋ **28.13** (*a*) What are the direction and magnitude of the magnetic field between the parallel current-carrying sheets of **Figure 28.26a**? What is the direction of $\vec{B}$ outside these sheets? (*b*) Repeat for the sheets of Figure 28.26*b*.

(b)

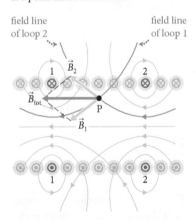

28.6 Solenoids and toroids

A **solenoid** is a long, tightly wound helical coil of wire (**Figure 28.27a**). In general, the diameter of the coil is much smaller than the length of the coil. When a current enters a solenoid at one end and exits at the other end, the solenoid generates a strong magnetic field. If a magnetic core is placed in the solenoid, the solenoid exerts a strong magnetic force on the core, turning electrical energy into motion. Solenoids are therefore often used in electrical valves and actuators. Because solenoids are generally very tightly wound, we can treat the windings of a solenoid as a stack of closely spaced current loops (Figure 28.27*b*).

Figure 28.28a shows that the magnetic field inside a long solenoid must be directed along the axis of the solenoid. Consider, for example, point P inside the solenoid. The figure shows the field lines of two of the loops, one on either side of P and equidistant from that point. The magnetic field contributions of the two loops give rise to a magnetic field that points along the axis of the solenoid. The same argument can be applied to any other pair of loops and to any other point inside the solenoid. Therefore the magnetic field everywhere inside a long solenoid must be directed parallel to the solenoid axis.

Figure 28.27 A solenoid is a tightly wound helical coil of wire.

(*a*) A solenoid

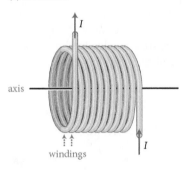

(*b*) A solenoid approximated as a stack of parallel current loops

Figure 28.28 The magnetic field of a solenoid that carries a current *I*.

(*a*) Cross section of a solenoid showing the contribution of two loops to the magnetic field at a point inside the solenoid

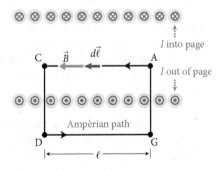

(*b*) Ampèrian path for calculating the magnetic field inside the solenoid

Figure 28.29 Magnetic field line pattern in a solenoid of finite length.

(a) Magnetic field of a solenoid, shown in cross section

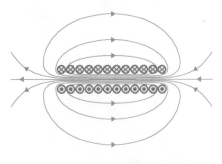

(b) The magnetic field of a solenoid

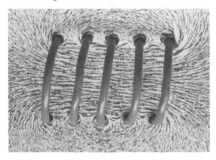

Figure 28.30 A toroid is a solenoid bent into a ring.

(a) Toroid

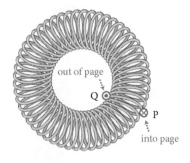

(b) Cross section showing magnetic field and a choice of Ampèrian path

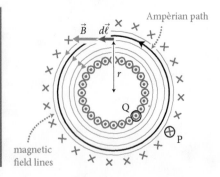

Because magnetic field lines form loops, all the lines that go through the solenoid must loop back from one end of the solenoid to the other. Because there is much more space available outside the solenoid, the density of the field lines as they loop back is much smaller than the field line density inside the solenoid. The longer the solenoid, the smaller the field line density in the immediate vicinity outside the solenoid. In the limit of an infinitely long solenoid, we can expect the magnetic field outside the solenoid to approach zero.

We can now use Ampère's law to determine the magnitude of the magnetic field when there is a current of magnitude I through the solenoid. Exploiting what we know about the direction of the magnetic field, we choose the rectangular path ACDG in Figure 28.28*b* as the Ampèrian path. Along AC, the magnetic field is parallel to the path, and so $\vec{B} \cdot d\vec{\ell} = B\,d\ell$. Along CD and GA, the magnetic field is perpendicular to the path, which means that these segments do not contribute: $\vec{B} \cdot d\vec{\ell} = 0$. Because the magnetic field is zero outside the solenoid, the segment DG also does not contribute. The line integral on the left side of Ampère's law (Eq. 28.1) thus becomes

$$\oint \vec{B} \cdot d\vec{\ell} = \int_A^C B\,d\ell. \tag{28.2}$$

The cylindrical symmetry of the solenoid requires the magnitude of the magnetic field to be constant along AC, so we can pull B out of the integral:

$$\oint \vec{B} \cdot d\vec{\ell} = \int_A^C B\,d\ell = B \int_A^C d\ell = B\ell, \tag{28.3}$$

where ℓ is the length of side AC.

What is the current encircled by the Ampèrian path? Each winding carries a current of magnitude I, but the path encircles more than one winding. If there are n windings per unit length, then $n\ell$ windings are encircled by the Ampèrian path, and the encircled current is I times the number of windings:

$$I_{\text{enc}} = n\ell I. \tag{28.4}$$

Substituting Eqs. 28.3 and 28.4 into Ampère's law (Eq. 28.1), we obtain

$$B\ell = \mu_0 n\ell I \tag{28.5}$$

$$B = \mu_0 n I \quad \text{(infinitely long solenoid).} \tag{28.6}$$

This result shows that the magnetic field inside the solenoid depends on the current through the windings and on the number of windings per unit length. The field within the solenoid is uniform—it does not depend on position inside the solenoid. Although Eq. 28.6 holds for an infinitely long solenoid, the result is pretty accurate even for a solenoid of finite length. For a solenoid that is at least four times as long as it is wide, the magnetic field is very weak outside the solenoid and approximately uniform and equal to the value given in Eq. 28.6 inside. An example of the magnetic field of a finite solenoid is shown in **Figure 28.29**. Note how the magnetic field pattern resembles that of the magnetic field around a bar magnet.

If a solenoid is bent into a circle so that its two ends are connected (**Figure 28.30***a*), we obtain a **toroid**. The magnetic field lines in the interior of a toroid (that is, the donut-shaped cavity enclosed by the coiled wire) close on themselves; thus they do not need to reconnect outside the toroid, as in the case of a solenoid. The entire magnetic field is contained inside the cavity. Symmetry requires the field lines to form circles inside the cavity; the field lines run in the direction of your right thumb when you curl the fingers of your right hand along the direction of the current through the windings. Figure 28.30b shows a few representative magnetic field lines in a cross section of the toroid where the current goes into the page on the outside rim of the toroid (as at point P, for example) and out of the page on the inside rim (as at point Q).

To determine the magnitude of the magnetic field, we apply Ampère's law (Eq. 28.1) to a circular path of radius r that coincides with a magnetic field line (Figure 28.30b). Because the field is tangential to the integration path, we have $\vec{B} \cdot d\vec{\ell} = B\,d\ell$. Furthermore symmetry requires the magnitude of the magnetic field to be the same all along the field line, and so we can pull B out of the integration:

$$\oint \vec{B} \cdot d\vec{\ell} = \oint B\,d\ell = B\oint dl = B(2\pi r). \qquad (28.7)$$

The Ampèrian path encircles one side of all of the windings, so if there are N windings, the encircled current is

$$I_{enc} = NI. \qquad (28.8)$$

Substituting these last two equations into Ampère's law (Eq. 28.1), we obtain

$$B = \mu_0 \frac{NI}{2\pi r} \quad \text{(toroid).} \qquad (28.9)$$

This result tells us that in contrast to a solenoid, the magnitude of the magnetic field in a toroid is not constant—it depends on the distance r to the axis through the center of the toroid.

Example 28.5 Square toroid

The toroid in **Figure 28.31** has 1000 windings carrying a current of 1.5 mA. Each winding is a square of side length 10 mm, and the toroid's inner radius is 10 mm. What is the magnitude of the magnetic field at the center of the winding squares?

Figure 28.31 Example 28.5

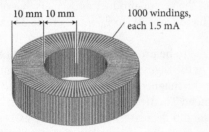

10 mm 10 mm

1000 windings, each 1.5 mA

❶ GETTING STARTED The fact that the windings are square does not change the magnetic field pattern. The magnetic field is still

circular as in Figure 28.30b, and the arguments given in the derivation of Eq. 28.9 still apply.

❷ DEVISE PLAN I can use Eq. 28.9 to determine the magnitude of the magnetic field.

❸ EXECUTE PLAN The distance from the center of the toroid to the center of each winding is 10 mm + 5 mm = 15 mm, so

$$B = (4\pi \times 10^{-7}\,\text{T}\cdot\text{m/A})\frac{(1000)(1.5 \times 10^{-3}\,\text{A})}{2\pi(0.015\,\text{m})}$$

$$= 2.0 \times 10^{-5}\,\text{T.} ✔$$

❹ EVALUATE RESULT The magnetic field magnitude I obtain is small—comparable to the magnitude of Earth's magnetic field at ground level—but the current through the toroid is very small, so my answer is not unreasonable.

28.14 Use Ampère's law to determine the magnetic field outside a toroid at a distance r from the center of the toroid (a) when r is greater than the toroid's outer radius and (b) when r is smaller than the toroid's inner radius.

Figure 28.32 The magnetic field at point P due to an arbitrarily shaped current-carrying wire cannot be obtained from Ampère's law. As shown in (c), the direction of $d\vec{B}_s$ can be found by taking the vector product of $d\vec{\ell}$, which has length $d\ell$ and direction given by the current through the wire, and $\hat{r}_{sP}$, which points from the segment to point P: $d\vec{\ell}_{sP} \times \hat{r}_{sP}$.

(a) Arbitrarily shaped current-carrying wire

(b) Small segment $d\vec{\ell}$ contributes magnetic field $d\vec{B}_s$ at point P

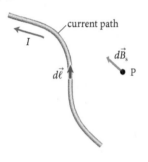

(c) Determining the direction of $d\vec{B}_s$

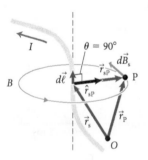

28.7 Magnetic fields due to currents

Ampère's law allows us to determine the magnetic field in only a few symmetrical situations involving current-carrying conductors. For any other situation, such as the one illustrated in **Figure 28.32a**, Ampère's law is of little help. Suppose, for example, that we are interested in determining the magnetic field at point P due to the current of magnitude I through this conductor. To calculate this field, we develop a procedure that parallels the procedure we developed in Section 23.7 to calculate the electric field of continuous charge distributions.

We begin by treating the wire as a current path and dividing it into small segments, each of length $d\ell$. For each segment we can define a vector $d\vec{\ell}$ that has length $d\ell$ and points in the direction of the current. One such vector segment is shown in Figure 28.32b. Let the magnetic field at P due to this segment be $d\vec{B}_s$. If we obtain an expression for the magnetic field due to the segment $d\vec{\ell}$ at an arbitrary position, we can determine the contributions of all the segments that make up the wire to the magnetic field at P and sum these to obtain $\vec{B}$. In the limit of infinitesimally small segments, this summation becomes an integral:

$$\vec{B} = \int_{\substack{\text{current} \\ \text{path}}} d\vec{B}_s. \tag{28.10}$$

where the integration is to be taken along the path followed by the current—that is, a path in the shape of the wire and in the direction of the current.

Before we can carry out the integration, we need to obtain an expression for the magnetic field $d\vec{B}_s$ of a current-carrying segment $d\vec{\ell}$. Because the magnetic field caused by one small segment is too feeble to measure, we cannot determine this field experimentally. From the expression for the magnetic field of a long straight current-carrying wire, however, it is possible to deduce what the field of a segment such as $d\vec{\ell}$ should be.

Let us first examine the direction of the magnetic field contribution $d\vec{B}_s$ due to a segment $d\vec{\ell}$ that is located such that $d\vec{\ell}$ is perpendicular to the vector $\vec{r}_{sP}$ pointing from $d\vec{\ell}$ to P ($\theta = 90°$, Figure 28.32c). The field lines of this segment are circles centered on the line of motion of the charge carriers causing the current through $d\vec{\ell}$ (see Figure 28.32a and c). Therefore the magnetic field at P due to $d\vec{\ell}$ is tangent to the circular field line through P. The direction of $d\vec{B}_s$ can be determined by associating the direction of $d\vec{B}_s$ with a vector product. To this end, we denote the unit vector pointing from the segment to P by $\hat{r}_{sP}$, as shown in Figure 28.32c. The direction of the vector-product $d\vec{\ell} \times \hat{r}_{sP}$ is that of $d\vec{B}_s$: If you curl the fingers of your right hand from $d\vec{\ell}$ to $\hat{r}_{sP}$ in Figure 28.32c, your thumb points in the direction of $d\vec{B}_s$.

We expect the magnitude dB_s to be proportional to the current I through the segment and to the length $d\ell$ of the segment. The greater the current or the longer the segment, the stronger the magnetic field. We also expect dB_s to depend on the distance r_{sP} to the segment. The field line picture for magnetic fields suggests that the magnetic field should decrease with distance r_{sP} as $1/r_{sP}^2$, just like the electric field. Finally, for any other segment than the one shown in Figure 28.32c we expect the magnetic field to depend on the angle θ between $d\vec{\ell}$ and $\vec{r}_{sP}$: For $\theta = 0$ (which is along the direction of the current) the magnetic field is zero, and for $\theta = 90°$ the magnetic field is maximum. The trigonometric function that fits this behavior is $\sin \theta$. Putting all this information in mathematical form, we obtain

$$dB_s = \frac{\mu_0}{4\pi} \frac{I \, d\ell \sin \theta}{r_{sP}^2} \quad (0 \leq \theta \leq \pi). \tag{28.11}$$

The proportionality factor $\mu_0/4\pi$ is obtained by deriving this expression from Ampère's law, but the mathematics is beyond the scope of this book. Instead, we use the reverse approach and show in Example 28.6 that Eq. 28.11 yields the correct result for a long straight current-carrying wire.

Incorporating what we know about the direction of $d\vec{B}_s$, we can also write Eq. 28.11 in vector form:

$$d\vec{B}_s = \frac{\mu_0}{4\pi} \frac{I\, d\vec{\ell} \times \hat{r}_{sP}}{r_{sP}^2} \quad \text{(constant current)}, \qquad (28.12)$$

where $\hat{r}_{sP}$ is the unit vector pointing along $\vec{r}_{sP}$ from s to P. Equation 28.12 is known as the **Biot-Savart law**. By substituting Eq. 28.12 into Eq. 28.10, we have a prescription for calculating the magnetic field produced by any constant current.

Example 28.6 Another look at the magnetic field generated by a long straight current-carrying wire

A long straight wire carries a current of magnitude I. Use the Biot-Savart law to derive an expression for the magnetic field $\vec{B}$ produced at point P a radial distance r from the wire.

❶ GETTING STARTED I begin by making a sketch of the wire (Figure 28.33). I arbitrarily orient the wire vertically and then choose the x axis along the direction of the wire. Because the magnetic field produced by the current has cylindrical symmetry, I can set the origin anywhere along the axis without loss of generality. For simplicity, I let the origin be at the height of point P. I assume the wire is of infinite length.

Figure 28.33

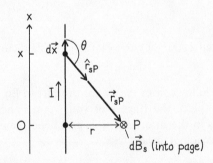

❷ DEVISE PLAN To use the Biot-Savart law (Eq. 28.12), I need to divide the wire into segments, determine the magnetic fields due to all segments, and then take the sum of the fields. In the limit of infinitesimally small segments, this sum becomes the integral given by Eq. 28.10 with the wire serving as the current path along which the integration is carried out.

I indicate one such segment in my sketch, calling it $d\vec{x}$. The magnetic field $d\vec{B}_s$ at P generated by this segment is given by Eq. 28.12. The unit vector $\hat{r}_{sP}$ in this equation points from $d\vec{x}$ to P, and the direction of $d\vec{B}_s$ is given by the right-hand current rule: I point my right thumb along the direction of the current and curl my fingers around the wire to determine the direction of $d\vec{B}_s$ (into the page at P in my sketch). Alternatively, I can use the vector product $d\vec{x} \times \hat{r}_{sP}$ to determine the direction of $d\vec{B}_s$: I line up the fingers of my right hand along $d\vec{x}$ in Figure 28.33 and curl them toward $\hat{r}_{sP}$. When I do this, my thumb points in

the direction of the magnetic field. Both methods yield the same result: $d\vec{B}_s$ points into the page.

Note that all the segments $d\vec{x}$ along the wire produce a magnetic field in the same direction. This means I can take the algebraic sum of the *magnitudes* of $d\vec{B}_s$ to determine the magnitude of the magnetic field at P. Then I can use Eq. 28.11 to express dB_s in terms of dx and integrate the resulting expression from $x = -\infty$ to $x = +\infty$ to determine the magnitude of the magnetic field at P.

❸ EXECUTE PLAN Because r_{sP} and θ in Eq. 28.11 both depend on x, I need to express them in terms of x before I can carry out the integration. By the Pythagorean theorem,

$$r_{sP}^2 = x^2 + r^2,$$

and remembering that $\sin\theta = \sin(180° - \theta)$, I write

$$\sin\theta = \frac{r}{r_{sP}} = \frac{r}{\sqrt{x^2 + r^2}}.$$

Substituting these last two results into Eq. 28.11 and using dx in place of $d\ell$, I get

$$dB_s = \frac{\mu_0 I}{4\pi} \frac{r}{\sqrt{x^2 + r^2}} \frac{1}{x^2 + r^2} dx = \frac{\mu_0 I r}{4\pi} \frac{dx}{[x^2 + r^2]^{3/2}},$$

and integrating this result over the length of the wire gives me

$$B = \frac{\mu_0 I r}{4\pi} \int_{-\infty}^{+\infty} \frac{dx}{[x^2 + r^2]^{3/2}} = \frac{\mu_0 I r}{4\pi} \left[\frac{1}{r^2} \frac{x}{[x^2 + r^2]^{1/2}} \right]_{x=-\infty}^{x=+\infty}$$

$$B = \frac{\mu_0 I}{2\pi r}. \checkmark$$

❹ EVALUATE RESULT This is identical to the result I obtained using Ampère's law in Example 28.3, a strong indication that my result here is correct.

28.15 Imagine a long straight wire of semi-infinite length, extending from $x = 0$ to $x = +\infty$, carrying a current of constant magnitude I. What is the magnitude of the magnetic field at a point P located a perpendicular distance d from the end of the wire that is at $x = 0$?

Figure 28.34 Calculating the magnetic force exerted by one current-carrying wire on another.

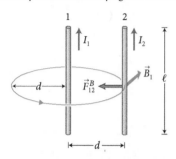

The expression for the magnitude of the magnetic field generated by a long straight wire obtained in Example 28.6 can lead us to an expression for the forces exerted by two current-carrying wires on each other. Consider the situation illustrated in **Figure 28.34**: Two parallel wires, of length ℓ and separated by a distance d, carry currents of magnitudes I_1 and I_2. To determine the magnetic force exerted by wire 1 on wire 2, we must first determine the magnitude and direction of the magnetic field generated by wire 1 at the location of wire 2 and then substitute this information into Eq. 27.8, which gives the magnetic force exerted on a straight current-carrying wire of length ℓ in a magnetic field $\vec{B}$:

$$\vec{F}^B = I\vec{\ell} \times \vec{B}, \tag{28.13}$$

where $\vec{\ell}$ is a vector whose magnitude is given by the length ℓ of the wire and whose direction is given by that of the current through the wire. Using the right-hand current rule, we see that the magnetic field $\vec{B}_1$ generated by wire 1 at the location of wire 2 points into the page (Figure 28.34). The result I obtained in Example 28.6 gives the magnitude of this field:

$$B_1 = \frac{\mu_0 I_1}{2\pi d}. \tag{28.14}$$

Because the magnetic field is perpendicular to $\vec{\ell}$, Eq. 28.13 yields for the magnitude of the magnetic force acting on wire 2

$$F_{12}^B = I_2 \ell B_1 \tag{28.15}$$

or, substituting Eq. 28.14,

$$F_{12}^B = \frac{\mu_0 \ell I_1 I_2}{2\pi d} \quad \text{(parallel straight wires carrying constant currents).} \tag{28.16}$$

The direction of this force follows from the vector product in Eq. 28.13: Using the right-hand force rule (place the fingers of your right hand along $\vec{\ell}$ in Figure 28.34, which points in the direction of I_2, and bend them toward $\vec{B}_1$), you can verify that the force points toward wire 1.

Example 28.7 The magnetic field generated by a circular arc of current-carrying wire

A wire bent into a circular arc of radius R subtending an angle ϕ carries a current of magnitude I (**Figure 28.35**). Use the Biot-Savart law to derive an expression for the magnitude of the magnetic field $\vec{B}$ produced at point P, located at the center of the arc.

Figure 28.35 Example 28.7

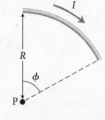

❶ **GETTING STARTED** I begin by evaluating what a small segment of length $d\ell$ along the arc contributes to the magnetic field. I therefore make a sketch showing one segment $d\vec{\ell}$ and the vector $\vec{r}_{sP}$ pointing from the segment to P (**Figure 28.36**). Using the right-hand vector product rule, I see that the direction of $d\vec{\ell} \times \vec{r}_{sP}$, and therefore the magnetic field $d\vec{B}_s$, are into the page at P.

Figure 28.36

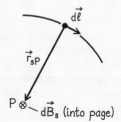

❷ **DEVISE PLAN** Because all segments contribute a magnetic field in the same direction, I can integrate Eq. 28.11 over the arc to obtain the magnitude of the magnetic field at P.

❸ **EXECUTE PLAN** Because $d\vec{\ell}$ and $\vec{r}_{sP}$ are always perpendicular to each other, I can write for Eq. 28.11

$$dB_s = \frac{\mu_0}{4\pi}\frac{I\,d\ell\,\sin 90°}{R^2} = \frac{\mu_0}{4\pi}\frac{I\,d\ell}{R^2},\qquad (1)$$

where I have substituted the radius R for the magnitude of $\vec{r}_{sP}$. To change the integration variable from $d\ell$ to the angular variable $d\phi$, I substitute $d\ell = R\,d\phi$ into Eq. 1:

$$dB_s = \frac{\mu_0 I}{4\pi R}\,d\phi$$

$$B = \int_{arc} dB_s = \frac{\mu_0 I}{4\pi R}\int_0^{\phi} d\phi$$

$$= \frac{\mu_0 I\phi}{4\pi R}.\checkmark$$

❹ **EVALUATE RESULT** My expression for B shows that at P the magnetic field magnitude is proportional to the current through the arc, as I expect. It is also proportional to the angle ϕ of the arc, also what I expect given that two such arcs should yield twice the magnetic field. Finally, B is inversely proportional to the arc radius R. That dependence on the radius makes sense because increasing the radius increases the distance between P and the arc, thus diminishing the magnetic field.

✋ **28.16** What is the magnitude of the magnetic field (*a*) at the center of a circular current loop of radius R and (*b*) at point P near the current loop in **Figure 28.37**? Both loops carry a current of constant magnitude I.

28.8 Magnetic field of a moving charged particle

Let us now use the Biot-Savart law to obtain an expression for the magnetic field caused by charged particles moving at constant velocity.* Consider first a straight wire carrying a current of magnitude I and aligned with the x axis, as in **Figure 28.38a**. The magnetic field generated at point P by a small segment $d\vec{x}$ is given by Eq. 28.12:

$$d\vec{B}_s = \frac{\mu_0}{4\pi}\frac{I\,d\vec{x}\times\hat{r}_{sP}}{r_{sP}^2},\qquad (28.17)$$

where $\hat{r}_{sP}$ is a unit vector pointing from the segment $d\vec{x}$ to the point at which we wish to determine the magnetic field, and r_{sP} is the distance between the segment and the point P. Suppose the segment contains an amount of charge dq. Let the charge carriers responsible for the current take a time interval dt to have displacement $d\vec{x}$ (**Figure 28.38b**). According to the definition of current (Eq. 27.2), we have

$$I \equiv \frac{dq}{dt},\qquad (28.18)$$

and so

$$I d\vec{x} = \frac{dq}{dt}d\vec{x} = dq\,\frac{d\vec{x}}{dt} = dq\,\vec{v},\qquad (28.19)$$

where $\vec{v}$ is the velocity at which the charge carriers move down the wire. In the limiting case where the segment $d\vec{x}$ contains just a single charge carrier carrying a charge q (**Figure 28.38c**), dq becomes q and

$$I d\vec{x} = q\vec{v}.\qquad (28.20)$$

*In the derivation that follows, we assume $v \ll c_0$ and ignore any relativistic effects as described in Sections 27.4 and 27.8.

Figure 28.37 A current loop carrying a current of constant magnitude I (Checkpoint 28.16).

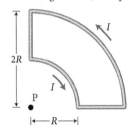

Figure 28.38 We use the Biot-Savart law to obtain an expression for the magnetic field caused by charged particles moving at constant velocity.

(*a*) Small segment of current-carrying wire causes magnetic field at point P

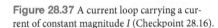

(*b*) Displacement of charge dq in time interval dt

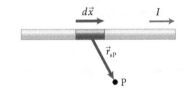

(*c*) Displacement of charged particle in time interval dt

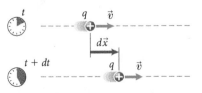

Substituting this result into Eq. 28.17, we obtain an expression for the magnetic field of a single moving charged particle:

$$\vec{B} = \frac{\mu_0}{4\pi} \frac{q\vec{v} \times \hat{r}_{pP}}{r_{pP}^2} \quad \text{(single particle),} \qquad (28.21)$$

where r_{pP} is the distance between the particle and P, and $\hat{r}_{pP}$ is the unit vector pointing from the particle to P.

Example 28.8 Magnetic field generated by a moving electron

An electron carrying a charge $-e = -1.60 \times 10^{-19}$ C moves in a straight line at a speed $v = 3.0 \times 10^7$ m/s. What are the magnitude and direction of the magnetic field caused by the electron at a point P 10 mm ahead of the electron and 20 mm away from its line of motion?

❶ GETTING STARTED I begin by drawing the moving electron and the point P at which I am to determine the magnetic field (Figure 28.39).

Figure 28.39

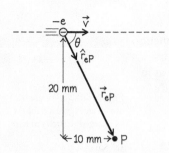

❷ DEVISE PLAN The magnetic field created by a moving charged particle is given by Eq. 28.21.

❸ EXECUTE PLAN The unit vector $\hat{r}_{eP}$ points from the electron to P, and so $\vec{v} \times \hat{r}_{eP}$ points into the page. Because the charge of the electron is negative, the magnetic field points in the opposite direction, out of the page. ✔

The magnitude of the vector $\vec{r}_{eP}$ in Figure 28.39 is $r_{eP} = \sqrt{(10 \text{ mm})^2 + (20 \text{ mm})^2} = 22.36$ mm, and $\sin \theta = (20 \text{ mm})/(22.36 \text{ mm}) = 0.8944$. Substituting these values into Eq. 28.11 thus yields for the magnitude of the magnetic field

$$B = \frac{(4\pi \times 10^{-7} \text{ T} \cdot \text{m/A})}{4\pi} \frac{(1.60 \times 10^{-19} \text{ C})(3.0 \times 10^7 \text{ m/s})(0.8944)}{(22.36 \times 10^{-3} \text{ m})^2}$$

$$= 8.6 \times 10^{-16} \text{ T.} ✔$$

❹ EVALUATE RESULT The magnetic field magnitude I obtained is much too small to be detected, but that's what I expect for a single electron. I can verify the direction of the magnetic field by applying the right-hand current rule to the current caused by the electron. Because the electron carries a negative charge, its motion to the right in Figure 28.39 causes a current to left. Pointing my right-hand thumb to the left in the figure, I observe that the fingers curl out of the page at P, in agreement with what I determined earlier for the direction of $\vec{B}$.

Figure 28.40 Magnetic interaction of two moving charged particles.

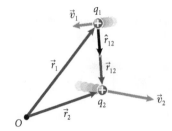

We can now combine the expression for the magnetic field caused by a moving charged particle (Eq. 28.21) with that for the magnetic force exerted on another moving charged particle (Eq. 27.19; $\vec{F}_P^B = q\vec{v} \times \vec{B}$) to determine the magnetic interaction between two moving charged particles. Consider the situation illustrated in Figure 28.40. The magnetic field caused by particle 1 at the location of particle 2 is given by Eq. 28.21:

$$\vec{B}_1(\vec{r}_2) = \frac{\mu_0}{4\pi} \frac{q_1\vec{v}_1 \times \hat{r}_{12}}{r_{12}^2}, \qquad (28.22)$$

where r_{12} is the magnitude of the vector $\vec{r}_{12}$ pointing from particle 1 to particle 2 and $\hat{r}_{12}$ is a unit vector pointing along $\vec{r}_{12}$. Substituting this expression into Eq. 27.19, we obtain for the magnetic force exerted by particle 1 on particle 2

$$\vec{F}_{12}^B = q_2\vec{v}_2 \times \vec{B}_1(\vec{r}_2) = \frac{\mu_0}{4\pi} \frac{q_1 q_2}{r_{12}^2} \vec{v}_2 \times (\vec{v}_1 \times \hat{r}_{12}). \qquad (28.23)$$

Notice the appearance of the double vector product: We must first take the vector product of $\vec{v}_1$ and $\hat{r}_{12}$ and then take the vector product of $\vec{v}_2$ and the result of the first vector product.

Equation 28.23 also shows that we need a moving charged particle in order to generate a magnetic field ($\vec{v}_1 \neq 0$) and another moving charged particle to "feel" that magnetic field ($\vec{v}_2 \neq 0$), in agreement with our discussion in the first part of this chapter.

In contrast to the electric force, the magnetic force does not satisfy Eq. 8.15, $\vec{F}_{12} = -\vec{F}_{21}$. To see this, consider the two positively charged moving particles shown in Figure 28.41. Particle 1, carrying a charge q_1, travels in the positive x direction, while particle 2, carrying charge q_2, travels in the positive y direction. The force exerted by 1 on 2 is given by Eq. 28.23. Applying the right-hand vector product rule, we determine that the vector product $\vec{v}_1 \times \hat{r}_{12}$ on the right side of Eq. 28.23 points out of the page. Applying the right-hand vector product rule again to the vector product of $\vec{v}_2$ and $\vec{v}_1 \times \hat{r}_{12}$, we see that $\vec{F}^B_{12}$ points in the positive x direction.* The force exerted by 2 on 1 is obtained by switching the subscripts 1 and 2 in Eq. 28.23:

$$\vec{F}^B_{21} = \frac{\mu_0}{4\pi} \frac{q_1 q_2}{r^2_{12}} \vec{v}_1 \times (\vec{v}_2 \times \hat{r}_{21}), \qquad (28.24)$$

where we have used that $q_1 q_2 = q_2 q_1$ and $r^2_{12} = r^2_{21}$. Applying the right-hand vector product rule twice, we find that $\vec{F}^B_{21}$ points in the positive y direction, and so the magnetic forces that the two particles exert on each other, while equal in magnitude, do not point in opposite directions, as we would expect from Eq. 8.15.

We derived Eq. 8.15 from the fact that the momentum of an isolated system of two particles is constant, which in turn follows from conservation of momentum, one of the most fundamental laws of physics. The electric and magnetic fields of isolated moving charged particles, however, are not constant, and as we shall see in Chapter 30, we can associate a flow of both momentum and energy with changing electric and magnetic fields. Therefore we need to account not only for the momenta of the two charged particles in Figure 28.41, but also for the momentum carried by their fields—which goes beyond the scope of this book. Even though Eq. 8.15 is not satisfied by the magnetic force, the momentum of the system comprising the two particles and their fields is still constant.

Substituting Coulomb's law and Eq. 28.23 into Eq. 27.20, $\vec{F}^{EB}_P = q(\vec{E} + \vec{v} \times \vec{B})$, we obtain an expression for the electromagnetic force that two moving charged particles exert on each other:

$$\vec{F}^{EB}_{12} = \frac{1}{4\pi\epsilon_0} \frac{q_1 q_2}{r^2_{12}} [\hat{r}_{12} + \mu_0 \epsilon_0 \vec{v}_2 \times (\vec{v}_1 \times \hat{r}_{12})]. \qquad (28.25)$$

By comparing the result obtained in Example 28.6 with Eq. 27.38 and substituting $k = 1/(4\pi\epsilon_0)$ (Eq. 24.7), we see that $\mu_0 = 1/(\epsilon_0 c^2_0)$ or $\mu_0 \epsilon_0 = 1/c^2_0$, where c_0 is the speed of light. Using this information, we can write Eq. 28.25 as

$$\vec{F}^{EB}_{12} = \frac{1}{4\pi\epsilon_0} \frac{q_1 q_2}{r^2_{12}} \left[\hat{r}_{12} + \frac{\vec{v}_2 \times (\vec{v}_1 \times \hat{r}_{12})}{c^2_0} \right] (v \ll c_0). \qquad (28.26)$$

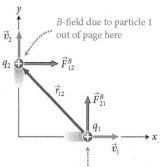

Figure 28.41 The forces that moving charged particles exert on each other do not necessarily point in opposite directions.

B-field due to particle 1 out of page here

B-field due to particle 2 into page here

---

*Here is another way you can determine the direction of the magnetic force exerted by particle 1 on particle 2. The motion of particle 1 corresponds to a current directed in the positive x direction. Using the right-hand current rule, it follows that the magnetic field of 1 points out of the xy plane at any location above the x axis. The magnetic field due to 1 at the location of 2 thus points out of the plane. The motion of particle 2 corresponds to a current directed in the positive y direction. To determine the direction of the magnetic force exerted on this "current," we can use the right-hand force rule. When you curl the fingers of your right hand from the positive y axis to the direction of $\vec{B}_1$ at particle 2 (out of the page), your thumb points in the positive x direction, as we found before.

This equation is the most general expression for the electromagnetic interaction between moving charged particles. Because in most applications the speeds of charged particles are significantly smaller than the speed of light, the second term in the square brackets in Eq. 28.26 is much smaller than the first term, which represents the electric contribution to the electromagnetic force from Coulomb's law.

28.17 Consider two protons 1 and 2, each carrying a charge $+e = 1.6 \times 10^{-19}$ C, separated by 1.0 mm moving at 3×10^5 m/s parallel to each other and perpendicular to their separation. (*a*) What is the direction of the magnetic force that each proton exerts on the other? (*b*) Determine the ratio of the magnitudes of the magnetic and electric forces that the two exert on each other.

Chapter Glossary

SI units of physical quantities are given in parentheses.

Ampère's law The line integral of the magnetic field along a closed path (called an *Ampèrian path*) is proportional to the current encircled by the path, I_{enc}:

$$\oint \vec{B} \cdot d\vec{\ell} = \mu_0 I_{enc}. \qquad (28.1)$$

In analogy to Gauss's law in electrostatics, Ampère's law allows us to determine the magnetic field for current distributions that exhibit planar, cylindrical, or toroidal symmetry.

Ampèrian path A closed path along which the magnetic field is integrated in Ampère's law.

Biot-Savart law An expression that gives the magnetic field at a point P due to a small segment $d\vec{\ell}$ of a wire carrying a current I:

$$d\vec{B}_s = \frac{\mu_0}{4\pi} \frac{I \, d\vec{\ell} \times \hat{r}_{sP}}{r_{sP}^2}, \qquad (28.12)$$

where $\hat{r}_{sP}$ is a unit vector pointing from the segment to the point at which the magnetic field is evaluated. The Biot-Savart law can be used to calculate the magnetic field of a current-carrying conductor of arbitrary shape by integration:

$$\vec{B} = \int_{\text{current path}} d\vec{B}_s. \qquad (28.10)$$

Current loop A current-carrying conductor in the shape of a loop. The magnetic field pattern of a current loop is similar to that of a magnetic dipole.

Magnetic constant μ_0 (T·m/A) A constant that relates the current encircled by an Ampèrian path and the line integral of the magnetic field along that path. In vacuum:

$$\mu_0 = 4\pi \times 10^{-7} \, \text{T·m/A}.$$

Magnetic dipole moment $\vec{\mu}$ (A·m²) A vector that points from the S pole to the N pole for a bar magnet or along the axis of a planar current loop in the direction given by the right-hand thumb when the fingers of that hand are curled along the direction of the current through the loop. In an external magnetic field, the magnetic dipole moment tends to align in the direction of the external magnetic field.

Solenoid A long, tightly wound helical coil of wire. The magnetic field of a current-carrying solenoid is similar to that of a bar magnet.

Toroid A solenoid bent into a circle. The magnetic field of a toroid is completely contained within the windings of the toroid.

29 Changing Magnetic Fields

CONCEPTS

QUANTITATIVE TOOLS

Up to now, the only electric fields we have encountered are those that arise from the presence of electrical charges. In this chapter we explore electric fields that accompany changing magnetic fields. As we saw in Chapter 27, constant magnetic fields exert forces on moving charge carriers. Now we discover that there is no fundamental difference between the interaction of changing magnetic fields with stationary charge carriers and the interaction of constant magnetic fields with moving charge carriers. We shall also learn that energy can be stored in magnetic fields. These ideas are harnessed in a wide variety of important applications, from electric motors and electrical power generation to electronic appliances and other everyday devices.

29.1 Moving conductors in magnetic fields

Let's examine what happens when a conducting rod moves with velocity $\vec{v}$ through a magnetic field of constant magnitude B, where $\vec{v}$ is perpendicular to $\vec{B}$ (**Figure 29.1**). Recall from Chapter 27 that a magnetic field exerts a force on moving charge carriers. That force is proportional to the field magnitude, the amount of charge, and the component of the velocity perpendicular to the field. The right-hand force rule gives the direction of the resulting force exerted on the charge carriers.

As the rod in Figure 29.1 moves to the right in the magnetic field, positive charge carriers in the rod experience a downward force. As a result, a positive charge accumulates at the lower end of the rod, leaving behind a negative charge at the other end. (If the charge carriers are negatively charged, they experience an upward force; the end result, however, is still as shown in Figure 29.1.)

Does charge continue to accumulate as the rod moves? No—as charge accumulates, the electric field between the oppositely charged ends of the rod resists further accumulation of charge. Once the opposing force due to the electric field is equal in magnitude to the force exerted by the

magnetic field on the charge carriers, the vector sum of the forces exerted on the charge carriers is zero, and no further accumulation of charge takes place. The motion of the rod through the magnetic field thus establishes a charge separation, the amount of which is determined by the magnitude of the force exerted by the magnetic field on the charge carriers.

29.1 What happens if the rod in Figure 29.1 moves to the left?

Now consider what happens if, instead of a conducting rod, a conducting rectangular loop moves into a magnetic field (**Figure 29.2**). As the right side of the loop enters the field, its charge carriers move in response to the magnetic force exerted on them. Rather than just accumulating at the ends of the right side, however, the carriers now have a closed path along which to travel. Thus, the magnetic force exerted on the carriers creates a current in the loop. (The carriers in the top and bottom sides of the loop also feel magnetic forces, as shown in Figure 29.2, but because these carriers can't move very far in the vertical direction, we can ignore the effects of these forces.)

The current that arises from moving a conducting loop through a magnetic field can be verified experimentally. **Figure 29.3** shows a simple experiment using a bar magnet and a current meter, which shows both the magnitude and the direction of the current through the wire. When the wire is at rest in the field of the magnet (Figure 29.3*a*), the meter indicates no current. Moving part of the wire to the right causes a counterclockwise current (Figure 29.3*b*); moving the same part of the wire to the left causes a clockwise current (Figure 29.3*c*).

The current that arises from the motion of charged particles relative to a magnetic field is called an **induced current** (as opposed to a "regular" current caused by a potential difference between, for instance, the terminals of a battery).

Figure 29.2 Direction of current in a rectangular loop moving into a magnetic field. Magnetic forces exerted on the charge carriers in the loop cause them to move around the loop.

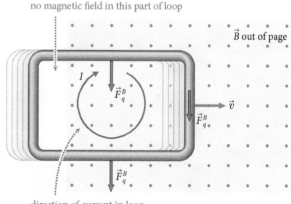

Figure 29.1 When a conducting rod moves in a magnetic field, charge carriers in the rod experience a magnetic force. In the configuration shown, the right-hand force rule tells us that a positive charge accumulates at the bottom of the rod.

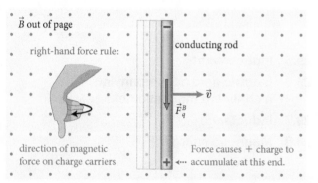

Figure 29.3 Experimental observation of induced current.

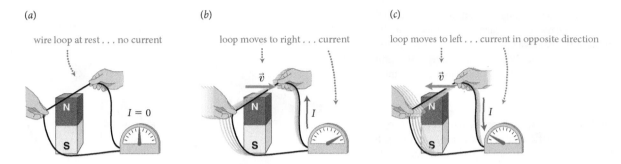

(a) wire loop at rest . . . no current

(b) loop moves to right . . . current

(c) loop moves to left . . . current in opposite direction

Example 29.1 A rectangular loop passing through a magnetic field

Consider a rectangular conducting loop traveling from left to right through a magnetic field as shown in **Figure 29.4**. During this motion, at what positions *a–e* is a current induced in the loop, and what is the current direction at each position?

❶ GETTING STARTED As the loop moves through the magnetic field, the charge carriers in the loop experience a magnetic force. For each of the five positions I need to determine whether or not the magnetic force causes the carriers to flow around the loop. To simplify the problem, I assume that only the positive charge carriers are free to move.

❷ DEVISE PLAN The direction of the magnetic force exerted on the charge carriers is given by the right-hand force rule. To see if there is a current in the loop, I'll sketch the loop in each position and determine the direction of the magnetic force exerted on the charge carriers in each side of the loop.

❸ EXECUTE PLAN No matter where along the loop perimeter the carriers are, they always move to the right with the loop, and so curling the fingers of my right hand from $\vec{v}$ to $\vec{B}$, I see that the magnetic force exerted on them is always toward the bottom of the page. I draw this force for each side of the loop that is inside the magnetic field (**Figure 29.5**). The magnetic forces exerted on the top and bottom sides of the loop cause no current through the loop because the charge carriers in these sides have essentially no mobility in the direction in which $\vec{F}^B$ is exerted.

In position (*a*), with no part of the loop in the magnetic field, no magnetic force is exerted on any of the carriers, and therefore no current is induced. When the right side of the loop passes into the magnetic field but the left edge has not yet

Figure 29.5

(a) (b) (c) (d) (e)

entered (*b*), the magnetic force exerted on the charge carriers in the right side of the loop causes a clockwise current around the loop. Once the entire loop is in the magnetic field (*c*), the magnetic force exerted on the carriers in the left side drives a counterclockwise current equal in magnitude to the clockwise current driven by the magnetic force exerted on the carriers in the right side. The combined effect is that no current is induced. In position (*d*), only the left side of the loop is in the field, and the magnetic force exerted on those charge carriers induces a counterclockwise current around the loop. When the loop has passed out of the magnetic field (*e*), once again no magnetic forces are exerted on the carriers and therefore no current is induced. ✔

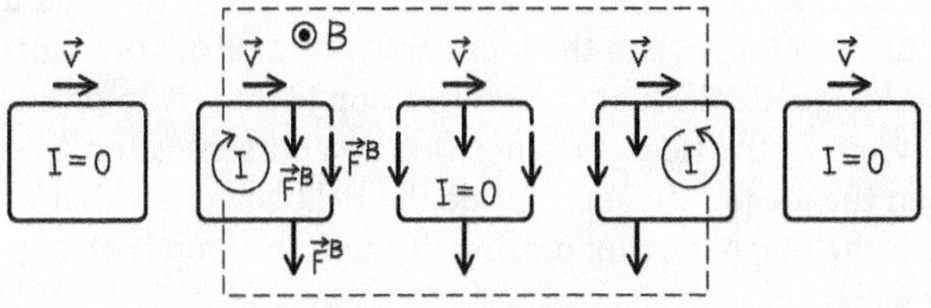

❹ EVALUATE RESULT In position (*b*) I have the same situation as in Figure 29.2, so it is reassuring that my answer is the same. If I ignore any dissipation of energy, the initial and final states are the same. Therefore it makes sense that the effect of moving the loop through the field is zero: The charge carriers first move clockwise as the loop moves into the field, then move counterclockwise as the loop moves out of the field.

CONCEPTS

Figure 29.4 Example 29.1.

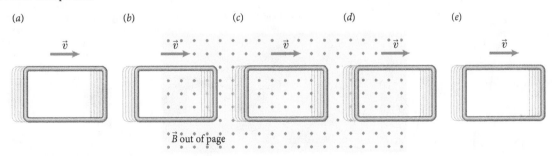

(a) (b) (c) (d) (e)

$\vec{B}$ out of page

I assumed that only the positive charge carriers can move. Had I assumed that the negative charge carriers are free to move, my answer would not have changed. Even though the direction of motion of the negative charge carriers is opposite the direction of motion of the positive carriers, the current is still in the directions shown in Figure 29.5 because current direction is *defined* as the direction in which positive charge carriers move.

✋ **29.2** In Example 29.1, suppose the loop is stationary and the source of the magnetic field is moved to the left such that their relative motion is the same. Do you expect there to be a current through the loop?

29.2 Faraday's law

As Example 29.1 indicates, a current is induced through a conducting loop when the loop moves into or out of a uniform magnetic field, not when the loop is entirely in or entirely out of the field. In other words, a current is induced when the area of the loop inside the field changes—that is, when the magnetic flux passing through the loop is changing with time.

Experiments show that the rate at which the magnetic flux through the loop changes affects the magnitude of the induced current. As illustrated in **Figure 29.6**, moving the loop through the magnetic field faster, which causes a greater rate of change in the magnetic flux through the loop, produces a greater induced current.

As we learned in Chapters 6 and 14, physical phenomena do not depend on the reference frame in which we observe them. Indeed, if the change in magnetic flux through a conducting loop is responsible for inducing a current, it should not matter whether the loop is moving in a constant magnetic field or the source of the magnetic field is moving and the loop is stationary. Only the motion of one relative to the other matters. Experiments confirm that the same current is measured in the loop no matter whether the loop moves to the right with speed v or the source of the magnetic field moves to the left with speed v (**Figure 29.7**). In fact, current is induced in a stationary conducting loop in any region where a magnetic field is changing, even if the

Figure 29.6 The speed at which a conducting loop is moved through a magnetic field affects the magnitude of the induced current.

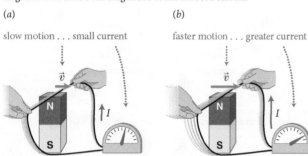

(a) slow motion . . . small current

(b) faster motion . . . greater current

Figure 29.7 A current can be induced by moving either the loop or the magnet.

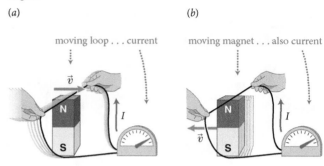

(a) moving loop . . . current

(b) moving magnet . . . also current

source of the magnetic field is stationary and the magnitude of the magnetic field is changing.

These experiments lead us to conclude:

> **A changing magnetic flux through a conducting loop induces a current in the loop.**

This statement is called **Faraday's law.** The process by which a changing magnetic flux causes charge carriers to move, establishing a separation of charge or inducing a current, is called **electromagnetic induction.**

Example 29.2 Changing magnetic flux

(*a*) In which of the four loops shown in **Figure 29.8** is there a current that is caused by a magnetic force? (*b*) In which situation(s) is a current induced in the loop?

❶ **GETTING STARTED** I begin by making a sketch of the magnetic field of the bar magnet, based on what I learned in Chapter 27 (Figure 27.13). I show a cross section of the loop placed above the north pole of the magnet (**Figure 29.9**).

❷ **DEVISE PLAN** Magnetic forces are exerted on charge carriers moving relative to a magnetic field, provided there is a nonzero component of the magnetic field perpendicular to the carriers' velocity. To answer part *a*, therefore, I must determine which of the four situations satisfies those conditions. For part *b* I know that a changing magnetic flux induces a current, and so for each situation I must establish whether or not the magnetic flux through the loop is changing.

❸ **EXECUTE PLAN** (*a*) Because the loop is stationary in situations 1 through 3, there cannot be a magnetic force in those situations. For situation 4, I draw vectors in Figure 29.9 indicating the

Figure 29.8 Example 29.2.

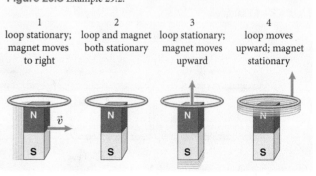

1	2	3	4
loop stationary; magnet moves to right	loop and magnet both stationary	loop stationary; magnet moves upward	loop moves upward; magnet stationary

Figure 29.9

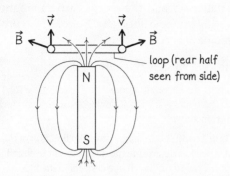

loop (rear half
seen from side)

directions of the loop velocity and of the magnetic field at the position of the ring and the velocity of the loop. Because the magnetic field lines spread outward from the north pole, the magnetic field is not parallel to the velocity, and so a magnetic force is exerted on the charge carriers. According to the right-hand force rule, the force is in opposite directions on the two sides of the loop, so this magnetic force creates a current in the loop. ✔

(*b*) Because the field of a bar magnet is not uniform, any relative motion between the magnet and the loop causes the magnetic flux through the loop to change. Therefore, the magnetic flux is changing in situations 1, 3, and 4, which means current is induced in the loop in these three cases. ✔

❹ **EVALUATE RESULT** That there is a current created by a magnetic force only in situation 4 makes sense because a magnetic force requires a nonzero velocity of the charge carriers and only in situation 4 is the loop moving. That there is an induced current in every situation except 2 makes sense because Faraday's law tells me that a current is induced whenever there is a changing magnetic flux through the loop—that is, whenever the loop and magnet move relative to each other.

✋ **29.3** Is a magnetic force exerted on the (stationary) charge carriers in the loop of wire held above the magnet in Figure 29.7*b*?

29.3 Electric fields accompany changing magnetic fields

Example 29.2 and Checkpoint 29.3 lead to a surprising conclusion: Although no magnetic force is exerted on the charge carriers in a stationary loop, a current is still induced! **Figure 29.10** shows this situation in more detail. Experiments show that as a magnetic field moves past a stationary conducting rod, a charge separation and hence a potential difference develop between the ends of the rod even though no magnetic force is exerted on stationary charge carriers.

The potential difference that develops between the ends of the rod shown in Figure 29.10 is the same as that which would develop if the magnetic field were stationary and the rod were moving to the right (recall Figure 29.1). Any motion of the rod relative to the magnetic field produces

Figure 29.10 A stationary conducting rod in a moving magnetic field develops a charge separation, but no magnetic force is exerted on the charge carriers because their speed in the rod is zero, which means $\vec{v} \times \vec{B} = \vec{0}$.

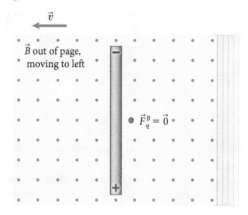

the same potential difference across the ends of the rod. By the same token, a current is induced in the stationary rectangular loop of **Figure 29.11** by the moving magnetic field, just as a current is induced in the rectangular loop moving through a constant magnetic field in Figure 29.2. In other words, whenever relative motion occurs between charge carriers and the source of a magnetic field, the resulting rearrangement of charge and potential difference is independent of the choice of inertial reference frame, as are any resulting currents.

What force causes the charge carriers to flow when the loop is stationary? We have encountered two types of forces that are exerted on charged particles: Electric and magnetic. Magnetic forces are exerted only on moving charged particles, so these forces cannot be responsible for the induced current in Figure 29.11. Only electric forces are exerted on stationary charged particles, so we must conclude that the force that causes charge to separate in Figure 29.10 is an electric force (because it behaves like one). Because electric forces are caused by electric fields, we are led to conclude:

A changing magnetic field is accompanied by an electric field.

Figure 29.11 A current is induced in a stationary conducting loop as a magnetic field moves past it when not all of the loop is in the field, even though no magnetic force is exerted on the charge carriers.

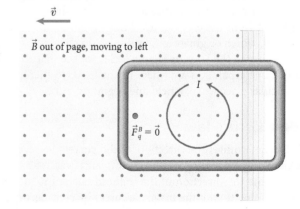

Figure 29.12 Examples 29.3 and 29.4.

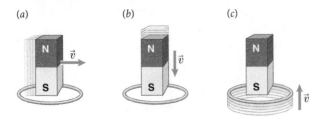

(a) (b) (c)

29.4 In Figure 29.1, charge accumulates at the ends of the moving rod until the amount at each end reaches an equilibrium value. Mechanical equilibrium is established when the magnetic force due to the motion of the rod counterbalances the electric force due to the charge separation. In Figure 29.10, what two forces determine the equilibrium state of the charge separation in the stationary rod?

We see that the nature of the force causing electromagnetic induction depends on the choice of reference frame. Forces that appear to be magnetic in one inertial reference frame are electric in another inertial reference frame. Indeed, the fact that our choice of reference frame affects whether a force is magnetic or electric is further evidence that electric and magnetic fields arise from fundamentally the same interaction (see Section 27.4). Because viewing the situation from a different inertial reference frame cannot alter the underlying interaction, magnetic and electric forces must be two manifestations of the same interaction.

29.5 As viewed from above, what is the direction of the induced current in situations 1 and 3 in Figure 29.8?

What do the field lines of the electric field that induces the current in Figure 29.11 look like? Electric field lines show the direction of the electric force exerted on a positively charged test particle. Because a counterclockwise current is induced in the loop (see position *d* in Example 29.1), we know that positive charge carriers in the loop experience a force directed counterclockwise all along the loop. This leads us to conclude that the field lines must form loops that close on themselves. The electric field lines that accompany a changing magnetic field are therefore very different from the electric field lines we have encountered thus far, which originate and terminate on charged objects. Returning to the loop in Figure 29.11, we see that there is no charge separation anywhere in the loop, so there is no particular place where a field line begins or ends.

29.4 Lenz's law

What determines the direction of the current induced by a changing magnetic flux? So far, the only way we have to determine the direction of the induced current is to work out the direction of the magnetic forces by considering the

situation from a frame of reference in which the magnetic field is stationary and the charge carriers are moving. As the following example illustrates, however, this procedure is cumbersome at best.

Example 29.3 What is the current direction?

For each of the three situations shown in **Figure 29.12**, is the induced current in the loop clockwise or counterclockwise as viewed from above?

❶ **GETTING STARTED** To determine the direction of the induced current, I need to determine the direction of the electric or magnetic force exerted on the charge carriers in the loop.

❷ **DEVISE PLAN** I can use the right-hand force rule to determine the direction of the magnetic forces (assuming positive charge carriers), and the only way I know to do this is to consider each situation from a reference frame in which the magnetic field is stationary and the loop is in motion. For (a) I choose a reference frame moving along with the magnet; in this reference frame the loop moves to the left (**Figure 29.13a**). Situations (b) and (c) are equivalent, so the direction of the induced current must be the same in both. I therefore need to consider only (c), where the magnetic field is stationary and the loop is in motion (Figure 29.13b). I can then use the right-hand force rule to determine the direction of the magnetic force at a couple of locations on the loop. Once I know these force directions, I can determine the direction of the induced current.

❸ **EXECUTE PLAN** I'll consider two locations on the loop: location P on the left and Q on the right. At each of these locations I draw arrows representing the velocity $\vec{v}$ of the (positive) charge carriers and the magnetic field $\vec{B}$. (All of the vectors I draw are in the vertical *xz* plane.)

(a) At location P on the loop in Figure 29.13a, the magnetic field points down and to the right, and the charge carriers in the loop are moving to the left at velocity $\vec{v}$. If I curl the fingers of my right hand from $\vec{v}$ to $\vec{B}$ at P, my thumb points out of the page, which means the magnetic force exerted on the charge carriers at P points out of the page. Thus at P the charge carriers are moving counterclockwise as viewed from above. At Q the magnetic field is up and to the left, and the charge carriers here are also moving to the left at velocity $\vec{v}$. When I curl the fingers of my right hand from $\vec{v}$ to $\vec{B}$, my thumb points into the page. This means the magnetic force at Q points into the page, moving the charge carriers in the loop counterclockwise as viewed from above. Thus the induced current is counterclockwise as seen from above. ✔

Figure 29.13

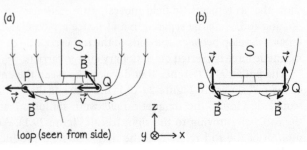

(a) (b)

loop (seen from side)

(b, c) At P in Figure 29.13b, the loop and its charge carriers are moving up at velocity $\vec{v}$, and the magnetic field has a vertical component and a component to the right in the plane of the page. Curling the fingers of my right hand from $\vec{v}$ to $\vec{B}$, I discover that the magnetic force exerted on the positive charge carriers points into the page. At Q it points out of the page, so the magnetic forces exerted on the charge carriers at P and Q induce a current that is clockwise as seen from above. ✔

❹ EVALUATE RESULT Comparing Figures 29.8 and 29.12, I note that Figure 29.12a is similar to situation 1 in Figure 29.8, and Figure 29.12b and c are similar to situations 3 and 4. In Example 29.2, I concluded that currents are induced in situations 1, 3, and 4, so in this example, too, I should expect currents in all three situations.

A simpler approach to determining the direction of an induced current follows from experimental observation:

The direction of an induced current through a conducting loop is always such that the magnetic flux produced by the induced current opposes the change in magnetic flux through the loop.

This principle is called **Lenz's law,** and the magnetic field resulting from the induced current is called the **induced magnetic field.**

Consider, for example, a conducting loop in a magnetic field that points up and is increasing in magnitude (**Figure 29.14a**). In order for the induced magnetic field $\vec{B}_{ind}$ to oppose the change in magnetic field, $\vec{B}_{ind}$ must point down, as shown in Figure 29.14b. What current direction produces a downward induced magnetic field? According to the right-hand dipole rule, the current must be clockwise as seen from above (Figure 29.14c). Indeed, this is what we found for a loop moving into a magnetic field in Example 29.1: As the magnetic flux through the loop increases, a clockwise current is induced.

Figure 29.14 Applying Lenz's law.

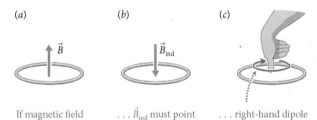

(a) If magnetic field points up and increases...

(b) ...$\vec{B}_{ind}$ must point down to oppose change, and...

(c) ...right-hand dipole rule says induced current is in this direction.

Example 29.4 Clockwise or counterclockwise?

Consider again the three situations of Figure 29.12. Use Lenz's law to determine the direction of the induced current in each loop.

❶ GETTING STARTED Using Lenz's law involves determining which direction for the induced current produces an induced

magnetic field opposing the change in magnetic flux through the loop. The magnetic field that passes through the loop points primarily up (toward the south pole of the magnet) and is strongest along the long axis of the magnet.

❷ DEVISE PLAN According to Lenz's law, the induced current must be directed in such a way as to create an induced magnetic field $\vec{B}_{ind}$ that counteracts the change in magnetic flux. If the magnetic flux that causes the induced current decreases, $\vec{B}_{ind}$ must be in the same direction as the magnetic field of the magnet so that $\vec{B}_{ind}$ increases the magnetic flux through the loop. If the magnetic flux that causes the induced current increases, $\vec{B}_{ind}$ must be in the direction opposite the direction of the magnet's magnetic field. Once I know the direction of $\vec{B}_{ind}$, I can use the right-hand dipole rule to determine the direction of the current.

❸ EXECUTE PLAN (a) As the magnet moves to the right, the magnitude of the magnetic field at the loop, and therefore the magnetic flux through the loop, decrease. Therefore $\vec{B}_{ind}$ is upward, in the direction of the magnet's magnetic field. To use the right-hand dipole rule, I recall from Section 28.3 that the direction of a magnetic field near the poles of a bar magnet is given by the magnetic dipole moment vector, which points from south pole to north pole. Thus I point my right thumb in this direction in Figure 29.12a and see that my fingers curl counterclockwise as viewed from above, telling me that this is the direction of the induced current. ✔

(b) Moving the magnet toward the loop causes the magnetic flux through the loop to increase. The induced magnetic field $\vec{B}_{ind}$ associated with the induced current must therefore point downward, opposite the direction of the field created by the magnet, to oppose the increase in magnetic flux through the loop. The induced current is therefore clockwise as viewed from above. ✔

(c) Moving the loop toward the magnet results in an increase in magnetic flux through the loop, as in part b. The induced current therefore must be clockwise as viewed from above to produce a downward-pointing induced magnetic field $\vec{B}_{ind}$ that opposes the increase in flux. ✔

❹ EVALUATE RESULT My answers are the same as in Example 29.3, which gives me confidence that they are correct.

✋ **29.6** When current is induced in a conducting loop by the motion of a nearby magnet, the induced magnetic field $\vec{B}_{ind}$ exerts a force on the magnet. (a) In Figure 29.12b, what is the direction of the force exerted on the magnet by $\vec{B}_{ind}$? (b) Suppose Lenz's law stated that the induced current *adds to* the change that produced it instead of opposing it. What would be the direction of the force exerted by $\vec{B}_{ind}$ on the magnet?

Checkpoint 29.6 gives us a clue to the reason for Lenz's law. In Figure 29.12b, let's consider the system made up of magnet and loop. If the induced current further increased the magnetic flux through the loop, the induced current would exert an attractive force on the magnet. This attractive force would pull the magnet closer to the loop, increasing the magnetic flux further and inducing even more current. In the process, the kinetic energy of the magnet and the energy associated with the current in the loop would increase, increasing the energy of the magnet-loop system without any work being done on it, violating the law of conservation of energy.

Conservation of energy therefore *requires* that an induced current oppose the change that created it. In Figure 29.12*b*, for example, the force exerted by the induced magnetic field $\vec{B}_{\text{ind}}$ resists the magnet moving closer to the loop. An alternative statement of Lenz's law therefore is:

An induced current is always in such a direction as to oppose the motion or change that caused it.

Because of this opposing force, an agent moving the magnet at constant speed toward the loop must do work on the magnet. Where does this energy go? As work is done on the system, the current through the loop—and therefore the induced magnetic field—increases. Just as we can associate electric potential energy with an electric field, we can associate **magnetic potential energy** with a magnetic field. In the absence of dissipation, the work done on the magnet-loop system therefore increases the magnetic potential energy of the system. The energy diagram in Figure 29.15 illustrates the conversion of mechanical work to magnetic potential energy.

Work must also be done to move a conducting loop into a magnetic field. When current is induced in the loop by its motion in the magnetic field, the magnetic field exerts a force on the current.

Figure 29.16 shows the directions of these magnetic forces exerted on each side of a rectangular conducting loop as it enters a magnetic field from the left. The vector sum of the forces points to the left. In order to move the loop through the field at constant speed, you must exert on

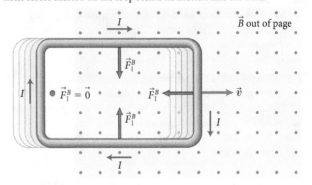

Figure 29.16 Direction of magnetic forces exerted by a magnetic field on each side of a rectangular conducting loop because of the current induced in the loop. As the loop moves into the field, the vector sum of the magnetic forces exerted on the loop resists its motion into the field.

the loop a constant force to the right that is equal in magnitude to the magnetic force (**Figure 29.17**). In doing so, you do work on the loop.

29.7 (*a*) After the left edge of the loop in Figure 29.17 enters the magnetic field, is the work required to continue pulling the loop through the field at constant speed *v* positive, negative, or zero? (*b*) As the right edge emerges from the magnetic field, is the work required positive, negative, or zero?

As you move a conductor relative to a magnetic field (or vice versa), the work you do transfers energy to the conductor, setting the charge carriers in motion. This energy associated with the induced current can then be converted to other forms of energy by some device placed in the current path, such as a lamp, toaster, or electric motor. In Figure 29.18, for example, a lamp is inserted in the current path and the work done on the system is converted to light energy and thermal energy in the lamp. Doing work by moving a loop through a magnetic field and thereby setting charge carriers in motion is the basic idea behind electric generators.

Figure 29.15 Energy diagram for a magnet-loop system when the magnet moves toward the conducting loop, inducing a current through the loop. Work must be done on the system to push the magnet at constant speed against the opposing induced magnetic field of the loop. This work causes the potential energy of the magnet-loop system to increase.

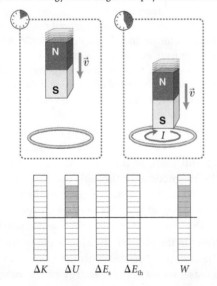

Figure 29.17 Work must be done on a conducting loop to move it into a magnetic field because magnetic forces exerted on the loop resist its motion when there is an induced current in the loop (shown in more detail in Figure 29.16).

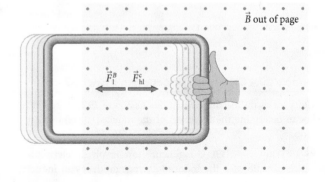

Figure 29.18 An induced current can be used to light a lamp. The work done on the magnet is converted to light and thermal energy in the filament of the lamp.

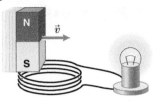

29.8 Which requires doing more work: moving a magnet toward a closed conducting loop or moving it toward a rod? Both motions are at constant speed.

It is not necessary to have a conducting object in the shape of a loop. If a bar magnet is moved toward an extended conducting object, such as a conducting sheet (**Figure 29.19**), circular currents are induced throughout the surface of the object. These currents are called **eddy currents** because they form eddy-like loops in the object's surface. (An *eddy* is the circular, whirlpool-like movement of water.)

Figure 29.19 The circular current loops induced in a conducting sheet by the motion of a nearby magnet are called eddy currents.

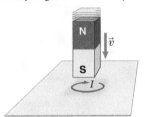

Because eddy currents dissipate energy, they can be used to convert kinetic energy (for example, the kinetic energy associated with the motion of a conductor relative to a magnet) to thermal energy. Dissipation of energy by eddy currents is used in braking systems, especially trains and roller coasters. In contrast to conventional friction brakes, eddy current brakes involve no physical contact and so there is no wear and tear and there are no brake pads to be replaced.

29.9 In **Figure 29.20**, a bar magnet moves parallel to a metal plate. (*a*) At the instant shown, does the magnitude of the magnetic flux increase, decrease, or stay the same through a small region around points P, Q, and R? (*b*) Are the eddy currents induced around these points clockwise, counterclockwise, or zero?

Figure 29.20 A bar magnet moves parallel to the surface of a conducting sheet, causing the local magnetic flux to change across the sheet surface and inducing eddy currents where the magnetic flux is changing (Checkpoint 29.9).

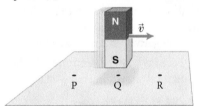

Self-quiz

1. A conducting rod moves through a magnetic field as shown in Figure 29.21. Which end of the bar, if any, becomes positively charged?

Figure 29.21
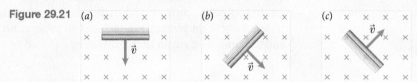

2. A conducting loop moves through a magnetic field as shown in Figure 29.22a–c. Which way does the current run in the loop at the instant shown in each figure?

Figure 29.22
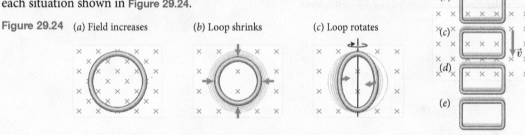

3. A conducting loop moves through a magnetic field at constant velocity as shown in Figure 29.23. For each case a–e, must the work done on the loop be positive, negative, or zero to keep the loop moving?

4. Using Faraday's law, determine whether charge carriers flow in the loop for each situation shown in Figure 29.24.

Figure 29.23

Figure 29.24   (a) Field increases   (b) Loop shrinks   (c) Loop rotates
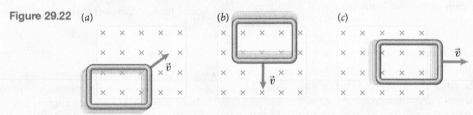

5. Using Lenz's law, determine the direction of the induced current, if any, at the instants shown in Figure 29.24.

Answers

1. The direction of the magnetic force on the positive charge carriers in the rod is given by the right-hand force rule: If you point the fingers of your right hand in the direction of $\vec{v}$ and then curl them toward $\vec{B}$ (into the page), your thumb points in the direction of the force. Applying this rule in the three situations shows that positive charge accumulates on the right end of the bar in (a) and (b) and on the left end of the bar in (c).

2. In (a) and (b), the magnitude of the magnetic flux through the loop increases into the page, so the induced magnetic field opposing this change must point out of the page. According to the right-hand dipole rule, an induced current must be counterclockwise to produce a magnetic field that points out of the page. In (c), the magnitude of the magnetic flux through the loop decreases into the page, so the induced current must be clockwise.

3. In (a), (c), and (e), the magnetic flux through the loop does not change and so no current is induced in the loop. The vector sum of the magnetic forces exerted on the loop is therefore zero, so no work is required to keep the loop moving at constant speed. In (b) and (d), the magnetic flux through the loop changes and a current is induced through the loop. According to Lenz's law the induced current is in such a direction as to oppose the motion or change that created it, so a positive amount of work is needed to keep the loop moving at constant speed.

4. Because the magnetic flux through the loop is changing in all three cases, charge carriers flow in all three cases.

5. (a) The magnetic field is directed into the page, and the magnitude of the magnetic flux is increasing into the page. The induced magnetic field opposing this change must therefore point out of the page, so according to the right-hand dipole rule, the induced current is counterclockwise. (b) Because the area of the loop decreases, the magnetic field is directed into the page and the magnitude of the magnetic flux decreases. The induced current is therefore clockwise. (c) At the instant shown in the figure, the magnitude of the magnetic flux through the loop decreases and so the induced current must be clockwise.

29.5 Induced emf

We first encountered the concept of emf in Chapter 26 in the context of batteries. Chemical reactions in a battery produce and maintain a separation of charge across the battery terminals; this separation results in a potential difference from one terminal to the other. If a light bulb is connected to the terminals, this potential difference drives a current through the light bulb. In Section 26.4, we defined the emf of a charge-separating device as the work done per unit charge by nonelectrostatic interactions in separating charge within the device. Because the charge separation across the ends of a rod caused by electromagnetic induction (Figure 29.10) is nonelectrostatic, we can associate an emf with this separation of charge. Emfs produced by electromagnetic induction are therefore called **induced emfs.**

Consider the setup illustrated in **Figure 29.25**. A conducting rod of length ℓ rests on two conducting rails connected to a light bulb. The rod is pulled to the right and moves at a constant speed v along the x axis. The motion of the rod changes the area of the magnetic field enclosed by the loop formed by the rod, rails, and wires attaching the bulb to the rails. Because of this area change, a current is induced through the loop. This current causes the light bulb to glow.

As we saw in Section 29.4, the magnetic field exerts a leftward force on the current-carrying rod. To move the rod at a constant speed v to the right, therefore, the agent pulling the rod must exert a rightward force of equal magnitude on it. In doing so, the external agent must do work on the rod, increasing the energy of the charge carriers in the loop. This energy is converted to light and thermal energy in the light bulb. We can obtain an expression for the induced emf in the loop by calculating the work done on the rod.

According to Eq. 27.4, the magnitude of the magnetic force exerted on a rod of length ℓ carrying a current I is $|I|\ell B$ when the rod is perpendicular to the magnetic field, so the contact force F_{ar}^c exerted by the agent on the rod required to pull it at constant speed must be of the same magnitude: $F_{ar}^c = |I|\ell B$. Because the work done on the rod is positive, it is equal to the product of the magnitude of the contact force exerted on it and the magnitude of the force displacement: $W_r \equiv F_{ar\,x}^c \Delta x_F = F_{ar}^c |\Delta x_F|$. In a time interval Δt, the rod is displaced a distance $\Delta x = v_x \Delta t$, so we have

$$W_r = F_{ar}^c |\Delta x_F| = B|I|\ell v \Delta t \qquad (29.1)$$

or, using Eq. 27.1, $I \equiv q/\Delta t$, to simplify,

$$W_r = B\ell v |q|, \qquad (29.2)$$

where q is the charge that passes through a given section of the rod in a time interval Δt. This result tells us how much energy is transferred to the current loop as the rod is pulled. Note that the force exerted by the magnetic field on the charge carriers in the rod transfers the mechanical work done on the rod to the charge carriers in the loop—the magnetic force itself does no work. (Because the direction of the magnetic force is always perpendicular to the motion of the charge carriers, the force cannot do any work on the carriers.)

If it bothers you that the external agent is doing the work and not the magnetic force that sets the charge carriers in motion, consider an analogy: Suppose you push yourself away from a wall starting from rest. The wall provides the force that accelerates you away from it. This force does *no work*, however, because the point of application does not move. In fact, no work at all is done on you. Internal (source) energy is converted to kinetic energy; the force exerted by the wall on you converts the internal energy to kinetic energy.

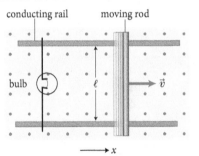

Figure 29.25 The emf that develops in a moving rod can be used to drive a current when that rod is part of a closed conducting loop. The rod is connected to the rest of the loop by sliding electrical contacts.

We can now calculate the magnitude of the induced emf, which is defined as the nonelectrostatic work per unit charge done on a charged particle (Eq. 26.7):

$$|\mathcal{E}_{\text{ind}}| \equiv \frac{W_{\text{r}}}{|q|} = B\ell v. \tag{29.3}$$

As shown in Figure 29.25, the induced emf in the closed conducting loop drives a current, just as a battery of the same emf would. This current has the same effect on the loop as a battery-driven current would have; in this case, it lights a light bulb. The magnitude of the resulting current—the **induced current**—depends on how much resistance the loop puts up to the motion of charge carriers. Experimentally we determine that the induced current through many conducting materials is proportional to the induced emf:

$$I_{\text{ind}} = \frac{\mathcal{E}_{\text{ind}}}{R}, \tag{29.4}$$

where the proportionality constant R is a measure of the *resistance* of the loop. The smaller the value of R, the greater the induced current for a given induced emf. We'll learn more about resistance in Chapter 31.

Equation 29.3 gives the magnitude of the induced emf in terms of v, B, and the length ℓ of the rod. In Section 29.2 we argued that the induced current—that is, the current that results from the induced emf—is due to a change in the magnetic flux enclosed by the loop. Let us therefore obtain an expression for the induced emf in terms of the magnetic flux Φ_B through the loop. In Section 27.6 we defined the magnetic flux through a loop by the surface integral (Eq. 27.10)

$$\Phi_B \equiv \int \vec{B} \cdot d\vec{A}, \tag{29.5}$$

where $d\vec{A}$ is the area vector of an infinitesimally small segment of a surface bounded by the loop.

Because the surface integral in Eq. 29.5 is independent of the choice of surface bounded by the loop (Eq. 27.15), we integrate over a flat surface bounded by the loop in Figure 29.25. Because the magnetic field is uniform, we can pull it out of the integral and so in Figure 29.25, if we let the vector $d\vec{A}$ point out of the page in the same direction as $\vec{B}$, $\Phi_B = BA$. When the moving side of the loop is displaced by Δx, the area enclosed by the expanding loop changes by $\Delta A = \ell \Delta x$. If this change occurs over a time interval Δt, the rate of change in the magnetic flux enclosed by the loop is thus

$$\frac{\Delta \Phi_B}{\Delta t} = \frac{B \Delta A}{\Delta t} = \frac{B\ell \Delta x}{\Delta t} = B\ell \frac{\Delta x}{\Delta t}, \tag{29.6}$$

so Eq. 29.3 can be written as

$$|\mathcal{E}_{\text{ind}}| = B\ell v = B\ell \left| \frac{\Delta x}{\Delta t} \right| = \left| \frac{\Delta \Phi_B}{\Delta t} \right|. \tag{29.7}$$

As we let the time interval Δt approach zero, this yields

$$\mathcal{E}_{\text{ind}} = -\frac{d\Phi_B}{dt}, \tag{29.8}$$

where we have added a negative sign to indicate that the direction of the induced emf is such that it drives a current that counteracts the change in magnetic flux, in accordance with Lenz's law. Equation 29.8, which relates a changing magnetic flux to an induced emf, is a quantitative statement of **Faraday's law.**

Note that Eq. 29.8 correctly gives a zero emf when applied to the situation shown in Figure 29.4c, in which the entire loop (not just one side) is moving and is in the magnetic field. We derived Eq. 29.8 for the special case of one side of a rectangular conducting loop moving in a magnetic field, but it is generally valid: It gives the value of the emf no matter what causes the magnetic flux through a conducting loop to change.

✋ **29.10** Sketch how the induced emf in the loop in Figure 29.4 varies as the loop moves through the five positions.

Equation 29.8 allows us to point out an important difference between potential difference and emf, both of which are related to the work per unit charge done on charge carriers—the former by electrostatic interactions, the latter by nonelectrostatic interactions. As we established in Chapter 25, potential difference depends only on starting and ending locations. It does not depend on the path followed to get from the starting location to the ending location. In other words, potential difference is path-independent. The work per unit charge done on charged particles by *nonelectrostatic* interactions, such as the induced emf in Figure 29.25, does depend on the path taken. Consider, for example, a rectangular conducting loop containing a light bulb and placed in a changing magnetic field (**Figure 29.26a**). The light bulb glows because the changing magnetic flux through the area enclosed by the loop causes an induced emf throughout the loop. If the area enclosed by the loop is smaller (Figure 29.26b), then the light bulb is less bright. The emf induced in the loop is now smaller because the smaller area causes a smaller rate of change in the magnetic flux. Had we established a potential difference across the light bulb by connecting it via wires to, say, the terminals of a battery, how much area the wires enclose would not affect the brightness of the bulb (**Figure 29.27**). We shall discuss this difference in path-dependence between potential difference and emf in more detail in the next section.

We can use an induced emf to establish a potential difference between two locations. Consider, for example, the conducting rod of Figure 29.1. As we saw in Section 29.1, the magnetic force exerted on the negative charge carriers in the moving rod drives them upward. A surplus of negative charge accumulates at the top of the rod, leaving a surplus of positive charge at the bottom. This separation of charge produces an electric field inside the rod. The negative charge carriers flow until the magnitude of the downward electric force exerted on one of them due to the electric field inside the rod is equal to the magnitude of the upward magnetic force (**Figure 29.28** on the next page):

$$F_q^E = F_q^B. \tag{29.9}$$

The magnitude of the magnetic force exerted on a charged particle of charge q moving perpendicular to a magnetic field is given by Eq. 27.18, $F_q^B \equiv |q|vB$, and the electric force is given by Eq. 23.6, $F_q^E = |q|E$. With this information, we can rewrite Eq. 29.9 as

$$|q|E = |q|vB. \tag{29.10}$$

Thus E, the magnitude of the electric field inside the rod, is

$$E = vB. \tag{29.11}$$

Figure 29.26 Path-dependence of an induced emf.

(a)

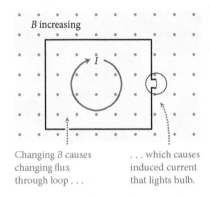

Changing *B* causes changing flux through loop . . .

. . . which causes induced current that lights bulb.

(b)

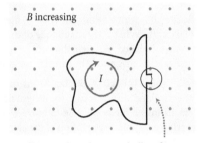

If loop encloses less area, bulb is dimmer.

Figure 29.27 The emf produced by a battery does not depend on enclosed area.

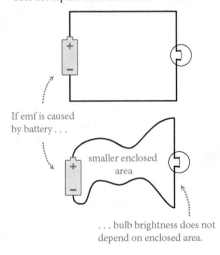

If emf is caused by battery . . .

smaller enclosed area

. . . bulb brightness does not depend on enclosed area.

QUANTITATIVE TOOLS

Figure 29.28 When a conducting rod moves in a magnetic field, the magnetic force exerted on the negative charge carriers in the rod causes a charge separation. This separation continues until the force exerted by the electric field resulting from the charge separation exactly counters the magnetic force exerted on the charge carriers.

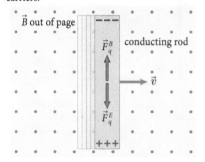

This electric field causes a potential difference between the top and the bottom of the rod. For a rod of length ℓ, the potential difference is (see Eq. 25.25)

$$V_{tb} = -\int_{top}^{bottom} \vec{E} \cdot \vec{d\ell} = \int_{top}^{bottom} E d\ell = E \int_{top}^{bottom} d\ell = E\ell, \quad (29.12)$$

where $\vec{E} \cdot \vec{d\ell} = -Ed\ell$ because the electric field points from the bottom to the top of the rod whereas $\vec{d\ell}$ points in the direction of integration. In addition we have pulled E out of the integral because it is constant throughout the rod. Substituting the magnitude of the electric field from Eq. 29.11, we obtain

$$V_{tb} = vB\ell. \quad (29.13)$$

Thus we see that the electric field in the rod establishes across the ends of the rod a potential difference that is equal in magnitude to the induced emf (Eq. 29.3).

Example 29.5 Airplane wing "battery"

A Boeing 747 with a wingspan of 60 m flies at a cruising speed of 850 km/h. What is the magnitude of the maximum potential difference induced between the two wingtips by Earth's magnetic field (the magnitude of which is roughly 0.50×10^{-4} T)?

❶ **GETTING STARTED** A Boeing 747 is made of metal and so, as the airplane flies through the magnetic field of Earth, the electrons in the metal are able to move when a magnetic force is exerted on them. This motion of the electrons causes a charge separation in the wings, and thus a potential difference is induced. To solve this problem, I must make several simplifying assumptions. I take the wings to be a 60-m conducting rod and assume the plane is flying horizontally through a vertical magnetic field. As a result, the problem reduces to a rod moving through a perpendicular magnetic field.

❷ **DEVISE PLAN** To calculate the induced potential difference between the ends of the "rod" formed by the wings, I can use Eq. 29.13.

❸ **EXECUTE PLAN**

$$V_{wings} = (8.50 \times 10^5 \text{ m/h})(1 \text{ h}/3600 \text{ s})(0.50 \times 10^{-4} \text{ T})(60 \text{ m})$$

$$= 0.71 \text{ T} \cdot \text{m}^2/\text{s} = 0.71 \text{ V}. ✔$$

❹ **EVALUATE RESULT** My calculation yields a potential difference on the order of the one from an AA battery, but that should not cause any problems in the aircraft. Also, the magnetic field is unlikely to be perpendicular to the velocity of the plane, as I assumed, and therefore the actual value of the potential difference is smaller than what I obtained.

Example 29.6 Generator

In an electric generator a solenoid that contains N windings each of area A is rotated at constant rotational speed ω in a uniform magnetic field of magnitude B (**Figure 29.29**). What is the emf induced in the solenoid?

Figure 29.29 Example 29.6.

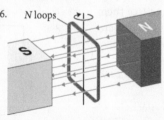

❶ **GETTING STARTED** As the solenoid rotates, the magnetic flux through it changes, and this changing flux causes an emf in the solenoid.

❷ **DEVISE PLAN** Equation 29.8 tells me that the emf induced in the solenoid equals the time rate of change of the magnetic flux through it. Therefore I must first determine how the magnetic flux varies as a function of time and then differentiate whatever expression I get to obtain the emf. I can determine the magnetic

flux through the solenoid by multiplying the magnetic flux through a single winding, $\Phi_B \equiv \vec{B} \cdot \vec{A}$ (Eq. 27.9), by the number of windings N. To determine the magnetic flux through a single winding, I sketch a top view of a winding in the magnetic field, indicating the directions of the magnetic field $\vec{B}$ and the area vector $\vec{A}$ (**Figure 29.30**). I let the plane of the winding be perpendicular to the direction of the magnetic field at $t = 0$.

Figure 29.30

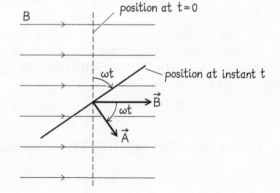

❸ EXECUTE PLAN As the solenoid rotates, the scalar product $\vec{B} \cdot \vec{A}$ changes with time. At instant t shown in my sketch, the angle between $\vec{A}$ and $\vec{B}$ is ωt, and so the magnetic flux through a single winding is $\Phi_B = \vec{B} \cdot \vec{A} = BA \cos \omega t$. Through the N windings of the solenoid the magnetic flux is $\Phi_B = NBA \cos \omega t$. Substituting this value into Eq. 29.8, I get

$$\mathcal{E}_{\text{ind}} = -\frac{d}{dt}(NBA \cos \omega t) = \omega NBA \sin \omega t. \checkmark$$

❹ EVALUATE RESULT My result shows that the emf oscillates sinusoidally. It is zero when $\omega t = n\pi$ ($n = 0, 1, 2, \ldots$) and maximum when $\omega t = n\pi + \frac{\pi}{2}$ ($n = 0, 1, 2, \ldots$). That result makes sense because the rate of change of the magnetic flux through the solenoid is zero when the area vector of the windings is parallel to the magnetic field ($\omega t = n\pi$) and maximum when the area vector is perpendicular to the magnetic field ($\omega t = n\pi + \frac{\pi}{2}$).

✋ **29.11** The expression I derived in Example 29.6 indicates that the emf becomes negative after the solenoid has rotated 180° and remains negative through the next 180° of rotation. However, the solenoid orientation looks the same when the solenoid has rotated 180° as when it started. Why does the emf have a different sign for half of the rotation?

29.6 Electric field accompanying a changing magnetic field

We saw in Section 29.3 that when the magnetic flux through a conducting loop changes, an electric force is exerted on the initially stationary charge carriers in the loop. This electric force is caused by the electric field that accompanies the changing magnetic field. Let's explore the properties of this electric field in more detail.

Consider the case of a conducting circular loop in a uniform magnetic field that has a circular cross section (**Figure 29.31a**). Suppose the magnitude of the magnetic field increases steadily over time. Because the conducting loop encloses an increasing magnetic flux directed out of the page, a clockwise current is induced in the loop. What do the electric field lines that are responsible for this current look like?

We found in Section 29.3 that the electric field lines that accompany a changing magnetic field form loops because there are no isolated charge carriers on which the electric field lines can begin or end. These loops must be circular and centered on the axis of the magnetic field because the electric field that accompanies the changing magnetic field must have the same cylindrical symmetry as the magnetic field (Figure 29.31b). Lenz's law tells us that the induced current is clockwise, and so the electric field must also be pointing clockwise. When the magnetic field does not exhibit cylindrical symmetry, the electric field lines are not circular, and in general it is difficult to determine the shape of the electric field lines.

✋ **29.12** What do the electric field lines look like when the magnitude of the magnetic field in Figure 29.31b (a) is held constant and (b) decreases steadily?

To determine the magnitude of the electric field that accompanies the changing magnetic field, let's think about its effect on charge carriers in the conducting loop. A current is induced in the loop because the electric field does work on the charge carriers in the loop. We can calculate the work done by the electric field using Eq. 25.1, which gives the work done by an electric field $\vec{E}$ on a particle carrying a charge q in moving it from point A to point B:

$$W_q(A \rightarrow B) = \int_A^B \vec{F}_q^E \cdot d\vec{\ell} = q \int_A^B \vec{E} \cdot d\vec{\ell}. \tag{29.14}$$

Figure 29.31 Electric field that accompanies an increasing cylindrical magnetic field.

(*a*) Conducting ring in cylindrical uniform magnetic field

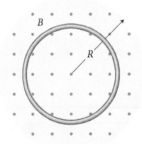

(*b*) Electric field that accompanies increasing magnetic field

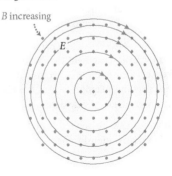

(Note that the electric field accompanying the changing magnetic field is *not* an electrostatic field, and thus the quantity calculated in Eq. 29.14 is *not* electrostatic work.)

The work done on a particle carrying a charge q as it travels around the closed path formed by the conducting loop is then

$$W_q(\text{closed path}) = q \oint \vec{E} \cdot d\vec{\ell}. \tag{29.15}$$

The work done by the electric field per unit charge is thus

$$\frac{W_q}{q} = \oint \vec{E} \cdot d\vec{\ell}. \tag{29.16}$$

This work per unit charge is the induced emf (Eq. 26.7), and so combining Eq. 29.16 with Eq. 29.8 gives us an expression that contains the electric field accompanying the changing magnetic field:

$$\oint \vec{E} \cdot d\vec{\ell} = -\frac{d\Phi_B}{dt}. \tag{29.17}$$

Equation 29.17 is an alternative statement of Faraday's law. Note that the magnetic flux appearing in this equation is a signed quantity. In order to obtain the correct direction of the electric field in Eq. 29.17, the magnetic flux must be taken to be positive when the magnetic field points in the direction of your right-hand thumb as you curl the fingers of your right hand in the direction of the integration path.

Keep in mind that the electric field in Eq. 29.17 is not an electrostatic field. Electrostatic fields originate directly from static charge distributions. Rather than originating from static charged objects, the electric field in Eq. 29.17 appears with a changing magnetic field and the electric field lines close on themselves. As we saw in Chapter 25, the work done by an electrostatic field on a charged particle moving around a closed path is zero, which means the line integral around a closed path on the left in Eq. 29.17 is always zero for an electrostatic field (see Eq. 25.32). However, as Eq. 29.17 shows, in a region where the magnetic field is changing, the line integral of an electric field around a closed path is not necessarily zero. This means that the electric field accompanying the changing magnetic field can do work on charged particles that travel along a closed path and the amount of work depends on the choice of path. We cannot define a potential difference between two points for such an electric field because it would have different values for different paths between those points.

Example 29.7 Electric field magnitude

Let the uniform cylindrical magnetic field in Figure 29.31 have a radius $R = 0.20$ m and increase at a steady rate of 0.050 T/s. What is the magnitude of the electric field at a radial distance $r = 0.10$ m from the center of the magnetic field?

❶ **GETTING STARTED** The changing magnetic field is accompanied by an electric field that has circular field lines pointing clockwise (Figure 29.31). Because of the circular symmetry, I know that the magnitude E of the electric field cannot vary along any given electric field line.

❷ **DEVISE PLAN** To solve this problem, I can use Eq. 29.17 to work out the relationship between E and the rate of change of

the magnetic field magnitude dB/dt. Because E is constant along any electric field line, it must be constant on *any* circular path centered on the axis of the cylindrical magnetic field. I therefore choose as my path of integration a clockwise circular path of radius $r = 0.10$ m centered on the long central axis of the magnetic field (**Figure 29.32**). Because $\vec{E}$ is always parallel to $d\vec{\ell}$, $\vec{E} \cdot d\vec{\ell} = E\,d\ell$ on the left side of Eq. 29.17, and because the magnitude of the electric field is the same everywhere on the path, I can take E out of the integral. To evaluate the right side, I must first calculate the magnetic flux through the circular integration path. Because the magnetic field is uniform, the magnetic flux is given by Eq. 27.9.

Figure 29.32

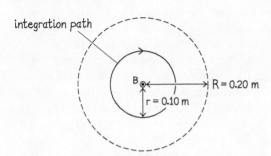

integration path

B $\odot$ ← R = 0.20 m

r = 0.10 m

③ **EXECUTE PLAN** The left side of Eq. 29.17 becomes

$$\oint \vec{E} \cdot d\vec{\ell} = 2\pi r E. \tag{1}$$

The magnetic flux through the area enclosed by the integration path is negative (curling the fingers of my right hand in the direction of the integration path makes my thumb point in the direction opposite the direction of $\vec{B}$), and so

$$\Phi_B = -BA = -\pi r^2 B.$$

Thus the right side of Eq. 29.17 becomes

$$-\frac{d\Phi_B}{dt} = +\pi r^2 \frac{dB}{dt}. \tag{2}$$

Substituting Eqs. 1 and 2 into Eq. 29.17 then yields

$$2\pi r E = \pi r^2 \frac{dB}{dt}$$

$$E = \tfrac{1}{2} r \frac{dB}{dt} = \frac{(0.10 \text{ m})(0.050 \text{ T/s})}{2}$$

$$= 2.5 \times 10^{-3} \text{ T} \cdot \text{m/s} = 2.5 \times 10^{-3} \text{ N/C.} \checkmark \tag{3}$$

④ **EVALUATE RESULT** Equation 3 shows that the magnitude of the electric field increases with radial distance r from the center of the magnetic field. Given the cylindrical symmetry of the situation, that's exactly what I expect.

Although the electric field lines and the conducting loop pictured in Figure 29.31 are circular, our derivation of Eq. 29.17 does not depend on the shape of the path of integration. This means we can use Eq. 29.17 to determine the line integral of the electric field around any closed path. For example, Figure 29.33 shows a square path of integration that encloses the same area (and hence the same magnetic flux) as the circular path I used to solve Example 29.7. Because the enclosed magnetic flux is the same, we know that the integral of the electric field around the square path must yield the same result as the integral around the round path, even though the electric field is no longer tangent to the path and no longer of the same magnitude all along the path.

29.13 In Example 29.7 what is the magnitude of the electric field at a distance of 0.30 m from the center of the magnetic field?

Figure 29.33 Square and circular paths of integration that enclose the same area and therefore enclose the same magnetic flux.

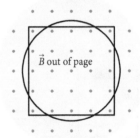

$\vec{B}$ out of page

29.7 Inductance

An interesting consequence of Faraday's law is that when a current through a conducting loop changes, the change induces an emf *in the loop itself*. To understand why this happens, consider Figure 29.34a (next page), in which a battery drives a current I through a solenoid. Because the current creates a magnetic field, a nonzero magnetic flux passes through the loops of the solenoid.

Now imagine the battery is suddenly disconnected, as in Figure 29.34b. Without the battery to supply an emf, the current, the magnitude of the magnetic field, and the magnitude of the magnetic flux all start to decrease. Faraday's law tells us, however, that the changing flux induces in the solenoid an emf that opposes the change in flux. Consequently, as the magnetic flux is decreasing, the induced emf causes an induced current that has the same direction as the original current. The magnetic flux associated with this induced current opposes the change in the magnetic flux.

The induced emf is proportional to the rate of change in the magnetic flux through the solenoid (Eq. 29.8). The magnetic flux is proportional to the magnitude B of the magnetic field, which is proportional to the current through the

Procedure: Calculating inductances

The inductance of a current-carrying device or current loop is a measure of the emf induced in the device or loop when current is changed. To determine the inductance of a particular device or current loop, follow these four steps.

1. Derive an expression for the magnitude of the magnetic field in the current-carrying device or current loop as a function of the current. Your expression should depend only on the current I and possibly—but not necessarily—the position within the device or current loop.

2. Calculate the magnetic flux Φ_B through the device or current loop. If the expression you derived in step 1

depends on position, you will have to integrate that expression over the volume of the device or circuit. Use symmetry to simplify the integral and divide the device into segments on which B is constant.

3. Substitute the resulting expression you obtained for Φ_B into Eq. 29.21. As you take the derivative with respect to time, keep in mind that only the current varies with respect to time, so you should end up with an expression that contains the derivative dI/dt on both sides of the equal sign.

4. Solve your expression for L after eliminating dI/dt.

Figure 29.34 Disconnecting a battery from a solenoid induces in the solenoid an emf that opposes the decrease in current through it.

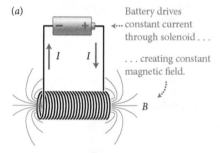

(a)

Battery drives constant current through solenoid . . .

. . . creating constant magnetic field.

B

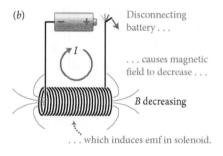

(b)

Disconnecting battery . . .

. . . causes magnetic field to decrease . . .

B decreasing

. . . which induces emf in solenoid.

loop (Eq. 28.6, $B = \mu_0 nI$). Thus the faster the current changes, the greater the magnitude of the induced emf:

$$\mathcal{E}_{\text{ind}} \propto \frac{d\Phi_B}{dt} \propto \frac{dB}{dt} \propto \frac{dI}{dt}. \tag{29.18}$$

We can define the **inductance** L of a loop or solenoid as the constant of proportionality between the emf and the rate of change of current:

$$\mathcal{E}_{\text{ind}} = -L\frac{dI}{dt}. \tag{29.19}$$

A large inductance leads to a large induced emf. The derived SI unit of inductance is the **henry:**

$$1\text{ H} \equiv 1\text{ V}\cdot\text{s/A} = 1\text{ kg}\cdot\text{m}^2/\text{C}^2. \tag{29.20}$$

A device that has an appreciable inductance is called an **inductor.** Inductors are used widely in electric circuits to even out variations in current.

The inductance describes how much change in magnetic flux is associated with a change in current for a particular loop or solenoid, as we can see by substituting Eq. 29.8 in the left side of Eq. 29.19:

$$\frac{d\Phi_B}{dt} = L\frac{dI}{dt}. \tag{29.21}$$

The inductance of a current-carrying device or circuit depends only on its geometry, but for most real devices, calculating the inductance is not simple. The general procedure for determining the inductance is described in the Procedure box on this page. In a few particularly simple cases, it is possible to derive an algebraic expression for the inductance, as in the following example.

Example 29.8 Inductance of a solenoid

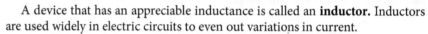

What is the inductance of a solenoid (Section 28.6) of length ℓ that has N windings, each of cross-sectional area A, when the current through the device is I?

① GETTING STARTED I begin by making a sketch of the solenoid (**Figure 29.35**). If I treat the solenoid as being infinitely long, I

know from Section 28.6 that the magnetic field is uniform inside the solenoid and zero outside of it. To calculate the inductance of the solenoid, I follow the steps in the Procedure box on this page.

Figure 29.35

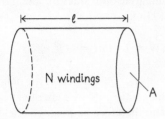

N windings

② **DEVISE PLAN** I first need to obtain an expression for the magnetic field magnitude B in terms of the current I. We derived this expression in Section 28.6 and found $B = \mu_0 n I$ (Eq. 28.6), where n is the number of windings per unit length $n = N/\ell$, and so $B = \mu_0 N I/\ell$. Knowing B, I can calculate the magnetic flux, substitute this result into Eq. 29.21, and then solve for the inductance L.

③ **EXECUTE PLAN** The magnetic flux through one winding is the area A enclosed by the winding multiplied by the magnetic field B inside the solenoid: $\Phi_B = BA$. For the solenoid, the magnetic flux is then

$$\Phi = NBA = \frac{\mu_0 N^2 I A}{\ell}.$$

Substituting this expression for the magnetic flux into Eq. 29.21 yields

$$\frac{d}{dt} \frac{\mu_0 N^2 I A}{\ell} = L \frac{dI}{dt}.$$

The only part of the left side that is time-dependent is the current I, so differentiating with respect to time and simplifying give me

$$L = \frac{\mu_0 N^2 A}{\ell}. ✔$$

④ **EVALUATE RESULT** My result shows that the inductance of a solenoid increases as the square of the number of windings N. This makes sense because both the magnetic field inside the solenoid and the magnetic flux increase with N. That inductance also depends on the area A enclosed by each winding also makes sense because increasing A increases the magnetic flux. Finally, that the inductance is inversely proportional to the length ℓ of the solenoid makes sense because the magnetic field is proportional to the number of turns per unit length.

29.14 A solenoid has 2760 windings of radius 50 mm and is 0.60 m long. If the current through the solenoid is increasing at a rate of 0.10 A/s, what is the magnitude of the induced emf?

29.8 Magnetic energy

We saw in Section 26.6 that work must be done to charge a capacitor and that this work increases the electric potential energy stored in the electric field of the capacitor. Likewise, work must be done on an inductor to establish the current through it because the change in current causes an induced emf that opposes this change; this work increases the **magnetic potential energy** stored in the magnetic field of the inductor.

In Chapter 26, we saw that electric potential energy can be attributed either to the configuration of charge in a system or to the electric field of this charge configuration. Likewise we can attribute the potential energy in an inductor either to the current through it or to the magnetic field caused by the current.

How much potential energy is stored in an inductor in which the current is I? To work this out, we calculate the work required to create a current I in an inductor. We begin by writing the work dW done on the inductor when an amount of charge dq moves through it. This work is the negative of the work done by the induced emf of the inductor:

$$dW = -\mathscr{E}_{\text{ind}}\, dq. \tag{29.22}$$

Using the definitions of current and Eq. 29.19, we obtain for the rate at which work is done

$$\frac{dW}{dt} = -\mathscr{E}_{\text{ind}} \frac{dq}{dt} = -\mathscr{E}_{\text{ind}} I = LI \frac{dI}{dt}, \tag{29.23}$$

and so $dW = LI\, dI$. We can then integrate both sides to obtain the work W done on the inductor to create a current I in it:

$$W = \int dW = L \int I\, dI = \tfrac{1}{2} LI^2. \tag{29.24}$$

To determine how much potential energy is stored in the magnetic field of an inductor, let's choose the zero of magnetic potential energy to be when there is no current through the inductor, and therefore no magnetic field. The magnetic potential energy stored in the inductor when there is a current through it is then equal to the work done to increase the current from zero to I:

$$U^B = \tfrac{1}{2} LI^2. \tag{29.25}$$

This is analogous to Eq. 26.4 for the electric potential energy stored in a capacitor, $U^E = \tfrac{1}{2} C V_{\text{cap}}^2$, with L taking the place of C and I taking the place of V_{cap}. Because this energy in the inductor is stored in the magnetic field, let's express the energy in terms of the magnetic field. We'll do this for the case of a long solenoid, but the result turns out to be generally applicable. Substituting the expression worked out in Example 29.8 for L into Eq. 29.25, we obtain

$$U^B = \tfrac{1}{2} \frac{\mu_0 N^2 A}{\ell} I^2, \tag{29.26}$$

where, as in Example 29.8, N is the number of windings, A is the area of each winding of the solenoid, and ℓ is the length of the solenoid.

In Chapter 26, we used the expression for the electric potential energy U^E to arrive at an expression for the energy density in an electric field: $u_E = \tfrac{1}{2} \epsilon_0 E^2$ (Eq. 26.6). Let us now use Eq. 29.26 to obtain an expression for the energy density in the magnetic field. Equation 28.6, $B = \mu_0 n I$, where $n = N/\ell$ is the number of windings per unit length, gives us the magnitude of the magnetic field inside a solenoid, and taking the square of this equation yields

$$B^2 = \frac{\mu_0^2 N^2 I^2}{\ell^2}. \tag{29.27}$$

Multiplying the right side of Eq. 29.26 by the factor $\mu_0 \ell / (\mu_0 \ell)$ gives

$$U^B = \tfrac{1}{2} \frac{\mu_0 N^2 A I^2}{\ell} \frac{\mu_0 \ell}{\mu_0 \ell} = \tfrac{1}{2} \frac{\mu_0^2 N^2 I^2}{\mu_0 \ell^2} A\ell = \tfrac{1}{2} \frac{B^2}{\mu_0} A\ell. \tag{29.28}$$

Because $A\ell$ is the volume of the region of magnetic field inside the solenoid, dividing the energy U^B by this volume gives us the **energy density** of the magnetic field:

$$u_B \equiv \tfrac{1}{2} \frac{B^2}{\mu_0}. \tag{29.29}$$

This expression is analogous to the expression for the energy density u_E in the electric field (Eq. 26.6), with B appearing instead of E and $1/\mu_0$ replacing ϵ_0. If the magnetic field is nonuniform, we must subdivide the volume of interest into small enough segments that B can be considered uniform within each segment, then apply Eq. 29.29 to each segment and take the sum of all the contributions. This amounts to integrating the energy density of the magnetic field over the volume containing the magnetic field:

$$U^B = \int u_B dV. \tag{29.30}$$

Example 29.9 Magnetic energy stored in a square toroid

Consider a toroid with square windings (**Figure 29.36**). The inner radius is $R = 60$ mm, each winding has width $w = 30$ mm, and there are 200 windings, each carrying a current of 1.5 mA. What is the magnetic potential energy stored in this toroid?

Figure 29.36 Example 29.9.

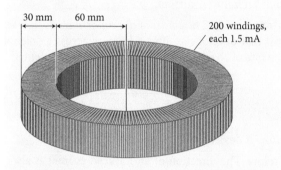

30 mm   60 mm

200 windings, each 1.5 mA

Figure 29.37

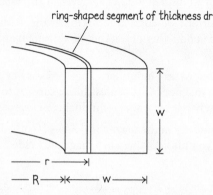

ring-shaped segment of thickness dr

❶ GETTING STARTED The current through the toroid causes a magnetic field inside the toroid. This magnetic field stores magnetic potential energy. I note that the width of the windings is not negligible relative to the toroid radius, and this tells me that I must assume a nonuniform magnetic field magnitude across the width of the windings. I therefore must use Eq. 29.30 to determine the amount of magnetic potential energy stored.

❷ DEVISE PLAN Before I can use Eq. 29.30, I must obtain the energy density of the magnetic field, which is given by Eq. 29.29. To calculate this energy density, I need to know the magnitude B of the magnetic field. Because the magnetic field is nonuniform, I must determine B by using Eq. 28.9, $B = \mu_0 NI/2\pi r$, where r is the radial distance from the center of the ring formed by the toroid. So to determine the magnetic potential energy stored, I must integrate Eq. 29.29 over the volume of the space enclosed by the windings of the toroid.

❸ EXECUTE PLAN Equation 29.29 gives me

$$U^B = \int u_B \, dV = \frac{1}{2\mu_0} \int B^2 \, dV = \frac{\mu_0 N^2 I^2}{8\pi^2} \int \frac{dV}{r^2}. \quad (1)$$

To integrate over the volume of the toroid, I divide the volume into ring-shaped segments (**Figure 29.37**) of height w, radius r, and thickness dr. The volume dV of each segment is equal to the product of the ring's circumference $2\pi r$, height w, and thickness dr: $dV = 2\pi r w \, dr$. Substituting this expression into Eq. 1 and integrating over r from the inner radius, $r = R$ of the toroid, to the outer radius, $r = R + w$, I get

$$U^B = \frac{\mu_0 N^2 I^2}{8\pi^2} \int_R^{R+w} \frac{2\pi w r \, dr}{r^2} = \frac{\mu_0 N^2 I^2 w}{4\pi} \int_R^{R+w} \frac{dr}{r}.$$

Working out the integration over r yields

$$U^B = \frac{\mu_0 N^2 I^2 w}{4\pi} \ln\left[1 + \frac{w}{R}\right]. \quad (2)$$

Substituting in the values for μ_0, N, I, R, and w, I obtain for the magnetic potential energy stored in the toroid

$$U^B = (10^{-7} \text{ T} \cdot \text{m/A})(200)^2 (1.5 \times 10^{-3} \text{A})^2$$

$$\times (30 \text{ mm}) \ln\left(1 + \frac{30 \text{ mm}}{60 \text{ mm}}\right)$$

$$= 1.1 \times 10^{-10} \text{ T} \cdot \text{m}^2 \cdot \text{A} = 1.1 \times 10^{-10} \text{ J}, \checkmark$$

where I have used $1 \text{ T} \cdot \text{m}^2 \cdot \text{A} = 1 \text{ [N}/(\text{A} \cdot \text{m})] \cdot \text{m}^2 \cdot \text{A} = 1 \text{ N} \cdot \text{m} = 1 \text{ J}$ (Eq. 27.6).

❹ EVALUATE RESULT The result I obtained is a very small amount of energy, but I have no idea (yet) how much energy is stored in a magnetic field. I can, however, compare my result with the result I would obtain in the limit where the width of each winding is much less than the toroid radius, $w \ll R$. In that limit the magnetic field is uniform inside the toroid, and its magnitude is given by Eq. 28.9, $B = \mu_0 NI/2\pi R$. The volume inside the windings of the toroid is equal to the toroid circumference, $2\pi R$, times the area of a winding, w^2: $V = 2\pi R w^2$. Substituting this expression for B in Eq. 29.29 and multiplying the energy density of the magnetic field by the volume V, I then get for the magnetic potential energy stored in the toroid

$$U^B = \frac{1}{2} \frac{\mu_0^2 N^2 I^2 / (2\pi R)^2}{\mu_0} (2\pi R w^2) = \frac{\mu_0 N^2 I^2 w^2}{4\pi R}$$

$$= \frac{\mu_0 N^2 I^2 w}{4\pi} \frac{w}{R}. \quad (3)$$

Because $\ln(1 + \epsilon) \approx \epsilon$ for $\epsilon \ll 1$, Eq. 2 is equal to Eq. 3 in the limit that $w \ll R$, giving me confidence in my integration result.

🖐 **29.15** How does the energy density of a 1.0-T magnetic field compare with the energy density of an 1.0-V/m electric field?

Chapter Glossary

SI units of physical quantities are given in parentheses.

Eddy current A circular current at the surface of an extended conducting object caused by a changing magnetic field.

Electromagnetic induction The process by which a changing magnetic flux causes charge carriers to move, inducing a charge separation or inducing a current.

Energy density of the magnetic field u_B (J/m^3) The energy per unit volume contained in a magnetic field:

$$u_B \equiv \frac{1}{2}\frac{B^2}{\mu_0}. \tag{29.29}$$

Faraday's law A changing magnetic flux induces an emf:

$$\mathscr{E}_{\text{ind}} = -\frac{d\Phi_B}{dt}. \tag{29.8}$$

Henry The derived SI unit of inductance:

$$1\ \text{H} \equiv 1\ \text{V} \cdot \text{s}/\text{A}. \tag{29.20}$$

Induced current (A) The current caused by a changing magnetic flux.

Induced emf $\mathscr{E}_{\text{ind}}$ (V) The work per unit charge done by electromagnetic induction in separating positive and negative charge carriers.

Induced magnetic field $\vec{B}_{\text{ind}}$ (T) The magnetic field produced by an induced current.

Inductance L (H) The constant of proportionality between the emf that develops around a loop or across a solenoid and the rate of change of current in that loop or solenoid:

$$\mathscr{E}_{\text{ind}} = -L\frac{dI}{dt}. \tag{29.19}$$

Inductor A device with an appreciable inductance.

Lenz's law The direction of an induced current is always such that the magnetic flux produced by the induced current opposes the change in the magnetic flux that induces the current.

Magnetic potential energy U^B (J) The form of potential energy associated with magnetic fields:

$$U^B = \int u_B dV. \tag{29.30}$$

30

Changing Electric Fields

CONCEPTS

QUANTITATIVE TOOLS

s we have seen in Section 29.3, electric fields accompany changing magnetic fields. Is the reverse true, too—do magnetic fields accompany changing electric fields? In this chapter we see that magnetic fields do indeed accompany changing electric fields. Consequently, a changing electric field can never occur without a magnetic field, and a changing magnetic field can never occur without an electric field. The interdependence of changing electric and magnetic fields gives rise to an oscillating form of changing fields called *electromagnetic waves*.

Electromagnetic waves are familiar to us as a wide range of phenomena: visible light, radio waves, and x-rays are all electromagnetic waves, the only difference being the frequency of oscillation of the electric and magnetic fields. We see our world by means of these waves, whether by using our eyes to observe our surroundings or by using x-ray diffraction to construct an image of a molecule or a material. Modern communications, from radio and television to mobile telephones, also make extensive use of electromagnetic waves. As we shall see, all these electromagnetic waves consist of changing electric and magnetic fields.

30.1 Magnetic fields accompany changing electric fields

In order to see that a magnetic field accompanies a changing electric field, let's revisit Ampère's law (see Section 28.5), which states that the line integral of the magnetic field along a closed path is proportional to the current encircled by the path (Eq. 28.1, $\oint \vec{B} \cdot d\vec{\ell} = \mu_0 I_{enc}$).

Figure 30.1 shows a current-carrying wire encircled by a closed path. The current encircled by the path is equal to the current through the wire, I. Another way to determine the encircled current is to consider any surface spanning the path and determine the current intercepted by that surface. For example, Figure 30.1 shows two different surfaces spanning the path. The current intercepted by either surface is I, the current encircled by the path.

✋ **30.1** Is the current intercepted by the surface equal to the current encircled by the closed path (*a*) in **Figure 30.2a** and (*b*) in **Figure 30.2b**?

Figure 30.2 Checkpoint 30.1.

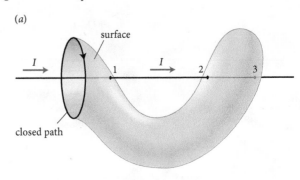

(*a*)

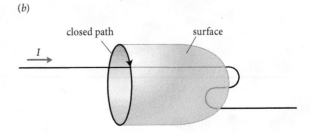

(*b*)

Checkpoint 30.1 shows that the current encircled by a closed path is equal to the current that is intercepted by any surface that spans the path, provided we keep track of the directions in which each interception takes place. Ampère's law can equally well be applied to the current encircled by a closed path and to the current intercepted by any surface spanning that closed path.

Now consider inserting a capacitor into our current-carrying wire while continuing to supply a constant current I to the wire. (That is, the capacitor is being charged.) Figure 30.3a again shows two surfaces A and B spanning the same closed path. The line integral of the magnetic field around the closed path does not depend on the choice of surface spanning the path. However, while the capacitor is charging, surface A is intercepted by a current I but

Figure 30.1 Current-carrying wire encircled by a closed path. Surfaces A and B both span the path. Surface A lies completely in the plane of the path. Surface B extends as a hemisphere whose rim is the path.

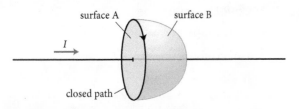

Figure 30.3 Capacitor being charged by a current-carrying wire. (*a*) The closed path of interest encircles the wire. Surface A intercepts the current, but surface B passes between the capacitor plates and does not intercept the current. (*b*) The closed path of interest lies between the capacitor plates. Surface A also lies between the plates and does not intercept the current, but surface B intercepts the current.

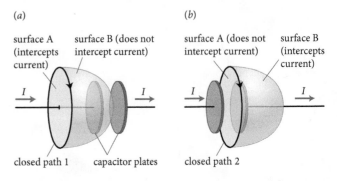

surface B, which passes between the capacitor plates, is not intercepted by any current. If we choose a closed path that lies between the capacitor plates (Figure 30.3b), a similar difficulty arises. Surface A intercepts no current, while surface B intercepts the current I.

In the case of a capacitor, therefore, the equivalence between encircled current and current intercepted by a surface spanning the encircling path doesn't hold. Surface B in Figure 30.3a would lead us to conclude that the line integral of the magnetic field around closed path 1 is zero. Because symmetry requires the magnetic field to always be tangent to the path and have the same magnitude all around the path, the line integral being zero means there is no magnetic field at the location of closed path 1 (even though the path encircles a current). Conversely, surface B in Figure 30.2b suggests there is a magnetic field at the location of closed path 2, even though that path encircles no current. Experiments do indeed confirm that there *is* a magnetic field in and around the gap between the plates of the charging capacitor. So only the surfaces that intersect the wires leading to the capacitor appear to provide the correct value of I_{enc} in Ampère's law for both closed paths in Figure 30.3.

Why must there be a magnetic field in and around the gap between the plates of the charging capacitor? Although there is no flow of charged particles between the plates of the capacitor, there *is* an electric field (**Figure 30.4**). Let us examine this electric field in more detail in the next checkpoint.

✋ **30.2** (a) While the capacitor of Figure 30.4 is being charged, is the current through the wire leading to or from the capacitor zero or nonzero? Is the electric field between the plates zero or nonzero? Is it constant or changing? (b) Answer the same questions for the capacitor fully charged.

The answers to Checkpoint 30.2 suggest that the magnetic field between the plates of the charging capacitor arises from the *changing* electric field. The current to the capacitor causes the electric field between the plates to

Figure 30.5 Parallels between (a) the electric field that accompanies a changing magnetic field and (b) the magnetic field that accompanies a changing electric field.

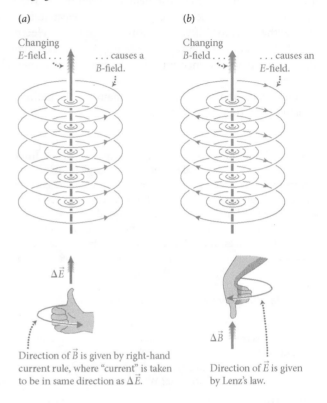

(a)

Changing E-field . . .   . . . causes a B-field.

$\Delta \vec{E}$

Direction of $\vec{B}$ is given by right-hand current rule, where "current" is taken to be in same direction as $\Delta \vec{E}$.

(b)

Changing B-field . . .   . . . causes an E-field.

$\Delta \vec{B}$

Direction of $\vec{E}$ is given by Lenz's law.

change, and the changing electric field between the capacitor plates acts in a way similar to the current that causes this change:

A changing electric field is accompanied by a magnetic field.

When the capacitor is fully charged, the current I into and out of the capacitor is zero, and there is no magnetic field surrounding the wires to the capacitor. Between the capacitor plates, the electric field is no longer changing, and the magnetic field is zero.

There are strong parallels between the electric field that accompanies a changing magnetic field and the magnetic field that accompanies a changing electric field, as **Figure 30.5** illustrates. Experiments show that the electric field lines that accompany a changing magnetic field form loops encircling the magnetic field, just as the magnetic field lines that accompany a changing electric field form loops encircling the electric field.

As we discussed in Section 27.3, the magnetic field surrounding a current-carrying wire forms loops that are clockwise when viewed looking along the direction of the current. The direction of these loops can be described by the right-hand current rule: Point the thumb of your right hand in the direction of the current, and your fingers curl

Figure 30.4 Capacitor being charged by a current-carrying wire. The electric field between the plates is shown. Closed path 1 encircles the current through the wire; closed path 2 encircles the electric field between the capacitor plates.

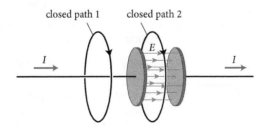

closed path 1   closed path 2

I E I

in the direction of the magnetic field. Similarly, the direction of the loops formed by the magnetic field lines that accompany a changing electric field are given by the right-hand current rule, taking the change in the electric field, $\Delta\vec{E}$, as the "current." If we take this change in the electric field into account in Figure 30.3, treating $\Delta\vec{E}$ like a current, the inconsistency we encountered before vanishes: Either a current or a change in the electric field, $\Delta\vec{E}$, is intercepted by the surface, and so for all surfaces spanning the paths we conclude that there is a magnetic field.

30.3 Consider disconnecting a charged capacitor from its source of current and allowing it to discharge (to release its charge into an external circuit). During discharge, the current reverses direction (relative to its direction when the capacitor was charging), but the electric field between the plates does not change direction. How does the direction of the magnetic field between the plates compare to the direction when the capacitor was charging? Does the right-hand current rule apply?

Example 30.1 Capacitor with dielectric

Consider a capacitor being charged with a constant current I and a dielectric between the plates. Is the magnitude of the magnetic field around a closed path spanning the capacitor (such as closed path 2 in Figure 30.4) any different from what it would be without the dielectric? Why or why not?

❶ **GETTING STARTED** I begin by making a two-dimensional sketch of the capacitor, indicating the position of the closed path (**Figure 30.6**).

Figure 30.6

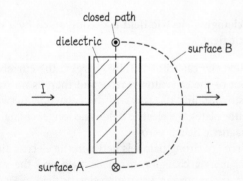

❷ **DEVISE PLAN** To determine the magnetic field magnitude at any position along the closed path, I need to examine the current and the changing electric field intercepted by a surface spanning the closed path.

❸ **EXECUTE PLAN** If I consider a flat surface through the closed path (surface A in Figure 30.6), the surface intersects the dielectric. While the capacitor is charging, the dielectric is being polarized: Negative charge carriers in the dielectric are displaced in one direction, and positive charge carriers are displaced in the opposite direction. This displacement of charge carriers

corresponds to a current. Surface A also intercepts a changing electric field. However, without further information about the capacitor, I can determine neither the current nor the electric field between the capacitor plates, which is affected by the presence of the dielectric. Surface A therefore doesn't permit me to compare the magnetic field magnitude to what it would be without the dielectric. I therefore draw another surface, making this surface loop around one of the capacitor plates (surface B in Figure 30.6). This surface intercepts only the wire leading from the capacitor, and I know that the current through the wire is unchanged by the presence of the dielectric. The fact that the current through this wire is unchanged tells me that the effective current in the region containing the dielectric is also unchanged. Therefore, the magnetic field must be the same as it would be without the dielectric. ✔

❹ **EVALUATE RESULT** Intuitively I expect the magnetic field magnitude around my closed path to be unchanged when the magnetic field magnitude around the wires attached to the capacitor is unchanged. The electric field between the capacitor plates gives rise to a displacement of charge carriers within the dielectric and thus affects the electric field between the capacitor plates, but apparently everything adds up to yield, for a given current through the capacitor, the same magnetic field magnitude outside the capacitor for a given current to the capacitor regardless of the presence or absence of the dielectric.

We now have a complete picture of what gives rise to electric and magnetic fields and on what kind of charged particle these fields exert forces. **Table 30.1** summarizes the properties of electric and magnetic fields. Note the remarkable symmetry between the two. Each type of field is produced by charged particles and accompanies a changing field of the other type. Electric fields are produced by charged particles either at rest or in motion, but magnetic fields are produced only by charged particles in motion. Likewise, any charged particle—at rest or in motion—is subject to a force in the presence of an electric field, but only charged particles in motion are subject to forces in a magnetic field.

Table 30.2 summarizes what we know about the field lines for electric and magnetic fields. The most striking difference between electric and magnetic fields is that magnetic field lines always form loops but electric field lines do

Table 30.1 Properties of electric and magnetic fields

	Electric field	Magnetic field
associated with	charged particle	moving charged particle
	changing magnetic field	changing electric field
exerts force on	any charged particle	moving charged particle

Table 30.2 Electric and magnetic field lines

	Electric field	Magnetic field
lines emanate from or terminate on	charged particle	–
loops encircle	–	moving charged particle
	changing magnetic field	changing electric field

not always form loops. This is a direct consequence of the difference in the sources of these fields. Magnetic field lines must form loops because there is no magnetic equivalent of electrical charge—no magnetic monopole (see Section 27.1). Instead, magnetic fields arise from current loops that act as magnetic dipoles.

Electric and magnetic field lines that accompany changing fields both form loops around the changing field. When particles serve as the field sources, however, the difference between magnetic and electric fields is evident: Electric field lines emanate or terminate from charged particles, while magnetic field lines always form loops around moving charged particles (currents).

We shall return to these ideas quantitatively in Section 30.5.

✋ **30.4** The neutron is a neutral particle that has a magnetic dipole moment. What does this nonzero magnetic dipole moment tell you about the structure of the neutron?

30.2 Fields of moving charged particles

We have seen that capacitors generate changing electric fields when charging or discharging. What else produces changing electric fields? One answer to this question is: changes in the motion of charged particles.

Before examining the electric fields of accelerating charged particles, let's consider the electric fields generated by charged particles moving at constant velocity. Figure 30.7 shows the electric field of a stationary charged particle and of the same particle moving at constant high speed. (By high speed, I mean a speed near enough the speed of light for relativistic effects to become important.)

The electric field of the stationary particle is spherically symmetrical; the electric field of the moving particle is still radial but definitely not spherically symmetrical. In this electric field, the field lines are sparse near the line along which the particle travels and are clustered together in the plane perpendicular to the motion. (This clustering is a relativistic effect and takes place for the same reason that objects moving at relativistic speeds appear shorter along the direction of motion, as discussed in Sections 14.3 and 14.6.) Consequently, the electric field created by the moving particle is strongest in that perpendicular plane. The faster the particle moves, the more the electric field lines bunch up in the transverse direction.

Keep in mind that as the particle moves at constant speed, the electric field lines move with it. At any instant, the electric field lines point directly away from the position of the particle *at that instant*. This means that as the particle moves, the electric field at a given position changes.

Because the particle in Figure 30.7b is moving, it is like a tiny current; it has a magnetic field that forms loops around its direction of travel, as shown in Figure 28.2b. The particle in Figure 30.7a does not have a magnetic field because it is at rest.

Now let's consider a particle that is initially at rest and then is suddenly set in motion. The electric field of this particle is shown at three successive instants in Figure 30.8 on the next page. Figure 30.8b and c show something we have not seen before: electric field lines that do not point directly away from the charged particle that is their source but instead are disrupted by sharp kinks. What is more, these kinks, which appear when the particle accelerates (just after Figure 30.8a), do not go away once the particle

Figure 30.7 Electric field line pattern of a charged particle (*a*) at rest and (*b*) moving to the right with speed v (v is a significant fraction of the speed of light). For the moving particle, the electric field lines cluster around the plane perpendicular to the direction of motion.

(*a*)

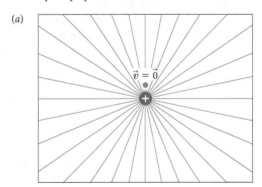

(*b*)

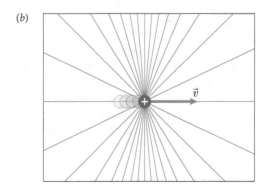

Figure 30.8 Electric field line pattern of a particle (*a*) initially at rest, (*b*) accelerating to speed *v*, and (*c*) moving at constant speed *v*. In (*b*) and (*c*), the ring of kinks in the electric field lines traveling outward from the particle corresponds to an electromagnetic wave pulse. Note that the speed *v* is smaller than the speed of the particle in Figure 30.7, indicated by the shorter arrow. Consequently, the electric field lines here are less sharply bunched around the vertical.

electric field line pattern of charged particle at rest:

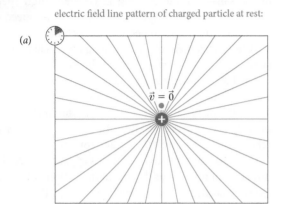

(*a*)

shortly after particle accelerates to constant speed:

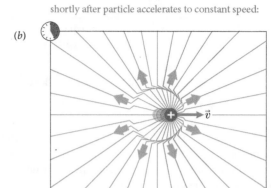

(*b*)

as particle continues at constant speed:

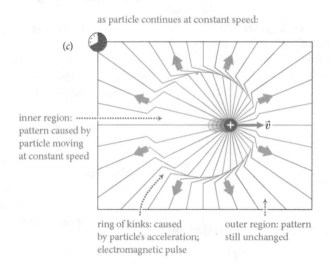

(*c*)

inner region: pattern caused by particle moving at constant speed

ring of kinks: caused by particle's acceleration; electromagnetic pulse

outer region: pattern still unchanged

reaches its final constant speed. Instead, they travel radially out from the location where the particle was when it started moving.

Where do these kinks in the electric field pattern come from? They arise because the electric field cannot change instantaneously everywhere in space to reflect changes in the source particle's motion. Remember that field lines extend infinitely far away from the particles that are their sources. If the electric field associated with a particle could change immediately everywhere in the universe when that particle changes its motion, then information about the change in motion would also be transmitted instantaneously throughout the entire universe. As we saw in Chapter 14, however, experiments show that such an instantaneous transmission of information does not happen. Changes in the electric field, and the information that these changes carry, travel at a finite (though very great) constant speed. In fact, in vacuum such changes always travel at the

same speed regardless of the details of the motion of the particles that produce them.

At distances that are too great for changes to reach in the time interval represented in Figure 30.8, the electric field line patterns in Figure 30.8*b* and *c* are still the same as the pattern of the stationary particle of Figure 30.8*a*. At distances that can be reached in that time interval, the electric field line patterns in parts *b* and *c* are those of the moving particle. Kinks form in order to connect these two patterns.

These kinks also form when a particle initially moving at constant velocity abruptly comes to a stop (**Figure 30.9**). The particle, initially moving at velocity *v*, stops just after the instant shown in Figure 30.9*e*. Part *f* shows the electric field line pattern of the stationary particle after some time interval has elapsed.

The electric field line density and consequently the magnitude of the electric field are much greater in the kinks than elsewhere. The energy density in the kinks region is

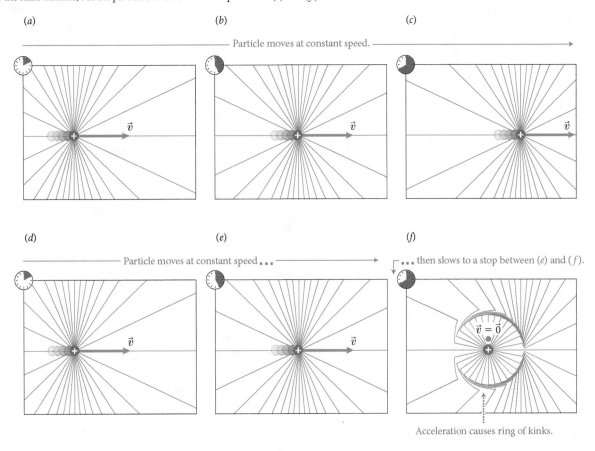

Figure 30.9 Electric field lines of a charged particle moving at some relativistic speed v. The upper diagrams show successive instants as the particle moves at constant velocity. The lower diagrams show the same instants, but the particle slows down to a stop between (e) and (f).

therefore greater than the energy density in other parts of the electric field. As the kinks move, they carry energy away from the particle. These kinks (and the energy carried by them) are one of the two parts of *electromagnetic waves*. As you might guess, kinks in magnetic field lines are the other part. Because changing electric fields are accompanied by changing magnetic fields (and vice versa), the two are always found together. An **electromagnetic wave** is thus a combined disturbance in an electric and a magnetic field that is propagating through space. Because a single isolated propagating disturbance is called a wave pulse (see Section 16.1), the kinks that appear in Figures 30.8 are 30.9 are *electromagnetic wave pulses*.

30.5 Estimate the final speed v of the charged particle in Figure 30.8 in terms of the speed of propagation c of the electromagnetic wave pulse produced by the particle's acceleration.

Let us now look at what effect an electromagnetic wave pulse has on a charged particle. Figure 30.10 on the next page shows the force exerted on a stationary charged test particle by the electric field of an accelerated charged particle. At

the first instant shown (Figure 30.10a), before the particle at the center of the panel is accelerated, the force exerted by the electric field on the test particle runs along the field line joining the two particles and points away from the center particle. At the second instant shown (Figure 30.10b), the center particle has been accelerated, and the wave pulse created by the acceleration has just reached the test particle. The force exerted on the test particle is no longer directed along the line joining the two particles but is directed along the kinks in the electric field lines. The force therefore has a component tangential to a circle centered on the original position of the accelerated particle at the center of the panel. (The exact direction of the force depends on the magnitude and duration of the acceleration of the accelerated particle.) Moreover, because the electric field line density is large in the region of the kinks, the force is large in magnitude.

At the final instant shown (Figure 30.10c), the wave pulse has traveled beyond the test particle and the force once again points away from the particle. The electric field line density is much smaller again, and so the magnitude of the force exerted on the test particle is again much smaller.

Figure 30.10 Force exerted on a stationary charged test particle by the electric field of an accelerated charged particle.

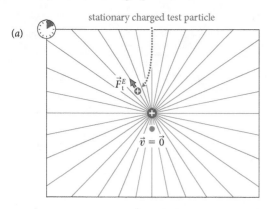

stationary charged test particle

(a)

(b)

(c)

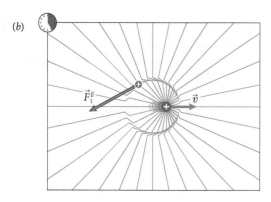

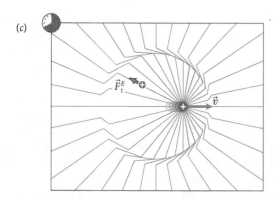

Example 30.2 Electromagnetic wave pulse

A particle carrying a negative charge is suddenly accelerated in a direction parallel to the long axis of a conducting rod, producing the electric field pattern shown in **Figure 30.11**. Does

Figure 30.11 Electric field of an accelerated particle near a conducting rod, before the wave pulse reaches the rod. (The electric field lines bend in near the conducting rod due to the rearrangement of charge carriers at the surface of the rod.)

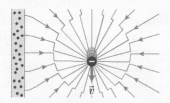

the electric field of the negatively charged particle create a current through the rod (*a*) at the instant shown in the figure, before the electromagnetic wave pulse created by the acceleration reaches the rod, and (*b*) at the instant the pulse reaches the rod? If you answer yes in either case, in which direction is the current through the rod?

❶ GETTING STARTED A current is created when charge carriers in the rod flow through the rod. For the carriers to flow, a force needs to be exerted on them.

❷ DEVISE PLAN To determine whether there is a current through the rod, I must determine if the electric field is oriented in such a way as to cause a flow of charge carriers through the rod. Even though in a metallic rod only electrons are free to move, I can pretend that only positive charge carriers are free to move because as I saw in Section 27.3, my answer is independent of the sign of the mobile charge carriers.

❸ EXECUTE PLAN (*a*) No. Before the pulse reaches the rod, the electric field is constant and so the rod is in electrostatic equilibrium. Therefore the electric field magnitude inside the rod is zero, so no charge carriers in the rod flow at the instant shown. ✔

(*b*) Yes. Once the pulse arrives at the rod, the electric field in the rod points downward, accelerating positively charged particles downward and causing a downward current. ✔

❹ EVALUATE RESULT Because the particle being accelerated is negatively charged, it makes sense that it pulls positive charge carriers in the rod along (with a delay caused by the time interval it takes the wave pulse to travel to the rod). In practice, electrons in the rod are accelerated upward, but the result is the same as what I describe for positive charge carriers.

30.6 In Figure 30.10, in which regions of space surrounding the accelerating particle does a magnetic field occur?

30.3 Oscillating dipoles and antennas

The wave pulse we have just considered is a brief, one-time, propagating disturbance in the electric field, analogous to the disturbance created when the end of a taut rope is suddenly displaced (as in Figure 16.2, for instance). Just as a harmonic wave can be generated on a rope by shaking the end of the rope back and forth in a sinusoidal fashion, a harmonic electromagnetic wave can be generated when a charged particle oscillates sinusoidally. **Figure 30.12** shows the electric field of a charged particle undergoing sinusoidal oscillation. This electric field consists of periodic kinks traveling away from the particle in a wavelike fashion.

In practice, isolated charged particles are not common. More often, positive and negative charged particles are present together, whether in individual atoms or in solid or liquid materials. Thus, displacing a positive particle leaves a negative particle behind, forming an electric dipole. Let us therefore consider the electric field pattern of an oscillating dipole.

Figure 30.12 Electric field of a sinusoidally oscillating charged particle.

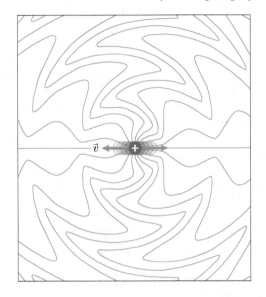

The electric field pattern of a stationary electric dipole with the positive charged particle above the negative charged particle is shown in **Figure 30.13a**. What about the electric field pattern of a stationary dipole made up of the same charged particles but with their positions switched, so that the dipole moment $\vec{p}$—which points from the negatively charged end to the positively charged end (see Section 23.4)—has reversed? The corresponding pattern of electric field lines has the same shape as shown in Figure 30.13a, but the directions of all the electric field lines are reversed.

Now let's work out the electric field pattern of an oscillating dipole, in which the two particles oscillate back and forth with a period T. We begin by considering the electric

field of a dipole that undergoes only a single reversal of its dipole moment (that is, one-half of a single oscillation) rather than oscillating continuously. The dipole starts out as shown in Figure 30.13a at instant $t = 0$. The charged particles that constitute the dipole then switch places in a time interval $T/2$ (half a cycle) and remain there.

Figure 30.13b shows the electric field pattern at instant $t = T$, after the dipole has been at rest in its new orientation for a time interval $T/2$. We can divide the space surrounding the dipole into the three regions shown. First consider the region sufficiently close to the dipole that the electric field is just the electric field of the stationary dipole in its new orientation. If we denote the speed at which changes in the electric field travel outward by c and the dipole has been stationary for a time interval $T/2$, this innermost region occupies a circle of radius $R = cT/2$. (The origin of our coordinate system is the center of the dipole, midway between the two particles.) Inside this circle, the electric field is that of the stationary dipole, the same shape as shown in Figure 30.13a but with the electric field line directions reversed.

Now consider the region sufficiently far away that no information about the motion of the dipole has reached it yet. This region lies outside a circle of radius $R = cT$. In this region, the electric field pattern is identical to that shown in Figure 30.13a, the electric field of the original dipole before it flipped over.

In the highlighted region of Figure 30.13b between these two circles, the electric field pattern is not dipolar. Because there are no charged particles in this region, the electric field lines cannot begin or end here. Instead, they must be connected to the electric field lines in the inner and outer regions. Consequently the electric field lines split into two disconnected sets: a set that emanates from the ends of the dipole and a set of loops detached from the dipole.

Figure 30.13 (a) Electric field of a stationary electric dipole in which the positive particle lies above the negative particle. (b) Electric field of the same dipole after the dipole moment has reversed and the charged particles have returned to rest.

(a) Electric field of stationary electric dipole

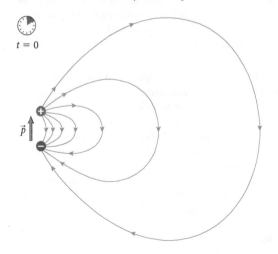

(b) Field shortly after dipole has reversed

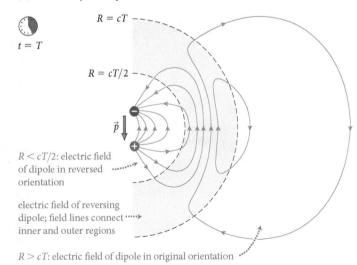

Figure 30.14 Snapshots of the electric field pattern of a sinusoidally oscillating dipole at time intervals of $T/8$ (where T is the period of oscillation).

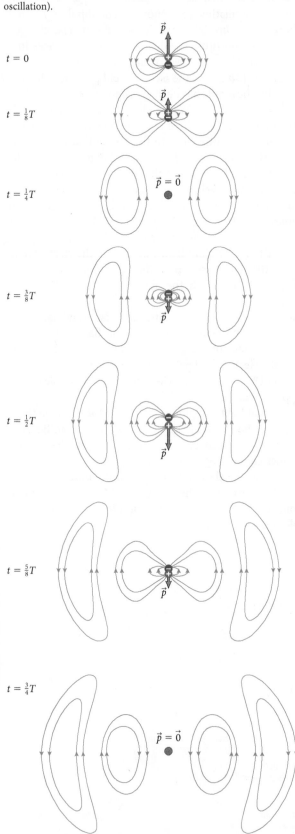

This electric field line pattern can be generalized to the case of a sinusoidally oscillating dipole (one that doesn't stop after half a cycle). **Figure 30.14** shows snapshots of the electric field pattern of such a dipole at time intervals of $T/8$. Just as with the single half-oscillation, we see a dipolar electric field near the dipole. Farther away, the electric field lines form loops.

Notice how these loops form every half-cycle (that is, at $t = T/4$ and $t = 3T/4$): As the charged particles of the dipole reach the origin during each oscillation, the electric field lines pinch off and the loops travel outward like puffs of smoke. This regular emission of looped electric field lines is a harmonic electromagnetic wave that travels away from the dipole horizontally left and right.

30.7 (*a*) If Figure 30.14 shows the oscillating electric field pattern at its actual size, estimate the wavelength of the electromagnetic wave. (*b*) If the wave is traveling at speed $c = 3 \times 10^8$ m/s, what is the wave frequency? (*c*) How long does one period last?

So far we have focused on the electric field pattern of this electromagnetic wave because it is natural to think about the electric field of a dipole. However, the changing electric field of the oscillating dipole is accompanied by a magnetic field. Consequently, the oscillation produces not only an electric field but also a magnetic field.

Example 30.3 Magnetic field pattern

Consider the electric field pattern of a sinusoidally oscillating dipole in Figure 30.14. (*a*) At $t = \frac{3}{4}T$, where along the horizontal axis bisecting the straight line connecting the two poles is the electric field increasing with time? Where is it decreasing? (*b*) Based on your answer to part *a*, what pattern of magnetic field lines do you expect in the horizontal plane that bisects the straight line connecting the two poles?

❶ **GETTING STARTED** Because the dipole oscillates sinusoidally, I expect the electric field to be a sinusoidally oscillating outward-traveling wave. The wave is three-dimensional, but the problem asks only about the electric field along the dipole's horizontal axis, so a one-dimensional treatment of this wave suffices. Because the wave is three-dimensional, the amplitude decreases as $1/r$ as the wave travels outward (see Section 17.1), but I'll ignore the decrease over the small distance over which the wave propagates in the figure.

❷ **DEVISE PLAN** I know from Chapter 16 that a one-dimensional sinusoidal wave can be represented by a sine function both in space and in time. (The wave function shows the value of the oscillating quantity as a function of position at a given instant in time, and the displacement curve shows the oscillating quantity as a function of time at a given position.) I can use the information shown in Figure 30.14 to draw the wave function for the electric field at $t = \frac{3}{4}T$. Once I have the wave function and know which way the wave is traveling, I can determine where the electric field increases. Because a changing electric field causes a magnetic field, I can use the information from part *a* to solve part *b*.

❸ **EXECUTE PLAN** (*a*) Because the problem asks for information at the instant $t = \frac{3}{4}T$, I begin by copying the right half of the bottom electric field pattern of Figure 30.14 (**Figure 30.15*a***). (The left half is simply the mirror image of the right half.) I draw a rightward-pointing horizontal axis through the center of the dipole and denote this as the z axis. I see that the electric field points downward parallel to the vertical axis (which I take to be the x axis) in the region between the dipole and the center of the first set of electric field loops. In the region between the centers of the first and the second set of loops, the electric field points upward. Because the electric field must vary sinusoidally, I can now sketch how its x component varies with position along the horizontal axis (Figure 30.15*b*).

Figure 30.15

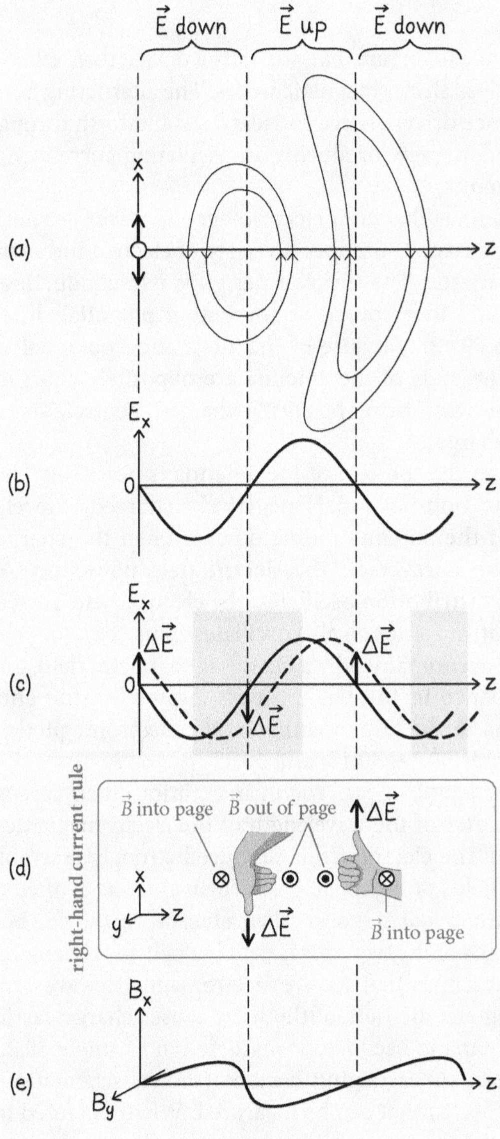

As the wave travels outward, the wave function of Figure 30.15*b* moves to the right (dashed curve in Figure 30.15*c*). The difference between the dashed and solid curves is the change in the electric field $\Delta\vec{E}$ (black arrows); $\Delta\vec{E}$ points down when the derivative of E_x with respect to z is positive (shaded regions) and up when it is negative (unshaded regions). ✔

(*b*) The direction of the magnetic field is determined by the right-hand current rule, taking the direction of the change in the electric field $\Delta\vec{E}$ as the "current." Pointing the thumb of my right hand down in the region where $\Delta\vec{E}$ points down and up where $\Delta\vec{E}$ points up (Figure 30.15*d*), I see from the way my fingers curl that the magnetic field lines form loops in the horizontal (yz) plane that are centered on the vertical black dashed lines, just like the electric field lines do. Consequently the magnetic field points out of the page when E_x is positive and into the page when E_x is negative. If I let the y axis point out of the page, the y component of the magnetic field must be positive when E_x is positive and negative when E_x is negative. I therefore draw a sinusoidally varying function for B_y as a function of position (Figure 30.15*e*). ✔

❹ **EVALUATE RESULT** My answer shows that the electric and magnetic fields have the same dependence on time but are oriented perpendicular to each other. Because the magnetic field changes, Faraday's law tells me that it is accompanied by an electric field. To analyze this electric field I can use an approach similar to the one I used to determine the magnetic field. **Figure 30.16*a*** shows the magnetic wave traveling outward. The difference between the dashed and solid curves is the change in the magnetic field $\Delta\vec{B}$. According to what I learned in Section 29.6, the direction of the electric field accompanying my changing magnetic field is given by Lenz's law and the right-hand dipole rule (Figure 30.16*b*). Consequently the electric field points into the page when B_y is positive and out of the page when B_y is negative. Because the x axis points into the page in this rendering (compare Figures 30.15*d* and 30.16*b*), the x component of the electric field must be positive when B_y is positive and negative when B_y is negative, as shown in Figure 30.16*c*. The electric field shown in Figure 30.16*c* is exactly the electric field I started out with in Figure 30.15*b*. In other words, the electric field yields the magnetic field and the magnetic field yields the electric field, and the two are entirely consistent with one another.

Figure 30.16

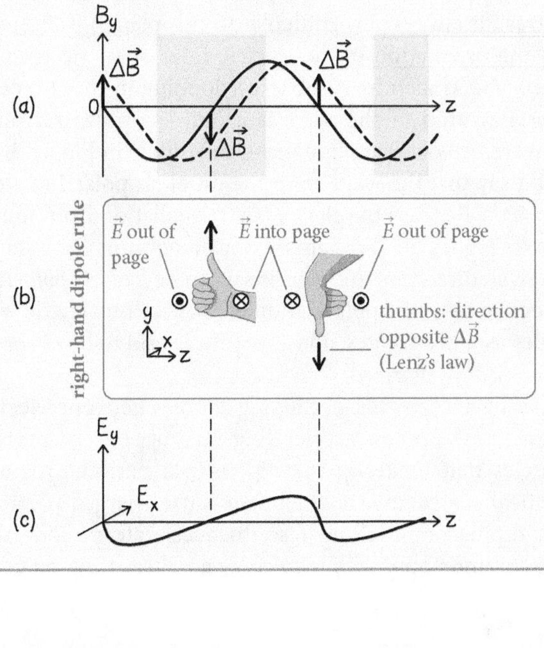

Figure 30.17 Electric and magnetic field pattern of oscillating dipole. The pink arrows indicate the direction of propagation of the electromagnetic wave pulse. For simplicity, only the fields in the xz and yz planes are shown.

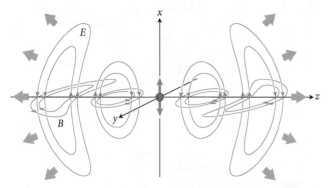

Figure 30.18 System of two antennas, one that emits electromagnetic waves and one that receives them. The emitting antenna is supplied with an oscillating current created by a source of alternating potential difference. An oscillating current is induced in the receiving antenna by the arriving electromagnetic wave.

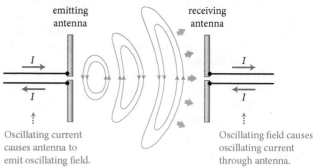

The solution to Example 30.3 suggests that the magnetic field line pattern is similar to the electric field line pattern, but perpendicular to it. **Figure 30.17** shows the combined electric and magnetic field pattern of an oscillating dipole. Traveling electromagnetic waves, like the one shown in Figure 30.17 are transverse waves (see Section 16.1) because both the magnetic field and the electric field are perpendicular to the direction of propagation. Also, the electric and magnetic fields propagate at the same frequency, and both reach their maxima (or minima) simultaneously; the electric and magnetic fields are therefore in phase with each other.

30.8 (*a*) At the origin of the graphs in Figure 30.15, the electric field is zero, but there is a current due to the motion of the charged particles that constitute the dipole. Is this current upward, downward, or zero at the instant shown in Figure 30.15? (*b*) Is this current (or the absence thereof) consistent with the magnetic field pattern shown in Figure 30.17?

In the wave shown in Figure 30.17, not only are the electric and magnetic fields perpendicular to each other, but throughout the entire wave the electric field has no component perpendicular to the xz plane. (The magnetic field, in contrast, is always perpendicular to this plane.) By convention the orientation of the electric field of an electromagnetic wave as seen by an observer looking in the direction of propagation of the wave is called the **polarization** of the wave. An observer looking at the dipole in Figure 30.17 would say that the wave from the dipole is polarized along the x axis. Because the electric field oscillation from the dipole in Figure 30.17 retains its orientation as it travels in any given direction, the wave is said to be *linearly polarized*. In certain cases the polarization of an electromagnetic wave rotates as it propagates, and the wave is said to be *circularly* or *elliptically polarized*.

We have seen that oscillating dipoles generate electromagnetic waves by accelerating the oppositely charged particles that make up the dipole in a periodic manner. Practically speaking, how can we cause charged particles, in a dipole or anything else, to accelerate periodically? One common approach is to apply an alternating potential difference to an *antenna*, which is a device that either emits or receives electromagnetic waves. The alternating potential difference drives charge carriers back and forth through the antenna, thereby producing an oscillating current through the antenna.

Antennas that emit electromagnetic waves are designed in many ways to produce a variety of electric and magnetic field patterns. The simplest design is two conducting rods connected to a source of alternating potential difference (**Figure 30.18**). Because of the alternating potential difference, the ends of the antenna are oppositely charged and cycle between being positively charged, neutral, and negatively charged.

When the top end of the antenna is positively charged and the bottom end is negatively charged, the electric field of the antenna points down. When the charge distribution is reversed, the electric field points up. As the charge distribution oscillates, the electric field adjacent to the emitting antenna also oscillates. This changing electric field is accompanied by a changing magnetic field, and the disturbance in the fields travels away from the emitting antenna in the same manner as the electromagnetic wave of Figures 30.14, 30.15, and 30.17.

If the length of each rod in an emitting antenna is exactly one-quarter of the wavelength of the electromagnetic wave emitted, the electric fields produced strongly resemble the dipole fields of Figure 30.17. Such an antenna is often called a *dipole antenna;* it is also called a *half-wave antenna* because the length of the two rods is equal to half a wavelength.

In antennas that receive electromagnetic waves, the oscillating electric field of the wave causes charge carriers in the antenna to oscillate, as discussed in Example 30.2. This produces an oscillating current (shown schematically in Figure 30.18) that can be measured. When operated in this mode, the antenna is said to be *receiving* a signal.

30.9 To maximize the magnitude of the current induced in a receiving antenna, should the antenna be oriented parallel or perpendicular to the polarization of the electromagnetic wave?

Self-quiz

1. Suppose the current shown in **Figure 30.19** discharges the capacitor. What are the directions of $\vec{E}$, $\Delta\vec{E}$, and $\vec{B}$ between the plates of the discharging capacitor?

Figure 30.19

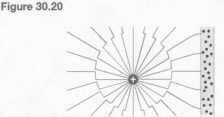

2. A positively charged particle creates the electric field shown in **Figure 30.20**. When the kinks in the electric field lines reach the rod, what is the direction of the current induced in the rod?

Figure 30.20

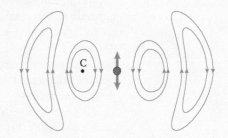

3. For the oscillating dipole of Figure 30.14, sketch the electric field pattern at $t = \frac{5}{4}T$.

4. In the electric field pattern for a sinusoidally oscillating dipole shown in **Figure 30.21**, what are (a) the direction of the change in the electric field $\Delta\vec{E}$ at point C as the electric field propagates and (b) the direction of the magnetic field loop near C?

Figure 30.21

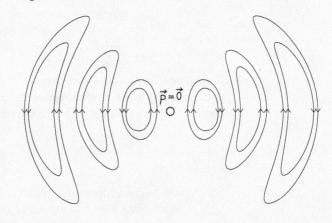

Answers

1. The current brings positive charge carriers to the left plate and removes them from the right plate. For the capacitor to discharge, the left plate must be negatively charged and the right one positively charged; this means $\vec{E}$ points left. The electric field decreases as the capacitor discharges, so $\Delta\vec{E}$ is to the right (just as the current is). The magnetic field lines are circular and centered on the axis of the capacitor in a direction given by the right-hand current rule with the thumb along the direction of $\Delta\vec{E}$. That is, the magnetic field lines are clockwise looking along the direction of the current.

2. Because the particle carries a positive charge, the electric field lines radiate outward. The electric field in the kinks therefore points up, and so the kinks induce an upward current through the rod.

3. See **Figure 30.22**. Because the loops move outward and a new pair of loops forms every half-period, the pattern now has three loops. Note in Figure 30.14 that the loops closest to the dipole at $t = \frac{1}{4}T$ and $t = \frac{3}{4}T$ have the same shape but opposite directions. Half a period after $t = \frac{3}{4}T$, at $t = \frac{5}{4}T$, the loops closest to the dipole again have the same shape (and the same direction as at $\frac{1}{4}T$). Likewise, at $t = \frac{5}{4}T$ the second closest loops curl in the direction opposite the direction at $\frac{3}{4}T$.

Figure 30.22

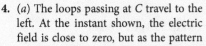

4. (a) The loops passing at C travel to the left. At the instant shown, the electric field is close to zero, but as the pattern moves to the left, the electric field lines point downward at C, so $\Delta\vec{E}$ is down. (b) The thumb of your right hand aligned in the direction of $\Delta\vec{E}$ makes your fingers curl in the direction of the magnetic field: clockwise viewed from the top.

Figure 30.23 Capacitor being charged by a current-carrying wire. Because surfaces A and B both span the closed path shown, either surface can be used to calculate the magnetic field around the path. Ampère's law must be the same in either case.

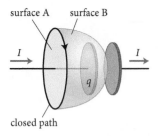

30.4 Displacement current

The work we did with Ampère's law in Sections 28.4 and 28.5 dealt only with the magnetic field associated with an electric current. As we saw in Section 30.1, though, magnetic fields also accompany changing electric fields, a phenomenon not covered by our Chapter 28 form of Ampère's law. Let us now see how the quantitative formulation of Ampère's law must be modified to account for the magnetic fields that accompany changing electric fields.

To do this, consider the charging capacitor in **Figure 30.23**. Ampère's law relates the integral of the magnetic field around a closed path to the current intercepted by a surface spanning the path (Eq. 28.1). Applying Ampère's law to surface A in Figure 30.23, we have

$$\oint \vec{B} \cdot d\vec{\ell} = \mu_0 I. \tag{30.1}$$

For surface B, however, the right-hand side of Eq. 30.1 is zero because the current is zero in the gap between the capacitor plates. As we discussed in Section 30.1, there is a change in the electric field between the capacitor plates, so surface B does not intercept a current, but it does intercept a change in electric flux. Let us therefore generalize Ampère's law by adding to the right side a term that depends on this change in electric flux.

We choose this term so that when we apply the generalized version of Ampère's law to the capacitor shown in Figure 30.23, for example, the magnetic field around the designated path is the same whether we calculate it from the current intercepted by surface A or from the change in electric flux $d\Phi_E/dt$ through surface B.

To obtain this generalizing term, let's determine a mathematical relationship between $d\Phi_E/dt$ through surface B and the current to the plates. First, note that the change in electric flux is related to the change in the charge q on the plates, which, in turn, is related to the current I to the plate. Consider the closed surface surrounding the left capacitor plate in Figure 30.23 made up by surfaces A and B combined. Applying Gauss's law to this closed surface, we find that the electric flux through it is

$$\oint_{A+B} \vec{E} \cdot d\vec{A} = \frac{q}{\epsilon_0}, \tag{30.2}$$

where q is the charge on the capacitor plate. Because the electric field is confined to the region between the plates, the electric flux through surface A is zero (Figure 30.23) and so

$$\oint_{A+B} \vec{E} \cdot d\vec{A} = \int_A \vec{E} \cdot d\vec{A} + \int_B \vec{E} \cdot d\vec{A} = \int_B \vec{E} \cdot d\vec{A} = \frac{q}{\epsilon_0}. \tag{30.3}$$

If we denote the electric flux through surface B by Φ_E, we see from Eqs. 30.3 that $q = \epsilon_0 \Phi_E$. The rate of change of the charge on the capacitor plates, dq/dt, is equal to the current supplied to the capacitor, so

$$I \equiv \frac{dq}{dt} = \epsilon_0 \frac{d\Phi_E}{dt}, \tag{30.4}$$

which is the relationship we were looking for.

If we substitute Eq. 30.4 in the right side of Eq. 30.1, we obtain

$$\oint \vec{B} \cdot d\vec{\ell} = \mu_0 \epsilon_0 \frac{d\Phi_E}{dt}. \tag{30.5}$$

We can now use this expression to determine the line integral of the magnetic field around the closed path in Figure 30.23 by evaluating the change in electric flux through surface B. Because the right side of Eq. 30.5 is equal to $\mu_0 I$, we obtain for $\oint \vec{B} \cdot d\vec{\ell}$ the same value we found using the original form of Ampère's law with the current intercepting surface A.

To account for both a current and a changing electric flux, we generalize Ampère's law as follows:

$$\oint \vec{B} \cdot d\vec{\ell} = \mu_0 I_{\text{int}} + \mu_0 \epsilon_0 \frac{d\Phi_E}{dt}. \qquad (30.6)$$

This equation holds for any surface spanning a closed path and is sometimes called the *Maxwell-Ampère law*, in honor of the Scottish physicist James Clerk Maxwell (1831–1879), who first introduced the additional term in Eq. 30.6. To reflect the fact that we must only include the current intercepted by the surface, not the current encircled by the integration path, we write I_{int} rather than I_{enc}.

The quantity on the right side of Eq. 30.4 is called the **displacement current:**

$$I_{\text{disp}} \equiv \epsilon_0 \frac{d\Phi_E}{dt}. \qquad (30.7)$$

As you can see from Eq. 30.4, the SI units of the displacement current are indeed those of a current. The name is somewhat misleading because the derivation holds for a capacitor in vacuum where no charged particles are present in the space between the plates. Even if the term is somewhat of a misnomer, it is still useful to associate the change in the electric field with a "current" to determine the direction of the magnetic field accompanying a changing electric field. As we argued in Section 30.1, the direction of this displacement current is the same as that of the change in the electric field $\Delta \vec{E}$. We can then use the right-hand current rule to determine the direction of the magnetic field from the displacement current (see, for example, Figure 30.5).

Using Eq. 30.7, we can write Eq. 30.6 in the form

$$\oint \vec{B} \cdot d\vec{\ell} = \mu_0 (I_{\text{int}} + I_{\text{disp}}). \qquad (30.8)$$

30.10 The parallel-plate capacitor in **Figure 30.24** is discharging so that the electric field between the plates *decreases*. What is the direction of the magnetic field (*a*) at point P above the plates and (*b*) at point S between the plates? Both P and S are on a line perpendicular to the axis of the capacitor.

Figure 30.24 Checkpoint 30.10.

P ●

decreasing *E*-field

S ●

Example 30.4 A bit of both

The parallel-plate capacitor in **Figure 30.25** has circular plates of radius R and is charged with a current of constant magnitude I. The surface is bounded by a circle that passes through point P and is centered on the wire leading to the left plate and perpendicular to that wire. The surface crosses the left plate in the middle so that the top half of the plate is on one side of the surface and the bottom half is on the other side. Use this surface and Eq. 30.6 to determine the magnitude of the magnetic field at point P, which is a distance $r = R$ from the capacitor's horizontal axis.

Figure 30.25 Example 30.4.

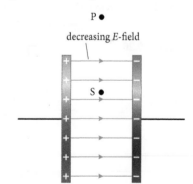

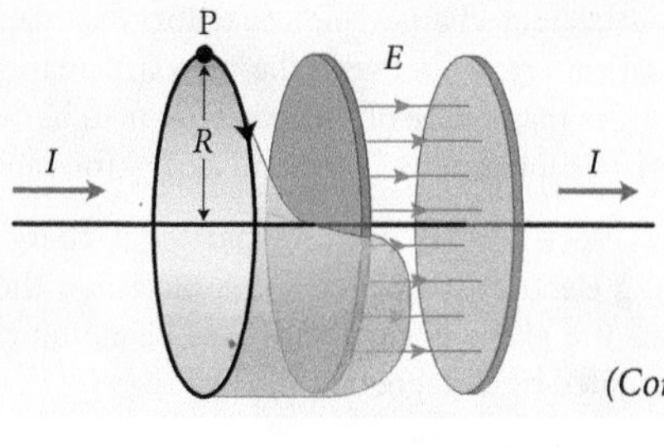

(Continued)

❶ GETTING STARTED The surface intercepts both a current (through the plate) and a changing electric field (between the plates). To apply Eq. 30.6, I must therefore determine both the current and the electric flux intercepted by the surface.

There also is a simple way to obtain the answer to this question: Ampère's law (Eq. 30.1). If I take the circle through point P centered on the wire and perpendicular to it as the integration path, then the integral on the left side of Eq. 30.1 is equal to the magnitude of the magnetic field times the circumference of the circle: $2\pi RB$. The path encircles the current I, so Ampère's law gives me $2\pi RB = \mu_0 I$ and $B = \mu_0 I/(2\pi R)$. Because the magnetic field magnitude cannot depend on the approach used to calculate it, I should obtain the same result using Eq. 30.6 and the surface in Figure 30.25.

❷ DEVISE PLAN The left side of Eq. 30.6 is identical to the left side of Eq. 30.1 and therefore equal to $2\pi RB$. To evaluate the right side of Eq. 30.6, I must determine both the ordinary current and the displacement current intercepted by the surface.

❸ EXECUTE PLAN If the surface intercepted the entire electric flux between the plates, the displacement current term on the right side of Eq. 30.6 would be equal to $\mu_0 I$. The surface intercepts only half of the electric flux, however, so the displacement current is

$$\epsilon_0 \frac{d\Phi_E}{dt} = \tfrac{1}{2} I.$$

Next I need to determine how much current is intercepted by the surface. To charge the plate uniformly, the current must carry charge carriers evenly to the two halves of the plate—half of the charge carriers go to the top half of the plate and the other half go to the bottom half of the plate (Figure 30.26).

Figure 30.26

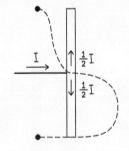

Because the top half of the plate is to the right of the surface, the current going to the top half of the plate must cross the surface. The current intercepted by the surface is thus $\tfrac{1}{2} I$, and the right side of Eq. 30.6 becomes

$$\mu_0 I_{int} + \mu_0\epsilon_0 \frac{d\Phi_E}{dt} = \mu_0(\tfrac{1}{2}I) + \mu_0(\tfrac{1}{2}I) = \mu_0 I. \qquad (1)$$

Substituting the right side of Eq. 1 into the right side of Eq. 30.6, I have $2\pi RB = \mu_0 I$ and so $B = \mu_0 I/(2\pi R)$, which is the same value I got using Eq. 30.1. ✔

❹ EVALUATE RESULT It's reassuring to see that the magnetic field magnitude at P does not depend on the choice of surface spanning the integration path. I can easily modify the argument above to show that any other surface that intercepts the left plate gives the same result. For example, a surface that intercepts one-quarter of the plate, as in Figure 30.27, intercepts one-quarter of the electric flux, and so the displacement current is only $I/4$. This surface intercepts the current twice: Where the surface intersects the wire, the current I crosses the surface from left to right, and where the surface intersects the plate, one-quarter of the current crosses the surface in the other direction, for a total contribution of $\tfrac{3}{4} I$. Again the sum of the ordinary and displacement currents intercepting the surface is I.

Figure 30.27

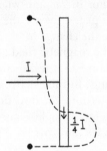

Example 30.5 Magnetic field in a capacitor

A parallel-plate capacitor has circular plates of radius $R = 0.10$ m and a plate separation distance $d = 0.10$ mm. While a current charges the capacitor, the magnitude of the potential difference between the plates increases by 10 V/μs. What is the magnitude of the magnetic field between the plates at a distance R from the horizontal axis of the capacitor?

❶ GETTING STARTED As the capacitor is charging, there is a changing electric flux between the plates, so the electric field between the plates is changing. This changing electric field is accompanied by a magnetic field.

❷ DEVISE PLAN Equation 30.6 relates the magnetic field to a changing electric flux. To work out the left side of Eq. 30.6, I chose a circular integration path centered on the horizontal axis of the capacitor so that I can exploit the circular symmetry of the problem. To work out the right side of Eq. 30.6, I need to determine the rate of change of the electric flux through an appropriate surface spanning the integration path. I choose the simplest possible surface: a flat surface parallel to the plates (Figure 30.28). To calculate the change in electric flux intercepted by this surface, I need to determine the magnitude of

Figure 30.28

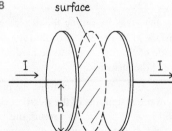

the (uniform) electric field between the plates. In Example 26.2 I determined that the magnitude of the electric field between the plates is related to the plate separation distance d and the magnitude of the potential difference between the plates: $V_{cap} = Ed$.

❸ **EXECUTE PLAN** Because the electric field between the plates is uniform and perpendicular to the plates, the electric flux Φ_E through the surface I chose is $\Phi_E = EA = E\pi R^2$, where $A = \pi R^2$ is also the area of the capacitor plates. The time rate of change of the electric flux is then given by

$$\frac{d\Phi_E}{dt} = \pi R^2 \frac{dE}{dt}.$$

Substituting this result in Eq. 30.6 and setting the current term equal to zero because no current is intercepted by the surface I've chosen, I get

$$\oint \vec{B} \cdot d\vec{\ell} = \mu_0 \epsilon_0 \pi R^2 \frac{dE}{dt}.$$

Around my integration path, the magnitude of the magnetic field is constant, and the left side of this expression simplifies to $2\pi R B$. Solving for B, I obtain

$$B = \frac{\mu_0 \epsilon_0 R}{2} \frac{dE}{dt}. \tag{1}$$

Because $V_{cap} = Ed$, I can write $E = V_{cap}/d$ and so

$$B = \frac{\mu_0 \epsilon_0 R}{2d} \frac{dV_{cap}}{dt} \tag{2}$$

$$B = (4\pi \times 10^{-7}\ \text{T}\cdot\text{m/A})(8.85 \times 10^{-12}\ \text{C}^2/\text{N}\cdot\text{m}^2)$$

$$\times \frac{0.10\ \text{m}}{2(0.10 \times 10^{-3}\ \text{m})} \frac{10\ \text{V}}{1.0 \times 10^{-6}\ \text{s}}$$

$$= 5.6 \times 10^{-8}\ \text{T},$$

where I have used the Eq. 25.16 definition of the volt, $1\ \text{V} \equiv 1\ \text{J/C} \equiv 1\ \text{N}\cdot\text{m/C}$ and the Eq. 27.3 definition of the ampere $1\ \text{C} \equiv 1\ \text{A}\cdot\text{s}$ to simplify the units. ✔

❹ **EVALUATE RESULT** The magnetic field magnitude I obtain is very small, in spite of the substantial rate at which the potential difference between the plates increases. I have no way of knowing whether my numerical result is reasonable or not, but what I can do to evaluate the result is use another method to obtain an expression for B. Because my flat surface intercepts all the electric flux, I know that the magnitude of the magnetic field should be the same at all positions a distance R from the current-carrying wire. I can obtain the current by solving Eq. 26.1, $q/V_{cap} = C$, for the charge q on the capacitor plate and then using the definition of current, $I \equiv dq/dt$:

$$I \equiv \frac{dq}{dt} = C\frac{dV_{cap}}{dt}.$$

Substituting the capacitance of a parallel-plate capacitor $C = \epsilon_0 A/d = \epsilon_0 \pi R^2/d$ (see Example 26.2) into this expression, and then substituting the result into the expression for the magnetic field around a current-carrying wire from Example 28.3, $B = \mu_0 I/(2\pi r)$ (setting $r = R$ for the distance to the wire), I get

$$B = \frac{\mu_0 I}{2\pi R} = \frac{\mu_0 \epsilon_0 R}{2d} \frac{dV_{cap}}{dt},$$

the same result I obtained in Eq. 2, as I expect.

Note that in Example 30.5 the rate of change of the electric field is very large (about $10^{11}\ \text{V}/(\text{m}\cdot\text{s})$), but the accompanying magnetic field is small. This is not the case for electric fields that accompany changing magnetic fields, as substantial emfs can be induced by the motion of ordinary magnets.

✋ **30.11** Consider again the parallel-plate capacitor of Figure 30.23. For circular plates of radius R, calculate the magnitude of the magnetic field a distance $r < R$ from the horizontal axis of the capacitor (*a*) between the plates and (*b*) a short distance to the right of the right plate.

Example 30.6 Displacement current in the presence of a dielectric

Suppose a slab of dielectric with dielectric constant κ is inserted between the plates of the capacitor in Figure 30.23 and the capacitor is charged with a current I, as considered in Example 30.1. How does Eq. 30.6 have to be modified to account for the dielectric?

❶ **GETTING STARTED** For a given amount of charge on the capacitor plates, the presence of a dielectric decreases the magnitude of the electric field between them. As I concluded in Example 30.1, however, the magnetic field surrounding the wires that lead to the capacitor wires is determined only by the current I through the wires and therefore cannot be affected by the insertion of the dielectric.

❷ **DEVISE PLAN** Given that the magnetic field surrounding the wires cannot be affected by the presence of the dielectric, the displacement current intercepted by a surface spanning a circular path around the wire and passing between the capacitor plates (Figure 30.29) should be equal to I, regardless of the

Figure 30.29

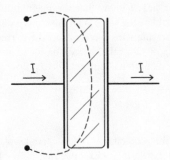

presence of the dielectric. By setting the displacement current through the surface in Figure 30.29 equal to I, I can determine how the right side of Eq. 30.6 needs to be modified to account for the presence of the dielectric.

❸ **EXECUTE PLAN** As the capacitor charges, the presence of the dielectric reduces the magnitude of the electric field by a factor $1/\kappa$ (see Eq. 26.15): $E = E_{\text{free}}/\kappa$. As a result, the rate of change of the electric field dE/dt and the rate of change in the electric flux intercepted by the surface $d\Phi_E/dt$ are also reduced by a factor $1/\kappa$. To compensate for this reduction, I need to multiply $d\Phi_E/dt$ by κ in order to make the right side of Eq. 30.4 equal to I again. Therefore Eq. 30.6 becomes

$$\oint \vec{B} \cdot d\vec{\ell} = \mu_0 I + \mu_0 \epsilon_0 \kappa \frac{d\Phi_E}{dt}. ✔$$

❹ **EVALUATE RESULT** My modification to Eq. 30.6 is identical to the modification we made to make Gauss's law work in dielectrics (Eq. 26.25): In both cases the term containing the electric field or electric flux includes the factor κ.

30.5 Maxwell's equations

With Maxwell's addition of the displacement current $\epsilon_0 d\Phi_E/dt$ to Ampère's law, $\oint \vec{B} \cdot d\vec{\ell} = \mu_0 I$, we now have a complete mathematical description of electric and magnetic phenomena and the relationship between the two. Let us summarize this description in the absence of a dielectric.

Electric and magnetic fields are *defined* by Eq. 27.20, which gives the force exerted on a charged particle moving in an electric field and a magnetic field:

$$\vec{F} = q(\vec{E} + \vec{v} \times \vec{B}). \tag{30.9}$$

Charged particles are the source of electrostatic fields, and electrostatic field lines always begin or end on charged particles. We can calculate electric fields from each individual charged particle if we wish, but when dealing with a distribution of charge that exhibits a certain symmetry (Section 24.4), it is most convenient to use Gauss's law (Eq. 24.8) to work out the electric field. Gauss's law tells us that the electric flux Φ_E through a closed surface (a Gaussian surface) is proportional to the charge enclosed by the surface:

$$\Phi_E \equiv \oint \vec{E} \cdot d\vec{A} = \frac{q_{\text{enc}}}{\epsilon_0}. \tag{30.10}$$

Figure 30.30*a* shows the electric field created by a charged particle, along with a Gaussian surface enclosing that particle.

Figure 30.30 Graphical representation of the physics behind Maxwell's equations, together with their mathematical expressions. (*a*) Electric field surrounding a charged particle and a Gaussian surface enclosing that particle. Gaussian surfaces can be used to relate the electric field to the enclosed charge. (*b*) Magnetic field surrounding a current-carrying wire and a closed surface intercepted by the wire; the integral of the magnetic field over a closed surface is always zero. (*c*) Electrostatic field and two closed paths through that field; the path integral of the electric field around either path must be zero. (*d*) Steady magnetic field surrounding a current and two closed paths through that field; the path integral of the magnetic field is proportional to the encircled current. (*e*) Changing magnetic field and two closed paths through it; an electric field accompanies the changing magnetic field; the path integral of this electric field around either path is nonzero. (*f*) Changing electric field and two closed paths through it; a magnetic field accompanies the changing electric field; the path integral of this magnetic field around either path is nonzero.

(*a*) Surface integral of electric field (Gauss's law)

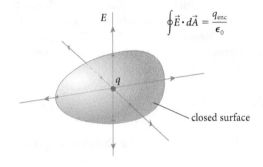

$$\oint \vec{E} \cdot d\vec{A} = \frac{q_{enc}}{\epsilon_0}$$

(*b*) Surface integral of magnetic field (Gauss's law for magnetism)

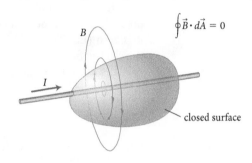

$$\oint \vec{B} \cdot d\vec{A} = 0$$

(*c*) Line integral of constant electric field

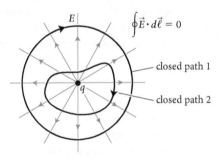

$$\oint \vec{E} \cdot d\vec{\ell} = 0$$

(*d*) Line integral of constant magnetic field (Ampère's law)

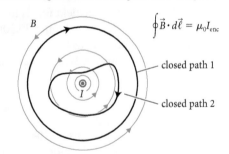

$$\oint \vec{B} \cdot d\vec{\ell} = \mu_0 I_{enc}$$

(*e*) Line integral of changing electric field (Faraday's law)

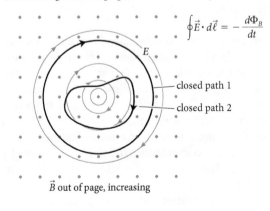

$$\oint \vec{E} \cdot d\vec{\ell} = -\frac{d\Phi_B}{dt}$$

$\vec{B}$ out of page, increasing

(*f*) Line integral of changing magnetic field (Maxwell's displacement current)

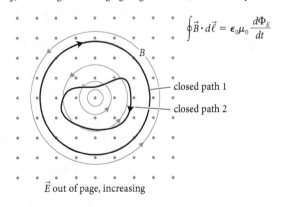

$$\oint \vec{B} \cdot d\vec{\ell} = \epsilon_0 \mu_0 \frac{d\Phi_E}{dt}$$

$\vec{E}$ out of page, increasing

QUANTITATIVE TOOLS

Magnetic fields are generated by moving charged particles, commonly in the form of currents. Unlike electric field lines, magnetic field lines always form loops. There are no isolated magnetic poles, only magnetic dipoles. Consequently, as we showed in Chapter 27, the magnetic flux through any closed surface is always zero (Eq. 27.11):

$$\Phi_B \equiv \oint \vec{B} \cdot d\vec{A} = 0. \tag{30.11}$$

Figure 30.30b shows the magnetic field surrounding a current-carrying wire, along with a closed surface intercepted by the wire.

For electrostatic fields, we showed in Chapter 25 that the path integral of the electric field around a closed path is zero, which means that when a charged object is moved around a closed path in an electrostatic field, the work done on it is zero (Eq. 25.32). This situation is represented in Figure 30.30c. However, the electric field accompanying a changing magnetic field does work on charged particles even when those particles travel around closed paths:

$$\oint \vec{E} \cdot d\vec{\ell} = -\frac{d\Phi_B}{dt}. \tag{30.12}$$

Figure 30.30e shows the electric field lines associated with a changing magnetic field. For such an electric field, no potential can be defined because the path integral of the electric field depends on the path chosen.

Finally, Ampère's law gives the line integral of the magnetic field produced by a current (Figure 30.30d). The magnetic field that accompanies a changing electric field forms loops around the direction of the change in electric field, as shown in Figure 30.30f. Combining these two contributions to the line integral of the magnetic field gives us Maxwell's generalization of Ampère's law, which is our Eq. 30.6, repeated here:

$$\oint \vec{B} \cdot d\vec{\ell} = \mu_0 I_{\text{int}} + \mu_0 \epsilon_0 \frac{d\Phi_E}{dt}. \tag{30.13}$$

Equations 30.10–30.13 are referred to as **Maxwell's equations** because Maxwell not only added the displacement current term to Ampère's law but also recognized the coherence and completeness of this set of equations. Together with conservation of charge, these four equations give a complete description of electromagnetic phenomena. In the presence of matter, these equations have to be modified to account for the effects of matter on electric and magnetic fields (see, for example, Example 30.6).

Maxwell's equations were developed from and subsequently verified by a vast body of experimental evidence. Equation 30.10 (Gauss's law) comes from the measured inverse-square dependence of the electric force on separation distance and the finding that, in the steady state, the interior of a hollow charged conductor carries no surplus charge. Equation 30.11 (Gauss's law for magnetism) states that isolated magnetic monopoles do not exist, and none have been detected to date, in spite of very sensitive experiments conducted to search for them. Equation 30.12, a quantitative statement of Faraday's law, comes from extensive experiments by Faraday and others on electromagnetic induction, and Eq. 30.13, Maxwell's generalization of Ampère's law, comes from measurements of the magnetic force between current-carrying wires and the observed properties of electromagnetic waves.

30.12 Suppose that isolated magnetic monopoles carrying a "magnetic charge" m did exist, and that the interaction between these monopoles depended on $1/r^2$, where r is the distance between two monopoles. How would you modify Maxwell's equations to account for these monopoles? Ignore any physical constants that may need to be added.

Example 30.7 Maxwell's equations in free space

What is the form of Maxwell's equations in a region of space that does not contain any charged particles?

① GETTING STARTED If there are no charged particles, there can be no accumulation of charge and no currents, which means $q_{enc} = 0$ and $I = 0$.

② DEVISE PLAN All I need to do is set q_{enc} and I equal to zero in Eqs. 30.10–30.13.

③ EXECUTE PLAN Setting $q_{enc} = 0$ in Eq. 30.10 and $I = 0$ in Eq. 30.13, I obtain the following form of Maxwell's equations:

$$\oint \vec{E} \cdot d\vec{A} = 0 \qquad (1)$$

$$\oint \vec{B} \cdot d\vec{A} = 0 \qquad (2)$$

$$\oint \vec{E} \cdot d\vec{\ell} = -\frac{d\Phi_B}{dt} \qquad (3)$$

$$\oint \vec{B} \cdot d\vec{\ell} = \mu_0 \epsilon_0 \frac{d\Phi_E}{dt}. \checkmark \qquad (4)$$

④ EVALUATE RESULT Maxwell's equations simplify greatly in the absence of charged particles (the only asymmetry is the sign difference between Eqs. 3 and 4, which comes from Lenz's law). Equations 1 and 2 state that both the electric and magnetic fluxes through a closed surface are zero in the absence of charged particles. Consequently, both electric and magnetic field lines must form loops. Equations 3 and 4 state that electric field line loops accompany changes in magnetic flux and magnetic field line loops accompany changes in electric flux.

✋ **30.13** As you saw in Section 30.3, the magnetic and electric fields in an electromagnetic wave are perpendicular to each other. How do Maxwell's equations in free space (Eqs. 1–4 of Example 30.7) express that perpendicular relationship?

30.6 Electromagnetic waves

From Maxwell's equations, we can derive the fundamental properties of electromagnetic waves. To begin, let's consider an electromagnetic wave pulse that arises from the sudden acceleration of a charged particle, as we discussed in the first part of this chapter (**Figure 30.31**). The magnitude of the electric field in the kinked part of the pulse in Figure 30.31 is essentially uniform and much greater than it is anywhere else. We shall consider the propagation of this wave pulse through a region of space containing no matter and no charged particles, so we can use the form of the Maxwell equations derived in Example 30.7. At great distances from the particle, only the transverse pulse is significant and we can ignore any other contributions to the electric field. This wave pulse is essentially a slab-like region of space that extends infinitely in the x and y directions and has a finite thickness in the z direction. Inside the slab, the electric field is uniform and has magnitude E; outside the slab, $E = 0$. We let the wave pulse move along the z axis (**Figure 30.32**) and denote its speed by c_0 (the subscript 0 indicates that

Figure 30.31 Electric field pattern of an accelerated charged particle. The kinks in the electric field pattern correspond to a transverse electric field pulse propagating away from the particle at speed c_0.

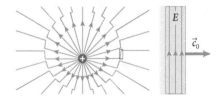

Figure 30.32 Perspective view of a planar electromagnetic wave pulse moving in the z direction. The electric field points in the x direction and has the same magnitude throughout an infinite plane parallel to the xy plane.

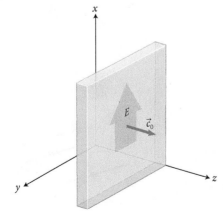

Figure 30.33 Side view of the planar electromagnetic wave pulse of Figure 30.32. The electric field inside the pulse is uniform except at the front and back surfaces, where it drops rapidly to zero.

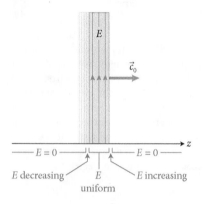

Figure 30.34 Displacement currents corresponding to the upwardly increasing electric field at the front surface and upwardly decreasing electric field at the back surface of the planar electromagnetic wave pulse of Figure 30.32.

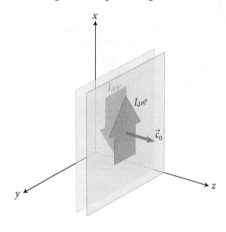

this speed is in vacuum). The magnitude of the electric field depends only on z, not on x and y. The wave pulse is an example of a **planar electromagnetic wave** because of the constant magnitude of the electric field in a plane normal to the direction of propagation.

What magnetic field pattern is associated with the electric field in the planar electromagnetic wave pulse in Figure 30.32? Viewing Figure 30.32 from the side (Figure 30.33), we see that the electric field is zero in front and in back of the pulse, nonzero and uniform inside the pulse, and changing at the front and back surfaces of the pulse. At the front surface, the electric field increases in the upward direction, corresponding to an upward displacement current I_{disp} (Figure 30.34). At the back surface of the pulse, the displacement current points down.

In Checkpoint 28.13b, you determined the magnetic field of two infinite planar sheets of oppositely directed current. The electric field in the planar electromagnetic wave pulse gives a similar arrangement of oppositely directed displacement currents. The magnetic field associated with this current distribution is uniform and points in the $+y$ direction (Figure 30.35). We now see that the planar electromagnetic wave pulse consists of uniform electric and magnetic fields that are perpendicular to each other and to the direction of propagation of the pulse, as we already concluded in Section 30.3. In fact, for a planar electromagnetic wave pulse, there is a right-hand relationship among the directions of $\vec{E}$, $\vec{B}$, and $\vec{c}_0$. If you curl the fingers of your right hand from the direction of $\vec{E}$ to the direction of $\vec{B}$, in Figure 30.35, your thumb points in the direction of propagation of the pulse. (This means that the vector product $\vec{E} \times \vec{B}$ yields a vector pointing in the direction of propagation of the electromagnetic wave pulse.)

To calculate the magnitude of the magnetic field in the planar electromagnetic wave pulse, we can use the version of Eq. 30.13 valid in a region of space that does not contain any charged particles (see Example 30.7):

$$\oint \vec{B} \cdot d\vec{\ell} = \mu_0 \epsilon_0 \frac{d\Phi_E}{dt}. \tag{30.14}$$

Let's begin by evaluating the left side of this equation. To exploit the fact that the magnetic field points in the $+y$ direction in the pulse, we choose the Ampèrian path in Figure 30.35. This rectangular path lies in the yz plane and has width ℓ in the y direction; side ad is inside the pulse and side fg is far off to the right in the positive z direction. We let the direction of the path be such that it coincides with the direction of the magnetic field, so that $\vec{B} \cdot d\vec{\ell} = B \, d\ell$. Only

Figure 30.35 Magnetic field associated with the planar electromagnetic wave pulse of Figure 30.32. The Ampèrian path in the yz plane can be used to calculate the magnitude of the magnetic field.

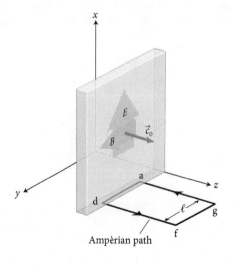

Ampèrian path

side ad of the rectangular path contributes to the line integral; the magnetic field is zero around side fg, and the two long sides are perpendicular to the magnetic field. Thus, the left side of Eq. 30.14 becomes

$$\oint \vec{B} \cdot d\vec{\ell} = B\ell. \tag{30.15}$$

The electric flux through the path is given by $\Phi_E = \vec{E} \cdot \vec{A} = EA$, where $\vec{A}$ is a surface area vector pointing in the $+x$ direction, as dictated by the choice of direction of the integration path (Appendix B). To determine the rate of change of the electric flux through the path, note that the planar electromagnetic wave pulse is moving to the right with speed c_0. Before the front surface of the pulse reaches side ad of the Ampèrian path, the electric flux Φ_E through the path is zero. In a time interval Δt after the front surface of the pulse reaches side ad, the pulse travels a distance $c_0 \Delta t$ into the rectangular path (Figure 30.36), and so at this instant the area over which the electric field is nonzero is $A = \ell c_0 \Delta t$. The electric flux through the path is then $\Phi_E = E\ell c_0 \Delta t$. The change in electric flux through the path during the interval Δt is thus

$$\Delta \Phi_E = E\ell c_0 \Delta t - 0, \tag{30.16}$$

and the rate of change in electric flux is

$$\frac{\Delta \Phi_E}{\Delta t} = E\ell c_0. \tag{30.17}$$

Substituting Eqs. 30.15 and 30.17 into Eq. 30.14 yields

$$B\ell = \mu_0 \epsilon_0 E\ell c_0 \tag{30.18}$$

or, solving for B, $\qquad B = \mu_0 \epsilon_0 E c_0. \tag{30.19}$

We have thus obtained a relationship between the magnitudes of the magnetic and electric fields in the planar electromagnetic wave pulse.

We can now use Faraday's law (Eq. 30.12) to obtain an additional relationship between these two transverse fields:

$$\oint \vec{E} \cdot d\vec{\ell} = -\frac{d\Phi_B}{dt}. \tag{30.20}$$

Let's begin by evaluating the left side of this equation. To exploit the fact that the electric field points in the $+x$ direction in the pulse, we choose the rectangular integration path in Figure 30.37. We let the direction of the path be such that it coincides with the direction of the electric field in the pulse, so that $\vec{E} \cdot d\vec{\ell} = E\, d\ell$. As in our derivation of the magnetic field, the only contribution to the line integral of the electric field around the rectangular path comes from the left side of the path:

$$\oint \vec{E} \cdot d\vec{\ell} = Ew. \tag{30.21}$$

To evaluate the right side of Eq. 30.20, we note that the geometry of this situation is the same as in our treatment of the electric pulse, except that now $\Phi_B = \vec{B} \cdot \vec{A} = -BA$, where $\vec{A}$ is a surface area vector pointing in the $-y$ direction,

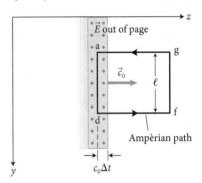

Figure 30.36 Top view of moving planar electromagnetic wave pulse of Figures 30.32–30.35, showing motion of the pulse through the Ampèrian path.

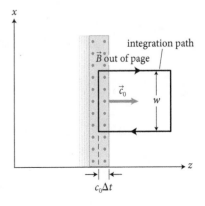

Figure 30.37 Top view of the planar electromagnetic wave pulse of Figures 30.32–30.35, showing motion of the pulse through an integration path lying in the xz plane.

as dictated by the choice of direction of the integration path (see Appendix B). In analogy to Eqs. 30.16 and 30.17, the rate at which the magnetic flux through the path changes is then given by

$$\frac{\Delta \Phi_B}{\Delta t} = -Bwc_0. \tag{30.22}$$

Substituting Eqs. 30.21 and 30.22 into Eq. 30.20 gives us

$$Ew = Bwc_0 \tag{30.23}$$

or

$$B = \frac{E}{c_0}. \tag{30.24}$$

We now have two different relationships between the magnitudes of the electric and magnetic fields: Eq. 30.19 comes from Maxwell's generalization of Ampère's law, and Eq. 30.24 comes from Faraday's law. Setting the right sides of these two equations equal, we get

$$\frac{E}{c_0} = \mu_0 \epsilon_0 \, E c_0. \tag{30.25}$$

This result implies that the speed of the planar electromagnetic wave pulse is

$$c_0 = \frac{1}{\sqrt{\epsilon_0 \mu_0}}. \tag{30.26}$$

Equation 30.26 tells us something surprising: The speed of the planar electromagnetic wave pulse in vacuum is determined by two fundamental constants, ϵ_0 and μ_0. The first, ϵ_0, is introduced in Coulomb's law (see Eqs. 22.1 and 24.7). The second, μ_0 (see Eq. 28.1), is set by the definition of the ampere. In 1862, when Maxwell first worked out the relationship expressed in Eq. 30.26, no one knew that light and electromagnetic waves were related. To evaluate c_0 in Eq. 30.26, Maxwell used the results of experiments made with electric circuits and obtained a value of $c_0 = 3 \times 10^8$ m/s, in excellent agreement with values obtained for the speed of light in vacuum. This agreement led Maxwell to the remarkable conclusion that light is an electromagnetic wave.

Nowadays, the speed of light is set to be exactly 299,792,458 m/s (see Section 1.3) to define the meter. Likewise μ_0 is set by the definition of the ampere (see Eq. 28.1). The value of $\epsilon_0 = 1/(c_0^2 \mu_0)$ as given in Eq. 24.7 is therefore also fixed.

Most electromagnetic waves have a more complex shape than the planar electromagnetic wave pulse we have used to arrive at Eq. 30.26. Through the superposition principle (see Section 16.3), however, we can superpose any number of planar electromagnetic wave pulses to obtain whatever planar electromagnetic wave shape interests us. The central property of these electromagnetic waves does not depend on shape: The electromagnetic wave pulse consists of electric and magnetic fields that are perpendicular to each other and to the direction of propagation of the pulse. The ratio of the magnitudes of the electric and magnetic fields is always given by Eq. 30.24. The field vectors $\vec{E}$ and $\vec{B}$ are always perpendicular, and they always travel at speed c_0 in a direction given by the vector product $\vec{E} \times \vec{B}$.

Mathematically, it is more convenient to build arbitrary wave shapes out of harmonic (sinusoidal) waves than out of rectangular wave pulses. **Figure 30.38** shows a planar electromagnetic wave for which the electric field varies sinusoidally in space. The field vectors for the electric field are shown embedded in rectangular slabs to emphasize that the electric field has the same magnitude

Figure 30.38 Perspective view of the electric field of a sinusoidal planar electromagnetic wave propagating in the z direction. The electric field vectors are embedded in rectangular slabs to emphasize that the electric field has the same magnitude everywhere in the plane of the slab. The magnitude of the electric field does vary from plane to plane along the z axis. The magnetic field (not shown) is uniform on planes parallel to the xy plane.

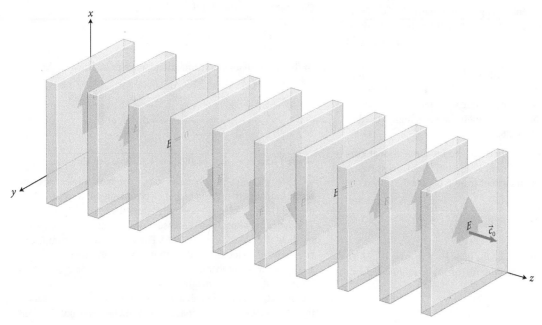

everywhere throughout the plane of a slab, not just on the z axis. The magnitude of this electric field depends only on z, not on x and y.

As we saw in Chapter 16, harmonic waves are characterized by a propagation speed c, a frequency f, and a wavelength λ, and these quantities are related by $c = f\lambda$. The remarkable thing about electromagnetic waves is that waves of all frequencies travel at the same constant speed c_0 in vacuum. Consequently, in vacuum, frequency and wavelength are inversely proportional to one another over a vast range of values.

Figure 30.39 shows the classification of electromagnetic waves as a function of wavelength and frequency. Extending over a span of nearly 20 orders of magnitude, the figure shows electromagnetic waves ranging from radio waves to gamma rays. Only a very small part of this range corresponds to what we are familiar with as "light." Our eyes are most sensitive to wavelengths between 430 nm and 690 nm, though we can see light somewhat outside this wavelength range if the light is sufficiently intense. However, waves outside the visible range are governed by exactly the same physics as visible light.

As we shall explore in more detail in Chapter 33, the frequency of an electromagnetic wave determines how the wave interacts with materials.

Figure 30.39 Classification of electromagnetic radiation as a function of frequency (top scale) and wavelength (bottom scale).

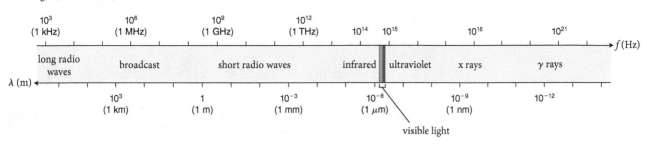

Example 30.8 Speed of light in a dielectric

At what speed does an electromagnetic wave pulse propagate through a dielectric for which the dielectric constant is κ?

1 GETTING STARTED In the presence of a dielectric, both Gauss's law and the displacement current are modified (Eq. 26.25 and Example 30.6, respectively). These changes affect the derivation I used to obtain the speed of electromagnetic waves in vacuum, Eq. 30.26.

2 DEVISE PLAN The modification of the displacement current by the dielectric changes Eq. 30.14. To obtain an expression for the speed at which the electromagnetic wave pulse propagates, I carry the modified expression through the same logic I used to go from Eq. 30.14 to Eq. 30.26.

3 EXECUTE PLAN Because there can be no conventional currents through a dielectric, I in Eq. 30.13 is still zero. Substituting the modified displacement current from Example 30.6, I get

$$\oint \vec{B} \cdot d\vec{\ell} = \mu_0 \epsilon_0 \kappa \frac{d\Phi_E}{dt}.$$

Carrying this expression through the logic leading from Eq. 30.14 to Eq. 30.19, I obtain the magnitude of the magnetic field in terms of the electric field:

$$B = \mu_0 \epsilon_0 \kappa \, Ec, \tag{1}$$

where the κ comes from the modification of the displacement current and where I have written c for the speed in the dielectric rather than c_0, our symbol for the speed in vacuum.

Because Faraday's law is unaffected by the presence of the dielectric, the only modification required in Eq. 30.24 is replacing c_0 with c. Solving Eq. 30.24 for E and substituting that expression for E into Eq. 1, I get

$$B = \mu_0 \epsilon_0 \kappa \, Bc^2,$$

so the speed of an electromagnetic wave pulse moving through a dielectric is

$$c = \frac{1}{\sqrt{\mu_0 \epsilon_0 \kappa}} = \frac{c_0}{\sqrt{\kappa}}, ✔$$

where I have used Eq. 30.26, $c_0 = 1/\sqrt{\epsilon_0 \mu_0}$, to simplify.

4 EVALUATE RESULT Because $\kappa > 1$ for most dielectrics, my result indicates that the speed of electromagnetic waves (including light) in a dielectric is smaller than their speed in vacuum. The dielectric constant is a measure of the reduction of the electric field (Eq. 26.15, $\kappa = E_{\text{free}}/E$). For a material that completely attenuates the electric field so that $E = 0$—in other words, for a conductor—κ is infinite and so my result yields $c = 0$. This means that an electromagnetic wave cannot propagate in such a material, a conclusion that agrees with the familiar observation that a conductor, such as a slab of metal, does not transmit visible light.

30.14 An electromagnetic wave with a wavelength of 600 nm in vacuum enters a dielectric for which $\kappa = 1.30$. What are the frequency and wavelength of the wave inside the dielectric?

30.7 Electromagnetic energy

Because electric and magnetic fields contain energy, energy is transported as an electromagnetic wave travels away from its source. Let us work out how much energy is transported by a planar electromagnetic wave. In Section 26.6, we calculated the energy density contained in an electric field in vacuum (Eq. 26.6):

$$u_E = \tfrac{1}{2}\epsilon_0 E^2. \tag{30.27}$$

Similarly, the energy density in a magnetic field in vacuum is given by Eq. 29.29:

$$u_B = \tfrac{1}{2}\frac{B^2}{\mu_0}. \tag{30.28}$$

The energy density in a combined electric and magnetic field is therefore

$$u = \tfrac{1}{2}\epsilon_0 E^2 + \tfrac{1}{2}\frac{B^2}{\mu_0}. \tag{30.29}$$

Because the magnitudes of the electric and magnetic fields in an electromagnetic wave are related by Eq. 30.24, we can rewrite Eq. 30.29 in terms of just the magnitude of the electric field. Using Eqs. 30.24 and 30.26, we get

$$u = \tfrac{1}{2}\epsilon_0 E^2 + \tfrac{1}{2}\frac{E^2}{c_0^2 \mu_0} = \tfrac{1}{2}\epsilon_0 E^2 + \tfrac{1}{2}\epsilon_0 E^2 = \epsilon_0 E^2. \qquad (30.30)$$

Comparing Eqs. 30.29 and 30.30, we see that in vacuum the electric and magnetic fields each contribute half of the energy density—the electric and magnetic energy densities are equal.

Alternatively, the energy density can be written in terms of only the magnitude of the magnetic field,

$$u = \frac{B^2}{\mu_0}, \qquad (30.31)$$

or in terms of the magnitudes of both the electric and the magnetic fields,

$$u = \sqrt{\frac{\epsilon_0}{\mu_0}}\, EB. \qquad (30.32)$$

Let us now calculate the rate at which energy flows through a certain area in an electromagnetic wave. Consider taking a slice of an electromagnetic wave normal to the direction of propagation (Figure 30.40). The slice has thickness dz and area A. The energy dU in this slice is the product of the energy density and the volume of the slice:

$$dU = uA\,dz. \qquad (30.33)$$

From Section 17.5, you know that the intensity S of a wave is defined as the energy flow per unit time (the power) across a unit area perpendicular to the direction of wave propagation.* Using Eq. 30.33, we can express this relationship in the form

$$S = \frac{1}{A}\frac{dU}{dt}. \qquad (30.34)$$

To determine the intensity of an electromagnetic wave, we substitute the expression for dU from Eq. 30.33 and recall that the wave travels at speed $dz/dt = c_0$. We can then rewrite Eq. 30.34 as

$$S = \frac{1}{A} uA \frac{dz}{dt} = uc_0 \qquad (30.35)$$

or, after substituting Eqs. 30.32 and 30.26,

$$S = \frac{1}{\mu_0} EB. \qquad (30.36)$$

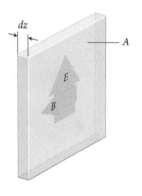

Figure 30.40 Perspective view of a slice of thickness dz of an electromagnetic wave, taken normal to the direction of propagation.

---

*In Chapter 17, we used I for intensity. That is fine in a discussion of mechanical waves, but now that we must deal with current I so frequently, we switch to the symbol S for intensity.

As the electromagnetic wave travels, energy travels with it in the direction of wave propagation. As we saw in Section 30.6, the propagation direction is the same as that of the vector product $\vec{E} \times \vec{B}$. So with Eq. 30.36 we can define a vector that fully describes energy flow in the electromagnetic wave:

$$\vec{S} \equiv \frac{1}{\mu_0} \vec{E} \times \vec{B}. \qquad (30.37)$$

This vector $\vec{S}$ is called the **Poynting vector,** after J. H. Poynting (1852–1914), the physicist who first defined this vector. The SI units of the Poynting vector are W/m^2 (Checkpoint 30.16). The Poynting vector represents the flow of energy in any combined electric and magnetic field, not just electromagnetic waves (Checkpoint 30.15), and the direction of $\vec{S}$ is the direction of energy flow.

When we describe electromagnetic waves, the magnitude S of the Poynting vector is called the *intensity of the electromagnetic wave* and, as noted in Eq. 30.34, is the instantaneous electromagnetic power (energy per unit time) crossing a unit area. For electromagnetic waves, this area is perpendicular to the direction of the vector product $\vec{E} \times \vec{B}$. To obtain the power P crossing a surface, we integrate the Poynting vector over the surface:

$$P = \int_{\text{surface}} \vec{S} \cdot d\vec{A}. \qquad (30.38)$$

The results of this section give expressions for the *instantaneous* values of electromagnetic energy density (Eqs. 30.30–32), intensity (Eq. 30.36), and power (Eq. 30.38). Often, however, it is useful to consider average values over some time interval. Because the average of the square of a sine function is $1/2$ (see Appendix B), the average value of $\sin^2 \omega t$ is $1/2$. So, although sinusoidally oscillating electric and magnetic fields average to zero, their squares average to $1/2$ of the square of their amplitudes: $(E^2)_{\text{av}} = \frac{1}{2}E_{\text{max}}^2$. The square roots of these average values are often referred to as the *root-mean-square values*, or rms values (see Eq. 19.21), and are represented by E_{rms} and B_{rms}:

$$E_{\text{rms}} \equiv \sqrt{(E^2)_{\text{av}}} \quad \text{and} \quad B_{\text{rms}} \equiv \sqrt{(B^2)_{\text{av}}}. \qquad (30.39)$$

For sinusoidal electromagnetic waves, energy density, intensity, and power are proportional to the squares of a sine function, and so the average values of these quantities are related to the rms values of the fields in the same manner that the instantaneous values of energy density, intensity, and power are related to the instantaneous values of the fields. For example, Eq. 30.36 yields

$$S_{\text{av}} = \frac{1}{\mu_0} E_{\text{rms}} B_{\text{rms}} \quad \text{(sinusoidal electromagnetic wave).} \qquad (30.40)$$

Example 30.9 Tanning fields

The average intensity S of the Sun's radiation at Earth's surface is approximately 1.0 kW/m^2. Assuming sinusoidal electromagnetic waves, what are the root-mean-square values of the electric and magnetic fields?

❶ GETTING STARTED The electric and magnetic fields in an electromagnetic wave each contribute half of the energy density of the wave (Eq. 30.29). That energy density is related to the wave's Poynting vector $\vec{S}$, whose magnitude S is the intensity of an electromagnetic wave.

❷ DEVISE PLAN I can rewrite Eq. 30.35 to obtain the average energy density u_{av} in terms of the average intensity S_{av}. I can then use Eqs. 30.30 and 30.31 to relate the average energy density to the rms values of the electric and magnetic fields.

❸ EXECUTE PLAN From Eq. 30.35, I have

$$u_{\text{av}} = \frac{S_{\text{av}}}{c_0} = \frac{1.0 \times 10^3 \text{ W/m}^2}{3.0 \times 10^8 \text{ m/s}} = 3.3 \times 10^{-6} \text{ J/m}^3.$$

To obtain the value of E_{rms}, I express the relationship between u_{av} and E_{rms} by analogy to Eq. 30.30:

$$u_{av} = \epsilon_0 E_{rms}^2$$

$$E_{rms} = \sqrt{\frac{u_{av}}{\epsilon_0}} = \sqrt{\frac{3.3 \times 10^{-6} \text{ J/m}^3}{8.85 \times 10^{-12} \text{ C}^2/\text{N} \cdot \text{m}^2}} = 6.1 \times 10^2 \text{ V/m}, ✔$$

where I have used the fact that $1 \text{ N/C} = 1 \text{ V/m}$.

For sinusoidal electromagnetic waves, the relationship between instantaneous values of the electric and magnetic field magnitudes holds for the rms values of the field magnitudes, so I can use Eq. 30.24 to determine the value of B_{rms}:

$$B_{rms} = \frac{E_{rms}}{c_0} = \frac{6.1 \times 10^2 \text{ N/C}}{3.0 \times 10^8 \text{ m/s}} = 2.0 \times 10^{-6} \text{ T}. ✔$$

④ **EVALUATE RESULT** As I expect based on the large value of c_0, the value of B_{rms} in teslas is much smaller than that of E_{rms} in volts per meter. This electric field magnitude is also not large, particularly compared with the electric fields often obtained in capacitors. (Applying a potential difference of just 1 V to the capacitor in the next example would produce an electric field magnitude of 10,000 V/m!) However, this electric field is greater than the typical atmospheric electric field magnitude of roughly 100 V/m (which is due to charged particles in the atmosphere rather than to the Sun's radiation).

✋ **30.15** Consider supplying a constant current to a parallel-plate capacitor in which the plates are circular. While the capacitor is charging, what is the direction of the Poynting vector at points that lie on the cylindrical surface surrounding the space between the capacitor plates? What does this Poynting vector represent?

Example 30.10 Capacitor power

A parallel-plate capacitor with circular plates of radius $R = 0.10$ m and separation distance $d = 0.10$ mm is charged by a constant current of 1.0 A. (*a*) What is the magnitude of the Poynting vector associated with the electric and magnetic fields at the edge of the space between the plates? (*b*) What is the rate at which electromagnetic energy is delivered to the cylindrical space between the plates?

① **GETTING STARTED** While the capacitor is charging, the charge on the plates and consequently the electric field between them and the energy stored in the capacitor are changing with time. The changing electric field is accompanied by a magnetic field. The electric field between the plates is uniform, and the magnetic field lines form circular loops centered on the horizontal axis of the capacitor, in a direction given by the right-hand current rule. At each point in the space between the plates, the electric and magnetic fields are perpendicular to each other, as shown in my sketch in **Figure 30.41**.

Figure 30.41

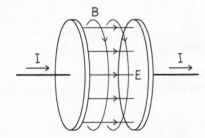

② **DEVISE PLAN** Because the electric and magnetic fields are perpendicular to each other, I can use Eq. 30.36 to calculate the magnitude of the Poynting vector. For this equation, I must

know E and B. I calculated the magnitude of the electric field between two capacitor plates in Example 26.2, and from Checkpoint 30.11*a* I have an expression for the magnitude of the magnetic field between the plates.

To obtain the rate at which electromagnetic energy is delivered to the capacitor, I must integrate the Poynting vector over the cylindrical surface surrounding the space between the plates, as given in Eq. 30.38. From Checkpoint 30.15, I know that the Poynting vector points radially inward toward the capacitor axis, and I also know that at any given instant, the magnitudes of the electric and magnetic fields do not vary on the cylindrical surface surrounding the space between the plates. To integrate the Poynting vector over that surface, I therefore need to multiply the expression for the magnitude of the Poynting vector at any point on the surface by the area of the surface.

③ **EXECUTE PLAN** (*a*) From Example 26.2 I know that the electric field between the capacitor plates is $E = q/(\epsilon_0 A)$, where $A = \pi R^2$ is the area of each plate. Because a constant current is supplied to the capacitor while the capacitor is charging, $q = I\Delta t$ (Eq. 27.1, where Δt is the time interval that has elapsed since the capacitor began charging), I can therefore rewrite my expression for E in the form

$$E = \frac{I\Delta t}{\epsilon_0 A} = \frac{I\Delta t}{\pi \epsilon_0 R^2}.$$

From the solution to Checkpoint 30.11*a* I have $B = \mu_0 Ir/2\pi R^2$, where r is the radial distance from the axis of the capacitor to the position where the magnetic field is measured and R is the radius of each plate. For this case, $r = R$, so

$$B = \frac{\mu_0 I}{2\pi R}.$$

(continued)

To obtain the magnitude of the Poynting vector, I substitute the above values for E and B into Eq. 30.36:

$$S = \frac{EB}{\mu_0} = \frac{I^2 \Delta t}{2\pi^2 \epsilon_0 R^3} = \frac{(1.0 \text{ A})^2 \Delta t}{2\pi^2 [8.85 \times 10^{-12} \text{ C}^2/(\text{N} \cdot \text{m}^2)](0.10 \text{ m})^3}$$

$$= [5.7 \times 10^{12} \text{ W}/(\text{m}^2 \cdot \text{s})]\Delta t. \text{ ✔}$$

(b) To calculate the rate at which electromagnetic energy is delivered to the cylindrical space between the capacitor plates, I multiply the expression for S by the area of the surface, which is the product of the circumference of the cylinder ($2\pi R$) and the height of the cylinder, which is the plate separation distance d:

$$P = 2\pi R d \frac{I^2 \Delta t}{2\pi^2 \epsilon_0 R^3} = \frac{I^2 d \Delta t}{\pi \epsilon_0 R^2}$$

$$= \frac{(1.0 \text{ A})^2 (1.0 \times 10^{-4} \text{ m}) \Delta t}{\pi [8.85 \times 10^{-12} \text{ C}^2/(\text{N} \cdot \text{m}^2)](0.10 \text{ m})^2}$$

$$= (3.6 \times 10^8 \text{ W/s})\Delta t. \text{ ✔}$$

❹ **EVALUATE RESULT** I have no way of gauging the reasonableness of my numerical results, but I can check the reasonableness of my expression for the power (which incorporates my result for part a) by using an alternative method to obtain that expression. The rate at which energy is delivered to the capacitor is the power delivered to it by the current. I know from Eq. 26.2 that the electric potential energy of the capacitor changes by an amount $dU^E = V_{\text{cap}}dq$ during charging. To get the rate at which the amount of energy stored in the capacitor changes with time, I take the derivative

$$P \equiv \frac{dU^E}{dt} = V_{\text{cap}}\frac{dq}{dt} = V_{\text{cap}}I = \frac{qI}{C} = \frac{I^2 \Delta t}{C}. \qquad (1)$$

For a parallel-plate capacitor, $C = \epsilon_0 A/d$, so with $A = \pi R^2$,

$$P = \frac{I^2 d \Delta t}{\epsilon_0 \pi R^2},$$

which is the same result I obtained for part b.

 30.16 Use Eq. 30.36 to show that the SI units of the Poynting vector are W/m^2.

Chapter Glossary

SI units of physical quantities are given in parentheses.

Displacement current (A) A current-like quantity that contributes to Ampère's law caused by a changing electric flux:

$$I_{\text{disp}} \equiv \epsilon_0 \frac{d\Phi_E}{dt}. \qquad (30.7)$$

Electromagnetic wave A wave disturbance that consists of combined electric and magnetic fields. In vacuum and dielectrics, the electric and magnetic fields in an electromagnetic wave are always perpendicular to each other and the direction of propagation is given by the vector product $\vec{E} \times \vec{B}$. In vacuum, electromagnetic waves always propagate at speed

$$c_0 = \frac{1}{\sqrt{\epsilon_0 \mu_0}}. \qquad (30.26)$$

Maxwell's equations Equations that together provide a complete description of electric and magnetic fields:

$$\Phi_E \equiv \oint \vec{E} \cdot d\vec{A} = \frac{q_{\text{enc}}}{\epsilon_0} \qquad (30.10)$$

$$\Phi_B \equiv \oint \vec{B} \cdot d\vec{A} = 0 \qquad (30.11)$$

$$\oint \vec{E} \cdot d\vec{\ell} = -\frac{d\Phi_B}{dt} \qquad (30.12)$$

$$\oint \vec{B} \cdot d\vec{\ell} = \mu_0 I + \mu_0 \epsilon_0 \frac{d\Phi_E}{dt}. \qquad (30.13)$$

Planar electromagnetic wave An electromagnetic wave in which the wavefronts (the surfaces of constant phase) are planes perpendicular to the direction of propagation. On these planes, the instantaneous magnitudes of the electric and magnetic fields both are uniform.

Polarization The property of an electromagnetic wave that describes the orientation of its electric field.

Poynting vector $\vec{S}$ (W/m^2) The vector that is associated with the flow of energy in electric and magnetic fields:

$$\vec{S} \equiv \frac{1}{\mu_0}\vec{E} \times \vec{B}. \qquad (30.37)$$

31

Electric Circuits

CONCEPTS

QUANTITATIVE TOOLS

Electric circuits surround us. We light our work and living spaces, start our cars, communicate with each other, and cook our food with electric circuits. Almost everything that consumes energy in our offices and homes is powered by electricity, whether by batteries or by the electric circuitry of the building.

Electrical devices are ubiquitous because electric circuits offer an extremely versatile means of producing and distributing electrical energy for a variety of tasks. Circuits are designed to control the flow of charge carriers—the current—in a device. The current then provides energy to the device, allowing it to perform its function.

In this chapter we explore the basic principles of electric circuits powered by sources of electric potential energy that maintain a constant potential difference, such as batteries. Such circuits are known as *direct-current circuits* or *DC circuits*. In the next chapter, we'll learn about *alternating-current (AC)* circuits, which run on time-varying potential differences, such as the electricity delivered to buildings by means of electrical power lines.

31.1 The basic circuit

One familiar way to make electricity do something useful is to connect a battery to a light bulb. To do so, we connect each terminal of the battery to one of the two contacts on the light bulb (**Figure 31.1a**). On standard light bulbs, the two contacts are the threaded metal casing and a metallic "foot" that is separated from the casing by an insulator (Figure 31.1*b*). Inside the bulb, each of these contacts is connected by a metal wire to one end of a tungsten *filament*, a very thin wire of tungsten wound in a tight coil. When connected as shown in Figure 31.1, the wires, contacts, and filament form a continuous conducting path from one terminal of the battery to the other, and charge carriers flow through the filament. The current in the filament causes the filament to become white hot and glow. (Tungsten is used for light bulb filaments because its very high melting temperature allows it to glow and not melt.)

Figure 31.1 A light bulb connected to a battery.

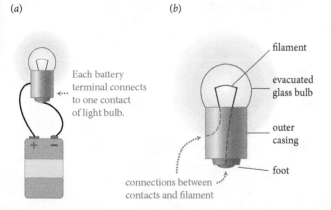

(a)   *(b)*

Each battery terminal connects to one contact of light bulb.

filament

evacuated glass bulb

outer casing

foot

connections between contacts and filament

Experiments with light bulbs show that:

1. A bulb doesn't glow—not even briefly—if only one contact is connected to a battery terminal.

2. In order for a bulb to glow, both contacts must be connected through a continuous conducting path to the terminals of a battery. If the path is broken, the bulb goes out. It doesn't matter where in the path the break occurs.

3. A glowing light bulb generates light and thermal energy.

4. The wires connecting a light bulb to a battery usually do not get hot.

5. A light bulb left connected to a battery over a long time interval eventually goes out because the battery runs down. Once this happens, not enough chemical energy remains in the battery to maintain a large enough potential difference to light the bulb.

In the rest of this chapter we explore the reasons for these observations. The first two observations can be summarized as follows: In order for the light bulb to glow, one of the bulb's contacts must be connected to the positive terminal of the battery and the bulb's other contact must be connected to the negative terminal of the battery. Because the battery maintains a potential difference between its terminals, charge carriers move from one terminal toward the other when a conducting path is provided between the terminals.

The arrangement shown in Figure 31.1 is an example of an **electric circuit**—an interconnection of electrical components (called *circuit elements*). Any closed conducting path through the circuit is called a **loop**. We shall first study electric circuits that have a single loop.

✋ **31.1** (*a*) Consider the system comprising the single-loop circuit shown in Figure 31.1*a*, including the light and thermal energy generated by the bulb. (*a*) Is this system closed? (*b*) Is the energy of the system constant? (*c*) Where do the light energy and the thermal energy come from?

Checkpoint 31.1 demonstrates an important feature of circuits: Energy is converted from electric potential energy to some other form. With this in mind, we can draw a general single-loop circuit (**Figure 31.2a**) consisting of a power source, a load, and wires that connect the load to the source. The **power source** provides electric potential energy to the rest of the circuit, usually by converting some form of energy to electric potential energy. The potential difference across the terminals of the power source drives a current in the circuit. The **load** in an electric circuit is all the circuit elements connected to the power source. In the load, the electric potential energy of the moving charge carriers is converted to other forms of energy, such as thermal or mechanical energy. The wires connecting the elements in a circuit are considered to be ideal; that is,

Figure 31.2 The energy conversions in a single-loop circuit and in a mechanical analog of the circuit.

(a) Schematic simple circuit

(b) Mechanical analog of circuit

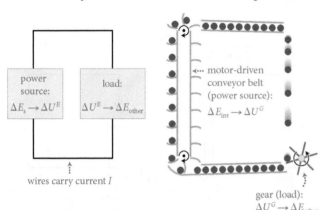

the wires serve to transport electric potential energy to the load, and we ignore the small amount of energy dissipated in the wires.

Figure 31.2b shows a mechanical analog of a single-loop electric circuit. The motor-driven conveyor belt is analogous to the power source in the electric circuit. The conveyor belt lifts the balls, converting source energy to gravitational potential energy. The balls then roll through a horizontal tube, losing only a tiny amount of gravitational potential energy in the process, in the same manner that charge carriers flow with very little loss of energy through wires. Next, the balls drop from the tube onto the gear, which corresponds to the load of an electric circuit. As the balls fall and make the gear turn, their gravitational potential energy is converted to mechanical energy. Finally, the balls roll down a second tube to the bottom of the conveyor belt, completing the cycle.

Notice two things about the mechanical circuit of Figure 31.2b. First, because the system is filled with balls, the balls have to move through the circuit one after the other. If the right end of the upper tube were blocked rather than open (the equivalent of the connecting wire in Figure 31.2a being disconnected from the load), no more balls could be pushed into it from the conveyor belt. There has to be a closed path through the system in order for the balls to move, just as charge carriers can flow through a circuit only when there is an unbroken conducting path.

The second thing to notice in Figure 31.2b is that there is no net transport of balls from the power source to the load. There are as many balls traveling from the power source to the load as the other way around. The balls travel around a closed path through the system, and if we watched one particular ball long enough, we would see it circle repeatedly around the circuit. The balls are simply vehicles transporting energy through the system.

We can therefore propose an operational description of a single-loop DC circuit in terms of energy conversion:

In a single-loop DC circuit, electric potential energy acquired by the carriers in the power source is converted to another form of energy in the load.

The power source in such an electric circuit doesn't have to be a battery—it can be anything that produces a constant potential difference, thereby driving a current around the circuit, such as a solar cell or an electric generator. Likewise, the load doesn't have to be a light bulb. A toaster acting as a circuit load converts electric potential energy to thermal energy, a loudspeaker acting as a load converts electric potential energy to mechanical energy in sound waves, and a motor load converts electric potential energy to mechanical energy.

Exercise 31.1 Solar fan

A solar cell, which converts solar energy (a form of source energy, see Section 7.4) to electric potential energy, is connected to a small fan. Represent this circuit by a diagram analogous to Figure 31.2a and describe the energy conversions taking place in this circuit.

SOLUTION The solar cell is the power source, and the fan is the load. Figure 31.3 shows my diagram for this circuit. The solar cell converts solar energy to electric potential energy, and the fan converts electric potential energy to another form of energy by setting the air in motion. I add these conversions to my diagram. ✔

Figure 31.3

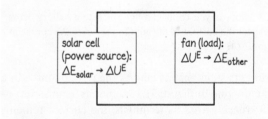

The various parts of electric circuits are commonly represented by graphical symbols rather than by such words as *power source* and *load*. Figure 31.4 shows the symbols for some common circuit elements. In Figure 31.5 these symbols are used to represent the bulb-and-battery circuit

Figure 31.4 Standard representations of common elements of electric circuits.

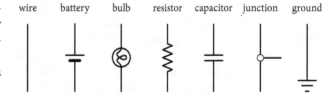

Figure 31.5 Circuit diagram of bulb-and-battery circuit of Figure 31.1.

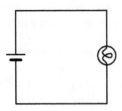

of Figure 31.1. Note that in the battery symbol the short, heavy line represents the negative terminal and the longer, thin line represents the positive terminal. A schematic representation of a circuit using the standard symbols shown in Figure 31.4 is called a **circuit diagram.**

✋ **31.2** Two wires connect the plates of a charged capacitor to the contacts of a light bulb. (*a*) Does this assembly constitute a circuit? If so, identify the power source and the load. (*b*) Does the bulb glow? (*c*) What energy conversions take place after the bulb and capacitor are connected?

31.2 Current and resistance

Why does the filament of a light bulb get hot and glow, but the wires connecting the bulb to the battery do not? And why does the battery run out of energy and the bulb eventually stop glowing? To answer these questions, we need to look at the motion of charge carriers through an electric circuit.

✋ **31.3** Suppose you connect a light bulb to a battery. How do you expect the current in the bulb to vary over the course of (*a*) a minute, (*b*) a few days?

Experiments show that as long as the power source in a circuit like the one in Figure 31.1 maintains a constant potential difference across its terminals, the current remains constant. Given that a battery connected to a bulb can maintain a constant potential difference for hours, the current in a circuit like the one in Figure 31.1 is constant over that same time interval. This current is established almost instantaneously after the circuit is completed (that is, after all the circuit elements are connected together), and it vanishes almost instantaneously when the circuit is broken. Other than the instants after the circuit is completed or broken, therefore, we have a **steady state** with constant current.

As we found in Section 25.3, positively charged particles tend to move toward regions of lower potential and negatively charged particles tends to move toward regions of higher potential. Indeed, this is why charge carriers flow through a closed circuit. For metal conductors, the mobile charge carriers are electrons; when a metallic conducting path is provided, electrons flow from the negative terminal of the power source to the positive terminal. In materials in which the mobile charge carriers are positively charged, the carriers travel in the opposite direction—from the positive

Figure 31.6 Because charge is conserved, in steady state it doesn't accumulate in the load or in any other part of the circuit. Hence, in steady state the current into any part of the circuit must be the same as the current out of that part.

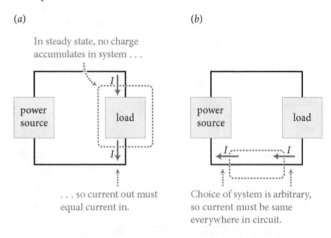

terminal of the power source to the negative terminal. The rate at which the charges on the carriers cross a section in the conductor is the current, and the sign of the carriers and the direction of their motion determine the direction of the current (see Sections 27.3 and 27.5). Remember, however, that for a given potential difference between two points, the current direction is from points of higher potential to points of lower potential regardless of the sign of the mobile charge carriers (see Section 27.3).

In steady state we can draw an important conclusion regarding currents in circuits. Consider, for example, the system comprising just the load in the circuit shown in Figure 31.6a. Because charge is conserved (see Section 22.3), the amount of charge inside this system can only change due to a flow of charge into or out of the system. In steady state the charge of the system is constant, and so the flow of charge carriers into the system must be equal to the flow of charge carriers out of it. Put differently, the current into the system must equal the current out of the system.

Because we can choose any system we like—as in Figure 31.6b, for example—we conclude:

In a steady state, the current is the same at all locations along a single-loop electric circuit.

We shall refer to this requirement as the **current continuity principle.** Figure 31.7 illustrates the principle for balls flowing through a tube. Because the tube is filled, if the flow of balls through the tube is steady, there is nowhere for the balls to pile up without changing the rate of flow. As a result, when one ball is pushed into the left end of the tube, one ball must come out at the right end. Likewise, if two balls are pushed into the left end at the same instant, two must come out at the right end.

The current continuity principle tells us something else about the operation of circuits: As electrons flow through a light bulb, they are not accumulating, or "used up," in the bulb. For every electron that goes into one end of the

Figure 31.7 The continuity principle.

If flow of balls through tube is steady:

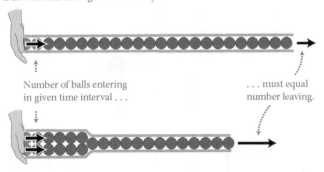

Number of balls entering in given time interval . . .

. . . must equal number leaving.

filament, one electron comes out the other end. What then is consumed (used up) by the bulb as it glows? In our discussion of the mechanical circuit of Figure 31.2*b*, we found that the balls moving through the circuit are not consumed anywhere. Instead, they act to transfer energy from the power source to the load. Before we begin our examination of the energy of the electrons in the electric circuit of Figure 31.2*a*, answer the next checkpoint.

31.4 Does an electron lose or gain electric potential energy (*a*) while moving inside a battery from the positive terminal to the negative terminal and (*b*) while moving through the rest of the circuit from the negative battery terminal to the positive terminal? (*c*) While flowing through the wire and load portions of the circuit, where do the electrons lose most of their energy?

As we saw in Section 26.4, *inside* a battery, chemical energy is converted to electric potential energy, and electrons at the negative terminal have greater potential energy than those at the positive terminal. In the rest of an electric circuit, electrons lose that same amount of electric potential energy as they move from the negative terminal through the load to the positive terminal, and this electric potential energy is converted to other forms of energy.

For a charged particle moving between two locations through the load, the difference in potential energy from one location to the other is equal to the potential difference between those two locations multiplied by the charge of the particle. Therefore, for each electron that moves through the load of a circuit, an amount of electric potential energy eV_{load} is converted to some other form of energy (light, say, or thermal energy, or mechanical energy), where V_{load} is the magnitude of the potential difference across the load.*

Thus, changes in the potential difference in an electric circuit show us where energy conversion is taking place. Experiments indicate that V_{load} is essentially equal to V_{batt}, while V_{wire} is negligible. In other words, essentially all of the energy conversion in the circuit takes place in the load and the source, and almost none takes place in the wires.

---

*When the subscript of potential refers to a device or circuit element (say, V_{device}), the symbol is taken to denote the *magnitude* of the potential difference across the device or circuit element.

Because we can ignore the energy dissipation that takes place in the wires:

> **Every point on any given wire is essentially at the same potential.**

Therefore we can consider the load in a circuit to be connected directly to the battery if the two are connected to each other through wires.

Example 31.2 Current and potential difference

In **Figure 31.8**, two light bulbs are connected to each other by a wire and the combination is connected to a battery. In steady state, bulb A glows brightly and bulb B glows dimly. If the magnitude of the potential difference across the battery is 9 V, what can you say about the magnitude of the potential difference across A?

Figure 31.8 Example 31.2.

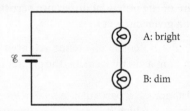

A: bright

B: dim

❶ **GETTING STARTED** The battery and light bulbs constitute a circuit in steady state. The electrons in this circuit gain electric potential energy inside the battery, and this electric potential energy is converted to light (a form of energy) and thermal energy in the bulbs. I assume that the ratio of light to thermal energy is the same in the two bulbs. Because bulb A glows more brightly than bulb B, bulb A must convert more electric potential energy than bulb B.

❷ **DEVISE PLAN** Because this circuit is in steady state, I know that the current is the same throughout the circuit, and therefore during any given time interval, the same number of electrons pass through both bulbs. For each electron that moves through the load of the circuit, an amount of electric potential energy eV_{load} is converted to light and thermal energy. Therefore the magnitude of the potential difference across each bulb is proportional to the amount of electric potential energy converted to light and thermal energy.

❸ **EXECUTE PLAN** The fact that A glows more brightly than B tells me that A is producing more light and thermal energy. Bulb A must therefore be converting more electric potential energy to these other forms, and the electrons must be losing more energy in A than in B. Because the electrons lose more energy in A than in B, the magnitude of the potential difference across A, V_A, must be greater than the magnitude of the potential difference across B, V_B. Because $V_A > V_B$ and because $V_A + V_B$ must equal V_{batt}, it must be true that V_A is more than half of V_{batt}, which means 9.0 V > V_A > 4.5 V. ✔

❹ **EVALUATE RESULT** It makes sense that the bulb that has the greater potential difference across it glows more brightly.

Example 31.2 illustrates an important point: The magnitudes of the potential difference across various circuit elements in an electric circuit need not be the same, even if the current in them is the same.

The two light bulbs in Figure 31.8 are said to be connected *in series:* There is only a single current path through them, and the charge carriers flow first through one and then through the other. As we have seen in Example 31.2:

> **The potential difference across circuit elements connected in series is equal to the sum of the individual potential differences across each circuit element.**

Experiments show that to obtain a particular current in a circuit element, different elements require widely different potential differences. As noted in passing in Section 29.5, for a given circuit element, the constant of proportionality between the potential difference across the element and the current in it is called the **resistance** of the circuit element:

> **The resistance of any element in an electric circuit is a measure of the potential difference across that element for a given current in it.**

The greater the potential difference required to obtain a certain current in a circuit element, the greater the element's resistance.

The resistance of a particular circuit element depends on the material of which it is made, its dimensions, and its temperature. The resistance of a combination of circuit elements connected in series is equal to the sum of the individual resistances of each element. Therefore, the more light bulbs we connect in series to a battery, the greater the resistance of the combination of light bulbs and the smaller the resulting current in the circuit.

✋ **31.5** (*a*) Which light bulb in Example 31.2 has the greater resistance? (*b*) Suppose you connect each bulb separately to the battery. Do you expect the current in bulb A to be greater than, equal to, or smaller than that in bulb B?

31.3 Junctions and multiple loops

Circuit elements can also be connected so that more than one conducting path is formed, as illustrated in Figure 31.9. Such circuits contain more than one loop and are called *multiloop circuits.* These circuits contain **junctions**—locations where more than two wires are connected together—and **branches**—conducting paths between two junctions that are not intercepted by another junction. The circuit in Figure 31.9, for example, contains two junctions (represented by open circles), three branches, and three loops.

The continuity principle permits us to draw some important conclusions about the currents in multiloop circuits. Let us begin by applying the current continuity principle to the circuit shown in Figure 31.10. Because we can select any system boundary along a branch, the current continuity principle requires the current to be the same throughout

Figure 31.9 Circuit diagram of two light bulbs connected to a battery.

This multiloop circuit has two junctions, . . .

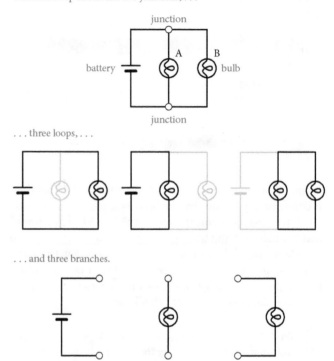

. . . three loops, . . .

. . . and three branches.

that branch. We can apply the same reasoning to the other two branches, so:

> **The current in each branch of a multiloop circuit is the same throughout that branch.**

This statement is known as the **branch rule.** We shall label the current in each branch according to the branch. In Figure 31.10, for example, the currents in the three branches are I_1, I_2, and I_3.

To examine how charge carriers flow at a junction, we draw a system boundary around a junction, as in **Figure 31.11a**. Because the charge inside the system is not changing and because charge is conserved, it follows that the flow of charge carriers into the system must be equal to the flow of charge carriers out of the system. Specifically, if a current I_1 goes into the junction and the other two wires carry currents I_2 and I_3 out of the junction, then $I_2 + I_3$ must

Figure 31.10 The branch rule: In each branch, the current must be the same throughout that branch.

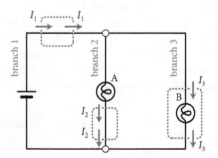

Figure 31.11 The continuity principle at a junction is illustrated by the flow of balls through a branched tube.

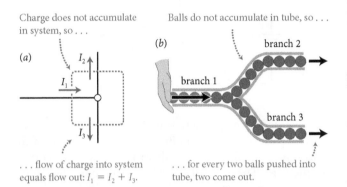

In steady state:

Charge does not accumulate in system, so . . .

Balls do not accumulate in tube, so . . .

(*a*)

(*b*)

branch 2

branch 1

branch 3

. . . flow of charge into system equals flow out: $I_1 = I_2 + I_3$.

. . . for every two balls pushed into tube, two come out.

equal I_1. The general statement of this principle is known as the **junction rule**:

> **The sum of the currents directed into a junction equals the sum of the currents directed out of the same junction.**

Figure 31.11*b* illustrates this principle for balls flowing through a junction of tubes. If we push four balls in at one end of branch 1, four balls have to come out from branches 2 and 3 combined. In the junction shown in the figure, branch 2 and branch 3 are equivalent and two balls are likely to come out from each. In general, however, the branches do not need to be equivalent, and in that case the number of balls going into branch 1 must be equal to the sum of the balls coming out of branches 2 and 3. Because the pushing in and the coming out occur during the same time interval, the "current" of balls through branch 1 equals the sum of the currents in branches 2 and 3, which is what we concluded for the currents in Figure 31.11*a*.

Let us return now to the circuit in Figure 31.9. Because the two bulbs are each connected to the same two junctions, they are said to be connected *in parallel* across the battery. Because each junction is at a given potential, we can conclude that for parallel circuit elements:

> **The potential differences across circuit elements connected in parallel are always equal.**

More specifically, for the circuit shown in Figure 31.9, the potential difference across the two light bulbs is equal to the emf of the battery.

In the next example we study the resistance of parallel combinations of circuit elements.

Example 31.3 Series versus parallel

Two identical light bulbs can be connected in parallel or in series to a battery to form a closed circuit. How do the magnitudes of the potential differences across each bulb and the current in the battery compare with those in a single-bulb circuit when the bulbs in the two-bulb circuit are connected in parallel? When they are connected in series?

❶ GETTING STARTED I begin by drawing circuit diagrams for the single-bulb circuit and for the parallel and series two-bulb circuits (**Figure 31.12**). To connect the bulbs in parallel to the battery, one contact of each bulb is connected directly to the positive terminal of the battery, and the other contact of each bulb is connected directly to the negative terminal of the battery. For the series circuit, one contact of bulb 1 is connected directly to the positive terminal of the battery, the other contact of bulb 1 is connected to one contact of bulb 2, and the other contact of bulb 2 is connected directly to the negative terminal of the battery.

Figure 31.12

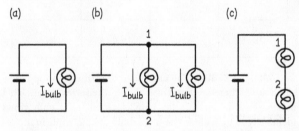

(a)   (b)   (c)

❷ DEVISE PLAN To determine the magnitude of the potential difference across each bulb, I need to analyze how the bulbs are connected to the battery. To determine the current in the battery, I must first determine the current through each bulb. To determine the current in each bulb, I use the fact that the resistance and potential difference determine the amount of current in each bulb. I can use the junction rule to determine the currents in the parallel circuit, and so I label the junctions 1 and 2 in Figure 31.12*b*. Because there is only one branch in the series circuit, I know that the current in the battery is the same as the current in the bulbs. Because the bulbs are identical, their resistances are identical too. I assume the wires have no resistance.

❸ EXECUTE PLAN Because in the parallel circuit one contact of each bulb is connected to the battery's positive terminal and the other contact of each bulb is connected to the negative terminal, the potential difference across each bulb is equal to the potential difference across the battery. By the same argument, the potential difference across the bulb in the single-bulb circuit is also equal to the potential difference across the battery. Therefore the potential difference across each bulb in the parallel circuit is the same as that across the bulb in the single-bulb circuit. ✔

Because the potential difference is the same across these three bulbs, the current must also be the same in all three bulbs. I'll denote this current by I_{bulb}. In the parallel circuit, the fact that the current in each bulb is I_{bulb} means that the current in the battery must be $2I_{bulb}$. (The current pathway at junction 1 of Figure 31.12*b* is just like the pathway shown in Figure 31.11, with $I_2 = I_3 = I_{bulb}$.) In the single-bulb circuit, the current must be the same at all locations in the circuit, so the current in the battery must be I_{bulb}. ✔

In the series circuit, the potential difference across the two-bulb combination is equal to the potential difference across the battery, which means the magnitude of the potential difference across each bulb must be half the magnitude of the potential difference across the battery. The potential difference across each bulb is thus equal to half the potential difference across the bulb in the single-bulb circuit. ✔

Because the potential difference across each bulb in series is half the potential difference across the battery, the current in each bulb must be half the current in the single-bulb circuit $I_{bulb}/2$. Because the series circuit in Figure 31.12c is a single-loop circuit, the current is the same at all locations. Therefore the current in the battery must also be $I_{bulb}/2$. The current in the battery in the single-bulb circuit is therefore twice that in the series two-bulb circuit. ✔

In tabular form my results are

	Parallel	Series	Single
V_{bulb}	V_{batt}	$V_{batt}/2$	V_{batt}
I_{batt}	$2I_{bulb}$	$I_{bulb}/2$	I_{bulb}

❹ **EVALUATE RESULT** The current in the battery in the parallel circuit is therefore four times that in the series circuit. That makes sense because in the parallel circuit each bulb glows identically to the bulb in the single-bulb circuit, while in the series circuit, the battery has to "push" the charge carriers through twice as much resistance. Therefore it makes sense that in the series circuit, both the potential difference across each bulb and the current in the battery are much smaller than they are in the parallel circuit.

✋ **31.6** In Figure 31.9, treat the parallel combination of two light bulbs as a single circuit element. Is the resistance of this element greater than, equal to, or smaller than the resistance of either bulb?

Checkpoint 31.6 highlights an important point about electric circuits: Adding circuit elements in parallel *lowers* the combined resistance and *increases* the current. How can adding elements with a certain resistance lower the combined resistance? The resolution to this apparent contradiction is that adding elements in parallel really amounts to adding paths through which charge carriers can flow, rather than adding resistance to existing paths in the circuit. If you are emptying the water out of a swimming pool, it empties faster if you have multiple hoses draining the water than if you drain the entire pool through a single hose.

Instead of connecting two light bulbs in parallel as in Figure 31.9, what if we connected a wire in parallel with a bulb as in **Figure 31.13**? Experimentally, we find that replacing bulb B of Figure 31.9 with a wire causes bulb A to stop glowing. Why does this happen? The potential difference across a branch is determined by the potential difference between the two junctions on either end of the branch, and therefore the potential difference across all branches between two junctions must be the same. Because the wire is made from a conducting material, the potential difference between its two ends is zero, and therefore the two junctions in Figure 31.13 are at the same potential. Consequently there is no current in the light bulb.

Figure 31.13 A wire that is connected to a battery *in parallel* to a light bulb constitutes a short circuit.

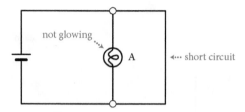

That doesn't mean there is no current in the circuit. As we'll discuss in more detail in Section 31.4, because of the wire's very small resistance, there is a very large current in the wire. Therefore, if the wire is left connected to the battery as in Figure 31.13, the battery quickly discharges through the wire and the wire heats up. A circuit branch with negligible resistance in parallel with an element is commonly called either a **short** or a **short circuit.**

The circuits in Figures 31.8, 31.10, and 31.12 are all drawn neatly with the circuit elements on a rectangular grid. Real circuits are rarely so neatly laid out, of course, and typically look more like the one shown in **Figure 31.14**. It takes time and concentration to look at the tangle of wires in Figure 31.14 and figure out whether or not the bulbs light up (they do) and identify the path taken by the charge carriers through the circuit.

To analyze such a circuit, therefore, it is helpful to draw a circuit diagram, with elements and wires arranged horizontally and vertically. The circuit diagram must accurately show the connections between elements that are present in the actual circuit, but wires that are not connected to each other should, as far as possible, be drawn so that they do not cross.

To draw a circuit diagram for the circuit in Figure 31.14, we begin by identifying the junctions and the branches connecting the junctions. Note that the two terminals of bulb A are connected to two wires each. Because the terminals of the bulb are also connected to each other via the filament inside the bulb, each of these terminals is a junction. There are three branches. One branch consists of the battery and the wires that connect the left and right contacts of light bulb A. The second branch connects the left contact of light bulb A through the filament to the right contact of light bulb A. The third branch connects the right contact of light bulb A through light bulbs B and C

Figure 31.14 Real circuits don't always look like circuit diagrams.

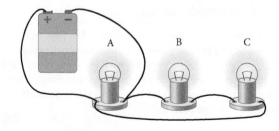

Figure 31.15 Circuit diagram for the circuit of Figure 31.14.

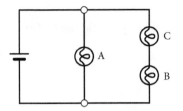

to the left contact of light bulb A. **Figure 31.15** shows a circuit diagram for the circuit in Figure 31.14. Junctions are marked by open circles. The battery and light bulbs have been replaced by the symbols introduced in Figure 31.4. The connecting wires between the circuit elements have been straightened out and replaced by lines.

It is not important that the branch of the circuit containing bulb A appears to the left of the branch containing bulbs B and C. The two-bulb branch could be shown to the left of the branch containing bulb A, and the diagram would still be correct. It is also not important that the branch that contains the battery is shown at the left. There is always more than one way to draw a circuit diagram for any given circuit; any diagram that correctly represents the connections between elements in the circuit is valid.

Exercise 31.4 Bulbs and batteries

Draw a circuit diagram for the arrangement shown in **Figure 31.16**.

Figure 31.16 Exercise 31.4.

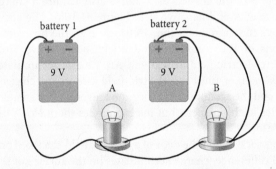

SOLUTION I begin by identifying the junctions and branches. I identify two junctions: one is the left terminal of bulb A and the other is the right terminal of bulb B. I can trace out three branches going from the left terminal of bulb A to the right terminal of bulb B. One branch includes battery 1. The second branch includes bulb A and battery 2, and the third branch includes just bulb B.

I begin by drawing the two junctions. Then I connect these two junctions by each of the three branches, making sure to get the directions of the batteries correct—positive terminal of battery 1 connected to the left terminals of bulbs A and B, and negative terminal of battery 1 connected to the positive terminal of battery 2 and bulb B. This yields the diagram shown in **Figure 31.17a**.

Figure 31.17

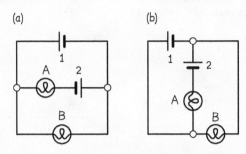

Note that the diagram I drew is not unique. Had I chosen to draw one junction above the other, I might have obtained the diagram shown in Figure 31.17b. By comparing the two diagrams, I see that by sliding battery 1 in Figure 31.17b to the leftmost vertical branch, sliding bulb B to the rightmost branch, and then rotating the diagram 90° clockwise, I obtain the diagram shown in Figure 31.17a.

31.7 If each battery in Figure 31.16 has an emf of 9 V, what is the magnitude of the potential difference across (*a*) bulb A and (*b*) bulb B? (*c*) If the two bulbs are identical, which one glows more brightly?

31.4 Electric fields in conductors

In Section 24.5 we found that the electric field is zero inside a conducting object in electrostatic equilibrium. A conductor through which charge carriers flow is *not* in electrostatic equilibrium, however. (Remember, *electrostatic* means that the arrangement of charged particles is fixed.) To keep charge carriers flowing through a conductor, we need an electric field inside it.

Let us examine more closely the electric field in current-carrying conductors. Consider connecting the two charged spheres of **Figure 31.18a** on the next page with a metal rod that is much thinner than the radius of the spheres. Charge carriers can now flow from one sphere to the other along this rod (Figure 31.18b). For simplicity, we assume that a power source (not shown in the figure) keeps the charge on each sphere and the potential difference between the two constant.

Initially, the electric field along the axis of the rod points from the positive sphere to the negative sphere. However, the electric field is stronger near the spheres, at A and C in Figure 31.18b, than in the middle, at B. Consequently, the charge carriers at A and C get pushed along horizontally more strongly than in the middle, causing positive charge carriers to accumulate between A and B and negative carriers to accumulate between B and C. This accumulation changes the electric field in the rod. For example, as charge carriers accumulate between A and B, the flow of carriers at A is reduced and that at B is enhanced. The carriers stop accumulating when the electric field is the same throughout

Figure 31.18 The source of the electric field in a current-carrying conductor. (For simplicity, we assume that the charge on each sphere is kept constant by an unseen power source.)

(*a*) Electric field around a pair of charged spheres

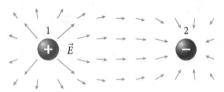

(*b*) Just after spheres are joined by conducting rod

Electric field in rod is smallest in middle (at B).

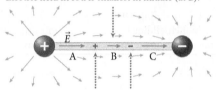

Unequal electric field in rod causes charge to accumulate.

(*c*) Steady state

Accumulation continues until field is equal throughout rod.

Figure 31.19 Electric field in a bent conductor that connects two charged plates.

(*a*) Just after charged plates are joined by bent conductor

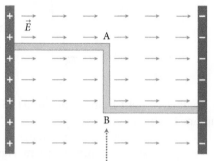

Electric field of capacitor pushes electrons in vertical segment to left.

(*b*) Steady state

Electric field causes charge to accumulate on surfaces of vertical segment.

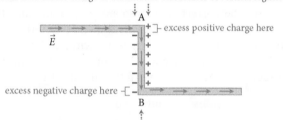

Excess charge at ends causes downward electric field in vertical segment.

the rod (Figure 31.18*c*). Thus, any unevenness in the electric field causes charge to accumulate, and the accumulation, in turn, affects the electric field. This feedback continues until the electric field is uniform and points along the conductor everywhere. In practice, this evening out takes place in an extremely short time interval (10^{-9} s).

✋ **31.8** Suppose the distance between the spheres in Figure 31.18 is ℓ and the potential difference between them is V_{12}. What is the magnitude of the electric field inside the connecting rod?

What is the electric field in a bent conductor, as opposed to the straight conductor of Figure 31.18? Consider the conductor connecting the charged plates in **Figure 31.19*a***. When the plates are first connected to each other via a long thin conductor, the only electric field present is that of the plates, and this field pushes electrons everywhere to the left. Because of this, electrons pile up on the left surface of vertical segment AB of the conductors, leaving behind positive ions on the right side (Figure 31.19*b*). The surface of the conductor accumulates charge in this manner until the horizontal component of the electric field is zero within segment AB. The electric field within AB then no longer pushes electrons to the left surface.

While charged particles accumulate on the two sides of segment AB, the corners at A and B also acquire a charge (positive at A and negative at B, Figure 31.19*b*). The accumulated charge reduces the electric field in the horizontal portions of the conductor and establishes a downward-pointing electric field in segment AB. Charge accumulates in this manner until the electric field due to the charged plates and the accumulated charge no longer pushes charged particles to the corners, but instead guides them along the conductor.

In general, when the ends of a wire are held at a fixed potential difference, charge accumulates on the surface of the conductor. The accumulation stops when the electric field due to the combined effects of the surface charge distribution and the applied potential difference has the same magnitude everywhere inside the conductor and points along a path through the conductor that is everywhere parallel to the sides of the conductor. Once this electric field is established, it no longer pushes charge carriers to the surface of the conductor but instead drives a steady flow of charge carriers through the conductor:

In a conductor of uniform cross section carrying a steady current, the electric field has the same magnitude everywhere in the conductor and is parallel to the walls of the conductor.

Keep in mind that because potential differences across conductors are generally small, the electric fields inside conductors are also generally small, even in the presence of substantial currents.

Example 31.5 Bending fields

Consider the three pieces of wire in **Figure 31.20**. All three are made of the same material and have identical circular cross sections. Conductor A has length ℓ; conductors B and C have length 2ℓ. With the wires kept in the configurations shown, a positively charged conducting plate is connected to the left end of each wire and a negatively charged conducting plate is connected to the right end of each wire. Rank the magnitudes of the steady-state electric field at P, Q, R, S, and T.

Figure 31.20 Examples 31.5.

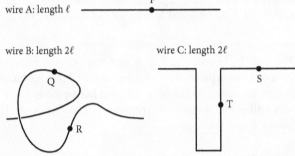

wire A: length ℓ

wire B: length 2ℓ wire C: length 2ℓ

❶ **GETTING STARTED** Once a steady flow of charge carriers has been established in each wire, the electric field in each wire has the same magnitude everywhere in that wire and is always parallel to the walls of the wire.

❷ **DEVISE PLAN** Because the electric field is uniform along each wire, I can use the result I obtained in Example 25.5a (Eq. 1): the magnitude of a uniform electric field between two points is equal to the magnitude of the potential difference between those points divided by the distance between those points.

❸ **EXECUTE PLAN** In each case the magnitude of the potential difference between the ends of the conductor is equal to the potential difference between the plates, which I'll denote by V_{plates}. The electric field magnitude in A is then V_{plates}/ℓ, and that in B and C is $V_{plates}/2\ell$. Therefore, if I use E_P for the electric field magnitude at P, I can say $E_P > E_Q = E_R = E_S = E_T$. ✔

❹ **EVALUATE RESULT** My result is not surprising because I can think of a wire as a load with very small resistance. A wire of length 2ℓ is equivalent to two wires of length ℓ connected in series. Based on the result I obtained in Example 31.3, I expect the current in the wire of length ℓ (analogous to the single-bulb case of Example 31.3) to be twice the current in the wire of length 2ℓ (analogous to the two-bulb series case). To obtain this greater current, I need a greater electric field, as my result shows.

✋ **31.9** Sketch the electric field lines inside the conductors in Figures 31.18 and 31.19.

Figure 31.21 (*a*) Conductor of circular cross section that decreases in radius by a factor of 2 at the midpoint. (b) Electric field lines in the conductor when it is connected between two charged objects (not shown).

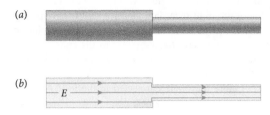

(*a*)

(*b*)

E

As Example 31.5 shows, the current in wire A is twice the current in wires B and C, which are twice as long. To get the same current in wires B and C as in A, the magnitude of the potential difference across wires B and C must be doubled. This means that the resistance of wires B and C is twice the resistance of wire A. Therefore, the resistance of a conductor is proportional to the length of the conductor.

So far, the only conductors we have considered have all had a uniform cross section. Suppose now that we have a conductor that has a circular cross section that decreases in radius by a factor of 2 in the middle (**Figure 31.21a**). How do the current and the electric field in this conductor compare in the wide and the narrow parts?

If we think of this conductor as two circuit elements (the wide part and the narrow part) connected in series, we see that the current must be the same through the entire conductor. At steady state, electric field lines inside a conductor in a circuit must satisfy the continuity principle, just as the current must. Therefore the electric field lines in the wide part of our conductor must all continue into the narrow part (Figure 31.21b). The density of the electric field lines, and consequently the magnitude of the electric field, must therefore increase by the same factor by which the area decreases. Because the cross-sectional area decreases by a factor of 4, the magnitude of the electric field must increase by a factor of 4.

Now consider the magnitude of the potential difference across each part of the conductor. In each part, the magnitude of the electric field is equal to the magnitude of the potential difference across the part divided by the length of the part (see Example 31.5). Because the electric field is four times greater in the narrow part and because the two parts have the same length, the potential difference across the narrow part must also be four times greater than the potential difference across the wide part.

To get the same current in both parts, four times as much potential difference must be applied to the narrow part! This means that the resistance of the narrow part is four times greater than the resistance of the wide part, even though both are made of the same material. Therefore, the resistance of a conductor not only depends on the material from which it is made and the length of the conductor, but also is inversely proportional to the cross-sectional area of the conductor.

Example 31.6 Electric fields in conductors

Figure 31.22 shows a rod made of three pieces of conducting material connected end to end. The pieces are of equal size but are made of different materials; pieces A and C have negligible resistance, but piece B has significant resistance. Consider connecting two oppositely charged plates to the ends of this rod. After the plates have been connected, how does the electric field in B compare with the electric field that existed between the plates before they were connected to the rod? Assume the charge on the plates is maintained.

Figure 31.22 Example 31.6.

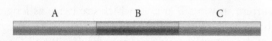

❶ **GETTING STARTED** I begin by making a sketch of the rod connected to the two charged plates (**Figure 31.23**).

Figure 31.23

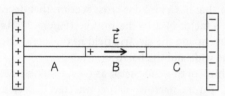

❷ **DEVISE PLAN** As I noted in Example 31.5 in using my result from Example 25.5a, the magnitude of the electric field in B is equal to the magnitude of the potential difference across B divided by the length of B. Likewise, the electric field magnitude between the plates is equal to the magnitude of the potential difference between the plates divided by the distance between them. To compare these two field magnitudes, I need to compare the magnitude of the potential difference across B with that across the plates.

❸ **EXECUTE PLAN** Because the resistance of A and C is negligible, the potential difference across A and C is negligibly small; the potential difference across B is therefore essentially the same as that between the plates. Because B is one-third as long as the distance between the plates, the electric field magnitude in B is roughly three times the electric field magnitude that existed between the plates before connection. ✔

❹ **EVALUATE RESULT** Because A and C have negligible resistance, I expect the charge to distribute itself along the rod as shown in Figure 31.23, with positive charge residing at the end of B nearer the positively charged plate and negative charge residing at the other end. This distribution of charge reduces the electric field in A and C and increases the electric field in B, as I concluded.

The rod in Figure 31.22 is a common circuit element called a **resistor,** a piece of conducting material that has non-negligible resistance (piece B) attached at both ends to pieces of wire called electrical *leads* (represented in Figure 31.22 by pieces A and C).

31.10 If a potential difference of 9 V is applied across the conductor in Figure 31.21, what is the magnitude of the potential difference (*a*) across the wide part and (*b*) across the narrow part?

Self-quiz

1. Why are none of the bulbs in **Figure 31.24** lit?

Figure 31.24

(a)   (b)   (c)

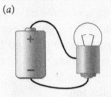

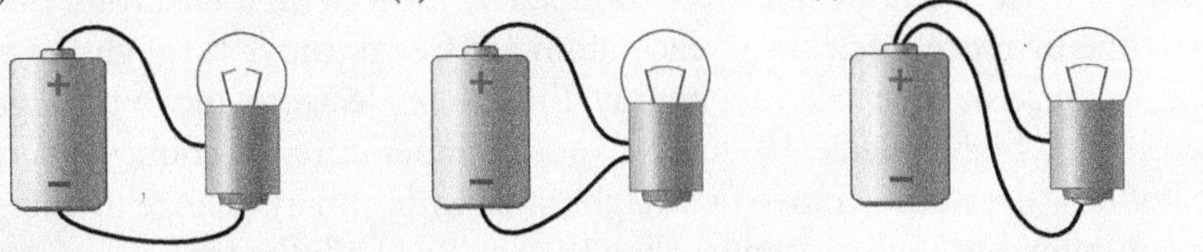

Figure 31.25

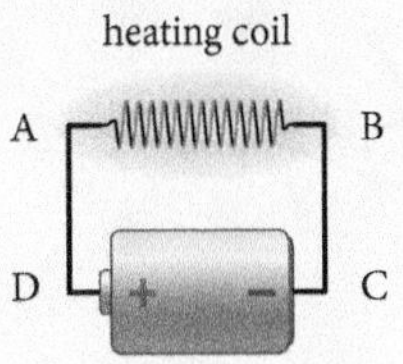

heating coil

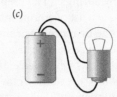

2. In **Figure 31.25**, identify the energy conversions that occur between points A and B, B and C, C and D, and D and A.

3. In **Figure 31.26**, bulb B is brighter than bulb C, which in turn is brighter than bulb A. Rank, largest first, (a) the magnitudes of the potential differences across the bulbs, (b) the currents in them, and (c) their resistances.

Figure 31.26

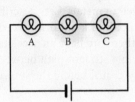

Figure 31.27

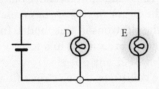

Figure 31.28

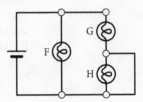

4. In **Figure 31.27**, bulb E is brighter than bulb D. Which bulb has (a) the greater potential difference across it, (b) more current in it, and (c) more resistance?

5. In **Figure 31.28**, which of the three bulbs F, G, and H light up?

6. A thick resistor and a thin resistor of the same length and material are connected in series, as shown in **Figure 31.29**. Which resistor has (a) the greater potential difference across it and (b) the greater resistance?

Figure 31.29

Answers

1. (a) Filament broken. (b) Battery connected to only one contact on bulb. (c) Bulb connected to only one battery terminal.

2. AB: electric potential energy converted to thermal energy; CD: chemical energy converted to electric potential energy; BC and DA: negligible conversion of energy.

3. (a) $V_B > V_C > V_A$ because the brightest bulb converts the most energy. (b) $I_B = I_C = I_A$ because the bulbs are in series and the current is the same throughout the circuit. (c) $R_B > R_C > R_A$ because, for a given current in a circuit element, R is proportional to V.

4. (a) $V_D = V_E$ because the bulbs are in parallel. (b) $I_E > I_D$ because E is brighter. (c) $R_D > R_E$ because less current passes through D.

5. F and G light up, but H does not because the wire in parallel with H carries all of the current through the right branch. (The wire "shorts out" H.)

6. (a) $V_{thin} > V_{thick}$ because the density of electric field lines through the thin resistor has to be greater, and if the lengths of the resistors are the same and E is greater, then V is greater. (b) $R_{thin} > R_{thick}$ because resistance is inversely proportional to cross-sectional area.

Figure 31.30 The effect of an applied electric field on the motion of a free electron through a lattice of ions.

(*a*) Motion in absence of an electric field

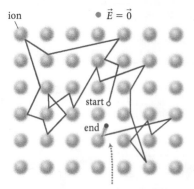

Electron's displacement is zero over long time interval.

(*b*) Motion with applied electric field

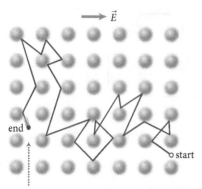

Electron undergoes displacement in direction opposite to electric field.

31.5 Resistance and Ohm's law

Recall from Section 22.3 that a metal consists of a lattice of positively charged ions through which electrons can move relatively freely. Although the lattice positions are fixed (meaning there is no motion of ions through the metal), the ions vibrate around their lattice positions, as if they were connected to those positions by springs. (The amplitude of the vibrations depends on temperature.) The ions consist of atomic nuclei surrounded by most of their electrons; the outermost one or more electrons of each atom are free to move through the entire lattice. Because of their thermal energy, these free electrons move through the lattice at very high speeds (10^5 m/s at room temperature, or about 0.1% of the speed of light), and they move in straight lines without any change in energy or momentum except in the instants when they collide with the ions.

In the absence of an electric field and over a time interval long enough for many collisions to take place, the displacement of an electron is very small in spite of its high speed (**Figure 31.30a**). This is true because each collision changes the direction of the electron's motion, randomizing the direction of the electron's velocity. Consequently, the average velocity of all the electrons is zero. However, when an electric field is applied, as in Figure 31.30b, the electric field causes the electrons to accelerate in the direction opposite to the electric field.

A quantitative description of the motion of charge carriers in a conductor was given by P. K. Drude in 1900, shortly after the discovery of the electron. His model (called the *Drude model*) applies remarkably well to metals. Let's describe the motion of the electrons in this model. In the presence of a uniform electric field $\vec{E}$, the electrons are subject to a force $-e\vec{E}$ and therefore have an acceleration $\vec{a} = -e\vec{E}/m_e$. For any one electron moving along a straight path between two consecutive collisions, the electron's final velocity $\vec{v}_f$ just before the second collision is

$$\vec{v}_f = \vec{v}_i + \vec{a}\Delta t = \vec{v}_i - \frac{e\vec{E}}{m_e}\Delta t, \qquad (31.1)$$

where $\vec{v}_i$ is the electron's initial velocity on that path (its velocity just after the first collision), Δt is the time interval the electron spends on that path (in other words, the time interval between collisions), e is the elementary charge (Eq. 22.3), $\vec{E}$ is the applied electric field, and m_e is the mass of the electron. The magnitude of $\vec{v}_i$ is roughly 10^5 m/s, as noted above, and the direction is determined by the first of the two collisions. Because of the high electron speed, the time interval Δt between collisions is extremely short—on the order of 10^{-14} s.

To calculate the average velocity of all the electrons, we take the average of Eq. 31.1 for all of the electrons:

$$(\vec{v}_f)_{av} = (\vec{v}_i)_{av} - \frac{e\vec{E}}{m_e}(\Delta t)_{av}. \qquad (31.2)$$

(I have assumed here that the electric field is either constant over time or takes a time interval much longer than Δt to change significantly.)

Even though the magnitude of $\vec{v}_i$ is quite large, its average value for all electrons is zero because the collisions produce a random distribution of the directions of the initial velocities. The resulting average velocity, called the **drift velocity** $\vec{v}_d$ of the electrons, is thus

$$\vec{v}_d = -\frac{e\vec{E}}{m_e}\tau, \qquad (31.3)$$

where $\tau \equiv (\Delta t)_{av}$ is the average time interval between collisions. (The value of τ depends on the number density, size, and charge of the lattice ions, and on temperature.) The magnitude of the drift velocity is called the *drift speed*. Equation 31.3 shows that the drift velocity of the electrons is in a direction opposite that of the electric field, and the drift speed is proportional to the electric field magnitude.

31.11 (*a*) Does the electric field do work on the electrons of Figure 31.30*b* as they accelerate between collisions? (*b*) On average, does the kinetic energy of the electrons increase as they drift through the lattice? (*c*) What do your answers to parts *a* and *b* imply about the energy in the lattice?

In Chapter 27 we found that the current in a conductor can be expressed in terms of the speed v of the charge carriers (Eq. 27.16). Because the speed of the charge carriers is what we are now calling the drift speed, we can write Eq. 27.16 in the form

$$I = nAqv_d, \tag{31.4}$$

where A is the cross-sectional area of the conductor, q is the charge on each charge carrier in the conductor, and n is the number of carriers per unit volume in the conductor. Because of the current's dependence on cross-sectional area, it is convenient to introduce the current per unit area, called the **current density,** whose magnitude is given by the magnitude of the current per unit area:

$$J \equiv \frac{|I|}{A} = n|q|v_d. \tag{31.5}$$

Because the drift velocity is a vector, the current density is a vector, too:

$$\vec{J} = nq\vec{v}_d. \tag{31.6}$$

The direction of the current density is the same as that of the drift velocity for positive charge carriers and opposite the direction of the drift velocity for negative charge carriers. Therefore the current density is always in the same direction as the current. The SI unit of current density is A/m^2.

Substituting the absolute value of the right side of Eq. 31.3 for v_d and e for $|q|$ in Eq. 31.5 yields

$$J = n(e)\left(\frac{eE}{m_e}\tau\right) = \frac{ne^2\tau}{m_e}E. \tag{31.7}$$

Equation 31.7 shows that the current density is proportional to the applied electric field. The proportionality constant σ is called the **conductivity** of the material of which the conductor is made:*

$$\sigma \equiv \frac{J}{E}. \tag{31.8}$$

*Note that conductivity is *not* the same as surface charge density, which is represented by the same symbol.

Table 31.1 Conductivities of various materials at room temperature A/(V · m)

Conductors	Silver	6.3×10^7
	Copper	5.9×10^7
	Aluminum	3.6×10^7
	Tungsten	1.8×10^7
	Nichrome	6.7×10^5
	Carbon	7.3×10^4
Semiconductors	Silicon	4×10^{-4}
	Germanium	2
Poor conductors	Seawater	4
Insulators	Pure water	4.0×10^{-6}
	Glass	10^{-12}

The SI unit of conductivity is equal to $(A/m^2)/(V/m) = A/(V \cdot m)$. The conductivity is a measure of a material's ability to conduct a current for a given applied electric field. The conductivity is a property of the material and is therefore the same for any piece of that material you might choose.

Table 31.1 gives the conductivities of some common materials at room temperature. Note that the first four materials, all metals, have very similar conductivities. Nichrome is an alloy of nickel and chromium used in heating elements because of its relatively low conductivity. Silicon is a *semiconductor;* its conductivity is intermediate between that of an electrical conductor and that of an electrical insulator. The conductivities of insulators (such as pure water and glass) are many orders of magnitude smaller than those of metals. Seawater has a greater conductivity than pure water because the ions dissolved in seawater serve as charge carriers for a current, just as the ions in the electrolyte of a battery do.

Comparing Eqs. 31.7 and 31.8 gives us an expression for the conductivity of a metal in terms of the average time interval between collisions, the mass and charge of the electrons, and the number density of electrons present in the material:

$$\sigma = \frac{ne^2\tau}{m_e}.\tag{31.9}$$

Because of the temperature dependence of τ, the conductivity σ depends on temperature and, for some materials, on the magnitude of the current in the material, but it is independent of the shape of the piece of material in which the current density is measured.

We obtained Eq. 31.9 for a metal, in which the charge carriers are electrons, but it applies to any system in which the charge carriers in a conducting material move freely between collisions. To generalize Eq. 31.9 for charge carriers other than electrons, simply substitute q^2 for e^2, and m_q for m_e, where q is the charge on the carriers and m_q is the mass of one carrier.

The conductivity describes how large a current density and hence current are created by an external electric field. Such an electric field is produced by applying a potential difference across a material. In Section 31.2, we looked at resistance as a way of relating the applied potential difference V across a circuit element to the resulting current in the element. In general, the resistance of any element is defined to be

$$R \equiv \frac{V}{I}.\tag{31.10}$$

The derived SI unit of resistance is the **ohm** ($1\ \Omega \equiv 1\ \text{V/A}$). Resistance is always positive, so in Eq. 31.10 the direction in which V and I are measured must be such that they both have the same algebraic sign. The resistance of most circuit elements is typically in the range of $10\ \Omega$ to $100{,}000\ \Omega$.

The concept of resistance is useful in describing conductors and other objects that provide a continuous conducting path for charge carriers, such as filaments, bulbs, and resistors. (It is not especially useful for other types of circuit elements, such as capacitors.) For some conducting materials, the resistance R of a piece of this material is fixed at a given temperature. For these materials, Eq. 31.10 indicates that the current in the conductor is proportional to the potential difference across it and inversely proportional to the resistance:

$$I = \frac{V}{R}. \tag{31.11}$$

Such materials are said to be *ohmic*. If we plot the magnitude of the current in a piece of this material as a function of the potential difference across it, the result is a straight line whose slope is equal to the inverse of the resistance R of the piece (**Figure 31.31**).

Equation 31.11 is often referred to as **Ohm's law.** It is important to keep two things in mind about Eq. 31.11. First, many materials and many circuit elements are not ohmic. For such materials and elements, a plot of current as a function of applied potential difference is not a straight line, but has a more complicated shape, indicating that R depends on the potential difference. Second, Eq. 31.11 is really a definition, not a law. It simply amounts to the observation that in certain materials, the current is proportional to the applied potential difference. In this chapter, we concern ourselves only with ohmic materials and circuit elements.

Let's now relate the resistance of a conductor to the conductivity of the material of which it is made. From Checkpoint 31.8 we know that the magnitude of the potential difference across a wire of length ℓ is given by (Eq. 25.25 and Example 25.5):

$$|V| = E\ell. \tag{31.12}$$

Combining Eqs. 31.5 and 31.8, we see that $J \equiv |I|/A = \sigma E$, so $E = |I|/(\sigma A)$. Substituting this expression into Eq. 31.12 and dropping the absolute-value symbol, we obtain a relationship between V and I:

$$V = \frac{I}{\sigma A}\ell = I\frac{\ell}{\sigma A}. \tag{31.13}$$

Substituting this expression into the definition of resistance, Eq. 31.10, we obtain for ohmic conductors,

$$R = \frac{\ell}{\sigma A}. \tag{31.14}$$

Equation 31.14 shows that the resistance of a conductor not only depends on the material from which it is made (through the conductivity σ) but also is proportional to the length ℓ of the conductor and inversely proportional to the cross-sectional area, as we had concluded in Section 31.4.

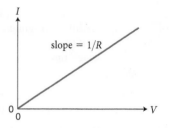

Figure 31.31 Current versus applied potential difference for an ohmic conductor. The current is proportional to the potential difference.

Example 31.7 Drifting electrons

Consider a piece of copper wire that is 10 m long and has a diameter of 1.0 mm. The number density of free electrons in copper is 8.4×10^{28} electrons/m^3. If the wire carries a current of 2.0 A, what are (a) the magnitude of the potential difference across the wire, (b) the drift speed of the electrons in the wire, and (c) the average time interval between collisions for the electrons?

❶ GETTING STARTED Ohm's law relates the potential difference across the wire to the current in it and its resistance. Even though the wire is made of copper, it has a finite resistance, which depends on the conductivity of copper and on the length and cross-sectional area of the wire. This resistance is caused by collisions between the electrons and copper ions in the wire, and these collisions limit the drift speed of the electrons through the wire. The drift speed is related to the average time interval between collisions, which is one of the parameters I need to calculate and appears in the expression for conductivity in the Drude model, Eq. 31.3, $v_d = eE\tau/m_e$.

❷ DEVISE PLAN The potential difference across the wire is related to the current by Eq. 31.11, but in order to use this equation, I must determine the wire's resistance, which is given by Eq. 31.14. I can look up the conductivity of copper in Table 31.1 ($\sigma = 5.9 \times 10^7$ A/V·m). To obtain the drift speed of the electrons, I can use Eq. 31.4, which contains the charge on the electron ($q = e = 1.6 \times 10^{-19}$ C), the current (given), the number density of the electrons (given), and the cross-sectional area of the wire, which I can calculate from the given diameter. To obtain the average time interval between collisions, I can use Eq. 31.9, which contains the mass of the electron ($m_e = 9.11 \times 10^{-31}$ kg).

❸ EXECUTE PLAN (a) I obtain the resistance of the wire from Eq. 31.14 and the cross-sectional area of the wire, $A = \pi r^2 = \pi(5.0 \times 10^{-4}$ m$)^2 = 7.9 \times 10^{-7}$ m^2:

$$R_{\text{wire}} = \frac{\ell}{\sigma A} = \frac{10 \text{ m}}{(5.9 \times 10^7 \text{ A/V} \cdot \text{m})(7.9 \times 10^{-7} \text{ m}^2)} = 0.21 \ \Omega.$$

Now that I have R_{wire}, I can obtain V_{wire} from Eq. 31.11:

$$V_{\text{wire}} = IR_{\text{wire}} = (2.0 \text{ A})(0.21 \ \Omega) = 0.42 \text{ V.} ✔$$

(b) I first obtain the drift speed of the electrons from Eq. 31.4:

$$v_d = \frac{I}{neA} = \frac{2.0 \text{ C/s}}{(8.4 \times 10^{28} \text{ m}^{-3})(1.6 \times 10^{-19} \text{ C})(7.9 \times 10^{-7} \text{ m}^2)}$$

$$= 1.9 \times 10^{-4} \text{ m/s.} ✔$$

(c) Solving Eq. 31.9 for τ, I get

$$\tau = \frac{m_e \sigma}{ne^2} = \frac{(9.11 \times 10^{-31} \text{ kg})(5.9 \times 10^7 \text{ A/V} \cdot \text{m})}{(8.4 \times 10^{28} \text{ m}^{-3})(1.6 \times 10^{-19} \text{ C})^2}$$

$$= 2.5 \times 10^{-14} \text{ s.} ✔$$

❹ EVALUATE RESULT Because the conductivity of copper is high, it makes sense that I obtain a small resistance and consequently a small potential difference across the wire even though it is 10 m long. My answer to part b shows that the drift speed of the electrons in the wire is very small. The drift speed indicates that it takes the electrons about 5 s to move just 1 mm, even though I know from experience that when I turn on a light switch, the light turns on instantly even if the bulb is meters away from the switch. The reason the light bulb comes on almost instantaneously, however, is that the current is the same throughout the circuit—all of the electrons throughout the circuit are set in motion almost simultaneously when I flip the switch.

How can my very small calculated value for drift speed be reasonable when the current is a significant 2.0 A = 2.0 C/s at any given location in the wire? The reason is that the number density of electrons in the wire is very high. Although it takes 5 s for an electron to move 1 mm, there are nearly 10^{20} electrons per cubic millimeter of the wire. This means that on the order of 10^{20} electrons, or 10 coulombs of charge, pass a given location in 5 s—hence the current is 2 C/s, or 2 A.

The very high number density of the electrons is also responsible for the extremely short time interval between collisions. If I imagine the 10^{20} electrons per cubic millimeter arranged on a cubic lattice where each cube of the lattice has a side length of 1.0 mm, there are about 10^7 electrons along each side of the cube, and so the average distance between them is about 10^{-10} m. As the electrons move at about 10^5 m/s (see Section 31.5), they cover this average distance in about 10^{-15} s, which is within an order of magnitude of what I obtained.

Figure 31.32 Circuit diagram for a battery connected in series with a single resistor.

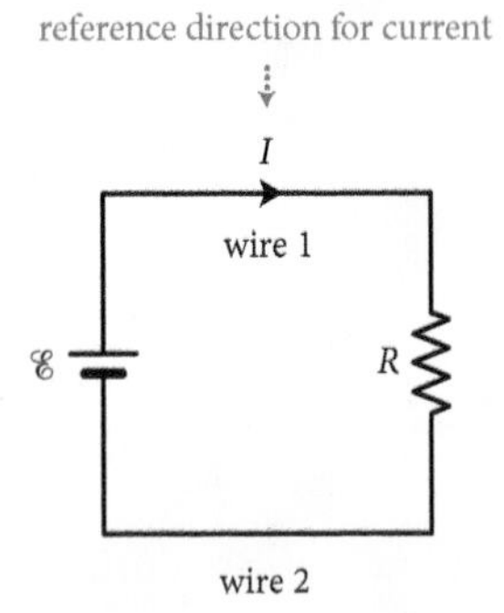

31.12 If the temperature of a metal is raised, the amplitude of the vibrations of the metal-lattice ions increases. (a) What effect, if any, do you expect these greater vibrations to have on the resistance of a piece of that metal? (b) What effect does running a current in a metal have on the temperature of the metal? (c) Make a graph of current versus potential difference, taking into account the effect you described in part b.

31.6 Single-loop circuits

In this section, we quantitatively analyze single-loop circuits. We begin with the simplest possible circuit: a single battery connected to a single resistor, obeying Ohm's law (Figure 31.32). The emf of the battery is $\mathscr{E}$, and the resistance of the resistor is R.

The emf of the battery establishes an electrostatic field that drives a current in the circuit. In steady state, this current is the same at all locations in the circuit, as we discussed in the first part of this chapter. For a single-loop circuit we can express this condition quantitatively as

$$I_{\text{battery}} = I_{\text{wire}} = I_{\text{resistor}} = I. \tag{31.15}$$

In Section 27.3 we called the direction of decreasing potential in a conducting object the *direction of current,* and we indicated this direction by an arrow labeled I next to the object carrying the current (Figure 27.18). When working with circuits, however, we generally won't know the direction of current in advance. To analyze the circuit we therefore need to choose a *reference direction for the current* in each circuit branch. We indicate this direction by an arrowhead on a wire in the circuit (Figure 31.32) and label that arrowhead with a symbol for the current in that branch. In a single-loop circuit, there are no junctions and therefore only one current. As the next checkpoint shows, the reference direction for the current and the direction of current need not be the same. As we shall see later, we obtain a positive value for I when the direction of current and the reference direction for the current are the same. When these two directions are opposite, we find that $I < 0$.

31.13 If $\mathscr{E}_1 < \mathscr{E}_2$ in Figure 31.33, is the direction of current the same as the reference direction for the current indicated in the diagram?

Let's next consider the energy transformations that occur in the circuit shown in Figure 31.34. Nonelectrostatic work done on the negative charge carriers as they travel from point a to point b through the source raises the electric potential energy of the charge carriers. In the load, this electric potential energy is converted to other forms of energy, such as thermal energy, mechanical energy, radiation energy:

$$\Delta E_{\text{other}} = W_{\text{nonelectrostatic}}(a \rightarrow b). \tag{31.16}$$

Considering just the load by itself, we also know that the amount of energy converted to other forms of energy must be equal to the work done by the electrostatic field on the charge carriers as they travel from point b to point a through the load:

$$\Delta E_{\text{other}} = W_{\text{electrostatic}}(b \rightarrow a), \tag{31.17}$$

so

$$W_{\text{nonelectrostatic}}(a \rightarrow b) = W_{\text{electrostatic}}(b \rightarrow a). \tag{31.18}$$

Using Eqs. 26.7 and 25.15 this becomes

$$q\mathscr{E} = -qV_{\text{ba}} \tag{31.19}$$

or

$$\mathscr{E} + V_{\text{ba}} = 0. \tag{31.20}$$

For circuits that contain many elements, Eq. 31.20 can be generalized by replacing each term by a sum. In that case the algebraic sum of the emfs and the potential differences around the loop is zero:

$$\sum \mathscr{E} + \sum V = 0 \quad \text{(steady state, around loop).} \tag{31.21}$$

Equation 31.21 is called the **loop rule.**

When evaluating the sum on the left in Eq. 31.21, we need to pay close attention to the signs of the emfs and potential differences. We begin by choosing a *direction of travel* around the loop. We denote this direction of travel by a curved

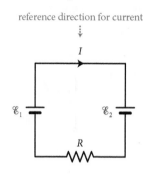

Figure 31.33 Checkpoint 31.13.

reference direction for current

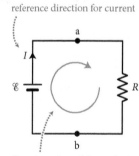

Figure 31.34 Choosing a direction of travel for analyzing a single-loop circuit.

reference direction for current

direction of travel for analyzing potential differences around loop

Figure 31.35 Potential differences across batteries, resistors, and capacitors.

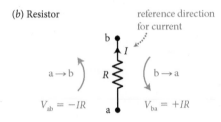

(a) Battery

direction of travel

$a \rightarrow b$ $\mathcal{E}$ $b \rightarrow a$

$V_{ab} = +\mathcal{E} < 0$ $V_{ba} = -\mathcal{E} > 0$

(b) Resistor

reference direction for current

$a \rightarrow b$ R $b \rightarrow a$

$V_{ab} = -IR$ $V_{ba} = +IR$

(c) Capacitor

$a \rightarrow b$ C $b \rightarrow a$

$V_{ab} = +q(t)/C > 0$ $V_{ba} = -q(t)/C < 0$

arrow at the center of the loop (Figure 31.34). As we shall see shortly, this choice does not affect the end result—we must just be sure to use the same direction for all elements in the loop. In single-loop circuits it makes sense to let the travel direction be the same as the reference direction for the current. In multiloop circuits, however, it is generally not possible to let both directions coincide.

Because the potential is the same everywhere along any of the wires, we only need to consider the potential differences across circuit elements: Each circuit element contributes one term to the sum. For a battery, the potential difference is positive when traveling from the negative terminal to the positive terminal because the potential of the positive terminal is higher than that of the negative terminal (**Figure 31.35a**). Conversely, the potential difference is negative when traveling in the opposite direction. If the battery is ideal, the magnitude of the potential difference is equal to the emf $\mathcal{E}$ of the battery (Eq. 26.8).

The sign of the potential difference across a resistor depends on both the choice of current reference direction and the choice of travel direction (Figure 31.35b). When traveling in the same direction as the reference direction for the current, the potential difference is $-IR$. To see why this is so, let us assume that in Figure 31.35b the current is positive (that is, the current is in the reference direction for the current). Because the direction of current is from higher potential to lower potential, we see that the potential at b must be lower than at a, $V_b < V_a$, and therefore $V_{ab} = V_b - V_a < 0$. Indeed when $I > 0$ the potential difference $V_{ab} = -IR$ is negative. When traveling in the opposite direction, the potential difference is $+IR$.

Next we examine the potential difference across a capacitor (Figure 31.35c). The potential difference is positive when traveling from the negatively charged plate to the positively charged one because the potential of the positive terminal is higher than that of the negative terminal; the potential difference is negative when traveling in the opposite direction. The magnitude of the potential difference can be obtained from Eq. 26.1: $C \equiv q/V_{cap}$, where C is the capacitance and q is the magnitude of the charge on the plates. It is important to keep in mind that the charge on the plate and the potential difference typically vary as a function of time as the capacitor charges or discharges. At any given instant, the magnitude of the potential difference across the capacitor is

$$V_{cap}(t) = \frac{q(t)}{C}, \tag{31.22}$$

where $q(t)$ is the magnitude of the charge on the plates at that instant. Because of this time-dependence, the current also depends on time. When solving Eq. 31.21 for circuits that contain capacitors, remember that the time-dependent current $I(t)$ and $q(t)$ are related by Eq. 27.2: $I(t) \equiv dq/dt$.

The Procedure box and **Table 31.2** on the next page summarize how to apply the loop rule to a single-loop circuit. Let us return to the circuit in Figure 31.34 and apply this procedure to obtain a relationship between the current and the emf. The figure already indicates a reference direction for the current (the arrowhead labeled I) and a direction of travel (the clockwise circular arrow). We shall begin our analysis at the bottom left corner. If we go clockwise, the first circuit element is the battery; because we are traveling from the negative to the positive terminal, the potential difference is $+\mathcal{E}$ (Figure 31.35a and Table 31.2). Next is the resistor; because we are traveling in the same direction as the reference direction for the current, the potential difference is $-IR$. Substituting these values in Eq. 31.21, we get

$$+\mathcal{E} - IR = 0 \tag{31.23}$$

or

$$I = \frac{\mathcal{E}}{R}. \tag{31.24}$$

Procedure: Applying the loop rule in single-loop circuits

When applying the loop rule to a single-loop circuit consisting of resistors, batteries, and capacitors, we need to make several choices in order to calculate the current or the potential difference across each circuit element.

1. Choose a reference direction for the current in the loop. (This direction is arbitrary and may or may not be the direction of current, but don't worry, things sort themselves out in step 4.) Indicate your chosen reference direction by an arrowhead, and label the arrowhead with the symbol for the current (I).
2. Choose a direction of travel around the loop. This choice is arbitrary and separate from the choice of the reference direction for the current in step 1. (You may

want to indicate the travel direction with a circular clockwise or counterclockwise arrow in the loop.)
3. Start traversing the loop in the direction chosen in step 2 from some arbitrary point on the loop. As you encounter circuit elements, each circuit element contributes a term to Eq. 31.21. Use Table 31.2 to determine the sign and value of each term. Add all terms to obtain the sum in Eq. 31.21. Make sure you traverse the loop completely.
4. Solve your expression for the desired quantity. If your solution indicates that $I < 0$, then the direction of current is opposite the reference direction you chose in step 1.

Table 31.2 Signs and values of potential differences across batteries and resistors (Figure 31.35)

Circuit element	Plus sign when traversing	Value
ideal battery	from − to +	$\mathcal{E}$
capacitor	from − to +	$q(t)/C$
resistor	opposite reference direction of current	IR

31.14 In the analysis of the circuit in Figure 31.34, we chose a clockwise reference direction for the current and a clockwise direction of travel. Redo the analysis using (a) a clockwise reference direction for the current and a counterclockwise direction of travel and (b) a counterclockwise reference direction for the current and a clockwise direction of travel.

Exercise 31.8 Series resistors

Consider the circuit shown in **Figure 31.36**, containing two resistors with resistances R_1 and R_2 and a battery with an emf $\mathcal{E}$. Determine the current in the circuit in terms of R_1, R_2, and $\mathcal{E}$.

Figure 31.36 Exercise 31.8.

SOLUTION I begin by making a diagram, labeling the points a, b, and c, and choosing a clockwise reference direction for the current and a clockwise direction of travel (**Figure 31.37**). If I begin at point a, as I travel from the negative to the positive terminal, the potential difference is $+\mathcal{E}$. As I travel across resistor 1, I am traveling in the same direction as the reference direction for the current, so the potential difference is $-IR_1$. Next I travel across resistor 2 in the same direction as the reference direction

Figure 31.37

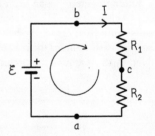

for the current, so the potential difference is $-IR_2$. I now add all these terms and set the sum to zero (Eq. 31.21):

$$+\mathcal{E} - IR_1 - IR_2 = 0.$$

Solving this equation for the current yields

$$I = \frac{\mathcal{E}}{R_1 + R_2}.$$

Comparing the result of Exercise 31.8 with Eq. 31.11, we see that the effect on the current of placing two (or more) resistors in series is equivalent to adding the resistances, as we had concluded in Section 31.2. In other words, the same current would exist in the circuit if we replaced the two resistors in series with a single resistor having resistance

$$R_{eq} = R_1 + R_2. \tag{31.25}$$

By following the same line of reasoning, I can show that for more than two resistors in series, this result generalizes to

$$R_{eq} = R_1 + R_2 + R_3 + \cdots \quad \text{(resistors in series)}. \tag{31.26}$$

The resistance that could be used to replace a combination of circuit elements without altering the current from the battery is often referred to as the **equivalent resistance** of that part of the circuit.

31.15 (*a*) In Figure 31.37, determine the potential difference between c and b by going counterclockwise from c to b. (*b*) For $\mathcal{E} = 9$ V, $R_1 = 10$ Ω, and $R_2 = 5$ Ω, calculate V_{cb} going counterclockwise from c to b.

Example 31.9 Internal resistance

Even the best batteries dissipate some energy, which means that not all the chemical energy is converted to electric potential energy. We can take this dissipation into account by modeling a nonideal battery as consisting of an ideal battery of emf $\mathcal{E}$ in series with a small resistor R_{batt}, often called the *internal resistance* of the battery. The effect of this internal resistance is that the potential difference across the battery terminals is smaller than $\mathcal{E}$ when there is a current. When a nonideal battery is used in a circuit where the load is a resistor of resistance R, is the potential difference across the battery terminals greatest when the resistance of the load is high or when it is low?

❶ **GETTING STARTED** I begin by drawing a circuit diagram for such a nonideal battery connected to a load of resistance R (**Figure 31.38**). I need to determine an expression that shows me how the relative values of R and the internal resistance R_{batt} affect the potential difference across the battery terminals. I arbitrarily choose clockwise both for the reference direction for the current and for my travel around the circuit.

Figure 31.38

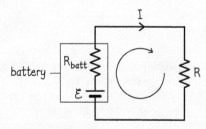

❷ **DEVISE PLAN** The potential difference across the nonideal battery terminals is the sum of the potential difference across the ideal battery, $\mathcal{E}$, and the potential difference across the internal

resistance, $-IR_{batt}$. This latter term means I need to obtain a value for the current. I note that my circuit is equivalent to the two resistors in series in the circuit shown in Figure 31.37, and the final expression I obtained in Exercise 31.8 lets me express the current in terms of these two resistances. Thus I should be able to use this result to write an expression relating R and R_{batt} to V_{batt}.

❸ **EXECUTE PLAN** I use the solution I obtained in Exercise 31.8 to calculate the current in the circuit:

$$I = \frac{\mathcal{E}}{R_{batt} + R}.$$

I can now substitute this expression for I in the $-IR_{batt}$ term in my expression for V_{batt}:

$$V_{batt} = \mathcal{E} - IR_{batt} = \mathcal{E}\left(1 - \frac{R_{batt}}{R_{batt} + R}\right) = \mathcal{E}\left(\frac{R}{R_{batt} + R}\right).$$

When $R \gg R_{batt}$, the effect of R_{batt} is small and the potential difference across the battery terminals is essentially what it is in an ideal battery. When R is small, and especially when it is comparable to R_{batt}, then R_{batt} reduces the potential difference across the terminals. Thus the potential difference across the terminals of a nonideal battery is greatest when the resistance of the load is high. ✔

❹ **EVALUATE RESULT** If $R \gg R_{batt}$, the resistance in the circuit is great and hence the current is small. Because the current is small, the potential difference due to the internal resistance is small ($-IR_{batt}$) and hence the potential difference across the terminals does not differ significantly from its ideal value. This conclusion agrees with what my expression for V_{batt} shows.

According to the model of Example 31.9, an ideal battery is simply a battery with $R_{batt} = 0$. For a high-quality nonideal battery, R_{batt} can be extremely small—at most a few ohms. Thus, for ordinary loads, which are typically hundreds of ohms or more, the battery's internal resistance is not important.

Electrical measuring instruments

Current, potential difference, and resistance can be measured with the following three electrical measuring devices. Often the three functions are combined into one instrument, called a *multimeter*.

Ammeter: To measure the current in a single-loop circuit (or in a branch of a multiloop circuit), an ammeter must be inserted into the loop or branch, as shown in Figure 31.39*a* or *c*. The ammeter then indicates the current passing through it.

Voltmeter: To measure the potential difference between two points, a voltmeter is connected to those two points

(Figure 31.39*b*). Unlike current measurement, a potential difference measurement does not require breaking the circuit.

Ohmmeter: A voltmeter and an ammeter can be combined to measure a resistance, as shown in Figure 31.39*c*. In practice resistance is measured using a device called an *ohmmeter*. Such a device puts a known potential difference across the resistor to be measured, measures the resulting current though the resistor, and then uses Ohm's law to determine the resistance. To measure the resistance of a circuit element, that element must be disconnected from the circuit.

Voltmeters and ammeters typically affect the circuit under observation. For example, if the resistance R_A of the ammeter in Figure 31.39*a* is nonzero, the current in the circuit is affected by the insertion of the ammeter. The resistance of ammeters must therefore be very small compared to the other resistances in the circuit; an *ideal ammeter* has zero resistance.

Similarly, the resistance R_V of the voltmeter in Figure 31.39*b* must be very great compared to other resistances in the circuit to prevent charge carriers from flowing through the voltmeter and changing the current in the circuit; an *ideal voltmeter* has infinite resistance.

In either case, if we know the resistance of the device, it is possible to correct for its effect on the circuit.

Figure 31.39 Electrical measuring instruments.

(*a*) Ammeter (*b*) Voltmeter (*c*) Ohmmeter

To measure current, we insert ammeter into circuit.

To measure potential difference between a and b, we connect voltmeter between those points.

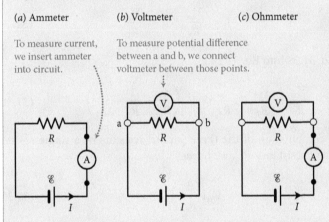

A number of measuring instruments exist to measure electrical quantities in electric circuits and circuit elements. The instrument to measure currents is called an *ammeter*; the instrument to measure potential differences is called a *voltmeter*. By measuring both the current in a resistor and the potential difference across it, one can determine the resistance using Ohm's law. An *ohmmeter* is used to accomplish this task: Connecting a resistor between the terminals of an ohmmeter yields a reading for the resistance. See the box above on Electrical measuring instruments for more information on these devices.

31.7 Multiloop circuits

The analysis of multiloop circuits follows the same principles we laid out in the preceding section, but we need to consider what happens in multiple branches, junctions, and loops. The first thing to note is that the branch rule requires that in steady state, the current in any branch is the same everywhere along that branch. Therefore, in a circuit containing M branches, there are M distinct currents. When analyzing multiloop circuits, you should therefore always begin by identifying all the branches and labeling the current in each of those branches.

As we saw in Section 31.3, the **junction rule** states that in steady state the sum of the currents going into a junction must be equal to the sum of the currents going out of that junction:

$$I_{in} = I_{out} \quad \text{(steady state).} \qquad (31.27)$$

Figure 31.40 Circuit diagram for three resistors connected in parallel across an ideal battery.

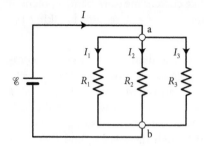

As an example of a simple multiloop circuit, let us begin by considering three resistors connected in parallel across an ideal battery, as shown in **Figure 31.40**. Our goal will be to determine the equivalent resistance of the parallel combination of resistors. The circuit in Figure 31.40 has four branches, two junctions, and six loops (Can you identify them all?). Because there are four branches, we begin by labeling the currents in the four branches: the current in the battery, I, and the currents I_1, I_2, and I_3 through the three resistors. At junction a in Figure 31.40 we have $I_{in} = I$ and $I_{out} = I_1 + I_2 + I_3$, so the junction rule yields

$$I = I_1 + I_2 + I_3. \tag{31.28}$$

We can follow the same reasoning at junction b, but as you can easily verify, we would obtain exactly the same equation.

The potential difference V_{ba} is the same across each of the three resistors, so we can write for the currents in the resistors

$$I_1 = \frac{V_{ba}}{R_1}, \quad I_2 = \frac{V_{ba}}{R_2}, \quad I_3 = \frac{V_{ba}}{R_3}. \tag{31.29}$$

Substituting Eq. 31.29 into Eq. 31.28 gives

$$I = \frac{V_{ba}}{R_1} + \frac{V_{ba}}{R_2} + \frac{V_{ba}}{R_3} = V_{ba}\left(\frac{1}{R_1} + \frac{1}{R_2} + \frac{1}{R_3}\right). \tag{31.30}$$

If we replace the combination of the three parallel resistors by a single resistor having an equivalent resistance R_{eq}, we have

$$I = V_{ba}\left(\frac{1}{R_{eq}}\right). \tag{31.31}$$

The value of R_{eq} needed to obtain the same current I in the battery as with the three resistors connected in parallel is obtained by comparing Eqs. 31.30 and 31.31:

$$\frac{1}{R_{eq}} = \frac{1}{R_1} + \frac{1}{R_2} + \frac{1}{R_3}, \tag{31.32}$$

which tells us that the reciprocal of the equivalent resistance of resistors connected in parallel equals the sum of the reciprocals of the individual resistances. Because the line of argument I used in deriving Eqs. 31.28–31.32 can be followed for any number of resistors, this statement is true for any number of resistors in parallel:

$$\frac{1}{R_{eq}} = \frac{1}{R_1} + \frac{1}{R_2} + \frac{1}{R_3} + \cdots \quad \text{(resistors in parallel).} \tag{31.33}$$

For two resistors in parallel, the equivalent resistance can also be written as

$$R_{eq} = \frac{R_1 R_2}{R_1 + R_2} \quad \text{(two resistors in parallel).} \tag{31.34}$$

Note that the equivalent resistance is smaller than either of the two individual resistances ($R_{eq} < R_1$ and $R_{eq} < R_2$). This is always true for resistors in parallel. As we discussed in Section 31.3, although it may seem paradoxical that combining multiple resistors reduces the equivalent resistance, combining resistors in parallel amounts to providing multiple paths for the current to follow. Viewed from this perspective, it is not surprising that increasing the number of branches in the circuit reduces the equivalent resistance of the circuit.

✋ **31.16** Let $\mathscr{E} = 9$ V, $R_1 = 3\ \Omega$, $R_2 = 10\ \Omega$, and $R_3 = 5\ \Omega$ in Figure 31.40. (*a*) What is the equivalent resistance of the three resistors? (*b*) What is the current in the battery?

The circuit in Figure 31.40 has more than one loop, but it is still a relatively simple circuit. To analyze more complex multiloop circuits, we need to derive a set of mathematical relationships among the unknown quantities (typically the currents in the various branches). Specifically, we need to have as many independent mathematical relationships as there are unknown quantities in the circuit.

One way to obtain a suitable set of equations is to apply the junction rule and the loop rule as many times as necessary to obtain a suitable number of equations. The junction rule and the loop rule are sometimes referred to as either *Kirchhoff's circuit rules* or *Kirchhoff's laws*, after the German physicist Gustav Kirchhoff (1824–1887). These rules do not contain any new physical principles, but instead are simply the application of principles we have already encountered—continuity (the junction rule) and Eq. 25.32 representing conservation of energy (the loop rule)—to an electric circuit in steady state.

Suppose we are interested in determining the currents in the circuit shown in Figure 31.41. Specifically, we want to determine the currents I_1, I_2, and I_3 in the three resistors. (The chosen reference directions in Figure 31.41 are arbitrary.) We wish to determine values for these currents in terms of the resistances R_1, R_2, and R_3 and the emfs of the batteries.

The circuit in Figure 31.41 has two junctions, three branches, and three loops. At junction b we have $I_{\text{in}} = I_1$ and $I_{\text{out}} = I_2 + I_3$, so, according to the junction rule,

$$I_1 = I_2 + I_3. \tag{31.35}$$

Applying the junction rule at f yields no new information because we would obtain the same equation. In general, it can be shown that in a circuit that contains N junctions, the junction rule yields $N - 1$ independent equations. This means that in a circuit with N junctions, we should apply the junction rule $N - 1$ times.

Next we need to use the loop rule. There are three unknown quantities in this problem (the three currents), so we need two equations in addition to Eq. 31.35 to determine the three currents. In other words, we need to apply the loop rule to two loops in the circuit.

First let us apply the loop rule to loop abdfea, choosing a clockwise travel direction around the loop:

$$V_{\text{ab}} + V_{\text{bd}} + V_{\text{df}} + V_{\text{fe}} + V_{\text{ea}} = 0. \tag{31.36}$$

Expressing the potential differences in this equation in terms of the emfs, resistances, and currents (see Table 32.2), we get

$$-I_1 R_1 - \mathscr{E}_2 - I_2 R_2 + 0 + \mathscr{E}_1 = 0. \tag{31.37}$$

Next we apply the loop rule to loop bcgfdb, going clockwise around the loop:

$$V_{\text{bc}} + V_{\text{cg}} + V_{\text{gf}} + V_{\text{fd}} + V_{\text{db}} = 0, \tag{31.38}$$

or

$$-I_3 R_3 + \mathscr{E}_3 + 0 + I_2 R_2 + \mathscr{E}_2 = 0. \tag{31.39}$$

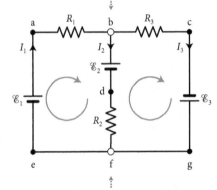

Figure 31.41 Using the junction and loop rules to determine the currents in the resistors in this multiloop circuit.

I_1 into junction; I_2 and I_3 out of junction

I_2 and I_3 into junction; I_1 out of junction

Procedure: Analyzing multiloop circuits

Here is a series of steps for calculating currents or potential differences in multiloop circuits.

1. Identify and label the junctions in the circuit.
2. Label the current in each branch of the circuit, arbitrarily assigning a direction to each current.
3. Apply the junction rule to all but one of the junctions. (The choice of which junctions to analyze is arbitrary; choose junctions that involve the quantities you are interested in calculating.)
4. Identify the loops in the circuit and apply the loop rule (see the Procedure box on page 831) enough times to obtain a suitable number of simultaneous equations relating the unknowns in the problem. The choice of loops is arbitrary, but every branch must be in at least one of the loops. Traverse each loop in whichever direction you prefer, but be sure you traverse each loop completely and stick with the direction of travel and with the chosen directions of the currents.

There are several simplifications you can make during your analysis.

1. Multiloop circuits can sometimes be simplified by replacing parallel or series combinations of resistors by their equivalent resistances. If you can reduce the circuit to a single loop, you can solve for the current in the source. You may then need to "unsimplify" and undo the resistor simplification to calculate the current or potential difference across a particular resistor.
2. In general when solving problems, you should solve equations analytically before substituting known numerical values. When solving the simultaneous equations you obtain for multiloop circuits, however, you can often simplify the algebra if you substitute the known numerical values earlier on.

We now have three equations (Eqs. 31.35, 31.37, and 31.39) involving three unknowns (I_1, I_2, and I_3). (As stated earlier, the resistances and emfs are not unknowns. We consider R_1, R_2, R_3, $\mathcal{E}_1$, $\mathcal{E}_2$, and $\mathcal{E}_3$ to have certain values, although we are not given numerical values for them. Other problems might treat one or more of the resistances or emfs as unknowns, in which case some additional information, such as the values of some of the currents, would be needed.) Equations 31.35, 31.37, and 31.39 can be solved using standard algebraic techniques. Because every circuit is different, we shall not solve this particular problem. Complete solutions to such problems are provided in the examples below and in the Practice Volume. All the physics of the circuit is expressed in Eqs. 31.35–31.39; the remainder of the solution simply involves algebra. The steps for analyzing multiloop circuits are summarized in the Procedure box above.

Example 31.10 Multiloop circuit

Consider the circuit shown in **Figure 31.42**. Determine the magnitude of the potential difference across R_1 if $\mathcal{E}_1 = \mathcal{E}_2 = 9.0$ V and $R_1 = R_2 = R_3 = 300\ \Omega$.

❶ **GETTING STARTED** I begin by drawing a circuit diagram for the circuit (**Figure 31.43**).

Figure 31.43

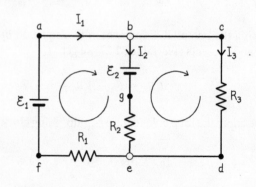

Figure 31.42 Example 31.10.

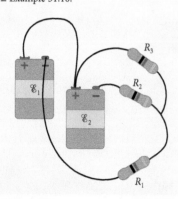

② DEVISE PLAN To solve this problem I follow the steps in the Procedure box on page 836. There are two junctions in the circuit (b, e), three loops (abgefa, bcdegb, abcdefa), and three branches (efab, egb, edcb). I label the currents in the three branches I_1, I_2, and I_3. If I apply the junction rule to one of the junctions and the loop rule to two of the loops, I will obtain three equations that I can solve for the three unknowns I_1, I_2, and I_3. I choose loops abgefa and bcdegb and a clockwise travel direction in each of these loops. Once I know the current I_1, I can use Ohm's law (Eq. 31.11) to compute the potential difference across the resistor of resistance R_1. Because the values of the emfs and resistances are all the same, I set $\mathscr{E}_1 = \mathscr{E}_2 = \mathscr{E}$ and $R_1 = R_2 = R_3 = R$ to simplify the calculation.

③ EXECUTE PLAN For junction b I have $I_{in} = I_1$ and $I_{out} = I_2 + I_3$, so applying the junction rule gives

$$I_1 = I_2 + I_3. \tag{1}$$

Applying the loop rule going clockwise around loop abgefa gives

$$0 - \mathscr{E} - I_2R - I_1R + \mathscr{E} = 0,$$

or
$$I_2 = -I_1. \tag{2}$$

Next I apply the loop rule going clockwise around loop bcdegb:

$$0 - I_3R + 0 + I_2R + \mathscr{E} = 0. \tag{3}$$

I now have three equations from which I can determine the three unknown currents.

Substituting Eq. 2 into Eq. 1 yields $I_1 = -I_1 + I_3$, or

$$I_3 = 2I_1. \tag{4}$$

Substituting this expression for I_3 and Eq. 2 into Eq. 3 yields

$$-2I_1R - I_1R + \mathscr{E} = 0.$$

Solving for I_1 then gives me

$$I_1 = \frac{\mathscr{E}}{3R}, \tag{5}$$

and substituting $\mathscr{E} = 9.0$ V and $R = 300$ Ω, I get

$$I_1 = \frac{9.0 \text{ V}}{900 \text{ Ω}} = 0.010 \text{ A}.$$

The magnitude of the potential difference across the resistor with resistance R_1 is therefore $|-I_1R_1| = (0.010 \text{ A})(300 \text{ Ω}) = 3.0$ V. ✔

④ EVALUATE RESULT I can use my expression for I_1 to determine the potential difference between positions b and e. Going through branch efab, I have $V_{eb} = V_{ef} + V_{fa} + V_{ab} = -I_1R + \mathscr{E} + 0$. Substituting my expression for I_1 (Eq. 5), I get $V_{eb} = \frac{2}{3}\mathscr{E} = +6.0$ V. Going through branch egb, I have $V_{eb} = V_{eg} + V_{gb} = +I_2R + \mathscr{E}$. Because $I_2 = -I_1$ (Eq. 2), this expression gives the same result, $V_{eb} = +6.0$ V. Going through branch edcb and using Eq. 4, I get $V_{eb} = 0 + I_3R + 0 = +2I_1R$. Substituting Eq. 5, I obtain $V_{eb} = \frac{2}{3}\mathscr{E} = +6.0$ V. Because I obtain the same value for the potential difference each time, I am confident that my result for I_1 is correct.

Example 31.11 Wheatstone bridge

The circuit shown in **Figure 31.44** includes a variable resistor, the resistance R_{var} of which can be adjusted, and a resistor of unknown value R. A circuit with such a network of resistors is called a *Wheatstone bridge* and can be used to determine the value of the unknown resistance R by adjusting the variable resistor so that the light bulb does not glow. The light bulb is initially glowing when R_{var} is set to 20 Ω, but it goes out when R_{var} is adjusted to 12 Ω. Determine the current in the battery when the light bulb is out.

Figure 31.44 Example 31.11.

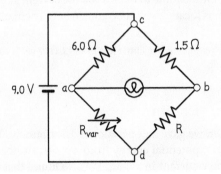

① GETTING STARTED The bulb glows when there is a current in it and stops glowing when the current in it is zero. In the latter case, it doesn't matter whether the bulb is connected or not, so I can omit it from the circuit and simplify the circuit diagram, as shown in **Figure 31.45a** on the next page. Because R_{var} is set to 12 Ω, I use that value rather than R_{var}. I use the label I_1 for the current in the branch containing the variable resistor, I_2 for the current in the branch containing R, and I_3 for the current in the battery, which is the current I need to determine.

② DEVISE PLAN In order for the current in the bulb to be zero, junctions a and b in Figure 31.44 must be at the same potential, and I must have $V_{ca} = V_{cb}$ and $V_{ad} = V_{bd}$. I can then use Ohm's law to express these two conditions in terms of the unknown resistance R and the currents I_1 and I_2, and solve the resulting set of equations for R. Once I know R, I can further simplify the circuit by replacing the sets of resistors connected in series in each branch by a single equivalent resistor. This yields the circuit shown in Figure 31.45b. To determine the current in the battery, I can then replace the two parallel resistors by a single equivalent resistor as in Figure 31.45c, and use Eq. 31.34 to determine its resistance.

Figure 31.45

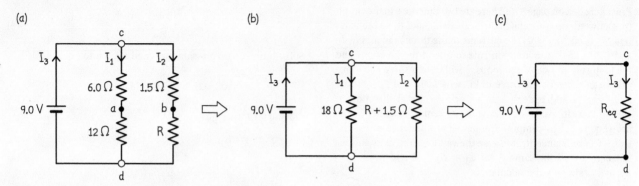

(a)   (b)   (c)

③ EXECUTE PLAN Rearranging the Ohm's law equation (Eq. 31.11) to $V = IR$ and applying it to the two conditions $V_{ca} = V_{cb}$ and $V_{ad} = V_{bd}$, I have, when I travel clockwise in the circuit,

$$-(6.0 \ \Omega)I_1 = -(1.5 \ \Omega)I_2 \tag{1}$$

$$-(12 \ \Omega)I_1 = -RI_2. \tag{2}$$

Solving Eq. 1 for I_1 gives me $I_1 = I_2/4$, and substituting this result into Eq. 2 and solving for R give me

$$R = 3.0 \ \Omega.$$

With this value of R the resistance in the rightmost branch in Figure 31.45b becomes 4.5 Ω. The equivalent resistance of the two parallel resistors is, from Eq. 31.34,

$$R_{eq} = \frac{(18 \ \Omega)(4.5 \ \Omega)}{18 \ \Omega + 4.5 \ \Omega} = 3.6 \ \Omega.$$

The current in the battery is then

$$I_3 = \frac{\mathscr{E}}{R_{eq}} = \frac{9.0 \ \text{V}}{3.6 \ \Omega} = 2.5 \ \text{A}. \ ✔$$

④ EVALUATE RESULT I can verify that $V_{ca} = V_{cb}$ by calculating I_1 and I_2 and substituting these values in Eq. 1. Applying the junction rule to junction c, I have $I_3 = I_1 + I_2$. I know that $I_1 = I_2/4$, so $I_2 = \frac{4}{5}I_3 = 2.0 \ \text{A}$ and $I_1 = \frac{1}{5}I_3 = 0.5 \ \text{A}$. Using these values, I get $V_{ca} = -(6.0 \ \Omega)(0.5 \ \text{A}) = -3.0 \ \text{V}$ and $V_{cb} = -(1.5 \ \Omega)(2.0 \ \text{A}) = -3.0 \ \text{V}$, so junctions a and b are at the same potential, as I expect when there is no current in the bulb.

Example 31.11 illustrates an alternative to our standard procedure for analyzing circuits. For circuits that contain many junctions and loops, it is sometimes easier to simplify the circuit by replacing part of it with a single equivalent resistance, as I did in Figure 31.45b of the example. When you do this, think about what question you ultimately want to answer about the circuit and about which parts of the circuit can be simplified without interfering with solving the problem. In Example 31.11, because what I wanted to determine was the current in the battery, I could replace everything outside the battery by a single equivalent resistor. However, if I had been asked to determine the currents at, say, junction b when the bulb was not glowing, I would have stopped simplifying the circuit at the stage shown in Figure 31.45b. Simplifying the circuit to just a single resistor and battery would have made it impossible to determine the current at b because the branch in which junction b is located would be gone from the circuit.

31.17 If R_{var} in Figure 31.44 is adjusted to a little less than 12 Ω, what is the direction of the current in the light bulb?

31.8 Power in electric circuits

At the beginning of this chapter, we gave an operational description of a single-loop circuit as converting electric potential energy from a power source to some other form of energy by driving a current in a load. We also found that the loop rule, which essentially embodies the idea of conservation of energy, is a powerful tool for analyzing circuits.

Let us examine how rapidly the power source in **Figure 31.46** can deliver energy to its load. From Chapter 25 we know that the electrostatic work done on a charge carrier carrying charge q in a time interval Δt as it passes through the load from point a to point b is the negative of the potential difference across the load multiplied by q (Eq. 25.17):

$$W_q(\text{a} \rightarrow \text{b}) = -qV_{\text{ab}}. \tag{31.40}$$

Because V_{ab} is negative, the work done by the power source on the charge carrier is positive, increasing its energy. While it travels through the load, however, this energy is converted to other forms, and so the amount of energy converted in the load is given by

$$\Delta E = -qV_{\text{ab}}. \tag{31.41}$$

To determine the rate at which energy is converted, we substitute this expression into the expression for the average power (time rate of change of energy, Eq. 9.29). Because we are considering a steady-state situation, average and instantaneous power are the same, so we can drop the subscript av:

$$P \equiv \frac{\Delta E}{\Delta t} = \frac{-qV_{\text{ab}}}{\Delta t} = -IV_{\text{ab}}. \tag{31.42}$$

The form of energy to which the electric potential energy in a circuit is converted depends on the type of load. When the load is a light bulb, for example, the conversion is from electric potential energy to light and thermal energy. When the load is an electric motor, the conversion is to kinetic energy as, say, the blades of a fan start rotating. Here we're interested in circuits in which the load is a resistor, in which case the conversion is to thermal energy. Substituting $V_{\text{ab}} = -IR$ in Eq. 31.42 gives us, for the rate at which energy is dissipated in a resistor,

$$P = I^2R. \tag{31.43}$$

It can also be useful to know the rate at which energy is dissipated in terms of the potential difference across the resistor. Substituting $-V_{\text{ab}}/R$ for I in Eq. 31.43 yields

$$P = \frac{V_{\text{ab}}^2}{R}. \tag{31.44}$$

Keep in mind that although Eq. 31.42 is valid for any electrical device, we have derived Eqs. 31.43 and 31.44 using Ohm's law for resistors, in which electric potential energy is converted to thermal energy.

We can apply a similar reasoning to the power source. Let's consider the case of a battery as the power source. When a charge carrier carrying charge q moves from the negative terminal to the positive terminal inside the battery, its electric potential energy increases at a rate given by

$$P = \frac{q\mathcal{E}}{\Delta t} = I\mathcal{E}. \tag{31.45}$$

For an ideal battery $\mathcal{E} = IR$ (Eq. 31.24), and so Eq. 31.45 becomes

$$P = I\mathcal{E} = I^2R. \tag{31.46}$$

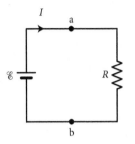

Figure 31.46 A battery delivers energy to a load connected to it.

Comparing Eqs. 31.46 and 31.43, we see that the rate at which electric potential energy increases in the battery is equal to the rate at which electric potential energy is dissipated in the load. For a nonideal battery, some energy is dissipated inside the battery. As we have seen in Example 31.9, we can account for this dissipation by attributing an internal resistance R_{batt} to the battery. This internal resistance decreases the potential difference across the terminals: $V_{ba} = \mathscr{E} - IR_{batt}$, so $\mathscr{E} = V_{ba} + IR_{batt}$. Substituting this expression and $V_{ba} = IR$ into Eq. 31.45 yields

$$P = I\mathscr{E} = IV_{ba} + I^2R_{batt} = I^2R + I^2R_{batt}. \tag{31.47}$$

So for a nonideal battery the rate at which chemical energy is converted is equal to the sum of the rates at which energy is dissipated in the load and inside the battery, as we would expect.

31.18 The SI units of power suggested by Eqs. 31.42 and 31.43 are $A \cdot V$ and $A^2 \cdot \Omega$, respectively. Show that these SI units are equivalent to the derived SI unit for power, the watt.

Example 31.12 Battery to battery

A 9.0-V and a 6.0-V battery are connected to each other (**Figure 31.47**). Each battery has an internal resistance of 0.25 Ω. At what rate is energy dissipated in the 6.0-V battery?

Figure 31.47 Example 31.12.

❶ **GETTING STARTED** I begin by drawing a circuit diagram for the two-battery combination, representing the internal resistance by two resistors in series with the batteries (**Figure 31.48**). I note that the negative terminals of the batteries are connected to each other, and the positive terminals are connected to each other. I arbitrarily choose clockwise both for the reference direction for the current and for my travel around the circuit.

Figure 31.48

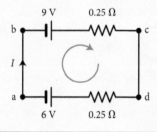

❷ **DEVISE PLAN** Because this circuit contains only one loop and no junctions, the current is the same everywhere in the circuit. To determine the current, I apply the loop rule and solve the resulting equation for the current. I can then apply Eq. 31.43 to determine the rate at which energy is dissipated in the 6.0-V battery.

❸ **EXECUTE PLAN** As I travel clockwise around the circuit starting at a, the loop rule yields

$$9.0 \text{ V} - I(0.25 \ \Omega) - I(0.25 \ \Omega) - 6.0 \text{ V} = 0$$

$$I = \frac{9.0 \text{ V} - 6.0 \text{ V}}{0.50 \ \Omega} = \frac{3.0 \text{ V}}{0.50 \ \Omega} = 6.0 \text{ A}.$$

The fact that I get a positive value for the current tells me that my assumed reference direction for the current (clockwise) is correct. Energy is dissipated in the 6.0-V battery as resistive losses in the internal resistance. Using Eq. 31.43 for the rate at which energy is dissipated in a resistor, I get

$$P = I^2R = (6.0 \text{ A})^2(0.25 \ \Omega) = 9.0 \text{ W}. ✔$$

❹ **EVALUATE RESULT** Even though the internal resistance is small, this power is substantial because the current is large. The current is large because there is very little resistance in the circuit. It's not a good idea to connect two batteries of different emf values in this manner!

31.19 (a) In Example 31.12, how would the answer change if we had chosen a counterclockwise travel direction around the circuit? (b) At what rate is energy dissipated in the 9-V battery?

Chapter Glossary

SI units of physical quantities are given in parentheses.

Branch The part of a circuit between two junctions that does not contain any junctions itself. In steady state, the current is the same at any location along a branch.

Branch rule The current in each branch of a multiloop circuit is the same throughout that branch.

Circuit diagram A schematic representation of an electric circuit, using standard symbols to represent circuit elements and straight lines to represent conducting connections between elements.

Conductivity σ (A/(V·m)) The proportionality constant relating current density to electric field in a conductor:

$$\sigma \equiv \frac{J}{E}. \tag{31.8}$$

The conductivity is a measure of a material's ability to conduct current. For a metal,

$$\sigma = \frac{ne^2\tau}{m_e}, \tag{31.9}$$

where n is the number density of the free electrons, e is the charge of the electron, m_e is the mass of the electron, and τ is the average time interval between collisions.

Current continuity principle In steady state, the current is the same at all locations along a single-loop electric circuit.

Current density $\vec{J}$ (A/m²) A vector whose magnitude represents the current per unit area through a conductor of cross-sectional area A:

$$J \equiv \frac{|I|}{A}. \tag{31.5}$$

Drift velocity (m/s) The average velocity that an electron attains in a conductor due to an electric field:

$$\vec{v}_d = -\frac{e\vec{E}}{m_e}\tau. \tag{31.3}$$

Electric circuit An interconnection of electrical elements.

Equivalent resistance The resistance that can replace a combination of circuit elements without altering the current from the battery. For resistors connected in series,

$$R_{eq} = R_1 + R_2 + R_3 + \cdots \quad \text{(resistors in series)}, \tag{31.26}$$

and for resistors connected in parallel,

$$\frac{1}{R_{eq}} = \frac{1}{R_1} + \frac{1}{R_2} + \frac{1}{R_3} + \cdots \quad \text{(resistors in parallel)}. \tag{31.33}$$

Junction A location in a circuit where more than two wires or other circuit elements are connected together.

Junction rule The sum of the currents going into a junction is equal to the sum of the currents coming out of the same junction.

Load A combination of circuit elements connected to a power source where electric potential energy is converted to other forms of energy.

Loop A closed conducting path in an electric circuit.

Loop rule In steady state, the sum of the emfs and the potential differences around any loop in an electric circuit is zero:

$$\sum \mathcal{E} + \sum V = 0 \quad \text{(steady state, around loop)}. \tag{31.21}$$

Ohm (Ω) The derived SI unit of resistance:

$$1\,\Omega \equiv 1\,\text{V/A}.$$

Ohm's law The current in a conductor between two points is directly proportional to the potential difference across the two points and inversely proportional to the resistance between them:

$$I = \frac{V}{R}. \tag{31.11}$$

Power source A circuit element that provides electric potential energy to the elements in an electric circuit by maintaining a potential difference between the two locations in the circuit in which it is connected.

Resistance R (Ω) The resistance of a circuit element is proportional to the potential difference that must be applied across it to obtain a current of 1 A in the load.

$$R \equiv \frac{V}{I}. \tag{31.10}$$

Resistor A conducting object that has nonnegligible constant resistance, usually attached to wires at either end for easy incorporation into a circuit.

Short circuit A branch in a circuit in parallel with a load that consists of only wire. Because of its negligible resistance, a short circuit diverts all the current away from the load.

Steady state A circuit is in steady state when the current has a constant value at all points in the circuit; at the instants when the current is established or cut off, the current is changing throughout the circuit and is not in steady state.

32

Electronics

*I*n the preceding chapter, we discussed electric circuits in which the current is steady. As noted in that chapter, the steady flow of charge carriers in one direction only is called *direct current*. Batteries and other devices that produce static electrical charge, such as van de Graaff generators, are sources of direct current. Although direct current has many uses, it has several limitations as well. For example, in order to produce substantial currents, direct-current sources must be quite large and are therefore cumbersome. More important, steady currents do not generate any electromagnetic waves, which can be used to transmit information and energy through space, as we saw in Chapter 30.

Because of these and many other factors, most electric and electronic circuits operate with **alternating currents** (abbreviated AC)—currents that periodically change direction. The current provided by household outlets in the United States, for instance, alternates in direction, completing 60 cycles per second (that is, with a frequency of 60 Hz), and the currents in computer circuits change direction billions of times per second. It is no understatement to say that contemporary society *depends* on alternating currents.

In this chapter we discuss the basics of both household currents and the electronics that lie at the heart of computers.

32.1 Alternating currents

We have already encountered one example of an electrical device that produces a changing current: a capacitor that is either charging or discharging. Let's consider what happens when we connect an inductor to a charged capacitor (**Figure 32.1**). A circuit that consists of an inductor and a capacitor is called an *LC* circuit. As soon as the two circuit elements are connected, positive charge carriers begin to flow clockwise through the circuit. The magnitude of the current increases from its initial value of zero (**Figure 32.2a** on next page) to a nonzero value (Figure 32.2*b–d*). The capacitor discharges through the inductor, and the current causes a magnetic field in the inductor. As the current in the inductor increases, the magnetic field also increases, causing an induced emf (see Section 29.7) that opposes this increase and prevents the current from increasing rapidly. Consequently, the capacitor discharges more slowly than it would if we had connected it to a wire.

Figure 32.1 What happens when we connect an inductor to a charged capacitor?

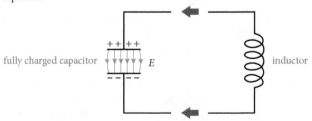

fully charged capacitor   inductor

E

++++++

↓↓↓↓↓↓

— — — —

32.1 (*a*) Just before the inductor is connected to the charged capacitor, what type of energy is contained in the system comprising the two elements? (*b*) Once the two elements are connected to each other, what happens to that energy? (*c*) Once the capacitor is completely discharged, in what form is the energy in the circuit?

As you saw in Checkpoint 32.1, when the capacitor is completely discharged, all of the energy in an *LC* circuit is contained in the magnetic field and this field reaches its maximum magnitude (Figure 32.2*c*). Because the magnetic energy is proportional to the square of the current in the inductor (Eq. 29.25), the current, too, reaches its maximum value at this instant. Once the magnetic field and the current reach their maximum values, the current begins to charge the capacitor in the opposite direction (Figure 32.2*d*), and the charge on the capacitor increases as the magnetic field in the inductor decreases. When the magnetic field in the inductor is zero, the current is also zero and the capacitor has again maximum charge but with the opposite polarity (Figure 32.2*e*). The process then repeats itself with the current in the opposite direction (Figure 32.2*f–h*) until the capacitor is restored to its starting configuration. Then the cycle begins again.

Figure 32.3 on the next page shows the time dependence of the electric potential energy U^E stored in the capacitor and the magnetic potential energy U^B stored in the inductor. In the absence of dissipation, the energy in the circuit, $U^E + U^B$, must stay constant. Therefore, when the capacitor is not charged and U^E drops to zero, U^B must reach its maximum value, U_{max}.

There is always some dissipation in a circuit. Resistance in the connecting wires gradually converts electrical energy to thermal energy. Consequently, the oscillations decay in the same manner as the damped mechanical oscillations we considered in Section 15.8. Resistance therefore plays the same role in oscillating circuits as damping does in mechanical oscillators.

Throughout this chapter we work with time-dependent potential differences and currents. To make the notation as concise as possible, we represent time-dependent quantities with lowercase letters. In other words, v_C is short for $V_C(t)$ and i is short for $I(t)$. We also need a symbol for the maximum value of an oscillating quantity—its *amplitude* (see Section 15.1). For this we use a capital letter without the time-dependent marker (t); thus V_C is the maximum value of the potential difference across a capacitor, and I is the maximum value of the current in a circuit.

Unlike their counterparts in DC circuits, the potential difference across the capacitor, v_C, and the current in the *LC* circuit, i, change sign periodically. So, when analyzing AC circuits, we must carefully define what we mean by the sign of these quantities. To analyze the *LC* circuit in Figure 32.2, for example, we choose a reference direction for the current i and let the potential difference v_C be positive when the top capacitor plate is at a higher potential than the bottom plate (**Figure 32.4a** on page 845). Note that both of these choices are arbitrary.

Figure 32.2 A series of "snapshots" showing what happens when we connect an inductor to a charged capacitor.

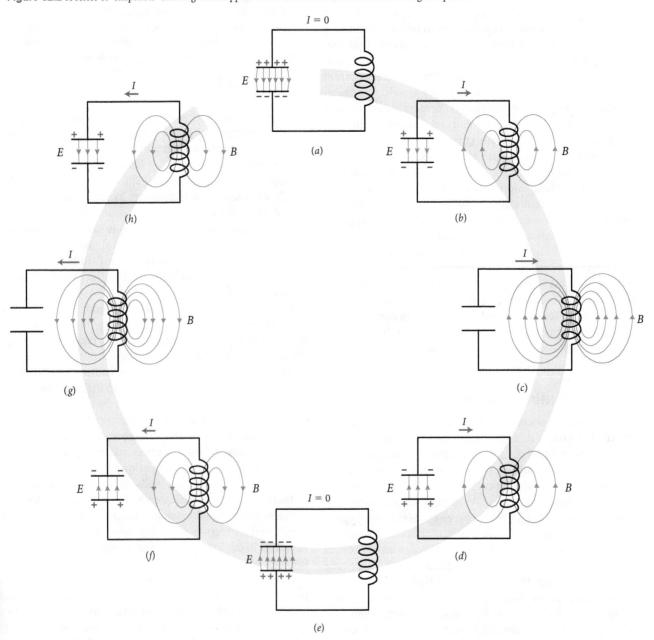

Figure 32.3 Time dependence of the electric potential energy U^E stored in the capacitor and the magnetic potential energy U^B stored in the inductor. In the absence of dissipation, the energy in the circuit, $U^E + U^B$, is a constant U_{max}.

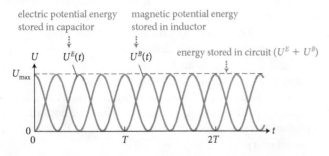

Figure 32.4b shows graphs for v_C and i with these choices. The potential difference across the capacitor v_C is initially positive, representing the situation at Figure 32.2a. During the first quarter cycle (Figure 32.2b), the capacitor is discharging and positive charge carriers travel away from the top plate of the capacitor in the chosen reference direction, and so i is positive. In the part of the cycle represented by Figure 32.2f, where the capacitor is again discharging, v_C is negative (because the top plate is negatively charged) and i is negative (because the direction of current is opposite the chosen reference direction), as shown in the time interval $\frac{1}{2}T < t < \frac{3}{4}T$ in Figure 32.4b. (See if you can work out the signs during the time intervals when the capacitor is

Figure 32.4 For the *LC* circuit shown in Figure 32.2, graphs of the time-dependent potential difference across the capacitor (defined to be positive when the top plate is at the higher potential) and the current in the circuit (defined to be positive when positive charge carriers travel away from top plate of the capacitor). One cycle is completed in a time interval *T* (the *period*).

(a)

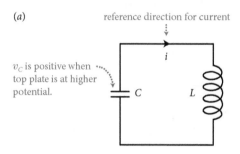

reference direction for current

v_C is positive when top plate is at higher potential.

(b)

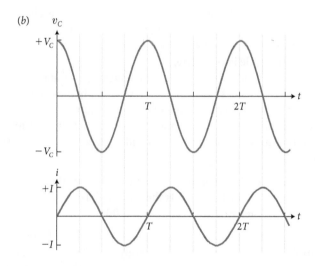

charging.) Both v_C and *i* vary sinusoidally in time, with v_C at its maximum when *i* is zero, and vice versa.

Because of dissipation, the *LC* circuit in Figure 32.1 is not a practical source of alternating current; instead, generators are widely used to produce sinusoidally alternating emfs in a circuit (see Example 29.6). The symbol for a source that generates a sinusoidally alternating potential difference or current is shown in **Figure 32.5**; such a source is called an **AC source**. The time-dependent emf an AC source produces across its terminals is designated $\mathscr{E}$, and its amplitude is designated $\mathscr{E}_{max}$.

Figure 32.5 Symbol that represents an AC source in an electric circuit. The AC source produces a sinusoidally varying emf $\mathscr{E}$ across its terminals.

$\mathscr{E}$

Exercise 32.1 AC source and resistor

Figure 32.6 shows a circuit consisting of an AC source and a resistor. The emf produced by the generator varies sinusoidally in time. Sketch the potential difference across the resistor as a function of time and the current in it as a function of time.

Figure 32.6 Exercise 32.1.

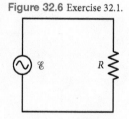

SOLUTION Ohm's law, the junction rule, and the loop rule (see Chapter 31) apply to alternating-current circuits just as they do to direct-current circuits. All I need to remember here is that the potential differences and currents are time dependent. Applying the loop rule to this circuit requires the time-dependent potential difference across the resistor v_R to equal the emf $\mathscr{E}$ of the AC source at every instant, so that the sum of the potential differences around the circuit is always zero. Consequently, v_R oscillates just as $\mathscr{E}$ oscillates, as shown in **Figure 32.7**; V_R is the maximum value of the potential difference across the resistor.

Figure 32.7

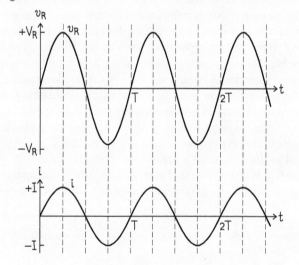

Ohm's law requires the time-dependent current *i* in the resistor to be proportional to v_R, which means that *i* also oscillates, with the current at its maximum when v_R is maximum. ✔

32.2 (*a*) Is energy dissipated in the resistor in the circuit of Figure 32.6? (*b*) If so, why doesn't the amplitude of the oscillations of v_R and *i* (shown in Figure 32.7) decrease with time?

32.2 AC circuits

The circuit discussed in Exercise 32.1 is an alternating current, or AC circuit. Such circuits exhibit more complex behavior when they contain elements that do not obey Ohm's law, so that the current is not proportional to the emf of the source. For example, let's consider the current in the circuit shown in **Figure 32.8** on the next page.

Figure 32.8 AC circuit with a capacitor connected to an AC source.

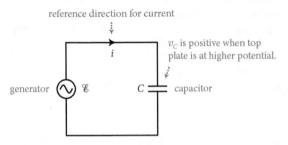

reference direction for current

v_C is positive when top plate is at higher potential.

generator $\mathcal{E}$ C capacitor

To analyze the circuit we choose a reference direction for the current i and let the potential difference v_C again be positive when the top capacitor plate is at a higher potential than the bottom plate (Figure 32.8). Because the capacitor is connected directly to the AC source, the time-dependent potential difference across the capacitor v_C equals the emf of the AC source at any instant. What is the current in the circuit? Let's begin considering what happens when the capacitor is uncharged. As v_C increases, the charge on the top plate of the capacitor increases. This means that positive charge carriers are moving toward the top plate, in the same direction as the chosen reference direction for the current, and so the current is positive (Figure 32.9a). When v_C reaches its maximum, the capacitor reaches its maximum charge and the current is instantaneously zero. As v_C decreases, the charge on the top plate of the capacitor decreases. Positive charge carriers now move away from the top plate and the current is negative (Figure 32.9b). At some instant the top plate becomes negatively charged (Figure 32.9c); v_C continues to decrease until it reaches its minimum value and the current is instantaneously zero. At that instant the capacitor again reaches its maximum charge but with the opposite

Figure 32.9 The charging and discharging of the capacitor in the circuit of Figure 32.8.

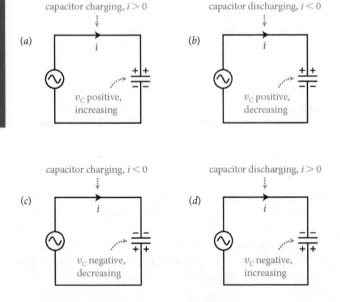

capacitor charging, $i > 0$

(a)

i

v_C positive, increasing

capacitor discharging, $i < 0$

(b)

i

v_C positive, decreasing

capacitor charging, $i < 0$

(c)

i

v_C negative, decreasing

capacitor discharging, $i > 0$

(d)

i

v_C negative, increasing

Figure 32.10 Time-dependent current in the circuit and potential difference across the capacitor for the circuit of Figure 32.9.

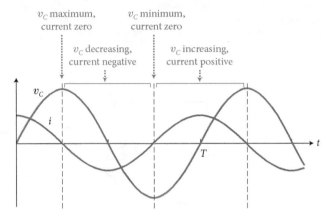

v_C maximum, current zero

v_C minimum, current zero

v_C decreasing, current negative

v_C increasing, current positive

v_C

i

T

t

polarity. As v_C begins to increase again, positive charge carriers flow toward the top plate and the current is positive again (Figure 32.9d). When both plates are uncharged again, the cycle is complete.

Figure 32.10 shows the time dependence of i and v_C in Figure 32.9. Note that i and v_C are not simply proportional to one another. Instead, the current maximum occurs one-quarter cycle before the potential difference maximum. For this reason, the current is said to *lead* the potential difference:

In an AC circuit that contains a capacitor, the current in the capacitor leads the potential difference by 90° (a quarter of an oscillation cycle).

To describe the time dependence of a sinusoidally oscillating quantity, we must specify both the angular frequency of oscillation ω and the instant at which the oscillating quantity equals zero. As discussed in Chapter 15, a sinusoidally time-dependent quantity (such as the circuit potential difference we are looking at here) can be written in the form $v = V\sin(\omega t + \phi_i)$. The argument of the sine, $\omega t + \phi_i$, is the *phase*. At $t = 0$ the phase is equal to the *initial phase* ϕ_i (Chapter 15). When the phase of an oscillating quantity is zero, $\omega t + \phi_i = 0$, the quantity is zero as well because $\sin(0) = 0$.

We can analyze phase differences in AC circuits with lots of algebra, but the underlying physics is much clearer (and the analysis much simpler!) if we use the phasor notation developed in Chapter 15 to describe oscillatory motion. Following the approach of Section 15.5, we can represent an oscillating potential difference v by a phasor rotating in a reference circle (**Figure 32.11**). Because the length of the phasor equals the amplitude (maximum value) of v, the phasor is labeled V. The phasor rotates counterclockwise at angular frequency ω. The magnitude of v at any instant is given by the vertical component of the phasor; as the phasor rotates, that component oscillates sinusoidally in time, as shown in Figure 32.11. The angle measured counterclockwise from the positive horizontal axis to the phasor is the phase $\omega t + \phi_i$.

Figure 32.11 Phasor representation of a sinusoidally varying potential difference v. The phasor rotates counterclockwise at the same angular frequency at which v oscillates. The instantaneous value of v equals the length of the vertical component of the phasor.

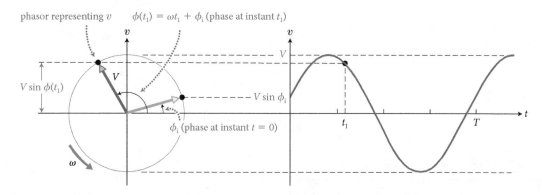

Example 32.2 Phasors

Consider the oscillating emf represented in the graph of **Figure 32.12**. Which of the phasors a–d, each shown at $t = 0$, correspond(s) to this oscillating emf?

Figure 32.12 Example 32.2.

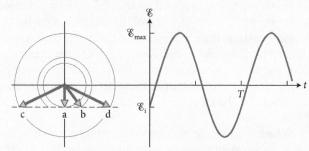

❶ **GETTING STARTED** I begin by observing from the graph that the emf is negative at instant $t = 0$ and increases until it reaches a maximum value $\mathcal{E}_{max}$.

❷ **DEVISE PLAN** To identify the correct phasor or phasors, I can use the following information: (1) the length of the phasor is equal to the amplitude of the oscillation, (2) the value of the emf at any instant corresponds to the vertical component of the phasor, and (3) the phasor rotates counterclockwise around the reference circle.

❸ **EXECUTE PLAN** The amplitudes of phasors a and b are too small and so I can rule these two out. The fact that the emf starts out negative at $t = 0$ and then increases tells me that the phasor representing it must be in the fourth quadrant (below the horizontal axis and to the right of the vertical axis), meaning the correct phasor must be d. ✔

❹ **EVALUATE RESULT** I can verify my answer by tracing out the projection of the phasor on the vertical axis as the phasors rotates counterclockwise. The initial value of the projection, initial phase, and amplitude all agree with the values of these variables represented in the graph.

32.3 Construct a phasor diagram for the time-dependent current and potential difference at $t = 0$ in the AC source-resistor circuit of Figure 32.6.

We can generalize the result of this checkpoint to represent i and v_R from Figure 32.7 at an arbitrary instant t_1. Because i and v_R are in phase for a resistor, the two phasors for i and v_R always have the same phase and so overlap (**Figure 32.13**). Note that the initial phase ϕ_i is zero because i and v_R are zero at $t = 0$ (at that instant both phasors point to the right along the horizontal axis).

The relative lengths of the I and V_R phasors are meaningless because the units of i and v_R are different. However, for circuits with multiple elements (resistors, inductors, or capacitors), the relative lengths of phasors showing the potential differences across different elements are meaningful and will prove very useful in analyzing the circuit.

Phasors are most useful when we need to represent quantities that are not in phase. **Figure 32.14** on the next page shows the phasor diagram that corresponds to Figure 32.10 (at the instant represented by Figure 32.9a). As the phasor diagram shows, the angle between V_C and I is 90°, and so the phase difference between the two phasors is $\pi/2$. Because the phasors rotate counterclockwise, we see that current phasor I is ahead of the potential difference phasor V_C, in agreement with our earlier conclusion that the current in a capacitor leads the potential difference across the capacitor.

Figure 32.13 Phasor diagram and graph showing time dependence of v_R and i from Figure 32.7.

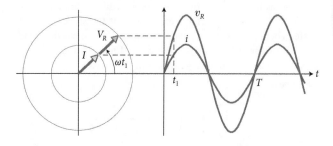

Figure 32.14 Phasor diagram and graph showing time dependence of i and v_C corresponding to Figure 32.10.

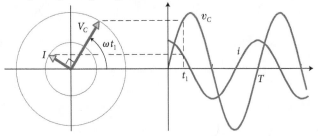

Example 32.3 Nonsinusoidal AC circuit

When a certain capacitor is connected to a nonsinusoidal source of emf as in **Figure 32.15a**, the emf varies in time as illustrated in Figure 32.15b. Sketch a graph showing the current in the circuit as a function of time.

Figure 32.15 Example 32.3.

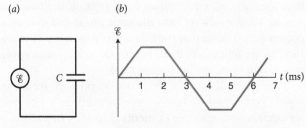

❶ **GETTING STARTED** From Figure 32.15b I see that the emf has five distinct parts during the time interval shown. During each part, the emf either is changing at a constant rate or is constant.

❷ **DEVISE PLAN** I know that the current is proportional to the rate at which the charge on the capacitor plates changes over time. I also know that the emf is proportional to the charge on the plates, and so the current is proportional to the derivative of the emf with respect to time.

❸ **EXECUTE PLAN** Between $t = 0$ and $t = 1$ ms, the emf increases at a constant rate, so $i = Cd\mathscr{E}/dt$ is constant and positive. Between $t = 1$ ms and $t = 2$ ms, the emf is constant, so $i = Cd\mathscr{E}/dt = 0$. Between $t = 2$ ms and $t = 4$ ms, the emf decreases at a constant rate, so $i = Cd\mathscr{E}/dt$ is constant and negative. Because the rate of decrease between $t = 2$ ms and $t = 4$ ms is the same as the rate of increase between $t = 0$ and $t = 1$ ms, the magnitude of the current between $t = 2$ ms and $t = 4$ ms should be the same as that between $t = 0$ and $t = 1$ ms. The current is zero again during the next millisecond ($t = 4$ ms to $t = 5$ ms) because here the emf is again constant. After $t = 5$ ms, the emf increases again at the same constant rate as between $t = 0$ and $t = 1$ ms, so the current has the same positive value as between $t = 0$ and $t = 1$ ms. The graph representing these current changes is shown in **Figure 32.16**. ✔

Figure 32.16

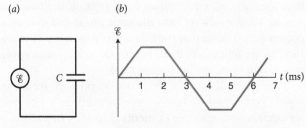

❹ **EVALUATE RESULT** When the current is positive, the emf is increasing; when the current is negative, the emf is decreasing; and when the current is zero, the emf is constant, as it should be.

Figure 32.17 AC circuit consisting of an inductor connected across the terminals of an AC source.

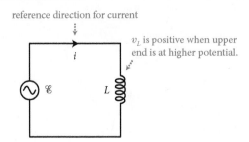

Now let's examine the behavior of an inductor connected to an AC generator (**Figure 32.17**). When the current in the circuit is changing, an emf is induced in the coil, in a direction to oppose this change (see Section 29.7). The potential difference between the ends of the inductor, which we'll denote by v_L, is proportional to the rate di/dt at which the current changes (Eq. 29.19). If the current is increasing in the reference direction for current indicated in Figure 32.17, the upper end of the inductor must be at a higher potential than the lower end to oppose the increase in current. If we take v_L to be positive when the upper end of the coil is at a higher potential, v_L must therefore be positive when the current is increasing in the reference direction for the current. This situation is represented in Figure 32.18a.

When the current reaches its maximum value in the cycle, v_L is instantaneously zero. After this instant, the current begins to decrease and the lower end of the inductor

Figure 32.18 Current and magnetic field oscillations through the inductor of Figure 32.17.

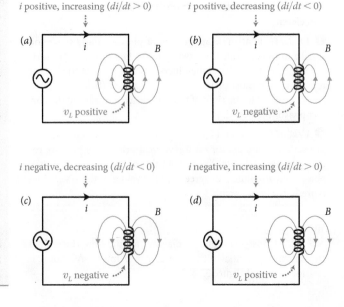

Figure 32.19 Graph of time-dependent current in the circuit and potential difference across the inductor for the circuit in Figure 32.17.

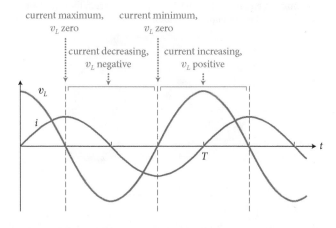

must be at a higher potential than the upper end to oppose this decrease in current. The potential difference v_L is now negative (Figure 32.18b). In the second half of the cycle, the current is in the opposite direction. As in the first part of the cycle, v_L has the same sign as di/dt (Figure 32.18c and d).

Figure 32.19 illustrates the time dependence of i and v_L in Figure 32.18. Note that the current maximum occurs one-quarter cycle after the potential difference maximum. For this reason, the current is said to *lag* the potential difference:

> **In an AC circuit that contains an inductor, the current in the inductor lags the potential difference by 90°.**

Figure 32.20 shows the phasor diagram that corresponds to Figure 32.19 (at the instant represented by Figure 32.18a). Just as with the capacitor, the angle between V_L and I is 90° and so the phase difference is $\pi/2$, but in this case the current phasor I is behind the potential difference phasor V_L, in agreement with our earlier conclusion that the current in an inductor lags the potential difference across the inductor.

32.4 What are the initial phases for the phasors in Figures 32.13 and 32.20?

Figure 32.20 Phasor diagram and graph showing time dependence of i and v_L corresponding to Figure 32.19.

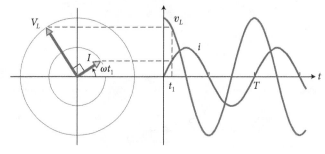

Figure 32.21 Schematic depiction of silicon, phosphorus, and boron atoms, shown as an inner core surrounded by valence electrons.

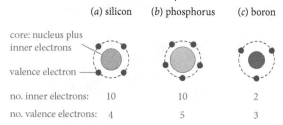

32.3 Semiconductors

Most modern electronic devices are made from a class of materials called **semiconductors**. Semiconductors have a limited supply of charge carriers that can move freely; consequently, their electrical conductivity is intermediate between that of conductors and that of insulators. Semiconductors are widely used in the manufacture of electronic devices such as transistors, diodes, and computer chips because their conductivity can be tailored chemically for particular applications layer by layer, even within a single piece of semiconductor.

Semiconductors are of two main types: intrinsic and extrinsic. *Intrinsic semiconductors* are chemically pure and have poor conductivity. *Extrinsic* or *doped semiconductors* are not chemically pure, have a conductivity that can be finely tuned, and are widely used in the microelectronics industry. The most widely used semiconductor is silicon, a nonmetallic element that makes up more than one-quarter of Earth's crust. **Figure 32.21a** shows a schematic of a silicon atom, which consists of a nucleus surrounded by fourteen electrons. Ten of these electrons are tightly bound to the nucleus—we'll refer to these electrons plus the nucleus as the *core* of the atom. The remaining outermost four electrons are called the atom's *valence electrons*. Each valence electron can form a covalent bond with a valence electron of another silicon atom. These bonds hold many identical silicon atoms together in a crystalline lattice (**Figure 32.22**).

Figure 32.22 Schematic of a crystalline lattice of silicon atoms, showing electrons participating in silicon-silicon bonds. (A real silicon crystal exists in three dimensions, and not all of the silicon-silicon bonds lie in a plane; this diagram illustrates only the essential idea that all of the valence electrons participate in covalent bonds.)

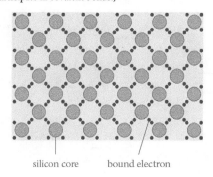

CONCEPTS

Figure 32.23 Schematic depiction of a crystalline lattice of silicon atoms doped with phosphorus atoms. The only charge carriers that are free to move in the crystal are the free electrons supplied by the phosphorus dopant atoms.

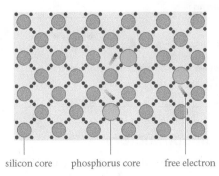

silicon core   phosphorus core   free electron

Figure 32.25 Schematic of crystalline lattice of silicon atoms with some boron atoms substituted for silicon, showing both bonding electrons and holes (missing electrons). The only free charge carriers in the crystal are the holes caused by the boron impurities.

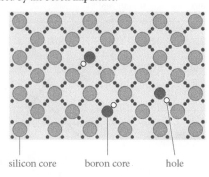

silicon core   boron core   hole

The electrons in a covalent bond are not free to move; consequently, pure silicon has a very low electrical conductivity because all of its valence electrons form covalent bonds.

In extrinsic silicon, other types of atoms, such as boron or phosphorus, replace some of the atoms in the silicon lattice, introducing freely moving charge carriers into the lattice. The substituted atoms are called either *impurities* or *dopants*. For example, phosphorus has five valence electrons (Figure 32.21*b*). Because the silicon lattice structure requires only four bonds from each atom, the fifth electron from a phosphorus atom dopant is not involved in a bond and is free to move through the solid (Figure 32.23).

If an electric field is applied to the doped semiconductor of Figure 32.23, the free electrons move, creating a current in the semiconductor (Figure 32.24). As free electrons leave the semiconductor from one side, other free electrons enter it on the opposite side. Because the semiconductor must remain electrically neutral, the number of free electrons in the semiconductor at any given instant is always the same and it is equal to the number of phosphorus atoms in the material.

If boron atoms, which have three valence electrons (Figure 32.21*c*), are substituted for some silicon atoms in a

silicon lattice, the "missing" fourth electron at each boron leaves behind what is called a **hole**—an incomplete bond (Figure 32.25). These holes behave like *positive* charge carriers and are free to move through the lattice (Figure 32.25). The holes therefore increase the ability of the silicon to conduct current, just as do the free electrons in phosphorus-doped silicon.

Keep in mind that the motion of holes involves electrons moving to fill existing holes, leaving new holes in the previous positions of the electrons (Figure 32.26). The boron

Figure 32.26 Sequence of four snapshots showing how holes "move" through a crystal by trading places with bonding electrons. In the presence of an electric field, holes move in the direction of the field (opposite to the directions in which the electrons move). To maintain continuity, free electrons from attached metal wires enter at the right, recombining with holes that accumulate there, and leave at the left.

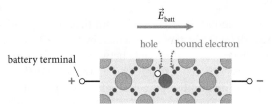

hole   bound electron

battery terminal

Electron jumps to position of hole…   …leaving new hole.

Second electron jumps to that hole…   …leaving new hole.

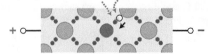

Effect is as though hole itself moves.

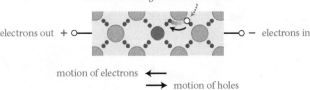

electrons out   electrons in

motion of electrons ←
→ motion of holes

Figure 32.24 In an applied electric field, the free electrons in a phosphorous-doped semiconductor are free to move in the direction opposite the field direction. Free electrons leave the semiconductor at the left, travel through the circuit wire, and enter the semiconductor at the right.

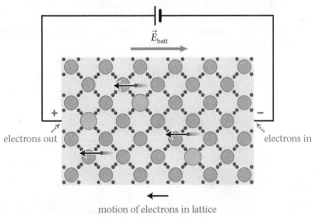

$\vec{E}_{batt}$

+   −

electrons out   electrons in

motion of electrons in lattice

cores do not move! In the presence of an electric field, the positively charged holes move in the direction of the field as the negatively charged electrons move in the opposite direction. If the semiconductor is attached to metal wires on either side, as in Figure 32.26, free electrons travel into the semiconductor from the right (eliminating holes that reach the right edge) and travel out of the semiconductor on the left (producing holes on the left edge). Electrons thus flow from right to left, making holes travel in the opposite direction. Unlike the electrons, however, the holes never leave the semiconductor.

Doped semiconductors are classified according to the nature of the dopant. In a *p-type* semiconductor, the dopant has fewer valence electrons than the host atoms, contributing positively charged holes as the free charge carriers (thus the *p* in the name). In an *n-type* semiconductor, the dopant has more valence electrons than the host atoms, contributing negatively charged electrons as the free charge carriers (thus the *n* in the name). Substituting as few as ten dopant atoms per million silicon atoms produces conductivities appropriate for most electronic devices.

✋ **32.5** Is a piece of *n*-type silicon positively charged, negatively charged, or neutral?

32.4 Diodes, transistors, and logic gates

Although tailoring the conductivity of a single piece of semiconductor can be a useful procedure, the most versatile semiconductor devices combine doped layers that have different types of charge carriers. The simplest such device is a **diode,** made by bringing a piece of *p*-type silicon into contact with a piece of *n*-type silicon (**Figure 32.27a**). Near the junction where the two pieces meet, free electrons from the *n*-type silicon wander into the *p*-type material, where they end up filling holes. This *recombination* process turns free electrons into bound electrons (that is, electrons not free to roam around in the material) and eliminates the holes. Likewise, some of the holes in the *p*-type silicon wander into the *n*-type silicon, where they recombine with free electrons.

As recombination events take place, a thin region containing no free charge carriers (neither free electrons nor holes), called the **depletion zone,** develops at the junction. Although there are no *free* charge carriers in this zone, the trapping of electrons on the *p*-side of the junction causes negative charge carriers that are nonmobile to accumulate there. Similarly, positive nonmobile charge carriers accumulate on the *n*-side of the junction. As a result, the depletion zone consists of a negatively charged region and a positively

(a) Pieces of *p*- and *n*-type doped silicon

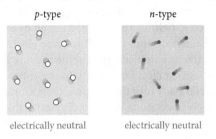

(b) When the two are put in contact, a diode is formed

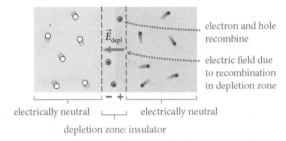

(c) Battery connected so as to produce electric field in *same* direction as electric field in depletion zone; diode blocks current

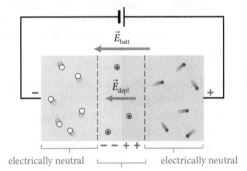

Electric field due to battery *broadens* depletion zone, so diode blocks current.

(d) Battery connected with the *opposite* polarity; diode conducts current

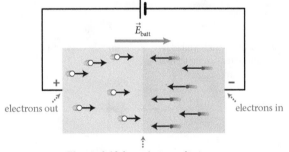

Electric field due to battery *eliminates* depletion zone, so diode conducts current.

Figure 32.27 How a diode transmits current in one direction but blocks it in the other. If the battery is connected as shown in part *d* and produces a sufficiently strong electric field to compensate for the field of the depletion zone, there is a steady flow of both electrons and holes. (Remember, though: The holes never leave the semiconductor. Only the electrons enter and leave the semiconductor.)

charged region, and an electric field points across the depletion zone from the *n*-side to the *p*-side (Figure 32.27*b*).

As this electric field in the depletion zone of the diode increases, it becomes more difficult for free electrons and holes to cross the junction and recombine because the electric field pushes free electrons back into the *n*-type silicon and pushes holes back into the *p*-type silicon. Consequently, the depletion zone stops growing. Typically this region is less than a micrometer wide. Because of the lack of free charge carriers in it,

the depletion zone acts as an electrical insulator.

If we now connect the *n*-side of this diode to the positive terminal of a battery and the *p*-side to the negative terminal, the battery produces across the diode an electric field that points in the same direction as the electric field in the depletion zone (Figure 32.27*c*). The electric field of the battery pulls free electrons in the *n*-type silicon toward the positive terminal and pulls holes in the *p*-type silicon toward the negative terminal, broadening the (nonconducting) depletion zone. Connecting the battery in this manner therefore causes no flow of charge carriers in the diode.

When the battery is connected in the opposite direction, however, the depletion zone narrows as the battery's electric field pushes free electrons and holes toward the junction (Figure 32.27*d*). When the magnitude of the applied electric field created by the battery equals that of the electric field across the depletion zone, both types of free charge carriers can reach the junction, resulting in a current in the device carried both by free electrons and by holes.

As Figure 32.27 shows, a diode conducts current in one direction only: from the *p*-type side to the *n*-type side. The symbol for a diode is shown in **Figure 32.28***a*; the triangle points in the direction in which the diode conducts current (from the *p*-side to the *n*-side).

32.6 In the diode of Figure 32.28*a*, which way do holes travel? Which way do electrons travel?

Figure 32.28 (*a*) Circuit symbol for a diode. (*b*) Schematic of a diode made using integrated-circuit technology.

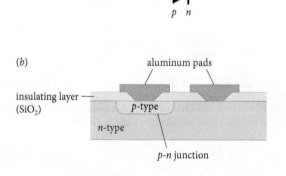

An *ideal diode* acts like a short circuit for current in the permitted direction and like an open circuit for current in the opposite direction. (That is not exactly how a diode behaves, but it's pretty close.)

32.7 Suppose a sinusoidally varying potential difference is applied across a diode connected in series with a resistor. Sketch the potential difference across the diode as a function of time, and then, on the same graph, sketch the current in the resistor as a function of time.

Example 32.4 Rectifier

Consider the arrangement of ideal diodes shown in **Figure 32.29**. This arrangement, called a *rectifier*, converts alternating current (AC) to direct current (DC). Sketch a graph showing, for a sinusoidally alternating source, the current in the resistor in the direction from b to c as a function of time.

Figure 32.29 Example 32.4.

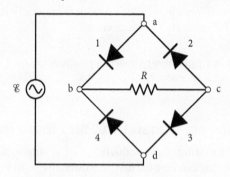

❶ **GETTING STARTED** Because the source is alternating, the current in the circuit periodically reverses direction. During part of the cycle the charge carriers creating the current flow clockwise through the source, and during another part of the cycle they flow counterclockwise. The diodes, however, conduct current in one direction only. I begin by making a sketch of the current between a and d, taking the direction from a to d to be positive (Figure 32.30*a*).

Figure 32.30

(*a*)

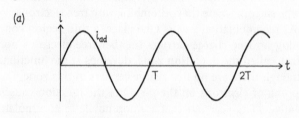

(*b*)

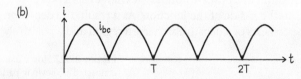

❷ **DEVISE PLAN** In an ideal diode, the charge carriers can flow only in the direction in which the triangle in the diode symbol points. I shall determine which diodes allow charge carriers to flow when the current direction is clockwise and when it is counterclockwise. I can then determine in each case which way the charge carriers flow through the resistor.

❸ **EXECUTE PLAN** When the current in the circuit is clockwise, only diodes 1 and 3 are conducting, so the current direction is abcd. When the current in the circuit is counterclockwise ($i_{ad} < 0$), only diodes 2 and 4 are conducting, so the current direction is dbca. At all instants, the current in the resistor points in the same direction: from b to c. This means that i_{bc} is positive regardless of whether i_{ad} is positive or negative. Whenever i_{ad} is negative, the diodes reverse the direction of the current in the resistor, so i_{bc} is always positive and my graph is as shown in Figure 32.30b. ✔

❹ **EVALUATE RESULT** The arrangement of diodes keeps the current from b to c always in the same direction, even though the current from a to d alternates in direction. It makes sense, then, that this arrangement of diodes is called a *rectifier*.

Figure 32.28b shows how a diode may be constructed as part of an integrated circuit (a computer chip, for example). An aluminum pad (part of the metal wire connecting the diode to the rest of the circuit) is in contact with a small *p*-type region of silicon, which is surrounded by a larger *n*-type region that is in contact with a second aluminum pad. The *p-n* junction forms at the interface between the *p*- and *n*-type regions. A thin layer of silicon oxide (SiO_2) insulates the aluminum from the underlying silicon except where electrical contact is needed. On a modern computer chip, the entire device is only a few micrometers wide.

Another important circuit element in modern electronics is the **transistor,** a device that allows current control that is more precise than the on/off control of a diode. A transistor consists of a thin layer of one type of doped semiconductor sandwiched between two layers of the opposite type of doped semiconductor. **Figure 32.31**, for example, shows an *npn-type bipolar transistor*—a thin layer of *p*-type silicon sandwiched between two thicker regions of *n*-type silicon.* If the *p*-type layer is thin, the depletion zone formed at the

left *p-n* junction merges with the depletion zone formed at the right *p-n* junction. The merged depletion zones form one wide depletion zone.

When a potential difference is applied across such a transistor (**Figure 32.32a**), the depletion zone across junction 1 disappears, but that across junction 2 grows, shifting the depleted region toward the positive terminal of the battery. While charge carriers can now cross junction 1 where the depletion zone has disappeared, the (shifted) depletion zone that still exists prohibits their movement, which means no current in the transistor. For historical reasons, the *n*-type region connected to the negative terminal is called the *emitter*, the *n*-type region connected to the positive terminal is called the *collector*, and the *p*-type layer is called the *base*. If the direction of the applied potential difference is reversed, the roles of the emitter and the collector are also reversed, and there is still no current in the transistor.

Figure 32.32 How an *npn*-type bipolar transistor works.

(*a*) Potential difference applied from collector to emitter only

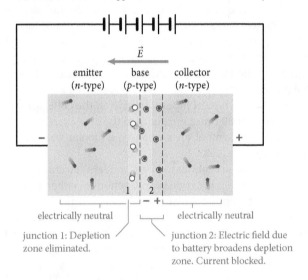

electrically neutral   electrically neutral

junction 1: Depletion   junction 2: Electric field due
zone eliminated.   to battery broadens depletion
                                                   zone. Current blocked.

(*b*) Potential difference also applied from base to emitter

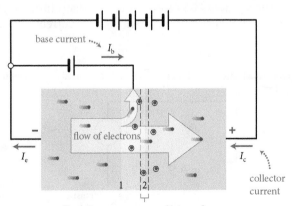

Depletion zone narrow; electrons have
enough kinetic energy to pass through it.

Figure 32.31 Schematic of an *npn*-type bipolar transistor, showing charge distribution and depletion zones for both *p-n* junctions.

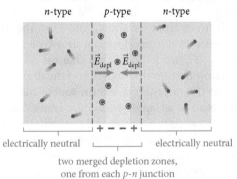

electrically neutral   electrically neutral

two merged depletion zones,
one from each *p-n* junction

---

*Transistors in which a thin layer of *n*-type silicon is sandwiched between pieces of *p*-type silicon, called *pnp*-type bipolar transistors, are also used.

Figure 32.33 Circuit symbol for an *npn*-type bipolar transistor.

npn-type bipolar transistor

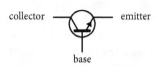

Figure 32.35 Schematic of an *npn*-type bipolar transistor made using integrated-circuit technology.

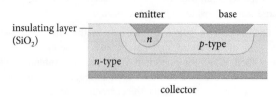

The situation changes drastically when, in addition to the potential difference between the emitter and the collector, a small potential difference is applied between the emitter and the base (Figure 32.32*b*). Adding this potential difference, called a *bias* or *bias potential difference,* makes the depletion zone much thinner than it is in Figure 32.32*a* because the formerly negatively charged region of this zone is brought to a positive potential, restoring mobile holes to that region. Because the emitter-base junction is conducting (remember, the depletion zone at junction 1 has disappeared), electrons now start flowing from the emitter toward the base. Once in the base, three things happen: (1) a small fraction of the electrons recombine with holes in the base, (2) electrons are attracted by the positive charge on the collector and have sufficient kinetic energy to pass straight through the very thin depletion zone, producing a collector current I_c, and (3) electrons diffuse through the base toward the positively charged end of the base, causing a small base current I_b. In a typical bipolar transistor, the collector current is 10 to 1000 times greater than the base current.

The circuit symbol for an *npn*-type bipolar transistor is shown in **Figure 32.33**.

Transistors are ubiquitous in modern electronics. In most applications, the transistor functions as either a switch or a current amplifier. If we consider I_b to be the input current and I_c the output current, the transistor acts as a switch in which I_b turns on and controls I_c. As a current amplifier, a small current I_b produces a much larger current I_c.

For electrical devices that draw large currents, it is useful to switch the device on and off with a mechanical switch wired in parallel with the device, rather than in series, so that the current in the device does not have to pass through the switch. **Figure 32.34** shows a circuit that utilizes such switching. When switch S is open, the base current is zero,

and so the collector current (and therefore the current in the device) is zero. When switch S is closed, the small current from base to emitter causes a large current from collector to emitter that turns on the motor.

32.8 In a bipolar transistor, what relationship, if any, exists among I_b, I_c, and the emitter current I_e?

Figure 32.35 shows how an *npn*-type bipolar transistor can be fabricated. A drawback of this type of transistor, however, is that a continuous small current through the base is required to make the transistor conducting. For this reason, another type of transistor, called the *field-effect transistor,* is used much more frequently. **Figure 32.36*a*** shows the configuration of one. Two *n*-type wells are made in a piece of *p*-type material. The *p*-type material between the two wells is covered with a nonconducting oxide layer (typically SiO_2) and then with a metal layer called the *gate.* The two *n*-type wells are called the *source* and the *drain* (the *n*-type well that is kept at a higher potential is the drain).

Because of the depletion zones between the *p*-type and *n*-type materials, no charge carriers can flow from the source to the drain (or vice versa). The nonconducting layer between the gate and the *p*-type material prevents charge carriers from traveling between the gate and the rest of the device.

If the gate is given a positive charge, as in Figure 32.36*b*, the (positively charged) holes just underneath the gate are pushed away, forming underneath the gate an additional depletion zone that connects the depletion zones around the two *n*-*p* junctions. If the positive charge is made large enough, electrons from the source and from the drain are pulled underneath the gate, forming an *n*-type channel below the gate (Figure 32.36*c*). This channel allows charge carriers to flow between the source and the drain. The gate thus controls the current between the source and the drain, just as the base in an *npn*-type bipolar transistor controls the current between the emitter and the collector. (The difference is that there is no current in the gate in a field-effect transistor.) Applying a positive charge to the gate is often referred to as putting a positive *bias* on the gate.

Figure 32.37*a* shows the circuit symbol for a field-effect transistor, and Figure 32.37*b* shows how this type of transistor can be realized in an integrated circuit. This type of transistor has two advantages over the bipolar transistor

Figure 32.34 Circuit in which a bipolar transistor is used to turn a motor on and off.

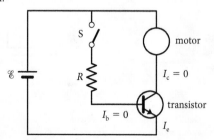

Figure 32.36 How a field-effect transistor works.

(*a*) Field-effect transistor with uncharged gate

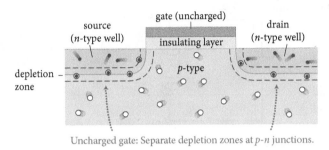

Uncharged gate: Separate depletion zones at *p-n* junctions.

(*b*) Small positive charge on gate attracts electrons to gate and extends depletion zone below gate

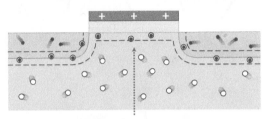

Small gate charge causes depletion zone to extend beneath gate.

(*c*) Large positive charge on gate attracts more electrons to gate and causes *n*-type channel, which connects source and drain

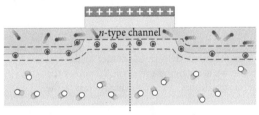

Strong gate charge pushes depletion zone away; conducting *n*-type channel now connnects source and drain.

Figure 32.37 (*a*) Circuit symbol for a field-effect transistor. (*b*) Schematic of a field-effect transistor made using integrated-circuit technology.

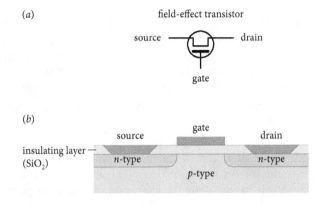

when both inputs are at positive potential with respect to ground. In an OR gate, the output potential is nonzero when either input potential is positive. The symbols used for these gates in circuit diagrams are shown in **Figure 32.38**; the inputs are on the left, and the output is on the right. In analyzing these circuits, we'll make the simplifying assumption that a transistor is just a switch that is open (off) when the potential of the gate is either at ground or negative with respect to ground and is closed (on) when the gate is at a positive potential.

Figure 32.38 Circuit symbols for AND and OR logic gates.

$$A \;\;\; B \;\;\; \boxed{AND} \;\; A \cap B \qquad A \;\;\; B \;\;\; \boxed{OR} \;\; A \cup B$$

✋ **32.9** Circuit diagrams for two logic gates are shown in **Figure 32.39**. Which is the AND gate, and which is the OR gate? Explain briefly how each one works.

shown in Figure 32.35. First, all the terminals in the field-effect transistor are on the same side of the chip, making fabrication in integrated circuits much easier. Second, the current between the source and the drain is controlled by the charge on the gate, allowing a potential difference rather than a current to be used to control the source-drain current. Because no current is leaving the gate, no energy is required to keep current flowing from the source to the drain.

Field-effect transistors are widely used in devices called *logic gates,* which are the building blocks of computer processors and memory. A logic gate takes two input signals and provides an output after performing a logic operation on the input signals. For example, in a so-called AND gate, the output potential is nonzero with respect to ground only

Figure 32.39 Checkpoint 32.9

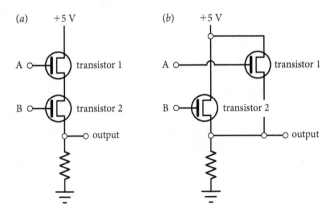

Self-quiz

1. At the instant shown in **Figure 32.40**, the potential difference across the capacitor is half its maximum value and the charge on the plates is increasing. Draw the direction of the current and sketch the magnetic field at this instant. Is the magnitude of current increasing or decreasing?

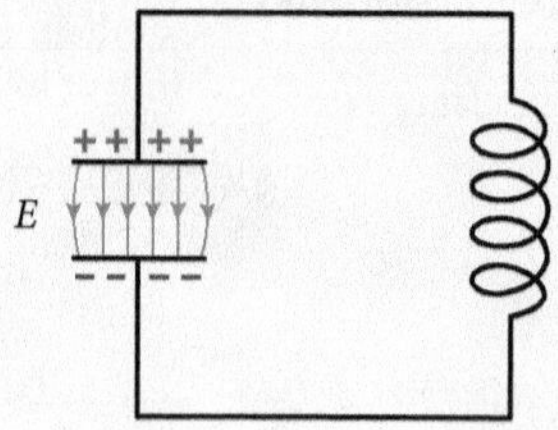

2. Construct a phasor diagram representing the current and potential difference in Figure 32.10 at $t = T/4$, $T/2$, and $3T/4$.

3. **Figure 32.41** shows the time-varying potential difference and current for the circuit of Figure 32.8. At the instant labeled t_a, what are the charge on the capacitor plates and the direction of the current?

Figure 32.40

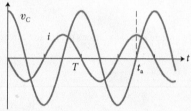

Figure 32.41

4. Is there any current in a diode connected as shown in **Figure 32.42**?

Figure 32.42

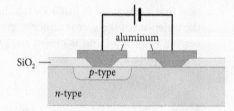

Answers:

1. Your sketch should show the current directed counterclockwise. The magnetic field in the center of the coil points up the page according to the right-hand dipole rule (assuming we are looking down on the top of the coil in Figure 32.40). Because the current is zero when the capacitor has maximum charge, the magnitude of the current is decreasing at the instant shown in Figure 32.40.

2. See **Figure 32.43**. At $t = T/4$ the potential difference phasor V_C points along the positive y axis because the potential difference reaches its maximum positive value at this instant, and the current phasor I points along the negative x axis because it leads the current by 90°. Each quarter cycle both phasors rotate 90° counterclockwise.

Figure 32.43

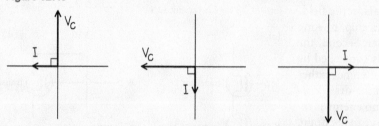

3. Because the potential difference across the capacitor is zero at instant t_a, the charge on the plates must be zero. The current is a maximum at this instant and is directed clockwise.

4. Yes. The holes in the p-type material move away from the positive terminal, and the electrons move toward it. According to Figure 32.27d, this flow shrinks the depletion zone, the charge carriers can flow, and so there is a current.

32.5 Reactance

Let us now develop a mathematical framework for analyzing alternating-current circuits. The instantaneous emf supplied by an AC source is customarily written as

$$\mathscr{E} = \mathscr{E}_{max} \sin \omega t, \qquad (32.1)$$

where $\mathscr{E}_{max}$ is the maximum value of the emf, typically called the *peak value* or *amplitude* (see Section 15.1), $\omega = 2\pi f$ is the angular frequency of oscillation in inverse seconds (Section 15.5), and f is the frequency in hertz. Most generators have frequencies of 50 Hz or 60 Hz. Audio circuits typically operate at kilohertz frequencies, radio transmitters at 10^8 Hz, for instance, and computer chips at 10^9 Hz. It's very important to remember to convert frequencies in hertz (cycles per second) to angular frequencies in s^{-1} when ω appears in the equations.

Note that the initial phase for the emf as written in Eq. 32.1 is zero. When we make this choice, the source emf serves as the reference for phase in the circuit.

Let's begin by revisiting the circuit from Exercise 32.1—a resistor connected to an AC source (Figure 32.44). At any instant, Ohm's law relates the potential difference across the resistor to the current in it, just as it does for DC circuits:

$$v_R = iR. \qquad (32.2)$$

The only difference between Eq. 32.2 and Ohm's law for DC circuits (Eq. 31.11) is that the potential difference and the current in Eq. 32.2 oscillate in time.

Applying the loop rule to this circuit gives the AC version of Eq. 31.23:

$$\mathscr{E} - iR = 0. \qquad (32.3)$$

Equations 32.2 and 32.3 show that the potential difference across the load equals the emf supplied by the source (as we would expect):

$$v_R = \mathscr{E} = \mathscr{E}_{max} \sin \omega t. \qquad (32.4)$$

✋ **32.10** (*a*) In Figure 32.44, is the potential at point a higher or lower than the potential at b when the current direction is clockwise through the circuit? (*b*) If we define such a current to be positive, is $\mathscr{E}$ positive or negative? Express v_R in terms of the potential at a and the potential at b. (*c*) Half a cycle later, when the current is negative, is $\mathscr{E}$ positive or negative? Express v_R again in terms of the potential at a and the potential at b.

Using Eqs. 32.2 and 32.4, we can write the current in the resistor as

$$i = \frac{v_R}{R} = \frac{\mathscr{E}_{max} \sin \omega t}{R} = I \sin \omega t, \qquad (32.5)$$

where $I = \mathscr{E}_{max}/R$ is the amplitude of the current. Note that the current and the potential difference both oscillate at angular frequency ω and are in phase, as we concluded in Exercise 32.1. If we write $v_R = V_R \sin \omega t$, we see that the amplitudes of the current and the potential difference satisfy the relationship

$$V_R = IR. \qquad (32.6)$$

Figure 32.45 shows the corresponding phasor diagram and time dependence of v_R and i.

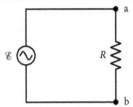

Figure 32.44 AC circuit consisting of a resistor connected across the terminals of an AC source.

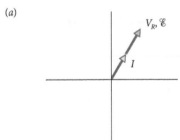

Figure 32.45 (*a*) Phasor diagram and (*b*) graph showing time dependence of i and v_R for the circuit shown in Figure 32.44.

(*a*)

(*b*)

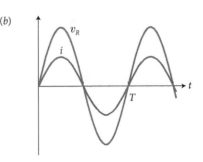

Exercise 32.5 AC circuit with two resistors

In **Figure 32.46**, the resistances are $R_1 = 100\ \Omega$ and $R_2 = 60\ \Omega$, the amplitude of the emf is $\mathscr{E}_{max} = 160$ V, and its frequency is 60 Hz. (*a*) What is the amplitude of the potential difference across each resistor? (*b*) What is the instantaneous potential difference across each resistor at $t = 50$ ms?

Figure 32.46 Exercise 32.5.

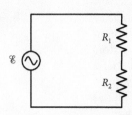

SOLUTION I analyze this circuit just as I would analyze a DC circuit containing two resistors, except now I must keep in mind that the current and potential differences are oscillating. The resistance of the load is

$$R_{load} = R_1 + R_2,$$

and the instantaneous current in the load is

$$i = \frac{\mathscr{E}}{R_{load}} = \frac{\mathscr{E}}{R_1 + R_2}.$$

(*a*) Because the current and the emf are in phase, they reach their maximum values at the same instant. As a result, the amplitude (maximum value) of the current is given by the amplitude of the emf divided by the resistance:

$$I = \frac{\mathscr{E}_{max}}{R_1 + R_2} = 1.0\ \text{A}.$$

The potential differences across the resistors are in phase with the current, and so I calculate the amplitude of the potential differences from the amplitude of the current using Eq. 32.6:

$$V_{R_1} = IR_1 = (1.0\ \text{A})(100\ \Omega) = 100\ \text{V}$$

$$V_{R_2} = IR_2 = (1.0\ \text{A})(60\ \Omega) = 60\ \text{V.} ✔$$

(*b*) I can use Eq. 32.5 to calculate the instantaneous value of the current:

$$i = (1.0\ \text{A})\sin(2\pi \cdot 60\ \text{Hz} \cdot 0.050\ \text{s}) = 0.$$

(In 50 ms, three full cycles at 60 Hz take place.) Because the current is zero at 50 ms, the potential differences v_{R_1} and v_{R_2} at 50 ms are also zero. ✔

Figure 32.47 AC circuit consisting of a capacitor connected across the terminals of an AC source.

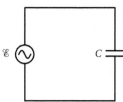

Next consider a capacitor connected to an AC source (**Figure 32.47**). Because the capacitor and the source are connected to each other, we have

$$\mathscr{E} = v_C, \tag{32.7}$$

and so the potential difference across the capacitor is

$$v_C = \mathscr{E}_{max} \sin \omega t = V_C \sin \omega t. \tag{32.8}$$

At any instant the potential difference across the capacitor and the charge on the upper plate are related by (see Eq. 26.1)

$$\frac{q}{v_C} = C, \tag{32.9}$$

where the potential difference v_C and the charge q on the plate oscillate in time. The charge on the upper capacitor plate is thus

$$q = Cv_C = CV_C \sin \omega t, \tag{32.10}$$

and the current is the rate of change of the charge on the plate:

$$i = \frac{dq}{dt} = \frac{d}{dt}(CV_C \sin \omega t) = \omega CV_C \cos \omega t. \tag{32.11}$$

Using the identity $\cos \alpha = \sin\left(\alpha + \frac{\pi}{2}\right)$, we can rewrite this as

$$i = \omega CV_C \sin\left(\omega t + \frac{\pi}{2}\right) = I \sin\left(\omega t + \frac{\pi}{2}\right). \tag{32.12}$$

We now see that v_C and i are not in phase: i reaches its maximum value one-quarter period before v_C reaches its maximum value (**Figure 32.48**), as we found in Section 32.2.

The current in the capacitor of Figure 32.47 is not simply proportional to the potential difference across the capacitor because the two are out of phase. However, the *amplitude* of the current is proportional to the amplitude of the potential difference: $I = \omega C V_C$. Rewriting this to express V_C in terms of I gives

$$V_C = \frac{I}{\omega C}. \tag{32.13}$$

Note how this expression differs from the expression for a circuit that consists of only an AC source and a resistor, $V_R = IR$ (Eq. 32.6), where R is the proportionality constant between V and I. In Eq. 32.13, the proportionality constant is no longer a resistance (though it still has units of ohms). In circuits that contain capacitors and/or inductors, we use the general name **reactance** for the proportionality constant between the potential difference amplitude and the current amplitude. From Eq. 32.13 we see that this proportionality constant for a circuit that contains a capacitor is $1/\omega C$, and we call this constant the *capacitive reactance X_C*:

$$X_C \equiv \frac{1}{\omega C}, \tag{32.14}$$

so Eq. 32.13 becomes
$$V_C = IX_C. \tag{32.15}$$

Reactance is a measure of the opposition of a circuit element to a change in current. Unlike resistance, reactance is frequency dependent. At low frequency, the capacitive reactance X_C is large, which means that the amplitude of the current is small for a given value of V_C. At zero frequency, the current $I = \omega C V_C$ is zero, as it should be. (There is no direct current in a capacitor because the capacitor is just like an open circuit!) The higher the frequency of the source, the smaller the capacitive reactance and the greater the current (the less the capacitor opposes the alternating current).

Often, when analyzing AC circuits, the only things we are interested in are the amplitudes of the currents and potential differences. The capacitive reactance allows us to calculate the amplitude of the current in the capacitor directly from the amplitude of the potential difference across it—in this case, the emf of the source.

It is conventional to write the current in an AC circuit in the form

$$i = I \sin(\omega t - \phi), \tag{32.16}$$

where ϕ is called the **phase constant.** The negative sign in front of the phase constant is chosen so that a positive ϕ corresponds to shifting the curve for the current to the right, in the positive direction along the time axis, and a negative ϕ corresponds to shifting the curve to the left, in the negative direction along this axis (Figure 32.49).

Figure 32.48 (*a*) Phasor diagram and (*b*) graph showing time dependence of i and v_C for the circuit of Figure 32.47. The phasor diagram shows the relative phase of i and v_C.

(*a*)
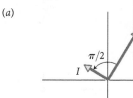

(*b*)
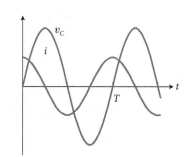

Figure 32.49 Positive and negative phase constant.

(*a*)
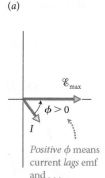
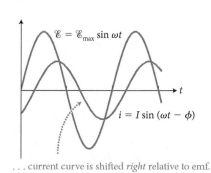

(*b*)
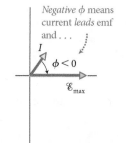
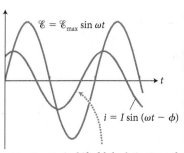

Figure 32.50 AC circuit consisting of an inductor connected across the terminals of an AC source.

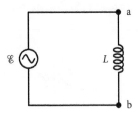

The phase constant represents the phase difference between the source emf and the current. It is measured from the current phasor to the source emf phasor with the counterclockwise direction being positive (Figure 32.49). As a result, when the current leads the source emf, ϕ is negative; when the current lags the source emf, ϕ is positive.

Comparing Eqs. 32.12 and Eq. 32.16, we see that for the capacitor-AC source circuit of Figure 32.47, $\phi = -\pi/2$, as shown in Figure 32.48a. The negative phase constant means that the current leads the source emf. The curve for i is shifted to the left relative to the curve for v_C, as shown in Figure 32.48b. As you can see from the figure, when the capacitor has maximum charge (v_C maximum), the current is zero because at that instant the current reverses direction as the capacitor begins discharging. The current reaches its maximum value when the capacitor is completely discharged ($v_C = 0$).

32.11 As in the *LC* circuit discussed in Section 32.1, the current in the circuit of Figure 32.47 oscillates. If we think of v_C as corresponding to the position of the simple harmonic oscillator described in Section 15.5, what property of the circuit of Figure 32.47 corresponds to the velocity of the oscillator?

Finally, consider an inductor connected to an AC source (**Figure 32.50**). Because the inductor and the source are connected to each other, we have

$$\mathcal{E} = v_L, \tag{32.17}$$

so the potential difference across the inductor is

$$v_L = \mathcal{E}_{max} \sin \omega t = V_L \sin \omega t. \tag{32.18}$$

In Chapter 29 we saw that a changing current in an inductor causes an induced emf (Eq. 29.19):

$$\mathcal{E}_{ind} = -L\frac{di}{dt}. \tag{32.19}$$

The negative sign in this expression means that the potential decreases across the inductor in the direction of increasing current. Consequently, in Figure 32.50, the potential at b is lower than the potential at a when the current is increasing clockwise around the circuit. However, for consistency with Eq. 32.3, we always measure the potential difference v_L from b to a, just as we did with the AC source-resistor circuit of Figure 32.44. Therefore the sign of the potential difference across the inductor is the opposite of the sign in Eq. 32.19:

$$v_L = L\frac{di}{dt}. \tag{32.20}$$

We obtain the current in the circuit by substituting Eq. 32.18 into Eq. 32.20:

$$L\frac{di}{dt} = V_L \sin \omega t \tag{32.21}$$

$$di = \frac{V_L}{L} \sin \omega t \, dt. \tag{32.22}$$

To obtain the current, we integrate this expression:

$$i = \frac{V_L}{L} \int \sin \omega t \, dt = -\frac{V_L}{\omega L} \cos \omega t. \tag{32.23}$$

The amplitude of the current is thus

$$I = \frac{V_L}{\omega L},\qquad(32.24)$$

and using the identity $\cos \omega t = -\sin(\omega t - \frac{\pi}{2})$, we get

$$i = I \sin\left(\omega t - \frac{\pi}{2}\right).\qquad(32.25)$$

The phase constant is $\phi = +\pi/2$, which means the current lags the source by 90°, as shown in Figure 32.51.

Just as we defined a capacitive reactance for a circuit that contains a capacitor, we define the *inductive reactance* X_L for a circuit that contains an inductor as the constant of proportionality between the amplitudes V_L and I in the circuit. From Eq. 32.24 we see that this proportionality constant is ωL:

$$X_L \equiv \omega L,\qquad(32.26)$$

so that

$$V_L = IX_L.\qquad(32.27)$$

Inductive reactance, like capacitive reactance, has units of ohms and depends on the frequency of the AC source. However, X_L increases with increasing frequency, so, at a given potential difference, the amplitude of the current is greatest at zero frequency and decreases as the frequency increases. This makes sense because for a constant current, an inductor is just a conducting wire and does not impede the current; as the frequency of the AC source increases, the emf induced across the inductor increases.

Figure 32.51 (*a*) Phasor diagram and (*b*) graph showing time dependence of i and v_L for the circuit of Figure 32.50. The phasor diagram shows the phase difference $\phi = \pi/2$ between i and v_L.

(*a*)

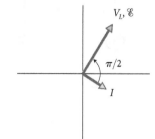

(*b*)

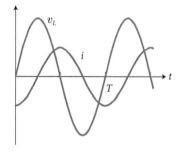

Example 32.6 Oscillating inductor

When a 3.0-H inductor is the only element in a circuit connected to a 60-Hz AC source that is delivering a maximum emf of 160 V, the current amplitude is I. When a capacitor is the only element in a circuit connected to the same source, what must the capacitance be in order to have the current amplitude again be I?

❶ **GETTING STARTED** I begin by identifying the information given in the problem statement: $\mathcal{E}_{max} = 160$ V, angular frequency $\omega = 2\pi(60 \text{ Hz})$, and inductance $L = 3.0$ H. The problem asks me to compare two circuits, one with an inductor connected to an AC source and the other with a capacitor connected to the same source. What I must determine is the capacitance value that makes the current amplitude the same in the two circuits.

❷ **DEVISE PLAN** For both circuits the potential difference across the load equals the source emf, so $\mathcal{E}_{max} = V_C = V_L$. I can use Eqs. 32.26 and 32.27 to get an expression for V_L in terms of I, from which I can express I in the inductor circuit in terms of V_L, ω, and L. Next I can use Eqs. 32.14 and 32.15 to get an expression for V_C in terms of I. I can then substitute this into my first expression for I and obtain an expression for C that contains only known quantities.

❸ **EXECUTE PLAN** Substituting the inductive reactance from Eq. 32.26, $X_L = \omega L$, into Eq. 32.27, $V_L = IX_L$, I get $V_L = I\omega L$, so the amplitude of the current is

$$I = \frac{V_L}{\omega L}.\qquad(1)$$

Substituting the capacitive reactance from Eq. 32.14, $X_C = 1/\omega C$, into Eq. 32.15, $V_C = IX_C$, I get

$$V_C = \frac{I}{\omega C}.\qquad(2)$$

Solving Eq. 2 for C and substituting Eq. 1 for I give

$$C = \frac{I}{\omega V_C} = \frac{V_L}{\omega^2 V_C L} = \frac{1}{\omega^2 L}$$

$$= \frac{1}{(2\pi \cdot 60 \text{ Hz})^2(3.0 \text{ H})} = 2.3 \times 10^{-6} \text{ F.}\ ✔$$

❹ **EVALUATE RESULT** To check my answer, I can calculate the inductive and capacitive reactances from Eqs. 32.26 and 32.14, respectively: $X_L = \omega L = (2\pi \cdot 60 \text{ Hz})(3.0 \text{ H}) = 1.1 \text{ k}\Omega$ and $X_C = 1/\omega C = 1/(2\pi \cdot 60 \text{ Hz})(2.3 \times 10^{-6} \text{ F}) = 1.1 \text{ k}\Omega$. The two are identical, as I expect given that they yield the same current amplitude for the same AC source.

QUANTITATIVE TOOLS

Figure 32.52 An *RC* series circuit, consisting of a resistor and a capacitor in series across the terminals of an AC source.

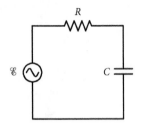

32.12 For the three circuits discussed in this section (AC source with resistor, capacitor, or inductor), sketch for a given emf amplitude (*a*) the resistance or reactance as a function of angular frequency ω and (*b*) the current amplitude in the circuit as a function of ω. Explain the meaning of each curve on your graphs.

32.6 *RC* and *RLC* series circuits

When an AC source is connected to multiple circuit elements, either in series or in parallel, applying the loop rule becomes more complicated than for DC circuits because we need to add several oscillating potential differences that may be out of phase with one another. For example, suppose we have a resistor and a capacitor in series with an AC source (**Figure 32.52**), known as an *RC series circuit*. The loop rule states that

$$\mathcal{E} = v_R + v_C. \tag{32.28}$$

To compute the sum on the right side of this equation, we must add potential differences that vary sinusoidally at the same angular frequency ω but are out of phase. The combined potential difference v of two potential differences v_1 and v_2 that oscillate at the same angular frequency is

$$v = V_1 \sin(\omega t + \phi_1) + V_2 \sin(\omega t + \phi_2), \tag{32.29}$$

where ϕ_1 and ϕ_2 are the initial phases of the two potential differences. Calculating this sum algebraically gets very messy, but using phasors to calculate it simplifies things greatly.

Figure 32.53*a* shows the phasors that correspond to the two terms on the right in Eq. 32.29. Recall that the instantaneous value of the quantity represented by a rotating phasor equals the vertical component of the phasor (see Figure 32.11). Therefore, v at any instant equals the sum of the vertical components of the phasors that represent v_1 and v_2. This sum is equal to the vertical component of the vector sum $V_1 + V_2$ of the phasors, as shown in Figure 32.53*b*.

Note that the combined potential difference v oscillates at the same angular frequency as v_1 and v_2. Consequently, the three phasors V_1, V_2, and $V_1 + V_2$ rotate as a unit at angular frequency ω, as shown in **Figure 32.54**. The phase relationship among the three phasors is constant, as is the phase relationship among the potential differences.

Figure 32.53 (*a*) Phasor diagram for a system of two oscillating potential differences v_1 and v_2. (b) Vector diagram indicating that the vertical component of the vector sum of the phasors equals the sum of the vertical components of the individual phasors.

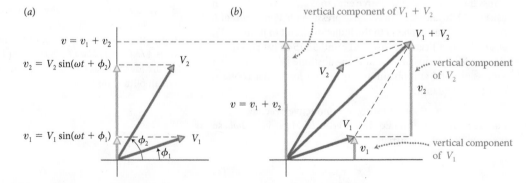

Figure 32.54 Phasor diagram and graph showing time dependence of v_1, v_2, and $v = v_1 + v_2$ from Figure 32.53. All three phasors rotate as a unit at angular frequency ω.

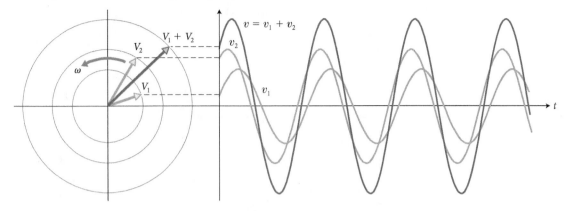

The next example shows how to apply these principles to a specific situation. To convince yourself that the phasor method is worthwhile, try adding the two original trigonometric functions algebraically after solving the problem using phasors!

Example 32.7 Adding phasors

Use phasors to determine the sum of the two oscillating potential differences $v_1 = (2.0 \text{ V}) \sin \omega t$ and $v_2 = (3.0 \text{ V}) \cos \omega t$.

❶ **GETTING STARTED** I begin by making a graph showing the time dependence of v_1 and v_2, and I draw the corresponding phasors V_1 and V_2 to the left of my graph (**Figure 32.55**). I add to my phasor diagram the phasor $V_1 + V_2$, which is the phasor that represents the potential difference sum $v_1 + v_2$ that I must determine. Using phasor $V_1 + V_2$, I can sketch the time dependence of the sum $v_1 + v_2$ by tracing out the projection of phasor $V_1 + V_2$ onto the vertical axis of my $V(\omega t)$ graph as this phasor rotates counterclockwise from the starting position I drew.

Figure 32.55

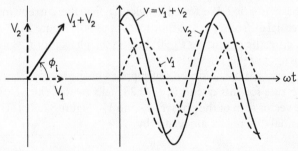

❷ **DEVISE PLAN** To obtain an algebraic expression for $v_1 + v_2$, I first write the oscillating potential differences in the form $v_1 = V_1 \sin(\omega t + \phi_1)$ and $v_2 = V_2 \sin(\omega t + \phi_2)$. Comparing these expressions with the given potential differences, I see that $V_1 = 2.0 \text{ V}$, $\phi_1 = 0$, and $V_2 = 3.0 \text{ V}$. In order to determine ϕ_2, I use the trigonometric identity $\cos(\omega t) = \sin(\omega t + \pi/2)$, and so my given information $v_2 = (3.0 \text{ V}) \cos \omega t = (3.0 \text{ V}) \sin(\omega t + \pi/2)$ tells me that $\phi_2 = \pi/2$. The sum $v_1 + v_2$ is a sinusoidally varying function that can be written as $v_1 + v_2 = A \sin(\omega t + \phi_i)$. The amplitude A is equal to the length of the phasor $V_1 + V_2$, and from my sketch I see that the initial phase ϕ_i is given by the angle between $V_1 + V_2$ and V_1.

❸ **EXECUTE PLAN** The length of the phasor $V_1 + V_2$ is given by the Pythagorean theorem applied to the right triangle containing ϕ_i in my phasor diagram:

$$A = \sqrt{V_1^2 + V_2^2} = \sqrt{(3.0 \text{ V})^2 + (2.0 \text{ V})^2}$$
$$= \sqrt{13 \text{ V}^2} = 3.6 \text{ V}.$$

The tangent of the angle between $V_1 + V_2$ and V_1 is then

$$\tan \phi_i = \frac{V_2}{V_1} = \frac{3.0 \text{ V}}{2.0 \text{ V}} = 1.5,$$

so $\phi_i = \tan^{-1}(1.5) = 56°$.

Now that I have determined A and ϕ_i, I can write the sum of the two potential differences as

$$v_1 + v_2 = (3.6 \text{ V}) \sin(\omega t + 56°). \checkmark$$

❹ **EVALUATE RESULT** The amplitude of the sinusoidal function I obtained is 3.6 V, which is greater than the larger of the two phasors I added, as I expect. My answer shows that the sum of the two potential differences reaches its maximum when $\omega t + \phi_i = 90°$, or when $\omega t = 90° - \phi_i = 34°$. This conclusion agrees with my phasor diagram: The phasor $V_1 + V_2$ reaches the vertical position after it rotates through an angle of $90° - \phi_i = 90° - 56° = 34°$. (I could also verify my answer by adding the two original sine functions algebraically, but the trigonometry needed in that approach is tedious.)

Figure 32.56 Steps involved in constructing a phasor diagram for the circuit in Figure 32.52. The diagram in part *d* indicates the phase of the current relative to the source emf.

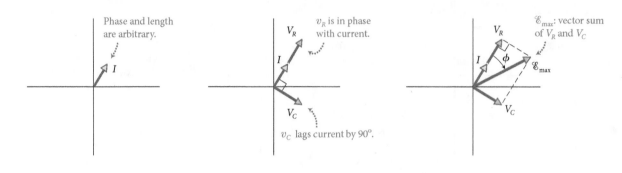

(*a*) Draw current phasor

(*b*) Add phasors for v_R and v_C

(*c*) Add phasor for emf

✋ **32.13** Suppose you need to add two potential differences that are oscillating at *different* angular frequencies—say, $2\sin(\omega t)$ and $3\cos(2\omega t)$. Can you use the phasor method described above to determine the sum? Why or why not?

Let us now return to the *RC* series circuit of Figure 32.52 and construct a phasor diagram in order to determine the amplitude and phase of the current in terms of the amplitude of the source emf and the resistance and capacitance of the circuit elements. From the current, we can calculate the potential differences across the circuit elements.

Because the circuit contains only one loop, the time-dependent current *i* is the same throughout. Therefore, we begin by drawing a phasor that represents *i* (Figure 32.56*a*). We are free to choose the phase of this phasor because we have not yet specified the phase of any of the potential differences in the circuit. Also, the length we draw for phasor *I* is unimportant because it is the only current phasor for this circuit.

Next, we draw the phasors for v_R and v_C, the potential differences across the resistor and capacitor, respectively. We must get the relative phases right, and the lengths of the phasors must also be appropriately proportioned. Because the current is in phase with v_R (Figure 32.45*a*), we draw the corresponding phasor as shown in Figure 32.56*b*; its length is $V_R = IR$.

What about the phasor for v_C? We found previously that the current in a capacitor leads the potential difference across the capacitor by 90° (Figure 32.48*a*), which means we must draw the phasor for v_C 90° behind the phasor for *i*, as it is in Figure 32.56*b*. The length of this phasor is $V_C = IX_C$.

Finally, we need to draw the phasor for the emf supplied by the source. Phasor addition with the loop rule for this circuit (Eq. 32.28) tells us that the phasor $\mathscr{E}_{max}$ for the emf is the vector sum of the phasors V_R and V_C (Figure 32.56*c*). The amplitudes of the potential differences are related by

$$\mathscr{E}_{max}^2 = V_R^2 + V_C^2. \tag{32.30}$$

If we substitute $V_R = IR$ (Eq. 32.6) and $V_C = IX_C$ (Eq. 32.15), this becomes

$$\mathscr{E}_{max}^2 = (IR)^2 + (IX_C)^2 = I^2(R^2 + X_C^2) = I^2\left(R^2 + \frac{1}{\omega^2 C^2}\right). \tag{32.31}$$

Solving for *I* gives
$$I = \frac{\mathscr{E}_{max}}{\sqrt{R^2 + 1/\omega^2 C^2}}. \tag{32.32}$$

Remembering that $\mathscr{E}_{max} = V_{load}$, we see that even though this load includes both resistive and reactive elements, I is still proportional to V_{load}! The constant of proportionality is called the **impedance** of the load and is denoted by Z:

$$I = \frac{\mathscr{E}_{max}}{Z}. \tag{32.33}$$

The impedance of the load is a property of the entire load. It is measured in ohms and depends on the frequency for any load that contains reactive elements.

Impedance plays the same role in AC circuits that resistance plays in DC circuits. In fact, Eq. 32.33 can be thought of as the equivalent of Ohm's law for AC circuits. Equation 32.32 shows that, for an RC series circuit, Z depends on both R and C:

$$Z_{RC} \equiv \sqrt{R^2 + 1/\omega^2 C^2} \quad \text{(RC series combination).} \tag{32.34}$$

To express V_R and V_C in terms of $\mathscr{E}_{max}$, R, C, and ω, we use Eq. 32.32:

$$V_R = IR = \frac{\mathscr{E}_{max}R}{\sqrt{R^2 + 1/\omega^2 C^2}} \tag{32.35}$$

$$V_C = IX_C = \frac{\mathscr{E}_{max}/\omega C}{\sqrt{R^2 + 1/\omega^2 C^2}}. \tag{32.36}$$

To calculate the phase constant ϕ, the geometry shown in Figure 32.56c gives us, with Eqs. 32.6 and 32.15,

$$\tan \phi = -\frac{V_C}{V_R} = -\frac{IX_C}{IR} = -\frac{1}{\omega RC} \tag{32.37}$$

or

$$\phi = \tan^{-1}\left(-\frac{1}{\omega RC}\right) \quad \text{(RC series circuit).} \tag{32.38}$$

The negative value of ϕ indicates that the current in an RC series circuit leads the emf, just as it does in an AC circuit with only a capacitor. As you can see in Figure 32.56c, however, the phase difference between the emf and the current in the RC series circuit is less than 90°.

Example 32.8 High-pass filter

A circuit that allows emfs in one angular-frequency range to pass through essentially unchanged but prevents emfs in other angular-frequency ranges from passing through is called a *filter*. Such a circuit is useful in a variety of electronic devices, including audio electronics. An example of a filter, called a *high-pass filter*, is shown in **Figure 32.57**. Emfs that have angular frequencies above a certain angular frequency, called the *cutoff angular frequency* ω_c, pass through to the two output terminals marked v_{out}, but the filter attenuates the amplitudes of emfs that have frequencies below the cutoff value. (*a*) Determine an expression that gives, in terms of R and C, the cutoff angular frequency ω_c at which $V_R = V_C$. (*b*) Determine the potential difference amplitude v_{out} across the output terminals for $\omega \gg \omega_c$ and for $\omega \ll \omega_c$.

Figure 32.57 Example 32.8.

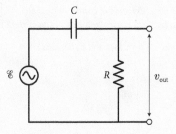

❶ **GETTING STARTED** This circuit is the same as the one in Figure 32.52, which I used to determine expressions for V_R (Eq. 32.35) and V_C (Eq. 32.36) in terms of R and C, so I can use

(Continued)

those results. From Figure 32.57 I see that the potential difference v_{out} is equal to the potential difference across the resistor, so $V_{out} = V_R$.

② DEVISE PLAN In order to determine the value of ω_c at which $V_R = V_C$, I equate the right sides of Eqs. 32.35 and 32.36. The resulting ω factor in my expression then is the cutoff value ω_c. For part *b*, I know that $V_{out} = V_R$. Therefore I can use Eq. 32.35 to determine V_{out} and then determine how V_{out} behaves in the limiting cases where $\omega \gg \omega_c$ and $\omega \ll \omega_c$.

③ EXECUTE PLAN (*a*) Equating the right sides of Eqs. 32.35 and 32.36, I get $R = 1/\omega C$. Solving for ω yields the desired cutoff angular frequency ω_c:

$$\omega_c = \frac{1}{RC}. ✔$$

(*b*) To obtain the values of V_{out} for $\omega \gg \omega_c$ and for $\omega \ll \omega_c$, I first rewrite Eq. 32.35 in a form that contains ω_c:

$$V_{out} = V_R = \frac{\mathscr{E}_{max}R}{\sqrt{R^2 + 1/\omega^2 C^2}}$$

$$= \frac{\mathscr{E}_{max}}{\sqrt{1 + 1/R^2\omega^2 C^2}} = \frac{\mathscr{E}_{max}}{\sqrt{1 + \omega_c^2/\omega^2}}. \quad (1)$$

For $\omega \gg \omega_c$, the second term in the square root vanishes and Eq. 1 reduces to $V_{out} = \mathscr{E}_{max}$. ✔

For $\omega \ll \omega_c$, the second term in the square root dominates, so I can ignore the first term. Equation 1 then becomes

$$V_{out} = V_R = \frac{\mathscr{E}_{max}}{\sqrt{1 + \omega_c^2/\omega^2}} \approx \frac{\mathscr{E}_{max}}{\sqrt{\omega_c^2/\omega^2}}$$

$$= \frac{\mathscr{E}_{max}\omega}{\omega_c} = \mathscr{E}_{max}\omega RC.$$

In the limit that the angular frequency ω approaches zero, V_{out} approaches zero as well. ✔

④ EVALUATE RESULT The name *high-pass filter* makes sense because this circuit allows emfs with an angular frequency higher than the cutoff angular frequency to pass through to the output but attenuates emfs of angular frequency lower than the cutoff angular frequency, preventing them from passing through to the output. It is the capacitor that does the actual passing or blocking. It blocks low-angular-frequency emfs because for these emfs the capacitive reactance, $X_C = 1/\omega C$, is very high. For high-angular-frequency emfs, X_C approaches zero, and so the capacitor passes the emf undiminished.

Figure 32.58 An *RLC* series circuit, consisting of a resistor, an inductor, and a capacitor in series across the terminals of an AC source.

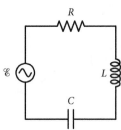

✋ **32.14** Interchange the resistor and the capacitor in Figure 32.57, and then show that the high-pass filter becomes a low-pass filter.

Filters can also be constructed by wiring an inductor and a resistor in series with an AC source. Such a circuit is called an *RL series circuit* and can be analyzed in exactly the manner we used to analyze an *RC* series circuit (see Example 32.9).

Finally, let's analyze an *RLC* series circuit: a resistor, a capacitor, and an inductor all in series with an AC source (**Figure 32.58**). As with the *RC* series circuit, the instantaneous current i is the same in all three elements, and the sum of all the potential differences equals the emf of the source:

$$\mathscr{E} = v_R + v_L + v_C. \quad (32.39)$$

The phasor diagram for this circuit is constructed in **Figure 32.59** for the case where $V_L > V_C$. As before, we begin with the phasors for i and v_R, and then note

Figure 32.59 Steps involved in constructing a phasor diagram for the *RLC* series circuit in Figure 32.58. The diagram in part *c* indicates the phase of the current relative to the source emf.

(*a*) Begin with phasors for i and v_R (in phase)

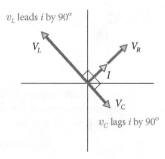

v_L leads i by 90°

v_C lags i by 90°

(*b*) Add V_C and V_L

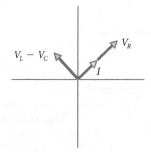

(*c*) Add $V_L - V_C$ and V_R to obtain $\mathscr{E}_{max}$

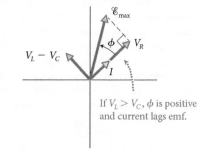

If $V_L > V_C$, ϕ is positive and current lags emf.

that v_C lags i by 90° and v_L leads i by 90° (Figure 32.59a). As a result, the phasors V_C and V_L can be added directly (Figure 32.59b). Finally, the loop rule (Eq. 32.39) requires the phasor for the emf to equal the vector sum of the phasors for the potential differences, as shown in Figure 32.59c. Consequently, the amplitudes V_R, V_L, and V_C must satisfy

$$\mathcal{E}_{max}^2 = V_R^2 + (V_L - V_C)^2. \tag{32.40}$$

Rewriting Eq. 32.40 in terms of I, R (from Eq. 32.6), X_L (from Eq. 32.27), and X_C (from Eq. 32.15) gives

$$\mathcal{E}_{max}^2 = I^2[R^2 + (X_L - X_C)^2] = I^2[R^2 + (\omega L - 1/\omega C)^2], \tag{32.41}$$

and thus
$$I = \frac{\mathcal{E}_{max}}{\sqrt{R^2 + (\omega L - 1/\omega C)^2}}. \tag{32.42}$$

The impedance of the *RLC* series combination (in other words, the constant of proportionality between I and $\mathcal{E}_{max}$) is therefore

$$Z_{RLC} \equiv \sqrt{R^2 + (\omega L - 1/\omega C)^2} \quad \text{(RLC series combination).} \tag{32.43}$$

Table 32.1 lists the impedances of various loads.

Figure 32.59c shows that the phase relationship between the current and the source emf depends on the relative magnitudes of V_L and V_C. The phase of the current relative to the emf is given by

$$\tan \phi = \frac{V_L - V_C}{V_R} = \frac{X_L - X_C}{R}$$

$$= \frac{\omega L - 1/\omega C}{R} \quad \text{(RLC series circuit).} \tag{32.44}$$

If $V_L > V_C$, as it is in Figure 32.59, ϕ is positive, meaning that the current lags the source emf. Here the inductor dominates the capacitor, and as a result the series combination of the inductor and capacitor behaves like an inductor. If $V_L < V_C$, ϕ is negative, the inductor-capacitor combination is dominated by the capacitor, and the current leads the source emf, just as in an *RC* series circuit.

In general, when analyzing AC series circuits, follow the procedure shown in the Procedure box on page 868.

Table 32.1 Impedances of various types of loads (all elements in series)

Load	Z
R	R
L	ωL
C	$1/\omega C$
RC	$\sqrt{R^2 + (1/\omega C)^2}$
RLC	$\sqrt{R^2 + (\omega L - 1/\omega C)^2}$

Note that impedances do not simply add the way resistances do. However, the impedance of any simpler load can be found from the impedance of the *RLC* combination; for example, $Z_{RC} = Z_{RLC}$ without the term containing L.

QUANTITATIVE TOOLS

Procedure: Analyzing AC series circuits

When analyzing AC series circuits, we generally know the properties of the various circuit elements (such as R, L, C, and $\mathscr{E}$) but not the potential differences across them. To determine these, follow this procedure:

1. To develop a feel for the problem and to help you evaluate the answer, construct a phasor diagram for the circuit.
2. Determine the impedance of the load using Eq. 32.43. If there is no inductor, then ignore the term containing L;

if there is no capacitor, ignore the term containing C; and so on.
3. To determine the amplitude of the current, in the circuit, you can now use Eq. 32.42; to determine the phase of the current relative to the emf, use Eq. 32.44.
4. Determine the amplitude of the potential difference across any reactive element using $V = XI$, where X is the reactance of that element. For a resistor, use $V = RI$.

Example 32.9 *RL* series circuit

Consider the circuit shown in **Figure 32.60**. (*a*) Determine the cutoff angular frequency ω_c and the phase constant at which $V_R = V_L$. (*b*) Can this circuit be used as a low-pass or high-pass filter?

Figure 32.60 Example 32.9.

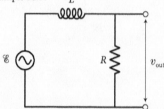

❶ GETTING STARTED This example is similar to Example 32.8, with the capacitor of that example replaced by an inductor here. As in Example 32.8, I see from the circuit diagram that the potential difference v_{out} is equal to the potential difference across the resistor, so $V_{out} = V_R$. I begin by drawing a phasor diagram for the circuit (**Figure 32.61**). I first draw phasors V_R and I, which I know from Figure 32.45*a* are in phase. I then add V_L, which leads I by 90° (Figure 32.51*a*). I make V_L have the same length as V_R because the problem asks about the circuit when $V_R = V_L$.

Figure 32.61

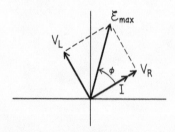

❷ DEVISE PLAN To determine the potential difference amplitudes V_L and V_R across the inductor and the resistor, I follow the procedure given in the Procedure box above. I then set these two amplitudes equal to each other in order to determine ω_c and the phase constant. To determine whether this circuit can be used as a low-pass or high-pass filter, I examine the behavior of V_{out} for $\omega \gg \omega_c$ and for $\omega \ll \omega_c$.

❸ EXECUTE PLAN (*a*) Ignoring the term containing C in Eq. 32.43 and substituting the result in Eq. 32.33, I get for the current amplitude

$$I = \frac{\mathscr{E}_{max}}{\sqrt{R^2 + (\omega L)^2}}.$$

I can now use Eq. 32.6 to calculate the amplitude of the potential difference across the resistor,

$$V_R = IR = \frac{\mathscr{E}_{max}R}{\sqrt{R^2 + (\omega L)^2}}, \tag{1}$$

and Eq. 32.27 to calculate the amplitude of the potential difference across the inductor,

$$V_L = IX_L = \frac{\mathscr{E}_{max}\omega L}{\sqrt{R^2 + (\omega L)^2}}, \tag{2}$$

where I have substituted ωL for X_L (Eq. 32.26). Equating the right sides of Eqs. 1 and 2 yields $R = \omega L$. Substituting ω_c for ω in this equation and solving for ω_c give me for the cutoff angular frequency value at which $V_R = V_L$:

$$\omega_c = \frac{R}{L}. \checkmark$$

To determine the phase constant for the condition $V_R = V_L$, I substitute V_R for V_L in Eq. 32.44 and set V_C equal to zero:

$$\tan \phi = \frac{V_L - V_C}{V_R} = \frac{V_R}{V_R} = 1,$$

so the phase constant is 45°. $\checkmark$

(*b*) Just as I did in Example 32.8, to obtain the limiting values of V_{out}, I first rewrite Eq. 1 in a form that contains ω_c:

$$V_{out} = V_R = \frac{\mathscr{E}_{max}R}{\sqrt{R^2 + (\omega L)^2}} = \frac{\mathscr{E}_{max}}{\sqrt{1 + (\omega L)^2/R^2}}$$

$$= \frac{\mathscr{E}_{max}}{\sqrt{1 + \omega^2/\omega_c^2}}. \tag{3}$$

For $\omega \ll \omega_c$, the second term in the square root vanishes and Eq. 3 reduces to $V_{out} = \mathcal{E}_{max}$. For $\omega \gg \omega_c$ the second term in the square root dominates and we can ignore the first term. Equation 3 then becomes

$$V_{out} = V_R = \frac{\mathcal{E}_{max}}{\sqrt{1+\omega^2/\omega_c^2}} \approx \frac{\mathcal{E}_{max}}{\sqrt{\omega^2/\omega_c^2}} = \frac{\mathcal{E}_{max}\omega_c}{\omega} = \frac{\mathcal{E}_{max}R}{\omega L}.$$

In the limit that the angular frequency ω becomes very large, V_{out} approaches zero. The circuit thus blocks high-frequency emfs and allows low-frequency ones to pass through to the output. Therefore it can be used as a low-pass filter. ✔

④ **EVALUATE RESULT** From my phasor diagram I see that the triangle that has V_R and V_L as two of its sides is an equilateral right-angle triangle, and so the phase constant ϕ must be 45°, as I obtained.

For part b, an emf is generated in the inductor whenever the current in it changes. This emf is proportional to the rate of change of the current in the inductor and it opposes the change in current (Eq. 29.19, $\mathcal{E}_{ind} = -L(di/dt)$). For a low-angular-frequency emf, di/dt is small, and the signal passes through the inductor essentially undiminished. For a high-angular-frequency signal, the inductive reactance $X_L = \omega L$ is high, and so the inductor essentially blocks the signal. It therefore makes sense that the arrangement in Figure 32.60 can serve as a low-pass filter.

Example 32.10 *RLC* series circuit

Consider an *RLC* circuit, such as the one shown in Figure 32.58. The source emf has amplitude 160 V and frequency 60 Hz. The resistance is $R = 50\ \Omega$ and the inductance is $L = 0.26$ H. If the amplitudes of the potential difference across the capacitor and the inductor are equal, what is the current in the circuit?

① **GETTING STARTED** I begin by drawing a phasor diagram for the circuit (**Figure 32.62**). I first draw phasors V_R and I, which are in phase, arbitrarily choosing the direction in which I draw them. I then add phasors V_C, which lags I by 90°, and V_L, which leads I by 90°.

Figure 32.62

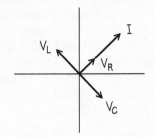

② **DEVISE PLAN** The current in the circuit depends on the impedance, which is given by Eq. 32.43. This equation contains C, however, and I am given no information about this variable. I am given, however, that $V_C = V_L$, and because both V_C and V_L are

proportional to the current, I should be able to determine the current without knowing the capacitance in the circuit.

③ **EXECUTE PLAN** I know from Eq. 32.24 that $V_L = I\omega L$, and I also know that $V_C = I/\omega C$ (Eq. 32.13). Because $V_L = V_C$ in this problem, I can equate the terms on the right in these two equations to obtain

$$\omega L = \frac{1}{\omega C}. \tag{1}$$

Substituting ωL for $1/\omega C$ in Eq. 32.43 then yields $Z_{RLC} = R$. Now I can use Eq. 32.33 to determine the current in the circuit:

$$I = \frac{\mathcal{E}_{max}}{Z_{RLC}} = \frac{\mathcal{E}_{max}}{R} = \frac{160\ \text{V}}{50\ \Omega} = 3.2\ \text{A}. ✔$$

④ **EVALUATE RESULT** When the amplitudes of the potential differences across the inductor and the capacitor are equal, the lengths of the phasors V_C and V_L are equal. Because the phasors point in opposite directions (Figure 32.62), they add to zero, and so the impedance in the circuit is due to the resistor only. This means the current is essentially given by Ohm's law, $I = V/R$, or, in the version I obtained here, $I = \mathcal{E}_{max}/R$.

✋ **32.15** (*a*) Calculate the maximum potential difference across each of the three circuit elements in Example 32.10. (*b*) Is the sum of the amplitudes V_R, V_L, and V_C equal to the amplitude of the source emf? Why or why not?

32.7 Resonance

Consider again the *RLC* series circuit of Figure 32.58. Suppose that the amplitude $\mathcal{E}_{max}$ of the source emf is held constant, but we vary its angular frequency. What happens to the amplitude of the current I and to the phase constant ϕ? Combining Eqs. 32.42 and Eq. 32.43, we can say

$$I = \frac{\mathcal{E}_{max}}{Z} = \frac{\mathcal{E}_{max}}{\sqrt{R^2 + (\omega L - 1/\omega C)^2}}. \tag{32.45}$$

Figure 32.63 Current and phase changes in the *RLC* circuit of Figure 32.58.

(*a*) Frequency dependence of the current at low, medium, and high *R*

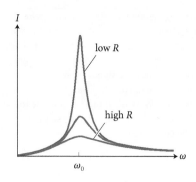

(*b*) Frequency dependence of current phase relative to source emf as function of angular frequency at low, medium, and high *R*

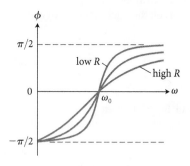

The current is at its maximum when the term in parentheses in the denominator is zero. When this term is zero, $V_L = V_C$ and the two potential differences are 180° out of phase. Therefore the effects of the inductor and the capacitor cancel each other and the circuit behaves as if only the resistor is present. The term in parentheses in Eq. 32.45 is zero when

$$\omega L = \frac{1}{\omega C}. \tag{32.46}$$

The angular frequency for which Eq. 32.46 is satisfied is called the **resonant angular frequency** ω_0 of the circuit:

$$\omega_0 = \frac{1}{\sqrt{LC}}. \tag{32.47}$$

The current amplitude and phase as a function of angular frequency are plotted in **Figure 32.63** for three values of *R* (with fixed values of $\mathcal{E}_{max}$, *L*, and *C*). Increasing or decreasing the angular frequency from ω_0 decreases the current amplitude. Changing *R* changes the maximum current that can be obtained and also changes how rapidly the current drops as the angular frequency increases or decreases from resonance.

Whenever an oscillating physical quantity has a peaked angular frequency dependence, the dependence is referred to as a *resonance curve*. The sharpness of the peak reflects the efficiency with which the source delivers energy to the system at or near resonance and depends on the amount of dissipation present in the system. A very tall, sharp peak corresponds to a system with low dissipation. In such a system, the source can pump an enormous amount of energy into the system at resonance. A short, broad peak corresponds to a system with high dissipation. Here, less energy goes into the system even at resonance, but that energy can be transferred in at angular frequencies farther from resonance. For the *RLC* series circuit, energy is dissipated via the resistor; high *R* values produce less current at ω_0 and a broader resonance curve, as Figure 32.63*a* shows.

Another system that exhibits resonance is a damped mechanical oscillator (see Section 15.8) driven by an external source. The damping in a mechanical oscillator is analogous to the resistance in the *RLC* series circuit.

32.16 How does the resonance curve in Figure 32.63 change if the value of *C* or *L* is changed?

In the *RLC* series circuit, the phase difference between the current and the driving emf also depends on the angular frequency of the AC source. The current can either lag or lead the emf (or be in phase with it), depending on the angular frequency. We found previously that the phase of the current relative to the source emf for an *RLC* series circuit is given by Eq. 32.44:

$$\tan \phi = \frac{\omega L - 1/\omega C}{R}. \tag{32.48}$$

Consider the limiting values of this expression for the relative phase by looking at the curves in Figure 32.63*b*. At resonance ($\omega = \omega_0$), $\phi = 0$ and the current and the source emf are in phase. When $\omega = 0$, $\tan \phi = -\infty$ and $\phi = -\pi/2$. When $\omega = \infty$, $\tan \phi = \infty$ and $\phi = +\pi/2$. Below resonance, $\phi < 0$, the capacitor provides the dominant contribution to the impedance, and the current leads the source emf. Above resonance, $\phi > 0$ and the inductor dominates, and the current lags the source emf.

32.17 In an *RLC* series circuit, you measure $V_R = 4.9$ V, $V_L = 6.7$ V, and $V_C = 2.5$ V. Is the angular frequency of the AC source above or below resonance?

32.8 Power in AC circuits

At the beginning of this chapter, we saw that in alternating-current circuits, the energy stored in capacitors and inductors can oscillate. Consequently, for part of each cycle, these elements put energy back into the source rather than taking up energy from the source. Thus, unlike what we see in DC circuits, the source in an AC circuit does not simply deliver energy steadily to the circuit. Let's take a closer look at how to determine the rate at which an AC source delivers energy to a load.

In general, the rate at which the source delivers energy to its load—in other words, the power of the source—is the time-dependent version of the result we found for DC circuits (Eq. 31.42):

$$p = iv_{\text{load}}. \tag{32.49}$$

Because the current and the emf oscillate, this power varies with time and in principle can be either positive or negative. Let's first consider a load that consists of just one resistor. Ohm's law tells us that the instantaneous energy delivered to the resistor is

$$p = iv_R = i^2R = I^2R \sin^2 \omega t. \tag{32.50}$$

The time dependence of the potential difference, current, and power are shown in **Figure 32.64**. Because the current and potential difference are in phase, the power is always positive, and so the source always delivers energy to the resistor. This makes sense because the resistor dissipates energy regardless of the current direction. Consequently, the rate at which energy is dissipated in the resistor (the *power at the resistor*) is always positive and oscillates at *twice* the angular frequency of the emf.

For most applications, we are interested in the time average of the power at the resistor. Using the trigonometric identity $\sin^2\alpha = \frac{1}{2}(1 - \cos 2\alpha)$, we can rewrite Eq. 32.50 as

$$p = I^2R[\tfrac{1}{2}(1 - \cos 2\,\omega t)] = \tfrac{1}{2}I^2R - \tfrac{1}{2}I^2R \cos 2\omega t. \tag{32.51}$$

The first term on the right is constant in time. The second term on the right averages to zero over a full cycle because the area under the positive half of the cosine is equal to the area under the negative half. As a result, for time intervals much longer than the period of oscillation, the time average of the power at the resistor is

$$P_{\text{av}} = \tfrac{1}{2}I^2R. \tag{32.52}$$

For a sinusoidally varying current, the *root-mean-square* or *rms* value of the current is (Eqs. 19.21 and 30.39)

$$I_{\text{rms}} \equiv \sqrt{(i^2)_{\text{av}}} = \sqrt{\tfrac{1}{2}I^2} = \frac{I}{\sqrt{2}}. \tag{32.53}$$

and so

$$P_{\text{av}} = I_{\text{rms}}^2 R. \tag{32.54}$$

The advantage of writing the average power in terms of the rms current is that Eq. 32.54 is completely analogous to the expression for the energy dissipated by a resistor connected to a DC source (Eq. 31.43). Similarly, we can introduce rms values of potential difference and source emf:

$$V_{\text{rms}} = \frac{V_R}{\sqrt{2}} \quad \text{and} \quad \mathscr{E}_{\text{rms}} = \frac{\mathscr{E}_{\text{max}}}{\sqrt{2}}. \tag{32.55}$$

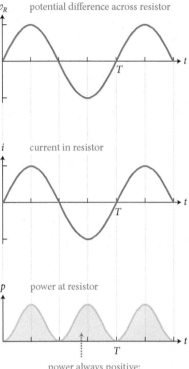

Figure 32.64 For an AC circuit consisting of a resistor connected across an AC source, time dependence of potential difference across the resistor, current in the resistor, and power at the resistor.

v_R potential difference across resistor

i current in resistor

p power at resistor

power always positive: energy always into resistor

Figure 32.65 For an AC circuit consisting of a capacitor connected across an AC source, time dependence of potential difference across the capacitor, current in the capacitor, and power at the capacitor.

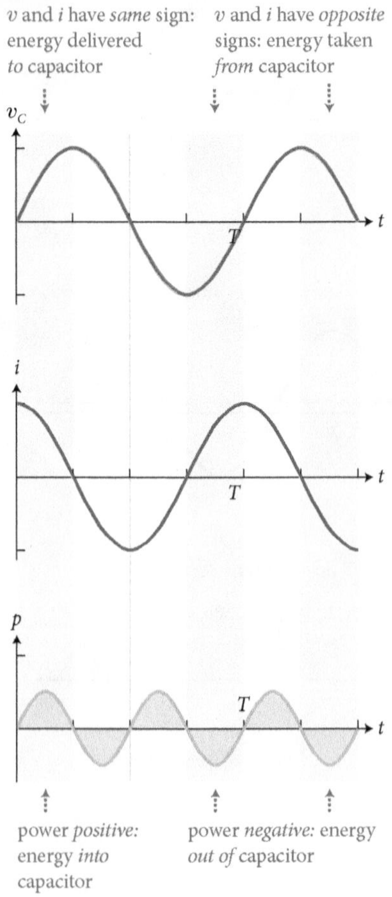

v and *i* have *same* sign: energy delivered *to* capacitor

v and *i* have *opposite* signs: energy taken *from* capacitor

power *positive*: energy *into* capacitor

power *negative*: energy *out of* capacitor

The rms value is a useful way to measure the average value of an oscillating current or emf because the strict time average of these quantities is zero. *Voltmeters* and *ammeters* typically measure the rms value of alternating potential differences and currents, respectively. Thus, for example, the wall potential difference in household electrical wiring in the United States is referred to as 120 V even though the amplitude is 170 V; the 120-V rating is the rms value.

Next let's look at a circuit made up of an AC source and a capacitor. How much power is delivered to the capacitor by the source? Now the current and the potential difference are out of phase (see Figure 32.48). Substituting from Eq. 32.8, $v_C = V_C \sin \omega t$, and Eq. 32.11, $i = \omega C V_C \cos \omega t$, into Eq. 32.49, we obtain

$$p = iv_C = (\omega C V_C \cos \omega t)(V_C \sin \omega t). \tag{32.56}$$

Using the trigonometric identity $\sin(2\alpha) = 2 \sin \alpha \cos \alpha$, we obtain

$$p = \tfrac{1}{2}\omega C V_C^2 \sin 2\omega t. \tag{32.57}$$

As in the resistor-only circuit, the power oscillates at twice the angular frequency of the source, but now the power is sometimes positive and sometimes negative, as shown in **Figure 32.65**. When v and i have the same sign, energy is transferred to the capacitor; when v and i have opposite signs, energy residing in the capacitor is transferred back to the source. The average power is zero, as it must be, because no energy is dissipated in a capacitor. The same is true for an inductor, except that in an inductor energy is stored as magnetic energy rather than electric energy; you can show this mathematically by converting Eqs. 32.56 and 32.57 to their V_L counterparts.

Finally, let's examine how power is delivered to the load in an *RLC* circuit. Although we could work out the power at each element, it's easier to consider the power for the entire load consisting of the *RLC* combination. The potential difference across the load is equal to the applied emf. Using Eq. 32.1, $\mathcal{E} = \mathcal{E}_{max} \sin \omega t$; and Eq. 32.16, $i = I \sin(\omega t - \phi)$; we can write the instantaneous power as

$$p = vi = \mathcal{E}_{max} I \sin \omega t \sin(\omega t - \phi). \tag{32.58}$$

Using the trigonometric identities $\sin(\alpha - \beta) = \sin \alpha \cos \beta - \cos \alpha \sin \beta$ to separate the ϕ and ω dependence and substituting $\sin \alpha \cos \alpha = \tfrac{1}{2} \sin 2\omega t$, we rewrite Eq. 32.58 in the form

$$p = \mathcal{E}_{max} I (\cos \phi \sin^2 \omega t - \sin \phi \sin \omega t \cos \omega t)$$

$$= \mathcal{E}_{max} I (\cos \phi \sin^2 \omega t - \sin \phi \tfrac{1}{2} \sin 2\omega t). \tag{32.59}$$

The time average of $\sin^2 \omega t$ is $1/2$, and the second term inside the parentheses averages to zero, leaving us with

$$P_{av} = \tfrac{1}{2} \mathcal{E}_{max} I \cos \phi. \tag{32.60}$$

Writing this result using rms values (Eqs. 32.54 and 32.55) gives

$$P_{av} = \tfrac{1}{2}(\sqrt{2}\mathcal{E}_{rms})(\sqrt{2}I_{rms})\cos \phi = \mathcal{E}_{rms} I_{rms} \cos \phi. \tag{32.61}$$

QUANTITATIVE TOOLS

We can rewrite this in a more physically insightful way if we note that $\mathscr{E}_{rms} = I_{rms}Z$ (Eq. 32.33) and note from Figure 32.59c that

$$\cos\phi = \frac{V_R}{\mathscr{E}_{max}} = \frac{RI}{ZI} = \frac{R}{Z}. \tag{32.62}$$

With these substitutions, Eq. 32.61 becomes

$$P_{av} = I_{rms}Z\,I_{rms}\frac{R}{Z} = I_{rms}^2 R. \tag{32.63}$$

This result tells us that all of the energy delivered to the circuit is dissipated as thermal energy in the resistor—as it must be, because neither the capacitor nor the inductor dissipates energy. This energy is dissipated at the same average rate as in a circuit made up of a single resistor connected to an AC source (Eq. 32.54).

The factor $\cos\phi$ that appears in Eqs. 32.60–32.62 is called the **power factor** which is a measure of the efficiency with which the source delivers energy to the load. At resonance, when the current and the emf are in phase ($\phi = 0$), the current and the power factor are greatest, and the maximum power possible is delivered to the load. At angular frequencies away from resonance, less power is delivered to the load.

32.18 Calculate the rate P_{av} at which energy is dissipated in the *RLC* series circuit of Example 32.10.

Chapter Glossary

SI units of physical quantities are given in parentheses.

AC source A power source that generates a sinusoidally alternating emf.

alternating current (AC) Current that periodically changes direction. Circuits in which the current is alternating are called *AC circuits*.

depletion zone A thin nonconducting region at the junction between *p*-doped and *n*-doped pieces of a semiconductor where the charge carriers have recombined and become immobile.

diode A circuit element that behaves like a one-way valve for current.

hole An incomplete bond in a semiconductor that behaves like a freely moving positive charge carrier.

impedance Z (Ω) The proportionality constant between the amplitudes of the potential difference and the current in any load connected to an AC source. The impedance for the load in an *RLC* series circuit is

$$Z_{RLC} \equiv \sqrt{R^2 + (\omega L - 1/\omega C)^2} \qquad (32.43)$$

phase constant ϕ (unitless) A scalar that represents the phase difference between the source emf and the current. When the current leads the source emf, ϕ is negative; when the current lags the source emf, ϕ is positive.

power factor $\cos \phi$ (unitless) A scalar factor that is a measure of the efficiency with which an AC source delivers energy to a load:

$$\cos \phi = \frac{V_R}{\mathscr{E}_{max}} = \frac{R}{Z}. \qquad (32.62)$$

reactance X (Ω) The proportionality constant between the amplitudes of the potential difference and the current in a capacitor or inductor connected to an AC source. The *capacitive reactance* is

$$X_C \equiv 1/\omega C, \qquad (32.14)$$

and the *inductive reactance* is

$$X_L \equiv \omega L. \qquad (32.26)$$

resonant angular frequency ω_0 (s^{-1}) In an *RLC* series circuit, the angular frequency at which the current is a maximum.

$$\omega_0 = \frac{1}{\sqrt{LC}}. \qquad (32.47)$$

In general, the resonant angular frequency of an oscillator of any kind is the angular frequency at which the maximum oscillation is obtained.

semiconductor A material that has a limited supply of charge carriers that can move freely and an electrical conductivity intermediate between that of conductors and that of insulators. An *intrinsic semiconductor* is made of atoms of one element only; a *doped/extrinsic semiconductor* contains trace amounts of atoms that alter the number of free electrons available and change the electronic properties. An *n-type* semiconductor has a surplus of valence electrons (relative to the number present in the original intrinsic semiconductor), which means it has some free electrons. A *p-type* semiconductor has a deficit of valence electrons, and so it has some free holes.

transistor A circuit element that behaves like a switch or a current amplifier.

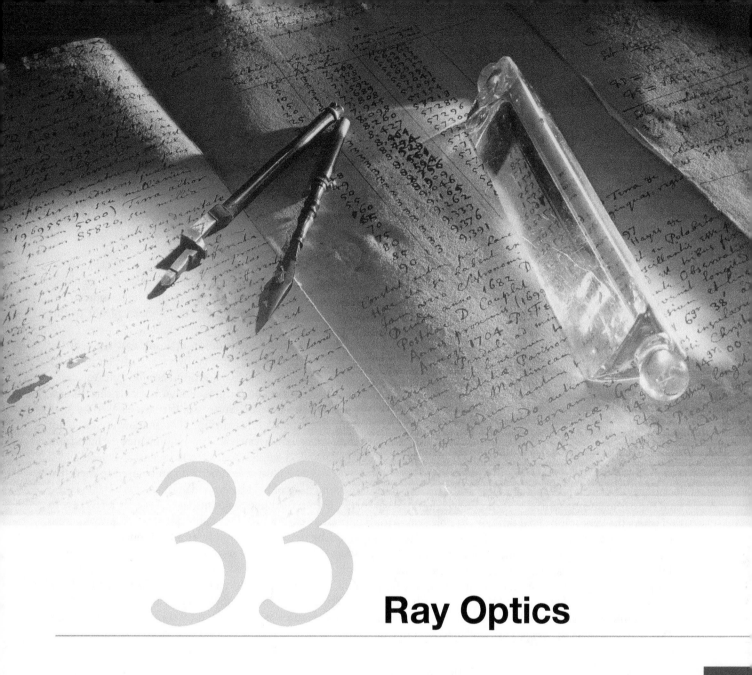

33

Ray Optics

CONCEPTS

QUANTITATIVE TOOLS

Y ou can read these words because this page reflects light toward you; your eyes intercept some of the reflected light, and the lenses of your eyes redirect it, forming an image of the page on the retina. Where does the light reflected from the page come from? Our primary source of light during the day is the Sun, and our secondary source is the brightness of the sky. Indoors and at night, our light sources are flames in candles, white-hot filaments in light bulbs, and glowing gases in fluorescent bulbs. The light from all these sources comes from the accelerated motion of electrons as this motion produces electromagnetic waves.

In Chapter 30 we studied the propagating electric and magnetic fields that constitute electromagnetic waves, and we learned that a narrow frequency range of these waves corresponds to what we know as visible light. In this chapter we continue to study light, particularly its propagation and its interactions with materials. We shall not consider the electric and magnetic fields individually, but instead think of the behavior of rays of light. Such behavior, which is called *ray optics,* was understood long before it was known that light is an electromagnetic wave.

33.1 Rays

If you pierce a small hole in a piece of cardboard and then hold the cardboard between a lamp and a screen, the position where the light transmitted through the hole strikes the screen lies on a straight line connecting the lamp and the hole (Figure 33.1). This observation suggests that we can think of a light source as made up of many straight beams that spread out in three dimensions from the source. Each beam travels in a straight line until it interacts with an object. That interaction changes the beam's direction of travel.

We can represent the propagation of light by drawing **rays:**

> A ray is a line that represents the direction in which light travels. A beam of light with a very small cross-sectional area approximately corresponds to a ray.

In order to see an object, our eyes form an image by collecting light that comes from the object. If the object is a

Figure 33.2 Rays emanating from a source of light.

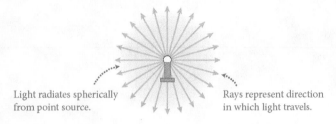

Light radiates spherically from point source.

Rays represent direction in which light travels.

light source, we see it by the light it emits. We can also see an object that is not a light source because such an object interacts with light that comes from a light source. The light is then redirected toward our eyes by means of this interaction.

When you stand outside on a sunny day, some of the rays from the Sun are blocked by your body while others travel in straight lines to the ground around you. You cast a shadow—a region on the ground that is darker than its surroundings because the Sun's rays that are blocked by your body do not strike this region. (The shadow region is not completely dark because it is still illuminated by light from the sky and by sunlight reflected from nearby objects.)

Figure 33.2 illustrates how rays can be used to represent the directions of light beams emanating from a light source. Just as with field line diagrams, we draw only a few rays to represent all the rays that could possibly be drawn; a ray could be drawn along any line radially outward from the source. Although most sources of light—the Sun, a flame, a light bulb—are extended, when the distance to the source is much greater than the extent of the source, we can treat that source as a *point source* (See Section 17.1). That is, we can treat the source as if all the light were emitted from a single point in space. In the first part of this chapter, we develop a feel for which rays to draw in a given situation.

33.1 Suppose a second bulb is added to the left of the one in Figure 33.1, as illustrated in Figure 33.3. What happens to (*a*) the brightness of the spot created on the screen by the first bulb and (*b*) the brightness at locations close to the vertical edges of the original shadow on the screen (points P, Q, R, and S)?

Figure 33.1 A light beam that is not disturbed travels in a straight line.

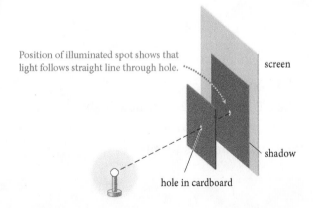

Position of illuminated spot shows that light follows straight line through hole.

screen

shadow

hole in cardboard

Figure 33.3 Checkpoint 33.1.

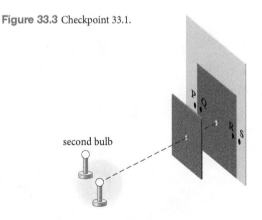

second bulb

Example 33.1 Light and shadow

An object that has a small aperture is placed between a light source and a screen, as shown in **Figure 33.4**. Which parts of the screen are in the shadow?

Figure 33.4 Example 33.1.

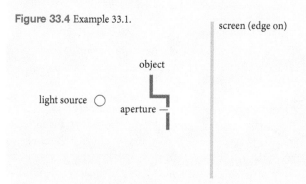

GETTING STARTED The rays emitted by the source radiate outward in all directions following straight paths. The shadow is cast because the object prevents some of the rays from reaching the screen (except for the rays that make it through the aperture).

DEVISE PLAN To locate the edges of the shadow, I draw straight lines from the source to the edges of the object (including the edges of the aperture) and extend these rays to the screen (**Figure 33.5**).

Figure 33.5

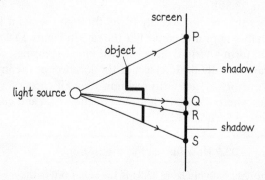

EXECUTE PLAN The top and bottom edges of the shadow correspond to the highest and lowest screen locations (P and S) reached by light rays that are not blocked by the object. The gap in the shadow between locations Q and R corresponds to light rays that pass through the aperture, which means that this region of the screen is not in shadow. ✔

EVALUATE RESULT A shadow that is taller than the object makes sense. Because the light rays from the source emanate in all directions, most of them reach the screen at angles other than 90°. This means that the distance from P to S must be greater than the object height. Indeed, I know from experience that the shadow cast by my hand gets larger as I move my hand closer to a lamp, increasing that angle.

✋ **33.2** Hold a piece of paper between your desk lamp (or any other source of light) and your desk or a wall. How does the sharpness of the edges of the shadow change as you move the paper closer to the bulb? Why does this happen?

33.2 Absorption, transmission, and reflection

Different materials interact differently with the light that strikes them, which is how you can visually distinguish wood from metal, fabric from skin, and a white piece of paper from a blue one. When light strikes an object, the light can be transmitted, absorbed, or reflected.

Transmitted light passes through a material. Objects that transmit light, such as a piece of glass, are said to be *transparent* (**Figure 33.6a**). In *translucent* materials, such as frosted glass, light rays are *transmitted diffusely*—that is, they are redirected in random directions as they pass through, so that the transmitted light does not come from a definite direction (Figure 33.6b). Because translucent materials scatter light in this manner, we cannot see objects clearly through them.

Absorbed light enters a material but never exits again. Objects that absorb most of the light that strikes them, such as a piece of wood, are said to be *opaque*. When light strikes such materials, the energy carried by the light is converted to some other form (usually thermal energy) and the light propagation stops.

Reflected light is any light that is redirected away from the surface of the material (**Figure 33.7** on the next page). Smooth surfaces reflect light *specularly*—that is, each ray bounces off the surface in such a way that the angle between it and the normal to the surface doesn't change (Figure 33.7a). The angle between the incoming ray and the normal to the surface is called the **angle of incidence** θ_i; the angle between the outgoing ray and the normal is called the **angle of reflection** θ_r.

Figure 33.6 We see objects clearly through a sheet of clear glass but diffusely through frosted glass.

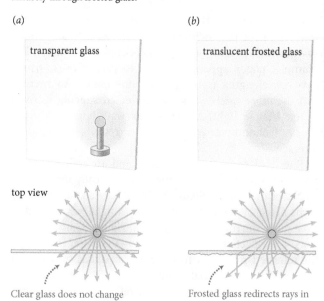

~~3.7 Light reflects specularly from a smooth surface, forming a ~~r image. From a rough surface, it reflects diffusely (in random ~~irections), so no image forms.

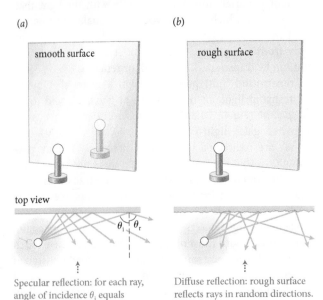

(a)

smooth surface

top view

θ_i θ_r

Specular reflection: for each ray, angle of incidence θ_i equals angle of reflection θ_r.

(b)

rough surface

Diffuse reflection: rough surface reflects rays in random directions.

Empirically we find:

For a ray striking a smooth surface, the angle of reflection is equal to the angle of incidence, and both angles are in the same plane.

This **law of reflection** holds at smooth surfaces for any angle of incidence.

Surfaces that are not smooth reflect light in many directions (Figure 33.7b). For such *diffuse reflection,* each ray obeys the law of reflection, but the direction of the surface normal varies over the surface and so the angle of reflection also varies.

How smooth is smooth? If the height and separation of irregularities on the surface are small relative to the wavelength of the incident light, the surface acts like a smooth surface and most light is reflected specularly. For example, paper appears smooth to microwaves, which have wavelengths ranging from millimeters to meters, and therefore microwaves are reflected specularly from paper. Visible light, however, has wavelengths of hundreds of nanometers, and so paper reflects visible light diffusely.

Rays that come from an object and are reflected from a smooth surface form an **image,** an optically formed duplicate of the object (Figure 33.7a). **Figure 33.8a** shows the paths taken by light rays emitted by a light bulb placed in front of a mirror. A diagram like Figure 33.8a showing just a few selected rays is called a **ray diagram.** If we trace the reflected rays back to the point at which they appear to intersect, we see that point is behind the mirror. Consequently, the brain interprets the reflected rays as having come from that point, creating the illusion that the light

Figure 33.8 Diagrams showing the paths taken by light rays that are produced by a bulb and reflected by a mirror into an observer's eye. The reflected rays appear to come from behind the mirror, forming an image behind the mirror.

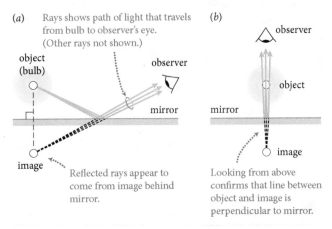

(a) Rays shows path of light that travels from bulb to observer's eye. (Other rays not shown.)

object (bulb)

observer

mirror

image

Reflected rays appear to come from image behind mirror.

(b)

observer

object

mirror

image

Looking from above confirms that line between object and image is perpendicular to mirror.

bulb is behind the mirror. The directions of the rays that reach the eyes of the observer are the same as if they had come from an object located behind the mirror.

Note that the image is located on the line through the object and perpendicular to the mirror, because if we look along that line, the image lies behind the object (Figure 33.8b).

Rays that do not actually travel through the point from which they appear to come, like the rays in Figure 33.8a, are said to form a *virtual image.* A *real image* is formed when the rays actually do intersect at the location of the image. (Flat mirrors cannot form real images; we'll encounter real images when we discuss lenses in Section 33.4 and curved mirrors in Section 33.7.)

Example 33.2 How far behind the mirror?

If the light bulb in Figure 33.8a is 1.0 m in front of the mirror, how far behind the mirror is the image?

❶ **GETTING STARTED** The location of the image is the location from which the rays reflected by the mirror appear to come—that is, the point at which they intersect. From Figure 33.8 I know that because the rays intersect directly behind the bulb, a line that passes through the bulb and is normal to the mirror passes through the image.

❷ **DEVISE PLAN** I can obtain the distance of the image behind the mirror by considering one ray that travels from the bulb to the observer and then tracing that ray back through the mirror to its intersection with the line that is perpendicular to the mirror and passes through the bulb.

❸ **EXECUTE PLAN** I begin by drawing a ray that travels from the bulb to the mirror and is reflected to the location of the observer (Figure 33.9). In my drawing, A denotes the bulb location, B denotes the point where the line connecting the bulb and its image intersects the mirror, and C denotes the point at which the ray that is reflected to the observer hits the mirror. According to the law of reflection, $\theta_r = \theta_i$.

Figure 33.9

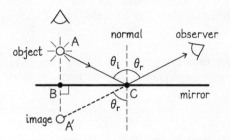

I now extend the reflected ray to behind the mirror (dashed line). I know that the image must lie somewhere along that dashed line and must also lie on the line that passes through the object and is perpendicular to the mirror. The image must therefore lie at the intersection of this line and the dashed ray extension; I denote that intersection point by A'.

To determine the distance BA', which is how far behind the mirror the image is, I note that angle A'CB is equal to $90° − θ_r$ and angle ACB is equal to $90° − θ_i$. Because $θ_i = θ_r$, angles A'CB and ACB are equal. Therefore triangles ABC and A'BC are congruent, and AB = BA'. That is, the image appears at the same distance behind the mirror as the object is in front of it: 1.0 m behind the mirror. ✔

❹ **EVALUATE RESULT** My result makes sense because I know from experience that as I walk toward a mirror, my image also approaches it.

✋ **33.3** If the observer in Figure 33.8 moves to a different position, does the location of the image change?

The colors of visible light we see correspond to different frequencies of electromagnetic waves. Red corresponds to the lowest frequency of the visible spectrum. As the frequency increases, the color changes to orange, yellow, green, blue, indigo, and finally violet (the highest frequency of the visible spectrum). The range of visible frequencies is quite small relative to the range of the complete electromagnetic spectrum, as **Figure 33.10** shows. Frequencies

Figure 33.10 Visible light makes up only a small part of the electromagnetic spectrum.

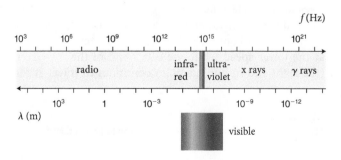

Figure 33.11 All colors of light pass through colorless glass (shown light blue for illustration purposes); orange glass transmits orange light and absorbs all colors of light except orange.

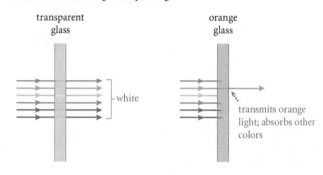

lower than the visible correspond to infrared radiation, and higher frequencies to ultraviolet. When a light source produces all the frequencies of the visible spectrum at roughly the same intensities, the emitted light appears white.

Different colors of light interact differently with different objects, affecting the color we perceive the object as being. Colorless materials, like a piece of ordinary window glass, transmit all colors of the visible spectrum. A piece of orange glass, on the other hand, transmits only the orange part of the visible spectrum. All other colors are absorbed in the glass (**Figure 33.11**). A red apple absorbs all colors of the visible spectrum except red, which is redirected to our eyes. Grass absorbs all colors except green, which is diffusively reflected at its surface.

Because light is a wave phenomenon, it is sometimes useful to represent the propagation of light with wavefronts, which we introduced in Section 17.1. Wavefronts are drawn perpendicular to the direction of propagation of the wave.* Because light rays point along the direction of propagation of the light, light wavefronts are perpendicular to light rays. **Figure 33.12a** shows the spherical

Figure 33.12 A point source of light produces spherical wavefronts; a beam of light contains planar wavefronts.

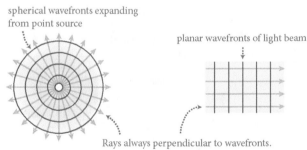

spherical wavefronts expanding from point source

planar wavefronts of light beam

Rays always perpendicular to wavefronts.

---

* For mechanical waves, the wavefronts are drawn at the locations of the wave crests, spaced by the wavelength of the wave. Because the wavelength of light is very short, the wavefronts of light cannot be represented to scale.

CONCEPTS

.13 (*a*) The reflection of wavefronts from a smooth surface ex-
the law of reflection. (*b*) The corresponding rays and their angles of
.cidence θ_i and reflection θ_r.

(*a*)

(*b*)

incident ray | reflected ray

$\theta_i | \theta_r$

C   B

A $\theta_i = \theta_r$ D

Distances AC and BD are equal,
so angle of incidence θ_i equals
angle of reflection θ_r.

Angles of incidence and reflection
for rays are measured from normal
to reflecting surface.

wavefronts for light coming from a point source, and Figure 33.12*b* shows the straight-line rays and wavefronts
corresponding to a planar electromagnetic wave. Note
that a planar wave is represented with rays that are parallel to one another because all the wavefronts are parallel
to one another.

By looking at how wavefronts behave, we can understand the law of reflection. When a light ray strikes a
smooth surface at an incidence angle $\theta_i \neq 0$ (**Figure 33.13**),
the left end of the first wavefront to reach the surface gets
there, at A, before the right end does. In the time interval
it takes the right end to reach the surface at D, the left end
has traveled back from the surface to C. The distance traveled by the right end toward the surface, BD, is the same as
that traveled by the left end away from the surface, AC, so
the angles BAD and CDA must be equal. The angle of incidence θ_i equals angle BAD. Likewise, the angle of reflection
θ_r equals angle CDA. So $\theta_i = \theta_r$.

So far we have treated the object (and consequently the
image) as a single point. **Figure 33.14** shows how images are
formed of extended objects. Each point on the object reflects (or emits) light rays, and the reflections of these rays

Figure 33.14 Paths taken by rays from more than one point on the
object, showing how extended images form.

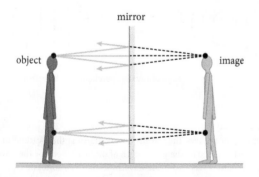

mirror

object

image

appear to come from a corresponding point on the image.
A flat mirror thus produces behind the mirror an exact
mirror image of the entire extended object.

✋ **33.4** In order for the person in Figure 33.14 to see a
complete image of himself, does the mirror need to be as tall
as he is?

33.3 Refraction and dispersion

As we found in Chapter 30, light propagates with speed
$c_0 = 3 \times 10^8 \text{ m/s}$ in vacuum. In air, the speed of light is
almost the same as that in vacuum. In a solid or liquid
medium, however, light propagates at a speed c that is
generally less than c_0.* In glass, for example, visible light
propagates at two-thirds of the speed of light in vacuum
(see Example 30.8).

How does this change in speed affect the propagation
of an electromagnetic wave? Recall from Chapter 16
that harmonic waves are characterized by both a wavelength λ and a frequency f and that the product of the
wavelength and frequency equals the wave's speed of
propagation (Eq. 16.10). The frequency of the wave must
remain the same because the oscillation frequency of
the electromagnetic field that makes up the wave is determined by the acceleration of charged particles at the
wave's source. The acceleration of the source does not
alter when the wave travels from one medium to another,
and thus the frequency of the traveling wave also cannot
change.

✋ **33.5** In vacuum, a particular light wave has a wavelength
of 400 nm. It then travels into a piece of glass, where its speed
decreases to two-thirds of its vacuum speed. What is the distance between the wavefronts in the glass?

As we found in Checkpoint 33.5, when rays of light
pass through the interface between vacuum and a transparent material, the wavefronts inside the material are
more closely spaced than they are in vacuum, due to the
lower speed of the wavefronts. **Figure 33.15** illustrates this
effect for wavefronts incident normal to the surface of the
material.

What if the wavefronts strike the transparent material
at an angle? In such a case, one end of the wavefront arrives at the surface before the other (**Figure 33.16**). Once
the end that reaches the surface first (this happens to be
the left end in Figure 33.16) enters the material, it travels
at the lower speed while the other end of the wavefront
(the right end in our example) continues to travel at the

*For visible light, c is less than c_0. For x rays, c can be greater than c_0.

CONCEPTS

Figure 33.15 Wavefronts for a ray traveling from vacuum into transparent glass in a direction normal to the glass surface.

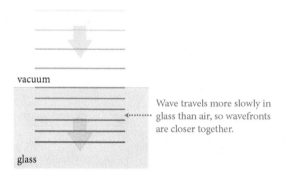

vacuum

Wave travels more slowly in glass than air, so wavefronts are closer together.

glass

vacuum speed. This means the distance AC traveled by the left end is less than the distance BD traveled by the right end during the same time interval (Figure 33.16a). Consequently, the wavefront CD in the material is no longer parallel to the wavefront AB that has not yet entered the material.

The direction of the ray associated with these wavefronts therefore changes on entering the material. As shown in Figure 33.16b, the angle of incidence θ_1, between the ray in vacuum and the normal to the interface between the two materials, is greater than the angle θ_2, between the ray in the material and the normal to that interface. This bending of light as it moves from one material into another is referred to as **refraction,** and the angle θ_2 between the refracted ray and the normal to the interface between the materials is called the **angle of refraction.** Whenever light is refracted, the angle between the ray and the normal is

Figure 33.16 (a) Refraction is explained by the behavior of wavefronts that cross at an angle into a transparent medium in which they travel more slowly. (b) Incident, reflected, and refracted rays, showing the angles of incidence θ_1 and refraction θ_2 (measured from the normal to the surface).

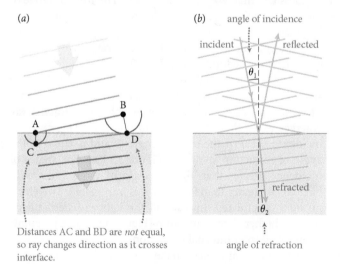

(a)

Distances AC and BD are *not* equal, so ray changes direction as it crosses interface.

(b) angle of incidence

angle of refraction

always greater in the material in which the light travels faster, so:

> When a light ray travels from one material into a second material where light travels more slowly, the ray bends toward the normal to the interface between the materials.

Generally, the speed of light decreases as the mass density of the material increases. Note also that, as shown in Figure 33.16b, both reflection and refraction take place at the interface between two media (or between vacuum and a medium).

The amount of bending depends on the angle of incidence and on the relative speeds in the two media. There is no bending for normal incidence (as we saw in Figure 33.15); the bending is less near normal incidence and becomes more pronounced as the angle of incidence increases. In Section 33.5, we'll work out a quantitative expression relating angles θ_1 and θ_2.

33.6 Suppose the ray in Figure 33.16 travels in the opposite direction—that is, from the denser medium to the less dense medium. If the angle of incidence is now θ_2, how does the angle of refraction compare with θ_1?

Because the relationship between the angles of incidence and refraction is completely determined by the speed of the wavefronts in the two media, the angles do not depend on which is the incident ray and which is the refracted ray. As shown in **Figure 33.17**, θ_1 and θ_2 have the same values whether θ_1 is the angle of incidence (Figure 33.17a) or the angle of refraction (Figure 33.17b). Keep in mind, however, that the reflected ray is always on the same side of the interface as the incident ray, and so the angle of reflection is *not* the same in Figure 33.17a and Figure 33.17b.

Figure 33.17 Because refraction is caused by the relative speeds of the wavefronts in two media, the angles of incidence and refraction do not depend on which is the incident ray and which the refracted ray.

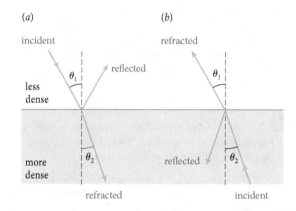

(a) *(b)*

CONCEPTS

.3 Crossing a slab

...sider a light ray incident on a parallel-sided slab of glass surrounded by air, as shown in Figure 33.17a. The ray travels all the way through the slab and emerges into air on the other side. In what direction does the ray emerge?

1 GETTING STARTED This problem involves two successive encounters of a light ray with interfaces between glass and air. At each interface, the ray is refracted. I need to determine the direction of the ray (its angle to the normal to the slab) after it crosses the lower interface of the slab represented in Figure 33.17a.

2 DEVISE PLAN Because I want to know the direction of the emerging ray, I construct an appropriate ray diagram. Figure 33.17a shows the direction of the ray inside the slab. I extend the ray through the slab to the lower interface (**Figure 33.18**) and draw the emerging ray, labeling its angle to the normal θ_{lower}. To determine this angle, I need to consider the refraction that occurs at the lower interface.

Figure 33.18

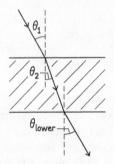

3 EXECUTE PLAN Because the two interfaces are parallel, their normals are also parallel, and so the angle at which the ray is incident on the lower interface is equal to the angle θ_2 at which it is refracted at the upper interface. I saw in Figure 33.17b that if the angle between the ray and the normal in the slab is θ_2, it doesn't matter whether the ray in the slab is the incident ray or the refracted ray; either way, the angle between the ray in the air and the normal is θ_1. Therefore $\theta_{lower} = \theta_1$. ✔

4 EVALUATE RESULT Crossing the lower interface from glass into air, the ray bends away from the normal, as it should because glass is denser than air. With $\theta_{lower} = \theta_1$, in fact, the ray emerges parallel to the original direction it had before entering the slab. This makes sense because the two air-glass interfaces are parallel. (Note, though, that the ray is shifted sideways by a small distance relative to its original path.)

🖐 **33.7** When the ray reflected from the bottom surface in Figure 33.18 reemerges from the top surface, how does the angle it makes with the normal compare with θ_1?

What range of refraction angles is possible? To answer this question, let's first consider the case where the ray

Figure 33.19 The range of possible refraction angles for a ray crossing into a medium of either higher or lower density.

Ray travels into higher-density medium at increasing angle of incidence

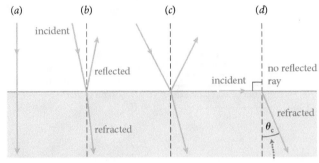

Ray travels into lower-density medium at increasing angle of incidence

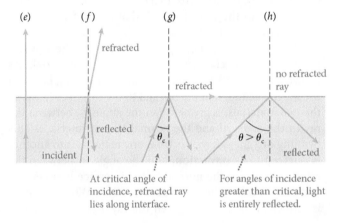

travels from a low-density medium into a denser medium (**Figure 33.19a–d**). Because the angle of incidence is always greater than the angle of refraction in this situation, as the angle of incidence approaches 90°, the angle of refraction remains less than 90° (Figure 33.19d). The full 90° range of incidence angles gives a range of refraction angles that is less than 90°.

Next consider the case where the ray travels from a high-density medium into a lower-density medium (Figure 34.18e–g). The angle of incidence is now less than the angle of refraction. Consequently, as the angle of incidence increases, it reaches a value for which the refracted ray emerges along the interface (Figure 33.19g). This angle of incidence is called the **critical angle** θ_c and is equal to the angle of refraction shown in Figure 33.19d. For angles of incidence greater than θ_c, the angle of refraction would have to be greater than 90°, which is impossible. Therefore, no light is refracted. Instead, all the light is *reflected* back into the higher-density medium (Figure 33.19h), a phenomenon called **total internal reflection.**

Several optical devices make use of total internal reflection to direct light. The glass prism shown in **Figure 33.20**

Figure 33.20 A prism can act as a perfect mirror by means of total internal reflection.

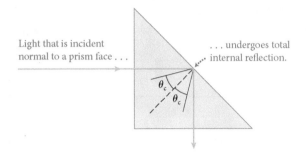

reflects light just as a mirror would. A light ray enters the prism's front surface at normal incidence. Because the back surface is slanted relative to the front surface, the angle at which the ray hits the back surface is less than 90°. This back-surface angle of incidence is greater than the critical angle for the glass, however, and so the light is totally reflected from the back surface. Such prisms are actually better mirrors than most regular mirrors; they reflect very close to 100% of the incident light, whereas mirrors are less reflective due to imperfections in the reflecting surface.

Optical fibers also guide light by means of total internal reflection. An optical fiber is a long, thin fiber made of a transparent material such as glass. If light shines into one end of the fiber at an angle greater than the critical angle, the light travels along the fiber through repeated total internal reflections, and essentially all of the light that entered the fiber emerges at the other end (Figure 33.21). Because very little light is lost as the light travels, only a faint glow comes from the rest of the fiber.

Figure 33.21 How optical fibers work.

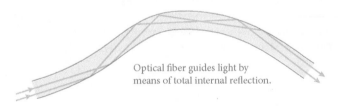

Optical fiber guides light by means of total internal reflection.

Figure 33.22 The phenomenon of dispersion, which results from the fact that the speed of light in a given medium (and hence the angle of refraction) depends slightly on the frequency of the light.

(*a*) Prism refracts light of single frequency

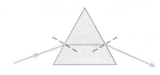

(*b*) Dispersion: different colors have different angles of refraction

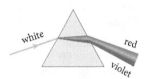

(*c*) Rainbows result from dispersion of sunlight by raindrops

Because the speed of light in any given medium depends slightly on the frequency of the light, the angle of refraction also depends on frequency. This phenomenon is called **dispersion** because it causes rays of different colors to separate—to be *dispersed*—when refracted. Prisms like the one shown in Figure 33.22 are designed to separate colors by the frequency dependence of the angle of refraction. In most media, high-frequency light travels more slowly than low-frequency light, and so high-frequency light bends more strongly toward the normal. The lowest frequency of visible light is red and the highest is violet, which means violet light bends the most, as the rainbow of Figure 33.22*c* shows.

Both rainbows and the brilliance of gems result from a combination of total internal reflection and dispersion. In a rainbow (Figure 33.22*c*), the combination of total internal reflection and dispersion means that we see different colors coming from water droplets at different viewing angles. Gems are cut with many internal surfaces from which total internal reflection takes place. Because the light is also dispersed, colorless gems such as diamonds shine with many distinct colors.

🖐 **33.8** Because of dispersion, the critical angle for total internal reflection in a given medium varies with frequency. Is the critical angle for a violet ray greater or less than that for a red ray?

CONCEPTS

iple

3.23 shows four ways in which a light ray can
vel between two locations A and B: directly, reflected
from a mirror, refracted through a glass slab, and refracted through a prism.* You could say that in each case
the ray reaches B because it is aimed properly from A.
However, an entirely different way of looking at the path
followed by the light was suggested by the French mathematician Pierre de Fermat (1601–1665) in a formulation
today known as **Fermat's principle:**

> The path taken by a light ray between two loca-
> tions is the path for which the time interval needed
> to travel between those locations is a minimum.

This principle may seem to imply that light always
travels in a straight line. However, the *quickest path*
between two locations is not necessarily the *shortest
distance* when the speed of light differs in different
regions.

Let's consider the four paths in Figure 33.23 using
Fermat's principle. In Figure 33.23*a*, the ray does follow a
straight path because the medium in which the ray travels is uniform. As a result, the quickest path is indeed the
shortest distance: a straight line from A to B.

In Figure 33.23*b*, the fact that the straight-line path
from A to B is blocked means that the ray must reflect
somewhere off the mirror in order to travel from A to B.
The path shown, which satisfies the law of reflection,
is the shortest distance from A to B involving reflection
from the mirror. Because the distance from A to the
reflection location P equals the distance between the
image location I and P, the straight line IB is equal in
length to the path traveled by the ray from A to B. Moving the reflection location to either side of P, so that the
angle of incidence does not equal the angle of reflection,

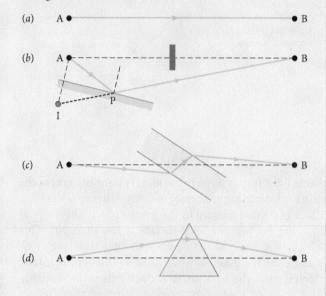

Figure 33.23 Ray diagrams illustrating the *quickest path* for a light ray
traveling from A to B for four situations.

increases the length of the path. Thus, Fermat's principle
implies the law of reflection.

When the ray must travel through some air and some
glass, as in Figure 33.22*c* and *d*, the quickest path is
not a straight line because the ray's speed in the glass is
only two-thirds of its speed in air. To minimize the time
interval needed to travel from A to B, the ray bends on
entering and exiting the glass. Such a bent path reduces
the distance traveled through the glass without increasing the distance traveled in air so much that it offsets the
amount of time saved. In Example 33.7, we shall see that
calculating the bending angles with Fermat's principle
gives the same result as with ray optics.

*Note that what distinguishes a glass slab from a glass prism is the way I use the terms: In a slab, the two opposite surfaces are parallel to each other;
in a prism, they are not.

33.4 Forming images

As shown in **Figure 33.24**, by combining two prisms and a
glass slab we can create a device that steers parallel light
rays toward each other. The rays through the center of the
device pass straight through, those through the top prism
are refracted downward, and those through the bottom
prism are refracted upward.

To bring all parallel incident rays to a single point, a
structure called a **lens** is used. A lens is designed with
curved surfaces so that the refraction of incident rays increases gradually as we move away from the center. To accomplish this, lenses are typically made with spherical surfaces, which are easy to manufacture.

Figure 33.25*a* shows a lens with *convex* spherical surfaces, where a *convex surface* is defined as one that curves
like the outside of a sphere. Rays parallel to the lens *axis*—a

Figure 33.24 A device that redirects parallel light rays toward each other.

Prism redirects ray.

Two prisms and a slab cause
parallel rays to converge.

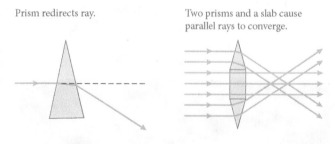

line perpendicular to the lens through its center—converge
through such a lens onto a single point called either the
focus or the **focal point.** A lens with convex surfaces is
therefore called a *converging lens*. The distance from the
center of the lens to the focus is called the **focal length** *f*.

Figure 33.25 Converging lens with convex spherical surfaces.

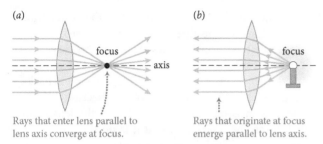

(a)

focus

axis

Rays that enter lens parallel to
lens axis converge at focus.

(b)

focus

Rays that originate at focus
emerge parallel to lens axis.

What if we place a light source at the focus of a lens, as in Figure 33.25b? As we saw in the preceding section, the path followed by a light ray is unaffected by reversing the direction of propagation of the ray, as long as the ray is not absorbed by the medium. So, if we place a light source at the focus of a lens, a beam of parallel rays emerges on the other side of the lens.

33.9 Sketch the wavefronts corresponding to all the rays in Figure 33.25a, both the parallel ones on the left and the refracted ones on the right.

We can reverse the direction of the rays through the lens of Figure 33.25, so that parallel rays enter from the right (Figure 33.26). The rays converge again at a focus that is the same distance f from the center of the lens. Thus, every lens has two foci, one on either side of the lens at the same distance f from it.

If we tilt the parallel rays a bit relative to the lens axis (Figure 33.27), the rays still converge at a distance f from the center, but the focus is no longer on the axis. Provided the parallel rays make only a small angle with the lens axis, they all converge at a point on a plane—called the *focal plane*—that is perpendicular to the axis a distance f from the lens. Rays that run near the lens axis—either parallel to it or at a small angle—are said to be *paraxial*.

Now that we know how parallel rays and rays that emanate from the focus of a lens are refracted by the lens, we can determine where images are formed. The image of a point on an object is formed where all the light rays emanating from that point converge. (These light rays then diverge from the location of the image; when they enter the eye, the brain interprets them as having come from the location of

Figure 33.26 A lens has two equivalent foci, one on each side, at equal distances from the center of the lens.

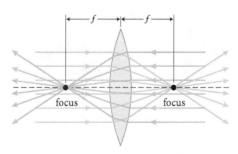

focus   focus

Figure 33.27 If the rays strike the lens at an angle, they no longer converge on the focus, but they still converge on a focal plane at the focal distance f.

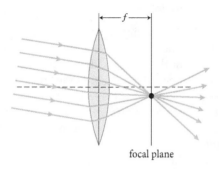

focal plane

the image.) An image of the entire object is made up of the images of all the individual points on the object.

To determine where the rays emanating from a point on an object converge, we don't need to draw all the rays. Instead, we draw three special ones, called **principal rays,** and see where they converge:

1. a ray that travels parallel to the lens axis before entering the lens,

2. a ray that passes through the center of the lens, and

3. a ray that passes through the focus that is on the same side of the lens as the object.

These three principal rays are shown in **Figure 33.28a** for the case where the object lies beyond the focus of the

Figure 33.28 The three principal rays for a spherical lens can be used to determine the location, size, and orientation of the image for a given object.

(a) The three principal rays

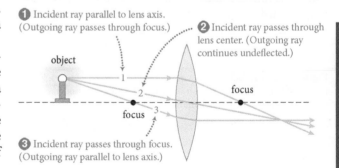

❶ Incident ray parallel to lens axis. (Outgoing ray passes through focus.)

❷ Incident ray passes through lens center. (Outgoing ray continues undeflected.)

object

focus

focus

❸ Incident ray passes through focus. (Outgoing ray parallel to lens axis.)

(b) Using principal rays to determine location and orientation of image

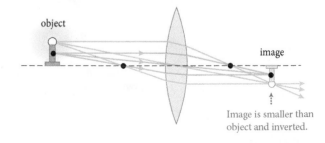

object

image

Image is smaller than object and inverted.

plified ray diagrams for lenses

rmine the location and orientation of an image
ned by a lens, follow this procedure:

1. Draw a horizontal line representing the lens axis (the line perpendicular to the lens through its center). In the center of the diagram, draw a vertical line representing the lens. Put a + above the line to represent a converging lens or a − to represent a diverging lens.
2. Put two dots on the axis on either side of the lens to represent the foci of the lens. The dots should be equidistant from the lens.
3. Represent the object by drawing an upward-pointing arrow from the axis at the appropriate relative distance from the lens. For example, if the distance from the object to the lens is twice the focal length of the lens, put the arrow twice as far from the lens as the dot you drew in step 2. The top of the arrow should be at about half the height of the lens.

4. From the top of the arrow representing the object draw two or three of the three *principal rays* listed in the Procedure box "Principal rays for lenses" on page 888.
5. The top of the image is at the point where the rays *that exit the lens* intersect (if they diverge, trace them backward to determine the point of intersection). If the intersection is on the opposite side of the lens from the object, the image is real; if it is on the same side, the image is virtual. Draw an arrow pointing from the axis to the intersection to represent the image (use a dashed arrow for a virtual image).

In general it is sufficient to draw two principal rays, but depending on the situation, some rays may be easier to draw than others. You can also use a third ray to verify that it, too, goes through the intersection. (If it doesn't, you have made a mistake.)

lens. We already know how rays 1 and 3 travel. Ray 1 passes through the focus on the other side of the lens, and ray 3 emerges from the lens parallel to the axis. As for ray 2, as long as it is paraxial, it passes straight through with negligible refraction (**Figure 33.29**). (Nonparaxial rays are shifted significantly; our treatment of lenses in this chapter is restricted to images formed by paraxial rays.) Ray 2 can therefore be drawn as traveling in a straight line through the center of the lens.

To determine the location and orientation of the image of an extended object, we work out the locations of the images of several points on the object. Figure 33.28b shows where the rays converge for two points on the object, and the extended image that can be inferred from these points. The image is smaller than the object and inverted. See the

Procedure box "Simplified ray diagrams for lenses" on this page for a general description of how to draw simplified ray diagrams for lenses.

33.10 Do you need to draw all three principal rays to determine the location of an image?

Example 33.4 Where is the image?

Consider the light bulb that is the object in Figure 33.28. If you move the bulb to the left, does the image shift left, shift right, or stay in the same place?

❶ **GETTING STARTED** Using the procedure for drawing simplified ray diagrams, I represent the lens, object, and image of Figure 33.28, and draw the bulb at its new position (**Figure 33.30a**).

Figure 33.30

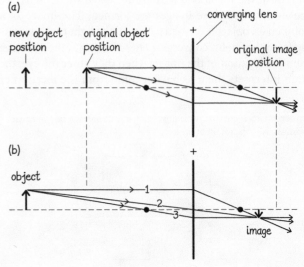

Figure 33.29 Paraxial and nonparaxial rays. For a paraxial ray or a lens that is not too thick, the refraction is so slight that we can consider this ray to be one uninterrupted straight line.

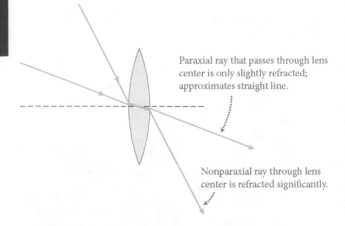

Paraxial ray that passes through lens center is only slightly refracted; approximates straight line.

Nonparaxial ray through lens center is refracted significantly.

❷ **DEVISE PLAN** I can determine which way the image shifts by drawing the principal rays for the light bulb in its new position.

❸ **EXECUTE PLAN** With the rays drawn (Figure 33.30b), I see that the image shifts left. ✔

❹ **EVALUATE RESULT** As I move the light bulb to the left, principal ray 1 remains the same, but principal ray 2 makes less of an angle with the lens axis than before. Consequently, the location at which these two rays intersect is closer to the lens than before, shifting the image to the left, as I concluded from the diagram.

Notice in Figure 33.30 that the image gets smaller as the object moves farther away from the lens. Conversely, moving the object closer to the lens makes the image larger. Indeed, one of the most common uses of lenses is to enlarge images. If the object is placed at the focus, no image forms because the rays all emerge from the lens parallel to each other—in other words, the rays do not converge.

Consider placing an object *between* the lens and its focus, as in **Figure 33.31**. In this configuration, principal ray 3 does not pass through the focus. Instead, it lies on the line that joins the focus to the point of interest on the object.

The image formed in the configuration of Figure 33.28 (object beyond focus) is real because the rays really do converge at the point where the image is formed. In contrast, the image in Figure 33.31 is virtual because the rays do not

actually converge at the point where the image is formed. (The extensions of these rays do cross the image point, however, and so an observer interprets the rays emerging from the lens as having traveled along straight lines from the location of the image, as indicated in Figure 33.31a.)

An important difference between real and virtual images is that if a screen is placed at the location of a real image, the image can be seen on the screen. Placing a screen at the location of a virtual image does not display the image because the light rays do not actually pass through the image location.

Figure 33.31a shows that, for this configuration of object and lens (object between focus and lens), the image is larger than the object and upright (unlike the image in Figure 33.28, which is inverted). A magnifying glass is designed to produce an enlarged, upright image of an object, which means that magnifying glasses are made with converging lenses and are held close to the object of interest (so that the object is between the lens and the focus).

✋ **33.11** As the object in Figure 33.31 is moved closer to the lens, does the size of the image increase, decrease, or stay the same?

Just as with electric circuits, it is convenient to use a simplified notation for ray diagrams. Figure 33.31b shows such a simplified version of the ray diagram of Figure 33.31a. Note that objects and real images are denoted by solid arrows and virtual images are denoted by dashed arrows.

Lenses can also be made with concave spherical surfaces rather than convex ones, where a *concave surface* is one that curves like the inside of a sphere. Such a lens is called a *diverging lens*, and **Figure 33.32** shows why: A series of parallel rays entering the lens are no longer parallel when they emerge. If we follow the path of the emerging rays back to the left side of the lens, we see that the diverging rays appear to all come from the same location on the left side. This location corresponds to the focus of a converging lens, but in a diverging lens it is a *virtual focus* rather than a real focus because the rays never actually travel through this location. (Just as for converging lenses, there is an equivalent focus on the other side of the lens.)

Figure 33.31 (a) When an object is located between the lens and the focus, the image is virtual and enlarged. (b) In a simplified ray diagram, the object and image are replaced with solid and dashed arrows, respectively, and the lens is replaced by a vertical line. (The + indicates a converging lens.)

(a) Ray diagram for an object located between the focus and the lens

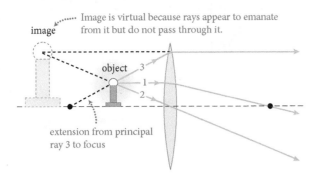

(b) Simplified version of ray diagram

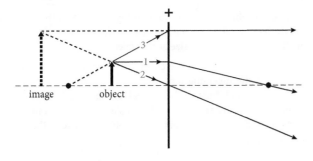

Figure 33.32 A diverging lens.

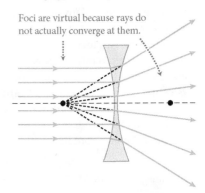

..icipal rays for lenses

..pagations of principal rays for converging and
..rging lenses are very similar. The description below
..olds for rays that travel from left to right.

Converging lens

1. A ray that travels parallel to the lens axis before entering the lens goes through the right focus after exiting
   the lens.
2. A ray that passes through the center of the lens continues undeflected.
3. A ray that passes through the left focus travels parallel
   to the lens axis after exiting the lens. If the object is

between the focus and the lens, this ray doesn't pass
through the focus but lies on the line from the focus to
the point where the ray originates.

Diverging lens

1. A ray that travels parallel to the lens axis before entering the lens continues along the line from the left
   focus to the point where the ray enters the lens.
2. A ray that passes through the center of the lens continues undeflected.
3. A ray that travels toward the right focus travels parallel to the lens axis after exiting the lens.

Figure 33.33 Ray diagram for an object outside the focus of a diverging lens.

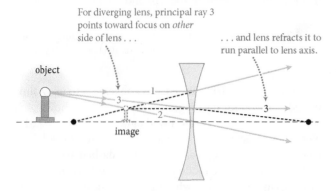

Figure 33.34 These lenses are all converging because each is thicker at the center than at the edges.

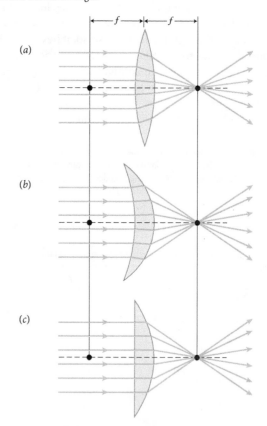

Figure 33.33 shows a ray diagram for a diverging lens. The same principal rays are drawn, but now ray 3 does not pass through the focus on the same side of the lens as the object. Instead, for diverging lenses ray 3 is drawn on the line that runs from the point where the ray originates to the focus on the other side of the lens; once refracted by the lens, this ray travels parallel to the lens axis. The general procedure for drawing ray diagrams for diverging lenses is still the same as the one for converging lenses (see the Procedure box "Simplified ray diagrams for lenses" on page 886), but the drawing of principal rays is a little bit different (see the Procedure box "Principal rays for lenses" above).

The lenses we have considered so far all have identical curved surfaces on each side. Many lenses have different surfaces, however. For example, it is possible to construct a converging lens with a certain focal length with two identical curved surfaces, two differently curved surfaces, or even a flat and a curved surface (**Figure 33.34**), as long as the lens is thicker at its center than at the edges. Regardless of the *radii of curvature* (that is, the radii of the spheres that best fit the surfaces), the lens has two foci, one on either side of the lens at the same distance f from it.

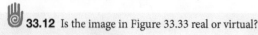 **33.12** Is the image in Figure 33.33 real or virtual?

Example 33.5 Demagnifying glass

Suppose the object in Figure 33.33 is placed between the focus and the lens. (*a*) Is the image real or virtual? (*b*) Is it larger than, smaller than, or the same size as the object?

❶ **GETTING STARTED** To sketch the situation (**Figure 33.35**), I represent the diverging lens as a vertical line with a minus sign above it, and draw the horizontal lens axis. I add the focal points at equal distances from the lens along its axis. Finally, I add a solid arrow, representing the object, between the left focal point and the lens.

Figure 33.35

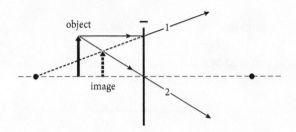

② DEVISE PLAN To locate the image and determine its size and whether it is real or virtual, I can draw the principal rays.

③ EXECUTE PLAN (*a*) I add to my diagram the principal rays coming from the tip of my object arrow. All I need is rays 1 and 2; I do not need ray 3 because the intersection of ray 2 and the dashed extension of ray 1 unambiguously determines the location of the tip of the image arrow. The dashed extension I had to draw for ray 1 tells me that the rays do not actually intersect at this location; they only appear to intersect here. Therefore the image is virtual. ✔

(*b*) My diagram tells me that the image is smaller than the object. ✔

④ EVALUATE RESULT Because a diverging lens spreads rays out rather than bringing them together, it makes sense that a virtual image will form. I also know from experience that in contrast to a converging lens, which magnifies images, diverging lenses create smaller images, as I found.

✋ 33.13 (*a*) Draw the third principal ray in **Figure 33.35**. Is there any position for the object in Figure 33.35 for which (*b*) the image is larger than the object and (*c*) the image is real?

~~~hy do you get a clear reflection from the surface of a lake on a calm day but little or no reflection from the surface on a windy day?

2. (*a*) As light travels from one medium into another, as shown in **Figure 33.36** ("fast" and "slow" refer to the wave speed in each medium), what happens to the wavelength of the light? (*b*) Draw the reflected and refracted rays at each surface.

Figure 33.36

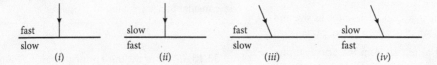

3. What is the difference between a real image and a virtual image?

4. In each situation in **Figure 33.37**, draw the three rays emanating from the top of the object and reflecting or refracting from the optical element shown. Show the image, and state whether it is real or virtual.

Figure 33.37

Answers:

1. On a calm day, the lake surface is smooth, and specular reflection is like that of a mirror. On a windy day, the surface is rough, which makes the reflection diffuse and prevents the formation of an image.

2. (*a*) The wavelength decreases when the wave travels more slowly in the second medium (*i* and *iii*) and increases when the wave travels faster in the second medium (*ii* and *iv*). (*b*) See **Figure 33.38**.

Figure 33.38

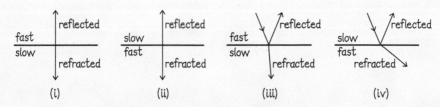

3. Real image: All rays actually pass through the location of the image, and the image can be seen on a screen placed at the image location. Virtual image: All rays do not pass through the location of the image (only the extensions of the rays do), and the image cannot be seen on a screen placed at the image location.

4. See **Figure 33.39**. For the lenses, the three principal rays can be used to locate the image; for the mirror, any rays and the law of reflection can be used to locate the image. The images are (*a*) real, (*b*) virtual, (*c*) virtual.

Figure 33.39

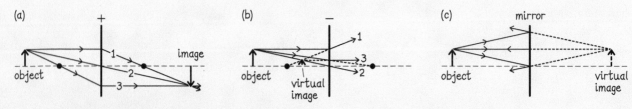

33.5 Snel's law

In the first part of this chapter, we saw that light refracts when it travels from one medium into another because the speed of light depends on the medium. The speed of light in a medium is specified by the **index of refraction:**

$$n \equiv \frac{c_0}{c}, \tag{33.1}$$

where c is the speed of light in the medium and c_0 is the speed of light in vacuum. (By definition, $n_{vacuum} = 1$; in air $n_{air} \approx 1$.) If a light wave of frequency f travels from one medium into another, the frequency doesn't change because the source determines the frequency (see also Checkpoint 16.10). The wavelength, however, does change; it is greater in the medium in which wave speed is greater.

The wavelength λ of the light is related to the wave speed and frequency, in the same manner that these quantities are related for harmonic waves (Eq. 16.10). In vacuum, for example,

$$\lambda = \frac{c_0}{f} \text{ (vacuum)}. \tag{33.2}$$

In a medium in which a wave has speed c_1, the wavelength λ_1 is given by

$$\lambda_1 = \frac{c_1}{f} = \frac{c_0/n_1}{f} = \frac{1}{n_1}\lambda, \tag{33.3}$$

where λ is the wavelength of the wave in vacuum and n_1 is the index of refraction of the medium. Thus, the wavelength decreases as the index of refraction increases. As discussed in Section 33.3, the amount of refraction a light wave undergoes varies somewhat with wavelength (see Figure 33.22) because different wavelengths of light travel at different speeds. Therefore the index of refraction depends on the wavelength. **Table 33.1** lists the indices of refraction for some common transparent materials at a wavelength of 589 nm.

Let us now work out the quantitative relationship between the angle of incidence and the angle of refraction. **Figure 33.40** shows wavefronts and one ray for a beam of light incident on the interface between medium 1 and medium 2 at angle θ_1 from the normal. The angle of refraction is θ_2. Using right triangles ABD and ACD, we can express angles θ_1 and θ_2 in terms of the wavelengths λ_1 and λ_2:

$$\sin \theta_1 = \frac{BD}{AD} = \frac{\lambda_1}{AD} \tag{33.4}$$

and

$$\sin \theta_2 = \frac{AC}{AD} = \frac{\lambda_2}{AD}. \tag{33.5}$$

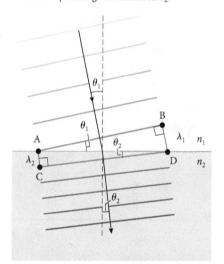

Figure 33.40 Relationship between angle of incidence θ_1 and angle of refraction θ_2.

Table 33.1 Indices of refraction for common transparent materials

Material	n (for $\lambda = 589$ nm)
Air (at standard temperature and pressure)	1.00029
Liquid water	1.33
Sugar solution (30%)	1.38
Sugar solution (80%)	1.49
Microscope cover slip glass	1.52
Sodium chloride (table salt)	1.54
Flint glass	1.65
Diamond	2.42

Combining these equations to eliminate AD and substituting $\lambda_1 = \lambda/n_1$ (Eq. 33.3), we get

$$\frac{\sin \theta_1}{\sin \theta_2} = \frac{\lambda_1}{\lambda_2} = \frac{\lambda/n_1}{\lambda/n_2} = \frac{n_2}{n_1}, \tag{33.6}$$

which can be written as

$$n_1 \sin \theta_1 = n_2 \sin \theta_2. \tag{33.7}$$

This relationship between the indices of refraction and the angles of incidence and refraction is called **Snel's law,** after the Dutch astronomer and mathematician Willebrord Snel van Royen (1580–1626).

Example 33.6 Bending 90°

A ray traveling through a medium for which the index of refraction is n_1 is incident on a medium for which the index of refraction is n_2. At what angle of incidence θ_1, expressed in terms of n_1 and n_2, must the ray strike the interface between the two media for the reflected and transmitted rays to be at right angles to each other?

❶ GETTING STARTED This problem involves both reflection and refraction at an interface between two media. To visualize the problem, I draw the incident, reflected, and refracted rays and indicate that the reflected and refracted rays are 90° apart (Figure 33.41).

Figure 33.41

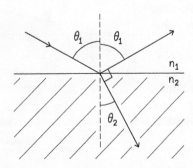

❷ DEVISE PLAN Snel's law (Eq. 33.7), the law of reflection, and the indices of refraction determine the paths taken by the

reflected and refracted rays. Therefore I need to use those relationships to obtain an expression that tells me the value of θ_1 that produces reflected and refracted rays oriented 90° to each other. To obtain θ_1 in terms of n_1 and n_2, I need to eliminate θ_2 from Eq. 33.7. To do so, I use the fact that the angles on the right side of the normal to the interface must add to 180°. Thus, with reflected and refracted rays forming a 90° angle, I can say $180° = \theta_1 + 90° + \theta_2$. Solving this expression for θ_2 gives $\theta_2 = 90° - \theta_1$, which I can substitute into Eq. 33.7.

❸ EXECUTE PLAN Substituting $\theta_2 = 90° - \theta_1$ into Eq. 33.7, I get

$$n_1 \sin \theta_1 = n_2 \sin(90° - \theta_1) = n_2 \cos \theta_1,$$

and isolating the terms that contain θ_1 gives

$$\frac{\sin \theta_1}{\cos \theta_1} = \tan \theta_1 = \frac{n_2}{n_1}$$

$$\theta_1 = \tan^{-1}\left(\frac{n_2}{n_1}\right). ✔$$

❹ EVALUATE RESULT My result says that θ_1 increases as n_2 increases. This makes sense because as n_2 increases, the refracted ray bends more, meaning that θ_2 becomes smaller. To keep the reflected and refracted rays perpendicular to each other, the angle of reflection must increase, and so θ_1 must also increase.

Figure 33.42 Critical angle for a ray traveling from a denser medium (n_2) to a less dense medium ($n_1 < n_2$).

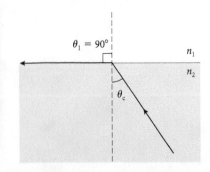

Earlier in this chapter, we found that for rays traveling from a denser medium to a less dense medium, we can define a critical angle of incidence θ_c such that the angle of refraction θ_1 is equal to 90° (Figure 33.42); beyond this critical angle θ_c, total internal reflection occurs. We can calculate the critical angle θ_c for an interface between two media with indices of refraction n_1 and n_2 ($n_2 > n_1$) by applying Snel's law (Eq. 33.7) and setting $\theta_1 = 90°$:

$$\frac{\sin \theta_1}{\sin \theta_2} = \frac{1}{\sin \theta_c} = \frac{n_2}{n_1}. \tag{33.8}$$

Solving for θ_c gives

$$\theta_c = \sin^{-1}\left(\frac{n_1}{n_2}\right). \tag{33.9}$$

Example 33.7 Fermat's principle

For a light ray that crosses the interface between medium 1 having index of refraction n_1 and medium 2 having index of refraction n_2, what relationship between θ_1 and θ_2 follows from Fermat's principle (page 884)?

❶ GETTING STARTED I begin with a diagram that shows the two media and a ray traveling from an arbitrary point A in the n_1 medium to an arbitrary point C in the n_2 medium (Figure 33.43). Fermat's principle states that the path the ray takes from A to C is the path for which the time interval needed for the motion is a minimum. Therefore this ray must cross the interface at a point B that makes the time interval a minimum.

Figure 33.43

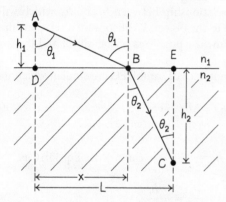

An alternative way to express this problem is: Given the locations of A and C, where must B lie so as to minimize the time interval needed to travel from A to C?

❷ DEVISE PLAN I add two more location labels to my drawing: D directly below A and lying on the interface, and E directly above C and lying on the interface. Doing so gives me two right-angle triangles that permit me to express the angles in terms of the distance traveled. I write h_1 for the distance AD and h_2 for the distance EC . I can think of the distance from D to B as unknown—I'll call it x—and the distance from D to E, which I'll call L, is fixed by the locations of A and C. My goal is to determine the value of x for which the travel time from A to C is minimized. Once I obtain x, I hope to obtain a relationship between θ_1 and θ_2.

❸ EXECUTE PLAN I begin by expressing the time interval Δt_{AC} the ray needs to travel from A to C in terms of the distances shown in Figure 33.43 and the speed of light in the two media:

$$\Delta t_{AC} = \Delta t_{AB} + \Delta t_{BC} = \frac{AB}{c_1} + \frac{BC}{c_2} = \frac{AB}{c_0/n_1} + \frac{BC}{c_0/n_2}. \quad (1)$$

Next I express AB and BC in terms of h_1, h_2, L, and x:

$$AB = \sqrt{h_1^2 + x^2}$$

$$BC = \sqrt{h_2^2 + (L - x)^2}.$$

Substituting these two expressions into Eq. 1 gives me

$$\Delta t_{AC} = \frac{\sqrt{h_1^2 + x^2}}{c_0/n_1} + \frac{\sqrt{h_2^2 + (L - x)^2}}{c_0/n_2}.$$

Except for x, all quantities in this expression are constants.

The path for which the time interval Δt_{AC} is a minimum—as it must be from Fermat's principle—is the path for which the derivative of Δt_{AC} with respect to x is zero:

$$\frac{d}{dx}(\Delta t_{AC}) = \frac{xn_1}{c_0\sqrt{h_1^2 + x^2}} - \frac{(L - x)n_2}{c_0\sqrt{h_2^2 + (L - x)^2}} = 0. \quad (2)$$

Solving this equation for x would tell me where the light ray crosses the interface, but I do not have values for L and x. However, the right triangles ADB and BEC in Figure 33.43 allow me to express these distances in terms of θ_1 and θ_2:

$$\sin \theta_1 = \frac{x}{\sqrt{h_1^2 + x^2}} \quad \text{and} \quad \sin \theta_2 = \frac{L - x}{\sqrt{h_2^2 + (L - x)^2}}. \quad (3)$$

I can now use these expressions to rewrite Eq. 2 in terms of θ_1 and θ_2. From Eq. 2 I obtain

$$\frac{xn_1}{c_0\sqrt{h_1^2 + x^2}} = \frac{(L - x)n_2}{c_0\sqrt{h_2^2 + (L - x)^2}}.$$

Canceling the c_0 factors that appear on both sides and substituting from Eq. 3, I get

$$n_1 \sin \theta_1 = n_2 \sin \theta_2. \checkmark$$

❹ EVALUATE RESULT My result is identical to Snel's law—which I derived by considering the effect of changing speed on the propagation of wavefronts. So Fermat's principle yields the same result as Snel's law, which I know to be correct.

Fermat's principle applies to all of ray optics, not only to refraction. As discussed in the box "Fermat's principle" on page 884, the law of reflection also follows from this principle.

✋ **33.14** We found in Example 33.3 that a light ray is refracted twice when it passes completely through a slab of transparent material (see Figure 33.18). The result of these two refractions is that the exiting ray is shifted sideways relative to the entering ray. Let the slab be in air with an index of refraction $n_1 = 1$. (a) Derive an expression for the distance (perpendicular to the ray) over which the ray is shifted sideways for an angle of incidence θ_1, slab thickness d, and slab index of refraction n_2. (b) Calculate the value of the shift for $\theta_1 = 30°$, $n_2 = 1.5$, $d = 0.010$ m.

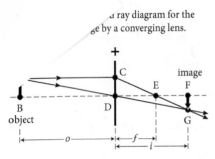

…a ray diagram for the …e by a converging lens.

33.6 Thin lenses and optical instruments

In the first part of this chapter, we found that converging lenses form images of objects by focusing the light rays emanating from those objects. Let us now work out quantitatively the location and size of such images. We shall restrict our discussion to lenses that are thin enough that we can ignore the type of effects shown in Figure 33.29. Such lenses are called *thin lenses*.

A simplified ray diagram of the image formed by a converging lens is shown in **Figure 33.44**. The focal length f of the lens is DE, the distance o from the lens to the object (also called the *object distance*) is BD, and the distance i from the lens to the image (also called the *image distance*) is DF. The height of the object is AB, and the height of the image is FG. Let us denote the height of the object by h_o and the height of the image by h_i. We choose the values of h_o and h_i to be positive for upright objects and images and negative for inverted objects and images. We want to obtain a relationship between h_i and h_o, which will tell us how large the image is relative to the object. We also want a relationship among f, i, and o, which will tell us how the positions of the object and the image are related.

We begin by noting that triangles ABD and DFG are similar, which means

$$\frac{AB}{DB} = \frac{FG}{DF}.$$ (33.10)

Because the image is inverted, $h_i = -FG$, we can rewrite Eq. 33.10 as

$$\frac{h_o}{o} = \frac{-h_i}{i}.$$ (33.11)

Rearranging this expression gives

$$-\frac{h_o}{h_i} = \frac{o}{i}.$$ (33.12)

In this case, the absolute value of the ratio of the object height to the image height equals the ratio of the object distance to the image distance.

Triangles CDE and EFG are also similar, which means

$$\frac{DE}{CD} = \frac{EF}{FG},$$ (33.13)

which can be written as

$$\frac{f}{h_o} = \frac{i-f}{-h_i}.$$ (33.14)

Using Eq. 33.12 to rewrite Eq. 33.14 in terms of f, o, and i gives us

$$\frac{f}{o} = \frac{i-f}{i} = 1 - \frac{f}{i}.$$ (33.15)

Dividing by f and rearranging terms yield

$$\frac{1}{f} = \frac{1}{o} + \frac{1}{i}.$$ (33.16)

This result is known as the **lens equation.**

It can be shown that Eq. 33.16 is generally true for either real or virtual images formed by either converging or diverging lenses, as long as we choose the signs of f, i, and o properly. For a converging lens, f is positive and o is positive if the object is in front of the lens. (This is always true for a single lens and for the first lens in a lens combination. For situations involving multiple lenses, however, it is possible that the object imaged by a secondary lens is on the opposite side of the lens from the side where the rays enter it—that is, the object is "behind the lens." In that case o is negative.) If the image is on the same side of the lens as the emerging light, the image is real and i is positive; if the image is on the opposite side of the lens from the emerging light, the image is virtual and i is negative.

For a diverging lens (a lens with concave surfaces), the focal length f is negative because the focus is virtual rather than real—that is, parallel rays appear to come from the same side of the lens where the light source is rather than converging on the other side of the lens. The same sign convention applies to o as for converging lenses. A single diverging lens always produces a virtual image (see Example 33.5), so i is always negative for such lenses.

The sign conventions for f, i, and o are similar for images formed by spherical mirrors, which are discussed in the next section. Table 33.2 summarizes these sign conventions.

The **magnification** of the image is defined as the ratio of the signed image height to the object height. Using Eq. 33.12, we get

$$M \equiv \frac{h_i}{h_o} = -\frac{i}{o}. \qquad (33.17)$$

We define M this way so that the magnification of upright images is positive and that of inverted images is negative. Examining Figures 33.30, 33.31, and 33.33, we can see that for a single lens, when the image distance and object distance are both positive, as in Figure 33.30, we obtain an inverted image, whereas when the image distance is negative, as in Figures 33.31 and 33.33, we obtain an upright image.

33.15 If the diverging lens in Figure 33.33 has a focal length of 80 mm and the object is located 100 mm from the lens, (a) what is the image distance and (b) how tall is the image relative to the object?

Table 33.2 Sign conventions for f, i, and o (positive = real; negative = virtual)

Sign	Lens	Mirror
$f > 0$	converging lens	converging mirror
$f < 0$	diverging lens	diverging mirror
$o > 0$	object in front[b] of lens	object in front of mirror
$o < 0$[a]	object behind lens	object behind mirror
$i > 0$	image behind lens	image in front of mirror
$i < 0$	image in front of lens	image behind mirror
$h_i > 0$	image upright	image upright
$h_i < 0$	image inverted	image inverted
$|M| > 1$	image larger than object	image larger than object
$|M| < 1$	image smaller than object	image smaller than object

[a] Encountered only with lens or mirror combinations.
[b] For both lenses and mirrors, "in front" means on the side where the rays originate; "behind" refers to the opposite side.

...nnot focus on an object that is closer than its near point (which represents
...ical lens's ability to change curvature). However, an external converging lens
...ying lens) makes it possible to see objects that are closer than the near point. It also
...n.

RETINAL IMAGE:

(a)

object

Object at eye's near point subtends angle θ_o of field of view.

eye

clear
and
sharp

near point

θ_o

0.25 m

(b)

object

For object closer to lens, image point is behind retina.

near point

clear
and
sharp

(c)

image   External lens forms virtual image that lies beyond near point, . . .

object   . . . subtends angle θ_i, . . .

clear
and
sharp

θ_i

near point

focal point

f

. . . and can be focused on retina by eye's lens.

The human eye focuses incoming light rays, forming an image on the retina of the eye (**Figure 33.45a**). One part of the eye is its lens, but unlike the lenses we have examined so far, the focal length of the eye's lens is variable, which allows us to see objects clearly over a wide range of distances. When the muscle around the lens is fully relaxed, the lens flattens out and the retina lies in the focal plane of the lens. Thus, light rays from distant objects focus onto the retina.

To the unaided eye, the largest (and thus most detailed) image of an object is observed when we bring the object as close as possible to the eye. However, there is a limit to how much the eye's lens can adjust. The *near point* is the closest object distance at which the eye can focus on the object comfortably. Typically, for an adult, the near point is about 0.25 m from the eye. An object positioned at the near point appears clear and sharp to the observer, as shown in Figure 33.45a. With age the distance between the near point and the eye tends to increase, and when an object is brought closer than the near point, the plane where the image is formed lies behind the retina and the image "seen" by the retina is blurry (Figure 33.45b). This situation can be corrected by an external lens that works in combination with your eye's lens to focus the image on the retina (Figure 33.45c).

An external converging lens properly placed between the object and the eye, as in Figure 33.45c, magnifies the object. To maximize the size of the image, the object is held near the focus of the external lens, and the lens is held as close as possible to the eye. The image formed by the external lens then serves as the object for the eye's lens. The image formed by the external lens is virtual and subtends an angle θ_i that is greater than the angle θ_o subtended by the object in Figure 33.45a, permitting the viewer to see finer details. The image is also outside the near point; if the object is placed exactly at the focus of the external lens, the image is at infinity and can be viewed comfortably. We can define the *angular magnification* produced by the lens as

$$M_\theta \equiv \left| \frac{\theta_i}{\theta_o} \right|. \tag{33.18}$$

For small angles and an object placed close to the focus of the external lens, as in Figure 33.45c, the angle θ_i subtended by the image can be expressed in terms of the object height h_o and the focal length f of the lens:

$$\theta_i \approx \tan \theta_i \approx \frac{h_o}{f} \quad \text{(object close to focus, small } \theta_i\text{).} \tag{33.19}$$

For small angles and an object placed at the eye's near point, as in Figure 33.45a, the angle subtended by the object is approximately

$$\theta_o \approx \tan \theta_o = \frac{h_o}{0.25 \text{ m}} \quad \text{(object close to near point, small } \theta_o\text{).} \tag{33.20}$$

Substituting Eqs. 33.19 and 33.20 into Eq. 33.18 gives an angular magnification of

$$M_\theta \approx \frac{0.25 \text{ m}}{f}. \tag{33.21}$$

This expression gives what is called either the *small-angle approximation* or the *paraxial approximation* to the angular magnification because it is obtained with the small-angle approximations of Eqs. 33.19 and 33.20. These approximations are good to within 1% for angles of $10°$ or less.

Lenses placed near the eye (in the form of eyeglasses) are used to correct vision for far-sighted or near-sighted eyes. The strength of eyeglass lenses (and of magnifying lenses, too) is commonly symbolized by d and measured in *diopters*:

$$d \equiv \frac{1 \text{ m}}{f}. \tag{33.22}$$

The *lens strength d*, like the lens focal length f, is positive for converging lenses and negative for diverging lenses. For example, a $+4$-diopter lens is a converging lens with a focal length of 0.25 m. Diverging lenses are typically used to correct nearsightedness, with lens strengths ranging from -0.5 to -4 diopters.

33.16 A single-lens magnifying glass used to examine photographic slides produces eightfold angular magnification. (*a*) What is the lens strength in diopters? (*b*) What is the focal length of the lens?

Many optical instruments combine two or more lenses to increase magnification. To trace rays through a combination of lenses, use the following procedure: The image formed by the first lens serves as the object for the second lens, the image formed by the second lens serves as the object for the third lens, and so on. **Figure 33.46** on the next page shows a ray diagram constructed in two steps for a combination of two lenses. Figure 33.46a shows the object, image, and rays for lens 1, and Figure 33.46b shows these elements for lens 2. Note that the rays from object 2 (which is the image formed by lens 1) are *not* the continuation of those used to locate image 1.

Figure 33.46 Two-step process for tracing rays through a combination of two lenses. When lenses are combined, the image of each lens serves as an object for the next lens.

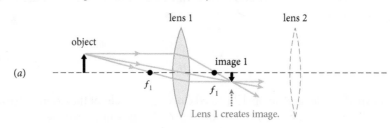

(a)

Lens 1 creates image.

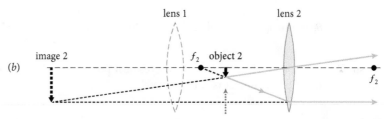

(b)

Image of lens 1 serves as object for lens 2, creating enlarged virtual image 2.

Example 33.8 Compound microscope

A compound microscope consists of two converging lenses, the *objective lens* and the *eyepiece lens*, positioned on a common optical axis (**Figure 33.47**). The objective lens is positioned to form a real, highly magnified image 1 of the sample being examined, and the eyepiece lens is positioned to form a virtual, further magnified image 2 of image 1. It is image 2 that the user sees. A knob on the microscope allows the user to move the objective lens upward and downward to change both the sample-objective lens distance and the distance between the two lenses. (a) How must the sample and the two lenses be positioned relative to one another so that the user sees a highly magnified, virtual image of the sample? (b) What is the overall magnification produced by the microscope?

Figure 33.47 Example 33.8.

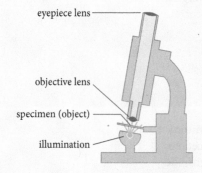

eyepiece lens

objective lens

specimen (object)

illumination

❶ GETTING STARTED I begin by examining Figure 33.46, which shows how, in a combination of two lenses 1 and 2, the image formed by lens 1 serves as the object for lens 2. The objective lens in a compound microscope corresponds to lens 1 in Figure 33.46, and the eyepiece lens corresponds to lens 2. Thus to keep things simple I refer to the objective lens as 1 and the eyepiece lens as 2.

❷ DEVISE PLAN To determine the relative positioning of the two lenses relative to each other, I must examine ray diagrams for various lens-sample distances and determine for which arrangement I get the greatest magnification. To determine the magnification M_1 of image 1, I can use the lens equation (Eq. 33.16) together with the relationship among magnification, image distance, and object distance (Eq. 33.17). The focal length of the lenses is fixed by their construction, and in operating a microscope, the observer can adjust the distance between the sample and the objective lens, so I can express this magnification in terms of f_1 and o_1. To determine the magnification of image 2, I recognize that lens 2 is used as a magnifying glass, so I can use Eq. 33.21, which gives the angular magnification $M_{\theta 2}$ produced by a simple magnifier. The overall magnification produced by the microscope is the product $M_1 M_{\theta 2}$.

❸ EXECUTE PLAN (a) In Figure 33.46, lens 1 produces an image that is smaller than the object. I am told that the image formed by lens 1 in a microscope is larger than the sample, and so I must choose a different sample position, one that yields an image 1 larger than the sample. I am also told that this image is real. Placing the sample just outside the focal point of lens 1 gives me an image 1 that is larger than the sample. My choices are to increase or decrease the sample-lens 1 distance. Drawing a ray diagram for each possibility, I see that moving the sample farther and farther from lens 1 makes the image smaller and smaller. Therefore I should position the sample closer to lens 1 than in Figure 33.46a. Should I choose a position inside or outside the lens focus? I know from the

Figure 33.48

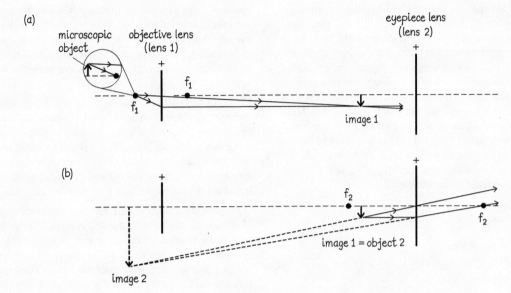

problem statement that this image is real, and I know from Figure 33.31 that an object inside the focus of a converging lens produces a virtual image. Thus my best choice is to adjust the sample-lens 1 distance so that the sample is just outside the lens focus (**Figure 33.48a**).

I am told that image 2 is virtual and larger than image 1. I know from Figure 33.31 that a converging lens produces a virtual, magnified image when the object is inside the lens focus. I again draw ray diagrams for various positions inside the focus and see that the greatest magnification is obtained when I adjust the distance from lens 1 to lens 2 to make image 1 fall just inside the focal point of lens 2, as shown in Figure 33.48b. ✔

(b) To determine M_1, I use the lens equation to write i_1 in terms of f_1 and o_1:

$$i_1 = \frac{1}{\left(\dfrac{1}{f_1} - \dfrac{1}{o_1}\right)}.$$

I substitute this expression into Eq. 33.17:

$$M_1 = -\frac{i_1}{o_1} = -\frac{1}{o_1} \times \frac{1}{\left(\dfrac{1}{f_1} - \dfrac{1}{o_1}\right)} = -\frac{1}{\dfrac{o_1}{f_1} - 1},$$

which tells me that the magnification M_1 produced by lens 1 is determined by the ratio o_1/f_1. Because I have made o_1 slightly larger than f_1 in order to produce a real image 1, the denominator is positive and therefore M_1 is negative.

The angular magnification produced by lens 2 is approximately

$$M_{\theta 2} = \frac{0.25 \text{ m}}{f_2}.$$

The overall magnification produced by the microscope is thus

$$M = M_1 M_{\theta 2} = \frac{-0.25 \text{ m}}{f_2\left(\dfrac{o_1}{f_1} - 1\right)}. ✔$$

❹ **EVALUATE RESULT** Figure 33.48a indicates that image 1 is inverted, making M_1 negative and giving me confidence in my expression for M_1. Figure 33.48b tells me that image 2 is upright relative to its object, and so $M_{\theta 2}$ is positive, which agrees with my result. Because image 2 is inverted relative to the sample, the overall magnification is negative, as my result shows.

33.17 (a) Consider replacing the objective lens in Fig. 33.48a with one that has a greater focal length, and moving the sample in order to keep it just outside the focal point of the lens. Does the image formed by the objective lens move closer to the objective lens, stay in the same place, or move farther from the objective lens? (b) In practice it is desirable for a microscope to be fairly compact. To keep the microscope compact, should the focal length of the objective lens be chosen to be short or long, or does it matter?

Example 33.9 Refracting telescope

A refracting telescope, like a compound microscope, contains two converging lenses, the objective lens and the eyepiece lens, positioned on a common optical axis (**Figure 33.49**). However, a telescope is designed to view large, very distant objects, whereas a microscope is used to view very small objects that are placed very close to the objective lens. Consequently, the arrangement of lenses in a telescope is different from the arrangement in a microscope. The telescope's objective lens is positioned to form a real image of very distant objects, and the eyepiece lens is positioned to form a virtual image of the image produced by the objective lens, to be viewed by an observer. (*a*) How should the lenses be arranged to accomplish this? (*b*) What is the overall magnification produced by the telescope?

Figure 33.49 Example 33.9.

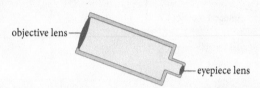

objective lens

eyepiece lens

❶ **GETTING STARTED** I begin by examining Figure 33.46 and then construct a similar ray diagram with the object at a very great distance from the lenses. As in Example 33.8, I use lens 1 to refer to the objective lens and lens 2 to refer to the eyepiece lens. Because the object is very far away, light rays from it enter lens 1 as parallel rays. These rays form an image 1 in the focal plane of lens 1 (**Figure 33.50***a*). Because I know the location of image 1, I need to draw only one principal ray. As in the microscope, lens 2 is used as a simple magnifier to view image 1.

❷ **DEVISE PLAN** Because the original object is very distant and the final image is viewed by the observer's eye, I can calculate the angular magnification of this image. Although I could calculate the angular magnification produced by each lens and

multiply them together, in this case it is simpler to determine the overall angular magnification because the angles θ_o and θ_i the object and image 2 subtend at the observer's eye are both very small. I can determine the overall angular magnification by taking the ratio of these angles while using the small-angle approximation.

❸ **EXECUTE PLAN** (*a*) In order for lens 2 to produce a magnified, virtual image of image 1, image 1 should be positioned just inside the focal plane of lens 2. If lens 2 is placed such that the image is at the focal plane of lens 2 (**Figure 33.50***b*), lens 2 forms an infinitely distant, virtual image that can be viewed comfortably by the observer's relaxed eye. As my diagram shows, the lenses are then arranged such that their foci coincide. ✔

(*b*) **Figure 33.51***a* shows the ray that passes through the foci of the lenses, labeled with the angles θ_o (subtended by the object) and θ_i (subtended by the image). Figure 33.51*b* shows the triangles I use to relate each of these angles to the height h_i of image 1 and the focal lengths of the lenses. The angular magnification is the ratio θ_o/θ_i. I can approximate these angles by

Figure 33.51

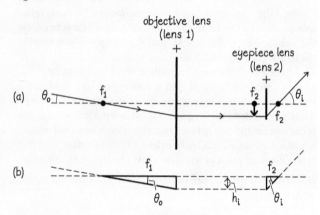

Figure 33.50

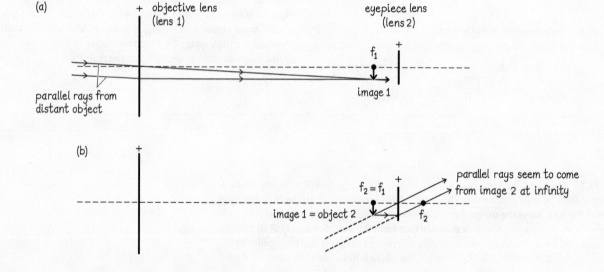

their tangents; substitute for the tangents of the angles in terms of h_i, f_1, and f_2; and simplify:

$$M_\theta = \left| \frac{\theta_i}{\theta_o} \right| \approx \left| \frac{\tan \theta_i}{\tan \theta_o} \right| = \frac{|h_i/f_2|}{|h_i/f_1|} = \left| \frac{f_1}{f_2} \right|. ✔$$

④ **EVALUATE RESULT** Figure 33.50 indicates that image 1 is inverted and that image 2 is upright relative to image 1, which means that image 2 is inverted relative to the distant object. This makes sense because the incoming ray in Figure 33.51 is angled downward but the outgoing ray is angled upward.

33.18 A telescope with a magnification of 22× has an eyepiece lens for which the focal length f_2 is 40.0 mm. (*a*) What is the focal length f_1 of the objective lens? (*b*) What is the length of the telescope?

33.7 Spherical mirrors

Just like lenses, spherical mirrors focus parallel rays (**Figure 33.52**). A concave mirror focuses rays to a point in front of the mirror, corresponding to a real focus; a diverging mirror makes rays diverge so that they appear to come from a point behind the mirror, corresponding to a virtual focus. Thus, just as with lenses, we can have both converging and diverging spherical mirrors. Unlike lenses, however, spherical mirrors have only a single focus.

To obtain the location of the focus of a converging mirror, we examine the reflection of the two rays shown in **Figure 33.53a**. Ray 1 comes in parallel to the axis of the mirror, striking the mirror at A, and ray 2 comes in along the axis of the mirror. The focus of the mirror is at D, where the two reflected rays cross, and the center of the sphere on which the surface of the mirror lies is at C. Line CA is therefore a radius of the sphere and perpendicular to the mirror surface. We denote the length of the radius by R. This distance is called the *radius of curvature*.

Ray 2 strikes the mirror at normal incidence and so is reflected back along the axis. Ray 1 is reflected through the focus at D. Consequently, the distances CD and AD in Figure 33.53a are equal. Dividing triangle ACD into two congruent right triangles by drawing a line perpendicular to the base from D, we see that CD = $(R/2)/\cos \theta_i$. For small θ, $\cos \theta \approx 1$, and so CD = $R/2$ and BD = $R -$ CD = $R/2$, independent of θ. Therefore, the focus, which is located at D because that is where the two reflected rays cross, lies halfway between the mirror and the center of the sphere, making the focal length

$$f = \frac{R}{2}. \tag{33.23}$$

As Figure 33.53*b* shows, the geometry and hence the position of the focus are exactly the same for diverging mirrors except that here the focus lies behind the mirror.

Figure 33.52 Spherical mirrors focus parallel rays just like lenses do.

(*a*) Concave spherical mirror

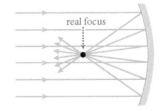

real focus

(*b*) Convex spherical mirror

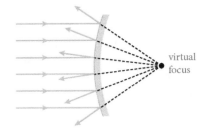

virtual focus

Figure 33.53 The two principal rays used to determine the focus of a spherical mirror.

(*a*) Concave spherical mirror

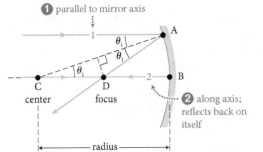

① parallel to mirror axis

C center
D focus

② along axis; reflects back on itself

radius

(*b*) Convex spherical mirror

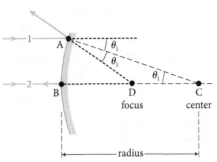

B
D focus
C center

radius

Procedure: Ray diagrams for spherical mirrors

Ray diagrams for spherical mirrors are very similar to those for lenses. This procedure is for rays traveling from the left to the right.

1. Draw a horizontal line representing the mirror axis. In the center of the diagram, draw a circular arc representing the mirror. A converging mirror curves toward the left; a diverging mirror curves toward the right.
2. Put a dot on the axis at the center of the circular arc and label it C. Add another dot on the axis, halfway between C and the mirror. This point is the focus. Label it f.
3. Represent the object by drawing an upward-pointing arrow from the axis at the appropriate relative distance to the left of the mirror. For example, if the distance from the object to a converging mirror is one-third the radius of curvature of the mirror, put the arrow a bit to the right of the focus. The top of the arrow should be at about half the height of the mirror.
4. From the top of the arrow representing the object draw two or three of the following three so-called *principal rays* listed in the Procedure box "Principal rays for spherical mirrors."
5. The top of the image is at the point where the rays that are reflected by the mirror intersect. If the intersection is on the left side of the lens, the image is real; if it is on the right side, the image is virtual. Draw an arrow pointing from the axis to the intersection to represent the image (use a dashed arrow for a virtual image).

To determine the location of images formed by mirrors, we follow the same ray-tracing procedure we used for lenses. The three principal rays emanating from a given point on the object are analogous: ray 1 approaching the mirror parallel to the mirror axis, ray 2 passing through the center C of the mirror, and ray 3 passing through the focus on its way to the mirror. Now, however, "center" refers to *the center of the sphere on which the mirror surface lies* rather than the center of the lens. **Figure 33.54** shows a ray diagram for an image formed by a converging mirror. Ray 1 is reflected through the focus, ray 2 strikes the mirror at normal incidence and thus reflects back on itself, and ray 3 is reflected parallel to the mirror axis. As Figure 33.54 shows, there is a fourth ray that can easily be drawn: A ray that hits the mirror on the axis is reflected back symmetrically about the axis. The procedures for drawing ray diagrams and principal rays for spherical mirrors are given in the Procedure boxes on this page.

Object distance, image distance, and focal length are measured from the surface of the mirror, and the relationship among o, i, and f is the same as that for lenses:

$$\frac{1}{f} = \frac{1}{o} + \frac{1}{i}. \tag{33.24}$$

Procedure: Principal rays for spherical mirrors

This description holds for rays that travel from left to right.

Converging mirror

1. A ray that travels parallel to the mirror axis before reaching the mirror goes through the focus after being reflected.
2. A ray that passes through the center of the sphere on which the mirror surface lies is reflected back onto itself. If the object is between the center and the mirror, this ray doesn't pass through the center but lies on the line from the center to the point at which the ray originates.
3. A ray that passes through the focus is reflected parallel to the axis. If the object is between the focus and the mirror, this ray doesn't pass through the focus but lies on the line from the focus to the point at which the ray originates.

Diverging mirror

1. A ray that travels parallel to the mirror axis before reaching the mirror is reflected along the line that goes through the focus and the point where the ray strikes the surface.
2. A ray that passes through the center of the sphere on which the mirror surface lies is reflected back onto itself.
3. A ray whose extension passes through the focus is reflected parallel to the axis.

For both converging and diverging mirrors, a ray that hits the mirror on the axis is reflected back symmetrically about the axis.

Figure 33.54 Principal ray diagram for an object outside the focus of a concave spherical mirror. The image is real, inverted, and smaller than the object.

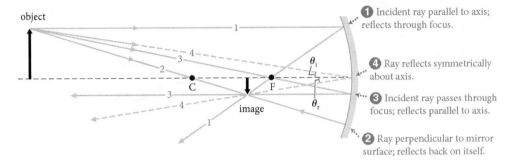

1 Incident ray parallel to axis; reflects through focus.

4 Ray reflects symmetrically about axis.

3 Incident ray passes through focus; reflects parallel to axis.

2 Ray perpendicular to mirror surface; reflects back on itself.

Just as for lenses, the focal length f for any spherical mirror is positive for a real focus and negative for a virtual focus, and o is positive when the object is in front of the mirror and negative when it is behind the mirror (that can happen only when the object is an image formed by another mirror or a lens). Similarly, i is positive for a real image and negative for a virtual image. With mirrors, however, a real image is located on the same side of the mirror as the object and a virtual image is located on the opposite side—the opposite of what happens with lenses. These sign conventions are summarized in Table 33.2. Finally, the relationship between object and image distances and heights for lenses (Eq. 33.12) also applies to mirrors, so that equation can be used to determine the size of the images formed by mirrors.

Example 33.10 Funny mirror

An object is placed 0.30 m in front of a converging mirror for which the radius of curvature is 1.0 m. (*a*) On which side of the mirror is the image? Is the image real or virtual? (*b*) If the object is 50 mm tall, what is the height of the image?

❶ **GETTING STARTED** To visualize the situation, I draw the mirror and its axis, and indicate its center of curvature C and its focal point f halfway between the mirror surface and C. I represent the object as a solid arrow (**Figure 33.55**).

Figure 33.55

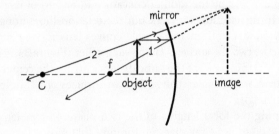

❷ **DEVISE PLAN** To locate the image and identify whether it is real or virtual, I can draw the principal rays. I can use Eq. 33.12 to obtain the image height from the object height and the image and object distances. I can determine the image distance from the focal length and the object distance using Eq. 33.24.

❸ **EXECUTE PLAN** (*a*) As Figure 33.55 shows, I need to draw only two principal rays because the intersection of two rays

unambiguously determines the position of the image. I draw ray 1 parallel to the mirror axis and reflecting through the focus. Because the object is between the mirror and the center of curvature, I draw ray 2 along the line defined by C and the tip of the object. Ray 2 does not pass through C until after it is reflected from the mirror, however, and I indicate this by adding an arrowhead pointing toward the mirror on the part of the ray to the right of the object and an arrowhead pointing away from the mirror on the part to the left of the object. My diagram shows that the rays do not actually meet but appear to come from a point behind the mirror. Therefore the image is behind the mirror and virtual. ✔

(*b*) I begin by determining the image distance i. The focal length of the mirror is half the radius of curvature, $f = 0.50$ m. Substituting this value and $o = 0.30$ m into Eq. 33.24 and solving for i give me $i = -0.75$ m. The negative sign indicates that the image is virtual and therefore behind the mirror; this is consistent with my result from part *a*. I then solve Eq. 33.12 for h_i (the signed image height) and substitute the values from this problem:

$$h_i = \frac{-i\,h_o}{o} = \frac{-(-0.75 \text{ m})(0.050 \text{ m})}{(0.30 \text{ m})} = 0.13 \text{ m.} ✔$$

❹ **EVALUATE RESULT** My ray diagram (Figure 33.55) indicates that the image is enlarged and upright, so h_i should be positive and greater than h_o. This agrees with my result.

Figure 33.56 Ray diagram for an image formed by a convex spherical mirror (object distance greater than focal length). The image is virtual, upright, and smaller than the object.

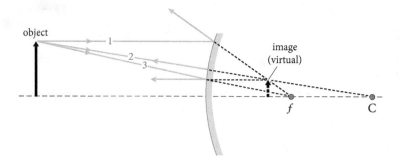

Example 33.10 shows that placing an object inside the focus of a converging mirror produces an upright, virtual image. Compare this with the situation in Figure 33.54, where a real, inverted image is formed for an object placed outside the mirror's focus. The same occurs with a converging lens: Placing the object inside the focus produces an upright, virtual image (Figure 33.31), while placing the object outside the focus produces a real, inverted image (Figure 33.30).

Diverging mirrors, like diverging lenses, always form virtual images when used alone because the light rays must diverge from the mirror surface. **Figure 33.56** shows a ray diagram for an image formed by a diverging mirror. The image is much smaller than the object, which allows a relatively large scene to be captured on a small mirror surface, and is upright. For these reasons, wide-angle rear-view mirrors on the passenger side of cars and trucks and wide-angle surveillance mirrors are typically convex. (A converging mirror also produces small images of distant objects, but the images are inverted, as Figure 33.54 shows.)

33.19 An object is placed 1.0 m in front of a diverging mirror for which the radius of curvature is 1.0 m. (*a*) Where is the image located relative to the mirror? Is the image real or virtual? (*b*) If the object is 0.30 m tall, what is the height of the image?

33.8 Lensmaker's formula

The focal length of a lens is determined by the refractive index n of the material of which the lens is made and by the radii of curvature R_1 and R_2 of its two surfaces (**Figure 33.57a**). In this section we work out the relationship among f, R_1, R_2, and n. In this analysis, we can think of a double-convex lens as two plano-convex lenses placed with the two flat surfaces facing each other (Figure 33.57b). Remember that both foci of a thin lens are the same distance f from the center of the lens. Because this is true, we can interchange the two surfaces of a thin lens without changing its focal length.

We begin by determining the focal lengths of the two plano-convex lenses in Figure 33.57b. **Figure 33.58** shows a ray diagram for light that passes through the right lens only. To calculate the focal length f_1 of this lens, consider a ray incident from the left that comes in parallel to the axis at a distance h above the axis. Because the ray is normal to the planar surface of the lens, it is not refracted at that surface. After passing through the lens, it strikes the curved surface at an angle θ_i measured from the normal to the curved surface and is refracted as it leaves the lens. The refracted ray emerges at an angle θ_r measured from the normal to the curved surface and crosses the lens axis a distance f_1 from the lens. Therefore the angle that the emerging ray makes with the lens axis is $\theta_r - \theta_i$.

Figure 33.57 Analysis of a double-convex lens.

Focal length of lens depends on lens's index of refraction n and on radii of curvature R_1, R_2 of lens faces.

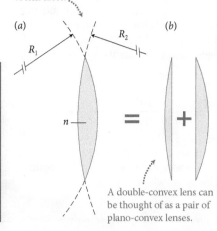

A double-convex lens can be thought of as a pair of plano-convex lenses.

Figure 33.58 Ray diagram for light passing through the right-hand plano-convex lens of Figure 33.57*b*.

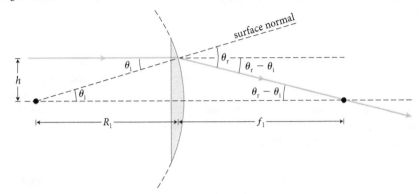

Applying Snel's law (Eq. 33.7) to this situation gives

$$n \sin \theta_{\mathrm{i}} = \sin \theta_{\mathrm{r}}. \qquad (33.25)$$

(We do not need to show an n on the right because the medium is air and $n_{\mathrm{air}} = 1$.) For paraxial rays, we can approximate the sines of the angles in Eq. 33.25 by the angles

$$n\theta_{\mathrm{i}} = \theta_{\mathrm{r}} \quad \text{(small angles)}. \qquad (33.26)$$

Using this relationship, we can express the angle between the emerging ray and the lens axis as

$$\theta_{\mathrm{r}} - \theta_{\mathrm{i}} = n\theta_{\mathrm{i}} - \theta_{\mathrm{i}} = (n - 1)\theta_{\mathrm{i}}. \qquad (33.27)$$

For small angles, the angles are approximately equal to their tangents, which means this relationship can be expressed as

$$\theta_{\mathrm{r}} - \theta_{\mathrm{i}} = (n - 1)\theta_{\mathrm{i}} \approx \frac{h}{f_1} \qquad (33.28)$$

and

$$\theta_{\mathrm{i}} \approx \frac{h}{R_1}. \qquad (33.29)$$

Substituting Eq. 33.29 into Eq. 33.28 and dividing both sides by h, we get

$$\frac{1}{f_1} = \frac{n - 1}{R_1}. \qquad (33.30)$$

We can follow the same procedure for the left lens in Figure 33.57*b*, using R_2 as our radius of curvature and a ray that originates at the left focus of that lens. The result analogous to Eq. 33.30 is

$$\frac{1}{f_2} = \frac{n - 1}{R_2}. \qquad (33.31)$$

Now let us determine the focal length of the lens combination by working out the lens equation (Eq. 33.16) for the combination, just as we did for the microscope and telescope in Section 33.6. Consider both the object and light source to be on the left side of the lens combination in Figure 33.57. First, the light from the object strikes the left lens from the left and forms an image someplace to the right of the right lens. The lens equation that relates the location of this object and image, in terms of the focal length f_2 calculated in Eq. 33.31, is

$$\frac{1}{o_2} + \frac{1}{i_2} = \frac{1}{f_2} = \frac{n-1}{R_2}. \tag{33.32}$$

The image formed by the left lens now serves as the (virtual) object for the right lens. Consequently, $o_1 = -i_2$. (The object for the right lens is virtual, and therefore the object distance o_1 is negative because the object is located on the right side of the lens and the illumination comes from the left side of the lens.) The lens equation for the right lens is thus

$$-\frac{1}{i_2} + \frac{1}{i_1} = \frac{1}{f_1} = \frac{n-1}{R_1}, \tag{33.33}$$

where the rightmost equality comes from Eq. 33.30.

Adding Eqs. 33.32 and 33.33 yields

$$\frac{1}{o_2} + \frac{1}{i_1} = (n-1)\left(\frac{1}{R_1} + \frac{1}{R_2}\right). \tag{33.34}$$

The lens equation for the lens as a whole is simply

$$\frac{1}{o} + \frac{1}{i} = \frac{1}{f}, \tag{33.35}$$

where f, o, and i are the focal length, object distance, and image distance of the lens combination, respectively. The object of the left lens is the actual object, and the image formed by the right lens is the final image, which means that in Eq. 33.35 $o = o_2$ and $i = i_1$. Comparing Eqs. 33.34 and 33.35 gives us the **lensmaker's formula** for the focal length of our lens combination:

$$\frac{1}{f} = (n-1)\left(\frac{1}{R_1} + \frac{1}{R_2}\right). \tag{33.36}$$

Our derivation was for a double-convex lens, but it can be shown that the lensmaker's formula applies to any thin lens, not just a double-convex lens. The radii of curvature are positive for convex surfaces, negative for concave surfaces, and infinity for planar surfaces. For a double-convex lens, f is positive. For a double-concave lens, f is negative (because $n > 1$ for any material used for lenses).

33.20 How should the lensmaker's formula be modified if a lens for which the index of refraction is n_1 is submerged in a medium for which the index of refraction is n_2?

Chapter Glossary

SI units of physical quantities are given in parentheses.

absorbed, reflected, and **transmitted light** Light that enters a material but never exits again, light that is redirected away from the surface of the material, and light that passes through a material, respectively.

angle of incidence θ_i The angle between a ray that is incident on a surface and the normal to that surface

angle of reflection θ_r The angle between a ray that is reflected from a surface and the normal to that surface.

angle of refraction θ The angle between a ray that is refracted after crossing the surface between one medium and another and the normal to that surface.

critical angle θ_c (unitless) The angle of incidence for which the angle of refraction equals 90° when a ray travels from a medium with an index of refraction n_2 to one with an index of refraction $n_1 < n_2$:

$$\theta_c = \sin^{-1}\left(\frac{n_1}{n_2}\right). \tag{33.9}$$

dispersion The spatial separation of waves of different wavelength caused by a frequency dependence of the wave speed.

Fermat's principle The path taken by a light ray between any two locations is the path for which the time interval needed to travel between those locations is a minimum.

focal length f (m) The distance f from the center of the lens or the surface of the mirror to the focus. The value of f is positive for a converging lens or mirror and negative for a diverging lens or mirror.

focus (also called **focal point**) The location where parallel rays come together. If the rays cross at the focus, the focus is *real*. If only the extensions of the rays cross at the focus, it is *virtual*.

image An optical likeness of an object produced by a lens or mirror. The image is at the point from which the rays emanating from the surface of the lens or mirror appear to originate. If the rays travel through the point from which they appear to come, the image is *real*; if they do not travel through that point, the image is *virtual*.

index of refraction n (unitless) The ratio of the speed of light in vacuum to the speed of light in a medium:

$$n \equiv \frac{c_0}{c}. \tag{33.1}$$

law of reflection The angle of reflection for a ray striking a smooth surface is equal to the angle of incidence, and both angles are in the same plane.

lens An optical element that redirects light in order to form images. A *converging lens* directs parallel incident rays to a single point on the other side of the lens. A *diverging lens* separates parallel incident rays in such a manner that they appear to all come from a single point on the side of the lens where the rays came from.

lens equation The equation that relates the object distance o, the image distance i, and the focal length f of a lens or mirror:

$$\frac{1}{f} = \frac{1}{o} + \frac{1}{i}. \tag{33.16}$$

lensmaker's formula The relationship among the focal length f of a lens, the refractive index n of the material of which the lens is made, and the radii of curvature R_1 and R_2 of its two surfaces:

$$\frac{1}{f} = (n - 1)\left(\frac{1}{R_1} + \frac{1}{R_2}\right). \tag{33.36}$$

magnification M (unitless) The ratio of the signed image height h_i ($h_i > 0$ for upright image, $h_i < 0$ for inverted image) to the object height h_o:

$$M \equiv \frac{h_i}{h_o} = -\frac{i}{o}. \tag{33.17}$$

The *angular magnification* is defined as the ratio of the angle θ_i subtended by the image and the angle θ_o subtended by the object:

$$M_\theta \equiv \left|\frac{\theta_i}{\theta_o}\right|. \tag{33.18}$$

Provided these angles are small and for an object that is placed close to both the focus of the lens and the eye's near point, the angular magnification is $M_\theta \approx (0.25 \text{ m}/f)$.

principal rays a set of rays that can be used in ray diagrams to determine the location, size, and orientation of images formed by lenses or spherical mirrors.

ray A line that represents the direction in which light travels. A beam of light with a very small cross-sectional area approximately corresponds to a ray.

ray diagram A diagram that shows just a few selected rays, typically the so-called *principal rays* (see the Procedure boxes on pages 888 and 902).

refraction The changing in direction of a ray when it travels from one medium to another.

Snel's law The relationship among the indices of refraction n_1 and n_2 of two materials and the angle of incidence θ_1 and angle of refraction θ_2 at the interface of the materials

$$n_1 \sin \theta_1 = n_2 \sin \theta_2. \tag{33.7}$$

total internal reflection Mirrorlike reflection that occurs when a ray traveling in a medium strikes the medium boundary at an angle greater than the critical angle. The ray is completely reflected back into the medium.

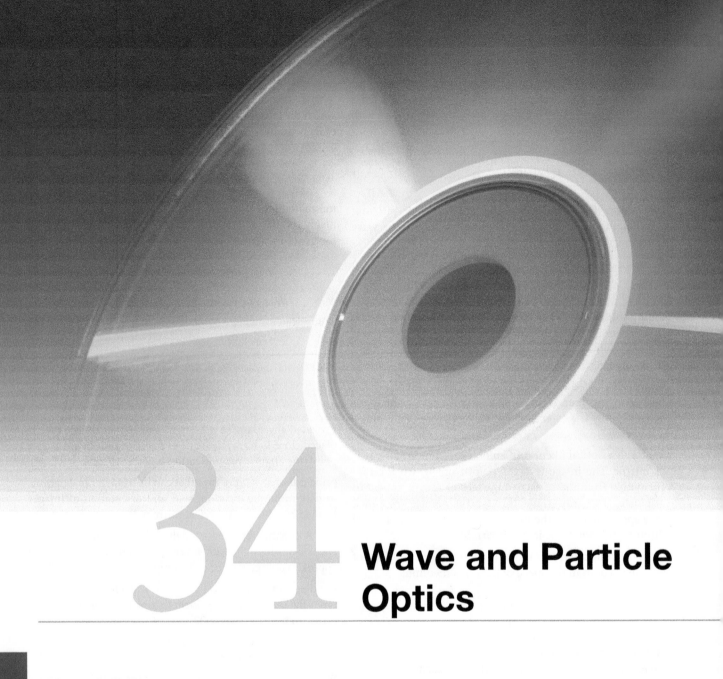

34 Wave and Particle Optics

In Chapter 33, we considered the propagation of light along a straight path. The chapter title, "Ray Optics," reflects the fact that we considered propagating light only in the simplest way—as straight-line motion. You know from Chapter 30, however, that light is an electromagnetic wave. This means that it must undergo interference and diffraction, just like any mechanical wave. As you will learn in this chapter, light waves can interfere with one another and diffract when they pass through small openings.

Another fact about light you will learn in this chapter is that it has a dual nature: It is a wave, yes, but also has the properties of a particle!

34.1 Diffraction of light

As we saw in Chapter 17, when a water wave strikes a barrier that has a small opening, the wave diffracts (spreads out) after it passes through the opening. Figure 34.1*a*, for example, shows surface water waves diffracting nearly circularly after they pass through an opening.

Given that light is a wave, as we discussed in Chapter 30, why don't we see light diffract in a similar fashion after it travels through, say, a window? As Figure 34.1*b* shows, after passing through a window, light continues to travel in a straight line, casting a sharp-edged shadow with no discernible diffraction.

The reason light does not diffract through a window is that the wavelength of the light is very much smaller than the size of the window. In Figure 34.1*a*, the wavelength of the water wave is about the same as the width of the opening, but the wavelength of the light in Figure 34.1*b* is about a million times smaller than the width of the window.

Diffraction is indeed observed with light waves but only when the width of the opening through which the light passes is not much greater in size than the wavelength of the light. Empirical evidence shows that diffraction occurs through openings approximately two orders of magnitude greater than the wavelength. Thus, visible light, with a wavelength on the order of 1 μm, diffracts through apertures up to hundreds of micrometers wide.

To understand diffraction, it is useful to consider the propagation of wavefronts. As discussed in Sections 17.1 and 33.2, a wavefront is a surface on which a wave spreading through space has constant phase. Wavefronts are everywhere perpendicular to the direction of propagation of the wave. By convention, wavefronts are drawn at the crests of the waves, which means the separation between adjacent wavefronts equals the wavelength (Figure 17.2). Although in principle wavefronts can take any shape, usually we consider only those that are either planar or spherical. Most sources of light, from light bulbs to stars, can be modeled as point sources—single points that produce concentric spherical wavefronts. As discussed in Section 17.1 (see especially Figure 17.7), far away from a point source, the radius of the spherical wavefronts is so great that the wavefronts are very nearly planar. As a result, distant point sources can be considered to be sources of planar waves. Lasers also produce planar waves, even very close to the source. For this reason, we can use a laser beam in seeing how electromagnetic waves behave. Keep in mind, however, that our analysis applies to any type of electromagnetic radiation, not just to laser beams.

Let us now determine under what conditions a planar wave spreads out as it propagates—in other words, under what conditions it undergoes diffraction. Figure 34.2 shows planar wavefronts from a beam of electromagnetic waves propagating to the right, with point Q located at the upper end of the wavefronts and point P located outside the region reached by the wavefronts. Because a wavelet (see Section 17.4) centered on Q radiates toward P along the line QP, we expect the beam to spread out as it propagates. However, such spreading is not observed. The reason is that the wavelets centered on points below Q also radiate toward P,

Figure 34.1 Water waves diffract when they pass through a gap, whereas light coming in through a window seems not to diffract—it forms a sharp-edged shadow. Notice that the gap in the breakwater is roughly as wide as the wavelength of the water waves.

(a)   *(b)*

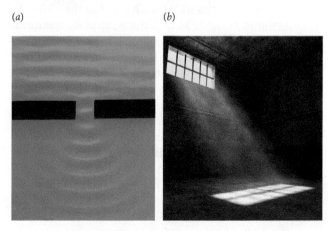

Figure 34.2 The reason we don't usually see diffraction for light beams, provided the beam is very much wider than the wavelength of the light waves.

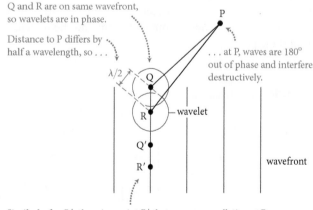

Q and R are on same wavefront, so wavelets are in phase.

Distance to P differs by half a wavelength, so . . .

$\lambda/2$

. . . at P, waves are 180° out of phase and interfere destructively.

wavelet

wavefront

Similarly, for Q′, there is a point R′ that causes cancellation at P.

and we need to sum the contributions of all these wavelets. Consider, for example, the wavelet centered on R, for which the distance PR is exactly half a wavelength longer than the distance PQ. Because Q and R lie on the same wavefront, they produce coherent wavelets. (You should review Section 17.4 if you do not see why this is true.) This means that at P the electric field part of the wavelet from Q is 180° out of phase with the electric field part of the wavelet from R. Thus the two electric fields interfere destructively at P. Because the same is true for the magnetic field part of the wavelets, there is complete destructive interference (see Section 16.3) at P.

In the same manner, for points Q′ lying below Q on the wavefront, we can find on the same wavefront a point R′ for which the distance R′P is exactly half a wavelength longer than the distance Q′P. Thus the fields from the wavelet traveling along R′P cancel those from the wavelet traveling along Q′P. If the wavefronts extend far enough below Q, we can always find points that cancel the radiation from any other point. As a result, we conclude that the light does not spread outside the beam; in other words, there is no diffraction.

✋ **34.1** Consider the point P located ahead of the wavefronts shown in **Figure 34.3**. Following the line of reasoning used in the preceding discussion, what can you say about the intensity (power/area, as defined in Eq. 30.34 and Section 17.5) at P once the wave fronts reach it? (Hint: Consider separately points R above, below, and on the ray through P.)

Checkpoint 34.1 demonstrates that the intensity of a planar wave is uniform as the wave propagates forward because the fields from individual wavelets cancel in any outward direction and reinforce only in the direction of propagation. Section 17.3 (especially Figure 17.20) shows that combining the waves from many adjacent point sources indeed produces a planar wave that propagates forward with very little spreading. This is true because the wavelets interfere destructively with one another in any direction other than the direction of propagation of the wavefronts.

The cancellation process we used in Figure 34.2 can be used only when the width of the laser beam, and hence the width of the wavefronts, is much greater than the wavelength of the light, so that all points on a wavefront can be paired with other points on that wavefront that are half a wavelength or more away from that point and the wavelets from these points cancel.

Figure 34.3 Checkpoint 34.1.

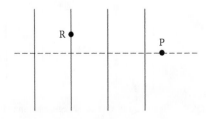

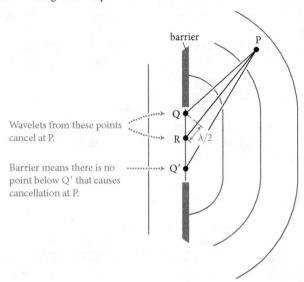

Figure 34.4 The reason we *do* see diffraction when a light beam is transmitted through a small aperture.

If the width of the beam is comparable to the wavelength of the light, not all points on the wavefronts can be paired in this manner. To see why this is so, let us place a barrier in front of our light source, as in **Figure 34.4**. You can think of this drawing as a bird's-eye view of a beam of light traveling to the right and running into a wall that has a gap in it. Light hitting the wall on either side of the gap cannot pass through. As each wavefront of the beam reaches the barrier, only the portion that hits the gap continues moving to the right. The width of each wavefront that passes through the gap is equal to the gap width.

Suppose the gap in Figure 34.4 has a width equal to 2λ. All radiation that reaches P from wavelets that originate above Q′ cancels, as indicated by the rays drawn from points Q and R in Figure 34.4; wavelets that originate at or below Q′ are not canceled at P because those wavelets lack corresponding wavelets at an appropriate distance below Q′. As a result, some of the light spreads out past the edges of the original path of the beam—the light is diffracted. Diffraction in this situation occurs for exactly the same reason as the diffraction of water waves pictured in Figure 34.1a. Diffraction of light occurs when a planar wave passes through an aperture that is only micrometers wide (much less than the width of a hair). As shown in Figure 34.1a, if the width of the aperture is equal to or less than the wavelength, the wavefronts coming from the aperture are spherical. If the aperture is a few wavelengths wide, the wavefronts are elongated right after they pass through the aperture, as shown in Figure 34.4.

An ordinary window, like that shown in Figure 34.1b, is effectively infinitely wide relative to the wavelength of the light, so only the light at the very edges of the window diffracts. In practice, not even diffraction from the edge of the window is observed because the edge is not perfectly smooth on a micrometer scale. However, it *is* possible to observe diffraction of light from the edge of a smooth razor blade, as shown in **Figure 34.5**.

Figure 34.5 Edge diffraction is not usually apparent for visible light because most edges are not smooth enough. However, it can be observed around the edge of a razor blade. The blade in this image is illuminated by a point source of monochromatic light.

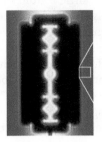

Figure 34.6 When the planar electromagnetic waves of a laser beam pass through a pair of narrow slits, what do we see on the screen?

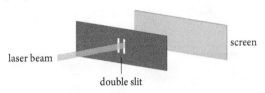

Example 34.1 Spreading out

Do you expect to be able to observe the diffraction of light through (*a*) the front door to your house; (*b*) the holes in a button; (*c*) the gaps between threads of the fabric of an umbrella?

❶ GETTING STARTED I expect to see noticeable diffraction through openings up to roughly two orders of magnitude times the wavelength of the light. I therefore need to estimate the width of each opening and determine the width-to-wavelength ratio.

❷ DEVISE PLAN To estimate the width of the front door and of the holes in the button, I can draw on my experience; to estimate the widths of the gaps between the threads of the fabric of an umbrella, I shall use an upper limit. Then I shall take the ratios of these widths to the wavelength of light in the middle of the visible range, 500 nm.

❸ EXECUTE PLAN (*a*) A door is about 1 m wide, so the ratio of the door's width to 500 nm is 2×10^6, much too great to see diffraction. ✔

(*b*) The holes in a button are about 1 mm in diameter, so the ratio of this width to 500 nm is 2×10^3, still too great to see diffraction. ✔

(*c*) I know that a human hair, which is less than 100 μm in diameter, cannot easily be threaded through a piece of fabric. Therefore I estimate the gaps between the threads in the fabric to be one-tenth of a hair diameter, or 10 μm at most. The ratio of a gap width to 500 nm is therefore 20 or less, and I expect to see diffraction through the gaps in the fabric. ✔

❹ EVALUATE RESULT I know from experience that I do not see diffraction through a doorway. I can check my answers to parts *b* and *c* by looking through the holes in a button and through an open umbrella. When I do so, I see no diffraction through the button but I can see diffraction through the umbrella if the fabric is dark. (This diffraction is particularly noticeable when I look at a streetlight at night through the umbrella.)

✋ **34.2** In discussing how a planar wave propagates, we could turn our earlier argument around and say that for each point Q in Figure 34.2 there is a point S somewhere on the wavefront that radiates toward P along a path exactly one wavelength longer than that from Q, and therefore there should be a nonzero intensity at P. What is wrong with this argument?

34.2 Diffraction gratings

What happens if instead of passing through a single small aperture, a planar electromagnetic wave strikes a barrier that contains two narrow slits at normal incidence, as shown in Figure 34.6? If the slit width is much less than the wavelength of the wave, the slits serve as two coherent point sources of electromagnetic waves of the same wavelength as the wave striking the barrier, and the waves from the two sources interfere with one another in the manner described in Section 17.3 for two adjacent sources of surface waves. The only difference is that electromagnetic waves are not confined to a planar surface but spread out in three dimensions. If the two slits are either round or square, the waves that emerge from them are spherical. If the slits are much taller than they are wide, as in Figure 34.6, the slits serve as lines of point sources and the waves that emerge from the slits have cylindrical wavefronts.

The crests of the waves from these two coherent point sources overlap in certain directions, as shown in Figure 34.7 (and Figure 17.13). Between these directions where there is overlap are directions along which the waves from the two sources cancel. If we place a screen to the right of the slits, a pattern of alternating bright and dark bands appears on the screen, as shown in Figure 34.8 on the next page. Such dark and bright bands are commonly called **interference fringes.** The bright fringes are labeled by a number called the **fringe order** *m*. The central bright fringe is the *zeroth-order bright fringe* (*m* = 0); around it are higher-order bright fringes (*m* = 1, 2, . . .). Note that the pattern is symmetrical about this zeroth-order bright fringe.

Figure 34.7 Interference between the diffracted waves emerging from the two slits.

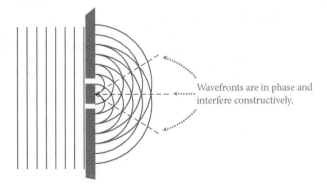

Wavefronts are in phase and interfere constructively.

Figure 34.8 The interference pattern produced when the laser beam of Figure 34.6 passes through a pair of slits and strikes the screen.

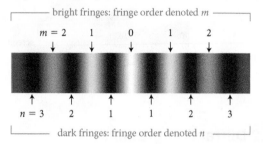

How do we determine the locations of the bright fringes? The central (zeroth-order, $m = 0$) bright fringe is simplest to locate. As shown in Figure 34.7, waves from the two sources interfere constructively along a perpendicular line running through the midpoint between the two slits. Because the waves coming from the two slits travel the same distance to reach any point along this perpendicular line, the waves arrive in phase with each other.

To locate the other bright interference fringes, we need to work out the directions in which the difference in path length between waves coming from the upper source and waves coming from the lower source is an integer number of wavelengths. In these directions, constructive interference between waves from the two sources produces bright fringes.

Let's consider two rays, one from each slit, that meet at the screen to form a fringe, as shown in **Figure 34.9a**. In general we shall take the distance from the sources to the screen to be much greater than the distance between the slits. Note from the figure how, even though the rays eventually meet at the screen, their paths are essentially

parallel when they emerge from the slits (Figure 34.9b). If we denote the angle between the nearly parallel rays and the normal to the barrier as θ, we can say that the difference in path length for waves emitted at angle θ from the two sources is $d \sin \theta$, where d is the distance between the slits (Figure 34.9b). When this path-length difference is equal to an integer multiple of the wavelength, constructive interference occurs. The central bright fringe corresponds to $d \sin \theta = 0$ or $\theta = 0$. Subsequent bright fringes are located at angles given by $d \sin \theta_m = \pm m\lambda$, where $m = 1, 2, \ldots$ denotes the order of the bright fringe. The plus and minus signs give rise to bright fringes on either side of the central maximum.

Likewise, in directions that correspond to path-length differences of an odd number of half-wavelengths, waves from the two sources interfere destructively, resulting in dark fringes. For these fringes, we use n to denote fringe order. The smallest angle at which destructive interference occurs corresponds to $d \sin \theta = \frac{1}{2}\lambda$. More generally, the angles at which dark fringes occur are given by $d \sin \theta_n = \pm(n - \frac{1}{2})\lambda$, where $n = 1, 2, \ldots$ denotes the order of the dark fringe. The dark fringes around the zeroth-order bright fringe are the first-order dark fringes.

Example 34.2 Two-slit diffraction grating

Coherent green light of wavelength 530 nm passes through two very narrow slits that are separated by 1.00 μm. (a) Where is the first-order bright fringe? (b) What is the angular separation between the $n = 1$ and $n = 2$ dark fringes?

❶ GETTING STARTED This problem involves interference between light rays passing through two closely spaced, very narrow slits, as shown in Figure 34.9.

Figure 34.9 Determining the path-length difference between two rays traveling to a point on a distant screen in a two-slit interference setup.

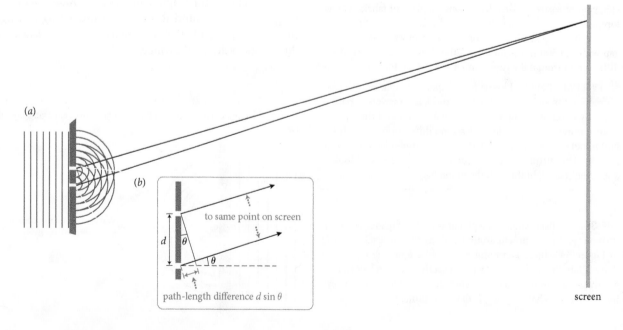

❷ DEVISE PLAN The angular locations of the centers of the bright fringes are given by the condition for constructive interference, $d \sin \theta_m = m\lambda$; the angular locations of the dark fringes are given by the condition for destructive interference, $d \sin \theta_n = (n - \frac{1}{2})\lambda$. (For both bright and dark fringes, I omit the $\pm$ signs because I'll only consider the fringes on one side of the central maximum.) I therefore need to use these relationships to calculate the angular locations of the fringes. To use the constructive and destructive interference conditions, I need to identify the appropriate values of m and n and then use them with the wavelength and the distance between slits to obtain the angular positions of the bright and dark fringes I am interested in.

❸ EXECUTE PLAN (*a*) The first-order bright fringe corresponds to $m = 1$, which means the center of the first-order bright fringe is located at the value θ_1 corresponding to $d \sin \theta_1 = \lambda$. Substituting for d and λ in this expression and solving for θ_1, I obtain

$$\theta_1 = \sin^{-1}\left(\frac{0.530 \ \mu m}{1.00 \ \mu m}\right) = 32.0°. ✔$$

(*b*) The two lowest-order dark fringes correspond to $n = 1$ and $n = 2$, and their centers occur at the angles corresponding to $d \sin \theta_1 = \lambda/2$ and $d \sin \theta_2 = 3\lambda/2$. Substituting and solving, I obtain

$$\theta_1 = \sin^{-1}\left(\frac{0.530 \ \mu m}{2 \times 1.00 \ \mu m}\right) = 15.4°$$

$$\theta_2 = \sin^{-1}\left(\frac{3 \times 0.530 \ \mu m}{2 \times 1.00 \ \mu m}\right) = 52.7°.$$

The angular separation is thus $52.7° - 15.4° = 37.3°. ✔$

❹ EVALUATE RESULT The center of the $m = 1$ bright fringe is roughly halfway between the centers of the $n = 1$ and $n = 2$ dark fringes, as I expect from Figure 34.8.

🖐 **34.3** Does the spacing of the bright fringes in the two-slit arrangement in Figure 34.6 increase, decrease, or stay the same if we (*a*) increase the spacing d of the slits, or (*b*) increase the wavelength λ of the light incident on the arrangement?

Now consider the effect of many equally spaced narrow slits in a barrier on which planar waves are incident normally (**Figure 34.10**). Once again, we can determine the fringe pattern produced at a given location on a distant screen by

Figure 34.10 Path-length difference for planar waves striking a barrier with multiple slits.

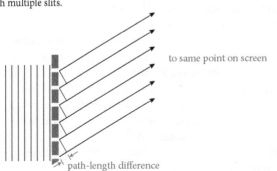

to same point on screen

path-length difference

Figure 34.11 Three coherent waves interfere destructively when each is out of phase with the other two by one-third of a cycle.

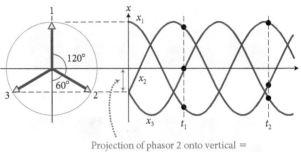

Projection of phasor 2 onto vertical = $\cos 60°$ times amplitude = half amplitude.

combining all the waves (one from each slit) that travel to that location. The condition for constructive interference among all the waves is equivalent to that for two slits. As can be seen from Figure 34.10, the path-length difference between each pair of adjacent waves is the same. This means that if the waves from two of the slits interfere constructively, the waves from *all* of the slits do the same. Bright fringes therefore appear at angles corresponding to $d \sin \theta_m = \pm m\lambda$, with d the separation between adjacent slits. The location of bright fringes therefore does not depend on the number of slits as long as the separation d between adjacent slits is the same for all slits.

🖐 **34.4** Suppose there are three slits in a barrier on which light is incident normally, with each slit separated from its neighbor by a distance d. Do the waves from all three slits cancel perfectly at angles given by $d \sin \theta = \pm(n - \frac{1}{2})\lambda$?

As Checkpoint 34.4 illustrates, the condition for complete destructive interference of all of the waves is not the same for three slits as for two. Instead, as **Figure 34.11** shows, three coherent waves cancel one another perfectly when each is out of phase with the other two by one-third of a cycle. The three waves also cancel one another perfectly when each is out of phase with the other two by two-thirds of a cycle. In that case, phasors 2 and 3 in Figure 34.11 are interchanged. So there are two dark fringes between each pair of bright fringes. In between these two dark fringes is a faint bright fringe (**Figure 34.12** on the next page). The brightest fringes are the *principal maxima* in the interference pattern. These correspond to constructive interference of the waves diffracted by all three slits. The fainter bright fringes are *secondary maxima*. At these locations the cancellation is not complete.

Four coherent waves cancel one another when adjacent waves are out of phase by one-fourth of a cycle. In the same manner, if there are N slits, the condition for complete destructive interference is that each wave must differ in phase by $1/N$ of a cycle from its immediate neighbors. Then the N waves are evenly distributed throughout one cycle of oscillation and add to zero. The condition for the path-length differences for the dark fringes is thus $d \sin \theta_k = \pm(k/N)\lambda$, where k is any integer that is *not* a whole-number multiple of N (because when $k/N = m$, we have constructive interference).

Figure 34.12 Interference pattern caused by the diffraction of a coherent beam of light through two, three, and eight narrow slits.

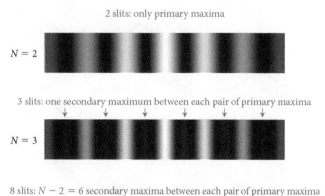

2 slits: only primary maxima

$N = 2$

3 slits: one secondary maximum between each pair of primary maxima

$N = 3$

8 slits: $N - 2 = 6$ secondary maxima between each pair of primary maxima

$N = 8$

Although the bright fringes are in the same location regardless of the number of slits, there are now $N - 1$ dark fringes between the bright fringes and $N - 2$ secondary maxima between each pair of principal maxima. As a result, as N increases, the bright fringes become narrower and brighter, as shown in Figure 34.12. (The brightness of the pattern corresponds to the intensity of the light striking the screen.)

✋ **34.5** Why does the brightness of the fringes increase as the number of slits increases?

The interference of a planar electromagnetic wave as it passes through many closely spaced narrow slits is due to the diffraction that occurs at the slits. A barrier that contains a very large number of such slits is therefore called a **diffraction grating.** Diffraction gratings can be either transmissive (such as the one shown in Figure 34.10) or reflective. Reflective diffraction gratings are made by engraving grooves to reflect light from a surface, as shown in Figure 34.13. The grooves on a music compact disc (opening picture in this chapter) form a reflective diffraction grating.

Why is the light reflected from the compact disc surface so colorful? You found in Checkpoint 34.3 that the position

Figure 34.13 Reflective diffraction grating.

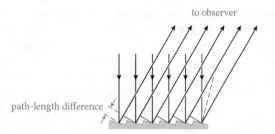

to observer

path-length difference

of interference fringes produced by light of a single color depends on the wavelength. When white light, which contains many different colors and therefore many different wavelengths, falls on a diffraction grating, the fringes for each wavelength are displaced from each other, producing a series of rainbows.

✋ **34.6** Suppose the light striking the reflective diffraction grating in Figure 34.13 is white light—that is, light consisting of all the colors of the rainbow. Red light has the longest wavelength of the colors that make up white light; violet has the shortest. (*a*) Is the angle at which first-order constructive interference occurs for violet light less than, equal to, or greater than the angle for red light? (*b*) Why are multiple rainbows visible in the reflected light?

Very precisely manufactured diffraction gratings have many uses in scientific equipment. They are most widely used to disperse visible, ultraviolet, or infrared light into its constituent wavelengths because the resulting spectrum provides information about the object that emitted the light. For example, astronomers often use a diffraction grating attached to a telescope to identify the wavelengths present in the light from stars, in order to understand the chemical composition of the stars or their distance from Earth.

Certain common objects can also function as diffraction gratings. For example, if you look through a piece of dark, finely woven, taut fabric at a point source of light (say, a distant street light through the fabric of a dark umbrella), you will see fringes.

✋ **34.7** Diffraction gratings used in astronomical instruments must be able to separate wavelengths that are quite close together. (*a*) To increase the ability to do this, should the separation between slits be made less or greater? (*b*) Does the width of the slits affect the diffraction pattern?

34.3 X-ray diffraction

The interference of very-short-wavelength electromagnetic radiation is widely used to study the structure of materials. **X rays** are electromagnetic waves that have wavelengths ranging from 0.01 nm to 10 nm, more than 100 times less than the wavelengths of visible light.

Figure 34.14 shows one way to generate X rays. Electrons are ejected from a heated cathode on the right and accelerated by a potential difference of several thousand volts toward a metal anode on the left. The electrons crash into the atoms that make up the anode; this inelastic collision decelerates the electrons very rapidly and gives the atoms a great deal of internal energy. The atoms then re-emit this energy in the form of x rays. In addition, the rapidly decelerating electrons radiate x rays, typically at a 90° angle to the path of the accelerated electrons.

Figure 34.14 Schematic diagram for a cathode ray tube x-ray emitter.

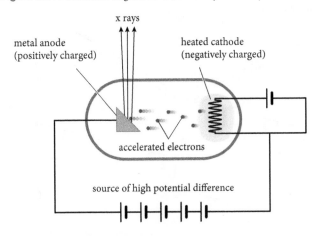

Figure 34.16 Two examples of crystal lattices.

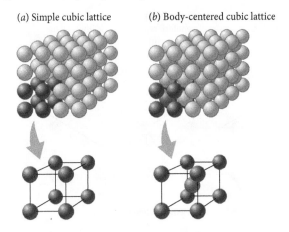

(a) Simple cubic lattice   (b) Body-centered cubic lattice

X rays can pass through many soft materials with low mass density that are opaque to visible light. For example, they pass through soft tissues in the human body but are strongly absorbed by bones and teeth. As a result, x rays are widely used to obtain photographic images of the skeleton (Figure 34.15). X-ray imaging of blood vessels or soft internal organs can be done by giving the patient a drug containing heavy atoms, such as iodine, because the heavy atoms absorb x rays. For example, one way cardiologists diagnose heart problems is to directly observe blood vessels in a patient's heart by injecting a heavy-atom drug into the patient's blood and taking x-ray movies of the beating heart as the blood is pumped through.

Because x-ray wavelengths are either shorter than or comparable to the typical distance between atoms in solid materials (0.1 nm to 1 nm), x-ray diffraction can be used to study atomic arrangements in solids. Many solids are *crystalline*, meaning that their atoms are arranged in a three-dimensional, regularly spaced grid called a *crystal lattice* (Figure 34.16). The lattice serves as a three-dimensional diffraction grating for x rays because the lattice spacing is comparable to the x-ray wavelength.

Consider what happens when x rays strike the top plane of atoms in a crystal lattice* at an angle θ (Figure 34.17). Each atom that is struck by the beam of x rays acts as the source of a wavelet emitting waves in all directions, much like the slits of the diffraction gratings we discussed in the preceding section. Waves emitted at $\theta' = \theta$ have the same path length and so they add constructively, yielding a strong reflected beam.

34.8 Considering only the top row of atoms in Figure 34.17, are there any other directions in which the x rays diffracted by the atoms interfere constructively?

As you saw in Checkpoint 34.8, x rays diffracted by the crystal in directions other than at angle $\theta' = \theta$ are much weaker than those diffracted at angle $\theta' = \theta$. We might therefore expect not to see any x-ray diffraction from crystals at other angles. However, a crystal consists of many planes of atoms, and some incident x rays penetrate into the

---

* The lattice shown is a so-called cubic lattice. Lattices can be much more complex than the cubic lattice, but the principles of diffraction are the same.

Figure 34.15 Bones and teeth absorb x rays, whereas soft tissues are nearly transparent to them.

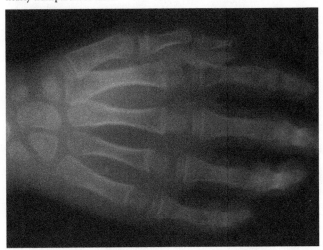

Figure 34.17 Diffraction of x rays by the atoms at the surface of a crystal lattice.

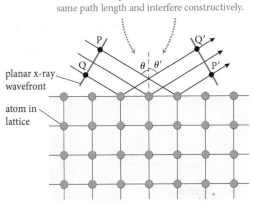

When these angles are the same, rays all have same path length and interfere constructively.

planar x-ray wavefront

atom in lattice

Figure 34.18 Interference of x rays diffracted by adjacent planes of a crystal.

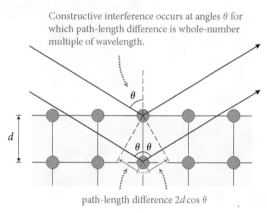

Constructive interference occurs at angles θ for which path-length difference is whole-number multiple of wavelength.

path-length difference $2d \cos \theta$

Figure 34.19 Constructive interference of x rays diffracted by two diagonal crystal planes.

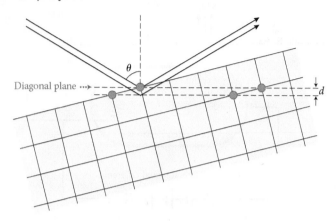

crystal. We thus need to take into account the diffraction of the x rays by the atoms in multiple crystal planes to determine the diffraction of the crystal as a whole.

Figure 34.18 shows the x rays diffracted by atoms in two adjacent planes of a crystal. For most angles of incidence, the waves diffracted by atoms in different planes differ in phase and so interfere destructively. However, when the difference in path length between rays diffracted by atoms in different planes is a whole-number multiple of the x rays' wavelength, the rays are in phase and interfere constructively. The path-length difference equals $2d \cos \theta$, where d is the distance between adjacent planes. Therefore the condition for constructive interference is $2d \cos \theta = m\lambda$. This condition is called the **Bragg condition,** after the father-and-son team of physicists who formulated it. Because the atomic spacing and the x-ray wavelength are fixed, crystals reflect x rays only at those angles for which $2d \cos \theta$ is an integer multiple of the x-ray wavelength.

The atoms in the lattice of a crystal define many different lattice planes. **Figure 34.19** shows two sets of lattice planes in a cubic crystal, one indicated by solid lines (planes parallel to the surface of the crystal) and the other indicated by dashed lines (diagonal planes). If we tilt the crystal so that the angle the incident x rays make with its surface is

different from the angle shown in Figure 34.18, a different set of planes with a different spacing can produce constructive interference of the diffracted waves.

By measuring the angles at which strong x-ray diffraction occurs, one can determine the arrangement of atoms in a crystalline solid. **Figure 34.20** shows how such a measurement is carried out. An x-ray source like that shown in Figure 34.14 is used to produce a beam of x rays of various wavelengths. The beam is then diffracted from a *crystal monochromator* (which is simply a crystal of known lattice spacing) positioned at an angle chosen so that the Bragg condition is satisfied for one desired wavelength of x rays. Because the other wavelengths in the original beam do not satisfy the Bragg condition, a *monochromatic* (single-wavelength) beam of x rays is diffracted from the monochromator to the sample.

The sample of crystalline material whose lattice is being studied is slowly rotated with respect to the monochromatic beam, and as this rotation takes place, the intensity of x rays diffracted from the sample is measured on a detector as a function of the angle α between the x rays and the sample surface (Figure 34.20b). This angle is often called the *Bragg angle α*. As this angle changes, the Bragg condition is

Figure 34.20 (*a*) Apparatus for studying x-ray diffraction from a crystalline solid. (*b*) Relationship between the incident angle θ and the Bragg angle α.

(*a*) Apparatus for studying x-ray diffraction from a crystalline solid

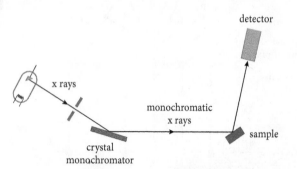

(*b*) Relationship between incident angle θ and Bragg angle α

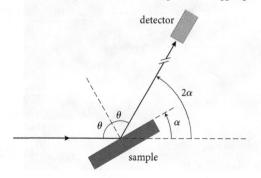

generally not satisfied. However, at specific Bragg angles, the various crystal planes in the sample satisfy the Bragg condition, producing a high intensity of x rays on the detector.

🖐 **34.9** Express the Bragg condition $2d \cos \theta = m\lambda$ in terms of the Bragg angle α between the x rays and the surface of the sample rather than the angle θ between the x rays and the surface normal.

Example 34.3 X-ray diffraction

Figure 34.21 shows diffracted x-ray intensity as a function of the Bragg angle α, obtained using x rays having a wavelength of 0.11 nm. (a) Without calculating values for the lattice spacing d, identify which of the two peaks corresponds to a greater distance between adjacent planes in the sample being studied. (b) Calculate the distance between adjacent planes corresponding to each peak.

Figure 34.21 Example 34.3.

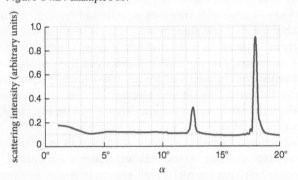

❶ **GETTING STARTED** The peaks in a graph of x ray intensity as a function of the Bragg angle α correspond to constructive interference and therefore values of α that satisfy the Bragg condition.

❷ **DEVISE PLAN** For part a, I can use the Bragg condition, together with the shape of the graph, to deduce which peak results from the greater d value. Because the graph gives intensity in terms of the Bragg angle, I shall need the Bragg condition expressed in terms of the Bragg angle.

For part b, I can solve this form of the Bragg condition for d and then insert my two given α values to determine the plane separation distance in each case. Looking at this form of the Bragg condition, I see that the Bragg angle α at which a peak occurs increases as m increases. Because this graph begins at $\theta = 0°$, the two peaks must correspond to $m = 1$ for the interference patterns when the crystal surface is oriented at the two Bragg angles I am working with.

❸ **EXECUTE PLAN** (a) From Checkpoint 34.9 I know that the Bragg condition in terms of the Bragg angle is $2d \sin \alpha = m\lambda$. In order for the product $2d \sin \alpha$ to remain constant, α must decrease as the distance d between adjacent planes increases. Therefore the peak at the smaller α value corresponds to the greater distance between planes. ✔

(b) To obtain d for each peak, I solve $2d \sin \alpha = m\lambda$ for d and then substitute $m = 1$ and the values for α and λ. For the short peak, $\alpha = 12.5°$, which gives $d = \lambda/(2 \sin \alpha) = (0.11 \text{ nm})/(2 \sin 12.5°) = 0.25$ nm. For the tall peak, $\alpha = 18°$, which gives $d = \lambda/2 \sin \alpha = (0.11 \text{ nm})/(2 \sin 18°) = 0.18$ nm. ✔

❹ **EVALUATE RESULT** The Bragg angle α at which constructive interference occurs decreases as the distance between planes increases. This is consistent with what I found previously for interference between two slits (Checkpoint 34.3), in which increasing the distance between slits also causes the angle between fringes to decrease. In general, the size of an interference pattern decreases as the distances between interfering sources increase. Finally, the smaller value of d multiplied by $\sqrt{2}$ gives the greater value of d, as it should for the distances between planes for the cubic lattice in Figure 34.19.

Many studies of crystal structure are done by passing x rays through the crystal rather than reflecting them from the various crystal planes. The crystal then acts like a three-dimensional transmissive diffraction grating for the x-ray beam. Instead of a single line of slits, the beam encounters many rows of slits. As a result, many rows of fringes usually called "spots" are formed. The experimental apparatus for such a measurement is shown in **Figure 34.22a**. From the

Figure 34.22 X-ray crystallography.

(a) Schematic apparatus for x-ray crystallography

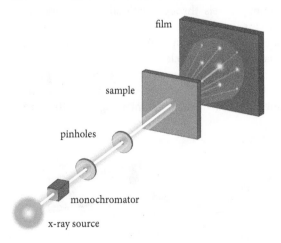

(b) X-ray diffraction pattern of diamond lattice

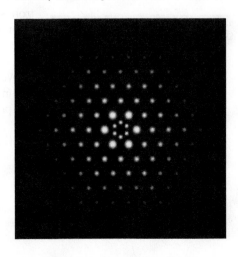

position and intensity of the spots in the resulting diffraction pattern (Figure 34.22b), we can deduce the arrangement of the atoms in the sample.

The earliest x-ray diffraction studies of crystals were done on simple crystals, such as metals that form cubic or other very simple lattices. However, x-ray diffraction has also been used to study the atomic structure of much more complicated molecules, by crystallizing solutions of such molecules. The double helical structure of DNA was determined using x-ray diffraction, also called *x-ray crystallography,* and crystallography is widely used today to determine the atomic structure of even more complicated biological molecules.

34.10 While comparing the x-ray diffraction patterns from two crystals, you determine that crystal A produces a pattern with more widely spaced diffraction spots than crystal B. Which crystal has the greater atomic spacing?

34.4 Matter waves

Patterns remarkably similar to x-ray diffraction patterns can be obtained by aiming a narrow beam of electrons at a crystal. **Figure 34.23** shows the pattern obtained by sending a beam of high-speed electrons through a solid crystalline sample in an instrument called an electron microscope. The shape of the pattern is similar to that obtained with x rays (Figure 34.22b), which tells us that electrons are also diffracted by crystals. This discovery, in turn, suggests that the electrons behave like waves because interference and diffraction are wave phenomena. Indeed, electrons have been found to exhibit interference in many other experiments. For example, a beam of electrons aimed at two very narrow slits produces an interference pattern similar to that of a beam of light aimed at two narrow slits (**Figure 34.24**).

Varying the speed of the electrons changes the spacing of the interference pattern, which indicates that the electron wavelength depends on speed.

Figure 34.23 Electron diffraction pattern for a diamond lattice. Notice the similarity to the x-ray pattern in Figure 34.22b.

34.11 Spots in an electron diffraction pattern, such as the one shown in Figure 34.23, move closer together as the speed of the electrons is increased. Does this mean the wavelength of the electrons increases or decreases with increasing speed?

Up to now we have considered electrons as being nearly pointlike particles, so it is surprising—to say the least—to discover that they also behave like waves. In fact, electrons can exhibit both particle behavior and wave behavior in a single experiment! This dual behavior was vividly demonstrated in an experiment done in 1989 in Japan using an apparatus similar to that shown in Figure 34.24. The number of electrons emitted by the source was kept very low, so as to ensure that at any given instant at most one electron was traveling from the source to the screen. The screen shows the place where each electron arrives as a bright dot, and at first these dots appear at seemingly random locations, as shown in **Figure 34.25a** and *b.* However, as more and more electrons reach the screen one after another, it becomes clear that the electron impacts are not randomly located (Figure 34.25c and *d*). Rather, they are arranged exactly in the two-slit interference pattern observed for a higher-intensity electron beam. Covering one slit and forcing the electrons to pass through the other makes the interference pattern disappear, just as it would for a light wave or a water wave.

Experiments show that after many electrons are allowed to pass one at a time through the apparatus, the number of electrons arriving on each small region of the screen per unit time is proportional to the intensity that would be observed at that region if light of the appropriate wavelength were shone on the slits. In other words, the *probability* of any single electron arriving at that small region in a fixed time interval corresponds to the intensity of the two-slit interference pattern at that location.

The observation of individual electrons arriving at the screen one after another indicates that the electrons are individual particles. However, if they are particles in the sense we think of for material objects, they must pass through either the right slit or the left slit, and so how can they produce an interference pattern?

The very counterintuitive conclusion is that each electron somehow travels through *both* slits simultaneously! Such a statement is not surprising for a wavefront hitting the two slits, but it goes completely against our intuitive notion of what we call a "particle."

We cannot explain the results of this experiment by concluding that electrons are *waves* because a classical wave could not produce the individual pinpoint images on the screen

Figure 34.24 Apparatus for observing two-slit interference with an electron beam.

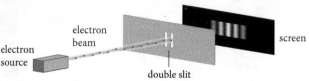

Figure 34.25 When we perform two-slit diffraction with a very weak electron beam, we can see the pattern build up over time. At first (*a*, *b*), the dots that mark electron impacts seem to be scattered randomly, but as more accumulate (*c*, *d*), the diffraction pattern becomes evident.

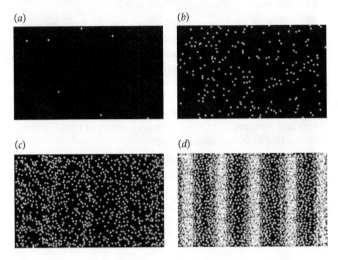

(*a*) (*b*)

(*c*) (*d*)

shown in, say, Figure 34.25*a*. If electrons were waves, the full diffraction pattern would be visible from the start (although it would be very faint). Scientists have concluded that electron behavior can be explained only if an electron has both particle properties and wave properties. This **wave-particle duality**—the possession of both wave properties and particle properties—has been observed not only for electrons but also for all other subatomic particles, for individual atoms, and, as we shall see in the next section, for light.

An expression for the wavelength of a particle, called the **de Broglie wavelength,** was proposed in 1924 by the French physicist Louis de Broglie* and confirmed experimentally a few years later. The de Broglie wavelength is inversely proportional to the momentum of the particle, $\lambda = h/p$, and the proportionality constant, called **Planck's constant,** is $h = 6.626 \times 10^{-34}$ J·s. That this constant is a very small number indicates that the wavelength of macroscopic objects is extremely small. Because waves exhibit diffraction and interference only on length scales comparable to their wavelength, the wave nature of matter has been observed only with subatomic particles, atoms, and molecules. Exercise 34.4 illustrates this point.

Exercise 34.4 Electron versus baseball

Calculate the de Broglie wavelength associated with (*a*) a 0.14-kg baseball thrown at 20 m/s and (*b*) an electron of mass 9.1×10^{-31} kg moving at 5.0×10^6 m/s.

SOLUTION I use the expression $\lambda = h/mv$ for the de Broglie wavelength.

(*a*)
$$\lambda_{\text{de Broglie, baseball}} = \frac{6.626 \times 10^{-34} \text{ J·s}}{(0.14 \text{ kg})(20 \text{ m/s})}$$
$$= 2.4 \times 10^{-34} \text{ m.} ✔$$

———————
* "de Broglie" is pronounced "duh-Br-uh-y," the "y" sounding as in "yikes."

This is 24 orders of magnitude less than the diameter of an atom!

(*b*)
$$\lambda_{\text{de Broglie, electron}} = \frac{6.626 \times 10^{-34} \text{ J·s}}{(9.1 \times 10^{-31} \text{ kg})(5.0 \times 10^6 \text{ m/s})}$$
$$= 1.5 \times 10^{-10} \text{ m} = 0.15 \text{ nm.} ✔$$

Electron diffraction takes place on distances comparable to the spacing between atoms, whereas a baseball would diffract only through apertures more than 10^{24} times smaller than the typical spacing between atoms. (But the diameter of a baseball is about 0.10 m, so such an experiment is not possible.)

34.12 How would the electron diffraction pattern in Figure 34.23 change if the electrons were traveling more slowly?

34.5 Photons

We have now found that particles have wave properties, but these properties are observed only when the particle wavelength is comparable to the size of the objects the particles interact with. Up until now, you have most probably thought of light as being solely a wave. However, the dual wave-particle nature of particles may lead you to wonder whether light, too, is not just a wave but also a particle.

34.13 Compare the two-slit interference pattern obtained with electrons (Figure 34.25) with the two-slit interference pattern obtained with light (Figure 34.8). If light and electrons exhibit a similar wave-particle duality, how might you modify the two-slit experiment with light shown in Figure 34.6 to observe light behaving like a particle?

Figure 34.26a on the next page shows the image obtained by shining a beam of light onto the sensor chip of a digital camera. Pixels in the center of the beam register more light, and therefore the image of these pixels is brighter. The pixels at the edge of the beam are dimmer than those at the center, and outside the beam the pixels are black. If we place in front of the sensor chip a filter that lets only 50% of the light through, the image darkens because the brightness measured by each camera pixel is cut in half (Figure 34.26*b*).

Suppose we keep adding such filters, cutting the beam intensity in half with each addition. Does the image keep getting proportionally darker? If you carry out the experiment, you will discover that it does not, for one of two reasons. The first reason is mechanical: If you use the sensor chip of a digital camera, it stops detecting below a certain level. As you decrease the intensity of the beam by adding more and more filters, the image first gets grainy and then turns black.

The second reason has to do with the fundamental nature of light. Even if you use an extremely sensitive detector, you will still see that once the beam becomes very weak, adding another filter does not simply halve the image intensity. Instead, as shown in Figure 34.26*c* and *d*, the image of the beam breaks up into individual point-like flashes of equal

Figure 34.26 Images formed by using the sensor of a digital camera to record increasingly faint beams of light for the same exposure period. The separate dots recorded for the faintest beams reveal the particle-like behavior of light.

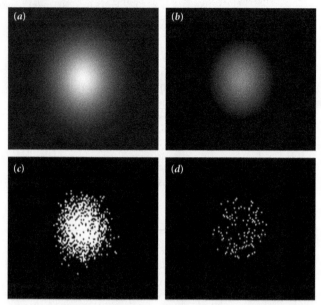

intensity, resembling the impacts of individual particles. The impacts may at first appear to be randomly distributed within the profile of the beam, but if you accumulate many of these impacts, you discover that the probability of observing a flash in a given location follows the intensity profile of the beam—the impacts are more likely to occur near the center of the beam. In fact, the probability of observing an impact in a particular location is proportional to the intensity of the beam at that location.

As the beam intensity is reduced, the individual impacts become separated in time. No two impacts occur at the same instant. Nor are any "half impacts" ever recorded. From these observations, we conclude that light indeed has particle properties as well as wave properties. The 'particles' of light are called **photons.** As we discuss further in Section 34.10, a photon represents the basic unit of a light wave and carries a certain amount of light energy. For a photon of frequency f, this energy equals hf, where h is again Planck's constant. Photons thus represent the quantum of electromagnetic energy—they cannot be subdivided.

Example 34.5 Photons from a light bulb

A 50-W incandescent light bulb emits about 5.0 W of visible light. (The rest is converted to thermal energy.) If a circular aperture 5.0 mm in diameter is placed 1.0 km away from the light bulb, approximately how many photons reach the aperture each second?

❶ GETTING STARTED This problem asks me to relate the power of light emitted by a light bulb to the rate at which photons pass through a certain area at a particular distance from the bulb. I can approximate the light bulb as a point source of light, meaning that its light is radiated uniformly in all directions. I can

also simplify the problem by assuming that all of the light has the same wavelength, and I choose 500 nm for that wavelength. (This is a significant simplification because real light bulbs emit all visible wavelengths of light as well as infrared.)

❷ DEVISE PLAN I begin by determining the intensity—the power per unit area—of light produced by the light bulb over a sphere of radius 1 km, which tells me the intensity at the aperture. I can then multiply this intensity by the area of the aperture to obtain the power passing through the aperture, and I multiply that by 1 s to calculate the energy passing through the aperture in 1 s. Finally I shall use the relationship between photon energy and wavelength to determine the number of photons corresponding to that amount of energy.

❸ EXECUTE PLAN The intensity at the aperture is given by the power emitted by the bulb divided by the surface area of a sphere of radius 1.0 km:

$$\frac{(5.0 \text{ W})}{(4\pi)(1.0 \times 10^3 \text{ m})^2} = 4.0 \times 10^{-7} \text{ J/m}^2 \cdot \text{s},$$

and therefore the amount of energy passing through a circular hole of radius 2.5 mm in 1.0 s is

$$(4.0 \times 10^{-7} \text{ J/m}^2 \cdot \text{s})[\pi(2.5 \times 10^{-3} \text{ m})^2] = 7.8 \times 10^{-12} \text{ J}.$$

The energy of a single photon of wavelength 500 nm is

$$E = hf = \frac{hc_0}{\lambda}$$

$$= \frac{(6.626 \times 10^{-34} \text{ J} \cdot \text{s})(3.00 \times 10^8 \text{ m/s})}{(500 \times 10^{-9} \text{ m})}$$

$$= 3.98 \times 10^{-19} \text{ J},$$

and so the number of photons that corresponds to the amount of energy passing through the hole is

$$\frac{7.8 \times 10^{-12} \text{ J}}{3.98 \times 10^{-19} \text{ J}} = 2.0 \times 10^7 \text{ photons.} \checkmark$$

❹ EVALUATE RESULT The number of photons I obtain is great even though I know from experience that the amount of light entering my eye 1.0 km from a 50-W bulb is exceedingly small. However, photons contain a vanishingly small amount of energy (about 10^{-19} J per photon), and so my answer is not implausible.

I assumed that all the photons have the same 500-nm wavelength. In reality the light bulb emits both longer- and shorter-wavelength photons. Suppose various wavelengths are equally distributed on both sides of 500 nm in the spectrum of the light emitted by the bulb. Photons with longer wavelengths have less energy and therefore there are more of them, but photons with shorter wavelengths have greater energy, and so there are fewer of them. The overall result is therefore about the same number of photons as the number I calculated using 500 nm as the only wavelength. To a first approximation, therefore, my result should be correct.

Figure 34.27 When we record a low-intensity beam of light with individual detectors, the beam acts like a stream of particles. Passing it through a double slit, however, causes an interference pattern to emerge.

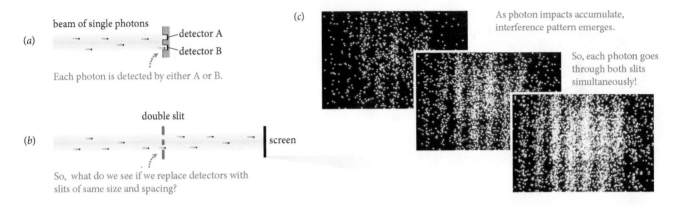

(a) beam of single photons
detector A
detector B
Each photon is detected by either A or B.

(b) double slit
screen
So, what do we see if we replace detectors with slits of same size and spacing?

(c) As photon impacts accumulate, interference pattern emerges.
So, each photon goes through both slits simultaneously!

What are photons? How can light be a wave *and* consist of photons that behave like particles? I cannot answer this question because no one really knows the answer. I can, however, describe how photons behave. As you will see, their behavior defies common sense—it is unlike anything we ever experience.

If we reduce the intensity of a light beam (or a beam of any other type of electromagnetic radiation) so greatly that the flashes from the impacts of individual photons are well separated in time and then aim this beam at two very small, very closely spaced detectors (Figure 34.27a), we see that each photon is detected by either one detector or the other. A simultaneous impact on both detectors is never recorded. (The detectors can be made as small as 1 μm wide and spaced by just a fraction of a micrometer.) This observation suggests that each photon takes a definite path—toward either detector A or detector B.

Now imagine replacing the two detectors in Figure 34.27a by two narrow slits of the same size as the detectors and placing a screen some distance back from the two slits (Figure 34.27b). The pattern on the screen initially looks like random impacts from photons that make it through either one slit or the other. As we accumulate the impacts of many photons, however, an interference pattern emerges (Figure 34.27c) that is identical to the one obtained by shining an intense beam of light on the two slits.

Notice that replacing the two detectors with two slits doesn't change anything about the beam or the photons contained in it. However, the experiment using the two detectors suggests that photons are particles detected by one or the other of the detectors; the experiment using the two slits indicates that each photon is a wave traveling through both slits simultaneously! So which is it? Do the photons travel through one slit only or through both? If we physically cover one of the slits (and therefore *force* the photons to go through the other slit), the interference pattern disappears. The pattern that emerges after accumulating many photons behind the slit corresponds to the diffraction pattern of light behind a single slit.

What this means is that photons behave as discrete particles when they are being detected. In transit, however, the wave nature of photons dictates their behavior. In other words, light behaves *both* like a wave and like a particle. It is impossible to explain the results of the above set of experiments by treating light as only a wave or only a particle; it must have qualities of both.

34.14 Figure 34.8 shows the interference pattern obtained by shining a strong laser beam on a pair of slits. If instead a very weak beam is shone on the same slits, so that the photons pass through the slits one photon at a time, at what angles is the probability of observing a particular photon the greatest?

Self-quiz

1. At point A in **Figure 34.28**, do the waves from the two slits add or cancel?

   Figure 34.28

   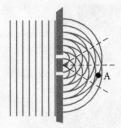

2. If the two sets of fringes shown in **Figure 34.29** were produced by the same diffraction grating, which set is the product of the longer-wavelength radiation?

   Figure 34.29

   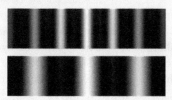

3. Coherent light of wavelength λ is normally incident on two slits separated by a distance d. What is the greatest possible fringe order?

4. Consider a proton and an electron moving at the same speed. Which has the longer wavelength?

5. Given the relationship between the energy E of a photon and its frequency f and the de Broglie expression relating momentum $p = mv$ and wavelength λ, determine the ratio E/p for a photon.

Answers:

1. At A, the crest of one wave overlaps the trough of the other wave, which means the waves cancel.
2. Because the wavelength is proportional to the sine of the angle the rays make with the normal to the diffraction grating, the fringes with the greater spacing were produced by the longer-wavelength radiation.
3. The fringe order is given by $d \sin \theta = \pm m\lambda$. Because the maximum value of $\sin \theta$ is 1, $d = m_{max}\lambda$ and so the maximum fringe order is $m_{max} = d/\lambda$. (Because m_{max} is an integer, you must truncate the value you obtain by dividing d by λ. For example, if $d/\lambda = 2.8$, then the greatest fringe order is 2.)
4. Because a proton has greater mass than an electron and because the de Broglie wavelength is inversely proportional to mass, $\lambda = h/mv$, the proton has a shorter wavelength than the electron.
5. The relationship between the energy of a photon and its frequency is $E = hf$ (see Section 34.5). The de Broglie wavelength is given by $\lambda = h/mv = h/p$, so $p = h/\lambda$. Therefore $E/p = hf/(h/\lambda) = f\lambda = c$, where c is the speed of light.

34.6 Multiple-slit interference

Let us now calculate the interference pattern produced by an electromagnetic wave normally incident on a barrier pierced by multiple closely spaced, very narrow slits. The width of each slit is much less than the wavelength of the radiation, so each slit serves as a point source of radiation. We begin by determining the pattern created when the barrier has just two very narrow slits.

As discussed earlier, the point sources corresponding to the two slits are in phase. Therefore the electric fields of the two waves that reach the screen differ in phase only due to any difference in the distance each wave travels from its slit to the screen. We shall refer to this difference as the *path-length difference* δs (**Figure 34.30a**). Taking the screen to be at position $x = 0$, we can write the electric fields of the waves traveling in the directions shown in Figure 34.30a as

$$E_1 = E_0 \sin \omega t \tag{34.1}$$

and
$$E_2 = E_0 \sin (\omega t + \phi), \tag{34.2}$$

where E_0 is the amplitude of the electric field and ϕ is a phase constant that is equal to the phase difference that results from the path-length difference between the two waves. The phase difference divided by 2π is the fraction of a cycle by which the two waves differ. This equals the path-length difference δs divided by the wavelength, or $\phi/2\pi = \delta s/\lambda$, and so

$$\phi = \frac{2\pi}{\lambda} d \sin \theta, \tag{34.3}$$

where d is the distance between the slits and θ is the angular position where the two rays meet on the screen.

We observe a bright fringe—a maximum in the intensity—when the two rays interfere constructively. This is the case when the phase difference ϕ equals an integer number times 2π:

$$\phi_m = \pm m(2\pi), \quad \text{for } m = 0, 1, 2, 3, \ldots. \tag{34.4}$$

Figure 34.30 (*a*) Interference of light diffracted by two very narrow slits. (*b*) We sum the phasors associated with the electric fields of the coherent light sources at slits 1 and 2.

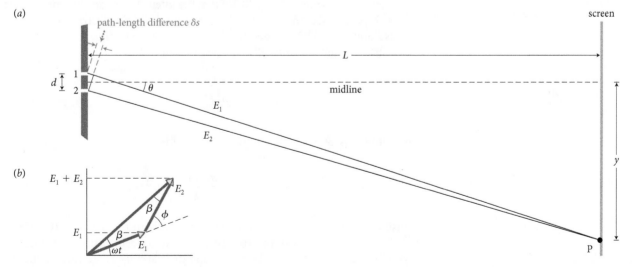

Combining Eqs. 34.3 and 34.4 and solving for $\sin \theta$ determine the angles θ_m for which bright fringes occur:

$$\sin \theta_m = \pm \frac{m\lambda}{d}, \quad \text{for } m = 0, 1, 2, 3, \ldots$$

$$\text{(bright interference fringes)}, \tag{34.5}$$

as we found in Section 34.2.

A dark fringe—a minimum in the intensity—occurs when the two rays interfere destructively. This is the case when the phase difference equals a odd number times π:

$$\phi_n = \pm (2n - 1)\pi, \quad \text{for } n = 1, 2, 3, \ldots \tag{34.6}$$

Substituting ϕ into Eq. 34.3 and solving for $\sin \theta$ determine the angles θ_n for which dark fringes occur:

$$\sin \theta_n = \pm \frac{(n - \frac{1}{2})}{d} \lambda, \quad \text{for } n = 1, 2, 3, \ldots$$

$$\text{(dark interference fringes)}. \tag{34.7}$$

To calculate the light intensity as a function of θ, we start by determining the sum of the electric fields E_1 and E_2 (Eqs. 34.1 and 34.2) using phasors. The amplitude of the sum of two sinusoidal quantities is equal to the vector sum of the phasors representing those two quantities (see Section 32.6). The phasors that represent E_1 and E_2 and their sum phasor are shown in Figure 34.30b. For the isosceles triangle formed by the E_1 and E_2 phasors, the exterior angle ϕ is equal to the sum of the two opposite interior angles β, and so $\beta = \frac{1}{2}\phi$. From this, we calculate the amplitude E_{12} of the sum of the two electric fields:

$$E_{12} = 2(E_0 \cos \beta) = 2(E_0 \cos \tfrac{1}{2}\phi). \tag{34.8}$$

Because this combined electric field oscillates at the same frequency as the incident wave, the time-dependent electric field at the screen is $E = E_{12} \sin \omega t$. At the central bright fringe, $\theta = 0$ and the two beams are in phase, so $\phi = 0$. The amplitude of E_{12} at the central bright fringe is therefore twice the amplitude of the incident wave, as we would expect.

We found in Chapter 30 (Eq. 30.36) that the intensity S of an electromagnetic wave is proportional to the product of the magnitudes of the electric and magnetic fields E and B: $S = EB/\mu_0$. Because B is proportional to E (Eq. 30.24, $E = Bc_0$), the intensity of light is commonly written in terms of E^2:

$$S = \frac{1}{\mu_0} EB = \frac{E(E/c_0)}{\mu_0}. \tag{34.9}$$

Substituting Eqs. 34.8 and 34.3 into Eq. 34.9 yields

$$S = \frac{4E_0^2}{\mu_0 c_0} \cos^2\left(\frac{\pi d \sin \theta}{\lambda}\right) \sin^2 \omega t. \tag{34.10}$$

Visible electromagnetic waves oscillate at such high frequencies (10^{14} Hz to 10^{15} Hz) that we ordinarily measure the time-averaged intensity. The time average of $\sin^2 \omega t$ is $\frac{1}{2}$. We can then write the time-averaged intensity of the interference pattern in terms of the time-averaged intensity of the incident wave:

$S_{0,\text{av}} = E_0^2/(2\mu_0 c_0)$. Thus substituting $\frac{1}{2}$ for $\sin^2 \omega t$ and $S_{0,\text{av}}(2\mu_0 c_0)$ for E_0^2 in Eq. 34.10 gives us

$$S_{\text{av}} = 4S_{0,\text{av}} \cos^2\!\left(\frac{\pi d \sin \theta}{\lambda}\right). \tag{34.11}$$

The maximum intensity of the interference pattern is *not* just the sum of the intensities of the two interfering waves—it is twice the sum! Because intensity is proportional to the square of the electric field, we must add the electric fields and then calculate the intensity from the square of the combined electric field.

✋ **34.15** How can the energy of the closed system made up of the two interfering waves remain constant (as the energy law states it must) if the maximum time-averaged intensity is four times the individual time-averaged intensities of the two waves?

Now let us work out how this pattern looks on the screen. If we consider only small angles θ, we can approximate $\sin \theta \approx \tan \theta = y/L$, where y is the position on the screen corresponding to the angle θ measured from the midline between the slits and L is the distance between the screen and the barrier that contains the slits, as shown in Figure 34.30*a*. (Positive y corresponds to positions above the midline; negative y to positions below the midline.) We can then write the time-averaged intensity as

$$S_{\text{av}} = 4S_{0,\text{av}} \cos^2 (\phi/2) \approx 4S_{0,\text{av}} \cos^2\!\left(\pi d \frac{y}{L\lambda}\right). \tag{34.12}$$

The intensity varies periodically with the phase difference ϕ. Near the central maximum, where θ is small, the intensity also varies periodically with y (**Figure 34.31**).

What is the distance D between adjacent intensity maxima of this pattern? Substituting y/L for $\sin \theta$ in Eq. 34.5, we obtain, for the positions of the maxima corresponding to any two values m and $m + 1$ of our order integer m,

$$m\lambda = d\frac{y_m}{L} \quad \text{and} \quad (m + 1)\lambda = d\frac{y_{m+1}}{L}. \tag{34.13}$$

Subtracting the first of these equation from the second yields

$$\lambda = \frac{d}{L}(y_{m+1} - y_m). \tag{34.14}$$

The distance between adjacent maxima is $y_{m+1} - y_m = D$, so

$$D = \frac{L}{d}\lambda. \tag{34.15}$$

This expression tells us that the distance between the maxima is proportional to the wavelength of the light. The interference pattern "magnifies" the wavelength by the factor L/d.

If there are more than two slits in the barrier, the analysis proceeds in the same fashion as before. **Figure 34.32** on the next page shows a wave diffracting through six slits, each separated from its immediate neighbors by distance d. The condition for a maximum in the intensity pattern is the same regardless of the number of slits because if the difference in path length for waves from slits 1 and 2 is λ, then the difference in path length for any pair of adjacent slits is also λ. The principal maxima therefore appear at those locations where

$$d \sin \theta_m = \pm m\lambda, \quad \text{for } m = 0, 1, 2, 3, \ldots$$
$$\text{(principal maxima)}, \tag{34.16}$$

just as we found for a two-slit barrier (Eq. 34.5).

The principal maxima occur at the same angles regardless of the number of slits. However, as discussed in Checkpoint 34.5, the *intensity* at the maxima increases with the number of slits. As we found in Section 34.2, as the number of

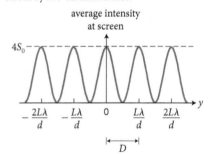

Figure 34.31 Average intensity produced on a screen by two-slit interference.

average intensity
at screen

$4S_0$

$-\dfrac{2L\lambda}{d}$ $-\dfrac{L\lambda}{d}$ 0 $\dfrac{L\lambda}{d}$ $\dfrac{2L\lambda}{d}$ y

D

Figure 34.32 Interference of light diffracted by six narrow slits.

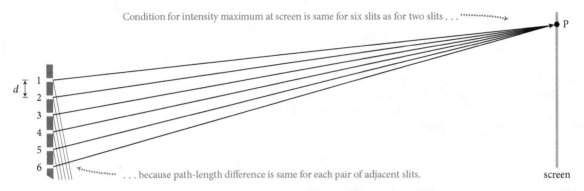

Condition for intensity maximum at screen is same for six slits as for two slits . . .

. . . because path-length difference is same for each pair of adjacent slits.

slits increases, the interference fringes also become narrower (**Figure 34.33**). This is so because the minima closest to the mth principal maximum occur for the ratio k/N that is as close as possible to m—namely, $(mN \pm 1)/N$. As the number of slits N becomes very great, the minima lie very close to m and so the interference pattern has extremely sharp maxima separated by broad dark regions with very faint secondary maxima.

Figure 34.33 Interference pattern produced by gratings with two, three, and eight slits.

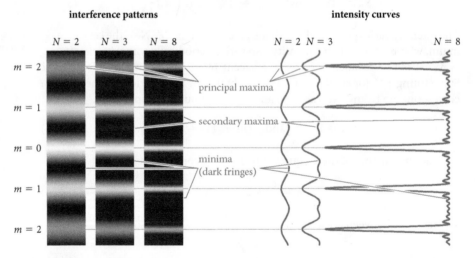

interference patterns

intensity curves

As discussed in Section 34.2, an important use of diffraction gratings is to disperse (separate) light into its constituent wavelengths in order to better understand the light source. The amount of information that can be obtained from a spectrum depends on whether the wavelengths of interest can be distinguished from one another in the spectrum (such wavelengths are said to be *resolved*).

Two wavelengths can be distinguished from each other if the principal maximum of one falls in the first dark region of the other, as shown in **Figure 34.34**. We found in Section 34.2 that, in general, a minimum occurs for interference through N slits when

$$d \sin \theta_{\text{min}} = \pm \frac{k}{N} \lambda, \quad \text{for integer } k \text{ that is not an integer multiple of } N$$

$$\text{(dark interference fringes).} \quad (34.17)$$

The angular position of the principal maxima increases with wavelength. Therefore, two wavelengths λ_1 and λ_2 can be distinguished from each other if the principal maximum for the longer wavelength falls at an angle greater than or equal to the angle for the $n = 1$ minimum for the shorter wavelength. As we

Figure 34.34 Two clearly separated bright fringes for light of different wavelength.

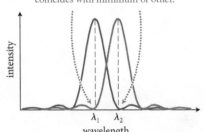

Principal maximum of each curve coincides with minimum of other.

intensity

λ_1 λ_2

wavelength

first explored in Checkpoint 34.7, the separation between the slits is critical in determining the smallest wavelength difference that can be distinguished (this wavelength difference is called the *resolution*).

Example 34.6 Resolution of wavelengths in a diffraction grating

An astronomer wishes to determine the relative heights of the intensity peaks for the bright fringes produced by two wavelengths of radiation emitted by sodium atoms. The wavelengths are 589.0 nm and 589.6 nm, and she uses a diffraction grating with 500.0 slits/mm to disperse the light collected by her telescope. (*a*) In which order are the intensity maxima for these two wavelengths farthest apart from each other: $m = 0$, $m = 1$, or $m = 2$? (*b*) If the part of the diffraction grating covered by the light is 4.000 mm wide, are the second-order principal maxima produced by these two spectral lines distinguishable from each other?

❶ GETTING STARTED This problem is about the overlapping diffraction patterns produced by two very similar wavelengths of light as the light passes through a diffraction grating. To answer the questions, I need to calculate the angular positions of the principal intensity maxima for each wavelength in three orders of diffraction. I also need to determine the angular position of the minimum adjacent to the second-order principal maximum for each wavelength, in order to determine whether the second-order principal maxima of the two wavelengths are distinguishable.

❷ DEVISE PLAN For part *a*, I can use Eq. 34.16 to locate the $m = 0$, 1, and 2 principal maxima for both wavelengths. For part *b*, I can locate the minimum adjacent to the second-order principal maximum for $\lambda = 589.0$ nm by applying the discussion following Eq. 34.17 and then compare that location with the location of the second-order principal maximum for $\lambda = 589.6$ nm.

❸ EXECUTE PLAN (*a*) I start by solving Eq. 34.16 for θ:

$$\theta_m = \pm \sin^{-1}\left(\frac{m\lambda}{d}\right). \tag{1}$$

This result indicates that the principal maxima are farthest apart for $m = 2$.

To calculate the positions of the principal maxima, I substitute the appropriate values of m, λ, and d in Eq. 1. The separation d between the slits is $1/(500 \text{ slits/mm}) = 2.000 \times 10^{-3}$ mm $= 2000$ nm. The $m = 0$ principal maximum occurs at $\theta = 0$ for any wavelength. The first-order principal maxima are located at

$$\theta_{589.0} = \pm \sin^{-1}\left(\frac{589.0 \text{ nm}}{2000 \text{ nm}}\right) = \pm 17.13°$$

$$\theta_{589.6} = \pm \sin^{-1}\left(\frac{589.6 \text{ nm}}{2000 \text{ nm}}\right) = \pm 17.15°,$$

and the second-order principal maxima are located at

$$\theta_{589.0} = \pm \sin^{-1}\left(\frac{2 \times 589.0 \text{ nm}}{2000 \text{ nm}}\right) = \pm 36.09°$$

$$\theta_{589.6} = \pm \sin^{-1}\left(\frac{2 \times 589.6 \text{ nm}}{2000 \text{ nm}}\right) = \pm 36.13°.$$

For these two wavelengths, the second-order principal maxima are the ones farthest apart from each other. ✔

(*b*) Using the information given in the discussion following Eq. 34.17, I can write that the condition for the minimum adjacent to the second-order principal maximum for $\lambda = 589.0$ nm is

$$d \sin \theta_{\min, 589.0} = \pm \lambda \frac{mN + 1}{N},$$

where I have replaced k in Eq. 34.17 by $mN + 1$ as explained in the discussion preceding that equation. Solving this expression for $\theta_{\min, 589.0}$ gives

$$\theta_{\min, 589.0} = \pm \sin^{-1}\left(\frac{\lambda}{d} \frac{mN + 1}{N}\right).$$

A region of the grating 4.000 mm wide contains 2000 slits. Substituting in the preceding expression, I get that, for $\lambda = 589.0$ nm, the minimum adjacent to the second-order principal maximum is at

$$\theta_{\min, 589.0} = \pm \sin^{-1}\left(\frac{589.0 \text{ nm}}{2000 \text{ nm}} \times \frac{4001}{2000}\right) = \pm 36.10°. ✔$$

This minimum lies between 36.09°, the second-order principal maximum for $\lambda = 589.0$ nm, and 36.13°, the second-order principal maximum for $\lambda = 589.6$ nm, telling me that the second-order principal maxima for these two wavelengths are distinguishable from each other.

❹ EVALUATE RESULT Equation 34.16 tells me that the angles at which principal maxima occur are equal to the inverse sine of integer multiples of λ/d. For small angles $\sin \theta_m \approx \theta_m$, and so the angles θ_m are approximately equally spaced (**Figure 34.35**). As the curve for $\sin \theta$ bends toward the horizontal, however, the distance between adjacent θ_m values increases, in agreement with the result I obtained.

Figure 34.35

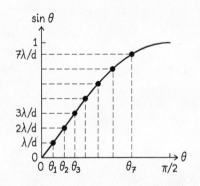

🖐 **34.16** As the above discussion indicates, the separation distance between the principal maxima for different wavelengths increases as the fringe order m increases. Can you obtain an arbitrarily great separation distance by going to extremely high orders?

Figure 34.36 Soap bubble.

34.7 Thin-film interference

A familiar manifestation of the interference of light is the rainbow of colors reflected from thin films such as soap bubbles (**Figure 34.36**). This type of interference occurs in transparent materials whose thickness is comparable to the wavelengths of visible light. When such a thin material (we refer to it as a *film*) is either suspended in air, as in a soap bubble, or supported on a much thicker material with a different index of refraction, as in an oil slick on a puddle of water, white light reflecting from both the front and back surfaces of the film interferes, in the same manner as x rays reflecting from adjacent layers of atoms in a crystal. This is shown schematically in **Figure 34.37**. The film thickness t and index of refraction n_b, together with the angle of incidence, determine the path-length difference and corresponding phase difference between the reflected beams, and therefore determine which colors undergo constructive interference and which undergo destructive interference in any given direction.

To identify the conditions for constructive and destructive interference, we begin by expressing the electric fields of the waves reflected from the two surfaces, just as we did for waves passing through two slits in the preceding section:

$$E_1 = E_0 \sin(\omega t) \tag{34.18}$$

$$E_2 = E_0 \sin(\omega t + \phi). \tag{34.19}$$

The phase difference ϕ is due to the path-length difference $2\Delta s$ and the effect of the reflections on the phases of each wave. The path-length difference gives rise to a phase difference

$$\phi_{\text{path}} = \frac{2\pi \Delta s}{\lambda_b} = \frac{2\pi(2t \cos\theta_b)}{\lambda/n_b} = \frac{4\pi n_b t \cos\theta_b}{\lambda}, \tag{34.20}$$

where the path-length difference Δs and the angle θ_b of the ray relative to the surface normal inside the film are shown in Figure 34.37, and $\lambda_b = \lambda/n_b$ is the wavelength of light inside the film (see Eq. 33.3). Because of refraction, expressing Δs in terms of the angle of incidence θ, instead of the angle in the film θ_b, involves Snel's law and produces a rather complicated result. However, for normal incidence $\cos\theta_b = 1$, and so the expression is greatly simplified.

To determine the phase difference due to the reflections from the two film surfaces, recall the discussion of mechanical waves at boundaries from Section 16.4. We saw there that when a wave pulse is launched in a first medium and travels into a second medium, the pulse is partially reflected and partially transmitted at the boundary. The reflected pulse is inverted relative to the incident pulse if the wave speed c_1 in the first medium is greater than the speed c_2 in the second medium. When $c_1 < c_2$, the incident pulse is not inverted upon reflection.

Let us now extend these ideas to sinusoidal electromagnetic waves. When the incident wave in medium 1 reflects at the boundary with medium 2 of greater index of refraction ($n_1 < n_2$), the speed is less in medium 2 ($c_2 < c_1$), and so the wave is inverted upon reflection. This inverting of the wave is equivalent to the wave undergoing a phase shift of π upon reflection. If instead $n_1 > n_2$, the wave is not inverted and there is no phase shift upon reflection.

A phase shift of π can occur at each of the two interfaces, depending on the refractive indices n_a, n_b, and n_c of the three media, as indicated in Figure 34.37. Thus the phase difference associated just with the reflections for the two reflected waves is

$$\phi_r = \phi_{r2} - \phi_{r1}, \tag{34.21}$$

where ϕ_{r1} and ϕ_{r2} are either π or 0, depending on the values of the indices of refraction. Therefore the total phase difference between the two reflected waves is

$$\phi = \frac{4\pi n_b t \cos\theta_b}{\lambda} + \phi_{r2} - \phi_{r1}. \tag{34.22}$$

Figure 34.37 Thin-film interference.

Wave is inverted at this interface if $n_b > n_a$.

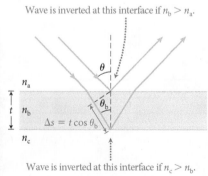

Wave is inverted at this interface if $n_c > n_b$.

Because ϕ_{r1} and ϕ_{r2} must each be π or 0, the effect of the two reflections on the phase cancels out when $\phi_{r1} = \phi_{r2}$, and the phase difference is due entirely to ϕ_{path}. If ϕ_{r1} is not equal to ϕ_{r2}, then ϕ differs from ϕ_{path} by π, causing the constructive and destructive interference conditions to be switched from what they would be due to just ϕ_{path}.

In our discussion of x-ray diffraction, the light source consisted of monochromatic x rays, and we found that constructive interference between adjacent planes takes place at only certain angles. In the current context—light reflections from thin films—we usually deal with light across the visible spectrum. Therefore ϕ depends on both wavelength and angle of incidence. For normal incidence, Eq. 34.22 simplifies to

$$\phi = \frac{4\pi n_b t}{\lambda} + \phi_{r2} - \phi_{r1} \quad \text{(normal incidence).} \qquad (34.23)$$

Then as before, constructive interference occurs when the phase difference corresponds to an integer number times 2π, and destructive when it corresponds to an odd number times π.

Example 34.7 Antireflective coating

Eyeglass lenses made of crown glass ($n = 1.52$) are given a thin coating of magnesium fluoride ($n = 1.38$) to minimize reflection of light from the lens surface. What is the minimum coating thickness for which reflection from the lens surface is minimized?

❶ GETTING STARTED Reflection is minimized at all values of the coating thickness that cause the waves reflected from the front coating surface to interfere destructively with those reflected from the back coating surface. My task therefore is to obtain the minimum thickness needed for destructive interference.

❷ DEVISE PLAN I see from Eq. 34.22 that the condition for destructive interference depends not only on the coating thickness t and index of refraction n_b but also on two variables I have no values for: the angle of incidence and the wavelength of the light. Thus I must make some simplifying assumptions to solve this problem. As in the text discussion, I shall consider only light normally incident on the coated lens and shall assume the lens and coating surfaces are flat, as shown in my sketch (**Figure 34.38**). I choose a representative visible wavelength, 500 nm—in the middle of the visible spectral range.

Figure 34.38

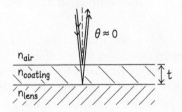

To obtain the condition for destructive interference, I can use Eq. 34.23 to determine the phase difference ϕ at normal incidence and equate it to an odd number times π radians. To work out the phase shifts ϕ_{r1} and ϕ_{r2} that occur at each reflection, I

note that at the air-coating interface the light reflects from a medium for which the index of refraction is greater than that of the medium through which the incident wave travels, and so this wave undergoes a π phase shift upon reflection: $\phi_{r1} = \pi$. The wave that reflects from the coating-lens interface likewise reflects from a medium for which the index of refraction is greater than that of the medium through which the incident wave travels and thus also undergoes a π phase shift ($\phi_{r2} = \pi$). Consequently, ϕ_{r1} and ϕ_{r2} in Eq. 34.23 cancel, leaving only the phase difference due to the path-length difference. The minimum thickness produces a phase difference corresponding to the smallest number of cycles that gives destructive interference—namely, half a cycle.

❸ EXECUTE PLAN To determine the thickness that produces destructive interference for 500 nm light, I express the phase difference using Eq. 34.23 and equate it to a half-cycle phase difference:

$$\frac{4\pi n_b t}{\lambda} + \phi_{r2} - \phi_{r2} = \pi.$$

Substituting $\phi_{r1} = \phi_{r2} = \pi$, solving for the thickness t, and substituting values give

$$t = \frac{\lambda}{4n_b} = \frac{500 \text{ nm}}{4(1.38)} = 90.6 \text{ nm.} ✔$$

❹ EVALUATE RESULT The coating is thinner than the wavelength of the light traveling through the coating. That makes sense because, to create destructive interference, the wave that travels through the coating and reflects from the coating-lens interface must travel only half a wavelength farther than the wave that reflects from the air-coating interface.

Figure 34.39 Diffraction through a narrow slit.

(*a*) Wavefront in slit treated as a series of point sources

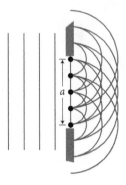

(*b*) Condition for first-order dark fringe: path lengths differ by $(\frac{1}{2}a)\sin\theta_1 = \lambda/2$

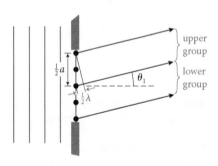

(*c*) Condition for second-order dark fringe: path lengths differ by $(\frac{1}{4}a)\sin\theta_2 = \lambda/2$

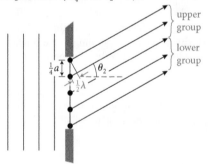

Figure 34.40 Intensity plot for a single-slit diffraction pattern.

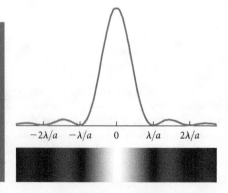

34.17 When oil spreads on water, bands of different colors are visible. What causes the different colors?

34.8 Diffraction at a single-slit barrier

Let us now describe quantitatively the diffraction by a single slit. Huygens' principle (see Section 17.4) allows us to describe the wavefront that reaches the opening as a series of point sources that emit spherical wavelets (**Figure 34.39a**). The wave beyond the slit is the superposition of all these spherical wavelets. In the original direction of propagation, all of the waves add in phase, and as a result the amplitude of the transmitted wave is the maximum possible: the sum of all the individual wave amplitudes. If we divide the rays representing the wavelets into two equal groups, as shown in Figure 34.39b, we can pair each ray in the upper half with a corresponding ray in the lower half. In Figure 34.39b, for example, we can pair the top ray in the upper group with the top ray in the bottom group. As we discussed in Section 34.2, these rays are essentially parallel when they emerge from the slits; as they travel and intersect at some location on a screen placed far to the right of the slits, the rays must travel different distances to reach that location. For rays traveling at an angle θ to the original propagation direction, the difference in path length from two such corresponding rays is $(a/2)\sin\theta$, where a is the width of the aperture. When the path lengths differ by half a wavelength, $(a/2)\sin\theta_1 = \lambda/2$, the rays interfere destructively, which means that the direction for the first-order dark fringe in a single-slit diffraction pattern is given by

$$\sin\theta_1 = \frac{\lambda}{a} \quad \text{(first-order dark diffraction fringe).} \quad (34.24)$$

Dividing the rays into four groups (Figure 34.39c) leads to a dark fringe (that is, a minimum in transmitted intensity) when $(a/4)\sin\theta = \frac{1}{2}\lambda$. Thus the direction for the second-order dark fringe is given by

$$\sin\theta_2 = 2\frac{\lambda}{a} \quad \text{(second-order dark diffraction fringe).} \quad (34.25)$$

The general condition for a dark fringe is thus

$$\sin\theta_n = \pm n\frac{\lambda}{a}, \quad n = 1, 2, 3, \ldots$$
$$\text{(dark diffraction fringes).} \quad (34.26)$$

Positive values of $\sin\theta_n$ correspond to dark fringes above the midline, while negative values correspond to dark fringes below the midline.

Calculation of the detailed intensity pattern is beyond the scope of this text, but we can examine some of the details of a representative pattern (**Figure 34.40**). In this single-slit diffraction pattern, the intensity of the first-order ($m = 1$) bright fringe is less than 5% of the intensity of the central ($m = 0$) bright fringe; the intensity of the second-order ($m = 2$) bright fringe is less than 2% of the central bright fringe intensity. In other words, most of the transmitted energy falls within the central peak. In general, in a single-slit diffraction pattern, the intensity is greatest at the central bright fringe and decreases rapidly with distance from the center of the pattern.

What if we want to calculate the linear positions of the dark diffraction fringes on a screen located a distance L away from a barrier containing a single slit (**Figure 34.41**) rather than the angular positions? For dark fringes located at small angles θ_n from the original direction of wave propagation, we can approximate $\tan \theta_n$ as $\sin \theta_n$:

$$y_n = L \tan \theta_n \approx L \sin \theta_n, \tag{34.27}$$

and so, from Eq. 34.26,

$$y_n = \pm n \frac{\lambda L}{a} \quad \text{(dark diffraction fringes).} \tag{34.28}$$

Figure 34.41 Calculating the positions of the dark fringes of a single-slit diffraction pattern.

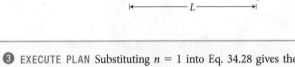

Example 34.8 Spreading light

Consider the diffraction pattern shown actual size in Figure 34.40. If the pattern was formed by light from a 623-nm (red) laser passing through a single narrow slit and the screen on which the pattern was cast was 1.0 m away from the slit, what is the slit width?

① GETTING STARTED This problem asks me to relate the diffraction pattern in Figure 34.40, which was produced by a setup such as the one shown in Figure 34.41, to the width of the slit that produced it. Thus I need to relate the slit width to fringes whose position I can calculate from the parameters of the problem and can also measure on the image. Because the only variable I know how to calculate is the positions of minima in the diffraction pattern, I can measure the distance between the two first-order minima and relate that distance to the slit width and the geometry of the setup.

② DEVISE PLAN The positions of the two first-order minima, in terms of wavelength λ, slit-to-screen distance L, and slit width a, are given by Eq. 34.28 with $n = 1$. Subtracting the two values I obtain from each other gives me an expression for the distance between these two minima in terms of λ, L, and a. I am given the values of λ and L, and my task is to determine a. Thus if I know the distance between the n_1 minima, I can calculate a. Because the image in Figure 34.40 is actual size, I can measure this distance directly. Then I can solve Eq. 34.28 for a and insert my known values.

③ EXECUTE PLAN Substituting $n = 1$ into Eq. 34.28 gives the linear positions of the two first-order minima:

$$y_1 = \pm \frac{\lambda L}{a},$$

so the distance w between the two minima is

$$w = \frac{2\lambda L}{a}. \tag{1}$$

I measure the distance between the centers of the two dark fringes on either side of the central bright fringe in Figure 34.40 to be 23 mm. Solving Eq. 1 for the slit width thus gives me

$$a = \frac{2\lambda L}{w} = \frac{2(623 \times 10^{-9} \text{ m})(1.0 \text{ m})}{23 \times 10^{-3} \text{ m}}$$

$$= 5.4 \times 10^{-5} \text{ m} = 0.054 \text{ mm} \quad ✔$$

④ EVALUATE RESULT The slit width is about a factor of 100 greater than the wavelength of the light, and that ratio is consistent with the general range of slit sizes that produce noticeable diffraction of visible light.

As long as λ is small relative to the slit width a, so that the small-angle approximation for θ_n is valid, most of the diffracted light intensity falls within this region defined by Eq. 1 in Example 34.8. Note that w increases with decreasing a: The narrower the slit, the more the wave spreads out after passing through the slit (**Figure 34.42**). If the slit width is equal to or less than the wavelength of the light, there are no dark fringes, as we found in Checkpoint 34.16. The wave simply spreads out in all directions behind the slit, which means the slit behaves as a point source.

✋ **34.18** Using Eq. 34.24, calculate the angle at which the first dark fringe occurs when (a) $a < \lambda$ and (b) $a \gg \lambda$. Interpret your results.

34.9 Circular apertures and limits of resolution

When light passes through a circular aperture, the symmetry of the aperture causes the resulting diffraction pattern also to be circular (**Figure 34.43** on the next page). A circular central bright fringe is surrounded by circular dark diffraction fringes and additional diffraction bright fringes. The central bright

Figure 34.42 The width of the central maximum in a diffraction pattern decreases as the slit is widened.

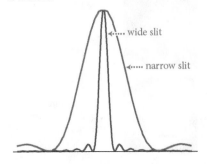

Figure 34.43 Diffraction pattern of light passing through a circular aperture.

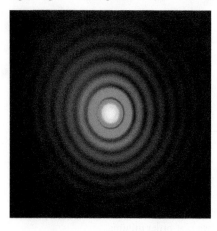

fringe is called the *Airy disk*, after the British astronomer and mathematician Sir George Airy, who developed the first detailed description of diffraction in 1835. Calculation of the circular diffraction pattern is rather involved, so all I shall do here is state the location of the first dark fringe. When light is diffracted by a circular aperture of diameter d, the first dark fringe occurs at angle θ_1 given by

$$\sin \theta_1 = 1.22 \frac{\lambda}{d}. \tag{34.29}$$

The result is similar to the one obtained in Eq. 34.24 for a slit of width a, except that now the sine of the angle is increased by a factor of 1.22. The increase in the angle for the same wavelength and aperture size can qualitatively be understood as follows: Equation 34.24 is obtained by considering the interference between wavelets coming from a slit whose width is a everywhere. A circular aperture has width d only across a diameter; the rest of the aperture is narrower. As the aperture gets narrower, diffraction through it becomes more pronounced. The factor of 1.22 quantitatively accounts for the varying horizontal width of the circular aperture.

The angular size of the Airy disk given in Eq. 34.29 determines the minimum angular separation of two point sources that can be distinguished by observing them with a (circular) lens—*regardless* of the magnification of the lens! To understand what this means, imagine imaging two distant, closely spaced point sources with a lens. These sources could be anything, from stars to organelles in a biological cell. The sources are not coherent, and so we can consider the Airy disks formed by the light from each source separately without considering interference between sources.

If there is overlap in the Airy disks of the images observed through the lens, it is difficult to tell whether there are two point sources or just one. Two objects being observed through a lens are just barely distinguishable when the center of one diffraction pattern is located at the first minimum of the other diffraction pattern. This happens when the angular separation between the two objects is at least the angle given in Eq. 34.29. If this is the case, we say that the two objects are *resolved*. This condition for distinguishability is called **Rayleigh's criterion.**

Because the diameter d of the lens is always much greater than the wavelength of the light, the angle in Eq. 34.29 is always small. Thus the minimum angular separation θ_r for which two sources can be resolved is approximately equal to the sine of the angle

$$\theta_r \approx \sin \theta_r = 1.22 \frac{\lambda}{d}. \tag{34.30}$$

Figure 34.44 The resolution of these two stars improves as the aperture is made larger.

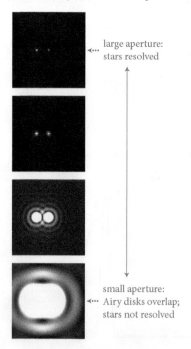

large aperture:
stars resolved

small aperture:
Airy disks overlap;
stars not resolved

Two objects that are separated by an angle equal to or greater than θ_r satisfy Rayleigh's criterion. For this reason, the closest two objects can be to each other and still be distinguished with an optical instrument such as a microscope or telescope depends not on the magnification but on the wavelength of the light and the size of the smallest aperture in the instrument.

Figure 34.44 shows the images of two stars obtained with a telescope. An aperture placed in front of the lens shows the effects of diffraction. When the opening of the aperture is small (bottom image in **Figure 34.44**), the images of the two stars are merged—the two stars cannot be resolved. As the aperture is opened, θ_r decreases and the two images separate cleanly.

Diffraction also determines the linear size of the images of point sources. In Chapter 33, we stated that the image of a point source formed on a screen by a lens is a point. **Figure 34.45a** shows parallel rays from a distant point source focused by a lens onto a screen placed in the focal plane of the lens. Without diffraction, the image formed by these rays would be an infinitesimally small point. In fact, because the aperture through which the light passes—the lens—has a

Figure 34.45 Analyzing the diffraction limit of a lens.

(a)

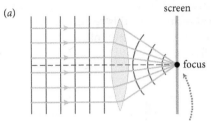

screen

focus

In absence of diffraction, lens would cause parallel rays to converge to point.

(b)

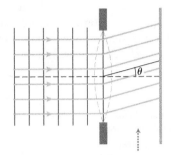

θ

But lens is an aperture, so light is also diffracted.

(c)

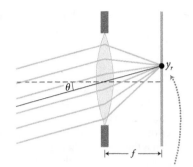

y_r

θ

f

Focusing of diffracted light. Angular size of Airy disk depends on both diffraction and focusing.

finite diameter, such rays do not focus to an infinitely small point, and the image formed is a diffraction pattern just like that shown in Figure 34.43. The angular size of the central bright fringe of this diffraction pattern—in other words, the Airy disk—is given by Eq. 34.29.

To calculate the radius of the Airy disk formed in the focal plane of the lens,* we must account for both diffraction and focusing. Figure 34.45b shows the diffraction of light through an aperture of the same diameter as the lens. Light that originally traveled parallel to the lens axis is diffracted by an angle θ. Now consider how the lens focuses the diffracted light. Parallel rays that make an angle θ with the lens axis are focused at a point located a distance y above the axis (Figure 34.45c, see also Figure 33.27). This distance y is given by $y = f \tan \theta$, where f is the focal length of the lens.

Our next step in determining the Airy disk radius is to calculate the distance y_r from the center of the disk to the first dark fringe in the diffraction pattern, which is found at the angle given by Eq. 34.29. In the small-angle limit, $\sin \theta \approx \tan \theta$ and so we can substitute $\sin \theta_r = y_r/f$ into Eq. 34.29, giving

$$y_r = 1.22 \frac{\lambda f}{d}. \qquad (34.31)$$

This expression gives the radius y_r of the Airy disk and the minimum size of the area to which light can be focused with light of wavelength λ by a lens of focal length f and diameter d. The best ratio of f/d that can be achieved with a lens is approximately unity, and so the smallest diameter to which light can be focused is about 2.5λ. This means that the smallest diameter of "points" in the resulting image is also 2.5λ.

This diffraction-determined minimum size of the features in an image is commonly called the *diffraction limit*. An industry in which the diffraction limit poses a serious problem is the manufacture of integrated circuits, such as computer chips. The transistors and logic gates described in Section 32.4 are produced by a series of processes known as *photolithography*, in which the semiconductor substrate is coated with a polymer that is sensitive to ultraviolet light. The polymer is then exposed to ultraviolet light in the pattern of the desired metal electrodes. This pattern of light is produced by illuminating a metal mask with holes in the shape of the electrodes and then imaging the resulting pattern of light onto the surface. Finally, the exposed polymer is dissolved with a chemical rinse, leaving the semiconductor surface exposed where metal is desired. The metal is then deposited in a subsequent step.

---

*The radius of the Airy disk is smallest when the screen is in the focal plane of the lens, and it increases as the screen is moved closer to or farther from the lens.

QUANTITATIVE TOOLS

The smallest electrode that can be made by this process is therefore determined by the diffraction limit for the ultraviolet light ($\lambda \leqslant 150$ nm) used to produce the pattern. With ordinary optical technology, this requires electrodes to be at least 300 nm wide. Many researchers are searching for other ways to produce electrodes that are not limited by diffraction.

Exercise 34.9 A point is not a point

A magnifying glass that has a focal length of 0.25 m and a diameter of 0.10 m is used to focus light of wavelength 623 nm. (*a*) What is the radius of the smallest Airy disk that can be produced by focusing light with this lens? (*b*) How large is the Airy disk formed when this lens focuses a laser beam that has a 2.0-mm diameter?

SOLUTION (*a*) The radius of the smallest Airy disk is given by Eq. 34.31 with d equal to the diameter of the lens:

$$y = 1.22\,\frac{(623 \times 10^{-9}\,\text{m})(0.25\,\text{m})}{0.10\,\text{m}}$$

$$= 1.9 \times 10^{-6}\,\text{m} = 1.9\,\mu\text{m}. ✔$$

The minimum spot size is about three times greater than the wavelength of the light.

(*b*) In this case the radius is given by Eq. 34.31 with d equal to the diameter of the laser beam. I must use the beam diameter because the beam does not make use of most of the area of the lens but effectively defines its own aperture. I therefore have

$$y = 1.22\,\frac{(623 \times 10^{-9}\,\text{m})(0.25\,\text{m})}{0.0020\,\text{m}}$$

$$= 9.5 \times 10^{-5}\,\text{m} = 95\,\mu\text{m}. ✔$$

This Airy disk is much bigger than the one found in part *a*. This Airy disk radius means the central bright fringe has a diameter of 190 μm = 0.19 mm, which is smaller than the original beam diameter due to the focusing of the lens. Because the laser beam is much smaller than the diameter of the lens, it cannot be effectively focused by this lens. The diffraction partially cancels the focusing.

Example 34.10 Blurry images

The widths of one pixel in the sensor for a digital camera is about 2.0 μm. If the camera lens has a diameter of 40 mm and a focal length of 30 mm, is the resolution of the resulting image limited by the lens or the sensor?

❶ GETTING STARTED This problem involves comparing, for the image formed by a digital camera, the resolution limit due to diffraction and the limit due to the size of the pixels in the sensor. The limit due to diffraction is the size of the image of a point source; the limit due to the sensor is the width of a single pixel. The greater limit determines the image resolution. The problem does not specify the wavelength of the light involved, but because the problem is concerned with forming images with visible light, I choose $\lambda = 500$ nm, near the center of the visible spectrum.

❷ DEVISE PLAN The diffraction-limited image of a point source is the Airy disk at the center of the diffraction pattern, so I can use Eq. 34.31 to obtain the radius of the Airy disk formed by the camera. I can then compare the diameter (not the radius) of that disk with the width of a pixel. Whichever is greater limits resolution possible for the image.

❸ EXECUTE PLAN Substituting the values given into Eq. 34.31, I obtain for the Airy disk radius

$$y = 1.22\,\frac{\lambda f}{d} = 1.22\,\frac{(0.500 \times 10^{-6}\,\text{m})(30 \times 10^{-3}\,\text{m})}{40 \times 10^{-3}\,\text{m}}$$

$$= 0.46 \times 10^{-6}\,\text{m} = 0.46\,\mu\text{m}.$$

The diameter of the Airy disk is thus $2y = 0.92\,\mu$m. This is significantly less than the pixel width, which means the resolution of the image is limited by pixel width rather than by diffraction. ✔

❹ EVALUATE RESULT My result for the radius of the Airy disk is reasonable because typically the diffraction-limited width of the image of a point source is comparable to the wavelength of light emitted by the source. (Note that if I had chosen $\lambda = 700$ nm, the longest visible wavelength, the Airy disk radius would increase only by a factor of $700/500 = 1.4$ and thus its diameter, 1.3 μm, would still not exceed the pixel width. If I had chosen a wavelength shorter than 500 nm, the Airy disk diameter would be less than my calculated value. So my conclusion is the same for any visible wavelength: The resolution is limited by the pixel width.).

34.19 Which of these three lenses offers (*a*) the highest resolution and (*b*) the lowest resolution: (*i*) $f = 10$ mm, $d = 8$ mm; (*ii*) $f = 15$ mm, $d = 10$ mm, (*iii*) $f = 20$ mm, $d = 18$ mm?

Figure 34.46 The photoelectric effect.

(a)

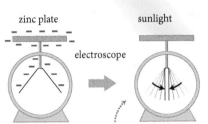

zinc plate

sunlight

electroscope

Sunlight discharges negatively charged zinc and electroscope.

(b)

Illuminating positively charged electroscope and ···▸ zinc plate has no effect.

(c)

glass (blocks UV)

Passing light through glass slide prevents discharging of negatively charged electroscope and zinc plate.

34.10 Photon energy and momentum

In Chapter 30 we described light as an electromagnetic wave that has a wavelength λ and a frequency f and moves in vacuum at speed c_0, such that

$$c_0 = \lambda f. \tag{34.32}$$

We also found that the energy density in the electromagnetic wave is proportional to the square of the amplitude of the electric field oscillation. In addition, the experiment described in Section 34.4 suggests that light has particle properties. If light always propagates at speed c_0 in vacuum, what determines its energy? And if light is a particle that has energy, shouldn't it also have momentum?

The answer to the first question is provided by the **photoelectric effect,** a surprising phenomenon that cannot be explained by thinking of light as a wave (**Figure 34.46**). Place a piece of metal, such as zinc, on an electroscope (see Section 22.3) that is negatively charged, as shown in Figure 34.46a. Some of the charge immediately moves to the zinc so that it, too, is negatively charged. If you then shine sunlight on the metal, the light discharges the zinc and the electroscope. If the electroscope is positively charged, however, as in Figure 34.46b, nothing happens when light shines on it. If we place a piece of glass in the beam of light (Figure 34.46c), nothing happens even if the zinc plate is negatively charged and the light is very intense.

What is going on? In the situations of Figure 34.46a and b, the light knocks electrons out of the zinc plate. When the plate is initially negatively charged, likecharge repulsion causes the ejected electrons to accelerate away from the plate. When the plate is initially positively charged, opposite-charge attraction causes the ejected electrons to be attracted back to the plate, so that the charge on the plate does not change. Ultraviolet radiation cannot pass through ordinary glass, and thus we conclude from the situation in Figure 34.46c that ultraviolet light is essential in order for electrons to be ejected.

The apparatus illustrated in **Figure 34.47** is used to study the photoelectric effect. It allows us to measure the energy of the ejected electrons while separately controlling either the wavelength or the intensity of the light. A zinc target T is placed in an evacuated quartz bulb (quartz is transparent to ultraviolet light), along with another metal electrode called the collector (C). A power supply is used to maintain a constant potential difference V_{CT} between the target and the collector. The current from the target to the collector is measured with an ammeter. If the target is kept at a negative potential relative to the collector, so that V_{CT} is negative, any electrons ejected from the target by the light are accelerated by the electric field and move to the collector. With negative V_{CT}, the current measured is proportional to the intensity of the light source, suggesting that ejecting each electron requires a certain amount of light energy.

If the potential difference V_{CT} is made slightly positive (so that the target is positive relative to the collector), there is a small current detected, but the electric field

Figure 34.47 Apparatus to study the photoelectric effect. The potential difference V_{CT} is positive when $V_T > V_C$.

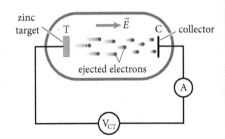

ultraviolet beam

zinc target

T

$\vec{E}$

C

collector

ejected electrons

A

V_{CT}

Figure 34.48 For the circuit in Figure 34.47, the current as a function of potential difference and the stopping potential as a function of the frequency of the incident light.

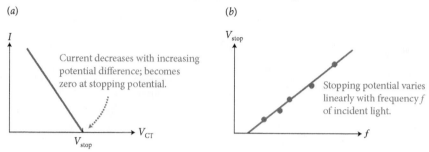

(a)

Current decreases with increasing potential difference; becomes zero at stopping potential.

(b)

Stopping potential varies linearly with frequency f of incident light.

between T and C now *slows down* any ejected electrons that initially move toward the collector. As V_{CT} increases, there is a certain value of V_{CT} at which the flow of electron stops completely, as shown in the graph of current I versus V_{CT} in **Figure 34.48a**. At this potential difference, called the *stopping potential difference,* the current is zero regardless of the intensity of the incident light. No matter how bright the light, there is no current between the target and the collector. This finding implies that the maximum kinetic energy with which the electrons leave the target does not depend on the intensity (and thus the incident power) of the light.

If we measure the kinetic energy of the ejected electrons (for $V_{CT} < V_{stop}$) directly, we discover that not all of them have the same kinetic energy. This happens because although the amount of energy absorbed from the photons is the same for all electrons, the energy required for the electron to make its way from its initial location in the target to the surface depends on the depth from which the electron is liberated. As a result, electrons released from the surface of the target have the maximum possible kinetic energy, which equals the amount of energy transferred to each electron by the light minus the energy required to liberate the electron from the metal.

For a given potential difference between the target and the collector, the electric field does work $-eV_{CT}$ on an electron as the electron moves from the target to the collector (see Eq. 25.17). The change in the electron's kinetic energy is thus

$$\Delta K = -eV_{CT}. \tag{34.33}$$

Given that the electrons just barely reach the collector at the stopping potential difference, we know that their final kinetic energy is zero, and so for these electrons $\Delta K = K_f - K_i = -K_i$. The maximum kinetic energy with which the electrons leave the target is thus

$$K_{max} = K_i = eV_{stop}. \tag{34.34}$$

Another clue to understanding the experiment in Figure 34.47 emerges when we change the frequency of the incident light and again measure the target-to-collector current as a function of V_{CT}. We observe that the stopping potential difference depends on the frequency; plotting this stopping potential difference as a function of the frequency of the light yields the results shown in Figure 34.48b.

34.20 (a) What does Figure 34.48b tell you about the relationship between the frequency of the incident light and the maximum kinetic energy of the ejected electrons? (b) What does the intercept of the line through the data points and the horizontal axis represent?

As Checkpoint 34.20, part *a* shows, the maximum kinetic energy of the ejected electrons depends on the frequency of the incident light, not its intensity. Electrons ejected by ultraviolet light, which has a higher frequency than visible light, have more kinetic energy than electrons ejected by visible light. (This is why putting glass in the beam of light in the experiment shown in Figure 34.46 essentially eliminates the effect. Although some electrons are ejected by the visible light, those electrons are liberated with less kinetic energy and are more likely to return to the target.) Furthermore, as you discovered in answering Checkpoint 34.20, part *b*, there is a certain minimum frequency of light below which electrons are not ejected at all, regardless of the intensity of the light.

Why does the photoelectric effect require us to think of light as a particle rather than as a wave? The stopping potential difference gives us the maximum kinetic energy with which electrons are released; the light must supply at least this much energy to the electrons in order to eject them. If light could be understood solely as a wave, the intensity of the wave, not its frequency, would determine the maximum amount of energy it could deliver to the electrons. Because the stopping potential difference depends not on light intensity but on frequency, we infer that light carries its energy in energy quanta and that the energy in each quantum is proportional to the frequency.

The photons described in Section 34.5 are these quanta. When an electron absorbs a photon, the electron acquires the photon's entire energy—the electron cannot absorb just part of a photon. The photon's energy frees the electron from the material and gives it additional kinetic energy. If we denote the minimum energy required to free the electron by E_0, we have

$$E_{photon} = hf = K_{max} + E_0, \tag{34.35}$$

where K_{max} is the maximum kinetic energy of the electron as it is ejected. The energy E_0, called the **work function** of the target metal, is a property of the metal that measures how tightly electrons are bound to the metal.

The value of Planck's constant h can be determined by using the relationship between V_{stop} and f given in Figure 34.48*b*. Substituting Eq. 34.34 into Eq. 34.35 and solving the result for V_{stop}, we get

$$V_{stop} = \frac{h}{e}f - \frac{E_0}{e}. \tag{34.36}$$

This result shows that V_{stop} depends linearly on f and that the slope of the line in Figure 34.48*b* is h/e. By measuring the slope in Figure 34.48*b* and dividing that slope by the charge e of the electron, one obtains $h = 6.626 \times 10^{-34}$ J·s, the value given in Section 34.4.

As discussed in Section 34.5 and expressed in Eq. 34.35, Planck's constant relates the energy and frequency of a photon, $E_{photon} = hf_{photon}$, and in Section 34.4 we learned the relationship between the momentum and the wavelength of an electron (or anything else that is ordinarily thought of as a particle): $\lambda_{electron} = h/p_{electron}$. Because of the wave-particle duality, we can apply this expression for wavelength to photons as well as electrons: $\lambda_{photon} = h/p_{photon}$. If we calculate the momentum of a photon from its wavelength using this expression and then substitute $\lambda = c_0/f$, we see that the momentum of a photon is proportional to its energy:

$$p_{photon} = \frac{h}{\lambda_{photon}} = \frac{hf_{photon}}{c_0} = \frac{E_{photon}}{c_0}. \tag{34.37}$$

If we substitute this result into the equation relating energy and momentum derived in Chapter 14 (Eq. 14.57)

$$E^2 - (c_0 p)^2 = (mc_0^2)^2, \tag{34.38}$$

the left side of this equation becomes zero, and so we see that photons have zero mass ($m_{photon} = 0$). We derived Eq. 34.38 for particles that have nonzero mass. Now we see that we can treat photons as massless "particles of light." While these particles have no mass, they do have both momentum and energy:

$$E_{photon} = hf_{photon} \tag{34.39}$$

$$p_{photon} = \frac{hf_{photon}}{c_0}, \tag{34.40}$$

and, unlike ordinary particles, they always move at the speed of light c_0. Remember also that the mass of a particle is associated with the internal energy of that particle (Eq. 14.54), so $m_{photon} = 0$ means that photons have no internal energy and therefore no internal structure.

Example 34.11 Photoelectric effect

Light of wavelength 380 nm strikes the metal target in Figure 34.47. As long as the potential difference V_{CT} between the target and the collector is no greater than +1.2 V, there is a current in the circuit. Determine the longest wavelength of light that can eject electrons from this metal.

① GETTING STARTED To solve this problem, I recognize that the longest wavelength of light that can eject electrons corresponds to the lowest-energy photon that can eject an electron; this energy is equal to the work function. I therefore need to use the idea of stopping potential difference to determine the work function from the information given in the problem.

② DEVISE PLAN The fact that the current is zero when $V_{CT} > +1.2$ V tells me that the stopping potential difference is 1.2 V. Equation 34.36 gives the relationship between photon frequency and stopping potential difference. I can use the relationship between photon frequency and wavelength to rewrite Eq. 34.36 in terms of wavelength. Finally, Eq. 34.35 shows that the lowest photon energy comes when $K_{max} = 0$ so that the lower energy equals the work function E_0. Therefore I must determine the wavelength of a photon that has an energy equal to the work function, and for this I can use the expression I developed for the relationship between photon energy and wavelength.

③ EXECUTE PLAN Solving Eq. 34.36 for E_0, then substituting c_0/λ for f and inserting numerical values, I obtain

$$\begin{aligned}
E_0 &= \frac{hc_0}{\lambda} - eV_{stop} \\
&= \frac{(6.626 \times 10^{-34} \text{ J} \cdot \text{s})(2.998 \times 10^8 \text{ m/s})}{380 \times 10^{-9} \text{ m}} \\
&\quad - (1.602 \times 10^{-19} \text{ C})(1.2 \text{ V}) \\
&= 3.3 \times 10^{-19} \text{ J},
\end{aligned}$$

where I have used the equality 1 V ≡ 1 J/C (Eq. 25.16). Because the longest wavelength that can eject electrons has energy equal to the work function, I solve $E_0 = hc_0/\lambda$ for λ and substitute the value of E_0 I just calculated to obtain this maximum wavelength:

$$\lambda = \frac{hc_0}{E_0} = \frac{(6.626 \times 10^{-34} \text{ J} \cdot \text{s})(2.998 \times 10^8 \text{ m/s})}{3.3 \times 10^{-19} \text{ J}} = 0.60 \ \mu\text{m}. ✔$$

④ EVALUATE RESULT This value for the longest wavelength that can eject electrons is greater than 380 nm, the wavelength corresponding to the stopping potential difference of 1.2 V, as it should be.

34.21 A photon enters a piece of glass for which the index of refraction is about 1.5. What happens to the photon's (a) speed, (b) frequency, (c) wavelength, and (d) energy?

Chapter Glossary

SI units of physical quantities are given in parentheses.

Bragg condition The condition under which x rays diffracted by planes of atoms in a crystal lattice interfere constructively. For x rays of wavelength λ, diffracting from a crystal lattice with spacing d between adjacent planes of atoms at an angle θ between the incident rays and the normal to the scattering planes, the condition states that $2d \cos \theta = m\lambda$.

de Broglie wavelength λ (m) The wavelength associated with the wave behavior of a particle, $\lambda = h/p$.

diffraction grating An optical component with a periodic structure of equally spaced slits or grooves that diffracts and splits light into several beams that travel in different directions. When a diffraction grating is made up of slits, light passes through it and it is called a *transmission diffraction grating*; when a diffraction grating is made up of grooves, light reflects from it and it is called a *reflection grating*. The so-called *principal maxima* in the intensity pattern created by a diffraction grating occur at angles given by

$$d \sin \theta_m = \pm m\lambda, \quad \text{for } m = 0, 1, 2, 3, \ldots \quad (34.16)$$

and minima occur at angles give by

$$d \sin \theta_{\min} = \pm \frac{k}{N}\lambda \quad (34.17)$$

for an integer k that is not an integer multiple of N.

fringe order m, or n (unitless) A number indexing interference fringes; the central bright fringe is called zeroth order ($m = 0$), and the index increases with distance from the central bright fringe. The dark fringes flanking the central bright fringe are first order ($n = 1$).

interference fringes A pattern of alternating bright and dark bands cast on a screen produced by coherent light passing through very small, closely spaced slits, apertures, or edges.

photoelectric effect The emission of electrons from matter as a consequence of their absorption of energy from electromagnetic radiation with photon energy greater than the work function.

photon The indivisible, discrete basic unit, or quantum, of light. A photon of frequency f_{photon} has energy

$$E_{\text{photon}} = hf_{\text{photon}} \quad (34.39)$$

and momentum

$$p_{\text{photon}} = \frac{hf_{\text{photon}}}{c_0}. \quad (34.40)$$

Planck's constant h (J·s) The fundamental constant that relates the energy of a photon to its frequency and also the de Broglie wavelength and momentum of a particle: $h = 6.626 \times 10^{-34}$ J·s.

Rayleigh's criterion Two features in the image formed by a lens can be visually separated (and are then said to be *resolved*) if they satisfy Rayleigh's criterion. For a lens of diameter d and light of wavelength λ, the minimum angular separation θ_r for which two sources can be resolved is

$$\theta_r \approx 1.22 \frac{\lambda}{d}. \quad (34.30)$$

wave-particle duality The possession of both wave properties and particle properties, observed both for all atomic-scale material particles and for photons.

work function E_0 (J) The minimum energy required to free an electron from the surface of a metal. This energy measures how tightly the electron is bound to the metal.

x rays Electromagnetic waves that have wavelengths ranging from 0.01 nm to 10 nm.

Appendix A

Notation

Notation used in this text, listed alphabetically, Greek letters first.
For information concerning superscripts and subscripts, see the explanation at the end of this table.

Symbol	Name of Quantity	Definition	Where Defined	SI units
α (alpha)	polarizability	scalar measure of amount of charge separation occurring in material due to external electric field	Eq. 23.24	$C^2 \cdot m/N$
α	Bragg angle	in x-ray diffraction, angle between incident x rays and sample surface	Section 34.3	degree, radian, or revolution
α_ϑ	(ϑ component of) rotational acceleration	rate at which rotational velocity ω_ϑ increases	Eq. 11.12	s^{-2}
β (beta)	sound intensity level	logarithmic scale for sound intensity, proportional to $\log(I/I_{th})$	Eq. 17.5	dB (not an SI unit)
γ (gamma)	Lorentz factor	factor indicating how much relativistic values deviate from nonrelativistic ones	Eq. 14.6	unitless
γ	surface tension	force per unit length exerted parallel to surface of liquid; energy per unit area required to increase surface area of liquid	Eq. 18.48	N/m
γ	heat capacity ratio	ratio of heat capacity at constant pressure to heat capacity at constant volume	Eq. 20.26	unitless
Δ	delta	change in	Eq. 2.4	
$\Delta\vec{r}$	displacement	vector from object's initial to final position	Eq. 2.8	m
$\Delta\vec{r}_F, \Delta x_F$	force displacement	displacement of point of application of a force	Eq. 9.7	m
Δt	interval of time	difference between final and initial instants	Table 2.2	s
Δt_{proper}	proper time interval	time interval between two events occurring at same position	Section 14.1	s
Δt_v	interval of time	time interval measured by observer moving at speed v with respect to events	Eq. 14.13	s
Δx	x component of displacement	difference between final and initial positions along x axis	Eq. 2.4	m
δ (delta)	delta	infinitesimally small amount of	Eq. 3.24	
ϵ_0 (epsilon)	electric constant	constant relating units of electrical charge to mechanical units	Eq. 24.7	$C^2/(N \cdot m^2)$
η (eta)	viscosity	measure of fluid's resistance to shear deformation	Eq. 18.38	$Pa \cdot s$
η	efficiency	ratio of work done by heat engine to thermal input of energy	Eq. 21.21	unitless
θ (theta)	angular coordinate	polar coordinate measuring angle between position vector and x axis	Eq. 10.2	degree, radian, or revolution
θ_c	contact angle	angle between solid surface and tangent to liquid surface at meeting point measured within liquid	Section 18.4	degree, radian, or revolution
θ_c	critical angle	angle of incidence greater than which total internal reflection occurs	Eq. 33.9	degree, radian, or revolution
θ_i	angle of incidence	angle between incident ray of light and normal to surface	Section 33.1	degree, radian, or revolution

Symbol	Name of Quantity	Definition	Where Defined	SI units
θ_i	angle subtended by image	angle subtended by image	Section 33.6	degree, radian, or revolution
θ_o	angle subtended by object	angle subtended by object	Section 33.6	degree, radian, or revolution
θ_r	angle of reflection	angle between reflected ray of light and normal to surface	Section 33.1	degree, radian, or revolution
θ_r	minimum resolving angle	smallest angular separation between objects that can be resolved by optical instrument with given aperture	Eq. 34.30	degree, radian, or revolution
ϑ (script theta)	rotational coordinate	for object traveling along circular path, arc length traveled divided by circle radius	Eq. 11.1	unitless
κ (kappa)	torsional constant	ratio of torque required to twist object to rotational displacement	Eq. 15.25	$N \cdot m$
κ	dielectric constant	factor by which potential difference across isolated capacitor is reduced by insertion of dielectric	Eq. 26.9	unitless
λ (lambda)	inertia per unit length	for uniform one-dimensional object, amount of inertia in a given length	Eq. 11.44	kg/m
λ	wavelength	minimum distance over which periodic wave repeats itself	Eq. 16.9	m
λ	linear charge density	amount of charge per unit length	Eq. 23.16	C/m
μ (mu)	reduced mass	product of two interacting objects' inertias divided by their sum	Eq. 6.39	kg
μ	linear mass density	mass per unit length	Eq. 16.25	kg/m
$\vec{\mu}$	magnetic dipole moment	vector pointing along direction of magnetic field of current loop, with magnitude equal to current times area of loop	Section 28.3	$A \cdot m^2$
μ_0	magnetic constant	constant relating units of electric current to mechanical units	Eq. 28.1	$T \cdot m/A$
μ_k	coefficient of kinetic friction	proportionality constant relating magnitudes of force of kinetic friction and normal force between two surfaces	Eq. 10.55	unitless
μ_s	coefficient of static friction	proportionality constant relating magnitudes of force of static friction and normal force between two surfaces	Eq. 10.46	unitless
ρ (rho)	mass density	amount of mass per unit volume	Eq. 1.4	kg/m^3
ρ	inertia per unit volume	for uniform three-dimensional object, amount of inertia in a given volume divided by that volume	Eq. 11.46	kg/m^3
ρ	(volume) charge density	amount of charge per unit volume	Eq. 23.18	C/m^3
σ (sigma)	inertia per unit area	for uniform two-dimensional object, inertia divided by area	Eq. 11.45	kg/m^2
σ	surface charge density	amount of charge per unit area	Eq. 23.17	C/m^2
σ	conductivity	ratio of current density to applied electric field	Eq. 31.8	$A/(V \cdot m)$
τ (tau)	torque	magnitude of axial vector describing ability of forces to change objects' rotational motion	Eq. 12.1	$N \cdot m$
τ	time constant	for damped oscillation, time for energy of oscillator to decrease by factor e^{-1}	Eq. 15.39	s
τ_ϑ	(ϑ component of) torque	ϑ component of axial vector describing ability of forces to change objects' rotational motion	Eq. 12.3	$N \cdot m$
Φ_E (phi, upper case)	electric flux	scalar product of electric field and area through which it passes	Eq. 24.1	$N \cdot m^2/C$

Symbol	Name of Quantity	Definition	Where Defined	SI units
Φ_B	magnetic flux	scalar product of magnetic field and area through which it passes	Eq. 27.10	Wb
ϕ (phi)	phase constant	phase difference between source emf and current in circuit	Eq. 32.16	unitless
$\phi(t)$	phase	time-dependent argument of sine function describing simple harmonic motion	Eq. 15.5	unitless
Ω (omega, upper case)	number of basic states	number of basic states corresponding to macrostate	Section 19.4, Eq. 19.1	unitless
ω (omega)	rotational speed	magnitude of rotational velocity	Eq. 11.7	s^{-1}
ω	angular frequency	for oscillation with period T, $2\pi/T$	Eq. 15.4	s^{-1}
ω_0	resonant angular frequency	angular frequency at which current in circuit is maximal	Eq. 32.47	s^{-1}
ω_ϑ	(ϑ component of) rotational velocity	rate at which rotational coordinate ϑ changes	Eq. 11.6	s^{-1}
A	area	length $\times$ width	Eq. 11.45	m^2
A	amplitude	magnitude of maximum displacement of oscillating object from equilibrium position	Eq. 15.6	m (for linear mechanical oscillation; unitless for rotational oscillation; various units for nonmechanical oscillation)
$\vec{A}$	area vector	vector with magnitude equal to area and direction normal to plane of area	Section 24.6	m^2
$\vec{a}$	acceleration	time rate of change in velocity	Section 3.1	m/s^2
$\vec{a}_{Ao}$	relative acceleration	value observer in reference frame A records for acceleration of object o in reference frame A	Eq. 6.11	m/s^2
a_c	magnitude of centripetal acceleration	acceleration required to make object follow circular trajectory	Eq. 11.15	m/s^2
a_r	radial component of acceleration	component of acceleration in radial direction	Eq. 11.16	m/s^2
a_t	tangential component of acceleration	component of acceleration tangent to trajectory; for circular motion at constant speed $a_t = 0$	Eq. 11.17	m/s^2
a_x	x component of acceleration	component of acceleration directed along x axis	Eq. 3.21	m/s^2
$\vec{B}$	magnetic field	vector field providing measure of magnetic interactions	Eq. 27.5	T
$\vec{B}_{ind}$	induced magnetic field	magnetic field produced by induced current	Section 29.4	T
b	damping coefficient	ratio of drag force on moving object to its speed	Eq. 15.34	kg/s
C	heat capacity per particle	ratio of energy transferred thermally per particle to change in temperature	Section 20.3	J/K
C	capacitance	ratio of magnitude of charge on one of a pair of oppositely charged conductors to magnitude of potential difference between them	Eq. 26.1	F
C_P	heat capacity per particle at constant pressure	ratio of energy transferred thermally per particle to change in temperature, while holding pressure constant	Eq. 20.20	J/K
C_V	heat capacity per particle at constant volume	ratio of energy transferred thermally per particle to change in temperature, while holding volume constant	Eq. 20.13	J/K
$COP_{cooling}$	coefficient of performance of cooling	ratio of thermal input of energy to work done on a heat pump	Eq. 21.27	unitless

Symbol	Name of Quantity	Definition	Where Defined	SI units
$COP_{heating}$	coefficient of performance of heating	ratio of thermal output of energy to work done on a heat pump	Eq. 21.25	unitless
c	shape factor	ratio of object's rotational inertia to mR^2; function of distribution of inertia within object	Table 11.3, Eq. 12.25	unitless
c	wave speed	speed at which mechanical wave travels through medium	Eq. 16.3	m/s
c	specific heat capacity	ratio of energy transferred thermally per unit mass to change in temperature	Section 20.3	$J/(K \cdot kg)$
c_0	speed of light in vacuum	speed of light in vacuum	Section 14.2	m/s
c_V	specific heat capacity at constant volume	ratio of energy transferred thermally per unit mass to change in temperature, while holding volume constant	Eq. 20.48	$J/(K \cdot kg)$
$\vec{D}$	displacement (of particle in wave)	displacement of particle from its equilibrium position	Eq. 16.1	m
d	diameter	diameter	Section 1.9	m
d	distance	distance between two locations	Eq. 2.5	m
d	degrees of freedom	number of ways particle can store thermal energy	Eq. 20.4	unitless
d	lens strength	1 m divided by focal length	Eq. 33.22	diopters
E	energy of system	sum of kinetic and internal energies of system	Table 1.1, Eq. 5.21	J
$\vec{E}$	electric field	vector field representing electric force per unit charge	Eq. 23.1	N/C
E_0	work function	minimum energy required to free electron from surface of metal	Eq. 34.35	J
E_{chem}	chemical energy	internal energy associated with object's chemical state	Eq. 5.27	J
E_{int}	internal energy of system	energy associated with an object's state	Eqs. 5.20, 14.54	J
E_{mech}	mechanical energy	sum of system's kinetic and potential energies	Eq. 7.9	J
E_s	source energy	incoherent energy used to produce other forms of energy	Eq. 7.7	J
E_{th}	thermal energy	internal energy associated with object's temperature	Eq. 5.27	J
$\mathscr{E}$	emf	in charge-separating device, nonelectrostatic work per unit charge done in separating positive and negative charge carriers	Eq. 26.7	V
$\mathscr{E}_{ind}$	induced emf	emf resulting from changing magnetic flux	Eqs. 29.3, 29.8	V
$\mathscr{E}_{max}$	amplitude of emf	amplitude of time-dependent emf produced by AC source	Section 32.1, Eq. 32.1	V
$\mathscr{E}_{rms}$	rms emf	root-mean-square emf	Eq. 32.55	V
e	coefficient of restitution	measure of amount of initial relative speed recovered after collision	Eq. 5.18	unitless
e	eccentricity	measure of deviation of conic section from circular	Section 13.7	unitless
e	elementary charge	magnitude of charge on electron	Eq. 22.3	C
$\vec{F}$	force	time rate of change of object's momentum	Eq. 8.2	N
$\vec{F}^B$	magnetic force	force exerted on electric current or moving charged particle by magnetic field	Eqs. 27.8, 27.19	N
$\vec{F}^b$	buoyant force	upward force exerted by fluid on submerged object	Eq. 18.12	N
$\vec{F}^c$	contact force	force between objects in physical contact	Section 8.5	N

Symbol	Name of Quantity	Definition	Where Defined	SI units
$\vec{F}^d$	drag force	force exerted by medium on object moving through medium	Eq. 15.34	N
$\vec{F}^E$	electric force	force exerted between electrically charged objects or on electrically charged objects by electric field	Eq. 22.1	N
$\vec{F}^{EB}$	electromagnetic force	force exerted on electrically charged objects by electric and magnetic fields	Eq. 27.20	N
$\vec{F}^f$	frictional force	force exerted on object due to friction between it and a second object or surface	Eq. 9.26	N
$\vec{F}^G$	gravitational force	force exerted by Earth or any object having mass on any other object having mass	Eqs. 8.16, 13.1	N
$\vec{F}^k$	force of kinetic friction	frictional force between two objects in relative motion	Section 10.4, Eq. 10.55	N
$\vec{F}^n$	normal force	force directed perpendicular to a surface	Section 10.4, Eq. 10.46	N
$\vec{F}^s$	force of static friction	frictional force between two objects not in relative motion	Section 10.4, Eq. 10.46	N
f	frequency	number of cycles per second of periodic motion	Eq. 15.2	Hz
f	focal length	distance from center of lens to focus	Section 33.4, Eq. 33.16	m
f_{beat}	beat frequency	frequency at which beats occur when waves of different frequency interfere	Eq. 17.8	Hz
G	gravitational constant	proportionality constant relating gravitational force between two objects to their masses and separation	Eq. 13.1	$N \cdot m^2 / kg^2$
g	magnitude of acceleration due to gravity	magnitude of acceleration of object in free fall near Earth's surface	Eq. 3.14	m/s^2
h	height	vertical distance	Eq. 10.26	m
h	Planck's constant	constant describing scale of quantum mechanics; relates photon energy to frequency and de Broglie wavelength to momentum of particle	Eq. 34.35	$J \cdot s$
I	rotational inertia	measure of object's resistance to change in its rotational velocity	Eq. 11.30	$kg \cdot m^2$
I	intensity	energy delivered by wave per unit time per unit area normal to direction of propagation	Eq. 17.1	W/m^2
I	(electric) current	rate at which charged particles cross a section of a conductor in a given direction	Eq. 27.2	A
I	amplitude of oscillating current	maximum value of oscillating current in circuit	Section 32.1, Eq. 32.5	A
I_{cm}	rotational inertia about center of mass	object's rotational inertia about an axis through its center of mass	Eq. 11.48	$kg \cdot m^2$
I_{disp}	displacement current	current-like quantity in Ampère's law caused by changing electric flux	Eq. 30.7	A
I_{enc}	enclosed current	current enclosed by Ampèrian path	Eq. 28.1	A
I_{ind}	induced current	current in loop caused by changing magnetic flux through loop	Eq. 29.4	A
I_{int}	intercepted current	current intercepted by surface spanning Ampèrian path	Eq. 30.6	A
I_{rms}	rms current	root-mean-square current	Eq. 32.53	A
I_{th}	intensity at threshold of hearing	minimum intensity audible to human ear	Eq. 17.4	W/m^2
i	time-dependent current	time-dependent current through circuit; $I(t)$	Section 32.1, Eq. 32.5	A
i	image distance	distance from lens to image	Section 33.6, Eq. 33.16	m

Symbol	Name of Quantity	Definition	Where Defined	SI units
$\hat{i}$	unit vector ("i hat")	vector for defining direction of x axis	Eq. 2.1	unitless
$\vec{J}$	impulse	amount of momentum transferred from environment to system	Eq. 4.18	$kg \cdot m/s$
$\vec{J}$	current density	current per unit area	Eq. 31.6	A/m^2
J_ϑ	rotational impulse	amount of angular momentum transferred from environment to system	Eq. 12.15	$kg \cdot m^2/s$
$\hat{j}$	unit vector	vector for defining direction of y axis	Eq. 10.4	unitless
K	kinetic energy	energy object has because of its translational motion	Eqs. 5.12, 14.51	J
K	surface current density	current per unit of sheet width	Section 28.5	A/m
K_{cm}	translational kinetic energy	kinetic energy associated with motion of center of mass of system	Eq. 6.32	J
K_{conv}	convertible kinetic energy	kinetic energy that can be converted to internal energy without changing system's momentum	Eq. 6.33	J
K_{rot}	rotational kinetic energy	energy object has due to its rotational motion	Eq. 11.31	J
k	spring constant	ratio of force exerted on spring to displacement of free end of spring	Eq. 8.18	N/m
k	wave number	number of wavelengths in 2π units of distance; for wave with wavelength λ, $2\pi/\lambda$	Eqs. 16.7, 16.11	m^{-1}
k	Coulomb's law constant	constant relating electrostatic force to charges and their separation distance	Eq. 22.5	$N \cdot m^2/C^2$
k_B	Boltzmann constant	constant relating thermal energy to absolute temperature	Eq. 19.39	J/K
L	inductance	negative of ratio of induced emf around loop to rate of change of current in loop	Eq. 29.19	H
L_ϑ	(ϑ component of) angular momentum	capacity of object to make other objects rotate	Eq. 11.34	$kg \cdot m^2/s$
L_m	specific transformation energy for melting	energy transferred thermally per unit mass required to melt substance	Eq. 20.55	J/kg
L_v	specific transformation energy for vaporization	energy transferred thermally per unit mass required to vaporize substance	Eq. 20.55	J/kg
ℓ	length	distance or extent in space	Table 1.1	m
ℓ_{proper}	proper length	length measured by observer at rest relative to object	Section 14.3	m
ℓ_v	length	measured length of object moving at speed v relative to observer	Eq. 14.28	m
M	magnification	ratio of signed image height to object height	Eq. 33.17	unitless
M_θ	angular magnification	ratio of angle subtended by image to angle subtended by object	Eq. 33.18	unitless
m	mass	amount of substance	Table 1.1, Eq. 13.1	kg
m	inertia	measure of object's resistance to change in its velocity	Eq. 4.2	kg
m	fringe order	number indexing bright interference fringes, counting from central, zeroth-order bright fringe	Section 34.2, Eq. 34.5	unitless
m_v	inertia	inertia of object moving at speed v relative to observer	Eq. 14.41	kg
N	number of objects	number of objects in sample	Eq. 1.3	unitless
N_A	Avogadro's number	number of particles in 1 mol of a substance	Eq. 1.2	unitless
n	number density	number of objects per unit volume	Eq. 1.3	m^{-3}

Symbol	Name of Quantity	Definition	Where Defined	SI units
n	windings per unit length	in a solenoid, number of windings per unit length	Eq. 28.4	unitless
n	index of refraction	ratio of speed of light in vacuum to speed of light in a medium	Eq. 33.1	unitless
n	fringe order	number indexing dark interference fringes, counting from central, zeroth-order bright fringe	Section 34.2, Eq. 34.7	unitless
O	origin	origin of coordinate system	Section 10.2	
o	object distance	distance from lens to object	Section 33.6, Eq. 33.16	m
P	power	time rate at which energy is transferred or converted	Eq. 9.30	W
P	pressure	force per unit area exerted by fluid	Eq. 18.1	Pa
P_{atm}	atmospheric pressure	average pressure in Earth's atmosphere at sea level	Eq. 18.3	Pa
P_{gauge}	gauge pressure	pressure measured as difference between absolute pressure and atmospheric pressure	Eq. 18.16	Pa
p	time-dependent power	time-dependent rate at which source delivers energy to load; $P(t)$	Eq. 32.49	W
$\vec{p}$	momentum	vector that is product of an object's inertia and velocity	Eq. 4.6	kg·m/s
$\vec{p}$	(electric) dipole moment	vector representing magnitude and direction of electric dipole, equal amounts of positive and negative charge separated by small distance	Eq. 23.9	C·m
$\vec{p}_{ind}$	induced dipole moment	dipole moment induced in material by external electric field	Eq. 23.24	C·m
p_x	x component of momentum	x component of momentum	Eq. 4.7	kg·m/s
Q	quality factor	for damped oscillation, number of cycles for energy of oscillator to decrease by factor $e^{-2\pi}$	Eq. 15.41	unitless
Q	volume flow rate	rate at which volume of fluid crosses section of tube	Eq. 18.25	m^3/s
Q	energy transferred thermally	energy transferred into system by thermal interactions	Eq. 20.1	J
Q_{in}	thermal input of energy	positive amount of energy transferred into system by thermal interactions	Sections 21.1, 21.5	J
Q_{out}	thermal output of energy	positive amount of energy transferred out of system by thermal interactions	Sections 21.1, 21.5	J
q	electrical charge	attribute responsible for electromagnetic interactions	Eq. 22.1	C
q_{enc}	enclosed charge	sum of all charge within a closed surface	Eq. 24.8	C
q_p	dipole charge	charge of positively charged pole of dipole	Section 23.6	C
R	radius	radius of an object	Eq. 11.47	m
R	resistance	ratio of applied potential difference to resulting current	Eqs. 29.4, 31.10	Ω
R_{eq}	equivalent resistance	resistance that could be used to replace combination of circuit elements	Eqs. 31.26, 31.33	Ω
r	radial coordinate	polar coordinate measuring distance from origin of coordinate system	Eq. 10.1	m
$\vec{r}$	position	vector for determining position	Eqs. 2.9, 10.4	m
$\hat{r}_{12}$	unit vector ("r hat")	unit vector pointing from tip of $\vec{r}_1$ to tip of $\vec{r}_2$	Eq. 22.6	unitless
$\vec{r}_{AB}$	relative position	position of observer B in reference frame of observer A	Eq. 6.3	m
$\vec{r}_{Ae}$	relative position	value observer in reference frame A records for position at which event e occurs	Eq. 6.3	m

Symbol	Name of Quantity	Definition	Where Defined	SI units
$\vec{r}_{cm}$	position of a system's center of mass	a fixed position in a system that is independent of choice of reference frame	Eq. 6.24	m
$\vec{r}_p$	dipole separation	position of positively charged particle relative to negatively charged particle in dipole	Section 23.6	m
$r_\perp$	lever arm distance *or* lever arm	perpendicular distance between rotation axis and line of action of a vector	Eq. 11.36	m
$\Delta\vec{r}$	displacement	vector from object's initial to final position	Eq. 2.8	m
$\Delta\vec{r}_F$	force displacement	displacement of point of application of a force	Eq. 9.7	m
S	entropy	logarithm of number of basic states	Eq. 19.4	unitless
S	intensity	intensity of electromagnetic wave	Eq. 30.36	W/m^2
$\vec{S}$	Poynting vector	vector representing flow of energy in combined electric and magnetic fields	Eq. 30.37	W/m^2
s	arc length	distance along circular path	Eq. 11.1	m
s^2	space-time interval	invariant measure of separation of events in space-time	Eq. 14.18	m^2
T	period	time interval needed for object in circular motion to complete one revolution	Eq. 11.20	s
T	absolute temperature	quantity related to rate of change of entropy with respect to thermal energy	Eq. 19.38	K
$\mathcal{T}$	tension	stress in object subject to opposing forces stretching the object	Section 8.6	N
t	instant in time	physical quantity that allows us to determine the sequence of related events	Table 1.1	s
t_{Ae}	instant in time	value observer A measures for instant at which event e occurs	Eq. 6.1	s
Δt	interval of time	difference between final and initial instants	Table 2.2	s
Δt_{proper}	proper time interval	time interval between two events occurring at same position	Section 14.1	s
Δt_v	interval of time	time interval between two events measured by observer moving at speed v relative to an observer for whom the events occur at the same position	Eq. 14.13	s
U	potential energy	energy stored in reversible changes to system's configuration state	Eq. 7.7	J
U^B	magnetic potential energy	potential energy stored in magnetic field	Eqs. 29.25, 29.30	J
U^E	electric potential energy	potential energy due to relative position of charged objects	Eq. 25.8	J
U^G	gravitational potential energy	potential energy due to relative position of gravitationally interacting objects	Eqs. 7.13, 13.14	J
u_B	energy density of magnetic field	energy per unit volume stored in magnetic field	Eq. 29.29	J/m^3
u_E	energy density of electric field	energy per unit volume stored in electric field	Eq. 26.6	J/m^3
V	volume	amount of space occupied by an object	Table 1.1	m^3
V_{AB}	potential difference	negative of electrostatic work per unit charge done on charged particle as it is moved from point A to point B	Eq. 25.15	V
V_{batt}	battery potential difference	magnitude of potential difference between terminals of battery	Eq. 25.19	V
V_C	amplitude of oscillating potential	maximum magnitude of potential across circuit element C	Section 32.1, Eq. 32.8	V
V_{disp}	displaced volume	volume of fluid displaced by submerged object	Eq. 18.12	m^3

Symbol	Name of Quantity	Definition	Where Defined	SI units
V_P	(electrostatic) potential	potential difference between conveniently chosen reference point of potential zero and point P	Eq. 25.30	V
V_{rms}	rms potential	root-mean-square potential difference	Eq. 32.55	V
V_{stop}	stopping potential	minimum potential difference required to stop flow of electrons from photoelectric effect	Eq. 34.34	V
$\mathcal{V}$	"volume" in velocity space	measure of range of velocities in three dimensions	Eq. 19.20	$(m/s)^3$
v	speed	magnitude of velocity	Table 1.1	m/s
$\vec{v}$	velocity	time rate of change in position	Eq. 2.23	m/s
$\vec{v}_{12}$	relative velocity	velocity of object 2 relative to object 1	Eq. 5.1	m/s
$\vec{v}_{AB}$	relative velocity	velocity of observer B in reference frame of observer A	Eq. 6.3	m/s
v_C	time-dependent potential	time-dependent potential across circuit element C; $V_C(t)$	Section 32.1, Eq. 32.8	V
$\vec{v}_{cm}$	velocity, center of mass	velocity of the center of mass of a system, equal to the velocity of the zero-momentum reference frame of the system	Eq. 6.26	m/s
$\vec{v}_d$	drift velocity	average velocity of electrons in conductor in presence of electric field	Eq. 31.3	m/s
v_{esc}	escape speed	minimum launch speed required for object to reach infinity	Eq. 13.23	m/s
v_r	radial component of velocity	for object moving along circular path, always zero	Eq. 11.18	m/s
v_{rms}	root-mean-square speed	square root of average of square of speed	Eq. 19.21	m/s
v_t	tangential component of velocity	for object in circular motion, rate at which arc length is swept out	Eq. 11.9	m/s
v_x	x component of velocity	component of velocity directed along x axis	Eq. 2.21	m/s
W	work	change in system's energy due to external forces exerted on system	Eqs. 9.1, 10.35	J
$W_{P \to Q}$	work	work done along path from P to Q	Eq. 13.12	J
W_{in}	mechanical input of energy	positive amount of mechanical work done on system	Section 21.1	J
W_{out}	mechanical output of energy	positive amount of mechanical work done by system	Section 21.1	J
W_q	electrostatic work	work done by electrostatic field on charged particle moving through field	Section 25.2, Eq. 25.17	J
X_C	capacitive reactance	ratio of potential difference amplitude to current amplitude for capacitor	Eq. 32.14	Ω
X_L	inductive reactance	ratio of potential difference amplitude to current amplitude for inductor	Eq. 32.26	Ω
x	position	position along x axis	Eq. 2.4	m
$x(t)$	position as function of time	position x at instant t	Section 2.3	m
Δx	x component of displacement	difference between final and initial positions along x axis	Eq. 2.4	m
Δx_F	force displacement	displacement of point of application of a force	Eq. 9.7	m
Z	impedance	(frequency-dependent) ratio of potential difference to current through circuit	Eq. 32.33	Ω
z	zero-momentum reference frame	reference frame in which system of interest has zero momentum	Eq. 6.23	

Math notation

Math notation	Name	Where introduced
$\equiv$	defined as	Eq. 1.3
$\approx$	approximately equal to	Section 1.9
Σ (sigma, upper case)	sum of	Eq. 3.25
$\int$	integral of	Eq. 3.27
$\parallel$	parallel	Section 10.2
$\perp$	perpendicular	Section 10.2
$\propto$	proportional to	Section 13.1
$\cdot$	scalar product of two vectors	Eq. 10.33
$\times$	vector product of two vectors	Eq. 12.35
$\dfrac{\partial f}{\partial x}$	partial derivative of f with respect to x	Eq. 16.47
$\vec{b}$	vector b	Eq. 2.2
$\lvert \vec{b} \rvert$ or b	magnitude of $\vec{b}$	Eq. 2.3
b_x	x component of $\vec{b}$	Eq. 2.2
$\vec{b}_x$	x component vector of $\vec{b}$	Eq. 10.5
$\hat{\imath}$	unit vector ("i hat")	Eq. 2.1
$\hat{r}_{12}$	unit vector ("r hat")	Eq. 22.6

Note concerning superscripts and subscripts

Superscripts are appended to forces and potential energies to indicate the type of force or energy. They may be found in the main list under F, for forces, and U, for potential energies. Uppercase superscripts are used for fundamental interactions.

Subscripts are used on many symbols to identify objects, reference frames, types (for example, of energy), and processes. Object identifiers may be numbers, letters, or groups of letters. Reference frames are indicated by capital letters. Object identifiers and reference frames can occur in pairs, indicating relative quantities. In this case, the main symbol describes a property of whatever is identified by the second subscript relative to that of the first. In the case of forces, the first subscript identifies the object that causes the force and the second identifies the object on which the force is exerted. Types and processes are identified in various ways; many are given in the main list. Here are some examples:

m_1	inertia of object 1
m_{ball}	inertia of ball
$\vec{v}_{\text{cm}}$	velocity of center of mass of system

$\vec{r}_{12}$	position of object 2 relative to object 1; $\vec{r}_{12} = \vec{r}_2 - \vec{r}_1$
$\vec{p}_1$	momentum of object 1
$\vec{p}_{Z2}$	momentum of object 2 as measured in zero-momentum reference frame
$\vec{v}_{AB}$	velocity of observer B as measured in reference frame of observer A
$\vec{v}_{Ao}$	velocity of object o as measured in reference frame A
$\vec{r}_{Ee}$	position of event e as measured in Earth reference frame
$\vec{F}^c_{pw}$	contact force exerted by person on wall
$\vec{F}^G_{Eb}$	gravitational force exerted by Earth on ball
E_{th}	thermal energy
K_{conv}	convertible kinetic energy
P_{av}	average power
a_c	centripetal acceleration
$W_{P \to Q}$	work done along path from P to Q

Initial and final conditions are identified by subscripts i and f, following other identifiers. For example:

$\vec{p}_{1i}$	initial momentum of object 1
$\vec{p}_{Z\text{ball},f}$	final momentum of ball as measured in zero-momentum reference frame

Italic subscripts are used to identify components of vectors. These include x, y, z, r (radial), t (tangential), and ϑ (angular, with respect to given axis). They are also used to enumerate collections, for example, as indices of summation, and to indicate that a subscript refers to another variable. Here are some examples:

r_x	x component of position
a_t	tangential component of acceleration
L_ϑ	ϑ component of angular momentum
$p_{Z\text{ball}\,y,f}$	final y component of momentum of ball as measured in zero-momentum reference frame
$\delta m_n r_n^2$	contribution to rotational inertia of extended object of small segment n, with inertia δm_n at position r_n
c_P	specific heat capacity at constant pressure
W_q	electrostatic work

Appendix B

Mathematics Review

1 Algebra

Factors

$$ax + bx + cx = (a + b + c)x$$
$$(a + b)^2 = a^2 + 2ab + b^2$$
$$(a - b)^2 = a^2 - 2ab + b^2$$
$$(a + b)(a - b) = a^2 - b^2$$

Fractions

$$\left(\frac{a}{b}\right)\left(\frac{c}{d}\right) = \frac{ac}{bd}$$
$$\left(\frac{a/b}{c/d}\right) = \frac{a}{b} \div \frac{c}{d} = \frac{a}{b} \cdot \frac{d}{c} = \frac{ad}{bc}$$
$$\left(\frac{1}{1/a}\right) = a$$

Exponents

$$a^n = \underbrace{a \times a \times a \times \cdots \times a}_{n \text{ factors}}$$

Any real number can be used as an exponent:

$$a^{-x} = \frac{1}{a^x}$$
$$a^0 = 1$$
$$a^1 = a$$
$$a^{1/2} = \sqrt{a}$$
$$a^{1/n} = \sqrt[n]{a}$$
$$a^x a^y = a^{x+y}$$
$$\frac{a^x}{a^y} = a^{x-y}$$
$$(a^x)^y = a^{x \cdot y}$$
$$a^x b^x = (ab)^x$$
$$\frac{a^x}{b^x} = \left(\frac{a}{b}\right)^x$$

Logarithms

Logarithm is the inverse function of the exponential function:

$$y = a^x \Leftrightarrow \log_a y = \log_a a^x = x \quad \text{and} \quad x = \log_a (a^x) = a^{\log_a x}$$

The two most common values for the base a are 10 (the common logarithm base) and e (the natural logarithm base).

$$y = e^x \Leftrightarrow \log_e y = \ln y = \ln e^x = x \quad \text{and} \quad x = \ln e^x = e^{\ln x}$$

Logarithm rules (valid for any base):

$$\ln (ab) = \ln (a) + \ln (b)$$
$$\ln \left(\frac{a}{b}\right) = \ln (a) - \ln (b)$$
$$\ln (a^n) = n \ln (a)$$
$$\ln 1 = 0$$

The expression $\ln (a + b)$ cannot be simplified.

Linear equations

A linear equation has the form $y = ax + b$, where a and b are constants. A graph of y versus x is a straight line. The value of a equals the slope of the line, and the value of b equals the value of y when x equals zero.

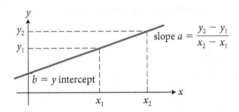

If $a = 0$, the line is horizontal. If $a > 0$, the line rises as x increases. If $a < 0$, the line falls as x increases. For any two values of x, say x_1 and x_2, the slope a can be calculated as

$$a = \frac{y_2 - y_1}{x_2 - x_1}$$

where y_1 and y_2 correspond to x_1 and x_2 (that is to say, $y_1 = ax_1 + b$ and $y_2 = ax_2 + b$).

Proportionality

If y is proportional to x (written $y \propto x$), then $y = ax$, where a is a constant. Proportionality is a subset of linearity. Because $y/x = a = $ constant for any corresponding x and y,

$$\frac{y_1}{x_1} = \frac{y_2}{x_2} \Leftrightarrow \frac{y_1}{y_2} = \frac{x_1}{x_2}.$$

Quadratic equation

The equation $ax^2 + bx + c = 0$ (the quadratic equation) has two solutions (called *roots*) for x:

$$x = \frac{-b \pm \sqrt{b^2 - 4ac}}{2a}$$

If $b^2 \geq 4ac$, the solutions are real numbers.

2 Geometry

Area and circumference for two-dimensional shapes

rectangle:
area $= ab$
circumference $= 2(a + b)$

parallelogram:
area $= bh$
circumference $= 2(a + b)$

triangle:
area $= \frac{1}{2}bh$
circumference $= a + b + c$

circle:
area $= \pi r^2$
circumference $= 2\pi r$

Volume and area for three-dimensional shapes

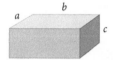

rectangular box:
volume $= abc$
area $= 2(a^2 + b^2 + c^2)$

sphere:
volume $= \frac{4}{3}\pi r^3$
area $= 4\pi r^2$

right circular cylinder:
volume $= \pi r^2 \ell$
area $= 2\pi r\ell + 2\pi r^2$

right circular cone:
volume $= \frac{1}{3}\pi r^2 h$
area $= \pi r^2 + \pi r \sqrt{r^2 + h^2}$

3 Trigonometry

Angle and arc length

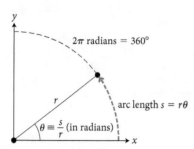

2π radians $= 360°$

arc length $s = r\theta$

$\theta \equiv \dfrac{s}{r}$ (in radians)

Right triangles

A right triangle is a triangle in which one of the angles is a right angle:

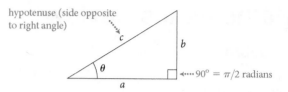

hypotenuse (side opposite to right angle)

$90° = \pi/2$ radians

Pythagorean theorem: $a^2 + b^2 = c^2 \Leftrightarrow c = \sqrt{a^2 + b^2}$

Trigonometric functions:

$$\sin\theta = \frac{b}{c} = \frac{\text{opposite side}}{\text{hypotenuse}}, \quad \theta = \sin^{-1}\left(\frac{b}{c}\right) = \arcsin\left(\frac{b}{c}\right)$$

$$\cos\theta = \frac{a}{c} = \frac{\text{adjacent side}}{\text{hypotenuse}}, \quad \theta = \cos^{-1}\left(\frac{a}{c}\right) = \arccos\left(\frac{a}{c}\right)$$

$$\tan\theta = \frac{b}{a} = \frac{\text{opposite side}}{\text{adjacent side}}, \quad \theta = \tan^{-1}\left(\frac{b}{a}\right) = \arctan\left(\frac{b}{a}\right)$$

General triangles

For any triangle, the following relationships hold:

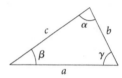

$$\alpha + \beta + \gamma = 180° = \pi \text{ rad}$$

Sine law: $\dfrac{\sin\alpha}{a} = \dfrac{\sin\beta}{b} = \dfrac{\sin\gamma}{c}$

Cosine law: $c^2 = a^2 + b^2 - 2ab\cos\gamma$

Identities

$$\tan\theta = \frac{\sin\theta}{\cos\theta}$$

$$\cot\theta = \frac{1}{\tan\theta} = \frac{\cos\theta}{\sin\theta}$$

$$\csc\theta = \frac{1}{\sin\theta}$$

$$\sec\theta = \frac{1}{\cos\theta}$$

Periodicity

$$\cos(\alpha + 2\pi) = \cos\alpha$$

$$\tan(\alpha + \pi) = \sin\alpha$$

Angle addition

$$\sin(\alpha \pm \beta) = \sin\alpha\cos\beta \pm \cos\alpha\sin\beta$$

$$\cos(\alpha \pm \beta) = \cos\alpha\cos\beta \mp \sin\alpha\sin\beta$$

Double angles

$$\sin(2\alpha) = 2\sin\alpha\cos\alpha$$

$$\cos(2\alpha) = \cos^2\alpha - \sin^2\alpha = 1 - 2\sin^2\alpha = 2\cos^2\alpha - 1$$

Other relations

$$\sin^2\alpha + \cos^2\alpha = 1$$

$$\sin(-\alpha) = -\sin\alpha$$

$$\cos(-\alpha) = \cos\alpha$$

$$\sin(\alpha \pm \pi) = -\sin\alpha$$

$$\cos(\alpha \pm \pi) = -\cos\alpha$$

$$\sin(\alpha \pm \pi/2) = \pm\cos\alpha$$

$$\cos(\alpha \pm \pi/2) = \mp\sin\alpha$$

The following graphs show $\sin\theta$, $\cos\theta$, and $\tan\theta$ as functions of θ:

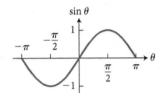
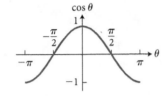

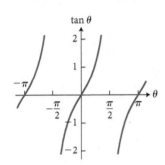

4 Vector algebra

A vector $\vec{A}$ in three-dimensional space can be written in terms of magnitudes A_x, A_y, and A_z of unit vectors $\hat{\imath}$, $\hat{\jmath}$, and $\hat{k}$, which have length 1 and lie along the x, y, and z axes:

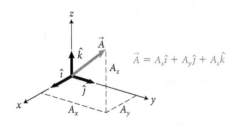

$$\vec{A} = A_x\hat{\imath} + A_y\hat{\jmath} + A_z\hat{k}$$

Dot products between vectors produce scalars:

$$\vec{A} \cdot \vec{B} = A_xB_x + A_yB_y + A_zB_z = |A||B|\cos\theta$$
$$(\theta \text{ is the angle between vectors } \vec{A} \text{ and } \vec{B})$$

Cross products between vectors produce vectors:

$$\vec{A} \times \vec{B} = (A_yB_z - A_zB_y)\hat{\imath} + (A_zB_x - A_xB_z)\hat{\jmath} + (A_xB_y - A_yB_x)\hat{k}$$

$$|\vec{A} \times \vec{B}| = |\vec{A}||\vec{B}|\sin\theta \; (\theta \text{ is the angle between vectors } \vec{A} \text{ and } \vec{B})$$

The direction of $\vec{A} \times \vec{B}$ is given by the right-hand rule (see Figure 12.44).

5 Calculus

In this section, x is a variable, and a and n are constants.

Derivatives

Geometrically, the derivative of a function $f(x)$ at $x = x_1$ is the slope of $f(x)$ at x_1:

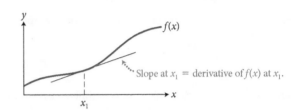

Derivatives of common functions

$$\frac{d}{dx}a = 0$$

$$\frac{d}{dx}x^n = nx^{n-1} \; (n \text{ need not be an integer})$$

$$\frac{d}{dx}\sin x = \cos x$$

$$\frac{d}{dx}\cos x = -\sin x$$

$$\frac{d}{dx}\tan x = \frac{1}{\cos^2 x}$$

$$\frac{d}{dx}e^{ax} = ae^{ax}$$

$$\frac{d}{dx}\ln(ax) = \frac{1}{x}$$

$$\frac{d}{dx}a^x = a^x\ln a$$

Derivatives of sums, products, and functions of functions

Constant times a function: $\dfrac{d}{dx}[a \cdot f(x)] = a \cdot \dfrac{d}{dx}f(x)$

Sum of functions: $\dfrac{d}{dx}[f(x) + g(x)] = \dfrac{d}{dx}f(x) + \dfrac{d}{dx}g(x)$

Product of functions:
$$\frac{d}{dx}[f(x) \cdot g(x)] = g(x)\frac{d}{dx}f(x) + f(x)\frac{d}{dx}g(x)$$

Quotient of functions: $\dfrac{d}{dx}\left[\dfrac{f(x)}{g(x)}\right] = \dfrac{g(x)\dfrac{d}{dx}f(x) - f(x)\dfrac{d}{dx}g(x)}{[g(x)]^2}$

Functions of functions (the chain rule): If f is a function of u, and u is a function of x, then

$$\frac{d[f(u)]}{du} \cdot \frac{d[u(x)]}{dx} = \frac{d[f(x)]}{dx}$$

Second and higher derivatives The second derivative of a function f with respect to x is the derivative of the derivative:

$$\frac{d^2 f(x)}{dx^2} = \frac{d}{dx}\left(\frac{d}{dx} f(x)\right)$$

Higher derivatives are defined similarly:

$$\frac{d^n f(x)}{dx^n} = \underbrace{\cdots \frac{d}{dx}\left(\frac{d}{dx}\left(\frac{d}{dx} f(x)\right)\right)}_{n \text{ uses of } \frac{d}{dx}} \quad \text{(where } n \text{ is a positive integer).}$$

Partial derivatives For functions of more than one variable, the partial derivative, written $\frac{\partial}{\partial x}$, is the derivative with respect to one variable; all other variables are treated as constants.

Integrals

Indefinite integrals Integration is the reverse of differentiation. An indefinite integral $\int f(x)dx$ is a function whose derivative is $f(x)$.

That is to say, $\frac{d}{dx}[\int f(x)dx] = f(x)$.

If $A(x)$ is an indefinite integral of $f(x)$, then so is $A(x) + C$, where C is any constant. Thus, it is customary when evaluating indefinite integrals to add a "constant of integration" C.

Definite integrals The definite integral of $f(x)$, written as $\int_{x1}^{x2} f(x)dx$, represents the sum of the area of contiguous rectangles that each intersect $f(x)$ at some point along one base and that each have another base coincident with the x axis over some part of the range between x_1 and x_2; the indefinite integral evaluates the sum in the limit of arbitrarily small rectangle bases. In other words, the indefinite integral gives the net area that lies under $f(x)$ but above the x axis between the boundaries x_1 and x_2.

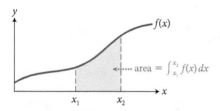

If $A(x)$ is any indefinite integral of $f(x)$, then the definite integral is given by $\int_{x1}^{x2} f(x)dx = A(x_2) - A(x_1) \equiv A(x)|_{x_1}^{x_2}$. The constant of integration C does not affect the value of definite integrals and thus can be ignored (i.e., set to zero) during evaluation.

Integration by parts $\int_a^b u \, dv$ is the area under the curve of $u(v)$. If $\int_a^b u \, dv$ is difficult to evaluate directly, it is sometimes easier to express the area under the curve as the area within part of a rectangle minus the area under the curve of $v(u)$. In other words:

$$\int_a^b u \, dv = uv|_a^b - \int_a^b v \, du.$$

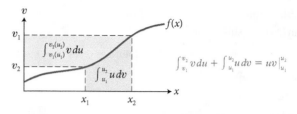

By choosing u and dv appropriately (both can be functions of x), this approach, called "integration by parts", can transform difficult integrals into easier ones.

Table of integrals In the following expressions, a and b are constants. An arbitrary constant of integration C can be added to the right-hand side.

$$\int x^n dx = \frac{1}{n+1} x^{n+1} \text{ (for } n \neq -1)$$

$$\int x^{-1} dx = \ln|x|$$

$$\int \frac{1}{a^2 + x^2} dx = \frac{1}{a} \tan^{-1}\frac{x}{a}$$

$$\int \frac{1}{(a^2 + x^2)^2} dx = \frac{1}{2a^3} \tan^{-1}\frac{x}{a} + \frac{x}{2a^2(x^2 + a^2)}$$

$$\int \frac{1}{\sqrt{\pm a^2 + x^2}} dx = \ln|x + \sqrt{\pm a^2 + x^2}|$$

$$\int \frac{1}{\sqrt{a^2 - x^2}} dx = \sin^{-1}\frac{x}{|a|} = \tan^{-1}\frac{x}{\sqrt{a^2 - x^2}}$$

$$\int \frac{x}{\sqrt{\pm a^2 - x^2}} dx = -\sqrt{\pm a^2 - x^2}$$

$$\int \frac{x}{\sqrt{\pm a^2 + x^2}} dx = \sqrt{\pm a^2 + x^2}$$

$$\int \frac{1}{(\pm a^2 + x^2)^{3/2}} dx = \frac{\pm x}{a^2\sqrt{\pm a^2 + x^2}}$$

$$\int \frac{x}{(a^2 + x^2)^{3/2}} dx = -\frac{1}{\sqrt{a^2 + x^2}}$$

$$\int \frac{1}{a + bx} dx = \frac{1}{b} \ln(a + bx)$$

$$\int \frac{1}{(a + bx)^2} dx = -\frac{1}{b(a + bx)}$$

$$\int \sin(ax)dx = -\frac{1}{a} \cos(ax)$$

$$\int \cos(ax)dx = \frac{1}{a} \sin(ax)$$

$$\int \tan(ax)dx = -\frac{1}{a} \ln(\cos ax)$$

$$\int \sin^2(ax)dx = \frac{x}{2} - \frac{\sin 2ax}{4a}$$

$$\int \cos^2(ax)dx = \frac{x}{2} + \frac{\sin 2ax}{4a}$$

$$\int x\sin(ax)dx = \frac{1}{a^2}\sin ax - \frac{1}{a}x\cos ax$$

$$\int x\cos(ax)dx = \frac{1}{a^2}\cos ax + \frac{1}{a}x\sin ax$$

$$\int e^{ax}dx = \frac{1}{a}e^{ax}$$

$$\int xe^{ax}dx = \frac{e^{ax}}{a^2}(ax - 1)$$

$$\int x^2 e^{ax}dx = \frac{x^2 e^{ax}}{a} - \frac{2}{a}\left[\frac{e^{ax}}{a^2}(ax - 1)\right]$$

$$\int \ln ax\, dx = x\ln(ax) - x$$

$$\int_0^\infty x^n e^{-ax}dx = \frac{n!}{a^{n+1}}$$

$$\int_0^\infty e^{-ax^2}dx = \frac{1}{2}\sqrt{\frac{\pi}{a}}$$

Line integrals. A *line integral* is an integral of a function that needs to be evaluated over a path (that is, a curve connecting two points in space). Consider, for example, the two-dimensional path C from point A to point B in the figure below. (The procedure described below is equally applicable in three dimensions.)

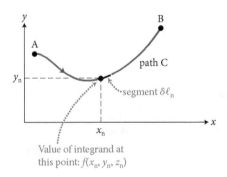

Value of integrand at this point: $f(x_n, y_n, z_n)$

The path from A to B is *directed*: At any point the direction along the path away from A and toward B is forward (positive). Suppose we have a function $f(x, y, z)$ defined everywhere along the path. The function can be either a scalar or a vector; we will first discuss line integrals of scalar functions. We divide the path between A and B into small segments of length $\delta\ell_n$, each segment small enough that we can consider it essentially straight and small enough that the value of the function $f(x, y, z)$ can be considered constant over that segment. We then calculate the product $f(x_n, y_n, z_n)\delta\ell_n$ for each segment. The line integral of the function $f(x, y, z)$ along path C is then given by the sum of all those products along the path in the limit of infinitesimally small segments:

$$\int_C f(x, y, z)d\ell = \lim_{\delta\ell\to 0}\sum_n f(x_n, y_n, z_n)\delta\ell_n.$$

To evaluate the integral on the right, we need to know the path C. Usually the path is specified in terms of the length parameter $\ell: x = x(\ell), y = y(\ell), z = z(\ell)$. The line integral can then be written as an ordinary definite integral:

$$\int_C f(x, y, z)d\ell = \int_A^B f[x(\ell), y(\ell), z(\ell)]d\ell.$$

Next we consider the line integral of a vector function. We consider the same path C from A to B, but now we consider a vector function $\vec{F}(x, y, z)$. Instead of taking infinitesimally small scalar segments $d\ell_n$ along the path, we take small vector segments $d\vec{\ell}_n$ along the path, of length $d\ell_n$ and whose direction is tangent to the path in the direction of the path from A to B:

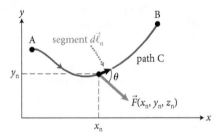

At each point we calculate the scalar product $\vec{F}(x_n, y_n, z_n)\cdot d\vec{\ell}_n$ and then sum these products over path C to obtain the line integral.

$$\int_C \vec{F}(x, y, z)\cdot d\vec{\ell}.$$

By writing out the scalar product, $\vec{F}(x, y, z)\cdot d\vec{\ell} = F(x, y, z)\cos\theta\, d\ell$, we can reduce the line integral of a vector function to that of a scalar function:

$$\int_C \vec{F}(x, y, z)\cdot d\vec{\ell} = \int_C F(x, y, z)\cos\theta\, d\ell.$$

In other words, we need to compute the line integral of the component of the vector $\vec{F}(x, y, z)$ along the tangent to the path.

If the path is closed—that is, the path returns to the starting point—we indicate that by putting a circle through the integration sign:

$$\oint_C \vec{F}(x, y, z)\cdot d\vec{\ell}.$$

Surface integrals. A *surface integral* is an integral of a function that needs to be evaluated over a surface. As with line integrals, the integrand of a surface integral can be a scalar or a vector function. We will only discuss the more general case of a vector function here.

The surface over which the integration is to be taken can be either *closed* or *open*. A closed surface, such as the surface of a sphere, divides space into two parts—an inside and an outside—and to get from one part to the other one has to go through the surface. An open surface does not have this property: For the surface S shown in the figure below, for example, one can go from one side of the surface to the other without passing through it.

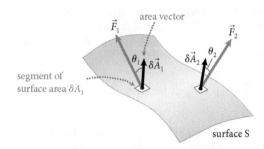

segment of surface area δA_1

surface S

Consider a vector function $\vec{F}(x, y, z)$. To calculate the surface integral of this function over surface S, we begin by dividing the surface into small segments of surface area δA_n, each segment being small enough that we can consider it to be essentially flat and small enough so that the function $\vec{F}(x, y, z)$ can be considered constant over the segment. We then define an *area vector* $\delta \vec{A}_n$ whose magnitude is equal to the surface area δA_n of the segment and whose direction is normal to that segment. For each segment we then calculate the scalar product of the area vector and the value $\vec{F}_n$ of the vector function at that location: $\vec{F}_n(x_n, y_n, z_n) \cdot \delta \vec{A}_n$. The surface integral of the vector function over the surface S is then given by the sum of all those products for all the segments that make up the surface:

$$\int_S \vec{F}(x, y, z) \cdot d\vec{A} = \lim_{\delta \vec{A}_n \to \infty} \sum_n \vec{F}_n(x_n, y_n, z_n) \cdot \delta \vec{A}_n.$$

If the surface is closed, we indicate that by putting a circle through the integration sign:

$$\oint_S \vec{F}(x, y, z) \cdot d\vec{A}$$

6 Complex numbers

A complex number $z = x + iy$ is defined in terms of its real part x and its imaginary part y. Both x and y are real numbers. i is Euler's constant, defined by the property $i^2 = -1$.

Each complex number z has a "complex conjugate" z^* which has the same real part but an imaginary part with opposite sign: $z = x + iy \Leftrightarrow z^* = x - iy$.

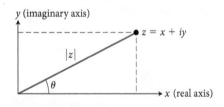

The real and imaginary parts can be expressed in terms of the complex number and its conjugate:

$$x = \tfrac{1}{2}(z + z^*)$$
$$y = \tfrac{1}{2}i(z - z^*)$$

A complex number is like a two-dimensional vector in a plane with a real axis and an imaginary axis. Thus, z can be described by a magnitude or length $|z|$ and an angle θ formed with the real axis (called the "phase angle"):

$$z = |z|(\cos \theta + i \sin \theta), \text{ where } |z| = \sqrt{zz^*} \text{ and}$$

$$\theta = \tan^{-1} \frac{y}{x} = \tan^{-1} \frac{i(z - z^*)}{(z + z^*)}.$$

Euler's formula says that $e^{i\theta} = \cos \theta + i \sin \theta$, allowing complex numbers to be written in the form $z = |z|e^{i\theta}$. This is a convenient form for expressing complex numbers. For example, it is easy to raise a complex number z to a power n: $z^n = |z|^n e^{in\theta}$.

7 Useful approximations

Binomial expansion

$$(1 + x)^n = 1 + nx + \frac{n(n-1)}{2}x^2 + \cdots$$

If $x \ll 1$, then $(1 + x)^n \approx 1 + nx$

Trigonometric expansions

$$\sin \alpha = \alpha - \frac{\alpha^3}{3!} + \frac{\alpha^5}{5!} - \frac{\alpha^7}{7!} + \cdots \quad (\alpha \text{ in rad})$$

$$\cos \alpha = 1 - \frac{\alpha^2}{2!} + \frac{\alpha^4}{4!} - \frac{\alpha^6}{6!} + \cdots \quad (\alpha \text{ in rad})$$

$$\tan \alpha = \alpha + \frac{1}{3}\alpha^3 + \frac{2}{15}\alpha^5 + \frac{17}{315}\alpha^7 + \cdots \quad (\alpha \text{ in rad})$$

If $\alpha \ll 1$ rad, then $\sin \alpha \approx \alpha$, $\cos \alpha \approx 1$, and $\tan \alpha \approx \alpha$.

Other useful expansions

$$\frac{1}{1 - x} = 1 + x + x^2 + x^3 + \cdots \text{ for } -1 < x < 1$$

$$e^x = 1 + x + \frac{1}{2}x^2 + \frac{1}{6}x^3 + \frac{1}{24}x^4 + \cdots$$

$$\ln(1 + x) = x - \frac{1}{2}x^2 + \frac{1}{3}x^3 - \frac{1}{4}x^4 + \cdots \quad \text{for } -1 < x < 1$$

$$\ln\left(\frac{1 + x}{1 - x}\right) = 2x + \frac{2}{3}x^3 + \frac{2}{5}x^5 - \frac{2}{7}x^7 + \cdots \text{ for } -1 < x < 1$$

Appendix C

SI Units, Useful Data, and Unit Conversion Factors

The seven base SI units

Unit	Abbreviation	Physical quantity
meter	m	length
kilogram	kg	mass
second	s	time
ampere	A	electric current
kelvin	K	thermodynamic temperature
mole	mol	amount of substance
candela	cd	luminous intensity

Some derived SI units

Unit	Abbreviation	Physical quantity	In terms of base units
newton	N	force	$kg \cdot m/s^2$
joule	J	energy	$kg \cdot m^2/s^2$
watt	W	power	$kg \cdot m^2/s^3$
pascal	Pa	pressure	$kg/m \cdot s^2$
hertz	Hz	frequency	s^{-1}
coulomb	C	electric charge	$A \cdot s$
volt	V	electric potential	$kg \cdot m^2/(A \cdot s^3)$
ohm	Ω	electric resistance	$kg \cdot m^2/(A^2 \cdot s^3)$
farad	F	capacitance	$A^2 \cdot s^4/(kg \cdot m^2)$
tesla	T	magnetic field	$kg/(A \cdot s^2)$
weber	Wb	magnetic flux	$kg \cdot m^2/(A \cdot s^2)$
henry	H	inductance	$kg \cdot m^2/(A^2 \cdot s^2)$

SI Prefixes

10^n	Prefix	Abbreviation	10^n	Prefix	Abbreviation
10^0	—	—			
10^3	kilo-	k	10^{-3}	milli-	m
10^6	mega-	M	10^{-6}	micro-	μ
10^9	giga-	G	10^{-9}	nano-	n
10^{12}	tera-	T	10^{-12}	pico-	p
10^{15}	peta-	P	10^{-15}	femto-	f
10^{18}	exa-	E	10^{-18}	atto-	a
10^{21}	zetta-	Z	10^{-21}	zepto-	z
10^{24}	yotta-	Y	10^{-24}	yocto-	y

Values of fundamental constants

Quantity	Symbol	Value
Speed of light in vacuum	c_0	$3.00 \times 10^8 \text{ m/s}$
Gravitational constant	G	$6.6738 \times 10^{-11} \text{ N} \cdot \text{m}^2/\text{kg}^2$
Avogadro's number	N_A	$6.0221413 \times 10^{23} \text{ mol}^{-1}$
Boltzmann's constant	k_B	$1.380 \times 10^{-23} \text{ J/K}$
Charge on electron	e	$1.60 \times 10^{-19} \text{ C}$
Electric constant	ϵ_0	$8.85418782 \times 10^{-12} \text{ C}^2/(\text{N} \cdot \text{m}^2)$
Magnetic constant	μ_0	$4\pi \times 10^{-7} \text{ T} \cdot \text{m/A}$
Planck's constant	h	$6.626 \times 10^{-34} \text{ J} \cdot \text{s}$
Electron mass	m_e	$9.11 \times 10^{-31} \text{ kg}$
Proton mass	m_p	$1.6726 \times 10^{-27} \text{ kg}$
Neutron mass	m_n	$1.6749 \times 10^{-27} \text{ kg}$
Atomic mass unit	amu	$1.6605 \times 10^{-27} \text{ kg}$

Other useful numbers

Number or quantity	Value
π	3.1415927
e	2.7182818
1 radian	$57.2957795°$
Absolute zero ($T = 0$)	$-273.15 \, °\text{C}$
Average acceleration g due to gravity near Earth's surface	9.8 m/s^2
Speed of sound in air at 20 °C	343 m/s
Density of dry air at atmospheric pressure and 20 °C	1.29 kg/m^3
Earth's mass	$5.97 \times 10^{24} \text{ kg}$
Earth's radius (mean)	$6.38 \times 10^6 \text{ m}$
Earth–Moon distance (mean)	$3.84 \times 10^8 \text{ m}$

Unit conversion factors

Length

1 in. = 2.54 cm (defined)

1 cm = 0.3937 in.

1 ft = 30.48 cm

1 m = 39.37 in. = 3.281 ft

1 mi = 5280 ft = 1.609 km

1 km = 0.6214 mi

1 nautical mile (U.S.) = 1.151 mi = 6076 ft = 1.852 km

1 fermi = 1 femtometer (fm) = 10^{-15} m

1 angstrom (Å) = 10^{-10} m = 0.1 nm

1 light − year (ly) = 9.461×10^{15} m

1 parsec = 3.26 ly = 3.09×10^{16} m

Volume

1 liter (L) = 1000 mL = 1000 cm^3 = 1.0×10^{-3} m^3
    = 1.057 qt (U.S.) = 61.02 $in.^3$

1 gal (U.S.) = 4 qt (U.S.) = 231 $in.^3$ = 3.785 L = 0.8327 gal (British)

1 quart (U.S.) = 2 pints (U.S.) = 946 mL

1 pint (British) = 1.20 pints (U.S.) = 568 mL

1 m^3 = 35.31 ft^3

Speed

1 mi/h = 1.4667 ft/s = 1.6093 km/h = 0.4470 m/s

1 km/h = 0.2778 m/s = 0.6214 mi/h

1 ft/s = 0.3048 m/s = 0.6818 mi/h = 1.0973 km/h

1 m/s = 3.281 ft/s = 3.600 km/h = 2.237 mi/h

1 knot = 1.151 mi/h = 0.5144 m/s

Angle

1 radian (rad) = 57.30° = 57°18'

1° = 0.01745 rad

1 rev/min (rpm) = 0.1047 rad/s

Time

1 day = 8.640×10^4 s

1 year = 365.242 days = 3.156×10^7 s

Mass

1 atomic mass unit (u) = 1.6605×10^{-27} kg

1 kg = 0.06852 slug

1 metric ton = 1000 kg

1 long ton = 2240 lbs = 1016 kg

1 short ton = 2000 lbs = 909.1 kg

1 kg has a weight of 2.20 lb where g = 9.80 m/s^2

Force

1 lb = 4.44822 N

1 N = 10^5 dyne = 0.2248 lb

Energy and work

1 J = 10^7 ergs = 0.7376 ft · lb

1 ft · lb = 1.356 J = 1.29×10^{-3} Btu = 3.24×10^{-4} kcal

1 kcal = 4.19×10^3 J = 3.97 Btu

1 eV = 1.6022×10^{-19} J

1 kWh = 3.600×10^6 J = 860 kcal

1 Btu = 1.056×10^3 J

Power

1 W = 1 J/s = 0.7376 ft · lb/s = 3.41 Btu/h

1 hp = 550 ft · lb/s = 746 W

1 kWh/day = 41.667 W

Pressure

1 atm = 1.01325 bar = 1.01325×10^5 N/m^2 = 14.7 $lb/in.^2$ = 760 torr

1 $lb/in.^2$ = 6.895×10^3 N/m^2

1 Pa = 1 N/m^2 = 1.450×10^{-4} $lb/in.^2$

Periodic Table of the Elements

Number of protons → 29
Symbol for element → **Cu**
63.546

Average atomic mass in g/mol. For elements having no stable isotope, value in parentheses is approximate atomic mass of longest-lived isotope.

Group	1	2	3	4	5	6	7	8	9	10	11	12	13	14	15	16	17	18
Period 1	1 **H** 1.008																	2 **He** 4.003
Period 2	3 **Li** 6.941	4 **Be** 9.012											5 **B** 10.811	6 **C** 12.011	7 **N** 14.007	8 **O** 15.999	9 **F** 18.998	10 **Ne** 20.180
Period 3	11 **Na** 22.990	12 **Mg** 24.305											13 **Al** 26.982	14 **Si** 28.086	15 **P** 30.974	16 **S** 32.065	17 **Cl** 35.453	18 **Ar** 39.948
Period 4	19 **K** 39.098	20 **Ca** 40.078	21 **Sc** 44.956	22 **Ti** 47.867	23 **V** 50.942	24 **Cr** 51.996	25 **Mn** 54.938	26 **Fe** 55.845	27 **Co** 58.933	28 **Ni** 58.693	29 **Cu** 63.546	30 **Zn** 65.409	31 **Ga** 69.723	32 **Ge** 72.64	33 **As** 74.922	34 **Se** 78.96	35 **Br** 79.904	36 **Kr** 83.798
Period 5	37 **Rb** 85.468	38 **Sr** 87.62	39 **Y** 88.906	40 **Zr** 91.224	41 **Nb** 92.906	42 **Mo** 95.94	43 **Tc** (98)	44 **Ru** 101.07	45 **Rh** 102.906	46 **Pd** 106.42	47 **Ag** 107.868	48 **Cd** 112.411	49 **In** 114.818	50 **Sn** 118.710	51 **Sb** 121.760	52 **Te** 127.60	53 **I** 126.904	54 **Xe** 131.293
Period 6	55 **Cs** 132.905	56 **Ba** 137.327	71 **Lu** 174.967	72 **Hf** 178.49	73 **Ta** 180.948	74 **W** 183.84	75 **Re** 186.207	76 **Os** 190.23	77 **Ir** 192.217	78 **Pt** 195.078	79 **Au** 196.967	80 **Hg** 200.59	81 **Tl** 204.383	82 **Pb** 207.2	83 **Bi** 208.980	84 **Po** (209)	85 **At** (210)	86 **Rn** (222)
Period 7	87 **Fr** (223)	88 **Ra** (226)	103 **Lr** (262)	104 **Rf** (261)	105 **Db** (262)	106 **Sg** (266)	107 **Bh** (264)	108 **Hs** (269)	109 **Mt** (268)	110 **Ds** (271)	111 **Rg** (272)	112 **Uub** (285)	113 **Uut** (284)	114 **Uuq** (289)	115 **Uup** (288)	116 **Uuh** (292)	117 **Uus** (294)	118 **Uuo** (?)

Lanthanoids

57 **La** 138.905	58 **Ce** 140.116	59 **Pr** 140.908	60 **Nd** 144.24	61 **Pm** (145)	62 **Sm** 150.36	63 **Eu** 151.964	64 **Gd** 157.25	65 **Tb** 158.925	66 **Dy** 162.500	67 **Ho** 164.930	68 **Er** 167.259	69 **Tm** 168.934	70 **Yb** 173.04

Actinoids

89 **Ac** (227)	90 **Th** (232)	91 **Pa** (231)	92 **U** (238)	93 **Np** (237)	94 **Pu** (244)	95 **Am** (243)	96 **Cm** (247)	97 **Bk** (247)	98 **Cf** (251)	99 **Es** (252)	100 **Fm** (257)	101 **Md** (258)	102 **No** (259)

Appendix D

Center of Mass of Extended Objects

We can apply the concept of center of mass to extended objects. Consider, for example, the object of inertia m in **Figure D.1**. If you imagine breaking down the object into many small segments of equal inertia δm, you can use Eq. 6.24 to compute the position of the center of mass:

$$x_{cm} = \frac{\delta m_1 x_1 + \delta m_2 x_2 + \cdots}{\delta m_1 + \delta m_2 + \cdots},$$

(D.1)

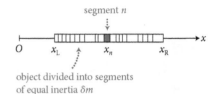

segment n

object divided into segments of equal inertia δm

Figure D.1

where x_n is the position of segment δm_n. Because the sum of the inertias of all segments is equal to the inertia m of the extended object, $\delta m_1 + \delta m_2 + \cdots = m$, we can write Eq. D.1 as

$$x_{cm} = \frac{1}{m}(\delta m_1 x_1 + \delta m_2 x_2 + \cdots) = \frac{1}{m}\sum_n (\delta m_n x_n)$$

(D.2)

To evaluate this sum for the extended object, we take the limit of this expression as $\delta m \to 0$. In this limit, the sum becomes an integral:

$$x_{cm} = \frac{1}{m}\lim_{\delta m \to 0}\sum_n (\delta m_n x_n) \equiv \frac{1}{m}\int_{\text{object}} x\, dm.$$

(D.3)

To evaluate the integral we need to know how the inertia is distributed over the object. Let the *inertia per unit length* of the object be $\lambda \equiv dm/dx$. In general λ is a function of position—that is, the inertia per unit length need not be the same at different locations along the object—and so $\lambda = \lambda(x)$ and $dm = \lambda(x)dx$. Substituting this expression for dm into Eq. D.3, we obtain

$$x_{cm} = \frac{1}{m}\int_{\text{object}} x\lambda(x)dx.$$

(D.4)

In this expression, the limits of integration should be taken to be the positions of the left and right ends of the object (x_L and x_R, respectively).

For example, let the object in Figure D.1 have a uniformly distributed inertia so that $\lambda(x)$ has the same value (the inertia of the extended object divided by the length of the object) everywhere:

$$\lambda(x) = \frac{m}{x_R - x_L}.$$

(D.5)

Because $\lambda(x)$ does not depend on x, we can pull it out of the integral in Eq. D.4 and so, substituting Eq. D.5 into Eq. D.4, we get

$$x_{cm} = \frac{\lambda(x)}{m}\int_{x_L}^{x_R} x\, dx = \frac{1}{x_R - x_L}\int_{x_L}^{x_R} x\, dx = \frac{1}{x_R - x_L}\left[\tfrac{1}{2}x^2\right]_{x_L}^{x_R}$$
$$= \frac{x_R^2 - x_L^2}{2(x_R - x_L)} = \frac{x_R + x_L}{2}.$$

(D.6)

In other words, the center of mass of the object is halfway between the ends of the object (at the center of the object), as we expect.

Appendix E

Derivation of the Lorentz Transformation Equations

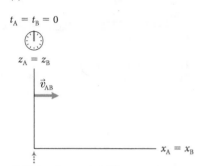

(a)

$t_A = t_B = 0$

$z_A = z_B$

$\vec{v}_{AB}$

$x_A = x_B$

Reference frames A and B overlap at $t_A = t_B = 0$.

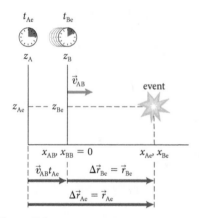

(b)

t_{Ae} t_{Be}

z_A z_B

$\vec{v}_{AB}$ event

z_{Ae} --- z_{Be}

$x_{AB}, x_{BB} = 0$ x_{Ae}, x_{Be}

$\vec{v}_{AB}t_{Ae}$ $\Delta\vec{r}_{Be} = \vec{r}_{Be}$

$\Delta\vec{r}_{Ae} = \vec{r}_{Ae}$

Figure E.1 Determining the space and time coordinates of an event from two reference frames that overlap at $t_A = t_B = 0$.

In this appendix we derive the Lorentz transformation equations (Eqs. 14.29–14.32), which relate position and time in two reference frames A and B. The origins of the reference frames coincide at $t_A = t_B = 0$, and the frames are moving at constant velocity $\vec{v}_{AB} = v_{ABx}\hat{\imath}$ in the x direction relative to each other (Figure E.1a).

Consider an event e that occurs at instant t_{Ae} and position x_{Ae} in reference frame A (Figure E.1b).* Our goal is to express t_{Be} and x_{Be}, the instant and the position at which the event occurs in reference frame B, in terms of t_{Ae} and x_{Ae}:

$$t_{Be} = f(x_{Ae}, t_{Ae}) \tag{E.1}$$

$$x_{Be} = g(x_{Ae}, t_{Ae}). \tag{E.2}$$

To obtain these relationships, we consider the space-time interval (Eq. 14.18) between the event in Figure E.1b and the event in Figure E.1a (this event being the overlapping of the two origins). The spatial separation of these two events is $\Delta x_{Ae} = x_{Ae} - 0 = x_{Ae}$ in reference frame A and $\Delta x_{Be} = x_{Be} - 0 = x_{Be}$ in reference frame B. The temporal separations are $\Delta t_{Ae} = t_{Ae} - 0 = t_{Ae}$ and $\Delta t_{Be} = t_{Be} - 0 = t_{Be}$. Because the space-time interval is an invariant, we have

$$(c_0 t_{Be})^2 - (x_{Be})^2 = (c_0 t_{Ae})^2 - (x_{Ae})^2. \tag{E.3}$$

To satisfy Eq. E.3, the relationships expressed in Eqs. E.1 and E.2 must be linear:

$$t_{Be} = D t_{Ae} + E x_{Ae} \tag{E.4}$$

$$x_{Be} = F t_{Ae} + G x_{Ae}, \tag{E.5}$$

where D, E, F, G are constants. To obtain relationships between these constants, consider the position of the origin of reference frame B at the instant illustrated in Figure E.1b. The origin of reference frame B moves at constant velocity $\vec{v}_{AB}$ relative to reference frame A, and so at t_{Ae} the position of B's origin in reference frame A is

$$x_{AB} = v_{ABx} t_{Ae}. \tag{E.6}$$

The position of B's origin in reference frame B is always zero, $x_{BB} = 0$.† According to Eq. E.5 we then have $x_{BB} = 0 = F t_{Ae} + G x_{AB}$, or substituting for x_{AB} from Eq. E.6, $0 = F t_{Ae} + G(v_{ABx} t_{Ae})$. Therefore

$$F = -v_{ABx} G. \tag{E.7}$$

Conversely, the origin of reference frame A is located at $x_{BA} = -v_{ABx} t_{Be}$ in reference frame B, and so, with $x_{AA} = 0$, Eqs. E.4 and E.5 become

$$t_{Be} = D t_{Ae} + 0 \tag{E.8}$$

$$-v_{ABx} t_{Be} = F t_{Ae} + 0. \tag{E.9}$$

Solving Eqs. E.8 and E.9 for F gives $F = -v_{ABx} D$, and so, from Eq. E.7,

$$D = G. \tag{E.10}$$

---

*Remember our subscript format: The capital letter refers to the reference frame; the lower case e is for "event." Thus the vector $\vec{r}_{Ae}$ represents observer **A**'s measurement of the position at which the event occurs.

†Do not be stymied by the unfamiliar subscript BB. It's the same format we've been using all along: The first B tells you the reference frame in which the measurement is made, the second B tells you what's being measured—in this case, the position of the origin of reference frame B.

Equations E.4 and E.5 then become

$$t_{Be} = Gt_{Ae} + Ex_{Ae} \tag{E.11}$$

$$x_{Be} = -v_{ABx}Gt_{Ae} + Gx_{Ae}. \tag{E.12}$$

Substituting Eqs. E.11 and E.12 into the left side of Eq. E.3 and using $v_{ABx}^2 = v^2$ yields

$$c_0^2 G^2 (1 - \frac{v^2}{c_0^2})t_{Ae}^2 + (c_0^2 EG + G^2 v_{ABx})2t_{Ae}x_{Ae} - (G^2 - c_0^2 E^2)x_{Ae}^2$$
$$= (c_0 t_{Ae})^2 - (x_{Ae})^2, \tag{E.13}$$

and then gather all the $(t_{Ae})^2$ terms, all the $x_{Ae}t_{Ae}$ terms, and all the $(t_{Ae})^2$ terms:

$$\left[c_0^2 G^2 \left(1 - \frac{v^2}{c_0^2} \right) - c_0^2 \right]t_{Ae}^2 + 2(c_0^2 EG + G^2 v_{ABx})x_{Ae}t_{Ae}$$
$$- (G^2 - c_0^2 E^2 - 1)x_{Ae}^2 = 0. \tag{E.14}$$

Because Eq. E.14 must hold for any value of t_{Ae} and x_{Ae}, the coefficient of each term must be zero. The coefficient of the t_{Ae}^2 term yields

$$c_0^2 G^2 \left(1 - \frac{v^2}{c_0^2} \right) = c_0^2. \tag{E.15}$$

Solving Eq. E.15 for G yields

$$G = \left(1 - \frac{v^2}{c_0^2} \right)^{-1/2} \equiv \gamma, \tag{E.16}$$

and so from Eq. E.7, $F = -v_{ABx}G$, we see that $F = -\gamma v_{ABx}$.

The coefficient of the $x_{Ae}t_{Ae}$ term in Eq. E.14 must also be zero. Substituting γ for G in that coefficient gives us

$$c_0^2 E\gamma + \gamma^2 v_{ABx} = 0 \tag{E.17}$$

$$E = -\gamma v_{ABx}/c_0^2. \tag{E.18}$$

Now we are ready to substitute for D, E, F, G in Eqs. E.4 and E.5—D from Eq. E.10, E from Eq. E.18, F from Eq. E.7, G from Eq. E.16:

$$t_{Be} = \gamma \left(t_{Ae} - \frac{1}{c_0^2} v_{ABx}x_{Ae} \right) \tag{E.19}$$

$$x_{Be} = \gamma(x_{Ae} - v_{ABx}t_{Ae}). \tag{E.20}$$

Because there is no length contraction in the directions perpendicular to the relative velocity $\vec{v}_{AB}$ of the two reference frames, we have

$$y_{Be} = y_{Ae} \tag{E.21}$$

$$z_{Be} = z_{Ae}. \tag{E.22}$$

Equations E.19–E.22 are the Lorentz transformation equations we wanted to derive.

Velocities in two reference frames

We can use Eqs. E.19 and E.20 to derive Eq. 14.33, the relationship between the x components of the velocities of an object o measured in two reference frames A and B. Let the x component be $v_{Aox} = dx_{Ao}/dt_A$ in reference frame A and $v_{Box} = dx_{Bo}/dt_B$ in reference frame B. We begin by writing Eqs. E.19 and E.20 in differential form:

$$dt_B = \gamma \left(dt_A - \frac{v_{ABx}}{c_0^2}dx_{Ao} \right) \tag{E.23}$$

$$dx_{Bo} = \gamma(dx_{Ao} - v_{ABx}dt_A), \tag{E.24}$$

$$dy_{Bo} = dy_{Ao}, \tag{E.25}$$

$$dz_{Bo} = dz_{Ao}, \tag{E.26}$$

where we have replaced the subscript e by o because we are considering an object rather than an event. Dividing Eq. E.24 by Eq. E.23 yields an expression for the x component of the velocity in reference frame B:

$$v_{Box} = \frac{dx_{Bo}}{dt_B} = \frac{\gamma(dx_{Ao} - v_{ABx}dt_A)}{\gamma\left(dt_A - \dfrac{v_{ABx}}{c_0^2}dx_{Ao}\right)} = \frac{(dx_{Ao} - v_{ABx}dt_A)}{\left(dt_A - \dfrac{v_{ABx}}{c_0^2}dx_{Ao}\right)}. \tag{E.27}$$

Finally, we divide the numerator and denominator of the rightmost term by dt_A to obtain Eq. 14.33:

$$v_{Box} = \frac{\left(\dfrac{dx_{Ao}}{dt_A} - v_{ABx}\right)}{\left(1 - \dfrac{v_{ABx}}{c_0^2}\dfrac{dx_{Ao}}{dt_A}\right)} = \frac{(v_{Aox} - v_{ABx})}{\left(1 - \dfrac{v_{ABx}}{c_0^2}v_{Aox}\right)}. \tag{E.28}$$

To find the y component of the velocity in reference frame B, we divide Eq. E.25 by Eq. E.23:

$$v_{Boy} = \frac{dy_{Bo}}{dt_B} = \frac{dy_{Ao}}{\gamma\left(dt_A - \dfrac{v_{ABx}}{c_0^2}dx_{Ao}\right)}. \tag{E.29}$$

Dividing numerator and denominator by dt_A then yields

$$v_{Boy} = \frac{v_{Aoy}}{\gamma\left(1 - \dfrac{v_{ABx}v_{Aox}}{c_0^2}\right)}. \tag{E.30}$$

For the z component of the velocity we obtain a similar equation with y replaced by z

$$v_{Boz} = \frac{v_{Aoz}}{\gamma\left(1 - \dfrac{v_{ABx}v_{Aox}}{c_0^2}\right)}. \tag{E.31}$$

$$\Delta p_{sx} = +0.38 \text{ kg} \cdot \text{m/s} - 0$$

Solutions to Checkpoints

Chapter 1

1.1 A nickel and a quarter. If you had to think about this for more than a few seconds, you probably unknowingly assumed that *neither* coin could be a nickel. This is an example of a hidden assumption.

1.2 That the player works with the new batteries does not necessarily mean the old batteries are dead. They could have been inserted incorrectly, or there could have been a bad contact that got fixed when the new batteries were inserted. That the player does not work with the new batteries does not necessarily mean the player is broken. Both the new and the old batteries could be inserted incorrectly, or both the new and the old batteries could be dead.

1.3 Two valuable goals are to become a better problem solver and to learn to apply the scientific method to any type of reasoning. Either skill is useful in any type of career. If you wrote "pass physics requirement" or "get into medical school," think again. With those goals, it is going to very difficult for you to derive much satisfaction or benefit from this course.

Here are some interesting responses I have collected over the years: to become a responsible citizen of the world, to understand modern technology; to bridge the gap between the sciences and the arts/humanities, to sharpen my mind.

1.4 Likely answers are that something about the case or cap has changed. The case opening may have expanded due to the heat, or the cap may have acquired a dent that prevents it from fitting over the opening. If you answered something along these lines, you assumed symmetry with respect to translation in both time and space: If the cap fits at home, it ought to fit elsewhere now and anytime in the future.

1.5 The distance from the North Pole to the equator is one-fourth of Earth's circumference c, which means $c = 4 \times (10,000,000 \text{ m}) = 40,000,000 \text{ m}$. Because $c = 2\pi R_E$, where R_E is Earth's radius, you have $R_E = c/2\pi = (40,000,000 \text{ m})/2\pi = 6,400,000 \text{ m}$.

1.6 The diameter of an atom is 10^{-10} m (see Figure 1.9); the diameter of an apple is about 10^{-1} m. Therefore the required magnification factor is $(10^{-1} \text{ m})/(10^{-10} \text{ m}) = 10^9$. With this magnification, the apple's diameter would be $10^9(10^{-1} \text{ m}) = 10^8$ m, which is about 10 times Earth's diameter.

1.7 (*a*) (*i*) Somebody switches the power off; the light bulb burns out. (*ii*) Thunder; tractor trailer passing dorm; stomach growling; dorm starting to collapse. (*iii*) You spent all the money in your account; you forgot to deposit the check your parents sent you last week; the bank made a mistake. (*b*) No. All these causes occur before the event. (*c*) One emotion you almost certainly experience is anxiety. Such events tend to be unsettling—we want to know the causes of things we observe.

1.8 A typical physics class lasts about 1 h, which is 3600 s, or order of magnitude 10^4 s. The number of sequential chemical reactions is thus $(10^4 \text{ s})(1 \text{ reaction}/10^{-13} \text{ s}) = 10^{17}$ (hundred million billion!) reactions.

1.9 Because Mike Masters imposes the most constraints, you begin by placing him in seat 1 (Figure S1.9). To satisfy the constraint of having men and women alternate, placing Mike here means the odd-numbered seats are for the men. Only two arrangements avoid putting Cyndi Ahlers next to Mike: Put her in 4 or in 6. If you put Cyndi in 4, Sylvia Masters has to go in 6 to avoid being next to her husband in 1. With Sylvia in 6, John Jones cannot be in 5 or 7, and so you have to put him in 3. Mary Jones then has to go in 8. The remaining empty seats—2, 5, and 7—leave you no choice: Bob Ahlers has to go in 7 because putting him in 5 would have him next to his wife in 4, and your spouse and you get 2 and 5. (Had you put Cyndi Ahlers in 6, you would have obtained the mirror image of the arrangement shown in Figure S1.9.)

Figure S1.9

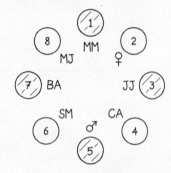

1.10 90 m. If you call the distance from the west end of the corridor to stop 1 x, the distance from stop 1 to the fountain is $2x$ (Figure S1.10). If you call the distance from stop 2 to the fountain $2y$, the distance from stop 2 to the east end of the corridor is y. Because the distance between the two stops is 60 m, you know that $2x + 2y = 60$ m. From the figure, you see that the total corridor length L is $L = x + 2x + 2y + y = 3x + 3y$. The rest is algebra: $2x + 2y = 60$ m, $x = (30 \text{ m} - y)$, $L = 3(30 \text{ m} - y) + 3y = 90$ m.

Figure S1.10

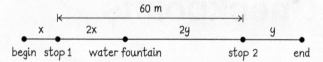

begin   stop 1   water fountain   stop 2   end

1.11 (a) $\ell = 150$ Mm (megameters) or 0.15 Gm (gigameters). (b) $t = 12$ ps (picoseconds). (c) 1.2 Mm/s. (d) 2.3 Mg.

1.12 The ratio on the left has inches in the numerator and millimeters in the denominator, meaning that the inches don't cancel out. In general, when converting from a given unit to a desired unit, you need to use a ratio that has the desired unit in the numerator and the given unit in the denominator.

1.13 (a) In 1 mol you have 6.022×10^{23} atoms, and when you pack this number of atoms into a cube, each side of the cube contains $\sqrt[3]{6.022 \times 10^{23}}$ atoms $= 8.445 \times 10^7$ atoms. The diameter of each atom is 10^{-10} m (see Figure S1.9), and so the length of each side of the cube is $(8.445 \times 10^7 \text{ atoms})(10^{-10} \text{ m/atom}) = 8.4 \times 10^{-3}$ m, which is about one-third of an inch. (b) Because 1 mol of carbon has a mass of 12×10^{-3} kg, Eq. 1.4 gives $V = m/\rho = (12 \times 10^{-3} \text{ kg})/(2.2 \times 10^3 \text{ kg/m}^3) = 5.5 \times 10^{-6}$ m^3. If you assume this value to be a cube, then the cube root of V gives the length of a side, 1.8×10^{-2} m, about twice the length for each side calculated in part a.

1.14 (a) The circumference is $2\pi R = 2\pi(27.3 \text{ mm}) = 172$ mm. You are allowed to report three significant digits because the radius R has three (the 2 in $2\pi R$ has an infinite number of significant digits, and π has as many as your calculator has digits). (b) $a + b + c = 12.3 + 3.241 + 55.74 = 71.3$ (12.3 has the fewest decimal places (one), and so the sum must be rounded to one decimal place). (c) $f = (4.00)^2/7 = 2.29$; $g = (3.00)^2/7 = 1.29$;

$f + g = (4.00)^2/7 + (3.00)^2/7 = 3.57$. Note that this result is not equal to the sum of the rounded values for f and g: $2.29 + 1.29 = 3.58$. Never use a rounded intermediate answer in a calculation!

1.15 Many answers are possible. One translation strategy would be to reason that water is like milk, and you know from experience that a quart of milk weighs about 2 lb. From Appendix [SI UNITS], you know that $1 \text{ qt} = 0.946 \times 10^{-3} \text{ m}^3 \approx 10^{-3} \text{ m}^3$ and that $2 \text{ lb} = 0.907 \text{ kg} \approx 1$ kg. Using Eq. 1.4, you thus obtain for the density of milk (and of water) $(1 \text{ kg})/(10^{-3} \text{ m}^3) = 10^3 \text{ kg/m}^3$.

One division strategy would be to estimate the mass and the volume of a glass of water and then use Eq. 1.4 to obtain the density. A glass of water is about 2 in. across and 4 in. tall, so $V \approx \pi[(1 \text{ in.})(2.54 \times 10^{-2} \text{ m/in.})]^2(4 \text{ in.})(2.54 \times 10^{-2} \text{ m/in.}) = 2 \times 10^{-4} \text{ m}^3$. Such a glass weighs about 0.25 lb, and so $m \approx (0.25 \text{ lb})(0.45 \text{ kg/lb}) = 0.11$ kg. Substituting V and m into Eq. 1.4 yields $\rho = m/V \approx (0.11 \text{ kg})/(2 \times 10^{-4} \text{ m}^3) = 5.5 \times 10^2 \text{ kg/m}^3$. The order of magnitude of both estimates is 10^3 kg/m^3 because $5.5 \times 10^2 \text{ kg/m}^3$ rounds to $10 \times 10^2 \text{ kg/m}^3 = 10^3 \text{ kg/m}^3$, which is the same answer we obtained before.

1.16 If we ignore the oceans, which constitute only a small layer at Earth's surface, the planet is mostly rock, which has a mass density higher than that of water—say, five times higher. As discussed in Section 1.9, the mass density of water is 1000 kg/m^3, meaning that rock should have a mass density of about $5 \times 10^3 \text{ kg/m}^3$. From Checkpoint 1.5 (which you worked out, didn't you?), you know that the Earth's radius is 6.4×10^6 m. Taking Earth to be a sphere, you obtain for its volume $V = \frac{4}{3}\pi R_E^3 = 1.1 \times 10^{21} \text{ m}^3$. Rearranging Eq. 1.4 to $m = \rho V$ gives you $m = (5 \times 10^3 \text{ kg/m}^3)(1.1 \times 10^{21} \text{ m}^3) = 5 \times 10^{24}$ kg, with order of magnitude 1025 kg. (The current best measurement of Earth's mass is 5.9736×10^{24} kg.)

Chapter 2

2.1 (a) See Table 2.1. (b) See Figure 2.2.

2.2 (a) Staying the same distance from the left edge in five sequential frames means I have stopped and am standing in one spot as time passes and the camera continues to roll. (b) I am again in the same spot in two frames, but now the frames are not sequential, and the data points in between are not aligned, telling you that in the time that passed between frames 7 and 14, I walked away from the spot and then came back to it. (c) Two vertically aligned points would mean that in one frame (which means one instant in time), I was at two different distances from the origin, something that could never happen.

2.3. (a) My image is 4.5 mm tall in the film clip. Estimating my height to be 1.8 m (6 ft), you get that 4.5 mm in the photograph corresponds to 1.8 m in the real world. So 1 mm in the photograph corresponds to $(1 \text{ mm in photo})(1.8 \text{ m in real world})/(4.5 \text{ mm in photo}) = 0.40$ m in real world. (b) In frame 1, I am 2.5 mm from the left edge of the frame; in frame 10, 12.0 mm. So the distance traveled between frames 1 and 10 is $(12.0 \text{ mm in photo}) - (2.5 \text{ mm in photo}) = 9.5$ mm in photo. This corresponds to an actual distance of $(9.5 \text{ mm in photo})(0.40 \text{ m in real world})/(1 \text{ mm in photo}) = 3.8$ m.

2.4. (a) $+3.4$ m in Figure 2.3; $+2.4$ m in Figure 2.4. (b) $+3.4 \text{ m} - (+1.0 \text{ m}) = +2.4$ m; $+2.4 \text{ m} - 0 = +2.4$ m.

2.5 (a) Because your final position is the same as your initial position, the x component of your displacement is zero. (b) Your distance traveled is 4 m. (c) No. In general the *distance traveled* and the x *component of the displacement* are two different quantities.

2.6 (a) A vertical line drawn from the first $x = +4.0$ m point down to the horizontal axis tells you that I was at that position at 1.7 s. The vertical line for the $x = +1$ m position is already drawn: It is the vertical axis. Therefore the length of time I took to travel this distance was $(1.7 \text{ s}) - (0) = 1.7$ s. (b) Vertical lines through the curve at $x = +2.0$ m and $x = +3.0$ m intersect the time axis at 0.6 s and 1.2 s, and so I traveled this distance in $(1.2 \text{ s}) - (0.6 \text{ s}) = 0.6 \text{ s}$. (c) At 0.8 s. (d) I was at $x = +2.5$ m for *zero* time (not even a split second, an infinitesimally small amount of time!) because it takes no time to move over a *point*. If this surprises you, look at your answers to parts *a* and *b*: To move from $+1.0$ m to $+4.0$ m took 1.7 s; from $+2.0$ m to $+3.0$ m it took 0.6 s. As we make the interval smaller and smaller, the time it takes to move over that interval gets smaller. If we consider just the point at $+2.5$ m, the distance shrinks to zero, and so does the time it takes to cross the point.

2.7 (a) Normal: $3.50 \text{ m} - 1.00 \text{ m} = 2.50 \text{ m}$; slow: $2.25 \text{ m} - 1.00 \text{ m} = 1.25$ m (half the distance, as one would expect). (b) Normal: $(2.50 \text{ m})/(1.50 \text{ s}) = 1.67$ m/s; slow: $(1.25 \text{ m})/(1.50 \text{ s}) = 0.83$ m/s (note units of meter per second). (c) Normal: $1.50 \text{ s} - 0.60 \text{ s} = 0.90$ s; slow: $3.00 \text{ s} - 1.20 \text{ s} = 1.80$ s (twice as long, as you expect). (d) Normal: $(1.50 \text{ m})/(0.90 \text{ s}) = 1.67$ m/s; slow: $(1.50 \text{ m})/(1.80 \text{ s}) = 0.83$ m/s (same as the values you calculated in part *b*).

2.8 (a) At $t = 1.3$ s. (b) $(3.4 \text{ m} - 1.0 \text{ m})/(1.3 \text{ s}) = 1.8$ m/s.

2.9 (a) The average speed is

$$\frac{\text{distance traveled}}{\text{time taken}} =$$

$$\frac{0.5 \text{ km}}{6.0 \text{ min}} \times \frac{1000 \text{ m}}{1.0 \text{ km}} \times \frac{1.0 \text{ min}}{60 \text{ s}} = 1.4 \text{ m/s}.$$

(b) The average speed is

$$\frac{\text{distance traveled}}{\text{time taken}} =$$

$$\frac{0.5 \text{ km}}{2.0 \text{ min}} \times \frac{1000 \text{ m}}{1.0 \text{ km}} \times \frac{1.0 \text{ min}}{60 \text{ s}} = 4.2 \text{ m/s}.$$

(c) $(1.4 \text{ m/s} + 4.2 \text{ m/s})/2 = 2.8$ m/s. This is not equal to the answer worked out in Example 2.3 because average speed is obtained not by averaging speeds over different segments of a trip but by dividing the distance traveled by the duration of the entire trip.

2.10 (a) Negative. The *x* component of the displacement is $-2.3 \text{ m} - (-1.2 \text{ m}) = -1.1$ m. Because the *x* component of the displacement is negative, the *x* component of the average velocity is negative. (b) Negative again. The *x* component of the displacement is $-1.2 \text{ m} - (+1.2 \text{ m}) = -2.4$ m, and so the *x* component of the average velocity is negative as well. (c) Yes. In both cases, the object moves in the direction of decreasing *x* and so must have a negative *x* component of the average velocity.

2.11 (*i*) and (*iv*) are scalars because no direction is required in order to specify these quantities completely; (*ii*) and (*iii*) are vectors because direction is required to specify them completely.

2.12 (a) Because my feet in Figure 2.18 are at $x = +4.8$ m, the *x* coordinate of the vector pointing from the origin to my feet is $x = +4.8$ m. (b) The vector $\vec{r}$ is obtained by multiplying the unit vector by $+4.8$ m: $\vec{r} = (+4.8 \text{ m})\vec{\imath}$. (c) It represents my position in the frame shown in Figure 2.18. (d) Yes. It is a vector that is 2.5 m long and points to the right (Figure S2.12). This vector could represent my position at an earlier instant, or it could be a vector representing the position of the right edge of the left garage door.

Figure S1.12

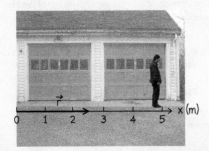

2.13 (a) $\Delta x = x_{\text{f}} - x_{\text{i}} = +4.8 \text{ m} - (+2.2 \text{ m}) = +2.6$ m. (b) $\Delta\vec{r} = (+2.6 \text{ m})\vec{\imath}$.

2.14 (a) Positive. If the object is initially at $x_{\text{i}} = -2$ m and moves to $x_{\text{f}} = -1$ m, then the *x* component of the displacement is $\Delta x = -1 \text{ m} - (-2 \text{ m}) = +1$ m. (b) No. Imagine any case where the object moves from a negative value of *x* to a positive value, and the displacement is again positive. Whenever positive *x* values are to the right of the origin, the *x* component of any displacement to the right is positive, regardless of the initial and final locations.

2.15 (a) See Figure S2.15a. (b) Yes, because addition is commutative: $\vec{a} + \vec{c} = \vec{c} + \vec{a}$; see Figure S2.15b. (c) No, because subtraction is not commutative: $\vec{a} - \vec{c} \neq \vec{c} - \vec{a}$; see Figure S2.15c.

Figure S2.15

(a)

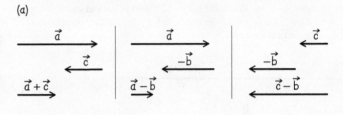

(b)   (c)

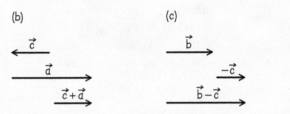

2.16 (a) The time interval between frames 6 and 17 is $\Delta t = t_{\text{f}} - t_{\text{i}} = 5.33 \text{ s} - 1.67 \text{ s} = 3.66$ s; the distance traveled is 2.0 m (from $x = +3.8$ m to $x = +4.8$ m and back). Average speed: $(2.0 \text{ m})/(3.66 \text{ s}) = 0.55$ m/s. *x* component of average velocity: $\Delta x = x_{\text{f}} - x_{\text{i}} = +3.8 \text{ m} - (+3.8 \text{ m}) = 0$, and so

$$v_{x,\text{av}} = \frac{\Delta x}{\Delta t} = \frac{0}{\Delta t} = 0.$$

Average velocity: $\vec{v}_{\text{av}} = v_{x,\text{av}}\vec{\imath} = (0)\vec{\imath} = \vec{0}$.

(b) The time interval between frames 1 and 17 is $\Delta t = t_{\text{f}} - t_{\text{i}} = 5.33 \text{ s} - 0 = 5.33$ s; the distance traveled is 4.8 m (3.8 m from $x = +1.0$ m to $x = +4.8$ m and then 1.0 m back to $x = +3.8$). Average speed: $(4.8 \text{ m})/(5.33 \text{ s}) = 0.90$ m/s. *x* component of average velocity: $\Delta x = x_{\text{f}} - x_{\text{i}} = +3.8 \text{ m} - (+1.0 \text{ m}) = +2.8$ m, and so

$$v_{x,\text{av}} = \frac{\Delta x}{\Delta t} = \frac{+2.8 \text{ m}}{5.33 \text{ s}} = +0.53 \text{ m/s}.$$

Average velocity: $\vec{v}_{\text{av}} = v_{x,\text{av}}\vec{\imath} = (+0.53 \text{ m/s})\vec{\imath}$.

2.17 From Figure 2.30 you see that, at $t = 2.0$ s, the object is at $x = +2.8$ m. Substituting $t_{\text{i}} = 2.0 \text{ s}$, $x_{\text{i}} = +2.8 \text{ m}$, and $v_x = +0.60$ m/s into Eq. 2.19 yields

$$x_{\text{f}} = +2.8 \text{ m} + (+0.60 \text{ m/s})(t_{\text{f}} - 2.0 \text{ s})$$

$$= +2.8 \text{ m} + (+0.60 \text{ m/s})t_{\text{f}} - (+0.60 \text{ m/s})(2.0 \text{ s})$$

$$= +2.8 \text{ m} + (+0.60 \text{ m/s})t_{\text{f}} - 1.2 \text{ m}$$

$$= +1.6 \text{ m} + (+0.60 \text{ m/s})t_{\text{f}},$$

which is identical to the result in Example 2.10.

2.18 (a) $x_2 = 0$, $x_9 = +0.314$ m;

$$t_2 = 2 \text{ intervals} \times \frac{0.0300 \text{ s}}{\text{interval}} = 0.0600 \text{ s}$$

$$t_9 = 9 \text{ intervals} \times \frac{0.0300 \text{ s}}{\text{interval}} = 0.2700 \text{ s}$$

$$v_{x,\text{av}} = \frac{\Delta x}{\Delta t} = \frac{+0.314 \text{ m}}{0.210 \text{ s}} = +1.50 \text{ m/s}.$$

(b) $x_8 = +0.250$ m, $t_8 = 0.240$ s, so

$$v_{x,\text{av}} = \frac{\Delta x}{\Delta t} = \frac{+0.250 \text{ m}}{0.180 \text{ s}} = +1.39 \text{ m/s}.$$

(c) See Table S2.18. (d) Decreases. Because the ball speeds up as it falls, averaging over a shorter time interval involves lower velocities, and therefore the average decreases (see also Figure 2.32). (e) See Figure S2.18. (f) Note that extrapolating the data to $\Delta t = 0$ does not yield a zero average velocity. This makes sense: The ball's velocity is nonzero before and after passing position 2, and so neither the average value nor the instantaneous value can be zero at position 2.

Table S2.18 Average velocities

Position	Δx (m)	Δt (s)	$\Delta x / \Delta t$ (m/s)
7	+0.197	0.1500	+1.31
6	+0.143	0.1200	+1.19
5	+0.102	0.0900	+1.13
4	+0.061	0.0600	+1.02
3	+0.028	0.0300	+0.93
2	0	0	–

Figure S2.18

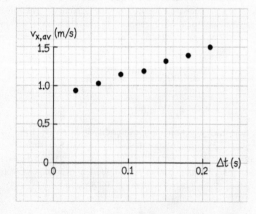

2.19 (a) Because the derivative with respect to time of a constant is zero, you have, from Eq. 2.22, $v_x = dx/dt = 0$. If the position is constant, then its x coordinate is also constant, and so Δx is always zero. Therefore $\Delta x / \Delta t = 0$ for any finite time interval Δt. Substituting $\Delta x / \Delta t = 0$ into Eq. 2.21 yields $v_x = 0$ because the limit of zero is zero. (b) If $x = ct$, then $v_x = dx/dt = d(ct)/dt = c\, dt/dt = c$. Over a finite time interval Δt, you have $x_i = ct_i$ and $x_f = c(t_i + \Delta t)$. Therefore $\Delta x = c(t_i + \Delta t) - ct_i = c\Delta t$, and so $\Delta x / \Delta t = c$. Because c does not depend on Δt, the limit in Eq. 2.21 has no effect and so $v_x = c$. This result makes sense because the relationship $x = ct$ means x changes by a constant amount c each second. Therefore the object travels the same distance in equal time intervals, and so its velocity must be constant.

Chapter 3

3.1 (a) The x component of the average acceleration is the change in the x component of the velocity divided by the time interval during which that change took place:

$$\frac{+5.0 \text{ m/s} - 0}{1.0 \text{ s}} = +5.0 \text{ m/s}^2.$$

(b)
$$\frac{+10 \text{ m/s} - (+5.0 \text{ m/s})}{2.0 \text{ s}} = +2.5 \text{ m/s}^2.$$

(c)
$$\frac{+10 \text{ m/s} - 0}{3.0 \text{ s}} = +3.3 \text{ m/s}^2.$$

3.2 (a) The minus signs on the two v_x values tell you that the car is traveling in the negative x direction. (b) $\Delta v_x = v_{x,f} - v_{x,i} = -2 \text{ m/s} - (-10 \text{ m/s}) = +8 \text{ m/s}$; because Δv_x is positive, the vector $\Delta \vec{v}$ points in the positive x direction. (c) $\vec{a}$ always points in the same direction as $\Delta \vec{v}$ and so points in the positive x direction. Consequently the x component of the acceleration is positive. (d) The acceleration points in the direction opposite the direction of the velocity, and so the car is slowing down.

3.3 (a) For both graphs, x decreases with time, which means both velocities are in the negative x direction, and so the x components of the velocities are negative. (b) The speeds are increasing in part a (more distance covered in equal time intervals as the $x(t)$ curve bends down; decreasing in part b as the curve bends up. (c) From part a you know that the velocity is always in the negative x direction; from part b you know that the speed is increasing in a and decreasing in b. Combining these results, you see that $\Delta \vec{v}$ points in the negative x direction in a and in the positive x direction in b, and so Δv_x is negative in a and positive in b. (d) The x components are negative in a and positive in b ($\vec{a}$ always points in the same direction as $\Delta \vec{v}$, and so the algebraic signs of Δv_x and a_x are the same).

3.4 (a) The car is moving in the negative x direction, and so $\vec{v}$ must point in that direction. To increase the car's speed, $\Delta \vec{v}$ must point in the negative x direction (Figure S3.4). Because $\vec{a}$ and $\Delta \vec{v}$ always point in the same direction, $\vec{a}$ also points in the negative x direction. (b) The figure shows a car slowing down in the negative x direction. Motion in the negative x direction implies a negative slope of the $x(t)$ curve, which eliminates the two graphs in Figure 3.2. Because the car slows down, it should cover increasingly smaller distances (smaller Δx for a given Δt at later instants t). This is so for Figure 3.3b.

Figure S3.4

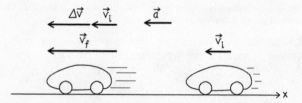

3.5 When the book and the paper are held side by side, the book falls faster. When the paper is on top of the book, they fall at same rate. The difference is due to air resistance, which slows down the light piece of paper but not the heavier book. When the paper is on top of the book, the paper is shielded from the flow of air and so feels no air resistance.

3.6 In Figure 3.6b you can see that Δx goes from -2.2 mm at $n = 2$ to -7.0 mm at $n = 10$: The displacement increases from one exposure to the next as n increases—the ball is accelerating. Because the magnitude of the displacement is increasing with time, the magnitude of the velocity must also be increasing.

3.7 (a) The speed increases because the object covers greater and greater distances in equal time intervals. (b) The x component of the velocity is negative because the motion is in the negative x direction. From part a you know that the magnitude of the velocity increases, and so the x component of the velocity becomes increasingly negative—it *decreases*. (c) Because the object speeds up in the negative x direction, the acceleration points in the negative x direction and so a_x is negative.

3.8 The ball's speed is about 20 m/s after 1 s and 30 m/s after 2 s. All your throwing has done is change the initial speed from zero (the initial speed if the ball is released from rest) to 10 m/s. Once the ball leaves your hand at this initial speed, the only thing affecting its motion is the acceleration due to gravity, which increases the speed at the rate of about 10 m/s².

3.9 See Figure S3.9. Because the rock reverses its motion, the motion diagram is split into two parts: one for the upward motion and the other for the downward motion. If you take up as the positive direction, the x component of the rock's acceleration is $a_x = -9.8$ m/s². With the origin chosen to be at the top of the cliff, the initial conditions are $t_i = 0$, $x_i = 0$, and $v_{x,i} = +15$ m/s. At the top of its trajectory, the rock reverses direction, and so the velocity is zero at that position. You do not know its position or the instant at which it reaches the top, and so your diagram must show $t_1 = ?$, $x_1 = ?$, and $v_{x,1} = 0$. The final conditions at the instant the rock reaches the water are $t_f = ?$, $x_f = -30$ m, and $v_{x,f} = ?$

Figure S3.9

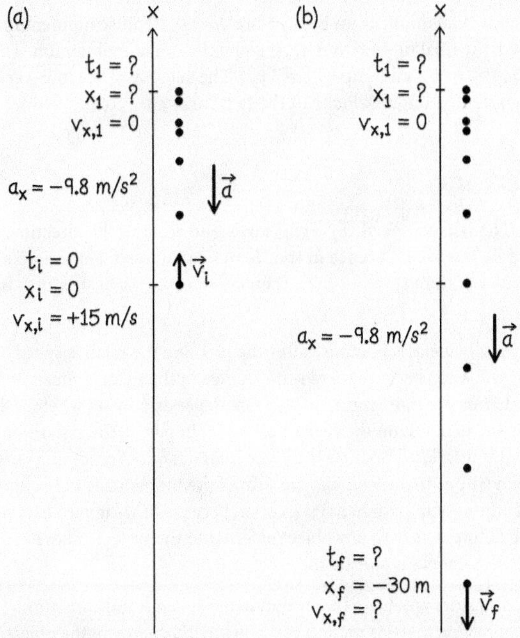

(a) (b)

$t_1 = ?$
$x_1 = ?$
$v_{x,1} = 0$

$a_x = -9.8$ m/s²

$t_i = 0$
$x_i = 0$
$v_{x,i} = +15$ m/s

$t_1 = ?$
$x_1 = ?$
$v_{x,1} = 0$

$a_x = -9.8$ m/s²

$t_f = ?$
$x_f = -30$ m
$v_{x,f} = ?$

3.10 From the motion diagram in Figure 3.11 you have $t_i = 0$, $x_i = 0$, and $v_{x,i} = 0$. Substituting these values into Eq. 3.9 gives $x_f = \frac{1}{2}a_x t_f^2$ or, solving for t_f: $t_f = \sqrt{2x_f/a_x}$. Substituting the values for x_f and a_x, you get $t_f = \sqrt{2(300 \text{ m})/(9.8 \text{ m/s}^2)} = 7.8$ s. Equation 3.10 then yields $v_{x,f} = v_{x,i} + a_x t_f = 0 + (9.8 \text{ m/s}^2)(7.8 \text{ s}) = 77$ m/s. (Alternatively, you can substitute the values for a_x and Δx into Eq. 3.13 to obtain the same result.)

3.11 Figure S3.11 shows the motion diagram for the new choice of x axis. The initial conditions are as before; the final conditions are $t_f = ?$, $x_f = +h$, and $v_{x,f} = ?$. The acceleration is now positive ($a_x = +g$) because it points in the same direction as the x axis, and so Eq. 3.9 is

$$h = 0 + 0 + \tfrac{1}{2}gt_f^2,$$

which yields the same result as before for t_f. The x component of the final velocity is $v_{x,f} = +gt_f$. This x component of the velocity is equal in magnitude to the velocity in Example 3.5 but opposite in sign (because the axis has been inverted).

Figure S3.11

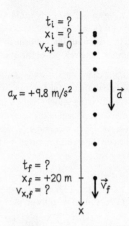

$t_i = ?$
$x_i = ?$
$v_{x,i} = 0$

$a_x = +9.8$ m/s²

$t_f = ?$
$x_f = +20$ m
$v_{x,f} = ?$

X

3.12 Applying Eq. 3.10 to the downward part of the motion, with $v_{x,i} = 0$, $a_x = -g$, and $t = v_{x,i}/g$, you get

$$v_{x,f} = 0 - g\frac{v_{x,i}}{g} = -v_{x,i}.$$

The stone hits the ground at the same speed it had when it was launched.

3.13 Figure S3.13 shows that any angle $180° > \theta > 90°$ gives the same result you would get for an angle of $180° - \theta$. All that matters is the angle the incline makes with the floor; once you go beyond 90°, the cart faces the other way, but that is immaterial. (Compare, for example, the 60° angle with $180° - 60° = 120°$.) Note that Eq. 3.20 properly accounts for this fact because $\sin \theta = \sin(180° - \theta)$. For example,

$$a_{120°} = g \sin(120°) = g \sin(180° - 120°) = a_{60°}.$$

Figure S3.13

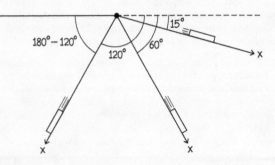

$180° - 120°$ $15°$

$120°$ $60°$

X   X   X

3.14 The first derivative, dropping the subscript f on t_f because it represents an (arbitrary) instant is,

$$\frac{dx_f}{dt} = \frac{d}{dt}(x_i + v_{x,i}t + \tfrac{1}{2}a_x t^2) = 0 + v_{x,i} + a_x t.$$

The expression on the right side is the x component of the velocity of an object moving at constant acceleration (Eq. 3.10). The second derivative yields the x component of the (constant) acceleration:

$$\frac{d^2 x_f}{dt^2} = \frac{d}{dt}\frac{dx_f}{dt} = \frac{d}{dt}(v_{x,i} + a_x t) = 0 + a_x.$$

Chapter 4

4.1 (a) Yes, because each of the $v_x(t)$ curves is a straight line (constant slope = constant acceleration). (b) The concrete, because the magnitude of the acceleration is largest when the magnitude of Δv_x for a given time interval is largest—that is, when the slope of the line is the steepest.

4.2 (a) $\Delta v_{1x} = v_{1x,f} - v_{1x,i} = +0.63\ \text{m/s} - (+0.14\ \text{m/s}) = +0.49\ \text{m/s}$.
(b) $\Delta v_{2x} = v_{2x,f} - v_{2x,i} = +0.14\ \text{m/s} - (+0.63\ \text{m/s}) = -0.49\ \text{m/s}$.
(c) The changes are equal in magnitude but opposite in sign, meaning that the velocity of one increases and the velocity of the other decreases.

4.3 No. The magnitude of the change in the x component of the velocity of the standard cart is always half that of the half cart. If you lower $v_{sx,i}$, the only thing that happens is that the velocity-versus-time graph shrinks in the vertical direction; the x component of the velocity of the standard cart still ends up being positive. If you increase $v_{sx,i}$, the graph expands in the vertical direction and again the x component of the final velocity of the cart is positive.

4.4 $\Delta v_{ux} = v_{ux,f} - v_{ux,i} = +0.21\ \text{m/s} - (0\ \text{m/s}) = +0.21\ \text{m/s}$;
$\Delta v_{sx} = v_{sx,f} - v_{sx,i} = -0.21\ \text{m/s} - (+0.42\ \text{m/s}) = -0.63\ \text{m/s}$.
Therefore $\Delta v_{ux}/\Delta v_{sx} = (+0.21\ \text{m/s})/(-0.63\ \text{m/s}) = -1/3$.

4.5 $\Delta v_{mx} = v_{mx,f} - v_{mx,i} = +0.07\ \text{m/s} - (0\ \text{m/s}) = +0.07\ \text{m/s}$;
$\Delta v_{px} = v_{px,f} - v_{px,i} = -0.20\ \text{m/s} - (+0.27\ \text{m/s}) = -0.47\ \text{m/s}$;
$\Delta v_{px}/\Delta v_{mx} = -6.7$.

4.6 Greater. Δv_{ux} is smaller than Δv_{sx}, which tells you that the cart of unknown inertia puts up more resistance to a change in its velocity and hence has more inertia.

4.7 (a) Extensive. The inertia of a double cart is equal to the sum of the inertias of the individual carts. (b) Intensive. The velocity of a double cart is equal to the velocity of each of the individual carts. (c) Extensive. The product of inertia and velocity of a double cart is equal to the double inertia times the velocity of the assembly: $2m_s v_d$; the sum of the product of inertia and velocity of each individual cart is equal to two times the single inertia times the velocity of the assembly: $2m_s v_d$. In general, the product of an extensive and an intensive quantity is extensive.

4.8 (a) Extensive. If I separate an amount of money into two piles, then the sum of the amounts of money in each pile equals the original sum of money. (b) Not extensive. The sum of the temperatures in each half of a room is not equal to the temperature of the entire room. (c) Not extensive. The sum of the humidities in each half of a greenhouse is not equal to the humidity of the entire greenhouse. (d) Extensive. The volume of a bathtub is equal to the sum of the volumes of each half of the bathtub.

4.9 (a) Yes. If, for example, our the new standard had only half the inertia of the French one, then the inertia in kilograms of all objects measured against the new standard would be twice as great as the values obtained with the French standard. (b) No, because we can't change the world just by choosing another standard. The only thing that matters in a collision is the *ratio* of the inertias. If we choose another standard, the numerical values of both inertias change by a certain factor, but their ratio is unchanged.

4.10 The x component of the change in momentum of the standard cart is positive, so the change in momentum $\Delta \vec{p}_s$ points in the positive x direction. The x component of the change in momentum of the red cart is negative and so $\Delta \vec{p}_r$ points in the negative x direction.

4.11 (a) Nonzero. The puck's velocity decreases, so its momentum changes. (b) Zero. The cart's velocity is constant. (c) Nonzero. The collision changes the cart's velocity. (d) Zero, see Eq. 4.9. (e) Nonzero. The cart's velocity changes.

4.12 (a) No. If you push with your legs against the sled, you are interacting with the sled and so you are not isolated. (b) Ignoring any interaction between sled and ice, however, you and the sled form an isolated system.

4.13 (a) The magnitude of the changes in momentum is $0.061\ \text{kg}\cdot\text{m/s}$, so the collision transfers $0.061\ \text{kg}\cdot\text{m/s}$. (b) Cart 1 gains that amount (Δp_{1x} is positive); cart 2 loses that same amount. So, momentum is transferred from cart 2 to cart 1.

4.14 (a) The magnitude of the impulse delivered to cart 1 is equal to the magnitude of the momentum change of cart 1, which we calculated in Checkpoint 4.13: $J = 0.061\ \text{kg}\cdot\text{m/s}$. (b) The impulse points in the same direction as the momentum change (in the positive x direction),

so $\vec{J} = (+0.061\ \text{kg}\cdot\text{m/s})\hat{\imath}$. (c) No. While momentum is conserved, the momentum of a system or object can change due to transfer of momentum.

4.15 (a) No, because the inertias are not equal: $\Delta v_{1x} = +0.17\ \text{m/s} - 0 = +0.17\ \text{m/s}$; $\Delta v_{2x} = -0.17\ \text{m/s} - (+0.34\ \text{m/s}) = -0.51\ \text{m/s}$.
(b) $m_1/m_2 = (0.36\ \text{kg})/(0.12\ \text{kg}) = 3$; $\Delta v_{2x}/\Delta v_{1x} = (-0.51\ \text{m/s})/(0.17\ \text{m/s}) = -3$. (c) $p_{1x,i} = m_1 v_{1x,i} = (0.36\ \text{kg})(0) = 0$; $p_{2x,i} = (0.12\ \text{kg})(+0.34\ \text{m/s}) = +0.041\ \text{kg}\cdot\text{m/s}$; $p_{1x,f} = (0.36\ \text{kg})(+0.17\ \text{m/s}) = +0.061\ \text{kg}\cdot\text{m/s}$; $p_{2x,f} = (0.12\ \text{kg})(-0.17\ \text{m/s}) = -0.020\ \text{kg}\cdot\text{m/s}$.
(d) $p_{1x,i} + p_{2x,i} = 0\ \text{kg}\cdot\text{m/s} + (+0.041\ \text{kg}\cdot\text{m/s}) = +0.041\ \text{kg}\cdot\text{m/s}$.
(e) $p_{1x,f} + p_{2x,f} = +0.061\ \text{kg}\cdot\text{m/s} + (-0.020\ \text{kg}\cdot\text{m/s}) = +0.041\ \text{kg}\cdot\text{m/s}$. (f) Yes. $\Delta p_{1x} = +0.061\ \text{kg}\cdot\text{m/s} - 0 = +0.061\ \text{kg}\cdot\text{m/s}$; $\Delta p_{2x} = -0.020\ \text{kg}\cdot\text{m/s} - (+0.041\ \text{kg}\cdot\text{m/s}) = -0.061\ \text{kg}\cdot\text{m/s}$. This is so because the momentum of the system does not change.

4.16 (a) Yes. Just as we do for the carts in Figure 4.3, we can ignore the interaction between the carts and the low-friction track, and so the two carts constitute an isolated system. (b) For an isolated system, the momentum law tells us that the momentum of the system cannot change.

4.17 There are two ways to solve this problem. The hard way is to apply the law of conservation of momentum to each collision separately. The easy way is to look at the momentum of the system at only two instants: before the first collision and after the final one. The system's momentum before the first collision is $mv_{x,i}$, where m is the inertia of a single railroad car. After all the cars collide, they move together at the same final velocity $v_{x,f}$. The system's final momentum is therefore $4mv_{x,f}$. Because momentum is conserved, the final momentum must equal the initial momentum. Therefore $4mv_{x,f} = mv_{x,i}$, and thus $v_{x,f} = \frac{1}{4}v_{x,i}$. The velocity of the four-car train is one-quarter the initial velocity of the first car.

Chapter 5

5.1 No. The relative speed $|\vec{v}_2 - \vec{v}_1|$ takes into account the directions of $\vec{v}_1$ and $\vec{v}_2$, but the difference in speeds $v_2 - v_1$ doesn't. For example, if $v_{1x} = +2\ \text{m/s}$ and $v_{2x} = -2\ \text{m/s}$, then $v_{12} = |v_{2x} - v_{1x}| = 4\ \text{m/s}$ but $v_2 - v_1 = 2\ \text{m/s} - 2\ \text{m/s} = 0$.

5.2 (a) Totally inelastic, because after the collision the relative speed is zero. (b) Elastic, because there is no change in the relative speed. (c) Whether or not momentum is constant depends only on whether the system is isolated, not on the type of collision. In part a, the system (ball + glove) is not isolated because the glove interacts with something outside the system (the outfielder). Thus the sum of the momenta is not constant. In part b, the system (two balls) is isolated because, if we ignore friction, neither ball interacts with any object external to the system. Therefore the sum of the momenta is constant.

5.3 Yes. Consider an object that is moving at speed v and consists of two pieces having inertias m_1 and m_2. The kinetic energy of the object, $\frac{1}{2}(m_1 + m_2)v^2$, is equal to the sum of the kinetic energies of its two parts, $\frac{1}{2}m_1 v^2 + \frac{1}{2}m_2 v^2$.

5.4 Denote the x component of the initial velocity of the moving cart $v_{x,i}$ and the x component of the final velocity of the combined carts $v_{x,f}$. Conservation of momentum requires that the two carts have a final momentum equal to the initial momentum, and so $mv_{x,i} = 2mv_{x,f}$. Thus $v_{x,f} = \frac{1}{2}v_{x,i}$. The initial kinetic energy is $\frac{1}{2}mv_{x,i}^2$; the final kinetic energy is $\frac{1}{2}(2m)v_{x,f}^2 = m(\frac{1}{2}v_{x,i})2 = \frac{1}{4}mv_{x,i}^2$. This is half the initial kinetic energy.

5.5 Increase. The system has some kinetic energy before the collision but zero kinetic energy after. This decrease in kinetic energy is accompanied by a change in the state of the dough (it changes shape). This change in state represents a change in the internal energy of the dough. If the sum of the internal and kinetic energies of the system is to remain constant, the internal energy must increase.

5.6 You could throw something against the cart, push it with your hand, blow against it, or ignite an explosive next to it. Changes in state or motion: The motion of the thrown object changes when it collides with the cart.

To push or blow, you have to move your muscles, which causes chemical reactions that change your chemical state. The ignition causes the chemical state of the explosive to change.

5.7 (*a*) No. The momentum of the system is zero before the spring expands and nonzero afterward. (*b*) No, because $\Delta p \neq 0$. There is an external interaction with the post that holds the left end of the spring fixed. Without this interaction, the expanding spring would not cause the cart to move.

5.8 See Figure S5.8. The car engine heats up, and so part of the chemical energy is converted to thermal energy. The sum of the changes ΔK and ΔE_{th} is equal in magnitude to the change ΔE_{chem}.

Figure S5.8

initial

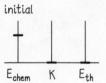

$E_{chem} \quad K \quad E_{th}$

final

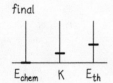

$E_{chem} \quad K \quad E_{th}$

change in motion: car accelerates
changes in state: • chemical state of fuel changes
                   • engine gets warm

5.9 (*a*) No. Changing the magnitude of an object's momentum means changing the object's speed, and this automatically means changing its kinetic energy. (*b*) Yes. If the change in kinetic energy is compensated by a change in the internal energy of the object, its energy remains unchanged. (*c*) Yes, for both parts. With a system of objects, it is possible to keep the kinetic energy constant while changing the magnitude of the momentum. Imagine two identical carts on a low-friction track moving away from each other in opposite directions at speed v. The momentum of the system is zero, and the kinetic energy of the system is mv^2. If the velocity of one of the two carts is reversed by some interaction, the system's kinetic energy remains unchanged but the momentum of the system becomes $2mv$. For the energy of the system, use the same reasoning you used to answer part *b*.

5.10 For a single object, if mv_x and mv_x^2 remain constant, then the product of these two quantities must also remain constant: $(mv_x)(mv_x^2) = m^2v_x^3 =$ constant. Because m does not change for a single, isolated object, $m^2v_x^3/m = mv_x^3$ must also be constant. Therefore mv_x^3 indeed remains constant for a single object.

 In an elastic collision between two objects of identical inertia m with velocities v_{1x} and v_{2x}, the system's momentum, $mv_{1x} + mv_{2x}$, remains constant. The system's kinetic energy is also constant, and so $mv_{1x}^2 + mv_{2x}^2$ is constant. The product of these two must also be constant: $m^2v_{1x}^3 + m^2v_{1x}^2v_{2x} + m^2v_{1x}v_{2x}^2 + m^2v_{2x}^3 = m(mv_{1x}^3 + mv_{2x}^3) + mv_{1x}v_{2x}(mv_{1x} + mv_{2x}) = $ constant. From this we see that $(mv_{1x}^3 + mv_{2x}^3)$ is constant if the second term, $mv_{1x}v_{2x}(mv_{1x} + mv_{2x})$, is constant. The part in parentheses is the system's momentum and is constant. The inertia m is also constant, so all we need to know is whether or not $v_{1x}v_{2x}$ is constant. Suppose $v_{1x,i}$ changes by Δv_{1x}. Conservation of momentum then requires $v_{2x,i}$ to change by $-\Delta v_{1x}$. So the product of the final velocities becomes $(v_{1x,i} + \Delta v_{1x})(v_{2x,i} - \Delta v_{1x}) = v_{1x,i}v_{2x,i} + \Delta v_{1x}(v_{2x,i} - v_{1x,i}) - (\Delta v_{1x})^2$. This can be equal to the product of the initial velocities, $v_{1x,i}v_{2x,i}$, only if the change in velocity Δv_{1x} is equal to $v_{2x,i} - v_{1x,i}$. This can indeed happen but, in general, need *not* be so. In general, therefore, mv_x^3 is not unchanged in an elastic collision.

5.11 No. If one object is initially at rest, then the other object must be moving in order for the collision to occur, and so the initial momentum of the system is nonzero. Because the system is isolated, its momentum is unchanged, and so the system's final momentum, too, is nonzero. Therefore the combination of the two objects has a nonzero velocity after the collision, which means the kinetic energy cannot be zero.

5.12 From Eq. 5.12 you know that $v = \sqrt{2K/m}$. Substituting 1.2×10^8 J for the kinetic energy K and 1200 kg for the inertia m gives $v = 4.5 \times 10^2$ m/s, which is more than 1000 mph! (Apparently it is not possible to convert all of the energy in gasoline to kinetic energy! You'll see why in Chapter 21.)

5.13 No. Cart 1 gets twice as much energy as cart 2: $K_{1f} = \frac{1}{2}m_1v_{1f}^2 = \frac{1}{2}(0.25 \text{ kg})(2.0 \text{ m/s})^2 = 0.50$ J, and $K_{2f} = \frac{1}{2}m_2v_{2f}^2 = \frac{1}{2}(0.50 \text{ kg})(1.0 \text{ m/s})^2 = 0.25$ J. The reason is that the system's final momentum needs to be zero, and so v_{1f} must be $2v_{2f}$. Because $m_2 = 2m_1$, you have $K_{2f} = \frac{1}{2}m_2v_{2f}^2 = \frac{1}{2}(2m_1)(\frac{1}{2}v_{1f})^2 = \frac{1}{4}m_1v_{1f}^2 = \frac{1}{2}K_{1f}$.

Chapter 6

6.1 Because $\vec{v}_{Eo} = \vec{v}_{EM} + \vec{v}_{Mo}$, you can write $v_{Mox} = v_{Eox} - v_{EMx}$. With $v_{EMx} = -3.0$ mm/frame, you have $v_{M1x} = 0 - (-3.0 \text{ mm/frame}) = +3.0$ mm/frame and $v_{M2x} = (3.6 \text{ mm/frame}) - (-3.0 \text{ mm/frame}) = +6.6$ mm/frame.

6.2 (*a*) There is no change in the state of either cart, and so the internal energy of either cart is not changing. From observer E's point of view, both velocities are constant and so both kinetic energies are constant. Observer M reaches the same conclusion: Both carts are moving at constant velocity and so their kinetic energies are constant. The observers agree that the energy of each cart is constant. (*b*) and (*c*) No state change or motion change in the environment surrounding each isolated system causes any state change or motion change inside the system, and so each system is closed.

6.3 (*a*) There is no change in the state of either cart and so no change in the internal energy. For observer E, cart 1 accelerates, and so its kinetic energy K and energy E increase; the velocity of cart 2 is constant, and so its kinetic energy K and energy E are constant. For observer M, cart 1 is at rest, and so its kinetic energy K and energy E are constant; cart 2 accelerates, and so its kinetic energy K and energy E are increasing. (*b*) For observer E, a change in state in the spring is responsible for the cart's acceleration, and so the system is not closed. For observer M, changes in both state and motion occur in the environment, but the expansion of the spring cannot account for the change in the kinetic energy of the entire environment (the universe minus cart 1), which means the energy changes don't add to zero and so the system is not closed. (*c*) For observer E, there are no changes in state or motion in the environment that are responsible for any changes in state or motion inside the system, and so the system is closed. For observer M, there are no changes in state or motion in the environment, and so it is closed. (*d*) For observer E, the cart 1 system is not closed and its energy increases; the cart 2 system is closed, and its energy is constant. The conservation of energy law holds for both systems. For observer M, the cart 1 system is not closed, and its energy remains constant; the cart 2 system is closed, and its energy is increasing. Neither observation agrees with the law of conservation of energy.

6.4 (*a*) $v_{Ecx,i} = +0.40$ m/s, $v_{Ecx,f} = +0.80$ m/s, $\Delta K = K_{Ec,f} - K_{Ec,i} = \frac{1}{2}(0.12 \text{ kg})(0.80 \text{ m/s})^2 - \frac{1}{2}(0.12 \text{ kg})(0.40 \text{ m/s})^2 = 29$ mJ. (*b*) $v_{Mcx,i} = -0.20$ m/s, $v_{Mcx,f} = +0.20$ m/s, $\Delta K = \frac{1}{2}(0.12 \text{ kg})(0.20 \text{ m/s})^2 - \frac{1}{2}(0.12 \text{ kg})(-0.20 \text{ m/s})^2 = 0$. (*c*) $v_{Mcx,i} = -0.40$ m/s, $v_{Mcx,f} = 0$, $\Delta K = 0 - \frac{1}{2}(0.12 \text{ kg})(-0.40 \text{ m/s})^2 = 10$ mJ.

6.5 Because the momentum of the isolated system is not changing: $m_1v_{E1x,i} + m_2v_{E2x,i} = (m_1 + m_2)v_{Ex,f}$, and so $v_{Ex,f} = (m_1v_{E1x,i} + m_2v_{E2x,i})/(m_1 + m_2) = +0.20$ m/s. In the moving reference frame, $v_{Mx,f} = +0.40$ m/s. See Table S6.5 on next page.

Table S6.5 Conversion of kinetic energy

	Inertia	v_x (m/s)			K (10^{-3} J)		
Cart	(kg)	before	after	Δv_x	before	after	ΔK
Earth reference frame							
1	0.36	0	+0.20	+0.20	0	7.2	+7.2
2	0.12	+0.80	+0.20	−0.60	38.4	2.4	−36
						$\Delta K =$	−29
Reference frame moving at −0.20 m/s relative to Earth							
1	0.36	+0.20	+0.40	+0.20	7.2	28.8	+21.6
2	0.12	+1.0	+0.40	−0.60	60	9.6	−50.4
						$\Delta K =$	−29

6.6 Because the laws of the universe are the same in all inertial reference frames, the coefficient of restitution must also be the same. Another way of seeing this: Relative velocities do not depend on the velocity of the reference frame (Figure 6.11), and so $e = v_{12f}/v_{12i}$ is also independent of the reference frame.

6.7 Less. In the Earth reference frame, $K_{E,i} = 0 + \frac{1}{2}(0.12$ kg$) \cdot (0.80$ m/s$)^2 = 38$ mJ. In the zero-momentum reference frame, $K_{Z,i} = \frac{1}{2}(0.36$ kg$)(0.20$ m/s$)^2 + \frac{1}{2}(0.12$ kg$)(0.60$ m/s$)^2 = 29$ mJ.

6.8 (a) If the jogger stays in place, he must be moving at the same speed as the belt but in the opposite direction. Therefore $v_{Bjx} = -2.0$ m/s, and so $r_{Bjx} = v_{Bjx}t = (-2.0$ m/s$)(10$ s$) = -20$ m. (b) Equation 6.3 yields $r_{Ejx} = r_{EBx} + r_{Bjx}$. The position of the belt relative to Earth is $r_{EBx} = v_{EBx}\Delta t$, and the position of the jogger relative to the belt is $r_{Bjx} = v_{Bjx}\Delta t$. Because $v_{Bjx} = -v_{EBx}$, you find that $r_{Ejx} = r_{EBx} + r_{Bjx} = v_{EBx}\Delta t + v_{Bjx}\Delta t = 0$.

6.9 If you define northward as the positive x direction, you have $v_{TPx} = +1.2$ m/s, $v_{ETx} = +3.1$ m/s, and $v_{Psx} = -0.5$ m/s. For the velocity of the spider relative to Earth, subscript cancellation gives $v_{Esx} = v_{EPx} + v_{Psx}$. For the velocity of the passenger relative to Earth, $v_{EPx} = v_{ETx} + v_{TPx}$, and so $v_{Esx} = (v_{ETx} + v_{TPx}) + v_{Psx} = (+3.1$ m/s$) + (+1.2$ m/s$) + (-0.5$ m/s$) = +3.8$ m/s. An observer standing by the side of the tracks sees the spider moving at 3.8 m/s northward.

6.10 (a) Substituting $m_1 = 3m_2 = 3m$ into Eq. 1 of Example 6.7 yields $x_{Acm} = (3mx_{A1} + mx_{A2})/(3m + m) = (3x_{A1} + x_{A2})/4$, one-quarter of the way from x_{A1} to x_{A2}. (b) $x_{Acm} = (m_1x_{A1} + m_2x_{A2} + m_3x_{A3})/(m_1 + m_2 + m_3)$. Substituting $m_1 = m_3 = 3m_2 = 3m$ and $x_{Acm} = x_{A2}$ gives $x_{A2} = (3mx_{A1} + mx_{A2} + 3mx_{A3})/(3m + m + 3m) = (3x_{A1} + x_{A2} + 3x_{A3})/7$. Then $x_{A3} = 2x_{A2} - x_{A1}$.

6.11 (a) Because $m_1 = 3m_2$, Eq. 6.26 gives: before the collision $v_{cmx} = [(m_1)(0) + (m_2)(+0.80$ m/s$)]/(m_1 + m_2) = +0.20$ m/s; after the collision $v_{cmx} = [(3m_2)(+0.40$ m/s$) + m_2(-0.40$ m/s$)]/(3m_2 + m_2) = +0.20$ m/s. This is the x component of the velocity where the two $v_x(t)$ curves intersect. (b) Yes. When two objects collide, the object that initially has greater velocity always ends up with less velocity after the collision (and vice versa). Therefore there always has to be a location at which the two objects have the same velocity. At that location $v_{1x} = v_{2x} = v_x$, and so

$$v_{cmx} = (m_1v_x + m_2v_x)/(m_1 + m_2)$$
$$= v_x(m_1 + m_2)/(m_1 + m_2) = v_x.$$

(c) From Eqs. 6.23 and 6.26, $v_{EZx} = v_{cmx} = +0.20$ m/s. From Eq. 6.8, $v_{E1x,i} = v_{EZx} + v_{Z1x,i}$, and so $v_{Z1x,i} = v_{E1x,i} - v_{EZx} = 0 - (+0.20$ m/s$) = -0.20$ m/s, $v_{Z2x,i} = +0.6$ m/s, $v_{Z1x,f} = +0.2$ m/s, and $v_{Z2x,f} = -0.6$ m/s. Then $p_{Zx,i} = (3m_2)(-0.20$ m/s$) + (m_2)(+0.60$ m/s$) = 0$ and $p_{Zx,f} = (3m_2)(+0.20$ m/s$) + (m_2)(-0.60$ m/s$) = 0$.

6.12

$$K_{conv} = (\tfrac{1}{2}m_1v_1^2 + \tfrac{1}{2}m_2v_2^2) - \tfrac{1}{2}(m_1 + m_2)v_{cm}^2$$

$$= (\tfrac{1}{2}m_1v_1^2 + \tfrac{1}{2}m_2v_2^2) - \tfrac{1}{2}\frac{(m_1v_{1x} + m_2v_{2x})^2}{m_1 + m_2}$$

$$= (\tfrac{1}{2}m_1v_1^2 + \tfrac{1}{2}m_2v_2^2) - \tfrac{1}{2}\frac{(m_1^2v_1^2 + 2m_1m_2v_{1x}v_{2x} + m_2^2v_2^2)}{m_1 + m_2}$$

$$= \tfrac{1}{2}\frac{(m_1v_1^2 + m_2v_2^2)(m_1 + m_2) - (m_1^2v_1^2 + 2m_1m_2v_{1x}v_{2x} + m_2^2v_2^2)}{m_1 + m_2}$$

$$= \tfrac{1}{2}\frac{(m_1m_2v_1^2 + m_1m_2v_2^2) - 2m_1m_2v_{1x}v_{2x}}{m_1 + m_2}$$

$$= \tfrac{1}{2}\frac{m_1m_2(v_1^2 - 2v_{1x}v_{2x} + v_2^2)}{m_1 + m_2} = \tfrac{1}{2}\left(\frac{m_1m_2}{m_1 + m_2}\right)(v_{2x} - v_{1x})^2$$

$$= \tfrac{1}{2}\left(\frac{m_1m_2}{m_1 + m_2}\right)v_{12}^2$$

6.13 (a) Because $\mu = m(0.5m)/(m + 0.5m) = m/3$, $K_{conv} = \frac{1}{2}(m/3)v^2$. Because $K_i = \frac{1}{2}mv^2$, the fraction that is convertible is $K_{conv}/K_i = 1/3$. (b) If it were converted, the momentum of the isolated two-object system would change.

6.14 (a) $v_{12i} = |0 - (+4.0$ m/s$)| = 4.0$ m/s and $v_{12f} = |(+1.5$ m/s$) - (-0.50$ m/s$)| = 2.0$ m/s, so $e = (2.0$ m/s$)/(4.0$ m/s$) = 0.50$. (b) See Table S6.14.

Table S6.14 Conversion of kinetic energy

	inertia	Velocity (m/s)			Kinetic energy (J)		
Object	(kg)	before	after	Δv_x	before	after	ΔK
Earth reference frame							
1	1.0	+4.0	−0.50	−4.5	8.0	0.13	−7.9
2	3.0	0	+1.5	+1.5	0	3.4	+3.4
						$\Delta K =$	−4.5
Reference frame moving at −1.0 m/s relative to Earth							
1	1.0	+5.0	+0.50	−4.5	12.5	0.13	−12.4
2	3.0	+1.0	+2.5	+1.5	1.5	9.4	+7.9
						$\Delta K =$	−4.5

Chapter 7

7.1 (a) The other object is Earth; this is an attractive interaction because it makes the ball and Earth accelerate toward each other. (b) Repulsive, because the interaction causes the two objects to accelerate away from each other.

7.2 (a) At $t = 30$ ms, the velocity-versus-time graph shows $v_{1x} = 0$ and $v_{2x} = +0.55$ m/s, so $p_{1x} = (0.12$ kg$)(0) = 0$ and $p_{2x} = (0.24$ kg$)(+0.55$ m/s$) = +0.13$ kg·m/s. At $t = 60$ ms, $v_{1x} = 0.37$ m/s and $v_{2x} = 0.37$ m/s, so $p_{1x} = (0.12$ kg$)(+0.37$ m/s$) = +0.044$ kg·m/s and $p_{2x} = (0.24$ kg$)(+0.37$ m/s$) = +0.089$ kg·m/s. At $t = 90$ ms, $v_{1x} = +0.73$ m/s and $v_{2x} = +0.18$ m/s, so $p_{1x} = +0.088$ kg·m/s and $p_{2x} = +0.043$ kg·m/s. (b) At 30 ms, $p_x = p_{1x} + p_{2x} = 0 + (+0.13$ kg·m/s$) = +0.13$ kg·m/s. At 60 ms, $p_x = (+0.044$ kg·m/s$) + (+0.089$ kg·m/s$) = +0.13$ kg·m/s. At 90 ms, $p_x = (+0.088$ kg·m/s$) + (+0.043$ kg·m/s$) = +0.13$ kg·m/s.

7.3 (a) With the velocity values from Checkpoint 7.2 and the formula $K = \frac{1}{2}mv^2$, the kinetic energies are: at $t = 30$ ms, $K_1 = 0$ and $K_2 = 0.036$ J; at $t = 60$ m/s, $K_1 = 0.0082$ J and $K_2 = 0.016$ J; at $t = 90$ ms, $K_1 = 0.032$ J and $K_2 = 0.0039$ J. (b) At 30 ms, $K = 0.036$ J; at 60 ms, $K = 0.025$ J; at 90 ms, $K = 0.036$ J.

7.4 (a) Zero, because the ball's speed is zero. (b) The momentum of the ball is constant only when the ball is isolated—in other words, when it is not interacting with anything. Ignoring the gravitational interaction with Earth because that interaction is much, much weaker than the ball-wall interaction, you can say that the ball is not interacting with anything before and after the collision, and so its momentum remains constant during these intervals. The ball interacts with the wall during the collision, and so now momentum is not constant for the system made up of only the ball; momentum is constant, however, for the system made up of ball plus wall.

7.5 As the spring is compressed, the kinetic energy of the cart is converted to potential energy in the deformed spring. Because the sum of the kinetic and potential energies must not change, any loss in kinetic energy must be made up by a gain in potential energy. Initially, all the energy of the system is kinetic and equal to $\frac{1}{2}mv_i^2$, where m is the inertia of the cart. In the middle drawing, the kinetic energy is zero, and so the potential energy is $\frac{1}{2}mv_i^2$.

7.6 (a) Let the direction in which the puck moves be the $+x$ direction. The x component of the puck's initial velocity is then $v_x = +8.0$ m/s, and the x component of its acceleration is $a_x = -1.0$ m/s^2. With that acceleration, the puck's speed is reduced by 1.0 m/s each second. So, after 4 s, the speed is half of its original value. Because kinetic energy depends on v^2, the puck's kinetic energy is reduced to one-quarter of its original value after 4 s. After 6 s, the speed is one-quarter of its original value and the kinetic energy is one-sixteenth. Based on these data, your sketches should look like Figure S7.6. (b) The kinetic energy is converted to thermal energy (both puck and ice get warmer).

Figure S7.6

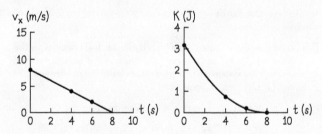

7.7 It should be classified as incoherent configuration energy. Chemical energy is associated with changes in the arrangement of atoms in molecules and so is configuration energy. In a chemical reaction, the orientation of the molecules and the kinetic energy given to the products are random, and so chemical energy is incoherent. It is not possible to make all products move in one direction and extract (coherent) kinetic energy. (The same arguments hold for nuclear energy, the form of incoherent configuration energy that corresponds to changes in nuclear configuration.)

7.8 Both of you, because the term *conserve* means one thing when used by physicists and another thing when used by environmentalists. To physicists, energy is always conserved, which means it is neither created out of nothing nor destroyed. When you leave the lights on, electrical energy is converted to light energy and thermal energy (lamps get hot). The sum of those forms of energy is equal to the electrical energy consumed by the lights. To environmentalists, *conserving energy* means reducing or eliminating the use of source energy. When you turn off the lights, less source energy (oil, gas, coal, or nuclear fuel) needs to be converted to electrical energy. The phrase *conserving energy* has nothing to do with the law of conservation of energy, and because Earth's supply of source energy is limited, you are right to turn off the lights.

7.9 (a) Elastic potential energy in the spring is converted to kinetic energy in the ball and in the spring as the spring expands. The interaction is nondissipative because it can run in reverse: The ball compresses the spring. (b) The ball gains speed as it changes its position relative to the ground, which means potential energy is converted to kinetic energy. The interaction is nondissipative because it can run in reverse: The ball is launched up. (c) Kinetic energy is converted to thermal energy: The bicycle slows down because of friction, which irreversibly dissipates the kinetic energy. The interaction is dissipative because the reverse motion—the bicycle accelerates backward without the pedals moving—is not possible. (d) Chemical energy in the gasoline is converted to kinetic energy (and thermal energy). The conversion is irreversible and therefore the interaction is dissipative. (If you played this situation in reverse, you would see a car slowing down backward. This is possible, but close inspection would reveal details that are not possible: The driver pressing the gas pedal instead of the brake pedal and fuel being added to the tank.)

7.10 (a) When you throw a ball, your momentum increases in the direction opposite the direction of the ball's momentum because the momentum of the isolated system comprising you and the ball must be constant. The throwing corresponds to an explosive separation ($e > 1$). The catching corresponds to a totally inelastic collision ($e = 0$). So, when your friend catches the ball, her momentum increases in the direction in which the ball was originally moving (away from you). (Normally these momentum changes aren't noticeable because of the friction between feet and ground, which is why this checkpoint has you standing on ice.) (b) The thrower moves back ("recoils") in the direction opposite the direction of the thrown ball (away from the catcher). The catcher recoils in the direction in which the ball was originally moving (away from the thrower). Therefore, each throw or catch increases the momentum of each of you in the direction away from the other, meaning that the interaction is repulsive. The ball serves as the carrier—the "gauge particle"—of the interaction. (c) Improbable as it may seem, there actually is a mechanical model for an attractive interaction based on an exchange of particles. Imagine two people standing within arm's reach of each other and extending their arms toward each other without touching. Person A holds a ball and throws it back toward himself; person B catches it before it hits person A, and throws it back toward herself. Repeat as needed. This is perhaps not easy to implement, but the throwing back and forth does result in an attraction.

7.11 (a) Once the vehicles lock together, they have identical velocities, the center-of-mass velocity. If you take the x component of the car's initial velocity as positive, you get from Eq. 6.26 $v_{cmx} = [(1000 \text{ kg})(+25 \text{ m/s}) + (2000 \text{ kg})(-25 \text{ m/s})]/[1000 \text{ kg} + 2000 \text{ kg}] = -8.3$ m/s. Therefore $\Delta v_c = (-8.3 \text{ m/s}) - (+25 \text{ m/s}) = -33$ m/s and $\Delta v_v = (-8.3 \text{ m/s}) - (-25 \text{ m/s}) = +17$ m/s. Because it has less inertia, the car has the larger change in velocity. Because both velocity changes occur during the same time interval, the acceleration is greater for the car, in agreement with Eq. 7.6. (b) $a_{cx,av} = (-33 \text{ m/s})/(0.20 \text{ s}) = -1.7 \times 10^2$ m/s^2 and $a_{cx,av} = (+17 \text{ m/s})/(0.20 \text{ s}) = +83$ m/s^2. The ratio is not exactly -2 because of the rounding of the velocity changes in part a.

7.12

$$\Delta K_E = \tfrac{1}{2}m_E v_{E,f}^2 - \tfrac{1}{2}m_E v_{E,i}^2 = \tfrac{1}{2}m_E(v_{E,f}^2 - v_{E,i}^2)$$

$$= \tfrac{1}{2}m_E[(v_{Ex,i} + \Delta v_{Ex})^2 - v_{E,i}^2]$$

$$= m_E v_{E,i}\Delta v_{Ex} + \tfrac{1}{2}m_E(\Delta v_{Ex})^2.$$

Rearranging terms and substituting Δp_{Ex} for $m_E \Delta v_{Ex}$, you get

$$\Delta K_E = v_{Ex,i}(m_E \Delta v_{Ex}) + \frac{1}{2}\frac{(m_E \Delta v_{Ex})^2}{m_E}$$

$$= v_{Ex,i}\Delta p_{Ex} + \frac{(\Delta p_{Ex})^2}{2m_E}.$$

The first term on the right of the final equals sign is zero because in the reference frame of Earth, $v_{Ex,i} = 0$. The second term to the right of the final equals sign is negligibly small compared to the corresponding term for the cart because the numerators are the same (conservation of momentum requires that $\Delta p_E = -\Delta p_c$) and $m_E \gg m_c$. Therefore $\Delta K_E \approx 0$, and the only change in kinetic energy in the Earth-cart system must be that of the cart.

7.13
$$\Delta U_{12} = U(x_2) - U(x_1)$$

$$\Delta U_{21} = U(x_1) - U(x_2) = -\Delta U_{12}$$

$$\Delta U_{1 \to 2 \to 1} = \Delta U_{12} + \Delta U_{21}$$

$$= [U(x_2) - U(x_1)] + [U(x_1) - U(x_2)]$$

$$= 0.$$

The potential energy change as an object moves in one direction has a sign, and the potential energy change when the object moves back to its original position has the opposite sign and the same magnitude. Thus the change in potential energy along a round trip is zero. This statement, trivial as it may seem, is another hallmark of potential energy. No other form of energy has this property.

7.14 As the ball moves upward, it slows down, reducing its kinetic energy. Because the Earth-ball system is closed, the gravitational potential energy of the system must therefore increase as the ball moves upward. The slowing down of the upward-moving ball means that the acceleration points down—the direction in which the gravitational potential energy of the Earth-ball system decreases. When the ball moves downward, it is again accelerated downward, its kinetic energy increases, and its potential energy decreases.

7.15 (a) It increases (see Checkpoint 7.14). (b) $\Delta U^G = (3.4 \text{ kg})(9.8 \text{ m/s}^2)$ $(1.0 \text{ m}) = 33 \text{ J}$. (c) Your arm muscles burn up some chemical energy as you raise the book.

7.16 No. Even though you may hear people say that the quantity expressed by Eq. 7.21 is "the gravitational potential energy of the ball," it is incorrect to do so. It is not the ball by itself that possesses the potential energy. The gravitational potential energy is a property of the configuration of the Earth-ball system. Put another way, the ball by itself is not a closed system—it does not fall by itself. Consequently, the energy of the ball by itself is not constant.

7.17 With $e = 0$, Eq. 7.26 gives $\Delta E_{th} = \frac{1}{2}\mu v_{12,i}^2$. The reduced inertia is $\mu = (1000 \text{ kg})(2000 \text{ kg})/(1000 \text{ kg} + 2000 \text{ kg}) = 667 \text{ kg}$, and the initial relative speed is $v_{12,i} = 50 \text{ m/s}$. Thus $\Delta E_{th} = \frac{1}{2}(667 \text{ kg})(50 \text{ m/s})^2 = 8.3 \times 10^5 \text{ J}$.

Chapter 8

8.1 Because the crate's velocity is constant, its momentum is also constant. So the change in its momentum is zero, and therefore the rate of change in its momentum is also zero. Surprised? Read on.

8.2 (a) As you set the crate in motion, the acceleration and the change in momentum point in the direction of travel. Because the vector sum of the forces exerted on the crate is equal to the time rate of change in its momentum, the vector sum points in the direction of travel. This makes sense: To make the crate move faster in a certain direction, you must push it in that direction. (b) When the crate is slowing down, all the signs are reversed: The acceleration points opposite the direction of travel, and so the change in momentum and the vector sum of the forces point opposite the direction of travel. Again, this makes sense because in order to stop the crate, you have to exert a force in the direction opposite its direction of travel. (c) When the crate is at rest—whether you are pushing against it or not—the vector sum of the forces exerted on it is zero, and so there is no direction involved.

8.3 (a) Because both collisions have the same initial and final velocities, it suffices to look at must one. To see whether the momentum of the

two-cart system is constant, we calculate the change in momentum of each cart:

$$\Delta v_{1x} = v_{1x,f} - v_{1x,i} = (+0.80 \text{ m/s}) - 0 = +0.80 \text{ m/s}$$

$$\Delta v_{2x} = v_{2x,f} - v_{2x,i} = (+0.20 \text{ m/s}) - (+0.60 \text{ m/s}) = -0.40 \text{ m/s}$$

$$\Delta p_{1x} = m_1 \Delta v_{1x} = (0.12 \text{ kg})(+0.80 \text{ m/s}) = +0.096 \text{ kg} \cdot \text{m/s}$$

$$\Delta p_{2x} = m_2 \Delta v_{2x} = (0.24 \text{ kg})(-0.40 \text{ m/s}) = -0.096 \text{ kg} \cdot \text{m/s}.$$

The total change in momentum is therefore zero, $\Delta p_{1x} + \Delta p_{2x} = 0$, and so the momentum of the system does not change. (b) Whether or not the collision is elastic can be determined by looking at the relative speeds. Before the collision $v_{12i} = |v_{1x,i} - v_{2x,i}| = |0 - (+0.60 \text{ m/s})| = 0.60 \text{ m/s}$; after the collision $v_{12f} = |v_{1x,f} - v_{2x,f}| = |(+0.80 \text{ m/s}) - (+0.20 \text{ m/s})| = 0.60 \text{ m/s}$. Therefore the relative speed is the same before and after the collision, $v_{12i} = v_{12f}$ and so $e \equiv v_{12f}/v_{12i} = 1$, and the collisions are indeed elastic. (You can obtain the same result by verifying that the kinetic energy of the two-cart system doesn't change.)

8.4 Yes. Because two colliding objects can be considered isolated, the momentum of the system remains unchanged regardless of the type of interaction. Therefore the changes in the momentum of the objects are equal in magnitude but opposite in direction, $\Delta \vec{p}_1 = -\Delta \vec{p}_2$, and so the objects exert equal forces on each other in opposite directions.

8.5 (a) The two magnitudes are equal. If that surprises you—how can something so small exert a force on something so large?—remember that any change in the book's momentum must be made up by an equal change of opposite sign in Earth's momentum, because the Earth-book system is isolated. (b) Yes. The fact that the book exerts a force $\vec{F}_{\text{by book on Earth}} = \Delta \vec{p}_E/\Delta t$ means Earth's momentum must be changing. Its velocity must therefore be changing, and so it is accelerating. Because its inertia is so great, however, the acceleration is negligible: The Earth-book system is isolated and so $\Delta \vec{p}_E + \Delta \vec{p}_b = m_E \Delta \vec{v}_E + m_b \Delta \vec{v}_b = \vec{0}$, so that

$$\Delta \vec{v}_E = \frac{m_b}{m_E} \Delta \vec{v}_b \approx \vec{0},$$

which means that Earth's acceleration during this interaction is vanishingly small.

8.6 (a) Because the collision is totally inelastic, the final velocities are the same: $v_{bx,f} = v_{mx,f}$. Because the mosquito-bus system is isolated, the final momentum of the system must be the same as the initial momentum. The x component of the initial momentum is

$$p_{x,i} = m_m v_{mx,i} + m_b v_{bx,i} = m_m(0) + m_b v_{bx,i}$$

$$= 0 + (10^4 \text{ kg})(+25 \text{ m/s}) = +2.5 \times 10^5 \text{ kg} \cdot \text{m/s}.$$

The x component of the final momentum is

$$p_{x,f} = m_m v_{mx,f} + m_b v_{bx,f} = (m_m + m_b)v_{x,f},$$

where $v_{x,f}$ is the x component of the common final velocity. Substituting inertia values gives

$$p_{x,f} = (10^{-4} \text{ kg} + 10^4 \text{ kg})v_{x,f} = p_{x,i} = +2.5 \times 10^5 \text{ kg} \cdot \text{m/s}$$

or, solving for $v_{x,f}$,

$$v_{x,f} = +25 \text{ m/s}.$$

It looks like the velocity of the bus hasn't changed at all. If you pretend that all the numerical values given are exact and keep more significant digits, however, you find that $v_{bx,f} = +24.99999975 \text{ m/s}$, which means the bus velocity changes by a minuscule amount: $\Delta v_{bx} = -0.00000025 \text{ m/s}$ (no one in the bus notices!). The x component of the change in the mosquito's velocity is $\Delta v_{mx} = +25 \text{ m/s}$.

(b) The x component of the change in the average acceleration of the bus is

$$a_{bx} = \Delta v_{bx}/\Delta t = (-0.00000025 \text{ m/s})/(0.005 \text{ s})$$

$$= -0.00005 \text{ m/s}^2.$$

That of the mosquito is

$$a_{mx} = \Delta v_{mx}/\Delta t = (+25 \text{ m/s})/(0.005 \text{ s}) = +5000 \text{ m/s}^2!$$

(c) The x component of the momentum changes are

$$\Delta p_{bx} = m_b \Delta v_{bx} = (10{,}000 \text{ kg})(-0.00000025 \text{ m/s})$$

$$= -0.0025 \text{ kg} \cdot \text{m/s}$$

and

$$\Delta p_{mx} = m_m \Delta v_{mx} = (10^{-4} \text{ kg})(+25 \text{ m/s})$$

$$= +0.0025 \text{ kg} \cdot \text{m/s}.$$

Notice that these changes in momentum are equal in magnitude and opposite in direction, so that the momentum of the isolated system doesn't change.

(d), (e)

$$(F_{\text{by bus on mos}})_x = \Delta p_{mx}/\Delta t = (+0.0025 \text{ kg} \cdot \text{m/s})/(0.005 \text{ s})$$

$$= +0.5 \text{ kg} \cdot \text{m/s}^2$$

$$(F_{\text{by mos on bus}})_x = \Delta p_{bx}/\Delta t = (-0.0025 \text{ kg} \cdot \text{m/s})/(0.005 \text{ s})$$

$$= -0.5 \text{ kg} \cdot \text{m/s}^2.$$

These two forces are equal in magnitude. Because of the large difference in inertia, however, the effect of this force on the bus is negligible, while on the mosquito the effect is enormous.

8.7 *Contact forces:* The only object in direct contact with the first magnet is the table, and so the table exerts a contact force on the magnet. There is no contact force with the other magnet because the two are not touching. Presumably the first magnet is also in contact with the surrounding air, and so there may also be a contact force exerted by the air on the magnet (in practice this force is negligible as long as the relative velocity of the magnet and the air is not too large). *Field forces:* Both gravitational and magnetic forces occur in this problem. There is a gravitational force exerted by Earth on the magnet and a magnetic force exerted by the second magnet. In principle, there is also a gravitational interaction between the two magnets and between the magnets and the table, but the forces due to these interactions are so small that we can safely ignore them.

8.8 No. Although the two forces are equal in magnitude and exerted in opposite directions, just as the two forces in an interaction pair are, they originate from different interactions and therefore are not part of the same interaction pair. The two forces in an interaction pair are exerted on different objects, whereas the two forces mentioned in this checkpoint are exerted on the same object (the book).

8.9 See Figure S8.9. During free fall, there is no contact force between the book and any other object. The only force exerted on the book is the gravitational force exerted by Earth (we ignore air resistance).

Figure S8.9

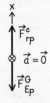

8.10 See Figure S8.10. The person is subject to one contact force: an upward force exerted by the ring (this is what keeps the person suspended). In addition, there is a downward gravitational force exerted by Earth.

Figure S8.10

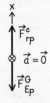

8.11 (a) See Figure S8.11a. As you accelerate the ball upward, the force exerted by your hand on the ball is larger than the force of gravity exerted by Earth on the ball. (b) See Figure S8.11b. Once the ball is released, the only force exerted on it is that exerted by Earth (ignoring air resistance). (c) The forces exerted on the ball are the same on all parts of the trajectory. Therefore your drawing for part c should look exactly like your drawing for part b. Although the direction of the ball's velocity reverses during flight, the force exerted by Earth always points downward.

Figure S8.11

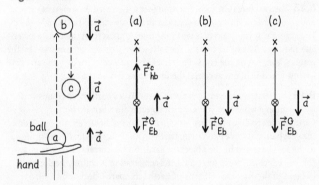

8.12 Figure S8.12 shows the free-body diagrams for the spring and the brick. The problem asks about the force exerted by the spring on the ceiling $\vec{F}^c_{sc}$, a force that forms an interaction pair with the force exerted by the ceiling on the spring. The brick is subject to a downward gravitational force exerted by Earth and an upward contact force exerted by the spring. Because the brick is at rest, we know that these two forces are equal in magnitude. The spring is subject to a downward gravitational force, a downward contact force exerted by the brick, and an upward contact force exerted by the ceiling. Because the spring is at rest we know that the forces must add up to zero and so the force exerted by the spring on the ceiling is equal to the combined gravitational forces exerted by Earth on the spring and the brick. (This result makes sense because the ceiling has to support both of them.)

Figure S8.12

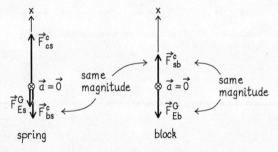

8.13 Figure S8.13 shows the free-body diagram for the rope. The tree now must exert a force of magnitude $2F$ to balance the forces exerted by the two people. This means a force of magnitude $2F$ is exerted on both ends of the rope, and so the tension is $T = 2F$—twice as large as when the two people pulled on opposite ends.

Figure S8.13

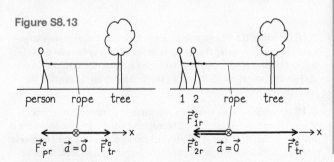

8.14 (*a*) The force you exert on your friend gives her a constant acceleration given by Eq. 8.7, where m is the inertia of your friend and her skates. Estimating m to be 60 kg (adjust for your own estimate), you have $a = (200\text{ N})/(60\text{ kg}) = 3.3\text{ m/s}^2$. Her displacement in the 2.0 s is $\Delta x = \frac{1}{2}a_x(\Delta t)^2 = \frac{1}{2}(+3.3\text{ m/s}^2)(2.0\text{ s})^2 = +6.7\text{ m}$. (*b*) When a person jumps off a wall, he is in free fall, and so the magnitude of his acceleration is 9.8 m/s^2. (*c*) When you put the person's inertia at 60 kg (adjust to your liking), the magnitude of the force exerted by Earth on the person is $F^G_{Ep} = (60\text{ kg})(9.8\text{ m/s}^2) = 5.9 \times 10^2\text{ N}$.

8.15 The two forces in an interaction pair are exerted on different objects and so never cancel when you consider the forces exerted on a single object. For example, if you push on a crate, the force you exert on the crate accelerates the crate. The other force in the interaction pair is a force exerted by the crate *on you*, not on the crate. The effect of this force exerted by the crate is to slow you down, not to cancel the force you exert on the crate.

8.16 (*a*) If you answered $m_E g$, think again! Does Earth have an upward acceleration g whenever you drop an object from a certain height? No; Earth doesn't budge. From the reciprocity of forces, however, you know that the gravitational force exerted by Earth on an object is equal in magnitude to the force exerted by the object on Earth. So the correct answer is $m_1 g$. (*b*) The effect this force has on Earth's acceleration is entirely negligible because of the enormous inertia of Earth (compare Checkpoint 8.6):

$$a_E = \frac{m_1}{m_E}g \approx 0.$$

If you are wondering how it is possible that the force exerted by Earth on an object and the force exerted by the object on Earth both depend on the inertia of the object only, then you are very perceptive. I shall explain this when we discuss gravity in Chapter 13.

8.17 (*a*) See Figure S8.17. Note that the $\vec{F}^G_{Em}$ vector arrow is shorter than the $\vec{F}^c_{fm}$ arrow. You drew your diagram that way, right? (*b*) Because the elevator is accelerating upward, so are you. Therefore the vector sum of all the forces acting on you must point upward. This means that the contact force exerted by the floor of the elevator on you is larger than the downward gravitational force exerted on you. From Eq. 8.8,

$$\sum F_x = ma_x = F^c_{fmx} + F^G_{Emx}$$

$$F^c_{fmx} = ma_x - F^G_{Emx} = ma_x - (-mg) = m(a_x + g).$$

For an inertia $m = 60$ kg (substitute your own inertia), this yields a force of $(60\text{ kg})[(+1\text{ m/s}^2) + (9.8\text{ m/s}^2)] = 6.5 \times 10^2\text{ N}$.

Figure S8.17

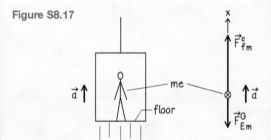

8.18 (*a*) Stiffer. If k is large, a small displacement $x - x_0$ requires a large force (Eq. 8.18). A spring that compresses only a little under a large load is stiff. (*b*) Steel. Steel compresses less (small Δx) than foam rubber (larger Δx) under equal loads. So the spring constant for steel is larger than that for foam rubber.

8.19 (*a*) Smaller. Both objects have acceleration g, but the feather has much less inertia than the brick. The magnitude of the gravitational force is $F^G_{Eo} = mg$ (Eq. 8.17), and so the gravitational force exerted on the feather is smaller than that exerted on the brick. (*b*) The momentum is the same for

both. The change in an object's momentum caused by a constant force—in other words, the impulse delivered to the object—is $\vec{F}\,\Delta t$. Because the same force is exerted on both objects for the same amount of time, $\Delta\vec{p}$ is the same for both. However the magnitude of $\Delta\vec{v}_{feather}$ is much larger than the magnitude of $\Delta\vec{v}_{brick}$ because of the difference in inertia.

8.20 (*a*) For a single object, the center-of-mass acceleration is equal to the particle's acceleration, $\vec{a}_{cm} = \vec{a}$, because the position of the center of mass coincides with the position x of the particle: $x_{cm} \equiv mx/m = x$ (Eq. 6.24). Finally for a single object there are no internal forces, so all forces are external and so $\sum \vec{F}_{ext} = \sum \vec{F}$. Consequently, Eqs. 8.46 and 8.47 are equivalent for a system made up of only one object. (*b*) Consider the motion of the system over a time interval $\Delta t = t_f - t_i$. If the center of mass initially moves at velocity $\vec{v}_{cm,i}$ at time t_i, then at the end of the time interval it has velocity

$$\vec{v}_{cm,f} = \vec{v}_{cm,i} + \vec{a}_{cm}\Delta t.$$

The change in the total momentum of the many-object system is thus

$$\Delta\vec{p} = \vec{p}_f - \vec{p}_i = m(\vec{v}_{cm,f} - \vec{v}_{cm,i})$$

$$= m(\vec{v}_{cm,i} + \vec{a}_{cm}\Delta t - \vec{v}_{cm,i}) = m\vec{a}_{cm}\Delta t.$$

After substituting Eq. 8.44 for the vector sum of the forces exerted on the system, you have your result:

$$\Delta\vec{p} = (\sum\vec{F}_{ext})\,\Delta t.$$

Chapter 9

9.1 (*a*) External, because the force is exerted from outside the system. (*b*) No, no, and no. Because the wall is hard, heavy, and anchored to the ground, it neither moves nor changes shape. Because there is no friction involved, the wall's temperature does not change. (*c*) No. Its kinetic energy doesn't change because it doesn't accelerate; its potential energy doesn't change because it doesn't change shape or deform; its thermal energy doesn't change because it doesn't heat up. (*d*) No. The energy of the wall remains the same, and so there is no transfer of energy from you to the wall, which means you do no work on the wall no matter how hard you push.

9.2 Both forces do work on the ball because for both the point of application is at the ball, and this point moves as you launch the ball.

9.3 (*a*) As the ball moves upward, its kinetic energy decreases and its internal energy doesn't change. So the change in the ball's energy (the sum of the kinetic and internal energies) is negative. This means the work done on the ball is negative. (*b*) The ball's kinetic energy now increases, causing a positive change in energy, and so the work done on the ball is positive.

9.4 The directions of the force and the force displacement are consistent with the answers obtained by analyzing energy changes. In part *a* the only force exerted on the ball (the gravitational force) is downward, but the displacement is upward, and so the work done on the ball is negative. In part *b* both vectors are directed downward, and so the work done on the ball is positive.

9.5 (*a*) Yes. There are no changes in motion or state in the environment. (*b*) Positive. As the block compresses the spring, the force exerted by the block on the spring is directed toward the wall and the point of application of that force moves toward the wall. Force and force displacement pointing in the same direction means positive work. (Also, the energy of the system increases as the spring is compressed, and so the work done on it must be positive.) (*c*) Zero. The point of contact between wall and spring is not moving, and so the force displacement is zero, meaning that the work done by the wall on the system is also zero. (Indeed, the system comprising spring and block is closed, as we concluded in part *a*.)

9.6 See Figure S9.6. The cart's kinetic energy decreases to zero, and there are no changes in the other forms of energy. The decrease in kinetic energy is the result of negative work done by the person on the cart. The force exerted by the person is directed to the left, and the force displacement is directed to the right, which is a second way of telling you that the work done by the person on the cart is negative.

Figure S9.6

9.7 See Figure S9.7. As the block comes to rest against the compressed spring, the block's kinetic energy decreases, so you must draw the kinetic energy bar below the baseline. The system has two changes in potential energy: a change in gravitational potential energy ΔU^G due to the change in the block-Earth configuration and a change in elastic potential energy ΔU_{spring} due to the compression of the spring. You should draw two potential energy bars, one for each kind. The gravitational potential energy decreases, and the potential energy of the spring increases. The thermal energy increases because of friction. The lengths of the bars add up to zero because the system is closed, so $W = 0$.

Figure S9.7

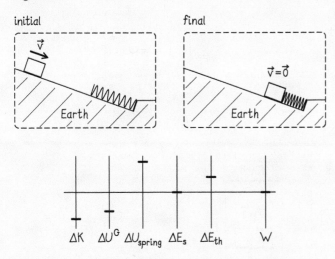

9.8 See Figure S9.8. The only energy in the system that changes is the kinetic energy, which decreases to zero. The work done on the system must therefore be negative. This negative work is the sum of the work done by the rope on the system and the work done by Earth on the system. The work done by the rope on the system is negative (force upward, force displacement downward), and the work done by Earth on the system is positive (force and force displacement both downward). The work done on the system is negative because, in order for the system to slow down, the magnitude of the force exerted by the rope must be larger than that exerted by Earth.

Figure S9.8

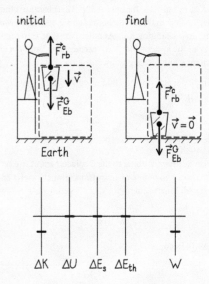

9.9 (a) See Figure S9.9. The kinetic energy remains unchanged, but energy is dissipated because of the friction between box and floor. No changes occur in potential energy. The changes in thermal energy can be attributed to work done on the system. The person exerts external forces both on the box (via the rope) and on the floor (via his feet). Only the force displacement of the force exerted on the box is nonzero. Because the force displacement and the force point in the same direction, the work done by the person on the system is positive. (b) Because the system now includes the person/rope, the system is closed, which means no work is done on it and the W bar in the energy diagram is blank. As in the system chosen in part a, you have $\Delta K = 0$, $\Delta U = 0$, and $\Delta E_{\text{th}} > 0$ (friction between box and floor increases E_{th} for the system). With the person/rope included as part of the system, the source energy bar must be filled in and it must be negative because muscle energy is used up as the person pulls.

Figure S9.9

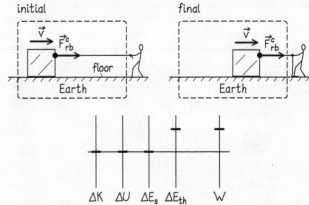

9.10 The x component of the impulse delivered by the (constant) gravitational force is given by Eq. 8.25, $\Delta p_x = F^G_{\text{Eb}x}\Delta t = -m_b g\Delta t$. Because the acceleration is constant, $\Delta v_x/\Delta t = a_{x,\text{av}} = -g$, you can say that $\Delta t = -\Delta v_x/g$ and thus

$$\Delta p_x = -m_b g(-\Delta v_x/g) = m_b\Delta v_x = m_b(v_{x,\text{f}} - v_{x,\text{i}}).$$

9.11 Because Δx_F is larger than Δx_{cm} (see Figure 9.19), W in Eq. 9.16 is larger than ΔK_{cm} in Eq. 9.14. Because $\Delta E = W$, this means that ΔE is larger than ΔK_{cm}. This makes sense because only part of the work done on the system goes into changing K_{cm}. Another part goes into increasing the internal energy of the system. $W = \Delta E$ accounts for both the change in K_{cm} and the change in internal energy.

9.12 For a single particle, the position of the center of mass is the position of the particle (Eq. 6.24):

$$x_{cm} \equiv \frac{mx}{m} = x,$$

so the force displacement is equal to the displacement of the center of mass: $\Delta x_{cm} = \Delta x_F$. The center-of-mass translational kinetic energy is simply the particle's kinetic energy:

$$K_{cm} = \tfrac{1}{2} m v_{cm}^2 = \tfrac{1}{2} m v^2.$$

Therefore Eq. 9.14 becomes

$$\Delta K = (\textstyle\sum F_x) \Delta x_F.$$

Because a particle has only kinetic energy, Eqs. 9.1 and 9.2 give $\Delta K = \Delta E = \Delta W$, so $\Delta W = (\sum F_x)\Delta x_F$, which is Eq. 9.9, the work equation for a single particle.

Now for reducing Eq. 9.18 to Eq. 9.9. Equation 9.18 is for a many-particle system. Because you want to use this equation for a single-particle system, there is only one particle experiencing the (external) force and so only one force displacement. That means you can take the factor Δx_{Fn} out of the summation in Eq. 9.18 and drop the subscript n. The equation is then identical to Eq. 9.9.

9.13 If the force is constant—in other words, if $F_x(x) = F_x$—it can be pulled out of the integral:

$$W = \int_{x_i}^{x_f} F_x(x)\,dx = F_x \int_{x_i}^{x_f} dx = F_x(x_f - x_i) = F_x\Delta x,$$

which is identical to Eq. 9.8.

9.14 See Figure S9.14. (*a*) This system is not closed, and the brick exerts a force on the system that has a nonzero force displacement. As it moves downward, the brick does positive work on the spring. This work ends up as elastic potential energy of the spring. (The force exerted by the floor on the spring does no work on the spring.) (*b*) The system is closed, so no

work is done on it. In addition, none of the four types of energy changes! The increase in the elastic potential energy of the spring is accompanied by a decrease in the gravitational potential energy of the brick. To show this, you should draw two separate potential energy bars.

9.15 mg is the magnitude of the constant force of gravity exerted by Earth *on the brick,* not a force exerted on the spring. The contact force between brick and spring is a variable force described by Hooke's law.

If you find it surprising that the brick doesn't push down on the spring with the same force as Earth pulls down on the brick, look at the free-body diagrams for the brick in Figure S9.15. At the instant it is released just above the uncompressed spring, the brick is subject to only the downward force of gravity; the spring exerts no upward force as long as it is not compressed (Figure S9.15a). Consequently the x component of the brick's initial acceleration is $a_x = -g$. As the brick moves downward, the upward force exerted by the spring on it increases according to Hooke's law (Eq. 8.20, $F_{sbx}^c = -k(x - x_o)$). In this problem you are dealing with the situation represented by the free-body diagram in Figure S9.15b. At the instant represented in Figure S9.15b, $\vec{F}_{sb}^c$ is nonzero but still smaller in magnitude than F_{Eb}^G (because the brick has not yet come to rest against the compressed spring):

$$F_{bs}^c < F_{Eb}^G = mg.$$

Answering this checkpoint requires knowing $\vec{F}_{bs}^c$, which forms an interaction pair with $\vec{F}_{sb}^c$: $\vec{F}_{bs}^c = -\vec{F}_{sb}^c$, so

$$F_{bs}^c < F_{Eb}^G = mg.$$

The difference between the two force magnitudes F_{bs}^c and mg is what makes the brick accelerate downward. (If you wonder where I get these brain twisters, the answer is simple: As I wrote the example, I was wondering about this question myself.)

Figure S9.15

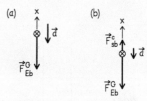

Figure S9.14

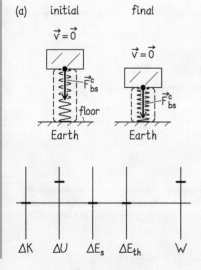

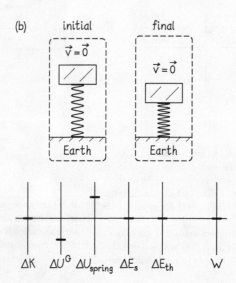

9.16 Although $\Delta x_{cm} = 0$ when the block comes to rest at its starting position, the distance it travels is nonzero: $d_{path} = 2(0.50 \text{ m}) = 1.0 \text{ m}$. All of the block's kinetic energy is converted to thermal energy, and so $\Delta E_{th} = K_i = \frac{1}{2}mv^2 = \frac{1}{2}(0.50 \text{ kg})(1.0 \text{ m/s})^2 = 0.25 \text{ J}$. From Eq. 9.28, you get $F_{sb}^f = \Delta E_{th}/d_{path} = (0.25 \text{ J})/(1.0 \text{ m}) = 0.25 \text{ N}$.

9.17 (a) Car: $(1.4 \times 10^8 \text{ J})/(1800 \text{ s}) = 78 \text{ kW}$; plane $(1.4 \times 10^8 \text{ J})/(1 \text{ s}) = 140 \text{ MW}$. (b) Equation 7.19 tells you that, to raise the load, the gravitational potential energy of the load-Earth system must be increased by $\Delta U^G = (10 \text{ kg})(10 \text{ m/s}^2)(50 \text{ m}) = 5000 \text{ J}$. The athlete can deliver $500 \text{ W} = 500 \text{ J/s}$ and so takes $(5000 \text{ J})/(500 \text{ J/s}) = 10 \text{ s}$. (c) $W_{\text{by engine}} = W_{\text{by athlete}}$ because the work done on the load does not depend on the rate at which it is done. In both cases, the work done on the load is equal to the change in the gravitational potential energy of the load-Earth system.

Chapter 10

10.1 (a) If your first reaction to this question was "In which reference frame?," you got the point! Before release, the ball is moving along with the cart, and so its initial velocity in the reference frame of the cart is zero. In the Earth reference frame, the ball's initial velocity is equal to that of the cart. (b) The speed is the magnitude of the velocity, which is the rate at which the displacement changes. Because of the cart's horizontal displacement, the ball's displacement from one frame of the film clip to the next is always greater in the Earth reference frame than in the cart's reference frame (see Figure 10.4). Therefore the ball's speed is higher in the Earth reference frame.

10.2 The ball's instantaneous velocity is the displacement over an infinitesimally small time interval. As Figure S10.2 shows, $\Delta \vec{r}$ approaches the tangent to the ball's trajectory as we make Δt smaller; $\vec{v}$ is thus tangent to the trajectory. Because of the downward curvature of the trajectory, $\vec{v}$ points above $\Delta \vec{r}$.

Figure S10.2

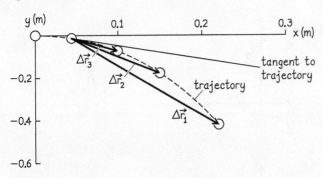

10.3 The suitcase is subject to a downward gravitational force and an upward contact force exerted by the belt. Because the suitcase's velocity is constant, its acceleration is zero, and so the vector sum of these forces must be zero. The two forces must therefore be equal in magnitude (Figure S10.3).

Figure S10.3

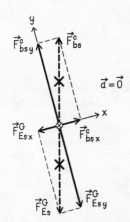

10.4 The cabinet participates in three interactions—with you, with the floor, and with Earth—and so is subject to three forces: a horizontal force due to your push $\vec{F}_{mc}^c$, a contact force $\vec{F}_{fc}^c$ exerted by the floor, and a downward force of gravity $\vec{F}_{Ec}^G$. Because the cabinet remains at rest, the vector sum of these forces must be zero. You know that your force is strictly horizontal and the gravitational force is strictly vertical. With these two forces being exerted on the cabinet, the only way for the vector sum of the forces exerted on the cabinet to be zero is for $\vec{F}_{fc}^c$ to have both a horizontal component $\vec{F}_{fcx}^c$ and a vertical component $\vec{F}_{fcy}^c$. The horizontal component of $\vec{F}_{fc}^c$ is therefore equal in magnitude to $\vec{F}_{mc}^c$, and the vertical component is equal in magnitude to $\vec{F}_{Ec}^G$. The free-body diagram for the file cabinet is shown in Figure S10.4a. (Alternatively, you can draw the two components of the contact force exerted by the floor as two separate forces, as shown in Figure S10.4b.)

Figure S10.4

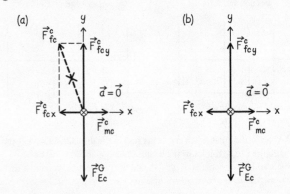

10.5 (a) The cabinet is subject to the same three forces as in Checkpoint 10.4. Because it moves at constant speed, the vector sum of these forces is still zero, and so the free-body diagram is as shown in Figure S10.5a. (b) The instant you stop pushing, $\vec{F}_{mc}^c$ disappears but $\vec{F}_{fcy}^c$ and $\vec{F}_{Ec}^G$ must be as before because there is no vertical motion. You know that the cabinet comes to rest when you stop pushing, and so there must be a horizontal force that slows it *after* you stop pushing. This force is $\vec{F}_{fcx}^c$. The free-body diagram for the cabinet is shown in Figure S10.5b. Once the cabinet is at rest, $\vec{F}_{fcx}^c$ vanishes.

Figure S10.5

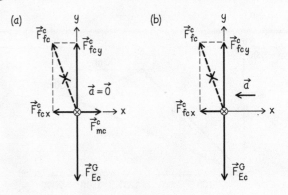

10.6 The motion of the two surfaces in contact relative to each other determines the type of friction: static friction between two surfaces not moving relative to each other and kinetic friction between two surfaces moving relative to each other. (a) Static, (b) kinetic, (c) kinetic, (d) kinetic, (e) static (as long as you are not slipping, your feet are at rest relative to the incline each time you put them on the ground while walking).

10.7 As the person accelerates, her kinetic energy increases. As you've just seen, this increase in kinetic energy is not due to work done on her by the force of static friction. The forces in the vertical direction add up to zero and so do no work either, so the work done on her is zero. What causes her to accelerate is that chemical energy in her muscles is converted to kinetic energy (and some of it is dissipated to thermal energy). So E_{th} increases and E_s decreases; the heights of the bars add up to zero (Figure S10.7). For the package, only its kinetic energy increases. Its shape and temperature remain the same, and so its internal energy does not change. The increase in kinetic energy is entirely due to work done by the conveyor belt on the package. The point of application of the force of friction exerted by the belt surface on the package is displaced in the direction of the force, so the work done on the package is positive, as Figure S10.7 shows.

Figure S10.7

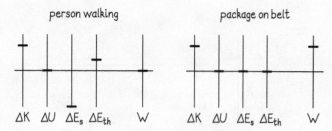

person walking   package on belt

ΔK ΔU ΔE_s ΔE_{th} W ΔK ΔU ΔE_s ΔE_{th} W

10.8 Your final displacement $\Delta \vec{r}$ is the sum of the displacements $\Delta \vec{r}_1$ and $\Delta \vec{r}_2$ in Figure S10.8. You know the length and angular coordinate of $\Delta \vec{r}_1$ and the x and y components of $\Delta \vec{r}_2$. To add the two vectors, you must first determine the x and y components of $\Delta \vec{r}_1$:

$$\Delta x_1 = \Delta r \cos \theta = (1500 \text{ m})(\cos 45°) = 1060 \text{ m}$$

$$\Delta y_1 = \Delta r \sin \theta = (1500 \text{ m})(\sin 45°) = 1060 \text{ m}.$$

Adding the components of $\Delta \vec{r}_1$ and $\Delta \vec{r}_2$ gives

$$\Delta x = \Delta x_1 + \Delta x_2 = 1060 \text{ m} + 700 \text{ m} = 1760 \text{ m}$$

$$\Delta y = \Delta y_1 + \Delta y_2 = 1060 \text{ m} - 300 \text{ m} = 760 \text{ m}.$$

The distance to the pier is thus

$$|\Delta \vec{r}| = \sqrt{\Delta x^2 + \Delta y^2} = 1.92 \times 10^3 \text{ m}.$$

Figure S10.8

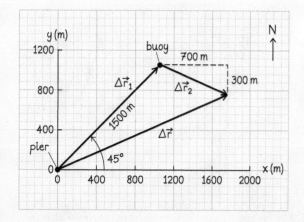

10.9 (a) According to Eqs. 10.3: $v_{x,i} = v_i \cos \theta$ and $v_{y,i} = v_i \sin \theta$. Substituting these values into the expressions we obtained in Examples 10.5 and 10.6 yields

$$x_f = \frac{2v_i^2 \sin \theta \cos \theta}{g} \text{ and } y_{top} = \frac{1}{2} \frac{v_i^2 \sin^2 \theta}{g}.$$

(b) For a given launch speed v_i, y_{top} has its greatest value when $\sin^2 \theta$ is maximal. This occurs when $\theta = 90°$, which makes sense because all of the projectile's speed goes into the upward motion when the projectile is launched straight up. (c) The greatest value of x_f is obtained when $\sin \theta \cos \theta = \frac{1}{2} \sin 2\theta$ is maximum. This occurs when $2\theta = 90°$ or $\theta = 45°$.

10.10 To answer this question, we need to compare the system's final and initial kinetic energies. Let the inertia of puck 1 be m; the inertia of puck 2 is then $2m$. The system's initial kinetic energy is $\frac{1}{2} m (1.8 \text{ m/s})^2 + \frac{1}{2} \cdot 2m(0.2 \text{ m/s})^2 = \frac{1}{2} m (3.3 \text{ m}^2/\text{s}^2)$. The final kinetic energy is $\frac{1}{2} m (0.8 \text{ m/s})^2 + \frac{1}{2} \cdot 2m(1.0 \text{ m/s})^2 = \frac{1}{2} m (2.6 \text{ m}^2/\text{s}^2)$. Because the system's final kinetic energy is not equal to the initial kinetic energy, the collision is not elastic.

10.11 To determine the sign of W, you need to know the sign of the scalar product for various relative positions of $\vec{F}$ and $\Delta \vec{r}$. To do so we choose a coordinate system that has its x axis aligned with $\Delta \vec{r}$ and then we place $\vec{F}$ in different positions (Figure S10.11). The scalar product is $F|\Delta \vec{r}| \cos \phi$, which can be written as $|\Delta \vec{r}| F \cos \phi = |\Delta \vec{r}| F_x$ and which is the product of the magnitude of $\Delta \vec{r}$ and the x component $\vec{F}$. Because $|\Delta \vec{r}|$ is always positive, the sign of the scalar product is determined by the algebraic sign of F_x. As you can see in Figure S10.11 F_x is positive when $\vec{F}$ is in quadrants I and IV and negative when $\vec{F}$ is in quadrants II and III. So W is positive when $0 < \phi < 90°$ or $270° < \phi < 360°$. For any other angle, F_x and W are negative. According to Section 9.2, W is positive when $\vec{F}$ and $\Delta \vec{r}$ point in the same direction and negative when they point in opposite directions. So Eq. 10.40 is in complete agreement with our earlier definition of when work is positive and when it is negative.

Figure S10.11

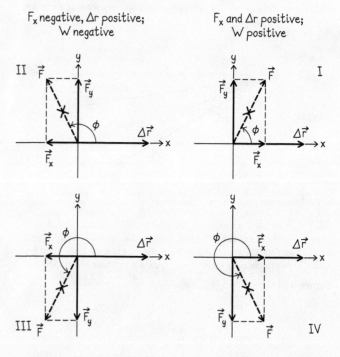

F_x negative, Δr positive; W negative

F_x and Δr positive; W positive

10.12 (a) That the brick slides more easily on ice tells you that the force of friction is smaller in the brick-ice case than in the brick-wood case, and so μ_s is smaller for brick on ice. (b) Substituting $\mu_s = 1$ into Eq. 10.46 gives $(F_{sr}^s)_{max} = F_{sr}^c$ for the maximum force of static friction exerted by the surface on the runner. This is the maximum horizontal force the surface can exert on the runner. Because the runner is not accelerating up or down, the magnitude of the normal force must be equal to the magnitude of the force of gravity, and so $F_{sr}^c = F_{Er}^G = mg$, where m is the runner's inertia. Substituting this value into the equation of motion $\Sigma F_x = ma_x$ shows that the maximum acceleration is $a_x = +g$.

10.13 (*a*) If $\mu_s = 1$, then $\theta_{max} = 45°$ according to Eq. 10.53. (*b*) All objects start sliding well before this angle is reached, telling you that generally $\mu_s < 1$.

10.14 (*a*) The free-body diagram for the bricks is given in Figure 10.42. At the point where the bricks begin to slide, F^G_{Ebx} is equal to the maximum value of F^s_{sb}, and so $F^G_{Ebx} + (F^s_{sb})_{max} = 0$. Because both F^G_{Ebx} and $(F^s_{sb})_{max}$ are proportional to the inertia m of the bricks, the angle at which $F^G_{Ebx} + (F^s_{sb})_{max} = 0$ is independent of m. The single and double bricks therefore begin sliding at the same instant. (You can also see this directly from Eq. 10.53, which shows that θ_{max} is independent of m). (*b*) Once the bricks begin to slide, kinetic friction replaces static friction. So the vector sum of the forces exerted on the bricks in the x direction is $\sum F_x = F^G_{Ebx} + F^k_{sb} = -mg \sin\theta + F^k_{sb}$. Using Eqs. 10.43 and 10.55, you can write this as $\sum F_x = -\mu_s F^c_{sb} + \mu_k F^c_{sb} = (\mu_k - \mu_s)F^c_{sb}$. The x component of the acceleration of the bricks is thus given by

$$ma_x = \mu_k mg \cos\theta - mg \sin\theta$$

and is independent of inertia: $a_x = \mu_k g \cos\theta - g \sin\theta$. The single and double brick therefore both slide down with the same acceleration. These results can also be justified as follows: In parts *a* and *b* all forces exerted on the double brick are twice as large, but the double brick, having twice the inertia, requires twice as large a force to reach the same acceleration. Consequently there cannot be any difference between the accelerations.

Chapter 11

11.1 If $|\vec{v}_f| > |\vec{v}_i|$, the vector $\vec{v}_f$ has a greater magnitude than the vector $\vec{v}_i$. As a result, the change in velocity $\Delta\vec{v}$ and $\vec{a}_{av}$ now point not toward the center of the circle but more in the direction of motion, as shown in Figure S11.1 (compare with Figure 11.8*b*).

Figure S11.1

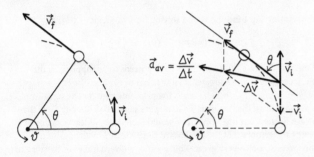

11.2 The centripetal acceleration is the same for both cubes, and so the force required to keep cube 1 going in a circle must be twice as large as the force required for cube 2. The maximum force of static friction, however, also scales with m—$(F^s_{tc})_{max} = \mu_s F^n_{tc} = \mu_s mg$—and so both cubes begin sliding at the same instant.

11.3 (*a*) Yes. In order to generate a vector sum of forces that points toward the center of the trajectory, the contact force exerted by the road surface on the bicycle (and that exerted by the bicycle on its rider) must be at an angle with the vertical. This can happen only when the bicycle leans into the curve. (*b*) No. The vertical component of $\vec{F}^c_{sb}$ is what prevents the bucket from falling to the ground: $F^c_{sbz} + F^G_{Eb} = 0$. If the rope is horizontal, then so is $\vec{F}^c_{sb}$ and consequently $F^c_{sbz} = 0$. Without this vertical component, the bucket cannot stay in the air.

11.4 (*a*) 3. The largest rotational inertia is obtained when the inertia of the pencil is as far as possible from the axis. For axis 3, half of the pencil is farther from the axis as for axis 2. (*b*) 1. For this axis all the inertia of the pencil is close to the axis.

11.5 (*a*) In 1.5 s the object covers an arc length of 4.5 m. Thus $\vartheta = (4.5\text{ m})/(2.0\text{ m}) = 2.3$. (*b*) The perimeter of the circle is $2\pi(2.0\text{ m})$. At 3.0 m/s, it therefore takes the object $2\pi(2.0\text{ m})/(3.0\text{ m/s}) = 4.2$ s to complete one revolution. (*c*) In 1.0 s the object moves through an arc of 3.0 m, and so the change in rotational coordinate in 1.0 s is $(3.0\text{ m})(2.0\text{ m}) = 1.5$. The object begins at a polar angle of 90°, which corresponds to a rotational coordinate of $(90°)(2\pi\text{ rad}/360°)/(1\text{ rad}) = \pi/2$. So the final rotational coordinate is $\pi/2 + 1.5 = 3.1$.

11.6 The faster you go, the larger the inward force required to make you go around the curve. When the maximum force $(F^s_{sp})_{max}$ is reached, $\tan\theta_{max} = \mu_s$ (Eq. 11.53) and Eq. 1 in Example 11.4 becomes

$$\mu_s mg = m\frac{v^2}{r}$$

$$v = \sqrt{\mu_s gr} = 6.6\text{ m/s},$$

where I have used the value $\mu_s \approx 1$ for the coefficient of static friction between skates and road.

11.7 (*a*) The speeds of B and C are equal, but the circumference of C's trajectory is twice as large, and so it takes C twice as long to complete one revolution. Therefore the ratio of the rotational velocities is 2. (*b*) Because of the r^2 dependence of rotational inertia (see Eq. 11.30), C's rotational inertia is four times as large as B's. Therefore the ratio of the rotational inertias is $\frac{1}{4}$.

11.8 Yes. Using Eq. 11.31, you can write for the final rotational kinetic energy:

$$\tfrac{1}{2}I_f\omega_f^2 = \tfrac{1}{2}(I_f\omega_f)\omega_f > \tfrac{1}{2}(I_i\omega_i)\omega_i = \tfrac{1}{2}I_i\omega_i^2.$$

Because his arms' centripetal acceleration must increase as he pulls them in, the force required to pull them in increases, requiring physical effort. Thus internal chemical energy in his body is converted to kinetic energy.

11.9 (*a*) Suppose the dumbbell has rotational velocity ω, with each puck moving at speed $v = r\omega = (l/2)\omega$. The kinetic energy of the dumbbell is equal to the sum of the kinetic energies of the two pucks:

$$K = 2(\tfrac{1}{2}mv^2) = mv^2.$$

Substituting the Eq. 11.11 value for v, you get

$$K = m(\tfrac{1}{2}l\omega)^2 = \tfrac{1}{4}ml^2\omega^2.$$

Knowing from Eq. 11.31 that $K = \tfrac{1}{2}I\omega^2$, you can say that

$$\tfrac{1}{4}ml^2\omega^2 = \tfrac{1}{2}I\omega^2.$$

Solve this expression for I and you see that, indeed, $I = \tfrac{1}{2}ml^2$. (*b*) No. The dumbbell rotates but has no momentum. Its axis is fixed to the table, so it cannot have any translational motion. Even though v is nonzero for each puck, $v_{dumbbell} = 0$ and therefore $p_{dumbbell} = 0$. Because the axis exerts a force on the dumbbell, the puck-dumbbell system is not isolated, and so it is not necessary for its momentum to be constant.

11.10 When the axis is at the end, you should choose the origin at that end, and so the integration boundaries in Example 11.9 become $x = 0$ and $x = l$. Therefore,

$$I = \frac{m}{l}\int_0^l x^2 dx = \frac{m}{l}\left[\frac{x^3}{3}\right]_0^l = \tfrac{1}{3}ml^2.$$

11.11 No, because drilling a hole removes material and its contribution to the rotational inertia, without changing the contribution of the remaining material: $I_{solid} = I_{removed} + I_{remaining} > I_{remaining}$. The m's in the two expressions for the rotational inertia are not the same: $m_{hollow} = \rho\pi l(R^2_{outer} - R^2_{inner})$, while $m_{solid} = \rho\pi lR^2_{outer}$. Substituting m_{hollow} into the expression for the rotational inertia of the hollow cylinder gives $I_{hollow} = \tfrac{1}{2}m_{hollow}(R^2_{outer} + R^2_{inner}) = \tfrac{1}{2}\rho\pi l(R^2_{outer} - R^2_{inner})(R^2_{outer} + R^2_{inner}) = \tfrac{1}{2}\rho\pi l(R^4_{outer} - R^4_{inner})$, which is *less* than $I_{solid} = \tfrac{1}{2}m_{solid}R^2_{outer} = \tfrac{1}{2}\rho\pi lR^4_{outer}$.

11.12 About an axis through the center of mass. I is larger about any other axis because the term md^2 in Eq. 11.53 is always positive.

Chapter 12

12.1 (a) The rod is subject to four forces: the force of gravity $\vec{F}^G_{Er}$ (which we can ignore because the inertia of the rod is negligible compared to m_1 and m_2), the downward contact forces $\vec{F}_1$ and $\vec{F}_2$ exerted by the two objects, and an upward force $\vec{F}^c_{pr}$ exerted by the pivot that holds the rod in place. Because the rod remains at rest, the vector sum of these forces must be zero, and so the upward force must be equal in magnitude to the sum of the downward forces (Figure S12.1). (b) The rod is still subject to the same forces. The rod is pinned down at the pivot, so it cannot accelerate. Therefore $\vec{a}_{cm} = \vec{0}$ as before, and the free-body diagram is unchanged. (The rod rotates to one side, pulling object 1 up and lowering object 2, but this is not reflected in the free-body diagram.) (c) $r_1/r_2 = m_2/m_1$. (d) The rod rotates so that the end where object 1 was shoots up and the end holding object 2 falls down. (e) The rod rotates so that the end of the rod where the object of inertia $2m_1$ is suspended goes down and object 2 is pulled up. (f) The end of the rod where object 1 is suspended goes down. (g) The difference is the speed at which the rod rotates into a vertical position: fast in part e and slowly in part f.

Figure S12.1

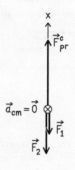

12.2 The seesaw remains at rest because the child causes a torque on the seesaw that is equal in magnitude to the torque you cause but tends to rotate the seesaw in the opposite direction.

12.3 (a) Because the torques caused by $\vec{F}_1$ and $\vec{F}_2$ cancel each other, the only way to prevent the rod from rotating is to make the torque caused by $\vec{F}_3$ zero. This can be achieved by aligning $\vec{F}_3$ along the long axis of the rod, either straight toward or straight away from the pivot. (b) The torque caused by $\vec{F}_1$ is $+r_{1\perp}F_1$; the torque caused by $\vec{F}_2$ is $-r_{2\perp}F_2 = -2r_{1\perp}F_2$. The lever arm distance of $\vec{F}_3$ is $r_{3\perp} = 3r_{1\perp}$, and so the torque caused by $\vec{F}_3$ is $+r_{3\perp}F_3 = +3r_{1\perp}F_1$ because $F_3 = F_1$. The sum of the torques is thus $+r_{1\perp}F_1 - 2r_{1\perp}F_2 + 3r_{1\perp}F_1$. Setting this sum to zero, I get $F_2 = 2F_1$. So, to balance the rod, you must increase the magnitude of $\vec{F}_2$ by a factor of 4.

12.4 If we neglect the very small effect of air resistance, the rotation of the wrench is steady. As you can verify by launching a stick or any other object, the object rotates in the same direction when it comes down. This shows that the rotational and translational motions are not coupled. The upward motion slows down because of the force of gravity, but this force does not affect the rotational motion because it does not cause a torque about the center of mass.

12.5 (a) The torque becomes greater because the lever arm distance of the force increases. See Figure S12.5a. (b) As the arm is raised, $\vec{F}^c_{mf}$ becomes more perpendicular to the forearm. See Figure S12.5b. This means that the torque caused by $\vec{F}^c_{mf}$ increases, and so the capacity to lift objects increases.

Figure S12.5

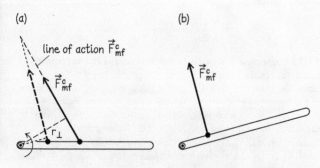

(a)

(b)

line of action $\vec{F}^c_{mf}$

$\vec{F}^c_{mf}$

$\vec{F}^c_{mf}$

$r_\perp$

12.6. The acceleration of the center of mass is the same regardless of where the force is exerted. The acceleration of the center of mass of a system of particles is *always* given by $\sum \vec{F}_{ext} = m\vec{a}_{cm}$. See Section 8.12.

12.7. (a) Because the rotation slows down, the length of $\vec{\omega}$ decreases and so $\vec{\omega}_f$ is shorter than $\vec{\omega}_i$. The vector $\Delta\vec{\omega} \equiv \vec{\omega}_f - \vec{\omega}_i$ points from the tip of $\vec{\omega}_i$ to the tip of $\vec{\omega}_f$, which means the direction of $\Delta\vec{\omega}$ is opposite the direction of $\vec{\omega}$ (Figure S12.7). (b) Yes, because rotational acceleration is defined as (Eq. 11.12)

$$\alpha_\vartheta \equiv \lim_{\Delta t \to 0} \frac{\Delta\omega_\vartheta}{\Delta t}.$$

Because $1/\Delta t$ is a scalar, $\vec{\alpha}$ points in the same direction as $\Delta\vec{\omega}$.

Figure S12.7

12.8. The fact that the forearm is not moving tells you that $\sum \tau_{ext} = 0$. This yields

$$\frac{\ell}{5}F^c_{mfy} - \frac{\ell}{2}F^G_{Ef} - \ell F^c_{bf} = 0.$$

Dividing both sides by $\ell/5$ and solving for F^c_{mfy} yield

$$F^c_{mfy} = \tfrac{5}{2}F^G_{Ef} + 5F^c_{bf}. \tag{1}$$

The condition that the vector sum of the forces exerted in the vertical direction must be zero yields

$$\sum F_y = F^c_{mfy} + F^c_{hfy} - F^G_{Ef} - F^c_{bf} = 0. \tag{2}$$

Substituting Eq. 1 into Eq. 2 and solving for F^c_{hfy} yields

$$|F^c_{hfy}| = \left|-\tfrac{3}{2}F^G_{Ef} - 4F^c_{bf}\right| = \tfrac{3}{2}F^G_{Ef} + 4F^c_{bf}.$$

Note that the magnitude of the force that needs to be supplied by the biceps muscle is *much* greater than the sum of F^G_{Ef} and F^c_{bf}: $F^c_{mf} > |F^c_{mfy}| = \tfrac{5}{2}F^G_{Ef} + 5F^c_{bf} \gg F^G_{Ef} + F^c_{bf}$. This is the price one pays for the versatility and mobility of the forearm: If a load were suspended directly from the biceps muscle, our lifting capacity would increase by at least a factor of 5! Note furthermore that the downward force exerted by the humerus on the forearm is also very great. This great downward force is necessary to counter the upward force exerted by the biceps muscle on the forearm. Without this downward force, the muscle would simply pull the forearm up rather than rotate it.

12.9. (a) No. A nonzero vector sum of forces would cause the disc's center of mass to accelerate, and we know that the disc remains in the same place, so $\vec{a}_{cm} = \vec{0}$. (b) No. Even though the *vector sum* of the forces exerted on the compact disc is zero, the individual forces cause a nonzero torque. The resulting nonzero torque is what gives the disc a rotational acceleration and increases its rotational kinetic energy.

12.10. (a) The cylindrical shell has greater rotational inertia than the solid cylinder (MR^2 versus $\frac{1}{2}MR^2$; see Table 11.3) and so, because $\tau_\vartheta = I\alpha_\vartheta$, it requires a greater torque to be accelerated rotationally. This torque is supplied by the force of static friction, so the frictional force exerted on the shell must be greater. (You can see this directly from Eq. 12.26: The shape factor c is greater for the cylindrical shell, and so F^s_{ro} is greater, too.) (b) The maximum value of the force of static friction is, from Eq. 10.46,

$$(F^s_{ro})_{max} = \mu_s F^n_{ro} = \mu_s mg \cos\theta.$$

Substituting this result into Eq. 12.26 yields

$$m_s\, mg\cos\theta = \frac{mg\sin\theta}{1 + c^{-1}}.$$

Solving this expression for θ, you obtain

$$\tan\theta = \mu_s(1 + c^{-1}).$$

For the cylindrical shell, $c = 1$, and so with $\mu_s = 1$, $\tan\theta = 2$ and $\theta = 63°$. For the solid cylinder, $c = \frac{1}{2}$, and so with $\mu_s = 1$, $\tan\theta = 3$ and $\theta = 72°$. Beyond these angles the cylinders no longer can roll without slipping. Note that both angles are larger than the 45° angle above which a block would begin to slip down a ramp. (See part a of Checkpoint 10.13.)

12.11. (a) Yes, the force of static friction does cause a torque on the wheel. The direction of this torque is counterclockwise in Figure 12.40a, opposite the direction of the torque caused by the chain. (b) Because the wheel's rotational acceleration is clockwise, the sum of the torques must be clockwise. Thus the clockwise torque caused by the chain must be greater than the counterclockwise torque caused by the frictional force.

12.12. Yes. The component $\vec{F}_\perp$ does work on the cylinder because its point of application is displaced along the circular arc. The arc length displacement is $\Delta s_F = r\Delta\vartheta$, and so the work done on the cylinder is $F_\perp\Delta s_F = F_\perp r\Delta\vartheta = \tau\Delta\vartheta$. This is equal to the change in the rotational kinetic energy in Eq. 12.31. So, for a rigid object subject to a constant torque, the quantity $t\Delta\vartheta$ is equal to the work done on the object.

12.13. $\vec{v} = \vec{\omega} \times \vec{r}$. Because $\vec{\omega}$ and $\vec{r}$ are orthogonal, the magnitude of this product is ωr, in agreement with Eq. 11.11. You can use the right-hand rule in Figure 12.44 to verify that the vector product of $\vec{\omega}$ and $\vec{r}$ indeed gives the right direction for $\vec{v}$.

Chapter 13

13.1 (a) Rotational speed is the change in the rotational position divided by the time interval during which that change takes place, and so the object with the smaller period has the greater rotational speed. Because the Moon takes less time to complete one revolution, it has the greater rotational speed. Rotations you encounter in daily life—a spinning disc, a spinning wheel, the hands of a clock—rotate significantly faster and thus have significantly greater rotational speeds. (b) The speed is the distance traveled in 1 rev ($2\pi R$) divided by the period. Moon: $[2\pi(3.84 \times 10^8 \text{ m})/(27.32 \text{ days})]/(86,400 \text{ s/day}) = 1.02 \times 10^3 \text{ m/s}$. Earth: $[2\pi(1.50 \times 10^{11} \text{ m})/(365.26 \text{ days})]/(86,400 \text{ s/day}) = 2.99 \times 10^4 \text{ m/s}$. Because of its much larger orbital radius, Earth's speed is considerably greater. (c) Moon: $a_c = v_M^2/R_M = (1.02 \times 10^3 \text{ m/s})^2/(3.84 \times 10^8 \text{ m}) = 0.00271 \text{ m/s}^2$. Earth: $(2.98 \times 10^4 \text{ m/s})^2/(1.50 \times 10^{11} \text{ m}) = 0.00595 \text{ m/s}^2$. ($d$) These accelerations are much smaller than the acceleration due to gravity at Earth's surface: $g = 9.8 \text{ m/s}^2$.

13.2 Earth's radius is about 6400 km, and so near the surface the distance to the center does not change much with altitude. For a plane 10 km above ground, the distance to the center of Earth is only $(10 \text{ km})/(6400 \text{ km}) = 0.16$ greater than that for an object on the ground. Because the gravitational force depends on the square of the distance, this increase results in a decrease in the strength of the force by a mere 2.4×10^{-6}!

13.3 (a) The light spreads out in a two-dimensional cone rather than a three-dimensional one, and the length of the straight line segment intersecting a two-dimensional wedge of light varies as $1/r$ (Figure S13.3), which means the force of gravity would exhibit a $1/r$ dependence. (b) The centripetal acceleration is still $a_c \propto R/T^2$. Because this acceleration is supplied by a gravitational force that is proportional to $1/r$, you have $R/T^2 \propto 1/R$ and so $T \propto R$.

Figure S13.3

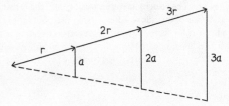

13.4 For requirement (i) to hold, you must be able to interchange the indices in the expression for the gravitational force without changing the force magnitude: $F_{12}^G = F_{21}^G$. Requirement (ii) means that if you double one mass, the force exerted on the object must double. (a) No. This dependence is consistent with requirement (i) because $m_1 + m_2 = m_2 + m_1$, but not with requirement (ii) because $m_1 + 2m_2 \neq 2(m_1 + m_2)$. ($b$) Only $m_1 m_2$ satisfies both requirements simultaneously. (c) If you take a very small piece m away, leaving $M - m$ behind, the attraction is proportional to $(M - m)m = Mm - m^2 \approx Mm$ (because the term m^2 is very small compared to Mm). So, the greater m is, the greater the force is. Symmetry thus suggests that the attraction is greatest when you divide the lump into two equal pieces. Mathematically you can prove this as follows: If you divide the lump into two pieces of masses cM and $(M - cM)$, where $0 < c < 1$, then the gravitational force is proportional to $(M - cM)cM = M^2(c - c^2)$. To find the value for c at which the force is maximum, you must set the first derivative of this expression with respect to c equal to zero: $d[M^2(c - c^2)]/dc = M^2(1 - 2c) = 0$, so $1 - 2c = 0$ or $c = \frac{1}{2}$.

13.5 For the two forces to be equal in magnitude, you need

$$\frac{m_1 m_2}{r_{12}^2} = \frac{m_1 m_E}{R_E^2}.$$

You calculated the right side in Example 13.1, so

$$\frac{m_1 m_2}{r_{12}^2} = \frac{(70 \text{ kg})^2}{r_{12}^2} = 1.0 \times 10^{13} \text{ kg}^2/\text{m}^2,$$

which yields

$$r_{12}^2 = \frac{(70 \text{ kg})^2}{1.0 \times 10^{13} \text{ kg}^2/\text{m}^2} = 4.9 \times 10^{-10} \text{ m}^2,$$

or $r_{12} = 2.2 \times 10^{-5}$ m. This is the required separation between the *centers* of the bodies. Because the width of a human body is many orders of magnitude greater than this value, it is not possible to verify this prediction.

13.6 (a) To find the center of mass, use Eq. 6.24: $x_{cm} = (m_1 x_1 + m_2 x_2)/(m_1 + m_2)$. Let 1 be the Sun and 2 Earth. If you place the Sun at the origin, $x_1 = 0$ and x_2 equals the radius of Earth's orbit. Using the values in Table 13.1, you find

$$x_{cm} = \frac{0 + (5.97 \times 10^{24} \text{ kg})(1.50 \times 10^{11} \text{ m})}{(2.0 \times 10^{30} \text{ kg}) + (5.97 \times 10^{24} \text{ kg})}$$

$$= \frac{8.96 \times 10^{35} \text{ kg} \cdot \text{m}}{2.0 \times 10^{30} \text{ kg}} = 4.5 \times 10^5 \text{ m}.$$

Compared with the radius of the Sun (7×10^8 m), this is a negligible distance: $x_{cm} = 0.00064 R_{Sun}$. (b) It speeds up because the force of gravity has a nonzero component parallel to the velocity, giving rise to an acceleration (Figure S13.6). The velocity keeps increasing until the comet reaches perihelion at the top of the ellipse. After passing that point, the comet slows down because the force of gravity causes an acceleration in the direction opposite the direction of the velocity. The slowing down continues until the comet reaches the point farthest from the Sun. At that point the velocity is a minimum.

Figure S13.6

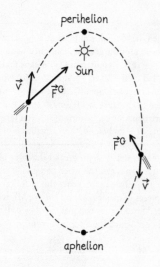

13.7 (*a*) The radius of Earth is about 6400 km, so the distance to Earth's center is increased by a factor of $(6400 \text{ km} + 300 \text{ km})/(6400 \text{ km}) = 1.05$. (*b*) Because the gravitational force changes as $1/r^2$, the force decreases by a factor of $1/(1.05)^2 = 0.91$. (*c*) The acceleration decreases by the same factor as the force: $0.91(9.8 \text{ m/s}^2) = 8.9 \text{ m/s}^2$. (*d*) The shuttle travels at such a high speed that the gravitational force exerted by Earth on it is just sufficient to provide its centripetal acceleration: $g = v^2/r$, with r the distance from the shuttle to the center of Earth.

13.8 (*a*) $F_{\text{Eb}}^G = mg = (1.0 \text{ kg})(9.8 \text{ m/s}^2) = 9.8 \text{ N}$. (*b*) The scale exerts an upward contact force of magnitude F_{sb}^c on the brick. The brick is at rest, and so you know that the vector sum of the forces exerted on it is zero. Choosing the *x* axis in the upward direction, we have $\sum F_x = F_{\text{sb}x}^c + F_{\text{Eb}x}^G = F_{\text{sb}}^c + (-F_{\text{Eb}}^G) = 0$, and so $F_{\text{sb}}^c = F_{\text{Eb}}^G$. Furthermore, the contact force exerted by the brick on the scale and that exerted by the scale on the brick form an interaction pair, which means they are equal in magnitude, so $F_{\text{bs}}^c = F_{\text{sb}}^c = F_{\text{Eb}}^G$. (*c*) Two. Step 1: $F_{\text{sb}x}^c + F_{\text{Eb}x}^G = 0$; step 2: $F_{\text{bs}}^c = F_{\text{sb}}^c$. (*d*) Yes, because, as you found in part *b*: $F_{\text{bs}}^c = F_{\text{Eb}}^G$. (If you think this question is trivial, hold off judgment until after doing Checkpoints 13.9 and 13.10.)

13.9 (*a*) Inside the orbiting shuttle, $g = 8.9 \text{ m/s}^2$ (see Checkpoint 13.7), and so $F_{\text{Eb}}^G = mg = (1.0 \text{ kg})(8.9 \text{ m/s}^2) = 8.9 \text{ N}$—almost the same as at the surface of Earth (see Checkpoint 13.8). (*b*) Zero. The brick floats: no force is necessary to support it because everything around it has the same downward acceleration. (*c*) No. The scale measures nothing, even though Earth pulls on the brick nearly as hard as it does on Earth's surface. (*d*) The answer is different because the reference frame of the shuttle is accelerating toward the Earth reference frame.

13.10 (*a*) The same. Because the elevator is moving at constant velocity, its acceleration (and that of the spring scale and the object) is zero. Consequently the vector sum of the forces exerted on the object is zero, as it is on the ground, and so $\sum F_x = F_{\text{so}x}^c + F_{\text{Eo}x}^G = 0$. (*b*) To

accelerate upward, the vector sum of the forces exerted on the object must be nonzero and point upward (Figure S13.10*a*). If $a_x = +0.5g$, $\sum F_x = ma_x = +\frac{1}{2}mg$. Consequently $\sum F_x = F_{\text{so}x}^c + F_{\text{Eo}x}^G = +\frac{1}{2}mg$, and so $F_{\text{so}x}^c = +\frac{1}{2}mg - F_{\text{Eo}}^G = +\frac{1}{2}mg - (-mg) = +\frac{3}{2}mg$. The scale reading is one and a half times what it is in a stationary elevator. (*c*) The object now has downward acceleration g and so is in free fall. The gravitational force exerted on it is thus the only force exerted on it (Figure S13.10*b*). The scale exerts no force on the object and so shows a zero reading.

Figure S13.10

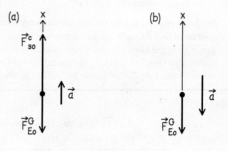

13.11 (*a*) In Checkpoint 13.7 you found that the shuttle's downward acceleration due to gravity is $g = 8.9 \text{ m/s}^2$. The vertical distance Δh through which the shuttle falls in a time interval Δt is $\Delta h = \frac{1}{2}gt^2 = \frac{1}{2}(8.9 \text{ m/s}^2)(1.0 \text{ s})^2 = 4.5 \text{ m}$. (*b*) If the shuttle remains 300 km above Earth, it is in a circular orbit, and so its centripetal acceleration is given by v^2/R, where v is its speed and R the radius of its orbit. Because the shuttle's acceleration is g, you have $v^2/R = g$, $v = \sqrt{gR} = \sqrt{(8.9 \text{ m/s}^2)(6.7 \times 10^6 \text{ m})} = \sqrt{5.7 \times 10^7 \text{ m}^2/\text{s}^2} = 7.5 \times 10^3 \text{ m/s}$.

13.12 (*a*) For the plane to be in free fall, it must have a downward acceleration of 9.8 m/s^2 (see Checkpoint 13.2). Starting with zero vertical velocity, a freely falling object drops in 40 s a distance $\Delta x = \frac{1}{2}gt^2 = \frac{1}{2}(9.8 \text{ m/s}^2)(40 \text{ s})^2 = 7.8 \text{ km}$. (*b*) The magnitude of the vertical speed is given by $v = gt = (9.8 \text{ m/s}^2) \times (40 \text{ s}) = 390 \text{ m/s}$. (*c*) You can't tell. The instantaneous velocity of an object in free fall can be in any direction. Remember that a ball launched upward is in free fall while moving up as well as while moving down. At all times, its acceleration has the same constant downward value (see Section 3.3). The answers to parts *a* and *b* show that the plane *must* have been moving upward before going into free fall. If it had started horizontally, it would have lost almost 8 km of altitude and be dangerously close to the ground (airplanes normally fly at an altitude of about 10 km). Worse, its downward velocity at the end of the free fall would have been an amazing 1400 km/h, which is well above the maximum airspeed for most airplanes.

13.13 (*a*) The plane's speed is $v = (900 \text{ km/h})(1000 \text{ m/km})/(3600 \text{ s/h}) = 250 \text{ m/s}$. Because the plane remains at a fixed altitude, it must follow a circular trajectory of radius $R \approx 6400 \text{ km}$. Its centripetal acceleration is thus $v^2/R = (250 \text{ m/s})^2/(6.4 \times 10^6 \text{ m}) = 9.8 \times 10^{-3} \text{ m/s}^2$. (*b*) See Figure S13.13. The centripetal acceleration of the plane is so small (about 0.1% of g) that you can safely ignore it, and so the free-body diagram is identical to what it would be if the plane were sitting on the ground. The plane is subject to two forces: a downward force of gravity and an upward force exerted by the air that is equal in magnitude to the downward force. (*c*) Because the airplane is not in free fall, its downward acceleration is only 0.1% of what it needs to be for the people to feel weightless.

Figure S13.13

Figure S13.18

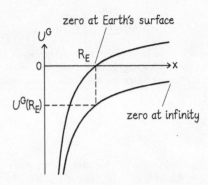

13.14 A passenger in a car that takes a right turn feels pushed outward from the turn, which means toward the left. To simulate this effect, the container should be tilted to the left.

13.15 (a) $\Delta t = \Delta x/v = (2.0 \text{ m})/(3.0 \times 10^8 \text{ m/s}) = 6.7 \times 10^{-9} \text{ s}$—just a few billionths of a second. (b) In the vertical direction you have $\Delta y = \frac{1}{2} a_y (\Delta t)^2$, $a_y = 2\Delta y/(\Delta t)^2$. To obtain a 1.0-mm deviation in the time interval obtained in part a, you need acceleration $a_y = 2(+1.0 \times 10^{-3} \text{ m})/(6.7 \times 10^{-9} \text{ s})^2 = +4.5 \times 10^{13} \text{ m/s}^2$. This is such a phenomenally large acceleration that you cannot expect to observe this effect. (c) In 0.0010 s, a light beam travels $(3.0 \times 10^8 \text{ m/s})(1.0 \times 10^{-3} \text{ s}) = 3.0 \times 10^5$ m. In that time interval, it falls like an object in free fall, and so $\Delta y = \frac{1}{2}(-9.8 \text{ m/s}^2)(1.0 \times 10^{-3} \text{ s})^2 = -4.9 \times 10^{-6}$ m—a tiny fraction of the diameter of a hair. Measuring this tiny displacement over a travel distance of 300 km is beyond current measurement accuracy.

13.16 (a) Zero. In the absence of both gravity and acceleration, the container and everything inside it are weightless. (b) The elevator moving at constant velocity has equal displacements in equal time intervals (Figure S13.16a). The path of the light pulse in the reference frame of the moving elevator is thus straight but at an angle (Figure S13.16b).

Figure S13.16

(a)

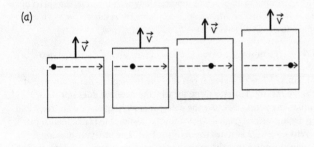

(b)

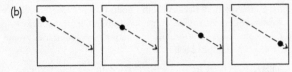

13.17 It increases because the gravitational force increases with decreasing distance r_{12}; see Eq. 13.1.

13.18 It yields values greater by an amount $|U(R_E)| = Gmm_E/R_E$. See Figure S13.18.

13.19 (a) If we let the star be at rest at the origin, then the initial kinetic energies of the star and the object are zero. The initial gravitational potential energy U^G of the closed star-object system is given by Eq. 13.14 and is negative. The initial mechanical energy of the system $E_{mech} = K + U^G$ is thus negative. If we consider the star-object system to be closed, then its mechanical energy is constant, and so it is always negative because there is no dissipation of energy. (b) Because $\vec{F}^G_{so}$ is directed toward the star's center, the object accelerates in a straight line toward that center. The minimum kinetic energy is zero (see part a). Because the object accelerates as it moves toward the star, the maximum kinetic energy is achieved when the object reaches the star's surface. Conservation of energy tells you that $E_{mech,i} = E_{mech,f}$. From part a, $E_{mech,i} = U^G_i = -GMm/r$. At the star's surface, you have $E_{mech,f} = K_f + U^G_f = K_f - GMm/R_s$. So $K_f = GMm(r - R_s)/(R_s r)$. (Because r is the distance from the object to the center of the star, you have $r > R_s$, and so $K_f > 0$, as it should be.) (c) Yes. The angular momentum is $r_\perp mv$ (Eq. 13.36). Because the motion of the object is directed straight toward the center of the star, the lever arm distance about the center of the star remains zero throughout the object's motion. The angular momentum is thus both zero and constant. (d) The answer depends on the magnitude of the launch velocity: If $K_i < |U^G_i|$, then $E_{mech} < 0$; if $K_i = |U^G_i|$, then $E_{mech} = 0$; and if $K_i > |U^G_i|$, then $E_{mech} > 0$.

13.20 (a) At r_{max}, $U^G = E_{mech}$ and so $K = 0$. This means $v = 0$ and so $\vec{L} = \vec{0}$. (b) Conservation of angular momentum requires $\vec{L}$ to remain zero. This can happen only if $\vec{r} \times m\vec{v}$ is zero (or if v remains zero, which is impossible, because then nothing holds the object in place). This vector product is zero when $\vec{r}$ and $\vec{v}$ point in the same direction—in other words, when the object's motion is restricted to a straight line. (c) No. If $\vec{L} \neq 0$, $\vec{v}$ can never point in the same direction as $\vec{r}$ and neither of these can ever become zero. This means the object can never move all the way out to r_{max}. Although moving out that far does not violate conservation of energy, it does not allow the object to keep enough angular momentum.

13.21 (a) For the planet to move in a circular trajectory of radius R at constant speed v, the star must exert a gravitational force on the planet to cause the centripetal acceleration $a_c = v^2/R$. The gravitational force is thus $GMm/R^2 = ma_c$, or

$$\frac{GMm}{R^2} = m\frac{v^2}{R}.$$

To obtain the kinetic energy $\frac{1}{2}mv^2$, multiply both sides by $\frac{1}{2}R$:

$$\frac{1}{2}\frac{GMm}{R} = \frac{1}{2}mv^2.$$

(b) The mechanical energy in the circular orbit is the sum of the kinetic energy you just determined and the gravitational potential energy from Eq. 13.11:

$$E = \tfrac{1}{2}\frac{GMm}{R} - \frac{GMm}{R} = -\tfrac{1}{2}\frac{GMm}{R}.$$

Note that this result is the negative of the kinetic energy. So, for an object in a circular orbit, the kinetic energy is half the magnitude of the system's potential energy. (c) For a circular orbit, $e = 0$, and so from Example 13.5 you know

$$\frac{2E_{mech}L^2}{G^2M^2m^3} = -1,$$

which means $E_{mech} = -\tfrac{1}{2}G^2M^2m^3/L^2$. Substituting L^2 from Example 13.5 with $e = 0$ and $a = R$ gives $E_{mech} = (-\tfrac{1}{2}G^2M^2m^3)/(GMm^2R) = -\tfrac{1}{2}GMm/R$. (d) The two results are the same.

13.22 (a) At Earth's surface, the distance to the planet's center is equal to Earth's radius: $r_i = R_E$, and so Eq. 13.23 gives

$$\tfrac{1}{2}mv_{esc}^2 = \frac{Gmm_E}{R_E}$$

or $v_{esc} = \sqrt{2Gm_E/R_E}$. (b) $v_{esc} = [2(6.6738 \times 10^{-11}\,\text{N}\cdot\text{m}^2/\text{kg}^2)(5.97 \times 10^{24}\,\text{kg})/(6.378 \times 10^6\,\text{m})]^{1/2} = 1.12 \times 10^4\,\text{m/s}$. (c) No. As long as an object is fired above the horizon (so it doesn't hit the ground), only the magnitude of the velocity matters. Once the object has enough kinetic energy to reach infinity ($E = 0$), it does so regardless of the direction in which it is fired.

13.23 (a) The integral has two parts:

$$(r^2 - R^2)\int_{r-R}^{r+R}\frac{1}{s^2}\,ds + \int_{r-R}^{r+R}ds.$$

The first term yields

$$(r^2 - R^2)\left[-\frac{1}{s}\right]_{r-R}^{r+R} = (r^2 - R^2)\left[-\frac{1}{r+R} + \frac{1}{r-R}\right].$$

Bringing both terms under a common denominator yields

$$(r^2 - R^2)\frac{(R-r)+(r+R)}{r^2 - R^2} = 2R.$$

The second term of the integral yields

$$[s]_{r-R}^{r+R} = (r+R) - (r-R) = 2R.$$

Adding both parts together yields $4R$. (b) Equations 13.24–13.27 still hold when m lies inside the shell. From Figure S13.23, we see that

$$R\cos\theta = r + s\cos(\pi - \alpha) = r - s\cos\alpha,$$

which yields the same relationship between α and θ as in Eq. 13.28. Equation 13.29 is also still valid and so we obtain the same result as in Eq. 13.33. (c) The integration boundaries are now from $R - r$ (at $\theta = 0$) to $R + r$ (at $\theta = \pi$). The first part of the integral then yields

$$(r^2 - R^2)\left[-\frac{1}{s}\right]_{R-r}^{R+r} = (r^2 - R^2)\left[-\frac{1}{R+r} + \frac{1}{R-r}\right].$$

Bringing both terms again under a common denominator gives

$$(r^2 - R^2)\frac{(r-R)+(R+r)}{R^2 - r^2} = -2r.$$

The second part of the integral yields

$$[s]_{R-r}^{R+r} = (R+r) - (R-r) = 2r.$$

Adding both parts together now yields zero. The gravitational attraction inside a uniform spherical shell is thus zero!

Figure S13.23

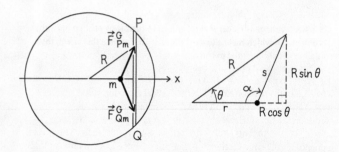

Chapter 14

14.1 (a) Because the light from event 1 reaches A before the light from event 2 reaches her, she observes event 1 first. (b) Because the light from event 1 and the light from event 2 reach B at the same instant, he observes the two events at the same instant.

14.2 (a) The light from each clock takes a finite time interval to travel from clock to observer. For clocks near the origin, this time interval is shorter than it is for clocks farther from the origin. This means (ii) is true: Nearby clocks display later time readings than distant clocks. (b) The clocks are synchronized and so by definition run at the same rate.

14.3 According to an observer moving along with the tube, the speed of the ball is l divided by Δt: $v_{Mb} = l/\Delta t$. Using the result $v_{Mb} = v_b + v_t$ from Example 14.3, you get $\Delta t = l/v_{Mb} = l/(v_b + v_t) = l/(2v)$.

14.4 (a) The observers agree that the signals do not reach the detector at the same instant because they both see that the blue light does not come on. (b) The observer at rest sees the two signals being emitted at the same instant. His reasoning for why the detector does not record simultaneity is that, because the detector is moving toward source 2 and away from source 1, signal 2 has a shorter distance to travel than signal 1. Therefore signal 2 reaches the detector first. The observer moving with the detector notes that the sources are equidistant from the detector but sees signal 2 arriving first. Because both signals travel at speed c_0 relative to him, he concludes that source 2 must have emitted its signal before source 1. The observers agree on the observation—no simultaneity—but have different explanations for it.

14.5 (a) Each signal takes $\Delta t = (5.0\,\text{m})/(3.00 \times 10^8\,\text{m/s}) = 1.7 \times 10^{-8}\,\text{s}$ to reach the detector. (b) $(20\,\text{m/s})(1.7 \times 10^{-8}\,\text{s}) = 3.3 \times 10^{-7}\,\text{m}$. (c) The travel time changes by $d/c_0 = (3.3 \times 10^{-7}\,\text{m})/(3.00 \times 10^8\,\text{m/s}) = 1.1 \times 10^{-15}\,\text{s} = 1.1\,\text{fs}$. (d) A 1.1-fs change in travel time is just barely measurable. (e) The detector moves $(1.0 \times 10^8\,\text{m/s})(1.7 \times 10^{-8}\,\text{s}) = 1.7\,\text{m}$. Light takes $(1.7\,\text{m})/(3.00 \times 10^8\,\text{m/s}) = 5.6 \times 10^{-9}\,\text{s}$ to travel this distance. Such a time interval can be measured, but note how fast the detector must be moving—$1.0 \times 10^8\,\text{m/s}$, about one-third the speed of light!—in order for the time interval to be long enough (if we can call $10^{-9}\,\text{s}$ "long") to be measurable.

14.6 (*a*) Greater than. Because the signal always travels at speed c_0, the longer distance the signal must travel according to B increases the duration of the cycle and therefore the period. (*b*) Increase. The faster the clock moves relative to B, the farther the mirror moves away from its initial position above the location where the signal was emitted in one period of a clock at rest and so the period increases.

14.7 (*a*) Slow, because the mechanical watch takes longer to complete 1 minute. (*b*) Fast.

14.8 Yes and no. By moving at high speed, you could slow down your biological clock and therefore slow down the rate at which your body ages, but *only as measured by an observer who is not in your reference frame.* From your perspective, you see your biological clock running at its "regular" rate (that is, a rate that is the same as what you would measure if you remained at rest on Earth). To you, there is no slowing down of your biological clock, regardless of your motion relative to any observer. So, although you could lengthen your life in the eyes of an observer, it would be impossible to do so from your own point of view.

14.9 Figure S14.9a shows the situation viewed from B's reference frame. B starts accelerating relative to A when her clocks read 3 o'clock. Figure S14.9b shows the situation viewed from A's reference frame. Because the unit is moving, the rear clock is ahead of the front clock, and so A sees the rear clock accelerate before the front clock, which means A sees B's unit getting shorter.

14.10 No. See Figure S14.10. To the observer on the ground, the front end of the pole reaches the exit (event 1) at 3 o'clock and the rear end reaches the entrance (event 2) at the same instant. To the observer moving along with the pole, these two events do not occur simultaneously. To this observer, the tunnel is in motion, and so clocks at the two ends are desynchronized, with the clock at the exit ahead of the clock at the entrance. Consequently, this observer sees the front end of the pole exit the tunnel before the rear end enters and concludes that the pole is longer than the tunnel.

Is this disagreement a problem (like the problem of the clock fitting between the pegs in Figure 14.19)? Suppose the tunnel is equipped with doors that can close or open instantly. Initially the entrance is open and the exit is closed. When the front of the pole reaches the exit, the exit door opens. When the rear of the pole reaches the entrance, the entrance door closes. According to the observer on the ground, these two events occur at the same instant, and at that instant the pole fits exactly between the two doors. For the observer moving along with the pole, the front of the pole

is at the exit before the rear is at the entrance, and so the exit door opens before the entrance door closes. The pole doesn't need to fit (and *can't* fit) in the tunnel.

Figure S14.10

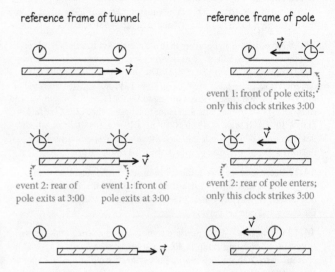

14.11 No, because the two observers measure different speeds for the object, and inertia depends on speed.

14.12 (*a*) When the spring is compressed, its internal energy increases, and so its mass increases. (*b*) When the coffee cools down, its internal energy decreases, and so its mass decreases.

14.13 (*a*) Because a particle has no internal energy, the mass is constant. Because the particle is accelerated, its kinetic energy increases. Therefore the energy of the system increases, and so the inertia increases. (*b*) As with any collision, we can consider the system of colliding objects to be closed, and so neither the energy nor the inertia can change. The internal energy must increase as the particle comes to rest in the target, and kinetic energy is converted to internal energy. Therefore the mass of the system increases.

Figure S14.9

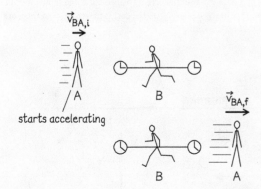

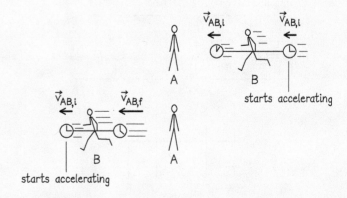

14.14 If you use Eq. 14.6, your calculator will round the result to 1 for all but d. A better approach is to use Eq. 14.11 to calculate $\frac{1}{2}v^2/c_o^2$ and then add 1. (a) $\frac{1}{2}(30 \text{ m/s})^2/(3.00 \times 10^8 \text{ m/s})^2 = 5.0 \times 10^{-15}$, $\gamma = 1 + 5.0 \times 10^{-15}$. (b) $\gamma = 1 + (3.5 \times 10^{-13})$. (c) $\gamma = 1 + (5.6 \times 10^{-10})$. (d) For this part you cannot use Eq. 14.11 because the speed is not small. Equation 14.6 gives $\gamma = 1/\sqrt{1 - (0.60)^2} = 1.25$.

14.15 From Eq. 14.17 you obtain $s^2 = [(3.00 \times 10^8 \text{ m/s})(5.0 \times 10^{-7} \text{ s})]^2 - (120 \text{ m})^2 = 8.1 \times 10^3 \text{ m}^2$. This value equals $(c_o\Delta t_{\text{proper}})^2$. So $(\Delta t_{\text{proper}})^2 = (8.1 \times 10^3 \text{ m}^2)/(3.00 \times 10^8 \text{ m/s})^2$ and $\Delta t_{\text{proper}} = 3.0 \times 10^{-7} \text{ s}$.

14.16 According to an observer in the car, the 1-km length of road travels past the car at 100 km/h = 28 m/s. According to this observer, the measured length l_v is l_{proper}/γ (Eq. 14.28). The amount by which the length contracts is $l_{\text{proper}} - l_{\text{proper}}/\gamma = l_{\text{proper}}(1 - 1/\gamma) = l_{\text{proper}}(\gamma - 1)/\gamma$. For low velocities, γ is close to 1. According to Eq. 14.11, $\gamma - 1 \approx \frac{1}{2}v^2/c_o^2$, and so $(\gamma - 1)/\gamma \approx \frac{1}{2}v^2/c_o^2$. The difference in length is thus $l_{\text{proper}}(\frac{1}{2}v^2/c_o^2) = (1.0 \times 10^3 \text{ m})\frac{1}{2}[(28 \text{ m/s})/(3.00 \times 10^8 \text{ m/s})]^2 = 4.3 \times 10^{-12} \text{ m}$, which is a few orders of magnitude less than the smallest distance that can be measured. For all practical purposes, the road is still 1.0 km long.

14.17 Because of symmetry, the collision seen by A is identical to that seen by B. The time interval measured by A between the throw and catch of particle 1 is therefore equal to the time interval measured by B between the throw and catch of particle 2.

14.18 (a) Because $\vec{v}_2$ has a nonzero component along the x axis, A sees $v_2 > v_1$. (b) According to Eq. 14.40, $m_{v2} > m_{v1}$. Hence A concludes that inertia increases with speed. (c) From the perspective of B, $v_1 > v_2$. In going from reference frame A to reference frame B, the roles of particles 1 and 2 are interchanged, and so in reference frame B, Eq. 14.40 reads $m_{v1} = \gamma m_{v2}$. According to B, $m_{v1} > m_{v2}$, and so B, too, concludes that inertia increases with speed.

14.19 You know that for the electron at rest, m_v and m in Eq. 14.41 are identical, and so you want to know what speed v yields $\gamma = 1000$. Substituting this value into Eq. 14.9 yields:

$$v^2/c_o^2 = 1 - 10^{-6}$$

$$v = \sqrt{1 - 10^{-6}}\,c_o = 0.9999995c_o.$$

14.20 (a) Because the final relative speed is zero, the collision is totally inelastic. (b) Using $\vec{p} = m\vec{v}$, you find $m_{\text{p}}v_{\text{px,i}} + m_{\text{e}}v_{\text{ex,i}} = 0$, $v_{\text{e,i}} = (m_{\text{p}}/m_{\text{e}})v_{\text{p,i}} = (1836)(0.6c_0) = 1102c_0$!

14.21 $K = (\gamma - 1)mc_o^2 \approx (\frac{1}{2}v^2/c_o^2)mc_o^2 = \frac{1}{2}mv^2$.

14.22 (a) See Figure S14.22. The collision converts kinetic energy to internal energy. (b) Kinetic energy is not constant because the final kinetic energy is zero. Mass is not constant because the increase in internal energy corresponds to an increase in mass. Energy is constant because the system is closed.

Figure S14.22

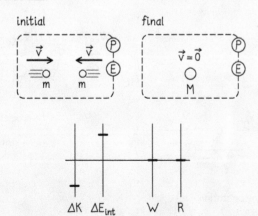

14.23 (a) See Figure S14.23. (b) The internal energy of the atom decreases because of the radiation emitted by it. The equivalence of mass and internal energy tells you that the mass of the atom decreases. (c) Because the atom is at rest, its inertia is equal to its mass. Because the mass decreases, the inertia also decreases.

Figure S14.23

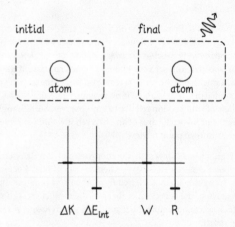

14.24 If you denote the initial energies as $E_{1\text{i}} = E_{2\text{i}} = E$, the energy law yields $E_{1\text{i}} + E_{2\text{i}} = 2E$. Because the momentum of the system is zero, we have from Eq. 14.57 $(2E)^2 = (300m_{\text{p}}c_o^2)^2$, or $E = 150m_{\text{p}}c_o^2$. From Eqs. 14.52 and 14.54, $K_{1\text{i}} = E - m_{\text{p}}c_o^2 = 149m_{\text{p}}c_o^2 = 2.2 \times 10^{-8}$ J, which is much less than the energy obtained in Example 14.13. (For this reason particle colliders use colliding beams rather than stationary targets.)

Chapter 15

15.1 (a) Two forces are exerted on the spring: one by the cart and the other by the post, and three forces are exerted on the cart: one by the spring, one by Earth, and one by the track. Figure S15.1 shows the free-body diagrams during compression (when the spring is stretched, the direction is reversed for the three horizontal forces in the diagrams). (b) None. The only forces that have a nonzero displacement along their line of action are the contact forces between the spring and the cart, but these are internal forces and so do no work on the system. (c) The contact force exerted by the cart on the spring points in the same direction as the force displacement, so the work done by this force on the spring is positive. This makes sense because, as the spring is compressed, its elastic potential energy increases. (d) The force exerted by the spring on the cart is directed in the direction opposite the force displacement, so the work done by this force on the cart is negative. Because $\vec{F}_{\text{cs}}^c = -\vec{F}_{\text{sc}}^c$ and the force displacement is the same for both forces, the work done by the spring on the cart is the negative of the work done by the cart on the spring.

Figure S15.1

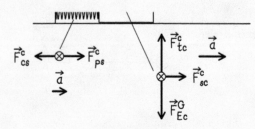

15.2 (*a*) Negative. Just before reaching the position where the spring is maximally stretched, the velocity is positive (points right), and just afterward, it is negative (points left). So the change in velocity and the acceleration are both negative. (*b*) It is greatest where the curvature of the $x(t)$ curve is greatest, when the spring is maximally stretched or compressed at the instants represented in Figure 15.2*a*, Figure 15.2*e*, and Figure 15.2*i*. It is smallest (zero) when the cart is at the equilibrium position at the instants represented in Figure 15.2*c* and 15.2*g*.

15.3 (*a*) *a*: tangential component of gravitational force, *b*: vertical component of elastic force in ruler, *c*: tangential component of gravitational force, *d*: vertical component of elastic force in string. (*b*) Gravitational potential energy in *a* and *c*, elastic potential energy in *b* and *d*.

15.4 (*a*) Factor of 4. From Chapter 9 you know that the potential energy of a spring is proportional to the square of the displacement from the equilibrium position. So, if the spring is compressed twice as much, the initial potential energy is four times as much. The initial kinetic energy is zero, so the energy is four times greater. (*b*) The initial compression determines the amplitude, and so the energy in the oscillator is proportional to the square of the amplitude.

15.5 (*a*) Upward, see Figure S15.5. (*b*) Downward. The two shaded regions in Figure S15.5 indicate that the shadow moves over increasingly smaller distances Δx during a given time interval Δt until it reaches the top. In other words, the shadow slows down. This means that the direction of the acceleration vector is opposite to that of the velocity vector.

Figure S15.5

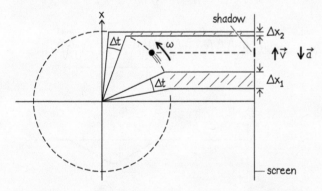

15.6 (*a*) A pure sinusoidal function requires just a single term in the Fourier series, so the spectrum consists of a single peak at frequency $f = 1/T$. The peak height A is the square of the amplitude of the function. (*b*) As T increases, $f = 1/T$ decreases, and the single peak in the spectrum shifts to a lower frequency.

15.7 (*a*) At $x = x_0$ and for large x, $\Sigma F_x = 0$, and so these are equilibrium positions. Only the position at x_0 is a stable equilibrium position. (At very large x, the equilibrium is unstable because the object will tend to accelerate in the negative x direction.) (*b*) The shape of the curve tells you that the magnitude of the restoring force is greater for a negative displacement from x_0 than for a equal positive displacement.

15.8 If you did not set your calculator to work with radians before getting the five values, you might have concluded that the small-angle approximation $\sin\theta \approx \theta$ is not correct! Because radians measure ratios of arc lengths and radii (see Section 11.4), $\sin\theta \approx \theta$ applies only if θ is expressed in radians. Thus your first step is to convert each angle from degrees to radians, and then (with your calculator set at "rad") get the sine values. As Table S15.8 shows, the approximation is correct to better than 1% up to rotational positions corresponding to polar angles of 10°.

Table S15.8 Small-angle approximation

polar angle θ in degrees	polar angle θ in radians	$\sin\theta$	error (%)
1	0.0174533	0.0174524	0.0051
5	0.0872665	0.0871557	0.1270
10	0.1745329	0.1736482	0.5095
20	0.3490659	0.3420201	2.0600

15.9 (*a*) If the length of the pendulum is increased, the displacement of the pendulum bob for a given angle increases, but the restoring force remains the same. So the restoring force for a given displacement is smaller and thus the period is longer (for a mathematical proof see Example 15.6). (*b*) The smaller acceleration of gravity on the Moon decreases the restoring force exerted on the pendulum, and so the period is increased. (*c*) Neither the mass nor the spring constant is affected by the smaller gravitational attraction on the Moon, and so the object's period is the same on the Moon and on Earth.

15.10 (*a*) x positive, v_x positive, a_x negative. (*b*) x negative, v_x negative, a_x positive.

15.11 (*a*) Substituting the maximum displacement $x = A$ into Eq. 15.15, you get $U = \frac{1}{2} m\omega^2 A^2$. (*b*) Yes. At maximum displacement, $K = 0$ and $E = U$.

15.12 (*a*) Yes. The velocity is given by the derivative of Eq. 15.6, as Eq. 15.7 shows. Because $\phi_i = \pi/2$, at $t = 0$, $v_x \propto \cos\pi/2 = 0$. (*b*) No. In Figure 15.26, for example, the cart moves to the left (negative x component of the velocity) even after crossing the position $x = 0$. Only after reaching $x = -A$ does the cart turn around and the x component of the velocity become positive.

15.13 (*a*) Because cart 2 remains in place after the first collision, the carts collide again when cart 1 returns to its initial position at $x = -15$ mm. Because the sine function is symmetrical about the maximum, this occurs after a time interval twice as long as that required to reach maximum compression. So $t = 2(0.17\text{ s}) = 0.34$ s. (*b*) Immediately after the second collision, cart 1 has zero velocity and then begins a new oscillation with an amplitude of 15 mm. Cart 2 moves away to the right at a constant speed of 0.10 m/s.

15.14 If the block is lifted above x_{eq}, Eq. 2 remains valid—the only difference now is that $x - x_{eq}$ is negative because x is on the other side of x_{eq}. The vector sum of the forces is then positive, reflecting the fact that the restoring force is now downward.

15.15 Decreasing the radius of the disk reduces its rotational inertia (which means it rotates more easily). Decreasing I increases ω (Eq. 15.29), and hence f increases (Eq. 15.4).

15.16 No. The oscillating rod experiment determines the acceleration due to gravity g, not G. The two are related (Eq. 13.4, $g = Gm_E/R_E^2$), but the relationship contains Earth's mass, which must be determined independently. [As you may remember, Earth's mass is determined from G, which is measured in the Cavendish experiment (see Section 13.5), not the other way around, see Example 13.3.]

15.17 Greater. The upward acceleration effectively increases g, which means that an object in your hand feels heavier. If g increases, the frequency of the pendulum increases too (see Example 15.6). An alternative way to see that f increases is to look at the free-body diagrams for the pendulum bob in an elevator at rest (Figure S15.17*a*, next page) and in an elevator that is accelerating upward (Figure S15.17*b*). For the elevator at rest (or moving at constant velocity), the upward vertical component of the tensile force exerted by the string on the bob and the downward force of gravity add approximately to zero (ignoring the small vertical acceleration due to the pendulum motion). When the elevator accelerates upward, the tensile force exerted by the string on the bob must increase so that the vertical component of $\vec{F}^c_{sb}$ becomes greater than the gravitational force. This causes the pendulum bob to accelerate along with the elevator. If $\vec{F}^c_{sb}$ becomes greater, however, the horizontal component also increases, and so the restoring force for a given displacement increases, which increases f.

Figure S15.17

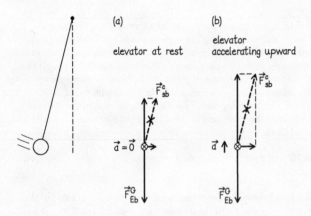

15.18 (*a*) The time-varying amplitude is given by $x_{max} = Ae^{-t/2\tau}$, where A is the initial amplitude. If the amplitude decreases by a factor of 3, $x_{max}/A = 1/3$, so $e^{-(4.0\,s)/2\tau} = 1/3$. Taking the natural logarithm (see Appendix B) of both sides, you get $-(4.0\,s)/2\tau = \ln(1/3) = /1.1$, $\tau = (-2.0\,s)/(-1.1) = 1.8$ s. (*b*) $Q = \omega t = 2\pi f\tau = 2\pi(262\,Hz)(1.8\,s) = 3.0 \times 10^3$.

Chapter 16

16.1 (*a*) See Figure S16.1. (*b*) The *x* component of the bead's velocity is always zero because the bead moves in only the *y* direction. The *y* component of the bead's velocity is positive on the leading edge of the pulse and negative on the trailing edge.

Figure S16.1

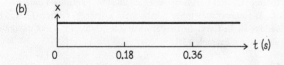

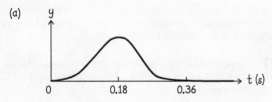

16.2 (*a*) Yes. Using a ruler, you can verify that the displacement of the pulse from one frame to the next is constant. (*b*) Between $t = 0$ and $t = 0.36$ s, the peak of the pulse moves over 16 beads, which corresponds to a displacement $\Delta x = (15)(5.0\,mm) = 75$ mm, and so the wave speed is $c = (75\,mm)/(0.36\,s) = 0.21$ m/s.

16.3 (*a*) 0.4 m. (*b*) 0.1 m. (*c*) 0. (*d*) Greater, because the pulse moves to the right and so the 1.5-m point on the string is about to be pulled upward.

16.4 (*a*) The point at $x = 0$ reaches its maximum height at $t = 0.50$ s; the point at $x = 1.0$ m reaches its maximum height at $t = 1.0$ s. So it takes the pulse 0.50 s to move 1.0 m. (*b*) $c = (1.0\,m)/(0.50\,s) = 2.0$ m/s.

16.5 (*a*) Figure 16.16*b* plots the horizontal displacement of each coil on the spring from its *equilibrium* position. The coils between $x = 0.20$ m and $x = 0.40$ m are displaced from their equilibrium position and so $x = 0.30$ m is the midpoint of the pulse. The coil originally at $x = 0.30$ m is displaced 0.05 m to the right, as shown by the slanted line in Figure 16.6*a*. So, even though this coil appears at $x = 0.35$ m in Figure 16.6*a*, it really is at the midpoint of the pulse. (*b*) Yes, but the displacement is now negative. See Figure S16.5.

Figure S16.5

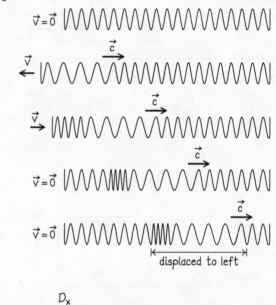

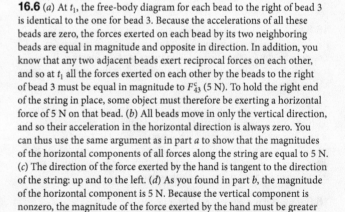

16.6 (*a*) At t_1, the free-body diagram for each bead to the right of bead 3 is identical to the one for bead 3. Because the accelerations of all these beads are zero, the forces exerted on each bead by its two neighboring beads are equal in magnitude and opposite in direction. In addition, you know that any two adjacent beads exert reciprocal forces on each other, and so at t_1 all the forces exerted on each other by the beads to the right of bead 3 must be equal in magnitude to F^c_{43} (5 N). To hold the right end of the string in place, some object must therefore be exerting a horizontal force of 5 N on that bead. (*b*) All beads move in only the vertical direction, and so their acceleration in the horizontal direction is always zero. You can thus use the same argument as in part *a* to show that the magnitudes of the horizontal components of all forces along the string are equal to 5 N. (*c*) The direction of the force exerted by the hand is tangent to the direction of the string: up and to the left. (*d*) As you found in part *b*, the magnitude of the horizontal component is 5 N. Because the vertical component is nonzero, the magnitude of the force exerted by the hand must be greater than 5 N. (*e*) The horizontal component keeps the string taut; together with the force exerted by the wall at the right end, it provides the tension in the string. The vertical component accelerates the first bead vertically upward.

16.7 (*a*) The shape of the pulse does not change as it propagates along the string, and so each bead executes the same motion as the pulse passes through it. This means that each bead has velocity $\vec{v}_{peak}$ and acceleration

$\vec{a}_{peak}$ when it is at its highest point, velocity $\vec{v}_{half}$ and acceleration $\vec{a}_{half}$ when it is halfway up to its highest point, and so on. Because the position of bead 4 at t_4 is the same as the position of bead 3 at t_3, $\vec{v}_4(t_4) = \vec{v}_3(t_3)$ and $\vec{a}_4(t_4) = \vec{a}_3(t_3)$. (*b*) They are the same. For each string, the motion of each point along the string is identical to the motion of the end you moved (provided the pulse doesn't change shape as it travels along the string). Because the end of A and the end of B move in identical fashion, all particles of *both* strings execute the same motion and so have equal velocities at equal displacements. (*c*) See Figure S16.7a. When the pulse on B has a displacement of $\Delta\vec{x}_B$, the displacement $\Delta\vec{x}_A$ of the pulse on A is twice as large because of the greater wave speed on string A. In addition—make sure you have not overlooked this—the pulse on A is twice as wide, for this reason: If, during the time interval it took your hand to move from initial position to maximum displacement, the pulse on A advanced a distance $\frac{1}{2}w_A$ (half the *width* of the pulse), then the pulse on B advanced only half as much: $\frac{1}{2}w_B = \frac{1}{2}(\frac{1}{2}w_A)$. (*d*) See Figure S16.7b. Each point carries out the same motion, taking the same amount of time to go up and then back down, and so the curves have the same shape. Because the pulse on B travels more slowly, the particle of B executes its motion at an instant t_B that is later than the instant t_A at which the particle of A executes its motion.

Figure S16.7

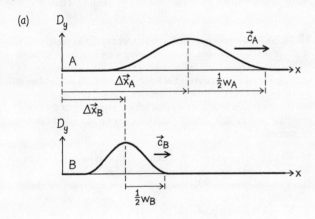

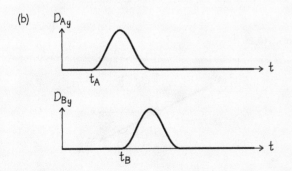

16.8 Because the tension in A is greater than that in B, the pulse travels faster on A than on B. This is the same as in Checkpoint 16.7, and so the sketches are the same. (*a*) See Figure S16.7a. (*b*) See Figure S16.7b.

16.9 See Figure S16.9. From the motion of the hand in Figure 16.9, you see that the end of the string first moved up, then down, and so on. Similarities: Your graph and the snapshot of the wave have identical shapes. Differences: In Figure S16.9, the beginning of the wave motion is on the left at $t = 0$; in the snapshots in Figure 16.9, however, the leading edge of the wave is on the right. So, in addition to your curve having a different scale, it is a mirror image of the wave of Figure 16.9.

Figure S16.9

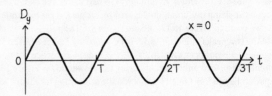

16.10 (*a*) Source: All particles of the string follow the motion of the source. (*b*) Properties of the string. (*c*) Both: The wavelength is determined by the wave speed and the period (see Figure 16.9). The wave speed is determined by the properties of the string, and the period is determined by the source. (*d*) Source: Because each string particle executes the same up-and-down motion as the source, the maximum speed is determined by the source.

16.11 Yes. Imagine putting some object in the path of the pulse. When the pulse displaces the string so that the string hits the object, the object is set in motion, and so the string has transferred momentum to the object.

16.12 (*a*) No. To see why, assume for simplicity that the pulses overlap at the midpoint of each string. In Figure 16.12, the maximum displacement at points away from the midpoint where the two pulses overlap is the peak displacement of the larger pulse. In the region of overlap near the midpoint, however, the maximum displacement is greater. In Figure 16.13, the maximum displacement at every point except where the pulses overlap is again the peak displacement of the larger pulse, but the maximum displacement at the midpoint is smaller than this. (*b*) See Figure S16.12a and b. In Figure S16.12a, the left peak shows how the point near the string's left end is displaced (just after $t = 0$) as the right-moving pulse in Figure 16.12 passes through the point. The right peak shows how, at some later instant, this same point is displaced by the left-moving pulse in Figure 16.12.

Figure S16.12

(a) Point near left end in Figure 16.12

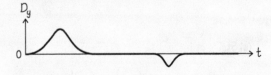

(b) Point near left end in Figure 16.13

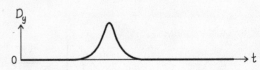

(c) String midpoint in Figure 16.12

(d) String midpoint in Figure 16.13

Figure S16.12*b* shows the same thing for the two pulses in Figure 16.13. (*c*) See Figure S16.12*c* and *d*. Assuming, as in part *a*, that the two pulses overlap at the string midpoint, the Figure S16.12*c* peak is the algebraic sum of the two peaks in Figure 16.12, and the Figure S16.12*d* peak is the algebraic sum of the two peaks in Figure 16.13. (*d*) Comparing Figure S16.12*a* and *c* tells you that a point near the string's left end is displaced twice but the midpoint is displaced only once, justifying the answer no for Figure 16.12. Comparing Figure S16.12*b* and *d* tells you the same thing for Figure 16.13.

16.13 (*a*) See Figure S16.13. At the midpoint, the two pulses always cause displacements of equal magnitude in opposite directions. (*b*) The velocity of a point along the string is equal to the derivative of the displacement of that point with respect to time. If the displacement is in the *y* direction, the *y* component of the velocity is $v_y = dD_y/dt$. Because the displacement is the sum of the individual displacements, velocity is the sum of the individual velocities.

Figure S16.13

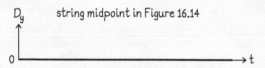

16.14 Because the string is straight, its elastic potential energy is zero, which means all the energy is kinetic energy. As Figure 16.16 shows, the pulse is in the process of inverting itself and, although the string is straight, every particle within the wave pulse is in motion.

16.15 (*a*) To see what happens, we construct the pulse shape by the superposition of the incident pulse and an imagined noninverted reflected pulse (Figure S16.15). At t_5 the peaks of both pulses have just reached the free end, and the leading edge of the reflected pulse overlaps the trailing end of the incident pulse. The velocities of particles of the leading edge of the reflected pulse are upward and those on the trailing edge of the incident pulse are downward, so the velocities of particles of the combined pulse are zero (see Checkpoint 16.13). The kinetic energy is zero. (*b*) Let the original pulse have elastic potential energy *U*; then $K = U$ (see Section 16.3) and $E = 2U$. If the string obeys Hooke's law, the potential energy is proportional to the square of the displacement. At t_5, when the displacement at the free end is equal to the sum of the amplitudes of the two pulses (that is, twice as great), the potential energy is four times as great. However, at t_5 the pulse is only half its original width, and so the potential energy is $2U$. (*c*) Yes, $E_1 = E_5 = 2U$.

Figure S16.15

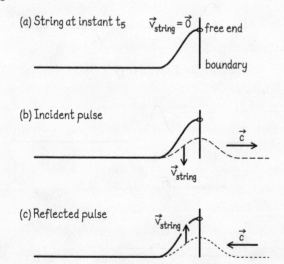

(a) String at instant t_5 $\vec{v}_{string} = \vec{0}$ free end
boundary

(b) Incident pulse $\vec{c}$
$\vec{v}_{string}$

(c) Reflected pulse $\vec{v}_{string}$ $\vec{c}$

16.16 (*a*) With $\mu_2 = 0$, the boundary becomes a free end, and so the reflected pulse has the same amplitude as the incident pulse and nothing is transmitted to string 2. (*b*) With $\mu_2 = \infty$, the boundary becomes fixed. Again the (inverted) reflected pulse has the same amplitude as the incident pulse and nothing is transmitted to string 2. (*c*) When $\mu_2 = \mu_1$, the strings become one: There is no boundary and so no reflection. The pulse travels unchanged from string 1 to string 2. (*d*) Because the wave speed is greater in string 2, the transmitted pulse is spread out more than the incident pulse (see Section 16.2).

16.17 (*a*) All but (*iii*), which cannot represent a traveling wave because the exponential factor e^{-x} is not a function of $(x \pm ct)$. (*b*) All three, by substituting $(x \pm ct)$ for each *x*.

16.18 (*a*) At $t = 0$, the string has zero velocity and so $K = 0$ and all the energy is stored as elastic potential energy. At $t = \frac{1}{8}T$, the string is moving vertically and is not in the equilibrium position, so the energy is distributed between kinetic and elastic potential. At $t = \frac{1}{4}T$, the string is horizontal, so $U = 0$. All the energy is kinetic. (*b*) Yes. Each particle of the string executes simple harmonic motion about its equilibrium position (but different points have different amplitudes—zero at the nodes, maximum at the antinodes). Because the energy of a simple harmonic oscillator is constant, the energy of any particle or any length of string is also constant. (*c*) No. The amount of energy f_1 carries rightward is equal to the amount f_2 carries leftward, so the combined flow of energy is zero.

16.19 (*a*) No. Suppose one amplitude is twice as large as the other. Half of the large-amplitude wave forms a standing wave with the other wave, but the remaining half remains a traveling wave. The superposition of a standing and a traveling wave is not a standing wave. (*b*) No. Because the wavelengths are not the same, the interference at any position is constructive at some instants and destructive at other instants, unlike the interference illustrated in Figure 16.37, which for a given position is the same at all instants.

16.20 (*a*) See Figure S16.20. Two contact forces act on A: F^c_{hA} and F^c_{BA}. If the gravitational force exerted on A is much smaller than these contact forces, we can ignore it: $\Sigma F_A = 0$. (*b*) The momentum does not change, which makes sense because A moves upward at constant velocity $\vec{v}$. (*c*) Yes, because the scalar product of the force and the force displacement is nonzero and positive: $dW = \vec{F} \cdot d\vec{r}_F$. (*d*) The energy increases because an increasing amount of string moves upward at velocity $\vec{v}$.

Figure S16.20

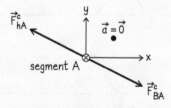

16.21 Greater (twice as great to be precise) because *two* waves travel away from the point you are shaking, one in each direction along the string. Another way to see this is by looking at the force you must exert to move the string. For a point in the middle, this force is twice as great as the force you must exert on an end because the pieces of string on either side of your hand each exert the same force as one end of the string. So the amount of work you do on the string increases by a factor of 2.

16.22 The frequency *f* (and therefore the angular frequency ω) is determined by the source that causes the wave. The wave number *k* is equal to ω divided by the wave speed *c*, which is determined by the tension and linear mass density of the string.

Chapter 17

17.1 (a) Because the wave speed c is constant, $R_2 = 2R_1$. (b) The circumference at R_1 is $2\pi R_1$, making the energy per unit length $E/2\pi R_1$; at R_2 this becomes $E/2\pi R_2 = E/4\pi R_1$. (c) The energy varies as $1/r$ because the wavefront circumference increases with r.

17.2 By reducing the amplitude of the source as time passes so that it always matches the decrease in amplitude of the outward traveling wave. Consequently the amplitude of the wave in Figure 17.1, though uniform over *space*, is not constant in time; it uniformly decreases over the entire wave pattern.

17.3 (a) Increase. The greater the spring constant, the faster any disturbance is passed from one bead to the next. The effect is the same as when the tension in a string carrying a transverse wave is increased (see Section 16.2). (b) Decrease. The greater mass slows down the transmission of the wave just as it does for a transverse pulse on a string of beads.

17.4 (a) Just as for a transverse wave, the x component of the velocity of each bead is dD_x/dt. If D_x is a sine function, the x component of the velocity is a cosine function. This means that the velocity wave is always one-quarter wavelength ahead of the displacement wave, as shown in Figure S17.14a. (b) By comparing Figure 17.9b and Figure S17.4a, you can see that the linear density is greatest where the velocity is maximum and smallest where the velocity is minimum. Figure S17.4b shows a plot of the linear density versus position.

17.5 Spherical. The shape of the wavefront is determined by the distance traveled by the wavefront. Because the wave speed is the same in all directions, a given wavefront reaches the same distance from the loudspeaker in all directions. (The sound is louder in the front than in the back because of differences in the wave amplitude, which is greater in front because of the way speakers are designed.)

Figure S17.4

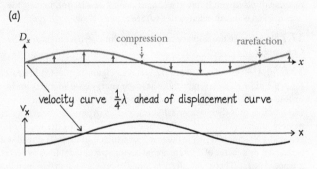

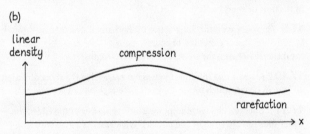

17.6 (a) Zero, because the displacement caused by each wavefront is zero. (b) The displacement at that point does not remain zero, however. A quarter period later, two crests (or troughs) meet at that point, and so the displacement at that point varies sinusoidally with an amplitude of $2A$ (Figure S17.6a). (c) Along the line that connects the half-filled circles, the waves generated by the two sources are $180°$ out of phase. Two

Figure S17.6

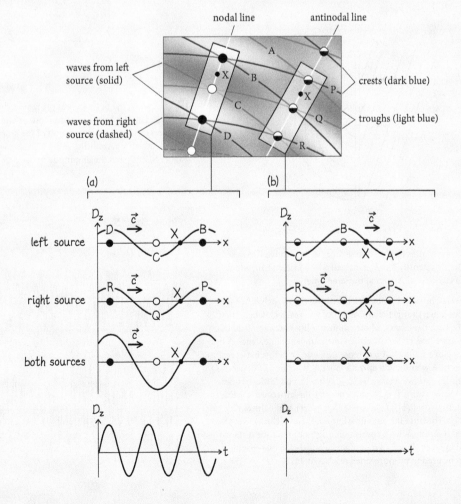

equal-amplitude sine waves that are 180° out of phase cancel, and so the medium displacement is zero at *all* points along the nodal line, not just the ones marked with half-filled circles. (*d*) See Figure S17.6*b*.

17.7 No, because the displacement at this point doesn't remain zero. See Checkpoint 17.6.

17.8 (*a*) $A + A = 2A$ in Figure 17.16*a*; A in Figure 17.16*b*. (*b*) Increase, for two reasons: There is now twice as much energy (two sources), and instead of radiating outward uniformly, the wave energy travels in only certain directions (between nodal lines). (*c*) $E \propto A^2$, and so the energy that passes P increases by a factor of 4. The energy delivered by the sources increases by a factor of 2, while the area over which the energy is spread out is reduced by a factor of 2 (there are just as many antinodal lines as there are nodal lines). These two effects multiply to produce a fourfold increase in the flow of energy at P.

17.9 There are six nodal lines above the horizontal straight line through S_1 and S_2 (and six below also, of course). So $2(d/\lambda) = 6$, $d/\lambda = 3$, $d = 3\lambda$; the distance between the sources is three wavelengths.

17.10 Because along this line, the two waves travel in the same direction and are in phase, which means they reinforce each other everywhere.

17.11 It does not change. Except for a little spreading at the edges, the wavefronts remain straight and parallel. What reaches a 10λ-wide line segment parallel to the row of sources at a distance R also reaches a similar segment at a distance $2R$.

17.12 See Figure S17.12. A narrow gap causes diffraction regardless of the orientation of the incident wavefronts because the incident waves cause the gap to become a point source. (For simplicity the reflections of the planar wavefronts from the back surface of the barrier have been omitted from the drawing.)

Figure S17.12

17.13 10 dB corresponds to 1 bel or one order of magnitude (because $\log 10 = 1$). So the intensity must be increased by a factor of 10, which means that ten clarinets must play at the same time.

17.14 (*a*) Three. The A and B lines fall exactly on top of each other at P, but as you move down they get out of step until you reach Q, where they match again. So each band has a whole number of lines between P and Q, but A has one more line. (If they had the same number of lines, each A line would fall on top of a B line. If A had more than one extra line between P and Q, the lines of A would catch up with those of B before Q.) Between Q and S the bands line up twice more, so band A has three additional lines between P and S. (*b*) Points P, Q, R, and S become dark (because the lines of the two patterns now fall right between each other), while the dark bands turn bright (because the lines now line up; Figure S17.14). Note that the moiré pattern moves down a considerable distance (half the distance between P and Q) even though B moved down only half a line spacing. This effect can be used to measure small displacements.

Figure S17.14

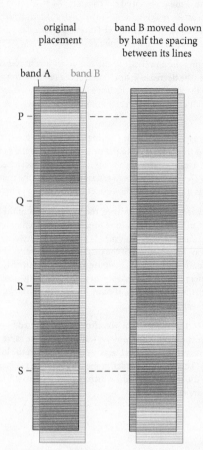

original placement band B moved down by half the spacing between its lines

band A   band B

17.15 Yes, but the beats are less pronounced than when the two amplitudes are equal. Let one wave have amplitude A and the other amplitude $2A$. The first wave produces beats with half of the second wave. The other half adds to the beat pattern, resulting in a oscillating wave with a varying amplitude (Figure S17.15), but the amplitude never goes all the way to zero.

Figure S17.15

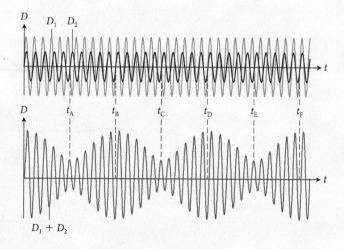

17.16 Equation 1 in Example 17.8 gives $f_s = f_1(c - v)/c$. Substituting the values for f_1, c, and v yields $f_s = 441$ Hz.

17.17 (a) With $v_o = \frac{1}{2}c$ and $v_s = 0$, Eq. 17.20 yields $f_o = \left[(c + \frac{1}{2}c)/(c - 0)\right]f_s = \frac{3}{2}f_s$. (b) With $v_o = 0$ and $v_s = \frac{1}{2}c$, $f_o = (c - 0)/(c - \frac{1}{2}c)f_s = 2f_s$. (c) The answers are different because there is a physical difference in the two situations. When the source moves, the wavefronts bunch up or spread out, changing the wavelength. When the observer moves, the wavelength, which is fixed by the wave speed in the stationary medium, does not change. Instead, the moving observer intercepts a different number of wavefronts per unit time. So it is not just the relative motion of source and observer that matters; the fact that their speeds are measured relative to the medium that transmits the wave also matters.

17.18 No. Shock waves are due to the piling up of wavefronts. When the source is stationary and the observer moves, the wavelength is unchanged. Setting $v_o = 2c$ with $v_s = 0$ in Eq. 17.20 yields $f_o = 3f_s$. This reflects the fact that the moving observer intercepts three times as many wavefronts per unit time as an observer at rest.

17.19 The cone angle is about 35°, and sin 35° = 0.57. From Eq. 17.22, you have $c/v_s = 0.57$, $v_s = 1.7c \approx 600$ m/s.

Chapter 18

18.1 (a) If the water is to remain at rest, the vector sum of the forces exerted on it must be zero. Because you initially exert a rightward 10-N force on the left piston, you must also exert a leftward 10-N force on the right piston. (b) Figure S18.1 shows free-body diagrams for the two pistons and the water. Because of the two interaction pairs marked in the figure, all six forces are equal in magnitude, so $\vec{F}_{pl}^c = \vec{F}_{wr}^c$. In other words, the force you exert on the left piston is transmitted by the water to the right piston.

Figure S18.1

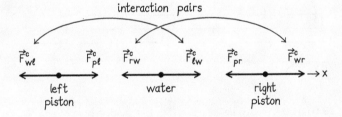

18.2 (a) Yes. The water is at rest, which means the downward force of gravity exerted on it must be balanced by an upward contact force exerted by the bottom of the cup. That the bottom exerts a force on the water tells you that the water must exert a force on the bottom. (b) Yes. If the sides of the cup were to disappear, the water would move radially outward. That the water remains in the cup tells you that the sides must exert a force on the water, and consequently the water must exert a force of equal magnitude on the cup in the opposite direction. (c) Positive. Because the water pushes outward on the surfaces of the cup, the surfaces must push inward on the water, putting it under compression. By definition, compression means positive pressure.

18.3 If you divide the liquid into three equal layers, each layer is subject to an identical force of gravity. The bottom layer supports the upper two, the middle layer supports only the top one, and the top one supports no other layer. If you make the division even finer, you can see that the pressure decreases linearly with height.

18.4 10 m. The pressure at the surface is 1.0×10^5 N/m², and that pressure doubles when the force of gravity exerted on a column of water of height h and cross-sectional area $A = 1$ m² has magnitude $F^G = mg = 1.0 \times 10^5$ N. The volume of this column is $V = hA$, its

mass is $m = \rho_{water}V = \rho_{water}\,hA$, and the magnitude of the force of gravity exerted on it is $mg = \rho_{water}\,hAg$. Thus $h = mg/(\rho_{water}\,Ag) = (1.0 \times 10^5$ N$)/[(1.0 \times 10^3$ kg/m³$)(1$ m²$)(9.8$ m/s²$)] = 10$ m.

18.5 (a) Nothing. Because the pressure in a liquid at rest decreases linearly with height, the pressure difference between the top and bottom of the brick is the same at every water depth. (b) Nothing. Pascal's principle tells you that a pressure change applied to a liquid is transmitted undiminished to every part of the liquid. So, increasing the pressure at the water surface increases the pressure at the top and at the bottom of the brick by the same amount, leaving the difference unchanged. (c) There is no effect because this pressure at the sides causes equal forces in opposite directions. (d) Yes, because the pressure in the atmosphere also decreases with height, although much less than in water. This force is upward.

18.6 The magnitude of the buoyant force exerted on the brick by the water is equal to the magnitude of the force of gravity exerted by Earth on the water displaced by the brick. Because the density of the brick is greater than the density of the water, the mass of the brick must be greater than the mass of the displaced water. Therefore the magnitude $m_{brick}g$ of the force of gravity on the brick is greater than the magnitude $m_{water}g$ of the force of gravity on the displaced water. The magnitude of the upward buoyant force exerted by the water on the brick is thus smaller than the magnitude of the downward force of gravity exerted on it, and so the brick sinks.

18.7 Equal. Because the pan-rock combination floats in both cases, the buoyant force exerted in both cases must be equal to the force of gravity exerted on the displaced water. You also know that the upward buoyant force on the combination must be equal to the downward force of gravity on the combination because the combination remains at rest. The mass of the pan-rock combination doesn't change when the rock is suspended (you can ignore the mass of the string), so the volume of water displaced doesn't change.

18.8 The same. The principle of relativity tells you that the laws of the universe are the same in all inertial reference frames. So, if you view the motion of an object moving at constant velocity through a stationary fluid from the reference frame of the object, you obtain the same situation—a laminar flow of fluid past a stationary object—as in Figure 18.18a.

18.9 As you can see in Figure 18.15, streamlines that pass over the roof of the car get closer together, indicating a higher flow speed and therefore lower pressure just above the roof. The air pressure inside the car, however, is unchanged and therefore greater than that outside, making the cloth roof bulge outward.

18.10 According to Pascal's principle, the pressure in the air in the balloon must be the same everywhere.

18.11 (a) Zero. The tensile forces due to surface tension are in the plane of the surface and therefore unable to compensate for any pressure difference across the surface when the surface is level. (b) Greater. The vector sum of the tensile forces exerted on a segment of the surface points toward the center of the drop. (c) It is higher. According to Laplace's law, the smaller radius of curvature of the small drop causes a greater pressure difference across the surface. Because both drops are surrounded by air at atmospheric pressure, the pressure must be greater inside the small drop.

18.12 Because the cylinder has a radius of curvature larger than that of the needle, the upward force exerted by the curved liquid surface on the cylinder is smaller than the upward force exerted on the needle. This smaller upward force is unable to support the (greater) downward force exerted by the cylinder on the surface.

18.13 The magnitude of the force exerted by the air on the cover is $F_{ac}^c = P_{atm}A = (101$ kPa$)(1000$ Pa$/1$ kPa$)(0.28$ m$)(0.22$ m$) = 6.2 \times 10^3$ N. The force of gravity exerted on the book is $F_{Eb}^G = (3.0$ kg$)(9.8$ m/s²$) = 29$ N. The force exerted by the atmosphere on the cover is more than 200 times greater than the gravitational force exerted by Earth on the book.

18.14 As the suction cup hits the ceiling, the bowl of the cup collapses, forcing out the air that was initially in the bowl. Consequently, the pressure in the space between the cup and the ceiling is lower than atmospheric pressure. As long as the seal between the cup edge and the ceiling is maintained, the air in the room underneath the cup exerts an upward force greater than the downward force exerted by the air inside the cup that holds the dart up against the ceiling.

18.15 That the object floats tells you that the buoyant force exerted on it must equal the force of gravity exerted on it, and so equating Eqs. 18.11 and 18.12 gives you $\rho_{o,av}V_o g = \rho_{water}V_{disp} g$. You are given that $V_{disp} = 0.80V_o$, and so $\rho_{o,av} = (\rho_{water}V_{disp})/V_o = 0.80\rho_{water}$ (see Example 18.2).

18.16 No. The pressure at the oil-water interface in the right leg must be the same as the pressure at a point that is at the same height in the left leg (Figure S18.16). Because the mass density of oil is smaller than that of water, the height of a column of oil required to cause a certain pressure difference is greater than the height of a column of water required to cause the same pressure difference. Therefore the top of the oil column must be higher than the top of the water column.

Figure S18.16

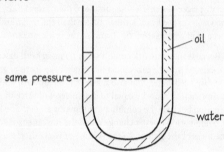

18.17 Because the atmosphere exerts a force on both pistons, it increases the pressure below them. When you add the effect of atmospheric pressure, the pressure in the liquid at the small piston is $P_1 = P_{atm} + F_1/A_1$ and the pressure at the large piston is $P_2 = P_{atm} + F_2/A_2$. Ignoring the term ρgh in Eq. 18.19 again, you have $P_1 \approx P_2$, and so $P_{atm} + F_1/A_1 = P_{atm} + F_2/A_2$. The atmospheric pressure cancels out, and you obtain the same result as in Eq. 18.21.

18.18 The continuity equation for an incompressible fluid (which means $\rho_1 = \rho_2$) gives $A_1v_1 = A_2v_2$ or $v_1 = v_2A_2/A_1 = v_2R_2^2/R_1^2$. Squaring all terms, subtracting v_2^2 from both sides, and dividing through by -1 give you $v_2^2 - v_1^2 = v_2^2(1 - R_2^4/R_1^4)$. With $R_1 = \frac{1}{2}(0.50\text{ m})$ and $R_2 = \frac{1}{2}(10.0\text{ mm})$, you have $(R_2/R_1)^4 = 1.6 \times 10^{-7}$, which is so small that you can ignore this term to get the result $v_2^2 - v_1^2 = v_2^2(1 - R_2^4/R_1^4) \approx v_2^2$. Thus the assumption that $v_1 \approx 0$ in Example 18.9 is excellent, and you can use the result in that example: $v_2 = \sqrt{2gd} = 2.4\text{ m/s}$.

18.19 A soap bubble consists of a thin film of soap solution surrounding a volume of air. Because the film has two surfaces, the force due to surface tension gains an additional factor 2. Thus Eq. 18.51 becomes $P_{in} - P_{out} = 4\gamma/R$.

Chapter 19

19.1 (a) Collisions with the leading edge of the pendulum slow it down, and collisions with the trailing edge speed it up. Because of the motion of the pendulum, nitrogen molecules that collide with the leading edge move faster relative to the pendulum than nitrogen molecules that collide with the trailing edge. Therefore there are more collisions with the leading edge than with the trailing edge, and the pendulum slows down. (b) When the pendulum is at rest, the average speed of the nitrogen molecules that hit the left edge is the same as the average speed of the molecules that hit the right edge, which means that the momentum transfers due to the collisions tend to balance out.

19.2 Each segment of the zigzag path would decrease in length, so the Brownian motion would be less pronounced and the maximum displacement of the grain from the initial position would be smaller.

19.3 7.

19.4 See Figure S19.4. (a) Five macrostates. (b) 16 basic states. (c) $4/16 = 1/4$. (d) The most likely macrostate is two heads because there are six basic states associated with this macrostate.

Figure S19.4

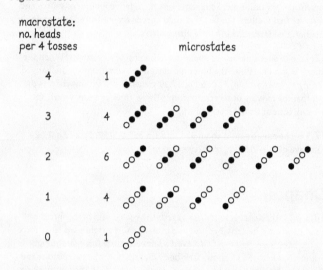

19.5 See Figure S19.5.

Figure S19.5

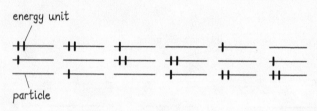

19.6 (a) 1/84, which is around 1%. (b) $\frac{28}{84} + \frac{21}{84} = \frac{49}{84}$, which is more than 50%.

19.7 (a) The pendulum has more than 50% of the energy when it has four, five, or six units of energy. There are $6 + 3 + 1 = 10$ basic states associated with these macrostates. The probability that the pendulum has more than 50% of the energy is then $10/84 = 0.12$. (b) Figure 19.7 shows that as the number of particles increases, the relative probability of any given fraction of energy being in the pendulum decreases.

19.8 (a) The average kinetic energies are equal because the energy of the gas is equally distributed among all the particles. (b) Because the two particles have the same average kinetic energy, $\frac{1}{2}m_1v_1^2 = \frac{1}{2}m_2v_2^2$. Therefore $v_1 = (m_2/m_1)^{1/2}v_2 = 2^{1/2}v_2$. Particle 1 has a greater average speed by a factor of $2^{1/2}$.

19.9. There are three ways of placing the two indistinguishable particles in the two compartments (Figure S19.9). The probability of finding one particle in each side is 1/3.

Figure S19.9

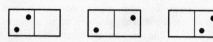

19.10. (*a*) Four basic states have all six particles in one compartment. They originate from different macrostates. Distribution A in Figure S19.10 has six particles in the top left compartment and is a macrostate. Distribution D is part of a different macrostate: no particles in the top left compartment. The three basic states in distribution D correspond to six particles in any of the other three compartments. There are 84 basic states, all equally probable, and so the probability of finding six particles in any compartment is $4/84 = 0.048$. (*b*) There are 12 basic states in which one compartment contains five particles: three for distribution B in Figure S19.10, three for distribution C, and six for distribution E. The probability of finding five particles in any compartment is thus $12/84 = 0.143$. Note that distribution E has six associated basic states, corresponding to the six ways one particle can be placed in one of three containers and five particles can be placed in another of the three containers.

Figure S19.10

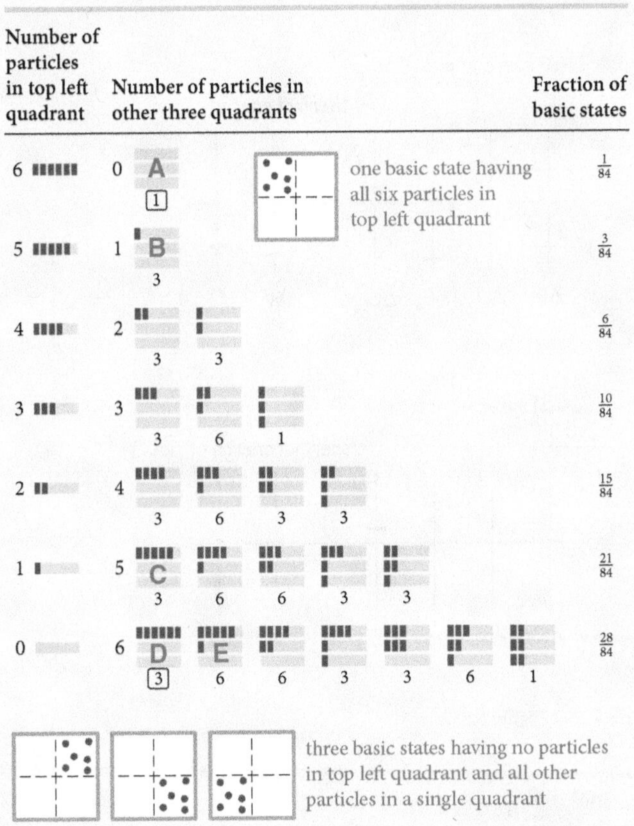

Number of particles in top left quadrant	Number of particles in other three quadrants		Fraction of basic states

19.11. The partition moves until the particles are uniformly distributed over the space enclosed by the two compartments. Because there are twice the number of particles in the left compartment, the partition moves to the right until the volume of the left compartment is twice the volume of the right compartment.

19.12. (*a*) All six energy units are initially in the pendulum, which means each collision moves one energy unit from the pendulum to one of the particles. The system then is in one of the three basic states depicted in the second of row of Figure 19.4. (*b*) Now there are three possibilities for how energy is transferred in any given collision: from the particle that initially has the one unit to either of the two other particles, from the particle that initially has the one unit to the pendulum, or from the pendulum to any one of the particles. When a unit is transferred from one particle to another, the amount of energy in the pendulum does not change, and so the macrostate does not change (that macrostate is represented in the second row in Figure 19.4). In a collision between the pendulum and a particle, the pendulum can either gain or lose a unit of energy, and so the

system ends up in the macrostates represented by the top and third rows in Figure 19.4. So, after one collision the system can be in any one of the top three macrostates in Figure 19.4. (*c*) The number of basic states in the top three macrostates is $1 + 3 + 6 = 10$.

19.13. (*a*) There are three accessible macrostates: three units in pendulum, two units in pendulum, one unit in pendulum. (*b*) To find the probability for each macrostate, you need to consider the number of basic states representing each: 10, 15, and 21, respectively, for a total of 46. The probabilities are thus $10/46$, $15/46$, and $21/46$.

19.14. (*a*) There are 21 ways that two energy units can be distributed among the six particles. Label the particles 1 through 6. Place one energy unit in particle 1. The second energy unit can be placed in any of the six particles, resulting in six basic states. Now place one energy unit in particle 2. The second energy unit can be placed in the five particles 2 through 6 but not in particle 1 because we already counted that basic state. Repeat this process for particles 3 through 6. The number of basic states is then $6 + 5 + 4 + 3 + 2 + 1 = 21$.

19.15. (*a*) There are 14 particles in A, so the average energy per particle in A is $1/14 = 0.07$ energy unit. The average energy per particle in B is $9/6 = 1.5$ energy units. (*b*) At equilibrium there are seven energy units in A, corresponding to an average energy per particle of $7/14 = 0.5$ energy unit. The average energy per particle in B is $3/6 = 0.5$ energy unit.

19.16. (*a*) Over time the ice cube heats up and melts, and the water cools. Because the system irreversibly evolves toward equilibrium, Ω must be increasing. (*b*) Because everything—cup, tea, air—is in equilibrium, Ω is at its maximum value and not changing. (*c*) Over time the oil separates and floats to the top. Because this separation constitutes an irreversible process (the oil doesn't spontaneously mix back into the water), Ω must be increasing. (*d*) Germination is an irreversible process, so Ω is increasing. (*e*) As far as we know, the universe evolves irreversibly, so Ω is increasing.

19.17. (*a*) The compartment size decreasing by a factor of 10 means that $M = 1000$. The number of basic states is then $\Omega_f = M^N = 1000^{10} = 10^{30}$. The initial value is $\Omega_i = 10^{20}$ (see Example 19.8), so the change in the number of basic states is $10^{30} - 10^{20}$, which for all practical purposes is equal to 10^{30}. This represents an enormous increase in the number of basic states. (*b*) $\ln \Omega_f = \ln (10^{30}) = 69$. The initial value is $\ln \Omega_i = 46$ (see Example 19.8), so the change is $69 - 46 = 23$.

19.18. (*a*) According to Eq. 19.4, an increase in entropy means the number of basic states has increased. (*b*) For an expanding gas, $V_f > V_i$, and so the $N \ln (V_f/V_i)$ term in Eq. 19.8 is positive, as expected for a system evolving toward equilibrium. (*c*) If the gas were to contract into a subvolume, $V_f < V_i$, and so the $N \ln (V_f/V_i)$ term is negative, meaning a decrease in entropy. Such a decrease would violate the entropy law.

19.19. Even though entropy depends on volume, the entropy change depends on a ratio of volumes (Eq. 19.8). Because the volume of a compartment is a fraction of the volume V of the box, V cancels when we take the ratio of volumes for each compartment.

19.20. (*a*) Adding up the x components of the velocities and dividing by 5 gives an x component of the average velocity of 0. (*b*) Adding up the speeds (the absolute values of the x components of the velocities) and dividing by 5 gives an average speed of 2.6 m/s. (*c*) Adding up the squares of the speeds, dividing by 5, and then taking the square root gives a rms speed of 3.0 m/s. Note that the rms speed is not equal to the average speed of the atoms: in a gas, the former is typically about 10% greater than the latter.

19.21. Putting all the numerical factors in the third term of Eq. 19.29 together gives

$$\left[2a \left(\frac{2E_{th}}{mN} \right)^{1/2} \right]^3 = \left[\left(\frac{8a^2 E_{th}}{mN} \right)^{1/2} \right]^3 = \left(\frac{8a^2}{mN} \right)^{3/2} (E_{th})^{3/2}.$$

19.22. (*a*) At thermal equilibrium there is equipartition of energy, and so, on average, the kinetic energy of all the atoms is the same. To the left of the dashed line, S_A is smaller and S_B is greater than at thermal equilibrium. According to Eq. 19.32, $E_{th,A}$ is therefore smaller than the equilibrium value and $E_{th,B}$ is greater than the equilibrium value. Consequently, to the left of the dashed line the average kinetic energy of the atoms in A is smaller than that in B. (*b*) At thermal equilibrium, both gases are equally hot. To the left of the dashed line gas B is hotter because the average kinetic energy of the atoms is greater. (*c*) At thermal equilibrium, $dS_A/dE_{th,A} = dS_B/dE_{th,B} = -dS_B/dE_{th,A}$ (Eq. 19.37). Figure 19.24 shows that the magnitude of the slope of curve S_A increases and the magnitude of the slope of curve S_B decreases when going left from the equilibrium point. Therefore, to the left of the equilibrium point $dS_A/dE_{th,A} > dS_B/dE_{th,B}$.

19.23. From Eq. 19.42 we have $E_{th,B}/E_{th,A} = N_B/N_A = 6/14 = 3/7$, and so 30% of the energy is in compartment B and 70% is in compartment A.

19.24. (*a*) Only the component of the momentum perpendicular to the wall changes. The atom's final momentum is away from the wall, and so the change in momentum is perpendicular to the wall and directed toward the interior of the container. (*b*) Because the system is isolated, the direction of the wall's change in momentum is opposite the direction of the atom's change in momentum. (*c*) As you saw in Section 8.7, the force exerted on an object is proportional to the object's change in momentum. The force exerted by the wall on the atom is therefore perpendicular to the wall and directed toward the interior of the container. The force exerted by the atom on the wall points in the opposite direction.

19.25. In deriving Eq. 19.20 we assumed that the system is in thermal equilibrium (see the beginning of Section 19.5), and so $T_A = T_B$. Multiplying the left and right sides of Eq. 19.20 by $k_B T_A$ gives $N_A k_B T_A/V_A = N_B k_B T_B/V_B$, or using the ideal gas law, $P_A = P_B$. In equilibrium, the pressure is the same on both sides of the partition, as we would expect.

19.26. (*a*) Neither. Because the pressures and volumes are equal, the thermal energies of the two gases are equal. (*b*) Neither. Because the two containers have the same number of atoms, the average kinetic energy is the same for all atoms. (*c*) Argon. The rms speed is proportional to the square root of the average kinetic energy divided by the mass (Eq. 19.53). Although the average kinetic energies of the two gases are equal, a helium atom has a smaller mass than an argon atom, so the rms speed of a helium atom is greater than that of an argon atom.

19.27. (*a*) No, because the system is not closed; as the gas cools, thermal energy leaves the system. The entropy law applies only to closed systems. (*b*) From Example 19.12a, we have $T_f = 145$ K, so $\Delta T = 145$ K $- 290$ K $= -145$ K, and so, from Eq. 19.50: $\Delta E_{th} = \frac{3}{2} N k_B \Delta T = \frac{3}{2}(10^{23})(1.38 \times 10^{-23} \text{ J/K})$ $(-145 \text{ K}) = -300$ J.

19.28. Decrease. From Eq. 19.61, because $T_f = T_i/2$ and $V_f = 2V_i$, we have

$$\Delta S = \tfrac{3}{2} N \ln \tfrac{1}{2} + N \ln 2.$$

Because $\ln \frac{1}{2} = -\ln 2$, this can be written as

$$\Delta S = -\tfrac{3}{2} N \ln 2 + N \ln 2 = -\tfrac{1}{2} N \ln 2 < 0,$$

and so the entropy decreases. When the temperature of the system is halved, the energy of the system decreases, and so the system is not closed.

Chapter 20

20.1. (*a*) Joules. (b) See Figure S20.1.

Figure S20.1

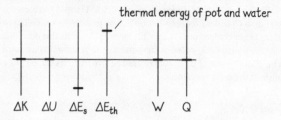

(i) system: water, pot, flame

thermal energy of pot and water

ΔK ΔU ΔE_s ΔE_{th} W Q

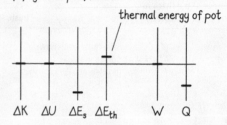

(ii) system: pot, flame

thermal energy of pot

ΔK ΔU ΔE_s ΔE_{th} W Q

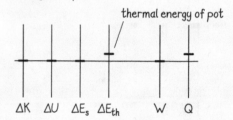

(iii) system: pot

thermal energy of pot

ΔK ΔU ΔE_s ΔE_{th} W Q

20.2. See Figure S20.2.

Figure S20.2

ΔK ΔU ΔE_s ΔE_{th} W Q

20.3. See Figure S20.3. The quasistatic process is possible because it depicts a succession of equilibrium states, all of which show an equipartition of space. The non-quasistatic process is not possible: As the first four frames show, it would require the gas molecules to "pull away" spontaneously from the piston, before the piston begins moving down in frame 2.

Figure S20.3

(*a*) Reverse quasistatic process

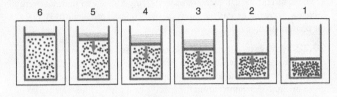

(*b*) Reverse non-quasistatic process

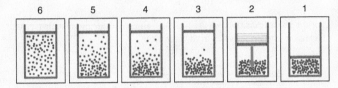

20.4. During a quasistatic process the system must remain near equilibrium at all instants during the process. (*a*) Not quasistatic; just after the balloon is popped, the pressure inside the container is not the same everywhere and so the system is not in equilibrium. (*b*) Quasistatic; even though the temperature of the coffee changes, the change is slow. The coffee therefore has a well-defined uniform temperature and is in equilibrium throughout the cooling process. (*c*) Quasistatic; as the gas pushes the piston outward, the pressure in the gas remains uniform and so the gas is in equilibrium. (*d*) Not quasistatic; the temperature of the ice cube and the hot water are different, and so the system is not in equilibrium.

20.5. (*a*) Using Eq. 2 from the "Temperature measurement scales" box, you find for $T_F = 60\,°F$:

$$T_C = \frac{5\,°C}{9\,°F}\,(60\,°F - 32\,°F) = 16\,°C.$$

Likewise, for $T_F = 80\,°F$, you obtain $T_C = 27\,°C$, making the range $16\,°C$ to $27\,°C$. By adding 273.15 to these values, you get the corresponding range in kelvins: 289 K to 300 K.

(*b*) Equation 1 from the box gives you

$$T_C = \frac{1\,°C}{1\,K}\,(2800\,K) - 273.15\,°C = 2500\,°C.$$

The expression I used in the solution to Exercise 20.2*b* gives

$$T_F = \frac{9\,°C}{5\,°F}\left[\frac{1\,°C}{1\,K}\,(2800\,K) - 273.15\,°C\right] + 32\,°F = 4600\,°F.$$

20.6. (*a*) There are many more water molecules in the pool than in the glass. Thermal energy is proportional to number of molecules, and so the pool has greater thermal energy. (*b*) The heat capacity of the pool is greater because the amount of thermal energy needed to increase the temperature of all the water in the pool by 1 K is greater than the amount needed to increase the temperature of all the water in the glass by 1 K. (*c*) The glass of warm water because thermal energy is proportional to temperature.

20.7. (*a*) All three. Unlike a hydrogen molecule, an ammonia molecule is not linear, so its rotational inertia is nonzero along the three orthogonal axes. (*b*) Because there are six degrees of freedom, three translational and three rotational, the thermal energy per molecule is $6(\frac{1}{2}k_BT) = 3k_BT$. (*c*) If you raise the temperature by $\Delta T = 1$ K, the increase in thermal energy per molecule is $\Delta E_{th} = 3k_B\Delta T$. The heat capacity per molecule is thus $\Delta E_{th}/\Delta T = 3k_B$. (*d*) It's within 10%. The experimental value listed in Table 20.3 is $3.37k_B$.

20.8. Less than. More energy must be transferred to the diatomic gas because it has more degrees of freedom, and so its heat capacity per particle is greater.

20.9. $P_1 > P_2$, $V_1 < V_2$, and $T_1 = T_2$ because the two states are on an isotherm. The entropy dependence on temperature and volume is given by Eq. 19.60, $S \propto \ln T^{3/2}V$. The temperatures are the same in the two states, and so the differences in entropy depend only on volumes. That $V_1 < V_2$ tells you that $S_1 < S_2$. Because $E_{th} \propto T$, the thermal energies of the two states are the same.

20.10. (*a*) The change in entropy is the same for the two processes (the same initial and final states and therefore the same initial and final temperatures and volumes). (*b*) No. Because the two processes have the same initial and final states, the change in temperature is the same for both. Hence the change in thermal energy ΔE_{th} is the same. This change is the sum of the work done on the gas and the energy transferred thermally to it: $\Delta E_{th} = W + Q$. Because the work done on the gas (the area under the path) is different for the two processes, the amount of energy transferred thermally must also be different.

20.11. See Table S20.11. The work done on the gas is positive during compression, negative during expansion, and zero for any isochoric process.

The thermal energy is proportional to the temperature (see Eq. 19.50). As you can see from Figure 20.25, the isentropic expansion brings the system to a lower isotherm, and so the thermal energy decreases; in the isochoric process, the temperature increases, and so $\Delta E_{th} > 0$; the isobaric process decreases the temperature, and so $\Delta E_{th} < 0$; and the isothermal compression doesn't change the temperature, so $\Delta E_{th} = 0$.

The isentropic process is adiabatic, so $Q = 0$. For the other processes, the sign of Q can be determined from ΔE_{th} and W, because the sum of the work done on the gas and the energy transferred thermally to it equals the change in thermal energy.

For the isentropic process the change in entropy is zero by definition. As you can see from Figure 20.25, the isochoric process brings the system to a higher isentrope and so $\Delta S > 0$. The isobaric and isothermal compressions bring the system to a lower isentrope, so $\Delta S < 0$. (Because the system is not closed, the entropy can decrease.)

20.12. Table 20.3 and Figure 20.15 tell you that at $T = 300$ K, the heat capacity per particle for hydrogen is $2.5k_B$ and so it has five degrees of freedom (three translational and two rotational). So, if we assume the gas can be treated as ideal, the thermal energy is $E_{th} = \frac{5}{2}Nk_BT$ (approximately N times 10^{-20} J).

Table S20.11 Checkpoint 20.11

Process	ΔE_{th}	W	Q	ΔS
isentropic expansion	negative	negative	0	0
isochoric pressure increase	positive	0	positive	positive
isobaric volume decrease	negative	positive	negative	negative
isothermal compression	0	positive	negative	negative

20.13. Process A. The vertical legs for both A and B are at constant volume, and Eq. 20.9 tells you that no work is done on the gas. In process A, the gas undergoes an isobaric decrease in volume at a higher pressure than process B. The volume change is the same in both cases. So, by Eq. 20.9 (which has ΔV no longer equal to zero), the work done on the gas during the isobaric process in process A is greater than in process B. (Note that the area under path A in Figure 20.28 is larger than the area under path B and the work done on the gas is positive, thus confirming our answer.)

20.14. (a) Process C. Because the initial and final states lie on an isotherm, $\Delta E_{th} = 0$ for all three processes. The work done on the system in each process is negative and equal in magnitude to the area under the path. Because $\Delta E_{th} = 0$, $Q = -W$ and so Q is equal to the area under the path, which is largest for process C and (b) smallest for process A.

20.15. Yes. If $\Delta T = 0$, then the temperature doesn't change, and if the temperature doesn't change, then Eq. 20.5 tells us that E_{th} doesn't change either, so $\Delta E_{th} = 0$.

20.16. Greater. Because $PV = Nk_BT$ and Nk_B is constant, the product PV is proportional to the temperature T of the state. The final temperature T_f is greater than the initial temperature T_i, and so $P_fV_f > P_iV_i$.

20.17. The isobaric process. Both processes have the same ΔE_{th} because the temperature change is the same. No work is done on the gas during the isochoric process, which means that in this process the energy transferred thermally goes entirely into raising the temperature. During the isobaric process, more energy must be transferred thermally into the gas during the expansion because the work done on the gas is negative.

20.18. Because the initial and final states lie on the same isotherm, the thermal energy of the gas doesn't change. From this you know that $\Delta E_{th} = W + Q = 0$ and so $W = -Q$. The work done on the gas is positive for both processes ($V_f < V_i$), and so the energy transferred thermally must be negative. (a) The work done on the gas is equal to the area under the path. From the paths in Figure 20.37 you see that the rectangular area under path A is larger than the area under path B, which means that the work done on the gas in process A is greater than the work done on the gas in process B. (b) Because W is greater for process A, the magnitude of Q must be greater for process A.

20.19. No. Equation 20.36 gives the change in entropy between two equilibrium states that are at the same temperature. During an isochoric process, the temperature changes, so we cannot use Eq. 20.36.

20.20. The same. Equation 20.34 is independent of the process. The entropy change depends on only the temperatures and volumes of the initial and final states. If the two combinations of processes have the same initial and final states, the change in entropy is the same in the two cases.

20.21. See Table S20.21. ΔS is positive when a process ends on a higher isentrope. To determine the sign of Q, we approximate the ideal gas process by a series of infinitesimal isentropic and isothermal segments. (We can do this for any process.) Along the isentropic segments, $Q = 0$. Along the isothermal segments, Q has the same sign as ΔS, which is positive. Therefore for any process for which $\Delta S > 0$, we know that Q must be positive too. The change in the thermal energy is the sum $Q + W$. Because isentropes

Table S20.21 Checkpoint 20.21

	Expansion	Compression
W	negative	positive
Q	positive	positive
ΔE_{th}	positive, negative, or zero	positive
ΔS	positive	positive

are steeper than isotherms, ending on a higher isentrope while compressing the gas means ending on a higher isotherm, and so the temperature T and the thermal energy both have to increase ($\Delta E_{th} > 0$). When the gas expands, its entropy can increase while its temperature and thermal energy increase, decrease, or stay the same. (a) $W < 0$ for expansion. (b) $W > 0$ for compression.

20.22. We have to calculate entropy changes associated with two processes: the cooling down of the water from 20 °C to 0.0 °C and the phase transition from water to ice. The entropy change during the cooling is given by Eq. 20.53. Substituting the specific heat capacity for water from Table 20.2, we obtain

$$\Delta S = \frac{(0.010 \text{ kg})(4181 \text{ J/kg·K})}{1.38 \times 10^{-23} \text{ J/K}} \ln\left(\frac{273}{293}\right) = -2.1 \times 10^{23}.$$

For the phase transition, we use Eq. 20.58 and substitute the specific transformation energy for water from Table 20.4:

$$\Delta S = -\frac{(0.010 \text{ kg})(3.34 \times 10^5 \text{ J/kg})}{(1.38 \times 10^{-23} \text{ J/K})(273 \text{ K})} = -8.9 \times 10^{23}.$$

The total entropy change is thus -11×10^{23}.

Chapter 21

21.1. (a) There is a change in the state of motion of the puck, which decreases its kinetic energy. Because the temperature of the puck and surface increases, their thermal energy increases. This is an irreversible process, so the entropy increases. (b) The spring relaxes from the compressed state, and the cart changes its state of motion. The configurational potential energy of the spring-cart system decreases, and the kinetic energy of the cart increases. The process is reversible, so the entropy does not change. (c) A chemical reaction changes the state of the explosives, releasing source energy. The projectile changes its state of motion, increasing its kinetic energy. The temperature and thermal energy of the cannon and projectile increase. Because the process is irreversible, the entropy increases. (d) The temperature of the hot object decreases, and therefore its thermal energy decreases. The temperature of the cold object increases, and therefore its thermal energy increases. This process is irreversible, so the entropy increases.

21.2. (a) See Figure S21.2. Energy is transferred thermally to the gas. This results in an increase in both the temperature and the volume of the gas. The expanding gas does work on the piston. (b) Both change. The temperature of the gas increases, so its thermal energy increases. The entropy increases because both the temperature and volume of the gas increase (Eq. 19.61).

Figure S21.2

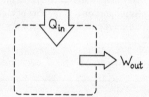

21.3. (a) Pushing the piston back to its original position requires us to do as much work on the device as the device has delivered to us (Eq. 20.30). Therefore once the piston is back to its original position $W = 0$, and because $\Delta E = 0$, $Q = 0$ too. No energy has been converted. (b) The volume and temperature of the gas are the same before expansion and after compression, so by Eq. 19.61 the entropy stays the same. (c) Yes, because the device keeps returning to its initial state.

21.4. (a) $\Delta E_R > 0$ (energy increases), $\Delta S_R > 0$ (entropy increases with energy), $\Delta S_R / \Delta E_R > 0$. (b) $\Delta E_R < 0$, $\Delta S_R < 0$, $\Delta S_R / \Delta E_R > 0$.

21.5. (a) Energy quality decreases because $\Delta S > 0$ means that energy is transferred to the right on the dS/dE axis, as in Figure 21.16b. (b) Energy quality remains the same because it remains at the same value of dS/dE.

21.6. (b), (c), and (f).

21.7. See Figure S21.7.

Figure S21.7

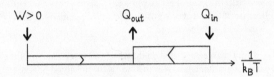

21.8. Because the process depicted in Figure 21.22a is reversible, $\Delta S_{env} = 0$, and so the two rectangles must have equal areas. The rectangle on the right is less long and therefore must have a greater height.

21.9. (a) Because you increase T_{in}, the label Q_{in} and the point where the two rectangles meet both move to the left. The length of the upgrade rectangle decreases, which means the magnitude of ΔS corresponding to the upgrade decreases. (b) Because the process is reversible, the magnitude of ΔS associated with the degradation also decreases. (c) Because the length of the degrade rectangle increases, the height must decrease; therefore Q_{out} decreases. That means, because $W = Q_{out} - Q_{in}$ and W is constant, Q_{in} decreases.

21.10. Lowering T_{out} requires a smaller thermal input of energy for the same work done on the environment, so the efficiency increases.

21.11. (a) In Figure 21.26, the Q_{in} arrow is roughly twice as wide as the W_{in} arrow. Because $W = W_{in}$, the coefficient of performance of cooling is $Q_{in}/W = 2$. (b) Because $Q_{out} = Q_{in} + W_{in} = 3W_{in} = 3W$, the coefficient of performance of heating is $Q_{out}/W = 3W/W = 3$.

21.12. (a) Now the area under the expansion path $1 \rightarrow 2$ is greater in magnitude than the area under the compression path $2 \rightarrow 1$, which means the negative work done on the system is greater than the positive work done on the system. So the work done on the system during the cycle is negative, and the magnitude is equal to the shaded area in Figure 21.28d. (b) The sign of the work done on the environment is the opposite of the sign of the work done on the system and so is positive. The magnitude is the same as in part a.

21.13. Although the transfer from low temperature to high temperature is an upgrade of the energy and therefore a decrease in entropy, the entropy increase from the degradation of the mechanical energy input is greater than the decrease.

21.14. Zero. The law of energy conservation requires that the energy of the closed system comprising the device and its environment cannot change. Because the energy of the steady device is not changing, $\Delta E = 0$ (Eq. 21.1), the energy of the environment cannot change either.

21.15. (a) Both negative. Because energy is transferred thermally out of the bar, the energy of the bar decreases and so its entropy decreases. (b) Energy is transferred thermally to the environment, and S_{env} increases. (c) The equilibration is irreversible, so ΔS_{env} must be greater than ΔS_{bar}, which means $S_{bar} + S_{env}$ increases.

21.16. (a) See Figure S21.16. (b) Because the transfer is to the left, the entropy change is negative. The height of the rectangle is Q_{in}. (Because

$W = -Q_{in}$ you could also write the height of the rectangle as $-W > 0$.) The rectangle stretches from the origin (0 on the $1/k_B T$ scale) to $1/k_B T_{in}$, so its length is $1/k_B T_{in}$. The entropy change is thus

$$\Delta S_{env} = -(Q_{in})\left(\frac{1}{k_B T_{in}}\right) = \frac{-Q_{in}}{k_B T_{in}}.$$

This process not allowed by the entropy law because $\Delta S_{env} < 0$.

Figure S21.16

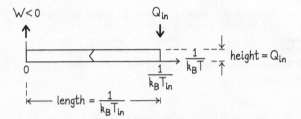

21.17. Yes, in theory. According to Eq. 21.20, Q_{out} goes to zero if T_{out} goes to zero. As T_{out} goes to zero, the length of the degrade rectangle becomes infinite, but the area of the rectangle can remain constant. Unfortunately there are no thermal reservoirs at $T = 0$.

21.18. (a) $\eta_{max} = (299\ K - 221\ K)/299\ K = 0.261$. (b) $\eta_{max} = (300\ K - 221\ K)/300 = 0.263$, so the efficiency increases by 0.002.

21.19. The blanket converts mechanical energy entirely to thermal energy, and so from Eq. 21.24, $COP_{heating} = Q_{out}/W = 1$, which is the smallest possible value.

21.20. (a) Equation 21.28 gives $COP_{heating} = COP_{cooling} + 1$. You know from Example 21.7 that $COP_{cooling}$ is 17. Therefore $COP_{heating}$ is 18. (b) You can use Eq. 21.24 to get the delivery rate: $dW/dt = (dQ_{out}/dt)/COP_{heating} = (500\ W)/18 = 28\ W$.

21.21. As the gas goes from state 1 to state 2 the entropy change of the gas is (Table 21.1):

$$\Delta S_{1 \rightarrow 2} = N \ln\left(\frac{V_2}{V_1}\right).$$

Substituting the values given then yields

$$\Delta S_{1 \rightarrow 2} = 5.10 \times 10^{23}.$$

During the two isentropic legs, the entropy doesn't change:

$$\Delta S_{2 \rightarrow 3} = 0$$
$$\Delta S_{4 \rightarrow 1} = 0.$$

The entropy change over the entire cycle is zero, so the entropy change as the gas goes from state 3 to state 4 must be the negative of the entropy change as the gas goes from state 1 to state 2:

$$\Delta S_{3 \rightarrow 4} = -\Delta S_{1 \rightarrow 2} = -5.10 \times 10^{23}.$$

21.22. (a) The efficiency of the jet engine is given by Eq. 21.44. Substituting T_1 and T_4 from Example 21.9, we get $\eta = 1 - (T_4/T_1) = 1 - (288\ K)/(760\ K) = 0.62$. (b) Equation 21.36 gives the Carnot efficiency: $\eta = (1520\ K - 288\ K)/(1520\ K) = 0.81$, which is significantly greater than the 0.62 efficiency of the engine running on the Brayton cycle. (Remember, however, that the Carnot cycle operates very slowly, which means the power is very low.)

Chapter 22

22.1. (*a*) You should observe that the tape and battery attract each other. (If the battery repels the tape, you can get rid of the repulsion by wiping the entire battery surface with your hand.) It makes no difference how you orient the battery or whether the battery is fresh or spent. Any wooden object should also attract the tape in the same way a battery does. (*b*) The tape and the power cord attract each other. Turning the lamp on or off doesn't make any difference. It does not appear that the power cord has an effect different from any other object.

22.2. (*a*) You should see that the two strips repel each other quite strongly. (*b*) No. Regardless of how you orient the strips—sticky sides facing, nonsticky sides facing, one sticky side facing one nonsticky side—the two always repel. As you bring them closer, they twist to avoid contact.

22.3. Paper-paper, no interaction; tape-paper, interaction; tape-tape, interaction.

22.4. (*a*) The strip should be attracted to your hand and the strip of paper and be repelled by the second charged strip. (*b*) The uncharged strip should remain motionless when you hold your hand close to it. (*c*) You should see no interaction between the uncharged hanging strip and the strip of paper, and there should be an attractive interaction between the two tape strips. This tape-tape interaction is just like the attractive interaction between a charged tape strip and any other uncharged object.

22.5. The recharged strip should be attracted to your hand and to a strip of paper and repelled by another charged tape strip.

22.6. (*a*) There is an attractive interaction between both types of strips and uncharged objects. (*b*) A third charged strip attracts strip B and repels strip T. (*c*) Yes. Strip T must be charged because it interacts with other charged strips and with uncharged objects. (*d*) Yes. Strip B must be charged because it interacts with other charged strips and with uncharged objects. (*e*) After discharging, neither strip B nor strip T interacts with uncharged objects, and both strips are attracted to a charged strip. (See the box "Troubleshooting B and T strips" if your experimental results do not agree with these.)

Troubleshooting B and T strips

If your experiment with B and T strips doesn't work as expected, check the following:

1. You must pull off the combination in step 2 of Figure 22.5 *very* slowly. (The amount of charge that builds up on the strips is roughly proportional to the speed at which you separate them.) Be sure to remove *all* charge before proceeding.

2. Separating the B and T strips, on the other hand, must be done fairly rapidly. (If you do it too fast, however, so much charge may build up on your strips that it becomes hard to prevent them from being attracted to your hands. If they curl around and touch your hands, you must start over.)

3. Avoid any air currents on the suspended strips.

4. If the humidity of the air is high, the strips may lose their charge rapidly; you may need to repeat the experiment in a drier environment.

22.7. See Figure 22.6.

22.8. (*a*) The comb attracts the uncharged paper strip and repels the (charged) tape strip. (*b*) The comb repels one strip (usually the B strip) and attracts the other. (*c*) The charge on the comb behaves the same way as the charge on the strip it repels (usually B). The charge on the comb is therefore the same type as the charge on the strip it repels.

22.9. If in Checkpoint 22.8 your comb repelled the B strip, your brand of tape produces B strips that carry a negative charge. If the comb attracted the B strip, your brand of tape produces B strips that carry a positive charge.

22.10. (*a*) The red marbles, all carrying a positive charge, exert repulsive forces on one another and so move as far away as possible from one another. (*b*) The blue marbles, all carrying a negative charge, also exert repulsive forces on one another and so move as far as possible from one another. (*c*) The red and blue marbles are attracted to each other and so form red-blue pairs. The (sum of the) charge on any given pair is zero and so does not repel the other pairs. Consequently the pairs do not spread out. (The positive and negative charges do not overlap completely—a red marble and a blue one cannot be at the same place—and so, as we shall see later, some residual interaction is left.) (*d*) Each blue marble becomes part of a red-blue pair; the leftover red marbles repel one another and spread out. (*e*) Because of the surplus of positively charged red marbles, the entire collection carries a positive charge.

22.11. (*a*) Any charge deposited on the metal rod spreads out over the entire conducting system, which in this case is the metal rod plus your body. Most of the charge therefore ends up on you. (*b*) Because rubber is an electrical insulator, any charge on the rod that is not in contact with your hand stays on the rod.

22.12. When you rub together surfaces made of the same material, friction does occur, which tells you that bonds do form between the surfaces. Because the two surfaces are made of the same material, however, there is no preferred direction to transfer charge—the same numbers of electrons are transferred in each direction.

22.13. (*a*) Before the charged rod is brought nearby, the electroscope is neutral, meaning it has as many positive charge carriers as negative charge carriers. No charge is transferred to the electroscope, and so even with the rod nearby, the electroscope still has as many positive charge carriers as negative charge carriers (although they are now separated) and is still neutral. (*b*) The magnitudes must be equal because otherwise the electroscope would not be neutral. (*c*) The negative charge carriers on the rod attract the positive charge carriers on the electroscope ball and repel the negative charge carriers on the leaves. (*d*) The positive charge on the ball is equal in magnitude to the negative charge on the leaves. Therefore, all other things being equal, the magnitudes of the forces are the same. However, the distance between leaves and rod is greater than the distance between ball and rod. Because the electric force decreases with increasing distance, the magnitude of the attractive force exerted on the ball will be *greater* than the magnitude of the repulsive force exerted on the leaves.

22.14. The electron cloud is attracted not only to the external positive charge but also to the positive nucleus. The attractive electric interaction between electron cloud and nucleus prevents the electron cloud from leaving the atom entirely.

22.15. (*a*) Attractive. When the external negatively charged body in Figure 22.21 is replaced by a positively charged body, the atoms in the paper still get polarized. The polarization is now in the other direction, however, so that the negative charge appears on the top in Figure 22.21, meaning the force is again attractive. The vector sum of the forces exerted by the charged object on the piece of paper is still attractive. (*b*) It makes no difference whether positively charged protons move up or negatively charged electrons move down. The effect is the same: a surplus of positive charge on the ball of the electroscope and a deficit of positive charge on its leaves (Figure S22.15). (*c*) No, because the outcome does not depend on the type of charge that moves. The experiment that demonstrates that it is indeed the electrons that flow in a metal—as I asserted in Section 22.3—is beyond the scope of this chapter. We will discuss it in Chapter 27.

Figure S22.15

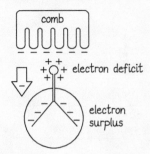

Outcome is same whether:

electrons move down   OR   protons move up

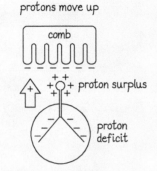

22.16. Greater. For simplicity we treat the spheres as particles so we can apply Coulomb's law. Before the spheres touch, the magnitude of the force exerted by them on each other is $k(q)(3q)/d^2 = 3kq^2/d^2$. When the spheres touch, the charge spreads out evenly over them, so each sphere carries a charge of $(q + 3q)/2 = 2q$. At separation d, therefore, the magnitude of the electric force between them is now $k(2q)(2q)/d^2 = 4kq^2/d^2$, which is greater than it was before.

22.17. Two particles carrying like charges repel each other. If you release them, they accelerate away from each other, gaining kinetic energy. Because the two particles form a closed system, this gain in kinetic energy must be due to a decrease in potential energy associated with the electric interaction. Thus the potential energy of a system consisting of two particles carrying like charges decreases as the separation of the two objects increases. Two particles carrying opposite charges attract each other and so accelerate toward each other. As with two particles carrying like charges, this acceleration means a gain in kinetic energy and a decrease in potential energy. The separation distance is decreasing, however, and we are asked about the change in the system's potential energy when the separation distance is *increasing*. We can reason that because the potential energy decreases as the separation distance between two oppositely charged particles decreases, it must increase as the separation distance increases.

22.18. (*a*) The positive charge carriers repel one another, making their separation greater than the distance between the centers of the spheres. Because Coulomb's law has a $1/r^2$ dependence, the greater separation results in a force of smaller magnitude. (*b*) If the two spheres carry opposite charges, the charge carriers attract one another, and so the distance between them becomes smaller than the distance between the centers of the spheres. Consequently, the magnitude of the attractive force is greater than that obtained from Coulomb's law.

22.19. Because particles 2 and 3 both interact attractively with particle 1, they must carry like charges, and so the interaction between 2 and 3 must be repulsive (Figure S22.19a). The vector sum of the forces exerted on 2

Figure S22.19

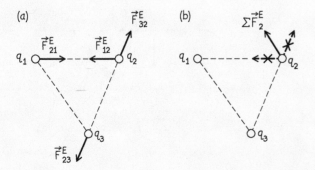

is the sum of a leftward horizontal force exerted by particle 1 and a force directed upward and to the right exerted by particle 3; see Figure S22.19b.

22.20. (*a*) The repulsive force $\vec{F}^E_{17}$ exerted by 1 on 7 points in the direction shown in Figure S22.20. To give 7 an acceleration to the right, the vector sum of the forces exerted on 7 must point to the right. This means that, in addition to $\vec{F}^E_{17}$, another sphere must exert a force $\vec{F}$ on 7 as indicated in the figure. Because the electric force is central (it always acts along the line connecting the two interacting objects), the only two spheres that can exert such a force are 2 and 5. (*b*) The magnitude F must be equal to $\vec{F}^E_{17}$ so that the components in the vertical direction add up to zero. So the magnitude of the charge on sphere 2 or 5 must be the same as that on 1 and 7. To give $\vec{F}$ the direction indicated in the figure, the charge must be negative if it is placed on 2 and positive if it is placed on 5.

Figure S22.20

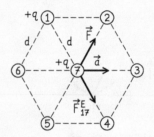

Chapter 23

23.1. (*a*) You know that the magnitude of the gravitational force exerted by Earth on an object of mass m_o is $F^G_{Eo} = Gm_Em_o/r^2_{Eo}$, where G is the gravitational constant and r_{Eo} is the Earth-object distance (Eq. 13.2). Because G and m_E are constants and r_{Eo} is the same for both objects, the magnitude of F^G_{Eo} is proportional to m_o, and thus $F^G_{E1}/F^G_{E2} = m_1/m_2$. (*b*) Yes, because mass is the only property of the objects in the gravitational force expression. Dividing by m_o removes that one property from the expression: $(Gm_Em_o/r^2_{Eo})/m_o = Gm_E/r^2_{Eo}$.

23.2. (*a*) The gravitational field at any location gives the gravitational force exerted on each kilogram of an object at that location. Because each kilogram of the satellite is subject to a force of 2.0 N, the satellite is subject to a force of magnitude $F^G_{Es} = (2.0 \text{ N/kg})(2000 \text{ kg}) = 4000$ N. (*b*) $F^G_{Eb} = (2.0 \text{ N/kg})(0.20 \text{ kg}) = 0.40$ N.

23.3. (*a*) The magnitudes are the same because the forces form an interaction pair. (*b*) Earth's gravitational field magnitude is greater than that of the ball because the gravitational pull exerted by Earth on objects is much greater than the pull of the ball on other ordinary objects. (*c*) The two forces are of equal magnitude because it is the *product* of the masses that determines the force magnitude: $F^G_{Eb} = Gm_Em_b/r^2_{Eb}$. The magnitude of the gravitational force exerted by a gravitational field on an object is equal to the product of the magnitude of the field at the location of the object and the mass of the object. Thus F^G_{Eb} is equal to the product of the magnitude g of Earth's gravitational field (proportional to m_E and large) and the (small) mass of the ball m_b. F^G_{bE} is equal to the product of the gravitational field of the ball (proportional to m_b and small) and the (large) mass of Earth m_E.

23.4. (*a*) No. $F^E_{pi} = kqq_i/d^2$ (Eq. 22.1) tells you that the magnitude of the force exerted by the particle on an object i is proportional to the charge on the object; because $q_1 \neq q_2$, the force magnitudes cannot be equal and therefore the forces cannot be equal. (*b*) The acceleration of the particle, from $\vec{a}_i = \vec{F}^E_{pi}/m_i$. (*c*) No, because the forces and mass of the particles are different (unless the difference in mass compensates for the difference in force). What is the same for both objects is $F^E_{pi}/q_i = kq/d^2$.

23.5. (*a*) Both forces point toward the particle because objects that carry charges of different types attract. (*b*) The electric field points in the same direction as the force exerted on a positively charged particle, so it points

Figure S23.7

Vector sum of forces on test particles

Electric field at points

Initial situation: $q_2 = q_1$

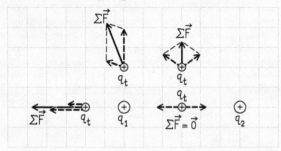

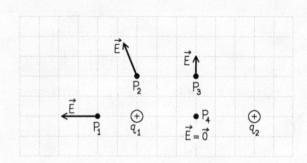

(a) q_2 is doubled: $q_2 = 2q_1$

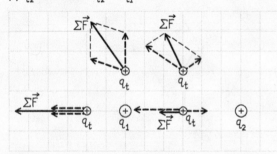

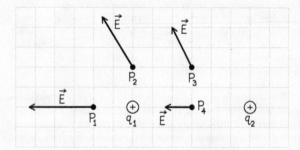

(b) Sign of q_2 is reversed: $q_2 = -q_1$

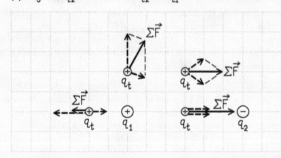

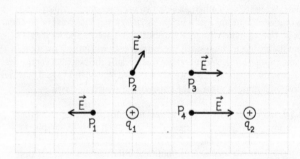

toward the particle. (c) Because objects that carry like charges repel, $\vec{F}_{p2}^E$ points away from the particle. The direction of the electric field, however, still points toward the particle. (d) The field points away from the particle because now the force exerted by the particle on a positively charged test particle is repulsive. (e) The two electric field magnitudes are the same because the magnitude of the force exerted on a test particle is proportional to the magnitude of the source charge and $|+q| = |-q|$.

23.6. The magnitude of the force is the product of the electric field magnitude $E = |\vec{E}|$ and the magnitude of the charge q. If q is positive, the direction of the force is the same as that of $\vec{E}$; if q is negative, the force points in the direction opposite the direction of $\vec{E}$.

23.7. (a) If q_2 is doubled, the force exerted by particle 2 doubles. Both the direction and the magnitude of $\sum \vec{F}^E$ change, and thus both the direction and the magnitude of $\vec{E}$ change (Figure S23.7a). (b) See Figure S23.7b.

23.8. (a) In the same direction as the electric field at that point, that is, to the left. (b) The magnitude becomes three times as great; the direction is the same. (c) Yes, because the force exerted by object 3 changes. This change affects both the direction and the magnitude of $\sum \vec{F}^E$. (d) Opposite the direction of the electric field at that point, that is, down the page.

23.9. The droplet is subject to a downward force $\vec{F}_d^G$ and a horizontal force $\vec{F}_d^E$. The vector sum of these forces is downward at an angle, as illustrated in Figure S23.9. Both forces are constant in magnitude, so their vector sum is also constant. Consequently the droplet has a constant acceleration along the direction of the vector sum. Because the droplet begins at rest, its trajectory is a straight line along this direction.

Figure S23.9

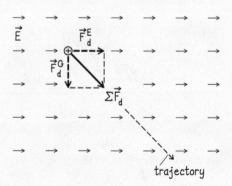

23.10. (*a*) The torque causes the molecule to rotate in the counterclockwise direction. (*b*) No. When the dipole axis that passes through the centers of the two particles is parallel to the electric field, the torque is zero because the two forces lie along the dipole axis and the lever arm of the torque is zero. When the axis is perpendicular to the electric field, the torque is a maximum.

23.11. (*a*) See Figure S23.11*a*. The vector sum of the forces—and therefore the center-of-mass acceleration—point to the left and down. (The force exerted on the negative end is greater than that on the positive end because the electric field is stronger there and both forces have a downward component.) (*b*) See Figure S23.11*b*. The acceleration now points to the right and down, and so the dipole begins by moving in that direction (*away* from the source of the electric field). The orientation of the dipole changes, however, because the forces cause a torque. As we saw in part *a*, as soon as the negative end of the dipole is more toward the left than the positive end, the dipole accelerates toward the left.

Figure S23.11

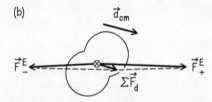

23.12. The magnitude of the electric field is $E = \sqrt{E_x^2 + E_y^2}$. Substituting the values we obtained for E_x and E_y in Example 23.3, we get $E = 1.9 \times 10^4$ N/C. The magnitude of the electric force is $F_e^E = eE = (1.6 \times 10^{-19} \text{ C})$ $(1.9 \times 10^4 \text{ N/C}) = 3.0 \times 10^{-15}$ N. This force causes an acceleration of magnitude $a_e = F_e^E/m_e = (3.0 \times 10^{-15} \text{ N})/(9.1 \times 10^{-31} \text{ kg}) = 3.3 \times 10^{15}$ m/s². As this example shows, the accumulation of just tens of microcoulombs causes a phenomenally large acceleration of the electron even when it is meters away.

23.13. Equations 23.10 and 23.13 show that the magnitude of the electric field of a dipole is proportional to the dipole moment magnitude p. So the dipole that creates the stronger field (dipole B) has the greater dipole moment. The magnitude of the dipole moment is the product of dipole separation d and dipole charge q_p, which means the greater dipole moment of B can be due either to greater d or to greater q_p. Because we cannot separate these two effects, we cannot say which dipole has the greater dipole charge.

23.14. To determine the electric field created by two parallel charged sheets, you must use the superposition principle. Figure S23.14*a* and *b* shows the electric fields of a positive and a negative sheet. In Figure S23.14*c* these two fields are added. (*a*) Between the sheets the electric fields are in the same direction, and so they reinforce, giving $E = 4k\pi\sigma$ between the sheets. (*b*) Outside the sheets, the electric fields are in opposite directions and therefore add to zero.

23.15. As you saw in Section 22.5, a direct consequence of the $1/r^2$ dependence in Coulomb's law is that the force exerted on a charged particle inside a hollow uniformly charged sphere is zero. This means that the electric field inside such a sphere is zero. For the electric field at some point P inside a uniformly charged sphere of radius R (Figure S23.15), the part of the sphere farther from the center than P does not contribute to the electric field at P. If the sphere carries a uniformly distributed charge q, the amount

Figure S23.14

(a) Electric field of positive sheet

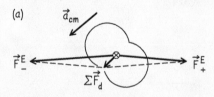

$E = 2k\pi\sigma$

(b) Electric field of negative sheet

$E = 2k\pi\sigma$

(c) Combined electric field

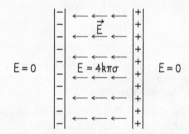

$E = 0$ $E = 4k\pi\sigma$ $E = 0$

of charge on the part of the sphere closer to the center than P is $(r^3/R^3)q$, and the electric field due to this charge is

$$E_r = k\frac{r^3}{R^3}\frac{q}{r^2} = k\frac{q}{R^3}r.$$

In other words, the electric field *increases* in proportion to the distance r from the center. At the center, it is exactly zero.

Figure S23.15

cross section of nested spheres

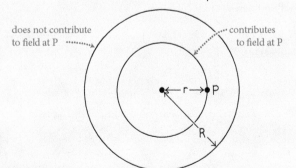

does not contribute to field at P   contributes to field at P

23.16. Yes. To see why, suppose the center of mass is off-center, as shown in Figure S23.16; everything else, including the magnitude of the charges and the dipole moment, is as before. Let the distance between the positive end and the center of mass be d_+. The force exerted on the positive end then causes a counterclockwise torque of magnitude $\tau_+ = r_\perp F_+^E = (d_+ \sin\theta)(q_p E)$, and the force exerted on the negative end causes a counterclockwise torque of magnitude $\tau_- = r_\perp F_-^E = [(d - d_+)\sin\theta](q_p E)$, where $d = |\vec{r}_p|$. Adding the torques yields

$$\sum\tau_\vartheta = (d_+ \sin\theta)(q_p E) + [(d - d_+)\sin\theta](q_p E)$$
$$= d\sin\theta(q_p E) = (q_p d) E \sin\theta = pE\sin\theta,$$

identical to the result in Eq. 23.20.

Figure S23.16

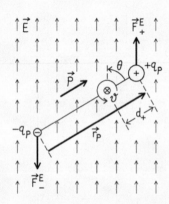

23.17. (a) Doubles the force (Eq. 23.23). (b) Doubles p and thus doubles the force. (c) Doubles p and thus doubles the force. (d) Reduces the force by a factor of $(\frac{1}{2})^3$ because of the $1/y^3$ dependence in Eq. 23.23.

23.18. The dipole moment $\vec{p}$ points in the same direction as the electric field that induces it, so α in Eq. 23.24 must be positive.

23.19. (a) Doubles the force (Eq. 22.1). (b) Quadruples the force because of the q^2 dependence in Eq. 23.26. (c) Doubling q_A in part b doubles the electric field created by A at the position of the induced dipole. This doubles the magnitude of p_{ind} (the stronger electric field increases the charge separation). This doubled dipole moment then interacts with a doubly strong electric field, increasing the force by a factor of 4. (d) No, because the polarization induced by the electric field is always along the electric field (Eq. 23.24). In other words, the induced dipole moment is parallel to the electric field, and the vector product of $\vec{p}$ and $\vec{E}$ is then zero.

Chapter 24

24.1 See Figure S24.1.

Figure S24.1

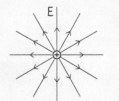
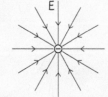

24.2 (a) No, because at the point of intersection the direction of the electric field would not be unique—it cannot be tangent to *both* intersecting lines at the same time. (b) No. Although two field lines that touch have the same tangent at the point where they touch, the two field lines would have different tangents on either side of that point, and so the direction of the electric field would again not be unique.

24.3 (a) As illustrated in Figure S24.3, the same 26 field lines pass through the surface of the hollow sphere. (b) The surface area of a sphere of radius R is $4\pi R^2$, so the number of field lines per unit area is $(26)/(4\pi R^2)$. (c) Again 26. (d) The number of field lines per unit area on the second sphere is $(26)/[4\pi(2R)^2] = (26)/(16\pi R^2)$, which is reduced by a factor of 4 from that for the sphere of radius R. (e) The electric field decreases as $1/r^2$, so doubling the distance reduces the electric field by a factor of 4.

Figure S24.3

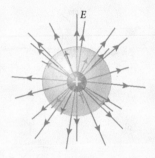

24.4 (a) No field lines cross the surface when it is parallel to the field lines. (b) If 16 field lines pass through an area of 1 m², then 8 field lines pass through an area of 0.5 m². (c) The number of field lines per unit area is $(8)/(0.5 \text{ m}^2) = 16 \text{ m}^{-2}$, or 16 field lines per square meter. (d) $(16)/(1 \text{ m}^2) = 16 \text{ m}^{-2}$, as above. The field line density is the same because the field is uniform.

24.5 (a) It remains the same—each field line still passes through the surface of the sphere. (b) Because neither the number of field lines nor the surface area of the sphere changes, the average number of field line crossings per unit surface area remains the same. (c) The electric field strength increases as the distance to the charged object decreases, so moving the sphere off-center increases the electric field strength on one side and decreases it on the opposite side. (d) No. The answer to part b gives the *average* field line density. Moving the sphere off-center increases the field line density on one side and decreases it on the other, so the average field line density can remain the same.

24.6 (a) Sixteen field lines emanate from the object, so 16 field lines pass through the surface of the sphere. (b) Eight field lines emanate from each of the objects, so 16 field lines pass through the surface of the sphere. (c) The amount of charge enclosed by the sphere is $(20 \text{ field lines})/(8/q \text{ field lines per unit charge}) = 2.5q$.

24.7 (a) Each field line that reenters the donut must also exit, contributing a value of $(+1) + (-1) = 0$ to the field line flux. Thus, regardless of how many field lines reenter the donut, the field line flux remains 6. (b) No. Regardless of the shape of the surface, each field line contributes no more and no less than $+1$ to the field line flux, so the field line flux is always equal to 6.

24.8 (a) The field line flux is zero because each field line that enters the sphere also exits the sphere, and so does not contribute to the field line flux. (b) No. Moving the particle changes the number of field lines that enter the sphere, but regardless of how many field lines enter the sphere, each of them must also leave the sphere, so the field line flux remains zero.

24.9 Five field lines emanating from the positively charged particle contribute a flux of $+5$ (the field line that points from the positive end to the negative end doesn't pass through the surface; see Figure S24.9). Five field lines terminating on the negatively charged particle contribute a flux of -5. This gives a flux of zero, which makes sense because the amount of charge enclosed by the box is $+q + (-q) = 0$. [Notice that this answer remains valid even if we make the box much larger so that all the curved field lines fit inside it. Then only two field lines pass through the box—one emanating from the positively charged particle, the other terminating on the negatively charged one. The flux is then still zero: $(+1) + (-1) = 0$.]

Figure S24.9

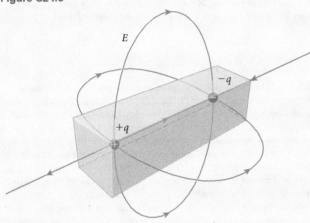

24.10 (a) Fifteen field lines go into the surface and three come out of it. The field line flux is thus $(-15) + (+3) = -12$. Twelve field lines emanate from the object carrying a charge of $+1$ C, so the charge inside the hidden region must be -1 C. Figure S24.10 shows the entire field line diagram; as you can see, the charge enclosed by the dashed line is indeed $(+3$ C$) + (-4$ C$) = -1$ C. (b) Going along the perimeter of the illustration, we note that four field lines leave the edge of the illustration and four enter it. The field line flux is thus $(+4) + (-4) = 0$. This makes sense because the charge enclosed within the diagram is $(+1$ C$) + (-1$ C$) = 0$.

Figure S24.10

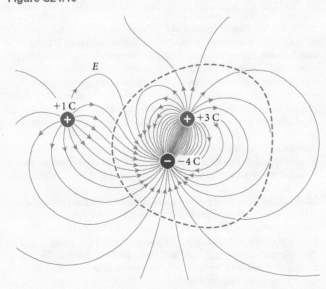

24.11 If there were a flux into the shell from the left and out of the shell on the right, then the situation would no longer have the required symmetry—if you were to rotate the sphere 180° about the vertical axis, the field line flux would be reversed and therefore the situation would be different. Because the situation is spherically symmetrical, this case is not possible.

24.12 (a) It stays the same because the field line density is the same everywhere on that surface. (b) It decreases because the field line spacing increases. (c) Zero. The field lines are all perpendicular to the wire, so no field lines pass through those surfaces.

24.13 (a) It stays the same because the field line density is the same everywhere along a plane parallel to the charged sheet. (b) It stays the same because the field line spacing doesn't change. (c) Positive, because the field lines cross from inside the surface to outside. (d) They are the same in magnitude and in sign. (e) Zero. The field lines are all perpendicular to the sheet, so no field lines pass through the curved surface.

24.14 (a) The field line flux is zero because the field is zero everywhere inside the conducting object. (b) According to the relationship between field line flux and enclosed charge derived in Section 24.3, the charge enclosed by the surface must be zero.

24.15 No. Imagine a Gaussian surface around the cavity, just inside the conducting object (Figure S24.15). Because $E = 0$ everywhere inside the object, the field line flux through this Gaussian surface is zero and so the charge enclosed by it must also be zero. So, the surplus charge on the object cannot reside on an inner surface.

Figure S24.15

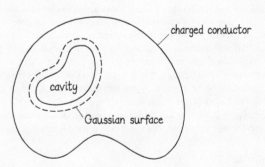

24.16 No. Field lines emanate from the positively charged particle and terminate on the negative charge carriers that line the cavity wall. Thus, "$E = 0$ everywhere inside a conducting object" means everywhere inside the bulk of the object. Cavities don't count!

24.17 (a) (i) Increases (more field lines intercepted); (ii) increases (greater field line density so more field lines intercepted); (iii) increases (more field lines intercepted). (b) (i) Increases (greater A yields greater Φ_E, by Eq. 24.1); (ii) increases (greater $|\vec{E}|$ yields greater Φ_E); (iii) increases (greater slope means smaller θ, which yields a greater $\cos\theta$).

24.18 (a) The area of the back surface is 1.0 m^2, so the magnitude of the corresponding area vector $\vec{A}_{\text{back}}$ is 1.0 m^2. From Figure S24.18 on the next page we see that $h/h_{\text{front}} = \cos\theta = \cos 30° = 0.87$, so the area of the front surface is larger by a factor of $h_{\text{front}}/h = 1.2$. The magnitude of the area vector $\vec{A}_{\text{front}}$ is thus 1.2 m^2. (b) We use Eq. 24.2 to calculate the electric fluxes. For the back surface we get $\Phi_E = EA_{\text{back}}\cos(180°) = (1.0\text{ N/C})(1.0\text{ m}^2)(-1) = -1.0\text{ N}\cdot\text{m}^2\text{/C}$; for the front surface we have $\Phi_E = EA_{\text{front}}\cos\theta = (1.0\text{ N/C})(1.2\text{ m}^2)(0.87) = 1.0\text{ N}\cdot\text{m}^2\text{/C}$.

Figure S24.18

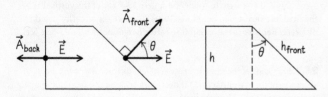

24.19 (a) At a distance r from a particle carrying a charge $+q$, the magnitude of the electric field is kq/r^2 (see Eq. 23.4). (b) Because the electric field is perpendicular to the sphere at all points and has the same constant magnitude, the electric flux is given by the product of the electric field and the area of the sphere: $\Phi_E = EA = E(4\pi r^2) = (kq/r^2)(4\pi r^2) = 4\pi kq$. (c) The enclosed charge is $+q$, so $\Phi_E = 4\pi kq_{enc}$. (d) No. If we double the radius r, the field decreases by a factor of 4 but the area of the sphere increases by a factor of 4. Thus the electric flux and its relationship to q_{enc} remain the same.

24.20 (a) With the modified Coulomb's law, the magnitude of the electric field at the surface of the sphere would be $kq/R^{2.00001}$. The electric flux thus becomes $\Phi_E = EA = E(4\pi R^2) = (kq/R^{2.00001})(4\pi R^2) = 4\pi kqR^{-0.00001}$. (b) $4\pi kqR^{-0.00001} = q_{enc}/\epsilon_0 = 4\pi kq_{enc}$ or $qR^{-0.00001} = q_{enc}$, which is not an equality. The cancellation of the radius R happens only when the dependence on r in Coulomb's law is *exactly* $1/r^2$.

24.21 Figure S24.21 shows how a wedge from q intersects the surface. The electric flux through intersection A_1 is equal to the electric flux through the intersection of the wedge with a sphere of radius r_1 centered on q. Likewise, the electric flux through A_2 is equal to the electric flux through the intersection of the wedge with a sphere of radius r_2 centered on q. As we saw in Checkpoint 24.19, the magnitudes of the electric flux through a sphere are independent of the radius. Because the intersections of the wedge with the two concentric spheres represent a fixed fraction of the total surface area of the spheres, the electric fluxes through these two intersections must also be the same. Consequently, the magnitudes of the electric fluxes through A_1 and A_2 are the same. Because the electric field points out of the sphere at A_1 and into the sphere at A_2, however, the algebraic signs of the electric fluxes are opposite: $\Phi_{E1} = -\Phi_{E2}$, so $\Phi_{E1} + \Phi_{E2} = 0$. Because this holds for *any* wedge, we learn that the electric flux through the surface due to q is zero when q is outside the surface.

Figure S24.21

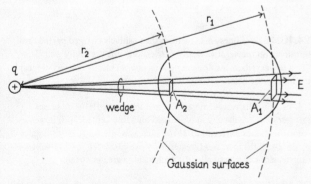

24.22 The electric field is also given by Eq. 24.15 because none of the arguments leading up to that result change: The field is still spherically symmetrical, and the enclosed charge is still $+q$.

24.23 No, because such a charge configuration does not have sufficient symmetry to make the calculation practical. A finite rod still has *rotational* symmetry (you can rotate it about its axis without changing the configuration), but it cannot be translated about its axis without changing the

configuration. For a finite rod the field changes as you move up and down parallel to the y axis in Figure 24.30. With only rotational symmetry, we cannot easily find a closed surface such that for the separate regions of the surface either the magnitude of the field is constant or the electric flux is zero.

24.24 (a) The charge spreads out over the outside surface area of the plate. If we ignore the edges of the plate, $\frac{1}{2}q$ spreads out over each side of surface area A, and so the surface charge density (charge per unit surface area) is $\frac{1}{2}q/A$ on each side. The magnitude of the field is given by Eq. 24.17: $E = \sigma/\epsilon_0 = (\frac{1}{2}q/A)/\epsilon_0 = q/(2\epsilon_0 A)$. (b) The surface charge density is q/A; the magnitude of the electric field is given by the solution of Exercise 24.8: $E = \sigma/(2\epsilon_0) = (q/A)/(2\epsilon_0) = q/(2\epsilon_0 A)$, which is the same result as in part a. The point to remember therefore is that the difference between the solutions of Exercise 24.8 and Eq. 24.17 arises solely from the difference in the surface charge that should be used in each of them.

Chapter 25

25.1 (a) The kinetic energies are the same because the two particles are subject to the same force and the force displacements are the same, so the work done on them is the same. (b) Particle 2 has the greater momentum. As you may recall from Chapter 8, the change in momentum of an object is given by the product of the force exerted on it and the time interval during which the force is exerted. Because particle 2 has greater mass, its acceleration is smaller, and so it takes longer to fall the same distance as particle 1. (c) Both have the same momentum (because the product of force and time interval is the same for both), but particle 1 has greater kinetic energy (because particle 1 has smaller mass, its acceleration is greater and so it undergoes a greater displacement than particle 2; therefore the work done on it is greater and it gains more kinetic energy). (d) The electric force exerted on each particle is equal to the product of the electric field and the charge on the particle. Thus, their accelerations are given by

$$\vec{a}_1 = \vec{F}_1^E/m_1 = q_1\vec{E}/m_1$$

$$\vec{a}_2 = \vec{F}_2^E/m_2 = q_2\vec{E}/m_2.$$

Equating these two accelerations yields $q_1/m_1 = q_2/m_2$ or, rearranging terms, $q_1/q_2 = m_1/m_2$. In other words, if we make the ratio of the charges on the particles equal to the ratio of their masses, their accelerations are the same.

25.2 (a) When the dipole is vertical, the electric potential energy has reached a minimum because the separation between the positive end of the dipole and the charged object is at a maximum and the distance between the negative end and the object is at a minimum. As the dipole rotates past the vertical, the electric potential energy increases. (b) When the dipole is vertical, the torque due to the electric field is zero. Because of its (rotational) kinetic energy, however, the dipole continues to rotate beyond this point and the torque due to the electric field reverses direction. The rotation of the dipole therefore slows down and (rotational) kinetic energy is converted back to electric potential energy. The dipole comes to rest when all of its kinetic energy has been converted—this happens when it is again horizontal (but now with the positive end to the right). The motion then reverses and the dipole continues to oscillate back and forth. (c) The same principles apply, but the oscillation would take place over different angles. In other words, if the dipole started at an angle of 45° to the vertical, then the oscillation would occur between $\pm 45°$ instead of $\pm 90°$ as in part b.

25.3 (a) The work done by the electric field on the particle is the product of the x components of the force exerted on it and the force displacement. The x component of the particle's displacement is $x_B - x_A$ and the x component of the force exerted on it is qE_x, so $W_{Ep}(A \rightarrow B) = qE_x(x_B - x_A)$. (b) As the particle is moved back to its initial position, the force displacement is reversed, so the work done by the electric field on the particle is the negative of the amount done in part a: $W_{Ep}(B \rightarrow A) = -qE_x(x_B - x_A)$. (c) There is no change in the particle's kinetic energy and the particle possesses no internal energy, so its energy does not change as the agent moves it back. Therefore the agent must do an amount of work on the particle that is the negative of the work done by the electric field on it calculated in part b: $W_{ap}(B \rightarrow A) = -W_{Ep}(B \rightarrow A) = +qE_x(x_B - x_A)$. (*d*) The sum of the work done by the agent and by the electric field on the particle is zero. (*e*) See Figure S25.3. (There are no changes in the particle's energies and so the total work done on the particle is zero.)

Figure S25.3

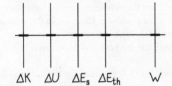

$$\Delta K \quad \Delta U \quad \Delta E_s \quad \Delta E_{th} \quad W$$

25.4 (*a*) The electrostatic work done on a particle as it moves along the gray path from A to C is equal to the electrostatic work done on it from A to B: $W_{Ep}(\text{gray path}) = W_{Ep}(A \rightarrow B)$. The work done from B to C is zero, as is that from C to B: $W_{Ep}(B \rightarrow C) = W_{Ep}(C \rightarrow B) = 0$. Thus, the electrostatic work done on the particle from C to B to A is just that done from B to A: $W_{Ep}(\text{CBA}) = W_{Ep}(B \rightarrow A)$. Because the electrostatic work done from B to A is the negative of that done from A to B (which is equal to W), the electrostatic work done from C to B to A must be $-W$. (*b*) Adding the result we found in part a to the electrostatic work done along the gray path, we calculate for the work along the closed path from A to C to B and back to A $W + (-W) = 0$.

25.5 (*a*) No. See Checkpoint 25.4. Although the electrostatic work done along the entire path is zero, that along segment AC is nonzero. (*b*) No. The electric field varies in magnitude (and therefore the electric force varies too) along the path AB. We must use integral calculus to calculate the work done by a variable force (see Section 9.7). (*c*) (*i*) If the charge on the particle is doubled, then the electric force exerted on the particle doubles and so the electrostatic work done on it doubles. (*ii*) The electric force does not depend on the mass m of the particle, so the electrostatic work done on it, too, is independent of m.

25.6 (*a*) Negative, because the electrostatic work done on a positively charged particle moving along any path from A to C is positive and the potential difference is the negative of the electrostatic work done per unit charge. (*b*) Zero, because the electrostatic work done on the particle is zero along any path from C to B. (*c*) Positive, because the electrostatic work done on the particle is negative (along the straight path from B to A the force and force displacement are in opposite directions). (*d*) The electrostatic work done on the particle is positive because the electric force and the force displacement are in the same direction. The potential difference is therefore negative. (*e*) The electrostatic work done on the particle is equal to the change in kinetic energy: $W_{Ep} = \Delta K$. This is the electrostatic work done on a particle carrying a charge q, and so the electrostatic work done per unit charge is $\Delta K/q$. The potential difference is the negative of this quantity: $-\Delta K/q$.

25.7 Yes. The surface of any sphere that is centered on the particle constitutes an equipotential surface. (The electric field is always perpendicular to such a surface, and so the electrostatic work done on a particle as it is moved along any path—regardless of its shape—that lies on such a surface is always zero.)

25.8 Zero. As we have seen in Section 24.5, the electric field inside a conducting object that is in electrostatic equilibrium is zero. The accumulation of positive and negative charge carriers on opposite ends of the sphere occurs precisely to cancel the effect of the external field of the charged rod anywhere inside the sphere: The electric field caused by the polarization of charge on the sphere exactly cancels the electric field of the rod. Therefore, because $\vec{E} = 0$ inside the conducting object, the electrostatic work done on a charged particle inside the object is zero, too. Consequently the potential difference between two points on or inside the object is zero.

25.9 (*a*) Along path CB, $r_i > r_f$, so the left side of Eq. 25.5 is negative. (*b*) Positive work must be done to push a positively charged particle toward another positively charged particle because the force the external agent doing the pushing must exert and the force displacement are in the same direction. Because the kinetic energy of particle 2 doesn't change, the total work done on 2 (the sum of the electrostatic work done by particle 1 on 2 and the work done by the external agent on 2) must be zero: $W_2 = W_{12} + W_{a2} = 0$. So if the electrostatic work done by particle 1 on 2 is negative ($W_{12} < 0$), then the work done by the external agent on 2 must be positive: $W_{a2} > 0$. (*c*) It we take both particles as our system, there is no longer any electrostatic work done on the system (the electric interaction is now internal). Consequently the work done by the external agent in moving 2 changes the electric potential energy of the two-particle system: $\Delta U^E = W_{a2}$. Using our answer to part b, we have $W_{a2} = -W_{12}(C \rightarrow B)$ or, using Eq. 25.4,

$$\Delta U^E = -W_{12}(C \rightarrow B) = -\frac{q_1 q_2}{4\pi\epsilon_0}\left[\frac{1}{r_C} - \frac{1}{r_B}\right].$$

Because $r_C > r_B$, this change in energy is positive.

25.10 Yes. Figure S25.10 shows particle 1 moving instead of particle 2 (compare with Figure 25.15). We follow the same derivation as in Eq. 25.2, substituting $\vec{F}_{21}^E$ for $\vec{F}_{12}^E$. Because the magnitudes of these two forces are the same, the right-hand side of Eq. 25.2 remains the same. The initial and final separations are also still the same and so we obtain the same end result (Eq. 25.4). Alternatively, you can look at Eq. 25.5 and note that the expression is symmetrical in q_1 and q_2 and that r_{12i} and r_{12f} are independent of which of the two charged particles is moved.

Figure S25.10

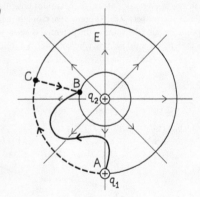

25.11 Yes. The electrostatic work done on 3 as it is brought in first is zero: $W_3 = 0$; the electrostatic work done on 1 as it is brought in next is (compare with Eq. 25.9)

$$W_{31} = -\frac{q_3 q_1}{4\pi\epsilon_0}\frac{1}{r_{31}}.$$

Likewise, the electrostatic work done on 2 when bringing in 2 is (compare with Eq. 25.12)

$$W_{12} + W_{32} = -\frac{q_1 q_2}{4\pi\epsilon_0}\frac{1}{r_{12}} - \frac{q_3 q_2}{4\pi\epsilon_0 r_{32}},$$

The total electrostatic work $W = W_{31} + W_{12} + W_{32}$ done on the system is still the same.

25.12 (*a*) Positive. The potential difference between A and B is $V_B - V_A$, so A is the initial point and B is the final point: $r_i = r_A$ and $r_f = r_B$. Because $r_A > r_B$, the right side of Eq. 25.20 is positive. (*b*) Along the straight path from A to B the angle between the electric force and the force displacement is between 90° and 180°, so the electrostatic work done on the particle is negative. This is consistent with the answer to part a; see Eq. 25.15.

25.13 The electric field of a charged particle is given by Eq. 23.4, $\vec{E}_s(P) = (kq_s/r_{sP}^2)\hat{r}_{sP}$ so Eq. 25.25 becomes

$$V_{AB} = -\int_A^B \vec{E} \cdot d\vec{\ell} = -\frac{q}{4\pi\epsilon_0}\int_A^B \frac{\hat{r}}{r^2} \cdot d\vec{\ell}.$$

The scalar product on the right-hand side is zero along a circular arc because $\hat{r}$ and $d\vec{\ell}$ are perpendicular. Along a radial line, $\hat{r}$ and $d\vec{\ell}$ are parallel, so $\hat{r} \cdot d\vec{\ell} = dr$. Therefore

$$V_{AB} = -\frac{q}{4\pi\epsilon_0}\int_{r_A}^{r_B} \frac{dr}{r^2} = \frac{q}{4\pi\epsilon_0}\left[\frac{1}{r}\right]_{r_A}^{r_B}$$

$$= \frac{q}{4\pi\epsilon_0}\left[\frac{1}{r_B} - \frac{1}{r_A}\right],$$

which is the same result we obtained in Eq. 25.20.

25.14 We begin by sketching some equipotentials for the electric field pattern shown in Figure 25.22. Because equipotentials are always perpendicular to field lines, we can sketch equipotential lines by always drawing them perpendicular to the field lines, as in Figure S25.14a. Based on the equipotential lines I have drawn, I choose a set of representative points to evaluate how the potential varies. Moving clockwise around the path from P, I note that the displacement is upward, whereas the direction of the electric field (and therefore the direction of the electric force exerted on a positively charged test particle) is downward. Thus, the electrostatic work done on the particle is negative and the potential difference between a point above the equipotential passing through P and P is positive. In other words, the potential *increases*. At point 1 we reach maximum potential; moving beyond 1, the potential decreases again. At point 2 we cross the equipotential through point P again (the value of the potential is again V_P). The potential continues to decrease until point 3, where it reaches a minimum value. The

Figure S25.14

(a)

(b)

potential then increases again until a local maximum at point 5 is reached. As we return to P, the potential reaches another local minimum at 7 before increasing again to the initial value. See Figure S25.14b.

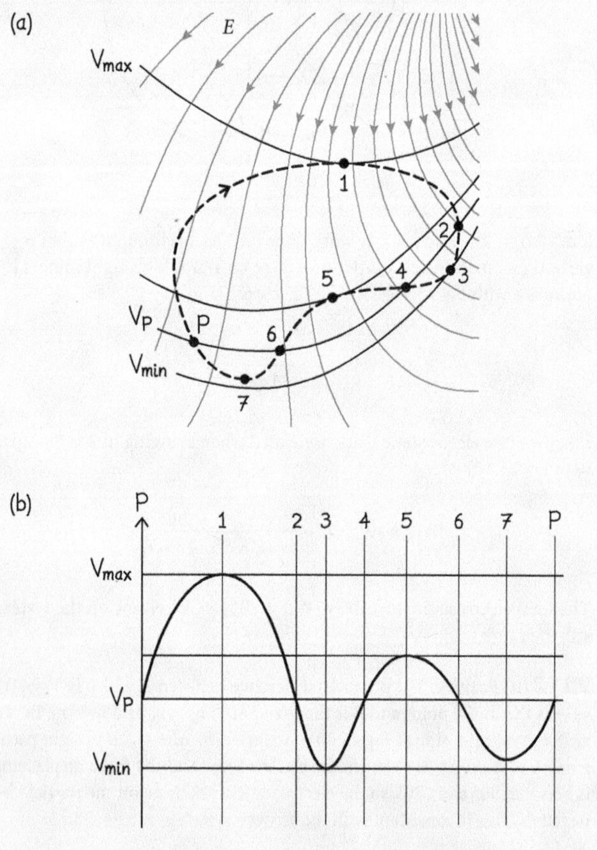

25.15 If the rod and the disk are positively charged and the zero of potential is at infinity, then both potentials should be *positive* (the electrostatic work done on a positively charged particle is negative as we bring it in from infinity). In the answer to Example 25.6, the factor in the logarithm is greater than 1 because $\sqrt{\ell^2 + d^2}$ is greater than d, so the numerator is always greater than the denominator. The logarithm is thus positive, yielding a positive V for positive q. Likewise, in the answer to Example 25.7, the factor in parentheses is always positive because $\sqrt{z^2 + R^2}$ is greater than $|z|$ so, for positive σ, the potential is positive.

25.16 In Example 25.5, we found for the potential between the plates, $V(a) = E(d - a)$, where a is the distance from the positive plate. Rewriting this result as a function of x, we have $V(x) = E(d - x)$. Because $V(x)$ is not a function of y and z, the partial derivatives of V with respect to y and z are zero. Thus $E_y = E_z = 0$. Therefore, the electric field must be in the positive x direction, and the component in that direction is

$$E_x = -\frac{\partial V}{\partial x} = -\frac{\partial}{\partial x}[E(d - x)]$$

$$= -\frac{\partial}{\partial x}(Ed) + \frac{\partial}{\partial x}(Ex) = 0 + E = E.$$

The electric field is in the x direction and of magnitude E, $\vec{E} = E\hat{\imath}$, in agreement with the situation shown in Figure 25.20.

25.17 The potential we obtained in Example 25.7 is a function of z only, so the x and y components of the electric field are zero. The z component is given by

$$E_z = -\frac{\partial V}{\partial z} = -\frac{\sigma}{2\epsilon_0}\frac{\partial}{\partial z}(\sqrt{z^2 + R^2} - |z|)$$

$$= -\frac{\sigma}{2\epsilon_0}\left(\frac{z}{\sqrt{z^2 + R^2}} - \frac{z}{|z|}\right).$$

Substituting $k = 1/(4\pi\epsilon_0)$, we get, for $z > 0$

$$E_z = 2k\pi\sigma\left[1 - \frac{z}{\sqrt{z^2 + R^2}}\right]$$

which is the same result we obtained by direct integration in Example 23.6 (see Eq. 2). Note how much easier and shorter the derivation of the electric field via the potential is compared to the direct integration of Section 23.7.

Chapter 26

26.1 (*i*) The force exerted on a unit positive charge at P is the vector sum of the electric forces due to all the charge carriers on the rod and the fur exerted on the unit charge. If the distribution of charge on the rod and the fur is the same as before, then doubling the charge doubles the magnitude of the individual electric forces without changing their direction. Consequently the direction of the electric field should be the same, but its magnitude should be twice as great. (*ii*) Because the electric field increases by a factor of 2, the potential difference (which is equal to the negative of the line integral of the electric field along any path between the two fixed points) increases by a factor of 2 as well. (*iii*) The potential energy stored in the system increases *by more than a factor of 2* because as charge is transferred from the fur to the rod, the charges already on the rod and the fur make the transfer more difficult because their field opposes the transfer. Thus, the transfer of each additional amount of charge requires more energy than the previous amount of charge. Put differently, the second half of the charge is transferred against a much greater potential difference than the first half.

26.2 The charged fur and rod attract each other, so you must do positive work on the rod-fur system to increase the separation of the rod and the fur. This work increases the electric potential energy of the rod-fur system.

26.3 See Figure S26.3. No external agent does any work on the system. The person, now part of the system, provides the energy from source energy.

Figure S26.3

$$\Delta K \quad \Delta U \quad \Delta E_s \quad \Delta E_{th} \quad W$$

26.4 (a) When the wires are disconnected, any charge already on the plates remains there, so the potential difference between the plates remains the same. (b) The charging of the capacitor continues until the potential difference across the capacitor is equal to that across the battery. Thus, if the battery has a greater potential difference, then the potential difference between the plates is greater too. This greater potential difference requires more charge on each plate.

26.5 (a) Connecting the capacitors to a 9-V battery means that the potential difference across each capacitor is 9 V. As we saw in Section 25.5, the potential difference across a parallel-plate capacitor is the product of the electric field E and the plate separation d (see Example 25.5). Therefore, the capacitor with the smaller plate separation has the greater electric field. Because the electric field of a charged plate is proportional to its surface charge density (see Exercise 24.8), the plates of the capacitor with the smaller plate separation carry the greater charge. (b) The size and shape of the plates don't enter into the expression for the electric field, so the smaller plates must have the same surface charge density as the larger ones. Consequently, the capacitor with the smaller plate area stores less charge.

26.6 (a) The amount of charge remains the same; there are no other conductors to or from which charge carriers can go. (b) The charge carriers in the conducting slab always rearrange themselves so as to cancel the electric field inside the slab (see Section 24.5). To do so, they must set up an electric field of the same magnitude as that inside the capacitor, but in the opposite direction. The charge carriers in the slab rearrange themselves at the surfaces as shown in Figure 26.9b. For the electric field due to the charge carriers at the surfaces of the slab to be equal in magnitude to that caused by the capacitor plates, the magnitude of the surface charge density on the slab surfaces must be the same as that on the plates. Because the surface area of the slab (or at least the part of the slab that is in the electric field) is the same as that of the capacitor plates, the magnitude of the charge on each side of the slab is the same as that on the capacitor plates. (c) Zero, because the electric field inside the slab is zero. (d) It decreases. Between the capacitor plates and the surfaces of the slab, the electric field is the same as before, but inside the slab the electric field is now zero. The electrostatic work done on a unit charge as it is moved from one plate to the other is therefore less than it was before because it is zero inside the slab. Therefore the potential difference is decreased.

26.7 (a) No. Regardless of where the slab is inserted, the field inside it is always zero, while the field between the capacitor plates and the slab is equal to what it was before the slab was inserted. In the extreme case that the slab makes contact with one of the plates, the charge carriers on the plate move through the slab, all the way to its opposite side, as illustrated in Figure S26.7a, effectively reducing the plate separation. The magnitude of the electric field, however, remains the same. (b) See Figure 26.7b. Note that the slope of $V(x)$ on either side of the slab does not depend on the position of the slab.

26.8 The direction of the electric field created by a given source is defined to coincide with the direction of the electric force exerted on a *positive* charge carrier (see Section 23.2). Because electrons carry a negative charge, the force exerted on them is in a direction opposite the direction of the electric field.

Figure S26.7

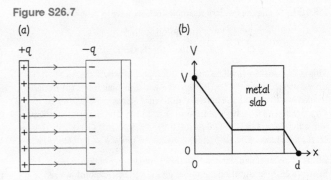

26.9 (a) The electric field is zero above the surface (so it has no direction) because the electric fields of the two surfaces cancel each other outside the dielectric, just like the fields of the two plates of a capacitor (see also Figure S23.14). (b) It points from the positively to the negatively charged surface—that is, opposite the direction of the electric field due to the capacitor plates.

26.10 (a) Zero, because the electric field due to the bound surface charge would be equal in magnitude and opposite in direction to that due to the free charge on the capacitor plates. In effect, the dielectric is like a conductor. (b) No. Suppose the bound charge accumulation at the surfaces becomes so great as to cancel the electric field due the free charge on the capacitor plates, as in part a. Because the electric field inside the dielectric is zero, there is nothing that would cause the material to polarize even further.

26.11 The capacitor with the dielectric, because the battery must do *additional* work on the charge carriers to increase the magnitude of the charge on each plate above that of the capacitor without the dielectric.

26.12 The electric field is in the direction of the flow of positive ions. (The electric force exerted on a positive charge carrier is in the same direction as the electric field.) Wait! Shouldn't the field be in the opposite direction, given the charge on the terminals and the fact that the electric field lines point away from positive charge carriers? The key to reconciling these two facts is to realize that the mechanism of the voltaic cell occurs in the small layers of electrolyte near the electrodes. There the field is in the opposite direction and chemical reactions push charge carriers against the electric field. The result is the pattern of charge and electric fields shown (not to scale) in Figure S26.12.

Figure S26.12

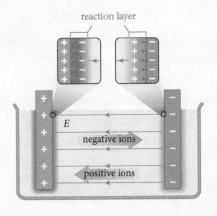

26.13 The work is negative because the system under consideration comprises the molecules undergoing chemical reactions, not the electrons. The molecules do positive work on the electrons, which are part of the molecules' environment, so the environment does *negative* work on the molecules.

26.14 A. The charge on each capacitor follows from Eq. 26.1: $q = C V_{cap}$. Thus, for the same V_{cap}, the capacitor with the greater capacitance C holds the greater charge q.

26.15 (a) Using the result of Example 26.2, we have

$$A = \frac{Cd}{\epsilon_0} = \frac{(1.0 \times 10^{-6}\,\text{F})(50 \times 10^{-6}\,\text{m})}{8.85 \times 10^{-12}\,\text{C}^2/(\text{N} \cdot \text{m}^2)} = 5.6\,\text{m}^2.$$

This is the area of a small room! A 1.0-F capacitor would therefore have a surface area of $5.6 \times 10^6\,\text{m}^2$—the size of a small town. (b) At the breakdown threshold, the potential difference across the capacitor is $V_{\text{cap}} = Ed = (3.0 \times 10^6\,\text{V/m})(50\,\mu\text{m}) = 150\,\text{V}$. The charge on the capacitor is then $q = C\,\Delta V = (1.0\,\mu\text{F})(150\,\text{V}) = 1.5 \times 10^{-4}\,\text{C}$. (c) This charge corresponds to $(1.5 \times 10^{-4}\,\text{C})/(1.6 \times 10^{-19}\,\text{C}) = 9.4 \times 10^{14}$ electrons.

26.16 The capacitance is given by the expression we obtained in Example 26.3. Substituting the values given in the checkpoint, we get

$$C = \frac{2\pi\epsilon_0\ell}{\ln(R_2/R_1)}$$

$$= \frac{2\pi[8.85 \times 10^{-12}\,\text{C}^2/(\text{N} \cdot \text{m}^2)](100\,\text{m})}{\ln[(2.0 \times 10^{-3}\,\text{m})/(2.0 \times 10^{-4}\,\text{m})]} = 2.4\,\text{nF}.$$

One nanofarad (1 nF) is 1×10^{-9} F, so this is a rather small capacitance. (A large capacitance is undesirable because it would allow charge carriers to "pile up" in the cable.)

26.17 (a) Example 26.4 gives an expression for the capacitance of a spherical capacitor. As R_2 approaches infinity, the R_1 in the denominator of this expression becomes negligible and so the capacitance reduces to

$$C = \lim_{R_2 \to \infty}\left[4\pi\epsilon_0\,\frac{R_1 R_2}{R_2 - R_1}\right] = 4\pi\epsilon_0\,\frac{R_1 R_2}{R_2} = 4\pi\epsilon_0 R_1.$$

(b) Substituting values for ϵ_0 and the radius R_1, we get

$$C = 4\pi[8.85 \times 10^{-12}\,\text{C}^2/(\text{N} \cdot \text{m}^2)](2.5\,\text{m})$$

$$= 2.8 \times 10^{-10}\,\text{F}.$$

(c) The electric field is maximum at the surface of the dome. From Eq. 24.15 we know that the electric field at the surface of a charged sphere is

$$E = \frac{1}{4\pi\epsilon_0}\frac{q}{R^2} = \frac{1}{4\pi\epsilon_0}\frac{q}{R}\frac{1}{R} = V_R\frac{1}{R},$$

where V_R is the potential at the surface of the sphere and the potential is zero at infinity. Therefore, $V_R = ER$ and the charge stored on the dome is

$$q = CV_{\text{cap}} = CER$$

$$= (2.8 \times 10^{-10}\,\text{F})(3.0 \times 10^6\,\text{V/m})(2.5\,\text{m})$$

$$= 2.1 \times 10^{-3}\,\text{C}.$$

26.18 (a) The potential difference across the capacitor is

$$V_{\text{cap}} = Ed = (3.0 \times 10^6\,\text{V/m})(50 \times 10^{-6}\,\text{m}) = 150\,\text{V}.$$

The energy stored in the capacitor then follows from Eq. 26.4:

$$U^E = \tfrac{1}{2}(1.0 \times 10^{-6}\,\text{F})(150\,\text{V})^2 = 1.1 \times 10^{-2}\,\text{F} \cdot \text{V}^2.$$

Because $1\,\text{F} = 1\,\text{C/V}$ and $1\,\text{V} = 1\,\text{J/C}$, we have $(1\,\text{F} \cdot \text{V}^2) = (1\,\text{J})$, so the energy stored in the capacitor is 11 mJ.

(b) The increase in gravitational potential energy is given by $\Delta U^G = mg\,\Delta y$, so $\Delta y = \Delta U^G/(mg) = (11\,\text{mJ})/[(2\,\text{kg})(10\,\text{m/s}^2)] = 0.6\,\text{mm}$.

26.19 (a) The surface charge density on each plate is given by $\sigma = q/A$, so the magnitude of the field due to one plate is

$$E_{\text{plate}} = \frac{\sigma}{2\epsilon_0} = \frac{q}{2\epsilon_0 A}.$$

Note that this is not the electric field between the capacitor plates. The field between the plates is twice as great because each plate contributes to that field. The force exerted by the electric field of one plate on the other plate is thus

$$F_{\text{Ep}} = qE_{\text{plate}} = \frac{q^2}{2\epsilon_0 A}.$$

(b) To increase the separation between the plates you must exert a force $\vec{F}$ of magnitude equal to the force exerted by the electric field on the plate calculated in part a. Because this force is constant, and because the force you exert must point in the same direction as the force displacement, we can write for the work you must do on the capacitor

$$W = F\Delta x_F = \frac{q^2 \Delta x}{2\epsilon_0 A}.$$

(c) The change in the electric potential energy is equal to the work done on the capacitor, which we determined in part b.

(d) The volume of the additional space is $A\,\Delta x$, so the energy stored in the electric field in this additional space is $u_E\,A\,\Delta x$. Because the electric field between the plates of the capacitor is twice the electric field of a single plate calculated in part a, the energy density of the electric field is

$$u_E = \tfrac{1}{2}\epsilon_0 E^2 = \tfrac{1}{2}\epsilon_0\left(\frac{q}{\epsilon_0 A}\right)^2 = \frac{q^2}{2\epsilon_0 A^2},$$

so the additional energy is

$$\Delta U^E = u_E\,A\Delta x = \frac{q^2}{2\epsilon_0 A^2}\,A\Delta x = \frac{q^2 \Delta x}{2\epsilon_0 A} = W.$$

26.20 (a) The energy stored in the capacitor is given by Eq. 26.4, so

$$U^E = \tfrac{1}{2}(10^{-4}\,\text{F})(300\,\text{V})^2 = 4.5\,\text{J}.$$

(b) The average power is given by the change in energy divided by the time interval over which the change takes place (Eq. 9.29):

$$P_{\text{av}} = \frac{4.5\,\text{J}}{1 \times 10^{-3}\,\text{s}} = 4.5 \times 10^3\,\text{W}.$$

This is a phenomenally large power (but it is delivered for only a very short amount of time)—much greater than the few watts that can be delivered by a typical battery. Because it requires a great amount of power to illuminate a large space for a short amount of time, all flash units use a capacitor that is charged by a battery between flash firings.

26.21 (a) Because $q_0 = q_{\text{free}} - q_{\text{bound}}$ (see Figure 26.29), we have $q_{\text{bound}} = q_{\text{free}} - q_0$. If we substitute Eq. 26.21, this becomes

$$q_{\text{bound}} = (\kappa - 1)q_0.$$

(b) The bound surface charge density on the dielectric is $\sigma_{\text{bound}} = q_{\text{bound}}/A$, where A is the area of the dielectric. Together with our answer to part a, this becomes

$$\sigma_{\text{bound}} = (\kappa - 1)\frac{q_0}{A} = (\kappa - 1)\epsilon_0 E,$$

where $E = q_0/(\epsilon_0 A)$ (Eq. 24.17).

26.22 (a) Because $1\,\text{V} = 1\,\text{J/C}$ and $1\,\text{J} = 1\,\text{N} \cdot \text{m}$, we have

$$\frac{\text{C}^2}{\text{N} \cdot \text{m}} = \frac{\text{C}^2}{\text{J}} = \frac{\text{C}}{\text{V}} = \text{F}.$$

(b) Similarly,

$$\text{F} \cdot \text{V}^2 = \frac{\text{C}}{\text{V}}\text{V}^2 = \text{C} \cdot \text{V} = \text{C}\left(\frac{\text{J}}{\text{C}}\right) = \text{J}.$$

26.23 Consider a cylindrical Gaussian surface of radius $r > R$. The electric flux through the surface of the cylinder is $\Phi = E(2\pi rL)$, where E is the electric field outside the insulation. The Gaussian surface encloses both bound and free charges. The enclosed bound charge is zero, however, because the surface encloses both the negative bound charge on the inside surface of the insulation and the positive bound charge on the outer surface. The free charge is still λL, so we obtain the same result as in Eq. 26.25 because $\kappa = 1$ for $r > R$.

Chapter 27

27.1 No. Whether a pole is N or S is determined by the way the magnet orients itself with respect to Earth's North Pole. The experiment illustrated in Figure 27.3 then *shows* that like poles repel and unlike poles attract.

27.2 (*a*) Yes, because the charged object always pulls charge carriers of the opposite type toward itself, exerting an attractive force on the near side of the neutral object and a repulsive force on the far side. Because the electric force decreases with increasing distance, the magnitude of the attractive force is greater than that of the repulsive force and so the sum of the forces is attractive. (*b*) A north pole is induced in the clip on the left and a south pole on the right because the interaction between opposite magnetic poles is attractive.

27.3 (*a*) See Figure S27.3. (*b*) No poles because there are no faces where only one type of elementary magnet is exposed. (*c*) If you broke the ring in half and put the exposed faces near paper clips and other magnets, you would find that the faces of the cuts are magnetic poles.

Figure S27.3

27.4 (*a*) By definition, the end that orients itself northward is the north pole of the needle. (*b*) The north pole of the magnet attracts the south pole of the needle and repels the north pole of the needle. Because the compass needle is free to rotate, the needle rotates so that its north-pointing end is as far away as possible from the north pole of the magnet. (*c*) No; if it were, it would repel the north pole of the compass needle. The earth's geographic North Pole is a magnetic south pole.

27.5 (*a*) The bar magnet's north pole attracts the needle's south pole and repels the needle's north pole. The two forces cause a torque on the needle that tends to align the needle with its north pole pointing away from the bar magnet's north pole. (*b*) Because the distance from the needle to the bar magnet's north pole is greater than the distance from the needle to the bar magnet's south pole, the magnitudes of the forces exerted by the north pole are smaller than those exerted by the south pole (Figure S27.5*a*). Consequently, the needle still aligns itself as in Figure 27.11*a*. (*c*) Once the needle has settled, we know that the torque on it must be zero. This means that the sum of the forces exerted by the poles of the bar magnet on each end of the needle must be aligned with the needle (Figure S27.5*b*).

Figure S27.5

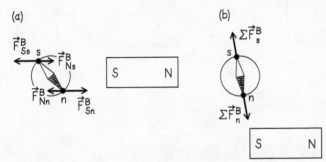

27.6 (*a*) Because elementary magnets have no physical extent, the magnetic field lines form loops that close on themselves (Figure S27.6). Because a particle cannot be cut in half, it must lie either inside or outside the closed surface. Consequently, each line that passes out through the closed surface must also pass back in through the surface. The straight field line entering at the bottom in Figure S27.6 is not an exception. It is balanced by the straight field line leaving at the top. The magnetic flux is zero. (*b*) No, because for this elementary magnet, too, each field line that leaves the surface must

eventually reenter it. (*c*) Zero. Because the magnetic flux must be zero for each elementary magnet making up a bar magnet, the magnetic flux of the magnet must also be zero. (*d*) No. Because an elementary magnet cannot be divided, there is always a whole number of elementary magnets inside the closed surface, so the magnetic flux is still zero.

Figure S27.6

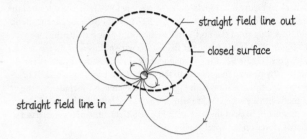

27.7 If we draw a closed surface around the north pole of the bar magnet (Figure S27.7), we see that all the field lines on the outside of the magnet are directed out of the closed surface. Because the magnetic flux through a closed surface must be zero, this means that the field lines inside the bar magnet point from the south pole to the north pole.

Figure S27.7

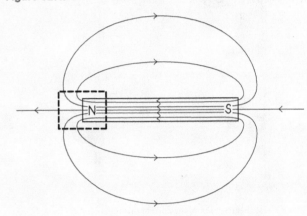

27.8 (*a*) No. If the rod were charged, the battery would have to carry an opposite charge, and charge carriers would start flowing back (see also Section 26.2). (*b*) No. Because it carries no surplus charge carriers, the rod cannot exert an electric force on another charged particle, and therefore there is no electric field due to the rod.

27.9 See Figure S27.9. If the needles align in a circular pattern, the magnetic field lines must be circular, curling in the direction that the north poles of the needles point.

Figure S27.9

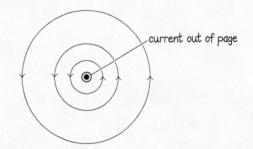

27.10 No. The circular magnetic field pattern generated by a current-carrying wire is unlike the field generated by any magnet or the elementary magnets of which any magnet is made. As we shall see in the next chapter, however, it is possible to use a current to generate a magnetic field pattern that approximates that of a bar magnet.

27.11 (a) Flipping the magnet in Figure 27.20b horizontally while keeping it behind the wire yields the situation in Figure S27.11a, with the magnetic field at the location of the wire pointing to the right. A field in the same direction can also be obtained by placing the magnet in front of the wire (Figure S27.11b). This is the same situation as shown in Figure 27.20b, but now seen from the back. Because the magnetic force exerted on the wire in Figure 27.20b is toward the front, it must be directed toward the back in Figure S27.11b. In other words, the force is directed *toward* the actual position of the magnet in Figure S27.11a. (b) Because the magnetic field lines loop from the north pole to the south pole outside the magnet (see Figure 27.13), the magnetic field at the wire points straight toward the south pole (Figure S27.11c). A magnetic field in the same direction can be obtained by placing the magnet to the right of the wire (Figure S27.11d). Comparing this situation with Figure 27.20b, you can conclude that the magnetic force exerted on the wire is now to the left. In other words, with the magnet in back of the wire as in Figure S27.11c, the magnet exerts a *sideways* force on the wire!

Figure S27.11

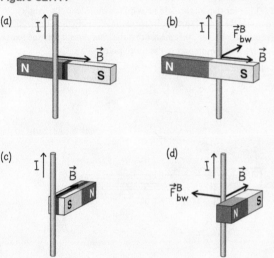

27.12 Reverse the direction of the current through rod 1 in Figure 27.23. As Figure S27.12a shows, at rod 1 the magnetic field due to the current through rod 2 is still out of the page. Placing the fingers of your right hand

Figure S27.12

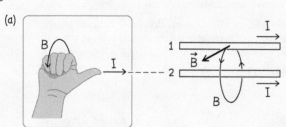

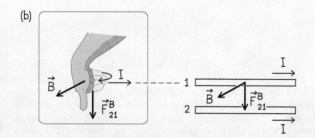

parallel to the current and curling them toward the magnetic field yield a downward-pointing thumb (Figure S27.12b), which means the magnetic force exerted by rod 2 on rod 1 is downward. The same reasoning tells you that the magnetic force exerted by rod 1 on rod 2 is upward. The two rods exert attractive forces on each other.

27.13 (a) From the point of view of observer M, the upper rod is moving to the left and the lower rod is at rest (Figure S27.13a.) Consequently, the upper rod's length is contracted and the lower one has length ℓ_{proper}. Because both rods carry the same charge, the shorter rod—that is, the upper one—has the greater charge density. (b) In Figure S27.13b, the charge on the lower rod has been adjusted so that the charge density matches that of the upper rod. (Because the lower rod's length is contracted according to observer E, the charge it carries is now smaller than the charge carried by the upper rod.) According to observer M, the upper rod, which is moving, is shorter than the lower rod. Thus observer M sees the shorter upper rod carrying more charge than the longer lower rod and says $\lambda_{\text{upper rod}} > \lambda_{\text{proper}}$ and $\lambda_{\text{lower rod}} < \lambda_{\text{proper}}$ (Figure S27.13c).

Figure S27.13

(a) M's reference frame

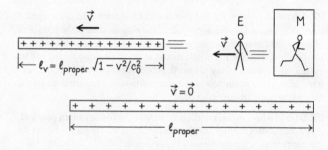

(b) Earth reference frame

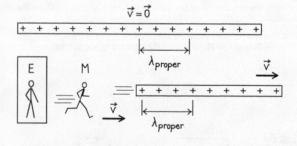

(c) M's reference frame

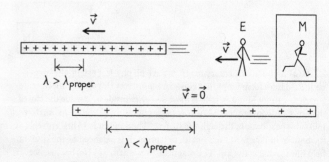

27.14 (a) The positive ions moving to the left correspond to a current to the left. (b) The negative electrons moving to the right correspond to a current to the left. (c) Yes. The positive particle moving to the right corresponds to a current to the right, and so the situation in Figure 27.27c corresponds to two parallel rods carrying currents in opposite directions. As you saw in Example 27.1, the force between two such rods is repulsive, in agreement with the repulsive force exerted by the rod on the particle. (d) The arguments for the wire remain unchanged: It appears electrically neutral to observer E and positively charged to observer M. The positively charged wire exerts an attractive electric force on the negatively charged particle. (e) Yes. Because a negative particle moving to the right corresponds to a current to the left, the situation is identical to that of two rods carrying currents in the same direction. As you saw in Checkpoint 27.12, the force between such rods is indeed attractive.

27.15 Example 27.2 shows that the current I is to the right in Figure 27.33. If this current is caused by negative charge carriers, the definition of current tells you that they must flow to the left.

27.16 (a) On the side through which the field enters, the field points into the cube but the area vector points outward, so $\theta = 180°$: $\Phi_B = AB \cos \theta = (1.0 \text{ m}^2)(1.0 \text{ T}) \cos 180° = -1.0 \text{ Wb}$. The minus sign reflects the fact that the magnetic field points *into* the cube. (b) Because as many field lines enter the cube as leave it, the magnetic flux is zero.

27.17 Because the magnitudes of the charge on the proton and the electron are the same, Eq. 27.23 tells you that the ratio of the radii of the paths is equal to the ratio of the masses: $R_p/R_e = m_p/m_e = (1.67 \times 10^{-27} \text{ kg})/ (9.11 \times 10^{-31} \text{ kg}) = 1.83 \times 10^3$.

27.18 (a) See Figure S27.18a For negative charge carriers, $\vec{F}_p^E$ points in the direction opposite the direction of the electric field lines. Because q changes sign, $\vec{F}_p^B = q\vec{v} \times \vec{B}$ (Eq. 27.19) reverses direction as well. The magnitudes of the two forces are unchanged, and so they still add up to zero. (b) See Figure S27.18b. Reversing the direction of motion of the charge carriers does not affect $\vec{F}_p^E$, but it does reverse the direction of $\vec{v}$ and thus the direction of $\vec{F}_p^B = q\vec{v} \times \vec{B}$. Both forces therefore point in the same direction and so do not cancel. (c) See Figure S27.18c. Because $\vec{F}_p^B$ is always perpendicular to both the magnetic field and the velocity $\vec{v}$, the direction of $\vec{F}_p^B$ no longer lines up with that of $\vec{F}_p^E$, and so the two forces do not cancel.

Figure S27.18

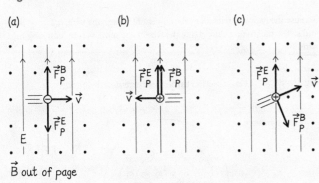

$\vec{B}$ out of page

27.19 (a) The charge distribution is similar to that of a charged capacitor, so E is given by $E = |V_{RL}|/w$ (see Example 25.5), with the width w of the strip substituted for the separation between the positively and negatively charged sides. (b) Because the electric and magnetic forces exerted on the charge carriers are equal in magnitude, we have from Eq. 27.26 $v = E/B = |V_{RL}|/(wB)$. (c) From Eq. 27.16, $n = I/(A|q|v)$. Substituting the result from part b, we obtain $n = (IwB)/(A|q|V_{RL})$. Because the charge

carriers carry an elementary charge e and because $A = wh$, with h the height of the strip, we have $n = IB/(eh|V_{RL}|)$.

27.20 The wire has cylindrical symmetry and so the magnetic field spreads out only in two dimensions. The electric force seen by observer M must therefore decrease as $1/r$ (see Section 24.4). Because the electric force seen by observer M is the same as the magnetic force seen by observer E, that force must also decrease as $1/r$.

Chapter 28

28.1. (a) See Figure S28.1a. (b) The two particles exert no magnetic forces on each other because it takes a moving charged particle to detect the magnetic field of another moving charged particle. (c) See Figure S28.1b. The forces exerted on the horizontal wire cause a torque that tends to align that wire with the vertical one. Conversely, the forces exerted on the vertical wire cause a torque in the opposite direction, tending to align that wire with the horizontal one.

Figure S28.1

(a)   (b)

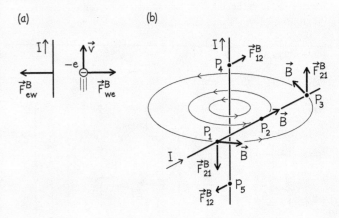

28.2. See Figure S28.2.

Figure S28.2

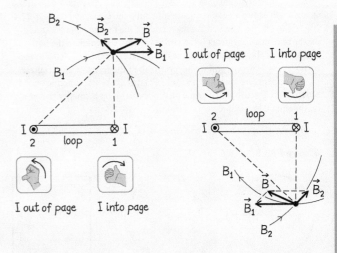

28.3. Comparing Figures 28.6b and 28.7, you see that the positively charged ring has a magnetic field similar to the field of a bar magnet, with its north pole up. The negatively charged ring thus has its north pole down, and this north pole is directly above the north pole of the positive ring. Because two north poles repel each other, the interaction is repulsive.

28.4. No. By definition the electric field points from positive charge carriers to negative charge carriers (see Section 23.2). The electric field along the axis passing through the poles of an electric dipole therefore points from the positive end to the negative end. The electric dipole moment, however, by definition points in the opposite direction (see Section 23.6).

28.5. (*a*) Because the magnetic field remains perpendicular to the horizontal sides of the loop, you know from Eq. 27.4 that the magnitude of the magnetic force exerted on each side is $F_{max}^B = |I|\ell B$, where ℓ is the length of each side of the loop. Because none of these quantities changes, the magnitude of the force exerted on the horizontal sides stays the same. (*b*) See Figure S28.5*a* and *b*. The lever arm of the force exerted on the horizontal sides become smaller as the loop rotates, and so the torque caused by these forces decreases. (*c*) In the initial position, the vertical sides are parallel to the magnetic field, which means the magnetic force exerted on them is zero (Eq. 27.7, $F^B = |I|\ell B \sin\theta$ with $\theta = 0$). After the loop has rotated 90°, the vertical sides are perpendicular to $\vec{B}$ and thus subject to an outward magnetic force. For $0 < \theta < 90°$, the magnitude of the force is again given by Eq. 27.7. As you can verify using the right-hand force rule, the forces exerted on the vertical sides are directed outward along the rotation axis and so cause no torque. (*d*) See Figure S28.5*c*. Because the top and bottom of the loop are now reversed, the direction of the torque reverses.

Figure S28.5

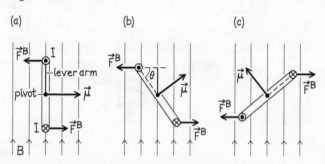

(a) (b) (c)

28.6. See Figure S28.6, where the circular loop is approximated by a series of vertical and horizontal segments. The vertical segments experience no force, but the horizontal ones do. All the horizontal segments add up to the same length as two sides of a square straddling the circle, and thus Eq. 27.4, $F_{max}^B = |I|\ell B$, tells you that the magnitude of the magnetic force exerted on the horizontal segments is the same as the force exerted on the horizontal side of the square loop. Some horizontal segments are closer to the rotation axis, however, and so the lever arms of the forces acting on these closer segments are smaller than the lever arms of the forces acting on the square loop. Therefore the torque on the circular loop is smaller than that on the square loop.

Figure S28.6

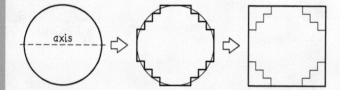

28.7. With the magnetic field on the left stronger than that on the right, $F_{left}^B > F_{right}^B$. Therefore the vector sum of the forces exerted on these two sides is nonzero and points upward. The effect is a clockwise rotation due to the torque and an upward acceleration due to the upward vector sum of the forces.

28.8. Increasing the current increases the magnitude of the magnetic field, and so the line integral increases.

28.9. (*a*) If the direction of the current is reversed, the direction of the magnetic field is reversed, and so the algebraic sign of the line integral is reversed. (*b*) Because the value of the line integral of the magnetic field around a closed path does not depend on the position of the wire inside the path, the second wire by itself gives rise to the same line integral as the first wire. Adding the second wire thus doubles the value of the line integral. (*c*) Reversing the current though the second wire flips the sign of the line integral. If its value was *C*, it is −*C* after the current is reversed. The sum of the two line integrals is thus $C + (−C) = 0$.

28.10. No. If the path tilts, you can break it down into arcs, radial segments, and *axial* segments, which are segments parallel to the wire. Because the magnetic field is perpendicular to the axial segments, they don't contribute to the line integral. For the same reason, of course, the radial segments contribute nothing. So the line integral again is that along all the arcs, which is equivalent to a single complete circular path.

28.11. (*a*) Reversing the direction of the current through wire 1 changes the sign of its contribution from negative to positive, so adding the contributions of wires 1 and 3 yields a positive value. (*b*) The path does not encircle wire 2, so reversing the current through wire 2 does not affect the answer to Exercise 28.2. (*c*) Reversing the direction of the path inverts the signs of all the contributions to the line integral, but because the integral is zero, the answer doesn't change.

28.12. See Figure S28.12. Inside the wire, use Ampèrian path 1 of radius $r < R$. For this path,

$$\oint \vec{B} \cdot d\vec{\ell} = B_{inside} \oint d\vec{\ell} = B_{inside} 2\pi r,$$

but the path encircles only part of the cross section and so encircles only part of the current *I* through the wire. The area of the cross section of the wire is πR^2, and the cross-sectional area of Ampèrian path 1 is πr^2, making the fraction of the wire cross section enclosed by the path $(\pi r^2)/(\pi R^2) = r^2/R^2$. Thus, the right side of Ampère's law is $\mu_0 I_{enc} = \mu_0 (r^2/R^2)I$. Ampère's law thus yields $B_{inside}(2\pi r) = \mu_0(r^2/R^2)I$, or $B_{inside} = \mu_0 Ir/2\pi R^2$. Outside the wire, use Ampèrian path 2 in Figure S28.12. For this path,

$$\oint \vec{B} \cdot d\vec{\ell} = B_{outside} 2\pi r.$$

Because the path encircles all of the current *I* through the wire, $\mu_0 I_{enc} = \mu_0 I$, and so the magnetic field outside the wire is the same as that for a long thin wire: $B_{outside} = \mu_0 I/2\pi r$.

Figure S28.12 Cross section through a current-carrying wire

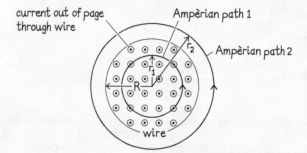

current out of page through wire

Ampèrian path 1

Ampèrian path 2

wire

28.13. (*a*) Figure S28.13 shows a head-on view of the magnetic fields of the two sheets separately and their sum. The magnetic fields add up to zero between the sheets. Below and above the sheets, the magnetic field points to the left and its magnitude is twice that of a single sheet: $2(\frac{1}{2}\mu_o K) = \mu_o K$. (*b*) With the current through the lower sheet reversed, its magnetic field is also reversed. Now the magnetic fields add up to zero below and above the sheets. Between them the magnetic field is to the right and its magnitude is twice that of a single sheet: $\mu_0 K$.

Figure S28.13

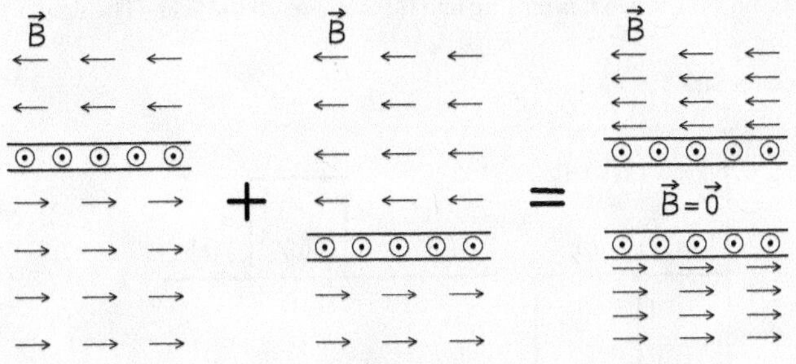

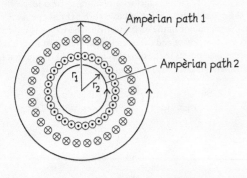

28.14. Outside the toroid, you use Ampèrian path 1 in Figure S28.14. The left side of Ampère's law is still given by Eq. 28.7, but the path now encircles both the current upward through the inside of each winding and the current downward through the outside of each winding. The encircled current is thus $NI - NI = 0$, making B outside the toroid zero. Inside the toroid's inner radius, use Ampèrian path 2. This path encloses no current, so here, too, $B = 0$.

Figure S28.14

Ampèrian path 1

Ampèrian path 2

28.15. For this wire the integration in Example 28.6 extends from $x = 0$ to $x = +\infty$:

$$B = \frac{\mu_0 I d}{4\pi} \int_0^{+\infty} \frac{dx}{[x^2 + d^2]^{3/2}}$$

$$= \frac{\mu_0 I d}{4\pi} \left[\frac{1}{d^2} \frac{x}{[x^2 + d^2]^{1/2}} \right]_{x=0}^{x=+\infty} = \frac{\mu_0 I}{4\pi d}.$$

28.16. (a) Using our result from Example 28.7 with $\phi = 2\pi$:

$$B = \frac{\mu_0 I(2\pi)}{4\pi R} = \frac{\mu_0 I}{2R}.$$

(b) The outer arc spans an angle $\phi = \pi/2$ and has a radius $2R$. Using our result from Example 28.7 again, it contributes a magnetic field

$$B_{\text{outer}} = \frac{\mu_0 I(\pi/2)}{4\pi(2R)} = \frac{\mu_0 I}{16R}.$$

The inner arc also spans an angle of $\phi = \pi/2$, but it has radius R, and the current through it runs in the opposite direction so the magnetic field

due to this arc is directed into the page. For this arc, the magnitude of the magnetic field is

$$B_{\text{inner}} = \frac{\mu_0 I(\pi/2)}{4\pi(R)} = \frac{\mu_0 I}{8R}.$$

The two straight segments do not contribute because they point straight toward P. Because B_{inner}, which points into the page, is greater than B_{outer}, which points out of the page, the magnetic field $\vec{B} = \vec{B}_{\text{inner}} + \vec{B}_{\text{outer}}$ is into the page rather than out of the page. The magnitude of the magnetic field is $B = \mu_0 I/8R - \mu_0 I/16R = \mu_0 I/16R$.

28.17. (a) The direction of the magnetic force is given by the vector products in Eq. 28.23. Figure S28.17a shows the vectors that appear in this expression. You must first evaluate $\vec{v}_1 \times \hat{r}_{12}$. The direction of the resulting vector is obtained by curling the fingers of your right hand from $\vec{v}_1$ to $\hat{r}_{12}$, which tells you $\vec{v}_1 \times \hat{r}_{12}$ points into the plane of the page (Figure S28.17b).

Figure S28.17

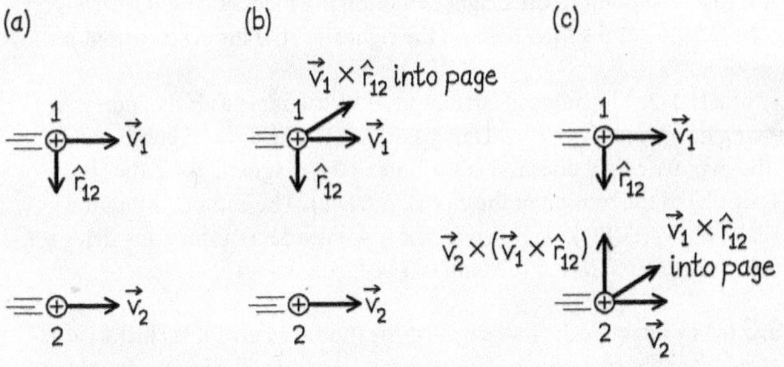

(a)   (b)   (c)

$\vec{v}_1 \times \hat{r}_{12}$ into page

$\vec{v}_2 \times (\vec{v}_1 \times \hat{r}_{12})$ $\vec{v}_1 \times \hat{r}_{12}$ into page

Next evaluate the vector product between $\vec{v}_2$ and the vector product you just found. The right-hand vector product rule tells you this yields a vector that points upward from proton 2 to proton 1. Thus, $\vec{F}^B_{12}$ points upward. Applying the same procedure to obtain $\vec{F}^B_{21}$, you find that it points downward, telling you that the magnetic interaction is attractive, like that of two parallel current-carrying wires.

(b) Because the angles involved in the vector products in Eq. 28.23 are all 90°,

$$F^B_{12} = \frac{\mu_0}{4\pi} \frac{q_1 q_2}{r_{12}^2} v_1 v_2.$$

The magnitude of the electric force is given by Coulomb's law:

$$F^E_{12} = \frac{1}{4\pi\epsilon_0} \frac{q_1 q_2}{r_{12}^2}.$$

The ratio is

$$\frac{F^B_{12}}{F^E_{12}} = \frac{\left(\dfrac{\mu_0}{4\pi}\right) \dfrac{q_1 q_2}{r_{12}^2} v_1 v_2}{\left(\dfrac{1}{4\pi\epsilon_0}\right) \dfrac{q_1 q_2}{r_{12}^2}} = \mu_0 \epsilon_0 v_1 v_2.$$

Substituting numerical values, you find that this ratio is 10^{-6}.

Chapter 29

29.1 The magnetic force exerted on the charge carriers points in the opposite direction—that is, upward rather than downward. As a result, positive charge accumulates at the upper end of the rod rather than the lower end.

29.2 Yes, in positions (b) and (d) in Figure 29.4. The only difference between moving the field and moving the loop is the observer's frame of reference. It would not make any sense for an observer in one reference frame to observe a current and an observer in another reference frame to see no current.

29.3 No, because magnetic forces are exerted only on moving charged particles, not on stationary ones.

29.4 There are two electric fields in Figure 29.10: the "regular" electrostatic field due to the charge separation and the electric field that accompanies the changing magnetic field due to the moving magnet. The charge distribution in the rod is in mechanical equilibrium when these two fields have the same magnitude and opposite directions, producing zero electric field inside the stationary rod, as we would expect.

29.5 Situation 1: The induced current must be the same as if the loop were moving to the left with the magnet stationary (the same relative motion). The upward component of the magnetic field exerts a force on a positive charge carrier that is directed into the page (toward the back of the loop). Because at this point in the magnet's motion the magnetic field is stronger on the left side of the loop than on the right side, the induced current is clockwise.

Situation 3: The induced current must be the same as if the loop were moving downward with the magnet stationary. The upward component of the magnetic field does not contribute to the magnetic force (because it's parallel to the motion of the charge carriers). The component pointing radially outward (looking down on the loop) produces forces that drive a clockwise current around the loop (viewed from above).

29.6 (a) In Example 29.4 we saw that the induced current is clockwise as viewed from above in Figure 29.12b. Inside the loop $\vec{B}_{ind}$ points down, and outside the loop $\vec{B}_{ind}$ points up. Remember that $\vec{B}_{ind}$ is stronger inside the loop than outside and that the direction of the force exerted by $\vec{B}_{ind}$ on the south pole is opposite the direction of $\vec{B}_{ind}$. Thus, the force exerted on the magnet by $\vec{B}_{ind}$ pushes the magnet upward, opposing the downward motion of the magnet that induced the current and field $\vec{B}_{ind}$.

(b) If Lenz's law indicated that the induced current and field should add to the change that produced them, the force exerted on the magnet by $\vec{B}_{ind}$ would accelerate the magnet downward more rapidly (and in the process increase the induced current).

29.7 (a) Zero. Once the loop is completely in the magnetic field, the magnetic flux enclosed by the loop is no longer changing. The loop's motion ceases to induce current, and the force that resists the motion vanishes, which means no further work is required to keep the loop moving at constant speed. (b) Positive. Once the right edge leaves the field, the magnetic flux enclosed by the loop begins to decrease. Lenz's law says that a force resists the magnetic flux decrease and therefore resists the loop's motion. Thus, the loop must be pulled out. The pulling force you exert points in the same direction as the motion, so the work done is positive.

29.8 More work must be done to move a magnet toward a closed conducting loop because the motion induces a current in the loop, and the work done on the magnet must provide the energy for this current. Because there cannot be a current through a rod, moving a magnet toward a rod induces only a static charge separation, which requires less work.

29.9 (a) The magnitude decreases through a small region around P because the magnet is moving away from P. It remains unchanged through a small region around Q because the magnet is directly overhead. It increases through a small region around R because the magnet is moving toward R. (b) The direction of any eddy current is such that the magnetic field associated with the current opposes the change in magnetic flux. A counterclockwise current (viewed from overhead) creates a field that points upward. Therefore the eddy current around P is counterclockwise

and that around R is clockwise. The magnetic flux through a small region around Q remains unchanged, so there are no eddy currents around Q.

29.10 See Figure S29.10 (with magnetic flux out of the page in Figure 29.4 chosen to be positive). The induced emf is zero when the loop is completely outside (a and e) or completely inside (c) the field. The emf has the same constant value when it is moving into (b) or out of (d) the field. (The sign

Figure S29.10

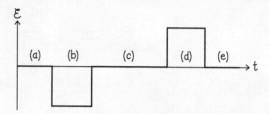

of the emf is different in b and d, but the magnitude is the same because $|\Delta\Phi_B/\Delta t|$ depends only on the speed of the loop and the magnitude of $\vec{B}$, both of which are constant.)

29.11 Although the solenoid orientation looks the same, the two sides have interchanged their positions. Because the two sides have changed positions, the sense of circulation has changed too. The emf thus has changed sign too.

29.12 (a) There is no change in the magnetic flux and thus no electric field. (b) They look like counterclockwise-pointing circular loops. By Lenz's law, the magnetic field of the induced current must oppose the decrease in the magnitude of the magnetic field. Therefore the induced current is counterclockwise and the electric field lines must form circles and the electric field points in the direction opposite the direction it has when the magnetic field is increasing.

29.13 For $r > R$, the left side of Eq. 29.17 is still given by Eq. 1 in Example 29.7, but the right side of Eq. 29.17 is $\pi R^2 \, dB/dt$. This gives

$$2\pi r E = \pi R^2 \frac{dB}{dt}$$
$$E = \frac{R^2}{2r}\frac{dB}{dt}.$$

Substituting $dB/dt = 0.050$ T/s, $R = 0.20$ m, and $r = 0.30$ m into this expression yields

$$E = \frac{(0.20\text{ m})^2(0.050\text{ T/s})}{2(0.30\text{ m})} = 3.3 \times 10^{-3}\text{ N/C}.$$

29.14 You can calculate the induced emf from the inductance of the solenoid (Example 29.8) and Eq. 29.19:

$$L = \frac{(4\pi \times 10^{-7}\text{ T}\cdot\text{m/A})(2760)^2\pi(50 \times 10^{-3}\text{ m})^2}{0.60\text{ m}} = 0.13\text{ H}$$

$$|\mathcal{E}_{ind}| = L\frac{dI}{dt} = (0.13\text{ H})(1.0 \times 10^{-1}\text{ A/s}) = 13 \times 10^{-3}\text{ V}.$$

29.15

$$u_B = \frac{B^2}{2\mu_0} = \frac{(1.0\text{ T})^2}{2(4\pi \times 10^{-7}\text{ T}\cdot\text{m/A})} = 4.0 \times 10^5\text{ J/m}^3$$

$$u_E = \tfrac{1}{2}\epsilon_0 E^2 = \tfrac{1}{2}(8.85 \times 10^{-12}\text{ C}^2/\text{N}\cdot\text{m}^2)(1.0\text{ V/m})^2 = 4.4 \times 10^{-12}\text{ J/m}^3.$$

The ratio of u_B to u_E is 9.0×10^{16}.

Chapter 30

30.1 (*a*) Yes. The wire is intercepted by the surface three times, but the directions of the intercepts are different. Because the magnetic field points in the same direction as the integration direction along the closed path, we can use the right-hand current rule (see the Procedure box in Chapter 28, point 4) to determine that intercepts 1 and 3 are positive and intercept 2 is negative. The direction of the current at intercepts 1 and 3 is from inside to outside; at intercept 2 it is from outside to inside. We can expect current intercepted going from inside to outside to contribute oppositely to current intercepted going from outside to inside. From Figure 30.1 we see that current intercepted going from inside to outside through surface B is positive, so the current intercepted by the surface is $+I$ at 1 and 3 and $-I$ at 2, for a total of I, which is equal to the current encircled by the path. (*b*) Yes. The current intercepted by the surface is I: $+I$ for the top intercept, $-I$ for the middle intercept, and $+I$ for the bottom intercept. Again, the current intercepted by the surface is equal to the current encircled by the path.

30.2 (*a*) While the capacitor is charging, there is a current through the wire leading to or from the capacitor, and the electric field between the plates is nonzero and changing. (*b*) Once the capacitor is fully charged, the current through the wire drops to zero. There is still a nonzero electric field between the plates, but that field is no longer changing.

30.3 Although the electric field direction does not change, the direction of the *change* $\Delta \vec{E}$ in the electric field is reversed because the electric field is now decreasing. Taking the direction of $\Delta \vec{E}$ as the "current," you see that the direction of the magnetic field is reversed from what it was when the unit was charging, and the right-hand current rule applies.

30.4 The magnetic dipole moment of the neutron indicates that it must have an internal structure consisting of charged particles that form current loops. Indeed, in Chapter 7 we learned that the neutron consists of one up and two down quarks (Figure 7.17). These quarks do indeed carry charge.

30.5 You can estimate v by comparing the distance the particle moves in the time interval between parts *b* and *c* of Figure 30.8 with the distance the kinks have moved in that interval as measured from the point where they originate, which is at the center of each panel. The kinks move roughly twice as far as the particle, which means $v \approx 0.5c$.

30.6 The electric field is changing at the kinks and in the region between the kinks and the particle. In these regions there is a magnetic field. Beyond the kinks, the magnetic field is zero because the electric field is static.

30.7 (*a*) In a time interval equal to the period T, the wave travels a distance equal to its wavelength. Because the wave travels horizontally, you should measure the distance traveled along the horizontal axis through the electric field pattern. Between $t = 0$ and $t = T/2$ the wave advances half a wavelength. The distance that the rightmost field line advances in this time interval is 16 mm, so the wavelength is 32 mm. (*b*) From Eq. 16.10, $f = c/\lambda = 9.4 \times 10^9$ Hz. (*c*) From Eq. 15.2, $T = 1/f = 1.1 \times 10^{-10}$ s, or 0.11 ns.

30.8. (*a*) Figure 30.15*a* shows the bottom electric field pattern of Figure 30.14, so the motion of the particles must be the same in those two cases. From Figure 30.14 you see that at $t = \frac{3}{4}T$, the positively charged particle is moving up and the negatively charged particle is moving down, which means the current direction is up. (Although the dipole moment is zero at this instant because the particles are in the same location, the current is not zero because they are both moving. The current is instantaneously zero when the particles are at the two extremes of their travel paths and thus have zero velocity for an instant.) (*b*) Yes, because the magnetic field lines closest to the dipole are in the proper direction for the magnetic field of an upward current: Immediately to the right of the dipole, the magnetic field lines point into the page; immediately to the left of the dipole, they point out of the page.

30.9 Parallel, because parallel to the polarization means parallel to the electric field. From Section 23.4, you know that an electric field exerts a force on charged particles (see Figure 23.15, for instance). Only electric forces with a component along the axis of the antenna can accelerate the charge carriers in the antenna in a direction that produces current (along the antenna's length), so an electric field with a component parallel to the antenna is needed.

30.10 Whether between the plates or outside the capacitor, the magnetic field accompanies a change in the electric field that points left (the electric field is decreasing). From Section 30.1, you know that the "current" you must use with the right-hand current rule for determining the direction of $\vec{B}$ also points left. If you point the thumb of your right hand to the left along the axis in Figure 30.24, your curled fingers point into the page above the axis of the capacitor and out of the page below the axis. This means $\vec{B}$ is into the page at both P and S.

30.11 (*a*) Use a surface that lies between the plates, like the one in Figure 30.28 but with a radius $r < R$. Then follow the approach used in Example 30.5, substituting r for R in Eq. 1:

$$B = \frac{\mu_0 \epsilon_0 r}{2} \frac{dE}{dt}.$$

To determine E, use Eq. 1 from Example 26.2 with $A = \pi R^2$ (because A is the area of each plate, not the area of the surface bounded by the path of integration). This substitution gives you

$$B = \frac{\mu_0 \epsilon_0 r}{2} \frac{d}{dt} \left(\frac{q}{\epsilon_0 \pi R^2} \right) = \frac{\mu_0 r}{2\pi R^2} \frac{dq}{dt}.$$

You know that dq/dt is the current, so you can write $B = \mu_0 I r/(2\pi R^2)$. Thus, the magnetic field between the plates is smaller closer to the axis of the plates. (*b*) To the right of the right plate (and, of course, to the left of the left plate), the magnetic field is simply that of a current-carrying wire at a distance r from the wire. The result we obtained in Example 28.3 gives you $B = \mu_0 I/(2\pi r)$.

30.12 Equation 30.10 relates the electric field to its source charge distributions and thus needs no modification. The right side of Eq. 30.11 is zero because there are no magnetic monopoles. If monopoles having a $1/r^2$ relationship did exist, Eq. 30.11 would have to be analogous to Eq. 30.10. This means the right side must be proportional to m_{enc} rather than zero. You would expect electric fields to form loops around monopole currents, in the same way that magnetic fields form loops around currents, so the right side of Eq. 30.12 must gain a term proportional to dm/dt. Equation 30.13 needs no modification because $\oint \vec{B} \cdot d\vec{\ell}$ for a static distribution of monopoles should be zero, just as $\oint \vec{E} \cdot d\vec{\ell}$ for a static charge distribution is zero.

30.13 In vacuum, the line integral of the electric field is proportional to the time rate of change of magnetic flux through the surface bounded by the path (Eq. 3), and the line integral of the magnetic field is proportional to the time rate of change of electric flux through the surface bounded by the path (Eq. 4). To satisfy these relationships, as shown in Figure 30.30*e* and *f*, the magnetic field and electric field must be perpendicular to each other. The magnetic field must form loops around the electric field, and the electric field must form loops around the magnetic field.

30.14 When an electromagnetic wave enters a dielectric medium, the electric field of the wave accelerates charged particles in the medium back and forth with the frequency of the wave. The accelerated particles radiate electromagnetic waves that propagate with the same frequency as the incoming wave. The electromagnetic wave in the medium is the combination of all of these waves and hence has the same frequency, $f = c_0/\lambda = (3.00 \times 10^8 \text{ m/s})/(600 \text{ nm}) = 5.00 \times 10^{14}$ Hz. The wavelength in the dielectric medium decreases because the speed of the wave in the medium is smaller than in vacuum:

$$\lambda_{new} = \frac{c}{f} = \frac{c_0}{f\sqrt{\kappa}} = \frac{\lambda}{\sqrt{\kappa}} = \frac{600 \text{ nm}}{\sqrt{1.30}} = 526 \text{ nm}.$$

30.15 The electric field points from one plate to the other, and the magnetic field forms loops around the electric field. (Figure 30.4 represents this situation.) The vector product of these two vectors points radially inward, toward the capacitor axis (Figure S30.15). The Poynting vector thus represents the flow of energy into the region between the capacitor plates, where energy is being stored in the electric field. This makes sense because the energy density associated with the electric field inside the cylindrical surface defined by the capacitor plates is increasing. So there must be a flow of electromagnetic energy into this region.

Figure S30.15

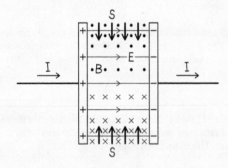

30.16 The SI units of μ_0 are $T \cdot m/A$; those of the electric field N/C and of the magnetic field T. Using Eq. 30.36, I thus get for the SI units of the Poynting vector:

$$\frac{[N/C][T]}{T \cdot m/A} = \frac{N \cdot A}{C \cdot m} = \frac{N \cdot A}{A \cdot s \cdot m} = \frac{N}{m \cdot s} = \frac{J}{m^2 \cdot s} = \frac{W}{m^2}.$$

Chapter 31

31.1 (*a*) Yes. In practice, the light and thermal energy travel away from the bulb, so you need either a flexible definition of the system or a container capable of keeping the light and thermal energy in a well-defined volume. (*b*) Yes. (*c*) The energy to produce light and thermal energy comes from the electric potential energy associated with the potential difference between the battery terminals. This energy comes from chemical reactions taking place inside the battery.

31.2 (*a*) Yes. The charged capacitor is the source, and the bulb is the load. (*b*) The bulb glows during the brief time interval when the capacitor is discharging. Electrons flow from the negatively charged capacitor plate through the connecting wires and the bulb to the positively charged plate until the capacitor plates are no longer charged. (*c*) Electric potential energy stored in the charged capacitor is converted to light and thermal energy in the bulb.

31.3 These answers are appropriate for a flashlight bulb connected to four D batteries that are fresh. For different batteries and different bulbs, the times may vary somewhat. (*a*) The current should stay constant for many minutes. (*b*) After 24 hours, the batteries are depleted and the current is zero.

31.4 (*a*) Gains electric potential energy. (*b*) Loses the same amount of electric potential energy. (*c*) In the load (they lose very little energy when flowing through the wires).

31.5 (*a*) That resistance is proportional to potential difference tells you that A has the greater resistance. (*b*) The greater resistance of bulb A means that a greater potential difference is required to obtain a given current in A. Thus, the same potential difference across both bulbs produces a smaller current in A than in B.

31.6 Smaller than. When the circuit load consists of the parallel combination of two bulbs, the current out of the battery is greater than the current out of the battery when the load consists of a single bulb. (To be exact, the current is twice as great with the parallel combination of bulbs as with the single bulb, as shown in Example 31.3.) The potential difference across the load is the same in both cases. Therefore, the resistance of the parallel combination of two bulbs is half the resistance of the single bulb.

31.7 (*a*) One contact of A is connected to the positive terminal of battery 1, and the other contact is connected to the negative terminal of battery 2. The negative terminal of battery 1 is connected to the positive terminal of battery 2. Therefore the potential difference across A equals the potential difference across the two-battery combination, which is the sum of the potential differences across each battery, or 18 V. (*b*) One contact of bulb B is connected to the positive terminal of battery 1, and the other contact of bulb B is connected to the negative terminal of the same battery. Therefore, the magnitude of the potential difference across B must be 9 V. (*c*) Bulb A glows more brightly because there is a greater potential difference across it.

31.8 Because the electric field is uniform throughout the rod, the magnitude of the electric field is equal to the magnitude of the potential difference across the rod divided by the length of the rod, $E = |V_{12}|/\ell$ (see Eq. 1 in Example 25.5).

31.9 See Figure S31.9. The electric field lines are parallel to the sides of the conductors everywhere. The electric field line density is also uniform, representing the uniform magnitude of the electric field.

Figure S31.9

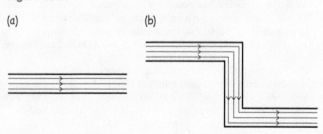

(*a*)

(*b*)

31.10 (*a*) You know from the text that $V_{wide} = V_{narrow}/4$. You also know that the magnitude of the potential difference across the entire conductor is 9 V. Therefore $V_{wide} + V_{narrow} = V_{wide} + 4V_{wide} = 9$ V, so $V_{wide} = 9$ V$/5 = 1.8$ V. (*b*) $V_{narrow} = 9$ V $- 1.8$ V $= 7.2$ V.

31.11 (*a*) Yes. The electric field does work on the electrons to accelerate them. (*b*) On average, the kinetic energy does not change over time because the average final velocity (drift velocity) does not depend on time. (*c*) Because the electric field does work on the electrons but their kinetic energy does not increase over time, some other form of energy in the system must increase. (As you'll see shortly, the increase takes the form of thermal energy.)

31.12 (*a*) Greater vibrations cause the ions to move around within a greater volume, increasing the probability of ion-electron collisions and decreasing the average time interval between collisions. Because conductivity is proportional to this time interval and resistance is inversely proportional to conductivity, decreasing the interval causes the resistance to increase. (*b*) A current causes the metal to heat up because the work done on the electrons by the applied electric field gets converted to thermal energy of the lattice ions. (*c*) See Figure S31.12. At low current, the heating is negligible. Because metals are ohmic at constant temperature, the curve is a line with slope $1/R$ at low current. At high current, which causes the metal to heat up, the resistance increases as the current increases, and so the line curves downward.

Figure S31.12

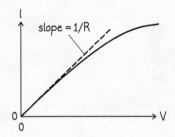

31.13 No. Because the positive terminals of the two batteries are connected to each other with a wire, the potential is the same at points a and b (Figure S31.13). So, if $\mathscr{E}_1 < \mathscr{E}_2$, then the potential at d must be greater than the potential at c. Because the direction of current is from high to low potential, the current is counterclockwise though the circuit, in the direction opposite the reference direction for the current indicated in the diagram.

Figure S31.13

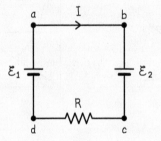

31.14 (a) See Figure S31.14a. If we go counterclockwise from the top, the first circuit element is the battery; because we are traveling from the positive to the negative terminal, the potential difference is $-\mathscr{E}$. Next is the resistor; because we are traveling in the opposite direction as the reference direction for the current, the potential difference is $+IR$. Substituting these values in Eq. 31.21, we get $-\mathscr{E} + IR = 0$ or $I = \mathscr{E}/R$, which is identical to the result we obtained in Eq. 31.24. We should get the same sign for the current because we have chosen the same reference direction for the current as in Figure 31.34. (b) See Figure S31.14b. If we go clockwise from the bottom, the first circuit element is the battery; because we are traveling from the negative to the positive terminal, the potential difference is again $\mathscr{E}$. Next is the resistor; because we are traveling in the opposite direction as the reference direction for the current, the potential difference is IR. Substituting these values in Eq. 31.21, we get $\mathscr{E} + IR = 0$ or $I = -\mathscr{E}/R$, which is opposite in sign to the result we obtained in Eq. 31.24. The negative sign means that the current direction is opposite the chosen reference direction for the current. That is, the current direction is not counterclockwise, but clockwise, in agreement with our earlier analysis.

Figure S31.14

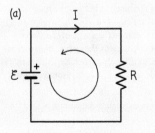

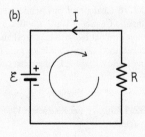

31.15 (a) The potential difference is now positive because you are moving closer to the positive terminal of the battery. The magnitudes of the current and resistance are the same, however, so the magnitude of the potential

difference must be the same. Thus, $V_{cb} = IR_1$. (b) From the solution I obtained in Exercise 31.8, $I = (9\ \text{V})/(10\ \Omega + 5\ \Omega) = 0.6\ \text{A}$. Therefore, $V_{cb} = (0.6\ \text{A})(10\ \Omega) = 6\ \text{V}$. This result makes sense because the resistance of R_1 is two-thirds of the resistance of the combination of R_1 and R_2, which means you expect the potential difference across R_1 to be two-thirds of 9 V, the potential difference across the combination.

31.16 (a) From Eq. 31.32, $R_{eq} = [1/(3\ \Omega) + 1/(10\ \Omega) + 1/(5\ \Omega)]^{-1} = 1.6\ \Omega$. (b) From Eq. 31.31, $I = (9\ \text{V})/(1.6\ \Omega) = 5.7\ \text{A}$.

31.17 Now the resistance in the path from a to d is smaller than it is when $R_{var} = 12\ \Omega$. More charge carriers now flow from a to d, so a small number of carriers flow from b to a through the bulb to increase the current from a to d.

31.18 Because 1 A = 1 C/s and 1 V = 1 J/C = 1 A·Ω,

$$1\ \text{A}\cdot\text{V} = 1\frac{\text{C}}{\text{s}}\cdot\frac{\text{J}}{\text{C}} = 1\ \text{J/s} = 1\ \text{W}$$

$$1\ \text{A}^2\cdot\Omega = 1\ \text{A}\cdot\text{A}\cdot\Omega = 1\ \text{A}\cdot\text{V} = 1\ \text{W}.$$

31.19 (a) If I travel counterclockwise around the circuit starting at a, the loop rule yields $6.0\ \text{V} + I(0.25\ \Omega) + I(0.25\ \Omega) - 9.0\ \text{V} = 0$. Solving this expression for the current again yields $I = 6.0\ \text{A}$, so the answer remains unchanged (as it should because nothing has physically changed). (b) Because the current is the same in both batteries and because they have the same internal resistance, the rate at which energy is dissipated is also the same: $P = 9.0\ \text{W}$.

Chapter 32

32.1 (a) Electric potential energy in the electric field of the capacitor. (b) As the capacitor discharges, that energy is converted to magnetic energy stored in the magnetic field in the inductor. (c) Magnetic energy in the inductor.

32.2 (a) Yes (see Eq. 31.43). (b) Energy is supplied by the AC source, and so as long as the AC source delivers energy at the rate at which it is dissipated by the resistor, the amplitudes of the current and potential difference oscillations remain constant. (Because the power is proportional to the current *squared*, the changing sign of the current does *not* imply that sometimes the resistor delivers energy to the AC source. We discuss this point in Section 32.8.)

32.3 See Figure S32.3. Because the current and potential difference are in phase with each other, the phasors V_R and I overlap as in Figure 32.13. Because v_R and i are both zero at $t = 0$ and increase as time goes on, the phasors must lie on the horizontal axis.

Figure S32.3

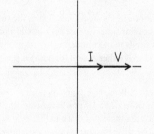

32.4 Figure 32.13: zero for both v_R and i. Figure 32.20: zero for i and $\pi/2$ for v_L.

32.5 A piece of doped silicon, whether n-type or p-type, is electrically neutral because the dopant atoms have the same number of protons as electrons. Although in a doped semiconductor there is either a surplus or a deficit of electrons relative to the number of electrons in a perfect silicon lattice, there is also an equal surplus or deficit of protons in the nuclei of the dopant atoms.

32.6 Holes travel from left to right (but only in the p-type region; in the n-type region they recombine with free electrons); simultaneously, electrons travel from right to left.

32.7 The current in the diode is in one direction only. If you assume the diode is attached in the circuit so that it conducts current only when $v_{\text{diode}} > 0$, your sketch looks like Figure S32.7a. If you assume the diode is attached so that it conducts current only when $v_{\text{diode}} < 0$, your sketch looks like Figure S32.7b.

Figure S32.7

(a)

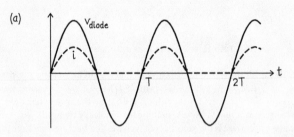

(b)

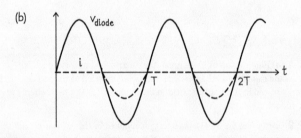

32.8 According to the junction rule, you must have $I_b + I_c = I_e$.

32.9 The AND gate is the one in Figure 32.39a. Both field-effect transistors function as switches and are closed (conducting) when their gates are positively charged. If the potential at A is positive with respect to ground, transistor 1 is conducting. If the potential at B is positive with respect to ground, transistor 2 is conducting. Because the transistors are connected in series, both must be conducting in order for the output potential to be different from ground, and so both A and B must be positive to obtain a nonzero output.

The OR gate is the one in Figure 32.39b. Again, the two transistors function as switches, but this time they are connected in parallel. Therefore only one of them needs to be conducting in order for the output potential to be nonzero, and so if either A or B (or both) is positive we obtain a nonzero output.

32.10 (a) Higher. (b) Because R is always positive, Eq. 32.2 tells you that v_R must be positive when i is positive. According to Eq. 32.3, $\mathcal{E}$ is positive, too. Given that v_R is positive and that $v_a > v_b$, you must have $v_R = v_a - v_b$. (c) Because the current is now negative, v_R and $\mathcal{E}$ must be negative, too. (Indeed, with the current now running counterclockwise, v_a is now smaller than v_b.) The potential difference across the resistor is still given by $v_R = v_a - v_b$.

32.11 The current in the circuit. The velocity of a simple harmonic oscillator is equal to the time derivative of its position. Because the potential difference v_C is proportional to the charge on the upper capacitor plate, the rate at which v_C changes—that is, the time derivative of v_C—is

proportional to the rate at which the amount of charge on the plate changes, dq/dt. This quantity is the current.

32.12 See Figure S32.12. The two horizontal lines mean that the resistance and the current in the resistor do not depend on angular frequency. The hyperbolic X_C curve shows that the capacitive reactance is inversely proportional to ω (Eq. 32.14). Consequently the amplitude of the current in the capacitor is directly proportional to ω, which means the I_C curve has a constant slope. The straight, positive-slope X_L curve means that the inductive reactance is directly proportional to ω (Eq. 32.26); consequently the current in the inductor is inversely proportional to ω, as shown by the hyperbolic I_L curve.

Figure S32.12

(a)   (b)

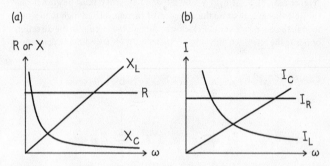

32.13 No. You cannot use the phasor method because the two phasors for the potential differences do not rotate as a unit with a constant phase difference.

32.14 We analyze this circuit as we did in Example 32.8, except that now $V_{\text{out}} = V_C$ instead of $V_{\text{out}} = V_R$. To take the high-ω and low-ω limits of V_{out}, we rewrite Eq. 32.36 as

$$V_{\text{out}} = V_C = \frac{\mathcal{E}_{\text{max}}}{\sqrt{\omega^2 R^2 C^2 + 1}}.$$

The low-ω limit of V_{out} is then $V_{\text{out}} = \mathcal{E}_{\text{max}}$, while the high-$\omega$ limit is $V_{\text{out}} = \mathcal{E}_{\text{max}}/\omega RC$, which approaches zero for large ω. This circuit therefore passes low-frequency signals essentially unchanged and attenuates high-frequency signals.

32.15 (a) The maximum potential differences can be found from the maximum current (using Eqs. 32.6, 32.13, and 32.24). Rounding off to two significant figures, you have $V_R = IR = 1.6 \times 10^2$ V, $V_L = I\omega L = 3.1 \times 10^2$ V, and $V_C = I/\omega C = 3.1 \times 10^2$ V.

(b) No. $(V_R + V_L + V_C) \gg \mathcal{E}_{\text{max}}$ because the potential differences are not in phase and are never simultaneously at their maxima.

32.16 Changing L or C while keeping R constant changes the resonant angular frequency (unless both L and C are changed in such a way that their product remains constant) and can change the shape of the $I(\omega)$ curve. Increasing C while keeping L fixed broadens the curve, and increasing L while keeping C fixed sharpens the curve. Because it depends only on R and on the applied emf, the amplitude of the current at resonance is not affected by the values of L or C.

32.17 Above, because when $V_L > V_C$, the inductor dominates the reactance.

32.18 We substitute the answers from Example 32.10 into Eq. 32.52:

$$P_{\text{av}} = I_{\text{rms}}^2 R = \frac{I^2}{(\sqrt{2})^2}R = \frac{(3.2 \text{ A})^2(50 \ \Omega)}{2} = 2.6 \times 10^2 \text{ W}.$$

Chapter 33

33.1 (*a*) See Figure S33.1. In considering whether the brightness of any location on the screen has changed, note that the distribution of light from the first bulb has not changed, which means the brightness of a particular location changes only if additional light is cast on that location by the second bulb. If the second bulb does not cast light on a particular location, the brightness does not change. (The fact that a given location would be shadowed if it were illuminated by only the second bulb does not decrease the brightness of the light from the first bulb.) The brightness of the spot created by the first bulb does not change because no light from the second bulb strikes this spot. (*b*) Locations P and Q are now brighter than before because some light from the second bulb strikes them. Locations R and S are unaffected because no light from the second bulb reaches them.

Figure S33.1

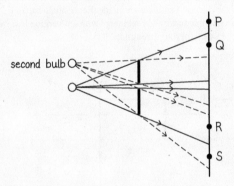

33.2 See Figure S33.2. The shadow edges become sharper as you move the paper farther from the bulb and fuzzier as you move the paper closer to the bulb. This happens because the bulb is not a point source of light. Rays from different parts of the bulb's surface pass by the edge of the paper at slightly different angles. When the paper is farther from the bulb, the difference in these angles is less and so the edge of the shadow is sharper. When the paper is close to the bulb, the angular size of the filament (as seen from the edge) is greater, which makes the edge of the shadow fuzzier.

Figure S33.2

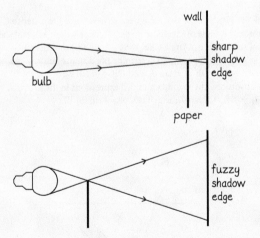

33.3 No. Figure S33.3 shows the reflected rays that reach the observer in the two locations. Because the angle of incidence always equals the angle of reflection when the reflecting surface is smooth, any ray from the object that reflects anywhere on the mirror can be traced back through the mirror to a location directly behind the object. Moving the observer changes only which subset of the reflected rays the observer sees and not where the rays appear to come from.

Figure S33.3

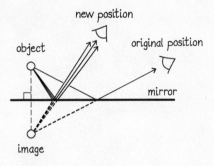

33.4 No. Figure S33.4 shows the rays that reflect into the person's eye from the highest and lowest points on his body. These rays strike the mirror at a height midway between where they originate on the person and the person's eye. So the mirror needs to extend only from halfway between the person's eye and the highest point on the person's body to halfway between the person's eye and the lowest point on the person's body. The ray from the highest point strikes the mirror at a height midway between the height h_1 in Figure S33.4 at which the ray originates and the height at which it strikes the eye, and the same is true for the ray from the lowest point. So the top of the mirror can be at a height that is half the height h_1 above the height of the eye, and the bottom of the mirror can be at a height that is half the height h_2 below the height of the eye. Because $h_1 + h_2$ is the person's height, this means the mirror needs to be only half this height. (However, the mirror must be positioned at the right height.)

Figure S33.4

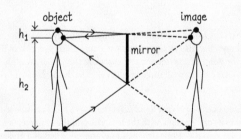

33.5 The wavefronts arrive at a given location in the glass at the same frequency that they arrive at any location in vacuum. However, they travel only two-thirds as fast in the glass as in vacuum. Sequential wavefronts arrive at the glass surface at instants separated by the period T of the wave ($T = 1/f$). If the wave traveled at the same speed in glass as it does in vacuum, then during one period, one wavefront would travel a distance into the glass equal to the vacuum wavelength (400 nm) before the next wavefront arrived at the surface of the glass, and the wavefronts would be separated by 400 nm. However, because the wavefronts travel at only two-thirds the speed of light in vacuum, the spacing between wavefronts in the glass must be two-thirds of the vacuum wavelength, or 267 nm.

33.6 The propagation of the wavefronts is exactly like that shown in Figure 33.16*a* with the direction of propagation reversed. Consequently, because the angle of incidence equals θ_2, the angle of refraction must equal θ_1.

33.7 The angle between the ray reflected from the bottom surface and the normal is θ_2, as shown in Figure S33.7 on the next page. This ray is now incident at the top surface and refracted as it emerges into the air. This is equivalent to the situation shown in Figure 33.17*b*, in which the ray originates in the glass and is refracted in the air. Therefore, as discussed in Example 33.3, in air the angle this refracted ray makes with the normal is θ_1. Note in Figure S33.7 that this refracted ray is shifted sideways relative to the ray *reflected* from the top surface.

Figure S33.7

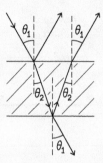

33.8 The critical angle corresponds to the angle of incidence for which the angle of refraction is 90°. When entering the medium, to obtain a given angle of refraction, the angle of incidence of a violet ray must be greater than that of a red ray; therefore the critical angle is smaller for a violet ray than for a red ray.

33.9 See Figure S33.9. The wavefronts are perpendicular to the light rays everywhere.

Figure S33.9

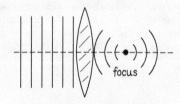

33.10 No. The image location can be found with any two of the three principal rays. Because all three intersect at one point, any two of them indicate the point of intersection and therefore the image location. It is, however, useful to draw all three rays in order to check that you have made no mistakes.

33.11 Decrease. See Figure S33.11. As the object moves closer to the lens, ray 3 makes a smaller angle with the lens axis and thus intercepts the lens closer to the axis. Ray 1 remains the same, and ray 2 makes a greater angle with the axis in order to pass through the center of the lens. As a result, the image point—the virtual intersection of the rays—is closer to the axis (as well as the lens), making the image smaller.

Figure S33.11

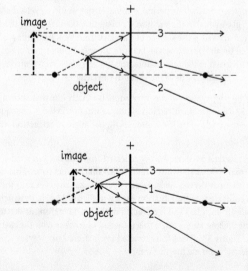

33.12 Virtual, because the rays do not actually intersect anywhere on the image. They merely appear to originate from a common point on the virtual image.

33.13 (a) See Figure S33.13a. You draw ray 3 by first drawing a line connecting the point of interest on the object with the focus on the side of the lens opposite the object. Run ray 3 along this line from the object to the lens surface; then change the ray direction so that it emerges parallel to the lens axis. (b) There is no position for which the image is larger than the object. To see why, all you need to consider is principal rays 1 and 2. When the object is moved, the direction of ray 1 does not change—only ray 2 changes direction. When the object is moved to the left, as in Figure S33.13b, ray 2 now intersects ray 1 closer to the lens axis and the image is smaller. Moving the object right, toward the lens, makes the image larger, but it cannot become larger than the object because on the left side of the lens where the image forms, ray 1 is always below the tip of the object. As shown in Figure 33.33, placing the object beyond the object-side focus also produces an image that is smaller than the object. (c) It is not possible to form a real image with a diverging lens because rays diverge when they go through it.

Figure S33.13

(a)

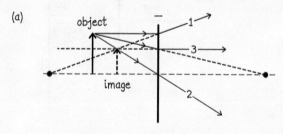

(b)

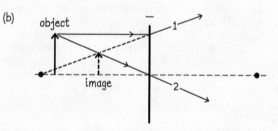

33.14 See Figure S33.14. (a) You need to determine the distance CE in terms of the angle of incidence θ_1, slab thickness $d = $ AD, and slab index of refraction n_2. Angle CBE $= 90° - \theta_1$, and so angle BEC $= \theta_1$. Therefore CE $=$ BE $\cos \theta_1$. To determine the distance BE, you can say DB $+$ BE $=$ DE, DE $= d \tan \theta_1$, and DB $= d \tan \theta_2$.

Substituting the second and third expressions in the first, you obtain $d \tan \theta_2 +$ BE $= d \tan \theta_1$, which can be solved for BE:

Figure S33.14

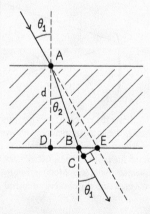

$$BE = d(\tan \theta_1 - \tan \theta_2) = d\left(\frac{\sin \theta_1}{\cos \theta_1} - \frac{\sin \theta_2}{\cos \theta_2}\right). \qquad (1)$$

Applying Snel's law (Eq. 33.7) to this case and solving Eq. 33.7 for $\sin \theta_2$ give you $\sin \theta_2 = n_1 \sin \theta_1 / n_2$. Next you can use the identity $\sin^2 \theta + \cos^2 \theta = 1$ to get an expression for $\cos \theta_2$:

$$\cos \theta_2 = \sqrt{1 - \sin^2 \theta_2} = \sqrt{1 - \left(\frac{n_1 \sin \theta_1}{n_2}\right)^2}.$$

You can substitute these expressions for $\sin \theta_2$ and $\cos \theta_2$ into Eq. 1:

$$BE = d\left[\frac{\sin \theta_1}{\cos \theta_1} - \frac{n_1 \sin \theta_1}{n_2 \sqrt{1 - \left(\frac{n_1 \sin \theta_1}{n_2}\right)^2}}\right],$$

which simplifies to

$$BE = d \sin \theta_1 \left(\frac{1}{\cos \theta_1} - \frac{n_1}{\sqrt{n_2^2 - n_1^2 \sin^2 \theta_1}}\right).$$

From this you can determine the distance CE:

$$CE = BE \cos \theta_1 = d \sin \theta_1 \left(1 - \frac{n_1 \cos \theta_1}{\sqrt{n_2^2 - n_1^2 \sin^2 \theta_1}}\right).$$

(b) $CE = (0.010 \text{ m})(0.50)\left(1 - \frac{(1)(0.87)}{\sqrt{(1.5)^2 - (1)^2(0.5)^2}}\right)$

$\qquad = 0.0019 \text{ m}.$

Thus, at an angle of incidence of 30°, in passing through a glass slab that is 10 mm thick (about three times the thickness of a typical windowpane), a light ray is shifted sideways by 2 mm (20% of the thickness of the slab).

33.15 (a) f is negative (lens is diverging), so $f = -80$ mm; o is positive (object is on same side as illumination), so $o = 100$ mm. Substituting these values into Eq. 33.16 gives

$$\frac{1}{i} = \frac{1}{-80 \text{ mm}} - \frac{1}{100 \text{ mm}} = -0.0225 \text{ mm}^{-1},$$

and solving for i gives $i = -44$ mm. The value of i is negative, as it should be for a virtual image, and the absolute value of i is less than o, which is consistent with Figure 33.33.

(b) Equation 33.17 gives

$$M = \frac{h_i}{h_o} = \frac{-i}{o} = \frac{-(-44 \text{ mm})}{100 \text{ mm}} = 0.44.$$

The image height is 44% of the object height.

33.16 (a) Solving Eq. 33.21 for f gives $f = (0.25 \text{ m})/M_\theta$, and substituting this result into Eq. 33.22 gives $d = 4M_\theta$. For $M_\theta = 8$, you have $d = +32$ diopters.

(b) $f = (0.25 \text{ m})/M_\theta = (0.25 \text{ m})/8 = 0.031 \text{ m} = 31 \text{ mm}$.

33.17 (a) The image moves farther from the objective lens. This can be seen either by constructing a ray diagram (Figure S33.17) or by considering the relationship among object distance, image distance, and focal length (Eq. 33.16). Solving Eq. 33.16 for i_1 gives

$$\frac{1}{i_1} = \frac{1}{f_1} - \frac{1}{o_1} = \frac{o_1 - f_1}{o_1 f_1} \text{ or } i_1 = \frac{o_1 f_1}{o_1 - f_1}.$$

Increasing both o_1 and f_1 will increase the numerator of the expression for i_1; keeping the sample "just outside the focal point" of the lens implies that the distance between the sample and the lens, $o_1 - f_1$, is kept at least roughly the same, so increasing the numerator while keeping the denominator constant increases i_1. (Practical limitations on the construction of

Figure S33.17

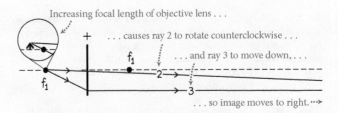

lenses actually allow the distance $o_1 - f_1$ to be smaller for shorter-focal-length lenses, so in fact, the denominator also increases as the focal length increases. However, this last effect is not a consequence of the simple thin-lens treatment but of ways that lenses are not ideal that are beyond the scope of this text.)

(b) Shorter focal lengths give a more compact microscope, because the first image must fall close to the focal point of the eyepiece lens, and thus the distance between the two lenses is roughly $i_1 + f_2$. From part a we know that i_1 depends on f_1, and so decreasing f_1 decreases i_1.

33.18 (a) Using the result of Example 33.9 gives you $f_1 = M_\theta f_2 = (22)(0.0400 \text{ m}) = 0.88 \text{ m}.$

(b) The length of the telescope is roughly the sum of the focal lengths of the two lenses (see Figure 33.50): $0.88 \text{ m} + 0.04 \text{ m} = 0.92 \text{ m}$. Because the eyepiece lens is by design a short-focal-length lens, the length of the telescope is determined primarily by the focal length of the objective lens.

33.19 (a) The ray diagram in Figure S33.19 shows that the image is behind the mirror and virtual. You calculate the image distance with Eq. 33.24. The focal length is half the radius of curvature, $C = 1.0$ m, and is negative because it is a virtual focus:

$$i = \left(\frac{1}{f} - \frac{1}{o}\right)^{-1} = \left(\frac{1}{-0.50} - \frac{1}{1.0}\right)^{-1} = -0.33 \text{ m}.$$

The negative value for i tells you that the image is located 0.33 m behind the mirror.

Figure S33.19

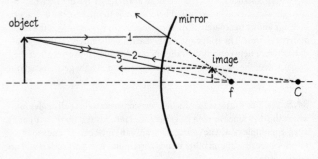

(b) $h_i = -\frac{h_o i}{o} = -\frac{(0.30 \text{ m})(-0.33 \text{ m})}{(1.0 \text{ m})} = 0.10 \text{ m}.$

The image is smaller than the object, and the positive value for h tells you that the image is upright.

33.20 The only change to the physics behind the derivation of the lensmaker's formula is that Eq. 33.25 becomes $n_1 \sin \theta_i = n_2 \sin \theta_r$. In the paraxial approximation, this result can be written as $(n_1/n_2)\theta_i = \theta_r$. Comparing this with Eq. 33.26, you see that the effect of submerging the lens is to substitute (n_1/n_2) for n. Once you make this change in the remainder of the derivation, the lensmaker's formula becomes

$$\frac{1}{f} = \left(\frac{n_1}{n_2} - 1\right)\left(\frac{1}{R_1} + \frac{1}{R_2}\right).$$

Chapter 34

34.1. Once the fronts reach P, the intensity at P is the same as the intensity at any point on any of the wavefronts. For any point R above or below the ray through P, you can locate a corresponding point R′ for which the distance R′P is exactly half a wavelength greater than the distance RP. Those points therefore do not contribute to the intensity at P. Only the point exactly on the ray (dashed line) contributes.

34.2. If you identify pairs of points that radiate in phase at P, then for each pair, there exists another pair that is exactly 180° out of phase with the first pair and thus cancels the radiation from the first pair.

34.3. (a) The spacing between fringes decreases as the separation between the slits increases because $\sin \theta$ is proportional to $1/d$. (b) The spacing between fringes increases as the wavelength increases because $\sin \theta$ is proportional to λ.

34.4. No. Two of the waves cancel each other perfectly, but the third wave is not canceled. The intensity observed on the screen, at angles given by $d \sin \theta = \pm(n - \frac{1}{2})\lambda$, is the intensity of one of the three waves.

34.5. Fringe brightness is determined by the intensity (power/area) that strikes the screen. Increasing the number of slits increases the amount of power that travels through the slits to the screen. In addition, increasing the number of slits causes the fringes to sharpen, with the result that the power that reaches the screen is concentrated into narrower fringes, decreasing the area and thus increasing the brightness further.

34.6. (a) As you found in Checkpoint 34.3, the angle of the fringes increases with wavelength, and so the angle is less for violet light than for red. White light is thus dispersed into a rainbow, with the violet end of the spectrum at smaller angles than the red. (b) Each rainbow corresponds to the fringes of a particular order for all colors, or said another way, each order produces its own rainbow.

34.7. (a) The spacing between fringes in a two-slit interference pattern increases with increasing wavelength and with decreasing distance between slits (see Checkpoint 34.3). The same is true for diffraction gratings: The separation distance between adjacent fringes is increased by decreasing the separation distance between adjacent slits. Therefore the distance between adjacent slits should be decreased if your aim is to increase the diffraction grating's ability to separate close wavelengths. (b) No, as long as the slits are very narrow. The fringe spacing is determined entirely by the separation distance between adjacent slits. (Slit width does, however, affect the brightness of the pattern by determining how much light gets through the diffraction grating.)

34.8. Yes. The diffracted x rays interfere constructively at all angles for which the difference in path length is either zero (as it is for $\theta' = \theta$) or an integer multiple of λ. The diffracted beam at $\theta' = \theta$ corresponds to the zeroth-order maximum; other beams correspond to higher orders and are much weaker.

34.9. From Figure 34.20b you can see that $2\theta + 2\alpha = \pi$, so $\theta = -\alpha + \frac{\pi}{2}$. Using trigonometry, I get $\cos \theta = \cos\left(-\alpha + \frac{\pi}{2}\right) = -\sin(-\alpha) = \sin \alpha$. Substituting this into the Bragg condition yields $2d \sin \alpha = m\lambda$.

34.10. Crystal B. Spots closer together correspond to greater distances between crystal planes and hence greater atomic spacing.

34.11. It decreases, because the spacing between spots is proportional to the wavelength.

34.12. If the electrons travel more slowly, their wavelength increases, and the diffraction spots spread farther apart.

34.13. You could perform the experiment shown in Figure 34.6 with a light source so weak that the light "particles" pass through the pair of slits only one at a time. If light indeed has particle properties, a source this weak would give you a pattern that initially is like the one in Figure 34.25a, showing where individual particles of light hit the screen.

34.14. The maximum probability is at the locations of greatest intensity in the interference pattern; these locations are at angles θ for which $\sin \theta$ is a multiple of λ/d.

34.15. Although the maximum time-averaged intensity is greater than the sum of the intensities of the original two beams, averaging this time-averaged intensity over the entire area filled by the bright and dark bands of the interference pattern gives just the sum of the intensities of the original two beams, $2S_{0,av}$ (because the average of the cosine squared is one-half).

34.16. No, because the higher-order bright fringes exist only as long as $m\lambda/d < 1$. For values of m such that $m\lambda/d > 1$, there are no bright fringes because these correspond to $\sin \theta_m > 1$, for which there is no angle θ_m.

34.17. The oil forms a thin film that causes thin-film interference. The thickness of the oil layer determines for which wavelength constructive interference occurs. If the thickness of the film varies spatially, different wavelengths interfere constructively at different locations.

34.18. (a) If $a < \lambda$, there are no dark fringes because that would require $\sin \theta > 1$. This indicates that no minimum exists, and the light coming through an aperture with a diameter less than the wavelength acts as a true point source as assumed in our original discussion of multiple-slit interference. When $a < \lambda$, the slit acts as a point source with the wavelets spreading out spherically from the slit. (b) For $a \gg \lambda$, the first dark fringe occurs very close to $\theta = 0$. This means there is essentially no diffraction, and the beam just propagates straight ahead. This illustrates that single-slit diffraction is observed primarily with slits from a few wavelengths wide to tens of wavelengths wide.

34.19. Resolution is determined by wavelength and the ratio f/d: $y_r = 1.22 \lambda f/d$. For a given wavelength, the greater f/d is, the larger the Airy disk. For the three lenses, f/d is (i) 1.3, (ii) 1.5, and (iii) 1.1. (a) The highest resolution is obtained with the smallest Airy disk, produced by lens iii. (b) The lowest resolution is obtained with the largest Airy disk, produced by lens ii. (The wavelength determines the exact size of the Airy disk; the comparisons made here assume the same wavelength for all three lenses.)

34.20. (a) The stopping potential difference V_{stop} is proportional to the frequency (Figure 34.48b) of the incident photons, and the maximum kinetic energy K_{max} of the ejected electrons is proportional to V_{stop} (Eq. 34.34). Therefore, K_{max} must be proportional to the frequency of the incident photons. (b) The minimum frequency the light can have and be able to eject electrons. Lower-frequency light does not eject electrons.

34.21. (a) The photon slows down because the speed of light is less in a medium that has an index of refraction of 1.5 than in air (index of refraction 1). (b) The frequency remains unchanged, as discussed in Chapter 33. (c) The wavelength decreases because the wavefronts travel less far in a given time interval. (d) The photon's energy does not change because the medium does not take away any energy. This is why photon energy must be expressed in terms of frequency rather than wavelength, because neither frequency nor energy depends on the medium, whereas wavelength does.

Credits

Index

Unit conversion factors

Length

1 in. = 2.54 cm (defined)

1 cm = 0.3937 in.

1 ft = 30.48 cm

1 m = 39.37 in. = 3.281 ft

1 mi = 5280 ft = 1.609 km

1 km = 0.6214 mi

1 nautical mile (U.S.) = 1.151 mi = 6076 ft = 1.852 km

1 fermi = 1 femtometer (fm) = 10^{-15} m

1 angstrom (Å) = 10^{-10} m = 0.1 nm

1 light − year (ly) = 9.461×10^{15} m

1 parsec = 3.26 ly = 3.09×10^{16} m

Volume

1 liter (L) = 1000 mL = 1000 cm^3 = 1.0×10^{-3} m^3
        = 1.057 qt (U.S.) = 61.02 $in.^3$

1 gal (U.S.) = 4 qt (U.S.) = 231 $in.^3$ = 3.785 L = 0.8327 gal (British)

1 quart (U.S.) = 2 pints (U.S.) = 946 mL

1 pint (British) = 1.20 pints (U.S.) = 568 mL

1 m^3 = 35.31 ft^3

Speed

1 mi/h = 1.4667 ft/s = 1.6093 km/h = 0.4470 m/s

1 km/h = 0.2778 m/s = 0.6214 mi/h

1 ft/s = 0.3048 m/s = 0.6818 mi/h = 1.0973 km/h

1 m/s = 3.281 ft/s = 3.600 km/h = 2.237 mi/h

1 knot = 1.151 mi/h = 0.5144 m/s

Angle

1 radian (rad) = 57.30° = 57°18'

1° = 0.01745 rad

1 rev/min (rpm) = 0.1047 rad/s

Time

1 day = 8.640×10^4 s

1 year = 365.242 days = 3.156×10^7 s

Mass

1 atomic mass unit (u) = 1.6605×10^{-27} kg

1 kg = 0.06852 slug

1 metric ton = 1000 kg

1 long ton = 2240 lbs = 1016 kg

1 short ton = 2000 lbs = 909.1 kg

1 kg has a weight of 2.20 lb where g = 9.80 m/s^2

Force

1 lb = 4.44822 N

1 N = 10^5 dyne = 0.2248 lb

Energy and work

1 J = 10^7 ergs = 0.7376 ft · lb

1 ft · lb = 1.356 J = 1.29×10^{-3} Btu = 3.24×10^{-4} kcal

1 kcal = 4.19×10^3 J = 3.97 Btu

1 eV = 1.6022×10^{-19} J

1 kWh = 3.600×10^6 J = 860 kcal

1 Btu = 1.056×10^3 J

Power

1 W = 1 J/s = 0.7376 ft · lb/s = 3.41 Btu/h

1 hp = 550 ft · lb/s = 746 W

1 kWh/day = 41.667 W

Pressure

1 atm = 1.01325 bar = 1.01325×10^5 N/m^2 = 14.7 $lb/in.^2$ = 760 torr

1 $lb/in.^2$ = 6.895×10^3 N/m^2

1 Pa = 1 N/m^2 = 1.450×10^{-4} $lb/in.^2$